ALTITUDE CORRECTION TABLES 0°-10°—SUN, STARS, PLANETS A3

App. Alt.	OCT.—MAR. SUN Lower Limb	OCT.—MAR. SUN Upper Limb	APR.—SEPT. SUN Lower Limb	APR.—SEPT. SUN Upper Limb	STARS PLANETS
° ′	′	′	′	′	′
0 00	−18·2	−50·5	−18·4	−50·2	−34·[illegible]
03	17·5	49·8	17·8	49·6	33·[illegible]
06	16·9	49·2	17·1	48·9	33·[illegible]
09	16·3	48·6	16·5	48·3	32·[illegible]
12	15·7	48·0	15·9	47·7	32·0
15	15·1	47·4	15·3	47·1	31·4
0 18	−14·5	−46·8	−14·8	−46·6	−30·8
21	14·0	46·3	14·2	46·0	30·3
24	13·5	45·8	13·7	45·5	29·8
27	12·9	45·2	13·2	45·0	29·2
30	12·4	44·7	12·7	44·5	28·7
33	11·9	44·2	12·2	44·0	28·2
0 36	−11·5	−43·8	−11·7	−43·5	−27·8
39	11·0	43·3	11·2	43·0	27·3
42	10·5	42·8	10·8	42·6	26·8
45	10·1	42·4	10·3	42·1	26·4
48	9·6	41·9	9·9	41·7	25·9
51	9·2	41·5	9·5	41·3	25·5
0 54	− 8·8	−41·1	− 9·1	−40·9	−25·1
0 57	8·4	40·7	8·7	40·5	24·7
1 00	8·0	40·3	8·3	40·1	24·3
03	7·7	40·0	7·9	39·7	24·0
06	7·3	39·6	7·5	39·3	23·6
09	6·9	39·2	7·2	39·0	23·2
1 12	− 6·6	−38·9	− 6·8	−38·6	−22·9
15	6·2	38·5	6·5	38·3	22·5
18	5·9	38·2	6·2	38·0	22·2
21	5·6	37·9	5·8	37·6	21·9
24	5·3	37·6	5·5	37·3	21·6
27	4·9	37·2	5·2	37·0	21·2
1 30	− 4·6	−36·9	− 4·9	−36·7	−20·9
35	4·2	36·5	4·4	36·2	20·5
40	3·7	36·0	4·0	35·8	20·0
45	3·2	35·5	3·5	35·3	19·5
50	2·8	35·1	3·1	34·9	19·1
1 55	2·4	34·7	2·6	34·4	18·7
2 00	− 2·0	−34·3	− 2·2	−34·0	−18·3
05	1·6	33·9	1·8	33·6	17·9
10	1·2	33·5	1·5	33·3	17·5
15	0·9	33·2	1·1	32·9	17·2
20	0·5	32·8	0·8	32·6	16·8
25	− 0·2	32·5	0·4	32·2	16·5
2 30	+ 0·2	−32·1	− 0·1	−31·9	−16·1
35	0·5	31·8	+ 0·2	31·6	15·8
40	0·8	31·5	0·5	31·3	15·5
45	1·1	31·2	0·8	31·0	15·2
50	1·4	30·9	1·1	30·7	14·9
2 55	1·6	30·7	1·4	30·4	14·7
3 00	+ 1·9	−30·4	+ 1·7	−30·1	−14·4
05	2·2	30·1	1·9	29·9	14·1
10	2·4	29·9	2·1	29·7	13·9
15	2·6	29·7	2·4	29·4	13·7
20	2·9	29·4	2·6	29·2	13·4
25	3·1	29·2	2·9	28·9	13·2
3 30	+ 3·3	−29·0	+ 3·1	−28·7	−13·0

App. Alt.	OCT.—MAR. SUN Lower Limb	OCT.—MAR. SUN Upper Limb	APR.—SEPT. SUN Lower Limb	APR.—SEPT. SUN Upper Limb	STARS PLANETS
° ′	′	′	′	′	′
[illegible]	[illegible]	[illegible]	[illegible]	−28·7	−13·0
[illegible]	[illegible]	[illegible]	[illegible]	28·5	12·7
[illegible]	[illegible]	[illegible]	[illegible]	28·3	12·5
[illegible]	[illegible]	[illegible]	[illegible]	28·1	12·3
[illegible]	4·2	[illegible]	3·9	27·9	12·1
3 55	4·4	27·9	4·1	27·7	11·9
4 00	+ 4·5	−27·8	+ 4·3	−27·5	−11·8
05	4·7	27·6	4·5	27·3	11·6
10	4·9	27·4	4·6	27·2	11·4
15	5·1	27·2	4·8	27·0	11·2
20	5·2	27·1	5·0	26·8	11·1
25	5·4	26·9	5·1	26·7	10·9
4 30	+ 5·6	−26·7	+ 5·3	−26·5	−10·7
35	5·7	26·6	5·5	26·3	10·6
40	5·9	26·4	5·6	26·2	10·4
45	6·0	26·3	5·8	26·0	10·3
50	6·2	26·1	5·9	25·9	10·1
4 55	6·3	26·0	6·0	25·8	10·0
5 00	+ 6·4	−25·9	+ 6·2	−25·6	− 9·9
05	6·6	25·7	6·3	25·5	9·7
10	6·7	25·6	6·4	25·4	9·6
15	6·8	25·5	6·6	25·2	9·5
20	6·9	25·4	6·7	25·1	9·4
25	7·1	25·2	6·8	25·0	9·2
5 30	+ 7·2	−25·1	+ 6·9	−24·9	− 9·1
35	7·3	25·0	7·0	24·8	9·0
40	7·4	24·9	7·2	24·6	8·9
45	7·5	24·8	7·3	24·5	8·8
50	7·6	24·7	7·4	24·4	8·7
5 55	7·7	24·6	7·5	24·3	8·6
6 00	+ 7·8	−24·5	+ 7·6	−24·2	− 8·5
10	8·0	24·3	7·8	24·0	8·3
20	8·2	24·1	8·0	23·8	8·1
30	8·4	23·9	8·1	23·7	7·9
40	8·6	23·7	8·3	23·5	7·7
50	8·7	23·6	8·5	23·3	7·6
7 00	+ 8·9	−23·4	+ 8·6	−23·2	− 7·4
10	9·1	23·2	8·8	23·0	7·2
20	9·2	23·1	9·0	22·8	7·1
30	9·3	23·0	9·1	22·7	7·0
40	9·5	22·8	9·2	22·6	6·8
7 50	9·6	22·7	9·4	22·4	6·7
8 00	+ 9·7	−22·6	+ 9·5	−22·3	− 6·6
10	9·9	22·4	9·6	22·2	6·4
20	10·0	22·3	9·7	22·1	6·3
30	10·1	22·2	9·8	22·0	6·2
40	10·2	22·1	10·0	21·8	6·1
8 50	10·3	22·0	10·1	21·7	6·0
9 00	+10·4	−21·9	+10·2	−21·6	− 5·9
10	10·5	21·8	10·3	21·5	5·8
20	10·6	21·7	10·4	21·4	5·7
30	10·7	21·6	10·5	21·3	5·6
40	10·8	21·5	10·6	21·2	5·5
9 50	10·9	21·4	10·6	21·2	5·4
10 00	+11·0	−21·3	+10·7	−21·1	− 5·3

Additional corrections for temperature and pressure are given on the following page.

For bubble sextant observations ignore dip and use the star corrections for Sun, planets and stars.

A4 ALTITUDE CORRECTION TABLES—ADDITIONAL CORRECTIONS

ADDITIONAL REFRACTION CORRECTIONS FOR NON-STANDARD CONDITIONS

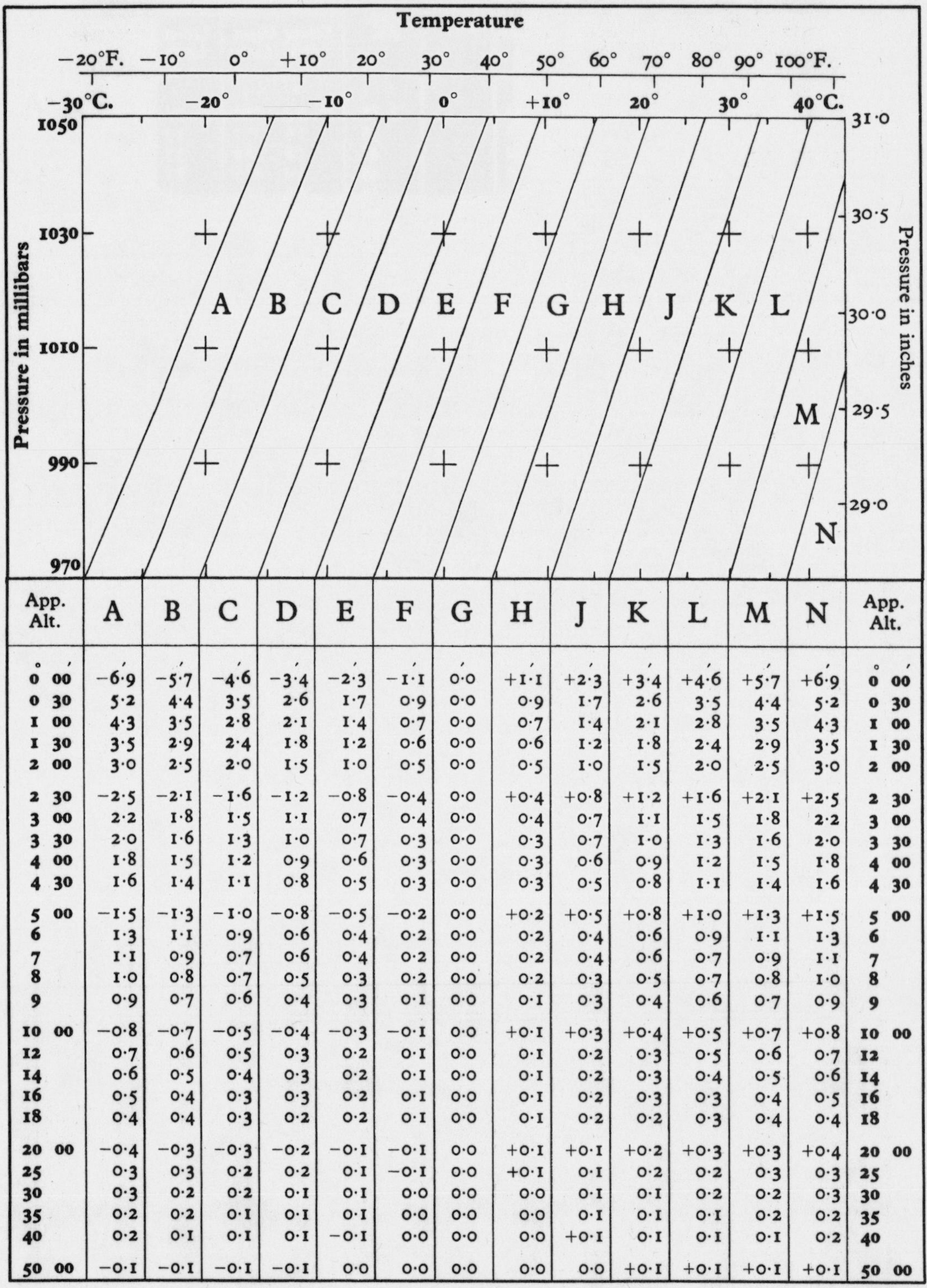

Temperature

−20°F. −10° 0° +10° 20° 30° 40° 50° 60° 70° 80° 90° 100°F.

−30°C. −20° −10° 0° +10° 20° 30° 40°C.

Pressure in millibars: 1050, 1030, 1010, 990, 970

Pressure in inches: 31·0, 30·5, 30·0, 29·5, 29·0

A B C D E F G H J K L M N

App. Alt.	A	B	C	D	E	F	G	H	J	K	L	M	N	App. Alt.
° ′	′	′	′	′	′	′	′	′	′	′	′	′	′	° ′
0 00	−6·9	−5·7	−4·6	−3·4	−2·3	−1·1	0·0	+1·1	+2·3	+3·4	+4·6	+5·7	+6·9	0 00
0 30	5·2	4·4	3·5	2·6	1·7	0·9	0·0	0·9	1·7	2·6	3·5	4·4	5·2	0 30
1 00	4·3	3·5	2·8	2·1	1·4	0·7	0·0	0·7	1·4	2·1	2·8	3·5	4·3	1 00
1 30	3·5	2·9	2·4	1·8	1·2	0·6	0·0	0·6	1·2	1·8	2·4	2·9	3·5	1 30
2 00	3·0	2·5	2·0	1·5	1·0	0·5	0·0	0·5	1·0	1·5	2·0	2·5	3·0	2 00
2 30	−2·5	−2·1	−1·6	−1·2	−0·8	−0·4	0·0	+0·4	+0·8	+1·2	+1·6	+2·1	+2·5	2 30
3 00	2·2	1·8	1·5	1·1	0·7	0·4	0·0	0·4	0·7	1·1	1·5	1·8	2·2	3 00
3 30	2·0	1·6	1·3	1·0	0·7	0·3	0·0	0·3	0·7	1·0	1·3	1·6	2·0	3 30
4 00	1·8	1·5	1·2	0·9	0·6	0·3	0·0	0·3	0·6	0·9	1·2	1·5	1·8	4 00
4 30	1·6	1·4	1·1	0·8	0·5	0·3	0·0	0·3	0·5	0·8	1·1	1·4	1·6	4 30
5 00	−1·5	−1·3	−1·0	−0·8	−0·5	−0·2	0·0	+0·2	+0·5	+0·8	+1·0	+1·3	+1·5	5 00
6	1·3	1·1	0·9	0·6	0·4	0·2	0·0	0·2	0·4	0·6	0·9	1·1	1·3	6
7	1·1	0·9	0·7	0·6	0·4	0·2	0·0	0·2	0·4	0·6	0·7	0·9	1·1	7
8	1·0	0·8	0·7	0·5	0·3	0·2	0·0	0·2	0·3	0·5	0·7	0·8	1·0	8
9	0·9	0·7	0·6	0·4	0·3	0·1	0·0	0·1	0·3	0·4	0·6	0·7	0·9	9
10 00	−0·8	−0·7	−0·5	−0·4	−0·3	−0·1	0·0	+0·1	+0·3	+0·4	+0·5	+0·7	+0·8	10 00
12	0·7	0·6	0·5	0·3	0·2	0·1	0·0	0·1	0·2	0·3	0·5	0·6	0·7	12
14	0·6	0·5	0·4	0·3	0·2	0·1	0·0	0·1	0·2	0·3	0·4	0·5	0·6	14
16	0·5	0·4	0·3	0·3	0·2	0·1	0·0	0·1	0·2	0·3	0·3	0·4	0·5	16
18	0·4	0·4	0·3	0·2	0·2	0·1	0·0	0·1	0·2	0·2	0·3	0·4	0·4	18
20 00	−0·4	−0·3	−0·3	−0·2	−0·1	−0·1	0·0	+0·1	+0·1	+0·2	+0·3	+0·3	+0·4	20 00
25	0·3	0·3	0·2	0·2	0·1	−0·1	0·0	+0·1	0·1	0·2	0·2	0·3	0·3	25
30	0·3	0·2	0·2	0·1	0·1	0·0	0·0	0·0	0·1	0·1	0·2	0·2	0·3	30
35	0·2	0·2	0·1	0·1	0·1	0·0	0·0	0·0	0·1	0·1	0·1	0·2	0·2	35
40	0·2	0·1	0·1	0·1	−0·1	0·0	0·0	0·0	+0·1	0·1	0·1	0·1	0·2	40
50 00	−0·1	−0·1	−0·1	−0·1	0·0	0·0	0·0	0·0	0·0	+0·1	+0·1	+0·1	+0·1	50 00

The graph is entered with arguments temperature and pressure to find a zone letter; using as arguments this zone letter and apparent altitude (sextant altitude corrected for dip), a correction is taken from the table. This correction is to be applied to the sextant altitude in addition to the corrections for standard conditions (for the Sun, stars and planets from page A2 and for the Moon from pages xxxiv and xxxv).

THE NAUTICAL ALMANAC

FOR THE YEAR

2003

WASHINGTON:
Issued by the
Nautical Almanac Office
United States
Naval Observatory
under the
authority of the
Secretary of the Navy

LONDON:
Issued by
Her Majesty's
Nautical Almanac Office
by order
of the
Secretary of State
for Defence

U.S. GOVERNMENT PRINTING OFFICE
WASHINGTON: 2002

UNITED STATES

For sale by the
U.S. GOVERNMENT PRINTING OFFICE
Superintendent of Documents
P. O. Box 371954
Pittsburgh, PA 15250-7954
phone: 1-202-512-1800
order online at http://bookstore.gpo.gov/

Printed by permission for official U. S. Government use.

PREFACE

The British and American editions of *The Nautical Almanac*, which are identical in content, are produced jointly by H. M. Nautical Almanac Office, Rutherford Appleton Laboratory, under the supervision of P. T. Wallace and C. Y. Hohenkerk, and by the Nautical Almanac Office, United States Naval Observatory, under the supervision of S. Howard and R. J. Miller, to the general requirements of the Royal Navy and of the United States Navy. The Almanac is printed separately in the United Kingdom and in the United States of America.

The data in this almanac, on written application, may be made available for reproduction. It can be made available in a form suitable for direct photographic reproduction, to the appropriate almanac-producing agency in any country; language changes in the headings of the ephemeral pages can be introduced, if desired, during reproduction. Under this arrangement, this almanac, with minor modifications and changes of language, has been adopted for the Brazilian, Chilean, Croatian, Danish, Greek, Indian, Indonesian and Mexican almanacs.

J.V. WOOD
Chief Executive, CCLRC
Rutherford Appleton Laboratory
Chilton, Didcot
Oxfordshire, OX11 0QX
United Kingdom

B.J. JARAMILLO
Captain, U.S. Navy
Superintendent, U.S. Naval Observatory
3450 Massachusetts Avenue NW
Washington, D.C. 20392-5420
U.S.A.

November 2001

For sale by the Superintendent of Documents, U.S. Government Printing Office
Internet: bookstore.gpo.gov Phone: toll free (866) 512-1800; DC area (202) 512-1800
Fax: (202) 512-2250 Mail: Stop SSOP, Washington, DC 20402-0001

ISBN 0-16-051042-2

RELIGIOUS CALENDARS

Epiphany	Jan. 6	Low Sunday	Apr. 27
Septuagesima Sunday	Feb. 16	Rogation Sunday	May 25
Quinquagesima Sunday	Mar. 2	Ascension Day—Holy Thursday	May 29
Ash Wednesday	Mar. 5	Whit Sunday—Pentecost	June 8
Quadragesima Sunday	Mar. 9	Trinity Sunday	June 15
Palm Sunday	Apr. 13	Corpus Christi	June 19
Good Friday	Apr. 18	First Sunday in Advent	Nov. 30
Easter Day	Apr. 20	Christmas Day (Thursday)	Dec. 25
First Day of Passover (Pesach)	Apr. 17	Day of Atonement (Yom Kippur)	Oct. 6
Feast of Weeks (Shavuot)	June 6	First day of Tabernacles (Succoth)	Oct. 11
Jewish New Year 5764 (Rosh Hashanah)	Sept. 27		
Islamic New Year (1424)	Mar. 5	Ramadân, First day of (tabular)	Oct. 27

The Jewish and Islamic dates above are tabular dates, which begin at sunset on the previous evening and end at sunset on the date tabulated. In practice, the dates of Islamic fasts and festivals are determined by an actual sighting of the appropriate new moon.

CIVIL CALENDAR—UNITED KINGDOM

Accession of Queen Elizabeth II	Feb. 6	Birthday of Prince Philip, Duke of Edinburgh	June 10
St David (Wales)	Mar. 1	The Queen's Official Birthday†	June 14
Commonwealth Day	Mar. 10	Remembrance Sunday	Nov. 9
St Patrick (Ireland)	Mar. 17	Birthday of the Prince of Wales	Nov. 14
Birthday of Queen Elizabeth II	Apr. 21	St Andrew (Scotland)	Nov. 30
St George (England)	Apr. 23		
Coronation Day	June 2		

PUBLIC HOLIDAYS

England and Wales—Jan. 1†, Apr. 18, Apr. 21, May 5†, May 26, Aug. 25, Dec. 25, Dec. 26
Northern Ireland—Jan. 1†, Mar. 17, Apr. 18, Apr. 21, May 5†, May 26, July 14†, Aug. 25, Dec. 25, Dec. 26
Scotland—Jan. 1, Jan. 2, Apr. 18, May 5, May 26†, Aug. 4, Dec. 25, Dec. 26†

CIVIL CALENDAR—UNITED STATES OF AMERICA

New Year's Day	Jan. 1	Labor Day	Sept. 1
Martin Luther King's Birthday	Jan. 20	Columbus Day	Oct. 13
Washington's Birthday	Feb. 17	Election Day (in certain States)	Nov. 4
Memorial Day	May 26	Veterans Day	Nov. 11
Independence Day	July 4	Thanksgiving Day	Nov. 27

†Dates subject to confirmation

PHASES OF THE MOON

New Moon	d	h	m	First Quarter	d	h	m	Full Moon	d	h	m	Last Quarter	d	h	m
Jan.	2	20	23	Jan.	10	13	15	Jan.	18	10	48	Jan.	25	08	33
Feb.	1	10	48	Feb.	9	11	11	Feb.	16	23	51	Feb.	23	16	46
Mar.	3	02	35	Mar.	11	07	15	Mar.	18	10	35	Mar.	25	01	51
Apr.	1	19	19	Apr.	9	23	40	Apr.	16	19	36	Apr.	23	12	18
May	1	12	15	May	9	11	53	May	16	03	36	May	23	00	31
May	31	04	20	June	7	20	28	June	14	11	16	June	21	14	45
June	29	18	39	July	7	02	32	July	13	19	21	July	21	07	01
July	29	06	53	Aug.	5	07	28	Aug.	12	04	48	Aug.	20	00	48
Aug.	27	17	26	Sept.	3	12	34	Sept.	10	16	36	Sept.	18	19	03
Sept.	26	03	09	Oct.	2	19	09	Oct.	10	07	27	Oct.	18	12	31
Oct.	25	12	50	Nov.	1	04	25	Nov.	9	01	13	Nov.	17	04	15
Nov.	23	22	59	Nov.	30	17	16	Dec.	8	20	37	Dec.	16	17	42
Dec.	23	09	43	Dec.	30	10	03								

DAYS OF THE WEEK AND DAYS OF THE YEAR

	JAN.	FEB.	MAR.	APR.	MAY	JUNE	JULY	AUG.	SEPT.	OCT.	NOV.	DEC.
Day	Wk Yr	Wk Yr	Wk Yr	Wk Yr	Wk Yr	Wk Yr	Wk Yr	Wk Yr	Wk Yr	Wk Yr	Wk Yr	Wk Yr
1	W. 1	Sa. 32	Sa. 60	Tu. 91	Th. 121	Su. 152	Tu. 182	F. 213	M. 244	W. 274	Sa. 305	M. 335
2	Th. 2	Su. 33	Su. 61	W. 92	F. 122	M. 153	W. 183	Sa. 214	Tu. 245	Th. 275	Su. 306	Tu. 336
3	F. 3	M. 34	M. 62	Th. 93	Sa. 123	Tu. 154	Th. 184	Su. 215	W. 246	F. 276	M. 307	W. 337
4	Sa. 4	Tu. 35	Tu. 63	F. 94	Su. 124	W. 155	F. 185	M. 216	Th. 247	Sa. 277	Tu. 308	Th. 338
5	Su. 5	W. 36	W. 64	Sa. 95	M. 125	Th. 156	Sa. 186	Tu. 217	F. 248	Su. 278	W. 309	F. 339
6	M. 6	Th. 37	Th. 65	Su. 96	Tu. 126	F. 157	Su. 187	W. 218	Sa. 249	M. 279	Th. 310	Sa. 340
7	Tu. 7	F. 38	F. 66	M. 97	W. 127	Sa. 158	M. 188	Th. 219	Su. 250	Tu. 280	F. 311	Su. 341
8	W. 8	Sa. 39	Sa. 67	Tu. 98	Th. 128	Su. 159	Tu. 189	F. 220	M. 251	W. 281	Sa. 312	M. 342
9	Th. 9	Su. 40	Su. 68	W. 99	F. 129	M. 160	W. 190	Sa. 221	Tu. 252	Th. 282	Su. 313	Tu. 343
10	F. 10	M. 41	M. 69	Th. 100	Sa. 130	Tu. 161	Th. 191	Su. 222	W. 253	F. 283	M. 314	W. 344
11	Sa. 11	Tu. 42	Tu. 70	F. 101	Su. 131	W. 162	F. 192	M. 223	Th. 254	Sa. 284	Tu. 315	Th. 345
12	Su. 12	W. 43	W. 71	Sa. 102	M. 132	Th. 163	Sa. 193	Tu. 224	F. 255	Su. 285	W. 316	F. 346
13	M. 13	Th. 44	Th. 72	Su. 103	Tu. 133	F. 164	Su. 194	W. 225	Sa. 256	M. 286	Th. 317	Sa. 347
14	Tu. 14	F. 45	F. 73	M. 104	W. 134	Sa. 165	M. 195	Th. 226	Su. 257	Tu. 287	F. 318	Su. 348
15	W. 15	Sa. 46	Sa. 74	Tu. 105	Th. 135	Su. 166	Tu. 196	F. 227	M. 258	W. 288	Sa. 319	M. 349
16	Th. 16	Su. 47	Su. 75	W. 106	F. 136	M. 167	W. 197	Sa. 228	Tu. 259	Th. 289	Su. 320	Tu. 350
17	F. 17	M. 48	M. 76	Th. 107	Sa. 137	Tu. 168	Th. 198	Su. 229	W. 260	F. 290	M. 321	W. 351
18	Sa. 18	Tu. 49	Tu. 77	F. 108	Su. 138	W. 169	F. 199	M. 230	Th. 261	Sa. 291	Tu. 322	Th. 352
19	Su. 19	W. 50	W. 78	Sa. 109	M. 139	Th. 170	Sa. 200	Tu. 231	F. 262	Su. 292	W. 323	F. 353
20	M. 20	Th. 51	Th. 79	Su. 110	Tu. 140	F. 171	Su. 201	W. 232	Sa. 263	M. 293	Th. 324	Sa. 354
21	Tu. 21	F. 52	F. 80	M. 111	W. 141	Sa. 172	M. 202	Th. 233	Su. 264	Tu. 294	F. 325	Su. 355
22	W. 22	Sa. 53	Sa. 81	Tu. 112	Th. 142	Su. 173	Tu. 203	F. 234	M. 265	W. 295	Sa. 326	M. 356
23	Th. 23	Su. 54	Su. 82	W. 113	F. 143	M. 174	W. 204	Sa. 235	Tu. 266	Th. 296	Su. 327	Tu. 357
24	F. 24	M. 55	M. 83	Th. 114	Sa. 144	Tu. 175	Th. 205	Su. 236	W. 267	F. 297	M. 328	W. 358
25	Sa. 25	Tu. 56	Tu. 84	F. 115	Su. 145	W. 176	F. 206	M. 237	Th. 268	Sa. 298	Tu. 329	Th. 359
26	Su. 26	W. 57	W. 85	Sa. 116	M. 146	Th. 177	Sa. 207	Tu. 238	F. 269	Su. 299	W. 330	F. 360
27	M. 27	Th. 58	Th. 86	Su. 117	Tu. 147	F. 178	Su. 208	W. 239	Sa. 270	M. 300	Th. 331	Sa. 361
28	Tu. 28	F. 59	F. 87	M. 118	W. 148	Sa. 179	M. 209	Th. 240	Su. 271	Tu. 301	F. 332	Su. 362
29	W. 29		Sa. 88	Tu. 119	Th. 149	Su. 180	Tu. 210	F. 241	M. 272	W. 302	Sa. 333	M. 363
30	Th. 30		Su. 89	W. 120	F. 150	M. 181	W. 211	Sa. 242	Tu. 273	Th. 303	Su. 334	Tu. 364
31	F. 31		M. 90		Sa. 151		Th. 212	Su. 243		F. 304		W. 365

ECLIPSES

There are two eclipses of the Sun and two of the Moon.

1. *A total eclipse of the Moon*, May 16. The eclipse begins at 02^{h} 03^{m} and ends at 05^{h} 17^{m}; the total phase begins at 03^{h} $14^{m}$and ends at 04^{h} 06^{m}. It is visible from part of Antarctica, Africa except the north-east and northern Madagascar, western Europe but excluding Scandinavia, southern Greenland, Atlantic Ocean, The Americas except north-west Canada and Alaska, and east Pacific Ocean.

2. *An annular eclipse of the Sun*, May 31. See map on page 6. The eclipse begins at 01^{h} 46^{m} and ends at 06^{h} 30^{m}; the annular phase begins at 03^{h} 45^{m} and ends at 04^{h} 31^{m}. The maximum duration of the annular phase is 3^{m} 37^{s}.

3. *A total eclipse of the Moon*, November 8-9. The eclipse begins on November 8 at 23^{h} 32^{m} and ends on November 9 at 03^{h} 05^{m}; the total phase on November 9 begins at 01^{h} 06^{m} and ends at 01^{h} 31^{m}. It is visible from western and central parts of Asia, west Indian Ocean, Africa, Europe, Atlantic Ocean, part of Antarctica, Greenland, The Americas except western Alaska, and east Pacific Ocean.

4. *A total eclipse of the Sun*, November 23-24. See map on page 7. The eclipse begins on November 23 at 20^{h} 46^{m} and ends on November 24 at 00^{h} 52^{m}; the total phase on November 23 begins at 22^{h} 19^{m} and ends at 23^{h} 19^{m}. The maximum duration of totality is 1^{m} 57^{s}.

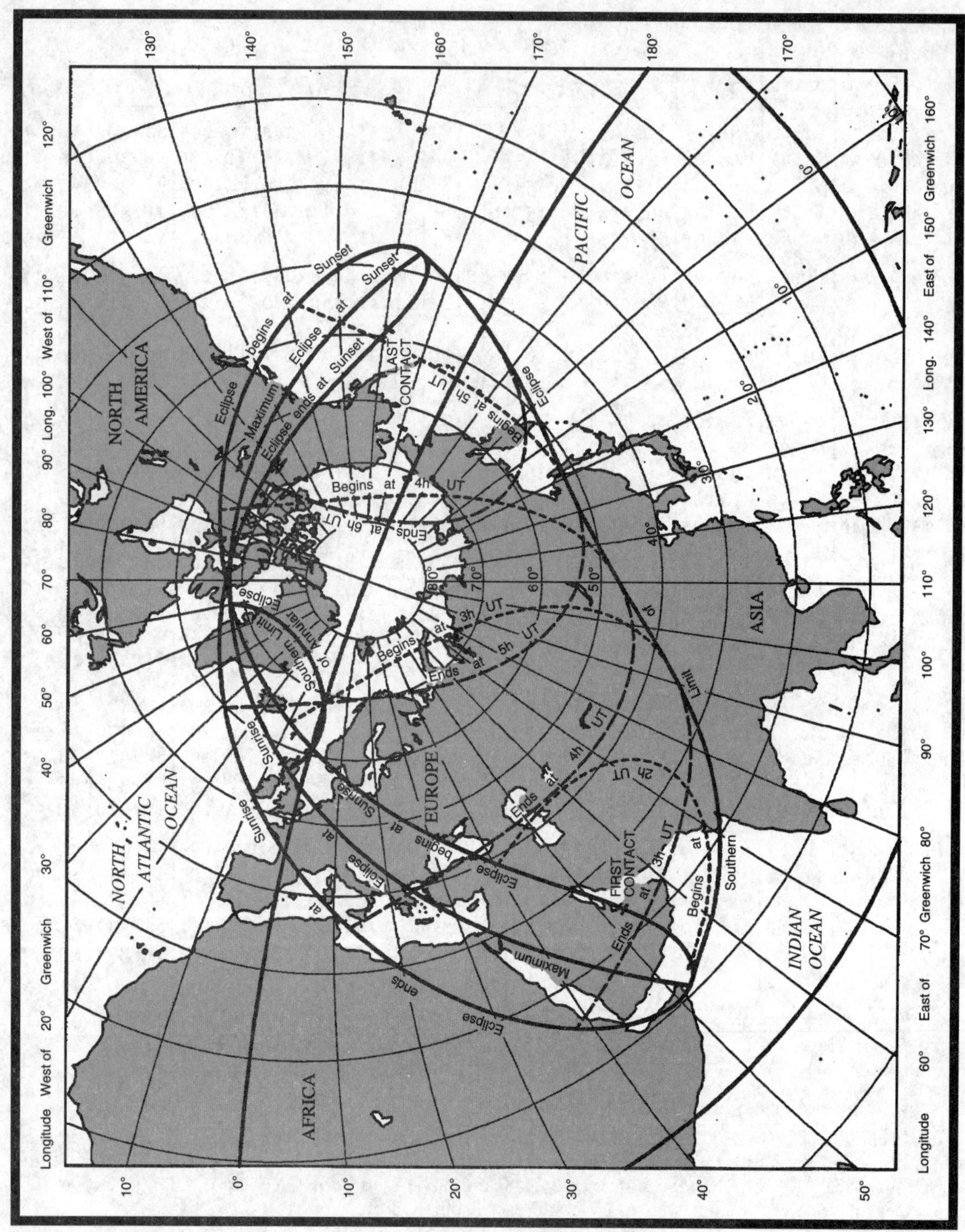

SOLAR ECLIPSE DIAGRAMS

The principal features shown on the above diagrams are: the paths of total and annular eclipses; the northern and southern limits of partial eclipse; the sunrise and sunset curves; dashed lines which show the times of beginning and end of partial eclipse at hourly intervals.

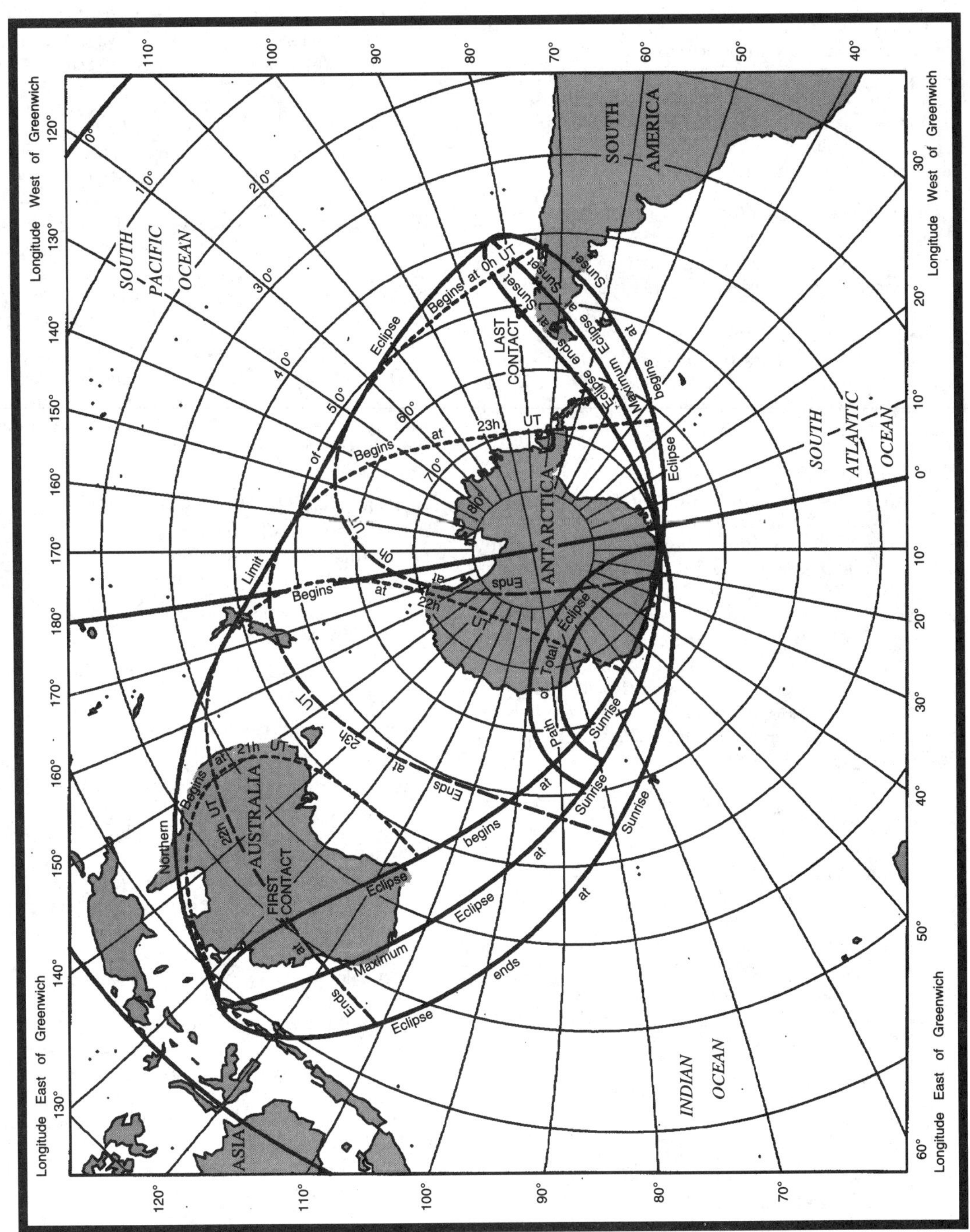

SOLAR ECLIPSE DIAGRAMS

Further details of the paths and times of central eclipse are given in *The Astronomical Almanac.*

VISIBILITY OF PLANETS

VENUS is a brilliant object in the morning sky from the beginning of the year until near the end of the second week of July when it becomes too close to the Sun for observation. During late September it reappears in the evening sky where it stays until the end of the year. Venus is in conjunction with Mercury on May 28 and June 21 and with Saturn on July 8.

MARS rises well after midnight in Libra at the beginning of the year, when it can only be seen in the morning sky. Its westward elongation gradually increases as it moves through Scorpius, Ophiuchus (passing 5° N of *Antares* on January 31), Sagittarius, Capricornus and into Aquarius in early June, when it can be seen for more than half the night. It is at opposition on August 28 when it can be seen throughout the night. It moves into Pisces in early December in which constellation it remains for the rest of the year.

JUPITER can be seen for most of the night in Cancer, its westward elongation gradually increases until it is at opposition on February 2 when it can be seen throughout the night. Its eastward elongation then gradually decreases as it passes into Leo at the very end of June where it can be seen only in the evening sky. In the second week of August it becomes too close to the Sun for observation until early September when it reappears in the morning sky in Leo in which constellation it remains for the rest of the year. Its westward elongation gradually increases and by mid-December it can be seen for more than half the night. Jupiter is in conjunction with Mercury on July 26.

SATURN is in Taurus at the beginning of the year. It can be seen for more than half the night until mid-March after which it can be seen only in the evening sky. Its eastward elongation gradually decreases and from mid-May passes into Orion. In early June it becomes too close to the Sun for observation, reappearing in the morning sky in mid-July in Gemini in which constellation it remains for the rest of the year. Its westward elongation gradually increases and from early October it can be seen for more than half the night. It is at opposition on December 31. Saturn is in conjunction with Venus on July 8.

MERCURY can only be seen low in the east before sunrise, or low in the west after sunset (about the time of beginning or end of civil twilight). It is visible in the mornings between the following approximate dates: January 18 (+1·8) to March 12 (−0·9), May 16 (+3·1) to June 28 (−1·5), and September 18 (+2·0) to October 13 (−1·2); the planet is brighter at the end of each period. It is visible in the evenings between the following approximate dates: January 1 (+0·1) to January 6 (+1·6), March 31 (−1·5) to April 28 (+2·6), July 13 (−1·3) to September 4 (+2·5) and November 10 (−0·6) to December 21 (+1·5); the planet is brighter at the beginning of each period. The figures in parentheses are the magnitudes.

Mercury transits the Sun's disk on May 7 from $05^h 13^m$ to $10^h 32^m$; the event is visible from Alaska, Melanesia, Australia, Asia, Arctic regions, Africa, Europe, Greenland, eastern Canada and north eastern S. America.

PLANET DIAGRAM

General Description. The diagram on the opposite page shows, in graphical form for any date during the year, the local mean time of meridian passage of the Sun, of the five planets Mercury, Venus, Mars, Jupiter, and Saturn, and of each 30° of SHA; intermediate lines corresponding to particular stars, may be drawn in by the user if desired. It is intended to provide a general picture of the availability of planets and stars for observation.

On each side of the line marking the time of meridian passage of the Sun a band, 45^m wide, is shaded to indicate that planets and most stars crossing the meridian within 45^m of the Sun are too close to the Sun for observation.

Method of use and interpretation. For any date the diagram provides immediately the local mean times of meridian passage of the Sun, planets and stars, and thus the following information:

(a) whether a planet or star is too close to the Sun for observation;

(b) some indication of its position in the sky, especially during twilight;

(c) the proximity of other planets.

When the meridian passage of an outer planet occurs at midnight the body is in opposition to the Sun and is visible all night; a planet may then be observable during both morning and evening twilights. As the time of meridian passage decreases, the body eventually ceases to be observable in the morning, but its altitude above the eastern horizon at sunset gradually increases; this continues until the body is on the meridian during evening twilight. From then onwards the body is observable above the western horizon and its altitude at sunset gradually decreases; eventually the body becomes too close to the Sun for observation. When the body again becomes visible it is seen low in the east during morning twilight; its altitude at sunrise increases until meridian passage occurs during morning twilight. Then, as the time of meridian passage decreases to 0^h, the body is observable in the west during morning twilight with a gradually decreasing altitude, until it once again reaches opposition.

DO NOT CONFUSE

Mercury with Venus from the end of the third week of May to late June when Venus is the brighter object, and with Jupiter in the second half of July when Jupiter is the brighter object.

LOCAL MEAN TIME OF MERIDIAN PASSAGE

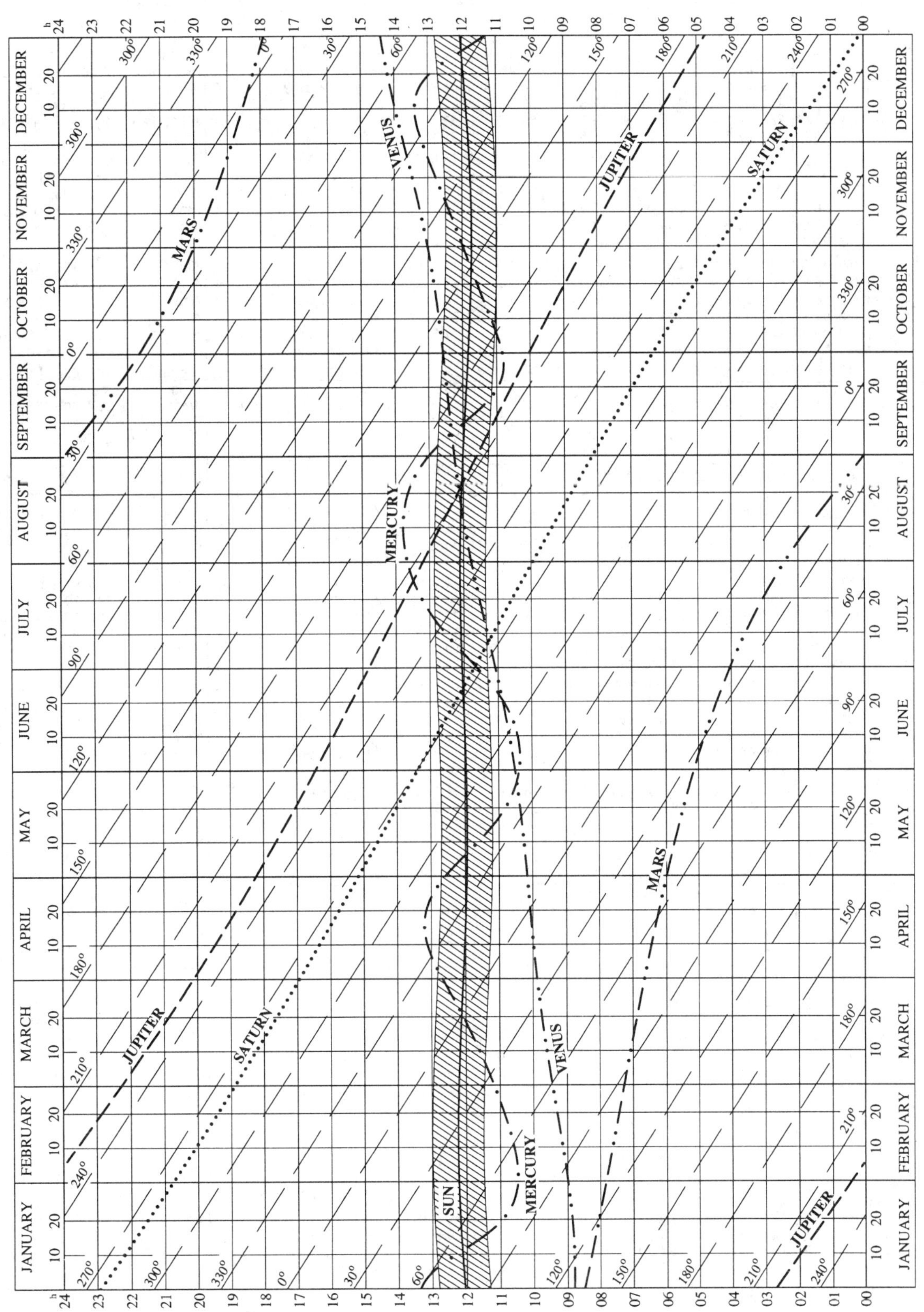

UT	ARIES	VENUS −4.5		MARS +1.5		JUPITER −2.5		SATURN −0.4	
d h	GHA ° ′	GHA ° ′	Dec ° ′	GHA ° ′	Dec ° ′	GHA ° ′	Dec ° ′	GHA ° ′	Dec ° ′
1 00	100 14.0	228 01.6	S15 14.8	232 55.3	S16 57.9	320 39.0	N16 30.6	16 13.6	N22 02.4
01	115 16.5	243 01.7	15.3	247 56.1	58.4	335 41.7	30.6	31 16.3	02.4
02	130 18.9	258 01.8	15.9	262 57.0	58.8	350 44.3	30.7	46 18.9	02.4
03	145 21.4	273 02.0	. . 16.4	277 57.8	. . 59.3	5 47.0	. . 30.8	61 21.6	. . 02.4
04	160 23.9	288 02.1	16.9	292 58.7	16 59.7	20 49.7	30.9	76 24.3	02.4
05	175 26.3	303 02.2	17.5	307 59.5	17 00.2	35 52.4	30.9	91 27.0	02.4
06	190 28.8	318 02.2	S15 18.0	323 00.4	S17 00.6	50 55.0	N16 31.0	106 29.6	N22 02.4
W 07	205 31.3	333 02.3	18.6	338 01.2	01.1	65 57.7	31.1	121 32.3	02.4
E 08	220 33.7	348 02.4	19.1	353 02.1	01.5	81 00.4	31.1	136 35.0	02.4
D 09	235 36.2	3 02.5	. . 19.7	8 02.9	. . 02.0	96 03.1	. . 31.2	151 37.6	. . 02.4
N 10	250 38.6	18 02.6	20.2	23 03.7	02.4	111 05.8	31.3	166 40.3	02.4
E 11	265 41.1	33 02.7	20.7	38 04.6	02.9	126 08.4	31.4	181 43.0	02.4
S 12	280 43.6	48 02.8	S15 21.3	53 05.4	S17 03.4	141 11.1	N16 31.4	196 45.7	N22 02.4
D 13	295 46.0	63 02.9	21.8	68 06.3	03.8	156 13.8	31.5	211 48.3	02.4
A 14	310 48.5	78 03.0	22.4	83 07.1	04.3	171 16.5	31.6	226 51.0	02.4
Y 15	325 51.0	93 03.1	. . 22.9	98 08.0	. . 04.7	186 19.1	. . 31.7	241 53.7	. . 02.4
16	340 53.4	108 03.2	23.5	113 08.8	05.2	201 21.8	31.7	256 56.4	02.4
17	355 55.9	123 03.3	24.0	128 09.7	05.6	216 24.5	31.8	271 59.0	02.3
18	10 58.4	138 03.3	S15 24.5	143 10.5	S17 06.1	231 27.2	N16 31.9	287 01.7	N22 02.3
19	26 00.8	153 03.4	25.1	158 11.4	06.5	246 29.9	32.0	302 04.4	02.3
20	41 03.3	168 03.5	25.6	173 12.2	07.0	261 32.5	32.0	317 07.0	02.3
21	56 05.8	183 03.6	. . 26.2	188 13.0	. . 07.4	276 35.2	. . 32.1	332 09.7	. . 02.3
22	71 08.2	198 03.7	26.7	203 13.9	07.9	291 37.9	32.2	347 12.4	02.3
23	86 10.7	213 03.7	27.3	218 14.7	08.3	306 40.6	32.3	2 15.1	02.3
2 00	101 13.1	228 03.8	S15 27.8	233 15.6	S17 08.8	321 43.3	N16 32.3	17 17.7	N22 02.3
01	116 15.6	243 03.9	28.3	248 16.4	09.2	336 45.9	32.4	32 20.4	02.3
02	131 18.1	258 04.0	28.9	263 17.3	09.7	351 48.6	32.5	47 23.1	02.3
03	146 20.5	273 04.0	. . 29.4	278 18.1	. . 10.1	6 51.3	. . 32.6	62 25.7	. . 02.3
04	161 23.0	288 04.1	30.0	293 18.9	10.5	21 54.0	32.6	77 28.4	02.3
05	176 25.5	303 04.2	30.5	308 19.8	11.0	36 56.7	32.7	92 31.1	02.3
06	191 27.9	318 04.3	S15 31.1	323 20.6	S17 11.4	51 59.4	N16 32.8	107 33.8	N22 02.3
07	206 30.4	333 04.3	31.6	338 21.5	11.9	67 02.0	32.9	122 36.4	02.3
T 08	221 32.9	348 04.4	32.1	353 22.3	12.3	82 04.7	32.9	137 39.1	02.3
H 09	236 35.3	3 04.5	. . 32.7	8 23.2	. . 12.8	97 07.4	. . 33.0	152 41.8	. . 02.3
U 10	251 37.8	18 04.5	33.2	23 24.0	13.2	112 10.1	33.1	167 44.4	02.3
R 11	266 40.3	33 04.6	33.8	38 24.8	13.7	127 12.8	33.2	182 47.1	02.3
S 12	281 42.7	48 04.7	S15 34.3	53 25.7	S17 14.1	142 15.5	N16 33.2	197 49.8	N22 02.3
D 13	296 45.2	63 04.7	34.9	68 26.5	14.6	157 18.1	33.3	212 52.5	02.3
A 14	311 47.6	78 04.8	35.4	83 27.4	15.0	172 20.8	33.4	227 55.1	02.3
Y 15	326 50.1	93 04.9	. . 36.0	98 28.2	. . 15.5	187 23.5	. . 33.5	242 57.8	. . 02.3
16	341 52.6	108 04.9	36.5	113 29.1	15.9	202 26.2	33.5	258 00.5	02.3
17	356 55.0	123 05.0	37.0	128 29.9	16.4	217 28.9	33.6	273 03.1	02.3
18	11 57.5	138 05.0	S15 37.6	143 30.7	S17 16.8	232 31.6	N16 33.7	288 05.8	N22 02.3
19	27 00.0	153 05.1	38.1	158 31.6	17.2	247 34.3	33.8	303 08.5	02.3
20	42 02.4	168 05.1	38.7	173 32.4	17.7	262 36.9	33.9	318 11.2	02.3
21	57 04.9	183 05.2	. . 39.2	188 33.3	. . 18.1	277 39.6	. . 33.9	333 13.8	. . 02.3
22	72 07.4	198 05.3	39.8	203 34.1	18.6	292 42.3	34.0	348 16.5	02.3
23	87 09.8	213 05.3	40.3	218 34.9	19.0	307 45.0	34.1	3 19.2	02.3
3 00	102 12.3	228 05.4	S15 40.8	233 35.8	S17 19.5	322 47.7	N16 34.2	18 21.8	N22 02.3
01	117 14.8	243 05.4	41.4	248 36.6	19.9	337 50.4	34.2	33 24.5	02.3
02	132 17.2	258 05.5	41.9	263 37.5	20.4	352 53.1	34.3	48 27.2	02.3
03	147 19.7	273 05.5	. . 42.5	278 38.3	. . 20.8	7 55.8	. . 34.4	63 29.8	. . 02.3
04	162 22.1	288 05.6	43.0	293 39.2	21.2	22 58.4	34.5	78 32.5	02.3
05	177 24.6	303 05.6	43.6	308 40.0	21.7	38 01.1	34.5	93 35.2	02.3
06	192 27.1	318 05.6	S15 44.1	323 40.8	S17 22.1	53 03.8	N16 34.6	108 37.9	N22 02.3
07	207 29.5	333 05.7	44.6	338 41.7	22.6	68 06.5	34.7	123 40.5	02.3
08	222 32.0	348 05.7	45.2	353 42.5	23.0	83 09.2	34.8	138 43.2	02.3
F 09	237 34.5	3 05.8	. . 45.7	8 43.4	. . 23.5	98 11.9	. . 34.9	153 45.9	. . 02.3
R 10	252 36.9	18 05.8	46.3	23 44.2	23.9	113 14.6	34.9	168 48.5	02.3
I 11	267 39.4	33 05.9	46.8	38 45.0	24.3	128 17.3	35.0	183 51.2	02.3
D 12	282 41.9	48 05.9	S15 47.4	53 45.9	S17 24.8	143 20.0	N16 35.1	198 53.9	N22 02.3
A 13	297 44.3	63 05.9	47.9	68 46.7	25.2	158 22.7	35.2	213 56.5	02.3
Y 14	312 46.8	78 06.0	48.4	83 47.6	25.7	173 25.3	35.2	228 59.2	02.3
15	327 49.3	93 06.0	. . 49.0	98 48.4	. . 26.1	188 28.0	. . 35.3	244 01.9	. . 02.3
16	342 51.7	108 06.0	49.5	113 49.2	26.5	203 30.7	35.4	259 04.5	02.2
17	357 54.2	123 06.1	50.1	128 50.1	27.0	218 33.4	35.5	274 07.2	02.2
18	12 56.6	138 06.1	S15 50.6	143 50.9	S17 27.4	233 36.1	N16 35.6	289 09.9	N22 02.2
19	27 59.1	153 06.1	51.2	158 51.7	27.9	248 38.8	35.6	304 12.5	02.2
20	43 01.6	168 06.2	51.7	173 52.6	28.3	263 41.5	35.7	319 15.2	02.2
21	58 04.0	183 06.2	. . 52.2	188 53.4	. . 28.7	278 44.2	. . 35.8	334 17.9	. . 02.2
22	73 06.5	198 06.2	52.8	203 54.3	29.2	293 46.9	35.9	349 20.5	02.2
23	88 09.0	213 06.2	53.3	218 55.1	29.6	308 49.6	36.0	4 23.2	02.2
Mer. Pass.	h m 17 12.3	v 0.1	d 0.5	v 0.8	d 0.4	v 2.7	d 0.1	v 2.7	d 0.0

STARS		
Name	SHA ° ′	Dec ° ′
Acamar	315 24.3	S40 17.8
Achernar	335 32.6	S57 13.7
Acrux	173 18.7	S63 06.6
Adhara	255 18.6	S28 58.5
Aldebaran	290 58.6	N16 30.9
Alioth	166 27.7	N55 56.4
Alkaid	153 05.4	N49 17.7
Al Na'ir	27 54.2	S46 57.1
Alnilam	275 54.4	S 1 12.0
Alphard	218 03.9	S 8 40.2
Alphecca	126 18.2	N26 42.2
Alpheratz	357 52.2	N29 06.5
Altair	62 16.6	N 8 52.5
Ankaa	353 23.7	S42 17.7
Antares	112 36.7	S26 26.3
Arcturus	146 03.4	N19 09.9
Atria	107 46.6	S69 01.8
Avior	234 20.9	S59 30.9
Bellatrix	278 40.5	N 6 21.1
Betelgeuse	271 09.9	N 7 24.5
Canopus	263 59.3	S52 41.8
Capella	280 46.2	N46 00.2
Deneb	49 37.6	N45 17.5
Denebola	182 41.9	N14 33.3
Diphda	349 04.1	S17 58.4
Dubhe	194 01.2	N61 43.9
Elnath	278 22.7	N28 36.6
Eltanin	90 50.5	N51 29.2
Enif	33 55.5	N 9 53.2
Fomalhaut	15 33.2	S29 36.7
Gacrux	172 10.3	S57 07.4
Gienah	176 00.8	S17 33.3
Hadar	149 00.1	S60 22.9
Hamal	328 10.0	N23 28.6
Kaus Aust.	83 55.1	S34 23.0
Kochab	137 20.0	N74 08.3
Markab	13 46.7	N15 13.2
Menkar	314 23.5	N 4 06.0
Menkent	148 17.6	S36 22.9
Miaplacidus	221 40.9	S69 43.5
Mirfak	308 51.9	N49 52.5
Nunki	76 08.9	S26 17.7
Peacock	53 32.6	S56 43.8
Pollux	243 37.4	N28 01.1
Procyon	245 08.0	N 5 13.1
Rasalhague	96 14.4	N12 33.4
Regulus	207 52.0	N11 57.2
Rigel	281 19.7	S 8 11.9
Rigil Kent.	140 03.6	S60 50.5
Sabik	102 22.3	S15 43.7
Schedar	349 50.1	N56 33.4
Shaula	96 33.5	S37 06.3
Sirius	258 40.7	S16 43.2
Spica	158 40.0	S11 10.5
Suhail	222 58.2	S43 26.5
Vega	80 45.0	N38 47.1
Zuben'ubi	137 14.8	S16 03.2

	SHA ° ′	Mer. Pass. h m
Venus	126 50.7	8 48
Mars	132 02.4	8 26
Jupiter	220 30.1	2 33
Saturn	276 04.6	22 47

2003 JANUARY 1, 2, 3 (WED., THURS., FRI.)

UT	SUN GHA	SUN Dec	MOON GHA	v	MOON Dec	d	HP
d h	° ′	° ′	° ′	′	° ′	′	′
1 00 (WEDNESDAY)	179 12.2	S23 03.1	205 28.6	5.3	S23 23.5	6.8	59.3
01	194 11.9	03.0	219 52.9	5.2	23 30.3	6.7	59.3
02	209 11.6	02.8	234 17.1	5.2	23 37.0	6.5	59.3
03	224 11.3	. . 02.6	248 41.3	5.2	23 43.5	6.3	59.3
04	239 11.0	02.4	263 05.5	5.1	23 49.8	6.2	59.2
05	254 10.7	02.2	277 29.6	5.1	23 56.0	6.1	59.2
06	269 10.4	S23 02.0	291 53.7	5.1	S24 02.1	5.8	59.2
07	284 10.1	01.8	306 17.8	5.0	24 07.9	5.7	59.2
08	299 09.8	01.6	320 41.8	4.9	24 13.6	5.6	59.2
09	314 09.6	. . 01.4	335 05.7	4.9	24 19.2	5.4	59.2
10	329 09.3	01.2	349 29.6	4.9	24 24.6	5.2	59.2
11	344 09.0	01.0	3 53.5	4.9	24 29.8	5.0	59.1
12	359 08.7	S23 00.8	18 17.4	4.8	S24 34.8	4.9	59.1
13	14 08.4	00.6	32 41.2	4.8	24 39.7	4.7	59.1
14	29 08.1	00.4	47 05.0	4.8	24 44.4	4.5	59.1
15	44 07.8	23 00.2	61 28.8	4.7	24 48.9	4.4	59.1
16	59 07.5	22 59.9	75 52.5	4.7	24 53.3	4.2	59.1
17	74 07.2	59.7	90 16.2	4.7	24 57.5	4.1	59.0
18	89 06.9	S22 59.5	104 39.9	4.7	S25 01.6	3.8	59.0
19	104 06.6	59.3	119 03.6	4.7	25 05.4	3.7	59.0
20	119 06.3	59.1	133 27.3	4.6	25 09.1	3.5	59.0
21	134 06.0	. . 58.9	147 50.9	4.6	25 12.6	3.4	59.0
22	149 05.7	58.7	162 14.5	4.7	25 16.0	3.2	58.9
23	164 05.4	58.5	176 38.2	4.6	25 19.2	3.0	58.9
2 00 (THURSDAY)	179 05.1	S22 58.3	191 01.8	4.6	S25 22.2	2.8	58.9
01	194 04.8	58.1	205 25.4	4.6	25 25.0	2.7	58.9
02	209 04.5	57.8	219 49.0	4.6	25 27.7	2.5	58.9
03	224 04.2	. . 57.6	234 12.6	4.6	25 30.2	2.3	58.8
04	239 03.9	57.4	248 36.2	4.6	25 32.5	2.1	58.8
05	254 03.6	57.2	262 59.8	4.6	25 34.6	2.0	58.8
06	269 03.3	S22 57.0	277 23.4	4.6	S25 36.6	1.8	58.8
07	284 03.1	56.8	291 47.0	4.7	25 38.4	1.6	58.8
08	299 02.8	56.5	306 10.7	4.6	25 40.0	1.5	58.7
09	314 02.5	. . 56.3	320 34.3	4.7	25 41.5	1.3	58.7
10	329 02.2	56.1	334 58.0	4.6	25 42.8	1.1	58.7
11	344 01.9	55.9	349 21.6	4.7	25 43.9	0.9	58.7
12	359 01.6	S22 55.7	3 45.3	4.7	S25 44.8	0.8	58.6
13	14 01.3	55.4	18 09.0	4.8	25 45.6	0.6	58.6
14	29 01.0	55.2	32 32.8	4.7	25 46.2	0.4	58.6
15	44 00.7	. . 55.0	46 56.5	4.8	25 46.6	0.3	58.6
16	59 00.4	54.8	61 20.3	4.8	25 46.9	0.0	58.6
17	74 00.1	54.5	75 44.1	4.8	25 46.9	0.1	58.5
18	88 59.8	S22 54.3	90 07.9	4.9	S25 46.8	0.2	58.5
19	103 59.5	54.1	104 31.8	4.9	25 46.6	0.4	58.5
20	118 59.2	53.8	118 55.7	5.0	25 46.2	0.6	58.5
21	133 59.0	. . 53.6	133 19.7	5.0	25 45.6	0.8	58.4
22	148 58.7	53.4	147 43.7	5.0	25 44.8	0.9	58.4
23	163 58.4	53.2	162 07.7	5.1	25 43.9	1.1	58.4
3 00 (FRIDAY)	178 58.1	S22 52.9	176 31.8	5.1	S25 42.8	1.3	58.4
01	193 57.8	52.7	190 55.9	5.2	25 41.5	1.4	58.3
02	208 57.5	52.5	205 20.1	5.2	25 40.1	1.6	58.3
03	223 57.2	. . 52.2	219 44.3	5.3	25 38.5	1.8	58.3
04	238 56.9	52.0	234 08.6	5.3	25 36.7	1.9	58.3
05	253 56.6	51.8	248 32.9	5.4	25 34.8	2.1	58.2
06	268 56.3	S22 51.5	262 57.3	5.4	S25 32.7	2.3	58.2
07	283 56.0	51.3	277 21.7	5.5	25 30.4	2.4	58.2
08	298 55.8	51.0	291 46.2	5.6	25 28.0	2.6	58.1
09	313 55.5	. . 50.8	306 10.8	5.6	25 25.4	2.7	58.1
10	328 55.2	50.6	320 35.4	5.7	25 22.7	2.9	58.1
11	343 54.9	50.3	335 00.1	5.8	25 19.8	3.0	58.1
12	358 54.6	S22 50.1	349 24.9	5.8	S25 16.8	3.2	58.0
13	13 54.3	49.8	3 49.7	5.9	25 13.6	3.4	58.0
14	28 54.0	49.6	18 14.6	6.0	25 10.2	3.5	58.0
15	43 53.7	. . 49.4	32 39.6	6.1	25 06.7	3.7	58.0
16	58 53.4	49.1	47 04.7	6.1	25 03.0	3.8	57.9
17	73 53.2	48.9	61 29.8	6.2	24 59.2	4.0	57.9
18	88 52.9	S22 48.6	75 55.0	6.3	S24 55.2	4.1	57.9
19	103 52.6	48.4	90 20.3	6.3	24 51.1	4.3	57.8
20	118 52.3	48.1	104 45.6	6.5	24 46.8	4.4	57.8
21	133 52.0	. . 47.9	119 11.1	6.5	24 42.4	4.6	57.8
22	148 51.7	47.6	133 36.6	6.6	24 37.8	4.7	57.8
23	163 51.4	47.4	148 02.2	6.7	S24 33.1	4.8	57.7
	SD 16.3	*d* 0.2	SD 16.1		16.0		15.8

Lat.	Twilight Naut.	Twilight Civil	Sunrise	Moonrise 1	Moonrise 2	Moonrise 3	Moonrise 4
°	h m	h m	h m	h m	h m	h m	h m
N 72	08 23	10 41	■	■	■	■	■
N 70	08 05	09 49	■	■	■	■	■
68	07 49	09 16	■	■	■	■	■
66	07 37	08 53	10 27	■	■	■	12 43
64	07 26	08 34	09 49	09 17	11 20	11 50	11 48
62	07 17	08 18	09 22	08 33	10 03	10 54	11 15
60	07 09	08 05	09 02	08 04	09 27	10 21	10 51
N 58	07 02	07 54	08 45	07 42	09 01	09 57	10 31
56	06 56	07 44	08 31	07 24	08 41	09 37	10 15
54	06 50	07 36	08 19	07 09	08 23	09 21	10 01
52	06 44	07 28	08 08	06 56	08 09	09 07	09 49
50	06 39	07 20	07 59	06 44	07 56	08 54	09 38
45	06 28	07 05	07 38	06 20	07 30	08 29	09 16
N 40	06 18	06 52	07 22	06 01	07 09	08 08	08 57
35	06 09	06 40	07 08	05 45	06 52	07 51	08 42
30	06 00	06 30	06 56	05 32	06 37	07 37	08 29
20	05 44	06 11	06 35	05 08	06 12	07 12	08 06
N 10	05 28	05 54	06 17	04 48	05 50	06 50	07 46
0	05 12	05 38	06 00	04 29	05 30	06 30	07 28
S 10	04 53	05 20	05 43	04 11	05 10	06 10	07 09
20	04 31	05 00	05 24	03 51	04 48	05 48	06 49
30	04 02	04 35	05 03	03 28	04 23	05 23	06 26
35	03 44	04 21	04 50	03 15	04 09	05 09	06 13
40	03 21	04 03	04 35	03 00	03 52	04 52	05 57
45	02 52	03 41	04 18	02 41	03 31	04 31	05 38
S 50	02 08	03 12	03 56	02 19	03 06	04 06	05 15
52	01 42	02 57	03 45	02 08	02 54	03 53	05 04
54	01 02	02 40	03 33	01 56	02 40	03 39	04 51
56	////	02 19	03 20	01 42	02 24	03 22	04 36
58	////	01 51	03 03	01 26	02 04	03 02	04 19
S 60	////	01 08	02 44	01 07	01 40	02 37	03 57

Lat.	Sunset	Twilight Civil	Twilight Naut.	Moonset 1	Moonset 2	Moonset 3	Moonset 4
°	h m	h m	h m	h m	h m	h m	h m
N 72	■	13 28	15 45	■	■	■	■
N 70	■	14 20	16 04	■	■	■	■
68	■	14 52	16 19	■	■	■	■
66	13 41	15 16	16 31	■	■	■	14 49
64	14 19	15 34	16 42	12 04	12 08	13 42	15 43
62	14 46	15 50	16 51	12 47	13 24	14 38	16 15
60	15 06	16 03	16 59	13 17	14 01	15 11	16 39
N 58	15 23	16 14	17 06	13 39	14 27	15 35	16 58
56	15 37	16 24	17 12	13 58	14 47	15 54	17 14
54	15 49	16 32	17 18	14 13	15 04	16 11	17 27
52	16 00	16 40	17 24	14 27	15 19	16 25	17 39
50	16 09	16 48	17 29	14 38	15 32	16 37	17 50
45	16 30	17 03	17 40	15 03	15 58	17 02	18 11
N 40	16 46	17 16	17 50	15 23	16 19	17 22	18 29
35	17 00	17 28	17 59	15 39	16 36	17 39	18 44
30	17 12	17 38	18 08	15 53	16 51	17 53	18 56
20	17 32	17 56	18 24	16 18	17 17	18 17	19 18
N 10	17 51	18 13	18 40	16 39	17 38	18 38	19 36
0	18 08	18 30	18 56	16 58	17 59	18 58	19 54
S 10	18 25	18 48	19 15	17 18	18 19	19 17	20 11
20	18 43	19 08	19 37	17 39	18 41	19 38	20 29
30	19 05	19 32	20 05	18 03	19 06	20 02	20 50
35	19 17	19 47	20 23	18 18	19 21	20 16	21 03
40	19 32	20 05	20 46	18 34	19 38	20 32	21 17
45	19 50	20 27	21 15	18 54	19 59	20 52	21 33
S 50	20 11	20 55	21 59	19 19	20 25	21 16	21 54
52	20 22	21 10	22 24	19 31	20 37	21 27	22 03
54	20 34	21 27	23 03	19 45	20 51	21 40	22 14
56	20 48	21 48	////	20 01	21 08	21 55	22 26
58	21 04	22 15	////	20 20	21 28	22 13	22 41
S 60	21 23	22 57	////	20 44	21 53	22 35	22 57

Day	SUN Eqn. of Time 00^h	SUN Eqn. of Time 12^h	SUN Mer. Pass.	MOON Mer. Pass. Upper	MOON Mer. Pass. Lower	MOON Age	MOON Phase
d	m s	m s	h m	h m	h m	d	%
1	03 10	03 25	12 03	10 44	23 14	28	2
2	03 39	03 53	12 04	11 44	24 14	29	0 ●
3	04 07	04 21	12 04	12 44	00 14	01	1

UT	ARIES	VENUS −4.5		MARS +1.5		JUPITER −2.5		SATURN −0.4	
	GHA	GHA	Dec	GHA	Dec	GHA	Dec	GHA	Dec
d h	° ′	° ′	° ′	° ′	° ′	° ′	° ′	° ′	° ′
4 00	103 11.4	228 06.3	S15 53.9	233 55.9	S17 30.1	323 52.3	N16 36.0	19 25.9	N22 02.2
01	118 13.9	243 06.3	54.4	248 56.8	30.5	338 55.0	36.1	34 28.6	02.2
02	133 16.4	258 06.3	55.0	263 57.6	30.9	353 57.7	36.2	49 31.2	02.2
03	148 18.8	273 06.3	. . 55.5	278 58.4	. . 31.4	9 00.4	. . 36.3	64 33.9	. . 02.2
04	163 21.3	288 06.4	56.0	293 59.3	31.8	24 03.1	36.3	79 36.6	02.2
05	178 23.7	303 06.4	56.6	309 00.1	32.3	39 05.8	36.4	94 39.2	02.2
06	193 26.2	318 06.4	S15 57.1	324 01.0	S17 32.7	54 08.4	N16 36.5	109 41.9	N22 02.2
07	208 28.7	333 06.4	57.7	339 01.8	33.1	69 11.1	36.6	124 44.6	02.2
S 08	223 31.1	348 06.4	58.2	354 02.6	33.6	84 13.8	36.7	139 47.2	02.2
A 09	238 33.6	3 06.5	. . 58.8	9 03.5	. . 34.0	99 16.5	. . 36.7	154 49.9	. . 02.2
T 10	253 36.1	18 06.5	59.3	24 04.3	34.4	114 19.2	36.8	169 52.6	02.2
U 11	268 38.5	33 06.5	15 59.8	39 05.1	34.9	129 21.9	36.9	184 55.2	02.2
R 12	283 41.0	48 06.5	S16 00.4	54 06.0	S17 35.3	144 24.6	N16 37.0	199 57.9	N22 02.2
D 13	298 43.5	63 06.5	00.9	69 06.8	35.8	159 27.3	37.1	215 00.6	02.2
A 14	313 45.9	78 06.5	01.5	84 07.7	36.2	174 30.0	37.1	230 03.2	02.2
Y 15	328 48.4	93 06.5	. . 02.0	99 08.5	. . 36.6	189 32.7	. . 37.2	245 05.9	. . 02.2
16	343 50.9	108 06.5	02.5	114 09.3	37.1	204 35.4	37.3	260 08.6	02.2
17	358 53.3	123 06.5	03.1	129 10.2	37.5	219 38.1	37.4	275 11.2	02.2
18	13 55.8	138 06.5	S16 03.6	144 11.0	S17 37.9	234 40.8	N16 37.5	290 13.9	N22 02.2
19	28 58.2	153 06.6	04.2	159 11.8	38.4	249 43.5	37.5	305 16.6	02.2
20	44 00.7	168 06.6	04.7	174 12.7	38.8	264 46.2	37.6	320 19.2	02.2
21	59 03.2	183 06.6	. . 05.3	189 13.5	. . 39.2	279 48.9	. . 37.7	335 21.9	. . 02.2
22	74 05.6	198 06.6	05.8	204 14.3	39.7	294 51.6	37.8	350 24.6	02.2
23	89 08.1	213 06.6	06.3	219 15.2	40.1	309 54.3	37.9	5 27.2	02.2
5 00	104 10.6	228 06.6	S16 06.9	234 16.0	S17 40.5	324 57.0	N16 37.9	20 29.9	N22 02.2
01	119 13.0	243 06.6	07.4	249 16.8	41.0	339 59.7	38.0	35 32.6	02.2
02	134 15.5	258 06.6	08.0	264 17.7	41.4	355 02.4	38.1	50 35.2	02.2
03	149 18.0	273 06.6	. . 08.5	279 18.5	. . 41.8	10 05.1	. . 38.2	65 37.9	. . 02.2
04	164 20.4	288 06.6	09.0	294 19.3	42.3	25 07.8	38.3	80 40.5	02.2
05	179 22.9	303 06.6	09.6	309 20.2	42.7	40 10.5	38.3	95 43.2	02.2
06	194 25.4	318 06.5	S16 10.1	324 21.0	S17 43.1	55 13.2	N16 38.4	110 45.9	N22 02.2
07	209 27.8	333 06.5	10.7	339 21.8	43.6	70 15.9	38.5	125 48.5	02.2
08	224 30.3	348 06.5	11.2	354 22.7	44.0	85 18.6	38.6	140 51.2	02.2
S 09	239 32.7	3 06.5	. . 11.7	9 23.5	. . 44.4	100 21.3	. . 38.7	155 53.9	. . 02.2
U 10	254 35.2	18 06.5	12.3	24 24.3	44.9	115 24.0	38.8	170 56.5	02.2
N 11	269 37.7	33 06.5	12.8	39 25.2	45.3	130 26.7	38.8	185 59.2	02.2
D 12	284 40.1	48 06.5	S16 13.4	54 26.0	S17 45.7	145 29.4	N16 38.9	201 01.9	N22 02.2
A 13	299 42.6	63 06.5	13.9	69 26.8	46.1	160 32.1	39.0	216 04.5	02.2
Y 14	314 45.1	78 06.5	14.4	84 27.7	46.6	175 34.8	39.1	231 07.2	02.2
15	329 47.5	93 06.4	. . 15.0	99 28.5	. . 47.0	190 37.6	. . 39.2	246 09.9	. . 02.2
16	344 50.0	108 06.4	15.5	114 29.3	47.4	205 40.3	39.2	261 12.5	02.2
17	359 52.5	123 06.4	16.1	129 30.2	47.9	220 43.0	39.3	276 15.2	02.2
18	14 54.9	138 06.4	S16 16.6	144 31.0	S17 48.3	235 45.7	N16 39.4	291 17.8	N22 02.2
19	29 57.4	153 06.4	17.1	159 31.8	48.7	250 48.4	39.5	306 20.5	02.2
20	44 59.9	168 06.4	17.7	174 32.7	49.2	265 51.1	39.6	321 23.2	02.2
21	60 02.3	183 06.3	. . 18.2	189 33.5	. . 49.6	280 53.8	. . 39.6	336 25.8	. . 02.1
22	75 04.8	198 06.3	18.7	204 34.3	50.0	295 56.5	39.7	351 28.5	02.1
23	90 07.2	213 06.3	19.3	219 35.2	50.4	310 59.2	39.8	6 31.2	02.1
6 00	105 09.7	228 06.3	S16 19.8	234 36.0	S17 50.9	326 01.9	N16 39.9	21 33.8	N22 02.1
01	120 12.2	243 06.2	20.4	249 36.8	51.3	341 04.6	40.0	36 36.5	02.1
02	135 14.6	258 06.2	20.9	264 37.7	51.7	356 07.3	40.1	51 39.2	02.1
03	150 17.1	273 06.2	. . 21.4	279 38.5	. . 52.2	11 10.0	. . 40.1	66 41.8	. . 02.1
04	165 19.6	288 06.2	22.0	294 39.3	52.6	26 12.7	40.2	81 44.5	02.1
05	180 22.0	303 06.1	22.5	309 40.2	53.0	41 15.4	40.3	96 47.1	02.1
06	195 24.5	318 06.1	S16 23.0	324 41.0	S17 53.4	56 18.1	N16 40.4	111 49.8	N22 02.1
07	210 27.0	333 06.1	23.6	339 41.8	53.9	71 20.9	40.5	126 52.5	02.1
08	225 29.4	348 06.0	24.1	354 42.6	54.3	86 23.6	40.6	141 55.1	02.1
M 09	240 31.9	3 06.0	. . 24.7	9 43.5	. . 54.7	101 26.3	. . 40.6	156 57.8	. . 02.1
O 10	255 34.3	18 06.0	25.2	24 44.3	55.1	116 29.0	40.7	172 00.5	02.1
N 11	270 36.8	33 05.9	25.7	39 45.1	55.6	131 31.7	40.8	187 03.1	02.1
D 12	285 39.3	48 05.9	S16 26.3	54 46.0	S17 56.0	146 34.4	N16 40.9	202 05.8	N22 02.1
A 13	300 41.7	63 05.9	26.8	69 46.8	56.4	161 37.1	41.0	217 08.4	02.1
Y 14	315 44.2	78 05.8	27.3	84 47.6	56.8	176 39.8	41.1	232 11.1	02.1
15	330 46.7	93 05.8	. . 27.9	99 48.5	. . 57.3	191 42.5	. . 41.1	247 13.8	. . 02.1
16	345 49.1	108 05.7	28.4	114 49.3	57.7	206 45.2	41.2	262 16.4	02.1
17	0 51.6	123 05.7	28.9	129 50.1	58.1	221 48.0	41.3	277 19.1	02.1
18	15 54.1	138 05.6	S16 29.5	144 50.9	S17 58.5	236 50.7	N16 41.4	292 21.8	N22 02.1
19	30 56.5	153 05.6	30.0	159 51.8	59.0	251 53.4	41.5	307 24.4	02.1
20	45 59.0	168 05.6	30.6	174 52.6	59.4	266 56.1	41.5	322 27.1	02.1
21	61 01.5	183 05.5	. . 31.1	189 53.4	17 59.8	281 58.8	. . 41.6	337 29.7	. . 02.1
22	76 03.9	198 05.5	31.6	204 54.3	18 00.2	297 01.5	41.7	352 32.4	02.1
23	91 06.4	213 05.4	32.2	219 55.1	S18 00.7	312 04.2	41.8	7 35.1	02.1
Mer. Pass.	h m 17 00.5	*v* 0.0	*d* 0.5	*v* 0.8	*d* 0.4	*v* 2.7	*d* 0.1	*v* 2.7	*d* 0.0

STARS

Name	SHA	Dec
	° ′	° ′
Acamar	315 24.3	S40 17.8
Achernar	335 32.6	S57 13.7
Acrux	173 18.7	S63 06.6
Adhara	255 18.6	S28 58.5
Aldebaran	290 58.6	N16 30.9
Alioth	166 27.7	N55 56.4
Alkaid	153 05.4	N49 17.7
Al Na'ir	27 54.2	S46 57.1
Alnilam	275 54.4	S 1 12.0
Alphard	218 03.9	S 8 40.2
Alphecca	126 18.2	N26 42.1
Alpheratz	357 52.2	N29 06.5
Altair	62 16.6	N 8 52.5
Ankaa	353 23.7	S42 17.7
Antares	112 36.7	S26 26.3
Arcturus	146 03.3	N19 09.9
Atria	107 46.5	S69 01.8
Avior	234 20.9	S59 31.0
Bellatrix	278 40.5	N 6 21.1
Betelgeuse	271 09.9	N 7 24.5
Canopus	263 59.3	S52 41.8
Capella	280 46.2	N46 00.2
Deneb	49 37.6	N45 17.5
Denebola	182 41.9	N14 33.3
Diphda	349 04.1	S17 58.4
Dubhe	194 01.2	N61 43.9
Elnath	278 22.7	N28 36.7
Eltanin	90 50.5	N51 29.2
Enif	33 55.5	N 9 53.2
Fomalhaut	15 33.2	S29 36.7
Gacrux	172 10.3	S57 07.4
Gienah	176 00.8	S17 33.4
Hadar	149 00.0	S60 22.9
Hamal	328 10.0	N23 28.6
Kaus Aust.	83 55.1	S34 23.0
Kochab	137 20.0	N74 08.3
Markab	13 46.7	N15 13.2
Menkar	314 23.5	N 4 06.0
Menkent	148 17.5	S36 22.9
Miaplacidus	221 40.9	S69 43.5
Mirfak	308 51.9	N49 52.5
Nunki	76 08.9	S26 17.7
Peacock	53 32.6	S56 43.7
Pollux	243 37.4	N28 01.1
Procyon	245 08.0	N 5 13.1
Rasalhague	96 14.4	N12 33.4
Regulus	207 52.0	N11 57.2
Rigel	281 19.7	S 8 11.9
Rigil Kent.	140 03.6	S60 50.5
Sabik	102 22.3	S15 43.7
Schedar	349 50.2	N56 33.4
Shaula	96 33.5	S37 06.3
Sirius	258 40.7	S16 43.2
Spica	158 40.0	S11 10.5
Suhail	222 58.2	S43 26.5
Vega	80 45.0	N38 47.1
Zuben'ubi	137 14.7	S16 03.2

	SHA	Mer. Pass.
	° ′	h m
Venus	123 56.0	8 48
Mars	130 05.4	8 22
Jupiter	220 46.4	2 20
Saturn	276 19.3	22 34

	UT	SUN GHA	SUN Dec	MOON GHA	v	MOON Dec	d	HP
	d h	° ′	° ′	° ′	′	° ′	′	′
	4 00	178 51.1	S22 47.1	162 27.9	6.8	S24 28.3	5.0	57.7
	01	193 50.8	46.9	176 53.7	6.8	24 23.3	5.2	57.7
	02	208 50.6	46.6	191 19.5	7.0	24 18.1	5.3	57.6
	03	223 50.3	. . 46.4	205 45.5	7.0	24 12.8	5.4	57.6
	04	238 50.0	46.1	220 11.5	7.2	24 07.4	5.6	57.6
	05	253 49.7	45.9	234 37.7	7.2	24 01.8	5.6	57.5
	06	268 49.4	S22 45.6	249 03.9	7.3	S23 56.2	5.9	57.5
	07	283 49.1	45.4	263 30.2	7.5	23 50.3	5.9	57.5
S	08	298 48.8	45.1	277 56.7	7.5	23 44.4	6.1	57.5
A	09	313 48.6	. . 44.8	292 23.2	7.6	23 38.3	6.3	57.4
T	10	328 48.3	44.6	306 49.8	7.7	23 32.0	6.3	57.4
U	11	343 48.0	44.3	321 16.5	7.8	23 25.7	6.5	57.4
R	12	358 47.7	S22 44.1	335 43.3	7.9	S23 19.2	6.6	57.3
D	13	13 47.4	43.8	350 10.2	8.0	23 12.6	6.8	57.3
A	14	28 47.1	43.5	4 37.2	8.1	23 05.8	6.8	57.3
Y	15	43 46.8	. . 43.3	19 04.3	8.2	22 59.0	7.0	57.3
	16	58 46.6	43.0	33 31.5	8.3	22 52.0	7.1	57.2
	17	73 46.3	42.7	47 58.8	8.3	22 44.9	7.3	57.2
	18	88 46.0	S22 42.5	62 26.1	8.5	S22 37.6	7.3	57.2
	19	103 45.7	42.2	76 53.6	8.6	22 30.3	7.5	57.1
	20	118 45.4	42.0	91 21.2	8.7	22 22.8	7.6	57.1
	21	133 45.1	. . 41.7	105 48.9	8.8	22 15.2	7.7	57.1
	22	148 44.9	41.4	120 16.7	9.0	22 07.5	7.8	57.0
	23	163 44.6	41.1	134 44.7	9.0	21 59.7	7.9	57.0
	5 00	178 44.3	S22 40.9	149 12.7	9.1	S21 51.8	8.0	57.0
	01	193 44.0	40.6	163 40.8	9.2	21 43.8	8.2	56.9
	02	208 43.7	40.3	178 09.0	9.3	21 35.6	8.2	56.9
	03	223 43.4	. . 40.1	192 37.3	9.4	21 27.4	8.4	56.9
	04	238 43.2	39.8	207 05.7	9.6	21 19.0	8.4	56.9
	05	253 42.9	39.5	221 34.3	9.6	21 10.6	8.6	56.8
	06	268 42.6	S22 39.2	236 02.9	9.7	S21 02.0	8.7	56.8
	07	283 42.3	39.0	250 31.6	9.9	20 53.3	8.7	56.8
	08	298 42.0	38.7	265 00.5	9.9	20 44.6	8.9	56.7
S	09	313 41.7	. . 38.4	279 29.4	10.0	20 35.7	8.9	56.7
U	10	328 41.5	38.1	293 58.4	10.2	20 26.8	9.1	56.7
N	11	343 41.2	37.9	308 27.6	10.2	20 17.7	9.1	56.6
D	12	358 40.9	S22 37.6	322 56.8	10.4	S20 08.6	9.3	56.6
A	13	13 40.6	37.3	337 26.2	10.5	19 59.3	9.3	56.6
Y	14	28 40.3	37.0	351 55.7	10.5	19 50.0	9.4	56.6
	15	43 40.1	. . 36.7	6 25.2	10.7	19 40.6	9.6	56.5
	16	58 39.8	36.5	20 54.9	10.7	19 31.0	9.6	56.5
	17	73 39.5	36.2	35 24.6	10.9	19 21.4	9.6	56.5
	18	88 39.2	S22 35.9	49 54.5	11.0	S19 11.8	9.8	56.4
	19	103 38.9	35.6	64 24.5	11.0	19 02.0	9.9	56.4
	20	118 38.7	35.3	78 54.5	11.2	18 52.1	9.9	56.4
	21	133 38.4	. . 35.0	93 24.7	11.3	18 42.2	10.0	56.3
	22	148 38.1	34.8	107 55.0	11.4	18 32.2	10.1	56.3
	23	163 37.8	34.5	122 25.4	11.4	18 22.1	10.2	56.3
	6 00	178 37.6	S22 34.2	136 55.8	11.6	S18 11.9	10.3	56.3
	01	193 37.3	33.9	151 26.4	11.6	18 01.6	10.3	56.2
	02	208 37.0	33.6	165 57.0	11.8	17 51.3	10.4	56.2
	03	223 36.7	. . 33.3	180 27.8	11.9	17 40.9	10.5	56.2
	04	238 36.4	33.0	194 58.7	11.9	17 30.4	10.5	56.1
	05	253 36.2	32.7	209 29.6	12.1	17 19.9	10.6	56.1
	06	268 35.9	S22 32.4	224 00.7	12.1	S17 09.3	10.7	56.1
	07	283 35.6	32.1	238 31.8	12.2	16 58.6	10.8	56.1
	08	298 35.3	31.8	253 03.0	12.4	16 47.8	10.8	56.0
M	09	313 35.1	. . 31.6	267 34.4	12.4	16 37.0	10.9	56.0
O	10	328 34.8	31.3	282 05.8	12.5	16 26.1	10.9	56.0
N	11	343 34.5	31.0	296 37.3	12.6	16 15.2	11.0	55.9
D	12	358 34.2	S22 30.7	311 08.9	12.7	S16 04.2	11.1	55.9
A	13	13 33.9	30.4	325 40.6	12.8	15 53.1	11.2	55.9
Y	14	28 33.7	30.1	340 12.4	12.9	15 41.9	11.2	55.9
	15	43 33.4	. . 29.8	354 44.3	13.0	15 30.7	11.2	55.8
	16	58 33.1	29.5	9 16.3	13.0	15 19.5	11.3	55.8
	17	73 32.8	29.2	23 48.3	13.1	15 08.2	11.4	55.8
	18	88 32.6	S22 28.9	38 20.4	13.3	S14 56.8	11.4	55.7
	19	103 32.3	28.6	52 52.7	13.3	14 45.4	11.5	55.7
	20	118 32.0	28.3	67 25.0	13.4	14 33.9	11.5	55.7
	21	133 31.7	. . 28.0	81 57.4	13.5	14 22.4	11.6	55.7
	22	148 31.5	27.6	96 29.9	13.5	14 10.8	11.6	55.6
	23	163 31.2	27.3	111 02.4	13.6	S13 59.2	11.7	55.6
		SD 16.3	*d* 0.3	SD		15.6	15.4	15.2

Lat.	Twilight Naut.	Twilight Civil	Sunrise	Moonrise 4	Moonrise 5	Moonrise 6	Moonrise 7
°	h m	h m	h m	h m	h m	h m	h m
N 72	08 20	10 31	■	■	■	13 24	12 33
N 70	08 02	09 43	■	■	14 12	12 46	12 15
68	07 47	09 13	11 32	■	12 51	12 19	12 00
66	07 35	08 50	10 21	12 43	12 12	11 58	11 48
64	07 25	08 32	09 45	11 48	11 45	11 42	11 38
62	07 16	08 17	09 20	11 15	11 24	11 28	11 29
60	07 08	08 04	09 00	10 51	11 07	11 16	11 22
N 58	07 01	07 53	08 44	10 31	10 52	11 06	11 15
56	06 55	07 44	08 30	10 15	10 40	10 57	11 09
54	06 49	07 35	08 18	10 01	10 29	10 49	11 04
52	06 44	07 27	08 07	09 49	10 19	10 42	10 59
50	06 39	07 20	07 58	09 38	10 11	10 35	10 55
45	06 28	07 05	07 38	09 16	09 52	10 21	10 45
N 40	06 18	06 52	07 22	08 57	09 37	10 10	10 37
35	06 09	06 41	07 09	08 42	09 24	10 00	10 30
30	06 01	06 30	06 57	08 29	09 13	09 51	10 24
20	05 45	06 12	06 36	08 06	08 54	09 36	10 14
N 10	05 29	05 56	06 18	07 46	08 37	09 23	10 05
0	05 13	05 39	06 02	07 28	08 21	09 11	09 56
S 10	04 55	05 22	05 45	07 09	08 05	08 58	09 47
20	04 33	05 02	05 26	06 49	07 48	08 45	09 38
30	04 05	04 38	05 05	06 26	07 29	08 29	09 27
35	03 47	04 23	04 53	06 13	07 17	08 20	09 21
40	03 25	04 06	04 38	05 57	07 04	08 10	09 14
45	02 55	03 44	04 21	05 38	06 48	07 58	09 06
S 50	02 13	03 16	03 59	05 15	06 29	07 43	08 56
52	01 48	03 01	03 49	05 04	06 20	07 37	08 51
54	01 11	02 45	03 37	04 51	06 10	07 29	08 46
56	////	02 24	03 24	04 36	05 58	07 20	08 40
58	////	01 57	03 08	04 19	05 44	07 11	08 34
S 60	////	01 18	02 49	03 57	05 28	06 59	08 26

Lat.	Sunset	Twilight Civil	Twilight Naut.	Moonset 4	Moonset 5	Moonset 6	Moonset 7
°	h m	h m	h m	h m	h m	h m	h m
N 72	■	13 40	15 51	■	■	17 41	20 04
N 70	■	14 28	16 09	■	15 11	18 17	20 20
68	12 39	14 58	16 24	■	16 31	18 43	20 33
66	13 50	15 21	16 36	14 49	17 09	19 02	20 44
64	14 26	15 39	16 46	15 43	17 35	19 18	20 53
62	14 51	15 54	16 55	16 15	17 55	19 30	21 00
60	15 11	16 07	17 03	16 39	18 11	19 41	21 07
N 58	15 27	16 18	17 10	16 58	18 25	19 51	21 12
56	15 41	16 27	17 16	17 14	18 37	19 59	21 17
54	15 53	16 36	17 22	17 27	18 47	20 06	21 22
52	16 03	16 44	17 27	17 39	18 56	20 12	21 26
50	16 13	16 51	17 32	17 50	19 05	20 18	21 29
45	16 33	17 06	17 43	18 11	19 22	20 31	21 37
N 40	16 49	17 19	17 53	18 29	19 36	20 41	21 44
35	17 02	17 30	18 02	18 44	19 48	20 50	21 49
30	17 14	17 40	18 10	18 56	19 58	20 58	21 54
20	17 34	17 58	18 26	19 18	20 16	21 11	22 02
N 10	17 52	18 15	18 41	19 36	20 31	21 22	22 10
0	18 09	18 31	18 57	19 54	20 45	21 33	22 17
S 10	18 26	18 49	19 16	20 11	21 00	21 43	22 23
20	18 44	19 09	19 37	20 29	21 15	21 55	22 31
30	19 05	19 33	20 05	20 50	21 32	22 07	22 39
35	19 18	19 47	20 23	21 03	21 42	22 15	22 43
40	19 32	20 05	20 45	21 17	21 53	22 23	22 49
45	19 49	20 26	21 14	21 33	22 06	22 32	22 55
S 50	20 11	20 54	21 57	21 54	22 22	22 44	23 02
52	20 21	21 08	22 21	22 03	22 30	22 49	23 05
54	20 33	21 25	22 57	22 14	22 38	22 55	23 09
56	20 46	21 45	////	22 26	22 47	23 02	23 13
58	21 02	22 12	////	22 41	22 58	23 09	23 17
S 60	21 20	22 50	////	22 57	23 10	23 17	23 22

Day	SUN Eqn. of Time 00^h	SUN Eqn. of Time 12^h	SUN Mer. Pass.	MOON Mer. Pass. Upper	MOON Mer. Pass. Lower	Age	Phase
d	m s	m s	h m	h m	h m	d	%
4	04 35	04 49	12 05	13 41	01 13	02	3
5	05 02	05 16	12 05	14 33	02 08	03	8
6	05 29	05 43	12 06	15 22	02 58	04	15

UT		ARIES	VENUS −4.5		MARS +1.5		JUPITER −2.5		SATURN −0.4	
		GHA	GHA	Dec	GHA	Dec	GHA	Dec	GHA	Dec
d	h	° ′	° ′	° ′	° ′	° ′	° ′	° ′	° ′	° ′
7	00	106 08.8	228 05.4	S16 32.7	234 55.9	S18 01.1	327 06.9	N16 41.9	22 37.7	N22 02.1
	01	121 11.3	243 05.3	33.2	249 56.7	01.5	342 09.7	42.0	37 40.4	02.1
	02	136 13.8	258 05.3	33.8	264 57.6	01.9	357 12.4	42.1	52 43.0	02.1
	03	151 16.2	273 05.2	. . 34.3	279 58.4	. . 02.3	12 15.1	. . 42.1	67 45.7	. . 02.1
	04	166 18.7	288 05.2	34.8	294 59.2	02.8	27 17.8	42.2	82 48.4	02.1
	05	181 21.2	303 05.1	35.4	310 00.1	03.2	42 20.5	42.3	97 51.0	02.1
	06	196 23.6	318 05.1	S16 35.9	325 00.9	S18 03.6	57 23.2	N16 42.4	112 53.7	N22 02.1
	07	211 26.1	333 05.0	36.4	340 01.7	04.0	72 25.9	42.5	127 56.3	02.1
T	08	226 28.6	348 04.9	37.0	355 02.5	04.5	87 28.6	42.6	142 59.0	02.1
U	09	241 31.0	3 04.9	. . 37.5	10 03.4	. . 04.9	102 31.4	. . 42.6	158 01.7	. . 02.1
E	10	256 33.5	18 04.8	38.0	25 04.2	05.3	117 34.1	42.7	173 04.3	02.1
	11	271 36.0	33 04.8	38.6	40 05.0	05.7	132 36.8	42.8	188 07.0	02.1
S	12	286 38.4	48 04.7	S16 39.1	55 05.9	S18 06.1	147 39.5	N16 42.9	203 09.6	N22 02.1
D	13	301 40.9	63 04.6	39.6	70 06.7	06.6	162 42.2	43.0	218 12.3	02.1
A	14	316 43.3	78 04.6	40.2	85 07.5	07.0	177 44.9	43.1	233 15.0	02.1
Y	15	331 45.8	93 04.5	. . 40.7	100 08.3	. . 07.4	192 47.7	. . 43.1	248 17.6	. . 02.1
	16	346 48.3	108 04.5	41.2	115 09.2	07.8	207 50.4	43.2	263 20.3	02.1
	17	1 50.7	123 04.4	41.8	130 10.0	08.2	222 53.1	43.3	278 22.9	02.1
	18	16 53.2	138 04.3	S16 42.3	145 10.8	S18 08.7	237 55.8	N16 43.4	293 25.6	N22 02.1
	19	31 55.7	153 04.3	42.8	160 11.6	09.1	252 58.5	43.5	308 28.3	02.1
	20	46 58.1	168 04.2	43.3	175 12.5	09.5	268 01.2	43.6	323 30.9	02.1
	21	62 00.6	183 04.1	. . 43.9	190 13.3	. . 09.9	283 04.0	. . 43.7	338 33.6	. . 02.1
	22	77 03.1	198 04.0	44.4	205 14.1	10.3	298 06.7	43.7	353 36.2	02.1
	23	92 05.5	213 04.0	44.9	220 14.9	10.7	313 09.4	43.8	8 38.9	02.1
8	00	107 08.0	228 03.9	S16 45.5	235 15.8	S18 11.2	328 12.1	N16 43.9	23 41.5	N22 02.1
	01	122 10.4	243 03.8	46.0	250 16.6	11.6	343 14.8	44.0	38 44.2	02.1
	02	137 12.9	258 03.8	46.5	265 17.4	12.0	358 17.5	44.1	53 46.9	02.1
	03	152 15.4	273 03.7	. . 47.1	280 18.2	. . 12.4	13 20.3	. . 44.2	68 49.5	. . 02.1
	04	167 17.8	288 03.6	47.6	295 19.1	12.8	28 23.0	44.3	83 52.2	02.1
	05	182 20.3	303 03.5	48.1	310 19.9	13.2	43 25.7	44.3	98 54.8	02.1
	06	197 22.8	318 03.5	S16 48.6	325 20.7	S18 13.7	58 28.4	N16 44.4	113 57.5	N22 02.1
W	07	212 25.2	333 03.4	49.2	340 21.5	14.1	73 31.1	44.5	129 00.1	02.1
E	08	227 27.7	348 03.3	49.7	355 22.4	14.5	88 33.9	44.6	144 02.8	02.1
D	09	242 30.2	3 03.2	. . 50.2	10 23.2	. . 14.9	103 36.6	. . 44.7	159 05.5	. . 02.1
N	10	257 32.6	18 03.1	50.8	25 24.0	15.3	118 39.3	44.8	174 08.1	02.0
E	11	272 35.1	33 03.0	51.3	40 24.8	15.7	133 42.0	44.9	189 10.8	02.0
S	12	287 37.6	48 03.0	S16 51.8	55 25.7	S18 16.1	148 44.8	N16 44.9	204 13.4	N22 02.0
D	13	302 40.0	63 02.9	52.3	70 26.5	16.6	163 47.5	45.0	219 16.1	02.0
A	14	317 42.5	78 02.8	52.9	85 27.3	17.0	178 50.2	45.1	234 18.7	02.0
Y	15	332 44.9	93 02.7	. . 53.4	100 28.1	. . 17.4	193 52.9	. . 45.2	249 21.4	. . 02.0
	16	347 47.4	108 02.6	53.9	115 29.0	17.8	208 55.6	45.3	264 24.1	02.0
	17	2 49.9	123 02.5	54.4	130 29.8	18.2	223 58.4	45.4	279 26.7	02.0
	18	17 52.3	138 02.4	S16 55.0	145 30.6	S18 18.6	239 01.1	N16 45.5	294 29.4	N22 02.0
	19	32 54.8	153 02.3	55.5	160 31.4	19.0	254 03.8	45.5	309 32.0	02.0
	20	47 57.3	168 02.3	56.0	175 32.3	19.5	269 06.5	45.6	324 34.7	02.0
	21	62 59.7	183 02.2	. . 56.6	190 33.1	. . 19.9	284 09.3	. . 45.7	339 37.3	. . 02.0
	22	78 02.2	198 02.1	57.1	205 33.9	20.3	299 12.0	45.8	354 40.0	02.0
	23	93 04.7	213 02.0	57.6	220 34.7	20.7	314 14.7	45.9	9 42.7	02.0
9	00	108 07.1	228 01.9	S16 58.1	235 35.5	S18 21.1	329 17.4	N16 46.0	24 45.3	N22 02.0
	01	123 09.6	243 01.8	58.7	250 36.4	21.5	344 20.1	46.1	39 48.0	02.0
	02	138 12.1	258 01.7	59.2	265 37.2	21.9	359 22.9	46.1	54 50.6	02.0
	03	153 14.5	273 01.6	16 59.7	280 38.0	. . 22.3	14 25.6	. . 46.2	69 53.3	. . 02.0
	04	168 17.0	288 01.5	17 00.2	295 38.8	22.7	29 28.3	46.3	84 55.9	02.0
	05	183 19.4	303 01.4	00.7	310 39.7	23.2	44 31.0	46.4	99 58.6	02.0
	06	198 21.9	318 01.3	S17 01.3	325 40.5	S18 23.6	59 33.8	N16 46.5	115 01.2	N22 02.0
	07	213 24.4	333 01.2	01.8	340 41.3	24.0	74 36.5	46.6	130 03.9	02.0
T	08	228 26.8	348 01.1	02.3	355 42.1	24.4	89 39.2	46.7	145 06.5	02.0
H	09	243 29.3	3 01.0	. . 02.8	10 42.9	. . 24.8	104 42.0	. . 46.8	160 09.2	. . 02.0
U	10	258 31.8	18 00.9	03.4	25 43.8	25.2	119 44.7	46.8	175 11.9	02.0
R	11	273 34.2	33 00.8	03.9	40 44.6	25.6	134 47.4	46.9	190 14.5	02.0
S	12	288 36.7	48 00.7	S17 04.4	55 45.4	S18 26.0	149 50.1	N16 47.0	205 17.2	N22 02.0
D	13	303 39.2	63 00.5	04.9	70 46.2	26.4	164 52.9	47.1	220 19.8	02.0
A	14	318 41.6	78 00.4	05.5	85 47.0	26.8	179 55.6	47.2	235 22.5	02.0
Y	15	333 44.1	93 00.3	. . 06.0	100 47.9	. . 27.3	194 58.3	. . 47.3	250 25.1	. . 02.0
	16	348 46.5	108 00.2	06.5	115 48.7	27.7	210 01.0	47.4	265 27.8	02.0
	17	3 49.0	123 00.1	07.0	130 49.5	28.1	225 03.8	47.5	280 30.4	02.0
	18	18 51.5	138 00.0	S17 07.5	145 50.3	S18 28.5	240 06.5	N16 47.5	295 33.1	N22 02.0
	19	33 53.9	152 59.9	08.1	160 51.1	28.9	255 09.2	47.6	310 35.7	02.0
	20	48 56.4	167 59.8	08.6	175 52.0	29.3	270 12.0	47.7	325 38.4	02.0
	21	63 58.9	182 59.6	. . 09.1	190 52.8	. . 29.7	285 14.7	. . 47.8	340 41.0	. . 02.0
	22	79 01.3	197 59.5	09.6	205 53.6	30.1	300 17.4	47.9	355 43.7	02.0
	23	94 03.8	212 59.4	10.1	220 54.4	30.5	315 20.1	48.0	10 46.4	02.0
Mer. Pass.		h m 16 48.7	*v* −0.1	*d* 0.5	*v* 0.8	*d* 0.4	*v* 2.7	*d* 0.1	*v* 2.7	*d* 0.0

STARS		
Name	SHA	Dec
	° ′	° ′
Acamar	315 24.3	S40 17.8
Achernar	335 32.6	S57 13.7
Acrux	173 18.7	S63 06.6
Adhara	255 18.6	S28 58.5
Aldebaran	290 58.6	N16 30.9
Alioth	166 27.7	N55 56.4
Alkaid	153 05.3	N49 17.7
Al Na'ir	27 54.2	S46 57.1
Alnilam	275 54.4	S 1 12.0
Alphard	218 03.9	S 8 40.2
Alphecca	126 18.2	N26 42.1
Alpheratz	357 52.2	N29 06.5
Altair	62 16.6	N 8 52.5
Ankaa	353 23.8	S42 17.7
Antares	112 36.7	S26 26.3
Arcturus	146 03.3	N19 09.9
Atria	107 46.5	S69 01.8
Avior	234 20.9	S59 31.0
Bellatrix	278 40.5	N 6 21.1
Betelgeuse	271 09.9	N 7 24.5
Canopus	263 59.3	S52 41.9
Capella	280 46.2	N46 00.2
Deneb	49 37.6	N45 17.4
Denebola	182 41.9	N14 33.3
Diphda	349 04.1	S17 58.4
Dubhe	194 01.1	N61 43.9
Elnath	278 22.7	N28 36.7
Eltanin	90 50.5	N51 29.2
Enif	33 55.5	N 9 53.2
Fomalhaut	15 33.2	S29 36.7
Gacrux	172 10.2	S57 07.5
Gienah	176 00.7	S17 33.4
Hadar	149 00.0	S60 22.9
Hamal	328 10.0	N23 28.6
Kaus Aust.	83 55.1	S34 23.0
Kochab	137 19.9	N74 08.3
Markab	13 46.7	N15 13.2
Menkar	314 23.5	N 4 06.0
Menkent	148 17.5	S36 22.9
Miaplacidus	221 40.8	S69 43.5
Mirfak	308 52.0	N49 52.5
Nunki	76 08.9	S26 17.7
Peacock	53 32.6	S56 43.7
Pollux	243 37.4	N28 01.1
Procyon	245 08.0	N 5 13.1
Rasalhague	96 14.4	N12 33.4
Regulus	207 52.0	N11 57.2
Rigel	281 19.7	S 8 11.9
Rigil Kent.	140 03.5	S60 50.5
Sabik	102 22.3	S15 43.7
Schedar	349 50.2	N56 33.4
Shaula	96 33.5	S37 06.3
Sirius	258 40.7	S16 43.2
Spica	158 40.0	S11 10.5
Suhail	222 58.2	S43 26.5
Vega	80 45.0	N38 47.1
Zuben'ubi	137 14.7	S16 03.2

	SHA	Mer. Pass.
	° ′	h m
Venus	120 55.9	8 48
Mars	128 07.8	8 18
Jupiter	221 04.1	2 07
Saturn	276 33.6	22 21

Day	UT (d h)	SUN GHA (° ′)	SUN Dec (° ′)	MOON GHA (° ′)	v (′)	MOON Dec (° ′)	d (′)	HP (′)
7	00	178 30.9	S22 27.0	125 35.0	13.8	S13 47.5	11.7	55.6
	01	193 30.7	26.7	140 07.8	13.8	13 35.8	11.8	55.6
	02	208 30.4	26.4	154 40.6	13.8	13 24.0	11.8	55.5
	03	223 30.1	. . 26.1	169 13.4	14.0	13 12.2	11.9	55.5
	04	238 29.8	25.8	183 46.4	14.0	13 00.3	11.9	55.5
	05	253 29.6	25.5	198 19.4	14.1	12 48.4	11.9	55.5
	06	268 29.3	S22 25.2	212 52.5	14.2	S12 36.5	12.0	55.4
	07	283 29.0	24.9	227 25.7	14.3	12 24.5	12.0	55.4
TUESDAY	08	298 28.7	24.6	241 59.0	14.3	12 12.5	12.1	55.4
	09	313 28.5	. . 24.2	256 32.3	14.4	12 00.4	12.1	55.4
	10	328 28.2	23.9	271 05.7	14.5	11 48.3	12.2	55.3
	11	343 27.9	23.6	285 39.2	14.5	11 36.1	12.1	55.3
	12	358 27.7	S22 23.3	300 12.7	14.6	S11 24.0	12.3	55.3
	13	13 27.4	23.0	314 46.3	14.7	11 11.7	12.2	55.3
	14	28 27.1	22.7	329 20.0	14.8	10 59.5	12.3	55.2
	15	43 26.9	. . 22.3	343 53.8	14.8	10 47.2	12.3	55.2
	16	58 26.6	22.0	358 27.6	14.8	10 34.9	12.4	55.2
	17	73 26.3	21.7	13 01.4	15.0	10 22.5	12.4	55.2
	18	88 26.0	S22 21.4	27 35.4	15.0	S10 10.1	12.4	55.2
	19	103 25.8	21.1	42 09.4	15.0	9 57.7	12.4	55.1
	20	118 25.5	20.7	56 43.4	15.1	9 45.3	12.5	55.1
	21	133 25.2	. . 20.4	71 17.5	15.2	9 32.8	12.5	55.1
	22	148 25.0	20.1	85 51.7	15.3	9 20.3	12.5	55.1
	23	163 24.7	19.8	100 26.0	15.3	9 07.8	12.6	55.0
8	00	178 24.4	S22 19.4	115 00.3	15.3	S 8 55.2	12.5	55.0
	01	193 24.2	19.1	129 34.6	15.4	8 42.7	12.6	55.0
	02	208 23.9	18.8	144 09.0	15.5	8 30.1	12.7	55.0
	03	223 23.6	. . 18.5	158 43.5	15.5	8 17.4	12.6	55.0
	04	238 23.4	18.1	173 18.0	15.6	8 04.8	12.7	54.9
	05	253 23.1	17.8	187 52.6	15.6	7 52.1	12.7	54.9
	06	268 22.8	S22 17.5	202 27.2	15.6	S 7 39.4	12.7	54.9
WEDNESDAY	07	283 22.6	17.1	217 01.8	15.7	7 26.7	12.7	54.9
	08	298 22.3	16.8	231 36.5	15.8	7 14.0	12.7	54.9
	09	313 22.0	. . 16.5	246 11.3	15.8	7 01.3	12.8	54.8
	10	328 21.8	16.2	260 46.1	15.9	6 48.5	12.8	54.8
	11	343 21.5	15.8	275 21.0	15.8	6 35.7	12.8	54.8
	12	358 21.2	S22 15.5	289 55.8	16.0	S 6 22.9	12.8	54.8
	13	13 21.0	15.1	304 30.8	16.0	6 10.1	12.8	54.8
	14	28 20.7	14.8	319 05.8	16.0	5 57.3	12.9	54.8
	15	43 20.4	. . 14.5	333 40.8	16.0	5 44.4	12.8	54.7
	16	58 20.2	14.1	348 15.8	16.1	5 31.6	12.9	54.7
	17	73 19.9	13.8	2 50.9	16.2	5 18.7	12.9	54.7
	18	88 19.6	S22 13.5	17 26.1	16.1	S 5 05.8	12.8	54.7
	19	103 19.4	13.1	32 01.2	16.3	4 53.0	12.9	54.7
	20	118 19.1	12.8	46 36.5	16.2	4 40.1	12.9	54.7
	21	133 18.8	. . 12.4	61 11.7	16.3	4 27.2	13.0	54.6
	22	148 18.6	12.1	75 47.0	16.3	4 14.2	12.9	54.6
	23	163 18.3	11.8	90 22.3	16.3	4 01.3	12.9	54.6
9	00	178 18.1	S22 11.4	104 57.6	16.4	S 3 48.4	13.0	54.6
	01	193 17.8	11.1	119 33.0	16.4	3 35.4	12.9	54.6
	02	208 17.5	10.7	134 08.4	16.4	3 22.5	12.9	54.6
	03	223 17.3	. . 10.4	148 43.8	16.4	3 09.6	13.0	54.5
	04	238 17.0	10.0	163 19.2	16.5	2 56.6	13.0	54.5
	05	253 16.7	09.7	177 54.7	16.5	2 43.6	12.9	54.5
	06	268 16.5	S22 09.3	192 30.2	16.5	S 2 30.7	13.0	54.5
THURSDAY	07	283 16.2	09.0	207 05.7	16.6	2 17.7	12.9	54.5
	08	298 16.0	08.6	221 41.3	16.5	2 04.8	13.0	54.5
	09	313 15.7	. . 08.3	236 16.8	16.6	1 51.8	13.0	54.5
	10	328 15.4	07.9	250 52.4	16.6	1 38.8	12.9	54.5
	11	343 15.2	07.6	265 28.0	16.6	1 25.9	13.0	54.4
	12	358 14.9	S22 07.2	280 03.6	16.7	S 1 12.9	13.0	54.4
	13	13 14.7	06.9	294 39.3	16.6	0 59.9	12.9	54.4
	14	28 14.4	06.5	309 14.9	16.7	0 47.0	13.0	54.4
	15	43 14.1	. . 06.2	323 50.6	16.7	0 34.0	12.9	54.4
	16	58 13.9	05.8	338 26.3	16.6	0 21.1	13.0	54.4
	17	73 13.6	05.5	353 01.9	16.7	S 0 08.1	13.0	54.4
	18	88 13.4	S22 05.1	7 37.6	16.8	N 0 04.9	12.9	54.4
	19	103 13.1	04.7	22 13.4	16.7	0 17.8	12.9	54.4
	20	118 12.9	04.4	36 49.1	16.7	0 30.7	13.0	54.4
	21	133 12.6	. . 04.0	51 24.8	16.7	0 43.7	12.9	54.3
	22	148 12.3	03.7	66 00.5	16.8	0 56.6	12.9	54.3
	23	163 12.1	03.3	80 36.3	16.7	N 1 09.5	12.9	54.3
		SD 16.3	d 0.3	SD 15.1		14.9		14.8

Lat. (°)	Twilight Naut. (h m)	Twilight Civil (h m)	Sunrise (h m)	Moonrise 7 (h m)	Moonrise 8 (h m)	Moonrise 9 (h m)	Moonrise 10 (h m)
N 72	08 16	10 20	▬	12 33	12 02	11 36	11 12
N 70	07 58	09 37	▬	12 15	11 53	11 35	11 17
68	07 44	09 08	11 11	12 00	11 46	11 33	11 21
66	07 33	08 46	10 14	11 48	11 40	11 32	11 24
64	07 23	08 29	09 41	11 38	11 34	11 31	11 27
62	07 14	08 15	09 16	11 29	11 30	11 30	11 30
60	07 07	08 02	08 57	11 22	11 26	11 29	11 32
N 58	07 00	07 52	08 42	11 15	11 22	11 28	11 34
56	06 54	07 42	08 28	11 09	11 19	11 27	11 36
54	06 48	07 34	08 16	11 04	11 16	11 27	11 37
52	06 43	07 26	08 06	10 59	11 13	11 26	11 39
50	06 38	07 19	07 57	10 55	11 11	11 26	11 40
45	06 28	07 04	07 38	10 45	11 06	11 25	11 43
N 40	06 18	06 52	07 22	10 37	11 01	11 24	11 46
35	06 09	06 41	07 09	10 30	10 58	11 23	11 48
30	06 01	06 31	06 57	10 24	10 54	11 22	11 50
20	05 46	06 13	06 37	10 14	10 48	11 21	11 53
N 10	05 31	05 57	06 19	10 05	10 43	11 20	11 56
0	05 15	05 41	06 03	09 56	10 38	11 19	11 59
S 10	04 57	05 23	05 46	09 47	10 34	11 18	12 02
20	04 35	05 04	05 28	09 38	10 28	11 17	12 05
30	04 07	04 40	05 07	09 27	10 22	11 16	12 08
35	03 50	04 26	04 55	09 21	10 19	11 15	12 11
40	03 28	04 09	04 41	09 14	10 15	11 14	12 13
45	03 00	03 47	04 24	09 06	10 11	11 13	12 16
S 50	02 18	03 20	04 03	08 56	10 05	11 12	12 19
52	01 55	03 06	03 53	08 51	10 03	11 12	12 20
54	01 21	02 50	03 41	08 46	10 00	11 11	12 22
56	////	02 30	03 28	08 40	09 57	11 11	12 24
58	////	02 05	03 13	08 34	09 53	11 10	12 26
S 60	////	01 28	02 55	08 26	09 49	11 09	12 28

Lat. (°)	Sunset (h m)	Twilight Civil (h m)	Twilight Naut. (h m)	Moonset 7 (h m)	Moonset 8 (h m)	Moonset 9 (h m)	Moonset 10 (h m)
N 72	▬	13 53	15 58	20 04	22 03	23 55	25 46
N 70	▬	14 37	16 15	20 20	22 09	23 53	25 36
68	13 02	15 05	16 29	20 33	22 14	23 51	25 28
66	14 00	15 27	16 41	20 44	22 19	23 50	25 21
64	14 33	15 45	16 51	20 53	22 22	23 49	25 15
62	14 57	15 59	16 59	21 00	22 25	23 48	25 10
60	15 16	16 11	17 07	21 07	22 28	23 47	25 06
N 58	15 32	16 22	17 13	21 12	22 30	23 47	25 02
56	15 45	16 31	17 20	21 17	22 33	23 46	24 59
54	15 57	16 40	17 25	21 22	22 34	23 45	24 56
52	16 07	16 47	17 30	21 26	22 36	23 45	24 53
50	16 16	16 54	17 35	21 29	22 38	23 44	24 51
45	16 36	17 09	17 46	21 37	22 41	23 43	24 46
N 40	16 51	17 22	17 55	21 44	22 44	23 43	24 41
35	17 05	17 33	18 04	21 49	22 46	23 42	24 37
30	17 16	17 42	18 12	21 54	22 48	23 41	24 34
20	17 36	18 00	18 27	22 02	22 52	23 40	24 28
N 10	17 54	18 16	18 43	22 10	22 55	23 39	24 23
0	18 10	18 33	18 59	22 17	22 58	23 38	24 18
S 10	18 27	18 50	19 17	22 23	23 01	23 37	24 13
20	18 45	19 09	19 38	22 31	23 04	23 36	24 08
30	19 06	19 33	20 05	22 39	23 08	23 35	24 03
35	19 18	19 47	20 23	22 43	23 10	23 34	23 59
40	19 32	20 04	20 45	22 49	23 12	23 34	23 56
45	19 49	20 25	21 13	22 55	23 14	23 33	23 51
S 50	20 10	20 53	21 54	23 02	23 17	23 32	23 46
52	20 20	21 06	22 17	23 05	23 19	23 31	23 44
54	20 31	21 22	22 49	23 09	23 20	23 31	23 41
56	20 44	21 42	////	23 13	23 22	23 30	23 39
58	20 59	22 07	////	23 17	23 24	23 30	23 35
S 60	21 17	22 42	////	23 22	23 26	23 29	23 32

Day	SUN Eqn. of Time 00^h (m s)	SUN Eqn. of Time 12^h (m s)	SUN Mer. Pass. (h m)	MOON Mer. Pass. Upper (h m)	MOON Mer. Pass. Lower (h m)	MOON Age (d)	MOON Phase (%)
7	05 56	06 09	12 06	16 06	03 44	05	22
8	06 22	06 35	12 07	16 48	04 28	06	31
9	06 47	07 00	12 07	17 29	05 09	07	40

UT d	UT h		ARIES GHA	VENUS −4.4 GHA	VENUS Dec	MARS +1.5 GHA	MARS Dec	JUPITER −2.5 GHA	JUPITER Dec	SATURN −0.3 GHA	SATURN Dec
			° ′	° ′	° ′	° ′	° ′	° ′	° ′	° ′	° ′
10	00		109 06.3	227 59.3	S17 10.7	235 55.2	S18 30.9	330 22.9	N16 48.1	25 49.0	N22 02.0
	01		124 08.7	242 59.2	11.2	250 56.1	31.3	345 25.6	48.2	40 51.7	02.0
	02		139 11.2	257 59.0	11.7	265 56.9	31.7	0 28.3	48.2	55 54.3	02.0
	03		154 13.7	272 58.9	. . 12.2	280 57.7	. . 32.1	15 31.1	. . 48.3	70 57.0	. . 02.0
	04		169 16.1	287 58.8	12.7	295 58.5	32.5	30 33.8	48.4	85 59.6	02.0
	05		184 18.6	302 58.7	13.2	310 59.3	32.9	45 36.5	48.5	101 02.3	02.0
	06		199 21.0	317 58.6	S17 13.8	326 00.2	S18 33.3	60 39.3	N16 48.6	116 04.9	N22 02.0
	07		214 23.5	332 58.4	14.3	341 01.0	33.8	75 42.0	48.7	131 07.6	02.0
	08		229 26.0	347 58.3	14.8	356 01.8	34.2	90 44.7	48.8	146 10.2	02.0
	09	F	244 28.4	2 58.2	. . 15.3	11 02.6	. . 34.6	105 47.4	. . 48.9	161 12.9	. . 02.0
	10	R	259 30.9	17 58.1	15.8	26 03.4	35.0	120 50.2	49.0	176 15.5	02.0
	11	I	274 33.4	32 57.9	16.3	41 04.3	35.4	135 52.9	49.0	191 18.2	02.0
	12	D	289 35.8	47 57.8	S17 16.9	56 05.1	S18 35.8	150 55.6	N16 49.1	206 20.8	N22 02.0
	13	A	304 38.3	62 57.7	17.4	71 05.9	36.2	165 58.4	49.2	221 23.5	02.0
	14	Y	319 40.8	77 57.5	17.9	86 06.7	36.6	181 01.1	49.3	236 26.1	02.0
	15		334 43.2	92 57.4	. . 18.4	101 07.5	. . 37.0	196 03.8	. . 49.4	251 28.8	. . 02.0
	16		349 45.7	107 57.3	18.9	116 08.3	37.4	211 06.6	49.5	266 31.4	02.0
	17		4 48.2	122 57.1	19.4	131 09.2	37.8	226 09.3	49.6	281 34.1	02.0
	18		19 50.6	137 57.0	S17 19.9	146 10.0	S18 38.2	241 12.0	N16 49.7	296 36.7	N22 02.0
	19		34 53.1	152 56.9	20.5	161 10.8	38.6	256 14.8	49.8	311 39.4	02.0
	20		49 55.5	167 56.7	21.0	176 11.6	39.0	271 17.5	49.8	326 42.0	02.0
	21		64 58.0	182 56.6	. . 21.5	191 12.4	. . 39.4	286 20.2	. . 49.9	341 44.7	. . 02.0
	22		80 00.5	197 56.4	22.0	206 13.2	39.8	301 23.0	50.0	356 47.3	02.0
	23		95 02.9	212 56.3	22.5	221 14.1	40.2	316 25.7	50.1	11 50.0	02.0
11	00		110 05.4	227 56.2	S17 23.0	236 14.9	S18 40.6	331 28.4	N16 50.2	26 52.6	N22 02.0
	01		125 07.9	242 56.0	23.5	251 15.7	41.0	346 31.2	50.3	41 55.3	02.0
	02		140 10.3	257 55.9	24.1	266 16.5	41.4	1 33.9	50.4	56 57.9	02.0
	03		155 12.8	272 55.7	. . 24.6	281 17.3	. . 41.8	16 36.6	. . 50.5	72 00.6	. . 02.0
	04		170 15.3	287 55.6	25.1	296 18.1	42.2	31 39.4	50.6	87 03.2	02.0
	05		185 17.7	302 55.4	25.6	311 19.0	42.6	46 42.1	50.6	102 05.9	02.0
	06		200 20.2	317 55.3	S17 26.1	326 19.8	S18 43.0	61 44.9	N16 50.7	117 08.5	N22 02.0
	07		215 22.6	332 55.1	26.6	341 20.6	43.4	76 47.6	50.8	132 11.2	02.0
	08	S	230 25.1	347 55.0	27.1	356 21.4	43.8	91 50.3	50.9	147 13.8	02.0
	09	A	245 27.6	2 54.8	. . 27.6	11 22.2	. . 44.2	106 53.1	. . 51.0	162 16.5	. . 02.0
	10	T	260 30.0	17 54.7	28.1	26 23.0	44.6	121 55.8	51.1	177 19.1	02.0
	11	U	275 32.5	32 54.5	28.6	41 23.8	45.0	136 58.5	51.2	192 21.8	02.0
	12	R	290 35.0	47 54.4	S17 29.2	56 24.7	S18 45.4	152 01.3	N16 51.3	207 24.4	N22 02.0
	13	D	305 37.4	62 54.2	29.7	71 25.5	45.8	167 04.0	51.4	222 27.1	01.9
	14	A	320 39.9	77 54.1	30.2	86 26.3	46.2	182 06.8	51.5	237 29.7	01.9
	15	Y	335 42.4	92 53.9	. . 30.7	101 27.1	. . 46.6	197 09.5	. . 51.5	252 32.4	. . 01.9
	16		350 44.8	107 53.8	31.2	116 27.9	47.0	212 12.2	51.6	267 35.0	01.9
	17		5 47.3	122 53.6	31.7	131 28.7	47.4	227 15.0	51.7	282 37.7	01.9
	18		20 49.8	137 53.5	S17 32.2	146 29.5	S18 47.8	242 17.7	N16 51.8	297 40.3	N22 01.9
	19		35 52.2	152 53.3	32.7	161 30.4	48.2	257 20.4	51.9	312 43.0	01.9
	20		50 54.7	167 53.1	33.2	176 31.2	48.6	272 23.2	52.0	327 45.6	01.9
	21		65 57.1	182 53.0	. . 33.7	191 32.0	. . 48.9	287 25.9	. . 52.1	342 48.2	. . 01.9
	22		80 59.6	197 52.8	34.2	206 32.8	49.3	302 28.7	52.2	357 50.9	01.9
	23		96 02.1	212 52.7	34.7	221 33.6	49.7	317 31.4	52.3	12 53.5	01.9
12	00		111 04.5	227 52.5	S17 35.2	236 34.4	S18 50.1	332 34.1	N16 52.4	27 56.2	N22 01.9
	01		126 07.0	242 52.3	35.7	251 35.2	50.5	347 36.9	52.4	42 58.8	01.9
	02		141 09.5	257 52.2	36.2	266 36.1	50.9	2 39.6	52.5	58 01.5	01.9
	03		156 11.9	272 52.0	. . 36.7	281 36.9	. . 51.3	17 42.4	. . 52.6	73 04.1	. . 01.9
	04		171 14.4	287 51.8	37.2	296 37.7	51.7	32 45.1	52.7	88 06.8	01.9
	05		186 16.9	302 51.7	37.7	311 38.5	52.1	47 47.8	52.8	103 09.4	01.9
	06		201 19.3	317 51.5	S17 38.3	326 39.3	S18 52.5	62 50.6	N16 52.9	118 12.1	N22 01.9
	07		216 21.8	332 51.3	38.8	341 40.1	52.9	77 53.3	53.0	133 14.7	01.9
	08		231 24.3	347 51.2	39.3	356 40.9	53.3	92 56.1	53.1	148 17.4	01.9
	09	S	246 26.7	2 51.0	. . 39.8	11 41.7	. . 53.7	107 58.8	. . 53.2	163 20.0	. . 01.9
	10	U	261 29.2	17 50.8	40.3	26 42.6	54.1	123 01.6	53.3	178 22.7	01.9
	11	N	276 31.6	32 50.6	40.8	41 43.4	54.5	138 04.3	53.4	193 25.3	01.9
	12	D	291 34.1	47 50.5	S17 41.3	56 44.2	S18 54.9	153 07.0	N16 53.4	208 27.9	N22 01.9
	13	A	306 36.6	62 50.3	41.8	71 45.0	55.2	168 09.8	53.5	223 30.6	01.9
	14	Y	321 39.0	77 50.1	42.3	86 45.8	55.6	183 12.5	53.6	238 33.2	01.9
	15		336 41.5	92 49.9	. . 42.8	101 46.6	. . 56.0	198 15.3	. . 53.7	253 35.9	. . 01.9
	16		351 44.0	107 49.8	43.3	116 47.4	56.4	213 18.0	53.8	268 38.5	01.9
	17		6 46.4	122 49.6	43.8	131 48.2	56.8	228 20.8	53.9	283 41.2	01.9
	18		21 48.9	137 49.4	S17 44.3	146 49.0	S18 57.2	243 23.5	N16 54.0	298 43.8	N22 01.9
	19		36 51.4	152 49.2	44.8	161 49.9	57.6	258 26.2	54.1	313 46.5	01.9
	20		51 53.8	167 49.0	45.3	176 50.7	58.0	273 29.0	54.2	328 49.1	01.9
	21		66 56.3	182 48.9	. . 45.8	191 51.5	. . 58.4	288 31.7	. . 54.3	343 51.7	. . 01.9
	22		81 58.7	197 48.7	46.2	206 52.3	58.8	303 34.5	54.4	358 54.4	01.9
	23		97 01.2	212 48.5	46.7	221 53.1	59.1	318 37.2	54.5	13 57.0	01.9
Mer. Pass.			h m 16 36.9	v −0.2	d 0.5	v 0.8	d 0.4	v 2.7	d 0.1	v 2.6	d 0.0

STARS

Name	SHA	Dec
	° ′	° ′
Acamar	315 24.3	S40 17.8
Achernar	335 32.6	S57 13.7
Acrux	173 18.6	S63 06.6
Adhara	255 18.6	S28 58.6
Aldebaran	290 58.6	N16 30.9
Alioth	166 27.6	N55 56.4
Alkaid	153 05.3	N49 17.7
Al Na'ir	27 54.2	S46 57.1
Alnilam	275 54.4	S 1 12.0
Alphard	218 03.9	S 8 40.2
Alphecca	126 18.1	N26 42.1
Alpheratz	357 52.2	N29 06.4
Altair	62 16.6	N 8 52.5
Ankaa	353 23.8	S42 17.7
Antares	112 36.7	S26 26.3
Arcturus	146 03.3	N19 09.9
Atria	107 46.5	S69 01.8
Avior	234 20.9	S59 31.0
Bellatrix	278 40.5	N 6 21.1
Betelgeuse	271 09.9	N 7 24.5
Canopus	263 59.3	S52 41.9
Capella	280 46.2	N46 00.2
Deneb	49 37.6	N45 17.4
Denebola	182 41.9	N14 33.3
Diphda	349 04.1	S17 58.4
Dubhe	194 01.1	N61 43.9
Elnath	278 22.7	N28 36.7
Eltanin	90 50.5	N51 29.1
Enif	33 55.5	N 9 53.2
Fomalhaut	15 33.2	S29 36.7
Gacrux	172 10.2	S57 07.5
Gienah	176 00.7	S17 33.4
Hadar	149 00.0	S60 22.9
Hamal	328 10.0	N23 28.6
Kaus Aust.	83 55.1	S34 23.0
Kochab	137 19.9	N74 08.3
Markab	13 46.7	N15 13.2
Menkar	314 23.5	N 4 06.0
Menkent	148 17.5	S36 22.9
Miaplacidus	221 40.8	S69 43.5
Mirfak	308 52.0	N49 52.5
Nunki	76 08.9	S26 17.7
Peacock	53 32.6	S56 43.7
Pollux	243 37.4	N28 01.2
Procyon	245 08.0	N 5 13.1
Rasalhague	96 14.4	N12 33.4
Regulus	207 52.0	N11 57.2
Rigel	281 19.7	S 8 11.9
Rigil Kent.	140 03.5	S60 50.5
Sabik	102 22.3	S15 43.7
Schedar	349 50.2	N56 33.4
Shaula	96 33.5	S37 06.3
Sirius	258 40.7	S16 43.2
Spica	158 40.0	S11 10.5
Suhail	222 58.2	S43 26.5
Vega	80 45.0	N38 47.0
Zuben'ubi	137 14.7	S16 03.2

	SHA	Mer. Pass.
	° ′	h m
Venus	117 50.8	8 48
Mars	126 09.5	8 15
Jupiter	221 23.0	1 54
Saturn	276 47.2	22 09

Day	UT (d h)	SUN GHA (° ′)	SUN Dec (° ′)	MOON GHA (° ′)	v (′)	MOON Dec (° ′)	d (′)	HP (′)
FRIDAY	10 00	178 11.8	S22 03.0	95 12.0	16.8	N 1 22.4	12.9	54.3
	01	193 11.6	02.6	109 47.8	16.7	1 35.3	12.9	54.3
	02	208 11.3	02.2	124 23.5	16.7	1 48.2	12.9	54.3
	03	223 11.1	. . 01.9	138 59.2	16.8	2 01.1	12.9	54.3
	04	238 10.8	01.5	153 35.0	16.7	2 14.0	12.8	54.3
	05	253 10.5	01.1	168 10.7	16.8	2 26.8	12.9	54.3
	06	268 10.3	S22 00.8	182 46.5	16.7	N 2 39.7	12.8	54.3
	07	283 10.0	00.4	197 22.2	16.8	2 52.5	12.8	54.3
	08	298 09.8	22 00.0	211 58.0	16.7	3 05.3	12.8	54.3
	09	313 09.5	21 59.7	226 33.7	16.7	3 18.1	12.8	54.3
	10	328 09.3	59.3	241 09.4	16.7	3 30.9	12.8	54.3
	11	343 09.0	58.9	255 45.1	16.7	3 43.7	12.8	54.3
	12	358 08.8	S21 58.6	270 20.8	16.7	N 3 56.5	12.7	54.3
	13	13 08.5	58.2	284 56.5	16.7	4 09.2	12.7	54.3
	14	28 08.3	57.8	299 32.2	16.7	4 21.9	12.7	54.2
	15	43 08.0	. . 57.4	314 07.9	16.7	4 34.6	12.7	54.2
	16	58 07.7	57.1	328 43.6	16.6	4 47.3	12.7	54.2
	17	73 07.5	56.7	343 19.2	16.7	5 00.0	12.7	54.2
	18	88 07.2	S21 56.3	357 54.9	16.6	N 5 12.7	12.6	54.2
	19	103 07.0	55.9	12 30.5	16.6	5 25.3	12.6	54.2
	20	118 06.7	55.6	27 06.1	16.6	5 37.9	12.6	54.2
	21	133 06.5	. . 55.2	41 41.7	16.5	5 50.5	12.6	54.2
	22	148 06.2	54.8	56 17.2	16.6	6 03.1	12.5	54.2
	23	163 06.0	54.4	70 52.8	16.5	6 15.6	12.6	54.2
SATURDAY	11 00	178 05.7	S21 54.1	85 28.3	16.5	N 6 28.2	12.5	54.2
	01	193 05.5	53.7	100 03.8	16.5	6 40.7	12.5	54.2
	02	208 05.2	53.3	114 39.3	16.4	6 53.2	12.4	54.2
	03	223 05.0	. . 52.9	129 14.7	16.4	7 05.6	12.4	54.2
	04	238 04.7	52.5	143 50.1	16.4	7 18.0	12.4	54.2
	05	253 04.5	52.1	158 25.5	16.4	7 30.4	12.4	54.2
	06	268 04.2	S21 51.8	173 00.9	16.3	N 7 42.8	12.4	54.2
	07	283 04.0	51.4	187 36.2	16.4	7 55.2	12.3	54.2
	08	298 03.7	51.0	202 11.6	16.2	8 07.5	12.3	54.2
	09	313 03.5	. . 50.6	216 46.8	16.3	8 19.8	12.2	54.2
	10	328 03.2	50.2	231 22.1	16.2	8 32.0	12.3	54.2
	11	343 03.0	49.8	245 57.3	16.2	8 44.3	12.2	54.2
	12	358 02.7	S21 49.4	260 32.5	16.1	N 8 56.5	12.2	54.2
	13	13 02.5	49.1	275 07.6	16.2	9 08.7	12.1	54.3
	14	28 02.2	48.7	289 42.8	16.0	9 20.8	12.1	54.3
	15	43 02.0	. . 48.3	304 17.8	16.1	9 32.9	12.1	54.3
	16	58 01.8	47.9	318 52.9	16.0	9 45.0	12.0	54.3
	17	73 01.5	47.5	333 27.9	15.9	9 57.0	12.0	54.3
	18	88 01.3	S21 47.1	348 02.8	15.9	N10 09.0	12.0	54.3
	19	103 01.0	46.7	2 37.7	15.9	10 21.0	11.9	54.3
	20	118 00.8	46.3	17 12.6	15.8	10 32.9	11.9	54.3
	21	133 00.5	. . 45.9	31 47.4	15.8	10 44.8	11.9	54.3
	22	148 00.3	45.5	46 22.2	15.8	10 56.7	11.8	54.3
	23	163 00.0	45.1	60 57.0	15.7	11 08.5	11.8	54.3
SUNDAY	12 00	177 59.8	S21 44.7	75 31.7	15.6	N11 20.3	11.7	54.3
	01	192 59.5	44.3	90 06.3	15.6	11 32.0	11.7	54.3
	02	207 59.3	43.9	104 40.9	15.6	11 43.7	11.7	54.3
	03	222 59.1	. . 43.5	119 15.5	15.5	11 55.4	11.6	54.3
	04	237 58.8	43.1	133 50.0	15.5	12 07.0	11.6	54.3
	05	252 58.6	42.7	148 24.5	15.4	12 18.6	11.5	54.4
	06	267 58.3	S21 42.3	162 58.9	15.3	N12 30.1	11.5	54.4
	07	282 58.1	41.9	177 33.2	15.3	12 41.6	11.4	54.4
	08	297 57.8	41.5	192 07.5	15.3	12 53.0	11.4	54.4
	09	312 57.6	. . 41.1	206 41.8	15.1	13 04.4	11.4	54.4
	10	327 57.4	40.7	221 15.9	15.2	13 15.8	11.3	54.4
	11	342 57.1	40.3	235 50.1	15.1	13 27.1	11.2	54.4
	12	357 56.9	S21 39.9	250 24.2	15.0	N13 38.3	11.2	54.4
	13	12 56.6	39.5	264 58.2	14.9	13 49.5	11.2	54.4
	14	27 56.4	39.1	279 32.1	14.9	14 00.7	11.1	54.4
	15	42 56.2	. . 38.7	294 06.0	14.9	14 11.8	11.0	54.5
	16	57 55.9	38.3	308 39.9	14.8	14 22.8	11.0	54.5
	17	72 55.7	37.9	323 13.7	14.7	14 33.8	11.0	54.5
	18	87 55.4	S21 37.5	337 47.4	14.6	N14 44.8	10.9	54.5
	19	102 55.2	37.1	352 21.0	14.6	14 55.7	10.8	54.5
	20	117 55.0	36.6	6 54.6	14.5	15 06.5	10.8	54.5
	21	132 54.7	. . 36.2	21 28.1	14.5	15 17.3	10.7	54.5
	22	147 54.5	35.8	36 01.6	14.4	15 28.0	10.7	54.5
	23	162 54.2	35.4	50 35.0	14.3	N15 38.7	10.6	54.5
		SD 16.3	*d* 0.4	SD 14.8		14.8		14.8

Lat.	Twilight Naut. (h m)	Twilight Civil (h m)	Sunrise (h m)	Moonrise 10 (h m)	Moonrise 11 (h m)	Moonrise 12 (h m)	Moonrise 13 (h m)
N 72	08 10	10 09	▬	11 12	10 45	10 10	09 03
N 70	07 54	09 30	▬	11 17	10 57	10 34	09 59
68	07 41	09 03	10 55	11 21	11 08	10 53	10 33
66	07 30	08 42	10 06	11 24	11 16	11 08	10 58
64	07 20	08 25	09 35	11 27	11 24	11 20	11 17
62	07 12	08 12	09 12	11 30	11 30	11 31	11 33
60	07 05	08 00	08 54	11 32	11 35	11 40	11 46
N 58	06 58	07 49	08 39	11 34	11 40	11 48	11 58
56	06 52	07 40	08 26	11 36	11 44	11 55	12 08
54	06 47	07 32	08 14	11 37	11 48	12 01	12 17
52	06 42	07 25	08 05	11 39	11 52	12 07	12 25
50	06 38	07 18	07 56	11 40	11 55	12 12	12 32
45	06 27	07 04	07 37	11 43	12 02	12 23	12 47
N 40	06 18	06 51	07 21	11 46	12 08	12 32	13 00
35	06 09	06 41	07 08	11 48	12 13	12 41	13 11
30	06 02	06 31	06 57	11 50	12 18	12 48	13 21
20	05 47	06 14	06 37	11 53	12 26	13 00	13 37
N 10	05 32	05 58	06 20	11 56	12 33	13 11	13 52
0	05 16	05 42	06 04	11 59	12 39	13 21	14 05
S 10	04 58	05 25	05 48	12 02	12 46	13 31	14 19
20	04 37	05 06	05 30	12 05	12 53	13 43	14 34
30	04 10	04 43	05 10	12 08	13 01	13 55	14 51
35	03 53	04 29	04 58	12 11	13 06	14 03	15 01
40	03 32	04 12	04 44	12 13	13 11	14 11	15 12
45	03 04	03 51	04 27	12 16	13 18	14 21	15 26
S 50	02 24	03 24	04 07	12 19	13 25	14 33	15 42
52	02 02	03 11	03 57	12 20	13 29	14 39	15 50
54	01 31	02 55	03 46	12 22	13 33	14 45	15 59
56	00 26	02 36	03 33	12 24	13 37	14 52	16 09
58	////	02 12	03 19	12 26	13 42	15 00	16 20
S 60	////	01 39	03 01	12 28	13 48	15 09	16 33

Lat.	Sunset (h m)	Twilight Civil (h m)	Twilight Naut. (h m)	Moonset 10 (h m)	Moonset 11 (h m)	Moonset 12 (h m)	Moonset 13 (h m)
N 72	▬	14 07	16 06	25 46	01 46	03 47	06 26
N 70	▬	14 46	16 22	25 36	01 36	03 25	05 31
68	13 21	15 13	16 36	25 28	01 28	03 08	04 58
66	14 10	15 34	16 47	25 21	01 21	02 55	04 35
64	14 41	15 51	16 56	25 15	01 15	02 44	04 16
62	15 04	16 05	17 04	25 10	01 10	02 34	04 01
60	15 22	16 16	17 11	25 06	01 06	02 26	03 49
N 58	15 37	16 27	17 18	25 02	01 02	02 19	03 38
56	15 50	16 36	17 24	24 59	00 59	02 13	03 29
54	16 02	16 44	17 29	24 56	00 56	02 07	03 20
52	16 11	16 51	17 34	24 53	00 53	02 02	03 13
50	16 20	16 58	17 38	24 51	00 51	01 58	03 06
45	16 39	17 12	17 49	24 46	00 46	01 48	02 52
N 40	16 54	17 24	17 58	24 41	00 41	01 40	02 40
35	17 07	17 35	18 06	24 37	00 37	01 33	02 31
30	17 19	17 45	18 14	24 34	00 34	01 27	02 22
20	17 38	18 02	18 29	24 28	00 28	01 17	02 07
N 10	17 55	18 18	18 44	24 23	00 23	01 08	01 54
0	18 12	18 34	19 00	24 18	00 18	00 59	01 42
S 10	18 28	18 51	19 17	24 13	00 13	00 51	01 30
20	18 45	19 09	19 38	24 08	00 08	00 42	01 17
30	19 06	19 32	20 05	24 03	00 03	00 31	01 03
35	19 17	19 47	20 22	23 59	24 26	00 26	00 54
40	19 31	20 03	20 43	23 56	24 19	00 19	00 45
45	19 48	20 24	21 11	23 51	24 11	00 11	00 34
S 50	20 08	20 50	21 50	23 46	24 02	00 02	00 20
52	20 18	21 04	22 12	23 44	23 58	24 14	00 14
54	20 29	21 19	22 42	23 41	23 53	24 07	00 07
56	20 41	21 38	23 38	23 39	23 48	23 59	24 14
58	20 56	22 01	////	23 35	23 42	23 51	24 02
S 60	21 13	22 34	////	23 32	23 36	23 41	23 49

Day (d)	SUN Eqn. of Time 00^h (m s)	SUN Eqn. of Time 12^h (m s)	SUN Mer. Pass. (h m)	MOON Mer. Pass. Upper (h m)	MOON Mer. Pass. Lower (h m)	MOON Age (d)	MOON Phase (%)
10	07 12	07 24	12 07	18 09	05 49	08	50
11	07 37	07 49	12 08	18 49	06 29	09	59
12	08 00	08 12	12 08	19 32	07 10	10	68

UT	ARIES	VENUS −4.4		MARS +1.4		JUPITER −2.5		SATURN −0.3	
d h	GHA ° ′	GHA ° ′	Dec ° ′	GHA ° ′	Dec ° ′	GHA ° ′	Dec ° ′	GHA ° ′	Dec ° ′
13 00 (MONDAY)	112 03.7	227 48.3	S17 47.2	236 53.9	S18 59.5	333 40.0	N16 54.5	28 59.7	N22 01.9
01	127 06.1	242 48.1	47.7	251 54.7	18 59.9	348 42.7	54.6	44 02.3	01.9
02	142 08.6	257 47.9	48.2	266 55.5	19 00.3	3 45.4	54.7	59 05.0	01.9
03	157 11.1	272 47.8	. . 48.7	281 56.3	. . 00.7	18 48.2	. . 54.8	74 07.6	. . 01.9
04	172 13.5	287 47.6	49.2	296 57.2	01.1	33 50.9	54.9	89 10.2	01.9
05	187 16.0	302 47.4	49.7	311 58.0	01.5	48 53.7	55.0	104 12.9	01.9
06	202 18.5	317 47.2	S17 50.2	326 58.8	S19 01.9	63 56.4	N16 55.1	119 15.5	N22 01.9
07	217 20.9	332 47.0	50.7	341 59.6	02.3	78 59.2	55.2	134 18.2	01.9
08	232 23.4	347 46.8	51.2	357 00.4	02.6	94 01.9	55.3	149 20.8	01.9
09	247 25.9	2 46.6	. . 51.7	12 01.2	. . 03.0	109 04.7	. . 55.4	164 23.5	. . 01.9
10	262 28.3	17 46.4	52.2	27 02.0	03.4	124 07.4	55.5	179 26.1	01.9
11	277 30.8	32 46.2	52.7	42 02.8	03.8	139 10.2	55.6	194 28.7	01.9
12	292 33.2	47 46.0	S17 53.2	57 03.6	S19 04.2	154 12.9	N16 55.6	209 31.4	N22 01.9
13	307 35.7	62 45.8	53.7	72 04.4	04.6	169 15.7	55.7	224 34.0	01.9
14	322 38.2	77 45.6	54.2	87 05.2	05.0	184 18.4	55.8	239 36.7	01.9
15	337 40.6	92 45.4	. . 54.6	102 06.1	. . 05.3	199 21.2	. . 55.9	254 39.3	. . 01.9
16	352 43.1	107 45.2	55.1	117 06.9	05.7	214 23.9	56.0	269 42.0	01.9
17	7 45.6	122 45.0	55.6	132 07.7	06.1	229 26.7	56.1	284 44.6	01.9
18	22 48.0	137 44.8	S17 56.1	147 08.5	S19 06.5	244 29.4	N16 56.2	299 47.2	N22 01.9
19	37 50.5	152 44.6	56.6	162 09.3	06.9	259 32.1	56.3	314 49.9	01.9
20	52 53.0	167 44.4	57.1	177 10.1	07.3	274 34.9	56.4	329 52.5	01.9
21	67 55.4	182 44.2	. . 57.6	192 10.9	. . 07.6	289 37.6	. . 56.5	344 55.2	. . 01.9
22	82 57.9	197 44.0	58.1	207 11.7	08.0	304 40.4	56.6	359 57.8	01.9
23	98 00.4	212 43.8	58.6	222 12.5	08.4	319 43.1	56.7	15 00.4	01.9
14 00 (TUESDAY)	113 02.8	227 43.6	S17 59.0	237 13.3	S19 08.8	334 45.9	N16 56.8	30 03.1	N22 01.9
01	128 05.3	242 43.4	17 59.5	252 14.1	09.2	349 48.6	56.8	45 05.7	01.9
02	143 07.7	257 43.2	18 00.0	267 14.9	09.6	4 51.4	56.9	60 08.4	01.9
03	158 10.2	272 43.0	. . 00.5	282 15.7	. . 09.9	19 54.1	. . 57.0	75 11.0	. . 01.9
04	173 12.7	287 42.8	01.0	297 16.5	10.3	34 56.9	57.1	90 13.6	01.9
05	188 15.1	302 42.6	01.5	312 17.4	10.7	49 59.6	57.2	105 16.3	01.9
06	203 17.6	317 42.4	S18 02.0	327 18.2	S19 11.1	65 02.4	N16 57.3	120 18.9	N22 01.9
07	218 20.1	332 42.1	02.4	342 19.0	11.5	80 05.1	57.4	135 21.6	01.9
08	233 22.5	347 41.9	02.9	357 19.8	11.9	95 07.9	57.5	150 24.2	01.9
09	248 25.0	2 41.7	. . 03.4	12 20.6	. . 12.2	110 10.6	. . 57.6	165 26.8	. . 01.9
10	263 27.5	17 41.5	03.9	27 21.4	12.6	125 13.4	57.7	180 29.5	01.9
11	278 29.9	32 41.3	04.4	42 22.2	13.0	140 16.1	57.8	195 32.1	01.9
12	293 32.4	47 41.1	S18 04.9	57 23.0	S19 13.4	155 18.9	N16 57.9	210 34.8	N22 01.9
13	308 34.9	62 40.9	05.4	72 23.8	13.8	170 21.7	58.0	225 37.4	01.9
14	323 37.3	77 40.6	05.8	87 24.6	14.1	185 24.4	58.1	240 40.0	01.9
15	338 39.8	92 40.4	. . 06.3	102 25.4	. . 14.5	200 27.2	. . 58.2	255 42.7	. . 01.9
16	353 42.2	107 40.2	06.8	117 26.2	14.9	215 29.9	58.2	270 45.3	01.9
17	8 44.7	122 40.0	07.3	132 27.0	15.3	230 32.7	58.3	285 48.0	01.9
18	23 47.2	137 39.8	S18 07.8	147 27.8	S19 15.7	245 35.4	N16 58.4	300 50.6	N22 01.9
19	38 49.6	152 39.5	08.2	162 28.6	16.0	260 38.2	58.5	315 53.2	01.9
20	53 52.1	167 39.3	08.7	177 29.4	16.4	275 40.9	58.6	330 55.9	01.9
21	68 54.6	182 39.1	. . 09.2	192 30.2	. . 16.8	290 43.7	. . 58.7	345 58.5	. . 01.9
22	83 57.0	197 38.9	09.7	207 31.0	17.2	305 46.4	58.8	1 01.1	01.9
23	98 59.5	212 38.6	10.2	222 31.9	17.5	320 49.2	58.9	16 03.8	01.9
15 00 (WEDNESDAY)	114 02.0	227 38.4	S18 10.6	237 32.7	S19 17.9	335 51.9	N16 59.0	31 06.4	N22 01.9
01	129 04.4	242 38.2	11.1	252 33.5	18.3	350 54.7	59.1	46 09.1	01.9
02	144 06.9	257 38.0	11.6	267 34.3	18.7	5 57.4	59.2	61 11.7	01.9
03	159 09.3	272 37.7	. . 12.1	282 35.1	. . 19.1	21 00.2	. . 59.3	76 14.3	. . 01.9
04	174 11.8	287 37.5	12.5	297 35.9	19.4	36 03.0	59.4	91 17.0	01.9
05	189 14.3	302 37.3	13.0	312 36.7	19.8	51 05.7	59.5	106 19.6	01.9
06	204 16.7	317 37.0	S18 13.5	327 37.5	S19 20.2	66 08.5	N16 59.6	121 22.2	N22 01.9
07	219 19.2	332 36.8	14.0	342 38.3	20.6	81 11.2	59.7	136 24.9	01.9
08	234 21.7	347 36.6	14.4	357 39.1	20.9	96 14.0	59.7	151 27.5	01.9
09	249 24.1	2 36.3	. . 14.9	12 39.9	. . 21.3	111 16.7	. . 59.8	166 30.1	. . 01.9
10	264 26.6	17 36.1	15.4	27 40.7	21.7	126 19.5	16 59.9	181 32.8	01.9
11	279 29.1	32 35.9	15.9	42 41.5	22.1	141 22.2	17 00.0	196 35.4	01.9
12	294 31.5	47 35.6	S18 16.3	57 42.3	S19 22.4	156 25.0	N17 00.1	211 38.1	N22 01.9
13	309 34.0	62 35.4	16.8	72 43.1	22.8	171 27.8	00.2	226 40.7	01.9
14	324 36.5	77 35.2	17.3	87 43.9	23.2	186 30.5	00.3	241 43.3	01.9
15	339 38.9	92 34.9	. . 17.7	102 44.7	. . 23.6	201 33.3	. . 00.4	256 46.0	. . 01.9
16	354 41.4	107 34.7	18.2	117 45.5	23.9	216 36.0	00.5	271 48.6	01.9
17	9 43.8	122 34.4	18.7	132 46.3	24.3	231 38.8	00.6	286 51.2	01.9
18	24 46.3	137 34.2	S18 19.2	147 47.1	S19 24.7	246 41.5	N17 00.7	301 53.9	N22 01.9
19	39 48.8	152 33.9	19.6	162 47.9	25.0	261 44.3	00.8	316 56.5	01.9
20	54 51.2	167 33.7	20.1	177 48.7	25.4	276 47.1	00.9	331 59.1	01.9
21	69 53.7	182 33.5	. . 20.6	192 49.5	. . 25.8	291 49.8	. . 01.0	347 01.8	. . 01.9
22	84 56.2	197 33.2	21.0	207 50.3	26.2	306 52.6	01.1	2 04.4	01.9
23	99 58.6	212 33.0	21.5	222 51.1	26.5	321 55.3	01.2	17 07.0	01.9
Mer. Pass.	h m 16 25.1	v −0.2	d 0.5	v 0.8	d 0.4	v 2.8	d 0.1	v 2.6	d 0.0

STARS

Name	SHA ° ′	Dec ° ′
Acamar	315 24.3	S40 17.8
Achernar	335 32.7	S57 13.7
Acrux	173 18.6	S63 06.6
Adhara	255 18.6	S28 58.6
Aldebaran	290 58.6	N16 30.9
Alioth	166 27.6	N55 56.4
Alkaid	153 05.3	N49 17.7
Al Na'ir	27 54.3	S46 57.1
Alnilam	275 54.4	S 1 12.0
Alphard	218 03.9	S 8 40.2
Alphecca	126 18.1	N26 42.1
Alpheratz	357 52.2	N29 06.4
Altair	62 16.6	N 8 52.4
Ankaa	353 23.8	S42 17.7
Antares	112 36.6	S26 26.3
Arcturus	146 03.3	N19 09.9
Atria	107 46.4	S69 01.8
Avior	234 20.9	S59 31.0
Bellatrix	278 40.5	N 6 21.1
Betelgeuse	271 09.9	N 7 24.5
Canopus	263 59.3	S52 41.9
Capella	280 46.2	N46 00.2
Deneb	49 37.6	N45 17.4
Denebola	182 41.8	N14 33.3
Diphda	349 04.1	S17 58.4
Dubhe	194 01.0	N61 43.9
Elnath	278 22.7	N28 36.7
Eltanin	90 50.5	N51 29.1
Enif	33 55.5	N 9 53.2
Fomalhaut	15 33.2	S29 36.7
Gacrux	172 10.1	S57 07.5
Gienah	176 00.7	S17 33.4
Hadar	148 59.9	S60 22.9
Hamal	328 10.0	N23 28.6
Kaus Aust.	83 55.1	S34 23.0
Kochab	137 19.8	N74 08.3
Markab	13 46.8	N15 13.2
Menkar	314 23.5	N 4 06.0
Menkent	148 17.4	S36 22.9
Miaplacidus	221 40.8	S69 43.6
Mirfak	308 52.0	N49 52.5
Nunki	76 08.8	S26 17.7
Peacock	53 32.6	S56 43.7
Pollux	243 37.4	N28 01.2
Procyon	245 08.0	N 5 13.1
Rasalhague	96 14.4	N12 33.3
Regulus	207 52.0	N11 57.1
Rigel	281 19.7	S 8 11.9
Rigil Kent.	140 03.4	S60 50.5
Sabik	102 22.3	S15 43.7
Schedar	349 50.2	N56 33.4
Shaula	96 33.5	S37 06.3
Sirius	258 40.7	S16 43.2
Spica	158 39.9	S11 10.5
Suhail	222 58.2	S43 26.5
Vega	80 45.0	N38 47.0
Zuben'ubi	137 14.7	S16 03.2

	SHA ° ′	Mer. Pass. h m
Venus	114 40.8	8 49
Mars	124 10.5	8 11
Jupiter	221 43.1	1 41
Saturn	277 00.3	21 56

d	UT h	SUN GHA ° ′	SUN Dec ° ′	MOON GHA ° ′	v ′	MOON Dec ° ′	d ′	HP ′
13	00	177 54.0	S21 35.0	65 08.3	14.3	N15 49.3	10.6	54.6
	01	192 53.8	34.6	79 41.6	14.2	15 59.9	10.5	54.6
	02	207 53.5	34.2	94 14.8	14.1	16 10.4	10.4	54.6
	03	222 53.3	. . 33.7	108 47.9	14.0	16 20.8	10.4	54.6
	04	237 53.1	33.3	123 20.9	14.0	16 31.2	10.3	54.6
	05	252 52.8	32.9	137 53.9	13.9	16 41.5	10.2	54.6
	06	267 52.6	S21 32.5	152 26.8	13.9	N16 51.7	10.2	54.6
	07	282 52.3	32.1	166 59.7	13.7	17 01.9	10.1	54.7
	08	297 52.1	31.6	181 32.4	13.7	17 12.0	10.1	54.7
MONDAY	09	312 51.9	. . 31.2	196 05.1	13.6	17 22.1	10.0	54.7
	10	327 51.6	30.8	210 37.7	13.6	17 32.1	9.9	54.7
	11	342 51.4	30.4	225 10.3	13.4	17 42.0	9.8	54.7
	12	357 51.2	S21 30.0	239 42.7	13.4	N17 51.8	9.8	54.7
	13	12 50.9	29.5	254 15.1	13.3	18 01.6	9.7	54.8
	14	27 50.7	29.1	268 47.4	13.3	18 11.3	9.6	54.8
	15	42 50.5	. . 28.7	283 19.7	13.1	18 20.9	9.6	54.8
	16	57 50.2	28.3	297 51.8	13.1	18 30.5	9.5	54.8
	17	72 50.0	27.8	312 23.9	13.0	18 40.0	9.4	54.8
	18	87 49.8	S21 27.4	326 55.9	12.9	N18 49.4	9.3	54.9
	19	102 49.5	27.0	341 27.8	12.9	18 58.7	9.3	54.9
	20	117 49.3	26.6	355 59.7	12.7	19 08.0	9.2	54.9
	21	132 49.1	. . 26.1	10 31.4	12.7	19 17.2	9.1	54.9
	22	147 48.8	25.7	25 03.1	12.6	19 26.3	9.0	54.9
	23	162 48.6	25.3	39 34.7	12.5	19 35.3	8.9	54.9
14	00	177 48.4	S21 24.8	54 06.2	12.4	N19 44.2	8.9	55.0
	01	192 48.1	24.4	68 37.6	12.4	19 53.1	8.8	55.0
	02	207 47.9	24.0	83 09.0	12.3	20 01.9	8.7	55.0
	03	222 47.7	. . 23.5	97 40.3	12.1	20 10.6	8.6	55.0
	04	237 47.4	23.1	112 11.4	12.1	20 19.2	8.5	55.0
	05	252 47.2	22.7	126 42.5	12.1	20 27.7	8.4	55.1
	06	267 47.0	S21 22.2	141 13.6	11.9	N20 36.1	8.4	55.1
	07	282 46.8	21.8	155 44.5	11.8	20 44.5	8.2	55.1
	08	297 46.5	21.3	170 15.3	11.8	20 52.7	8.2	55.1
TUESDAY	09	312 46.3	. . 20.9	184 46.1	11.7	21 00.9	8.1	55.1
	10	327 46.1	20.5	199 16.8	11.6	21 09.0	7.9	55.2
	11	342 45.8	20.0	213 47.4	11.5	21 16.9	7.9	55.2
	12	357 45.6	S21 19.6	228 17.9	11.4	N21 24.8	7.8	55.2
	13	12 45.4	19.2	242 48.3	11.3	21 32.6	7.7	55.2
	14	27 45.2	18.7	257 18.6	11.3	21 40.3	7.6	55.3
	15	42 44.9	. . 18.3	271 48.9	11.2	21 47.9	7.5	55.3
	16	57 44.7	17.8	286 19.1	11.0	21 55.4	7.4	55.3
	17	72 44.5	17.4	300 49.1	11.0	22 02.8	7.3	55.3
	18	87 44.3	S21 16.9	315 19.1	10.9	N22 10.1	7.2	55.3
	19	102 44.0	16.5	329 49.0	10.9	22 17.3	7.1	55.4
	20	117 43.8	16.0	344 18.9	10.7	22 24.4	7.0	55.4
	21	132 43.6	. . 15.6	358 48.6	10.6	22 31.4	6.9	55.4
	22	147 43.4	15.2	13 18.2	10.6	22 38.3	6.8	55.4
	23	162 43.1	14.7	27 47.8	10.5	22 45.1	6.7	55.5
15	00	177 42.9	S21 14.3	42 17.3	10.4	N22 51.8	6.6	55.5
	01	192 42.7	13.8	56 46.7	10.3	22 58.4	6.4	55.5
	02	207 42.5	13.4	71 16.0	10.2	23 04.8	6.4	55.5
	03	222 42.2	. . 12.9	85 45.2	10.2	23 11.2	6.2	55.6
	04	237 42.0	12.5	100 14.4	10.0	23 17.4	6.2	55.6
	05	252 41.8	12.0	114 43.4	10.0	23 23.6	6.0	55.6
	06	267 41.6	S21 11.5	129 12.4	9.9	N23 29.6	5.9	55.6
	07	282 41.3	11.1	143 41.3	9.8	23 35.5	5.8	55.7
	08	297 41.1	10.6	158 10.1	9.7	23 41.3	5.7	55.7
WEDNESDAY	09	312 40.9	. . 10.2	172 38.8	9.7	23 47.0	5.5	55.7
	10	327 40.7	09.7	187 07.5	9.5	23 52.5	5.5	55.7
	11	342 40.5	09.3	201 36.0	9.5	23 58.0	5.3	55.8
	12	357 40.2	S21 08.8	216 04.5	9.4	N24 03.3	5.2	55.8
	13	12 40.0	08.4	230 32.9	9.3	24 08.5	5.0	55.8
	14	27 39.8	07.9	245 01.2	9.3	24 13.5	5.0	55.8
	15	42 39.6	. . 07.4	259 29.5	9.2	24 18.5	4.8	55.9
	16	57 39.4	07.0	273 57.7	9.0	24 23.3	4.7	55.9
	17	72 39.1	06.5	288 25.7	9.1	24 28.0	4.6	55.9
	18	87 38.9	S21 06.1	302 53.8	8.9	N24 32.6	4.5	55.9
	19	102 38.7	05.6	317 21.7	8.9	24 37.1	4.3	56.0
	20	117 38.5	05.1	331 49.6	8.7	24 41.4	4.2	56.0
	21	132 38.3	. . 04.7	346 17.3	8.8	24 45.6	4.1	56.0
	22	147 38.0	04.2	0 45.1	8.6	24 49.7	3.9	56.0
	23	162 37.8	03.7	15 12.7	8.6	N24 53.6	3.8	56.1
		SD 16.3	*d* 0.4	SD 14.9		15.0		15.2

Lat. °	Twilight Naut. h m	Twilight Civil h m	Sunrise h m	Moonrise 13 h m	Moonrise 14 h m	Moonrise 15 h m	Moonrise 16 h m
N 72	08 04	09 57	■	09 03	□	□	□
N 70	07 49	09 22	■	09 59	□	□	□
68	07 36	08 57	10 40	10 33	09 54	□	□
66	07 26	08 37	09 57	10 58	10 43	10 05	□
64	07 17	08 21	09 29	11 17	11 14	11 13	11 19
62	07 09	08 08	09 07	11 33	11 38	11 49	12 14
60	07 02	07 57	08 50	11 46	11 57	12 15	12 47
N 58	06 56	07 47	08 35	11 58	12 13	12 35	13 12
56	06 51	07 38	08 23	12 08	12 26	12 52	13 31
54	06 45	07 30	08 12	12 17	12 38	13 07	13 48
52	06 41	07 23	08 02	12 25	12 48	13 19	14 02
50	06 36	07 17	07 54	12 32	12 57	13 30	14 14
45	06 26	07 03	07 35	12 47	13 17	13 54	14 40
N 40	06 17	06 51	07 21	13 00	13 33	14 12	15 00
35	06 09	06 40	07 08	13 11	13 46	14 28	15 17
30	06 02	06 31	06 57	13 21	13 58	14 42	15 32
20	05 47	06 14	06 38	13 37	14 19	15 05	15 57
N 10	05 33	05 59	06 21	13 52	14 36	15 25	16 18
0	05 17	05 43	06 05	14 05	14 53	15 44	16 38
S 10	05 00	05 27	05 49	14 19	15 10	16 03	16 59
20	04 39	05 08	05 32	14 34	15 28	16 24	17 20
30	04 13	04 46	05 12	14 51	15 49	16 47	17 46
35	03 56	04 32	05 01	15 01	16 01	17 01	18 00
40	03 36	04 15	04 47	15 12	16 15	17 17	18 18
45	03 09	03 55	04 31	15 26	16 32	17 37	18 38
S 50	02 31	03 29	04 11	15 42	16 53	18 01	19 05
52	02 10	03 16	04 02	15 50	17 03	18 13	19 17
54	01 41	03 01	03 51	15 59	17 14	18 26	19 32
56	00 54	02 43	03 39	16 09	17 27	18 42	19 49
58	////	02 21	03 25	16 20	17 42	19 01	20 09
S 60	////	01 50	03 08	16 33	18 00	19 24	20 35

Lat. °	Sunset h m	Twilight Civil h m	Twilight Naut. h m	Moonset 13 h m	Moonset 14 h m	Moonset 15 h m	Moonset 16 h m
N 72	■	14 22	16 15	06 26	□	□	□
N 70	■	14 57	16 30	05 31	□	□	□
68	13 39	15 22	16 42	04 58	07 14	□	□
66	14 21	15 42	16 53	04 35	06 25	08 48	□
64	14 50	15 57	17 02	04 16	05 55	07 40	09 26
62	15 11	16 11	17 09	04 01	05 32	07 05	08 31
60	15 29	16 22	17 16	03 49	05 14	06 39	07 58
N 58	15 43	16 32	17 22	03 38	04 59	06 19	07 34
56	15 56	16 40	17 28	03 29	04 46	06 03	07 14
54	16 06	16 48	17 33	03 20	04 35	05 49	06 58
52	16 16	16 55	17 38	03 13	04 25	05 36	06 44
50	16 25	17 02	17 42	03 06	04 16	05 26	06 32
45	16 43	17 16	17 52	02 52	03 57	05 03	06 07
N 40	16 58	17 27	18 01	02 40	03 42	04 45	05 47
35	17 10	17 38	18 09	02 31	03 30	04 30	05 30
30	17 21	17 47	18 17	02 22	03 19	04 17	05 15
20	17 40	18 04	18 31	02 07	03 00	03 54	04 51
N 10	17 57	18 19	18 45	01 54	02 43	03 35	04 30
0	18 13	18 35	19 01	01 42	02 28	03 17	04 10
S 10	18 28	18 51	19 18	01 30	02 13	02 59	03 50
20	18 45	19 10	19 38	01 17	01 56	02 40	03 29
30	19 05	19 32	20 04	01 03	01 38	02 18	03 05
35	19 17	19 46	20 21	00 54	01 27	02 05	02 50
40	19 30	20 02	20 42	00 45	01 15	01 50	02 34
45	19 46	20 22	21 08	00 34	01 00	01 33	02 14
S 50	20 06	20 48	21 46	00 20	00 42	01 11	01 49
52	20 15	21 01	22 06	00 14	00 34	01 01	01 37
54	20 26	21 15	22 34	00 07	00 25	00 49	01 23
56	20 38	21 33	23 18	24 14	00 14	00 36	01 07
58	20 52	21 55	////	24 02	00 02	00 20	00 48
S 60	21 08	22 25	////	23 49	24 02	00 02	00 25

Day	SUN Eqn. of Time 00^h	SUN Eqn. of Time 12^h	SUN Mer. Pass.	MOON Mer. Pass. Upper	MOON Mer. Pass. Lower	MOON Age	MOON Phase
d	m s	m s	h m	h m	h m	d	%
13	08 24	08 35	12 09	20 17	07 54	11	77
14	08 46	08 57	12 09	21 05	08 40	12	84
15	09 08	09 19	12 09	21 57	09 30	13	91

	UT	ARIES	VENUS −4.4		MARS +1.4		JUPITER −2.6		SATURN −0.3	
	d h	GHA ° ′	GHA ° ′	Dec ° ′	GHA ° ′	Dec ° ′	GHA ° ′	Dec ° ′	GHA ° ′	Dec ° ′
	16 00	115 01.1	227 32.7	S18 22.0	237 51.9	S19 26.9	336 58.1	N17 01.3	32 09.7	N22 01.9
	01	130 03.6	242 32.5	22.4	252 52.7	27.3	352 00.8	01.4	47 12.3	01.9
	02	145 06.0	257 32.2	22.9	267 53.5	27.6	7 03.6	01.4	62 14.9	01.9
	03	160 08.5	272 32.0	. . 23.4	282 54.3	. . 28.0	22 06.4	. . 01.5	77 17.6	. . 01.9
	04	175 11.0	287 31.7	23.8	297 55.1	28.4	37 09.1	01.6	92 20.2	01.9
	05	190 13.4	302 31.5	24.3	312 55.9	28.8	52 11.9	01.7	107 22.8	01.9
	06	205 15.9	317 31.2	S18 24.8	327 56.7	S19 29.1	67 14.6	N17 01.8	122 25.5	N22 01.9
	07	220 18.3	332 31.0	25.2	342 57.5	29.5	82 17.4	01.9	137 28.1	01.9
T	08	235 20.8	347 30.7	25.7	357 58.3	29.9	97 20.2	02.0	152 30.7	01.9
H	09	250 23.3	2 30.5	. . 26.2	12 59.1	. . 30.2	112 22.9	. . 02.1	167 33.4	. . 01.9
U	10	265 25.7	17 30.2	26.6	27 59.9	30.6	127 25.7	02.2	182 36.0	01.9
R	11	280 28.2	32 29.9	27.1	43 00.7	31.0	142 28.4	02.3	197 38.6	01.9
S	12	295 30.7	47 29.7	S18 27.5	58 01.5	S19 31.3	157 31.2	N17 02.4	212 41.3	N22 01.9
D	13	310 33.1	62 29.4	28.0	73 02.3	31.7	172 34.0	02.5	227 43.9	01.9
A	14	325 35.6	77 29.2	28.5	88 03.1	32.1	187 36.7	02.6	242 46.5	01.9
Y	15	340 38.1	92 28.9	. . 28.9	103 03.9	. . 32.4	202 39.5	. . 02.7	257 49.2	. . 01.9
	16	355 40.5	107 28.7	29.4	118 04.7	32.8	217 42.2	02.8	272 51.8	01.9
	17	10 43.0	122 28.4	29.8	133 05.5	33.2	232 45.0	02.9	287 54.4	01.9
	18	25 45.5	137 28.1	S18 30.3	148 06.3	S19 33.6	247 47.8	N17 03.0	302 57.1	N22 01.9
	19	40 47.9	152 27.9	30.8	163 07.1	33.9	262 50.5	03.1	317 59.7	01.9
	20	55 50.4	167 27.6	31.2	178 07.9	34.3	277 53.3	03.2	333 02.3	01.9
	21	70 52.8	182 27.3	. . 31.7	193 08.7	. . 34.6	292 56.0	. . 03.3	348 04.9	. . 01.9
	22	85 55.3	197 27.1	32.1	208 09.5	35.0	307 58.8	03.4	3 07.6	01.9
	23	100 57.8	212 26.8	32.6	223 10.3	35.4	323 01.6	03.4	18 10.2	01.9
	17 00	116 00.2	227 26.5	S18 33.1	238 11.1	S19 35.7	338 04.3	N17 03.5	33 12.8	N22 01.9
	01	131 02.7	242 26.3	33.5	253 11.9	36.1	353 07.1	03.6	48 15.5	01.9
	02	146 05.2	257 26.0	34.0	268 12.7	36.5	8 09.9	03.7	63 18.1	01.9
	03	161 07.6	272 25.7	. . 34.4	283 13.5	. . 36.8	23 12.6	. . 03.8	78 20.7	. . 01.9
	04	176 10.1	287 25.5	34.9	298 14.3	37.2	38 15.4	03.9	93 23.4	01.9
	05	191 12.6	302 25.2	35.3	313 15.1	37.6	53 18.2	04.0	108 26.0	01.9
	06	206 15.0	317 24.9	S18 35.8	328 15.9	S19 37.9	68 20.9	N17 04.1	123 28.6	N22 01.9
	07	221 17.5	332 24.6	36.2	343 16.7	38.3	83 23.7	04.2	138 31.2	01.9
	08	236 20.0	347 24.4	36.7	358 17.5	38.7	98 26.4	04.3	153 33.9	01.9
F	09	251 22.4	2 24.1	. . 37.1	13 18.3	. . 39.0	113 29.2	. . 04.4	168 36.5	. . 01.9
R	10	266 24.9	17 23.8	37.6	28 19.1	39.4	128 32.0	04.5	183 39.1	01.9
I	11	281 27.3	32 23.5	38.0	43 19.9	39.8	143 34.7	04.6	198 41.8	01.9
D	12	296 29.8	47 23.3	S18 38.5	58 20.7	S19 40.1	158 37.5	N17 04.7	213 44.4	N22 01.9
A	13	311 32.3	62 23.0	38.9	73 21.5	40.5	173 40.3	04.8	228 47.0	01.9
Y	14	326 34.7	77 22.7	39.4	88 22.3	40.8	188 43.0	04.9	243 49.6	01.9
	15	341 37.2	92 22.4	. . 39.8	103 23.1	. . 41.2	203 45.8	. . 05.0	258 52.3	. . 01.9
	16	356 39.7	107 22.2	40.3	118 23.9	41.6	218 48.6	05.1	273 54.9	01.9
	17	11 42.1	122 21.9	40.7	133 24.7	41.9	233 51.3	05.2	288 57.5	01.9
	18	26 44.6	137 21.6	S18 41.2	148 25.5	S19 42.3	248 54.1	N17 05.3	304 00.2	N22 01.9
	19	41 47.1	152 21.3	41.6	163 26.3	42.6	263 56.9	05.4	319 02.8	01.9
	20	56 49.5	167 21.0	42.1	178 27.0	43.0	278 59.6	05.5	334 05.4	01.9
	21	71 52.0	182 20.7	. . 42.5	193 27.8	. . 43.4	294 02.4	. . 05.6	349 08.0	. . 01.9
	22	86 54.5	197 20.5	43.0	208 28.6	43.7	309 05.2	05.7	4 10.7	01.9
	23	101 56.9	212 20.2	43.4	223 29.4	44.1	324 07.9	05.8	19 13.3	01.9
	18 00	116 59.4	227 19.9	S18 43.9	238 30.2	S19 44.4	339 10.7	N17 05.9	34 15.9	N22 01.9
	01	132 01.8	242 19.6	44.3	253 31.0	44.8	354 13.4	05.9	49 18.5	01.9
	02	147 04.3	257 19.3	44.8	268 31.8	45.2	9 16.2	06.0	64 21.2	01.9
	03	162 06.8	272 19.0	. . 45.2	283 32.6	. . 45.5	24 19.0	. . 06.1	79 23.8	. . 01.9
	04	177 09.2	287 18.7	45.6	298 33.4	45.9	39 21.7	06.2	94 26.4	01.9
	05	192 11.7	302 18.4	46.1	313 34.2	46.2	54 24.5	06.3	109 29.1	01.9
	06	207 14.2	317 18.1	S18 46.5	328 35.0	S19 46.6	69 27.3	N17 06.4	124 31.7	N22 01.9
	07	222 16.6	332 17.9	47.0	343 35.8	47.0	84 30.1	06.5	139 34.3	01.9
S	08	237 19.1	347 17.6	47.4	358 36.6	47.3	99 32.8	06.6	154 36.9	01.9
A	09	252 21.6	2 17.3	. . 47.8	13 37.4	. . 47.7	114 35.6	. . 06.7	169 39.6	. . 01.9
T	10	267 24.0	17 17.0	48.3	28 38.2	48.0	129 38.4	06.8	184 42.2	01.9
U	11	282 26.5	32 16.7	48.7	43 39.0	48.4	144 41.1	06.9	199 44.8	01.9
R	12	297 28.9	47 16.4	S18 49.2	58 39.8	S19 48.7	159 43.9	N17 07.0	214 47.4	N22 01.9
D	13	312 31.4	62 16.1	49.6	73 40.6	49.1	174 46.7	07.1	229 50.1	01.9
A	14	327 33.9	77 15.8	50.0	88 41.3	49.5	189 49.4	07.2	244 52.7	01.9
Y	15	342 36.3	92 15.5	. . 50.5	103 42.1	. . 49.8	204 52.2	. . 07.3	259 55.3	. . 01.9
	16	357 38.8	107 15.2	50.9	118 42.9	50.2	219 55.0	07.4	274 57.9	01.9
	17	12 41.3	122 14.9	51.3	133 43.7	50.5	234 57.7	07.5	290 00.6	01.9
	18	27 43.7	137 14.6	S18 51.8	148 44.5	S19 50.9	250 00.5	N17 07.6	305 03.2	N22 01.9
	19	42 46.2	152 14.3	52.2	163 45.3	51.2	265 03.3	07.7	320 05.8	01.9
	20	57 48.7	167 14.0	52.6	178 46.1	51.6	280 06.0	07.8	335 08.4	01.9
	21	72 51.1	182 13.7	. . 53.1	193 46.9	. . 51.9	295 08.8	. . 07.9	350 11.1	. . 01.9
	22	87 53.6	197 13.4	53.5	208 47.7	52.3	310 11.6	08.0	5 13.7	01.9
	23	102 56.1	212 13.1	53.9	223 48.5	52.6	325 14.4	08.1	20 16.3	01.9
	Mer. Pass.	h m 16 13.3	*v* −0.3	*d* 0.5	*v* 0.8	*d* 0.4	*v* 2.8	*d* 0.1	*v* 2.6	*d* 0.0

STARS

Name	SHA ° ′	Dec ° ′
Acamar	315 24.3	S40 17.8
Achernar	335 32.7	S57 13.7
Acrux	173 18.5	S63 06.6
Adhara	255 18.6	S28 58.6
Aldebaran	290 58.6	N16 30.9
Alioth	166 27.6	N55 56.4
Alkaid	153 05.2	N49 17.7
Al Na'ir	27 54.3	S46 57.1
Alnilam	275 54.4	S 1 12.0
Alphard	218 03.9	S 8 40.2
Alphecca	126 18.1	N26 42.1
Alpheratz	357 52.2	N29 06.4
Altair	62 16.6	N 8 52.4
Ankaa	353 23.8	S42 17.7
Antares	112 36.6	S26 26.3
Arcturus	146 03.2	N19 09.9
Atria	107 46.4	S69 01.8
Avior	234 20.8	S59 31.0
Bellatrix	278 40.5	N 6 21.1
Betelgeuse	271 09.9	N 7 24.5
Canopus	263 59.3	S52 41.9
Capella	280 46.2	N46 00.2
Deneb	49 37.6	N45 17.4
Denebola	182 41.8	N14 33.3
Diphda	349 04.2	S17 58.4
Dubhe	194 01.0	N61 43.9
Elnath	278 22.7	N28 36.7
Eltanin	90 50.4	N51 29.1
Enif	33 55.5	N 9 53.2
Fomalhaut	15 33.2	S29 36.7
Gacrux	172 10.1	S57 07.5
Gienah	176 00.7	S17 33.4
Hadar	148 59.9	S60 22.9
Hamal	328 10.0	N23 28.6
Kaus Aust.	83 55.1	S34 23.0
Kochab	137 19.7	N74 08.3
Markab	13 46.8	N15 13.2
Menkar	314 23.5	N 4 06.0
Menkent	148 17.4	S36 22.9
Miaplacidus	221 40.8	S69 43.6
Mirfak	308 52.0	N49 52.5
Nunki	76 08.8	S26 17.7
Peacock	53 32.6	S56 43.7
Pollux	243 37.3	N28 01.2
Procyon	245 08.0	N 5 13.0
Rasalhague	96 14.4	N12 33.3
Regulus	207 51.9	N11 57.1
Rigel	281 19.7	S 8 12.0
Rigil Kent.	140 03.4	S60 50.5
Sabik	102 22.2	S15 43.7
Schedar	349 50.3	N56 33.4
Shaula	96 33.4	S37 06.3
Sirius	258 40.6	S16 43.2
Spica	158 39.9	S11 10.6
Suhail	222 58.1	S43 26.5
Vega	80 44.9	N38 47.0
Zuben'ubi	137 14.6	S16 03.2

	SHA ° ′	Mer. Pass. h m
Venus	111 26.3	8 50
Mars	122 10.9	8 07
Jupiter	222 04.1	1 27
Saturn	277 12.6	21 43

UT	SUN GHA	SUN Dec	MOON GHA	v	MOON Dec	d	HP
d h	° ′	° ′	° ′	′	° ′	′	′
16 00	177 37.6	S21 03.3	29 40.3	8.5	N24 57.4	3.7	56.1
01	192 37.4	02.8	44 07.8	8.4	25 01.1	3.5	56.1
02	207 37.2	02.3	58 35.2	8.3	25 04.6	3.4	56.1
03	222 37.0	· · 01.9	73 02.5	8.3	25 08.0	3.3	56.2
04	237 36.7	01.4	87 29.8	8.3	25 11.3	3.2	56.2
05	252 36.5	00.9	101 57.1	8.1	25 14.5	3.0	56.2
06	267 36.3	S21 00.5	116 24.2	8.1	N25 17.5	2.8	56.2
07	282 36.1	21 00.0	130 51.3	8.0	25 20.3	2.7	56.3
T 08	297 35.9	20 59.5	145 18.3	8.0	25 23.0	2.6	56.3
H 09	312 35.7	· · 59.1	159 45.3	7.9	25 25.6	2.5	56.3
U 10	327 35.4	58.6	174 12.2	7.9	25 28.1	2.3	56.4
R 11	342 35.2	58.1	188 39.1	7.7	25 30.4	2.1	56.4
S 12	357 35.0	S20 57.6	203 05.8	7.8	N25 32.5	2.1	56.4
D 13	12 34.8	57.2	217 32.6	7.7	25 34.6	1.8	56.4
A 14	27 34.6	56.7	231 59.3	7.6	25 36.4	1.8	56.5
Y 15	42 34.4	· · 56.2	246 25.9	7.5	25 38.2	1.6	56.5
16	57 34.2	55.7	260 52.4	7.6	25 39.8	1.4	56.5
17	72 34.0	55.3	275 19.0	7.4	25 41.2	1.3	56.5
18	87 33.7	S20 54.8	289 45.4	7.4	N25 42.5	1.2	56.6
19	102 33.5	54.3	304 11.8	7.4	25 43.7	1.0	56.6
20	117 33.3	53.8	318 38.2	7.3	25 44.7	0.8	56.6
21	132 33.1	· · 53.3	333 04.5	7.3	25 45.5	0.7	56.7
22	147 32.9	52.9	347 30.8	7.3	25 46.2	0.6	56.7
23	162 32.7	52.4	1 57.1	7.2	25 46.8	0.4	56.7
17 00	177 32.5	S20 51.9	16 23.3	7.1	N25 47.2	0.2	56.7
01	192 32.3	51.4	30 49.4	7.1	25 47.4	0.1	56.8
02	207 32.1	50.9	45 15.5	7.1	25 47.5	0.0	56.8
03	222 31.8	· · 50.4	59 41.6	7.1	25 47.5	0.2	56.8
04	237 31.6	50.0	74 07.7	7.0	25 47.3	0.3	56.8
05	252 31.4	49.5	88 33.7	7.0	25 47.0	0.5	56.9
06	267 31.2	S20 49.0	102 59.7	6.9	N25 46.5	0.7	56.9
07	282 31.0	48.5	117 25.6	6.9	25 45.8	0.8	56.9
08	297 30.8	48.0	131 51.5	6.9	25 45.0	1.0	56.9
F 09	312 30.6	· · 47.5	146 17.4	6.9	25 44.0	1.1	57.0
R 10	327 30.4	47.0	160 43.3	6.8	25 42.9	1.2	57.0
I 11	342 30.2	46.5	175 09.1	6.9	25 41.7	1.5	57.0
D 12	357 30.0	S20 46.1	189 35.0	6.8	N25 40.2	1.5	57.1
A 13	12 29.8	45.6	204 00.8	6.8	25 38.7	1.8	57.1
Y 14	27 29.6	45.1	218 26.6	6.7	25 36.9	1.8	57.1
15	42 29.4	· · 44.6	232 52.3	6.8	25 35.1	2.1	57.1
16	57 29.2	44.1	247 18.1	6.7	25 33.0	2.2	57.2
17	72 29.0	43.6	261 43.8	6.8	25 30.8	2.3	57.2
18	87 28.7	S20 43.1	276 09.6	6.7	N25 28.5	2.5	57.2
19	102 28.5	42.6	290 35.3	6.7	25 26.0	2.7	57.2
20	117 28.3	42.1	305 01.0	6.7	25 23.3	2.8	57.3
21	132 28.1	· · 41.6	319 26.7	6.7	25 20.5	3.0	57.3
22	147 27.9	41.1	333 52.4	6.7	25 17.5	3.1	57.3
23	162 27.7	40.6	348 18.1	6.7	25 14.4	3.3	57.3
18 00	177 27.5	S20 40.1	2 43.8	6.7	N25 11.1	3.4	57.4
01	192 27.3	39.6	17 09.5	6.7	25 07.7	3.6	57.4
02	207 27.1	39.1	31 35.2	6.7	25 04.1	3.7	57.4
03	222 26.9	· · 38.6	46 00.9	6.7	25 00.4	3.9	57.4
04	237 26.7	38.1	60 26.6	6.7	24 56.5	4.1	57.5
05	252 26.5	37.6	74 52.3	6.8	24 52.4	4.2	57.5
06	267 26.3	S20 37.1	89 18.1	6.7	N24 48.2	4.3	57.5
07	282 26.1	36.6	103 43.8	6.7	24 43.9	4.6	57.5
S 08	297 25.9	36.1	118 09.5	6.8	24 39.3	4.6	57.6
A 09	312 25.7	· · 35.6	132 35.3	6.8	24 34.7	4.8	57.6
T 10	327 25.5	35.1	147 01.1	6.8	24 29.9	5.0	57.6
U 11	342 25.3	34.6	161 26.9	6.8	24 24.9	5.1	57.6
R 12	357 25.1	S20 34.1	175 52.7	6.8	N24 19.8	5.3	57.7
D 13	12 24.9	33.6	190 18.5	6.8	24 14.5	5.4	57.7
A 14	27 24.7	33.1	204 44.3	6.9	24 09.1	5.6	57.7
Y 15	42 24.5	· · 32.6	219 10.2	6.9	24 03.5	5.7	57.7
16	57 24.3	32.0	233 36.1	6.9	23 57.8	5.9	57.8
17	72 24.1	31.5	248 02.0	7.0	23 51.9	6.0	57.8
18	87 23.9	S20 31.0	262 28.0	6.9	N23 45.9	6.2	57.8
19	102 23.7	30.5	276 53.9	7.0	23 39.7	6.3	57.8
20	117 23.5	30.0	291 19.9	7.1	23 33.4	6.5	57.9
21	132 23.3	· · 29.5	305 46.0	7.0	23 26.9	6.6	57.9
22	147 23.1	29.0	320 12.0	7.1	23 20.3	6.8	57.9
23	162 22.9	28.5	334 38.1	7.2	N23 13.5	6.9	57.9
	SD 16.3	*d* 0.5	SD 15.4		15.5		15.7

Lat.	Twilight Naut.	Twilight Civil	Sunrise	Moonrise 16	17	18	19
°	h m	h m	h m	h m	h m	h m	h m
N 72	07 57	09 45	■	▭	▭	▭	▭
N 70	07 43	09 13	11 51	▭	▭	▭	▭
68	07 31	08 50	10 25	▭	▭	▭	14 27
66	07 21	08 31	09 48	▭	▭	▭	15 22
64	07 13	08 16	09 22	11 19	12 08	13 57	15 54
62	07 06	08 04	09 02	12 14	13 09	14 36	16 18
60	06 59	07 53	08 45	12 47	13 43	15 03	16 37
N 58	06 53	07 43	08 31	13 12	14 08	15 24	16 53
56	06 48	07 35	08 19	13 31	14 28	15 41	17 06
54	06 43	07 28	08 09	13 48	14 44	15 56	17 18
52	06 39	07 21	08 00	14 02	14 58	16 08	17 28
50	06 34	07 15	07 51	14 14	15 11	16 20	17 37
45	06 25	07 01	07 34	14 40	15 36	16 43	17 56
N 40	06 16	06 50	07 19	15 00	15 57	17 02	18 12
35	06 09	06 40	07 07	15 17	16 14	17 17	18 25
30	06 01	06 31	06 56	15 32	16 29	17 31	18 36
20	05 47	06 14	06 38	15 57	16 53	17 54	18 55
N 10	05 33	05 59	06 22	16 18	17 15	18 13	19 12
0	05 18	05 44	06 06	16 38	17 35	18 32	19 28
S 10	05 02	05 28	05 51	16 59	17 55	18 50	19 43
20	04 42	05 10	05 34	17 20	18 17	19 10	20 00
30	04 16	04 48	05 15	17 46	18 41	19 33	20 19
35	04 00	04 35	05 04	18 00	18 56	19 46	20 30
40	03 40	04 19	04 51	18 18	19 13	20 01	20 43
45	03 14	04 00	04 35	18 38	19 33	20 20	20 58
S 50	02 37	03 34	04 16	19 05	19 59	20 42	21 16
52	02 18	03 22	04 07	19 17	20 11	20 53	21 24
54	01 52	03 08	03 56	19 32	20 25	21 05	21 34
56	01 13	02 50	03 45	19 49	20 41	21 19	21 44
58	////	02 29	03 31	20 09	21 01	21 35	21 57
S 60	////	02 02	03 15	20 35	21 25	21 54	22 11

Lat.	Sunset	Twilight Civil	Twilight Naut.	Moonset 16	17	18	19
°	h m	h m	h m	h m	h m	h m	h m
N 72	■	14 36	16 24	▭	▭	▭	▭
N 70	12 30	15 08	16 38	▭	▭	▭	▭
68	13 55	15 31	16 50	▭	▭	▭	12 14
66	14 33	15 49	16 59	▭	▭	▭	11 18
64	14 59	16 04	17 08	09 26	10 34	10 45	10 45
62	15 19	16 17	17 15	08 31	09 33	10 06	10 20
60	15 35	16 28	17 21	07 58	08 59	09 38	10 00
N 58	15 49	16 37	17 27	07 34	08 34	09 17	09 44
56	16 01	16 45	17 32	07 14	08 14	08 59	09 30
54	16 11	16 53	17 37	06 58	07 58	08 44	09 18
52	16 21	17 00	17 42	06 44	07 43	08 31	09 07
50	16 29	17 06	17 46	06 32	07 31	08 20	08 58
45	16 47	17 19	17 55	06 07	07 05	07 56	08 38
N 40	17 01	17 31	18 04	05 47	06 45	07 36	08 21
35	17 13	17 41	18 12	05 30	06 27	07 20	08 07
30	17 24	17 50	18 19	05 15	06 13	07 06	07 55
20	17 42	18 06	18 33	04 51	05 47	06 42	07 34
N 10	17 58	18 21	18 47	04 30	05 26	06 22	07 16
0	18 14	18 36	19 01	04 10	05 05	06 02	06 59
S 10	18 29	18 52	19 18	03 50	04 45	05 43	06 42
20	18 46	19 10	19 38	03 29	04 23	05 22	06 23
30	19 05	19 31	20 03	03 05	03 58	04 58	06 02
35	19 16	19 45	20 20	02 50	03 43	04 43	05 49
40	19 29	20 00	20 40	02 34	03 26	04 27	05 34
45	19 44	20 20	21 05	02 14	03 05	04 07	05 17
S 50	20 03	20 45	21 41	01 49	02 39	03 42	04 55
52	20 12	20 57	22 00	01 37	02 26	03 30	04 45
54	20 23	21 11	22 26	01 23	02 11	03 16	04 33
56	20 34	21 28	23 03	01 07	01 54	03 00	04 20
58	20 47	21 49	////	00 48	01 34	02 40	04 04
S 60	21 03	22 15	////	00 25	01 08	02 16	03 45

Day	SUN Eqn. of Time 00^h	SUN Eqn. of Time 12^h	SUN Mer. Pass.	MOON Mer. Pass. Upper	MOON Mer. Pass. Lower	Age	Phase
d	m s	m s	h m	h m	h m	d %	
16	09 29	09 39	12 10	22 52	10 24	14 96	○
17	09 50	10 00	12 10	23 49	11 20	15 99	
18	10 09	10 19	12 10	24 45	12 17	16 100	

	UT	ARIES	VENUS −4.4		MARS +1.4		JUPITER −2.6		SATURN −0.3	
		GHA	GHA	Dec	GHA	Dec	GHA	Dec	GHA	Dec
	d h	° ′	° ′	° ′	° ′	° ′	° ′	° ′	° ′	° ′
	19 00	117 58.5	227 12.8	S18 54.4	238 49.3	S19 53.0	340 17.1	N17 08.2	35 18.9	N22 01.9
	01	133 01.0	242 12.5	54.8	253 50.1	53.4	355 19.9	08.3	50 21.5	01.9
	02	148 03.4	257 12.2	55.2	268 50.9	53.7	10 22.7	08.4	65 24.2	01.9
	03	163 05.9	272 11.8	. . 55.7	283 51.7	. . 54.1	25 25.4	. . 08.5	80 26.8	. . 01.9
	04	178 08.4	287 11.5	56.1	298 52.4	54.4	40 28.2	08.6	95 29.4	01.9
	05	193 10.8	302 11.2	56.5	313 53.2	54.8	55 31.0	08.7	110 32.0	01.9
	06	208 13.3	317 10.9	S18 57.0	328 54.0	S19 55.1	70 33.7	N17 08.8	125 34.7	N22 01.9
	07	223 15.8	332 10.6	57.4	343 54.8	55.5	85 36.5	08.9	140 37.3	01.9
	08	238 18.2	347 10.3	57.8	358 55.6	55.8	100 39.3	09.0	155 39.9	01.9
S	09	253 20.7	2 10.0	. . 58.2	13 56.4	. . 56.2	115 42.1	. . 09.1	170 42.5	. . 01.9
U	10	268 23.2	17 09.7	58.7	28 57.2	56.5	130 44.8	09.1	185 45.1	01.9
N	11	283 25.6	32 09.4	59.1	43 58.0	56.9	145 47.6	09.2	200 47.8	01.9
D	12	298 28.1	47 09.0	S18 59.5	58 58.8	S19 57.2	160 50.4	N17 09.3	215 50.4	N22 01.9
A	13	313 30.6	62 08.7	19 00.0	73 59.6	57.6	175 53.1	09.4	230 53.0	01.9
Y	14	328 33.0	77 08.4	00.4	89 00.4	57.9	190 55.9	09.5	245 55.6	01.9
	15	343 35.5	92 08.1	. . 00.8	104 01.1	. . 58.3	205 58.7	. . 09.6	260 58.3	. . 01.9
	16	358 37.9	107 07.8	01.2	119 01.9	58.6	221 01.5	09.7	276 00.9	01.9
	17	13 40.4	122 07.4	01.6	134 02.7	59.0	236 04.2	09.8	291 03.5	01.9
	18	28 42.9	137 07.1	S19 02.1	149 03.5	S19 59.3	251 07.0	N17 09.9	306 06.1	N22 01.9
	19	43 45.3	152 06.8	02.5	164 04.3	19 59.7	266 09.8	10.0	321 08.7	01.9
	20	58 47.8	167 06.5	02.9	179 05.1	20 00.0	281 12.6	10.1	336 11.4	01.9
	21	73 50.3	182 06.2	. . 03.3	194 05.9	. . 00.4	296 15.3	. . 10.2	351 14.0	. . 01.9
	22	88 52.7	197 05.8	03.8	209 06.7	00.7	311 18.1	10.3	6 16.6	01.9
	23	103 55.2	212 05.5	04.2	224 07.5	01.1	326 20.9	10.4	21 19.2	01.9
	20 00	118 57.7	227 05.2	S19 04.6	239 08.3	S20 01.4	341 23.6	N17 10.5	36 21.8	N22 01.9
	01	134 00.1	242 04.9	05.0	254 09.0	01.8	356 26.4	10.6	51 24.5	01.9
	02	149 02.6	257 04.5	05.4	269 09.8	02.1	11 29.2	10.7	66 27.1	01.9
	03	164 05.1	272 04.2	. . 05.8	284 10.6	. . 02.4	26 32.0	. . 10.8	81 29.7	. . 01.9
	04	179 07.5	287 03.9	06.3	299 11.4	02.8	41 34.7	10.9	96 32.3	01.9
	05	194 10.0	302 03.6	06.7	314 12.2	03.1	56 37.5	11.0	111 34.9	01.9
	06	209 12.4	317 03.2	S19 07.1	329 13.0	S20 03.5	71 40.3	N17 11.1	126 37.5	N22 01.9
	07	224 14.9	332 02.9	07.5	344 13.8	03.8	86 43.1	11.2	141 40.2	01.9
	08	239 17.4	347 02.6	07.9	359 14.6	04.2	101 45.8	11.3	156 42.8	01.9
M	09	254 19.8	2 02.2	. . 08.3	14 15.4	. . 04.5	116 48.6	. . 11.4	171 45.4	. . 01.9
O	10	269 22.3	17 01.9	08.8	29 16.1	04.9	131 51.4	11.5	186 48.0	01.9
N	11	284 24.8	32 01.6	09.2	44 16.9	05.2	146 54.2	11.6	201 50.6	01.9
D	12	299 27.2	47 01.2	S19 09.6	59 17.7	S20 05.6	161 56.9	N17 11.7	216 53.3	N22 01.9
A	13	314 29.7	62 00.9	10.0	74 18.5	05.9	176 59.7	11.8	231 55.9	01.9
Y	14	329 32.2	77 00.6	10.4	89 19.3	06.2	192 02.5	11.9	246 58.5	01.9
	15	344 34.6	92 00.2	. . 10.8	104 20.1	. . 06.6	207 05.3	. . 12.0	262 01.1	. . 01.9
	16	359 37.1	106 59.9	11.2	119 20.9	06.9	222 08.0	12.1	277 03.7	01.9
	17	14 39.6	121 59.6	11.6	134 21.7	07.3	237 10.8	12.2	292 06.3	01.9
	18	29 42.0	136 59.2	S19 12.0	149 22.4	S20 07.6	252 13.6	N17 12.3	307 09.0	N22 01.9
	19	44 44.5	151 58.9	12.5	164 23.2	08.0	267 16.4	12.4	322 11.6	01.9
	20	59 46.9	166 58.5	12.9	179 24.0	08.3	282 19.1	12.5	337 14.2	01.9
	21	74 49.4	181 58.2	. . 13.3	194 24.8	. . 08.6	297 21.9	. . 12.6	352 16.8	. . 01.9
	22	89 51.9	196 57.9	13.7	209 25.6	09.0	312 24.7	12.7	7 19.4	01.9
	23	104 54.3	211 57.5	14.1	224 26.4	09.3	327 27.5	12.8	22 22.0	01.9
	21 00	119 56.8	226 57.2	S19 14.5	239 27.2	S20 09.7	342 30.3	N17 12.9	37 24.7	N22 01.9
	01	134 59.3	241 56.8	14.9	254 27.9	10.0	357 33.0	13.0	52 27.3	01.9
	02	150 01.7	256 56.5	15.3	269 28.7	10.3	12 35.8	13.1	67 29.9	01.9
	03	165 04.2	271 56.1	. . 15.7	284 29.5	. . 10.7	27 38.6	. . 13.2	82 32.5	. . 01.9
	04	180 06.7	286 55.8	16.1	299 30.3	11.0	42 41.4	13.3	97 35.1	01.9
	05	195 09.1	301 55.4	16.5	314 31.1	11.4	57 44.1	13.4	112 37.7	01.9
	06	210 11.6	316 55.1	S19 16.9	329 31.9	S20 11.7	72 46.9	N17 13.5	127 40.3	N22 01.9
	07	225 14.0	331 54.8	17.3	344 32.7	12.0	87 49.7	13.6	142 43.0	01.9
T	08	240 16.5	346 54.4	17.7	359 33.4	12.4	102 52.5	13.7	157 45.6	01.9
	09	255 19.0	1 54.1	. . 18.1	14 34.2	. . 12.7	117 55.2	. . 13.8	172 48.2	. . 01.9
U	10	270 21.4	16 53.7	18.5	29 35.0	13.1	132 58.0	13.9	187 50.8	01.9
E	11	285 23.9	31 53.4	18.9	44 35.8	13.4	148 00.8	14.0	202 53.4	01.9
S	12	300 26.4	46 53.0	S19 19.3	59 36.6	S20 13.7	163 03.6	N17 14.1	217 56.0	N22 01.9
D	13	315 28.8	61 52.7	19.7	74 37.4	14.1	178 06.4	14.2	232 58.6	01.9
A	14	330 31.3	76 52.3	20.1	89 38.2	14.4	193 09.1	14.3	248 01.3	01.9
Y	15	345 33.8	91 51.9	. . 20.5	104 38.9	. . 14.8	208 11.9	. . 14.4	263 03.9	. . 01.9
	16	0 36.2	106 51.6	20.9	119 39.7	15.1	223 14.7	14.5	278 06.5	01.9
	17	15 38.7	121 51.2	21.3	134 40.5	15.4	238 17.5	14.6	293 09.1	01.9
	18	30 41.2	136 50.9	S19 21.7	149 41.3	S20 15.8	253 20.3	N17 14.7	308 11.7	N22 01.9
	19	45 43.6	151 50.5	22.1	164 42.1	16.1	268 23.0	14.8	323 14.3	01.9
	20	60 46.1	166 50.2	22.5	179 42.9	16.4	283 25.8	14.9	338 16.9	01.9
	21	75 48.5	181 49.8	. . 22.9	194 43.6	. . 16.8	298 28.6	. . 15.0	353 19.6	. . 01.9
	22	90 51.0	196 49.4	23.3	209 44.4	17.1	313 31.4	15.0	8 22.2	01.9
	23	105 53.5	211 49.1	23.6	224 45.2	17.4	328 34.2	15.1	23 24.8	01.9
	Mer. Pass.	h m 16 01.5	v −0.3	d 0.4	v 0.8	d 0.3	v 2.8	d 0.1	v 2.6	d 0.0

STARS

Name	SHA	Dec
	° ′	° ′
Acamar	315 24.3	S40 17.8
Achernar	335 32.7	S57 13.7
Acrux	173 18.5	S63 06.6
Adhara	255 18.6	S28 58.6
Aldebaran	290 58.6	N16 30.9
Alioth	166 27.5	N55 56.4
Alkaid	153 05.2	N49 17.6
Al Na'ir	27 54.3	S46 57.1
Alnilam	275 54.4	S 1 12.0
Alphard	218 03.8	S 8 40.3
Alphecca	126 18.1	N26 42.1
Alpheratz	357 52.2	N29 06.4
Altair	62 16.6	N 8 52.4
Ankaa	353 23.8	S42 17.7
Antares	112 36.6	S26 26.3
Arcturus	146 03.2	N19 09.9
Atria	107 46.3	S69 01.8
Avior	234 20.8	S59 31.1
Bellatrix	278 40.5	N 6 21.1
Betelgeuse	271 09.9	N 7 24.5
Canopus	263 59.3	S52 41.9
Capella	280 46.2	N46 00.2
Deneb	49 37.6	N45 17.4
Denebola	182 41.8	N14 33.3
Diphda	349 04.2	S17 58.4
Dubhe	194 01.0	N61 43.9
Elnath	278 22.7	N28 36.7
Eltanin	90 50.4	N51 29.1
Enif	33 55.5	N 9 53.2
Fomalhaut	15 33.2	S29 36.6
Gacrux	172 10.1	S57 07.5
Gienah	176 00.6	S17 33.4
Hadar	148 59.8	S60 22.9
Hamal	328 10.0	N23 28.6
Kaus Aust.	83 55.0	S34 23.0
Kochab	137 19.7	N74 08.3
Markab	13 46.8	N15 13.2
Menkar	314 23.5	N 4 06.0
Menkent	148 17.4	S36 22.9
Miaplacidus	221 40.8	S69 43.6
Mirfak	308 52.0	N49 52.5
Nunki	76 08.8	S26 17.7
Peacock	53 32.6	S56 43.7
Pollux	243 37.3	N28 01.2
Procyon	245 08.0	N 5 13.0
Rasalhague	96 14.3	N12 33.3
Regulus	207 51.9	N11 57.1
Rigel	281 19.7	S 8 12.0
Rigil Kent.	140 03.4	S60 50.6
Sabik	102 22.2	S15 43.8
Schedar	349 50.3	N56 33.4
Shaula	96 33.4	S37 06.3
Sirius	258 40.6	S16 43.2
Spica	158 39.9	S11 10.6
Suhail	222 58.1	S43 26.6
Vega	80 44.9	N38 47.0
Zuben'ubi	137 14.6	S16 03.2

	SHA	Mer. Pass.
	° ′	h m
Venus	108 07.5	8 52
Mars	120 10.6	8 03
Jupiter	222 26.0	1 14
Saturn	277 24.2	21 31

Day	UT (h)	SUN GHA (° ′)	SUN Dec (° ′)	MOON GHA (° ′)	v (′)	MOON Dec (° ′)	d (′)	HP (′)
19 SUNDAY	00	177 22.8	S20 27.9	349 04.3	7.1	N23 06.6	7.0	58.0
	01	192 22.6	27.4	3 30.4	7.2	22 59.6	7.2	58.0
	02	207 22.4	26.9	17 56.6	7.3	22 52.4	7.3	58.0
	03	222 22.2	. . 26.4	32 22.9	7.2	22 45.1	7.5	58.0
	04	237 22.0	25.9	46 49.1	7.4	22 37.6	7.6	58.0
	05	252 21.8	25.4	61 15.5	7.3	22 30.0	7.8	58.1
	06	267 21.6	S20 24.8	75 41.8	7.4	N22 22.2	7.9	58.1
	07	282 21.4	24.3	90 08.2	7.5	22 14.3	8.0	58.1
	08	297 21.2	23.8	104 34.7	7.4	22 06.3	8.2	58.1
	09	312 21.0	. . 23.3	119 01.1	7.6	21 58.1	8.3	58.2
	10	327 20.8	22.8	133 27.7	7.5	21 49.8	8.4	58.2
	11	342 20.6	22.2	147 54.2	7.7	21 41.4	8.6	58.2
	12	357 20.4	S20 21.7	162 20.9	7.6	N21 32.8	8.7	58.2
	13	12 20.2	21.2	176 47.5	7.8	21 24.1	8.9	58.2
	14	27 20.1	20.7	191 14.3	7.7	21 15.2	8.9	58.3
	15	42 19.9	. . 20.1	205 41.0	7.8	21 06.3	9.2	58.3
	16	57 19.7	19.6	220 07.8	7.9	20 57.1	9.2	58.3
	17	72 19.5	19.1	234 34.7	7.9	20 47.9	9.4	58.3
	18	87 19.3	S20 18.6	249 01.6	8.0	N20 38.5	9.4	58.3
	19	102 19.1	18.0	263 28.6	8.0	20 29.1	9.7	58.4
	20	117 18.9	17.5	277 55.6	8.1	20 19.4	9.7	58.4
	21	132 18.7	. . 17.0	292 22.7	8.1	20 09.7	9.9	58.4
	22	147 18.5	16.5	306 49.8	8.2	19 59.8	10.0	58.4
	23	162 18.3	15.9	321 17.0	8.2	19 49.8	10.1	58.4
20 MONDAY	00	177 18.2	S20 15.4	335 44.2	8.3	N19 39.7	10.2	58.5
	01	192 18.0	14.9	350 11.5	8.3	19 29.5	10.4	58.5
	02	207 17.8	14.3	4 38.8	8.4	19 19.1	10.4	58.5
	03	222 17.6	. . 13.8	19 06.2	8.5	19 08.7	10.6	58.5
	04	237 17.4	13.3	33 33.7	8.5	18 58.1	10.7	58.5
	05	252 17.2	12.7	48 01.2	8.5	18 47.4	10.8	58.5
	06	267 17.0	S20 12.2	62 28.7	8.7	N18 36.6	11.0	58.6
	07	282 16.9	11.7	76 56.4	8.6	18 25.6	11.0	58.6
	08	297 16.7	11.1	91 24.0	8.8	18 14.6	11.2	58.6
	09	312 16.5	. . 10.6	105 51.8	8.7	18 03.4	11.2	58.6
	10	327 16.3	10.0	120 19.5	8.9	17 52.2	11.4	58.6
	11	342 16.1	09.5	134 47.4	8.9	17 40.8	11.5	58.6
	12	357 15.9	S20 09.0	149 15.3	8.9	N17 29.3	11.6	58.7
	13	12 15.7	08.4	163 43.2	9.1	17 17.7	11.6	58.7
	14	27 15.6	07.9	178 11.3	9.0	17 06.1	11.8	58.7
	15	42 15.4	. . 07.4	192 39.3	9.1	16 54.3	11.9	58.7
	16	57 15.2	06.8	207 07.4	9.2	16 42.4	12.0	58.7
	17	72 15.0	06.3	221 35.6	9.3	16 30.4	12.1	58.7
	18	87 14.8	S20 05.7	236 03.9	9.3	N16 18.3	12.2	58.7
	19	102 14.7	05.2	250 32.2	9.3	16 06.1	12.3	58.8
	20	117 14.5	04.6	265 00.5	9.4	15 53.8	12.3	58.8
	21	132 14.3	. . 04.1	279 28.9	9.5	15 41.5	12.5	58.8
	22	147 14.1	03.5	293 57.4	9.5	15 29.0	12.6	58.8
	23	162 13.9	03.0	308 25.9	9.5	15 16.4	12.6	58.8
21 TUESDAY	00	177 13.7	S20 02.5	322 54.4	9.7	N15 03.8	12.8	58.8
	01	192 13.6	01.9	337 23.1	9.6	14 51.0	12.8	58.8
	02	207 13.4	01.4	351 51.7	9.8	14 38.2	12.9	58.9
	03	222 13.2	. . 00.8	6 20.5	9.8	14 25.3	13.0	58.9
	04	237 13.0	20 00.3	20 49.3	9.8	14 12.3	13.1	58.9
	05	252 12.9	19 59.7	35 18.1	9.9	13 59.2	13.1	58.9
	06	267 12.7	S19 59.2	49 47.0	9.9	N13 46.1	13.3	58.9
	07	282 12.5	58.6	64 15.9	10.0	13 32.8	13.3	58.9
	08	297 12.3	58.1	78 44.9	10.1	13 19.5	13.4	58.9
	09	312 12.1	. . 57.5	93 14.0	10.1	13 06.1	13.5	58.9
	10	327 12.0	57.0	107 43.1	10.1	12 52.6	13.5	59.0
	11	342 11.8	56.4	122 12.2	10.2	12 39.1	13.7	59.0
	12	357 11.6	S19 55.8	136 41.4	10.2	N12 25.4	13.7	59.0
	13	12 11.4	55.3	151 10.6	10.3	12 11.7	13.7	59.0
	14	27 11.3	54.7	165 39.9	10.4	11 58.0	13.9	59.0
	15	42 11.1	. . 54.2	180 09.3	10.4	11 44.1	13.9	59.0
	16	57 10.9	53.6	194 38.7	10.4	11 30.2	13.9	59.0
	17	72 10.7	53.1	209 08.1	10.5	11 16.3	14.1	59.0
	18	87 10.6	S19 52.5	223 37.6	10.5	N11 02.2	14.1	59.0
	19	102 10.4	51.9	238 07.1	10.5	10 48.1	14.1	59.1
	20	117 10.2	51.4	252 36.6	10.7	10 34.0	14.2	59.1
	21	132 10.0	. . 50.8	267 06.3	10.6	10 19.8	14.3	59.1
	22	147 09.9	50.3	281 35.9	10.7	10 05.5	14.4	59.1
	23	162 09.7	49.7	296 05.6	10.7	N 9 51.1	14.3	59.1
		SD 16.3	*d* 0.5	SD 15.9		16.0		16.1

Lat.	Twilight Naut. (h m)	Twilight Civil (h m)	Sunrise (h m)	Moonrise 19 (h m)	Moonrise 20 (h m)	Moonrise 21 (h m)	Moonrise 22 (h m)
N 72	07 49	09 32	■	□	15 44	18 45	21 07
N 70	07 36	09 04	11 09	□	16 36	19 03	21 14
68	07 25	08 42	10 12	14 27	17 08	19 18	21 19
66	07 16	08 25	09 39	15 22	17 31	19 30	21 23
64	07 08	08 11	09 14	15 54	17 49	19 40	21 26
62	07 02	07 59	08 56	16 18	18 04	19 48	21 29
60	06 56	07 49	08 40	16 37	18 16	19 55	21 32
N 58	06 50	07 40	08 27	16 53	18 27	20 01	21 34
56	06 45	07 32	08 15	17 06	18 36	20 07	21 36
54	06 41	07 25	08 06	17 18	18 44	20 12	21 38
52	06 36	07 18	07 57	17 28	18 52	20 16	21 40
50	06 32	07 12	07 49	17 37	18 58	20 20	21 42
45	06 23	06 59	07 32	17 56	19 12	20 29	21 45
N 40	06 15	06 48	07 18	18 12	19 24	20 36	21 48
35	06 08	06 39	07 06	18 25	19 33	20 42	21 50
30	06 01	06 30	06 56	18 36	19 42	20 48	21 52
20	05 47	06 14	06 38	18 55	19 57	20 57	21 56
N 10	05 34	06 00	06 22	19 12	20 09	21 05	21 59
0	05 20	05 45	06 07	19 28	20 21	21 13	22 02
S 10	05 03	05 30	05 52	19 43	20 33	21 20	22 05
20	04 44	05 12	05 36	20 00	20 46	21 28	22 08
30	04 19	04 51	05 18	20 19	21 00	21 37	22 12
35	04 03	04 38	05 07	20 30	21 08	21 43	22 14
40	03 44	04 23	04 54	20 43	21 18	21 49	22 17
45	03 19	04 04	04 39	20 58	21 29	21 55	22 19
S 50	02 44	03 40	04 21	21 16	21 42	22 04	22 23
52	02 26	03 28	04 12	21 24	21 48	22 07	22 24
54	02 02	03 14	04 02	21 34	21 55	22 12	22 26
56	01 29	02 58	03 51	21 44	22 02	22 16	22 27
58	////	02 38	03 38	21 57	22 11	22 21	22 29
S 60	////	02 13	03 23	22 11	22 20	22 27	22 32

Lat.	Sunset (h m)	Twilight Civil (h m)	Twilight Naut. (h m)	Moonset 19 (h m)	Moonset 20 (h m)	Moonset 21 (h m)	Moonset 22 (h m)
N 72	■	14 50	16 34	□	12 53	11 42	11 05
N 70	13 14	15 19	16 47	□	11 59	11 21	10 56
68	14 11	15 41	16 58	12 14	11 26	11 04	10 48
66	14 44	15 58	17 06	11 18	11 02	10 51	10 42
64	15 08	16 12	17 14	10 45	10 42	10 39	10 36
62	15 27	16 24	17 21	10 20	10 26	10 30	10 31
60	15 43	16 34	17 27	10 00	10 13	10 21	10 27
N 58	15 56	16 43	17 32	09 44	10 02	10 14	10 23
56	16 07	16 51	17 37	09 30	09 52	10 07	10 20
54	16 17	16 58	17 42	09 18	09 43	10 01	10 17
52	16 26	17 04	17 46	09 07	09 35	09 56	10 14
50	16 34	17 10	17 50	08 58	09 27	09 51	10 11
45	16 50	17 23	17 59	08 38	09 12	09 41	10 06
N 40	17 04	17 34	18 07	08 21	08 59	09 32	10 01
35	17 16	17 43	18 14	08 07	08 48	09 24	09 57
30	17 26	17 52	18 21	07 55	08 39	09 18	09 53
20	17 44	18 08	18 35	07 34	08 22	09 06	09 47
N 10	18 00	18 22	18 48	07 16	08 07	08 56	09 42
0	18 15	18 37	19 02	06 59	07 54	08 46	09 36
S 10	18 29	18 52	19 18	06 42	07 40	08 36	09 31
20	18 45	19 09	19 38	06 23	07 25	08 26	09 25
30	19 04	19 30	20 02	06 02	07 08	08 13	09 19
35	19 15	19 43	20 18	05 49	06 57	08 06	09 15
40	19 27	19 59	20 37	05 34	06 46	07 58	09 10
45	19 42	20 17	21 02	05 17	06 32	07 49	09 05
S 50	20 00	20 41	21 36	04 55	06 15	07 37	08 59
52	20 09	20 53	21 54	04 45	06 07	07 32	08 56
54	20 19	21 06	22 17	04 33	05 58	07 26	08 53
56	20 30	21 22	22 49	04 20	05 48	07 19	08 49
58	20 43	21 41	////	04 04	05 37	07 12	08 46
S 60	20 57	22 06	////	03 45	05 24	07 03	08 41

Day	SUN Eqn. of Time 00^h (m s)	SUN Eqn. of Time 12^h (m s)	SUN Mer. Pass. (h m)	MOON Mer. Pass. Upper (h m)	MOON Mer. Pass. Lower (h m)	MOON Age (d)	MOON Phase (%)
19	10 29	10 38	12 11	00 45	13 13	17	99
20	10 47	10 56	12 11	01 41	14 08	18	95
21	11 05	11 13	12 11	02 34	14 59	19	89

UT d h	ARIES GHA ° ′	VENUS −4.3 GHA ° ′	VENUS Dec ° ′	MARS +1.3 GHA ° ′	MARS Dec ° ′	JUPITER −2.6 GHA ° ′	JUPITER Dec ° ′	SATURN −0.3 GHA ° ′	SATURN Dec ° ′
22 00 Wednesday	120 55.9	226 48.7	S19 24.0	239 46.0	S20 17.8	343 36.9	N17 15.2	38 27.4	N22 01.9
01	135 58.4	241 48.4	24.4	254 46.8	18.1	358 39.7	15.3	53 30.0	01.9
02	151 00.9	256 48.0	24.8	269 47.6	18.4	13 42.5	15.4	68 32.6	01.9
03	166 03.3	271 47.6	25.2	284 48.3	18.8	28 45.3	15.5	83 35.2	01.9
04	181 05.8	286 47.3	25.6	299 49.1	19.1	43 48.1	15.6	98 37.8	01.9
05	196 08.3	301 46.9	26.0	314 49.9	19.4	58 50.8	15.7	113 40.4	01.9
06	211 10.7	316 46.5	S19 26.4	329 50.7	S20 19.8	73 53.6	N17 15.8	128 43.1	N22 01.9
07	226 13.2	331 46.2	26.8	344 51.5	20.1	88 56.4	15.9	143 45.7	01.9
08	241 15.6	346 45.8	27.1	359 52.3	20.4	103 59.2	16.0	158 48.3	01.9
09	256 18.1	1 45.4	27.5	14 53.0	20.8	119 02.0	16.1	173 50.9	01.9
10	271 20.6	16 45.1	27.9	29 53.8	21.1	134 04.7	16.2	188 53.5	01.9
11	286 23.0	31 44.7	28.3	44 54.6	21.4	149 07.5	16.3	203 56.1	01.9
12	301 25.5	46 44.3	S19 28.7	59 55.4	S20 21.8	164 10.3	N17 16.4	218 58.7	N22 01.9
13	316 28.0	61 44.0	29.1	74 56.2	22.1	179 13.1	16.5	234 01.3	01.9
14	331 30.4	76 43.6	29.5	89 56.9	22.4	194 15.9	16.6	249 03.9	01.9
15	346 32.9	91 43.2	29.8	104 57.7	22.8	209 18.6	16.7	264 06.5	01.9
16	1 35.4	106 42.9	30.2	119 58.5	23.1	224 21.4	16.8	279 09.2	01.9
17	16 37.8	121 42.5	30.6	134 59.3	23.4	239 24.2	16.9	294 11.8	01.9
18	31 40.3	136 42.1	S19 31.0	150 00.1	S20 23.8	254 27.0	N17 17.0	309 14.4	N22 01.9
19	46 42.8	151 41.7	31.4	165 00.9	24.1	269 29.8	17.1	324 17.0	01.9
20	61 45.2	166 41.4	31.7	180 01.6	24.4	284 32.6	17.2	339 19.6	01.9
21	76 47.7	181 41.0	32.1	195 02.4	24.8	299 35.3	17.3	354 22.2	01.9
22	91 50.1	196 40.6	32.5	210 03.2	25.1	314 38.1	17.4	9 24.8	01.9
23	106 52.6	211 40.2	32.9	225 04.0	25.4	329 40.9	17.5	24 27.4	01.9
23 00 Thursday	121 55.1	226 39.9	S19 33.2	240 04.8	S20 25.7	344 43.7	N17 17.6	39 30.0	N22 01.9
01	136 57.5	241 39.5	33.6	255 05.5	26.1	359 46.5	17.7	54 32.6	01.9
02	152 00.0	256 39.1	34.0	270 06.3	26.4	14 49.3	17.8	69 35.2	01.9
03	167 02.5	271 38.7	34.4	285 07.1	26.7	29 52.0	17.9	84 37.8	01.9
04	182 04.9	286 38.3	34.7	300 07.9	27.0	44 54.8	18.0	99 40.5	01.9
05	197 07.4	301 38.0	35.1	315 08.7	27.4	59 57.6	18.1	114 43.1	01.9
06	212 09.9	316 37.6	S19 35.5	330 09.4	S20 27.7	75 00.4	N17 18.2	129 45.7	N22 01.9
07	227 12.3	331 37.2	35.9	345 10.2	28.0	90 03.2	18.3	144 48.3	01.9
08	242 14.8	346 36.8	36.2	0 11.0	28.4	105 06.0	18.4	159 50.9	01.9
09	257 17.3	1 36.4	36.6	15 11.8	28.7	120 08.7	18.5	174 53.5	01.9
10	272 19.7	16 36.0	37.0	30 12.6	29.0	135 11.5	18.6	189 56.1	01.9
11	287 22.2	31 35.6	37.3	45 13.3	29.3	150 14.3	18.7	204 58.7	01.9
12	302 24.6	46 35.3	S19 37.7	60 14.1	S20 29.7	165 17.1	N17 18.8	220 01.3	N22 01.9
13	317 27.1	61 34.9	38.1	75 14.9	30.0	180 19.9	18.9	235 03.9	01.9
14	332 29.6	76 34.5	38.4	90 15.7	30.3	195 22.7	19.0	250 06.5	01.9
15	347 32.0	91 34.1	38.8	105 16.5	30.6	210 25.4	19.1	265 09.1	01.9
16	2 34.5	106 33.7	39.2	120 17.2	31.0	225 28.2	19.2	280 11.7	01.9
17	17 37.0	121 33.3	39.5	135 18.0	31.3	240 31.0	19.3	295 14.3	01.9
18	32 39.4	136 32.9	S19 39.9	150 18.8	S20 31.6	255 33.8	N17 19.4	310 16.9	N22 01.9
19	47 41.9	151 32.5	40.3	165 19.6	31.9	270 36.6	19.5	325 19.5	01.9
20	62 44.4	166 32.1	40.6	180 20.3	32.3	285 39.4	19.6	340 22.2	01.9
21	77 46.8	181 31.8	41.0	195 21.1	32.6	300 42.2	19.7	355 24.8	01.9
22	92 49.3	196 31.4	41.4	210 21.9	32.9	315 44.9	19.8	10 27.4	01.9
23	107 51.7	211 31.0	41.7	225 22.7	33.2	330 47.7	19.9	25 30.0	01.9
24 00 Friday	122 54.2	226 30.6	S19 42.1	240 23.5	S20 33.5	345 50.5	N17 20.0	40 32.6	N22 01.9
01	137 56.7	241 30.2	42.4	255 24.2	33.9	0 53.3	20.1	55 35.2	01.9
02	152 59.1	256 29.8	42.8	270 25.0	34.2	15 56.1	20.2	70 37.8	01.9
03	168 01.6	271 29.4	43.2	285 25.8	34.5	30 58.9	20.3	85 40.4	01.9
04	183 04.1	286 29.0	43.5	300 26.6	34.8	46 01.7	20.4	100 43.0	01.9
05	198 06.5	301 28.6	43.9	315 27.3	35.2	61 04.4	20.5	115 45.6	01.9
06	213 09.0	316 28.2	S19 44.2	330 28.1	S20 35.5	76 07.2	N17 20.6	130 48.2	N22 01.9
07	228 11.5	331 27.8	44.6	345 28.9	35.8	91 10.0	20.7	145 50.8	01.9
08	243 13.9	346 27.4	44.9	0 29.7	36.1	106 12.8	20.8	160 53.4	01.9
09	258 16.4	1 27.0	45.3	15 30.4	36.4	121 15.6	20.9	175 56.0	01.9
10	273 18.9	16 26.6	45.6	30 31.2	36.8	136 18.4	21.0	190 58.6	01.9
11	288 21.3	31 26.2	46.0	45 32.0	37.1	151 21.2	21.1	206 01.2	01.9
12	303 23.8	46 25.8	S19 46.4	60 32.8	S20 37.4	166 23.9	N17 21.2	221 03.8	N22 01.9
13	318 26.2	61 25.4	46.7	75 33.6	37.7	181 26.7	21.3	236 06.4	01.9
14	333 28.7	76 25.0	47.1	90 34.3	38.0	196 29.5	21.4	251 09.0	01.9
15	348 31.2	91 24.6	47.4	105 35.1	38.4	211 32.3	21.5	266 11.6	02.0
16	3 33.6	106 24.2	47.8	120 35.9	38.7	226 35.1	21.6	281 14.2	02.0
17	18 36.1	121 23.8	48.1	135 36.7	39.0	241 37.9	21.7	296 16.8	02.0
18	33 38.6	136 23.3	S19 48.5	150 37.4	S20 39.3	256 40.7	N17 21.8	311 19.4	N22 02.0
19	48 41.0	151 22.9	48.8	165 38.2	39.6	271 43.5	21.9	326 22.0	02.0
20	63 43.5	166 22.5	49.2	180 39.0	39.9	286 46.2	22.0	341 24.6	02.0
21	78 46.0	181 22.1	49.5	195 39.8	40.3	301 49.0	22.1	356 27.2	02.0
22	93 48.4	196 21.7	49.8	210 40.5	40.6	316 51.8	22.2	11 29.8	02.0
23	108 50.9	211 21.3	50.2	225 41.3	40.9	331 54.6	22.3	26 32.4	02.0
Mer. Pass. (h m)	15 49.7	v −0.4	d 0.4	v 0.8	d 0.3	v 2.8	d 0.1	v 2.6	d 0.0

STARS

Name	SHA ° ′	Dec ° ′
Acamar	315 24.4	S40 17.8
Achernar	335 32.7	S57 13.7
Acrux	173 18.5	S63 06.6
Adhara	255 18.6	S28 58.6
Aldebaran	290 58.6	N16 30.9
Alioth	166 27.5	N55 56.4
Alkaid	153 05.2	N49 17.6
Al Na'ir	27 54.3	S46 57.1
Alnilam	275 54.4	S 1 12.0
Alphard	218 03.8	S 8 40.3
Alphecca	126 18.0	N26 42.1
Alpheratz	357 52.2	N29 06.4
Altair	62 16.6	N 8 52.4
Ankaa	353 23.8	S42 17.7
Antares	112 36.6	S26 26.3
Arcturus	146 03.2	N19 09.9
Atria	107 46.3	S69 01.8
Avior	234 20.8	S59 31.1
Bellatrix	278 40.5	N 6 21.1
Betelgeuse	271 09.9	N 7 24.5
Canopus	263 59.3	S52 41.9
Capella	280 46.2	N46 00.2
Deneb	49 37.6	N45 17.4
Denebola	182 41.8	N14 33.2
Diphda	349 04.2	S17 58.4
Dubhe	194 00.9	N61 43.9
Elnath	278 22.7	N28 36.7
Eltanin	90 50.4	N51 29.1
Enif	33 55.5	N 9 53.2
Fomalhaut	15 33.2	S29 36.6
Gacrux	172 10.0	S57 07.5
Gienah	176 00.6	S17 33.4
Hadar	148 59.8	S60 23.0
Hamal	328 10.1	N23 28.6
Kaus Aust.	83 55.0	S34 23.0
Kochab	137 19.6	N74 08.3
Markab	13 46.8	N15 13.2
Menkar	314 23.5	N 4 06.0
Menkent	148 17.3	S36 22.9
Miaplacidus	221 40.8	S69 43.6
Mirfak	308 52.0	N49 52.5
Nunki	76 08.8	S26 17.7
Peacock	53 32.6	S56 43.7
Pollux	243 37.3	N28 01.2
Procyon	245 08.0	N 5 13.0
Rasalhague	96 14.3	N12 33.3
Regulus	207 51.9	N11 57.1
Rigel	281 19.7	S 8 12.0
Rigil Kent.	140 03.3	S60 50.6
Sabik	102 22.2	S15 43.8
Schedar	349 50.3	N56 33.4
Shaula	96 33.4	S37 06.3
Sirius	258 40.6	S16 43.2
Spica	158 39.9	S11 10.6
Suhail	222 58.1	S43 26.6
Vega	80 44.9	N38 47.0
Zuben'ubi	137 14.6	S16 03.2

	SHA ° ′	Mer. Pass. h m
Venus	104 44.8	8 54
Mars	118 09.7	7 59
Jupiter	222 48.6	1 01
Saturn	277 35.0	21 18

UT d	UT h	SUN GHA	SUN Dec	MOON GHA	v	MOON Dec	d	HP
d	h	° ′	° ′	° ′	′	° ′	′	′
22	00	177 09.5	S19 49.1	310 35.3	10.8	N 9 36.8	14.5	59.1
	01	192 09.4	48.6	325 05.1	10.8	9 22.3	14.5	59.1
	02	207 09.2	48.0	339 34.9	10.9	9 07.8	14.5	59.1
	03	222 09.0	. . 47.5	354 04.8	10.8	8 53.3	14.6	59.1
	04	237 08.8	46.9	8 34.6	11.0	8 38.7	14.7	59.1
	05	252 08.7	46.3	23 04.6	10.9	8 24.0	14.7	59.1
	06	267 08.5	S19 45.8	37 34.5	11.0	N 8 09.3	14.7	59.1
W	07	282 08.3	45.2	52 04.5	11.0	7 54.6	14.8	59.1
E	08	297 08.2	44.6	66 34.5	11.1	7 39.8	14.8	59.2
D	09	312 08.0	. . 44.1	81 04.6	11.1	7 25.0	14.9	59.2
N	10	327 07.8	43.5	95 34.7	11.1	7 10.1	14.9	59.2
E	11	342 07.6	42.9	110 04.8	11.2	6 55.2	15.0	59.2
S	12	357 07.5	S19 42.4	124 35.0	11.1	N 6 40.2	15.0	59.2
D	13	12 07.3	41.8	139 05.1	11.3	6 25.2	15.0	59.2
A	14	27 07.1	41.2	153 35.4	11.2	6 10.2	15.0	59.2
Y	15	42 07.0	. . 40.6	168 05.6	11.3	5 55.2	15.1	59.2
	16	57 06.8	40.1	182 35.9	11.2	5 40.1	15.1	59.2
	17	72 06.6	39.5	197 06.1	11.3	5 25.0	15.2	59.2
	18	87 06.5	S19 38.9	211 36.4	11.4	N 5 09.8	15.2	59.2
	19	102 06.3	38.3	226 06.8	11.3	4 54.6	15.2	59.2
	20	117 06.1	37.8	240 37.1	11.4	4 39.4	15.2	59.2
	21	132 06.0	. . 37.2	255 07.5	11.4	4 24.2	15.3	59.2
	22	147 05.8	36.6	269 37.9	11.4	4 08.9	15.2	59.2
	23	162 05.6	36.0	284 08.3	11.4	3 53.7	15.3	59.2
23	00	177 05.5	S19 35.5	298 38.7	11.5	N 3 38.4	15.4	59.2
	01	192 05.3	34.9	313 09.2	11.5	3 23.0	15.3	59.2
	02	207 05.2	34.3	327 39.7	11.4	3 07.7	15.4	59.2
	03	222 05.0	. . 33.7	342 10.1	11.5	2 52.3	15.3	59.2
	04	237 04.8	33.2	356 40.6	11.5	2 37.0	15.4	59.3
	05	252 04.7	32.6	11 11.1	11.5	2 21.6	15.4	59.3
	06	267 04.5	S19 32.0	25 41.6	11.6	N 2 06.2	15.4	59.3
	07	282 04.3	31.4	40 12.2	11.5	1 50.8	15.4	59.3
T	08	297 04.2	30.8	54 42.7	11.5	1 35.4	15.5	59.3
H	09	312 04.0	. . 30.2	69 13.2	11.6	1 19.9	15.4	59.3
U	10	327 03.9	29.7	83 43.8	11.5	1 04.5	15.4	59.3
R	11	342 03.7	29.1	98 14.3	11.6	0 49.1	15.5	59.3
S	12	357 03.5	S19 28.5	112 44.9	11.5	N 0 33.6	15.4	59.3
D	13	12 03.4	27.9	127 15.4	11.6	0 18.2	15.5	59.3
A	14	27 03.2	27.3	141 46.0	11.6	N 0 02.7	15.4	59.3
Y	15	42 03.1	. . 26.7	156 16.6	11.5	S 0 12.7	15.5	59.3
	16	57 02.9	26.1	170 47.1	11.6	0 28.2	15.4	59.3
	17	72 02.7	25.6	185 17.7	11.5	0 43.6	15.5	59.3
	18	87 02.6	S19 25.0	199 48.2	11.6	S 0 59.1	15.4	59.3
	19	102 02.4	24.4	214 18.8	11.5	1 14.5	15.5	59.3
	20	117 02.3	23.8	228 49.3	11.6	1 30.0	15.4	59.3
	21	132 02.1	. . 23.2	243 19.9	11.5	1 45.4	15.4	59.3
	22	147 01.9	22.6	257 50.4	11.5	2 00.8	15.4	59.3
	23	162 01.8	22.0	272 20.9	11.5	2 16.2	15.4	59.3
24	00	177 01.6	S19 21.4	286 51.4	11.5	S 2 31.6	15.4	59.3
	01	192 01.5	20.8	301 21.9	11.5	2 47.0	15.3	59.3
	02	207 01.3	20.2	315 52.4	11.5	3 02.3	15.4	59.3
	03	222 01.2	. . 19.6	330 22.9	11.5	3 17.7	15.3	59.3
	04	237 01.0	19.0	344 53.4	11.4	3 33.0	15.3	59.3
	05	252 00.9	18.5	359 23.8	11.5	3 48.3	15.3	59.3
	06	267 00.7	S19 17.9	13 54.3	11.4	S 4 03.6	15.2	59.3
	07	282 00.5	17.3	28 24.7	11.4	4 18.8	15.3	59.3
	08	297 00.4	16.7	42 55.1	11.3	4 34.1	15.2	59.3
F	09	312 00.2	. . 16.1	57 25.4	11.4	4 49.3	15.2	59.3
R	10	327 00.1	15.5	71 55.8	11.3	5 04.5	15.2	59.3
I	11	341 59.9	14.9	86 26.1	11.3	5 19.7	15.1	59.3
D	12	356 59.8	S19 14.3	100 56.4	11.3	S 5 34.8	15.1	59.3
A	13	11 59.6	13.7	115 26.7	11.3	5 49.9	15.1	59.3
Y	14	26 59.5	13.1	129 57.0	11.2	6 05.0	15.0	59.3
	15	41 59.3	. . 12.5	144 27.2	11.2	6 20.0	15.0	59.3
	16	56 59.2	11.9	158 57.4	11.2	6 35.0	15.0	59.3
	17	71 59.0	11.3	173 27.6	11.1	6 50.0	14.9	59.3
	18	86 58.9	S19 10.7	187 57.7	11.1	S 7 04.9	14.9	59.3
	19	101 58.7	10.1	202 27.8	11.1	7 19.8	14.8	59.2
	20	116 58.6	09.4	216 57.9	11.1	7 34.6	14.8	59.2
	21	131 58.4	. . 08.8	231 28.0	11.0	7 49.4	14.8	59.2
	22	146 58.3	08.2	245 58.0	11.0	8 04.2	14.7	59.2
	23	161 58.1	07.6	260 28.0	10.9	S 8 18.9	14.7	59.2
		SD 16.3	d 0.6	SD		16.1	16.1	16.1

Lat.	Twilight Naut.	Twilight Civil	Sunrise	Moonrise 22	Moonrise 23	Moonrise 24	Moonrise 25
°	h m	h m	h m	h m	h m	h m	h m
N 72	07 40	09 20	▬	21 07	23 23	25 42	01 42
N 70	07 29	08 54	10 44	21 14	23 19	25 27	01 27
68	07 19	08 34	09 58	21 19	23 16	25 15	01 15
66	07 11	08 18	09 29	21 23	23 13	25 06	01 06
64	07 03	08 05	09 07	21 26	23 11	24 58	00 58
62	06 57	07 54	08 49	21 29	23 10	24 51	00 51
60	06 52	07 44	08 34	21 32	23 08	24 45	00 45
N 58	06 46	07 36	08 22	21 34	23 07	24 39	00 39
56	06 42	07 28	08 11	21 36	23 05	24 35	00 35
54	06 38	07 21	08 02	21 38	23 04	24 31	00 31
52	06 34	07 15	07 53	21 40	23 03	24 27	00 27
50	06 30	07 10	07 46	21 42	23 03	24 24	00 24
45	06 21	06 57	07 29	21 45	23 01	24 16	00 16
N 40	06 14	06 47	07 16	21 48	22 59	24 10	00 10
35	06 07	06 38	07 05	21 50	22 58	24 05	00 05
30	06 00	06 29	06 55	21 52	22 56	24 01	00 01
20	05 47	06 14	06 38	21 56	22 54	23 53	24 53
N 10	05 35	06 00	06 22	21 59	22 53	23 46	24 41
0	05 21	05 46	06 08	22 02	22 51	23 40	24 31
S 10	05 05	05 31	05 54	22 05	22 49	23 34	24 20
20	04 46	05 14	05 38	22 08	22 48	23 27	24 09
30	04 22	04 54	05 20	22 12	22 46	23 20	23 56
35	04 07	04 41	05 10	22 14	22 45	23 16	23 49
40	03 48	04 27	04 58	22 17	22 43	23 11	23 41
45	03 24	04 08	04 43	22 19	22 42	23 05	23 31
S 50	02 51	03 45	04 25	22 23	22 40	22 59	23 20
52	02 34	03 34	04 17	22 24	22 40	22 56	23 14
54	02 12	03 21	04 08	22 26	22 39	22 53	23 08
56	01 43	03 06	03 57	22 27	22 38	22 49	23 02
58	00 50	02 47	03 45	22 29	22 37	22 45	22 55
S 60	////	02 24	03 31	22 32	22 36	22 41	22 47

Lat.	Sunset	Twilight Civil	Twilight Naut.	Moonset 22	Moonset 23	Moonset 24	Moonset 25
°	h m	h m	h m	h m	h m	h m	h m
N 72	▬	15 05	16 44	11 05	10 36	10 07	09 32
N 70	13 41	15 31	16 56	10 56	10 35	10 14	09 50
68	14 26	15 51	17 06	10 48	10 34	10 20	10 04
66	14 56	16 06	17 14	10 42	10 33	10 25	10 15
64	15 18	16 20	17 21	10 36	10 32	10 29	10 25
62	15 35	16 31	17 27	10 31	10 32	10 33	10 34
60	15 50	16 40	17 33	10 27	10 31	10 36	10 41
N 58	16 02	16 49	17 38	10 23	10 31	10 39	10 47
56	16 13	16 56	17 42	10 20	10 31	10 41	10 53
54	16 22	17 03	17 47	10 17	10 30	10 44	10 58
52	16 31	17 09	17 50	10 14	10 30	10 46	11 03
50	16 38	17 14	17 54	10 11	10 30	10 48	11 07
45	16 55	17 27	18 02	10 06	10 29	10 52	11 16
N 40	17 08	17 37	18 10	10 01	10 28	10 55	11 24
35	17 19	17 46	18 17	09 57	10 28	10 59	11 30
30	17 29	17 54	18 24	09 53	10 27	11 01	11 36
20	17 46	18 09	18 36	09 47	10 27	11 06	11 46
N 10	18 01	18 23	18 49	09 42	10 26	11 10	11 55
0	18 15	18 37	19 03	09 36	10 25	11 14	12 04
S 10	18 30	18 52	19 18	09 31	10 25	11 18	12 12
20	18 45	19 09	19 37	09 25	10 24	11 22	12 21
30	19 03	19 29	20 01	09 19	10 23	11 27	12 32
35	19 13	19 42	20 16	09 15	10 22	11 30	12 38
40	19 25	19 56	20 35	09 10	10 22	11 33	12 45
45	19 40	20 14	20 58	09 05	10 21	11 37	12 53
S 50	19 57	20 37	21 30	08 59	10 20	11 41	13 03
52	20 05	20 48	21 48	08 56	10 20	11 43	13 07
54	20 15	21 01	22 09	08 53	10 19	11 45	13 12
56	20 25	21 16	22 37	08 49	10 19	11 48	13 18
58	20 37	21 34	23 24	08 46	10 18	11 50	13 24
S 60	20 51	21 57	////	08 41	10 18	11 54	13 31

Day	SUN Eqn. of Time 00^h	SUN Eqn. of Time 12^h	SUN Mer. Pass.	MOON Mer. Pass. Upper	MOON Mer. Pass. Lower	MOON Age	MOON Phase
d	m s	m s	h m	h m	h m	d %	
22	11 22	11 30	12 11	03 24	15 49	20 80	
23	11 38	11 46	12 12	04 14	16 38	21 71	
24	11 53	12 01	12 12	05 02	17 27	22 60	

UT	ARIES	VENUS −4.3		MARS +1.3		JUPITER −2.6		SATURN −0.2	
	GHA	GHA	Dec	GHA	Dec	GHA	Dec	GHA	Dec
d h	° ′	° ′	° ′	° ′	° ′	° ′	° ′	° ′	° ′
25 00	123 53.4	226 20.9	S19 50.5	240 42.1	S20 41.2	346 57.4	N17 22.4	41 35.0	N22 02.0
01	138 55.8	241 20.5	50.9	255 42.9	41.5	2 00.2	22.5	56 37.6	02.0
02	153 58.3	256 20.1	51.2	270 43.6	41.8	17 03.0	22.6	71 40.2	02.0
03	169 00.7	271 19.6	. . 51.6	285 44.4	. . 42.2	32 05.8	. . 22.7	86 42.8	. . 02.0
04	184 03.2	286 19.2	51.9	300 45.2	42.5	47 08.5	22.8	101 45.4	02.0
05	199 05.7	301 18.8	52.2	315 46.0	42.8	62 11.3	22.9	116 48.0	02.0
06	214 08.1	316 18.4	S19 52.6	330 46.7	S20 43.1	77 14.1	N17 23.0	131 50.6	N22 02.0
07	229 10.6	331 18.0	52.9	345 47.5	43.4	92 16.9	23.1	146 53.2	02.0
S 08	244 13.1	346 17.6	53.3	0 48.3	43.7	107 19.7	23.2	161 55.8	02.0
A 09	259 15.5	1 17.2	. . 53.6	15 49.1	. . 44.0	122 22.5	. . 23.3	176 58.4	. . 02.0
T 10	274 18.0	16 16.7	53.9	30 49.8	44.3	137 25.3	23.4	192 01.0	02.0
U 11	289 20.5	31 16.3	54.3	45 50.6	44.7	152 28.1	23.5	207 03.6	02.0
R 12	304 22.9	46 15.9	S19 54.6	60 51.4	S20 45.0	167 30.9	N17 23.6	222 06.2	N22 02.0
D 13	319 25.4	61 15.5	55.0	75 52.1	45.3	182 33.6	23.7	237 08.8	02.0
A 14	334 27.8	76 15.1	55.3	90 52.9	45.6	197 36.4	23.8	252 11.4	02.0
Y 15	349 30.3	91 14.6	. . 55.6	105 53.7	. . 45.9	212 39.2	. . 23.9	267 14.0	. . 02.0
16	4 32.8	106 14.2	56.0	120 54.5	46.2	227 42.0	24.0	282 16.6	02.0
17	19 35.2	121 13.8	56.3	135 55.2	46.5	242 44.8	24.1	297 19.2	02.0
18	34 37.7	136 13.4	S19 56.6	150 56.0	S20 46.8	257 47.6	N17 24.2	312 21.8	N22 02.0
19	49 40.2	151 12.9	56.9	165 56.8	47.2	272 50.4	24.3	327 24.4	02.0
20	64 42.6	166 12.5	57.3	180 57.6	47.5	287 53.2	24.4	342 27.0	02.0
21	79 45.1	181 12.1	. . 57.6	195 58.3	. . 47.8	302 56.0	. . 24.5	357 29.6	. . 02.0
22	94 47.6	196 11.7	57.9	210 59.1	48.1	317 58.7	24.6	12 32.2	02.0
23	109 50.0	211 11.2	58.3	225 59.9	48.4	333 01.5	24.7	27 34.8	02.0
26 00	124 52.5	226 10.8	S19 58.6	241 00.6	S20 48.7	348 04.3	N17 24.8	42 37.4	N22 02.0
01	139 55.0	241 10.4	58.9	256 01.4	49.0	3 07.1	24.9	57 40.0	02.0
02	154 57.4	256 10.0	59.2	271 02.2	49.3	18 09.9	25.0	72 42.6	02.0
03	169 59.9	271 09.5	. . 59.6	286 03.0	. . 49.6	33 12.7	. . 25.1	87 45.1	. . 02.0
04	185 02.3	286 09.1	19 59.9	301 03.7	49.9	48 15.5	25.2	102 47.7	02.0
05	200 04.8	301 08.7	20 00.2	316 04.5	50.3	63 18.3	25.3	117 50.3	02.0
06	215 07.3	316 08.2	S20 00.5	331 05.3	S20 50.6	78 21.1	N17 25.4	132 52.9	N22 02.0
07	230 09.7	331 07.8	00.9	346 06.1	50.9	93 23.9	25.5	147 55.5	02.0
08	245 12.2	346 07.4	01.2	1 06.8	51.2	108 26.7	25.6	162 58.1	02.0
S 09	260 14.7	1 06.9	. . 01.5	16 07.6	. . 51.5	123 29.4	. . 25.7	178 00.7	. . 02.0
U 10	275 17.1	16 06.5	01.8	31 08.4	51.8	138 32.2	25.8	193 03.3	02.0
N 11	290 19.6	31 06.1	02.2	46 09.1	52.1	153 35.0	25.9	208 05.9	02.0
D 12	305 22.1	46 05.6	S20 02.5	61 09.9	S20 52.4	168 37.8	N17 26.0	223 08.5	N22 02.0
A 13	320 24.5	61 05.2	02.8	76 10.7	52.7	183 40.6	26.1	238 11.1	02.0
Y 14	335 27.0	76 04.8	03.1	91 11.4	53.0	198 43.4	26.2	253 13.7	02.0
15	350 29.5	91 04.3	. . 03.4	106 12.2	. . 53.3	213 46.2	. . 26.3	268 16.3	. . 02.0
16	5 31.9	106 03.9	03.7	121 13.0	53.6	228 49.0	26.4	283 18.9	02.0
17	20 34.4	121 03.5	04.1	136 13.8	53.9	243 51.8	26.5	298 21.5	02.0
18	35 36.8	136 03.0	S20 04.4	151 14.5	S20 54.2	258 54.6	N17 26.6	313 24.1	N22 02.0
19	50 39.3	151 02.6	04.7	166 15.3	54.5	273 57.4	26.7	328 26.6	02.0
20	65 41.8	166 02.1	05.0	181 16.1	54.8	289 00.1	26.8	343 29.2	02.0
21	80 44.2	181 01.7	. . 05.3	196 16.8	. . 55.1	304 02.9	. . 26.9	358 31.8	. . 02.0
22	95 46.7	196 01.3	05.6	211 17.6	55.4	319 05.7	27.0	13 34.4	02.0
23	110 49.2	211 00.8	05.9	226 18.4	55.8	334 08.5	27.1	28 37.0	02.0
27 00	125 51.6	226 00.4	S20 06.2	241 19.2	S20 56.1	349 11.3	N17 27.2	43 39.6	N22 02.0
01	140 54.1	240 59.9	06.6	256 19.9	56.4	4 14.1	27.3	58 42.2	02.0
02	155 56.6	255 59.5	06.9	271 20.7	56.7	19 16.9	27.4	73 44.8	02.0
03	170 59.0	270 59.1	. . 07.2	286 21.5	. . 57.0	34 19.7	. . 27.5	88 47.4	. . 02.0
04	186 01.5	285 58.6	07.5	301 22.2	57.3	49 22.5	27.6	103 50.0	02.0
05	201 03.9	300 58.2	07.8	316 23.0	57.6	64 25.3	27.7	118 52.6	02.0
06	216 06.4	315 57.7	S20 08.1	331 23.8	S20 57.9	79 28.1	N17 27.8	133 55.2	N22 02.0
07	231 08.9	330 57.3	08.4	346 24.5	58.2	94 30.9	27.9	148 57.7	02.0
08	246 11.3	345 56.8	08.7	1 25.3	58.5	109 33.7	28.0	164 00.3	02.0
M 09	261 13.8	0 56.4	. . 09.0	16 26.1	. . 58.8	124 36.4	. . 28.1	179 02.9	. . 02.1
O 10	276 16.3	15 55.9	09.3	31 26.8	59.1	139 39.2	28.2	194 05.5	02.1
N 11	291 18.7	30 55.5	09.6	46 27.6	59.4	154 42.0	28.3	209 08.1	02.1
D 12	306 21.2	45 55.0	S20 09.9	61 28.4	S20 59.7	169 44.8	N17 28.4	224 10.7	N22 02.1
A 13	321 23.7	60 54.6	10.2	76 29.1	21 00.0	184 47.6	28.5	239 13.3	02.1
Y 14	336 26.1	75 54.1	10.5	91 29.9	00.3	199 50.4	28.6	254 15.9	02.1
15	351 28.6	90 53.7	. . 10.8	106 30.7	. . 00.6	214 53.2	. . 28.8	269 18.5	. . 02.1
16	6 31.1	105 53.2	11.1	121 31.5	00.9	229 56.0	28.9	284 21.1	02.1
17	21 33.5	120 52.8	11.4	136 32.2	01.2	244 58.8	29.0	299 23.6	02.1
18	36 36.0	135 52.3	S20 11.7	151 33.0	S21 01.5	260 01.6	N17 29.1	314 26.2	N22 02.1
19	51 38.4	150 51.9	12.0	166 33.8	01.8	275 04.4	29.2	329 28.8	02.1
20	66 40.9	165 51.4	12.3	181 34.5	02.1	290 07.2	29.3	344 31.4	02.1
21	81 43.4	180 51.0	. . 12.6	196 35.3	. . 02.4	305 10.0	. . 29.4	359 34.0	. . 02.1
22	96 45.8	195 50.5	12.9	211 36.1	02.7	320 12.8	29.5	14 36.6	02.1
23	111 48.3	210 50.0	13.2	226 36.8	03.0	335 15.6	29.6	29 39.2	02.1
Mer. Pass.	h m 15 37.9	v −0.4	d 0.3	v 0.8	d 0.3	v 2.8	d 0.1	v 2.6	d 0.0

STARS

Name	SHA	Dec
	° ′	° ′
Acamar	315 24.4	S40 17.9
Achernar	335 32.7	S57 13.7
Acrux	173 18.4	S63 06.7
Adhara	255 18.6	S28 58.6
Aldebaran	290 58.6	N16 30.9
Alioth	166 27.4	N55 56.4
Alkaid	153 05.1	N49 17.6
Al Na'ir	27 54.3	S46 57.0
Alnilam	275 54.4	S 1 12.0
Alphard	218 03.8	S 8 40.3
Alphecca	126 18.0	N26 42.1
Alpheratz	357 52.3	N29 06.4
Altair	62 16.6	N 8 52.4
Ankaa	353 23.8	S42 17.7
Antares	112 36.5	S26 26.3
Arcturus	146 03.2	N19 09.9
Atria	107 46.2	S69 01.8
Avior	234 20.8	S59 31.1
Bellatrix	278 40.5	N 6 21.1
Betelgeuse	271 09.9	N 7 24.4
Canopus	263 59.3	S52 41.9
Capella	280 46.2	N46 00.2
Deneb	49 37.6	N45 17.3
Denebola	182 41.8	N14 33.2
Diphda	349 04.2	S17 58.4
Dubhe	194 00.9	N61 43.9
Elnath	278 22.7	N28 36.7
Eltanin	90 50.4	N51 29.1
Enif	33 55.5	N 9 53.2
Fomalhaut	15 33.2	S29 36.6
Gacrux	172 10.0	S57 07.5
Gienah	176 00.6	S17 33.4
Hadar	148 59.7	S60 23.0
Hamal	328 10.1	N23 28.6
Kaus Aust.	83 55.0	S34 23.0
Kochab	137 19.5	N74 08.3
Markab	13 46.8	N15 13.2
Menkar	314 23.6	N 4 06.0
Menkent	148 17.3	S36 22.9
Miaplacidus	221 40.8	S69 43.6
Mirfak	308 52.0	N49 52.5
Nunki	76 08.8	S26 17.7
Peacock	53 32.6	S56 43.7
Pollux	243 37.3	N28 01.2
Procyon	245 08.0	N 5 13.0
Rasalhague	96 14.3	N12 33.3
Regulus	207 51.9	N11 57.1
Rigel	281 19.7	S 8 12.0
Rigil Kent.	140 03.3	S60 50.6
Sabik	102 22.2	S15 43.8
Schedar	349 50.3	N56 33.4
Shaula	96 33.4	S37 06.3
Sirius	258 40.7	S16 43.3
Spica	158 39.8	S11 10.6
Suhail	222 58.1	S43 26.6
Vega	80 44.9	N38 47.0
Zuben'ubi	137 14.6	S16 03.2

	SHA	Mer. Pass.
	° ′	h m
Venus	101 18.3	8 56
Mars	116 08.2	7 56
Jupiter	223 11.8	0 48
Saturn	277 44.9	21 06

	UT	SUN GHA	SUN Dec	MOON GHA	v	MOON Dec	d	HP
	d h	° ′	° ′	° ′	′	° ′	′	′
	25 00	176 58.0	S19 07.0	274 57.9	10.9	S 8 33.6	14.6	59.2
	01	191 57.8	06.4	289 27.8	10.9	8 48.2	14.6	59.2
	02	206 57.7	05.8	303 57.7	10.8	9 02.8	14.5	59.2
	03	221 57.5	. . 05.2	318 27.5	10.8	9 17.3	14.5	59.2
	04	236 57.4	04.6	332 57.3	10.7	9 31.8	14.4	59.2
	05	251 57.2	04.0	347 27.0	10.7	9 46.2	14.4	59.2
	06	266 57.1	S19 03.4	1 56.7	10.7	S10 00.6	14.3	59.2
	07	281 56.9	02.8	16 26.4	10.6	10 14.9	14.3	59.2
S	08	296 56.8	02.1	30 56.0	10.6	10 29.2	14.2	59.2
A	09	311 56.7	. . 01.5	45 25.6	10.5	10 43.4	14.1	59.2
T	10	326 56.5	00.9	59 55.1	10.5	10 57.5	14.1	59.2
U	11	341 56.4	19 00.3	74 24.6	10.4	11 11.6	14.0	59.2
R	12	356 56.2	S18 59.7	88 54.0	10.4	S11 25.6	13.9	59.2
D	13	11 56.1	59.1	103 23.4	10.3	11 39.5	13.9	59.2
A	14	26 55.9	58.5	117 52.7	10.3	11 53.4	13.8	59.2
Y	15	41 55.8	. . 57.8	132 22.0	10.2	12 07.2	13.8	59.2
	16	56 55.6	57.2	146 51.2	10.2	12 21.0	13.7	59.2
	17	71 55.5	56.6	161 20.4	10.1	12 34.7	13.6	59.2
	18	86 55.4	S18 56.0	175 49.5	10.1	S12 48.3	13.5	59.1
	19	101 55.2	55.4	190 18.6	10.0	13 01.8	13.5	59.1
	20	116 55.1	54.8	204 47.6	10.0	13 15.3	13.4	59.1
	21	131 54.9	. . 54.1	219 16.6	9.9	13 28.7	13.3	59.1
	22	146 54.8	53.5	233 45.5	9.8	13 42.0	13.2	59.1
	23	161 54.6	52.9	248 14.3	9.8	13 55.2	13.2	59.1
	26 00	176 54.5	S18 52.3	262 43.1	9.7	S14 08.4	13.0	59.1
	01	191 54.4	51.6	277 11.8	9.7	14 21.4	13.0	59.1
	02	206 54.2	51.0	291 40.5	9.6	14 34.4	12.9	59.1
	03	221 54.1	. . 50.4	306 09.1	9.6	14 47.3	12.9	59.1
	04	236 53.9	49.8	320 37.7	9.4	15 00.2	12.7	59.1
	05	251 53.8	49.2	335 06.1	9.5	15 12.9	12.6	59.1
	06	266 53.7	S18 48.5	349 34.6	9.3	S15 25.5	12.6	59.1
	07	281 53.5	47.9	4 02.9	9.4	15 38.1	12.5	59.1
	08	296 53.4	47.3	18 31.3	9.2	15 50.6	12.3	59.1
S	09	311 53.3	. . 46.6	32 59.5	9.2	16 02.9	12.3	59.0
U	10	326 53.1	46.0	47 27.7	9.1	16 15.2	12.2	59.0
N	11	341 53.0	45.4	61 55.8	9.0	16 27.4	12.1	59.0
D	12	356 52.8	S18 44.8	76 23.8	9.0	S16 39.5	12.0	59.0
A	13	11 52.7	44.1	90 51.8	9.0	16 51.5	11.9	59.0
Y	14	26 52.6	43.5	105 19.8	8.8	17 03.4	11.8	59.0
	15	41 52.4	. . 42.9	119 47.6	8.8	17 15.2	11.7	59.0
	16	56 52.3	42.2	134 15.4	8.7	17 26.9	11.6	59.0
	17	71 52.2	41.6	148 43.1	8.7	17 38.5	11.4	59.0
	18	86 52.0	S18 41.0	163 10.8	8.6	S17 49.9	11.4	59.0
	19	101 51.9	40.3	177 38.4	8.5	18 01.3	11.3	59.0
	20	116 51.8	39.7	192 05.9	8.5	18 12.6	11.2	59.0
	21	131 51.6	. . 39.1	206 33.4	8.4	18 23.8	11.0	59.0
	22	146 51.5	38.4	221 00.8	8.3	18 34.8	10.9	58.9
	23	161 51.4	37.8	235 28.1	8.3	18 45.7	10.9	58.9
	27 00	176 51.2	S18 37.2	249 55.4	8.2	S18 56.6	10.7	58.9
	01	191 51.1	36.5	264 22.6	8.1	19 07.3	10.6	58.9
	02	206 51.0	35.9	278 49.7	8.0	19 17.9	10.5	58.9
	03	221 50.8	. . 35.3	293 16.7	8.0	19 28.4	10.3	58.9
	04	236 50.7	34.6	307 43.7	8.0	19 38.7	10.3	58.9
	05	251 50.6	34.0	322 10.7	7.8	19 49.0	10.1	58.9
	06	266 50.4	S18 33.3	336 37.5	7.8	S19 59.1	10.0	58.9
	07	281 50.3	32.7	351 04.3	7.8	20 09.1	9.9	58.9
	08	296 50.2	32.1	5 31.1	7.6	20 19.0	9.8	58.8
M	09	311 50.1	. . 31.4	19 57.7	7.6	20 28.8	9.6	58.8
O	10	326 49.9	30.8	34 24.3	7.6	20 38.4	9.5	58.8
N	11	341 49.8	30.1	48 50.9	7.4	20 47.9	9.4	58.8
D	12	356 49.7	S18 29.5	63 17.3	7.4	S20 57.3	9.2	58.8
A	13	11 49.5	28.8	77 43.7	7.4	21 06.5	9.1	58.8
Y	14	26 49.4	28.2	92 10.1	7.3	21 15.6	9.0	58.8
	15	41 49.3	. . 27.6	106 36.4	7.2	21 24.6	8.9	58.8
	16	56 49.2	26.9	121 02.6	7.1	21 33.5	8.7	58.8
	17	71 49.0	26.3	135 28.7	7.1	21 42.2	8.6	58.8
	18	86 48.9	S18 25.6	149 54.8	7.0	S21 50.8	8.4	58.7
	19	101 48.8	25.0	164 20.8	7.0	21 59.2	8.4	58.7
	20	116 48.7	24.3	178 46.8	6.9	22 07.6	8.1	58.7
	21	131 48.5	. . 23.7	193 12.7	6.9	22 15.7	8.1	58.7
	22	146 48.4	23.0	207 38.6	6.8	22 23.8	7.9	58.7
	23	161 48.3	22.4	222 04.4	6.7	S22 31.7	7.7	58.7
		SD 16.3	d 0.6	SD 16.1		16.1		16.0

Lat.	Twilight Naut.	Twilight Civil	Sunrise	Moonrise 25	26	27	28
°	h m	h m	h m	h m	h m	h m	h m
N 72	07 31	09 07	11 49	01 42	04 27	■	■
N 70	07 21	08 43	10 23	01 27	03 49	■	■
68	07 12	08 25	09 45	01 15	03 23	05 53	■
66	07 04	08 10	09 19	01 06	03 03	05 11	07 57
64	06 58	07 58	08 58	00 58	02 47	04 43	06 44
62	06 52	07 48	08 42	00 51	02 34	04 21	06 08
60	06 47	07 39	08 28	00 45	02 23	04 04	05 43
N 58	06 42	07 31	08 17	00 39	02 14	03 49	05 22
56	06 38	07 24	08 06	00 35	02 06	03 37	05 06
54	06 34	07 18	07 57	00 31	01 58	03 26	04 52
52	06 31	07 12	07 49	00 27	01 52	03 17	04 39
50	06 27	07 07	07 42	00 24	01 46	03 08	04 28
45	06 19	06 55	07 27	00 16	01 33	02 50	04 06
N 40	06 12	06 45	07 14	00 10	01 23	02 36	03 48
35	06 06	06 36	07 03	00 05	01 14	02 23	03 32
30	05 59	06 28	06 54	00 01	01 06	02 13	03 19
20	05 47	06 14	06 37	24 53	00 53	01 54	02 57
N 10	05 35	06 00	06 23	24 41	00 41	01 39	02 38
0	05 22	05 47	06 09	24 31	00 31	01 24	02 20
S 10	05 07	05 33	05 55	24 20	00 20	01 09	02 02
20	04 49	05 17	05 40	24 09	00 09	00 54	01 43
30	04 25	04 57	05 23	23 56	24 36	00 36	01 22
35	04 11	04 45	05 13	23 49	24 26	00 26	01 09
40	03 53	04 30	05 01	23 41	24 15	00 15	00 54
45	03 30	04 13	04 47	23 31	24 01	00 01	00 37
S 50	02 59	03 51	04 30	23 20	23 45	24 16	00 16
52	02 42	03 40	04 23	23 14	23 37	24 06	00 06
54	02 22	03 28	04 14	23 08	23 28	23 55	24 32
56	01 56	03 14	04 04	23 02	23 19	23 42	24 17
58	01 16	02 56	03 52	22 55	23 08	23 27	23 58
S 60	////	02 35	03 39	22 47	22 55	23 10	23 36

Lat.	Sunset	Twilight Civil	Twilight Naut.	Moonset 25	26	27	28
°	h m	h m	h m	h m	h m	h m	h m
N 72	12 37	15 19	16 55	09 32	08 36	■	■
N 70	14 03	15 43	17 06	09 50	09 16	■	■
68	14 41	16 01	17 14	10 04	09 43	09 05	■
66	15 07	16 15	17 22	10 15	10 04	09 48	09 01
64	15 28	16 28	17 28	10 25	10 21	10 18	10 14
62	15 44	16 38	17 34	10 34	10 36	10 40	10 51
60	15 57	16 47	17 39	10 41	10 48	10 58	11 17
N 58	16 09	16 55	17 43	10 47	10 58	11 14	11 38
56	16 19	17 02	17 48	10 53	11 07	11 27	11 55
54	16 28	17 08	17 51	10 58	11 15	11 38	12 09
52	16 36	17 14	17 55	11 03	11 23	11 48	12 22
50	16 43	17 19	17 58	11 07	11 29	11 57	12 33
45	16 59	17 31	18 06	11 16	11 44	12 16	12 57
N 40	17 11	17 40	18 13	11 24	11 55	12 32	13 15
35	17 22	17 49	18 20	11 30	12 05	12 45	13 31
30	17 32	17 57	18 26	11 36	12 14	12 57	13 45
20	17 48	18 11	18 38	11 46	12 30	13 17	14 08
N 10	18 02	18 25	18 50	11 55	12 43	13 34	14 29
0	18 16	18 38	19 03	12 04	12 56	13 50	14 48
S 10	18 30	18 52	19 18	12 12	13 08	14 07	15 07
20	18 45	19 08	19 36	12 21	13 22	14 24	15 27
30	19 02	19 28	19 59	12 32	13 38	14 44	15 51
35	19 12	19 40	20 14	12 38	13 47	14 56	16 04
40	19 23	19 54	20 31	12 45	13 57	15 10	16 20
45	19 37	20 11	20 54	12 53	14 10	15 26	16 40
S 50	19 54	20 33	21 25	13 03	14 25	15 46	17 04
52	20 01	20 43	21 41	13 07	14 32	15 56	17 15
54	20 10	20 55	22 00	13 12	14 40	16 07	17 29
56	20 20	21 10	22 25	13 18	14 49	16 19	17 44
58	20 31	21 26	23 02	13 24	14 59	16 33	18 02
S 60	20 44	21 47	////	13 31	15 10	16 50	18 24

Day	SUN Eqn. of Time 00^h	SUN Eqn. of Time 12^h	SUN Mer. Pass.	MOON Mer. Pass. Upper	MOON Mer. Pass. Lower	MOON Age	MOON Phase
d	m s	m s	h m	h m	h m	d %	
25	12 08	12 15	12 12	05 52	18 17	23 48	
26	12 22	12 28	12 12	06 43	19 10	24 37	
27	12 35	12 41	12 13	07 37	20 05	25 27	

UT d	UT h	ARIES GHA	VENUS −4.3 GHA	VENUS Dec	MARS +1.3 GHA	MARS Dec	JUPITER −2.6 GHA	JUPITER Dec	SATURN −0.2 GHA	SATURN Dec
		° ′	° ′	° ′	° ′	° ′	° ′	° ′	° ′	° ′
28 TUESDAY	00	126 50.8	225 49.6	S20 13.5	241 37.6	S21 03.2	350 18.3	N17 29.7	44 41.8	N22 02.1
	01	141 53.2	240 49.1	13.8	256 38.4	03.5	5 21.1	29.8	59 44.3	02.1
	02	156 55.7	255 48.7	14.1	271 39.1	03.8	20 23.9	29.9	74 46.9	02.1
	03	171 58.2	270 48.2	. . 14.4	286 39.9	. . 04.1	35 26.7	. . 30.0	89 49.5	. . 02.1
	04	187 00.6	285 47.8	14.6	301 40.7	04.4	50 29.5	30.1	104 52.1	02.1
	05	202 03.1	300 47.3	14.9	316 41.4	04.7	65 32.3	30.2	119 54.7	02.1
	06	217 05.6	315 46.8	S20 15.2	331 42.2	S21 05.0	80 35.1	N17 30.3	134 57.3	N22 02.1
	07	232 08.0	330 46.4	15.5	346 43.0	05.3	95 37.9	30.4	149 59.9	02.1
	08	247 10.5	345 45.9	15.8	1 43.7	05.6	110 40.7	30.5	165 02.4	02.1
	09	262 12.9	0 45.4	. . 16.1	16 44.5	. . 05.9	125 43.5	. . 30.6	180 05.0	. . 02.1
	10	277 15.4	15 45.0	16.4	31 45.3	06.2	140 46.3	30.7	195 07.6	02.1
	11	292 17.9	30 44.5	16.7	46 46.0	06.5	155 49.1	30.8	210 10.2	02.1
	12	307 20.3	45 44.1	S20 16.9	61 46.8	S21 06.8	170 51.9	N17 30.9	225 12.8	N22 02.1
	13	322 22.8	60 43.6	17.2	76 47.6	07.1	185 54.7	31.0	240 15.4	02.1
	14	337 25.3	75 43.1	17.5	91 48.3	07.4	200 57.5	31.1	255 18.0	02.1
	15	352 27.7	90 42.7	. . 17.8	106 49.1	. . 07.7	216 00.3	. . 31.2	270 20.5	. . 02.1
	16	7 30.2	105 42.2	18.1	121 49.9	08.0	231 03.1	31.3	285 23.1	02.1
	17	22 32.7	120 41.7	18.3	136 50.6	08.2	246 05.8	31.4	300 25.7	02.1
	18	37 35.1	135 41.3	S20 18.6	151 51.4	S21 08.5	261 08.6	N17 31.5	315 28.3	N22 02.1
	19	52 37.6	150 40.8	18.9	166 52.1	08.8	276 11.4	31.6	330 30.9	02.1
	20	67 40.1	165 40.3	19.2	181 52.9	09.1	291 14.2	31.7	345 33.5	02.1
	21	82 42.5	180 39.9	. . 19.5	196 53.7	. . 09.4	306 17.0	. . 31.8	0 36.0	. . 02.1
	22	97 45.0	195 39.4	19.7	211 54.4	09.7	321 19.8	31.9	15 38.6	02.1
	23	112 47.4	210 38.9	20.0	226 55.2	10.0	336 22.6	32.0	30 41.2	02.1
29 WEDNESDAY	00	127 49.9	225 38.4	S20 20.3	241 56.0	S21 10.3	351 25.4	N17 32.1	45 43.8	N22 02.1
	01	142 52.4	240 38.0	20.6	256 56.7	10.6	6 28.2	32.2	60 46.4	02.1
	02	157 54.8	255 37.5	20.8	271 57.5	10.9	21 31.0	32.3	75 49.0	02.1
	03	172 57.3	270 37.0	. . 21.1	286 58.3	. . 11.2	36 33.8	. . 32.4	90 51.5	. . 02.1
	04	187 59.8	285 36.6	21.4	301 59.0	11.4	51 36.6	32.5	105 54.1	02.1
	05	203 02.2	300 36.1	21.6	316 59.8	11.7	66 39.4	32.6	120 56.7	02.1
	06	218 04.7	315 35.6	S20 21.9	332 00.6	S21 12.0	81 42.2	N17 32.7	135 59.3	N22 02.1
	07	233 07.2	330 35.1	22.2	347 01.3	12.3	96 45.0	32.8	151 01.9	02.1
	08	248 09.6	345 34.7	22.5	2 02.1	12.6	111 47.8	32.9	166 04.5	02.1
	09	263 12.1	0 34.2	. . 22.7	17 02.8	. . 12.9	126 50.6	. . 33.0	181 07.0	. . 02.2
	10	278 14.6	15 33.7	23.0	32 03.6	13.2	141 53.4	33.1	196 09.6	02.2
	11	293 17.0	30 33.2	23.3	47 04.4	13.5	156 56.2	33.2	211 12.2	02.2
	12	308 19.5	45 32.7	S20 23.5	62 05.1	S21 13.7	171 59.0	N17 33.3	226 14.8	N22 02.2
	13	323 21.9	60 32.3	23.8	77 05.9	14.0	187 01.8	33.4	241 17.4	02.2
	14	338 24.4	75 31.8	24.1	92 06.7	14.3	202 04.5	33.5	256 19.9	02.2
	15	353 26.9	90 31.3	. . 24.3	107 07.4	. . 14.6	217 07.3	. . 33.6	271 22.5	. . 02.2
	16	8 29.3	105 30.8	24.6	122 08.2	14.9	232 10.1	33.7	286 25.1	02.2
	17	23 31.8	120 30.4	24.8	137 09.0	15.2	247 12.9	33.8	301 27.7	02.2
	18	38 34.3	135 29.9	S20 25.1	152 09.7	S21 15.5	262 15.7	N17 33.9	316 30.3	N22 02.2
	19	53 36.7	150 29.4	25.4	167 10.5	15.7	277 18.5	34.0	331 32.8	02.2
	20	68 39.2	165 28.9	25.6	182 11.2	16.0	292 21.3	34.1	346 35.4	02.2
	21	83 41.7	180 28.4	. . 25.9	197 12.0	. . 16.3	307 24.1	. . 34.2	1 38.0	. . 02.2
	22	98 44.1	195 27.9	26.1	212 12.8	16.6	322 26.9	34.3	16 40.6	02.2
	23	113 46.6	210 27.5	26.4	227 13.5	16.9	337 29.7	34.4	31 43.2	02.2
30 THURSDAY	00	128 49.1	225 27.0	S20 26.6	242 14.3	S21 17.2	352 32.5	N17 34.5	46 45.7	N22 02.2
	01	143 51.5	240 26.5	26.9	257 15.1	17.4	7 35.3	34.6	61 48.3	02.2
	02	158 54.0	255 26.0	27.2	272 15.8	17.7	22 38.1	34.7	76 50.9	02.2
	03	173 56.4	270 25.5	. . 27.4	287 16.6	. . 18.0	37 40.9	. . 34.8	91 53.5	. . 02.2
	04	188 58.9	285 25.0	27.7	302 17.3	18.3	52 43.7	34.9	106 56.0	02.2
	05	204 01.4	300 24.5	27.9	317 18.1	18.6	67 46.5	35.0	121 58.6	02.2
	06	219 03.8	315 24.1	S20 28.2	332 18.9	S21 18.9	82 49.3	N17 35.1	137 01.2	N22 02.2
	07	234 06.3	330 23.6	28.4	347 19.6	19.1	97 52.1	35.2	152 03.8	02.2
	08	249 08.8	345 23.1	28.7	2 20.4	19.4	112 54.9	35.3	167 06.4	02.2
	09	264 11.2	0 22.6	. . 28.9	17 21.2	. . 19.7	127 57.7	. . 35.4	182 08.9	. . 02.2
	10	279 13.7	15 22.1	29.2	32 21.9	20.0	143 00.5	35.5	197 11.5	02.2
	11	294 16.2	30 21.6	29.4	47 22.7	20.3	158 03.3	35.6	212 14.1	02.2
	12	309 18.6	45 21.1	S20 29.7	62 23.4	S21 20.5	173 06.1	N17 35.7	227 16.7	N22 02.2
	13	324 21.1	60 20.6	29.9	77 24.2	20.8	188 08.9	35.8	242 19.2	02.2
	14	339 23.5	75 20.1	30.1	92 25.0	21.1	203 11.7	35.9	257 21.8	02.2
	15	354 26.0	90 19.6	. . 30.4	107 25.7	. . 21.4	218 14.5	. . 36.0	272 24.4	. . 02.2
	16	9 28.5	105 19.1	30.6	122 26.5	21.7	233 17.3	36.1	287 27.0	02.2
	17	24 30.9	120 18.7	30.9	137 27.2	21.9	248 20.1	36.2	302 29.5	02.2
	18	39 33.4	135 18.2	S20 31.1	152 28.0	S21 22.2	263 22.8	N17 36.3	317 32.1	N22 02.2
	19	54 35.9	150 17.7	31.4	167 28.8	22.5	278 25.6	36.4	332 34.7	02.2
	20	69 38.3	165 17.2	31.6	182 29.5	22.8	293 28.4	36.5	347 37.3	02.2
	21	84 40.8	180 16.7	. . 31.8	197 30.3	. . 23.0	308 31.2	. . 36.6	2 39.8	. . 02.2
	22	99 43.3	195 16.2	32.1	212 31.0	23.3	323 34.0	36.7	17 42.4	02.2
	23	114 45.7	210 15.7	32.3	227 31.8	23.6	338 36.8	36.8	32 45.0	02.2
Mer. Pass.		h m 15 26.1	v −0.5	d 0.3	v 0.8	d 0.3	v 2.8	d 0.1	v 2.6	d 0.0

STARS

Name	SHA	Dec
	° ′	° ′
Acamar	315 24.4	S40 17.9
Achernar	335 32.8	S57 13.7
Acrux	173 18.4	S63 06.7
Adhara	255 18.6	S28 58.6
Aldebaran	290 58.6	N16 30.9
Alioth	166 27.4	N55 56.4
Alkaid	153 05.1	N49 17.6
Al Na'ir	27 54.3	S46 57.0
Alnilam	275 54.4	S 1 12.1
Alphard	218 03.8	S 8 40.3
Alphecca	126 18.0	N26 42.1
Alpheratz	357 52.3	N29 06.4
Altair	62 16.5	N 8 52.4
Ankaa	353 23.8	S42 17.7
Antares	112 36.5	S26 26.3
Arcturus	146 03.1	N19 09.8
Atria	107 46.2	S69 01.8
Avior	234 20.8	S59 31.1
Bellatrix	278 40.5	N 6 21.1
Betelgeuse	271 09.9	N 7 24.4
Canopus	263 59.3	S52 42.0
Capella	280 46.2	N46 00.2
Deneb	49 37.6	N45 17.3
Denebola	182 41.7	N14 33.2
Diphda	349 04.2	S17 58.4
Dubhe	194 00.9	N61 43.9
Elnath	278 22.7	N28 36.7
Eltanin	90 50.4	N51 29.0
Enif	33 55.5	N 9 53.2
Fomalhaut	15 33.2	S29 36.6
Gacrux	172 10.0	S57 07.5
Gienah	176 00.6	S17 33.4
Hadar	148 59.7	S60 23.0
Hamal	328 10.1	N23 28.6
Kaus Aust.	83 55.0	S34 23.0
Kochab	137 19.5	N74 08.3
Markab	13 46.8	N15 13.2
Menkar	314 23.6	N 4 06.0
Menkent	148 17.3	S36 22.9
Miaplacidus	221 40.7	S69 43.6
Mirfak	308 52.1	N49 52.5
Nunki	76 08.8	S26 17.7
Peacock	53 32.5	S56 43.7
Pollux	243 37.3	N28 01.2
Procyon	245 07.9	N 5 13.0
Rasalhague	96 14.3	N12 33.3
Regulus	207 51.9	N11 57.1
Rigel	281 19.7	S 8 12.0
Rigil Kent.	140 03.2	S60 50.6
Sabik	102 22.2	S15 43.8
Schedar	349 50.3	N56 33.4
Shaula	96 33.3	S37 06.3
Sirius	258 40.7	S16 43.3
Spica	158 39.8	S11 10.6
Suhail	222 58.1	S43 26.6
Vega	80 44.9	N38 47.0
Zuben'ubi	137 14.5	S16 03.2

	SHA	Mer. Pass.
	° ′	h m
Venus	97 48.5	8 58
Mars	114 06.1	7 52
Jupiter	223 35.5	0 34
Saturn	277 53.9	20 53

	UT	SUN GHA	SUN Dec	MOON GHA	v	Dec	d	HP
	d h	° ′	° ′	° ′	′	° ′	′	′
	28 00	176 48.2	S18 21.7	236 30.1	6.7	S22 39.4	7.6	58.7
	01	191 48.0	21.1	250 55.8	6.6	22 47.0	7.5	58.7
	02	206 47.9	20.4	265 21.4	6.6	22 54.5	7.3	58.6
	03	221 47.8	. . 19.8	279 47.0	6.5	23 01.8	7.2	58.6
	04	236 47.7	19.1	294 12.5	6.5	23 09.0	7.0	58.6
	05	251 47.5	18.5	308 38.0	6.4	23 16.0	6.9	58.6
	06	266 47.4	S18 17.8	323 03.4	6.4	S23 22.9	6.7	58.6
	07	281 47.3	17.2	337 28.8	6.3	23 29.6	6.6	58.6
TUESDAY	08	296 47.2	16.5	351 54.1	6.2	23 36.2	6.5	58.6
	09	311 47.1	. . 15.9	6 19.3	6.3	23 42.7	6.2	58.6
	10	326 46.9	15.2	20 44.6	6.1	23 48.9	6.2	58.5
	11	341 46.8	14.5	35 09.7	6.2	23 55.1	6.0	58.5
	12	356 46.7	S18 13.9	49 34.9	6.1	S24 01.1	5.8	58.5
	13	11 46.6	13.2	64 00.0	6.0	24 06.9	5.7	58.5
	14	26 46.5	12.6	78 25.0	6.0	24 12.6	5.5	58.5
	15	41 46.3	. . 11.9	92 50.0	6.0	24 18.1	5.3	58.5
	16	56 46.2	11.3	107 15.0	5.9	24 23.4	5.2	58.5
	17	71 46.1	10.6	121 39.9	5.9	24 28.6	5.1	58.4
	18	86 46.0	S18 09.9	136 04.8	5.9	S24 33.7	4.9	58.4
	19	101 45.9	09.3	150 29.7	5.8	24 38.6	4.7	58.4
	20	116 45.7	08.6	164 54.5	5.8	24 43.3	4.6	58.4
	21	131 45.6	. . 07.9	179 19.3	5.8	24 47.9	4.4	58.4
	22	146 45.5	07.3	193 44.1	5.7	24 52.3	4.3	58.4
	23	161 45.4	06.6	208 08.8	5.7	24 56.6	4.0	58.4
	29 00	176 45.3	S18 06.0	222 33.5	5.7	S25 00.6	4.0	58.3
	01	191 45.2	05.3	236 58.2	5.7	25 04.6	3.8	58.3
	02	206 45.0	04.6	251 22.9	5.6	25 08.4	3.6	58.3
	03	221 44.9	. . 04.0	265 47.5	5.6	25 12.0	3.4	58.3
	04	236 44.8	03.3	280 12.1	5.6	25 15.4	3.3	58.3
	05	251 44.7	02.6	294 36.7	5.6	25 18.7	3.1	58.3
	06	266 44.6	S18 02.0	309 01.3	5.6	S25 21.8	3.0	58.3
WEDNESDAY	07	281 44.5	01.3	323 25.9	5.6	25 24.8	2.8	58.2
	08	296 44.4	00.6	337 50.5	5.5	25 27.6	2.6	58.2
	09	311 44.2	18 00.0	352 15.0	5.6	25 30.2	2.5	58.2
	10	326 44.1	17 59.3	6 39.6	5.5	25 32.7	2.3	58.2
	11	341 44.0	58.6	21 04.1	5.5	25 35.0	2.1	58.2
	12	356 43.9	S17 57.9	35 28.6	5.6	S25 37.1	2.0	58.2
	13	11 43.8	57.3	49 53.2	5.5	25 39.1	1.8	58.1
	14	26 43.7	56.6	64 17.7	5.5	25 40.9	1.6	58.1
	15	41 43.6	. . 55.9	78 42.2	5.5	25 42.5	1.5	58.1
	16	56 43.5	55.3	93 06.7	5.6	25 44.0	1.3	58.1
	17	71 43.4	54.6	107 31.3	5.5	25 45.3	1.2	58.1
	18	86 43.2	S17 53.9	121 55.8	5.6	S25 46.5	1.0	58.1
	19	101 43.1	53.2	136 20.4	5.5	25 47.5	0.8	58.0
	20	116 43.0	52.6	150 44.9	5.6	25 48.3	0.6	58.0
	21	131 42.9	. . 51.9	165 09.5	5.6	25 48.9	0.5	58.0
	22	146 42.8	51.2	179 34.1	5.6	25 49.4	0.4	58.0
	23	161 42.7	50.5	193 58.7	5.6	25 49.8	0.1	58.0
	30 00	176 42.6	S17 49.9	208 23.3	5.6	S25 49.9	0.0	57.9
	01	191 42.5	49.2	222 47.9	5.7	25 49.9	0.1	57.9
	02	206 42.4	48.5	237 12.6	5.7	25 49.8	0.4	57.9
	03	221 42.3	. . 47.8	251 37.3	5.7	25 49.4	0.5	57.9
	04	236 42.2	47.1	266 02.0	5.7	25 48.9	0.6	57.9
	05	251 42.1	46.5	280 26.7	5.8	25 48.3	0.8	57.9
	06	266 42.0	S17 45.8	294 51.5	5.8	S25 47.5	1.0	57.8
THURSDAY	07	281 41.9	45.1	309 16.3	5.8	25 46.5	1.1	57.8
	08	296 41.7	44.4	323 41.1	5.9	25 45.4	1.3	57.8
	09	311 41.6	. . 43.7	338 06.0	5.9	25 44.1	1.5	57.8
	10	326 41.5	43.1	352 30.9	5.9	25 42.6	1.6	57.8
	11	341 41.4	42.4	6 55.8	6.0	25 41.0	1.8	57.7
	12	356 41.3	S17 41.7	21 20.8	6.1	S25 39.2	1.9	57.7
	13	11 41.2	41.0	35 45.9	6.0	25 37.3	2.1	57.7
	14	26 41.1	40.3	50 10.9	6.1	25 35.2	2.3	57.7
	15	41 41.0	. . 39.6	64 36.0	6.2	25 32.9	2.4	57.7
	16	56 40.9	38.9	79 01.2	6.2	25 30.5	2.5	57.6
	17	71 40.8	38.3	93 26.4	6.3	25 28.0	2.8	57.6
	18	86 40.7	S17 37.6	107 51.7	6.3	S25 25.2	2.8	57.6
	19	101 40.6	36.9	122 17.0	6.4	25 22.4	3.1	57.6
	20	116 40.5	36.2	136 42.4	6.4	25 19.3	3.1	57.6
	21	131 40.4	. . 35.5	151 07.8	6.5	25 16.2	3.4	57.5
	22	146 40.3	34.8	165 33.3	6.5	25 12.8	3.5	57.5
	23	161 40.2	34.1	179 58.8	6.6	S25 09.3	3.6	57.5
		SD 16.3	*d* 0.7	SD 15.9		15.8		15.7

Lat.	Twilight Naut.	Twilight Civil	Sunrise	Moonrise 28	Moonrise 29	Moonrise 30	Moonrise 31
°	h m	h m	h m	h m	h m	h m	h m
N 72	07 21	08 54	11 00	■	■	■	■
N 70	07 12	08 32	10 05	■	■	■	■
68	07 04	08 16	09 32	■	■	■	■
66	06 58	08 03	09 08	07 57	■	■	■
64	06 52	07 51	08 50	06 44	08 49	10 03	10 01
62	06 47	07 42	08 34	06 08	07 45	08 50	09 20
60	06 42	07 33	08 22	05 43	07 11	08 15	08 52
N 58	06 38	07 26	08 11	05 22	06 45	07 49	08 30
56	06 34	07 19	08 01	05 06	06 25	07 28	08 12
54	06 30	07 13	07 53	04 52	06 09	07 11	07 57
52	06 27	07 08	07 45	04 39	05 54	06 57	07 44
50	06 24	07 03	07 38	04 28	05 42	06 44	07 33
45	06 17	06 52	07 24	04 06	05 16	06 18	07 08
N 40	06 10	06 43	07 12	03 48	04 56	05 57	06 49
35	06 04	06 35	07 01	03 32	04 39	05 40	06 33
30	05 58	06 27	06 52	03 19	04 24	05 25	06 19
20	05 47	06 13	06 36	02 57	03 59	04 59	05 55
N 10	05 35	06 01	06 23	02 38	03 38	04 38	05 34
0	05 22	05 48	06 10	02 20	03 18	04 17	05 15
S 10	05 08	05 34	05 56	02 02	02 58	03 57	04 56
20	04 51	05 19	05 42	01 43	02 37	03 35	04 35
30	04 28	05 00	05 26	01 22	02 13	03 10	04 11
35	04 14	04 48	05 16	01 09	01 59	02 55	03 56
40	03 57	04 34	05 05	00 54	01 42	02 38	03 40
45	03 35	04 18	04 52	00 37	01 22	02 17	03 20
S 50	03 06	03 57	04 36	00 16	00 58	01 51	02 56
52	02 51	03 47	04 28	00 06	00 46	01 38	02 44
54	02 32	03 35	04 20	24 32	00 32	01 24	02 30
56	02 09	03 22	04 10	24 17	00 17	01 07	02 14
58	01 36	03 06	03 59	23 58	24 47	00 47	01 56
S 60	////	02 46	03 47	23 36	24 22	00 22	01 32

Lat.	Sunset	Twilight Civil	Twilight Naut.	Moonset 28	Moonset 29	Moonset 30	Moonset 31
°	h m	h m	h m	h m	h m	h m	h m
N 72	13 28	15 34	17 06	■	■	■	■
N 70	14 22	15 55	17 15	■	■	■	■
68	14 55	16 11	17 23	■	■	■	■
66	15 19	16 25	17 30	09 01	■	■	■
64	15 37	16 36	17 35	10 14	10 12	11 02	13 03
62	15 52	16 45	17 40	10 51	11 16	12 14	13 43
60	16 05	16 54	17 45	11 17	11 51	12 50	14 11
N 58	16 16	17 01	17 49	11 38	12 16	13 15	14 33
56	16 26	17 07	17 53	11 55	12 37	13 36	14 50
54	16 34	17 13	17 56	12 09	12 53	13 53	15 05
52	16 42	17 19	18 00	12 22	13 08	14 07	15 18
50	16 48	17 24	18 03	12 33	13 20	14 20	15 29
45	17 03	17 34	18 10	12 57	13 46	14 46	15 53
N 40	17 15	17 44	18 16	13 15	14 07	15 06	16 11
35	17 25	17 52	18 22	13 31	14 24	15 24	16 27
30	17 34	17 59	18 28	13 45	14 39	15 38	16 40
20	17 50	18 13	18 40	14 08	15 04	16 03	17 03
N 10	18 04	18 26	18 51	14 29	15 26	16 25	17 23
0	18 17	18 38	19 04	14 48	15 46	16 45	17 42
S 10	18 30	18 52	19 18	15 07	16 07	17 05	18 00
20	18 44	19 07	19 35	15 27	16 28	17 27	18 20
30	19 00	19 26	19 57	15 51	16 54	17 51	18 42
35	19 10	19 38	20 11	16 04	17 08	18 06	18 55
40	19 21	19 51	20 28	16 20	17 26	18 23	19 10
45	19 34	20 07	20 50	16 40	17 46	18 43	19 28
S 50	19 49	20 28	21 19	17 04	18 12	19 08	19 50
52	19 57	20 38	21 34	17 15	18 25	19 20	20 01
54	20 05	20 49	21 51	17 29	18 39	19 34	20 12
56	20 15	21 03	22 14	17 44	18 56	19 50	20 26
58	20 25	21 18	22 45	18 02	19 16	20 09	20 42
S 60	20 37	21 37	23 48	18 24	19 41	20 32	21 00

Day	SUN Eqn. of Time 00^h	SUN Eqn. of Time 12^h	SUN Mer. Pass.	MOON Mer. Pass. Upper	MOON Mer. Pass. Lower	Age	Phase
d	m s	m s	h m	h m	h m	d	%
28	12 47	12 53	12 13	08 34	21 03	26	17
29	12 59	13 04	12 13	09 32	22 02	27	10
30	13 09	13 14	12 13	10 31	23 00	28	4

UT d	h	ARIES GHA	VENUS −4.3 GHA	VENUS Dec	MARS +1.3 GHA	MARS Dec	JUPITER −2.6 GHA	JUPITER Dec	SATURN −0.2 GHA	SATURN Dec
		° ′	° ′	° ′	° ′	° ′	° ′	° ′	° ′	° ′
31	00	129 48.2	225 15.2	S20 32.5	242 32.6	S21 23.9	353 39.6	N17 36.9	47 47.6	N22 02.3
	01	144 50.7	240 14.7	32.8	257 33.3	24.2	8 42.4	37.0	62 50.1	02.3
	02	159 53.1	255 14.2	33.0	272 34.1	24.4	23 45.2	37.1	77 52.7	02.3
	03	174 55.6	270 13.7	33.3	287 34.8	24.7	38 48.0	37.2	92 55.3	02.3
	04	189 58.0	285 13.2	33.5	302 35.6	25.0	53 50.8	37.3	107 57.9	02.3
	05	205 00.5	300 12.7	33.7	317 36.4	25.3	68 53.6	37.4	123 00.4	02.3
	06	220 03.0	315 12.2	S20 34.0	332 37.1	S21 25.5	83 56.4	N17 37.5	138 03.0	N22 02.3
	07	235 05.4	330 11.7	34.2	347 37.9	25.8	98 59.2	37.6	153 05.6	02.3
	08	250 07.9	345 11.2	34.4	2 38.6	26.1	114 02.0	37.7	168 08.1	02.3
F	09	265 10.4	0 10.7	34.6	17 39.4	26.4	129 04.8	37.8	183 10.7	02.3
R	10	280 12.8	15 10.2	34.9	32 40.2	26.6	144 07.6	37.9	198 13.3	02.3
I	11	295 15.3	30 09.7	35.1	47 40.9	26.9	159 10.4	38.0	213 15.9	02.3
D	12	310 17.8	45 09.2	S20 35.3	62 41.7	S21 27.2	174 13.2	N17 38.1	228 18.4	N22 02.3
A	13	325 20.2	60 08.7	35.6	77 42.4	27.4	189 16.0	38.2	243 21.0	02.3
Y	14	340 22.7	75 08.2	35.8	92 43.2	27.7	204 18.8	38.3	258 23.6	02.3
	15	355 25.2	90 07.7	36.0	107 44.0	28.0	219 21.6	38.4	273 26.1	02.3
	16	10 27.6	105 07.2	36.2	122 44.7	28.3	234 24.4	38.5	288 28.7	02.3
	17	25 30.1	120 06.7	36.4	137 45.5	28.5	249 27.2	38.6	303 31.3	02.3
	18	40 32.5	135 06.1	S20 36.7	152 46.2	S21 28.8	264 30.0	N17 38.7	318 33.9	N22 02.3
	19	55 35.0	150 05.6	36.9	167 47.0	29.1	279 32.8	38.8	333 36.4	02.3
	20	70 37.5	165 05.1	37.1	182 47.7	29.3	294 35.6	38.9	348 39.0	02.3
	21	85 39.9	180 04.6	37.3	197 48.5	29.6	309 38.4	39.0	3 41.6	02.3
	22	100 42.4	195 04.1	37.6	212 49.3	29.9	324 41.2	39.1	18 44.1	02.3
	23	115 44.9	210 03.6	37.8	227 50.0	30.2	339 44.0	39.2	33 46.7	02.3
1	00	130 47.3	225 03.1	S20 38.0	242 50.8	S21 30.4	354 46.8	N17 39.3	48 49.3	N22 02.3
	01	145 49.8	240 02.6	38.2	257 51.5	30.7	9 49.6	39.4	63 51.8	02.3
	02	160 52.3	255 02.1	38.4	272 52.3	31.0	24 52.4	39.5	78 54.4	02.3
	03	175 54.7	270 01.6	38.6	287 53.0	31.2	39 55.2	39.6	93 57.0	02.3
	04	190 57.2	285 01.1	38.8	302 53.8	31.5	54 58.0	39.7	108 59.6	02.3
	05	205 59.7	300 00.5	39.1	317 54.6	31.8	70 00.8	39.8	124 02.1	02.3
	06	221 02.1	315 00.0	S20 39.3	332 55.3	S21 32.0	85 03.6	N17 39.9	139 04.7	N22 02.3
S	07	236 04.6	329 59.5	39.5	347 56.1	32.3	100 06.4	40.0	154 07.3	02.3
	08	251 07.0	344 59.0	39.7	2 56.8	32.6	115 09.2	40.1	169 09.8	02.3
A	09	266 09.5	359 58.5	39.9	17 57.6	32.8	130 12.0	40.2	184 12.4	02.4
T	10	281 12.0	14 58.0	40.1	32 58.3	33.1	145 14.8	40.3	199 15.0	02.4
U	11	296 14.4	29 57.5	40.3	47 59.1	33.4	160 17.6	40.4	214 17.5	02.4
R	12	311 16.9	44 56.9	S20 40.5	62 59.9	S21 33.6	175 20.4	N17 40.5	229 20.1	N22 02.4
D	13	326 19.4	59 56.4	40.7	78 00.6	33.9	190 23.1	40.6	244 22.7	02.4
A	14	341 21.8	74 55.9	40.9	93 01.4	34.2	205 25.9	40.7	259 25.2	02.4
Y	15	356 24.3	89 55.4	41.1	108 02.1	34.4	220 28.7	40.8	274 27.8	02.4
	16	11 26.8	104 54.9	41.3	123 02.9	34.7	235 31.5	40.9	289 30.4	02.4
	17	26 29.2	119 54.4	41.6	138 03.6	35.0	250 34.3	41.0	304 32.9	02.4
	18	41 31.7	134 53.8	S20 41.8	153 04.4	S21 35.2	265 37.1	N17 41.1	319 35.5	N22 02.4
	19	56 34.2	149 53.3	42.0	168 05.2	35.5	280 39.9	41.2	334 38.1	02.4
	20	71 36.6	164 52.8	42.2	183 05.9	35.8	295 42.7	41.3	349 40.6	02.4
	21	86 39.1	179 52.3	42.4	198 06.7	36.0	310 45.5	41.4	4 43.2	02.4
	22	101 41.5	194 51.8	42.6	213 07.4	36.3	325 48.3	41.5	19 45.8	02.4
	23	116 44.0	209 51.2	42.8	228 08.2	36.6	340 51.1	41.6	34 48.3	02.4
2	00	131 46.5	224 50.7	S20 42.9	243 08.9	S21 36.8	355 53.9	N17 41.7	49 50.9	N22 02.4
	01	146 48.9	239 50.2	43.1	258 09.7	37.1	10 56.7	41.8	64 53.5	02.4
	02	161 51.4	254 49.7	43.3	273 10.4	37.4	25 59.5	41.9	79 56.0	02.4
	03	176 53.9	269 49.2	43.5	288 11.2	37.6	41 02.3	42.0	94 58.6	02.4
	04	191 56.3	284 48.6	43.7	303 12.0	37.9	56 05.1	42.1	110 01.1	02.4
	05	206 58.8	299 48.1	43.9	318 12.7	38.1	71 07.9	42.2	125 03.7	02.4
	06	222 01.3	314 47.6	S20 44.1	333 13.5	S21 38.4	86 10.7	N17 42.3	140 06.3	N22 02.4
	07	237 03.7	329 47.1	44.3	348 14.2	38.7	101 13.5	42.4	155 08.8	02.4
	08	252 06.2	344 46.5	44.5	3 15.0	38.9	116 16.3	42.5	170 11.4	02.4
S	09	267 08.6	359 46.0	44.7	18 15.7	39.2	131 19.1	42.6	185 14.0	02.4
U	10	282 11.1	14 45.5	44.9	33 16.5	39.4	146 21.9	42.7	200 16.5	02.4
N	11	297 13.6	29 45.0	45.1	48 17.2	39.7	161 24.7	42.8	215 19.1	02.4
D	12	312 16.0	44 44.4	S20 45.2	63 18.0	S21 40.0	176 27.5	N17 42.9	230 21.7	N22 02.4
A	13	327 18.5	59 43.9	45.4	78 18.8	40.2	191 30.3	43.0	245 24.2	02.4
Y	14	342 21.0	74 43.4	45.6	93 19.5	40.5	206 33.1	43.1	260 26.8	02.4
	15	357 23.4	89 42.8	45.8	108 20.3	40.7	221 35.9	43.2	275 29.3	02.4
	16	12 25.9	104 42.3	46.0	123 21.0	41.0	236 38.7	43.3	290 31.9	02.5
	17	27 28.4	119 41.8	46.2	138 21.8	41.3	251 41.5	43.4	305 34.5	02.5
	18	42 30.8	134 41.3	S20 46.4	153 22.5	S21 41.5	266 44.3	N17 43.5	320 37.0	N22 02.5
	19	57 33.3	149 40.7	46.5	168 23.3	41.8	281 47.1	43.6	335 39.6	02.5
	20	72 35.8	164 40.2	46.7	183 24.0	42.0	296 49.9	43.7	350 42.1	02.5
	21	87 38.2	179 39.7	46.9	198 24.8	42.3	311 52.7	43.8	5 44.7	02.5
	22	102 40.7	194 39.1	47.1	213 25.5	42.5	326 55.5	43.9	20 47.3	02.5
	23	117 43.1	209 38.6	47.2	228 26.3	42.8	341 58.3	44.0	35 49.8	02.5
Mer. Pass.		h m 15 14.3	v −0.5	d 0.2	v 0.8	d 0.3	v 2.8	d 0.1	v 2.6	d 0.0

STARS

Name	SHA	Dec
	° ′	° ′
Acamar	315 24.4	S40 17.9
Achernar	335 32.8	S57 13.6
Acrux	173 18.3	S63 06.7
Adhara	255 18.6	S28 58.6
Aldebaran	290 58.6	N16 30.9
Alioth	166 27.4	N55 56.4
Alkaid	153 05.1	N49 17.6
Al Na'ir	27 54.2	S46 57.0
Alnilam	275 54.4	S 1 12.1
Alphard	218 03.8	S 8 40.3
Alphecca	126 18.0	N26 42.0
Alpheratz	357 52.3	N29 06.4
Altair	62 16.5	N 8 52.4
Ankaa	353 23.9	S42 17.7
Antares	112 36.5	S26 26.3
Arcturus	146 03.1	N19 09.8
Atria	107 46.1	S69 01.8
Avior	234 20.8	S59 31.1
Bellatrix	278 40.5	N 6 21.1
Betelgeuse	271 09.9	N 7 24.4
Canopus	263 59.3	S52 42.0
Capella	280 46.2	N46 00.2
Deneb	49 37.6	N45 17.3
Denebola	182 41.7	N14 33.2
Diphda	349 04.2	S17 58.4
Dubhe	194 00.8	N61 43.9
Elnath	278 22.7	N28 36.7
Eltanin	90 50.3	N51 29.0
Enif	33 55.5	N 9 53.2
Fomalhaut	15 33.2	S29 36.6
Gacrux	172 09.9	S57 07.6
Gienah	176 00.6	S17 33.5
Hadar	148 59.6	S60 23.0
Hamal	328 10.1	N23 28.6
Kaus Aust.	83 55.0	S34 23.0
Kochab	137 19.4	N74 08.3
Markab	13 46.8	N15 13.2
Menkar	314 23.6	N 4 06.0
Menkent	148 17.3	S36 22.9
Miaplacidus	221 40.7	S69 43.7
Mirfak	308 52.1	N49 52.5
Nunki	76 08.7	S26 17.7
Peacock	53 32.5	S56 43.6
Pollux	243 37.3	N28 01.2
Procyon	245 07.9	N 5 13.0
Rasalhague	96 14.3	N12 33.3
Regulus	207 51.9	N11 57.1
Rigel	281 19.7	S 8 12.0
Rigil Kent.	140 03.2	S60 50.6
Sabik	102 22.1	S15 43.8
Schedar	349 50.4	N56 33.4
Shaula	96 33.3	S37 06.3
Sirius	258 40.7	S16 43.3
Spica	158 39.8	S11 10.6
Suhail	222 58.1	S43 26.6
Vega	80 44.9	N38 46.9
Zuben'ubi	137 14.5	S16 03.2

	SHA	Mer. Pass.
	° ′	h m
Venus	94 15.8	9 00
Mars	112 03.4	7 48
Jupiter	223 59.4	0 21
Saturn	278 01.9	20 41

UT (d h)	SUN GHA ° ′	SUN Dec ° ′	MOON GHA ° ′	v ′	MOON Dec ° ′	d ′	HP ′
31 (FRIDAY) 00	176 40.1	S17 33.4	194 24.4	6.7	S25 05.7	3.8	57.5
01	191 40.0	32.8	208 50.1	6.7	25 01.9	3.9	57.5
02	206 39.9	32.1	223 15.8	6.8	24 58.0	4.1	57.4
03	221 39.8	. . 31.4	237 41.6	6.9	24 53.9	4.2	57.4
04	236 39.7	30.7	252 07.5	6.9	24 49.7	4.4	57.4
05	251 39.6	30.0	266 33.4	7.0	24 45.3	4.5	57.4
06	266 39.5	S17 29.3	280 59.4	7.1	S24 40.8	4.7	57.4
07	281 39.4	28.6	295 25.5	7.1	24 36.1	4.8	57.3
08	296 39.3	27.9	309 51.6	7.2	24 31.3	4.9	57.3
09	311 39.2	. . 27.2	324 17.8	7.3	24 26.4	5.1	57.3
10	326 39.1	26.5	338 44.1	7.4	24 21.3	5.2	57.3
11	341 39.1	25.8	353 10.5	7.4	24 16.1	5.4	57.3
12	356 39.0	S17 25.1	7 36.9	7.5	S24 10.7	5.5	57.2
13	11 38.9	24.4	22 03.4	7.6	24 05.2	5.6	57.2
14	26 38.8	23.7	36 30.0	7.7	23 59.6	5.8	57.2
15	41 38.7	. . 23.0	50 56.7	7.7	23 53.8	5.9	57.2
16	56 38.6	22.3	65 23.4	7.9	23 47.9	6.0	57.1
17	71 38.5	21.6	79 50.3	7.9	23 41.9	6.2	57.1
18	86 38.4	S17 20.9	94 17.2	8.0	S23 35.7	6.3	57.1
19	101 38.3	20.2	108 44.2	8.1	23 29.4	6.4	57.1
20	116 38.2	19.5	123 11.3	8.1	23 23.0	6.6	57.1
21	131 38.1	. . 18.8	137 38.4	8.3	23 16.4	6.7	57.0
22	146 38.0	18.1	152 05.7	8.3	23 09.7	6.8	57.0
23	161 37.9	17.4	166 33.0	8.5	23 02.9	6.9	57.0
1 (SATURDAY) 00	176 37.8	S17 16.7	181 00.5	8.5	S22 56.0	7.1	57.0
01	191 37.8	16.0	195 28.0	8.6	22 48.9	7.1	56.9
02	206 37.7	15.3	209 55.6	8.7	22 41.8	7.3	56.9
03	221 37.6	. . 14.6	224 23.3	8.8	22 34.5	7.4	56.9
04	236 37.5	13.9	238 51.1	8.8	22 27.1	7.6	56.9
05	251 37.4	13.2	253 18.9	9.0	22 19.5	7.6	56.8
06	266 37.3	S17 12.5	267 46.9	9.0	S22 11.9	7.8	56.8
07	281 37.2	11.8	282 14.9	9.2	22 04.1	7.9	56.8
08	296 37.1	11.1	296 43.1	9.2	21 56.2	8.0	56.8
09	311 37.0	. . 10.4	311 11.3	9.3	21 48.2	8.1	56.8
10	326 37.0	09.7	325 39.6	9.5	21 40.1	8.2	56.7
11	341 36.9	08.9	340 08.1	9.5	21 31.9	8.3	56.7
12	356 36.8	S17 08.2	354 36.6	9.6	S21 23.6	8.4	56.7
13	11 36.7	07.5	9 05.2	9.7	21 15.2	8.6	56.7
14	26 36.6	06.8	23 33.9	9.8	21 06.6	8.6	56.6
15	41 36.5	. . 06.1	38 02.7	9.8	20 58.0	8.7	56.6
16	56 36.4	05.4	52 31.5	10.0	20 49.3	8.9	56.6
17	71 36.4	04.7	67 00.5	10.1	20 40.4	8.9	56.6
18	86 36.3	S17 04.0	81 29.6	10.2	S20 31.5	9.1	56.5
19	101 36.2	03.2	95 58.8	10.2	20 22.4	9.1	56.5
20	116 36.1	02.5	110 28.0	10.4	20 13.3	9.2	56.5
21	131 36.0	. . 01.8	124 57.4	10.4	20 04.1	9.4	56.5
22	146 35.9	01.1	139 26.8	10.5	19 54.7	9.4	56.5
23	161 35.9	17 00.4	153 56.3	10.7	19 45.3	9.5	56.4
2 (SUNDAY) 00	176 35.8	S16 59.7	168 26.0	10.7	S19 35.8	9.6	56.4
01	191 35.7	59.0	182 55.7	10.8	19 26.2	9.7	56.4
02	206 35.6	58.2	197 25.5	10.9	19 16.5	9.8	56.4
03	221 35.5	. . 57.5	211 55.4	11.0	19 06.7	9.9	56.3
04	236 35.4	56.8	226 25.4	11.1	18 56.8	9.9	56.3
05	251 35.4	56.1	240 55.5	11.2	18 46.9	10.1	56.3
06	266 35.3	S16 55.4	255 25.7	11.3	S18 36.8	10.1	56.3
07	281 35.2	54.7	269 56.0	11.4	18 26.7	10.2	56.2
08	296 35.1	53.9	284 26.4	11.4	18 16.5	10.3	56.2
09	311 35.0	. . 53.2	298 56.8	11.6	18 06.2	10.4	56.2
10	326 35.0	52.5	313 27.4	11.6	17 55.8	10.4	56.2
11	341 34.9	51.8	327 58.0	11.7	17 45.4	10.5	56.1
12	356 34.8	S16 51.0	342 28.7	11.9	S17 34.9	10.6	56.1
13	11 34.7	50.3	356 59.6	11.9	17 24.3	10.7	56.1
14	26 34.7	49.6	11 30.5	12.0	17 13.6	10.8	56.1
15	41 34.6	. . 48.9	26 01.5	12.0	17 02.8	10.8	56.1
16	56 34.5	48.2	40 32.5	12.2	16 52.0	10.9	56.0
17	71 34.4	47.4	55 03.7	12.3	16 41.1	10.9	56.0
18	86 34.4	S16 46.7	69 35.0	12.3	S16 30.2	11.0	56.0
19	101 34.3	46.0	84 06.3	12.5	16 19.2	11.1	56.0
20	116 34.2	45.3	98 37.8	12.5	16 08.1	11.2	55.9
21	131 34.1	. . 44.5	113 09.3	12.6	15 56.9	11.2	55.9
22	146 34.1	43.8	127 40.9	12.7	15 45.7	11.3	55.9
23	161 34.0	43.1	142 12.6	12.7	S15 34.4	11.3	55.9
	SD 16.3	*d* 0.7	SD 15.6		15.4		15.3

Lat. °	Twilight Naut. h m	Twilight Civil h m	Sunrise h m	Moonrise 31 h m	Moonrise 1 h m	Moonrise 2 h m	Moonrise 3 h m
N 72	07 11	08 40	10 31	■	■	12 23	11 01
N 70	07 03	08 21	09 48	■	■	11 17	10 37
68	06 56	08 07	09 19	■	11 33	10 40	10 18
66	06 50	07 54	08 58	■	10 32	10 14	10 03
64	06 45	07 44	08 41	10 01	09 58	09 54	09 50
62	06 41	07 35	08 27	09 20	09 32	09 37	09 39
60	06 37	07 27	08 15	08 52	09 12	09 23	09 30
N 58	06 33	07 21	08 05	08 30	08 56	09 12	09 22
56	06 30	07 15	07 56	08 12	08 42	09 01	09 15
54	06 26	07 09	07 48	07 57	08 29	08 52	09 09
52	06 23	07 04	07 41	07 44	08 19	08 44	09 03
50	06 20	06 59	07 34	07 33	08 09	08 36	08 58
45	06 14	06 49	07 20	07 08	07 49	08 20	08 46
N 40	06 08	06 40	07 09	06 49	07 32	08 07	08 37
35	06 02	06 33	06 59	06 33	07 18	07 56	08 29
30	05 57	06 26	06 51	06 19	07 06	07 46	08 22
20	05 46	06 13	06 36	05 55	06 45	07 30	08 09
N 10	05 35	06 01	06 22	05 34	06 27	07 15	07 58
0	05 23	05 48	06 10	05 15	06 10	07 01	07 48
S 10	05 09	05 35	05 57	04 56	05 53	06 47	07 38
20	04 53	05 20	05 44	04 35	05 34	06 32	07 27
30	04 32	05 02	05 28	04 11	05 13	06 15	07 14
35	04 18	04 51	05 19	03 56	05 01	06 05	07 07
40	04 02	04 38	05 08	03 40	04 46	05 53	06 58
45	03 41	04 23	04 56	03 20	04 29	05 39	06 48
S 50	03 13	04 03	04 41	02 56	04 08	05 23	06 36
52	02 59	03 53	04 34	02 44	03 58	05 15	06 31
54	02 42	03 42	04 26	02 30	03 46	05 06	06 25
56	02 21	03 30	04 17	02 14	03 33	04 56	06 18
58	01 53	03 15	04 07	01 56	03 18	04 45	06 10
S 60	01 06	02 57	03 55	01 32	03 00	04 31	06 01

Lat. °	Sunset h m	Twilight Civil h m	Twilight Naut. h m	Moonset 31 h m	Moonset 1 h m	Moonset 2 h m	Moonset 3 h m
N 72	13 58	15 48	17 18	■	■	14 22	17 20
N 70	14 41	16 07	17 26	■	■	15 26	17 43
68	15 09	16 22	17 32	■	13 25	16 01	18 00
66	15 30	16 34	17 38	■	14 25	16 26	18 13
64	15 47	16 44	17 43	13 03	14 59	16 46	18 25
62	16 01	16 53	17 47	13 43	15 23	17 01	18 34
60	16 13	17 01	17 51	14 11	15 43	17 14	18 42
N 58	16 23	17 07	17 55	14 33	15 59	17 25	18 49
56	16 32	17 13	17 58	14 50	16 12	17 35	18 56
54	16 40	17 19	18 02	15 05	16 24	17 44	19 01
52	16 47	17 24	18 04	15 18	16 34	17 51	19 06
50	16 53	17 28	18 07	15 29	16 43	17 58	19 11
45	17 07	17 39	18 14	15 53	17 03	18 13	19 21
N 40	17 19	17 47	18 20	16 11	17 18	18 25	19 29
35	17 28	17 55	18 25	16 27	17 31	18 35	19 36
30	17 37	18 02	18 31	16 40	17 43	18 43	19 42
20	17 52	18 15	18 41	17 03	18 02	18 59	19 52
N 10	18 05	18 27	18 52	17 23	18 19	19 12	20 01
0	18 17	18 39	19 04	17 42	18 35	19 24	20 09
S 10	18 30	18 52	19 17	18 00	18 50	19 36	20 18
20	18 43	19 06	19 34	18 20	19 07	19 49	20 27
30	18 58	19 24	19 55	18 42	19 26	20 03	20 37
35	19 08	19 35	20 08	18 55	19 37	20 12	20 42
40	19 18	19 48	20 25	19 10	19 49	20 21	20 49
45	19 30	20 03	20 45	19 28	20 04	20 33	20 56
S 50	19 45	20 23	21 12	19 50	20 22	20 46	21 05
52	19 52	20 33	21 26	20 01	20 30	20 52	21 10
54	20 00	20 43	21 43	20 12	20 40	20 59	21 14
56	20 09	20 56	22 03	20 26	20 50	21 07	21 19
58	20 19	21 10	22 30	20 42	21 02	21 15	21 25
S 60	20 30	21 28	23 12	21 00	21 16	21 25	21 31

Day	SUN Eqn. of Time 00^h	SUN Eqn. of Time 12^h	SUN Mer. Pass.	MOON Mer. Pass. Upper	MOON Mer. Pass. Lower	Age	Phase
d	m s	m s	h m	h m	h m	d %	
31	13 19	13 24	12 13	11 28	23 56	29 1	●
1	13 28	13 33	12 14	12 22	24 48	00 0	
2	13 37	13 41	12 14	13 12	00 48	01 1	

UT (d h)	ARIES GHA ° ′	VENUS −4.3 GHA ° ′	VENUS Dec ° ′	MARS +1.2 GHA ° ′	MARS Dec ° ′	JUPITER −2.6 GHA ° ′	JUPITER Dec ° ′	SATURN −0.2 GHA ° ′	SATURN Dec ° ′
3 00 (MONDAY)	132 45.6	224 38.1	S20 47.4	243 27.0	S21 43.1	357 01.1	N17 44.1	50 52.4	N22 02.5
01	147 48.1	239 37.5	47.6	258 27.8	43.3	12 03.9	44.2	65 54.9	02.5
02	162 50.5	254 37.0	47.8	273 28.6	43.6	27 06.7	44.3	80 57.5	02.5
03	177 53.0	269 36.5	. . 48.0	288 29.3	. . 43.8	42 09.5	. . 44.4	96 00.1	. . 02.5
04	192 55.5	284 35.9	48.1	303 30.1	44.1	57 12.3	44.5	111 02.6	02.5
05	207 57.9	299 35.4	48.3	318 30.8	44.3	72 15.1	44.6	126 05.2	02.5
06	223 00.4	314 34.9	S20 48.5	333 31.6	S21 44.6	87 17.9	N17 44.7	141 07.7	N22 02.5
07	238 02.9	329 34.3	48.6	348 32.3	44.8	102 20.7	44.8	156 10.3	02.5
08	253 05.3	344 33.8	48.8	3 33.1	45.1	117 23.5	44.9	171 12.9	02.5
09	268 07.8	359 33.3	. . 49.0	18 33.8	. . 45.3	132 26.3	. . 45.0	186 15.4	. . 02.5
10	283 10.3	14 32.7	49.1	33 34.6	45.6	147 29.1	45.1	201 18.0	02.5
11	298 12.7	29 32.2	49.3	48 35.3	45.9	162 31.9	45.2	216 20.5	02.5
12	313 15.2	44 31.6	S20 49.5	63 36.1	S21 46.1	177 34.7	N17 45.3	231 23.1	N22 02.5
13	328 17.6	59 31.1	49.6	78 36.8	46.4	192 37.5	45.4	246 25.7	02.5
14	343 20.1	74 30.6	49.8	93 37.6	46.6	207 40.3	45.5	261 28.2	02.5
15	358 22.6	89 30.0	. . 50.0	108 38.3	. . 46.9	222 43.1	. . 45.6	276 30.8	. . 02.5
16	13 25.0	104 29.5	50.1	123 39.1	47.1	237 45.9	45.7	291 33.3	02.5
17	28 27.5	119 29.0	50.3	138 39.8	47.4	252 48.7	45.8	306 35.9	02.5
18	43 30.0	134 28.4	S20 50.5	153 40.6	S21 47.6	267 51.5	N17 45.9	321 38.4	N22 02.5
19	58 32.4	149 27.9	50.6	168 41.4	47.9	282 54.3	46.0	336 41.0	02.6
20	73 34.9	164 27.3	50.8	183 42.1	48.1	297 57.1	46.1	351 43.6	02.6
21	88 37.4	179 26.8	. . 50.9	198 42.9	. . 48.4	312 59.9	. . 46.2	6 46.1	. . 02.6
22	103 39.8	194 26.2	51.1	213 43.6	48.6	328 02.7	46.3	21 48.7	02.6
23	118 42.3	209 25.7	51.3	228 44.4	48.9	343 05.5	46.4	36 51.2	02.6
4 00 (TUESDAY)	133 44.7	224 25.2	S20 51.4	243 45.1	S21 49.1	358 08.3	N17 46.5	51 53.8	N22 02.6
01	148 47.2	239 24.6	51.6	258 45.9	49.4	13 11.0	46.6	66 56.3	02.6
02	163 49.7	254 24.1	51.7	273 46.6	49.6	28 13.8	46.7	81 58.9	02.6
03	178 52.1	269 23.5	. . 51.9	288 47.4	. . 49.9	43 16.6	. . 46.8	97 01.4	. . 02.6
04	193 54.6	284 23.0	52.0	303 48.1	50.1	58 19.4	46.9	112 04.0	02.6
05	208 57.1	299 22.4	52.2	318 48.9	50.4	73 22.2	47.0	127 06.5	02.6
06	223 59.5	314 21.9	S20 52.3	333 49.6	S21 50.6	88 25.0	N17 47.1	142 09.1	N22 02.6
07	239 02.0	329 21.3	52.5	348 50.4	50.9	103 27.8	47.2	157 11.7	02.6
08	254 04.5	344 20.8	52.6	3 51.1	51.1	118 30.6	47.3	172 14.2	02.6
09	269 06.9	359 20.3	. . 52.8	18 51.9	. . 51.4	133 33.4	. . 47.4	187 16.8	. . 02.6
10	284 09.4	14 19.7	52.9	33 52.6	51.6	148 36.2	47.5	202 19.3	02.6
11	299 11.9	29 19.2	53.1	48 53.4	51.8	163 39.0	47.6	217 21.9	02.6
12	314 14.3	44 18.6	S20 53.2	63 54.1	S21 52.1	178 41.8	N17 47.7	232 24.4	N22 02.6
13	329 16.8	59 18.1	53.4	78 54.9	52.3	193 44.6	47.8	247 27.0	02.6
14	344 19.2	74 17.5	53.5	93 55.6	52.6	208 47.4	47.9	262 29.5	02.6
15	359 21.7	89 17.0	. . 53.6	108 56.4	. . 52.8	223 50.2	. . 48.0	277 32.1	. . 02.6
16	14 24.2	104 16.4	53.8	123 57.1	53.1	238 53.0	48.1	292 34.6	02.6
17	29 26.6	119 15.9	53.9	138 57.9	53.3	253 55.8	48.2	307 37.2	02.6
18	44 29.1	134 15.3	S20 54.1	153 58.6	S21 53.6	268 58.6	N17 48.3	322 39.7	N22 02.6
19	59 31.6	149 14.8	54.2	168 59.4	53.8	284 01.4	48.4	337 42.3	02.6
20	74 34.0	164 14.2	54.3	184 00.1	54.1	299 04.2	48.5	352 44.8	02.6
21	89 36.5	179 13.7	. . 54.5	199 00.9	. . 54.3	314 07.0	. . 48.6	7 47.4	. . 02.7
22	104 39.0	194 13.1	54.6	214 01.6	54.5	329 09.8	48.7	22 49.9	02.7
23	119 41.4	209 12.6	54.8	229 02.4	54.8	344 12.6	48.8	37 52.5	02.7
5 00 (WEDNESDAY)	134 43.9	224 12.0	S20 54.9	244 03.1	S21 55.0	359 15.4	N17 48.9	52 55.0	N22 02.7
01	149 46.3	239 11.5	55.0	259 03.9	55.3	14 18.2	49.0	67 57.6	02.7
02	164 48.8	254 10.9	55.2	274 04.6	55.5	29 21.0	49.0	83 00.1	02.7
03	179 51.3	269 10.3	. . 55.3	289 05.4	. . 55.7	44 23.8	. . 49.1	98 02.7	. . 02.7
04	194 53.7	284 09.8	55.4	304 06.1	56.0	59 26.6	49.2	113 05.2	02.7
05	209 56.2	299 09.2	55.5	319 06.9	56.2	74 29.4	49.3	128 07.8	02.7
06	224 58.7	314 08.7	S20 55.7	334 07.6	S21 56.5	89 32.2	N17 49.4	143 10.3	N22 02.7
07	240 01.1	329 08.1	55.8	349 08.4	56.7	104 35.0	49.5	158 12.9	02.7
08	255 03.6	344 07.6	55.9	4 09.1	57.0	119 37.8	49.6	173 15.4	02.7
09	270 06.1	359 07.0	. . 56.1	19 09.9	. . 57.2	134 40.6	. . 49.7	188 18.0	. . 02.7
10	285 08.5	14 06.5	56.2	34 10.6	57.4	149 43.4	49.8	203 20.5	02.7
11	300 11.0	29 05.9	56.3	49 11.4	57.7	164 46.2	49.9	218 23.1	02.7
12	315 13.5	44 05.3	S20 56.4	64 12.1	S21 57.9	179 49.0	N17 50.0	233 25.6	N22 02.7
13	330 15.9	59 04.8	56.6	79 12.9	58.2	194 51.8	50.1	248 28.2	02.7
14	345 18.4	74 04.2	56.7	94 13.6	58.4	209 54.6	50.2	263 30.7	02.7
15	0 20.8	89 03.7	. . 56.8	109 14.4	. . 58.6	224 57.4	. . 50.3	278 33.3	. . 02.7
16	15 23.3	104 03.1	56.9	124 15.1	58.9	240 00.2	50.4	293 35.8	02.7
17	30 25.8	119 02.5	57.0	139 15.9	59.1	255 03.0	50.5	308 38.4	02.7
18	45 28.2	134 02.0	S20 57.2	154 16.6	S21 59.3	270 05.8	N17 50.6	323 40.9	N22 02.7
19	60 30.7	149 01.4	57.3	169 17.4	59.6	285 08.6	50.7	338 43.5	02.7
20	75 33.2	164 00.9	57.4	184 18.1	21 59.8	300 11.4	50.8	353 46.0	02.7
21	90 35.6	179 00.3	. . 57.5	199 18.9	22 00.1	315 14.2	. . 50.9	8 48.6	. . 02.8
22	105 38.1	193 59.7	57.6	214 19.6	00.3	330 17.0	51.0	23 51.1	02.8
23	120 40.6	208 59.2	57.7	229 20.4	00.5	345 19.8	51.1	38 53.7	02.8
Mer. Pass.	h m 15 02.5	v −0.5	d 0.1	v 0.8	d 0.2	v 2.8	d 0.1	v 2.6	d 0.0

STARS Name	SHA ° ′	Dec ° ′
Acamar	315 24.4	S40 17.
Achernar	335 32.8	S57 13.
Acrux	173 18.3	S63 06.
Adhara	255 18.6	S28 58.
Aldebaran	290 58.6	N16 30.
Alioth	166 27.3	N55 56.
Alkaid	153 05.0	N49 17.
Al Na'ir	27 54.2	S46 57.
Alnilam	275 54.4	S 1 12.
Alphard	218 03.8	S 8 40.
Alphecca	126 17.9	N26 42.
Alpheratz	357 52.3	N29 06.
Altair	62 16.5	N 8 52.
Ankaa	353 23.9	S42 17.
Antares	112 36.5	S26 26.
Arcturus	146 03.1	N19 09.8
Atria	107 46.0	S69 01.
Avior	234 20.8	S59 31.
Bellatrix	278 40.6	N 6 21.
Betelgeuse	271 09.9	N 7 24.
Canopus	263 59.3	S52 42.
Capella	280 46.3	N46 00.2
Deneb	49 37.6	N45 17.3
Denebola	182 41.7	N14 33.2
Diphda	349 04.2	S17 58.4
Dubhe	194 00.8	N61 43.9
Elnath	278 22.7	N28 36.7
Eltanin	90 50.3	N51 29.0
Enif	33 55.5	N 9 53.2
Fomalhaut	15 33.2	S29 36.6
Gacrux	172 09.9	S57 07.6
Gienah	176 00.5	S17 33.5
Hadar	148 59.6	S60 23.0
Hamal	328 10.1	N23 28.6
Kaus Aust.	83 54.9	S34 23.0
Kochab	137 19.3	N74 08.2
Markab	13 46.8	N15 13.2
Menkar	314 23.6	N 4 06.0
Menkent	148 17.2	S36 22.9
Miaplacidus	221 40.7	S69 43.7
Mirfak	308 52.1	N49 52.5
Nunki	76 08.7	S26 17.7
Peacock	53 32.5	S56 43.6
Pollux	243 37.3	N28 01.2
Procyon	245 07.9	N 5 13.0
Rasalhague	96 14.2	N12 33.3
Regulus	207 51.8	N11 57.1
Rigel	281 19.7	S 8 12.0
Rigil Kent.	140 03.1	S60 50.6
Sabik	102 22.1	S15 43.8
Schedar	349 50.4	N56 33.4
Shaula	96 33.3	S37 06.3
Sirius	258 40.7	S16 43.3
Spica	158 39.8	S11 10.6
Suhail	222 58.1	S43 26.7
Vega	80 44.8	N38 46.9
Zuben'ubi	137 14.5	S16 03.2

	SHA ° ′	Mer. Pass h m
Venus	90 40.4	9 03
Mars	110 00.4	7 45
Jupiter	224 23.5	0 07
Saturn	278 09.0	20 29

	UT	SUN GHA	SUN Dec	MOON GHA	v	MOON Dec	d	HP
	d h	° ′	° ′	° ′	′	° ′	′	′
	3 00	176 33.9	S16 42.3	156 44.3	12.9	S15 23.1	11.4	55.8
	01	191 33.8	41.6	171 16.2	12.9	15 11.7	11.5	55.8
	02	206 33.8	40.9	185 48.1	13.0	15 00.2	11.5	55.8
	03	221 33.7	. . 40.2	200 20.1	13.1	14 48.7	11.6	55.8
	04	236 33.6	39.4	214 52.2	13.2	14 37.1	11.6	55.8
	05	251 33.5	38.7	229 24.4	13.3	14 25.5	11.7	55.7
	06	266 33.5	S16 38.0	243 56.7	13.3	S14 13.8	11.7	55.7
	07	281 33.4	37.2	258 29.0	13.4	14 02.1	11.8	55.7
	08	296 33.3	36.5	273 01.4	13.5	13 50.3	11.9	55.7
M	09	311 33.3	. . 35.8	287 33.9	13.6	13 38.4	11.9	55.6
O	10	326 33.2	35.0	302 06.5	13.6	13 26.5	11.9	55.6
N	11	341 33.1	34.3	316 39.1	13.7	13 14.6	12.0	55.6
D	12	356 33.1	S16 33.6	331 11.8	13.8	S13 02.6	12.0	55.6
A	13	11 33.0	32.8	345 44.6	13.9	12 50.6	12.1	55.6
Y	14	26 32.9	32.1	0 17.5	13.9	12 38.5	12.2	55.5
	15	41 32.8	. . 31.4	14 50.4	14.0	12 26.3	12.1	55.5
	16	56 32.8	30.6	29 23.4	14.1	12 14.2	12.2	55.5
	17	71 32.7	29.9	43 56.5	14.2	12 02.0	12.3	55.5
	18	86 32.6	S16 29.1	58 29.7	14.2	S11 49.7	12.3	55.4
	19	101 32.6	28.4	73 02.9	14.3	11 37.4	12.3	55.4
	20	116 32.5	27.7	87 36.2	14.3	11 25.1	12.4	55.4
	21	131 32.4	. . 26.9	102 09.5	14.4	11 12.7	12.4	55.4
	22	146 32.4	26.2	116 42.9	14.5	11 00.3	12.4	55.4
	23	161 32.3	25.5	131 16.4	14.6	10 47.9	12.5	55.3
	4 00	176 32.2	S16 24.7	145 50.0	14.6	S10 35.4	12.5	55.3
	01	191 32.2	24.0	160 23.6	14.6	10 22.9	12.6	55.3
	02	206 32.1	23.2	174 57.2	14.8	10 10.3	12.5	55.3
	03	221 32.1	. . 22.5	189 31.0	14.8	9 57.8	12.6	55.3
	04	236 32.0	21.7	204 04.8	14.8	9 45.2	12.7	55.2
	05	251 31.9	21.0	218 38.6	14.9	9 32.5	12.6	55.2
	06	266 31.9	S16 20.3	233 12.5	15.0	S 9 19.9	12.7	55.2
	07	281 31.8	19.5	247 46.5	15.0	9 07.2	12.8	55.2
T	08	296 31.7	18.8	262 20.5	15.1	8 54.4	12.7	55.2
U	09	311 31.7	. . 18.0	276 54.6	15.2	8 41.7	12.8	55.1
E	10	326 31.6	17.3	291 28.8	15.2	8 28.9	12.8	55.1
S	11	341 31.6	16.5	306 03.0	15.2	8 16.1	12.8	55.1
D	12	356 31.5	S16 15.8	320 37.2	15.3	S 8 03.3	12.8	55.1
A	13	11 31.4	15.0	335 11.5	15.4	7 50.5	12.9	55.1
Y	14	26 31.4	14.3	349 45.9	15.4	7 37.6	12.9	55.0
	15	41 31.3	. . 13.6	4 20.3	15.4	7 24.7	12.9	55.0
	16	56 31.3	12.8	18 54.7	15.5	7 11.8	12.9	55.0
	17	71 31.2	12.1	33 29.2	15.5	6 58.9	12.9	55.0
	18	86 31.1	S16 11.3	48 03.7	15.6	S 6 46.0	13.0	55.0
	19	101 31.1	10.6	62 38.3	15.7	6 33.0	13.0	54.9
	20	116 31.0	09.8	77 13.0	15.6	6 20.0	12.9	54.9
	21	131 31.0	. . 09.1	91 47.6	15.8	6 07.1	13.0	54.9
	22	146 30.9	08.3	106 22.4	15.7	5 54.1	13.1	54.9
	23	161 30.8	07.6	120 57.1	15.8	5 41.0	13.0	54.9
	5 00	176 30.8	S16 06.8	135 31.9	15.9	S 5 28.0	13.0	54.9
	01	191 30.7	06.1	150 06.8	15.9	5 15.0	13.1	54.8
	02	206 30.7	05.3	164 41.7	15.9	5 01.9	13.0	54.8
	03	221 30.6	. . 04.5	179 16.6	16.0	4 48.9	13.1	54.8
	04	236 30.6	03.8	193 51.6	16.0	4 35.8	13.1	54.8
	05	251 30.5	03.0	208 26.6	16.0	4 22.7	13.1	54.8
	06	266 30.5	S16 02.3	223 01.6	16.1	S 4 09.6	13.1	54.8
W	07	281 30.4	01.5	237 36.7	16.1	3 56.5	13.1	54.7
E	08	296 30.4	00.8	252 11.8	16.1	3 43.4	13.1	54.7
D	09	311 30.3	16 00.0	266 46.9	16.1	3 30.3	13.1	54.7
N	10	326 30.2	15 59.3	281 22.0	16.2	3 17.2	13.1	54.7
	11	341 30.2	58.5	295 57.2	16.3	3 04.1	13.1	54.7
E	12	356 30.1	S15 57.7	310 32.5	16.2	S 2 51.0	13.2	54.7
S	13	11 30.1	57.0	325 07.7	16.3	2 37.8	13.1	54.6
D	14	26 30.0	56.2	339 43.0	16.3	2 24.7	13.1	54.6
A	15	41 30.0	. . 55.5	354 18.3	16.3	2 11.6	13.1	54.6
Y	16	56 29.9	54.7	8 53.6	16.4	1 58.5	13.2	54.6
	17	71 29.9	53.9	23 29.0	16.3	1 45.3	13.1	54.6
	18	86 29.8	S15 53.2	38 04.3	16.4	S 1 32.2	13.1	54.6
	19	101 29.8	52.4	52 39.7	16.4	1 19.1	13.1	54.6
	20	116 29.7	51.7	67 15.1	16.5	1 06.0	13.1	54.5
	21	131 29.7	. . 50.9	81 50.6	16.4	0 52.9	13.2	54.5
	22	146 29.6	50.1	96 26.0	16.5	0 39.7	13.1	54.5
	23	161 29.6	49.4	111 01.5	16.5	S 0 26.6	13.1	54.5
		SD 16.3	d 0.7	SD 15.1		15.0		14.9

Lat.	Twilight Naut.	Twilight Civil	Sunrise	Moonrise 3	Moonrise 4	Moonrise 5	Moonrise 6
°	h m	h m	h m	h m	h m	h m	h m
N 72	07 00	08 27	10 07	11 01	10 24	09 57	09 31
N 70	06 53	08 10	09 31	10 37	10 12	09 52	09 34
68	06 48	07 57	09 06	10 18	10 02	09 48	09 36
66	06 43	07 46	08 47	10 03	09 53	09 45	09 37
64	06 38	07 36	08 32	09 50	09 46	09 43	09 39
62	06 34	07 28	08 19	09 39	09 40	09 40	09 40
60	06 31	07 21	08 08	09 30	09 35	09 38	09 41
N 58	06 28	07 15	07 58	09 22	09 30	09 36	09 42
56	06 25	07 09	07 50	09 15	09 26	09 35	09 43
54	06 22	07 04	07 43	09 09	09 22	09 33	09 44
52	06 19	07 00	07 36	09 03	09 18	09 32	09 44
50	06 17	06 55	07 30	08 58	09 15	09 31	09 45
45	06 11	06 46	07 17	08 46	09 08	09 28	09 46
N 40	06 05	06 38	07 06	08 37	09 02	09 26	09 48
35	06 00	06 30	06 57	08 29	08 57	09 24	09 49
30	05 55	06 24	06 49	08 22	08 53	09 22	09 50
20	05 45	06 12	06 35	08 09	08 45	09 19	09 51
N 10	05 35	06 00	06 22	07 58	08 38	09 16	09 53
0	05 24	05 49	06 10	07 48	08 32	09 14	09 54
S 10	05 11	05 37	05 58	07 38	08 26	09 11	09 55
20	04 55	05 22	05 46	07 27	08 19	09 08	09 57
30	04 35	05 05	05 31	07 14	08 11	09 05	09 59
35	04 22	04 55	05 22	07 07	08 06	09 04	10 00
40	04 06	04 42	05 12	06 58	08 01	09 02	10 01
45	03 46	04 28	05 00	06 48	07 55	08 59	10 02
S 50	03 20	04 09	04 46	06 36	07 48	08 57	10 04
52	03 07	04 00	04 40	06 31	07 44	08 55	10 05
54	02 51	03 49	04 32	06 25	07 41	08 54	10 05
56	02 32	03 38	04 24	06 18	07 36	08 52	10 06
58	02 08	03 24	04 14	06 10	07 32	08 51	10 07
S 60	01 32	03 07	04 03	06 01	07 27	08 49	10 09

Lat.	Sunset	Twilight Civil	Twilight Naut.	Moonset 3	Moonset 4	Moonset 5	Moonset 6
°	h m	h m	h m	h m	h m	h m	h m
N 72	14 22	16 02	17 29	17 20	19 27	21 22	23 14
N 70	14 58	16 19	17 36	17 43	19 37	21 23	23 07
68	15 23	16 32	17 42	18 00	19 45	21 24	23 02
66	15 42	16 43	17 46	18 13	19 52	21 25	22 57
64	15 57	16 53	17 51	18 25	19 57	21 26	22 53
62	16 10	17 01	17 54	18 34	20 02	21 27	22 50
60	16 21	17 08	17 58	18 42	20 06	21 27	22 47
N 58	16 30	17 14	18 01	18 49	20 10	21 28	22 44
56	16 39	17 19	18 04	18 56	20 13	21 28	22 42
54	16 46	17 24	18 07	19 01	20 16	21 28	22 40
52	16 53	17 29	18 09	19 06	20 19	21 29	22 38
50	16 59	17 33	18 12	19 11	20 21	21 29	22 36
45	17 11	17 43	18 18	19 21	20 26	21 30	22 32
N 40	17 22	17 51	18 23	19 29	20 30	21 30	22 29
35	17 31	17 58	18 28	19 36	20 34	21 31	22 27
30	17 39	18 04	18 33	19 42	20 37	21 31	22 24
20	17 53	18 16	18 43	19 52	20 43	21 32	22 20
N 10	18 06	18 28	18 53	20 01	20 48	21 33	22 17
0	18 17	18 39	19 04	20 09	20 52	21 33	22 13
S 10	18 29	18 51	19 17	20 18	20 57	21 34	22 10
20	18 42	19 05	19 32	20 27	21 01	21 34	22 06
30	18 57	19 22	19 53	20 37	21 07	21 35	22 02
35	19 05	19 32	20 05	20 42	21 10	21 35	22 00
40	19 15	19 45	20 21	20 49	21 13	21 36	21 58
45	19 26	19 59	20 40	20 56	21 17	21 36	21 55
S 50	19 41	20 18	21 06	21 05	21 22	21 37	21 51
52	19 47	20 27	21 19	21 10	21 24	21 37	21 49
54	19 54	20 37	21 34	21 14	21 26	21 37	21 48
56	20 03	20 48	21 53	21 19	21 29	21 37	21 46
58	20 12	21 02	22 16	21 25	21 32	21 38	21 43
S 60	20 23	21 18	22 50	21 31	21 35	21 38	21 41

Day	SUN Eqn. of Time 00^h	SUN Eqn. of Time 12^h	SUN Mer. Pass.	MOON Mer. Pass. Upper	MOON Mer. Pass. Lower	MOON Age	MOON Phase
d	m s	m s	h m	h m	h m	d %	
3	13 44	13 48	12 14	13 59	01 36	02 5	
4	13 51	13 54	12 14	14 42	02 21	03 10	
5	13 57	13 59	12 14	15 23	03 03	04 16	

UT d h	ARIES GHA ° ′	VENUS −4.2 GHA ° ′	VENUS Dec ° ′	MARS +1.2 GHA ° ′	MARS Dec ° ′	JUPITER −2.6 GHA ° ′	JUPITER Dec ° ′	SATURN −0.2 GHA ° ′	SATURN Dec ° ′
6 00 THURSDAY	135 43.0	223 58.6	S20 57.9	244 21.1	S22 00.8	0 22.5	N17 51.2	53 56.2	N22 02.8
01	150 45.5	238 58.1	58.0	259 21.9	01.0	15 25.3	51.3	68 58.8	02.8
02	165 47.9	253 57.5	58.1	274 22.6	01.2	30 28.1	51.4	84 01.3	02.8
03	180 50.4	268 56.9	. . 58.2	289 23.3	. . 01.5	45 30.9	. . 51.5	99 03.8	. . 02.8
04	195 52.9	283 56.4	58.3	304 24.1	01.7	60 33.7	51.6	114 06.4	02.8
05	210 55.3	298 55.8	58.4	319 24.8	01.9	75 36.5	51.7	129 08.9	02.8
06	225 57.8	313 55.2	S20 58.5	334 25.6	S22 02.2	90 39.3	N17 51.8	144 11.5	N22 02.8
07	241 00.3	328 54.7	58.6	349 26.3	02.4	105 42.1	51.9	159 14.0	02.8
08	256 02.7	343 54.1	58.7	4 27.1	02.6	120 44.9	52.0	174 16.6	02.8
09	271 05.2	358 53.5	. . 58.8	19 27.8	. . 02.9	135 47.7	. . 52.1	189 19.1	. . 02.8
10	286 07.7	13 53.0	58.9	34 28.6	03.1	150 50.5	52.2	204 21.7	02.8
11	301 10.1	28 52.4	59.0	49 29.3	03.3	165 53.3	52.3	219 24.2	02.8
12	316 12.6	43 51.8	S20 59.1	64 30.1	S22 03.6	180 56.1	N17 52.4	234 26.7	N22 02.8
13	331 15.1	58 51.3	59.2	79 30.8	03.8	195 58.9	52.5	249 29.3	02.8
14	346 17.5	73 50.7	59.3	94 31.6	04.0	211 01.7	52.6	264 31.8	02.8
15	1 20.0	88 50.1	. . 59.4	109 32.3	. . 04.3	226 04.5	. . 52.7	279 34.4	. . 02.8
16	16 22.4	103 49.6	59.5	124 33.1	04.5	241 07.3	52.8	294 36.9	02.8
17	31 24.9	118 49.0	59.6	139 33.8	04.7	256 10.1	52.9	309 39.5	02.8
18	46 27.4	133 48.4	S20 59.7	154 34.6	S22 05.0	271 12.9	N17 53.0	324 42.0	N22 02.8
19	61 29.8	148 47.9	59.8	169 35.3	05.2	286 15.7	53.1	339 44.5	02.9
20	76 32.3	163 47.3	20 59.9	184 36.0	05.4	301 18.5	53.2	354 47.1	02.9
21	91 34.8	178 46.7	21 00.0	199 36.8	. . 05.6	316 21.3	. . 53.3	9 49.6	. . 02.9
22	106 37.2	193 46.2	00.1	214 37.5	05.9	331 24.1	53.4	24 52.2	02.9
23	121 39.7	208 45.6	00.2	229 38.3	06.1	346 26.9	53.5	39 54.7	02.9
7 00 FRIDAY	136 42.2	223 45.0	S21 00.3	244 39.0	S22 06.3	1 29.7	N17 53.6	54 57.2	N22 02.9
01	151 44.6	238 44.4	00.4	259 39.8	06.6	16 32.5	53.6	69 59.8	02.9
02	166 47.1	253 43.9	00.5	274 40.5	06.8	31 35.3	53.7	85 02.3	02.9
03	181 49.6	268 43.3	. . 00.6	289 41.3	. . 07.0	46 38.1	. . 53.8	100 04.9	. . 02.9
04	196 52.0	283 42.7	00.7	304 42.0	07.2	61 40.9	53.9	115 07.4	02.9
05	211 54.5	298 42.1	00.7	319 42.8	07.5	76 43.7	54.0	130 10.0	02.9
06	226 56.9	313 41.6	S21 00.8	334 43.5	S22 07.7	91 46.4	N17 54.1	145 12.5	N22 02.9
07	241 59.4	328 41.0	00.9	349 44.3	07.9	106 49.2	54.2	160 15.0	02.9
08	257 01.9	343 40.4	01.0	4 45.0	08.1	121 52.0	54.3	175 17.6	02.9
09	272 04.3	358 39.9	. . 01.1	19 45.7	. . 08.4	136 54.8	. . 54.4	190 20.1	. . 02.9
10	287 06.8	13 39.3	01.2	34 46.5	08.6	151 57.6	54.5	205 22.7	02.9
11	302 09.3	28 38.7	01.2	49 47.2	08.8	167 00.4	54.6	220 25.2	02.9
12	317 11.7	43 38.1	S21 01.3	64 48.0	S22 09.1	182 03.2	N17 54.7	235 27.7	N22 02.9
13	332 14.2	58 37.6	01.4	79 48.7	09.3	197 06.0	54.8	250 30.3	02.9
14	347 16.7	73 37.0	01.5	94 49.5	09.5	212 08.8	54.9	265 32.8	02.9
15	2 19.1	88 36.4	. . 01.6	109 50.2	. . 09.7	227 11.6	. . 55.0	280 35.3	. . 02.9
16	17 21.6	103 35.8	01.6	124 51.0	09.9	242 14.4	55.1	295 37.9	02.9
17	32 24.0	118 35.2	01.7	139 51.7	10.2	257 17.2	55.2	310 40.4	03.0
18	47 26.5	133 34.7	S21 01.8	154 52.5	S22 10.4	272 20.0	N17 55.3	325 43.0	N22 03.0
19	62 29.0	148 34.1	01.9	169 53.2	10.6	287 22.8	55.4	340 45.5	03.0
20	77 31.4	163 33.5	01.9	184 53.9	10.8	302 25.6	55.5	355 48.0	03.0
21	92 33.9	178 32.9	. . 02.0	199 54.7	. . 11.1	317 28.4	. . 55.6	10 50.6	. . 03.0
22	107 36.4	193 32.4	02.1	214 55.4	11.3	332 31.2	55.7	25 53.1	03.0
23	122 38.8	208 31.8	02.2	229 56.2	11.5	347 34.0	55.8	40 55.6	03.0
8 00 SATURDAY	137 41.3	223 31.2	S21 02.2	244 56.9	S22 11.7	2 36.8	N17 55.9	55 58.2	N22 03.0
01	152 43.8	238 30.6	02.3	259 57.7	12.0	17 39.6	56.0	71 00.7	03.0
02	167 46.2	253 30.0	02.4	274 58.4	12.2	32 42.4	56.1	86 03.2	03.0
03	182 48.7	268 29.5	. . 02.4	289 59.2	. . 12.4	47 45.2	. . 56.2	101 05.8	. . 03.0
04	197 51.2	283 28.9	02.5	304 59.9	12.6	62 47.9	56.3	116 08.3	03.0
05	212 53.6	298 28.3	02.6	320 00.6	12.8	77 50.7	56.4	131 10.9	03.0
06	227 56.1	313 27.7	S21 02.6	335 01.4	S22 13.1	92 53.5	N17 56.5	146 13.4	N22 03.0
07	242 58.5	328 27.1	02.7	350 02.1	13.3	107 56.3	56.5	161 15.9	03.0
08	258 01.0	343 26.6	02.7	5 02.9	13.5	122 59.1	56.6	176 18.5	03.0
09	273 03.5	358 26.0	. . 02.8	20 03.6	. . 13.7	138 01.9	. . 56.7	191 21.0	. . 03.0
10	288 05.9	13 25.4	02.9	35 04.4	13.9	153 04.7	56.8	206 23.5	03.0
11	303 08.4	28 24.8	02.9	50 05.1	14.2	168 07.5	56.9	221 26.1	03.0
12	318 10.9	43 24.2	S21 03.0	65 05.8	S22 14.4	183 10.3	N17 57.0	236 28.6	N22 03.0
13	333 13.3	58 23.6	03.0	80 06.6	14.6	198 13.1	57.1	251 31.1	03.1
14	348 15.8	73 23.1	03.1	95 07.3	14.8	213 15.9	57.2	266 33.7	03.1
15	3 18.3	88 22.5	. . 03.2	110 08.1	. . 15.0	228 18.7	. . 57.3	281 36.2	. . 03.1
16	18 20.7	103 21.9	03.2	125 08.8	15.2	243 21.5	57.4	296 38.7	03.1
17	33 23.2	118 21.3	03.3	140 09.6	15.5	258 24.3	57.5	311 41.3	03.1
18	48 25.7	133 20.7	S21 03.3	155 10.3	S22 15.7	273 27.1	N17 57.6	326 43.8	N22 03.1
19	63 28.1	148 20.1	03.4	170 11.0	15.9	288 29.9	57.7	341 46.3	03.1
20	78 30.6	163 19.5	03.4	185 11.8	16.1	303 32.7	57.8	356 48.9	03.1
21	93 33.0	178 19.0	. . 03.5	200 12.5	. . 16.3	318 35.5	. . 57.9	11 51.4	. . 03.1
22	108 35.5	193 18.4	03.5	215 13.3	16.5	333 38.2	58.0	26 53.9	03.1
23	123 38.0	208 17.8	03.6	230 14.0	16.8	348 41.0	58.1	41 56.5	03.1
Mer. Pass.	h m 14 50.8	v −0.6	d 0.1	v 0.7	d 0.2	v 2.8	d 0.1	v 2.5	d 0.0

STARS

Name	SHA ° ′	Dec ° ′
Acamar	315 24.4	S40 17.9
Achernar	335 32.8	S57 13.6
Acrux	173 18.3	S63 06.7
Adhara	255 18.6	S28 58.7
Aldebaran	290 58.6	N16 30.9
Alioth	166 27.3	N55 56.4
Alkaid	153 05.0	N49 17.6
Al Na'ir	27 54.3	S46 57.0
Alnilam	275 54.4	S 1 12.1
Alphard	218 03.8	S 8 40.3
Alphecca	126 17.9	N26 42.0
Alpheratz	357 52.3	N29 06.4
Altair	62 16.5	N 8 52.4
Ankaa	353 23.9	S42 17.7
Antares	112 36.4	S26 26.3
Arcturus	146 03.1	N19 09.8
Atria	107 46.0	S69 01.8
Avior	234 20.9	S59 31.2
Bellatrix	278 40.6	N 6 21.1
Betelgeuse	271 09.9	N 7 24.4
Canopus	263 59.4	S52 42.0
Capella	280 46.3	N46 00.2
Deneb	49 37.6	N45 17.3
Denebola	182 41.7	N14 33.2
Diphda	349 04.2	S17 58.4
Dubhe	194 00.8	N61 44.0
Elnath	278 22.7	N28 36.7
Eltanin	90 50.3	N51 29.0
Enif	33 55.5	N 9 53.2
Fomalhaut	15 33.2	S29 36.6
Gacrux	172 09.9	S57 07.6
Gienah	176 00.5	S17 33.5
Hadar	148 59.6	S60 23.0
Hamal	328 10.1	N23 28.6
Kaus Aust.	83 54.9	S34 23.0
Kochab	137 19.3	N74 08.2
Markab	13 46.8	N15 13.2
Menkar	314 23.6	N 4 06.0
Menkent	148 17.2	S36 23.0
Miaplacidus	221 40.7	S69 43.7
Mirfak	308 52.1	N49 52.5
Nunki	76 08.7	S26 17.7
Peacock	53 32.5	S56 43.6
Pollux	243 37.3	N28 01.2
Procyon	245 08.0	N 5 13.0
Rasalhague	96 14.2	N12 33.3
Regulus	207 51.8	N11 57.1
Rigel	281 19.7	S 8 12.0
Rigil Kent.	140 03.1	S60 50.6
Sabik	102 22.1	S15 43.8
Schedar	349 50.4	N56 33.3
Shaula	96 33.3	S37 06.3
Sirius	258 40.7	S16 43.3
Spica	158 39.7	S11 10.6
Suhail	222 58.1	S43 26.7
Vega	80 44.8	N38 46.9
Zuben'ubi	137 14.5	S16 03.3

	SHA ° ′	Mer. Pass. h m
Venus	87 02.9	9 05
Mars	107 56.9	7 41
Jupiter	224 47.5	23 50
Saturn	278 15.1	20 17

Day	UT	SUN GHA	SUN Dec	MOON GHA	v	MOON Dec	d	HP
	d h	° ′	° ′	° ′	′	° ′	′	′
THURSDAY	6 00	176 29.5	S15 48.6	125 37.0	16.5	S 0 13.5	13.1	54.5
	01	191 29.5	47.9	140 12.5	16.5	S 0 00.4	13.1	54.5
	02	206 29.4	47.1	154 48.0	16.5	N 0 12.7	13.0	54.5
	03	221 29.4	. . 46.3	169 23.5	16.6	0 25.7	13.1	54.5
	04	236 29.4	45.6	183 59.1	16.5	0 38.8	13.1	54.4
	05	251 29.3	44.8	198 34.6	16.6	0 51.9	13.0	54.4
	06	266 29.3	S15 44.0	213 10.2	16.6	N 1 04.9	13.1	54.4
	07	281 29.2	43.3	227 45.8	16.5	1 18.0	13.0	54.4
	08	296 29.2	42.5	242 21.3	16.6	1 31.0	13.0	54.4
	09	311 29.1	. . 41.7	256 56.9	16.6	1 44.0	13.1	54.4
	10	326 29.1	41.0	271 32.5	16.6	1 57.1	13.0	54.4
	11	341 29.0	40.2	286 08.1	16.6	2 10.1	12.9	54.4
	12	356 29.0	S15 39.4	300 43.7	16.6	N 2 23.0	13.0	54.4
	13	11 29.0	38.6	315 19.3	16.7	2 36.0	13.0	54.4
	14	26 28.9	37.9	329 55.0	16.6	2 49.0	12.9	54.3
	15	41 28.9	. . 37.1	344 30.6	16.6	3 01.9	12.9	54.3
	16	56 28.8	36.3	359 06.2	16.6	3 14.8	12.9	54.3
	17	71 28.8	35.6	13 41.8	16.6	3 27.7	12.9	54.3
	18	86 28.7	S15 34.8	28 17.4	16.6	N 3 40.6	12.9	54.3
	19	101 28.7	34.0	42 53.0	16.6	3 53.5	12.9	54.3
	20	116 28.7	33.2	57 28.6	16.6	4 06.4	12.8	54.3
	21	131 28.6	. . 32.5	72 04.2	16.6	4 19.2	12.8	54.3
	22	146 28.6	31.7	86 39.8	16.6	4 32.0	12.8	54.3
	23	161 28.5	30.9	101 15.4	16.6	4 44.8	12.8	54.3
FRIDAY	7 00	176 28.5	S15 30.2	115 51.0	16.5	N 4 57.6	12.7	54.3
	01	191 28.5	29.4	130 26.5	16.6	5 10.3	12.8	54.3
	02	206 28.4	28.6	145 02.1	16.5	5 23.1	12.7	54.3
	03	221 28.4	. . 27.8	159 37.6	16.6	5 35.8	12.7	54.3
	04	236 28.3	27.1	174 13.2	16.5	5 48.5	12.6	54.2
	05	251 28.3	26.3	188 48.7	16.5	6 01.1	12.7	54.2
	06	266 28.3	S15 25.5	203 24.2	16.5	N 6 13.8	12.6	54.2
	07	281 28.2	24.7	217 59.7	16.5	6 26.4	12.5	54.2
	08	296 28.2	23.9	232 35.2	16.4	6 38.9	12.6	54.2
	09	311 28.2	. . 23.2	247 10.6	16.5	6 51.5	12.5	54.2
	10	326 28.1	22.4	261 46.1	16.4	7 04.0	12.5	54.2
	11	341 28.1	21.6	276 21.5	16.4	7 16.5	12.5	54.2
	12	356 28.1	S15 20.8	290 56.9	16.4	N 7 29.0	12.4	54.2
	13	11 28.0	20.0	305 32.3	16.3	7 41.4	12.5	54.2
	14	26 28.0	19.3	320 07.6	16.4	7 53.9	12.3	54.2
	15	41 28.0	. . 18.5	334 43.0	16.3	8 06.2	12.4	54.2
	16	56 27.9	17.7	349 18.3	16.3	8 18.6	12.3	54.2
	17	71 27.9	16.9	3 53.6	16.2	8 30.9	12.3	54.2
	18	86 27.9	S15 16.1	18 28.8	16.2	N 8 43.2	12.2	54.2
	19	101 27.8	15.4	33 04.0	16.2	8 55.4	12.2	54.2
	20	116 27.8	14.6	47 39.2	16.2	9 07.6	12.2	54.2
	21	131 27.8	. . 13.8	62 14.4	16.2	9 19.8	12.2	54.2
	22	146 27.7	13.0	76 49.6	16.1	9 32.0	12.1	54.2
	23	161 27.7	12.2	91 24.7	16.1	9 44.1	12.1	54.2
SATURDAY	8 00	176 27.7	S15 11.4	105 59.8	16.0	N 9 56.2	12.0	54.2
	01	191 27.6	10.6	120 34.8	16.0	10 08.2	12.0	54.2
	02	206 27.6	09.9	135 09.8	16.0	10 20.2	12.0	54.2
	03	221 27.6	. . 09.1	149 44.8	16.0	10 32.2	11.9	54.2
	04	236 27.5	08.3	164 19.8	15.9	10 44.1	11.9	54.2
	05	251 27.5	07.5	178 54.7	15.8	10 56.0	11.8	54.2
	06	266 27.5	S15 06.7	193 29.5	15.9	N11 07.8	11.8	54.2
	07	281 27.5	05.9	208 04.4	15.8	11 19.6	11.8	54.2
	08	296 27.4	05.1	222 39.2	15.7	11 31.4	11.7	54.2
	09	311 27.4	. . 04.3	237 13.9	15.7	11 43.1	11.6	54.2
	10	326 27.4	03.6	251 48.6	15.7	11 54.7	11.7	54.2
	11	341 27.3	02.8	266 23.3	15.6	12 06.4	11.5	54.2
	12	356 27.3	S15 02.0	280 57.9	15.6	N12 17.9	11.6	54.2
	13	11 27.3	01.2	295 32.5	15.5	12 29.5	11.5	54.2
	14	26 27.3	15 00.4	310 07.0	15.5	12 41.0	11.4	54.2
	15	41 27.2	14 59.6	324 41.5	15.5	12 52.4	11.4	54.2
	16	56 27.2	58.8	339 16.0	15.4	13 03.8	11.4	54.3
	17	71 27.2	58.0	353 50.4	15.3	13 15.2	11.3	54.3
	18	86 27.2	S14 57.2	8 24.7	15.3	N13 26.5	11.2	54.3
	19	101 27.1	56.4	22 59.0	15.3	13 37.7	11.2	54.3
	20	116 27.1	55.6	37 33.3	15.1	13 48.9	11.1	54.3
	21	131 27.1	. . 54.8	52 07.4	15.2	14 00.0	11.1	54.3
	22	146 27.1	54.0	66 41.6	15.1	14 11.1	11.1	54.3
	23	161 27.0	53.2	81 15.7	15.0	N14 22.2	11.0	54.3
		SD 16.2	d 0.8	SD 14.8		14.8		14.8

Lat.	Twilight Naut.	Twilight Civil	Sunrise	Moonrise 6	Moonrise 7	Moonrise 8	Moonrise 9
°	h m	h m	h m	h m	h m	h m	h m
N 72	06 49	08 13	09 46	09 31	09 06	08 35	07 47
N 70	06 43	07 58	09 16	09 34	09 15	08 54	08 24
68	06 39	07 47	08 54	09 36	09 23	09 09	08 51
66	06 35	07 37	08 36	09 37	09 29	09 21	09 11
64	06 31	07 28	08 22	09 39	09 35	09 31	09 28
62	06 28	07 21	08 10	09 40	09 40	09 40	09 41
60	06 25	07 15	08 00	09 41	09 44	09 48	09 53
N 58	06 22	07 09	07 52	09 42	09 48	09 55	10 03
56	06 19	07 04	07 44	09 43	09 51	10 01	10 12
54	06 17	06 59	07 37	09 44	09 54	10 06	10 20
52	06 15	06 55	07 31	09 44	09 57	10 11	10 27
50	06 13	06 51	07 25	09 45	10 00	10 15	10 34
45	06 07	06 42	07 13	09 46	10 05	10 25	10 47
N 40	06 03	06 35	07 03	09 48	10 10	10 33	10 59
35	05 58	06 28	06 54	09 49	10 14	10 40	11 09
30	05 54	06 22	06 47	09 50	10 17	10 46	11 17
20	05 44	06 11	06 33	09 51	10 23	10 57	11 32
N 10	05 35	06 00	06 22	09 53	10 29	11 06	11 46
0	05 24	05 49	06 11	09 54	10 34	11 15	11 58
S 10	05 12	05 38	05 59	09 55	10 39	11 24	12 11
20	04 57	05 24	05 47	09 57	10 45	11 34	12 24
30	04 38	05 08	05 33	09 59	10 52	11 45	12 40
35	04 25	04 58	05 25	10 00	10 55	11 51	12 49
40	04 11	04 46	05 16	10 01	11 00	11 59	12 59
45	03 52	04 32	05 05	10 02	11 05	12 07	13 11
S 50	03 27	04 15	04 52	10 04	11 11	12 18	13 26
52	03 15	04 06	04 45	10 05	11 13	12 23	13 33
54	03 01	03 57	04 38	10 05	11 16	12 28	13 41
56	02 43	03 46	04 31	10 06	11 20	12 34	13 50
58	02 21	03 33	04 22	10 07	11 24	12 41	13 59
S 60	01 51	03 18	04 12	10 09	11 28	12 48	14 11

Lat.	Sunset	Twilight Civil	Twilight Naut.	Moonset 6	Moonset 7	Moonset 8	Moonset 9
°	h m	h m	h m	h m	h m	h m	h m
N 72	14 44	16 17	17 41	23 14	25 11	01 11	03 27
N 70	15 14	16 31	17 47	23 07	24 54	00 54	02 51
68	15 36	16 43	17 51	23 02	24 40	00 40	02 26
66	15 53	16 53	17 55	22 57	24 30	00 30	02 06
64	16 07	17 01	17 59	22 53	24 21	00 21	01 51
62	16 19	17 08	18 02	22 50	24 13	00 13	01 38
60	16 29	17 15	18 05	22 47	24 06	00 06	01 27
N 58	16 38	17 20	18 07	22 44	24 00	00 00	01 18
56	16 45	17 25	18 10	22 42	23 55	25 10	01 10
54	16 52	17 30	18 12	22 40	23 51	25 03	01 03
52	16 58	17 34	18 14	22 38	23 47	24 56	00 56
50	17 04	17 38	18 16	22 36	23 43	24 50	00 50
45	17 16	17 47	18 21	22 32	23 35	24 38	00 38
N 40	17 26	17 54	18 26	22 29	23 28	24 28	00 28
35	17 34	18 01	18 31	22 27	23 22	24 19	00 19
30	17 42	18 07	18 35	22 24	23 17	24 11	00 11
20	17 55	18 18	18 44	22 20	23 09	23 58	24 49
N 10	18 07	18 28	18 53	22 17	23 01	23 46	24 34
0	18 18	18 39	19 04	22 13	22 54	23 36	24 20
S 10	18 29	18 51	19 16	22 10	22 47	23 25	24 06
20	18 41	19 04	19 31	22 06	22 39	23 13	23 51
30	18 54	19 20	19 50	22 02	22 31	23 01	23 33
35	19 02	19 30	20 02	22 00	22 26	22 53	23 23
40	19 12	19 41	20 17	21 58	22 20	22 45	23 12
45	19 23	19 55	20 35	21 55	22 14	22 35	22 59
S 50	19 36	20 12	20 59	21 51	22 06	22 23	22 43
52	19 42	20 21	21 11	21 49	22 02	22 17	22 35
54	19 49	20 30	21 25	21 48	21 59	22 11	22 27
56	19 56	20 41	21 42	21 46	21 54	22 05	22 17
58	20 05	20 53	22 04	21 43	21 50	21 57	22 07
S 60	20 15	21 08	22 32	21 41	21 44	21 48	21 55

Day	SUN Eqn. of Time 00^h	SUN Eqn. of Time 12^h	SUN Mer. Pass.	MOON Mer. Pass. Upper	MOON Mer. Pass. Lower	MOON Age	MOON Phase
d	m s	m s	h m	h m	h m	d %	
6	14 02	14 04	12 14	16 04	03 44	05 24	
7	14 06	14 08	12 14	16 44	04 24	06 32	
8	14 09	14 11	12 14	17 25	05 04	07 41	

UT	ARIES	VENUS −4.2		MARS +1.2		JUPITER −2.6		SATURN −0.1	
	GHA	GHA	Dec	GHA	Dec	GHA	Dec	GHA	Dec
d h	° ′	° ′	° ′	° ′	° ′	° ′	° ′	° ′	° ′
9 00	138 40.4	223 17.2	S21 03.6	245 14.8	S22 17.0	3 43.8	N17 58.2	56 59.0	N22 03.1
01	153 42.9	238 16.6	03.7	260 15.5	17.2	18 46.6	58.3	72 01.5	03.1
02	168 45.4	253 16.0	03.7	275 16.2	17.4	33 49.4	58.4	87 04.0	03.1
03	183 47.8	268 15.4	. . 03.8	290 17.0	. . 17.6	48 52.2	. . 58.5	102 06.6	. . 03.1
04	198 50.3	283 14.8	03.8	305 17.7	17.8	63 55.0	58.6	117 09.1	03.1
05	213 52.8	298 14.3	03.8	320 18.5	18.0	78 57.8	58.7	132 11.6	03.1
06	228 55.2	313 13.7	S21 03.9	335 19.2	S22 18.2	94 00.6	N17 58.8	147 14.2	N22 03.1
07	243 57.7	328 13.1	03.9	350 20.0	18.5	109 03.4	58.8	162 16.7	03.1
S 08	259 00.1	343 12.5	04.0	5 20.7	18.7	124 06.2	58.9	177 19.2	03.1
U 09	274 02.6	358 11.9	. . 04.0	20 21.4	. . 18.9	139 09.0	. . 59.0	192 21.8	. . 03.2
N 10	289 05.1	13 11.3	04.0	35 22.2	19.1	154 11.8	59.1	207 24.3	03.2
D 11	304 07.5	28 10.7	04.1	50 22.9	19.3	169 14.6	59.2	222 26.8	03.2
A 12	319 10.0	43 10.1	S21 04.1	65 23.7	S22 19.5	184 17.4	N17 59.3	237 29.3	N22 03.2
Y 13	334 12.5	58 09.5	04.1	80 24.4	19.7	199 20.2	59.4	252 31.9	03.2
14	349 14.9	73 08.9	04.2	95 25.2	19.9	214 22.9	59.5	267 34.4	03.2
15	4 17.4	88 08.4	. . 04.2	110 25.9	. . 20.2	229 25.7	. . 59.6	282 36.9	. . 03.2
16	19 19.9	103 07.8	04.2	125 26.6	20.4	244 28.5	59.7	297 39.5	03.2
17	34 22.3	118 07.2	04.3	140 27.4	20.6	259 31.3	59.8	312 42.0	03.2
18	49 24.8	133 06.6	S21 04.3	155 28.1	S22 20.8	274 34.1	N17 59.9	327 44.5	N22 03.2
19	64 27.3	148 06.0	04.3	170 28.9	21.0	289 36.9	18 00.0	342 47.0	03.2
20	79 29.7	163 05.4	04.4	185 29.6	21.2	304 39.7	00.1	357 49.6	03.2
21	94 32.2	178 04.8	. . 04.4	200 30.3	. . 21.4	319 42.5	. . 00.2	12 52.1	. . 03.2
22	109 34.6	193 04.2	04.4	215 31.1	21.6	334 45.3	00.3	27 54.6	03.2
23	124 37.1	208 03.6	04.4	230 31.8	21.8	349 48.1	00.4	42 57.2	03.2
10 00	139 39.6	223 03.0	S21 04.5	245 32.6	S22 22.0	4 50.9	N18 00.5	57 59.7	N22 03.2
01	154 42.0	238 02.4	04.5	260 33.3	22.2	19 53.7	00.6	73 02.2	03.2
02	169 44.5	253 01.8	04.5	275 34.1	22.4	34 56.5	00.7	88 04.7	03.2
03	184 47.0	268 01.2	. . 04.5	290 34.8	. . 22.7	49 59.2	. . 00.7	103 07.3	. . 03.3
04	199 49.4	283 00.6	04.6	305 35.5	22.9	65 02.0	00.8	118 09.8	03.3
05	214 51.9	298 00.0	04.6	320 36.3	23.1	80 04.8	00.9	133 12.3	03.3
06	229 54.4	312 59.4	S21 04.6	335 37.0	S22 23.3	95 07.6	N18 01.0	148 14.8	N22 03.3
07	244 56.8	327 58.8	04.6	350 37.8	23.5	110 10.4	01.1	163 17.4	03.3
08	259 59.3	342 58.2	04.6	5 38.5	23.7	125 13.2	01.2	178 19.9	03.3
M 09	275 01.8	357 57.7	. . 04.6	20 39.2	. . 23.9	140 16.0	. . 01.3	193 22.4	. . 03.3
O 10	290 04.2	12 57.1	04.7	35 40.0	24.1	155 18.8	01.4	208 24.9	03.3
N 11	305 06.7	27 56.5	04.7	50 40.7	24.3	170 21.6	01.5	223 27.5	03.3
D 12	320 09.1	42 55.9	S21 04.7	65 41.5	S22 24.5	185 24.4	N18 01.6	238 30.0	N22 03.3
A 13	335 11.6	57 55.3	04.7	80 42.2	24.7	200 27.2	01.7	253 32.5	03.3
Y 14	350 14.1	72 54.7	04.7	95 42.9	24.9	215 29.9	01.8	268 35.0	03.3
15	5 16.5	87 54.1	. . 04.7	110 43.7	. . 25.1	230 32.7	. . 01.9	283 37.6	. . 03.3
16	20 19.0	102 53.5	04.7	125 44.4	25.3	245 35.5	02.0	298 40.1	03.3
17	35 21.5	117 52.9	04.7	140 45.2	25.5	260 38.3	02.1	313 42.6	03.3
18	50 23.9	132 52.3	S21 04.7	155 45.9	S22 25.7	275 41.1	N18 02.2	328 45.1	N22 03.3
19	65 26.4	147 51.7	04.8	170 46.6	25.9	290 43.9	02.3	343 47.6	03.3
20	80 28.9	162 51.1	04.8	185 47.4	26.1	305 46.7	02.4	358 50.2	03.3
21	95 31.3	177 50.5	. . 04.8	200 48.1	. . 26.3	320 49.5	. . 02.4	13 52.7	. . 03.4
22	110 33.8	192 49.9	04.8	215 48.9	26.5	335 52.3	02.5	28 55.2	03.4
23	125 36.2	207 49.3	04.8	230 49.6	26.7	350 55.1	02.6	43 57.7	03.4
11 00	140 38.7	222 48.7	S21 04.8	245 50.3	S22 26.9	5 57.9	N18 02.7	59 00.3	N22 03.4
01	155 41.2	237 48.1	04.8	260 51.1	27.1	21 00.6	02.8	74 02.8	03.4
02	170 43.6	252 47.5	04.8	275 51.8	27.3	36 03.4	02.9	89 05.3	03.4
03	185 46.1	267 46.9	. . 04.8	290 52.6	. . 27.5	51 06.2	. . 03.0	104 07.8	. . 03.4
04	200 48.6	282 46.3	04.8	305 53.3	27.7	66 09.0	03.1	119 10.3	03.4
05	215 51.0	297 45.7	04.8	320 54.0	27.9	81 11.8	03.2	134 12.9	03.4
06	230 53.5	312 45.1	S21 04.8	335 54.8	S22 28.1	96 14.6	N18 03.3	149 15.4	N22 03.4
07	245 56.0	327 44.5	04.8	350 55.5	28.3	111 17.4	03.4	164 17.9	03.4
T 08	260 58.4	342 43.8	04.8	5 56.3	28.5	126 20.2	03.5	179 20.4	03.4
U 09	276 00.9	357 43.2	. . 04.7	20 57.0	. . 28.7	141 23.0	. . 03.6	194 22.9	. . 03.4
E 10	291 03.4	12 42.6	04.7	35 57.7	28.9	156 25.8	03.7	209 25.5	03.4
S 11	306 05.8	27 42.0	04.7	50 58.5	29.1	171 28.5	03.8	224 28.0	03.4
D 12	321 08.3	42 41.4	S21 04.7	65 59.2	S22 29.3	186 31.3	N18 03.8	239 30.5	N22 03.4
A 13	336 10.7	57 40.8	04.7	80 59.9	29.5	201 34.1	03.9	254 33.0	03.4
Y 14	351 13.2	72 40.2	04.7	96 00.7	29.7	216 36.9	04.0	269 35.5	03.4
15	6 15.7	87 39.6	. . 04.7	111 01.4	. . 29.9	231 39.7	. . 04.1	284 38.1	. . 03.5
16	21 18.1	102 39.0	04.7	126 02.2	30.1	246 42.5	04.2	299 40.6	03.5
17	36 20.6	117 38.4	04.7	141 02.9	30.3	261 45.3	04.3	314 43.1	03.5
18	51 23.1	132 37.8	S21 04.6	156 03.6	S22 30.5	276 48.1	N18 04.4	329 45.6	N22 03.5
19	66 25.5	147 37.2	04.6	171 04.4	30.7	291 50.9	04.5	344 48.1	03.5
20	81 28.0	162 36.6	04.6	186 05.1	30.9	306 53.6	04.6	359 50.6	03.5
21	96 30.5	177 36.0	. . 04.6	201 05.9	. . 31.1	321 56.4	. . 04.7	14 53.2	. . 03.5
22	111 32.9	192 35.4	04.6	216 06.6	31.3	336 59.2	04.8	29 55.7	03.5
23	126 35.4	207 34.8	04.6	231 07.3	31.5	352 02.0	04.9	44 58.2	03.5
Mer. Pass.	h m 14 39.0	v −0.6	d 0.0	v 0.7	d 0.2	v 2.8	d 0.1	v 2.5	d 0.0

STARS

Name	SHA	Dec
	° ′	° ′
Acamar	315 24.5	S40 17.9
Achernar	335 32.9	S57 13.6
Acrux	173 18.2	S63 06.7
Adhara	255 18.6	S28 58.7
Aldebaran	290 58.6	N16 30.9
Alioth	166 27.3	N55 56.4
Alkaid	153 05.0	N49 17.6
Al Na'ir	27 54.2	S46 57.0
Alnilam	275 54.5	S 1 12.1
Alphard	218 03.8	S 8 40.3
Alphecca	126 17.9	N26 42.0
Alpheratz	357 52.3	N29 06.4
Altair	62 16.5	N 8 52.4
Ankaa	353 23.9	S42 17.7
Antares	112 36.4	S26 26.3
Arcturus	146 03.0	N19 09.8
Atria	107 45.9	S69 01.8
Avior	234 20.9	S59 31.2
Bellatrix	278 40.6	N 6 21.1
Betelgeuse	271 09.9	N 7 24.4
Canopus	263 59.4	S52 42.0
Capella	280 46.3	N46 00.2
Deneb	49 37.6	N45 17.3
Denebola	182 41.7	N14 33.2
Diphda	349 04.2	S17 58.4
Dubhe	194 00.8	N61 44.0
Elnath	278 22.7	N28 36.7
Eltanin	90 50.3	N51 29.0
Enif	33 55.5	N 9 53.1
Fomalhaut	15 33.2	S29 36.6
Gacrux	172 09.9	S57 07.6
Gienah	176 00.5	S17 33.5
Hadar	148 59.5	S60 23.0
Hamal	328 10.1	N23 28.6
Kaus Aust.	83 54.9	S34 23.0
Kochab	137 19.2	N74 08.2
Markab	13 46.8	N15 13.2
Menkar	314 23.6	N 4 06.0
Menkent	148 17.2	S36 23.0
Miaplacidus	221 40.7	S69 43.7
Mirfak	308 52.1	N49 52.5
Nunki	76 08.7	S26 17.7
Peacock	53 32.5	S56 43.6
Pollux	243 37.3	N28 01.2
Procyon	245 08.0	N 5 13.0
Rasalhague	96 14.2	N12 33.3
Regulus	207 51.8	N11 57.1
Rigel	281 19.7	S 8 12.0
Rigil Kent.	140 03.1	S60 50.6
Sabik	102 22.1	S15 43.8
Schedar	349 50.4	N56 33.3
Shaula	96 33.2	S37 06.3
Sirius	258 40.7	S16 43.3
Spica	158 39.7	S11 10.6
Suhail	222 58.1	S43 26.7
Vega	80 44.8	N38 46.9
Zuben'ubi	137 14.5	S16 03.3

	SHA	Mer. Pass.
	° ′	h m
Venus	83 23.4	9 08
Mars	105 53.0	7 37
Jupiter	225 11.3	23 36
Saturn	278 20.1	20 05

UT	SUN GHA	SUN Dec	MOON GHA	v	MOON Dec	d	HP
d h	° ′	° ′	° ′	′	° ′	′	′
9 00	176 27.0	S14 52.5	95 49.7	15.0	N14 33.2	10.9	54.3
01	191 27.0	51.7	110 23.7	14.9	14 44.1	10.9	54.3
02	206 27.0	50.9	124 57.6	14.9	14 55.0	10.8	54.3
03	221 27.0	. . 50.1	139 31.5	14.8	15 05.8	10.8	54.3
04	236 26.9	49.3	154 05.3	14.7	15 16.6	10.7	54.3
05	251 26.9	48.5	168 39.0	14.7	15 27.3	10.6	54.3
06	266 26.9	S14 47.7	183 12.7	14.7	N15 37.9	10.6	54.4
07	281 26.9	46.9	197 46.4	14.5	15 48.5	10.6	54.4
08	296 26.9	46.1	212 19.9	14.6	15 59.1	10.4	54.4
S 09	311 26.8	. . 45.3	226 53.5	14.4	16 09.5	10.4	54.4
U 10	326 26.8	44.5	241 26.9	14.4	16 19.9	10.4	54.4
N 11	341 26.8	43.7	256 00.3	14.3	16 30.3	10.3	54.4
D 12	356 26.8	S14 42.9	270 33.6	14.3	N16 40.6	10.2	54.4
A 13	11 26.8	42.1	285 06.9	14.2	16 50.8	10.2	54.4
Y 14	26 26.7	41.3	299 40.1	14.1	17 01.0	10.0	54.4
15	41 26.7	. . 40.5	314 13.2	14.1	17 11.0	10.1	54.5
16	56 26.7	39.7	328 46.3	13.9	17 21.1	9.9	54.5
17	71 26.7	38.9	343 19.2	14.0	17 31.0	9.9	54.5
18	86 26.7	S14 38.1	357 52.2	13.8	N17 40.9	9.8	54.5
19	101 26.7	37.2	12 25.0	13.8	17 50.7	9.8	54.5
20	116 26.6	36.4	26 57.8	13.7	18 00.5	9.7	54.5
21	131 26.6	. . 35.6	41 30.5	13.7	18 10.2	9.6	54.5
22	146 26.6	34.8	56 03.2	13.5	18 19.8	9.5	54.6
23	161 26.6	34.0	70 35.7	13.5	18 29.3	9.5	54.6
10 00	176 26.6	S14 33.2	85 08.2	13.5	N18 38.8	9.4	54.6
01	191 26.6	32.4	99 40.7	13.3	18 48.2	9.3	54.6
02	206 26.6	31.6	114 13.0	13.3	18 57.5	9.2	54.6
03	221 26.5	. . 30.8	128 45.3	13.2	19 06.7	9.2	54.6
04	236 26.5	30.0	143 17.5	13.1	19 15.9	9.1	54.6
05	251 26.5	29.2	157 49.6	13.1	19 25.0	9.0	54.7
06	266 26.5	S14 28.4	172 21.7	13.0	N19 34.0	8.9	54.7
07	281 26.5	27.6	186 53.7	12.9	19 42.9	8.9	54.7
08	296 26.5	26.8	201 25.6	12.8	19 51.8	8.8	54.7
M 09	311 26.5	. . 25.9	215 57.4	12.8	20 00.6	8.6	54.7
O 10	326 26.5	25.1	230 29.2	12.6	20 09.2	8.6	54.7
N 11	341 26.5	24.3	245 00.8	12.6	20 17.8	8.6	54.8
D 12	356 26.4	S14 23.5	259 32.4	12.5	N20 26.4	8.4	54.8
A 13	11 26.4	22.7	274 03.9	12.5	20 34.8	8.3	54.8
Y 14	26 26.4	21.9	288 35.4	12.3	20 43.1	8.3	54.8
15	41 26.4	. . 21.1	303 06.7	12.3	20 51.4	8.2	54.8
16	56 26.4	20.3	317 38.0	12.2	20 59.6	8.1	54.9
17	71 26.4	19.5	332 09.2	12.1	21 07.7	8.0	54.9
18	86 26.4	S14 18.6	346 40.3	12.0	N21 15.7	7.9	54.9
19	101 26.4	17.8	1 11.3	12.0	21 23.6	7.8	54.9
20	116 26.4	17.0	15 42.3	11.8	21 31.4	7.7	54.9
21	131 26.4	. . 16.2	30 13.1	11.8	21 39.1	7.6	55.0
22	146 26.4	15.4	44 43.9	11.7	21 46.7	7.5	55.0
23	161 26.4	14.6	59 14.6	11.6	21 54.2	7.5	55.0
11 00	176 26.4	S14 13.8	73 45.2	11.6	N22 01.7	7.3	55.0
01	191 26.3	12.9	88 15.8	11.4	22 09.0	7.3	55.0
02	206 26.3	12.1	102 46.2	11.4	22 16.3	7.1	55.1
03	221 26.3	. . 11.3	117 16.6	11.3	22 23.4	7.0	55.1
04	236 26.3	10.5	131 46.9	11.2	22 30.4	7.0	55.1
05	251 26.3	09.7	146 17.1	11.1	22 37.4	6.8	55.1
06	266 26.3	S14 08.8	160 47.2	11.1	N22 44.2	6.8	55.2
07	281 26.3	08.0	175 17.3	10.9	22 51.0	6.6	55.2
T 08	296 26.3	07.2	189 47.2	10.9	22 57.6	6.5	55.2
U 09	311 26.3	. . 06.4	204 17.1	10.8	23 04.1	6.4	55.2
E 10	326 26.3	05.6	218 46.9	10.7	23 10.5	6.4	55.3
S 11	341 26.3	04.7	233 16.6	10.6	23 16.9	6.2	55.3
D 12	356 26.3	S14 03.9	247 46.2	10.6	N23 23.1	6.1	55.3
A 13	11 26.3	03.1	262 15.8	10.4	23 29.2	6.0	55.3
Y 14	26 26.3	02.3	276 45.2	10.4	23 35.2	5.8	55.4
15	41 26.3	. . 01.5	291 14.6	10.3	23 41.0	5.8	55.4
16	56 26.3	14 00.6	305 43.9	10.2	23 46.8	5.7	55.4
17	71 26.3	13 59.8	320 13.1	10.2	23 52.5	5.5	55.4
18	86 26.3	S13 59.0	334 42.3	10.0	N23 58.0	5.4	55.5
19	101 26.3	58.2	349 11.3	10.0	24 03.4	5.3	55.5
20	116 26.3	57.3	3 40.3	9.9	24 08.7	5.2	55.5
21	131 26.3	. . 56.5	18 09.2	9.8	24 13.9	5.1	55.5
22	146 26.3	55.7	32 38.0	9.7	24 19.0	5.0	55.6
23	161 26.3	54.9	47 06.7	9.7	N24 24.0	4.8	55.6
	SD 16.2	*d* 0.8	SD 14.8		14.9		15.1

Lat.	Twilight Naut.	Twilight Civil	Sunrise	Moonrise 9	Moonrise 10	Moonrise 11	Moonrise 12
°	h m	h m	h m	h m	h m	h m	h m
N 72	06 37	08 00	09 26	07 47	▭	▭	▭
N 70	06 33	07 47	09 00	08 24	07 17	▭	▭
68	06 29	07 36	08 41	08 51	08 22	▭	▭
66	06 26	07 27	08 25	09 11	08 58	08 35	▭
64	06 23	07 20	08 13	09 28	09 24	09 22	09 21
62	06 21	07 13	08 02	09 41	09 45	09 52	10 09
60	06 18	07 08	07 53	09 53	10 01	10 15	10 39
N 58	06 16	07 02	07 45	10 03	10 15	10 34	11 03
56	06 14	06 58	07 37	10 12	10 27	10 49	11 22
54	06 12	06 54	07 31	10 20	10 38	11 03	11 37
52	06 10	06 50	07 25	10 27	10 47	11 14	11 51
50	06 08	06 46	07 20	10 34	10 56	11 25	12 03
45	06 04	06 38	07 09	10 47	11 14	11 47	12 28
N 40	06 00	06 32	07 00	10 59	11 29	12 05	12 48
35	05 56	06 25	06 52	11 09	11 41	12 20	13 05
30	05 52	06 20	06 44	11 17	11 52	12 33	13 19
20	05 43	06 09	06 32	11 32	12 11	12 55	13 44
N 10	05 34	06 00	06 21	11 46	12 28	13 14	14 05
0	05 25	05 49	06 11	11 58	12 44	13 33	14 25
S 10	05 13	05 39	06 00	12 11	12 59	13 51	14 45
20	04 59	05 26	05 49	12 24	13 16	14 10	15 06
30	04 41	05 11	05 36	12 40	13 36	14 33	15 31
35	04 29	05 01	05 28	12 49	13 47	14 47	15 46
40	04 15	04 50	05 20	12 59	14 00	15 02	16 03
45	03 57	04 37	05 09	13 11	14 16	15 21	16 23
S 50	03 34	04 21	04 57	13 26	14 35	15 44	16 49
52	03 23	04 13	04 51	13 33	14 44	15 55	17 02
54	03 09	04 04	04 45	13 41	14 55	16 08	17 16
56	02 54	03 53	04 37	13 50	15 06	16 22	17 33
58	02 34	03 42	04 29	13 59	15 20	16 40	17 53
S 60	02 08	03 28	04 20	14 11	15 36	17 01	18 19

Lat.	Sunset	Twilight Civil	Twilight Naut.	Moonset 9	Moonset 10	Moonset 11	Moonset 12
°	h m	h m	h m	h m	h m	h m	h m
N 72	15 04	16 31	17 53	03 27	▭	▭	▭
N 70	15 29	16 43	17 57	02 51	05 31	▭	▭
68	15 49	16 54	18 01	02 26	04 27	▭	▭
66	16 04	17 02	18 04	02 06	03 52	05 54	▭
64	16 17	17 10	18 07	01 51	03 26	05 08	06 56
62	16 28	17 16	18 09	01 38	03 07	04 39	06 08
60	16 37	17 22	18 11	01 27	02 51	04 16	05 38
N 58	16 45	17 27	18 13	01 18	02 38	03 58	05 15
56	16 52	17 31	18 16	01 10	02 26	03 43	04 56
54	16 58	17 36	18 17	01 03	02 16	03 30	04 41
52	17 04	17 39	18 19	00 56	02 07	03 18	04 27
50	17 09	17 43	18 21	00 50	01 59	03 08	04 15
45	17 20	17 51	18 25	00 38	01 42	02 47	03 51
N 40	17 29	17 57	18 29	00 28	01 28	02 30	03 31
35	17 37	18 03	18 33	00 19	01 17	02 16	03 15
30	17 44	18 09	18 37	00 11	01 06	02 03	03 01
20	17 57	18 19	18 45	24 49	00 49	01 42	02 37
N 10	18 07	18 29	18 54	24 34	00 34	01 24	02 17
0	18 18	18 39	19 04	24 20	00 20	01 07	01 57
S 10	18 28	18 50	19 15	24 06	00 06	00 50	01 38
20	18 39	19 02	19 29	23 51	24 32	00 32	01 17
30	18 52	19 17	19 47	23 33	24 11	00 11	00 54
35	19 00	19 26	19 59	23 23	23 59	24 40	00 40
40	19 08	19 37	20 13	23 12	23 45	24 24	00 24
45	19 18	19 50	20 30	22 59	23 28	24 05	00 05
S 50	19 31	20 07	20 52	22 43	23 08	23 41	24 25
52	19 36	20 15	21 04	22 35	22 58	23 29	24 12
54	19 43	20 23	21 17	22 27	22 48	23 16	23 57
56	19 50	20 33	21 32	22 17	22 35	23 01	23 40
58	19 58	20 45	21 51	22 07	22 21	22 44	23 20
S 60	20 07	20 58	22 16	21 55	22 05	22 22	22 54

Day	SUN Eqn. of Time 00^h	SUN Eqn. of Time 12^h	SUN Mer. Pass.	MOON Mer. Pass. Upper	MOON Mer. Pass. Lower	Age	Phase
d	m s	m s	h m	h m	h m	d %	
9	14 12	14 13	12 14	18 09	05 47	08 50	
10	14 14	14 14	12 14	18 55	06 32	09 60	
11	14 15	14 15	12 14	19 45	07 19	10 69	

UT d	h	ARIES GHA	VENUS −4.2 GHA	VENUS Dec	MARS +1.1 GHA	MARS Dec	JUPITER −2.6 GHA	JUPITER Dec	SATURN −0.1 GHA	SATURN Dec
		° ′	° ′	° ′	° ′	° ′	° ′	° ′	° ′	° ′
12	00	141 37.9	222 34.2	S21 04.5	246 08.1	S22 31.7	7 04.8	N18 05.0	60 00.7	N22 03.5
	01	156 40.3	237 33.6	04.5	261 08.8	31.9	22 07.6	05.1	75 03.2	03.5
	02	171 42.8	252 33.0	04.5	276 09.5	32.0	37 10.4	05.2	90 05.7	03.5
	03	186 45.2	267 32.3	04.5	291 10.3	32.2	52 13.2	05.2	105 08.3	03.5
	04	201 47.7	282 31.7	04.4	306 11.0	32.4	67 15.9	05.3	120 10.8	03.5
	05	216 50.2	297 31.1	04.4	321 11.8	32.6	82 18.7	05.4	135 13.3	03.5
WEDNESDAY	06	231 52.6	312 30.5	S21 04.4	336 12.5	S22 32.8	97 21.5	N18 05.5	150 15.8	N22 03.5
	07	246 55.1	327 29.9	04.4	351 13.2	33.0	112 24.3	05.6	165 18.3	03.6
	08	261 57.6	342 29.3	04.3	6 14.0	33.2	127 27.1	05.7	180 20.8	03.6
	09	277 00.0	357 28.7	04.3	21 14.7	33.4	142 29.9	05.8	195 23.3	03.6
	10	292 02.5	12 28.1	04.3	36 15.4	33.6	157 32.7	05.9	210 25.9	03.6
	11	307 05.0	27 27.5	04.2	51 16.2	33.8	172 35.5	06.0	225 28.4	03.6
	12	322 07.4	42 26.9	S21 04.2	66 16.9	S22 34.0	187 38.2	N18 06.1	240 30.9	N22 03.6
	13	337 09.9	57 26.3	04.2	81 17.7	34.1	202 41.0	06.2	255 33.4	03.6
	14	352 12.4	72 25.6	04.1	96 18.4	34.3	217 43.8	06.3	270 35.9	03.6
	15	7 14.8	87 25.0	04.1	111 19.1	34.5	232 46.6	06.4	285 38.4	03.6
	16	22 17.3	102 24.4	04.1	126 19.9	34.7	247 49.4	06.4	300 40.9	03.6
	17	37 19.7	117 23.8	04.0	141 20.6	34.9	262 52.2	06.5	315 43.5	03.6
	18	52 22.2	132 23.2	S21 04.0	156 21.3	S22 35.1	277 55.0	N18 06.6	330 46.0	N22 03.6
	19	67 24.7	147 22.6	04.0	171 22.1	35.3	292 57.8	06.7	345 48.5	03.6
	20	82 27.1	162 22.0	03.9	186 22.8	35.5	308 00.5	06.8	0 51.0	03.6
	21	97 29.6	177 21.4	03.9	201 23.6	35.7	323 03.3	06.9	15 53.5	03.6
	22	112 32.1	192 20.8	03.8	216 24.3	35.8	338 06.1	07.0	30 56.0	03.6
	23	127 34.5	207 20.1	03.8	231 25.0	36.0	353 08.9	07.1	45 58.5	03.7
13	00	142 37.0	222 19.5	S21 03.7	246 25.8	S22 36.2	8 11.7	N18 07.2	61 01.0	N22 03.7
	01	157 39.5	237 18.9	03.7	261 26.5	36.4	23 14.5	07.3	76 03.6	03.7
	02	172 41.9	252 18.3	03.7	276 27.2	36.6	38 17.3	07.4	91 06.1	03.7
	03	187 44.4	267 17.7	03.6	291 28.0	36.8	53 20.0	07.5	106 08.6	03.7
	04	202 46.8	282 17.1	03.6	306 28.7	37.0	68 22.8	07.5	121 11.1	03.7
	05	217 49.3	297 16.5	03.5	321 29.4	37.1	83 25.6	07.6	136 13.6	03.7
THURSDAY	06	232 51.8	312 15.9	S21 03.5	336 30.2	S22 37.3	98 28.4	N18 07.7	151 16.1	N22 03.7
	07	247 54.2	327 15.2	03.4	351 30.9	37.5	113 31.2	07.8	166 18.6	03.7
	08	262 56.7	342 14.6	03.4	6 31.7	37.7	128 34.0	07.9	181 21.1	03.7
	09	277 59.2	357 14.0	03.3	21 32.4	37.9	143 36.7	08.0	196 23.6	03.7
	10	293 01.6	12 13.4	03.2	36 33.1	38.1	158 39.5	08.1	211 26.2	03.7
	11	308 04.1	27 12.8	03.2	51 33.9	38.3	173 42.3	08.2	226 28.7	03.7
	12	323 06.6	42 12.2	S21 03.1	66 34.6	S22 38.4	188 45.1	N18 08.3	241 31.2	N22 03.7
	13	338 09.0	57 11.6	03.1	81 35.3	38.6	203 47.9	08.4	256 33.7	03.7
	14	353 11.5	72 10.9	03.0	96 36.1	38.8	218 50.7	08.5	271 36.2	03.7
	15	8 14.0	87 10.3	03.0	111 36.8	39.0	233 53.5	08.5	286 38.7	03.8
	16	23 16.4	102 09.7	02.9	126 37.5	39.2	248 56.2	08.6	301 41.2	03.8
	17	38 18.9	117 09.1	02.8	141 38.3	39.3	263 59.0	08.7	316 43.7	03.8
	18	53 21.3	132 08.5	S21 02.8	156 39.0	S22 39.5	279 01.8	N18 08.8	331 46.2	N22 03.8
	19	68 23.8	147 07.9	02.7	171 39.7	39.7	294 04.6	08.9	346 48.7	03.8
	20	83 26.3	162 07.2	02.7	186 40.5	39.9	309 07.4	09.0	1 51.2	03.8
	21	98 28.7	177 06.6	02.6	201 41.2	40.1	324 10.2	09.1	16 53.7	03.8
	22	113 31.2	192 06.0	02.5	216 42.0	40.2	339 12.9	09.2	31 56.2	03.8
	23	128 33.7	207 05.4	02.5	231 42.7	40.4	354 15.7	09.3	46 58.8	03.8
14	00	143 36.1	222 04.8	S21 02.4	246 43.4	S22 40.6	9 18.5	N18 09.4	62 01.3	N22 03.8
	01	158 38.6	237 04.2	02.3	261 44.2	40.8	24 21.3	09.5	77 03.8	03.8
	02	173 41.1	252 03.5	02.3	276 44.9	41.0	39 24.1	09.5	92 06.3	03.8
	03	188 43.5	267 02.9	02.2	291 45.6	41.1	54 26.9	09.6	107 08.8	03.8
	04	203 46.0	282 02.3	02.1	306 46.4	41.3	69 29.6	09.7	122 11.3	03.8
	05	218 48.5	297 01.7	02.0	321 47.1	41.5	84 32.4	09.8	137 13.8	03.8
FRIDAY	06	233 50.9	312 01.1	S21 02.0	336 47.8	S22 41.7	99 35.2	N18 09.9	152 16.3	N22 03.9
	07	248 53.4	327 00.4	01.9	351 48.6	41.9	114 38.0	10.0	167 18.8	03.9
	08	263 55.8	341 59.8	01.8	6 49.3	42.0	129 40.8	10.1	182 21.3	03.9
	09	278 58.3	356 59.2	01.7	21 50.0	42.2	144 43.5	10.2	197 23.8	03.9
	10	294 00.8	11 58.6	01.7	36 50.8	42.4	159 46.3	10.3	212 26.3	03.9
	11	309 03.2	26 58.0	01.6	51 51.5	42.6	174 49.1	10.4	227 28.8	03.9
	12	324 05.7	41 57.3	S21 01.5	66 52.2	S22 42.7	189 51.9	N18 10.5	242 31.3	N22 03.9
	13	339 08.2	56 56.7	01.4	81 53.0	42.9	204 54.7	10.5	257 33.8	03.9
	14	354 10.6	71 56.1	01.3	96 53.7	43.1	219 57.5	10.6	272 36.3	03.9
	15	9 13.1	86 55.5	01.3	111 54.4	43.3	235 00.2	10.7	287 38.8	03.9
	16	24 15.6	101 54.9	01.2	126 55.2	43.4	250 03.0	10.8	302 41.3	03.9
	17	39 18.0	116 54.2	01.1	141 55.9	43.6	265 05.8	10.9	317 43.8	03.9
	18	54 20.5	131 53.6	S21 01.0	156 56.6	S22 43.8	280 08.6	N18 11.0	332 46.3	N22 03.9
	19	69 23.0	146 53.0	00.9	171 57.4	44.0	295 11.4	11.1	347 48.8	03.9
	20	84 25.4	161 52.4	00.8	186 58.1	44.1	310 14.1	11.2	2 51.4	03.9
	21	99 27.9	176 51.8	00.8	201 58.8	44.3	325 16.9	11.3	17 53.9	04.0
	22	114 30.3	191 51.1	00.7	216 59.6	44.5	340 19.7	11.3	32 56.4	04.0
	23	129 32.8	206 50.5	00.6	232 00.3	44.7	355 22.5	11.4	47 58.9	04.0
Mer. Pass.		h m 14 27.2	v −0.6	d 0.1	v 0.7	d 0.2	v 2.8	d 0.1	v 2.5	d 0.0

STARS

Name	SHA	Dec
	° ′	° ′
Acamar	315 24.5	S40 17.9
Achernar	335 32.9	S57 13.6
Acrux	173 18.2	S63 06.7
Adhara	255 18.6	S28 58.7
Aldebaran	290 58.6	N16 30.9
Alioth	166 27.2	N55 56.4
Alkaid	153 04.9	N49 17.6
Al Na'ir	27 54.2	S46 57.0
Alnilam	275 54.5	S 1 12.1
Alphard	218 03.8	S 8 40.3
Alphecca	126 17.9	N26 42.0
Alpheratz	357 52.3	N29 06.4
Altair	62 16.5	N 8 52.4
Ankaa	353 23.9	S42 17.7
Antares	112 36.4	S26 26.3
Arcturus	146 03.0	N19 09.8
Atria	107 45.9	S69 01.8
Avior	234 20.9	S59 31.2
Bellatrix	278 40.6	N 6 21.1
Betelgeuse	271 09.9	N 7 24.4
Canopus	263 59.4	S52 42.0
Capella	280 46.3	N46 00.2
Deneb	49 37.6	N45 17.3
Denebola	182 41.6	N14 33.2
Diphda	349 04.2	S17 58.4
Dubhe	194 00.7	N61 44.0
Elnath	278 22.7	N28 36.7
Eltanin	90 50.2	N51 29.0
Enif	33 55.5	N 9 53.1
Fomalhaut	15 33.2	S29 36.6
Gacrux	172 09.8	S57 07.6
Gienah	176 00.5	S17 33.5
Hadar	148 59.5	S60 23.0
Hamal	328 10.1	N23 28.6
Kaus Aust.	83 54.9	S34 23.0
Kochab	137 19.1	N74 08.2
Markab	13 46.8	N15 13.1
Menkar	314 23.6	N 4 06.0
Menkent	148 17.2	S36 23.0
Miaplacidus	221 40.8	S69 43.7
Mirfak	308 52.2	N49 52.5
Nunki	76 08.7	S26 17.7
Peacock	53 32.5	S56 43.6
Pollux	243 37.3	N28 01.2
Procyon	245 08.0	N 5 13.0
Rasalhague	96 14.2	N12 33.3
Regulus	207 51.8	N11 57.1
Rigel	281 19.7	S 8 12.0
Rigil Kent.	140 03.0	S60 50.6
Sabik	102 22.0	S15 43.8
Schedar	349 50.4	N56 33.3
Shaula	96 33.2	S37 06.3
Sirius	258 40.7	S16 43.3
Spica	158 39.7	S11 10.6
Suhail	222 58.1	S43 26.7
Vega	80 44.8	N38 46.9
Zuben'ubi	137 14.4	S16 03.3

	SHA	Mer. Pass.
	° ′	h m
Venus	79 42.5	9 11
Mars	103 48.8	7 34
Jupiter	225 34.7	23 23
Saturn	278 24.1	19 53

UT d	UT h	SUN GHA	SUN Dec	MOON GHA	v	MOON Dec	d	HP
		° ′	° ′	° ′	′	° ′	′	′
12 WEDNESDAY	00	176 26.3	S13 54.0	61 35.4	9.5	N24 28.8	4.7	55.6
	01	191 26.3	53.2	76 03.9	9.5	24 33.5	4.6	55.7
	02	206 26.3	52.4	90 32.4	9.5	24 38.1	4.5	55.7
	03	221 26.3	. . 51.6	105 00.9	9.3	24 42.6	4.3	55.7
	04	236 26.3	50.7	119 29.2	9.3	24 46.9	4.2	55.7
	05	251 26.3	49.9	133 57.5	9.1	24 51.1	4.1	55.8
	06	266 26.3	S13 49.1	148 25.6	9.1	N24 55.2	4.0	55.8
	07	281 26.3	48.3	162 53.7	9.1	24 59.2	3.9	55.8
	08	296 26.3	47.4	177 21.8	8.9	25 03.1	3.7	55.9
	09	311 26.4	. . 46.6	191 49.7	8.9	25 06.8	3.6	55.9
	10	326 26.4	45.8	206 17.6	8.8	25 10.4	3.4	55.9
	11	341 26.4	44.9	220 45.4	8.8	25 13.8	3.3	55.9
	12	356 26.4	S13 44.1	235 13.2	8.6	N25 17.1	3.2	56.0
	13	11 26.4	43.3	249 40.8	8.6	25 20.3	3.1	56.0
	14	26 26.4	42.4	264 08.4	8.6	25 23.4	2.9	56.0
	15	41 26.4	. . 41.6	278 36.0	8.4	25 26.3	2.8	56.1
	16	56 26.4	40.8	293 03.4	8.4	25 29.1	2.6	56.1
	17	71 26.4	39.9	307 30.8	8.4	25 31.7	2.6	56.1
	18	86 26.4	S13 39.1	321 58.2	8.2	N25 34.3	2.3	56.2
	19	101 26.4	38.3	336 25.4	8.2	25 36.6	2.3	56.2
	20	116 26.4	37.4	350 52.6	8.1	25 38.9	2.1	56.2
	21	131 26.4	. . 36.6	5 19.7	8.1	25 41.0	1.9	56.2
	22	146 26.5	35.8	19 46.8	8.0	25 42.9	1.9	56.3
	23	161 26.5	34.9	34 13.8	8.0	25 44.8	1.6	56.3
13 THURSDAY	00	176 26.5	S13 34.1	48 40.8	7.9	N25 46.4	1.6	56.3
	01	191 26.5	33.3	63 07.7	7.8	25 48.0	1.4	56.4
	02	206 26.5	32.4	77 34.5	7.8	25 49.4	1.2	56.4
	03	221 26.5	. . 31.6	92 01.3	7.7	25 50.6	1.1	56.4
	04	236 26.5	30.8	106 28.0	7.7	25 51.7	1.0	56.5
	05	251 26.5	29.9	120 54.7	7.6	25 52.7	0.8	56.5
	06	266 26.5	S13 29.1	135 21.3	7.5	N25 53.5	0.7	56.5
	07	281 26.6	28.3	149 47.8	7.5	25 54.2	0.5	56.6
	08	296 26.6	27.4	164 14.3	7.5	25 54.7	0.4	56.6
	09	311 26.6	. . 26.6	178 40.8	7.4	25 55.1	0.2	56.6
	10	326 26.6	25.7	193 07.2	7.4	25 55.3	0.1	56.7
	11	341 26.6	24.9	207 33.6	7.3	25 55.4	0.1	56.7
	12	356 26.6	S13 24.1	221 59.9	7.3	N25 55.3	0.2	56.7
	13	11 26.6	23.2	236 26.2	7.2	25 55.1	0.4	56.8
	14	26 26.7	22.4	250 52.4	7.2	25 54.7	0.5	56.8
	15	41 26.7	. . 21.5	265 18.6	7.1	25 54.2	0.7	56.8
	16	56 26.7	20.7	279 44.7	7.2	25 53.5	0.9	56.9
	17	71 26.7	19.9	294 10.9	7.0	25 52.6	0.9	56.9
	18	86 26.7	S13 19.0	308 36.9	7.1	N25 51.7	1.2	56.9
	19	101 26.7	18.2	323 03.0	7.0	25 50.5	1.3	57.0
	20	116 26.7	17.3	337 29.0	7.0	25 49.2	1.4	57.0
	21	131 26.8	. . 16.5	351 55.0	6.9	25 47.8	1.6	57.0
	22	146 26.8	15.6	6 20.9	6.9	25 46.2	1.7	57.1
	23	161 26.8	14.8	20 46.8	6.9	25 44.5	2.0	57.1
14 FRIDAY	00	176 26.8	S13 14.0	35 12.7	6.9	N25 42.5	2.0	57.1
	01	191 26.8	13.1	49 38.6	6.8	25 40.5	2.2	57.2
	02	206 26.9	12.3	64 04.4	6.9	25 38.3	2.4	57.2
	03	221 26.9	. . 11.4	78 30.3	6.8	25 35.9	2.5	57.2
	04	236 26.9	10.6	92 56.1	6.7	25 33.4	2.7	57.3
	05	251 26.9	09.7	107 21.8	6.8	25 30.7	2.9	57.3
	06	266 26.9	S13 08.9	121 47.6	6.7	N25 27.8	2.9	57.3
	07	281 27.0	08.0	136 13.3	6.8	25 24.9	3.2	57.4
	08	296 27.0	07.2	150 39.1	6.7	25 21.7	3.3	57.4
	09	311 27.0	. . 06.3	165 04.8	6.7	25 18.4	3.5	57.4
	10	326 27.0	05.5	179 30.5	6.7	25 14.9	3.6	57.5
	11	341 27.0	04.6	193 56.2	6.7	25 11.3	3.8	57.5
	12	356 27.1	S13 03.8	208 21.9	6.6	N25 07.5	3.9	57.5
	13	11 27.1	02.9	222 47.5	6.7	25 03.6	4.1	57.6
	14	26 27.1	02.1	237 13.2	6.7	24 59.5	4.2	57.6
	15	41 27.1	. . 01.2	251 38.9	6.6	24 55.3	4.5	57.6
	16	56 27.2	13 00.4	266 04.5	6.7	24 50.8	4.5	57.7
	17	71 27.2	12 59.5	280 30.2	6.7	24 46.3	4.7	57.7
	18	86 27.2	S12 58.7	294 55.9	6.6	N24 41.6	4.9	57.7
	19	101 27.2	57.8	309 21.5	6.7	24 36.7	5.0	57.8
	20	116 27.3	57.0	323 47.2	6.7	24 31.7	5.2	57.8
	21	131 27.3	. . 56.1	338 12.9	6.7	24 26.5	5.3	57.8
	22	146 27.3	55.3	352 38.6	6.6	24 21.2	5.5	57.9
	23	161 27.3	54.4	7 04.2	6.7	N24 15.7	5.7	57.9
		SD 16.2	*d* 0.8	SD	15.3		15.5	15.7

Lat.	Twilight Naut.	Twilight Civil	Sunrise	Moonrise 12	Moonrise 13	Moonrise 14	Moonrise 15
°	h m	h m	h m	h m	h m	h m	h m
N 72	06 25	07 46	09 08	▭	▭	▭	▭
N 70	06 22	07 35	08 46	▭	▭	▭	▭
68	06 19	07 25	08 28	▭	▭	▭	▭
66	06 17	07 18	08 14	▭	▭	▭	12 24
64	06 15	07 11	08 03	09 21	09 37	11 11	13 11
62	06 13	07 05	07 53	10 09	10 48	12 02	13 41
60	06 11	07 00	07 45	10 39	11 23	12 33	14 04
N 58	06 10	06 56	07 37	11 03	11 49	12 57	14 22
56	06 08	06 52	07 31	11 22	12 09	13 15	14 37
54	06 06	06 48	07 25	11 37	12 26	13 31	14 50
52	06 05	06 45	07 20	11 51	12 41	13 45	15 02
50	06 03	06 41	07 15	12 03	12 53	13 57	15 12
45	06 00	06 34	07 05	12 28	13 20	14 22	15 33
N 40	05 56	06 28	06 56	12 48	13 40	14 42	15 50
35	05 53	06 23	06 49	13 05	13 58	14 58	16 05
30	05 49	06 18	06 42	13 19	14 12	15 12	16 17
20	05 42	06 08	06 31	13 44	14 38	15 37	16 38
N 10	05 34	05 59	06 20	14 05	15 00	15 57	16 57
0	05 25	05 50	06 11	14 25	15 20	16 17	17 14
S 10	05 14	05 39	06 01	14 45	15 40	16 36	17 31
20	05 01	05 28	05 51	15 06	16 02	16 57	17 49
30	04 44	05 13	05 38	15 31	16 28	17 21	18 10
35	04 33	05 05	05 31	15 46	16 43	17 35	18 22
40	04 19	04 54	05 23	16 03	17 00	17 52	18 36
45	04 03	04 42	05 14	16 23	17 21	18 11	18 53
S 50	03 41	04 26	05 02	16 49	17 47	18 35	19 13
52	03 31	04 19	04 57	17 02	18 00	18 47	19 23
54	03 18	04 11	04 51	17 16	18 15	19 00	19 34
56	03 04	04 01	04 44	17 33	18 32	19 15	19 46
58	02 46	03 50	04 37	17 53	18 52	19 34	20 00
S 60	02 24	03 38	04 28	18 19	19 19	19 56	20 16

Lat.	Sunset	Twilight Civil	Twilight Naut.	Moonset 12	Moonset 13	Moonset 14	Moonset 15
°	h m	h m	h m	h m	h m	h m	h m
N 72	15 22	16 44	18 06	▭	▭	▭	▭
N 70	15 44	16 55	18 08	▭	▭	▭	▭
68	16 02	17 04	18 11	▭	▭	▭	▭
66	16 15	17 12	18 13	▭	▭	▭	09 45
64	16 27	17 19	18 15	06 56	08 33	08 58	08 58
62	16 36	17 24	18 17	06 08	07 23	08 07	08 27
60	16 45	17 29	18 18	05 38	06 47	07 36	08 04
N 58	16 52	17 34	18 20	05 15	06 22	07 12	07 45
56	16 58	17 38	18 21	04 56	06 01	06 53	07 29
54	17 04	17 41	18 23	04 41	05 44	06 36	07 16
52	17 09	17 45	18 24	04 27	05 30	06 23	07 04
50	17 14	17 48	18 26	04 15	05 17	06 10	06 53
45	17 24	17 55	18 29	03 51	04 51	05 45	06 31
N 40	17 33	18 01	18 33	03 31	04 31	05 25	06 13
35	17 40	18 06	18 36	03 15	04 13	05 08	05 58
30	17 47	18 11	18 40	03 01	03 58	04 54	05 45
20	17 58	18 21	18 47	02 37	03 33	04 29	05 22
N 10	18 08	18 30	18 55	02 17	03 11	04 07	05 03
0	18 18	18 39	19 04	01 57	02 51	03 47	04 44
S 10	18 27	18 49	19 14	01 38	02 31	03 27	04 26
20	18 38	19 00	19 27	01 17	02 09	03 05	04 06
30	18 50	19 15	19 44	00 54	01 44	02 40	03 43
35	18 57	19 23	19 55	00 40	01 29	02 25	03 29
40	19 05	19 33	20 08	00 24	01 11	02 08	03 13
45	19 14	19 46	20 24	00 05	00 51	01 48	02 55
S 50	19 25	20 01	20 46	24 25	00 25	01 21	02 31
52	19 31	20 08	20 56	24 12	00 12	01 09	02 20
54	19 36	20 16	21 08	23 57	24 54	00 54	02 07
56	19 43	20 26	21 22	23 40	24 37	00 37	01 52
58	19 50	20 36	21 39	23 20	24 17	00 17	01 34
S 60	19 58	20 49	22 01	22 54	23 51	25 12	01 12

Day	SUN Eqn. of Time 00^h	SUN Eqn. of Time 12^h	SUN Mer. Pass.	MOON Mer. Pass. Upper	MOON Mer. Pass. Lower	MOON Age	MOON Phase
d	m s	m s	h m	h m	h m	d %	
12	14 15	14 15	12 14	20 38	08 11	11 78	
13	14 14	14 14	12 14	21 34	09 05	12 86	
14	14 13	14 12	12 14	22 31	10 02	13 92	

UT d h	ARIES GHA	VENUS −4.2 GHA	VENUS Dec	MARS +1.1 GHA	MARS Dec	JUPITER −2.6 GHA	JUPITER Dec	SATURN −0.1 GHA	SATURN Dec
	° ′	° ′	° ′	° ′	° ′	° ′	° ′	° ′	° ′
15 00 (SATURDAY)	144 35.3	221 49.9	S21 00.5	247 01.1	S22 44.8	10 25.3	N18 11.5	63 01.4	N22 04.0
01	159 37.7	236 49.3	00.4	262 01.8	45.0	25 28.0	11.6	78 03.9	04.0
02	174 40.2	251 48.7	00.3	277 02.5	45.2	40 30.8	11.7	93 06.4	04.0
03	189 42.7	266 48.0	00.2	292 03.3	45.3	55 33.6	11.8	108 08.9	04.0
04	204 45.1	281 47.4	00.1	307 04.0	45.5	70 36.4	11.9	123 11.4	04.0
05	219 47.6	296 46.8	21 00.0	322 04.7	45.7	85 39.2	12.0	138 13.9	04.0
06	234 50.1	311 46.2	S20 59.9	337 05.5	S22 45.9	100 41.9	N18 12.1	153 16.4	N22 04.0
07	249 52.5	326 45.5	59.8	352 06.2	46.0	115 44.7	12.2	168 18.9	04.0
08	264 55.0	341 44.9	59.7	7 06.9	46.2	130 47.5	12.2	183 21.4	04.0
09	279 57.5	356 44.3	59.6	22 07.7	46.4	145 50.3	12.3	198 23.9	04.0
10	294 59.9	11 43.7	59.5	37 08.4	46.5	160 53.1	12.4	213 26.4	04.0
11	310 02.4	26 43.0	59.4	52 09.1	46.7	175 55.8	12.5	228 28.9	04.0
12	325 04.8	41 42.4	S20 59.3	67 09.9	S22 46.9	190 58.6	N18 12.6	243 31.4	N22 04.1
13	340 07.3	56 41.8	59.2	82 10.6	47.0	206 01.4	12.7	258 33.9	04.1
14	355 09.8	71 41.2	59.1	97 11.3	47.2	221 04.2	12.8	273 36.4	04.1
15	10 12.2	86 40.6	59.0	112 12.1	47.4	236 07.0	12.9	288 38.9	04.1
16	25 14.7	101 39.9	58.9	127 12.8	47.5	251 09.7	13.0	303 41.4	04.1
17	40 17.2	116 39.3	58.8	142 13.5	47.7	266 12.5	13.0	318 43.9	04.1
18	55 19.6	131 38.7	S20 58.7	157 14.3	S22 47.9	281 15.3	N18 13.1	333 46.4	N22 04.1
19	70 22.1	146 38.1	58.6	172 15.0	48.0	296 18.1	13.2	348 48.9	04.1
20	85 24.6	161 37.4	58.5	187 15.7	48.2	311 20.8	13.3	3 51.3	04.1
21	100 27.0	176 36.8	58.4	202 16.4	48.4	326 23.6	13.4	18 53.8	04.1
22	115 29.5	191 36.2	58.2	217 17.2	48.5	341 26.4	13.5	33 56.3	04.1
23	130 31.9	206 35.6	58.1	232 17.9	48.7	356 29.2	13.6	48 58.8	04.1
16 00 (SUNDAY)	145 34.4	221 34.9	S20 58.0	247 18.6	S22 48.9	11 32.0	N18 13.7	64 01.3	N22 04.1
01	160 36.9	236 34.3	57.9	262 19.4	49.0	26 34.7	13.7	79 03.8	04.1
02	175 39.3	251 33.7	57.8	277 20.1	49.2	41 37.5	13.8	94 06.3	04.2
03	190 41.8	266 33.1	57.7	292 20.8	49.4	56 40.3	13.9	109 08.8	04.2
04	205 44.3	281 32.4	57.5	307 21.6	49.5	71 43.1	14.0	124 11.3	04.2
05	220 46.7	296 31.8	57.4	322 22.3	49.7	86 45.8	14.1	139 13.8	04.2
06	235 49.2	311 31.2	S20 57.3	337 23.0	S22 49.9	101 48.6	N18 14.2	154 16.3	N22 04.2
07	250 51.7	326 30.5	57.2	352 23.8	50.0	116 51.4	14.3	169 18.8	04.2
08	265 54.1	341 29.9	57.1	7 24.5	50.2	131 54.2	14.4	184 21.3	04.2
09	280 56.6	356 29.3	56.9	22 25.2	50.4	146 56.9	14.5	199 23.8	04.2
10	295 59.1	11 28.7	56.8	37 26.0	50.5	161 59.7	14.5	214 26.3	04.2
11	311 01.5	26 28.0	56.7	52 26.7	50.7	177 02.5	14.6	229 28.8	04.2
12	326 04.0	41 27.4	S20 56.6	67 27.4	S22 50.8	192 05.3	N18 14.7	244 31.3	N22 04.2
13	341 06.4	56 26.8	56.4	82 28.2	51.0	207 08.0	14.8	259 33.8	04.2
14	356 08.9	71 26.2	56.3	97 28.9	51.2	222 10.8	14.9	274 36.3	04.2
15	11 11.4	86 25.5	56.2	112 29.6	51.3	237 13.6	15.0	289 38.8	04.3
16	26 13.8	101 24.9	56.0	127 30.4	51.5	252 16.4	15.1	304 41.3	04.3
17	41 16.3	116 24.3	55.9	142 31.1	51.6	267 19.1	15.2	319 43.7	04.3
18	56 18.8	131 23.6	S20 55.8	157 31.8	S22 51.8	282 21.9	N18 15.2	334 46.2	N22 04.3
19	71 21.2	146 23.0	55.7	172 32.6	52.0	297 24.7	15.3	349 48.7	04.3
20	86 23.7	161 22.4	55.5	187 33.3	52.1	312 27.5	15.4	4 51.2	04.3
21	101 26.2	176 21.8	55.4	202 34.0	52.3	327 30.2	15.5	19 53.7	04.3
22	116 28.6	191 21.1	55.2	217 34.8	52.4	342 33.0	15.6	34 56.2	04.3
23	131 31.1	206 20.5	55.1	232 35.5	52.6	357 35.8	15.7	49 58.7	04.3
17 00 (MONDAY)	146 33.6	221 19.9	S20 55.0	247 36.2	S22 52.8	12 38.6	N18 15.8	65 01.2	N22 04.3
01	161 36.0	236 19.2	54.8	262 36.9	52.9	27 41.3	15.8	80 03.7	04.3
02	176 38.5	251 18.6	54.7	277 37.7	53.1	42 44.1	15.9	95 06.2	04.3
03	191 40.9	266 18.0	54.6	292 38.4	53.2	57 46.9	16.0	110 08.7	04.3
04	206 43.4	281 17.4	54.4	307 39.1	53.4	72 49.7	16.1	125 11.2	04.3
05	221 45.9	296 16.7	54.3	322 39.9	53.5	87 52.4	16.2	140 13.6	04.4
06	236 48.3	311 16.1	S20 54.1	337 40.6	S22 53.7	102 55.2	N18 16.3	155 16.1	N22 04.4
07	251 50.8	326 15.5	54.0	352 41.3	53.9	117 58.0	16.4	170 18.6	04.4
08	266 53.3	341 14.8	53.8	7 42.1	54.0	133 00.8	16.5	185 21.1	04.4
09	281 55.7	356 14.2	53.7	22 42.8	54.2	148 03.5	16.5	200 23.6	04.4
10	296 58.2	11 13.6	53.5	37 43.5	54.3	163 06.3	16.6	215 26.1	04.4
11	312 00.7	26 12.9	53.4	52 44.3	54.5	178 09.1	16.7	230 28.6	04.4
12	327 03.1	41 12.3	S20 53.2	67 45.0	S22 54.6	193 11.8	N18 16.8	245 31.1	N22 04.4
13	342 05.6	56 11.7	53.1	82 45.7	54.8	208 14.6	16.9	260 33.6	04.4
14	357 08.0	71 11.1	52.9	97 46.4	54.9	223 17.4	17.0	275 36.1	04.4
15	12 10.5	86 10.4	52.8	112 47.2	55.1	238 20.2	17.1	290 38.5	04.4
16	27 13.0	101 09.8	52.6	127 47.9	55.2	253 22.9	17.1	305 41.0	04.4
17	42 15.4	116 09.2	52.5	142 48.6	55.4	268 25.7	17.2	320 43.5	04.4
18	57 17.9	131 08.5	S20 52.3	157 49.4	S22 55.6	283 28.5	N18 17.3	335 46.0	N22 04.5
19	72 20.4	146 07.9	52.2	172 50.1	55.7	298 31.2	17.4	350 48.5	04.5
20	87 22.8	161 07.3	52.0	187 50.8	55.9	313 34.0	17.5	5 51.0	04.5
21	102 25.3	176 06.6	51.8	202 51.6	56.0	328 36.8	17.6	20 53.5	04.5
22	117 27.8	191 06.0	51.7	217 52.3	56.2	343 39.6	17.7	35 56.0	04.5
23	132 30.2	206 05.4	51.5	232 53.0	56.3	358 42.3	17.7	50 58.4	04.5
Mer. Pass. (h m)	14 15.4	v −0.6	d 0.1	v 0.7	d 0.2	v 2.8	d 0.1	v 2.5	d 0.0

STARS

Name	SHA	Dec
	° ′	° ′
Acamar	315 24.5	S40 17.8
Achernar	335 32.9	S57 13.6
Acrux	173 18.2	S63 06.8
Adhara	255 18.7	S28 58.7
Aldebaran	290 58.7	N16 30.9
Alioth	166 27.2	N55 56.4
Alkaid	153 04.9	N49 17.6
Al Na'ir	27 54.2	S46 57.0
Alnilam	275 54.5	S 1 12.1
Alphard	218 03.8	S 8 40.3
Alphecca	126 17.8	N26 42.0
Alpheratz	357 52.3	N29 06.4
Altair	62 16.5	N 8 52.4
Ankaa	353 23.9	S42 17.7
Antares	112 36.4	S26 26.3
Arcturus	146 03.0	N19 09.8
Atria	107 45.8	S69 01.8
Avior	234 20.9	S59 31.2
Bellatrix	278 40.6	N 6 21.1
Betelgeuse	271 09.9	N 7 24.4
Canopus	263 59.4	S52 42.0
Capella	280 46.3	N46 00.2
Deneb	49 37.6	N45 17.2
Denebola	182 41.6	N14 33.2
Diphda	349 04.2	S17 58.4
Dubhe	194 00.7	N61 44.0
Elnath	278 22.8	N28 36.7
Eltanin	90 50.2	N51 29.0
Enif	33 55.4	N 9 53.1
Fomalhaut	15 33.2	S29 36.6
Gacrux	172 09.8	S57 07.6
Gienah	176 00.5	S17 33.5
Hadar	148 59.4	S60 23.0
Hamal	328 10.1	N23 28.6
Kaus Aust.	83 54.8	S34 23.0
Kochab	137 19.1	N74 08.2
Markab	13 46.8	N15 13.1
Menkar	314 23.6	N 4 06.0
Menkent	148 17.1	S36 23.0
Miaplacidus	221 40.8	S69 43.8
Mirfak	308 52.2	N49 52.5
Nunki	76 08.6	S26 17.7
Peacock	53 32.4	S56 43.6
Pollux	243 37.3	N28 01.2
Procyon	245 08.0	N 5 13.0
Rasalhague	96 14.2	N12 33.2
Regulus	207 51.8	N11 57.1
Rigel	281 19.7	S 8 12.0
Rigil Kent.	140 03.0	S60 50.6
Sabik	102 22.0	S15 43.8
Schedar	349 50.5	N56 33.3
Shaula	96 33.2	S37 06.3
Sirius	258 40.7	S16 43.3
Spica	158 39.7	S11 10.7
Suhail	222 58.1	S43 26.7
Vega	80 44.8	N38 46.9
Zuben'ubi	137 14.4	S16 03.3

	SHA	Mer. Pass.
	° ′	h m
Venus	76 00.5	9 14
Mars	101 44.2	7 30
Jupiter	225 57.5	23 10
Saturn	278 26.9	19 41

UT		SUN GHA	SUN Dec	MOON GHA	*v*	Dec	*d*	HP
d h		° ′	° ′	° ′	′	° ′	′	′
15 00		176 27.4	S12 53.6	21 29.9	6.8	N24 10.0	5.8	57.9
01		191 27.4	52.7	35 55.7	6.7	24 04.2	5.9	58.0
02		206 27.4	51.9	50 21.4	6.7	23 58.3	6.2	58.0
03		221 27.4	. . 51.0	64 47.1	6.8	23 52.1	6.2	58.0
04		236 27.5	50.2	79 12.9	6.7	23 45.9	6.4	58.1
05		251 27.5	49.3	93 38.6	6.8	23 39.5	6.6	58.1
06		266 27.5	S12 48.5	108 04.4	6.8	N23 32.9	6.7	58.1
07		281 27.5	47.6	122 30.2	6.8	23 26.2	6.9	58.2
08	S	296 27.6	46.7	136 56.0	6.9	23 19.3	7.0	58.2
09	A	311 27.6	. . 45.9	151 21.9	6.9	23 12.3	7.2	58.2
10	T	326 27.6	45.0	165 47.8	6.8	23 05.1	7.3	58.3
11	U	341 27.7	44.2	180 13.6	7.0	22 57.8	7.4	58.3
12	R	356 27.7	S12 43.3	194 39.6	6.9	N22 50.4	7.7	58.3
13	D	11 27.7	42.5	209 05.5	7.0	22 42.7	7.7	58.4
14	A	26 27.8	41.6	223 31.5	7.0	22 35.0	7.9	58.4
15	Y	41 27.8	. . 40.7	237 57.5	7.0	22 27.1	8.1	58.4
16		56 27.8	39.9	252 23.5	7.1	22 19.0	8.2	58.5
17		71 27.8	39.0	266 49.6	7.0	22 10.8	8.3	58.5
18		86 27.9	S12 38.2	281 15.6	7.2	N22 02.5	8.5	58.5
19		101 27.9	37.3	295 41.8	7.1	21 54.0	8.6	58.6
20		116 27.9	36.5	310 07.9	7.2	21 45.4	8.8	58.6
21		131 28.0	. . 35.6	324 34.1	7.2	21 36.6	8.9	58.6
22		146 28.0	34.7	339 00.3	7.3	21 27.7	9.1	58.6
23		161 28.0	33.9	353 26.6	7.3	21 18.6	9.2	58.7
16 00		176 28.1	S12 33.0	7 52.9	7.3	N21 09.4	9.3	58.7
01		191 28.1	32.1	22 19.2	7.4	21 00.1	9.4	58.7
02		206 28.1	31.3	36 45.6	7.4	20 50.7	9.6	58.8
03		221 28.2	. . 30.4	51 12.0	7.5	20 41.1	9.8	58.8
04		236 28.2	29.6	65 38.5	7.5	20 31.3	9.9	58.8
05		251 28.3	28.7	80 05.0	7.5	20 21.4	10.0	58.8
06		266 28.3	S12 27.8	94 31.5	7.6	N20 11.4	10.1	58.9
07		281 28.3	27.0	108 58.1	7.6	20 01.3	10.3	58.9
08		296 28.4	26.1	123 24.7	7.7	19 51.0	10.4	58.9
09	S	311 28.4	. . 25.2	137 51.4	7.7	19 40.6	10.5	59.0
10	U	326 28.4	24.4	152 18.1	7.7	19 30.1	10.6	59.0
11	N	341 28.5	23.5	166 44.8	7.8	19 19.5	10.8	59.0
12	D	356 28.5	S12 22.6	181 11.6	7.9	N19 08.7	10.9	59.0
13	A	11 28.5	21.8	195 38.5	7.9	18 57.8	11.0	59.1
14	Y	26 28.6	20.9	210 05.4	7.9	18 46.8	11.2	59.1
15		41 28.6	. . 20.0	224 32.3	8.0	18 35.6	11.2	59.1
16		56 28.7	19.2	238 59.3	8.0	18 24.4	11.4	59.1
17		71 28.7	18.3	253 26.3	8.1	18 13.0	11.5	59.2
18		86 28.7	S12 17.4	267 53.4	8.1	N18 01.5	11.7	59.2
19		101 28.8	16.6	282 20.5	8.2	17 49.8	11.7	59.2
20		116 28.8	15.7	296 47.7	8.2	17 38.1	11.9	59.2
21		131 28.9	. . 14.8	311 14.9	8.3	17 26.2	11.9	59.3
22		146 28.9	14.0	325 42.2	8.3	17 14.3	12.1	59.3
23		161 28.9	13.1	340 09.5	8.4	17 02.2	12.2	59.3
17 00		176 29.0	S12 12.2	354 36.9	8.4	N16 50.0	12.3	59.3
01		191 29.0	11.4	9 04.3	8.5	16 37.7	12.4	59.4
02		206 29.1	10.5	23 31.8	8.5	16 25.3	12.5	59.4
03		221 29.1	. . 09.6	37 59.3	8.5	16 12.8	12.6	59.4
04		236 29.1	08.8	52 26.8	8.6	16 00.2	12.8	59.4
05		251 29.2	07.9	66 54.4	8.7	15 47.4	12.8	59.4
06		266 29.2	S12 07.0	81 22.1	8.7	N15 34.6	12.9	59.5
07		281 29.3	06.1	95 49.8	8.7	15 21.7	13.0	59.5
08		296 29.3	05.3	110 17.5	8.8	15 08.7	13.2	59.5
09	M	311 29.4	. . 04.4	124 45.3	8.9	14 55.5	13.2	59.5
10	O	326 29.4	03.5	139 13.2	8.9	14 42.3	13.3	59.6
11	N	341 29.5	02.6	153 41.1	8.9	14 29.0	13.4	59.6
12	D	356 29.5	S12 01.8	168 09.0	9.0	N14 15.6	13.5	59.6
13	A	11 29.5	00.9	182 37.0	9.0	14 02.1	13.6	59.6
14	Y	26 29.6	12 00.0	197 05.0	9.1	13 48.5	13.7	59.6
15		41 29.6	11 59.1	211 33.1	9.2	13 34.8	13.7	59.6
16		56 29.7	58.3	226 01.3	9.1	13 21.1	13.9	59.7
17		71 29.7	57.4	240 29.4	9.3	13 07.2	13.9	59.7
18		86 29.8	S11 56.5	254 57.7	9.2	N12 53.3	14.0	59.7
19		101 29.8	55.6	269 25.9	9.3	12 39.3	14.1	59.7
20		116 29.9	54.8	283 54.2	9.4	12 25.2	14.2	59.7
21		131 29.9	. . 53.9	298 22.6	9.4	12 11.0	14.2	59.8
22		146 30.0	53.0	312 51.0	9.4	11 56.8	14.4	59.8
23		161 30.0	52.1	327 19.4	9.5	N11 42.4	14.4	59.8
		SD 16.2	*d* 0.9	SD 15.9		16.1		16.2

Lat.	Twilight Naut.	Twilight Civil	Sunrise	Moonrise 15	16	17	18
°	h m	h m	h m	h m	h m	h m	h m
N 72	06 12	07 32	08 50	▭	▭	15 50	18 26
N 70	06 11	07 22	08 31	▭	13 23	16 17	18 37
68	06 09	07 15	08 16	▭	14 15	16 37	18 45
66	06 08	07 08	08 03	12 24	14 47	16 53	18 52
64	06 07	07 02	07 53	13 11	15 10	17 06	18 58
62	06 05	06 57	07 44	13 41	15 29	17 17	19 03
60	06 04	06 53	07 37	14 04	15 44	17 26	19 07
N 58	06 03	06 49	07 30	14 22	15 57	17 34	19 11
56	06 02	06 45	07 24	14 37	16 08	17 41	19 14
54	06 01	06 42	07 19	14 50	16 17	17 47	19 17
52	06 00	06 39	07 14	15 02	16 26	17 53	19 20
50	05 59	06 36	07 10	15 12	16 33	17 58	19 22
45	05 56	06 30	07 00	15 33	16 50	18 09	19 28
N 40	05 53	06 25	06 52	15 50	17 03	18 18	19 32
35	05 50	06 20	06 45	16 05	17 14	18 25	19 36
30	05 47	06 15	06 39	16 17	17 24	18 32	19 39
20	05 41	06 06	06 29	16 38	17 41	18 44	19 45
N 10	05 33	05 58	06 20	16 57	17 56	18 54	19 50
0	05 25	05 50	06 11	17 14	18 09	19 03	19 55
S 10	05 15	05 40	06 02	17 31	18 23	19 12	20 00
20	05 03	05 29	05 52	17 49	18 38	19 22	20 05
30	04 46	05 16	05 41	18 10	18 54	19 34	20 10
35	04 36	05 08	05 34	18 22	19 04	19 40	20 14
40	04 24	04 58	05 27	18 36	19 14	19 48	20 17
45	04 08	04 47	05 18	18 53	19 27	19 56	20 22
S 50	03 48	04 32	05 07	19 13	19 43	20 06	20 27
52	03 38	04 25	05 03	19 23	19 50	20 11	20 29
54	03 27	04 18	04 57	19 34	19 58	20 16	20 32
56	03 13	04 09	04 51	19 46	20 07	20 22	20 34
58	02 57	03 59	04 44	20 00	20 17	20 28	20 37
S 60	02 38	03 47	04 36	20 16	20 28	20 36	20 41

Lat.	Sunset	Twilight Civil	Twilight Naut.	Moonset 15	16	17	18
°	h m	h m	h m	h m	h m	h m	h m
N 72	15 40	16 58	18 18	▭	▭	10 14	09 29
N 70	15 59	17 08	18 20	▭	10 46	09 45	09 16
68	16 14	17 15	18 21	▭	09 52	09 23	09 05
66	16 26	17 22	18 22	09 45	09 20	09 06	08 56
64	16 36	17 27	18 23	08 58	08 55	08 52	08 48
62	16 45	17 32	18 24	08 27	08 36	08 40	08 42
60	16 53	17 36	18 25	08 04	08 20	08 29	08 36
N 58	16 59	17 40	18 26	07 45	08 06	08 20	08 31
56	17 05	17 44	18 27	07 29	07 54	08 12	08 26
54	17 10	17 47	18 28	07 16	07 44	08 05	08 22
52	17 15	17 50	18 29	07 04	07 35	07 59	08 18
50	17 19	17 53	18 30	06 53	07 27	07 53	08 15
45	17 29	17 59	18 33	06 31	07 09	07 40	08 07
N 40	17 36	18 04	18 36	06 13	06 54	07 30	08 01
35	17 43	18 09	18 39	05 58	06 42	07 21	07 56
30	17 49	18 14	18 42	05 45	06 31	07 13	07 51
20	18 00	18 22	18 48	05 22	06 12	06 59	07 42
N 10	18 09	18 30	18 55	05 03	05 56	06 47	07 35
0	18 17	18 39	19 03	04 44	05 41	06 35	07 28
S 10	18 26	18 48	19 13	04 26	05 25	06 24	07 21
20	18 36	18 59	19 25	04 06	05 08	06 11	07 13
30	18 47	19 12	19 41	03 43	04 49	05 57	07 04
35	18 53	19 20	19 51	03 29	04 38	05 48	06 59
40	19 01	19 29	20 04	03 13	04 25	05 39	06 53
45	19 09	19 41	20 19	02 55	04 09	05 27	06 46
S 50	19 20	19 55	20 39	02 31	03 50	05 13	06 38
52	19 25	20 02	20 48	02 20	03 41	05 07	06 34
54	19 30	20 09	21 00	02 07	03 30	05 00	06 30
56	19 36	20 18	21 12	01 52	03 19	04 51	06 25
58	19 42	20 27	21 28	01 34	03 05	04 42	06 20
S 60	19 50	20 39	21 47	01 12	02 50	04 32	06 14

Day	SUN Eqn. of Time 00^h	12^h	Mer. Pass.	MOON Mer. Pass. Upper	Lower	Age	Phase
d	m s	m s	h m	h m	h m	d %	
15	14 11	14 09	12 14	23 27	10 59	14 97	○
16	14 08	14 06	12 14	24 22	11 55	15 99	
17	14 04	14 02	12 14	00 22	12 49	16 99	

	UT	ARIES	VENUS −4.2		MARS +1.1		JUPITER −2.5		SATURN −0.1	
		GHA	GHA	Dec	GHA	Dec	GHA	Dec	GHA	Dec
	d h	° ′	° ′	° ′	° ′	° ′	° ′	° ′	° ′	° ′
	18 00	147 32.7	221 04.7	S20 51.4	247 53.8	S22 56.5	13 45.1	N18 17.8	66 00.9	N22 04.5
	01	162 35.2	236 04.1	51.2	262 54.5	56.6	28 47.9	17.9	81 03.4	04.5
	02	177 37.6	251 03.5	51.0	277 55.2	56.8	43 50.6	18.0	96 05.9	04.5
	03	192 40.1	266 02.8	50.9	292 55.9	56.9	58 53.4	18.1	111 08.4	04.5
	04	207 42.5	281 02.2	50.7	307 56.7	57.1	73 56.2	18.2	126 10.9	04.5
	05	222 45.0	296 01.6	50.5	322 57.4	57.2	88 59.0	18.3	141 13.4	04.5
	06	237 47.5	311 01.0	S20 50.4	337 58.1	S22 57.4	104 01.7	N18 18.3	156 15.8	N22 04.5
	07	252 49.9	326 00.3	50.2	352 58.9	57.5	119 04.5	18.4	171 18.3	04.6
T	08	267 52.4	340 59.7	50.0	7 59.6	57.7	134 07.3	18.5	186 20.8	04.6
U	09	282 54.9	355 59.1	49.9	23 00.3	57.8	149 10.0	18.6	201 23.3	04.6
E	10	297 57.3	10 58.4	49.7	38 01.1	58.0	164 12.8	18.7	216 25.8	04.6
S	11	312 59.8	25 57.8	49.5	53 01.8	58.1	179 15.6	18.8	231 28.3	04.6
D	12	328 02.3	40 57.2	S20 49.3	68 02.5	S22 58.3	194 18.3	N18 18.9	246 30.8	N22 04.6
A	13	343 04.7	55 56.5	49.2	83 03.2	58.4	209 21.1	18.9	261 33.2	04.6
Y	14	358 07.2	70 55.9	49.0	98 04.0	58.5	224 23.9	19.0	276 35.7	04.6
	15	13 09.7	85 55.3	48.8	113 04.7	58.7	239 26.6	19.1	291 38.2	04.6
	16	28 12.1	100 54.6	48.6	128 05.4	58.8	254 29.4	19.2	306 40.7	04.6
	17	43 14.6	115 54.0	48.5	143 06.2	59.0	269 32.2	19.3	321 43.2	04.6
	18	58 17.0	130 53.4	S20 48.3	158 06.9	S22 59.1	284 34.9	N18 19.4	336 45.7	N22 04.6
	19	73 19.5	145 52.7	48.1	173 07.6	59.3	299 37.7	19.4	351 48.1	04.6
	20	88 22.0	160 52.1	47.9	188 08.3	59.4	314 40.5	19.5	6 50.6	04.7
	21	103 24.4	175 51.5	47.7	203 09.1	59.6	329 43.3	19.6	21 53.1	04.7
	22	118 26.9	190 50.8	47.6	218 09.8	59.7	344 46.0	19.7	36 55.6	04.7
	23	133 29.4	205 50.2	47.4	233 10.5	22 59.9	359 48.8	19.8	51 58.1	04.7
	19 00	148 31.8	220 49.6	S20 47.2	248 11.3	S23 00.0	14 51.6	N18 19.9	67 00.5	N22 04.7
	01	163 34.3	235 48.9	47.0	263 12.0	00.1	29 54.3	20.0	82 03.0	04.7
	02	178 36.8	250 48.3	46.8	278 12.7	00.3	44 57.1	20.0	97 05.5	04.7
	03	193 39.2	265 47.7	46.6	293 13.4	00.4	59 59.9	20.1	112 08.0	04.7
	04	208 41.7	280 47.0	46.4	308 14.2	00.6	75 02.6	20.2	127 10.5	04.7
	05	223 44.1	295 46.4	46.2	323 14.9	00.7	90 05.4	20.3	142 12.9	04.7
	06	238 46.6	310 45.7	S20 46.1	338 15.6	S23 00.9	105 08.2	N18 20.4	157 15.4	N22 04.7
W	07	253 49.1	325 45.1	45.9	353 16.4	01.0	120 10.9	20.5	172 17.9	04.7
E	08	268 51.5	340 44.5	45.7	8 17.1	01.1	135 13.7	20.5	187 20.4	04.8
D	09	283 54.0	355 43.8	45.5	23 17.8	01.3	150 16.4	20.6	202 22.9	04.8
N	10	298 56.5	10 43.2	45.3	38 18.6	01.4	165 19.2	20.7	217 25.3	04.8
E	11	313 58.9	25 42.6	45.1	53 19.3	01.6	180 22.0	20.8	232 27.8	04.8
S	12	329 01.4	40 41.9	S20 44.9	68 20.0	S23 01.7	195 24.7	N18 20.9	247 30.3	N22 04.8
D	13	344 03.9	55 41.3	44.7	83 20.7	01.8	210 27.5	21.0	262 32.8	04.8
A	14	359 06.3	70 40.7	44.5	98 21.5	02.0	225 30.3	21.0	277 35.3	04.8
Y	15	14 08.8	85 40.0	44.3	113 22.2	02.1	240 33.0	21.1	292 37.7	04.8
	16	29 11.3	100 39.4	44.1	128 22.9	02.3	255 35.8	21.2	307 40.2	04.8
	17	44 13.7	115 38.8	43.9	143 23.7	02.4	270 38.6	21.3	322 42.7	04.8
	18	59 16.2	130 38.1	S20 43.7	158 24.4	S23 02.5	285 41.3	N18 21.4	337 45.2	N22 04.8
	19	74 18.6	145 37.5	43.5	173 25.1	02.7	300 44.1	21.5	352 47.7	04.8
	20	89 21.1	160 36.9	43.3	188 25.8	02.8	315 46.9	21.5	7 50.1	04.8
	21	104 23.6	175 36.2	43.1	203 26.6	02.9	330 49.6	21.6	22 52.6	04.9
	22	119 26.0	190 35.6	42.9	218 27.3	03.1	345 52.4	21.7	37 55.1	04.9
	23	134 28.5	205 35.0	42.6	233 28.0	03.2	0 55.1	21.8	52 57.6	04.9
	20 00	149 31.0	220 34.3	S20 42.4	248 28.7	S23 03.4	15 57.9	N18 21.9	68 00.0	N22 04.9
	01	164 33.4	235 33.7	42.2	263 29.5	03.5	31 00.7	22.0	83 02.5	04.9
	02	179 35.9	250 33.0	42.0	278 30.2	03.6	46 03.4	22.0	98 05.0	04.9
	03	194 38.4	265 32.4	41.8	293 30.9	03.8	61 06.2	22.1	113 07.5	04.9
	04	209 40.8	280 31.8	41.6	308 31.7	03.9	76 09.0	22.2	128 09.9	04.9
	05	224 43.3	295 31.1	41.4	323 32.4	04.0	91 11.7	22.3	143 12.4	04.9
	06	239 45.7	310 30.5	S20 41.2	338 33.1	S23 04.2	106 14.5	N18 22.4	158 14.9	N22 04.9
	07	254 48.2	325 29.9	40.9	353 33.8	04.3	121 17.2	22.5	173 17.4	04.9
T	08	269 50.7	340 29.2	40.7	8 34.6	04.4	136 20.0	22.5	188 19.8	04.9
H	09	284 53.1	355 28.6	40.5	23 35.3	04.6	151 22.8	22.6	203 22.3	05.0
U	10	299 55.6	10 28.0	40.3	38 36.0	04.7	166 25.5	22.7	218 24.8	05.0
R	11	314 58.1	25 27.3	40.1	53 36.8	04.8	181 28.3	22.8	233 27.3	05.0
S	12	330 00.5	40 26.7	S20 39.8	68 37.5	S23 05.0	196 31.1	N18 22.9	248 29.7	N22 05.0
D	13	345 03.0	55 26.0	39.6	83 38.2	05.1	211 33.8	22.9	263 32.2	05.0
A	14	0 05.5	70 25.4	39.4	98 38.9	05.2	226 36.6	23.0	278 34.7	05.0
Y	15	15 07.9	85 24.8	39.2	113 39.7	05.4	241 39.3	23.1	293 37.2	05.0
	16	30 10.4	100 24.1	39.0	128 40.4	05.5	256 42.1	23.2	308 39.6	05.0
	17	45 12.9	115 23.5	38.7	143 41.1	05.6	271 44.9	23.3	323 42.1	05.0
	18	60 15.3	130 22.9	S20 38.5	158 41.8	S23 05.8	286 47.6	N18 23.4	338 44.6	N22 05.0
	19	75 17.8	145 22.2	38.3	173 42.6	05.9	301 50.4	23.4	353 47.1	05.0
	20	90 20.2	160 21.6	38.0	188 43.3	06.0	316 53.1	23.5	8 49.5	05.0
	21	105 22.7	175 21.0	37.8	203 44.0	06.2	331 55.9	23.6	23 52.0	05.1
	22	120 25.2	190 20.3	37.6	218 44.8	06.3	346 58.7	23.7	38 54.5	05.1
	23	135 27.6	205 19.7	37.3	233 45.5	06.4	2 01.4	23.8	53 56.9	05.1
	Mer. Pass.	h m 14 03.6	*v* −0.6	*d* 0.2	*v* 0.7	*d* 0.1	*v* 2.8	*d* 0.1	*v* 2.5	*d* 0.0

STARS

Name	SHA	Dec
	° ′	° ′
Acamar	315 24.5	S40 17.8
Achernar	335 32.9	S57 13.6
Acrux	173 18.2	S63 06.8
Adhara	255 18.7	S28 58.7
Aldebaran	290 58.7	N16 30.9
Alioth	166 27.2	N55 56.4
Alkaid	153 04.9	N49 17.6
Al Na'ir	27 54.2	S46 57.0
Alnilam	275 54.5	S 1 12.1
Alphard	218 03.8	S 8 40.3
Alphecca	126 17.8	N26 42.0
Alpheratz	357 52.3	N29 06.4
Altair	62 16.4	N 8 52.4
Ankaa	353 23.9	S42 17.7
Antares	112 36.3	S26 26.3
Arcturus	146 03.0	N19 09.8
Atria	107 45.8	S69 01.8
Avior	234 20.9	S59 31.2
Bellatrix	278 40.6	N 6 21.1
Betelgeuse	271 09.9	N 7 24.4
Canopus	263 59.4	S52 42.0
Capella	280 46.3	N46 00.2
Deneb	49 37.5	N45 17.2
Denebola	182 41.6	N14 33.2
Diphda	349 04.2	S17 58.4
Dubhe	194 00.7	N61 44.0
Elnath	278 22.8	N28 36.7
Eltanin	90 50.2	N51 29.0
Enif	33 55.4	N 9 53.1
Fomalhaut	15 33.2	S29 36.6
Gacrux	172 09.8	S57 07.6
Gienah	176 00.5	S17 33.5
Hadar	148 59.4	S60 23.0
Hamal	328 10.2	N23 28.6
Kaus Aust.	83 54.8	S34 23.0
Kochab	137 19.0	N74 08.2
Markab	13 46.8	N15 13.1
Menkar	314 23.6	N 4 06.0
Menkent	148 17.1	S36 23.0
Miaplacidus	221 40.8	S69 43.8
Mirfak	308 52.2	N49 52.5
Nunki	76 08.6	S26 17.7
Peacock	53 32.4	S56 43.6
Pollux	243 37.3	N28 01.2
Procyon	245 08.0	N 5 13.0
Rasalhague	96 14.1	N12 33.2
Regulus	207 51.8	N11 57.1
Rigel	281 19.8	S 8 12.0
Rigil Kent.	140 02.9	S60 50.6
Sabik	102 22.0	S15 43.8
Schedar	349 50.5	N56 33.3
Shaula	96 33.1	S37 06.3
Sirius	258 40.7	S16 43.3
Spica	158 39.7	S11 10.7
Suhail	222 58.1	S43 26.7
Vega	80 44.8	N38 46.9
Zuben'ubi	137 14.4	S16 03.3

	SHA	Mer. Pass.
	° ′	h m
Venus	72 17.7	9 17
Mars	99 39.4	7 27
Jupiter	226 19.7	22 56
Saturn	278 28.7	19 29

Day	UT (h)	SUN GHA ° ′	SUN Dec ° ′	MOON GHA ° ′	v ′	MOON Dec ° ′	d ′	HP ′
18	00	176 30.1	S11 51.3	341 47.9	9.5	N11 28.0	14.4	59.8
	01	191 30.1	50.4	356 16.4	9.6	11 13.6	14.6	59.8
	02	206 30.2	49.5	10 45.0	9.6	10 59.0	14.6	59.8
	03	221 30.2	48.6	25 13.6	9.6	10 44.4	14.7	59.8
	04	236 30.3	47.7	39 42.2	9.7	10 29.7	14.7	59.9
	05	251 30.3	46.9	54 10.9	9.8	10 15.0	14.8	59.9
	06	266 30.4	S11 46.0	68 39.7	9.7	N10 00.2	14.9	59.9
	07	281 30.4	45.1	83 08.4	9.8	9 45.3	15.0	59.9
T	08	296 30.5	44.2	97 37.2	9.9	9 30.3	15.0	59.9
U	09	311 30.5	43.3	112 06.1	9.8	9 15.3	15.0	59.9
E	10	326 30.6	42.5	126 34.9	10.0	9 00.3	15.1	59.9
S	11	341 30.6	41.6	141 03.9	9.9	8 45.2	15.2	59.9
D	12	356 30.7	S11 40.7	155 32.8	10.0	N 8 30.0	15.2	60.0
A	13	11 30.7	39.8	170 01.8	10.0	8 14.8	15.3	60.0
Y	14	26 30.8	38.9	184 30.8	10.0	7 59.5	15.3	60.0
	15	41 30.8	38.1	198 59.8	10.1	7 44.2	15.3	60.0
	16	56 30.9	37.2	213 28.9	10.1	7 28.9	15.5	60.0
	17	71 30.9	36.3	227 58.0	10.2	7 13.4	15.4	60.0
	18	86 31.0	S11 35.4	242 27.2	10.1	N 6 58.0	15.5	60.0
	19	101 31.0	34.5	256 56.3	10.2	6 42.5	15.5	60.0
	20	116 31.1	33.6	271 25.5	10.3	6 27.0	15.6	60.0
	21	131 31.1	32.8	285 54.8	10.2	6 11.4	15.6	60.0
	22	146 31.2	31.9	300 24.0	10.3	5 55.8	15.7	60.0
	23	161 31.3	31.0	314 53.3	10.3	5 40.1	15.7	60.0
19	00	176 31.3	S11 30.1	329 22.6	10.3	N 5 24.4	15.7	60.1
	01	191 31.4	29.2	343 51.9	10.4	5 08.7	15.7	60.1
	02	206 31.4	28.3	358 21.3	10.4	4 53.0	15.8	60.1
	03	221 31.5	27.4	12 50.7	10.4	4 37.2	15.8	60.1
	04	236 31.5	26.6	27 20.1	10.4	4 21.4	15.9	60.1
	05	251 31.6	25.7	41 49.5	10.4	4 05.5	15.8	60.1
	06	266 31.6	S11 24.8	56 18.9	10.5	N 3 49.7	15.9	60.1
W	07	281 31.7	23.9	70 48.4	10.4	3 33.8	15.9	60.1
E	08	296 31.8	23.0	85 17.8	10.5	3 17.9	15.9	60.1
D	09	311 31.8	22.1	99 47.3	10.5	3 02.0	15.9	60.1
N	10	326 31.9	21.2	114 16.8	10.6	2 46.1	16.0	60.1
E	11	341 31.9	20.3	128 46.4	10.5	2 30.1	16.0	60.1
S	12	356 32.0	S11 19.5	143 15.9	10.5	N 2 14.1	15.9	60.1
D	13	11 32.1	18.6	157 45.4	10.6	1 58.2	16.0	60.1
A	14	26 32.1	17.7	172 15.0	10.6	1 42.2	16.0	60.1
Y	15	41 32.2	16.8	186 44.6	10.5	1 26.2	16.0	60.1
	16	56 32.2	15.9	201 14.1	10.6	1 10.2	16.0	60.1
	17	71 32.3	15.0	215 43.7	10.6	0 54.2	16.1	60.1
	18	86 32.4	S11 14.1	230 13.3	10.6	N 0 38.1	16.0	60.1
	19	101 32.4	13.2	244 42.9	10.6	0 22.1	16.0	60.1
	20	116 32.5	12.3	259 12.5	10.6	N 0 06.1	16.0	60.1
	21	131 32.5	11.4	273 42.1	10.7	S 0 09.9	16.0	60.1
	22	146 32.6	10.5	288 11.8	10.6	0 25.9	16.0	60.1
	23	161 32.7	09.7	302 41.4	10.6	0 41.9	16.0	60.1
20	00	176 32.7	S11 08.8	317 11.0	10.6	S 0 57.9	16.0	60.1
	01	191 32.8	07.9	331 40.6	10.6	1 13.9	16.0	60.1
	02	206 32.9	07.0	346 10.2	10.6	1 29.9	16.0	60.1
	03	221 32.9	06.1	0 39.8	10.7	1 45.9	16.0	60.1
	04	236 33.0	05.2	15 09.5	10.6	2 01.9	15.9	60.1
	05	251 33.0	04.3	29 39.1	10.6	2 17.8	15.9	60.1
	06	266 33.1	S11 03.4	44 08.7	10.6	S 2 33.7	16.0	60.1
T	07	281 33.2	02.5	58 38.3	10.6	2 49.7	15.9	60.1
H	08	296 33.2	01.6	73 07.9	10.6	3 05.6	15.8	60.1
U	09	311 33.3	11 00.7	87 37.5	10.5	3 21.4	15.9	60.1
R	10	326 33.4	10 59.8	102 07.0	10.6	3 37.3	15.8	60.0
S	11	341 33.4	58.9	116 36.6	10.5	3 53.1	15.8	60.0
D	12	356 33.5	S10 58.0	131 06.1	10.6	S 4 08.9	15.8	60.0
A	13	11 33.6	57.1	145 35.7	10.5	4 24.7	15.7	60.0
Y	14	26 33.6	56.2	160 05.2	10.5	4 40.4	15.8	60.0
	15	41 33.7	55.3	174 34.7	10.5	4 56.2	15.7	60.0
	16	56 33.8	54.4	189 04.2	10.5	5 11.9	15.6	60.0
	17	71 33.8	53.5	203 33.7	10.5	5 27.5	15.6	60.0
	18	86 33.9	S10 52.6	218 03.2	10.4	S 5 43.1	15.6	60.0
	19	101 34.0	51.7	232 32.6	10.5	5 58.7	15.5	60.0
	20	116 34.0	50.8	247 02.1	10.4	6 14.2	15.6	60.0
	21	131 34.1	49.9	261 31.5	10.4	6 29.8	15.4	60.0
	22	146 34.2	49.0	276 00.9	10.3	6 45.2	15.4	60.0
	23	161 34.2	48.1	290 30.2	10.4	S 7 00.6	15.4	59.9
		SD 16.2	d 0.9	SD 16.3		16.4		16.4

Lat.	Twilight Naut. (h m)	Twilight Civil (h m)	Sunrise (h m)	Moonrise 18 (h m)	Moonrise 19 (h m)	Moonrise 20 (h m)	Moonrise 21 (h m)
N 72	05 59	07 18	08 33	18 26	20 48	23 10	25 48
N 70	05 59	07 10	08 16	18 37	20 48	22 59	25 19
68	05 59	07 03	08 03	18 45	20 48	22 50	24 58
66	05 58	06 58	07 52	18 52	20 47	22 42	24 41
64	05 58	06 53	07 43	18 58	20 47	22 36	24 28
62	05 57	06 49	07 35	19 03	20 47	22 31	24 16
60	05 57	06 45	07 28	19 07	20 47	22 26	24 07
N 58	05 56	06 42	07 22	19 11	20 47	22 22	23 59
56	05 56	06 39	07 17	19 14	20 47	22 19	23 51
54	05 55	06 36	07 12	19 17	20 46	22 15	23 45
52	05 54	06 33	07 08	19 20	20 46	22 13	23 39
50	05 53	06 31	07 04	19 22	20 46	22 10	23 34
45	05 51	06 26	06 56	19 28	20 46	22 04	23 23
N 40	05 49	06 21	06 48	19 32	20 46	22 00	23 13
35	05 47	06 16	06 42	19 36	20 46	21 56	23 06
30	05 44	06 12	06 37	19 39	20 46	21 52	22 59
20	05 39	06 05	06 27	19 45	20 46	21 46	22 47
N 10	05 32	05 57	06 19	19 50	20 45	21 41	22 37
0	05 25	05 49	06 10	19 55	20 45	21 36	22 27
S 10	05 16	05 41	06 02	20 00	20 45	21 31	22 18
20	05 04	05 31	05 53	20 05	20 45	21 26	22 08
30	04 49	05 18	05 43	20 10	20 45	21 20	21 56
35	04 39	05 11	05 37	20 14	20 45	21 17	21 50
40	04 28	05 02	05 30	20 17	20 45	21 13	21 43
45	04 13	04 51	05 22	20 22	20 45	21 09	21 34
S 50	03 55	04 38	05 13	20 27	20 45	21 04	21 24
52	03 45	04 31	05 08	20 29	20 45	21 01	21 19
54	03 35	04 24	05 03	20 32	20 45	20 59	21 14
56	03 23	04 16	04 58	20 34	20 45	20 56	21 08
58	03 08	04 07	04 52	20 37	20 45	20 53	21 02
S 60	02 51	03 57	04 45	20 41	20 45	20 50	20 55

Lat.	Sunset (h m)	Twilight Civil (h m)	Twilight Naut. (h m)	Moonset 18 (h m)	Moonset 19 (h m)	Moonset 20 (h m)	Moonset 21 (h m)
N 72	15 56	17 12	18 31	09 29	08 56	08 26	07 53
N 70	16 13	17 20	18 31	09 16	08 53	08 31	08 07
68	16 26	17 26	18 31	09 05	08 49	08 35	08 19
66	16 37	17 31	18 31	08 56	08 47	08 38	08 28
64	16 46	17 36	18 31	08 48	08 44	08 40	08 36
62	16 54	17 40	18 32	08 42	08 42	08 43	08 43
60	17 00	17 44	18 32	08 36	08 41	08 45	08 49
N 58	17 06	17 47	18 33	08 31	08 39	08 47	08 55
56	17 12	17 50	18 33	08 26	08 38	08 48	09 00
54	17 16	17 53	18 34	08 22	08 36	08 50	09 04
52	17 21	17 55	18 35	08 18	08 35	08 51	09 08
50	17 24	17 58	18 35	08 15	08 34	08 52	09 11
45	17 33	18 03	18 37	08 07	08 32	08 55	09 19
N 40	17 40	18 08	18 39	08 01	08 30	08 57	09 26
35	17 46	18 12	18 41	07 56	08 28	08 59	09 31
30	17 51	18 16	18 44	07 51	08 26	09 01	09 36
20	18 01	18 23	18 49	07 42	08 24	09 04	09 45
N 10	18 09	18 31	18 55	07 35	08 21	09 07	09 53
0	18 17	18 38	19 03	07 28	08 19	09 09	10 00
S 10	18 25	18 47	19 12	07 21	08 16	09 12	10 07
20	18 34	18 57	19 23	07 13	08 14	09 14	10 15
30	18 44	19 09	19 38	07 04	08 11	09 17	10 24
35	18 50	19 16	19 48	06 59	08 09	09 19	10 29
40	18 57	19 25	19 59	06 53	08 07	09 21	10 35
45	19 05	19 36	20 13	06 46	08 05	09 23	10 42
S 50	19 14	19 49	20 32	06 38	08 02	09 26	10 50
52	19 18	19 55	20 41	06 34	08 01	09 27	10 54
54	19 23	20 02	20 51	06 30	07 59	09 29	10 58
56	19 29	20 10	21 03	06 25	07 58	09 30	11 02
58	19 35	20 19	21 17	06 20	07 56	09 32	11 08
S 60	19 41	20 29	21 34	06 14	07 54	09 34	11 14

Day	SUN Eqn. of Time 00^h	SUN Eqn. of Time 12^h	SUN Mer. Pass.	MOON Mer. Pass. Upper	MOON Mer. Pass. Lower	MOON Age	MOON Phase
d	m s	m s	h m	h m	h m	d	%
18	14 00	13 57	12 14	01 15	13 41	17	97
19	13 55	13 52	12 14	02 07	14 32	18	91
20	13 49	13 46	12 14	02 57	15 22	19	84

UT	ARIES	VENUS −4.2		MARS +1.0		JUPITER −2.5		SATURN −0.1	
d h	GHA ° ′	GHA ° ′	Dec ° ′	GHA ° ′	Dec ° ′	GHA ° ′	Dec ° ′	GHA ° ′	Dec ° ′
21 00	150 30.1	220 19.0	S20 37.1	248 46.2	S23 06.5	17 04.2	N18 23.8	68 59.4	N22 05.1
01	165 32.6	235 18.4	36.9	263 46.9	06.7	32 06.9	23.9	84 01.9	05.1
02	180 35.0	250 17.8	36.6	278 47.7	06.8	47 09.7	24.0	99 04.4	05.1
03	195 37.5	265 17.1	36.4	293 48.4	06.9	62 12.4	24.1	114 06.8	05.1
04	210 40.0	280 16.5	36.2	308 49.1	07.1	77 15.2	24.2	129 09.3	05.1
05	225 42.4	295 15.9	35.9	323 49.8	07.2	92 18.0	24.2	144 11.8	05.1
06	240 44.9	310 15.2	S20 35.7	338 50.6	S23 07.3	107 20.7	N18 24.3	159 14.2	N22 05.1
07	255 47.3	325 14.6	35.5	353 51.3	07.4	122 23.5	24.4	174 16.7	05.1
08	270 49.8	340 14.0	35.2	8 52.0	07.6	137 26.2	24.5	189 19.2	05.2
F 09	285 52.3	355 13.3	35.0	23 52.8	07.7	152 29.0	24.6	204 21.6	05.2
R 10	300 54.7	10 12.7	34.7	38 53.5	07.8	167 31.8	24.7	219 24.1	05.2
I 11	315 57.2	25 12.0	34.5	53 54.2	08.0	182 34.5	24.7	234 26.6	05.2
D 12	330 59.7	40 11.4	S20 34.2	68 54.9	S23 08.1	197 37.3	N18 24.8	249 29.1	N22 05.2
A 13	346 02.1	55 10.8	34.0	83 55.7	08.2	212 40.0	24.9	264 31.5	05.2
Y 14	1 04.6	70 10.1	33.7	98 56.4	08.3	227 42.8	25.0	279 34.0	05.2
15	16 07.1	85 09.5	33.5	113 57.1	08.5	242 45.5	25.1	294 36.5	05.2
16	31 09.5	100 08.9	33.2	128 57.8	08.6	257 48.3	25.1	309 38.9	05.2
17	46 12.0	115 08.2	33.0	143 58.6	08.7	272 51.0	25.2	324 41.4	05.2
18	61 14.5	130 07.6	S20 32.7	158 59.3	S23 08.8	287 53.8	N18 25.3	339 43.9	N22 05.2
19	76 16.9	145 06.9	32.5	174 00.0	09.0	302 56.6	25.4	354 46.3	05.2
20	91 19.4	160 06.3	32.2	189 00.7	09.1	317 59.3	25.5	9 48.8	05.3
21	106 21.8	175 05.7	32.0	204 01.5	09.2	333 02.1	25.5	24 51.3	05.3
22	121 24.3	190 05.0	31.7	219 02.2	09.3	348 04.8	25.6	39 53.7	05.3
23	136 26.8	205 04.4	31.5	234 02.9	09.4	3 07.6	25.7	54 56.2	05.3
22 00	151 29.2	220 03.8	S20 31.2	249 03.7	S23 09.6	18 10.3	N18 25.8	69 58.7	N22 05.3
01	166 31.7	235 03.1	31.0	264 04.4	09.7	33 13.1	25.9	85 01.1	05.3
02	181 34.2	250 02.5	30.7	279 05.1	09.8	48 15.8	25.9	100 03.6	05.3
03	196 36.6	265 01.8	30.4	294 05.8	09.9	63 18.6	26.0	115 06.1	05.3
04	211 39.1	280 01.2	30.2	309 06.6	10.1	78 21.4	26.1	130 08.5	05.3
05	226 41.6	295 00.6	29.9	324 07.3	10.2	93 24.1	26.2	145 11.0	05.3
06	241 44.0	309 59.9	S20 29.7	339 08.0	S23 10.3	108 26.9	N18 26.3	160 13.5	N22 05.3
S 07	256 46.5	324 59.3	29.4	354 08.7	10.4	123 29.6	26.3	175 15.9	05.4
08	271 49.0	339 58.7	29.1	9 09.5	10.5	138 32.4	26.4	190 18.4	05.4
A 09	286 51.4	354 58.0	28.9	24 10.2	10.7	153 35.1	26.5	205 20.9	05.4
T 10	301 53.9	9 57.4	28.6	39 10.9	10.8	168 37.9	26.6	220 23.3	05.4
U 11	316 56.3	24 56.7	28.3	54 11.6	10.9	183 40.6	26.6	235 25.8	05.4
R 12	331 58.8	39 56.1	S20 28.1	69 12.4	S23 11.0	198 43.4	N18 26.7	250 28.2	N22 05.4
D 13	347 01.3	54 55.5	27.8	84 13.1	11.1	213 46.1	26.8	265 30.7	05.4
A 14	2 03.7	69 54.8	27.5	99 13.8	11.2	228 48.9	26.9	280 33.2	05.4
Y 15	17 06.2	84 54.2	27.2	114 14.5	11.4	243 51.6	27.0	295 35.6	05.4
16	32 08.7	99 53.6	27.0	129 15.3	11.5	258 54.4	27.0	310 38.1	05.4
17	47 11.1	114 52.9	26.7	144 16.0	11.6	273 57.1	27.1	325 40.6	05.4
18	62 13.6	129 52.3	S20 26.4	159 16.7	S23 11.7	288 59.9	N18 27.2	340 43.0	N22 05.4
19	77 16.1	144 51.6	26.1	174 17.4	11.8	304 02.6	27.3	355 45.5	05.5
20	92 18.5	159 51.0	25.9	189 18.2	11.9	319 05.4	27.4	10 47.9	05.5
21	107 21.0	174 50.4	25.6	204 18.9	12.1	334 08.1	27.4	25 50.4	05.5
22	122 23.4	189 49.7	25.3	219 19.6	12.2	349 10.9	27.5	40 52.9	05.5
23	137 25.9	204 49.1	25.0	234 20.4	12.3	4 13.6	27.6	55 55.3	05.5
23 00	152 28.4	219 48.5	S20 24.7	249 21.1	S23 12.4	19 16.4	N18 27.7	70 57.8	N22 05.5
01	167 30.8	234 47.8	24.5	264 21.8	12.5	34 19.1	27.7	86 00.3	05.5
02	182 33.3	249 47.2	24.2	279 22.5	12.6	49 21.9	27.8	101 02.7	05.5
03	197 35.8	264 46.5	23.9	294 23.3	12.8	64 24.6	27.9	116 05.2	05.5
04	212 38.2	279 45.9	23.6	309 24.0	12.9	79 27.4	28.0	131 07.6	05.5
05	227 40.7	294 45.3	23.3	324 24.7	13.0	94 30.1	28.1	146 10.1	05.5
06	242 43.2	309 44.6	S20 23.0	339 25.4	S23 13.1	109 32.9	N18 28.1	161 12.6	N22 05.6
07	257 45.6	324 44.0	22.8	354 26.2	13.2	124 35.6	28.2	176 15.0	05.6
08	272 48.1	339 43.4	22.5	9 26.9	13.3	139 38.4	28.3	191 17.5	05.6
S 09	287 50.6	354 42.7	22.2	24 27.6	13.4	154 41.1	28.4	206 19.9	05.6
U 10	302 53.0	9 42.1	21.9	39 28.3	13.5	169 43.9	28.4	221 22.4	05.6
N 11	317 55.5	24 41.4	21.6	54 29.1	13.7	184 46.6	28.5	236 24.9	05.6
D 12	332 57.9	39 40.8	S20 21.3	69 29.8	S23 13.8	199 49.4	N18 28.6	251 27.3	N22 05.6
A 13	348 00.4	54 40.2	21.0	84 30.5	13.9	214 52.1	28.7	266 29.8	05.6
Y 14	3 02.9	69 39.5	20.7	99 31.2	14.0	229 54.9	28.8	281 32.2	05.6
15	18 05.3	84 38.9	20.4	114 32.0	14.1	244 57.6	28.8	296 34.7	05.6
16	33 07.8	99 38.3	20.1	129 32.7	14.2	260 00.4	28.9	311 37.1	05.6
17	48 10.3	114 37.6	19.8	144 33.4	14.3	275 03.1	29.0	326 39.6	05.7
18	63 12.7	129 37.0	S20 19.5	159 34.1	S23 14.4	290 05.9	N18 29.1	341 42.1	N22 05.7
19	78 15.2	144 36.4	19.2	174 34.9	14.5	305 08.6	29.1	356 44.5	05.7
20	93 17.7	159 35.7	18.9	189 35.6	14.6	320 11.4	29.2	11 47.0	05.7
21	108 20.1	174 35.1	18.6	204 36.3	14.8	335 14.1	29.3	26 49.4	05.7
22	123 22.6	189 34.4	18.3	219 37.0	14.9	350 16.8	29.4	41 51.9	05.7
23	138 25.1	204 33.8	18.0	234 37.8	15.0	5 19.6	29.5	56 54.3	05.7
Mer. Pass.	h m 13 51.8	*v* −0.6	*d* 0.3	*v* 0.7	*d* 0.1	*v* 2.8	*d* 0.1	*v* 2.5	*d* 0.0

STARS Name	SHA ° ′	Dec ° ′
Acamar	315 24.5	S40 17.8
Achernar	335 32.9	S57 13.6
Acrux	173 18.1	S63 06.8
Adhara	255 18.7	S28 58.7
Aldebaran	290 58.7	N16 30.9
Alioth	166 27.2	N55 56.4
Alkaid	153 04.9	N49 17.6
Al Na'ir	27 54.2	S46 56.9
Alnilam	275 54.5	S 1 12.1
Alphard	218 03.8	S 8 40.4
Alphecca	126 17.8	N26 42.0
Alpheratz	357 52.3	N29 06.3
Altair	62 16.4	N 8 52.4
Ankaa	353 23.9	S42 17.6
Antares	112 36.3	S26 26.3
Arcturus	146 03.0	N19 09.8
Atria	107 45.7	S69 01.8
Avior	234 20.9	S59 31.3
Bellatrix	278 40.6	N 6 21.1
Betelgeuse	271 09.9	N 7 24.4
Canopus	263 59.5	S52 42.1
Capella	280 46.4	N46 00.2
Deneb	49 37.5	N45 17.2
Denebola	182 41.6	N14 33.2
Diphda	349 04.2	S17 58.4
Dubhe	194 00.7	N61 44.0
Elnath	278 22.8	N28 36.7
Eltanin	90 50.2	N51 29.0
Enif	33 55.4	N 9 53.1
Fomalhaut	15 33.2	S29 36.6
Gacrux	172 09.8	S57 07.7
Gienah	176 00.4	S17 33.5
Hadar	148 59.4	S60 23.1
Hamal	328 10.2	N23 28.6
Kaus Aust.	83 54.8	S34 23.0
Kochab	137 18.9	N74 08.3
Markab	13 46.8	N15 13.1
Menkar	314 23.6	N 4 06.0
Menkent	148 17.1	S36 23.0
Miaplacidus	221 40.8	S69 43.8
Mirfak	308 52.2	N49 52.5
Nunki	76 08.6	S26 17.7
Peacock	53 32.4	S56 43.6
Pollux	243 37.3	N28 01.2
Procyon	245 08.0	N 5 13.0
Rasalhague	96 14.1	N12 33.2
Regulus	207 51.8	N11 57.1
Rigel	281 19.8	S 8 12.0
Rigil Kent.	140 02.9	S60 50.6
Sabik	102 22.0	S15 43.8
Schedar	349 50.5	N56 33.3
Shaula	96 33.1	S37 06.3
Sirius	258 40.7	S16 43.3
Spica	158 39.6	S11 10.7
Suhail	222 58.1	S43 26.7
Vega	80 44.7	N38 46.9
Zuben'ubi	137 14.4	S16 03.3

	SHA ° ′	Mer. Pass. h m
Venus	68 34.5	9 20
Mars	97 34.4	7 23
Jupiter	226 41.1	22 43
Saturn	278 29.4	19 17

	UT (d h)	SUN GHA ° ′	SUN Dec ° ′	MOON GHA ° ′	MOON v ′	MOON Dec ° ′	MOON d ′	MOON HP ′
	21 00	176 34.3	S10 47.2	304 59.6	10.3	S 7 16.0	15.3	59.9
	01	191 34.4	46.3	319 28.9	10.3	7 31.3	15.3	59.9
	02	206 34.4	45.4	333 58.2	10.2	7 46.6	15.3	59.9
	03	221 34.5	. . 44.5	348 27.4	10.3	8 01.9	15.1	59.9
	04	236 34.6	43.6	2 56.7	10.2	8 17.0	15.2	59.9
	05	251 34.7	42.7	17 25.9	10.2	8 32.2	15.0	59.9
	06	266 34.7	S10 41.8	31 55.1	10.1	S 8 47.2	15.1	59.9
	07	281 34.8	40.9	46 24.2	10.2	9 02.3	14.9	59.9
	08	296 34.9	40.0	60 53.4	10.0	9 17.2	14.9	59.8
F	09	311 34.9	. . 39.1	75 22.4	10.1	9 32.1	14.9	59.8
R	10	326 35.0	38.2	89 51.5	10.0	9 47.0	14.7	59.8
I	11	341 35.1	37.3	104 20.5	10.0	10 01.7	14.8	59.8
D	12	356 35.2	S10 36.4	118 49.5	10.0	S10 16.5	14.6	59.8
A	13	11 35.2	35.5	133 18.5	9.9	10 31.1	14.6	59.8
Y	14	26 35.3	34.6	147 47.4	9.9	10 45.7	14.5	59.8
	15	41 35.4	. . 33.7	162 16.3	9.8	11 00.2	14.5	59.8
	16	56 35.4	32.8	176 45.1	9.8	11 14.7	14.4	59.7
	17	71 35.5	31.9	191 13.9	9.8	11 29.1	14.3	59.7
	18	86 35.6	S10 31.0	205 42.7	9.7	S11 43.4	14.2	59.7
	19	101 35.7	30.1	220 11.4	9.7	11 57.6	14.2	59.7
	20	116 35.7	29.2	234 40.1	9.7	12 11.8	14.1	59.7
	21	131 35.8	. . 28.3	249 08.8	9.6	12 25.9	14.0	59.7
	22	146 35.9	27.4	263 37.4	9.5	12 39.9	13.9	59.7
	23	161 36.0	26.5	278 05.9	9.6	12 53.8	13.9	59.6
	22 00	176 36.0	S10 25.6	292 34.5	9.4	S13 07.7	13.7	59.6
	01	191 36.1	24.7	307 02.9	9.5	13 21.4	13.7	59.6
	02	206 36.2	23.8	321 31.4	9.4	13 35.1	13.6	59.6
	03	221 36.3	. . 22.8	335 59.8	9.3	13 48.7	13.5	59.6
	04	236 36.3	21.9	350 28.1	9.3	14 02.2	13.4	59.6
	05	251 36.4	21.0	4 56.4	9.2	14 15.6	13.4	59.5
	06	266 36.5	S10 20.1	19 24.6	9.2	S14 29.0	13.2	59.5
	07	281 36.6	19.2	33 52.8	9.2	14 42.2	13.2	59.5
S	08	296 36.7	18.3	48 21.0	9.1	14 55.4	13.0	59.5
A	09	311 36.7	. . 17.4	62 49.1	9.1	15 08.4	13.0	59.5
T	10	326 36.8	16.5	77 17.2	9.0	15 21.4	12.9	59.5
U	11	341 36.9	15.6	91 45.2	8.9	15 34.3	12.8	59.4
R	12	356 37.0	S10 14.7	106 13.1	8.9	S15 47.1	12.6	59.4
D	13	11 37.1	13.8	120 41.0	8.9	15 59.7	12.6	59.4
A	14	26 37.1	12.8	135 08.9	8.8	16 12.3	12.5	59.4
Y	15	41 37.2	. . 11.9	149 36.7	8.7	16 24.8	12.3	59.4
	16	56 37.3	11.0	164 04.4	8.7	16 37.1	12.3	59.3
	17	71 37.4	10.1	178 32.1	8.7	16 49.4	12.2	59.3
	18	86 37.5	S10 09.2	192 59.8	8.6	S17 01.6	12.0	59.3
	19	101 37.5	08.3	207 27.4	8.5	17 13.6	12.0	59.3
	20	116 37.6	07.4	221 54.9	8.5	17 25.6	11.8	59.3
	21	131 37.7	. . 06.5	236 22.4	8.4	17 37.4	11.7	59.3
	22	146 37.8	05.5	250 49.8	8.4	17 49.1	11.6	59.2
	23	161 37.9	04.6	265 17.2	8.4	18 00.7	11.5	59.2
	23 00	176 37.9	S10 03.7	279 44.6	8.2	S18 12.2	11.4	59.2
	01	191 38.0	02.8	294 11.8	8.3	18 23.6	11.3	59.2
	02	206 38.1	01.9	308 39.1	8.1	18 34.9	11.1	59.2
	03	221 38.2	. . 01.0	323 06.2	8.1	18 46.0	11.1	59.1
	04	236 38.3	10 00.1	337 33.3	8.1	18 57.1	10.9	59.1
	05	251 38.4	9 59.1	352 00.4	8.0	19 08.0	10.8	59.1
	06	266 38.4	S 9 58.2	6 27.4	8.0	S19 18.8	10.7	59.1
	07	281 38.5	57.3	20 54.4	7.9	19 29.5	10.5	59.1
	08	296 38.6	56.4	35 21.3	7.8	19 40.0	10.4	59.0
S	09	311 38.7	. . 55.5	49 48.1	7.8	19 50.4	10.3	59.0
U	10	326 38.8	54.6	64 14.9	7.8	20 00.7	10.2	59.0
N	11	341 38.9	53.7	78 41.7	7.7	20 10.9	10.1	59.0
D	12	356 38.9	S 9 52.7	93 08.4	7.6	S20 21.0	9.9	59.0
A	13	11 39.0	51.8	107 35.0	7.6	20 30.9	9.8	58.9
Y	14	26 39.1	50.9	122 01.6	7.5	20 40.7	9.6	58.9
	15	41 39.2	. . 50.0	136 28.1	7.5	20 50.3	9.6	58.9
	16	56 39.3	49.1	150 54.6	7.5	20 59.9	9.4	58.9
	17	71 39.4	48.2	165 21.1	7.3	21 09.3	9.2	58.9
	18	86 39.5	S 9 47.2	179 47.4	7.4	S21 18.5	9.2	58.8
	19	101 39.5	46.3	194 13.8	7.3	21 27.7	8.9	58.8
	20	116 39.6	45.4	208 40.1	7.2	21 36.6	8.9	58.8
	21	131 39.7	. . 44.5	223 06.3	7.2	21 45.5	8.7	58.8
	22	146 39.8	43.6	237 32.5	7.1	21 54.2	8.6	58.8
	23	161 39.9	42.6	251 58.6	7.1	S22 02.8	8.4	58.7
		SD 16.2	*d* 0.9	SD 16.3		16.2		16.1

Lat. °	Twilight Naut. h m	Twilight Civil h m	Sunrise h m	Moonrise 21 h m	Moonrise 22 h m	Moonrise 23 h m	Moonrise 24 h m
N 72	05 45	07 04	08 17	25 48	01 48	■	■
N 70	05 47	06 57	08 02	25 19	01 19	04 22	■
68	05 48	06 52	07 51	24 58	00 58	03 22	■
66	05 48	06 48	07 41	24 41	00 41	02 48	05 17
64	05 49	06 44	07 33	24 28	00 28	02 24	04 25
62	05 49	06 40	07 26	24 16	00 16	02 04	03 53
60	05 49	06 37	07 20	24 07	00 07	01 49	03 29
N 58	05 49	06 34	07 15	23 59	25 35	01 35	03 10
56	05 49	06 32	07 10	23 51	25 24	01 24	02 55
54	05 49	06 30	07 06	23 45	25 14	01 14	02 41
52	05 48	06 27	07 02	23 39	25 05	01 05	02 29
50	05 48	06 25	06 58	23 34	24 58	00 58	02 19
45	05 47	06 21	06 51	23 23	24 41	00 41	01 57
N 40	05 45	06 17	06 44	23 13	24 27	00 27	01 40
35	05 44	06 13	06 39	23 06	24 16	00 16	01 25
30	05 42	06 10	06 34	22 59	24 06	00 06	01 13
20	05 37	06 03	06 25	22 47	23 49	24 51	00 51
N 10	05 32	05 56	06 17	22 37	23 34	24 33	00 33
0	05 25	05 49	06 10	22 27	23 20	24 16	00 16
S 10	05 16	05 41	06 03	22 18	23 07	23 58	24 53
20	05 06	05 32	05 55	22 08	22 52	23 40	24 32
30	04 52	05 21	05 45	21 56	22 36	23 19	24 08
35	04 43	05 14	05 40	21 50	22 26	23 07	23 54
40	04 32	05 06	05 34	21 43	22 15	22 53	23 38
45	04 18	04 56	05 27	21 34	22 03	22 37	23 19
S 50	04 01	04 43	05 18	21 24	21 47	22 17	22 55
52	03 52	04 38	05 14	21 19	21 40	22 07	22 43
54	03 43	04 31	05 09	21 14	21 32	21 57	22 30
56	03 32	04 24	05 04	21 08	21 24	21 45	22 15
58	03 18	04 15	04 59	21 02	21 14	21 31	21 57
S 60	03 03	04 06	04 53	20 55	21 03	21 15	21 36

Lat. °	Sunset h m	Twilight Civil h m	Twilight Naut. h m	Moonset 21 h m	Moonset 22 h m	Moonset 23 h m	Moonset 24 h m
N 72	16 12	17 25	18 44	07 53	07 05	■	■
N 70	16 27	17 32	18 42	08 07	07 36	06 26	■
68	16 38	17 37	18 41	08 19	07 59	07 27	■
66	16 47	17 41	18 40	08 28	08 17	08 02	07 30
64	16 55	17 45	18 40	08 36	08 32	08 28	08 23
62	17 02	17 48	18 40	08 43	08 45	08 48	08 55
60	17 08	17 51	18 39	08 49	08 55	09 04	09 20
N 58	17 13	17 54	18 39	08 55	09 05	09 18	09 39
56	17 18	17 56	18 39	09 00	09 13	09 31	09 55
54	17 22	17 58	18 39	09 04	09 20	09 41	10 09
52	17 26	18 00	18 40	09 08	09 27	09 51	10 21
50	17 30	18 02	18 40	09 11	09 33	09 59	10 32
45	17 37	18 07	18 41	09 19	09 46	10 17	10 55
N 40	17 43	18 11	18 42	09 26	09 57	10 32	11 13
35	17 49	18 14	18 44	09 31	10 06	10 44	11 28
30	17 54	18 18	18 46	09 36	10 14	10 55	11 42
20	18 02	18 24	18 50	09 45	10 28	11 14	12 04
N 10	18 10	18 31	18 56	09 53	10 40	11 31	12 24
0	18 17	18 38	19 02	10 00	10 52	11 46	12 43
S 10	18 24	18 46	19 10	10 07	11 04	12 02	13 01
20	18 32	18 55	19 21	10 15	11 16	12 19	13 21
30	18 41	19 06	19 35	10 24	11 31	12 38	13 44
35	18 46	19 13	19 44	10 29	11 39	12 49	13 58
40	18 53	19 21	19 54	10 35	11 49	13 02	14 13
45	19 00	19 30	20 08	10 42	12 00	13 17	14 32
S 50	19 08	19 42	20 25	10 50	12 14	13 36	14 55
52	19 12	19 48	20 33	10 54	12 20	13 45	15 07
54	19 16	19 54	20 42	10 58	12 27	13 56	15 20
56	19 21	20 02	20 53	11 02	12 35	14 07	15 34
58	19 27	20 10	21 06	11 08	12 44	14 20	15 52
S 60	19 33	20 19	21 21	11 14	12 55	14 36	16 13

Day	SUN Eqn. of Time 00h m s	SUN Eqn. of Time 12h m s	SUN Mer. Pass. h m	MOON Mer. Pass. Upper h m	MOON Mer. Pass. Lower h m	MOON Age d	MOON Phase %
21	13 43	13 40	12 14	03 48	16 13	20	74
22	13 36	13 32	12 14	04 40	17 06	21	64
23	13 28	13 24	12 13	05 33	18 01	22	52

UT	ARIES	VENUS −4.1		MARS +1.0		JUPITER −2.5		SATURN −0.1	
	GHA	GHA	Dec	GHA	Dec	GHA	Dec	GHA	Dec
d h	° ′	° ′	° ′	° ′	° ′	° ′	° ′	° ′	° ′
24 00 (MONDAY)	153 27.5	219 33.2	S20 17.7	249 38.5	S23 15.1	20 22.3	N18 29.5	71 56.8	N22 05.7
01	168 30.0	234 32.5	17.4	264 39.2	15.2	35 25.1	29.6	86 59.3	05.7
02	183 32.4	249 31.9	17.1	279 39.9	15.3	50 27.8	29.7	102 01.7	05.7
03	198 34.9	264 31.3	. . 16.8	294 40.7	. . 15.4	65 30.6	. . 29.8	117 04.2	. . 05.8
04	213 37.4	279 30.6	16.5	309 41.4	15.5	80 33.3	29.8	132 06.6	05.8
05	228 39.8	294 30.0	16.2	324 42.1	15.6	95 36.1	29.9	147 09.1	05.8
06	243 42.3	309 29.3	S20 15.9	339 42.8	S23 15.7	110 38.8	N18 30.0	162 11.5	N22 05.8
07	258 44.8	324 28.7	15.5	354 43.6	15.8	125 41.6	30.1	177 14.0	05.8
08	273 47.2	339 28.1	15.2	9 44.3	15.9	140 44.3	30.1	192 16.4	05.8
09	288 49.7	354 27.4	. . 14.9	24 45.0	. . 16.0	155 47.0	. . 30.2	207 18.9	. . 05.8
10	303 52.2	9 26.8	14.6	39 45.7	16.1	170 49.8	30.3	222 21.4	05.8
11	318 54.6	24 26.2	14.3	54 46.5	16.2	185 52.5	30.4	237 23.8	05.8
12	333 57.1	39 25.5	S20 14.0	69 47.2	S23 16.3	200 55.3	N18 30.4	252 26.3	N22 05.8
13	348 59.6	54 24.9	13.7	84 47.9	16.5	215 58.0	30.5	267 28.7	05.8
14	4 02.0	69 24.3	13.3	99 48.6	16.6	231 00.8	30.6	282 31.2	05.9
15	19 04.5	84 23.6	. . 13.0	114 49.4	. . 16.7	246 03.5	. . 30.7	297 33.6	. . 05.9
16	34 06.9	99 23.0	12.7	129 50.1	16.8	261 06.2	30.7	312 36.1	05.9
17	49 09.4	114 22.3	12.4	144 50.8	16.9	276 09.0	30.8	327 38.5	05.9
18	64 11.9	129 21.7	S20 12.1	159 51.5	S23 17.0	291 11.7	N18 30.9	342 41.0	N22 05.9
19	79 14.3	144 21.1	11.7	174 52.3	17.1	306 14.5	31.0	357 43.4	05.9
20	94 16.8	159 20.4	11.4	189 53.0	17.2	321 17.2	31.0	12 45.9	05.9
21	109 19.3	174 19.8	. . 11.1	204 53.7	. . 17.3	336 19.9	. . 31.1	27 48.3	. . 05.9
22	124 21.7	189 19.2	10.8	219 54.4	17.4	351 22.7	31.2	42 50.8	05.9
23	139 24.2	204 18.5	10.4	234 55.2	17.5	6 25.4	31.3	57 53.2	05.9
25 00 (TUESDAY)	154 26.7	219 17.9	S20 10.1	249 55.9	S23 17.6	21 28.2	N18 31.3	72 55.7	N22 06.0
01	169 29.1	234 17.3	09.8	264 56.6	17.7	36 30.9	31.4	87 58.1	06.0
02	184 31.6	249 16.6	09.4	279 57.3	17.8	51 33.7	31.5	103 00.6	06.0
03	199 34.0	264 16.0	. . 09.1	294 58.1	. . 17.9	66 36.4	. . 31.6	118 03.0	. . 06.0
04	214 36.5	279 15.3	08.8	309 58.8	18.0	81 39.1	31.6	133 05.5	06.0
05	229 39.0	294 14.7	08.4	324 59.5	18.1	96 41.9	31.7	148 07.9	06.0
06	244 41.4	309 14.1	S20 08.1	340 00.2	S23 18.2	111 44.6	N18 31.8	163 10.4	N22 06.0
07	259 43.9	324 13.4	07.8	355 01.0	18.3	126 47.3	31.9	178 12.8	06.0
08	274 46.4	339 12.8	07.4	10 01.7	18.4	141 50.1	31.9	193 15.3	06.0
09	289 48.8	354 12.2	. . 07.1	25 02.4	. . 18.5	156 52.8	. . 32.0	208 17.7	. . 06.0
10	304 51.3	9 11.5	06.8	40 03.1	18.6	171 55.6	32.1	223 20.2	06.0
11	319 53.8	24 10.9	06.4	55 03.9	18.7	186 58.3	32.2	238 22.6	06.1
12	334 56.2	39 10.3	S20 06.1	70 04.6	S23 18.8	202 01.0	N18 32.2	253 25.1	N22 06.1
13	349 58.7	54 09.6	05.7	85 05.3	18.9	217 03.8	32.3	268 27.5	06.1
14	5 01.2	69 09.0	05.4	100 06.0	18.9	232 06.5	32.4	283 30.0	06.1
15	20 03.6	84 08.4	. . 05.1	115 06.8	. . 19.0	247 09.3	. . 32.5	298 32.4	. . 06.1
16	35 06.1	99 07.7	04.7	130 07.5	19.1	262 12.0	32.5	313 34.9	06.1
17	50 08.5	114 07.1	04.4	145 08.2	19.2	277 14.7	32.6	328 37.3	06.1
18	65 11.0	129 06.5	S20 04.0	160 08.9	S23 19.3	292 17.5	N18 32.7	343 39.8	N22 06.1
19	80 13.5	144 05.8	03.7	175 09.7	19.4	307 20.2	32.8	358 42.2	06.1
20	95 15.9	159 05.2	03.3	190 10.4	19.5	322 22.9	32.8	13 44.7	06.1
21	110 18.4	174 04.6	. . 03.0	205 11.1	. . 19.6	337 25.7	. . 32.9	28 47.1	. . 06.2
22	125 20.9	189 03.9	02.6	220 11.8	19.7	352 28.4	33.0	43 49.6	06.2
23	140 23.3	204 03.3	02.3	235 12.6	19.8	7 31.1	33.1	58 52.0	06.2
26 00 (WEDNESDAY)	155 25.8	219 02.7	S20 01.9	250 13.3	S23 19.9	22 33.9	N18 33.1	73 54.5	N22 06.2
01	170 28.3	234 02.0	01.6	265 14.0	20.0	37 36.6	33.2	88 56.9	06.2
02	185 30.7	249 01.4	01.2	280 14.7	20.1	52 39.4	33.3	103 59.3	06.2
03	200 33.2	264 00.7	. . 00.9	295 15.5	. . 20.2	67 42.1	. . 33.3	119 01.8	. . 06.2
04	215 35.7	279 00.1	00.5	310 16.2	20.3	82 44.8	33.4	134 04.2	06.2
05	230 38.1	293 59.5	20 00.1	325 16.9	20.4	97 47.6	33.5	149 06.7	06.2
06	245 40.6	308 58.8	S19 59.8	340 17.6	S23 20.4	112 50.3	N18 33.6	164 09.1	N22 06.2
07	260 43.0	323 58.2	59.4	355 18.4	20.5	127 53.0	33.6	179 11.6	06.3
08	275 45.5	338 57.6	59.1	10 19.1	20.6	142 55.8	33.7	194 14.0	06.3
09	290 48.0	353 56.9	. . 58.7	25 19.8	. . 20.7	157 58.5	. . 33.8	209 16.5	. . 06.3
10	305 50.4	8 56.3	58.3	40 20.5	20.8	173 01.2	33.9	224 18.9	06.3
11	320 52.9	23 55.7	58.0	55 21.2	20.9	188 04.0	33.9	239 21.3	06.3
12	335 55.4	38 55.0	S19 57.6	70 22.0	S23 21.0	203 06.7	N18 34.0	254 23.8	N22 06.3
13	350 57.8	53 54.4	57.3	85 22.7	21.1	218 09.4	34.1	269 26.2	06.3
14	6 00.3	68 53.8	56.9	100 23.4	21.2	233 12.2	34.1	284 28.7	06.3
15	21 02.8	83 53.1	. . 56.5	115 24.1	. . 21.3	248 14.9	. . 34.2	299 31.1	. . 06.3
16	36 05.2	98 52.5	56.2	130 24.9	21.3	263 17.6	34.3	314 33.6	06.3
17	51 07.7	113 51.9	55.8	145 25.6	21.4	278 20.4	34.4	329 36.0	06.4
18	66 10.2	128 51.3	S19 55.4	160 26.3	S23 21.5	293 23.1	N18 34.4	344 38.4	N22 06.4
19	81 12.6	143 50.6	55.0	175 27.0	21.6	308 25.8	34.5	359 40.9	06.4
20	96 15.1	158 50.0	54.7	190 27.8	21.7	323 28.5	34.6	14 43.3	06.4
21	111 17.5	173 49.4	. . 54.3	205 28.5	. . 21.8	338 31.3	. . 34.6	29 45.8	. . 06.4
22	126 20.0	188 48.7	53.9	220 29.2	21.9	353 34.0	34.7	44 48.2	06.4
23	141 22.5	203 48.1	53.6	235 29.9	22.0	8 36.7	34.8	59 50.7	06.4
Mer. Pass.	h m 13 40.0	*v* −0.6	*d* 0.3	*v* 0.7	*d* 0.1	*v* 2.7	*d* 0.1	*v* 2.4	*d* 0.0

STARS

Name	SHA	Dec
	° ′	° ′
Acamar	315 24.5	S40 17.8
Achernar	335 33.0	S57 13.6
Acrux	173 18.1	S63 06.8
Adhara	255 18.7	S28 58.7
Aldebaran	290 58.7	N16 30.9
Alioth	166 27.1	N55 56.4
Alkaid	153 04.8	N49 17.6
Al Na'ir	27 54.2	S46 56.9
Alnilam	275 54.5	S 1 12.1
Alphard	218 03.8	S 8 40.4
Alphecca	126 17.8	N26 42.0
Alpheratz	357 52.3	N29 06.3
Altair	62 16.4	N 8 52.3
Ankaa	353 23.9	S42 17.6
Antares	112 36.3	S26 26.3
Arcturus	146 02.9	N19 09.8
Atria	107 45.6	S69 01.8
Avior	234 20.9	S59 31.3
Bellatrix	278 40.6	N 6 21.1
Betelgeuse	271 10.0	N 7 24.4
Canopus	263 59.5	S52 42.1
Capella	280 46.4	N46 00.2
Deneb	49 37.5	N45 17.2
Denebola	182 41.6	N14 33.2
Diphda	349 04.2	S17 58.4
Dubhe	194 00.7	N61 44.0
Elnath	278 22.8	N28 36.7
Eltanin	90 50.1	N51 28.9
Enif	33 55.4	N 9 53.1
Fomalhaut	15 33.2	S29 36.6
Gacrux	172 09.7	S57 07.7
Gienah	176 00.4	S17 33.5
Hadar	148 59.3	S60 23.1
Hamal	328 10.2	N23 28.6
Kaus Aust.	83 54.8	S34 23.0
Kochab	137 18.9	N74 08.3
Markab	13 46.8	N15 13.1
Menkar	314 23.7	N 4 06.0
Menkent	148 17.1	S36 23.0
Miaplacidus	221 40.8	S69 43.8
Mirfak	308 52.2	N49 52.5
Nunki	76 08.6	S26 17.7
Peacock	53 32.4	S56 43.5
Pollux	243 37.3	N28 01.2
Procyon	245 08.0	N 5 13.0
Rasalhague	96 14.1	N12 33.2
Regulus	207 51.8	N11 57.1
Rigel	281 19.8	S 8 12.0
Rigil Kent.	140 02.9	S60 50.6
Sabik	102 21.9	S15 43.8
Schedar	349 50.5	N56 33.3
Shaula	96 33.1	S37 06.3
Sirius	258 40.7	S16 43.3
Spica	158 39.6	S11 10.7
Suhail	222 58.1	S43 26.8
Vega	80 44.7	N38 46.9
Zuben'ubi	137 14.3	S16 03.3

	SHA	Mer. Pass.
	° ′	h m
Venus	64 51.2	9 23
Mars	95 29.2	7 20
Jupiter	227 01.5	22 30
Saturn	278 29.0	19 05

	UT	SUN GHA	SUN Dec	MOON GHA	v	MOON Dec	d	HP
	d h	° ′	° ′	° ′	′	° ′	′	′
	24 00	176 40.0	S 9 41.7	266 24.7	7.1	S22 11.2	8.3	58.7
	01	191 40.1	40.8	280 50.8	7.0	22 19.5	8.2	58.7
	02	206 40.2	39.9	295 16.8	6.9	22 27.7	8.0	58.7
	03	221 40.2	. . 39.0	309 42.7	6.9	22 35.7	7.9	58.6
	04	236 40.3	38.0	324 08.6	6.9	22 43.6	7.7	58.6
	05	251 40.4	37.1	338 34.5	6.8	22 51.3	7.6	58.6
	06	266 40.5	S 9 36.2	353 00.3	6.8	S22 58.9	7.4	58.6
	07	281 40.6	35.3	7 26.1	6.8	23 06.3	7.3	58.6
	08	296 40.7	34.4	21 51.9	6.7	23 13.6	7.1	58.5
MONDAY	09	311 40.8	. . 33.4	36 17.6	6.6	23 20.7	7.0	58.5
	10	326 40.9	32.5	50 43.2	6.7	23 27.7	6.9	58.5
	11	341 41.0	31.6	65 08.9	6.5	23 34.6	6.7	58.5
	12	356 41.1	S 9 30.7	79 34.4	6.6	S23 41.3	6.5	58.5
	13	11 41.1	29.7	94 00.0	6.5	23 47.8	6.4	58.4
	14	26 41.2	28.8	108 25.5	6.5	23 54.2	6.2	58.4
	15	41 41.3	. . 27.9	122 51.0	6.4	24 00.4	6.1	58.4
	16	56 41.4	27.0	137 16.4	6.5	24 06.5	6.0	58.4
	17	71 41.5	26.0	151 41.9	6.3	24 12.5	5.7	58.3
	18	86 41.6	S 9 25.1	166 07.2	6.4	S24 18.2	5.7	58.3
	19	101 41.7	24.2	180 32.6	6.3	24 23.9	5.4	58.3
	20	116 41.8	23.3	194 57.9	6.3	24 29.3	5.4	58.3
	21	131 41.9	. . 22.3	209 23.2	6.3	24 34.7	5.1	58.3
	22	146 42.0	21.4	223 48.5	6.3	24 39.8	5.0	58.2
	23	161 42.1	20.5	238 13.8	6.2	24 44.8	4.9	58.2
	25 00	176 42.2	S 9 19.6	252 39.0	6.2	S24 49.7	4.7	58.2
	01	191 42.3	18.6	267 04.2	6.2	24 54.4	4.5	58.2
	02	206 42.4	17.7	281 29.4	6.2	24 58.9	4.4	58.2
	03	221 42.5	. . 16.8	295 54.6	6.1	25 03.3	4.2	58.1
	04	236 42.5	15.9	310 19.7	6.1	25 07.5	4.0	58.1
	05	251 42.6	14.9	324 44.8	6.2	25 11.5	3.9	58.1
	06	266 42.7	S 9 14.0	339 10.0	6.1	S25 15.4	3.8	58.1
	07	281 42.8	13.1	353 35.1	6.1	25 19.2	3.6	58.0
TUESDAY	08	296 42.9	12.2	8 00.2	6.1	25 22.8	3.4	58.0
	09	311 43.0	. . 11.2	22 25.3	6.0	25 26.2	3.2	58.0
	10	326 43.1	10.3	36 50.3	6.1	25 29.4	3.1	58.0
	11	341 43.2	09.4	51 15.4	6.1	25 32.5	3.0	58.0
	12	356 43.3	S 9 08.4	65 40.5	6.0	S25 35.5	2.7	57.9
	13	11 43.4	07.5	80 05.5	6.1	25 38.2	2.7	57.9
	14	26 43.5	06.6	94 30.6	6.1	25 40.9	2.4	57.9
	15	41 43.6	. . 05.7	108 55.7	6.0	25 43.3	2.3	57.9
	16	56 43.7	04.7	123 20.7	6.1	25 45.6	2.1	57.8
	17	71 43.8	03.8	137 45.8	6.0	25 47.7	2.0	57.8
	18	86 43.9	S 9 02.9	152 10.8	6.1	S25 49.7	1.8	57.8
	19	101 44.0	01.9	166 35.9	6.1	25 51.5	1.7	57.8
	20	116 44.1	01.0	181 01.0	6.1	25 53.2	1.5	57.8
	21	131 44.2	9 00.1	195 26.1	6.1	25 54.7	1.3	57.7
	22	146 44.3	8 59.1	209 51.2	6.1	25 56.0	1.2	57.7
	23	161 44.4	58.2	224 16.3	6.1	25 57.2	1.0	57.7
	26 00	176 44.5	S 8 57.3	238 41.4	6.2	S25 58.2	0.8	57.7
	01	191 44.6	56.3	253 06.6	6.1	25 59.0	0.7	57.7
	02	206 44.7	55.4	267 31.7	6.2	25 59.7	0.5	57.6
	03	221 44.8	. . 54.5	281 56.9	6.2	26 00.2	0.4	57.6
	04	236 44.9	53.5	296 22.1	6.2	26 00.6	0.2	57.6
	05	251 45.0	52.6	310 47.3	6.3	26 00.8	0.1	57.6
	06	266 45.1	S 8 51.7	325 12.6	6.2	S26 00.9	0.2	57.5
WEDNESDAY	07	281 45.2	50.7	339 37.8	6.3	26 00.7	0.2	57.5
	08	296 45.3	49.8	354 03.1	6.4	26 00.5	0.5	57.5
	09	311 45.4	. . 48.9	8 28.5	6.3	26 00.0	0.5	57.5
	10	326 45.5	47.9	22 53.8	6.4	25 59.5	0.8	57.5
	11	341 45.6	47.0	37 19.2	6.4	25 58.7	0.9	57.4
	12	356 45.7	S 8 46.1	51 44.6	6.5	S25 57.8	1.0	57.4
	13	11 45.8	45.1	66 10.1	6.5	25 56.8	1.3	57.4
	14	26 45.9	44.2	80 35.6	6.5	25 55.5	1.3	57.4
	15	41 46.0	. . 43.3	95 01.1	6.6	25 54.2	1.5	57.4
	16	56 46.1	42.3	109 26.7	6.6	25 52.7	1.7	57.3
	17	71 46.2	41.4	123 52.3	6.6	25 51.0	1.9	57.3
	18	86 46.3	S 8 40.5	138 17.9	6.7	S25 49.1	1.9	57.3
	19	101 46.4	39.5	152 43.6	6.8	25 47.2	2.2	57.3
	20	116 46.6	38.6	167 09.4	6.8	25 45.0	2.3	57.2
	21	131 46.7	. . 37.7	181 35.2	6.8	25 42.7	2.4	57.2
	22	146 46.8	36.7	196 01.0	6.9	25 40.3	2.6	57.2
	23	161 46.9	35.8	210 26.9	7.0	S25 37.7	2.8	57.2
		SD 16.2	d 0.9	SD 15.9		15.8		15.6

Lat.	Twilight Naut.	Twilight Civil	Sunrise	Moonrise 24	Moonrise 25	Moonrise 26	Moonrise 27
°	h m	h m	h m	h m	h m	h m	h m
N 72	05 31	06 49	08 00	▬	▬	▬	▬
N 70	05 34	06 45	07 48	▬	▬	▬	▬
68	05 36	06 41	07 38	▬	▬	▬	▬
66	05 38	06 37	07 30	05 17	▬	▬	▬
64	05 39	06 34	07 23	04 25	06 33	▬	08 19
62	05 40	06 31	07 17	03 53	05 35	06 51	07 28
60	05 41	06 29	07 11	03 29	05 01	06 13	06 56
N 58	05 42	06 27	07 07	03 10	04 37	05 46	06 33
56	05 42	06 25	07 03	02 55	04 17	05 25	06 13
54	05 42	06 23	06 59	02 41	04 01	05 07	05 57
52	05 42	06 21	06 56	02 29	03 47	04 52	05 43
50	05 42	06 20	06 52	02 19	03 35	04 39	05 31
45	05 42	06 16	06 46	01 57	03 09	04 13	05 06
N 40	05 41	06 13	06 40	01 40	02 49	03 52	04 46
35	05 40	06 10	06 35	01 25	02 32	03 34	04 29
30	05 39	06 07	06 31	01 13	02 18	03 19	04 14
20	05 35	06 01	06 23	00 51	01 53	02 54	03 50
N 10	05 30	05 55	06 16	00 33	01 32	02 31	03 28
0	05 25	05 49	06 10	00 16	01 13	02 11	03 08
S 10	05 17	05 42	06 03	24 53	00 53	01 50	02 48
20	05 07	05 34	05 56	24 32	00 32	01 28	02 27
30	04 54	05 23	05 48	24 08	00 08	01 03	02 02
35	04 46	05 17	05 43	23 54	24 48	00 48	01 47
40	04 36	05 09	05 37	23 38	24 31	00 31	01 30
45	04 23	05 00	05 31	23 19	24 10	00 10	01 10
S 50	04 07	04 49	05 23	22 55	23 44	24 44	00 44
52	03 59	04 44	05 19	22 43	23 31	24 32	00 32
54	03 51	04 38	05 15	22 30	23 17	24 18	00 18
56	03 40	04 31	05 11	22 15	23 00	24 01	00 01
58	03 28	04 23	05 06	21 57	22 39	23 41	24 59
S 60	03 14	04 15	05 01	21 36	22 14	23 16	24 39

Lat.	Sunset	Twilight Civil	Twilight Naut.	Moonset 24	Moonset 25	Moonset 26	Moonset 27
°	h m	h m	h m	h m	h m	h m	h m
N 72	16 28	17 39	18 57	▬	▬	▬	▬
N 70	16 40	17 44	18 54	▬	▬	▬	▬
68	16 50	17 47	18 52	▬	▬	▬	▬
66	16 58	17 51	18 50	07 30	▬	▬	▬
64	17 05	17 54	18 49	08 23	08 15	▬	10 29
62	17 11	17 56	18 47	08 55	09 14	09 59	11 20
60	17 16	17 58	18 47	09 20	09 47	10 37	11 52
N 58	17 21	18 00	18 46	09 39	10 12	11 04	12 15
56	17 25	18 02	18 45	09 55	10 32	11 25	12 34
54	17 28	18 04	18 45	10 09	10 49	11 43	12 50
52	17 32	18 06	18 45	10 21	11 03	11 58	13 04
50	17 35	18 07	18 45	10 32	11 15	12 10	13 16
45	17 41	18 11	18 45	10 55	11 41	12 37	13 41
N 40	17 47	18 14	18 46	11 13	12 02	12 58	14 00
35	17 52	18 17	18 47	11 28	12 19	13 16	14 17
30	17 56	18 20	18 48	11 42	12 34	13 31	14 31
20	18 03	18 26	18 51	12 04	12 59	13 56	14 55
N 10	18 10	18 31	18 56	12 24	13 20	14 18	15 16
0	18 16	18 37	19 02	12 43	13 41	14 39	15 35
S 10	18 23	18 44	19 09	13 01	14 01	14 59	15 54
20	18 30	18 52	19 19	13 21	14 23	15 21	16 14
30	18 38	19 02	19 31	13 44	14 48	15 46	16 38
35	18 43	19 09	19 39	13 58	15 02	16 01	16 52
40	18 48	19 16	19 49	14 13	15 20	16 18	17 08
45	18 55	19 25	20 02	14 32	15 40	16 39	17 26
S 50	19 02	19 36	20 17	14 55	16 06	17 05	17 50
52	19 06	19 41	20 25	15 07	16 19	17 17	18 01
54	19 10	19 47	20 34	15 20	16 33	17 32	18 14
56	19 14	19 54	20 44	15 34	16 50	17 48	18 29
58	19 19	20 01	20 55	15 52	17 10	18 08	18 46
S 60	19 24	20 09	21 09	16 13	17 36	18 33	19 06

Day	SUN Eqn. of Time 00^h	SUN Eqn. of Time 12^h	SUN Mer. Pass.	MOON Mer. Pass. Upper	MOON Mer. Pass. Lower	Age	Phase
d	m s	m s	h m	h m	h m	d %	
24	13 20	13 16	12 13	06 29	18 58	23 41	
25	13 12	13 07	12 13	07 27	19 56	24 31	
26	13 02	12 57	12 13	08 25	20 53	25 21	

Day	UT d h	ARIES GHA ° ′	VENUS −4.1 GHA ° ′	VENUS Dec ° ′	MARS +1.0 GHA ° ′	MARS Dec ° ′	JUPITER −2.5 GHA ° ′	JUPITER Dec ° ′	SATURN +0.0 GHA ° ′	SATURN Dec ° ′
THURSDAY	27 00	156 24.9	218 47.5	S19 53.2	250 30.7	S23 22.0	23 39.5	N18 34.9	74 53.1	N22 06.4
	01	171 27.4	233 46.8	52.8	265 31.4	22.1	38 42.2	34.9	89 55.5	06.4
	02	186 29.9	248 46.2	52.4	280 32.1	22.2	53 44.9	35.0	104 58.0	06.4
	03	201 32.3	263 45.6	. . 52.0	295 32.8	. . 22.3	68 47.7	. . 35.1	120 00.4	. . 06.5
	04	216 34.8	278 44.9	51.7	310 33.6	22.4	83 50.4	35.1	135 02.9	06.5
	05	231 37.3	293 44.3	51.3	325 34.3	22.5	98 53.1	35.2	150 05.3	06.5
	06	246 39.7	308 43.7	S19 50.9	340 35.0	S23 22.6	113 55.8	N18 35.3	165 07.7	N22 06.5
	07	261 42.2	323 43.0	50.5	355 35.7	22.6	128 58.6	35.4	180 10.2	06.5
	08	276 44.7	338 42.4	50.1	10 36.5	22.7	144 01.3	35.4	195 12.6	06.5
	09	291 47.1	353 41.8	. . 49.8	25 37.2	. . 22.8	159 04.0	. . 35.5	210 15.1	. . 06.5
	10	306 49.6	8 41.1	49.4	40 37.9	22.9	174 06.8	35.6	225 17.5	06.5
	11	321 52.0	23 40.5	49.0	55 38.6	23.0	189 09.5	35.6	240 19.9	06.5
	12	336 54.5	38 39.9	S19 48.6	70 39.4	S23 23.1	204 12.2	N18 35.7	255 22.4	N22 06.5
	13	351 57.0	53 39.2	48.2	85 40.1	23.1	219 14.9	35.8	270 24.8	06.6
	14	6 59.4	68 38.6	47.8	100 40.8	23.2	234 17.7	35.9	285 27.2	06.6
	15	22 01.9	83 38.0	. . 47.4	115 41.5	. . 23.3	249 20.4	. . 35.9	300 29.7	. . 06.6
	16	37 04.4	98 37.4	47.0	130 42.3	23.4	264 23.1	36.0	315 32.1	06.6
	17	52 06.8	113 36.7	46.6	145 43.0	23.5	279 25.9	36.1	330 34.6	06.6
	18	67 09.3	128 36.1	S19 46.3	160 43.7	S23 23.5	294 28.6	N18 36.1	345 37.0	N22 06.6
	19	82 11.8	143 35.5	45.9	175 44.4	23.6	309 31.3	36.2	0 39.4	06.6
	20	97 14.2	158 34.8	45.5	190 45.2	23.7	324 34.0	36.3	15 41.9	06.6
	21	112 16.7	173 34.2	. . 45.1	205 45.9	. . 23.8	339 36.8	. . 36.3	30 44.3	. . 06.6
	22	127 19.1	188 33.6	44.7	220 46.6	23.9	354 39.5	36.4	45 46.7	06.7
	23	142 21.6	203 32.9	44.3	235 47.3	23.9	9 42.2	36.5	60 49.2	06.7
FRIDAY	28 00	157 24.1	218 32.3	S19 43.9	250 48.0	S23 24.0	24 44.9	N18 36.6	75 51.6	N22 06.7
	01	172 26.5	233 31.7	43.5	265 48.8	24.1	39 47.7	36.6	90 54.1	06.7
	02	187 29.0	248 31.1	43.1	280 49.5	24.2	54 50.4	36.7	105 56.5	06.7
	03	202 31.5	263 30.4	. . 42.7	295 50.2	. . 24.3	69 53.1	. . 36.8	120 58.9	. . 06.7
	04	217 33.9	278 29.8	42.3	310 50.9	24.3	84 55.8	36.8	136 01.4	06.7
	05	232 36.4	293 29.2	41.9	325 51.7	24.4	99 58.5	36.9	151 03.8	06.7
	06	247 38.9	308 28.5	S19 41.5	340 52.4	S23 24.5	115 01.3	N18 37.0	166 06.2	N22 06.7
	07	262 41.3	323 27.9	41.1	355 53.1	24.6	130 04.0	37.0	181 08.7	06.7
	08	277 43.8	338 27.3	40.6	10 53.8	24.6	145 06.7	37.1	196 11.1	06.8
	09	292 46.3	353 26.7	. . 40.2	25 54.6	. . 24.7	160 09.4	. . 37.2	211 13.5	. . 06.8
	10	307 48.7	8 26.0	39.8	40 55.3	24.8	175 12.2	37.3	226 16.0	06.8
	11	322 51.2	23 25.4	39.4	55 56.0	24.9	190 14.9	37.3	241 18.4	06.8
	12	337 53.6	38 24.8	S19 39.0	70 56.7	S23 24.9	205 17.6	N18 37.4	256 20.8	N22 06.8
	13	352 56.1	53 24.1	38.6	85 57.5	25.0	220 20.3	37.5	271 23.3	06.8
	14	7 58.6	68 23.5	38.2	100 58.2	25.1	235 23.1	37.5	286 25.7	06.8
	15	23 01.0	83 22.9	. . 37.8	115 58.9	. . 25.2	250 25.8	. . 37.6	301 28.1	. . 06.8
	16	38 03.5	98 22.3	37.4	130 59.6	25.2	265 28.5	37.7	316 30.6	06.8
	17	53 06.0	113 21.6	36.9	146 00.4	25.3	280 31.2	37.7	331 33.0	06.9
	18	68 08.4	128 21.0	S19 36.5	161 01.1	S23 25.4	295 33.9	N18 37.8	346 35.4	N22 06.9
	19	83 10.9	143 20.4	36.1	176 01.8	25.5	310 36.7	37.9	1 37.9	06.9
	20	98 13.4	158 19.8	35.7	191 02.5	25.5	325 39.4	37.9	16 40.3	06.9
	21	113 15.8	173 19.1	. . 35.3	206 03.3	. . 25.6	340 42.1	. . 38.0	31 42.7	. . 06.9
	22	128 18.3	188 18.5	34.9	221 04.0	25.7	355 44.8	38.1	46 45.2	06.9
	23	143 20.8	203 17.9	34.4	236 04.7	25.8	10 47.5	38.1	61 47.6	06.9
SATURDAY	1 00	158 23.2	218 17.3	S19 34.0	251 05.4	S23 25.8	25 50.3	N18 38.2	76 50.0	N22 06.9
	01	173 25.7	233 16.6	33.6	266 06.2	25.9	40 53.0	38.3	91 52.4	06.9
	02	188 28.1	248 16.0	33.2	281 06.9	26.0	55 55.7	38.3	106 54.9	06.9
	03	203 30.6	263 15.4	. . 32.7	296 07.6	. . 26.0	70 58.4	. . 38.4	121 57.3	. . 07.0
	04	218 33.1	278 14.7	32.3	311 08.3	26.1	86 01.1	38.5	136 59.7	07.0
	05	233 35.5	293 14.1	31.9	326 09.1	26.2	101 03.8	38.6	152 02.2	07.0
	06	248 38.0	308 13.5	S19 31.5	341 09.8	S23 26.2	116 06.6	N18 38.6	167 04.6	N22 07.0
	07	263 40.5	323 12.9	31.0	356 10.5	26.3	131 09.3	38.7	182 07.0	07.0
	08	278 42.9	338 12.2	30.6	11 11.2	26.4	146 12.0	38.8	197 09.5	07.0
	09	293 45.4	353 11.6	. . 30.2	26 12.0	. . 26.5	161 14.7	. . 38.8	212 11.9	. . 07.0
	10	308 47.9	8 11.0	29.7	41 12.7	26.5	176 17.4	38.9	227 14.3	07.0
	11	323 50.3	23 10.4	29.3	56 13.4	26.6	191 20.1	39.0	242 16.7	07.0
	12	338 52.8	38 09.7	S19 28.9	71 14.1	S23 26.7	206 22.9	N18 39.0	257 19.2	N22 07.1
	13	353 55.2	53 09.1	28.4	86 14.9	26.7	221 25.6	39.1	272 21.6	07.1
	14	8 57.7	68 08.5	28.0	101 15.6	26.8	236 28.3	39.2	287 24.0	07.1
	15	24 00.2	83 07.9	. . 27.6	116 16.3	. . 26.9	251 31.0	. . 39.2	302 26.5	. . 07.1
	16	39 02.6	98 07.2	27.1	131 17.0	26.9	266 33.7	39.3	317 28.9	07.1
	17	54 05.1	113 06.6	26.7	146 17.7	27.0	281 36.4	39.4	332 31.3	07.1
	18	69 07.6	128 06.0	S19 26.2	161 18.5	S23 27.1	296 39.2	N18 39.4	347 33.7	N22 07.1
	19	84 10.0	143 05.4	25.8	176 19.2	27.1	311 41.9	39.5	2 36.2	07.1
	20	99 12.5	158 04.8	25.4	191 19.9	27.2	326 44.6	39.6	17 38.6	07.1
	21	114 15.0	173 04.1	. . 24.9	206 20.6	. . 27.3	341 47.3	. . 39.6	32 41.0	. . 07.2
	22	129 17.4	188 03.5	24.5	221 21.4	27.3	356 50.0	39.7	47 43.4	07.2
	23	144 19.9	203 02.9	24.0	236 22.1	27.4	11 52.7	39.8	62 45.9	07.2
	Mer. Pass.	h m 13 28.2	*v* −0.6	*d* 0.4	*v* 0.7	*d* 0.1	*v* 2.7	*d* 0.1	*v* 2.4	*d* 0.0

STARS

Name	SHA ° ′	Dec ° ′
Acamar	315 24.6	S40 17.8
Achernar	335 33.0	S57 13.6
Acrux	173 18.1	S63 06.8
Adhara	255 18.7	S28 58.7
Aldebaran	290 58.7	N16 30.9
Alioth	166 27.1	N55 56.4
Alkaid	153 04.8	N49 17.6
Al Na'ir	27 54.2	S46 56.9
Alnilam	275 54.5	S 1 12.1
Alphard	218 03.8	S 8 40.4
Alphecca	126 17.7	N26 42.0
Alpheratz	357 52.3	N29 06.3
Altair	62 16.4	N 8 52.3
Ankaa	353 23.9	S42 17.6
Antares	112 36.3	S26 26.3
Arcturus	146 02.9	N19 09.8
Atria	107 45.6	S69 01.8
Avior	234 20.9	S59 31.3
Bellatrix	278 40.6	N 6 21.1
Betelgeuse	271 10.0	N 7 24.4
Canopus	263 59.5	S52 42.1
Capella	280 46.4	N46 00.2
Deneb	49 37.5	N45 17.2
Denebola	182 41.6	N14 33.2
Diphda	349 04.2	S17 58.4
Dubhe	194 00.6	N61 44.0
Elnath	278 22.8	N28 36.7
Eltanin	90 50.1	N51 28.9
Enif	33 55.4	N 9 53.1
Fomalhaut	15 33.2	S29 36.6
Gacrux	172 09.7	S57 07.7
Gienah	176 00.4	S17 33.6
Hadar	148 59.3	S60 23.1
Hamal	328 10.2	N23 28.6
Kaus Aust.	83 54.7	S34 23.0
Kochab	137 18.8	N74 08.3
Markab	13 46.8	N15 13.1
Menkar	314 23.7	N 4 06.0
Menkent	148 17.0	S36 23.0
Miaplacidus	221 40.8	S69 43.8
Mirfak	308 52.2	N49 52.5
Nunki	76 08.5	S26 17.7
Peacock	53 32.3	S56 43.5
Pollux	243 37.3	N28 01.2
Procyon	245 08.0	N 5 13.0
Rasalhague	96 14.1	N12 33.2
Regulus	207 51.8	N11 57.1
Rigel	281 19.8	S 8 12.0
Rigil Kent.	140 02.8	S60 50.7
Sabik	102 21.9	S15 43.8
Schedar	349 50.5	N56 33.3
Shaula	96 33.0	S37 06.3
Sirius	258 40.7	S16 43.3
Spica	158 39.6	S11 10.7
Suhail	222 58.1	S43 26.8
Vega	80 44.7	N38 46.8
Zuben'ubi	137 14.3	S16 03.3

	SHA ° ′	Mer. Pass. h m
Venus	61 08.2	9 26
Mars	93 24.0	7 16
Jupiter	227 20.9	22 17
Saturn	278 27.5	18 53

Day	UT (h)	SUN GHA (° ′)	SUN Dec (° ′)	MOON GHA (° ′)	v (′)	MOON Dec (° ′)	d (′)	HP (′)
27 THURSDAY	00	176 47.0	S 8 34.8	224 52.9	7.0	S25 34.9	2.8	57.2
	01	191 47.1	33.9	239 18.9	7.0	25 32.1	3.1	57.1
	02	206 47.2	33.0	253 44.9	7.1	25 29.0	3.2	57.1
	03	221 47.3	. . 32.0	268 11.0	7.2	25 25.8	3.3	57.1
	04	236 47.4	31.1	282 37.2	7.2	25 22.5	3.5	57.1
	05	251 47.5	30.2	297 03.4	7.3	25 19.0	3.6	57.1
	06	266 47.6	S 8 29.2	311 29.7	7.3	S25 15.4	3.8	57.0
	07	281 47.7	28.3	325 56.0	7.4	25 11.6	3.9	57.0
	08	296 47.8	27.3	340 22.4	7.5	25 07.7	4.1	57.0
	09	311 47.9	. . 26.4	354 48.9	7.5	25 03.6	4.2	57.0
	10	326 48.0	25.5	9 15.4	7.6	24 59.4	4.3	56.9
	11	341 48.1	24.5	23 42.0	7.7	24 55.1	4.5	56.9
	12	356 48.3	S 8 23.6	38 08.7	7.7	S24 50.6	4.6	56.9
	13	11 48.4	22.6	52 35.4	7.8	24 46.0	4.8	56.9
	14	26 48.5	21.7	67 02.2	7.9	24 41.2	4.9	56.9
	15	41 48.6	. . 20.8	81 29.1	7.9	24 36.3	5.0	56.8
	16	56 48.7	19.8	95 56.0	8.0	24 31.3	5.2	56.8
	17	71 48.8	18.9	110 23.0	8.1	24 26.1	5.3	56.8
	18	86 48.9	S 8 17.9	124 50.1	8.1	S24 20.8	5.4	56.8
	19	101 49.0	17.0	139 17.2	8.3	24 15.4	5.6	56.8
	20	116 49.1	16.1	153 44.5	8.3	24 09.8	5.7	56.7
	21	131 49.2	. . 15.1	168 11.8	8.3	24 04.1	5.8	56.7
	22	146 49.4	14.2	182 39.1	8.5	23 58.3	6.0	56.7
	23	161 49.5	13.2	197 06.6	8.5	23 52.3	6.1	56.7
28 FRIDAY	00	176 49.6	S 8 12.3	211 34.1	8.6	S23 46.2	6.2	56.7
	01	191 49.7	11.3	226 01.7	8.7	23 40.0	6.3	56.6
	02	206 49.8	10.4	240 29.4	8.8	23 33.7	6.5	56.6
	03	221 49.9	. . 09.5	254 57.2	8.8	23 27.2	6.6	56.6
	04	236 50.0	08.5	269 25.0	8.9	23 20.6	6.7	56.6
	05	251 50.1	07.6	283 52.9	9.0	23 13.9	6.8	56.6
	06	266 50.3	S 8 06.6	298 20.9	9.1	S23 07.1	7.0	56.5
	07	281 50.4	05.7	312 49.0	9.2	23 00.1	7.0	56.5
	08	296 50.5	04.7	327 17.2	9.2	22 53.1	7.2	56.5
	09	311 50.6	. . 03.8	341 45.4	9.4	22 45.9	7.3	56.5
	10	326 50.7	02.9	356 13.8	9.4	22 38.6	7.5	56.5
	11	341 50.8	01.9	10 42.2	9.5	22 31.1	7.5	56.4
	12	356 50.9	S 8 01.0	25 10.7	9.6	S22 23.6	7.6	56.4
	13	11 51.0	8 00.0	39 39.3	9.6	22 16.0	7.8	56.4
	14	26 51.2	7 59.1	54 07.9	9.8	22 08.2	7.9	56.4
	15	41 51.3	. . 58.1	68 36.7	9.8	22 00.3	8.0	56.4
	16	56 51.4	57.2	83 05.5	10.0	21 52.3	8.1	56.3
	17	71 51.5	56.2	97 34.5	10.0	21 44.2	8.2	56.3
	18	86 51.6	S 7 55.3	112 03.5	10.1	S21 36.0	8.3	56.3
	19	101 51.7	54.3	126 32.6	10.1	21 27.7	8.4	56.3
	20	116 51.9	53.4	141 01.7	10.3	21 19.3	8.5	56.3
	21	131 52.0	. . 52.5	155 31.0	10.3	21 10.8	8.6	56.2
	22	146 52.1	51.5	170 00.3	10.5	21 02.2	8.7	56.2
	23	161 52.2	50.6	184 29.8	10.5	20 53.5	8.9	56.2
1 SATURDAY	00	176 52.3	S 7 49.6	198 59.3	10.6	S20 44.6	8.9	56.2
	01	191 52.4	48.7	213 28.9	10.7	20 35.7	9.0	56.2
	02	206 52.5	47.7	227 58.6	10.8	20 26.7	9.1	56.1
	03	221 52.7	. . 46.8	242 28.4	10.8	20 17.6	9.2	56.1
	04	236 52.8	45.8	256 58.2	11.0	20 08.4	9.3	56.1
	05	251 52.9	44.9	271 28.2	11.0	19 59.1	9.4	56.1
	06	266 53.0	S 7 43.9	285 58.2	11.1	S19 49.7	9.5	56.1
	07	281 53.1	43.0	300 28.3	11.2	19 40.2	9.6	56.0
	08	296 53.3	42.0	314 58.5	11.3	19 30.6	9.7	56.0
	09	311 53.4	. . 41.1	329 28.8	11.4	19 20.9	9.7	56.0
	10	326 53.5	40.1	343 59.2	11.4	19 11.2	9.9	56.0
	11	341 53.6	39.2	358 29.6	11.6	19 01.3	9.9	56.0
	12	356 53.7	S 7 38.2	13 00.2	11.6	S18 51.4	10.0	55.9
	13	11 53.9	37.3	27 30.8	11.7	18 41.4	10.1	55.9
	14	26 54.0	36.3	42 01.5	11.8	18 31.3	10.2	55.9
	15	41 54.1	. . 35.4	56 32.3	11.9	18 21.1	10.2	55.9
	16	56 54.2	34.4	71 03.2	11.9	18 10.9	10.4	55.9
	17	71 54.3	33.5	85 34.1	12.1	18 00.5	10.4	55.8
	18	86 54.5	S 7 32.5	100 05.2	12.1	S17 50.1	10.5	55.8
	19	101 54.6	31.6	114 36.3	12.2	17 39.6	10.5	55.8
	20	116 54.7	30.6	129 07.5	12.3	17 29.1	10.7	55.8
	21	131 54.8	. . 29.7	143 38.8	12.3	17 18.4	10.7	55.8
	22	146 54.9	28.7	158 10.1	12.5	17 07.7	10.8	55.7
	23	161 55.1	27.8	172 41.6	12.5	S16 56.9	10.9	55.7
		SD 16.2	*d* 0.9	SD 15.5		15.4		15.2

Lat.	Twilight Naut. (h m)	Twilight Civil (h m)	Sunrise (h m)	Moonrise 27 (h m)	Moonrise 28 (h m)	Moonrise 1 (h m)	Moonrise 2 (h m)
N 72	05 17	06 35	07 44	▬	▬	▬	09 33
N 70	05 21	06 32	07 34	▬	▬	10 01	09 01
68	05 25	06 29	07 25	▬	▬	09 06	08 37
66	05 27	06 26	07 18	▬	08 59	08 33	08 19
64	05 30	06 24	07 12	08 19	08 13	08 08	08 04
62	05 31	06 22	07 07	07 28	07 43	07 49	07 51
60	05 33	06 21	07 03	06 56	07 20	07 33	07 40
N 58	05 34	06 19	06 59	06 33	07 01	07 19	07 31
56	05 35	06 18	06 55	06 13	06 46	07 07	07 22
54	05 36	06 16	06 52	05 57	06 32	06 57	07 15
52	05 36	06 15	06 49	05 43	06 21	06 48	07 08
50	05 37	06 14	06 46	05 31	06 10	06 39	07 02
45	05 37	06 11	06 41	05 06	05 48	06 22	06 49
N 40	05 37	06 08	06 36	04 46	05 30	06 07	06 38
35	05 37	06 06	06 31	04 29	05 16	05 55	06 29
30	05 36	06 04	06 28	04 14	05 03	05 44	06 21
20	05 33	05 59	06 21	03 50	04 40	05 26	06 06
N 10	05 29	05 54	06 15	03 28	04 21	05 10	05 54
0	05 24	05 48	06 09	03 08	04 03	04 54	05 42
S 10	05 17	05 42	06 03	02 48	03 45	04 39	05 31
20	05 09	05 35	05 57	02 27	03 25	04 23	05 18
30	04 57	05 26	05 50	02 02	03 03	04 04	05 03
35	04 49	05 20	05 46	01 47	02 50	03 53	04 55
40	04 40	05 13	05 41	01 30	02 34	03 40	04 45
45	04 28	05 05	05 35	01 10	02 16	03 25	04 34
S 50	04 13	04 55	05 28	00 44	01 53	03 07	04 20
52	04 06	04 50	05 25	00 32	01 43	02 58	04 14
54	03 58	04 44	05 21	00 18	01 30	02 48	04 07
56	03 49	04 38	05 18	00 01	01 16	02 37	03 59
58	03 38	04 31	05 13	24 59	00 59	02 24	03 50
S 60	03 25	04 23	05 08	24 39	00 39	02 09	03 39

Lat.	Sunset (h m)	Twilight Civil (h m)	Twilight Naut. (h m)	Moonset 27 (h m)	Moonset 28 (h m)	Moonset 1 (h m)	Moonset 2 (h m)
N 72	16 43	17 53	19 11	▬	▬	▬	14 36
N 70	16 53	17 56	19 06	▬	▬	12 29	15 06
68	17 01	17 58	19 03	▬	▬	13 22	15 28
66	17 08	18 00	19 00	▬	11 43	13 54	15 45
64	17 14	18 02	18 57	10 29	12 29	14 18	15 59
62	17 19	18 04	18 55	11 20	12 58	14 37	16 11
60	17 24	18 06	18 54	11 52	13 20	14 52	16 21
N 58	17 28	18 07	18 53	12 15	13 39	15 05	16 29
56	17 31	18 09	18 52	12 34	13 54	15 16	16 37
54	17 34	18 10	18 51	12 50	14 07	15 26	16 44
52	17 37	18 11	18 50	13 04	14 18	15 34	16 50
50	17 40	18 12	18 50	13 16	14 28	15 42	16 55
45	17 45	18 15	18 49	13 41	14 49	15 58	17 07
N 40	17 50	18 17	18 49	14 00	15 06	16 12	17 16
35	17 54	18 20	18 49	14 17	15 20	16 23	17 24
30	17 58	18 22	18 50	14 31	15 32	16 33	17 31
20	18 05	18 27	18 52	14 55	15 53	16 50	17 44
N 10	18 10	18 32	18 56	15 16	16 11	17 04	17 54
0	18 16	18 37	19 01	15 35	16 28	17 18	18 04
S 10	18 22	18 43	19 08	15 54	16 45	17 31	18 14
20	18 28	18 50	19 16	16 14	17 03	17 46	18 24
30	18 35	18 59	19 28	16 38	17 23	18 02	18 36
35	18 39	19 05	19 35	16 52	17 35	18 11	18 43
40	18 44	19 11	19 44	17 08	17 48	18 22	18 51
45	18 49	19 20	19 56	17 26	18 04	18 35	19 00
S 50	18 56	19 30	20 10	17 50	18 24	18 50	19 10
52	18 59	19 34	20 17	18 01	18 33	18 57	19 15
54	19 02	19 40	20 25	18 14	18 44	19 05	19 21
56	19 06	19 45	20 35	18 29	18 55	19 13	19 27
58	19 10	19 52	20 45	18 46	19 09	19 23	19 33
S 60	19 15	20 00	20 58	19 06	19 24	19 35	19 41

Day	SUN Eqn. of Time 00^h (m s)	SUN Eqn. of Time 12^h (m s)	SUN Mer. Pass. (h m)	MOON Mer. Pass. Upper (h m)	MOON Mer. Pass. Lower (h m)	MOON Age (d)	MOON Phase (%)
27	12 52	12 47	12 13	09 22	21 49	26	13
28	12 42	12 37	12 13	10 16	22 41	27	7
1	12 31	12 25	12 12	11 06	23 30	28	3

UT		ARIES	VENUS −4.1		MARS +0.9		JUPITER −2.5		SATURN +0.0	
d	h	GHA ° ′	GHA ° ′	Dec ° ′	GHA ° ′	Dec ° ′	GHA ° ′	Dec ° ′	GHA ° ′	Dec ° ′
2 SUNDAY	00	159 22.4	218 02.3	S19 23.6	251 22.8	S23 27.4	26 55.4	N18 39.8	77 48.3	N22 07.2
	01	174 24.8	233 01.6	23.1	266 23.5	27.5	41 58.2	39.9	92 50.7	07.2
	02	189 27.3	248 01.0	22.7	281 24.3	27.6	57 00.9	40.0	107 53.1	07.2
	03	204 29.7	263 00.4	22.2	296 25.0	27.6	72 03.6	40.0	122 55.6	07.2
	04	219 32.2	277 59.8	21.8	311 25.7	27.7	87 06.3	40.1	137 58.0	07.2
	05	234 34.7	292 59.2	21.3	326 26.4	27.8	102 09.0	40.2	153 00.4	07.2
	06	249 37.1	307 58.5	S19 20.9	341 27.2	S23 27.8	117 11.7	N18 40.2	168 02.8	N22 07.3
	07	264 39.6	322 57.9	20.4	356 27.9	27.9	132 14.4	40.3	183 05.3	07.3
	08	279 42.1	337 57.3	20.0	11 28.6	28.0	147 17.1	40.4	198 07.7	07.3
	09	294 44.5	352 56.7	19.5	26 29.3	28.0	162 19.8	40.4	213 10.1	07.3
	10	309 47.0	7 56.0	19.1	41 30.1	28.1	177 22.6	40.5	228 12.5	07.3
	11	324 49.5	22 55.4	18.6	56 30.8	28.1	192 25.3	40.5	243 15.0	07.3
	12	339 51.9	37 54.8	S19 18.2	71 31.5	S23 28.2	207 28.0	N18 40.6	258 17.4	N22 07.3
	13	354 54.4	52 54.2	17.7	86 32.2	28.3	222 30.7	40.7	273 19.8	07.3
	14	9 56.9	67 53.6	17.3	101 33.0	28.3	237 33.4	40.7	288 22.2	07.3
	15	24 59.3	82 52.9	16.8	116 33.7	28.4	252 36.1	40.8	303 24.7	07.4
	16	40 01.8	97 52.3	16.3	131 34.4	28.4	267 38.8	40.9	318 27.1	07.4
	17	55 04.2	112 51.7	15.9	146 35.1	28.5	282 41.5	40.9	333 29.5	07.4
	18	70 06.7	127 51.1	S19 15.4	161 35.9	S23 28.6	297 44.2	N18 41.0	348 31.9	N22 07.4
	19	85 09.2	142 50.5	15.0	176 36.6	28.6	312 46.9	41.1	3 34.3	07.4
	20	100 11.6	157 49.8	14.5	191 37.3	28.7	327 49.7	41.1	18 36.8	07.4
	21	115 14.1	172 49.2	14.0	206 38.0	28.7	342 52.4	41.2	33 39.2	07.4
	22	130 16.6	187 48.6	13.6	221 38.8	28.8	357 55.1	41.3	48 41.6	07.4
	23	145 19.0	202 48.0	13.1	236 39.5	28.8	12 57.8	41.3	63 44.0	07.4
3 MONDAY	00	160 21.5	217 47.4	S19 12.6	251 40.2	S23 28.9	28 00.5	N18 41.4	78 46.4	N22 07.5
	01	175 24.0	232 46.7	12.2	266 40.9	29.0	43 03.2	41.5	93 48.9	07.5
	02	190 26.4	247 46.1	11.7	281 41.7	29.0	58 05.9	41.5	108 51.3	07.5
	03	205 28.9	262 45.5	11.2	296 42.4	29.1	73 08.6	41.6	123 53.7	07.5
	04	220 31.3	277 44.9	10.7	311 43.1	29.1	88 11.3	41.6	138 56.1	07.5
	05	235 33.8	292 44.3	10.3	326 43.8	29.2	103 14.0	41.7	153 58.5	07.5
	06	250 36.3	307 43.7	S19 09.8	341 44.6	S23 29.2	118 16.7	N18 41.8	169 01.0	N22 07.5
	07	265 38.7	322 43.0	09.3	356 45.3	29.3	133 19.4	41.8	184 03.4	07.5
	08	280 41.2	337 42.4	08.8	11 46.0	29.4	148 22.1	41.9	199 05.8	07.5
	09	295 43.7	352 41.8	08.4	26 46.7	29.4	163 24.8	42.0	214 08.2	07.6
	10	310 46.1	7 41.2	07.9	41 47.5	29.5	178 27.5	42.0	229 10.6	07.6
	11	325 48.6	22 40.6	07.4	56 48.2	29.5	193 30.2	42.1	244 13.1	07.6
	12	340 51.1	37 40.0	S19 06.9	71 48.9	S23 29.6	208 33.0	N18 42.2	259 15.5	N22 07.6
	13	355 53.5	52 39.3	06.5	86 49.6	29.6	223 35.7	42.2	274 17.9	07.6
	14	10 56.0	67 38.7	06.0	101 50.4	29.7	238 38.4	42.3	289 20.3	07.6
	15	25 58.5	82 38.1	05.5	116 51.1	29.7	253 41.1	42.3	304 22.7	07.6
	16	41 00.9	97 37.5	05.0	131 51.8	29.8	268 43.8	42.4	319 25.1	07.6
	17	56 03.4	112 36.9	04.5	146 52.5	29.8	283 46.5	42.5	334 27.6	07.6
	18	71 05.8	127 36.3	S19 04.0	161 53.3	S23 29.9	298 49.2	N18 42.5	349 30.0	N22 07.7
	19	86 08.3	142 35.6	03.5	176 54.0	29.9	313 51.9	42.6	4 32.4	07.7
	20	101 10.8	157 35.0	03.1	191 54.7	30.0	328 54.6	42.7	19 34.8	07.7
	21	116 13.2	172 34.4	02.6	206 55.4	30.0	343 57.3	42.7	34 37.2	07.7
	22	131 15.7	187 33.8	02.1	221 56.2	30.1	359 00.0	42.8	49 39.6	07.7
	23	146 18.2	202 33.2	01.6	236 56.9	30.1	14 02.7	42.9	64 42.1	07.7
4 TUESDAY	00	161 20.6	217 32.6	S19 01.1	251 57.6	S23 30.2	29 05.4	N18 42.9	79 44.5	N22 07.7
	01	176 23.1	232 32.0	00.6	266 58.3	30.2	44 08.1	43.0	94 46.9	07.7
	02	191 25.6	247 31.3	19 00.1	281 59.1	30.3	59 10.8	43.0	109 49.3	07.8
	03	206 28.0	262 30.7	18 59.6	296 59.8	30.3	74 13.5	43.1	124 51.7	07.8
	04	221 30.5	277 30.1	59.1	312 00.5	30.4	89 16.2	43.2	139 54.1	07.8
	05	236 32.9	292 29.5	58.6	327 01.3	30.4	104 18.9	43.2	154 56.6	07.8
	06	251 35.4	307 28.9	S18 58.1	342 02.0	S23 30.5	119 21.6	N18 43.3	169 59.0	N22 07.8
	07	266 37.9	322 28.3	57.6	357 02.7	30.5	134 24.3	43.3	185 01.4	07.8
	08	281 40.3	337 27.7	57.1	12 03.4	30.6	149 27.0	43.4	200 03.8	07.8
	09	296 42.8	352 27.1	56.6	27 04.2	30.6	164 29.7	43.5	215 06.2	07.8
	10	311 45.3	7 26.4	56.1	42 04.9	30.7	179 32.4	43.5	230 08.6	07.8
	11	326 47.7	22 25.8	55.6	57 05.6	30.7	194 35.1	43.6	245 11.0	07.9
	12	341 50.2	37 25.2	S18 55.1	72 06.3	S23 30.8	209 37.8	N18 43.7	260 13.5	N22 07.9
	13	356 52.7	52 24.6	54.6	87 07.1	30.8	224 40.5	43.7	275 15.9	07.9
	14	11 55.1	67 24.0	54.1	102 07.8	30.9	239 43.2	43.8	290 18.3	07.9
	15	26 57.6	82 23.4	53.6	117 08.5	30.9	254 45.9	43.8	305 20.7	07.9
	16	42 00.1	97 22.8	53.1	132 09.2	30.9	269 48.6	43.9	320 23.1	07.9
	17	57 02.5	112 22.2	52.6	147 10.0	31.0	284 51.3	44.0	335 25.5	07.9
	18	72 05.0	127 21.6	S18 52.1	162 10.7	S23 31.0	299 54.0	N18 44.0	350 27.9	N22 07.9
	19	87 07.4	142 20.9	51.6	177 11.4	31.1	314 56.7	44.1	5 30.3	07.9
	20	102 09.9	157 20.3	51.1	192 12.1	31.1	329 59.4	44.1	20 32.7	08.0
	21	117 12.4	172 19.7	50.6	207 12.9	31.2	345 02.0	44.2	35 35.2	08.0
	22	132 14.8	187 19.1	50.1	222 13.6	31.2	0 04.7	44.3	50 37.6	08.0
	23	147 17.3	202 18.5	49.6	237 14.3	31.3	15 07.4	44.3	65 40.0	08.0
Mer. Pass.		h m 13 16.4	*v* −0.6	*d* 0.5	*v* 0.7	*d* 0.1	*v* 2.7	*d* 0.1	*v* 2.4	*d* 0.0

STARS

Name	SHA ° ′	Dec ° ′
Acamar	315 24.6	S40 17.8
Achernar	335 33.0	S57 13.6
Acrux	173 18.1	S63 06.9
Adhara	255 18.7	S28 58.7
Aldebaran	290 58.7	N16 30.9
Alioth	166 27.1	N55 56.4
Alkaid	153 04.8	N49 17.6
Al Na'ir	27 54.2	S46 56.9
Alnilam	275 54.5	S 1 12.1
Alphard	218 03.8	S 8 40.4
Alphecca	126 17.7	N26 42.0
Alpheratz	357 52.3	N29 06.3
Altair	62 16.4	N 8 52.3
Ankaa	353 23.9	S42 17.6
Antares	112 36.2	S26 26.3
Arcturus	146 02.9	N19 09.8
Atria	107 45.5	S69 01.8
Avior	234 21.0	S59 31.3
Bellatrix	278 40.7	N 6 21.1
Betelgeuse	271 10.0	N 7 24.4
Canopus	263 59.5	S52 42.1
Capella	280 46.4	N46 00.3
Deneb	49 37.5	N45 17.2
Denebola	182 41.6	N14 33.2
Diphda	349 04.2	S17 58.4
Dubhe	194 00.6	N61 44.0
Elnath	278 22.8	N28 36.7
Eltanin	90 50.1	N51 28.9
Enif	33 55.4	N 9 53.1
Fomalhaut	15 33.2	S29 36.6
Gacrux	172 09.7	S57 07.7
Gienah	176 00.4	S17 33.6
Hadar	148 59.3	S60 23.1
Hamal	328 10.2	N23 28.6
Kaus Aust.	83 54.7	S34 23.0
Kochab	137 18.8	N74 08.3
Markab	13 46.8	N15 13.1
Menkar	314 23.7	N 4 06.0
Menkent	148 17.0	S36 23.0
Miaplacidus	221 40.9	S69 43.9
Mirfak	308 52.3	N49 52.5
Nunki	76 08.5	S26 17.7
Peacock	53 32.3	S56 43.5
Pollux	243 37.3	N28 01.2
Procyon	245 08.0	N 5 13.0
Rasalhague	96 14.0	N12 33.2
Regulus	207 51.8	N11 57.1
Rigel	281 19.8	S 8 12.0
Rigil Kent.	140 02.8	S60 50.7
Sabik	102 21.9	S15 43.8
Schedar	349 50.5	N56 33.3
Shaula	96 33.0	S37 06.3
Sirius	258 40.7	S16 43.3
Spica	158 39.6	S11 10.7
Suhail	222 58.1	S43 26.8
Vega	80 44.7	N38 46.8
Zuben'ubi	137 14.3	S16 03.3

	SHA ° ′	Mer. Pass. h m
Venus	57 25.9	9 29
Mars	91 18.7	7 13
Jupiter	227 39.0	22 04
Saturn	278 25.0	18 42

UT	SUN GHA	SUN Dec	MOON GHA	v	MOON Dec	d	HP
d h	° ′	° ′	° ′	′	° ′	′	′
2 00	176 55.2	S 7 26.8	187 13.1	12.6	S16 46.0	10.9	55.7
01	191 55.3	25.9	201 44.7	12.7	16 35.1	11.0	55.7
02	206 55.4	24.9	216 16.4	12.7	16 24.1	11.1	55.7
03	221 55.5	. . 24.0	230 48.1	12.9	16 13.0	11.1	55.6
04	236 55.7	23.0	245 20.0	12.9	16 01.9	11.2	55.6
05	251 55.8	22.1	259 51.9	13.0	15 50.7	11.3	55.6
06	266 55.9	S 7 21.1	274 23.9	13.0	S15 39.4	11.3	55.6
07	281 56.0	20.2	288 55.9	13.2	15 28.1	11.4	55.6
08	296 56.2	19.2	303 28.1	13.2	15 16.7	11.4	55.6
S 09	311 56.3	. . 18.2	318 00.3	13.3	15 05.3	11.6	55.5
U 10	326 56.4	17.3	332 32.6	13.3	14 53.7	11.5	55.5
N 11	341 56.5	16.3	347 04.9	13.5	14 42.2	11.7	55.5
D 12	356 56.7	S 7 15.4	1 37.4	13.5	S14 30.5	11.6	55.5
A 13	11 56.8	14.4	16 09.9	13.5	14 18.9	11.8	55.5
Y 14	26 56.9	13.5	30 42.4	13.7	14 07.1	11.8	55.4
15	41 57.0	. . 12.5	45 15.1	13.7	13 55.3	11.8	55.4
16	56 57.2	11.6	59 47.8	13.8	13 43.5	11.9	55.4
17	71 57.3	10.6	74 20.6	13.9	13 31.6	12.0	55.4
18	86 57.4	S 7 09.7	88 53.5	13.9	S13 19.6	12.0	55.4
19	101 57.5	08.7	103 26.4	14.0	13 07.6	12.0	55.4
20	116 57.7	07.7	117 59.4	14.0	12 55.6	12.1	55.3
21	131 57.8	. . 06.8	132 32.4	14.1	12 43.5	12.2	55.3
22	146 57.9	05.8	147 05.5	14.2	12 31.3	12.2	55.3
23	161 58.0	04.9	161 38.7	14.3	12 19.1	12.2	55.3
3 00	176 58.2	S 7 03.9	176 12.0	14.3	S1[illegible] 06.9	12.3	55.3
01	191 58.3	03.0	190 45.3	14.4	11 54.6	12.3	55.2
02	206 58.4	02.0	205 18.7	14.4	11 42.3	12.4	55.2
03	221 58.6	. . 01.1	219 52.1	14.5	11 29.9	12.4	55.2
04	236 58.7	7 00.1	234 25.6	14.6	11 17.5	12.4	55.2
05	251 58.8	6 59.1	248 59.2	14.6	11 05.1	12.5	55.2
06	266 58.9	S 6 58.2	263 32.8	14.7	S10 52.6	12.5	55.2
07	281 59.1	57.2	278 06.5	14.7	10 40.1	12.5	55.1
08	296 59.2	56.3	292 40.2	14.8	10 27.6	12.6	55.1
M 09	311 59.3	. . 55.3	307 14.0	14.8	10 15.0	12.7	55.1
O 10	326 59.5	54.3	321 47.8	14.9	10 02.3	12.6	55.1
N 11	341 59.6	53.4	336 21.7	15.0	9 49.7	12.7	55.1
D 12	356 59.7	S 6 52.4	350 55.7	15.0	S 9 37.0	12.7	55.1
A 13	11 59.8	51.5	5 29.7	15.1	9 24.3	12.8	55.0
Y 14	27 00.0	50.5	20 03.8	15.1	9 11.5	12.8	55.0
15	42 00.1	. . 49.6	34 37.9	15.1	8 58.7	12.8	55.0
16	57 00.2	48.6	49 12.0	15.3	8 45.9	12.8	55.0
17	72 00.4	47.6	63 46.3	15.2	8 33.1	12.9	55.0
18	87 00.5	S 6 46.7	78 20.5	15.3	S 8 20.2	12.9	55.0
19	102 00.6	45.7	92 54.8	15.4	8 07.3	12.9	54.9
20	117 00.8	44.8	107 29.2	15.4	7 54.4	12.9	54.9
21	132 00.9	. . 43.8	122 03.6	15.4	7 41.5	13.0	54.9
22	147 01.0	42.8	136 38.0	15.5	7 28.5	13.0	54.9
23	162 01.2	41.9	151 12.5	15.6	7 15.5	13.0	54.9
4 00	177 01.3	S 6 40.9	165 47.1	15.6	S 7 02.5	13.0	54.9
01	192 01.4	40.0	180 21.7	15.6	6 49.5	13.0	54.9
02	207 01.5	39.0	194 56.3	15.6	6 36.5	13.1	54.8
03	222 01.7	. . 38.0	209 30.9	15.8	6 23.4	13.1	54.8
04	237 01.8	37.1	224 05.7	15.7	6 10.3	13.1	54.8
05	252 01.9	36.1	238 40.4	15.8	5 57.2	13.1	54.8
06	267 02.1	S 6 35.2	253 15.2	15.8	S 5 44.1	13.1	54.8
07	282 02.2	34.2	267 50.0	15.9	5 31.0	13.2	54.8
T 08	297 02.3	33.2	282 24.9	15.8	5 17.8	13.1	54.7
U 09	312 02.5	. . 32.3	296 59.7	16.0	5 04.7	13.2	54.7
E 10	327 02.6	31.3	311 34.7	15.9	4 51.5	13.2	54.7
S 11	342 02.7	30.3	326 09.6	16.0	4 38.3	13.2	54.7
D 12	357 02.9	S 6 29.4	340 44.6	16.1	S 4 25.1	13.2	54.7
A 13	12 03.0	28.4	355 19.7	16.0	4 11.9	13.2	54.7
Y 14	27 03.2	27.5	9 54.7	16.1	3 58.7	13.2	54.7
15	42 03.3	. . 26.5	24 29.8	16.1	3 45.5	13.2	54.6
16	57 03.4	25.5	39 04.9	16.2	3 32.3	13.2	54.6
17	72 03.6	24.6	53 40.1	16.1	3 19.1	13.3	54.6
18	87 03.7	S 6 23.6	68 15.2	16.2	S 3 05.8	13.2	54.6
19	102 03.8	22.6	82 50.4	16.2	2 52.6	13.3	54.6
20	117 04.0	21.7	97 25.6	16.3	2 39.3	13.2	54.6
21	132 04.1	. . 20.7	112 00.9	16.3	2 26.1	13.3	54.6
22	147 04.2	19.8	126 36.2	16.3	2 12.8	13.2	54.6
23	162 04.4	18.8	141 11.5	16.3	S 1 59.6	13.3	54.5
	SD 16.2	d 1.0	SD	15.1	15.0		14.9

Lat.	Twilight Naut.	Twilight Civil	Sunrise	Moonrise 2	Moonrise 3	Moonrise 4	Moonrise 5
°	h m	h m	h m	h m	h m	h m	h m
N 72	05 02	06 20	07 29	09 33	08 48	08 18	07 51
N 70	05 08	06 18	07 20	09 01	08 32	08 10	07 51
68	05 13	06 17	07 13	08 37	08 19	08 04	07 51
66	05 17	06 16	07 07	08 19	08 08	07 59	07 51
64	05 20	06 14	07 02	08 04	07 59	07 55	07 51
62	05 22	06 13	06 58	07 51	07 52	07 51	07 51
60	05 24	06 12	06 54	07 40	07 45	07 48	07 51
N 58	05 26	06 11	06 51	07 31	07 39	07 45	07 51
56	05 28	06 10	06 48	07 22	07 33	07 43	07 51
54	05 29	06 10	06 45	07 15	07 29	07 40	07 51
52	05 30	06 09	06 43	07 08	07 24	07 38	07 51
50	05 31	06 08	06 40	07 02	07 20	07 36	07 51
45	05 32	06 06	06 35	06 49	07 12	07 32	07 50
N 40	05 33	06 04	06 31	06 38	07 05	07 28	07 50
35	05 33	06 02	06 28	06 29	06 58	07 25	07 50
30	05 33	06 00	06 24	06 21	06 53	07 22	07 50
20	05 31	05 57	06 19	06 06	06 43	07 18	07 50
N 10	05 28	05 52	06 14	05 54	06 35	07 13	07 50
0	05 24	05 48	06 09	05 42	06 27	07 09	07 50
S 10	05 18	05 43	06 04	05 31	06 19	07 05	07 50
20	05 10	05 36	05 58	05 18	06 11	07 01	07 50
30	04 59	05 28	05 52	05 03	06 01	06 56	07 50
35	04 52	05 23	05 48	04 55	05 55	06 53	07 50
40	04 44	05 16	05 44	04 45	05 49	06 50	07 50
45	04 33	05 09	05 39	04 34	05 41	06 46	07 50
S 50	04 19	05 00	05 33	04 20	05 32	06 42	07 50
52	04 13	04 56	05 30	04 14	05 28	06 40	07 50
54	04 05	04 51	05 27	04 07	05 23	06 38	07 50
56	03 57	04 45	05 24	03 59	05 18	06 35	07 50
58	03 47	04 39	05 20	03 50	05 12	06 32	07 50
S 60	03 35	04 32	05 16	03 39	05 06	06 29	07 50

Lat.	Sunset	Twilight Civil	Twilight Naut.	Moonset 2	Moonset 3	Moonset 4	Moonset 5
°	h m	h m	h m	h m	h m	h m	h m
N 72	16 57	18 06	19 25	14 36	16 53	18 51	20 44
N 70	17 06	18 08	19 19	15 06	17 06	18 55	20 40
68	17 13	18 09	19 14	15 28	17 18	18 59	20 37
66	17 18	18 10	19 10	15 45	17 27	19 02	20 34
64	17 23	18 11	19 06	15 59	17 34	19 04	20 32
62	17 28	18 12	19 04	16 11	17 41	19 06	20 30
60	17 31	18 13	19 01	16 21	17 46	19 08	20 29
N 58	17 35	18 14	18 59	16 29	17 51	19 10	20 27
56	17 37	18 15	18 58	16 37	17 55	19 11	20 26
54	17 40	18 16	18 56	16 44	17 59	19 13	20 25
52	17 42	18 16	18 55	16 50	18 03	19 14	20 24
50	17 45	18 17	18 54	16 55	18 06	19 15	20 23
45	17 49	18 19	18 53	17 07	18 13	19 17	20 20
N 40	17 53	18 21	18 52	17 16	18 19	19 19	20 19
35	17 57	18 22	18 52	17 24	18 23	19 21	20 17
30	18 00	18 24	18 52	17 31	18 28	19 22	20 16
20	18 06	18 28	18 53	17 44	18 35	19 25	20 13
N 10	18 11	18 32	18 56	17 54	18 42	19 27	20 11
0	18 15	18 36	19 00	18 04	18 48	19 29	20 09
S 10	18 20	18 41	19 06	18 14	18 54	19 31	20 08
20	18 25	18 48	19 14	18 24	19 00	19 33	20 05
30	18 32	18 56	19 24	18 36	19 07	19 36	20 03
35	18 35	19 01	19 31	18 43	19 11	19 37	20 02
40	18 39	19 07	19 39	18 51	19 16	19 38	20 00
45	18 44	19 14	19 50	19 00	19 21	19 40	19 59
S 50	18 50	19 23	20 03	19 10	19 27	19 42	19 56
52	18 52	19 27	20 10	19 15	19 30	19 43	19 55
54	18 55	19 32	20 17	19 21	19 33	19 44	19 54
56	18 59	19 37	20 25	19 27	19 37	19 45	19 53
58	19 02	19 43	20 35	19 33	19 41	19 47	19 52
S 60	19 06	19 50	20 46	19 41	19 45	19 48	19 51

Day	SUN Eqn. of Time 00^h	SUN Eqn. of Time 12^h	SUN Mer. Pass.	MOON Mer. Pass. Upper	MOON Mer. Pass. Lower	MOON Age	MOON Phase
d	m s	m s	h m	h m	h m	d	%
2	12 20	12 14	12 12	11 53	24 16	29	1
3	12 08	12 01	12 12	12 37	00 16	00	0
4	11 55	11 49	12 12	13 19	00 59	01	2

UT	ARIES	VENUS −4.1		MARS +0.9		JUPITER −2.5		SATURN +0.0	
	GHA	GHA	Dec	GHA	Dec	GHA	Dec	GHA	Dec
d h	° ′	° ′	° ′	° ′	° ′	° ′	° ′	° ′	° ′
WEDNESDAY									
5 00	162 19.8	217 17.9	S18 49.0	252 15.0	S23 31.3	30 10.1	N18 44.4	80 42.4	N22 08.0
01	177 22.2	232 17.3	48.5	267 15.8	31.3	45 12.8	44.4	95 44.8	08.0
02	192 24.7	247 16.7	48.0	282 16.5	31.4	60 15.5	44.5	110 47.2	08.0
03	207 27.2	262 16.1	. . 47.5	297 17.2	. . 31.4	75 18.2	. . 44.6	125 49.6	. . 08.0
04	222 29.6	277 15.5	47.0	312 17.9	31.5	90 20.9	44.6	140 52.0	08.1
05	237 32.1	292 14.8	46.5	327 18.7	31.5	105 23.6	44.7	155 54.4	08.1
06	252 34.5	307 14.2	S18 45.9	342 19.4	S23 31.5	120 26.3	N18 44.8	170 56.9	N22 08.1
07	267 37.0	322 13.6	45.4	357 20.1	31.6	135 29.0	44.8	185 59.3	08.1
08	282 39.5	337 13.0	44.9	12 20.9	31.6	150 31.7	44.9	201 01.7	08.1
09	297 41.9	352 12.4	. . 44.4	27 21.6	. . 31.7	165 34.4	. . 44.9	216 04.1	. . 08.1
10	312 44.4	7 11.8	43.9	42 22.3	31.7	180 37.1	45.0	231 06.5	08.1
11	327 46.9	22 11.2	43.3	57 23.0	31.7	195 39.8	45.0	246 08.9	08.1
12	342 49.3	37 10.6	S18 42.8	72 23.8	S23 31.8	210 42.5	N18 45.1	261 11.3	N22 08.1
13	357 51.8	52 10.0	42.3	87 24.5	31.8	225 45.1	45.2	276 13.7	08.2
14	12 54.3	67 09.4	41.8	102 25.2	31.9	240 47.8	45.2	291 16.1	08.2
15	27 56.7	82 08.8	. . 41.2	117 25.9	. . 31.9	255 50.5	. . 45.3	306 18.5	. . 08.2
16	42 59.2	97 08.2	40.7	132 26.7	31.9	270 53.2	45.3	321 20.9	08.2
17	58 01.7	112 07.6	40.2	147 27.4	32.0	285 55.9	45.4	336 23.3	08.2
18	73 04.1	127 07.0	S18 39.7	162 28.1	S23 32.0	300 58.6	N18 45.5	351 25.7	N22 08.2
19	88 06.6	142 06.4	39.1	177 28.8	32.0	316 01.3	45.5	6 28.2	08.2
20	103 09.0	157 05.7	38.6	192 29.6	32.1	331 04.0	45.6	21 30.6	08.2
21	118 11.5	172 05.1	. . 38.1	207 30.3	. . 32.1	346 06.7	. . 45.6	36 33.0	. . 08.3
22	133 14.0	187 04.5	37.5	222 31.0	32.2	1 09.4	45.7	51 35.4	08.3
23	148 16.4	202 03.9	37.0	237 31.7	32.2	16 12.0	45.8	66 37.8	08.3
THURSDAY									
6 00	163 18.9	217 03.3	S18 36.5	252 32.5	S23 32.2	31 14.7	N18 45.8	81 40.2	N22 08.3
01	178 21.4	232 02.7	35.9	267 33.2	32.3	46 17.4	45.9	96 42.6	08.3
02	193 23.8	247 02.1	35.4	282 33.9	32.3	61 20.1	45.9	111 45.0	08.3
03	208 26.3	262 01.5	. . 34.8	297 34.7	. . 32.3	76 22.8	. . 46.0	126 47.4	. . 08.3
04	223 28.8	277 00.9	34.3	312 35.4	32.4	91 25.5	46.1	141 49.8	08.3
05	238 31.2	292 00.3	33.8	327 36.1	32.4	106 28.2	46.1	156 52.2	08.3
06	253 33.7	306 59.7	S18 33.2	342 36.8	S23 32.4	121 30.9	N18 46.2	171 54.6	N22 08.4
07	268 36.2	321 59.1	32.7	357 37.6	32.5	136 33.5	46.2	186 57.0	08.4
08	283 38.6	336 58.5	32.1	12 38.3	32.5	151 36.2	46.3	201 59.4	08.4
09	298 41.1	351 57.9	. . 31.6	27 39.0	. . 32.5	166 38.9	. . 46.3	217 01.8	. . 08.4
10	313 43.5	6 57.3	31.1	42 39.7	32.6	181 41.6	46.4	232 04.2	08.4
11	328 46.0	21 56.7	30.5	57 40.5	32.6	196 44.3	46.5	247 06.6	08.4
12	343 48.5	36 56.1	S18 30.0	72 41.2	S23 32.6	211 47.0	N18 46.5	262 09.0	N22 08.4
13	358 50.9	51 55.5	29.4	87 41.9	32.7	226 49.7	46.6	277 11.4	08.4
14	13 53.4	66 54.9	28.9	102 42.7	32.7	241 52.3	46.6	292 13.8	08.5
15	28 55.9	81 54.3	. . 28.3	117 43.4	. . 32.7	256 55.0	. . 46.7	307 16.2	. . 08.5
16	43 58.3	96 53.7	27.8	132 44.1	32.8	271 57.7	46.7	322 18.6	08.5
17	59 00.8	111 53.1	27.2	147 44.8	32.8	287 00.4	46.8	337 21.0	08.5
18	74 03.3	126 52.5	S18 26.7	162 45.6	S23 32.8	302 03.1	N18 46.9	352 23.5	N22 08.5
19	89 05.7	141 51.9	26.1	177 46.3	32.8	317 05.8	46.9	7 25.9	08.5
20	104 08.2	156 51.3	25.6	192 47.0	32.9	332 08.4	47.0	22 28.3	08.5
21	119 10.6	171 50.7	. . 25.0	207 47.7	. . 32.9	347 11.1	. . 47.0	37 30.7	. . 08.5
22	134 13.1	186 50.1	24.4	222 48.5	32.9	2 13.8	47.1	52 33.1	08.6
23	149 15.6	201 49.5	23.9	237 49.2	33.0	17 16.5	47.1	67 35.5	08.6
FRIDAY									
7 00	164 18.0	216 48.9	S18 23.3	252 49.9	S23 33.0	32 19.2	N18 47.2	82 37.9	N22 08.6
01	179 20.5	231 48.3	22.8	267 50.7	33.0	47 21.8	47.3	97 40.3	08.6
02	194 23.0	246 47.7	22.2	282 51.4	33.0	62 24.5	47.3	112 42.7	08.6
03	209 25.4	261 47.1	. . 21.7	297 52.1	. . 33.1	77 27.2	. . 47.4	127 45.1	. . 08.6
04	224 27.9	276 46.5	21.1	312 52.8	33.1	92 29.9	47.4	142 47.5	08.6
05	239 30.4	291 45.9	20.5	327 53.6	33.1	107 32.6	47.5	157 49.9	08.6
06	254 32.8	306 45.3	S18 20.0	342 54.3	S23 33.2	122 35.3	N18 47.5	172 52.3	N22 08.7
07	269 35.3	321 44.7	19.4	357 55.0	33.2	137 37.9	47.6	187 54.7	08.7
08	284 37.8	336 44.1	18.8	12 55.7	33.2	152 40.6	47.6	202 57.1	08.7
09	299 40.2	351 43.5	. . 18.3	27 56.5	. . 33.2	167 43.3	. . 47.7	217 59.5	. . 08.7
10	314 42.7	6 42.9	17.7	42 57.2	33.3	182 46.0	47.8	233 01.9	08.7
11	329 45.1	21 42.3	17.1	57 57.9	33.3	197 48.6	47.8	248 04.3	08.7
12	344 47.6	36 41.7	S18 16.6	72 58.7	S23 33.3	212 51.3	N18 47.9	263 06.6	N22 08.7
13	359 50.1	51 41.1	16.0	87 59.4	33.3	227 54.0	47.9	278 09.0	08.7
14	14 52.5	66 40.5	15.4	103 00.1	33.4	242 56.7	48.0	293 11.4	08.7
15	29 55.0	81 39.9	. . 14.9	118 00.8	. . 33.4	257 59.4	. . 48.0	308 13.8	. . 08.8
16	44 57.5	96 39.3	14.3	133 01.6	33.4	273 02.0	48.1	323 16.2	08.8
17	59 59.9	111 38.8	13.7	148 02.3	33.4	288 04.7	48.1	338 18.6	08.8
18	75 02.4	126 38.2	S18 13.1	163 03.0	S23 33.5	303 07.4	N18 48.2	353 21.0	N22 08.8
19	90 04.9	141 37.6	12.6	178 03.8	33.5	318 10.1	48.3	8 23.4	08.8
20	105 07.3	156 37.0	12.0	193 04.5	33.5	333 12.7	48.3	23 25.8	08.8
21	120 09.8	171 36.4	. . 11.4	208 05.2	. . 33.5	348 15.4	. . 48.4	38 28.2	. . 08.8
22	135 12.2	186 35.8	10.8	223 05.9	33.5	3 18.1	48.4	53 30.6	08.8
23	150 14.7	201 35.2	10.3	238 06.7	33.6	18 20.8	48.5	68 33.0	08.9
Mer. Pass.	h m 13 04.6	v −0.6	d 0.5	v 0.7	d 0.0	v 2.7	d 0.1	v 2.4	d 0.0

STARS

Name	SHA	Dec
	° ′	° ′
Acamar	315 24.6	S40 17.8
Achernar	335 33.0	S57 13.5
Acrux	173 18.1	S63 06.9
Adhara	255 18.7	S28 58.7
Aldebaran	290 58.7	N16 30.9
Alioth	166 27.1	N55 56.4
Alkaid	153 04.8	N49 17.7
Al Na'ir	27 54.2	S46 56.9
Alnilam	275 54.5	S 1 12.1
Alphard	218 03.8	S 8 40.4
Alphecca	126 17.7	N26 42.0
Alpheratz	357 52.3	N29 06.3
Altair	62 16.4	N 8 52.3
Ankaa	353 23.9	S42 17.6
Antares	112 36.2	S26 26.3
Arcturus	146 02.9	N19 09.8
Atria	107 45.5	S69 01.8
Avior	234 21.0	S59 31.3
Bellatrix	278 40.7	N 6 21.1
Betelgeuse	271 10.0	N 7 24.4
Canopus	263 59.6	S52 42.1
Capella	280 46.4	N46 00.3
Deneb	49 37.5	N45 17.2
Denebola	182 41.6	N14 33.2
Diphda	349 04.3	S17 58.4
Dubhe	194 00.6	N61 44.1
Elnath	278 22.8	N28 36.7
Eltanin	90 50.0	N51 28.9
Enif	33 55.4	N 9 53.1
Fomalhaut	15 33.2	S29 36.6
Gacrux	172 09.7	S57 07.7
Gienah	176 00.4	S17 33.6
Hadar	148 59.2	S60 23.1
Hamal	328 10.2	N23 28.6
Kaus Aust.	83 54.7	S34 23.0
Kochab	137 18.7	N74 08.3
Markab	13 46.8	N15 13.1
Menkar	314 23.7	N 4 06.0
Menkent	148 17.0	S36 23.1
Miaplacidus	221 40.9	S69 43.9
Mirfak	308 52.3	N49 52.5
Nunki	76 08.5	S26 17.6
Peacock	53 32.3	S56 43.5
Pollux	243 37.4	N28 01.2
Procyon	245 08.0	N 5 13.0
Rasalhague	96 14.0	N12 33.2
Regulus	207 51.8	N11 57.1
Rigel	281 19.8	S 8 12.0
Rigil Kent.	140 02.8	S60 50.7
Sabik	102 21.9	S15 43.8
Schedar	349 50.5	N56 33.2
Shaula	96 33.0	S37 06.3
Sirius	258 40.8	S16 43.3
Spica	158 39.6	S11 10.7
Suhail	222 58.1	S43 26.8
Vega	80 44.6	N38 46.8
Zuben'ubi	137 14.3	S16 03.3

	SHA	Mer. Pass.
	° ′	h m
Venus	53 44.4	9 32
Mars	89 13.6	7 09
Jupiter	227 55.8	21 51
Saturn	278 21.3	18 30

UT		SUN GHA	SUN Dec	MOON GHA	v	MOON Dec	d	HP
d	h	° ′	° ′	° ′	′	° ′	′	′
5	00	177 04.5	S 6 17.8	155 46.8	16.3	S 1 46.3	13.2	54.5
	01	192 04.6	16.9	170 22.1	16.4	1 33.1	13.3	54.5
	02	207 04.8	15.9	184 57.5	16.3	1 19.8	13.2	54.5
	03	222 04.9	. . 14.9	199 32.8	16.4	1 06.6	13.3	54.5
	04	237 05.1	14.0	214 08.2	16.4	0 53.3	13.2	54.5
	05	252 05.2	13.0	228 43.6	16.5	0 40.1	13.3	54.5
	06	267 05.3	S 6 12.0	243 19.1	16.4	S 0 26.8	13.2	54.5
WEDNESDAY	07	282 05.5	11.1	257 54.5	16.5	0 13.6	13.2	54.4
	08	297 05.6	10.1	272 30.0	16.4	S 0 00.4	13.3	54.4
	09	312 05.8	. . 09.1	287 05.4	16.5	N 0 12.9	13.2	54.4
	10	327 05.9	08.2	301 40.9	16.5	0 26.1	13.2	54.4
	11	342 06.0	07.2	316 16.4	16.5	0 39.3	13.2	54.4
	12	357 06.2	S 6 06.2	330 51.9	16.5	N 0 52.5	13.2	54.4
	13	12 06.3	05.3	345 27.4	16.5	1 05.7	13.2	54.4
	14	27 06.4	04.3	0 02.9	16.6	1 18.9	13.2	54.4
	15	42 06.6	. . 03.3	14 38.5	16.5	1 32.1	13.1	54.4
	16	57 06.7	02.4	29 14.0	16.5	1 45.2	13.2	54.4
	17	72 06.9	01.4	43 49.5	16.6	1 58.4	13.1	54.3
	18	87 07.0	S 6 00.4	58 25.1	16.5	N 2 11.5	13.1	54.3
	19	102 07.1	5 59.5	73 00.6	16.6	2 24.6	13.1	54.3
	20	117 07.3	58.5	87 36.2	16.6	2 37.7	13.1	54.3
	21	132 07.4	. . 57.5	102 11.8	16.5	2 50.8	13.1	54.3
	22	147 07.6	56.6	116 47.3	16.6	3 03.9	13.1	54.3
	23	162 07.7	55.6	131 22.9	16.5	3 17.0	13.0	54.3
6	00	177 07.9	S 5 54.6	145 58.4	16.6	N 3 30.0	13.1	54.3
	01	192 08.0	53.7	160 34.0	16.5	3 43.1	13.0	54.3
	02	207 08.1	52.7	175 09.5	16.6	3 56.1	13.0	54.3
	03	222 08.3	. . 51.7	189 45.1	16.5	4 09.1	12.9	54.2
	04	237 08.4	50.8	204 20.6	16.6	4 22.0	13.0	54.2
	05	252 08.6	49.8	218 56.2	16.5	4 35.0	12.9	54.2
	06	267 08.7	S 5 48.8	233 31.7	16.5	N 4 47.9	12.9	54.2
	07	282 08.8	47.9	248 07.2	16.6	5 00.8	12.9	54.2
THURSDAY	08	297 09.0	46.9	262 42.8	16.5	5 13.7	12.9	54.2
	09	312 09.1	. . 45.9	277 18.3	16.5	5 26.6	12.8	54.2
	10	327 09.3	45.0	291 53.8	16.5	5 39.4	12.8	54.2
	11	342 09.4	44.0	306 29.3	16.4	5 52.2	12.8	54.2
	12	357 09.6	S 5 43.0	321 04.7	16.5	N 6 05.0	12.8	54.2
	13	12 09.7	42.1	335 40.2	16.4	6 17.8	12.7	54.2
	14	27 09.9	41.1	350 15.6	16.5	6 30.5	12.7	54.2
	15	42 10.0	. . 40.1	4 51.1	16.4	6 43.2	12.7	54.2
	16	57 10.1	39.1	19 26.5	16.4	6 55.9	12.7	54.2
	17	72 10.3	38.2	34 01.9	16.4	7 08.6	12.6	54.2
	18	87 10.4	S 5 37.2	48 37.3	16.4	N 7 21.2	12.6	54.1
	19	102 10.6	36.2	63 12.7	16.3	7 33.8	12.5	54.1
	20	117 10.7	35.3	77 48.0	16.4	7 46.3	12.6	54.1
	21	132 10.9	. . 34.3	92 23.4	16.3	7 58.9	12.5	54.1
	22	147 11.0	33.3	106 58.7	16.3	8 11.4	12.4	54.1
	23	162 11.2	32.4	121 34.0	16.2	8 23.8	12.5	54.1
7	00	177 11.3	S 5 31.4	136 09.2	16.3	N 8 36.3	12.4	54.1
	01	192 11.4	30.4	150 44.5	16.2	8 48.7	12.3	54.1
	02	207 11.6	29.4	165 19.7	16.2	9 01.0	12.3	54.1
	03	222 11.7	. . 28.5	179 54.9	16.2	9 13.3	12.3	54.1
	04	237 11.9	27.5	194 30.1	16.1	9 25.6	12.3	54.1
	05	252 12.0	26.5	209 05.2	16.1	9 37.9	12.2	54.1
	06	267 12.2	S 5 25.6	223 40.3	16.1	N 9 50.1	12.2	54.1
	07	282 12.3	24.6	238 15.4	16.1	10 02.3	12.1	54.1
	08	297 12.5	23.6	252 50.5	16.0	10 14.4	12.1	54.1
FRIDAY	09	312 12.6	. . 22.6	267 25.5	16.0	10 26.5	12.1	54.1
	10	327 12.8	21.7	282 00.5	16.0	10 38.6	12.0	54.1
	11	342 12.9	20.7	296 35.5	15.9	10 50.6	11.9	54.1
	12	357 13.1	S 5 19.7	311 10.4	15.9	N11 02.5	12.0	54.1
	13	12 13.2	18.8	325 45.3	15.9	11 14.5	11.8	54.1
	14	27 13.4	17.8	340 20.2	15.8	11 26.3	11.9	54.1
	15	42 13.5	. . 16.8	354 55.0	15.8	11 38.2	11.8	54.1
	16	57 13.7	15.8	9 29.8	15.7	11 50.0	11.7	54.1
	17	72 13.8	14.9	24 04.5	15.8	12 01.7	11.7	54.1
	18	87 14.0	S 5 13.9	38 39.3	15.6	N12 13.4	11.7	54.1
	19	102 14.1	12.9	53 13.9	15.7	12 25.1	11.6	54.1
	20	117 14.3	11.9	67 48.6	15.6	12 36.7	11.6	54.1
	21	132 14.4	. . 11.0	82 23.2	15.5	12 48.3	11.5	54.1
	22	147 14.6	10.0	96 57.7	15.6	12 59.8	11.4	54.1
	23	162 14.7	09.0	111 32.3	15.4	N13 11.2	11.4	54.1
		SD 16.1	d 1.0	SD 14.8		14.8		14.7

Lat.	Twilight Naut.	Twilight Civil	Sunrise	Moonrise 5	Moonrise 6	Moonrise 7	Moonrise 8
°	h m	h m	h m	h m	h m	h m	h m
N 72	04 47	06 05	07 13	07 51	07 26	06 56	06 16
N 70	04 54	06 05	07 06	07 51	07 32	07 12	06 45
68	05 00	06 05	07 00	07 51	07 38	07 24	07 07
66	05 05	06 05	06 56	07 51	07 43	07 34	07 24
64	05 09	06 04	06 52	07 51	07 47	07 43	07 38
62	05 13	06 04	06 48	07 51	07 50	07 50	07 50
60	05 16	06 04	06 45	07 51	07 53	07 56	08 01
N 58	05 18	06 03	06 42	07 51	07 56	08 02	08 10
56	05 20	06 03	06 40	07 51	07 59	08 07	08 17
54	05 22	06 03	06 38	07 51	08 01	08 12	08 24
52	05 23	06 02	06 36	07 51	08 03	08 16	08 31
50	05 24	06 02	06 34	07 51	08 05	08 20	08 37
45	05 27	06 01	06 30	07 50	08 09	08 28	08 49
N 40	05 28	06 00	06 27	07 50	08 12	08 35	08 59
35	05 29	05 58	06 24	07 50	08 15	08 41	09 08
30	05 29	05 57	06 21	07 50	08 18	08 46	09 16
20	05 29	05 54	06 16	07 50	08 22	08 55	09 30
N 10	05 27	05 51	06 12	07 50	08 26	09 03	09 42
0	05 23	05 47	06 08	07 50	08 30	09 11	09 53
S 10	05 18	05 43	06 04	07 50	08 34	09 19	10 04
20	05 11	05 37	05 59	07 50	08 38	09 27	10 16
30	05 02	05 30	05 54	07 50	08 43	09 36	10 30
35	04 55	05 25	05 51	07 50	08 46	09 42	10 39
40	04 47	05 20	05 47	07 50	08 49	09 48	10 48
45	04 38	05 13	05 43	07 50	08 53	09 56	10 59
S 50	04 25	05 05	05 38	07 50	08 57	10 05	11 12
52	04 19	05 01	05 36	07 50	08 59	10 09	11 19
54	04 12	04 57	05 33	07 50	09 02	10 13	11 26
56	04 05	04 52	05 31	07 50	09 04	10 18	11 33
58	03 56	04 47	05 27	07 50	09 07	10 24	11 42
S 60	03 45	04 40	05 24	07 50	09 10	10 30	11 52

Lat.	Sunset	Twilight Civil	Twilight Naut.	Moonset 5	Moonset 6	Moonset 7	Moonset 8
°	h m	h m	h m	h m	h m	h m	h m
N 72	17 12	18 20	19 39	20 44	22 39	24 46	00 46
N 70	17 19	18 20	19 31	20 40	22 26	24 19	00 19
68	17 24	18 20	19 25	20 37	22 16	23 58	25 53
66	17 29	18 20	19 20	20 34	22 07	23 43	25 24
64	17 32	18 20	19 15	20 32	22 00	23 30	25 03
62	17 36	18 20	19 12	20 30	21 54	23 19	24 46
60	17 39	18 20	19 09	20 29	21 48	23 09	24 32
N 58	17 41	18 21	19 06	20 27	21 44	23 01	24 20
56	17 44	18 21	19 04	20 26	21 40	22 54	24 10
54	17 46	18 21	19 02	20 25	21 36	22 48	24 01
52	17 48	18 22	19 01	20 24	21 33	22 42	23 53
50	17 50	18 22	18 59	20 23	21 30	22 37	23 45
45	17 53	18 23	18 57	20 20	21 23	22 26	23 30
N 40	17 57	18 24	18 55	20 19	21 18	22 17	23 17
35	18 00	18 25	18 54	20 17	21 13	22 09	23 06
30	18 02	18 26	18 54	20 16	21 09	22 02	22 57
20	18 07	18 29	18 54	20 13	21 02	21 51	22 41
N 10	18 11	18 32	18 56	20 11	20 56	21 41	22 27
0	18 15	18 35	19 00	20 09	20 50	21 31	22 14
S 10	18 19	18 40	19 04	20 08	20 44	21 21	22 01
20	18 23	18 45	19 11	20 05	20 38	21 11	21 47
30	18 28	18 52	19 21	20 03	20 31	21 00	21 31
35	18 31	18 57	19 27	20 02	20 27	20 53	21 22
40	18 35	19 02	19 34	20 00	20 22	20 46	21 12
45	18 39	19 08	19 44	19 59	20 17	20 37	21 00
S 50	18 43	19 16	19 56	19 56	20 11	20 27	20 45
52	18 46	19 20	20 02	19 55	20 08	20 22	20 38
54	18 48	19 24	20 09	19 54	20 05	20 17	20 30
56	18 51	19 29	20 16	19 53	20 01	20 11	20 22
58	18 54	19 34	20 25	19 52	19 58	20 04	20 12
S 60	18 57	19 41	20 35	19 51	19 53	19 57	20 02

Day	SUN Eqn. of Time 00^h	SUN Eqn. of Time 12^h	SUN Mer. Pass.	MOON Mer. Pass. Upper	MOON Mer. Pass. Lower	MOON Age	MOON Phase
d	m s	m s	h m	h m	h m	d %	
5	11 42	11 36	12 12	14 00	01 40	02 6	
6	11 29	11 22	12 11	14 40	02 20	03 11	
7	11 15	11 08	12 11	15 21	03 00	04 17	

UT		ARIES	VENUS −4.1		MARS +0.8		JUPITER −2.5		SATURN +0.0	
	d h	GHA ° ′	GHA ° ′	Dec ° ′	GHA ° ′	Dec ° ′	GHA ° ′	Dec ° ′	GHA ° ′	Dec ° ′
	8 00	165 17.2	216 34.6	S18 09.7	253 07.4	S23 33.6	33 23.4	N18 48.5	83 35.4	N22 08.9
	01	180 19.6	231 34.0	09.1	268 08.1	33.6	48 26.1	48.6	98 37.8	08.9
	02	195 22.1	246 33.4	08.5	283 08.8	33.6	63 28.8	48.6	113 40.2	08.9
	03	210 24.6	261 32.8	. . 07.9	298 09.6	. . 33.6	78 31.5	. . 48.7	128 42.6	. . 08.9
	04	225 27.0	276 32.2	07.4	313 10.3	33.7	93 34.1	48.8	143 45.0	08.9
	05	240 29.5	291 31.6	06.8	328 11.0	33.7	108 36.8	48.8	158 47.4	08.9
	06	255 32.0	306 31.0	S18 06.2	343 11.8	S23 33.7	123 39.5	N18 48.9	173 49.8	N22 08.9
	07	270 34.4	321 30.5	05.6	358 12.5	33.7	138 42.2	48.9	188 52.2	09.0
S	08	285 36.9	336 29.9	05.0	13 13.2	33.7	153 44.8	49.0	203 54.6	09.0
A	09	300 39.4	351 29.3	. . 04.4	28 13.9	. . 33.8	168 47.5	. . 49.0	218 57.0	. . 09.0
T	10	315 41.8	6 28.7	03.9	43 14.7	33.8	183 50.2	49.1	233 59.4	09.0
U	11	330 44.3	21 28.1	03.3	58 15.4	33.8	198 52.9	49.1	249 01.7	09.0
R	12	345 46.7	36 27.5	S18 02.7	73 16.1	S23 33.8	213 55.5	N18 49.2	264 04.1	N22 09.0
D	13	0 49.2	51 26.9	02.1	88 16.9	33.8	228 58.2	49.2	279 06.5	09.0
A	14	15 51.7	66 26.3	01.5	103 17.6	33.8	244 00.9	49.3	294 08.9	09.0
Y	15	30 54.1	81 25.7	. . 00.9	118 18.3	. . 33.9	259 03.5	. . 49.3	309 11.3	. . 09.1
	16	45 56.6	96 25.1	18 00.3	133 19.1	33.9	274 06.2	49.4	324 13.7	09.1
	17	60 59.1	111 24.6	17 59.7	148 19.8	33.9	289 08.9	49.4	339 16.1	09.1
	18	76 01.5	126 24.0	S17 59.1	163 20.5	S23 33.9	304 11.5	N18 49.5	354 18.5	N22 09.1
	19	91 04.0	141 23.4	58.5	178 21.2	33.9	319 14.2	49.6	9 20.9	09.1
	20	106 06.5	156 22.8	57.9	193 22.0	33.9	334 16.9	49.6	24 23.3	09.1
	21	121 08.9	171 22.2	. . 57.3	208 22.7	. . 34.0	349 19.6	. . 49.7	39 25.7	. . 09.1
	22	136 11.4	186 21.6	56.7	223 23.4	34.0	4 22.2	49.7	54 28.1	09.1
	23	151 13.9	201 21.0	56.1	238 24.2	34.0	19 24.9	49.8	69 30.5	09.2
	9 00	166 16.3	216 20.4	S17 55.5	253 24.9	S23 34.0	34 27.6	N18 49.8	84 32.8	N22 09.2
	01	181 18.8	231 19.9	54.9	268 25.6	34.0	49 30.2	49.9	99 35.2	09.2
	02	196 21.2	246 19.3	54.3	283 26.3	34.0	64 32.9	49.9	114 37.6	09.2
	03	211 23.7	261 18.7	. . 53.7	298 27.1	. . 34.0	79 35.6	. . 50.0	129 40.0	. . 09.2
	04	226 26.2	276 18.1	53.1	313 27.8	34.1	94 38.2	50.0	144 42.4	09.2
	05	241 28.6	291 17.5	52.5	328 28.5	34.1	109 40.9	50.1	159 44.8	09.2
	06	256 31.1	306 16.9	S17 51.9	343 29.3	S23 34.1	124 43.6	N18 50.1	174 47.2	N22 09.2
	07	271 33.6	321 16.3	51.3	358 30.0	34.1	139 46.2	50.2	189 49.6	09.3
	08	286 36.0	336 15.8	50.7	13 30.7	34.1	154 48.9	50.2	204 52.0	09.3
S	09	301 38.5	351 15.2	. . 50.1	28 31.4	. . 34.1	169 51.6	. . 50.3	219 54.4	. . 09.3
U	10	316 41.0	6 14.6	49.5	43 32.2	34.1	184 54.2	50.3	234 56.7	09.3
N	11	331 43.4	21 14.0	48.9	58 32.9	34.1	199 56.9	50.4	249 59.1	09.3
D	12	346 45.9	36 13.4	S17 48.3	73 33.6	S23 34.1	214 59.6	N18 50.4	265 01.5	N22 09.3
A	13	1 48.3	51 12.8	47.6	88 34.4	34.2	230 02.2	50.5	280 03.9	09.3
Y	14	16 50.8	66 12.3	47.0	103 35.1	34.2	245 04.9	50.5	295 06.3	09.3
	15	31 53.3	81 11.7	. . 46.4	118 35.8	. . 34.2	260 07.5	. . 50.6	310 08.7	. . 09.4
	16	46 55.7	96 11.1	45.8	133 36.6	34.2	275 10.2	50.6	325 11.1	09.4
	17	61 58.2	111 10.5	45.2	148 37.3	34.2	290 12.9	50.7	340 13.5	09.4
	18	77 00.7	126 09.9	S17 44.6	163 38.0	S23 34.2	305 15.5	N18 50.8	355 15.8	N22 09.4
	19	92 03.1	141 09.3	44.0	178 38.7	34.2	320 18.2	50.8	10 18.2	09.4
	20	107 05.6	156 08.8	43.3	193 39.5	34.2	335 20.9	50.9	25 20.6	09.4
	21	122 08.1	171 08.2	. . 42.7	208 40.2	. . 34.2	350 23.5	. . 50.9	40 23.0	. . 09.4
	22	137 10.5	186 07.6	42.1	223 40.9	34.2	5 26.2	51.0	55 25.4	09.5
	23	152 13.0	201 07.0	41.5	238 41.7	34.2	20 28.8	51.0	70 27.8	09.5
	10 00	167 15.5	216 06.4	S17 40.9	253 42.4	S23 34.3	35 31.5	N18 51.1	85 30.2	N22 09.5
	01	182 17.9	231 05.9	40.2	268 43.1	34.3	50 34.2	51.1	100 32.5	09.5
	02	197 20.4	246 05.3	39.6	283 43.9	34.3	65 36.8	51.2	115 34.9	09.5
	03	212 22.8	261 04.7	. . 39.0	298 44.6	. . 34.3	80 39.5	. . 51.2	130 37.3	. . 09.5
	04	227 25.3	276 04.1	38.4	313 45.3	34.3	95 42.2	51.3	145 39.7	09.5
	05	242 27.8	291 03.5	37.7	328 46.0	34.3	110 44.8	51.3	160 42.1	09.5
	06	257 30.2	306 03.0	S17 37.1	343 46.8	S23 34.3	125 47.5	N18 51.4	175 44.5	N22 09.6
	07	272 32.7	321 02.4	36.5	358 47.5	34.3	140 50.1	51.4	190 46.9	09.6
	08	287 35.2	336 01.8	35.9	13 48.2	34.3	155 52.8	51.5	205 49.2	09.6
M	09	302 37.6	351 01.2	. . 35.2	28 49.0	. . 34.3	170 55.4	. . 51.5	220 51.6	. . 09.6
O	10	317 40.1	6 00.7	34.6	43 49.7	34.3	185 58.1	51.6	235 54.0	09.6
N	11	332 42.6	21 00.1	34.0	58 50.4	34.3	201 00.8	51.6	250 56.4	09.6
D	12	347 45.0	35 59.5	S17 33.3	73 51.2	S23 34.3	216 03.4	N18 51.7	265 58.8	N22 09.6
A	13	2 47.5	50 58.9	32.7	88 51.9	34.3	231 06.1	51.7	281 01.2	09.6
Y	14	17 50.0	65 58.3	32.1	103 52.6	34.3	246 08.7	51.8	296 03.5	09.7
	15	32 52.4	80 57.8	. . 31.4	118 53.4	. . 34.3	261 11.4	. . 51.8	311 05.9	. . 09.7
	16	47 54.9	95 57.2	30.8	133 54.1	34.3	276 14.0	51.9	326 08.3	09.7
	17	62 57.3	110 56.6	30.2	148 54.8	34.3	291 16.7	51.9	341 10.7	09.7
	18	77 59.8	125 56.0	S17 29.5	163 55.5	S23 34.3	306 19.4	N18 52.0	356 13.1	N22 09.7
	19	93 02.3	140 55.5	28.9	178 56.3	34.3	321 22.0	52.0	11 15.4	09.7
	20	108 04.7	155 54.9	28.2	193 57.0	34.3	336 24.7	52.1	26 17.8	09.7
	21	123 07.2	170 54.3	. . 27.6	208 57.7	. . 34.3	351 27.3	. . 52.1	41 20.2	. . 09.7
	22	138 09.7	185 53.7	27.0	223 58.5	34.3	6 30.0	52.1	56 22.6	09.8
	23	153 12.1	200 53.2	26.3	238 59.2	34.3	21 32.6	52.2	71 25.0	09.8
Mer. Pass.		h m 12 52.8	v −0.6	d 0.6	v 0.7	d 0.0	v 2.7	d 0.1	v 2.4	d 0.0

STARS		
Name	SHA ° ′	Dec ° ′
Acamar	315 24.6	S40 17.8
Achernar	335 33.0	S57 13.5
Acrux	173 18.0	S63 06.9
Adhara	255 18.7	S28 58.7
Aldebaran	290 58.8	N16 30.9
Alioth	166 27.0	N55 56.4
Alkaid	153 04.7	N49 17.7
Al Na'ir	27 54.2	S46 56.9
Alnilam	275 54.6	S 1 12.1
Alphard	218 03.8	S 8 40.4
Alphecca	126 17.7	N26 42.0
Alpheratz	357 52.3	N29 06.3
Altair	62 16.3	N 8 52.3
Ankaa	353 23.9	S42 17.6
Antares	112 36.2	S26 26.3
Arcturus	146 02.9	N19 09.8
Atria	107 45.4	S69 01.8
Avior	234 21.0	S59 31.3
Bellatrix	278 40.7	N 6 21.1
Betelgeuse	271 10.0	N 7 24.4
Canopus	263 59.6	S52 42.1
Capella	280 46.4	N46 00.3
Deneb	49 37.5	N45 17.2
Denebola	182 41.6	N14 33.2
Diphda	349 04.3	S17 58.4
Dubhe	194 00.6	N61 44.1
Elnath	278 22.9	N28 36.7
Eltanin	90 50.0	N51 28.9
Enif	33 55.4	N 9 53.1
Fomalhaut	15 33.2	S29 36.5
Gacrux	172 09.7	S57 07.7
Gienah	176 00.4	S17 33.6
Hadar	148 59.2	S60 23.1
Hamal	328 10.2	N23 28.6
Kaus Aust.	83 54.7	S34 23.0
Kochab	137 18.7	N74 08.3
Markab	13 46.8	N15 13.1
Menkar	314 23.7	N 4 06.0
Menkent	148 17.0	S36 23.1
Miaplacidus	221 40.9	S69 43.9
Mirfak	308 52.3	N49 52.5
Nunki	76 08.5	S26 17.6
Peacock	53 32.3	S56 43.5
Pollux	243 37.4	N28 01.2
Procyon	245 08.0	N 5 13.0
Rasalhague	96 14.0	N12 33.2
Regulus	207 51.8	N11 57.1
Rigel	281 19.8	S 8 12.0
Rigil Kent.	140 02.7	S60 50.7
Sabik	102 21.9	S15 43.8
Schedar	349 50.5	N56 33.2
Shaula	96 33.0	S37 06.3
Sirius	258 40.8	S16 43.3
Spica	158 39.6	S11 10.7
Suhail	222 58.1	S43 26.8
Vega	80 44.6	N38 46.8
Zuben'ubi	137 14.3	S16 03.3

	SHA ° ′	Mer. Pass. h m
Venus	50 04.1	9 35
Mars	87 08.6	7 06
Jupiter	228 11.2	21 38
Saturn	278 16.5	18 19

UT d	h	SUN GHA ° ′	SUN Dec ° ′	MOON GHA ° ′	v ′	MOON Dec ° ′	d ′	HP ′
8 SATURDAY	00	177 14.9	S 5 08.1	126 06.7	15.5	N13 22.6	11.4	54.1
	01	192 15.0	07.1	140 41.2	15.3	13 34.0	11.3	54.1
	02	207 15.2	06.1	155 15.5	15.4	13 45.3	11.3	54.1
	03	222 15.3	. . 05.1	169 49.9	15.3	13 56.6	11.2	54.1
	04	237 15.5	04.2	184 24.2	15.2	14 07.8	11.1	54.1
	05	252 15.6	03.2	198 58.4	15.2	14 18.9	11.1	54.1
	06	267 15.8	S 5 02.2	213 32.6	15.2	N14 30.0	11.0	54.1
	07	282 15.9	01.2	228 06.8	15.1	14 41.0	11.0	54.1
	08	297 16.1	5 00.3	242 40.9	15.0	14 52.0	10.9	54.1
	09	312 16.2	4 59.3	257 14.9	15.0	15 02.9	10.9	54.1
	10	327 16.4	58.3	271 48.9	15.0	15 13.8	10.8	54.1
	11	342 16.5	57.3	286 22.9	14.9	15 24.6	10.8	54.1
	12	357 16.7	S 4 56.4	300 56.8	14.8	N15 35.4	10.6	54.1
	13	12 16.8	55.4	315 30.6	14.8	15 46.0	10.7	54.1
	14	27 17.0	54.4	330 04.4	14.8	15 56.7	10.5	54.1
	15	42 17.1	. . 53.4	344 38.2	14.7	16 07.2	10.5	54.2
	16	57 17.3	52.5	359 11.9	14.6	16 17.7	10.5	54.2
	17	72 17.4	51.5	13 45.5	14.6	16 28.2	10.3	54.2
	18	87 17.6	S 4 50.5	28 19.1	14.5	N16 38.5	10.3	54.2
	19	102 17.7	49.5	42 52.6	14.4	16 48.8	10.3	54.2
	20	117 17.9	48.6	57 26.0	14.5	16 59.1	10.2	54.2
	21	132 18.0	. . 47.6	71 59.5	14.3	17 09.3	10.1	54.2
	22	147 18.2	46.6	86 32.8	14.3	17 19.4	10.0	54.2
	23	162 18.4	45.6	101 06.1	14.2	17 29.4	10.0	54.2
9 SUNDAY	00	177 18.5	S 4 44.6	115 39.3	14.2	N17 39.4	9.9	54.2
	01	192 18.7	43.7	130 12.5	14.1	17 49.3	9.8	54.2
	02	207 18.8	42.7	144 45.6	14.0	17 59.1	9.8	54.2
	03	222 19.0	. . 41.7	159 18.6	14.0	18 08.9	9.7	54.2
	04	237 19.1	40.7	173 51.6	13.9	18 18.6	9.6	54.3
	05	252 19.3	39.8	188 24.5	13.8	18 28.2	9.5	54.3
	06	267 19.4	S 4 38.8	202 57.3	13.8	N18 37.7	9.5	54.3
	07	282 19.6	37.8	217 30.1	13.8	18 47.2	9.4	54.3
	08	297 19.7	36.8	232 02.9	13.6	18 56.6	9.3	54.3
	09	312 19.9	. . 35.9	246 35.5	13.6	19 05.9	9.3	54.3
	10	327 20.1	34.9	261 08.1	13.5	19 15.2	9.1	54.3
	11	342 20.2	33.9	275 40.6	13.5	19 24.3	9.1	54.3
	12	357 20.4	S 4 32.9	290 13.1	13.4	N19 33.4	9.0	54.3
	13	12 20.5	31.9	304 45.5	13.3	19 42.4	8.9	54.3
	14	27 20.7	31.0	319 17.8	13.2	19 51.3	8.9	54.4
	15	42 20.8	. . 30.0	333 50.0	13.2	20 00.2	8.8	54.4
	16	57 21.0	29.0	348 22.2	13.1	20 09.0	8.6	54.4
	17	72 21.2	28.0	2 54.3	13.1	20 17.6	8.6	54.4
	18	87 21.3	S 4 27.1	17 26.4	13.0	N20 26.2	8.5	54.4
	19	102 21.5	26.1	31 58.4	12.9	20 34.7	8.5	54.4
	20	117 21.6	25.1	46 30.3	12.8	20 43.2	8.3	54.4
	21	132 21.8	. . 24.1	61 02.1	12.8	20 51.5	8.3	54.5
	22	147 21.9	23.1	75 33.9	12.7	20 59.8	8.1	54.5
	23	162 22.1	22.2	90 05.6	12.6	21 07.9	8.1	54.5
10 MONDAY	00	177 22.3	S 4 21.2	104 37.2	12.5	N21 16.0	8.0	54.5
	01	192 22.4	20.2	119 08.7	12.5	21 24.0	7.9	54.5
	02	207 22.6	19.2	133 40.2	12.4	21 31.9	7.8	54.5
	03	222 22.7	. . 18.3	148 11.6	12.4	21 39.7	7.7	54.5
	04	237 22.9	17.3	162 43.0	12.2	21 47.4	7.7	54.6
	05	252 23.1	16.3	177 14.2	12.2	21 55.1	7.5	54.6
	06	267 23.2	S 4 15.3	191 45.4	12.1	N22 02.6	7.4	54.6
	07	282 23.4	14.3	206 16.5	12.0	22 10.0	7.3	54.6
	08	297 23.5	13.4	220 47.5	12.0	22 17.3	7.3	54.6
	09	312 23.7	. . 12.4	235 18.5	11.9	22 24.6	7.1	54.6
	10	327 23.8	11.4	249 49.4	11.8	22 31.7	7.1	54.7
	11	342 24.0	10.4	264 20.2	11.8	22 38.8	6.9	54.7
	12	357 24.2	S 4 09.4	278 51.0	11.6	N22 45.7	6.9	54.7
	13	12 24.3	08.5	293 21.6	11.6	22 52.6	6.7	54.7
	14	27 24.5	07.5	307 52.2	11.5	22 59.3	6.7	54.7
	15	42 24.6	. . 06.5	322 22.7	11.5	23 06.0	6.5	54.8
	16	57 24.8	05.5	336 53.2	11.4	23 12.5	6.4	54.8
	17	72 25.0	04.5	351 23.6	11.2	23 18.9	6.4	54.8
	18	87 25.1	S 4 03.6	5 53.8	11.3	N23 25.3	6.2	54.8
	19	102 25.3	02.6	20 24.1	11.1	23 31.5	6.1	54.8
	20	117 25.5	01.6	34 54.2	11.1	23 37.6	6.0	54.9
	21	132 25.6	4 00.6	49 24.3	11.0	23 43.6	5.9	54.9
	22	147 25.8	3 59.6	63 54.3	10.9	23 49.5	5.8	54.9
	23	162 25.9	S 3 58.7	78 24.2	10.9	N23 55.3	5.7	54.9
		SD 16.1	*d* 1.0	SD 14.8		14.8		14.9

Lat. °	Twilight Naut. h m	Twilight Civil h m	Sunrise h m	Moonrise 8 h m	Moonrise 9 h m	Moonrise 10 h m	Moonrise 11 h m
N 72	04 30	05 50	06 57	06 16	▭	▭	▭
N 70	04 40	05 52	06 52	06 45	06 00	▭	▭
68	04 48	05 53	06 48	07 07	06 43	05 35	▭
66	04 54	05 53	06 44	07 24	07 12	06 53	▭
64	04 59	05 54	06 41	07 38	07 34	07 30	07 25
62	05 03	05 55	06 39	07 50	07 52	07 56	08 07
60	05 07	05 55	06 36	08 01	08 07	08 17	08 35
N 58	05 10	05 55	06 34	08 10	08 20	08 34	08 57
56	05 12	05 55	06 32	08 17	08 31	08 49	09 15
54	05 15	05 55	06 31	08 24	08 40	09 01	09 31
52	05 16	05 56	06 29	08 31	08 49	09 12	09 44
50	05 18	05 56	06 28	08 37	08 57	09 22	09 55
45	05 21	05 55	06 25	08 49	09 13	09 43	10 20
N 40	05 24	05 55	06 22	08 59	09 27	10 00	10 39
35	05 25	05 54	06 20	09 08	09 39	10 14	10 55
30	05 26	05 54	06 18	09 16	09 49	10 26	11 09
20	05 26	05 52	06 14	09 30	10 07	10 48	11 33
N 10	05 25	05 50	06 11	09 42	10 22	11 06	11 54
0	05 23	05 47	06 07	09 53	10 37	11 24	12 14
S 10	05 18	05 43	06 04	10 04	10 52	11 41	12 33
20	05 12	05 38	06 00	10 16	11 07	12 00	12 54
30	05 04	05 32	05 56	10 30	11 26	12 22	13 19
35	04 58	05 28	05 54	10 39	11 36	12 35	13 33
40	04 51	05 23	05 51	10 48	11 49	12 50	13 50
45	04 42	05 18	05 47	10 59	12 03	13 07	14 10
S 50	04 31	05 10	05 43	11 12	12 21	13 29	14 35
52	04 25	05 07	05 41	11 19	12 29	13 40	14 48
54	04 19	05 03	05 39	11 26	12 39	13 52	15 02
56	04 12	04 59	05 37	11 33	12 50	14 06	15 18
58	04 04	04 54	05 34	11 42	13 02	14 22	15 38
S 60	03 55	04 48	05 32	11 52	13 16	14 41	16 02

Lat. °	Sunset h m	Twilight Civil h m	Twilight Naut. h m	Moonset 8 h m	Moonset 9 h m	Moonset 10 h m	Moonset 11 h m
N 72	17 26	18 34	19 54	00 46	▭	▭	▭
N 70	17 31	18 32	19 44	00 19	02 35	▭	▭
68	17 35	18 31	19 36	25 53	01 53	04 36	▭
66	17 39	18 30	19 30	25 24	01 24	03 19	▭
64	17 42	18 29	19 24	25 03	01 03	02 43	04 29
62	17 44	18 28	19 20	24 46	00 46	02 17	03 48
60	17 46	18 28	19 16	24 32	00 32	01 56	03 20
N 58	17 48	18 27	19 13	24 20	00 20	01 40	02 58
56	17 50	18 27	19 10	24 10	00 10	01 26	02 40
54	17 52	18 27	19 08	24 01	00 01	01 14	02 26
52	17 53	18 27	19 06	23 53	25 03	01 03	02 13
50	17 55	18 27	19 04	23 45	24 54	00 54	02 01
45	17 57	18 27	19 01	23 30	24 34	00 34	01 38
N 40	18 00	18 27	18 58	23 17	24 18	00 18	01 19
35	18 02	18 27	18 57	23 06	24 04	00 04	01 03
30	18 04	18 28	18 56	22 57	23 53	24 49	00 49
20	18 08	18 30	18 55	22 41	23 33	24 26	00 26
N 10	18 11	18 32	18 56	22 27	23 15	24 06	00 06
0	18 14	18 35	18 59	22 14	22 59	23 48	24 39
S 10	18 17	18 38	19 03	22 01	22 43	23 29	24 19
20	18 21	18 43	19 09	21 47	22 26	23 09	23 57
30	18 25	18 49	19 17	21 31	22 06	22 46	23 32
35	18 27	18 53	19 23	21 22	21 55	22 33	23 17
40	18 30	18 57	19 29	21 12	21 42	22 17	23 00
45	18 33	19 03	19 38	21 00	21 26	21 59	22 40
S 50	18 37	19 10	19 49	20 45	21 07	21 36	22 14
52	18 39	19 13	19 54	20 38	20 58	21 25	22 02
54	18 41	19 17	20 00	20 30	20 48	21 13	21 47
56	18 43	19 21	20 07	20 22	20 37	20 59	21 31
58	18 45	19 26	20 15	20 12	20 24	20 42	21 11
S 60	18 48	19 31	20 24	20 02	20 09	20 22	20 46

Day	SUN Eqn. of Time 00^h	SUN Eqn. of Time 12^h	SUN Mer. Pass.	MOON Mer. Pass. Upper	MOON Mer. Pass. Lower	MOON Age	MOON Phase
d	m s	m s	h m	h m	h m	d %	
8	11 01	10 54	12 11	16 03	03 42	05 25	
9	10 46	10 39	12 11	16 48	04 25	06 33	
10	10 31	10 24	12 10	17 36	05 11	07 42	

UT		ARIES	VENUS −4.1		MARS +0.8		JUPITER −2.4		SATURN +0.0	
		GHA	GHA	Dec	GHA	Dec	GHA	Dec	GHA	Dec
d	h	° ′	° ′	° ′	° ′	° ′	° ′	° ′	° ′	° ′
11 TUESDAY	00	168 14.6	215 52.6	S17 25.7	253 59.9	S23 34.3	36 35.3	N18 52.2	86 27.4	N22 09.8
	01	183 17.1	230 52.0	25.0	269 00.7	34.3	51 37.9	52.3	101 29.7	09.8
	02	198 19.5	245 51.4	24.4	284 01.4	34.3	66 40.6	52.3	116 32.1	09.8
	03	213 22.0	260 50.9	. . 23.7	299 02.1	. . 34.3	81 43.2	. . 52.4	131 34.5	. . 09.8
	04	228 24.4	275 50.3	23.1	314 02.9	34.3	96 45.9	52.4	146 36.9	09.8
	05	243 26.9	290 49.7	22.5	329 03.6	34.3	111 48.6	52.5	161 39.3	09.9
	06	258 29.4	305 49.2	S17 21.8	344 04.3	S23 34.3	126 51.2	N18 52.5	176 41.6	N22 09.9
	07	273 31.8	320 48.6	21.2	359 05.1	34.3	141 53.9	52.6	191 44.0	09.9
	08	288 34.3	335 48.0	20.5	14 05.8	34.3	156 56.5	52.6	206 46.4	09.9
	09	303 36.8	350 47.4	. . 19.9	29 06.5	. . 34.3	171 59.2	. . 52.7	221 48.8	. . 09.9
	10	318 39.2	5 46.9	19.2	44 07.3	34.3	187 01.8	52.7	236 51.2	09.9
	11	333 41.7	20 46.3	18.6	59 08.0	34.3	202 04.5	52.8	251 53.5	09.9
	12	348 44.2	35 45.7	S17 17.9	74 08.7	S23 34.3	217 07.1	N18 52.8	266 55.9	N22 09.9
	13	3 46.6	50 45.2	17.3	89 09.4	34.3	232 09.8	52.9	281 58.3	10.0
	14	18 49.1	65 44.6	16.6	104 10.2	34.3	247 12.4	52.9	297 00.7	10.0
	15	33 51.6	80 44.0	. . 15.9	119 10.9	. . 34.3	262 15.1	. . 53.0	312 03.0	. . 10.0
	16	48 54.0	95 43.5	15.3	134 11.6	34.3	277 17.7	53.0	327 05.4	10.0
	17	63 56.5	110 42.9	14.6	149 12.4	34.3	292 20.4	53.1	342 07.8	10.0
	18	78 58.9	125 42.3	S17 14.0	164 13.1	S23 34.3	307 23.0	N18 53.1	357 10.2	N22 10.0
	19	94 01.4	140 41.7	13.3	179 13.8	34.3	322 25.7	53.2	12 12.6	10.0
	20	109 03.9	155 41.2	12.7	194 14.6	34.3	337 28.3	53.2	27 14.9	10.1
	21	124 06.3	170 40.6	. . 12.0	209 15.3	. . 34.3	352 31.0	. . 53.2	42 17.3	. . 10.1
	22	139 08.8	185 40.0	11.3	224 16.0	34.3	7 33.6	53.3	57 19.7	10.1
	23	154 11.3	200 39.5	10.7	239 16.8	34.3	22 36.2	53.3	72 22.1	10.1
12 WEDNESDAY	00	169 13.7	215 38.9	S17 10.0	254 17.5	S23 34.2	37 38.9	N18 53.4	87 24.4	N22 10.1
	01	184 16.2	230 38.3	09.3	269 18.2	34.2	52 41.5	53.4	102 26.8	10.1
	02	199 18.7	245 37.8	08.7	284 19.0	34.2	67 44.2	53.5	117 29.2	10.1
	03	214 21.1	260 37.2	. . 08.0	299 19.7	. . 34.2	82 46.8	. . 53.5	132 31.6	. . 10.1
	04	229 23.6	275 36.6	07.3	314 20.4	34.2	97 49.5	53.6	147 33.9	10.2
	05	244 26.1	290 36.1	06.7	329 21.2	34.2	112 52.1	53.6	162 36.3	10.2
	06	259 28.5	305 35.5	S17 06.0	344 21.9	S23 34.2	127 54.8	N18 53.7	177 38.7	N22 10.2
	07	274 31.0	320 34.9	05.3	359 22.6	34.2	142 57.4	53.7	192 41.1	10.2
	08	289 33.4	335 34.4	04.7	14 23.4	34.2	158 00.1	53.8	207 43.4	10.2
	09	304 35.9	350 33.8	. . 04.0	29 24.1	. . 34.2	173 02.7	. . 53.8	222 45.8	. . 10.2
	10	319 38.4	5 33.3	03.3	44 24.8	34.2	188 05.3	53.8	237 48.2	10.2
	11	334 40.8	20 32.7	02.7	59 25.6	34.1	203 08.0	53.9	252 50.6	10.3
	12	349 43.3	35 32.1	S17 02.0	74 26.3	S23 34.1	218 10.6	N18 53.9	267 52.9	N22 10.3
	13	4 45.8	50 31.6	01.3	89 27.0	34.1	233 13.3	54.0	282 55.3	10.3
	14	19 48.2	65 31.0	00.6	104 27.8	34.1	248 15.9	54.0	297 57.7	10.3
	15	34 50.7	80 30.4	17 00.0	119 28.5	. . 34.1	263 18.6	. . 54.1	313 00.1	. . 10.3
	16	49 53.2	95 29.9	16 59.3	134 29.2	34.1	278 21.2	54.1	328 02.4	10.3
	17	64 55.6	110 29.3	58.6	149 30.0	34.1	293 23.8	54.2	343 04.8	10.3
	18	79 58.1	125 28.8	S16 57.9	164 30.7	S23 34.1	308 26.5	N18 54.2	358 07.2	N22 10.3
	19	95 00.6	140 28.2	57.3	179 31.4	34.1	323 29.1	54.3	13 09.5	10.4
	20	110 03.0	155 27.6	56.6	194 32.2	34.0	338 31.8	54.3	28 11.9	10.4
	21	125 05.5	170 27.1	. . 55.9	209 32.9	. . 34.0	353 34.4	. . 54.3	43 14.3	. . 10.4
	22	140 07.9	185 26.5	55.2	224 33.6	34.0	8 37.1	54.4	58 16.7	10.4
	23	155 10.4	200 25.9	54.5	239 34.4	34.0	23 39.7	54.4	73 19.0	10.4
13 THURSDAY	00	170 12.9	215 25.4	S16 53.9	254 35.1	S23 34.0	38 42.3	N18 54.5	88 21.4	N22 10.4
	01	185 15.3	230 24.8	53.2	269 35.8	34.0	53 45.0	54.5	103 23.8	10.4
	02	200 17.8	245 24.3	52.5	284 36.6	34.0	68 47.6	54.6	118 26.1	10.5
	03	215 20.3	260 23.7	. . 51.8	299 37.3	. . 33.9	83 50.2	. . 54.6	133 28.5	. . 10.5
	04	230 22.7	275 23.2	51.1	314 38.0	33.9	98 52.9	54.7	148 30.9	10.5
	05	245 25.2	290 22.6	50.4	329 38.8	33.9	113 55.5	54.7	163 33.3	10.5
	06	260 27.7	305 22.0	S16 49.7	344 39.5	S23 33.9	128 58.2	N18 54.7	178 35.6	N22 10.5
	07	275 30.1	320 21.5	49.0	359 40.2	33.9	144 00.8	54.8	193 38.0	10.5
	08	290 32.6	335 20.9	48.4	14 41.0	33.9	159 03.4	54.8	208 40.4	10.5
	09	305 35.0	350 20.4	. . 47.7	29 41.7	. . 33.8	174 06.1	. . 54.9	223 42.7	. . 10.5
	10	320 37.5	5 19.8	47.0	44 42.4	33.8	189 08.7	54.9	238 45.1	10.6
	11	335 40.0	20 19.2	46.3	59 43.2	33.8	204 11.3	55.0	253 47.5	10.6
	12	350 42.4	35 18.7	S16 45.6	74 43.9	S23 33.8	219 14.0	N18 55.0	268 49.8	N22 10.6
	13	5 44.9	50 18.1	44.9	89 44.6	33.8	234 16.6	55.0	283 52.2	10.6
	14	20 47.4	65 17.6	44.2	104 45.4	33.8	249 19.3	55.1	298 54.6	10.6
	15	35 49.8	80 17.0	. . 43.5	119 46.1	. . 33.7	264 21.9	. . 55.1	313 56.9	. . 10.6
	16	50 52.3	95 16.5	42.8	134 46.8	33.7	279 24.5	55.2	328 59.3	10.6
	17	65 54.8	110 15.9	42.1	149 47.6	33.7	294 27.2	55.2	344 01.7	10.7
	18	80 57.2	125 15.4	S16 41.4	164 48.3	S23 33.7	309 29.8	N18 55.3	359 04.0	N22 10.7
	19	95 59.7	140 14.8	40.7	179 49.1	33.7	324 32.4	55.3	14 06.4	10.7
	20	111 02.2	155 14.3	40.0	194 49.8	33.6	339 35.1	55.3	29 08.8	10.7
	21	126 04.6	170 13.7	. . 39.3	209 50.5	. . 33.6	354 37.7	. . 55.4	44 11.1	. . 10.7
	22	141 07.1	185 13.1	38.6	224 51.3	33.6	9 40.3	55.4	59 13.5	10.7
	23	156 09.5	200 12.6	37.9	239 52.0	33.6	24 43.0	55.5	74 15.9	10.7
Mer. Pass.		h m 12 41.0	*v* −0.6	*d* 0.7	*v* 0.7	*d* 0.0	*v* 2.6	*d* 0.0	*v* 2.4	*d* 0.0

STARS

Name	SHA	Dec
	° ′	° ′
Acamar	315 24.6	S40 17.8
Achernar	335 33.0	S57 13.5
Acrux	173 18.0	S63 06.9
Adhara	255 18.8	S28 58.8
Aldebaran	290 58.8	N16 30.9
Alioth	166 27.0	N55 56.5
Alkaid	153 04.7	N49 17.7
Al Na'ir	27 54.1	S46 56.9
Alnilam	275 54.6	S 1 12.1
Alphard	218 03.8	S 8 40.4
Alphecca	126 17.7	N26 42.0
Alpheratz	357 52.3	N29 06.3
Altair	62 16.3	N 8 52.3
Ankaa	353 23.9	S42 17.6
Antares	112 36.2	S26 26.3
Arcturus	146 02.8	N19 09.8
Atria	107 45.3	S69 01.8
Avior	234 21.0	S59 31.3
Bellatrix	278 40.7	N 6 21.1
Betelgeuse	271 10.0	N 7 24.4
Canopus	263 59.6	S52 42.1
Capella	280 46.5	N46 00.2
Deneb	49 37.4	N45 17.2
Denebola	182 41.5	N14 33.2
Diphda	349 04.3	S17 58.4
Dubhe	194 00.6	N61 44.1
Elnath	278 22.9	N28 36.7
Eltanin	90 50.0	N51 28.9
Enif	33 55.4	N 9 53.1
Fomalhaut	15 33.2	S29 36.5
Gacrux	172 09.6	S57 07.8
Gienah	176 00.4	S17 33.6
Hadar	148 59.2	S60 23.1
Hamal	328 10.2	N23 28.6
Kaus Aust.	83 54.6	S34 23.0
Kochab	137 18.6	N74 08.3
Markab	13 46.8	N15 13.1
Menkar	314 23.7	N 4 06.0
Menkent	148 16.9	S36 23.1
Miaplacidus	221 40.9	S69 43.9
Mirfak	308 52.3	N49 52.5
Nunki	76 08.5	S26 17.6
Peacock	53 32.2	S56 43.5
Pollux	243 37.4	N28 01.2
Procyon	245 08.0	N 5 13.0
Rasalhague	96 14.0	N12 33.2
Regulus	207 51.8	N11 57.1
Rigel	281 19.8	S 8 12.0
Rigil Kent.	140 02.7	S60 50.7
Sabik	102 21.8	S15 43.8
Schedar	349 50.5	N56 33.2
Shaula	96 32.9	S37 06.3
Sirius	258 40.8	S16 43.3
Spica	158 39.5	S11 10.7
Suhail	222 58.1	S43 26.8
Vega	80 44.6	N38 46.8
Zuben'ubi	137 14.2	S16 03.3

	SHA	Mer. Pass.
	° ′	h m
Venus	46 25.2	9 38
Mars	85 03.8	7 02
Jupiter	228 25.2	21 26
Saturn	278 10.7	18 08

Day	UT	SUN GHA	SUN Dec	MOON GHA	v	MOON Dec	d	HP
	d h	° ′	° ′	° ′	′	° ′	′	′
TUESDAY	11 00	177 26.1	S 3 57.7	92 54.1	10.7	N24 01.0	5.6	54.9
	01	192 26.3	56.7	107 23.8	10.7	24 06.6	5.5	55.0
	02	207 26.4	55.7	121 53.5	10.7	24 12.1	5.3	55.0
	03	222 26.6	. . 54.7	136 23.2	10.5	24 17.4	5.2	55.0
	04	237 26.7	53.7	150 52.7	10.5	24 22.6	5.2	55.0
	05	252 26.9	52.8	165 22.2	10.4	24 27.8	5.0	55.1
	06	267 27.1	S 3 51.8	179 51.6	10.4	N24 32.8	4.9	55.1
	07	282 27.2	50.8	194 21.0	10.2	24 37.7	4.7	55.1
	08	297 27.4	49.8	208 50.2	10.2	24 42.4	4.7	55.1
	09	312 27.6	. . 48.8	223 19.4	10.2	24 47.1	4.5	55.2
	10	327 27.7	47.9	237 48.6	10.0	24 51.6	4.5	55.2
	11	342 27.9	46.9	252 17.6	10.0	24 56.1	4.3	55.2
	12	357 28.1	S 3 45.9	266 46.6	9.9	N25 00.4	4.1	55.2
	13	12 28.2	44.9	281 15.5	9.9	25 04.5	4.1	55.3
	14	27 28.4	43.9	295 44.4	9.7	25 08.6	3.9	55.3
	15	42 28.5	. . 43.0	310 13.1	9.8	25 12.5	3.8	55.3
	16	57 28.7	42.0	324 41.9	9.6	25 16.3	3.7	55.3
	17	72 28.9	41.0	339 10.5	9.6	25 20.0	3.6	55.4
	18	87 29.0	S 3 40.0	353 39.1	9.5	N25 23.6	3.4	55.4
	19	102 29.2	39.0	8 07.6	9.4	25 27.0	3.3	55.4
	20	117 29.4	38.0	22 36.0	9.4	25 30.3	3.2	55.4
	21	132 29.5	. . 37.1	37 04.4	9.3	25 33.5	3.0	55.5
	22	147 29.7	36.1	51 32.7	9.3	25 36.5	3.0	55.5
	23	162 29.9	35.1	66 01.0	9.2	25 39.5	2.8	55.5
WEDNESDAY	12 00	177 30.0	S 3 34.1	80 29.2	9.1	N25 42.3	2.6	55.6
	01	192 30.2	33.1	94 57.3	9.0	25 44.9	2.5	55.6
	02	207 30.4	32.1	109 25.3	9.0	25 47.4	2.4	55.6
	03	222 30.5	. . 31.2	123 53.3	9.0	25 49.8	2.3	55.6
	04	237 30.7	30.2	138 21.3	8.9	25 52.1	2.1	55.7
	05	252 30.9	29.2	152 49.2	8.8	25 54.2	2.0	55.7
	06	267 31.0	S 3 28.2	167 17.0	8.8	N25 56.2	1.9	55.7
	07	282 31.2	27.2	181 44.8	8.7	25 58.1	1.7	55.8
	08	297 31.4	26.2	196 12.5	8.6	25 59.8	1.6	55.8
	09	312 31.5	. . 25.3	210 40.1	8.6	26 01.4	1.4	55.8
	10	327 31.7	24.3	225 07.7	8.6	26 02.8	1.3	55.9
	11	342 31.9	23.3	239 35.3	8.4	26 04.1	1.2	55.9
	12	357 32.0	S 3 22.3	254 02.7	8.5	N26 05.3	1.0	55.9
	13	12 32.2	21.3	268 30.2	8.4	26 06.3	0.9	56.0
	14	27 32.4	20.3	282 57.6	8.3	26 07.2	0.8	56.0
	15	42 32.5	. . 19.4	297 24.9	8.3	26 08.0	0.6	56.0
	16	57 32.7	18.4	311 52.2	8.2	26 08.6	0.4	56.0
	17	72 32.9	17.4	326 19.4	8.2	26 09.0	0.4	56.1
	18	87 33.0	S 3 16.4	340 46.6	8.2	N26 09.4	0.1	56.1
	19	102 33.2	15.4	355 13.8	8.1	26 09.5	0.1	56.1
	20	117 33.4	14.4	9 40.9	8.1	26 09.6	0.1	56.2
	21	132 33.5	. . 13.5	24 08.0	8.0	26 09.5	0.3	56.2
	22	147 33.7	12.5	38 35.0	8.0	26 09.2	0.4	56.2
	23	162 33.9	11.5	53 02.0	7.9	26 08.8	0.5	56.3
THURSDAY	13 00	177 34.0	S 3 10.5	67 28.9	7.9	N26 08.3	0.7	56.3
	01	192 34.2	09.5	81 55.8	7.9	26 07.6	0.9	56.3
	02	207 34.4	08.5	96 22.7	7.8	26 06.7	0.9	56.4
	03	222 34.5	. . 07.6	110 49.5	7.8	26 05.8	1.2	56.4
	04	237 34.7	06.6	125 16.3	7.7	26 04.6	1.3	56.5
	05	252 34.9	05.6	139 43.0	7.8	26 03.3	1.4	56.5
	06	267 35.0	S 3 04.6	154 09.8	7.7	N26 01.9	1.6	56.5
	07	282 35.2	03.6	168 36.5	7.6	26 00.3	1.7	56.6
	08	297 35.4	02.6	183 03.1	7.7	25 58.6	1.9	56.6
	09	312 35.6	. . 01.6	197 29.8	7.6	25 56.7	2.0	56.6
	10	327 35.7	3 00.7	211 56.4	7.5	25 54.7	2.2	56.7
	11	342 35.9	2 59.7	226 22.9	7.6	25 52.5	2.3	56.7
	12	357 36.1	S 2 58.7	240 49.5	7.5	N25 50.2	2.5	56.7
	13	12 36.2	57.7	255 16.0	7.5	25 47.7	2.6	56.8
	14	27 36.4	56.7	269 42.5	7.5	25 45.1	2.8	56.8
	15	42 36.6	. . 55.7	284 09.0	7.5	25 42.3	2.9	56.8
	16	57 36.7	54.8	298 35.5	7.4	25 39.4	3.1	56.9
	17	72 36.9	53.8	313 01.9	7.5	25 36.3	3.2	56.9
	18	87 37.1	S 2 52.8	327 28.4	7.4	N25 33.1	3.4	57.0
	19	102 37.3	51.8	341 54.8	7.4	25 29.7	3.5	57.0
	20	117 37.4	50.8	356 21.2	7.4	25 26.2	3.7	57.0
	21	132 37.6	. . 49.8	10 47.6	7.3	25 22.5	3.9	57.1
	22	147 37.8	48.8	25 13.9	7.4	25 18.6	4.0	57.1
	23	162 37.9	47.9	39 40.3	7.4	N25 14.6	4.1	57.1
		SD 16.1	*d* 1.0	SD 15.0		15.2		15.5

Lat.	Twilight Naut.	Twilight Civil	Sunrise	Moonrise 11	Moonrise 12	Moonrise 13	Moonrise 14
°	h m	h m	h m	h m	h m	h m	h m
N 72	04 14	05 35	06 42	▭	▭	▭	▭
N 70	04 25	05 38	06 38	▭	▭	▭	▭
68	04 34	05 40	06 35	▭	▭	▭	▭
66	04 42	05 42	06 33	▭	▭	▭	▭
64	04 48	05 44	06 31	07 25	07 18	08 15	10 19
62	04 53	05 45	06 29	08 07	08 33	09 29	10 59
60	04 57	05 46	06 27	08 35	09 08	10 05	11 27
N 58	05 01	05 47	06 26	08 57	09 34	10 31	11 48
56	05 04	05 48	06 25	09 15	09 55	10 52	12 06
54	05 07	05 48	06 23	09 31	10 12	11 09	12 20
52	05 10	05 49	06 22	09 44	10 26	11 23	12 33
50	05 12	05 49	06 21	09 55	10 39	11 36	12 45
45	05 16	05 50	06 19	10 20	11 05	12 02	13 08
N 40	05 19	05 50	06 17	10 39	11 26	12 22	13 27
35	05 21	05 50	06 16	10 55	11 44	12 40	13 43
30	05 23	05 50	06 14	11 09	11 59	12 55	13 56
20	05 24	05 49	06 11	11 33	12 24	13 20	14 19
N 10	05 24	05 48	06 09	11 54	12 46	13 41	14 39
0	05 22	05 46	06 07	12 14	13 07	14 02	14 58
S 10	05 19	05 43	06 04	12 33	13 27	14 22	15 16
20	05 13	05 39	06 01	12 54	13 49	14 44	15 36
30	05 06	05 34	05 58	13 19	14 15	15 09	15 59
35	05 01	05 31	05 56	13 33	14 30	15 24	16 12
40	04 55	05 27	05 54	13 50	14 48	15 41	16 28
45	04 47	05 22	05 51	14 10	15 09	16 01	16 46
S 50	04 36	05 15	05 48	14 35	15 36	16 27	17 09
52	04 32	05 13	05 47	14 48	15 49	16 40	17 19
54	04 26	05 09	05 45	15 02	16 04	16 54	17 32
56	04 20	05 05	05 43	15 18	16 21	17 11	17 46
58	04 12	05 01	05 41	15 38	16 43	17 31	18 02
S 60	04 04	04 57	05 39	16 02	17 10	17 56	18 21

Lat.	Sunset	Twilight Civil	Twilight Naut.	Moonset 11	Moonset 12	Moonset 13	Moonset 14
°	h m	h m	h m	h m	h m	h m	h m
N 72	17 40	18 47	20 10	▭	▭	▭	▭
N 70	17 43	18 44	19 58	▭	▭	▭	▭
68	17 46	18 42	19 48	▭	▭	▭	▭
66	17 49	18 40	19 40	▭	▭	▭	▭
64	17 51	18 38	19 34	04 29	06 25	07 22	07 15
62	17 52	18 36	19 29	03 48	05 11	06 08	06 35
60	17 54	18 35	19 24	03 20	04 35	05 32	06 07
N 58	17 55	18 34	19 20	02 58	04 09	05 06	05 45
56	17 56	18 33	19 17	02 40	03 49	04 45	05 27
54	17 57	18 33	19 14	02 26	03 32	04 28	05 12
52	17 58	18 32	19 11	02 13	03 17	04 13	04 59
50	17 59	18 32	19 09	02 01	03 05	04 01	04 47
45	18 01	18 31	19 05	01 38	02 39	03 34	04 23
N 40	18 03	18 30	19 02	01 19	02 18	03 14	04 03
35	18 05	18 30	18 59	01 03	02 01	02 56	03 47
30	18 06	18 30	18 58	00 49	01 46	02 41	03 33
20	18 09	18 31	18 56	00 26	01 21	02 16	03 09
N 10	18 11	18 32	18 56	00 06	00 59	01 54	02 48
0	18 13	18 34	18 58	24 39	00 39	01 33	02 29
S 10	18 16	18 37	19 01	24 19	00 19	01 12	02 09
20	18 18	18 40	19 06	23 57	24 50	00 50	01 48
30	18 21	18 45	19 13	23 32	24 25	00 25	01 23
35	18 23	18 48	19 18	23 17	24 09	00 09	01 09
40	18 25	18 52	19 24	23 00	23 52	24 52	00 52
45	18 28	18 57	19 32	22 40	23 31	24 32	00 32
S 50	18 31	19 03	19 42	22 14	23 04	24 07	00 07
52	18 32	19 06	19 47	22 02	22 51	23 54	25 10
54	18 34	19 09	19 52	21 47	22 36	23 40	24 58
56	18 35	19 13	19 58	21 31	22 18	23 24	24 45
58	18 37	19 17	20 05	21 11	21 57	23 04	24 29
S 60	18 39	19 21	20 14	20 46	21 29	22 39	24 10

Day	SUN Eqn. of Time 00^h	SUN Eqn. of Time 12^h	SUN Mer. Pass.	MOON Mer. Pass. Upper	MOON Mer. Pass. Lower	MOON Age	MOON Phase
d	m s	m s	h m	h m	h m	d	%
11	10 16	10 08	12 10	18 26	06 01	08	52
12	10 00	09 52	12 10	19 20	06 53	09	62
13	09 44	09 36	12 10	20 15	07 47	10	71

	UT (d h)	ARIES GHA	VENUS −4.0 GHA	VENUS Dec	MARS +0.8 GHA	MARS Dec	JUPITER −2.4 GHA	JUPITER Dec	SATURN +0.0 GHA	SATURN Dec
		° ′	° ′	° ′	° ′	° ′	° ′	° ′	° ′	° ′
	14 00	171 12.0	215 12.0	S16 37.2	254 52.7	S23 33.6	39 45.6	N18 55.5	89 18.2	N22 10.8
	01	186 14.5	230 11.5	36.5	269 53.5	33.5	54 48.2	55.6	104 20.6	10.8
	02	201 16.9	245 10.9	35.8	284 54.2	33.5	69 50.9	55.6	119 23.0	10.8
	03	216 19.4	260 10.4	35.1	299 54.9	33.5	84 53.5	55.6	134 25.3	10.8
	04	231 21.9	275 09.8	34.4	314 55.7	33.5	99 56.1	55.7	149 27.7	10.8
	05	246 24.3	290 09.3	33.7	329 56.4	33.5	114 58.7	55.7	164 30.1	10.8
	06	261 26.8	305 08.7	S16 33.0	344 57.1	S23 33.4	130 01.4	N18 55.8	179 32.4	N22 10.8
	07	276 29.3	320 08.2	32.3	359 57.9	33.4	145 04.0	55.8	194 34.8	10.8
	08	291 31.7	335 07.6	31.6	14 58.6	33.4	160 06.6	55.8	209 37.2	10.9
F	09	306 34.2	350 07.1	30.9	29 59.3	33.4	175 09.3	55.9	224 39.5	10.9
R	10	321 36.7	5 06.5	30.1	45 00.1	33.3	190 11.9	55.9	239 41.9	10.9
I	11	336 39.1	20 06.0	29.4	60 00.8	33.3	205 14.5	56.0	254 44.3	10.9
D	12	351 41.6	35 05.4	S16 28.7	75 01.6	S23 33.3	220 17.2	N18 56.0	269 46.6	N22 10.9
A	13	6 44.0	50 04.9	28.0	90 02.3	33.3	235 19.8	56.1	284 49.0	10.9
Y	14	21 46.5	65 04.3	27.3	105 03.0	33.2	250 22.4	56.1	299 51.4	10.9
	15	36 49.0	80 03.8	26.6	120 03.8	33.2	265 25.0	56.1	314 53.7	11.0
	16	51 51.4	95 03.2	25.9	135 04.5	33.2	280 27.7	56.2	329 56.1	11.0
	17	66 53.9	110 02.7	25.1	150 05.2	33.2	295 30.3	56.2	344 58.4	11.0
	18	81 56.4	125 02.1	S16 24.4	165 06.0	S23 33.1	310 32.9	N18 56.3	0 00.8	N22 11.0
	19	96 58.8	140 01.6	23.7	180 06.7	33.1	325 35.5	56.3	15 03.2	11.0
	20	112 01.3	155 01.0	23.0	195 07.4	33.1	340 38.2	56.3	30 05.5	11.0
	21	127 03.8	170 00.5	22.3	210 08.2	33.1	355 40.8	56.4	45 07.9	11.0
	22	142 06.2	185 00.0	21.5	225 08.9	33.0	10 43.4	56.4	60 10.3	11.1
	23	157 08.7	199 59.4	20.8	240 09.7	33.0	25 46.0	56.5	75 12.6	11.1
	15 00	172 11.2	214 58.9	S16 20.1	255 10.4	S23 33.0	40 48.7	N18 56.5	90 15.0	N22 11.1
	01	187 13.6	229 58.3	19.4	270 11.1	32.9	55 51.3	56.5	105 17.3	11.1
	02	202 16.1	244 57.8	18.7	285 11.9	32.9	70 53.9	56.6	120 19.7	11.1
	03	217 18.5	259 57.2	17.9	300 12.6	32.9	85 56.5	56.6	135 22.1	11.1
	04	232 21.0	274 56.7	17.2	315 13.3	32.9	100 59.2	56.7	150 24.4	11.1
	05	247 23.5	289 56.1	16.5	330 14.1	32.8	116 01.8	56.7	165 26.8	11.2
	06	262 25.9	304 55.6	S16 15.8	345 14.8	S23 32.8	131 04.4	N18 56.7	180 29.2	N22 11.2
S	07	277 28.4	319 55.1	15.0	0 15.5	32.8	146 07.0	56.8	195 31.5	11.2
A	08	292 30.9	334 54.5	14.3	15 16.3	32.7	161 09.7	56.8	210 33.9	11.2
T	09	307 33.3	349 54.0	13.6	30 17.0	32.7	176 12.3	56.9	225 36.2	11.2
U	10	322 35.8	4 53.4	12.8	45 17.8	32.7	191 14.9	56.9	240 38.6	11.2
R	11	337 38.3	19 52.9	12.1	60 18.5	32.6	206 17.5	56.9	255 40.9	11.2
D	12	352 40.7	34 52.3	S16 11.4	75 19.2	S23 32.6	221 20.1	N18 57.0	270 43.3	N22 11.2
A	13	7 43.2	49 51.8	10.6	90 20.0	32.6	236 22.8	57.0	285 45.7	11.3
Y	14	22 45.6	64 51.3	09.9	105 20.7	32.6	251 25.4	57.1	300 48.0	11.3
	15	37 48.1	79 50.7	09.2	120 21.4	32.5	266 28.0	57.1	315 50.4	11.3
	16	52 50.6	94 50.2	08.4	135 22.2	32.5	281 30.6	57.1	330 52.7	11.3
	17	67 53.0	109 49.6	07.7	150 22.9	32.5	296 33.2	57.2	345 55.1	11.3
	18	82 55.5	124 49.1	S16 07.0	165 23.7	S23 32.4	311 35.9	N18 57.2	0 57.5	N22 11.3
	19	97 58.0	139 48.6	06.2	180 24.4	32.4	326 38.5	57.3	15 59.8	11.3
	20	113 00.4	154 48.0	05.5	195 25.1	32.4	341 41.1	57.3	31 02.2	11.4
	21	128 02.9	169 47.5	04.8	210 25.9	32.3	356 43.7	57.3	46 04.5	11.4
	22	143 05.4	184 46.9	04.0	225 26.6	32.3	11 46.3	57.4	61 06.9	11.4
	23	158 07.8	199 46.4	03.3	240 27.3	32.3	26 49.0	57.4	76 09.2	11.4
	16 00	173 10.3	214 45.9	S16 02.5	255 28.1	S23 32.2	41 51.6	N18 57.4	91 11.6	N22 11.4
	01	188 12.8	229 45.3	01.8	270 28.8	32.2	56 54.2	57.5	106 14.0	11.4
	02	203 15.2	244 44.8	01.0	285 29.6	32.1	71 56.8	57.5	121 16.3	11.4
	03	218 17.7	259 44.2	16 00.3	300 30.3	32.1	86 59.4	57.6	136 18.7	11.5
	04	233 20.1	274 43.7	15 59.6	315 31.0	32.1	102 02.0	57.6	151 21.0	11.5
	05	248 22.6	289 43.2	58.8	330 31.8	32.0	117 04.7	57.6	166 23.4	11.5
	06	263 25.1	304 42.6	S15 58.1	345 32.5	S23 32.0	132 07.3	N18 57.7	181 25.7	N22 11.5
	07	278 27.5	319 42.1	57.3	0 33.2	32.0	147 09.9	57.7	196 28.1	11.5
	08	293 30.0	334 41.6	56.6	15 34.0	31.9	162 12.5	57.7	211 30.5	11.5
S	09	308 32.5	349 41.0	55.8	30 34.7	31.9	177 15.1	57.8	226 32.8	11.5
U	10	323 34.9	4 40.5	55.1	45 35.5	31.9	192 17.7	57.8	241 35.2	11.6
N	11	338 37.4	19 40.0	54.3	60 36.2	31.8	207 20.3	57.9	256 37.5	11.6
D	12	353 39.9	34 39.4	S15 53.6	75 36.9	S23 31.8	222 23.0	N18 57.9	271 39.9	N22 11.6
A	13	8 42.3	49 38.9	52.8	90 37.7	31.7	237 25.6	57.9	286 42.2	11.6
Y	14	23 44.8	64 38.4	52.1	105 38.4	31.7	252 28.2	58.0	301 44.6	11.6
	15	38 47.3	79 37.8	51.3	120 39.2	31.7	267 30.8	58.0	316 46.9	11.6
	16	53 49.7	94 37.3	50.6	135 39.9	31.6	282 33.4	58.0	331 49.3	11.6
	17	68 52.2	109 36.8	49.8	150 40.6	31.6	297 36.0	58.1	346 51.6	11.7
	18	83 54.6	124 36.2	S15 49.0	165 41.4	S23 31.5	312 38.6	N18 58.1	1 54.0	N22 11.7
	19	98 57.1	139 35.7	48.3	180 42.1	31.5	327 41.2	58.2	16 56.4	11.7
	20	113 59.6	154 35.2	47.5	195 42.9	31.5	342 43.8	58.2	31 58.7	11.7
	21	129 02.0	169 34.6	46.8	210 43.6	31.4	357 46.5	58.2	47 01.1	11.7
	22	144 04.5	184 34.1	46.0	225 44.3	31.4	12 49.1	58.3	62 03.4	11.7
	23	159 07.0	199 33.6	45.3	240 45.1	31.3	27 51.7	58.3	77 05.8	11.7
	Mer. Pass. (h m)	12 29.2	v −0.5	d 0.7	v 0.7	d 0.0	v 2.6	d 0.0	v 2.4	d 0.0

STARS

Name	SHA	Dec
	° ′	° ′
Acamar	315 24.6	S40 17.8
Achernar	335 33.1	S57 13.5
Acrux	173 18.0	S63 06.9
Adhara	255 18.8	S28 58.8
Aldebaran	290 58.8	N16 30.9
Alioth	166 27.0	N55 56.5
Alkaid	153 04.7	N49 17.7
Al Na'ir	27 54.1	S46 56.9
Alnilam	275 54.6	S 1 12.1
Alphard	218 03.8	S 8 40.4
Alphecca	126 17.6	N26 42.0
Alpheratz	357 52.3	N29 06.3
Altair	62 16.3	N 8 52.3
Ankaa	353 23.9	S42 17.6
Antares	112 36.1	S26 26.4
Arcturus	146 02.8	N19 09.8
Atria	107 45.3	S69 01.8
Avior	234 21.1	S59 31.3
Bellatrix	278 40.7	N 6 21.1
Betelgeuse	271 10.0	N 7 24.4
Canopus	263 59.6	S52 42.1
Capella	280 46.5	N46 00.3
Deneb	49 37.4	N45 17.1
Denebola	182 41.5	N14 33.2
Diphda	349 04.3	S17 58.4
Dubhe	194 00.6	N61 44.1
Elnath	278 22.9	N28 36.7
Eltanin	90 49.9	N51 28.9
Enif	33 55.4	N 9 53.1
Fomalhaut	15 33.1	S29 36.5
Gacrux	172 09.6	S57 07.8
Gienah	176 00.4	S17 33.6
Hadar	148 59.2	S60 23.2
Hamal	328 10.2	N23 28.6
Kaus Aust.	83 54.6	S34 23.0
Kochab	137 18.6	N74 08.3
Markab	13 46.8	N15 13.1
Menkar	314 23.7	N 4 06.0
Menkent	148 16.9	S36 23.1
Miaplacidus	221 41.0	S69 43.9
Mirfak	308 52.3	N49 52.5
Nunki	76 08.4	S26 17.6
Peacock	53 32.2	S56 43.5
Pollux	243 37.4	N28 01.2
Procyon	245 08.0	N 5 13.0
Rasalhague	96 14.0	N12 33.2
Regulus	207 51.8	N11 57.1
Rigel	281 19.9	S 8 12.0
Rigil Kent.	140 02.7	S60 50.7
Sabik	102 21.8	S15 43.8
Schedar	349 50.6	N56 33.2
Shaula	96 32.9	S37 06.3
Sirius	258 40.8	S16 43.3
Spica	158 39.5	S11 10.7
Suhail	222 58.2	S43 26.8
Vega	80 44.6	N38 46.8
Zuben'ubi	137 14.2	S16 03.3

	SHA	Mer. Pass
	° ′	h m
Venus	42 47.7	9 40
Mars	82 59.2	6 59
Jupiter	228 37.5	21 13
Saturn	278 03.8	17 56

UT d h		SUN GHA	SUN Dec	MOON GHA	v	MOON Dec	d	HP
		° ′	° ′	° ′	′	° ′	′	′
14 FRIDAY	00	177 38.1	S 2 46.9	54 06.7	7.3	N25 10.5	4.3	57.2
	01	192 38.3	45.9	68 33.0	7.3	25 06.2	4.5	57.2
	02	207 38.5	44.9	82 59.3	7.4	25 01.7	4.6	57.3
	03	222 38.6	. . 43.9	97 25.7	7.3	24 57.1	4.7	57.3
	04	237 38.8	42.9	111 52.0	7.3	24 52.4	4.9	57.3
	05	252 39.0	41.9	126 18.3	7.3	24 47.5	5.1	57.4
	06	267 39.1	S 2 41.0	140 44.6	7.4	N24 42.4	5.2	57.4
	07	282 39.3	40.0	155 11.0	7.3	24 37.2	5.3	57.5
	08	297 39.5	39.0	169 37.3	7.3	24 31.9	5.6	57.5
	09	312 39.7	. . 38.0	184 03.6	7.3	24 26.3	5.6	57.5
	10	327 39.8	37.0	198 29.9	7.3	24 20.7	5.8	57.6
	11	342 40.0	36.0	212 56.2	7.4	24 14.9	6.0	57.6
	12	357 40.2	S 2 35.0	227 22.6	7.3	N24 08.9	6.1	57.6
	13	12 40.4	34.1	241 48.9	7.3	24 02.8	6.3	57.7
	14	27 40.5	33.1	256 15.2	7.4	23 56.5	6.4	57.7
	15	42 40.7	. . 32.1	270 41.6	7.3	23 50.1	6.6	57.8
	16	57 40.9	31.1	285 07.9	7.4	23 43.5	6.7	57.8
	17	72 41.1	30.1	299 34.3	7.4	23 36.8	6.9	57.8
	18	87 41.2	S 2 29.1	314 00.7	7.4	N23 29.9	7.0	57.9
	19	102 41.4	28.1	328 27.1	7.4	23 22.9	7.1	57.9
	20	117 41.6	27.2	342 53.5	7.4	23 15.8	7.3	58.0
	21	132 41.7	. . 26.2	357 19.9	7.4	23 08.5	7.5	58.0
	22	147 41.9	25.2	11 46.3	7.5	23 01.0	7.6	58.0
	23	162 42.1	24.2	26 12.8	7.4	22 53.4	7.8	58.1
15 SATURDAY	00	177 42.3	S 2 23.2	40 39.2	7.5	N22 45.6	7.9	58.1
	01	192 42.4	22.2	55 05.7	7.5	22 37.7	8.0	58.1
	02	207 42.6	21.2	69 32.2	7.6	22 29.7	8.2	58.2
	03	222 42.8	. . 20.2	83 58.8	7.5	22 21.5	8.3	58.2
	04	237 43.0	19.3	98 25.3	7.6	22 13.2	8.5	58.3
	05	252 43.1	18.3	112 51.9	7.5	22 04.7	8.6	58.3
	06	267 43.3	S 2 17.3	127 18.4	7.7	N21 56.1	8.8	58.3
	07	282 43.5	16.3	141 45.1	7.6	21 47.3	8.9	58.4
	08	297 43.7	15.3	156 11.7	7.6	21 38.4	9.0	58.4
	09	312 43.8	. . 14.3	170 38.3	7.7	21 29.4	9.2	58.5
	10	327 44.0	13.3	185 05.0	7.7	21 20.2	9.3	58.5
	11	342 44.2	12.4	199 31.7	7.8	21 10.9	9.5	58.5
	12	357 44.4	S 2 11.4	213 58.5	7.7	N21 01.4	9.6	58.6
	13	12 44.5	10.4	228 25.2	7.8	20 51.8	9.7	58.6
	14	27 44.7	09.4	242 52.0	7.8	20 42.1	9.9	58.7
	15	42 44.9	. . 08.4	257 18.8	7.9	20 32.2	10.0	58.7
	16	57 45.1	07.4	271 45.7	7.8	20 22.2	10.1	58.7
	17	72 45.3	06.4	286 12.5	7.9	20 12.1	10.3	58.8
	18	87 45.4	S 2 05.4	300 39.4	8.0	N20 01.8	10.4	58.8
	19	102 45.6	04.5	315 06.4	7.9	19 51.4	10.5	58.8
	20	117 45.8	03.5	329 33.3	8.0	19 40.9	10.7	58.9
	21	132 46.0	. . 02.5	344 00.3	8.1	19 30.2	10.8	58.9
	22	147 46.1	01.5	358 27.4	8.0	19 19.4	10.9	59.0
	23	162 46.3	2 00.5	12 54.4	8.1	19 08.5	11.1	59.0
16 SUNDAY	00	177 46.5	S 1 59.5	27 21.5	8.1	N18 57.4	11.1	59.0
	01	192 46.7	58.5	41 48.6	8.2	18 46.3	11.4	59.1
	02	207 46.8	57.5	56 15.8	8.2	18 34.9	11.4	59.1
	03	222 47.0	. . 56.6	70 43.0	8.2	18 23.5	11.5	59.1
	04	237 47.2	55.6	85 10.2	8.3	18 12.0	11.7	59.2
	05	252 47.4	54.6	99 37.5	8.2	18 00.3	11.8	59.2
	06	267 47.6	S 1 53.6	114 04.7	8.4	N17 48.5	11.9	59.2
	07	282 47.7	52.6	128 32.1	8.3	17 36.6	12.1	59.3
	08	297 47.9	51.6	142 59.4	8.4	17 24.5	12.1	59.3
	09	312 48.1	. . 50.6	157 26.8	8.4	17 12.4	12.3	59.3
	10	327 48.3	49.6	171 54.2	8.5	17 00.1	12.4	59.4
	11	342 48.4	48.7	186 21.7	8.5	16 47.7	12.5	59.4
	12	357 48.6	S 1 47.7	200 49.2	8.5	N16 35.2	12.6	59.4
	13	12 48.8	46.7	215 16.7	8.6	16 22.6	12.7	59.5
	14	27 49.0	45.7	229 44.3	8.6	16 09.9	12.8	59.5
	15	42 49.2	. . 44.7	244 11.9	8.6	15 57.1	13.0	59.6
	16	57 49.3	43.7	258 39.5	8.6	15 44.1	13.0	59.6
	17	72 49.5	42.7	273 07.1	8.7	15 31.1	13.2	59.6
	18	87 49.7	S 1 41.7	287 34.8	8.8	N15 17.9	13.3	59.6
	19	102 49.9	40.8	302 02.6	8.7	15 04.6	13.3	59.7
	20	117 50.0	39.8	316 30.3	8.8	14 51.3	13.5	59.7
	21	132 50.2	. . 38.8	330 58.1	8.8	14 37.8	13.6	59.7
	22	147 50.4	37.8	345 25.9	8.9	14 24.2	13.6	59.8
	23	162 50.6	36.8	359 53.8	8.9	N14 10.6	13.8	59.8
		SD 16.1	d 1.0	SD 15.7		16.0		16.2

Lat.	Twilight Naut.	Twilight Civil	Sunrise	Moonrise 14	Moonrise 15	Moonrise 16	Moonrise 17
°	h m	h m	h m	h m	h m	h m	h m
N 72	03 56	05 19	06 26	□	□	12 19	15 23
N 70	04 10	05 24	06 24	□	□	13 07	15 40
68	04 21	05 28	06 23	□	10 52	13 38	15 54
66	04 30	05 31	06 21	□	11 47	14 01	16 05
64	04 37	05 33	06 20	10 19	12 20	14 18	16 14
62	04 43	05 35	06 19	10 59	12 44	14 33	16 21
60	04 48	05 37	06 18	11 27	13 03	14 45	16 28
N 58	04 52	05 38	06 17	11 48	13 19	14 55	16 34
56	04 56	05 40	06 17	12 06	13 32	15 04	16 39
54	05 00	05 41	06 16	12 20	13 43	15 12	16 44
52	05 03	05 42	06 15	12 33	13 54	15 20	16 48
50	05 05	05 43	06 15	12 45	14 03	15 26	16 52
45	05 10	05 44	06 14	13 08	14 22	15 40	17 00
N 40	05 14	05 45	06 12	13 27	14 37	15 51	17 06
35	05 17	05 46	06 11	13 43	14 50	16 01	17 12
30	05 19	05 47	06 11	13 56	15 02	16 09	17 17
20	05 21	05 47	06 09	14 19	15 21	16 23	17 26
N 10	05 22	05 46	06 07	14 39	15 38	16 36	17 34
0	05 21	05 45	06 06	14 58	15 53	16 48	17 41
S 10	05 19	05 43	06 04	15 16	16 09	16 59	17 48
20	05 15	05 40	06 02	15 36	16 26	17 12	17 55
30	05 08	05 36	06 00	15 59	16 45	17 26	18 04
35	05 04	05 33	05 59	16 12	16 56	17 34	18 09
40	04 58	05 30	05 57	16 28	17 08	17 43	18 14
45	04 51	05 26	05 55	16 46	17 23	17 54	18 21
S 50	04 42	05 20	05 53	17 09	17 41	18 07	18 29
52	04 37	05 18	05 52	17 19	17 50	18 13	18 32
54	04 32	05 15	05 51	17 32	17 59	18 20	18 36
56	04 27	05 12	05 50	17 46	18 10	18 27	18 40
58	04 20	05 08	05 48	18 02	18 22	18 35	18 45
S 60	04 13	05 04	05 47	18 21	18 36	18 45	18 50

Lat.	Sunset	Twilight Civil	Twilight Naut.	Moonset 14	Moonset 15	Moonset 16	Moonset 17
°	h m	h m	h m	h m	h m	h m	h m
N 72	17 54	19 01	20 26	□	□	09 12	08 01
N 70	17 56	18 57	20 12	□	□	08 21	07 41
68	17 57	18 53	20 00	□	08 40	07 49	07 26
66	17 58	18 49	19 51	□	07 44	07 25	07 13
64	17 59	18 47	19 43	07 15	07 11	07 06	07 02
62	18 00	18 45	19 37	06 35	06 46	06 51	06 53
60	18 01	18 43	19 32	06 07	06 26	06 38	06 45
N 58	18 02	18 41	19 27	05 45	06 10	06 26	06 38
56	18 03	18 40	19 23	05 27	05 56	06 16	06 31
54	18 03	18 38	19 20	05 12	05 44	06 08	06 26
52	18 04	18 37	19 17	04 59	05 33	06 00	06 21
50	18 04	18 36	19 14	04 47	05 24	05 53	06 16
45	18 05	18 35	19 09	04 23	05 03	05 37	06 06
N 40	18 06	18 33	19 05	04 03	04 47	05 25	05 57
35	18 07	18 32	19 02	03 47	04 33	05 14	05 50
30	18 08	18 32	19 00	03 33	04 21	05 04	05 44
20	18 09	18 31	18 57	03 09	04 00	04 48	05 32
N 10	18 11	18 32	18 56	02 48	03 42	04 33	05 22
0	18 12	18 33	18 57	02 29	03 24	04 19	05 13
S 10	18 14	18 35	18 59	02 09	03 07	04 06	05 04
20	18 16	18 38	19 03	01 48	02 49	03 51	04 53
30	18 18	18 42	19 10	01 23	02 27	03 34	04 42
35	18 19	18 44	19 14	01 09	02 14	03 24	04 35
40	18 20	18 47	19 19	00 52	02 00	03 12	04 27
45	18 22	18 51	19 26	00 32	01 42	02 58	04 18
S 50	18 24	18 56	19 35	00 07	01 20	02 42	04 06
52	18 25	18 59	19 39	25 10	01 10	02 34	04 01
54	18 26	19 02	19 44	24 58	00 58	02 25	03 55
56	18 27	19 05	19 50	24 45	00 45	02 15	03 49
58	18 29	19 08	19 56	24 29	00 29	02 04	03 42
S 60	18 30	19 12	20 03	24 10	00 10	01 50	03 34

Day	SUN Eqn. of Time 00^h	SUN Eqn. of Time 12^h	SUN Mer. Pass.	MOON Mer. Pass. Upper	MOON Mer. Pass. Lower	MOON Age	MOON Phase
d	m s	m s	h m	h m	h m	d %	
14	09 28	09 20	12 09	21 11	08 43	11 80	
15	09 11	09 03	12 09	22 06	09 39	12 88	
16	08 54	08 46	12 09	23 00	10 34	13 95	

UT d h	ARIES GHA ° ′	VENUS −4.0 GHA ° ′	VENUS Dec ° ′	MARS +0.7 GHA ° ′	MARS Dec ° ′	JUPITER −2.4 GHA ° ′	JUPITER Dec ° ′	SATURN +0.0 GHA ° ′	SATURN Dec ° ′
17 00 (MONDAY)	174 09.4	214 33.0	S15 44.5	255 45.8	S23 31.3	42 54.3	N18 58.3	92 08.1	N22 11.8
01	189 11.9	229 32.5	43.7	270 46.6	31.3	57 56.9	58.4	107 10.5	11.8
02	204 14.4	244 32.0	43.0	285 47.3	31.2	72 59.5	58.4	122 12.8	11.8
03	219 16.8	259 31.4	. . 42.2	300 48.0	. . 31.2	88 02.1	. . 58.4	137 15.2	. . 11.8
04	234 19.3	274 30.9	41.5	315 48.8	31.1	103 04.7	58.5	152 17.5	11.8
05	249 21.7	289 30.4	40.7	330 49.5	31.1	118 07.3	58.5	167 19.9	11.8
06	264 24.2	304 29.9	S15 39.9	345 50.3	S23 31.0	133 09.9	N18 58.5	182 22.2	N22 11.8
07	279 26.7	319 29.3	39.2	0 51.0	31.0	148 12.5	58.6	197 24.6	11.9
08	294 29.1	334 28.8	38.4	15 51.7	31.0	163 15.2	58.6	212 26.9	11.9
09	309 31.6	349 28.3	. . 37.6	30 52.5	. . 30.9	178 17.8	. . 58.7	227 29.3	. . 11.9
10	324 34.1	4 27.8	36.9	45 53.2	30.9	193 20.4	58.7	242 31.6	11.9
11	339 36.5	19 27.2	36.1	60 54.0	30.8	208 23.0	58.7	257 34.0	11.9
12	354 39.0	34 26.7	S15 35.3	75 54.7	S23 30.8	223 25.6	N18 58.8	272 36.3	N22 11.9
13	9 41.5	49 26.2	34.5	90 55.4	30.7	238 28.2	58.8	287 38.7	11.9
14	24 43.9	64 25.6	33.8	105 56.2	30.7	253 30.8	58.8	302 41.0	12.0
15	39 46.4	79 25.1	. . 33.0	120 56.9	. . 30.6	268 33.4	. . 58.9	317 43.4	. . 12.0
16	54 48.9	94 24.6	32.2	135 57.7	30.6	283 36.0	58.9	332 45.7	12.0
17	69 51.3	109 24.1	31.5	150 58.4	30.5	298 38.6	58.9	347 48.1	12.0
18	84 53.8	124 23.5	S15 30.7	165 59.1	S23 30.5	313 41.2	N18 59.0	2 50.4	N22 12.0
19	99 56.2	139 23.0	29.9	180 59.9	30.5	328 43.8	59.0	17 52.8	12.0
20	114 58.7	154 22.5	29.1	196 00.6	30.4	343 46.4	59.0	32 55.1	12.0
21	130 01.2	169 22.0	. . 28.4	211 01.4	. . 30.4	358 49.0	. . 59.1	47 57.5	. . 12.1
22	145 03.6	184 21.4	27.6	226 02.1	30.3	13 51.6	59.1	62 59.8	12.1
23	160 06.1	199 20.9	26.8	241 02.8	30.3	28 54.2	59.1	78 02.2	12.1
18 00 (TUESDAY)	175 08.6	214 20.4	S15 26.0	256 03.6	S23 30.2	43 56.8	N18 59.2	93 04.5	N22 12.1
01	190 11.0	229 19.9	25.2	271 04.3	30.2	58 59.4	59.2	108 06.9	12.1
02	205 13.5	244 19.4	24.5	286 05.1	30.1	74 02.0	59.2	123 09.2	12.1
03	220 16.0	259 18.8	. . 23.7	301 05.8	. . 30.1	89 04.6	. . 59.3	138 11.6	. . 12.1
04	235 18.4	274 18.3	22.9	316 06.5	30.0	104 07.2	59.3	153 13.9	12.2
05	250 20.9	289 17.8	22.1	331 07.3	30.0	119 09.8	59.3	168 16.2	12.2
06	265 23.3	304 17.3	S15 21.3	346 08.0	S23 29.9	134 12.4	N18 59.4	183 18.6	N22 12.2
07	280 25.8	319 16.7	20.6	1 08.8	29.9	149 15.0	59.4	198 20.9	12.2
08	295 28.3	334 16.2	19.8	16 09.5	29.8	164 17.6	59.4	213 23.3	12.2
09	310 30.7	349 15.7	. . 19.0	31 10.3	. . 29.8	179 20.2	. . 59.5	228 25.6	. . 12.2
10	325 33.2	4 15.2	18.2	46 11.0	29.7	194 22.8	59.5	243 28.0	12.2
11	340 35.7	19 14.7	17.4	61 11.7	29.7	209 25.4	59.5	258 30.3	12.3
12	355 38.1	34 14.1	S15 16.6	76 12.5	S23 29.6	224 28.0	N18 59.6	273 32.7	N22 12.3
13	10 40.6	49 13.6	15.8	91 13.2	29.6	239 30.6	59.6	288 35.0	12.3
14	25 43.1	64 13.1	15.0	106 14.0	29.5	254 33.2	59.6	303 37.4	12.3
15	40 45.5	79 12.6	. . 14.3	121 14.7	. . 29.5	269 35.8	. . 59.7	318 39.7	. . 12.3
16	55 48.0	94 12.1	13.5	136 15.5	29.4	284 38.4	59.7	333 42.0	12.3
17	70 50.5	109 11.6	12.7	151 16.2	29.3	299 41.0	59.7	348 44.4	12.3
18	85 52.9	124 11.0	S15 11.9	166 16.9	S23 29.3	314 43.6	N18 59.8	3 46.7	N22 12.4
19	100 55.4	139 10.5	11.1	181 17.7	29.2	329 46.2	59.8	18 49.1	12.4
20	115 57.8	154 10.0	10.3	196 18.4	29.2	344 48.8	59.8	33 51.4	12.4
21	131 00.3	169 09.5	. . 09.5	211 19.2	. . 29.1	359 51.4	. . 59.9	48 53.8	. . 12.4
22	146 02.8	184 09.0	08.7	226 19.9	29.1	14 54.0	59.9	63 56.1	12.4
23	161 05.2	199 08.5	07.9	241 20.6	29.0	29 56.6	18 59.9	78 58.5	12.4
19 00 (WEDNESDAY)	176 07.7	214 07.9	S15 07.1	256 21.4	S23 29.0	44 59.2	N19 00.0	94 00.8	N22 12.4
01	191 10.2	229 07.4	06.3	271 22.1	28.9	60 01.8	00.0	109 03.1	12.5
02	206 12.6	244 06.9	05.5	286 22.9	28.9	75 04.4	00.0	124 05.5	12.5
03	221 15.1	259 06.4	. . 04.7	301 23.6	. . 28.8	90 07.0	. . 00.1	139 07.8	. . 12.5
04	236 17.6	274 05.9	03.9	316 24.4	28.7	105 09.6	00.1	154 10.2	12.5
05	251 20.0	289 05.4	03.1	331 25.1	28.7	120 12.1	00.1	169 12.5	12.5
06	266 22.5	304 04.9	S15 02.3	346 25.8	S23 28.6	135 14.7	N19 00.2	184 14.9	N22 12.5
07	281 24.9	319 04.3	01.5	1 26.6	28.6	150 17.3	00.2	199 17.2	12.6
08	296 27.4	334 03.8	15 00.7	16 27.3	28.5	165 19.9	00.2	214 19.5	12.6
09	311 29.9	349 03.3	14 59.9	31 28.1	. . 28.5	180 22.5	. . 00.2	229 21.9	. . 12.6
10	326 32.3	4 02.8	59.1	46 28.8	28.4	195 25.1	00.3	244 24.2	12.6
11	341 34.8	19 02.3	58.3	61 29.6	28.3	210 27.7	00.3	259 26.6	12.6
12	356 37.3	34 01.8	S14 57.5	76 30.3	S23 28.3	225 30.3	N19 00.3	274 28.9	N22 12.6
13	11 39.7	49 01.3	56.7	91 31.1	28.2	240 32.9	00.4	289 31.2	12.6
14	26 42.2	64 00.8	55.9	106 31.8	28.2	255 35.5	00.4	304 33.6	12.7
15	41 44.7	79 00.2	. . 55.1	121 32.5	. . 28.1	270 38.1	. . 00.4	319 35.9	. . 12.7
16	56 47.1	93 59.7	54.3	136 33.3	28.0	285 40.6	00.5	334 38.3	12.7
17	71 49.6	108 59.2	53.5	151 34.0	28.0	300 43.2	00.5	349 40.6	12.7
18	86 52.1	123 58.7	S14 52.6	166 34.8	S23 27.9	315 45.8	N19 00.5	4 42.9	N22 12.7
19	101 54.5	138 58.2	51.8	181 35.5	27.9	330 48.4	00.6	19 45.3	12.7
20	116 57.0	153 57.7	51.0	196 36.3	27.8	345 51.0	00.6	34 47.6	12.7
21	131 59.4	168 57.2	. . 50.2	211 37.0	. . 27.7	0 53.6	. . 00.6	49 50.0	. . 12.8
22	147 01.9	183 56.7	49.4	226 37.8	27.7	15 56.2	00.6	64 52.3	12.8
23	162 04.4	198 56.2	48.6	241 38.5	27.6	30 58.8	00.7	79 54.6	12.8
Mer. Pass.	12 h 17.4 m	v −0.5	d 0.8	v 0.7	d 0.1	v 2.6	d 0.0	v 2.3	d 0.0

STARS

Name	SHA ° ′	Dec ° ′
Acamar	315 24.6	S40 17.8
Achernar	335 33.1	S57 13.5
Acrux	173 18.0	S63 06.9
Adhara	255 18.8	S28 58.8
Aldebaran	290 58.8	N16 30.9
Alioth	166 27.0	N55 56.5
Alkaid	153 04.7	N49 17.7
Al Na'ir	27 54.1	S46 56.8
Alnilam	275 54.6	S 1 12.1
Alphard	218 03.8	S 8 40.4
Alphecca	126 17.6	N26 42.0
Alpheratz	357 52.3	N29 06.3
Altair	62 16.3	N 8 52.3
Ankaa	353 23.9	S42 17.5
Antares	112 36.1	S26 26.4
Arcturus	146 02.8	N19 09.8
Atria	107 45.2	S69 01.8
Avior	234 21.1	S59 31.4
Bellatrix	278 40.7	N 6 21.1
Betelgeuse	271 10.0	N 7 24.4
Canopus	263 59.7	S52 42.1
Capella	280 46.5	N46 00.3
Deneb	49 37.4	N45 17.1
Denebola	182 41.5	N14 33.2
Diphda	349 04.3	S17 58.4
Dubhe	194 00.6	N61 44.1
Elnath	278 22.9	N28 36.7
Eltanin	90 49.9	N51 28.9
Enif	33 55.4	N 9 53.1
Fomalhaut	15 33.1	S29 36.5
Gacrux	172 09.6	S57 07.8
Gienah	176 00.4	S17 33.6
Hadar	148 59.1	S60 23.2
Hamal	328 10.2	N23 28.6
Kaus Aust.	83 54.6	S34 23.0
Kochab	137 18.5	N74 08.3
Markab	13 46.7	N15 13.1
Menkar	314 23.7	N 4 06.0
Menkent	148 16.9	S36 23.1
Miaplacidus	221 41.0	S69 43.9
Mirfak	308 52.4	N49 52.5
Nunki	76 08.4	S26 17.6
Peacock	53 32.2	S56 43.5
Pollux	243 37.4	N28 01.2
Procyon	245 08.0	N 5 13.0
Rasalhague	96 13.9	N12 33.2
Regulus	207 51.8	N11 57.1
Rigel	281 19.9	S 8 12.0
Rigil Kent.	140 02.6	S60 50.7
Sabik	102 21.8	S15 43.8
Schedar	349 50.6	N56 33.2
Shaula	96 32.9	S37 06.3
Sirius	258 40.8	S16 43.3
Spica	158 39.5	S11 10.7
Suhail	222 58.2	S43 26.8
Vega	80 44.5	N38 46.8
Zuben'ubi	137 14.2	S16 03.3

	SHA ° ′	Mer. Pass. h m
Venus	39 11.8	9 43
Mars	80 55.0	6 55
Jupiter	228 48.3	21 01
Saturn	277 55.9	17 45

Day	UT	SUN GHA	SUN Dec	MOON GHA	v	MOON Dec	d	HP
	d h	° ′	° ′	° ′	′	° ′	′	′
MONDAY	17 00	177 50.8	S 1 35.8	14 21.7	8.9	N13 56.8	13.8	59.8
	01	192 50.9	34.8	28 49.6	9.0	13 43.0	14.0	59.9
	02	207 51.1	33.8	43 17.6	9.0	13 29.0	14.1	59.9
	03	222 51.3	. . 32.9	57 45.6	9.0	13 14.9	14.1	59.9
	04	237 51.5	31.9	72 13.6	9.0	13 00.8	14.2	60.0
	05	252 51.7	30.9	86 41.6	9.1	12 46.6	14.4	60.0
	06	267 51.8	S 1 29.9	101 09.7	9.1	N12 32.2	14.4	60.0
	07	282 52.0	28.9	115 37.8	9.2	12 17.8	14.5	60.0
	08	297 52.2	27.9	130 06.0	9.1	12 03.3	14.5	60.1
	09	312 52.4	. . 26.9	144 34.1	9.2	11 48.8	14.7	60.1
	10	327 52.6	25.9	159 02.3	9.2	11 34.1	14.7	60.1
	11	342 52.7	25.0	173 30.5	9.3	11 19.4	14.9	60.2
	12	357 52.9	S 1 24.0	187 58.8	9.3	N11 04.5	14.9	60.2
	13	12 53.1	23.0	202 27.1	9.3	10 49.6	14.9	60.2
	14	27 53.3	22.0	216 55.4	9.3	10 34.7	15.1	60.2
	15	42 53.5	. . 21.0	231 23.7	9.4	10 19.6	15.1	60.3
	16	57 53.6	20.0	245 52.1	9.3	10 04.5	15.2	60.3
	17	72 53.8	19.0	260 20.4	9.4	9 49.3	15.3	60.3
	18	87 54.0	S 1 18.0	274 48.8	9.5	N 9 34.0	15.3	60.3
	19	102 54.2	17.1	289 17.3	9.4	9 18.7	15.4	60.4
	20	117 54.4	16.1	303 45.7	9.5	9 03.3	15.5	60.4
	21	132 54.6	. . 15.1	318 14.2	9.5	8 47.8	15.5	60.4
	22	147 54.7	14.1	332 42.7	9.5	8 32.3	15.6	60.4
	23	162 54.9	13.1	347 11.2	9.5	8 16.7	15.7	60.4
TUESDAY	18 00	177 55.1	S 1 12.1	1 39.7	9.6	N 8 01.0	15.7	60.5
	01	192 55.3	11.1	16 08.3	9.6	7 45.3	15.7	60.5
	02	207 55.5	10.1	30 36.9	9.5	7 29.6	15.9	60.5
	03	222 55.6	. . 09.1	45 05.4	9.7	7 13.7	15.8	60.5
	04	237 55.8	08.2	59 34.1	9.6	6 57.9	16.0	60.6
	05	252 56.0	07.2	74 02.7	9.6	6 41.9	15.9	60.6
	06	267 56.2	S 1 06.2	88 31.3	9.7	N 6 26.0	16.0	60.6
	07	282 56.4	05.2	103 00.0	9.6	6 10.0	16.1	60.6
	08	297 56.5	04.2	117 28.6	9.7	5 53.9	16.1	60.6
	09	312 56.7	. . 03.2	131 57.3	9.7	5 37.8	16.2	60.6
	10	327 56.9	02.2	146 26.0	9.7	5 21.6	16.2	60.7
	11	342 57.1	01.2	160 54.7	9.7	5 05.4	16.2	60.7
	12	357 57.3	S 1 00.3	175 23.4	9.7	N 4 49.2	16.3	60.7
	13	12 57.5	0 59.3	189 52.1	9.8	4 32.9	16.3	60.7
	14	27 57.6	58.3	204 20.9	9.7	4 16.6	16.3	60.7
	15	42 57.8	. . 57.3	218 49.6	9.8	4 00.3	16.4	60.7
	16	57 58.0	56.3	233 18.4	9.7	3 43.9	16.3	60.8
	17	72 58.2	55.3	247 47.1	9.8	3 27.6	16.5	60.8
	18	87 58.4	S 0 54.3	262 15.9	9.7	N 3 11.1	16.4	60.8
	19	102 58.6	53.3	276 44.6	9.8	2 54.7	16.5	60.8
	20	117 58.7	52.4	291 13.4	9.8	2 38.2	16.5	60.8
	21	132 58.9	. . 51.4	305 42.2	9.7	2 21.7	16.5	60.8
	22	147 59.1	50.4	320 10.9	9.8	2 05.2	16.5	60.8
	23	162 59.3	49.4	334 39.7	9.8	1 48.7	16.6	60.8
WEDNESDAY	19 00	177 59.5	S 0 48.4	349 08.5	9.7	N 1 32.1	16.5	60.8
	01	192 59.7	47.4	3 37.2	9.8	1 15.6	16.6	60.9
	02	207 59.8	46.4	18 06.0	9.8	0 59.0	16.6	60.9
	03	223 00.0	. . 45.4	32 34.8	9.7	0 42.4	16.6	60.9
	04	238 00.2	44.4	47 03.5	9.8	0 25.8	16.6	60.9
	05	253 00.4	43.5	61 32.3	9.7	N 0 09.2	16.6	60.9
	06	268 00.6	S 0 42.5	76 01.0	9.8	S 0 07.4	16.6	60.9
	07	283 00.8	41.5	90 29.8	9.7	0 24.0	16.6	60.9
	08	298 00.9	40.5	104 58.5	9.7	0 40.6	16.6	60.9
	09	313 01.1	. . 39.5	119 27.2	9.7	0 57.2	16.6	60.9
	10	328 01.3	38.5	133 55.9	9.7	1 13.8	16.6	60.9
	11	343 01.5	37.5	148 24.6	9.7	1 30.4	16.6	60.9
	12	358 01.7	S 0 36.5	162 53.3	9.7	S 1 47.0	16.6	60.9
	13	13 01.9	35.6	177 22.0	9.7	2 03.6	16.6	60.9
	14	28 02.0	34.6	191 50.7	9.6	2 20.2	16.5	60.9
	15	43 02.2	. . 33.6	206 19.3	9.6	2 36.7	16.6	60.9
	16	58 02.4	32.6	220 47.9	9.6	2 53.3	16.5	60.9
	17	73 02.6	31.6	235 16.5	9.6	3 09.8	16.5	60.9
	18	88 02.8	S 0 30.6	249 45.1	9.6	S 3 26.3	16.5	60.9
	19	103 03.0	29.6	264 13.7	9.6	3 42.8	16.5	60.9
	20	118 03.1	28.6	278 42.3	9.5	3 59.3	16.4	60.9
	21	133 03.3	. . 27.6	293 10.8	9.5	4 15.7	16.4	60.9
	22	148 03.5	26.7	307 39.3	9.5	4 32.1	16.4	60.9
	23	163 03.7	25.7	322 07.8	9.5	S 4 48.5	16.4	60.9
		SD 16.1	*d* 1.0	SD	16.4	16.5		16.6

Lat.	Twilight Naut.	Twilight Civil	Sunrise	Moonrise 17	Moonrise 18	Moonrise 19	Moonrise 20
°	h m	h m	h m	h m	h m	h m	h m
N 72	03 37	05 04	06 11	15 23	17 53	20 19	22 54
N 70	03 54	05 10	06 11	15 40	17 58	20 12	22 33
68	04 07	05 15	06 10	15 54	18 01	20 07	22 18
66	04 17	05 19	06 10	16 05	18 04	20 03	22 05
64	04 25	05 22	06 10	16 14	18 07	20 00	21 55
62	04 32	05 25	06 09	16 21	18 09	19 57	21 46
60	04 38	05 28	06 09	16 28	18 11	19 54	21 38
N 58	04 44	05 30	06 09	16 34	18 13	19 52	21 32
56	04 48	05 32	06 09	16 39	18 14	19 49	21 26
54	04 52	05 33	06 09	16 44	18 15	19 48	21 21
52	04 55	05 35	06 08	16 48	18 17	19 46	21 16
50	04 58	05 36	06 08	16 52	18 18	19 44	21 12
45	05 05	05 39	06 08	17 00	18 20	19 41	21 03
N 40	05 09	05 41	06 08	17 06	18 22	19 38	20 55
35	05 13	05 42	06 07	17 12	18 24	19 36	20 49
30	05 15	05 43	06 07	17 17	18 26	19 34	20 43
20	05 19	05 44	06 06	17 26	18 28	19 31	20 33
N 10	05 20	05 45	06 06	17 34	18 31	19 27	20 25
0	05 20	05 44	06 05	17 41	18 33	19 25	20 17
S 10	05 19	05 43	06 04	17 48	18 35	19 22	20 10
20	05 15	05 41	06 03	17 55	18 37	19 19	20 02
30	05 10	05 38	06 02	18 04	18 40	19 16	19 52
35	05 06	05 36	06 01	18 09	18 42	19 14	19 47
40	05 01	05 33	06 00	18 14	18 43	19 12	19 41
45	04 55	05 30	05 59	18 21	18 45	19 09	19 34
S 50	04 47	05 25	05 58	18 29	18 48	19 07	19 26
52	04 43	05 23	05 57	18 32	18 49	19 05	19 22
54	04 39	05 21	05 57	18 36	18 50	19 04	19 18
56	04 34	05 18	05 56	18 40	18 52	19 02	19 14
58	04 28	05 15	05 55	18 45	18 53	19 01	19 09
S 60	04 21	05 12	05 54	18 50	18 55	18 59	19 03

Lat.	Sunset	Twilight Civil	Twilight Naut.	Moonset 17	Moonset 18	Moonset 19	Moonset 20
°	h m	h m	h m	h m	h m	h m	h m
N 72	18 08	19 16	20 43	08 01	07 23	06 51	06 18
N 70	18 08	19 09	20 26	07 41	07 15	06 52	06 28
68	18 08	19 04	20 13	07 26	07 08	06 52	06 36
66	18 08	18 59	20 02	07 13	07 02	06 53	06 43
64	18 08	18 56	19 53	07 02	06 58	06 53	06 49
62	18 08	18 53	19 46	06 53	06 53	06 54	06 54
60	18 09	18 50	19 40	06 45	06 50	06 54	06 58
N 58	18 09	18 48	19 34	06 38	06 46	06 54	07 02
56	18 09	18 46	19 30	06 31	06 43	06 54	07 05
54	18 09	18 44	19 26	06 26	06 41	06 55	07 08
52	18 09	18 43	19 22	06 21	06 38	06 55	07 11
50	18 09	18 41	19 19	06 16	06 36	06 55	07 14
45	18 09	18 38	19 13	06 06	06 31	06 55	07 20
N 40	18 09	18 36	19 08	05 57	06 27	06 56	07 24
35	18 10	18 35	19 04	05 50	06 24	06 56	07 28
30	18 10	18 34	19 02	05 44	06 21	06 56	07 32
20	18 10	18 32	18 58	05 32	06 15	06 57	07 38
N 10	18 11	18 32	18 56	05 22	06 10	06 57	07 44
0	18 11	18 32	18 56	05 13	06 05	06 57	07 49
S 10	18 12	18 33	18 58	05 04	06 01	06 58	07 55
20	18 13	18 35	19 01	04 53	05 56	06 58	08 00
30	18 14	18 38	19 06	04 42	05 50	06 58	08 07
35	18 15	18 40	19 09	04 35	05 46	06 58	08 11
40	18 15	18 43	19 14	04 27	05 42	06 58	08 15
45	18 16	18 46	19 20	04 18	05 38	06 59	08 20
S 50	18 18	18 50	19 28	04 06	05 32	06 59	08 26
52	18 18	18 52	19 32	04 01	05 30	06 59	08 28
54	18 19	18 54	19 36	03 55	05 27	06 59	08 31
56	18 19	18 57	19 41	03 49	05 24	06 59	08 35
58	18 20	18 59	19 47	03 42	05 21	06 59	08 39
S 60	18 21	19 03	19 53	03 34	05 17	06 59	08 43

Day	SUN Eqn. of Time 00^h	SUN Eqn. of Time 12^h	SUN Mer. Pass.	MOON Mer. Pass. Upper	MOON Mer. Pass. Lower	Age	Phase
d	m s	m s	h m	h m	h m	d %	
17	08 37	08 29	12 08	23 53	11 27	14 99	
18	08 20	08 11	12 08	24 45	12 19	15 100	○
19	08 02	07 54	12 08	00 45	13 11	16 98	

UT	ARIES	VENUS −4.0		MARS +0.7		JUPITER −2.4		SATURN +0.1		STARS		
	GHA	GHA	Dec	GHA	Dec	GHA	Dec	GHA	Dec	Name	SHA	Dec
d h	° ′	° ′	° ′	° ′	° ′	° ′	° ′	° ′	° ′		° ′	° ′
20 00	177 06.8	213 55.7	S14 47.8	256 39.2	S23 27.6	46 01.3	N19 00.7	94 57.0	N22 12.8	Acamar	315 24.7	S40 17.
01	192 09.3	228 55.2	46.9	271 40.0	27.5	61 03.9	00.7	109 59.3	12.8	Achernar	335 33.1	S57 13.
02	207 11.8	243 54.6	46.1	286 40.7	27.4	76 06.5	00.8	125 01.6	12.8	Acrux	173 18.0	S63 07.
03	222 14.2	258 54.1	45.3	301 41.5	27.4	91 09.1	00.8	140 04.0	12.8	Adhara	255 18.8	S28 58.
04	237 16.7	273 53.6	44.5	316 42.2	27.3	106 11.7	00.8	155 06.3	12.9	Aldebaran	290 58.8	N16 30.
05	252 19.2	288 53.1	43.7	331 43.0	27.2	121 14.3	00.8	170 08.7	12.9			
06	267 21.6	303 52.6	S14 42.9	346 43.7	S23 27.2	136 16.9	N19 00.9	185 11.0	N22 12.9	Alioth	166 27.0	N55 56.
T 07	282 24.1	318 52.1	42.0	1 44.5	27.1	151 19.4	00.9	200 13.3	12.9	Alkaid	153 04.7	N49 17.
H 08	297 26.5	333 51.6	41.2	16 45.2	27.1	166 22.0	00.9	215 15.7	12.9	Al Na'ir	27 54.1	S46 56.
U 09	312 29.0	348 51.1	40.4	31 45.9	27.0	181 24.6	01.0	230 18.0	12.9	Alnilam	275 54.6	S 1 12.
R 10	327 31.5	3 50.6	39.6	46 46.7	26.9	196 27.2	01.0	245 20.3	12.9	Alphard	218 03.8	S 8 40.
S 11	342 33.9	18 50.1	38.8	61 47.4	26.9	211 29.8	01.0	260 22.7	13.0			
D 12	357 36.4	33 49.6	S14 37.9	76 48.2	S23 26.8	226 32.4	N19 01.0	275 25.0	N22 13.0	Alphecca	126 17.6	N26 42.
A 13	12 38.9	48 49.1	37.1	91 48.9	26.7	241 34.9	01.1	290 27.4	13.0	Alpheratz	357 52.3	N29 06.
Y 14	27 41.3	63 48.6	36.3	106 49.7	26.7	256 37.5	01.1	305 29.7	13.0	Altair	62 16.3	N 8 52.
15	42 43.8	78 48.1	35.5	121 50.4	26.6	271 40.1	01.1	320 32.0	13.0	Ankaa	353 23.9	S42 17.
16	57 46.3	93 47.6	34.6	136 51.2	26.5	286 42.7	01.2	335 34.4	13.0	Antares	112 36.1	S26 26.
17	72 48.7	108 47.1	33.8	151 51.9	26.5	301 45.3	01.2	350 36.7	13.0			
18	87 51.2	123 46.6	S14 33.0	166 52.7	S23 26.4	316 47.9	N19 01.2	5 39.0	N22 13.1	Arcturus	146 02.8	N19 09.
19	102 53.7	138 46.1	32.2	181 53.4	26.3	331 50.4	01.2	20 41.4	13.1	Atria	107 45.2	S69 01.
20	117 56.1	153 45.6	31.3	196 54.2	26.3	346 53.0	01.3	35 43.7	13.1	Avior	234 21.1	S59 31.
21	132 58.6	168 45.1	30.5	211 54.9	26.2	1 55.6	01.3	50 46.0	13.1	Bellatrix	278 40.7	N 6 21.
22	148 01.0	183 44.6	29.7	226 55.6	26.1	16 58.2	01.3	65 48.4	13.1	Betelgeuse	271 10.1	N 7 24.
23	163 03.5	198 44.1	28.8	241 56.4	26.1	32 00.8	01.4	80 50.7	13.1			
21 00	178 06.0	213 43.6	S14 28.0	256 57.1	S23 26.0	47 03.3	N19 01.4	95 53.0	N22 13.2	Canopus	263 59.7	S52 42.
01	193 08.4	228 43.1	27.2	271 57.9	25.9	62 05.9	01.4	110 55.4	13.2	Capella	280 46.5	N46 00.
02	208 10.9	243 42.6	26.3	286 58.6	25.8	77 08.5	01.4	125 57.7	13.2	Deneb	49 37.4	N45 17.
03	223 13.4	258 42.1	25.5	301 59.4	25.8	92 11.1	01.5	141 00.0	13.2	Denebola	182 41.5	N14 33.
04	238 15.8	273 41.6	24.7	317 00.1	25.7	107 13.6	01.5	156 02.4	13.2	Diphda	349 04.3	S17 58.
05	253 18.3	288 41.1	23.8	332 00.9	25.6	122 16.2	01.5	171 04.7	13.2			
06	268 20.8	303 40.6	S14 23.0	347 01.6	S23 25.6	137 18.8	N19 01.5	186 07.0	N22 13.2	Dubhe	194 00.6	N61 44.
07	283 23.2	318 40.1	22.2	2 02.4	25.5	152 21.4	01.6	201 09.4	13.3	Elnath	278 22.9	N28 36.
08	298 25.7	333 39.6	21.3	17 03.1	25.4	167 23.9	01.6	216 11.7	13.3	Eltanin	90 49.9	N51 28.
F 09	313 28.2	348 39.1	20.5	32 03.9	25.4	182 26.5	01.6	231 14.0	13.3	Enif	33 55.3	N 9 53.
R 10	328 30.6	3 38.6	19.6	47 04.6	25.3	197 29.1	01.7	246 16.4	13.3	Fomalhaut	15 33.1	S29 36.
I 11	343 33.1	18 38.1	18.8	62 05.4	25.2	212 31.7	01.7	261 18.7	13.3			
D 12	358 35.5	33 37.6	S14 18.0	77 06.1	S23 25.1	227 34.2	N19 01.7	276 21.0	N22 13.3	Gacrux	172 09.6	S57 07.
A 13	13 38.0	48 37.1	17.1	92 06.8	25.1	242 36.8	01.7	291 23.4	13.3	Gienah	176 00.4	S17 33.
Y 14	28 40.5	63 36.6	16.3	107 07.6	25.0	257 39.4	01.8	306 25.7	13.4	Hadar	148 59.1	S60 23.
15	43 42.9	78 36.1	15.4	122 08.3	24.9	272 42.0	01.8	321 28.0	13.4	Hamal	328 10.2	N23 28.
16	58 45.4	93 35.6	14.6	137 09.1	24.8	287 44.5	01.8	336 30.4	13.4	Kaus Aust.	83 54.6	S34 23.
17	73 47.9	108 35.1	13.7	152 09.8	24.8	302 47.1	01.8	351 32.7	13.4			
18	88 50.3	123 34.6	S14 12.9	167 10.6	S23 24.7	317 49.7	N19 01.9	6 35.0	N22 13.4	Kochab	137 18.5	N74 08.
19	103 52.8	138 34.1	12.1	182 11.3	24.6	332 52.3	01.9	21 37.3	13.4	Markab	13 46.7	N15 13.
20	118 55.3	153 33.6	11.2	197 12.1	24.6	347 54.8	01.9	36 39.7	13.4	Menkar	314 23.7	N 4 06.
21	133 57.7	168 33.1	10.4	212 12.8	24.5	2 57.4	01.9	51 42.0	13.5	Menkent	148 16.9	S36 23.
22	149 00.2	183 32.6	09.5	227 13.6	24.4	18 00.0	02.0	66 44.3	13.5	Miaplacidus	221 41.0	S69 43.
23	164 02.6	198 32.1	08.7	242 14.3	24.3	33 02.6	02.0	81 46.7	13.5			
22 00	179 05.1	213 31.7	S14 07.8	257 15.1	S23 24.3	48 05.1	N19 02.0	96 49.0	N22 13.5	Mirfak	308 52.4	N49 52.
01	194 07.6	228 31.2	07.0	272 15.8	24.2	63 07.7	02.0	111 51.3	13.5	Nunki	76 08.4	S26 17.
02	209 10.0	243 30.7	06.1	287 16.6	24.1	78 10.3	02.1	126 53.7	13.5	Peacock	53 32.1	S56 43.
03	224 12.5	258 30.2	05.3	302 17.3	24.0	93 12.8	02.1	141 56.0	13.6	Pollux	243 37.4	N28 01.
04	239 15.0	273 29.7	04.4	317 18.1	23.9	108 15.4	02.1	156 58.3	13.6	Procyon	245 08.1	N 5 13.
05	254 17.4	288 29.2	03.6	332 18.8	23.9	123 18.0	02.1	172 00.6	13.6			
06	269 19.9	303 28.7	S14 02.7	347 19.6	S23 23.8	138 20.5	N19 02.2	187 03.0	N22 13.6	Rasalhague	96 13.9	N12 33.
07	284 22.4	318 28.2	01.9	2 20.3	23.7	153 23.1	02.2	202 05.3	13.6	Regulus	207 51.8	N11 57.
S 08	299 24.8	333 27.7	01.0	17 21.1	23.6	168 25.7	02.2	217 07.6	13.6	Rigel	281 19.9	S 8 12.
A 09	314 27.3	348 27.2	14 00.1	32 21.8	23.6	183 28.2	02.2	232 09.9	13.6	Rigil Kent.	140 02.6	S60 50.
T 10	329 29.8	3 26.7	13 59.3	47 22.6	23.5	198 30.8	02.3	247 12.3	13.7	Sabik	102 21.8	S15 43.
U 11	344 32.2	18 26.3	58.4	62 23.3	23.4	213 33.4	02.3	262 14.6	13.7			
R 12	359 34.7	33 25.8	S13 57.6	77 24.1	S23 23.3	228 35.9	N19 02.3	277 16.9	N22 13.7	Schedar	349 50.6	N56 33.
D 13	14 37.1	48 25.3	56.7	92 24.8	23.2	243 38.5	02.3	292 19.3	13.7	Shaula	96 32.9	S37 06.
A 14	29 39.6	63 24.8	55.9	107 25.6	23.2	258 41.1	02.4	307 21.6	13.7	Sirius	258 40.8	S16 43.
Y 15	44 42.1	78 24.3	55.0	122 26.3	23.1	273 43.6	02.4	322 23.9	13.7	Spica	158 39.5	S11 10.
16	59 44.5	93 23.8	54.1	137 27.1	23.0	288 46.2	02.4	337 26.2	13.7	Suhail	222 58.2	S43 26.
17	74 47.0	108 23.3	53.3	152 27.8	22.9	303 48.8	02.4	352 28.6	13.8			
18	89 49.5	123 22.8	S13 52.4	167 28.6	S23 22.8	318 51.3	N19 02.5	7 30.9	N22 13.8	Vega	80 44.5	N38 46.
19	104 51.9	138 22.4	51.6	182 29.3	22.8	333 53.9	02.5	22 33.2	13.8	Zuben'ubi	137 14.2	S16 03.
20	119 54.4	153 21.9	50.7	197 30.1	22.7	348 56.5	02.5	37 35.5	13.8		SHA	Mer. Pass
21	134 56.9	168 21.4	49.8	212 30.8	22.6	3 59.0	02.5	52 37.9	13.8		° ′	h m
22	149 59.3	183 20.9	49.0	227 31.6	22.5	19 01.6	02.6	67 40.2	13.8	Venus	35 37.6	9 45
23	165 01.8	198 20.4	48.1	242 32.3	22.4	34 04.2	02.6	82 42.5	13.9	Mars	78 51.2	6 52
	h m									Jupiter	228 57.4	20 48
Mer. Pass.	12 05.6	*v* −0.5	*d* 0.8	*v* 0.7	*d* 0.1	*v* 2.6	*d* 0.0	*v* 2.3	*d* 0.0	Saturn	277 47.1	17 34

UT	SUN GHA	SUN Dec	MOON GHA	MOON v	MOON Dec	MOON d	MOON HP
d h	° ′	° ′	° ′	′	° ′	′	′
20 00	178 03.9	S 0 24.7	336 36.3	9.5	S 5 04.9	16.3	60.9
01	193 04.1	23.7	351 04.8	9.4	5 21.2	16.3	60.9
02	208 04.3	22.7	5 33.2	9.4	5 37.5	16.3	60.9
03	223 04.4	. . 21.7	20 01.6	9.4	5 53.8	16.2	60.9
04	238 04.6	20.7	34 30.0	9.3	6 10.0	16.2	60.9
05	253 04.8	19.7	48 58.3	9.3	6 26.2	16.1	60.9
06	268 05.0	S 0 18.8	63 26.6	9.3	S 6 42.3	16.1	60.9
07	283 05.2	17.8	77 54.9	9.3	6 58.4	16.1	60.9
T 08	298 05.4	16.8	92 23.2	9.2	7 14.5	16.0	60.9
H 09	313 05.6	. . 15.8	106 51.4	9.2	7 30.5	16.0	60.9
U 10	328 05.7	14.8	121 19.6	9.1	7 46.5	15.9	60.9
R 11	343 05.9	13.8	135 47.7	9.2	8 02.4	15.8	60.9
S 12	358 06.1	S 0 12.8	150 15.9	9.0	S 8 18.2	15.9	60.9
D 13	13 06.3	11.8	164 43.9	9.1	8 34.1	15.7	60.9
A 14	28 06.5	10.9	179 12.0	9.0	8 49.8	15.7	60.9
Y 15	43 06.7	. . 09.9	193 40.0	9.0	9 05.5	15.6	60.8
16	58 06.9	08.9	208 08.0	9.0	9 21.1	15.6	60.8
17	73 07.0	07.9	222 36.0	8.9	9 36.7	15.5	60.8
18	88 07.2	S 0 06.9	237 03.9	8.8	S 9 52.2	15.5	60.8
19	103 07.4	05.9	251 31.7	8.9	10 07.7	15.3	60.8
20	118 07.6	04.9	265 59.6	8.7	10 23.0	15.3	60.8
21	133 07.8	. . 03.9	280 27.3	8.8	10 38.3	15.3	60.8
22	148 08.0	03.0	294 55.1	8.7	10 53.6	15.1	60.8
23	163 08.2	02.0	309 22.8	8.7	11 08.7	15.1	60.7
21 00	178 08.3	S 0 01.0	323 50.5	8.6	S11 23.8	15.0	60.7
01	193 08.5	00.0	338 18.1	8.5	11 38.8	15.0	60.7
02	208 08.7	N 01.0	352 45.6	8.6	11 53.8	14.8	60.7
03	223 08.9	. . 02.0	7 13.2	8.5	12 08.6	14.8	60.7
04	238 09.1	03.0	21 40.7	8.4	12 23.4	14.7	60.7
05	253 09.3	04.0	36 08.1	8.4	12 38.1	14.6	60.7
06	268 09.5	N 0 04.9	50 35.5	8.3	S12 52.7	14.5	60.6
07	283 09.6	05.9	65 02.8	8.3	13 07.2	14.5	60.6
08	298 09.8	06.9	79 30.1	8.3	13 21.7	14.3	60.6
F 09	313 10.0	. . 07.9	93 57.4	8.2	13 36.0	14.3	60.6
R 10	328 10.2	08.9	108 24.6	8.2	13 50.3	14.1	60.6
I 11	343 10.4	09.9	122 51.8	8.1	14 04.4	14.1	60.6
D 12	358 10.6	N 0 10.9	137 18.9	8.0	S14 18.5	14.0	60.5
A 13	13 10.8	11.9	151 45.9	8.0	14 32.5	13.8	60.5
Y 14	28 10.9	12.8	166 12.9	8.0	14 46.3	13.8	60.5
15	43 11.1	. . 13.8	180 39.9	7.9	15 00.1	13.7	60.5
16	58 11.3	14.8	195 06.8	7.9	15 13.8	13.5	60.5
17	73 11.5	15.8	209 33.7	7.8	15 27.3	13.5	60.4
18	88 11.7	N 0 16.8	224 00.5	7.7	S15 40.8	13.4	60.4
19	103 11.9	17.8	238 27.2	7.7	15 54.2	13.2	60.4
20	118 12.1	18.8	252 53.9	7.7	16 07.4	13.2	60.4
21	133 12.3	. . 19.8	267 20.6	7.6	16 20.6	13.0	60.4
22	148 12.4	20.7	281 47.2	7.6	16 33.6	12.9	60.3
23	163 12.6	21.7	296 13.8	7.5	16 46.5	12.8	60.3
22 00	178 12.8	N 0 22.7	310 40.3	7.4	S16 59.3	12.7	60.3
01	193 13.0	23.7	325 06.7	7.4	17 12.0	12.5	60.3
02	208 13.2	24.7	339 33.1	7.3	17 24.5	12.5	60.2
03	223 13.4	. . 25.7	353 59.4	7.3	17 37.0	12.3	60.2
04	238 13.6	26.7	8 25.7	7.3	17 49.3	12.2	60.2
05	253 13.8	27.7	22 52.0	7.2	18 01.5	12.1	60.2
06	268 13.9	N 0 28.6	37 18.2	7.1	S18 13.6	12.0	60.1
S 07	283 14.1	29.6	51 44.3	7.1	18 25.6	11.8	60.1
A 08	298 14.3	30.6	66 10.4	7.0	18 37.4	11.7	60.1
T 09	313 14.5	. . 31.6	80 36.4	7.0	18 49.1	11.6	60.1
U 10	328 14.7	32.6	95 02.4	7.0	19 00.7	11.4	60.0
R 11	343 14.9	33.6	109 28.4	6.9	19 12.1	11.4	60.0
D 12	358 15.1	N 0 34.6	123 54.3	6.8	S19 23.5	11.2	60.0
A 13	13 15.3	35.5	138 20.1	6.8	19 34.7	11.0	60.0
Y 14	28 15.4	36.5	152 45.9	6.7	19 45.7	10.9	59.9
15	43 15.6	. . 37.5	167 11.6	6.7	19 56.6	10.8	59.9
16	58 15.8	38.5	181 37.3	6.6	20 07.4	10.6	59.9
17	73 16.0	39.5	196 02.9	6.6	20 18.0	10.6	59.9
18	88 16.2	N 0 40.5	210 28.5	6.6	S20 28.6	10.3	59.8
19	103 16.4	41.5	224 54.1	6.5	20 38.9	10.2	59.8
20	118 16.6	42.5	239 19.6	6.4	20 49.1	10.1	59.8
21	133 16.8	. . 43.4	253 45.0	6.4	20 59.2	10.0	59.8
22	148 16.9	44.4	268 10.4	6.4	21 09.2	9.8	59.7
23	163 17.1	45.4	282 35.8	6.3	S21 19.0	9.6	59.7
	SD 16.1	*d* 1.0	SD 16.6		16.5		16.3

Lat.	Twilight Naut.	Twilight Civil	Sunrise	Moonrise 20	Moonrise 21	Moonrise 22	Moonrise 23
°	h m	h m	h m	h m	h m	h m	h m
N 72	03 18	04 47	05 56	22 54	26 41	02 41	■
N 70	03 37	04 55	05 57	22 33	25 19	01 19	■
68	03 52	05 02	05 58	22 18	24 41	00 41	■
66	04 04	05 07	05 58	22 05	24 14	00 14	02 40
64	04 14	05 12	05 59	21 55	23 54	25 59	01 59
62	04 22	05 15	06 00	21 46	23 38	25 32	01 32
60	04 29	05 19	06 00	21 38	23 24	25 10	01 10
N 58	04 35	05 21	06 01	21 32	23 13	24 53	00 53
56	04 40	05 24	06 01	21 26	23 03	24 39	00 39
54	04 44	05 26	06 01	21 21	22 54	24 26	00 26
52	04 48	05 28	06 02	21 16	22 46	24 15	00 15
50	04 52	05 30	06 02	21 12	22 39	24 05	00 05
45	04 59	05 33	06 02	21 03	22 24	23 45	25 01
N 40	05 04	05 36	06 03	20 55	22 12	23 28	24 41
35	05 08	05 38	06 03	20 49	22 02	23 15	24 25
30	05 12	05 39	06 03	20 43	21 53	23 03	24 11
20	05 16	05 42	06 04	20 33	21 37	22 42	23 47
N 10	05 19	05 43	06 04	20 25	21 24	22 25	23 26
0	05 19	05 43	06 04	20 17	21 12	22 08	23 07
S 10	05 19	05 43	06 04	20 10	20 59	21 52	22 48
20	05 16	05 42	06 04	20 02	20 46	21 35	22 27
30	05 12	05 40	06 04	19 52	20 32	21 15	22 04
35	05 09	05 38	06 04	19 47	20 23	21 04	21 50
40	05 05	05 36	06 03	19 41	20 14	20 51	21 34
45	04 59	05 34	06 03	19 34	20 02	20 35	21 15
S 50	04 52	05 30	06 03	19 26	19 49	20 16	20 52
52	04 49	05 29	06 02	19 22	19 42	20 08	20 41
54	04 45	05 27	06 02	19 18	19 36	19 58	20 28
56	04 40	05 25	06 02	19 14	19 28	19 47	20 14
58	04 35	05 22	06 02	19 09	19 19	19 34	19 57
S 60	04 30	05 20	06 01	19 03	19 10	19 19	19 36

Lat.	Sunset	Twilight Civil	Twilight Naut.	Moonset 20	Moonset 21	Moonset 22	Moonset 23
°	h m	h m	h m	h m	h m	h m	h m
N 72	18 22	19 30	21 02	06 18	05 36	03 45	■
N 70	18 20	19 22	20 41	06 28	05 59	05 10	■
68	18 19	19 15	20 26	06 36	06 17	05 49	■
66	18 18	19 10	20 13	06 43	06 32	06 17	05 51
64	18 17	19 05	20 03	06 49	06 44	06 39	06 32
62	18 16	19 01	19 55	06 54	06 54	06 56	07 01
60	18 16	18 58	19 48	06 58	07 03	07 10	07 23
N 58	18 15	18 55	19 42	07 02	07 11	07 23	07 41
56	18 15	18 52	19 36	07 05	07 18	07 34	07 56
54	18 14	18 50	19 32	07 08	07 24	07 43	08 09
52	18 14	18 48	19 28	07 11	07 30	07 52	08 20
50	18 14	18 46	19 24	07 14	07 35	07 59	08 31
45	18 13	18 42	19 17	07 20	07 46	08 16	08 52
N 40	18 13	18 40	19 11	07 24	07 55	08 29	09 10
35	18 12	18 37	19 07	07 28	08 03	08 41	09 24
30	18 12	18 36	19 03	07 32	08 10	08 51	09 37
20	18 11	18 33	18 59	07 38	08 22	09 09	09 59
N 10	18 11	18 32	18 56	07 44	08 33	09 24	10 18
0	18 11	18 31	18 55	07 49	08 43	09 38	10 36
S 10	18 10	18 31	18 56	07 55	08 53	09 53	10 54
20	18 10	18 32	18 58	08 00	09 04	10 08	11 13
30	18 10	18 34	19 02	08 07	09 16	10 26	11 35
35	18 11	18 36	19 05	08 11	09 23	10 37	11 48
40	18 11	18 38	19 09	08 15	09 32	10 49	12 03
45	18 11	18 40	19 14	08 20	09 41	11 03	12 22
S 50	18 11	18 43	19 21	08 26	09 53	11 20	12 44
52	18 11	18 45	19 25	08 28	09 59	11 28	12 55
54	18 11	18 47	19 28	08 31	10 05	11 38	13 07
56	18 12	18 49	19 33	08 35	10 11	11 48	13 21
58	18 12	18 51	19 37	08 39	10 19	12 00	13 37
S 60	18 12	18 53	19 43	08 43	10 28	12 14	13 57

Day	SUN Eqn. of Time 00^h	SUN Eqn. of Time 12^h	SUN Mer. Pass.	MOON Mer. Pass. Upper	MOON Mer. Pass. Lower	MOON Age	MOON Phase
d	m s	m s	h m	h m	h m	d %	
20	07 45	07 36	12 08	01 37	14 03	17 94	
21	07 27	07 18	12 07	02 30	14 57	18 87	
22	07 09	07 00	12 07	03 25	15 53	19 78	

UT		ARIES	VENUS −4.0		MARS +0.6		JUPITER −2.4		SATURN +0.1	
		GHA	GHA	Dec	GHA	Dec	GHA	Dec	GHA	Dec
d	h	° ′	° ′	° ′	° ′	° ′	° ′	° ′	° ′	° ′
23	00	180 04.3	213 19.9	S13 47.2	257 33.1	S23 22.4	49 06.7	N19 02.6	97 44.8	N22 13.9
	01	195 06.7	228 19.4	46.4	272 33.8	22.3	64 09.3	02.6	112 47.2	13.9
	02	210 09.2	243 19.0	45.5	287 34.6	22.2	79 11.9	02.7	127 49.5	13.9
	03	225 11.6	258 18.5	. . 44.6	302 35.3	. . 22.1	94 14.4	. . 02.7	142 51.8	. . 13.9
	04	240 14.1	273 18.0	43.8	317 36.1	22.0	109 17.0	02.7	157 54.1	13.9
	05	255 16.6	288 17.5	42.9	332 36.8	21.9	124 19.5	02.7	172 56.5	13.9
	06	270 19.0	303 17.0	S13 42.0	347 37.6	S23 21.9	139 22.1	N19 02.7	187 58.8	N22 14.0
	07	285 21.5	318 16.5	41.2	2 38.3	21.8	154 24.7	02.8	203 01.1	14.0
	08	300 24.0	333 16.1	40.3	17 39.1	21.7	169 27.2	02.8	218 03.4	14.0
S	09	315 26.4	348 15.6	. . 39.4	32 39.8	. . 21.6	184 29.8	. . 02.8	233 05.8	. . 14.0
U	10	330 28.9	3 15.1	38.5	47 40.6	21.5	199 32.3	02.8	248 08.1	14.0
N	11	345 31.4	18 14.6	37.7	62 41.3	21.4	214 34.9	02.9	263 10.4	14.0
D	12	0 33.8	33 14.1	S13 36.8	77 42.1	S23 21.4	229 37.5	N19 02.9	278 12.7	N22 14.0
A	13	15 36.3	48 13.6	35.9	92 42.8	21.3	244 40.0	02.9	293 15.0	14.1
Y	14	30 38.7	63 13.2	35.1	107 43.6	21.2	259 42.6	02.9	308 17.4	14.1
	15	45 41.2	78 12.7	. . 34.2	122 44.3	. . 21.1	274 45.1	. . 02.9	323 19.7	. . 14.1
	16	60 43.7	93 12.2	33.3	137 45.1	21.0	289 47.7	03.0	338 22.0	14.1
	17	75 46.1	108 11.7	32.4	152 45.8	20.9	304 50.3	03.0	353 24.3	14.1
	18	90 48.6	123 11.2	S13 31.5	167 46.6	S23 20.8	319 52.8	N19 03.0	8 26.7	N22 14.1
	19	105 51.1	138 10.8	30.7	182 47.4	20.7	334 55.4	03.0	23 29.0	14.2
	20	120 53.5	153 10.3	29.8	197 48.1	20.7	349 57.9	03.1	38 31.3	14.2
	21	135 56.0	168 09.8	. . 28.9	212 48.9	. . 20.6	5 00.5	. . 03.1	53 33.6	. . 14.2
	22	150 58.5	183 09.3	28.0	227 49.6	20.5	20 03.0	03.1	68 35.9	14.2
	23	166 00.9	198 08.9	27.1	242 50.4	20.4	35 05.6	03.1	83 38.3	14.2
24	00	181 03.4	213 08.4	S13 26.3	257 51.1	S23 20.3	50 08.1	N19 03.1	98 40.6	N22 14.2
	01	196 05.9	228 07.9	25.4	272 51.9	20.2	65 10.7	03.2	113 42.9	14.2
	02	211 08.3	243 07.4	24.5	287 52.6	20.1	80 13.3	03.2	128 45.2	14.3
	03	226 10.8	258 06.9	. . 23.6	302 53.4	. . 20.0	95 15.8	. . 03.2	143 47.5	. . 14.3
	04	241 13.2	273 06.5	22.7	317 54.1	20.0	110 18.4	03.2	158 49.9	14.3
	05	256 15.7	288 06.0	21.8	332 54.9	19.9	125 20.9	03.2	173 52.2	14.3
	06	271 18.2	303 05.5	S13 21.0	347 55.6	S23 19.8	140 23.5	N19 03.3	188 54.5	N22 14.3
	07	286 20.6	318 05.0	20.1	2 56.4	19.7	155 26.0	03.3	203 56.8	14.3
	08	301 23.1	333 04.6	19.2	17 57.1	19.6	170 28.6	03.3	218 59.1	14.4
M	09	316 25.6	348 04.1	. . 18.3	32 57.9	. . 19.5	185 31.1	. . 03.3	234 01.5	. . 14.4
O	10	331 28.0	3 03.6	17.4	47 58.7	19.4	200 33.7	03.3	249 03.8	14.4
N	11	346 30.5	18 03.1	16.5	62 59.4	19.3	215 36.2	03.4	264 06.1	14.4
D	12	1 33.0	33 02.7	S13 15.6	78 00.2	S23 19.2	230 38.8	N19 03.4	279 08.4	N22 14.4
A	13	16 35.4	48 02.2	14.7	93 00.9	19.1	245 41.3	03.4	294 10.7	14.4
Y	14	31 37.9	63 01.7	13.9	108 01.7	19.0	260 43.9	03.4	309 13.0	14.4
	15	46 40.4	78 01.3	. . 13.0	123 02.4	. . 18.9	275 46.4	. . 03.4	324 15.4	. . 14.5
	16	61 42.8	93 00.8	12.1	138 03.2	18.9	290 49.0	03.5	339 17.7	14.5
	17	76 45.3	108 00.3	11.2	153 03.9	18.8	305 51.5	03.5	354 20.0	14.5
	18	91 47.7	122 59.8	S13 10.3	168 04.7	S23 18.7	320 54.1	N19 03.5	9 22.3	N22 14.5
	19	106 50.2	137 59.4	09.4	183 05.4	18.6	335 56.6	03.5	24 24.6	14.5
	20	121 52.7	152 58.9	08.5	198 06.2	18.5	350 59.2	03.5	39 27.0	14.5
	21	136 55.1	167 58.4	. . 07.6	213 07.0	. . 18.4	6 01.7	. . 03.6	54 29.3	. . 14.6
	22	151 57.6	182 58.0	06.7	228 07.7	18.3	21 04.3	03.6	69 31.6	14.6
	23	167 00.1	197 57.5	05.8	243 08.5	18.2	36 06.8	03.6	84 33.9	14.6
25	00	182 02.5	212 57.0	S13 04.9	258 09.2	S23 18.1	51 09.4	N19 03.6	99 36.2	N22 14.6
	01	197 05.0	227 56.5	04.0	273 10.0	18.0	66 11.9	03.6	114 38.5	14.6
	02	212 07.5	242 56.1	03.1	288 10.7	17.9	81 14.5	03.7	129 40.8	14.6
	03	227 09.9	257 55.6	. . 02.2	303 11.5	. . 17.8	96 17.0	. . 03.7	144 43.2	. . 14.6
	04	242 12.4	272 55.1	01.3	318 12.2	17.7	111 19.6	03.7	159 45.5	14.7
	05	257 14.9	287 54.7	13 00.4	333 13.0	17.6	126 22.1	03.7	174 47.8	14.7
	06	272 17.3	302 54.2	S12 59.5	348 13.8	S23 17.5	141 24.6	N19 03.7	189 50.1	N22 14.7
	07	287 19.8	317 53.7	58.6	3 14.5	17.4	156 27.2	03.8	204 52.4	14.7
T	08	302 22.2	332 53.3	57.7	18 15.3	17.3	171 29.7	03.8	219 54.7	14.7
U	09	317 24.7	347 52.8	. . 56.8	33 16.0	. . 17.2	186 32.3	. . 03.8	234 57.1	. . 14.7
E	10	332 27.2	2 52.3	55.9	48 16.8	17.1	201 34.8	03.8	249 59.4	14.8
S	11	347 29.6	17 51.9	55.0	63 17.5	17.0	216 37.4	03.8	265 01.7	14.8
D	12	2 32.1	32 51.4	S12 54.1	78 18.3	S23 16.9	231 39.9	N19 03.8	280 04.0	N22 14.8
A	13	17 34.6	47 50.9	53.2	93 19.1	16.8	246 42.4	03.9	295 06.3	14.8
Y	14	32 37.0	62 50.5	52.3	108 19.8	16.7	261 45.0	03.9	310 08.6	14.8
	15	47 39.5	77 50.0	. . 51.4	123 20.6	. . 16.6	276 47.5	. . 03.9	325 10.9	. . 14.8
	16	62 42.0	92 49.5	50.5	138 21.3	16.5	291 50.1	03.9	340 13.2	14.8
	17	77 44.4	107 49.1	49.6	153 22.1	16.4	306 52.6	03.9	355 15.6	14.9
	18	92 46.9	122 48.6	S12 48.6	168 22.8	S23 16.3	321 55.2	N19 03.9	10 17.9	N22 14.9
	19	107 49.4	137 48.2	47.7	183 23.6	16.2	336 57.7	04.0	25 20.2	14.9
	20	122 51.8	152 47.7	46.8	198 24.4	16.1	352 00.2	04.0	40 22.5	14.9
	21	137 54.3	167 47.2	. . 45.9	213 25.1	. . 16.0	7 02.8	. . 04.0	55 24.8	. . 14.9
	22	152 56.7	182 46.8	45.0	228 25.9	15.9	22 05.3	04.0	70 27.1	14.9
	23	167 59.2	197 46.3	44.1	243 26.6	15.8	37 07.9	04.0	85 29.4	15.0
Mer. Pass.		h m 11 53.8	*v* −0.5	*d* 0.9	*v* 0.8	*d* 0.1	*v* 2.6	*d* 0.0	*v* 2.3	*d* 0.0

STARS

Name	SHA	Dec
	° ′	° ′
Acamar	315 24.7	S40 17.8
Achernar	335 33.1	S57 13.5
Acrux	173 18.0	S63 07.0
Adhara	255 18.8	S28 58.8
Aldebaran	290 58.8	N16 30.9
Alioth	166 27.0	N55 56.5
Alkaid	153 04.6	N49 17.7
Al Na'ir	27 54.1	S46 56.8
Alnilam	275 54.6	S 1 12.1
Alphard	218 03.8	S 8 40.4
Alphecca	126 17.6	N26 42.0
Alpheratz	357 52.3	N29 06.3
Altair	62 16.2	N 8 52.3
Ankaa	353 23.9	S42 17.5
Antares	112 36.1	S26 26.4
Arcturus	146 02.8	N19 09.8
Atria	107 45.1	S69 01.8
Avior	234 21.1	S59 31.4
Bellatrix	278 40.7	N 6 21.1
Betelgeuse	271 10.1	N 7 24.4
Canopus	263 59.7	S52 42.1
Capella	280 46.5	N46 00.2
Deneb	49 37.3	N45 17.1
Denebola	182 41.5	N14 33.2
Diphda	349 04.3	S17 58.4
Dubhe	194 00.6	N61 44.1
Elnath	278 22.9	N28 36.7
Eltanin	90 49.9	N51 28.9
Enif	33 55.3	N 9 53.1
Fomalhaut	15 33.1	S29 36.5
Gacrux	172 09.6	S57 07.8
Gienah	176 00.3	S17 33.6
Hadar	148 59.1	S60 23.2
Hamal	328 10.2	N23 28.5
Kaus Aust.	83 54.5	S34 23.0
Kochab	137 18.4	N74 08.3
Markab	13 46.7	N15 13.1
Menkar	314 23.7	N 4 06.0
Menkent	148 16.9	S36 23.1
Miaplacidus	221 41.1	S69 44.0
Mirfak	308 52.4	N49 52.4
Nunki	76 08.4	S26 17.6
Peacock	53 32.1	S56 43.5
Pollux	243 37.4	N28 01.2
Procyon	245 08.1	N 5 13.0
Rasalhague	96 13.9	N12 33.2
Regulus	207 51.8	N11 57.1
Rigel	281 19.9	S 8 12.0
Rigil Kent.	140 02.6	S60 50.8
Sabik	102 21.7	S15 43.8
Schedar	349 50.6	N56 33.2
Shaula	96 32.8	S37 06.3
Sirius	258 40.8	S16 43.4
Spica	158 39.5	S11 10.7
Suhail	222 58.2	S43 26.9
Vega	80 44.5	N38 46.8
Zuben'ubi	137 14.2	S16 03.4

	SHA	Mer. Pass.
	° ′	h m
Venus	32 05.0	9 48
Mars	76 47.7	6 48
Jupiter	229 04.8	20 36
Saturn	277 37.2	17 23

UT		SUN GHA	SUN Dec	MOON GHA	v	MOON Dec	d	HP
d	h	° ′	° ′	° ′	′	° ′	′	′
23	00	178 17.3	N 0 46.4	297 01.1	6.3	S21 28.6	9.5	59.7
	01	193 17.5	47.4	311 26.4	6.2	21 38.1	9.4	59.6
	02	208 17.7	48.4	325 51.6	6.2	21 47.5	9.2	59.6
	03	223 17.9	. . 49.4	340 16.8	6.1	21 56.7	9.0	59.6
	04	238 18.1	50.3	354 41.9	6.1	22 05.7	8.9	59.6
	05	253 18.3	51.3	9 07.0	6.1	22 14.6	8.8	59.5
	06	268 18.4	N 0 52.3	23 32.1	6.0	S22 23.4	8.6	59.5
	07	283 18.6	53.3	37 57.1	6.0	22 32.0	8.4	59.5
	08	298 18.8	54.3	52 22.1	6.0	22 40.4	8.3	59.4
S	09	313 19.0	. . 55.3	66 47.1	5.9	22 48.7	8.2	59.4
U	10	328 19.2	56.3	81 12.0	5.9	22 56.9	8.0	59.4
N	11	343 19.4	57.2	95 36.9	5.8	23 04.9	7.8	59.3
D	12	358 19.6	N 0 58.2	110 01.7	5.9	S23 12.7	7.7	59.3
A	13	13 19.8	0 59.2	124 26.6	5.7	23 20.4	7.5	59.3
Y	14	28 19.9	1 00.2	138 51.3	5.8	23 27.9	7.3	59.3
	15	43 20.1	. . 01.2	153 16.1	5.7	23 35.2	7.2	59.2
	16	58 20.3	02.2	167 40.8	5.7	23 42.4	7.1	59.2
	17	73 20.5	03.2	182 05.5	5.7	23 49.5	6.8	59.2
	18	88 20.7	N 1 04.1	196 30.2	5.7	S23 56.3	6.7	59.1
	19	103 20.9	05.1	210 54.9	5.6	24 03.0	6.6	59.1
	20	118 21.1	06.1	225 19.5	5.6	24 09.6	6.4	59.1
	21	133 21.3	. . 07.1	239 44.1	5.6	24 16.0	6.2	59.0
	22	148 21.5	08.1	254 08.7	5.5	24 22.2	6.1	59.0
	23	163 21.6	09.1	268 33.2	5.6	24 28.3	5.9	59.0
24	00	178 21.8	N 1 10.1	282 57.8	5.5	S24 34.2	5.7	58.9
	01	193 22.0	11.0	297 22.3	5.5	24 39.9	5.6	58.9
	02	208 22.2	12.0	311 46.8	5.5	24 45.5	5.4	58.9
	03	223 22.4	. . 13.0	326 11.3	5.5	24 50.9	5.2	58.8
	04	238 22.6	14.0	340 35.8	5.5	24 56.1	5.1	58.8
	05	253 22.8	15.0	355 00.3	5.4	25 01.2	4.9	58.8
	06	268 23.0	N 1 16.0	9 24.7	5.5	S25 06.1	4.7	58.8
	07	283 23.2	16.9	23 49.2	5.4	25 10.8	4.6	58.7
	08	298 23.3	17.9	38 13.6	5.5	25 15.4	4.4	58.7
M	09	313 23.5	. . 18.9	52 38.1	5.4	25 19.8	4.2	58.7
O	10	328 23.7	19.9	67 02.5	5.5	25 24.0	4.1	58.6
N	11	343 23.9	20.9	81 27.0	5.4	25 28.1	3.9	58.6
D	12	358 24.1	N 1 21.9	95 51.4	5.4	S25 32.0	3.7	58.6
A	13	13 24.3	22.9	110 15.8	5.5	25 35.7	3.6	58.5
Y	14	28 24.5	23.8	124 40.3	5.4	25 39.3	3.4	58.5
	15	43 24.7	. . 24.8	139 04.7	5.5	25 42.7	3.2	58.5
	16	58 24.9	25.8	153 29.2	5.5	25 45.9	3.1	58.4
	17	73 25.0	26.8	167 53.7	5.4	25 49.0	2.9	58.4
	18	88 25.2	N 1 27.8	182 18.1	5.5	S25 51.9	2.7	58.4
	19	103 25.4	28.8	196 42.6	5.5	25 54.6	2.5	58.3
	20	118 25.6	29.7	211 07.1	5.5	25 57.1	2.4	58.3
	21	133 25.8	. . 30.7	225 31.6	5.6	25 59.5	2.3	58.3
	22	148 26.0	31.7	239 56.2	5.5	26 01.8	2.0	58.2
	23	163 26.2	32.7	254 20.7	5.6	26 03.8	1.9	58.2
25	00	178 26.4	N 1 33.7	268 45.3	5.6	S26 05.7	1.7	58.2
	01	193 26.5	34.7	283 09.9	5.6	26 07.4	1.6	58.2
	02	208 26.7	35.6	297 34.5	5.6	26 09.0	1.4	58.1
	03	223 26.9	. . 36.6	311 59.1	5.7	26 10.4	1.2	58.1
	04	238 27.1	37.6	326 23.8	5.7	26 11.6	1.0	58.1
	05	253 27.3	38.6	340 48.5	5.7	26 12.6	0.9	58.0
	06	268 27.5	N 1 39.6	355 13.2	5.8	S26 13.5	0.8	58.0
	07	283 27.7	40.6	9 38.0	5.8	26 14.3	0.5	58.0
T	08	298 27.9	41.5	24 02.8	5.8	26 14.8	0.4	57.9
U	09	313 28.1	. . 42.5	38 27.6	5.9	26 15.2	0.3	57.9
E	10	328 28.2	43.5	52 52.5	5.9	26 15.5	0.0	57.9
S	11	343 28.4	44.5	67 17.4	5.9	26 15.5	0.1	57.8
D	12	358 28.6	N 1 45.5	81 42.3	6.0	S26 15.4	0.2	57.8
A	13	13 28.8	46.5	96 07.3	6.0	26 15.2	0.4	57.8
Y	14	28 29.0	47.4	110 32.3	6.1	26 14.8	0.6	57.7
	15	43 29.2	. . 48.4	124 57.4	6.1	26 14.2	0.8	57.7
	16	58 29.4	49.4	139 22.5	6.2	26 13.4	0.8	57.7
	17	73 29.6	50.4	153 47.7	6.2	26 12.6	1.1	57.7
	18	88 29.8	N 1 51.4	168 12.9	6.2	S26 11.5	1.2	57.6
	19	103 29.9	52.4	182 38.1	6.4	26 10.3	1.4	57.6
	20	118 30.1	53.3	197 03.5	6.3	26 08.9	1.5	57.6
	21	133 30.3	. . 54.3	211 28.8	6.5	26 07.4	1.7	57.5
	22	148 30.5	55.3	225 54.3	6.4	26 05.7	1.8	57.5
	23	163 30.7	56.3	240 19.7	6.6	S26 03.9	2.0	57.5
		SD 16.1	*d* 1.0	SD 16.2		16.0		15.8

Lat.	Twilight Naut.	Twilight Civil	Sunrise	Moonrise 23	Moonrise 24	Moonrise 25	Moonrise 26
°	h m	h m	h m	h m	h m	h m	h m
N 72	02 56	04 31	05 40	■	■	■	■
N 70	03 19	04 41	05 43	■	■	■	■
68	03 37	04 49	05 45	■	■	■	■
66	03 50	04 55	05 47	02 40	■	■	■
64	04 02	05 01	05 48	01 59	04 14	■	06 51
62	04 11	05 05	05 50	01 32	03 22	04 51	05 38
60	04 19	05 09	05 51	01 10	02 50	04 11	05 02
N 58	04 25	05 13	05 52	00 53	02 26	03 43	04 36
56	04 31	05 16	05 53	00 39	02 07	03 22	04 16
54	04 36	05 18	05 54	00 26	01 51	03 04	03 59
52	04 41	05 21	05 55	00 15	01 38	02 49	03 44
50	04 45	05 23	05 55	00 05	01 26	02 36	03 32
45	04 53	05 27	05 57	25 01	01 01	02 09	03 05
N 40	04 59	05 31	05 58	24 41	00 41	01 47	02 44
35	05 04	05 34	05 59	24 25	00 25	01 30	02 27
30	05 08	05 36	06 00	24 11	00 11	01 14	02 12
20	05 13	05 39	06 01	23 47	24 49	00 49	01 47
N 10	05 17	05 41	06 02	23 26	24 26	00 26	01 25
0	05 18	05 42	06 03	23 07	24 06	00 06	01 04
S 10	05 19	05 43	06 04	22 48	23 45	24 43	00 43
20	05 17	05 43	06 05	22 27	23 23	24 21	00 21
30	05 14	05 42	06 06	22 04	22 58	23 56	24 57
35	05 11	05 41	06 06	21 50	22 43	23 41	24 43
40	05 08	05 39	06 06	21 34	22 25	23 24	24 27
45	05 03	05 38	06 07	21 15	22 04	23 03	24 08
S 50	04 57	05 35	06 07	20 52	21 38	22 36	23 43
52	04 54	05 34	06 08	20 41	21 26	22 23	23 32
54	04 51	05 33	06 08	20 28	21 11	22 09	23 19
56	04 47	05 31	06 08	20 14	20 54	21 51	23 03
58	04 43	05 29	06 08	19 57	20 34	21 30	22 45
S 60	04 38	05 27	06 09	19 36	20 08	21 04	22 23

Lat.	Sunset	Twilight Civil	Twilight Naut.	Moonset 23	Moonset 24	Moonset 25	Moonset 26
°	h m	h m	h m	h m	h m	h m	h m
N 72	18 35	19 45	21 22	■	■	■	■
N 70	18 32	19 35	20 58	■	■	■	■
68	18 30	19 27	20 40	■	■	■	■
66	18 28	19 20	20 25	05 51	■	■	■
64	18 26	19 14	20 14	06 32	06 21	■	07 49
62	18 24	19 09	20 04	07 01	07 13	07 48	09 01
60	18 23	19 05	19 56	07 23	07 45	08 28	09 37
N 58	18 22	19 02	19 49	07 41	08 09	08 56	10 02
56	18 21	18 58	19 43	07 56	08 29	09 17	10 23
54	18 20	18 56	19 38	08 09	08 45	09 35	10 40
52	18 19	18 53	19 33	08 20	08 59	09 50	10 54
50	18 19	18 51	19 29	08 31	09 11	10 03	11 07
45	18 17	18 46	19 21	08 52	09 37	10 30	11 33
N 40	18 16	18 43	19 15	09 10	09 57	10 52	11 53
35	18 15	18 40	19 09	09 24	10 14	11 10	12 10
30	18 14	18 37	19 05	09 37	10 28	11 25	12 25
20	18 12	18 34	19 00	09 59	10 53	11 51	12 50
N 10	18 11	18 32	18 56	10 18	11 15	12 13	13 11
0	18 10	18 30	18 54	10 36	11 35	12 34	13 31
S 10	18 09	18 30	18 54	10 54	11 55	12 55	13 51
20	18 08	18 30	18 55	11 13	12 17	13 17	14 12
30	18 07	18 31	18 58	11 35	12 42	13 43	14 37
35	18 06	18 32	19 01	11 48	12 56	13 58	14 51
40	18 06	18 33	19 04	12 03	13 13	14 15	15 08
45	18 05	18 34	19 09	12 22	13 34	14 36	15 27
S 50	18 05	18 37	19 14	12 44	14 00	15 03	15 52
52	18 04	18 38	19 17	12 55	14 12	15 16	16 04
54	18 04	18 39	19 21	13 07	14 27	15 31	16 17
56	18 04	18 41	19 24	13 21	14 43	15 48	16 33
58	18 03	18 42	19 28	13 37	15 04	16 09	16 52
S 60	18 03	18 44	19 33	13 57	15 29	16 36	17 14

Day	SUN Eqn. of Time 00^h	SUN Eqn. of Time 12^h	SUN Mer. Pass.	MOON Mer. Pass. Upper	MOON Mer. Pass. Lower	MOON Age	MOON Phase
d	m s	m s	h m	h m	h m	d %	
23	06 51	06 42	12 07	04 22	16 51	20 67	
24	06 33	06 24	12 06	05 21	17 50	21 56	
25	06 15	06 06	12 06	06 20	18 49	22 46	

Day	UT	ARIES	VENUS −4.0		MARS +0.6		JUPITER −2.3		SATURN +0.1		STARS		
	d h	GHA ° ′	GHA ° ′	Dec ° ′	GHA ° ′	Dec ° ′	GHA ° ′	Dec ° ′	GHA ° ′	Dec ° ′	Name	SHA ° ′	Dec ° ′
WEDNESDAY	26 00	183 01.7	212 45.8	S12 43.2	258 27.4	S23 15.7	52 10.4	N19 04.1	100 31.7	N22 15.0	Acamar	315 24.7	S40 17.8
	01	198 04.1	227 45.4	42.3	273 28.1	15.6	67 12.9	04.1	115 34.1	15.0	Achernar	335 33.1	S57 13.4
	02	213 06.6	242 44.9	41.4	288 28.9	15.5	82 15.5	04.1	130 36.4	15.0	Acrux	173 18.0	S63 07.0
	03	228 09.1	257 44.5	. . 40.4	303 29.7	. . 15.4	97 18.0	. . 04.1	145 38.7	. . 15.0	Adhara	255 18.8	S28 58.8
	04	243 11.5	272 44.0	39.5	318 30.4	15.3	112 20.6	04.1	160 41.0	15.0	Aldebaran	290 58.8	N16 30.9
	05	258 14.0	287 43.5	38.6	333 31.2	15.2	127 23.1	04.1	175 43.3	15.0			
	06	273 16.5	302 43.1	S12 37.7	348 31.9	S23 15.1	142 25.6	N19 04.2	190 45.6	N22 15.1	Alioth	166 27.0	N55 56.5
	07	288 18.9	317 42.6	36.8	3 32.7	15.0	157 28.2	04.2	205 47.9	15.1	Alkaid	153 04.6	N49 17.7
	08	303 21.4	332 42.2	35.9	18 33.5	14.9	172 30.7	04.2	220 50.2	15.1	Al Na'ir	27 54.0	S46 56.8
	09	318 23.8	347 41.7	. . 34.9	33 34.2	. . 14.8	187 33.2	. . 04.2	235 52.5	. . 15.1	Alnilam	275 54.6	S 1 12.1
	10	333 26.3	2 41.2	34.0	48 35.0	14.7	202 35.8	04.2	250 54.9	15.1	Alphard	218 03.8	S 8 40.4
	11	348 28.8	17 40.8	33.1	63 35.7	14.6	217 38.3	04.2	265 57.2	15.1			
	12	3 31.2	32 40.3	S12 32.2	78 36.5	S23 14.5	232 40.8	N19 04.2	280 59.5	N22 15.2	Alphecca	126 17.6	N26 42.0
	13	18 33.7	47 39.9	31.3	93 37.3	14.4	247 43.4	04.3	296 01.8	15.2	Alpheratz	357 52.3	N29 06.3
	14	33 36.2	62 39.4	30.3	108 38.0	14.3	262 45.9	04.3	311 04.1	15.2	Altair	62 16.2	N 8 52.3
	15	48 38.6	77 38.9	. . 29.4	123 38.8	. . 14.2	277 48.4	. . 04.3	326 06.4	. . 15.2	Ankaa	353 23.9	S42 17.5
	16	63 41.1	92 38.5	28.5	138 39.5	14.1	292 51.0	04.3	341 08.7	15.2	Antares	112 36.0	S26 26.4
	17	78 43.6	107 38.0	27.6	153 40.3	14.0	307 53.5	04.3	356 11.0	15.2			
	18	93 46.0	122 37.6	S12 26.6	168 41.1	S23 13.9	322 56.0	N19 04.3	11 13.3	N22 15.2	Arcturus	146 02.8	N19 09.8
	19	108 48.5	137 37.1	25.7	183 41.8	13.8	337 58.6	04.4	26 15.6	15.3	Atria	107 45.0	S69 01.8
	20	123 51.0	152 36.7	24.8	198 42.6	13.7	353 01.1	04.4	41 17.9	15.3	Avior	234 21.2	S59 31.4
	21	138 53.4	167 36.2	. . 23.9	213 43.3	. . 13.5	8 03.6	. . 04.4	56 20.2	. . 15.3	Bellatrix	278 40.8	N 6 21.1
	22	153 55.9	182 35.7	22.9	228 44.1	13.4	23 06.2	04.4	71 22.6	15.3	Betelgeuse	271 10.1	N 7 24.4
	23	168 58.3	197 35.3	22.0	243 44.9	13.3	38 08.7	04.4	86 24.9	15.3			
THURSDAY	27 00	184 00.8	212 34.8	S12 21.1	258 45.6	S23 13.2	53 11.2	N19 04.4	101 27.2	N22 15.3	Canopus	263 59.8	S52 42.1
	01	199 03.3	227 34.4	20.2	273 46.4	13.1	68 13.8	04.4	116 29.5	15.4	Capella	280 46.5	N46 00.2
	02	214 05.7	242 33.9	19.2	288 47.1	13.0	83 16.3	04.5	131 31.8	15.4	Deneb	49 37.3	N45 17.1
	03	229 08.2	257 33.5	. . 18.3	303 47.9	. . 12.9	98 18.8	. . 04.5	146 34.1	. . 15.4	Denebola	182 41.5	N14 33.2
	04	244 10.7	272 33.0	17.4	318 48.7	12.8	113 21.4	04.5	161 36.4	15.4	Diphda	349 04.3	S17 58.4
	05	259 13.1	287 32.6	16.4	333 49.4	12.7	128 23.9	04.5	176 38.7	15.4			
	06	274 15.6	302 32.1	S12 15.5	348 50.2	S23 12.6	143 26.4	N19 04.5	191 41.0	N22 15.4	Dubhe	194 00.6	N61 44.1
	07	289 18.1	317 31.7	14.6	3 50.9	12.5	158 28.9	04.5	206 43.3	15.5	Elnath	278 22.9	N28 36.7
	08	304 20.5	332 31.2	13.7	18 51.7	12.3	173 31.5	04.5	221 45.6	15.5	Eltanin	90 49.8	N51 28.9
	09	319 23.0	347 30.8	. . 12.7	33 52.5	. . 12.2	188 34.0	. . 04.6	236 47.9	. . 15.5	Enif	33 55.3	N 9 53.1
	10	334 25.5	2 30.3	11.8	48 53.2	12.1	203 36.5	04.6	251 50.2	15.5	Fomalhaut	15 33.1	S29 36.5
	11	349 27.9	17 29.9	10.9	63 54.0	12.0	218 39.1	04.6	266 52.5	15.5			
	12	4 30.4	32 29.4	S12 09.9	78 54.8	S23 11.9	233 41.6	N19 04.6	281 54.8	N22 15.5	Gacrux	172 09.6	S57 07.9
	13	19 32.8	47 29.0	09.0	93 55.5	11.8	248 44.1	04.6	296 57.1	15.5	Gienah	176 00.3	S17 33.6
	14	34 35.3	62 28.5	08.0	108 56.3	11.7	263 46.6	04.6	311 59.5	15.6	Hadar	148 59.1	S60 23.2
	15	49 37.8	77 28.1	. . 07.1	123 57.0	. . 11.6	278 49.2	. . 04.6	327 01.8	. . 15.6	Hamal	328 10.2	N23 28.5
	16	64 40.2	92 27.6	06.2	138 57.8	11.5	293 51.7	04.7	342 04.1	15.6	Kaus Aust.	83 54.5	S34 23.0
	17	79 42.7	107 27.2	05.2	153 58.6	11.3	308 54.2	04.7	357 06.4	15.6			
	18	94 45.2	122 26.7	S12 04.3	168 59.3	S23 11.2	323 56.7	N19 04.7	12 08.7	N22 15.6	Kochab	137 18.4	N74 08.3
	19	109 47.6	137 26.3	03.4	184 00.1	11.1	338 59.3	04.7	27 11.0	15.6	Markab	13 46.7	N15 13.1
	20	124 50.1	152 25.8	02.4	199 00.9	11.0	354 01.8	04.7	42 13.3	15.7	Menkar	314 23.7	N 4 06.0
	21	139 52.6	167 25.4	. . 01.5	214 01.6	. . 10.9	9 04.3	. . 04.7	57 15.6	. . 15.7	Menkent	148 16.9	S36 23.1
	22	154 55.0	182 24.9	12 00.5	229 02.4	10.8	24 06.8	04.7	72 17.9	15.7	Miaplacidus	221 41.1	S69 44.0
	23	169 57.5	197 24.5	11 59.6	244 03.1	10.7	39 09.4	04.8	87 20.2	15.7			
FRIDAY	28 00	185 00.0	212 24.0	S11 58.7	259 03.9	S23 10.6	54 11.9	N19 04.8	102 22.5	N22 15.7	Mirfak	308 52.4	N49 52.4
	01	200 02.4	227 23.6	57.7	274 04.7	10.4	69 14.4	04.8	117 24.8	15.7	Nunki	76 08.3	S26 17.6
	02	215 04.9	242 23.1	56.8	289 05.4	10.3	84 16.9	04.8	132 27.1	15.7	Peacock	53 32.0	S56 43.4
	03	230 07.3	257 22.7	. . 55.8	304 06.2	. . 10.2	99 19.4	. . 04.8	147 29.4	. . 15.8	Pollux	243 37.4	N28 01.2
	04	245 09.8	272 22.2	54.9	319 07.0	10.1	114 22.0	04.8	162 31.7	15.8	Procyon	245 08.1	N 5 13.0
	05	260 12.3	287 21.8	53.9	334 07.7	10.0	129 24.5	04.8	177 34.0	15.8			
	06	275 14.7	302 21.3	S11 53.0	349 08.5	S23 09.9	144 27.0	N19 04.8	192 36.3	N22 15.8	Rasalhague	96 13.9	N12 33.2
	07	290 17.2	317 20.9	52.0	4 09.3	09.7	159 29.5	04.9	207 38.6	15.8	Regulus	207 51.8	N11 57.1
	08	305 19.7	332 20.5	51.1	19 10.0	09.6	174 32.1	04.9	222 40.9	15.8	Rigel	281 19.9	S 8 12.0
	09	320 22.1	347 20.0	. . 50.1	34 10.8	. . 09.5	189 34.6	. . 04.9	237 43.2	. . 15.9	Rigil Kent.	140 02.5	S60 50.8
	10	335 24.6	2 19.6	49.2	49 11.6	09.4	204 37.1	04.9	252 45.5	15.9	Sabik	102 21.7	S15 43.8
	11	350 27.1	17 19.1	48.3	64 12.3	09.3	219 39.6	04.9	267 47.8	15.9			
	12	5 29.5	32 18.7	S11 47.3	79 13.1	S23 09.2	234 42.1	N19 04.9	282 50.1	N22 15.9	Schedar	349 50.6	N56 33.2
	13	20 32.0	47 18.2	46.4	94 13.9	09.0	249 44.6	04.9	297 52.4	15.9	Shaula	96 32.8	S37 06.3
	14	35 34.4	62 17.8	45.4	109 14.6	08.9	264 47.2	04.9	312 54.7	15.9	Sirius	258 40.9	S16 43.4
	15	50 36.9	77 17.4	. . 44.5	124 15.4	. . 08.8	279 49.7	. . 04.9	327 57.0	. . 16.0	Spica	158 39.5	S11 10.7
	16	65 39.4	92 16.9	43.5	139 16.1	08.7	294 52.2	05.0	342 59.3	16.0	Suhail	222 58.2	S43 26.9
	17	80 41.8	107 16.5	42.6	154 16.9	08.6	309 54.7	05.0	358 01.6	16.0			
	18	95 44.3	122 16.0	S11 41.6	169 17.7	S23 08.5	324 57.2	N19 05.0	13 03.9	N22 16.0	Vega	80 44.5	N38 46.8
	19	110 46.8	137 15.6	40.6	184 18.4	08.3	339 59.8	05.0	28 06.2	16.0	Zuben'ubi	137 14.1	S16 03.4
	20	125 49.2	152 15.1	39.7	199 19.2	08.2	355 02.3	05.0	43 08.5	16.0		SHA	Mer. Pass.
	21	140 51.7	167 14.7	. . 38.7	214 20.0	. . 08.1	10 04.8	. . 05.0	58 10.8	. . 16.0		° ′	h m
	22	155 54.2	182 14.3	37.8	229 20.7	08.0	25 07.3	05.0	73 13.1	16.1	Venus	28 34.0	9 50
	23	170 56.6	197 13.8	36.8	244 21.5	07.9	40 09.8	05.0	88 15.4	16.1	Mars	74 44.8	6 45
	Mer. Pass.	h m 11 42.0	*v* −0.5	*d* 0.9	*v* 0.8	*d* 0.1	*v* 2.5	*d* 0.0	*v* 2.3	*d* 0.0	Jupiter	229 10.4	20 24
											Saturn	277 26.4	17 12

	UT	SUN GHA	SUN Dec	MOON GHA	*v*	MOON Dec	*d*	HP
d	h	° ′	° ′	° ′	′	° ′	′	′
26	00	178 30.9	N 1 57.3	254 45.3	6.6	S26 01.9	2.2	57.4
	01	193 31.1	58.3	269 10.9	6.6	25 59.7	2.3	57.4
	02	208 31.3	1 59.2	283 36.5	6.8	25 57.4	2.5	57.4
	03	223 31.5	2 00.2	298 02.3	6.7	25 54.9	2.6	57.3
	04	238 31.7	01.2	312 28.0	6.9	25 52.3	2.7	57.3
	05	253 31.8	02.2	326 53.9	6.9	25 49.6	2.9	57.3
	06	268 32.0	N 2 03.2	341 19.8	7.0	S25 46.7	3.1	57.3
W	07	283 32.2	04.1	355 45.8	7.0	25 43.6	3.2	57.2
E	08	298 32.4	05.1	10 11.8	7.1	25 40.4	3.4	57.2
D	09	313 32.6	. . 06.1	24 37.9	7.2	25 37.0	3.5	57.2
N	10	328 32.8	07.1	39 04.1	7.3	25 33.5	3.6	57.1
E	11	343 33.0	08.1	53 30.4	7.3	25 29.9	3.8	57.1
S	12	358 33.2	N 2 09.1	67 56.7	7.4	S25 26.1	3.9	57.1
D	13	13 33.4	10.0	82 23.1	7.5	25 22.2	4.1	57.1
A	14	28 33.5	11.0	96 49.6	7.6	25 18.1	4.2	57.0
Y	15	43 33.7	. . 12.0	111 16.2	7.6	25 13.9	4.4	57.0
	16	58 33.9	13.0	125 42.8	7.7	25 09.5	4.5	57.0
	17	73 34.1	14.0	140 09.5	7.8	25 05.0	4.6	56.9
	18	88 34.3	N 2 14.9	154 36.3	7.8	S25 00.4	4.8	56.9
	19	103 34.5	15.9	169 03.1	8.0	24 55.6	4.9	56.9
	20	118 34.7	16.9	183 30.1	8.0	24 50.7	5.0	56.9
	21	133 34.9	. . 17.9	197 57.1	8.1	24 45.7	5.2	56.8
	22	148 35.1	18.9	212 24.2	8.2	24 40.5	5.3	56.8
	23	163 35.2	19.8	226 51.4	8.3	24 35.2	5.5	56.8
27	00	178 35.4	N 2 20.8	241 18.7	8.3	S24 29.7	5.5	56.8
	01	193 35.6	21.8	255 46.0	8.5	24 24.2	5.7	56.7
	02	208 35.8	22.8	270 13.5	8.5	24 18.5	5.9	56.7
	03	223 36.0	. . 23.8	284 41.0	8.6	24 12.6	5.9	56.7
	04	238 36.2	24.7	299 08.6	8.7	24 06.7	6.1	56.6
	05	253 36.4	25.7	313 36.3	8.7	24 00.6	6.2	56.6
	06	268 36.6	N 2 26.7	328 04.0	8.9	S23 54.4	6.4	56.6
	07	283 36.8	27.7	342 31.9	8.9	23 48.0	6.4	56.6
T	08	298 36.9	28.7	356 59.8	9.1	23 41.6	6.6	56.5
H	09	313 37.1	. . 29.6	11 27.9	9.1	23 35.0	6.7	56.5
U	10	328 37.3	30.6	25 56.0	9.2	23 28.3	6.8	56.5
R	11	343 37.5	31.6	40 24.2	9.3	23 21.5	6.9	56.5
S	12	358 37.7	N 2 32.6	54 52.5	9.4	S23 14.6	7.1	56.4
D	13	13 37.9	33.6	69 20.9	9.5	23 07.5	7.2	56.4
A	14	28 38.1	34.5	83 49.4	9.5	23 00.3	7.3	56.4
Y	15	43 38.3	. . 35.5	98 17.9	9.7	22 53.0	7.4	56.4
	16	58 38.5	36.5	112 46.6	9.7	22 45.6	7.5	56.3
	17	73 38.6	37.5	127 15.3	9.9	22 38.1	7.6	56.3
	18	88 38.8	N 2 38.5	141 44.2	9.9	S22 30.5	7.7	56.3
	19	103 39.0	39.4	156 13.1	10.0	22 22.8	7.9	56.3
	20	118 39.2	40.4	170 42.1	10.1	22 14.9	7.9	56.2
	21	133 39.4	. . 41.4	185 11.2	10.2	22 07.0	8.1	56.2
	22	148 39.6	42.4	199 40.4	10.3	21 58.9	8.1	56.2
	23	163 39.8	43.3	214 09.7	10.4	21 50.8	8.3	56.2
28	00	178 40.0	N 2 44.3	228 39.1	10.4	S21 42.5	8.4	56.1
	01	193 40.1	45.3	243 08.5	10.6	21 34.1	8.4	56.1
	02	208 40.3	46.3	257 38.1	10.6	21 25.7	8.6	56.1
	03	223 40.5	. . 47.3	272 07.7	10.7	21 17.1	8.7	56.1
	04	238 40.7	48.2	286 37.4	10.8	21 08.4	8.7	56.0
	05	253 40.9	49.2	301 07.2	10.9	20 59.7	8.9	56.0
	06	268 41.1	N 2 50.2	315 37.1	11.0	S20 50.8	9.0	56.0
	07	283 41.3	51.2	330 07.1	11.1	20 41.8	9.0	56.0
	08	298 41.5	52.1	344 37.2	11.2	20 32.8	9.2	55.9
F	09	313 41.7	. . 53.1	359 07.4	11.2	20 23.6	9.2	55.9
R	10	328 41.8	54.1	13 37.6	11.4	20 14.4	9.3	55.9
I	11	343 42.0	55.1	28 08.0	11.4	20 05.1	9.5	55.9
D	12	358 42.2	N 2 56.1	42 38.4	11.5	S19 55.6	9.5	55.9
A	13	13 42.4	57.0	57 08.9	11.7	19 46.1	9.6	55.8
Y	14	28 42.6	58.0	71 39.6	11.6	19 36.5	9.7	55.8
	15	43 42.8	2 59.0	86 10.2	11.8	19 26.8	9.7	55.8
	16	58 43.0	3 00.0	100 41.0	11.9	19 17.1	9.9	55.8
	17	73 43.2	00.9	115 11.9	11.9	19 07.2	9.9	55.7
	18	88 43.4	N 3 01.9	129 42.8	12.1	S18 57.3	10.0	55.7
	19	103 43.5	02.9	144 13.9	12.1	18 47.3	10.1	55.7
	20	118 43.7	03.9	158 45.0	12.2	18 37.2	10.2	55.7
	21	133 43.9	. . 04.8	173 16.2	12.3	18 27.0	10.3	55.7
	22	148 44.1	05.8	187 47.5	12.3	18 16.7	10.3	55.6
	23	163 44.3	06.8	202 18.8	12.5	S18 06.4	10.4	55.6
		SD 16.1	*d* 1.0	SD		15.6	15.4	15.2

Lat.	Twilight Naut.	Twilight Civil	Sunrise	Moonrise 26	Moonrise 27	Moonrise 28	Moonrise 29
°	h m	h m	h m	h m	h m	h m	h m
N 72	02 32	04 14	05 25	■	■	■	08 13
N 70	03 00	04 26	05 29	■	■	■	07 29
68	03 21	04 35	05 32	■	■	07 38	06 59
66	03 36	04 43	05 35	■	07 43	06 54	06 37
64	03 49	04 50	05 38	06 51	06 32	06 24	06 19
62	04 00	04 55	05 40	05 38	05 56	06 02	06 04
60	04 08	05 00	05 42	05 02	05 30	05 44	05 52
N 58	04 16	05 04	05 44	04 36	05 09	05 29	05 41
56	04 23	05 08	05 45	04 16	04 52	05 15	05 31
54	04 28	05 11	05 46	03 59	04 37	05 04	05 23
52	04 33	05 14	05 48	03 44	04 25	04 54	05 15
50	04 38	05 16	05 49	03 32	04 14	04 45	05 09
45	04 47	05 22	05 51	03 05	03 50	04 26	04 54
N 40	04 54	05 26	05 53	02 44	03 32	04 10	04 42
35	05 00	05 29	05 55	02 27	03 16	03 57	04 32
30	05 04	05 32	05 56	02 12	03 02	03 45	04 23
20	05 11	05 36	05 58	01 47	02 39	03 25	04 07
N 10	05 15	05 39	06 00	01 25	02 19	03 08	03 53
0	05 18	05 42	06 02	01 04	02 00	02 52	03 40
S 10	05 19	05 43	06 04	00 43	01 41	02 35	03 27
20	05 18	05 44	06 06	00 21	01 20	02 18	03 13
30	05 16	05 43	06 07	24 57	00 57	01 58	02 57
35	05 14	05 43	06 08	24 43	00 43	01 46	02 48
40	05 11	05 42	06 09	24 27	00 27	01 32	02 37
45	05 07	05 41	06 11	24 08	00 08	01 16	02 24
S 50	05 02	05 40	06 12	23 43	24 56	00 56	02 09
52	05 00	05 39	06 13	23 32	24 46	00 46	02 02
54	04 57	05 38	06 13	23 19	24 35	00 35	01 54
56	04 54	05 37	06 14	23 03	24 23	00 23	01 44
58	04 50	05 36	06 15	22 45	24 09	00 09	01 34
S 60	04 45	05 34	06 16	22 23	23 52	25 22	01 22

Lat.	Sunset	Twilight Civil	Twilight Naut.	Moonset 26	Moonset 27	Moonset 28	Moonset 29
°	h m	h m	h m	h m	h m	h m	h m
N 72	18 49	20 01	21 45	■	■	■	11 50
N 70	18 45	19 48	21 16	■	■	■	12 33
68	18 41	19 38	20 54	■	■	10 43	13 01
66	18 38	19 30	20 38	■	08 51	11 27	13 22
64	18 35	19 23	20 25	07 49	10 02	11 55	13 39
62	18 33	19 18	20 14	09 01	10 38	12 17	13 53
60	18 31	19 13	20 05	09 37	11 03	12 35	14 04
N 58	18 29	19 08	19 57	10 02	11 23	12 49	14 14
56	18 27	19 05	19 50	10 23	11 40	13 02	14 23
54	18 26	19 01	19 44	10 40	11 54	13 13	14 31
52	18 24	18 58	19 39	10 54	12 06	13 22	14 37
50	18 23	18 56	19 35	11 07	12 17	13 31	14 44
45	18 21	18 50	19 25	11 33	12 40	13 49	14 57
N 40	18 19	18 46	19 18	11 53	12 58	14 03	15 08
35	18 17	18 42	19 12	12 10	13 13	14 16	15 17
30	18 15	18 39	19 07	12 25	13 26	14 26	15 25
20	18 13	18 35	19 01	12 50	13 48	14 45	15 39
N 10	18 11	18 32	18 56	13 11	14 07	15 01	15 51
0	18 09	18 29	18 53	13 31	14 25	15 15	16 02
S 10	18 07	18 28	18 52	13 51	14 43	15 30	16 13
20	18 05	18 27	18 53	14 12	15 02	15 46	16 25
30	18 03	18 27	18 55	14 37	15 23	16 03	16 38
35	18 02	18 27	18 57	14 51	15 36	16 14	16 46
40	18 01	18 28	18 59	15 08	15 50	16 25	16 55
45	18 00	18 29	19 03	15 27	16 08	16 39	17 05
S 50	17 58	18 30	19 08	15 52	16 29	16 56	17 17
52	17 57	18 31	19 10	16 04	16 39	17 04	17 23
54	17 57	18 32	19 13	16 17	16 50	17 12	17 29
56	17 56	18 33	19 16	16 33	17 03	17 22	17 36
58	17 55	18 34	19 20	16 52	17 17	17 33	17 44
S 60	17 54	18 35	19 24	17 14	17 35	17 46	17 52

Day	SUN Eqn. of Time 00h	SUN Eqn. of Time 12h	SUN Mer. Pass.	MOON Mer. Pass. Upper	MOON Mer. Pass. Lower	MOON Age	MOON Phase
d	m s	m s	h m	h m	h m	d %	
26	05 57	05 48	12 06	07 18	19 45	23 35	
27	05 39	05 30	12 05	08 12	20 39	24 25	◑
28	05 21	05 11	12 05	09 04	21 28	25 17	

UT	ARIES	VENUS −4.0		MARS +0.5		JUPITER −2.3		SATURN +0.1		STARS		
	GHA	GHA	Dec	GHA	Dec	GHA	Dec	GHA	Dec	Name	SHA	Dec
d h	° ′	° ′	° ′	° ′	° ′	° ′	° ′	° ′	° ′		° ′	° ′
29 00	185 59.1	212 13.4	S11 35.9	259 22.3	S23 07.7	55 12.3	N19 05.0	103 17.7	N22 16.1	Acamar	315 24.7	S40 17.7
01	201 01.6	227 12.9	34.9	274 23.0	07.6	70 14.8	05.1	118 20.0	16.1	Achernar	335 33.1	S57 13.4
02	216 04.0	242 12.5	34.0	289 23.8	07.5	85 17.4	05.1	133 22.3	16.1	Acrux	173 18.0	S63 07.0
03	231 06.5	257 12.1	. . 33.0	304 24.6	. . 07.4	100 19.9	. . 05.1	148 24.6	. . 16.1	Adhara	255 18.9	S28 58.8
04	246 08.9	272 11.6	32.0	319 25.3	07.2	115 22.4	05.1	163 26.9	16.2	Aldebaran	290 58.8	N16 30.9
05	261 11.4	287 11.2	31.1	334 26.1	07.1	130 24.9	05.1	178 29.2	16.2			
06	276 13.9	302 10.8	S11 30.1	349 26.9	S23 07.0	145 27.4	N19 05.1	193 31.5	N22 16.2	Alioth	166 27.0	N55 56.5
07	291 16.3	317 10.3	29.2	4 27.7	06.9	160 29.9	05.1	208 33.8	16.2	Alkaid	153 04.6	N49 17.7
S 08	306 18.8	332 09.9	28.2	19 28.4	06.8	175 32.4	05.1	223 36.1	16.2	Al Na'ir	27 54.0	S46 56.8
A 09	321 21.3	347 09.4	. . 27.2	34 29.2	. . 06.6	190 34.9	. . 05.1	238 38.4	. . 16.2	Alnilam	275 54.6	S 1 12.1
T 10	336 23.7	2 09.0	26.3	49 30.0	06.5	205 37.5	05.1	253 40.7	16.3	Alphard	218 03.8	S 8 40.4
U 11	351 26.2	17 08.6	25.3	64 30.7	06.4	220 40.0	05.2	268 43.0	16.3			
R 12	6 28.7	32 08.1	S11 24.4	79 31.5	S23 06.3	235 42.5	N19 05.2	283 45.3	N22 16.3	Alphecca	126 17.5	N26 42.0
D 13	21 31.1	47 07.7	23.4	94 32.3	06.1	250 45.0	05.2	298 47.6	16.3	Alpheratz	357 52.3	N29 06.3
A 14	36 33.6	62 07.3	22.4	109 33.0	06.0	265 47.5	05.2	313 49.9	16.3	Altair	62 16.2	N 8 52.3
Y 15	51 36.0	77 06.8	. . 21.5	124 33.8	. . 05.9	280 50.0	. . 05.2	328 52.2	. . 16.3	Ankaa	353 23.9	S42 17.5
16	66 38.5	92 06.4	20.5	139 34.6	05.8	295 52.5	05.2	343 54.5	16.3	Antares	112 36.0	S26 26.4
17	81 41.0	107 06.0	19.5	154 35.3	05.6	310 55.0	05.2	358 56.8	16.4			
18	96 43.4	122 05.5	S11 18.6	169 36.1	S23 05.5	325 57.5	N19 05.2	13 59.0	N22 16.4	Arcturus	146 02.7	N19 09.8
19	111 45.9	137 05.1	17.6	184 36.9	05.4	341 00.0	05.2	29 01.3	16.4	Atria	107 45.0	S69 01.8
20	126 48.4	152 04.7	16.6	199 37.6	05.3	356 02.6	05.2	44 03.6	16.4	Avior	234 21.2	S59 31.4
21	141 50.8	167 04.2	. . 15.7	214 38.4	. . 05.1	11 05.1	. . 05.2	59 05.9	. . 16.4	Bellatrix	278 40.8	N 6 21.1
22	156 53.3	182 03.8	14.7	229 39.2	05.0	26 07.6	05.3	74 08.2	16.4	Betelgeuse	271 10.1	N 7 24.4
23	171 55.8	197 03.4	13.7	244 40.0	04.9	41 10.1	05.3	89 10.5	16.5			
30 00	186 58.2	212 02.9	S11 12.8	259 40.7	S23 04.8	56 12.6	N19 05.3	104 12.8	N22 16.5	Canopus	263 59.8	S52 42.1
01	202 00.7	227 02.5	11.8	274 41.5	04.6	71 15.1	05.3	119 15.1	16.5	Capella	280 46.6	N46 00.2
02	217 03.2	242 02.1	10.8	289 42.3	04.5	86 17.6	05.3	134 17.4	16.5	Deneb	49 37.3	N45 17.1
03	232 05.6	257 01.6	. . 09.8	304 43.0	. . 04.4	101 20.1	. . 05.3	149 19.7	. . 16.5	Denebola	182 41.5	N14 33.2
04	247 08.1	272 01.2	08.9	319 43.8	04.3	116 22.6	05.3	164 22.0	16.5	Diphda	349 04.2	S17 58.3
05	262 10.5	287 00.8	07.9	334 44.6	04.1	131 25.1	05.3	179 24.3	16.6			
06	277 13.0	302 00.3	S11 06.9	349 45.3	S23 04.0	146 27.6	N19 05.3	194 26.6	N22 16.6	Dubhe	194 00.6	N61 44.2
07	292 15.5	316 59.9	06.0	4 46.1	03.9	161 30.1	05.3	209 28.9	16.6	Elnath	278 22.9	N28 36.7
08	307 17.9	331 59.5	05.0	19 46.9	03.7	176 32.6	05.3	224 31.2	16.6	Eltanin	90 49.8	N51 28.9
S 09	322 20.4	346 59.0	. . 04.0	34 47.7	. . 03.6	191 35.1	. . 05.3	239 33.5	. . 16.6	Enif	33 55.3	N 9 53.1
U 10	337 22.9	1 58.6	03.0	49 48.4	03.5	206 37.6	05.4	254 35.8	16.6	Fomalhaut	15 33.1	S29 36.5
N 11	352 25.3	16 58.2	02.1	64 49.2	03.4	221 40.1	05.4	269 38.0	16.6			
D 12	7 27.8	31 57.8	S11 01.1	79 50.0	S23 03.2	236 42.6	N19 05.4	284 40.3	N22 16.7	Gacrux	172 09.6	S57 07.9
A 13	22 30.3	46 57.3	11 00.1	94 50.7	03.1	251 45.1	05.4	299 42.6	16.7	Gienah	176 00.3	S17 33.6
Y 14	37 32.7	61 56.9	10 59.1	109 51.5	03.0	266 47.6	05.4	314 44.9	16.7	Hadar	148 59.0	S60 23.2
15	52 35.2	76 56.5	. . 58.2	124 52.3	. . 02.8	281 50.1	. . 05.4	329 47.2	. . 16.7	Hamal	328 10.2	N23 28.5
16	67 37.7	91 56.0	57.2	139 53.1	02.7	296 52.6	05.4	344 49.5	16.7	Kaus Aust.	83 54.5	S34 23.0
17	82 40.1	106 55.6	56.2	154 53.8	02.6	311 55.1	05.4	359 51.8	16.7			
18	97 42.6	121 55.2	S10 55.2	169 54.6	S23 02.4	326 57.6	N19 05.4	14 54.1	N22 16.8	Kochab	137 18.3	N74 08.3
19	112 45.0	136 54.8	54.2	184 55.4	02.3	342 00.1	05.4	29 56.4	16.8	Markab	13 46.7	N15 13.1
20	127 47.5	151 54.3	53.3	199 56.1	02.2	357 02.6	05.4	44 58.7	16.8	Menkar	314 23.8	N 4 06.0
21	142 50.0	166 53.9	. . 52.3	214 56.9	. . 02.0	12 05.1	. . 05.4	60 01.0	. . 16.8	Menkent	148 16.8	S36 23.1
22	157 52.4	181 53.5	51.3	229 57.7	01.9	27 07.6	05.4	75 03.3	16.8	Miaplacidus	221 41.1	S69 44.0
23	172 54.9	196 53.1	50.3	244 58.5	01.8	42 10.1	05.4	90 05.5	16.8			
31 00	187 57.4	211 52.6	S10 49.3	259 59.2	S23 01.6	57 12.6	N19 05.4	105 07.8	N22 16.9	Mirfak	308 52.4	N49 52.4
01	202 59.8	226 52.2	48.3	275 00.0	01.5	72 15.1	05.5	120 10.1	16.9	Nunki	76 08.3	S26 17.6
02	218 02.3	241 51.8	47.4	290 00.8	01.4	87 17.6	05.5	135 12.4	16.9	Peacock	53 32.0	S56 43.4
03	233 04.8	256 51.4	. . 46.4	305 01.6	. . 01.2	102 20.1	. . 05.5	150 14.7	. . 16.9	Pollux	243 37.4	N28 01.2
04	248 07.2	271 50.9	45.4	320 02.3	01.1	117 22.6	05.5	165 17.0	16.9	Procyon	245 08.1	N 5 13.0
05	263 09.7	286 50.5	44.4	335 03.1	01.0	132 25.1	05.5	180 19.3	16.9			
06	278 12.1	301 50.1	S10 43.4	350 03.9	S23 00.8	147 27.6	N19 05.5	195 21.6	N22 17.0	Rasalhague	96 13.8	N12 33.2
07	293 14.6	316 49.7	42.4	5 04.7	00.7	162 30.1	05.5	210 23.9	17.0	Regulus	207 51.8	N11 57.1
08	308 17.1	331 49.2	41.5	20 05.4	00.6	177 32.6	05.5	225 26.1	17.0	Rigel	281 19.9	S 8 12.0
M 09	323 19.5	346 48.8	. . 40.5	35 06.2	. . 00.4	192 35.1	. . 05.5	240 28.4	. . 17.0	Rigil Kent.	140 02.5	S60 50.8
O 10	338 22.0	1 48.4	39.5	50 07.0	00.3	207 37.6	05.5	255 30.7	17.0	Sabik	102 21.7	S15 43.8
N 11	353 24.5	16 48.0	38.5	65 07.8	00.2	222 40.1	05.5	270 33.0	17.0			
D 12	8 26.9	31 47.6	S10 37.5	80 08.5	S23 00.0	237 42.6	N19 05.5	285 35.3	N22 17.0	Schedar	349 50.6	N56 33.2
A 13	23 29.4	46 47.1	36.5	95 09.3	22 59.9	252 45.1	05.5	300 37.6	17.1	Shaula	96 32.8	S37 06.3
Y 14	38 31.9	61 46.7	35.5	110 10.1	59.8	267 47.6	05.5	315 39.9	17.1	Sirius	258 40.9	S16 43.3
15	53 34.3	76 46.3	. . 34.5	125 10.9	. . 59.6	282 50.1	. . 05.5	330 42.2	. . 17.1	Spica	158 39.5	S11 10.7
16	68 36.8	91 45.9	33.5	140 11.6	59.5	297 52.6	05.5	345 44.5	17.1	Suhail	222 58.2	S43 26.9
17	83 39.3	106 45.5	32.5	155 12.4	59.3	312 55.1	05.5	0 46.7	17.1			
18	98 41.7	121 45.0	S10 31.6	170 13.2	S22 59.2	327 57.6	N19 05.5	15 49.0	N22 17.1	Vega	80 44.4	N38 46.8
19	113 44.2	136 44.6	30.6	185 14.0	59.1	343 00.1	05.6	30 51.3	17.2	Zuben'ubi	137 14.1	S16 03.4
20	128 46.6	151 44.2	29.6	200 14.7	58.9	358 02.5	05.6	45 53.6	17.2		SHA	Mer. Pass.
21	143 49.1	166 43.8	. . 28.6	215 15.5	. . 58.8	13 05.0	. . 05.6	60 55.9	. . 17.2		° ′	h m
22	158 51.6	181 43.4	27.6	230 16.3	58.7	28 07.5	05.6	75 58.2	17.2	Venus	25 04.7	9 52
23	173 54.0	196 42.9	26.6	245 17.1	58.5	43 10.0	05.6	91 00.5	17.2	Mars	72 42.5	6 41
Mer. Pass.	h m 11 30.2	*v* −0.4	*d* 1.0	*v* 0.8	*d* 0.1	*v* 2.5	*d* 0.0	*v* 2.3	*d* 0.0	Jupiter	229 14.4	20 12
										Saturn	277 14.6	17 01

UT		SUN GHA	SUN Dec	MOON GHA	MOON v	MOON Dec	MOON d	MOON HP
d	h	° ′	° ′	° ′	′	° ′	′	′
29	00	178 44.5	N 3 07.8	216 50.3	12.5	S17 56.0	10.5	55.6
	01	193 44.7	08.8	231 21.8	12.6	17 45.5	10.6	55.6
	02	208 44.9	09.7	245 53.4	12.7	17 34.9	10.6	55.6
	03	223 45.0	. . 10.7	260 25.1	12.8	17 24.3	10.7	55.5
	04	238 45.2	11.7	274 56.9	12.9	17 13.6	10.8	55.5
	05	253 45.4	12.7	289 28.8	12.9	17 02.8	10.8	55.5
	06	268 45.6	N 3 13.6	304 00.7	13.0	S16 52.0	11.0	55.5
	07	283 45.8	14.6	318 32.7	13.1	16 41.0	10.9	55.5
S	08	298 46.0	15.6	333 04.8	13.1	16 30.1	11.1	55.4
A	09	313 46.2	. . 16.6	347 36.9	13.3	16 19.0	11.1	55.4
T	10	328 46.4	17.5	2 09.2	13.3	16 07.9	11.2	55.4
U	11	343 46.6	18.5	16 41.5	13.4	15 56.7	11.2	55.4
R	12	358 46.7	N 3 19.5	31 13.9	13.4	S15 45.5	11.3	55.4
D	13	13 46.9	20.5	45 46.3	13.6	15 34.2	11.4	55.3
A	14	28 47.1	21.4	60 18.9	13.6	15 22.8	11.4	55.3
Y	15	43 47.3	. . 22.4	74 51.5	13.7	15 11.4	11.5	55.3
	16	58 47.5	23.4	89 24.2	13.7	14 59.9	11.5	55.3
	17	73 47.7	24.4	103 56.9	13.8	14 48.4	11.6	55.3
	18	88 47.9	N 3 25.3	118 29.7	13.9	S14 36.8	11.7	55.2
	19	103 48.1	26.3	133 02.6	14.0	14 25.1	11.7	55.2
	20	118 48.2	27.3	147 35.6	14.0	14 13.4	11.7	55.2
	21	133 48.4	. . 28.2	162 08.6	14.1	14 01.7	11.9	55.2
	22	148 48.6	29.2	176 41.7	14.2	13 49.8	11.8	55.2
	23	163 48.8	30.2	191 14.9	14.2	13 38.0	11.9	55.2
30	00	178 49.0	N 3 31.2	205 48.1	14.3	S13 26.1	12.0	55.1
	01	193 49.2	32.1	220 21.4	14.3	13 14.1	12.0	55.1
	02	208 49.4	33.1	234 54.7	14.5	13 02.1	12.1	55.1
	03	223 49.6	. . 34.1	249 28.2	14.4	12 50.0	12.1	55.1
	04	238 49.7	35.1	264 01.6	14.6	12 37.9	12.1	55.1
	05	253 49.9	36.0	278 35.2	14.6	12 25.8	12.2	55.0
	06	268 50.1	N 3 37.0	293 08.8	14.7	S12 13.6	12.3	55.0
	07	283 50.3	38.0	307 42.5	14.7	12 01.3	12.3	55.0
	08	298 50.5	39.0	322 16.2	14.8	11 49.0	12.3	55.0
S	09	313 50.7	. . 39.9	336 50.0	14.8	11 36.7	12.4	55.0
U	10	328 50.9	40.9	351 23.8	14.9	11 24.3	12.4	55.0
N	11	343 51.1	41.9	5 57.7	15.0	11 11.9	12.4	54.9
D	12	358 51.2	N 3 42.8	20 31.7	15.0	S10 59.5	12.5	54.9
A	13	13 51.4	43.8	35 05.7	15.0	10 47.0	12.5	54.9
Y	14	28 51.6	44.8	49 39.7	15.1	10 34.5	12.6	54.9
	15	43 51.8	. . 45.8	64 13.8	15.2	10 21.9	12.6	54.9
	16	58 52.0	46.7	78 48.0	15.2	10 09.3	12.6	54.9
	17	73 52.2	47.7	93 22.2	15.3	9 56.7	12.7	54.9
	18	88 52.4	N 3 48.7	107 56.5	15.3	S 9 44.0	12.7	54.8
	19	103 52.6	49.6	122 30.8	15.4	9 31.3	12.7	54.8
	20	118 52.7	50.6	137 05.2	15.4	9 18.6	12.8	54.8
	21	133 52.9	. . 51.6	151 39.6	15.4	9 05.8	12.7	54.8
	22	148 53.1	52.6	166 14.0	15.6	8 53.1	12.9	54.8
	23	163 53.3	53.5	180 48.6	15.5	8 40.2	12.8	54.8
31	00	178 53.5	N 3 54.5	195 23.1	15.6	S 8 27.4	12.9	54.8
	01	193 53.7	55.5	209 57.7	15.6	8 14.5	12.9	54.7
	02	208 53.9	56.4	224 32.3	15.7	8 01.6	12.9	54.7
	03	223 54.1	. . 57.4	239 07.0	15.8	7 48.7	12.9	54.7
	04	238 54.2	58.4	253 41.8	15.7	7 35.8	13.0	54.7
	05	253 54.4	3 59.3	268 16.5	15.8	7 22.8	13.0	54.7
	06	268 54.6	N 4 00.3	282 51.3	15.9	S 7 09.8	13.0	54.7
	07	283 54.8	01.3	297 26.2	15.8	6 56.8	13.1	54.7
	08	298 55.0	02.3	312 01.0	16.0	6 43.7	13.0	54.6
M	09	313 55.2	. . 03.2	326 36.0	15.9	6 30.7	13.1	54.6
O	10	328 55.4	04.2	341 10.9	16.0	6 17.6	13.1	54.6
N	11	343 55.5	05.2	355 45.9	16.0	6 04.5	13.1	54.6
D	12	358 55.7	N 4 06.1	10 20.9	16.1	S 5 51.4	13.1	54.6
A	13	13 55.9	07.1	24 56.0	16.1	5 38.3	13.2	54.6
Y	14	28 56.1	08.1	39 31.1	16.1	5 25.1	13.1	54.6
	15	43 56.3	. . 09.0	54 06.2	16.2	5 12.0	13.2	54.5
	16	58 56.5	10.0	68 41.4	16.1	4 58.8	13.2	54.5
	17	73 56.7	11.0	83 16.5	16.3	4 45.6	13.2	54.5
	18	88 56.8	N 4 11.9	97 51.8	16.2	S 4 32.4	13.2	54.5
	19	103 57.0	12.9	112 27.0	16.3	4 19.2	13.2	54.5
	20	118 57.2	13.9	127 02.3	16.2	4 06.0	13.3	54.5
	21	133 57.4	. . 14.8	141 37.5	16.4	3 52.7	13.2	54.5
	22	148 57.6	15.8	156 12.9	16.3	3 39.5	13.3	54.5
	23	163 57.8	16.8	170 48.2	16.4	S 3 26.2	13.2	54.5
		SD 16.0	d 1.0	SD 15.1		15.0		14.9

Lat.	Twilight Naut.	Twilight Civil	Sunrise	Moonrise 29	30	31	1
°	h m	h m	h m	h m	h m	h m	h m
N 72	02 05	03 56	05 09	08 13	07 14	06 40	06 12
N 70	02 40	04 10	05 15	07 29	06 54	06 30	06 10
68	03 04	04 22	05 20	06 59	06 38	06 22	06 08
66	03 22	04 31	05 24	06 37	06 25	06 15	06 06
64	03 36	04 38	05 27	06 19	06 14	06 09	06 05
62	03 48	04 45	05 30	06 04	06 05	06 04	06 03
60	03 58	04 50	05 33	05 52	05 56	06 00	06 02
N 58	04 06	04 55	05 35	05 41	05 49	05 56	06 01
56	04 14	04 59	05 37	05 31	05 43	05 52	06 00
54	04 20	05 03	05 39	05 23	05 37	05 49	05 59
52	04 26	05 07	05 41	05 15	05 32	05 46	05 59
50	04 31	05 10	05 42	05 09	05 28	05 43	05 58
45	04 41	05 16	05 45	04 54	05 17	05 38	05 56
N 40	04 49	05 21	05 48	04 42	05 09	05 33	05 55
35	04 55	05 25	05 50	04 32	05 02	05 29	05 54
30	05 00	05 28	05 52	04 23	04 55	05 25	05 53
20	05 08	05 34	05 56	04 07	04 44	05 19	05 51
N 10	05 13	05 38	05 59	03 53	04 34	05 13	05 50
0	05 17	05 41	06 01	03 40	04 25	05 07	05 48
S 10	05 18	05 43	06 04	03 27	04 16	05 02	05 47
20	05 19	05 44	06 06	03 13	04 06	04 56	05 45
30	05 18	05 45	06 09	02 57	03 54	04 50	05 44
35	05 16	05 45	06 11	02 48	03 48	04 46	05 43
40	05 14	05 45	06 12	02 37	03 40	04 42	05 42
45	05 11	05 45	06 14	02 24	03 31	04 37	05 40
S 50	05 07	05 45	06 17	02 09	03 21	04 31	05 39
52	05 05	05 44	06 18	02 02	03 16	04 28	05 38
54	05 03	05 44	06 19	01 54	03 10	04 25	05 37
56	05 00	05 43	06 20	01 44	03 04	04 21	05 37
58	04 57	05 42	06 22	01 34	02 57	04 18	05 36
S 60	04 53	05 42	06 23	01 22	02 49	04 13	05 35

Lat.	Sunset	Twilight Civil	Twilight Naut.	Moonset 29	30	31	1
°	h m	h m	h m	h m	h m	h m	h m
N 72	19 03	20 17	22 13	11 50	14 21	16 23	18 17
N 70	18 57	20 02	21 35	12 33	14 40	16 31	18 16
68	18 52	19 50	21 10	13 01	14 54	16 37	18 15
66	18 47	19 41	20 51	13 22	15 05	16 42	18 15
64	18 44	19 33	20 36	13 39	15 15	16 46	18 14
62	18 41	19 26	20 24	13 53	15 23	16 49	18 14
60	18 38	19 20	20 13	14 04	15 30	16 53	18 13
N 58	18 35	19 16	20 05	14 14	15 36	16 55	18 13
56	18 33	19 11	19 57	14 23	15 42	16 58	18 12
54	18 31	19 07	19 51	14 31	15 46	17 00	18 12
52	18 30	19 04	19 45	14 37	15 51	17 02	18 12
50	18 28	19 01	19 40	14 44	15 55	17 04	18 11
45	18 25	18 54	19 29	14 57	16 03	17 08	18 11
N 40	18 22	18 49	19 21	15 08	16 10	17 11	18 10
35	18 19	18 45	19 15	15 17	16 16	17 14	18 10
30	18 17	18 41	19 09	15 25	16 21	17 16	18 09
20	18 14	18 36	19 01	15 39	16 30	17 20	18 09
N 10	18 11	18 32	18 56	15 51	16 38	17 24	18 08
0	18 08	18 29	18 53	16 02	16 46	17 27	18 08
S 10	18 05	18 26	18 51	16 13	16 53	17 31	18 07
20	18 03	18 25	18 50	16 25	17 01	17 34	18 06
30	18 00	18 23	18 51	16 38	17 09	17 38	18 06
35	17 58	18 23	18 52	16 46	17 15	17 41	18 05
40	17 56	18 23	18 54	16 55	17 20	17 43	18 05
45	17 54	18 23	18 57	17 05	17 27	17 46	18 04
S 50	17 52	18 24	19 01	17 17	17 35	17 50	18 04
52	17 50	18 24	19 03	17 23	17 38	17 51	18 03
54	17 49	18 24	19 05	17 29	17 42	17 53	18 03
56	17 48	18 25	19 08	17 36	17 46	17 55	18 03
58	17 46	18 25	19 11	17 44	17 51	17 57	18 02
S 60	17 45	18 26	19 15	17 52	17 57	17 59	18 02

Day	SUN Eqn. of Time 00^h	SUN Eqn. of Time 12^h	SUN Mer. Pass.	MOON Mer. Pass. Upper	MOON Mer. Pass. Lower	MOON Age	MOON Phase
d	m s	m s	h m	h m	h m	d	%
29	05 02	04 53	12 05	09 51	22 14	26	10
30	04 44	04 35	12 05	10 35	22 57	27	5
31	04 26	04 17	12 04	11 17	23 38	28	2

	UT	ARIES	VENUS −4.0		MARS +0.5		JUPITER −2.3		SATURN +0.1	
	d h	GHA ° ′	GHA ° ′	Dec ° ′	GHA ° ′	Dec ° ′	GHA ° ′	Dec ° ′	GHA ° ′	Dec ° ′
TUESDAY	1 00	188 56.5	211 42.5	S10 25.6	260 17.8	S22 58.4	58 12.5	N19 05.6	106 02.7	N22 17.2
	01	203 59.0	226 42.1	24.6	275 18.6	58.2	73 15.0	05.6	121 05.0	17.3
	02	219 01.4	241 41.7	23.6	290 19.4	58.1	88 17.5	05.6	136 07.3	17.3
	03	234 03.9	256 41.3	. . 22.6	305 20.2	. . 58.0	103 20.0	. . 05.6	151 09.6	. . 17.3
	04	249 06.4	271 40.9	21.6	320 20.9	57.8	118 22.5	05.6	166 11.9	17.3
	05	264 08.8	286 40.4	20.6	335 21.7	57.7	133 25.0	05.6	181 14.2	17.3
	06	279 11.3	301 40.0	S10 19.6	350 22.5	S22 57.5	148 27.4	N19 05.6	196 16.5	N22 17.3
	07	294 13.7	316 39.6	18.6	5 23.3	57.4	163 29.9	05.6	211 18.7	17.4
	08	309 16.2	331 39.2	17.6	20 24.1	57.3	178 32.4	05.6	226 21.0	17.4
	09	324 18.7	346 38.8	. . 16.6	35 24.8	. . 57.1	193 34.9	. . 05.6	241 23.3	. . 17.4
	10	339 21.1	1 38.4	15.6	50 25.6	57.0	208 37.4	05.6	256 25.6	17.4
	11	354 23.6	16 37.9	14.6	65 26.4	56.8	223 39.9	05.6	271 27.9	17.4
	12	9 26.1	31 37.5	S10 13.6	80 27.2	S22 56.7	238 42.4	N19 05.6	286 30.2	N22 17.4
	13	24 28.5	46 37.1	12.6	95 28.0	56.5	253 44.9	05.6	301 32.4	17.4
	14	39 31.0	61 36.7	11.6	110 28.7	56.4	268 47.3	05.6	316 34.7	17.5
	15	54 33.5	76 36.3	. . 10.6	125 29.5	. . 56.3	283 49.8	. . 05.6	331 37.0	. . 17.5
	16	69 35.9	91 35.9	09.6	140 30.3	56.1	298 52.3	05.6	346 39.3	17.5
	17	84 38.4	106 35.5	08.6	155 31.1	56.0	313 54.8	05.6	1 41.6	17.5
	18	99 40.9	121 35.1	S10 07.6	170 31.8	S22 55.8	328 57.3	N19 05.6	16 43.9	N22 17.5
	19	114 43.3	136 34.6	06.6	185 32.6	55.7	343 59.8	05.6	31 46.1	17.5
	20	129 45.8	151 34.2	05.6	200 33.4	55.5	359 02.2	05.6	46 48.4	17.6
	21	144 48.2	166 33.8	. . 04.6	215 34.2	. . 55.4	14 04.7	. . 05.6	61 50.7	. . 17.6
	22	159 50.7	181 33.4	03.6	230 35.0	55.3	29 07.2	05.6	76 53.0	17.6
	23	174 53.2	196 33.0	02.6	245 35.7	55.1	44 09.7	05.6	91 55.3	17.6
WEDNESDAY	2 00	189 55.6	211 32.6	S10 01.6	260 36.5	S22 55.0	59 12.2	N19 05.7	106 57.6	N22 17.6
	01	204 58.1	226 32.2	10 00.5	275 37.3	54.8	74 14.7	05.7	121 59.8	17.6
	02	220 00.6	241 31.8	9 59.5	290 38.1	54.7	89 17.1	05.7	137 02.1	17.7
	03	235 03.0	256 31.3	. . 58.5	305 38.9	. . 54.5	104 19.6	. . 05.7	152 04.4	. . 17.7
	04	250 05.5	271 30.9	57.5	320 39.6	54.4	119 22.1	05.7	167 06.7	17.7
	05	265 08.0	286 30.5	56.5	335 40.4	54.2	134 24.6	05.7	182 09.0	17.7
	06	280 10.4	301 30.1	S 9 55.5	350 41.2	S22 54.1	149 27.1	N19 05.7	197 11.2	N22 17.7
	07	295 12.9	316 29.7	54.5	5 42.0	53.9	164 29.5	05.7	212 13.5	17.7
	08	310 15.3	331 29.3	53.5	20 42.8	53.8	179 32.0	05.7	227 15.8	17.8
	09	325 17.8	346 28.9	. . 52.5	35 43.6	. . 53.7	194 34.5	. . 05.7	242 18.1	. . 17.8
	10	340 20.3	1 28.5	51.5	50 44.3	53.5	209 37.0	05.7	257 20.4	17.8
	11	355 22.7	16 28.1	50.4	65 45.1	53.4	224 39.5	05.7	272 22.6	17.8
	12	10 25.2	31 27.7	S 9 49.4	80 45.9	S22 53.2	239 41.9	N19 05.7	287 24.9	N22 17.8
	13	25 27.7	46 27.3	48.4	95 46.7	53.1	254 44.4	05.7	302 27.2	17.8
	14	40 30.1	61 26.9	47.4	110 47.5	52.9	269 46.9	05.7	317 29.5	17.8
	15	55 32.6	76 26.4	. . 46.4	125 48.2	. . 52.8	284 49.4	. . 05.7	332 31.8	. . 17.9
	16	70 35.1	91 26.0	45.4	140 49.0	52.6	299 51.8	05.7	347 34.0	17.9
	17	85 37.5	106 25.6	44.4	155 49.8	52.5	314 54.3	05.7	2 36.3	17.9
	18	100 40.0	121 25.2	S 9 43.3	170 50.6	S22 52.3	329 56.8	N19 05.7	17 38.6	N22 17.9
	19	115 42.5	136 24.8	42.3	185 51.4	52.2	344 59.3	05.7	32 40.9	17.9
	20	130 44.9	151 24.4	41.3	200 52.2	52.0	0 01.8	05.7	47 43.2	17.9
	21	145 47.4	166 24.0	. . 40.3	215 52.9	. . 51.9	15 04.2	. . 05.7	62 45.4	. . 18.0
	22	160 49.8	181 23.6	39.3	230 53.7	51.7	30 06.7	05.7	77 47.7	18.0
	23	175 52.3	196 23.2	38.2	245 54.5	51.6	45 09.2	05.7	92 50.0	18.0
THURSDAY	3 00	190 54.8	211 22.8	S 9 37.2	260 55.3	S22 51.4	60 11.7	N19 05.7	107 52.3	N22 18.0
	01	205 57.2	226 22.4	36.2	275 56.1	51.3	75 14.1	05.7	122 54.5	18.0
	02	220 59.7	241 22.0	35.2	290 56.9	51.1	90 16.6	05.7	137 56.8	18.0
	03	236 02.2	256 21.6	. . 34.2	305 57.6	. . 51.0	105 19.1	. . 05.7	152 59.1	. . 18.1
	04	251 04.6	271 21.2	33.1	320 58.4	50.8	120 21.5	05.7	168 01.4	18.1
	05	266 07.1	286 20.8	32.1	335 59.2	50.7	135 24.0	05.7	183 03.7	18.1
	06	281 09.6	301 20.4	S 9 31.1	351 00.0	S22 50.5	150 26.5	N19 05.7	198 05.9	N22 18.1
	07	296 12.0	316 20.0	30.1	6 00.8	50.4	165 29.0	05.7	213 08.2	18.1
	08	311 14.5	331 19.6	29.1	21 01.6	50.2	180 31.4	05.7	228 10.5	18.1
	09	326 16.9	346 19.2	. . 28.0	36 02.4	. . 50.0	195 33.9	. . 05.7	243 12.8	. . 18.2
	10	341 19.4	1 18.8	27.0	51 03.1	49.9	210 36.4	05.7	258 15.0	18.2
	11	356 21.9	16 18.4	26.0	66 03.9	49.7	225 38.8	05.7	273 17.3	18.2
	12	11 24.3	31 18.0	S 9 25.0	81 04.7	S22 49.6	240 41.3	N19 05.7	288 19.6	N22 18.2
	13	26 26.8	46 17.6	23.9	96 05.5	49.4	255 43.8	05.7	303 21.9	18.2
	14	41 29.3	61 17.2	22.9	111 06.3	49.3	270 46.3	05.7	318 24.1	18.2
	15	56 31.7	76 16.8	. . 21.9	126 07.1	. . 49.1	285 48.7	. . 05.7	333 26.4	. . 18.2
	16	71 34.2	91 16.4	20.9	141 07.9	49.0	300 51.2	05.7	348 28.7	18.3
	17	86 36.7	106 16.0	19.8	156 08.6	48.8	315 53.7	05.7	3 31.0	18.3
	18	101 39.1	121 15.6	S 9 18.8	171 09.4	S22 48.7	330 56.1	N19 05.7	18 33.2	N22 18.3
	19	116 41.6	136 15.2	17.8	186 10.2	48.5	345 58.6	05.7	33 35.5	18.3
	20	131 44.1	151 14.8	16.7	201 11.0	48.3	1 01.1	05.7	48 37.8	18.3
	21	146 46.5	166 14.4	. . 15.7	216 11.8	. . 48.2	16 03.5	. . 05.7	63 40.1	. . 18.3
	22	161 49.0	181 14.0	14.7	231 12.6	48.0	31 06.0	05.7	78 42.3	18.4
	23	176 51.4	196 13.6	13.7	246 13.4	47.9	46 08.5	05.7	93 44.6	18.4
	Mer. Pass.	h m 11 18.4	*v* −0.4	*d* 1.0	*v* 0.8	*d* 0.1	*v* 2.5	*d* 0.0	*v* 2.3	*d* 0.0

STARS

Name	SHA ° ′	Dec ° ′
Acamar	315 24.7	S40 17.7
Achernar	335 33.1	S57 13.4
Acrux	173 18.0	S63 07.0
Adhara	255 18.9	S28 58.8
Aldebaran	290 58.8	N16 30.9
Alioth	166 27.0	N55 56.5
Alkaid	153 04.6	N49 17.7
Al Na'ir	27 54.0	S46 56.8
Alnilam	275 54.7	S 1 12.1
Alphard	218 03.8	S 8 40.4
Alphecca	126 17.5	N26 42.0
Alpheratz	357 52.3	N29 06.3
Altair	62 16.2	N 8 52.3
Ankaa	353 23.9	S42 17.5
Antares	112 36.0	S26 26.4
Arcturus	146 02.7	N19 09.8
Atria	107 45.0	S69 01.8
Avior	234 21.2	S59 31.4
Bellatrix	278 40.8	N 6 21.1
Betelgeuse	271 10.1	N 7 24.4
Canopus	263 59.8	S52 42.1
Capella	280 46.6	N46 00.2
Deneb	49 37.3	N45 17.1
Denebola	182 41.5	N14 33.2
Diphda	349 04.2	S17 58.3
Dubhe	194 00.6	N61 44.2
Elnath	278 23.0	N28 36.7
Eltanin	90 49.8	N51 28.9
Enif	33 55.3	N 9 53.1
Fomalhaut	15 33.1	S29 36.5
Gacrux	172 09.6	S57 07.9
Gienah	176 00.3	S17 33.6
Hadar	148 59.0	S60 23.2
Hamal	328 10.3	N23 28.5
Kaus Aust.	83 54.4	S34 23.0
Kochab	137 18.3	N74 08.4
Markab	13 46.7	N15 13.1
Menkar	314 23.8	N 4 06.0
Menkent	148 16.8	S36 23.2
Miaplacidus	221 41.2	S69 44.0
Mirfak	308 52.4	N49 52.4
Nunki	76 08.3	S26 17.6
Peacock	53 32.0	S56 43.4
Pollux	243 37.5	N28 01.2
Procyon	245 08.1	N 5 13.0
Rasalhague	96 13.8	N12 33.2
Regulus	207 51.8	N11 57.1
Rigel	281 19.9	S 8 12.0
Rigil Kent.	140 02.5	S60 50.8
Sabik	102 21.7	S15 43.8
Schedar	349 50.6	N56 33.1
Shaula	96 32.7	S37 06.4
Sirius	258 40.9	S16 43.3
Spica	158 39.5	S11 10.7
Suhail	222 58.2	S43 26.9
Vega	80 44.4	N38 46.8
Zuben'ubi	137 14.1	S16 03.4

	SHA ° ′	Mer. Pass. h m
Venus	21 36.9	9 54
Mars	70 40.9	6 37
Jupiter	229 16.5	20 00
Saturn	277 01.9	16 50

Day	UT (d h)	SUN GHA (° ′)	SUN Dec (° ′)	MOON GHA (° ′)	v (′)	MOON Dec (° ′)	d (′)	HP (′)
TUESDAY	1 00	178 58.0	N 4 17.7	185 23.6	16.4	S 3 13.0	13.3	54.4
	01	193 58.2	18.7	199 59.0	16.4	2 59.7	13.3	54.4
	02	208 58.3	19.7	214 34.4	16.4	2 46.4	13.3	54.4
	03	223 58.5	. . 20.6	229 09.8	16.4	2 33.1	13.2	54.4
	04	238 58.7	21.6	243 45.2	16.5	2 19.9	13.3	54.4
	05	253 58.9	22.6	258 20.7	16.5	2 06.6	13.3	54.4
	06	268 59.1	N 4 23.5	272 56.2	16.5	S 1 53.3	13.3	54.4
	07	283 59.3	24.5	287 31.7	16.5	1 40.0	13.3	54.4
	08	298 59.5	25.5	302 07.2	16.6	1 26.7	13.3	54.4
	09	313 59.6	. . 26.4	316 42.8	16.5	1 13.4	13.3	54.3
	10	328 59.8	27.4	331 18.3	16.6	1 00.1	13.3	54.3
	11	344 00.0	28.4	345 53.9	16.5	0 46.8	13.3	54.3
	12	359 00.2	N 4 29.3	0 29.4	16.6	S 0 33.5	13.3	54.3
	13	14 00.4	30.3	15 05.0	16.6	0 20.2	13.3	54.3
	14	29 00.6	31.3	29 40.6	16.6	S 0 06.9	13.3	54.3
	15	44 00.8	. . 32.2	44 16.2	16.7	N 0 06.4	13.2	54.3
	16	59 00.9	33.2	58 51.9	16.6	0 19.6	13.3	54.3
	17	74 01.1	34.2	73 27.5	16.6	0 32.9	13.3	54.3
	18	89 01.3	N 4 35.1	88 03.1	16.7	N 0 46.2	13.3	54.3
	19	104 01.5	36.1	102 38.8	16.6	0 59.5	13.2	54.2
	20	119 01.7	37.1	117 14.4	16.6	1 12.7	13.3	54.2
	21	134 01.9	. . 38.0	131 50.0	16.7	1 26.0	13.2	54.2
	22	149 02.0	39.0	146 25.7	16.6	1 39.2	13.2	54.2
	23	164 02.2	40.0	161 01.3	16.7	1 52.4	13.2	54.2
WEDNESDAY	2 00	179 02.4	N 4 40.9	175 37.0	16.7	N 2 05.6	13.2	54.2
	01	194 02.6	41.9	190 12.7	16.6	2 18.8	13.2	54.2
	02	209 02.8	42.9	204 48.3	16.7	2 32.0	13.2	54.2
	03	224 03.0	. . 43.8	219 24.0	16.6	2 45.2	13.2	54.2
	04	239 03.2	44.8	233 59.6	16.7	2 58.4	13.1	54.2
	05	254 03.3	45.7	248 35.3	16.6	3 11.5	13.2	54.2
	06	269 03.5	N 4 46.7	263 10.9	16.7	N 3 24.7	13.1	54.2
	07	284 03.7	47.7	277 46.6	16.6	3 37.8	13.1	54.2
	08	299 03.9	48.6	292 22.2	16.6	3 50.9	13.1	54.1
	09	314 04.1	. . 49.6	306 57.8	16.7	4 04.0	13.1	54.1
	10	329 04.3	50.6	321 33.5	16.6	4 17.1	13.0	54.1
	11	344 04.4	51.5	336 09.1	16.6	4 30.1	13.0	54.1
	12	359 04.6	N 4 52.5	350 44.7	16.6	N 4 43.1	13.0	54.1
	13	14 04.8	53.4	5 20.3	16.6	4 56.1	13.0	54.1
	14	29 05.0	54.4	19 55.9	16.5	5 09.1	13.0	54.1
	15	44 05.2	. . 55.4	34 31.4	16.6	5 22.1	12.9	54.1
	16	59 05.4	56.3	49 07.0	16.6	5 35.0	13.0	54.1
	17	74 05.6	57.3	63 42.6	16.5	5 48.0	12.9	54.1
	18	89 05.7	N 4 58.3	78 18.1	16.5	N 6 00.9	12.8	54.1
	19	104 05.9	4 59.2	92 53.6	16.5	6 13.7	12.9	54.1
	20	119 06.1	5 00.2	107 29.1	16.5	6 26.6	12.8	54.1
	21	134 06.3	. . 01.1	122 04.6	16.5	6 39.4	12.8	54.1
	22	149 06.5	02.1	136 40.1	16.5	6 52.2	12.8	54.1
	23	164 06.7	03.1	151 15.6	16.4	7 05.0	12.7	54.1
THURSDAY	3 00	179 06.8	N 5 04.0	165 51.0	16.4	N 7 17.7	12.7	54.0
	01	194 07.0	05.0	180 26.4	16.4	7 30.4	12.7	54.0
	02	209 07.2	05.9	195 01.8	16.4	7 43.1	12.6	54.0
	03	224 07.4	. . 06.9	209 37.2	16.3	7 55.7	12.7	54.0
	04	239 07.6	07.9	224 12.5	16.4	8 08.4	12.5	54.0
	05	254 07.8	08.8	238 47.9	16.3	8 20.9	12.6	54.0
	06	269 07.9	N 5 09.8	253 23.2	16.3	N 8 33.5	12.5	54.0
	07	284 08.1	10.7	267 58.5	16.2	8 46.0	12.5	54.0
	08	299 08.3	11.7	282 33.7	16.3	8 58.5	12.4	54.0
	09	314 08.5	. . 12.7	297 09.0	16.2	9 10.9	12.5	54.0
	10	329 08.7	13.6	311 44.2	16.1	9 23.4	12.3	54.0
	11	344 08.9	14.6	326 19.3	16.2	9 35.7	12.4	54.0
	12	359 09.0	N 5 15.5	340 54.5	16.1	N 9 48.1	12.3	54.0
	13	14 09.2	16.5	355 29.6	16.1	10 00.4	12.2	54.0
	14	29 09.4	17.4	10 04.7	16.0	10 12.6	12.3	54.0
	15	44 09.6	. . 18.4	24 39.7	16.1	10 24.9	12.1	54.0
	16	59 09.8	19.4	39 14.8	16.0	10 37.0	12.2	54.0
	17	74 10.0	20.3	53 49.8	15.9	10 49.2	12.1	54.0
	18	89 10.1	N 5 21.3	68 24.7	15.9	N11 01.3	12.0	54.0
	19	104 10.3	22.2	82 59.6	15.9	11 13.3	12.0	54.0
	20	119 10.5	23.2	97 34.5	15.9	11 25.3	12.0	54.0
	21	134 10.7	. . 24.1	112 09.4	15.8	11 37.3	11.9	54.0
	22	149 10.9	25.1	126 44.2	15.8	11 49.2	11.9	54.0
	23	164 11.0	26.1	141 19.0	15.7	N12 01.1	11.8	54.0
		SD 16.0	*d* 1.0	SD 14.8		14.7		14.7

Lat.	Twilight Naut.	Twilight Civil	Sunrise	Moonrise 1	Moonrise 2	Moonrise 3	Moonrise 4
°	h m	h m	h m	h m	h m	h m	h m
N 72	01 31	03 38	04 53	06 12	05 46	05 18	04 42
N 70	02 17	03 54	05 01	06 10	05 51	05 30	05 06
68	02 45	04 08	05 07	06 08	05 54	05 40	05 24
66	03 06	04 18	05 12	06 06	05 57	05 48	05 39
64	03 23	04 27	05 17	06 05	06 00	05 55	05 51
62	03 36	04 34	05 20	06 03	06 02	06 01	06 01
60	03 47	04 41	05 24	06 02	06 04	06 07	06 10
N 58	03 57	04 46	05 27	06 01	06 06	06 11	06 18
56	04 05	04 51	05 29	06 00	06 08	06 16	06 25
54	04 12	04 56	05 32	05 59	06 09	06 19	06 31
52	04 18	04 59	05 34	05 59	06 10	06 23	06 37
50	04 24	05 03	05 36	05 58	06 12	06 26	06 42
45	04 35	05 10	05 40	05 56	06 14	06 33	06 53
N 40	04 44	05 16	05 43	05 55	06 17	06 38	07 02
35	04 51	05 21	05 46	05 54	06 18	06 43	07 10
30	04 57	05 25	05 49	05 53	06 20	06 48	07 17
20	05 05	05 31	05 53	05 51	06 23	06 56	07 29
N 10	05 11	05 36	05 57	05 50	06 26	07 02	07 40
0	05 16	05 40	06 00	05 48	06 28	07 09	07 50
S 10	05 18	05 43	06 04	05 47	06 31	07 15	08 00
20	05 20	05 45	06 07	05 45	06 34	07 22	08 11
30	05 19	05 47	06 11	05 44	06 37	07 30	08 24
35	05 18	05 48	06 13	05 43	06 39	07 35	08 31
40	05 17	05 48	06 15	05 42	06 41	07 40	08 39
45	05 15	05 49	06 18	05 40	06 43	07 46	08 49
S 50	05 12	05 49	06 21	05 39	06 46	07 53	09 01
52	05 10	05 49	06 23	05 38	06 48	07 57	09 07
54	05 08	05 49	06 25	05 37	06 49	08 01	09 13
56	05 06	05 49	06 26	05 37	06 51	08 05	09 20
58	05 03	05 49	06 28	05 36	06 53	08 09	09 28
S 60	05 00	05 49	06 30	05 35	06 55	08 15	09 36

Lat.	Sunset	Twilight Civil	Twilight Naut.	Moonset 1	Moonset 2	Moonset 3	Moonset 4
°	h m	h m	h m	h m	h m	h m	h m
N 72	19 17	20 34	22 49	18 17	20 11	22 13	24 50
N 70	19 09	20 16	21 57	18 16	20 01	21 52	23 57
68	19 03	20 03	21 27	18 15	19 54	21 35	23 25
66	18 57	19 52	21 05	18 15	19 47	21 22	23 02
64	18 53	19 43	20 48	18 14	19 42	21 11	22 44
62	18 49	19 35	20 34	18 14	19 37	21 02	22 29
60	18 45	19 28	20 22	18 13	19 33	20 54	22 16
N 58	18 42	19 23	20 13	18 13	19 29	20 47	22 06
56	18 39	19 18	20 04	18 12	19 26	20 41	21 56
54	18 37	19 13	19 57	18 12	19 23	20 35	21 48
52	18 35	19 09	19 51	18 12	19 21	20 30	21 41
50	18 33	19 06	19 45	18 11	19 19	20 26	21 34
45	18 28	18 58	19 34	18 11	19 14	20 16	21 20
N 40	18 25	18 52	19 25	18 10	19 09	20 09	21 09
35	18 22	18 47	19 17	18 10	19 06	20 02	20 59
30	18 19	18 43	19 11	18 09	19 03	19 56	20 50
20	18 14	18 37	19 02	18 09	18 57	19 46	20 35
N 10	18 11	18 32	18 56	18 08	18 52	19 37	20 23
0	18 07	18 28	18 52	18 08	18 48	19 28	20 11
S 10	18 04	18 25	18 49	18 07	18 43	19 20	19 59
20	18 00	18 22	18 48	18 06	18 38	19 11	19 46
30	17 56	18 20	18 48	18 06	18 33	19 01	19 32
35	17 54	18 19	18 48	18 05	18 30	18 56	19 23
40	17 51	18 18	18 50	18 05	18 26	18 49	19 14
45	17 48	18 18	18 52	18 04	18 22	18 41	19 03
S 50	17 45	18 17	18 55	18 04	18 18	18 32	18 49
52	17 44	18 17	18 56	18 03	18 15	18 28	18 43
54	17 42	18 17	18 58	18 03	18 13	18 24	18 36
56	17 40	18 17	19 00	18 03	18 10	18 19	18 29
58	17 38	18 17	19 03	18 02	18 07	18 13	18 20
S 60	17 36	18 17	19 05	18 02	18 04	18 07	18 11

Day	SUN Eqn. of Time 00^h	SUN Eqn. of Time 12^h	SUN Mer. Pass.	MOON Mer. Pass. Upper	MOON Mer. Pass. Lower	MOON Age	MOON Phase
d	m s	m s	h m	h m	h m	d	%
1	04 09	04 00	12 04	11 58	24 18	29	0
2	03 51	03 42	12 04	12 38	00 18	01	1
3	03 33	03 24	12 03	13 19	00 58	02	3

	UT	ARIES	VENUS −4.0		MARS +0.4		JUPITER −2.3		SATURN +0.1	
	d h	GHA ° ′	GHA ° ′	Dec ° ′	GHA ° ′	Dec ° ′	GHA ° ′	Dec ° ′	GHA ° ′	Dec ° ′
	4 00	191 53.9	211 13.2	S 9 12.6	261 14.1	S22 47.7	61 10.9	N19 05.6	108 46.9	N22 18.4
	01	206 56.4	226 12.8	11.6	276 14.9	47.6	76 13.4	05.6	123 49.2	18.4
	02	221 58.8	241 12.4	10.6	291 15.7	47.4	91 15.9	05.6	138 51.4	18.4
	03	237 01.3	256 12.0	. . 09.5	306 16.5	. . 47.3	106 18.3	. . 05.6	153 53.7	. . 18.4
	04	252 03.8	271 11.6	08.5	321 17.3	47.1	121 20.8	05.6	168 56.0	18.5
	05	267 06.2	286 11.2	07.5	336 18.1	46.9	136 23.3	05.6	183 58.3	18.5
	06	282 08.7	301 10.8	S 9 06.4	351 18.9	S22 46.8	151 25.7	N19 05.6	199 00.5	N22 18.5
	07	297 11.2	316 10.4	05.4	6 19.7	46.6	166 28.2	05.6	214 02.8	18.5
	08	312 13.6	331 10.0	04.4	21 20.5	46.5	181 30.6	05.6	229 05.1	18.5
F	09	327 16.1	346 09.6	. . 03.3	36 21.2	. . 46.3	196 33.1	. . 05.6	244 07.3	. . 18.5
R	10	342 18.6	1 09.2	02.3	51 22.0	46.1	211 35.6	05.6	259 09.6	18.6
I	11	357 21.0	16 08.8	01.3	66 22.8	46.0	226 38.0	05.6	274 11.9	18.6
D	12	12 23.5	31 08.4	S 9 00.2	81 23.6	S22 45.8	241 40.5	N19 05.6	289 14.2	N22 18.6
A	13	27 25.9	46 08.0	8 59.2	96 24.4	45.7	256 43.0	05.6	304 16.4	18.6
Y	14	42 28.4	61 07.6	58.2	111 25.2	45.5	271 45.4	05.6	319 18.7	18.6
	15	57 30.9	76 07.2	. . 57.1	126 26.0	. . 45.3	286 47.9	. . 05.6	334 21.0	. . 18.6
	16	72 33.3	91 06.9	56.1	141 26.8	45.2	301 50.3	05.6	349 23.2	18.7
	17	87 35.8	106 06.5	55.0	156 27.6	45.0	316 52.8	05.6	4 25.5	18.7
	18	102 38.3	121 06.1	S 8 54.0	171 28.4	S22 44.9	331 55.3	N19 05.6	19 27.8	N22 18.7
	19	117 40.7	136 05.7	53.0	186 29.1	44.7	346 57.7	05.6	34 30.1	18.7
	20	132 43.2	151 05.3	51.9	201 29.9	44.5	2 00.2	05.6	49 32.3	18.7
	21	147 45.7	166 04.9	. . 50.9	216 30.7	. . 44.4	17 02.6	. . 05.6	64 34.6	. . 18.7
	22	162 48.1	181 04.5	49.8	231 31.5	44.2	32 05.1	05.6	79 36.9	18.7
	23	177 50.6	196 04.1	48.8	246 32.3	44.1	47 07.6	05.6	94 39.1	18.8
	5 00	192 53.0	211 03.7	S 8 47.8	261 33.1	S22 43.9	62 10.0	N19 05.6	109 41.4	N22 18.8
	01	207 55.5	226 03.3	46.7	276 33.9	43.7	77 12.5	05.6	124 43.7	18.8
	02	222 58.0	241 02.9	45.7	291 34.7	43.6	92 14.9	05.6	139 45.9	18.8
	03	238 00.4	256 02.5	. . 44.6	306 35.5	. . 43.4	107 17.4	. . 05.6	154 48.2	. . 18.8
	04	253 02.9	271 02.2	43.6	321 36.3	43.2	122 19.8	05.6	169 50.5	18.8
	05	268 05.4	286 01.8	42.6	336 37.1	43.1	137 22.3	05.5	184 52.8	18.9
	06	283 07.8	301 01.4	S 8 41.5	351 37.8	S22 42.9	152 24.8	N19 05.5	199 55.0	N22 18.9
S	07	298 10.3	316 01.0	40.5	6 38.6	42.7	167 27.2	05.5	214 57.3	18.9
	08	313 12.8	331 00.6	39.4	21 39.4	42.6	182 29.7	05.5	229 59.6	18.9
A	09	328 15.2	346 00.2	. . 38.4	36 40.2	. . 42.4	197 32.1	. . 05.5	245 01.8	. . 18.9
T	10	343 17.7	0 59.8	37.3	51 41.0	42.3	212 34.6	05.5	260 04.1	18.9
U	11	358 20.2	15 59.4	36.3	66 41.8	42.1	227 37.0	05.5	275 06.4	19.0
R	12	13 22.6	30 59.0	S 8 35.2	81 42.6	S22 41.9	242 39.5	N19 05.5	290 08.6	N22 19.0
D	13	28 25.1	45 58.7	34.2	96 43.4	41.8	257 41.9	05.5	305 10.9	19.0
A	14	43 27.5	60 58.3	33.1	111 44.2	41.6	272 44.4	05.5	320 13.2	19.0
Y	15	58 30.0	75 57.9	. . 32.1	126 45.0	. . 41.4	287 46.8	. . 05.5	335 15.4	. . 19.0
	16	73 32.5	90 57.5	31.0	141 45.8	41.3	302 49.3	05.5	350 17.7	19.0
	17	88 34.9	105 57.1	30.0	156 46.6	41.1	317 51.7	05.5	5 20.0	19.1
	18	103 37.4	120 56.7	S 8 28.9	171 47.4	S22 40.9	332 54.2	N19 05.5	20 22.2	N22 19.1
	19	118 39.9	135 56.3	27.9	186 48.2	40.8	347 56.6	05.5	35 24.5	19.1
	20	133 42.3	150 55.9	26.8	201 49.0	40.6	2 59.1	05.5	50 26.8	19.1
	21	148 44.8	165 55.6	. . 25.8	216 49.8	. . 40.4	18 01.5	. . 05.5	65 29.0	. . 19.1
	22	163 47.3	180 55.2	24.7	231 50.5	40.3	33 04.0	05.5	80 31.3	19.1
	23	178 49.7	195 54.8	23.7	246 51.3	40.1	48 06.4	05.4	95 33.6	19.2
	6 00	193 52.2	210 54.4	S 8 22.6	261 52.1	S22 39.9	63 08.9	N19 05.4	110 35.8	N22 19.2
	01	208 54.7	225 54.0	21.6	276 52.9	39.7	78 11.3	05.4	125 38.1	19.2
	02	223 57.1	240 53.6	20.5	291 53.7	39.6	93 13.8	05.4	140 40.4	19.2
	03	238 59.6	255 53.2	. . 19.5	306 54.5	. . 39.4	108 16.2	. . 05.4	155 42.6	. . 19.2
	04	254 02.0	270 52.9	18.4	321 55.3	39.2	123 18.7	05.4	170 44.9	19.2
	05	269 04.5	285 52.5	17.4	336 56.1	39.1	138 21.1	05.4	185 47.2	19.2
	06	284 07.0	300 52.1	S 8 16.3	351 56.9	S22 38.9	153 23.6	N19 05.4	200 49.4	N22 19.3
	07	299 09.4	315 51.7	15.3	6 57.7	38.7	168 26.0	05.4	215 51.7	19.3
	08	314 11.9	330 51.3	14.2	21 58.5	38.6	183 28.5	05.4	230 54.0	19.3
S	09	329 14.4	345 50.9	. . 13.2	36 59.3	. . 38.4	198 30.9	. . 05.4	245 56.2	. . 19.3
U	10	344 16.8	0 50.6	12.1	52 00.1	38.2	213 33.4	05.4	260 58.5	19.3
N	11	359 19.3	15 50.2	11.0	67 00.9	38.1	228 35.8	05.4	276 00.7	19.3
D	12	14 21.8	30 49.8	S 8 10.0	82 01.7	S22 37.9	243 38.3	N19 05.4	291 03.0	N22 19.4
A	13	29 24.2	45 49.4	08.9	97 02.5	37.7	258 40.7	05.4	306 05.3	19.4
Y	14	44 26.7	60 49.0	07.9	112 03.3	37.5	273 43.2	05.3	321 07.5	19.4
	15	59 29.1	75 48.7	. . 06.8	127 04.1	. . 37.4	288 45.6	. . 05.3	336 09.8	. . 19.4
	16	74 31.6	90 48.3	05.8	142 04.9	37.2	303 48.0	05.3	351 12.1	19.4
	17	89 34.1	105 47.9	04.7	157 05.7	37.0	318 50.5	05.3	6 14.3	19.4
	18	104 36.5	120 47.5	S 8 03.6	172 06.5	S22 36.9	333 52.9	N19 05.3	21 16.6	N22 19.5
	19	119 39.0	135 47.1	02.6	187 07.3	36.7	348 55.4	05.3	36 18.9	19.5
	20	134 41.5	150 46.8	01.5	202 08.1	36.5	3 57.8	05.3	51 21.1	19.5
	21	149 43.9	165 46.4	8 00.5	217 08.9	. . 36.3	19 00.3	. . 05.3	66 23.4	. . 19.5
	22	164 46.4	180 46.0	7 59.4	232 09.7	36.2	34 02.7	05.3	81 25.6	19.5
	23	179 48.9	195 45.6	S 7 58.3	247 10.5	36.0	49 05.1	05.3	96 27.9	19.5
	Mer. Pass.	h m 11 06.6	v −0.4	d 1.0	v 0.8	d 0.2	v 2.5	d 0.0	v 2.3	d 0.0

STARS

Name	SHA ° ′	Dec ° ′
Acamar	315 24.7	S40 17.7
Achernar	335 33.1	S57 13.4
Acrux	173 18.0	S63 07.0
Adhara	255 18.9	S28 58.8
Aldebaran	290 58.9	N16 30.9
Alioth	166 27.0	N55 56.6
Alkaid	153 04.6	N49 17.8
Al Na'ir	27 54.0	S46 56.8
Alnilam	275 54.7	S 1 12.1
Alphard	218 03.8	S 8 40.4
Alphecca	126 17.5	N26 42.0
Alpheratz	357 52.3	N29 06.3
Altair	62 16.2	N 8 52.3
Ankaa	353 23.9	S42 17.5
Antares	112 36.0	S26 26.4
Arcturus	146 02.7	N19 09.8
Atria	107 44.9	S69 01.8
Avior	234 21.3	S59 31.4
Bellatrix	278 40.8	N 6 21.1
Betelgeuse	271 10.1	N 7 24.4
Canopus	263 59.8	S52 42.1
Capella	280 46.6	N46 00.2
Deneb	49 37.3	N45 17.1
Denebola	182 41.5	N14 33.2
Diphda	349 04.2	S17 58.3
Dubhe	194 00.7	N61 44.2
Elnath	278 23.0	N28 36.7
Eltanin	90 49.7	N51 28.9
Enif	33 55.3	N 9 53.1
Fomalhaut	15 33.1	S29 36.5
Gacrux	172 09.6	S57 07.9
Gienah	176 00.3	S17 33.6
Hadar	148 59.0	S60 23.3
Hamal	328 10.3	N23 28.5
Kaus Aust.	83 54.4	S34 23.0
Kochab	137 18.3	N74 08.4
Markab	13 46.7	N15 13.1
Menkar	314 23.8	N 4 06.0
Menkent	148 16.8	S36 23.2
Miaplacidus	221 41.2	S69 44.0
Mirfak	308 52.4	N49 52.4
Nunki	76 08.3	S26 17.6
Peacock	53 31.9	S56 43.4
Pollux	243 37.5	N28 01.2
Procyon	245 08.1	N 5 13.0
Rasalhague	96 13.8	N12 33.2
Regulus	207 51.8	N11 57.1
Rigel	281 20.0	S 8 12.0
Rigil Kent.	140 02.5	S60 50.8
Sabik	102 21.6	S15 43.8
Schedar	349 50.5	N56 33.1
Shaula	96 32.7	S37 06.4
Sirius	258 40.9	S16 43.3
Spica	158 39.5	S11 10.7
Suhail	222 58.3	S43 26.9
Vega	80 44.4	N38 46.8
Zuben'ubi	137 14.1	S16 03.4
	SHA ° ′	**Mer. Pass.** h m
Venus	18 10.7	9 56
Mars	68 40.0	6 33
Jupiter	229 17.0	19 48
Saturn	276 48.4	16 39

UT		SUN GHA	SUN Dec	MOON GHA	v	MOON Dec	d	HP
d	h	° ′	° ′	° ′	′	° ′	′	′
4	00	179 11.2	N 5 27.0	155 53.7	15.7	N12 12.9	11.8	54.0
	01	194 11.4	28.0	170 28.4	15.7	12 24.7	11.8	54.0
	02	209 11.6	28.9	185 03.1	15.6	12 36.5	11.7	54.0
	03	224 11.8	. . 29.9	199 37.7	15.6	12 48.2	11.6	54.0
	04	239 12.0	30.8	214 12.3	15.6	12 59.8	11.6	54.0
	05	254 12.1	31.8	228 46.9	15.5	13 11.4	11.5	54.0
	06	269 12.3	N 5 32.8	243 21.4	15.4	N13 22.9	11.5	54.0
	07	284 12.5	33.7	257 55.8	15.5	13 34.4	11.4	54.0
	08	299 12.7	34.7	272 30.3	15.3	13 45.8	11.4	54.0
F	09	314 12.9	. . 35.6	287 04.6	15.4	13 57.2	11.3	54.0
R	10	329 13.0	36.6	301 39.0	15.2	14 08.5	11.3	54.0
I	11	344 13.2	37.5	316 13.2	15.3	14 19.8	11.2	54.0
D	12	359 13.4	N 5 38.5	330 47.5	15.2	N14 31.0	11.2	54.0
A	13	14 13.6	39.4	345 21.7	15.1	14 42.2	11.1	54.0
Y	14	29 13.8	40.4	359 55.8	15.1	14 53.3	11.0	54.0
	15	44 14.0	. . 41.3	14 29.9	15.0	15 04.3	11.0	54.0
	16	59 14.1	42.3	29 03.9	15.0	15 15.3	11.0	54.0
	17	74 14.3	43.3	43 37.9	15.0	15 26.3	10.8	54.0
	18	89 14.5	N 5 44.2	58 11.9	14.9	N15 37.1	10.8	54.0
	19	104 14.7	45.2	72 45.8	14.8	15 47.9	10.8	54.0
	20	119 14.9	46.1	87 19.6	14.8	15 58.7	10.7	54.0
	21	134 15.0	. . 47.1	101 53.4	14.8	16 09.4	10.6	54.0
	22	149 15.2	48.0	116 27.2	14.6	16 20.0	10.5	54.0
	23	164 15.4	49.0	131 00.8	14.7	16 30.5	10.5	54.0
5	00	179 15.6	N 5 49.9	145 34.5	14.6	N16 41.0	10.4	54.0
	01	194 15.8	50.9	160 08.1	14.5	16 51.4	10.4	54.0
	02	209 15.9	51.8	174 41.6	14.4	17 01.8	10.3	54.0
	03	224 16.1	. . 52.8	189 15.0	14.5	17 12.1	10.2	54.0
	04	239 16.3	53.7	203 48.5	14.3	17 22.3	10.1	54.0
	05	254 16.5	54.7	218 21.8	14.3	17 32.4	10.1	54.0
	06	269 16.7	N 5 55.6	232 55.1	14.3	N17 42.5	10.0	54.0
	07	284 16.8	56.6	247 28.4	14.1	17 52.5	10.0	54.0
S	08	299 17.0	57.5	262 01.5	14.2	18 02.5	9.8	54.1
A	09	314 17.2	. . 58.5	276 34.7	14.0	18 12.3	9.8	54.1
T	10	329 17.4	5 59.4	291 07.7	14.1	18 22.1	9.8	54.1
U	11	344 17.6	6 00.4	305 40.8	13.9	18 31.9	9.6	54.1
R	12	359 17.7	N 6 01.3	320 13.7	13.9	N18 41.5	9.6	54.1
D	13	14 17.9	02.3	334 46.6	13.8	18 51.1	9.5	54.1
A	14	29 18.1	03.2	349 19.4	13.8	19 00.6	9.4	54.1
Y	15	44 18.3	. . 04.2	3 52.2	13.7	19 10.0	9.3	54.1
	16	59 18.5	05.1	18 24.9	13.6	19 19.3	9.3	54.1
	17	74 18.6	06.1	32 57.5	13.6	19 28.6	9.2	54.1
	18	89 18.8	N 6 07.0	47 30.1	13.6	N19 37.8	9.1	54.1
	19	104 19.0	08.0	62 02.7	13.4	19 46.9	9.0	54.1
	20	119 19.2	08.9	76 35.1	13.4	19 55.9	8.9	54.1
	21	134 19.4	. . 09.9	91 07.5	13.3	20 04.8	8.9	54.1
	22	149 19.5	10.8	105 39.8	13.3	20 13.7	8.8	54.1
	23	164 19.7	11.8	120 12.1	13.2	20 22.5	8.6	54.2
6	00	179 19.9	N 6 12.7	134 44.3	13.2	N20 31.1	8.6	54.2
	01	194 20.1	13.7	149 16.5	13.0	20 39.7	8.6	54.2
	02	209 20.2	14.6	163 48.5	13.0	20 48.3	8.4	54.2
	03	224 20.4	. . 15.6	178 20.5	13.0	20 56.7	8.3	54.2
	04	239 20.6	16.5	192 52.5	12.9	21 05.0	8.3	54.2
	05	254 20.8	17.5	207 24.4	12.8	21 13.3	8.1	54.2
	06	269 21.0	N 6 18.4	221 56.2	12.7	N21 21.4	8.1	54.2
	07	284 21.1	19.4	236 27.9	12.7	21 29.5	8.0	54.2
	08	299 21.3	20.3	250 59.6	12.6	21 37.5	7.9	54.2
S	09	314 21.5	. . 21.3	265 31.2	12.6	21 45.4	7.8	54.3
U	10	329 21.7	22.2	280 02.8	12.5	21 53.2	7.7	54.3
N	11	344 21.9	23.1	294 34.3	12.4	22 00.9	7.6	54.3
D	12	359 22.0	N 6 24.1	309 05.7	12.4	N22 08.5	7.5	54.3
A	13	14 22.2	25.0	323 37.1	12.2	22 16.0	7.4	54.3
Y	14	29 22.4	26.0	338 08.3	12.3	22 23.4	7.3	54.3
	15	44 22.6	. . 26.9	352 39.6	12.1	22 30.7	7.2	54.3
	16	59 22.7	27.9	7 10.7	12.1	22 37.9	7.2	54.3
	17	74 22.9	28.8	21 41.8	12.0	22 45.1	7.0	54.4
	18	89 23.1	N 6 29.8	36 12.8	12.0	N22 52.1	6.9	54.4
	19	104 23.3	30.7	50 43.8	11.9	22 59.0	6.8	54.4
	20	119 23.5	31.7	65 14.7	11.8	23 05.8	6.8	54.4
	21	134 23.6	. . 32.6	79 45.5	11.8	23 12.6	6.6	54.4
	22	149 23.8	33.5	94 16.3	11.7	23 19.2	6.5	54.4
	23	164 24.0	34.5	108 47.0	11.6	N23 25.7	6.4	54.4
		SD 16.0	*d* 1.0	SD 14.7		14.7		14.8

Lat.	Twilight Naut.	Twilight Civil	Sunrise	Moonrise 4	Moonrise 5	Moonrise 6	Moonrise 7
°	h m	h m	h m	h m	h m	h m	h m
N 72	00 33	03 19	04 37	04 42	03 34	□	□
N 70	01 50	03 38	04 47	05 06	04 29	□	□
68	02 26	03 53	04 54	05 24	05 02	04 20	□
66	02 50	04 05	05 01	05 39	05 27	05 09	04 19
64	03 09	04 16	05 06	05 51	05 46	05 41	05 34
62	03 24	04 24	05 11	06 01	06 02	06 04	06 11
60	03 36	04 31	05 15	06 10	06 15	06 23	06 37
N 58	03 47	04 38	05 18	06 18	06 26	06 39	06 57
56	03 56	04 43	05 21	06 25	06 36	06 52	07 14
54	04 04	04 48	05 24	06 31	06 45	07 04	07 29
52	04 10	04 52	05 27	06 37	06 53	07 14	07 42
50	04 16	04 56	05 29	06 42	07 00	07 23	07 53
45	04 29	05 04	05 34	06 53	07 16	07 43	08 16
N 40	04 39	05 11	05 39	07 02	07 28	07 59	08 35
35	04 46	05 17	05 42	07 10	07 39	08 12	08 51
30	04 53	05 21	05 45	07 17	07 49	08 24	09 05
20	05 03	05 28	05 51	07 29	08 05	08 44	09 28
N 10	05 10	05 34	05 55	07 40	08 20	09 02	09 48
0	05 15	05 39	06 00	07 50	08 33	09 19	10 07
S 10	05 18	05 43	06 04	08 00	08 47	09 36	10 26
20	05 20	05 46	06 08	08 11	09 02	09 54	10 47
30	05 21	05 49	06 13	08 24	09 19	10 14	11 11
35	05 21	05 50	06 15	08 31	09 29	10 27	11 25
40	05 20	05 51	06 19	08 39	09 40	10 41	11 41
45	05 19	05 53	06 22	08 49	09 53	10 58	12 01
S 50	05 16	05 54	06 26	09 01	10 10	11 18	12 25
52	05 15	05 54	06 28	09 07	10 18	11 28	12 37
54	05 14	05 55	06 30	09 13	10 26	11 40	12 51
56	05 12	05 55	06 32	09 20	10 36	11 53	13 07
58	05 10	05 56	06 35	09 28	10 47	12 08	13 25
S 60	05 08	05 56	06 38	09 36	11 00	12 25	13 49

Lat.	Sunset	Twilight Civil	Twilight Naut.	Moonset 4	Moonset 5	Moonset 6	Moonset 7
°	h m	h m	h m	h m	h m	h m	h m
N 72	19 32	20 52	////	24 50	00 50	□	□
N 70	19 22	20 31	22 24	23 57	□	□	□
68	19 14	20 16	21 46	23 25	25 40	01 40	□
66	19 07	20 03	21 19	23 02	24 52	00 52	03 21
64	19 02	19 52	21 00	22 44	24 22	00 22	02 07
62	18 57	19 44	20 44	22 29	23 59	25 31	01 31
60	18 52	19 36	20 32	22 16	23 41	25 05	01 05
N 58	18 49	19 30	20 21	22 06	23 25	24 45	00 45
56	18 45	19 24	20 12	21 56	23 13	24 28	00 28
54	18 43	19 19	20 04	21 48	23 01	24 14	00 14
52	18 40	19 15	19 57	21 41	22 52	24 02	00 02
50	18 37	19 11	19 51	21 34	22 43	23 51	24 55
45	18 32	19 02	19 38	21 20	22 24	23 28	24 30
N 40	18 28	18 55	19 28	21 09	22 09	23 10	24 09
35	18 24	18 50	19 20	20 59	21 57	22 55	23 52
30	18 21	18 45	19 13	20 50	21 45	22 42	23 38
20	18 15	18 37	19 03	20 35	21 27	22 19	23 13
N 10	18 10	18 32	18 56	20 23	21 10	22 00	22 52
0	18 06	18 27	18 51	20 11	20 55	21 42	22 32
S 10	18 02	18 23	18 47	19 59	20 40	21 24	22 12
20	17 57	18 19	18 45	19 46	20 24	21 05	21 50
30	17 52	18 16	18 44	19 32	20 05	20 43	21 26
35	17 50	18 15	18 44	19 23	19 54	20 30	21 11
40	17 47	18 14	18 45	19 14	19 42	20 15	20 54
45	17 43	18 12	18 46	19 03	19 27	19 57	20 34
S 50	17 39	18 11	18 48	18 49	19 10	19 36	20 09
52	17 37	18 10	18 49	18 43	19 02	19 25	19 57
54	17 35	18 10	18 51	18 36	18 52	19 14	19 43
56	17 32	18 09	18 52	18 29	18 42	19 00	19 27
58	17 30	18 09	18 54	18 20	18 30	18 45	19 08
S 60	17 27	18 08	18 56	18 11	18 17	18 26	18 44

Day	SUN Eqn. of Time 00^h	SUN Eqn. of Time 12^h	SUN Mer. Pass.	MOON Mer. Pass. Upper	MOON Mer. Pass. Lower	Age	Phase
d	m s	m s	h m	h m	h m	d %	
4	03 15	03 07	12 03	14 00	01 39	03 6	
5	02 58	02 49	12 03	14 44	02 22	04 12	
6	02 41	02 32	12 03	15 30	03 07	05 19	

UT	ARIES	VENUS −4.0		MARS +0.4		JUPITER −2.3		SATURN +0.1	
	GHA	GHA	Dec	GHA	Dec	GHA	Dec	GHA	Dec
d h	° ′	° ′	° ′	° ′	° ′	° ′	° ′	° ′	° ′
7 00 (MONDAY)	194 51.3	210 45.2	S 7 57.3	262 11.3	S22 35.8	64 07.6	N19 05.3	111 30.2	N22 19.6
01	209 53.8	225 44.9	56.2	277 12.1	35.6	79 10.0	05.3	126 32.4	19.6
02	224 56.3	240 44.5	55.2	292 12.9	35.5	94 12.5	05.2	141 34.7	19.6
03	239 58.7	255 44.1	54.1	307 13.7	35.3	109 14.9	05.2	156 37.0	19.6
04	255 01.2	270 43.7	53.0	322 14.5	35.1	124 17.4	05.2	171 39.2	19.6
05	270 03.6	285 43.3	52.0	337 15.3	34.9	139 19.8	05.2	186 41.5	19.6
06	285 06.1	300 43.0	S 7 50.9	352 16.1	S22 34.8	154 22.2	N19 05.2	201 43.7	N22 19.7
07	300 08.6	315 42.6	49.8	7 16.9	34.6	169 24.7	05.2	216 46.0	19.7
08	315 11.0	330 42.2	48.8	22 17.7	34.4	184 27.1	05.2	231 48.3	19.7
09	330 13.5	345 41.8	47.7	37 18.5	34.2	199 29.5	05.2	246 50.5	19.7
10	345 16.0	0 41.5	46.6	52 19.3	34.1	214 32.0	05.2	261 52.8	19.7
11	0 18.4	15 41.1	45.6	67 20.1	33.9	229 34.4	05.2	276 55.0	19.7
12	15 20.9	30 40.7	S 7 44.5	82 20.9	S22 33.7	244 36.9	N19 05.2	291 57.3	N22 19.7
13	30 23.4	45 40.3	43.4	97 21.7	33.5	259 39.3	05.1	306 59.6	19.8
14	45 25.8	60 40.0	42.4	112 22.5	33.4	274 41.7	05.1	322 01.8	19.8
15	60 28.3	75 39.6	41.3	127 23.3	33.2	289 44.2	05.1	337 04.1	19.8
16	75 30.8	90 39.2	40.2	142 24.1	33.0	304 46.6	05.1	352 06.3	19.8
17	90 33.2	105 38.8	39.2	157 24.9	32.8	319 49.0	05.1	7 08.6	19.8
18	105 35.7	120 38.4	S 7 38.1	172 25.7	S22 32.6	334 51.5	N19 05.1	22 10.9	N22 19.8
19	120 38.1	135 38.1	37.0	187 26.5	32.5	349 53.9	05.1	37 13.1	19.9
20	135 40.6	150 37.7	36.0	202 27.3	32.3	4 56.4	05.1	52 15.4	19.9
21	150 43.1	165 37.3	34.9	217 28.1	32.1	19 58.8	05.1	67 17.6	19.9
22	165 45.5	180 37.0	33.8	232 28.9	31.9	35 01.2	05.1	82 19.9	19.9
23	180 48.0	195 36.6	32.8	247 29.7	31.8	50 03.7	05.0	97 22.2	19.9
8 00 (TUESDAY)	195 50.5	210 36.2	S 7 31.7	262 30.5	S22 31.6	65 06.1	N19 05.0	112 24.4	N22 19.9
01	210 52.9	225 35.8	30.6	277 31.3	31.4	80 08.5	05.0	127 26.7	20.0
02	225 55.4	240 35.5	29.5	292 32.1	31.2	95 11.0	05.0	142 28.9	20.0
03	240 57.9	255 35.1	28.5	307 32.9	31.0	110 13.4	05.0	157 31.2	20.0
04	256 00.3	270 34.7	27.4	322 33.7	30.9	125 15.8	05.0	172 33.4	20.0
05	271 02.8	285 34.3	26.3	337 34.5	30.7	140 18.2	05.0	187 35.7	20.0
06	286 05.3	300 34.0	S 7 25.3	352 35.3	S22 30.5	155 20.7	N19 05.0	202 38.0	N22 20.0
07	301 07.7	315 33.6	24.2	7 36.1	30.3	170 23.1	05.0	217 40.2	20.1
08	316 10.2	330 33.2	23.1	22 36.9	30.1	185 25.5	04.9	232 42.5	20.1
09	331 12.6	345 32.9	22.0	37 37.7	29.9	200 28.0	04.9	247 44.7	20.1
10	346 15.1	0 32.5	21.0	52 38.5	29.8	215 30.4	04.9	262 47.0	20.1
11	1 17.6	15 32.1	19.9	67 39.4	29.6	230 32.8	04.9	277 49.2	20.1
12	16 20.0	30 31.7	S 7 18.8	82 40.2	S22 29.4	245 35.3	N19 04.9	292 51.5	N22 20.1
13	31 22.5	45 31.4	17.7	97 41.0	29.2	260 37.7	04.9	307 53.8	20.2
14	46 25.0	60 31.0	16.7	112 41.8	29.0	275 40.1	04.9	322 56.0	20.2
15	61 27.4	75 30.6	15.6	127 42.6	28.9	290 42.6	04.9	337 58.3	20.2
16	76 29.9	90 30.3	14.5	142 43.4	28.7	305 45.0	04.9	353 00.5	20.2
17	91 32.4	105 29.9	13.4	157 44.2	28.5	320 47.4	04.8	8 02.8	20.2
18	106 34.8	120 29.5	S 7 12.3	172 45.0	S22 28.3	335 49.8	N19 04.8	23 05.0	N22 20.2
19	121 37.3	135 29.2	11.3	187 45.8	28.1	350 52.3	04.8	38 07.3	20.3
20	136 39.7	150 28.8	10.2	202 46.6	27.9	5 54.7	04.8	53 09.5	20.3
21	151 42.2	165 28.4	09.1	217 47.4	27.8	20 57.1	04.8	68 11.8	20.3
22	166 44.7	180 28.0	08.0	232 48.2	27.6	35 59.5	04.8	83 14.0	20.3
23	181 47.1	195 27.7	07.0	247 49.0	27.4	51 02.0	04.8	98 16.3	20.3
9 00 (WEDNESDAY)	196 49.6	210 27.3	S 7 05.9	262 49.8	S22 27.2	66 04.4	N19 04.8	113 18.6	N22 20.3
01	211 52.1	225 26.9	04.8	277 50.6	27.0	81 06.8	04.7	128 20.8	20.3
02	226 54.5	240 26.6	03.7	292 51.5	26.8	96 09.2	04.7	143 23.1	20.4
03	241 57.0	255 26.2	02.6	307 52.3	26.6	111 11.7	04.7	158 25.3	20.4
04	256 59.5	270 25.8	01.5	322 53.1	26.5	126 14.1	04.7	173 27.6	20.4
05	272 01.9	285 25.5	7 00.5	337 53.9	26.3	141 16.5	04.7	188 29.8	20.4
06	287 04.4	300 25.1	S 6 59.4	352 54.7	S22 26.1	156 18.9	N19 04.7	203 32.1	N22 20.4
07	302 06.9	315 24.7	58.3	7 55.5	25.9	171 21.4	04.7	218 34.3	20.4
08	317 09.3	330 24.4	57.2	22 56.3	25.7	186 23.8	04.7	233 36.6	20.5
09	332 11.8	345 24.0	56.1	37 57.1	25.5	201 26.2	04.6	248 38.8	20.5
10	347 14.2	0 23.6	55.1	52 57.9	25.3	216 28.6	04.6	263 41.1	20.5
11	2 16.7	15 23.3	54.0	67 58.7	25.2	231 31.1	04.6	278 43.3	20.5
12	17 19.2	30 22.9	S 6 52.9	82 59.5	S22 25.0	246 33.5	N19 04.6	293 45.6	N22 20.5
13	32 21.6	45 22.5	51.8	98 00.3	24.8	261 35.9	04.6	308 47.9	20.5
14	47 24.1	60 22.2	50.7	113 01.2	24.6	276 38.3	04.6	323 50.1	20.6
15	62 26.6	75 21.8	49.6	128 02.0	24.4	291 40.7	04.6	338 52.4	20.6
16	77 29.0	90 21.5	48.5	143 02.8	24.2	306 43.2	04.5	353 54.6	20.6
17	92 31.5	105 21.1	47.5	158 03.6	24.0	321 45.6	04.5	8 56.9	20.6
18	107 34.0	120 20.7	S 6 46.4	173 04.4	S22 23.8	336 48.0	N19 04.5	23 59.1	N22 20.6
19	122 36.4	135 20.4	45.3	188 05.2	23.7	351 50.4	04.5	39 01.4	20.6
20	137 38.9	150 20.0	44.2	203 06.0	23.5	6 52.8	04.5	54 03.6	20.7
21	152 41.4	165 19.6	43.1	218 06.8	23.3	21 55.3	04.5	69 05.9	20.7
22	167 43.8	180 19.3	42.0	233 07.6	23.1	36 57.7	04.5	84 08.1	20.7
23	182 46.3	195 18.9	40.9	248 08.5	22.9	52 00.1	04.4	99 10.4	20.7
Mer. Pass. (h m)	10 54.8	*v* −0.4	*d* 1.1	*v* 0.8	*d* 0.2	*v* 2.4	*d* 0.0	*v* 2.3	*d* 0.0

STARS

Name	SHA	Dec
	° ′	° ′
Acamar	315 24.7	S40 17.7
Achernar	335 33.1	S57 13.4
Acrux	173 18.0	S63 07.1
Adhara	255 18.9	S28 58.8
Aldebaran	290 58.9	N16 30.9
Alioth	166 26.9	N55 56.6
Alkaid	153 04.6	N49 17.8
Al Na'ir	27 54.0	S46 56.8
Alnilam	275 54.7	S 1 12.1
Alphard	218 03.8	S 8 40.4
Alphecca	126 17.5	N26 42.0
Alpheratz	357 52.3	N29 06.2
Altair	62 16.1	N 8 52.3
Ankaa	353 23.9	S42 17.4
Antares	112 35.9	S26 26.4
Arcturus	146 02.7	N19 09.8
Atria	107 44.8	S69 01.9
Avior	234 21.3	S59 31.4
Bellatrix	278 40.8	N 6 21.1
Betelgeuse	271 10.1	N 7 24.4
Canopus	263 59.9	S52 42.1
Capella	280 46.6	N46 00.2
Deneb	49 37.2	N45 17.1
Denebola	182 41.5	N14 33.2
Diphda	349 04.2	S17 58.3
Dubhe	194 00.7	N61 44.2
Elnath	278 23.0	N28 36.7
Eltanin	90 49.7	N51 28.9
Enif	33 55.2	N 9 53.1
Fomalhaut	15 33.0	S29 36.4
Gacrux	172 09.6	S57 07.9
Gienah	176 00.3	S17 33.7
Hadar	148 59.0	S60 23.3
Hamal	328 10.3	N23 28.5
Kaus Aust.	83 54.4	S34 23.0
Kochab	137 18.2	N74 08.4
Markab	13 46.7	N15 13.1
Menkar	314 23.8	N 4 06.0
Menkent	148 16.8	S36 23.2
Miaplacidus	221 41.3	S69 44.0
Mirfak	308 52.4	N49 52.4
Nunki	76 08.2	S26 17.6
Peacock	53 31.9	S56 43.4
Pollux	243 37.5	N28 01.2
Procyon	245 08.1	N 5 13.0
Rasalhague	96 13.8	N12 33.2
Regulus	207 51.8	N11 57.1
Rigel	281 20.0	S 8 12.0
Rigil Kent.	140 02.5	S60 50.8
Sabik	102 21.6	S15 43.8
Schedar	349 50.5	N56 33.1
Shaula	96 32.7	S37 06.4
Sirius	258 40.9	S16 43.3
Spica	158 39.5	S11 10.8
Suhail	222 58.3	S43 26.9
Vega	80 44.4	N38 46.8
Zuben'ubi	137 14.1	S16 03.4

	SHA	Mer. Pass.
	° ′	h m
Venus	14 45.7	9 58
Mars	66 40.0	6 30
Jupiter	229 15.6	19 36
Saturn	276 33.9	16 28

Day	UT (d h)	SUN GHA (° ′)	SUN Dec (° ′)	MOON GHA (° ′)	MOON v (′)	MOON Dec (° ′)	MOON d (′)	MOON HP (′)
MONDAY	7 00	179 24.2	N 6 35.4	123 17.6	11.5	N23 32.1	6.3	54.4
	01	194 24.3	36.4	137 48.1	11.5	23 38.4	6.2	54.5
	02	209 24.5	37.3	152 18.6	11.5	23 44.6	6.1	54.5
	03	224 24.7	. . 38.3	166 49.1	11.3	23 50.7	6.0	54.5
	04	239 24.9	39.2	181 19.4	11.3	23 56.7	5.9	54.5
	05	254 25.0	40.1	195 49.7	11.3	24 02.6	5.7	54.5
	06	269 25.2	N 6 41.1	210 20.0	11.1	N24 08.3	5.7	54.5
	07	284 25.4	42.0	224 50.1	11.1	24 14.0	5.5	54.6
	08	299 25.6	43.0	239 20.2	11.1	24 19.5	5.5	54.6
	09	314 25.7	. . 43.9	253 50.3	10.9	24 25.0	5.3	54.6
	10	329 25.9	44.9	268 20.2	11.0	24 30.3	5.2	54.6
	11	344 26.1	45.8	282 50.2	10.8	24 35.5	5.1	54.6
	12	359 26.3	N 6 46.7	297 20.0	10.8	N24 40.6	5.0	54.6
	13	14 26.4	47.7	311 49.8	10.7	24 45.6	4.9	54.7
	14	29 26.6	48.6	326 19.5	10.7	24 50.5	4.7	54.7
	15	44 26.8	. . 49.6	340 49.2	10.6	24 55.2	4.6	54.7
	16	59 27.0	50.5	355 18.8	10.5	24 59.8	4.6	54.7
	17	74 27.2	51.4	9 48.3	10.5	25 04.4	4.4	54.7
	18	89 27.3	N 6 52.4	24 17.8	10.4	N25 08.8	4.2	54.8
	19	104 27.5	53.3	38 47.2	10.4	25 13.0	4.2	54.8
	20	119 27.7	54.3	53 16.6	10.3	25 17.2	4.0	54.8
	21	134 27.9	. . 55.2	67 45.9	10.2	25 21.2	3.9	54.8
	22	149 28.0	56.1	82 15.1	10.2	25 25.1	3.8	54.8
	23	164 28.2	57.1	96 44.3	10.1	25 28.9	3.7	54.9
TUESDAY	8 00	179 28.4	N 6 58.0	111 13.4	10.1	N25 32.6	3.6	54.9
	01	194 28.6	58.9	125 42.5	10.0	25 36.2	3.4	54.9
	02	209 28.7	6 59.9	140 11.5	10.0	25 39.6	3.3	54.9
	03	224 28.9	7 00.8	154 40.5	9.9	25 42.9	3.2	54.9
	04	239 29.1	01.8	169 09.4	9.8	25 46.1	3.0	55.0
	05	254 29.2	02.7	183 38.2	9.8	25 49.1	2.9	55.0
	06	269 29.4	N 7 03.6	198 07.0	9.8	N25 52.0	2.8	55.0
	07	284 29.6	04.6	212 35.8	9.6	25 54.8	2.7	55.0
	08	299 29.8	05.5	227 04.4	9.7	25 57.5	2.5	55.0
	09	314 29.9	. . 06.4	241 33.1	9.6	26 00.0	2.4	55.1
	10	329 30.1	07.4	256 01.7	9.5	26 02.4	2.3	55.1
	11	344 30.3	08.3	270 30.2	9.5	26 04.7	2.2	55.1
	12	359 30.5	N 7 09.3	284 58.7	9.4	N26 06.9	2.0	55.1
	13	14 30.6	10.2	299 27.1	9.4	26 08.9	1.9	55.2
	14	29 30.8	11.1	313 55.5	9.4	26 10.8	1.7	55.2
	15	44 31.0	. . 12.1	328 23.9	9.3	26 12.5	1.6	55.2
	16	59 31.2	13.0	342 52.2	9.2	26 14.1	1.5	55.2
	17	74 31.3	13.9	357 20.4	9.3	26 15.6	1.4	55.3
	18	89 31.5	N 7 14.9	11 48.7	9.1	N26 17.0	1.2	55.3
	19	104 31.7	15.8	26 16.8	9.2	26 18.2	1.1	55.3
	20	119 31.9	16.7	40 45.0	9.1	26 19.3	0.9	55.3
	21	134 32.0	. . 17.7	55 13.1	9.0	26 20.2	0.9	55.4
	22	149 32.2	18.6	69 41.1	9.0	26 21.1	0.6	55.4
	23	164 32.4	19.5	84 09.1	9.0	26 21.7	0.6	55.4
WEDNESDAY	9 00	179 32.5	N 7 20.5	98 37.1	8.9	N26 22.3	0.4	55.4
	01	194 32.7	21.4	113 05.0	8.9	26 22.7	0.2	55.5
	02	209 32.9	22.3	127 32.9	8.9	26 22.9	0.2	55.5
	03	224 33.1	. . 23.3	142 00.8	8.8	26 23.1	0.1	55.5
	04	239 33.2	24.2	156 28.6	8.8	26 23.0	0.1	55.6
	05	254 33.4	25.1	170 56.4	8.8	26 22.9	0.3	55.6
	06	269 33.6	N 7 26.1	185 24.2	8.7	N26 22.6	0.4	55.6
	07	284 33.7	27.0	199 51.9	8.7	26 22.2	0.6	55.6
	08	299 33.9	27.9	214 19.6	8.7	26 21.6	0.7	55.7
	09	314 34.1	. . 28.9	228 47.3	8.6	26 20.9	0.9	55.7
	10	329 34.3	29.8	243 14.9	8.6	26 20.0	1.0	55.7
	11	344 34.4	30.7	257 42.5	8.6	26 19.0	1.1	55.8
	12	359 34.6	N 7 31.7	272 10.1	8.6	N26 17.9	1.3	55.8
	13	14 34.8	32.6	286 37.7	8.5	26 16.6	1.4	55.8
	14	29 34.9	33.5	301 05.2	8.6	26 15.2	1.6	55.9
	15	44 35.1	. . 34.4	315 32.8	8.5	26 13.6	1.7	55.9
	16	59 35.3	35.4	330 00.3	8.4	26 11.9	1.8	55.9
	17	74 35.5	36.3	344 27.7	8.5	26 10.1	2.0	55.9
	18	89 35.6	N 7 37.2	358 55.2	8.4	N26 08.1	2.1	56.0
	19	104 35.8	38.2	13 22.6	8.5	26 06.0	2.3	56.0
	20	119 36.0	39.1	27 50.1	8.4	26 03.7	2.4	56.0
	21	134 36.1	. . 40.0	42 17.5	8.4	26 01.3	2.6	56.1
	22	149 36.3	40.9	56 44.9	8.3	25 58.7	2.7	56.1
	23	164 36.5	41.9	71 12.2	8.4	N25 56.0	2.9	56.1
		SD 16.0	*d* 0.9	SD 14.9		15.0		15.2

Lat.	Twilight Naut. (h m)	Twilight Civil (h m)	Sunrise (h m)	Moonrise 7 (h m)	Moonrise 8 (h m)	Moonrise 9 (h m)	Moonrise 10 (h m)
N 72	////	02 58	04 21	□	□	□	□
N 70	01 16	03 21	04 32	□	□	□	□
68	02 03	03 39	04 41	□	□	□	□
66	02 33	03 52	04 49	04 19	□	□	□
64	02 55	04 04	04 55	05 34	05 21	□	07 29
62	03 11	04 13	05 01	06 11	06 27	07 08	08 25
60	03 25	04 22	05 06	06 37	07 02	07 48	08 58
N 58	03 37	04 29	05 10	06 57	07 28	08 15	09 22
56	03 47	04 35	05 14	07 14	07 48	08 36	09 42
54	03 55	04 40	05 17	07 29	08 05	08 54	09 58
52	04 03	04 45	05 20	07 42	08 19	09 09	10 12
50	04 09	04 50	05 23	07 53	08 32	09 22	10 24
45	04 23	04 59	05 29	08 16	08 58	09 49	10 49
N 40	04 33	05 06	05 34	08 35	09 18	10 10	11 09
35	04 42	05 12	05 38	08 51	09 36	10 28	11 26
30	04 49	05 18	05 42	09 05	09 51	10 43	11 41
20	05 00	05 26	05 48	09 28	10 16	11 09	12 05
N 10	05 08	05 32	05 54	09 48	10 38	11 31	12 26
0	05 14	05 38	05 59	10 07	10 58	11 52	12 46
S 10	05 18	05 43	06 04	10 26	11 19	12 12	13 05
20	05 21	05 47	06 09	10 47	11 41	12 35	13 26
30	05 23	05 51	06 15	11 11	12 07	13 00	13 51
35	05 23	05 52	06 18	11 25	12 22	13 16	14 05
40	05 23	05 54	06 22	11 41	12 39	13 33	14 21
45	05 22	05 56	06 26	12 01	13 01	13 54	14 41
S 50	05 21	05 58	06 31	12 25	13 27	14 21	15 05
52	05 20	05 59	06 33	12 37	13 40	14 34	15 17
54	05 19	06 00	06 36	12 51	13 56	14 49	15 31
56	05 18	06 01	06 38	13 07	14 13	15 07	15 46
58	05 17	06 02	06 41	13 25	14 35	15 28	16 05
S 60	05 15	06 03	06 45	13 49	15 02	15 56	16 27

Lat.	Sunset (h m)	Twilight Civil (h m)	Twilight Naut. (h m)	Moonset 7 (h m)	Moonset 8 (h m)	Moonset 9 (h m)	Moonset 10 (h m)
N 72	19 46	21 11	////	□	□	□	□
N 70	19 35	20 47	23 02	□	□	□	□
68	19 25	20 29	22 07	□	□	□	□
66	19 17	20 14	21 36	03 21	□	□	□
64	19 10	20 02	21 13	02 07	04 05	□	05 39
62	19 05	19 53	20 56	01 31	02 59	04 07	04 43
60	19 00	19 44	20 41	01 05	02 24	03 28	04 10
N 58	18 55	19 37	20 29	00 45	01 59	03 01	03 45
56	18 52	19 31	20 19	00 28	01 39	02 39	03 26
54	18 48	19 25	20 11	00 14	01 22	02 22	03 09
52	18 45	19 20	20 03	00 02	01 08	02 07	02 55
50	18 42	19 16	19 56	24 55	00 55	01 54	02 43
45	18 36	19 06	19 42	24 30	00 30	01 27	02 17
N 40	18 31	18 59	19 31	24 09	00 09	01 06	01 57
35	18 26	18 52	19 23	23 52	24 48	00 48	01 39
30	18 23	18 47	19 15	23 38	24 33	00 33	01 25
20	18 16	18 38	19 04	23 13	24 07	00 07	01 00
N 10	18 10	18 32	18 56	22 52	23 45	24 38	00 38
0	18 05	18 26	18 50	22 32	23 24	24 17	00 17
S 10	18 00	18 21	18 46	22 12	23 03	23 57	24 53
20	17 55	18 17	18 43	21 50	22 41	23 35	24 33
30	17 49	18 13	18 41	21 26	22 15	23 10	24 10
35	17 46	18 11	18 40	21 11	21 59	22 55	23 56
40	17 42	18 09	18 40	20 54	21 42	22 37	23 40
45	17 37	18 07	18 41	20 34	21 20	22 16	23 21
S 50	17 32	18 05	18 42	20 09	20 53	21 50	22 57
52	17 30	18 04	18 43	19 57	20 40	21 37	22 46
54	17 27	18 03	18 44	19 43	20 25	21 22	22 33
56	17 25	18 02	18 45	19 27	20 07	21 04	22 17
58	17 21	18 01	18 46	19 08	19 46	20 43	21 59
S 60	17 18	18 00	18 48	18 44	19 18	20 16	21 37

Day	SUN Eqn. of Time 00^h (m s)	SUN Eqn. of Time 12^h (m s)	SUN Mer. Pass. (h m)	MOON Mer. Pass. Upper (h m)	MOON Mer. Pass. Lower (h m)	MOON Age (d)	MOON %	Phase
7	02 24	02 15	12 02	16 19	03 55	06	27	
8	02 07	01 58	12 02	17 11	04 45	07	35	
9	01 50	01 42	12 02	18 04	05 38	08	45	

UT d	h	ARIES GHA	VENUS −4.0 GHA	Dec	MARS +0.3 GHA	Dec	JUPITER −2.2 GHA	Dec	SATURN +0.1 GHA	Dec
		° ′	° ′	° ′	° ′	° ′	° ′	° ′	° ′	° ′
10	00	197 48.7	210 18.5	S 6 39.8	263 09.3	S22 22.7	67 02.5	N19 04.4	114 12.6	N22 20.7
	01	212 51.2	225 18.2	38.8	278 10.1	22.5	82 04.9	04.4	129 14.9	20.7
	02	227 53.7	240 17.8	37.7	293 10.9	22.3	97 07.3	04.4	144 17.1	20.8
	03	242 56.1	255 17.5	. . 36.6	308 11.7	. . 22.1	112 09.8	. . 04.4	159 19.4	. . 20.8
	04	257 58.6	270 17.1	35.5	323 12.5	21.9	127 12.2	04.4	174 21.6	20.8
	05	273 01.1	285 16.7	34.4	338 13.3	21.8	142 14.6	04.4	189 23.9	20.8
	06	288 03.5	300 16.4	S 6 33.3	353 14.1	S22 21.6	157 17.0	N19 04.3	204 26.1	N22 20.8
	07	303 06.0	315 16.0	32.2	8 15.0	21.4	172 19.4	04.3	219 28.4	20.8
T	08	318 08.5	330 15.6	31.1	23 15.8	21.2	187 21.8	04.3	234 30.6	20.8
H	09	333 10.9	345 15.3	. . 30.0	38 16.6	. . 21.0	202 24.2	. . 04.3	249 32.9	. . 20.9
U	10	348 13.4	0 14.9	28.9	53 17.4	20.8	217 26.7	04.3	264 35.1	20.9
R	11	3 15.9	15 14.6	27.9	68 18.2	20.6	232 29.1	04.3	279 37.4	20.9
S	12	18 18.3	30 14.2	S 6 26.8	83 19.0	S22 20.4	247 31.5	N19 04.2	294 39.6	N22 20.9
D	13	33 20.8	45 13.8	25.7	98 19.8	20.2	262 33.9	04.2	309 41.9	20.9
A	14	48 23.2	60 13.5	24.6	113 20.6	20.0	277 36.3	04.2	324 44.1	20.9
Y	15	63 25.7	75 13.1	. . 23.5	128 21.5	. . 19.8	292 38.7	. . 04.2	339 46.4	. . 21.0
	16	78 28.2	90 12.8	22.4	143 22.3	19.6	307 41.1	04.2	354 48.6	21.0
	17	93 30.6	105 12.4	21.3	158 23.1	19.4	322 43.6	04.2	9 50.9	21.0
	18	108 33.1	120 12.0	S 6 20.2	173 23.9	S22 19.2	337 46.0	N19 04.2	24 53.1	N22 21.0
	19	123 35.6	135 11.7	19.1	188 24.7	19.1	352 48.4	04.1	39 55.4	21.0
	20	138 38.0	150 11.3	18.0	203 25.5	18.9	7 50.8	04.1	54 57.6	21.0
	21	153 40.5	165 11.0	. . 16.9	218 26.4	. . 18.7	22 53.2	. . 04.1	69 59.8	. . 21.1
	22	168 43.0	180 10.6	15.8	233 27.2	18.5	37 55.6	04.1	85 02.1	21.1
	23	183 45.4	195 10.3	14.7	248 28.0	18.3	52 58.0	04.1	100 04.3	21.1
11	00	198 47.9	210 09.9	S 6 13.6	263 28.8	S22 18.1	68 00.4	N19 04.1	115 06.6	N22 21.1
	01	213 50.3	225 09.5	12.5	278 29.6	17.9	83 02.8	04.0	130 08.8	21.1
	02	228 52.8	240 09.2	11.4	293 30.4	17.7	98 05.2	04.0	145 11.1	21.1
	03	243 55.3	255 08.8	. . 10.3	308 31.2	. . 17.5	113 07.7	. . 04.0	160 13.3	. . 21.2
	04	258 57.7	270 08.5	09.2	323 32.1	17.3	128 10.1	04.0	175 15.6	21.2
	05	274 00.2	285 08.1	08.1	338 32.9	17.1	143 12.5	04.0	190 17.8	21.2
	06	289 02.7	300 07.7	S 6 07.0	353 33.7	S22 16.9	158 14.9	N19 04.0	205 20.1	N22 21.2
	07	304 05.1	315 07.4	05.9	8 34.5	16.7	173 17.3	03.9	220 22.3	21.2
	08	319 07.6	330 07.0	04.8	23 35.3	16.5	188 19.7	03.9	235 24.6	21.2
F	09	334 10.1	345 06.7	. . 03.7	38 36.2	. . 16.3	203 22.1	. . 03.9	250 26.8	. . 21.3
R	10	349 12.5	0 06.3	02.6	53 37.0	16.1	218 24.5	03.9	265 29.0	21.3
I	11	4 15.0	15 06.0	01.5	68 37.8	15.9	233 26.9	03.9	280 31.3	21.3
D	12	19 17.5	30 05.6	S 6 00.4	83 38.6	S22 15.7	248 29.3	N19 03.8	295 33.5	N22 21.3
A	13	34 19.9	45 05.3	5 59.3	98 39.4	15.5	263 31.7	03.8	310 35.8	21.3
Y	14	49 22.4	60 04.9	58.2	113 40.2	15.3	278 34.1	03.8	325 38.0	21.3
	15	64 24.8	75 04.5	. . 57.1	128 41.1	. . 15.1	293 36.5	. . 03.8	340 40.3	. . 21.4
	16	79 27.3	90 04.2	56.0	143 41.9	14.9	308 38.9	03.8	355 42.5	21.4
	17	94 29.8	105 03.8	54.9	158 42.7	14.7	323 41.3	03.8	10 44.8	21.4
	18	109 32.2	120 03.5	S 5 53.8	173 43.5	S22 14.5	338 43.7	N19 03.7	25 47.0	N22 21.4
	19	124 34.7	135 03.1	52.7	188 44.3	14.3	353 46.1	03.7	40 49.3	21.4
	20	139 37.2	150 02.8	51.6	203 45.2	14.1	8 48.6	03.7	55 51.5	21.4
	21	154 39.6	165 02.4	. . 50.5	218 46.0	. . 13.9	23 51.0	. . 03.7	70 53.7	. . 21.4
	22	169 42.1	180 02.1	49.4	233 46.8	13.7	38 53.4	03.7	85 56.0	21.5
	23	184 44.6	195 01.7	48.3	248 47.6	13.5	53 55.8	03.6	100 58.2	21.5
12	00	199 47.0	210 01.4	S 5 47.2	263 48.4	S22 13.3	68 58.2	N19 03.6	116 00.5	N22 21.5
	01	214 49.5	225 01.0	46.1	278 49.3	13.1	84 00.6	03.6	131 02.7	21.5
	02	229 52.0	240 00.7	45.0	293 50.1	12.9	99 03.0	03.6	146 05.0	21.5
	03	244 54.4	255 00.3	. . 43.9	308 50.9	. . 12.7	114 05.4	. . 03.6	161 07.2	. . 21.5
	04	259 56.9	269 59.9	42.8	323 51.7	12.5	129 07.8	03.6	176 09.4	21.6
	05	274 59.3	284 59.6	41.7	338 52.5	12.3	144 10.2	03.5	191 11.7	21.6
	06	290 01.8	299 59.2	S 5 40.6	353 53.4	S22 12.1	159 12.6	N19 03.5	206 13.9	N22 21.6
	07	305 04.3	314 58.9	39.5	8 54.2	11.9	174 15.0	03.5	221 16.2	21.6
S	08	320 06.7	329 58.5	38.4	23 55.0	11.7	189 17.4	03.5	236 18.4	21.6
A	09	335 09.2	344 58.2	. . 37.3	38 55.8	. . 11.5	204 19.8	. . 03.5	251 20.7	. . 21.6
T	10	350 11.7	359 57.8	36.2	53 56.6	11.3	219 22.2	03.4	266 22.9	21.7
U	11	5 14.1	14 57.5	35.0	68 57.5	11.1	234 24.6	03.4	281 25.1	21.7
R	12	20 16.6	29 57.1	S 5 33.9	83 58.3	S22 10.9	249 27.0	N19 03.4	296 27.4	N22 21.7
D	13	35 19.1	44 56.8	32.8	98 59.1	10.7	264 29.4	03.4	311 29.6	21.7
A	14	50 21.5	59 56.4	31.7	113 59.9	10.5	279 31.7	03.4	326 31.9	21.7
Y	15	65 24.0	74 56.1	. . 30.6	129 00.8	. . 10.3	294 34.1	. . 03.3	341 34.1	. . 21.7
	16	80 26.4	89 55.7	29.5	144 01.6	10.1	309 36.5	03.3	356 36.3	21.8
	17	95 28.9	104 55.4	28.4	159 02.4	09.9	324 38.9	03.3	11 38.6	21.8
	18	110 31.4	119 55.0	S 5 27.3	174 03.2	S22 09.7	339 41.3	N19 03.3	26 40.8	N22 21.8
	19	125 33.8	134 54.7	26.2	189 04.1	09.5	354 43.7	03.3	41 43.1	21.8
	20	140 36.3	149 54.3	25.1	204 04.9	09.3	9 46.1	03.2	56 45.3	21.8
	21	155 38.8	164 54.0	. . 24.0	219 05.7	. . 09.1	24 48.5	. . 03.2	71 47.5	. . 21.8
	22	170 41.2	179 53.6	22.8	234 06.5	08.9	39 50.9	03.2	86 49.8	21.9
	23	185 43.7	194 53.3	21.7	249 07.4	08.7	54 53.3	03.2	101 52.0	21.9
Mer. Pass.		h m 10 43.0	*v* −0.4	*d* 1.1	*v* 0.8	*d* 0.2	*v* 2.4	*d* 0.0	*v* 2.2	*d* 0.0

STARS

Name	SHA	Dec
	° ′	° ′
Acamar	315 24.7	S40 17.7
Achernar	335 33.1	S57 13.4
Acrux	173 18.0	S63 07.1
Adhara	255 18.9	S28 58.8
Aldebaran	290 58.9	N16 30.9
Alioth	166 26.9	N55 56.6
Alkaid	153 04.6	N49 17.8
Al Na'ir	27 53.9	S46 56.7
Alnilam	275 54.7	S 1 12.1
Alphard	218 03.9	S 8 40.4
Alphecca	126 17.5	N26 42.0
Alpheratz	357 52.3	N29 06.2
Altair	62 16.1	N 8 52.3
Ankaa	353 23.9	S42 17.4
Antares	112 35.9	S26 26.4
Arcturus	146 02.7	N19 09.8
Atria	107 44.8	S69 01.9
Avior	234 21.3	S59 31.4
Bellatrix	278 40.8	N 6 21.1
Betelgeuse	271 10.1	N 7 24.4
Canopus	263 59.9	S52 42.1
Capella	280 46.6	N46 00.2
Deneb	49 37.2	N45 17.1
Denebola	182 41.5	N14 33.2
Diphda	349 04.2	S17 58.3
Dubhe	194 00.7	N61 44.2
Elnath	278 23.0	N28 36.7
Eltanin	90 49.7	N51 28.9
Enif	33 55.2	N 9 53.1
Fomalhaut	15 33.0	S29 36.4
Gacrux	172 09.6	S57 07.9
Gienah	176 00.3	S17 33.7
Hadar	148 59.0	S60 23.3
Hamal	328 10.3	N23 28.5
Kaus Aust.	83 54.4	S34 23.0
Kochab	137 18.2	N74 08.4
Markab	13 46.7	N15 13.1
Menkar	314 23.8	N 4 06.0
Menkent	148 16.8	S36 23.2
Miaplacidus	221 41.3	S69 44.0
Mirfak	308 52.5	N49 52.4
Nunki	76 08.2	S26 17.6
Peacock	53 31.9	S56 43.4
Pollux	243 37.5	N28 01.2
Procyon	245 08.1	N 5 13.0
Rasalhague	96 13.8	N12 33.2
Regulus	207 51.8	N11 57.1
Rigel	281 20.0	S 8 12.0
Rigil Kent.	140 02.4	S60 50.9
Sabik	102 21.6	S15 43.8
Schedar	349 50.5	N56 33.1
Shaula	96 32.7	S37 06.4
Sirius	258 40.9	S16 43.3
Spica	158 39.4	S11 10.8
Suhail	222 58.3	S43 26.9
Vega	80 44.3	N38 46.8
Zuben'ubi	137 14.1	S16 03.4

	SHA	Mer. Pass.
	° ′	h m
Venus	11 22.0	10 00
Mars	64 40.9	6 26
Jupiter	229 12.5	19 25
Saturn	276 18.7	16 17

UT		SUN GHA	SUN Dec	MOON GHA	v	MOON Dec	d	HP
d	h	° ′	° ′	° ′	′	° ′	′	′
10	00	179 36.7	N 7 42.8	85 39.6	8.4	N25 53.1	2.9	56.2
	01	194 36.8	43.7	100 07.0	8.3	25 50.2	3.2	56.2
	02	209 37.0	44.7	114 34.3	8.3	25 47.0	3.3	56.2
	03	224 37.2	. . 45.6	129 01.6	8.3	25 43.7	3.4	56.3
	04	239 37.3	46.5	143 28.9	8.4	25 40.3	3.6	56.3
	05	254 37.5	47.4	157 56.3	8.3	25 36.7	3.7	56.3
	06	269 37.7	N 7 48.4	172 23.6	8.3	N25 33.0	3.9	56.4
	07	284 37.8	49.3	186 50.9	8.3	25 29.1	4.0	56.4
T	08	299 38.0	50.2	201 18.2	8.3	25 25.1	4.1	56.4
H	09	314 38.2	. . 51.1	215 45.5	8.3	25 21.0	4.3	56.5
U	10	329 38.3	52.1	230 12.8	8.2	25 16.7	4.5	56.5
R	11	344 38.5	53.0	244 40.0	8.3	25 12.2	4.6	56.5
S	12	359 38.7	N 7 53.9	259 07.3	8.3	N25 07.6	4.7	56.6
D	13	14 38.9	54.8	273 34.6	8.3	25 02.9	4.9	56.6
A	14	29 39.0	55.8	288 01.9	8.3	24 58.0	5.0	56.6
Y	15	44 39.2	. . 56.7	302 29.2	8.3	24 53.0	5.1	56.7
	16	59 39.4	57.6	316 56.5	8.3	24 47.9	5.3	56.7
	17	74 39.5	58.5	331 23.8	8.3	24 42.6	5.5	56.8
	18	89 39.7	N 7 59.5	345 51.1	8.3	N24 37.1	5.6	56.8
	19	104 39.9	8 00.4	0 18.4	8.3	24 31.5	5.7	56.8
	20	119 40.0	01.3	14 45.7	8.4	24 25.8	5.9	56.9
	21	134 40.2	. . 02.2	29 13.1	8.3	24 19.9	6.0	56.9
	22	149 40.4	03.2	43 40.4	8.3	24 13.9	6.2	56.9
	23	164 40.5	04.1	58 07.7	8.4	24 07.7	6.3	57.0
11	00	179 40.7	N 8 05.0	72 35.1	8.3	N24 01.4	6.5	57.0
	01	194 40.9	05.9	87 02.4	8.4	23 54.9	6.5	57.1
	02	209 41.0	06.9	101 29.8	8.4	23 48.4	6.8	57.1
	03	224 41.2	. . 07.8	115 57.2	8.4	23 41.6	6.9	57.1
	04	239 41.4	08.7	130 24.6	8.4	23 34.7	7.0	57.2
	05	254 41.5	09.6	144 52.0	8.4	23 27.7	7.1	57.2
	06	269 41.7	N 8 10.5	159 19.4	8.5	N23 20.6	7.3	57.2
	07	284 41.9	11.5	173 46.9	8.4	23 13.3	7.5	57.3
	08	299 42.0	12.4	188 14.3	8.5	23 05.8	7.5	57.3
F	09	314 42.2	. . 13.3	202 41.8	8.5	22 58.3	7.7	57.4
R	10	329 42.4	14.2	217 09.3	8.5	22 50.6	7.9	57.4
I	11	344 42.5	15.1	231 36.8	8.6	22 42.7	8.0	57.4
D	12	359 42.7	N 8 16.1	246 04.4	8.5	N22 34.7	8.1	57.5
A	13	14 42.9	17.0	260 31.9	8.6	22 26.6	8.3	57.5
Y	14	29 43.0	17.9	274 59.5	8.6	22 18.3	8.4	57.6
	15	44 43.2	. . 18.8	289 27.1	8.6	22 09.9	8.5	57.6
	16	59 43.4	19.7	303 54.7	8.6	22 01.4	8.7	57.6
	17	74 43.5	20.7	318 22.3	8.6	21 52.7	8.8	57.7
	18	89 43.7	N 8 21.6	332 49.9	8.7	N21 43.9	8.9	57.7
	19	104 43.9	22.5	347 17.6	8.7	21 35.0	9.1	57.7
	20	119 44.0	23.4	1 45.3	8.7	21 25.9	9.2	57.8
	21	134 44.2	. . 24.3	16 13.0	8.8	21 16.7	9.3	57.8
	22	149 44.4	25.2	30 40.8	8.7	21 07.4	9.5	57.9
	23	164 44.5	26.2	45 08.5	8.8	20 57.9	9.6	57.9
12	00	179 44.7	N 8 27.1	59 36.3	8.8	N20 48.3	9.7	57.9
	01	194 44.8	28.0	74 04.1	8.9	20 38.6	9.9	58.0
	02	209 45.0	28.9	88 32.0	8.8	20 28.7	9.9	58.0
	03	224 45.2	. . 29.8	102 59.8	8.9	20 18.8	10.2	58.1
	04	239 45.3	30.7	117 27.7	8.9	20 08.6	10.2	58.1
	05	254 45.5	31.7	131 55.6	9.0	19 58.4	10.4	58.2
	06	269 45.7	N 8 32.6	146 23.6	8.9	N19 48.0	10.5	58.2
	07	284 45.8	33.5	160 51.5	9.0	19 37.5	10.6	58.2
S	08	299 46.0	34.4	175 19.5	9.0	19 26.9	10.7	58.3
A	09	314 46.2	. . 35.3	189 47.5	9.1	19 16.2	10.9	58.3
T	10	329 46.3	36.2	204 15.6	9.0	19 05.3	11.0	58.4
U	11	344 46.5	37.1	218 43.6	9.1	18 54.3	11.1	58.4
R	12	359 46.6	N 8 38.1	233 11.7	9.1	N18 43.2	11.2	58.4
D	13	14 46.8	39.0	247 39.8	9.2	18 32.0	11.4	58.5
A	14	29 47.0	39.9	262 08.0	9.1	18 20.6	11.4	58.5
Y	15	44 47.1	. . 40.8	276 36.1	9.2	18 09.2	11.6	58.6
	16	59 47.3	41.7	291 04.3	9.3	17 57.6	11.7	58.6
	17	74 47.5	42.6	305 32.6	9.2	17 45.9	11.8	58.6
	18	89 47.6	N 8 43.5	320 00.8	9.3	N17 34.1	12.0	58.7
	19	104 47.8	44.4	334 29.1	9.3	17 22.1	12.0	58.7
	20	119 47.9	45.4	348 57.4	9.3	17 10.1	12.2	58.8
	21	134 48.1	. . 46.3	3 25.7	9.3	16 57.9	12.3	58.8
	22	149 48.3	47.2	17 54.0	9.4	16 45.6	12.3	58.8
	23	164 48.4	48.1	32 22.4	9.4	N16 33.3	12.5	58.9
		SD 16.0	d 0.9	SD 15.4		15.7		15.9

Lat.	Twilight Naut.	Twilight Civil	Sunrise	Moonrise 10	11	12	13
°	h m	h m	h m	h m	h m	h m	h m
N 72	////	02 36	04 04	▭	▭	▭	12 07
N 70	////	03 03	04 18	▭	▭	09 37	12 35
68	01 37	03 23	04 28	▭	▭	10 34	12 56
66	02 14	03 39	04 37	▭	08 43	11 07	13 12
64	02 39	03 52	04 45	07 29	09 33	11 31	13 26
62	02 58	04 03	04 51	08 25	10 04	11 50	13 37
60	03 14	04 12	04 57	08 58	10 27	12 05	13 46
N 58	03 26	04 20	05 02	09 22	10 46	12 18	13 54
56	03 37	04 27	05 06	09 42	11 01	12 30	14 02
54	03 47	04 33	05 10	09 58	11 15	12 39	14 08
52	03 55	04 38	05 13	10 12	11 26	12 48	14 14
50	04 02	04 43	05 17	10 24	11 37	12 56	14 19
45	04 17	04 53	05 23	10 49	11 58	13 13	14 30
N 40	04 28	05 01	05 29	11 09	12 16	13 26	14 39
35	04 38	05 08	05 34	11 26	12 30	13 37	14 47
30	04 45	05 14	05 38	11 41	12 43	13 47	14 54
20	04 57	05 23	05 46	12 05	13 04	14 05	15 06
N 10	05 06	05 31	05 52	12 26	13 23	14 19	15 16
0	05 13	05 37	05 58	12 46	13 40	14 33	15 26
S 10	05 18	05 42	06 04	13 05	13 57	14 47	15 35
20	05 22	05 47	06 10	13 26	14 15	15 02	15 45
30	05 24	05 52	06 16	13 51	14 37	15 18	15 57
35	05 25	05 55	06 20	14 05	14 49	15 28	16 04
40	05 26	05 57	06 25	14 21	15 03	15 39	16 11
45	05 26	06 00	06 29	14 41	15 20	15 52	16 20
S 50	05 25	06 03	06 35	15 05	15 40	16 08	16 30
52	05 25	06 04	06 38	15 17	15 50	16 15	16 35
54	05 24	06 05	06 41	15 31	16 01	16 23	16 40
56	05 24	06 07	06 44	15 46	16 13	16 32	16 46
58	05 23	06 08	06 48	16 05	16 27	16 42	16 53
S 60	05 22	06 10	06 52	16 27	16 44	16 54	17 00

Lat.	Sunset	Twilight Civil	Twilight Naut.	Moonset 10	11	12	13
°	h m	h m	h m	h m	h m	h m	h m
N 72	20 01	21 32	////	▭	▭	▭	06 43
N 70	19 48	21 04	////	▭	▭	07 20	06 13
68	19 36	20 43	22 33	▭	▭	06 22	05 50
66	19 27	20 26	21 54	▭	06 20	05 48	05 32
64	19 20	20 13	21 27	05 39	05 29	05 23	05 18
62	19 13	20 02	21 07	04 43	04 57	05 03	05 05
60	19 07	19 52	20 51	04 10	04 34	04 47	04 55
N 58	19 02	19 44	20 38	03 45	04 14	04 33	04 45
56	18 58	19 37	20 27	03 26	03 58	04 21	04 37
54	18 54	19 31	20 17	03 09	03 45	04 10	04 30
52	18 50	19 26	20 09	02 55	03 33	04 01	04 23
50	18 47	19 21	20 02	02 43	03 22	03 53	04 17
45	18 40	19 10	19 47	02 17	02 59	03 35	04 05
N 40	18 34	19 02	19 35	01 57	02 41	03 20	03 54
35	18 29	18 55	19 25	01 39	02 26	03 08	03 45
30	18 24	18 49	19 18	01 25	02 13	02 57	03 37
20	18 17	18 39	19 05	01 00	01 50	02 38	03 22
N 10	18 10	18 32	18 56	00 38	01 30	02 21	03 10
0	18 04	18 25	18 50	00 17	01 12	02 05	02 58
S 10	17 59	18 20	18 44	24 53	00 53	01 50	02 46
20	17 52	18 15	18 40	24 33	00 33	01 33	02 34
30	17 46	18 10	18 37	24 10	00 10	01 13	02 19
35	17 42	18 07	18 36	23 56	25 02	01 02	02 10
40	17 37	18 04	18 36	23 40	24 49	00 49	02 01
45	17 32	18 02	18 36	23 21	24 33	00 33	01 49
S 50	17 26	17 59	18 36	22 57	24 14	00 14	01 35
52	17 23	17 57	18 36	22 46	24 04	00 04	01 28
54	17 20	17 56	18 37	22 33	23 54	25 21	01 21
56	17 17	17 54	18 37	22 17	23 42	25 13	01 13
58	17 13	17 53	18 38	21 59	23 28	25 03	01 03
S 60	17 09	17 51	18 39	21 37	23 12	24 53	00 53

Day	SUN Eqn. of Time 00^h	SUN Eqn. of Time 12^h	SUN Mer. Pass.	MOON Mer. Pass. Upper	MOON Mer. Pass. Lower	Age	Phase
d	m s	m s	h m	h m	h m	d %	
10	01 34	01 26	12 01	18 59	06 32	09 55	
11	01 18	01 10	12 01	19 53	07 26	10 66	
12	01 02	00 54	12 01	20 46	08 19	11 76	

UT		ARIES	VENUS −3.9		MARS +0.3		JUPITER −2.2		SATURN +0.1	
		GHA	GHA	Dec	GHA	Dec	GHA	Dec	GHA	Dec
d	h	° ′	° ′	° ′	° ′	° ′	° ′	° ′	° ′	° ′
13	00	200 46.2	209 52.9	S 5 20.6	264 08.2	S22 08.5	69 55.7	N19 03.2	116 54.3	N22 21.9
	01	215 48.6	224 52.6	19.5	279 09.0	08.2	84 58.1	03.1	131 56.5	21.9
	02	230 51.1	239 52.2	18.4	294 09.8	08.0	100 00.5	03.1	146 58.7	21.9
	03	245 53.6	254 51.9	. . 17.3	309 10.7	. . 07.8	115 02.9	. . 03.1	162 01.0	. . 21.9
	04	260 56.0	269 51.5	16.2	324 11.5	07.6	130 05.3	03.1	177 03.2	21.9
	05	275 58.5	284 51.2	15.1	339 12.3	07.4	145 07.7	03.1	192 05.5	22.0
	06	291 00.9	299 50.8	S 5 13.9	354 13.1	S22 07.2	160 10.1	N19 03.0	207 07.7	N22 22.0
	07	306 03.4	314 50.5	12.8	9 14.0	07.0	175 12.5	03.0	222 09.9	22.0
	08	321 05.9	329 50.1	11.7	24 14.8	06.8	190 14.8	03.0	237 12.2	22.0
S	09	336 08.3	344 49.8	. . 10.6	39 15.6	. . 06.6	205 17.2	. . 03.0	252 14.4	. . 22.0
U	10	351 10.8	359 49.4	09.5	54 16.4	06.4	220 19.6	02.9	267 16.7	22.0
N	11	6 13.3	14 49.1	08.4	69 17.3	06.2	235 22.0	02.9	282 18.9	22.1
D	12	21 15.7	29 48.8	S 5 07.3	84 18.1	S22 06.0	250 24.4	N19 02.9	297 21.1	N22 22.1
A	13	36 18.2	44 48.4	06.1	99 18.9	05.8	265 26.8	02.9	312 23.4	22.1
Y	14	51 20.7	59 48.1	05.0	114 19.7	05.6	280 29.2	02.9	327 25.6	22.1
	15	66 23.1	74 47.7	. . 03.9	129 20.6	. . 05.3	295 31.6	. . 02.8	342 27.8	. . 22.1
	16	81 25.6	89 47.4	02.8	144 21.4	05.1	310 34.0	02.8	357 30.1	22.1
	17	96 28.1	104 47.0	01.7	159 22.2	04.9	325 36.3	02.8	12 32.3	22.2
	18	111 30.5	119 46.7	S 5 00.6	174 23.1	S22 04.7	340 38.7	N19 02.8	27 34.6	N22 22.2
	19	126 33.0	134 46.3	4 59.5	189 23.9	04.5	355 41.1	02.7	42 36.8	22.2
	20	141 35.4	149 46.0	58.3	204 24.7	04.3	10 43.5	02.7	57 39.0	22.2
	21	156 37.9	164 45.6	. . 57.2	219 25.5	. . 04.1	25 45.9	. . 02.7	72 41.3	. . 22.2
	22	171 40.4	179 45.3	56.1	234 26.4	03.9	40 48.3	02.7	87 43.5	22.2
	23	186 42.8	194 44.9	55.0	249 27.2	03.7	55 50.7	02.7	102 45.7	22.3
14	00	201 45.3	209 44.6	S 4 53.9	264 28.0	S22 03.5	70 53.1	N19 02.6	117 48.0	N22 22.3
	01	216 47.8	224 44.3	52.7	279 28.9	03.2	85 55.4	02.6	132 50.2	22.3
	02	231 50.2	239 43.9	51.6	294 29.7	03.0	100 57.8	02.6	147 52.5	22.3
	03	246 52.7	254 43.6	. . 50.5	309 30.5	. . 02.8	116 00.2	. . 02.6	162 54.7	. . 22.3
	04	261 55.2	269 43.2	49.4	324 31.3	02.6	131 02.6	02.5	177 56.9	22.3
	05	276 57.6	284 42.9	48.3	339 32.2	02.4	146 05.0	02.5	192 59.2	22.3
	06	292 00.1	299 42.5	S 4 47.1	354 33.0	S22 02.2	161 07.4	N19 02.5	208 01.4	N22 22.4
	07	307 02.5	314 42.2	46.0	9 33.8	02.0	176 09.8	02.5	223 03.6	22.4
	08	322 05.0	329 41.8	44.9	24 34.7	01.8	191 12.1	02.4	238 05.9	22.4
M	09	337 07.5	344 41.5	. . 43.8	39 35.5	. . 01.6	206 14.5	. . 02.4	253 08.1	. . 22.4
O	10	352 09.9	359 41.2	42.7	54 36.3	01.3	221 16.9	02.4	268 10.3	22.4
N	11	7 12.4	14 40.8	41.5	69 37.2	01.1	236 19.3	02.4	283 12.6	22.4
D	12	22 14.9	29 40.5	S 4 40.4	84 38.0	S22 00.9	251 21.7	N19 02.4	298 14.8	N22 22.5
A	13	37 17.3	44 40.1	39.3	99 38.8	00.7	266 24.0	02.3	313 17.0	22.5
Y	14	52 19.8	59 39.8	38.2	114 39.6	00.5	281 26.4	02.3	328 19.3	22.5
	15	67 22.3	74 39.4	. . 37.1	129 40.5	. . 00.3	296 28.8	. . 02.3	343 21.5	. . 22.5
	16	82 24.7	89 39.1	35.9	144 41.3	22 00.1	311 31.2	02.3	358 23.7	22.5
	17	97 27.2	104 38.8	34.8	159 42.1	21 59.8	326 33.6	02.2	13 26.0	22.5
	18	112 29.7	119 38.4	S 4 33.7	174 43.0	S21 59.6	341 36.0	N19 02.2	28 28.2	N22 22.6
	19	127 32.1	134 38.1	32.6	189 43.8	59.4	356 38.3	02.2	43 30.4	22.6
	20	142 34.6	149 37.7	31.4	204 44.6	59.2	11 40.7	02.2	58 32.7	22.6
	21	157 37.0	164 37.4	. . 30.3	219 45.5	. . 59.0	26 43.1	. . 02.1	73 34.9	. . 22.6
	22	172 39.5	179 37.0	29.2	234 46.3	58.8	41 45.5	02.1	88 37.1	22.6
	23	187 42.0	194 36.7	28.1	249 47.1	58.6	56 47.9	02.1	103 39.4	22.6
15	00	202 44.4	209 36.4	S 4 27.0	264 48.0	S21 58.3	71 50.2	N19 02.1	118 41.6	N22 22.7
	01	217 46.9	224 36.0	25.8	279 48.8	58.1	86 52.6	02.0	133 43.8	22.7
	02	232 49.4	239 35.7	24.7	294 49.6	57.9	101 55.0	02.0	148 46.1	22.7
	03	247 51.8	254 35.3	. . 23.6	309 50.5	. . 57.7	116 57.4	. . 02.0	163 48.3	. . 22.7
	04	262 54.3	269 35.0	22.5	324 51.3	57.5	131 59.7	02.0	178 50.5	22.7
	05	277 56.8	284 34.7	21.3	339 52.1	57.3	147 02.1	01.9	193 52.8	22.7
	06	292 59.2	299 34.3	S 4 20.2	354 53.0	S21 57.0	162 04.5	N19 01.9	208 55.0	N22 22.7
	07	308 01.7	314 34.0	19.1	9 53.8	56.8	177 06.9	01.9	223 57.2	22.8
T	08	323 04.1	329 33.6	17.9	24 54.7	56.6	192 09.2	01.9	238 59.5	22.8
U	09	338 06.6	344 33.3	. . 16.8	39 55.5	. . 56.4	207 11.6	. . 01.8	254 01.7	. . 22.8
E	10	353 09.1	359 33.0	15.7	54 56.3	56.2	222 14.0	01.8	269 03.9	22.8
S	11	8 11.5	14 32.6	14.6	69 57.2	56.0	237 16.4	01.8	284 06.2	22.8
D	12	23 14.0	29 32.3	S 4 13.4	84 58.0	S21 55.7	252 18.7	N19 01.8	299 08.4	N22 22.8
	13	38 16.5	44 31.9	12.3	99 58.8	55.5	267 21.1	01.7	314 10.6	22.9
A	14	53 18.9	59 31.6	11.2	114 59.7	55.3	282 23.5	01.7	329 12.9	22.9
Y	15	68 21.4	74 31.3	. . 10.1	130 00.5	. . 55.1	297 25.9	. . 01.7	344 15.1	. . 22.9
	16	83 23.9	89 30.9	08.9	145 01.3	54.9	312 28.2	01.7	359 17.3	22.9
	17	98 26.3	104 30.6	07.8	160 02.2	54.7	327 30.6	01.6	14 19.5	22.9
	18	113 28.8	119 30.2	S 4 06.7	175 03.0	S21 54.4	342 33.0	N19 01.6	29 21.8	N22 22.9
	19	128 31.3	134 29.9	05.5	190 03.9	54.2	357 35.4	01.6	44 24.0	23.0
	20	143 33.7	149 29.6	04.4	205 04.7	54.0	12 37.7	01.5	59 26.2	23.0
	21	158 36.2	164 29.2	. . 03.3	220 05.5	. . 53.8	27 40.1	. . 01.5	74 28.5	. . 23.0
	22	173 38.6	179 28.9	02.2	235 06.4	53.6	42 42.5	01.5	89 30.7	23.0
	23	188 41.1	194 28.5	01.0	250 07.2	53.3	57 44.8	01.5	104 32.9	23.0
Mer. Pass.		h m 10 31.3	v −0.3	d 1.1	v 0.8	d 0.2	v 2.4	d 0.0	v 2.2	d 0.0

STARS

Name	SHA	Dec
	° ′	° ′
Acamar	315 24.7	S40 17.7
Achernar	335 33.1	S57 13.3
Acrux	173 18.0	S63 07.1
Adhara	255 18.9	S28 58.8
Aldebaran	290 58.9	N16 30.9
Alioth	166 26.9	N55 56.6
Alkaid	153 04.6	N49 17.8
Al Na'ir	27 53.9	S46 56.7
Alnilam	275 54.7	S 1 12.1
Alphard	218 03.9	S 8 40.4
Alphecca	126 17.5	N26 42.0
Alpheratz	357 52.3	N29 06.2
Altair	62 16.1	N 8 52.3
Ankaa	353 23.9	S42 17.4
Antares	112 35.9	S26 26.4
Arcturus	146 02.7	N19 09.8
Atria	107 44.7	S69 01.9
Avior	234 21.3	S59 31.4
Bellatrix	278 40.8	N 6 21.1
Betelgeuse	271 10.2	N 7 24.4
Canopus	263 59.9	S52 42.1
Capella	280 46.6	N46 00.2
Deneb	49 37.2	N45 17.1
Denebola	182 41.5	N14 33.2
Diphda	349 04.2	S17 58.3
Dubhe	194 00.7	N61 44.2
Elnath	278 23.0	N28 36.7
Eltanin	90 49.6	N51 28.9
Enif	33 55.2	N 9 53.1
Fomalhaut	15 33.0	S29 36.4
Gacrux	172 09.6	S57 07.9
Gienah	176 00.3	S17 33.7
Hadar	148 59.0	S60 23.3
Hamal	328 10.3	N23 28.5
Kaus Aust.	83 54.3	S34 23.0
Kochab	137 18.2	N74 08.4
Markab	13 46.6	N15 13.1
Menkar	314 23.8	N 4 06.0
Menkent	148 16.8	S36 23.2
Miaplacidus	221 41.3	S69 44.0
Mirfak	308 52.5	N49 52.4
Nunki	76 08.2	S26 17.6
Peacock	53 31.8	S56 43.4
Pollux	243 37.5	N28 01.2
Procyon	245 08.1	N 5 13.0
Rasalhague	96 13.7	N12 33.2
Regulus	207 51.8	N11 57.1
Rigel	281 20.0	S 8 12.0
Rigil Kent.	140 02.4	S60 50.9
Sabik	102 21.6	S15 43.8
Schedar	349 50.5	N56 33.1
Shaula	96 32.6	S37 06.4
Sirius	258 40.9	S16 43.3
Spica	158 39.4	S11 10.8
Suhail	222 58.3	S43 26.9
Vega	80 44.3	N38 46.8
Zuben'ubi	137 14.1	S16 03.4

	SHA	Mer. Pass.
	° ′	h m
Venus	7 59.3	10 01
Mars	62 42.7	6 22
Jupiter	229 07.8	19 13
Saturn	276 02.7	16 06

	UT	SUN GHA	SUN Dec	MOON GHA	v	MOON Dec	d	HP
	d h	° ′	° ′	° ′	′	° ′	′	′
	13 00	179 48.6	N 8 49.0	46 50.8	9.4	N16 20.8	12.6	58.9
	01	194 48.8	49.9	61 19.2	9.5	16 08.2	12.8	59.0
	02	209 48.9	50.8	75 47.7	9.4	15 55.4	12.8	59.0
	03	224 49.1	· · 51.7	90 16.1	9.5	15 42.6	12.9	59.0
	04	239 49.2	52.6	104 44.6	9.5	15 29.7	13.0	59.1
	05	254 49.4	53.5	119 13.1	9.6	15 16.7	13.2	59.1
	06	269 49.6	N 8 54.5	133 41.7	9.5	N15 03.5	13.2	59.2
	07	284 49.7	55.4	148 10.2	9.6	14 50.3	13.4	59.2
	08	299 49.9	56.3	162 38.8	9.6	14 36.9	13.4	59.2
S	09	314 50.0	· · 57.2	177 07.4	9.7	14 23.5	13.5	59.3
U	10	329 50.2	58.1	191 36.1	9.6	14 10.0	13.7	59.3
N	11	344 50.4	59.0	206 04.7	9.7	13 56.3	13.7	59.3
D	12	359 50.5	N 8 59.9	220 33.4	9.7	N13 42.6	13.8	59.4
A	13	14 50.7	9 00.8	235 02.1	9.7	13 28.8	14.0	59.4
Y	14	29 50.8	01.7	249 30.8	9.7	13 14.8	14.0	59.5
	15	44 51.0	· · 02.6	263 59.5	9.7	13 00.8	14.1	59.5
	16	59 51.2	03.5	278 28.2	9.8	12 46.7	14.2	59.5
	17	74 51.3	04.4	292 57.0	9.8	12 32.5	14.3	59.6
	18	89 51.5	N 9 05.3	307 25.8	9.8	N12 18.2	14.3	59.6
	19	104 51.6	06.2	321 54.6	9.8	12 03.9	14.5	59.7
	20	119 51.8	07.2	336 23.4	9.8	11 49.4	14.5	59.7
	21	134 52.0	· · 08.1	350 52.2	9.9	11 34.9	14.7	59.7
	22	149 52.1	09.0	5 21.1	9.8	11 20.2	14.7	59.8
	23	164 52.3	09.9	19 49.9	9.9	11 05.5	14.8	59.8
	14 00	179 52.4	N 9 10.8	34 18.8	9.9	N10 50.7	14.9	59.8
	01	194 52.6	11.7	48 47.7	9.9	10 35.8	14.9	59.9
	02	209 52.7	12.6	63 16.6	9.9	10 20.9	15.0	59.9
	03	224 52.9	· · 13.5	77 45.5	9.9	10 05.9	15.1	59.9
	04	239 53.1	14.4	92 14.4	9.9	9 50.8	15.2	60.0
	05	254 53.2	15.3	106 43.3	10.0	9 35.6	15.3	60.0
	06	269 53.4	N 9 16.2	121 12.3	9.9	N 9 20.3	15.3	60.0
	07	284 53.5	17.1	135 41.2	10.0	9 05.0	15.4	60.1
	08	299 53.7	18.0	150 10.2	9.9	8 49.6	15.4	60.1
M	09	314 53.9	· · 18.9	164 39.1	10.0	8 34.2	15.6	60.1
O	10	329 54.0	19.8	179 08.1	10.0	8 18.6	15.6	60.2
N	11	344 54.2	20.7	193 37.1	9.9	8 03.0	15.6	60.2
D	12	359 54.3	N 9 21.6	208 06.0	10.0	N 7 47.4	15.7	60.2
A	13	14 54.5	22.5	222 35.0	10.0	7 31.7	15.8	60.3
Y	14	29 54.6	23.4	237 04.0	10.0	7 15.9	15.9	60.3
	15	44 54.8	· · 24.3	251 33.0	10.0	7 00.0	15.8	60.3
	16	59 54.9	25.2	266 02.0	10.0	6 44.2	16.0	60.4
	17	74 55.1	26.1	280 31.0	9.9	6 28.2	16.0	60.4
	18	89 55.3	N 9 27.0	294 59.9	10.0	N 6 12.2	16.1	60.4
	19	104 55.4	27.9	309 28.9	10.0	5 56.1	16.1	60.5
	20	119 55.6	28.8	323 57.9	10.0	5 40.0	16.1	60.5
	21	134 55.7	· · 29.7	338 26.9	10.0	5 23.9	16.2	60.5
	22	149 55.9	30.6	352 55.9	9.9	5 07.7	16.3	60.6
	23	164 56.0	31.5	7 24.8	10.0	4 51.4	16.3	60.6
	15 00	179 56.2	N 9 32.4	21 53.8	9.9	N 4 35.1	16.3	60.6
	01	194 56.3	33.3	36 22.7	10.0	4 18.8	16.4	60.6
	02	209 56.5	34.2	50 51.7	9.9	4 02.4	16.4	60.7
	03	224 56.7	· · 35.1	65 20.6	9.9	3 46.0	16.4	60.7
	04	239 56.8	36.0	79 49.5	10.0	3 29.6	16.5	60.7
	05	254 57.0	36.9	94 18.5	9.9	3 13.1	16.5	60.7
	06	269 57.1	N 9 37.8	108 47.4	9.9	N 2 56.6	16.6	60.8
	07	284 57.3	38.7	123 16.3	9.8	2 40.0	16.6	60.8
	08	299 57.4	39.6	137 45.1	9.9	2 23.4	16.6	60.8
T	09	314 57.6	· · 40.5	152 14.0	9.8	2 06.8	16.6	60.8
U	10	329 57.7	41.3	166 42.8	9.9	1 50.2	16.7	60.9
E	11	344 57.9	42.2	181 11.7	9.8	1 33.5	16.7	60.9
S	12	359 58.0	N 9 43.1	195 40.5	9.8	N 1 16.8	16.7	60.9
D	13	14 58.2	44.0	210 09.3	9.7	1 00.1	16.7	60.9
A	14	29 58.3	44.9	224 38.0	9.8	0 43.4	16.7	61.0
Y	15	44 58.5	· · 45.8	239 06.8	9.7	0 26.7	16.8	61.0
	16	59 58.7	46.7	253 35.5	9.7	N 0 09.9	16.7	61.0
	17	74 58.8	47.6	268 04.2	9.7	S 0 06.8	16.8	61.0
	18	89 59.0	N 9 48.5	282 32.9	9.7	S 0 23.6	16.8	61.0
	19	104 59.1	49.4	297 01.6	9.6	0 40.4	16.8	61.1
	20	119 59.3	50.3	311 30.2	9.6	0 57.2	16.8	61.1
	21	134 59.4	· · 51.2	325 58.8	9.6	1 14.0	16.8	61.1
	22	149 59.6	52.1	340 27.4	9.5	1 30.8	16.8	61.1
	23	164 59.7	53.0	354 55.9	9.5	S 1 47.6	16.8	61.1
		SD 16.0	d 0.9	SD 16.2		16.4		16.6

Lat.	Twilight Naut.	Twilight Civil	Sunrise	Moonrise 13	Moonrise 14	Moonrise 15	Moonrise 16
°	h m	h m	h m	h m	h m	h m	h m
N 72	////	02 11	03 48	12 07	14 46	17 12	19 43
N 70	////	02 44	04 03	12 35	14 57	17 12	19 31
68	01 03	03 08	04 16	12 56	15 05	17 12	19 21
66	01 53	03 26	04 26	13 12	15 12	17 11	19 13
64	02 23	03 40	04 34	13 26	15 18	17 11	19 06
62	02 45	03 52	04 42	13 37	15 23	17 11	19 00
60	03 02	04 02	04 48	13 46	15 28	17 11	18 55
N 58	03 16	04 11	04 53	13 54	15 32	17 10	18 51
56	03 28	04 18	04 58	14 02	15 35	17 10	18 47
54	03 38	04 25	05 03	14 08	15 38	17 10	18 44
52	03 47	04 31	05 07	14 14	15 41	17 10	18 41
50	03 55	04 36	05 10	14 19	15 44	17 10	18 38
45	04 11	04 48	05 18	14 30	15 49	17 10	18 32
N 40	04 23	04 57	05 25	14 39	15 54	17 09	18 27
35	04 33	05 04	05 30	14 47	15 58	17 09	18 22
30	04 42	05 10	05 35	14 54	16 01	17 09	18 19
20	04 55	05 21	05 43	15 06	16 07	17 09	18 12
N 10	05 04	05 29	05 50	15 16	16 12	17 09	18 07
0	05 12	05 36	05 57	15 26	16 17	17 09	18 01
S 10	05 18	05 42	06 04	15 35	16 22	17 09	17 56
20	05 23	05 48	06 10	15 45	16 27	17 09	17 51
30	05 26	05 54	06 18	15 57	16 33	17 08	17 45
35	05 28	05 57	06 23	16 04	16 36	17 08	17 41
40	05 29	06 00	06 28	16 11	16 40	17 08	17 37
45	05 29	06 04	06 33	16 20	16 45	17 08	17 33
S 50	05 30	06 07	06 40	16 30	16 50	17 08	17 27
52	05 30	06 09	06 43	16 35	16 52	17 08	17 25
54	05 30	06 11	06 47	16 40	16 55	17 08	17 22
56	05 29	06 13	06 50	16 46	16 58	17 08	17 19
58	05 29	06 15	06 55	16 53	17 01	17 08	17 16
S 60	05 29	06 17	06 59	17 00	17 05	17 08	17 12

Lat.	Sunset	Twilight Civil	Twilight Naut.	Moonset 13	Moonset 14	Moonset 15	Moonset 16
°	h m	h m	h m	h m	h m	h m	h m
N 72	20 17	21 57	////	06 43	05 54	05 20	04 48
N 70	20 01	21 22	////	06 13	05 41	05 16	04 52
68	19 48	20 57	23 12	05 50	05 30	05 13	04 56
66	19 37	20 38	22 14	05 32	05 20	05 10	05 00
64	19 29	20 23	21 42	05 18	05 13	05 08	05 03
62	19 21	20 11	21 19	05 05	05 06	05 06	05 05
60	19 15	20 01	21 02	04 55	05 00	05 04	05 07
N 58	19 09	19 52	20 47	04 45	04 55	05 02	05 09
56	19 04	19 44	20 35	04 37	04 50	05 01	05 11
54	18 59	19 37	20 25	04 30	04 46	04 59	05 13
52	18 55	19 31	20 15	04 23	04 42	04 58	05 14
50	18 52	19 26	20 08	04 17	04 38	04 57	05 15
45	18 44	19 14	19 51	04 05	04 31	04 55	05 18
N 40	18 37	19 05	19 38	03 54	04 24	04 53	05 21
35	18 31	18 57	19 28	03 45	04 19	04 51	05 23
30	18 26	18 51	19 20	03 37	04 14	04 49	05 25
20	18 18	18 40	19 07	03 22	04 05	04 46	05 28
N 10	18 10	18 32	18 57	03 10	03 57	04 44	05 31
0	18 04	18 25	18 49	02 58	03 50	04 41	05 33
S 10	17 57	18 18	18 43	02 46	03 43	04 39	05 36
20	17 50	18 12	18 38	02 34	03 35	04 36	05 39
30	17 42	18 06	18 34	02 19	03 26	04 33	05 42
35	17 38	18 03	18 33	02 10	03 20	04 32	05 44
40	17 33	18 00	18 31	02 01	03 14	04 30	05 46
45	17 27	17 57	18 31	01 49	03 07	04 27	05 48
S 50	17 20	17 53	18 30	01 35	02 59	04 24	05 51
52	17 17	17 51	18 30	01 28	02 55	04 23	05 53
54	17 13	17 49	18 30	01 21	02 51	04 22	05 54
56	17 09	17 47	18 30	01 13	02 46	04 20	05 56
58	17 05	17 45	18 30	01 03	02 40	04 18	05 58
S 60	17 00	17 43	18 31	00 53	02 34	04 16	06 00

Day	SUN Eqn. of Time 00^h	SUN Eqn. of Time 12^h	SUN Mer. Pass.	MOON Mer. Pass. Upper	MOON Mer. Pass. Lower	MOON Age	MOON Phase
d	m s	m s	h m	h m	h m	d	%
13	00 46	00 38	12 01	21 38	09 12	12	85
14	00 31	00 23	12 00	22 29	10 04	13	92
15	00 16	00 08	12 00	23 21	10 55	14	97

UT d h		ARIES GHA ° ′	VENUS −3.9 GHA ° ′	VENUS Dec ° ′	MARS +0.2 GHA ° ′	MARS Dec ° ′	JUPITER −2.2 GHA ° ′	JUPITER Dec ° ′	SATURN +0.1 GHA ° ′	SATURN Dec ° ′
16 00		203 43.6	209 28.2	S 3 59.9	265 08.0	S21 53.1	72 47.2	N19 01.4	119 35.2	N22 23.0
01		218 46.0	224 27.9	58.8	280 08.9	52.9	87 49.6	01.4	134 37.4	23.1
02		233 48.5	239 27.5	57.6	295 09.7	52.7	102 52.0	01.4	149 39.6	23.1
03		248 51.0	254 27.2	. . 56.5	310 10.6	. . 52.5	117 54.3	. . 01.4	164 41.8	. . 23.1
04		263 53.4	269 26.8	55.4	325 11.4	52.2	132 56.7	01.3	179 44.1	23.1
05		278 55.9	284 26.5	54.2	340 12.2	52.0	147 59.1	01.3	194 46.3	23.1
06		293 58.4	299 26.2	S 3 53.1	355 13.1	S21 51.8	163 01.4	N19 01.3	209 48.5	N22 23.1
07	W	309 00.8	314 25.8	52.0	10 13.9	51.6	178 03.8	01.3	224 50.8	23.1
08	E	324 03.3	329 25.5	50.8	25 14.8	51.3	193 06.2	01.2	239 53.0	23.2
09	D	339 05.7	344 25.2	. . 49.7	40 15.6	. . 51.1	208 08.5	. . 01.2	254 55.2	. . 23.2
10	N	354 08.2	359 24.8	48.6	55 16.4	50.9	223 10.9	01.2	269 57.4	23.2
11	E	9 10.7	14 24.5	47.4	70 17.3	50.7	238 13.3	01.1	284 59.7	23.2
12	S	24 13.1	29 24.1	S 3 46.3	85 18.1	S21 50.5	253 15.6	N19 01.1	300 01.9	N22 23.2
13	D	39 15.6	44 23.8	45.2	100 19.0	50.2	268 18.0	01.1	315 04.1	23.2
14	A	54 18.1	59 23.5	44.0	115 19.8	50.0	283 20.4	01.1	330 06.4	23.3
15	Y	69 20.5	74 23.1	. . 42.9	130 20.6	. . 49.8	298 22.7	. . 01.0	345 08.6	. . 23.3
16		84 23.0	89 22.8	41.8	145 21.5	49.6	313 25.1	01.0	0 10.8	23.3
17		99 25.5	104 22.5	40.6	160 22.3	49.3	328 27.5	01.0	15 13.0	23.3
18		114 27.9	119 22.1	S 3 39.5	175 23.2	S21 49.1	343 29.8	N19 00.9	30 15.3	N22 23.3
19		129 30.4	134 21.8	38.4	190 24.0	48.9	358 32.2	00.9	45 17.5	23.3
20		144 32.9	149 21.5	37.2	205 24.8	48.7	13 34.6	00.9	60 19.7	23.4
21		159 35.3	164 21.1	. . 36.1	220 25.7	. . 48.4	28 36.9	. . 00.9	75 21.9	. . 23.4
22		174 37.8	179 20.8	35.0	235 26.5	48.2	43 39.3	00.8	90 24.2	23.4
23		189 40.2	194 20.5	33.8	250 27.4	48.0	58 41.7	00.8	105 26.4	23.4
17 00		204 42.7	209 20.1	S 3 32.7	265 28.2	S21 47.8	73 44.0	N19 00.8	120 28.6	N22 23.4
01		219 45.2	224 19.8	31.6	280 29.1	47.5	88 46.4	00.7	135 30.8	23.4
02		234 47.6	239 19.4	30.4	295 29.9	47.3	103 48.7	00.7	150 33.1	23.4
03		249 50.1	254 19.1	. . 29.3	310 30.7	. . 47.1	118 51.1	. . 00.7	165 35.3	. . 23.5
04		264 52.6	269 18.8	28.2	325 31.6	46.9	133 53.5	00.7	180 37.5	23.5
05		279 55.0	284 18.4	27.0	340 32.4	46.6	148 55.8	00.6	195 39.7	23.5
06		294 57.5	299 18.1	S 3 25.9	355 33.3	S21 46.4	163 58.2	N19 00.6	210 42.0	N22 23.5
07	T	310 00.0	314 17.8	24.7	10 34.1	46.2	179 00.6	00.6	225 44.2	23.5
08	H	325 02.4	329 17.4	23.6	25 35.0	46.0	194 02.9	00.5	240 46.4	23.5
09	U	340 04.9	344 17.1	. . 22.5	40 35.8	. . 45.7	209 05.3	. . 00.5	255 48.6	. . 23.6
10	R	355 07.4	359 16.8	21.3	55 36.7	45.5	224 07.6	00.5	270 50.9	23.6
11	S	10 09.8	14 16.4	20.2	70 37.5	45.3	239 10.0	00.5	285 53.1	23.6
12	D	25 12.3	29 16.1	S 3 19.1	85 38.3	S21 45.0	254 12.4	N19 00.4	300 55.3	N22 23.6
13	A	40 14.7	44 15.8	17.9	100 39.2	44.8	269 14.7	00.4	315 57.5	23.6
14	Y	55 17.2	59 15.4	16.8	115 40.0	44.6	284 17.1	00.4	330 59.8	23.6
15		70 19.7	74 15.1	. . 15.6	130 40.9	. . 44.4	299 19.4	. . 00.3	346 02.0	. . 23.7
16		85 22.1	89 14.8	14.5	145 41.7	44.1	314 21.8	00.3	1 04.2	23.7
17		100 24.6	104 14.4	13.4	160 42.6	43.9	329 24.1	00.3	16 06.4	23.7
18		115 27.1	119 14.1	S 3 12.2	175 43.4	S21 43.7	344 26.5	N19 00.2	31 08.7	N22 23.7
19		130 29.5	134 13.8	11.1	190 44.3	43.5	359 28.9	00.2	46 10.9	23.7
20		145 32.0	149 13.4	09.9	205 45.1	43.2	14 31.2	00.2	61 13.1	23.7
21		160 34.5	164 13.1	. . 08.8	220 46.0	. . 43.0	29 33.6	. . 00.2	76 15.3	. . 23.7
22		175 36.9	179 12.8	07.7	235 46.8	42.8	44 35.9	00.1	91 17.6	23.8
23		190 39.4	194 12.4	06.5	250 47.7	42.5	59 38.3	00.1	106 19.8	23.8
18 00		205 41.8	209 12.1	S 3 05.4	265 48.5	S21 42.3	74 40.6	N19 00.1	121 22.0	N22 23.8
01		220 44.3	224 11.8	04.2	280 49.3	42.1	89 43.0	00.0	136 24.2	23.8
02		235 46.8	239 11.4	03.1	295 50.2	41.8	104 45.3	00.0	151 26.5	23.8
03		250 49.2	254 11.1	. . 02.0	310 51.0	. . 41.6	119 47.7	19 00.0	166 28.7	. . 23.8
04		265 51.7	269 10.8	3 00.8	325 51.9	41.4	134 50.1	18 59.9	181 30.9	23.9
05		280 54.2	284 10.4	2 59.7	340 52.7	41.2	149 52.4	59.9	196 33.1	23.9
06		295 56.6	299 10.1	S 2 58.5	355 53.6	S21 40.9	164 54.8	N18 59.9	211 35.3	N22 23.9
07		310 59.1	314 09.8	57.4	10 54.4	40.7	179 57.1	59.8	226 37.6	23.9
08		326 01.6	329 09.4	56.3	25 55.3	40.5	194 59.5	59.8	241 39.8	23.9
09	F	341 04.0	344 09.1	. . 55.1	40 56.1	. . 40.2	210 01.8	. . 59.8	256 42.0	. . 23.9
10	R	356 06.5	359 08.8	54.0	55 57.0	40.0	225 04.2	59.7	271 44.2	23.9
11	I	11 09.0	14 08.4	52.8	70 57.8	39.8	240 06.5	59.7	286 46.4	24.0
12	D	26 11.4	29 08.1	S 2 51.7	85 58.7	S21 39.5	255 08.9	N18 59.7	301 48.7	N22 24.0
13	A	41 13.9	44 07.8	50.5	100 59.5	39.3	270 11.2	59.7	316 50.9	24.0
14	Y	56 16.3	59 07.4	49.4	116 00.4	39.1	285 13.6	59.6	331 53.1	24.0
15		71 18.8	74 07.1	. . 48.3	131 01.2	. . 38.8	300 15.9	. . 59.6	346 55.3	. . 24.0
16		86 21.3	89 06.8	47.1	146 02.1	38.6	315 18.3	59.6	1 57.6	24.0
17		101 23.7	104 06.4	46.0	161 02.9	38.4	330 20.6	59.5	16 59.8	24.1
18		116 26.2	119 06.1	S 2 44.8	176 03.8	S21 38.1	345 23.0	N18 59.5	32 02.0	N22 24.1
19		131 28.7	134 05.8	43.7	191 04.6	37.9	0 25.3	59.5	47 04.2	24.1
20		146 31.1	149 05.5	42.5	206 05.5	37.7	15 27.7	59.4	62 06.4	24.1
21		161 33.6	164 05.1	. . 41.4	221 06.3	. . 37.4	30 30.0	. . 59.4	77 08.7	. . 24.1
22		176 36.1	179 04.8	40.2	236 07.2	37.2	45 32.4	59.4	92 10.9	24.1
23		191 38.5	194 04.5	39.1	251 08.0	37.0	60 34.7	59.3	107 13.1	24.2
Mer. Pass.		h m 10 19.5	v −0.3	d 1.1	v 0.8	d 0.2	v 2.4	d 0.0	v 2.2	d 0.0

STARS Name	SHA ° ′	Dec ° ′
Acamar	315 24.7	S40 17.7
Achernar	335 33.1	S57 13.3
Acrux	173 18.0	S63 07.1
Adhara	255 19.0	S28 58.8
Aldebaran	290 58.9	N16 30.9
Alioth	166 26.9	N55 56.6
Alkaid	153 04.6	N49 17.8
Al Na'ir	27 53.9	S46 56.7
Alnilam	275 54.7	S 1 12.1
Alphard	218 03.9	S 8 40.4
Alphecca	126 17.4	N26 42.1
Alpheratz	357 52.2	N29 06.2
Altair	62 16.1	N 8 52.3
Ankaa	353 23.9	S42 17.4
Antares	112 35.9	S26 26.4
Arcturus	146 02.7	N19 09.8
Atria	107 44.7	S69 01.9
Avior	234 21.4	S59 31.4
Bellatrix	278 40.8	N 6 21.1
Betelgeuse	271 10.2	N 7 24.4
Canopus	263 59.9	S52 42.1
Capella	280 46.7	N46 00.2
Deneb	49 37.1	N45 17.1
Denebola	182 41.5	N14 33.3
Diphda	349 04.2	S17 58.3
Dubhe	194 00.7	N61 44.2
Elnath	278 23.0	N28 36.7
Eltanin	90 49.6	N51 28.9
Enif	33 55.2	N 9 53.1
Fomalhaut	15 33.0	S29 36.4
Gacrux	172 09.6	S57 08.0
Gienah	176 00.3	S17 33.7
Hadar	148 58.9	S60 23.3
Hamal	328 10.3	N23 28.5
Kaus Aust.	83 54.3	S34 23.0
Kochab	137 18.2	N74 08.4
Markab	13 46.6	N15 13.1
Menkar	314 23.8	N 4 06.0
Menkent	148 16.8	S36 23.2
Miaplacidus	221 41.4	S69 44.0
Mirfak	308 52.5	N49 52.4
Nunki	76 08.2	S26 17.6
Peacock	53 31.8	S56 43.4
Pollux	243 37.5	N28 01.2
Procyon	245 08.2	N 5 13.0
Rasalhague	96 13.7	N12 33.2
Regulus	207 51.9	N11 57.1
Rigel	281 20.0	S 8 12.0
Rigil Kent.	140 02.4	S60 50.9
Sabik	102 21.6	S15 43.8
Schedar	349 50.5	N56 33.1
Shaula	96 32.6	S37 06.4
Sirius	258 41.0	S16 43.3
Spica	158 39.4	S11 10.8
Suhail	222 58.3	S43 26.9
Vega	80 44.3	N38 46.8
Zuben'ubi	137 14.0	S16 03.4

	SHA ° ′	Mer. Pass. h m
Venus	4 37.4	10 03
Mars	60 45.5	6 18
Jupiter	229 01.3	19 02
Saturn	275 45.9	15 56

UT		SUN GHA	SUN Dec	MOON GHA	v	MOON Dec	d	HP
d	h	° ′	° ′	° ′	′	° ′	′	′
16 WEDNESDAY	00	179 59.9	N 9 53.8	9 24.4	9.5	S 2 04.4	16.8	61.2
	01	195 00.0	54.7	23 52.9	9.4	2 21.2	16.8	61.2
	02	210 00.2	55.6	38 21.3	9.5	2 38.0	16.8	61.2
	03	225 00.3	. . 56.5	52 49.8	9.3	2 54.8	16.8	61.2
	04	240 00.5	57.4	67 18.1	9.4	3 11.6	16.8	61.2
	05	255 00.6	58.3	81 46.5	9.3	3 28.4	16.7	61.2
	06	270 00.8	N 9 59.2	96 14.8	9.3	S 3 45.1	16.8	61.2
	07	285 00.9	10 00.1	110 43.1	9.2	4 01.9	16.7	61.3
	08	300 01.1	01.0	125 11.3	9.2	4 18.6	16.7	61.3
	09	315 01.2	. . 01.8	139 39.5	9.2	4 35.3	16.7	61.3
	10	330 01.4	02.7	154 07.7	9.1	4 52.0	16.7	61.3
	11	345 01.5	03.6	168 35.8	9.1	5 08.7	16.6	61.3
	12	0 01.7	N10 04.5	183 03.9	9.0	S 5 25.3	16.6	61.3
	13	15 01.8	05.4	197 31.9	9.0	5 41.9	16.6	61.3
	14	30 02.0	06.3	211 59.9	8.9	5 58.5	16.6	61.3
	15	45 02.1	. . 07.2	226 27.8	8.9	6 15.1	16.5	61.3
	16	60 02.3	08.1	240 55.7	8.9	6 31.6	16.5	61.3
	17	75 02.4	08.9	255 23.6	8.8	6 48.1	16.5	61.4
	18	90 02.6	N10 09.8	269 51.4	8.7	S 7 04.6	16.4	61.4
	19	105 02.7	10.7	284 19.1	8.7	7 21.0	16.4	61.4
	20	120 02.9	11.6	298 46.8	8.7	7 37.4	16.3	61.4
	21	135 03.0	. . 12.5	313 14.5	8.6	7 53.7	16.3	61.4
	22	150 03.2	13.4	327 42.1	8.5	8 10.0	16.2	61.4
	23	165 03.3	14.3	342 09.6	8.5	8 26.2	16.2	61.4
17 THURSDAY	00	180 03.5	N10 15.1	356 37.1	8.5	S 8 42.4	16.1	61.4
	01	195 03.6	16.0	11 04.6	8.4	8 58.5	16.1	61.4
	02	210 03.8	16.9	25 32.0	8.3	9 14.6	16.1	61.4
	03	225 03.9	. . 17.8	39 59.3	8.3	9 30.7	15.9	61.4
	04	240 04.0	18.7	54 26.6	8.2	9 46.6	15.9	61.4
	05	255 04.2	19.6	68 53.8	8.2	10 02.5	15.9	61.4
	06	270 04.3	N10 20.4	83 21.0	8.1	S10 18.4	15.8	61.4
	07	285 04.5	21.3	97 48.1	8.0	10 34.2	15.7	61.4
	08	300 04.6	22.2	112 15.1	8.0	10 49.9	15.6	61.4
	09	315 04.8	. . 23.1	126 42.1	8.0	11 05.5	15.6	61.4
	10	330 04.9	24.0	141 09.1	7.8	11 21.1	15.5	61.4
	11	345 05.1	24.8	155 35.9	7.8	11 36.6	15.4	61.4
	12	0 05.2	N10 25.7	170 02.7	7.8	S11 52.0	15.4	61.4
	13	15 05.4	26.6	184 29.5	7.7	12 07.4	15.3	61.4
	14	30 05.5	27.5	198 56.2	7.6	12 22.7	15.2	61.4
	15	45 05.7	. . 28.4	213 22.8	7.6	12 37.9	15.1	61.4
	16	60 05.8	29.2	227 49.4	7.5	12 53.0	15.0	61.4
	17	75 05.9	30.1	242 15.9	7.4	13 08.0	14.9	61.4
	18	90 06.1	N10 31.0	256 42.3	7.4	S13 22.9	14.9	61.3
	19	105 06.2	31.9	271 08.7	7.3	13 37.8	14.8	61.3
	20	120 06.4	32.8	285 35.0	7.2	13 52.6	14.6	61.3
	21	135 06.5	. . 33.6	300 01.2	7.2	14 07.2	14.6	61.3
	22	150 06.7	34.5	314 27.4	7.1	14 21.8	14.5	61.3
	23	165 06.8	35.4	328 53.5	7.0	14 36.3	14.3	61.3
18 FRIDAY	00	180 07.0	N10 36.3	343 19.5	7.0	S14 50.6	14.3	61.3
	01	195 07.1	37.1	357 45.5	6.9	15 04.9	14.2	61.3
	02	210 07.2	38.0	12 11.4	6.9	15 19.1	14.0	61.3
	03	225 07.4	. . 38.9	26 37.3	6.8	15 33.1	14.0	61.3
	04	240 07.5	39.8	41 03.1	6.7	15 47.1	13.9	61.2
	05	255 07.7	40.6	55 28.8	6.6	16 01.0	13.7	61.2
	06	270 07.8	N10 41.5	69 54.4	6.6	S16 14.7	13.6	61.2
	07	285 08.0	42.4	84 20.0	6.5	16 28.3	13.5	61.2
	08	300 08.1	43.3	98 45.5	6.5	16 41.8	13.4	61.2
	09	315 08.2	. . 44.1	113 11.0	6.3	16 55.2	13.3	61.2
	10	330 08.4	45.0	127 36.3	6.4	17 08.5	13.2	61.2
	11	345 08.5	45.9	142 01.7	6.2	17 21.7	13.0	61.1
	12	0 08.7	N10 46.8	156 26.9	6.2	S17 34.7	12.9	61.1
	13	15 08.8	47.6	170 52.1	6.1	17 47.6	12.8	61.1
	14	30 08.9	48.5	185 17.2	6.0	18 00.4	12.6	61.1
	15	45 09.1	. . 49.4	199 42.2	6.0	18 13.0	12.6	61.1
	16	60 09.2	50.2	214 07.2	6.0	18 25.6	12.4	61.0
	17	75 09.4	51.1	228 32.2	5.8	18 38.0	12.2	61.0
	18	90 09.5	N10 52.0	242 57.0	5.8	S18 50.2	12.1	61.0
	19	105 09.6	52.9	257 21.8	5.7	19 02.3	12.0	61.0
	20	120 09.8	53.7	271 46.5	5.7	19 14.3	11.9	61.0
	21	135 09.9	. . 54.6	286 11.2	5.6	19 26.2	11.7	60.9
	22	150 10.1	55.5	300 35.8	5.6	19 37.9	11.6	60.9
	23	165 10.2	56.3	315 00.4	5.4	S19 49.5	11.4	60.9
		SD 16.0	d 0.9	SD 16.7		16.7		16.7

Lat.	Twilight Naut.	Twilight Civil	Sunrise	Moonrise 16	Moonrise 17	Moonrise 18	Moonrise 19
°	h m	h m	h m	h m	h m	h m	h m
N 72	////	01 42	03 30	19 43	22 39	■	■
N 70	////	02 24	03 48	19 31	22 04	■	■
68	////	02 51	04 02	19 21	21 40	24 28	00 28
66	01 28	03 12	04 14	19 13	21 22	23 44	■
64	02 05	03 28	04 24	19 06	21 07	23 14	25 32
62	02 30	03 41	04 32	19 00	20 54	22 52	24 51
60	02 50	03 52	04 39	18 55	20 44	22 35	24 23
N 58	03 05	04 02	04 45	18 51	20 35	22 20	24 02
56	03 18	04 10	04 51	18 47	20 27	22 07	23 44
54	03 29	04 18	04 56	18 44	20 20	21 56	23 30
52	03 39	04 24	05 00	18 41	20 13	21 47	23 17
50	03 47	04 30	05 04	18 38	20 08	21 38	23 05
45	04 05	04 42	05 13	18 32	19 56	21 20	22 42
N 40	04 18	04 52	05 20	18 27	19 46	21 05	22 23
35	04 29	05 00	05 26	18 22	19 37	20 53	22 08
30	04 38	05 07	05 32	18 19	19 30	20 42	21 54
20	04 52	05 18	05 41	18 12	19 17	20 24	21 31
N 10	05 03	05 28	05 49	18 07	19 06	20 08	21 12
0	05 11	05 35	05 56	18 01	18 56	19 53	20 53
S 10	05 18	05 42	06 04	17 56	18 46	19 39	20 35
20	05 23	05 49	06 11	17 51	18 35	19 23	20 16
30	05 28	05 56	06 20	17 45	18 23	19 06	19 54
35	05 30	05 59	06 25	17 41	18 16	18 55	19 41
40	05 31	06 03	06 30	17 37	18 08	18 44	19 26
45	05 33	06 07	06 37	17 33	17 59	18 30	19 08
S 50	05 34	06 12	06 45	17 27	17 48	18 14	18 46
52	05 34	06 14	06 48	17 25	17 43	18 06	18 36
54	05 35	06 16	06 52	17 22	17 38	17 58	18 25
56	05 35	06 18	06 56	17 19	17 32	17 48	18 11
58	05 35	06 21	07 01	17 16	17 25	17 37	17 56
S 60	05 35	06 24	07 06	17 12	17 17	17 25	17 38

Lat.	Sunset	Twilight Civil	Twilight Naut.	Moonset 16	Moonset 17	Moonset 18	Moonset 19
°	h m	h m	h m	h m	h m	h m	h m
N 72	20 33	22 27	////	04 48	04 11	03 12	■
N 70	20 14	21 41	////	04 52	04 26	03 49	■
68	20 00	21 12	////	04 56	04 39	04 15	03 29
66	19 48	20 51	22 40	05 00	04 49	04 35	04 15
64	19 38	20 34	21 59	05 03	04 57	04 52	04 45
62	19 29	20 20	21 33	05 05	05 05	05 06	05 08
60	19 22	20 09	21 13	05 07	05 11	05 17	05 27
N 58	19 16	19 59	20 56	05 09	05 17	05 27	05 42
56	19 10	19 51	20 43	05 11	05 22	05 36	05 55
54	19 05	19 43	20 32	05 13	05 27	05 44	06 07
52	19 00	19 37	20 22	05 14	05 31	05 51	06 17
50	18 56	19 31	20 13	05 15	05 35	05 58	06 26
45	18 47	19 18	19 56	05 18	05 43	06 12	06 46
N 40	18 40	19 08	19 42	05 21	05 50	06 23	07 02
35	18 34	19 00	19 31	05 23	05 56	06 33	07 15
30	18 28	18 53	19 22	05 25	06 02	06 42	07 27
20	18 19	18 41	19 08	05 28	06 11	06 57	07 47
N 10	18 11	18 32	18 57	05 31	06 19	07 10	08 05
0	18 03	18 24	18 48	05 33	06 27	07 23	08 22
S 10	17 56	18 17	18 42	05 36	06 35	07 35	08 38
20	17 48	18 10	18 36	05 39	06 43	07 49	08 56
30	17 39	18 03	18 31	05 42	06 52	08 04	09 17
35	17 34	18 00	18 29	05 44	06 58	08 13	09 29
40	17 28	17 56	18 27	05 46	07 04	08 24	09 43
45	17 22	17 52	18 26	05 48	07 11	08 36	09 59
S 50	17 14	17 47	18 24	05 51	07 20	08 51	10 20
52	17 10	17 45	18 24	05 53	07 24	08 58	10 30
54	17 06	17 43	18 24	05 54	07 29	09 05	10 41
56	17 02	17 40	18 23	05 56	07 34	09 14	10 53
58	16 57	17 37	18 23	05 58	07 40	09 24	11 08
S 60	16 52	17 35	18 23	06 00	07 46	09 35	11 25

Day	SUN Eqn. of Time 00^h	SUN Eqn. of Time 12^h	SUN Mer. Pass.	MOON Mer. Pass. Upper	MOON Mer. Pass. Lower	MOON Age	MOON Phase
d	m s	m s	h m	h m	h m	d %	
16	00 01	00 06	12 00	24 14	11 47	15 100	○
17	00 14	00 21	12 00	00 14	12 41	16 99	
18	00 28	00 34	11 59	01 09	13 38	17 96	

UT d h	ARIES GHA ° ′	VENUS −3.9 GHA ° ′	VENUS Dec ° ′	MARS +0.2 GHA ° ′	MARS Dec ° ′	JUPITER −2.2 GHA ° ′	JUPITER Dec ° ′	SATURN +0.1 GHA ° ′	SATURN Dec ° ′
19 00	206 41.0	209 04.1	S 2 38.0	266 08.9	S21 36.7	75 37.1	N18 59.3	122 15.3	N22 24.2
01	221 43.5	224 03.8	36.8	281 09.8	36.5	90 39.4	59.3	137 17.5	24.2
02	236 45.9	239 03.5	35.7	296 10.6	36.3	105 41.8	59.2	152 19.7	24.2
03	251 48.4	254 03.1	34.5	311 11.5	36.0	120 44.1	59.2	167 22.0	24.2
04	266 50.8	269 02.8	33.4	326 12.3	35.8	135 46.5	59.2	182 24.2	24.2
05	281 53.3	284 02.5	32.2	341 13.2	35.6	150 48.8	59.1	197 26.4	24.2
06	296 55.8	299 02.1	S 2 31.1	356 14.0	S21 35.3	165 51.2	N18 59.1	212 28.6	N22 24.3
S 07	311 58.2	314 01.8	29.9	11 14.9	35.1	180 53.5	59.1	227 30.8	24.3
A 08	327 00.7	329 01.5	28.8	26 15.7	34.9	195 55.9	59.0	242 33.1	24.3
T 09	342 03.2	344 01.2	27.6	41 16.6	34.6	210 58.2	59.0	257 35.3	24.3
U 10	357 05.6	359 00.8	26.5	56 17.4	34.4	226 00.5	59.0	272 37.5	24.3
R 11	12 08.1	14 00.5	25.3	71 18.3	34.1	241 02.9	58.9	287 39.7	24.3
D 12	27 10.6	29 00.2	S 2 24.2	86 19.1	S21 33.9	256 05.2	N18 58.9	302 41.9	N22 24.4
A 13	42 13.0	43 59.8	23.1	101 20.0	33.7	271 07.6	58.9	317 44.2	24.4
Y 14	57 15.5	58 59.5	21.9	116 20.9	33.4	286 09.9	58.8	332 46.4	24.4
15	72 18.0	73 59.2	20.8	131 21.7	33.2	301 12.3	58.8	347 48.6	24.4
16	87 20.4	88 58.8	19.6	146 22.6	33.0	316 14.6	58.8	2 50.8	24.4
17	102 22.9	103 58.5	18.5	161 23.4	32.7	331 16.9	58.7	17 53.0	24.4
18	117 25.3	118 58.2	S 2 17.3	176 24.3	S21 32.5	346 19.3	N18 58.7	32 55.2	N22 24.4
19	132 27.8	133 57.9	16.2	191 25.1	32.2	1 21.6	58.7	47 57.5	24.5
20	147 30.3	148 57.5	15.0	206 26.0	32.0	16 24.0	58.6	62 59.7	24.5
21	162 32.7	163 57.2	13.9	221 26.8	31.8	31 26.3	58.6	78 01.9	24.5
22	177 35.2	178 56.9	12.7	236 27.7	31.5	46 28.7	58.6	93 04.1	24.5
23	192 37.7	193 56.5	11.6	251 28.6	31.3	61 31.0	58.5	108 06.3	24.5
20 00	207 40.1	208 56.2	S 2 10.4	266 29.4	S21 31.1	76 33.3	N18 58.5	123 08.5	N22 24.5
01	222 42.6	223 55.9	09.3	281 30.3	30.8	91 35.7	58.5	138 10.8	24.6
02	237 45.1	238 55.6	08.1	296 31.1	30.6	106 38.0	58.4	153 13.0	24.6
03	252 47.5	253 55.2	07.0	311 32.0	30.3	121 40.4	58.4	168 15.2	24.6
04	267 50.0	268 54.9	05.8	326 32.9	30.1	136 42.7	58.3	183 17.4	24.6
05	282 52.4	283 54.6	04.7	341 33.7	29.9	151 45.0	58.3	198 19.6	24.6
06	297 54.9	298 54.2	S 2 03.5	356 34.6	S21 29.6	166 47.4	N18 58.3	213 21.8	N22 24.6
07	312 57.4	313 53.9	02.4	11 35.4	29.4	181 49.7	58.2	228 24.0	24.6
08	327 59.8	328 53.6	01.2	26 36.3	29.1	196 52.1	58.2	243 26.3	24.7
S 09	343 02.3	343 53.3	2 00.1	41 37.1	28.9	211 54.4	58.2	258 28.5	24.7
U 10	358 04.8	358 52.9	1 58.9	56 38.0	28.7	226 56.7	58.1	273 30.7	24.7
N 11	13 07.2	13 52.6	57.8	71 38.9	28.4	241 59.1	58.1	288 32.9	24.7
D 12	28 09.7	28 52.3	S 1 56.6	86 39.7	S21 28.2	257 01.4	N18 58.1	303 35.1	N22 24.7
A 13	43 12.2	43 51.9	55.5	101 40.6	27.9	272 03.7	58.0	318 37.3	24.7
Y 14	58 14.6	58 51.6	54.3	116 41.4	27.7	287 06.1	58.0	333 39.6	24.8
15	73 17.1	73 51.3	53.2	131 42.3	27.5	302 08.4	58.0	348 41.8	24.8
16	88 19.6	88 51.0	52.0	146 43.2	27.2	317 10.7	57.9	3 44.0	24.8
17	103 22.0	103 50.6	50.9	161 44.0	27.0	332 13.1	57.9	18 46.2	24.8
18	118 24.5	118 50.3	S 1 49.7	176 44.9	S21 26.7	347 15.4	N18 57.9	33 48.4	N22 24.8
19	133 26.9	133 50.0	48.6	191 45.8	26.5	2 17.7	57.8	48 50.6	24.8
20	148 29.4	148 49.6	47.4	206 46.6	26.2	17 20.1	57.8	63 52.8	24.9
21	163 31.9	163 49.3	46.3	221 47.5	26.0	32 22.4	57.7	78 55.0	24.9
22	178 34.3	178 49.0	45.1	236 48.3	25.8	47 24.8	57.7	93 57.3	24.9
23	193 36.8	193 48.7	43.9	251 49.2	25.5	62 27.1	57.7	108 59.5	24.9
21 00	208 39.3	208 48.3	S 1 42.8	266 50.1	S21 25.3	77 29.4	N18 57.6	124 01.7	N22 24.9
01	223 41.7	223 48.0	41.6	281 50.9	25.0	92 31.8	57.6	139 03.9	24.9
02	238 44.2	238 47.7	40.5	296 51.8	24.8	107 34.1	57.6	154 06.1	24.9
03	253 46.7	253 47.3	39.3	311 52.6	24.5	122 36.4	57.5	169 08.3	25.0
04	268 49.1	268 47.0	38.2	326 53.5	24.3	137 38.7	57.5	184 10.5	25.0
05	283 51.6	283 46.7	37.0	341 54.4	24.1	152 41.1	57.4	199 12.8	25.0
06	298 54.1	298 46.4	S 1 35.9	356 55.2	S21 23.8	167 43.4	N18 57.4	214 15.0	N22 25.0
07	313 56.5	313 46.0	34.7	11 56.1	23.6	182 45.7	57.4	229 17.2	25.0
08	328 59.0	328 45.7	33.6	26 57.0	23.3	197 48.1	57.3	244 19.4	25.0
M 09	344 01.4	343 45.4	32.4	41 57.8	23.1	212 50.4	57.3	259 21.6	25.1
O 10	359 03.9	358 45.1	31.3	56 58.7	22.8	227 52.7	57.3	274 23.8	25.1
N 11	14 06.4	13 44.7	30.1	71 59.6	22.6	242 55.1	57.2	289 26.0	25.1
D 12	29 08.8	28 44.4	S 1 28.9	87 00.4	S21 22.3	257 57.4	N18 57.2	304 28.2	N22 25.1
A 13	44 11.3	43 44.1	27.8	102 01.3	22.1	272 59.7	57.2	319 30.4	25.1
Y 14	59 13.8	58 43.7	26.6	117 02.2	21.8	288 02.1	57.1	334 32.7	25.1
15	74 16.2	73 43.4	25.5	132 03.0	21.6	303 04.4	57.1	349 34.9	25.1
16	89 18.7	88 43.1	24.3	147 03.9	21.4	318 06.7	57.0	4 37.1	25.2
17	104 21.2	103 42.8	23.2	162 04.8	21.1	333 09.0	57.0	19 39.3	25.2
18	119 23.6	118 42.4	S 1 22.0	177 05.6	S21 20.9	348 11.4	N18 57.0	34 41.5	N22 25.2
19	134 26.1	133 42.1	20.9	192 06.5	20.6	3 13.7	56.9	49 43.7	25.2
20	149 28.6	148 41.8	19.7	207 07.4	20.4	18 16.0	56.9	64 45.9	25.2
21	164 31.0	163 41.5	18.6	222 08.2	20.1	33 18.3	56.8	79 48.1	25.2
22	179 33.5	178 41.1	17.4	237 09.1	19.9	48 20.7	56.8	94 50.3	25.3
23	194 35.9	193 40.8	16.2	252 10.0	19.6	63 23.0	56.8	109 52.6	25.3
Mer. Pass.	h m 10 07.7	*v* −0.3	*d* 1.2	*v* 0.9	*d* 0.2	*v* 2.3	*d* 0.0	*v* 2.2	*d* 0.0

STARS

Name	SHA ° ′	Dec ° ′
Acamar	315 24.7	S40 17.7
Achernar	335 33.1	S57 13.3
Acrux	173 18.0	S63 07.1
Adhara	255 19.0	S28 58.8
Aldebaran	290 58.9	N16 30.9
Alioth	166 27.0	N55 56.6
Alkaid	153 04.6	N49 17.8
Al Na'ir	27 53.9	S46 56.7
Alnilam	275 54.7	S 1 12.1
Alphard	218 03.9	S 8 40.4
Alphecca	126 17.4	N26 42.1
Alpheratz	357 52.2	N29 06.2
Altair	62 16.0	N 8 52.4
Ankaa	353 23.9	S42 17.4
Antares	112 35.9	S26 26.4
Arcturus	146 02.7	N19 09.8
Atria	107 44.7	S69 01.9
Avior	234 21.4	S59 31.4
Bellatrix	278 40.8	N 6 21.1
Betelgeuse	271 10.2	N 7 24.4
Canopus	264 00.0	S52 42.1
Capella	280 46.7	N46 00.2
Deneb	49 37.1	N45 17.1
Denebola	182 41.5	N14 33.3
Diphda	349 04.2	S17 58.3
Dubhe	194 00.7	N61 44.2
Elnath	278 23.0	N28 36.7
Eltanin	90 49.6	N51 29.0
Enif	33 55.2	N 9 53.1
Fomalhaut	15 33.0	S29 36.4
Gacrux	172 09.6	S57 08.0
Gienah	176 00.3	S17 33.7
Hadar	148 58.9	S60 23.3
Hamal	328 10.2	N23 28.5
Kaus Aust.	83 54.3	S34 23.0
Kochab	137 18.2	N74 08.5
Markab	13 46.6	N15 13.1
Menkar	314 23.8	N 4 06.0
Menkent	148 16.8	S36 23.2
Miaplacidus	221 41.4	S69 44.0
Mirfak	308 52.5	N49 52.4
Nunki	76 08.1	S26 17.6
Peacock	53 31.8	S56 43.4
Pollux	243 37.5	N28 01.2
Procyon	245 08.2	N 5 13.0
Rasalhague	96 13.7	N12 33.2
Regulus	207 51.9	N11 57.1
Rigel	281 20.0	S 8 12.0
Rigil Kent.	140 02.4	S60 50.9
Sabik	102 21.5	S15 43.8
Schedar	349 50.5	N56 33.1
Shaula	96 32.6	S37 06.4
Sirius	258 41.0	S16 43.3
Spica	158 39.4	S11 10.8
Suhail	222 58.3	S43 26.9
Vega	80 44.3	N38 46.8
Zuben'ubi	137 14.0	S16 03.4

	SHA ° ′	Mer. Pass h m
Venus	1 16.1	10 04
Mars	58 49.3	6 14
Jupiter	228 53.2	18 51
Saturn	275 28.4	15 45

UT	SUN GHA	SUN Dec	MOON GHA	v	MOON Dec	d	HP
d h	° ′	° ′	° ′	′	° ′	′	′
19 00	180 10.3	N10 57.2	329 24.8	5.5	S20 00.9	11.3	60.9
01	195 10.5	58.1	343 49.3	5.3	20 12.2	11.1	60.9
02	210 10.6	59.0	358 13.6	5.3	20 23.3	11.0	60.8
03	225 10.8	10 59.8	12 37.9	5.3	20 34.3	10.9	60.8
04	240 10.9	11 00.7	27 02.2	5.2	20 45.2	10.7	60.8
05	255 11.0	01.6	41 26.4	5.1	20 55.9	10.5	60.8
06	270 11.2	N11 02.4	55 50.5	5.1	S21 06.4	10.4	60.7
S 07	285 11.3	03.3	70 14.6	5.0	21 16.8	10.2	60.7
A 08	300 11.5	04.2	84 38.6	5.0	21 27.0	10.1	60.7
T 09	315 11.6 . .	05.0	99 02.6	4.9	21 37.1	9.9	60.6
U 10	330 11.7	05.9	113 26.5	4.9	21 47.0	9.8	60.6
R 11	345 11.9	06.8	127 50.4	4.8	21 56.8	9.6	60.6
D 12	0 12.0	N11 07.6	142 14.2	4.7	S22 06.4	9.4	60.6
A 13	15 12.1	08.5	156 37.9	4.8	22 15.8	9.3	60.5
Y 14	30 12.3	09.4	171 01.7	4.6	22 25.1	9.1	60.5
15	45 12.4 . .	10.2	185 25.3	4.7	22 34.2	9.0	60.5
16	60 12.6	11.1	199 49.0	4.6	22 43.2	8.8	60.5
17	75 12.7	11.9	214 12.6	4.5	22 52.0	8.6	60.4
18	90 12.8	N11 12.8	228 36.1	4.5	S23 00.6	8.4	60.4
19	105 13.0	13.7	242 59.6	4.5	23 09.0	8.3	60.4
20	120 13.1	14.5	257 23.1	4.4	23 17.3	8.1	60.3
21	135 13.2 . .	15.4	271 46.5	4.4	23 25.4	8.0	60.3
22	150 13.4	16.3	286 09.9	4.3	23 33.4	7.8	60.3
23	165 13.5	17.1	300 33.2	4.4	23 41.2	7.6	60.2
20 00	180 13.6	N11 18.0	314 56.6	4.3	S23 48.8	7.4	60.2
01	195 13.8	18.8	329 19.9	4.2	23 56.2	7.2	60.2
02	210 13.9	19.7	343 43.1	4.3	24 03.4	7.1	60.1
03	225 14.0 . .	20.6	358 06.4	4.2	24 10.5	6.9	60.1
04	240 14.2	21.4	12 29.6	4.2	24 17.4	6.7	60.1
05	255 14.3	22.3	26 52.8	4.1	24 24.1	6.6	60.0
06	270 14.4	N11 23.1	41 15.9	4.2	S24 30.7	6.4	60.0
07	285 14.6	24.0	55 39.1	4.1	24 37.1	6.2	60.0
08	300 14.7	24.9	70 02.2	4.1	24 43.3	6.0	59.9
S 09	315 14.8 . .	25.7	84 25.3	4.1	24 49.3	5.8	59.9
U 10	330 15.0	26.6	98 48.4	4.1	24 55.1	5.7	59.9
N 11	345 15.1	27.4	113 11.5	4.0	25 00.8	5.5	59.8
D 12	0 15.2	N11 28.3	127 34.5	4.1	S25 06.3	5.3	59.8
A 13	15 15.4	29.2	141 57.6	4.1	25 11.6	5.1	59.8
Y 14	30 15.5	30.0	156 20.7	4.0	25 16.7	4.9	59.7
15	45 15.6 . .	30.9	170 43.7	4.1	25 21.6	4.8	59.7
16	60 15.8	31.7	185 06.8	4.0	25 26.4	4.6	59.7
17	75 15.9	32.6	199 29.8	4.0	25 31.0	4.3	59.6
18	90 16.0	N11 33.4	213 52.8	4.1	S25 35.3	4.3	59.6
19	105 16.2	34.3	228 15.9	4.0	25 39.6	4.0	59.6
20	120 16.3	35.2	242 38.9	4.1	25 43.6	3.8	59.5
21	135 16.4 . .	36.0	257 02.0	4.1	25 47.4	3.7	59.5
22	150 16.6	36.9	271 25.1	4.1	25 51.1	3.5	59.4
23	165 16.7	37.7	285 48.2	4.0	25 54.6	3.3	59.4
21 00	180 16.8	N11 38.6	300 11.2	4.2	S25 57.9	3.1	59.4
01	195 17.0	39.4	314 34.4	4.1	26 01.0	3.0	59.3
02	210 17.1	40.3	328 57.5	4.1	26 04.0	2.7	59.3
03	225 17.2 . .	41.1	343 20.6	4.2	26 06.7	2.6	59.3
04	240 17.3	42.0	357 43.8	4.2	26 09.3	2.4	59.2
05	255 17.5	42.8	12 07.0	4.2	26 11.7	2.2	59.2
06	270 17.6	N11 43.7	26 30.2	4.3	S26 13.9	2.1	59.1
07	285 17.7	44.5	40 53.5	4.3	26 16.0	1.8	59.1
08	300 17.9	45.4	55 16.8	4.3	26 17.8	1.7	59.1
M 09	315 18.0 . .	46.2	69 40.1	4.3	26 19.5	1.5	59.0
O 10	330 18.1	47.1	84 03.4	4.4	26 21.0	1.3	59.0
N 11	345 18.2	47.9	98 26.8	4.4	26 22.3	1.2	59.0
D 12	0 18.4	N11 48.8	112 50.2	4.5	S26 23.5	1.0	58.9
A 13	15 18.5	49.6	127 13.7	4.5	26 24.5	0.7	58.9
Y 14	30 18.6	50.5	141 37.2	4.5	26 25.2	0.7	58.8
15	45 18.8 . .	51.3	156 00.7	4.6	26 25.9	0.4	58.8
16	60 18.9	52.2	170 24.3	4.7	26 26.3	0.3	58.8
17	75 19.0	53.0	184 48.0	4.7	26 26.6	0.0	58.7
18	90 19.1	N11 53.9	199 11.7	4.7	S26 26.6	0.0	58.7
19	105 19.3	54.7	213 35.4	4.8	26 26.6	0.3	58.7
20	120 19.4	55.6	227 59.2	4.9	26 26.3	0.4	58.6
21	135 19.5 . .	56.4	242 23.1	4.9	26 25.9	0.6	58.6
22	150 19.6	57.3	256 47.0	5.0	26 25.3	0.8	58.5
23	165 19.8	58.1	271 11.0	5.0	S26 24.5	1.0	58.5
	SD 15.9	d 0.9	SD 16.5		16.3		16.1

Lat.	Twilight Naut.	Twilight Civil	Sunrise	Moonrise 19	Moonrise 20	Moonrise 21	Moonrise 22
°	h m	h m	h m	h m	h m	h m	h m
N 72	////	01 03	03 12	■	■	■	■
N 70	////	02 01	03 33	■	■	■	■
68	////	02 34	03 49	00 28	■	■	■
66	00 55	02 57	04 02	■	■	■	■
64	01 45	03 16	04 13	25 32	01 32	■	■
62	02 15	03 30	04 22	24 51	00 51	02 38	03 44
60	02 37	03 43	04 30	24 23	00 23	01 58	03 03
N 58	02 54	03 53	04 37	24 02	00 02	01 31	02 35
56	03 09	04 02	04 43	23 44	25 09	01 09	02 14
54	03 21	04 10	04 49	23 30	24 52	00 52	01 56
52	03 31	04 17	04 54	23 17	24 37	00 37	01 40
50	03 40	04 23	04 58	23 05	24 24	00 24	01 27
45	03 59	04 37	05 08	22 42	23 57	25 00	01 00
N 40	04 13	04 47	05 16	22 23	23 36	24 38	00 38
35	04 25	04 56	05 22	22 08	23 18	24 21	00 21
30	04 34	05 04	05 28	21 54	23 03	24 05	00 05
20	04 49	05 16	05 39	21 31	22 37	23 39	24 35
N 10	05 01	05 26	05 47	21 12	22 15	23 17	24 14
0	05 10	05 35	05 56	20 53	21 55	22 56	23 54
S 10	05 18	05 42	06 04	20 35	21 34	22 35	23 34
20	05 24	05 50	06 12	20 16	21 13	22 12	23 13
30	05 29	05 57	06 22	19 54	20 47	21 46	22 48
35	05 32	06 02	06 27	19 41	20 33	21 31	22 34
40	05 34	06 06	06 33	19 26	20 16	21 13	22 17
45	05 36	06 11	06 41	19 08	19 55	20 52	21 57
S 50	05 38	06 16	06 49	18 46	19 30	20 25	21 31
52	05 39	06 18	06 53	18 36	19 17	20 12	21 19
54	05 40	06 21	06 57	18 25	19 03	19 57	21 05
56	05 40	06 24	07 02	18 11	18 47	19 39	20 48
58	05 41	06 27	07 08	17 56	18 27	19 17	20 29
S 60	05 42	06 30	07 14	17 38	18 03	18 50	20 04

Lat.	Sunset	Twilight Civil	Twilight Naut.	Moonset 19	Moonset 20	Moonset 21	Moonset 22
°	h m	h m	h m	h m	h m	h m	h m
N 72	20 50	23 12	////	■	■	■	■
N 70	20 28	22 04	////	■	■	■	■
68	20 12	21 29	////	03 29	■	■	■
66	19 58	21 04	23 18	04 15	■	■	■
64	19 47	20 45	22 19	04 45	04 34	■	■
62	19 38	20 30	21 47	05 08	05 15	05 38	06 39
60	19 29	20 18	21 24	05 27	05 44	06 18	07 19
N 58	19 22	20 07	21 06	05 42	06 06	06 45	07 47
56	19 16	19 58	20 51	05 55	06 24	07 07	08 08
54	19 10	19 49	20 39	06 07	06 39	07 25	08 26
52	19 05	19 42	20 28	06 17	06 52	07 40	08 41
50	19 01	19 36	20 19	06 26	07 04	07 53	08 55
45	18 51	19 22	20 00	06 46	07 28	08 20	09 22
N 40	18 43	19 11	19 46	07 02	07 48	08 42	09 43
35	18 36	19 02	19 34	07 15	08 04	08 59	10 00
30	18 30	18 55	19 24	07 27	08 18	09 15	10 16
20	18 20	18 42	19 09	07 47	08 42	09 41	10 41
N 10	18 11	18 32	18 57	08 05	09 03	10 03	11 03
0	18 02	18 23	18 48	08 22	09 23	10 24	11 24
S 10	17 54	18 16	18 40	08 38	09 42	10 45	11 45
20	17 45	18 08	18 34	08 56	10 03	11 07	12 07
30	17 36	18 00	18 28	09 17	10 27	11 33	12 32
35	17 30	17 56	18 26	09 29	10 42	11 49	12 47
40	17 24	17 52	18 23	09 43	10 58	12 06	13 04
45	17 17	17 47	18 21	09 59	11 18	12 28	13 25
S 50	17 08	17 41	18 19	10 20	11 43	12 55	13 51
52	17 04	17 39	18 18	10 30	11 55	13 08	14 03
54	17 00	17 36	18 17	10 41	12 09	13 23	14 17
56	16 55	17 33	18 16	10 53	12 25	13 41	14 34
58	16 49	17 30	18 16	11 08	12 45	14 02	14 54
S 60	16 43	17 27	18 15	11 25	13 09	14 30	15 19

Day	SUN Eqn. of Time 00ʰ	SUN Eqn. of Time 12ʰ	SUN Mer. Pass.	MOON Mer. Pass. Upper	MOON Mer. Pass. Lower	MOON Age	MOON Phase
d	m s	m s	h m	h m	h m	d	%
19	00 41	00 48	11 59	02 07	14 37	18	90
20	00 54	01 01	11 59	03 08	15 39	19	81
21	01 07	01 13	11 59	04 09	16 40	20	71

UT d	UT h	ARIES GHA	VENUS −3.9 GHA	VENUS Dec	MARS +0.1 GHA	MARS Dec	JUPITER −2.2 GHA	JUPITER Dec	SATURN +0.1 GHA	SATURN Dec
d	h	° ′	° ′	° ′	° ′	° ′	° ′	° ′	° ′	° ′
22 TUESDAY	00	209 38.4	208 40.5	S 1 15.1	267 10.8	S21 19.4	78 25.3	N18 56.7	124 54.8	N22 25.3
	01	224 40.9	223 40.2	13.9	282 11.7	19.1	93 27.6	56.7	139 57.0	25.3
	02	239 43.3	238 39.8	12.8	297 12.6	18.9	108 30.0	56.7	154 59.2	25.3
	03	254 45.8	253 39.5	. . 11.6	312 13.4	. . 18.6	123 32.3	. . 56.6	170 01.4	. . 25.3
	04	269 48.3	268 39.2	10.5	327 14.3	18.4	138 34.6	56.6	185 03.6	25.3
	05	284 50.7	283 38.9	09.3	342 15.2	18.1	153 36.9	56.5	200 05.8	25.4
	06	299 53.2	298 38.5	S 1 08.1	357 16.0	S21 17.9	168 39.3	N18 56.5	215 08.0	N22 25.4
	07	314 55.7	313 38.2	07.0	12 16.9	17.6	183 41.6	56.5	230 10.2	25.4
	08	329 58.1	328 37.9	05.8	27 17.8	17.4	198 43.9	56.4	245 12.4	25.4
	09	345 00.6	343 37.5	. . 04.7	42 18.6	. . 17.1	213 46.2	. . 56.4	260 14.6	. . 25.4
	10	0 03.1	358 37.2	03.5	57 19.5	16.9	228 48.6	56.3	275 16.9	25.4
	11	15 05.5	13 36.9	02.4	72 20.4	16.7	243 50.9	56.3	290 19.1	25.5
	12	30 08.0	28 36.6	S 1 01.2	87 21.3	S21 16.4	258 53.2	N18 56.3	305 21.3	N22 25.5
	13	45 10.4	43 36.2	1 00.0	102 22.1	16.2	273 55.5	56.2	320 23.5	25.5
	14	60 12.9	58 35.9	0 58.9	117 23.0	15.9	288 57.9	56.2	335 25.7	25.5
	15	75 15.4	73 35.6	. 57.7	132 23.9	. . 15.7	304 00.2	. . 56.1	350 27.9	. . 25.5
	16	90 17.8	88 35.3	56.6	147 24.7	15.4	319 02.5	56.1	5 30.1	25.5
	17	105 20.3	103 34.9	55.4	162 25.6	15.2	334 04.8	56.1	20 32.3	25.5
	18	120 22.8	118 34.6	S 0 54.3	177 26.5	S21 14.9	349 07.1	N18 56.0	35 34.5	N22 25.6
	19	135 25.2	133 34.3	53.1	192 27.4	14.7	4 09.5	56.0	50 36.7	25.6
	20	150 27.7	148 34.0	51.9	207 28.2	14.4	19 11.8	55.9	65 38.9	25.6
	21	165 30.2	163 33.6	. . 50.8	222 29.1	. . 14.1	34 14.1	. . 55.9	80 41.1	. . 25.6
	22	180 32.6	178 33.3	49.6	237 30.0	13.9	49 16.4	55.9	95 43.3	25.6
	23	195 35.1	193 33.0	48.5	252 30.8	13.6	64 18.7	55.8	110 45.6	25.6
23 WEDNESDAY	00	210 37.5	208 32.7	S 0 47.3	267 31.7	S21 13.4	79 21.0	N18 55.8	125 47.8	N22 25.7
	01	225 40.0	223 32.3	46.1	282 32.6	13.1	94 23.4	55.7	140 50.0	25.7
	02	240 42.5	238 32.0	45.0	297 33.5	12.9	109 25.7	55.7	155 52.2	25.7
	03	255 44.9	253 31.7	. . 43.8	312 34.3	. . 12.6	124 28.0	. . 55.7	170 54.4	. . 25.7
	04	270 47.4	268 31.4	42.7	327 35.2	12.4	139 30.3	55.6	185 56.6	25.7
	05	285 49.9	283 31.0	41.5	342 36.1	12.1	154 32.6	55.6	200 58.8	25.7
	06	300 52.3	298 30.7	S 0 40.4	357 37.0	S21 11.9	169 35.0	N18 55.5	216 01.0	N22 25.7
	07	315 54.8	313 30.4	39.2	12 37.8	11.6	184 37.3	55.5	231 03.2	25.8
	08	330 57.3	328 30.1	38.0	27 38.7	11.4	199 39.6	55.5	246 05.4	25.8
	09	345 59.7	343 29.7	. . 36.9	42 39.6	. . 11.1	214 41.9	. . 55.4	261 07.6	. . 25.8
	10	1 02.2	358 29.4	35.7	57 40.5	10.9	229 44.2	55.4	276 09.8	25.8
	11	16 04.7	13 29.1	34.6	72 41.3	10.6	244 46.5	55.3	291 12.0	25.8
	12	31 07.1	28 28.8	S 0 33.4	87 42.2	S21 10.4	259 48.8	N18 55.3	306 14.2	N22 25.8
	13	46 09.6	43 28.4	32.2	102 43.1	10.1	274 51.2	55.3	321 16.4	25.8
	14	61 12.0	58 28.1	31.1	117 44.0	09.9	289 53.5	55.2	336 18.6	25.9
	15	76 14.5	73 27.8	. . 29.9	132 44.8	. . 09.6	304 55.8	. . 55.2	351 20.8	. . 25.9
	16	91 17.0	88 27.4	28.8	147 45.7	09.4	319 58.1	55.1	6 23.1	25.9
	17	106 19.4	103 27.1	27.6	162 46.6	09.1	335 00.4	55.1	21 25.3	25.9
	18	121 21.9	118 26.8	S 0 26.4	177 47.5	S21 08.8	350 02.7	N18 55.0	36 27.5	N22 25.9
	19	136 24.4	133 26.5	25.3	192 48.4	08.6	5 05.0	55.0	51 29.7	25.9
	20	151 26.8	148 26.1	24.1	207 49.2	08.3	20 07.4	55.0	66 31.9	26.0
	21	166 29.3	163 25.8	. . 23.0	222 50.1	. . 08.1	35 09.7	. . 54.9	81 34.1	. . 26.0
	22	181 31.8	178 25.5	21.8	237 51.0	07.8	50 12.0	54.9	96 36.3	26.0
	23	196 34.2	193 25.2	20.6	252 51.9	07.6	65 14.3	54.8	111 38.5	26.0
24 THURSDAY	00	211 36.7	208 24.8	S 0 19.5	267 52.7	S21 07.3	80 16.6	N18 54.8	126 40.7	N22 26.0
	01	226 39.2	223 24.5	18.3	282 53.6	07.1	95 18.9	54.7	141 42.9	26.0
	02	241 41.6	238 24.2	17.1	297 54.5	06.8	110 21.2	54.7	156 45.1	26.0
	03	256 44.1	253 23.9	. . 16.0	312 55.4	. . 06.5	125 23.5	. . 54.7	171 47.3	. . 26.1
	04	271 46.5	268 23.5	14.8	327 56.3	06.3	140 25.8	54.6	186 49.5	26.1
	05	286 49.0	283 23.2	13.7	342 57.1	06.0	155 28.2	54.6	201 51.7	26.1
	06	301 51.5	298 22.9	S 0 12.5	357 58.0	S21 05.8	170 30.5	N18 54.5	216 53.9	N22 26.1
	07	316 53.9	313 22.6	11.3	12 58.9	05.5	185 32.8	54.5	231 56.1	26.1
	08	331 56.4	328 22.2	10.2	27 59.8	05.3	200 35.1	54.4	246 58.3	26.1
	09	346 58.9	343 21.9	. . 09.0	43 00.7	. . 05.0	215 37.4	. . 54.4	262 00.5	. . 26.2
	10	2 01.3	358 21.6	07.9	58 01.5	04.7	230 39.7	54.4	277 02.7	26.2
	11	17 03.8	13 21.3	06.7	73 02.4	04.5	245 42.0	54.3	292 04.9	26.2
	12	32 06.3	28 20.9	S 0 05.5	88 03.3	S21 04.2	260 44.3	N18 54.3	307 07.1	N22 26.2
	13	47 08.7	43 20.6	04.4	103 04.2	04.0	275 46.6	54.2	322 09.3	26.2
	14	62 11.2	58 20.3	03.2	118 05.1	03.7	290 48.9	54.2	337 11.5	26.2
	15	77 13.7	73 20.0	. . 02.0	133 06.0	. . 03.5	305 51.2	. . 54.1	352 13.7	. . 26.2
	16	92 16.1	88 19.6	S 00.9	148 06.8	03.2	320 53.5	54.1	7 15.9	26.3
	17	107 18.6	103 19.3	N 00.3	163 07.7	02.9	335 55.8	54.1	22 18.1	26.3
	18	122 21.0	118 19.0	N 0 01.4	178 08.6	S21 02.7	350 58.2	N18 54.0	37 20.3	N22 26.3
	19	137 23.5	133 18.7	02.6	193 09.5	02.4	6 00.5	54.0	52 22.5	26.3
	20	152 26.0	148 18.3	03.8	208 10.4	02.2	21 02.8	53.9	67 24.7	26.3
	21	167 28.4	163 18.0	. . 04.9	223 11.3	. . 01.9	36 05.1	. . 53.9	82 26.9	. . 26.3
	22	182 30.9	178 17.7	06.1	238 12.1	01.7	51 07.4	53.8	97 29.1	26.3
	23	197 33.4	193 17.4	07.2	253 13.0	01.4	66 09.7	53.8	112 31.3	26.4
Mer. Pass.		h m 9 55.9	*v* −0.3	*d* 1.2	*v* 0.9	*d* 0.3	*v* 2.3	*d* 0.0	*v* 2.2	*d* 0.0

STARS Name	SHA ° ′	Dec ° ′
Acamar	315 24.7	S40 17.6
Achernar	335 33.1	S57 13.3
Acrux	173 18.0	S63 07.1
Adhara	255 19.0	S28 58.7
Aldebaran	290 58.9	N16 30.9
Alioth	166 26.9	N55 56.6
Alkaid	153 04.6	N49 17.8
Al Na'ir	27 53.8	S46 56.7
Alnilam	275 54.7	S 1 12.1
Alphard	218 03.9	S 8 40.4
Alphecca	126 17.4	N26 42.1
Alpheratz	357 52.2	N29 06.2
Altair	62 16.0	N 8 52.4
Ankaa	353 23.8	S42 17.4
Antares	112 35.8	S26 26.4
Arcturus	146 02.7	N19 09.9
Atria	107 44.6	S69 01.9
Avior	234 21.4	S59 31.4
Bellatrix	278 40.8	N 6 21.1
Betelgeuse	271 10.2	N 7 24.4
Canopus	264 00.0	S52 42.1
Capella	280 46.7	N46 00.2
Deneb	49 37.1	N45 17.1
Denebola	182 41.5	N14 33.3
Diphda	349 04.2	S17 58.3
Dubhe	194 00.8	N61 44.3
Elnath	278 23.0	N28 36.7
Eltanin	90 49.6	N51 29.0
Enif	33 55.1	N 9 53.1
Fomalhaut	15 33.0	S29 36.4
Gacrux	172 09.6	S57 08.0
Gienah	176 00.3	S17 33.7
Hadar	148 58.9	S60 23.4
Hamal	328 10.2	N23 28.5
Kaus Aust.	83 54.2	S34 23.0
Kochab	137 18.1	N74 08.5
Markab	13 46.6	N15 13.1
Menkar	314 23.8	N 4 06.0
Menkent	148 16.8	S36 23.2
Miaplacidus	221 41.5	S69 44.1
Mirfak	308 52.5	N49 52.4
Nunki	76 08.1	S26 17.6
Peacock	53 31.7	S56 43.4
Pollux	243 37.6	N28 01.2
Procyon	245 08.2	N 5 13.0
Rasalhague	96 13.7	N12 33.2
Regulus	207 51.9	N11 57.1
Rigel	281 20.0	S 8 12.0
Rigil Kent.	140 02.4	S60 50.9
Sabik	102 21.5	S15 43.8
Schedar	349 50.5	N56 33.1
Shaula	96 32.6	S37 06.4
Sirius	258 41.0	S16 43.3
Spica	158 39.4	S11 10.8
Suhail	222 58.4	S43 26.9
Vega	80 44.2	N38 46.8
Zuben'ubi	137 14.0	S16 03.4

	SHA ° ′	Mer. Pass. h m
Venus	357 55.1	10 06
Mars	56 54.2	6 10
Jupiter	228 43.5	18 40
Saturn	275 10.2	15 35

UT d h	SUN GHA ° ′	SUN Dec ° ′	MOON GHA ° ′	v ′	MOON Dec ° ′	d ′	HP ′
22 00	180 19.9	N11 59.0	285 35.0	5.1	S26 23.5	1.1	58.5
01	195 20.0	11 59.8	299 59.1	5.2	26 22.4	1.3	58.4
02	210 20.2	12 00.7	314 23.3	5.2	26 21.1	1.4	58.4
03	225 20.3	. . 01.5	328 47.5	5.3	26 19.7	1.6	58.3
04	240 20.4	02.3	343 11.8	5.4	26 18.1	1.8	58.3
05	255 20.5	03.2	357 36.2	5.5	26 16.3	2.0	58.3
06	270 20.7	N12 04.0	12 00.7	5.5	S26 14.3	2.1	58.2
T 07	285 20.8	04.9	26 25.2	5.6	26 12.2	2.3	58.2
U 08	300 20.9	05.7	40 49.8	5.7	26 09.9	2.4	58.2
E 09	315 21.0	. . 06.6	55 14.5	5.7	26 07.5	2.6	58.1
S 10	330 21.1	07.4	69 39.2	5.8	26 04.9	2.7	58.1
D 11	345 21.3	08.3	84 04.0	5.9	26 02.2	3.0	58.0
A 12	0 21.4	N12 09.1	98 28.9	6.0	S25 59.2	3.0	58.0
Y 13	15 21.5	09.9	112 53.9	6.1	25 56.2	3.2	58.0
14	30 21.6	10.8	127 19.0	6.2	25 53.0	3.4	57.9
15	45 21.8	. . 11.6	141 44.2	6.2	25 49.6	3.6	57.9
16	60 21.9	12.5	156 09.4	6.3	25 46.0	3.7	57.9
17	75 22.0	13.3	170 34.7	6.5	25 42.3	3.8	57.8
18	90 22.1	N12 14.1	185 00.2	6.5	S25 38.5	4.0	57.8
19	105 22.3	15.0	199 25.7	6.6	25 34.5	4.1	57.7
20	120 22.4	15.8	213 51.3	6.7	25 30.4	4.3	57.7
21	135 22.5	. . 16.7	228 17.0	6.8	25 26.1	4.4	57.7
22	150 22.6	17.5	242 42.8	6.8	25 21.7	4.6	57.6
23	165 22.7	18.3	257 08.6	7.0	25 17.1	4.7	57.6
23 00	180 22.9	N12 19.2	271 34.6	7.1	S25 12.4	4.9	57.6
01	195 23.0	20.0	286 00.7	7.1	25 07.5	5.0	57.5
02	210 23.1	20.8	300 26.8	7.3	25 02.5	5.2	57.5
03	225 23.2	. . 21.7	314 53.1	7.4	24 57.3	5.2	57.4
04	240 23.3	22.5	329 19.5	7.4	24 52.1	5.5	57.4
05	255 23.5	23.4	343 45.9	7.6	24 46.6	5.5	57.4
06	270 23.6	N12 24.2	358 12.5	7.6	S24 41.1	5.7	57.3
W 07	285 23.7	25.0	12 39.1	7.8	24 35.4	5.8	57.3
E 08	300 23.8	25.9	27 05.9	7.8	24 29.6	6.0	57.3
D 09	315 23.9	. . 26.7	41 32.7	8.0	24 23.6	6.1	57.2
N 10	330 24.1	27.5	55 59.7	8.0	24 17.5	6.2	57.2
E 11	345 24.2	28.4	70 26.7	8.2	24 11.3	6.3	57.2
S 12	0 24.3	N12 29.2	84 53.9	8.3	S24 05.0	6.5	57.1
D 13	15 24.4	30.0	99 21.2	8.3	23 58.5	6.6	57.1
A 14	30 24.5	30.9	113 48.5	8.5	23 51.9	6.7	57.1
Y 15	45 24.6	. . 31.7	128 16.0	8.6	23 45.2	6.9	57.0
16	60 24.8	32.5	142 43.6	8.6	23 38.3	7.0	57.0
17	75 24.9	33.4	157 11.2	8.8	23 31.3	7.0	56.9
18	90 25.0	N12 34.2	171 39.0	8.9	S23 24.3	7.2	56.9
19	105 25.1	35.0	186 06.9	9.0	23 17.1	7.4	56.9
20	120 25.2	35.9	200 34.9	9.1	23 09.7	7.4	56.8
21	135 25.4	. . 36.7	215 03.0	9.1	23 02.3	7.6	56.8
22	150 25.5	37.5	229 31.1	9.3	22 54.7	7.6	56.8
23	165 25.6	38.3	243 59.4	9.4	22 47.1	7.8	56.7
24 00	180 25.7	N12 39.2	258 27.8	9.6	S22 39.3	7.9	56.7
01	195 25.8	40.0	272 56.4	9.6	22 31.4	8.0	56.7
02	210 25.9	40.8	287 25.0	9.7	22 23.4	8.1	56.6
03	225 26.0	. . 41.7	301 53.7	9.8	22 15.3	8.2	56.6
04	240 26.2	42.5	316 22.5	9.9	22 07.1	8.3	56.6
05	255 26.3	43.3	330 51.4	10.0	21 58.8	8.5	56.5
06	270 26.4	N12 44.1	345 20.4	10.2	S21 50.3	8.5	56.5
T 07	285 26.5	45.0	359 49.6	10.2	21 41.8	8.6	56.5
H 08	300 26.6	45.8	14 18.8	10.3	21 33.2	8.7	56.4
U 09	315 26.7	. . 46.6	28 48.1	10.5	21 24.5	8.9	56.4
R 10	330 26.9	47.4	43 17.6	10.5	21 15.6	8.9	56.4
S 11	345 27.0	48.3	57 47.1	10.6	21 06.7	9.0	56.4
D 12	0 27.1	N12 49.1	72 16.7	10.8	S20 57.7	9.1	56.3
A 13	15 27.2	49.9	86 46.5	10.8	20 48.6	9.2	56.3
Y 14	30 27.3	50.7	101 16.3	11.0	20 39.4	9.3	56.3
15	45 27.4	. . 51.6	115 46.3	11.0	20 30.1	9.4	56.2
16	60 27.5	52.4	130 16.3	11.2	20 20.7	9.5	56.2
17	75 27.6	53.2	144 46.5	11.2	20 11.2	9.5	56.2
18	90 27.8	N12 54.0	159 16.7	11.3	S20 01.7	9.7	56.1
19	105 27.9	54.9	173 47.0	11.5	19 52.0	9.7	56.1
20	120 28.0	55.7	188 17.5	11.5	19 42.3	9.9	56.1
21	135 28.1	. . 56.5	202 48.0	11.6	19 32.4	9.9	56.1
22	150 28.2	57.3	217 18.6	11.8	19 22.5	10.0	56.0
23	165 28.3	58.1	231 49.4	11.8	S19 12.5	10.0	56.0
	SD 15.9	*d* 0.8	SD 15.8		15.6		15.3

Lat. °	Twilight Naut. h m	Twilight Civil h m	Sunrise h m	Moonrise 22 h m	Moonrise 23 h m	Moonrise 24 h m	Moonrise 25 h m
N 72	////	////	02 53	■	■	■	07 21
N 70	////	01 34	03 18	■	■	■	06 02
68	////	02 15	03 36	■	■	06 34	05 23
66	////	02 42	03 51	■	■	05 19	04 56
64	01 22	03 03	04 03	■	04 54	04 42	04 35
62	01 59	03 19	04 13	03 44	04 08	04 16	04 18
60	02 24	03 33	04 22	03 03	03 38	03 55	04 04
N 58	02 43	03 44	04 29	02 35	03 15	03 38	03 52
56	02 59	03 54	04 36	02 14	02 56	03 23	03 41
54	03 12	04 03	04 42	01 56	02 41	03 11	03 32
52	03 23	04 10	04 47	01 40	02 27	03 00	03 23
50	03 33	04 17	04 52	01 27	02 15	02 50	03 16
45	03 53	04 31	05 03	01 00	01 50	02 29	02 59
N 40	04 08	04 43	05 11	00 38	01 30	02 12	02 46
35	04 21	04 52	05 19	00 21	01 14	01 58	02 35
30	04 31	05 00	05 25	00 05	00 59	01 46	02 25
20	04 47	05 14	05 36	24 35	00 35	01 24	02 08
N 10	04 59	05 25	05 46	24 14	00 14	01 06	01 52
0	05 09	05 34	05 55	23 54	24 48	00 48	01 38
S 10	05 18	05 42	06 04	23 34	24 31	00 31	01 24
20	05 25	05 51	06 13	23 13	24 12	00 12	01 09
30	05 31	05 59	06 24	22 48	23 51	24 51	00 51
35	05 34	06 04	06 30	22 34	23 38	24 41	00 41
40	05 37	06 09	06 36	22 17	23 23	24 29	00 29
45	05 40	06 14	06 44	21 57	23 06	24 15	00 15
S 50	05 43	06 20	06 54	21 31	22 44	23 58	25 11
52	05 44	06 23	06 58	21 19	22 34	23 50	25 05
54	05 45	06 26	07 03	21 05	22 22	23 41	24 59
56	05 46	06 29	07 08	20 48	22 08	23 31	24 52
58	05 47	06 33	07 14	20 29	21 53	23 19	24 44
S 60	05 48	06 37	07 21	20 04	21 34	23 06	24 35

Lat. °	Sunset h m	Twilight Civil h m	Twilight Naut. h m	Moonset 22 h m	Moonset 23 h m	Moonset 24 h m	Moonset 25 h m
N 72	21 08	////	////	■	■	■	08 38
N 70	20 43	22 31	////	■	■	■	09 56
68	20 24	21 47	////	■	■	07 41	10 33
66	20 09	21 18	////	■	■	08 55	10 59
64	19 56	20 57	22 42	■	07 29	09 31	11 19
62	19 46	20 40	22 03	06 39	08 14	09 57	11 35
60	19 37	20 26	21 36	07 19	08 44	10 17	11 49
N 58	19 29	20 15	21 16	07 47	09 07	10 33	12 00
56	19 22	20 04	21 00	08 08	09 25	10 47	12 10
54	19 16	19 56	20 47	08 26	09 40	10 59	12 19
52	19 11	19 48	20 35	08 41	09 53	11 10	12 26
50	19 06	19 41	20 25	08 55	10 05	11 19	12 33
45	18 55	19 26	20 05	09 22	10 29	11 39	12 48
N 40	18 46	19 15	19 49	09 43	10 48	11 55	13 00
35	18 39	19 05	19 37	10 00	11 04	12 08	13 11
30	18 32	18 57	19 26	10 16	11 18	12 20	13 20
20	18 21	18 43	19 10	10 41	11 42	12 40	13 35
N 10	18 11	18 32	18 58	11 03	12 02	12 57	13 48
0	18 02	18 23	18 48	11 24	12 21	13 13	14 01
S 10	17 53	18 14	18 39	11 45	12 39	13 29	14 13
20	17 43	18 06	18 32	12 07	12 59	13 46	14 26
30	17 33	17 57	18 25	12 32	13 22	14 05	14 41
35	17 27	17 53	18 22	12 47	13 36	14 16	14 50
40	17 20	17 48	18 19	13 04	13 51	14 29	15 00
45	17 12	17 42	18 16	13 25	14 09	14 44	15 11
S 50	17 02	17 36	18 13	13 51	14 32	15 02	15 25
52	16 58	17 33	18 12	14 03	14 43	15 11	15 31
54	16 53	17 30	18 11	14 17	14 55	15 20	15 38
56	16 48	17 26	18 10	14 34	15 09	15 31	15 46
58	16 42	17 23	18 09	14 54	15 25	15 43	15 55
S 60	16 35	17 19	18 07	15 19	15 44	15 57	16 05

Day	SUN Eqn. of Time 00^h	SUN Eqn. of Time 12^h	SUN Mer. Pass.	MOON Mer. Pass. Upper	MOON Mer. Pass. Lower	MOON Age	MOON Phase
d	m s	m s	h m	h m	h m	d %	
22	01 19	01 25	11 59	05 10	17 39	21 61	
23	01 31	01 37	11 58	06 07	18 35	22 50	◑
24	01 43	01 48	11 58	07 01	19 26	23 40	

UT	ARIES	VENUS −3.9		MARS +0.1		JUPITER −2.1		SATURN +0.1		STARS		
	GHA	GHA	Dec	GHA	Dec	GHA	Dec	GHA	Dec	Name	SHA	Dec
d h	° ′	° ′	° ′	° ′	° ′	° ′	° ′	° ′	° ′		° ′	° ′
25 00	212 35.8	208 17.0	N 0 08.4	268 13.9	S21 01.1	81 12.0	N18 53.8	127 33.5	N22 26.4	Acamar	315 24.7	S40 17.6
01	227 38.3	223 16.7	09.6	283 14.8	00.9	96 14.3	53.7	142 35.7	26.4	Achernar	335 33.1	S57 13.3
02	242 40.8	238 16.4	10.7	298 15.7	00.6	111 16.6	53.7	157 37.9	26.4	Acrux	173 18.0	S63 07.2
03	257 43.2	253 16.1	. . 11.9	313 16.6	. . 00.4	126 18.9	. . 53.6	172 40.1	. . 26.4	Adhara	255 19.0	S28 58.7
04	272 45.7	268 15.7	13.1	328 17.5	21 00.1	141 21.2	53.6	187 42.3	26.4	Aldebaran	290 58.9	N16 30.9
05	287 48.1	283 15.4	14.2	343 18.3	20 59.8	156 23.5	53.5	202 44.5	26.5			
F 06	302 50.6	298 15.1	N 0 15.4	358 19.2	S20 59.6	171 25.8	N18 53.5	217 46.7	N22 26.5	Alioth	166 27.0	N55 56.6
R 07	317 53.1	313 14.8	16.6	13 20.1	59.3	186 28.1	53.4	232 48.9	26.5	Alkaid	153 04.6	N49 17.8
I 08	332 55.5	328 14.4	17.7	28 21.0	59.1	201 30.4	53.4	247 51.1	26.5	Al Na'ir	27 53.8	S46 56.7
D 09	347 58.0	343 14.1	. . 18.9	43 21.9	. . 58.8	216 32.7	. . 53.3	262 53.3	. . 26.5	Alnilam	275 54.7	S 1 12.1
A 10	3 00.5	358 13.8	20.0	58 22.8	58.5	231 35.0	53.3	277 55.5	26.5	Alphard	218 03.9	S 8 40.4
Y 11	18 02.9	13 13.5	21.2	73 23.7	58.3	246 37.3	53.3	292 57.7	26.5			
12	33 05.4	28 13.1	N 0 22.4	88 24.5	S20 58.0	261 39.6	N18 53.2	307 59.9	N22 26.6	Alphecca	126 17.4	N26 42.1
13	48 07.9	43 12.8	23.5	103 25.4	57.7	276 41.9	53.2	323 02.1	26.6	Alpheratz	357 52.2	N29 06.2
14	63 10.3	58 12.5	24.7	118 26.3	57.5	291 44.2	53.1	338 04.3	26.6	Altair	62 16.0	N 8 52.4
15	78 12.8	73 12.2	. . 25.9	133 27.2	. . 57.2	306 46.5	. . 53.1	353 06.5	. . 26.6	Ankaa	353 23.8	S42 17.4
16	93 15.3	88 11.8	27.0	148 28.1	57.0	321 48.8	53.0	8 08.7	26.6	Antares	112 35.8	S26 26.4
17	108 17.7	103 11.5	28.2	163 29.0	56.7	336 51.1	53.0	23 10.9	26.6			
18	123 20.2	118 11.2	N 0 29.3	178 29.9	S20 56.4	351 53.4	N18 52.9	38 13.1	N22 26.6	Arcturus	146 02.7	N19 09.9
19	138 22.6	133 10.9	30.5	193 30.8	56.2	6 55.7	52.9	53 15.3	26.7	Atria	107 44.6	S69 01.9
20	153 25.1	148 10.5	31.7	208 31.7	55.9	21 58.0	52.8	68 17.5	26.7	Avior	234 21.5	S59 31.4
21	168 27.6	163 10.2	. . 32.8	223 32.5	. . 55.7	37 00.3	. . 52.8	83 19.7	. . 26.7	Bellatrix	278 40.9	N 6 21.1
22	183 30.0	178 09.9	34.0	238 33.4	55.4	52 02.6	52.8	98 21.9	26.7	Betelgeuse	271 10.2	N 7 24.4
23	198 32.5	193 09.6	35.2	253 34.3	55.1	67 04.9	52.7	113 24.1	26.7			
26 00	213 35.0	208 09.2	N 0 36.3	268 35.2	S20 54.9	82 07.2	N18 52.7	128 26.3	N22 26.7	Canopus	264 00.0	S52 42.1
01	228 37.4	223 08.9	37.5	283 36.1	54.6	97 09.5	52.6	143 28.5	26.7	Capella	280 46.7	N46 00.2
02	243 39.9	238 08.6	38.7	298 37.0	54.3	112 11.8	52.6	158 30.7	26.8	Deneb	49 37.1	N45 17.1
03	258 42.4	253 08.3	. . 39.8	313 37.9	. . 54.1	127 14.1	. . 52.5	173 32.9	. . 26.8	Denebola	182 41.5	N14 33.3
04	273 44.8	268 07.9	41.0	328 38.8	53.8	142 16.4	52.5	188 35.1	26.8	Diphda	349 04.2	S17 58.3
05	288 47.3	283 07.6	42.2	343 39.7	53.5	157 18.7	52.4	203 37.3	26.8			
06	303 49.8	298 07.3	N 0 43.3	358 40.6	S20 53.3	172 21.0	N18 52.4	218 39.5	N22 26.8	Dubhe	194 00.8	N61 44.3
S 07	318 52.2	313 07.0	44.5	13 41.5	53.0	187 23.3	52.3	233 41.7	26.8	Elnath	278 23.0	N28 36.7
A 08	333 54.7	328 06.6	45.6	28 42.3	52.8	202 25.6	52.3	248 43.9	26.9	Eltanin	90 49.5	N51 29.0
T 09	348 57.1	343 06.3	. . 46.8	43 43.2	. . 52.5	217 27.8	. . 52.2	263 46.1	. . 26.9	Enif	33 55.1	N 9 53.1
U 10	3 59.6	358 06.0	48.0	58 44.1	52.2	232 30.1	52.2	278 48.3	26.9	Fomalhaut	15 32.9	S29 36.4
R 11	19 02.1	13 05.7	49.1	73 45.0	52.0	247 32.4	52.2	293 50.5	26.9			
D 12	34 04.5	28 05.3	N 0 50.3	88 45.9	S20 51.7	262 34.7	N18 52.1	308 52.7	N22 26.9	Gacrux	172 09.6	S57 08.0
A 13	49 07.0	43 05.0	51.5	103 46.8	51.4	277 37.0	52.1	323 54.9	26.9	Gienah	176 00.3	S17 33.7
Y 14	64 09.5	58 04.7	52.6	118 47.7	51.2	292 39.3	52.0	338 57.1	26.9	Hadar	148 58.9	S60 23.4
15	79 11.9	73 04.3	. . 53.8	133 48.6	. . 50.9	307 41.6	. . 52.0	353 59.3	. . 27.0	Hamal	328 10.2	N23 28.5
16	94 14.4	88 04.0	55.0	148 49.5	50.6	322 43.9	51.9	9 01.5	27.0	Kaus Aust.	83 54.2	S34 23.0
17	109 16.9	103 03.7	56.1	163 50.4	50.4	337 46.2	51.9	24 03.7	27.0			
18	124 19.3	118 03.4	N 0 57.3	178 51.3	S20 50.1	352 48.5	N18 51.8	39 05.9	N22 27.0	Kochab	137 18.1	N74 08.5
19	139 21.8	133 03.0	58.5	193 52.2	49.8	7 50.8	51.8	54 08.1	27.0	Markab	13 46.6	N15 13.1
20	154 24.2	148 02.7	0 59.6	208 53.1	49.6	22 53.1	51.7	69 10.2	27.0	Menkar	314 23.8	N 4 06.0
21	169 26.7	163 02.4	1 00.8	223 54.0	. . 49.3	37 55.4	. . 51.7	84 12.4	. . 27.0	Menkent	148 16.8	S36 23.2
22	184 29.2	178 02.1	01.9	238 54.9	49.0	52 57.6	51.6	99 14.6	27.1	Miaplacidus	221 41.5	S69 44.1
23	199 31.6	193 01.7	03.1	253 55.8	48.8	67 59.9	51.6	114 16.8	27.1			
27 00	214 34.1	208 01.4	N 1 04.3	268 56.7	S20 48.5	83 02.2	N18 51.5	129 19.0	N22 27.1	Mirfak	308 52.5	N49 52.3
01	229 36.6	223 01.1	05.4	283 57.6	48.2	98 04.5	51.5	144 21.2	27.1	Nunki	76 08.1	S26 17.6
02	244 39.0	238 00.8	06.6	298 58.5	48.0	113 06.8	51.4	159 23.4	27.1	Peacock	53 31.7	S56 43.4
03	259 41.5	253 00.4	. . 07.8	313 59.4	. . 47.7	128 09.1	. . 51.4	174 25.6	. . 27.1	Pollux	243 37.6	N28 01.2
04	274 44.0	268 00.1	08.9	329 00.2	47.4	143 11.4	51.3	189 27.8	27.1	Procyon	245 08.2	N 5 13.0
05	289 46.4	282 59.8	10.1	344 01.1	47.2	158 13.7	51.3	204 30.0	27.2			
06	304 48.9	297 59.5	N 1 11.3	359 02.0	S20 46.9	173 16.0	N18 51.2	219 32.2	N22 27.2	Rasalhague	96 13.7	N12 33.2
07	319 51.4	312 59.1	12.4	14 02.9	46.6	188 18.2	51.2	234 34.4	27.2	Regulus	207 51.9	N11 57.1
08	334 53.8	327 58.8	13.6	29 03.8	46.4	203 20.5	51.1	249 36.6	27.2	Rigel	281 20.0	S 8 12.0
S 09	349 56.3	342 58.5	. . 14.8	44 04.7	. . 46.1	218 22.8	. . 51.1	264 38.8	. . 27.2	Rigil Kent.	140 02.4	S60 50.9
U 10	4 58.7	357 58.2	15.9	59 05.6	45.8	233 25.1	51.1	279 41.0	27.2	Sabik	102 21.5	S15 43.8
N 11	20 01.2	12 57.8	17.1	74 06.5	45.6	248 27.4	51.0	294 43.2	27.3			
D 12	35 03.7	27 57.5	N 1 18.3	89 07.4	S20 45.3	263 29.7	N18 51.0	309 45.4	N22 27.3	Schedar	349 50.4	N56 33.1
A 13	50 06.1	42 57.2	19.4	104 08.3	45.0	278 32.0	50.9	324 47.5	27.3	Shaula	96 32.5	S37 06.4
Y 14	65 08.6	57 56.8	20.6	119 09.2	44.8	293 34.3	50.9	339 49.7	27.3	Sirius	258 41.0	S16 43.3
15	80 11.1	72 56.5	. . 21.7	134 10.1	. . 44.5	308 36.5	. . 50.8	354 51.9	. . 27.3	Spica	158 39.4	S11 10.8
16	95 13.5	87 56.2	22.9	149 11.0	44.2	323 38.8	50.8	9 54.1	27.3	Suhail	222 58.4	S43 26.9
17	110 16.0	102 55.9	24.1	164 11.9	44.0	338 41.1	50.7	24 56.3	27.3			
18	125 18.5	117 55.5	N 1 25.2	179 12.8	S20 43.7	353 43.4	N18 50.7	39 58.5	N22 27.4	Vega	80 44.2	N38 46.9
19	140 20.9	132 55.2	26.4	194 13.7	43.4	8 45.7	50.6	55 00.7	27.4	Zuben'ubi	137 14.0	S16 03.4
20	155 23.4	147 54.9	27.6	209 14.6	43.1	23 48.0	50.6	70 02.9	27.4		SHA	Mer. Pass
21	170 25.8	162 54.6	. . 28.7	224 15.5	. . 42.9	38 50.2	. . 50.5	85 05.1	. . 27.4		° ′	h m
22	185 28.3	177 54.2	29.9	239 16.4	42.6	53 52.5	50.5	100 07.3	27.4	Venus	354 34.3	10 08
23	200 30.8	192 53.9	31.1	254 17.3	42.3	68 54.8	50.4	115 09.5	27.4	Mars	55 00.2	6 05
Mer. Pass.	h m 9 44.1	*v* −0.3	*d* 1.2	*v* 0.9	*d* 0.3	*v* 2.3	*d* 0.0	*v* 2.2	*d* 0.0	Jupiter	228 32.2	18 29
										Saturn	274 51.4	15 24

UT	SUN GHA	SUN Dec	MOON GHA	v	MOON Dec	d	HP
d h	° ′	° ′	° ′	′	° ′	′	′
25 00	180 28.4	N12 59.0	246 20.2	11.9	S19 02.5	10.2	56.0
01	195 28.5	12 59.8	260 51.1	12.0	18 52.3	10.2	55.9
02	210 28.6	13 00.6	275 22.1	12.1	18 42.1	10.3	55.9
03	225 28.8	. . 01.4	289 53.2	12.2	18 31.8	10.4	55.9
04	240 28.9	02.2	304 24.4	12.3	18 21.4	10.4	55.9
05	255 29.0	03.1	318 55.7	12.4	18 11.0	10.6	55.8
06	270 29.1	N13 03.9	333 27.1	12.5	S18 00.4	10.6	55.8
07	285 29.2	04.7	347 58.6	12.5	17 49.8	10.7	55.8
F 08	300 29.3	05.5	2 30.1	12.7	17 39.1	10.7	55.7
R 09	315 29.4	. . 06.3	17 01.8	12.7	17 28.4	10.8	55.7
I 10	330 29.5	07.2	31 33.5	12.9	17 17.6	10.9	55.7
D 11	345 29.6	08.0	46 05.4	12.9	17 06.7	10.9	55.7
A 12	0 29.7	N13 08.8	60 37.3	13.0	S16 55.8	11.1	55.6
Y 13	15 29.8	09.6	75 09.3	13.1	16 44.7	11.0	55.6
14	30 30.0	10.4	89 41.4	13.2	16 33.7	11.2	55.6
15	45 30.1	. . 11.2	104 13.6	13.2	16 22.5	11.2	55.6
16	60 30.2	12.0	118 45.8	13.3	16 11.3	11.3	55.5
17	75 30.3	12.9	133 18.1	13.5	16 00.0	11.3	55.5
18	90 30.4	N13 13.7	147 50.6	13.5	S15 48.7	11.4	55.5
19	105 30.5	14.5	162 23.1	13.6	15 37.3	11.4	55.5
20	120 30.6	15.3	176 55.7	13.6	15 25.9	11.5	55.4
21	135 30.7	. . 16.1	191 28.3	13.8	15 14.4	11.6	55.4
22	150 30.8	16.9	206 01.1	13.8	15 02.8	11.6	55.4
23	165 30.9	17.7	220 33.9	13.9	14 51.2	11.7	55.4
26 00	180 31.0	N13 18.6	235 06.8	13.9	S14 39.5	11.7	55.3
01	195 31.1	19.4	249 39.7	14.1	14 27.8	11.8	55.3
02	210 31.2	20.2	264 12.8	14.1	14 16.0	11.8	55.3
03	225 31.3	. . 21.0	278 45.9	14.2	14 04.2	11.9	55.3
04	240 31.4	21.8	293 19.1	14.3	13 52.3	11.9	55.2
05	255 31.5	22.6	307 52.4	14.3	13 40.4	12.0	55.2
06	270 31.7	N13 23.4	322 25.7	14.4	S13 28.4	12.0	55.2
S 07	285 31.8	24.2	336 59.1	14.5	13 16.4	12.0	55.2
A 08	300 31.9	25.0	351 32.6	14.5	13 04.4	12.1	55.2
T 09	315 32.0	. . 25.8	6 06.1	14.6	12 52.3	12.2	55.1
U 10	330 32.1	26.6	20 39.7	14.7	12 40.1	12.2	55.1
R 11	345 32.2	27.5	35 13.4	14.7	12 27.9	12.2	55.1
D 12	0 32.3	N13 28.3	49 47.1	14.8	S12 15.7	12.3	55.1
A 13	15 32.4	29.1	64 20.9	14.9	12 03.4	12.3	55.1
Y 14	30 32.5	29.9	78 54.8	14.9	11 51.1	12.4	55.0
15	45 32.6	. . 30.7	93 28.7	15.0	11 38.7	12.4	55.0
16	60 32.7	31.5	108 02.7	15.1	11 26.3	12.4	55.0
17	75 32.8	32.3	122 36.8	15.1	11 13.9	12.5	55.0
18	90 32.9	N13 33.1	137 10.9	15.1	S11 01.4	12.5	54.9
19	105 33.0	33.9	151 45.0	15.3	10 48.9	12.6	54.9
20	120 33.1	34.7	166 19.3	15.2	10 36.3	12.6	54.9
21	135 33.2	. . 35.5	180 53.5	15.4	10 23.7	12.6	54.9
22	150 33.3	36.3	195 27.9	15.4	10 11.1	12.6	54.9
23	165 33.4	37.1	210 02.3	15.4	9 58.5	12.7	54.9
27 00	180 33.5	N13 37.9	224 36.7	15.5	S 9 45.8	12.7	54.8
01	195 33.6	38.7	239 11.2	15.5	9 33.1	12.7	54.8
02	210 33.7	39.5	253 45.7	15.6	9 20.4	12.8	54.8
03	225 33.8	. . 40.3	268 20.3	15.7	9 07.6	12.8	54.8
04	240 33.9	41.1	282 55.0	15.7	8 54.8	12.8	54.8
05	255 34.0	41.9	297 29.7	15.7	8 42.0	12.9	54.7
06	270 34.1	N13 42.7	312 04.4	15.8	S 8 29.1	12.9	54.7
07	285 34.2	43.5	326 39.2	15.8	8 16.2	12.9	54.7
08	300 34.3	44.3	341 14.0	15.9	8 03.3	12.9	54.7
S 09	315 34.4	. . 45.1	355 48.9	15.9	7 50.4	12.9	54.7
U 10	330 34.5	45.9	10 23.8	16.0	7 37.5	13.0	54.7
N 11	345 34.6	46.7	24 58.8	16.0	7 24.5	13.0	54.6
D 12	0 34.7	N13 47.5	39 33.8	16.0	S 7 11.5	13.0	54.6
A 13	15 34.8	48.3	54 08.8	16.1	6 58.5	13.0	54.6
Y 14	30 34.9	49.1	68 43.9	16.1	6 45.5	13.1	54.6
15	45 35.0	. . 49.9	83 19.0	16.1	6 32.4	13.1	54.6
16	60 35.1	50.7	97 54.1	16.2	6 19.3	13.1	54.6
17	75 35.2	51.5	112 29.3	16.3	6 06.2	13.1	54.6
18	90 35.3	N13 52.3	127 04.6	16.2	S 5 53.1	13.1	54.5
19	105 35.4	53.1	141 39.8	16.3	5 40.0	13.1	54.5
20	120 35.5	53.9	156 15.1	16.3	5 26.9	13.2	54.5
21	135 35.6	. . 54.7	170 50.4	16.4	5 13.7	13.1	54.5
22	150 35.6	55.5	185 25.8	16.4	5 00.6	13.2	54.5
23	165 35.7	56.3	200 01.2	16.4	S 4 47.4	13.2	54.5
	SD 15.9	d 0.8	SD 15.2		15.0		14.9

Lat.	Twilight Naut.	Twilight Civil	Sunrise	Moonrise 25	Moonrise 26	Moonrise 27	Moonrise 28
°	h m	h m	h m	h m	h m	h m	h m
N 72	////	////	02 33	07 21	05 43	05 04	04 34
N 70	////	00 59	03 02	06 02	05 18	04 51	04 30
68	////	01 54	03 22	05 23	04 58	04 40	04 26
66	////	02 27	03 39	04 56	04 42	04 32	04 22
64	00 51	02 50	03 52	04 35	04 29	04 24	04 19
62	01 41	03 08	04 04	04 18	04 18	04 18	04 17
60	02 10	03 23	04 13	04 04	04 09	04 12	04 15
N 58	02 32	03 35	04 21	03 52	04 01	04 07	04 13
56	02 49	03 46	04 29	03 41	03 53	04 03	04 11
54	03 03	03 55	04 35	03 32	03 47	03 59	04 09
52	03 15	04 04	04 41	03 23	03 41	03 55	04 08
50	03 26	04 11	04 46	03 16	03 36	03 52	04 06
45	03 47	04 26	04 58	02 59	03 24	03 45	04 03
N 40	04 04	04 38	05 07	02 46	03 14	03 39	04 01
35	04 17	04 49	05 15	02 35	03 06	03 34	03 59
30	04 28	04 57	05 22	02 25	02 59	03 29	03 57
20	04 45	05 12	05 34	02 08	02 46	03 21	03 54
N 10	04 58	05 23	05 45	01 52	02 35	03 14	03 51
0	05 09	05 33	05 54	01 38	02 24	03 07	03 48
S 10	05 18	05 42	06 04	01 24	02 14	03 01	03 45
20	05 26	05 52	06 14	01 09	02 02	02 53	03 43
30	05 33	06 01	06 26	00 51	01 49	02 45	03 39
35	05 36	06 06	06 32	00 41	01 42	02 41	03 37
40	05 40	06 12	06 39	00 29	01 33	02 35	03 35
45	05 43	06 18	06 48	00 15	01 23	02 29	03 33
S 50	05 47	06 25	06 58	25 11	01 11	02 21	03 30
52	05 48	06 28	07 03	25 05	01 05	02 18	03 28
54	05 50	06 31	07 08	24 59	00 59	02 14	03 27
56	05 51	06 35	07 14	24 52	00 52	02 10	03 25
58	05 53	06 39	07 21	24 44	00 44	02 05	03 23
S 60	05 54	06 44	07 28	24 35	00 35	02 00	03 21

Lat.	Sunset	Twilight Civil	Twilight Naut.	Moonset 25	Moonset 26	Moonset 27	Moonset 28
°	h m	h m	h m	h m	h m	h m	h m
N 72	21 28	////	////	08 38	11 50	13 58	15 53
N 70	20 58	23 12	////	09 56	12 14	14 08	15 55
68	20 36	22 07	////	10 33	12 32	14 17	15 56
66	20 19	21 33	////	10 59	12 46	14 24	15 57
64	20 06	21 09	23 18	11 19	12 58	14 30	15 58
62	19 54	20 50	22 20	11 35	13 07	14 35	15 59
60	19 44	20 35	21 49	11 49	13 16	14 39	16 00
N 58	19 36	20 22	21 27	12 00	13 23	14 43	16 00
56	19 28	20 11	21 09	12 10	13 30	14 46	16 01
54	19 22	20 02	20 54	12 19	13 35	14 49	16 01
52	19 16	19 54	20 42	12 26	13 40	14 52	16 02
50	19 10	19 46	20 31	12 33	13 45	14 54	16 02
45	18 59	19 30	20 10	12 48	13 55	15 00	16 03
N 40	18 49	19 18	19 53	13 00	14 03	15 04	16 04
35	18 41	19 08	19 40	13 11	14 10	15 08	16 04
30	18 34	18 59	19 29	13 20	14 17	15 12	16 05
20	18 22	18 44	19 11	13 35	14 27	15 17	16 06
N 10	18 11	18 33	18 58	13 48	14 37	15 22	16 07
0	18 01	18 22	18 47	14 01	14 45	15 27	16 07
S 10	17 52	18 13	18 38	14 13	14 54	15 32	16 08
20	17 41	18 04	18 30	14 26	15 03	15 37	16 09
30	17 30	17 54	18 23	14 41	15 13	15 42	16 10
35	17 23	17 49	18 19	14 50	15 19	15 45	16 10
40	17 16	17 44	18 15	15 00	15 26	15 49	16 11
45	17 07	17 37	18 12	15 11	15 34	15 53	16 11
S 50	16 57	17 30	18 08	15 25	15 43	15 58	16 12
52	16 52	17 27	18 07	15 31	15 47	16 00	16 12
54	16 47	17 24	18 05	15 38	15 52	16 03	16 13
56	16 41	17 20	18 04	15 46	15 57	16 06	16 13
58	16 34	17 16	18 02	15 55	16 03	16 09	16 14
S 60	16 27	17 11	18 00	16 05	16 09	16 12	16 14

Day	SUN Eqn. of Time 00^h	SUN Eqn. of Time 12^h	SUN Mer. Pass.	MOON Mer. Pass. Upper	MOON Mer. Pass. Lower	MOON Age	MOON Phase
d	m s	m s	h m	h m	h m	d %	
25	01 53	01 59	11 58	07 50	20 13	24 30	
26	02 04	02 09	11 58	08 35	20 56	25 22	
27	02 14	02 19	11 58	09 17	21 38	26 14	

UT d h	ARIES GHA	VENUS −3.9 GHA	VENUS Dec	MARS +0.0 GHA	MARS Dec	JUPITER −2.1 GHA	JUPITER Dec	SATURN +0.1 GHA	SATURN Dec
	° ′	° ′	° ′	° ′	° ′	° ′	° ′	° ′	° ′
28 00 (MONDAY)	215 33.2	207 53.6	N 1 32.2	269 18.3	S20 42.1	83 57.1	N18 50.4	130 11.7	N22 27.4
01	230 35.7	222 53.3	33.4	284 19.2	41.8	98 59.4	50.3	145 13.9	27.5
02	245 38.2	237 52.9	34.6	299 20.1	41.5	114 01.7	50.3	160 16.1	27.5
03	260 40.6	252 52.6	. . 35.7	314 21.0	. . 41.3	129 03.9	. . 50.2	175 18.2	. . 27.5
04	275 43.1	267 52.3	36.9	329 21.9	41.0	144 06.2	50.2	190 20.4	27.5
05	290 45.6	282 51.9	38.1	344 22.8	40.7	159 08.5	50.1	205 22.6	27.5
06	305 48.0	297 51.6	N 1 39.2	359 23.7	S20 40.4	174 10.8	N18 50.1	220 24.8	N22 27.5
07	320 50.5	312 51.3	40.4	14 24.6	40.2	189 13.1	50.0	235 27.0	27.5
08	335 53.0	327 51.0	41.5	29 25.5	39.9	204 15.3	50.0	250 29.2	27.6
09	350 55.4	342 50.6	. . 42.7	44 26.4	. . 39.6	219 17.6	. . 49.9	265 31.4	. . 27.6
10	5 57.9	357 50.3	43.9	59 27.3	39.4	234 19.9	49.9	280 33.6	27.6
11	21 00.3	12 50.0	45.0	74 28.2	39.1	249 22.2	49.8	295 35.8	27.6
12	36 02.8	27 49.7	N 1 46.2	89 29.1	S20 38.8	264 24.5	N18 49.8	310 38.0	N22 27.6
13	51 05.3	42 49.3	47.4	104 30.0	38.5	279 26.7	49.7	325 40.2	27.6
14	66 07.7	57 49.0	48.5	119 30.9	38.3	294 29.0	49.7	340 42.3	27.6
15	81 10.2	72 48.7	. . 49.7	134 31.8	. . 38.0	309 31.3	. . 49.6	355 44.5	. . 27.7
16	96 12.7	87 48.3	50.9	149 32.7	37.7	324 33.6	49.6	10 46.7	27.7
17	111 15.1	102 48.0	52.0	164 33.6	37.5	339 35.9	49.5	25 48.9	27.7
18	126 17.6	117 47.7	N 1 53.2	179 34.5	S20 37.2	354 38.1	N18 49.4	40 51.1	N22 27.7
19	141 20.1	132 47.4	54.4	194 35.5	36.9	9 40.4	49.4	55 53.3	27.7
20	156 22.5	147 47.0	55.5	209 36.4	36.6	24 42.7	49.3	70 55.5	27.7
21	171 25.0	162 46.7	. . 56.7	224 37.3	. . 36.4	39 45.0	. . 49.3	85 57.7	. . 27.7
22	186 27.4	177 46.4	57.9	239 38.2	36.1	54 47.2	49.2	100 59.9	27.8
23	201 29.9	192 46.0	1 59.0	254 39.1	35.8	69 49.5	49.2	116 02.1	27.8
29 00 (TUESDAY)	216 32.4	207 45.7	N 2 00.2	269 40.0	S20 35.5	84 51.8	N18 49.1	131 04.2	N22 27.8
01	231 34.8	222 45.4	01.4	284 40.9	35.3	99 54.1	49.1	146 06.4	27.8
02	246 37.3	237 45.1	02.5	299 41.8	35.0	114 56.3	49.0	161 08.6	27.8
03	261 39.8	252 44.7	. . 03.7	314 42.7	. . 34.7	129 58.6	. . 49.0	176 10.8	. . 27.8
04	276 42.2	267 44.4	04.8	329 43.6	34.4	145 00.9	48.9	191 13.0	27.8
05	291 44.7	282 44.1	06.0	344 44.5	34.2	160 03.2	48.9	206 15.2	27.9
06	306 47.2	297 43.7	N 2 07.2	359 45.5	S20 33.9	175 05.4	N18 48.8	221 17.4	N22 27.9
07	321 49.6	312 43.4	08.3	14 46.4	33.6	190 07.7	48.8	236 19.6	27.9
08	336 52.1	327 43.1	09.5	29 47.3	33.3	205 10.0	48.7	251 21.7	27.9
09	351 54.6	342 42.8	. . 10.7	44 48.2	. . 33.1	220 12.3	. . 48.7	266 23.9	. . 27.9
10	6 57.0	357 42.4	11.8	59 49.1	32.8	235 14.5	48.6	281 26.1	27.9
11	21 59.5	12 42.1	13.0	74 50.0	32.5	250 16.8	48.6	296 28.3	27.9
12	37 01.9	27 41.8	N 2 14.2	89 50.9	S20 32.2	265 19.1	N18 48.5	311 30.5	N22 28.0
13	52 04.4	42 41.4	15.3	104 51.8	32.0	280 21.4	48.5	326 32.7	28.0
14	67 06.9	57 41.1	16.5	119 52.7	31.7	295 23.6	48.4	341 34.9	28.0
15	82 09.3	72 40.8	. . 17.7	134 53.7	. . 31.4	310 25.9	. . 48.4	356 37.1	. . 28.0
16	97 11.8	87 40.5	18.8	149 54.6	31.1	325 28.2	48.3	11 39.3	28.0
17	112 14.3	102 40.1	20.0	164 55.5	30.9	340 30.4	48.2	26 41.4	28.0
18	127 16.7	117 39.8	N 2 21.2	179 56.4	S20 30.6	355 32.7	N18 48.2	41 43.6	N22 28.0
19	142 19.2	132 39.5	22.3	194 57.3	30.3	10 35.0	48.1	56 45.8	28.1
20	157 21.7	147 39.1	23.5	209 58.2	30.0	25 37.3	48.1	71 48.0	28.1
21	172 24.1	162 38.8	. . 24.6	224 59.1	. . 29.8	40 39.5	. . 48.0	86 50.2	. . 28.1
22	187 26.6	177 38.5	25.8	240 00.1	29.5	55 41.8	48.0	101 52.4	28.1
23	202 29.1	192 38.1	27.0	255 01.0	29.2	70 44.1	47.9	116 54.6	28.1
30 00 (WEDNESDAY)	217 31.5	207 37.8	N 2 28.1	270 01.9	S20 28.9	85 46.3	N18 47.9	131 56.7	N22 28.1
01	232 34.0	222 37.5	29.3	285 02.8	28.7	100 48.6	47.8	146 58.9	28.1
02	247 36.4	237 37.2	30.5	300 03.7	28.4	115 50.9	47.8	162 01.1	28.2
03	262 38.9	252 36.8	. . 31.6	315 04.6	. . 28.1	130 53.1	. . 47.7	177 03.3	. . 28.2
04	277 41.4	267 36.5	32.8	330 05.6	27.8	145 55.4	47.7	192 05.5	28.2
05	292 43.8	282 36.2	34.0	345 06.5	27.5	160 57.7	47.6	207 07.7	28.2
06	307 46.3	297 35.8	N 2 35.1	0 07.4	S20 27.3	175 59.9	N18 47.6	222 09.9	N22 28.2
07	322 48.8	312 35.5	36.3	15 08.3	27.0	191 02.2	47.5	237 12.0	28.2
08	337 51.2	327 35.2	37.4	30 09.2	26.7	206 04.5	47.4	252 14.2	28.2
09	352 53.7	342 34.8	. . 38.6	45 10.1	. . 26.4	221 06.7	. . 47.4	267 16.4	. . 28.3
10	7 56.2	357 34.5	39.8	60 11.1	26.2	236 09.0	47.3	282 18.6	28.3
11	22 58.6	12 34.2	40.9	75 12.0	25.9	251 11.3	47.3	297 20.8	28.3
12	38 01.1	27 33.9	N 2 42.1	90 12.9	S20 25.6	266 13.5	N18 47.2	312 23.0	N22 28.3
13	53 03.5	42 33.5	43.3	105 13.8	25.3	281 15.8	47.2	327 25.2	28.3
14	68 06.0	57 33.2	44.4	120 14.7	25.0	296 18.1	47.1	342 27.3	28.3
15	83 08.5	72 32.9	. . 45.6	135 15.6	. . 24.8	311 20.3	. . 47.1	357 29.5	. . 28.3
16	98 10.9	87 32.5	46.8	150 16.6	24.5	326 22.6	47.0	12 31.7	28.4
17	113 13.4	102 32.2	47.9	165 17.5	24.2	341 24.9	47.0	27 33.9	28.4
18	128 15.9	117 31.9	N 2 49.1	180 18.4	S20 23.9	356 27.1	N18 46.9	42 36.1	N22 28.4
19	143 18.3	132 31.5	50.2	195 19.3	23.7	11 29.4	46.8	57 38.3	28.4
20	158 20.8	147 31.2	51.4	210 20.3	23.4	26 31.6	46.8	72 40.4	28.4
21	173 23.3	162 30.9	. . 52.6	225 21.2	. . 23.1	41 33.9	. . 46.7	87 42.6	. . 28.4
22	188 25.7	177 30.5	53.7	240 22.1	22.8	56 36.2	46.7	102 44.8	28.4
23	203 28.2	192 30.2	54.9	255 23.0	22.5	71 38.4	46.6	117 47.0	28.5
Mer. Pass.	h m 9 32.3	*v* −0.3	*d* 1.2	*v* 0.9	*d* 0.3	*v* 2.3	*d* 0.1	*v* 2.2	*d* 0.0

STARS

Name	SHA	Dec
	° ′	° ′
Acamar	315 24.7	S40 17.6
Achernar	335 33.1	S57 13.2
Acrux	173 18.0	S63 07.2
Adhara	255 19.0	S28 58.7
Aldebaran	290 58.9	N16 30.9
Alioth	166 27.0	N55 56.7
Alkaid	153 04.6	N49 17.9
Al Na'ir	27 53.8	S46 56.7
Alnilam	275 54.7	S 1 12.0
Alphard	218 03.9	S 8 40.4
Alphecca	126 17.4	N26 42.1
Alpheratz	357 52.2	N29 06.2
Altair	62 16.0	N 8 52.4
Ankaa	353 23.8	S42 17.3
Antares	112 35.8	S26 26.4
Arcturus	146 02.7	N19 09.9
Atria	107 44.5	S69 01.9
Avior	234 21.5	S59 31.4
Bellatrix	278 40.9	N 6 21.1
Betelgeuse	271 10.2	N 7 24.4
Canopus	264 00.0	S52 42.0
Capella	280 46.7	N46 00.2
Deneb	49 37.0	N45 17.1
Denebola	182 41.5	N14 33.3
Diphda	349 04.2	S17 58.2
Dubhe	194 00.8	N61 44.1
Elnath	278 23.1	N28 36.7
Eltanin	90 49.5	N51 29.0
Enif	33 55.1	N 9 53.1
Fomalhaut	15 32.9	S29 36.4
Gacrux	172 09.6	S57 08.0
Gienah	176 00.3	S17 33.7
Hadar	148 58.9	S60 23.4
Hamal	328 10.2	N23 28.5
Kaus Aust.	83 54.2	S34 23.0
Kochab	137 18.1	N74 08.5
Markab	13 46.6	N15 13.1
Menkar	314 23.8	N 4 06.0
Menkent	148 16.8	S36 23.2
Miaplacidus	221 41.6	S69 44.1
Mirfak	308 52.5	N49 52.3
Nunki	76 08.1	S26 17.6
Peacock	53 31.6	S56 43.4
Pollux	243 37.6	N28 01.2
Procyon	245 08.2	N 5 13.0
Rasalhague	96 13.6	N12 33.3
Regulus	207 51.9	N11 57.1
Rigel	281 20.0	S 8 12.0
Rigil Kent.	140 02.4	S60 50.9
Sabik	102 21.5	S15 43.8
Schedar	349 50.4	N56 33.0
Shaula	96 32.5	S37 06.4
Sirius	258 41.0	S16 43.3
Spica	158 39.4	S11 10.8
Suhail	222 58.4	S43 26.7
Vega	80 44.2	N38 46.9
Zuben'ubi	137 14.0	S16 03.4

	SHA	Mer. Pass.
	° ′	h m
Venus	351 13.3	10 09
Mars	53 07.6	6 01
Jupiter	228 19.4	18 18
Saturn	274 31.9	15 13

UT d h	SUN GHA ° ′	SUN Dec ° ′	MOON GHA ° ′	MOON *v* ′	MOON Dec ° ′	MOON *d* ′	MOON HP ′
28 00	180 35.8	N13 57.1	214 36.6	16.4	S 4 34.2	13.2	54.5
01	195 35.9	57.9	229 12.0	16.5	4 21.0	13.2	54.4
02	210 36.0	58.6	243 47.5	16.5	4 07.8	13.3	54.4
03	225 36.1	13 59.4	258 23.0	16.5	3 54.5	13.2	54.4
04	240 36.2	14 00.2	272 58.5	16.5	3 41.3	13.2	54.4
05	255 36.3	01.0	287 34.0	16.6	3 28.1	13.3	54.4
06	270 36.4	N14 01.8	302 09.6	16.6	S 3 14.8	13.2	54.4
07	285 36.5	02.6	316 45.2	16.6	3 01.6	13.3	54.4
08	300 36.6	03.4	331 20.8	16.6	2 48.3	13.3	54.4
M 09	315 36.7	. . 04.2	345 56.4	16.6	2 35.0	13.2	54.3
O 10	330 36.8	05.0	0 32.0	16.7	2 21.8	13.3	54.3
N 11	345 36.9	05.8	15 07.7	16.7	2 08.5	13.3	54.3
D 12	0 37.0	N14 06.5	29 43.4	16.7	S 1 55.2	13.3	54.3
A 13	15 37.1	07.3	44 19.1	16.7	1 41.9	13.3	54.3
Y 14	30 37.1	08.1	58 54.8	16.7	1 28.6	13.3	54.3
15	45 37.2	. . 08.9	73 30.5	16.7	1 15.3	13.2	54.3
16	60 37.3	09.7	88 06.2	16.8	1 02.1	13.3	54.3
17	75 37.4	10.5	102 42.0	16.8	0 48.8	13.3	54.3
18	90 37.5	N14 11.3	117 17.8	16.7	S 0 35.5	13.3	54.2
19	105 37.6	12.1	131 53.5	16.8	0 22.2	13.3	54.2
20	120 37.7	12.8	146 29.3	16.8	S 0 08.9	13.3	54.2
21	135 37.8	. . 13.6	161 05.1	16.8	N 0 04.4	13.2	54.2
22	150 37.9	14.4	175 40.9	16.8	0 17.6	13.3	54.2
23	165 38.0	15.2	190 16.7	16.8	0 30.9	13.3	54.2
29 00	180 38.1	N14 16.0	204 52.5	16.8	N 0 44.2	13.2	54.2
01	195 38.1	16.8	219 28.3	16.8	0 57.4	13.3	54.2
02	210 38.2	17.5	234 04.1	16.8	1 10.7	13.2	54.2
03	225 38.3	. . 18.3	248 39.9	16.9	1 23.9	13.3	54.2
04	240 38.4	19.1	263 15.8	16.8	1 37.2	13.2	54.1
05	255 38.5	19.9	277 51.6	16.8	1 50.4	13.2	54.1
06	270 38.6	N14 20.7	292 27.4	16.8	N 2 03.6	13.2	54.1
07	285 38.7	21.4	307 03.2	16.8	2 16.8	13.2	54.1
T 08	300 38.8	22.2	321 39.0	16.8	2 30.0	13.2	54.1
U 09	315 38.9	. . 23.0	336 14.8	16.8	2 43.2	13.2	54.1
E 10	330 38.9	23.8	350 50.6	16.8	2 56.4	13.1	54.1
S 11	345 39.0	24.6	5 26.4	16.8	3 09.5	13.2	54.1
D 12	0 39.1	N14 25.3	20 02.2	16.8	N 3 22.7	13.1	54.1
A 13	15 39.2	26.1	34 38.0	16.8	3 35.8	13.1	54.1
Y 14	30 39.3	26.9	49 13.8	16.8	3 48.9	13.1	54.1
15	45 39.4	. . 27.7	63 49.6	16.8	4 02.0	13.1	54.1
16	60 39.5	28.5	78 25.4	16.7	4 15.1	13.1	54.1
17	75 39.5	29.2	93 01.1	16.8	4 28.2	13.0	54.1
18	90 39.6	N14 30.0	107 36.9	16.7	N 4 41.2	13.1	54.0
19	105 39.7	30.8	122 12.6	16.7	4 54.3	13.0	54.0
20	120 39.8	31.6	136 48.3	16.7	5 07.3	13.0	54.0
21	135 39.9	. . 32.3	151 24.0	16.7	5 20.3	12.9	54.0
22	150 40.0	33.1	165 59.7	16.7	5 33.2	13.0	54.0
23	165 40.1	33.9	180 35.4	16.6	5 46.2	12.9	54.0
30 00	180 40.1	N14 34.7	195 11.0	16.7	N 5 59.1	12.9	54.0
01	195 40.2	35.4	209 46.7	16.6	6 12.0	12.9	54.0
02	210 40.3	36.2	224 22.3	16.6	6 24.9	12.8	54.0
03	225 40.4	. . 37.0	238 57.9	16.6	6 37.7	12.8	54.0
04	240 40.5	37.7	253 33.5	16.5	6 50.5	12.8	54.0
05	255 40.6	38.5	268 09.0	16.6	7 03.3	12.8	54.0
06	270 40.6	N14 39.3	282 44.6	16.5	N 7 16.1	12.8	54.0
W 07	285 40.7	40.1	297 20.1	16.5	7 28.9	12.7	54.0
E 08	300 40.8	40.8	311 55.6	16.5	7 41.6	12.7	54.0
D 09	315 40.9	. . 41.6	326 31.1	16.4	7 54.3	12.6	54.0
N 10	330 41.0	42.4	341 06.5	16.4	8 06.9	12.6	54.0
E 11	345 41.0	43.1	355 41.9	16.4	8 19.5	12.6	54.0
S 12	0 41.1	N14 43.9	10 17.3	16.4	N 8 32.1	12.6	54.0
D 13	15 41.2	44.7	24 52.7	16.3	8 44.7	12.5	54.0
A 14	30 41.3	45.4	39 28.0	16.3	8 57.2	12.5	54.0
Y 15	45 41.4	. . 46.2	54 03.3	16.3	9 09.7	12.5	54.0
16	60 41.5	47.0	68 38.6	16.3	9 22.2	12.4	54.0
17	75 41.5	47.7	83 13.9	16.2	9 34.6	12.4	54.0
18	90 41.6	N14 48.5	97 49.1	16.2	N 9 47.0	12.4	54.0
19	105 41.7	49.3	112 24.3	16.1	9 59.4	12.3	54.0
20	120 41.8	50.0	126 59.4	16.1	10 11.7	12.3	53.9
21	135 41.9	. . 50.8	141 34.5	16.1	10 24.0	12.2	53.9
22	150 41.9	51.6	156 09.6	16.0	10 36.2	12.2	53.9
23	165 42.0	52.3	170 44.6	16.1	N10 48.4	12.2	53.9
	SD 15.9	*d* 0.8	SD 14.8		14.7		14.7

Lat.	Twilight Naut.	Twilight Civil	Sunrise	Moonrise 28	Moonrise 29	Moonrise 30	Moonrise 1
°	h m	h m	h m	h m	h m	h m	h m
N 72	////	////	02 12	04 34	04 08	03 41	03 08
N 70	////	////	02 45	04 30	04 10	03 50	03 27
68	////	01 31	03 09	04 26	04 12	03 58	03 42
66	////	02 10	03 27	04 22	04 13	04 04	03 54
64	////	02 37	03 42	04 19	04 14	04 10	04 05
62	01 20	02 57	03 54	04 17	04 15	04 14	04 14
60	01 55	03 13	04 05	04 15	04 16	04 19	04 21
N 58	02 20	03 27	04 14	04 13	04 17	04 22	04 28
56	02 39	03 38	04 22	04 11	04 18	04 26	04 34
54	02 55	03 48	04 29	04 09	04 19	04 29	04 39
52	03 08	03 57	04 35	04 08	04 19	04 31	04 44
50	03 19	04 05	04 41	04 06	04 20	04 34	04 49
45	03 42	04 21	04 53	04 03	04 21	04 39	04 58
N 40	03 59	04 34	05 03	04 01	04 22	04 44	05 06
35	04 13	04 45	05 12	03 59	04 23	04 48	05 13
30	04 24	04 54	05 19	03 57	04 24	04 51	05 20
20	04 42	05 09	05 32	03 54	04 26	04 57	05 30
N 10	04 56	05 22	05 44	03 51	04 27	05 03	05 40
0	05 08	05 33	05 54	03 48	04 28	05 08	05 49
S 10	05 18	05 43	06 04	03 45	04 29	05 13	05 58
20	05 26	05 52	06 15	03 43	04 31	05 19	06 07
30	05 34	06 03	06 27	03 39	04 32	05 25	06 19
35	05 38	06 08	06 34	03 37	04 33	05 29	06 25
40	05 42	06 14	06 42	03 35	04 34	05 33	06 32
45	05 46	06 21	06 52	03 33	04 35	05 38	06 41
S 50	05 51	06 29	07 03	03 30	04 37	05 44	06 51
52	05 53	06 32	07 08	03 28	04 38	05 47	06 56
54	05 54	06 36	07 14	03 27	04 38	05 50	07 02
56	05 56	06 40	07 20	03 25	04 39	05 53	07 08
58	05 58	06 45	07 27	03 23	04 40	05 57	07 14
S 60	06 01	06 50	07 35	03 21	04 41	06 01	07 22

Lat.	Sunset	Twilight Civil	Twilight Naut.	Moonset 28	Moonset 29	Moonset 30	Moonset 1
°	h m	h m	h m	h m	h m	h m	h m
N 72	21 49	////	////	15 53	17 46	19 44	22 03
N 70	21 14	////	////	15 55	17 39	19 27	21 26
68	20 49	22 32	////	15 56	17 34	19 14	21 00
66	20 30	21 50	////	15 57	17 29	19 03	20 41
64	20 15	21 22	////	15 58	17 25	18 54	20 25
62	20 02	21 01	22 42	15 59	17 22	18 46	20 13
60	19 52	20 44	22 04	16 00	17 19	18 40	20 02
N 58	19 43	20 30	21 38	16 00	17 17	18 34	19 52
56	19 34	20 18	21 18	16 01	17 15	18 29	19 44
54	19 27	20 08	21 02	16 01	17 13	18 24	19 37
52	19 21	19 59	20 49	16 02	17 11	18 20	19 30
50	19 15	19 51	20 37	16 02	17 09	18 16	19 24
45	19 02	19 35	20 14	16 03	17 06	18 08	19 12
N 40	18 52	19 21	19 57	16 04	17 03	18 02	19 01
35	18 43	19 10	19 43	16 04	17 00	17 56	18 53
30	18 36	19 01	19 31	16 05	16 58	17 51	18 45
20	18 23	18 46	19 13	16 06	16 54	17 42	18 31
N 10	18 11	18 33	18 59	16 07	16 50	17 35	18 20
0	18 01	18 22	18 47	16 07	16 47	17 27	18 09
S 10	17 50	18 12	18 37	16 08	16 44	17 20	17 58
20	17 39	18 02	18 28	16 09	16 41	17 13	17 47
30	17 27	17 52	18 20	16 10	16 37	17 04	17 34
35	17 20	17 46	18 16	16 10	16 35	16 59	17 26
40	17 12	17 40	18 12	16 11	16 32	16 54	17 18
45	17 02	17 33	18 08	16 11	16 29	16 48	17 08
S 50	16 51	17 25	18 03	16 12	16 26	16 40	16 56
52	16 46	17 22	18 01	16 12	16 24	16 36	16 50
54	16 40	17 18	18 00	16 13	16 22	16 33	16 44
56	16 34	17 14	17 58	16 13	16 20	16 28	16 38
58	16 27	17 09	17 55	16 14	16 18	16 24	16 30
S 60	16 19	17 04	17 53	16 14	16 16	16 18	16 21

Day	SUN Eqn. of Time 00^h	SUN Eqn. of Time 12^h	SUN Mer. Pass.	MOON Mer. Pass. Upper	MOON Mer. Pass. Lower	MOON Age	MOON Phase
d	m s	m s	h m	h m	h m	d %	
28	02 23	02 28	11 58	09 58	22 18	27 8	●
29	02 32	02 36	11 57	10 38	22 58	28 4	
30	02 40	02 44	11 57	11 18	23 38	29 1	

	UT	ARIES	VENUS −3.9		MARS +0.0		JUPITER −2.1		SATURN +0.1		STARS		
		GHA	GHA	Dec	GHA	Dec	GHA	Dec	GHA	Dec	Name	SHA	Dec
	d h	° ′	° ′	° ′	° ′	° ′	° ′	° ′	° ′	° ′		° ′	° ′
	1 00	218 30.7	207 29.9	N 2 56.1	270 23.9	S20 22.3	86 40.7	N18 46.6	132 49.2	N22 28.5	Acamar	315 24.7	S40 17.
	01	233 33.1	222 29.5	57.2	285 24.9	22.0	101 43.0	46.5	147 51.4	28.5	Achernar	335 33.1	S57 13.
	02	248 35.6	237 29.2	58.4	300 25.8	21.7	116 45.2	46.5	162 53.6	28.5	Acrux	173 18.0	S63 07.
	03	263 38.0	252 28.9	2 59.6	315 26.7	. . 21.4	131 47.5	. . 46.4	177 55.7	. . 28.5	Adhara	255 19.0	S28 58.
	04	278 40.5	267 28.5	3 00.7	330 27.6	21.1	146 49.7	46.3	192 57.9	28.5	Aldebaran	290 58.9	N16 30.
	05	293 43.0	282 28.2	01.9	345 28.5	20.9	161 52.0	46.3	208 00.1	28.5			
	06	308 45.4	297 27.9	N 3 03.0	0 29.5	S20 20.6	176 54.3	N18 46.2	223 02.3	N22 28.6	Alioth	166 27.0	N55 56.
	07	323 47.9	312 27.6	04.2	15 30.4	20.3	191 56.5	46.2	238 04.5	28.6	Alkaid	153 04.6	N49 17.
T	08	338 50.4	327 27.2	05.4	30 31.3	20.0	206 58.8	46.1	253 06.6	28.6	Al Na'ir	27 53.8	S46 56.
H	09	353 52.8	342 26.9	. . 06.5	45 32.2	. . 19.7	222 01.0	. . 46.1	268 08.8	. . 28.6	Alnilam	275 54.7	S 1 12.
U	10	8 55.3	357 26.6	07.7	60 33.2	19.4	237 03.3	46.0	283 11.0	28.6	Alphard	218 03.9	S 8 40.
R	11	23 57.8	12 26.2	08.9	75 34.1	19.2	252 05.6	45.9	298 13.2	28.6			
S	12	39 00.2	27 25.9	N 3 10.0	90 35.0	S20 18.9	267 07.8	N18 45.9	313 15.4	N22 28.6	Alphecca	126 17.4	N26 42.
D	13	54 02.7	42 25.6	11.2	105 35.9	18.6	282 10.1	45.8	328 17.6	28.6	Alpheratz	357 52.2	N29 06.
A	14	69 05.1	57 25.2	12.3	120 36.9	18.3	297 12.3	45.8	343 19.7	28.7	Altair	62 16.0	N 8 52.
Y	15	84 07.6	72 24.9	. . 13.5	135 37.8	. . 18.0	312 14.6	. . 45.7	358 21.9	. . 28.7	Ankaa	353 23.8	S42 17.
	16	99 10.1	87 24.6	14.7	150 38.7	17.8	327 16.9	45.7	13 24.1	28.7	Antares	112 35.8	S26 26.
	17	114 12.5	102 24.2	15.8	165 39.6	17.5	342 19.1	45.6	28 26.3	28.7			
	18	129 15.0	117 23.9	N 3 17.0	180 40.6	S20 17.2	357 21.4	N18 45.6	43 28.5	N22 28.7	Arcturus	146 02.7	N19 09.
	19	144 17.5	132 23.6	18.2	195 41.5	16.9	12 23.6	45.5	58 30.7	28.7	Atria	107 44.5	S69 01.
	20	159 19.9	147 23.2	19.3	210 42.4	16.6	27 25.9	45.4	73 32.8	28.7	Avior	234 21.5	S59 31.
	21	174 22.4	162 22.9	. . 20.5	225 43.4	. . 16.3	42 28.1	. . 45.4	88 35.0	. . 28.8	Bellatrix	278 40.9	N 6 21.
	22	189 24.9	177 22.6	21.6	240 44.3	16.1	57 30.4	45.3	103 37.2	28.8	Betelgeuse	271 10.2	N 7 24.
	23	204 27.3	192 22.2	22.8	255 45.2	15.8	72 32.6	45.3	118 39.4	28.8			
	2 00	219 29.8	207 21.9	N 3 24.0	270 46.1	S20 15.5	87 34.9	N18 45.2	133 41.6	N22 28.8	Canopus	264 00.1	S52 42.
	01	234 32.3	222 21.5	25.1	285 47.1	15.2	102 37.2	45.2	148 43.7	28.8	Capella	280 46.7	N46 00.
	02	249 34.7	237 21.2	26.3	300 48.0	14.9	117 39.4	45.1	163 45.9	28.8	Deneb	49 37.0	N45 17.
	03	264 37.2	252 20.9	. . 27.4	315 48.9	. . 14.6	132 41.7	. . 45.0	178 48.1	. . 28.8	Denebola	182 41.6	N14 33.
	04	279 39.6	267 20.5	28.6	330 49.9	14.4	147 43.9	45.0	193 50.3	28.9	Diphda	349 04.2	S17 58.
	05	294 42.1	282 20.2	29.8	345 50.8	14.1	162 46.2	44.9	208 52.5	28.9			
	06	309 44.6	297 19.9	N 3 30.9	0 51.7	S20 13.8	177 48.4	N18 44.9	223 54.6	N22 28.9	Dubhe	194 00.8	N61 44.
	07	324 47.0	312 19.5	32.1	15 52.7	13.5	192 50.7	44.8	238 56.8	28.9	Elnath	278 23.1	N28 36.
	08	339 49.5	327 19.2	33.2	30 53.6	13.2	207 52.9	44.7	253 59.0	28.9	Eltanin	90 49.5	N51 29.
F	09	354 52.0	342 18.9	. . 34.4	45 54.5	. . 12.9	222 55.2	. . 44.7	269 01.2	. . 28.9	Enif	33 55.1	N 9 53.
R	10	9 54.4	357 18.5	35.6	60 55.4	12.7	237 57.4	44.6	284 03.4	28.9	Fomalhaut	15 32.9	S29 36.
I	11	24 56.9	12 18.2	36.7	75 56.4	12.4	252 59.7	44.6	299 05.5	29.0			
D	12	39 59.4	27 17.9	N 3 37.9	90 57.3	S20 12.1	268 01.9	N18 44.5	314 07.7	N22 29.0	Gacrux	172 09.6	S57 08.
A	13	55 01.8	42 17.5	39.1	105 58.2	11.8	283 04.2	44.5	329 09.9	29.0	Gienah	176 00.3	S17 33.
Y	14	70 04.3	57 17.2	40.2	120 59.2	11.5	298 06.5	44.4	344 12.1	29.0	Hadar	148 58.9	S60 23.
	15	85 06.8	72 16.9	. . 41.4	136 00.1	. . 11.2	313 08.7	. . 44.3	359 14.3	. . 29.0	Hamal	328 10.2	N23 28.
	16	100 09.2	87 16.5	42.5	151 01.0	11.0	328 11.0	44.3	14 16.4	29.0	Kaus Aust.	83 54.2	S34 23.
	17	115 11.7	102 16.2	43.7	166 02.0	10.7	343 13.2	44.2	29 18.6	29.0			
	18	130 14.1	117 15.8	N 3 44.9	181 02.9	S20 10.4	358 15.5	N18 44.2	44 20.8	N22 29.0	Kochab	137 18.1	N74 08.
	19	145 16.6	132 15.5	46.0	196 03.8	10.1	13 17.7	44.1	59 23.0	29.1	Markab	13 46.5	N15 13.
	20	160 19.1	147 15.2	47.2	211 04.8	09.8	28 20.0	44.0	74 25.2	29.1	Menkar	314 23.8	N 4 06.
	21	175 21.5	162 14.8	. . 48.3	226 05.7	. . 09.5	43 22.2	. . 44.0	89 27.3	. . 29.1	Menkent	148 16.8	S36 23.
	22	190 24.0	177 14.5	49.5	241 06.6	09.2	58 24.5	43.9	104 29.5	29.1	Miaplacidus	221 41.6	S69 44.
	23	205 26.5	192 14.2	50.7	256 07.6	09.0	73 26.7	43.9	119 31.7	29.1			
	3 00	220 28.9	207 13.8	N 3 51.8	271 08.5	S20 08.7	88 28.9	N18 43.8	134 33.9	N22 29.1	Mirfak	308 52.5	N49 52.
	01	235 31.4	222 13.5	53.0	286 09.4	08.4	103 31.2	43.7	149 36.0	29.1	Nunki	76 08.0	S26 17.
	02	250 33.9	237 13.2	54.1	301 10.4	08.1	118 33.4	43.7	164 38.2	29.2	Peacock	53 31.6	S56 43.
	03	265 36.3	252 12.8	. . 55.3	316 11.3	. . 07.8	133 35.7	. . 43.6	179 40.4	. . 29.2	Pollux	243 37.6	N28 01.
	04	280 38.8	267 12.5	56.4	331 12.3	07.5	148 37.9	43.6	194 42.6	29.2	Procyon	245 08.2	N 5 13.
	05	295 41.2	282 12.1	57.6	346 13.2	07.2	163 40.2	43.5	209 44.8	29.2			
	06	310 43.7	297 11.8	N 3 58.8	1 14.1	S20 07.0	178 42.4	N18 43.5	224 46.9	N22 29.2	Rasalhague	96 13.6	N12 33.
	07	325 46.2	312 11.5	3 59.9	16 15.1	06.7	193 44.7	43.4	239 49.1	29.2	Regulus	207 51.9	N11 57.
S	08	340 48.6	327 11.1	4 01.1	31 16.0	06.4	208 46.9	43.3	254 51.3	29.2	Rigel	281 20.0	S 8 12.
A	09	355 51.1	342 10.8	. . 02.2	46 16.9	. . 06.1	223 49.2	. . 43.3	269 53.5	. . 29.3	Rigil Kent.	140 02.4	S60 51.
T	10	10 53.6	357 10.5	03.4	61 17.9	05.8	238 51.4	43.2	284 55.6	29.3	Sabik	102 21.5	S15 43.
U	11	25 56.0	12 10.1	04.6	76 18.8	05.5	253 53.7	43.2	299 57.8	29.3			
R	12	40 58.5	27 09.8	N 4 05.7	91 19.8	S20 05.2	268 55.9	N18 43.1	315 00.0	N22 29.3	Schedar	349 50.4	N56 33.
D	13	56 01.0	42 09.4	06.9	106 20.7	04.9	283 58.2	43.0	330 02.2	29.3	Shaula	96 32.5	S37 06.
A	14	71 03.4	57 09.1	08.0	121 21.6	04.7	299 00.4	43.0	345 04.4	29.3	Sirius	258 41.0	S16 43.
Y	15	86 05.9	72 08.8	. . 09.2	136 22.6	. . 04.4	314 02.6	. . 42.9	0 06.5	. . 29.3	Spica	158 39.4	S11 10.
	16	101 08.4	87 08.4	10.3	151 23.5	04.1	329 04.9	42.9	15 08.7	29.3	Suhail	222 58.4	S43 26.
	17	116 10.8	102 08.1	11.5	166 24.5	03.8	344 07.1	42.8	30 10.9	29.4			
	18	131 13.3	117 07.7	N 4 12.7	181 25.4	S20 03.5	359 09.4	N18 42.7	45 13.1	N22 29.4	Vega	80 44.2	N38 46.
	19	146 15.7	132 07.4	13.8	196 26.3	03.2	14 11.6	42.7	60 15.2	29.4	Zuben'ubi	137 14.0	S16 03.
	20	161 18.2	147 07.1	15.0	211 27.3	02.9	29 13.9	42.6	75 17.4	29.4		SHA	Mer.Pas
	21	176 20.7	162 06.7	. . 16.1	226 28.2	. . 02.6	44 16.1	. . 42.5	90 19.6	. . 29.4		° ′	h m
	22	191 23.1	177 06.4	17.3	241 29.2	02.4	59 18.4	42.5	105 21.8	29.4	Venus	347 52.1	10 11
	23	206 25.6	192 06.0	18.5	256 30.1	02.1	74 20.6	42.4	120 23.9	29.4	Mars	51 16.4	5 57
		h m									Jupiter	228 05.1	18 07
	Mer. Pass.	9 20.5	v −0.3	d 1.2	v 0.9	d 0.3	v 2.3	d 0.1	v 2.2	d 0.0	Saturn	274 11.8	15 03

UT		SUN GHA	SUN Dec	MOON GHA	v	MOON Dec	d	HP
d h		° ′	° ′	° ′	′	° ′	′	′
1 00		180 42.1	N14 53.1	185 19.7	15.9	N11 00.6	12.1	53.9
01		195 42.2	53.9	199 54.6	16.0	11 12.7	12.0	53.9
02		210 42.2	54.6	214 29.6	15.9	11 24.7	12.1	53.9
03		225 42.3	. . 55.4	229 04.5	15.8	11 36.8	12.0	53.9
04		240 42.4	56.1	243 39.3	15.8	11 48.8	11.9	53.9
05		255 42.5	56.9	258 14.1	15.8	12 00.7	11.9	53.9
06		270 42.6	N14 57.7	272 48.9	15.7	N12 12.6	11.9	53.9
07		285 42.6	58.4	287 23.6	15.7	12 24.5	11.8	53.9
08	THURSDAY	300 42.7	59.2	301 58.3	15.7	12 36.3	11.7	53.9
09		315 42.8	14 59.9	316 33.0	15.6	12 48.0	11.7	53.9
10		330 42.9	15 00.7	331 07.6	15.6	12 59.7	11.7	53.9
11		345 42.9	01.5	345 42.2	15.5	13 11.4	11.6	53.9
12		0 43.0	N15 02.2	0 16.7	15.5	N13 23.0	11.6	53.9
13		15 43.1	03.0	14 51.2	15.4	13 34.6	11.5	53.9
14		30 43.2	03.7	29 25.6	15.4	13 46.1	11.4	53.9
15		45 43.2	. . 04.5	44 00.0	15.3	13 57.5	11.4	53.9
16		60 43.3	05.3	58 34.3	15.3	14 08.9	11.4	53.9
17		75 43.4	06.0	73 08.6	15.2	14 20.3	11.3	53.9
18		90 43.5	N15 06.8	87 42.8	15.2	N14 31.6	11.2	53.9
19		105 43.5	07.5	102 17.0	15.1	14 42.8	11.2	53.9
20		120 43.6	08.3	116 51.1	15.1	14 54.0	11.1	53.9
21		135 43.7	. . 09.0	131 25.2	15.1	15 05.1	11.1	54.0
22		150 43.8	09.8	145 59.3	14.9	15 16.2	11.0	54.0
23		165 43.8	10.5	160 33.2	15.0	15 27.2	10.9	54.0
2 00		180 43.9	N15 11.3	175 07.2	14.8	N15 38.1	10.9	54.0
01		195 44.0	12.0	189 41.0	14.9	15 49.0	10.9	54.0
02		210 44.1	12.8	204 14.9	14.7	15 59.9	10.7	54.0
03		225 44.1	. . 13.5	218 48.6	14.8	16 10.6	10.7	54.0
04		240 44.2	14.3	233 22.4	14.6	16 21.3	10.7	54.0
05		255 44.3	15.0	247 56.0	14.6	16 32.0	10.5	54.0
06		270 44.3	N15 15.8	262 29.6	14.6	N16 42.5	10.5	54.0
07		285 44.4	16.5	277 03.2	14.5	16 53.0	10.5	54.0
08		300 44.5	17.3	291 36.7	14.4	17 03.5	10.4	54.0
09	FRIDAY	315 44.6	. . 18.0	306 10.1	14.4	17 13.9	10.3	54.0
10		330 44.6	18.8	320 43.5	14.3	17 24.2	10.2	54.0
11		345 44.7	19.5	335 16.8	14.2	17 34.4	10.2	54.0
12		0 44.8	N15 20.3	349 50.0	14.2	N17 44.6	10.0	54.0
13		15 44.8	21.0	4 23.2	14.2	17 54.6	10.1	54.0
14		30 44.9	21.8	18 56.4	14.1	18 04.7	9.9	54.0
15		45 45.0	. . 22.5	33 29.5	14.0	18 14.6	9.9	54.0
16		60 45.0	23.3	48 02.5	14.0	18 24.5	9.8	54.0
17		75 45.1	24.0	62 35.5	13.9	18 34.3	9.7	54.0
18		90 45.2	N15 24.8	77 08.4	13.8	N18 44.0	9.7	54.0
19		105 45.3	25.5	91 41.2	13.8	18 53.7	9.6	54.0
20		120 45.3	26.3	106 14.0	13.7	19 03.3	9.5	54.0
21		135 45.4	. . 27.0	120 46.7	13.7	19 12.8	9.4	54.0
22		150 45.5	27.7	135 19.4	13.5	19 22.2	9.3	54.1
23		165 45.5	28.5	149 51.9	13.6	19 31.5	9.3	54.1
3 00		180 45.6	N15 29.2	164 24.5	13.4	N19 40.8	9.2	54.1
01		195 45.7	30.0	178 56.9	13.4	19 50.0	9.1	54.1
02		210 45.7	30.7	193 29.3	13.4	19 59.1	9.0	54.1
03		225 45.8	. . 31.5	208 01.7	13.3	20 08.1	8.9	54.1
04		240 45.9	32.2	222 34.0	13.2	20 17.0	8.8	54.1
05		255 45.9	32.9	237 06.2	13.1	20 25.8	8.8	54.1
06		270 46.0	N15 33.7	251 38.3	13.1	N20 34.6	8.7	54.1
07		285 46.1	34.4	266 10.4	13.0	20 43.3	8.6	54.1
08	SATURDAY	300 46.1	35.2	280 42.4	13.0	20 51.9	8.5	54.1
09		315 46.2	. . 35.9	295 14.4	12.9	21 00.4	8.4	54.1
10		330 46.3	36.6	309 46.3	12.8	21 08.8	8.3	54.1
11		345 46.3	37.4	324 18.1	12.8	21 17.1	8.2	54.1
12		0 46.4	N15 38.1	338 49.9	12.7	N21 25.3	8.2	54.1
13		15 46.4	38.8	353 21.6	12.6	21 33.5	8.0	54.2
14		30 46.5	39.6	7 53.2	12.6	21 41.5	8.0	54.2
15		45 46.6	. . 40.3	22 24.8	12.5	21 49.5	7.8	54.2
16		60 46.6	41.1	36 56.3	12.4	21 57.3	7.8	54.2
17		75 46.7	41.8	51 27.7	12.4	22 05.1	7.7	54.2
18		90 46.8	N15 42.5	65 59.1	12.3	N22 12.8	7.6	54.2
19		105 46.8	43.3	80 30.4	12.2	22 20.4	7.4	54.2
20		120 46.9	44.0	95 01.6	12.2	22 27.8	7.4	54.2
21		135 47.0	. . 44.7	109 32.8	12.1	22 35.2	7.3	54.2
22		150 47.0	45.5	124 03.9	12.1	22 42.5	7.2	54.2
23		165 47.1	46.2	138 35.0	11.9	N22 49.7	7.0	54.2
		SD 15.9	d 0.7	SD 14.7		14.7		14.8

Lat.	Twilight Naut.	Twilight Civil	Sunrise	Moonrise 1	Moonrise 2	Moonrise 3	Moonrise 4
°	h m	h m	h m	h m	h m	h m	h m
N 72	////	////	01 48	03 08	02 17	▭	▭
N 70	////	////	02 28	03 27	02 55	01 36	▭
68	////	01 01	02 55	03 42	03 22	02 50	▭
66	////	01 52	03 15	03 54	03 43	03 27	02 56
64	////	02 23	03 32	04 05	04 00	03 54	03 47
62	00 53	02 45	03 45	04 14	04 13	04 15	04 19
60	01 39	03 03	03 57	04 21	04 25	04 32	04 43
N 58	02 08	03 18	04 06	04 28	04 35	04 46	05 02
56	02 29	03 30	04 15	04 34	04 44	04 58	05 18
54	02 46	03 41	04 23	04 39	04 52	05 09	05 32
52	03 00	03 50	04 29	04 44	05 00	05 19	05 44
50	03 12	03 59	04 36	04 49	05 06	05 27	05 54
45	03 36	04 16	04 49	04 58	05 20	05 46	06 17
N 40	03 54	04 30	05 00	05 06	05 32	06 01	06 35
35	04 09	04 42	05 09	05 13	05 42	06 13	06 50
30	04 21	04 51	05 17	05 20	05 50	06 24	07 03
20	04 40	05 07	05 31	05 30	06 05	06 44	07 26
N 10	04 55	05 21	05 43	05 40	06 19	07 00	07 45
0	05 07	05 32	05 54	05 49	06 31	07 16	08 04
S 10	05 18	05 43	06 05	05 58	06 44	07 32	08 22
20	05 27	05 53	06 16	06 07	06 57	07 49	08 42
30	05 36	06 04	06 29	06 19	07 13	08 09	09 05
35	05 41	06 11	06 37	06 25	07 22	08 20	09 19
40	05 45	06 17	06 45	06 32	07 33	08 34	09 35
45	05 50	06 25	06 55	06 41	07 45	08 50	09 53
S 50	05 55	06 33	07 07	06 51	08 00	09 09	10 17
52	05 57	06 37	07 13	06 56	08 07	09 18	10 28
54	05 59	06 41	07 19	07 02	08 15	09 29	10 41
56	06 01	06 46	07 26	07 08	08 24	09 41	10 56
58	06 04	06 51	07 34	07 14	08 34	09 55	11 14
S 60	06 07	06 56	07 42	07 22	08 45	10 11	11 36

Lat.	Sunset	Twilight Civil	Twilight Naut.	Moonset 1	Moonset 2	Moonset 3	Moonset 4
°	h m	h m	h m	h m	h m	h m	h m
N 72	22 14	////	////	22 03	▭	▭	▭
N 70	21 31	////	////	21 26	24 17	00 17	▭
68	21 03	23 05	////	21 00	23 04	▭	▭
66	20 41	22 07	////	20 41	22 28	24 36	00 36
64	20 25	21 35	////	20 25	22 02	23 46	25 40
62	20 11	21 12	23 11	20 13	21 42	23 15	24 46
60	19 59	20 53	22 19	20 02	21 26	22 51	24 13
N 58	19 49	20 38	21 50	19 52	21 12	22 33	23 49
56	19 40	20 25	21 28	19 44	21 01	22 17	23 30
54	19 33	20 14	21 10	19 37	20 50	22 04	23 14
52	19 26	20 05	20 56	19 30	20 41	21 52	23 00
50	19 20	19 56	20 44	19 24	20 33	21 42	22 48
45	19 06	19 39	20 19	19 12	20 16	21 20	22 23
N 40	18 55	19 25	20 01	19 01	20 02	21 03	22 03
35	18 46	19 13	19 46	18 53	19 50	20 49	21 47
30	18 38	19 03	19 33	18 45	19 40	20 36	21 32
20	18 24	18 47	19 14	18 31	19 22	20 15	21 08
N 10	18 12	18 34	18 59	18 20	19 07	19 56	20 47
0	18 00	18 22	18 47	18 09	18 53	19 39	20 28
S 10	17 49	18 11	18 36	17 58	18 38	19 22	20 08
20	17 38	18 00	18 27	17 47	18 23	19 03	19 47
30	17 24	17 49	18 18	17 34	18 06	18 42	19 23
35	17 17	17 43	18 13	17 26	17 56	18 30	19 09
40	17 08	17 36	18 08	17 18	17 45	18 16	18 53
45	16 58	17 29	18 04	17 08	17 31	17 59	18 33
S 50	16 46	17 20	17 59	16 56	17 15	17 38	18 09
52	16 40	17 16	17 56	16 50	17 07	17 29	17 57
54	16 34	17 12	17 54	16 44	16 59	17 18	17 44
56	16 27	17 07	17 52	16 38	16 49	17 05	17 29
58	16 20	17 02	17 49	16 30	16 39	16 51	17 11
S 60	16 11	16 57	17 46	16 21	16 26	16 34	16 49

Day	SUN Eqn. of Time 00^h	SUN Eqn. of Time 12^h	SUN Mer. Pass.	MOON Mer. Pass. Upper	MOON Mer. Pass. Lower	MOON Age	MOON Phase
d	m s	m s	h m	h m	h m	d %	
1	02 48	02 52	11 57	11 59	24 20	30 0	
2	02 55	02 59	11 57	12 42	00 20	01 1	●
3	03 02	03 05	11 57	13 27	01 04	02 4	

UT d h	ARIES GHA ° ′	VENUS −3.9 GHA ° ′	Dec ° ′	MARS −0.1 GHA ° ′	Dec ° ′	JUPITER −2.1 GHA ° ′	Dec ° ′	SATURN +0.1 GHA ° ′	Dec ° ′
SUNDAY 4 00	221 28.1	207 05.7	N 4 19.6	271 31.0	S20 01.8	89 22.8	N18 42.4	135 26.1	N22 29.5
01	236 30.5	222 05.4	20.8	286 32.0	01.5	104 25.1	42.3	150 28.3	29.5
02	251 33.0	237 05.0	21.9	301 32.9	01.2	119 27.3	42.2	165 30.5	29.5
03	266 35.5	252 04.7	. . 23.1	316 33.9	. . 00.9	134 29.6	. . 42.2	180 32.6	. . 29.5
04	281 37.9	267 04.3	24.2	331 34.8	00.6	149 31.8	42.1	195 34.8	29.5
05	296 40.4	282 04.0	25.4	346 35.8	00.3	164 34.0	42.1	210 37.0	29.5
06	311 42.9	297 03.7	N 4 26.5	1 36.7	S20 00.0	179 36.3	N18 42.0	225 39.2	N22 29.5
07	326 45.3	312 03.3	27.7	16 37.6	19 59.8	194 38.5	41.9	240 41.3	29.5
08	341 47.8	327 03.0	28.9	31 38.6	59.5	209 40.8	41.9	255 43.5	29.6
09	356 50.2	342 02.6	. . 30.0	46 39.5	. . 59.2	224 43.0	. . 41.8	270 45.7	. . 29.6
10	11 52.7	357 02.3	31.2	61 40.5	58.9	239 45.2	41.8	285 47.9	29.6
11	26 55.2	12 02.0	32.3	76 41.4	58.6	254 47.5	41.7	300 50.0	29.6
12	41 57.6	27 01.6	N 4 33.5	91 42.4	S19 58.3	269 49.7	N18 41.6	315 52.2	N22 29.6
13	57 00.1	42 01.3	34.6	106 43.3	58.0	284 52.0	41.6	330 54.4	29.6
14	72 02.6	57 00.9	35.8	121 44.3	57.7	299 54.2	41.5	345 56.6	29.6
15	87 05.0	72 00.6	. . 36.9	136 45.2	. . 57.4	314 56.4	. . 41.4	0 58.7	. . 29.7
16	102 07.5	87 00.2	38.1	151 46.2	57.2	329 58.7	41.4	16 00.9	29.7
17	117 10.0	101 59.9	39.3	166 47.1	56.9	345 00.9	41.3	31 03.1	29.7
18	132 12.4	116 59.6	N 4 40.4	181 48.0	S19 56.6	0 03.1	N18 41.3	46 05.3	N22 29.7
19	147 14.9	131 59.2	41.6	196 49.0	56.3	15 05.4	41.2	61 07.4	29.7
20	162 17.4	146 58.9	42.7	211 49.9	56.0	30 07.6	41.1	76 09.6	29.7
21	177 19.8	161 58.5	. . 43.9	226 50.9	. . 55.7	45 09.9	. . 41.1	91 11.8	. . 29.7
22	192 22.3	176 58.2	45.0	241 51.8	55.4	60 12.1	41.0	106 14.0	29.7
23	207 24.7	191 57.8	46.2	256 52.8	55.1	75 14.3	40.9	121 16.1	29.8
MONDAY 5 00	222 27.2	206 57.5	N 4 47.3	271 53.7	S19 54.8	90 16.6	N18 40.9	136 18.3	N22 29.8
01	237 29.7	221 57.2	48.5	286 54.7	54.5	105 18.8	40.8	151 20.5	29.8
02	252 32.1	236 56.8	49.6	301 55.6	54.2	120 21.0	40.8	166 22.7	29.8
03	267 34.6	251 56.5	. . 50.8	316 56.6	. . 54.0	135 23.3	. . 40.7	181 24.8	. . 29.8
04	282 37.1	266 56.1	52.0	331 57.5	53.7	150 25.5	40.6	196 27.0	29.8
05	297 39.5	281 55.8	53.1	346 58.5	53.4	165 27.7	40.6	211 29.2	29.8
06	312 42.0	296 55.4	N 4 54.3	1 59.4	S19 53.1	180 30.0	N18 40.5	226 31.3	N22 29.9
07	327 44.5	311 55.1	55.4	17 00.4	52.8	195 32.2	40.4	241 33.5	29.9
08	342 46.9	326 54.7	56.6	32 01.3	52.5	210 34.4	40.4	256 35.7	29.9
09	357 49.4	341 54.4	. . 57.7	47 02.3	. . 52.2	225 36.7	. . 40.3	271 37.9	. . 29.9
10	12 51.9	356 54.1	4 58.9	62 03.2	51.9	240 38.9	40.2	286 40.0	29.9
11	27 54.3	11 53.7	5 00.0	77 04.2	51.6	255 41.1	40.2	301 42.2	29.9
12	42 56.8	26 53.4	N 5 01.2	92 05.1	S19 51.3	270 43.4	N18 40.1	316 44.4	N22 29.9
13	57 59.2	41 53.0	02.3	107 06.1	51.0	285 45.6	40.1	331 46.6	29.9
14	73 01.7	56 52.7	03.5	122 07.1	50.7	300 47.8	40.0	346 48.7	30.0
15	88 04.2	71 52.3	. . 04.6	137 08.0	. . 50.5	315 50.1	. . 39.9	1 50.9	. . 30.0
16	103 06.6	86 52.0	05.8	152 09.0	50.2	330 52.3	39.9	16 53.1	30.0
17	118 09.1	101 51.6	06.9	167 09.9	49.9	345 54.5	39.8	31 55.2	30.0
18	133 11.6	116 51.3	N 5 08.1	182 10.9	S19 49.6	0 56.8	N18 39.7	46 57.4	N22 30.0
19	148 14.0	131 50.9	09.2	197 11.8	49.3	15 59.0	39.7	61 59.6	30.0
20	163 16.5	146 50.6	10.4	212 12.8	49.0	31 01.2	39.6	77 01.8	30.0
21	178 19.0	161 50.3	. . 11.5	227 13.7	. . 48.7	46 03.4	. . 39.5	92 03.9	. . 30.0
22	193 21.4	176 49.9	12.7	242 14.7	48.4	61 05.7	39.5	107 06.1	30.1
23	208 23.9	191 49.6	13.8	257 15.6	48.1	76 07.9	39.4	122 08.3	30.1
TUESDAY 6 00	223 26.3	206 49.2	N 5 15.0	272 16.6	S19 47.8	91 10.1	N18 39.3	137 10.4	N22 30.1
01	238 28.8	221 48.9	16.1	287 17.6	47.5	106 12.4	39.3	152 12.6	30.1
02	253 31.3	236 48.5	17.3	302 18.5	47.2	121 14.6	39.2	167 14.8	30.1
03	268 33.7	251 48.2	. . 18.4	317 19.5	. . 46.9	136 16.8	. . 39.2	182 16.9	. . 30.1
04	283 36.2	266 47.8	19.6	332 20.4	46.6	151 19.1	39.1	197 19.1	30.1
05	298 38.7	281 47.5	20.7	347 21.4	46.3	166 21.3	39.0	212 21.3	30.2
06	313 41.1	296 47.1	N 5 21.9	2 22.3	S19 46.1	181 23.5	N18 39.0	227 23.5	N22 30.2
07	328 43.6	311 46.8	23.0	17 23.3	45.8	196 25.7	38.9	242 25.6	30.2
08	343 46.1	326 46.4	24.2	32 24.3	45.5	211 28.0	38.8	257 27.8	30.2
09	358 48.5	341 46.1	. . 25.3	47 25.2	. . 45.2	226 30.2	. . 38.8	272 30.0	. . 30.2
10	13 51.0	356 45.7	26.5	62 26.2	44.9	241 32.4	38.7	287 32.1	30.2
11	28 53.5	11 45.4	27.6	77 27.1	44.6	256 34.6	38.6	302 34.3	30.2
12	43 55.9	26 45.0	N 5 28.8	92 28.1	S19 44.3	271 36.9	N18 38.6	317 36.5	N22 30.2
13	58 58.4	41 44.7	29.9	107 29.1	44.0	286 39.1	38.5	332 38.7	30.3
14	74 00.8	56 44.3	31.1	122 30.0	43.7	301 41.3	38.4	347 40.8	30.3
15	89 03.3	71 44.0	. . 32.2	137 31.0	. . 43.4	316 43.5	. . 38.4	2 43.0	. . 30.3
16	104 05.8	86 43.6	33.4	152 31.9	43.1	331 45.8	38.3	17 45.2	30.3
17	119 08.2	101 43.3	34.5	167 32.9	42.8	346 48.0	38.2	32 47.3	30.3
18	134 10.7	116 42.9	N 5 35.7	182 33.9	S19 42.5	1 50.2	N18 38.2	47 49.5	N22 30.3
19	149 13.2	131 42.6	36.8	197 34.8	42.2	16 52.4	38.1	62 51.7	30.3
20	164 15.6	146 42.2	38.0	212 35.8	41.9	31 54.7	38.0	77 53.8	30.3
21	179 18.1	161 41.9	. . 39.1	227 36.7	. . 41.6	46 56.9	. . 38.0	92 56.0	. . 30.4
22	194 20.6	176 41.5	40.3	242 37.7	41.3	61 59.1	37.9	107 58.2	30.4
23	209 23.0	191 41.2	41.4	257 38.7	41.0	77 01.3	37.8	123 00.3	30.4
Mer. Pass. (h m)	9 08.7	v −0.3	d 1.2	v 1.0	d 0.3	v 2.2	d 0.1	v 2.2	d 0.0

STARS Name	SHA ° ′	Dec ° ′
Acamar	315 24.7	S40 17.
Achernar	335 33.1	S57 13.
Acrux	173 18.0	S63 07.
Adhara	255 19.0	S28 58.
Aldebaran	290 58.9	N16 30.
Alioth	166 27.0	N55 56.
Alkaid	153 04.6	N49 17.
Al Na'ir	27 53.7	S46 56.
Alnilam	275 54.8	S 1 12.
Alphard	218 03.9	S 8 40.
Alphecca	126 17.4	N26 42.
Alpheratz	357 52.1	N29 06.
Altair	62 15.9	N 8 52.
Ankaa	353 23.8	S42 17.
Antares	112 35.8	S26 26.
Arcturus	146 02.7	N19 09.
Atria	107 44.4	S69 01.
Avior	234 21.6	S59 31.
Bellatrix	278 40.9	N 6 21.
Betelgeuse	271 10.2	N 7 24.5
Canopus	264 00.1	S52 42.0
Capella	280 46.7	N46 00.
Deneb	49 37.0	N45 17.
Denebola	182 41.6	N14 33.
Diphda	349 04.1	S17 58.
Dubhe	194 00.8	N61 44.
Elnath	278 23.1	N28 36.
Eltanin	90 49.5	N51 29.0
Enif	33 55.1	N 9 53.
Fomalhaut	15 32.9	S29 36.
Gacrux	172 09.6	S57 08.0
Gienah	176 00.3	S17 33.7
Hadar	148 58.9	S60 23.
Hamal	328 10.2	N23 28.5
Kaus Aust.	83 54.2	S34 23.0
Kochab	137 18.1	N74 08.5
Markab	13 46.5	N15 13.
Menkar	314 23.8	N 4 06.0
Menkent	148 16.7	S36 23.
Miaplacidus	221 41.7	S69 44.
Mirfak	308 52.5	N49 52.
Nunki	76 08.0	S26 17.6
Peacock	53 31.6	S56 43.4
Pollux	243 37.6	N28 01.
Procyon	245 08.2	N 5 13.0
Rasalhague	96 13.6	N12 33.
Regulus	207 51.9	N11 57.
Rigel	281 20.0	S 8 11.
Rigil Kent.	140 02.3	S60 51.0
Sabik	102 21.4	S15 43.8
Schedar	349 50.4	N56 33.0
Shaula	96 32.5	S37 06.4
Sirius	258 41.0	S16 43.
Spica	158 39.4	S11 10.8
Suhail	222 58.4	S43 26.
Vega	80 44.1	N38 46.
Zuben'ubi	137 14.0	S16 03.4

	SHA ° ′	Mer. Pass. h m
Venus	344 30.3	10 12
Mars	49 26.5	5 52
Jupiter	227 49.4	17 56
Saturn	273 51.1	14 53

UT	SUN GHA	SUN Dec	MOON GHA	v	MOON Dec	d	HP
d h	° ′	° ′	° ′	′	° ′	′	′
4 00	180 47.1	N15 46.9	153 05.9	12.0	N22 56.7	7.0	54.3
01	195 47.2	47.6	167 36.9	11.8	23 03.7	6.9	54.3
02	210 47.3	48.4	182 07.7	11.8	23 10.6	6.8	54.3
03	225 47.3	. . 49.1	196 38.5	11.7	23 17.4	6.6	54.3
04	240 47.4	49.8	211 09.2	11.7	23 24.0	6.6	54.3
05	255 47.4	50.6	225 39.9	11.6	23 30.6	6.5	54.3
06	270 47.5	N15 51.3	240 10.5	11.5	N23 37.1	6.3	54.3
07	285 47.6	52.0	254 41.0	11.5	23 43.4	6.3	54.3
08	300 47.6	52.8	269 11.5	11.4	23 49.7	6.1	54.3
09	315 47.7	. . 53.5	283 41.9	11.4	23 55.8	6.0	54.4
10	330 47.7	54.2	298 12.3	11.3	24 01.8	6.0	54.4
11	345 47.8	54.9	312 42.6	11.2	24 07.8	5.8	54.4
12	0 47.9	N15 55.7	327 12.8	11.2	N24 13.6	5.7	54.4
13	15 47.9	56.4	341 43.0	11.1	24 19.3	5.5	54.4
14	30 48.0	57.1	356 13.1	11.0	24 24.8	5.5	54.4
15	45 48.0	. . 57.8	10 43.1	11.0	24 30.3	5.4	54.4
16	60 48.1	58.6	25 13.1	11.0	24 35.7	5.2	54.4
17	75 48.2	15 59.3	39 43.1	10.8	24 40.9	5.2	54.5
18	90 48.2	N16 00.0	54 12.9	10.8	N24 46.1	5.0	54.5
19	105 48.3	00.7	68 42.7	10.8	24 51.1	4.9	54.5
20	120 48.3	01.5	83 12.5	10.7	24 56.0	4.8	54.5
21	135 48.4	. . 02.2	97 42.2	10.6	25 00.8	4.6	54.5
22	150 48.4	02.9	112 11.8	10.6	25 05.4	4.6	54.5
23	165 48.5	03.6	126 41.4	10.6	25 10.0	4.4	54.5
5 00	180 48.6	N16 04.3	141 11.0	10.4	N25 14.4	4.3	54.6
01	195 48.6	05.1	155 40.4	10.5	25 18.7	4.2	54.6
02	210 48.7	05.8	170 09.9	10.3	25 22.9	4.1	54.6
03	225 48.7	. . 06.5	184 39.2	10.4	25 27.0	3.9	54.6
04	240 48.8	07.2	199 08.6	10.2	25 30.9	3.8	54.6
05	255 48.8	07.9	213 37.8	10.3	25 34.7	3.7	54.6
06	270 48.9	N16 08.7	228 07.1	10.1	N25 38.4	3.6	54.6
07	285 48.9	09.4	242 36.2	10.1	25 42.0	3.4	54.7
08	300 49.0	10.1	257 05.3	10.1	25 45.4	3.4	54.7
09	315 49.0	. . 10.8	271 34.4	10.0	25 48.8	3.2	54.7
10	330 49.1	11.5	286 03.4	10.0	25 52.0	3.0	54.7
11	345 49.2	12.2	300 32.4	9.9	25 55.0	3.0	54.7
12	0 49.2	N16 13.0	315 01.3	9.9	N25 58.0	2.8	54.7
13	15 49.3	13.7	329 30.2	9.9	26 00.8	2.7	54.8
14	30 49.3	14.4	343 59.1	9.8	26 03.5	2.6	54.8
15	45 49.4	. . 15.1	358 27.9	9.7	26 06.1	2.4	54.8
16	60 49.4	15.8	12 56.6	9.7	26 08.5	2.3	54.8
17	75 49.5	16.5	27 25.3	9.7	26 10.8	2.2	54.8
18	90 49.5	N16 17.2	41 54.0	9.6	N26 13.0	2.0	54.8
19	105 49.6	17.9	56 22.6	9.6	26 15.0	1.9	54.9
20	120 49.6	18.7	70 51.2	9.6	26 16.9	1.8	54.9
21	135 49.7	. . 19.4	85 19.8	9.5	26 18.7	1.7	54.9
22	150 49.7	20.1	99 48.3	9.5	26 20.4	1.5	54.9
23	165 49.8	20.8	114 16.8	9.4	26 21.9	1.4	54.9
6 00	180 49.8	N16 21.5	128 45.2	9.5	N26 23.3	1.2	55.0
01	195 49.9	22.2	143 13.7	9.3	26 24.5	1.1	55.0
02	210 49.9	22.9	157 42.0	9.4	26 25.6	1.0	55.0
03	225 50.0	. . 23.6	172 10.4	9.3	26 26.6	0.9	55.0
04	240 50.0	24.3	186 38.7	9.3	26 27.5	0.7	55.0
05	255 50.1	25.0	201 07.0	9.3	26 28.2	0.6	55.1
06	270 50.1	N16 25.7	215 35.3	9.2	N26 28.8	0.4	55.1
07	285 50.2	26.5	230 03.5	9.2	26 29.2	0.3	55.1
08	300 50.2	27.2	244 31.7	9.2	26 29.5	0.2	55.1
09	315 50.3	. . 27.9	258 59.9	9.1	26 29.7	0.0	55.1
10	330 50.3	28.6	273 28.0	9.2	26 29.7	0.1	55.2
11	345 50.4	29.3	287 56.2	9.1	26 29.6	0.2	55.2
12	0 50.4	N16 30.0	302 24.3	9.1	N26 29.4	0.4	55.2
13	15 50.5	30.7	316 52.4	9.0	26 29.0	0.5	55.2
14	30 50.5	31.4	331 20.4	9.1	26 28.5	0.7	55.3
15	45 50.6	. . 32.1	345 48.5	9.0	26 27.8	0.7	55.3
16	60 50.6	32.8	0 16.5	9.0	26 27.1	1.0	55.3
17	75 50.7	33.5	14 44.5	9.0	26 26.1	1.0	55.3
18	90 50.7	N16 34.2	29 12.5	9.0	N26 25.1	1.2	55.3
19	105 50.7	34.9	43 40.5	9.0	26 23.9	1.4	55.4
20	120 50.8	35.6	58 08.5	8.9	26 22.5	1.5	55.4
21	135 50.8	. . 36.3	72 36.4	9.0	26 21.0	1.6	55.4
22	150 50.9	37.0	87 04.4	8.9	26 19.4	1.7	55.4
23	165 50.9	37.7	101 32.3	8.9	N26 17.7	1.9	55.5
	SD 15.9	d 0.7	SD 14.8		14.9		15.0

Lat.	Twilight Naut.	Twilight Civil	Sunrise	Moonrise 4	Moonrise 5	Moonrise 6	Moonrise 7
°	h m	h m	h m	h m	h m	h m	h m
N 72	////	////	01 19	▭	▭	▭	▭
N 70	////	////	02 09	▭	▭	▭	▭
68	////	////	02 41	▭	▭	▭	▭
66	////	01 32	03 04	02 56	▭	▭	▭
64	////	02 08	03 21	03 47	03 36	▭	04 36
62	////	02 34	03 36	04 19	04 30	05 00	06 05
60	01 22	02 53	03 49	04 43	05 03	05 40	06 42
N 58	01 55	03 09	03 59	05 02	05 27	06 08	07 08
56	02 18	03 23	04 08	05 18	05 47	06 29	07 28
54	02 37	03 34	04 16	05 32	06 03	06 47	07 46
52	02 52	03 44	04 24	05 44	06 17	07 02	08 00
50	03 05	03 53	04 30	05 54	06 30	07 15	08 13
45	03 31	04 12	04 44	06 17	06 55	07 42	08 39
N 40	03 50	04 26	04 56	06 35	07 15	08 04	09 00
35	04 05	04 38	05 06	06 50	07 32	08 22	09 17
30	04 18	04 49	05 14	07 03	07 47	08 37	09 32
20	04 38	05 06	05 29	07 26	08 12	09 03	09 57
N 10	04 54	05 20	05 42	07 45	08 34	09 25	10 19
0	05 07	05 32	05 53	08 04	08 54	09 46	10 39
S 10	05 18	05 43	06 05	08 22	09 14	10 07	11 00
20	05 28	05 54	06 17	08 42	09 36	10 29	11 21
30	05 38	06 06	06 31	09 05	10 01	10 56	11 46
35	05 43	06 13	06 39	09 19	10 16	11 11	12 01
40	05 48	06 20	06 48	09 35	10 34	11 29	12 18
45	05 53	06 28	06 59	09 53	10 55	11 50	12 39
S 50	05 59	06 37	07 12	10 17	11 21	12 18	13 04
52	06 01	06 41	07 18	10 28	11 34	12 31	13 17
54	06 04	06 46	07 24	10 41	11 49	12 46	13 31
56	06 06	06 51	07 32	10 56	12 06	13 04	13 48
58	06 09	06 57	07 40	11 14	12 27	13 26	14 08
S 60	06 12	07 03	07 49	11 36	12 54	13 55	14 32

Lat.	Sunset	Twilight Civil	Twilight Naut.	Moonset 4	Moonset 5	Moonset 6	Moonset 7
°	h m	h m	h m	h m	h m	h m	h m
N 72	22 45	////	////	▭	▭	▭	▭
N 70	21 49	////	////	▭	▭	▭	▭
68	21 16	////	////	▭	▭	▭	▭
66	20 53	22 28	////	00 36	▭	▭	▭
64	20 34	21 49	////	25 40	01 40	▭	04 19
62	20 19	21 23	////	24 46	00 46	02 03	02 49
60	20 07	21 03	22 38	24 13	00 13	01 24	02 12
N 58	19 56	20 46	22 02	23 49	24 56	00 56	01 46
56	19 46	20 33	21 38	23 30	24 35	00 35	01 25
54	19 38	20 21	21 19	23 14	24 17	00 17	01 08
52	19 31	20 11	21 03	23 00	24 02	00 02	00 53
50	19 24	20 02	20 50	22 48	23 49	24 40	00 40
45	19 10	19 43	20 24	22 23	23 22	24 14	00 14
N 40	18 58	19 28	20 04	22 03	23 01	23 53	24 39
35	18 48	19 16	19 49	21 47	22 43	23 35	24 23
30	18 40	19 05	19 36	21 32	22 28	23 20	24 09
20	18 25	18 48	19 15	21 08	22 02	22 55	23 45
N 10	18 12	18 34	19 00	20 47	21 40	22 32	23 24
0	18 00	18 22	18 47	20 28	21 19	22 12	23 05
S 10	17 48	18 10	18 35	20 08	20 58	21 51	22 45
20	17 36	17 59	18 25	19 47	20 36	21 29	22 24
30	17 22	17 47	18 15	19 23	20 10	21 03	22 00
35	17 14	17 40	18 10	19 09	19 55	20 47	21 46
40	17 05	17 33	18 05	18 53	19 37	20 30	21 29
45	16 54	17 25	18 00	18 33	19 16	20 08	21 09
S 50	16 41	17 16	17 54	18 09	18 49	19 41	20 44
52	16 35	17 11	17 52	17 57	18 36	19 28	20 32
54	16 28	17 07	17 49	17 44	18 21	19 12	20 18
56	16 21	17 02	17 46	17 29	18 04	18 54	20 01
58	16 13	16 56	17 43	17 11	17 43	18 32	19 42
S 60	16 03	16 50	17 40	16 49	17 16	18 04	19 17

Day	SUN Eqn. of Time 00^h	SUN Eqn. of Time 12^h	SUN Mer. Pass.	MOON Mer. Pass. Upper	MOON Mer. Pass. Lower	MOON Age	MOON Phase
d	m s	m s	h m	h m	h m	d	%
4	03 08	03 11	11 57	14 16	01 51	03	8
5	03 14	03 17	11 57	15 06	02 41	04	14
6	03 19	03 22	11 57	15 59	03 32	05	21

UT	ARIES	VENUS −3.9		MARS −0.2		JUPITER −2.1		SATURN +0.1		STARS		
d h	GHA ° ′	GHA ° ′	Dec ° ′	GHA ° ′	Dec ° ′	GHA ° ′	Dec ° ′	GHA ° ′	Dec ° ′	Name	SHA ° ′	Dec ° ′
7 00 WEDNESDAY	224 25.5	206 40.8	N 5 42.6	272 39.6	S19 40.7	92 03.6	N18 37.8	138 02.5	N22 30.4	Acamar	315 24.7	S40 17.
01	239 28.0	221 40.5	43.7	287 40.6	40.4	107 05.8	37.7	153 04.7	30.4	Achernar	335 33.0	S57 13.
02	254 30.4	236 40.1	44.8	302 41.6	40.2	122 08.0	37.6	168 06.8	30.4	Acrux	173 18.0	S63 07.
03	269 32.9	251 39.8	. . 46.0	317 42.5	. . 39.9	137 10.2	. . 37.6	183 09.0	. . 30.4	Adhara	255 19.0	S28 58.
04	284 35.3	266 39.4	47.1	332 43.5	39.6	152 12.4	37.5	198 11.2	30.4	Aldebaran	290 58.9	N16 30.
05	299 37.8	281 39.1	48.3	347 44.5	39.3	167 14.7	37.4	213 13.4	30.5			
06	314 40.3	296 38.7	N 5 49.4	2 45.4	S19 39.0	182 16.9	N18 37.4	228 15.5	N22 30.5	Alioth	166 27.0	N55 56.
07	329 42.7	311 38.4	50.6	17 46.4	38.7	197 19.1	37.3	243 17.7	30.5	Alkaid	153 04.6	N49 17.
08	344 45.2	326 38.0	51.7	32 47.4	38.4	212 21.3	37.2	258 19.9	30.5	Al Na'ir	27 53.7	S46 56.
09	359 47.7	341 37.7	. . 52.9	47 48.3	. . 38.1	227 23.5	. . 37.2	273 22.0	. . 30.5	Alnilam	275 54.8	S 1 12.
10	14 50.1	356 37.3	54.0	62 49.3	37.8	242 25.8	37.1	288 24.2	30.5	Alphard	218 03.9	S 8 40.
11	29 52.6	11 36.9	55.2	77 50.3	37.5	257 28.0	37.0	303 26.4	30.5			
12	44 55.1	26 36.6	N 5 56.3	92 51.2	S19 37.2	272 30.2	N18 37.0	318 28.5	N22 30.5	Alphecca	126 17.4	N26 42.
13	59 57.5	41 36.2	57.4	107 52.2	36.9	287 32.4	36.9	333 30.7	30.6	Alpheratz	357 52.1	N29 06.
14	75 00.0	56 35.9	58.6	122 53.2	36.6	302 34.6	36.8	348 32.9	30.6	Altair	62 15.9	N 8 52.
15	90 02.5	71 35.5	5 59.7	137 54.1	. . 36.3	317 36.9	. . 36.8	3 35.0	. . 30.6	Ankaa	353 23.8	S42 17.
16	105 04.9	86 35.2	6 00.9	152 55.1	36.0	332 39.1	36.7	18 37.2	30.6	Antares	112 35.8	S26 26.
17	120 07.4	101 34.8	02.0	167 56.1	35.7	347 41.3	36.6	33 39.4	30.6			
18	135 09.8	116 34.5	N 6 03.2	182 57.0	S19 35.4	2 43.5	N18 36.6	48 41.5	N22 30.6	Arcturus	146 02.7	N19 09.
19	150 12.3	131 34.1	04.3	197 58.0	35.1	17 45.7	36.5	63 43.7	30.6	Atria	107 44.4	S69 02.
20	165 14.8	146 33.8	05.4	212 59.0	34.8	32 48.0	36.4	78 45.9	30.7	Avior	234 21.6	S59 31.
21	180 17.2	161 33.4	. . 06.6	227 59.9	. . 34.5	47 50.2	. . 36.4	93 48.0	. . 30.7	Bellatrix	278 40.9	N 6 21.
22	195 19.7	176 33.0	07.7	243 00.9	34.2	62 52.4	36.3	108 50.2	30.7	Betelgeuse	271 10.2	N 7 24.
23	210 22.2	191 32.7	08.9	258 01.9	33.9	77 54.6	36.2	123 52.4	30.7			
8 00 THURSDAY	225 24.6	206 32.3	N 6 10.0	273 02.8	S19 33.6	92 56.8	N18 36.2	138 54.5	N22 30.7	Canopus	264 00.1	S52 42.
01	240 27.1	221 32.0	11.2	288 03.8	33.3	107 59.0	36.1	153 56.7	30.7	Capella	280 46.7	N46 00.
02	255 29.6	236 31.6	12.3	303 04.8	33.0	123 01.3	36.0	168 58.9	30.7	Deneb	49 36.9	N45 17.
03	270 32.0	251 31.3	. . 13.4	318 05.8	. . 32.7	138 03.5	. . 36.0	184 01.0	. . 30.7	Denebola	182 41.6	N14 33.
04	285 34.5	266 30.9	14.6	333 06.7	32.4	153 05.7	35.9	199 03.2	30.8	Diphda	349 04.1	S17 58.
05	300 37.0	281 30.6	15.7	348 07.7	32.1	168 07.9	35.8	214 05.4	30.8			
06	315 39.4	296 30.2	N 6 16.9	3 08.7	S19 31.8	183 10.1	N18 35.7	229 07.5	N22 30.8	Dubhe	194 00.9	N61 44.
07	330 41.9	311 29.8	18.0	18 09.6	31.5	198 12.3	35.7	244 09.7	30.8	Elnath	278 23.1	N28 36.
08	345 44.3	326 29.5	19.1	33 10.6	31.2	213 14.5	35.6	259 11.8	30.8	Eltanin	90 49.4	N51 29.
09	0 46.8	341 29.1	. . 20.3	48 11.6	. . 30.9	228 16.8	. . 35.5	274 14.0	. . 30.8	Enif	33 55.0	N 9 53.
10	15 49.3	356 28.8	21.4	63 12.6	30.6	243 19.0	35.5	289 16.2	30.8	Fomalhaut	15 32.8	S29 36.
11	30 51.7	11 28.4	22.6	78 13.5	30.3	258 21.2	35.4	304 18.3	30.8			
12	45 54.2	26 28.1	N 6 23.7	93 14.5	S19 30.0	273 23.4	N18 35.3	319 20.5	N22 30.9	Gacrux	172 09.6	S57 08.
13	60 56.7	41 27.7	24.8	108 15.5	29.7	288 25.6	35.3	334 22.7	30.9	Gienah	176 00.3	S17 33.
14	75 59.1	56 27.3	26.0	123 16.5	29.4	303 27.8	35.2	349 24.8	30.9	Hadar	148 58.9	S60 23.
15	91 01.6	71 27.0	. . 27.1	138 17.4	. . 29.1	318 30.0	. . 35.1	4 27.0	. . 30.9	Hamal	328 10.2	N23 28.
16	106 04.1	86 26.6	28.3	153 18.4	28.8	333 32.3	35.1	19 29.2	30.9	Kaus Aust.	83 54.1	S34 23.
17	121 06.5	101 26.3	29.4	168 19.4	28.5	348 34.5	35.0	34 31.3	30.9			
18	136 09.0	116 25.9	N 6 30.5	183 20.4	S19 28.2	3 36.7	N18 34.9	49 33.5	N22 30.9	Kochab	137 18.1	N74 08.
19	151 11.4	131 25.5	31.7	198 21.3	27.9	18 38.9	34.9	64 35.7	30.9	Markab	13 46.5	N15 13.
20	166 13.9	146 25.2	32.8	213 22.3	27.6	33 41.1	34.8	79 37.8	31.0	Menkar	314 23.8	N 4 06.
21	181 16.4	161 24.8	. . 34.0	228 23.3	. . 27.3	48 43.3	. . 34.7	94 40.0	. . 31.0	Menkent	148 16.7	S36 23.
22	196 18.8	176 24.5	35.1	243 24.3	27.0	63 45.5	34.6	109 42.2	31.0	Miaplacidus	221 41.7	S69 44.
23	211 21.3	191 24.1	36.2	258 25.2	26.7	78 47.7	34.6	124 44.3	31.0			
9 00 FRIDAY	226 23.8	206 23.7	N 6 37.4	273 26.2	S19 26.4	93 49.9	N18 34.5	139 46.5	N22 31.0	Mirfak	308 52.4	N49 52.
01	241 26.2	221 23.4	38.5	288 27.2	26.1	108 52.2	34.4	154 48.6	31.0	Nunki	76 08.0	S26 17.
02	256 28.7	236 23.0	39.6	303 28.2	25.8	123 54.4	34.4	169 50.8	31.0	Peacock	53 31.5	S56 43.
03	271 31.2	251 22.7	. . 40.8	318 29.2	. . 25.5	138 56.6	. . 34.3	184 53.0	. . 31.0	Pollux	243 37.6	N28 01.
04	286 33.6	266 22.3	41.9	333 30.1	25.2	153 58.8	34.2	199 55.1	31.0	Procyon	245 08.2	N 5 13.
05	301 36.1	281 21.9	43.0	348 31.1	24.9	169 01.0	34.2	214 57.3	31.1			
06	316 38.6	296 21.6	N 6 44.2	3 32.1	S19 24.6	184 03.2	N18 34.1	229 59.5	N22 31.1	Rasalhague	96 13.6	N12 33.
07	331 41.0	311 21.2	45.3	18 33.1	24.3	199 05.4	34.0	245 01.6	31.1	Regulus	207 51.9	N11 57.
08	346 43.5	326 20.8	46.5	33 34.1	24.0	214 07.6	33.9	260 03.8	31.1	Rigel	281 20.0	S 8 11.
09	1 45.9	341 20.5	. . 47.6	48 35.0	. . 23.7	229 09.8	. . 33.9	275 06.0	. . 31.1	Rigil Kent.	140 02.3	S60 51.
10	16 48.4	356 20.1	48.7	63 36.0	23.4	244 12.0	33.8	290 08.1	31.1	Sabik	102 21.4	S15 43.
11	31 50.9	11 19.8	49.9	78 37.0	23.1	259 14.2	33.7	305 10.3	31.1			
12	46 53.3	26 19.4	N 6 51.0	93 38.0	S19 22.8	274 16.4	N18 33.7	320 12.4	N22 31.1	Schedar	349 50.3	N56 33.
13	61 55.8	41 19.0	52.1	108 39.0	22.5	289 18.7	33.6	335 14.6	31.2	Shaula	96 32.4	S37 06.
14	76 58.3	56 18.7	53.3	123 39.9	22.2	304 20.9	33.5	350 16.8	31.2	Sirius	258 41.0	S16 43.
15	92 00.7	71 18.3	. . 54.4	138 40.9	. . 21.9	319 23.1	. . 33.4	5 18.9	. . 31.2	Spica	158 39.4	S11 10.
16	107 03.2	86 17.9	55.5	153 41.9	21.6	334 25.3	33.4	20 21.1	31.2	Suhail	222 58.4	S43 26.
17	122 05.7	101 17.6	56.7	168 42.9	21.3	349 27.5	33.3	35 23.3	31.2			
18	137 08.1	116 17.2	N 6 57.8	183 43.9	S19 21.0	4 29.7	N18 33.2	50 25.4	N22 31.2	Vega	80 44.1	N38 46.
19	152 10.6	131 16.8	6 58.9	198 44.9	20.7	19 31.9	33.2	65 27.6	31.2	Zuben'ubi	137 14.0	S16 03.
20	167 13.1	146 16.5	7 00.1	213 45.8	20.4	34 34.1	33.1	80 29.7	31.2			
21	182 15.5	161 16.1	. . 01.2	228 46.8	. . 20.1	49 36.3	. . 33.0	95 31.9	. . 31.3			
22	197 18.0	176 15.7	02.3	243 47.8	19.8	64 38.5	32.9	110 34.1	31.3			
23	212 20.4	191 15.4	03.5	258 48.8	19.5	79 40.7	32.9	125 36.2	31.3			
Mer. Pass.	h m 8 56.9	*v* −0.4	*d* 1.1	*v* 1.0	*d* 0.3	*v* 2.2	*d* 0.1	*v* 2.2	*d* 0.0			

	SHA ° ′	Mer. Pass. h m
Venus	341 07.7	10 14
Mars	47 38.2	5 47
Jupiter	227 32.2	17 46
Saturn	273 29.9	14 42

UT	SUN GHA	SUN Dec	MOON GHA	v	MOON Dec	d	HP
d h	° ′	° ′	° ′	′	° ′	′	′
7 00	180 51.0	N16 38.4	116 00.2	9.0	N26 15.8	2.1	55.5
01	195 51.0	39.1	130 28.2	8.9	26 13.7	2.2	55.5
02	210 51.1	39.8	144 56.1	8.9	26 11.5	2.3	55.5
03	225 51.1	. . 40.5	159 24.0	8.9	26 09.2	2.4	55.6
04	240 51.1	41.2	173 51.9	8.9	26 06.8	2.6	55.6
05	255 51.2	41.9	188 19.8	8.9	26 04.2	2.8	55.6
06	270 51.2	N16 42.6	202 47.7	8.9	N26 01.4	2.8	55.6
07	285 51.3	43.3	217 15.6	8.9	25 58.6	3.1	55.7
08	300 51.3	43.9	231 43.5	8.9	25 55.5	3.1	55.7
09	315 51.4	. . 44.6	246 11.4	8.9	25 52.4	3.3	55.7
10	330 51.4	45.3	260 39.3	8.9	25 49.1	3.4	55.7
11	345 51.4	46.0	275 07.2	8.9	25 45.7	3.6	55.8
12	0 51.5	N16 46.7	289 35.1	8.9	N25 42.1	3.7	55.8
13	15 51.5	47.4	304 03.0	8.9	25 38.4	3.9	55.8
14	30 51.6	48.1	318 30.9	8.9	25 34.5	3.9	55.8
15	45 51.6	. . 48.8	332 58.8	8.9	25 30.6	4.2	55.9
16	60 51.7	49.5	347 26.7	8.9	25 26.4	4.2	55.9
17	75 51.7	50.2	1 54.6	9.0	25 22.2	4.4	55.9
18	90 51.7	N16 50.9	16 22.6	8.9	N25 17.8	4.6	55.9
19	105 51.8	51.6	30 50.5	9.0	25 13.2	4.7	56.0
20	120 51.8	52.2	45 18.5	9.0	25 08.5	4.8	56.0
21	135 51.9	. . 52.9	59 46.5	8.9	25 03.7	4.9	56.0
22	150 51.9	53.6	74 14.4	9.0	24 58.8	5.1	56.1
23	165 51.9	54.3	88 42.4	9.1	24 53.7	5.2	56.1
8 00	180 52.0	N16 55.0	103 10.5	9.0	N24 48.5	5.4	56.1
01	195 52.0	55.7	117 38.5	9.0	24 43.1	5.5	56.1
02	210 52.0	56.4	132 06.5	9.1	24 37.6	5.6	56.2
03	225 52.1	. . 57.0	146 34.6	9.1	24 32.0	5.8	56.2
04	240 52.1	57.7	161 02.7	9.1	24 26.2	5.9	56.2
05	255 52.2	58.4	175 30.8	9.1	24 20.3	6.1	56.3
06	270 52.2	N16 59.1	189 58.9	9.1	N24 14.2	6.1	56.3
07	285 52.2	16 59.8	204 27.0	9.2	24 08.1	6.4	56.3
08	300 52.3	17 00.5	218 55.2	9.2	24 01.7	6.4	56.4
09	315 52.3	. . 01.1	233 23.4	9.2	23 55.3	6.6	56.4
10	330 52.3	01.8	247 51.6	9.2	23 48.7	6.7	56.4
11	345 52.4	02.5	262 19.8	9.2	23 42.0	6.8	56.5
12	0 52.4	N17 03.2	276 48.0	9.3	N23 35.2	7.0	56.5
13	15 52.5	03.9	291 16.3	9.3	23 28.2	7.1	56.5
14	30 52.5	04.5	305 44.6	9.3	23 21.1	7.3	56.5
15	45 52.5	. . 05.2	320 12.9	9.3	23 13.8	7.3	56.6
16	60 52.6	05.9	334 41.2	9.4	23 06.5	7.5	56.6
17	75 52.6	06.6	349 09.6	9.4	22 59.0	7.7	56.6
18	90 52.6	N17 07.3	3 38.0	9.4	N22 51.3	7.7	56.7
19	105 52.7	07.9	18 06.4	9.4	22 43.6	7.9	56.7
20	120 52.7	08.6	32 34.8	9.5	22 35.7	8.0	56.7
21	135 52.7	. . 09.3	47 03.3	9.5	22 27.7	8.2	56.8
22	150 52.8	10.0	61 31.8	9.5	22 19.5	8.2	56.8
23	165 52.8	10.6	76 00.3	9.6	22 11.3	8.4	56.8
9 00	180 52.8	N17 11.3	90 28.9	9.6	N22 02.9	8.6	56.9
01	195 52.9	12.0	104 57.5	9.6	21 54.3	8.6	56.9
02	210 52.9	12.7	119 26.1	9.6	21 45.7	8.8	56.9
03	225 52.9	. . 13.3	133 54.7	9.7	21 36.9	8.9	57.0
04	240 53.0	14.0	148 23.4	9.7	21 28.0	9.0	57.0
05	255 53.0	14.7	162 52.1	9.7	21 19.0	9.2	57.0
06	270 53.0	N17 15.3	177 20.8	9.8	N21 09.8	9.2	57.1
07	285 53.1	16.0	191 49.6	9.7	21 00.6	9.4	57.1
08	300 53.1	16.7	206 18.3	9.9	20 51.2	9.5	57.1
09	315 53.1	. . 17.4	220 47.2	9.8	20 41.7	9.7	57.2
10	330 53.2	18.0	235 16.0	9.9	20 32.0	9.7	57.2
11	345 53.2	18.7	249 44.9	9.9	20 22.3	9.9	57.2
12	0 53.2	N17 19.4	264 13.8	9.9	N20 12.4	10.0	57.3
13	15 53.2	20.0	278 42.7	10.0	20 02.4	10.1	57.3
14	30 53.3	20.7	293 11.7	10.0	19 52.3	10.2	57.4
15	45 53.3	. . 21.4	307 40.7	10.0	19 42.1	10.4	57.4
16	60 53.3	22.0	322 09.7	10.1	19 31.7	10.4	57.4
17	75 53.4	22.7	336 38.8	10.1	19 21.3	10.6	57.5
18	90 53.4	N17 23.4	351 07.9	10.1	N19 10.7	10.7	57.5
19	105 53.4	24.0	5 37.0	10.1	19 00.0	10.8	57.5
20	120 53.5	24.7	20 06.1	10.2	18 49.2	10.9	57.6
21	135 53.5	. . 25.4	34 35.3	10.2	18 38.3	11.0	57.6
22	150 53.5	26.0	49 04.5	10.3	18 27.3	11.1	57.6
23	165 53.5	26.7	63 33.8	10.2	N18 16.2	11.3	57.7
	SD 15.9	d 0.7	SD		15.2	15.4	15.6

Lat.	Twilight Naut.	Twilight Civil	Sunrise	Moonrise 7	Moonrise 8	Moonrise 9	Moonrise 10
°	h m	h m	h m	h m	h m	h m	h m
N 72	////	////	00 35	▭	▭	▭	08 52
N 70	////	////	01 49	▭	▭	▭	09 39
68	////	////	02 26	▭	▭	07 34	10 09
66	////	01 08	02 52	▭	▭	08 24	10 31
64	////	01 53	03 11	04 36	06 58	08 55	10 48
62	////	02 22	03 27	06 05	07 37	09 19	11 02
60	01 00	02 43	03 41	06 42	08 04	09 37	11 14
N 58	01 41	03 01	03 52	07 08	08 25	09 52	11 24
56	02 08	03 15	04 02	07 28	08 42	10 05	11 33
54	02 28	03 27	04 11	07 46	08 57	10 17	11 41
52	02 45	03 38	04 18	08 00	09 10	10 27	11 48
50	02 58	03 48	04 25	08 13	09 21	10 36	11 55
45	03 26	04 07	04 40	08 39	09 44	10 54	12 08
N 40	03 46	04 23	04 52	09 00	10 02	11 10	12 19
35	04 02	04 35	05 03	09 17	10 18	11 22	12 29
30	04 15	04 46	05 12	09 32	10 31	11 34	12 37
20	04 36	05 04	05 27	09 57	10 54	11 53	12 51
N 10	04 53	05 19	05 41	10 19	11 14	12 09	13 04
0	05 06	05 31	05 53	10 39	11 32	12 24	13 15
S 10	05 18	05 43	06 05	11 00	11 51	12 40	13 27
20	05 29	05 55	06 18	11 21	12 10	12 56	13 39
30	05 39	06 08	06 33	11 46	12 33	13 15	13 53
35	05 45	06 15	06 42	12 01	12 46	13 26	14 01
40	05 50	06 23	06 51	12 18	13 01	13 38	14 10
45	05 56	06 31	07 03	12 39	13 19	13 53	14 21
S 50	06 02	06 41	07 16	13 04	13 41	14 10	14 34
52	06 05	06 46	07 23	13 17	13 52	14 19	14 40
54	06 08	06 51	07 30	13 31	14 04	14 28	14 46
56	06 11	06 56	07 37	13 48	14 18	14 38	14 53
58	06 15	07 02	07 46	14 08	14 34	14 50	15 02
S 60	06 18	07 09	07 56	14 32	14 53	15 04	15 11

Lat.	Sunset	Twilight Civil	Twilight Naut.	Moonset 7	Moonset 8	Moonset 9	Moonset 10
°	h m	h m	h m	h m	h m	h m	h m
N 72	▭	▭	▭	▭	▭	▭	05 35
N 70	22 10	////	////	▭	▭	▭	04 47
68	21 31	////	////	▭	▭	05 03	04 15
66	21 04	22 54	////	▭	▭	04 12	03 52
64	20 44	22 04	////	04 19	03 48	03 39	03 33
62	20 28	21 34	////	02 49	03 08	03 15	03 18
60	20 14	21 12	23 01	02 12	02 41	02 56	03 05
N 58	20 02	20 54	22 16	01 46	02 19	02 40	02 54
56	19 52	20 40	21 48	01 25	02 02	02 27	02 44
54	19 44	20 27	21 27	01 08	01 47	02 15	02 35
52	19 36	20 16	21 10	00 53	01 34	02 04	02 28
50	19 29	20 07	20 56	00 40	01 22	01 55	02 21
45	19 14	19 47	20 29	00 14	00 58	01 35	02 06
N 40	19 01	19 31	20 08	24 39	00 39	01 19	01 53
35	18 51	19 18	19 52	24 23	00 23	01 05	01 42
30	18 42	19 07	19 38	24 09	00 09	00 53	01 33
20	18 26	18 49	19 17	23 45	24 32	00 32	01 17
N 10	18 12	18 35	19 00	23 24	24 14	00 14	01 02
0	18 00	18 22	18 47	23 05	23 57	24 49	00 49
S 10	17 48	18 10	18 35	22 45	23 40	24 35	00 35
20	17 34	17 57	18 24	22 24	23 22	24 21	00 21
30	17 20	17 45	18 13	22 00	23 01	24 04	00 04
35	17 11	17 38	18 08	21 46	22 48	23 54	25 01
40	17 01	17 30	18 02	21 29	22 34	23 43	24 53
45	16 50	17 21	17 56	21 09	22 17	23 29	24 44
S 50	16 36	17 11	17 50	20 44	21 56	23 13	24 33
52	16 30	17 07	17 47	20 32	21 45	23 05	24 28
54	16 23	17 02	17 44	20 18	21 34	22 56	24 22
56	16 15	16 56	17 41	20 01	21 20	22 47	24 16
58	16 06	16 50	17 38	19 42	21 05	22 35	24 08
S 60	15 56	16 43	17 34	19 17	20 47	22 22	24 00

Day	SUN Eqn. of Time 00^h	SUN Eqn. of Time 12^h	SUN Mer. Pass.	MOON Mer. Pass. Upper	MOON Mer. Pass. Lower	MOON Age	MOON Phase
d	m s	m s	h m	h m	h m	d %	
7	03 24	03 26	11 57	16 52	04 25	06 30	
8	03 28	03 30	11 57	17 45	05 19	07 40	
9	03 31	03 33	11 56	18 37	06 11	08 50	

UT d	h	ARIES GHA ° ′	VENUS −3.9 GHA ° ′	Dec ° ′	MARS −0.2 GHA ° ′	Dec ° ′	JUPITER −2.0 GHA ° ′	Dec ° ′	SATURN +0.1 GHA ° ′	Dec ° ′
10	00	227 22.9	206 15.0	N 7 04.6	273 49.8	S19 19.2	94 42.9	N18 32.8	140 38.4	N22 31.3
	01	242 25.4	221 14.6	05.7	288 50.8	18.9	109 45.1	32.7	155 40.6	31.3
	02	257 27.8	236 14.3	06.8	303 51.7	18.6	124 47.3	32.7	170 42.7	31.3
	03	272 30.3	251 13.9	. . 08.0	318 52.7	. . 18.3	139 49.5	. . 32.6	185 44.9	. . 31.3
	04	287 32.8	266 13.5	09.1	333 53.7	18.0	154 51.7	32.5	200 47.0	31.3
	05	302 35.2	281 13.2	10.2	348 54.7	17.7	169 53.9	32.4	215 49.2	31.4
	06	317 37.7	296 12.8	N 7 11.4	3 55.7	S19 17.4	184 56.1	N18 32.4	230 51.4	N22 31.4
	07	332 40.2	311 12.4	12.5	18 56.7	17.1	199 58.3	32.3	245 53.5	31.4
S	08	347 42.6	326 12.1	13.6	33 57.7	16.8	215 00.5	32.2	260 55.7	31.4
A	09	2 45.1	341 11.7	. . 14.8	48 58.7	. . 16.5	230 02.7	. . 32.2	275 57.8	. . 31.4
T	10	17 47.5	356 11.3	15.9	63 59.6	16.2	245 04.9	32.1	291 00.0	31.4
U	11	32 50.0	11 11.0	17.0	79 00.6	15.9	260 07.1	32.0	306 02.2	31.4
R	12	47 52.5	26 10.6	N 7 18.1	94 01.6	S19 15.6	275 09.3	N18 31.9	321 04.3	N22 31.4
D	13	62 54.9	41 10.2	19.3	109 02.6	15.3	290 11.5	31.9	336 06.5	31.5
A	14	77 57.4	56 09.9	20.4	124 03.6	15.0	305 13.7	31.8	351 08.6	31.5
Y	15	92 59.9	71 09.5	. . 21.5	139 04.6	. . 14.7	320 15.9	. . 31.7	6 10.8	. . 31.5
	16	108 02.3	86 09.1	22.7	154 05.6	14.4	335 18.1	31.6	21 13.0	31.5
	17	123 04.8	101 08.8	23.8	169 06.6	14.1	350 20.3	31.6	36 15.1	31.5
	18	138 07.3	116 08.4	N 7 24.9	184 07.6	S19 13.8	5 22.5	N18 31.5	51 17.3	N22 31.5
	19	153 09.7	131 08.0	26.0	199 08.6	13.5	20 24.7	31.4	66 19.4	31.5
	20	168 12.2	146 07.6	27.2	214 09.5	13.2	35 26.9	31.4	81 21.6	31.5
	21	183 14.7	161 07.3	. . 28.3	229 10.5	. . 12.9	50 29.1	. . 31.3	96 23.8	. . 31.5
	22	198 17.1	176 06.9	29.4	244 11.5	12.6	65 31.3	31.2	111 25.9	31.6
	23	213 19.6	191 06.5	30.6	259 12.5	12.3	80 33.5	31.1	126 28.1	31.6
11	00	228 22.0	206 06.2	N 7 31.7	274 13.5	S19 12.0	95 35.7	N18 31.1	141 30.2	N22 31.6
	01	243 24.5	221 05.8	32.8	289 14.5	11.7	110 37.9	31.0	156 32.4	31.6
	02	258 27.0	236 05.4	33.9	304 15.5	11.3	125 40.1	30.9	171 34.6	31.6
	03	273 29.4	251 05.0	. . 35.1	319 16.5	. . 11.0	140 42.3	. . 30.8	186 36.7	. . 31.6
	04	288 31.9	266 04.7	36.2	334 17.5	10.7	155 44.5	30.8	201 38.9	31.6
	05	303 34.4	281 04.3	37.3	349 18.5	10.4	170 46.7	30.7	216 41.0	31.6
	06	318 36.8	296 03.9	N 7 38.4	4 19.5	S19 10.1	185 48.9	N18 30.6	231 43.2	N22 31.7
	07	333 39.3	311 03.6	39.6	19 20.5	09.8	200 51.1	30.5	246 45.4	31.7
	08	348 41.8	326 03.2	40.7	34 21.5	09.5	215 53.3	30.5	261 47.5	31.7
S	09	3 44.2	341 02.8	. . 41.8	49 22.5	. . 09.2	230 55.5	. . 30.4	276 49.7	. . 31.7
U	10	18 46.7	356 02.4	42.9	64 23.5	08.9	245 57.7	30.3	291 51.8	31.7
N	11	33 49.2	11 02.1	44.0	79 24.4	08.6	260 59.9	30.2	306 54.0	31.7
D	12	48 51.6	26 01.7	N 7 45.2	94 25.4	S19 08.3	276 02.1	N18 30.2	321 56.1	N22 31.7
A	13	63 54.1	41 01.3	46.3	109 26.4	08.0	291 04.3	30.1	336 58.3	31.7
Y	14	78 56.5	56 00.9	47.4	124 27.4	07.7	306 06.5	30.0	352 00.5	31.7
	15	93 59.0	71 00.6	. . 48.5	139 28.4	. . 07.4	321 08.7	. . 30.0	7 02.6	. . 31.8
	16	109 01.5	86 00.2	49.7	154 29.4	07.1	336 10.9	29.9	22 04.8	31.8
	17	124 03.9	100 59.8	50.8	169 30.4	06.8	351 13.1	29.8	37 06.9	31.8
	18	139 06.4	115 59.4	N 7 51.9	184 31.4	S19 06.5	6 15.3	N18 29.7	52 09.1	N22 31.8
	19	154 08.9	130 59.1	53.0	199 32.4	06.2	21 17.5	29.7	67 11.3	31.8
	20	169 11.3	145 58.7	54.1	214 33.4	05.9	36 19.7	29.6	82 13.4	31.8
	21	184 13.8	160 58.3	. . 55.3	229 34.4	. . 05.6	51 21.8	. . 29.5	97 15.6	. . 31.8
	22	199 16.3	175 57.9	56.4	244 35.4	05.3	66 24.0	29.4	112 17.7	31.8
	23	214 18.7	190 57.5	57.5	259 36.4	05.0	81 26.2	29.4	127 19.9	31.9
12	00	229 21.2	205 57.2	N 7 58.6	274 37.4	S19 04.6	96 28.4	N18 29.3	142 22.0	N22 31.9
	01	244 23.6	220 56.8	7 59.7	289 38.4	04.3	111 30.6	29.2	157 24.2	31.9
	02	259 26.1	235 56.4	8 00.9	304 39.4	04.0	126 32.8	29.1	172 26.4	31.9
	03	274 28.6	250 56.0	. . 02.0	319 40.4	. . 03.7	141 35.0	. . 29.1	187 28.5	. . 31.9
	04	289 31.0	265 55.7	03.1	334 41.4	03.4	156 37.2	29.0	202 30.7	31.9
	05	304 33.5	280 55.3	04.2	349 42.4	03.1	171 39.4	28.9	217 32.8	31.9
	06	319 36.0	295 54.9	N 8 05.3	4 43.4	S19 02.8	186 41.6	N18 28.8	232 35.0	N22 31.9
	07	334 38.4	310 54.5	06.4	19 44.4	02.5	201 43.8	28.8	247 37.1	31.9
	08	349 40.9	325 54.1	07.6	34 45.4	02.2	216 46.0	28.7	262 39.3	32.0
M	09	4 43.4	340 53.8	. . 08.7	49 46.4	. . 01.9	231 48.1	. . 28.6	277 41.5	. . 32.0
O	10	19 45.8	355 53.4	09.8	64 47.4	01.6	246 50.3	28.5	292 43.6	32.0
N	11	34 48.3	10 53.0	10.9	79 48.4	01.3	261 52.5	28.4	307 45.8	32.0
D	12	49 50.8	25 52.6	N 8 12.0	94 49.4	S19 01.0	276 54.7	N18 28.4	322 47.9	N22 32.0
A	13	64 53.2	40 52.2	13.1	109 50.4	00.7	291 56.9	28.3	337 50.1	32.0
Y	14	79 55.7	55 51.9	14.3	124 51.5	00.4	306 59.1	28.2	352 52.2	32.0
	15	94 58.1	70 51.5	. . 15.4	139 52.5	19 00.1	322 01.3	. . 28.1	7 54.4	. . 32.0
	16	110 00.6	85 51.1	16.5	154 53.5	18 59.8	337 03.5	28.1	22 56.6	32.0
	17	125 03.1	100 50.7	17.6	169 54.5	59.4	352 05.7	28.0	37 58.7	32.1
	18	140 05.5	115 50.3	N 8 18.7	184 55.5	S18 59.1	7 07.8	N18 27.9	53 00.9	N22 32.1
	19	155 08.0	130 50.0	19.8	199 56.5	58.8	22 10.0	27.8	68 03.0	32.1
	20	170 10.5	145 49.6	21.0	214 57.5	58.5	37 12.2	27.8	83 05.2	32.1
	21	185 12.9	160 49.2	. . 22.1	229 58.5	. . 58.2	52 14.4	. . 27.7	98 07.3	. . 32.1
	22	200 15.3	175 48.8	23.2	244 59.5	57.9	67 16.6	27.6	113 09.5	32.1
	23	215 17.9	190 48.4	24.3	260 00.5	57.6	82 18.8	27.5	128 11.6	32.1
Mer. Pass.		h m 8 45.1	*v* −0.4	*d* 1.1	*v* 1.0	*d* 0.3	*v* 2.2	*d* 0.1	*v* 2.2	*d* 0.0

STARS

Name	SHA ° ′	Dec °
Acamar	315 24.7	S40 17.
Achernar	335 33.0	S57 13.
Acrux	173 18.1	S63 07.
Adhara	255 19.1	S28 58.
Aldebaran	290 58.9	N16 30.
Alioth	166 27.0	N55 56.
Alkaid	153 04.6	N49 17.
Al Na'ir	27 53.7	S46 56.
Alnilam	275 54.8	S 1 12.
Alphard	218 04.0	S 8 40.
Alphecca	126 17.4	N26 42.
Alpheratz	357 52.1	N29 06.
Altair	62 15.9	N 8 52.
Ankaa	353 23.7	S42 17.
Antares	112 35.7	S26 26.
Arcturus	146 02.7	N19 09.
Atria	107 44.4	S69 02.
Avior	234 21.6	S59 31.
Bellatrix	278 40.9	N 6 21.
Betelgeuse	271 10.2	N 7 24.
Canopus	264 00.1	S52 42.
Capella	280 46.7	N46 00.
Deneb	49 36.9	N45 17.
Denebola	182 41.6	N14 33.
Diphda	349 04.1	S17 58.
Dubhe	194 00.9	N61 44.
Elnath	278 23.1	N28 36.
Eltanin	90 49.4	N51 29.
Enif	33 55.0	N 9 53.
Fomalhaut	15 32.8	S29 36.
Gacrux	172 09.6	S57 08.
Gienah	176 00.4	S17 33.
Hadar	148 58.9	S60 23
Hamal	328 10.2	N23 28.
Kaus Aust.	83 54.1	S34 23.
Kochab	137 18.1	N74 08.
Markab	13 46.5	N15 13.
Menkar	314 23.7	N 4 06.
Menkent	148 16.7	S36 23.
Miaplacidus	221 41.7	S69 44.
Mirfak	308 52.4	N49 52.
Nunki	76 08.0	S26 17.
Peacock	53 31.5	S56 43.
Pollux	243 37.6	N28 01.
Procyon	245 08.2	N 5 13.
Rasalhague	96 13.6	N12 33
Regulus	207 51.9	N11 57.
Rigel	281 20.0	S 8 11.
Rigil Kent.	140 02.3	S60 51.
Sabik	102 21.4	S15 43.
Schedar	349 50.3	N56 33.
Shaula	96 32.4	S37 06.
Sirius	258 41.0	S16 43.
Spica	158 39.4	S11 10.
Suhail	222 58.5	S43 26.
Vega	80 44.1	N38 46.
Zuben'ubi	137 14.0	S16 03.

	SHA ° ′	Mer. Pas h m
Venus	337 44.1	10 16
Mars	45 51.5	5 43
Jupiter	227 13.7	17 35
Saturn	273 08.2	14 32

UT	SUN GHA	SUN Dec	MOON GHA	v	MOON Dec	d	HP
d h	° ′	° ′	° ′	′	° ′	′	′
10 00 (SATURDAY)	180 53.6	N17 27.3	78 03.0	10.3	N18 04.9	11.3	57.7
01	195 53.6	28.0	92 32.3	10.3	17 53.6	11.5	57.7
02	210 53.6	28.7	107 01.6	10.4	17 42.1	11.5	57.8
03	225 53.6	. . 29.3	121 31.0	10.4	17 30.6	11.7	57.8
04	240 53.7	30.0	136 00.4	10.4	17 18.9	11.8	57.9
05	255 53.7	30.6	150 29.8	10.4	17 07.1	11.9	57.9
06	270 53.7	N17 31.3	164 59.2	10.5	N16 55.2	11.9	57.9
07	285 53.7	32.0	179 28.7	10.4	16 43.3	12.1	58.0
08	300 53.8	32.6	193 58.1	10.6	16 31.2	12.2	58.0
09	315 53.8	. . 33.3	208 27.7	10.5	16 19.0	12.3	58.0
10	330 53.8	33.9	222 57.2	10.6	16 06.7	12.4	58.1
11	345 53.8	34.6	237 26.8	10.5	15 54.3	12.5	58.1
12	0 53.9	N17 35.2	251 56.3	10.7	N15 41.8	12.6	58.2
13	15 53.9	35.9	266 26.0	10.6	15 29.2	12.7	58.2
14	30 53.9	36.6	280 55.6	10.6	15 16.5	12.7	58.2
15	45 53.9	. . 37.2	295 25.2	10.7	15 03.8	12.9	58.3
16	60 54.0	37.9	309 54.9	10.7	14 50.9	13.0	58.3
17	75 54.0	38.5	324 24.6	10.8	14 37.9	13.1	58.3
18	90 54.0	N17 39.2	338 54.4	10.7	N14 24.8	13.1	58.4
19	105 54.0	39.8	353 24.1	10.8	14 11.7	13.3	58.4
20	120 54.1	40.5	7 53.9	10.8	13 58.4	13.3	58.5
21	135 54.1	. . 41.1	22 23.7	10.8	13 45.1	13.4	58.5
22	150 54.1	41.8	36 53.5	10.8	13 31.7	13.6	58.5
23	165 54.1	42.4	51 23.3	10.8	13 18.1	13.6	58.6
11 00 (SUNDAY)	180 54.2	N17 43.1	65 53.1	10.9	N13 04.5	13.7	58.6
01	195 54.2	43.7	80 23.0	10.8	12 50.8	13.7	58.6
02	210 54.2	44.4	94 52.8	10.9	12 37.1	13.9	58.7
03	225 54.2	. . 45.0	109 22.7	10.9	12 23.2	13.9	58.7
04	240 54.2	45.7	123 52.6	11.0	12 09.3	14.1	58.8
05	255 54.3	46.3	138 22.6	10.9	11 55.2	14.1	58.8
06	270 54.3	N17 47.0	152 52.5	10.9	N11 41.1	14.2	58.8
07	285 54.3	47.6	167 22.4	11.0	11 26.9	14.2	58.9
08	300 54.3	48.3	181 52.4	11.0	11 12.7	14.4	58.9
09	315 54.3	. . 48.9	196 22.4	10.9	10 58.3	14.4	58.9
10	330 54.4	49.5	210 52.3	11.0	10 43.9	14.5	59.0
11	345 54.4	50.2	225 22.3	11.0	10 29.4	14.6	59.0
12	0 54.4	N17 50.8	239 52.3	11.0	N10 14.8	14.6	59.1
13	15 54.4	51.5	254 22.3	11.0	10 00.2	14.7	59.1
14	30 54.4	52.1	268 52.3	11.0	9 45.5	14.8	59.1
15	45 54.4	. . 52.8	283 22.3	11.0	9 30.7	14.8	59.2
16	60 54.5	53.4	297 52.3	11.1	9 15.9	15.0	59.2
17	75 54.5	54.0	312 22.4	11.0	9 00.9	15.0	59.2
18	90 54.5	N17 54.7	326 52.4	11.0	N 8 45.9	15.0	59.3
19	105 54.5	55.3	341 22.4	11.0	8 30.9	15.1	59.3
20	120 54.5	56.0	355 52.4	11.1	8 15.8	15.2	59.3
21	135 54.6	. . 56.6	10 22.5	11.0	8 00.6	15.2	59.4
22	150 54.6	57.2	24 52.5	11.0	7 45.4	15.3	59.4
23	165 54.6	57.9	39 22.5	11.0	7 30.1	15.4	59.5
12 00 (MONDAY)	180 54.6	N17 58.5	53 52.5	11.0	N 7 14.7	15.4	59.5
01	195 54.6	59.2	68 22.5	11.1	6 59.3	15.5	59.5
02	210 54.6	17 59.8	82 52.6	11.0	6 43.8	15.5	59.6
03	225 54.6	18 00.4	97 22.6	11.0	6 28.3	15.6	59.6
04	240 54.7	01.1	111 52.6	10.9	6 12.7	15.7	59.6
05	255 54.7	01.7	126 22.5	11.0	5 57.0	15.6	59.7
06	270 54.7	N18 02.3	140 52.5	11.0	N 5 41.4	15.8	59.7
07	285 54.7	03.0	155 22.5	11.0	5 25.6	15.8	59.7
08	300 54.7	03.6	169 52.5	10.9	5 09.8	15.8	59.8
09	315 54.7	. . 04.2	184 22.4	10.9	4 54.0	15.9	59.8
10	330 54.7	04.9	198 52.3	11.0	4 38.1	15.9	59.8
11	345 54.8	05.5	213 22.3	10.9	4 22.2	16.0	59.9
12	0 54.8	N18 06.1	227 52.2	10.8	N 4 06.2	16.0	59.9
13	15 54.8	06.8	242 22.0	10.9	3 50.2	16.0	59.9
14	30 54.8	07.4	256 51.9	10.9	3 34.2	16.1	60.0
15	45 54.8	. . 08.0	271 21.8	10.8	3 18.1	16.1	60.0
16	60 54.8	08.6	285 51.6	10.8	3 02.0	16.2	60.0
17	75 54.8	09.3	300 21.4	10.8	2 45.8	16.2	60.1
18	90 54.8	N18 09.9	314 51.2	10.7	N 2 29.6	16.2	60.1
19	105 54.9	10.5	329 20.9	10.8	2 13.4	16.2	60.1
20	120 54.9	11.2	343 50.7	10.7	1 57.2	16.3	60.2
21	135 54.9	. . 11.8	358 20.4	10.7	1 40.9	16.3	60.2
22	150 54.9	12.4	12 50.1	10.6	1 24.6	16.4	60.2
23	165 54.9	13.0	27 19.7	10.6	N 1 08.2	16.3	60.3
	SD 15.9	d 0.6	SD 15.8		16.1		16.3

Lat.	Twilight Naut.	Twilight Civil	Sunrise	Moonrise 10	Moonrise 11	Moonrise 12	Moonrise 13
°	h m	h m	h m	h m	h m	h m	h m
N 72	▭	▭	▭	08 52	11 46	14 11	16 35
N 70	////	////	01 27	09 39	12 04	14 16	16 29
68	////	////	02 11	10 09	12 17	14 20	16 24
66	////	00 32	02 40	10 31	12 28	14 24	16 21
64	////	01 37	03 01	10 48	12 38	14 26	16 17
62	////	02 10	03 19	11 02	12 45	14 29	16 15
60	00 29	02 34	03 33	11 14	12 52	14 31	16 12
N 58	01 27	02 52	03 45	11 24	12 58	14 33	16 10
56	01 57	03 08	03 56	11 33	13 03	14 35	16 08
54	02 19	03 21	04 05	11 41	13 08	14 36	16 06
52	02 37	03 32	04 13	11 48	13 12	14 37	16 05
50	02 52	03 42	04 21	11 55	13 16	14 39	16 04
45	03 21	04 03	04 36	12 08	13 24	14 41	16 01
N 40	03 42	04 19	04 49	12 19	13 31	14 43	15 58
35	03 59	04 32	05 00	12 29	13 37	14 45	15 56
30	04 13	04 44	05 10	12 37	13 42	14 47	15 54
20	04 35	05 02	05 26	12 51	13 50	14 50	15 51
N 10	04 52	05 18	05 40	13 04	13 58	14 53	15 48
0	05 06	05 31	05 53	13 15	14 05	14 55	15 46
S 10	05 18	05 44	06 06	13 27	14 12	14 57	15 43
20	05 30	05 56	06 19	13 39	14 20	15 00	15 40
30	05 41	06 10	06 35	13 53	14 29	15 03	15 38
35	05 47	06 17	06 44	14 01	14 34	15 05	15 36
40	05 53	06 25	06 54	14 10	14 39	15 07	15 34
45	05 59	06 35	07 06	14 21	14 46	15 09	15 32
S 50	06 06	06 45	07 20	14 34	14 53	15 11	15 29
52	06 09	06 50	07 27	14 40	14 57	15 13	15 28
54	06 12	06 55	07 35	14 46	15 01	15 14	15 27
56	06 16	07 01	07 43	14 53	15 05	15 15	15 25
58	06 20	07 08	07 52	15 02	15 10	15 17	15 24
S 60	06 24	07 15	08 03	15 11	15 15	15 19	15 22

Lat.	Sunset	Twilight Civil	Twilight Naut.	Moonset 10	Moonset 11	Moonset 12	Moonset 13
°	h m	h m	h m	h m	h m	h m	h m
N 72	▭	▭	▭	05 35	04 27	03 49	03 17
N 70	22 33	////	////	04 47	04 07	03 40	03 17
68	21 46	////	////	04 15	03 52	03 34	03 17
66	21 16	23 45	////	03 52	03 39	03 28	03 17
64	20 54	22 21	////	03 33	03 28	03 23	03 18
62	20 36	21 46	////	03 18	03 19	03 18	03 18
60	20 21	21 22	23 46	03 05	03 11	03 15	03 18
N 58	20 09	21 02	22 31	02 54	03 04	03 11	03 18
56	19 58	20 47	21 59	02 44	02 57	03 08	03 18
54	19 49	20 33	21 36	02 35	02 52	03 05	03 18
52	19 41	20 22	21 18	02 28	02 47	03 03	03 18
50	19 33	20 12	21 03	02 21	02 42	03 01	03 18
45	19 17	19 51	20 33	02 06	02 32	02 56	03 18
N 40	19 04	19 34	20 12	01 53	02 23	02 51	03 18
35	18 53	19 21	19 55	01 42	02 16	02 48	03 18
30	18 44	19 09	19 41	01 33	02 10	02 44	03 19
20	18 27	18 51	19 18	01 17	01 58	02 39	03 19
N 10	18 13	18 35	19 01	01 02	01 49	02 34	03 19
0	18 00	18 22	18 47	00 49	01 39	02 29	03 19
S 10	17 47	18 09	18 34	00 35	01 30	02 24	03 19
20	17 33	17 56	18 23	00 21	01 20	02 19	03 19
30	17 17	17 43	18 11	00 04	01 08	02 13	03 19
35	17 08	17 35	18 06	25 01	01 01	02 09	03 19
40	16 58	17 27	18 00	24 53	00 53	02 05	03 19
45	16 46	17 18	17 53	24 44	00 44	02 01	03 19
S 50	16 32	17 07	17 46	24 33	00 33	01 55	03 19
52	16 25	17 02	17 43	24 28	00 28	01 52	03 19
54	16 17	16 57	17 40	24 22	00 22	01 49	03 19
56	16 09	16 51	17 36	24 16	00 16	01 46	03 18
58	16 00	16 44	17 32	24 08	00 08	01 43	03 18
S 60	15 49	16 37	17 28	24 00	00 00	01 39	03 18

Day	SUN Eqn. of Time 00^h	SUN Eqn. of Time 12^h	SUN Mer. Pass.	MOON Mer. Pass. Upper	MOON Mer. Pass. Lower	MOON Age	MOON Phase
d	m s	m s	h m	h m	h m	d %	
10	03 34	03 35	11 56	19 27	07 02	09 61	
11	03 37	03 38	11 56	20 17	07 52	10 72	◐
12	03 38	03 39	11 56	21 07	08 42	11 81	

UT		ARIES	VENUS −3.9		MARS −0.3		JUPITER −2.0		SATURN +0.1	
		GHA	GHA	Dec	GHA	Dec	GHA	Dec	GHA	Dec
	d h	° ′	° ′	° ′	° ′	° ′	° ′	° ′	° ′	° ′
	13 00	230 20.3	205 48.0	N 8 25.4	275 01.5	S18 57.3	97 21.0	N18 27.5	143 13.8	N22 32.1
	01	245 22.8	220 47.7	26.5	290 02.5	57.0	112 23.2	27.4	158 16.0	32.1
	02	260 25.2	235 47.3	27.6	305 03.5	56.7	127 25.3	27.3	173 18.1	32.2
	03	275 27.7	250 46.9	. . 28.7	320 04.5	. . 56.4	142 27.5	. . 27.2	188 20.3	. . 32.2
	04	290 30.2	265 46.5	29.9	335 05.5	56.1	157 29.7	27.1	203 22.4	32.2
	05	305 32.6	280 46.1	31.0	350 06.6	55.8	172 31.9	27.1	218 24.6	32.2
	06	320 35.1	295 45.7	N 8 32.1	5 07.6	S18 55.5	187 34.1	N18 27.0	233 26.7	N22 32.2
	07	335 37.6	310 45.3	33.2	20 08.6	55.2	202 36.3	26.9	248 28.9	32.2
TUESDAY	08	350 40.0	325 45.0	34.3	35 09.6	54.8	217 38.4	26.8	263 31.0	32.2
	09	5 42.5	340 44.6	. . 35.4	50 10.6	. . 54.5	232 40.6	. . 26.8	278 33.2	. . 32.2
	10	20 45.0	355 44.2	36.5	65 11.6	54.2	247 42.8	26.7	293 35.3	32.2
	11	35 47.4	10 43.8	37.6	80 12.6	53.9	262 45.0	26.6	308 37.5	32.3
	12	50 49.9	25 43.4	N 8 38.7	95 13.6	S18 53.6	277 47.2	N18 26.5	323 39.7	N22 32.3
	13	65 52.4	40 43.0	39.8	110 14.6	53.3	292 49.4	26.4	338 41.8	32.3
	14	80 54.8	55 42.6	41.0	125 15.7	53.0	307 51.6	26.4	353 44.0	32.3
	15	95 57.3	70 42.2	. . 42.1	140 16.7	. . 52.7	322 53.7	. . 26.3	8 46.1	. . 32.3
	16	110 59.7	85 41.9	43.2	155 17.7	52.4	337 55.9	26.2	23 48.3	32.3
	17	126 02.2	100 41.5	44.3	170 18.7	52.1	352 58.1	26.1	38 50.4	32.3
	18	141 04.7	115 41.1	N 8 45.4	185 19.7	S18 51.8	8 00.3	N18 26.1	53 52.6	N22 32.3
	19	156 07.1	130 40.7	46.5	200 20.7	51.5	23 02.5	26.0	68 54.7	32.4
	20	171 09.6	145 40.3	47.6	215 21.7	51.2	38 04.6	25.9	83 56.9	32.4
	21	186 12.1	160 39.9	. . 48.7	230 22.7	. . 50.8	53 06.8	. . 25.8	98 59.0	. . 32.4
	22	201 14.5	175 39.5	49.8	245 23.8	50.5	68 09.0	25.7	114 01.2	32.4
	23	216 17.0	190 39.1	50.9	260 24.8	50.2	83 11.2	25.7	129 03.3	32.4
	14 00	231 19.5	205 38.7	N 8 52.0	275 25.8	S18 49.9	98 13.4	N18 25.6	144 05.5	N22 32.4
	01	246 21.9	220 38.3	53.1	290 26.8	49.6	113 15.5	25.5	159 07.7	32.4
	02	261 24.4	235 38.0	54.2	305 27.8	49.3	128 17.7	25.4	174 09.8	32.4
	03	276 26.9	250 37.6	. . 55.3	320 28.8	. . 49.0	143 19.9	. . 25.4	189 12.0	. . 32.4
	04	291 29.3	265 37.2	56.4	335 29.9	48.7	158 22.1	25.3	204 14.1	32.4
	05	306 31.8	280 36.8	57.5	350 30.9	48.4	173 24.3	25.2	219 16.3	32.5
	06	321 34.2	295 36.4	N 8 58.6	5 31.9	S18 48.1	188 26.4	N18 25.1	234 18.4	N22 32.5
WEDNESDAY	07	336 36.7	310 36.0	8 59.7	20 32.9	47.8	203 28.6	25.0	249 20.6	32.5
	08	351 39.2	325 35.6	9 00.9	35 33.9	47.5	218 30.8	25.0	264 22.7	32.5
	09	6 41.6	340 35.2	. . 02.0	50 34.9	. . 47.1	233 33.0	. . 24.9	279 24.9	. . 32.5
	10	21 44.1	355 34.8	03.1	65 36.0	46.8	248 35.2	24.8	294 27.0	32.5
	11	36 46.6	10 34.4	04.2	80 37.0	46.5	263 37.3	24.7	309 29.2	32.5
	12	51 49.0	25 34.0	N 9 05.3	95 38.0	S18 46.2	278 39.5	N18 24.6	324 31.3	N22 32.5
	13	66 51.5	40 33.6	06.4	110 39.0	45.9	293 41.7	24.6	339 33.5	32.5
	14	81 54.0	55 33.2	07.5	125 40.0	45.6	308 43.9	24.5	354 35.6	32.6
	15	96 56.4	70 32.8	. . 08.6	140 41.1	. . 45.3	323 46.1	. . 24.4	9 37.8	. . 32.6
	16	111 58.9	85 32.5	09.7	155 42.1	45.0	338 48.2	24.3	24 39.9	32.6
	17	127 01.3	100 32.1	10.8	170 43.1	44.7	353 50.4	24.2	39 42.1	32.6
	18	142 03.8	115 31.7	N 9 11.9	185 44.1	S18 44.4	8 52.6	N18 24.2	54 44.2	N22 32.6
	19	157 06.3	130 31.3	13.0	200 45.1	44.1	23 54.8	24.1	69 46.4	32.6
	20	172 08.7	145 30.9	14.1	215 46.2	43.7	38 56.9	24.0	84 48.5	32.6
	21	187 11.2	160 30.5	. . 15.2	230 47.2	. . 43.4	53 59.1	. . 23.9	99 50.7	. . 32.6
	22	202 13.7	175 30.1	16.3	245 48.2	43.1	69 01.3	23.8	114 52.9	32.6
	23	217 16.1	190 29.7	17.4	260 49.2	42.8	84 03.5	23.8	129 55.0	32.7
	15 00	232 18.6	205 29.3	N 9 18.5	275 50.2	S18 42.5	99 05.6	N18 23.7	144 57.2	N22 32.7
	01	247 21.1	220 28.9	19.6	290 51.3	42.2	114 07.8	23.6	159 59.3	32.7
	02	262 23.5	235 28.5	20.6	305 52.3	41.9	129 10.0	23.5	175 01.5	32.7
	03	277 26.0	250 28.1	. . 21.7	320 53.3	. . 41.6	144 12.2	. . 23.4	190 03.6	. . 32.7
	04	292 28.5	265 27.7	22.8	335 54.3	41.3	159 14.3	23.4	205 05.8	32.7
	05	307 30.9	280 27.3	23.9	350 55.4	41.0	174 16.5	23.3	220 07.9	32.7
	06	322 33.4	295 26.9	N 9 25.0	5 56.4	S18 40.6	189 18.7	N18 23.2	235 10.1	N22 32.7
	07	337 35.8	310 26.5	26.1	20 57.4	40.3	204 20.9	23.1	250 12.2	32.7
THURSDAY	08	352 38.3	325 26.1	27.2	35 58.4	40.0	219 23.0	23.0	265 14.4	32.8
	09	7 40.8	340 25.7	. . 28.3	50 59.5	. . 39.7	234 25.2	. . 23.0	280 16.5	. . 32.8
	10	22 43.2	355 25.3	29.4	66 00.5	39.4	249 27.4	22.9	295 18.7	32.8
	11	37 45.7	10 24.9	30.5	81 01.5	39.1	264 29.5	22.8	310 20.8	32.8
	12	52 48.2	25 24.5	N 9 31.6	96 02.5	S18 38.8	279 31.7	N18 22.7	325 23.0	N22 32.8
	13	67 50.6	40 24.1	32.7	111 03.6	38.5	294 33.9	22.6	340 25.1	32.8
	14	82 53.1	55 23.7	33.8	126 04.6	38.2	309 36.1	22.6	355 27.3	32.8
	15	97 55.6	70 23.3	. . 34.9	141 05.6	. . 37.9	324 38.2	. . 22.5	10 29.4	. . 32.8
	16	112 58.0	85 22.9	36.0	156 06.7	37.5	339 40.4	22.4	25 31.6	32.8
	17	128 00.5	100 22.5	37.1	171 07.7	37.2	354 42.6	22.3	40 33.7	32.8
	18	143 03.0	115 22.1	N 9 38.2	186 08.7	S18 36.9	9 44.7	N18 22.2	55 35.9	N22 32.9
	19	158 05.4	130 21.7	39.2	201 09.7	36.6	24 46.9	22.1	70 38.0	32.9
	20	173 07.9	145 21.3	40.3	216 10.8	36.3	39 49.1	22.1	85 40.2	32.9
	21	188 10.3	160 20.9	. . 41.4	231 11.8	. . 36.0	54 51.3	. . 22.0	100 42.3	. . 32.9
	22	203 12.8	175 20.5	42.5	246 12.8	35.7	69 53.4	21.9	115 44.5	32.9
	23	218 15.3	190 20.1	43.6	261 13.9	35.4	84 55.6	21.8	130 46.6	32.9
	Mer. Pass.	h m 8 33.3	v −0.4	d 1.1	v 1.0	d 0.3	v 2.2	d 0.1	v 2.2	d 0.0

STARS

Name	SHA	Dec
	° ′	° ′
Acamar	315 24.7	S40 17.5
Achernar	335 33.0	S57 13.2
Acrux	173 18.1	S63 07.2
Adhara	255 19.1	S28 58.7
Aldebaran	290 58.9	N16 30.9
Alioth	166 27.0	N55 56.7
Alkaid	153 04.6	N49 17.9
Al Na'ir	27 53.6	S46 56.6
Alnilam	275 54.8	S 1 12.0
Alphard	218 04.0	S 8 40.4
Alphecca	126 17.4	N26 42.1
Alpheratz	357 52.1	N29 06.2
Altair	62 15.9	N 8 52.4
Ankaa	353 23.7	S42 17.3
Antares	112 35.7	S26 26.4
Arcturus	146 02.7	N19 09.9
Atria	107 44.3	S69 02.0
Avior	234 21.6	S59 31.4
Bellatrix	278 40.9	N 6 21.1
Betelgeuse	271 10.2	N 7 24.5
Canopus	264 00.1	S52 42.0
Capella	280 46.7	N46 00.2
Deneb	49 36.9	N45 17.1
Denebola	182 41.6	N14 33.3
Diphda	349 04.1	S17 58.2
Dubhe	194 00.9	N61 44.3
Elnath	278 23.1	N28 36.7
Eltanin	90 49.4	N51 29.0
Enif	33 55.0	N 9 53.2
Fomalhaut	15 32.8	S29 36.3
Gacrux	172 09.7	S57 08.1
Gienah	176 00.4	S17 33.7
Hadar	148 58.9	S60 23.4
Hamal	328 10.2	N23 28.5
Kaus Aust.	83 54.1	S34 23.0
Kochab	137 18.1	N74 08.6
Markab	13 46.5	N15 13.1
Menkar	314 23.7	N 4 06.1
Menkent	148 16.7	S36 23.3
Miaplacidus	221 41.8	S69 44.1
Mirfak	308 52.4	N49 52.3
Nunki	76 07.9	S26 17.6
Peacock	53 31.4	S56 43.4
Pollux	243 37.6	N28 01.2
Procyon	245 08.3	N 5 13.0
Rasalhague	96 13.6	N12 33.3
Regulus	207 51.9	N11 57.2
Rigel	281 20.0	S 8 11.9
Rigil Kent.	140 02.3	S60 51.0
Sabik	102 21.4	S15 43.8
Schedar	349 50.3	N56 33.0
Shaula	96 32.4	S37 06.4
Sirius	258 41.0	S16 43.3
Spica	158 39.4	S11 10.8
Suhail	222 58.5	S43 26.9
Vega	80 44.1	N38 46.9
Zuben'ubi	137 14.0	S16 03.4

	SHA	Mer. Pass.
	° ′	h m
Venus	334 19.3	10 18
Mars	44 06.3	5 38
Jupiter	226 53.9	17 25
Saturn	272 46.0	14 22

	UT	SUN GHA	SUN Dec	MOON GHA	MOON v	MOON Dec	MOON d	MOON HP
	d h	° ′	° ′	° ′	′	° ′	′	′
	13 00	180 54.9	N18 13.7	41 49.3	10.6	N 0 51.9	16.4	60.3
	01	195 54.9	14.3	56 18.9	10.6	0 35.5	16.4	60.3
	02	210 54.9	14.9	70 48.5	10.5	0 19.1	16.4	60.3
	03	225 54.9 . .	15.5	85 18.0	10.5	N 0 02.7	16.4	60.4
	04	240 55.0	16.1	99 47.5	10.5	S 0 13.7	16.5	60.4
	05	255 55.0	16.8	114 17.0	10.4	0 30.2	16.5	60.4
	06	270 55.0	N18 17.4	128 46.4	10.4	S 0 46.7	16.4	60.5
	07	285 55.0	18.0	143 15.8	10.3	1 03.1	16.5	60.5
T	08	300 55.0	18.6	157 45.1	10.4	1 19.6	16.5	60.5
U	09	315 55.0 . .	19.2	172 14.5	10.2	1 36.1	16.5	60.5
	10	330 55.0	19.9	186 43.7	10.2	1 52.6	16.5	60.6
E	11	345 55.0	20.5	201 12.9	10.2	2 09.1	16.5	60.6
S	12	0 55.0	N18 21.1	215 42.1	10.2	S 2 25.6	16.6	60.6
D	13	15 55.0	21.7	230 11.3	10.1	2 42.2	16.5	60.7
A	14	30 55.0	22.3	244 40.4	10.0	2 58.7	16.5	60.7
Y	15	45 55.0 . .	23.0	259 09.4	10.0	3 15.2	16.5	60.7
	16	60 55.0	23.6	273 38.4	10.0	3 31.7	16.5	60.7
	17	75 55.0	24.2	288 07.4	9.9	3 48.2	16.5	60.8
	18	90 55.1	N18 24.8	302 36.3	9.8	S 4 04.7	16.5	60.8
	19	105 55.1	25.4	317 05.1	9.8	4 21.2	16.5	60.8
	20	120 55.1	26.0	331 33.9	9.7	4 37.7	16.4	60.8
	21	135 55.1 . .	26.6	346 02.6	9.7	4 54.1	16.5	60.8
	22	150 55.1	27.3	0 31.3	9.7	5 10.6	16.4	60.9
	23	165 55.1	27.9	15 00.0	9.5	5 27.0	16.4	60.9
	14 00	180 55.1	N18 28.5	29 28.5	9.5	S 5 43.4	16.4	60.9
	01	195 55.1	29.1	43 57.0	9.5	5 59.8	16.4	60.9
	02	210 55.1	29.7	58 25.5	9.4	6 16.2	16.3	61.0
	03	225 55.1 . .	30.3	72 53.9	9.3	6 32.5	16.3	61.0
	04	240 55.1	30.9	87 22.2	9.3	6 48.8	16.3	61.0
	05	255 55.1	31.5	101 50.5	9.2	7 05.1	16.3	61.0
	06	270 55.1	N18 32.1	116 18.7	9.2	S 7 21.4	16.2	61.0
W	07	285 55.1	32.7	130 46.9	9.0	7 37.6	16.2	61.0
E	08	300 55.1	33.4	145 14.9	9.1	7 53.8	16.1	61.1
D	09	315 55.1 . .	34.0	159 43.0	8.9	8 09.9	16.1	61.1
N	10	330 55.1	34.6	174 10.9	8.9	8 26.0	16.1	61.1
E	11	345 55.1	35.2	188 38.8	8.8	8 42.1	16.0	61.1
S	12	0 55.1	N18 35.8	203 06.6	8.7	S 8 58.1	16.0	61.1
D	13	15 55.1	36.4	217 34.3	8.7	9 14.1	16.0	61.1
A	14	30 55.1	37.0	232 02.0	8.6	9 30.1	15.9	61.2
Y	15	45 55.1 . .	37.6	246 29.6	8.5	9 46.0	15.8	61.2
	16	60 55.1	38.2	260 57.1	8.5	10 01.8	15.8	61.2
	17	75 55.1	38.8	275 24.6	8.4	10 17.6	15.7	61.2
	18	90 55.1	N18 39.4	289 52.0	8.3	S10 33.3	15.7	61.2
	19	105 55.1	40.0	304 19.3	8.2	10 49.0	15.6	61.2
	20	120 55.1	40.6	318 46.5	8.1	11 04.6	15.5	61.2
	21	135 55.1 . .	41.2	333 13.6	8.1	11 20.1	15.5	61.2
	22	150 55.1	41.8	347 40.7	8.0	11 35.6	15.4	61.3
	23	165 55.1	42.4	2 07.7	7.9	11 51.0	15.4	61.3
	15 00	180 55.1	N18 43.0	16 34.6	7.8	S12 06.4	15.3	61.3
	01	195 55.1	43.6	31 01.4	7.8	12 21.7	15.2	61.3
	02	210 55.1	44.2	45 28.2	7.7	12 36.9	15.1	61.3
	03	225 55.1 . .	44.8	59 54.9	7.6	12 52.0	15.1	61.3
	04	240 55.1	45.4	74 21.5	7.5	13 07.1	14.9	61.3
	05	255 55.1	46.0	88 48.0	7.4	13 22.0	14.9	61.3
	06	270 55.1	N18 46.6	103 14.4	7.3	S13 36.9	14.8	61.3
T	07	285 55.1	47.2	117 40.7	7.3	13 51.7	14.8	61.3
H	08	300 55.1	47.8	132 07.0	7.2	14 06.5	14.6	61.3
U	09	315 55.1 . .	48.4	146 33.2	7.1	14 21.1	14.5	61.3
R	10	330 55.1	49.0	160 59.3	7.0	14 35.6	14.5	61.3
S	11	345 55.1	49.5	175 25.3	6.9	14 50.1	14.4	61.3
D	12	0 55.1	N18 50.1	189 51.2	6.8	S15 04.5	14.2	61.3
A	13	15 55.1	50.7	204 17.0	6.8	15 18.7	14.2	61.3
Y	14	30 55.1	51.3	218 42.8	6.7	15 32.9	14.1	61.3
	15	45 55.1 . .	51.9	233 08.5	6.5	15 47.0	13.9	61.3
	16	60 55.1	52.5	247 34.0	6.5	16 00.9	13.9	61.3
	17	75 55.1	53.1	261 59.5	6.4	16 14.8	13.7	61.3
	18	90 55.0	N18 53.7	276 24.9	6.4	S16 28.5	13.7	61.3
	19	105 55.0	54.3	290 50.3	6.2	16 42.2	13.5	61.3
	20	120 55.0	54.8	305 15.5	6.1	16 55.7	13.4	61.3
	21	135 55.0 . .	55.4	319 40.6	6.1	17 09.1	13.3	61.3
	22	150 55.0	56.0	334 05.7	6.0	17 22.4	13.2	61.3
	23	165 55.0	56.6	348 30.7	5.9	S17 35.6	13.1	61.3
		SD 15.9	*d* 0.6	SD 16.5		16.7		16.7

Lat.	Twilight Naut.	Twilight Civil	Sunrise	Moonrise 13	Moonrise 14	Moonrise 15	Moonrise 16
°	h m	h m	h m	h m	h m	h m	h m
N 72	□	□	□	16 35	19 11	23 12	■
N 70	////	////	00 59	16 29	18 51	21 43	■
68	////	////	01 55	16 24	18 36	21 04	■
66	////	////	02 28	16 21	18 23	20 37	23 16
64	////	01 18	02 52	16 17	18 13	20 17	22 31
62	////	01 57	03 10	16 15	18 05	20 01	22 02
60	////	02 24	03 26	16 12	17 57	19 47	21 40
N 58	01 10	02 44	03 39	16 10	17 51	19 35	21 22
56	01 46	03 01	03 50	16 08	17 45	19 25	21 07
54	02 11	03 15	04 00	16 06	17 40	19 17	20 54
52	02 30	03 27	04 08	16 05	17 35	19 09	20 43
50	02 45	03 37	04 16	16 04	17 31	19 02	20 33
45	03 16	03 59	04 33	16 01	17 22	18 47	20 12
N 40	03 38	04 16	04 46	15 58	17 15	18 35	19 55
35	03 56	04 30	04 58	15 56	17 09	18 24	19 41
30	04 10	04 42	05 08	15 54	17 03	18 15	19 29
20	04 33	05 01	05 25	15 51	16 54	18 00	19 08
N 10	04 51	05 17	05 39	15 48	16 46	17 46	18 50
0	05 06	05 31	05 53	15 46	16 38	17 34	18 33
S 10	05 19	05 44	06 06	15 43	16 31	17 22	18 17
20	05 31	05 57	06 21	15 40	16 23	17 09	17 59
30	05 43	06 12	06 37	15 38	16 14	16 54	17 39
35	05 49	06 19	06 46	15 36	16 09	16 45	17 28
40	05 55	06 28	06 57	15 34	16 03	16 36	17 15
45	06 02	06 38	07 09	15 32	15 56	16 25	16 59
S 50	06 10	06 49	07 25	15 29	15 49	16 11	16 40
52	06 13	06 54	07 32	15 28	15 45	16 05	16 31
54	06 17	07 00	07 40	15 27	15 41	15 58	16 21
56	06 20	07 06	07 49	15 25	15 37	15 50	16 10
58	06 25	07 13	07 59	15 24	15 32	15 42	15 57
S 60	06 29	07 21	08 10	15 22	15 26	15 32	15 42

Lat.	Sunset	Twilight Civil	Twilight Naut.	Moonset 13	Moonset 14	Moonset 15	Moonset 16
°	h m	h m	h m	h m	h m	h m	h m
N 72	□	□	□	03 17	02 44	{02 01 / 23 59}	■
N 70	23 05	////	////	03 17	02 53	02 23	01 30
68	22 02	////	////	03 17	03 00	02 41	02 11
66	21 28	////	////	03 17	03 07	02 55	02 39
64	21 03	22 41	////	03 18	03 12	03 07	03 00
62	20 44	21 59	////	03 18	03 17	03 17	03 18
60	20 29	21 31	////	03 18	03 21	03 26	03 32
N 58	20 15	21 11	22 48	03 18	03 25	03 33	03 45
56	20 04	20 54	22 10	03 18	03 28	03 40	03 56
54	19 54	20 39	21 45	03 18	03 31	03 46	04 06
52	19 45	20 27	21 25	03 18	03 34	03 52	04 14
50	19 37	20 17	21 09	03 18	03 36	03 57	04 22
45	19 21	19 55	20 38	03 18	03 42	04 08	04 39
N 40	19 07	19 38	20 15	03 18	03 46	04 17	04 52
35	18 55	19 24	19 58	03 18	03 50	04 25	05 04
30	18 45	19 12	19 43	03 19	03 54	04 32	05 14
20	18 28	18 52	19 20	03 19	04 00	04 44	05 32
N 10	18 14	18 36	19 02	03 19	04 05	04 54	05 47
0	18 00	18 22	18 47	03 19	04 10	05 04	06 02
S 10	17 46	18 08	18 34	03 19	04 15	05 14	06 16
20	17 32	17 55	18 22	03 19	04 21	05 25	06 32
30	17 16	17 41	18 10	03 19	04 27	05 37	06 50
35	17 06	17 33	18 04	03 19	04 30	05 44	07 01
40	16 55	17 24	17 57	03 19	04 34	05 53	07 13
45	16 43	17 15	17 50	03 19	04 39	06 02	07 27
S 50	16 27	17 03	17 42	03 19	04 45	06 14	07 45
52	16 20	16 58	17 39	03 19	04 47	06 19	07 53
54	16 12	16 52	17 35	03 19	04 50	06 25	08 02
56	16 04	16 46	17 32	03 18	04 53	06 32	08 13
58	15 53	16 39	17 27	03 18	04 57	06 39	08 25
S 60	15 42	16 31	17 23	03 18	05 01	06 48	08 39

Day	SUN Eqn. of Time 00^h	SUN Eqn. of Time 12^h	SUN Mer. Pass.	MOON Mer. Pass. Upper	MOON Mer. Pass. Lower	MOON Age	MOON Phase
d	m s	m s	h m	h m	h m	d %	
13	03 40	03 40	11 56	21 58	09 32	12 90	
14	03 40	03 40	11 56	22 51	10 24	13 96	○
15	03 40	03 40	11 56	23 48	11 19	14 99	

UT	ARIES	VENUS −3.9		MARS −0.3		JUPITER −2.0		SATURN +0.1	
d h	GHA ° ′	GHA ° ′	Dec ° ′	GHA ° ′	Dec ° ′	GHA ° ′	Dec ° ′	GHA ° ′	Dec ° ′
16 00	233 17.7	205 19.6	N 9 44.7	276 14.9	S18 35.1	99 57.8	N18 21.7	145 48.8	N22 32.9
01	248 20.2	220 19.2	45.8	291 15.9	34.8	114 59.9	21.7	160 50.9	32.9
02	263 22.7	235 18.8	46.9	306 17.0	34.4	130 02.1	21.6	175 53.1	32.9
03	278 25.1	250 18.4	. . 48.0	321 18.0	. . 34.1	145 04.3	. . 21.5	190 55.2	. . 33.0
04	293 27.6	265 18.0	49.1	336 19.0	33.8	160 06.4	21.4	205 57.4	33.0
05	308 30.1	280 17.6	50.1	351 20.1	33.5	175 08.6	21.3	220 59.5	33.0
06	323 32.5	295 17.2	N 9 51.2	6 21.1	S18 33.2	190 10.8	N18 21.2	236 01.7	N22 33.0
07	338 35.0	310 16.8	52.3	21 22.1	32.9	205 12.9	21.2	251 03.8	33.0
08	353 37.4	325 16.4	53.4	36 23.2	32.6	220 15.1	21.1	266 06.0	33.0
F 09	8 39.9	340 16.0	. . 54.5	51 24.2	. . 32.3	235 17.3	. . 21.0	281 08.1	. . 33.0
R 10	23 42.4	355 15.6	55.6	66 25.2	32.0	250 19.4	20.9	296 10.3	33.0
I 11	38 44.8	10 15.2	56.7	81 26.3	31.6	265 21.6	20.8	311 12.4	33.0
D 12	53 47.3	25 14.8	N 9 57.7	96 27.3	S18 31.3	280 23.8	N18 20.7	326 14.5	N22 33.0
A 13	68 49.8	40 14.4	58.8	111 28.3	31.0	295 25.9	20.7	341 16.7	33.1
Y 14	83 52.2	55 13.9	9 59.9	126 29.4	30.7	310 28.1	20.6	356 18.8	33.1
15	98 54.7	70 13.5	10 01.0	141 30.4	. . 30.4	325 30.3	. . 20.5	11 21.0	. . 33.1
16	113 57.2	85 13.1	02.1	156 31.4	30.1	340 32.4	20.4	26 23.1	33.1
17	128 59.6	100 12.7	03.2	171 32.5	29.8	355 34.6	20.3	41 25.3	33.1
18	144 02.1	115 12.3	N10 04.3	186 33.5	S18 29.5	10 36.8	N18 20.2	56 27.4	N22 33.1
19	159 04.6	130 11.9	05.3	201 34.5	29.2	25 38.9	20.2	71 29.6	33.1
20	174 07.0	145 11.5	06.4	216 35.6	28.8	40 41.1	20.1	86 31.7	33.1
21	189 09.5	160 11.1	. . 07.5	231 36.6	. . 28.5	55 43.3	. . 20.0	101 33.9	. . 33.1
22	204 11.9	175 10.7	08.6	246 37.7	28.2	70 45.4	19.9	116 36.0	33.1
23	219 14.4	190 10.2	09.7	261 38.7	27.9	85 47.6	19.8	131 38.2	33.2
17 00	234 16.9	205 09.8	N10 10.7	276 39.7	S18 27.6	100 49.8	N18 19.7	146 40.3	N22 33.2
01	249 19.3	220 09.4	11.8	291 40.8	27.3	115 51.9	19.7	161 42.5	33.2
02	264 21.8	235 09.0	12.9	306 41.8	27.0	130 54.1	19.6	176 44.6	33.2
03	279 24.3	250 08.6	. . 14.0	321 42.8	. . 26.7	145 56.3	. . 19.5	191 46.8	. . 33.2
04	294 26.7	265 08.2	15.1	336 43.9	26.3	160 58.4	19.4	206 48.9	33.2
05	309 29.2	280 07.8	16.1	351 44.9	26.0	176 00.6	19.3	221 51.1	33.2
06	324 31.7	295 07.3	N10 17.2	6 46.0	S18 25.7	191 02.7	N18 19.2	236 53.2	N22 33.2
07	339 34.1	310 06.9	18.3	21 47.0	25.4	206 04.9	19.2	251 55.4	33.2
S 08	354 36.6	325 06.5	19.4	36 48.1	25.1	221 07.1	19.1	266 57.5	33.2
A 09	9 39.1	340 06.1	. . 20.5	51 49.1	. . 24.8	236 09.2	. . 19.0	281 59.6	. . 33.3
T 10	24 41.5	355 05.7	21.5	66 50.1	24.5	251 11.4	18.9	297 01.8	33.3
U 11	39 44.0	10 05.3	22.6	81 51.2	24.2	266 13.6	18.8	312 03.9	33.3
R 12	54 46.4	25 04.9	N10 23.7	96 52.2	S18 23.8	281 15.7	N18 18.7	327 06.1	N22 33.3
D 13	69 48.9	40 04.4	24.8	111 53.3	23.5	296 17.9	18.7	342 08.2	33.3
A 14	84 51.4	55 04.0	25.8	126 54.3	23.2	311 20.0	18.6	357 10.4	33.3
Y 15	99 53.8	70 03.6	. . 26.9	141 55.4	. . 22.9	326 22.2	. . 18.5	12 12.5	. . 33.3
16	114 56.3	85 03.2	28.0	156 56.4	22.6	341 24.4	18.4	27 14.7	33.3
17	129 58.8	100 02.8	29.1	171 57.4	22.3	356 26.5	18.3	42 16.8	33.3
18	145 01.2	115 02.3	N10 30.1	186 58.5	S18 22.0	11 28.7	N18 18.2	57 19.0	N22 33.3
19	160 03.7	130 01.9	31.2	201 59.5	21.7	26 30.8	18.1	72 21.1	33.4
20	175 06.2	145 01.5	32.3	217 00.6	21.4	41 33.0	18.1	87 23.3	33.4
21	190 08.6	160 01.1	. . 33.4	232 01.6	. . 21.0	56 35.2	. . 18.0	102 25.4	. . 33.4
22	205 11.1	175 00.7	34.4	247 02.7	20.7	71 37.3	17.9	117 27.6	33.4
23	220 13.6	190 00.2	35.5	262 03.7	20.4	86 39.5	17.8	132 29.7	33.4
18 00	235 16.0	204 59.8	N10 36.6	277 04.8	S18 20.1	101 41.6	N18 17.7	147 31.8	N22 33.4
01	250 18.5	219 59.4	37.6	292 05.8	19.8	116 43.8	17.6	162 34.0	33.4
02	265 20.9	234 59.0	38.7	307 06.9	19.5	131 45.9	17.6	177 36.1	33.4
03	280 23.4	249 58.6	. . 39.8	322 07.9	. . 19.2	146 48.1	. . 17.5	192 38.3	. . 33.4
04	295 25.9	264 58.1	40.9	337 09.0	18.9	161 50.3	17.4	207 40.4	33.4
05	310 28.3	279 57.7	41.9	352 10.0	18.5	176 52.4	17.3	222 42.6	33.5
06	325 30.8	294 57.3	N10 43.0	7 11.1	S18 18.2	191 54.6	N18 17.2	237 44.7	N22 33.5
07	340 33.3	309 56.9	44.1	22 12.1	17.9	206 56.7	17.1	252 46.9	33.5
08	355 35.7	324 56.4	45.1	37 13.1	17.6	221 58.9	17.0	267 49.0	33.5
S 09	10 38.2	339 56.0	. . 46.2	52 14.2	. . 17.3	237 01.0	. . 17.0	282 51.1	. . 33.5
U 10	25 40.7	354 55.6	47.3	67 15.2	17.0	252 03.2	16.9	297 53.3	33.5
N 11	40 43.1	9 55.2	48.3	82 16.3	16.7	267 05.4	16.8	312 55.4	33.5
D 12	55 45.6	24 54.7	N10 49.4	97 17.4	S18 16.4	282 07.5	N18 16.7	327 57.6	N22 33.5
A 13	70 48.1	39 54.3	50.5	112 18.4	16.0	297 09.7	16.6	342 59.7	33.5
Y 14	85 50.5	54 53.9	51.5	127 19.5	15.7	312 11.8	16.5	358 01.9	33.5
15	100 53.0	69 53.5	. . 52.6	142 20.5	. . 15.4	327 14.0	. . 16.4	13 04.0	. . 33.6
16	115 55.4	84 53.0	53.7	157 21.6	15.1	342 16.1	16.3	28 06.2	33.6
17	130 57.9	99 52.6	54.7	172 22.6	14.8	357 18.3	16.3	43 08.3	33.6
18	146 00.4	114 52.2	N10 55.8	187 23.7	S18 14.5	12 20.4	N18 16.2	58 10.5	N22 33.6
19	161 02.8	129 51.8	56.9	202 24.7	14.2	27 22.6	16.1	73 12.6	33.6
20	176 05.3	144 51.3	57.9	217 25.8	13.8	42 24.8	16.0	88 14.7	33.6
21	191 07.8	159 50.9	10 59.0	232 26.8	. . 13.5	57 26.9	. . 15.9	103 16.9	. . 33.6
22	206 10.2	174 50.5	11 00.1	247 27.9	13.2	72 29.1	15.8	118 19.0	33.6
23	221 12.7	189 50.0	N11 01.1	262 28.9	12.9	87 31.2	15.7	133 21.2	33.6
Mer. Pass.	h m 8 21.5	*v* −0.4	*d* 1.1	*v* 1.0	*d* 0.3	*v* 2.2	*d* 0.1	*v* 2.1	*d* 0.0

STARS Name	SHA ° ′	Dec ° ′
Acamar	315 24.7	S40 17.5
Achernar	335 33.0	S57 13.1
Acrux	173 18.1	S63 07.2
Adhara	255 19.1	S28 58.7
Aldebaran	290 58.9	N16 30.9
Alioth	166 27.0	N55 56.7
Alkaid	153 04.6	N49 17.9
Al Na'ir	27 53.6	S46 56.6
Alnilam	275 54.8	S 1 12.0
Alphard	218 04.0	S 8 40.4
Alphecca	126 17.3	N26 42.2
Alpheratz	357 52.1	N29 06.2
Altair	62 15.8	N 8 52.4
Ankaa	353 23.7	S42 17.2
Antares	112 35.7	S26 26.4
Arcturus	146 02.7	N19 09.9
Atria	107 44.3	S69 02.0
Avior	234 21.7	S59 31.4
Bellatrix	278 40.9	N 6 21.1
Betelgeuse	271 10.2	N 7 24.5
Canopus	264 00.1	S52 42.0
Capella	280 46.7	N46 00.2
Deneb	49 36.9	N45 17.2
Denebola	182 41.6	N14 33.3
Diphda	349 04.1	S17 58.2
Dubhe	194 00.9	N61 44.3
Elnath	278 23.1	N28 36.7
Eltanin	90 49.4	N51 29.1
Enif	33 55.0	N 9 53.2
Fomalhaut	15 32.8	S29 36.3
Gacrux	172 09.7	S57 08.1
Gienah	176 00.4	S17 33.7
Hadar	148 58.9	S60 23.5
Hamal	328 10.2	N23 28.5
Kaus Aust.	83 54.1	S34 23.0
Kochab	137 18.2	N74 08.6
Markab	13 46.4	N15 13.1
Menkar	314 23.7	N 4 06.1
Menkent	148 16.7	S36 23.3
Miaplacidus	221 41.8	S69 44.1
Mirfak	308 52.4	N49 52.3
Nunki	76 07.9	S26 17.6
Peacock	53 31.4	S56 43.3
Pollux	243 37.6	N28 01.2
Procyon	245 08.3	N 5 13.0
Rasalhague	96 13.5	N12 33.3
Regulus	207 51.9	N11 57.2
Rigel	281 20.0	S 8 11.9
Rigil Kent.	140 02.3	S60 51.0
Sabik	102 21.4	S15 43.8
Schedar	349 50.3	N56 33.0
Shaula	96 32.4	S37 06.4
Sirius	258 41.1	S16 43.3
Spica	158 39.4	S11 10.8
Suhail	222 58.5	S43 26.9
Vega	80 44.1	N38 46.9
Zuben'ubi	137 14.0	S16 03.4

	SHA ° ′	Mer. Pass. h m
Venus	330 53.0	10 20
Mars	42 22.9	5 33
Jupiter	226 32.9	17 14
Saturn	272 23.4	14 11

Day	UT (d h)	SUN GHA	SUN Dec	MOON GHA	v	MOON Dec	d	HP
		° ′	° ′	° ′	′	° ′	′	′
16	00	180 55.0	N18 57.2	2 55.6	5.8	S17 48.7	12.9	61.3
	01	195 55.0	57.8	17 20.4	5.7	18 01.6	12.8	61.3
	02	210 55.0	58.4	31 45.1	5.6	18 14.4	12.7	61.3
	03	225 55.0	. . 58.9	46 09.7	5.6	18 27.1	12.6	61.3
	04	240 55.0	18 59.5	60 34.3	5.4	18 39.7	12.4	61.3
	05	255 55.0	19 00.1	74 58.7	5.4	18 52.1	12.3	61.3
	06	270 55.0	N19 00.7	89 23.1	5.3	S19 04.4	12.2	61.3
	07	285 54.9	01.3	103 47.4	5.2	19 16.6	12.0	61.3
	08	300 54.9	01.9	118 11.6	5.2	19 28.6	11.9	61.3
F	09	315 54.9	. . 02.4	132 35.8	5.0	19 40.5	11.8	61.3
R	10	330 54.9	03.0	146 59.8	5.0	19 52.3	11.6	61.2
I	11	345 54.9	03.6	161 23.8	4.9	20 03.9	11.4	61.2
D	12	0 54.9	N19 04.2	175 47.7	4.8	S20 15.3	11.4	61.2
A	13	15 54.9	04.8	190 11.5	4.7	20 26.7	11.1	61.2
Y	14	30 54.9	05.3	204 35.2	4.7	20 37.8	11.1	61.2
	15	45 54.9	. . 05.9	218 58.9	4.6	20 48.9	10.8	61.2
	16	60 54.9	06.5	233 22.5	4.5	20 59.7	10.8	61.2
	17	75 54.8	07.1	247 46.0	4.4	21 10.5	10.5	61.2
	18	90 54.8	N19 07.6	262 09.4	4.4	S21 21.0	10.5	61.1
	19	105 54.8	08.2	276 32.8	4.2	21 31.5	10.2	61.1
	20	120 54.8	08.8	290 56.0	4.2	21 41.7	10.1	61.1
	21	135 54.8	. . 09.4	305 19.2	4.2	21 51.8	9.9	61.1
	22	150 54.8	09.9	319 42.4	4.1	22 01.7	9.8	61.1
	23	165 54.8	10.5	334 05.5	4.0	22 11.5	9.6	61.1
17	00	180 54.8	N19 11.1	348 28.5	3.9	S22 21.1	9.5	61.0
	01	195 54.7	11.6	2 51.4	3.9	22 30.6	9.3	61.0
	02	210 54.7	12.2	17 14.3	3.8	22 39.9	9.1	61.0
	03	225 54.7	. . 12.8	31 37.1	3.8	22 49.0	8.9	61.0
	04	240 54.7	13.4	45 59.9	3.7	22 57.9	8.8	61.0
	05	255 54.7	13.9	60 22.6	3.6	23 06.7	8.6	60.9
	06	270 54.7	N19 14.5	74 45.2	3.6	S23 15.3	8.4	60.9
	07	285 54.7	15.1	89 07.8	3.5	23 23.7	8.2	60.9
S	08	300 54.6	15.6	103 30.3	3.5	23 31.9	8.1	60.9
A	09	315 54.6	. . 16.2	117 52.8	3.4	23 40.0	7.9	60.9
T	10	330 54.6	16.8	132 15.2	3.4	23 47.9	7.7	60.8
U	11	345 54.6	17.3	146 37.6	3.3	23 55.6	7.6	60.8
R	12	0 54.6	N19 17.9	160 59.9	3.3	S24 03.2	7.3	60.8
D	13	15 54.6	18.5	175 22.2	3.3	24 10.5	7.2	60.8
A	14	30 54.5	19.0	189 44.5	3.2	24 17.7	7.0	60.7
Y	15	45 54.5	. . 19.6	204 06.7	3.1	24 24.7	6.8	60.7
	16	60 54.5	20.1	218 28.8	3.2	24 31.5	6.6	60.7
	17	75 54.5	20.7	232 51.0	3.1	24 38.1	6.4	60.7
	18	90 54.5	N19 21.3	247 13.1	3.0	S24 44.5	6.3	60.6
	19	105 54.5	21.8	261 35.1	3.1	24 50.8	6.1	60.6
	20	120 54.4	22.4	275 57.2	3.0	24 56.9	5.8	60.6
	21	135 54.4	. . 23.0	290 19.2	3.0	25 02.7	5.7	60.6
	22	150 54.4	23.5	304 41.2	2.9	25 08.4	5.5	60.5
	23	165 54.4	24.1	319 03.1	3.0	25 13.9	5.3	60.5
18	00	180 54.4	N19 24.6	333 25.1	2.9	S25 19.2	5.1	60.5
	01	195 54.3	25.2	347 47.0	2.9	25 24.3	5.0	60.4
	02	210 54.3	25.7	2 08.9	2.9	25 29.3	4.7	60.4
	03	225 54.3	. . 26.3	16 30.8	2.9	25 34.0	4.6	60.4
	04	240 54.3	26.9	30 52.7	2.9	25 38.6	4.3	60.4
	05	255 54.3	27.4	45 14.6	2.9	25 42.9	4.2	60.3
	06	270 54.2	N19 28.0	59 36.5	2.9	S25 47.1	4.0	60.3
	07	285 54.2	28.5	73 58.4	2.9	25 51.1	3.7	60.3
	08	300 54.2	29.1	88 20.3	2.8	25 54.8	3.6	60.2
S	09	315 54.2	. . 29.6	102 42.1	2.9	25 58.4	3.4	60.2
U	10	330 54.2	30.2	117 04.0	2.9	26 01.8	3.2	60.2
N	11	345 54.1	30.7	131 25.9	2.9	26 05.0	3.0	60.1
D	12	0 54.1	N19 31.3	145 47.8	3.0	S26 08.0	2.8	60.1
A	13	15 54.1	31.8	160 09.8	2.9	26 10.8	2.7	60.1
Y	14	30 54.1	32.4	174 31.7	3.0	26 13.5	2.4	60.0
	15	45 54.1	. . 32.9	188 53.7	3.0	26 15.9	2.2	60.0
	16	60 54.0	33.5	203 15.7	3.0	26 18.1	2.1	60.0
	17	75 54.0	34.0	217 37.7	3.0	26 20.2	1.8	59.9
	18	90 54.0	N19 34.6	231 59.7	3.1	S26 22.0	1.7	59.9
	19	105 54.0	35.1	246 21.8	3.0	26 23.7	1.5	59.9
	20	120 53.9	35.7	260 43.8	3.2	26 25.2	1.3	59.8
	21	135 53.9	. . 36.2	275 06.0	3.1	26 26.5	1.1	59.8
	22	150 53.9	36.8	289 28.1	3.2	26 27.6	0.9	59.7
	23	165 53.9	37.3	303 50.3	3.3	S26 28.5	0.7	59.7
		SD 15.8	*d* 0.6	SD 16.7		16.6		16.4

Lat.	Twilight Naut.	Twilight Civil	Sunrise	Moonrise 16	Moonrise 17	Moonrise 18	Moonrise 19
°	h m	h m	h m	h m	h m	h m	h m
N 72	□	□	□	■	■	■	■
N 70	□	□	□	■	■	■	■
68	////	////	01 38	■	■	■	■
66	////	////	02 16	23 16	■	■	■
64	////	00 56	02 42	22 31	25 05	01 05	■
62	////	01 44	03 02	22 02	24 01	00 01	01 34
60	////	02 14	03 19	21 40	23 27	24 51	00 51
N 58	00 50	02 36	03 33	21 22	23 02	24 22	00 22
56	01 34	02 54	03 44	21 07	22 42	24 00	00 00
54	02 02	03 09	03 55	20 54	22 25	23 41	24 36
52	02 22	03 21	04 04	20 43	22 11	23 26	24 22
50	02 39	03 32	04 12	20 33	21 59	23 12	24 09
45	03 11	03 55	04 29	20 12	21 33	22 45	23 43
N 40	03 35	04 13	04 43	19 55	21 13	22 23	23 22
35	03 53	04 27	04 55	19 41	20 56	22 05	23 05
30	04 08	04 39	05 06	19 29	20 41	21 49	22 49
20	04 32	05 00	05 23	19 08	20 17	21 23	22 24
N 10	04 50	05 16	05 39	18 50	19 56	21 00	22 02
0	05 05	05 31	05 53	18 33	19 36	20 39	21 41
S 10	05 19	05 45	06 07	18 17	19 16	20 18	21 21
20	05 32	05 58	06 22	17 59	18 55	19 56	20 59
30	05 44	06 13	06 39	17 39	18 31	19 30	20 33
35	05 51	06 21	06 49	17 28	18 17	19 15	20 18
40	05 58	06 31	07 00	17 15	18 01	18 57	20 01
45	06 05	06 41	07 13	16 59	17 42	18 36	19 40
S 50	06 13	06 53	07 29	16 40	17 18	18 09	19 13
52	06 17	06 58	07 36	16 31	17 07	17 56	19 00
54	06 21	07 04	07 45	16 21	16 54	17 41	18 45
56	06 25	07 11	07 54	16 10	16 39	17 23	18 28
58	06 29	07 18	08 04	15 57	16 21	17 02	18 07
S 60	06 34	07 27	08 17	15 42	16 00	16 35	17 40

Lat.	Sunset	Twilight Civil	Twilight Naut.	Moonset 16	Moonset 17	Moonset 18	Moonset 19
°	h m	h m	h m	h m	h m	h m	h m
N 72	□	□	□	■	■	■	■
N 70	□	□	□	01 30	■	■	■
68	22 20	////	////	02 11	■	■	■
66	21 40	////	////	02 39	02 06	■	■
64	21 13	23 05	////	03 00	02 52	02 30	■
62	20 52	22 12	////	03 18	03 22	03 34	04 15
60	20 36	21 41	////	03 32	03 45	04 09	04 58
N 58	20 22	21 19	23 10	03 45	04 03	04 34	05 27
56	20 10	21 01	22 22	03 56	04 19	04 55	05 49
54	19 59	20 45	21 53	04 06	04 32	05 12	06 08
52	19 50	20 33	21 32	04 14	04 44	05 26	06 23
50	19 42	20 21	21 15	04 22	04 55	05 39	06 37
45	19 24	19 59	20 43	04 39	05 17	06 05	07 04
N 40	19 10	19 41	20 19	04 52	05 35	06 26	07 26
35	18 58	19 26	20 00	05 04	05 50	06 43	07 44
30	18 47	19 14	19 45	05 14	06 03	06 58	08 00
20	18 30	18 53	19 21	05 32	06 25	07 24	08 26
N 10	18 14	18 37	19 03	05 47	06 45	07 46	08 48
0	18 00	18 22	18 47	06 02	07 03	08 06	09 09
S 10	17 46	18 08	18 34	06 16	07 21	08 27	09 30
20	17 31	17 54	18 21	06 32	07 41	08 49	09 53
30	17 14	17 39	18 08	06 50	08 04	09 14	10 19
35	17 04	17 31	18 02	07 01	08 17	09 29	10 34
40	16 53	17 22	17 55	07 13	08 32	09 47	10 52
45	16 39	17 12	17 47	07 27	08 51	10 08	11 13
S 50	16 23	16 59	17 39	07 45	09 14	10 34	11 40
52	16 16	16 54	17 35	07 53	09 25	10 47	11 53
54	16 08	16 48	17 31	08 02	09 38	11 02	12 08
56	15 58	16 41	17 27	08 13	09 52	11 20	12 26
58	15 48	16 34	17 23	08 25	10 09	11 41	12 47
S 60	15 35	16 25	17 18	08 39	10 30	12 07	13 14

Day	SUN Eqn. of Time 00^h	SUN Eqn. of Time 12^h	SUN Mer. Pass.	MOON Mer. Pass. Upper	MOON Mer. Pass. Lower	MOON Age	MOON Phase
d	m s	m s	h m	h m	h m	d %	
16	03 40	03 40	11 56	24 48	12 18	15 100	
17	03 39	03 38	11 56	00 48	13 19	16 97	○
18	03 38	03 37	11 56	01 51	14 23	17 92	

UT	ARIES	VENUS −3.9		MARS −0.4		JUPITER −2.0		SATURN +0.1	
d h	GHA ° ′	GHA ° ′	Dec ° ′	GHA ° ′	Dec ° ′	GHA ° ′	Dec ° ′	GHA ° ′	Dec ° ′
19 00	236 15.2	204 49.6	N11 02.2	277 30.0	S18 12.6	102 33.4	N18 15.7	148 23.3	N22 33.6
01	251 17.6	219 49.2	03.2	292 31.0	12.3	117 35.5	15.6	163 25.5	33.7
02	266 20.1	234 48.8	04.3	307 32.1	12.0	132 37.7	15.5	178 27.6	33.7
03	281 22.6	249 48.3	. . 05.4	322 33.2	. . 11.7	147 39.8	. . 15.4	193 29.7	. . 33.7
04	296 25.0	264 47.9	06.4	337 34.2	11.3	162 42.0	15.3	208 31.9	33.7
05	311 27.5	279 47.5	07.5	352 35.3	11.0	177 44.1	15.2	223 34.0	33.7
06	326 29.9	294 47.0	N11 08.5	7 36.3	S18 10.7	192 46.3	N18 15.1	238 36.2	N22 33.7
07	341 32.4	309 46.6	09.6	22 37.4	10.4	207 48.4	15.0	253 38.3	33.7
08	356 34.9	324 46.2	10.7	37 38.4	10.1	222 50.6	15.0	268 40.5	33.7
M 09	11 37.3	339 45.7	. . 11.7	52 39.5	. . 09.8	237 52.7	. . 14.9	283 42.6	. . 33.7
O 10	26 39.8	354 45.3	12.8	67 40.6	09.5	252 54.9	14.8	298 44.8	33.7
N 11	41 42.3	9 44.9	13.8	82 41.6	09.1	267 57.0	14.7	313 46.9	33.7
D 12	56 44.7	24 44.4	N11 14.9	97 42.7	S18 08.8	282 59.2	N18 14.6	328 49.0	N22 33.8
A 13	71 47.2	39 44.0	15.9	112 43.7	08.5	298 01.3	14.5	343 51.2	33.8
Y 14	86 49.7	54 43.6	17.0	127 44.8	08.2	313 03.5	14.4	358 53.3	33.8
15	101 52.1	69 43.1	. . 18.1	142 45.9	. . 07.9	328 05.6	. . 14.3	13 55.5	. . 33.8
16	116 54.6	84 42.7	19.1	157 46.9	07.6	343 07.8	14.3	28 57.6	33.8
17	131 57.1	99 42.3	20.2	172 48.0	07.3	358 09.9	14.2	43 59.8	33.8
18	146 59.5	114 41.8	N11 21.2	187 49.0	S18 07.0	13 12.1	N18 14.1	59 01.9	N22 33.8
19	162 02.0	129 41.4	22.3	202 50.1	06.6	28 14.2	14.0	74 04.0	33.8
20	177 04.4	144 41.0	23.3	217 51.2	06.3	43 16.4	13.9	89 06.2	33.8
21	192 06.9	159 40.5	. . 24.4	232 52.2	. . 06.0	58 18.5	. . 13.8	104 08.3	. . 33.8
22	207 09.4	174 40.1	25.4	247 53.3	05.7	73 20.7	13.7	119 10.5	33.9
23	222 11.8	189 39.6	26.5	262 54.4	05.4	88 22.8	13.6	134 12.6	33.9
20 00	237 14.3	204 39.2	N11 27.5	277 55.4	S18 05.1	103 25.0	N18 13.6	149 14.7	N22 33.9
01	252 16.8	219 38.8	28.6	292 56.5	04.8	118 27.1	13.5	164 16.9	33.9
02	267 19.2	234 38.3	29.6	307 57.6	04.4	133 29.3	13.4	179 19.0	33.9
03	282 21.7	249 37.9	. . 30.7	322 58.6	. . 04.1	148 31.4	. . 13.3	194 21.2	. . 33.9
04	297 24.2	264 37.5	31.7	337 59.7	03.8	163 33.6	13.2	209 23.3	33.9
05	312 26.6	279 37.0	32.8	353 00.7	03.5	178 35.7	13.1	224 25.5	33.9
06	327 29.1	294 36.6	N11 33.8	8 01.8	S18 03.2	193 37.9	N18 13.0	239 27.6	N22 33.9
07	342 31.6	309 36.1	34.9	23 02.9	02.9	208 40.0	12.9	254 29.7	33.9
T 08	357 34.0	324 35.7	35.9	38 03.9	02.6	223 42.1	12.8	269 31.9	33.9
U 09	12 36.5	339 35.3	. . 37.0	53 05.0	. . 02.3	238 44.3	. . 12.8	284 34.0	. . 34.0
E 10	27 38.9	354 34.8	38.0	68 06.1	01.9	253 46.4	12.7	299 36.2	34.0
S 11	42 41.4	9 34.4	39.1	83 07.2	01.6	268 48.6	12.6	314 38.3	34.0
D 12	57 43.9	24 33.9	N11 40.1	98 08.2	S18 01.3	283 50.7	N18 12.5	329 40.4	N22 34.0
A 13	72 46.3	39 33.5	41.2	113 09.3	01.0	298 52.9	12.4	344 42.6	34.0
Y 14	87 48.8	54 33.0	42.2	128 10.4	00.7	313 55.0	12.3	359 44.7	34.0
15	102 51.3	69 32.6	. . 43.3	143 11.4	. . 00.4	328 57.2	. . 12.2	14 46.9	. . 34.0
16	117 53.7	84 32.2	44.3	158 12.5	18 00.1	343 59.3	12.1	29 49.0	34.0
17	132 56.2	99 31.7	45.4	173 13.6	17 59.7	359 01.5	12.0	44 51.2	34.0
18	147 58.7	114 31.3	N11 46.4	188 14.6	S17 59.4	14 03.6	N18 11.9	59 53.3	N22 34.0
19	163 01.1	129 30.8	47.5	203 15.7	59.1	29 05.7	11.9	74 55.4	34.0
20	178 03.6	144 30.4	48.5	218 16.8	58.8	44 07.9	11.8	89 57.6	34.1
21	193 06.0	159 29.9	. . 49.5	233 17.8	. . 58.5	59 10.0	. . 11.7	104 59.7	. . 34.1
22	208 08.5	174 29.5	50.6	248 18.9	58.2	74 12.2	11.6	120 01.9	34.1
23	223 11.0	189 29.0	51.6	263 20.0	57.9	89 14.3	11.5	135 04.0	34.1
21 00	238 13.4	204 28.6	N11 52.7	278 21.1	S17 57.5	104 16.5	N18 11.4	150 06.1	N22 34.1
01	253 15.9	219 28.1	53.7	293 22.1	57.2	119 18.6	11.3	165 08.3	34.1
02	268 18.4	234 27.7	54.7	308 23.2	56.9	134 20.7	11.2	180 10.4	34.1
03	283 20.8	249 27.2	. . 55.8	323 24.3	. . 56.6	149 22.9	. . 11.1	195 12.6	. . 34.1
04	298 23.3	264 26.8	56.8	338 25.4	56.3	164 25.0	11.0	210 14.7	34.1
05	313 25.8	279 26.3	57.9	353 26.4	56.0	179 27.2	11.0	225 16.8	34.1
06	328 28.2	294 25.9	N11 58.9	8 27.5	S17 55.7	194 29.3	N18 10.9	240 19.0	N22 34.1
W 07	343 30.7	309 25.4	11 59.9	23 28.6	55.3	209 31.4	10.8	255 21.1	34.2
E 08	358 33.2	324 25.0	12 01.0	38 29.7	55.0	224 33.6	10.7	270 23.3	34.2
D 09	13 35.6	339 24.5	. . 02.0	53 30.7	. . 54.7	239 35.7	. . 10.6	285 25.4	. . 34.2
N 10	28 38.1	354 24.1	03.1	68 31.8	54.4	254 37.9	10.5	300 27.5	34.2
E 11	43 40.5	9 23.6	04.1	83 32.9	54.1	269 40.0	10.4	315 29.7	34.2
S 12	58 43.0	24 23.2	N12 05.1	98 34.0	S17 53.8	284 42.2	N18 10.3	330 31.8	N22 34.2
D 13	73 45.5	39 22.7	06.2	113 35.0	53.5	299 44.3	10.2	345 34.0	34.2
A 14	88 47.9	54 22.3	07.2	128 36.1	53.2	314 46.4	10.1	0 36.1	34.2
Y 15	103 50.4	69 21.8	. . 08.2	143 37.2	. . 52.8	329 48.6	. . 10.0	15 38.2	. . 34.2
16	118 52.9	84 21.4	09.3	158 38.3	52.5	344 50.7	10.0	30 40.4	34.2
17	133 55.3	99 20.9	10.3	173 39.4	52.2	359 52.8	09.9	45 42.5	34.2
18	148 57.8	114 20.5	N12 11.3	188 40.4	S17 51.9	14 55.0	N18 09.8	60 44.7	N22 34.3
19	164 00.3	129 20.0	12.4	203 41.5	51.6	29 57.1	09.7	75 46.8	34.3
20	179 02.7	144 19.6	13.4	218 42.6	51.3	44 59.3	09.6	90 48.9	34.3
21	194 05.2	159 19.1	. . 14.4	233 43.7	. . 51.0	60 01.4	. . 09.5	105 51.1	. . 34.3
22	209 07.7	174 18.7	15.5	248 44.8	50.6	75 03.5	09.4	120 53.2	34.3
23	224 10.1	189 18.2	16.5	263 45.8	50.3	90 05.7	09.3	135 55.3	34.3
Mer. Pass.	h m 8 09.7	*v* −0.4	*d* 1.0	*v* 1.1	*d* 0.3	*v* 2.1	*d* 0.1	*v* 2.1	*d* 0.0

STARS Name	SHA ° ′	Dec ° ′
Acamar	315 24.7	S40 17.5
Achernar	335 33.0	S57 13.1
Acrux	173 18.1	S63 07.3
Adhara	255 19.1	S28 58.7
Aldebaran	290 58.9	N16 30.9
Alioth	166 27.0	N55 56.7
Alkaid	153 04.6	N49 17.9
Al Na'ir	27 53.6	S46 56.6
Alnilam	275 54.8	S 1 12.0
Alphard	218 04.0	S 8 40.4
Alphecca	126 17.3	N26 42.2
Alpheratz	357 52.0	N29 06.2
Altair	62 15.8	N 8 52.4
Ankaa	353 23.7	S42 17.2
Antares	112 35.7	S26 26.4
Arcturus	146 02.6	N19 09.9
Atria	107 44.3	S69 02.0
Avior	234 21.7	S59 31.4
Bellatrix	278 40.9	N 6 21.1
Betelgeuse	271 10.2	N 7 24.5
Canopus	264 00.1	S52 42.0
Capella	280 46.7	N46 00.1
Deneb	49 36.8	N45 17.2
Denebola	182 41.6	N14 33.3
Diphda	349 04.0	S17 58.2
Dubhe	194 01.0	N61 44.3
Elnath	278 23.1	N28 36.7
Eltanin	90 49.4	N51 29.1
Enif	33 54.9	N 9 53.2
Fomalhaut	15 32.7	S29 36.3
Gacrux	172 09.7	S57 08.1
Gienah	176 00.4	S17 33.7
Hadar	148 58.9	S60 23.5
Hamal	328 10.1	N23 28.5
Kaus Aust.	83 54.0	S34 23.0
Kochab	137 18.2	N74 08.6
Markab	13 46.4	N15 13.1
Menkar	314 23.7	N 4 06.1
Menkent	148 16.7	S36 23.3
Miaplacidus	221 41.9	S69 44.1
Mirfak	308 52.4	N49 52.3
Nunki	76 07.9	S26 17.6
Peacock	53 31.4	S56 43.3
Pollux	243 37.6	N28 01.2
Procyon	245 08.3	N 5 13.0
Rasalhague	96 13.5	N12 33.3
Regulus	207 51.9	N11 57.2
Rigel	281 20.0	S 8 11.9
Rigil Kent.	140 02.3	S60 51.0
Sabik	102 21.4	S15 43.8
Schedar	349 50.2	N56 33.0
Shaula	96 32.4	S37 06.4
Sirius	258 41.1	S16 43.3
Spica	158 39.4	S11 10.8
Suhail	222 58.5	S43 26.9
Vega	80 44.0	N38 46.9
Zuben'ubi	137 14.0	S16 03.4

	SHA ° ′	Mer. Pass h m
Venus	327 24.9	10 22
Mars	40 41.1	5 28
Jupiter	226 10.7	17 04
Saturn	272 00.4	14 01

	UT	SUN GHA	SUN Dec	MOON GHA	v	MOON Dec	d	HP
	d h	° ′	° ′	° ′	′	° ′	′	′
	19 00	180 53.8	N19 37.9	318 12.6	3.3	S26 29.2	0.5	59.7
	01	195 53.8	38.4	332 34.9	3.3	26 29.7	0.4	59.6
	02	210 53.8	38.9	346 57.2	3.4	26 30.1	0.1	59.6
	03	225 53.8	. . 39.5	1 19.6	3.5	26 30.2	0.0	59.6
	04	240 53.7	40.0	15 42.1	3.5	26 30.2	0.3	59.5
	05	255 53.7	40.6	30 04.6	3.5	26 29.9	0.4	59.5
	06	270 53.7	N19 41.1	44 27.1	3.6	S26 29.5	0.5	59.5
	07	285 53.7	41.6	58 49.7	3.7	26 29.0	0.8	59.4
	08	300 53.6	42.2	73 12.4	3.7	26 28.2	1.0	59.4
M	09	315 53.6	. . 42.7	87 35.1	3.8	26 27.2	1.1	59.3
O	10	330 53.6	43.3	101 57.9	3.9	26 26.1	1.3	59.3
N	11	345 53.5	43.8	116 20.8	3.9	26 24.8	1.5	59.3
D	12	0 53.5	N19 44.3	130 43.7	4.1	S26 23.3	1.7	59.2
A	13	15 53.5	44.9	145 06.8	4.0	26 21.6	1.8	59.2
Y	14	30 53.5	45.4	159 29.8	4.2	26 19.8	2.1	59.1
	15	45 53.4	. . 46.0	173 53.0	4.2	26 17.7	2.2	59.1
	16	60 53.4	46.5	188 16.2	4.4	26 15.5	2.3	59.1
	17	75 53.4	47.0	202 39.6	4.4	26 13.2	2.6	59.0
	18	90 53.3	N19 47.6	217 03.0	4.5	S26 10.6	2.7	59.0
	19	105 53.3	48.1	231 26.5	4.5	26 07.9	2.9	59.0
	20	120 53.3	48.6	245 50.0	4.7	26 05.0	3.1	58.9
	21	135 53.3	. . 49.2	260 13.7	4.8	26 01.9	3.2	58.9
	22	150 53.2	49.7	274 37.5	4.8	25 58.7	3.4	58.8
	23	165 53.2	50.2	289 01.3	5.0	25 55.3	3.6	58.8
	20 00	180 53.2	N19 50.7	303 25.3	5.0	S25 51.7	3.7	58.8
	01	195 53.1	51.3	317 49.3	5.1	25 48.0	3.9	58.7
	02	210 53.1	51.8	332 13.4	5.2	25 44.1	4.0	58.7
	03	225 53.1	. . 52.3	346 37.6	5.4	25 40.1	4.2	58.6
	04	240 53.0	52.9	1 02.0	5.4	25 35.9	4.4	58.6
	05	255 53.0	53.4	15 26.4	5.5	25 31.5	4.5	58.6
	06	270 53.0	N19 53.9	29 50.9	5.6	S25 27.0	4.7	58.5
	07	285 52.9	54.4	44 15.5	5.8	25 22.3	4.8	58.5
T	08	300 52.9	55.0	58 40.3	5.8	25 17.5	5.0	58.4
U	09	315 52.9	. . 55.5	73 05.1	6.0	25 12.5	5.1	58.4
E	10	330 52.8	56.0	87 30.1	6.0	25 07.4	5.3	58.4
S	11	345 52.8	56.5	101 55.1	6.2	25 02.1	5.4	58.3
D	12	0 52.8	N19 57.1	116 20.3	6.3	S24 56.7	5.6	58.3
A	13	15 52.7	57.6	130 45.6	6.3	24 51.1	5.7	58.2
Y	14	30 52.7	58.1	145 10.9	6.5	24 45.4	5.9	58.2
	15	45 52.7	. . 58.6	159 36.4	6.6	24 39.5	6.0	58.2
	16	60 52.6	59.2	174 02.0	6.8	24 33.5	6.2	58.1
	17	75 52.6	19 59.7	188 27.8	6.8	24 27.3	6.3	58.1
	18	90 52.6	N20 00.2	202 53.6	6.9	S24 21.0	6.4	58.0
	19	105 52.5	00.7	217 19.5	7.1	24 14.6	6.5	58.0
	20	120 52.5	01.2	231 45.6	7.2	24 08.1	6.7	57.9
	21	135 52.5	. . 01.8	246 11.8	7.3	24 01.4	6.9	57.9
	22	150 52.4	02.3	260 38.1	7.4	23 54.5	6.9	57.9
	23	165 52.4	02.8	275 04.5	7.5	23 47.6	7.1	57.8
	21 00	180 52.3	N20 03.3	289 31.0	7.6	S23 40.5	7.2	57.8
	01	195 52.3	03.8	303 57.6	7.8	23 33.3	7.3	57.8
	02	210 52.3	04.3	318 24.4	7.9	23 26.0	7.5	57.7
	03	225 52.2	. . 04.9	332 51.3	7.9	23 18.5	7.6	57.7
	04	240 52.2	05.4	347 18.2	8.2	23 10.9	7.7	57.6
	05	255 52.2	05.9	1 45.4	8.2	23 03.2	7.8	57.6
	06	270 52.1	N20 06.4	16 12.6	8.3	S22 55.4	8.0	57.6
W	07	285 52.1	06.9	30 39.9	8.5	22 47.4	8.0	57.5
E	08	300 52.0	07.4	45 07.4	8.6	22 39.4	8.2	57.5
D	09	315 52.0	. . 07.9	59 35.0	8.7	22 31.2	8.3	57.4
N	10	330 52.0	08.4	74 02.7	8.8	22 22.9	8.4	57.4
E	11	345 51.9	08.9	88 30.5	8.9	22 14.5	8.5	57.4
S	12	0 51.9	N20 09.5	102 58.4	9.1	S22 06.0	8.6	57.3
D	13	15 51.8	10.0	117 26.5	9.2	21 57.4	8.7	57.3
A	14	30 51.8	10.5	131 54.7	9.2	21 48.7	8.9	57.2
Y	15	45 51.8	. . 11.0	146 22.9	9.4	21 39.8	8.9	57.2
	16	60 51.7	11.5	160 51.3	9.6	21 30.9	9.1	57.2
	17	75 51.7	12.0	175 19.9	9.6	21 21.8	9.1	57.1
	18	90 51.6	N20 12.5	189 48.5	9.8	S21 12.7	9.2	57.1
	19	105 51.6	13.0	204 17.3	9.8	21 03.5	9.4	57.1
	20	120 51.6	13.5	218 46.1	10.0	20 54.1	9.4	57.0
	21	135 51.5	. . 14.0	233 15.1	10.1	20 44.7	9.5	57.0
	22	150 51.5	14.5	247 44.2	10.2	20 35.2	9.6	56.9
	23	165 51.4	15.0	262 13.4	10.4	S20 25.6	9.8	56.9
		SD 15.8	*d* 0.5	SD 16.1		15.9		15.6

Lat.	Twilight Naut.	Twilight Civil	Sunrise	Moonrise 19	Moonrise 20	Moonrise 21	Moonrise 22
°	h m	h m	h m	h m	h m	h m	h m
N 72	□	□	□	■	■	■	■
N 70	□	□	□	■	■	■	04 53
68	////	////	01 19	■	■	■	03 52
66	////	////	02 04	■	■	03 53	03 16
64	////	00 22	02 33	■	03 23	03 00	02 51
62	////	01 31	02 55	01 34	02 15	02 27	02 31
60	////	02 04	03 12	00 51	01 39	02 03	02 14
N 58	00 20	02 28	03 27	00 22	01 14	01 43	02 00
56	01 22	02 47	03 39	00 00	00 53	01 27	01 48
54	01 53	03 03	03 50	24 36	00 36	01 13	01 38
52	02 16	03 16	04 00	24 22	00 22	01 01	01 28
50	02 33	03 28	04 08	24 09	00 09	00 50	01 20
45	03 07	03 51	04 26	23 43	24 28	00 28	01 02
N 40	03 31	04 10	04 41	23 22	24 09	00 09	00 47
35	03 50	04 25	04 53	23 05	23 54	24 35	00 35
30	04 06	04 38	05 04	22 49	23 41	24 24	00 24
20	04 30	04 59	05 22	22 24	23 18	24 05	00 05
N 10	04 49	05 16	05 38	22 02	22 58	23 48	24 33
0	05 05	05 31	05 53	21 41	22 40	23 33	24 21
S 10	05 19	05 45	06 08	21 21	22 21	23 17	24 09
20	05 33	06 00	06 23	20 59	22 01	23 01	23 57
30	05 46	06 15	06 41	20 33	21 38	22 41	23 42
35	05 53	06 24	06 51	20 18	21 24	22 30	23 33
40	06 00	06 33	07 02	20 01	21 09	22 17	23 24
45	06 08	06 44	07 16	19 40	20 50	22 02	23 12
S 50	06 17	06 56	07 33	19 13	20 26	21 43	22 58
52	06 20	07 02	07 41	19 00	20 15	21 34	22 51
54	06 25	07 09	07 49	18 45	20 02	21 24	22 44
56	06 29	07 16	07 59	18 28	19 47	21 12	22 36
58	06 34	07 23	08 10	18 07	19 30	20 59	22 27
S 60	06 39	07 32	08 23	17 40	19 08	20 43	22 16

Lat.	Sunset	Twilight Civil	Twilight Naut.	Moonset 19	Moonset 20	Moonset 21	Moonset 22
°	h m	h m	h m	h m	h m	h m	h m
N 72	□	□	□	■	■	■	■
N 70	□	□	□	■	■	■	06 53
68	22 40	////	////	■	■	■	07 54
66	21 53	////	////	■	■	06 04	08 28
64	21 23	////	////	■	04 35	06 57	08 53
62	21 00	22 26	////	04 15	05 42	07 29	09 12
60	20 42	21 51	////	04 58	06 17	07 52	09 28
N 58	20 28	21 27	////	05 27	06 43	08 11	09 41
56	20 15	21 07	22 34	05 49	07 03	08 27	09 52
54	20 04	20 51	22 02	06 08	07 20	08 40	10 02
52	19 54	20 38	21 39	06 23	07 34	08 52	10 11
50	19 46	20 26	21 21	06 37	07 46	09 02	10 19
45	19 27	20 02	20 47	07 04	08 12	09 24	10 36
N 40	19 13	19 44	20 22	07 26	08 32	09 41	10 49
35	19 00	19 29	20 03	07 44	08 49	09 56	11 01
30	18 49	19 16	19 48	08 00	09 04	10 09	11 11
20	18 31	18 55	19 23	08 26	09 29	10 30	11 28
N 10	18 15	18 37	19 04	08 48	09 50	10 48	11 43
0	18 00	18 22	18 48	09 09	10 10	11 06	11 56
S 10	17 45	18 08	18 33	09 30	10 29	11 23	12 10
20	17 30	17 53	18 20	09 53	10 50	11 41	12 25
30	17 12	17 38	18 07	10 19	11 15	12 02	12 41
35	17 02	17 29	18 00	10 34	11 29	12 14	12 51
40	16 50	17 20	17 53	10 52	11 45	12 28	13 02
45	16 37	17 09	17 45	11 13	12 05	12 44	13 15
S 50	16 20	16 56	17 36	11 40	12 29	13 04	13 30
52	16 12	16 50	17 32	11 53	12 41	13 14	13 37
54	16 03	16 44	17 28	12 08	12 54	13 24	13 45
56	15 53	16 37	17 23	12 26	13 09	13 36	13 54
58	15 42	16 29	17 19	12 47	13 27	13 50	14 04
S 60	15 29	16 20	17 13	13 14	13 49	14 06	14 16

Day	SUN Eqn. of Time 00^h	SUN Eqn. of Time 12^h	SUN Mer. Pass.	MOON Mer. Pass. Upper	MOON Mer. Pass. Lower	Age	Phase
d	m s	m s	h m	h m	h m	d %	
19	03 35	03 34	11 56	02 54	15 25	18 85	
20	03 33	03 31	11 56	03 56	16 25	19 76	
21	03 29	03 28	11 57	04 53	17 19	20 66	

	UT	ARIES	VENUS −3.9		MARS −0.5		JUPITER −2.0		SATURN +0.1		STARS		
		GHA	GHA	Dec	GHA	Dec	GHA	Dec	GHA	Dec	Name	SHA	Dec
	d h	° ′	° ′	° ′	° ′	° ′	° ′	° ′	° ′	° ′		° ′	° ′
	22 00	239 12.6	204 17.7	N12 17.5	278 46.9	S17 50.0	105 07.8	N18 09.2	150 57.5	N22 34.3	Acamar	315 24.7	S40 17.5
	01	254 15.0	219 17.3	18.6	293 48.0	49.7	120 10.0	09.1	165 59.6	34.3	Achernar	335 32.9	S57 13.1
	02	269 17.5	234 16.8	19.6	308 49.1	49.4	135 12.1	09.0	181 01.8	34.3	Acrux	173 18.1	S63 07.3
	03	284 20.0	249 16.4	. . 20.6	323 50.2	. . 49.1	150 14.2	. . 08.9	196 03.9	. . 34.3	Adhara	255 19.1	S28 58.7
	04	299 22.4	264 15.9	21.6	338 51.3	48.8	165 16.4	08.9	211 06.0	34.3	Aldebaran	290 58.9	N16 30.9
	05	314 24.9	279 15.5	22.7	353 52.3	48.4	180 18.5	08.8	226 08.2	34.4			
	06	329 27.4	294 15.0	N12 23.7	8 53.4	S17 48.1	195 20.6	N18 08.7	241 10.3	N22 34.4	Alioth	166 27.0	N55 56.7
	07	344 29.8	309 14.5	24.7	23 54.5	47.8	210 22.8	08.6	256 12.5	34.4	Alkaid	153 04.6	N49 18.0
T	08	359 32.3	324 14.1	25.7	38 55.6	47.5	225 24.9	08.5	271 14.6	34.4	Al Na'ir	27 53.5	S46 56.6
H	09	14 34.8	339 13.6	. . 26.8	53 56.7	. . 47.2	240 27.0	. . 08.4	286 16.7	. . 34.4	Alnilam	275 54.8	S 1 12.0
U	10	29 37.2	354 13.2	27.8	68 57.8	46.9	255 29.2	08.3	301 18.9	34.4	Alphard	218 04.0	S 8 40.4
R	11	44 39.7	9 12.7	28.8	83 58.8	46.6	270 31.3	08.2	316 21.0	34.4			
S	12	59 42.2	24 12.2	N12 29.8	98 59.9	S17 46.2	285 33.4	N18 08.1	331 23.1	N22 34.4	Alphecca	126 17.3	N26 42.2
D	13	74 44.6	39 11.8	30.9	114 01.0	45.9	300 35.6	08.0	346 25.3	34.4	Alpheratz	357 52.0	N29 06.3
A	14	89 47.1	54 11.3	31.9	129 02.1	45.6	315 37.7	07.9	1 27.4	34.4	Altair	62 15.8	N 8 52.4
Y	15	104 49.5	69 10.9	. . 32.9	144 03.2	. . 45.3	330 39.8	. . 07.8	16 29.6	. . 34.4	Ankaa	353 23.6	S42 17.2
	16	119 52.0	84 10.4	33.9	159 04.3	45.0	345 42.0	07.7	31 31.7	34.4	Antares	112 35.7	S26 26.4
	17	134 54.5	99 09.9	35.0	174 05.4	44.7	0 44.1	07.7	46 33.8	34.5			
	18	149 56.9	114 09.5	N12 36.0	189 06.5	S17 44.4	15 46.3	N18 07.6	61 36.0	N22 34.5	Arcturus	146 02.6	N19 09.9
	19	164 59.4	129 09.0	37.0	204 07.5	44.0	30 48.4	07.5	76 38.1	34.5	Atria	107 44.2	S69 02.0
	20	180 01.9	144 08.5	38.0	219 08.6	43.7	45 50.5	07.4	91 40.2	34.5	Avior	234 21.7	S59 31.4
	21	195 04.3	159 08.1	. . 39.0	234 09.7	. . 43.4	60 52.6	. . 07.3	106 42.4	. . 34.5	Bellatrix	278 40.9	N 6 21.2
	22	210 06.8	174 07.6	40.1	249 10.8	43.1	75 54.8	07.2	121 44.5	34.5	Betelgeuse	271 10.2	N 7 24.5
	23	225 09.3	189 07.1	41.1	264 11.9	42.8	90 56.9	07.1	136 46.7	34.5			
	23 00	240 11.7	204 06.7	N12 42.1	279 13.0	S17 42.5	105 59.0	N18 07.0	151 48.8	N22 34.5	Canopus	264 00.2	S52 42.0
	01	255 14.2	219 06.2	43.1	294 14.1	42.2	121 01.2	06.9	166 50.9	34.5	Capella	280 46.7	N46 00.1
	02	270 16.6	234 05.7	44.1	309 15.2	41.8	136 03.3	06.8	181 53.1	34.5	Deneb	49 36.8	N45 17.2
	03	285 19.1	249 05.3	. . 45.2	324 16.3	. . 41.5	151 05.4	. . 06.7	196 55.2	. . 34.5	Denebola	182 41.6	N14 33.3
	04	300 21.6	264 04.8	46.2	339 17.4	41.2	166 07.6	06.6	211 57.3	34.5	Diphda	349 04.0	S17 58.1
	05	315 24.0	279 04.3	47.2	354 18.5	40.9	181 09.7	06.5	226 59.5	34.6			
	06	330 26.5	294 03.9	N12 48.2	9 19.5	S17 40.6	196 11.8	N18 06.4	242 01.6	N22 34.6	Dubhe	194 01.0	N61 44.3
	07	345 29.0	309 03.4	49.2	24 20.6	40.3	211 14.0	06.3	257 03.8	34.6	Elnath	278 23.1	N28 36.7
	08	0 31.4	324 02.9	50.2	39 21.7	40.0	226 16.1	06.3	272 05.9	34.6	Eltanin	90 49.4	N51 29.1
F	09	15 33.9	339 02.5	. . 51.3	54 22.8	. . 39.7	241 18.2	. . 06.2	287 08.0	. . 34.6	Enif	33 54.9	N 9 53.2
R	10	30 36.4	354 02.0	52.3	69 23.9	39.3	256 20.4	06.1	302 10.2	34.6	Fomalhaut	15 32.7	S29 36.3
I	11	45 38.8	9 01.5	53.3	84 25.0	39.0	271 22.5	06.0	317 12.3	34.6			
D	12	60 41.3	24 01.1	N12 54.3	99 26.1	S17 38.7	286 24.6	N18 05.9	332 14.4	N22 34.6	Gacrux	172 09.7	S57 08.1
A	13	75 43.8	39 00.6	55.3	114 27.2	38.4	301 26.7	05.8	347 16.6	34.6	Gienah	176 00.4	S17 33.7
Y	14	90 46.2	54 00.1	56.3	129 28.3	38.1	316 28.9	05.7	2 18.7	34.6	Hadar	148 58.9	S60 23.5
	15	105 48.7	68 59.6	. . 57.3	144 29.4	. . 37.8	331 31.0	. . 05.6	17 20.8	. . 34.6	Hamal	328 10.1	N23 28.5
	16	120 51.1	83 59.2	58.3	159 30.5	37.5	346 33.1	05.5	32 23.0	34.6	Kaus Aust.	83 54.0	S34 23.0
	17	135 53.6	98 58.7	12 59.3	174 31.6	37.1	1 35.3	05.4	47 25.1	34.7			
	18	150 56.1	113 58.2	N13 00.4	189 32.7	S17 36.8	16 37.4	N18 05.3	62 27.3	N22 34.7	Kochab	137 18.2	N74 08.6
	19	165 58.5	128 57.8	01.4	204 33.8	36.5	31 39.5	05.2	77 29.4	34.7	Markab	13 46.4	N15 13.1
	20	181 01.0	143 57.3	02.4	219 34.9	36.2	46 41.6	05.1	92 31.5	34.7	Menkar	314 23.7	N 4 06.1
	21	196 03.5	158 56.8	. . 03.4	234 36.0	. . 35.9	61 43.8	. . 05.0	107 33.7	. . 34.7	Menkent	148 16.7	S36 23.3
	22	211 05.9	173 56.3	04.4	249 37.1	35.6	76 45.9	04.9	122 35.8	34.7	Miaplacidus	221 41.9	S69 44.1
	23	226 08.4	188 55.9	05.4	264 38.2	35.3	91 48.0	04.8	137 37.9	34.7			
	24 00	241 10.9	203 55.4	N13 06.4	279 39.3	S17 34.9	106 50.2	N18 04.7	152 40.1	N22 34.7	Mirfak	308 52.4	N49 52.3
	01	256 13.3	218 54.9	07.4	294 40.4	34.6	121 52.3	04.7	167 42.2	34.7	Nunki	76 07.9	S26 17.6
	02	271 15.8	233 54.4	08.4	309 41.5	34.3	136 54.4	04.6	182 44.3	34.7	Peacock	53 31.3	S56 43.3
	03	286 18.3	248 53.9	. . 09.4	324 42.6	. . 34.0	151 56.5	. . 04.5	197 46.5	. . 34.7	Pollux	243 37.7	N28 01.2
	04	301 20.7	263 53.5	10.4	339 43.7	33.7	166 58.7	04.4	212 48.6	34.7	Procyon	245 08.3	N 5 13.0
	05	316 23.2	278 53.0	11.4	354 44.8	33.4	182 00.8	04.3	227 50.7	34.8			
	06	331 25.6	293 52.5	N13 12.4	9 45.9	S17 33.1	197 02.9	N18 04.2	242 52.9	N22 34.8	Rasalhague	96 13.5	N12 33.3
	07	346 28.1	308 52.0	13.4	24 47.0	32.8	212 05.0	04.1	257 55.0	34.8	Regulus	207 52.0	N11 57.2
S	08	1 30.6	323 51.6	14.4	39 48.1	32.4	227 07.2	04.0	272 57.2	34.8	Rigel	281 20.0	S 8 11.9
A	09	16 33.0	338 51.1	. . 15.4	54 49.2	. . 32.1	242 09.3	. . 03.9	287 59.3	. . 34.8	Rigil Kent.	140 02.3	S60 51.0
T	10	31 35.5	353 50.6	16.4	69 50.3	31.8	257 11.4	03.8	303 01.4	34.8	Sabik	102 21.3	S15 43.8
U	11	46 38.0	8 50.1	17.4	84 51.4	31.5	272 13.5	03.7	318 03.6	34.8			
R	12	61 40.4	23 49.6	N13 18.4	99 52.5	S17 31.2	287 15.7	N18 03.6	333 05.7	N22 34.8	Schedar	349 50.2	N56 33.0
D	13	76 42.9	38 49.2	19.4	114 53.6	30.9	302 17.8	03.5	348 07.8	34.8	Shaula	96 32.3	S37 06.4
A	14	91 45.4	53 48.7	20.4	129 54.7	30.6	317 19.9	03.4	3 10.0	34.8	Sirius	258 41.1	S16 43.3
Y	15	106 47.8	68 48.2	. . 21.4	144 55.8	. . 30.2	332 22.0	. . 03.3	18 12.1	. . 34.8	Spica	158 39.4	S11 10.8
	16	121 50.3	83 47.7	22.4	159 57.0	29.9	347 24.2	03.2	33 14.2	34.8	Suhail	222 58.5	S43 26.9
	17	136 52.7	98 47.2	23.4	174 58.1	29.6	2 26.3	03.1	48 16.4	34.9			
	18	151 55.2	113 46.7	N13 24.4	189 59.2	S17 29.3	17 28.4	N18 03.0	63 18.5	N22 34.9	Vega	80 44.0	N38 47.0
	19	166 57.7	128 46.3	25.4	205 00.3	29.0	32 30.5	02.9	78 20.6	34.9	Zuben'ubi	137 13.9	S16 03.4
	20	182 00.1	143 45.8	26.4	220 01.4	28.7	47 32.7	02.8	93 22.8	34.9			
	21	197 02.6	158 45.3	. . 27.4	235 02.5	. . 28.4	62 34.8	. . 02.7	108 24.9	. . 34.9			
	22	212 05.1	173 44.8	28.4	250 03.6	28.0	77 36.9	02.6	123 27.0	34.9			
	23	227 07.5	188 44.3	29.4	265 04.7	27.7	92 39.0	02.5	138 29.2	34.9			
	Mer. Pass.	h m 7 57.9	*v* −0.5	*d* 1.0	*v* 1.1	*d* 0.3	*v* 2.1	*d* 0.1	*v* 2.1	*d* 0.0			

	SHA	Mer. Pass.
	° ′	h m
Venus	323 55.0	10 24
Mars	39 01.3	5 23
Jupiter	225 47.3	16 54
Saturn	271 37.1	13 51

	UT	SUN GHA	SUN Dec	MOON GHA	v	MOON Dec	d	HP
	d h	° ′	° ′	° ′	′	° ′	′	′
	22 00	180 51.4	N20 15.5	276 42.8	10.4	S20 15.8	9.8	56.9
	01	195 51.4	16.0	291 12.2	10.6	20 06.0	9.8	56.8
	02	210 51.3	16.5	305 41.8	10.6	19 56.2	10.0	56.8
	03	225 51.3	. . 17.0	320 11.4	10.8	19 46.2	10.1	56.8
	04	240 51.2	17.5	334 41.2	10.9	19 36.1	10.1	56.7
	05	255 51.2	18.0	349 11.1	11.0	19 26.0	10.3	56.7
	06	270 51.1	N20 18.5	3 41.1	11.1	S19 15.7	10.3	56.6
	07	285 51.1	19.0	18 11.2	11.2	19 05.4	10.4	56.6
T	08	300 51.0	19.5	32 41.4	11.4	18 55.0	10.5	56.6
H	09	315 51.0	. . 20.0	47 11.8	11.4	18 44.5	10.5	56.5
U	10	330 51.0	20.5	61 42.2	11.5	18 34.0	10.6	56.5
R	11	345 50.9	21.0	76 12.7	11.7	18 23.4	10.7	56.5
S	12	0 50.9	N20 21.5	90 43.4	11.7	S18 12.7	10.8	56.4
D	13	15 50.8	22.0	105 14.1	11.9	18 01.9	10.9	56.4
A	14	30 50.8	22.5	119 45.0	11.9	17 51.0	10.9	56.4
Y	15	45 50.7	. . 23.0	134 15.9	12.1	17 40.1	11.0	56.3
	16	60 50.7	23.5	148 47.0	12.2	17 29.1	11.0	56.3
	17	75 50.6	24.0	163 18.2	12.2	17 18.1	11.2	56.3
	18	90 50.6	N20 24.5	177 49.4	12.4	S17 06.9	11.2	56.2
	19	105 50.5	24.9	192 20.8	12.4	16 55.7	11.2	56.2
	20	120 50.5	25.4	206 52.2	12.6	16 44.5	11.3	56.2
	21	135 50.4	. . 25.9	221 23.8	12.6	16 33.2	11.4	56.1
	22	150 50.4	26.4	235 55.4	12.8	16 21.8	11.5	56.1
	23	165 50.4	26.9	250 27.2	12.8	16 10.3	11.5	56.1
	23 00	180 50.3	N20 27.4	264 59.0	13.0	S15 58.8	11.5	56.0
	01	195 50.3	27.9	279 31.0	13.0	15 47.3	11.7	56.0
	02	210 50.2	28.4	294 03.0	13.1	15 35.6	11.6	56.0
	03	225 50.2	. . 28.9	308 35.1	13.2	15 24.0	11.8	55.9
	04	240 50.1	29.3	323 07.3	13.3	15 12.2	11.8	55.9
	05	255 50.1	29.8	337 39.6	13.4	15 00.4	11.8	55.9
	06	270 50.0	N20 30.3	352 12.0	13.5	S14 48.6	11.9	55.8
	07	285 50.0	30.8	6 44.5	13.6	14 36.7	11.9	55.8
	08	300 49.9	31.3	21 17.1	13.6	14 24.8	12.0	55.8
F	09	315 49.9	. . 31.8	35 49.7	13.8	14 12.8	12.1	55.8
R	10	330 49.8	32.2	50 22.5	13.8	14 00.7	12.1	55.7
I	11	345 49.8	32.7	64 55.3	13.9	13 48.6	12.1	55.7
D	12	0 49.7	N20 33.2	79 28.2	14.0	S13 36.5	12.2	55.7
A	13	15 49.7	33.7	94 01.2	14.0	13 24.3	12.2	55.6
Y	14	30 49.6	34.2	108 34.2	14.2	13 12.1	12.3	55.6
	15	45 49.6	. . 34.6	123 07.4	14.2	12 59.8	12.3	55.6
	16	60 49.5	35.1	137 40.6	14.3	12 47.5	12.3	55.6
	17	75 49.4	35.6	152 13.9	14.3	12 35.2	12.4	55.5
	18	90 49.4	N20 36.1	166 47.2	14.5	S12 22.8	12.5	55.5
	19	105 49.3	36.5	181 20.7	14.5	12 10.3	12.4	55.5
	20	120 49.3	37.0	195 54.2	14.6	11 57.9	12.5	55.4
	21	135 49.2	. . 37.5	210 27.8	14.7	11 45.4	12.6	55.4
	22	150 49.2	38.0	225 01.5	14.7	11 32.8	12.6	55.4
	23	165 49.1	38.4	239 35.2	14.8	11 20.2	12.6	55.4
	24 00	180 49.1	N20 38.9	254 09.0	14.9	S11 07.6	12.6	55.3
	01	195 49.0	39.4	268 42.9	14.9	10 55.0	12.7	55.3
	02	210 49.0	39.9	283 16.8	15.0	10 42.3	12.7	55.3
	03	225 48.9	. . 40.3	297 50.8	15.0	10 29.6	12.8	55.3
	04	240 48.9	40.8	312 24.8	15.2	10 16.8	12.7	55.2
	05	255 48.8	41.3	326 59.0	15.2	10 04.1	12.8	55.2
	06	270 48.8	N20 41.7	341 33.2	15.2	S 9 51.3	12.9	55.2
S	07	285 48.7	42.2	356 07.4	15.3	9 38.4	12.8	55.2
	08	300 48.6	42.7	10 41.7	15.4	9 25.6	12.9	55.1
A	09	315 48.6	. . 43.1	25 16.1	15.4	9 12.7	12.9	55.1
T	10	330 48.5	43.6	39 50.5	15.5	8 59.8	13.0	55.1
U	11	345 48.5	44.1	54 25.0	15.5	8 46.8	12.9	55.1
R	12	0 48.4	N20 44.5	68 59.5	15.6	S 8 33.9	13.0	55.0
D	13	15 48.4	45.0	83 34.1	15.6	8 20.9	13.0	55.0
A	14	30 48.3	45.5	98 08.7	15.7	8 07.9	13.0	55.0
Y	15	45 48.2	. . 45.9	112 43.4	15.8	7 54.9	13.1	55.0
	16	60 48.2	46.4	127 18.2	15.8	7 41.8	13.0	55.0
	17	75 48.1	46.9	141 53.0	15.8	7 28.8	13.1	54.9
	18	90 48.1	N20 47.3	156 27.8	15.9	S 7 15.7	13.1	54.9
	19	105 48.0	47.8	171 02.7	15.9	7 02.6	13.1	54.9
	20	120 48.0	48.2	185 37.6	16.0	6 49.5	13.2	54.9
	21	135 47.9	. . 48.7	200 12.6	16.0	6 36.3	13.1	54.8
	22	150 47.8	49.2	214 47.6	16.1	6 23.2	13.2	54.8
	23	165 47.8	49.6	229 22.7	16.1	S 6 10.0	13.2	54.8
		SD 15.8	*d* 0.5	SD	15.4		15.2	15.0

Lat.	Twilight Naut.	Twilight Civil	Sunrise	Moonrise 22	Moonrise 23	Moonrise 24	Moonrise 25
°	h m	h m	h m	h m	h m	h m	h m
N 72	▭	▭	▭	■	04 16	03 29	02 57
N 70	▭	▭	▭	04 53	03 43	03 12	02 49
68	////	////	00 57	03 52	03 19	02 59	02 43
66	////	////	01 52	03 16	03 00	02 48	02 38
64	////	////	02 24	02 51	02 45	02 39	02 34
62	////	01 17	02 47	02 31	02 32	02 31	02 30
60	////	01 55	03 06	02 14	02 21	02 24	02 27
N 58	////	02 21	03 21	02 00	02 11	02 18	02 24
56	01 09	02 41	03 34	01 48	02 02	02 13	02 21
54	01 44	02 58	03 46	01 38	01 55	02 08	02 19
52	02 09	03 12	03 56	01 28	01 48	02 04	02 17
50	02 28	03 24	04 05	01 20	01 42	02 00	02 15
45	03 03	03 48	04 23	01 02	01 29	01 51	02 10
N 40	03 28	04 07	04 39	00 47	01 18	01 44	02 07
35	03 48	04 23	04 51	00 35	01 08	01 37	02 03
30	04 04	04 36	05 03	00 24	01 00	01 32	02 01
20	04 29	04 58	05 22	00 05	00 46	01 22	01 56
N 10	04 49	05 15	05 38	24 33	00 33	01 14	01 52
0	05 05	05 31	05 53	24 21	00 21	01 06	01 48
S 10	05 20	05 46	06 08	24 09	00 09	00 58	01 43
20	05 34	06 01	06 24	23 57	24 49	00 49	01 39
30	05 47	06 17	06 42	23 42	24 39	00 39	01 34
35	05 55	06 26	06 53	23 33	24 34	00 34	01 31
40	06 02	06 35	07 05	23 24	24 27	00 27	01 28
45	06 10	06 47	07 19	23 12	24 19	00 19	01 24
S 50	06 20	07 00	07 36	22 58	24 10	00 10	01 20
52	06 24	07 06	07 45	22 51	24 06	00 06	01 18
54	06 28	07 13	07 54	22 44	24 01	00 01	01 15
56	06 33	07 20	08 04	22 36	23 56	25 13	01 13
58	06 38	07 28	08 16	22 27	23 50	25 10	01 10
S 60	06 44	07 37	08 29	22 16	23 44	25 07	01 07

Lat.	Sunset	Twilight Civil	Twilight Naut.	Moonset 22	Moonset 23	Moonset 24	Moonset 25
°	h m	h m	h m	h m	h m	h m	h m
N 72	▭	▭	▭	■	09 11	11 30	13 28
N 70	▭	▭	▭	06 53	09 42	11 44	13 33
68	23 04	////	////	07 54	10 05	11 55	13 36
66	22 05	////	////	08 28	10 22	12 04	13 39
64	21 32	////	////	08 53	10 37	12 12	13 42
62	21 08	22 41	////	09 12	10 48	12 18	13 44
60	20 49	22 01	////	09 28	10 58	12 24	13 46
N 58	20 33	21 34	////	09 41	11 07	12 29	13 47
56	20 20	21 14	22 48	09 52	11 15	12 33	13 49
54	20 09	20 57	22 11	10 02	11 21	12 37	13 50
52	19 59	20 43	21 46	10 11	11 28	12 41	13 52
50	19 49	20 31	21 27	10 19	11 33	12 44	13 53
45	19 31	20 06	20 51	10 36	11 45	12 51	13 55
N 40	19 15	19 47	20 26	10 49	11 54	12 57	13 57
35	19 02	19 31	20 06	11 01	12 03	13 02	13 59
30	18 51	19 18	19 50	11 11	12 10	13 06	14 00
20	18 32	18 56	19 24	11 28	12 22	13 14	14 03
N 10	18 16	18 38	19 05	11 43	12 33	13 20	14 05
0	18 00	18 22	18 48	11 56	12 43	13 26	14 07
S 10	17 45	18 08	18 33	12 10	12 53	13 32	14 09
20	17 29	17 53	18 20	12 25	13 03	13 38	14 11
30	17 11	17 36	18 06	12 41	13 15	13 46	14 14
35	17 00	17 28	17 59	12 51	13 22	13 50	14 15
40	16 48	17 18	17 51	13 02	13 30	13 54	14 16
45	16 34	17 06	17 43	13 15	13 39	14 00	14 18
S 50	16 16	16 53	17 33	13 30	13 50	14 06	14 20
52	16 08	16 47	17 29	13 37	13 55	14 09	14 21
54	15 59	16 40	17 25	13 45	14 00	14 12	14 22
56	15 49	16 33	17 20	13 54	14 06	14 16	14 24
58	15 37	16 25	17 15	14 04	14 13	14 20	14 25
S 60	15 24	16 16	17 09	14 16	14 21	14 24	14 26

Day	SUN Eqn. of Time 00^h	SUN Eqn. of Time 12^h	SUN Mer. Pass.	MOON Mer. Pass. Upper	MOON Mer. Pass. Lower	MOON Age	Phase
d	m s	m s	h m	h m	h m	d %	
22	03 26	03 24	11 57	05 45	18 09	21 55	
23	03 21	03 19	11 57	06 32	18 54	22 45	
24	03 16	03 14	11 57	07 16	19 37	23 36	

UT		ARIES	VENUS −3.9		MARS −0.5		JUPITER −1.9		SATURN +0.1		STARS		
		GHA	GHA	Dec	GHA	Dec	GHA	Dec	GHA	Dec	Name	SHA	Dec
d	h	° ′	° ′	° ′	° ′	° ′	° ′	° ′	° ′	° ′		° ′	° ′
25	00	242 10.0	203 43.8	N13 30.4	280 05.8	S17 27.4	107 41.2	N18 02.4	153 31.3	N22 34.9	Acamar	315 24.7	S40 17.5
	01	257 12.5	218 43.4	31.4	295 06.9	27.1	122 43.3	02.4	168 33.4	34.9	Achernar	335 32.9	S57 13.1
	02	272 14.9	233 42.9	32.4	310 08.0	26.8	137 45.4	02.3	183 35.6	34.9	Acrux	173 18.1	S63 07.3
	03	287 17.4	248 42.4	. . 33.4	325 09.2	. . 26.5	152 47.5	. . 02.2	198 37.7	. . 34.9	Adhara	255 19.1	S28 58.7
	04	302 19.9	263 41.9	34.4	340 10.3	26.2	167 49.6	02.1	213 39.8	34.9	Aldebaran	290 58.9	N16 30.9
	05	317 22.3	278 41.4	35.4	355 11.4	25.9	182 51.8	02.0	228 42.0	34.9			
	06	332 24.8	293 40.9	N13 36.4	10 12.5	S17 25.5	197 53.9	N18 01.9	243 44.1	N22 35.0	Alioth	166 27.1	N55 56.8
	07	347 27.2	308 40.4	37.4	25 13.6	25.2	212 56.0	01.8	258 46.2	35.0	Alkaid	153 04.6	N49 18.0
	08	2 29.7	323 39.9	38.3	40 14.7	24.9	227 58.1	01.7	273 48.4	35.0	Al Na'ir	27 53.5	S46 56.6
S	09	17 32.2	338 39.5	. . 39.3	55 15.8	. . 24.6	243 00.2	. . 01.6	288 50.5	. . 35.0	Alnilam	275 54.8	S 1 12.0
U	10	32 34.6	353 39.0	40.3	70 16.9	24.3	258 02.4	01.5	303 52.6	35.0	Alphard	218 04.0	S 8 40.4
N	11	47 37.1	8 38.5	41.3	85 18.1	24.0	273 04.5	01.4	318 54.8	35.0			
D	12	62 39.6	23 38.0	N13 42.3	100 19.2	S17 23.7	288 06.6	N18 01.3	333 56.9	N22 35.0	Alphecca	126 17.3	N26 42.2
A	13	77 42.0	38 37.5	43.3	115 20.3	23.4	303 08.7	01.2	348 59.0	35.0	Alpheratz	357 52.0	N29 06.3
Y	14	92 44.5	53 37.0	44.3	130 21.4	23.0	318 10.8	01.1	4 01.2	35.0	Altair	62 15.8	N 8 52.4
	15	107 47.0	68 36.5	. . 45.3	145 22.5	. . 22.7	333 13.0	. . 01.0	19 03.3	. . 35.0	Ankaa	353 23.6	S42 17.2
	16	122 49.4	83 36.0	46.2	160 23.6	22.4	348 15.1	00.9	34 05.4	35.0	Antares	112 35.7	S26 26.4
	17	137 51.9	98 35.5	47.2	175 24.8	22.1	3 17.2	00.8	49 07.6	35.0			
	18	152 54.3	113 35.0	N13 48.2	190 25.9	S17 21.8	18 19.3	N18 00.7	64 09.7	N22 35.0	Arcturus	146 02.7	N19 09.9
	19	167 56.8	128 34.5	49.2	205 27.0	21.5	33 21.4	00.6	79 11.8	35.1	Atria	107 44.2	S69 02.0
	20	182 59.3	143 34.0	50.2	220 28.1	21.2	48 23.6	00.5	94 14.0	35.1	Avior	234 21.8	S59 31.4
	21	198 01.7	158 33.5	. . 51.2	235 29.2	. . 20.9	63 25.7	. . 00.4	109 16.1	. . 35.1	Bellatrix	278 40.9	N 6 21.2
	22	213 04.2	173 33.1	52.1	250 30.3	20.5	78 27.8	00.3	124 18.2	35.1	Betelgeuse	271 10.2	N 7 24.5
	23	228 06.7	188 32.6	53.1	265 31.5	20.2	93 29.9	00.2	139 20.4	35.1			
26	00	243 09.1	203 32.1	N13 54.1	280 32.6	S17 19.9	108 32.0	N18 00.1	154 22.5	N22 35.1	Canopus	264 00.2	S52 41.9
	01	258 11.6	218 31.6	55.1	295 33.7	19.6	123 34.1	18 00.0	169 24.6	35.1	Capella	280 46.7	N46 00.1
	02	273 14.1	233 31.1	56.1	310 34.8	19.3	138 36.3	17 59.9	184 26.8	35.1	Deneb	49 36.8	N45 17.2
	03	288 16.5	248 30.6	. . 57.0	325 35.9	. . 19.0	153 38.4	. . 59.8	199 28.9	. . 35.1	Denebola	182 41.6	N14 33.3
	04	303 19.0	263 30.1	58.0	340 37.1	18.7	168 40.5	59.7	214 31.0	35.1	Diphda	349 04.0	S17 58.1
	05	318 21.5	278 29.6	13 59.0	355 38.2	18.4	183 42.6	59.6	229 33.2	35.1			
	06	333 23.9	293 29.1	N14 00.0	10 39.3	S17 18.0	198 44.7	N17 59.5	244 35.3	N22 35.1	Dubhe	194 01.0	N61 44.3
	07	348 26.4	308 28.6	01.0	25 40.4	17.7	213 46.8	59.4	259 37.4	35.1	Elnath	278 23.1	N28 36.7
	08	3 28.8	323 28.1	01.9	40 41.6	17.4	228 49.0	59.3	274 39.6	35.1	Eltanin	90 49.3	N51 29.1
M	09	18 31.3	338 27.6	. . 02.9	55 42.7	. . 17.1	243 51.1	. . 59.2	289 41.7	. . 35.2	Enif	33 54.9	N 9 53.2
O	10	33 33.8	353 27.1	03.9	70 43.8	16.8	258 53.2	59.1	304 43.8	35.2	Fomalhaut	15 32.7	S29 36.3
N	11	48 36.2	8 26.6	04.9	85 44.9	16.5	273 55.3	59.0	319 46.0	35.2			
D	12	63 38.7	23 26.1	N14 05.8	100 46.1	S17 16.2	288 57.4	N17 58.9	334 48.1	N22 35.2	Gacrux	172 09.7	S57 08.1
A	13	78 41.2	38 25.6	06.8	115 47.2	15.9	303 59.5	58.8	349 50.2	35.2	Gienah	176 00.4	S17 33.7
Y	14	93 43.6	53 25.1	07.8	130 48.3	15.5	319 01.6	58.7	4 52.4	35.2	Hadar	148 58.9	S60 23.5
	15	108 46.1	68 24.6	. . 08.8	145 49.4	. . 15.2	334 03.8	. . 58.6	19 54.5	. . 35.2	Hamal	328 10.1	N23 28.5
	16	123 48.6	83 24.1	09.7	160 50.6	14.9	349 05.9	58.5	34 56.6	35.2	Kaus Aust.	83 54.0	S34 23.0
	17	138 51.0	98 23.6	10.7	175 51.7	14.6	4 08.0	58.4	49 58.8	35.2			
	18	153 53.5	113 23.1	N14 11.7	190 52.8	S17 14.3	19 10.1	N17 58.3	65 00.9	N22 35.2	Kochab	137 18.2	N74 08.6
	19	168 56.0	128 22.6	12.6	205 53.9	14.0	34 12.2	58.2	80 03.0	35.2	Markab	13 46.4	N15 13.2
	20	183 58.4	143 22.1	13.6	220 55.1	13.7	49 14.3	58.1	95 05.1	35.2	Menkar	314 23.7	N 4 06.1
	21	199 00.9	158 21.6	. . 14.6	235 56.2	. . 13.4	64 16.4	. . 58.0	110 07.3	. . 35.2	Menkent	148 16.7	S36 23.3
	22	214 03.3	173 21.0	15.5	250 57.3	13.0	79 18.6	57.9	125 09.4	35.3	Miaplacidus	221 42.0	S69 44.1
	23	229 05.8	188 20.5	16.5	265 58.5	12.7	94 20.7	57.8	140 11.5	35.3			
27	00	244 08.3	203 20.0	N14 17.5	280 59.6	S17 12.4	109 22.8	N17 57.7	155 13.7	N22 35.3	Mirfak	308 52.4	N49 52.3
	01	259 10.7	218 19.5	18.5	296 00.7	12.1	124 24.9	57.6	170 15.8	35.3	Nunki	76 07.8	S26 17.6
	02	274 13.2	233 19.0	19.4	311 01.8	11.8	139 27.0	57.5	185 17.9	35.3	Peacock	53 31.3	S56 43.3
	03	289 15.7	248 18.5	. . 20.4	326 03.0	. . 11.5	154 29.1	. . 57.4	200 20.1	. . 35.3	Pollux	243 37.7	N28 01.2
	04	304 18.1	263 18.0	21.3	341 04.1	11.2	169 31.2	57.3	215 22.2	35.3	Procyon	245 08.3	N 5 13.0
	05	319 20.6	278 17.5	22.3	356 05.2	10.9	184 33.3	57.2	230 24.3	35.3			
	06	334 23.1	293 17.0	N14 23.3	11 06.4	S17 10.6	199 35.5	N17 57.1	245 26.5	N22 35.3	Rasalhague	96 13.5	N12 33.3
	07	349 25.5	308 16.5	24.2	26 07.5	10.2	214 37.6	57.0	260 28.6	35.3	Regulus	207 52.0	N11 57.2
T	08	4 28.0	323 16.0	25.2	41 08.6	09.9	229 39.7	56.9	275 30.7	35.3	Rigel	281 20.0	S 8 11.9
U	09	19 30.4	338 15.5	. . 26.2	56 09.8	. . 09.6	244 41.8	. . 56.8	290 32.9	. . 35.3	Rigil Kent.	140 02.3	S60 51.1
E	10	34 32.9	353 14.9	27.1	71 10.9	09.3	259 43.9	56.7	305 35.0	35.3	Sabik	102 21.3	S15 43.8
S	11	49 35.4	8 14.4	28.1	86 12.0	09.0	274 46.0	56.6	320 37.1	35.3			
D	12	64 37.8	23 13.9	N14 29.0	101 13.2	S17 08.7	289 48.1	N17 56.5	335 39.2	N22 35.4	Schedar	349 50.2	N56 33.0
A	13	79 40.3	38 13.4	30.0	116 14.3	08.4	304 50.2	56.4	350 41.4	35.4	Shaula	96 32.3	S37 06.4
Y	14	94 42.8	53 12.9	31.0	131 15.4	08.1	319 52.3	56.3	5 43.5	35.4	Sirius	258 41.1	S16 43.3
	15	109 45.2	68 12.4	. . 31.9	146 16.6	. . 07.7	334 54.4	. . 56.2	20 45.6	. . 35.4	Spica	158 39.4	S11 10.8
	16	124 47.7	83 11.9	32.9	161 17.7	07.4	349 56.6	56.1	35 47.8	35.4	Suhail	222 58.6	S43 26.9
	17	139 50.2	98 11.4	33.8	176 18.9	07.1	4 58.7	56.0	50 49.9	35.4			
	18	154 52.6	113 10.8	N14 34.8	191 20.0	S17 06.8	20 00.8	N17 55.9	65 52.0	N22 35.4	Vega	80 44.0	N38 47.0
	19	169 55.1	128 10.3	35.8	206 21.1	06.5	35 02.9	55.8	80 54.2	35.4	Zuben'ubi	137 14.0	S16 03.4
	20	184 57.6	143 09.8	36.7	221 22.3	06.2	50 05.0	55.7	95 56.3	35.4		SHA	Mer. Pass.
	21	200 00.0	158 09.3	. . 37.7	236 23.4	. . 05.9	65 07.1	. . 55.6	110 58.4	. . 35.4		° ′	h m
	22	215 02.5	173 08.8	38.6	251 24.6	05.6	80 09.2	55.5	126 00.5	35.4	Venus	320 22.9	10 26
	23	230 04.9	188 08.3	39.6	266 25.7	05.3	95 11.3	55.4	141 02.7	35.4	Mars	37 23.4	5 17
Mer. Pass.		h m 7 46.1	v −0.5	d 1.0	v 1.1	d 0.3	v 2.1	d 0.1	v 2.1	d 0.0	Jupiter	225 22.9	16 44
											Saturn	271 13.4	13 41

UT	SUN GHA	SUN Dec	MOON GHA	v	MOON Dec	d	HP
d h	° ′	° ′	° ′	′	° ′	′	′
SUNDAY 25 00	180 47.7	N20 50.1	243 57.8	16.1	S 5 56.8	13.2	54.8
01	195 47.7	50.5	258 32.9	16.2	5 43.6	13.2	54.8
02	210 47.6	51.0	273 08.1	16.2	5 30.4	13.2	54.7
03	225 47.5	. . 51.5	287 43.3	16.3	5 17.2	13.2	54.7
04	240 47.5	51.9	302 18.6	16.3	5 04.0	13.3	54.7
05	255 47.4	52.4	316 53.9	16.3	4 50.7	13.2	54.7
06	270 47.4	N20 52.8	331 29.2	16.4	S 4 37.5	13.3	54.7
07	285 47.3	53.3	346 04.6	16.4	4 24.2	13.3	54.7
08	300 47.2	53.7	0 40.0	16.4	4 10.9	13.3	54.6
09	315 47.2	. . 54.2	15 15.4	16.4	3 57.6	13.2	54.6
10	330 47.1	54.6	29 50.8	16.5	3 44.4	13.3	54.6
11	345 47.1	55.1	44 26.3	16.5	3 31.1	13.3	54.6
12	0 47.0	N20 55.5	59 01.8	16.6	S 3 17.8	13.3	54.6
13	15 46.9	56.0	73 37.4	16.5	3 04.5	13.3	54.5
14	30 46.9	56.4	88 12.9	16.6	2 51.2	13.3	54.5
15	45 46.8	. . 56.9	102 48.5	16.6	2 37.9	13.4	54.5
16	60 46.7	57.3	117 24.1	16.6	2 24.5	13.3	54.5
17	75 46.7	57.8	131 59.7	16.7	2 11.2	13.3	54.5
18	90 46.6	N20 58.2	146 35.4	16.7	S 1 57.9	13.3	54.5
19	105 46.6	58.7	161 11.1	16.7	1 44.6	13.3	54.5
20	120 46.5	59.1	175 46.8	16.7	1 31.3	13.3	54.4
21	135 46.4	20 59.6	190 22.5	16.7	1 18.0	13.4	54.4
22	150 46.4	21 00.0	204 58.2	16.7	1 04.6	13.3	54.4
23	165 46.3	00.4	219 33.9	16.8	0 51.3	13.3	54.4
MONDAY 26 00	180 46.2	N21 00.9	234 09.7	16.8	S 0 38.0	13.3	54.4
01	195 46.2	01.3	248 45.5	16.7	0 24.7	13.3	54.4
02	210 46.1	01.8	263 21.2	16.8	S 0 11.4	13.3	54.4
03	225 46.0	. . 02.2	277 57.0	16.8	N 0 01.9	13.3	54.3
04	240 46.0	02.7	292 32.8	16.9	0 15.2	13.3	54.3
05	255 45.9	03.1	307 08.7	16.8	0 28.5	13.2	54.3
06	270 45.8	N21 03.5	321 44.5	16.8	N 0 41.7	13.3	54.3
07	285 45.8	04.0	336 20.3	16.8	0 55.0	13.3	54.3
08	300 45.7	04.4	350 56.1	16.9	1 08.3	13.2	54.3
09	315 45.6	. . 04.9	5 32.0	16.8	1 21.5	13.3	54.3
10	330 45.6	05.3	20 07.8	16.9	1 34.8	13.2	54.3
11	345 45.5	05.7	34 43.7	16.8	1 48.0	13.3	54.2
12	0 45.4	N21 06.2	49 19.5	16.9	N 2 01.3	13.2	54.2
13	15 45.4	06.6	63 55.4	16.8	2 14.5	13.2	54.2
14	30 45.3	07.0	78 31.2	16.9	2 27.7	13.2	54.2
15	45 45.2	. . 07.5	93 07.1	16.8	2 40.9	13.1	54.2
16	60 45.2	07.9	107 42.9	16.9	2 54.0	13.2	54.2
17	75 45.1	08.3	122 18.8	16.8	3 07.2	13.1	54.2
18	90 45.0	N21 08.8	136 54.6	16.8	N 3 20.3	13.2	54.2
19	105 45.0	09.2	151 30.4	16.9	3 33.5	13.1	54.2
20	120 44.9	09.6	166 06.3	16.8	3 46.6	13.1	54.2
21	135 44.8	. . 10.1	180 42.1	16.8	3 59.7	13.1	54.2
22	150 44.8	10.5	195 17.9	16.8	4 12.8	13.0	54.1
23	165 44.7	10.9	209 53.7	16.8	4 25.8	13.1	54.1
TUESDAY 27 00	180 44.6	N21 11.3	224 29.5	16.8	N 4 38.9	13.0	54.1
01	195 44.6	11.8	239 05.3	16.8	4 51.9	13.0	54.1
02	210 44.5	12.2	253 41.1	16.7	5 04.9	13.0	54.1
03	225 44.4	. . 12.6	268 16.8	16.8	5 17.9	13.0	54.1
04	240 44.3	13.0	282 52.6	16.7	5 30.9	12.9	54.1
05	255 44.3	13.5	297 28.3	16.7	5 43.8	12.9	54.1
06	270 44.2	N21 13.9	312 04.0	16.7	N 5 56.7	12.9	54.1
07	285 44.1	14.3	326 39.7	16.7	6 09.6	12.9	54.1
08	300 44.1	14.7	341 15.4	16.7	6 22.5	12.8	54.1
09	315 44.0	. . 15.2	355 51.1	16.6	6 35.3	12.9	54.1
10	330 43.9	15.6	10 26.7	16.6	6 48.2	12.8	54.1
11	345 43.8	16.0	25 02.3	16.6	7 01.0	12.7	54.1
12	0 43.8	N21 16.4	39 37.9	16.6	N 7 13.7	12.8	54.0
13	15 43.7	16.8	54 13.5	16.5	7 26.5	12.7	54.0
14	30 43.6	17.3	68 49.0	16.6	7 39.2	12.7	54.0
15	45 43.6	. . 17.7	83 24.6	16.5	7 51.9	12.6	54.0
16	60 43.5	18.1	98 00.1	16.4	8 04.5	12.6	54.0
17	75 43.4	18.5	112 35.5	16.5	8 17.1	12.6	54.0
18	90 43.3	N21 18.9	127 11.0	16.4	N 8 29.7	12.6	54.0
19	105 43.3	19.4	141 46.4	16.4	8 42.3	12.5	54.0
20	120 43.2	19.8	156 21.8	16.4	8 54.8	12.5	54.0
21	135 43.1	. . 20.2	170 57.2	16.3	9 07.3	12.5	54.0
22	150 43.0	20.6	185 32.5	16.3	9 19.8	12.4	54.0
23	165 43.0	21.0	200 07.8	16.3	N 9 32.2	12.4	54.0
	SD 15.8	d 0.4	SD 14.9		14.8		14.7

Lat.	Twilight Naut.	Twilight Civil	Sunrise	Moonrise 25	26	27	28
°	h m	h m	h m	h m	h m	h m	h m
N 72	▭	▭	▭	02 57	02 30	02 03	01 33
N 70	▭	▭	▭	02 49	02 29	02 10	01 48
68	////	////	00 25	02 43	02 29	02 15	02 00
66	////	////	01 39	02 38	02 29	02 20	02 10
64	////	////	02 15	02 34	02 29	02 24	02 19
62	////	01 01	02 40	02 30	02 29	02 27	02 26
60	////	01 46	03 00	02 27	02 29	02 31	02 33
N 58	////	02 14	03 16	02 24	02 29	02 33	02 39
56	00 55	02 35	03 30	02 21	02 29	02 36	02 44
54	01 36	02 53	03 42	02 19	02 28	02 38	02 48
52	02 02	03 07	03 52	02 17	02 28	02 40	02 52
50	02 23	03 20	04 01	02 15	02 28	02 42	02 56
45	02 59	03 45	04 21	02 10	02 28	02 46	03 05
N 40	03 26	04 05	04 37	02 07	02 28	02 49	03 11
35	03 46	04 21	04 50	02 03	02 28	02 52	03 17
30	04 02	04 35	05 01	02 01	02 28	02 55	03 23
20	04 28	04 57	05 21	01 56	02 28	03 00	03 32
N 10	04 49	05 15	05 38	01 52	02 28	03 04	03 40
0	05 05	05 31	05 53	01 48	02 28	03 07	03 48
S 10	05 20	05 46	06 09	01 43	02 28	03 11	03 56
20	05 35	06 02	06 25	01 39	02 28	03 15	04 04
30	05 49	06 18	06 44	01 34	02 28	03 20	04 13
35	05 56	06 27	06 55	01 31	02 28	03 23	04 19
40	06 04	06 38	07 07	01 28	02 28	03 26	04 25
45	06 13	06 49	07 22	01 24	02 27	03 30	04 33
S 50	06 23	07 03	07 40	01 20	02 27	03 34	04 42
52	06 27	07 09	07 49	01 18	02 27	03 36	04 46
54	06 32	07 16	07 58	01 15	02 27	03 39	04 50
56	06 37	07 24	08 09	01 13	02 27	03 41	04 55
58	06 42	07 33	08 21	01 10	02 27	03 44	05 01
S 60	06 48	07 42	08 35	01 07	02 27	03 47	05 07

Lat.	Sunset	Twilight Civil	Twilight Naut.	Moonset 25	26	27	28
°	h m	h m	h m	h m	h m	h m	h m
N 72	▭	▭	▭	13 28	15 21	17 16	19 25
N 70	▭	▭	▭	13 33	15 17	17 03	18 57
68	▭	▭	▭	13 36	15 14	16 53	18 36
66	22 18	////	////	13 39	15 11	16 44	18 20
64	21 41	////	////	13 42	15 09	16 37	18 07
62	21 15	22 58	////	13 44	15 07	16 31	17 56
60	20 55	22 11	////	13 46	15 06	16 26	17 47
N 58	20 39	21 42	////	13 47	15 04	16 21	17 39
56	20 25	21 20	23 04	13 49	15 03	16 17	17 31
54	20 13	21 02	22 20	13 50	15 02	16 13	17 25
52	20 03	20 48	21 53	13 52	15 01	16 10	17 19
50	19 53	20 35	21 33	13 53	15 00	16 07	17 14
45	19 34	20 09	20 56	13 55	14 58	16 00	17 03
N 40	19 18	19 49	20 29	13 57	14 56	15 55	16 54
35	19 04	19 33	20 09	13 59	14 54	15 50	16 46
30	18 53	19 20	19 52	14 00	14 53	15 46	16 39
20	18 33	18 57	19 26	14 03	14 51	15 39	16 28
N 10	18 16	18 39	19 06	14 05	14 49	15 33	16 17
0	18 01	18 23	18 49	14 07	14 47	15 27	16 08
S 10	17 45	18 07	18 33	14 09	14 45	15 21	15 58
20	17 28	17 52	18 19	14 11	14 43	15 15	15 48
30	17 10	17 35	18 05	14 14	14 41	15 08	15 36
35	16 59	17 26	17 57	14 15	14 39	15 04	15 30
40	16 46	17 16	17 49	14 16	14 38	14 59	15 22
45	16 31	17 04	17 41	14 18	14 36	14 54	15 13
S 50	16 13	16 50	17 31	14 20	14 34	14 48	15 03
52	16 05	16 44	17 26	14 21	14 33	14 45	14 58
54	15 56	16 37	17 22	14 22	14 32	14 42	14 53
56	15 45	16 29	17 17	14 24	14 31	14 38	14 47
58	15 33	16 21	17 11	14 25	14 30	14 35	14 40
S 60	15 18	16 11	17 05	14 26	14 28	14 30	14 33

Day	SUN Eqn. of Time 00^h	Eqn. of Time 12^h	Mer. Pass.	MOON Mer. Pass. Upper	Mer. Pass. Lower	Age	Phase
d	m s	m s	h m	h m	h m	d %	
25	03 11	03 08	11 57	07 57	20 17	24 27	
26	03 05	03 02	11 57	08 37	20 57	25 19	
27	02 59	02 55	11 57	09 17	21 37	26 12	

	UT	ARIES	VENUS −3.9		MARS −0.6		JUPITER −1.9		SATURN +0.1		STARS		
		GHA	GHA	Dec	GHA	Dec	GHA	Dec	GHA	Dec	Name	SHA	Dec
	d h	° ′	° ′	° ′	° ′	° ′	° ′	° ′	° ′	° ′		° ′	° ′
	28 00	245 07.4	203 07.8	N14 40.5	281 26.8	S17 04.9	110 13.4	N17 55.3	156 04.8	N22 35.4	Acamar	315 24.7	S40 17.4
	01	260 09.9	218 07.2	41.5	296 28.0	04.6	125 15.5	55.2	171 06.9	35.4	Achernar	335 32.9	S57 13.1
	02	275 12.3	233 06.7	42.4	311 29.1	04.3	140 17.6	55.1	186 09.1	35.4	Acrux	173 18.2	S63 07.3
	03	290 14.8	248 06.2	. . 43.4	326 30.3	. . 04.0	155 19.7	. . 55.0	201 11.2	. . 35.5	Adhara	255 19.1	S28 58.7
	04	305 17.3	263 05.7	44.3	341 31.4	03.7	170 21.9	54.9	216 13.3	35.5	Aldebaran	290 58.9	N16 30.9
	05	320 19.7	278 05.2	45.3	356 32.5	03.4	185 24.0	54.8	231 15.5	35.5			
	06	335 22.2	293 04.6	N14 46.2	11 33.7	S17 03.1	200 26.1	N17 54.7	246 17.6	N22 35.5	Alioth	166 27.1	N55 56.8
W	07	350 24.7	308 04.1	47.2	26 34.8	02.8	215 28.2	54.6	261 19.7	35.5	Alkaid	153 04.6	N49 18.0
E	08	5 27.1	323 03.6	48.1	41 36.0	02.5	230 30.3	54.5	276 21.8	35.5	Al Na'ir	27 53.5	S46 56.6
D	09	20 29.6	338 03.1	. . 49.1	56 37.1	. . 02.2	245 32.4	. . 54.4	291 24.0	. . 35.5	Alnilam	275 54.8	S 1 12.0
N	10	35 32.1	353 02.6	50.0	71 38.3	01.8	260 34.5	54.3	306 26.1	35.5	Alphard	218 04.0	S 8 40.4
E	11	50 34.5	8 02.0	51.0	86 39.4	01.5	275 36.6	54.2	321 28.2	35.5			
S	12	65 37.0	23 01.5	N14 51.9	101 40.5	S17 01.2	290 38.7	N17 54.1	336 30.4	N22 35.5	Alphecca	126 17.3	N26 42.2
D	13	80 39.4	38 01.0	52.9	116 41.7	00.9	305 40.8	54.0	351 32.5	35.5	Alpheratz	357 52.0	N29 06.3
A	14	95 41.9	53 00.5	53.8	131 42.8	00.6	320 42.9	53.9	6 34.6	35.5	Altair	62 15.8	N 8 52.5
Y	15	110 44.4	68 00.0	. . 54.8	146 44.0	. . 00.3	335 45.0	. . 53.8	21 36.7	. . 35.5	Ankaa	353 23.6	S42 17.2
	16	125 46.8	82 59.4	55.7	161 45.1	17 00.0	350 47.1	53.7	36 38.9	35.5	Antares	112 35.7	S26 26.4
	17	140 49.3	97 58.9	56.6	176 46.3	16 59.7	5 49.2	53.6	51 41.0	35.5			
	18	155 51.8	112 58.4	N14 57.6	191 47.4	S16 59.4	20 51.3	N17 53.5	66 43.1	N22 35.6	Arcturus	146 02.7	N19 09.9
	19	170 54.2	127 57.9	58.5	206 48.6	59.1	35 53.4	53.4	81 45.3	35.6	Atria	107 44.2	S69 02.1
	20	185 56.7	142 57.3	14 59.5	221 49.7	58.7	50 55.5	53.3	96 47.4	35.6	Avior	234 21.8	S59 31.4
	21	200 59.2	157 56.8	15 00.4	236 50.9	. . 58.4	65 57.6	. . 53.2	111 49.5	. . 35.6	Bellatrix	278 40.9	N 6 21.2
	22	216 01.6	172 56.3	01.3	251 52.0	58.1	80 59.8	53.1	126 51.6	35.6	Betelgeuse	271 10.2	N 7 24.5
	23	231 04.1	187 55.7	02.3	266 53.2	57.8	96 01.9	53.0	141 53.8	35.6			
	29 00	246 06.5	202 55.2	N15 03.2	281 54.3	S16 57.5	111 04.0	N17 52.9	156 55.9	N22 35.6	Canopus	264 00.2	S52 41.9
	01	261 09.0	217 54.7	04.2	296 55.5	57.2	126 06.1	52.8	171 58.0	35.6	Capella	280 46.7	N46 00.1
	02	276 11.5	232 54.2	05.1	311 56.6	56.9	141 08.2	52.7	187 00.2	35.6	Deneb	49 36.8	N45 17.2
	03	291 13.9	247 53.6	. . 06.0	326 57.8	. . 56.6	156 10.3	. . 52.6	202 02.3	. . 35.6	Denebola	182 41.6	N14 33.3
	04	306 16.4	262 53.1	07.0	341 58.9	56.3	171 12.4	52.5	217 04.4	35.6	Diphda	349 04.0	S17 58.1
	05	321 18.9	277 52.6	07.9	357 00.1	56.0	186 14.5	52.4	232 06.5	35.6			
	06	336 21.3	292 52.0	N15 08.8	12 01.2	S16 55.6	201 16.6	N17 52.3	247 08.7	N22 35.6	Dubhe	194 01.0	N61 44.3
	07	351 23.8	307 51.5	09.8	27 02.4	55.3	216 18.7	52.2	262 10.8	35.6	Elnath	278 23.1	N28 36.6
T	08	6 26.3	322 51.0	10.7	42 03.6	55.0	231 20.8	52.1	277 12.9	35.6	Eltanin	90 49.3	N51 29.1
H	09	21 28.7	337 50.5	. . 11.6	57 04.7	. . 54.7	246 22.9	. . 51.9	292 15.1	. . 35.7	Enif	33 54.9	N 9 53.2
U	10	36 31.2	352 49.9	12.6	72 05.9	54.4	261 25.0	51.8	307 17.2	35.7	Fomalhaut	15 32.7	S29 36.2
R	11	51 33.7	7 49.4	13.5	87 07.0	54.1	276 27.1	51.7	322 19.3	35.7			
S	12	66 36.1	22 48.9	N15 14.4	102 08.2	S16 53.8	291 29.2	N17 51.6	337 21.4	N22 35.7	Gacrux	172 09.7	S57 08.1
D	13	81 38.6	37 48.3	15.4	117 09.3	53.5	306 31.3	51.5	352 23.6	35.7	Gienah	176 00.4	S17 33.7
A	14	96 41.0	52 47.8	16.3	132 10.5	53.2	321 33.4	51.4	7 25.7	35.7	Hadar	148 58.9	S60 23.5
Y	15	111 43.5	67 47.3	. . 17.2	147 11.6	. . 52.9	336 35.5	. . 51.3	22 27.8	. . 35.7	Hamal	328 10.1	N23 28.5
	16	126 46.0	82 46.7	18.2	162 12.8	52.6	351 37.6	51.2	37 30.0	35.7	Kaus Aust.	83 54.0	S34 23.0
	17	141 48.4	97 46.2	19.1	177 14.0	52.2	6 39.7	51.1	52 32.1	35.7			
	18	156 50.9	112 45.7	N15 20.0	192 15.1	S16 51.9	21 41.8	N17 51.0	67 34.2	N22 35.7	Kochab	137 18.2	N74 08.6
	19	171 53.4	127 45.1	20.9	207 16.3	51.6	36 43.9	50.9	82 36.3	35.7	Markab	13 46.3	N15 13.2
	20	186 55.8	142 44.6	21.9	222 17.4	51.3	51 46.0	50.8	97 38.5	35.7	Menkar	314 23.7	N 4 06.1
	21	201 58.3	157 44.0	. . 22.8	237 18.6	. . 51.0	66 48.1	. . 50.7	112 40.6	. . 35.7	Menkent	148 16.7	S36 23.3
	22	217 00.8	172 43.5	23.7	252 19.8	50.7	81 50.2	50.6	127 42.7	35.7	Miaplacidus	221 42.0	S69 44.1
	23	232 03.2	187 43.0	24.6	267 20.9	50.4	96 52.3	50.5	142 44.8	35.7			
	30 00	247 05.7	202 42.4	N15 25.6	282 22.1	S16 50.1	111 54.4	N17 50.4	157 47.0	N22 35.7	Mirfak	308 52.4	N49 52.3
	01	262 08.2	217 41.9	26.5	297 23.2	49.8	126 56.5	50.3	172 49.1	35.8	Nunki	76 07.8	S26 17.6
	02	277 10.6	232 41.3	27.4	312 24.4	49.5	141 58.6	50.2	187 51.2	35.8	Peacock	53 31.2	S56 43.3
	03	292 13.1	247 40.8	. . 28.3	327 25.6	. . 49.2	157 00.7	. . 50.1	202 53.4	. . 35.8	Pollux	243 37.7	N28 01.2
	04	307 15.5	262 40.3	29.3	342 26.7	48.9	172 02.8	50.0	217 55.5	35.8	Procyon	245 08.3	N 5 13.0
	05	322 18.0	277 39.7	30.2	357 27.9	48.5	187 04.9	49.9	232 57.6	35.8			
	06	337 20.5	292 39.2	N15 31.1	12 29.1	S16 48.2	202 07.0	N17 49.8	247 59.7	N22 35.8	Rasalhague	96 13.5	N12 33.3
	07	352 22.9	307 38.6	32.0	27 30.2	47.9	217 09.1	49.7	263 01.9	35.8	Regulus	207 52.0	N11 57.2
	08	7 25.4	322 38.1	32.9	42 31.4	47.6	232 11.2	49.6	278 04.0	35.8	Rigel	281 20.0	S 8 11.9
F	09	22 27.9	337 37.6	. . 33.9	57 32.6	. . 47.3	247 13.3	. . 49.4	293 06.1	. . 35.8	Rigil Kent.	140 02.3	S60 51.1
R	10	37 30.3	352 37.0	34.8	72 33.7	47.0	262 15.4	49.3	308 08.2	35.8	Sabik	102 21.3	S15 43.8
I	11	52 32.8	7 36.5	35.7	87 34.9	46.7	277 17.4	49.2	323 10.4	35.8			
D	12	67 35.3	22 35.9	N15 36.6	102 36.1	S16 46.4	292 19.5	N17 49.1	338 12.5	N22 35.8	Schedar	349 50.1	N56 33.0
A	13	82 37.7	37 35.4	37.5	117 37.2	46.1	307 21.6	49.0	353 14.6	35.8	Shaula	96 32.3	S37 06.4
Y	14	97 40.2	52 34.8	38.4	132 38.4	45.8	322 23.7	48.9	8 16.7	35.8	Sirius	258 41.1	S16 43.3
	15	112 42.6	67 34.3	. . 39.3	147 39.6	. . 45.5	337 25.8	. . 48.8	23 18.9	. . 35.8	Spica	158 39.4	S11 10.8
	16	127 45.1	82 33.8	40.3	162 40.7	45.2	352 27.9	48.7	38 21.0	35.8	Suhail	222 58.6	S43 26.9
	17	142 47.6	97 33.2	41.2	177 41.9	44.9	7 30.0	48.6	53 23.1	35.8			
	18	157 50.0	112 32.7	N15 42.1	192 43.1	S16 44.6	22 32.1	N17 48.5	68 25.3	N22 35.9	Vega	80 44.0	N38 47.0
	19	172 52.5	127 32.1	43.0	207 44.2	44.2	37 34.2	48.4	83 27.4	35.9	Zuben'ubi	137 14.0	S16 03.4
	20	187 55.0	142 31.6	43.9	222 45.4	43.9	52 36.3	48.3	98 29.5	35.9			
	21	202 57.4	157 31.0	. . 44.8	237 46.6	. . 43.6	67 38.4	. . 48.2	113 31.6	. . 35.9			
	22	217 59.9	172 30.5	45.7	252 47.7	43.3	82 40.5	48.1	128 33.8	35.9			
	23	233 02.4	187 29.9	46.6	267 48.9	43.0	97 42.6	48.0	143 35.9	35.9			
	Mer. Pass.	h m 7 34.3	v −0.5	d 0.9	v 1.2	d 0.3	v 2.1	d 0.1	v 2.1	d 0.0			

	SHA	Mer. Pass.
	° ′	h m
Venus	316 48.7	10 29
Mars	35 47.8	5 12
Jupiter	224 57.4	16 33
Saturn	270 49.4	13 30

UT	SUN GHA	SUN Dec	MOON GHA	v	MOON Dec	d	HP
d h	° ′	° ′	° ′	′	° ′	′	′
28 00	180 42.9	N21 21.4	214 43.1	16.2	N 9 44.6	12.3	54.0
01	195 42.8	21.8	229 18.3	16.2	9 56.9	12.4	54.0
02	210 42.7	22.2	243 53.5	16.2	10 09.3	12.2	54.0
03	225 42.7	. . 22.7	258 28.7	16.1	10 21.5	12.3	54.0
04	240 42.6	23.1	273 03.8	16.1	10 33.8	12.2	54.0
05	255 42.5	23.5	287 38.9	16.1	10 46.0	12.1	54.0
06	270 42.4	N21 23.9	302 14.0	16.0	N10 58.1	12.2	54.0
W 07	285 42.4	24.3	316 49.0	16.0	11 10.3	12.0	54.0
E 08	300 42.3	24.7	331 24.0	15.9	11 22.3	12.1	54.0
D 09	315 42.2	. . 25.1	345 58.9	15.9	11 34.4	12.0	54.0
N 10	330 42.1	25.5	0 33.8	15.9	11 46.4	11.9	54.0
E 11	345 42.0	25.9	15 08.7	15.8	11 58.3	11.9	54.0
S 12	0 42.0	N21 26.3	29 43.5	15.7	N12 10.2	11.9	54.0
D 13	15 41.9	26.7	44 18.2	15.8	12 22.1	11.8	54.0
A 14	30 41.8	27.1	58 53.0	15.7	12 33.9	11.8	54.0
Y 15	45 41.7	. . 27.5	73 27.7	15.6	12 45.7	11.8	54.0
16	60 41.7	27.9	88 02.3	15.6	12 57.4	11.7	54.0
17	75 41.6	28.3	102 36.9	15.5	13 09.1	11.6	54.0
18	90 41.5	N21 28.7	117 11.4	15.5	N13 20.7	11.6	54.0
19	105 41.4	29.1	131 45.9	15.5	13 32.3	11.5	54.0
20	120 41.3	29.5	146 20.4	15.4	13 43.8	11.5	54.0
21	135 41.3	. . 29.9	160 54.8	15.3	13 55.3	11.4	54.0
22	150 41.2	30.3	175 29.1	15.3	14 06.7	11.4	54.0
23	165 41.1	30.7	190 03.4	15.3	14 18.1	11.3	54.0
29 00	180 41.0	N21 31.1	204 37.7	15.2	N14 29.4	11.2	54.0
01	195 40.9	31.5	219 11.9	15.1	14 40.6	11.2	54.0
02	210 40.9	31.9	233 46.0	15.1	14 51.8	11.2	54.0
03	225 40.8	. . 32.3	248 20.1	15.0	15 03.0	11.1	54.0
04	240 40.7	32.7	262 54.1	15.0	15 14.1	11.0	54.0
05	255 40.6	33.1	277 28.1	14.9	15 25.1	11.0	54.0
06	270 40.5	N21 33.5	292 02.0	14.9	N15 36.1	10.9	54.0
07	285 40.5	33.9	306 35.9	14.8	15 47.0	10.8	54.0
T 08	300 40.4	34.3	321 09.7	14.8	15 57.8	10.8	54.0
H 09	315 40.3	. . 34.7	335 43.5	14.7	16 08.6	10.8	54.0
U 10	330 40.2	35.1	350 17.2	14.6	16 19.4	10.6	54.0
R 11	345 40.1	35.5	4 50.8	14.6	16 30.0	10.6	54.0
S 12	0 40.1	N21 35.9	19 24.4	14.5	N16 40.6	10.6	54.0
D 13	15 40.0	36.2	33 57.9	14.5	16 51.2	10.4	54.0
A 14	30 39.9	36.6	48 31.4	14.4	17 01.6	10.4	54.0
Y 15	45 39.8	. . 37.0	63 04.8	14.3	17 12.0	10.4	54.0
16	60 39.7	37.4	77 38.1	14.3	17 22.4	10.3	54.1
17	75 39.6	37.8	92 11.4	14.3	17 32.7	10.2	54.1
18	90 39.6	N21 38.2	106 44.7	14.1	N17 42.9	10.1	54.1
19	105 39.5	38.6	121 17.8	14.1	17 53.0	10.0	54.1
20	120 39.4	38.9	135 50.9	14.0	18 03.0	10.0	54.1
21	135 39.3	. . 39.3	150 23.9	14.0	18 13.0	9.9	54.1
22	150 39.2	39.7	164 56.9	13.9	18 22.9	9.9	54.1
23	165 39.1	40.1	179 29.8	13.9	18 32.8	9.8	54.1
30 00	180 39.1	N21 40.5	194 02.7	13.7	N18 42.6	9.7	54.1
01	195 39.0	40.9	208 35.4	13.8	18 52.3	9.6	54.1
02	210 38.9	41.2	223 08.2	13.6	19 01.9	9.5	54.1
03	225 38.8	. . 41.6	237 40.8	13.6	19 11.4	9.5	54.1
04	240 38.7	42.0	252 13.4	13.5	19 20.9	9.4	54.1
05	255 38.6	42.4	266 45.9	13.4	19 30.3	9.3	54.1
06	270 38.5	N21 42.8	281 18.3	13.4	N19 39.6	9.2	54.1
07	285 38.5	43.1	295 50.7	13.3	19 48.8	9.1	54.1
08	300 38.4	43.5	310 23.0	13.3	19 57.9	9.1	54.1
F 09	315 38.3	. . 43.9	324 55.3	13.2	20 07.0	8.9	54.2
R 10	330 38.2	44.3	339 27.5	13.1	20 15.9	8.9	54.2
I 11	345 38.1	44.6	353 59.6	13.0	20 24.8	8.8	54.2
D 12	0 38.0	N21 45.0	8 31.6	13.0	N20 33.6	8.8	54.2
A 13	15 37.9	45.4	23 03.6	12.9	20 42.4	8.6	54.2
Y 14	30 37.9	45.8	37 35.5	12.9	20 51.0	8.5	54.2
15	45 37.8	. . 46.1	52 07.4	12.8	20 59.5	8.5	54.2
16	60 37.7	46.5	66 39.2	12.7	21 08.0	8.4	54.2
17	75 37.6	46.9	81 10.9	12.6	21 16.4	8.2	54.2
18	90 37.5	N21 47.2	95 42.5	12.6	N21 24.6	8.2	54.2
19	105 37.4	47.6	110 14.1	12.5	21 32.8	8.1	54.2
20	120 37.3	48.0	124 45.6	12.4	21 40.9	8.0	54.2
21	135 37.2	. . 48.3	139 17.0	12.4	21 48.9	7.9	54.3
22	150 37.2	48.7	153 48.4	12.3	21 56.8	7.8	54.3
23	165 37.1	49.1	168 19.7	12.2	N22 04.6	7.7	54.3
	SD 15.8	*d* 0.4	SD 14.7		14.7		14.8

Lat.	Twilight Naut.	Twilight Civil	Sunrise	Moonrise 28	Moonrise 29	Moonrise 30	Moonrise 31
°	h m	h m	h m	h m	h m	h m	h m
N 72	▭	▭	▭	01 33	00 51	▭	▭
N 70	▭	▭	▭	01 48	01 20	00 31	▭
68	▭	▭	▭	02 00	01 42	01 16	▭
66	////	////	01 27	02 10	02 00	01 46	01 23
64	////	////	02 07	02 19	02 14	02 09	02 02
62	////	00 42	02 34	02 26	02 26	02 27	02 30
60	////	01 36	02 55	02 33	02 36	02 42	02 51
N 58	////	02 07	03 12	02 39	02 45	02 55	03 09
56	00 38	02 30	03 26	02 44	02 53	03 06	03 23
54	01 28	02 48	03 38	02 48	03 00	03 16	03 36
52	01 56	03 03	03 49	02 52	03 07	03 24	03 47
50	02 18	03 17	03 59	02 56	03 13	03 32	03 57
45	02 56	03 43	04 19	03 05	03 25	03 49	04 18
N 40	03 23	04 03	04 35	03 11	03 36	04 03	04 36
35	03 44	04 20	04 49	03 17	03 45	04 15	04 50
30	04 01	04 33	05 00	03 23	03 52	04 25	05 03
20	04 28	04 56	05 20	03 32	04 06	04 43	05 24
N 10	04 48	05 15	05 38	03 40	04 18	04 59	05 43
0	05 06	05 31	05 54	03 48	04 30	05 14	06 01
S 10	05 21	05 47	06 10	03 56	04 41	05 29	06 18
20	05 36	06 03	06 27	04 04	04 53	05 45	06 37
30	05 50	06 20	06 46	04 13	05 07	06 03	07 00
35	05 58	06 29	06 57	04 19	05 16	06 14	07 13
40	06 06	06 40	07 10	04 25	05 25	06 26	07 28
45	06 15	06 52	07 25	04 33	05 36	06 41	07 46
S 50	06 26	07 06	07 43	04 42	05 50	06 59	08 08
52	06 30	07 13	07 52	04 46	05 56	07 08	08 19
54	06 35	07 20	08 02	04 50	06 03	07 17	08 31
56	06 40	07 28	08 13	04 55	06 11	07 28	08 45
58	06 46	07 37	08 26	05 01	06 20	07 41	09 01
S 60	06 52	07 47	08 41	05 07	06 30	07 55	09 21

Lat.	Sunset	Twilight Civil	Twilight Naut.	Moonset 28	Moonset 29	Moonset 30	Moonset 31
°	h m	h m	h m	h m	h m	h m	h m
N 72	▭	▭	▭	19 25	▭	▭	▭
N 70	▭	▭	▭	18 57	21 16	▭	▭
68	▭	▭	▭	18 36	20 33	▭	▭
66	22 32	////	////	18 20	20 04	22 02	▭
64	21 50	////	////	18 07	19 42	21 24	23 14
62	21 22	23 20	////	17 56	19 25	20 57	22 30
60	21 01	22 21	////	17 47	19 10	20 36	22 00
N 58	20 44	21 49	////	17 39	18 58	20 19	21 38
56	20 30	21 26	23 24	17 31	18 48	20 05	21 20
54	20 17	21 08	22 29	17 25	18 39	19 52	21 05
52	20 06	20 52	22 00	17 19	18 30	19 42	20 51
50	19 57	20 39	21 38	17 14	18 23	19 32	20 40
45	19 36	20 12	20 59	17 03	18 07	19 12	20 16
N 40	19 20	19 52	20 32	16 54	17 54	18 56	19 57
35	19 06	19 35	20 11	16 46	17 44	18 42	19 41
30	18 55	19 22	19 54	16 39	17 34	18 30	19 27
20	18 34	18 59	19 27	16 28	17 18	18 10	19 03
N 10	18 17	18 40	19 07	16 17	17 04	17 52	18 43
0	18 01	18 23	18 49	16 08	16 51	17 36	18 24
S 10	17 45	18 08	18 34	15 58	16 37	17 20	18 05
20	17 28	17 52	18 19	15 48	16 23	17 02	17 45
30	17 09	17 35	18 04	15 36	16 08	16 42	17 22
35	16 57	17 25	17 56	15 30	15 58	16 31	17 08
40	16 45	17 14	17 48	15 22	15 48	16 18	16 53
45	16 29	17 02	17 39	15 13	15 36	16 02	16 34
S 50	16 11	16 48	17 29	15 03	15 21	15 43	16 11
52	16 02	16 42	17 24	14 58	15 14	15 34	16 00
54	15 52	16 34	17 19	14 53	15 06	15 24	15 47
56	15 41	16 26	17 14	14 47	14 58	15 12	15 33
58	15 29	16 17	17 08	14 40	14 48	14 59	15 16
S 60	15 14	16 07	17 02	14 33	14 37	14 44	14 56

Day	SUN Eqn. of Time 00^h	SUN Eqn. of Time 12^h	SUN Mer. Pass.	MOON Mer. Pass. Upper	MOON Mer. Pass. Lower	MOON Age	MOON Phase
d	m s	m s	h m	h m	h m	d %	
28	02 52	02 48	11 57	09 58	22 19	27 6	
29	02 44	02 40	11 57	10 40	23 02	28 3	
30	02 36	02 32	11 57	11 25	23 48	29 0	

	UT	ARIES	VENUS −3.9		MARS −0.7		JUPITER −1.9		SATURN +0.1		STARS		
		GHA	GHA	Dec	GHA	Dec	GHA	Dec	GHA	Dec	Name	SHA	Dec
	d h	° ′	° ′	° ′	° ′	° ′	° ′	° ′	° ′	° ′		° ′	° ′
	31 00	248 04.8	202 29.4	N15 47.5	282 50.1	S16 42.7	112 44.7	N17 47.9	158 38.0	N22 35.9	Acamar	315 24.7	S40 17.4
	01	263 07.3	217 28.8	48.5	297 51.3	42.4	127 46.8	47.8	173 40.1	35.9	Achernar	335 32.9	S57 13.0
	02	278 09.8	232 28.3	49.4	312 52.4	42.1	142 48.9	47.7	188 42.3	35.9	Acrux	173 18.2	S63 07.3
	03	293 12.2	247 27.7	. . 50.3	327 53.6	. . 41.8	157 51.0	. . 47.5	203 44.4	. . 35.9	Adhara	255 19.1	S28 58.6
	04	308 14.7	262 27.2	51.2	342 54.8	41.5	172 53.1	47.4	218 46.5	35.9	Aldebaran	290 58.9	N16 30.9
	05	323 17.1	277 26.6	52.1	357 56.0	41.2	187 55.2	47.3	233 48.6	35.9			
	06	338 19.6	292 26.1	N15 53.0	12 57.1	S16 40.9	202 57.2	N17 47.2	248 50.8	N22 35.9	Alioth	166 27.1	N55 56.8
	07	353 22.1	307 25.5	53.9	27 58.3	40.6	217 59.3	47.1	263 52.9	35.9	Alkaid	153 04.6	N49 18.0
S	08	8 24.5	322 25.0	54.8	42 59.5	40.3	233 01.4	47.0	278 55.0	35.9	Al Na'ir	27 53.4	S46 56.6
A	09	23 27.0	337 24.4	. . 55.7	58 00.7	. . 40.0	248 03.5	. . 46.9	293 57.1	. . 35.9	Alnilam	275 54.8	S 1 12.0
T	10	38 29.5	352 23.9	56.6	73 01.8	39.6	263 05.6	46.8	308 59.3	35.9	Alphard	218 04.0	S 8 40.4
U	11	53 31.9	7 23.3	57.5	88 03.0	39.3	278 07.7	46.7	324 01.4	36.0			
R	12	68 34.4	22 22.7	N15 58.4	103 04.2	S16 39.0	293 09.8	N17 46.6	339 03.5	N22 36.0	Alphecca	126 17.3	N26 42.2
D	13	83 36.9	37 22.2	15 59.3	118 05.4	38.7	308 11.9	46.5	354 05.6	36.0	Alpheratz	357 51.9	N29 06.3
A	14	98 39.3	52 21.6	16 00.2	133 06.6	38.4	323 14.0	46.4	9 07.8	36.0	Altair	62 15.7	N 8 52.5
Y	15	113 41.8	67 21.1	. . 01.1	148 07.7	. . 38.1	338 16.1	. . 46.3	24 09.9	. . 36.0	Ankaa	353 23.6	S42 17.2
	16	128 44.3	82 20.5	02.0	163 08.9	37.8	353 18.2	46.2	39 12.0	36.0	Antares	112 35.7	S26 26.4
	17	143 46.7	97 20.0	02.9	178 10.1	37.5	8 20.3	46.1	54 14.1	36.0			
	18	158 49.2	112 19.4	N16 03.8	193 11.3	S16 37.2	23 22.3	N17 46.0	69 16.3	N22 36.0	Arcturus	146 02.7	N19 10.0
	19	173 51.6	127 18.8	04.7	208 12.5	36.9	38 24.4	45.8	84 18.4	36.0	Atria	107 44.2	S69 02.1
	20	188 54.1	142 18.3	05.6	223 13.6	36.6	53 26.5	45.7	99 20.5	36.0	Avior	234 21.8	S59 31.4
	21	203 56.6	157 17.7	. . 06.5	238 14.8	. . 36.3	68 28.6	. . 45.6	114 22.6	. . 36.0	Bellatrix	278 40.9	N 6 21.2
	22	218 59.0	172 17.2	07.4	253 16.0	36.0	83 30.7	45.5	129 24.8	36.0	Betelgeuse	271 10.2	N 7 24.5
	23	234 01.5	187 16.6	08.2	268 17.2	35.7	98 32.8	45.4	144 26.9	36.0			
	1 00	249 04.0	202 16.0	N16 09.1	283 18.4	S16 35.4	113 34.9	N17 45.3	159 29.0	N22 36.0	Canopus	264 00.2	S52 41.9
	01	264 06.4	217 15.5	10.0	298 19.6	35.1	128 37.0	45.2	174 31.1	36.0	Capella	280 46.7	N46 00.1
	02	279 08.9	232 14.9	10.9	313 20.7	34.8	143 39.1	45.1	189 33.3	36.0	Deneb	49 36.7	N45 17.2
	03	294 11.4	247 14.4	. . 11.8	328 21.9	. . 34.5	158 41.2	. . 45.0	204 35.4	. . 36.0	Denebola	182 41.6	N14 33.3
	04	309 13.8	262 13.8	12.7	343 23.1	34.2	173 43.2	44.9	219 37.5	36.0	Diphda	349 04.0	S17 58.1
	05	324 16.3	277 13.2	13.6	358 24.3	33.8	188 45.3	44.8	234 39.6	36.1			
	06	339 18.8	292 12.7	N16 14.5	13 25.5	S16 33.5	203 47.4	N17 44.7	249 41.8	N22 36.1	Dubhe	194 01.1	N61 44.3
	07	354 21.2	307 12.1	15.4	28 26.7	33.2	218 49.5	44.6	264 43.9	36.1	Elnath	278 23.1	N28 36.6
	08	9 23.7	322 11.5	16.3	43 27.9	32.9	233 51.6	44.4	279 46.0	36.1	Eltanin	90 49.3	N51 29.1
S	09	24 26.1	337 11.0	. . 17.1	58 29.0	. . 32.6	248 53.7	. . 44.3	294 48.1	. . 36.1	Enif	33 54.8	N 9 53.2
U	10	39 28.6	352 10.4	18.0	73 30.2	32.3	263 55.8	44.2	309 50.3	36.1	Fomalhaut	15 32.6	S29 36.2
N	11	54 31.1	7 09.9	18.9	88 31.4	32.0	278 57.9	44.1	324 52.4	36.1			
D	12	69 33.5	22 09.3	N16 19.8	103 32.6	S16 31.7	294 00.0	N17 44.0	339 54.5	N22 36.1	Gacrux	172 09.7	S57 08.1
A	13	84 36.0	37 08.7	20.7	118 33.8	31.4	309 02.0	43.9	354 56.6	36.1	Gienah	176 00.4	S17 33.7
Y	14	99 38.5	52 08.2	21.6	133 35.0	31.1	324 04.1	43.8	9 58.8	36.1	Hadar	148 58.9	S60 23.5
	15	114 40.9	67 07.6	. . 22.4	148 36.2	. . 30.8	339 06.2	. . 43.7	25 00.9	. . 36.1	Hamal	328 10.1	N23 28.5
	16	129 43.4	82 07.0	23.3	163 37.4	30.5	354 08.3	43.6	40 03.0	36.1	Kaus Aust.	83 54.0	S34 23.0
	17	144 45.9	97 06.5	24.2	178 38.6	30.2	9 10.4	43.5	55 05.1	36.1			
	18	159 48.3	112 05.9	N16 25.1	193 39.7	S16 29.9	24 12.5	N17 43.4	70 07.3	N22 36.1	Kochab	137 18.3	N74 08.7
	19	174 50.8	127 05.3	26.0	208 40.9	29.6	39 14.6	43.3	85 09.4	36.1	Markab	13 46.3	N15 13.2
	20	189 53.3	142 04.7	26.8	223 42.1	29.3	54 16.6	43.1	100 11.5	36.1	Menkar	314 23.7	N 4 06.1
	21	204 55.7	157 04.2	. . 27.7	238 43.3	. . 29.0	69 18.7	. . 43.0	115 13.6	. . 36.1	Menkent	148 16.7	S36 23.3
	22	219 58.2	172 03.6	28.6	253 44.5	28.7	84 20.8	42.9	130 15.8	36.1	Miaplacidus	221 42.0	S69 44.1
	23	235 00.6	187 03.0	29.5	268 45.7	28.4	99 22.9	42.8	145 17.9	36.1			
	2 00	250 03.1	202 02.5	N16 30.3	283 46.9	S16 28.1	114 25.0	N17 42.7	160 20.0	N22 36.2	Mirfak	308 52.3	N49 52.3
	01	265 05.6	217 01.9	31.2	298 48.1	27.8	129 27.1	42.6	175 22.1	36.2	Nunki	76 07.8	S26 17.6
	02	280 08.0	232 01.3	32.1	313 49.3	27.5	144 29.2	42.5	190 24.3	36.2	Peacock	53 31.2	S56 43.3
	03	295 10.5	247 00.7	. . 33.0	328 50.5	. . 27.2	159 31.2	. . 42.4	205 26.4	. . 36.2	Pollux	243 37.7	N28 01.2
	04	310 13.0	262 00.2	33.8	343 51.7	26.9	174 33.3	42.3	220 28.5	36.2	Procyon	245 08.3	N 5 13.0
	05	325 15.4	276 59.6	34.7	358 52.9	26.6	189 35.4	42.2	235 30.6	36.2			
	06	340 17.9	291 59.0	N16 35.6	13 54.1	S16 26.3	204 37.5	N17 42.1	250 32.7	N22 36.2	Rasalhague	96 13.5	N12 33.3
	07	355 20.4	306 58.4	36.5	28 55.3	26.0	219 39.6	41.9	265 34.9	36.2	Regulus	207 52.0	N11 57.2
	08	10 22.8	321 57.9	37.3	43 56.5	25.7	234 41.7	41.8	280 37.0	36.2	Rigel	281 20.0	S 8 11.9
M	09	25 25.3	336 57.3	. . 38.2	58 57.7	. . 25.4	249 43.7	. . 41.7	295 39.1	. . 36.2	Rigil Kent.	140 02.3	S60 51.1
O	10	40 27.8	351 56.7	39.1	73 58.9	25.0	264 45.8	41.6	310 41.2	36.2	Sabik	102 21.3	S15 43.8
N	11	55 30.2	6 56.1	39.9	89 00.1	24.7	279 47.9	41.5	325 43.4	36.2			
D	12	70 32.7	21 55.6	N16 40.8	104 01.3	S16 24.4	294 50.0	N17 41.4	340 45.5	N22 36.2	Schedar	349 50.1	N56 33.0
A	13	85 35.1	36 55.0	41.7	119 02.5	24.1	309 52.1	41.3	355 47.6	36.2	Shaula	96 32.3	S37 06.4
Y	14	100 37.6	51 54.4	42.5	134 03.7	23.8	324 54.2	41.2	10 49.7	36.2	Sirius	258 41.1	S16 43.2
	15	115 40.1	66 53.8	. . 43.4	149 04.9	. . 23.5	339 56.2	. . 41.1	25 51.9	. . 36.2	Spica	158 39.4	S11 10.8
	16	130 42.5	81 53.2	44.3	164 06.1	23.2	354 58.3	41.0	40 54.0	36.2	Suhail	222 58.6	S43 26.9
	17	145 45.0	96 52.7	45.1	179 07.3	22.9	10 00.4	40.8	55 56.1	36.2			
	18	160 47.5	111 52.1	N16 46.0	194 08.5	S16 22.6	25 02.5	N17 40.7	70 58.2	N22 36.2	Vega	80 44.0	N38 47.0
	19	175 49.9	126 51.5	46.8	209 09.7	22.3	40 04.6	40.6	86 00.3	36.2	Zuben'ubi	137 13.9	S16 03.4
	20	190 52.4	141 50.9	47.7	224 10.9	22.0	55 06.7	40.5	101 02.5	36.3		SHA	Mer. Pass.
	21	205 54.9	156 50.3	. . 48.6	239 12.1	. . 21.7	70 08.7	. . 40.4	116 04.6	. . 36.3		° ′	h m
	22	220 57.3	171 49.8	49.4	254 13.3	21.4	85 10.8	40.3	131 06.7	36.3	Venus	313 12.1	10 31
	23	235 59.8	186 49.2	50.3	269 14.5	21.1	100 12.9	40.2	146 08.8	36.3	Mars	34 14.4	5 06
		h m									Jupiter	224 30.9	16 23
	Mer. Pass.	7 22.5	*v* −0.6	*d* 0.9	*v* 1.2	*d* 0.3	*v* 2.1	*d* 0.1	*v* 2.1	*d* 0.0	Saturn	270 25.1	13 20

	UT	SUN GHA	SUN Dec	MOON GHA	v	MOON Dec	d	HP
	d h	° ′	° ′	° ′	′	° ′	′	′
	31 00	180 37.0	N21 49.4					
	01	195 36.9	49.8					
	02	210 36.8	50.2					
	03	225 36.7	. . 50.5					
	04	240 36.6	50.9					
	05	255 36.5	51.3					
	06	270 36.4	N21 51.6	269 56.9	11.7	N22 56.6	7.0	54.3
	07	285 36.4	52.0	284 27.6	11.7	23 03.6	6.9	54.4
S	08	300 36.3	52.3	298 58.3	11.7	23 10.5	6.8	54.4
A	09	315 36.2	. . 52.7	313 29.0	11.5	23 17.3	6.7	54.4
T	10	330 36.1	53.1	327 59.5	11.5	23 24.0	6.6	54.4
U	11	345 36.0	53.4	342 30.0	11.4	23 30.6	6.5	54.4
R	12	0 35.9	N21 53.8	357 00.4	11.4	N23 37.1	6.4	54.4
D	13	15 35.8	54.1	11 30.8	11.3	23 43.5	6.3	54.4
A	14	30 35.7	54.5	26 01.1	11.2	23 49.8	6.1	54.4
Y	15	45 35.6	. . 54.9	40 31.3	11.2	23 55.9	6.1	54.4
	16	60 35.5	55.2	55 01.5	11.1	24 02.0	5.9	54.4
	17	75 35.4	55.6	69 31.6	11.0	24 07.9	5.9	54.5
	18	90 35.3	N21 55.9	84 01.6	11.0	N24 13.8	5.7	54.5
	19	105 35.3	56.3	98 31.6	10.9	24 19.5	5.6	54.5
	20	120 35.2	56.6	113 01.5	10.9	24 25.1	5.5	54.5
	21	135 35.1	. . 57.0	127 31.4	10.8	24 30.6	5.4	54.5
	22	150 35.0	57.3	142 01.2	10.7	24 36.0	5.2	54.5
	23	165 34.9	57.7	156 30.9	10.7	24 41.2	5.2	54.5
	1 00	180 34.8	N21 58.0	171 00.6	10.6	N24 46.4	5.0	54.5
	01	195 34.7	58.4	185 30.2	10.5	24 51.4	4.9	54.6
	02	210 34.6	58.7	199 59.7	10.5	24 56.3	4.8	54.6
	03	225 34.5	. . 59.1	214 29.2	10.5	25 01.1	4.7	54.6
	04	240 34.4	59.4	228 58.7	10.4	25 05.8	4.6	54.6
	05	255 34.3	21 59.8	243 28.1	10.3	25 10.4	4.4	54.6
	06	270 34.2	N22 00.1	257 57.4	10.3	N25 14.8	4.3	54.6
	07	285 34.1	00.5	272 26.7	10.2	25 19.1	4.2	54.6
	08	300 34.0	00.8	286 55.9	10.2	25 23.3	4.1	54.7
S	09	315 33.9	. . 01.2	301 25.1	10.1	25 27.4	3.9	54.7
U	10	330 33.8	01.5	315 54.2	10.0	25 31.3	3.9	54.7
N	11	345 33.8	01.8	330 23.2	10.0	25 35.2	3.7	54.7
D	12	0 33.7	N22 02.2	344 52.2	10.0	N25 38.9	3.5	54.7
A	13	15 33.6	02.5	359 21.2	9.9	25 42.4	3.5	54.7
Y	14	30 33.5	02.9	13 50.1	9.9	25 45.9	3.3	54.7
	15	45 33.4	. . 03.2	28 19.0	9.8	25 49.2	3.2	54.8
	16	60 33.3	03.5	42 47.8	9.8	25 52.4	3.0	54.8
	17	75 33.2	03.9	57 16.6	9.7	25 55.4	3.0	54.8
	18	90 33.1	N22 04.2	71 45.3	9.7	N25 58.4	2.8	54.8
	19	105 33.0	04.6	86 14.0	9.6	26 01.2	2.7	54.8
	20	120 32.9	04.9	100 42.6	9.6	26 03.9	2.5	54.8
	21	135 32.8	. . 05.2	115 11.2	9.6	26 06.4	2.4	54.8
	22	150 32.7	05.6	129 39.8	9.5	26 08.8	2.3	54.9
	23	165 32.6	05.9	144 08.3	9.4	26 11.1	2.1	54.9
	2 00	180 32.5	N22 06.2	158 36.7	9.5	N26 13.2	2.1	54.9
	01	195 32.4	06.6	173 05.2	9.4	26 15.3	1.8	54.9
	02	210 32.3	06.9	187 33.6	9.3	26 17.1	1.8	54.9
	03	225 32.2	. . 07.2	202 01.9	9.4	26 18.9	1.6	54.9
	04	240 32.1	07.6	216 30.3	9.3	26 20.5	1.5	54.9
	05	255 32.0	07.9	230 58.6	9.2	26 22.0	1.4	55.0
	06	270 31.9	N22 08.2	245 26.8	9.3	N26 23.4	1.2	55.0
	07	285 31.8	08.6	259 55.1	9.2	26 24.6	1.0	55.0
	08	300 31.7	08.9	274 23.3	9.2	26 25.6	1.0	55.0
M	09	315 31.6	. . 09.2	288 51.5	9.1	26 26.6	0.8	55.0
O	10	330 31.5	09.5	303 19.6	9.1	26 27.4	0.7	55.0
N	11	345 31.4	09.9	317 47.7	9.1	26 28.1	0.5	55.1
D	12	0 31.3	N22 10.2	332 15.8	9.1	N26 28.6	0.4	55.1
A	13	15 31.2	10.5	346 43.9	9.1	26 29.0	0.3	55.1
Y	14	30 31.1	10.8	1 12.0	9.0	26 29.3	0.1	55.1
	15	45 31.0	. . 11.2	15 40.0	9.0	26 29.4	0.0	55.1
	16	60 30.9	11.5	30 08.0	9.0	26 29.4	0.2	55.2
	17	75 30.8	11.8	44 36.0	9.0	26 29.2	0.3	55.2
	18	90 30.7	N22 12.1	59 04.0	8.9	N26 28.9	0.4	55.2
	19	105 30.6	12.5	73 31.9	9.0	26 28.5	0.6	55.2
	20	120 30.5	12.8	87 59.9	8.9	26 27.9	0.7	55.2
	21	135 30.4	. . 13.1	102 27.8	8.9	26 27.2	0.8	55.2
	22	150 30.3	13.4	116 55.7	8.9	26 26.4	1.0	55.3
	23	165 30.2	13.7	131 23.6	8.9	N26 25.4	1.1	55.3
		SD 15.8	d 0.3	SD 14.8		14.9		15.0

An annular eclipse of the Sun occurs on this date. See page 5.

Lat.	Twilight Naut.	Twilight Civil	Sunrise	Moonrise 31	Moonrise 1	Moonrise 2	Moonrise 3
°	h m	h m	h m	h m	h m	h m	h m
N 72	▭	▭	▭	▭	▭	▭	▭
N 70	▭	▭	▭	▭	▭	▭	▭
68	▭	▭	▭	▭	▭	▭	▭
66	////	////	01 14	01 23	▭	▭	▭
64	////	////	01 59	02 02	01 53	▭	▭
62	////	00 09	02 28	02 30	02 38	03 01	03 54
60	////	01 27	02 50	02 51	03 08	03 39	04 33
N 58	////	02 01	03 08	03 09	03 31	04 06	05 00
56	00 08	02 25	03 22	03 23	03 49	04 27	05 21
54	01 20	02 44	03 35	03 36	04 05	04 45	05 39
52	01 51	03 00	03 46	03 47	04 18	05 00	05 54
50	02 13	03 14	03 56	03 57	04 30	05 13	06 07
45	02 53	03 41	04 17	04 18	04 55	05 39	06 33
N 40	03 21	04 01	04 33	04 36	05 14	06 00	06 54
35	03 42	04 18	04 47	04 50	05 31	06 18	07 12
30	04 00	04 32	04 59	05 03	05 45	06 33	07 27
20	04 27	04 56	05 20	05 24	06 09	06 59	07 53
N 10	04 48	05 15	05 38	05 43	06 31	07 21	08 15
0	05 06	05 32	05 54	06 01	06 50	07 42	08 35
S 10	05 22	05 48	06 10	06 18	07 10	08 03	08 56
20	05 37	06 04	06 28	06 37	07 31	08 26	09 18
30	05 52	06 21	06 47	07 00	07 56	08 52	09 44
35	06 00	06 31	06 59	07 13	08 11	09 07	09 59
40	06 08	06 42	07 12	07 28	08 28	09 25	10 16
45	06 18	06 54	07 28	07 46	08 48	09 46	10 37
S 50	06 28	07 09	07 47	08 08	09 14	10 14	11 04
52	06 33	07 16	07 56	08 19	09 27	10 27	11 16
54	06 38	07 23	08 06	08 31	09 41	10 42	11 31
56	06 43	07 31	08 17	08 45	09 58	11 00	11 48
58	06 49	07 41	08 30	09 01	10 18	11 22	12 09
S 60	06 56	07 51	08 46	09 21	10 43	11 51	12 35

Lat.	Sunset	Twilight Civil	Twilight Naut.	Moonset 31	Moonset 1	Moonset 2	Moonset 3
°	h m	h m	h m	h m	h m	h m	h m
N 72	▭	▭	▭	▭	▭	▭	▭
N 70	▭	▭	▭	▭	▭	▭	▭
68	▭	▭	▭	▭	▭	▭	▭
66	22 45	////	////	▭	▭	▭	▭
64	21 59	////	////	23 14	▭	▭	▭
62	21 29	////	////	22 30	23 54	24 51	00 51
60	21 07	22 31	////	22 00	23 16	24 12	00 12
N 58	20 49	21 56	////	21 38	22 49	23 45	24 23
56	20 34	21 32	////	21 20	22 28	23 24	24 04
54	20 21	21 12	22 38	21 05	22 11	23 06	23 49
52	20 10	20 56	22 06	20 51	21 56	22 51	23 35
50	20 00	20 43	21 43	20 40	21 43	22 38	23 23
45	19 39	20 15	21 03	20 16	21 17	22 11	22 58
N 40	19 22	19 55	20 35	19 57	20 56	21 50	22 38
35	19 08	19 38	20 13	19 41	20 38	21 32	22 21
30	18 56	19 23	19 56	19 27	20 23	21 17	22 07
20	18 36	19 00	19 29	19 03	19 58	20 51	21 42
N 10	18 18	18 41	19 07	18 43	19 36	20 29	21 21
0	18 01	18 24	18 50	18 24	19 15	20 08	21 01
S 10	17 45	18 08	18 34	18 05	18 55	19 47	20 41
20	17 28	17 51	18 19	17 45	18 33	19 24	20 20
30	17 08	17 34	18 04	17 22	18 07	18 58	19 55
35	16 56	17 24	17 55	17 08	17 52	18 43	19 40
40	16 43	17 13	17 47	16 53	17 35	18 25	19 23
45	16 28	17 01	17 38	16 34	17 14	18 04	19 02
S 50	16 09	16 46	17 27	16 11	16 48	17 36	18 36
52	16 00	16 39	17 22	16 00	16 35	17 23	18 23
54	15 49	16 32	17 17	15 47	16 21	17 08	18 09
56	15 38	16 24	17 12	15 33	16 04	16 49	17 52
58	15 25	16 14	17 06	15 16	15 43	16 27	17 31
S 60	15 09	16 04	16 59	14 56	15 18	15 59	17 05

Day	SUN Eqn. of Time 00ʰ	SUN Eqn. of Time 12ʰ	SUN Mer. Pass.	MOON Mer. Pass. Upper	MOON Mer. Pass. Lower	MOON Age	MOON Phase
d	m s	m s	h m	h m	h m	d	%
31	02 28	02 24	11 58	12 12	24 37	00	0
1	02 19	02 15	11 58	13 03	00 37	01	2 ●
2	02 10	02 05	11 58	13 55	01 29	02	5

UT		ARIES	VENUS −3.9		MARS −0.8		JUPITER −1.9		SATURN +0.1	
		GHA	GHA	Dec	GHA	Dec	GHA	Dec	GHA	Dec
d	h	° ′	° ′	° ′	° ′	° ′	° ′	° ′	° ′	° ′
3 (TUESDAY)	00	251 02.2	201 48.6	N16 51.1	284 15.7	S16 20.8	115 15.0	N17 40.1	161 11.0	N22 36.3
	01	266 04.7	216 48.0	52.0	299 16.9	20.5	130 17.1	40.0	176 13.1	36.3
	02	281 07.2	231 47.4	52.9	314 18.1	20.2	145 19.1	39.9	191 15.2	36.3
	03	296 09.6	246 46.8	. . 53.7	329 19.4	. . 19.9	160 21.2	. . 39.7	206 17.3	. . 36.3
	04	311 12.1	261 46.3	54.6	344 20.6	19.6	175 23.3	39.6	221 19.5	36.3
	05	326 14.6	276 45.7	55.4	359 21.8	19.3	190 25.4	39.5	236 21.6	36.3
	06	341 17.0	291 45.1	N16 56.3	14 23.0	S16 19.0	205 27.5	N17 39.4	251 23.7	N22 36.3
	07	356 19.5	306 44.5	57.1	29 24.2	18.7	220 29.5	39.3	266 25.8	36.3
	08	11 22.0	321 43.9	58.0	44 25.4	18.4	235 31.6	39.2	281 27.9	36.3
	09	26 24.4	336 43.3	. . 58.8	59 26.6	. . 18.1	250 33.7	. . 39.1	296 30.1	. . 36.3
	10	41 26.9	351 42.7	16 59.7	74 27.8	17.8	265 35.8	39.0	311 32.2	36.3
	11	56 29.4	6 42.2	17 00.5	89 29.0	17.5	280 37.9	38.9	326 34.3	36.3
	12	71 31.8	21 41.6	N17 01.4	104 30.2	S16 17.2	295 39.9	N17 38.7	341 36.4	N22 36.3
	13	86 34.3	36 41.0	02.2	119 31.5	16.9	310 42.0	38.6	356 38.6	36.3
	14	101 36.7	51 40.4	03.1	134 32.7	16.6	325 44.1	38.5	11 40.7	36.3
	15	116 39.2	66 39.8	. . 03.9	149 33.9	. . 16.3	340 46.2	. . 38.4	26 42.8	. . 36.3
	16	131 41.7	81 39.2	04.8	164 35.1	16.0	355 48.3	38.3	41 44.9	36.3
	17	146 44.1	96 38.6	05.6	179 36.3	15.7	10 50.3	38.2	56 47.0	36.3
	18	161 46.6	111 38.0	N17 06.5	194 37.5	S16 15.4	25 52.4	N17 38.1	71 49.2	N22 36.4
	19	176 49.1	126 37.4	07.3	209 38.7	15.1	40 54.5	38.0	86 51.3	36.4
	20	191 51.5	141 36.8	08.2	224 40.0	14.8	55 56.6	37.9	101 53.4	36.4
	21	206 54.0	156 36.2	. . 09.0	239 41.2	. . 14.5	70 58.6	. . 37.7	116 55.5	. . 36.4
	22	221 56.5	171 35.7	09.8	254 42.4	14.2	86 00.7	37.6	131 57.6	36.4
	23	236 58.9	186 35.1	10.7	269 43.6	13.9	101 02.8	37.5	146 59.8	36.4
4 (WEDNESDAY)	00	252 01.4	201 34.5	N17 11.5	284 44.8	S16 13.6	116 04.9	N17 37.4	162 01.9	N22 36.4
	01	267 03.9	216 33.9	12.4	299 46.0	13.3	131 06.9	37.3	177 04.0	36.4
	02	282 06.3	231 33.3	13.2	314 47.3	13.0	146 09.0	37.2	192 06.1	36.4
	03	297 08.8	246 32.7	. . 14.0	329 48.5	. . 12.7	161 11.1	. . 37.1	207 08.3	. . 36.4
	04	312 11.2	261 32.1	14.9	344 49.7	12.4	176 13.2	37.0	222 10.4	36.4
	05	327 13.7	276 31.5	15.7	359 50.9	12.1	191 15.3	36.8	237 12.5	36.4
	06	342 16.2	291 30.9	N17 16.6	14 52.1	S16 11.8	206 17.3	N17 36.7	252 14.6	N22 36.4
	07	357 18.6	306 30.3	17.4	29 53.4	11.5	221 19.4	36.6	267 16.7	36.4
	08	12 21.1	321 29.7	18.2	44 54.6	11.3	236 21.5	36.5	282 18.9	36.4
	09	27 23.6	336 29.1	. . 19.1	59 55.8	. . 11.0	251 23.6	. . 36.4	297 21.0	. . 36.4
	10	42 26.0	351 28.5	19.9	74 57.0	10.7	266 25.6	36.3	312 23.1	36.4
	11	57 28.5	6 27.9	20.7	89 58.3	10.4	281 27.7	36.2	327 25.2	36.4
	12	72 31.0	21 27.3	N17 21.6	104 59.5	S16 10.1	296 29.8	N17 36.1	342 27.4	N22 36.4
	13	87 33.4	36 26.7	22.4	120 00.7	09.8	311 31.9	35.9	357 29.5	36.4
	14	102 35.9	51 26.1	23.2	135 01.9	09.5	326 33.9	35.8	12 31.6	36.4
	15	117 38.4	66 25.5	. . 24.1	150 03.2	. . 09.2	341 36.0	. . 35.7	27 33.7	. . 36.4
	16	132 40.8	81 24.9	24.9	165 04.4	08.9	356 38.1	35.6	42 35.8	36.5
	17	147 43.3	96 24.3	25.7	180 05.6	08.6	11 40.2	35.5	57 38.0	36.5
	18	162 45.7	111 23.7	N17 26.5	195 06.8	S16 08.3	26 42.2	N17 35.4	72 40.1	N22 36.5
	19	177 48.2	126 23.1	27.4	210 08.1	08.0	41 44.3	35.3	87 42.2	36.5
	20	192 50.7	141 22.5	28.2	225 09.3	07.7	56 46.4	35.2	102 44.3	36.5
	21	207 53.1	156 21.9	. . 29.0	240 10.5	. . 07.4	71 48.4	. . 35.0	117 46.4	. . 36.5
	22	222 55.6	171 21.3	29.8	255 11.7	07.1	86 50.5	34.9	132 48.6	36.5
	23	237 58.1	186 20.7	30.7	270 13.0	06.8	101 52.6	34.8	147 50.7	36.5
5 (THURSDAY)	00	253 00.5	201 20.1	N17 31.5	285 14.2	S16 06.5	116 54.7	N17 34.7	162 52.8	N22 36.5
	01	268 03.0	216 19.5	32.3	300 15.4	06.2	131 56.7	34.6	177 54.9	36.5
	02	283 05.5	231 18.8	33.1	315 16.7	05.9	146 58.8	34.5	192 57.0	36.5
	03	298 07.9	246 18.2	. . 33.9	330 17.9	. . 05.6	162 00.9	. . 34.4	207 59.2	. . 36.5
	04	313 10.4	261 17.6	34.8	345 19.1	05.3	177 03.0	34.3	223 01.3	36.5
	05	328 12.9	276 17.0	35.6	0 20.4	05.0	192 05.0	34.1	238 03.4	36.5
	06	343 15.3	291 16.4	N17 36.4	15 21.6	S16 04.7	207 07.1	N17 34.0	253 05.5	N22 36.5
	07	358 17.8	306 15.8	37.2	30 22.8	04.4	222 09.2	33.9	268 07.6	36.5
	08	13 20.2	321 15.2	38.0	45 24.1	04.1	237 11.2	33.8	283 09.8	36.5
	09	28 22.7	336 14.6	. . 38.9	60 25.3	. . 03.8	252 13.3	. . 33.7	298 11.9	. . 36.5
	10	43 25.2	351 14.0	39.7	75 26.5	03.5	267 15.4	33.6	313 14.0	36.5
	11	58 27.6	6 13.4	40.5	90 27.8	03.3	282 17.5	33.5	328 16.1	36.5
	12	73 30.1	21 12.8	N17 41.3	105 29.0	S16 03.0	297 19.5	N17 33.3	343 18.2	N22 36.5
	13	88 32.6	36 12.1	42.1	120 30.2	02.7	312 21.6	33.2	358 20.4	36.5
	14	103 35.0	51 11.5	42.9	135 31.5	02.4	327 23.7	33.1	13 22.5	36.5
	15	118 37.5	66 10.9	. . 43.7	150 32.7	. . 02.1	342 25.7	. . 33.0	28 24.6	. . 36.5
	16	133 40.0	81 10.3	44.5	165 34.0	01.8	357 27.8	32.9	43 26.7	36.5
	17	148 42.4	96 09.7	45.4	180 35.2	01.5	12 29.9	32.8	58 28.8	36.6
	18	163 44.9	111 09.1	N17 46.2	195 36.4	S16 01.2	27 31.9	N17 32.7	73 31.0	N22 36.6
	19	178 47.3	126 08.5	47.0	210 37.7	00.9	42 34.0	32.5	88 33.1	36.6
	20	193 49.8	141 07.8	47.8	225 38.9	00.6	57 36.1	32.4	103 35.2	36.6
	21	208 52.3	156 07.2	. . 48.6	240 40.1	. . 00.3	72 38.2	. . 32.3	118 37.3	. . 36.6
	22	223 54.7	171 06.6	49.4	255 41.4	16 00.0	87 40.2	32.2	133 39.4	36.6
	23	238 57.2	186 06.0	50.2	270 42.6	S15 59.7	102 42.3	32.1	148 41.6	36.6
Mer. Pass.		h m 7 10.7	*v* −0.6	*d* 0.8	*v* 1.2	*d* 0.3	*v* 2.1	*d* 0.1	*v* 2.1	*d* 0.0

STARS

Name	SHA	Dec
	° ′	° ′
Acamar	315 24.6	S40 17.4
Achernar	335 32.8	S57 13.0
Acrux	173 18.2	S63 07.3
Adhara	255 19.1	S28 58.6
Aldebaran	290 58.9	N16 30.9
Alioth	166 27.1	N55 56.8
Alkaid	153 04.6	N49 18.0
Al Na'ir	27 53.4	S46 56.6
Alnilam	275 54.7	S 1 12.0
Alphard	218 04.0	S 8 40.4
Alphecca	126 17.3	N26 42.2
Alpheratz	357 51.9	N29 06.3
Altair	62 15.7	N 8 52.5
Ankaa	353 23.5	S42 17.2
Antares	112 35.7	S26 26.4
Arcturus	146 02.7	N19 10.0
Atria	107 44.2	S69 02.1
Avior	234 21.8	S59 31.4
Bellatrix	278 40.9	N 6 21.2
Betelgeuse	271 10.2	N 7 24.5
Canopus	264 00.2	S52 41.9
Capella	280 46.7	N46 00.1
Deneb	49 36.7	N45 17.2
Denebola	182 41.6	N14 33.3
Diphda	349 03.9	S17 58.1
Dubhe	194 01.1	N61 44.3
Elnath	278 23.0	N28 36.6
Eltanin	90 49.3	N51 29.2
Enif	33 54.8	N 9 53.2
Fomalhaut	15 32.6	S29 36.2
Gacrux	172 09.8	S57 08.1
Gienah	176 00.4	S17 33.7
Hadar	148 58.9	S60 23.5
Hamal	328 10.0	N23 28.5
Kaus Aust.	83 53.9	S34 23.0
Kochab	137 18.3	N74 08.7
Markab	13 46.3	N15 13.2
Menkar	314 23.7	N 4 06.1
Menkent	148 16.7	S36 23.3
Miaplacidus	221 42.1	S69 44.1
Mirfak	308 52.3	N49 52.3
Nunki	76 07.8	S26 17.6
Peacock	53 31.2	S56 43.3
Pollux	243 37.7	N28 01.2
Procyon	245 08.3	N 5 13.1
Rasalhague	96 13.5	N12 33.4
Regulus	207 52.0	N11 57.2
Rigel	281 20.0	S 8 11.9
Rigil Kent.	140 02.3	S60 51.1
Sabik	102 21.3	S15 43.8
Schedar	349 50.1	N56 33.0
Shaula	96 32.3	S37 06.4
Sirius	258 41.1	S16 43.2
Spica	158 39.4	S11 10.8
Suhail	222 58.6	S43 26.9
Vega	80 43.9	N38 47.0
Zuben'ubi	137 13.9	S16 03.4

	SHA	Mer. Pass.
	° ′	h m
Venus	309 33.1	10 34
Mars	32 43.4	5 01
Jupiter	224 03.5	16 13
Saturn	270 00.5	13 10

Day	UT (d h)	SUN GHA (° ′)	SUN Dec (° ′)	MOON GHA (° ′)	v (′)	MOON Dec (° ′)	d (′)	HP (′)
TUESDAY	3 00	180 30.1	N22 14.1	145 51.5	8.9	N26 24.3	1.3	55.3
	01	195 30.0	14.4	160 19.4	8.9	26 23.0	1.4	55.3
	02	210 29.9	14.7	174 47.3	8.8	26 21.6	1.5	55.3
	03	225 29.8	. . 15.0	189 15.1	8.9	26 20.1	1.7	55.4
	04	240 29.7	15.3	203 43.0	8.9	26 18.4	1.8	55.4
	05	255 29.6	15.6	218 10.9	8.8	26 16.6	2.0	55.4
	06	270 29.5	N22 15.9	232 38.7	8.9	N26 14.6	2.1	55.4
	07	285 29.4	16.3	247 06.6	8.9	26 12.5	2.3	55.4
	08	300 29.3	16.6	261 34.5	8.8	26 10.2	2.3	55.4
	09	315 29.2	. . 16.9	276 02.3	8.9	26 07.9	2.6	55.5
	10	330 29.1	17.2	290 30.2	8.8	26 05.3	2.6	55.5
	11	345 29.0	17.5	304 58.0	8.9	26 02.7	2.8	55.5
	12	0 28.9	N22 17.8	319 25.9	8.9	N25 59.9	2.9	55.5
	13	15 28.8	18.1	333 53.8	8.9	25 57.0	3.1	55.5
	14	30 28.7	18.4	348 21.7	8.9	25 53.9	3.2	55.6
	15	45 28.6	. . 18.7	2 49.6	8.9	25 50.7	3.4	55.6
	16	60 28.5	19.0	17 17.5	8.9	25 47.3	3.5	55.6
	17	75 28.4	19.4	31 45.4	8.9	25 43.8	3.6	55.6
	18	90 28.3	N22 19.7	46 13.3	8.9	N25 40.2	3.8	55.7
	19	105 28.2	20.0	60 41.2	9.0	25 36.4	3.9	55.7
	20	120 28.1	20.3	75 09.2	9.0	25 32.5	4.0	55.7
	21	135 28.0	. . 20.6	89 37.2	8.9	25 28.5	4.2	55.7
	22	150 27.9	20.9	104 05.1	9.0	25 24.3	4.3	55.7
	23	165 27.7	21.2	118 33.1	9.1	25 20.0	4.5	55.8
WEDNESDAY	4 00	180 27.6	N22 21.5	133 01.2	9.0	N25 15.5	4.6	55.8
	01	195 27.5	21.8	147 29.2	9.0	25 10.9	4.7	55.8
	02	210 27.4	22.1	161 57.2	9.1	25 06.2	4.8	55.8
	03	225 27.3	. . 22.4	176 25.3	9.1	25 01.4	5.0	55.8
	04	240 27.2	22.7	190 53.4	9.1	24 56.4	5.2	55.9
	05	255 27.1	23.0	205 21.5	9.2	24 51.2	5.2	55.9
	06	270 27.0	N22 23.3	219 49.7	9.1	N24 46.0	5.4	55.9
	07	285 26.9	23.6	234 17.8	9.2	24 40.6	5.6	55.9
	08	300 26.8	23.9	248 46.0	9.3	24 35.0	5.6	56.0
	09	315 26.7	. . 24.2	263 14.3	9.2	24 29.4	5.8	56.0
	10	330 26.6	24.5	277 42.5	9.3	24 23.6	6.0	56.0
	11	345 26.5	24.8	292 10.8	9.3	24 17.6	6.0	56.0
	12	0 26.4	N22 25.0	306 39.1	9.3	N24 11.6	6.2	56.0
	13	15 26.3	25.3	321 07.4	9.4	24 05.4	6.4	56.1
	14	30 26.2	25.6	335 35.8	9.4	23 59.0	6.4	56.1
	15	45 26.1	. . 25.9	350 04.2	9.4	23 52.6	6.6	56.1
	16	60 25.9	26.2	4 32.6	9.4	23 46.0	6.7	56.1
	17	75 25.8	26.5	19 01.0	9.5	23 39.3	6.9	56.2
	18	90 25.7	N22 26.8	33 29.5	9.5	N23 32.4	7.0	56.2
	19	105 25.6	27.1	47 58.0	9.6	23 25.4	7.1	56.2
	20	120 25.5	27.4	62 26.6	9.6	23 18.3	7.2	56.2
	21	135 25.4	. . 27.7	76 55.2	9.6	23 11.1	7.3	56.3
	22	150 25.3	27.9	91 23.8	9.7	23 03.8	7.5	56.3
	23	165 25.2	28.2	105 52.5	9.7	22 56.3	7.6	56.3
THURSDAY	5 00	180 25.1	N22 28.5	120 21.2	9.7	N22 48.7	7.8	56.3
	01	195 25.0	28.8	134 49.9	9.8	22 40.9	7.8	56.4
	02	210 24.9	29.1	149 18.7	9.8	22 33.1	8.0	56.4
	03	225 24.8	. . 29.4	163 47.5	9.8	22 25.1	8.1	56.4
	04	240 24.6	29.7	178 16.3	9.9	22 17.0	8.2	56.4
	05	255 24.5	29.9	192 45.2	9.9	22 08.8	8.4	56.5
	06	270 24.4	N22 30.2	207 14.1	10.0	N22 00.4	8.4	56.5
	07	285 24.3	30.5	221 43.1	10.0	21 52.0	8.6	56.5
	08	300 24.2	30.8	236 12.1	10.0	21 43.4	8.7	56.5
	09	315 24.1	. . 31.1	250 41.1	10.1	21 34.7	8.8	56.6
	10	330 24.0	31.3	265 10.2	10.1	21 25.9	9.0	56.6
	11	345 23.9	31.6	279 39.3	10.2	21 16.9	9.0	56.6
	12	0 23.8	N22 31.9	294 08.5	10.2	N21 07.9	9.2	56.6
	13	15 23.7	32.2	308 37.7	10.2	20 58.7	9.3	56.7
	14	30 23.5	32.4	323 06.9	10.3	20 49.4	9.4	56.7
	15	45 23.4	. . 32.7	337 36.2	10.3	20 40.0	9.5	56.7
	16	60 23.3	33.0	352 05.5	10.3	20 30.5	9.6	56.7
	17	75 23.2	33.3	6 34.8	10.4	20 20.9	9.8	56.8
	18	90 23.1	N22 33.5	21 04.2	10.5	N20 11.1	9.8	56.8
	19	105 23.0	33.8	35 33.7	10.4	20 01.3	10.0	56.8
	20	120 22.9	34.1	50 03.1	10.6	19 51.3	10.1	56.8
	21	135 22.8	. . 34.4	64 32.7	10.5	19 41.2	10.2	56.9
	22	150 22.7	34.6	79 02.2	10.6	19 31.0	10.3	56.9
	23	165 22.6	34.9	93 31.8	10.6	N19 20.7	10.4	56.9
		SD 15.8	*d* 0.3	SD	15.1	15.3		15.4

Lat.	Twilight Naut. (h m)	Twilight Civil (h m)	Sunrise (h m)	Moonrise 3 (h m)	Moonrise 4 (h m)	Moonrise 5 (h m)	Moonrise 6 (h m)
N 72	▭	▭	▭	▭	▭	▭	▭
N 70	▭	▭	▭	▭	▭	▭	06 54
68	▭	▭	▭	▭	▭	▭	07 35
66	////	////	01 01	▭	▭	05 53	08 03
64	////	////	01 52	▭	04 33	06 32	08 24
62	////	////	02 23	03 54	05 20	06 59	08 40
60	////	01 19	02 46	04 33	05 50	07 20	08 54
N 58	////	01 55	03 04	05 00	06 12	07 37	09 06
56	////	02 21	03 19	05 21	06 31	07 51	09 16
54	01 12	02 41	03 33	05 39	06 46	08 03	09 25
52	01 46	02 57	03 44	05 54	07 00	08 14	09 33
50	02 10	03 11	03 54	06 07	07 11	08 24	09 41
45	02 51	03 39	04 15	06 33	07 36	08 44	09 56
N 40	03 19	04 00	04 32	06 54	07 55	09 00	10 08
35	03 41	04 17	04 47	07 12	08 11	09 14	10 19
30	03 59	04 32	04 59	07 27	08 25	09 26	10 28
20	04 27	04 55	05 20	07 53	08 49	09 47	10 44
N 10	04 48	05 15	05 38	08 15	09 09	10 04	10 58
0	05 06	05 32	05 55	08 35	09 29	10 21	11 11
S 10	05 22	05 49	06 11	08 56	09 48	10 37	11 24
20	05 38	06 05	06 29	09 18	10 08	10 54	11 37
30	05 53	06 23	06 49	09 44	10 31	11 14	11 53
35	06 01	06 33	07 01	09 59	10 45	11 26	12 02
40	06 10	06 44	07 14	10 16	11 01	11 39	12 12
45	06 20	06 57	07 30	10 37	11 20	11 55	12 24
S 50	06 31	07 12	07 49	11 04	11 43	12 14	12 39
52	06 36	07 19	07 59	11 16	11 55	12 23	12 45
54	06 41	07 26	08 09	11 31	12 07	12 34	12 53
56	06 46	07 35	08 21	11 48	12 22	12 45	13 01
58	06 53	07 44	08 34	12 09	12 39	12 58	13 10
S 60	06 59	07 55	08 50	12 35	13 00	13 13	13 21

Lat.	Sunset (h m)	Twilight Civil (h m)	Twilight Naut. (h m)	Moonset 3 (h m)	Moonset 4 (h m)	Moonset 5 (h m)	Moonset 6 (h m)
N 72	▭	▭	▭	▭	▭	▭	▭
N 70	▭	▭	▭	▭	▭	▭	03 20
68	▭	▭	▭	▭	▭	▭	02 38
66	23 00	////	////	▭	▭	02 34	02 10
64	22 07	////	////	▭	02 04	01 54	01 48
62	21 35	////	////	00 51	01 17	01 26	01 30
60	21 12	22 40	////	00 12	00 46	01 05	01 15
N 58	20 53	22 03	////	24 23	00 23	00 47	01 02
56	20 38	21 37	////	24 04	00 04	00 32	00 51
54	20 25	21 17	22 47	23 49	24 19	00 19	00 42
52	20 13	21 00	22 12	23 35	24 08	00 08	00 33
50	20 03	20 46	21 48	23 23	23 58	24 25	00 25
45	19 42	20 18	21 06	22 58	23 37	24 09	00 09
N 40	19 24	19 57	20 38	22 38	23 20	23 55	24 26
35	19 10	19 40	20 16	22 21	23 05	23 43	24 17
30	18 58	19 25	19 58	22 07	22 52	23 33	24 10
20	18 37	19 01	19 30	21 42	22 30	23 15	23 57
N 10	18 19	18 42	19 08	21 21	22 12	23 00	23 45
0	18 02	18 24	18 50	21 01	21 54	22 45	23 34
S 10	17 45	18 08	18 34	20 41	21 36	22 30	23 23
20	17 28	17 51	18 19	20 20	21 17	22 14	23 12
30	17 07	17 34	18 03	19 55	20 54	21 56	22 58
35	16 56	17 23	17 55	19 40	20 41	21 45	22 50
40	16 42	17 12	17 46	19 23	20 26	21 33	22 41
45	16 26	17 00	17 36	19 02	20 08	21 18	22 31
S 50	16 07	16 45	17 26	18 36	19 45	21 00	22 18
52	15 57	16 38	17 21	18 23	19 34	20 51	22 12
54	15 47	16 30	17 15	18 09	19 22	20 42	22 05
56	15 35	16 21	17 10	17 52	19 07	20 31	21 57
58	15 22	16 12	17 04	17 31	18 51	20 18	21 49
S 60	15 06	16 01	16 57	17 05	18 30	20 04	21 39

Day	SUN Eqn. of Time 00^h (m s)	SUN Eqn. of Time 12^h (m s)	SUN Mer. Pass. (h m)	MOON Mer. Pass. Upper (h m)	MOON Mer. Pass. Lower (h m)	MOON Age (d)	MOON Phase (%)
3	02 01	01 56	11 58	14 48	02 22	03	10
4	01 51	01 46	11 58	15 41	03 15	04	17
5	01 41	01 35	11 58	16 33	04 07	05	26

UT	ARIES	VENUS −3.9		MARS −0.8		JUPITER −1.9		SATURN +0.1	
	GHA	GHA	Dec	GHA	Dec	GHA	Dec	GHA	Dec
d h	° ′	° ′	° ′	° ′	° ′	° ′	° ′	° ′	° ′
6 00	253 59.7	201 05.4	N17 51.0	285 43.9	S15 59.4	117 44.4	N17 32.0	163 43.7	N22 36.6
01	269 02.1	216 04.8	51.8	300 45.1	59.1	132 46.4	31.9	178 45.8	36.6
02	284 04.6	231 04.1	52.6	315 46.4	58.8	147 48.5	31.7	193 47.9	36.6
03	299 07.1	246 03.5	. . 53.4	330 47.6	. . 58.5	162 50.6	. . 31.6	208 50.0	. . 36.6
04	314 09.5	261 02.9	54.2	345 48.8	58.3	177 52.6	31.5	223 52.2	36.6
05	329 12.0	276 02.3	55.0	0 50.1	58.0	192 54.7	31.4	238 54.3	36.6
06	344 14.5	291 01.7	N17 55.8	15 51.3	S15 57.7	207 56.8	N17 31.3	253 56.4	N22 36.6
07	359 16.9	306 01.0	56.6	30 52.6	57.4	222 58.8	31.2	268 58.5	36.6
08	14 19.4	321 00.4	57.4	45 53.8	57.1	238 00.9	31.0	284 00.6	36.6
F 09	29 21.8	335 59.8	. . 58.2	60 55.1	. . 56.8	253 03.0	. . 30.9	299 02.8	. . 36.6
R 10	44 24.3	350 59.2	59.0	75 56.3	56.5	268 05.0	30.8	314 04.9	36.6
I 11	59 26.8	5 58.6	17 59.8	90 57.6	56.2	283 07.1	30.7	329 07.0	36.6
D 12	74 29.2	20 57.9	N18 00.6	105 58.8	S15 55.9	298 09.2	N17 30.6	344 09.1	N22 36.6
A 13	89 31.7	35 57.3	01.4	121 00.1	55.6	313 11.2	30.5	359 11.2	36.6
Y 14	104 34.2	50 56.7	02.2	136 01.3	55.3	328 13.3	30.4	14 13.4	36.6
15	119 36.6	65 56.1	. . 03.0	151 02.6	. . 55.0	343 15.4	. . 30.2	29 15.5	. . 36.6
16	134 39.1	80 55.4	03.8	166 03.8	54.7	358 17.4	30.1	44 17.6	36.6
17	149 41.6	95 54.8	04.6	181 05.1	54.5	13 19.5	30.0	59 19.7	36.6
18	164 44.0	110 54.2	N18 05.3	196 06.3	S15 54.2	28 21.6	N17 29.9	74 21.8	N22 36.6
19	179 46.5	125 53.6	06.1	211 07.6	53.9	43 23.6	29.8	89 24.0	36.6
20	194 49.0	140 52.9	06.9	226 08.8	53.6	58 25.7	29.7	104 26.1	36.7
21	209 51.4	155 52.3	. . 07.7	241 10.1	. . 53.3	73 27.8	. . 29.5	119 28.2	. . 36.7
22	224 53.9	170 51.7	08.5	256 11.3	53.0	88 29.8	29.4	134 30.3	36.7
23	239 56.3	185 51.1	09.3	271 12.6	52.7	103 31.9	29.3	149 32.4	36.7
7 00	254 58.8	200 50.4	N18 10.1	286 13.8	S15 52.4	118 34.0	N17 29.2	164 34.6	N22 36.7
01	270 01.3	215 49.8	10.9	301 15.1	52.1	133 36.0	29.1	179 36.7	36.7
02	285 03.7	230 49.2	11.6	316 16.3	51.8	148 38.1	29.0	194 38.8	36.7
03	300 06.2	245 48.5	. . 12.4	331 17.6	. . 51.5	163 40.1	. . 28.8	209 40.9	. . 36.7
04	315 08.7	260 47.9	13.2	346 18.9	51.3	178 42.2	28.7	224 43.0	36.7
05	330 11.1	275 47.3	14.0	1 20.1	51.0	193 44.3	28.6	239 45.2	36.7
06	345 13.6	290 46.6	N18 14.8	16 21.4	S15 50.7	208 46.3	N17 28.5	254 47.3	N22 36.7
07	0 16.1	305 46.0	15.5	31 22.6	50.4	223 48.4	28.4	269 49.4	36.7
S 08	15 18.5	320 45.4	16.3	46 23.9	50.1	238 50.5	28.3	284 51.5	36.7
A 09	30 21.0	335 44.7	. . 17.1	61 25.1	. . 49.8	253 52.5	. . 28.1	299 53.6	. . 36.7
T 10	45 23.4	350 44.1	17.9	76 26.4	49.5	268 54.6	28.0	314 55.7	36.7
U 11	60 25.9	5 43.5	18.7	91 27.7	49.2	283 56.7	27.9	329 57.9	36.7
R 12	75 28.4	20 42.8	N18 19.4	106 28.9	S15 48.9	298 58.7	N17 27.8	345 00.0	N22 36.7
D 13	90 30.8	35 42.2	20.2	121 30.2	48.7	314 00.8	27.7	0 02.1	36.7
A 14	105 33.3	50 41.6	21.0	136 31.4	48.4	329 02.8	27.6	15 04.2	36.7
Y 15	120 35.8	65 40.9	. . 21.8	151 32.7	. . 48.1	344 04.9	. . 27.4	30 06.3	. . 36.7
16	135 38.2	80 40.3	22.5	166 34.0	47.8	359 07.0	27.3	45 08.5	36.7
17	150 40.7	95 39.7	23.3	181 35.2	47.5	14 09.0	27.2	60 10.6	36.7
18	165 43.2	110 39.0	N18 24.1	196 36.5	S15 47.2	29 11.1	N17 27.1	75 12.7	N22 36.7
19	180 45.6	125 38.4	24.8	211 37.8	46.9	44 13.2	27.0	90 14.8	36.7
20	195 48.1	140 37.8	25.6	226 39.0	46.6	59 15.2	26.9	105 16.9	36.7
21	210 50.6	155 37.1	. . 26.4	241 40.3	. . 46.3	74 17.3	. . 26.7	120 19.0	. . 36.7
22	225 53.0	170 36.5	27.1	256 41.6	46.1	89 19.3	26.6	135 21.2	36.7
23	240 55.5	185 35.8	27.9	271 42.8	45.8	104 21.4	26.5	150 23.3	36.7
8 00	255 57.9	200 35.2	N18 28.7	286 44.1	S15 45.5	119 23.5	N17 26.4	165 25.4	N22 36.7
01	271 00.4	215 34.6	29.4	301 45.4	45.2	134 25.5	26.3	180 27.5	36.7
02	286 02.9	230 33.9	30.2	316 46.6	44.9	149 27.6	26.2	195 29.6	36.7
03	301 05.3	245 33.3	. . 31.0	331 47.9	. . 44.6	164 29.6	. . 26.0	210 31.8	. . 36.8
04	316 07.8	260 32.6	31.7	346 49.2	44.3	179 31.7	25.9	225 33.9	36.8
05	331 10.3	275 32.0	32.5	1 50.4	44.0	194 33.8	25.8	240 36.0	36.8
06	346 12.7	290 31.3	N18 33.3	16 51.7	S15 43.8	209 35.8	N17 25.7	255 38.1	N22 36.8
07	1 15.2	305 30.7	34.0	31 53.0	43.5	224 37.9	25.6	270 40.2	36.8
08	16 17.7	320 30.1	34.8	46 54.2	43.2	239 39.9	25.4	285 42.3	36.8
S 09	31 20.1	335 29.4	. . 35.5	61 55.5	. . 42.9	254 42.0	. . 25.3	300 44.5	. . 36.8
U 10	46 22.6	350 28.8	36.3	76 56.8	42.6	269 44.1	25.2	315 46.6	36.8
N 11	61 25.1	5 28.1	37.0	91 58.1	42.3	284 46.1	25.1	330 48.7	36.8
D 12	76 27.5	20 27.5	N18 37.8	106 59.3	S15 42.0	299 48.2	N17 25.0	345 50.8	N22 36.8
A 13	91 30.0	35 26.8	38.6	122 00.6	41.8	314 50.2	24.9	0 52.9	36.8
Y 14	106 32.4	50 26.2	39.3	137 01.9	41.5	329 52.3	24.7	15 55.1	36.8
15	121 34.9	65 25.5	. . 40.1	152 03.1	. . 41.2	344 54.4	. . 24.6	30 57.2	. . 36.8
16	136 37.4	80 24.9	40.8	167 04.4	40.9	359 56.4	24.5	45 59.3	36.8
17	151 39.8	95 24.2	41.6	182 05.7	40.6	14 58.5	24.4	61 01.4	36.8
18	166 42.3	110 23.6	N18 42.3	197 07.0	S15 40.3	30 00.5	N17 24.3	76 03.5	N22 36.8
19	181 44.8	125 22.9	43.1	212 08.3	40.0	45 02.6	24.1	91 05.6	36.8
20	196 47.2	140 22.3	43.8	227 09.5	39.8	60 04.6	24.0	106 07.8	36.8
21	211 49.7	155 21.6	. . 44.6	242 10.8	. . 39.5	75 06.7	. . 23.9	121 09.9	. . 36.8
22	226 52.2	170 21.0	45.3	257 12.1	39.2	90 08.8	23.8	136 12.0	36.8
23	241 54.6	185 20.3	46.1	272 13.4	38.9	105 10.8	23.7	151 14.1	36.8
Mer. Pass.	h m 6 58.9	v −0.6	d 0.8	v 1.3	d 0.3	v 2.1	d 0.1	v 2.1	d 0.0

STARS

Name	SHA	Dec
	° ′	° ′
Acamar	315 24.6	S40 17.4
Achernar	335 32.8	S57 13.0
Acrux	173 18.2	S63 07.3
Adhara	255 19.1	S28 58.6
Aldebaran	290 58.8	N16 30.9
Alioth	166 27.1	N55 56.8
Alkaid	153 04.6	N49 18.0
Al Na'ir	27 53.4	S46 56.6
Alnilam	275 54.7	S 1 12.0
Alphard	218 04.0	S 8 40.4
Alphecca	126 17.3	N26 42.2
Alpheratz	357 51.9	N29 06.3
Altair	62 15.7	N 8 52.5
Ankaa	353 23.5	S42 17.1
Antares	112 35.6	S26 26.4
Arcturus	146 02.7	N19 10.0
Atria	107 44.1	S69 02.1
Avior	234 21.8	S59 31.4
Bellatrix	278 40.8	N 6 21.2
Betelgeuse	271 10.2	N 7 24.5
Canopus	264 00.2	S52 41.9
Capella	280 46.7	N46 00.1
Deneb	49 36.7	N45 17.2
Denebola	182 41.6	N14 33.3
Diphda	349 03.9	S17 58.1
Dubhe	194 01.1	N61 44.3
Elnath	278 23.0	N28 36.6
Eltanin	90 49.3	N51 29.2
Enif	33 54.8	N 9 53.2
Fomalhaut	15 32.6	S29 36.2
Gacrux	172 09.8	S57 08.1
Gienah	176 00.4	S17 33.7
Hadar	148 58.9	S60 23.5
Hamal	328 10.0	N23 28.5
Kaus Aust.	83 53.9	S34 23.0
Kochab	137 18.3	N74 08.7
Markab	13 46.3	N15 13.2
Menkar	314 23.6	N 4 06.1
Menkent	148 16.7	S36 23.3
Miaplacidus	221 42.1	S69 44.0
Mirfak	308 52.3	N49 52.3
Nunki	76 07.8	S26 17.6
Peacock	53 31.1	S56 43.4
Pollux	243 37.7	N28 01.2
Procyon	245 08.3	N 5 13.1
Rasalhague	96 13.5	N12 33.4
Regulus	207 52.0	N11 57.2
Rigel	281 20.0	S 8 11.9
Rigil Kent.	140 02.3	S60 51.1
Sabik	102 21.3	S15 43.8
Schedar	349 50.0	N56 33.0
Shaula	96 32.3	S37 06.4
Sirius	258 41.1	S16 43.2
Spica	158 39.4	S11 10.8
Suhail	222 58.6	S43 26.9
Vega	80 43.9	N38 47.0
Zuben'ubi	137 13.9	S16 03.4

	SHA	Mer. Pass.
	° ′	h m
Venus	305 51.6	10 37
Mars	31 15.0	4 55
Jupiter	223 35.1	16 04
Saturn	269 35.7	13 00

UT	SUN GHA	SUN Dec	MOON GHA	v	MOON Dec	d	HP
d h	° ′	° ′	° ′	′	° ′	′	′
6 00	180 22.4	N22 35.2	108 01.4	10.7	N19 10.3	10.5	57.0
01	195 22.3	35.4	122 31.1	10.7	18 59.8	10.6	57.0
02	210 22.2	35.7	137 00.8	10.8	18 49.2	10.7	57.0
03	225 22.1	. . 36.0	151 30.6	10.8	18 38.5	10.8	57.0
04	240 22.0	36.2	166 00.4	10.8	18 27.7	10.9	57.1
05	255 21.9	36.5	180 30.2	10.9	18 16.8	11.1	57.1
06	270 21.8	N22 36.8	195 00.1	10.9	N18 05.7	11.1	57.1
07	285 21.7	37.0	209 30.0	10.9	17 54.6	11.2	57.2
08	300 21.5	37.3	223 59.9	11.0	17 43.4	11.4	57.2
09	315 21.4	. . 37.5	238 29.9	11.0	17 32.0	11.4	57.2
10	330 21.3	37.8	252 59.9	11.1	17 20.6	11.5	57.2
11	345 21.2	38.1	267 30.0	11.1	17 09.1	11.6	57.3
12	0 21.1	N22 38.3	282 00.1	11.1	N16 57.5	11.8	57.3
13	15 21.0	38.6	296 30.2	11.2	16 45.7	11.8	57.3
14	30 20.9	38.9	311 00.4	11.2	16 33.9	11.9	57.4
15	45 20.8	. . 39.1	325 30.6	11.2	16 22.0	12.0	57.4
16	60 20.6	39.4	340 00.8	11.3	16 10.0	12.1	57.4
17	75 20.5	39.6	354 31.1	11.3	15 57.9	12.2	57.4
18	90 20.4	N22 39.9	9 01.4	11.3	N15 45.7	12.2	57.5
19	105 20.3	40.1	23 31.7	11.4	15 33.5	12.4	57.5
20	120 20.2	40.4	38 02.1	11.4	15 21.1	12.5	57.5
21	135 20.1	. . 40.6	52 32.5	11.4	15 08.6	12.5	57.6
22	150 20.0	40.9	67 02.9	11.5	14 56.1	12.6	57.6
23	165 19.8	41.2	81 33.4	11.5	14 43.5	12.7	57.6
7 00	180 19.7	N22 41.4	96 03.9	11.5	N14 30.8	12.8	57.6
01	195 19.6	41.7	110 34.4	11.6	14 18.0	12.9	57.7
02	210 19.5	41.9	125 05.0	11.5	14 05.1	13.0	57.7
03	225 19.4	. . 42.2	139 35.5	11.6	13 52.1	13.0	57.7
04	240 19.3	42.4	154 06.1	11.7	13 39.1	13.2	57.8
05	255 19.1	42.7	168 36.8	11.6	13 25.9	13.2	57.8
06	270 19.0	N22 42.9	183 07.4	11.7	N13 12.7	13.3	57.8
07	285 18.9	43.2	197 38.1	11.7	12 59.4	13.3	57.9
08	300 18.8	43.4	212 08.8	11.8	12 46.1	13.5	57.9
09	315 18.7	. . 43.6	226 39.6	11.7	12 32.6	13.5	57.9
10	330 18.6	43.9	241 10.3	11.8	12 19.1	13.6	57.9
11	345 18.5	44.1	255 41.1	11.8	12 05.5	13.7	58.0
12	0 18.3	N22 44.4	270 11.9	11.8	N11 51.8	13.7	58.0
13	15 18.2	44.6	284 42.7	11.9	11 38.1	13.8	58.0
14	30 18.1	44.9	299 13.6	11.8	11 24.3	13.9	58.1
15	45 18.0	. . 45.1	313 44.4	11.9	11 10.4	13.9	58.1
16	60 17.9	45.3	328 15.3	11.9	10 56.5	14.1	58.1
17	75 17.8	45.6	342 46.2	11.9	10 42.4	14.1	58.2
18	90 17.6	N22 45.8	357 17.1	11.9	N10 28.3	14.1	58.2
19	105 17.5	46.1	11 48.0	12.0	10 14.2	14.2	58.2
20	120 17.4	46.3	26 19.0	11.9	10 00.0	14.3	58.2
21	135 17.3	. . 46.5	40 49.9	12.0	9 45.7	14.4	58.3
22	150 17.2	46.8	55 20.9	11.9	9 31.3	14.4	58.3
23	165 17.1	47.0	69 51.8	12.0	9 16.9	14.4	58.3
8 00	180 16.9	N22 47.3	84 22.8	12.0	N 9 02.5	14.6	58.4
01	195 16.8	47.5	98 53.8	12.0	8 47.9	14.6	58.4
02	210 16.7	47.7	113 24.8	12.0	8 33.3	14.6	58.4
03	225 16.6	. . 48.0	127 55.8	12.0	8 18.7	14.7	58.5
04	240 16.5	48.2	142 26.8	12.1	8 04.0	14.8	58.5
05	255 16.3	48.4	156 57.9	12.0	7 49.2	14.8	58.5
06	270 16.2	N22 48.7	171 28.9	12.0	N 7 34.4	14.9	58.6
07	285 16.1	48.9	185 59.9	12.0	7 19.5	14.9	58.6
08	300 16.0	49.1	200 30.9	12.1	7 04.6	15.0	58.6
09	315 15.9	. . 49.3	215 02.0	12.0	6 49.6	15.0	58.6
10	330 15.8	49.6	229 33.0	12.0	6 34.6	15.0	58.7
11	345 15.6	49.8	244 04.0	12.0	6 19.6	15.2	58.7
12	0 15.5	N22 50.0	258 35.0	12.1	N 6 04.4	15.1	58.7
13	15 15.4	50.3	273 06.1	12.0	5 49.3	15.2	58.8
14	30 15.3	50.5	287 37.1	12.0	5 34.1	15.3	58.8
15	45 15.2	. . 50.7	302 08.1	12.0	5 18.8	15.3	58.8
16	60 15.0	50.9	316 39.1	12.0	5 03.5	15.3	58.9
17	75 14.9	51.2	331 10.1	12.0	4 48.2	15.4	58.9
18	90 14.8	N22 51.4	345 41.1	11.9	N 4 32.8	15.4	58.9
19	105 14.7	51.6	0 12.0	12.0	4 17.4	15.5	58.9
20	120 14.6	51.8	14 43.0	11.9	4 01.9	15.5	59.0
21	135 14.4	. . 52.0	29 13.9	12.0	3 46.4	15.5	59.0
22	150 14.3	52.3	43 44.9	11.9	3 30.9	15.5	59.0
23	165 14.2	52.5	58 15.8	11.9	N 3 15.4	15.6	59.1
	SD 15.8	*d* 0.2	SD 15.6		15.8		16.0

Lat.	Twilight Naut.	Twilight Civil	Sunrise	Moonrise 6	Moonrise 7	Moonrise 8	Moonrise 9
°	h m	h m	h m	h m	h m	h m	h m
N 72	▭	▭	▭	▭	09 03	11 28	13 45
N 70	▭	▭	▭	06 54	09 27	11 38	13 44
68	▭	▭	▭	07 35	09 45	11 45	13 43
66	////	////	00 47	08 03	09 59	11 51	13 43
64	////	////	01 45	08 24	10 11	11 56	13 42
62	////	////	02 18	08 40	10 21	12 01	13 41
60	////	01 11	02 42	08 54	10 29	12 05	13 41
N 58	////	01 51	03 01	09 06	10 37	12 08	13 41
56	////	02 17	03 17	09 16	10 43	12 11	13 40
54	01 05	02 38	03 30	09 25	10 49	12 14	13 40
52	01 41	02 55	03 42	09 33	10 54	12 16	13 40
50	02 06	03 09	03 53	09 41	10 59	12 18	13 39
45	02 49	03 37	04 14	09 56	11 09	12 23	13 39
N 40	03 18	03 59	04 31	10 08	11 17	12 27	13 38
35	03 40	04 17	04 46	10 19	11 25	12 31	13 38
30	03 58	04 31	04 58	10 28	11 31	12 34	13 38
20	04 26	04 55	05 20	10 44	11 42	12 39	13 37
N 10	04 48	05 15	05 38	10 58	11 51	12 43	13 36
0	05 07	05 33	05 55	11 11	12 00	12 48	13 36
S 10	05 23	05 49	06 12	11 24	12 09	12 52	13 36
20	05 39	06 06	06 30	11 37	12 18	12 57	13 35
30	05 54	06 24	06 50	11 53	12 28	13 02	13 35
35	06 03	06 34	07 02	12 02	12 34	13 05	13 35
40	06 12	06 46	07 16	12 12	12 41	13 08	13 34
45	06 22	06 59	07 32	12 24	12 49	13 12	13 34
S 50	06 33	07 14	07 52	12 39	12 59	13 17	13 34
52	06 38	07 21	08 02	12 45	13 03	13 19	13 33
54	06 43	07 29	08 12	12 53	13 08	13 21	13 33
56	06 49	07 38	08 24	13 01	13 13	13 24	13 33
58	06 55	07 47	08 38	13 10	13 19	13 26	13 33
S 60	07 02	07 58	08 55	13 21	13 26	13 30	13 33

Lat.	Sunset	Twilight Civil	Twilight Naut.	Moonset 6	Moonset 7	Moonset 8	Moonset 9
°	h m	h m	h m	h m	h m	h m	h m
N 72	▭	▭	▭	▭	02 57	02 14	01 42
N 70	▭	▭	▭	03 20	02 31	02 02	01 39
68	▭	▭	▭	02 38	02 11	01 52	01 36
66	23 16	////	////	02 10	01 55	01 44	01 34
64	22 14	////	////	01 48	01 42	01 37	01 32
62	21 41	////	////	01 30	01 31	01 31	01 30
60	21 16	22 49	////	01 15	01 21	01 25	01 29
N 58	20 57	22 08	////	01 02	01 13	01 21	01 27
56	20 41	21 41	////	00 51	01 05	01 17	01 26
54	20 28	21 21	22 55	00 42	00 59	01 13	01 25
52	20 16	21 04	22 18	00 33	00 53	01 09	01 24
50	20 05	20 49	21 52	00 25	00 47	01 06	01 23
45	19 44	20 21	21 09	00 09	00 36	00 59	01 21
N 40	19 26	19 59	20 40	24 26	00 26	00 53	01 20
35	19 12	19 41	20 18	24 17	00 17	00 48	01 18
30	18 59	19 26	20 00	24 10	00 10	00 44	01 17
20	18 38	19 02	19 31	23 57	24 36	00 36	01 15
N 10	18 19	18 42	19 09	23 45	24 29	00 29	01 13
0	18 02	18 25	18 51	23 34	24 23	00 23	01 11
S 10	17 46	18 08	18 34	23 23	24 16	00 16	01 09
20	17 28	17 51	18 19	23 12	24 09	00 09	01 06
30	17 07	17 33	18 03	22 58	24 01	00 01	01 04
35	16 55	17 23	17 55	22 50	23 56	25 03	01 03
40	16 41	17 12	17 46	22 41	23 51	25 01	01 01
45	16 25	16 59	17 36	22 31	23 44	24 59	00 59
S 50	16 05	16 43	17 24	22 18	23 37	24 57	00 57
52	15 56	16 36	17 19	22 12	23 33	24 56	00 56
54	15 45	16 28	17 14	22 05	23 29	24 54	00 54
56	15 33	16 20	17 08	21 57	23 25	24 53	00 53
58	15 19	16 10	17 02	21 49	23 20	24 52	00 52
S 60	15 03	15 59	16 55	21 39	23 14	24 50	00 50

Day	SUN Eqn. of Time 00^h	SUN Eqn. of Time 12^h	SUN Mer. Pass.	MOON Mer. Pass. Upper	MOON Mer. Pass. Lower	MOON Age	MOON Phase
d	m s	m s	h m	h m	h m	d	%
6	01 30	01 25	11 59	17 23	04 58	06	36
7	01 19	01 14	11 59	18 11	05 47	07	46
8	01 08	01 02	11 59	18 59	06 35	08	57

	UT	ARIES	VENUS −3.9		MARS −0.9		JUPITER −1.9		SATURN +0.0		STARS		
	d h	GHA ° ′	GHA ° ′	Dec ° ′	GHA ° ′	Dec ° ′	GHA ° ′	Dec ° ′	GHA ° ′	Dec ° ′	Name	SHA ° ′	Dec ° ′
	9 00	256 57.1	200 19.7	N18 46.8	287 14.6	S15 38.6	120 12.9	N17 23.5	166 16.2	N22 36.8	Acamar	315 24.6	S40 17.4
	01	271 59.5	215 19.0	47.6	302 15.9	38.3	135 14.9	23.4	181 18.4	36.8	Achernar	335 32.8	S57 13.0
	02	287 02.0	230 18.4	48.3	317 17.2	38.1	150 17.0	23.3	196 20.5	36.8	Acrux	173 18.3	S63 07.3
	03	302 04.5	245 17.7	49.0	332 18.5	37.8	165 19.0	23.2	211 22.6	36.8	Adhara	255 19.1	S28 58.6
	04	317 06.9	260 17.1	49.8	347 19.8	37.5	180 21.1	23.1	226 24.7	36.8	Aldebaran	290 58.8	N16 30.9
	05	332 09.4	275 16.4	50.5	2 21.0	37.2	195 23.2	22.9	241 26.8	36.8			
	06	347 11.9	290 15.8	N18 51.3	17 22.3	S15 36.9	210 25.2	N17 22.8	256 28.9	N22 36.8	Alioth	166 27.2	N55 56.8
	07	2 14.3	305 15.1	52.0	32 23.6	36.6	225 27.3	22.7	271 31.1	36.8	Alkaid	153 04.7	N49 18.0
	08	17 16.8	320 14.5	52.7	47 24.9	36.4	240 29.3	22.6	286 33.2	36.8	Al Na'ir	27 53.3	S46 56.6
M	09	32 19.3	335 13.8	53.5	62 26.2	36.1	255 31.4	22.5	301 35.3	36.8	Alnilam	275 54.7	S 1 12.0
O	10	47 21.7	350 13.2	54.2	77 27.5	35.8	270 33.4	22.4	316 37.4	36.8	Alphard	218 04.0	S 8 40.4
N	11	62 24.2	5 12.5	55.0	92 28.7	35.5	285 35.5	22.2	331 39.5	36.8			
D	12	77 26.7	20 11.8	N18 55.7	107 30.0	S15 35.2	300 37.5	N17 22.1	346 41.6	N22 36.8	Alphecca	126 17.3	N26 42.2
A	13	92 29.1	35 11.2	56.4	122 31.3	34.9	315 39.6	22.0	1 43.8	36.8	Alpheratz	357 51.9	N29 06.3
Y	14	107 31.6	50 10.5	57.2	137 32.6	34.7	330 41.7	21.9	16 45.9	36.8	Altair	62 15.7	N 8 52.5
	15	122 34.0	65 09.9	57.9	152 33.9	34.4	345 43.7	21.8	31 48.0	36.8	Ankaa	353 23.5	S42 17.1
	16	137 36.5	80 09.2	58.6	167 35.2	34.1	0 45.8	21.6	46 50.1	36.9	Antares	112 35.6	S26 26.4
	17	152 39.0	95 08.5	18 59.4	182 36.5	33.8	15 47.8	21.5	61 52.2	36.9			
	18	167 41.4	110 07.9	N19 00.1	197 37.8	S15 33.5	30 49.9	N17 21.4	76 54.3	N22 36.9	Arcturus	146 02.7	N19 10.0
	19	182 43.9	125 07.2	00.8	212 39.0	33.2	45 51.9	21.3	91 56.5	36.9	Atria	107 44.1	S69 02.1
	20	197 46.4	140 06.6	01.5	227 40.3	33.0	60 54.0	21.2	106 58.6	36.9	Avior	234 21.9	S59 31.3
	21	212 48.8	155 05.9	02.3	242 41.6	32.7	75 56.0	21.0	122 00.7	36.9	Bellatrix	278 40.8	N 6 21.2
	22	227 51.3	170 05.2	03.0	257 42.9	32.4	90 58.1	20.9	137 02.8	36.9	Betelgeuse	271 10.2	N 7 24.5
	23	242 53.8	185 04.6	03.7	272 44.2	32.1	106 00.1	20.8	152 04.9	36.9			
	10 00	257 56.2	200 03.9	N19 04.5	287 45.5	S15 31.8	121 02.2	N17 20.7	167 07.0	N22 36.9	Canopus	264 00.2	S52 41.9
	01	272 58.7	215 03.3	05.2	302 46.8	31.6	136 04.3	20.5	182 09.2	36.9	Capella	280 46.7	N46 00.1
	02	288 01.2	230 02.6	05.9	317 48.1	31.3	151 06.3	20.4	197 11.3	36.9	Deneb	49 36.7	N45 17.2
	03	303 03.6	245 01.9	06.6	332 49.4	31.0	166 08.4	20.3	212 13.4	36.9	Denebola	182 41.6	N14 33.3
	04	318 06.1	260 01.3	07.4	347 50.7	30.7	181 10.4	20.2	227 15.5	36.9	Diphda	349 03.9	S17 58.1
	05	333 08.5	275 00.6	08.1	2 52.0	30.4	196 12.5	20.1	242 17.6	36.9			
	06	348 11.0	289 59.9	N19 08.8	17 53.3	S15 30.2	211 14.5	N17 19.9	257 19.7	N22 36.9	Dubhe	194 01.1	N61 44.3
	07	3 13.5	304 59.3	09.5	32 54.6	29.9	226 16.6	19.8	272 21.9	36.9	Elnath	278 23.0	N28 36.6
T	08	18 15.9	319 58.6	10.2	47 55.9	29.6	241 18.6	19.7	287 24.0	36.9	Eltanin	90 49.3	N51 29.2
U	09	33 18.4	334 57.9	10.9	62 57.2	29.3	256 20.7	19.6	302 26.1	36.9	Enif	33 54.8	N 9 53.2
E	10	48 20.9	349 57.3	11.7	77 58.5	29.0	271 22.7	19.5	317 28.2	36.9	Fomalhaut	15 32.6	S29 36.2
S	11	63 23.3	4 56.6	12.4	92 59.7	28.8	286 24.8	19.3	332 30.3	36.9			
D	12	78 25.8	19 55.9	N19 13.1	108 01.0	S15 28.5	301 26.8	N17 19.2	347 32.4	N22 36.9	Gacrux	172 09.8	S57 08.1
A	13	93 28.3	34 55.3	13.8	123 02.3	28.2	316 28.9	19.1	2 34.6	36.9	Gienah	176 00.4	S17 33.7
Y	14	108 30.7	49 54.6	14.5	138 03.6	27.9	331 30.9	19.0	17 36.7	36.9	Hadar	148 59.0	S60 23.5
	15	123 33.2	64 53.9	15.2	153 04.9	27.6	346 33.0	18.9	32 38.8	36.9	Hamal	328 10.0	N23 28.5
	16	138 35.6	79 53.2	15.9	168 06.2	27.4	1 35.0	18.7	47 40.9	36.9	Kaus Aust.	83 53.9	S34 23.0
	17	153 38.1	94 52.6	16.7	183 07.5	27.1	16 37.1	18.6	62 43.0	36.9			
	18	168 40.6	109 51.9	N19 17.4	198 08.8	S15 26.8	31 39.1	N17 18.5	77 45.1	N22 36.9	Kochab	137 18.3	N74 08.7
	19	183 43.0	124 51.2	18.1	213 10.2	26.5	46 41.2	18.4	92 47.3	36.9	Markab	13 46.2	N15 13.2
	20	198 45.5	139 50.6	18.8	228 11.5	26.2	61 43.2	18.2	107 49.4	36.9	Menkar	314 23.6	N 4 06.1
	21	213 48.0	154 49.9	19.5	243 12.8	26.0	76 45.3	18.1	122 51.5	36.9	Menkent	148 16.7	S36 23.3
	22	228 50.4	169 49.2	20.2	258 14.1	25.7	91 47.3	18.0	137 53.6	36.9	Miaplacidus	221 42.2	S69 44.0
	23	243 52.9	184 48.5	20.9	273 15.4	25.4	106 49.4	17.9	152 55.7	36.9			
	11 00	258 55.4	199 47.9	N19 21.6	288 16.7	S15 25.1	121 51.4	N17 17.8	167 57.8	N22 36.9	Mirfak	308 52.3	N49 52.2
	01	273 57.8	214 47.2	22.3	303 18.0	24.9	136 53.5	17.6	183 00.0	36.9	Nunki	76 07.8	S26 17.6
	02	289 00.3	229 46.5	23.0	318 19.3	24.6	151 55.5	17.5	198 02.1	36.9	Peacock	53 31.1	S56 43.4
	03	304 02.8	244 45.8	23.7	333 20.6	24.3	166 57.6	17.4	213 04.2	36.9	Pollux	243 37.7	N28 01.2
	04	319 05.2	259 45.2	24.4	348 21.9	24.0	181 59.6	17.3	228 06.3	36.9	Procyon	245 08.3	N 5 13.1
	05	334 07.7	274 44.5	25.1	3 23.2	23.8	197 01.7	17.1	243 08.4	36.9			
	06	349 10.1	289 43.8	N19 25.8	18 24.5	S15 23.5	212 03.7	N17 17.0	258 10.5	N22 36.9	Rasalhague	96 13.4	N12 33.4
W	07	4 12.6	304 43.1	26.5	33 25.8	23.2	227 05.8	16.9	273 12.7	36.9	Regulus	207 52.0	N11 57.2
E	08	19 15.1	319 42.5	27.2	48 27.1	22.9	242 07.8	16.8	288 14.8	36.9	Rigel	281 20.0	S 8 11.8
D	09	34 17.5	334 41.8	27.9	63 28.4	22.6	257 09.9	16.7	303 16.9	36.9	Rigil Kent.	140 02.4	S60 51.1
N	10	49 20.0	349 41.1	28.6	78 29.8	22.4	272 11.9	16.5	318 19.0	36.9	Sabik	102 21.3	S15 43.8
E	11	64 22.5	4 40.4	29.3	93 31.1	22.1	287 14.0	16.4	333 21.1	36.9			
S	12	79 24.9	19 39.7	N19 30.0	108 32.4	S15 21.8	302 16.0	N17 16.3	348 23.2	N22 36.9	Schedar	349 50.0	N56 33.0
D	13	94 27.4	34 39.1	30.7	123 33.7	21.5	317 18.1	16.2	3 25.4	36.9	Shaula	96 32.3	S37 06.4
A	14	109 29.9	49 38.4	31.4	138 35.0	21.3	332 20.1	16.0	18 27.5	36.9	Sirius	258 41.1	S16 43.2
	15	124 32.3	64 37.7	32.1	153 36.3	21.0	347 22.2	15.9	33 29.6	36.9	Spica	158 39.5	S11 10.8
Y	16	139 34.8	79 37.0	32.8	168 37.6	20.7	2 24.2	15.8	48 31.7	36.9	Suhail	222 58.6	S43 26.9
	17	154 37.3	94 36.3	33.5	183 38.9	20.4	17 26.3	15.7	63 33.8	36.9			
	18	169 39.7	109 35.7	N19 34.1	198 40.3	S15 20.2	32 28.3	N17 15.6	78 35.9	N22 37.0	Vega	80 43.9	N38 47.0
	19	184 42.2	124 35.0	34.8	213 41.6	19.9	47 30.3	15.4	93 38.0	37.0	Zuben'ubi	137 13.9	S16 03.4
	20	199 44.6	139 34.3	35.5	228 42.9	19.6	62 32.4	15.3	108 40.2	37.0			
	21	214 47.1	154 33.6	36.2	243 44.2	19.3	77 34.4	15.2	123 42.3	37.0			
	22	229 49.6	169 32.9	36.9	258 45.5	19.1	92 36.5	15.1	138 44.4	37.0			
	23	244 52.0	184 32.2	37.6	273 46.8	18.8	107 38.5	14.9	153 46.5	37.0			
	Mer. Pass.	h m 6 47.1	*v* −0.7	*d* 0.7	*v* 1.3	*d* 0.3	*v* 2.1	*d* 0.1	*v* 2.1	*d* 0.0			

	SHA ° ′	Mer. Pass. h m
Venus	302 07.7	10 40
Mars	29 49.3	4 49
Jupiter	223 06.0	15 54
Saturn	269 10.8	12 50

UT	SUN GHA	SUN Dec	MOON GHA	v	MOON Dec	d	HP
d h	° ′	° ′	° ′	′	° ′	′	′
9 00	180 14.1	N22 52.7	72 46.7	11.9	N 2 59.8	15.7	59.1
01	195 14.0	52.9	87 17.6	11.8	2 44.1	15.6	59.1
02	210 13.8	53.1	101 48.4	11.9	2 28.5	15.7	59.2
03	225 13.7	. . 53.3	116 19.3	11.8	2 12.8	15.7	59.2
04	240 13.6	53.6	130 50.1	11.8	1 57.1	15.7	59.2
05	255 13.5	53.8	145 20.9	11.8	1 41.4	15.8	59.2
06	270 13.4	N22 54.0	159 51.7	11.7	N 1 25.6	15.8	59.3
07	285 13.2	54.2	174 22.4	11.8	1 09.8	15.8	59.3
M 08	300 13.1	54.4	188 53.2	11.7	0 54.0	15.8	59.3
O 09	315 13.0	. . 54.6	203 23.9	11.6	0 38.2	15.8	59.4
N 10	330 12.9	54.8	217 54.5	11.7	0 22.4	15.9	59.4
D 11	345 12.8	55.1	232 25.2	11.6	N 0 06.5	15.9	59.4
A 12	0 12.6	N22 55.3	246 55.8	11.5	S 0 09.4	15.9	59.4
Y 13	15 12.5	55.5	261 26.3	11.6	0 25.3	15.9	59.5
14	30 12.4	55.7	275 56.9	11.5	0 41.2	15.9	59.5
15	45 12.3	. . 55.9	290 27.4	11.5	0 57.1	15.9	59.5
16	60 12.2	56.1	304 57.9	11.4	1 13.0	16.0	59.6
17	75 12.0	56.3	319 28.3	11.4	1 29.0	15.9	59.6
18	90 11.9	N22 56.5	333 58.7	11.3	S 1 44.9	16.0	59.6
19	105 11.8	56.7	348 29.0	11.3	2 00.9	15.9	59.6
20	120 11.7	56.9	2 59.3	11.3	2 16.8	16.0	59.7
21	135 11.5	. . 57.1	17 29.6	11.2	2 32.8	16.0	59.7
22	150 11.4	57.3	31 59.8	11.2	2 48.8	15.9	59.7
23	165 11.3	57.5	46 30.0	11.1	3 04.7	16.0	59.7
10 00	180 11.2	N22 57.7	61 00.1	11.1	S 3 20.7	16.0	59.8
01	195 11.1	57.9	75 30.2	11.0	3 36.7	15.9	59.8
02	210 10.9	58.1	90 00.2	11.0	3 52.6	16.0	59.8
03	225 10.8	. . 58.3	104 30.2	11.0	4 08.6	15.9	59.9
04	240 10.7	58.5	119 00.2	10.8	4 24.5	15.9	59.9
05	255 10.6	58.7	133 30.0	10.9	4 40.4	16.0	59.9
06	270 10.4	N22 58.9	147 59.9	10.7	S 4 56.4	15.9	59.9
07	285 10.3	59.1	162 29.6	10.7	5 12.3	15.9	60.0
T 08	300 10.2	59.3	176 59.3	10.7	5 28.2	15.8	60.0
U 09	315 10.1	. . 59.5	191 29.0	10.6	5 44.0	15.9	60.0
E 10	330 09.9	59.7	205 58.6	10.5	5 59.9	15.8	60.0
S 11	345 09.8	22 59.9	220 28.1	10.5	6 15.7	15.9	60.1
D 12	0 09.7	N23 00.1	234 57.6	10.4	S 6 31.6	15.8	60.1
A 13	15 09.6	00.3	249 27.0	10.3	6 47.4	15.7	60.1
Y 14	30 09.5	00.5	263 56.3	10.3	7 03.1	15.8	60.1
15	45 09.3	. . 00.7	278 25.6	10.2	7 18.9	15.7	60.1
16	60 09.2	00.9	292 54.8	10.2	7 34.6	15.7	60.2
17	75 09.1	01.1	307 24.0	10.1	7 50.3	15.7	60.2
18	90 09.0	N23 01.3	321 53.1	10.0	S 8 06.0	15.6	60.2
19	105 08.8	01.4	336 22.1	9.9	8 21.6	15.6	60.2
20	120 08.7	01.6	350 51.0	9.9	8 37.2	15.6	60.3
21	135 08.6	. . 01.8	5 19.9	9.8	8 52.8	15.5	60.3
22	150 08.5	02.0	19 48.7	9.7	9 08.3	15.4	60.3
23	165 08.3	02.2	34 17.4	9.6	9 23.7	15.5	60.3
11 00	180 08.2	N23 02.4	48 46.0	9.6	S 9 39.2	15.4	60.3
01	195 08.1	02.6	63 14.6	9.5	9 54.6	15.3	60.4
02	210 08.0	02.7	77 43.1	9.4	10 09.9	15.3	60.4
03	225 07.8	. . 02.9	92 11.5	9.3	10 25.2	15.3	60.4
04	240 07.7	03.1	106 39.8	9.3	10 40.5	15.2	60.4
05	255 07.6	03.3	121 08.1	9.1	10 55.7	15.1	60.4
06	270 07.5	N23 03.5	135 36.2	9.1	S11 10.8	15.1	60.4
W 07	285 07.3	03.7	150 04.3	9.0	11 25.9	15.0	60.5
E 08	300 07.2	03.8	164 32.3	8.9	11 40.9	15.0	60.5
D 09	315 07.1	. . 04.0	179 00.2	8.9	11 55.9	14.9	60.5
N 10	330 07.0	04.2	193 28.1	8.7	12 10.8	14.9	60.5
E 11	345 06.8	04.4	207 55.8	8.7	12 25.7	14.7	60.5
S 12	0 06.7	N23 04.5	222 23.5	8.5	S12 40.4	14.8	60.5
D 13	15 06.6	04.7	236 51.0	8.5	12 55.2	14.6	60.6
A 14	30 06.5	04.9	251 18.5	8.4	13 09.8	14.6	60.6
Y 15	45 06.3	. . 05.1	265 45.9	8.3	13 24.4	14.5	60.6
16	60 06.2	05.2	280 13.2	8.2	13 38.9	14.4	60.6
17	75 06.1	05.4	294 40.4	8.2	13 53.3	14.4	60.6
18	90 06.0	N23 05.6	309 07.6	8.0	S14 07.7	14.2	60.6
19	105 05.8	05.8	323 34.6	7.9	14 21.9	14.2	60.7
20	120 05.7	05.9	338 01.5	7.9	14 36.1	14.1	60.7
21	135 05.6	. . 06.1	352 28.4	7.7	14 50.2	14.1	60.7
22	150 05.4	06.3	6 55.1	7.7	15 04.3	13.9	60.7
23	165 05.3	06.4	21 21.8	7.5	S15 18.2	13.8	60.7
	SD 15.8	d 0.2	SD 16.2		16.4		16.5

Lat.	Twilight Naut.	Twilight Civil	Sunrise	Moonrise 9	10	11	12
°	h m	h m	h m	h m	h m	h m	h m
N 72	□	□	□	13 45	16 08	18 57	■
N 70	□	□	□	13 44	15 56	18 24	■
68	□	□	□	13 43	15 46	18 00	20 44
66	////	////	00 31	13 43	15 38	17 42	20 01
64	////	////	01 40	13 42	15 31	17 27	19 33
62	////	////	02 15	13 41	15 25	17 15	19 11
60	////	01 03	02 39	13 41	15 20	17 05	18 54
N 58	////	01 47	02 59	13 41	15 16	16 56	18 39
56	////	02 14	03 15	13 40	15 12	16 48	18 27
54	00 58	02 35	03 29	13 40	15 09	16 41	18 16
52	01 38	02 53	03 41	13 40	15 06	16 35	18 07
50	02 04	03 07	03 51	13 39	15 03	16 29	17 58
45	02 47	03 36	04 13	13 39	14 57	16 17	17 40
N 40	03 17	03 58	04 31	13 38	14 51	16 07	17 26
35	03 40	04 16	04 46	13 38	14 47	15 59	17 14
30	03 58	04 31	04 58	13 38	14 43	15 52	17 03
20	04 26	04 55	05 20	13 37	14 37	15 39	16 45
N 10	04 48	05 15	05 38	13 36	14 31	15 28	16 29
0	05 07	05 33	05 56	13 36	14 26	15 18	16 15
S 10	05 24	05 50	06 13	13 36	14 21	15 08	16 00
20	05 40	06 07	06 31	13 35	14 15	14 58	15 45
30	05 56	06 25	06 52	13 35	14 09	14 46	15 27
35	06 04	06 36	07 04	13 35	14 05	14 39	15 17
40	06 13	06 47	07 18	13 34	14 01	14 31	15 06
45	06 23	07 00	07 34	13 34	13 57	14 22	14 52
S 50	06 35	07 16	07 54	13 34	13 51	14 11	14 36
52	06 40	07 23	08 04	13 33	13 49	14 06	14 28
54	06 45	07 31	08 15	13 33	13 46	14 01	14 20
56	06 51	07 40	08 27	13 33	13 43	13 55	14 11
58	06 58	07 50	08 41	13 33	13 40	13 48	14 00
S 60	07 05	08 01	08 58	13 33	13 36	13 41	13 48

Lat.	Sunset	Twilight Civil	Twilight Naut.	Moonset 9	10	11	12
°	h m	h m	h m	h m	h m	h m	h m
N 72	□	□	□	01 42	01 11	00 35 / 23 38	■
N 70	□	□	□	01 39	01 16	00 50	00 14
68	□	□	□	01 36	01 20	01 03	00 39 / 23 55
66	23 34	////	////	01 34	01 24	01 13	00 59
64	22 20	////	////	01 32	01 27	01 21	01 15
62	21 45	////	////	01 30	01 29	01 29	01 29
60	21 20	22 57	////	01 29	01 32	01 35	01 40
N 58	21 00	22 13	////	01 27	01 34	01 41	01 50
56	20 44	21 45	////	01 26	01 36	01 46	01 59
54	20 30	21 24	23 03	01 25	01 37	01 51	02 07
52	20 18	21 07	22 22	01 24	01 39	01 55	02 14
50	20 08	20 52	21 56	01 23	01 40	01 59	02 21
45	19 46	20 23	21 12	01 21	01 43	02 07	02 34
N 40	19 28	20 01	20 42	01 20	01 46	02 14	02 46
35	19 13	19 43	20 19	01 18	01 48	02 20	02 55
30	19 01	19 28	20 01	01 17	01 50	02 25	03 04
20	18 39	19 03	19 32	01 15	01 53	02 34	03 19
N 10	18 20	18 43	19 10	01 13	01 56	02 43	03 32
0	18 03	18 25	18 52	01 11	01 59	02 50	03 44
S 10	17 46	18 09	18 35	01 09	02 02	02 58	03 57
20	17 28	17 52	18 19	01 06	02 05	03 06	04 10
30	17 07	17 33	18 03	01 04	02 09	03 16	04 25
35	16 55	17 23	17 54	01 03	02 11	03 21	04 34
40	16 41	17 11	17 45	01 01	02 13	03 27	04 44
45	16 24	16 58	17 35	00 59	02 15	03 34	04 56
S 50	16 04	16 42	17 24	00 57	02 19	03 43	05 11
52	15 55	16 35	17 19	00 56	02 20	03 47	05 18
54	15 44	16 27	17 13	00 54	02 22	03 52	05 26
56	15 31	16 18	17 07	00 53	02 23	03 57	05 34
58	15 17	16 08	17 01	00 52	02 25	04 02	05 44
S 60	15 00	15 57	16 53	00 50	02 27	04 09	05 55

Day	SUN Eqn. of Time 00^h	Eqn. of Time 12^h	Mer. Pass.	MOON Mer. Pass. Upper	Mer. Pass. Lower	Age	Phase
d	m s	m s	h m	h m	h m	d %	
9	00 57	00 51	11 59	19 48	07 23	09 68	
10	00 45	00 39	11 59	20 38	08 12	10 79	
11	00 33	00 27	12 00	21 31	09 04	11 88	

	UT (d h)	ARIES GHA	VENUS −3.9 GHA	VENUS Dec	MARS −1.0 GHA	MARS Dec	JUPITER −1.9 GHA	JUPITER Dec	SATURN +0.0 GHA	SATURN Dec
		° ′	° ′	° ′	° ′	° ′	° ′	° ′	° ′	° ′
	12 00	259 54.5	199 31.5	N19 38.3	288 48.2	S15 18.5	122 40.6	N17 14.8	168 48.6	N22 37.0
	01	274 57.0	214 30.9	38.9	303 49.5	18.2	137 42.6	14.7	183 50.7	37.0
	02	289 59.4	229 30.2	39.6	318 50.8	18.0	152 44.7	14.6	198 52.9	37.0
	03	305 01.9	244 29.5	. . 40.3	333 52.1	. . 17.7	167 46.7	. . 14.4	213 55.0	. . 37.0
	04	320 04.4	259 28.8	41.0	348 53.4	17.4	182 48.8	14.3	228 57.1	37.0
	05	335 06.8	274 28.1	41.7	3 54.8	17.2	197 50.8	14.2	243 59.2	37.0
	06	350 09.3	289 27.4	N19 42.3	18 56.1	S15 16.9	212 52.9	N17 14.1	259 01.3	N22 37.0
	07	5 11.7	304 26.7	43.0	33 57.4	16.6	227 54.9	13.9	274 03.4	37.0
T	08	20 14.2	319 26.0	43.7	48 58.7	16.3	242 56.9	13.8	289 05.5	37.0
H	09	35 16.7	334 25.4	. . 44.4	64 00.0	. . 16.1	257 59.0	. . 13.7	304 07.7	. . 37.0
U	10	50 19.1	349 24.7	45.0	79 01.4	15.8	273 01.0	13.6	319 09.8	37.0
R	11	65 21.6	4 24.0	45.7	94 02.7	15.5	288 03.1	13.5	334 11.9	37.0
S	12	80 24.1	19 23.3	N19 46.4	109 04.0	S15 15.2	303 05.1	N17 13.3	349 14.0	N22 37.0
D	13	95 26.5	34 22.6	47.1	124 05.3	15.0	318 07.2	13.2	4 16.1	37.0
A	14	110 29.0	49 21.9	47.7	139 06.7	14.7	333 09.2	13.1	19 18.2	37.0
Y	15	125 31.5	64 21.2	. . 48.4	154 08.0	. . 14.4	348 11.3	. . 13.0	34 20.4	. . 37.0
	16	140 33.9	79 20.5	49.1	169 09.3	14.2	3 13.3	12.8	49 22.5	37.0
	17	155 36.4	94 19.8	49.7	184 10.7	13.9	18 15.3	12.7	64 24.6	37.0
	18	170 38.9	109 19.1	N19 50.4	199 12.0	S15 13.6	33 17.4	N17 12.6	79 26.7	N22 37.0
	19	185 41.3	124 18.4	51.1	214 13.3	13.4	48 19.4	12.5	94 28.8	37.0
	20	200 43.8	139 17.7	51.7	229 14.6	13.1	63 21.5	12.3	109 30.9	37.0
	21	215 46.2	154 17.0	. . 52.4	244 16.0	. . 12.8	78 23.5	. . 12.2	124 33.0	. . 37.0
	22	230 48.7	169 16.3	53.1	259 17.3	12.5	93 25.6	12.1	139 35.2	37.0
	23	245 51.2	184 15.7	53.7	274 18.6	12.3	108 27.6	12.0	154 37.3	37.0
	13 00	260 53.6	199 15.0	N19 54.4	289 20.0	S15 12.0	123 29.6	N17 11.8	169 39.4	N22 37.0
	01	275 56.1	214 14.3	55.1	304 21.3	11.7	138 31.7	11.7	184 41.5	37.0
	02	290 58.6	229 13.6	55.7	319 22.6	11.5	153 33.7	11.6	199 43.6	37.0
	03	306 01.0	244 12.9	. . 56.4	334 24.0	. . 11.2	168 35.8	. . 11.5	214 45.7	. . 37.0
	04	321 03.5	259 12.2	57.0	349 25.3	10.9	183 37.8	11.3	229 47.9	37.0
	05	336 06.0	274 11.5	57.7	4 26.6	10.7	198 39.9	11.2	244 50.0	37.0
	06	351 08.4	289 10.8	N19 58.3	19 28.0	S15 10.4	213 41.9	N17 11.1	259 52.1	N22 37.0
	07	6 10.9	304 10.1	59.0	34 29.3	10.1	228 43.9	11.0	274 54.2	37.0
	08	21 13.4	319 09.4	19 59.7	49 30.6	09.8	243 46.0	10.8	289 56.3	37.0
F	09	36 15.8	334 08.7	20 00.3	64 32.0	. . 09.6	258 48.0	. . 10.7	304 58.4	. . 37.0
R	10	51 18.3	349 08.0	01.0	79 33.3	09.3	273 50.1	10.6	320 00.5	37.0
I	11	66 20.7	4 07.3	01.6	94 34.6	09.0	288 52.1	10.5	335 02.7	37.0
D	12	81 23.2	19 06.6	N20 02.3	109 36.0	S15 08.8	303 54.1	N17 10.3	350 04.8	N22 37.0
A	13	96 25.7	34 05.9	02.9	124 37.3	08.5	318 56.2	10.2	5 06.9	37.0
Y	14	111 28.1	49 05.2	03.6	139 38.7	08.2	333 58.2	10.1	20 09.0	37.0
	15	126 30.6	64 04.5	. . 04.2	154 40.0	. . 08.0	349 00.3	. . 10.0	35 11.1	. . 37.0
	16	141 33.1	79 03.7	04.9	169 41.3	07.7	4 02.3	09.8	50 13.2	37.0
	17	156 35.5	94 03.0	05.5	184 42.7	07.4	19 04.3	09.7	65 15.3	37.0
	18	171 38.0	109 02.3	N20 06.2	199 44.0	S15 07.2	34 06.4	N17 09.6	80 17.5	N22 37.0
	19	186 40.5	124 01.6	06.8	214 45.4	06.9	49 08.4	09.5	95 19.6	37.0
	20	201 42.9	139 00.9	07.4	229 46.7	06.6	64 10.5	09.3	110 21.7	37.0
	21	216 45.4	154 00.2	. . 08.1	244 48.1	. . 06.4	79 12.5	. . 09.2	125 23.8	. . 37.0
	22	231 47.9	168 59.5	08.7	259 49.4	06.1	94 14.5	09.1	140 25.9	37.0
	23	246 50.3	183 58.8	09.4	274 50.7	05.8	109 16.6	09.0	155 28.0	37.0
	14 00	261 52.8	198 58.1	N20 10.0	289 52.1	S15 05.6	124 18.6	N17 08.8	170 30.1	N22 37.0
	01	276 55.2	213 57.4	10.6	304 53.4	05.3	139 20.7	08.7	185 32.3	37.0
	02	291 57.7	228 56.7	11.3	319 54.8	05.0	154 22.7	08.6	200 34.4	37.0
	03	307 00.2	243 56.0	. . 11.9	334 56.1	. . 04.8	169 24.7	. . 08.4	215 36.5	. . 37.0
	04	322 02.6	258 55.3	12.6	349 57.5	04.5	184 26.8	08.3	230 38.6	37.0
	05	337 05.1	273 54.6	13.2	4 58.8	04.2	199 28.8	08.2	245 40.7	37.0
	06	352 07.6	288 53.8	N20 13.8	20 00.2	S15 04.0	214 30.9	N17 08.1	260 42.8	N22 37.0
	07	7 10.0	303 53.1	14.5	35 01.5	03.7	229 32.9	07.9	275 44.9	37.0
S	08	22 12.5	318 52.4	15.1	50 02.9	03.5	244 34.9	07.8	290 47.1	37.0
A	09	37 15.0	333 51.7	. . 15.7	65 04.2	. . 03.2	259 37.0	. . 07.7	305 49.2	. . 37.0
T	10	52 17.4	348 51.0	16.4	80 05.6	02.9	274 39.0	07.6	320 51.3	37.0
U	11	67 19.9	3 50.3	17.0	95 06.9	02.7	289 41.0	07.4	335 53.4	37.0
R	12	82 22.4	18 49.6	N20 17.6	110 08.3	S15 02.4	304 43.1	N17 07.3	350 55.5	N22 37.0
D	13	97 24.8	33 48.9	18.2	125 09.6	02.1	319 45.1	07.2	5 57.6	37.0
A	14	112 27.3	48 48.1	18.9	140 11.0	01.9	334 47.2	07.1	20 59.7	37.0
Y	15	127 29.7	63 47.4	. . 19.5	155 12.3	. . 01.6	349 49.2	. . 06.9	36 01.9	. . 37.0
	16	142 32.2	78 46.7	20.1	170 13.7	01.3	4 51.2	06.8	51 04.0	37.0
	17	157 34.7	93 46.0	20.7	185 15.1	01.1	19 53.3	06.7	66 06.1	37.0
	18	172 37.1	108 45.3	N20 21.4	200 16.4	S15 00.8	34 55.3	N17 06.5	81 08.2	N22 37.0
	19	187 39.6	123 44.6	22.0	215 17.8	00.6	49 57.3	06.4	96 10.3	37.0
	20	202 42.1	138 43.9	22.6	230 19.1	00.3	64 59.4	06.3	111 12.4	37.0
	21	217 44.5	153 43.1	. . 23.2	245 20.5	15 00.0	80 01.4	. . 06.2	126 14.5	. . 37.0
	22	232 47.0	168 42.4	23.8	260 21.8	14 59.8	95 03.5	06.0	141 16.7	37.0
	23	247 49.5	183 41.7	24.5	275 23.2	S14 59.5	110 05.5	05.9	156 18.8	37.0
	Mer. Pass.	h m 6 35.3	v −0.7	d 0.7	v 1.3	d 0.3	v 2.0	d 0.1	v 2.1	d 0.0

STARS

Name	SHA	Dec
	° ′	° ′
Acamar	315 24.6	S40 17.4
Achernar	335 32.8	S57 13.0
Acrux	173 18.3	S63 07.3
Adhara	255 19.1	S28 58.6
Aldebaran	290 58.8	N16 30.9
Alioth	166 27.2	N55 56.8
Alkaid	153 04.7	N49 18.0
Al Na'ir	27 53.3	S46 56.6
Alnilam	275 54.7	S 1 12.0
Alphard	218 04.0	S 8 40.4
Alphecca	126 17.3	N26 42.3
Alpheratz	357 51.8	N29 06.3
Altair	62 15.7	N 8 52.5
Ankaa	353 23.5	S42 17.1
Antares	112 35.6	S26 26.4
Arcturus	146 02.7	N19 10.0
Atria	107 44.1	S69 02.1
Avior	234 21.9	S59 31.3
Bellatrix	278 40.8	N 6 21.2
Betelgeuse	271 10.2	N 7 24.5
Canopus	264 00.2	S52 41.9
Capella	280 46.7	N46 00.1
Deneb	49 36.6	N45 17.3
Denebola	182 41.7	N14 33.3
Diphda	349 03.9	S17 58.1
Dubhe	194 01.2	N61 44.3
Elnath	278 23.0	N28 36.6
Eltanin	90 49.3	N51 29.2
Enif	33 54.7	N 9 53.3
Fomalhaut	15 32.5	S29 36.2
Gacrux	172 09.8	S57 08.1
Gienah	176 00.4	S17 33.7
Hadar	148 59.0	S60 23.6
Hamal	328 10.0	N23 28.5
Kaus Aust.	83 53.9	S34 23.0
Kochab	137 18.4	N74 08.7
Markab	13 46.2	N15 13.2
Menkar	314 23.6	N 4 06.1
Menkent	148 16.8	S36 23.3
Miaplacidus	221 42.2	S69 44.0
Mirfak	308 52.3	N49 52.2
Nunki	76 07.7	S26 17.6
Peacock	53 31.1	S56 43.4
Pollux	243 37.7	N28 01.2
Procyon	245 08.3	N 5 13.1
Rasalhague	96 13.4	N12 33.4
Regulus	207 52.0	N11 57.2
Rigel	281 20.0	S 8 11.8
Rigil Kent.	140 02.4	S60 51.1
Sabik	102 21.3	S15 43.8
Schedar	349 50.0	N56 33.0
Shaula	96 32.2	S37 06.4
Sirius	258 41.1	S16 43.2
Spica	158 39.5	S11 10.8
Suhail	222 58.6	S43 26.9
Vega	80 43.9	N38 47.1
Zuben'ubi	137 13.9	S16 03.4

	SHA	Mer. Pass.
	° ′	h m
Venus	298 21.3	10 44
Mars	28 26.3	4 42
Jupiter	222 36.0	15 44
Saturn	268 45.8	12 40

UT (d h)	SUN GHA (° ′)	SUN Dec (° ′)	MOON GHA (° ′)	v (′)	MOON Dec (° ′)	d (′)	HP (′)
12 00 THURSDAY	180 05.2	N23 06.6	35 48.3	7.5	S15 32.0	13.8	60.7
01	195 05.1	06.8	50 14.8	7.4	15 45.8	13.6	60.7
02	210 04.9	06.9	64 41.2	7.3	15 59.4	13.6	60.7
03	225 04.8	. . 07.1	79 07.5	7.1	16 13.0	13.5	60.7
04	240 04.7	07.3	93 33.6	7.1	16 26.5	13.3	60.7
05	255 04.6	07.4	107 59.7	7.0	16 39.8	13.3	60.8
06	270 04.4	N23 07.6	122 25.7	6.9	S16 53.1	13.2	60.8
07	285 04.3	07.8	136 51.6	6.8	17 06.3	13.0	60.8
08	300 04.2	07.9	151 17.4	6.7	17 19.3	13.0	60.8
09	315 04.1	. . 08.1	165 43.1	6.6	17 32.3	12.8	60.8
10	330 03.9	08.2	180 08.7	6.5	17 45.1	12.7	60.8
11	345 03.8	08.4	194 34.2	6.4	17 57.8	12.7	60.8
12	0 03.7	N23 08.6	208 59.6	6.3	S18 10.5	12.5	60.8
13	15 03.5	08.7	223 24.9	6.2	18 23.0	12.3	60.8
14	30 03.4	08.9	237 50.1	6.1	18 35.3	12.3	60.8
15	45 03.3	. . 09.0	252 15.2	6.0	18 47.6	12.1	60.8
16	60 03.2	09.2	266 40.2	5.9	18 59.7	12.0	60.8
17	75 03.0	09.4	281 05.1	5.8	19 11.7	11.9	60.8
18	90 02.9	N23 09.5	295 29.9	5.7	S19 23.6	11.8	60.8
19	105 02.8	09.7	309 54.6	5.6	19 35.4	11.6	60.8
20	120 02.6	09.8	324 19.2	5.6	19 47.0	11.5	60.8
21	135 02.5	. . 10.0	338 43.8	5.4	19 58.5	11.4	60.8
22	150 02.4	10.1	353 08.2	5.3	20 09.9	11.2	60.8
23	165 02.3	10.3	7 32.5	5.3	20 21.1	11.1	60.8
13 00 FRIDAY	180 02.1	N23 10.4	21 56.8	5.1	S20 32.2	10.9	60.8
01	195 02.0	10.6	36 20.9	5.1	20 43.1	10.9	60.8
02	210 01.9	10.7	50 45.0	4.9	20 54.0	10.6	60.8
03	225 01.7	. . 10.9	65 08.9	4.9	21 04.6	10.5	60.8
04	240 01.6	11.0	79 32.8	4.8	21 15.1	10.4	60.8
05	255 01.5	11.2	93 56.6	4.7	21 25.5	10.2	60.8
06	270 01.4	N23 11.3	108 20.3	4.6	S21 35.7	10.1	60.8
07	285 01.2	11.5	122 43.9	4.5	21 45.8	9.9	60.8
08	300 01.1	11.6	137 07.4	4.5	21 55.7	9.8	60.8
09	315 01.0	. . 11.8	151 30.9	4.3	22 05.5	9.6	60.8
10	330 00.8	11.9	165 54.2	4.3	22 15.1	9.4	60.8
11	345 00.7	12.0	180 17.5	4.2	22 24.5	9.3	60.8
12	0 00.6	N23 12.2	194 40.7	4.1	S22 33.8	9.2	60.8
13	15 00.5	12.3	209 03.8	4.0	22 43.0	8.9	60.8
14	30 00.3	12.5	223 26.8	3.9	22 51.9	8.8	60.8
15	45 00.2	. . 12.6	237 49.7	3.9	23 00.7	8.7	60.8
16	60 00.1	12.8	252 12.6	3.8	23 09.4	8.4	60.8
17	74 59.9	12.9	266 35.4	3.7	23 17.8	8.4	60.8
18	89 59.8	N23 13.0	280 58.1	3.7	S23 26.2	8.1	60.8
19	104 59.7	13.2	295 20.8	3.6	23 34.3	7.9	60.7
20	119 59.6	13.3	309 43.4	3.5	23 42.2	7.8	60.7
21	134 59.4	. . 13.4	324 05.9	3.4	23 50.0	7.6	60.7
22	149 59.3	13.6	338 28.3	3.4	23 57.6	7.5	60.7
23	164 59.2	13.7	352 50.7	3.3	24 05.1	7.2	60.7
14 00 SATURDAY	179 59.0	N23 13.8	7 13.0	3.3	S24 12.3	7.1	60.7
01	194 58.9	14.0	21 35.3	3.2	24 19.4	6.9	60.7
02	209 58.8	14.1	35 57.5	3.2	24 26.3	6.7	60.7
03	224 58.6	. . 14.2	50 19.7	3.1	24 33.0	6.6	60.6
04	239 58.5	14.4	64 41.8	3.0	24 39.6	6.3	60.6
05	254 58.4	14.5	79 03.8	3.0	24 45.9	6.2	60.6
06	269 58.3	N23 14.6	93 25.8	3.0	S24 52.1	6.0	60.6
07	284 58.1	14.8	107 47.8	2.9	24 58.1	5.8	60.6
08	299 58.0	14.9	122 09.7	2.8	25 03.9	5.6	60.6
09	314 57.9	. . 15.0	136 31.5	2.9	25 09.5	5.4	60.6
10	329 57.7	15.1	150 53.4	2.7	25 14.9	5.2	60.5
11	344 57.6	15.3	165 15.1	2.8	25 20.1	5.1	60.5
12	359 57.5	N23 15.4	179 36.9	2.7	S25 25.2	4.8	60.5
13	14 57.3	15.5	193 58.6	2.7	25 30.0	4.7	60.5
14	29 57.2	15.6	208 20.3	2.7	25 34.7	4.5	60.5
15	44 57.1	. . 15.8	222 42.0	2.6	25 39.2	4.3	60.5
16	59 56.9	15.9	237 03.6	2.6	25 43.5	4.0	60.4
17	74 56.8	16.0	251 25.2	2.6	25 47.5	3.9	60.4
18	89 56.7	N23 16.1	265 46.8	2.6	S25 51.4	3.7	60.4
19	104 56.6	16.3	280 08.4	2.6	25 55.1	3.5	60.4
20	119 56.4	16.4	294 30.0	2.6	25 58.6	3.4	60.4
21	134 56.3	. . 16.5	308 51.6	2.5	26 02.0	3.1	60.3
22	149 56.2	16.6	323 13.1	2.6	26 05.1	2.9	60.3
23	164 56.0	16.7	337 34.7	2.5	S26 08.0	2.7	60.3
	SD 15.8	*d* 0.1	SD 16.6		16.6		16.5

Lat. (°)	Twilight Naut. (h m)	Twilight Civil (h m)	Sunrise (h m)	Moonrise 12 (h m)	Moonrise 13 (h m)	Moonrise 14 (h m)	Moonrise 15 (h m)
N 72	□	□	□	■	■	■	■
N 70	□	□	□	■	■	■	■
68	□	□	□	20 44	■	■	■
66	□	□	□	20 01	■	■	■
64	////	////	01 36	19 33	21 52	■	■
62	////	////	02 12	19 11	21 12	23 03	24 11
60	////	00 57	02 37	18 54	20 44	22 23	23 31
N 58	////	01 43	02 57	18 39	20 23	21 55	23 03
56	////	02 12	03 14	18 27	20 06	21 34	22 41
54	00 52	02 34	03 28	18 16	19 51	21 16	22 23
52	01 35	02 51	03 40	18 07	19 38	21 01	22 08
50	02 02	03 06	03 51	17 58	19 27	20 48	21 54
45	02 46	03 36	04 13	17 40	19 03	20 21	21 27
N 40	03 16	03 58	04 31	17 26	18 45	20 00	21 05
35	03 39	04 16	04 45	17 14	18 29	19 42	20 47
30	03 58	04 31	04 58	17 03	18 16	19 27	20 32
20	04 26	04 56	05 20	16 45	17 53	19 01	20 06
N 10	04 49	05 16	05 39	16 29	17 33	18 39	19 43
0	05 08	05 34	05 56	16 15	17 15	18 18	19 22
S 10	05 24	05 51	06 14	16 00	16 57	17 58	19 01
20	05 40	06 08	06 32	15 45	16 38	17 36	18 39
30	05 57	06 27	06 53	15 27	16 15	17 11	18 13
35	06 05	06 37	07 05	15 17	16 02	16 56	17 57
40	06 15	06 49	07 19	15 06	15 48	16 39	17 40
45	06 25	07 02	07 36	14 52	15 30	16 18	17 18
S 50	06 36	07 18	07 56	14 36	15 09	15 53	16 51
52	06 42	07 25	08 06	14 28	14 58	15 40	16 38
54	06 47	07 33	08 17	14 20	14 47	15 26	16 23
56	06 53	07 42	08 29	14 11	14 34	15 10	16 05
58	07 00	07 52	08 44	14 00	14 19	14 50	15 43
S 60	07 07	08 04	09 01	13 48	14 01	14 26	15 16

Lat. (°)	Sunset (h m)	Twilight Civil (h m)	Twilight Naut. (h m)	Moonset 12 (h m)	Moonset 13 (h m)	Moonset 14 (h m)	Moonset 15 (h m)
N 72	□	□	□	■	■	■	■
N 70	□	□	□	00 14	■	■	■
68	□	□	□	00 39 / 23 55	■	■	■
66	□	□	□	00 59	00 39	■	■
64	22 25	////	////	01 15	01 08	00 57	■
62	21 49	////	////	01 29	01 31	01 38	02 01
60	21 23	23 05	////	01 40	01 49	02 06	02 41
N 58	21 03	22 17	////	01 50	02 05	02 28	03 09
56	20 47	21 48	////	01 59	02 18	02 46	03 31
54	20 33	21 27	23 09	02 07	02 29	03 01	03 49
52	20 20	21 09	22 26	02 14	02 39	03 14	04 04
50	20 10	20 54	21 59	02 21	02 48	03 26	04 17
45	19 47	20 25	21 14	02 34	03 08	03 50	04 44
N 40	19 30	20 02	20 44	02 46	03 24	04 10	05 06
35	19 15	19 44	20 21	02 55	03 37	04 26	05 24
30	19 02	19 29	20 02	03 04	03 49	04 40	05 39
20	18 40	19 04	19 34	03 19	04 09	05 04	06 05
N 10	18 21	18 44	19 11	03 32	04 26	05 25	06 28
0	18 04	18 26	18 52	03 44	04 43	05 45	06 49
S 10	17 46	18 09	18 35	03 57	04 59	06 04	07 10
20	17 28	17 52	18 19	04 10	05 17	06 26	07 32
30	17 07	17 33	18 03	04 25	05 38	06 50	07 59
35	16 55	17 23	17 54	04 34	05 50	07 04	08 14
40	16 41	17 11	17 45	04 44	06 03	07 21	08 32
45	16 24	16 58	17 35	04 56	06 20	07 41	08 53
S 50	16 04	16 42	17 23	05 11	06 40	08 06	09 20
52	15 54	16 35	17 18	05 18	06 50	08 18	09 33
54	15 43	16 26	17 13	05 26	07 01	08 32	09 49
56	15 30	16 18	17 07	05 34	07 14	08 48	10 06
58	15 16	16 07	17 00	05 44	07 28	09 07	10 28
S 60	14 59	15 56	16 52	05 55	07 45	09 31	10 56

Day	SUN Eqn. of Time 00^h	SUN Eqn. of Time 12^h	SUN Mer. Pass.	MOON Mer. Pass. Upper	MOON Mer. Pass. Lower	MOON Age	MOON Phase
d	m s	m s	h m	h m	h m	d %	
12	00 21	00 15	12 00	22 29	09 59	12 94	
13	00 09	00 03	12 00	23 30	10 59	13 99	○
14	00 04	00 10	12 00	24 34	12 02	14 100	

UT	ARIES	VENUS −3.9		MARS −1.1		JUPITER −1.8		SATURN +0.0		STARS		
	GHA	GHA	Dec	GHA	Dec	GHA	Dec	GHA	Dec	Name	SHA	Dec
d h	° ′	° ′	° ′	° ′	° ′	° ′	° ′	° ′	° ′		° ′	° ′
15 00	262 51.9	198 41.0	N20 25.1	290 24.6	S14 59.2	125 07.5	N17 05.8	171 20.9	N22 37.0	Acamar	315 24.6	S40 17.3
01	277 54.4	213 40.3	25.7	305 25.9	59.0	140 09.6	05.7	186 23.0	37.0	Achernar	335 32.7	S57 13.0
02	292 56.9	228 39.5	26.3	320 27.3	58.7	155 11.6	05.5	201 25.1	37.0	Acrux	173 18.3	S63 07.3
03	307 59.3	243 38.8	. . 26.9	335 28.6	. . 58.5	170 13.6	. . 05.4	216 27.2	. . 37.0	Adhara	255 19.1	S28 58.6
04	323 01.8	258 38.1	27.5	350 30.0	58.2	185 15.7	05.3	231 29.3	37.0	Aldebaran	290 58.8	N16 31.0
05	338 04.2	273 37.4	28.2	5 31.4	57.9	200 17.7	05.1	246 31.5	37.0			
06	353 06.7	288 36.7	N20 28.8	20 32.7	S14 57.7	215 19.7	N17 05.0	261 33.6	N22 37.0	Alioth	166 27.2	N55 56.8
07	8 09.2	303 35.9	29.4	35 34.1	57.4	230 21.8	04.9	276 35.7	37.0	Alkaid	153 04.7	N49 18.0
08	23 11.6	318 35.2	30.0	50 35.5	57.2	245 23.8	04.8	291 37.8	37.0	Al Na'ir	27 53.3	S46 56.5
S 09	38 14.1	333 34.5	. . 30.6	65 36.8	. . 56.9	260 25.8	. . 04.6	306 39.9	. . 37.0	Alnilam	275 54.7	S 1 12.0
U 10	53 16.6	348 33.8	31.2	80 38.2	56.6	275 27.9	04.5	321 42.0	37.0	Alphard	218 04.0	S 8 40.4
N 11	68 19.0	3 33.0	31.8	95 39.6	56.4	290 29.9	04.4	336 44.1	37.0			
D 12	83 21.5	18 32.3	N20 32.4	110 40.9	S14 56.1	305 31.9	N17 04.3	351 46.3	N22 37.0	Alphecca	126 17.3	N26 42.3
A 13	98 24.0	33 31.6	33.0	125 42.3	55.9	320 34.0	04.1	6 48.4	37.0	Alpheratz	357 51.8	N29 06.3
Y 14	113 26.4	48 30.9	33.6	140 43.7	55.6	335 36.0	04.0	21 50.5	37.0	Altair	62 15.6	N 8 52.5
15	128 28.9	63 30.1	. . 34.2	155 45.0	. . 55.3	350 38.0	. . 03.9	36 52.6	. . 37.0	Ankaa	353 23.4	S42 17.1
16	143 31.4	78 29.4	34.8	170 46.4	55.1	5 40.1	03.7	51 54.7	37.0	Antares	112 35.6	S26 26.4
17	158 33.8	93 28.7	35.4	185 47.8	54.8	20 42.1	03.6	66 56.8	37.0			
18	173 36.3	108 28.0	N20 36.0	200 49.1	S14 54.6	35 44.1	N17 03.5	81 58.9	N22 37.0	Arcturus	146 02.7	N19 10.0
19	188 38.7	123 27.2	36.6	215 50.5	54.3	50 46.2	03.4	97 01.1	37.0	Atria	107 44.1	S69 02.1
20	203 41.2	138 26.5	37.2	230 51.9	54.1	65 48.2	03.2	112 03.2	37.0	Avior	234 21.9	S59 31.3
21	218 43.7	153 25.8	. . 37.8	245 53.3	. . 53.8	80 50.2	. . 03.1	127 05.3	. . 37.0	Bellatrix	278 40.8	N 6 21.2
22	233 46.1	168 25.1	38.4	260 54.6	53.5	95 52.3	03.0	142 07.4	37.0	Betelgeuse	271 10.2	N 7 24.5
23	248 48.6	183 24.3	39.0	275 56.0	53.3	110 54.3	02.8	157 09.5	37.0			
16 00	263 51.1	198 23.6	N20 39.6	290 57.4	S14 53.0	125 56.3	N17 02.7	172 11.6	N22 37.0	Canopus	264 00.2	S52 41.8
01	278 53.5	213 22.9	40.2	305 58.7	52.8	140 58.4	02.6	187 13.7	37.0	Capella	280 46.6	N46 00.1
02	293 56.0	228 22.1	40.8	321 00.1	52.5	156 00.4	02.5	202 15.8	37.0	Deneb	49 36.6	N45 17.3
03	308 58.5	243 21.4	. . 41.4	336 01.5	. . 52.3	171 02.4	. . 02.3	217 18.0	. . 37.0	Denebola	182 41.7	N14 33.3
04	324 00.9	258 20.7	42.0	351 02.9	52.0	186 04.5	02.2	232 20.1	37.0	Diphda	349 03.8	S17 58.1
05	339 03.4	273 19.9	42.6	6 04.3	51.7	201 06.5	02.1	247 22.2	37.0			
06	354 05.9	288 19.2	N20 43.2	21 05.6	S14 51.5	216 08.5	N17 01.9	262 24.3	N22 37.0	Dubhe	194 01.2	N61 44.3
07	9 08.3	303 18.5	43.7	36 07.0	51.2	231 10.6	01.8	277 26.4	37.0	Elnath	278 23.0	N28 36.6
08	24 10.8	318 17.7	44.3	51 08.4	51.0	246 12.6	01.7	292 28.5	37.0	Eltanin	90 49.3	N51 29.2
M 09	39 13.2	333 17.0	. . 44.9	66 09.8	. . 50.7	261 14.6	. . 01.5	307 30.6	. . 37.0	Enif	33 54.7	N 9 53.3
O 10	54 15.7	348 16.3	45.5	81 11.1	50.5	276 16.7	01.4	322 32.8	37.0	Fomalhaut	15 32.5	S29 36.2
N 11	69 18.2	3 15.5	46.1	96 12.5	50.2	291 18.7	01.3	337 34.9	37.0			
D 12	84 20.6	18 14.8	N20 46.7	111 13.9	S14 50.0	306 20.7	N17 01.2	352 37.0	N22 37.0	Gacrux	172 09.8	S57 08.1
A 13	99 23.1	33 14.1	47.2	126 15.3	49.7	321 22.8	01.0	7 39.1	37.0	Gienah	176 00.4	S17 33.7
Y 14	114 25.6	48 13.3	47.8	141 16.7	49.4	336 24.8	00.9	22 41.2	37.0	Hadar	148 59.0	S60 23.6
15	129 28.0	63 12.6	. . 48.4	156 18.1	. . 49.2	351 26.8	. . 00.8	37 43.3	. . 37.0	Hamal	328 10.0	N23 28.6
16	144 30.5	78 11.9	49.0	171 19.4	48.9	6 28.8	00.6	52 45.4	37.0	Kaus Aust.	83 53.9	S34 23.0
17	159 33.0	93 11.1	49.6	186 20.8	48.7	21 30.9	00.5	67 47.5	37.0			
18	174 35.4	108 10.4	N20 50.1	201 22.2	S14 48.4	36 32.9	N17 00.4	82 49.7	N22 37.0	Kochab	137 18.4	N74 08.7
19	189 37.9	123 09.7	50.7	216 23.6	48.2	51 34.9	00.3	97 51.8	37.0	Markab	13 46.2	N15 13.2
20	204 40.3	138 08.9	51.3	231 25.0	47.9	66 37.0	00.1	112 53.9	37.0	Menkar	314 23.6	N 4 06.1
21	219 42.8	153 08.2	. . 51.9	246 26.4	. . 47.7	81 39.0	17 00.0	127 56.0	. . 37.0	Menkent	148 16.7	S36 23.3
22	234 45.3	168 07.4	52.4	261 27.8	47.4	96 41.0	16 59.9	142 58.1	37.0	Miaplacidus	221 42.2	S69 44.0
23	249 47.7	183 06.7	53.0	276 29.1	47.2	111 43.1	59.7	158 00.2	37.0			
17 00	264 50.2	198 06.0	N20 53.6	291 30.5	S14 46.9	126 45.1	N16 59.6	173 02.3	N22 37.0	Mirfak	308 52.2	N49 52.2
01	279 52.7	213 05.2	54.2	306 31.9	46.7	141 47.1	59.5	188 04.5	37.0	Nunki	76 07.7	S26 17.6
02	294 55.1	228 04.5	54.7	321 33.3	46.4	156 49.1	59.3	203 06.6	37.0	Peacock	53 31.0	S56 43.4
03	309 57.6	243 03.7	. . 55.3	336 34.7	. . 46.2	171 51.2	. . 59.2	218 08.7	. . 37.0	Pollux	243 37.7	N28 01.2
04	325 00.1	258 03.0	55.9	351 36.1	45.9	186 53.2	59.1	233 10.8	37.0	Procyon	245 08.3	N 5 13.1
05	340 02.5	273 02.3	56.4	6 37.5	45.7	201 55.2	59.0	248 12.9	37.0			
06	355 05.0	288 01.5	N20 57.0	21 38.9	S14 45.4	216 57.3	N16 58.8	263 15.0	N22 37.0	Rasalhague	96 13.4	N12 33.4
07	10 07.5	303 00.8	57.6	36 40.3	45.2	231 59.3	58.7	278 17.1	37.0	Regulus	207 52.0	N11 57.2
T 08	25 09.9	318 00.0	58.1	51 41.7	44.9	247 01.3	58.6	293 19.2	37.0	Rigel	281 20.0	S 8 11.8
U 09	40 12.4	332 59.3	. . 58.7	66 43.1	. . 44.7	262 03.3	. . 58.4	308 21.4	. . 37.0	Rigil Kent.	140 02.4	S60 51.1
E 10	55 14.8	347 58.5	59.2	81 44.5	44.4	277 05.4	58.3	323 23.5	37.0	Sabik	102 21.3	S15 43.8
S 11	70 17.3	2 57.8	20 59.8	96 45.9	44.2	292 07.4	58.2	338 25.6	37.0			
D 12	85 19.8	17 57.0	N21 00.4	111 47.3	S14 43.9	307 09.4	N16 58.0	353 27.7	N22 37.0	Schedar	349 49.9	N56 33.0
A 13	100 22.2	32 56.3	00.9	126 48.6	43.7	322 11.5	57.9	8 29.8	37.0	Shaula	96 32.2	S37 06.4
Y 14	115 24.7	47 55.6	01.5	141 50.0	43.4	337 13.5	57.8	23 31.9	37.0	Sirius	258 41.1	S16 43.2
15	130 27.2	62 54.8	. . 02.0	156 51.4	. . 43.2	352 15.5	. . 57.6	38 34.0	. . 37.0	Spica	158 39.5	S11 10.8
16	145 29.6	77 54.1	02.6	171 52.8	42.9	7 17.5	57.5	53 36.2	37.0	Suhail	222 58.6	S43 26.9
17	160 32.1	92 53.3	03.1	186 54.2	42.7	22 19.6	57.4	68 38.3	37.0			
18	175 34.6	107 52.6	N21 03.7	201 55.6	S14 42.4	37 21.6	N16 57.3	83 40.4	N22 37.0	Vega	80 43.9	N38 47.1
19	190 37.0	122 51.8	04.2	216 57.0	42.2	52 23.6	57.1	98 42.5	37.0	Zuben'ubi	137 13.9	S16 03.4
20	205 39.5	137 51.1	04.8	231 58.4	41.9	67 25.6	57.0	113 44.6	37.0		SHA	Mer. Pass
21	220 42.0	152 50.3	. . 05.3	246 59.8	. . 41.7	82 27.7	. . 56.9	128 46.7	. . 37.0		° ′	h m
22	235 44.4	167 49.6	05.9	262 01.3	41.4	97 29.7	56.7	143 48.8	37.0	Venus	294 32.5	10 47
23	250 46.9	182 48.8	06.4	277 02.7	41.2	112 31.7	56.6	158 50.9	37.0	Mars	27 06.3	4 36
										Jupiter	222 05.3	15 34
Mer. Pass.	h m 6 23.5	v −0.7	d 0.6	v 1.4	d 0.3	v 2.0	d 0.1	v 2.1	d 0.0	Saturn	268 20.6	12 29

	UT d h	SUN GHA ° ′	SUN Dec ° ′	MOON GHA ° ′	v ′	MOON Dec ° ′	d ′	HP ′
	15 00	179 55.9	N23 16.9	351 56.2	2.6	S26 10.7	2.6	60.3
	01	194 55.8	17.0	6 17.8	2.5	26 13.3	2.3	60.2
	02	209 55.6	17.1	20 39.3	2.6	26 15.6	2.2	60.2
	03	224 55.5	. . 17.2	35 00.9	2.6	26 17.8	1.9	60.2
	04	239 55.4	17.3	49 22.5	2.6	26 19.7	1.8	60.2
	05	254 55.2	17.4	63 44.1	2.6	26 21.5	1.5	60.1
	06	269 55.1	N23 17.5	78 05.7	2.6	S26 23.0	1.4	60.1
	07	284 55.0	17.6	92 27.3	2.6	26 24.4	1.2	60.1
	08	299 54.8	17.8	106 48.9	2.7	26 25.6	0.9	60.1
S	09	314 54.7	. . 17.9	121 10.6	2.7	26 26.5	0.8	60.0
U	10	329 54.6	18.0	135 32.3	2.7	26 27.3	0.6	60.0
N	11	344 54.4	18.1	149 54.0	2.7	26 27.9	0.4	60.0
D	12	359 54.3	N23 18.2	164 15.7	2.8	S26 28.3	0.2	60.0
A	13	14 54.2	18.3	178 37.5	2.8	26 28.5	0.0	59.9
Y	14	29 54.0	18.4	192 59.3	2.9	26 28.5	0.2	59.9
	15	44 53.9	. . 18.5	207 21.2	2.9	26 28.3	0.3	59.9
	16	59 53.8	18.6	221 43.1	3.0	26 28.0	0.6	59.8
	17	74 53.7	18.7	236 05.1	3.0	26 27.4	0.8	59.8
	18	89 53.5	N23 18.8	250 27.1	3.0	S26 26.6	0.9	59.8
	19	104 53.4	18.9	264 49.1	3.1	26 25.7	1.1	59.8
	20	119 53.3	19.0	279 11.2	3.2	26 24.6	1.3	59.7
	21	134 53.1	. . 19.1	293 33.4	3.2	26 23.3	1.6	59.7
	22	149 53.0	19.2	307 55.6	3.3	26 21.7	1.6	59.7
	23	164 52.9	19.3	322 17.9	3.4	26 20.1	1.9	59.6
	16 00	179 52.7	N23 19.4	336 40.3	3.4	S26 18.2	2.1	59.6
	01	194 52.6	19.5	351 02.7	3.5	26 16.1	2.2	59.6
	02	209 52.5	19.6	5 25.2	3.5	26 13.9	2.4	59.5
	03	224 52.3	. . 19.7	19 47.7	3.7	26 11.5	2.6	59.5
	04	239 52.2	19.8	34 10.4	3.7	26 08.9	2.8	59.5
	05	254 52.1	19.9	48 33.1	3.8	26 06.1	3.0	59.4
	06	269 51.9	N23 20.0	62 55.9	3.8	S26 03.1	3.1	59.4
	07	284 51.8	20.1	77 18.7	4.0	26 00.0	3.3	59.4
	08	299 51.7	20.2	91 41.7	4.0	25 56.7	3.5	59.3
M	09	314 51.5	. . 20.3	106 04.7	4.1	25 53.2	3.7	59.3
O	10	329 51.4	20.4	120 27.8	4.3	25 49.5	3.8	59.3
N	11	344 51.3	20.5	134 51.1	4.3	25 45.7	4.0	59.2
D	12	359 51.1	N23 20.6	149 14.4	4.3	S25 41.7	4.2	59.2
A	13	14 51.0	20.7	163 37.7	4.5	25 37.5	4.3	59.2
Y	14	29 50.9	20.8	178 01.2	4.6	25 33.2	4.5	59.1
	15	44 50.7	. . 20.9	192 24.8	4.7	25 28.7	4.7	59.1
	16	59 50.6	21.0	206 48.5	4.8	25 24.0	4.8	59.1
	17	74 50.5	21.0	221 12.3	4.9	25 19.2	5.0	59.0
	18	89 50.3	N23 21.1	235 36.2	4.9	S25 14.2	5.2	59.0
	19	104 50.2	21.2	250 00.1	5.1	25 09.0	5.3	59.0
	20	119 50.1	21.3	264 24.2	5.2	25 03.7	5.5	58.9
	21	134 49.9	. . 21.4	278 48.4	5.3	24 58.2	5.6	58.9
	22	149 49.8	21.5	293 12.7	5.4	24 52.6	5.8	58.9
	23	164 49.7	21.6	307 37.1	5.5	24 46.8	5.9	58.8
	17 00	179 49.5	N23 21.6	322 01.6	5.7	S24 40.9	6.1	58.8
	01	194 49.4	21.7	336 26.3	5.7	24 34.8	6.2	58.7
	02	209 49.3	21.8	350 51.0	5.9	24 28.6	6.4	58.7
	03	224 49.1	. . 21.9	5 15.9	5.9	24 22.2	6.5	58.7
	04	239 49.0	22.0	19 40.8	6.1	24 15.7	6.7	58.6
	05	254 48.9	22.0	34 05.9	6.2	24 09.0	6.8	58.6
	06	269 48.7	N23 22.1	48 31.1	6.3	S24 02.2	7.0	58.6
	07	284 48.6	22.2	62 56.4	6.4	23 55.2	7.0	58.5
T	08	299 48.5	22.3	77 21.8	6.6	23 48.2	7.3	58.5
U	09	314 48.3	. . 22.3	91 47.4	6.6	23 40.9	7.3	58.5
E	10	329 48.2	22.4	106 13.0	6.8	23 33.6	7.5	58.4
S	11	344 48.1	22.5	120 38.8	6.9	23 26.1	7.7	58.4
D	12	359 47.9	N23 22.6	135 04.7	7.1	S23 18.4	7.7	58.3
A	13	14 47.8	22.6	149 30.8	7.1	23 10.7	7.9	58.3
Y	14	29 47.6	22.7	163 56.9	7.3	23 02.8	8.1	58.3
	15	44 47.5	. . 22.8	178 23.2	7.4	22 54.7	8.1	58.2
	16	59 47.4	22.9	192 49.6	7.5	22 46.6	8.3	58.2
	17	74 47.2	22.9	207 16.1	7.6	22 38.3	8.4	58.1
	18	89 47.1	N23 23.0	221 42.7	7.7	S22 29.9	8.5	58.1
	19	104 47.0	23.1	236 09.4	7.9	22 21.4	8.6	58.1
	20	119 46.8	23.1	250 36.3	8.0	22 12.8	8.7	58.0
	21	134 46.7	. . 23.2	265 03.3	8.1	22 04.1	8.9	58.0
	22	149 46.6	23.3	279 30.4	8.3	21 55.2	9.0	58.0
	23	164 46.4	23.3	293 57.7	8.3	S21 46.2	9.0	57.9
		SD 15.8	*d* 0.1	SD	16.3		16.1	15.9

Lat.	Twilight Naut.	Twilight Civil	Sunrise	Moonrise 15	Moonrise 16	Moonrise 17	Moonrise 18
°	h m	h m	h m	h m	h m	h m	h m
N 72	□	□	□	■	■	■	■
N 70	□	□	□	■	■	■	■
68	□	□	□	■	■	■	02 34
66	□	□	□	■	■	■	01 39
64	////	////	01 33	■	■	01 17	01 06
62	////	////	02 10	24 11	00 11	00 34	00 41
60	////	00 52	02 36	23 31	24 05	00 05	00 21
N 58	////	01 41	02 56	23 03	23 43	24 05	00 05
56	////	02 11	03 13	22 41	23 24	23 51	24 09
54	00 48	02 33	03 27	22 23	23 09	23 39	24 00
52	01 33	02 51	03 39	22 08	22 56	23 29	23 52
50	02 00	03 06	03 50	21 54	22 44	23 19	23 45
45	02 46	03 35	04 13	21 27	22 19	22 59	23 30
N 40	03 16	03 58	04 31	21 05	22 00	22 43	23 17
35	03 39	04 16	04 46	20 47	21 43	22 29	23 07
30	03 58	04 31	04 59	20 32	21 29	22 17	22 57
20	04 27	04 56	05 20	20 06	21 05	21 56	22 41
N 10	04 49	05 16	05 39	19 43	20 44	21 38	22 27
0	05 08	05 34	05 57	19 22	20 24	21 21	22 13
S 10	05 25	05 51	06 14	19 01	20 04	21 04	22 00
20	05 41	06 09	06 33	18 39	19 43	20 46	21 45
30	05 58	06 28	06 54	18 13	19 19	20 25	21 29
35	06 06	06 38	07 06	17 57	19 04	20 12	21 19
40	06 16	06 50	07 20	17 40	18 48	19 58	21 08
45	06 26	07 03	07 37	17 18	18 27	19 41	20 55
S 50	06 38	07 19	07 58	16 51	18 02	19 20	20 38
52	06 43	07 27	08 08	16 38	17 50	19 10	20 31
54	06 49	07 35	08 19	16 23	17 36	18 59	20 22
56	06 55	07 44	08 31	16 05	17 20	18 46	20 13
58	07 02	07 54	08 46	15 43	17 00	18 30	20 02
S 60	07 09	08 06	09 03	15 16	16 36	18 12	19 50

Lat.	Sunset	Twilight Civil	Twilight Naut.	Moonset 15	Moonset 16	Moonset 17	Moonset 18
°	h m	h m	h m	h m	h m	h m	h m
N 72	□	□	□	■	■	■	■
N 70	□	□	□	■	■	■	■
68	□	□	□	■	■	■	04 50
66	□	□	□	■	■	■	05 43
64	22 29	////	////	■	■	04 08	06 16
62	21 52	////	////	02 01	03 07	04 51	06 39
60	21 26	23 10	////	02 41	03 47	05 19	06 58
N 58	21 05	22 21	////	03 09	04 15	05 41	07 14
56	20 48	21 51	////	03 31	04 37	05 59	07 27
54	20 34	21 29	23 14	03 49	04 54	06 14	07 38
52	20 22	21 11	22 29	04 04	05 10	06 27	07 48
50	20 11	20 56	22 01	04 17	05 23	06 38	07 57
45	19 49	20 26	21 16	04 44	05 50	07 02	08 16
N 40	19 31	20 04	20 45	05 06	06 11	07 21	08 32
35	19 16	19 45	20 22	05 24	06 29	07 37	08 45
30	19 03	19 30	20 04	05 39	06 44	07 50	08 56
20	18 41	19 05	19 35	06 05	07 10	08 13	09 15
N 10	18 22	18 45	19 12	06 28	07 32	08 33	09 31
0	18 04	18 27	18 53	06 49	07 52	08 52	09 47
S 10	17 47	18 10	18 36	07 10	08 13	09 10	10 02
20	17 28	17 52	18 20	07 32	08 35	09 30	10 18
30	17 07	17 34	18 03	07 59	09 00	09 52	10 37
35	16 55	17 23	17 55	08 14	09 15	10 06	10 47
40	16 41	17 11	17 45	08 32	09 32	10 21	10 59
45	16 24	16 58	17 35	08 53	09 53	10 39	11 14
S 50	16 03	16 42	17 23	09 20	10 18	11 01	11 31
52	15 53	16 34	17 18	09 33	10 31	11 11	11 39
54	15 42	16 26	17 12	09 49	10 45	11 23	11 48
56	15 30	16 17	17 06	10 06	11 02	11 37	11 59
58	15 15	16 07	17 00	10 28	11 21	11 52	12 10
S 60	14 58	15 55	16 52	10 56	11 46	12 11	12 24

Day	SUN Eqn. of Time 00^h	SUN Eqn. of Time 12^h	SUN Mer. Pass.	MOON Mer. Pass. Upper	MOON Mer. Pass. Lower	Age	Phase
d	m s	m s	h m	h m	h m	d %	
15	00 16	00 22	12 00	00 34	13 06	15 98	○
16	00 29	00 35	12 01	01 37	14 08	16 94	
17	00 42	00 48	12 01	02 38	15 07	17 88	

UT	ARIES	VENUS −3.9		MARS −1.1		JUPITER −1.8		SATURN +0.0	
	GHA	GHA	Dec	GHA	Dec	GHA	Dec	GHA	Dec
d h	° ′	° ′	° ′	° ′	° ′	° ′	° ′	° ′	° ′
18 00 (WEDNESDAY)	265 49.3	197 48.1	N21 07.0	292 04.1	S14 40.9	127 33.7	N16 56.5	173 53.1	N22 37.0
01	280 51.8	212 47.3	07.5	307 05.5	40.7	142 35.8	56.3	188 55.2	37.0
02	295 54.3	227 46.6	08.1	322 06.9	40.4	157 37.8	56.2	203 57.3	37.0
03	310 56.7	242 45.8	. . 08.6	337 08.3	. . 40.2	172 39.8	. . 56.1	218 59.4	. . 37.0
04	325 59.2	257 45.1	09.2	352 09.7	39.9	187 41.9	55.9	234 01.5	37.0
05	341 01.7	272 44.3	09.7	7 11.1	39.7	202 43.9	55.8	249 03.6	37.0
06	356 04.1	287 43.6	N21 10.3	22 12.5	S14 39.4	217 45.9	N16 55.7	264 05.7	N22 37.0
07	11 06.6	302 42.8	10.8	37 13.9	39.2	232 47.9	55.5	279 07.8	37.0
08	26 09.1	317 42.1	11.3	52 15.3	39.0	247 50.0	55.4	294 10.0	37.0
09	41 11.5	332 41.3	. . 11.9	67 16.7	. . 38.7	262 52.0	. . 55.3	309 12.1	. . 37.0
10	56 14.0	347 40.5	12.4	82 18.1	38.5	277 54.0	55.2	324 14.2	37.0
11	71 16.5	2 39.8	12.9	97 19.5	38.2	292 56.0	55.0	339 16.3	37.0
12	86 18.9	17 39.0	N21 13.5	112 21.0	S14 38.0	307 58.1	N16 54.9	354 18.4	N22 37.0
13	101 21.4	32 38.3	14.0	127 22.4	37.7	323 00.1	54.8	9 20.5	37.0
14	116 23.8	47 37.5	14.5	142 23.8	37.5	338 02.1	54.6	24 22.6	37.0
15	131 26.3	62 36.8	. . 15.1	157 25.2	. . 37.2	353 04.1	. . 54.5	39 24.7	. . 37.0
16	146 28.8	77 36.0	15.6	172 26.6	37.0	8 06.1	54.4	54 26.9	37.0
17	161 31.2	92 35.3	16.1	187 28.0	36.8	23 08.2	54.2	69 29.0	37.0
18	176 33.7	107 34.5	N21 16.7	202 29.4	S14 36.5	38 10.2	N16 54.1	84 31.1	N22 37.0
19	191 36.2	122 33.7	17.2	217 30.9	36.3	53 12.2	54.0	99 33.2	37.0
20	206 38.6	137 33.0	17.7	232 32.3	36.0	68 14.2	53.8	114 35.3	37.0
21	221 41.1	152 32.2	. . 18.3	247 33.7	. . 35.8	83 16.3	. . 53.7	129 37.4	. . 37.0
22	236 43.6	167 31.5	18.8	262 35.1	35.5	98 18.3	53.6	144 39.5	37.0
23	251 46.0	182 30.7	19.3	277 36.5	35.3	113 20.3	53.4	159 41.6	37.0
19 00 (THURSDAY)	266 48.5	197 29.9	N21 19.8	292 38.0	S14 35.1	128 22.3	N16 53.3	174 43.8	N22 37.0
01	281 51.0	212 29.2	20.3	307 39.4	34.8	143 24.4	53.2	189 45.9	37.0
02	296 53.4	227 28.4	20.9	322 40.8	34.6	158 26.4	53.0	204 48.0	37.0
03	311 55.9	242 27.7	. . 21.4	337 42.2	. . 34.3	173 28.4	. . 52.9	219 50.1	. . 37.0
04	326 58.3	257 26.9	21.9	352 43.6	34.1	188 30.4	52.8	234 52.2	37.0
05	342 00.8	272 26.1	22.4	7 45.1	33.9	203 32.4	52.6	249 54.3	37.0
06	357 03.3	287 25.4	N21 22.9	22 46.5	S14 33.6	218 34.5	N16 52.5	264 56.4	N22 37.0
07	12 05.7	302 24.6	23.5	37 47.9	33.4	233 36.5	52.4	279 58.5	37.0
08	27 08.2	317 23.8	24.0	52 49.3	33.1	248 38.5	52.2	295 00.7	37.0
09	42 10.7	332 23.1	. . 24.5	67 50.8	. . 32.9	263 40.5	. . 52.1	310 02.8	. . 37.0
10	57 13.1	347 22.3	25.0	82 52.2	32.7	278 42.6	52.0	325 04.9	37.0
11	72 15.6	2 21.5	25.5	97 53.6	32.4	293 44.6	51.8	340 07.0	37.0
12	87 18.1	17 20.8	N21 26.0	112 55.0	S14 32.2	308 46.6	N16 51.7	355 09.1	N22 37.0
13	102 20.5	32 20.0	26.5	127 56.5	31.9	323 48.6	51.6	10 11.2	37.0
14	117 23.0	47 19.3	27.0	142 57.9	31.7	338 50.6	51.4	25 13.3	37.0
15	132 25.4	62 18.5	. . 27.6	157 59.3	. . 31.5	353 52.7	. . 51.3	40 15.4	. . 37.0
16	147 27.9	77 17.7	28.1	173 00.8	31.2	8 54.7	51.2	55 17.6	37.0
17	162 30.4	92 16.9	28.6	188 02.2	31.0	23 56.7	51.0	70 19.7	37.0
18	177 32.8	107 16.2	N21 29.1	203 03.6	S14 30.7	38 58.7	N16 50.9	85 21.8	N22 37.0
19	192 35.3	122 15.4	29.6	218 05.1	30.5	54 00.7	50.8	100 23.9	37.0
20	207 37.8	137 14.6	30.1	233 06.5	30.3	69 02.8	50.6	115 26.0	37.0
21	222 40.2	152 13.9	. . 30.6	248 07.9	. . 30.0	84 04.8	. . 50.5	130 28.1	. . 37.0
22	237 42.7	167 13.1	31.1	263 09.4	29.8	99 06.8	50.4	145 30.2	37.0
23	252 45.2	182 12.3	31.6	278 10.8	29.6	114 08.8	50.2	160 32.3	37.0
20 00 (FRIDAY)	267 47.6	197 11.6	N21 32.1	293 12.2	S14 29.3	129 10.8	N16 50.1	175 34.5	N22 37.0
01	282 50.1	212 10.8	32.6	308 13.7	29.1	144 12.9	50.0	190 36.6	37.0
02	297 52.6	227 10.0	33.1	323 15.1	28.8	159 14.9	49.8	205 38.7	37.0
03	312 55.0	242 09.3	. . 33.6	338 16.5	. . 28.6	174 16.9	. . 49.7	220 40.8	. . 37.0
04	327 57.5	257 08.5	34.1	353 18.0	28.4	189 18.9	49.6	235 42.9	37.0
05	342 59.9	272 07.7	34.6	8 19.4	28.1	204 20.9	49.4	250 45.0	37.0
06	358 02.4	287 06.9	N21 35.0	23 20.9	S14 27.9	219 23.0	N16 49.3	265 47.1	N22 37.0
07	13 04.9	302 06.2	35.5	38 22.3	27.7	234 25.0	49.2	280 49.2	37.0
08	28 07.3	317 05.4	36.0	53 23.7	27.4	249 27.0	49.0	295 51.4	37.0
09	43 09.8	332 04.6	. . 36.5	68 25.2	. . 27.2	264 29.0	. . 48.9	310 53.5	. . 37.0
10	58 12.3	347 03.8	37.0	83 26.6	27.0	279 31.0	48.8	325 55.6	37.0
11	73 14.7	2 03.1	37.5	98 28.1	26.7	294 33.1	48.6	340 57.7	37.0
12	88 17.2	17 02.3	N21 38.0	113 29.5	S14 26.5	309 35.1	N16 48.5	355 59.8	N22 37.0
13	103 19.7	32 01.5	38.5	128 31.0	26.3	324 37.1	48.4	11 01.9	37.0
14	118 22.1	47 00.7	39.0	143 32.4	26.0	339 39.1	48.2	26 04.0	37.0
15	133 24.6	62 00.0	. . 39.4	158 33.9	. . 25.8	354 41.1	. . 48.1	41 06.1	. . 37.0
16	148 27.1	76 59.2	39.9	173 35.3	25.6	9 43.1	47.9	56 08.3	37.0
17	163 29.5	91 58.4	40.4	188 36.7	25.3	24 45.2	47.8	71 10.4	37.0
18	178 32.0	106 57.6	N21 40.9	203 38.2	S14 25.1	39 47.2	N16 47.7	86 12.5	N22 37.0
19	193 34.4	121 56.9	41.4	218 39.6	24.9	54 49.2	47.5	101 14.6	36.9
20	208 36.9	136 56.1	41.8	233 41.1	24.6	69 51.2	47.4	116 16.7	36.9
21	223 39.4	151 55.3	. . 42.3	248 42.5	. . 24.4	84 53.2	. . 47.3	131 18.8	. . 36.9
22	238 41.8	166 54.5	42.8	263 44.0	24.2	99 55.3	47.1	146 20.9	36.9
23	253 44.3	181 53.7	43.3	278 45.4	23.9	114 57.3	47.0	161 23.0	36.9
Mer. Pass.	h m 6 11.7	v −0.8	d 0.5	v 1.4	d 0.2	v 2.0	d 0.1	v 2.1	d 0.0

STARS

Name	SHA	Dec
	° ′	° ′
Acamar	315 24.5	S40 17.
Achernar	335 32.7	S57 13.
Acrux	173 18.3	S63 07.
Adhara	255 19.1	S28 58.
Aldebaran	290 58.8	N16 31.
Alioth	166 27.2	N55 56.
Alkaid	153 04.7	N49 18.
Al Na'ir	27 53.2	S46 56.
Alnilam	275 54.7	S 1 11.
Alphard	218 04.1	S 8 40.
Alphecca	126 17.3	N26 42.
Alpheratz	357 51.8	N29 06.
Altair	62 15.6	N 8 52.
Ankaa	353 23.4	S42 17.
Antares	112 35.6	S26 26.
Arcturus	146 02.7	N19 10.
Atria	107 44.1	S69 02.
Avior	234 21.9	S59 31.
Bellatrix	278 40.8	N 6 21.
Betelgeuse	271 10.2	N 7 24.
Canopus	264 00.2	S52 41.
Capella	280 46.6	N46 00.
Deneb	49 36.6	N45 17.
Denebola	182 41.7	N14 33.
Diphda	349 03.8	S17 58.
Dubhe	194 01.2	N61 44.
Elnath	278 23.0	N28 36.
Eltanin	90 49.3	N51 29.
Enif	33 54.7	N 9 53.
Fomalhaut	15 32.5	S29 36.
Gacrux	172 09.8	S57 08.
Gienah	176 00.4	S17 33.
Hadar	148 59.0	S60 23.
Hamal	328 09.9	N23 28
Kaus Aust.	83 53.8	S34 23.
Kochab	137 18.5	N74 08.
Markab	13 46.2	N15 13.
Menkar	314 23.6	N 4 06.
Menkent	148 16.8	S36 23.
Miaplacidus	221 42.3	S69 44.
Mirfak	308 52.2	N49 52.
Nunki	76 07.7	S26 17.
Peacock	53 31.0	S56 43.
Pollux	243 37.7	N28 01.
Procyon	245 08.3	N 5 13.
Rasalhague	96 13.4	N12 33
Regulus	207 52.0	N11 57
Rigel	281 20.0	S 8 11
Rigil Kent.	140 02.4	S60 51
Sabik	102 21.2	S15 43
Schedar	349 49.9	N56 33
Shaula	96 32.2	S37 06
Sirius	258 41.1	S16 43
Spica	158 39.5	S11 10
Suhail	222 58.7	S43 26
Vega	80 43.9	N38 47
Zuben'ubi	137 13.9	S16 03

	SHA	Mer. Pas
	° ′	h m
Venus	290 41.5	10 51
Mars	25 49.5	4 29
Jupiter	221 33.8	15 24
Saturn	267 55.3	12 19

UT	SUN GHA	SUN Dec	MOON GHA	v	MOON Dec	d	HP
d h	° ′	° ′	° ′	′	° ′	′	′
18 00	179 46.3	N23 23.4	308 25.0	8.5	S21 37.2	9.2	57.9
01	194 46.2	23.5	322 52.5	8.6	21 28.0	9.3	57.8
02	209 46.0	23.5	337 20.1	8.8	21 18.7	9.4	57.8
03	224 45.9	. . 23.6	351 47.9	8.8	21 09.3	9.6	57.8
04	239 45.8	23.7	6 15.7	9.0	20 59.7	9.6	57.7
05	254 45.6	23.7	20 43.7	9.1	20 50.1	9.7	57.7
06	269 45.5	N23 23.8	35 11.8	9.2	S20 40.4	9.8	57.7
W 07	284 45.4	23.8	49 40.0	9.4	20 30.6	9.9	57.6
E 08	299 45.2	23.9	64 08.4	9.4	20 20.7	10.0	57.6
D 09	314 45.1	. . 24.0	78 36.8	9.6	20 10.7	10.1	57.5
N 10	329 45.0	24.0	93 05.4	9.7	20 00.6	10.2	57.5
E 11	344 44.8	24.1	107 34.1	9.8	19 50.4	10.3	57.5
S 12	359 44.7	N23 24.1	122 02.9	10.0	S19 40.1	10.4	57.4
D 13	14 44.5	24.2	136 31.9	10.0	19 29.7	10.5	57.4
A 14	29 44.4	24.3	151 00.9	10.2	19 19.2	10.5	57.4
Y 15	44 44.3	. . 24.3	165 30.1	10.3	19 08.7	10.7	57.3
16	59 44.1	24.4	179 59.4	10.4	18 58.0	10.7	57.3
17	74 44.0	24.4	194 28.8	10.5	18 47.3	10.8	57.2
18	89 43.9	N23 24.5	208 58.3	10.6	S18 36.5	10.9	57.2
19	104 43.7	24.5	223 27.9	10.8	18 25.6	11.0	57.2
20	119 43.6	24.6	237 57.7	10.9	18 14.6	11.0	57.1
21	134 43.5	. . 24.6	252 27.6	10.9	18 03.6	11.1	57.1
22	149 43.3	24.7	266 57.5	11.1	17 52.5	11.2	57.1
23	164 43.2	24.7	281 27.6	11.2	17 41.3	11.3	57.0
19 00	179 43.1	N23 24.8	295 57.8	11.3	S17 30.0	11.3	57.0
01	194 42.9	24.8	310 28.1	11.5	17 18.7	11.4	56.9
02	209 42.8	24.9	324 58.6	11.5	17 07.3	11.5	56.9
03	224 42.7	. . 24.9	339 29.1	11.6	16 55.8	11.5	56.9
04	239 42.5	25.0	353 59.7	11.8	16 44.3	11.6	56.8
05	254 42.4	25.0	8 30.5	11.8	16 32.7	11.7	56.8
06	269 42.3	N23 25.0	23 01.3	12.0	S16 21.0	11.7	56.8
T 07	284 42.1	25.1	37 32.3	12.0	16 09.3	11.8	56.7
H 08	299 42.0	25.1	52 03.3	12.2	15 57.5	11.9	56.7
U 09	314 41.8	. . 25.2	66 34.5	12.3	15 45.6	11.9	56.7
R 10	329 41.7	25.2	81 05.8	12.3	15 33.7	12.0	56.6
S 11	344 41.6	25.3	95 37.1	12.5	15 21.7	12.0	56.6
D 12	359 41.4	N23 25.3	110 08.6	12.6	S15 09.7	12.1	56.6
A 13	14 41.3	25.3	124 40.2	12.6	14 57.6	12.1	56.5
Y 14	29 41.2	25.4	139 11.8	12.8	14 45.5	12.2	56.5
15	44 41.0	. . 25.4	153 43.6	12.8	14 33.3	12.3	56.5
16	59 40.9	25.5	168 15.4	13.0	14 21.0	12.2	56.4
17	74 40.8	25.5	182 47.4	13.0	14 08.8	12.4	56.4
18	89 40.6	N23 25.5	197 19.4	13.1	S13 56.4	12.4	56.3
19	104 40.5	25.6	211 51.5	13.3	13 44.0	12.4	56.3
20	119 40.4	25.6	226 23.8	13.3	13 31.6	12.5	56.3
21	134 40.2	. . 25.6	240 56.1	13.4	13 19.1	12.5	56.2
22	149 40.1	25.7	255 28.5	13.5	13 06.6	12.6	56.2
23	164 39.9	25.7	270 01.0	13.6	12 54.0	12.6	56.2
20 00	179 39.8	N23 25.7	284 33.6	13.6	S12 41.4	12.6	56.1
01	194 39.7	25.8	299 06.2	13.8	12 28.8	12.7	56.1
02	209 39.5	25.8	313 39.0	13.8	12 16.1	12.7	56.1
03	224 39.4	. . 25.8	328 11.8	13.9	12 03.4	12.8	56.1
04	239 39.3	25.8	342 44.7	14.0	11 50.6	12.8	56.0
05	254 39.1	25.9	357 17.7	14.1	11 37.8	12.8	56.0
06	269 39.0	N23 25.9	11 50.8	14.1	S11 25.0	12.9	56.0
F 07	284 38.9	25.9	26 23.9	14.3	11 12.1	12.9	55.9
R 08	299 38.7	25.9	40 57.2	14.3	10 59.2	12.9	55.9
I 09	314 38.6	. . 26.0	55 30.5	14.3	10 46.3	13.0	55.9
D 10	329 38.5	26.0	70 03.8	14.5	10 33.3	12.9	55.8
A 11	344 38.3	26.0	84 37.3	14.5	10 20.4	13.1	55.8
Y 12	359 38.2	N23 26.0	99 10.8	14.6	S10 07.3	13.0	55.8
13	14 38.0	26.1	113 44.4	14.7	9 54.3	13.1	55.7
14	29 37.9	26.1	128 18.1	14.7	9 41.2	13.1	55.7
15	44 37.8	. . 26.1	142 51.8	14.8	9 28.1	13.1	55.7
16	59 37.6	26.1	157 25.6	14.9	9 15.0	13.1	55.7
17	74 37.5	26.1	171 59.5	14.9	9 01.9	13.2	55.6
18	89 37.4	N23 26.2	186 33.4	15.0	S 8 48.7	13.2	55.6
19	104 37.2	26.2	201 07.4	15.1	8 35.5	13.2	55.6
20	119 37.1	26.2	215 41.5	15.1	8 22.3	13.2	55.5
21	134 37.0	. . 26.2	230 15.6	15.2	8 09.1	13.3	55.5
22	149 36.8	26.2	244 49.8	15.2	7 55.8	13.3	55.5
23	164 36.7	26.2	259 24.0	15.3	S 7 42.5	13.2	55.5
	SD 15.8	*d* 0.0	SD	15.6		15.4	15.2

Lat.	Twilight Naut.	Twilight Civil	Sunrise	Moonrise 18	19	20	21
°	h m	h m	h m	h m	h m	h m	h m
N 72	▭	▭	▭	■	03 05	01 56	01 19
N 70	▭	▭	▭	■	02 14	01 34	01 09
68	▭	▭	▭	02 34	01 41	01 18	01 01
66	▭	▭	▭	01 39	01 17	01 04	00 54
64	////	////	01 31	01 06	00 59	00 53	00 47
62	////	////	02 09	00 41	00 43	00 43	00 42
60	////	00 49	02 35	00 21	00 30	00 35	00 38
N 58	////	01 40	02 56	00 05	00 19	00 27	00 34
56	////	02 10	03 13	24 09	00 09	00 21	00 30
54	00 45	02 32	03 27	24 00	00 00	00 15	00 27
52	01 32	02 50	03 39	23 52	24 10	00 10	00 24
50	02 00	03 06	03 50	23 45	24 05	00 05	00 21
45	02 46	03 35	04 13	23 30	23 54	24 15	00 15
N 40	03 16	03 58	04 31	23 17	23 46	24 10	00 10
35	03 39	04 16	04 46	23 07	23 38	24 06	00 06
30	03 58	04 31	04 59	22 57	23 32	24 02	00 02
20	04 27	04 56	05 21	22 41	23 20	23 55	24 28
N 10	04 50	05 17	05 40	22 27	23 10	23 50	24 27
0	05 09	05 35	05 58	22 13	23 00	23 44	24 25
S 10	05 26	05 52	06 15	22 00	22 51	23 39	24 24
20	05 42	06 10	06 34	21 45	22 41	23 33	24 22
30	05 59	06 28	06 55	21 29	22 29	23 26	24 21
35	06 07	06 39	07 07	21 19	22 22	23 22	24 20
40	06 17	06 51	07 21	21 08	22 14	23 18	24 18
45	06 27	07 04	07 38	20 55	22 05	23 12	24 17
S 50	06 39	07 20	07 59	20 38	21 54	23 06	24 15
52	06 44	07 28	08 09	20 31	21 49	23 03	24 15
54	06 50	07 36	08 20	20 22	21 43	23 00	24 14
56	06 56	07 45	08 33	20 13	21 37	22 57	24 13
58	07 03	07 55	08 47	20 02	21 30	22 53	24 12
S 60	07 10	08 07	09 05	19 50	21 22	22 48	24 11

Lat.	Sunset	Twilight Civil	Twilight Naut.	Moonset 18	19	20	21
°	h m	h m	h m	h m	h m	h m	h m
N 72	▭	▭	▭	■	06 07	08 52	10 58
N 70	▭	▭	▭	■	06 56	09 11	11 06
68	▭	▭	▭	04 50	07 27	09 26	11 12
66	▭	▭	▭	05 43	07 50	09 38	11 17
64	22 32	////	////	06 16	08 07	09 48	11 21
62	21 54	////	////	06 39	08 22	09 56	11 25
60	21 27	23 14	////	06 58	08 34	10 04	11 28
N 58	21 07	22 22	////	07 14	08 44	10 10	11 31
56	20 50	21 52	////	07 27	08 53	10 16	11 34
54	20 36	21 30	23 18	07 38	09 01	10 21	11 36
52	20 23	21 12	22 31	07 48	09 09	10 25	11 38
50	20 12	20 57	22 03	07 57	09 15	10 29	11 40
45	19 50	20 27	21 17	08 16	09 29	10 38	11 44
N 40	19 32	20 05	20 46	08 32	09 40	10 45	11 47
35	19 17	19 46	20 23	08 45	09 50	10 51	11 50
30	19 04	19 31	20 04	08 56	09 58	10 57	11 53
20	18 42	19 06	19 35	09 15	10 12	11 06	11 57
N 10	18 23	18 46	19 13	09 31	10 25	11 14	12 01
0	18 05	18 27	18 54	09 47	10 36	11 22	12 04
S 10	17 47	18 10	18 37	10 02	10 48	11 29	12 08
20	17 29	17 53	18 20	10 18	11 00	11 37	12 11
30	17 08	17 34	18 04	10 37	11 14	11 46	12 15
35	16 55	17 23	17 55	10 47	11 22	11 51	12 18
40	16 41	17 12	17 46	10 59	11 31	11 57	12 20
45	16 24	16 58	17 35	11 14	11 41	12 04	12 23
S 50	16 03	16 42	17 24	11 31	11 54	12 12	12 27
52	15 54	16 35	17 18	11 39	12 00	12 16	12 29
54	15 42	16 26	17 13	11 48	12 06	12 20	12 31
56	15 30	16 17	17 06	11 59	12 13	12 24	12 33
58	15 15	16 07	17 00	12 10	12 21	12 29	12 35
S 60	14 57	15 55	16 52	12 24	12 30	12 34	12 37

Day	SUN Eqn. of Time 00^h	SUN Eqn. of Time 12^h	SUN Mer. Pass.	MOON Mer. Pass. Upper	MOON Mer. Pass. Lower	Age	Phase
d	m s	m s	h m	h m	h m	d %	
18	00 55	01 01	12 01	03 34	16 00	18 80	◑
19	01 07	01 14	12 01	04 25	16 48	19 71	
20	01 20	01 27	12 01	05 11	17 33	20 61	

UT	ARIES	VENUS −3.9		MARS −1.2		JUPITER −1.8		SATURN +0.0	
d h	GHA ° ′	GHA ° ′	Dec ° ′	GHA ° ′	Dec ° ′	GHA ° ′	Dec ° ′	GHA ° ′	Dec ° ′
21 00	268 46.8	196 53.0	N21 43.7	293 46.9	S14 23.7	129 59.3	N16 46.9	176 25.2	N22 36.9
01	283 49.2	211 52.2	44.2	308 48.4	23.5	145 01.3	46.7	191 27.3	36.9
02	298 51.7	226 51.4	44.7	323 49.8	23.3	160 03.3	46.6	206 29.4	36.9
03	313 54.2	241 50.6	45.2	338 51.3	23.0	175 05.3	46.5	221 31.5	36.9
04	328 56.6	256 49.8	45.6	353 52.7	22.8	190 07.4	46.3	236 33.6	36.9
05	343 59.1	271 49.1	46.1	8 54.2	22.6	205 09.4	46.2	251 35.7	36.9
06	359 01.5	286 48.3	N21 46.6	23 55.6	S14 22.3	220 11.4	N16 46.1	266 37.8	N22 36.9
07	14 04.0	301 47.5	47.0	38 57.1	22.1	235 13.4	45.9	281 39.9	36.9
S 08	29 06.5	316 46.7	47.5	53 58.5	21.9	250 15.4	45.8	296 42.0	36.9
A 09	44 08.9	331 45.9	48.0	69 00.0	21.7	265 17.4	45.6	311 44.2	36.9
T 10	59 11.4	346 45.1	48.4	84 01.5	21.4	280 19.4	45.5	326 46.3	36.9
U 11	74 13.9	1 44.4	48.9	99 02.9	21.2	295 21.5	45.4	341 48.4	36.9
R 12	89 16.3	16 43.6	N21 49.3	114 04.4	S14 21.0	310 23.5	N16 45.2	356 50.5	N22 36.9
D 13	104 18.8	31 42.8	49.8	129 05.8	20.7	325 25.5	45.1	11 52.6	36.9
A 14	119 21.3	46 42.0	50.3	144 07.3	20.5	340 27.5	45.0	26 54.7	36.9
Y 15	134 23.7	61 41.2	50.7	159 08.8	20.3	355 29.5	44.8	41 56.8	36.9
16	149 26.2	76 40.4	51.2	174 10.2	20.1	10 31.5	44.7	56 58.9	36.9
17	164 28.7	91 39.6	51.6	189 11.7	19.8	25 33.6	44.6	72 01.1	36.9
18	179 31.1	106 38.9	N21 52.1	204 13.2	S14 19.6	40 35.6	N16 44.4	87 03.2	N22 36.9
19	194 33.6	121 38.1	52.5	219 14.6	19.4	55 37.6	44.3	102 05.3	36.9
20	209 36.0	136 37.3	53.0	234 16.1	19.2	70 39.6	44.2	117 07.4	36.9
21	224 38.5	151 36.5	53.4	249 17.6	18.9	85 41.6	44.0	132 09.5	36.9
22	239 41.0	166 35.7	53.9	264 19.0	18.7	100 43.6	43.9	147 11.6	36.9
23	254 43.4	181 34.9	54.3	279 20.5	18.5	115 45.6	43.7	162 13.7	36.9
22 00	269 45.9	196 34.1	N21 54.8	294 22.0	S14 18.3	130 47.7	N16 43.6	177 15.8	N22 36.9
01	284 48.4	211 33.3	55.2	309 23.4	18.0	145 49.7	43.5	192 18.0	36.9
02	299 50.8	226 32.6	55.7	324 24.9	17.8	160 51.7	43.3	207 20.1	36.9
03	314 53.3	241 31.8	56.1	339 26.4	17.6	175 53.7	43.2	222 22.2	36.9
04	329 55.8	256 31.0	56.6	354 27.9	17.4	190 55.7	43.1	237 24.3	36.9
05	344 58.2	271 30.2	57.0	9 29.3	17.1	205 57.7	42.9	252 26.4	36.9
06	0 00.7	286 29.4	N21 57.5	24 30.8	S14 16.9	220 59.7	N16 42.8	267 28.5	N22 36.9
07	15 03.1	301 28.6	57.9	39 32.3	16.7	236 01.7	42.7	282 30.6	36.9
08	30 05.6	316 27.8	58.3	54 33.7	16.5	251 03.8	42.5	297 32.7	36.9
S 09	45 08.1	331 27.0	58.8	69 35.2	16.2	266 05.8	42.4	312 34.8	36.9
U 10	60 10.5	346 26.2	59.2	84 36.7	16.0	281 07.8	42.2	327 37.0	36.9
N 11	75 13.0	1 25.4	21 59.7	99 38.2	15.8	296 09.8	42.1	342 39.1	36.9
D 12	90 15.5	16 24.6	N22 00.1	114 39.7	S14 15.6	311 11.8	N16 42.0	357 41.2	N22 36.9
A 13	105 17.9	31 23.8	00.5	129 41.1	15.4	326 13.8	41.8	12 43.3	36.9
Y 14	120 20.4	46 23.0	01.0	144 42.6	15.1	341 15.8	41.7	27 45.4	36.9
15	135 22.9	61 22.3	01.4	159 44.1	14.9	356 17.8	41.6	42 47.5	36.9
16	150 25.3	76 21.5	01.8	174 45.6	14.7	11 19.9	41.4	57 49.6	36.9
17	165 27.8	91 20.7	02.3	189 47.1	14.5	26 21.9	41.3	72 51.7	36.9
18	180 30.3	106 19.9	N22 02.7	204 48.5	S14 14.3	41 23.9	N16 41.1	87 53.9	N22 36.9
19	195 32.7	121 19.1	03.1	219 50.0	14.0	56 25.9	41.0	102 56.0	36.9
20	210 35.2	136 18.3	03.5	234 51.5	13.8	71 27.9	40.9	117 58.1	36.8
21	225 37.6	151 17.5	04.0	249 53.0	13.6	86 29.9	40.7	133 00.2	36.8
22	240 40.1	166 16.7	04.4	264 54.5	13.4	101 31.9	40.6	148 02.3	36.8
23	255 42.6	181 15.9	04.8	279 56.0	13.2	116 33.9	40.5	163 04.4	36.8
23 00	270 45.0	196 15.1	N22 05.2	294 57.4	S14 12.9	131 36.0	N16 40.3	178 06.5	N22 36.8
01	285 47.5	211 14.3	05.7	309 58.9	12.7	146 38.0	40.2	193 08.6	36.8
02	300 50.0	226 13.5	06.1	325 00.4	12.5	161 40.0	40.0	208 10.7	36.8
03	315 52.4	241 12.7	06.5	340 01.9	12.3	176 42.0	39.9	223 12.9	36.8
04	330 54.9	256 11.9	06.9	355 03.4	12.1	191 44.0	39.8	238 15.0	36.8
05	345 57.4	271 11.1	07.3	10 04.9	11.8	206 46.0	39.6	253 17.1	36.8
06	0 59.8	286 10.3	N22 07.8	25 06.4	S14 11.6	221 48.0	N16 39.5	268 19.2	N22 36.8
07	16 02.3	301 09.5	08.2	40 07.9	11.4	236 50.0	39.3	283 21.3	36.8
08	31 04.8	316 08.7	08.6	55 09.4	11.2	251 52.0	39.2	298 23.4	36.8
M 09	46 07.2	331 07.9	09.0	70 10.9	11.0	266 54.0	39.1	313 25.5	36.8
O 10	61 09.7	346 07.1	09.4	85 12.4	10.8	281 56.1	38.9	328 27.6	36.8
N 11	76 12.1	1 06.3	09.8	100 13.8	10.6	296 58.1	38.8	343 29.8	36.8
D 12	91 14.6	16 05.5	N22 10.2	115 15.3	S14 10.3	312 00.1	N16 38.7	358 31.9	N22 36.8
A 13	106 17.1	31 04.7	10.6	130 16.8	10.1	327 02.1	38.5	13 34.0	36.8
Y 14	121 19.5	46 03.9	11.1	145 18.3	09.9	342 04.1	38.4	28 36.1	36.8
15	136 22.0	61 03.1	11.5	160 19.8	09.7	357 06.1	38.2	43 38.2	36.8
16	151 24.5	76 02.3	11.9	175 21.3	09.5	12 08.1	38.1	58 40.3	36.8
17	166 26.9	91 01.5	12.3	190 22.8	09.3	27 10.1	38.0	73 42.4	36.8
18	181 29.4	106 00.7	N22 12.7	205 24.3	S14 09.1	42 12.1	N16 37.8	88 44.5	N22 36.8
19	196 31.9	120 59.9	13.1	220 25.8	08.8	57 14.1	37.7	103 46.7	36.8
20	211 34.3	135 59.1	13.5	235 27.3	08.6	72 16.1	37.5	118 48.8	36.8
21	226 36.8	150 58.3	13.9	250 28.8	08.4	87 18.2	37.4	133 50.9	36.8
22	241 39.2	165 57.4	14.3	265 30.3	08.2	102 20.2	37.3	148 53.0	36.8
23	256 41.7	180 56.6	14.7	280 31.8	08.0	117 22.2	37.1	163 55.1	36.8
Mer. Pass.	h m 6 00.0	v −0.8	d 0.4	v 1.5	d 0.2	v 2.0	d 0.1	v 2.1	d 0.0

STARS

Name	SHA ° ′	Dec ° ′
Acamar	315 24.5	S40 17.3
Achernar	335 32.7	S57 12.9
Acrux	173 18.4	S63 07.3
Adhara	255 19.1	S28 58.6
Aldebaran	290 58.8	N16 31.0
Alioth	166 27.2	N55 56.8
Alkaid	153 04.7	N49 18.0
Al Na'ir	27 53.2	S46 56.5
Alnilam	275 54.7	S 1 11.9
Alphard	218 04.1	S 8 40.3
Alphecca	126 17.3	N26 42.3
Alpheratz	357 51.8	N29 06.3
Altair	62 15.6	N 8 52.5
Ankaa	353 23.4	S42 17.1
Antares	112 35.6	S26 26.5
Arcturus	146 02.7	N19 10.0
Atria	107 44.1	S69 02.2
Avior	234 21.9	S59 31.3
Bellatrix	278 40.8	N 6 21.2
Betelgeuse	271 10.2	N 7 24.5
Canopus	264 00.2	S52 41.8
Capella	280 46.6	N46 00.1
Deneb	49 36.6	N45 17.3
Denebola	182 41.7	N14 33.3
Diphda	349 03.8	S17 58.0
Dubhe	194 01.2	N61 44.3
Elnath	278 23.0	N28 36.6
Eltanin	90 49.3	N51 29.3
Enif	33 54.7	N 9 53.3
Fomalhaut	15 32.5	S29 36.2
Gacrux	172 09.9	S57 08.2
Gienah	176 00.4	S17 33.7
Hadar	148 59.0	S60 23.6
Hamal	328 09.9	N23 28.6
Kaus Aust.	83 53.8	S34 23.0
Kochab	137 18.5	N74 08.7
Markab	13 46.1	N15 13.2
Menkar	314 23.5	N 4 06.2
Menkent	148 16.8	S36 23.3
Miaplacidus	221 42.3	S69 44.6
Mirfak	308 52.2	N49 52.2
Nunki	76 07.7	S26 17.6
Peacock	53 31.0	S56 43.4
Pollux	243 37.7	N28 01.2
Procyon	245 08.3	N 5 13.
Rasalhague	96 13.4	N12 33.
Regulus	207 52.0	N11 57.2
Rigel	281 20.0	S 8 11.8
Rigil Kent.	140 02.4	S60 51.
Sabik	102 21.2	S15 43.
Schedar	349 49.9	N56 33.
Shaula	96 32.2	S37 06.
Sirius	258 41.1	S16 43.
Spica	158 39.5	S11 10.
Suhail	222 58.7	S43 26.
Vega	80 43.9	N38 47.
Zuben'ubi	137 14.0	S16 03.

	SHA ° ′	Mer. Pass. h m
Venus	286 48.2	10 54
Mars	24 36.1	4 22
Jupiter	221 01.8	15 15
Saturn	267 29.9	12 09

UT	SUN GHA	SUN Dec	MOON GHA	v	MOON Dec	d	HP
d h	° ′	° ′	° ′	′	° ′	′	′
21 00	179 36.5	N23 26.3	273 58.3	15.4	S 7 29.3	13.4	55.4
01	194 36.4	26.3	288 32.7	15.4	7 15.9	13.3	55.4
02	209 36.3	26.3	303 07.1	15.5	7 02.6	13.3	55.4
03	224 36.1	. . 26.3	317 41.6	15.5	6 49.3	13.4	55.4
04	239 36.0	26.3	332 16.1	15.6	6 35.9	13.3	55.3
05	254 35.9	26.3	346 50.7	15.6	6 22.6	13.4	55.3
06	269 35.7	N23 26.3	1 25.3	15.7	S 6 09.2	13.4	55.3
07	284 35.6	26.3	16 00.0	15.7	5 55.8	13.4	55.2
08	299 35.5	26.3	30 34.7	15.8	5 42.4	13.4	55.2
09	314 35.3	. . 26.4	45 09.5	15.8	5 29.0	13.4	55.2
10	329 35.2	26.4	59 44.3	15.8	5 15.6	13.4	55.2
11	344 35.1	26.4	74 19.1	15.9	5 02.2	13.4	55.1
12	359 34.9	N23 26.4	88 54.0	16.0	S 4 48.8	13.5	55.1
13	14 34.8	26.4	103 29.0	15.9	4 35.3	13.4	55.1
14	29 34.6	26.4	118 03.9	16.1	4 21.9	13.5	55.1
15	44 34.5	. . 26.4	132 39.0	16.0	4 08.4	13.4	55.1
16	59 34.4	26.4	147 14.0	16.1	3 55.0	13.5	55.0
17	74 34.2	26.4	161 49.1	16.2	3 41.5	13.4	55.0
18	89 34.1	N23 26.4	176 24.3	16.1	S 3 28.1	13.5	55.0
19	104 34.0	26.4	190 59.4	16.2	3 14.6	13.4	55.0
20	119 33.8	26.4	205 34.6	16.3	3 01.2	13.5	54.9
21	134 33.7	. . 26.4	220 09.9	16.2	2 47.7	13.5	54.9
22	149 33.6	26.4	234 45.1	16.3	2 34.2	13.4	54.9
23	164 33.4	26.4	249 20.4	16.4	2 20.8	13.5	54.9
22 00	179 33.3	N23 26.4	263 55.8	16.3	S 2 07.3	13.5	54.9
01	194 33.1	26.4	278 31.1	16.4	1 53.8	13.4	54.8
02	209 33.0	26.4	293 06.5	16.4	1 40.4	13.5	54.8
03	224 32.9	. . 26.4	307 41.9	16.4	1 26.9	13.4	54.8
04	239 32.7	26.4	322 17.3	16.5	1 13.5	13.5	54.8
05	254 32.6	26.4	336 52.8	16.5	1 00.0	13.4	54.8
06	269 32.5	N23 26.4	351 28.3	16.5	S 0 46.6	13.4	54.7
07	284 32.3	26.3	6 03.8	16.5	0 33.2	13.5	54.7
08	299 32.2	26.3	20 39.3	16.5	0 19.7	13.4	54.7
09	314 32.1	. . 26.3	35 14.8	16.6	S 0 06.3	13.4	54.7
10	329 31.9	26.3	49 50.4	16.5	N 0 07.1	13.4	54.7
11	344 31.8	26.3	64 25.9	16.6	0 20.5	13.4	54.6
12	359 31.7	N23 26.3	79 01.5	16.6	N 0 33.9	13.4	54.6
13	14 31.5	26.3	93 37.1	16.6	0 47.3	13.3	54.6
14	29 31.4	26.3	108 12.7	16.7	1 00.6	13.4	54.6
15	44 31.2	. . 26.3	122 48.4	16.6	1 14.0	13.3	54.6
16	59 31.1	26.2	137 24.0	16.6	1 27.3	13.4	54.6
17	74 31.0	26.2	151 59.6	16.7	1 40.7	13.3	54.5
18	89 30.8	N23 26.2	166 35.3	16.6	N 1 54.0	13.3	54.5
19	104 30.7	26.2	181 10.9	16.7	2 07.3	13.3	54.5
20	119 30.6	26.2	195 46.6	16.7	2 20.6	13.3	54.5
21	134 30.4	. . 26.2	210 22.3	16.7	2 33.9	13.2	54.5
22	149 30.3	26.1	224 58.0	16.6	2 47.1	13.3	54.5
23	164 30.2	26.1	239 33.6	16.7	3 00.4	13.2	54.5
23 00	179 30.0	N23 26.1	254 09.3	16.7	N 3 13.6	13.2	54.4
01	194 29.9	26.1	268 45.0	16.7	3 26.8	13.2	54.4
02	209 29.8	26.1	283 20.7	16.6	3 40.0	13.2	54.4
03	224 29.6	. . 26.0	297 56.3	16.7	3 53.2	13.1	54.4
04	239 29.5	26.0	312 32.0	16.7	4 06.3	13.2	54.4
05	254 29.4	26.0	327 07.7	16.7	4 19.5	13.1	54.4
06	269 29.2	N23 26.0	341 43.4	16.6	N 4 32.6	13.1	54.4
07	284 29.1	25.9	356 19.0	16.7	4 45.7	13.0	54.4
08	299 28.9	25.9	10 54.7	16.6	4 58.7	13.1	54.3
09	314 28.8	. . 25.9	25 30.3	16.7	5 11.8	13.0	54.3
10	329 28.7	25.9	40 06.0	16.6	5 24.8	13.0	54.3
11	344 28.5	25.8	54 41.6	16.6	5 37.8	13.0	54.3
12	359 28.4	N23 25.8	69 17.2	16.6	N 5 50.8	12.9	54.3
13	14 28.3	25.8	83 52.8	16.6	6 03.7	12.9	54.3
14	29 28.1	25.7	98 28.4	16.6	6 16.6	12.9	54.3
15	44 28.0	. . 25.7	113 04.0	16.5	6 29.5	12.9	54.3
16	59 27.9	25.7	127 39.5	16.6	6 42.4	12.8	54.3
17	74 27.7	25.6	142 15.1	16.5	6 55.2	12.9	54.3
18	89 27.6	N23 25.6	156 50.6	16.5	N 7 08.1	12.7	54.2
19	104 27.5	25.6	171 26.1	16.5	7 20.8	12.8	54.2
20	119 27.3	25.5	186 01.6	16.5	7 33.6	12.7	54.2
21	134 27.2	. . 25.5	200 37.1	16.4	7 46.3	12.7	54.2
22	149 27.1	25.5	215 12.5	16.4	7 59.0	12.6	54.2
23	164 26.9	25.4	229 47.9	16.4	N 8 11.6	12.7	54.2
	SD 15.8	*d* 0.0	SD	15.0	14.9		14.8

Lat.	Twilight Naut.	Twilight Civil	Sunrise	Moonrise 21	22	23	24
°	h m	h m	h m	h m	h m	h m	h m
N 72	▭	▭	▭	01 19	00 51	{00 24 / 23 56}	23 19
N 70	▭	▭	▭	01 09	00 48	00 28	{00 08 / 23 43}
68	▭	▭	▭	01 01	00 46	00 32	00 18
66	▭	▭	▭	00 54	00 44	00 35	00 26
64	////	////	01 31	00 47	00 42	00 38	00 33
62	////	////	02 09	00 42	00 41	00 40	00 39
60	////	00 49	02 36	00 38	00 40	00 42	00 44
N 58	////	01 40	02 56	00 34	00 39	00 44	00 49
56	////	02 10	03 13	00 30	00 38	00 45	00 53
54	00 45	02 33	03 27	00 27	00 37	00 47	00 57
52	01 32	02 51	03 40	00 24	00 36	00 48	01 00
50	02 00	03 06	03 51	00 21	00 35	00 49	01 03
45	02 46	03 36	04 13	00 15	00 34	00 52	01 10
N 40	03 17	03 58	04 31	00 10	00 32	00 54	01 16
35	03 40	04 17	04 46	00 06	00 31	00 56	01 21
30	03 59	04 32	04 59	00 02	00 30	00 57	01 25
20	04 28	04 57	05 22	24 28	00 28	01 00	01 33
N 10	04 50	05 18	05 41	24 27	00 27	01 03	01 39
0	05 09	05 36	05 58	24 25	00 25	01 06	01 46
S 10	05 26	05 53	06 16	24 24	00 24	01 08	01 52
20	05 43	06 10	06 34	24 22	00 22	01 11	01 59
30	05 59	06 29	06 56	24 21	00 21	01 14	02 07
35	06 08	06 40	07 08	24 20	00 20	01 16	02 12
40	06 17	06 52	07 22	24 18	00 18	01 18	02 17
45	06 28	07 05	07 39	24 17	00 17	01 20	02 23
S 50	06 40	07 21	08 00	24 15	00 15	01 23	02 30
52	06 45	07 29	08 10	24 15	00 15	01 24	02 34
54	06 51	07 37	08 21	24 14	00 14	01 26	02 37
56	06 57	07 46	08 33	24 13	00 13	01 28	02 42
58	07 04	07 56	08 48	24 12	00 12	01 29	02 46
S 60	07 11	08 08	09 06	24 11	00 11	01 31	02 52

Lat.	Sunset	Twilight Civil	Twilight Naut.	Moonset 21	22	23	24
°	h m	h m	h m	h m	h m	h m	h m
N 72	▭	▭	▭	10 58	12 54	14 48	16 50
N 70	▭	▭	▭	11 06	12 53	14 38	16 28
68	▭	▭	▭	11 12	12 52	14 30	16 12
66	▭	▭	▭	11 17	12 51	14 24	15 59
64	22 33	////	////	11 21	12 51	14 18	15 48
62	21 54	////	////	11 25	12 50	14 14	15 38
60	21 28	23 15	////	11 28	12 50	14 10	15 31
N 58	21 07	22 23	////	11 31	12 49	14 06	15 24
56	20 51	21 53	////	11 34	12 49	14 03	15 18
54	20 36	21 31	23 19	11 36	12 49	14 00	15 12
52	20 24	21 13	22 31	11 38	12 48	13 58	15 07
50	20 13	20 58	22 03	11 40	12 48	13 55	15 03
45	19 50	20 28	21 18	11 44	12 48	13 50	14 53
N 40	19 32	20 05	20 47	11 47	12 47	13 46	14 45
35	19 17	19 47	20 24	11 50	12 47	13 43	14 39
30	19 04	19 32	20 05	11 53	12 46	13 40	14 33
20	18 42	19 07	19 36	11 57	12 46	13 34	14 23
N 10	18 23	18 46	19 13	12 01	12 45	13 29	14 14
0	18 06	18 28	18 54	12 04	12 45	13 25	14 05
S 10	17 48	18 11	18 37	12 08	12 44	13 20	13 57
20	17 30	17 54	18 21	12 11	12 44	13 16	13 48
30	17 08	17 35	18 05	12 15	12 43	13 10	13 38
35	16 56	17 24	17 56	12 18	12 43	13 07	13 33
40	16 42	17 12	17 46	12 20	12 42	13 04	13 26
45	16 25	16 59	17 36	12 23	12 42	13 00	13 18
S 50	16 04	16 43	17 24	12 27	12 41	12 55	13 09
52	15 54	16 35	17 19	12 29	12 41	12 53	13 05
54	15 43	16 27	17 13	12 31	12 40	12 50	13 01
56	15 30	16 18	17 07	12 33	12 40	12 48	12 56
58	15 16	16 08	17 00	12 35	12 40	12 45	12 50
S 60	14 58	15 56	16 53	12 37	12 39	12 41	12 44

Day	SUN Eqn. of Time 00^h	SUN Eqn. of Time 12^h	SUN Mer. Pass.	MOON Mer. Pass. Upper	MOON Mer. Pass. Lower	Age	Phase
d	m s	m s	h m	h m	h m	d %	
21	01 34	01 40	12 02	05 54	18 15	21 51	
22	01 47	01 53	12 02	06 35	18 55	22 42	
23	02 00	02 06	12 02	07 15	19 35	23 32	

UT	ARIES	VENUS −3.9		MARS −1.3		JUPITER −1.8		SATURN +0.0	
d h	GHA ° ′	GHA ° ′	Dec ° ′	GHA ° ′	Dec ° ′	GHA ° ′	Dec ° ′	GHA ° ′	Dec ° ′
24 00	271 44.2	195 55.8	N22 15.1	295 33.4	S14 07.8	132 24.2	N16 37.0	178 57.2	N22 36.8
01	286 46.6	210 55.0	15.5	310 34.9	07.6	147 26.2	36.9	193 59.3	36.8
02	301 49.1	225 54.2	15.9	325 36.4	07.4	162 28.2	36.7	209 01.4	36.8
03	316 51.6	240 53.4	. . 16.3	340 37.9	. . 07.1	177 30.2	. . 36.6	224 03.5	. . 36.8
04	331 54.0	255 52.6	16.7	355 39.4	06.9	192 32.2	36.4	239 05.7	36.8
05	346 56.5	270 51.8	17.0	10 40.9	06.7	207 34.2	36.3	254 07.8	36.8
06	1 59.0	285 51.0	N22 17.4	25 42.4	S14 06.5	222 36.2	N16 36.2	269 09.9	N22 36.8
07	17 01.4	300 50.2	17.8	40 43.9	06.3	237 38.2	36.0	284 12.0	36.7
T 08	32 03.9	315 49.4	18.2	55 45.4	06.1	252 40.2	35.9	299 14.1	36.7
U 09	47 06.4	330 48.6	. . 18.6	70 46.9	. . 05.9	267 42.3	. . 35.7	314 16.2	. . 36.7
E 10	62 08.8	345 47.8	19.0	85 48.4	05.7	282 44.3	35.6	329 18.3	36.7
S 11	77 11.3	0 46.9	19.4	100 49.9	05.5	297 46.3	35.5	344 20.4	36.7
D 12	92 13.7	15 46.1	N22 19.8	115 51.5	S14 05.3	312 48.3	N16 35.3	359 22.6	N22 36.7
A 13	107 16.2	30 45.3	20.1	130 53.0	05.0	327 50.3	35.2	14 24.7	36.7
Y 14	122 18.7	45 44.5	20.5	145 54.5	04.8	342 52.3	35.0	29 26.8	36.7
15	137 21.1	60 43.7	. . 20.9	160 56.0	. . 04.6	357 54.3	. . 34.9	44 28.9	. . 36.7
16	152 23.6	75 42.9	21.3	175 57.5	04.4	12 56.3	34.8	59 31.0	36.7
17	167 26.1	90 42.1	21.7	190 59.0	04.2	27 58.3	34.6	74 33.1	36.7
18	182 28.5	105 41.3	N22 22.0	206 00.6	S14 04.0	43 00.3	N16 34.5	89 35.2	N22 36.7
19	197 31.0	120 40.5	22.4	221 02.1	03.8	58 02.3	34.3	104 37.3	36.7
20	212 33.5	135 39.6	22.8	236 03.6	03.6	73 04.3	34.2	119 39.5	36.7
21	227 35.9	150 38.8	. . 23.2	251 05.1	. . 03.4	88 06.3	. . 34.1	134 41.6	. . 36.7
22	242 38.4	165 38.0	23.5	266 06.6	03.2	103 08.3	33.9	149 43.7	36.7
23	257 40.9	180 37.2	23.9	281 08.2	03.0	118 10.3	33.8	164 45.8	36.7
25 00	272 43.3	195 36.4	N22 24.3	296 09.7	S14 02.8	133 12.3	N16 33.6	179 47.9	N22 36.7
01	287 45.8	210 35.6	24.7	311 11.2	02.6	148 14.3	33.5	194 50.0	36.7
02	302 48.2	225 34.8	25.0	326 12.7	02.4	163 16.4	33.4	209 52.1	36.7
03	317 50.7	240 33.9	. . 25.4	341 14.3	. . 02.2	178 18.4	. . 33.2	224 54.2	. . 36.7
04	332 53.2	255 33.1	25.8	356 15.8	01.9	193 20.4	33.1	239 56.3	36.7
05	347 55.6	270 32.3	26.1	11 17.3	01.7	208 22.4	32.9	254 58.5	36.7
06	2 58.1	285 31.5	N22 26.5	26 18.8	S14 01.5	223 24.4	N16 32.8	270 00.6	N22 36.7
W 07	18 00.6	300 30.7	26.8	41 20.4	01.3	238 26.4	32.7	285 02.7	36.7
E 08	33 03.0	315 29.9	27.2	56 21.9	01.1	253 28.4	32.5	300 04.8	36.7
D 09	48 05.5	330 29.0	. . 27.6	71 23.4	. . 00.9	268 30.4	. . 32.4	315 06.9	. . 36.7
N 10	63 08.0	345 28.2	27.9	86 25.0	00.7	283 32.4	32.2	330 09.0	36.7
E 11	78 10.4	0 27.4	28.3	101 26.5	00.5	298 34.4	32.1	345 11.1	36.7
S 12	93 12.9	15 26.6	N22 28.6	116 28.0	S14 00.3	313 36.4	N16 32.0	0 13.2	N22 36.7
D 13	108 15.3	30 25.8	29.0	131 29.6	14 00.1	328 38.4	31.8	15 15.4	36.7
A 14	123 17.8	45 25.0	29.4	146 31.1	13 59.9	343 40.4	31.7	30 17.5	36.7
Y 15	138 20.3	60 24.1	. . 29.7	161 32.6	. . 59.7	358 42.4	. . 31.5	45 19.6	. . 36.6
16	153 22.7	75 23.3	30.1	176 34.2	59.5	13 44.4	31.4	60 21.7	36.6
17	168 25.2	90 22.5	30.4	191 35.7	59.3	28 46.4	31.2	75 23.8	36.6
18	183 27.7	105 21.7	N22 30.8	206 37.2	S13 59.1	43 48.4	N16 31.1	90 25.9	N22 36.6
19	198 30.1	120 20.9	31.1	221 38.8	58.9	58 50.4	31.0	105 28.0	36.6
20	213 32.6	135 20.0	31.5	236 40.3	58.7	73 52.4	30.8	120 30.1	36.6
21	228 35.1	150 19.2	. . 31.8	251 41.8	. . 58.5	88 54.4	. . 30.7	135 32.2	. . 36.6
22	243 37.5	165 18.4	32.2	266 43.4	58.3	103 56.4	30.5	150 34.4	36.6
23	258 40.0	180 17.6	32.5	281 44.9	58.1	118 58.4	30.4	165 36.5	36.6
26 00	273 42.5	195 16.8	N22 32.9	296 46.5	S13 57.9	134 00.4	N16 30.3	180 38.6	N22 36.6
01	288 44.9	210 15.9	33.2	311 48.0	57.7	149 02.4	30.1	195 40.7	36.6
02	303 47.4	225 15.1	33.5	326 49.5	57.5	164 04.4	30.0	210 42.8	36.6
03	318 49.8	240 14.3	. . 33.9	341 51.1	. . 57.3	179 06.4	. . 29.8	225 44.9	. . 36.6
04	333 52.3	255 13.5	34.2	356 52.6	57.1	194 08.4	29.7	240 47.0	36.6
05	348 54.8	270 12.6	34.6	11 54.2	56.9	209 10.5	29.5	255 49.1	36.6
06	3 57.2	285 11.8	N22 34.9	26 55.7	S13 56.7	224 12.5	N16 29.4	270 51.3	N22 36.6
07	18 59.7	300 11.0	35.2	41 57.3	56.5	239 14.5	29.3	285 53.4	36.6
T 08	34 02.2	315 10.2	35.6	56 58.8	56.3	254 16.5	29.1	300 55.5	36.6
H 09	49 04.6	330 09.3	. . 35.9	72 00.4	. . 56.2	269 18.5	. . 29.0	315 57.6	. . 36.6
U 10	64 07.1	345 08.5	36.2	87 01.9	56.0	284 20.5	28.8	330 59.7	36.6
R 11	79 09.6	0 07.7	36.6	102 03.5	55.8	299 22.5	28.7	346 01.8	36.6
S 12	94 12.0	15 06.9	N22 36.9	117 05.0	S13 55.6	314 24.5	N16 28.6	1 03.9	N22 36.6
D 13	109 14.5	30 06.0	37.2	132 06.6	55.4	329 26.5	28.4	16 06.0	36.6
A 14	124 17.0	45 05.2	37.6	147 08.1	55.2	344 28.5	28.3	31 08.1	36.6
Y 15	139 19.4	60 04.4	. . 37.9	162 09.7	. . 55.0	359 30.5	. . 28.1	46 10.3	. . 36.6
16	154 21.9	75 03.6	38.2	177 11.2	54.8	14 32.5	28.0	61 12.4	36.6
17	169 24.3	90 02.7	38.5	192 12.8	54.6	29 34.5	27.8	76 14.5	36.6
18	184 26.8	105 01.9	N22 38.9	207 14.3	S13 54.4	44 36.5	N16 27.7	91 16.6	N22 36.6
19	199 29.3	120 01.1	39.2	222 15.9	54.2	59 38.5	27.6	106 18.7	36.5
20	214 31.7	135 00.3	39.5	237 17.5	54.0	74 40.5	27.4	121 20.8	36.5
21	229 34.2	149 59.4	. . 39.8	252 19.0	. . 53.8	89 42.5	. . 27.3	136 22.9	. . 36.5
22	244 36.7	164 58.6	40.2	267 20.6	53.6	104 44.5	27.1	151 25.0	36.5
23	259 39.1	179 57.8	40.5	282 22.1	53.4	119 46.5	27.0	166 27.2	36.5
Mer. Pass.	h m 5 48.2	v −0.8	d 0.4	v 1.5	d 0.2	v 2.0	d 0.1	v 2.1	d 0.0

STARS Name	SHA ° ′	Dec ° ′
Acamar	315 24.5	S40 17
Achernar	335 32.6	S57 12
Acrux	173 18.4	S63 07
Adhara	255 19.1	S28 58
Aldebaran	290 58.8	N16 31
Alioth	166 27.3	N55 56
Alkaid	153 04.7	N49 18
Al Na'ir	27 53.2	S46 56
Alnilam	275 54.7	S 1 11
Alphard	218 04.1	S 8 40
Alphecca	126 17.3	N26 42
Alpheratz	357 51.7	N29 06
Altair	62 15.6	N 8 52
Ankaa	353 23.3	S42 17
Antares	112 35.6	S26 26
Arcturus	146 02.7	N19 10
Atria	107 44.1	S69 02
Avior	234 21.9	S59 31
Bellatrix	278 40.8	N 6 21
Betelgeuse	271 10.2	N 7 24
Canopus	264 00.2	S52 41
Capella	280 46.6	N46 00
Deneb	49 36.6	N45 17
Denebola	182 41.7	N14 33
Diphda	349 03.8	S17 58
Dubhe	194 01.2	N61 44
Elnath	278 23.0	N28 36
Eltanin	90 49.3	N51 29
Enif	33 54.7	N 9 53
Fomalhaut	15 32.4	S29 36
Gacrux	172 09.9	S57 08
Gienah	176 00.5	S17 3
Hadar	148 59.0	S60 2
Hamal	328 09.9	N23 28
Kaus Aust.	83 53.8	S34 2
Kochab	137 18.5	N74 08
Markab	13 46.1	N15 1
Menkar	314 23.5	N 4 0
Menkent	148 16.8	S36 2
Miaplacidus	221 42.3	S69 44
Mirfak	308 52.2	N49 5
Nunki	76 07.7	S26 1
Peacock	53 30.9	S56 4
Pollux	243 37.7	N28 0
Procyon	245 08.3	N 5 1
Rasalhague	96 13.4	N12 3
Regulus	207 52.0	N11 5
Rigel	281 20.0	S 8 1
Rigil Kent.	140 02.4	S60 5
Sabik	102 21.2	S15 4
Schedar	349 49.8	N56 3
Shaula	96 32.2	S37 0
Sirius	258 41.1	S16 4
Spica	158 39.5	S11 1
Suhail	222 58.7	S43 2
Vega	80 43.9	N38 4
Zuben'ubi	137 14.0	S16 0

	SHA ° ′	Mer. Pa[ss.] h m
Venus	282 53.1	10 5
Mars	23 26.4	4 1
Jupiter	220 29.0	15 0
Saturn	267 04.6	11 5

UT	SUN GHA	SUN Dec	MOON GHA	v	MOON Dec	d	HP
d h	° ′	° ′	° ′	′	° ′	′	′
4 00	179 26.8	N23 25.4	244 23.3	16.4	N 8 24.3	12.6	54.2
01	194 26.6	25.4	258 58.7	16.4	8 36.9	12.5	54.2
02	209 26.5	25.3	273 34.1	16.3	8 49.4	12.5	54.2
03	224 26.4	. . 25.3	288 09.4	16.3	9 01.9	12.5	54.2
04	239 26.2	25.2	302 44.7	16.3	9 14.4	12.5	54.2
05	254 26.1	25.2	317 20.0	16.2	9 26.9	12.4	54.2
06	269 26.0	N23 25.2	331 55.2	16.2	N 9 39.3	12.4	54.2
07	284 25.8	25.1	346 30.4	16.2	9 51.7	12.3	54.2
08	299 25.7	25.1	1 05.6	16.1	10 04.0	12.3	54.2
09	314 25.6	. . 25.0	15 40.7	16.2	10 16.3	12.3	54.1
10	329 25.4	25.0	30 15.9	16.0	10 28.6	12.2	54.1
11	344 25.3	24.9	44 50.9	16.1	10 40.8	12.2	54.1
12	359 25.2	N23 24.9	59 26.0	16.0	N10 53.0	12.1	54.1
13	14 25.0	24.8	74 01.0	16.0	11 05.1	12.1	54.1
14	29 24.9	24.8	88 36.0	15.9	11 17.2	12.0	54.1
15	44 24.8	. . 24.7	103 10.9	15.9	11 29.2	12.0	54.1
16	59 24.6	24.7	117 45.8	15.9	11 41.2	12.0	54.1
17	74 24.5	24.6	132 20.7	15.8	11 53.2	11.9	54.1
18	89 24.4	N23 24.6	146 55.5	15.8	N12 05.1	11.9	54.1
19	104 24.2	24.5	161 30.3	15.7	12 17.0	11.8	54.1
20	119 24.1	24.5	176 05.0	15.7	12 28.8	11.8	54.1
21	134 24.0	. . 24.4	190 39.7	15.6	12 40.6	11.8	54.1
22	149 23.8	24.4	205 14.3	15.7	12 52.4	11.6	54.1
23	164 23.7	24.3	219 49.0	15.5	13 04.0	11.7	54.1
5 00	179 23.5	N23 24.3	234 23.5	15.5	N13 15.7	11.6	54.1
01	194 23.4	24.2	248 58.0	15.5	13 27.3	11.5	54.1
02	209 23.3	24.2	263 32.5	15.4	13 38.8	11.5	54.1
03	224 23.1	. . 24.1	278 06.9	15.4	13 50.3	11.4	54.1
04	239 23.0	24.1	292 41.3	15.3	14 01.7	11.4	54.1
05	254 22.9	24.0	307 15.6	15.3	14 13.1	11.3	54.1
06	269 22.7	N23 23.9	321 49.9	15.2	N14 24.4	11.3	54.1
07	284 22.6	23.9	336 24.1	15.2	14 35.7	11.2	54.1
08	299 22.5	23.8	350 58.3	15.1	14 46.9	11.1	54.1
09	314 22.3	. . 23.8	5 32.4	15.0	14 58.0	11.1	54.1
10	329 22.2	23.7	20 06.4	15.0	15 09.1	11.1	54.1
11	344 22.1	23.6	34 40.4	15.0	15 20.2	11.0	54.1
12	359 21.9	N23 23.6	49 14.4	14.9	N15 31.2	10.9	54.1
13	14 21.8	23.5	63 48.3	14.8	15 42.1	10.9	54.1
14	29 21.7	23.4	78 22.1	14.8	15 53.0	10.8	54.1
15	44 21.5	. . 23.4	92 55.9	14.7	16 03.8	10.7	54.1
16	59 21.4	23.3	107 29.6	14.7	16 14.5	10.7	54.1
17	74 21.3	23.2	122 03.3	14.6	16 25.2	10.6	54.1
18	89 21.1	N23 23.2	136 36.9	14.6	N16 35.8	10.5	54.1
19	104 21.0	23.1	151 10.5	14.5	16 46.3	10.5	54.1
20	119 20.9	23.0	165 44.0	14.4	16 56.8	10.4	54.1
21	134 20.7	. . 23.0	180 17.4	14.4	17 07.2	10.4	54.2
22	149 20.6	22.9	194 50.8	14.3	17 17.6	10.3	54.2
23	164 20.5	22.8	209 24.1	14.2	17 27.9	10.2	54.2
6 00	179 20.3	N23 22.8	223 57.3	14.2	N17 38.1	10.1	54.2
01	194 20.2	22.7	238 30.5	14.1	17 48.2	10.1	54.2
02	209 20.1	22.6	253 03.6	14.1	17 58.3	10.0	54.2
03	224 19.9	. . 22.5	267 36.7	14.0	18 08.3	10.0	54.2
04	239 19.8	22.5	282 09.7	13.9	18 18.3	9.8	54.2
05	254 19.7	22.4	296 42.6	13.8	18 28.1	9.8	54.2
06	269 19.5	N23 22.3	311 15.4	13.8	N18 37.9	9.7	54.2
07	284 19.4	22.2	325 48.2	13.8	18 47.6	9.7	54.2
08	299 19.3	22.2	340 21.0	13.6	18 57.3	9.5	54.2
09	314 19.1	. . 22.1	354 53.6	13.6	19 06.8	9.5	54.2
10	329 19.0	22.0	9 26.2	13.5	19 16.3	9.4	54.2
11	344 18.9	21.9	23 58.7	13.5	19 25.7	9.3	54.2
12	359 18.7	N23 21.8	38 31.2	13.4	N19 35.0	9.3	54.2
13	14 18.6	21.8	53 03.6	13.3	19 44.3	9.2	54.2
14	29 18.5	21.7	67 35.9	13.2	19 53.5	9.0	54.3
15	44 18.4	. . 21.6	82 08.1	13.2	20 02.5	9.1	54.3
16	59 18.2	21.5	96 40.3	13.1	20 11.6	8.9	54.3
17	74 18.1	21.4	111 12.4	13.1	20 20.5	8.8	54.3
18	89 18.0	N23 21.3	125 44.5	12.9	N20 29.3	8.8	54.3
19	104 17.8	21.3	140 16.4	12.9	20 38.1	8.6	54.3
20	119 17.7	21.2	154 48.3	12.9	20 46.7	8.6	54.3
21	134 17.6	. . 21.1	169 20.2	12.7	20 55.3	8.5	54.3
22	149 17.4	21.0	183 51.9	12.7	21 03.8	8.4	54.3
23	164 17.3	20.9	198 23.6	12.6	N21 12.2	8.3	54.3
	SD 15.8	d 0.1	SD 14.8		14.7		14.8

Lat.	Twilight Naut.	Twilight Civil	Sunrise	Moonrise 24	Moonrise 25	Moonrise 26	Moonrise 27
°	h m	h m	h m	h m	h m	h m	h m
N 72	▭	▭	▭	23 19	22 10	▭	▭
N 70	▭	▭	▭	{00 08 / 23 43}	23 05	▭	▭
68	▭	▭	▭	00 18	{00 01 / 23 39}	22 55	▭
66	▭	▭	▭	00 26	00 16	{00 03 / 23 45}	22 47
64	////	////	01 33	00 33	00 28	00 23	00 17
62	////	////	02 11	00 39	00 38	00 38	00 40
60	////	00 51	02 37	00 44	00 47	00 52	00 59
N 58	////	01 42	02 57	00 49	00 55	01 03	01 15
56	////	02 12	03 14	00 53	01 02	01 13	01 29
54	00 47	02 34	03 28	00 57	01 08	01 22	01 40
52	01 34	02 52	03 41	01 00	01 14	01 30	01 51
50	02 01	03 07	03 52	01 03	01 19	01 37	02 00
45	02 47	03 37	04 14	01 10	01 30	01 52	02 20
N 40	03 18	03 59	04 32	01 16	01 39	02 05	02 36
35	03 41	04 18	04 47	01 21	01 47	02 16	02 49
30	03 59	04 33	05 00	01 25	01 54	02 26	03 01
20	04 29	04 58	05 22	01 33	02 06	02 42	03 22
N 10	04 51	05 18	05 41	01 39	02 17	02 57	03 39
0	05 10	05 36	05 59	01 46	02 27	03 10	03 56
S 10	05 27	05 53	06 16	01 52	02 37	03 24	04 13
20	05 43	06 11	06 35	01 59	02 48	03 39	04 31
30	06 00	06 30	06 56	02 07	03 01	03 56	04 52
35	06 09	06 40	07 08	02 12	03 08	04 06	05 04
40	06 18	06 52	07 23	02 17	03 16	04 17	05 18
45	06 28	07 06	07 39	02 23	03 26	04 31	05 35
S 50	06 40	07 22	08 00	02 30	03 38	04 47	05 56
52	06 45	07 29	08 10	02 34	03 44	04 55	06 06
54	06 51	07 37	08 21	02 37	03 50	05 04	06 18
56	06 57	07 46	08 34	02 42	03 57	05 13	06 31
58	07 04	07 57	08 48	02 46	04 05	05 25	06 46
S 60	07 11	08 08	09 06	02 52	04 13	05 38	07 04

Lat.	Sunset	Twilight Civil	Twilight Naut.	Moonset 24	Moonset 25	Moonset 26	Moonset 27
°	h m	h m	h m	h m	h m	h m	h m
N 72	▭	▭	▭	16 50	19 28	▭	▭
N 70	▭	▭	▭	16 28	18 35	▭	▭
68	▭	▭	▭	16 12	18 02	20 19	▭
66	▭	▭	▭	15 59	17 39	19 30	22 09
64	22 32	////	////	15 48	17 21	19 00	20 47
62	21 54	////	////	15 38	17 06	18 37	20 10
60	21 28	23 13	////	15 31	16 53	18 18	19 44
N 58	21 07	22 23	////	15 24	16 43	18 03	19 23
56	20 51	21 53	////	15 18	16 33	17 50	19 07
54	20 36	21 31	23 17	15 12	16 25	17 39	18 52
52	20 24	21 13	22 31	15 07	16 18	17 29	18 40
50	20 13	20 58	22 03	15 03	16 11	17 20	18 29
45	19 51	20 28	21 18	14 53	15 57	17 02	18 06
N 40	19 33	20 06	20 47	14 45	15 46	16 47	17 48
35	19 18	19 47	20 24	14 39	15 36	16 34	17 33
30	19 05	19 32	20 06	14 33	15 27	16 23	17 19
20	18 43	19 07	19 37	14 23	15 12	16 04	16 57
N 10	18 24	18 47	19 14	14 14	15 00	15 47	16 38
0	18 06	18 29	18 55	14 05	14 48	15 32	16 19
S 10	17 49	18 12	18 38	13 57	14 36	15 17	16 01
20	17 30	17 54	18 22	13 48	14 23	15 01	15 42
30	17 09	17 35	18 05	13 38	14 08	14 42	15 20
35	16 57	17 25	17 57	13 33	14 00	14 31	15 07
40	16 43	17 13	17 47	13 26	13 51	14 19	14 52
45	16 26	17 00	17 37	13 18	13 40	14 04	14 34
S 50	16 05	16 43	17 25	13 09	13 26	13 47	14 13
52	15 55	16 36	17 20	13 05	13 20	13 38	14 02
54	15 44	16 28	17 14	13 01	13 13	13 29	13 50
56	15 31	16 19	17 08	12 56	13 06	13 19	13 37
58	15 17	16 09	17 01	12 50	12 57	13 07	13 21
S 60	14 59	15 57	16 54	12 44	12 47	12 53	13 03

Day	SUN Eqn. of Time 00^h	SUN Eqn. of Time 12^h	SUN Mer. Pass.	MOON Mer. Pass. Upper	MOON Mer. Pass. Lower	MOON Age	MOON Phase
d	m s	m s	h m	h m	h m	d %	
24	02 13	02 19	12 02	07 55	20 16	24 24	
25	02 26	02 32	12 03	08 37	20 59	25 16	
26	02 38	02 45	12 03	09 21	21 44	26 10	

	UT	ARIES	VENUS −3.9		MARS −1.4		JUPITER −1.8		SATURN +0.0		STARS		
		GHA	GHA	Dec	GHA	Dec	GHA	Dec	GHA	Dec	Name	SHA	Dec
	d h	° ′	° ′	° ′	° ′	° ′	° ′	° ′	° ′	° ′		° ′	°
	27 00	274 41.6	194 56.9	N22 40.8	297 23.7	S13 53.3	134 48.5	N16 26.8	181 29.3	N22 36.5	Acamar	315 24.5	S40 17.
	01	289 44.1	209 56.1	41.1	312 25.3	53.1	149 50.5	26.7	196 31.4	36.5	Achernar	335 32.6	S57 12.
	02	304 46.5	224 55.3	41.4	327 26.8	52.9	164 52.5	26.6	211 33.5	36.5	Acrux	173 18.4	S63 07.
	03	319 49.0	239 54.5	41.7	342 28.4	52.7	179 54.5	26.4	226 35.6	36.5	Adhara	255 19.1	S28 58.
	04	334 51.5	254 53.6	42.0	357 29.9	52.5	194 56.5	26.3	241 37.7	36.5	Aldebaran	290 58.7	N16 31.
	05	349 53.9	269 52.8	42.4	12 31.5	52.3	209 58.5	26.1	256 39.8	36.5			
	06	4 56.4	284 52.0	N22 42.7	27 33.1	S13 52.1	225 00.5	N16 26.0	271 41.9	N22 36.5	Alioth	166 27.3	N55 56
	07	19 58.8	299 51.1	43.0	42 34.6	51.9	240 02.5	25.8	286 44.1	36.5	Alkaid	153 04.7	N49 18.
	08	35 01.3	314 50.3	43.3	57 36.2	51.7	255 04.5	25.7	301 46.2	36.5	Al Na'ir	27 53.1	S46 56.
F	09	50 03.8	329 49.5	43.6	72 37.8	51.5	270 06.5	25.6	316 48.3	36.5	Alnilam	275 54.7	S 1 11.
R	10	65 06.2	344 48.6	43.9	87 39.3	51.4	285 08.5	25.4	331 50.4	36.5	Alphard	218 04.1	S 8 40.
I	11	80 08.7	359 47.8	44.2	102 40.9	51.2	300 10.5	25.3	346 52.5	36.5			
D	12	95 11.2	14 47.0	N22 44.5	117 42.5	S13 51.0	315 12.5	N16 25.1	1 54.6	N22 36.5	Alphecca	126 17.3	N26 42.
A	13	110 13.6	29 46.1	44.8	132 44.1	50.8	330 14.5	25.0	16 56.7	36.5	Alpheratz	357 51.7	N29 06.
Y	14	125 16.1	44 45.3	45.1	147 45.6	50.6	345 16.5	24.8	31 58.8	36.5	Altair	62 15.6	N 8 52.
	15	140 18.6	59 44.5	45.4	162 47.2	50.4	0 18.5	24.7	47 00.9	36.5	Ankaa	353 23.3	S42 17.
	16	155 21.0	74 43.6	45.7	177 48.8	50.2	15 20.5	24.6	62 03.1	36.5	Antares	112 35.6	S26 26
	17	170 23.5	89 42.8	46.0	192 50.3	50.0	30 22.5	24.4	77 05.2	36.5			
	18	185 26.0	104 42.0	N22 46.3	207 51.9	S13 49.9	45 24.5	N16 24.3	92 07.3	N22 36.5	Arcturus	146 02.7	N19 10
	19	200 28.4	119 41.1	46.6	222 53.5	49.7	60 26.4	24.1	107 09.4	36.5	Atria	107 44.1	S69 02
	20	215 30.9	134 40.3	46.9	237 55.1	49.5	75 28.4	24.0	122 11.5	36.4	Avior	234 22.0	S59 31
	21	230 33.3	149 39.5	47.2	252 56.7	49.3	90 30.4	23.8	137 13.6	36.4	Bellatrix	278 40.8	N 6 21
	22	245 35.8	164 38.6	47.5	267 58.2	49.1	105 32.4	23.7	152 15.7	36.4	Betelgeuse	271 10.2	N 7 24
	23	260 38.3	179 37.8	47.8	282 59.8	48.9	120 34.4	23.5	167 17.8	36.4			
	28 00	275 40.7	194 37.0	N22 48.1	298 01.4	S13 48.8	135 36.4	N16 23.4	182 20.0	N22 36.4	Canopus	264 00.2	S52 41.
	01	290 43.2	209 36.1	48.4	313 03.0	48.6	150 38.4	23.3	197 22.1	36.4	Capella	280 46.6	N46 00
	02	305 45.7	224 35.3	48.7	328 04.6	48.4	165 40.4	23.1	212 24.2	36.4	Deneb	49 36.5	N45 17.
	03	320 48.1	239 34.5	48.9	343 06.1	48.2	180 42.4	23.0	227 26.3	36.4	Denebola	182 41.7	N14 33.
	04	335 50.6	254 33.6	49.2	358 07.7	48.0	195 44.4	22.8	242 28.4	36.4	Diphda	349 03.8	S17 58
	05	350 53.1	269 32.8	49.5	13 09.3	47.8	210 46.4	22.7	257 30.5	36.4			
	06	5 55.5	284 31.9	N22 49.8	28 10.9	S13 47.7	225 48.4	N16 22.5	272 32.6	N22 36.4	Dubhe	194 01.3	N61 44
S	07	20 58.0	299 31.1	50.1	43 12.5	47.5	240 50.4	22.4	287 34.7	36.4	Elnath	278 23.0	N28 36
	08	36 00.4	314 30.3	50.4	58 14.1	47.3	255 52.4	22.2	302 36.8	36.4	Eltanin	90 49.3	N51 29
A	09	51 02.9	329 29.4	50.6	73 15.6	47.1	270 54.4	22.1	317 39.0	36.4	Enif	33 54.6	N 9 53
T	10	66 05.4	344 28.6	50.9	88 17.2	46.9	285 56.4	22.0	332 41.1	36.4	Fomalhaut	15 32.4	S29 36
U	11	81 07.8	359 27.8	51.2	103 18.8	46.7	300 58.4	21.8	347 43.2	36.4			
R	12	96 10.3	14 26.9	N22 51.5	118 20.4	S13 46.6	316 00.4	N16 21.7	2 45.3	N22 36.4	Gacrux	172 09.9	S57 08
D	13	111 12.8	29 26.1	51.8	133 22.0	46.4	331 02.4	21.5	17 47.4	36.4	Gienah	176 00.5	S17 33
A	14	126 15.2	44 25.2	52.0	148 23.6	46.2	346 04.4	21.4	32 49.5	36.4	Hadar	148 59.0	S60 23
Y	15	141 17.7	59 24.4	52.3	163 25.2	46.0	1 06.4	21.2	47 51.6	36.4	Hamal	328 09.9	N23 28
	16	156 20.2	74 23.6	52.6	178 26.8	45.8	16 08.4	21.1	62 53.7	36.4	Kaus Aust.	83 53.8	S34 23
	17	171 22.6	89 22.7	52.8	193 28.4	45.7	31 10.4	20.9	77 55.9	36.4			
	18	186 25.1	104 21.9	N22 53.1	208 30.0	S13 45.5	46 12.4	N16 20.8	92 58.0	N22 36.4	Kochab	137 18.6	N74 08
	19	201 27.6	119 21.0	53.4	223 31.6	45.3	61 14.4	20.7	108 00.1	36.3	Markab	13 46.1	N15 13
	20	216 30.0	134 20.2	53.7	238 33.2	45.1	76 16.4	20.5	123 02.2	36.3	Menkar	314 23.5	N 4 06
	21	231 32.5	149 19.4	53.9	253 34.8	45.0	91 18.4	20.4	138 04.3	36.3	Menkent	148 16.8	S36 23
	22	246 34.9	164 18.5	54.2	268 36.4	44.8	106 20.4	20.2	153 06.4	36.3	Miaplacidus	221 42.4	S69 44
	23	261 37.4	179 17.7	54.5	283 38.0	44.6	121 22.3	20.1	168 08.5	36.3			
	29 00	276 39.9	194 16.8	N22 54.7	298 39.6	S13 44.4	136 24.3	N16 19.9	183 10.6	N22 36.3	Mirfak	308 52.1	N49 52
	01	291 42.3	209 16.0	55.0	313 41.2	44.3	151 26.3	19.8	198 12.8	36.3	Nunki	76 07.7	S26 17
	02	306 44.8	224 15.1	55.2	328 42.8	44.1	166 28.3	19.6	213 14.9	36.3	Peacock	53 30.9	S56 43
	03	321 47.3	239 14.3	55.5	343 44.4	43.9	181 30.3	19.5	228 17.0	36.3	Pollux	243 37.7	N28 01
	04	336 49.7	254 13.5	55.8	358 46.0	43.7	196 32.3	19.3	243 19.1	36.3	Procyon	245 08.3	N 5 13
	05	351 52.2	269 12.6	56.0	13 47.6	43.6	211 34.3	19.2	258 21.2	36.3			
	06	6 54.7	284 11.8	N22 56.3	28 49.2	S13 43.4	226 36.3	N16 19.1	273 23.3	N22 36.3	Rasalhague	96 13.4	N12 33
	07	21 57.1	299 10.9	56.5	43 50.8	43.2	241 38.3	18.9	288 25.4	36.3	Regulus	207 52.0	N11 57
	08	36 59.6	314 10.1	56.8	58 52.4	43.0	256 40.3	18.8	303 27.5	36.3	Rigel	281 19.9	S 8 11
S	09	52 02.1	329 09.2	57.0	73 54.0	42.9	271 42.3	18.6	318 29.7	36.3	Rigil Kent.	140 02.4	S60 51
U	10	67 04.5	344 08.4	57.3	88 55.6	42.7	286 44.3	18.5	333 31.8	36.3	Sabik	102 21.2	S15 43
N	11	82 07.0	359 07.6	57.5	103 57.2	42.5	301 46.3	18.3	348 33.9	36.3			
D	12	97 09.4	14 06.7	N22 57.8	118 58.8	S13 42.3	316 48.3	N16 18.2	3 36.0	N22 36.3	Schedar	349 49.8	N56 33
A	13	112 11.9	29 05.9	58.0	134 00.4	42.2	331 50.3	18.0	18 38.1	36.3	Shaula	96 32.2	S37 06
Y	14	127 14.4	44 05.0	58.3	149 02.0	42.0	346 52.3	17.9	33 40.2	36.3	Sirius	258 41.0	S16 43
	15	142 16.8	59 04.2	58.5	164 03.7	41.8	1 54.3	17.7	48 42.3	36.3	Spica	158 39.5	S11 10
	16	157 19.3	74 03.3	58.8	179 05.3	41.7	16 56.2	17.6	63 44.4	36.2	Suhail	222 58.7	S43 26
	17	172 21.8	89 02.5	59.0	194 06.9	41.5	31 58.2	17.5	78 46.6	36.2			
	18	187 24.2	104 01.6	N22 59.3	209 08.5	S13 41.3	47 00.2	N16 17.3	93 48.7	N22 36.2	Vega	80 43.9	N38 47
	19	202 26.7	119 00.8	59.5	224 10.1	41.1	62 02.2	17.2	108 50.8	36.2	Zuben'ubi	137 14.0	S16 03
	20	217 29.2	133 59.9	22 59.7	239 11.7	41.0	77 04.2	17.0	123 52.9	36.2		SHA	Mer.Pass
	21	232 31.6	148 59.1	23 00.0	254 13.4	40.8	92 06.2	16.9	138 55.0	36.2		° ′	h m
	22	247 34.1	163 58.3	00.2	269 15.0	40.6	107 08.2	16.7	153 57.1	36.2	Venus	278 56.2	11 02
	23	262 36.6	178 57.4	00.5	284 16.6	40.5	122 10.2	16.6	168 59.2	36.2	Mars	22 20.7	4 07
		h m									Jupiter	219 55.7	14 56
	Mer. Pass.	5 36.4	v −0.8	d 0.3	v 1.6	d 0.2	v 2.0	d 0.1	v 2.1	d 0.0	Saturn	266 39.2	11 49

	UT	SUN GHA	SUN Dec	MOON GHA	v	MOON Dec	d	HP
	d h	° ′	° ′	° ′	′	° ′	′	′
	27 00	179 17.2	N23 20.8	212 55.2	12.5	N21 20.5	8.2	54.3
	01	194 17.0	20.7	227 26.7	12.5	21 28.7	8.1	54.4
	02	209 16.9	20.6	241 58.2	12.4	21 36.8	8.1	54.4
	03	224 16.8	. . 20.5	256 29.6	12.3	21 44.9	7.9	54.4
	04	239 16.6	20.5	271 00.9	12.3	21 52.8	7.9	54.4
	05	254 16.5	20.4	285 32.2	12.2	22 00.7	7.7	54.4
	06	269 16.4	N23 20.3	300 03.4	12.1	N22 08.4	7.7	54.4
	07	284 16.2	20.2	314 34.5	12.0	22 16.1	7.5	54.4
	08	299 16.1	20.1	329 05.5	12.0	22 23.6	7.5	54.4
F	09	314 16.0	. . 20.0	343 36.5	11.9	22 31.1	7.4	54.4
R	10	329 15.8	19.9	358 07.4	11.8	22 38.5	7.2	54.5
I	11	344 15.7	19.8	12 38.2	11.8	22 45.7	7.2	54.5
D	12	359 15.6	N23 19.7	27 09.0	11.7	N22 52.9	7.1	54.5
A	13	14 15.5	19.6	41 39.7	11.6	23 00.0	6.9	54.5
Y	14	29 15.3	19.5	56 10.3	11.5	23 06.9	6.9	54.5
	15	44 15.2	. . 19.4	70 40.8	11.5	23 13.8	6.7	54.5
	16	59 15.1	19.3	85 11.3	11.4	23 20.5	6.7	54.5
	17	74 14.9	19.2	99 41.7	11.3	23 27.2	6.5	54.5
	18	89 14.8	N23 19.1	114 12.0	11.3	N23 33.7	6.4	54.5
	19	104 14.7	19.0	128 42.3	11.2	23 40.1	6.3	54.6
	20	119 14.5	18.9	143 12.5	11.1	23 46.4	6.3	54.6
	21	134 14.4	. . 18.8	157 42.6	11.0	23 52.7	6.1	54.6
	22	149 14.3	18.7	172 12.6	11.0	23 58.8	6.0	54.6
	23	164 14.2	18.6	186 42.6	10.9	24 04.8	5.8	54.6
	28 00	179 14.0	N23 18.5	201 12.5	10.9	N24 10.6	5.8	54.6
	01	194 13.9	18.4	215 42.4	10.7	24 16.4	5.7	54.6
	02	209 13.8	18.3	230 12.1	10.8	24 22.1	5.5	54.7
	03	224 13.6	. . 18.1	244 41.9	10.6	24 27.6	5.4	54.7
	04	239 13.5	18.0	259 11.5	10.6	24 33.0	5.3	54.7
	05	254 13.4	17.9	273 41.1	10.5	24 38.3	5.2	54.7
	06	269 13.2	N23 17.8	288 10.6	10.5	N24 43.5	5.1	54.7
	07	284 13.1	17.7	302 40.1	10.4	24 48.6	5.0	54.7
S	08	299 13.0	17.6	317 09.5	10.3	24 53.6	4.8	54.7
A	09	314 12.9	. . 17.5	331 38.8	10.2	24 58.4	4.7	54.8
T	10	329 12.7	17.4	346 08.0	10.2	25 03.1	4.6	54.8
U	11	344 12.6	17.3	0 37.2	10.2	25 07.7	4.5	54.8
R	12	359 12.5	N23 17.1	15 06.4	10.1	N25 12.2	4.4	54.8
D	13	14 12.3	17.0	29 35.5	10.0	25 16.6	4.2	54.8
A	14	29 12.2	16.9	44 04.5	10.0	25 20.8	4.1	54.8
Y	15	44 12.1	. . 16.8	58 33.5	9.9	25 24.9	4.0	54.8
	16	59 11.9	16.7	73 02.4	9.8	25 28.9	3.8	54.9
	17	74 11.8	16.6	87 31.2	9.8	25 32.7	3.8	54.9
	18	89 11.7	N23 16.4	102 00.0	9.7	N25 36.5	3.6	54.9
	19	104 11.6	16.3	116 28.7	9.7	25 40.1	3.4	54.9
	20	119 11.4	16.2	130 57.4	9.7	25 43.5	3.4	54.9
	21	134 11.3	. . 16.1	145 26.1	9.5	25 46.9	3.2	54.9
	22	149 11.2	16.0	159 54.6	9.6	25 50.1	3.1	55.0
	23	164 11.0	15.8	174 23.2	9.4	25 53.2	3.0	55.0
	29 00	179 10.9	N23 15.7	188 51.6	9.5	N25 56.2	2.8	55.0
	01	194 10.8	15.6	203 20.1	9.4	25 59.0	2.7	55.0
	02	209 10.7	15.5	217 48.5	9.3	26 01.7	2.6	55.0
	03	224 10.5	. . 15.3	232 16.8	9.3	26 04.3	2.4	55.0
	04	239 10.4	15.2	246 45.1	9.2	26 06.7	2.3	55.1
	05	254 10.3	15.1	261 13.3	9.2	26 09.0	2.2	55.1
	06	269 10.2	N23 15.0	275 41.5	9.2	N26 11.2	2.0	55.1
	07	284 10.0	14.8	290 09.7	9.1	26 13.2	1.9	55.1
	08	299 09.9	14.7	304 37.8	9.1	26 15.1	1.8	55.1
S	09	314 09.8	. . 14.6	319 05.9	9.0	26 16.9	1.6	55.1
U	10	329 09.6	14.4	333 33.9	9.0	26 18.5	1.5	55.2
N	11	344 09.5	14.3	348 01.9	9.0	26 20.0	1.3	55.2
D	12	359 09.4	N23 14.2	2 29.9	8.9	N26 21.3	1.2	55.2
A	13	14 09.3	14.0	16 57.8	8.9	26 22.5	1.1	55.2
Y	14	29 09.1	13.9	31 25.7	8.9	26 23.6	1.0	55.2
	15	44 09.0	. . 13.8	45 53.6	8.8	26 24.6	0.8	55.2
	16	59 08.9	13.6	60 21.4	8.8	26 25.4	0.6	55.3
	17	74 08.8	13.5	74 49.2	8.8	26 26.0	0.5	55.3
	18	89 08.6	N23 13.4	89 17.0	8.7	N26 26.5	0.4	55.3
	19	104 08.5	13.2	103 44.7	8.8	26 26.9	0.3	55.3
	20	119 08.4	13.1	118 12.5	8.7	26 27.2	0.1	55.3
	21	134 08.2	. . 13.0	132 40.2	8.6	26 27.3	0.1	55.4
	22	149 08.1	12.8	147 07.8	8.7	26 27.2	0.2	55.4
	23	164 08.0	12.7	161 35.5	8.6	N26 27.0	0.3	55.4
		SD 15.8	d 0.1	SD 14.8		14.9		15.0

Lat.	Twilight Naut.	Twilight Civil	Sunrise	Moonrise 27	Moonrise 28	Moonrise 29	Moonrise 30
°	h m	h m	h m	h m	h m	h m	h m
N 72	□	□	□	□	□	□	□
N 70	□	□	□	□	□	□	□
68	□	□	□	□	□	□	□
66	□	□	□	22 47	□	□	□
64	////	////	01 36	00 17	00 10 / 23 52	□	□
62	////	////	02 13	00 40	00 47	01 03	01 46
60	////	00 56	02 39	00 59	01 13	01 39	02 25
N 58	////	01 44	02 59	01 15	01 34	02 05	02 53
56	////	02 14	03 16	01 29	01 51	02 25	03 14
54	00 51	02 36	03 30	01 40	02 06	02 42	03 32
52	01 36	02 53	03 42	01 51	02 19	02 57	03 47
50	02 03	03 08	03 53	02 00	02 30	03 09	04 00
45	02 48	03 38	04 15	02 20	02 53	03 36	04 27
N 40	03 19	04 00	04 33	02 36	03 12	03 56	04 48
35	03 42	04 19	04 48	02 49	03 28	04 14	05 06
30	04 00	04 34	05 01	03 01	03 42	04 29	05 21
20	04 29	04 59	05 23	03 22	04 06	04 54	05 47
N 10	04 52	05 19	05 42	03 39	04 26	05 16	06 10
0	05 11	05 37	05 59	03 56	04 45	05 37	06 30
S 10	05 28	05 54	06 17	04 13	05 04	05 58	06 51
20	05 44	06 11	06 35	04 31	05 25	06 20	07 13
30	06 00	06 30	06 56	04 52	05 49	06 45	07 39
35	06 09	06 41	07 09	05 04	06 03	07 01	07 55
40	06 18	06 52	07 23	05 18	06 20	07 18	08 12
45	06 28	07 06	07 39	05 35	06 39	07 40	08 34
S 50	06 40	07 22	08 00	05 56	07 04	08 07	09 00
52	06 45	07 29	08 10	06 06	07 16	08 20	09 14
54	06 51	07 37	08 21	06 18	07 30	08 35	09 29
56	06 57	07 46	08 33	06 31	07 46	08 53	09 46
58	07 04	07 56	08 48	06 46	08 05	09 14	10 08
S 60	07 11	08 08	09 05	07 04	08 28	09 42	10 35

Lat.	Sunset	Twilight Civil	Twilight Naut.	Moonset 27	Moonset 28	Moonset 29	Moonset 30
°	h m	h m	h m	h m	h m	h m	h m
N 72	□	□	□	□	□	□	□
N 70	□	□	□	□	□	□	□
68	□	□	□	□	□	□	□
66	□	□	□	22 09	□	□	□
64	22 30	////	////	20 47	22 50	□	□
62	21 53	////	////	20 10	21 39	22 48	23 22
60	21 27	23 09	////	19 44	21 04	22 08	22 49
N 58	21 07	22 21	////	19 23	20 38	21 40	22 24
56	20 50	21 52	////	19 07	20 18	21 19	22 05
54	20 36	21 30	23 14	18 52	20 01	21 01	21 48
52	20 24	21 13	22 30	18 40	19 47	20 46	21 34
50	20 13	20 58	22 03	18 29	19 34	20 33	21 22
45	19 51	20 28	21 18	18 06	19 09	20 06	20 56
N 40	19 33	20 06	20 47	17 48	18 48	19 45	20 35
35	19 18	19 48	20 24	17 33	18 31	19 27	20 18
30	19 05	19 32	20 06	17 19	18 16	19 12	20 04
20	18 43	19 08	19 37	16 57	17 51	18 46	19 38
N 10	18 24	18 47	19 14	16 38	17 30	18 23	19 17
0	18 07	18 29	18 56	16 19	17 10	18 02	18 56
S 10	17 50	18 12	18 39	16 01	16 50	17 42	18 36
20	17 31	17 55	18 23	15 42	16 28	17 19	18 14
30	17 10	17 36	18 06	15 20	16 04	16 53	17 48
35	16 58	17 26	17 58	15 07	15 49	16 38	17 33
40	16 44	17 14	17 48	14 52	15 32	16 20	17 16
45	16 27	17 01	17 38	14 34	15 12	15 59	16 55
S 50	16 06	16 45	17 26	14 13	14 47	15 31	16 28
52	15 57	16 37	17 21	14 02	14 34	15 18	16 15
54	15 46	16 29	17 15	13 50	14 20	15 03	16 00
56	15 33	16 20	17 09	13 37	14 04	14 45	15 43
58	15 18	16 10	17 03	13 21	13 45	14 23	15 21
S 60	15 01	15 59	16 55	13 03	13 21	13 55	14 54

Day	SUN Eqn. of Time 00^h	SUN Eqn. of Time 12^h	SUN Mer. Pass.	MOON Mer. Pass. Upper	MOON Mer. Pass. Lower	MOON Age	MOON Phase
d	m s	m s	h m	h m	h m	d	%
27	02 51	02 57	12 03	10 08	22 32	27	5
28	03 04	03 10	12 03	10 57	23 23	28	2 ●
29	03 16	03 22	12 03	11 50	24 16	29	0

UT (d h)	ARIES GHA	VENUS −3.9 GHA	VENUS Dec	MARS −1.5 GHA	MARS Dec	JUPITER −1.8 GHA	JUPITER Dec	SATURN +0.0 GHA	SATURN Dec
d h	° ′	° ′	° ′	° ′	° ′	° ′	° ′	° ′	° ′
30 00 (MONDAY)	277 39.0	193 56.6	N23 00.7	299 18.2	S13 40.3	137 12.2	N16 16.4	184 01.3	N22 36.2
01	292 41.5	208 55.7	00.9	314 19.8	40.1	152 14.2	16.3	199 03.5	36.2
02	307 43.9	223 54.9	01.2	329 21.5	40.0	167 16.2	16.1	214 05.6	36.2
03	322 46.4	238 54.0	. . 01.4	344 23.1	. . 39.8	182 18.2	. . 16.0	229 07.7	. . 36.2
04	337 48.9	253 53.2	01.6	359 24.7	39.6	197 20.2	15.8	244 09.8	36.2
05	352 51.3	268 52.3	01.9	14 26.3	39.5	212 22.2	15.7	259 11.9	36.2
06	7 53.8	283 51.5	N23 02.1	29 28.0	S13 39.3	227 24.1	N16 15.5	274 14.0	N22 36.2
07	22 56.3	298 50.6	02.3	44 29.6	39.1	242 26.1	15.4	289 16.1	36.2
08	37 58.7	313 49.8	02.5	59 31.2	39.0	257 28.1	15.3	304 18.2	36.2
09	53 01.2	328 48.9	. . 02.8	74 32.8	. . 38.8	272 30.1	. . 15.1	319 20.4	. . 36.2
10	68 03.7	343 48.1	03.0	89 34.5	38.6	287 32.1	15.0	334 22.5	36.2
11	83 06.1	358 47.2	03.2	104 36.1	38.5	302 34.1	14.8	349 24.6	36.2
12	98 08.6	13 46.4	N23 03.4	119 37.7	S13 38.3	317 36.1	N16 14.7	4 26.7	N22 36.1
13	113 11.1	28 45.5	03.7	134 39.4	38.1	332 38.1	14.5	19 28.8	36.1
14	128 13.5	43 44.7	03.9	149 41.0	38.0	347 40.1	14.4	34 30.9	36.1
15	143 16.0	58 43.8	. . 04.1	164 42.6	. . 37.8	2 42.1	. . 14.2	49 33.0	. . 36.1
16	158 18.4	73 43.0	04.3	179 44.3	37.6	17 44.1	14.1	64 35.1	36.1
17	173 20.9	88 42.1	04.5	194 45.9	37.5	32 46.0	13.9	79 37.3	36.1
18	188 23.4	103 41.3	N23 04.7	209 47.5	S13 37.3	47 48.0	N16 13.8	94 39.4	N22 36.1
19	203 25.8	118 40.4	05.0	224 49.2	37.2	62 50.0	13.6	109 41.5	36.1
20	218 28.3	133 39.6	05.2	239 50.8	37.0	77 52.0	13.5	124 43.6	36.1
21	233 30.8	148 38.7	. . 05.4	254 52.4	. . 36.8	92 54.0	. . 13.3	139 45.7	. . 36.1
22	248 33.2	163 37.9	05.6	269 54.1	36.7	107 56.0	13.2	154 47.8	36.1
23	263 35.7	178 37.0	05.8	284 55.7	36.5	122 58.0	13.0	169 49.9	36.1
1 00 (TUESDAY)	278 38.2	193 36.2	N23 06.0	299 57.4	S13 36.3	138 00.0	N16 12.9	184 52.0	N22 36.1
01	293 40.6	208 35.3	06.2	314 59.0	36.2	153 02.0	12.8	199 54.2	36.1
02	308 43.1	223 34.5	06.4	330 00.6	36.0	168 04.0	12.6	214 56.3	36.1
03	323 45.6	238 33.6	. . 06.6	345 02.3	. . 35.9	183 05.9	. . 12.5	229 58.4	. . 36.1
04	338 48.0	253 32.7	06.8	0 03.9	35.7	198 07.9	12.3	245 00.5	36.1
05	353 50.5	268 31.9	07.0	15 05.6	35.6	213 09.9	12.2	260 02.6	36.1
06	8 52.9	283 31.0	N23 07.2	30 07.2	S13 35.4	228 11.9	N16 12.0	275 04.7	N22 36.1
07	23 55.4	298 30.2	07.4	45 08.9	35.2	243 13.9	11.9	290 06.8	36.0
08	38 57.9	313 29.3	07.6	60 10.5	35.1	258 15.9	11.7	305 08.9	36.0
09	54 00.3	328 28.5	. . 07.8	75 12.2	. . 34.9	273 17.9	. . 11.6	320 11.1	. . 36.0
10	69 02.8	343 27.6	08.0	90 13.8	34.8	288 19.9	11.4	335 13.2	36.0
11	84 05.3	358 26.8	08.2	105 15.5	34.6	303 21.9	11.3	350 15.3	36.0
12	99 07.7	13 25.9	N23 08.4	120 17.1	S13 34.5	318 23.9	N16 11.1	5 17.4	N22 36.0
13	114 10.2	28 25.1	08.6	135 18.8	34.3	333 25.8	11.0	20 19.5	36.0
14	129 12.7	43 24.2	08.8	150 20.4	34.1	348 27.8	10.8	35 21.6	36.0
15	144 15.1	58 23.4	. . 09.0	165 22.1	. . 34.0	3 29.8	. . 10.7	50 23.7	. . 36.0
16	159 17.6	73 22.5	09.2	180 23.7	33.8	18 31.8	10.5	65 25.8	36.0
17	174 20.0	88 21.6	09.4	195 25.4	33.7	33 33.8	10.4	80 28.0	36.0
18	189 22.5	103 20.8	N23 09.6	210 27.0	S13 33.5	48 35.8	N16 10.2	95 30.1	N22 36.0
19	204 25.0	118 19.9	09.7	225 28.7	33.4	63 37.8	10.1	110 32.2	36.0
20	219 27.4	133 19.1	09.9	240 30.4	33.2	78 39.8	09.9	125 34.3	36.0
21	234 29.9	148 18.2	. . 10.1	255 32.0	. . 33.1	93 41.7	. . 09.8	140 36.4	. . 36.0
22	249 32.4	163 17.4	10.3	270 33.7	32.9	108 43.7	09.6	155 38.5	36.0
23	264 34.8	178 16.5	10.5	285 35.3	32.8	123 45.7	09.5	170 40.6	36.0
2 00 (WEDNESDAY)	279 37.3	193 15.6	N23 10.7	300 37.0	S13 32.6	138 47.7	N16 09.3	185 42.8	N22 36.0
01	294 39.8	208 14.8	10.8	315 38.7	32.4	153 49.7	09.2	200 44.9	35.9
02	309 42.2	223 13.9	11.0	330 40.3	32.3	168 51.7	09.0	215 47.0	35.9
03	324 44.7	238 13.1	. . 11.2	345 42.0	. . 32.1	183 53.7	. . 08.9	230 49.1	. . 35.9
04	339 47.2	253 12.2	11.4	0 43.7	32.0	198 55.7	08.7	245 51.2	35.9
05	354 49.6	268 11.4	11.5	15 45.3	31.8	213 57.6	08.6	260 53.3	35.9
06	9 52.1	283 10.5	N23 11.7	30 47.0	S13 31.7	228 59.6	N16 08.5	275 55.4	N22 35.9
07	24 54.5	298 09.6	11.9	45 48.7	31.5	244 01.6	08.3	290 57.5	35.9
08	39 57.0	313 08.8	12.1	60 50.3	31.4	259 03.6	08.2	305 59.7	35.9
09	54 59.5	328 07.9	. . 12.2	75 52.0	. . 31.2	274 05.6	. . 08.0	321 01.8	. . 35.9
10	70 01.9	343 07.1	12.4	90 53.7	31.1	289 07.6	07.9	336 03.9	35.9
11	85 04.4	358 06.2	12.6	105 55.3	30.9	304 09.6	07.7	351 06.0	35.9
12	100 06.9	13 05.3	N23 12.7	120 57.0	S13 30.8	319 11.6	N16 07.6	6 08.1	N22 35.9
13	115 09.3	28 04.5	12.9	135 58.7	30.7	334 13.5	07.4	21 10.2	35.9
14	130 11.8	43 03.6	13.1	151 00.4	30.5	349 15.5	07.3	36 12.3	35.9
15	145 14.3	58 02.8	. . 13.2	166 02.0	. . 30.4	4 17.5	. . 07.1	51 14.4	. . 35.9
16	160 16.7	73 01.9	13.4	181 03.7	30.2	19 19.5	07.0	66 16.6	35.9
17	175 19.2	88 01.1	13.5	196 05.4	30.1	34 21.5	06.8	81 18.7	35.9
18	190 21.7	103 00.2	N23 13.7	211 07.1	S13 29.9	49 23.5	N16 06.7	96 20.8	N22 35.8
19	205 24.1	117 59.3	13.9	226 08.8	29.8	64 25.5	06.5	111 22.9	35.8
20	220 26.6	132 58.5	14.0	241 10.4	29.6	79 27.4	06.4	126 25.0	35.8
21	235 29.0	147 57.6	. . 14.2	256 12.1	. . 29.5	94 29.4	. . 06.2	141 27.1	. . 35.8
22	250 31.5	162 56.7	14.3	271 13.8	29.3	109 31.4	06.1	156 29.2	35.8
23	265 34.0	177 55.9	14.5	286 15.5	29.2	124 33.4	05.9	171 31.4	35.8
Mer. Pass. (h m)	5 24.6	v −0.9	d 0.2	v 1.7	d 0.2	v 2.0	d 0.1	v 2.1	d 0.0

STARS

Name	SHA ° ′	Dec ° ′
Acamar	315 24.5	S40 17.3
Achernar	335 32.6	S57 12.9
Acrux	173 18.4	S63 07.3
Adhara	255 19.1	S28 58.5
Aldebaran	290 58.7	N16 31.0
Alioth	166 27.3	N55 56.8
Alkaid	153 04.8	N49 18.1
Al Na'ir	27 53.1	S46 56.5
Alnilam	275 54.7	S 1 11.9
Alphard	218 04.1	S 8 40.3
Alphecca	126 17.3	N26 42.3
Alpheratz	357 51.7	N29 06.4
Altair	62 15.6	N 8 52.6
Ankaa	353 23.3	S42 17.1
Antares	112 35.6	S26 26.5
Arcturus	146 02.7	N19 10.0
Atria	107 44.1	S69 02.2
Avior	234 22.0	S59 31.3
Bellatrix	278 40.8	N 6 21.2
Betelgeuse	271 10.1	N 7 24.5
Canopus	264 00.2	S52 41.8
Capella	280 46.6	N46 00.1
Deneb	49 36.5	N45 17.4
Denebola	182 41.7	N14 33.3
Diphda	349 03.7	S17 58.0
Dubhe	194 01.3	N61 44.3
Elnath	278 22.9	N28 36.6
Eltanin	90 49.3	N51 29.3
Enif	33 54.6	N 9 53.3
Fomalhaut	15 32.4	S29 36.2
Gacrux	172 09.9	S57 08.2
Gienah	176 00.5	S17 33.7
Hadar	148 59.0	S60 23.6
Hamal	328 09.8	N23 28.6
Kaus Aust.	83 53.8	S34 23.0
Kochab	137 18.6	N74 08.8
Markab	13 46.1	N15 13.3
Menkar	314 23.5	N 4 06.2
Menkent	148 16.8	S36 23.4
Miaplacidus	221 42.4	S69 44.0
Mirfak	308 52.1	N49 52.2
Nunki	76 07.6	S26 17.6
Peacock	53 30.9	S56 43.4
Pollux	243 37.7	N28 01.2
Procyon	245 08.3	N 5 13.1
Rasalhague	96 13.4	N12 33.4
Regulus	207 52.0	N11 57.2
Rigel	281 19.9	S 8 11.8
Rigil Kent.	140 02.4	S60 51.2
Sabik	102 21.2	S15 43.8
Schedar	349 49.7	N56 33.0
Shaula	96 32.2	S37 06.5
Sirius	258 41.0	S16 43.1
Spica	158 39.5	S11 10.8
Suhail	222 58.7	S43 26.8
Vega	80 43.9	N38 47.2
Zuben'ubi	137 14.0	S16 03.4

	SHA ° ′	Mer. Pass h m
Venus	274 58.0	11 06
Mars	21 19.2	4 00
Jupiter	219 21.8	14 46
Saturn	266 13.9	11 39

UT	SUN GHA	SUN Dec	MOON GHA	v	MOON Dec	d	HP
d h	° ′	° ′	° ′	′	° ′	′	′
30 00	179 07.9	N23 12.5	176 03.1	8.6	N26 26.7	0.4	55.4
01	194 07.7	12.4	190 30.7	8.6	26 26.3	0.6	55.4
02	209 07.6	12.3	204 58.3	8.6	26 25.7	0.8	55.4
03	224 07.5	. . 12.1	219 25.9	8.6	26 24.9	0.9	55.5
04	239 07.4	12.0	233 53.5	8.5	26 24.0	1.0	55.5
05	254 07.2	11.8	248 21.0	8.5	26 23.0	1.2	55.5
06	269 07.1	N23 11.7	262 48.5	8.6	N26 21.8	1.3	55.5
07	284 07.0	11.5	277 16.1	8.5	26 20.5	1.4	55.5
08	299 06.9	11.4	291 43.6	8.5	26 19.1	1.6	55.6
M 09	314 06.7	. . 11.2	306 11.1	8.5	26 17.5	1.8	55.6
O 10	329 06.6	11.1	320 38.6	8.5	26 15.7	1.9	55.6
N 11	344 06.5	11.0	335 06.1	8.5	26 13.8	2.0	55.6
D 12	359 06.4	N23 10.8	349 33.6	8.5	N26 11.8	2.2	55.6
A 13	14 06.2	10.7	4 01.1	8.5	26 09.6	2.3	55.7
Y 14	29 06.1	10.5	18 28.6	8.5	26 07.3	2.4	55.7
15	44 06.0	. . 10.4	32 56.1	8.5	26 04.9	2.6	55.7
16	59 05.9	10.2	47 23.6	8.5	26 02.3	2.8	55.7
17	74 05.7	10.1	61 51.1	8.5	25 59.5	2.8	55.7
18	89 05.6	N23 09.9	76 18.6	8.5	N25 56.7	3.1	55.8
19	104 05.5	09.7	90 46.1	8.5	25 53.6	3.1	55.8
20	119 05.4	09.6	105 13.6	8.5	25 50.5	3.3	55.8
21	134 05.2	. . 09.4	119 41.1	8.6	25 47.2	3.5	55.8
22	149 05.1	09.3	134 08.7	8.5	25 43.7	3.6	55.8
23	164 05.0	09.1	148 36.2	8.6	25 40.1	3.7	55.8
1 00	179 04.9	N23 09.0	163 03.8	8.5	N25 36.4	3.9	55.9
01	194 04.8	08.8	177 31.3	8.6	25 32.5	4.0	55.9
02	209 04.6	08.7	191 58.9	8.6	25 28.5	4.1	55.9
03	224 04.5	. . 08.5	206 26.5	8.6	25 24.4	4.3	55.9
04	239 04.4	08.3	220 54.1	8.7	25 20.1	4.5	55.9
05	254 04.3	08.2	235 21.8	8.6	25 15.6	4.5	56.0
06	269 04.1	N23 08.0	249 49.4	8.7	N25 11.1	4.7	56.0
07	284 04.0	07.8	264 17.1	8.7	25 06.4	4.9	56.0
T 08	299 03.9	07.7	278 44.8	8.7	25 01.5	5.0	56.0
U 09	314 03.8	. . 07.5	293 12.5	8.8	24 56.5	5.1	56.0
E 10	329 03.6	07.4	307 40.3	8.7	24 51.4	5.3	56.1
E 11	344 03.5	07.2	322 08.0	8.8	24 46.1	5.4	56.1
S 12	359 03.4	N23 07.0	336 35.8	8.9	N24 40.7	5.5	56.1
D 13	14 03.3	06.9	351 03.7	8.8	24 35.2	5.7	56.1
A 14	29 03.2	06.7	5 31.5	8.9	24 29.5	5.8	56.2
Y 15	44 03.0	. . 06.5	19 59.4	8.9	24 23.7	5.9	56.2
16	59 02.9	06.4	34 27.3	9.0	24 17.8	6.1	56.2
17	74 02.8	06.2	48 55.3	8.9	24 11.7	6.2	56.2
18	89 02.7	N23 06.0	63 23.2	9.0	N24 05.5	6.4	56.2
19	104 02.5	05.9	77 51.2	9.1	23 59.1	6.4	56.3
20	119 02.4	05.7	92 19.3	9.1	23 52.7	6.7	56.3
21	134 02.3	. . 05.5	106 47.4	9.1	23 46.0	6.7	56.3
22	149 02.2	05.3	121 15.5	9.1	23 39.3	6.9	56.3
23	164 02.1	05.2	135 43.6	9.2	23 32.4	7.0	56.3
2 00	179 01.9	N23 05.0	150 11.8	9.3	N23 25.4	7.1	56.4
01	194 01.8	04.8	164 40.1	9.2	23 18.3	7.3	56.4
02	209 01.7	04.6	179 08.3	9.3	23 11.0	7.3	56.4
03	224 01.6	. . 04.5	193 36.6	9.4	23 03.7	7.6	56.4
04	239 01.5	04.3	208 05.0	9.4	22 56.1	7.6	56.4
05	254 01.3	04.1	222 33.4	9.4	22 48.5	7.8	56.5
06	269 01.2	N23 03.9	237 01.8	9.4	N22 40.7	7.9	56.5
W 07	284 01.1	03.8	251 30.2	9.6	22 32.8	8.0	56.5
E 08	299 01.0	03.6	265 58.8	9.5	22 24.8	8.1	56.5
D 09	314 00.9	. . 03.4	280 27.3	9.6	22 16.7	8.3	56.5
N 10	329 00.7	03.2	294 55.9	9.7	22 08.4	8.4	56.6
E 11	344 00.6	03.0	309 24.6	9.6	22 00.0	8.5	56.6
S 12	359 00.5	N23 02.8	323 53.2	9.8	N21 51.5	8.6	56.6
D 13	14 00.4	02.7	338 22.0	9.7	21 42.9	8.8	56.6
A 14	29 00.3	02.5	352 50.7	9.9	21 34.1	8.8	56.6
Y 15	44 00.1	. . 02.3	7 19.6	9.8	21 25.3	9.0	56.7
16	59 00.0	02.1	21 48.4	10.0	21 16.3	9.1	56.7
17	73 59.9	01.9	36 17.4	9.9	21 07.2	9.2	56.7
18	88 59.8	N23 01.7	50 46.3	10.0	N20 58.0	9.4	56.7
19	103 59.7	01.6	65 15.3	10.1	20 48.6	9.4	56.8
20	118 59.6	01.4	79 44.4	10.1	20 39.2	9.6	56.8
21	133 59.4	. . 01.2	94 13.5	10.1	20 29.6	9.7	56.8
22	148 59.3	01.0	108 42.6	10.2	20 19.9	9.8	56.8
23	163 59.2	00.8	123 11.8	10.3	N20 10.1	9.9	56.8
	SD 15.8	d 0.2	SD 15.2		15.3		15.4

Lat.	Twilight Naut.	Twilight Civil	Sunrise	Moonrise 30	Moonrise 1	Moonrise 2	Moonrise 3
°	h m	h m	h m	h m	h m	h m	h m
N 72	▭	▭	▭	▭	▭	▭	▭
N 70	▭	▭	▭	▭	▭	▭	04 06
68	▭	▭	▭	▭	▭	▭	05 06
66	////	////	00 13	▭	▭	03 20	05 40
64	////	////	01 40	▭	02 07	04 10	06 04
62	////	////	02 16	01 46	03 04	04 41	06 23
60	////	01 02	02 42	02 25	03 37	05 05	06 39
N 58	////	01 48	03 02	02 53	04 01	05 23	06 52
56	////	02 16	03 18	03 14	04 20	05 39	07 04
54	00 57	02 38	03 32	03 32	04 37	05 52	07 14
52	01 39	02 56	03 44	03 47	04 51	06 04	07 22
50	02 06	03 10	03 55	04 00	05 03	06 14	07 30
45	02 50	03 40	04 17	04 27	05 28	06 36	07 47
N 40	03 20	04 02	04 35	04 48	05 48	06 53	08 01
35	03 43	04 20	04 49	05 06	06 05	07 07	08 12
30	04 02	04 35	05 02	05 21	06 19	07 20	08 22
20	04 30	05 00	05 24	05 47	06 43	07 41	08 39
N 10	04 53	05 20	05 43	06 10	07 05	08 00	08 54
0	05 11	05 38	06 00	06 30	07 24	08 17	09 08
S 10	05 28	05 55	06 17	06 51	07 44	08 34	09 22
20	05 44	06 12	06 36	07 13	08 05	08 53	09 37
30	06 00	06 30	06 57	07 39	08 29	09 14	09 54
35	06 09	06 41	07 09	07 55	08 43	09 26	10 04
40	06 18	06 52	07 23	08 12	09 00	09 40	10 15
45	06 28	07 06	07 39	08 34	09 19	09 57	10 28
S 50	06 40	07 21	08 00	09 00	09 44	10 18	10 44
52	06 45	07 29	08 09	09 14	09 56	10 27	10 51
54	06 51	07 37	08 20	09 29	10 09	10 38	10 59
56	06 57	07 46	08 33	09 46	10 25	10 51	11 09
58	07 03	07 56	08 47	10 08	10 43	11 05	11 19
S 60	07 11	08 07	09 04	10 35	11 05	11 22	11 31

Lat.	Sunset	Twilight Civil	Twilight Naut.	Moonset 30	Moonset 1	Moonset 2	Moonset 3
°	h m	h m	h m	h m	h m	h m	h m
N 72	▭	▭	▭	▭	▭	▭	▭
N 70	▭	▭	▭	▭	▭	▭	02 02
68	▭	▭	▭	▭	▭	▭	01 01
66	23 44	////	////	▭	▭	00 58	00 26
64	22 26	////	////	▭	00 19	00 07	(00 01 / 23 55)
62	21 51	////	////	23 22	23 35	23 41	23 42
60	21 25	23 04	////	22 49	23 12	23 24	23 31
N 58	21 05	22 19	////	22 24	22 52	23 10	23 22
56	20 49	21 50	////	22 05	22 36	22 58	23 13
54	20 35	21 29	23 09	21 48	22 23	22 47	23 06
52	20 23	21 11	22 27	21 34	22 11	22 38	22 59
50	20 13	20 57	22 01	21 22	22 00	22 30	22 53
45	19 50	20 28	21 17	20 56	21 38	22 12	22 40
N 40	19 33	20 05	20 47	20 35	21 19	21 57	22 29
35	19 18	19 48	20 24	20 18	21 04	21 44	22 19
30	19 05	19 33	20 06	20 04	20 51	21 33	22 11
20	18 43	19 08	19 37	19 38	20 28	21 14	21 57
N 10	18 25	18 48	19 15	19 17	20 08	20 58	21 44
0	18 07	18 30	18 56	18 56	19 50	20 42	21 32
S 10	17 50	18 13	18 39	18 36	19 31	20 26	21 20
20	17 32	17 56	18 23	18 14	19 11	20 09	21 07
30	17 11	17 37	18 07	17 48	18 48	19 50	20 52
35	16 59	17 27	17 59	17 33	18 34	19 38	20 43
40	16 45	17 15	17 49	17 16	18 18	19 25	20 33
45	16 28	17 02	17 39	16 55	17 59	19 09	20 22
S 50	16 08	16 46	17 28	16 28	17 36	18 50	20 07
52	15 58	16 39	17 23	16 15	17 24	18 41	20 01
54	15 47	16 31	17 17	16 00	17 11	18 30	19 53
56	15 35	16 22	17 11	15 43	16 56	18 18	19 45
58	15 21	16 12	17 04	15 21	16 38	18 05	19 35
S 60	15 04	16 01	16 57	14 54	16 16	17 48	19 24

Day	SUN Eqn. of Time 00^h	SUN Eqn. of Time 12^h	SUN Mer. Pass.	MOON Mer. Pass. Upper	MOON Mer. Pass. Lower	Age	Phase
d	m s	m s	h m	h m	h m	d	%
30	03 28	03 34	12 04	12 43	00 16	01	1
1	03 40	03 46	12 04	13 37	01 10	02	3
2	03 52	03 58	12 04	14 30	02 04	03	8

	UT d h	ARIES GHA	VENUS −3.9 GHA	VENUS Dec	MARS −1.5 GHA	MARS Dec	JUPITER −1.8 GHA	JUPITER Dec	SATURN +0.0 GHA	SATURN Dec
		° ′	° ′	° ′	° ′	° ′	° ′	° ′	° ′	° ′
	3 00	280 36.4	192 55.0	N23 14.6	301 17.2	S13 29.0	139 35.4	N16 05.8	186 33.5	N22 35.8
	01	295 38.9	207 54.2	14.8	316 18.9	28.9	154 37.4	05.6	201 35.6	35.8
	02	310 41.4	222 53.3	14.9	331 20.5	28.8	169 39.4	05.5	216 37.7	35.8
	03	325 43.8	237 52.4	15.1	346 22.2	28.6	184 41.3	05.3	231 39.8	35.8
	04	340 46.3	252 51.6	15.2	1 23.9	28.5	199 43.3	05.2	246 41.9	35.8
	05	355 48.8	267 50.7	15.4	16 25.6	28.3	214 45.3	05.0	261 44.0	35.8
	06	10 51.2	282 49.9	N23 15.5	31 27.3	S13 28.2	229 47.3	N16 04.9	276 46.2	N22 35.8
	07	25 53.7	297 49.0	15.7	46 29.0	28.1	244 49.3	04.7	291 48.3	35.8
T	08	40 56.2	312 48.1	15.8	61 30.7	27.9	259 51.3	04.6	306 50.4	35.8
H	09	55 58.6	327 47.3	16.0	76 32.4	27.8	274 53.3	04.4	321 52.5	35.8
U	10	71 01.1	342 46.4	16.1	91 34.1	27.6	289 55.2	04.3	336 54.6	35.8
R	11	86 03.5	357 45.5	16.2	106 35.7	27.5	304 57.2	04.1	351 56.7	35.7
S	12	101 06.0	12 44.7	N23 16.4	121 37.4	S13 27.3	319 59.2	N16 04.0	6 58.8	N22 35.7
D	13	116 08.5	27 43.8	16.5	136 39.1	27.2	335 01.2	03.8	22 00.9	35.7
A	14	131 10.9	42 43.0	16.6	151 40.8	27.1	350 03.2	03.7	37 03.1	35.7
Y	15	146 13.4	57 42.1	16.8	166 42.5	26.9	5 05.2	03.5	52 05.2	35.7
	16	161 15.9	72 41.2	16.9	181 44.2	26.8	20 07.1	03.4	67 07.3	35.7
	17	176 18.3	87 40.4	17.0	196 45.9	26.7	35 09.1	03.2	82 09.4	35.7
	18	191 20.8	102 39.5	N23 17.2	211 47.6	S13 26.5	50 11.1	N16 03.1	97 11.5	N22 35.7
	19	206 23.3	117 38.6	17.3	226 49.3	26.4	65 13.1	02.9	112 13.6	35.7
	20	221 25.7	132 37.8	17.4	241 51.0	26.2	80 15.1	02.8	127 15.7	35.7
	21	236 28.2	147 36.9	17.6	256 52.7	26.1	95 17.1	02.6	142 17.9	35.7
	22	251 30.6	162 36.0	17.7	271 54.4	26.0	110 19.0	02.5	157 20.0	35.7
	23	266 33.1	177 35.2	17.8	286 56.1	25.8	125 21.0	02.3	172 22.1	35.7
	4 00	281 35.6	192 34.3	N23 17.9	301 57.8	S13 25.7	140 23.0	N16 02.2	187 24.2	N22 35.7
	01	296 38.0	207 33.5	18.1	316 59.6	25.6	155 25.0	02.0	202 26.3	35.7
	02	311 40.5	222 32.6	18.2	332 01.3	25.4	170 27.0	01.8	217 28.4	35.7
	03	326 43.0	237 31.7	18.3	347 03.0	25.3	185 29.0	01.7	232 30.5	35.6
	04	341 45.4	252 30.9	18.4	2 04.7	25.2	200 30.9	01.5	247 32.7	35.6
	05	356 47.9	267 30.0	18.5	17 06.4	25.0	215 32.9	01.4	262 34.8	35.6
	06	11 50.4	282 29.1	N23 18.7	32 08.1	S13 24.9	230 34.9	N16 01.2	277 36.9	N22 35.6
	07	26 52.8	297 28.3	18.8	47 09.8	24.8	245 36.9	01.1	292 39.0	35.6
	08	41 55.3	312 27.4	18.9	62 11.5	24.6	260 38.9	00.9	307 41.1	35.6
F	09	56 57.8	327 26.5	19.0	77 13.2	24.5	275 40.9	00.8	322 43.2	35.6
R	10	72 00.2	342 25.7	19.1	92 15.0	24.4	290 42.8	00.6	337 45.3	35.6
I	11	87 02.7	357 24.8	19.2	107 16.7	24.2	305 44.8	00.5	352 47.5	35.6
D	12	102 05.1	12 23.9	N23 19.3	122 18.4	S13 24.1	320 46.8	N16 00.3	7 49.6	N22 35.6
A	13	117 07.6	27 23.1	19.4	137 20.1	24.0	335 48.8	00.2	22 51.7	35.6
Y	14	132 10.1	42 22.2	19.6	152 21.8	23.8	350 50.8	16 00.0	37 53.8	35.6
	15	147 12.5	57 21.3	19.7	167 23.5	23.7	5 52.7	15 59.9	52 55.9	35.6
	16	162 15.0	72 20.5	19.8	182 25.3	23.6	20 54.7	59.7	67 58.0	35.6
	17	177 17.5	87 19.6	19.9	197 27.0	23.5	35 56.7	59.6	83 00.1	35.6
	18	192 19.9	102 18.7	N23 20.0	212 28.7	S13 23.3	50 58.7	N15 59.4	98 02.3	N22 35.5
	19	207 22.4	117 17.9	20.1	227 30.4	23.2	66 00.7	59.3	113 04.4	35.5
	20	222 24.9	132 17.0	20.2	242 32.2	23.1	81 02.7	59.1	128 06.5	35.5
	21	237 27.3	147 16.1	20.3	257 33.9	22.9	96 04.6	59.0	143 08.6	35.5
	22	252 29.8	162 15.3	20.4	272 35.6	22.8	111 06.6	58.8	158 10.7	35.5
	23	267 32.3	177 14.4	20.5	287 37.3	22.7	126 08.6	58.7	173 12.8	35.5
	5 00	282 34.7	192 13.5	N23 20.6	302 39.1	S13 22.6	141 10.6	N15 58.5	188 14.9	N22 35.5
	01	297 37.2	207 12.7	20.7	317 40.8	22.4	156 12.6	58.4	203 17.1	35.5
	02	312 39.6	222 11.8	20.7	332 42.5	22.3	171 14.5	58.2	218 19.2	35.5
	03	327 42.1	237 10.9	20.8	347 44.3	22.2	186 16.5	58.1	233 21.3	35.5
	04	342 44.6	252 10.1	20.9	2 46.0	22.1	201 18.5	57.9	248 23.4	35.5
	05	357 47.0	267 09.2	21.0	17 47.7	21.9	216 20.5	57.8	263 25.5	35.5
	06	12 49.5	282 08.3	N23 21.1	32 49.5	S13 21.8	231 22.5	N15 57.6	278 27.6	N22 35.5
	07	27 52.0	297 07.5	21.2	47 51.2	21.7	246 24.4	57.4	293 29.7	35.5
S	08	42 54.4	312 06.6	21.3	62 52.9	21.6	261 26.4	57.3	308 31.9	35.5
A	09	57 56.9	327 05.7	21.4	77 54.7	21.4	276 28.4	57.1	323 34.0	35.4
T	10	72 59.4	342 04.9	21.5	92 56.4	21.3	291 30.4	57.0	338 36.1	35.4
U	11	88 01.8	357 04.0	21.5	107 58.1	21.2	306 32.4	56.8	353 38.2	35.4
R	12	103 04.3	12 03.1	N23 21.6	122 59.9	S13 21.1	321 34.3	N15 56.7	8 40.3	N22 35.4
D	13	118 06.7	27 02.3	21.7	138 01.6	20.9	336 36.3	56.5	23 42.4	35.4
A	14	133 09.2	42 01.4	21.8	153 03.4	20.8	351 38.3	56.4	38 44.5	35.4
Y	15	148 11.7	57 00.5	21.9	168 05.1	20.7	6 40.3	56.2	53 46.7	35.4
	16	163 14.1	71 59.6	21.9	183 06.8	20.6	21 42.3	56.1	68 48.8	35.4
	17	178 16.6	86 58.8	22.0	198 08.6	20.5	36 44.2	55.9	83 50.9	35.4
	18	193 19.1	101 57.9	N23 22.1	213 10.3	S13 20.3	51 46.2	N15 55.8	98 53.0	N22 35.4
	19	208 21.5	116 57.0	22.2	228 12.1	20.2	66 48.2	55.6	113 55.1	35.4
	20	223 24.0	131 56.2	22.2	243 13.8	20.1	81 50.2	55.5	128 57.2	35.4
	21	238 26.5	146 55.3	22.3	258 15.6	20.0	96 52.2	55.3	143 59.3	35.4
	22	253 28.9	161 54.4	22.4	273 17.3	19.9	111 54.1	55.2	159 01.5	35.4
	23	268 31.4	176 53.6	22.4	288 19.1	19.7	126 56.1	55.0	174 03.6	35.3
	Mer. Pass.	h m 5 12.8	*v* −0.9	*d* 0.1	*v* 1.7	*d* 0.1	*v* 2.0	*d* 0.2	*v* 2.1	*d* 0.0

STARS

Name	SHA	Dec
	° ′	° ′
Acamar	315 24.4	S40 17.3
Achernar	335 32.5	S57 12.9
Acrux	173 18.5	S63 07.3
Adhara	255 19.1	S28 58.5
Aldebaran	290 58.7	N16 31.0
Alioth	166 27.3	N55 56.8
Alkaid	153 04.8	N49 18.1
Al Na'ir	27 53.1	S46 56.5
Alnilam	275 54.7	S 1 11.9
Alphard	218 04.1	S 8 40.3
Alphecca	126 17.3	N26 42.3
Alpheratz	357 51.6	N29 06.4
Altair	62 15.6	N 8 52.6
Ankaa	353 23.2	S42 17.1
Antares	112 35.6	S26 26.5
Arcturus	146 02.7	N19 10.0
Atria	107 44.1	S69 02.2
Avior	234 22.0	S59 31.2
Bellatrix	278 40.7	N 6 21.2
Betelgeuse	271 10.1	N 7 24.5
Canopus	264 00.2	S52 41.7
Capella	280 46.5	N46 00.1
Deneb	49 36.5	N45 17.4
Denebola	182 41.7	N14 33.3
Diphda	349 03.7	S17 58.0
Dubhe	194 01.3	N61 44.3
Elnath	278 22.9	N28 36.6
Eltanin	90 49.3	N51 29.3
Enif	33 54.6	N 9 53.3
Fomalhaut	15 32.3	S29 36.2
Gacrux	172 10.0	S57 08.2
Gienah	176 00.5	S17 33.7
Hadar	148 59.1	S60 23.6
Hamal	328 09.8	N23 28.6
Kaus Aust.	83 53.8	S34 23.0
Kochab	137 18.7	N74 08.8
Markab	13 46.0	N15 13.3
Menkar	314 23.5	N 4 06.2
Menkent	148 16.8	S36 23.4
Miaplacidus	221 42.4	S69 44.0
Mirfak	308 52.1	N49 52.2
Nunki	76 07.6	S26 17.6
Peacock	53 30.9	S56 43.4
Pollux	243 37.7	N28 01.2
Procyon	245 08.3	N 5 13.1
Rasalhague	96 13.4	N12 33.4
Regulus	207 52.0	N11 57.2
Rigel	281 19.9	S 8 11.8
Rigil Kent.	140 02.5	S60 51.2
Sabik	102 21.2	S15 43.8
Schedar	349 49.7	N56 33.1
Shaula	96 32.2	S37 06.5
Sirius	258 41.0	S16 43.1
Spica	158 39.5	S11 10.8
Suhail	222 58.7	S43 26.8
Vega	80 43.9	N38 47.2
Zuben'ubi	137 14.0	S16 03.4

	SHA	Mer. Pass
	° ′	h m
Venus	270 58.7	11 10
Mars	20 22.3	3 52
Jupiter	218 47.4	14 37
Saturn	265 48.6	11 29

UT		SUN GHA	SUN Dec	MOON GHA	v	MOON Dec	d	HP
d	h	° ′	° ′	° ′	′	° ′	′	′
3 THURSDAY	00	178 59.1	N23 00.6	137 41.1	10.3	N20 00.2	10.0	56.9
	01	193 59.0	00.4	152 10.4	10.3	19 50.2	10.1	56.9
	02	208 58.8	00.2	166 39.7	10.4	19 40.1	10.2	56.9
	03	223 58.7	23 00.0	181 09.1	10.4	19 29.9	10.3	56.9
	04	238 58.6	22 59.8	195 38.5	10.5	19 19.6	10.5	56.9
	05	253 58.5	59.6	210 08.0	10.5	19 09.1	10.5	57.0
	06	268 58.4	N22 59.4	224 37.5	10.6	N18 58.6	10.7	57.0
	07	283 58.3	59.3	239 07.1	10.6	18 47.9	10.7	57.0
	08	298 58.1	59.1	253 36.7	10.7	18 37.2	10.9	57.0
	09	313 58.0	. . 58.9	268 06.4	10.7	18 26.3	10.9	57.1
	10	328 57.9	58.7	282 36.1	10.8	18 15.4	11.1	57.1
	11	343 57.8	58.5	297 05.9	10.8	18 04.3	11.1	57.1
	12	358 57.7	N22 58.3	311 35.7	10.9	N17 53.2	11.3	57.1
	13	13 57.6	58.1	326 05.6	10.9	17 41.9	11.3	57.1
	14	28 57.4	57.9	340 35.5	10.9	17 30.6	11.5	57.2
	15	43 57.3	. . 57.7	355 05.4	11.0	17 19.1	11.5	57.2
	16	58 57.2	57.5	9 35.4	11.1	17 07.6	11.7	57.2
	17	73 57.1	57.3	24 05.5	11.1	16 55.9	11.7	57.2
	18	88 57.0	N22 57.1	38 35.6	11.1	N16 44.2	11.8	57.2
	19	103 56.9	56.9	53 05.7	11.2	16 32.4	11.9	57.3
	20	118 56.8	56.6	67 35.9	11.2	16 20.5	12.0	57.3
	21	133 56.6	. . 56.4	82 06.1	11.3	16 08.5	12.1	57.3
	22	148 56.5	56.2	96 36.4	11.3	15 56.4	12.2	57.3
	23	163 56.4	56.0	111 06.7	11.3	15 44.2	12.3	57.4
4 FRIDAY	00	178 56.3	N22 55.8	125 37.0	11.4	N15 31.9	12.3	57.4
	01	193 56.2	55.6	140 07.4	11.4	15 19.6	12.5	57.4
	02	208 56.1	55.4	154 37.8	11.5	15 07.1	12.5	57.4
	03	223 55.9	. . 55.2	169 08.3	11.5	14 54.6	12.6	57.4
	04	238 55.8	55.0	183 38.8	11.6	14 42.0	12.7	57.5
	05	253 55.7	54.8	198 09.4	11.6	14 29.3	12.7	57.5
	06	268 55.6	N22 54.6	212 40.0	11.6	N14 16.6	12.9	57.5
	07	283 55.5	54.4	227 10.6	11.7	14 03.7	12.9	57.5
	08	298 55.4	54.1	241 41.3	11.7	13 50.8	13.0	57.6
	09	313 55.3	. . 53.9	256 12.0	11.8	13 37.8	13.1	57.6
	10	328 55.2	53.7	270 42.8	11.7	13 24.7	13.1	57.6
	11	343 55.0	53.5	285 13.5	11.9	13 11.6	13.3	57.6
	12	358 54.9	N22 53.3	299 44.4	11.8	N12 58.3	13.3	57.6
	13	13 54.8	53.1	314 15.2	11.9	12 45.0	13.3	57.7
	14	28 54.7	52.8	328 46.1	11.9	12 31.7	13.5	57.7
	15	43 54.6	. . 52.6	343 17.0	12.0	12 18.2	13.5	57.7
	16	58 54.5	52.4	357 48.0	12.0	12 04.7	13.6	57.7
	17	73 54.4	52.2	12 19.0	12.0	11 51.1	13.6	57.7
	18	88 54.3	N22 52.0	26 50.0	12.0	N11 37.5	13.7	57.8
	19	103 54.1	51.8	41 21.0	12.1	11 23.8	13.8	57.8
	20	118 54.0	51.5	55 52.1	12.1	11 10.0	13.9	57.8
	21	133 53.9	. . 51.3	70 23.2	12.1	10 56.1	13.9	57.8
	22	148 53.8	51.1	84 54.3	12.2	10 42.2	13.9	57.9
	23	163 53.7	50.9	99 25.5	12.2	10 28.3	14.1	57.9
5 SATURDAY	00	178 53.6	N22 50.6	113 56.7	12.2	N10 14.2	14.0	57.9
	01	193 53.5	50.4	128 27.9	12.2	10 00.2	14.2	57.9
	02	208 53.4	50.2	142 59.1	12.2	9 46.0	14.2	57.9
	03	223 53.3	. . 50.0	157 30.3	12.3	9 31.8	14.2	58.0
	04	238 53.1	49.7	172 01.6	12.3	9 17.6	14.4	58.0
	05	253 53.0	49.5	186 32.9	12.3	9 03.2	14.3	58.0
	06	268 52.9	N22 49.3	201 04.2	12.4	N 8 48.9	14.5	58.0
	07	283 52.8	49.0	215 35.6	12.3	8 34.4	14.4	58.1
	08	298 52.7	48.8	230 06.9	12.4	8 20.0	14.5	58.1
	09	313 52.6	. . 48.6	244 38.3	12.4	8 05.5	14.6	58.1
	10	328 52.5	48.4	259 09.7	12.4	7 50.9	14.6	58.1
	11	343 52.4	48.1	273 41.1	12.4	7 36.3	14.7	58.1
	12	358 52.3	N22 47.9	288 12.5	12.4	N 7 21.6	14.7	58.2
	13	13 52.2	47.7	302 43.9	12.4	7 06.9	14.8	58.2
	14	28 52.0	47.4	317 15.3	12.5	6 52.1	14.8	58.2
	15	43 51.9	. . 47.2	331 46.8	12.4	6 37.3	14.8	58.2
	16	58 51.8	47.0	346 18.2	12.5	6 22.5	14.9	58.2
	17	73 51.7	46.7	0 49.7	12.5	6 07.6	14.9	58.3
	18	88 51.6	N22 46.5	15 21.2	12.5	N 5 52.7	15.0	58.3
	19	103 51.5	46.3	29 52.7	12.5	5 37.7	15.0	58.3
	20	118 51.4	46.0	44 24.2	12.4	5 22.7	15.0	58.3
	21	133 51.3	. . 45.8	58 55.6	12.5	5 07.7	15.1	58.4
	22	148 51.2	45.5	73 27.1	12.5	4 52.6	15.1	58.4
	23	163 51.1	45.3	87 58.6	12.5	N 4 37.5	15.2	58.4
		SD 15.8	d 0.2	SD 15.6		15.7		15.8

Lat.	Twilight Naut.	Twilight Civil	Sunrise	Moonrise 3	Moonrise 4	Moonrise 5	Moonrise 6
°	h m	h m	h m	h m	h m	h m	h m
N 72	▭	▭	▭	▭	06 29	09 00	11 15
N 70	▭	▭	▭	04 06	06 59	09 12	11 17
68	▭	▭	▭	05 06	07 22	09 22	11 18
66	////	////	00 38	05 40	07 39	09 31	11 20
64	////	////	01 46	06 04	07 53	09 38	11 21
62	////	////	02 20	06 23	08 04	09 43	11 22
60	////	01 10	02 45	06 39	08 14	09 49	11 23
N 58	////	01 52	03 04	06 52	08 23	09 53	11 23
56	////	02 20	03 21	07 04	08 30	09 57	11 24
54	01 04	02 41	03 34	07 14	08 37	10 01	11 25
52	01 43	02 58	03 46	07 22	08 43	10 04	11 25
50	02 09	03 13	03 57	07 30	08 48	10 07	11 26
45	02 53	03 42	04 19	07 47	09 00	10 13	11 27
N 40	03 22	04 04	04 36	08 01	09 09	10 18	11 28
35	03 45	04 21	04 51	08 12	09 18	10 23	11 28
30	04 03	04 36	05 03	08 22	09 25	10 27	11 29
20	04 31	05 01	05 25	08 39	09 37	10 34	11 30
N 10	04 54	05 21	05 44	08 54	09 48	10 40	11 31
0	05 12	05 38	06 01	09 08	09 58	10 45	11 33
S 10	05 29	05 55	06 18	09 22	10 08	10 51	11 34
20	05 44	06 12	06 36	09 37	10 18	10 57	11 35
30	06 00	06 30	06 56	09 54	10 30	11 04	11 36
35	06 09	06 41	07 09	10 04	10 37	11 08	11 37
40	06 18	06 52	07 22	10 15	10 45	11 12	11 38
45	06 28	07 05	07 39	10 28	10 54	11 17	11 39
S 50	06 39	07 21	07 59	10 44	11 05	11 23	11 40
52	06 44	07 28	08 08	10 51	11 10	11 26	11 40
54	06 50	07 36	08 19	10 59	11 16	11 29	11 41
56	06 56	07 45	08 31	11 09	11 22	11 32	11 41
58	07 02	07 54	08 45	11 19	11 28	11 36	11 42
S 60	07 09	08 06	09 02	11 31	11 36	11 40	11 43

Lat.	Sunset	Twilight Civil	Twilight Naut.	Moonset 3	Moonset 4	Moonset 5	Moonset 6
°	h m	h m	h m	h m	h m	h m	h m
N 72	▭	▭	▭	▭	01 25	00 36	{00 02 / 23 32}
N 70	▭	▭	▭	02 02	00 53	{00 21 / 23 57}	23 35
68	▭	▭	▭	01 01	00 29	{00 09 / 23 52}	23 37
66	23 25	////	////	00 26	{00 10 / 23 58}	23 48	23 38
64	22 21	////	////	{00 01 / 23 55}	23 50	23 45	23 40
62	21 47	////	////	23 42	23 42	23 42	23 41
60	21 23	22 57	////	23 31	23 36	23 39	23 42
N 58	21 03	22 15	////	23 22	23 30	23 37	23 43
56	20 48	21 48	////	23 13	23 25	23 35	23 44
54	20 34	21 27	23 03	23 06	23 20	23 33	23 45
52	20 22	21 10	22 24	22 59	23 16	23 31	23 45
50	20 12	20 55	21 59	22 53	23 12	23 30	23 46
45	19 50	20 27	21 16	22 40	23 04	23 26	23 48
N 40	19 32	20 05	20 46	22 29	22 57	23 23	23 49
35	19 18	19 47	20 24	22 19	22 51	23 21	23 50
30	19 05	19 32	20 05	22 11	22 46	23 19	23 51
20	18 44	19 08	19 37	21 57	22 37	23 15	23 52
N 10	18 25	18 48	19 15	21 44	22 28	23 11	23 54
0	18 08	18 30	18 57	21 32	22 20	23 08	23 55
S 10	17 51	18 14	18 40	21 20	22 13	23 04	23 56
20	17 33	17 57	18 24	21 07	22 04	23 01	23 58
30	17 12	17 39	18 08	20 52	21 54	22 57	23 59
35	17 00	17 28	18 00	20 43	21 49	22 54	24 00
40	16 46	17 17	17 51	20 33	21 42	22 51	24 01
45	16 30	17 04	17 41	20 22	21 35	22 48	24 02
S 50	16 10	16 48	17 29	20 07	21 26	22 44	24 04
52	16 01	16 41	17 24	20 01	21 21	22 42	24 04
54	15 50	16 33	17 19	19 53	21 17	22 40	24 05
56	15 38	16 24	17 13	19 45	21 11	22 38	24 06
58	15 24	16 15	17 07	19 35	21 06	22 36	24 07
S 60	15 07	16 03	17 00	19 24	20 59	22 33	24 08

Day	SUN Eqn. of Time 00^h	SUN Eqn. of Time 12^h	SUN Mer. Pass.	MOON Mer. Pass. Upper	MOON Mer. Pass. Lower	MOON Age	MOON Phase
d	m s	m s	h m	h m	h m	d %	
3	04 03	04 09	12 04	15 20	02 55	04 14	
4	04 15	04 20	12 04	16 09	03 45	05 23	
5	04 25	04 31	12 05	16 57	04 33	06 32	

UT d h		ARIES GHA ° ′	VENUS −3.9 GHA ° ′	VENUS Dec ° ′	MARS −1.6 GHA ° ′	MARS Dec ° ′	JUPITER −1.8 GHA ° ′	JUPITER Dec ° ′	SATURN +0.1 GHA ° ′	SATURN Dec ° ′
6 00		283 33.9	191 52.7	N23 22.5	303 20.8	S13 19.6	141 58.1	N15 54.9	189 05.7	N22 35.3
01		298 36.3	206 51.8	22.6	318 22.6	19.5	157 00.1	54.7	204 07.8	35.3
02		313 38.8	221 51.0	22.6	333 24.3	19.4	172 02.1	54.5	219 09.9	35.3
03		328 41.2	236 50.1	. . 22.7	348 26.1	. . 19.3	187 04.0	. . 54.4	234 12.0	. . 35.3
04		343 43.7	251 49.2	22.8	3 27.8	19.2	202 06.0	54.2	249 14.2	35.3
05		358 46.2	266 48.3	22.8	18 29.6	19.0	217 08.0	54.1	264 16.3	35.3
06		13 48.6	281 47.5	N23 22.9	33 31.3	S13 18.9	232 10.0	N15 53.9	279 18.4	N22 35.3
07		28 51.1	296 46.6	22.9	48 33.1	18.8	247 12.0	53.8	294 20.5	35.3
08		43 53.6	311 45.7	23.0	63 34.9	18.7	262 13.9	53.6	309 22.6	35.3
09	S	58 56.0	326 44.9	. . 23.1	78 36.6	. . 18.6	277 15.9	. . 53.5	324 24.7	. . 35.3
10	U	73 58.5	341 44.0	23.1	93 38.4	18.5	292 17.9	53.3	339 26.8	35.3
11	N	89 01.0	356 43.1	23.2	108 40.1	18.4	307 19.9	53.2	354 29.0	35.3
12	D	104 03.4	11 42.2	N23 23.2	123 41.9	S13 18.2	322 21.8	N15 53.0	9 31.1	N22 35.3
13	A	119 05.9	26 41.4	23.3	138 43.7	18.1	337 23.8	52.9	24 33.2	35.2
14	Y	134 08.4	41 40.5	23.3	153 45.4	18.0	352 25.8	52.7	39 35.3	35.2
15		149 10.8	56 39.6	. . 23.4	168 47.2	. . 17.9	7 27.8	. . 52.5	54 37.4	. . 35.2
16		164 13.3	71 38.8	23.4	183 49.0	17.8	22 29.8	52.4	69 39.5	35.2
17		179 15.7	86 37.9	23.5	198 50.7	17.7	37 31.7	52.2	84 41.7	35.2
18		194 18.2	101 37.0	N23 23.5	213 52.5	S13 17.6	52 33.7	N15 52.1	99 43.8	N22 35.2
19		209 20.7	116 36.2	23.6	228 54.3	17.5	67 35.7	51.9	114 45.9	35.2
20		224 23.1	131 35.3	23.6	243 56.0	17.4	82 37.7	51.8	129 48.0	35.2
21		239 25.6	146 34.4	. . 23.6	258 57.8	. . 17.2	97 39.6	. . 51.6	144 50.1	. . 35.2
22		254 28.1	161 33.5	23.7	273 59.6	17.1	112 41.6	51.5	159 52.2	35.2
23		269 30.5	176 32.7	23.7	289 01.4	17.0	127 43.6	51.3	174 54.3	35.2
7 00		284 33.0	191 31.8	N23 23.8	304 03.1	S13 16.9	142 45.6	N15 51.2	189 56.5	N22 35.2
01		299 35.5	206 30.9	23.8	319 04.9	16.8	157 47.6	51.0	204 58.6	35.2
02		314 37.9	221 30.1	23.8	334 06.7	16.7	172 49.5	50.9	220 00.7	35.2
03		329 40.4	236 29.2	. . 23.9	349 08.5	. . 16.6	187 51.5	. . 50.7	235 02.8	. . 35.1
04		344 42.8	251 28.3	23.9	4 10.2	16.5	202 53.5	50.5	250 04.9	35.1
05		359 45.3	266 27.4	23.9	19 12.0	16.4	217 55.5	50.4	265 07.0	35.1
06		14 47.8	281 26.6	N23 24.0	34 13.8	S13 16.3	232 57.4	N15 50.2	280 09.2	N22 35.1
07		29 50.2	296 25.7	24.0	49 15.6	16.2	247 59.4	50.1	295 11.3	35.1
08		44 52.7	311 24.8	24.0	64 17.4	16.1	263 01.4	49.9	310 13.4	35.1
09	M	59 55.2	326 24.0	. . 24.1	79 19.1	. . 16.0	278 03.4	. . 49.8	325 15.5	. . 35.1
10	O	74 57.6	341 23.1	24.1	94 20.9	15.9	293 05.3	49.6	340 17.6	35.1
11	N	90 00.1	356 22.2	24.1	109 22.7	15.7	308 07.3	49.5	355 19.7	35.1
12	D	105 02.6	11 21.3	N23 24.1	124 24.5	S13 15.6	323 09.3	N15 49.3	10 21.9	N22 35.1
13	A	120 05.0	26 20.5	24.2	139 26.3	15.5	338 11.3	49.2	25 24.0	35.1
14	Y	135 07.5	41 19.6	24.2	154 28.1	15.4	353 13.2	49.0	40 26.1	35.1
15		150 10.0	56 18.7	. . 24.2	169 29.9	. . 15.3	8 15.2	. . 48.8	55 28.2	. . 35.1
16		165 12.4	71 17.8	24.2	184 31.6	15.2	23 17.2	48.7	70 30.3	35.0
17		180 14.9	86 17.0	24.2	199 33.4	15.1	38 19.2	48.5	85 32.4	35.0
18		195 17.3	101 16.1	N23 24.2	214 35.2	S13 15.0	53 21.1	N15 48.4	100 34.5	N22 35.0
19		210 19.8	116 15.2	24.3	229 37.0	14.9	68 23.1	48.2	115 36.7	35.0
20		225 22.3	131 14.4	24.3	244 38.8	14.8	83 25.1	48.1	130 38.8	35.0
21		240 24.7	146 13.5	. . 24.3	259 40.6	. . 14.7	98 27.1	. . 47.9	145 40.9	. . 35.0
22		255 27.2	161 12.6	24.3	274 42.4	14.6	113 29.1	47.8	160 43.0	35.0
23		270 29.7	176 11.7	24.3	289 44.2	14.5	128 31.0	47.6	175 45.1	35.0
8 00		285 32.1	191 10.9	N23 24.3	304 46.0	S13 14.4	143 33.0	N15 47.4	190 47.2	N22 35.0
01		300 34.6	206 10.0	24.3	319 47.8	14.3	158 35.0	47.3	205 49.4	35.0
02		315 37.1	221 09.1	24.3	334 49.6	14.2	173 37.0	47.1	220 51.5	35.0
03		330 39.5	236 08.3	. . 24.3	349 51.4	. . 14.1	188 38.9	. . 47.0	235 53.6	. . 35.0
04		345 42.0	251 07.4	24.4	4 53.2	14.0	203 40.9	46.8	250 55.7	35.0
05		0 44.5	266 06.5	24.4	19 55.0	13.9	218 42.9	46.7	265 57.8	34.9
06		15 46.9	281 05.6	N23 24.4	34 56.8	S13 13.8	233 44.9	N15 46.5	280 59.9	N22 34.9
07		30 49.4	296 04.8	24.4	49 58.6	13.7	248 46.8	46.4	296 02.1	34.9
08		45 51.8	311 03.9	24.4	65 00.4	13.6	263 48.8	46.2	311 04.2	34.9
09	T	60 54.3	326 03.0	. . 24.4	80 02.2	. . 13.5	278 50.8	. . 46.0	326 06.3	. . 34.9
10	U	75 56.8	341 02.1	24.4	95 04.0	13.4	293 52.7	45.9	341 08.4	34.9
11	E	90 59.2	356 01.3	24.4	110 05.8	13.4	308 54.7	45.7	356 10.5	34.9
12	S	106 01.7	11 00.4	N23 24.4	125 07.6	S13 13.3	323 56.7	N15 45.6	11 12.6	N22 34.9
13	D	121 04.2	25 59.5	24.3	140 09.4	13.2	338 58.7	45.4	26 14.8	34.9
14	A	136 06.6	40 58.6	24.3	155 11.3	13.1	354 00.6	45.3	41 16.9	34.9
15	Y	151 09.1	55 57.8	. . 24.3	170 13.1	. . 13.0	9 02.6	. . 45.1	56 19.0	. . 34.9
16		166 11.6	70 56.9	24.3	185 14.9	12.9	24 04.6	45.0	71 21.1	34.9
17		181 14.0	85 56.0	24.3	200 16.7	12.8	39 06.6	44.8	86 23.2	34.8
18		196 16.5	100 55.2	N23 24.3	215 18.5	S13 12.7	54 08.5	N15 44.6	101 25.3	N22 34.8
19		211 18.9	115 54.3	24.3	230 20.3	12.6	69 10.5	44.5	116 27.5	34.8
20		226 21.4	130 53.4	24.3	245 22.1	12.5	84 12.5	44.3	131 29.6	34.8
21		241 23.9	145 52.5	. . 24.3	260 24.0	. . 12.4	99 14.5	. . 44.2	146 31.7	. . 34.8
22		256 26.3	160 51.7	24.2	275 25.8	12.3	114 16.4	44.0	161 33.8	34.8
23		271 28.8	175 50.8	24.2	290 27.6	12.2	129 18.4	43.9	176 35.9	34.8
Mer. Pass.		h m 5 01.0	*v* −0.9	*d* 0.0	*v* 1.8	*d* 0.1	*v* 2.0	*d* 0.2	*v* 2.1	*d* 0.0

STARS

Name	SHA ° ′	Dec ° ′
Acamar	315 24.4	S40 17.2
Achernar	335 32.5	S57 12.9
Acrux	173 18.5	S63 07.3
Adhara	255 19.1	S28 58.5
Aldebaran	290 58.7	N16 31.0
Alioth	166 27.3	N55 56.8
Alkaid	153 04.8	N49 18.1
Al Na'ir	27 53.1	S46 56.5
Alnilam	275 54.6	S 1 11.9
Alphard	218 04.1	S 8 40.3
Alphecca	126 17.4	N26 42.3
Alpheratz	357 51.6	N29 06.4
Altair	62 15.6	N 8 52.6
Ankaa	353 23.2	S42 17.1
Antares	112 35.6	S26 26.5
Arcturus	146 02.7	N19 10.0
Atria	107 44.1	S69 02.2
Avior	234 22.0	S59 31.2
Bellatrix	278 40.7	N 6 21.2
Betelgeuse	271 10.1	N 7 24.5
Canopus	264 00.2	S52 41.7
Capella	280 46.5	N46 00.1
Deneb	49 36.5	N45 17.4
Denebola	182 41.7	N14 33.4
Diphda	349 03.7	S17 58.0
Dubhe	194 01.3	N61 44.3
Elnath	278 22.9	N28 36.6
Eltanin	90 49.3	N51 29.3
Enif	33 54.6	N 9 53.3
Fomalhaut	15 32.3	S29 36.1
Gacrux	172 10.0	S57 08.1
Gienah	176 00.5	S17 33.7
Hadar	148 59.1	S60 23.6
Hamal	328 09.8	N23 28.6
Kaus Aust.	83 53.8	S34 23.0
Kochab	137 18.7	N74 08.8
Markab	13 46.0	N15 13.3
Menkar	314 23.4	N 4 06.2
Menkent	148 16.8	S36 23.4
Miaplacidus	221 42.4	S69 43.9
Mirfak	308 52.0	N49 52.2
Nunki	76 07.6	S26 17.6
Peacock	53 30.8	S56 43.4
Pollux	243 37.7	N28 01.2
Procyon	245 08.3	N 5 13.1
Rasalhague	96 13.4	N12 33.5
Regulus	207 52.1	N11 57.2
Rigel	281 19.9	S 8 11.8
Rigil Kent.	140 02.5	S60 51.2
Sabik	102 21.2	S15 43.8
Schedar	349 49.7	N56 33.1
Shaula	96 32.2	S37 06.5
Sirius	258 41.0	S16 43.1
Spica	158 39.5	S11 10.7
Suhail	222 58.7	S43 26.8
Vega	80 43.9	N38 47.2
Zuben'ubi	137 14.0	S16 03.4

	SHA ° ′	Mer. Pass. h m
Venus	266 58.8	11 15
Mars	19 30.1	3 43
Jupiter	218 12.6	14 27
Saturn	265 23.5	11 19

UT		SUN GHA	SUN Dec	MOON GHA	v	MOON Dec	d	HP
d	h	° ′	° ′	° ′	′	° ′	′	′
6	00	178 51.0	N22 45.1	102 30.1	12.5	N 4 22.3	15.1	58.4
	01	193 50.9	44.8	117 01.6	12.5	4 07.2	15.3	58.4
	02	208 50.7	44.6	131 33.1	12.5	3 51.9	15.2	58.5
	03	223 50.6	. . 44.3	146 04.6	12.5	3 36.7	15.2	58.5
	04	238 50.5	44.1	160 36.1	12.5	3 21.5	15.3	58.5
	05	253 50.4	43.8	175 07.6	12.4	3 06.2	15.4	58.5
	06	268 50.3	N22 43.6	189 39.0	12.5	N 2 50.8	15.3	58.5
	07	283 50.2	43.4	204 10.5	12.5	2 35.5	15.4	58.6
	08	298 50.1	43.1	218 42.0	12.4	2 20.1	15.3	58.6
S	09	313 50.0	. . 42.9	233 13.4	12.5	2 04.8	15.4	58.6
U	10	328 49.9	42.6	247 44.9	12.4	1 49.4	15.5	58.6
N	11	343 49.8	42.4	262 16.3	12.4	1 33.9	15.4	58.6
D	12	358 49.7	N22 42.1	276 47.7	12.4	N 1 18.5	15.5	58.7
A	13	13 49.6	41.9	291 19.1	12.4	1 03.0	15.4	58.7
Y	14	28 49.5	41.6	305 50.5	12.3	0 47.6	15.5	58.7
	15	43 49.4	. . 41.4	320 21.8	12.4	0 32.1	15.5	58.7
	16	58 49.3	41.1	334 53.2	12.3	0 16.6	15.6	58.7
	17	73 49.2	40.9	349 24.5	12.3	N 0 01.0	15.5	58.8
	18	88 49.1	N22 40.6	3 55.8	12.3	S 0 14.5	15.5	58.8
	19	103 49.0	40.4	18 27.1	12.2	0 30.0	15.6	58.8
	20	118 48.8	40.1	32 58.3	12.3	0 45.6	15.5	58.8
	21	133 48.7	. . 39.8	47 29.6	12.2	1 01.1	15.6	58.9
	22	148 48.6	39.6	62 00.8	12.1	1 16.7	15.5	58.9
	23	163 48.5	39.3	76 31.9	12.2	1 32.2	15.6	58.9
7	00	178 48.4	N22 39.1	91 03.1	12.1	S 1 47.8	15.6	58.9
	01	193 48.3	38.8	105 34.2	12.1	2 03.4	15.5	58.9
	02	208 48.2	38.6	120 05.3	12.1	2 18.9	15.6	59.0
	03	223 48.1	. . 38.3	134 36.4	12.0	2 34.5	15.5	59.0
	04	238 48.0	38.0	149 07.4	12.0	2 50.0	15.6	59.0
	05	253 47.9	37.8	163 38.4	11.9	3 05.6	15.6	59.0
	06	268 47.8	N22 37.5	178 09.3	11.9	S 3 21.2	15.5	59.0
	07	283 47.7	37.3	192 40.2	11.9	3 36.7	15.5	59.0
	08	298 47.6	37.0	207 11.1	11.8	3 52.2	15.6	59.1
M	09	313 47.5	. . 36.7	221 41.9	11.8	4 07.8	15.5	59.1
O	10	328 47.4	36.5	236 12.7	11.8	4 23.3	15.5	59.1
N	11	343 47.3	36.2	250 43.5	11.7	4 38.8	15.5	59.1
D	12	358 47.2	N22 35.9	265 14.2	11.6	S 4 54.3	15.5	59.1
A	13	13 47.1	35.7	279 44.8	11.6	5 09.8	15.4	59.2
Y	14	28 47.0	35.4	294 15.4	11.6	5 25.2	15.5	59.2
	15	43 46.9	. . 35.1	308 46.0	11.5	5 40.7	15.4	59.2
	16	58 46.8	34.9	323 16.5	11.5	5 56.1	15.4	59.2
	17	73 46.7	34.6	337 47.0	11.4	6 11.5	15.4	59.2
	18	88 46.6	N22 34.3	352 17.4	11.4	S 6 26.9	15.3	59.3
	19	103 46.5	34.1	6 47.8	11.3	6 42.2	15.3	59.3
	20	118 46.4	33.8	21 18.1	11.2	6 57.5	15.4	59.3
	21	133 46.3	. . 33.5	35 48.3	11.2	7 12.9	15.2	59.3
	22	148 46.2	33.3	50 18.5	11.2	7 28.1	15.3	59.3
	23	163 46.1	33.0	64 48.7	11.0	7 43.4	15.2	59.3
8	00	178 46.0	N22 32.7	79 18.7	11.1	S 7 58.6	15.2	59.4
	01	193 45.9	32.4	93 48.8	10.9	8 13.8	15.1	59.4
	02	208 45.8	32.2	108 18.7	10.9	8 28.9	15.1	59.4
	03	223 45.7	. . 31.9	122 48.6	10.9	8 44.0	15.1	59.4
	04	238 45.6	31.6	137 18.5	10.7	8 59.1	15.0	59.4
	05	253 45.5	31.3	151 48.2	10.7	9 14.1	15.0	59.4
	06	268 45.4	N22 31.1	166 17.9	10.7	S 9 29.1	15.0	59.5
	07	283 45.3	30.8	180 47.6	10.5	9 44.1	14.9	59.5
T	08	298 45.2	30.5	195 17.1	10.5	9 59.0	14.9	59.5
U	09	313 45.1	. . 30.2	209 46.6	10.5	10 13.9	14.8	59.5
E	10	328 45.0	29.9	224 16.1	10.3	10 28.7	14.7	59.5
S	11	343 44.9	29.7	238 45.4	10.3	10 43.4	14.8	59.5
D	12	358 44.8	N22 29.4	253 14.7	10.2	S10 58.2	14.6	59.6
A	13	13 44.7	29.1	267 43.9	10.1	11 12.8	14.6	59.6
Y	14	28 44.6	28.8	282 13.0	10.1	11 27.4	14.6	59.6
	15	43 44.5	. . 28.5	296 42.1	10.0	11 42.0	14.5	59.6
	16	58 44.4	28.2	311 11.1	9.9	11 56.5	14.4	59.6
	17	73 44.3	28.0	325 40.0	9.8	12 10.9	14.4	59.6
	18	88 44.2	N22 27.7	340 08.8	9.8	S12 25.3	14.4	59.6
	19	103 44.1	27.4	354 37.6	9.6	12 39.7	14.2	59.7
	20	118 44.1	27.1	9 06.2	9.6	12 53.9	14.2	59.7
	21	133 44.0	. . 26.8	23 34.8	9.5	13 08.1	14.1	59.7
	22	148 43.9	26.5	38 03.3	9.4	13 22.2	14.1	59.7
	23	163 43.8	26.2	52 31.7	9.4	S13 36.3	14.0	59.7
		SD 15.8	*d* 0.3	SD		16.0	16.1	16.2

Lat.	Twilight Naut.	Twilight Civil	Sunrise	Moonrise 6	Moonrise 7	Moonrise 8	Moonrise 9
°	h m	h m	h m	h m	h m	h m	h m
N 72	□	□	□	11 15	13 30	16 00	■
N 70	□	□	□	11 17	13 22	15 38	18 26
68	□	□	□	11 18	13 16	15 20	17 43
66	////	////	00 55	11 20	13 10	15 06	17 14
64	////	////	01 53	11 21	13 05	14 55	16 52
62	////	////	02 25	11 22	13 02	14 45	16 35
60	////	01 18	02 49	11 23	12 58	14 37	16 21
N 58	////	01 58	03 08	11 23	12 55	14 30	16 09
56	////	02 24	03 24	11 24	12 52	14 24	15 58
54	01 12	02 45	03 37	11 25	12 50	14 18	15 49
52	01 48	03 01	03 49	11 25	12 48	14 13	15 41
50	02 13	03 16	03 59	11 26	12 46	14 09	15 34
45	02 55	03 44	04 21	11 27	12 42	13 59	15 18
N 40	03 24	04 05	04 38	11 28	12 38	13 51	15 06
35	03 47	04 23	04 52	11 28	12 35	13 44	14 55
30	04 05	04 38	05 05	11 29	12 32	13 38	14 46
20	04 33	05 02	05 26	11 30	12 28	13 27	14 30
N 10	04 54	05 21	05 44	11 31	12 24	13 18	14 16
0	05 13	05 39	06 01	11 33	12 20	13 10	14 03
S 10	05 29	05 55	06 18	11 34	12 17	13 02	13 50
20	05 45	06 12	06 36	11 35	12 13	12 53	13 37
30	06 00	06 30	06 56	11 36	12 09	12 43	13 21
35	06 09	06 40	07 08	11 37	12 06	12 37	13 12
40	06 18	06 51	07 22	11 38	12 03	12 31	13 02
45	06 27	07 04	07 38	11 39	12 00	12 24	12 51
S 50	06 38	07 20	07 57	11 40	11 56	12 15	12 37
52	06 43	07 27	08 07	11 40	11 55	12 11	12 30
54	06 49	07 34	08 17	11 41	11 53	12 06	12 23
56	06 54	07 43	08 29	11 41	11 51	12 01	12 15
58	07 01	07 53	08 43	11 42	11 49	11 56	12 06
S 60	07 08	08 04	08 59	11 43	11 46	11 50	11 56

Lat.	Sunset	Twilight Civil	Twilight Naut.	Moonset 6	Moonset 7	Moonset 8	Moonset 9
°	h m	h m	h m	h m	h m	h m	h m
N 72	□	□	□	{00 02 / 23 32}	23 00	22 16	■
N 70	□	□	□	23 35	23 11	22 41	21 45
68	□	□	□	23 37	23 20	23 01	22 30
66	23 10	////	////	23 38	23 28	23 16	23 00
64	22 15	////	////	23 40	23 35	23 29	23 23
62	21 43	////	////	23 41	23 40	23 40	23 41
60	21 20	22 49	////	23 42	23 45	23 50	23 56
N 58	21 01	22 11	////	23 43	23 50	23 58	24 09
56	20 45	21 45	////	23 44	23 54	24 05	00 05
54	20 32	21 24	22 56	23 45	23 57	24 12	00 12
52	20 20	21 08	22 20	23 45	24 00	00 00	00 18
50	20 10	20 54	21 56	23 46	24 03	00 03	00 23
45	19 49	20 25	21 14	23 48	24 10	00 10	00 35
N 40	19 32	20 04	20 45	23 49	24 15	00 15	00 44
35	19 17	19 47	20 23	23 50	24 20	00 20	00 53
30	19 05	19 32	20 05	23 51	24 24	00 24	01 00
20	18 44	19 08	19 37	23 52	24 31	00 31	01 13
N 10	18 25	18 48	19 15	23 54	24 38	00 38	01 24
0	18 09	18 31	18 57	23 55	24 44	00 44	01 35
S 10	17 52	18 14	18 41	23 56	24 50	00 50	01 45
20	17 34	17 58	18 25	23 58	24 56	00 56	01 57
30	17 14	17 40	18 09	23 59	25 03	01 03	02 10
35	17 02	17 30	18 01	24 00	00 00	01 08	02 17
40	16 48	17 18	17 52	24 01	00 01	01 12	02 26
45	16 32	17 06	17 43	24 02	00 02	01 18	02 36
S 50	16 12	16 50	17 32	24 04	00 04	01 25	02 49
52	16 03	16 43	17 27	24 04	00 04	01 28	02 54
54	15 53	16 36	17 21	24 05	00 05	01 31	03 01
56	15 41	16 27	17 16	24 06	00 06	01 35	03 08
58	15 27	16 17	17 09	24 07	00 07	01 40	03 16
S 60	15 11	16 06	17 02	24 08	00 08	01 44	03 25

Day	SUN Eqn. of Time 00^h	SUN Eqn. of Time 12^h	SUN Mer. Pass.	MOON Mer. Pass. Upper	MOON Mer. Pass. Lower	Age	Phase
d	m s	m s	h m	h m	h m	d %	
6	04 36	04 41	12 05	17 44	05 20	07 43	
7	04 46	04 51	12 05	18 32	06 08	08 55	
8	04 56	05 01	12 05	19 22	06 57	09 66	

	UT	ARIES	VENUS −3.9		MARS −1.7		JUPITER −1.8		SATURN +0.1	
		GHA	GHA	Dec	GHA	Dec	GHA	Dec	GHA	Dec
	d h	° ′	° ′	° ′	° ′	° ′	° ′	° ′	° ′	° ′
	9 00	286 31.3	190 49.9	N23 24.2	305 29.4	S13 12.1	144 20.4	N15 43.7	191 38.0	N22 34.8
	01	301 33.7	205 49.0	24.2	320 31.2	12.1	159 22.4	43.5	206 40.2	34.8
	02	316 36.2	220 48.2	24.2	335 33.1	12.0	174 24.3	43.4	221 42.3	34.8
	03	331 38.7	235 47.3	. . 24.1	350 34.9	. . 11.9	189 26.3	. . 43.2	236 44.4	. . 34.8
	04	346 41.1	250 46.4	24.1	5 36.7	11.8	204 28.3	43.1	251 46.5	34.8
	05	1 43.6	265 45.6	24.1	20 38.5	11.7	219 30.2	42.9	266 48.6	34.8
	06	16 46.1	280 44.7	N23 24.1	35 40.4	S13 11.6	234 32.2	N15 42.8	281 50.7	N22 34.7
W	07	31 48.5	295 43.8	24.0	50 42.2	11.5	249 34.2	42.6	296 52.9	34.7
E	08	46 51.0	310 42.9	24.0	65 44.0	11.4	264 36.2	42.4	311 55.0	34.7
D	09	61 53.4	325 42.1	. . 24.0	80 45.9	. . 11.3	279 38.1	. . 42.3	326 57.1	. . 34.7
N	10	76 55.9	340 41.2	24.0	95 47.7	11.3	294 40.1	42.1	341 59.2	34.7
E	11	91 58.4	355 40.3	23.9	110 49.5	11.2	309 42.1	42.0	357 01.3	34.7
S	12	107 00.8	10 39.4	N23 23.9	125 51.4	S13 11.1	324 44.1	N15 41.8	12 03.4	N22 34.7
D	13	122 03.3	25 38.6	23.9	140 53.2	11.0	339 46.0	41.7	27 05.6	34.7
A	14	137 05.8	40 37.7	23.8	155 55.0	10.9	354 48.0	41.5	42 07.7	34.7
Y	15	152 08.2	55 36.8	. . 23.8	170 56.9	. . 10.8	9 50.0	. . 41.4	57 09.8	. . 34.7
	16	167 10.7	70 35.9	23.7	185 58.7	10.8	24 51.9	41.2	72 11.9	34.7
	17	182 13.2	85 35.1	23.7	201 00.5	10.7	39 53.9	41.0	87 14.0	34.7
	18	197 15.6	100 34.2	N23 23.7	216 02.4	S13 10.6	54 55.9	N15 40.9	102 16.2	N22 34.6
	19	212 18.1	115 33.3	23.6	231 04.2	10.5	69 57.9	40.7	117 18.3	34.6
	20	227 20.6	130 32.5	23.6	246 06.1	10.4	84 59.8	40.6	132 20.4	34.6
	21	242 23.0	145 31.6	. . 23.5	261 07.9	. . 10.3	100 01.8	. . 40.4	147 22.5	. . 34.6
	22	257 25.5	160 30.7	23.5	276 09.7	10.3	115 03.8	40.2	162 24.6	34.6
	23	272 27.9	175 29.8	23.4	291 11.6	10.2	130 05.7	40.1	177 26.7	34.6
	10 00	287 30.4	190 29.0	N23 23.4	306 13.4	S13 10.1	145 07.7	N15 39.9	192 28.9	N22 34.6
	01	302 32.9	205 28.1	23.4	321 15.3	10.0	160 09.7	39.8	207 31.0	34.6
	02	317 35.3	220 27.2	23.3	336 17.1	09.9	175 11.7	39.6	222 33.1	34.6
	03	332 37.8	235 26.3	. . 23.3	351 19.0	. . 09.9	190 13.6	. . 39.5	237 35.2	. . 34.6
	04	347 40.3	250 25.5	23.2	6 20.8	09.8	205 15.6	39.3	252 37.3	34.6
	05	2 42.7	265 24.6	23.1	21 22.7	09.7	220 17.6	39.1	267 39.4	34.5
	06	17 45.2	280 23.7	N23 23.1	36 24.5	S13 09.6	235 19.5	N15 39.0	282 41.6	N22 34.5
T	07	32 47.7	295 22.9	23.0	51 26.4	09.5	250 21.5	38.8	297 43.7	34.5
H	08	47 50.1	310 22.0	23.0	66 28.2	09.5	265 23.5	38.7	312 45.8	34.5
U	09	62 52.6	325 21.1	. . 22.9	81 30.1	. . 09.4	280 25.5	. . 38.5	327 47.9	. . 34.5
R	10	77 55.0	340 20.2	22.9	96 31.9	09.3	295 27.4	38.4	342 50.0	34.5
S	11	92 57.5	355 19.4	22.8	111 33.8	09.2	310 29.4	38.2	357 52.2	34.5
D	12	108 00.0	10 18.5	N23 22.7	126 35.7	S13 09.2	325 31.4	N15 38.0	12 54.3	N22 34.5
A	13	123 02.4	25 17.6	22.7	141 37.5	09.1	340 33.3	37.9	27 56.4	34.5
Y	14	138 04.9	40 16.7	22.6	156 39.4	09.0	355 35.3	37.7	42 58.5	34.5
	15	153 07.4	55 15.9	. . 22.5	171 41.2	. . 08.9	10 37.3	. . 37.6	58 00.6	. . 34.5
	16	168 09.8	70 15.0	22.5	186 43.1	08.9	25 39.2	37.4	73 02.7	34.5
	17	183 12.3	85 14.1	22.4	201 45.0	08.8	40 41.2	37.3	88 04.9	34.4
	18	198 14.8	100 13.2	N23 22.3	216 46.8	S13 08.7	55 43.2	N15 37.1	103 07.0	N22 34.4
	19	213 17.2	115 12.4	22.3	231 48.7	08.6	70 45.2	36.9	118 09.1	34.4
	20	228 19.7	130 11.5	22.2	246 50.6	08.6	85 47.1	36.8	133 11.2	34.4
	21	243 22.2	145 10.6	. . 22.1	261 52.4	. . 08.5	100 49.1	. . 36.6	148 13.3	. . 34.4
	22	258 24.6	160 09.8	22.1	276 54.3	08.4	115 51.1	36.5	163 15.5	34.4
	23	273 27.1	175 08.9	22.0	291 56.2	08.3	130 53.0	36.3	178 17.6	34.4
	11 00	288 29.5	190 08.0	N23 21.9	306 58.0	S13 08.3	145 55.0	N15 36.1	193 19.7	N22 34.4
	01	303 32.0	205 07.1	21.8	321 59.9	08.2	160 57.0	36.0	208 21.8	34.4
	02	318 34.5	220 06.3	21.7	337 01.8	08.1	175 58.9	35.8	223 23.9	34.4
	03	333 36.9	235 05.4	. . 21.7	352 03.7	. . 08.1	191 00.9	. . 35.7	238 26.0	. . 34.4
	04	348 39.4	250 04.5	21.6	7 05.5	08.0	206 02.9	35.5	253 28.2	34.4
	05	3 41.9	265 03.7	21.5	22 07.4	07.9	221 04.9	35.3	268 30.3	34.3
	06	18 44.3	280 02.8	N23 21.4	37 09.3	S13 07.8	236 06.8	N15 35.2	283 32.4	N22 34.3
	07	33 46.8	295 01.9	21.3	52 11.2	07.8	251 08.8	35.0	298 34.5	34.3
	08	48 49.3	310 01.0	21.3	67 13.0	07.7	266 10.8	34.9	313 36.6	34.3
F	09	63 51.7	325 00.2	. . 21.2	82 14.9	. . 07.6	281 12.7	. . 34.7	328 38.8	. . 34.3
R	10	78 54.2	339 59.3	21.1	97 16.8	07.6	296 14.7	34.6	343 40.9	34.3
I	11	93 56.7	354 58.4	21.0	112 18.7	07.5	311 16.7	34.4	358 43.0	34.3
D	12	108 59.1	9 57.5	N23 20.9	127 20.6	S13 07.4	326 18.6	N15 34.2	13 45.1	N22 34.3
A	13	124 01.6	24 56.7	20.8	142 22.4	07.4	341 20.6	34.1	28 47.2	34.3
Y	14	139 04.0	39 55.8	20.7	157 24.3	07.3	356 22.6	33.9	43 49.4	34.3
	15	154 06.5	54 54.9	. . 20.6	172 26.2	. . 07.2	11 24.5	. . 33.8	58 51.5	. . 34.3
	16	169 09.0	69 54.1	20.5	187 28.1	07.2	26 26.5	33.6	73 53.6	34.2
	17	184 11.4	84 53.2	20.4	202 30.0	07.1	41 28.5	33.4	88 55.7	34.2
	18	199 13.9	99 52.3	N23 20.3	217 31.9	S13 07.1	56 30.5	N15 33.3	103 57.8	N22 34.2
	19	214 16.4	114 51.4	20.2	232 33.8	07.0	71 32.4	33.1	118 59.9	34.2
	20	229 18.8	129 50.6	20.1	247 35.7	06.9	86 34.4	33.0	134 02.1	34.2
	21	244 21.3	144 49.7	. . 20.0	262 37.5	. . 06.9	101 36.4	. . 32.8	149 04.2	. . 34.2
	22	259 23.8	159 48.8	19.9	277 39.4	06.8	116 38.3	32.6	164 06.3	34.2
	23	274 26.2	174 48.0	19.8	292 41.3	06.7	131 40.3	32.5	179 08.4	34.2
	Mer. Pass.	h m 4 49.2	*v* −0.9	*d* 0.1	*v* 1.9	*d* 0.1	*v* 2.0	*d* 0.2	*v* 2.1	*d* 0.0

STARS

Name	SHA	Dec
	° ′	° ′
Acamar	315 24.4	S40 17.2
Achernar	335 32.5	S57 12.9
Acrux	173 18.5	S63 07.3
Adhara	255 19.1	S28 58.5
Aldebaran	290 58.7	N16 31.0
Alioth	166 27.4	N55 56.8
Alkaid	153 04.8	N49 18.1
Al Na'ir	27 53.0	S46 56.5
Alnilam	275 54.6	S 1 11.9
Alphard	218 04.1	S 8 40.3
Alphecca	126 17.4	N26 42.3
Alpheratz	357 51.6	N29 06.4
Altair	62 15.5	N 8 52.6
Ankaa	353 23.2	S42 17.0
Antares	112 35.6	S26 26.5
Arcturus	146 02.7	N19 10.0
Atria	107 44.1	S69 02.2
Avior	234 22.0	S59 31.2
Bellatrix	278 40.7	N 6 21.2
Betelgeuse	271 10.1	N 7 24.5
Canopus	264 00.2	S52 41.7
Capella	280 46.5	N46 00.1
Deneb	49 36.5	N45 17.4
Denebola	182 41.7	N14 33.4
Diphda	349 03.7	S17 58.0
Dubhe	194 01.3	N61 44.3
Elnath	278 22.9	N28 36.6
Eltanin	90 49.3	N51 29.3
Enif	33 54.6	N 9 53.4
Fomalhaut	15 32.3	S29 36.1
Gacrux	172 10.0	S57 08.1
Gienah	176 00.5	S17 33.7
Hadar	148 59.1	S60 23.6
Hamal	328 09.8	N23 28.6
Kaus Aust.	83 53.8	S34 23.0
Kochab	137 18.8	N74 08.8
Markab	13 46.0	N15 13.3
Menkar	314 23.4	N 4 06.2
Menkent	148 16.8	S36 23.4
Miaplacidus	221 42.5	S69 43.9
Mirfak	308 52.0	N49 52.2
Nunki	76 07.6	S26 17.6
Peacock	53 30.8	S56 43.4
Pollux	243 37.6	N28 01.2
Procyon	245 08.3	N 5 13.1
Rasalhague	96 13.4	N12 33.5
Regulus	207 52.1	N11 57.2
Rigel	281 19.9	S 8 11.7
Rigil Kent.	140 02.5	S60 51.2
Sabik	102 21.2	S15 43.8
Schedar	349 49.6	N56 33.1
Shaula	96 32.2	S37 06.5
Sirius	258 41.0	S16 43.1
Spica	158 39.5	S11 10.7
Suhail	222 58.7	S43 26.8
Vega	80 43.9	N38 47.2
Zuben'ubi	137 14.0	S16 03.4

	SHA	Mer. Pass.
	° ′	h m
Venus	262 58.6	11 19
Mars	18 43.0	3 35
Jupiter	217 37.3	14 18
Saturn	264 58.5	11 09

UT d	UT h	SUN GHA ° ′	SUN Dec ° ′	MOON GHA ° ′	MOON v ′	MOON Dec ° ′	MOON d ′	MOON HP ′
9 Wednesday	00	178 43.7	N22 25.9	67 00.1	9.2	S13 50.3	13.9	59.7
	01	193 43.6	25.7	81 28.3	9.2	14 04.2	13.8	59.7
	02	208 43.5	25.4	95 56.5	9.1	14 18.0	13.8	59.8
	03	223 43.4	. . 25.1	110 24.6	8.9	14 31.8	13.7	59.8
	04	238 43.3	24.8	124 52.5	8.9	14 45.5	13.6	59.8
	05	253 43.2	24.5	139 20.4	8.9	14 59.1	13.5	59.8
	06	268 43.1	N22 24.2	153 48.3	8.7	S15 12.6	13.4	59.8
	07	283 43.0	23.9	168 16.0	8.6	15 26.0	13.4	59.8
	08	298 42.9	23.6	182 43.6	8.5	15 39.4	13.3	59.8
	09	313 42.8	. . 23.3	197 11.1	8.5	15 52.7	13.2	59.8
	10	328 42.7	23.0	211 38.6	8.4	16 05.9	13.1	59.8
	11	343 42.6	22.7	226 06.0	8.2	16 19.0	13.0	59.9
	12	358 42.5	N22 22.4	240 33.2	8.2	S16 32.0	12.9	59.9
	13	13 42.5	22.1	255 00.4	8.1	16 44.9	12.8	59.9
	14	28 42.4	21.8	269 27.5	7.9	16 57.7	12.7	59.9
	15	43 42.3	. . 21.5	283 54.4	7.9	17 10.4	12.6	59.9
	16	58 42.2	21.2	298 21.3	7.8	17 23.0	12.5	59.9
	17	73 42.1	20.9	312 48.1	7.7	17 35.5	12.4	59.9
	18	88 42.0	N22 20.6	327 14.8	7.6	S17 47.9	12.4	59.9
	19	103 41.9	20.3	341 41.4	7.5	18 00.3	12.2	59.9
	20	118 41.8	20.0	356 07.9	7.5	18 12.5	12.1	59.9
	21	133 41.7	. . 19.7	10 34.4	7.3	18 24.6	11.9	60.0
	22	148 41.6	19.4	25 00.7	7.2	18 36.5	11.9	60.0
	23	163 41.5	19.1	39 26.9	7.1	18 48.4	11.8	60.0
10 Thursday	00	178 41.4	N22 18.8	53 53.0	7.0	S19 00.2	11.6	60.0
	01	193 41.4	18.5	68 19.0	7.0	19 11.8	11.6	60.0
	02	208 41.3	18.2	82 45.0	6.8	19 23.4	11.4	60.0
	03	223 41.2	. . 17.9	97 10.8	6.7	19 34.8	11.3	60.0
	04	238 41.1	17.6	111 36.5	6.7	19 46.1	11.1	60.0
	05	253 41.0	17.3	126 02.2	6.5	19 57.2	11.1	60.0
	06	268 40.9	N22 16.9	140 27.7	6.5	S20 08.3	10.9	60.0
	07	283 40.8	16.6	154 53.2	6.3	20 19.2	10.8	60.0
	08	298 40.7	16.3	169 18.5	6.3	20 30.0	10.6	60.0
	09	313 40.6	. . 16.0	183 43.8	6.2	20 40.6	10.6	60.0
	10	328 40.6	15.7	198 09.0	6.0	20 51.2	10.3	60.0
	11	343 40.5	15.4	212 34.0	6.0	21 01.5	10.3	60.0
	12	358 40.4	N22 15.1	226 59.0	5.9	S21 11.8	10.1	60.0
	13	13 40.3	14.8	241 23.9	5.8	21 21.9	10.0	60.0
	14	28 40.2	14.4	255 48.7	5.7	21 31.9	9.9	60.0
	15	43 40.1	. . 14.1	270 13.4	5.6	21 41.8	9.7	60.0
	16	58 40.0	13.8	284 38.0	5.5	21 51.5	9.5	60.0
	17	73 39.9	13.5	299 02.5	5.5	22 01.0	9.4	60.0
	18	88 39.9	N22 13.2	313 27.0	5.3	S22 10.4	9.3	60.0
	19	103 39.8	12.9	327 51.3	5.3	22 19.7	9.1	60.1
	20	118 39.7	12.5	342 15.6	5.1	22 28.8	9.0	60.1
	21	133 39.6	. . 12.2	356 39.7	5.1	22 37.8	8.8	60.1
	22	148 39.5	11.9	11 03.8	5.0	22 46.6	8.7	60.1
	23	163 39.4	11.6	25 27.8	4.9	22 55.3	8.5	60.1
11 Friday	00	178 39.3	N22 11.3	39 51.7	4.8	S23 03.8	8.3	60.1
	01	193 39.2	10.9	54 15.5	4.8	23 12.1	8.2	60.1
	02	208 39.2	10.6	68 39.3	4.7	23 20.3	8.1	60.0
	03	223 39.1	. . 10.3	83 03.0	4.6	23 28.4	7.9	60.0
	04	238 39.0	10.0	97 26.6	4.5	23 36.3	7.7	60.0
	05	253 38.9	09.6	111 50.1	4.4	23 44.0	7.5	60.0
	06	268 38.8	N22 09.3	126 13.5	4.4	S23 51.5	7.4	60.0
	07	283 38.7	09.0	140 36.9	4.3	23 58.9	7.2	60.0
	08	298 38.7	08.7	155 00.2	4.2	24 06.1	7.1	60.0
	09	313 38.6	. . 08.3	169 23.4	4.1	24 13.2	6.9	60.0
	10	328 38.5	08.0	183 46.5	4.1	24 20.1	6.7	60.0
	11	343 38.4	07.7	198 09.6	4.0	24 26.8	6.5	60.0
	12	358 38.3	N22 07.4	212 32.6	4.0	S24 33.3	6.4	60.0
	13	13 38.2	07.0	226 55.6	3.9	24 39.7	6.2	60.0
	14	28 38.2	06.7	241 18.5	3.8	24 45.9	6.0	60.0
	15	43 38.1	. . 06.4	255 41.3	3.8	24 51.9	5.9	60.0
	16	58 38.0	06.0	270 04.1	3.7	24 57.8	5.7	60.0
	17	73 37.9	05.7	284 26.8	3.7	25 03.5	5.5	60.0
	18	88 37.8	N22 05.4	298 49.5	3.6	S25 09.0	5.3	60.0
	19	103 37.7	05.0	313 12.1	3.6	25 14.3	5.1	60.0
	20	118 37.7	04.7	327 34.7	3.5	25 19.4	5.0	60.0
	21	133 37.6	. . 04.4	341 57.2	3.4	25 24.4	4.7	60.0
	22	148 37.5	04.0	356 19.6	3.5	25 29.1	4.6	60.0
	23	163 37.4	03.7	10 42.1	3.4	S25 33.7	4.4	59.9
		SD 15.8	*d* 0.3	SD 16.3		16.4		16.4

Lat. °	Twilight Naut. h m	Twilight Civil h m	Sunrise h m	Moonrise 9 h m	Moonrise 10 h m	Moonrise 11 h m	Moonrise 12 h m
N 72	□	□	□	■	■	■	■
N 70	□	□	□	18 26	■	■	■
68	□	□	□	17 43	■	■	■
66	////	////	01 10	17 14	19 48	■	■
64	////	////	02 00	16 52	19 01	21 31	■
62	////	////	02 31	16 35	18 31	20 26	21 57
60	////	01 27	02 54	16 21	18 08	19 52	21 14
N 58	////	02 04	03 12	16 09	17 50	19 27	20 45
56	////	02 29	03 27	15 58	17 35	19 07	20 23
54	01 20	02 49	03 40	15 49	17 22	18 50	20 05
52	01 54	03 05	03 52	15 41	17 10	18 36	19 49
50	02 17	03 19	04 02	15 34	17 00	18 23	19 36
45	02 58	03 46	04 23	15 18	16 39	17 58	19 08
N 40	03 27	04 08	04 40	15 06	16 22	17 37	18 47
35	03 49	04 25	04 54	14 55	16 08	17 21	18 29
30	04 06	04 39	05 06	14 46	15 56	17 06	18 13
20	04 34	05 03	05 27	14 30	15 35	16 41	17 47
N 10	04 55	05 22	05 45	14 16	15 17	16 20	17 24
0	05 13	05 39	06 02	14 03	15 00	16 00	17 03
S 10	05 29	05 56	06 18	13 50	14 43	15 41	16 42
20	05 45	06 12	06 36	13 37	14 25	15 20	16 20
30	06 00	06 30	06 56	13 21	14 05	14 56	15 54
35	06 08	06 40	07 07	13 12	13 53	14 42	15 39
40	06 17	06 51	07 21	13 02	13 40	14 26	15 21
45	06 26	07 03	07 36	12 51	13 24	14 06	15 00
S 50	06 37	07 18	07 56	12 37	13 05	13 42	14 33
52	06 42	07 25	08 05	12 30	12 55	13 31	14 20
54	06 47	07 33	08 15	12 23	12 45	13 18	14 05
56	06 53	07 41	08 27	12 15	12 34	13 02	13 47
58	06 59	07 50	08 40	12 06	12 21	12 45	13 26
S 60	07 06	08 01	08 56	11 56	12 05	12 23	13 00

Lat. °	Sunset h m	Twilight Civil h m	Twilight Naut. h m	Moonset 9 h m	Moonset 10 h m	Moonset 11 h m	Moonset 12 h m
N 72	□	□	□	■	■	■	■
N 70	□	□	□	21 45	■	■	■
68	□	□	□	22 30	■	■	■
66	22 56	////	////	23 00	22 27	■	■
64	22 08	////	////	23 23	23 14	22 53	■
62	21 38	////	////	23 41	23 45	23 58	24 40
60	21 16	22 41	////	23 56	24 08	00 08	00 33
N 58	20 58	22 05	////	24 09	00 09	00 27	00 59
56	20 43	21 41	////	00 05	00 20	00 43	01 19
54	20 30	21 21	22 48	00 12	00 30	00 57	01 36
52	20 18	21 05	22 15	00 18	00 39	01 09	01 50
50	20 08	20 51	21 52	00 23	00 47	01 19	02 03
45	19 47	20 24	21 11	00 35	01 04	01 42	02 29
N 40	19 31	20 03	20 43	00 44	01 18	01 59	02 50
35	19 16	19 46	20 22	00 53	01 30	02 15	03 07
30	19 04	19 31	20 04	01 00	01 41	02 28	03 22
20	18 43	19 08	19 37	01 13	01 59	02 50	03 48
N 10	18 26	18 48	19 15	01 24	02 15	03 10	04 10
0	18 09	18 31	18 57	01 35	02 29	03 28	04 30
S 10	17 52	18 15	18 41	01 45	02 44	03 47	04 51
20	17 35	17 59	18 26	01 57	03 00	04 06	05 13
30	17 15	17 41	18 11	02 10	03 19	04 29	05 38
35	17 03	17 31	18 03	02 17	03 30	04 43	05 54
40	16 50	17 20	17 54	02 26	03 42	04 58	06 11
45	16 34	17 08	17 44	02 36	03 57	05 17	06 32
S 50	16 15	16 53	17 34	02 49	04 15	05 40	06 59
52	16 06	16 46	17 29	02 54	04 23	05 51	07 11
54	15 56	16 38	17 24	03 01	04 33	06 04	07 26
56	15 44	16 30	17 18	03 08	04 44	06 18	07 44
58	15 31	16 21	17 12	03 16	04 56	06 36	08 05
S 60	15 15	16 10	17 05	03 25	05 11	06 57	08 31

Day	SUN Eqn. of Time 00^h	SUN Eqn. of Time 12^h	SUN Mer. Pass.	MOON Mer. Pass. Upper	MOON Mer. Pass. Lower	MOON Age	MOON Phase
d	m s	m s	h m	h m	h m	d	%
9	05 05	05 10	12 05	20 16	07 49	10	76
10	05 14	05 18	12 05	21 14	08 44	11	86
11	05 22	05 27	12 05	22 15	09 44	12	93

UT d	UT h	ARIES GHA ° ′	VENUS −3.9 GHA ° ′	VENUS Dec ° ′	MARS −1.8 GHA ° ′	MARS Dec ° ′	JUPITER −1.8 GHA ° ′	JUPITER Dec ° ′	SATURN +0.1 GHA ° ′	SATURN Dec ° ′
12	00	289 28.7	189 47.1	N23 19.7	307 43.2	S13 06.7	146 42.3	N15 32.3	194 10.5	N22 34.2
	01	304 31.2	204 46.2	19.6	322 45.1	06.6	161 44.2	32.2	209 12.7	34.2
	02	319 33.6	219 45.3	19.5	337 47.0	06.6	176 46.2	32.0	224 14.8	34.2
	03	334 36.1	234 44.5	. . 19.4	352 48.9	. . 06.5	191 48.2	. . 31.8	239 16.9	. . 34.1
	04	349 38.5	249 43.6	19.3	7 50.8	06.4	206 50.1	31.7	254 19.0	34.1
	05	4 41.0	264 42.7	19.2	22 52.7	06.4	221 52.1	31.5	269 21.1	34.1
	06	19 43.5	279 41.9	N23 19.1	37 54.6	S13 06.3	236 54.1	N15 31.4	284 23.3	N22 34.1
	07	34 45.9	294 41.0	19.0	52 56.5	06.3	251 56.0	31.2	299 25.4	34.1
S	08	49 48.4	309 40.1	18.8	67 58.4	06.2	266 58.0	31.0	314 27.5	34.1
A	09	64 50.9	324 39.2	. . 18.7	83 00.3	. . 06.1	282 00.0	. . 30.9	329 29.6	. . 34.1
T	10	79 53.3	339 38.4	18.6	98 02.2	06.1	297 01.9	30.7	344 31.7	34.1
U	11	94 55.8	354 37.5	18.5	113 04.2	06.0	312 03.9	30.6	359 33.9	34.1
R	12	109 58.3	9 36.6	N23 18.4	128 06.1	S13 06.0	327 05.9	N15 30.4	14 36.0	N22 34.1
D	13	125 00.7	24 35.8	18.2	143 08.0	05.9	342 07.8	30.2	29 38.1	34.1
A	14	140 03.2	39 34.9	18.1	158 09.9	05.9	357 09.8	30.1	44 40.2	34.0
Y	15	155 05.7	54 34.0	. . 18.0	173 11.8	. . 05.8	12 11.8	. . 29.9	59 42.3	. . 34.0
	16	170 08.1	69 33.1	17.9	188 13.7	05.7	27 13.7	29.8	74 44.5	34.0
	17	185 10.6	84 32.3	17.8	203 15.6	05.7	42 15.7	29.6	89 46.6	34.0
	18	200 13.0	99 31.4	N23 17.6	218 17.5	S13 05.6	57 17.7	N15 29.4	104 48.7	N22 34.0
	19	215 15.5	114 30.5	17.5	233 19.4	05.6	72 19.6	29.3	119 50.8	34.0
	20	230 18.0	129 29.7	17.4	248 21.4	05.5	87 21.6	29.1	134 52.9	34.0
	21	245 20.4	144 28.8	. . 17.2	263 23.3	. . 05.5	102 23.6	. . 29.0	149 55.1	. . 34.0
	22	260 22.9	159 27.9	17.1	278 25.2	05.4	117 25.5	28.8	164 57.2	34.0
	23	275 25.4	174 27.1	17.0	293 27.1	05.4	132 27.5	28.6	179 59.3	34.0
13	00	290 27.8	189 26.2	N23 16.8	308 29.0	S13 05.3	147 29.5	N15 28.5	195 01.4	N22 33.9
	01	305 30.3	204 25.3	16.7	323 31.0	05.3	162 31.4	28.3	210 03.5	33.9
	02	320 32.8	219 24.4	16.6	338 32.9	05.2	177 33.4	28.2	225 05.7	33.9
	03	335 35.2	234 23.6	. . 16.4	353 34.8	. . 05.2	192 35.4	. . 28.0	240 07.8	. . 33.9
	04	350 37.7	249 22.7	16.3	8 36.7	05.1	207 37.3	27.8	255 09.9	33.9
	05	5 40.2	264 21.8	16.2	23 38.7	05.1	222 39.3	27.7	270 12.0	33.9
	06	20 42.6	279 21.0	N23 16.0	38 40.6	S13 05.0	237 41.3	N15 27.5	285 14.1	N22 33.9
	07	35 45.1	294 20.1	15.9	53 42.5	05.0	252 43.2	27.4	300 16.3	33.9
	08	50 47.5	309 19.2	15.7	68 44.5	04.9	267 45.2	27.2	315 18.4	33.9
S	09	65 50.0	324 18.4	. . 15.6	83 46.4	. . 04.9	282 47.2	. . 27.0	330 20.5	. . 33.9
U	10	80 52.5	339 17.5	15.4	98 48.3	04.8	297 49.1	26.9	345 22.6	33.9
N	11	95 54.9	354 16.6	15.3	113 50.2	04.8	312 51.1	26.7	0 24.7	33.8
D	12	110 57.4	9 15.7	N23 15.2	128 52.2	S13 04.7	327 53.1	N15 26.6	15 26.9	N22 33.8
A	13	125 59.9	24 14.9	15.0	143 54.1	04.7	342 55.0	26.4	30 29.0	33.8
Y	14	141 02.3	39 14.0	14.9	158 56.1	04.6	357 57.0	26.2	45 31.1	33.8
	15	156 04.8	54 13.1	. . 14.7	173 58.0	. . 04.6	12 59.0	. . 26.1	60 33.2	. . 33.8
	16	171 07.3	69 12.3	14.6	188 59.9	04.5	28 00.9	25.9	75 35.3	33.8
	17	186 09.7	84 11.4	14.4	204 01.9	04.5	43 02.9	25.8	90 37.5	33.8
	18	201 12.2	99 10.5	N23 14.2	219 03.8	S13 04.5	58 04.9	N15 25.6	105 39.6	N22 33.8
	19	216 14.7	114 09.7	14.1	234 05.8	04.4	73 06.8	25.4	120 41.7	33.8
	20	231 17.1	129 08.8	13.9	249 07.7	04.4	88 08.8	25.3	135 43.8	33.8
	21	246 19.6	144 07.9	. . 13.8	264 09.6	. . 04.3	103 10.7	. . 25.1	150 45.9	. . 33.7
	22	261 22.0	159 07.1	13.6	279 11.6	04.3	118 12.7	24.9	165 48.1	33.7
	23	276 24.5	174 06.2	13.5	294 13.5	04.2	133 14.7	24.8	180 50.2	33.7
14	00	291 27.0	189 05.3	N23 13.3	309 15.5	S13 04.2	148 16.6	N15 24.6	195 52.3	N22 33.7
	01	306 29.4	204 04.5	13.1	324 17.4	04.2	163 18.6	24.5	210 54.4	33.7
	02	321 31.9	219 03.6	13.0	339 19.4	04.1	178 20.6	24.3	225 56.6	33.7
	03	336 34.4	234 02.7	. . 12.8	354 21.3	. . 04.1	193 22.5	. . 24.1	240 58.7	. . 33.7
	04	351 36.8	249 01.9	12.6	9 23.3	04.0	208 24.5	24.0	256 00.8	33.7
	05	6 39.3	264 01.0	12.5	24 25.2	04.0	223 26.5	23.8	271 02.9	33.7
	06	21 41.8	279 00.1	N23 12.3	39 27.2	S13 04.0	238 28.4	N15 23.7	286 05.0	N22 33.7
	07	36 44.2	293 59.3	12.1	54 29.1	03.9	253 30.4	23.5	301 07.2	33.7
	08	51 46.7	308 58.4	12.0	69 31.1	03.9	268 32.4	23.3	316 09.3	33.6
M	09	66 49.2	323 57.5	. . 11.8	84 33.1	. . 03.8	283 34.3	. . 23.2	331 11.4	. . 33.6
O	10	81 51.6	338 56.7	11.6	99 35.0	03.8	298 36.3	23.0	346 13.5	33.6
N	11	96 54.1	353 55.8	11.4	114 37.0	03.8	313 38.2	22.8	1 15.6	33.6
D	12	111 56.5	8 54.9	N23 11.3	129 38.9	S13 03.7	328 40.2	N15 22.7	16 17.8	N22 33.6
A	13	126 59.0	23 54.1	11.1	144 40.9	03.7	343 42.2	22.5	31 19.9	33.6
Y	14	142 01.5	38 53.2	10.9	159 42.9	03.7	358 44.1	22.4	46 22.0	33.6
	15	157 03.9	53 52.3	. . 10.7	174 44.8	. . 03.6	13 46.1	. . 22.2	61 24.1	. . 33.6
	16	172 06.4	68 51.5	10.5	189 46.8	03.6	28 48.1	22.0	76 26.3	33.6
	17	187 08.9	83 50.6	10.4	204 48.8	03.6	43 50.0	21.9	91 28.4	33.6
	18	202 11.3	98 49.7	N23 10.2	219 50.7	S13 03.5	58 52.0	N15 21.7	106 30.5	N22 33.5
	19	217 13.8	113 48.9	10.0	234 52.7	03.5	73 54.0	21.5	121 32.6	33.5
	20	232 16.3	128 48.0	09.8	249 54.7	03.5	88 55.9	21.4	136 34.7	33.5
	21	247 18.7	143 47.1	. . 09.6	264 56.6	. . 03.4	103 57.9	. . 21.2	151 36.9	. . 33.5
	22	262 21.2	158 46.3	09.4	279 58.6	03.4	118 59.8	21.1	166 39.0	33.5
	23	277 23.6	173 45.4	09.2	295 00.6	03.4	134 01.8	20.9	181 41.1	33.5
Mer. Pass.	h m	4 37.4	*v* −0.9	*d* 0.1	*v* 1.9	*d* 0.0	*v* 2.0	*d* 0.2	*v* 2.1	*d* 0.0

STARS

Name	SHA ° ′	Dec ° ′
Acamar	315 24.4	S40 17.2
Achernar	335 32.4	S57 12.9
Acrux	173 18.5	S63 07.3
Adhara	255 19.1	S28 58.5
Aldebaran	290 58.6	N16 31.0
Alioth	166 27.4	N55 56.8
Alkaid	153 04.8	N49 18.1
Al Na'ir	27 53.0	S46 56.5
Alnilam	275 54.6	S 1 11.9
Alphard	218 04.1	S 8 40.3
Alphecca	126 17.4	N26 42.3
Alpheratz	357 51.6	N29 06.4
Altair	62 15.5	N 8 52.6
Ankaa	353 23.2	S42 17.0
Antares	112 35.6	S26 26.5
Arcturus	146 02.7	N19 10.0
Atria	107 44.1	S69 02.2
Avior	234 22.0	S59 31.2
Bellatrix	278 40.7	N 6 21.2
Betelgeuse	271 10.1	N 7 24.6
Canopus	264 00.2	S52 41.7
Capella	280 46.5	N46 00.0
Deneb	49 36.5	N45 17.4
Denebola	182 41.7	N14 33.4
Diphda	349 03.6	S17 58.0
Dubhe	194 01.3	N61 44.3
Elnath	278 22.9	N28 36.6
Eltanin	90 49.3	N51 29.4
Enif	33 54.5	N 9 53.4
Fomalhaut	15 32.3	S29 36.1
Gacrux	172 10.0	S57 08.1
Gienah	176 00.5	S17 33.7
Hadar	148 59.1	S60 23.6
Hamal	328 09.7	N23 28.6
Kaus Aust.	83 53.8	S34 23.0
Kochab	137 18.8	N74 08.8
Markab	13 46.0	N15 13.3
Menkar	314 23.4	N 4 06.2
Menkent	148 16.8	S36 23.4
Miaplacidus	221 42.5	S69 43.9
Mirfak	308 52.0	N49 52.2
Nunki	76 07.6	S26 17.6
Peacock	53 30.8	S56 43.4
Pollux	243 37.6	N28 01.2
Procyon	245 08.2	N 5 13.1
Rasalhague	96 13.4	N12 33.5
Regulus	207 52.1	N11 57.2
Rigel	281 19.9	S 8 11.7
Rigil Kent.	140 02.5	S60 51.2
Sabik	102 21.2	S15 43.8
Schedar	349 49.6	N56 33.1
Shaula	96 32.2	S37 06.5
Sirius	258 41.0	S16 43.1
Spica	158 39.5	S11 10.7
Suhail	222 58.7	S43 26.8
Vega	80 43.8	N38 47.2
Zuben'ubi	137 14.0	S16 03.4

	SHA ° ′	Mer. Pass. h m
Venus	258 58.4	11 23
Mars	18 01.2	3 26
Jupiter	217 01.6	14 08
Saturn	264 33.6	10 58

d	h	SUN GHA	SUN Dec	MOON GHA	v	MOON Dec	d	HP
		° ′	° ′	° ′	′	° ′	′	′
12	00	178 37.3	N22 03.4	25 04.5	3.3	S25 38.1	4.3	59.9
	01	193 37.2	03.0	39 26.8	3.3	25 42.4	4.0	59.9
	02	208 37.2	02.7	53 49.1	3.3	25 46.4	3.9	59.9
	03	223 37.1	. . 02.3	68 11.4	3.3	25 50.3	3.6	59.9
	04	238 37.0	02.0	82 33.7	3.2	25 53.9	3.5	59.9
	05	253 36.9	01.7	96 55.9	3.2	25 57.4	3.3	59.9
	06	268 36.8	N22 01.3	111 18.1	3.2	S26 00.7	3.1	59.9
	07	283 36.8	01.0	125 40.3	3.1	26 03.8	2.9	59.9
S	08	298 36.7	00.6	140 02.4	3.1	26 06.7	2.8	59.8
A	09	313 36.6	22 00.3	154 24.5	3.2	26 09.5	2.5	59.8
T	10	328 36.5	21 59.9	168 46.7	3.1	26 12.0	2.4	59.8
U	11	343 36.5	59.6	183 08.8	3.1	26 14.4	2.1	59.8
R	12	358 36.4	N21 59.3	197 30.9	3.1	S26 16.5	2.0	59.8
D	13	13 36.3	58.9	211 53.0	3.0	26 18.5	1.8	59.8
A	14	28 36.2	58.6	226 15.0	3.1	26 20.3	1.6	59.8
Y	15	43 36.1	. . 58.2	240 37.1	3.1	26 21.9	1.4	59.8
	16	58 36.1	57.9	254 59.2	3.1	26 23.3	1.2	59.7
	17	73 36.0	57.5	269 21.3	3.1	26 24.5	1.0	59.7
	18	88 35.9	N21 57.2	283 43.4	3.1	S26 25.5	0.9	59.7
	19	103 35.8	56.8	298 05.5	3.1	26 26.4	0.6	59.7
	20	118 35.7	56.5	312 27.6	3.1	26 27.0	0.5	59.7
	21	133 35.7	. . 56.1	326 49.7	3.2	26 27.5	0.3	59.7
	22	148 35.6	55.8	341 11.9	3.2	26 27.8	0.0	59.6
	23	163 35.5	55.4	355 34.1	3.1	26 27.8	0.1	59.6
13	00	178 35.4	N21 55.1	9 56.2	3.3	S26 27.7	0.3	59.6
	01	193 35.4	54.7	24 18.5	3.2	26 27.4	0.4	59.6
	02	208 35.3	54.4	38 40.7	3.3	26 27.0	0.7	59.6
	03	223 35.2	. . 54.0	53 03.0	3.3	26 26.3	0.9	59.6
	04	238 35.1	53.6	67 25.3	3.3	26 25.4	1.0	59.5
	05	253 35.1	53.3	81 47.6	3.4	26 24.4	1.2	59.5
	06	268 35.0	N21 52.9	96 10.0	3.4	S26 23.2	1.5	59.5
	07	283 34.9	52.6	110 32.4	3.5	26 21.7	1.6	59.5
	08	298 34.8	52.2	124 54.9	3.5	26 20.1	1.7	59.5
S	09	313 34.8	. . 51.9	139 17.4	3.6	26 18.4	2.0	59.4
U	10	328 34.7	51.5	153 40.0	3.6	26 16.4	2.2	59.4
N	11	343 34.6	51.1	168 02.6	3.6	26 14.2	2.3	59.4
D	12	358 34.5	N21 50.8	182 25.2	3.8	S26 11.9	2.5	59.4
A	13	13 34.5	50.4	196 48.0	3.7	26 09.4	2.7	59.4
Y	14	28 34.4	50.1	211 10.7	3.9	26 06.7	2.9	59.3
	15	43 34.3	. . 49.7	225 33.6	3.9	26 03.8	3.0	59.3
	16	58 34.2	49.3	239 56.5	3.9	26 00.8	3.3	59.3
	17	73 34.2	49.0	254 19.4	4.1	25 57.5	3.4	59.3
	18	88 34.1	N21 48.6	268 42.5	4.1	S25 54.1	3.5	59.2
	19	103 34.0	48.2	283 05.6	4.2	25 50.6	3.8	59.2
	20	118 34.0	47.9	297 28.8	4.2	25 46.8	3.9	59.2
	21	133 33.9	. . 47.5	311 52.0	4.3	25 42.9	4.1	59.2
	22	148 33.8	47.1	326 15.3	4.5	25 38.8	4.3	59.1
	23	163 33.7	46.8	340 38.8	4.4	25 34.5	4.4	59.1
14	00	178 33.7	N21 46.4	355 02.2	4.6	S25 30.1	4.6	59.1
	01	193 33.6	46.0	9 25.8	4.7	25 25.5	4.8	59.1
	02	208 33.5	45.7	23 49.5	4.7	25 20.7	4.9	59.0
	03	223 33.5	. . 45.3	38 13.2	4.8	25 15.8	5.1	59.0
	04	238 33.4	44.9	52 37.0	4.9	25 10.7	5.3	59.0
	05	253 33.3	44.5	67 00.9	5.1	25 05.4	5.4	59.0
	06	268 33.2	N21 44.2	81 25.0	5.1	S25 00.0	5.6	58.9
	07	283 33.2	43.8	95 49.1	5.1	24 54.4	5.7	58.9
	08	298 33.1	43.4	110 13.2	5.3	24 48.7	5.9	58.9
M	09	313 33.0	. . 43.1	124 37.5	5.4	24 42.8	6.1	58.9
O	10	328 33.0	42.7	139 01.9	5.5	24 36.7	6.2	58.8
N	11	343 32.9	42.3	153 26.4	5.6	24 30.5	6.4	58.8
D	12	358 32.8	N21 41.9	167 51.0	5.7	S24 24.1	6.5	58.8
A	13	13 32.8	41.5	182 15.7	5.8	24 17.6	6.6	58.7
Y	14	28 32.7	41.2	196 40.5	5.9	24 11.0	6.8	58.7
	15	43 32.6	. . 40.8	211 05.4	5.9	24 04.2	7.0	58.7
	16	58 32.6	40.4	225 30.3	6.1	23 57.2	7.1	58.7
	17	73 32.5	40.0	239 55.4	6.3	23 50.1	7.2	58.6
	18	88 32.4	N21 39.7	254 20.7	6.3	S23 42.9	7.4	58.6
	19	103 32.4	39.3	268 46.0	6.4	23 35.5	7.5	58.6
	20	118 32.3	38.9	283 11.4	6.5	23 28.0	7.7	58.5
	21	133 32.2	. . 38.5	297 36.9	6.7	23 20.3	7.8	58.5
	22	148 32.2	38.1	312 02.6	6.7	23 12.5	7.9	58.5
	23	163 32.1	37.7	326 28.3	6.9	S23 04.6	8.1	58.5
		SD 15.8	*d* 0.4	SD 16.3		16.2		16.0

Lat.	Twilight Naut.	Twilight Civil	Sunrise	Moonrise 12	13	14	15
°	h m	h m	h m	h m	h m	h m	h m
N 72	▭	▭	▭	■	■	■	■
N 70	▭	▭	▭	■	■	■	■
68	▭	▭	▭	■	■	■	■
66	////	////	01 24	■	■	■	{00 09 / 23 36}
64	////	////	02 08	■	23 40	23 20	23 12
62	////	00 24	02 37	21 57	22 37	22 49	22 53
60	////	01 37	02 59	21 14	22 03	22 26	22 37
N 58	////	02 10	03 16	20 45	21 37	22 07	22 24
56	00 23	02 34	03 31	20 23	21 17	21 51	22 12
54	01 29	02 53	03 44	20 05	21 00	21 37	22 02
52	02 00	03 09	03 55	19 49	20 46	21 25	21 53
50	02 22	03 22	04 05	19 36	20 33	21 15	21 45
45	03 02	03 49	04 25	19 08	20 07	20 53	21 28
N 40	03 30	04 10	04 42	18 47	19 46	20 35	21 13
35	03 51	04 27	04 56	18 29	19 29	20 19	21 01
30	04 08	04 41	05 08	18 13	19 14	20 06	20 50
20	04 35	05 04	05 28	17 47	18 49	19 44	20 32
N 10	04 56	05 23	05 46	17 24	18 27	19 24	20 16
0	05 14	05 40	06 02	17 03	18 06	19 06	20 01
S 10	05 30	05 56	06 18	16 42	17 46	18 47	19 46
20	05 44	06 12	06 35	16 20	17 24	18 28	19 30
30	06 00	06 29	06 55	15 54	16 58	18 05	19 11
35	06 07	06 39	07 06	15 39	16 43	17 51	19 00
40	06 16	06 49	07 19	15 21	16 26	17 36	18 47
45	06 25	07 02	07 35	15 00	16 05	17 17	18 32
S 50	06 36	07 16	07 54	14 33	15 38	16 54	18 14
52	06 40	07 23	08 03	14 20	15 25	16 43	18 05
54	06 45	07 30	08 13	14 05	15 11	16 30	17 55
56	06 51	07 38	08 24	13 47	14 53	16 16	17 44
58	06 56	07 48	08 37	13 26	14 32	15 58	17 32
S 60	07 03	07 58	08 52	13 00	14 06	15 37	17 17

Lat.	Sunset	Twilight Civil	Twilight Naut.	Moonset 12	13	14	15
°	h m	h m	h m	h m	h m	h m	h m
N 72	▭	▭	▭	■	■	■	■
N 70	▭	▭	▭	■	■	■	■
68	▭	▭	▭	■	■	■	■
66	22 43	////	////	■	■	■	02 42
64	22 01	////	////	■	■	01 08	03 30
62	21 33	23 35	////	24 40	00 40	02 10	04 00
60	21 11	22 32	////	00 33	01 23	02 45	04 23
N 58	20 54	21 59	////	00 59	01 52	03 10	04 42
56	20 39	21 36	23 38	01 19	02 14	03 30	04 57
54	20 27	21 17	22 40	01 36	02 32	03 46	05 10
52	20 16	21 02	22 10	01 50	02 48	04 00	05 21
50	20 06	20 48	21 48	02 03	03 01	04 13	05 32
45	19 46	20 22	21 09	02 29	03 29	04 38	05 53
N 40	19 29	20 01	20 41	02 50	03 50	04 58	06 10
35	19 15	19 44	20 20	03 07	04 08	05 15	06 24
30	19 03	19 30	20 03	03 22	04 24	05 30	06 37
20	18 43	19 07	19 36	03 48	04 50	05 54	06 58
N 10	18 26	18 48	19 15	04 10	05 13	06 15	07 16
0	18 09	18 32	18 57	04 30	05 34	06 35	07 33
S 10	17 53	18 16	18 42	04 51	05 54	06 55	07 50
20	17 36	18 00	18 27	05 13	06 17	07 16	08 07
30	17 17	17 43	18 12	05 38	06 43	07 40	08 28
35	17 05	17 33	18 04	05 54	06 58	07 54	08 40
40	16 52	17 22	17 56	06 11	07 16	08 10	08 53
45	16 37	17 10	17 47	06 32	07 37	08 29	09 09
S 50	16 18	16 55	17 36	06 59	08 04	08 53	09 29
52	16 09	16 49	17 32	07 11	08 17	09 05	09 38
54	15 59	16 41	17 27	07 26	08 32	09 18	09 49
56	15 48	16 33	17 21	07 44	08 49	09 33	10 00
58	15 35	16 24	17 15	08 05	09 10	09 50	10 14
S 60	15 20	16 14	17 09	08 31	09 37	10 12	10 29

Day	SUN Eqn. of Time 00^h	SUN Eqn. of Time 12^h	SUN Mer. Pass.	MOON Mer. Pass. Upper	MOON Mer. Pass. Lower	Age	Phase
d	m s	m s	h m	h m	h m	d %	
12	05 31	05 34	12 06	23 19	10 47	13 98	
13	05 38	05 42	12 06	24 21	11 50	14 100	○
14	05 45	05 49	12 06	00 21	12 51	15 99	

UT		ARIES	VENUS −3.9		MARS −1.9		JUPITER −1.8		SATURN +0.1		STARS		
d	h	GHA	GHA	Dec	GHA	Dec	GHA	Dec	GHA	Dec	Name	SHA	Dec
		° ′	° ′	° ′	° ′	° ′	° ′	° ′	° ′	° ′		° ′	° ′
15 TUESDAY	00	292 26.1	188 44.5	N23 09.0	310 02.6	S13 03.3	149 03.8	N15 20.7	196 43.2	N22 33.5	Acamar	315 24.3	S40 17.2
	01	307 28.6	203 43.7	08.9	325 04.5	03.3	164 05.7	20.6	211 45.4	33.5	Achernar	335 32.4	S57 12.9
	02	322 31.0	218 42.8	08.7	340 06.5	03.3	179 07.7	20.4	226 47.5	33.5	Acrux	173 18.6	S63 07.3
	03	337 33.5	233 41.9	08.5	355 08.5	03.2	194 09.7	20.2	241 49.6	33.5	Adhara	255 19.1	S28 58.5
	04	352 36.0	248 41.1	08.3	10 10.5	03.2	209 11.6	20.1	256 51.7	33.4	Aldebaran	290 58.6	N16 31.0
	05	7 38.4	263 40.2	08.1	25 12.4	03.2	224 13.6	19.9	271 53.8	33.4			
	06	22 40.9	278 39.3	N23 07.9	40 14.4	S13 03.1	239 15.6	N15 19.8	286 56.0	N22 33.4	Alioth	166 27.4	N55 56.8
	07	37 43.4	293 38.5	07.7	55 16.4	03.1	254 17.5	19.6	301 58.1	33.4	Alkaid	153 04.8	N49 18.1
	08	52 45.8	308 37.6	07.5	70 18.4	03.1	269 19.5	19.4	317 00.2	33.4	Al Na'ir	27 53.0	S46 56.5
	09	67 48.3	323 36.8	07.3	85 20.4	03.1	284 21.4	19.3	332 02.3	33.4	Alnilam	275 54.6	S 1 11.9
	10	82 50.8	338 35.9	07.1	100 22.4	03.0	299 23.4	19.1	347 04.5	33.4	Alphard	218 04.1	S 8 40.3
	11	97 53.2	353 35.0	06.9	115 24.3	03.0	314 25.4	18.9	2 06.6	33.4			
	12	112 55.7	8 34.2	N23 06.7	130 26.3	S13 03.0	329 27.3	N15 18.8	17 08.7	N22 33.4	Alphecca	126 17.4	N26 42.3
	13	127 58.1	23 33.3	06.5	145 28.3	03.0	344 29.3	18.6	32 10.8	33.4	Alpheratz	357 51.5	N29 06.4
	14	143 00.6	38 32.4	06.3	160 30.3	02.9	359 31.3	18.5	47 12.9	33.3	Altair	62 15.5	N 8 52.6
	15	158 03.1	53 31.6	06.0	175 32.3	02.9	14 33.2	18.3	62 15.1	33.3	Ankaa	353 23.1	S42 17.0
	16	173 05.5	68 30.7	05.8	190 34.3	02.9	29 35.2	18.1	77 17.2	33.3	Antares	112 35.6	S26 26.5
	17	188 08.0	83 29.8	05.6	205 36.3	02.9	44 37.1	18.0	92 19.3	33.3			
	18	203 10.5	98 29.0	N23 05.4	220 38.3	S13 02.8	59 39.1	N15 17.8	107 21.4	N22 33.3	Arcturus	146 02.7	N19 10.0
	19	218 12.9	113 28.1	05.2	235 40.3	02.8	74 41.1	17.6	122 23.6	33.3	Atria	107 44.1	S69 02.3
	20	233 15.4	128 27.3	05.0	250 42.3	02.8	89 43.0	17.5	137 25.7	33.3	Avior	234 22.0	S59 31.2
	21	248 17.9	143 26.4	04.8	265 44.3	02.8	104 45.0	17.3	152 27.8	33.3	Bellatrix	278 40.7	N 6 21.3
	22	263 20.3	158 25.5	04.6	280 46.3	02.7	119 46.9	17.2	167 29.9	33.3	Betelgeuse	271 10.1	N 7 24.6
	23	278 22.8	173 24.7	04.3	295 48.3	02.7	134 48.9	17.0	182 32.0	33.3			
16 WEDNESDAY	00	293 25.3	188 23.8	N23 04.1	310 50.3	S13 02.7	149 50.9	N15 16.8	197 34.2	N22 33.2	Canopus	264 00.2	S52 41.7
	01	308 27.7	203 22.9	03.9	325 52.3	02.7	164 52.8	16.7	212 36.3	33.2	Capella	280 46.4	N46 00.0
	02	323 30.2	218 22.1	03.7	340 54.3	02.7	179 54.8	16.5	227 38.4	33.2	Deneb	49 36.4	N45 17.4
	03	338 32.6	233 21.2	03.5	355 56.3	02.6	194 56.8	16.3	242 40.5	33.2	Denebola	182 41.7	N14 33.4
	04	353 35.1	248 20.4	03.2	10 58.3	02.6	209 58.7	16.2	257 42.7	33.2	Diphda	349 03.6	S17 58.0
	05	8 37.6	263 19.5	03.0	26 00.3	02.6	225 00.7	16.0	272 44.8	33.2			
	06	23 40.0	278 18.6	N23 02.8	41 02.3	S13 02.6	240 02.6	N15 15.8	287 46.9	N22 33.2	Dubhe	194 01.4	N61 44.3
	07	38 42.5	293 17.8	02.6	56 04.3	02.6	255 04.6	15.7	302 49.0	33.2	Elnath	278 22.8	N28 36.6
	08	53 45.0	308 16.9	02.3	71 06.3	02.5	270 06.6	15.5	317 51.2	33.2	Eltanin	90 49.3	N51 29.4
	09	68 47.4	323 16.1	02.1	86 08.3	02.5	285 08.5	15.4	332 53.3	33.1	Enif	33 54.5	N 9 53.4
	10	83 49.9	338 15.2	01.9	101 10.4	02.5	300 10.5	15.2	347 55.4	33.1	Fomalhaut	15 32.2	S29 36.1
	11	98 52.4	353 14.3	01.6	116 12.4	02.5	315 12.4	15.0	2 57.5	33.1			
	12	113 54.8	8 13.5	N23 01.4	131 14.4	S13 02.5	330 14.4	N15 14.9	17 59.6	N22 33.1	Gacrux	172 10.0	S57 08.1
	13	128 57.3	23 12.6	01.2	146 16.4	02.5	345 16.4	14.7	33 01.8	33.1	Gienah	176 00.5	S17 33.7
	14	143 59.8	38 11.8	00.9	161 18.4	02.4	0 18.3	14.5	48 03.9	33.1	Hadar	148 59.2	S60 23.6
	15	159 02.2	53 10.9	00.7	176 20.4	02.4	15 20.3	14.4	63 06.0	33.1	Hamal	328 09.7	N23 28.6
	16	174 04.7	68 10.0	00.5	191 22.5	02.4	30 22.2	14.2	78 08.1	33.1	Kaus Aust.	83 53.8	S34 23.1
	17	189 07.1	83 09.2	00.2	206 24.5	02.4	45 24.2	14.0	93 10.3	33.1			
	18	204 09.6	98 08.3	N23 00.0	221 26.5	S13 02.4	60 26.2	N15 13.9	108 12.4	N22 33.1	Kochab	137 18.9	N74 08.8
	19	219 12.1	113 07.5	22 59.7	236 28.5	02.4	75 28.1	13.7	123 14.5	33.0	Markab	13 46.0	N15 13.3
	20	234 14.5	128 06.6	59.5	251 30.6	02.4	90 30.1	13.5	138 16.6	33.0	Menkar	314 23.4	N 4 06.2
	21	249 17.0	143 05.7	59.3	266 32.6	02.4	105 32.0	13.4	153 18.8	33.0	Menkent	148 16.8	S36 23.4
	22	264 19.5	158 04.9	59.0	281 34.6	02.3	120 34.0	13.2	168 20.9	33.0	Miaplacidus	221 42.5	S69 43.9
	23	279 21.9	173 04.0	58.8	296 36.6	02.3	135 36.0	13.1	183 23.0	33.0			
17 THURSDAY	00	294 24.4	188 03.2	N22 58.5	311 38.7	S13 02.3	150 37.9	N15 12.9	198 25.1	N22 33.0	Mirfak	308 51.9	N49 52.2
	01	309 26.9	203 02.3	58.3	326 40.7	02.3	165 39.9	12.7	213 27.3	33.0	Nunki	76 07.6	S26 17.6
	02	324 29.3	218 01.5	58.0	341 42.7	02.3	180 41.8	12.6	228 29.4	33.0	Peacock	53 30.8	S56 43.4
	03	339 31.8	233 00.6	57.8	356 44.8	02.3	195 43.8	12.4	243 31.5	33.0	Pollux	243 37.6	N28 01.2
	04	354 34.2	247 59.7	57.5	11 46.8	02.3	210 45.8	12.2	258 33.6	32.9	Procyon	245 08.2	N 5 13.1
	05	9 36.7	262 58.9	57.3	26 48.8	02.3	225 47.7	12.1	273 35.8	32.9			
	06	24 39.2	277 58.0	N22 57.0	41 50.9	S13 02.3	240 49.7	N15 11.9	288 37.9	N22 32.9	Rasalhague	96 13.4	N12 33.5
	07	39 41.6	292 57.2	56.8	56 52.9	02.3	255 51.6	11.7	303 40.0	32.9	Regulus	207 52.1	N11 57.2
	08	54 44.1	307 56.3	56.5	71 54.9	02.3	270 53.6	11.6	318 42.1	32.9	Rigel	281 19.8	S 8 11.7
	09	69 46.6	322 55.5	56.2	86 57.0	02.2	285 55.6	11.4	333 44.2	32.9	Rigil Kent.	140 02.5	S60 51.2
	10	84 49.0	337 54.6	56.0	101 59.0	02.2	300 57.5	11.2	348 46.4	32.9	Sabik	102 21.2	S15 43.8
	11	99 51.5	352 53.7	55.7	117 01.1	02.2	315 59.5	11.1	3 48.5	32.9			
	12	114 54.0	7 52.9	N22 55.5	132 03.1	S13 02.2	331 01.4	N15 10.9	18 50.6	N22 32.9	Schedar	349 49.6	N56 33.1
	13	129 56.4	22 52.0	55.2	147 05.2	02.2	346 03.4	10.8	33 52.7	32.9	Shaula	96 32.2	S37 06.5
	14	144 58.9	37 51.2	54.9	162 07.2	02.2	1 05.4	10.6	48 54.9	32.8	Sirius	258 41.0	S16 43.1
	15	160 01.4	52 50.3	54.7	177 09.3	02.2	16 07.3	10.4	63 57.0	32.8	Spica	158 39.5	S11 10.7
	16	175 03.8	67 49.5	54.4	192 11.3	02.2	31 09.3	10.3	78 59.1	32.8	Suhail	222 58.7	S43 26.8
	17	190 06.3	82 48.6	54.1	207 13.3	02.2	46 11.2	10.1	94 01.2	32.8			
	18	205 08.7	97 47.8	N22 53.9	222 15.4	S13 02.2	61 13.2	N15 09.9	109 03.4	N22 32.8	Vega	80 43.8	N38 47.2
	19	220 11.2	112 46.9	53.6	237 17.5	02.2	76 15.2	09.8	124 05.5	32.8	Zuben'ubi	137 14.0	S16 03.4
	20	235 13.7	127 46.0	53.3	252 19.5	02.2	91 17.1	09.6	139 07.6	32.8		SHA	Mer. Pass.
	21	250 16.1	142 45.2	53.1	267 21.6	02.2	106 19.1	09.4	154 09.7	32.8		° ′	h m
	22	265 18.6	157 44.3	52.8	282 23.6	02.2	121 21.0	09.3	169 11.9	32.8	Venus	254 58.6	11 27
	23	280 21.1	172 43.5	52.5	297 25.7	02.2	136 23.0	09.1	184 14.0	32.7	Mars	17 25.0	3 16
		h m									Jupiter	216 25.6	13 59
Mer. Pass.		4 25.6	*v* −0.9	*d* 0.2	*v* 2.0	*d* 0.0	*v* 2.0	*d* 0.2	*v* 2.1	*d* 0.0	Saturn	264 08.9	10 48

Day	UT (d h)	SUN GHA (° ′)	SUN Dec (° ′)	MOON GHA (° ′)	MOON v (′)	MOON Dec (° ′)	MOON d (′)	MOON HP (′)
TUESDAY	15 00	178 32.0	N21 37.4	340 54.2	7.0	S22 56.5	8.2	58.4
	01	193 32.0	37.0	355 20.2	7.1	22 48.3	8.3	58.4
	02	208 31.9	36.6	9 46.3	7.2	22 40.0	8.4	58.4
	03	223 31.8	. . 36.2	24 12.5	7.3	22 31.6	8.6	58.3
	04	238 31.8	35.8	38 38.8	7.4	22 23.0	8.7	58.3
	05	253 31.7	35.4	53 05.2	7.6	22 14.3	8.8	58.3
	06	268 31.6	N21 35.0	67 31.8	7.6	S22 05.5	9.0	58.2
	07	283 31.6	34.7	81 58.4	7.8	21 56.5	9.1	58.2
	08	298 31.5	34.3	96 25.2	7.9	21 47.4	9.1	58.2
	09	313 31.4	. . 33.9	110 52.1	8.0	21 38.3	9.3	58.1
	10	328 31.4	33.5	125 19.1	8.1	21 29.0	9.4	58.1
	11	343 31.3	33.1	139 46.2	8.3	21 19.6	9.6	58.1
	12	358 31.2	N21 32.7	154 13.5	8.4	S21 10.0	9.6	58.1
	13	13 31.2	32.3	168 40.9	8.4	21 00.4	9.7	58.0
	14	28 31.1	31.9	183 08.3	8.6	20 50.7	9.9	58.0
	15	43 31.1	. . 31.5	197 35.9	8.7	20 40.8	9.9	58.0
	16	58 31.0	31.1	212 03.6	8.9	20 30.9	10.1	57.9
	17	73 30.9	30.7	226 31.5	8.9	20 20.8	10.1	57.9
	18	88 30.9	N21 30.3	240 59.4	9.0	S20 10.7	10.3	57.9
	19	103 30.8	29.9	255 27.4	9.2	20 00.4	10.4	57.8
	20	118 30.7	29.6	269 55.6	9.3	19 50.0	10.4	57.8
	21	133 30.7	. . 29.2	284 23.9	9.4	19 39.6	10.6	57.8
	22	148 30.6	28.8	298 52.3	9.5	19 29.0	10.6	57.7
	23	163 30.6	28.4	313 20.8	9.7	19 18.4	10.7	57.7
WEDNESDAY	16 00	178 30.5	N21 28.0	327 49.5	9.7	S19 07.7	10.9	57.7
	01	193 30.4	27.6	342 18.2	9.9	18 56.8	10.9	57.6
	02	208 30.4	27.2	356 47.1	9.9	18 45.9	11.0	57.6
	03	223 30.3	. . 26.8	11 16.0	10.1	18 34.9	11.1	57.6
	04	238 30.3	26.4	25 45.1	10.2	18 23.8	11.1	57.5
	05	253 30.2	26.0	40 14.3	10.3	18 12.7	11.3	57.5
	06	268 30.1	N21 25.6	54 43.6	10.4	S18 01.4	11.3	57.5
	07	283 30.1	25.1	69 13.0	10.6	17 50.1	11.4	57.4
	08	298 30.0	24.7	83 42.6	10.6	17 38.7	11.5	57.4
	09	313 30.0	. . 24.3	98 12.2	10.8	17 27.2	11.6	57.4
	10	328 29.9	23.9	112 42.0	10.8	17 15.6	11.6	57.3
	11	343 29.8	23.5	127 11.8	11.0	17 04.0	11.7	57.3
	12	358 29.8	N21 23.1	141 41.8	11.1	S16 52.3	11.8	57.3
	13	13 29.7	22.7	156 11.9	11.1	16 40.5	11.8	57.2
	14	28 29.7	22.3	170 42.0	11.3	16 28.7	12.0	57.2
	15	43 29.6	. . 21.9	185 12.3	11.4	16 16.7	11.9	57.2
	16	58 29.5	21.5	199 42.7	11.5	16 04.8	12.1	57.1
	17	73 29.5	21.1	214 13.2	11.6	15 52.7	12.1	57.1
	18	88 29.4	N21 20.7	228 43.8	11.7	S15 40.6	12.2	57.1
	19	103 29.4	20.3	243 14.5	11.8	15 28.4	12.2	57.0
	20	118 29.3	19.8	257 45.3	11.9	15 16.2	12.3	57.0
	21	133 29.3	. . 19.4	272 16.2	12.0	15 03.9	12.4	57.0
	22	148 29.2	19.0	286 47.2	12.1	14 51.5	12.4	56.9
	23	163 29.1	18.6	301 18.3	12.2	14 39.1	12.4	56.9
THURSDAY	17 00	178 29.1	N21 18.2	315 49.5	12.3	S14 26.7	12.5	56.9
	01	193 29.0	17.8	330 20.8	12.4	14 14.2	12.6	56.8
	02	208 29.0	17.4	344 52.2	12.5	14 01.6	12.6	56.8
	03	223 28.9	. . 16.9	359 23.7	12.5	13 49.0	12.7	56.8
	04	238 28.9	16.5	13 55.2	12.7	13 36.3	12.7	56.7
	05	253 28.8	16.1	28 26.9	12.8	13 23.6	12.8	56.7
	06	268 28.8	N21 15.7	42 58.7	12.8	S13 10.8	12.8	56.7
	07	283 28.7	15.3	57 30.5	13.0	12 58.0	12.8	56.6
	08	298 28.7	14.9	72 02.5	13.0	12 45.2	12.9	56.6
	09	313 28.6	. . 14.4	86 34.5	13.1	12 32.3	13.0	56.6
	10	328 28.5	14.0	101 06.6	13.2	12 19.3	13.0	56.6
	11	343 28.5	13.6	115 38.8	13.3	12 06.3	13.0	56.5
	12	358 28.4	N21 13.2	130 11.1	13.4	S11 53.3	13.0	56.5
	13	13 28.4	12.7	144 43.5	13.4	11 40.3	13.1	56.5
	14	28 28.3	12.3	159 15.9	13.6	11 27.2	13.1	56.4
	15	43 28.3	. . 11.9	173 48.5	13.6	11 14.1	13.2	56.4
	16	58 28.2	11.5	188 21.1	13.7	11 00.9	13.2	56.4
	17	73 28.2	11.1	202 53.8	13.8	10 47.7	13.2	56.3
	18	88 28.1	N21 10.6	217 26.6	13.8	S10 34.5	13.3	56.3
	19	103 28.1	10.2	231 59.4	14.0	10 21.2	13.3	56.3
	20	118 28.0	09.8	246 32.4	14.0	10 07.9	13.3	56.2
	21	133 28.0	. . 09.3	261 05.4	14.1	9 54.6	13.3	56.2
	22	148 27.9	08.9	275 38.5	14.1	9 41.3	13.4	56.2
	23	163 27.9	08.5	290 11.6	14.3	S 9 27.9	13.4	56.1
		SD 15.8	*d* 0.4	SD 15.8		15.6		15.4

Lat. (°)	Twilight Naut. (h m)	Twilight Civil (h m)	Sunrise (h m)	Moonrise 15 (h m)	Moonrise 16 (h m)	Moonrise 17 (h m)	Moonrise 18 (h m)
N 72	□	□	□	■	■	(00 28 / 23 43)	23 12
N 70	□	□	□	■	(00 58 / 23 59)	23 29	23 07
68	□	□	□	■	(00 08 / 23 37)	23 18	23 02
66	////	////	01 38	(00 09 / 23 36)	23 20	23 08	22 58
64	////	////	02 17	23 12	23 06	23 00	22 55
62	////	00 54	02 44	22 53	22 54	22 53	22 52
60	////	01 47	03 05	22 37	22 44	22 47	22 50
N 58	////	02 17	03 21	22 24	22 35	22 42	22 48
56	00 49	02 40	03 36	22 12	22 27	22 37	22 46
54	01 38	02 58	03 48	22 02	22 20	22 33	22 44
52	02 06	03 13	03 58	21 53	22 13	22 29	22 42
50	02 27	03 26	04 08	21 45	22 08	22 25	22 41
45	03 05	03 52	04 28	21 28	21 55	22 18	22 37
N 40	03 32	04 12	04 44	21 13	21 45	22 11	22 35
35	03 53	04 29	04 58	21 01	21 36	22 06	22 32
30	04 10	04 42	05 09	20 50	21 28	22 01	22 30
20	04 37	05 05	05 29	20 32	21 14	21 52	22 26
N 10	04 57	05 24	05 46	20 16	21 02	21 44	22 23
0	05 14	05 40	06 02	20 01	20 51	21 37	22 20
S 10	05 30	05 56	06 18	19 46	20 40	21 30	22 17
20	05 44	06 11	06 35	19 30	20 28	21 22	22 14
30	05 59	06 28	06 54	19 11	20 14	21 14	22 10
35	06 06	06 38	07 05	19 00	20 06	21 08	22 08
40	06 15	06 48	07 18	18 47	19 57	21 03	22 06
45	06 24	07 00	07 33	18 32	19 46	20 56	22 03
S 50	06 34	07 14	07 51	18 14	19 32	20 48	22 00
52	06 38	07 21	08 00	18 05	19 26	20 44	21 58
54	06 43	07 28	08 09	17 55	19 19	20 40	21 56
56	06 48	07 35	08 20	17 44	19 12	20 35	21 54
58	06 54	07 44	08 33	17 32	19 03	20 30	21 52
S 60	07 00	07 54	08 48	17 17	18 53	20 24	21 50

Lat. (°)	Sunset (h m)	Twilight Civil (h m)	Twilight Naut. (h m)	Moonset 15 (h m)	Moonset 16 (h m)	Moonset 17 (h m)	Moonset 18 (h m)
N 72	□	□	□	■	■	06 01	08 20
N 70	□	□	□	■	03 47	06 28	08 32
68	□	□	□	■	04 36	06 49	08 41
66	22 31	////	////	02 42	05 07	07 04	08 49
64	21 53	////	////	03 30	05 30	07 17	08 55
62	21 27	23 12	////	04 00	05 48	07 28	09 01
60	21 06	22 23	////	04 23	06 03	07 37	09 06
N 58	20 49	21 53	////	04 42	06 15	07 45	09 10
56	20 35	21 31	23 17	04 57	06 26	07 52	09 13
54	20 23	21 13	22 32	05 10	06 36	07 58	09 17
52	20 13	20 58	22 04	05 21	06 44	08 04	09 20
50	20 03	20 45	21 43	05 32	06 52	08 09	09 23
45	19 44	20 19	21 06	05 53	07 08	08 20	09 28
N 40	19 28	19 59	20 39	06 10	07 21	08 29	09 33
35	19 14	19 43	20 18	06 24	07 32	08 36	09 37
30	19 02	19 29	20 02	06 37	07 41	08 43	09 41
20	18 43	19 07	19 35	06 58	07 58	08 54	09 47
N 10	18 25	18 48	19 15	07 16	08 12	09 04	09 53
0	18 10	18 32	18 58	07 33	08 25	09 14	09 58
S 10	17 54	18 16	18 42	07 50	08 39	09 23	10 03
20	17 37	18 01	18 28	08 07	08 53	09 33	10 09
30	17 18	17 44	18 13	08 28	09 09	09 44	10 15
35	17 07	17 35	18 06	08 40	09 18	09 50	10 18
40	16 54	17 24	17 58	08 53	09 28	09 57	10 22
45	16 39	17 12	17 49	09 09	09 40	10 05	10 27
S 50	16 21	16 58	17 39	09 29	09 55	10 15	10 32
52	16 13	16 52	17 34	09 38	10 02	10 20	10 34
54	16 03	16 45	17 30	09 49	10 10	10 25	10 37
56	15 52	16 37	17 24	10 00	10 18	10 30	10 40
58	15 40	16 28	17 19	10 14	10 28	10 37	10 43
S 60	15 25	16 18	17 13	10 29	10 38	10 44	10 47

Day (d)	SUN Eqn. of Time 00^h (m s)	SUN Eqn. of Time 12^h (m s)	SUN Mer. Pass. (h m)	MOON Mer. Pass. Upper (h m)	MOON Mer. Pass. Lower (h m)	MOON Age (d)	MOON Phase (%)
15	05 52	05 55	12 06	01 19	13 47	16	96
16	05 58	06 01	12 06	02 13	14 38	17	91
17	06 04	06 06	12 06	03 02	15 26	18	84

UT d h	ARIES GHA	VENUS −3.9 GHA	VENUS Dec	MARS −2.0 GHA	MARS Dec	JUPITER −1.7 GHA	JUPITER Dec	SATURN +0.1 GHA	SATURN Dec
	° ′	° ′	° ′	° ′	° ′	° ′	° ′	° ′	° ′
18 00 FRIDAY	295 23.5	187 42.6	N22 52.2	312 27.7	S13 02.2	151 25.0	N15 08.9	199 16.1	N22 32.7
01	310 26.0	202 41.8	52.0	327 29.8	02.2	166 26.9	08.8	214 18.2	32.7
02	325 28.5	217 40.9	51.7	342 31.9	02.2	181 28.9	08.6	229 20.4	32.7
03	340 30.9	232 40.1	. . 51.4	357 33.9	. . 02.2	196 30.8	. . 08.4	244 22.5	. . 32.7
04	355 33.4	247 39.2	51.1	12 36.0	02.2	211 32.8	08.3	259 24.6	32.7
05	10 35.9	262 38.4	50.8	27 38.0	02.2	226 34.7	08.1	274 26.7	32.7
06	25 38.3	277 37.5	N22 50.6	42 40.1	S13 02.2	241 36.7	N15 07.9	289 28.9	N22 32.7
07	40 40.8	292 36.7	50.3	57 42.2	02.2	256 38.7	07.8	304 31.0	32.7
08	55 43.2	307 35.8	50.0	72 44.2	02.2	271 40.6	07.6	319 33.1	32.6
09	70 45.7	322 34.9	. . 49.7	87 46.3	. . 02.2	286 42.6	. . 07.4	334 35.2	. . 32.6
10	85 48.2	337 34.1	49.4	102 48.4	02.2	301 44.5	07.3	349 37.4	32.6
11	100 50.6	352 33.2	49.1	117 50.5	02.2	316 46.5	07.1	4 39.5	32.6
12	115 53.1	7 32.4	N22 48.8	132 52.5	S13 02.2	331 48.5	N15 07.0	19 41.6	N22 32.6
13	130 55.6	22 31.5	48.5	147 54.6	02.2	346 50.4	06.8	34 43.8	32.6
14	145 58.0	37 30.7	48.3	162 56.7	02.2	1 52.4	06.6	49 45.9	32.6
15	161 00.5	52 29.8	. . 48.0	177 58.7	. . 02.2	16 54.3	. . 06.5	64 48.0	. . 32.6
16	176 03.0	67 29.0	47.7	193 00.8	02.3	31 56.3	06.3	79 50.1	32.6
17	191 05.4	82 28.1	47.4	208 02.9	02.3	46 58.2	06.1	94 52.3	32.5
18	206 07.9	97 27.3	N22 47.1	223 05.0	S13 02.3	62 00.2	N15 06.0	109 54.4	N22 32.5
19	221 10.3	112 26.4	46.8	238 07.1	02.3	77 02.2	05.8	124 56.5	32.5
20	236 12.8	127 25.6	46.5	253 09.1	02.3	92 04.1	05.6	139 58.6	32.5
21	251 15.3	142 24.7	. . 46.2	268 11.2	. . 02.3	107 06.1	. . 05.5	155 00.8	. . 32.5
22	266 17.7	157 23.9	45.9	283 13.3	02.3	122 08.0	05.3	170 02.9	32.5
23	281 20.2	172 23.0	45.6	298 15.4	02.3	137 10.0	05.1	185 05.0	32.5
19 00 SATURDAY	296 22.7	187 22.2	N22 45.3	313 17.5	S13 02.3	152 11.9	N15 05.0	200 07.1	N22 32.5
01	311 25.1	202 21.3	45.0	328 19.6	02.3	167 13.9	04.8	215 09.3	32.5
02	326 27.6	217 20.5	44.7	343 21.7	02.3	182 15.9	04.6	230 11.4	32.4
03	341 30.1	232 19.6	. . 44.4	358 23.8	. . 02.4	197 17.8	. . 04.5	245 13.5	. . 32.4
04	356 32.5	247 18.8	44.0	13 25.9	02.4	212 19.8	04.3	260 15.6	32.4
05	11 35.0	262 17.9	43.7	28 27.9	02.4	227 21.7	04.1	275 17.8	32.4
06	26 37.5	277 17.1	N22 43.4	43 30.0	S13 02.4	242 23.7	N15 04.0	290 19.9	N22 32.4
07	41 39.9	292 16.3	43.1	58 32.1	02.4	257 25.6	03.8	305 22.0	32.4
08	56 42.4	307 15.4	42.8	73 34.2	02.4	272 27.6	03.6	320 24.1	32.4
09	71 44.8	322 14.6	. . 42.5	88 36.3	. . 02.4	287 29.6	. . 03.5	335 26.3	. . 32.4
10	86 47.3	337 13.7	42.2	103 38.4	02.5	302 31.5	03.3	350 28.4	32.4
11	101 49.8	352 12.9	41.9	118 40.5	02.5	317 33.5	03.1	5 30.5	32.3
12	116 52.2	7 12.0	N22 41.5	133 42.6	S13 02.5	332 35.4	N15 03.0	20 32.7	N22 32.3
13	131 54.7	22 11.2	41.2	148 44.7	02.5	347 37.4	02.8	35 34.8	32.3
14	146 57.2	37 10.3	40.9	163 46.8	02.5	2 39.3	02.6	50 36.9	32.3
15	161 59.6	52 09.5	. . 40.6	178 48.9	. . 02.5	17 41.3	. . 02.5	65 39.0	. . 32.3
16	177 02.1	67 08.6	40.3	193 51.1	02.6	32 43.3	02.3	80 41.2	32.3
17	192 04.6	82 07.8	39.9	208 53.2	02.6	47 45.2	02.1	95 43.3	32.3
18	207 07.0	97 06.9	N22 39.6	223 55.3	S13 02.6	62 47.2	N15 02.0	110 45.4	N22 32.3
19	222 09.5	112 06.1	39.3	238 57.4	02.6	77 49.1	01.8	125 47.5	32.3
20	237 11.9	127 05.3	39.0	253 59.5	02.6	92 51.1	01.6	140 49.7	32.2
21	252 14.4	142 04.4	. . 38.6	269 01.6	. . 02.7	107 53.0	. . 01.5	155 51.8	. . 32.2
22	267 16.9	157 03.6	38.3	284 03.7	02.7	122 55.0	01.3	170 53.9	32.2
23	282 19.3	172 02.7	38.0	299 05.8	02.7	137 56.9	01.1	185 56.1	32.2
20 00 SUNDAY	297 21.8	187 01.9	N22 37.6	314 08.0	S13 02.7	152 58.9	N15 01.0	200 58.2	N22 32.2
01	312 24.3	202 01.0	37.3	329 10.1	02.7	168 00.9	00.8	216 00.3	32.2
02	327 26.7	217 00.2	37.0	344 12.2	02.8	183 02.8	00.6	231 02.4	32.2
03	342 29.2	231 59.3	. . 36.6	359 14.3	. . 02.8	198 04.8	. . 00.5	246 04.6	. . 32.2
04	357 31.7	246 58.5	36.3	14 16.4	02.8	213 06.7	00.3	261 06.7	32.2
05	12 34.1	261 57.7	36.0	29 18.6	02.8	228 08.7	00.1	276 08.8	32.1
06	27 36.6	276 56.8	N22 35.6	44 20.7	S13 02.9	243 10.6	N15 00.0	291 10.9	N22 32.1
07	42 39.1	291 56.0	35.3	59 22.8	02.9	258 12.6	14 59.8	306 13.1	32.1
08	57 41.5	306 55.1	34.9	74 24.9	02.9	273 14.5	59.6	321 15.2	32.1
09	72 44.0	321 54.3	. . 34.6	89 27.1	. . 02.9	288 16.5	. . 59.5	336 17.3	. . 32.1
10	87 46.4	336 53.5	34.3	104 29.2	03.0	303 18.5	59.3	351 19.5	32.1
11	102 48.9	351 52.6	33.9	119 31.3	03.0	318 20.4	59.1	6 21.6	32.1
12	117 51.4	6 51.8	N22 33.6	134 33.5	S13 03.0	333 22.4	N14 59.0	21 23.7	N22 32.1
13	132 53.8	21 50.9	33.2	149 35.6	03.0	348 24.3	58.8	36 25.8	32.0
14	147 56.3	36 50.1	32.9	164 37.7	03.1	3 26.3	58.6	51 28.0	32.0
15	162 58.8	51 49.2	. . 32.5	179 39.9	. . 03.1	18 28.2	. . 58.5	66 30.1	. . 32.0
16	178 01.2	66 48.4	32.2	194 42.0	03.1	33 30.2	58.3	81 32.2	32.0
17	193 03.7	81 47.6	31.8	209 44.1	03.1	48 32.1	58.1	96 34.4	32.0
18	208 06.2	96 46.7	N22 31.5	224 46.3	S13 03.2	63 34.1	N14 57.9	111 36.5	N22 32.0
19	223 08.6	111 45.9	31.1	239 48.4	03.2	78 36.1	57.8	126 38.6	32.0
20	238 11.1	126 45.0	30.8	254 50.6	03.2	93 38.0	57.6	141 40.7	32.0
21	253 13.6	141 44.2	. . 30.4	269 52.7	. . 03.3	108 40.0	. . 57.4	156 42.9	. . 32.0
22	268 16.0	156 43.4	30.1	284 54.8	03.3	123 41.9	57.3	171 45.0	31.9
23	283 18.5	171 42.5	29.7	299 57.0	03.3	138 43.9	57.1	186 47.1	31.9
Mer. Pass.	h m 4 13.8	*v* −0.8	*d* 0.3	*v* 2.1	*d* 0.0	*v* 2.0	*d* 0.2	*v* 2.1	*d* 0.0

STARS Name	SHA	Dec
	° ′	° ′
Acamar	315 24.3	S40 17.2
Achernar	335 32.3	S57 12.9
Acrux	173 18.6	S63 07.3
Adhara	255 19.1	S28 58.4
Aldebaran	290 58.6	N16 31.0
Alioth	166 27.4	N55 56.8
Alkaid	153 04.9	N49 18.1
Al Na'ir	27 53.0	S46 56.5
Alnilam	275 54.6	S 1 11.9
Alphard	218 04.1	S 8 40.3
Alphecca	126 17.4	N26 42.3
Alpheratz	357 51.5	N29 06.4
Altair	62 15.5	N 8 52.6
Ankaa	353 23.1	S42 17.0
Antares	112 35.6	S26 26.5
Arcturus	146 02.8	N19 10.0
Atria	107 44.2	S69 02.3
Avior	234 22.0	S59 31.2
Bellatrix	278 40.7	N 6 21.3
Betelgeuse	271 10.1	N 7 24.6
Canopus	264 00.2	S52 41.7
Capella	280 46.4	N46 00.0
Deneb	49 36.4	N45 17.4
Denebola	182 41.7	N14 33.4
Diphda	349 03.6	S17 57.9
Dubhe	194 01.4	N61 44.3
Elnath	278 22.8	N28 36.6
Eltanin	90 49.3	N51 29.4
Enif	33 54.5	N 9 53.4
Fomalhaut	15 32.2	S29 36.1
Gacrux	172 10.1	S57 08.1
Gienah	176 00.5	S17 33.7
Hadar	148 59.2	S60 23.6
Hamal	328 09.7	N23 28.6
Kaus Aust.	83 53.8	S34 23.1
Kochab	137 18.9	N74 08.8
Markab	13 45.9	N15 13.3
Menkar	314 23.3	N 4 06.2
Menkent	148 16.9	S36 23.4
Miaplacidus	221 42.5	S69 43.9
Mirfak	308 51.9	N49 52.2
Nunki	76 07.6	S26 17.6
Peacock	53 30.8	S56 43.4
Pollux	243 37.6	N28 01.2
Procyon	245 08.2	N 5 13.1
Rasalhague	96 13.4	N12 33.5
Regulus	207 52.1	N11 57.2
Rigel	281 19.8	S 8 11.7
Rigil Kent.	140 02.6	S60 51.2
Sabik	102 21.2	S15 43.8
Schedar	349 49.5	N56 33.1
Shaula	96 32.2	S37 06.5
Sirius	258 41.0	S16 43.1
Spica	158 39.5	S11 10.7
Suhail	222 58.7	S43 26.7
Vega	80 43.8	N38 47.2
Zuben'ubi	137 14.0	S16 03.4
	SHA	Mer. Pass
	° ′	h m
Venus	250 59.5	11 31
Mars	16 54.8	3 06
Jupiter	215 49.3	13 49
Saturn	263 44.5	10 38

UT	SUN GHA	SUN Dec	MOON GHA	*v*	MOON Dec	*d*	HP
d h	° ′	° ′	° ′	′	° ′	′	′
18 00	178 27.8	N21 08.1	304 44.9	14.3	S 9 14.5	13.4	56.1
01	193 27.8	07.6	319 18.2	14.3	9 01.1	13.4	56.1
02	208 27.7	07.2	333 51.5	14.5	8 47.7	13.5	56.1
03	223 27.7	. . 06.8	348 25.0	14.5	8 34.2	13.5	56.0
04	238 27.6	06.3	2 58.5	14.5	8 20.7	13.5	56.0
05	253 27.6	05.9	17 32.0	14.7	8 07.2	13.5	56.0
06	268 27.5	N21 05.5	32 05.7	14.7	S 7 53.7	13.6	55.9
07	283 27.5	05.0	46 39.4	14.7	7 40.1	13.5	55.9
08	298 27.4	04.6	61 13.1	14.8	7 26.6	13.6	55.9
F 09	313 27.4	. . 04.2	75 46.9	14.9	7 13.0	13.6	55.9
R 10	328 27.3	03.7	90 20.8	15.0	6 59.4	13.6	55.8
I 11	343 27.3	03.3	104 54.8	15.0	6 45.8	13.6	55.8
D 12	358 27.2	N21 02.9	119 28.8	15.0	S 6 32.2	13.6	55.8
A 13	13 27.2	02.4	134 02.8	15.1	6 18.6	13.7	55.7
Y 14	28 27.1	02.0	148 36.9	15.2	6 04.9	13.6	55.7
15	43 27.1	. . 01.5	163 11.1	15.2	5 51.3	13.7	55.7
16	58 27.0	01.1	177 45.3	15.3	5 37.6	13.6	55.7
17	73 27.0	00.7	192 19.6	15.3	5 24.0	13.7	55.6
18	88 26.9	N21 00.2	206 53.9	15.4	S 5 10.3	13.7	55.6
19	103 26.9	20 59.8	221 28.3	15.4	4 56.6	13.7	55.6
20	118 26.9	59.3	236 02.7	15.4	4 42.9	13.7	55.6
21	133 26.8	. . 58.9	250 37.1	15.5	4 29.2	13.7	55.5
22	148 26.8	58.5	265 11.6	15.6	4 15.5	13.7	55.5
23	163 26.7	58.0	279 46.2	15.6	4 01.8	13.7	55.5
19 00	178 26.7	N20 57.6	294 20.8	15.6	S 3 48.1	13.7	55.4
01	193 26.6	57.1	308 55.4	15.7	3 34.4	13.7	55.4
02	208 26.6	56.7	323 30.1	15.7	3 20.7	13.7	55.4
03	223 26.5	. . 56.2	338 04.8	15.8	3 07.0	13.7	55.4
04	238 26.5	55.8	352 39.6	15.8	2 53.3	13.7	55.3
05	253 26.4	55.3	7 14.4	15.8	2 39.6	13.7	55.3
06	268 26.4	N20 54.9	21 49.2	15.9	S 2 25.9	13.7	55.3
07	283 26.4	54.4	36 24.1	15.9	2 12.2	13.7	55.3
S 08	298 26.3	54.0	50 59.0	15.9	1 58.5	13.7	55.3
A 09	313 26.3	. . 53.6	65 33.9	16.0	1 44.8	13.6	55.2
T 10	328 26.2	53.1	80 08.9	16.0	1 31.2	13.7	55.2
U 11	343 26.2	52.7	94 43.9	16.0	1 17.5	13.7	55.2
R 12	358 26.1	N20 52.2	109 18.9	16.1	S 1 03.8	13.6	55.2
D 13	13 26.1	51.7	123 54.0	16.1	0 50.2	13.7	55.1
A 14	28 26.1	51.3	138 29.1	16.1	0 36.5	13.6	55.1
Y 15	43 26.0	. . 50.8	153 04.2	16.1	0 22.9	13.7	55.1
16	58 26.0	50.4	167 39.3	16.2	S 0 09.2	13.6	55.1
17	73 25.9	49.9	182 14.5	16.1	N 0 04.4	13.6	55.0
18	88 25.9	N20 49.5	196 49.6	16.2	N 0 18.0	13.6	55.0
19	103 25.9	49.0	211 24.8	16.3	0 31.6	13.6	55.0
20	118 25.8	48.6	226 00.1	16.2	0 45.2	13.6	55.0
21	133 25.8	. . 48.1	240 35.3	16.3	0 58.8	13.5	55.0
22	148 25.7	47.7	255 10.6	16.2	1 12.3	13.5	54.9
23	163 25.7	47.2	269 45.8	16.3	1 25.8	13.6	54.9
20 00	178 25.7	N20 46.7	284 21.1	16.3	N 1 39.4	13.5	54.9
01	193 25.6	46.3	298 56.4	16.3	1 52.9	13.5	54.9
02	208 25.6	45.8	313 31.7	16.4	2 06.4	13.5	54.9
03	223 25.5	. . 45.4	328 07.1	16.3	2 19.9	13.4	54.8
04	238 25.5	44.9	342 42.4	16.4	2 33.3	13.5	54.8
05	253 25.5	44.4	357 17.8	16.3	2 46.8	13.4	54.8
06	268 25.4	N20 44.0	11 53.1	16.4	N 3 00.2	13.4	54.8
07	283 25.4	43.5	26 28.5	16.3	3 13.6	13.3	54.8
08	298 25.3	43.0	41 03.8	16.4	3 26.9	13.4	54.8
S 09	313 25.3	. . 42.6	55 39.2	16.4	3 40.3	13.3	54.7
U 10	328 25.3	42.1	70 14.6	16.4	3 53.6	13.3	54.7
N 11	343 25.2	41.6	84 50.0	16.4	4 06.9	13.3	54.7
D 12	358 25.2	N20 41.2	99 25.4	16.4	N 4 20.2	13.3	54.7
A 13	13 25.2	40.7	114 00.8	16.4	4 33.5	13.2	54.7
Y 14	28 25.1	40.2	128 36.2	16.3	4 46.7	13.2	54.7
15	43 25.1	. . 39.8	143 11.5	16.4	4 59.9	13.2	54.6
16	58 25.0	39.3	157 46.9	16.4	5 13.1	13.2	54.6
17	73 25.0	38.8	172 22.3	16.4	5 26.3	13.1	54.6
18	88 25.0	N20 38.4	186 57.7	16.4	N 5 39.4	13.1	54.6
19	103 24.9	37.9	201 33.1	16.3	5 52.5	13.1	54.6
20	118 24.9	37.4	216 08.4	16.4	6 05.6	13.0	54.6
21	133 24.9	. . 37.0	230 43.8	16.3	6 18.6	13.1	54.6
22	148 24.8	36.5	245 19.1	16.4	6 31.7	12.9	54.5
23	163 24.8	36.0	259 54.5	16.3	N 6 44.6	13.0	54.5
	SD 15.8	*d* 0.5	SD	15.2		15.0	14.9

Lat.	Twilight Naut.	Twilight Civil	Sunrise	Moonrise 18	Moonrise 19	Moonrise 20	Moonrise 21
°	h m	h m	h m	h m	h m	h m	h m
N 72	▭	▭	▭	23 12	22 45	22 17	21 45
N 70	▭	▭	▭	23 07	22 47	22 26	22 03
68	////	////	00 40	23 02	22 48	22 34	22 18
66	////	////	01 51	22 58	22 49	22 40	22 30
64	////	////	02 26	22 55	22 50	22 45	22 40
62	////	01 12	02 51	22 52	22 51	22 50	22 49
60	////	01 56	03 11	22 50	22 52	22 54	22 56
N 58	////	02 24	03 27	22 48	22 53	22 57	23 03
56	01 06	02 46	03 40	22 46	22 53	23 01	23 09
54	01 47	03 03	03 52	22 44	22 54	23 04	23 14
52	02 13	03 18	04 02	22 42	22 54	23 06	23 19
50	02 33	03 30	04 11	22 41	22 55	23 09	23 24
45	03 09	03 55	04 31	22 37	22 56	23 14	23 33
N 40	03 35	04 15	04 46	22 35	22 57	23 18	23 41
35	03 56	04 31	05 00	22 32	22 57	23 22	23 48
30	04 12	04 44	05 11	22 30	22 58	23 26	23 54
20	04 38	05 06	05 30	22 26	22 59	23 32	24 05
N 10	04 58	05 25	05 47	22 23	23 00	23 37	24 14
0	05 15	05 41	06 03	22 20	23 01	23 42	24 23
S 10	05 30	05 56	06 18	22 17	23 02	23 47	24 32
20	05 44	06 11	06 34	22 14	23 04	23 52	24 41
30	05 58	06 27	06 53	22 10	23 05	23 59	24 52
35	06 05	06 36	07 04	22 08	23 06	24 02	00 02
40	06 13	06 46	07 16	22 06	23 07	24 06	00 06
45	06 22	06 58	07 31	22 03	23 08	24 11	00 11
S 50	06 31	07 12	07 48	22 00	23 09	24 17	00 17
52	06 36	07 18	07 57	21 58	23 09	24 19	00 19
54	06 40	07 25	08 06	21 56	23 10	24 22	00 22
56	06 45	07 32	08 16	21 54	23 11	24 26	00 26
58	06 50	07 41	08 29	21 52	23 11	24 29	00 29
S 60	06 56	07 50	08 42	21 50	23 12	24 33	00 33

Lat.	Sunset	Twilight Civil	Twilight Naut.	Moonset 18	Moonset 19	Moonset 20	Moonset 21
°	h m	h m	h m	h m	h m	h m	h m
N 72	▭	▭	▭	08 20	10 21	12 16	14 15
N 70	▭	▭	▭	08 32	10 23	12 10	13 59
68	23 21	////	////	08 41	10 25	12 05	13 46
66	22 18	////	////	08 49	10 26	12 01	13 35
64	21 44	////	////	08 55	10 28	11 57	13 26
62	21 20	22 55	////	09 01	10 29	11 54	13 19
60	21 01	22 13	////	09 06	10 30	11 51	13 12
N 58	20 45	21 46	////	09 10	10 30	11 49	13 07
56	20 31	21 25	23 02	09 13	10 31	11 47	13 02
54	20 20	21 08	22 23	09 17	10 32	11 45	12 57
52	20 09	20 54	21 58	09 20	10 32	11 43	12 53
50	20 00	20 42	21 38	09 23	10 33	11 42	12 50
45	19 41	20 17	21 02	09 28	10 34	11 38	12 42
N 40	19 26	19 57	20 36	09 33	10 35	11 35	12 35
35	19 13	19 41	20 16	09 37	10 36	11 33	12 29
30	19 01	19 28	20 00	09 41	10 37	11 31	12 24
20	18 42	19 06	19 34	09 47	10 38	11 27	12 16
N 10	18 25	18 48	19 14	09 53	10 39	11 24	12 08
0	18 10	18 32	18 58	09 58	10 40	11 21	12 01
S 10	17 55	18 17	18 43	10 03	10 41	11 18	11 55
20	17 38	18 02	18 29	10 09	10 42	11 15	11 47
30	17 20	17 46	18 15	10 15	10 43	11 11	11 39
35	17 09	17 36	18 07	10 18	10 44	11 09	11 34
40	16 57	17 26	18 00	10 22	10 45	11 06	11 29
45	16 42	17 15	17 51	10 27	10 45	11 04	11 22
S 50	16 25	17 01	17 42	10 32	10 46	11 00	11 15
52	16 16	16 55	17 37	10 34	10 47	10 59	11 11
54	16 07	16 48	17 33	10 37	10 47	10 57	11 07
56	15 56	16 41	17 28	10 40	10 48	10 55	11 03
58	15 44	16 32	17 23	10 43	10 49	10 53	10 59
S 60	15 31	16 23	17 17	10 47	10 49	10 51	10 53

Day	SUN Eqn. of Time 00^h	SUN Eqn. of Time 12^h	SUN Mer. Pass.	MOON Mer. Pass. Upper	MOON Mer. Pass. Lower	MOON Age	MOON Phase
d	m s	m s	h m	h m	h m	d	%
18	06 09	06 11	12 06	03 48	16 09	19	76
19	06 13	06 15	12 06	04 30	16 51	20	67
20	06 17	06 19	12 06	05 11	17 31	21	58

UT	ARIES	VENUS −3.9		MARS −2.1		JUPITER −1.7		SATURN +0.1		STARS		
	GHA	GHA	Dec	GHA	Dec	GHA	Dec	GHA	Dec	Name	SHA	Dec
d h	° ′	° ′	° ′	° ′	° ′	° ′	° ′	° ′	° ′		° ′	° ′
21 00 (MONDAY)	298 20.9	186 41.7	N22 29.3	314 59.1	S13 03.4	153 45.8	N14 56.9	201 49.3	N22 31.9	Acamar	315 24.3	S40 17.2
01	313 23.4	201 40.9	29.0	330 01.3	03.4	168 47.8	56.8	216 51.4	31.9	Achernar	335 32.3	S57 12.9
02	328 25.9	216 40.0	28.6	345 03.4	03.4	183 49.7	56.6	231 53.5	31.9	Acrux	173 18.6	S63 07.3
03	343 28.3	231 39.2	28.3	0 05.6	03.5	198 51.7	56.4	246 55.6	31.9	Adhara	255 19.1	S28 58.4
04	358 30.8	246 38.3	27.9	15 07.7	03.5	213 53.6	56.3	261 57.8	31.9	Aldebaran	290 58.6	N16 31.0
05	13 33.3	261 37.5	27.5	30 09.9	03.5	228 55.6	56.1	276 59.9	31.9			
06	28 35.7	276 36.7	N22 27.2	45 12.0	S13 03.6	243 57.6	N14 55.9	292 02.0	N22 31.8	Alioth	166 27.4	N55 56.8
07	43 38.2	291 35.8	26.8	60 14.2	03.6	258 59.5	55.8	307 04.2	31.8	Alkaid	153 04.9	N49 18.1
08	58 40.7	306 35.0	26.4	75 16.4	03.6	274 01.5	55.6	322 06.3	31.8	Al Na'ir	27 52.9	S46 56.6
09	73 43.1	321 34.2	26.1	90 18.5	03.7	289 03.4	55.4	337 08.4	31.8	Alnilam	275 54.6	S 1 11.9
10	88 45.6	336 33.3	25.7	105 20.7	03.7	304 05.4	55.3	352 10.5	31.8	Alphard	218 04.1	S 8 40.3
11	103 48.0	351 32.5	25.3	120 22.8	03.7	319 07.3	55.1	7 12.7	31.8			
12	118 50.5	6 31.7	N22 24.9	135 25.0	S13 03.8	334 09.3	N14 54.9	22 14.8	N22 31.8	Alphecca	126 17.4	N26 42.4
13	133 53.0	21 30.8	24.6	150 27.2	03.8	349 11.2	54.8	37 16.9	31.8	Alpheratz	357 51.5	N29 06.4
14	148 55.4	36 30.0	24.2	165 29.3	03.9	4 13.2	54.6	52 19.1	31.8	Altair	62 15.5	N 8 52.6
15	163 57.9	51 29.2	23.8	180 31.5	03.9	19 15.1	54.4	67 21.2	31.7	Ankaa	353 23.1	S42 17.0
16	179 00.4	66 28.3	23.4	195 33.7	03.9	34 17.1	54.2	82 23.3	31.7	Antares	112 35.6	S26 26.5
17	194 02.8	81 27.5	23.1	210 35.8	04.0	49 19.0	54.1	97 25.4	31.7			
18	209 05.3	96 26.7	N22 22.7	225 38.0	S13 04.0	64 21.0	N14 53.9	112 27.6	N22 31.7	Arcturus	146 02.8	N19 10.0
19	224 07.8	111 25.8	22.3	240 40.2	04.1	79 23.0	53.7	127 29.7	31.7	Atria	107 44.2	S69 02.3
20	239 10.2	126 25.0	21.9	255 42.3	04.1	94 24.9	53.6	142 31.8	31.7	Avior	234 22.0	S59 31.2
21	254 12.7	141 24.2	21.5	270 44.5	04.1	109 26.9	53.4	157 34.0	31.7	Bellatrix	278 40.7	N 6 21.3
22	269 15.2	156 23.3	21.2	285 46.7	04.2	124 28.8	53.2	172 36.1	31.7	Betelgeuse	271 10.0	N 7 24.6
23	284 17.6	171 22.5	20.8	300 48.9	04.2	139 30.8	53.1	187 38.2	31.6			
22 00 (TUESDAY)	299 20.1	186 21.7	N22 20.4	315 51.0	S13 04.3	154 32.7	N14 52.9	202 40.4	N22 31.6	Canopus	264 00.1	S52 41.6
01	314 22.5	201 20.8	20.0	330 53.2	04.3	169 34.7	52.7	217 42.5	31.6	Capella	280 46.4	N46 00.0
02	329 25.0	216 20.0	19.6	345 55.4	04.4	184 36.6	52.6	232 44.6	31.6	Deneb	49 36.4	N45 17.5
03	344 27.5	231 19.2	19.2	0 57.6	04.4	199 38.6	52.4	247 46.7	31.6	Denebola	182 41.7	N14 33.4
04	359 29.9	246 18.3	18.8	15 59.8	04.5	214 40.5	52.2	262 48.9	31.6	Diphda	349 03.6	S17 57.9
05	14 32.4	261 17.5	18.4	31 01.9	04.5	229 42.5	52.1	277 51.0	31.6			
06	29 34.9	276 16.7	N22 18.0	46 04.1	S13 04.5	244 44.4	N14 51.9	292 53.1	N22 31.6	Dubhe	194 01.4	N61 44.3
07	44 37.3	291 15.8	17.6	61 06.3	04.6	259 46.4	51.7	307 55.3	31.5	Elnath	278 22.8	N28 36.6
08	59 39.8	306 15.0	17.2	76 08.5	04.6	274 48.3	51.5	322 57.4	31.5	Eltanin	90 49.3	N51 29.4
09	74 42.3	321 14.2	16.9	91 10.7	04.7	289 50.3	51.4	337 59.5	31.5	Enif	33 54.5	N 9 53.4
10	89 44.7	336 13.4	16.5	106 12.9	04.7	304 52.3	51.2	353 01.7	31.5	Fomalhaut	15 32.2	S29 36.1
11	104 47.2	351 12.5	16.1	121 15.1	04.8	319 54.2	51.0	8 03.8	31.5			
12	119 49.7	6 11.7	N22 15.7	136 17.3	S13 04.8	334 56.2	N14 50.9	23 05.9	N22 31.5	Gacrux	172 10.1	S57 08.1
13	134 52.1	21 10.9	15.3	151 19.5	04.9	349 58.1	50.7	38 08.0	31.5	Gienah	176 00.5	S17 33.6
14	149 54.6	36 10.0	14.8	166 21.7	04.9	5 00.1	50.5	53 10.2	31.5	Hadar	148 59.2	S60 23.6
15	164 57.0	51 09.2	14.4	181 23.9	05.0	20 02.0	50.4	68 12.3	31.5	Hamal	328 09.7	N23 28.6
16	179 59.5	66 08.4	14.0	196 26.1	05.0	35 04.0	50.2	83 14.4	31.4	Kaus Aust.	83 53.8	S34 23.1
17	195 02.0	81 07.6	13.6	211 28.3	05.1	50 05.9	50.0	98 16.6	31.4			
18	210 04.4	96 06.7	N22 13.2	226 30.5	S13 05.1	65 07.9	N14 49.9	113 18.7	N22 31.4	Kochab	137 19.0	N74 08.8
19	225 06.9	111 05.9	12.8	241 32.7	05.2	80 09.8	49.7	128 20.8	31.4	Markab	13 45.9	N15 13.4
20	240 09.4	126 05.1	12.4	256 34.9	05.2	95 11.8	49.5	143 23.0	31.4	Menkar	314 23.3	N 4 06.2
21	255 11.8	141 04.3	12.0	271 37.1	05.3	110 13.7	49.3	158 25.1	31.4	Menkent	148 16.9	S36 23.4
22	270 14.3	156 03.4	11.6	286 39.3	05.3	125 15.7	49.2	173 27.2	31.4	Miaplacidus	221 42.5	S69 43.9
23	285 16.8	171 02.6	11.2	301 41.5	05.4	140 17.6	49.0	188 29.4	31.4			
23 00 (WEDNESDAY)	300 19.2	186 01.8	N22 10.8	316 43.7	S13 05.4	155 19.6	N14 48.8	203 31.5	N22 31.3	Mirfak	308 51.9	N49 52.2
01	315 21.7	201 01.0	10.3	331 45.9	05.5	170 21.5	48.7	218 33.6	31.3	Nunki	76 07.6	S26 17.6
02	330 24.1	216 00.1	09.9	346 48.1	05.6	185 23.5	48.5	233 35.7	31.3	Peacock	53 30.7	S56 43.4
03	345 26.6	230 59.3	09.5	1 50.3	05.6	200 25.4	48.3	248 37.9	31.3	Pollux	243 37.6	N28 01.2
04	0 29.1	245 58.5	09.1	16 52.5	05.7	215 27.4	48.2	263 40.0	31.3	Procyon	245 08.2	N 5 13.1
05	15 31.5	260 57.7	08.7	31 54.7	05.7	230 29.3	48.0	278 42.1	31.3			
06	30 34.0	275 56.8	N22 08.3	46 57.0	S13 05.8	245 31.3	N14 47.8	293 44.3	N22 31.3	Rasalhague	96 13.4	N12 33.5
07	45 36.5	290 56.0	07.8	61 59.2	05.8	260 33.3	47.6	308 46.4	31.3	Regulus	207 52.1	N11 57.2
08	60 38.9	305 55.2	07.4	77 01.4	05.9	275 35.2	47.5	323 48.5	31.2	Rigel	281 19.8	S 8 11.7
09	75 41.4	320 54.4	07.0	92 03.6	06.0	290 37.2	47.3	338 50.7	31.2	Rigil Kent.	140 02.6	S60 51.2
10	90 43.9	335 53.5	06.6	107 05.8	06.0	305 39.1	47.1	353 52.8	31.2	Sabik	102 21.2	S15 43.8
11	105 46.3	350 52.7	06.1	122 08.1	06.1	320 41.1	47.0	8 54.9	31.2			
12	120 48.8	5 51.9	N22 05.7	137 10.3	S13 06.1	335 43.0	N14 46.8	23 57.1	N22 31.2	Schedar	349 49.5	N56 33.1
13	135 51.3	20 51.1	05.3	152 12.5	06.2	350 45.0	46.6	38 59.2	31.2	Shaula	96 32.2	S37 06.5
14	150 53.7	35 50.3	04.9	167 14.7	06.2	5 46.9	46.5	54 01.3	31.2	Sirius	258 41.0	S16 43.1
15	165 56.2	50 49.4	04.4	182 17.0	06.3	20 48.9	46.3	69 03.5	31.2	Spica	158 39.5	S11 10.7
16	180 58.6	65 48.6	04.0	197 19.2	06.4	35 50.8	46.1	84 05.6	31.1	Suhail	222 58.7	S43 26.7
17	196 01.1	80 47.8	03.6	212 21.4	06.4	50 52.8	45.9	99 07.7	31.1			
18	211 03.6	95 47.0	N22 03.1	227 23.7	S13 06.5	65 54.7	N14 45.8	114 09.9	N22 31.1	Vega	80 43.9	N38 47.3
19	226 06.0	110 46.2	02.7	242 25.9	06.6	80 56.7	45.6	129 12.0	31.1	Zuben'ubi	137 14.0	S16 03.4
20	241 08.5	125 45.3	02.3	257 28.1	06.6	95 58.6	45.4	144 14.1	31.1		SHA	Mer. Pass.
21	256 11.0	140 44.5	01.8	272 30.4	06.7	111 00.6	45.3	159 16.3	31.1		° ′	h m
22	271 13.4	155 43.7	01.4	287 32.6	06.7	126 02.5	45.1	174 18.4	31.1	Venus	247 01.6	11 35
23	286 15.9	170 42.9	00.9	302 34.8	06.8	141 04.5	44.9	189 20.5	31.1	Mars	16 31.0	2 56
Mer. Pass.	h m 4 02.0	*v* −0.8	*d* 0.4	*v* 2.2	*d* 0.0	*v* 2.0	*d* 0.2	*v* 2.1	*d* 0.0	Jupiter	215 12.6	13 40
										Saturn	263 20.3	10 28

	UT	SUN GHA	SUN Dec	MOON GHA	v	MOON Dec	d	HP
	d h	° ′	° ′	° ′	′	° ′	′	′
MONDAY	21 00	178 24.8	N20 35.5	274 29.8	16.3	N 6 57.6	12.9	54.5
	01	193 24.7	35.1	289 05.1	16.3	7 10.5	12.9	54.5
	02	208 24.7	34.6	303 40.4	16.3	7 23.4	12.9	54.5
	03	223 24.7	. . 34.1	318 15.7	16.3	7 36.3	12.8	54.5
	04	238 24.6	33.6	332 51.0	16.2	7 49.1	12.8	54.5
	05	253 24.6	33.2	347 26.2	16.3	8 01.9	12.7	54.5
	06	268 24.6	N20 32.7	2 01.5	16.2	N 8 14.6	12.7	54.4
	07	283 24.5	32.2	16 36.7	16.2	8 27.3	12.7	54.4
	08	298 24.5	31.7	31 11.9	16.2	8 40.0	12.7	54.4
	09	313 24.5	. . 31.3	45 47.1	16.2	8 52.7	12.6	54.4
	10	328 24.4	30.8	60 22.3	16.1	9 05.3	12.5	54.4
	11	343 24.4	30.3	74 57.4	16.1	9 17.8	12.5	54.4
	12	358 24.4	N20 29.8	89 32.5	16.1	N 9 30.3	12.5	54.4
	13	13 24.3	29.3	104 07.6	16.1	9 42.8	12.5	54.4
	14	28 24.3	28.9	118 42.7	16.0	9 55.3	12.4	54.4
	15	43 24.3	. . 28.4	133 17.7	16.0	10 07.7	12.3	54.4
	16	58 24.3	27.9	147 52.7	16.0	10 20.0	12.4	54.3
	17	73 24.2	27.4	162 27.7	16.0	10 32.4	12.3	54.3
	18	88 24.2	N20 26.9	177 02.7	15.9	N10 44.7	12.2	54.3
	19	103 24.2	26.4	191 37.6	15.9	10 56.9	12.2	54.3
	20	118 24.1	25.9	206 12.5	15.9	11 09.1	12.1	54.3
	21	133 24.1	. . 25.5	220 47.4	15.8	11 21.2	12.1	54.3
	22	148 24.1	25.0	235 22.2	15.8	11 33.3	12.1	54.3
	23	163 24.0	24.5	249 57.0	15.8	11 45.4	12.0	54.3
TUESDAY	22 00	178 24.0	N20 24.0	264 31.8	15.7	N11 57.4	11.9	54.3
	01	193 24.0	23.5	279 06.5	15.7	12 09.3	12.0	54.3
	02	208 24.0	23.0	293 41.2	15.6	12 21.3	11.8	54.3
	03	223 23.9	. . 22.5	308 15.8	15.7	12 33.1	11.8	54.3
	04	238 23.9	22.0	322 50.5	15.5	12 44.9	11.8	54.3
	05	253 23.9	21.6	337 25.0	15.6	12 56.7	11.7	54.3
	06	268 23.9	N20 21.1	351 59.6	15.5	N13 08.4	11.7	54.3
	07	283 23.8	20.6	6 34.1	15.4	13 20.1	11.6	54.3
	08	298 23.8	20.1	21 08.5	15.4	13 31.7	11.5	54.3
	09	313 23.8	. . 19.6	35 42.9	15.4	13 43.2	11.5	54.2
	10	328 23.8	19.1	50 17.3	15.3	13 54.7	11.5	54.2
	11	343 23.7	18.6	64 51.6	15.3	14 06.2	11.4	54.2
	12	358 23.7	N20 18.1	79 25.9	15.2	N14 17.6	11.3	54.2
	13	13 23.7	17.6	94 00.1	15.2	14 28.9	11.3	54.2
	14	28 23.6	17.1	108 34.3	15.1	14 40.2	11.2	54.2
	15	43 23.6	. . 16.6	123 08.4	15.1	14 51.4	11.1	54.2
	16	58 23.6	16.1	137 42.5	15.1	15 02.5	11.2	54.2
	17	73 23.6	15.6	152 16.6	14.9	15 13.7	11.0	54.2
	18	88 23.6	N20 15.1	166 50.5	15.0	N15 24.7	11.0	54.2
	19	103 23.5	14.6	181 24.5	14.9	15 35.7	10.9	54.2
	20	118 23.5	14.1	195 58.4	14.8	15 46.6	10.9	54.2
	21	133 23.5	. . 13.6	210 32.2	14.8	15 57.5	10.8	54.2
	22	148 23.5	13.1	225 06.0	14.7	16 08.3	10.7	54.2
	23	163 23.4	12.6	239 39.7	14.7	16 19.0	10.7	54.2
WEDNESDAY	23 00	178 23.4	N20 12.1	254 13.4	14.6	N16 29.7	10.6	54.2
	01	193 23.4	11.6	268 47.0	14.5	16 40.3	10.5	54.2
	02	208 23.4	11.1	283 20.5	14.5	16 50.8	10.5	54.2
	03	223 23.3	. . 10.6	297 54.0	14.5	17 01.3	10.4	54.2
	04	238 23.3	10.1	312 27.5	14.4	17 11.7	10.3	54.2
	05	253 23.3	09.6	327 00.9	14.3	17 22.0	10.3	54.2
	06	268 23.3	N20 09.1	341 34.2	14.2	N17 32.3	10.2	54.2
	07	283 23.3	08.6	356 07.4	14.2	17 42.5	10.1	54.3
	08	298 23.2	08.1	10 40.6	14.2	17 52.6	10.1	54.3
	09	313 23.2	. . 07.6	25 13.8	14.1	18 02.7	10.0	54.3
	10	328 23.2	07.1	39 46.9	14.0	18 12.7	9.9	54.3
	11	343 23.2	06.6	54 19.9	13.9	18 22.6	9.8	54.3
	12	358 23.2	N20 06.1	68 52.8	13.9	N18 32.4	9.8	54.3
	13	13 23.1	05.5	83 25.7	13.9	18 42.2	9.7	54.3
	14	28 23.1	05.0	97 58.6	13.7	18 51.9	9.6	54.3
	15	43 23.1	. . 04.5	112 31.3	13.7	19 01.5	9.5	54.3
	16	58 23.1	04.0	127 04.0	13.7	19 11.0	9.5	54.3
	17	73 23.1	03.5	141 36.7	13.5	19 20.5	9.4	54.3
	18	88 23.0	N20 03.0	156 09.2	13.5	N19 29.9	9.3	54.3
	19	103 23.0	02.5	170 41.7	13.5	19 39.2	9.2	54.3
	20	118 23.0	02.0	185 14.2	13.3	19 48.4	9.1	54.3
	21	133 23.0	. . 01.4	199 46.5	13.3	19 57.5	9.1	54.3
	22	148 23.0	00.9	214 18.8	13.2	20 06.6	9.0	54.3
	23	163 23.0	00.4	228 51.0	13.2	N20 15.6	8.9	54.3
		SD 15.8	d 0.5	SD 14.8		14.8		14.8

Lat.	Twilight Naut.	Twilight Civil	Sunrise	Moonrise 21	Moonrise 22	Moonrise 23	Moonrise 24
°	h m	h m	h m	h m	h m	h m	h m
N 72	□	□	□	21 45	20 54	□	□
N 70	□	□	□	22 03	21 32	20 17	□
68	////	////	01 10	22 18	21 58	21 26	□
66	////	////	02 03	22 30	22 19	22 03	21 33
64	////	////	02 35	22 40	22 35	22 29	22 23
62	////	01 28	02 58	22 49	22 49	22 50	22 54
60	////	02 06	03 17	22 56	23 00	23 07	23 18
N 58	////	02 32	03 32	23 03	23 10	23 21	23 37
56	01 21	02 52	03 45	23 09	23 19	23 33	23 53
54	01 56	03 08	03 56	23 14	23 27	23 44	24 06
52	02 20	03 22	04 06	23 19	23 34	23 53	24 18
50	02 38	03 34	04 15	23 24	23 41	24 02	00 02
45	03 13	03 59	04 34	23 33	23 55	24 20	00 20
N 40	03 39	04 18	04 49	23 41	24 06	00 06	00 35
35	03 58	04 33	05 02	23 48	24 16	00 16	00 48
30	04 14	04 46	05 13	23 54	24 25	00 25	00 59
20	04 39	05 08	05 32	24 05	00 05	00 40	01 18
N 10	04 59	05 25	05 48	24 14	00 14	00 53	01 35
0	05 15	05 41	06 03	24 23	00 23	01 05	01 50
S 10	05 30	05 55	06 18	24 32	00 32	01 18	02 06
20	05 43	06 10	06 34	24 41	00 41	01 31	02 23
30	05 57	06 26	06 52	24 52	00 52	01 47	02 43
35	06 04	06 35	07 02	00 02	00 59	01 56	02 54
40	06 11	06 45	07 14	00 06	01 06	02 06	03 07
45	06 20	06 56	07 28	00 11	01 14	02 18	03 23
S 50	06 29	07 09	07 45	00 17	01 25	02 33	03 43
52	06 33	07 15	07 53	00 19	01 30	02 40	03 52
54	06 37	07 21	08 02	00 22	01 35	02 48	04 02
56	06 42	07 28	08 12	00 26	01 41	02 57	04 14
58	06 47	07 36	08 24	00 29	01 47	03 07	04 28
S 60	06 52	07 46	08 37	00 33	01 55	03 18	04 44

Lat.	Sunset	Twilight Civil	Twilight Naut.	Moonset 21	Moonset 22	Moonset 23	Moonset 24
°	h m	h m	h m	h m	h m	h m	h m
N 72	□	□	□	14 15	16 34	□	□
N 70	□	□	□	13 59	15 58	18 44	□
68	22 56	////	////	13 46	15 33	17 36	□
66	22 06	////	////	13 35	15 13	17 00	19 08
64	21 35	////	////	13 26	14 58	16 35	18 19
62	21 13	22 40	////	13 19	14 45	16 15	17 48
60	20 54	22 04	////	13 12	14 35	15 59	17 25
N 58	20 39	21 39	////	13 07	14 25	15 45	17 06
56	20 27	21 19	22 48	13 02	14 17	15 34	16 51
54	20 15	21 03	22 15	12 57	14 10	15 24	16 37
52	20 06	20 49	21 51	12 53	14 04	15 15	16 26
50	19 57	20 38	21 33	12 50	13 58	15 07	16 16
45	19 38	20 13	20 58	12 42	13 45	14 50	15 54
N 40	19 23	19 55	20 33	12 35	13 35	14 36	15 37
35	19 11	19 39	20 14	12 29	13 26	14 24	15 22
30	19 00	19 26	19 58	12 24	13 19	14 14	15 10
20	18 41	19 05	19 33	12 16	13 05	13 56	14 49
N 10	18 25	18 47	19 14	12 08	12 54	13 41	14 30
0	18 10	18 32	18 58	12 01	12 43	13 27	14 13
S 10	17 55	18 17	18 43	11 55	12 32	13 13	13 56
20	17 39	18 03	18 30	11 47	12 21	12 57	13 37
30	17 21	17 47	18 16	11 39	12 08	12 40	13 16
35	17 11	17 38	18 09	11 34	12 01	12 30	13 04
40	16 59	17 29	18 02	11 29	11 52	12 19	12 50
45	16 45	17 18	17 54	11 22	11 42	12 06	12 33
S 50	16 28	17 05	17 45	11 15	11 31	11 49	12 13
52	16 20	16 59	17 41	11 11	11 25	11 42	12 03
54	16 11	16 52	17 36	11 07	11 19	11 33	11 52
56	16 01	16 45	17 32	11 03	11 12	11 24	11 40
58	15 50	16 37	17 27	10 59	11 05	11 13	11 26
S 60	15 36	16 28	17 21	10 53	10 57	11 01	11 09

Day	SUN Eqn. of Time 00^h	SUN Eqn. of Time 12^h	SUN Mer. Pass.	MOON Mer. Pass. Upper	MOON Mer. Pass. Lower	MOON Age	MOON Phase
d	m s	m s	h m	h m	h m	d	%
21	06 21	06 22	12 06	05 52	18 12	22	48
22	06 24	06 25	12 06	06 33	18 54	23	39
23	06 26	06 27	12 06	07 16	19 38	24	30

	UT	ARIES	VENUS −3.9		MARS −2.1		JUPITER −1.7		SATURN +0.1	
		GHA	GHA	Dec	GHA	Dec	GHA	Dec	GHA	Dec
	d h	° ′	° ′	° ′	° ′	° ′	° ′	° ′	° ′	° ′
	24 00	301 18.4	185 42.1	N22 00.5	317 37.1	S13 06.9	156 06.4	N14 44.7	204 22.7	N22 31.0
	01	316 20.8	200 41.2	22 00.1	332 39.3	06.9	171 08.4	44.6	219 24.8	31.0
	02	331 23.3	215 40.4	21 59.6	347 41.6	07.0	186 10.3	44.4	234 26.9	31.0
	03	346 25.8	230 39.6	. . 59.2	2 43.8	. . 07.1	201 12.3	. . 44.2	249 29.0	. . 31.0
	04	1 28.2	245 38.8	58.7	17 46.1	07.1	216 14.2	44.1	264 31.2	31.0
	05	16 30.7	260 38.0	58.3	32 48.3	07.2	231 16.2	43.9	279 33.3	31.0
	06	31 33.1	275 37.2	N21 57.8	47 50.5	S13 07.3	246 18.1	N14 43.7	294 35.4	N22 31.0
	07	46 35.6	290 36.3	57.4	62 52.8	07.3	261 20.1	43.6	309 37.6	31.0
T	08	61 38.1	305 35.5	56.9	77 55.0	07.4	276 22.0	43.4	324 39.7	30.9
H	09	76 40.5	320 34.7	. . 56.5	92 57.3	. . 07.5	291 24.0	. . 43.2	339 41.8	. . 30.9
U	10	91 43.0	335 33.9	56.0	107 59.6	07.6	306 25.9	43.0	354 44.0	30.9
R	11	106 45.5	350 33.1	55.6	123 01.8	07.6	321 27.9	42.9	9 46.1	30.9
S	12	121 47.9	5 32.3	N21 55.1	138 04.1	S13 07.7	336 29.8	N14 42.7	24 48.2	N22 30.9
D	13	136 50.4	20 31.5	54.7	153 06.3	07.8	351 31.8	42.5	39 50.4	30.9
A	14	151 52.9	35 30.6	54.2	168 08.6	07.8	6 33.7	42.4	54 52.5	30.9
Y	15	166 55.3	50 29.8	. . 53.7	183 10.8	. . 07.9	21 35.7	. . 42.2	69 54.6	. . 30.9
	16	181 57.8	65 29.0	53.3	198 13.1	08.0	36 37.6	42.0	84 56.8	30.8
	17	197 00.3	80 28.2	52.8	213 15.4	08.1	51 39.6	41.8	99 58.9	30.8
	18	212 02.7	95 27.4	N21 52.4	228 17.6	S13 08.1	66 41.5	N14 41.7	115 01.0	N22 30.8
	19	227 05.2	110 26.6	51.9	243 19.9	08.2	81 43.5	41.5	130 03.2	30.8
	20	242 07.6	125 25.8	51.4	258 22.2	08.3	96 45.4	41.3	145 05.3	30.8
	21	257 10.1	140 25.0	. . 51.0	273 24.4	. . 08.3	111 47.4	. . 41.2	160 07.4	. . 30.8
	22	272 12.6	155 24.1	50.5	288 26.7	08.4	126 49.3	41.0	175 09.6	30.8
	23	287 15.0	170 23.3	50.0	303 29.0	08.5	141 51.3	40.8	190 11.7	30.8
	25 00	302 17.5	185 22.5	N21 49.6	318 31.2	S13 08.6	156 53.2	N14 40.6	205 13.8	N22 30.7
	01	317 20.0	200 21.7	49.1	333 33.5	08.6	171 55.2	40.5	220 16.0	30.7
	02	332 22.4	215 20.9	48.6	348 35.8	08.7	186 57.1	40.3	235 18.1	30.7
	03	347 24.9	230 20.1	. . 48.2	3 38.1	. . 08.8	201 59.1	. . 40.1	250 20.3	. . 30.7
	04	2 27.4	245 19.3	47.7	18 40.3	08.9	217 01.0	40.0	265 22.4	30.7
	05	17 29.8	260 18.5	47.2	33 42.6	09.0	232 03.0	39.8	280 24.5	30.7
	06	32 32.3	275 17.7	N21 46.8	48 44.9	S13 09.0	247 04.9	N14 39.6	295 26.7	N22 30.7
	07	47 34.7	290 16.9	46.3	63 47.2	09.1	262 06.9	39.4	310 28.8	30.6
	08	62 37.2	305 16.1	45.8	78 49.5	09.2	277 08.8	39.3	325 30.9	30.6
F	09	77 39.7	320 15.2	. . 45.3	93 51.7	. . 09.3	292 10.8	. . 39.1	340 33.1	. . 30.6
R	10	92 42.1	335 14.4	44.8	108 54.0	09.4	307 12.7	38.9	355 35.2	30.6
I	11	107 44.6	350 13.6	44.4	123 56.3	09.4	322 14.7	38.8	10 37.3	30.6
D	12	122 47.1	5 12.8	N21 43.9	138 58.6	S13 09.5	337 16.6	N14 38.6	25 39.5	N22 30.6
A	13	137 49.5	20 12.0	43.4	154 00.9	09.6	352 18.6	38.4	40 41.6	30.6
Y	14	152 52.0	35 11.2	42.9	169 03.2	09.7	7 20.5	38.2	55 43.7	30.6
	15	167 54.5	50 10.4	. . 42.4	184 05.5	. . 09.8	22 22.5	. . 38.1	70 45.9	. . 30.5
	16	182 56.9	65 09.6	42.0	199 07.8	09.8	37 24.4	37.9	85 48.0	30.5
	17	197 59.4	80 08.8	41.5	214 10.1	09.9	52 26.4	37.7	100 50.1	30.5
	18	213 01.9	95 08.0	N21 41.0	229 12.3	S13 10.0	67 28.3	N14 37.6	115 52.3	N22 30.5
	19	228 04.3	110 07.2	40.5	244 14.6	10.1	82 30.3	37.4	130 54.4	30.5
	20	243 06.8	125 06.4	40.0	259 16.9	10.2	97 32.2	37.2	145 56.5	30.5
	21	258 09.2	140 05.6	. . 39.5	274 19.2	. . 10.3	112 34.2	. . 37.0	160 58.7	. . 30.5
	22	273 11.7	155 04.8	39.0	289 21.5	10.4	127 36.1	36.9	176 00.8	30.5
	23	288 14.2	170 04.0	38.5	304 23.8	10.4	142 38.1	36.7	191 02.9	30.4
	26 00	303 16.6	185 03.2	N21 38.0	319 26.2	S13 10.5	157 40.0	N14 36.5	206 05.1	N22 30.4
	01	318 19.1	200 02.4	37.5	334 28.5	10.6	172 42.0	36.3	221 07.2	30.4
	02	333 21.6	215 01.6	37.0	349 30.8	10.7	187 43.9	36.2	236 09.4	30.4
	03	348 24.0	230 00.8	. . 36.5	4 33.1	. . 10.8	202 45.9	. . 36.0	251 11.5	. . 30.4
	04	3 26.5	245 00.0	36.0	19 35.4	10.9	217 47.8	35.8	266 13.6	30.4
	05	18 29.0	259 59.2	35.5	34 37.7	11.0	232 49.8	35.7	281 15.8	30.4
	06	33 31.4	274 58.4	N21 35.0	49 40.0	S13 11.1	247 51.7	N14 35.5	296 17.9	N22 30.3
	07	48 33.9	289 57.6	34.5	64 42.3	11.1	262 53.7	35.3	311 20.0	30.3
S	08	63 36.4	304 56.8	34.0	79 44.6	11.2	277 55.6	35.1	326 22.2	30.3
A	09	78 38.8	319 56.0	. . 33.5	94 46.9	. . 11.3	292 57.6	. . 35.0	341 24.3	. . 30.3
T	10	93 41.3	334 55.2	33.0	109 49.3	11.4	307 59.5	34.8	356 26.4	30.3
U	11	108 43.7	349 54.4	32.5	124 51.6	11.5	323 01.5	34.6	11 28.6	30.3
R	12	123 46.2	4 53.6	N21 32.0	139 53.9	S13 11.6	338 03.4	N14 34.4	26 30.7	N22 30.3
D	13	138 48.7	19 52.8	31.5	154 56.2	11.7	353 05.4	34.3	41 32.8	30.3
A	14	153 51.1	34 52.0	31.0	169 58.5	11.8	8 07.3	34.1	56 35.0	30.2
Y	15	168 53.6	49 51.2	. . 30.5	185 00.9	. . 11.9	23 09.2	. . 33.9	71 37.1	. . 30.2
	16	183 56.1	64 50.4	30.0	200 03.2	12.0	38 11.2	33.8	86 39.3	30.2
	17	198 58.5	79 49.6	29.5	215 05.5	12.1	53 13.1	33.6	101 41.4	30.2
	18	214 01.0	94 48.8	N21 29.0	230 07.8	S13 12.2	68 15.1	N14 33.4	116 43.5	N22 30.2
	19	229 03.5	109 48.0	28.4	245 10.2	12.3	83 17.0	33.2	131 45.7	30.2
	20	244 05.9	124 47.2	27.9	260 12.5	12.4	98 19.0	33.1	146 47.8	30.2
	21	259 08.4	139 46.4	. . 27.4	275 14.8	. . 12.4	113 20.9	. . 32.9	161 49.9	. . 30.2
	22	274 10.9	154 45.6	26.9	290 17.2	12.5	128 22.9	32.7	176 52.1	30.1
	23	289 13.3	169 44.8	26.4	305 19.5	12.6	143 24.8	32.5	191 54.2	30.1
	Mer. Pass.	h m 3 50.2	*v* −0.8	*d* 0.5	*v* 2.3	*d* 0.1	*v* 1.9	*d* 0.2	*v* 2.1	*d* 0.0

STARS

Name	SHA	Dec
	° ′	° ′
Acamar	315 24.3	S40 17.
Achernar	335 32.3	S57 12.
Acrux	173 18.6	S63 07.
Adhara	255 19.0	S28 58.
Aldebaran	290 58.6	N16 31.
Alioth	166 27.4	N55 56.
Alkaid	153 04.9	N49 18.
Al Na'ir	27 52.9	S46 56.
Alnilam	275 54.5	S 1 11.
Alphard	218 04.1	S 8 40.
Alphecca	126 17.4	N26 42.
Alpheratz	357 51.5	N29 06.
Altair	62 15.5	N 8 52.
Ankaa	353 23.1	S42 17.
Antares	112 35.6	S26 26.
Arcturus	146 02.8	N19 10.
Atria	107 44.2	S69 02.
Avior	234 22.0	S59 31.
Bellatrix	278 40.6	N 6 21.
Betelgeuse	271 10.0	N 7 24.
Canopus	264 00.1	S52 41.
Capella	280 46.4	N46 00.
Deneb	49 36.4	N45 17.
Denebola	182 41.7	N14 33.
Diphda	349 03.5	S17 57.
Dubhe	194 01.4	N61 44.
Elnath	278 22.8	N28 36.
Eltanin	90 49.3	N51 29.
Enif	33 54.5	N 9 53.
Fomalhaut	15 32.2	S29 36.
Gacrux	172 10.1	S57 08.
Gienah	176 00.5	S17 33.
Hadar	148 59.2	S60 23.
Hamal	328 09.6	N23 28.
Kaus Aust.	83 53.8	S34 23.
Kochab	137 19.1	N74 08.
Markab	13 45.9	N15 13.
Menkar	314 23.3	N 4 06.
Menkent	148 16.9	S36 23.
Miaplacidus	221 42.6	S69 43.
Mirfak	308 51.8	N49 52.
Nunki	76 07.6	S26 17.
Peacock	53 30.7	S56 43.
Pollux	243 37.6	N28 01.
Procyon	245 08.2	N 5 13.
Rasalhague	96 13.4	N12 33.
Regulus	207 52.1	N11 57.
Rigel	281 19.8	S 8 11.
Rigil Kent.	140 02.6	S60 51.
Sabik	102 21.2	S15 43.
Schedar	349 49.5	N56 33.
Shaula	96 32.2	S37 06.
Sirius	258 41.0	S16 43.
Spica	158 39.6	S11 10.
Suhail	222 58.7	S43 26.
Vega	80 43.9	N38 47.
Zuben'ubi	137 14.0	S16 03.

	SHA	Mer. Pass.
	° ′	h m
Venus	243 05.0	11 39
Mars	16 13.7	2 45
Jupiter	214 35.7	13 31
Saturn	262 56.3	10 18

UT d h	SUN GHA ° ′	SUN Dec ° ′	MOON GHA ° ′	v ′	MOON Dec ° ′	d ′	HP ′
24 00	178 22.9	N19 59.9	243 23.2	13.1	N20 24.5	8.8	54.3
01	193 22.9	59.4	257 55.3	13.0	20 33.3	8.7	54.4
02	208 22.9	58.9	272 27.3	13.0	20 42.0	8.6	54.4
03	223 22.9	. . 58.3	286 59.3	12.9	20 50.6	8.6	54.4
04	238 22.9	57.8	301 31.2	12.8	20 59.2	8.4	54.4
05	253 22.9	57.3	316 03.0	12.7	21 07.6	8.4	54.4
06	268 22.8	N19 56.8	330 34.7	12.7	N21 16.0	8.3	54.4
07	283 22.8	56.3	345 06.4	12.6	21 24.3	8.2	54.4
T 08	298 22.8	55.8	359 38.0	12.5	21 32.5	8.1	54.4
H 09	313 22.8	. . 55.2	14 09.5	12.5	21 40.6	8.0	54.4
U 10	328 22.8	54.7	28 41.0	12.3	21 48.6	7.9	54.4
R 11	343 22.8	54.2	43 12.3	12.4	21 56.5	7.8	54.4
S 12	358 22.8	N19 53.7	57 43.7	12.2	N22 04.3	7.8	54.5
D 13	13 22.7	53.1	72 14.9	12.2	22 12.1	7.6	54.5
A 14	28 22.7	52.6	86 46.1	12.1	22 19.7	7.5	54.5
Y 15	43 22.7	. . 52.1	101 17.2	12.0	22 27.2	7.4	54.5
16	58 22.7	51.6	115 48.2	11.9	22 34.6	7.4	54.5
17	73 22.7	51.0	130 19.1	11.9	22 42.0	7.2	54.5
18	88 22.7	N19 50.5	144 50.0	11.8	N22 49.2	7.2	54.5
19	103 22.7	50.0	159 20.8	11.8	22 56.4	7.0	54.5
20	118 22.7	49.5	173 51.6	11.6	23 03.4	6.9	54.6
21	133 22.6	. . 48.9	188 22.2	11.6	23 10.3	6.9	54.6
22	148 22.6	48.4	202 52.8	11.5	23 17.2	6.7	54.6
23	163 22.6	47.9	217 23.3	11.5	23 23.9	6.6	54.6
25 00	178 22.6	N19 47.3	231 53.8	11.3	N23 30.5	6.5	54.6
01	193 22.6	46.8	246 24.1	11.3	23 37.0	6.4	54.6
02	208 22.6	46.3	260 54.4	11.3	23 43.4	6.3	54.6
03	223 22.6	. . 45.8	275 24.7	11.1	23 49.7	6.2	54.6
04	238 22.6	45.2	289 54.8	11.1	23 55.9	6.1	54.7
05	253 22.6	44.7	304 24.9	11.0	24 02.0	6.0	54.7
06	268 22.6	N19 44.2	318 54.9	11.0	N24 08.0	5.8	54.7
07	283 22.5	43.6	333 24.9	10.9	24 13.8	5.8	54.7
08	298 22.5	43.1	347 54.8	10.8	24 19.6	5.6	54.7
F 09	313 22.5	. . 42.6	2 24.6	10.7	24 25.2	5.5	54.7
R 10	328 22.5	42.0	16 54.3	10.7	24 30.7	5.5	54.7
I 11	343 22.5	41.5	31 24.0	10.6	24 36.2	5.2	54.8
D 12	358 22.5	N19 40.9	45 53.6	10.5	N24 41.4	5.2	54.8
A 13	13 22.5	40.4	60 23.1	10.5	24 46.6	5.1	54.8
Y 14	28 22.5	39.9	74 52.6	10.4	24 51.7	4.9	54.8
15	43 22.5	. . 39.3	89 22.0	10.3	24 56.6	4.8	54.8
16	58 22.5	38.8	103 51.3	10.2	25 01.4	4.7	54.8
17	73 22.5	38.3	118 20.5	10.2	25 06.1	4.6	54.9
18	88 22.5	N19 37.7	132 49.7	10.2	N25 10.7	4.5	54.9
19	103 22.5	37.2	147 18.9	10.1	25 15.2	4.3	54.9
20	118 22.5	36.6	161 48.0	10.0	25 19.5	4.2	54.9
21	133 22.4	. . 36.1	176 17.0	9.9	25 23.7	4.1	54.9
22	148 22.4	35.5	190 45.9	9.9	25 27.8	4.0	54.9
23	163 22.4	35.0	205 14.8	9.8	25 31.8	3.9	55.0
26 00	178 22.4	N19 34.5	219 43.6	9.8	N25 35.7	3.7	55.0
01	193 22.4	33.9	234 12.4	9.7	25 39.4	3.6	55.0
02	208 22.4	33.4	248 41.1	9.6	25 43.0	3.4	55.0
03	223 22.4	. . 32.8	263 09.7	9.6	25 46.4	3.4	55.0
04	238 22.4	32.3	277 38.3	9.5	25 49.8	3.2	55.1
05	253 22.4	31.7	292 06.8	9.5	25 53.0	3.0	55.1
06	268 22.4	N19 31.2	306 35.3	9.4	N25 56.0	3.0	55.1
07	283 22.4	30.6	321 03.7	9.4	25 59.0	2.8	55.1
S 08	298 22.4	30.1	335 32.1	9.3	26 01.8	2.7	55.1
A 09	313 22.4	. . 29.5	350 00.4	9.2	26 04.5	2.5	55.2
T 10	328 22.4	29.0	4 28.6	9.2	26 07.0	2.4	55.2
U 11	343 22.4	28.4	18 56.8	9.2	26 09.4	2.3	55.2
R 12	358 22.4	N19 27.9	33 25.0	9.1	N26 11.7	2.2	55.2
D 13	13 22.4	27.3	47 53.1	9.0	26 13.9	2.0	55.2
A 14	28 22.4	26.8	62 21.1	9.1	26 15.9	1.9	55.3
Y 15	43 22.4	. . 26.2	76 49.2	8.9	26 17.8	1.7	55.3
16	58 22.4	25.7	91 17.1	8.9	26 19.5	1.6	55.3
17	73 22.4	25.1	105 45.0	8.9	26 21.1	1.5	55.3
18	88 22.4	N19 24.6	120 12.9	8.8	N26 22.6	1.3	55.3
19	103 22.4	24.0	134 40.7	8.8	26 23.9	1.2	55.4
20	118 22.4	23.5	149 08.5	8.8	26 25.1	1.1	55.4
21	133 22.4	. . 22.9	163 36.3	8.7	26 26.2	0.9	55.4
22	148 22.4	22.4	178 04.0	8.7	26 27.1	0.7	55.4
23	163 22.4	21.8	192 31.7	8.6	N26 27.8	0.7	55.4
	SD 15.8	*d* 0.5	SD 14.8		14.9		15.0

Lat.	Twilight Naut.	Twilight Civil	Sunrise	Moonrise 24	Moonrise 25	Moonrise 26	Moonrise 27
°	h m	h m	h m	h m	h m	h m	h m
N 72	▭	▭	▭	▭	▭	▭	▭
N 70	▭	▭	▭	▭	▭	▭	▭
68	////	////	01 32	▭	▭	▭	▭
66	////	////	02 16	21 33	▭	▭	▭
64	////	00 37	02 44	22 23	22 11	▭	23 21
62	////	01 43	03 06	22 54	23 05	23 36	24 43
60	////	02 16	03 23	23 18	23 38	24 16	00 16
N 58	00 34	02 40	03 38	23 37	24 02	00 02	00 44
56	01 34	02 59	03 50	23 53	24 22	00 22	01 05
54	02 04	03 14	04 01	24 06	00 06	00 38	01 23
52	02 27	03 27	04 11	24 18	00 18	00 52	01 38
50	02 44	03 39	04 19	00 02	00 29	01 04	01 51
45	03 18	04 02	04 37	00 20	00 51	01 30	02 18
N 40	03 42	04 20	04 52	00 35	01 09	01 50	02 40
35	04 01	04 35	05 04	00 48	01 24	02 07	02 57
30	04 16	04 48	05 14	00 59	01 37	02 22	03 13
20	04 41	05 09	05 33	01 18	02 00	02 47	03 39
N 10	05 00	05 26	05 48	01 35	02 20	03 09	04 01
0	05 15	05 41	06 03	01 50	02 38	03 29	04 22
S 10	05 29	05 55	06 17	02 06	02 57	03 49	04 43
20	05 43	06 09	06 33	02 23	03 16	04 11	05 05
30	05 56	06 25	06 50	02 43	03 39	04 36	05 31
35	06 02	06 33	07 00	02 54	03 53	04 51	05 47
40	06 09	06 42	07 12	03 07	04 09	05 09	06 05
45	06 17	06 53	07 25	03 23	04 28	05 30	06 26
S 50	06 26	07 05	07 42	03 43	04 51	05 56	06 54
52	06 30	07 11	07 49	03 52	05 03	06 09	07 07
54	06 34	07 17	07 58	04 02	05 15	06 24	07 22
56	06 38	07 24	08 07	04 14	05 30	06 41	07 40
58	06 43	07 32	08 18	04 28	05 48	07 02	08 02
S 60	06 48	07 41	08 31	04 44	06 10	07 29	08 31

Lat.	Sunset	Twilight Civil	Twilight Naut.	Moonset 24	Moonset 25	Moonset 26	Moonset 27
°	h m	h m	h m	h m	h m	h m	h m
N 72	▭	▭	▭	▭	▭	▭	▭
N 70	▭	▭	▭	▭	▭	▭	▭
68	22 35	////	////	▭	▭	▭	▭
66	21 54	////	////	19 08	▭	▭	▭
64	21 26	23 23	////	18 19	20 14	▭	22 47
62	21 05	22 26	////	17 48	19 20	20 39	21 25
60	20 48	21 54	////	17 25	18 48	19 59	20 49
N 58	20 34	21 31	23 28	17 06	18 24	19 31	20 22
56	20 22	21 13	22 35	16 51	18 05	19 10	20 02
54	20 11	20 58	22 06	16 37	17 48	18 52	19 44
52	20 02	20 45	21 44	16 26	17 35	18 37	19 30
50	19 53	20 33	21 27	16 16	17 23	18 24	19 17
45	19 35	20 10	20 54	15 54	16 58	17 57	18 50
N 40	19 21	19 52	20 30	15 37	16 38	17 36	18 29
35	19 09	19 37	20 12	15 22	16 21	17 18	18 12
30	18 58	19 25	19 56	15 10	16 07	17 03	17 57
20	18 40	19 04	19 32	14 49	15 43	16 37	17 31
N 10	18 25	18 47	19 13	14 30	15 22	16 15	17 09
0	18 10	18 32	18 58	14 13	15 02	15 54	16 48
S 10	17 56	18 18	18 44	13 56	14 43	15 33	16 27
20	17 40	18 04	18 31	13 37	14 22	15 11	16 05
30	17 23	17 49	18 18	13 16	13 58	14 45	15 39
35	17 13	17 40	18 11	13 04	13 44	14 30	15 24
40	17 02	17 31	18 04	12 50	13 27	14 12	15 06
45	16 48	17 20	17 56	12 33	13 08	13 51	14 44
S 50	16 32	17 08	17 48	12 13	12 44	13 24	14 17
52	16 24	17 02	17 44	12 03	12 32	13 11	14 04
54	16 16	16 56	17 40	11 52	12 19	12 56	13 49
56	16 06	16 49	17 36	11 40	12 04	12 39	13 31
58	15 55	16 42	17 31	11 26	11 45	12 18	13 09
S 60	15 43	16 33	17 26	11 09	11 23	11 51	12 40

Day	SUN Eqn. of Time 00^h	SUN Eqn. of Time 12^h	SUN Mer. Pass.	MOON Mer. Pass. Upper	MOON Mer. Pass. Lower	MOON Age	MOON Phase
d	m s	m s	h m	h m	h m	d %	
24	06 28	06 29	12 06	08 02	20 25	25 21	
25	06 30	06 30	12 06	08 50	21 15	26 14	
26	06 30	06 30	12 07	09 41	22 08	27 8	

UT	ARIES	VENUS −3.9		MARS −2.2		JUPITER −1.7		SATURN +0.1	
	GHA	GHA	Dec	GHA	Dec	GHA	Dec	GHA	Dec
d h	° ′	° ′	° ′	° ′	° ′	° ′	° ′	° ′	° ′
27 00 SUNDAY	304 15.8	184 44.0	N21 25.8	320 21.8	S13 12.7	158 26.8	N14 32.4	206 56.4	N22 30.1
01	319 18.2	199 43.2	25.3	335 24.2	12.8	173 28.7	32.2	221 58.5	30.1
02	334 20.7	214 42.4	24.8	350 26.5	12.9	188 30.7	32.0	237 00.6	30.1
03	349 23.2	229 41.6	. . 24.3	5 28.9	. . 13.0	203 32.6	. . 31.9	252 02.8	. . 30.1
04	4 25.6	244 40.8	23.8	20 31.2	13.1	218 34.6	31.7	267 04.9	30.1
05	19 28.1	259 40.1	23.2	35 33.5	13.2	233 36.5	31.5	282 07.0	30.0
06	34 30.6	274 39.3	N21 22.7	50 35.9	S13 13.3	248 38.5	N14 31.3	297 09.2	N22 30.0
07	49 33.0	289 38.5	22.2	65 38.2	13.4	263 40.4	31.2	312 11.3	30.0
08	64 35.5	304 37.7	21.6	80 40.6	13.5	278 42.4	31.0	327 13.4	30.0
09	79 38.0	319 36.9	. . 21.1	95 42.9	. . 13.6	293 44.3	. . 30.8	342 15.6	. . 30.0
10	94 40.4	334 36.1	20.6	110 45.3	13.7	308 46.3	30.6	357 17.7	30.0
11	109 42.9	349 35.3	20.1	125 47.6	13.8	323 48.2	30.5	12 19.9	30.0
12	124 45.4	4 34.5	N21 19.5	140 50.0	S13 13.9	338 50.2	N14 30.3	27 22.0	N22 30.0
13	139 47.8	19 33.7	19.0	155 52.3	14.0	353 52.1	30.1	42 24.1	29.9
14	154 50.3	34 32.9	18.4	170 54.7	14.1	8 54.0	29.9	57 26.3	29.9
15	169 52.7	49 32.2	. . 17.9	185 57.0	. . 14.3	23 56.0	. . 29.8	72 28.4	. . 29.9
16	184 55.2	64 31.4	17.4	200 59.4	14.4	38 57.9	29.6	87 30.6	29.9
17	199 57.7	79 30.6	16.8	216 01.8	14.5	53 59.9	29.4	102 32.7	29.9
18	215 00.1	94 29.8	N21 16.3	231 04.1	S13 14.6	69 01.8	N14 29.2	117 34.8	N22 29.9
19	230 02.6	109 29.0	15.8	246 06.5	14.7	84 03.8	29.1	132 37.0	29.9
20	245 05.1	124 28.2	15.2	261 08.8	14.8	99 05.7	28.9	147 39.1	29.8
21	260 07.5	139 27.4	. . 14.7	276 11.2	. . 14.9	114 07.7	. . 28.7	162 41.2	. . 29.8
22	275 10.0	154 26.6	14.1	291 13.6	15.0	129 09.6	28.6	177 43.4	29.8
23	290 12.5	169 25.9	13.6	306 15.9	15.1	144 11.6	28.4	192 45.5	29.8
28 00 MONDAY	305 14.9	184 25.1	N21 13.0	321 18.3	S13 15.2	159 13.5	N14 28.2	207 47.7	N22 29.8
01	320 17.4	199 24.3	12.5	336 20.7	15.3	174 15.5	28.0	222 49.8	29.8
02	335 19.8	214 23.5	11.9	351 23.0	15.4	189 17.4	27.9	237 51.9	29.8
03	350 22.3	229 22.7	. . 11.4	6 25.4	. . 15.5	204 19.4	. . 27.7	252 54.1	. . 29.7
04	5 24.8	244 21.9	10.8	21 27.8	15.6	219 21.3	27.5	267 56.2	29.7
05	20 27.2	259 21.2	10.3	36 30.2	15.8	234 23.3	27.3	282 58.4	29.7
06	35 29.7	274 20.4	N21 09.7	51 32.5	S13 15.9	249 25.2	N14 27.2	298 00.5	N22 29.7
07	50 32.2	289 19.6	09.2	66 34.9	16.0	264 27.1	27.0	313 02.6	29.7
08	65 34.6	304 18.8	08.6	81 37.3	16.1	279 29.1	26.8	328 04.8	29.7
09	80 37.1	319 18.0	. . 08.1	96 39.7	. . 16.2	294 31.0	. . 26.6	343 06.9	. . 29.7
10	95 39.6	334 17.2	07.5	111 42.1	16.3	309 33.0	26.5	358 09.0	29.7
11	110 42.0	349 16.5	07.0	126 44.5	16.4	324 34.9	26.3	13 11.2	29.6
12	125 44.5	4 15.7	N21 06.4	141 46.8	S13 16.5	339 36.9	N14 26.1	28 13.3	N22 29.6
13	140 47.0	19 14.9	05.8	156 49.2	16.6	354 38.8	25.9	43 15.5	29.6
14	155 49.4	34 14.1	05.3	171 51.6	16.8	9 40.8	25.8	58 17.6	29.6
15	170 51.9	49 13.3	. . 04.7	186 54.0	. . 16.9	24 42.7	. . 25.6	73 19.7	. . 29.6
16	185 54.3	64 12.6	04.2	201 56.4	17.0	39 44.7	25.4	88 21.9	29.6
17	200 56.8	79 11.8	03.6	216 58.8	17.1	54 46.6	25.2	103 24.0	29.6
18	215 59.3	94 11.0	N21 03.0	232 01.2	S13 17.2	69 48.6	N14 25.1	118 26.2	N22 29.5
19	231 01.7	109 10.2	02.5	247 03.6	17.3	84 50.5	24.9	133 28.3	29.5
20	246 04.2	124 09.4	01.9	262 06.0	17.5	99 52.4	24.7	148 30.4	29.5
21	261 06.7	139 08.7	. . 01.3	277 08.4	. . 17.6	114 54.4	. . 24.5	163 32.6	. . 29.5
22	276 09.1	154 07.9	00.8	292 10.8	17.7	129 56.3	24.4	178 34.7	29.5
23	291 11.6	169 07.1	21 00.2	307 13.2	17.8	144 58.3	24.2	193 36.9	29.5
29 00 TUESDAY	306 14.1	184 06.3	N20 59.6	322 15.6	S13 17.9	160 00.2	N14 24.0	208 39.0	N22 29.5
01	321 16.5	199 05.6	59.0	337 18.0	18.0	175 02.2	23.8	223 41.1	29.4
02	336 19.0	214 04.8	58.5	352 20.4	18.2	190 04.1	23.7	238 43.3	29.4
03	351 21.5	229 04.0	. . 57.9	7 22.8	. . 18.3	205 06.1	. . 23.5	253 45.4	. . 29.4
04	6 23.9	244 03.2	57.3	22 25.2	18.4	220 08.0	23.3	268 47.6	29.4
05	21 26.4	259 02.5	56.7	37 27.6	18.5	235 10.0	23.2	283 49.7	29.4
06	36 28.8	274 01.7	N20 56.2	52 30.0	S13 18.6	250 11.9	N14 23.0	298 51.8	N22 29.4
07	51 31.3	289 00.9	55.6	67 32.4	18.8	265 13.9	22.8	313 54.0	29.4
08	66 33.8	304 00.1	55.0	82 34.8	18.9	280 15.8	22.6	328 56.1	29.4
09	81 36.2	318 59.4	. . 54.4	97 37.2	. . 19.0	295 17.7	. . 22.5	343 58.3	. . 29.3
10	96 38.7	333 58.6	53.8	112 39.6	19.1	310 19.7	22.3	359 00.4	29.3
11	111 41.2	348 57.8	53.3	127 42.1	19.3	325 21.6	22.1	14 02.5	29.3
12	126 43.6	3 57.1	N20 52.7	142 44.5	S13 19.4	340 23.6	N14 21.9	29 04.7	N22 29.3
13	141 46.1	18 56.3	52.1	157 46.9	19.5	355 25.5	21.8	44 06.8	29.3
14	156 48.6	33 55.5	51.5	172 49.3	19.6	10 27.5	21.6	59 09.0	29.3
15	171 51.0	48 54.7	. . 50.9	187 51.7	. . 19.7	25 29.4	. . 21.4	74 11.1	. . 29.3
16	186 53.5	63 54.0	50.3	202 54.2	19.9	40 31.4	21.2	89 13.3	29.2
17	201 56.0	78 53.2	49.7	217 56.6	20.0	55 33.3	21.1	104 15.4	29.2
18	216 58.4	93 52.4	N20 49.1	232 59.0	S13 20.1	70 35.3	N14 20.9	119 17.5	N22 29.2
19	232 00.9	108 51.7	48.6	248 01.4	20.2	85 37.2	20.7	134 19.7	29.2
20	247 03.3	123 50.9	48.0	263 03.9	20.4	100 39.1	20.5	149 21.8	29.2
21	262 05.8	138 50.1	. . 47.4	278 06.3	. . 20.5	115 41.1	. . 20.3	164 24.0	. . 29.2
22	277 08.3	153 49.4	46.8	293 08.7	20.6	130 43.0	20.2	179 26.1	29.2
23	292 10.7	168 48.6	46.2	308 11.2	20.8	145 45.0	20.0	194 28.2	29.1
Mer. Pass.	h m 3 38.4	*v* −0.8	*d* 0.6	*v* 2.4	*d* 0.1	*v* 1.9	*d* 0.2	*v* 2.1	*d* 0.0

STARS Name	SHA	Dec
	° ′	° ′
Acamar	315 24.2	S40 17.2
Achernar	335 32.2	S57 12.8
Acrux	173 18.7	S63 07.3
Adhara	255 19.0	S28 58.4
Aldebaran	290 58.5	N16 31.0
Alioth	166 27.5	N55 56.8
Alkaid	153 04.9	N49 18.1
Al Na'ir	27 52.9	S46 56.6
Alnilam	275 54.5	S 1 11.8
Alphard	218 04.1	S 8 40.3
Alphecca	126 17.4	N26 42.4
Alpheratz	357 51.5	N29 06.5
Altair	62 15.5	N 8 52.6
Ankaa	353 23.0	S42 17.0
Antares	112 35.6	S26 26.5
Arcturus	146 02.8	N19 10.0
Atria	107 44.2	S69 02.3
Avior	234 22.0	S59 31.1
Bellatrix	278 40.6	N 6 21.3
Betelgeuse	271 10.0	N 7 24.6
Canopus	264 00.1	S52 41.6
Capella	280 46.3	N46 00.0
Deneb	49 36.4	N45 17.5
Denebola	182 41.8	N14 33.4
Diphda	349 03.5	S17 57.9
Dubhe	194 01.4	N61 44.2
Elnath	278 22.8	N28 36.6
Eltanin	90 49.3	N51 29.4
Enif	33 54.5	N 9 53.4
Fomalhaut	15 32.2	S29 36.1
Gacrux	172 10.1	S57 08.1
Gienah	176 00.5	S17 33.6
Hadar	148 59.2	S60 23.6
Hamal	328 09.6	N23 28.7
Kaus Aust.	83 53.8	S34 23.1
Kochab	137 19.1	N74 08.8
Markab	13 45.9	N15 13.4
Menkar	314 23.3	N 4 06.3
Menkent	148 16.9	S36 23.3
Miaplacidus	221 42.6	S69 43.9
Mirfak	308 51.8	N49 52.2
Nunki	76 07.6	S26 17.6
Peacock	53 30.7	S56 43.5
Pollux	243 37.6	N28 01.2
Procyon	245 08.2	N 5 13.1
Rasalhague	96 13.4	N12 33.5
Regulus	207 52.1	N11 57.2
Rigel	281 19.8	S 8 11.7
Rigil Kent.	140 02.6	S60 51.2
Sabik	102 21.2	S15 43.8
Schedar	349 49.4	N56 33.1
Shaula	96 32.2	S37 06.5
Sirius	258 40.9	S16 43.0
Spica	158 39.6	S11 10.7
Suhail	222 58.7	S43 26.7
Vega	80 43.9	N38 47.3
Zuben'ubi	137 14.0	S16 03.4

	SHA	Mer. Pass.
	° ′	h m
Venus	239 10.2	11 43
Mars	16 03.4	2 34
Jupiter	213 58.6	13 21
Saturn	262 32.7	10 07

UT	SUN GHA	SUN Dec	MOON GHA	v	MOON Dec	d	HP
d h	° ′	° ′	° ′	′	° ′	′	′
27 00	178 22.4	N19 21.3	206 59.3	8.6	N26 28.5	0.5	55.5
01	193 22.4	20.7	221 26.9	8.6	26 29.0	0.3	55.5
02	208 22.4	20.1	235 54.5	8.5	26 29.3	0.2	55.5
03	223 22.4	. . 19.6	250 22.0	8.5	26 29.5	0.1	55.5
04	238 22.4	19.0	264 49.5	8.5	26 29.6	0.1	55.6
05	253 22.4	18.5	279 17.0	8.5	26 29.5	0.2	55.6
06	268 22.4	N19 17.9	293 44.5	8.4	N26 29.3	0.4	55.6
07	283 22.4	17.3	308 11.9	8.4	26 28.9	0.5	55.6
08	298 22.4	16.8	322 39.3	8.4	26 28.4	0.6	55.6
09	313 22.4	. . 16.2	337 06.7	8.3	26 27.8	0.8	55.7
10	328 22.4	15.7	351 34.0	8.3	26 27.0	0.9	55.7
11	343 22.4	15.1	6 01.3	8.4	26 26.1	1.1	55.7
12	358 22.4	N19 14.5	20 28.7	8.2	N26 25.0	1.3	55.7
13	13 22.4	14.0	34 55.9	8.3	26 23.7	1.3	55.8
14	28 22.4	13.4	49 23.2	8.3	26 22.4	1.5	55.8
15	43 22.4	. . 12.8	63 50.5	8.2	26 20.9	1.7	55.8
16	58 22.4	12.3	78 17.7	8.2	26 19.2	1.8	55.8
17	73 22.5	11.7	92 44.9	8.3	26 17.4	2.0	55.8
18	88 22.5	N19 11.1	107 12.2	8.2	N26 15.4	2.1	55.9
19	103 22.5	10.6	121 39.4	8.2	26 13.3	2.2	55.9
20	118 22.5	10.0	136 06.6	8.2	26 11.1	2.4	55.9
21	133 22.5	. . 09.4	150 33.8	8.2	26 08.7	2.5	55.9
22	148 22.5	08.9	165 01.0	8.1	26 06.2	2.7	56.0
23	163 22.5	08.3	179 28.1	8.2	26 03.5	2.8	56.0
28 00	178 22.5	N19 07.7	193 55.3	8.2	N26 00.7	3.0	56.0
01	193 22.5	07.1	208 22.5	8.2	25 57.7	3.1	56.0
02	208 22.5	06.6	222 49.7	8.2	25 54.6	3.3	56.0
03	223 22.5	. . 06.0	237 16.9	8.1	25 51.3	3.4	56.1
04	238 22.5	05.4	251 44.0	8.2	25 47.9	3.5	56.1
05	253 22.5	04.9	266 11.2	8.2	25 44.4	3.7	56.1
06	268 22.6	N19 04.3	280 38.4	8.2	N25 40.7	3.9	56.1
07	283 22.6	03.7	295 05.6	8.2	25 36.8	4.0	56.2
08	298 22.6	03.1	309 32.8	8.2	25 32.8	4.1	56.2
09	313 22.6	. . 02.6	324 00.0	8.3	25 28.7	4.3	56.2
10	328 22.6	02.0	338 27.3	8.2	25 24.4	4.4	56.2
11	343 22.6	01.4	352 54.5	8.2	25 20.0	4.6	56.3
12	358 22.6	N19 00.8	7 21.7	8.3	N25 15.4	4.7	56.3
13	13 22.6	19 00.3	21 49.0	8.3	25 10.7	4.8	56.3
14	28 22.6	18 59.7	36 16.3	8.3	25 05.9	5.0	56.3
15	43 22.6	. . 59.1	50 43.6	8.3	25 00.9	5.2	56.4
16	58 22.7	58.5	65 10.9	8.3	24 55.7	5.2	56.4
17	73 22.7	57.9	79 38.2	8.4	24 50.5	5.5	56.4
18	88 22.7	N18 57.4	94 05.6	8.4	N24 45.0	5.5	56.4
19	103 22.7	56.8	108 33.0	8.4	24 39.5	5.7	56.4
20	118 22.7	56.2	123 00.4	8.4	24 33.8	5.9	56.5
21	133 22.7	. . 55.6	137 27.8	8.4	24 27.9	6.0	56.5
22	148 22.7	55.0	151 55.2	8.5	24 21.9	6.1	56.5
23	163 22.7	54.5	166 22.7	8.5	24 15.8	6.2	56.5
29 00	178 22.8	N18 53.9	180 50.2	8.6	N24 09.6	6.4	56.6
01	193 22.8	53.3	195 17.8	8.5	24 03.2	6.6	56.6
02	208 22.8	52.7	209 45.3	8.6	23 56.6	6.7	56.6
03	223 22.8	. . 52.1	224 12.9	8.6	23 49.9	6.8	56.6
04	238 22.8	51.5	238 40.5	8.7	23 43.1	6.9	56.7
05	253 22.8	50.9	253 08.2	8.7	23 36.2	7.1	56.7
06	268 22.8	N18 50.4	267 35.9	8.7	N23 29.1	7.2	56.7
07	283 22.9	49.8	282 03.6	8.7	23 21.9	7.4	56.7
08	298 22.9	49.2	296 31.3	8.8	23 14.5	7.5	56.8
09	313 22.9	. . 48.6	310 59.1	8.9	23 07.0	7.6	56.8
10	328 22.9	48.0	325 27.0	8.8	22 59.4	7.7	56.8
11	343 22.9	47.4	339 54.8	8.9	22 51.7	7.9	56.8
12	358 22.9	N18 46.8	354 22.7	9.0	N22 43.8	8.0	56.8
13	13 23.0	46.2	8 50.7	9.0	22 35.8	8.2	56.9
14	28 23.0	45.6	23 18.7	9.0	22 27.6	8.2	56.9
15	43 23.0	. . 45.1	37 46.7	9.1	22 19.4	8.4	56.9
16	58 23.0	44.5	52 14.8	9.1	22 11.0	8.6	56.9
17	73 23.0	43.9	66 42.9	9.1	22 02.4	8.6	57.0
18	88 23.1	N18 43.3	81 11.0	9.2	N21 53.8	8.8	57.0
19	103 23.1	42.7	95 39.2	9.3	21 45.0	8.9	57.0
20	118 23.1	42.1	110 07.5	9.2	21 36.1	9.0	57.0
21	133 23.1	. . 41.5	124 35.7	9.4	21 27.1	9.2	57.1
22	148 23.1	40.9	139 04.1	9.3	21 17.9	9.3	57.1
23	163 23.1	40.3	153 32.4	9.4	N21 08.6	9.3	57.1
	SD 15.8	*d* 0.6	SD 15.2		15.3		15.5

Lat.	Twilight Naut.	Twilight Civil	Sunrise	Moonrise 27	Moonrise 28	Moonrise 29	Moonrise 30
°	h m	h m	h m	h m	h m	h m	h m
N 72	▭	▭	▭	▭	▭	▭	▭
N 70	////	////	00 22	▭	▭	▭	▭
68	////	////	01 51	▭	▭	▭	02 23
66	////	////	02 28	▭	▭	▭	03 09
64	////	01 09	02 54	23 21	25 40	01 40	03 39
62	////	01 56	03 14	24 43	00 43	02 18	04 01
60	////	02 26	03 30	00 16	01 19	02 44	04 19
N 58	01 03	02 48	03 44	00 44	01 45	03 05	04 34
56	01 46	03 05	03 56	01 05	02 06	03 22	04 47
54	02 13	03 20	04 06	01 23	02 23	03 36	04 58
52	02 34	03 32	04 15	01 38	02 38	03 49	05 08
50	02 50	03 43	04 23	01 51	02 50	04 00	05 16
45	03 22	04 06	04 40	02 18	03 16	04 23	05 35
N 40	03 45	04 23	04 54	02 40	03 37	04 41	05 50
35	04 04	04 38	05 06	02 57	03 54	04 57	06 02
30	04 19	04 50	05 16	03 13	04 09	05 10	06 13
20	04 42	05 10	05 34	03 39	04 34	05 33	06 32
N 10	05 00	05 27	05 49	04 01	04 56	05 52	06 48
0	05 16	05 41	06 03	04 22	05 16	06 11	07 03
S 10	05 29	05 55	06 17	04 43	05 37	06 29	07 18
20	05 42	06 08	06 32	05 05	05 58	06 48	07 34
30	05 54	06 23	06 48	05 31	06 23	07 10	07 53
35	06 00	06 31	06 58	05 47	06 38	07 24	08 03
40	06 07	06 40	07 09	06 05	06 55	07 39	08 16
45	06 14	06 50	07 22	06 26	07 16	07 56	08 30
S 50	06 22	07 02	07 38	06 54	07 41	08 18	08 47
52	06 26	07 07	07 45	07 07	07 54	08 29	08 55
54	06 30	07 13	07 53	07 22	08 08	08 41	09 04
56	06 34	07 20	08 02	07 40	08 24	08 54	09 15
58	06 38	07 27	08 13	08 02	08 44	09 10	09 26
S 60	06 43	07 35	08 25	08 31	09 08	09 29	09 40

Lat.	Sunset	Twilight Civil	Twilight Naut.	Moonset 27	Moonset 28	Moonset 29	Moonset 30
°	h m	h m	h m	h m	h m	h m	h m
N 72	▭	▭	▭	▭	▭	▭	▭
N 70	23 28	////	////	▭	▭	▭	23 18
68	22 18	////	////	▭	▭	23 32	22 48
66	21 42	////	////	▭	▭	22 45	22 26
64	21 17	22 57	////	22 47	22 22	22 14	22 08
62	20 57	22 13	////	21 25	21 44	21 51	21 53
60	20 41	21 45	////	20 49	21 17	21 32	21 41
N 58	20 28	21 23	23 04	20 22	20 56	21 16	21 30
56	20 16	21 06	22 24	20 02	20 38	21 03	21 20
54	20 06	20 52	21 57	19 44	20 23	20 51	21 12
52	19 57	20 39	21 37	19 30	20 10	20 41	21 04
50	19 49	20 28	21 21	19 17	19 59	20 32	20 57
45	19 32	20 06	20 50	18 50	19 35	20 12	20 43
N 40	19 18	19 49	20 27	18 29	19 16	19 56	20 31
35	19 07	19 35	20 09	18 12	19 00	19 43	20 20
30	18 56	19 23	19 54	17 57	18 46	19 31	20 11
20	18 39	19 03	19 31	17 31	18 22	19 11	19 55
N 10	18 24	18 46	19 12	17 09	18 02	18 53	19 41
0	18 10	18 32	18 57	16 48	17 42	18 36	19 28
S 10	17 56	18 18	18 44	16 27	17 23	18 19	19 14
20	17 42	18 05	18 32	16 05	17 02	18 01	19 00
30	17 25	17 50	18 19	15 39	16 38	17 40	18 44
35	17 15	17 42	18 13	15 24	16 24	17 28	18 34
40	17 04	17 33	18 06	15 06	16 07	17 14	18 23
45	16 51	17 23	17 59	14 44	15 47	16 57	18 10
S 50	16 36	17 12	17 51	14 17	15 22	16 36	17 54
52	16 28	17 06	17 48	14 04	15 10	16 25	17 46
54	16 20	17 00	17 44	13 49	14 56	16 14	17 38
56	16 11	16 54	17 40	13 31	14 40	16 01	17 28
58	16 01	16 47	17 35	13 09	14 20	15 46	17 17
S 60	15 49	16 39	17 31	12 40	13 56	15 28	17 05

Day	SUN Eqn. of Time 00^h	SUN Eqn. of Time 12^h	SUN Mer. Pass.	MOON Mer. Pass. Upper	MOON Mer. Pass. Lower	MOON Age	MOON Phase
d	m s	m s	h m	h m	h m	d %	
27	06 30	06 30	12 07	10 35	23 02	28 4	
28	06 30	06 30	12 06	11 29	23 57	29 1	
29	06 29	06 28	12 06	12 23	24 50	00 0	

UT	ARIES	VENUS −3.9		MARS −2.3		JUPITER −1.7		SATURN +0.1		STARS		
	GHA	GHA	Dec	GHA	Dec	GHA	Dec	GHA	Dec	Name	SHA	Dec
d h	° ′	° ′	° ′	° ′	° ′	° ′	° ′	° ′	° ′		° ′	°
30 00	307 13.2	183 47.8	N20 45.6	323 13.6	S13 20.9	160 46.9	N14 19.8	209 30.4	N22 29.1	Acamar	315 24.2	S40 17
01	322 15.7	198 47.1	45.0	338 16.0	21.0	175 48.9	19.6	224 32.5	29.1	Achernar	335 32.2	S57 12
02	337 18.1	213 46.3	44.4	353 18.5	21.1	190 50.8	19.5	239 34.7	29.1	Acrux	173 18.7	S63 07
03	352 20.6	228 45.5	. . 43.8	8 20.9	. . 21.3	205 52.8	. . 19.3	254 36.8	. . 29.1	Adhara	255 19.0	S28 58
04	7 23.1	243 44.8	43.2	23 23.3	21.4	220 54.7	19.1	269 39.0	29.1	Aldebaran	290 58.5	N16 31
05	22 25.5	258 44.0	42.6	38 25.8	21.5	235 56.7	18.9	284 41.1	29.1			
06	37 28.0	273 43.2	N20 42.0	53 28.2	S13 21.7	250 58.6	N14 18.8	299 43.2	N22 29.0	Alioth	166 27.5	N55 56
W 07	52 30.5	288 42.5	41.4	68 30.7	21.8	266 00.5	18.6	314 45.4	29.0	Alkaid	153 04.9	N49 18
E 08	67 32.9	303 41.7	40.8	83 33.1	21.9	281 02.5	18.4	329 47.5	29.0	Al Na'ir	27 52.9	S46 56
D 09	82 35.4	318 41.0	. . 40.2	98 35.6	. . 22.1	296 04.4	. . 18.2	344 49.7	. . 29.0	Alnilam	275 54.5	S 1 11
N 10	97 37.8	333 40.2	39.6	113 38.0	22.2	311 06.4	18.1	359 51.8	29.0	Alphard	218 04.1	S 8 40
E 11	112 40.3	348 39.4	38.9	128 40.5	22.3	326 08.3	17.9	14 53.9	29.0			
S 12	127 42.8	3 38.7	N20 38.3	143 42.9	S13 22.5	341 10.3	N14 17.7	29 56.1	N22 29.0	Alphecca	126 17.4	N26 42
D 13	142 45.2	18 37.9	37.7	158 45.4	22.6	356 12.2	17.5	44 58.2	28.9	Alpheratz	357 51.4	N29 06
A 14	157 47.7	33 37.1	37.1	173 47.8	22.7	11 14.2	17.4	60 00.4	28.9	Altair	62 15.5	N 8 52
Y 15	172 50.2	48 36.4	. . 36.5	188 50.3	. . 22.9	26 16.1	. . 17.2	75 02.5	. . 28.9	Ankaa	353 23.0	S42 17
16	187 52.6	63 35.6	35.9	203 52.7	23.0	41 18.0	17.0	90 04.7	28.9	Antares	112 35.7	S26 26
17	202 55.1	78 34.9	35.3	218 55.2	23.1	56 20.0	16.8	105 06.8	28.9			
18	217 57.6	93 34.1	N20 34.7	233 57.6	S13 23.3	71 21.9	N14 16.7	120 09.0	N22 28.9	Arcturus	146 02.8	N19 10
19	233 00.0	108 33.3	34.0	249 00.1	23.4	86 23.9	16.5	135 11.1	28.9	Atria	107 44.2	S69 02
20	248 02.5	123 32.6	33.4	264 02.6	23.5	101 25.8	16.3	150 13.2	28.9	Avior	234 22.0	S59 31
21	263 04.9	138 31.8	. . 32.8	279 05.0	. . 23.7	116 27.8	. . 16.1	165 15.4	. . 28.8	Bellatrix	278 40.6	N 6 21
22	278 07.4	153 31.1	32.2	294 07.5	23.8	131 29.7	16.0	180 17.5	28.8	Betelgeuse	271 10.0	N 7 24
23	293 09.9	168 30.3	31.6	309 09.9	24.0	146 31.7	15.8	195 19.7	28.8			
31 00	308 12.3	183 29.6	N20 30.9	324 12.4	S13 24.1	161 33.6	N14 15.6	210 21.8	N22 28.8	Canopus	264 00.1	S52 41.
01	323 14.8	198 28.8	30.3	339 14.9	24.2	176 35.5	15.4	225 24.0	28.8	Capella	280 46.3	N46 00
02	338 17.3	213 28.0	29.7	354 17.3	24.4	191 37.5	15.3	240 26.1	28.8	Deneb	49 36.4	N45 17.
03	353 19.7	228 27.3	. . 29.1	9 19.8	. . 24.5	206 39.4	. . 15.1	255 28.2	. . 28.8	Denebola	182 41.8	N14 33.
04	8 22.2	243 26.5	28.4	24 22.3	24.6	221 41.4	14.9	270 30.4	28.7	Diphda	349 03.5	S17 57.
05	23 24.7	258 25.8	27.8	39 24.8	24.8	236 43.3	14.7	285 32.5	28.7			
06	38 27.1	273 25.0	N20 27.2	54 27.2	S13 24.9	251 45.3	N14 14.5	300 34.7	N22 28.7	Dubhe	194 01.4	N61 44.
T 07	53 29.6	288 24.3	26.6	69 29.7	25.1	266 47.2	14.4	315 36.8	28.7	Elnath	278 22.7	N28 36.
H 08	68 32.1	303 23.5	25.9	84 32.2	25.2	281 49.2	14.2	330 39.0	28.7	Eltanin	90 49.3	N51 29.
U 09	83 34.5	318 22.8	. . 25.3	99 34.7	. . 25.4	296 51.1	. . 14.0	345 41.1	. . 28.7	Enif	33 54.5	N 9 53.
R 10	98 37.0	333 22.0	24.7	114 37.2	25.5	311 53.0	13.8	0 43.3	28.7	Fomalhaut	15 32.1	S29 36.
11	113 39.4	348 21.3	24.0	129 39.6	25.6	326 55.0	13.7	15 45.4	28.6			
S 12	128 41.9	3 20.5	N20 23.4	144 42.1	S13 25.8	341 56.9	N14 13.5	30 47.5	N22 28.6	Gacrux	172 10.1	S57 08.
D 13	143 44.4	18 19.8	22.8	159 44.6	25.9	356 58.9	13.3	45 49.7	28.6	Gienah	176 00.5	S17 33.
A 14	158 46.8	33 19.0	22.1	174 47.1	26.1	12 00.8	13.1	60 51.8	28.6	Hadar	148 59.3	S60 23.
Y 15	173 49.3	48 18.3	. . 21.5	189 49.6	. . 26.2	27 02.8	. . 13.0	75 54.0	. . 28.6	Hamal	328 09.6	N23 28.
16	188 51.8	63 17.5	20.9	204 52.1	26.4	42 04.7	12.8	90 56.1	28.6	Kaus Aust.	83 53.8	S34 23.
17	203 54.2	78 16.8	20.2	219 54.5	26.5	57 06.7	12.6	105 58.3	28.6			
18	218 56.7	93 16.0	N20 19.6	234 57.0	S13 26.6	72 08.6	N14 12.4	121 00.4	N22 28.5	Kochab	137 19.2	N74 08.
19	233 59.2	108 15.3	18.9	249 59.5	26.8	87 10.5	12.3	136 02.6	28.5	Markab	13 45.9	N15 13.
20	249 01.6	123 14.5	18.3	265 02.0	26.9	102 12.5	12.1	151 04.7	28.5	Menkar	314 23.3	N 4 06.
21	264 04.1	138 13.8	. . 17.7	280 04.5	. . 27.1	117 14.4	. . 11.9	166 06.8	. . 28.5	Menkent	148 16.9	S36 23.
22	279 06.6	153 13.0	17.0	295 07.0	27.2	132 16.4	11.7	181 09.0	28.5	Miaplacidus	221 42.6	S69 43.
23	294 09.0	168 12.3	16.4	310 09.5	27.4	147 18.3	11.5	196 11.1	28.5			
1 00	309 11.5	183 11.5	N20 15.7	325 12.0	S13 27.5	162 20.3	N14 11.4	211 13.3	N22 28.5	Mirfak	308 51.8	N49 52.
01	324 13.9	198 10.8	15.1	340 14.5	27.7	177 22.2	11.2	226 15.4	28.4	Nunki	76 07.6	S26 17.
02	339 16.4	213 10.0	14.4	355 17.0	27.8	192 24.1	11.0	241 17.6	28.4	Peacock	53 30.7	S56 43.
03	354 18.9	228 09.3	. . 13.8	10 19.5	. . 28.0	207 26.1	. . 10.8	256 19.7	. . 28.4	Pollux	243 37.6	N28 01.
04	9 21.3	243 08.5	13.1	25 22.0	28.1	222 28.0	10.7	271 21.9	28.4	Procyon	245 08.2	N 5 13.
05	24 23.8	258 07.8	12.5	40 24.5	28.3	237 30.0	10.5	286 24.0	28.4			
06	39 26.3	273 07.0	N20 11.8	55 27.0	S13 28.4	252 31.9	N14 10.3	301 26.2	N22 28.4	Rasalhague	96 13.4	N12 33.
07	54 28.7	288 06.3	11.2	70 29.5	28.6	267 33.9	10.1	316 28.3	28.4	Regulus	207 52.1	N11 57.
08	69 31.2	303 05.6	10.5	85 32.1	28.7	282 35.8	10.0	331 30.4	28.3	Rigel	281 19.8	S 8 11.
F 09	84 33.7	318 04.8	. . 09.9	100 34.6	. . 28.9	297 37.8	. . 09.8	346 32.6	. . 28.3	Rigil Kent.	140 02.7	S60 51.
R 10	99 36.1	333 04.1	09.2	115 37.1	29.0	312 39.7	09.6	1 34.7	28.3	Sabik	102 21.2	S15 43.
I 11	114 38.6	348 03.3	08.5	130 39.6	29.2	327 41.6	09.4	16 36.9	28.3			
D 12	129 41.0	3 02.6	N20 07.9	145 42.1	S13 29.3	342 43.6	N14 09.2	31 39.0	N22 28.3	Schedar	349 49.4	N56 33.
A 13	144 43.5	18 01.8	07.2	160 44.6	29.5	357 45.5	09.1	46 41.2	28.3	Shaula	96 32.2	S37 06.
Y 14	159 46.0	33 01.1	06.6	175 47.1	29.6	12 47.5	08.9	61 43.3	28.2	Sirius	258 40.9	S16 43.
15	174 48.4	48 00.4	. . 05.9	190 49.7	. . 29.8	27 49.4	. . 08.7	76 45.5	. . 28.2	Spica	158 39.6	S11 10.
16	189 50.9	62 59.6	05.2	205 52.2	29.9	42 51.4	08.5	91 47.6	28.2	Suhail	222 58.7	S43 26.
17	204 53.4	77 58.9	04.6	220 54.7	30.1	57 53.3	08.4	106 49.8	28.2			
18	219 55.8	92 58.1	N20 03.9	235 57.2	S13 30.3	72 55.2	N14 08.2	121 51.9	N22 28.2	Vega	80 43.9	N38 47.
19	234 58.3	107 57.4	03.3	250 59.8	30.4	87 57.2	08.0	136 54.1	28.2	Zuben'ubi	137 14.0	S16 03.
20	250 00.8	122 56.7	02.6	266 02.3	30.6	102 59.1	07.8	151 56.2	28.2		SHA	Mer. Pas
21	265 03.2	137 55.9	. . 01.9	281 04.8	. . 30.7	118 01.1	. . 07.6	166 58.3	. . 28.1		° ′	h m
22	280 05.7	152 55.2	01.3	296 07.3	30.9	133 03.0	07.5	182 00.5	28.1	Venus	235 17.2	11 47
23	295 08.2	167 54.5	00.6	311 09.9	31.0	148 05.0	07.3	197 02.6	28.1	Mars	16 00.1	2 23
	h m									Jupiter	213 21.3	13 12
Mer. Pass.	3 26.6	v −0.8	d 0.6	v 2.5	d 0.1	v 1.9	d 0.2	v 2.1	d 0.0	Saturn	262 09.5	9 57

	UT	SUN GHA	SUN Dec	MOON GHA	v	MOON Dec	d	HP
	d h	° ′	° ′	° ′	′	° ′	′	′
	30 00	178 23.2	N18 39.7	168 00.8	9.5	N20 59.3	9.6	57.1
	01	193 23.2	39.1	182 29.3	9.5	20 49.7	9.6	57.1
	02	208 23.2	38.5	196 57.8	9.6	20 40.1	9.7	57.2
	03	223 23.2	. . 37.9	211 26.4	9.6	20 30.4	9.9	57.2
	04	238 23.3	37.3	225 55.0	9.6	20 20.5	10.0	57.2
	05	253 23.3	36.7	240 23.6	9.7	20 10.5	10.1	57.2
	06	268 23.3	N18 36.1	254 52.3	9.8	N20 00.4	10.2	57.3
W	07	283 23.3	35.5	269 21.1	9.7	19 50.2	10.4	57.3
E	08	298 23.3	34.9	283 49.8	9.9	19 39.8	10.4	57.3
	09	313 23.4	. . 34.3	298 18.7	9.9	19 29.4	10.5	57.3
D	10	328 23.4	33.7	312 47.6	9.9	19 18.9	10.7	57.3
N	11	343 23.4	33.1	327 16.5	10.0	19 08.2	10.8	57.4
E	12	358 23.4	N18 32.5	341 45.5	10.0	N18 57.4	10.9	57.4
S	13	13 23.5	31.9	356 14.5	10.1	18 46.5	11.0	57.4
D	14	28 23.5	31.3	10 43.6	10.1	18 35.5	11.0	57.4
A	15	43 23.5	. . 30.7	25 12.7	10.2	18 24.5	11.2	57.5
Y	16	58 23.5	30.1	39 41.9	10.2	18 13.3	11.3	57.5
	17	73 23.6	29.5	54 11.1	10.3	18 02.0	11.5	57.5
	18	88 23.6	N18 28.9	68 40.4	10.3	N17 50.5	11.5	57.5
	19	103 23.6	28.3	83 09.7	10.4	17 39.0	11.6	57.5
	20	118 23.6	27.7	97 39.1	10.4	17 27.4	11.7	57.6
	21	133 23.7	. . 27.1	112 08.5	10.5	17 15.7	11.8	57.6
	22	148 23.7	26.5	126 38.0	10.5	17 03.9	11.9	57.6
	23	163 23.7	25.9	141 07.5	10.6	16 52.0	12.0	57.6
	31 00	178 23.7	N18 25.2	155 37.1	10.6	N16 40.0	12.1	57.6
	01	193 23.8	24.6	170 06.7	10.6	16 27.9	12.2	57.7
	02	208 23.8	24.0	184 36.3	10.7	16 15.7	12.2	57.7
	03	223 23.8	. . 23.4	199 06.0	10.8	16 03.5	12.4	57.7
	04	238 23.8	22.8	213 35.8	10.8	15 51.1	12.5	57.7
	05	253 23.9	22.2	228 05.6	10.8	15 38.6	12.5	57.7
	06	268 23.9	N18 21.6	242 35.4	10.9	N15 26.1	12.7	57.8
	07	283 23.9	21.0	257 05.3	10.9	15 13.4	12.7	57.8
T	08	298 24.0	20.4	271 35.2	11.0	15 00.7	12.8	57.8
H	09	313 24.0	. . 19.7	286 05.2	11.0	14 47.9	12.9	57.8
U	10	328 24.0	19.1	300 35.2	11.1	14 35.0	13.0	57.8
R	11	343 24.0	18.5	315 05.3	11.1	14 22.0	13.0	57.9
S	12	358 24.1	N18 17.9	329 35.4	11.2	N14 09.0	13.2	57.9
D	13	13 24.1	17.3	344 05.6	11.1	13 55.8	13.2	57.9
A	14	28 24.1	16.7	358 35.7	11.3	13 42.6	13.3	57.9
Y	15	43 24.2	. . 16.1	13 06.0	11.3	13 29.3	13.3	57.9
	16	58 24.2	15.4	27 36.3	11.3	13 16.0	13.5	58.0
	17	73 24.2	14.8	42 06.6	11.3	13 02.5	13.5	58.0
	18	88 24.3	N18 14.2	56 36.9	11.4	N12 49.0	13.6	58.0
	19	103 24.3	13.6	71 07.3	11.5	12 35.4	13.7	58.0
	20	118 24.3	13.0	85 37.8	11.4	12 21.7	13.7	58.0
	21	133 24.3	. . 12.3	100 08.2	11.5	12 08.0	13.8	58.1
	22	148 24.4	11.7	114 38.7	11.6	11 54.2	13.9	58.1
	23	163 24.4	11.1	129 09.3	11.6	11 40.3	13.9	58.1
	1 00	178 24.4	N18 10.5	143 39.9	11.6	N11 26.4	14.0	58.1
	01	193 24.5	09.9	158 10.5	11.6	11 12.4	14.1	58.1
	02	208 24.5	09.2	172 41.1	11.7	10 58.3	14.1	58.1
	03	223 24.5	. . 08.6	187 11.8	11.7	10 44.2	14.2	58.2
	04	238 24.6	08.0	201 42.5	11.8	10 30.0	14.3	58.2
	05	253 24.6	07.4	216 13.3	11.7	10 15.7	14.3	58.2
	06	268 24.7	N18 06.7	230 44.0	11.9	N10 01.4	14.4	58.2
	07	283 24.7	06.1	245 14.9	11.8	9 47.0	14.4	58.2
	08	298 24.7	05.5	259 45.7	11.9	9 32.6	14.5	58.2
F	09	313 24.8	. . 04.9	274 16.6	11.9	9 18.1	14.5	58.3
R	10	328 24.8	04.2	288 47.5	11.9	9 03.6	14.6	58.3
I	11	343 24.8	03.6	303 18.4	11.9	8 49.0	14.6	58.3
D	12	358 24.9	N18 03.0	317 49.3	12.0	N 8 34.4	14.7	58.3
A	13	13 24.9	02.4	332 20.3	12.0	8 19.7	14.8	58.3
Y	14	28 24.9	01.7	346 51.3	12.0	8 04.9	14.8	58.3
	15	43 25.0	. . 01.1	1 22.3	12.1	7 50.1	14.8	58.4
	16	58 25.0	18 00.5	15 53.4	12.0	7 35.3	14.9	58.4
	17	73 25.0	17 59.8	30 24.4	12.1	7 20.4	14.9	58.4
	18	88 25.1	N17 59.2	44 55.5	12.1	N 7 05.5	14.9	58.4
	19	103 25.1	58.6	59 26.6	12.1	6 50.6	15.1	58.4
	20	118 25.2	57.9	73 57.7	12.2	6 35.5	15.0	58.4
	21	133 25.2	. . 57.3	88 28.9	12.1	6 20.5	15.1	58.5
	22	148 25.2	56.7	103 00.0	12.2	6 05.4	15.1	58.5
	23	163 25.3	56.0	117 31.2	12.2	N 5 50.3	15.2	58.5
		SD 15.8	d 0.6	SD 15.6		15.8		15.9

Lat.	Twilight Naut.	Twilight Civil	Sunrise	Moonrise 30	Moonrise 31	Moonrise 1	Moonrise 2
°	h m	h m	h m	h m	h m	h m	h m
N 72	▭	▭	▭	▭	03 44	06 30	08 49
N 70	////	////	01 12	▭	04 26	06 47	08 54
68	////	////	02 07	02 23	04 54	07 00	08 58
66	////	////	02 40	03 09	05 15	07 10	09 01
64	////	01 30	03 03	03 39	05 31	07 19	09 04
62	////	02 09	03 22	04 01	05 45	07 26	09 06
60	////	02 35	03 37	04 19	05 56	07 33	09 08
N 58	01 22	02 56	03 50	04 34	06 06	07 38	09 10
56	01 57	03 12	04 01	04 47	06 15	07 43	09 11
54	02 22	03 26	04 11	04 58	06 23	07 48	09 13
52	02 41	03 38	04 19	05 08	06 29	07 52	09 14
50	02 56	03 48	04 27	05 16	06 36	07 55	09 15
45	03 27	04 10	04 44	05 35	06 49	08 03	09 18
N 40	03 49	04 26	04 57	05 50	07 00	08 10	09 20
35	04 06	04 40	05 08	06 02	07 09	08 15	09 22
30	04 21	04 52	05 18	06 13	07 17	08 20	09 23
20	04 43	05 11	05 35	06 32	07 31	08 29	09 26
N 10	05 01	05 27	05 49	06 48	07 43	08 36	09 28
0	05 16	05 41	06 03	07 03	07 54	08 43	09 31
S 10	05 29	05 54	06 16	07 18	08 05	08 50	09 33
20	05 41	06 07	06 30	07 34	08 17	08 57	09 35
30	05 52	06 21	06 46	07 53	08 31	09 05	09 38
35	05 58	06 29	06 56	08 03	08 38	09 10	09 40
40	06 05	06 37	07 06	08 16	08 47	09 16	09 42
45	06 11	06 47	07 19	08 30	08 58	09 22	09 44
S 50	06 19	06 58	07 34	08 47	09 10	09 29	09 46
52	06 22	07 03	07 41	08 55	09 16	09 33	09 47
54	06 26	07 09	07 48	09 04	09 22	09 36	09 49
56	06 29	07 15	07 57	09 15	09 29	09 40	09 50
58	06 33	07 22	08 07	09 26	09 37	09 45	09 52
S 60	06 38	07 29	08 18	09 40	09 46	09 50	09 53

Lat.	Sunset	Twilight Civil	Twilight Naut.	Moonset 30	Moonset 31	Moonset 1	Moonset 2
°	h m	h m	h m	h m	h m	h m	h m
N 72	▭	▭	▭	▭	{00 01 / 22 59}	22 22	21 51
N 70	22 52	////	////	23 18	22 40	22 14	21 52
68	22 01	////	////	22 48	22 25	22 08	21 52
66	21 30	////	////	22 26	22 13	22 02	21 52
64	21 07	22 37	////	22 08	22 02	21 57	21 52
62	20 49	22 00	////	21 53	21 54	21 53	21 52
60	20 34	21 35	////	21 41	21 46	21 49	21 52
N 58	20 21	21 15	22 46	21 30	21 39	21 46	21 52
56	20 10	20 59	22 12	21 20	21 33	21 43	21 52
54	20 01	20 46	21 49	21 12	21 28	21 41	21 53
52	19 52	20 34	21 30	21 04	21 23	21 38	21 53
50	19 45	20 24	21 15	20 57	21 18	21 36	21 53
45	19 29	20 02	20 45	20 43	21 08	21 31	21 53
N 40	19 15	19 46	20 23	20 31	21 00	21 27	21 53
35	19 04	19 32	20 06	20 20	20 53	21 24	21 53
30	18 54	19 20	19 52	20 11	20 47	21 21	21 53
20	18 38	19 01	19 29	19 55	20 36	21 15	21 53
N 10	18 23	18 45	19 12	19 41	20 27	21 10	21 53
0	18 10	18 32	18 57	19 28	20 18	21 06	21 53
S 10	17 57	18 19	18 44	19 14	20 08	21 01	21 53
20	17 43	18 06	18 32	19 00	19 59	20 56	21 53
30	17 27	17 52	18 21	18 44	19 47	20 50	21 53
35	17 17	17 44	18 15	18 34	19 41	20 47	21 53
40	17 07	17 36	18 09	18 23	19 33	20 43	21 53
45	16 55	17 26	18 02	18 10	19 24	20 39	21 53
S 50	16 40	17 15	17 54	17 54	19 14	20 33	21 53
52	16 33	17 10	17 51	17 46	19 09	20 31	21 53
54	16 25	17 05	17 48	17 38	19 03	20 28	21 53
56	16 16	16 58	17 44	17 28	18 57	20 25	21 53
58	16 07	16 52	17 40	17 17	18 50	20 22	21 53
S 60	15 55	16 44	17 36	17 05	18 42	20 18	21 53

Day	SUN Eqn. of Time 00^h	SUN Eqn. of Time 12^h	SUN Mer. Pass.	MOON Mer. Pass. Upper	MOON Mer. Pass. Lower	MOON Age	MOON Phase
d	m s	m s	h m	h m	h m	d %	
30	06 27	06 26	12 06	13 16	00 50	01 2	
31	06 25	06 24	12 06	14 06	01 41	02 6	
1	06 22	06 21	12 06	14 54	02 30	03 12	

UT d	UT h	ARIES GHA	VENUS −3.9 GHA	VENUS Dec	MARS −2.4 GHA	MARS Dec	JUPITER −1.7 GHA	JUPITER Dec	SATURN +0.1 GHA	SATURN Dec
		° ′	° ′	° ′	° ′	° ′	° ′	° ′	° ′	° ′
2	00	310 10.6	182 53.7	N19 59.9	326 12.4	S13 31.2	163 06.9	N14 07.1	212 04.8	N22 28.1
	01	325 13.1	197 53.0	59.2	341 14.9	31.3	178 08.8	06.9	227 06.9	28.1
	02	340 15.5	212 52.3	58.6	356 17.5	31.5	193 10.8	06.8	242 09.1	28.1
	03	355 18.0	227 51.5	. . 57.9	11 20.0	. . 31.7	208 12.7	. . 06.6	257 11.2	. . 28.1
	04	10 20.5	242 50.8	57.2	26 22.5	31.8	223 14.7	06.4	272 13.4	28.0
	05	25 22.9	257 50.0	56.5	41 25.1	32.0	238 16.6	06.2	287 15.5	28.0
	06	40 25.4	272 49.3	N19 55.9	56 27.6	S13 32.1	253 18.6	N14 06.1	302 17.7	N22 28.0
S	07	55 27.9	287 48.6	55.2	71 30.2	32.3	268 20.5	05.9	317 19.8	28.0
A	08	70 30.3	302 47.8	54.5	86 32.7	32.5	283 22.4	05.7	332 22.0	28.0
T	09	85 32.8	317 47.1	. . 53.8	101 35.2	. . 32.6	298 24.4	. . 05.5	347 24.1	. . 28.0
U	10	100 35.3	332 46.4	53.2	116 37.8	32.8	313 26.3	05.3	2 26.3	28.0
R	11	115 37.7	347 45.7	52.5	131 40.3	32.9	328 28.3	05.2	17 28.4	27.9
D	12	130 40.2	2 44.9	N19 51.8	146 42.9	S13 33.1	343 30.2	N14 05.0	32 30.6	N22 27.9
A	13	145 42.6	17 44.2	51.1	161 45.4	33.3	358 32.2	04.8	47 32.7	27.9
Y	14	160 45.1	32 43.5	50.4	176 48.0	33.4	13 34.1	04.6	62 34.9	27.9
	15	175 47.6	47 42.7	. . 49.7	191 50.5	. . 33.6	28 36.0	. . 04.4	77 37.0	. . 27.9
	16	190 50.0	62 42.0	49.1	206 53.1	33.8	43 38.0	04.3	92 39.2	27.9
	17	205 52.5	77 41.3	48.4	221 55.6	33.9	58 39.9	04.1	107 41.3	27.9
	18	220 55.0	92 40.5	N19 47.7	236 58.2	S13 34.1	73 41.9	N14 03.9	122 43.5	N22 27.8
	19	235 57.4	107 39.8	47.0	252 00.8	34.2	88 43.8	03.7	137 45.6	27.8
	20	250 59.9	122 39.1	46.3	267 03.3	34.4	103 45.8	03.6	152 47.8	27.8
	21	266 02.4	137 38.4	. . 45.6	282 05.9	. . 34.6	118 47.7	. . 03.4	167 49.9	. . 27.8
	22	281 04.8	152 37.6	44.9	297 08.4	34.7	133 49.6	03.2	182 52.0	27.8
	23	296 07.3	167 36.9	44.2	312 11.0	34.9	148 51.6	03.0	197 54.2	27.8
3	00	311 09.8	182 36.2	N19 43.5	327 13.6	S13 35.1	163 53.5	N14 02.8	212 56.3	N22 27.8
	01	326 12.2	197 35.5	42.8	342 16.1	35.2	178 55.5	02.7	227 58.5	27.7
	02	341 14.7	212 34.7	42.1	357 18.7	35.4	193 57.4	02.5	243 00.6	27.7
	03	356 17.1	227 34.0	. . 41.4	12 21.3	. . 35.6	208 59.4	. . 02.3	258 02.8	. . 27.7
	04	11 19.6	242 33.3	40.7	27 23.8	35.7	224 01.3	02.1	273 04.9	27.7
	05	26 22.1	257 32.6	40.0	42 26.4	35.9	239 03.2	02.0	288 07.1	27.7
	06	41 24.5	272 31.8	N19 39.3	57 29.0	S13 36.1	254 05.2	N14 01.8	303 09.2	N22 27.7
	07	56 27.0	287 31.1	38.6	72 31.5	36.2	269 07.1	01.6	318 11.4	27.6
S	08	71 29.5	302 30.4	37.9	87 34.1	36.4	284 09.1	01.4	333 13.5	27.6
U	09	86 31.9	317 29.7	. . 37.2	102 36.7	. . 36.6	299 11.0	. . 01.2	348 15.7	. . 27.6
N	10	101 34.4	332 28.9	36.5	117 39.3	36.7	314 13.0	01.1	3 17.8	27.6
D	11	116 36.9	347 28.2	35.8	132 41.8	36.9	329 14.9	00.9	18 20.0	27.6
A	12	131 39.3	2 27.5	N19 35.1	147 44.4	S13 37.1	344 16.8	N14 00.7	33 22.1	N22 27.6
Y	13	146 41.8	17 26.8	34.4	162 47.0	37.3	359 18.8	00.5	48 24.3	27.6
	14	161 44.3	32 26.1	33.7	177 49.6	37.4	14 20.7	00.3	63 26.4	27.5
	15	176 46.7	47 25.3	. . 33.0	192 52.2	. . 37.6	29 22.7	. . 00.2	78 28.6	. . 27.5
	16	191 49.2	62 24.6	32.3	207 54.8	37.8	44 24.6	14 00.0	93 30.7	27.5
	17	206 51.6	77 23.9	31.6	222 57.3	37.9	59 26.5	13 59.8	108 32.9	27.5
	18	221 54.1	92 23.2	N19 30.9	237 59.9	S13 38.1	74 28.5	N13 59.6	123 35.0	N22 27.5
	19	236 56.6	107 22.5	30.2	253 02.5	38.3	89 30.4	59.5	138 37.2	27.5
	20	251 59.0	122 21.8	29.4	268 05.1	38.5	104 32.4	59.3	153 39.3	27.5
	21	267 01.5	137 21.0	. . 28.7	283 07.7	. . 38.6	119 34.3	. . 59.1	168 41.5	. . 27.4
	22	282 04.0	152 20.3	28.0	298 10.3	38.8	134 36.3	58.9	183 43.6	27.4
	23	297 06.4	167 19.6	27.3	313 12.9	39.0	149 38.2	58.7	198 45.8	27.4
4	00	312 08.9	182 18.9	N19 26.6	328 15.5	S13 39.2	164 40.1	N13 58.6	213 47.9	N22 27.4
	01	327 11.4	197 18.2	25.9	343 18.1	39.3	179 42.1	58.4	228 50.1	27.4
	02	342 13.8	212 17.5	25.1	358 20.7	39.5	194 44.0	58.2	243 52.3	27.4
	03	357 16.3	227 16.7	. . 24.4	13 23.3	. . 39.7	209 46.0	. . 58.0	258 54.4	. . 27.3
	04	12 18.7	242 16.0	23.7	28 25.9	39.9	224 47.9	57.8	273 56.6	27.3
	05	27 21.2	257 15.3	23.0	43 28.5	40.0	239 49.9	57.7	288 58.7	27.3
	06	42 23.7	272 14.6	N19 22.3	58 31.1	S13 40.2	254 51.8	N13 57.5	304 00.9	N22 27.3
	07	57 26.1	287 13.9	21.5	73 33.7	40.4	269 53.7	57.3	319 03.0	27.3
	08	72 28.6	302 13.2	20.8	88 36.3	40.6	284 55.7	57.1	334 05.2	27.3
M	09	87 31.1	317 12.5	. . 20.1	103 38.9	. . 40.7	299 57.6	. . 56.9	349 07.3	. . 27.3
O	10	102 33.5	332 11.8	19.4	118 41.5	40.9	314 59.6	56.8	4 09.5	27.2
N	11	117 36.0	347 11.1	18.6	133 44.1	41.1	330 01.5	56.6	19 11.6	27.2
D	12	132 38.5	2 10.3	N19 17.9	148 46.7	S13 41.3	345 03.4	N13 56.4	34 13.8	N22 27.2
A	13	147 40.9	17 09.6	17.2	163 49.4	41.5	0 05.4	56.2	49 15.9	27.2
Y	14	162 43.4	32 08.9	16.4	178 52.0	41.6	15 07.3	56.1	64 18.1	27.2
	15	177 45.9	47 08.2	. . 15.7	193 54.6	. . 41.8	30 09.3	. . 55.9	79 20.2	. . 27.2
	16	192 48.3	62 07.5	15.0	208 57.2	42.0	45 11.2	55.7	94 22.4	27.2
	17	207 50.8	77 06.8	14.2	223 59.8	42.2	60 13.2	55.5	109 24.5	27.1
	18	222 53.2	92 06.1	N19 13.5	239 02.4	S13 42.3	75 15.1	N13 55.3	124 26.7	N22 27.1
	19	237 55.7	107 05.4	12.8	254 05.1	42.5	90 17.0	55.2	139 28.8	27.1
	20	252 58.2	122 04.7	12.0	269 07.7	42.7	105 19.0	55.0	154 31.0	27.1
	21	268 00.6	137 04.0	. . 11.3	284 10.3	. . 42.9	120 20.9	. . 54.8	169 33.1	. . 27.1
	22	283 03.1	152 03.3	10.6	299 12.9	43.1	135 22.9	54.6	184 35.3	27.1
	23	298 05.6	167 02.6	09.8	314 15.6	43.3	150 24.8	54.4	199 37.4	27.0
Mer. Pass.		h m 3 14.8	*v* −0.7	*d* 0.7	*v* 2.6	*d* 0.2	*v* 1.9	*d* 0.2	*v* 2.2	*d* 0.0

STARS

Name	SHA	Dec
	° ′	° ′
Acamar	315 24.2	S40 17.2
Achernar	335 32.2	S57 12.8
Acrux	173 18.7	S63 07.3
Adhara	255 19.0	S28 58.4
Aldebaran	290 58.5	N16 31.0
Alioth	166 27.5	N55 56.8
Alkaid	153 04.9	N49 18.1
Al Na'ir	27 52.9	S46 56.6
Alnilam	275 54.5	S 1 11.8
Alphard	218 04.1	S 8 40.3
Alphecca	126 17.4	N26 42.4
Alpheratz	357 51.4	N29 06.5
Altair	62 15.5	N 8 52.7
Ankaa	353 23.0	S42 17.0
Antares	112 35.7	S26 26.5
Arcturus	146 02.8	N19 10.0
Atria	107 44.3	S69 02.3
Avior	234 22.0	S59 31.1
Bellatrix	278 40.6	N 6 21.3
Betelgeuse	271 10.0	N 7 24.6
Canopus	264 00.1	S52 41.6
Capella	280 46.3	N46 00.0
Deneb	49 36.4	N45 17.5
Denebola	182 41.8	N14 33.4
Diphda	349 03.5	S17 57.9
Dubhe	194 01.4	N61 44.2
Elnath	278 22.7	N28 36.6
Eltanin	90 49.3	N51 29.4
Enif	33 54.5	N 9 53.4
Fomalhaut	15 32.1	S29 36.1
Gacrux	172 10.2	S57 08.1
Gienah	176 00.6	S17 33.6
Hadar	148 59.3	S60 23.6
Hamal	328 09.6	N23 28.7
Kaus Aust.	83 53.8	S34 23.1
Kochab	137 19.2	N74 08.8
Markab	13 45.8	N15 13.4
Menkar	314 23.2	N 4 06.3
Menkent	148 16.9	S36 23.3
Miaplacidus	221 42.6	S69 43.8
Mirfak	308 51.7	N49 52.3
Nunki	76 07.6	S26 17.6
Peacock	53 30.7	S56 43.5
Pollux	243 37.6	N28 01.2
Procyon	245 08.2	N 5 13.1
Rasalhague	96 13.4	N12 33.5
Regulus	207 52.1	N11 57.2
Rigel	281 19.7	S 8 11.7
Rigil Kent.	140 02.7	S60 51.2
Sabik	102 21.3	S15 43.8
Schedar	349 49.4	N56 33.2
Shaula	96 32.2	S37 06.5
Sirius	258 40.9	S16 43.0
Spica	158 39.6	S11 10.7
Suhail	222 58.7	S43 26.7
Vega	80 43.9	N38 47.3
Zuben'ubi	137 14.0	S16 03.4

	SHA	Mer. Pass
	° ′	h m
Venus	231 26.4	11 50
Mars	16 03.8	2 11
Jupiter	212 43.8	13 03
Saturn	261 46.6	9 47

UT	SUN GHA	SUN Dec	MOON GHA	v	MOON Dec	d	HP
d h	° ′	° ′	° ′	′	° ′	′	′
2 00	178 25.3	N17 55.4	132 02.4	12.2	N 5 35.1	15.1	58.5
01	193 25.4	54.8	146 33.6	12.2	5 20.0	15.3	58.5
02	208 25.4	54.1	161 04.8	12.2	5 04.7	15.2	58.5
03	223 25.4	. . 53.5	175 36.0	12.3	4 49.5	15.3	58.5
04	238 25.5	52.9	190 07.3	12.2	4 34.2	15.3	58.6
05	253 25.5	52.2	204 38.5	12.3	4 18.9	15.4	58.6
06	268 25.6	N17 51.6	219 09.8	12.2	N 4 03.5	15.3	58.6
07	283 25.6	51.0	233 41.0	12.3	3 48.2	15.4	58.6
08	298 25.6	50.3	248 12.3	12.3	3 32.8	15.4	58.6
09	313 25.7	. . 49.7	262 43.6	12.3	3 17.4	15.5	58.6
10	328 25.7	49.1	277 14.9	12.2	3 01.9	15.5	58.6
11	343 25.8	48.4	291 46.1	12.3	2 46.4	15.4	58.7
12	358 25.8	N17 47.8	306 17.4	12.3	N 2 31.0	15.5	58.7
13	13 25.9	47.1	320 48.7	12.3	2 15.5	15.6	58.7
14	28 25.9	46.5	335 20.0	12.2	1 59.9	15.5	58.7
15	43 25.9	. . 45.9	349 51.2	12.3	1 44.4	15.5	58.7
16	58 26.0	45.2	4 22.5	12.3	1 28.9	15.6	58.7
17	73 26.0	44.6	18 53.8	12.2	1 13.3	15.6	58.7
18	88 26.1	N17 43.9	33 25.0	12.3	N 0 57.7	15.6	58.8
19	103 26.1	43.3	47 56.3	12.3	0 42.1	15.6	58.8
20	118 26.2	42.6	62 27.6	12.2	0 26.5	15.6	58.8
21	133 26.2	. . 42.0	76 58.8	12.2	N 0 10.9	15.6	58.8
22	148 26.3	41.3	91 30.0	12.2	S 0 04.7	15.6	58.8
23	163 26.3	40.7	106 01.2	12.3	0 20.3	15.6	58.8
3 00	178 26.3	N17 40.1	120 32.5	12.1	S 0 35.9	15.6	58.8
01	193 26.4	39.4	135 03.6	12.2	0 51.5	15.6	58.8
02	208 26.4	38.8	149 34.8	12.2	1 07.1	15.7	58.8
03	223 26.5	. . 38.1	164 06.0	12.1	1 22.8	15.6	58.9
04	238 26.5	37.5	178 37.1	12.2	1 38.4	15.6	58.9
05	253 26.6	36.8	193 08.3	12.1	1 54.0	15.6	58.9
06	268 26.6	N17 36.2	207 39.4	12.1	S 2 09.6	15.6	58.9
07	283 26.7	35.5	222 10.5	12.0	2 25.2	15.6	58.9
08	298 26.7	34.9	236 41.5	12.1	2 40.8	15.6	58.9
09	313 26.8	. . 34.2	251 12.6	12.0	2 56.4	15.6	58.9
10	328 26.8	33.6	265 43.6	12.0	3 12.0	15.6	58.9
11	343 26.9	32.9	280 14.6	12.0	3 27.6	15.5	58.9
12	358 26.9	N17 32.3	294 45.6	11.9	S 3 43.1	15.6	59.0
13	13 27.0	31.6	309 16.5	11.9	3 58.7	15.5	59.0
14	28 27.0	31.0	323 47.4	11.9	4 14.2	15.5	59.0
15	43 27.1	. . 30.3	338 18.3	11.8	4 29.7	15.5	59.0
16	58 27.1	29.7	352 49.1	11.8	4 45.2	15.5	59.0
17	73 27.2	29.0	7 19.9	11.8	5 00.7	15.4	59.0
18	88 27.2	N17 28.4	21 50.7	11.8	S 5 16.1	15.4	59.0
19	103 27.3	27.7	36 21.5	11.7	5 31.5	15.5	59.0
20	118 27.3	27.0	50 52.2	11.7	5 47.0	15.3	59.0
21	133 27.4	. . 26.4	65 22.9	11.6	6 02.3	15.4	59.0
22	148 27.4	25.7	79 53.5	11.6	6 17.7	15.3	59.1
23	163 27.5	25.1	94 24.1	11.6	6 33.0	15.3	59.1
4 00	178 27.5	N17 24.4	108 54.7	11.5	S 6 48.3	15.3	59.1
01	193 27.6	23.8	123 25.2	11.4	7 03.6	15.2	59.1
02	208 27.6	23.1	137 55.6	11.5	7 18.8	15.2	59.1
03	223 27.7	. . 22.4	152 26.1	11.3	7 34.0	15.2	59.1
04	238 27.7	21.8	166 56.4	11.4	7 49.2	15.1	59.1
05	253 27.8	21.1	181 26.8	11.3	8 04.3	15.1	59.1
06	268 27.9	N17 20.5	195 57.1	11.2	S 8 19.4	15.1	59.1
07	283 27.9	19.8	210 27.3	11.2	8 34.5	15.0	59.1
08	298 28.0	19.1	224 57.5	11.1	8 49.5	14.9	59.1
09	313 28.0	. . 18.5	239 27.6	11.1	9 04.4	15.0	59.1
10	328 28.1	17.8	253 57.7	11.0	9 19.4	14.8	59.2
11	343 28.1	17.1	268 27.7	11.0	9 34.2	14.9	59.2
12	358 28.2	N17 16.5	282 57.7	10.9	S 9 49.1	14.8	59.2
13	13 28.2	15.8	297 27.6	10.9	10 03.9	14.7	59.2
14	28 28.3	15.2	311 57.5	10.8	10 18.6	14.7	59.2
15	43 28.4	. . 14.5	326 27.3	10.7	10 33.3	14.6	59.2
16	58 28.4	13.8	340 57.0	10.7	10 47.9	14.6	59.2
17	73 28.5	13.2	355 26.7	10.7	11 02.5	14.5	59.2
18	88 28.5	N17 12.5	9 56.4	10.5	S11 17.0	14.5	59.2
19	103 28.6	11.8	24 25.9	10.5	11 31.5	14.4	59.2
20	118 28.6	11.2	38 55.4	10.5	11 45.9	14.3	59.2
21	133 28.7	. . 10.5	53 24.9	10.3	12 00.2	14.3	59.2
22	148 28.8	09.8	67 54.2	10.3	12 14.5	14.2	59.2
23	163 28.8	09.2	82 23.5	10.3	S12 28.7	14.2	59.2
	SD 15.8	*d* 0.7	SD 16.0		16.1		16.1

Lat.	Twilight Naut.	Twilight Civil	Sunrise	Moonrise 2	Moonrise 3	Moonrise 4	Moonrise 5
°	h m	h m	h m	h m	h m	h m	h m
N 72	▭	▭	▭	08 49	11 04	13 28	16 33
N 70	////	////	01 39	08 54	10 59	13 10	15 41
68	////	////	02 23	08 58	10 55	12 56	15 10
66	////	00 46	02 51	09 01	10 51	12 45	14 46
64	////	01 48	03 13	09 04	10 48	12 35	14 28
62	////	02 21	03 30	09 06	10 45	12 27	14 14
60	00 42	02 45	03 44	09 08	10 43	12 21	14 01
N 58	01 38	03 03	03 56	09 10	10 41	12 15	13 51
56	02 08	03 19	04 07	09 11	10 39	12 09	13 42
54	02 30	03 32	04 16	09 13	10 38	12 05	13 34
52	02 48	03 43	04 24	09 14	10 36	12 00	13 26
50	03 03	03 53	04 31	09 15	10 35	11 57	13 20
45	03 31	04 13	04 47	09 18	10 32	11 48	13 06
N 40	03 52	04 29	05 00	09 20	10 30	11 41	12 55
35	04 09	04 43	05 10	09 22	10 28	11 36	12 45
30	04 23	04 54	05 20	09 23	10 26	11 31	12 37
20	04 45	05 13	05 36	09 26	10 23	11 22	12 22
N 10	05 02	05 28	05 50	09 28	10 21	11 14	12 10
0	05 16	05 41	06 03	09 31	10 18	11 07	11 58
S 10	05 28	05 54	06 15	09 33	10 16	11 00	11 46
20	05 39	06 06	06 29	09 35	10 13	10 53	11 34
30	05 50	06 19	06 44	09 38	10 11	10 44	11 20
35	05 56	06 26	06 53	09 40	10 09	10 39	11 12
40	06 02	06 34	07 03	09 42	10 07	10 34	11 03
45	06 08	06 44	07 15	09 44	10 05	10 28	10 53
S 50	06 15	06 54	07 29	09 46	10 03	10 20	10 40
52	06 18	06 59	07 36	09 47	10 02	10 17	10 35
54	06 21	07 04	07 43	09 49	10 00	10 13	10 28
56	06 25	07 10	07 51	09 50	09 59	10 09	10 21
58	06 28	07 16	08 01	09 52	09 58	10 04	10 13
S 60	06 32	07 23	08 11	09 53	09 56	09 59	10 04

Lat.	Sunset	Twilight Civil	Twilight Naut.	Moonset 2	Moonset 3	Moonset 4	Moonset 5
°	h m	h m	h m	h m	h m	h m	h m
N 72	▭	▭	▭	21 51	21 20	20 41	19 24
N 70	22 26	////	////	21 52	21 29	21 01	20 18
68	21 46	////	////	21 52	21 36	21 17	20 51
66	21 18	23 14	////	21 52	21 42	21 30	21 16
64	20 57	22 19	////	21 52	21 47	21 41	21 35
62	20 40	21 48	////	21 52	21 51	21 51	21 51
60	20 27	21 25	23 19	21 52	21 55	21 59	22 04
N 58	20 15	21 07	22 30	21 52	21 59	22 06	22 16
56	20 04	20 52	22 01	21 52	22 02	22 12	22 26
54	19 55	20 39	21 40	21 53	22 05	22 18	22 35
52	19 47	20 28	21 23	21 53	22 07	22 23	22 43
50	19 40	20 18	21 08	21 53	22 09	22 28	22 50
45	19 25	19 58	20 40	21 53	22 14	22 38	23 05
N 40	19 12	19 42	20 19	21 53	22 19	22 47	23 18
35	19 01	19 29	20 03	21 53	22 22	22 54	23 29
30	18 52	19 18	19 49	21 53	22 26	23 00	23 39
20	18 36	19 00	19 27	21 53	22 31	23 12	23 55
N 10	18 22	18 45	19 10	21 53	22 36	23 22	24 10
0	18 10	18 31	18 57	21 53	22 41	23 31	24 23
S 10	17 57	18 19	18 44	21 53	22 46	23 40	24 37
20	17 44	18 07	18 33	21 53	22 51	23 50	24 52
30	17 28	17 54	18 22	21 53	22 57	24 02	00 02
35	17 20	17 46	18 17	21 53	23 00	24 08	00 08
40	17 10	17 38	18 11	21 53	23 04	24 16	00 16
45	16 58	17 29	18 05	21 53	23 08	24 25	00 25
S 50	16 44	17 19	17 58	21 53	23 14	24 36	00 36
52	16 37	17 14	17 55	21 53	23 16	24 41	00 41
54	16 30	17 09	17 52	21 53	23 19	24 46	00 46
56	16 22	17 03	17 49	21 53	23 22	24 53	00 53
58	16 13	16 57	17 45	21 53	23 25	25 00	01 00
S 60	16 02	16 50	17 41	21 53	23 29	25 08	01 08

Day	SUN Eqn. of Time 00^h	SUN Eqn. of Time 12^h	SUN Mer. Pass.	MOON Mer. Pass. Upper	MOON Mer. Pass. Lower	MOON Age	MOON Phase
d	m s	m s	h m	h m	h m	d %	
2	06 19	06 17	12 06	15 42	03 18	04 20	
3	06 15	06 12	12 06	16 30	04 06	05 30	
4	06 10	06 07	12 06	17 19	04 54	06 41	

d	h	ARIES GHA	VENUS −3.9 GHA	VENUS Dec	MARS −2.5 GHA	MARS Dec	JUPITER −1.7 GHA	JUPITER Dec	SATURN +0.1 GHA	SATURN Dec
		° ′	° ′	° ′	° ′	° ′	° ′	° ′	° ′	° ′
5 TUESDAY	00	313 08.0	182 01.9	N19 09.1	329 18.2	S13 43.4	165 26.7	N13 54.3	214 39.6	N22 27.0
	01	328 10.5	197 01.2	08.3	344 20.8	43.6	180 28.7	54.1	229 41.8	27.0
	02	343 13.0	212 00.5	07.6	359 23.4	43.8	195 30.6	53.9	244 43.9	27.0
	03	358 15.4	226 59.8	06.9	14 26.1	44.0	210 32.6	53.7	259 46.1	27.0
	04	13 17.9	241 59.1	06.1	29 28.7	44.2	225 34.5	53.5	274 48.2	27.0
	05	28 20.3	256 58.3	05.4	44 31.3	44.4	240 36.4	53.4	289 50.4	27.0
	06	43 22.8	271 57.6	N19 04.6	59 34.0	S13 44.5	255 38.4	N13 53.2	304 52.5	N22 26.9
	07	58 25.3	286 56.9	03.9	74 36.6	44.7	270 40.3	53.0	319 54.7	26.9
	08	73 27.7	301 56.2	03.1	89 39.2	44.9	285 42.3	52.8	334 56.8	26.9
	09	88 30.2	316 55.5	02.4	104 41.9	45.1	300 44.2	52.6	349 59.0	26.9
	10	103 32.7	331 54.8	01.6	119 44.5	45.3	315 46.2	52.5	5 01.1	26.9
	11	118 35.1	346 54.1	00.9	134 47.2	45.5	330 48.1	52.3	20 03.3	26.9
	12	133 37.6	1 53.4	N19 00.1	149 49.8	S13 45.6	345 50.0	N13 52.1	35 05.4	N22 26.8
	13	148 40.1	16 52.8	18 59.4	164 52.5	45.8	0 52.0	51.9	50 07.6	26.8
	14	163 42.5	31 52.1	58.6	179 55.1	46.0	15 53.9	51.7	65 09.7	26.8
	15	178 45.0	46 51.4	57.9	194 57.7	46.2	30 55.9	51.6	80 11.9	26.8
	16	193 47.5	61 50.7	57.1	210 00.4	46.4	45 57.8	51.4	95 14.1	26.8
	17	208 49.9	76 50.0	56.4	225 03.0	46.6	60 59.7	51.2	110 16.2	26.8
	18	223 52.4	91 49.3	N18 55.6	240 05.7	S13 46.8	76 01.7	N13 51.0	125 18.4	N22 26.8
	19	238 54.8	106 48.6	54.8	255 08.3	47.0	91 03.6	50.8	140 20.5	26.7
	20	253 57.3	121 47.9	54.1	270 11.0	47.1	106 05.6	50.7	155 22.7	26.7
	21	268 59.8	136 47.2	53.3	285 13.7	47.3	121 07.5	50.5	170 24.8	26.7
	22	284 02.2	151 46.5	52.6	300 16.3	47.5	136 09.4	50.3	185 27.0	26.7
	23	299 04.7	166 45.8	51.8	315 19.0	47.7	151 11.4	50.1	200 29.1	26.7
6 WEDNESDAY	00	314 07.2	181 45.1	N18 51.0	330 21.6	S13 47.9	166 13.3	N13 49.9	215 31.3	N22 26.7
	01	329 09.6	196 44.4	50.3	345 24.3	48.1	181 15.3	49.8	230 33.5	26.6
	02	344 12.1	211 43.7	49.5	0 26.9	48.3	196 17.2	49.6	245 35.6	26.6
	03	359 14.6	226 43.0	48.7	15 29.6	48.5	211 19.2	49.4	260 37.8	26.6
	04	14 17.0	241 42.3	48.0	30 32.3	48.7	226 21.1	49.2	275 39.9	26.6
	05	29 19.5	256 41.6	47.2	45 34.9	48.9	241 23.0	49.0	290 42.1	26.6
	06	44 22.0	271 41.0	N18 46.4	60 37.6	S13 49.1	256 25.0	N13 48.9	305 44.2	N22 26.6
	07	59 24.4	286 40.3	45.7	75 40.3	49.2	271 26.9	48.7	320 46.4	26.6
	08	74 26.9	301 39.6	44.9	90 42.9	49.4	286 28.9	48.5	335 48.5	26.5
	09	89 29.3	316 38.9	44.1	105 45.6	49.6	301 30.8	48.3	350 50.7	26.5
	10	104 31.8	331 38.2	43.4	120 48.3	49.8	316 32.7	48.1	5 52.9	26.5
	11	119 34.3	346 37.5	42.6	135 51.0	50.0	331 34.7	48.0	20 55.0	26.5
	12	134 36.7	1 36.8	N18 41.8	150 53.6	S13 50.2	346 36.6	N13 47.8	35 57.2	N22 26.5
	13	149 39.2	16 36.1	41.0	165 56.3	50.4	1 38.6	47.6	50 59.3	26.5
	14	164 41.7	31 35.4	40.3	180 59.0	50.6	16 40.5	47.4	66 01.5	26.4
	15	179 44.1	46 34.8	39.5	196 01.7	50.8	31 42.4	47.2	81 03.6	26.4
	16	194 46.6	61 34.1	38.7	211 04.3	51.0	46 44.4	47.1	96 05.8	26.4
	17	209 49.1	76 33.4	37.9	226 07.0	51.2	61 46.3	46.9	111 07.9	26.4
	18	224 51.5	91 32.7	N18 37.2	241 09.7	S13 51.4	76 48.3	N13 46.7	126 10.1	N22 26.4
	19	239 54.0	106 32.0	36.4	256 12.4	51.6	91 50.2	46.5	141 12.3	26.4
	20	254 56.5	121 31.3	35.6	271 15.1	51.8	106 52.1	46.3	156 14.4	26.4
	21	269 58.9	136 30.7	34.8	286 17.8	52.0	121 54.1	46.2	171 16.6	26.3
	22	285 01.4	151 30.0	34.0	301 20.4	52.2	136 56.0	46.0	186 18.7	26.3
	23	300 03.8	166 29.3	33.2	316 23.1	52.4	151 58.0	45.8	201 20.9	26.3
7 THURSDAY	00	315 06.3	181 28.6	N18 32.5	331 25.8	S13 52.5	166 59.9	N13 45.6	216 23.0	N22 26.3
	01	330 08.8	196 27.9	31.7	346 28.5	52.7	182 01.8	45.4	231 25.2	26.3
	02	345 11.2	211 27.2	30.9	1 31.2	52.9	197 03.8	45.2	246 27.4	26.3
	03	0 13.7	226 26.6	30.1	16 33.9	53.1	212 05.7	45.1	261 29.5	26.2
	04	15 16.2	241 25.9	29.3	31 36.6	53.3	227 07.7	44.9	276 31.7	26.2
	05	30 18.6	256 25.2	28.5	46 39.3	53.5	242 09.6	44.7	291 33.8	26.2
	06	45 21.1	271 24.5	N18 27.7	61 42.0	S13 53.7	257 11.5	N13 44.5	306 36.0	N22 26.2
	07	60 23.6	286 23.8	26.9	76 44.7	53.9	272 13.5	44.3	321 38.1	26.2
	08	75 26.0	301 23.2	26.2	91 47.4	54.1	287 15.4	44.2	336 40.3	26.2
	09	90 28.5	316 22.5	25.4	106 50.1	54.3	302 17.4	44.0	351 42.5	26.2
	10	105 30.9	331 21.8	24.6	121 52.8	54.5	317 19.3	43.8	6 44.6	26.1
	11	120 33.4	346 21.1	23.8	136 55.5	54.7	332 21.3	43.6	21 46.8	26.1
	12	135 35.9	1 20.5	N18 23.0	151 58.2	S13 54.9	347 23.2	N13 43.4	36 48.9	N22 26.1
	13	150 38.3	16 19.8	22.2	167 00.9	55.1	2 25.1	43.3	51 51.1	26.1
	14	165 40.8	31 19.1	21.4	182 03.6	55.3	17 27.1	43.1	66 53.3	26.1
	15	180 43.3	46 18.4	20.6	197 06.3	55.5	32 29.0	42.9	81 55.4	26.1
	16	195 45.7	61 17.8	19.8	212 09.0	55.7	47 31.0	42.7	96 57.6	26.0
	17	210 48.2	76 17.1	19.0	227 11.7	55.9	62 32.9	42.5	111 59.7	26.0
	18	225 50.7	91 16.4	N18 18.2	242 14.4	S13 56.1	77 34.8	N13 42.4	127 01.9	N22 26.0
	19	240 53.1	106 15.7	17.4	257 17.2	56.3	92 36.8	42.2	142 04.0	26.0
	20	255 55.6	121 15.1	16.6	272 19.9	56.5	107 38.7	42.0	157 06.2	26.0
	21	270 58.1	136 14.4	15.8	287 22.6	56.7	122 40.7	41.8	172 08.4	26.0
	22	286 00.5	151 13.7	15.0	302 25.3	56.9	137 42.6	41.6	187 10.5	25.9
	23	301 03.0	166 13.0	14.2	317 28.0	57.2	152 44.5	41.4	202 12.7	25.9
Mer. Pass.		h m 3 03.0	v −0.7	d 0.8	v 2.7	d 0.2	v 1.9	d 0.2	v 2.2	d 0.0

STARS

Name	SHA	Dec
	° ′	° ′
Acamar	315 24.1	S40 17.1
Achernar	335 32.1	S57 12.8
Acrux	173 18.7	S63 07.3
Adhara	255 19.0	S28 58.4
Aldebaran	290 58.5	N16 31.0
Alioth	166 27.5	N55 56.8
Alkaid	153 05.0	N49 18.1
Al Na'ir	27 52.8	S46 56.6
Alnilam	275 54.5	S 1 11.8
Alphard	218 04.1	S 8 40.3
Alphecca	126 17.5	N26 42.4
Alpheratz	357 51.4	N29 06.5
Altair	62 15.5	N 8 52.7
Ankaa	353 22.9	S42 17.0
Antares	112 35.7	S26 26.5
Arcturus	146 02.8	N19 10.6
Atria	107 44.3	S69 02.3
Avior	234 22.0	S59 31.1
Bellatrix	278 40.6	N 6 21.3
Betelgeuse	271 10.0	N 7 24.6
Canopus	264 00.1	S52 41.6
Capella	280 46.3	N46 00.0
Deneb	49 36.4	N45 17.5
Denebola	182 41.8	N14 33.4
Diphda	349 03.4	S17 57.9
Dubhe	194 01.4	N61 44.2
Elnath	278 22.7	N28 36.6
Eltanin	90 49.4	N51 29.5
Enif	33 54.4	N 9 53.5
Fomalhaut	15 32.1	S29 36.
Gacrux	172 10.2	S57 08.
Gienah	176 00.6	S17 33.6
Hadar	148 59.3	S60 23.6
Hamal	328 09.5	N23 28.
Kaus Aust.	83 53.8	S34 23.
Kochab	137 19.3	N74 08.
Markab	13 45.8	N15 13.
Menkar	314 23.2	N 4 06.
Menkent	148 16.9	S36 23.
Miaplacidus	221 42.6	S69 43.
Mirfak	308 51.7	N49 52.
Nunki	76 07.6	S26 17.
Peacock	53 30.7	S56 43.
Pollux	243 37.5	N28 01.
Procyon	245 08.2	N 5 13.
Rasalhague	96 13.4	N12 33.
Regulus	207 52.1	N11 57.
Rigel	281 19.7	S 8 11.
Rigil Kent.	140 02.7	S60 51.
Sabik	102 21.3	S15 43.
Schedar	349 49.3	N56 33.
Shaula	96 32.2	S37 06.
Sirius	258 40.9	S16 43.
Spica	158 39.6	S11 10.
Suhail	222 58.7	S43 26.
Vega	80 43.9	N38 47.
Zuben'ubi	137 14.1	S16 03.

	SHA	Mer. Pass.
	° ′	h m
Venus	227 37.9	11 54
Mars	16 14.5	1 58
Jupiter	212 06.2	12 53
Saturn	261 24.1	9 37

UT	SUN GHA	SUN Dec	MOON GHA	v	MOON Dec	d	HP
d h	° ′	° ′	° ′	′	° ′	′	′
5 00	178 28.9	N17 08.5	96 52.8	10.2	S12 42.9	14.0	59.2
01	193 28.9	07.8	111 22.0	10.1	12 56.9	14.1	59.2
02	208 29.0	07.1	125 51.1	10.0	13 11.0	13.9	59.3
03	223 29.0	. . 06.5	140 20.1	10.0	13 24.9	13.9	59.3
04	238 29.1	05.8	154 49.1	9.9	13 38.8	13.8	59.3
05	253 29.2	05.1	169 18.0	9.8	13 52.6	13.7	59.3
06	268 29.2	N17 04.5	183 46.8	9.7	S14 06.3	13.7	59.3
07	283 29.3	03.8	198 15.5	9.7	14 20.0	13.6	59.3
08	298 29.4	03.1	212 44.2	9.6	14 33.6	13.5	59.3
09	313 29.4	. . 02.4	227 12.8	9.5	14 47.1	13.4	59.3
10	328 29.5	01.8	241 41.3	9.5	15 00.5	13.3	59.3
11	343 29.5	01.1	256 09.8	9.4	15 13.8	13.3	59.3
12	358 29.6	N17 00.4	270 38.2	9.2	S15 27.1	13.1	59.3
13	13 29.7	16 59.7	285 06.4	9.3	15 40.2	13.1	59.3
14	28 29.7	59.1	299 34.7	9.1	15 53.3	13.0	59.3
15	43 29.8	. . 58.4	314 02.8	9.1	16 06.3	12.9	59.3
16	58 29.9	57.7	328 30.9	8.9	16 19.2	12.8	59.3
17	73 29.9	57.0	342 58.8	8.9	16 32.0	12.7	59.3
18	88 30.0	N16 56.4	357 26.7	8.9	S16 44.7	12.7	59.3
19	103 30.0	55.7	11 54.6	8.7	16 57.4	12.5	59.3
20	118 30.1	55.0	26 22.3	8.7	17 09.9	12.4	59.3
21	133 30.2	. . 54.3	40 50.0	8.5	17 22.3	12.3	59.3
22	148 30.2	53.6	55 17.5	8.5	17 34.6	12.3	59.3
23	163 30.3	53.0	69 45.0	8.4	17 46.9	12.1	59.3
6 00	178 30.4	N16 52.3	84 12.4	8.4	S17 59.0	12.0	59.3
01	193 30.4	51.6	98 39.8	8.2	18 11.0	11.9	59.3
02	208 30.5	50.9	113 07.0	8.2	18 22.9	11.8	59.3
03	223 30.6	. . 50.2	127 34.2	8.0	18 34.7	11.7	59.3
04	238 30.6	49.6	142 01.2	8.0	18 46.4	11.6	59.3
05	253 30.7	48.9	156 28.2	7.9	18 58.0	11.5	59.3
06	268 30.8	N16 48.2	170 55.1	7.8	S19 09.5	11.4	59.3
07	283 30.8	47.5	185 21.9	7.8	19 20.9	11.2	59.4
08	298 30.9	46.8	199 48.7	7.6	19 32.1	11.2	59.4
09	313 31.0	. . 46.1	214 15.3	7.6	19 43.3	11.0	59.4
10	328 31.0	45.5	228 41.9	7.5	19 54.3	10.9	59.4
11	343 31.1	44.8	243 08.4	7.4	20 05.2	10.7	59.4
12	358 31.2	N16 44.1	257 34.8	7.3	S20 15.9	10.7	59.4
13	13 31.2	43.4	272 01.1	7.2	20 26.6	10.5	59.4
14	28 31.3	42.7	286 27.3	7.1	20 37.1	10.4	59.4
15	43 31.4	. . 42.0	300 53.4	7.1	20 47.5	10.3	59.4
16	58 31.4	41.3	315 19.5	7.0	20 57.8	10.1	59.4
17	73 31.5	40.6	329 45.5	6.8	21 07.9	10.0	59.4
18	88 31.6	N16 40.0	344 11.3	6.8	S21 17.9	9.9	59.4
19	103 31.7	39.3	358 37.1	6.8	21 27.8	9.8	59.4
20	118 31.7	38.6	13 02.9	6.6	21 37.6	9.6	59.4
21	133 31.8	. . 37.9	27 28.5	6.6	21 47.2	9.5	59.4
22	148 31.9	37.2	41 54.1	6.4	21 56.7	9.3	59.3
23	163 31.9	36.5	56 19.5	6.4	22 06.0	9.2	59.3
7 00	178 32.0	N16 35.8	70 44.9	6.3	S22 15.2	9.1	59.3
01	193 32.1	35.1	85 10.2	6.3	22 24.3	8.9	59.3
02	208 32.2	34.4	99 35.5	6.1	22 33.2	8.7	59.3
03	223 32.2	. . 33.7	114 00.6	6.1	22 41.9	8.7	59.3
04	238 32.3	33.0	128 25.7	6.0	22 50.6	8.5	59.3
05	253 32.4	32.3	142 50.7	5.9	22 59.1	8.3	59.3
06	268 32.5	N16 31.7	157 15.6	5.9	S23 07.4	8.2	59.3
07	283 32.5	31.0	171 40.5	5.7	23 15.6	8.0	59.3
08	298 32.6	30.3	186 05.2	5.7	23 23.6	7.9	59.3
09	313 32.7	. . 29.6	200 29.9	5.7	23 31.5	7.7	59.3
10	328 32.7	28.9	214 54.6	5.5	23 39.2	7.6	59.3
11	343 32.8	28.2	229 19.1	5.5	23 46.8	7.4	59.3
12	358 32.9	N16 27.5	243 43.6	5.4	S23 54.2	7.3	59.3
13	13 33.0	26.8	258 08.0	5.3	24 01.5	7.1	59.3
14	28 33.1	26.1	272 32.3	5.3	24 08.6	6.9	59.3
15	43 33.1	. . 25.4	286 56.6	5.2	24 15.5	6.8	59.3
16	58 33.2	24.7	301 20.8	5.2	24 22.3	6.7	59.3
17	73 33.3	24.0	315 45.0	5.1	24 29.0	6.4	59.3
18	88 33.4	N16 23.3	330 09.1	5.0	S24 35.4	6.3	59.3
19	103 33.4	22.6	344 33.1	5.0	24 41.7	6.2	59.3
20	118 33.5	21.9	358 57.1	4.9	24 47.9	5.9	59.3
21	133 33.6	. . 21.2	13 21.0	4.8	24 53.8	5.9	59.3
22	148 33.7	20.5	27 44.8	4.8	24 59.7	5.6	59.3
23	163 33.7	19.8	42 08.6	4.7	S25 05.3	5.5	59.3
	SD 15.8	*d* 0.7	SD 16.2		16.2		16.2

Lat.	Twilight Naut.	Twilight Civil	Sunrise	Moonrise 5	Moonrise 6	Moonrise 7	Moonrise 8
°	h m	h m	h m	h m	h m	h m	h m
N 72	////	////	00 49	16 33	■	■	■
N 70	////	////	02 01	15 41	■	■	■
68	////	////	02 37	15 10	18 11	■	■
66	////	01 19	03 03	14 46	17 06	■	■
64	////	02 04	03 22	14 28	16 31	18 47	■
62	////	02 33	03 38	14 14	16 05	18 00	19 42
60	01 12	02 54	03 51	14 01	15 46	17 29	18 59
N 58	01 52	03 11	04 03	13 51	15 30	17 06	18 31
56	02 19	03 26	04 12	13 42	15 16	16 48	18 09
54	02 39	03 38	04 21	13 34	15 04	16 32	17 51
52	02 55	03 48	04 29	13 26	14 54	16 19	17 35
50	03 09	03 58	04 36	13 20	14 44	16 07	17 22
45	03 36	04 17	04 50	13 06	14 25	15 43	16 55
N 40	03 56	04 33	05 02	12 55	14 09	15 23	16 33
35	04 12	04 45	05 13	12 45	13 56	15 07	16 15
30	04 25	04 56	05 22	12 37	13 44	14 53	16 00
20	04 46	05 14	05 37	12 22	13 25	14 29	15 34
N 10	05 02	05 28	05 50	12 10	13 08	14 09	15 12
0	05 16	05 41	06 02	11 58	12 52	13 50	14 51
S 10	05 27	05 53	06 15	11 46	12 37	13 31	14 30
20	05 38	06 05	06 27	11 34	12 20	13 11	14 08
30	05 48	06 17	06 42	11 20	12 01	12 48	13 42
35	05 54	06 24	06 50	11 12	11 50	12 35	13 27
40	05 59	06 31	07 00	11 03	11 38	12 19	13 10
45	06 05	06 40	07 11	10 53	11 23	12 01	12 49
S 50	06 11	06 50	07 24	10 40	11 05	11 38	12 23
52	06 14	06 54	07 31	10 35	10 57	11 27	12 10
54	06 17	06 59	07 38	10 28	10 48	11 15	11 55
56	06 20	07 04	07 45	10 21	10 37	11 01	11 38
58	06 23	07 10	07 54	10 13	10 25	10 45	11 18
S 60	06 26	07 17	08 04	10 04	10 12	10 25	10 52

Lat.	Sunset	Twilight Civil	Twilight Naut.	Moonset 5	Moonset 6	Moonset 7	Moonset 8
°	h m	h m	h m	h m	h m	h m	h m
N 72	23 07	////	////	19 24	■	■	■
N 70	22 05	////	////	20 18	■	■	■
68	21 31	////	////	20 51	19 45	■	■
66	21 06	22 45	////	21 16	20 51	■	■
64	20 47	22 04	////	21 35	21 27	21 13	■
62	20 32	21 36	////	21 51	21 53	22 01	22 27
60	20 19	21 15	22 54	22 04	22 14	22 32	23 09
N 58	20 08	20 58	22 16	22 16	22 31	22 55	23 38
56	19 58	20 44	21 51	22 26	22 45	23 14	24 00
54	19 50	20 32	21 31	22 35	22 57	23 30	24 18
52	19 42	20 22	21 15	22 43	23 08	23 44	24 34
50	19 35	20 13	21 02	22 50	23 18	23 56	24 47
45	19 21	19 54	20 35	23 05	23 39	24 21	00 21
N 40	19 09	19 39	20 15	23 18	23 56	24 41	00 41
35	18 59	19 26	19 59	23 29	24 10	00 10	00 58
30	18 50	19 15	19 46	23 39	24 22	00 22	01 13
20	18 35	18 58	19 25	23 55	24 44	00 44	01 37
N 10	18 22	18 44	19 09	24 10	00 10	01 02	01 59
0	18 09	18 31	18 56	24 23	00 23	01 19	02 19
S 10	17 57	18 19	18 44	24 37	00 37	01 37	02 39
20	17 45	18 08	18 34	24 52	00 52	01 56	03 00
30	17 30	17 55	18 24	00 02	01 09	02 17	03 25
35	17 22	17 48	18 19	00 08	01 19	02 30	03 40
40	17 12	17 41	18 13	00 16	01 30	02 44	03 57
45	17 01	17 33	18 08	00 25	01 43	03 02	04 17
S 50	16 48	17 23	18 02	00 36	02 00	03 24	04 43
52	16 42	17 18	17 59	00 41	02 08	03 34	04 56
54	16 35	17 14	17 56	00 46	02 16	03 46	05 10
56	16 27	17 08	17 53	00 53	02 26	03 59	05 27
58	16 19	17 02	17 50	01 00	02 37	04 15	05 47
S 60	16 09	16 56	17 46	01 08	02 50	04 34	06 12

Day	SUN Eqn. of Time 00^h	SUN Eqn. of Time 12^h	SUN Mer. Pass.	MOON Mer. Pass. Upper	MOON Mer. Pass. Lower	MOON Age	MOON Phase
d	m s	m s	h m	h m	h m	d %	
5	06 05	06 02	12 06	18 11	05 44	07 52	
6	05 59	05 55	12 06	19 06	06 38	08 64	
7	05 52	05 49	12 06	20 04	07 35	09 74	

UT	ARIES	VENUS −3.9		MARS −2.6		JUPITER −1.7		SATURN +0.1		STARS		
	GHA	GHA	Dec	GHA	Dec	GHA	Dec	GHA	Dec	Name	SHA	Dec
d h	° ′	° ′	° ′	° ′	° ′	° ′	° ′	° ′	° ′		° ′	° ′
8 00	316 05.4	181 12.4	N18 13.4	332 30.7	S13 57.4	167 46.5	N13 41.3	217 14.8	N22 25.9	Acamar	315 24.1	S40 17.1
01	331 07.9	196 11.7	12.6	347 33.5	57.6	182 48.4	41.1	232 17.0	25.9	Achernar	335 32.1	S57 12.8
02	346 10.4	211 11.0	11.7	2 36.2	57.8	197 50.4	40.9	247 19.2	25.9	Acrux	173 18.8	S63 07.3
03	1 12.8	226 10.4	. . 10.9	17 38.9	. . 58.0	212 52.3	. . 40.7	262 21.3	. . 25.9	Adhara	255 19.0	S28 58.4
04	16 15.3	241 09.7	10.1	32 41.6	58.2	227 54.2	40.5	277 23.5	25.9	Aldebaran	290 58.4	N16 31.0
05	31 17.8	256 09.0	09.3	47 44.4	58.4	242 56.2	40.4	292 25.6	25.8			
06	46 20.2	271 08.4	N18 08.5	62 47.1	S13 58.6	257 58.1	N13 40.2	307 27.8	N22 25.8	Alioth	166 27.5	N55 56.8
07	61 22.7	286 07.7	07.7	77 49.8	58.8	273 00.1	40.0	322 30.0	25.8	Alkaid	153 05.0	N49 18.1
08	76 25.2	301 07.0	06.9	92 52.6	59.0	288 02.0	39.8	337 32.1	25.8	Al Na'ir	27 52.8	S46 56.6
F 09	91 27.6	316 06.4	. . 06.1	107 55.3	. . 59.2	303 03.9	. . 39.6	352 34.3	. . 25.8	Alnilam	275 54.4	S 1 11.8
R 10	106 30.1	331 05.7	05.3	122 58.0	59.4	318 05.9	39.4	7 36.4	25.8	Alphard	218 04.1	S 8 40.3
I 11	121 32.6	346 05.0	04.4	138 00.7	59.6	333 07.8	39.3	22 38.6	25.7			
D 12	136 35.0	1 04.4	N18 03.6	153 03.5	S13 59.8	348 09.8	N13 39.1	37 40.8	N22 25.7	Alphecca	126 17.5	N26 42.4
A 13	151 37.5	16 03.7	02.8	168 06.2	14 00.0	3 11.7	38.9	52 42.9	25.7	Alpheratz	357 51.4	N29 06.5
Y 14	166 39.9	31 03.0	02.0	183 09.0	00.2	18 13.6	38.7	67 45.1	25.7	Altair	62 15.5	N 8 52.7
15	181 42.4	46 02.4	. . 01.2	198 11.7	. . 00.4	33 15.6	. . 38.5	82 47.2	. . 25.7	Ankaa	353 22.9	S42 17.0
16	196 44.9	61 01.7	18 00.3	213 14.4	00.6	48 17.5	38.4	97 49.4	25.7	Antares	112 35.7	S26 26.5
17	211 47.3	76 01.0	17 59.5	228 17.2	00.8	63 19.5	38.2	112 51.6	25.7			
18	226 49.8	91 00.4	N17 58.7	243 19.9	S14 01.1	78 21.4	N13 38.0	127 53.7	N22 25.6	Arcturus	146 02.8	N19 10.0
19	241 52.3	105 59.7	57.9	258 22.7	01.3	93 23.3	37.8	142 55.9	25.6	Atria	107 44.3	S69 02.3
20	256 54.7	120 59.1	57.1	273 25.4	01.5	108 25.3	37.6	157 58.0	25.6	Avior	234 22.0	S59 31.1
21	271 57.2	135 58.4	. . 56.2	288 28.2	. . 01.7	123 27.2	. . 37.4	173 00.2	. . 25.6	Bellatrix	278 40.5	N 6 21.3
22	286 59.7	150 57.7	55.4	303 30.9	01.9	138 29.2	37.3	188 02.4	25.6	Betelgeuse	271 09.9	N 7 24.6
23	302 02.1	165 57.1	54.6	318 33.6	02.1	153 31.1	37.1	203 04.5	25.6			
9 00	317 04.6	180 56.4	N17 53.7	333 36.4	S14 02.3	168 33.0	N13 36.9	218 06.7	N22 25.5	Canopus	264 00.0	S52 41.6
01	332 07.1	195 55.8	52.9	348 39.1	02.5	183 35.0	36.7	233 08.8	25.5	Capella	280 46.2	N46 00.0
02	347 09.5	210 55.1	52.1	3 41.9	02.7	198 36.9	36.5	248 11.0	25.5	Deneb	49 36.4	N45 17.6
03	2 12.0	225 54.4	. . 51.3	18 44.7	. . 02.9	213 38.9	. . 36.4	263 13.2	. . 25.5	Denebola	182 41.8	N14 33.4
04	17 14.4	240 53.8	50.4	33 47.4	03.2	228 40.8	36.2	278 15.3	25.5	Diphda	349 03.4	S17 57.9
05	32 16.9	255 53.1	49.6	48 50.2	03.4	243 42.7	36.0	293 17.5	25.5			
06	47 19.4	270 52.5	N17 48.8	63 52.9	S14 03.6	258 44.7	N13 35.8	308 19.7	N22 25.4	Dubhe	194 01.4	N61 44.2
S 07	62 21.8	285 51.8	47.9	78 55.7	03.8	273 46.6	35.6	323 21.8	25.4	Elnath	278 22.7	N28 36.6
A 08	77 24.3	300 51.2	47.1	93 58.4	04.0	288 48.6	35.4	338 24.0	25.4	Eltanin	90 49.4	N51 29.5
T 09	92 26.8	315 50.5	. . 46.3	109 01.2	. . 04.2	303 50.5	. . 35.3	353 26.1	. . 25.4	Enif	33 54.4	N 9 53.5
U 10	107 29.2	330 49.8	45.4	124 04.0	04.4	318 52.4	35.1	8 28.3	25.4	Fomalhaut	15 32.1	S29 36.1
R 11	122 31.7	345 49.2	44.6	139 06.7	04.6	333 54.4	34.9	23 30.5	25.4			
D 12	137 34.2	0 48.5	N17 43.8	154 09.5	S14 04.8	348 56.3	N13 34.7	38 32.6	N22 25.4	Gacrux	172 10.2	S57 08.1
A 13	152 36.6	15 47.9	42.9	169 12.3	05.1	3 58.3	34.5	53 34.8	25.3	Gienah	176 00.6	S17 33.6
Y 14	167 39.1	30 47.2	42.1	184 15.0	05.3	19 00.2	34.4	68 37.0	25.3	Hadar	148 59.3	S60 23.
15	182 41.6	45 46.6	. . 41.2	199 17.8	. . 05.5	34 02.1	. . 34.2	83 39.1	. . 25.3	Hamal	328 09.5	N23 28.
16	197 44.0	60 45.9	40.4	214 20.6	05.7	49 04.1	34.0	98 41.3	25.3	Kaus Aust.	83 53.8	S34 23.
17	212 46.5	75 45.3	39.5	229 23.3	05.9	64 06.0	33.8	113 43.4	25.3			
18	227 48.9	90 44.6	N17 38.7	244 26.1	S14 06.1	79 08.0	N13 33.6	128 45.6	N22 25.3	Kochab	137 19.3	N74 08.
19	242 51.4	105 44.0	37.9	259 28.9	06.3	94 09.9	33.4	143 47.8	25.2	Markab	13 45.8	N15 13.
20	257 53.9	120 43.3	37.0	274 31.7	06.5	109 11.8	33.3	158 49.9	25.2	Menkar	314 23.2	N 4 06.
21	272 56.3	135 42.7	. . 36.2	289 34.4	. . 06.8	124 13.8	. . 33.1	173 52.1	. . 25.2	Menkent	148 16.9	S36 23.
22	287 58.8	150 42.0	35.3	304 37.2	07.0	139 15.7	32.9	188 54.3	25.2	Miaplacidus	221 42.6	S69 43.
23	303 01.3	165 41.4	34.5	319 40.0	07.2	154 17.6	32.7	203 56.4	25.2			
10 00	318 03.7	180 40.7	N17 33.6	334 42.8	S14 07.4	169 19.6	N13 32.5	218 58.6	N22 25.2	Mirfak	308 51.7	N49 52.
01	333 06.2	195 40.1	32.8	349 45.5	07.6	184 21.5	32.3	234 00.8	25.1	Nunki	76 07.6	S26 17.
02	348 08.7	210 39.4	31.9	4 48.3	07.8	199 23.5	32.2	249 02.9	25.1	Peacock	53 30.7	S56 43.
03	3 11.1	225 38.8	. . 31.1	19 51.1	. . 08.1	214 25.4	. . 32.0	264 05.1	. . 25.1	Pollux	243 37.5	N28 01.
04	18 13.6	240 38.1	30.2	34 53.9	08.3	229 27.3	31.8	279 07.2	25.1	Procyon	245 08.1	N 5 13.
05	33 16.1	255 37.5	29.4	49 56.7	08.5	244 29.3	31.6	294 09.4	25.1			
06	48 18.5	270 36.8	N17 28.5	64 59.5	S14 08.7	259 31.2	N13 31.4	309 11.6	N22 25.1	Rasalhague	96 13.4	N12 33.
07	63 21.0	285 36.2	27.7	80 02.2	08.9	274 33.2	31.3	324 13.7	25.0	Regulus	207 52.0	N11 57.
08	78 23.4	300 35.6	26.8	95 05.0	09.1	289 35.1	31.1	339 15.9	25.0	Rigel	281 19.7	S 8 11.
S 09	93 25.9	315 34.9	. . 26.0	110 07.8	. . 09.4	304 37.0	. . 30.9	354 18.1	. . 25.0	Rigil Kent.	140 02.7	S60 51.
U 10	108 28.4	330 34.3	25.1	125 10.6	09.6	319 39.0	30.7	9 20.2	25.0	Sabik	102 21.3	S15 43.
N 11	123 30.8	345 33.6	24.2	140 13.4	09.8	334 40.9	30.5	24 22.4	25.0			
D 12	138 33.3	0 33.0	N17 23.4	155 16.2	S14 10.0	349 42.9	N13 30.3	39 24.6	N22 25.0	Schedar	349 49.3	N56 33.
A 13	153 35.8	15 32.3	22.5	170 19.0	10.2	4 44.8	30.2	54 26.7	25.0	Shaula	96 32.2	S37 06.
Y 14	168 38.2	30 31.7	21.7	185 21.8	10.4	19 46.7	30.0	69 28.9	24.9	Sirius	258 40.9	S16 43.
15	183 40.7	45 31.1	. . 20.8	200 24.6	. . 10.7	34 48.7	. . 29.8	84 31.1	. . 24.9	Spica	158 39.6	S11 10.
16	198 43.2	60 30.4	19.9	215 27.4	10.9	49 50.6	29.6	99 33.2	24.9	Suhail	222 58.7	S43 26.
17	213 45.6	75 29.8	19.1	230 30.2	11.1	64 52.6	29.4	114 35.4	24.9			
18	228 48.1	90 29.1	N17 18.2	245 33.0	S14 11.3	79 54.5	N13 29.2	129 37.5	N22 24.9	Vega	80 43.9	N38 47.
19	243 50.5	105 28.5	17.4	260 35.8	11.5	94 56.4	29.1	144 39.7	24.9	Zuben'ubi	137 14.1	S16 03.
20	258 53.0	120 27.9	16.5	275 38.6	11.7	109 58.4	28.9	159 41.9	24.8		SHA	Mer. Pass
21	273 55.5	135 27.2	. . 15.6	290 41.4	. . 12.0	125 00.3	. . 28.7	174 44.0	. . 24.8		° ′	h m
22	288 57.9	150 26.6	14.8	305 44.2	12.2	140 02.3	28.5	189 46.2	24.8	Venus	223 51.8	11 57
23	304 00.4	165 25.9	13.9	320 47.0	12.4	155 04.2	28.3	204 48.4	24.8	Mars	16 31.8	1 45
	h m									Jupiter	211 28.4	12 44
Mer. Pass.	2 51.2	*v* −0.7	*d* 0.8	*v* 2.8	*d* 0.2	*v* 1.9	*d* 0.2	*v* 2.2	*d* 0.0	Saturn	261 02.1	9 26

UT	SUN GHA	SUN Dec	MOON GHA	v	MOON Dec	d	HP
d h	° ′	° ′	° ′	′	° ′	′	′
8 00	178 33.8	N16 19.1	56 32.3	4.7	S25 10.8	5.3	59.3
01	193 33.9	18.4	70 56.0	4.7	25 16.1	5.1	59.2
02	208 34.0	17.7	85 19.7	4.6	25 21.2	5.0	59.2
03	223 34.1	. . 17.0	99 43.3	4.5	25 26.2	4.7	59.2
04	238 34.1	16.3	114 06.8	4.5	25 30.9	4.7	59.2
05	253 34.2	15.6	128 30.3	4.5	25 35.6	4.4	59.2
06	268 34.3	N16 14.9	142 53.8	4.4	S25 40.0	4.3	59.2
07	283 34.4	14.1	157 17.2	4.3	25 44.3	4.1	59.2
08	298 34.5	13.4	171 40.5	4.4	25 48.4	3.9	59.2
F 09	313 34.5	. . 12.7	186 03.9	4.3	25 52.3	3.7	59.2
R 10	328 34.6	12.0	200 27.2	4.3	25 56.0	3.6	59.2
I 11	343 34.7	11.3	214 50.5	4.2	25 59.6	3.4	59.2
D 12	358 34.8	N16 10.6	229 13.7	4.2	S26 03.0	3.2	59.2
A 13	13 34.9	09.9	243 36.9	4.2	26 06.2	3.0	59.2
Y 14	28 34.9	09.2	258 00.1	4.2	26 09.2	2.9	59.1
15	43 35.0	. . 08.5	272 23.3	4.1	26 12.1	2.7	59.1
16	58 35.1	07.8	286 46.4	4.1	26 14.8	2.5	59.1
17	73 35.2	07.1	301 09.5	4.1	26 17.3	2.3	59.1
18	88 35.3	N16 06.4	315 32.6	4.1	S26 19.6	2.1	59.1
19	103 35.4	05.6	329 55.7	4.1	26 21.7	2.0	59.1
20	118 35.4	04.9	344 18.8	4.1	26 23.7	1.7	59.1
21	133 35.5	. . 04.2	358 41.9	4.0	26 25.4	1.6	59.1
22	148 35.6	03.5	13 04.9	4.1	26 27.0	1.5	59.1
23	163 35.7	02.8	27 28.0	4.0	26 28.5	1.2	59.1
9 00	178 35.8	N16 02.1	41 51.0	4.1	S26 29.7	1.0	59.0
01	193 35.9	01.4	56 14.1	4.0	26 30.7	0.9	59.0
02	208 35.9	16 00.6	70 37.1	4.1	26 31.6	0.7	59.0
03	223 36.0	15 59.9	85 00.2	4.1	26 32.3	0.5	59.0
04	238 36.1	59.2	99 23.3	4.0	26 32.8	0.3	59.0
05	253 36.2	58.5	113 46.3	4.1	26 33.1	0.2	59.0
06	268 36.3	N15 57.8	128 09.4	4.1	S26 33.3	0.0	59.0
S 07	283 36.4	57.1	142 32.5	4.1	26 33.3	0.3	59.0
08	298 36.5	56.4	156 55.6	4.1	26 33.0	0.4	59.0
A 09	313 36.5	. . 55.6	171 18.7	4.2	26 32.6	0.5	58.9
T 10	328 36.6	54.9	185 41.9	4.1	26 32.1	0.8	58.9
U 11	343 36.7	54.2	200 05.0	4.2	26 31.3	0.9	58.9
R 12	358 36.8	N15 53.5	214 28.2	4.2	S26 30.4	1.1	58.9
D 13	13 36.9	52.8	228 51.4	4.3	26 29.3	1.3	58.9
A 14	28 37.0	52.0	243 14.7	4.3	26 28.0	1.5	58.9
Y 15	43 37.1	. . 51.3	257 38.0	4.3	26 26.5	1.6	58.9
16	58 37.2	50.6	272 01.3	4.3	26 24.9	1.9	58.8
17	73 37.2	49.9	286 24.6	4.4	26 23.0	2.0	58.8
18	88 37.3	N15 49.2	300 48.0	4.5	S26 21.0	2.1	58.8
19	103 37.4	48.4	315 11.5	4.4	26 18.9	2.4	58.8
20	118 37.5	47.7	329 34.9	4.5	26 16.5	2.5	58.8
21	133 37.6	. . 47.0	343 58.4	4.6	26 14.0	2.7	58.8
22	148 37.7	46.3	358 22.0	4.6	26 11.3	2.9	58.8
23	163 37.8	45.5	12 45.6	4.7	26 08.4	3.0	58.7
10 00	178 37.9	N15 44.8	27 09.3	4.7	S26 05.4	3.3	58.7
01	193 38.0	44.1	41 33.0	4.8	26 02.1	3.4	58.7
02	208 38.1	43.4	55 56.8	4.8	25 58.7	3.5	58.7
03	223 38.2	. . 42.6	70 20.6	4.9	25 55.2	3.7	58.7
04	238 38.2	41.9	84 44.5	5.0	25 51.5	3.9	58.7
05	253 38.3	41.2	99 08.5	5.0	25 47.6	4.1	58.6
06	268 38.4	N15 40.5	113 32.5	5.1	S25 43.5	4.2	58.6
07	283 38.5	39.7	127 56.6	5.1	25 39.3	4.4	58.6
S 08	298 38.6	39.0	142 20.7	5.2	25 34.9	4.6	58.6
U 09	313 38.7	. . 38.3	156 44.9	5.3	25 30.3	4.7	58.6
N 10	328 38.8	37.6	171 09.2	5.4	25 25.6	4.9	58.6
D 11	343 38.9	36.8	185 33.6	5.4	25 20.7	5.0	58.5
A 12	358 39.0	N15 36.1	199 58.0	5.5	S25 15.7	5.2	58.5
Y 13	13 39.1	35.4	214 22.5	5.6	25 10.5	5.4	58.5
14	28 39.2	34.6	228 47.1	5.7	25 05.1	5.5	58.5
15	43 39.3	. . 33.9	243 11.8	5.7	24 59.6	5.7	58.5
16	58 39.4	33.2	257 36.5	5.8	24 53.9	5.8	58.4
17	73 39.5	32.4	272 01.3	6.0	24 48.1	6.0	58.4
18	88 39.6	N15 31.7	286 26.3	5.9	S24 42.1	6.1	58.4
19	103 39.6	31.0	300 51.2	6.1	24 36.0	6.3	58.4
20	118 39.7	30.3	315 16.3	6.2	24 29.7	6.4	58.4
21	133 39.8	. . 29.5	329 41.5	6.2	24 23.3	6.6	58.3
22	148 39.9	28.8	344 06.7	6.4	24 16.7	6.7	58.3
23	163 40.0	28.1	358 32.1	6.4	S24 10.0	6.9	58.3
	SD 15.8	d 0.7	SD 16.1		16.0		15.9

Lat.	Twilight Naut.	Twilight Civil	Sunrise	Moonrise 8	Moonrise 9	Moonrise 10	Moonrise 11
°	h m	h m	h m	h m	h m	h m	h m
N 72	////	////	01 30	▬	▬	▬	▬
N 70	////	////	02 20	▬	▬	▬	▬
68	////	////	02 51	▬	▬	▬	22 45
66	////	01 42	03 14	▬	▬	▬	21 58
64	////	02 19	03 32	▬	▬	21 38	21 27
62	////	02 44	03 46	19 42	20 39	20 58	21 03
60	01 32	03 03	03 59	18 59	19 59	20 30	20 45
N 58	02 05	03 19	04 09	18 31	19 32	20 08	20 29
56	02 28	03 33	04 18	18 09	19 10	19 50	20 16
54	02 47	03 44	04 26	17 51	18 52	19 35	20 04
52	03 02	03 54	04 33	17 35	18 37	19 22	19 54
50	03 15	04 03	04 40	17 22	18 24	19 10	19 44
45	03 40	04 21	04 54	16 55	17 56	18 46	19 25
N 40	04 00	04 36	05 05	16 33	17 35	18 27	19 09
35	04 15	04 48	05 15	16 15	17 17	18 11	18 55
30	04 27	04 58	05 23	16 00	17 02	17 56	18 43
20	04 47	05 15	05 38	15 34	16 36	17 32	18 23
N 10	05 03	05 29	05 50	15 12	16 13	17 12	18 05
0	05 16	05 41	06 02	14 51	15 52	16 52	17 49
S 10	05 27	05 52	06 14	14 30	15 31	16 33	17 32
20	05 37	06 03	06 26	14 08	15 09	16 12	17 14
30	05 46	06 15	06 40	13 42	14 43	15 48	16 53
35	05 51	06 21	06 47	13 27	14 27	15 33	16 41
40	05 56	06 28	06 56	13 10	14 10	15 17	16 27
45	06 01	06 36	07 07	12 49	13 48	14 57	16 11
S 50	06 07	06 45	07 20	12 23	13 21	14 32	15 50
52	06 09	06 49	07 25	12 10	13 08	14 20	15 40
54	06 11	06 54	07 32	11 55	12 53	14 06	15 29
56	06 14	06 59	07 39	11 38	12 35	13 50	15 17
58	06 17	07 04	07 47	11 18	12 13	13 31	15 02
S 60	06 20	07 10	07 56	10 52	11 45	13 07	14 44

Lat.	Sunset	Twilight Civil	Twilight Naut.	Moonset 8	Moonset 9	Moonset 10	Moonset 11
°	h m	h m	h m	h m	h m	h m	h m
N 72	22 32	////	////	▬	▬	▬	▬
N 70	21 46	////	////	▬	▬	▬	▬
68	21 16	23 30	////	▬	▬	▬	▬
66	20 54	22 23	////	▬	▬	▬	▬
64	20 37	21 49	////	▬	▬	▬	00 44
62	20 23	21 24	23 34	22 27	23 38	25 23	01 23
60	20 11	21 05	22 34	23 09	24 18	00 18	01 51
N 58	20 01	20 50	22 03	23 38	24 45	00 45	02 12
56	19 52	20 37	21 40	24 00	00 00	01 07	02 30
54	19 44	20 26	21 22	24 18	00 18	01 25	02 44
52	19 37	20 16	21 07	24 34	00 34	01 40	02 57
50	19 30	20 07	20 55	24 47	00 47	01 53	03 08
45	19 16	19 49	20 30	00 21	01 15	02 19	03 32
N 40	19 05	19 35	20 11	00 41	01 37	02 41	03 50
35	18 56	19 23	19 56	00 58	01 55	02 58	04 06
30	18 47	19 13	19 43	01 13	02 10	03 13	04 19
20	18 33	18 56	19 23	01 37	02 36	03 39	04 42
N 10	18 20	18 42	19 08	01 59	02 59	04 01	05 02
0	18 09	18 30	18 55	02 19	03 20	04 21	05 20
S 10	17 58	18 19	18 45	02 39	03 41	04 42	05 38
20	17 46	18 08	18 35	03 00	04 04	05 03	05 57
30	17 32	17 57	18 25	03 25	04 30	05 29	06 19
35	17 24	17 50	18 21	03 40	04 45	05 43	06 32
40	17 15	17 43	18 16	03 57	05 03	06 00	06 47
45	17 05	17 36	18 11	04 17	05 25	06 21	07 05
S 50	16 52	17 27	18 05	04 43	05 52	06 46	07 26
52	16 46	17 23	18 03	04 56	06 05	06 59	07 37
54	16 40	17 18	18 00	05 10	06 21	07 13	07 48
56	16 33	17 13	17 58	05 27	06 38	07 29	08 01
58	16 25	17 08	17 55	05 47	07 00	07 48	08 17
S 60	16 16	17 02	17 52	06 12	07 28	08 13	08 35

Day	SUN Eqn. of Time 00^h	SUN Eqn. of Time 12^h	SUN Mer. Pass.	MOON Mer. Pass. Upper	MOON Mer. Pass. Lower	MOON Age	MOON Phase
d	m s	m s	h m	h m	h m	d %	
8	05 45	05 41	12 06	21 05	08 35	10 84	
9	05 37	05 33	12 06	22 07	09 36	11 91	
10	05 29	05 24	12 05	23 06	10 37	12 96	

UT	ARIES	VENUS −3.9		MARS −2.6		JUPITER −1.7		SATURN +0.1	
	GHA	GHA	Dec	GHA	Dec	GHA	Dec	GHA	Dec
d h	° ′	° ′	° ′	° ′	° ′	° ′	° ′	° ′	° ′
11 00	319 02.9	180 25.3	N17 13.0	335 49.8	S14 12.6	170 06.1	N13 28.1	219 50.5	N22 24.8
01	334 05.3	195 24.7	12.2	350 52.6	12.8	185 08.1	28.0	234 52.7	24.8
02	349 07.8	210 24.0	11.3	5 55.4	13.1	200 10.0	27.8	249 54.9	24.7
03	4 10.3	225 23.4	10.4	20 58.3	13.3	215 12.0	27.6	264 57.0	24.7
04	19 12.7	240 22.8	09.5	36 01.1	13.5	230 13.9	27.4	279 59.2	24.7
05	34 15.2	255 22.1	08.7	51 03.9	13.7	245 15.8	27.2	295 01.4	24.7
06	49 17.7	270 21.5	N17 07.8	66 06.7	S14 14.0	260 17.8	N13 27.0	310 03.5	N22 24.7
07	64 20.1	285 20.9	06.9	81 09.5	14.2	275 19.7	26.9	325 05.7	24.7
08	79 22.6	300 20.2	06.0	96 12.3	14.4	290 21.6	26.7	340 07.9	24.6
M 09	94 25.0	315 19.6	05.2	111 15.1	14.6	305 23.6	26.5	355 10.0	24.6
O 10	109 27.5	330 19.0	04.3	126 18.0	14.8	320 25.5	26.3	10 12.2	24.6
N 11	124 30.0	345 18.3	03.4	141 20.8	15.1	335 27.5	26.1	25 14.4	24.6
D 12	139 32.4	0 17.7	N17 02.5	156 23.6	S14 15.3	350 29.4	N13 25.9	40 16.5	N22 24.6
A 13	154 34.9	15 17.1	01.7	171 26.4	15.5	5 31.3	25.8	55 18.7	24.6
Y 14	169 37.4	30 16.4	17 00.8	186 29.3	15.7	20 33.3	25.6	70 20.9	24.6
15	184 39.8	45 15.8	16 59.9	201 32.1	16.0	35 35.2	25.4	85 23.0	24.5
16	199 42.3	60 15.2	59.0	216 34.9	16.2	50 37.2	25.2	100 25.2	24.5
17	214 44.8	75 14.6	58.1	231 37.7	16.4	65 39.1	25.0	115 27.4	24.5
18	229 47.2	90 13.9	N16 57.3	246 40.6	S14 16.6	80 41.0	N13 24.8	130 29.5	N22 24.5
19	244 49.7	105 13.3	56.4	261 43.4	16.8	95 43.0	24.7	145 31.7	24.5
20	259 52.2	120 12.7	55.5	276 46.2	17.1	110 44.9	24.5	160 33.9	24.5
21	274 54.6	135 12.0	54.6	291 49.1	17.3	125 46.9	24.3	175 36.0	24.4
22	289 57.1	150 11.4	53.7	306 51.9	17.5	140 48.8	24.1	190 38.2	24.4
23	304 59.5	165 10.8	52.8	321 54.7	17.7	155 50.7	23.9	205 40.4	24.4
12 00	320 02.0	180 10.2	N16 51.9	336 57.6	S14 18.0	170 52.7	N13 23.7	220 42.5	N22 24.4
01	335 04.5	195 09.5	51.0	352 00.4	18.2	185 54.6	23.6	235 44.7	24.4
02	350 06.9	210 08.9	50.2	7 03.2	18.4	200 56.6	23.4	250 46.9	24.4
03	5 09.4	225 08.3	49.3	22 06.1	18.6	215 58.5	23.2	265 49.1	24.3
04	20 11.9	240 07.7	48.4	37 08.9	18.9	231 00.4	23.0	280 51.2	24.3
05	35 14.3	255 07.0	47.5	52 11.8	19.1	246 02.4	22.8	295 53.4	24.3
06	50 16.8	270 06.4	N16 46.6	67 14.6	S14 19.3	261 04.3	N13 22.6	310 55.6	N22 24.3
07	65 19.3	285 05.8	45.7	82 17.4	19.5	276 06.3	22.5	325 57.7	24.3
T 08	80 21.7	300 05.2	44.8	97 20.3	19.8	291 08.2	22.3	340 59.9	24.3
U 09	95 24.2	315 04.6	43.9	112 23.1	20.0	306 10.1	22.1	356 02.1	24.2
E 10	110 26.7	330 03.9	43.0	127 26.0	20.2	321 12.1	21.9	11 04.2	24.2
S 11	125 29.1	345 03.3	42.1	142 28.8	20.4	336 14.0	21.7	26 06.4	24.2
D 12	140 31.6	0 02.7	N16 41.2	157 31.7	S14 20.7	351 15.9	N13 21.5	41 08.6	N22 24.2
A 13	155 34.0	15 02.1	40.3	172 34.5	20.9	6 17.9	21.4	56 10.7	24.2
Y 14	170 36.5	30 01.5	39.4	187 37.4	21.1	21 19.8	21.2	71 12.9	24.2
15	185 39.0	45 00.8	38.5	202 40.2	21.3	36 21.8	21.0	86 15.1	24.1
16	200 41.4	60 00.2	37.6	217 43.1	21.6	51 23.7	20.8	101 17.3	24.1
17	215 43.9	74 59.6	36.7	232 46.0	21.8	66 25.6	20.6	116 19.4	24.1
18	230 46.4	89 59.0	N16 35.8	247 48.8	S14 22.0	81 27.6	N13 20.4	131 21.6	N22 24.1
19	245 48.8	104 58.4	34.9	262 51.7	22.3	96 29.5	20.3	146 23.8	24.1
20	260 51.3	119 57.7	34.0	277 54.5	22.5	111 31.5	20.1	161 25.9	24.1
21	275 53.8	134 57.1	33.1	292 57.4	22.7	126 33.4	19.9	176 28.1	24.1
22	290 56.2	149 56.5	32.2	308 00.2	22.9	141 35.3	19.7	191 30.3	24.0
23	305 58.7	164 55.9	31.3	323 03.1	23.2	156 37.3	19.5	206 32.4	24.0
13 00	321 01.1	179 55.3	N16 30.4	338 06.0	S14 23.4	171 39.2	N13 19.3	221 34.6	N22 24.0
01	336 03.6	194 54.7	29.5	353 08.8	23.6	186 41.2	19.1	236 36.8	24.0
02	351 06.1	209 54.1	28.6	8 11.7	23.9	201 43.1	19.0	251 39.0	24.0
03	6 08.5	224 53.4	27.7	23 14.6	24.1	216 45.0	18.8	266 41.1	24.0
04	21 11.0	239 52.8	26.7	38 17.4	24.3	231 47.0	18.6	281 43.3	23.9
05	36 13.5	254 52.2	25.8	53 20.3	24.5	246 48.9	18.4	296 45.5	23.9
06	51 15.9	269 51.6	N16 24.9	68 23.2	S14 24.8	261 50.8	N13 18.2	311 47.6	N22 23.9
W 07	66 18.4	284 51.0	24.0	83 26.0	25.0	276 52.8	18.0	326 49.8	23.9
E 08	81 20.9	299 50.4	23.1	98 28.9	25.2	291 54.7	17.9	341 52.0	23.9
D 09	96 23.3	314 49.8	22.2	113 31.8	25.5	306 56.7	17.7	356 54.2	23.9
N 10	111 25.8	329 49.2	21.3	128 34.7	25.7	321 58.6	17.5	11 56.3	23.8
E 11	126 28.3	344 48.6	20.3	143 37.5	25.9	337 00.5	17.3	26 58.5	23.8
S 12	141 30.7	359 48.0	N16 19.4	158 40.4	S14 26.1	352 02.5	N13 17.1	42 00.7	N22 23.8
D 13	156 33.2	14 47.3	18.5	173 43.3	26.4	7 04.4	16.9	57 02.8	23.8
A 14	171 35.6	29 46.7	17.6	188 46.2	26.6	22 06.4	16.8	72 05.0	23.8
Y 15	186 38.1	44 46.1	16.7	203 49.1	26.8	37 08.3	16.6	87 07.2	23.8
16	201 40.6	59 45.5	15.8	218 51.9	27.1	52 10.2	16.4	102 09.4	23.7
17	216 43.0	74 44.9	14.8	233 54.8	27.3	67 12.2	16.2	117 11.5	23.7
18	231 45.5	89 44.3	N16 13.9	248 57.7	S14 27.5	82 14.1	N13 16.0	132 13.7	N22 23.7
19	246 48.0	104 43.7	13.0	264 00.6	27.8	97 16.1	15.8	147 15.9	23.7
20	261 50.4	119 43.1	12.1	279 03.5	28.0	112 18.0	15.6	162 18.0	23.7
21	276 52.9	134 42.5	11.1	294 06.4	28.2	127 19.9	15.5	177 20.2	23.7
22	291 55.4	149 41.9	10.2	309 09.3	28.5	142 21.9	15.3	192 22.4	23.6
23	306 57.8	164 41.3	09.3	324 12.1	28.7	157 23.8	15.1	207 24.6	23.6
Mer. Pass.	h m 2 39.4	*v* −0.6	*d* 0.9	*v* 2.9	*d* 0.2	*v* 1.9	*d* 0.2	*v* 2.2	*d* 0.0

STARS

Name	SHA	Dec
	° ′	° ′
Acamar	315 24.1	S40 17.
Achernar	335 32.1	S57 12.
Acrux	173 18.8	S63 07.
Adhara	255 18.9	S28 58.
Aldebaran	290 58.4	N16 31.
Alioth	166 27.5	N55 56.
Alkaid	153 05.0	N49 18.
Al Na'ir	27 52.8	S46 56.
Alnilam	275 54.4	S 1 11.
Alphard	218 04.0	S 8 40.
Alphecca	126 17.5	N26 42.
Alpheratz	357 51.4	N29 06.
Altair	62 15.5	N 8 52.
Ankaa	353 22.9	S42 17.
Antares	112 35.7	S26 26.
Arcturus	146 02.8	N19 10.
Atria	107 44.3	S69 02.
Avior	234 22.0	S59 31.
Bellatrix	278 40.5	N 6 21.
Betelgeuse	271 09.9	N 7 24.
Canopus	264 00.0	S52 41.
Capella	280 46.2	N46 00.
Deneb	49 36.4	N45 17.
Denebola	182 41.8	N14 33.
Diphda	349 03.4	S17 57.
Dubhe	194 01.4	N61 44.
Elnath	278 22.6	N28 36.
Eltanin	90 49.4	N51 29.
Enif	33 54.4	N 9 53.
Fomalhaut	15 32.1	S29 36.
Gacrux	172 10.2	S57 08.
Gienah	176 00.6	S17 33.
Hadar	148 59.4	S60 23.
Hamal	328 09.5	N23 28.
Kaus Aust.	83 53.8	S34 23.
Kochab	137 19.4	N74 08.
Markab	13 45.8	N15 13.
Menkar	314 23.2	N 4 06.
Menkent	148 16.9	S36 23.
Miaplacidus	221 42.6	S69 43.
Mirfak	308 51.6	N49 52.
Nunki	76 07.6	S26 17.
Peacock	53 30.7	S56 43.
Pollux	243 37.5	N28 01.
Procyon	245 08.1	N 5 13.
Rasalhague	96 13.4	N12 33.
Regulus	207 52.0	N11 57.
Rigel	281 19.7	S 8 11.
Rigil Kent.	140 02.8	S60 51.
Sabik	102 21.3	S15 43.
Schedar	349 49.3	N56 33.
Shaula	96 32.2	S37 06.
Sirius	258 40.9	S16 43
Spica	158 39.6	S11 10.
Suhail	222 58.7	S43 26.
Vega	80 43.9	N38 47.
Zuben'ubi	137 14.1	S16 03.

	SHA	Mer. Pas
	° ′	h m
Venus	220 08.2	12 00
Mars	16 55.6	1 32
Jupiter	210 50.7	12 35
Saturn	260 40.5	9 16

UT	SUN GHA	SUN Dec	MOON GHA	v	MOON Dec	d	HP
d h	° ′	° ′	° ′	′	° ′	′	′
11 00	178 40.1	N15 27.3	12 57.5	6.6	S24 03.1	7.0	58.3
01	193 40.2	26.6	27 23.1	6.6	23 56.1	7.2	58.3
02	208 40.3	25.8	41 48.7	6.7	23 48.9	7.3	58.2
03	223 40.4	. . 25.1	56 14.4	6.8	23 41.6	7.4	58.2
04	238 40.5	24.4	70 40.2	6.9	23 34.2	7.6	58.2
05	253 40.6	23.6	85 06.1	7.0	23 26.6	7.7	58.2
06	268 40.7	N15 22.9	99 32.1	7.1	S23 18.9	7.9	58.2
07	283 40.8	22.2	113 58.2	7.2	23 11.0	8.0	58.1
08	298 40.9	21.4	128 24.4	7.3	23 03.0	8.1	58.1
09	313 41.0	. . 20.7	142 50.7	7.4	22 54.9	8.2	58.1
10	328 41.1	19.9	157 17.1	7.5	22 46.7	8.4	58.1
11	343 41.2	19.2	171 43.6	7.6	22 38.3	8.5	58.0
12	358 41.3	N15 18.5	186 10.2	7.7	S22 29.8	8.6	58.0
13	13 41.4	17.7	200 36.9	7.8	22 21.2	8.8	58.0
14	28 41.5	17.0	215 03.7	7.9	22 12.4	8.9	58.0
15	43 41.6	. . 16.2	229 30.6	8.0	22 03.5	8.9	58.0
16	58 41.7	15.5	243 57.6	8.1	21 54.6	9.2	57.9
17	73 41.8	14.8	258 24.7	8.2	21 45.4	9.2	57.9
18	88 41.9	N15 14.0	272 51.9	8.3	S21 36.2	9.3	57.9
19	103 42.0	13.3	287 19.2	8.4	21 26.9	9.5	57.9
20	118 42.1	12.5	301 46.6	8.5	21 17.4	9.6	57.8
21	133 42.2	. . 11.8	316 14.1	8.6	21 07.8	9.7	57.8
22	148 42.3	11.1	330 41.7	8.8	20 58.1	9.8	57.8
23	163 42.4	10.3	345 09.5	8.8	20 48.3	9.9	57.8
12 00	178 42.5	N15 09.6	359 37.3	8.9	S20 38.4	10.0	57.7
01	193 42.6	08.8	14 05.2	9.1	20 28.4	10.1	57.7
02	208 42.7	08.1	28 33.3	9.1	20 18.3	10.2	57.7
03	223 42.8	. . 07.3	43 01.4	9.3	20 08.1	10.4	57.7
04	238 42.9	06.6	57 29.7	9.3	19 57.7	10.4	57.6
05	253 43.0	05.8	71 58.0	9.5	19 47.3	10.5	57.6
06	268 43.1	N15 05.1	86 26.5	9.6	S19 36.8	10.6	57.6
07	283 43.2	04.3	100 55.1	9.6	19 26.2	10.8	57.6
08	298 43.4	03.6	115 23.7	9.8	19 15.4	10.8	57.5
09	313 43.5	. . 02.8	129 52.5	9.9	19 04.6	10.9	57.5
10	328 43.6	02.1	144 21.4	9.9	18 53.7	11.0	57.5
11	343 43.7	01.3	158 50.3	10.1	18 42.7	11.1	57.5
12	358 43.8	N15 00.6	173 19.4	10.2	S18 31.6	11.1	57.4
13	13 43.9	14 59.8	187 48.6	10.3	18 20.5	11.3	57.4
14	28 44.0	59.1	202 17.9	10.4	18 09.2	11.3	57.4
15	43 44.1	. . 58.3	216 47.3	10.5	17 57.9	11.5	57.4
16	58 44.2	57.6	231 16.8	10.6	17 46.4	11.5	57.3
17	73 44.3	56.8	245 46.4	10.6	17 34.9	11.6	57.3
18	88 44.4	N14 56.1	260 16.0	10.8	S17 23.3	11.6	57.3
19	103 44.5	55.3	274 45.8	10.9	17 11.7	11.8	57.3
20	118 44.6	54.6	289 15.7	11.0	16 59.9	11.8	57.2
21	133 44.7	. . 53.8	303 45.7	11.1	16 48.1	11.9	57.2
22	148 44.8	53.1	318 15.8	11.2	16 36.2	12.0	57.2
23	163 44.9	52.3	332 46.0	11.3	16 24.2	12.0	57.2
13 00	178 45.1	N14 51.6	347 16.3	11.3	S16 12.2	12.1	57.1
01	193 45.2	50.8	1 46.6	11.5	16 00.1	12.2	57.1
02	208 45.3	50.1	16 17.1	11.6	15 47.9	12.2	57.1
03	223 45.4	. . 49.3	30 47.7	11.6	15 35.7	12.3	57.0
04	238 45.5	48.6	45 18.3	11.8	15 23.4	12.4	57.0
05	253 45.6	47.8	59 49.1	11.8	15 11.0	12.4	57.0
06	268 45.7	N14 47.0	74 19.9	12.0	S14 58.6	12.5	57.0
07	283 45.8	46.3	88 50.9	12.0	14 46.1	12.6	56.9
08	298 45.9	45.5	103 21.9	12.1	14 33.5	12.6	56.9
09	313 46.0	. . 44.8	117 53.0	12.2	14 20.9	12.6	56.9
10	328 46.2	44.0	132 24.2	12.3	14 08.3	12.7	56.9
11	343 46.3	43.2	146 55.5	12.4	13 55.6	12.8	56.8
12	358 46.4	N14 42.5	161 26.9	12.5	S13 42.8	12.8	56.8
13	13 46.5	41.7	175 58.4	12.6	13 30.0	12.9	56.8
14	28 46.6	41.0	190 30.0	12.6	13 17.1	13.0	56.8
15	43 46.7	. . 40.2	205 01.6	12.7	13 04.1	12.9	56.7
16	58 46.8	39.4	219 33.3	12.8	12 51.2	13.0	56.7
17	73 46.9	38.7	234 05.1	12.9	12 38.2	13.1	56.7
18	88 47.1	N14 37.9	248 37.0	13.0	S12 25.1	13.1	56.6
19	103 47.2	37.2	263 09.0	13.1	12 12.0	13.2	56.6
20	118 47.3	36.4	277 41.1	13.1	11 58.8	13.2	56.6
21	133 47.4	. . 35.6	292 13.2	13.2	11 45.6	13.2	56.6
22	148 47.5	34.9	306 45.4	13.3	11 32.4	13.3	56.5
23	163 47.6	34.1	321 17.7	13.4	S11 19.1	13.3	56.5
	SD 15.8	*d* 0.7	SD 15.8		15.7		15.5

Lat.	Twilight Naut.	Twilight Civil	Sunrise	Moonrise 11	Moonrise 12	Moonrise 13	Moonrise 14
°	h m	h m	h m	h m	h m	h m	h m
N 72	////	////	01 58	■	23 12	22 10	21 35
N 70	////	////	02 38	■	22 29	21 51	21 26
68	////	01 11	03 05	22 45	21 59	21 36	21 19
66	////	02 02	03 25	21 58	21 37	21 24	21 13
64	////	02 32	03 41	21 27	21 19	21 13	21 08
62	01 03	02 55	03 54	21 03	21 05	21 04	21 03
60	01 49	03 12	04 06	20 45	20 52	20 57	20 59
N 58	02 17	03 27	04 15	20 29	20 42	20 50	20 56
56	02 38	03 39	04 24	20 16	20 32	20 44	20 53
54	02 55	03 50	04 32	20 04	20 24	20 38	20 50
52	03 09	03 59	04 38	19 54	20 16	20 33	20 48
50	03 21	04 08	04 44	19 44	20 09	20 29	20 45
45	03 45	04 25	04 57	19 25	19 55	20 19	20 40
N 40	04 03	04 39	05 08	19 09	19 43	20 11	20 36
35	04 18	04 50	05 17	18 55	19 32	20 04	20 32
30	04 30	05 00	05 25	18 43	19 23	19 58	20 29
20	04 49	05 16	05 39	18 23	19 08	19 47	20 23
N 10	05 03	05 29	05 51	18 05	18 54	19 38	20 18
0	05 15	05 40	06 02	17 49	18 41	19 29	20 13
S 10	05 26	05 51	06 13	17 32	18 28	19 20	20 09
20	05 35	06 01	06 24	17 14	18 14	19 10	20 04
30	05 44	06 12	06 37	16 53	17 58	18 59	19 58
35	05 48	06 18	06 44	16 41	17 48	18 53	19 54
40	05 53	06 25	06 53	16 27	17 38	18 46	19 51
45	05 57	06 32	07 03	16 11	17 25	18 37	19 46
S 50	06 02	06 40	07 14	15 50	17 10	18 27	19 41
52	06 04	06 44	07 20	15 40	17 02	18 22	19 38
54	06 06	06 48	07 26	15 29	16 54	18 17	19 36
56	06 08	06 53	07 33	15 17	16 45	18 11	19 33
58	06 11	06 58	07 40	15 02	16 35	18 04	19 29
S 60	06 13	07 03	07 49	14 44	16 23	17 57	19 26

Lat.	Sunset	Twilight Civil	Twilight Naut.	Moonset 11	Moonset 12	Moonset 13	Moonset 14
°	h m	h m	h m	h m	h m	h m	h m
N 72	22 05	////	////	■	■	02 56	05 36
N 70	21 28	////	////	■	■	03 38	05 53
68	21 02	22 49	////	■	01 34	04 06	06 06
66	20 43	22 04	////	■	02 20	04 26	06 17
64	20 27	21 34	////	00 44	02 50	04 43	06 26
62	20 14	21 13	22 58	01 23	03 13	04 56	06 33
60	20 03	20 56	22 17	01 51	03 31	05 08	06 40
N 58	19 53	20 41	21 50	02 12	03 46	05 18	06 45
56	19 45	20 29	21 30	02 30	03 58	05 26	06 50
54	19 37	20 19	21 13	02 44	04 09	05 34	06 55
52	19 31	20 09	21 00	02 57	04 19	05 41	06 59
50	19 25	20 01	20 48	03 08	04 28	05 47	07 03
45	19 12	19 44	20 24	03 32	04 46	06 00	07 11
N 40	19 01	19 31	20 06	03 50	05 01	06 11	07 17
35	18 52	19 19	19 52	04 06	05 14	06 20	07 23
30	18 44	19 10	19 40	04 19	05 25	06 28	07 28
20	18 31	18 54	19 21	04 42	05 43	06 41	07 36
N 10	18 19	18 41	19 07	05 02	05 59	06 53	07 44
0	18 08	18 30	18 55	05 20	06 14	07 04	07 50
S 10	17 58	18 19	18 45	05 38	06 29	07 15	07 57
20	17 46	18 09	18 35	05 57	06 45	07 27	08 04
30	17 34	17 58	18 27	06 19	07 03	07 40	08 13
35	17 26	17 52	18 22	06 32	07 13	07 48	08 17
40	17 18	17 46	18 18	06 47	07 25	07 56	08 23
45	17 08	17 39	18 14	07 05	07 39	08 06	08 29
S 50	16 56	17 31	18 09	07 26	07 56	08 18	08 36
52	16 51	17 27	18 07	07 37	08 04	08 24	08 39
54	16 45	17 23	18 05	07 48	08 12	08 30	08 43
56	16 38	17 18	18 03	08 01	08 22	08 36	08 47
58	16 31	17 14	18 00	08 17	08 33	08 44	08 51
S 60	16 23	17 08	17 58	08 35	08 46	08 52	08 56

Day	SUN Eqn. of Time 00h	SUN Eqn. of Time 12h	SUN Mer. Pass.	MOON Mer. Pass. Upper	MOON Mer. Pass. Lower	MOON Age	MOON Phase
d	m s	m s	h m	h m	h m	d %	
11	05 20	05 15	12 05	24 02	11 34	13 99	
12	05 10	05 05	12 05	00 02	12 28	14 100	○
13	05 00	04 55	12 05	00 53	13 17	15 98	

UT	ARIES	VENUS −3.9		MARS −2.7		JUPITER −1.7		SATURN +0.1		STARS		
	GHA	GHA	Dec	GHA	Dec	GHA	Dec	GHA	Dec	Name	SHA	Dec
d h	° ′	° ′	° ′	° ′	° ′	° ′	° ′	° ′	° ′		° ′	°
14 00	322 00.3	179 40.7	N16 08.4	339 15.0	S14 28.9	172 25.7	N13 14.9	222 26.7	N22 23.6	Acamar	315 24.1	S40 17
01	337 02.8	194 40.1	07.4	354 17.9	29.1	187 27.7	14.7	237 28.9	23.6	Achernar	335 32.0	S57 12
02	352 05.2	209 39.5	06.5	9 20.8	29.4	202 29.6	14.5	252 31.1	23.6	Acrux	173 18.8	S63 07
03	7 07.7	224 38.9	. . 05.6	24 23.7	. . 29.6	217 31.6	. . 14.4	267 33.3	. . 23.6	Adhara	255 18.9	S28 58
04	22 10.1	239 38.3	04.7	39 26.6	29.8	232 33.5	14.2	282 35.4	23.5	Aldebaran	290 58.4	N16 31
05	37 12.6	254 37.7	03.7	54 29.5	30.1	247 35.4	14.0	297 37.6	23.5			
06	52 15.1	269 37.1	N16 02.8	69 32.4	S14 30.3	262 37.4	N13 13.8	312 39.8	N22 23.5	Alioth	166 27.6	N55 56
07	67 17.5	284 36.5	01.9	84 35.3	30.5	277 39.3	13.6	327 41.9	23.5	Alkaid	153 05.0	N49 18
T 08	82 20.0	299 35.9	00.9	99 38.2	30.8	292 41.3	13.4	342 44.1	23.5	Al Na'ir	27 52.8	S46 56
H 09	97 22.5	314 35.3	16 00.0	114 41.1	. . 31.0	307 43.2	. . 13.2	357 46.3	. . 23.5	Alnilam	275 54.4	S 1 11
U 10	112 24.9	329 34.7	15 59.1	129 44.0	31.2	322 45.1	13.1	12 48.5	23.4	Alphard	218 04.0	S 8 40
R 11	127 27.4	344 34.1	58.1	144 46.9	31.5	337 47.1	12.9	27 50.6	23.4			
S 12	142 29.9	359 33.5	N15 57.2	159 49.8	S14 31.7	352 49.0	N13 12.7	42 52.8	N22 23.4	Alphecca	126 17.5	N26 42
D 13	157 32.3	14 32.9	56.3	174 52.7	31.9	7 51.0	12.5	57 55.0	23.4	Alpheratz	357 51.3	N29 06
A 14	172 34.8	29 32.3	55.3	189 55.6	32.2	22 52.9	12.3	72 57.2	23.4	Altair	62 15.5	N 8 52
Y 15	187 37.2	44 31.7	. . 54.4	204 58.5	. . 32.4	37 54.8	. . 12.1	87 59.3	. . 23.4	Ankaa	353 22.9	S42 17
16	202 39.7	59 31.1	53.4	220 01.4	32.6	52 56.8	12.0	103 01.5	23.3	Antares	112 35.7	S26 26
17	217 42.2	74 30.5	52.5	235 04.3	32.9	67 58.7	11.8	118 03.7	23.3			
18	232 44.6	89 29.9	N15 51.6	250 07.2	S14 33.1	83 00.6	N13 11.6	133 05.9	N22 23.3	Arcturus	146 02.9	N19 10
19	247 47.1	104 29.3	50.6	265 10.1	33.3	98 02.6	11.4	148 08.0	23.3	Atria	107 44.4	S69 02
20	262 49.6	119 28.7	49.7	280 13.1	33.6	113 04.5	11.2	163 10.2	23.3	Avior	234 22.0	S59 31
21	277 52.0	134 28.1	. . 48.7	295 16.0	. . 33.8	128 06.5	. . 11.0	178 12.4	. . 23.3	Bellatrix	278 40.5	N 6 21
22	292 54.5	149 27.5	47.8	310 18.9	34.0	143 08.4	10.8	193 14.6	23.2	Betelgeuse	271 09.9	N 7 24
23	307 57.0	164 26.9	46.8	325 21.8	34.3	158 10.3	10.7	208 16.7	23.2			
15 00	322 59.4	179 26.3	N15 45.9	340 24.7	S14 34.5	173 12.3	N13 10.5	223 18.9	N22 23.2	Canopus	264 00.0	S52 41
01	338 01.9	194 25.7	45.0	355 27.6	34.7	188 14.2	10.3	238 21.1	23.2	Capella	280 46.2	N46 00
02	353 04.4	209 25.2	44.0	10 30.6	35.0	203 16.2	10.1	253 23.3	23.2	Deneb	49 36.4	N45 17
03	8 06.8	224 24.6	. . 43.1	25 33.5	. . 35.2	218 18.1	. . 09.9	268 25.4	. . 23.2	Denebola	182 41.8	N14 33
04	23 09.3	239 24.0	42.1	40 36.4	35.4	233 20.0	09.7	283 27.6	23.1	Diphda	349 03.4	S17 57
05	38 11.7	254 23.4	41.2	55 39.3	35.7	248 22.0	09.6	298 29.8	23.1			
06	53 14.2	269 22.8	N15 40.2	70 42.2	S14 35.9	263 23.9	N13 09.4	313 32.0	N22 23.1	Dubhe	194 01.4	N61 44
07	68 16.7	284 22.2	39.3	85 45.2	36.1	278 25.9	09.2	328 34.1	23.1	Elnath	278 22.6	N28 36
08	83 19.1	299 21.6	38.3	100 48.1	36.4	293 27.8	09.0	343 36.3	23.1	Eltanin	90 49.4	N51 29
F 09	98 21.6	314 21.0	. . 37.4	115 51.0	. . 36.6	308 29.7	. . 08.8	358 38.5	. . 23.1	Enif	33 54.4	N 9 53
R 10	113 24.1	329 20.4	36.4	130 53.9	36.9	323 31.7	08.6	13 40.7	23.1	Fomalhaut	15 32.1	S29 36
I 11	128 26.5	344 19.9	35.5	145 56.9	37.1	338 33.6	08.4	28 42.8	23.0			
D 12	143 29.0	359 19.3	N15 34.5	160 59.8	S14 37.3	353 35.5	N13 08.3	43 45.0	N22 23.0	Gacrux	172 10.2	S57 08
A 13	158 31.5	14 18.7	33.6	176 02.7	37.6	8 37.5	08.1	58 47.2	23.0	Gienah	176 00.6	S17 33
Y 14	173 33.9	29 18.1	32.6	191 05.6	37.8	23 39.4	07.9	73 49.4	23.0	Hadar	148 59.4	S60 23
15	188 36.4	44 17.5	. . 31.6	206 08.6	. . 38.0	38 41.4	. . 07.7	88 51.6	. . 23.0	Hamal	328 09.5	N23 28
16	203 38.8	59 16.9	30.7	221 11.5	38.3	53 43.3	07.5	103 53.7	23.0	Kaus Aust.	83 53.8	S34 23
17	218 41.3	74 16.3	29.7	236 14.4	38.5	68 45.2	07.3	118 55.9	22.9			
18	233 43.8	89 15.8	N15 28.8	251 17.4	S14 38.7	83 47.2	N13 07.1	133 58.1	N22 22.9	Kochab	137 19.4	N74 08
19	248 46.2	104 15.2	27.8	266 20.3	39.0	98 49.1	07.0	149 00.3	22.9	Markab	13 45.8	N15 13
20	263 48.7	119 14.6	26.8	281 23.3	39.2	113 51.1	06.8	164 02.4	22.9	Menkar	314 23.1	N 4 06
21	278 51.2	134 14.0	. . 25.9	296 26.2	. . 39.4	128 53.0	. . 06.6	179 04.6	. . 22.9	Menkent	148 17.0	S36 23
22	293 53.6	149 13.4	24.9	311 29.1	39.7	143 54.9	06.4	194 06.8	22.9	Miaplacidus	221 42.6	S69 43
23	308 56.1	164 12.8	24.0	326 32.1	39.9	158 56.9	06.2	209 09.0	22.8			
16 00	323 58.6	179 12.3	N15 23.0	341 35.0	S14 40.2	173 58.8	N13 06.0	224 11.1	N22 22.8	Mirfak	308 51.6	N49 52
01	339 01.0	194 11.7	22.0	356 38.0	40.4	189 00.7	05.8	239 13.3	22.8	Nunki	76 07.6	S26 17
02	354 03.5	209 11.1	21.1	11 40.9	40.6	204 02.7	05.7	254 15.5	22.8	Peacock	53 30.7	S56 43
03	9 06.0	224 10.5	. . 20.1	26 43.8	. . 40.9	219 04.6	. . 05.5	269 17.7	. . 22.8	Pollux	243 37.5	N28 01
04	24 08.4	239 09.9	19.1	41 46.8	41.1	234 06.6	05.3	284 19.9	22.8	Procyon	245 08.1	N 5 13
05	39 10.9	254 09.4	18.2	56 49.7	41.3	249 08.5	05.1	299 22.0	22.7			
06	54 13.3	269 08.8	N15 17.2	71 52.7	S14 41.6	264 10.4	N13 04.9	314 24.2	N22 22.7	Rasalhague	96 13.5	N12 33
07	69 15.8	284 08.2	16.2	86 55.6	41.8	279 12.4	04.7	329 26.4	22.7	Regulus	207 52.0	N11 57
S 08	84 18.3	299 07.6	15.3	101 58.6	42.0	294 14.3	04.5	344 28.6	22.7	Rigel	281 19.7	S 8 11
A 09	99 20.7	314 07.0	. . 14.3	117 01.5	. . 42.3	309 16.3	. . 04.4	359 30.8	. . 22.7	Rigil Kent.	140 02.8	S60 51
T 10	114 23.2	329 06.5	13.3	132 04.5	42.5	324 18.2	04.2	14 32.9	22.7	Sabik	102 21.3	S15 43
U 11	129 25.7	344 05.9	12.4	147 07.4	42.8	339 20.1	04.0	29 35.1	22.6			
R 12	144 28.1	359 05.3	N15 11.4	162 10.4	S14 43.0	354 22.1	N13 03.8	44 37.3	N22 22.6	Schedar	349 49.3	N56 33
D 13	159 30.6	14 04.7	10.4	177 13.3	43.2	9 24.0	03.6	59 39.5	22.6	Shaula	96 32.2	S37 06
A 14	174 33.1	29 04.2	09.4	192 16.3	43.5	24 26.0	03.4	74 41.6	22.6	Sirius	258 40.8	S16 43
Y 15	189 35.5	44 03.6	. . 08.5	207 19.2	. . 43.7	39 27.9	. . 03.2	89 43.8	. . 22.6	Spica	158 39.6	S11 10
16	204 38.0	59 03.0	07.5	222 22.2	43.9	54 29.8	03.1	104 46.0	22.6	Suhail	222 58.7	S43 26
17	219 40.4	74 02.4	06.5	237 25.2	44.2	69 31.8	02.9	119 48.2	22.5			
18	234 42.9	89 01.9	N15 05.6	252 28.1	S14 44.4	84 33.7	N13 02.7	134 50.4	N22 22.5	Vega	80 43.9	N38 47
19	249 45.4	104 01.3	04.6	267 31.1	44.6	99 35.6	02.5	149 52.5	22.5	Zuben'ubi	137 14.1	S16 03
20	264 47.8	119 00.7	03.6	282 34.0	44.9	114 37.6	02.3	164 54.7	22.5		SHA	Mer. Pass.
21	279 50.3	134 00.2	. . 02.6	297 37.0	. . 45.1	129 39.5	. . 02.1	179 56.9	. . 22.5		° ′	h m
22	294 52.8	148 59.6	01.6	312 40.0	45.4	144 41.5	01.9	194 59.1	22.5	Venus	216 26.9	12 03
23	309 55.2	163 59.0	00.7	327 42.9	45.6	159 43.4	01.8	210 01.3	22.4	Mars	17 25.3	1 18
	h m									Jupiter	210 12.9	12 26
Mer. Pass.	2 27.6	v −0.6	d 1.0	v 2.9	d 0.2	v 1.9	d 0.2	v 2.2	d 0.0	Saturn	260 19.5	9 05

Day	UT (d h)	SUN GHA (° ′)	SUN Dec (° ′)	MOON GHA (° ′)	*v* (′)	MOON Dec (° ′)	*d* (′)	HP (′)
	14 00	178 47.7	N14 33.3	335 50.1	13.5	S11 05.8	13.4	56.5
	01	193 47.8	32.6	350 22.6	13.5	10 52.4	13.3	56.5
	02	208 48.0	31.8	4 55.1	13.6	10 39.1	13.5	56.4
	03	223 48.1	. . 31.0	19 27.7	13.7	10 25.6	13.4	56.4
	04	238 48.2	30.3	34 00.4	13.7	10 12.2	13.5	56.4
	05	253 48.3	29.5	48 33.1	13.8	9 58.7	13.5	56.4
	06	268 48.4	N14 28.8	63 05.9	13.9	S 9 45.2	13.6	56.3
	07	283 48.5	28.0	77 38.8	13.9	9 31.6	13.5	56.3
T	08	298 48.7	27.2	92 11.7	14.1	9 18.1	13.6	56.3
H	09	313 48.8	. . 26.4	106 44.8	14.1	9 04.5	13.6	56.2
U	10	328 48.9	25.7	121 17.9	14.1	8 50.9	13.7	56.2
R	11	343 49.0	24.9	135 51.0	14.2	8 37.2	13.7	56.2
S	12	358 49.1	N14 24.1	150 24.2	14.3	S 8 23.5	13.7	56.2
D	13	13 49.2	23.4	164 57.5	14.4	8 09.8	13.7	56.1
A	14	28 49.4	22.6	179 30.9	14.4	7 56.1	13.7	56.1
Y	15	43 49.5	. . 21.8	194 04.3	14.4	7 42.4	13.8	56.1
	16	58 49.6	21.1	208 37.7	14.6	7 28.6	13.8	56.1
	17	73 49.7	20.3	223 11.3	14.5	7 14.8	13.8	56.0
	18	88 49.8	N14 19.5	237 44.8	14.7	S 7 01.0	13.8	56.0
	19	103 49.9	18.7	252 18.5	14.7	6 47.2	13.8	56.0
	20	118 50.1	18.0	266 52.2	14.7	6 33.4	13.8	56.0
	21	133 50.2	. . 17.2	281 25.9	14.8	6 19.6	13.9	55.9
	22	148 50.3	16.4	295 59.7	14.9	6 05.7	13.8	55.9
	23	163 50.4	15.7	310 33.6	14.9	5 51.9	13.9	55.9
	15 00	178 50.5	N14 14.9	325 07.5	15.0	S 5 38.0	13.9	55.9
	01	193 50.7	14.1	339 41.5	15.0	5 24.1	13.9	55.8
	02	208 50.8	13.3	354 15.5	15.1	5 10.2	13.9	55.8
	03	223 50.9	. . 12.6	8 49.6	15.1	4 56.3	13.9	55.8
	04	238 51.0	11.8	23 23.7	15.1	4 42.4	13.9	55.8
	05	253 51.1	11.0	37 57.8	15.2	4 28.5	13.9	55.7
	06	268 51.3	N14 10.2	52 32.0	15.3	S 4 14.6	13.9	55.7
	07	283 51.4	09.5	67 06.3	15.3	4 00.7	14.0	55.7
	08	298 51.5	08.7	81 40.6	15.3	3 46.7	13.9	55.7
F	09	313 51.6	. . 07.9	96 14.9	15.4	3 32.8	13.9	55.6
R	10	328 51.7	07.1	110 49.3	15.4	3 18.9	13.9	55.6
I	11	343 51.9	06.3	125 23.7	15.4	3 05.0	14.0	55.6
D	12	358 52.0	N14 05.6	139 58.1	15.5	S 2 51.0	13.9	55.6
A	13	13 52.1	04.8	154 32.6	15.5	2 37.1	13.9	55.5
Y	14	28 52.2	04.0	169 07.1	15.6	2 23.2	13.9	55.5
	15	43 52.4	. . 03.2	183 41.7	15.6	2 09.3	14.0	55.5
	16	58 52.5	02.5	198 16.3	15.6	1 55.3	13.9	55.5
	17	73 52.6	01.7	212 50.9	15.7	1 41.4	13.9	55.4
	18	88 52.7	N14 00.9	227 25.6	15.7	S 1 27.5	13.9	55.4
	19	103 52.9	14 00.1	242 00.3	15.7	1 13.6	13.9	55.4
	20	118 53.0	13 59.3	256 35.0	15.7	0 59.7	13.9	55.4
	21	133 53.1	. . 58.5	271 09.7	15.8	0 45.8	13.8	55.4
	22	148 53.2	57.8	285 44.5	15.8	0 32.0	13.9	55.3
	23	163 53.4	57.0	300 19.3	15.8	0 18.1	13.9	55.3
	16 00	178 53.5	N13 56.2	314 54.1	15.9	S 0 04.2	13.8	55.3
	01	193 53.6	55.4	329 29.0	15.9	N 0 09.6	13.8	55.3
	02	208 53.7	54.6	344 03.9	15.9	0 23.4	13.8	55.2
	03	223 53.9	. . 53.8	358 38.8	15.9	0 37.2	13.9	55.2
	04	238 54.0	53.1	13 13.7	15.9	0 51.1	13.7	55.2
	05	253 54.1	52.3	27 48.6	16.0	1 04.8	13.8	55.2
	06	268 54.2	N13 51.5	42 23.6	16.0	N 1 18.6	13.8	55.2
	07	283 54.4	50.7	56 58.6	16.0	1 32.4	13.7	55.1
S	08	298 54.5	49.9	71 33.6	16.0	1 46.1	13.7	55.1
A	09	313 54.6	. . 49.1	86 08.6	16.0	1 59.8	13.7	55.1
T	10	328 54.7	48.3	100 43.6	16.0	2 13.5	13.7	55.1
U	11	343 54.9	47.6	115 18.6	16.1	2 27.2	13.7	55.1
R	12	358 55.0	N13 46.8	129 53.7	16.1	N 2 40.9	13.6	55.0
D	13	13 55.1	46.0	144 28.8	16.0	2 54.5	13.7	55.0
A	14	28 55.3	45.2	159 03.8	16.1	3 08.2	13.6	55.0
Y	15	43 55.4	. . 44.4	173 38.9	16.1	3 21.8	13.5	55.0
	16	58 55.5	43.6	188 14.0	16.1	3 35.3	13.6	55.0
	17	73 55.6	42.8	202 49.1	16.1	3 48.9	13.5	54.9
	18	88 55.8	N13 42.0	217 24.2	16.1	N 4 02.4	13.5	54.9
	19	103 55.9	41.2	231 59.3	16.2	4 15.9	13.5	54.9
	20	118 56.0	40.5	246 34.5	16.1	4 29.4	13.4	54.9
	21	133 56.2	. . 39.7	261 09.6	16.1	4 42.8	13.5	54.9
	22	148 56.3	38.9	275 44.7	16.1	4 56.3	13.4	54.9
	23	163 56.4	38.1	290 19.8	16.1	N 5 09.7	13.3	54.8
		SD 15.8	*d* 0.8	SD	15.3	15.1		15.0

Lat. (°)	Twilight Naut. (h m)	Twilight Civil (h m)	Sunrise (h m)	Moonrise 14 (h m)	Moonrise 15 (h m)	Moonrise 16 (h m)	Moonrise 17 (h m)
N 72	////	////	02 21	21 35	21 06	20 38	20 08
N 70	////	////	02 54	21 26	21 05	20 44	20 22
68	////	01 40	03 17	21 19	21 04	20 49	20 34
66	////	02 18	03 36	21 13	21 03	20 54	20 44
64	////	02 45	03 50	21 08	21 03	20 57	20 52
62	01 28	03 05	04 03	21 03	21 02	21 01	20 59
60	02 04	03 21	04 13	20 59	21 02	21 03	21 05
N 58	02 28	03 35	04 22	20 56	21 01	21 06	21 11
56	02 47	03 46	04 30	20 53	21 01	21 08	21 16
54	03 03	03 56	04 37	20 50	21 00	21 10	21 20
52	03 16	04 05	04 43	20 48	21 00	21 12	21 24
50	03 27	04 13	04 49	20 45	21 00	21 14	21 28
45	03 49	04 29	05 01	20 40	20 59	21 17	21 36
N 40	04 07	04 42	05 11	20 36	20 58	21 20	21 43
35	04 20	04 53	05 20	20 32	20 58	21 23	21 48
30	04 32	05 02	05 27	20 29	20 58	21 25	21 54
20	04 50	05 17	05 40	20 23	20 57	21 30	22 02
N 10	05 04	05 29	05 51	20 18	20 56	21 33	22 10
0	05 15	05 40	06 01	20 13	20 56	21 37	22 18
S 10	05 25	05 50	06 11	20 09	20 55	21 40	22 25
20	05 33	05 59	06 22	20 04	20 55	21 44	22 33
30	05 41	06 09	06 34	19 58	20 54	21 48	22 43
35	05 45	06 15	06 41	19 54	20 53	21 51	22 48
40	05 49	06 21	06 49	19 51	20 53	21 54	22 54
45	05 53	06 28	06 58	19 46	20 53	21 57	23 01
S 50	05 57	06 35	07 09	19 41	20 52	22 01	23 10
52	05 59	06 39	07 14	19 38	20 52	22 03	23 14
54	06 01	06 42	07 20	19 36	20 51	22 05	23 18
56	06 02	06 46	07 26	19 33	20 51	22 08	23 23
58	06 04	06 51	07 33	19 29	20 51	22 10	23 29
S 60	06 07	06 56	07 41	19 26	20 50	22 13	23 35

Lat. (°)	Sunset (h m)	Twilight Civil (h m)	Twilight Naut. (h m)	Moonset 14 (h m)	Moonset 15 (h m)	Moonset 16 (h m)	Moonset 17 (h m)
N 72	21 42	////	////	05 36	07 44	09 42	11 40
N 70	21 11	23 46	////	05 53	07 49	09 39	11 28
68	20 48	22 23	////	06 06	07 54	09 37	11 18
66	20 31	21 46	////	06 17	07 58	09 35	11 10
64	20 17	21 21	23 48	06 26	08 01	09 33	11 03
62	20 05	21 01	22 34	06 33	08 04	09 31	10 57
60	19 54	20 46	22 01	06 40	08 07	09 30	10 52
N 58	19 46	20 32	21 38	06 45	08 09	09 29	10 48
56	19 38	20 21	21 20	06 50	08 11	09 28	10 44
54	19 31	20 11	21 04	06 55	08 12	09 27	10 40
52	19 25	20 03	20 52	06 59	08 14	09 26	10 37
50	19 19	19 55	20 41	07 03	08 15	09 25	10 34
45	19 07	19 39	20 19	07 11	08 18	09 24	10 28
N 40	18 57	19 26	20 02	07 17	08 21	09 23	10 23
35	18 49	19 16	19 48	07 23	08 23	09 21	10 19
30	18 42	19 07	19 37	07 28	08 25	09 20	10 15
20	18 29	18 52	19 19	07 36	08 28	09 19	10 08
N 10	18 18	18 40	19 05	07 44	08 31	09 17	10 02
0	18 08	18 29	18 54	07 50	08 34	09 15	09 56
S 10	17 58	18 19	18 44	07 57	08 36	09 14	09 51
20	17 47	18 10	18 36	08 04	08 39	09 12	09 45
30	17 35	18 00	18 28	08 13	08 42	09 11	09 38
35	17 28	17 55	18 24	08 17	08 44	09 09	09 35
40	17 21	17 49	18 21	08 23	08 46	09 08	09 30
45	17 12	17 42	18 17	08 29	08 48	09 07	09 25
S 50	17 01	17 35	18 13	08 36	08 51	09 05	09 19
52	16 56	17 31	18 11	08 39	08 52	09 05	09 17
54	16 50	17 28	18 09	08 43	08 54	09 04	09 14
56	16 44	17 24	18 08	08 47	08 55	09 03	09 10
58	16 37	17 19	18 06	08 51	08 57	09 02	09 07
S 60	16 29	17 14	18 04	08 56	08 59	09 01	09 03

Day	SUN Eqn. of Time 00^h	SUN Eqn. of Time 12^h	SUN Mer. Pass.	MOON Mer. Pass. Upper	MOON Mer. Pass. Lower	MOON Age	MOON Phase
d	m s	m s	h m	h m	h m	d	%
14	04 49	04 44	12 05	01 40	14 02	16	94
15	04 38	04 32	12 05	02 24	14 45	17	88
16	04 26	04 20	12 04	03 06	15 26	18	81

UT	ARIES	VENUS −3.9		MARS −2.8		JUPITER −1.7		SATURN +0.1		STARS		
	GHA	GHA	Dec	GHA	Dec	GHA	Dec	GHA	Dec	Name	SHA	Dec
d h	° ′	° ′	° ′	° ′	° ′	° ′	° ′	° ′	° ′		° ′	° ′
17 00	324 57.7	178 58.4	N14 59.7	342 45.9	S14 45.8	174 45.3	N13 01.6	225 03.4	N22 22.4	Acamar	315 24.0	S40 17.
01	340 00.2	193 57.9	58.7	357 48.8	46.1	189 47.3	01.4	240 05.6	22.4	Achernar	335 32.0	S57 12.
02	355 02.6	208 57.3	57.7	12 51.8	46.3	204 49.2	01.2	255 07.8	22.4	Acrux	173 18.8	S63 07.
03	10 05.1	223 56.7	. . 56.7	27 54.8	. . 46.5	219 51.2	. . 01.0	270 10.0	. . 22.4	Adhara	255 18.9	S28 58.
04	25 07.6	238 56.2	55.8	42 57.7	46.8	234 53.1	00.8	285 12.2	22.4	Aldebaran	290 58.4	N16 31.
05	40 10.0	253 55.6	54.8	58 00.7	47.0	249 55.0	00.6	300 14.3	22.3			
06	55 12.5	268 55.0	N14 53.8	73 03.7	S14 47.3	264 57.0	N13 00.5	315 16.5	N22 22.3	Alioth	166 27.6	N55 56.
07	70 14.9	283 54.5	52.8	88 06.7	47.5	279 58.9	00.3	330 18.7	22.3	Alkaid	153 05.0	N49 18.
08	85 17.4	298 53.9	51.8	103 09.6	47.7	295 00.8	13 00.1	345 20.9	22.3	Al Na'ir	27 52.8	S46 56.
S 09	100 19.9	313 53.3	. . 50.8	118 12.6	. . 48.0	310 02.8	12 59.9	0 23.1	. . 22.3	Alnilam	275 54.4	S 1 11.
U 10	115 22.3	328 52.8	49.8	133 15.6	48.2	325 04.7	59.7	15 25.3	22.3	Alphard	218 04.0	S 8 40.
N 11	130 24.8	343 52.2	48.9	148 18.6	48.4	340 06.7	59.5	30 27.4	22.2			
D 12	145 27.3	358 51.6	N14 47.9	163 21.5	S14 48.7	355 08.6	N12 59.3	45 29.6	N22 22.2	Alphecca	126 17.5	N26 42.
A 13	160 29.7	13 51.1	46.9	178 24.5	48.9	10 10.5	59.2	60 31.8	22.2	Alpheratz	357 51.3	N29 06.
Y 14	175 32.2	28 50.5	45.9	193 27.5	49.2	25 12.5	59.0	75 34.0	22.2	Altair	62 15.5	N 8 52.
15	190 34.7	43 49.9	. . 44.9	208 30.5	. . 49.4	40 14.4	. . 58.8	90 36.2	. . 22.2	Ankaa	353 22.9	S42 17.
16	205 37.1	58 49.4	43.9	223 33.4	49.6	55 16.4	58.6	105 38.3	22.2	Antares	112 35.7	S26 26.
17	220 39.6	73 48.8	42.9	238 36.4	49.9	70 18.3	58.4	120 40.5	22.1			
18	235 42.1	88 48.2	N14 41.9	253 39.4	S14 50.1	85 20.2	N12 58.2	135 42.7	N22 22.1	Arcturus	146 02.9	N19 10.
19	250 44.5	103 47.7	40.9	268 42.4	50.3	100 22.2	58.0	150 44.9	22.1	Atria	107 44.4	S69 02.
20	265 47.0	118 47.1	39.9	283 45.4	50.6	115 24.1	57.9	165 47.1	22.1	Avior	234 21.9	S59 31.
21	280 49.4	133 46.6	. . 38.9	298 48.3	. . 50.8	130 26.1	. . 57.7	180 49.3	. . 22.1	Bellatrix	278 40.5	N 6 21.
22	295 51.9	148 46.0	37.9	313 51.3	51.1	145 28.0	57.5	195 51.4	22.1	Betelgeuse	271 09.9	N 7 24.
23	310 54.4	163 45.4	36.9	328 54.3	51.3	160 29.9	57.3	210 53.6	22.0			
18 00	325 56.8	178 44.9	N14 35.9	343 57.3	S14 51.5	175 31.9	N12 57.1	225 55.8	N22 22.0	Canopus	264 00.0	S52 41.
01	340 59.3	193 44.3	34.9	359 00.3	51.8	190 33.8	56.9	240 58.0	22.0	Capella	280 46.1	N46 00.
02	356 01.8	208 43.8	33.9	14 03.3	52.0	205 35.7	56.7	256 00.2	22.0	Deneb	49 36.4	N45 17.
03	11 04.2	223 43.2	. . 32.9	29 06.3	. . 52.2	220 37.7	. . 56.5	271 02.4	. . 22.0	Denebola	182 41.8	N14 33.
04	26 06.7	238 42.6	31.9	44 09.3	52.5	235 39.6	56.4	286 04.5	22.0	Diphda	349 03.4	S17 57.
05	41 09.2	253 42.1	30.9	59 12.3	52.7	250 41.6	56.2	301 06.7	21.9			
06	56 11.6	268 41.5	N14 29.9	74 15.2	S14 53.0	265 43.5	N12 56.0	316 08.9	N22 21.9	Dubhe	194 01.5	N61 44.
07	71 14.1	283 41.0	28.9	89 18.2	53.2	280 45.4	55.8	331 11.1	21.9	Elnath	278 22.6	N28 36.
08	86 16.5	298 40.4	27.9	104 21.2	53.4	295 47.4	55.6	346 13.3	21.9	Eltanin	90 49.4	N51 29.
M 09	101 19.0	313 39.9	. . 26.9	119 24.2	. . 53.7	310 49.3	. . 55.4	1 15.5	. . 21.9	Enif	33 54.4	N 9 53.
O 10	116 21.5	328 39.3	25.9	134 27.2	53.9	325 51.3	55.2	16 17.6	21.9	Fomalhaut	15 32.1	S29 36.
N 11	131 23.9	343 38.7	24.9	149 30.2	54.1	340 53.2	55.1	31 19.8	21.8			
D 12	146 26.4	358 38.2	N14 23.9	164 33.2	S14 54.4	355 55.1	N12 54.9	46 22.0	N22 21.8	Gacrux	172 10.2	S57 08.
A 13	161 28.9	13 37.6	22.9	179 36.2	54.6	10 57.1	54.7	61 24.2	21.8	Gienah	176 00.6	S17 33.
Y 14	176 31.3	28 37.1	21.9	194 39.2	54.9	25 59.0	54.5	76 26.4	21.8	Hadar	148 59.4	S60 23.
15	191 33.8	43 36.5	. . 20.9	209 42.2	. . 55.1	41 00.9	. . 54.3	91 28.6	. . 21.8	Hamal	328 09.4	N23 28.
16	206 36.3	58 36.0	19.9	224 45.2	55.3	56 02.9	54.1	106 30.7	21.8	Kaus Aust.	83 53.8	S34 23.
17	221 38.7	73 35.4	18.9	239 48.2	55.6	71 04.8	53.9	121 32.9	21.7			
18	236 41.2	88 34.9	N14 17.9	254 51.2	S14 55.8	86 06.8	N12 53.8	136 35.1	N22 21.7	Kochab	137 19.5	N74 08.
19	251 43.7	103 34.3	16.9	269 54.2	56.0	101 08.7	53.6	151 37.3	21.7	Markab	13 45.8	N15 13.
20	266 46.1	118 33.8	15.9	284 57.2	56.3	116 10.6	53.4	166 39.5	21.7	Menkar	314 23.1	N 4 06.
21	281 48.6	133 33.2	. . 14.8	300 00.2	. . 56.5	131 12.6	. . 53.2	181 41.7	. . 21.7	Menkent	148 17.0	S36 23.
22	296 51.0	148 32.7	13.8	315 03.2	56.8	146 14.5	53.0	196 43.9	21.6	Miaplacidus	221 42.6	S69 43.
23	311 53.5	163 32.1	12.8	330 06.2	57.0	161 16.5	52.8	211 46.0	21.6			
19 00	326 56.0	178 31.6	N14 11.8	345 09.3	S14 57.2	176 18.4	N12 52.6	226 48.2	N22 21.6	Mirfak	308 51.6	N49 52.
01	341 58.4	193 31.0	10.8	0 12.3	57.5	191 20.3	52.4	241 50.4	21.6	Nunki	76 07.6	S26 17.
02	357 00.9	208 30.5	09.8	15 15.3	57.7	206 22.3	52.3	256 52.6	21.6	Peacock	53 30.7	S56 43.
03	12 03.4	223 29.9	. . 08.8	30 18.3	. . 57.9	221 24.2	. . 52.1	271 54.8	. . 21.6	Pollux	243 37.5	N28 01.
04	27 05.8	238 29.4	07.7	45 21.3	58.2	236 26.2	51.9	286 57.0	21.5	Procyon	245 08.1	N 5 13.
05	42 08.3	253 28.8	06.7	60 24.3	58.4	251 28.1	51.7	301 59.2	21.5			
06	57 10.8	268 28.3	N14 05.7	75 27.3	S14 58.7	266 30.0	N12 51.5	317 01.3	N22 21.5	Rasalhague	96 13.5	N12 33.
07	72 13.2	283 27.7	04.7	90 30.3	58.9	281 32.0	51.3	332 03.5	21.5	Regulus	207 52.0	N11 57.
T 08	87 15.7	298 27.2	03.7	105 33.3	59.1	296 33.9	51.1	347 05.7	21.5	Rigel	281 19.6	S 8 11.
U 09	102 18.2	313 26.6	. . 02.7	120 36.4	. . 59.4	311 35.8	. . 50.9	2 07.9	. . 21.5	Rigil Kent.	140 02.8	S60 51.
E 10	117 20.6	328 26.1	01.6	135 39.4	59.6	326 37.8	50.8	17 10.1	21.4	Sabik	102 21.3	S15 43.
S 11	132 23.1	343 25.6	14 00.6	150 42.4	14 59.8	341 39.7	50.6	32 12.3	21.4			
D 12	147 25.5	358 25.0	N13 59.6	165 45.4	S15 00.1	356 41.7	N12 50.4	47 14.5	N22 21.4	Schedar	349 49.2	N56 33.
A 13	162 28.0	13 24.5	58.6	180 48.4	00.3	11 43.6	50.2	62 16.6	21.4	Shaula	96 32.2	S37 06.
Y 14	177 30.5	28 23.9	57.5	195 51.5	00.6	26 45.5	50.0	77 18.8	21.4	Sirius	258 40.8	S16 43.
15	192 32.9	43 23.4	. . 56.5	210 54.5	. . 00.8	41 47.5	. . 49.8	92 21.0	. . 21.4	Spica	158 39.6	S11 10.
16	207 35.4	58 22.8	55.5	225 57.5	01.0	56 49.4	49.6	107 23.2	21.3	Suhail	222 58.7	S43 26.
17	222 37.9	73 22.3	54.5	241 00.5	01.3	71 51.4	49.5	122 25.4	21.3			
18	237 40.3	88 21.8	N13 53.4	256 03.5	S15 01.5	86 53.3	N12 49.3	137 27.6	N22 21.3	Vega	80 43.9	N38 47.
19	252 42.8	103 21.2	52.4	271 06.6	01.7	101 55.2	49.1	152 29.8	21.3	Zuben'ubi	137 14.1	S16 03.
20	267 45.3	118 20.7	51.4	286 09.6	02.0	116 57.2	48.9	167 32.0	21.3		SHA	Mer. Pass
21	282 47.7	133 20.1	. . 50.4	301 12.6	. . 02.2	131 59.1	. . 48.7	182 34.1	. . 21.3		° ′	h m
22	297 50.2	148 19.6	49.3	316 15.6	02.5	147 01.1	48.5	197 36.3	21.2	Venus	212 48.0	12 05
23	312 52.6	163 19.1	48.3	331 18.7	02.7	162 03.0	48.3	212 38.5	21.2	Mars	18 00.5	1 04
	h m									Jupiter	209 35.0	12 16
Mer. Pass.	2 15.8	v −0.6	d 1.0	v 3.0	d 0.2	v 1.9	d 0.2	v 2.2	d 0.0	Saturn	259 59.0	8 55

UT		SUN GHA	SUN Dec	MOON GHA	v	MOON Dec	d	HP
d	h	° ′	° ′	° ′	′	° ′	′	′
17	00	178 56.5	N13 37.3	304 54.9	16.2	N 5 23.0	13.4	54.8
	01	193 56.7	36.5	319 30.1	16.1	5 36.4	13.3	54.8
	02	208 56.8	35.7	334 05.2	16.1	5 49.7	13.2	54.8
	03	223 56.9	. . 34.9	348 40.3	16.1	6 02.9	13.3	54.8
	04	238 57.1	34.1	3 15.4	16.1	6 16.2	13.2	54.8
	05	253 57.2	33.3	17 50.5	16.1	6 29.4	13.2	54.7
	06	268 57.3	N13 32.5	32 25.6	16.1	N 6 42.6	13.1	54.7
	07	283 57.5	31.7	47 00.7	16.1	6 55.7	13.1	54.7
	08	298 57.6	30.9	61 35.8	16.1	7 08.8	13.1	54.7
S	09	313 57.7	. . 30.1	76 10.9	16.1	7 21.9	13.1	54.7
U	10	328 57.9	29.3	90 46.0	16.0	7 35.0	13.0	54.7
N	11	343 58.0	28.5	105 21.0	16.1	7 48.0	12.9	54.6
D	12	358 58.1	N13 27.8	119 56.1	16.0	N 8 00.9	13.0	54.6
A	13	13 58.3	27.0	134 31.1	16.0	8 13.9	12.9	54.6
Y	14	28 58.4	26.2	149 06.1	16.1	8 26.8	12.8	54.6
	15	43 58.5	. . 25.4	163 41.2	15.9	8 39.6	12.8	54.6
	16	58 58.7	24.6	178 16.1	16.0	8 52.4	12.8	54.6
	17	73 58.8	23.8	192 51.1	16.0	9 05.2	12.7	54.6
	18	88 58.9	N13 23.0	207 26.1	15.9	N 9 17.9	12.7	54.5
	19	103 59.1	22.2	222 01.0	15.9	9 30.6	12.7	54.5
	20	118 59.2	21.4	236 35.9	15.9	9 43.3	12.6	54.5
	21	133 59.3	. . 20.6	251 10.8	15.9	9 55.9	12.5	54.5
	22	148 59.5	19.8	265 45.7	15.9	10 08.4	12.5	54.5
	23	163 59.6	19.0	280 20.6	15.8	10 20.9	12.5	54.5
18	00	178 59.7	N13 18.2	294 55.4	15.8	N10 33.4	12.4	54.5
	01	193 59.9	17.4	309 30.2	15.8	10 45.8	12.4	54.5
	02	209 00.0	16.6	324 05.0	15.7	10 58.2	12.4	54.5
	03	224 00.1	. . 15.8	338 39.7	15.8	11 10.6	12.2	54.4
	04	239 00.3	15.0	353 14.5	15.7	11 22.8	12.3	54.4
	05	254 00.4	14.2	7 49.2	15.6	11 35.1	12.2	54.4
	06	269 00.6	N13 13.3	22 23.8	15.7	N11 47.3	12.1	54.4
	07	284 00.7	12.5	36 58.5	15.6	11 59.4	12.1	54.4
	08	299 00.8	11.7	51 33.1	15.6	12 11.5	12.0	54.4
M	09	314 01.0	. . 10.9	66 07.7	15.5	12 23.5	12.0	54.4
O	10	329 01.1	10.1	80 42.2	15.5	12 35.5	12.0	54.4
N	11	344 01.2	09.3	95 16.7	15.5	12 47.5	11.8	54.4
D	12	359 01.4	N13 08.5	109 51.2	15.4	N12 59.3	11.9	54.4
A	13	14 01.5	07.7	124 25.6	15.4	13 11.2	11.7	54.4
Y	14	29 01.7	06.9	139 00.0	15.4	13 22.9	11.7	54.3
	15	44 01.8	. . 06.1	153 34.4	15.3	13 34.6	11.7	54.3
	16	59 01.9	05.3	168 08.7	15.3	13 46.3	11.6	54.3
	17	74 02.1	04.5	182 43.0	15.3	13 57.9	11.6	54.3
	18	89 02.2	N13 03.7	197 17.3	15.2	N14 09.5	11.4	54.3
	19	104 02.4	02.9	211 51.5	15.2	14 20.9	11.5	54.3
	20	119 02.5	02.1	226 25.7	15.1	14 32.4	11.3	54.3
	21	134 02.6	. . 01.3	240 59.8	15.1	14 43.7	11.4	54.3
	22	149 02.8	13 00.4	255 33.9	15.0	14 55.1	11.2	54.3
	23	164 02.9	12 59.6	270 07.9	15.0	15 06.3	11.2	54.3
19	00	179 03.1	N12 58.8	284 41.9	15.0	N15 17.5	11.1	54.3
	01	194 03.2	58.0	299 15.9	14.9	15 28.6	11.1	54.3
	02	209 03.3	57.2	313 49.8	14.8	15 39.7	11.0	54.3
	03	224 03.5	. . 56.4	328 23.6	14.9	15 50.7	10.9	54.3
	04	239 03.6	55.6	342 57.5	14.7	16 01.6	10.9	54.3
	05	254 03.8	54.8	357 31.2	14.7	16 12.5	10.8	54.3
	06	269 03.9	N12 54.0	12 04.9	14.7	N16 23.3	10.7	54.3
	07	284 04.0	53.1	26 38.6	14.6	16 34.0	10.7	54.3
T	08	299 04.2	52.3	41 12.2	14.6	16 44.7	10.6	54.3
U	09	314 04.3	. . 51.5	55 45.8	14.5	16 55.3	10.5	54.3
E	10	329 04.5	50.7	70 19.3	14.5	17 05.8	10.5	54.3
S	11	344 04.6	49.9	84 52.8	14.4	17 16.3	10.4	54.3
D	12	359 04.8	N12 49.1	99 26.2	14.3	N17 26.7	10.3	54.3
A	13	14 04.9	48.3	113 59.5	14.3	17 37.0	10.3	54.3
Y	14	29 05.0	47.4	128 32.8	14.3	17 47.3	10.2	54.3
	15	44 05.2	. . 46.6	143 06.1	14.2	17 57.5	10.1	54.3
	16	59 05.3	45.8	157 39.3	14.1	18 07.6	10.0	54.3
	17	74 05.5	45.0	172 12.4	14.1	18 17.6	10.0	54.3
	18	89 05.6	N12 44.2	186 45.5	14.0	N18 27.6	9.9	54.3
	19	104 05.8	43.4	201 18.5	14.0	18 37.5	9.8	54.3
	20	119 05.9	42.5	215 51.5	13.9	18 47.3	9.7	54.3
	21	134 06.1	. . 41.7	230 24.4	13.8	18 57.0	9.7	54.3
	22	149 06.2	40.9	244 57.2	13.8	19 06.7	9.5	54.3
	23	164 06.3	40.1	259 30.0	13.7	N19 16.2	9.5	54.3
		SD 15.8	*d* 0.8	SD 14.9		14.8		14.8

Lat.	Twilight Naut.	Twilight Civil	Sunrise	Moonrise 17	Moonrise 18	Moonrise 19	Moonrise 20
°	h m	h m	h m	h m	h m	h m	h m
N 72	////	////	02 41	20 08	19 26	▭	▭
N 70	////	01 07	03 09	20 22	19 54	19 06	▭
68	////	02 02	03 30	20 34	20 16	19 49	18 23
66	////	02 34	03 46	20 44	20 33	20 19	19 55
64	00 59	02 57	04 00	20 52	20 47	20 41	20 34
62	01 48	03 15	04 11	20 59	20 58	20 59	21 01
60	02 17	03 30	04 20	21 05	21 09	21 13	21 22
N 58	02 39	03 42	04 28	21 11	21 17	21 26	21 39
56	02 56	03 53	04 36	21 16	21 25	21 37	21 54
54	03 10	04 02	04 42	21 20	21 32	21 47	22 06
52	03 22	04 10	04 48	21 24	21 38	21 56	22 18
50	03 33	04 18	04 53	21 28	21 44	22 03	22 27
45	03 54	04 33	05 05	21 36	21 56	22 20	22 48
N 40	04 10	04 45	05 14	21 43	22 07	22 34	23 05
35	04 23	04 55	05 22	21 48	22 16	22 45	23 20
30	04 34	05 04	05 29	21 54	22 23	22 56	23 32
20	04 51	05 18	05 41	22 02	22 37	23 14	23 54
N 10	05 04	05 29	05 51	22 10	22 49	23 29	24 13
0	05 15	05 39	06 01	22 18	23 00	23 44	24 30
S 10	05 24	05 48	06 10	22 25	23 11	23 58	24 48
20	05 31	05 57	06 20	22 33	23 23	24 14	00 14
30	05 38	06 07	06 31	22 43	23 37	24 32	00 32
35	05 42	06 12	06 38	22 48	23 45	24 43	00 43
40	05 45	06 17	06 45	22 54	23 55	24 55	00 55
45	05 49	06 23	06 53	23 01	24 05	00 05	01 10
S 50	05 52	06 30	07 03	23 10	24 19	00 19	01 28
52	05 53	06 33	07 08	23 14	24 25	00 25	01 36
54	05 55	06 36	07 13	23 18	24 32	00 32	01 46
56	05 56	06 40	07 19	23 23	24 40	00 40	01 57
58	05 58	06 44	07 25	23 29	24 48	00 48	02 09
S 60	05 59	06 48	07 32	23 35	24 58	00 58	02 24

Lat.	Sunset	Twilight Civil	Twilight Naut.	Moonset 17	Moonset 18	Moonset 19	Moonset 20
°	h m	h m	h m	h m	h m	h m	h m
N 72	21 22	////	////	11 40	13 50	▭	▭
N 70	20 55	22 49	////	11 28	13 23	15 41	▭
68	20 35	22 00	////	11 18	13 03	14 59	18 00
66	20 19	21 30	////	11 10	12 47	14 31	16 29
64	20 06	21 08	22 59	11 03	12 35	14 10	15 51
62	19 55	20 50	22 15	10 57	12 24	13 53	15 25
60	19 46	20 36	21 47	10 52	12 15	13 39	15 04
N 58	19 38	20 24	21 26	10 48	12 07	13 27	14 48
56	19 31	20 13	21 10	10 44	12 00	13 16	14 33
54	19 24	20 04	20 56	10 40	11 54	13 07	14 21
52	19 19	19 56	20 44	10 37	11 48	12 59	14 11
50	19 14	19 49	20 34	10 34	11 43	12 52	14 01
45	19 02	19 34	20 13	10 28	11 32	12 37	13 41
N 40	18 53	19 22	19 57	10 23	11 23	12 24	13 25
35	18 45	19 12	19 44	10 19	11 16	12 13	13 11
30	18 39	19 04	19 33	10 15	11 09	12 04	13 00
20	18 27	18 50	19 16	10 08	10 57	11 48	12 39
N 10	18 17	18 38	19 04	10 02	10 47	11 34	12 22
0	18 07	18 28	18 53	09 56	10 38	11 21	12 06
S 10	17 58	18 19	18 44	09 51	10 29	11 08	11 50
20	17 48	18 11	18 37	09 45	10 19	10 54	11 32
30	17 37	18 02	18 30	09 38	10 07	10 38	11 13
35	17 31	17 57	18 26	09 35	10 01	10 29	11 01
40	17 23	17 51	18 23	09 30	09 53	10 19	10 48
45	17 15	17 45	18 20	09 25	09 45	10 07	10 32
S 50	17 05	17 39	18 17	09 19	09 35	09 52	10 13
52	17 00	17 36	18 15	09 17	09 30	09 45	10 04
54	16 55	17 32	18 14	09 14	09 25	09 38	09 54
56	16 50	17 29	18 13	09 10	09 19	09 29	09 43
58	16 43	17 25	18 11	09 07	09 12	09 20	09 30
S 60	16 36	17 21	18 10	09 03	09 05	09 09	09 15

Day	SUN Eqn. of Time 00^h	SUN Eqn. of Time 12^h	SUN Mer. Pass.	MOON Mer. Pass. Upper	MOON Mer. Pass. Lower	MOON Age	MOON Phase
d	m s	m s	h m	h m	h m	d	%
17	04 14	04 08	12 04	03 47	16 07	19	73
18	04 01	03 55	12 04	04 28	16 49	20	64
19	03 48	03 41	12 04	05 10	17 32	21	55

UT d	h	ARIES GHA ° ′	VENUS −3.9 GHA ° ′	Dec ° ′	MARS −2.8 GHA ° ′	Dec ° ′	JUPITER −1.7 GHA ° ′	Dec ° ′	SATURN +0.1 GHA ° ′	Dec ° ′
20 WEDNESDAY	00	327 55.1	178 18.5	N13 47.3	346 21.7	S15 02.9	177 04.9	N12 48.1	227 40.7	N22 21.2
	01	342 57.6	193 18.0	46.2	1 24.7	03.2	192 06.9	48.0	242 42.9	21.2
	02	358 00.0	208 17.4	45.2	16 27.8	03.4	207 08.8	47.8	257 45.1	21.2
	03	13 02.5	223 16.9	44.2	31 30.8	03.6	222 10.7	47.6	272 47.3	21.2
	04	28 05.0	238 16.4	43.2	46 33.8	03.9	237 12.7	47.4	287 49.5	21.1
	05	43 07.4	253 15.8	42.1	61 36.8	04.1	252 14.6	47.2	302 51.6	21.1
	06	58 09.9	268 15.3	N13 41.1	76 39.9	S15 04.3	267 16.6	N12 47.0	317 53.8	N22 21.1
	07	73 12.4	283 14.8	40.1	91 42.9	04.6	282 18.5	46.8	332 56.0	21.1
	08	88 14.8	298 14.2	39.0	106 45.9	04.8	297 20.4	46.6	347 58.2	21.1
	09	103 17.3	313 13.7	38.0	121 49.0	05.0	312 22.4	46.5	3 00.4	21.1
	10	118 19.8	328 13.1	36.9	136 52.0	05.3	327 24.3	46.3	18 02.6	21.0
	11	133 22.2	343 12.6	35.9	151 55.1	05.5	342 26.3	46.1	33 04.8	21.0
	12	148 24.7	358 12.1	N13 34.9	166 58.1	S15 05.8	357 28.2	N12 45.9	48 07.0	N22 21.0
	13	163 27.1	13 11.5	33.8	182 01.1	06.0	12 30.1	45.7	63 09.2	21.0
	14	178 29.6	28 11.0	32.8	197 04.2	06.2	27 32.1	45.5	78 11.3	21.0
	15	193 32.1	43 10.5	31.8	212 07.2	06.5	42 34.0	45.3	93 13.5	21.0
	16	208 34.5	58 09.9	30.7	227 10.2	06.7	57 36.0	45.1	108 15.7	20.9
	17	223 37.0	73 09.4	29.7	242 13.3	06.9	72 37.9	45.0	123 17.9	20.9
	18	238 39.5	88 08.9	N13 28.6	257 16.3	S15 07.2	87 39.8	N12 44.8	138 20.1	N22 20.9
	19	253 41.9	103 08.4	27.6	272 19.4	07.4	102 41.8	44.6	153 22.3	20.9
	20	268 44.4	118 07.8	26.5	287 22.4	07.6	117 43.7	44.4	168 24.5	20.9
	21	283 46.9	133 07.3	25.5	302 25.5	07.9	132 45.6	44.2	183 26.7	20.9
	22	298 49.3	148 06.8	24.5	317 28.5	08.1	147 47.6	44.0	198 28.9	20.8
	23	313 51.8	163 06.2	23.4	332 31.5	08.3	162 49.5	43.8	213 31.1	20.8
21 THURSDAY	00	328 54.3	178 05.7	N13 22.4	347 34.6	S15 08.6	177 51.5	N12 43.6	228 33.3	N22 20.8
	01	343 56.7	193 05.2	21.3	2 37.6	08.8	192 53.4	43.5	243 35.4	20.8
	02	358 59.2	208 04.6	20.3	17 40.7	09.0	207 55.3	43.3	258 37.6	20.8
	03	14 01.6	223 04.1	19.2	32 43.7	09.3	222 57.3	43.1	273 39.8	20.8
	04	29 04.1	238 03.6	18.2	47 46.8	09.5	237 59.2	42.9	288 42.0	20.7
	05	44 06.6	253 03.1	17.1	62 49.8	09.8	253 01.2	42.7	303 44.2	20.7
	06	59 09.0	268 02.5	N13 16.1	77 52.9	S15 10.0	268 03.1	N12 42.5	318 46.4	N22 20.7
	07	74 11.5	283 02.0	15.0	92 55.9	10.2	283 05.0	42.3	333 48.6	20.7
	08	89 14.0	298 01.5	14.0	107 59.0	10.5	298 07.0	42.1	348 50.8	20.7
	09	104 16.4	313 01.0	12.9	123 02.0	10.7	313 08.9	42.0	3 53.0	20.7
	10	119 18.9	328 00.4	11.9	138 05.1	10.9	328 10.9	41.8	18 55.2	20.6
	11	134 21.4	342 59.9	10.8	153 08.1	11.2	343 12.8	41.6	33 57.4	20.6
	12	149 23.8	357 59.4	N13 09.8	168 11.2	S15 11.4	358 14.7	N12 41.4	48 59.6	N22 20.6
	13	164 26.3	12 58.9	08.7	183 14.3	11.6	13 16.7	41.2	64 01.7	20.6
	14	179 28.7	27 58.3	07.7	198 17.3	11.9	28 18.6	41.0	79 03.9	20.6
	15	194 31.2	42 57.8	06.6	213 20.4	12.1	43 20.6	40.8	94 06.1	20.6
	16	209 33.7	57 57.3	05.6	228 23.4	12.3	58 22.5	40.6	109 08.3	20.5
	17	224 36.1	72 56.8	04.5	243 26.5	12.6	73 24.4	40.5	124 10.5	20.5
	18	239 38.6	87 56.3	N13 03.4	258 29.5	S15 12.8	88 26.4	N12 40.3	139 12.7	N22 20.5
	19	254 41.1	102 55.7	02.4	273 32.6	13.0	103 28.3	40.1	154 14.9	20.5
	20	269 43.5	117 55.2	01.3	288 35.7	13.3	118 30.2	39.9	169 17.1	20.5
	21	284 46.0	132 54.7	13 00.3	303 38.7	13.5	133 32.2	39.7	184 19.3	20.5
	22	299 48.5	147 54.2	12 59.2	318 41.8	13.7	148 34.1	39.5	199 21.5	20.4
	23	314 50.9	162 53.7	58.2	333 44.8	13.9	163 36.1	39.3	214 23.7	20.4
22 FRIDAY	00	329 53.4	177 53.1	N12 57.1	348 47.9	S15 14.2	178 38.0	N12 39.1	229 25.9	N22 20.4
	01	344 55.9	192 52.6	56.0	3 51.0	14.4	193 39.9	39.0	244 28.1	20.4
	02	359 58.3	207 52.1	55.0	18 54.0	14.6	208 41.9	38.8	259 30.3	20.4
	03	15 00.8	222 51.6	53.9	33 57.1	14.9	223 43.8	38.6	274 32.4	20.3
	04	30 03.2	237 51.1	52.9	49 00.2	15.1	238 45.8	38.4	289 34.6	20.3
	05	45 05.7	252 50.5	51.8	64 03.2	15.3	253 47.7	38.2	304 36.8	20.3
	06	60 08.2	267 50.0	N12 50.7	79 06.3	S15 15.6	268 49.6	N12 38.0	319 39.0	N22 20.3
	07	75 10.6	282 49.5	49.7	94 09.4	15.8	283 51.6	37.8	334 41.2	20.3
	08	90 13.1	297 49.0	48.6	109 12.4	16.0	298 53.5	37.6	349 43.4	20.3
	09	105 15.6	312 48.5	47.5	124 15.5	16.3	313 55.5	37.5	4 45.6	20.2
	10	120 18.0	327 48.0	46.5	139 18.6	16.5	328 57.4	37.3	19 47.8	20.2
	11	135 20.5	342 47.5	45.4	154 21.6	16.7	343 59.3	37.1	34 50.0	20.2
	12	150 23.0	357 46.9	N12 44.3	169 24.7	S15 17.0	359 01.3	N12 36.9	49 52.2	N22 20.2
	13	165 25.4	12 46.4	43.3	184 27.8	17.2	14 03.2	36.7	64 54.4	20.2
	14	180 27.9	27 45.9	42.2	199 30.9	17.4	29 05.2	36.5	79 56.6	20.2
	15	195 30.4	42 45.4	41.1	214 33.9	17.6	44 07.1	36.3	94 58.8	20.1
	16	210 32.8	57 44.9	40.1	229 37.0	17.9	59 09.0	36.1	110 01.0	20.1
	17	225 35.3	72 44.4	39.0	244 40.1	18.1	74 11.0	36.0	125 03.2	20.1
	18	240 37.7	87 43.9	N12 37.9	259 43.1	S15 18.3	89 12.9	N12 35.8	140 05.4	N22 20.1
	19	255 40.2	102 43.4	36.8	274 46.2	18.6	104 14.9	35.6	155 07.6	20.1
	20	270 42.7	117 42.8	35.8	289 49.3	18.8	119 16.8	35.4	170 09.8	20.1
	21	285 45.1	132 42.3	34.7	304 52.4	19.0	134 18.7	35.2	185 12.0	20.0
	22	300 47.6	147 41.8	33.6	319 55.4	19.3	149 20.7	35.0	200 14.2	20.0
	23	315 50.1	162 41.3	32.5	334 58.5	19.5	164 22.6	34.8	215 16.3	20.0
Mer. Pass.		h m 2 04.0	v −0.5	d 1.1	v 3.1	d 0.2	v 1.9	d 0.2	v 2.2	d 0.0

STARS

Name	SHA ° ′	Dec ° ′
Acamar	315 24.0	S40 17.1
Achernar	335 32.0	S57 12.9
Acrux	173 18.8	S63 07.2
Adhara	255 18.9	S28 58.3
Aldebaran	290 58.4	N16 31.1
Alioth	166 27.6	N55 56.7
Alkaid	153 05.0	N49 18.0
Al Na'ir	27 52.8	S46 56.6
Alnilam	275 54.4	S 1 11.8
Alphard	218 04.0	S 8 40.2
Alphecca	126 17.5	N26 42.4
Alpheratz	357 51.3	N29 06.6
Altair	62 15.5	N 8 52.7
Ankaa	353 22.8	S42 17.0
Antares	112 35.7	S26 26.5
Arcturus	146 02.9	N19 10.0
Atria	107 44.5	S69 02.3
Avior	234 21.9	S59 31.0
Bellatrix	278 40.5	N 6 21.3
Betelgeuse	271 09.9	N 7 24.6
Canopus	263 59.9	S52 41.5
Capella	280 46.1	N46 00.0
Deneb	49 36.4	N45 17.6
Denebola	182 41.8	N14 33.3
Diphda	349 03.3	S17 57.9
Dubhe	194 01.5	N61 44.1
Elnath	278 22.6	N28 36.7
Eltanin	90 49.5	N51 29.5
Enif	33 54.4	N 9 53.5
Fomalhaut	15 32.0	S29 36.1
Gacrux	172 10.3	S57 08.1
Gienah	176 00.6	S17 33.6
Hadar	148 59.4	S60 23.6
Hamal	328 09.4	N23 28.7
Kaus Aust.	83 53.8	S34 23.1
Kochab	137 19.6	N74 08.8
Markab	13 45.8	N15 13.5
Menkar	314 23.1	N 4 06.3
Menkent	148 17.0	S36 23.3
Miaplacidus	221 42.6	S69 43.7
Mirfak	308 51.5	N49 52.3
Nunki	76 07.6	S26 17.6
Peacock	53 30.7	S56 43.5
Pollux	243 37.5	N28 01.2
Procyon	245 08.1	N 5 13.1
Rasalhague	96 13.5	N12 33.5
Regulus	207 52.0	N11 57.2
Rigel	281 19.6	S 8 11.6
Rigil Kent.	140 02.9	S60 51.2
Sabik	102 21.3	S15 43.8
Schedar	349 49.2	N56 33.2
Shaula	96 32.2	S37 06.5
Sirius	258 40.8	S16 43.0
Spica	158 39.6	S11 10.7
Suhail	222 58.7	S43 26.6
Vega	80 43.9	N38 47.4
Zuben'ubi	137 14.1	S16 03.4

	SHA ° ′	Mer. Pass. h m
Venus	209 11.5	12 08
Mars	18 40.3	0 50
Jupiter	208 57.2	12 07
Saturn	259 39.0	8 45

UT	SUN GHA	SUN Dec	MOON GHA	v	MOON Dec	d	HP
d h	° ′	° ′	° ′	′	° ′	′	′
20 00	179 06.5	N12 39.3	274 02.7	13.7	N19 25.7	9.5	54.3
01	194 06.6	38.5	288 35.4	13.6	19 35.2	9.3	54.3
02	209 06.8	37.6	303 08.0	13.5	19 44.5	9.2	54.3
03	224 06.9	. . 36.8	317 40.5	13.5	19 53.7	9.2	54.3
04	239 07.1	36.0	332 13.0	13.4	20 02.9	9.1	54.3
05	254 07.2	35.2	346 45.4	13.3	20 12.0	9.0	54.3
06	269 07.4	N12 34.4	1 17.7	13.3	N20 21.0	8.9	54.3
07	284 07.5	33.5	15 50.0	13.2	20 29.9	8.9	54.3
08	299 07.7	32.7	30 22.2	13.2	20 38.8	8.7	54.3
09	314 07.8	. . 31.9	44 54.4	13.1	20 47.5	8.7	54.3
10	329 08.0	31.1	59 26.5	13.0	20 56.2	8.6	54.3
11	344 08.1	30.3	73 58.5	13.0	21 04.8	8.4	54.3
12	359 08.3	N12 29.4	88 30.5	12.9	N21 13.2	8.4	54.3
13	14 08.4	28.6	103 02.4	12.8	21 21.6	8.3	54.3
14	29 08.6	27.8	117 34.2	12.8	21 29.9	8.3	54.3
15	44 08.7	. . 27.0	132 06.0	12.7	21 38.2	8.1	54.4
16	59 08.8	26.1	146 37.7	12.6	21 46.3	8.0	54.4
17	74 09.0	25.3	161 09.3	12.5	21 54.3	8.0	54.4
18	89 09.1	N12 24.5	175 40.8	12.5	N22 02.3	7.8	54.4
19	104 09.3	23.7	190 12.3	12.5	22 10.1	7.8	54.4
20	119 09.4	22.8	204 43.8	12.3	22 17.9	7.6	54.4
21	134 09.6	. . 22.0	219 15.1	12.3	22 25.5	7.6	54.4
22	149 09.7	21.2	233 46.4	12.2	22 33.1	7.4	54.4
23	164 09.9	20.4	248 17.6	12.2	22 40.5	7.4	54.4
21 00	179 10.0	N12 19.5	262 48.8	12.1	N22 47.9	7.2	54.4
01	194 10.2	18.7	277 19.9	12.0	22 55.1	7.2	54.4
02	209 10.3	17.9	291 50.9	11.9	23 02.3	7.1	54.5
03	224 10.5	. . 17.0	306 21.8	11.9	23 09.4	6.9	54.5
04	239 10.6	16.2	320 52.7	11.8	23 16.3	6.9	54.5
05	254 10.8	15.4	335 23.5	11.8	23 23.2	6.7	54.5
06	269 11.0	N12 14.6	349 54.3	11.6	N23 29.9	6.7	54.5
07	284 11.1	13.7	4 24.9	11.6	23 36.6	6.5	54.5
08	299 11.3	12.9	18 55.5	11.6	23 43.1	6.4	54.5
09	314 11.4	. . 12.1	33 26.1	11.4	23 49.5	6.4	54.5
10	329 11.6	11.2	47 56.5	11.4	23 55.9	6.2	54.5
11	344 11.7	10.4	62 26.9	11.4	24 02.1	6.1	54.6
12	359 11.9	N12 09.6	76 57.3	11.2	N24 08.2	6.0	54.6
13	14 12.0	08.8	91 27.5	11.2	24 14.2	5.9	54.6
14	29 12.2	07.9	105 57.7	11.1	24 20.1	5.8	54.6
15	44 12.3	. . 07.1	120 27.8	11.1	24 25.9	5.7	54.6
16	59 12.5	06.3	134 57.9	11.0	24 31.6	5.5	54.6
17	74 12.6	05.4	149 27.9	10.9	24 37.1	5.5	54.6
18	89 12.8	N12 04.6	163 57.8	10.9	N24 42.6	5.3	54.7
19	104 12.9	03.8	178 27.7	10.8	24 47.9	5.2	54.7
20	119 13.1	02.9	192 57.5	10.7	24 53.1	5.1	54.7
21	134 13.2	. . 02.1	207 27.2	10.6	24 58.2	5.0	54.7
22	149 13.4	01.3	221 56.8	10.6	25 03.2	4.9	54.7
23	164 13.6	12 00.4	236 26.4	10.6	25 08.1	4.7	54.7
22 00	179 13.7	N11 59.6	250 56.0	10.4	N25 12.8	4.7	54.8
01	194 13.9	58.8	265 25.4	10.4	25 17.5	4.5	54.8
02	209 14.0	57.9	279 54.8	10.4	25 22.0	4.4	54.8
03	224 14.2	. . 57.1	294 24.2	10.2	25 26.4	4.2	54.8
04	239 14.3	56.2	308 53.4	10.3	25 30.6	4.2	54.8
05	254 14.5	55.4	323 22.7	10.1	25 34.8	4.0	54.8
06	269 14.6	N11 54.6	337 51.8	10.1	N25 38.8	3.9	54.9
07	284 14.8	53.7	352 20.9	10.0	25 42.7	3.8	54.9
08	299 15.0	52.9	6 49.9	10.0	25 46.5	3.6	54.9
09	314 15.1	. . 52.1	21 18.9	9.9	25 50.1	3.6	54.9
10	329 15.3	51.2	35 47.8	9.9	25 53.7	3.4	54.9
11	344 15.4	50.4	50 16.7	9.7	25 57.1	3.2	55.0
12	359 15.6	N11 49.5	64 45.4	9.8	N26 00.3	3.2	55.0
13	14 15.7	48.7	79 14.2	9.7	26 03.5	3.0	55.0
14	29 15.9	47.9	93 42.9	9.6	26 06.5	2.9	55.0
15	44 16.1	. . 47.0	108 11.5	9.6	26 09.4	2.7	55.0
16	59 16.2	46.2	122 40.1	9.5	26 12.1	2.7	55.1
17	74 16.4	45.3	137 08.6	9.4	26 14.8	2.5	55.1
18	89 16.5	N11 44.5	151 37.0	9.4	N26 17.3	2.3	55.1
19	104 16.7	43.7	166 05.4	9.4	26 19.6	2.3	55.1
20	119 16.9	42.8	180 33.8	9.3	26 21.9	2.1	55.1
21	134 17.0	. . 42.0	195 02.1	9.3	26 24.0	1.9	55.2
22	149 17.2	41.1	209 30.4	9.2	26 25.9	1.8	55.2
23	164 17.3	40.3	223 58.6	9.1	N26 27.7	1.7	55.2
	SD 15.8	*d* 0.8	SD 14.8		14.9		15.0

Lat.	Twilight Naut.	Twilight Civil	Sunrise	Moonrise 20	Moonrise 21	Moonrise 22	Moonrise 23
°	h m	h m	h m	h m	h m	h m	h m
N 72	////	////	03 00	▭	▭	▭	▭
N 70	////	01 40	03 24	▭	▭	▭	▭
68	////	02 21	03 42	18 23	▭	▭	▭
66	////	02 48	03 57	19 55	▭	▭	▭
64	01 28	03 09	04 09	20 34	20 23	▭	▭
62	02 05	03 25	04 19	21 01	21 08	21 27	22 18
60	02 30	03 38	04 27	21 22	21 37	22 06	22 59
N 58	02 49	03 50	04 35	21 39	22 00	22 34	23 27
56	03 05	04 00	04 42	21 54	22 18	22 55	23 48
54	03 18	04 08	04 48	22 06	22 34	23 13	24 06
52	03 29	04 16	04 53	22 18	22 47	23 28	24 21
50	03 39	04 23	04 58	22 27	22 59	23 41	24 35
45	03 58	04 37	05 08	22 48	23 24	24 08	00 08
N 40	04 14	04 48	05 17	23 05	23 43	24 29	00 29
35	04 26	04 58	05 24	23 20	24 00	00 00	00 47
30	04 36	05 06	05 31	23 32	24 14	00 14	01 02
20	04 52	05 19	05 42	23 54	24 38	00 38	01 28
N 10	05 04	05 29	05 51	24 13	00 13	01 00	01 50
0	05 14	05 39	06 00	24 30	00 30	01 19	02 11
S 10	05 22	05 47	06 09	24 48	00 48	01 39	02 32
20	05 29	05 55	06 18	00 14	01 07	02 01	02 55
30	05 36	06 04	06 28	00 32	01 29	02 25	03 21
35	05 39	06 08	06 34	00 43	01 42	02 40	03 37
40	05 41	06 13	06 41	00 55	01 57	02 57	03 55
45	05 44	06 18	06 48	01 10	02 15	03 17	04 16
S 50	05 47	06 24	06 58	01 28	02 37	03 43	04 44
52	05 48	06 27	07 02	01 36	02 48	03 56	04 57
54	05 49	06 30	07 07	01 46	03 00	04 10	05 13
56	05 50	06 33	07 12	01 57	03 14	04 27	05 31
58	05 51	06 37	07 18	02 09	03 30	04 47	05 53
S 60	05 52	06 40	07 24	02 24	03 50	05 13	06 23

Lat.	Sunset	Twilight Civil	Twilight Naut.	Moonset 20	Moonset 21	Moonset 22	Moonset 23
°	h m	h m	h m	h m	h m	h m	h m
N 72	21 02	23 57	////	▭	▭	▭	▭
N 70	20 39	22 18	////	▭	▭	▭	▭
68	20 22	21 40	////	18 00	▭	▭	▭
66	20 07	21 15	23 58	16 29	▭	▭	▭
64	19 56	20 55	22 32	15 51	17 42	▭	▭
62	19 46	20 39	21 57	15 25	16 58	18 25	19 26
60	19 37	20 26	21 33	15 04	16 29	17 46	18 45
N 58	19 30	20 15	21 15	14 48	16 07	17 19	18 17
56	19 23	20 05	21 00	14 33	15 49	16 58	17 55
54	19 18	19 57	20 47	14 21	15 34	16 41	17 37
52	19 12	19 49	20 36	14 11	15 20	16 26	17 22
50	19 08	19 43	20 26	14 01	15 09	16 13	17 09
45	18 57	19 29	20 07	13 41	14 45	15 46	16 42
N 40	18 49	19 17	19 52	13 25	14 26	15 25	16 20
35	18 42	19 08	19 40	13 11	14 10	15 07	16 02
30	18 35	19 00	19 30	13 00	13 56	14 52	15 47
20	18 25	18 47	19 14	12 39	13 33	14 27	15 21
N 10	18 15	18 37	19 02	12 22	13 12	14 05	14 58
0	18 07	18 28	18 52	12 06	12 54	13 44	14 37
S 10	17 58	18 19	18 44	11 50	12 35	13 24	14 16
20	17 49	18 11	18 37	11 32	12 15	13 02	13 53
30	17 39	18 03	18 31	11 13	11 51	12 36	13 27
35	17 33	17 59	18 28	11 01	11 38	12 21	13 12
40	17 26	17 54	18 26	10 48	11 22	12 04	12 54
45	17 19	17 49	18 23	10 32	11 04	11 43	12 32
S 50	17 09	17 43	18 21	10 13	10 41	11 17	12 04
52	17 05	17 40	18 20	10 04	10 30	11 04	11 51
54	17 01	17 37	18 19	09 54	10 17	10 49	11 35
56	16 55	17 34	18 18	09 43	10 03	10 32	11 17
58	16 50	17 31	18 17	09 30	09 46	10 12	10 54
S 60	16 43	17 27	18 16	09 15	09 26	09 46	10 25

Day	SUN Eqn. of Time 00^h	SUN Eqn. of Time 12^h	SUN Mer. Pass.	MOON Mer. Pass. Upper	MOON Mer. Pass. Lower	MOON Age	MOON Phase
d	m s	m s	h m	h m	h m	d	%
20	03 34	03 27	12 03	05 55	18 18	22	46
21	03 20	03 13	12 03	06 42	19 06	23	36
22	03 05	02 58	12 03	07 32	19 58	24	27

d	h	ARIES GHA	VENUS −3.9 GHA	VENUS Dec	MARS −2.8 GHA	MARS Dec	JUPITER −1.7 GHA	JUPITER Dec	SATURN +0.2 GHA	SATURN Dec
		° ′	° ′	° ′	° ′	° ′	° ′	° ′	° ′	° ′
23	00	330 52.5	177 40.8	N12 31.5	350 01.6	S15 19.7	179 24.6	N12 34.6	230 18.5	N22 20.0
	01	345 55.0	192 40.3	30.4	5 04.7	19.9	194 26.5	34.4	245 20.7	20.0
	02	0 57.5	207 39.8	29.3	20 07.8	20.2	209 28.4	34.3	260 22.9	20.0
	03	15 59.9	222 39.3	. . 28.2	35 10.8	. . 20.4	224 30.4	. . 34.1	275 25.1	. . 19.9
	04	31 02.4	237 38.8	27.2	50 13.9	20.6	239 32.3	33.9	290 27.3	19.9
	05	46 04.9	252 38.3	26.1	65 17.0	20.9	254 34.3	33.7	305 29.5	19.9
	06	61 07.3	267 37.8	N12 25.0	80 20.1	S15 21.1	269 36.2	N12 33.5	320 31.7	N22 19.9
	07	76 09.8	282 37.3	23.9	95 23.2	21.3	284 38.1	33.3	335 33.9	19.9
S	08	91 12.2	297 36.7	22.9	110 26.2	21.5	299 40.1	33.1	350 36.1	19.9
A	09	106 14.7	312 36.2	. . 21.8	125 29.3	. . 21.8	314 42.0	. . 32.9	5 38.3	. . 19.8
T	10	121 17.2	327 35.7	20.7	140 32.4	22.0	329 44.0	32.8	20 40.5	19.8
U	11	136 19.6	342 35.2	19.6	155 35.5	22.2	344 45.9	32.6	35 42.7	19.8
R	12	151 22.1	357 34.7	N12 18.5	170 38.6	S15 22.4	359 47.8	N12 32.4	50 44.9	N22 19.8
D	13	166 24.6	12 34.2	17.4	185 41.7	22.7	14 49.8	32.2	65 47.1	19.8
A	14	181 27.0	27 33.7	16.4	200 44.7	22.9	29 51.7	32.0	80 49.3	19.8
Y	15	196 29.5	42 33.2	. . 15.3	215 47.8	. . 23.1	44 53.7	. . 31.8	95 51.5	. . 19.7
	16	211 32.0	57 32.7	14.2	230 50.9	23.3	59 55.6	31.6	110 53.7	19.7
	17	226 34.4	72 32.2	13.1	245 54.0	23.6	74 57.5	31.4	125 55.9	19.7
	18	241 36.9	87 31.7	N12 12.0	260 57.1	S15 23.8	89 59.5	N12 31.2	140 58.1	N22 19.7
	19	256 39.4	102 31.2	10.9	276 00.2	24.0	105 01.4	31.1	156 00.3	19.7
	20	271 41.8	117 30.7	09.9	291 03.3	24.2	120 03.4	30.9	171 02.5	19.7
	21	286 44.3	132 30.2	. . 08.8	306 06.4	. . 24.5	135 05.3	. . 30.7	186 04.7	. . 19.6
	22	301 46.7	147 29.7	07.7	321 09.5	24.7	150 07.2	30.5	201 06.9	19.6
	23	316 49.2	162 29.2	06.6	336 12.5	24.9	165 09.2	30.3	216 09.1	19.6
24	00	331 51.7	177 28.7	N12 05.5	351 15.6	S15 25.1	180 11.1	N12 30.1	231 11.3	N22 19.6
	01	346 54.1	192 28.2	04.4	6 18.7	25.4	195 13.1	29.9	246 13.5	19.6
	02	1 56.6	207 27.7	03.3	21 21.8	25.6	210 15.0	29.7	261 15.7	19.6
	03	16 59.1	222 27.2	. . 02.2	36 24.9	. . 25.8	225 16.9	. . 29.5	276 17.9	. . 19.5
	04	32 01.5	237 26.7	01.1	51 28.0	26.0	240 18.9	29.4	291 20.1	19.5
	05	47 04.0	252 26.2	12 00.1	66 31.1	26.3	255 20.8	29.2	306 22.3	19.5
	06	62 06.5	267 25.7	N11 59.0	81 34.2	S15 26.5	270 22.8	N12 29.0	321 24.5	N22 19.5
	07	77 08.9	282 25.2	57.9	96 37.3	26.7	285 24.7	28.8	336 26.7	19.5
	08	92 11.4	297 24.7	56.8	111 40.4	26.9	300 26.6	28.6	351 28.9	19.5
S	09	107 13.8	312 24.2	. . 55.7	126 43.5	. . 27.2	315 28.6	. . 28.4	6 31.1	. . 19.4
U	10	122 16.3	327 23.7	54.6	141 46.6	27.4	330 30.5	28.2	21 33.3	19.4
N	11	137 18.8	342 23.2	53.5	156 49.7	27.6	345 32.5	28.0	36 35.5	19.4
D	12	152 21.2	357 22.7	N11 52.4	171 52.8	S15 27.8	0 34.4	N12 27.8	51 37.7	N22 19.4
A	13	167 23.7	12 22.2	51.3	186 55.9	28.0	15 36.3	27.7	66 39.9	19.4
Y	14	182 26.2	27 21.7	50.2	201 59.0	28.3	30 38.3	27.5	81 42.1	19.3
	15	197 28.6	42 21.3	. . 49.1	217 02.1	. . 28.5	45 40.2	. . 27.3	96 44.3	. . 19.3
	16	212 31.1	57 20.8	48.0	232 05.2	28.7	60 42.2	27.1	111 46.5	19.3
	17	227 33.6	72 20.3	46.9	247 08.3	28.9	75 44.1	26.9	126 48.7	19.3
	18	242 36.0	87 19.8	N11 45.8	262 11.4	S15 29.2	90 46.0	N12 26.7	141 50.9	N22 19.3
	19	257 38.5	102 19.3	44.7	277 14.5	29.4	105 48.0	26.5	156 53.1	19.3
	20	272 41.0	117 18.8	43.6	292 17.6	29.6	120 49.9	26.3	171 55.3	19.2
	21	287 43.4	132 18.3	. . 42.5	307 20.7	. . 29.8	135 51.9	. . 26.1	186 57.5	. . 19.2
	22	302 45.9	147 17.8	41.4	322 23.8	30.0	150 53.8	26.0	201 59.7	19.2
	23	317 48.3	162 17.3	40.3	337 26.9	30.3	165 55.7	25.8	217 01.9	19.2
25	00	332 50.8	177 16.8	N11 39.2	352 30.0	S15 30.5	180 57.7	N12 25.6	232 04.1	N22 19.2
	01	347 53.3	192 16.3	38.1	7 33.1	30.7	195 59.6	25.4	247 06.3	19.2
	02	2 55.7	207 15.8	37.0	22 36.2	30.9	211 01.6	25.2	262 08.5	19.1
	03	17 58.2	222 15.4	. . 35.9	37 39.3	. . 31.1	226 03.5	. . 25.0	277 10.7	. . 19.1
	04	33 00.7	237 14.9	34.8	52 42.4	31.4	241 05.4	24.8	292 12.9	19.1
	05	48 03.1	252 14.4	33.7	67 45.5	31.6	256 07.4	24.6	307 15.1	19.1
	06	63 05.6	267 13.9	N11 32.6	82 48.6	S15 31.8	271 09.3	N12 24.4	322 17.3	N22 19.1
	07	78 08.1	282 13.4	31.5	97 51.7	32.0	286 11.3	24.3	337 19.5	19.1
	08	93 10.5	297 12.9	30.4	112 54.8	32.2	301 13.2	24.1	352 21.7	19.0
M	09	108 13.0	312 12.4	. . 29.3	127 57.9	. . 32.4	316 15.1	. . 23.9	7 23.9	. . 19.0
O	10	123 15.5	327 11.9	28.2	143 01.0	32.7	331 17.1	23.7	22 26.1	19.0
N	11	138 17.9	342 11.5	27.0	158 04.1	32.9	346 19.0	23.5	37 28.3	19.0
D	12	153 20.4	357 11.0	N11 25.9	173 07.2	S15 33.1	1 21.0	N12 23.3	52 30.5	N22 19.0
A	13	168 22.8	12 10.5	24.8	188 10.3	33.3	16 22.9	23.1	67 32.8	19.0
Y	14	183 25.3	27 10.0	23.7	203 13.4	33.5	31 24.8	22.9	82 35.0	18.9
	15	198 27.8	42 09.5	. . 22.6	218 16.5	. . 33.7	46 26.8	. . 22.7	97 37.2	. . 18.9
	16	213 30.2	57 09.0	21.5	233 19.7	34.0	61 28.7	22.6	112 39.4	18.9
	17	228 32.7	72 08.5	20.4	248 22.8	34.2	76 30.7	22.4	127 41.6	18.9
	18	243 35.2	87 08.1	N11 19.3	263 25.9	S15 34.4	91 32.6	N12 22.2	142 43.8	N22 18.9
	19	258 37.6	102 07.6	18.2	278 29.0	34.6	106 34.5	22.0	157 46.0	18.9
	20	273 40.1	117 07.1	17.0	293 32.1	34.8	121 36.5	21.8	172 48.2	18.8
	21	288 42.6	132 06.6	. . 15.9	308 35.2	. . 35.0	136 38.4	. . 21.6	187 50.4	. . 18.8
	22	303 45.0	147 06.1	14.8	323 38.3	35.2	151 40.4	21.4	202 52.6	18.8
	23	318 47.5	162 05.7	13.7	338 41.4	35.5	166 42.3	21.2	217 54.8	18.8
Mer. Pass.		h m 1 52.2	v −0.5	d 1.1	v 3.1	d 0.2	v 1.9	d 0.2	v 2.2	d 0.0

STARS

Name	SHA	Dec
	° ′	° ′
Acamar	315 24.0	S40 17.1
Achernar	335 31.9	S57 12.9
Acrux	173 18.9	S63 07.2
Adhara	255 18.9	S28 58.3
Aldebaran	290 58.3	N16 31.1
Alioth	166 27.6	N55 56.7
Alkaid	153 05.1	N49 18.0
Al Na'ir	27 52.8	S46 56.6
Alnilam	275 54.3	S 1 11.8
Alphard	218 04.0	S 8 40.2
Alphecca	126 17.5	N26 42.4
Alpheratz	357 51.3	N29 06.6
Altair	62 15.5	N 8 52.7
Ankaa	353 22.8	S42 17.0
Antares	112 35.7	S26 26.5
Arcturus	146 02.9	N19 10.0
Atria	107 44.5	S69 02.3
Avior	234 21.9	S59 31.0
Bellatrix	278 40.4	N 6 21.3
Betelgeuse	271 09.8	N 7 24.6
Canopus	263 59.9	S52 41.5
Capella	280 46.1	N46 00.0
Deneb	49 36.4	N45 17.6
Denebola	182 41.8	N14 33.3
Diphda	349 03.3	S17 57.9
Dubhe	194 01.5	N61 44.1
Elnath	278 22.5	N28 36.7
Eltanin	90 49.5	N51 29.5
Enif	33 54.4	N 9 53.5
Fomalhaut	15 32.0	S29 36.1
Gacrux	172 10.3	S57 08.6
Gienah	176 00.6	S17 33.6
Hadar	148 59.5	S60 23.6
Hamal	328 09.4	N23 28.7
Kaus Aust.	83 53.8	S34 23.1
Kochab	137 19.6	N74 08.8
Markab	13 45.8	N15 13.9
Menkar	314 23.1	N 4 06.3
Menkent	148 17.0	S36 23.3
Miaplacidus	221 42.6	S69 43.7
Mirfak	308 51.5	N49 52.3
Nunki	76 07.6	S26 17.6
Peacock	53 30.7	S56 43.6
Pollux	243 37.4	N28 01.2
Procyon	245 08.1	N 5 13.1
Rasalhague	96 13.5	N12 33.6
Regulus	207 52.0	N11 57.2
Rigel	281 19.6	S 8 11.6
Rigil Kent.	140 02.9	S60 51.2
Sabik	102 21.3	S15 43.8
Schedar	349 49.2	N56 33.1
Shaula	96 32.3	S37 06.5
Sirius	258 40.8	S16 43.0
Spica	158 39.6	S11 10.1
Suhail	222 58.7	S43 26.6
Vega	80 43.9	N38 47.4
Zuben'ubi	137 14.1	S16 03.4

	SHA	Mer. Pass
	° ′	h m
Venus	205 37.0	12 10
Mars	19 24.0	0 35
Jupiter	208 19.4	11 58
Saturn	259 19.6	8 34

	UT	SUN GHA	SUN Dec	MOON GHA	v	MOON Dec	d	HP
	d h	° ′	° ′	° ′	′	° ′	′	′
	23 00	179 17.5	N11 39.5	238 26.7	9.2	N26 29.4	1.6	55.2
	01	194 17.6	38.6	252 54.9	9.0	26 31.0	1.4	55.2
	02	209 17.8	37.8	267 22.9	9.1	26 32.4	1.3	55.3
	03	224 18.0	. . 36.9	281 51.0	8.9	26 33.7	1.2	55.3
	04	239 18.1	36.1	296 18.9	9.0	26 34.9	1.0	55.3
	05	254 18.3	35.2	310 46.9	8.9	26 35.9	0.8	55.3
	06	269 18.4	N11 34.4	325 14.8	8.8	N26 36.7	0.8	55.4
	07	284 18.6	33.5	339 42.6	8.9	26 37.5	0.6	55.4
S	08	299 18.8	32.7	354 10.5	8.7	26 38.1	0.4	55.4
A	09	314 18.9	. . 31.9	8 38.2	8.8	26 38.5	0.3	55.4
T	10	329 19.1	31.0	23 06.0	8.7	26 38.8	0.2	55.5
U	11	344 19.3	30.2	37 33.7	8.7	26 39.0	0.0	55.5
R	12	359 19.4	N11 29.3	52 01.4	8.6	N26 39.0	0.1	55.5
D	13	14 19.6	28.5	66 29.0	8.6	26 38.9	0.2	55.5
A	14	29 19.7	27.6	80 56.6	8.6	26 38.7	0.4	55.6
Y	15	44 19.9	. . 26.8	95 24.2	8.5	26 38.3	0.6	55.6
	16	59 20.1	25.9	109 51.7	8.6	26 37.7	0.6	55.6
	17	74 20.2	25.1	124 19.3	8.4	26 37.1	0.9	55.6
	18	89 20.4	N11 24.2	138 46.7	8.5	N26 36.2	0.9	55.7
	19	104 20.6	23.4	153 14.2	8.5	26 35.3	1.1	55.7
	20	119 20.7	22.5	167 41.7	8.4	26 34.2	1.3	55.7
	21	134 20.9	. . 21.7	182 09.1	8.4	26 32.9	1.4	55.7
	22	149 21.0	20.8	196 36.5	8.3	26 31.5	1.5	55.8
	23	164 21.2	20.0	211 03.8	8.4	26 30.0	1.7	55.8
	24 00	179 21.4	N11 19.1	225 31.2	8.3	N26 28.3	1.9	55.8
	01	194 21.5	18.3	239 58.5	8.3	26 26.4	1.9	55.8
	02	209 21.7	17.4	254 25.8	8.4	26 24.5	2.2	55.9
	03	224 21.9	. . 16.6	268 53.2	8.2	26 22.3	2.2	55.9
	04	239 22.0	15.7	283 20.4	8.3	26 20.1	2.5	55.9
	05	254 22.2	14.9	297 47.7	8.3	26 17.6	2.5	56.0
	06	269 22.4	N11 14.0	312 15.0	8.2	N26 15.1	2.7	56.0
	07	284 22.5	13.2	326 42.2	8.3	26 12.4	2.9	56.0
	08	299 22.7	12.3	341 09.5	8.2	26 09.5	3.0	56.0
S	09	314 22.9	. . 11.5	355 36.7	8.2	26 06.5	3.1	56.1
U	10	329 23.0	10.6	10 03.9	8.3	26 03.4	3.3	56.1
N	11	344 23.2	09.8	24 31.2	8.2	26 00.1	3.5	56.1
D	12	359 23.4	N11 08.9	38 58.4	8.2	N25 56.6	3.6	56.1
A	13	14 23.5	08.0	53 25.6	8.2	25 53.0	3.7	56.2
Y	14	29 23.7	07.2	67 52.8	8.2	25 49.3	3.9	56.2
	15	44 23.9	. . 06.3	82 20.0	8.2	25 45.4	4.0	56.2
	16	59 24.0	05.5	96 47.2	8.2	25 41.4	4.2	56.3
	17	74 24.2	04.6	111 14.4	8.3	25 37.2	4.3	56.3
	18	89 24.4	N11 03.8	125 41.7	8.2	N25 32.9	4.5	56.3
	19	104 24.5	02.9	140 08.9	8.2	25 28.4	4.6	56.3
	20	119 24.7	02.1	154 36.1	8.2	25 23.8	4.7	56.4
	21	134 24.9	. . 01.2	169 03.3	8.3	25 19.1	4.9	56.4
	22	149 25.0	11 00.3	183 30.6	8.2	25 14.2	5.1	56.4
	23	164 25.2	10 59.5	197 57.8	8.3	25 09.1	5.2	56.5
	25 00	179 25.4	N10 58.6	212 25.1	8.3	N25 03.9	5.3	56.5
	01	194 25.5	57.8	226 52.4	8.3	24 58.6	5.5	56.5
	02	209 25.7	56.9	241 19.7	8.3	24 53.1	5.6	56.5
	03	224 25.9	. . 56.1	255 47.0	8.3	24 47.5	5.8	56.6
	04	239 26.0	55.2	270 14.3	8.3	24 41.7	5.9	56.6
	05	254 26.2	54.3	284 41.6	8.3	24 35.8	6.0	56.6
	06	269 26.4	N10 53.5	299 08.9	8.4	N24 29.8	6.2	56.7
	07	284 26.5	52.6	313 36.3	8.4	24 23.6	6.4	56.7
	08	299 26.7	51.8	328 03.7	8.4	24 17.2	6.5	56.7
M	09	314 26.9	. . 50.9	342 31.1	8.4	24 10.7	6.6	56.8
O	10	329 27.1	50.0	356 58.5	8.5	24 04.1	6.7	56.8
N	11	344 27.2	49.2	11 26.0	8.4	23 57.4	6.9	56.8
D	12	359 27.4	N10 48.3	25 53.4	8.5	N23 50.5	7.1	56.8
A	13	14 27.6	47.4	40 20.9	8.6	23 43.4	7.2	56.9
Y	14	29 27.7	46.6	54 48.5	8.5	23 36.2	7.3	56.9
	15	44 27.9	. . 45.7	69 16.0	8.6	23 28.9	7.4	56.9
	16	59 28.1	44.9	83 43.6	8.6	23 21.5	7.6	57.0
	17	74 28.3	44.0	98 11.2	8.6	23 13.9	7.8	57.0
	18	89 28.4	N10 43.1	112 38.8	8.7	N23 06.1	7.8	57.0
	19	104 28.6	42.3	127 06.5	8.7	22 58.3	8.0	57.1
	20	119 28.8	41.4	141 34.2	8.7	22 50.3	8.2	57.1
	21	134 28.9	. . 40.5	156 01.9	8.7	22 42.1	8.2	57.1
	22	149 29.1	39.7	170 29.6	8.8	22 33.9	8.5	57.1
	23	164 29.3	38.8	184 57.4	8.8	N22 25.4	8.5	57.2
		SD 15.8	*d* 0.9	SD	15.1	15.3		15.5

Lat.	Twilight Naut.	Twilight Civil	Sunrise	Moonrise 23	24	25	26
°	h m	h m	h m	h m	h m	h m	h m
N 72	////	01 08	03 17	▭	▭	▭	▭
N 70	////	02 05	03 38	▭	▭	▭	▭
68	////	02 38	03 54	▭	▭	▭	▭
66	00 59	03 02	04 07	▭	▭	▭	00 20
64	01 50	03 20	04 18	▭	22 55	25 01	01 01
62	02 19	03 35	04 27	22 18	23 45	25 29	01 29
60	02 41	03 47	04 35	22 59	24 17	00 17	01 50
N 58	02 59	03 57	04 41	23 27	24 40	00 40	02 07
56	03 13	04 06	04 48	23 48	24 59	00 59	02 22
54	03 25	04 14	04 53	24 06	00 06	01 15	02 35
52	03 35	04 21	04 58	24 21	00 21	01 29	02 46
50	03 44	04 27	05 02	24 35	00 35	01 41	02 55
45	04 03	04 41	05 12	00 08	01 02	02 05	03 16
N 40	04 17	04 51	05 20	00 29	01 23	02 25	03 32
35	04 29	05 00	05 26	00 47	01 41	02 41	03 46
30	04 38	05 07	05 32	01 02	01 56	02 55	03 58
20	04 53	05 20	05 42	01 28	02 22	03 19	04 19
N 10	05 05	05 30	05 51	01 50	02 44	03 40	04 37
0	05 14	05 38	05 59	02 11	03 05	04 00	04 53
S 10	05 21	05 46	06 07	02 32	03 26	04 19	05 10
20	05 27	05 53	06 15	02 55	03 48	04 39	05 28
30	05 33	06 01	06 25	03 21	04 14	05 03	05 48
35	05 35	06 05	06 30	03 37	04 29	05 17	05 59
40	05 37	06 09	06 36	03 55	04 47	05 33	06 13
45	05 39	06 13	06 43	04 16	05 08	05 52	06 29
S 50	05 41	06 19	06 52	04 44	05 35	06 16	06 48
52	05 42	06 21	06 56	04 57	05 48	06 28	06 57
54	05 43	06 24	07 00	05 13	06 03	06 41	07 08
56	05 43	06 26	07 05	05 31	06 21	06 56	07 19
58	05 44	06 29	07 10	05 53	06 42	07 13	07 33
S 60	05 44	06 33	07 16	06 23	07 09	07 35	07 48

Lat.	Sunset	Twilight Civil	Twilight Naut.	Moonset 23	24	25	26
°	h m	h m	h m	h m	h m	h m	h m
N 72	20 44	22 44	////	▭	▭	▭	▭
N 70	20 24	21 53	////	▭	▭	▭	21 56
68	20 08	21 22	////	▭	▭	▭	21 13
66	19 56	21 00	22 55	▭	▭	21 12	20 44
64	19 45	20 42	22 10	▭	20 43	20 30	20 22
62	19 36	20 28	21 42	19 26	19 52	20 01	20 05
60	19 29	20 16	21 20	18 45	19 20	19 39	19 50
N 58	19 22	20 06	21 04	18 17	18 57	19 21	19 37
56	19 16	19 57	20 50	17 55	18 37	19 06	19 26
54	19 11	19 49	20 38	17 37	18 21	18 53	19 16
52	19 06	19 42	20 28	17 22	18 07	18 42	19 07
50	19 02	19 36	20 19	17 09	17 55	18 31	19 00
45	18 52	19 23	20 01	16 42	17 30	18 10	18 43
N 40	18 45	19 13	19 47	16 20	17 10	17 52	18 29
35	18 38	19 04	19 35	16 02	16 53	17 38	18 17
30	18 32	18 57	19 26	15 47	16 38	17 25	18 07
20	18 22	18 45	19 11	15 21	16 13	17 03	17 49
N 10	18 14	18 35	19 00	14 58	15 51	16 43	17 33
0	18 06	18 27	18 51	14 37	15 31	16 25	17 19
S 10	17 58	18 19	18 44	14 16	15 11	16 07	17 04
20	17 50	18 12	18 38	13 53	14 49	15 48	16 48
30	17 40	18 05	18 33	13 27	14 24	15 25	16 29
35	17 35	18 01	18 30	13 12	14 09	15 12	16 18
40	17 29	17 57	18 28	12 54	13 52	14 56	16 06
45	17 22	17 52	18 26	12 32	13 31	14 38	15 51
S 50	17 14	17 47	18 25	12 04	13 04	14 15	15 33
52	17 10	17 45	18 24	11 51	12 51	14 04	15 24
54	17 06	17 42	18 23	11 35	12 36	13 51	15 14
56	17 01	17 40	18 23	11 17	12 19	13 37	15 03
58	16 56	17 37	18 22	10 54	11 58	13 20	14 51
S 60	16 50	17 33	18 22	10 25	11 31	12 59	14 36

Day	SUN Eqn. of Time 00^h	SUN Eqn. of Time 12^h	SUN Mer. Pass.	MOON Mer. Pass. Upper	MOON Mer. Pass. Lower	Age	Phase
d	m s	m s	h m	h m	h m	d %	
23	02 50	02 43	12 03	08 24	20 51	25 19	
24	02 35	02 27	12 02	09 18	21 45	26 12	
25	02 19	02 11	12 02	10 13	22 39	27 6	

UT		ARIES	VENUS −3.9		MARS −2.9		JUPITER −1.7		SATURN +0.2	
d h	Day	GHA ° ′	GHA ° ′	Dec ° ′	GHA ° ′	Dec ° ′	GHA ° ′	Dec ° ′	GHA ° ′	Dec ° ′
26 00		333 50.0	177 05.2	N11 12.6	353 44.5	S15 35.7	181 44.2	N12 21.0	232 57.0	N22 18.8
01		348 52.4	192 04.7	11.5	8 47.6	35.9	196 46.2	20.9	247 59.2	18.8
02		3 54.9	207 04.2	10.4	23 50.8	36.1	211 48.1	20.7	263 01.4	18.7
03		18 57.3	222 03.7	. . 09.2	38 53.9	. . 36.3	226 50.1	. . 20.5	278 03.6	. . 18.7
04		33 59.8	237 03.3	08.1	53 57.0	36.5	241 52.0	20.3	293 05.8	18.7
05		49 02.3	252 02.8	07.0	69 00.1	36.7	256 53.9	20.1	308 08.0	18.7
06		64 04.7	267 02.3	N11 05.9	84 03.2	S15 37.0	271 55.9	N12 19.9	323 10.2	N22 18.7
07		79 07.2	282 01.8	04.8	99 06.3	37.2	286 57.8	19.7	338 12.4	18.7
08	T	94 09.7	297 01.3	03.6	114 09.4	37.4	301 59.8	19.5	353 14.6	18.6
09	U	109 12.1	312 00.9	. . 02.5	129 12.6	. . 37.6	317 01.7	. . 19.3	8 16.9	. . 18.6
10	E	124 14.6	327 00.4	01.4	144 15.7	37.8	332 03.7	19.2	23 19.1	18.6
11	S	139 17.1	341 59.9	11 00.3	159 18.8	38.0	347 05.6	19.0	38 21.3	18.6
12	D	154 19.5	356 59.4	N10 59.2	174 21.9	S15 38.2	2 07.5	N12 18.8	53 23.5	N22 18.6
13	A	169 22.0	11 58.9	58.0	189 25.0	38.4	17 09.5	18.6	68 25.7	18.6
14	Y	184 24.4	26 58.5	56.9	204 28.1	38.6	32 11.4	18.4	83 27.9	18.5
15		199 26.9	41 58.0	. . 55.8	219 31.3	. . 38.9	47 13.4	. . 18.2	98 30.1	. . 18.5
16		214 29.4	56 57.5	54.7	234 34.4	39.1	62 15.3	18.0	113 32.3	18.5
17		229 31.8	71 57.0	53.6	249 37.5	39.3	77 17.2	17.8	128 34.5	18.5
18		244 34.3	86 56.6	N10 52.4	264 40.6	S15 39.5	92 19.2	N12 17.6	143 36.7	N22 18.5
19		259 36.8	101 56.1	51.3	279 43.7	39.7	107 21.1	17.5	158 38.9	18.5
20		274 39.2	116 55.6	50.2	294 46.8	39.9	122 23.1	17.3	173 41.1	18.4
21		289 41.7	131 55.1	. . 49.0	309 50.0	. . 40.1	137 25.0	. . 17.1	188 43.3	. . 18.4
22		304 44.2	146 54.7	47.9	324 53.1	40.3	152 26.9	16.9	203 45.5	18.4
23		319 46.6	161 54.2	46.8	339 56.2	40.5	167 28.9	16.7	218 47.8	18.4
27 00		334 49.1	176 53.7	N10 45.7	354 59.3	S15 40.7	182 30.8	N12 16.5	233 50.0	N22 18.4
01		349 51.6	191 53.3	44.5	10 02.4	40.9	197 32.8	16.3	248 52.2	18.4
02		4 54.0	206 52.8	43.4	25 05.6	41.1	212 34.7	16.1	263 54.4	18.3
03		19 56.5	221 52.3	. . 42.3	40 08.7	. . 41.3	227 36.6	. . 15.9	278 56.6	. . 18.3
04		34 58.9	236 51.8	41.2	55 11.8	41.6	242 38.6	15.7	293 58.8	18.3
05		50 01.4	251 51.4	40.0	70 14.9	41.8	257 40.5	15.6	309 01.0	18.3
06		65 03.9	266 50.9	N10 38.9	85 18.0	S15 42.0	272 42.5	N12 15.4	324 03.2	N22 18.3
07	W	80 06.3	281 50.4	37.8	100 21.2	42.2	287 44.4	15.2	339 05.4	18.2
08	E	95 08.8	296 50.0	36.6	115 24.3	42.4	302 46.4	15.0	354 07.6	18.2
09	D	110 11.3	311 49.5	. . 35.5	130 27.4	. . 42.6	317 48.3	. . 14.8	9 09.8	. . 18.2
10	N	125 13.7	326 49.0	34.4	145 30.5	42.8	332 50.2	14.6	24 12.1	18.2
11	E	140 16.2	341 48.6	33.2	160 33.6	43.0	347 52.2	14.4	39 14.3	18.2
12	S	155 18.7	356 48.1	N10 32.1	175 36.8	S15 43.2	2 54.1	N12 14.2	54 16.5	N22 18.2
13	D	170 21.1	11 47.6	31.0	190 39.9	43.4	17 56.1	14.0	69 18.7	18.1
14	A	185 23.6	26 47.2	29.8	205 43.0	43.6	32 58.0	13.9	84 20.9	18.1
15	Y	200 26.1	41 46.7	. . 28.7	220 46.1	. . 43.8	47 59.9	. . 13.7	99 23.1	. . 18.1
16		215 28.5	56 46.2	27.6	235 49.3	44.0	63 01.9	13.5	114 25.3	18.1
17		230 31.0	71 45.8	26.4	250 52.4	44.2	78 03.8	13.3	129 27.5	18.1
18		245 33.4	86 45.3	N10 25.3	265 55.5	S15 44.4	93 05.8	N12 13.1	144 29.7	N22 18.1
19		260 35.9	101 44.8	24.1	280 58.6	44.6	108 07.7	12.9	159 31.9	18.0
20		275 38.4	116 44.4	23.0	296 01.7	44.8	123 09.6	12.7	174 34.2	18.0
21		290 40.8	131 43.9	. . 21.9	311 04.9	. . 45.0	138 11.6	. . 12.5	189 36.4	. . 18.0
22		305 43.3	146 43.4	20.7	326 08.0	45.2	153 13.5	12.3	204 38.6	18.0
23		320 45.8	161 43.0	19.6	341 11.1	45.4	168 15.5	12.1	219 40.8	18.0
28 00		335 48.2	176 42.5	N10 18.5	356 14.2	S15 45.6	183 17.4	N12 12.0	234 43.0	N22 18.0
01		350 50.7	191 42.0	17.3	11 17.4	45.8	198 19.4	11.8	249 45.2	17.9
02		5 53.2	206 41.6	16.2	26 20.5	46.0	213 21.3	11.6	264 47.4	17.9
03		20 55.6	221 41.1	. . 15.0	41 23.6	. . 46.2	228 23.2	. . 11.4	279 49.6	. . 17.9
04		35 58.1	236 40.6	13.9	56 26.7	46.4	243 25.2	11.2	294 51.8	17.9
05		51 00.5	251 40.2	12.7	71 29.9	46.6	258 27.1	11.0	309 54.1	17.9
06		66 03.0	266 39.7	N10 11.6	86 33.0	S15 46.8	273 29.1	N12 10.8	324 56.3	N22 17.9
07		81 05.5	281 39.3	10.5	101 36.1	47.0	288 31.0	10.6	339 58.5	17.8
08	T	96 07.9	296 38.8	09.3	116 39.2	47.2	303 32.9	10.4	355 00.7	17.8
09	H	111 10.4	311 38.3	. . 08.2	131 42.4	. . 47.4	318 34.9	. . 10.3	10 02.9	. . 17.8
10	U	126 12.9	326 37.9	07.0	146 45.5	47.6	333 36.8	10.1	25 05.1	17.8
11	R	141 15.3	341 37.4	05.9	161 48.6	47.8	348 38.8	09.9	40 07.3	17.8
12	S	156 17.8	356 37.0	N10 04.7	176 51.8	S15 48.0	3 40.7	N12 09.7	55 09.5	N22 17.8
13	D	171 20.3	11 36.5	03.6	191 54.9	48.2	18 42.7	09.5	70 11.8	17.7
14	A	186 22.7	26 36.0	02.4	206 58.0	48.4	33 44.6	09.3	85 14.0	17.7
15	Y	201 25.2	41 35.6	. . 01.3	222 01.1	. . 48.6	48 46.5	. . 09.1	100 16.2	. . 17.7
16		216 27.7	56 35.1	10 00.2	237 04.3	48.8	63 48.5	08.9	115 18.4	17.7
17		231 30.1	71 34.7	9 59.0	252 07.4	49.0	78 50.4	08.7	130 20.6	17.7
18		246 32.6	86 34.2	N 9 57.9	267 10.5	S15 49.2	93 52.4	N12 08.5	145 22.8	N22 17.7
19		261 35.0	101 33.7	56.7	282 13.6	49.3	108 54.3	08.4	160 25.0	17.6
20		276 37.5	116 33.3	55.6	297 16.8	49.5	123 56.3	08.2	175 27.3	17.6
21		291 40.0	131 32.8	. . 54.4	312 19.9	. . 49.7	138 58.2	. . 08.0	190 29.5	. . 17.6
22		306 42.4	146 32.4	53.3	327 23.0	49.9	154 00.1	07.8	205 31.7	17.6
23		321 44.9	161 31.9	52.1	342 26.2	50.1	169 02.1	07.6	220 33.9	17.6
Mer. Pass.		h m 1 40.5	v −0.5	d 1.1	v 3.1	d 0.2	v 1.9	d 0.2	v 2.2	d 0.0

STARS

Name	SHA ° ′	Dec ° ′
Acamar	315 24.0	S40 17.1
Achernar	335 31.9	S57 12.9
Acrux	173 18.9	S63 07.2
Adhara	255 18.9	S28 58.3
Aldebaran	290 58.3	N16 31.1
Alioth	166 27.6	N55 56.7
Alkaid	153 05.1	N49 18.0
Al Na'ir	27 52.8	S46 56.6
Alnilam	275 54.3	S 1 11.8
Alphard	218 04.0	S 8 40.2
Alphecca	126 17.5	N26 42.4
Alpheratz	357 51.3	N29 06.6
Altair	62 15.5	N 8 52.7
Ankaa	353 22.8	S42 17.0
Antares	112 35.7	S26 26.5
Arcturus	146 02.9	N19 10.0
Atria	107 44.5	S69 02.4
Avior	234 21.9	S59 31.0
Bellatrix	278 40.4	N 6 21.3
Betelgeuse	271 09.8	N 7 24.6
Canopus	263 59.9	S52 41.5
Capella	280 46.0	N46 00.0
Deneb	49 36.4	N45 17.7
Denebola	182 41.8	N14 33.3
Diphda	349 03.3	S17 57.9
Dubhe	194 01.4	N61 44.1
Elnath	278 22.5	N28 36.7
Eltanin	90 49.5	N51 29.5
Enif	33 54.4	N 9 53.5
Fomalhaut	15 32.0	S29 36.1
Gacrux	172 10.3	S57 08.0
Gienah	176 00.6	S17 33.6
Hadar	148 59.5	S60 23.6
Hamal	328 09.4	N23 28.8
Kaus Aust.	83 53.8	S34 23.1
Kochab	137 19.7	N74 08.7
Markab	13 45.7	N15 13.5
Menkar	314 23.0	N 4 06.3
Menkent	148 17.0	S36 23.3
Miaplacidus	221 42.5	S69 43.7
Mirfak	308 51.5	N49 52.3
Nunki	76 07.6	S26 17.6
Peacock	53 30.7	S56 43.6
Pollux	243 37.4	N28 01.2
Procyon	245 08.1	N 5 13.2
Rasalhague	96 13.5	N12 33.6
Regulus	207 52.0	N11 57.2
Rigel	281 19.6	S 8 11.6
Rigil Kent.	140 02.9	S60 51.2
Sabik	102 21.3	S15 43.8
Schedar	349 49.2	N56 33.3
Shaula	96 32.3	S37 06.5
Sirius	258 40.8	S16 43.0
Spica	158 39.6	S11 10.7
Suhail	222 58.7	S43 26.6
Vega	80 44.0	N38 47.4
Zuben'ubi	137 14.1	S16 03.4

	SHA ° ′	Mer. Pass h m
Venus	202 04.6	12 13
Mars	20 10.2	0 20
Jupiter	207 41.7	11 48
Saturn	259 00.9	8 23

UT	SUN GHA	SUN Dec	MOON GHA	v	MOON Dec	d	HP
d h	° ′	° ′	° ′	′	° ′	′	′
26 00	179 29.5	N10 38.0	199 25.2	8.9	N22 16.9	8.7	57.2
01	194 29.6	37.1	213 53.1	8.9	22 08.2	8.8	57.2
02	209 29.8	36.2	228 21.0	8.9	21 59.4	8.9	57.3
03	224 30.0	. . 35.4	242 48.9	8.9	21 50.5	9.1	57.3
04	239 30.1	34.5	257 16.8	9.0	21 41.4	9.2	57.3
05	254 30.3	33.6	271 44.8	9.0	21 32.2	9.3	57.3
06	269 30.5	N10 32.8	286 12.8	9.1	N21 22.9	9.4	57.4
07	284 30.7	31.9	300 40.9	9.1	21 13.5	9.6	57.4
T 08	299 30.8	31.0	315 09.0	9.1	21 03.9	9.7	57.4
U 09	314 31.0	. . 30.2	329 37.1	9.2	20 54.2	9.8	57.5
E 10	329 31.2	29.3	344 05.3	9.2	20 44.4	10.0	57.5
S 11	344 31.4	28.4	358 33.5	9.3	20 34.4	10.0	57.5
D 12	359 31.5	N10 27.5	13 01.8	9.3	N20 24.4	10.2	57.6
A 13	14 31.7	26.7	27 30.1	9.3	20 14.2	10.3	57.6
Y 14	29 31.9	25.8	41 58.4	9.4	20 03.9	10.5	57.6
15	44 32.1	. . 24.9	56 26.8	9.4	19 53.4	10.5	57.6
16	59 32.2	24.1	70 55.2	9.5	19 42.9	10.7	57.7
17	74 32.4	23.2	85 23.7	9.5	19 32.2	10.8	57.7
18	89 32.6	N10 22.3	99 52.2	9.5	N19 21.4	10.9	57.7
19	104 32.8	21.5	114 20.7	9.6	19 10.5	11.0	57.8
20	119 32.9	20.6	128 49.3	9.6	18 59.5	11.2	57.8
21	134 33.1	. . 19.7	143 17.9	9.7	18 48.3	11.2	57.8
22	149 33.3	18.8	157 46.6	9.7	18 37.1	11.4	57.8
23	164 33.5	18.0	172 15.3	9.7	18 25.7	11.4	57.9
27 00	179 33.6	N10 17.1	186 44.0	9.8	N18 14.3	11.6	57.9
01	194 33.8	16.2	201 12.8	9.8	18 02.7	11.7	57.9
02	209 34.0	15.4	215 41.6	9.9	17 51.0	11.8	57.9
03	224 34.2	. . 14.5	230 10.5	9.9	17 39.2	11.9	58.0
04	239 34.4	13.6	244 39.4	10.0	17 27.3	12.0	58.0
05	254 34.5	12.7	259 08.4	9.9	17 15.3	12.1	58.0
06	269 34.7	N10 11.9	273 37.3	10.1	N17 03.2	12.2	58.1
W 07	284 34.9	11.0	288 06.4	10.1	16 51.0	12.4	58.1
E 08	299 35.1	10.1	302 35.5	10.1	16 38.6	12.4	58.1
D 09	314 35.2	. . 09.2	317 04.6	10.1	16 26.2	12.5	58.1
N 10	329 35.4	08.4	331 33.7	10.2	16 13.7	12.6	58.2
E 11	344 35.6	07.5	346 02.9	10.3	16 01.1	12.7	58.2
S 12	359 35.8	N10 06.6	0 32.2	10.3	N15 48.4	12.8	58.2
D 13	14 36.0	05.7	15 01.5	10.3	15 35.6	12.9	58.2
A 14	29 36.1	04.9	29 30.8	10.3	15 22.7	13.0	58.3
Y 15	44 36.3	. . 04.0	44 00.1	10.4	15 09.7	13.1	58.3
16	59 36.5	03.1	58 29.5	10.5	14 56.6	13.2	58.3
17	74 36.7	02.2	72 59.0	10.4	14 43.4	13.3	58.3
18	89 36.9	N10 01.4	87 28.4	10.6	N14 30.1	13.3	58.4
19	104 37.0	10 00.5	101 58.0	10.5	14 16.8	13.5	58.4
20	119 37.2	9 59.6	116 27.5	10.6	14 03.3	13.5	58.4
21	134 37.4	. . 58.7	130 57.1	10.6	13 49.8	13.6	58.4
22	149 37.6	57.9	145 26.7	10.7	13 36.2	13.7	58.5
23	164 37.8	57.0	159 56.4	10.7	13 22.5	13.8	58.5
28 00	179 37.9	N 9 56.1	174 26.1	10.7	N13 08.7	13.9	58.5
01	194 38.1	55.2	188 55.8	10.8	12 54.8	13.9	58.5
02	209 38.3	54.3	203 25.6	10.8	12 40.9	14.0	58.6
03	224 38.5	. . 53.5	217 55.4	10.9	12 26.9	14.1	58.6
04	239 38.7	52.6	232 25.3	10.8	12 12.8	14.2	58.6
05	254 38.8	51.7	246 55.1	10.9	11 58.6	14.3	58.6
06	269 39.0	N 9 50.8	261 25.0	11.0	N11 44.3	14.3	58.6
T 07	284 39.2	49.9	275 55.0	10.9	11 30.0	14.4	58.7
H 08	299 39.4	49.1	290 24.9	11.0	11 15.6	14.4	58.7
U 09	314 39.6	. . 48.2	304 54.9	11.1	11 01.2	14.5	58.7
R 10	329 39.7	47.3	319 25.0	11.0	10 46.7	14.6	58.7
S 11	344 39.9	46.4	333 55.0	11.1	10 32.1	14.7	58.8
D 12	359 40.1	N 9 45.5	348 25.1	11.1	N10 17.4	14.7	58.8
A 13	14 40.3	44.6	2 55.2	11.2	10 02.7	14.8	58.8
Y 14	29 40.5	43.8	17 25.4	11.1	9 47.9	14.8	58.8
15	44 40.7	. . 42.9	31 55.5	11.2	9 33.1	14.9	58.8
16	59 40.8	42.0	46 25.7	11.3	9 18.2	15.0	58.9
17	74 41.0	41.1	60 56.0	11.2	9 03.2	15.0	58.9
18	89 41.2	N 9 40.2	75 26.2	11.3	N 8 48.2	15.1	58.9
19	104 41.4	39.4	89 56.5	11.3	8 33.1	15.1	58.9
20	119 41.6	38.5	104 26.8	11.3	8 18.0	15.2	58.9
21	134 41.8	. . 37.6	118 57.1	11.3	8 02.8	15.3	59.0
22	149 41.9	36.7	133 27.4	11.4	7 47.5	15.2	59.0
23	164 42.1	35.8	147 57.8	11.4	N 7 32.3	15.4	59.0
	SD 15.9	*d* 0.9	SD 15.7		15.9		16.0

Lat.	Twilight Naut.	Twilight Civil	Sunrise	Moonrise 26	Moonrise 27	Moonrise 28	Moonrise 29
°	h m	h m	h m	h m	h m	h m	h m
N 72	////	01 45	03 33	□	□	03 44	06 13
N 70	////	02 26	03 51	□	01 28	04 07	06 22
68	////	02 54	04 05	□	02 09	04 25	06 28
66	01 30	03 14	04 17	00 20	02 37	04 39	06 34
64	02 08	03 31	04 27	01 01	02 58	04 50	06 39
62	02 33	03 44	04 35	01 29	03 15	05 00	06 43
60	02 52	03 55	04 42	01 50	03 29	05 08	06 46
N 58	03 08	04 05	04 48	02 07	03 41	05 15	06 49
56	03 21	04 13	04 53	02 22	03 51	05 21	06 52
54	03 32	04 20	04 58	02 35	04 00	05 27	06 54
52	03 42	04 27	05 03	02 46	04 08	05 32	06 57
50	03 50	04 32	05 07	02 55	04 15	05 37	06 59
45	04 07	04 45	05 15	03 16	04 30	05 47	07 03
N 40	04 21	04 54	05 22	03 32	04 43	05 55	07 07
35	04 31	05 02	05 29	03 46	04 54	06 02	07 10
30	04 40	05 09	05 34	03 58	05 03	06 08	07 12
20	04 54	05 21	05 43	04 19	05 19	06 18	07 17
N 10	05 05	05 30	05 51	04 37	05 33	06 28	07 21
0	05 13	05 37	05 58	04 53	05 46	06 36	07 25
S 10	05 20	05 44	06 06	05 10	05 58	06 45	07 29
20	05 25	05 51	06 13	05 28	06 12	06 54	07 33
30	05 30	05 57	06 22	05 48	06 28	07 04	07 38
35	05 31	06 01	06 26	05 59	06 37	07 10	07 41
40	05 33	06 05	06 32	06 13	06 47	07 17	07 44
45	05 34	06 08	06 38	06 29	06 59	07 25	07 47
S 50	05 35	06 13	06 46	06 48	07 13	07 34	07 52
52	05 36	06 15	06 49	06 57	07 20	07 38	07 54
54	05 36	06 17	06 53	07 08	07 28	07 43	07 56
56	05 36	06 19	06 57	07 19	07 36	07 48	07 58
58	05 36	06 22	07 02	07 33	07 45	07 54	08 01
S 60	05 36	06 24	07 07	07 48	07 56	08 00	08 03

Lat.	Sunset	Twilight Civil	Twilight Naut.	Moonset 26	Moonset 27	Moonset 28	Moonset 29
°	h m	h m	h m	h m	h m	h m	h m
N 72	20 26	22 10	////	□	21 29	20 46	20 12
N 70	20 09	21 32	////	21 56	21 04	20 34	20 10
68	19 55	21 05	23 44	21 13	20 45	20 25	20 08
66	19 44	20 46	22 25	20 44	20 29	20 17	20 06
64	19 35	20 30	21 51	20 22	20 16	20 10	20 05
62	19 27	20 17	21 27	20 05	20 05	20 05	20 04
60	19 20	20 06	21 08	19 50	19 56	20 00	20 02
N 58	19 14	19 57	20 53	19 37	19 47	19 55	20 01
56	19 08	19 49	20 40	19 26	19 40	19 51	20 00
54	19 04	19 42	20 29	19 16	19 33	19 47	20 00
52	18 59	19 35	20 20	19 07	19 28	19 44	19 59
50	18 56	19 30	20 12	19 00	19 22	19 41	19 58
45	18 47	19 18	19 55	18 43	19 10	19 35	19 57
N 40	18 40	19 08	19 42	18 29	19 01	19 29	19 55
35	18 34	19 00	19 31	18 17	18 52	19 24	19 54
30	18 29	18 53	19 22	18 07	18 45	19 20	19 53
20	18 20	18 42	19 09	17 49	18 32	19 12	19 52
N 10	18 12	18 33	18 58	17 33	18 21	19 06	19 50
0	18 05	18 26	18 50	17 19	18 10	19 00	19 48
S 10	17 58	18 19	18 44	17 04	17 59	18 53	19 47
20	17 50	18 13	18 38	16 48	17 47	18 47	19 45
30	17 42	18 06	18 34	16 29	17 34	18 39	19 43
35	17 37	18 03	18 32	16 18	17 26	18 34	19 42
40	17 32	17 59	18 31	16 06	17 17	18 29	19 41
45	17 26	17 55	18 30	15 51	17 07	18 23	19 39
S 50	17 18	17 51	18 29	15 33	16 54	18 16	19 38
52	17 15	17 49	18 28	15 24	16 48	18 12	19 37
54	17 11	17 47	18 28	15 14	16 41	18 08	19 36
56	17 07	17 45	18 28	15 03	16 34	18 04	19 35
58	17 02	17 43	18 28	14 51	16 25	18 00	19 34
S 60	16 57	17 40	18 28	14 36	16 16	17 54	19 32

Day	SUN Eqn. of Time 00^h	SUN Eqn. of Time 12^h	SUN Mer. Pass.	MOON Mer. Pass. Upper	MOON Mer. Pass. Lower	MOON Age	MOON Phase
d	m s	m s	h m	h m	h m	d	%
26	02 03	01 54	12 02	11 06	23 32	28	2
27	01 46	01 37	12 02	11 58	24 23	29	0
28	01 29	01 20	12 01	12 48	00 23	01	1

UT d h		ARIES GHA	VENUS −3.9 GHA	VENUS Dec	MARS −2.9 GHA	MARS Dec	JUPITER −1.7 GHA	JUPITER Dec	SATURN +0.2 GHA	SATURN Dec	STARS Name	SHA	Dec
		° ′	° ′	° ′	° ′	° ′	° ′	° ′	° ′	° ′		° ′	° ′
29 00		336 47.4	176 31.5	N 9 51.0	357 29.3	S15 50.3	184 04.0	N12 07.4	235 36.1	N22 17.6	Acamar	315 23.9	S40 17.1
01		351 49.8	191 31.0	49.8	12 32.4	50.5	199 06.0	07.2	250 38.3	17.5	Achernar	335 31.9	S57 12.9
02		6 52.3	206 30.6	48.7	27 35.5	50.7	214 07.9	07.0	265 40.5	17.5	Acrux	173 18.9	S63 07.2
03		21 54.8	221 30.1	47.5	42 38.7	50.9	229 09.8	06.8	280 42.8	17.5	Adhara	255 18.8	S28 58.3
04		36 57.2	236 29.6	46.3	57 41.8	51.1	244 11.8	06.7	295 45.0	17.5	Aldebaran	290 58.3	N16 31.1
05		51 59.7	251 29.2	45.2	72 44.9	51.3	259 13.7	06.5	310 47.2	17.5			
06		67 02.1	266 28.7	N 9 44.0	87 48.0	S15 51.5	274 15.7	N12 06.3	325 49.4	N22 17.5	Alioth	166 27.6	N55 56.7
07		82 04.6	281 28.3	42.9	102 51.2	51.6	289 17.6	06.1	340 51.6	17.4	Alkaid	153 05.1	N49 18.0
08		97 07.1	296 27.8	41.7	117 54.3	51.8	304 19.6	05.9	355 53.8	17.4	Al Na'ir	27 52.8	S46 56.6
09	F	112 09.5	311 27.4	40.6	132 57.4	52.0	319 21.5	05.7	10 56.0	17.4	Alnilam	275 54.3	S 1 11.8
10	R	127 12.0	326 26.9	39.4	148 00.6	52.2	334 23.4	05.5	25 58.3	17.4	Alphard	218 04.0	S 8 40.2
11	I	142 14.5	341 26.5	38.3	163 03.7	52.4	349 25.4	05.3	41 00.5	17.4			
12	D	157 16.9	356 26.0	N 9 37.1	178 06.8	S15 52.6	4 27.3	N12 05.1	56 02.7	N22 17.4	Alphecca	126 17.6	N26 42.4
13	A	172 19.4	11 25.6	36.0	193 09.9	52.8	19 29.3	04.9	71 04.9	17.3	Alpheratz	357 51.3	N29 06.6
14	Y	187 21.9	26 25.1	34.8	208 13.1	53.0	34 31.2	04.8	86 07.1	17.3	Altair	62 15.5	N 8 52.7
15		202 24.3	41 24.7	33.6	223 16.2	53.1	49 33.2	04.6	101 09.3	17.3	Ankaa	353 22.8	S42 17.0
16		217 26.8	56 24.2	32.5	238 19.3	53.3	64 35.1	04.4	116 11.6	17.3	Antares	112 35.8	S26 26.5
17		232 29.3	71 23.8	31.3	253 22.5	53.5	79 37.0	04.2	131 13.8	17.3			
18		247 31.7	86 23.3	N 9 30.2	268 25.6	S15 53.7	94 39.0	N12 04.0	146 16.0	N22 17.3	Arcturus	146 02.9	N19 10.0
19		262 34.2	101 22.9	29.0	283 28.7	53.9	109 40.9	03.8	161 18.2	17.2	Atria	107 44.6	S69 02.4
20		277 36.6	116 22.4	27.8	298 31.8	54.1	124 42.9	03.6	176 20.4	17.2	Avior	234 21.9	S59 31.0
21		292 39.1	131 22.0	26.7	313 35.0	54.3	139 44.8	03.4	191 22.6	17.2	Bellatrix	278 40.4	N 6 21.3
22		307 41.6	146 21.5	25.5	328 38.1	54.4	154 46.8	03.2	206 24.9	17.2	Betelgeuse	271 09.8	N 7 24.6
23		322 44.0	161 21.1	24.4	343 41.2	54.6	169 48.7	03.0	221 27.1	17.2			
30 00		337 46.5	176 20.6	N 9 23.2	358 44.4	S15 54.8	184 50.6	N12 02.9	236 29.3	N22 17.2	Canopus	263 59.9	S52 41.5
01		352 49.0	191 20.2	22.0	13 47.5	55.0	199 52.6	02.7	251 31.5	17.1	Capella	280 46.0	N46 00.0
02		7 51.4	206 19.7	20.9	28 50.6	55.2	214 54.5	02.5	266 33.7	17.1	Deneb	49 36.4	N45 17.7
03		22 53.9	221 19.3	19.7	43 53.8	55.3	229 56.5	02.3	281 36.0	17.1	Denebola	182 41.8	N14 33.3
04		37 56.4	236 18.8	18.5	58 56.9	55.5	244 58.4	02.1	296 38.2	17.1	Diphda	349 03.3	S17 57.9
05		52 58.8	251 18.4	17.4	74 00.0	55.7	260 00.4	01.9	311 40.4	17.1			
06		68 01.3	266 17.9	N 9 16.2	89 03.1	S15 55.9	275 02.3	N12 01.7	326 42.6	N22 17.1	Dubhe	194 01.4	N61 44.1
07		83 03.7	281 17.5	15.0	104 06.3	56.1	290 04.2	01.5	341 44.8	17.0	Elnath	278 22.5	N28 36.7
08	S	98 06.2	296 17.0	13.9	119 09.4	56.3	305 06.2	01.3	356 47.0	17.0	Eltanin	90 49.5	N51 29.5
09	A	113 08.7	311 16.6	12.7	134 12.5	56.4	320 08.1	01.1	11 49.3	17.0	Enif	33 54.4	N 9 53.5
10	T	128 11.1	326 16.2	11.6	149 15.7	56.6	335 10.1	01.0	26 51.5	17.0	Fomalhaut	15 32.0	S29 36.1
11	U	143 13.6	341 15.7	10.4	164 18.8	56.8	350 12.0	00.8	41 53.7	17.0			
12	R	158 16.1	356 15.3	N 9 09.2	179 21.9	S15 57.0	5 14.0	N12 00.6	56 55.9	N22 17.0	Gacrux	172 10.3	S57 08.0
13	D	173 18.5	11 14.8	08.0	194 25.0	57.1	20 15.9	00.4	71 58.1	16.9	Gienah	176 00.6	S17 33.6
14	A	188 21.0	26 14.4	06.9	209 28.2	57.3	35 17.8	00.2	87 00.4	16.9	Hadar	148 59.5	S60 23.6
15	Y	203 23.5	41 13.9	05.7	224 31.3	57.5	50 19.8	12 00.0	102 02.6	16.9	Hamal	328 09.4	N23 28.8
16		218 25.9	56 13.5	04.5	239 34.4	57.7	65 21.7	11 59.8	117 04.8	16.9	Kaus Aust.	83 53.8	S34 23.1
17		233 28.4	71 13.1	03.4	254 37.6	57.9	80 23.7	59.6	132 07.0	16.9			
18		248 30.9	86 12.6	N 9 02.2	269 40.7	S15 58.0	95 25.6	N11 59.4	147 09.2	N22 16.8	Kochab	137 19.7	N74 08.7
19		263 33.3	101 12.2	9 01.0	284 43.8	58.2	110 27.6	59.2	162 11.5	16.8	Markab	13 45.7	N15 13.5
20		278 35.8	116 11.7	8 59.9	299 46.9	58.4	125 29.5	59.1	177 13.7	16.8	Menkar	314 23.0	N 4 06.3
21		293 38.2	131 11.3	58.7	314 50.1	58.6	140 31.4	58.9	192 15.9	16.8	Menkent	148 17.0	S36 23.3
22		308 40.7	146 10.8	57.5	329 53.2	58.7	155 33.4	58.7	207 18.1	16.8	Miaplacidus	221 42.5	S69 43.7
23		323 43.2	161 10.4	56.3	344 56.3	58.9	170 35.3	58.5	222 20.3	16.8			
31 00		338 45.6	176 10.0	N 8 55.2	359 59.5	S15 59.1	185 37.3	N11 58.3	237 22.6	N22 16.7	Mirfak	308 51.4	N49 52.3
01		353 48.1	191 09.5	54.0	15 02.6	59.2	200 39.2	58.1	252 24.8	16.7	Nunki	76 07.6	S26 17.6
02		8 50.6	206 09.1	52.8	30 05.7	59.4	215 41.2	57.9	267 27.0	16.7	Peacock	53 30.7	S56 43.6
03		23 53.0	221 08.6	51.7	45 08.8	59.6	230 43.1	57.7	282 29.2	16.7	Pollux	243 37.4	N28 01.2
04		38 55.5	236 08.2	50.5	60 12.0	59.8	245 45.0	57.5	297 31.4	16.7	Procyon	245 08.0	N 5 13.2
05		53 58.0	251 07.8	49.3	75 15.1	15 59.9	260 47.0	57.3	312 33.7	16.7			
06		69 00.4	266 07.3	N 8 48.1	90 18.2	S16 00.1	275 48.9	N11 57.2	327 35.9	N22 16.6	Rasalhague	96 13.5	N12 33.6
07		84 02.9	281 06.9	47.0	105 21.3	00.3	290 50.9	57.0	342 38.1	16.6	Regulus	207 52.0	N11 57.2
08		99 05.3	296 06.5	45.8	120 24.5	00.4	305 52.8	56.8	357 40.3	16.6	Rigel	281 19.5	S 8 11.6
09	S	114 07.8	311 06.0	44.6	135 27.6	00.6	320 54.8	56.6	12 42.6	16.6	Rigil Kent.	140 02.9	S60 51.2
10	U	129 10.3	326 05.6	43.4	150 30.7	00.8	335 56.7	56.4	27 44.8	16.6	Sabik	102 21.3	S15 43.8
11	N	144 12.7	341 05.1	42.3	165 33.9	01.0	350 58.7	56.2	42 47.0	16.6			
12	D	159 15.2	356 04.7	N 8 41.1	180 37.0	S16 01.1	6 00.6	N11 56.0	57 49.2	N22 16.5	Schedar	349 49.1	N56 33.3
13	A	174 17.7	11 04.3	39.9	195 40.1	01.3	21 02.5	55.8	72 51.4	16.5	Shaula	96 32.3	S37 06.5
14	Y	189 20.1	26 03.8	38.7	210 43.2	01.5	36 04.5	55.6	87 53.7	16.5	Sirius	258 40.8	S16 43.0
15		204 22.6	41 03.4	37.5	225 46.4	01.6	51 06.4	55.4	102 55.9	16.5	Spica	158 39.7	S11 10.7
16		219 25.1	56 03.0	36.4	240 49.5	01.8	66 08.4	55.3	117 58.1	16.5	Suhail	222 58.7	S43 26.6
17		234 27.5	71 02.5	35.2	255 52.6	02.0	81 10.3	55.1	133 00.3	16.5			
18		249 30.0	86 02.1	N 8 34.0	270 55.7	S16 02.1	96 12.3	N11 54.9	148 02.6	N22 16.4	Vega	80 44.0	N38 47.4
19		264 32.5	101 01.7	32.8	285 58.9	02.3	111 14.2	54.7	163 04.8	16.4	Zuben'ubi	137 14.1	S16 03.4
20		279 34.9	116 01.2	31.6	301 02.0	02.5	126 16.2	54.5	178 07.0	16.4			
21		294 37.4	131 00.8	30.5	316 05.1	02.6	141 18.1	54.3	193 09.2	16.4			
22		309 39.8	146 00.4	29.3	331 08.2	02.8	156 20.0	54.1	208 11.5	16.4			
23		324 42.3	160 59.9	28.1	346 11.4	02.9	171 22.0	53.9	223 13.7	16.4			
Mer. Pass. (h m)		1 28.7	v −0.4	d 1.2	v 3.1	d 0.2	v 1.9	d 0.2	v 2.2	d 0.0			

	SHA ° ′	Mer. Pass h m
Venus	198 34.1	12 15
Mars	20 57.9	0 05
Jupiter	207 04.1	11 39
Saturn	258 42.8	8 13

UT	SUN GHA	SUN Dec	MOON GHA	*v*	MOON Dec	*d*	HP
d h	° ′	° ′	° ′	′	° ′	′	′
29 00	179 42.3	N 9 34.9	162 28.2	11.4	N 7 16.9	15.3	59.0
01	194 42.5	34.0	176 58.6	11.4	7 01.6	15.5	59.0
02	209 42.7	33.2	191 29.0	11.4	6 46.1	15.4	59.0
03	224 42.9	. . 32.3	205 59.4	11.4	6 30.7	15.5	59.1
04	239 43.1	31.4	220 29.8	11.5	6 15.2	15.6	59.1
05	254 43.2	30.5	235 00.3	11.5	5 59.6	15.6	59.1
06	269 43.4	N 9 29.6	249 30.8	11.5	N 5 44.0	15.6	59.1
07	284 43.6	28.7	264 01.3	11.5	5 28.4	15.7	59.1
08	299 43.8	27.8	278 31.8	11.5	5 12.7	15.7	59.1
09	314 44.0	. . 27.0	293 02.3	11.5	4 57.0	15.7	59.2
10	329 44.2	26.1	307 32.8	11.5	4 41.3	15.8	59.2
11	344 44.4	25.2	322 03.3	11.6	4 25.5	15.8	59.2
12	359 44.5	N 9 24.3	336 33.9	11.5	N 4 09.7	15.8	59.2
13	14 44.7	23.4	351 04.4	11.6	3 53.9	15.8	59.2
14	29 44.9	22.5	5 35.0	11.5	3 38.1	15.9	59.2
15	44 45.1	. . 21.6	20 05.5	11.6	3 22.2	15.9	59.2
16	59 45.3	20.7	34 36.1	11.6	3 06.3	15.9	59.3
17	74 45.5	19.8	49 06.7	11.5	2 50.4	16.0	59.3
18	89 45.7	N 9 19.0	63 37.2	11.6	N 2 34.4	16.0	59.3
19	104 45.8	18.1	78 07.8	11.6	2 18.4	15.9	59.3
20	119 46.0	17.2	92 38.4	11.5	2 02.5	16.0	59.3
21	134 46.2	. . 16.3	107 08.9	11.6	1 46.5	16.0	59.3
22	149 46.4	15.4	121 39.5	11.6	1 30.5	16.1	59.3
23	164 46.6	14.5	136 10.1	11.5	1 14.4	16.0	59.3
30 00	179 46.8	N 9 13.6	150 40.6	11.6	N 0 58.4	16.1	59.4
01	194 47.0	12.7	165 11.2	11.5	0 42.3	16.0	59.4
02	209 47.2	11.8	179 41.7	11.6	0 26.3	16.1	59.4
03	224 47.4	. . 10.9	194 12.3	11.5	N 0 10.2	16.1	59.4
04	239 47.5	10.0	208 42.8	11.5	S 0 05.9	16.0	59.4
05	254 47.7	09.1	223 13.3	11.5	0 21.9	16.1	59.4
06	269 47.9	N 9 08.3	237 43.8	11.5	S 0 38.0	16.1	59.4
07	284 48.1	07.4	252 14.3	11.5	0 54.1	16.1	59.4
08	299 48.3	06.5	266 44.8	11.5	1 10.2	16.1	59.4
09	314 48.5	. . 05.6	281 15.3	11.5	1 26.3	16.0	59.5
10	329 48.7	04.7	295 45.8	11.4	1 42.3	16.1	59.5
11	344 48.9	03.8	310 16.2	11.4	1 58.4	16.1	59.5
12	359 49.1	N 9 02.9	324 46.6	11.5	S 2 14.5	16.0	59.5
13	14 49.2	02.0	339 17.1	11.4	2 30.5	16.1	59.5
14	29 49.4	01.1	353 47.5	11.3	2 46.6	16.0	59.5
15	44 49.6	9 00.2	8 17.8	11.4	3 02.6	16.0	59.5
16	59 49.8	8 59.3	22 48.2	11.3	3 18.6	16.0	59.5
17	74 50.0	58.4	37 18.5	11.3	3 34.6	16.0	59.5
18	89 50.2	N 8 57.5	51 48.8	11.3	S 3 50.6	16.0	59.5
19	104 50.4	56.6	66 19.1	11.2	4 06.6	16.0	59.5
20	119 50.6	55.7	80 49.3	11.3	4 22.6	15.9	59.5
21	134 50.8	. . 54.8	95 19.6	11.2	4 38.5	15.9	59.5
22	149 51.0	53.9	109 49.8	11.1	4 54.4	15.9	59.5
23	164 51.2	53.0	124 19.9	11.2	5 10.3	15.9	59.5
31 00	179 51.4	N 8 52.1	138 50.1	11.1	S 5 26.2	15.8	59.6
01	194 51.5	51.2	153 20.2	11.1	5 42.0	15.8	59.6
02	209 51.7	50.3	167 50.3	11.0	5 57.8	15.8	59.6
03	224 51.9	. . 49.4	182 20.3	11.0	6 13.6	15.7	59.6
04	239 52.1	48.5	196 50.3	11.0	6 29.3	15.7	59.6
05	254 52.3	47.7	211 20.3	10.9	6 45.0	15.7	59.6
06	269 52.5	N 8 46.8	225 50.2	10.9	S 7 00.7	15.7	59.6
07	284 52.7	45.9	240 20.1	10.9	7 16.4	15.6	59.6
08	299 52.9	45.0	254 50.0	10.8	7 32.0	15.5	59.6
09	314 53.1	. . 44.1	269 19.8	10.8	7 47.5	15.6	59.6
10	329 53.3	43.2	283 49.6	10.7	8 03.1	15.4	59.6
11	344 53.5	42.3	298 19.3	10.7	8 18.5	15.5	59.6
12	359 53.7	N 8 41.4	312 49.0	10.7	S 8 34.0	15.4	59.6
13	14 53.9	40.5	327 18.7	10.6	8 49.4	15.3	59.6
14	29 54.1	39.6	341 48.3	10.5	9 04.7	15.3	59.6
15	44 54.2	. . 38.7	356 17.8	10.6	9 20.0	15.2	59.6
16	59 54.4	37.8	10 47.4	10.4	9 35.2	15.2	59.6
17	74 54.6	36.8	25 16.8	10.5	9 50.4	15.2	59.6
18	89 54.8	N 8 35.9	39 46.3	10.3	S10 05.6	15.0	59.6
19	104 55.0	35.0	54 15.6	10.3	10 20.6	15.1	59.6
20	119 55.2	34.1	68 44.9	10.3	10 35.7	14.9	59.6
21	134 55.4	. . 33.2	83 14.2	10.2	10 50.6	14.9	59.6
22	149 55.6	32.3	97 43.4	10.2	11 05.5	14.9	59.6
23	164 55.8	31.4	112 12.6	10.1	S11 20.4	14.7	59.6
	SD 15.9	*d* 0.9	SD 16.1		16.2		16.2

Lat.	Twilight Naut.	Twilight Civil	Sunrise	Moonrise 29	30	31	1
°	h m	h m	h m	h m	h m	h m	h m
N 72	////	02 12	03 49	06 13	08 32	10 55	13 44
N 70	////	02 45	04 04	06 22	08 30	10 42	13 08
68	01 03	03 09	04 17	06 28	08 28	10 31	12 43
66	01 54	03 27	04 27	06 34	08 27	10 22	12 24
64	02 24	03 41	04 35	06 39	08 26	10 15	12 08
62	02 46	03 53	04 43	06 43	08 25	10 08	11 55
60	03 03	04 03	04 49	06 46	08 24	10 03	11 45
N 58	03 17	04 12	04 54	06 49	08 23	09 58	11 35
56	03 29	04 19	04 59	06 52	08 22	09 54	11 27
54	03 39	04 26	05 04	06 54	08 22	09 50	11 20
52	03 48	04 32	05 08	06 57	08 21	09 47	11 14
50	03 56	04 37	05 11	06 59	08 21	09 44	11 08
45	04 12	04 48	05 19	07 03	08 20	09 37	10 56
N 40	04 24	04 57	05 25	07 07	08 19	09 31	10 45
35	04 34	05 05	05 31	07 10	08 18	09 27	10 37
30	04 42	05 11	05 36	07 12	08 17	09 23	10 29
20	04 55	05 21	05 44	07 17	08 16	09 15	10 16
N 10	05 05	05 30	05 51	07 21	08 15	09 09	10 05
0	05 12	05 37	05 57	07 25	08 14	09 03	09 54
S 10	05 18	05 43	06 04	07 29	08 13	08 58	09 44
20	05 23	05 48	06 11	07 33	08 12	08 52	09 33
30	05 26	05 54	06 18	07 38	08 11	08 45	09 21
35	05 28	05 57	06 23	07 41	08 11	08 41	09 14
40	05 29	06 00	06 27	07 44	08 10	08 37	09 05
45	05 29	06 03	06 33	07 47	08 09	08 32	08 56
S 50	05 30	06 07	06 40	07 52	08 08	08 26	08 45
52	05 30	06 09	06 43	07 54	08 08	08 23	08 40
54	05 29	06 10	06 46	07 56	08 08	08 20	08 34
56	05 29	06 12	06 50	07 58	08 07	08 17	08 28
58	05 29	06 14	06 54	08 01	08 07	08 13	08 21
S 60	05 28	06 16	06 58	08 03	08 06	08 09	08 13

Lat.	Sunset	Twilight Civil	Twilight Naut.	Moonset 29	30	31	1
°	h m	h m	h m	h m	h m	h m	h m
N 72	20 09	21 43	////	20 12	19 41	19 04	18 04
N 70	19 54	21 12	////	20 10	19 47	19 20	18 42
68	19 42	20 49	22 47	20 08	19 51	19 33	19 09
66	19 32	20 32	22 02	20 06	19 56	19 44	19 30
64	19 24	20 18	21 34	20 05	19 59	19 53	19 47
62	19 17	20 06	21 13	20 04	20 02	20 01	20 01
60	19 11	19 56	20 56	20 02	20 05	20 08	20 13
N 58	19 06	19 48	20 42	20 01	20 07	20 14	20 23
56	19 01	19 41	20 31	20 00	20 10	20 20	20 32
54	18 57	19 34	20 21	20 00	20 12	20 24	20 40
52	18 53	19 28	20 12	19 59	20 13	20 29	20 47
50	18 49	19 23	20 05	19 58	20 15	20 33	20 54
45	18 42	19 12	19 49	19 57	20 19	20 42	21 08
N 40	18 35	19 03	19 37	19 55	20 22	20 49	21 20
35	18 30	18 56	19 27	19 54	20 24	20 55	21 30
30	18 25	18 50	19 19	19 53	20 26	21 01	21 38
20	18 17	18 40	19 06	19 52	20 30	21 11	21 54
N 10	18 10	18 32	18 57	19 50	20 34	21 19	22 07
0	18 04	18 25	18 49	19 48	20 37	21 27	22 20
S 10	17 58	18 19	18 44	19 47	20 41	21 35	22 32
20	17 51	18 13	18 39	19 45	20 44	21 44	22 46
30	17 44	18 08	18 36	19 43	20 48	21 54	23 01
35	17 39	18 05	18 34	19 42	20 51	22 00	23 11
40	17 35	18 02	18 33	19 41	20 53	22 06	23 21
45	17 29	17 59	18 33	19 39	20 56	22 14	23 33
S 50	17 23	17 55	18 33	19 38	21 00	22 23	23 48
52	17 20	17 54	18 33	19 37	21 02	22 28	23 55
54	17 16	17 52	18 33	19 36	21 03	22 33	24 03
56	17 13	17 50	18 34	19 35	21 06	22 38	24 12
58	17 09	17 49	18 34	19 34	21 08	22 44	24 22
S 60	17 04	17 47	18 35	19 32	21 10	22 50	24 33

Day	SUN Eqn. of Time 00^h	SUN Eqn. of Time 12^h	SUN Mer. Pass.	MOON Mer. Pass. Upper	MOON Mer. Pass. Lower	MOON Age	MOON Phase
d	m s	m s	h m	h m	h m	d %	
29	01 11	01 02	12 01	13 37	01 13	02 4	
30	00 53	00 44	12 01	14 26	02 01	03 10	
31	00 35	00 26	12 00	15 15	02 50	04 18	

Day	UT d h	ARIES GHA ° ′	VENUS −3.9 GHA ° ′	VENUS Dec ° ′	MARS −2.8 GHA ° ′	MARS Dec ° ′	JUPITER −1.7 GHA ° ′	JUPITER Dec ° ′	SATURN +0.1 GHA ° ′	SATURN Dec ° ′
	1 00	339 44.8	175 59.5	N 8 26.9	1 14.5	S16 03.1	186 23.9	N11 53.7	238 15.9	N22 16.3
	01	354 47.2	190 59.1	25.7	16 17.6	03.3	201 25.9	53.5	253 18.1	16.3
	02	9 49.7	205 58.6	24.5	31 20.7	03.4	216 27.8	53.4	268 20.4	16.3
	03	24 52.2	220 58.2	. . 23.4	46 23.9	. . 03.6	231 29.8	. . 53.2	283 22.6	. . 16.3
	04	39 54.6	235 57.8	22.2	61 27.0	03.8	246 31.7	53.0	298 24.8	16.3
	05	54 57.1	250 57.3	21.0	76 30.1	03.9	261 33.6	52.8	313 27.0	16.3
	06	69 59.6	265 56.9	N 8 19.8	91 33.2	S16 04.1	276 35.6	N11 52.6	328 29.3	N22 16.2
	07	85 02.0	280 56.5	18.6	106 36.4	04.2	291 37.5	52.4	343 31.5	16.2
	08	100 04.5	295 56.0	17.4	121 39.5	04.4	306 39.5	52.2	358 33.7	16.2
MONDAY	09	115 07.0	310 55.6	. . 16.3	136 42.6	. . 04.6	321 41.4	. . 52.0	13 35.9	. . 16.2
	10	130 09.4	325 55.2	15.1	151 45.7	04.7	336 43.4	51.8	28 38.2	16.2
	11	145 11.9	340 54.7	13.9	166 48.9	04.9	351 45.3	51.6	43 40.4	16.2
	12	160 14.3	355 54.3	N 8 12.7	181 52.0	S16 05.0	6 47.3	N11 51.5	58 42.6	N22 16.1
	13	175 16.8	10 53.9	11.5	196 55.1	05.2	21 49.2	51.3	73 44.8	16.1
	14	190 19.3	25 53.5	10.3	211 58.2	05.3	36 51.1	51.1	88 47.1	16.1
	15	205 21.7	40 53.0	. . 09.1	227 01.3	. . 05.5	51 53.1	. . 50.9	103 49.3	. . 16.1
	16	220 24.2	55 52.6	07.9	242 04.5	05.7	66 55.0	50.7	118 51.5	16.1
	17	235 26.7	70 52.2	06.8	257 07.6	05.8	81 57.0	50.5	133 53.7	16.1
	18	250 29.1	85 51.7	N 8 05.6	272 10.7	S16 06.0	96 58.9	N11 50.3	148 56.0	N22 16.0
	19	265 31.6	100 51.3	04.4	287 13.8	06.1	112 00.9	50.1	163 58.2	16.0
	20	280 34.1	115 50.9	03.2	302 17.0	06.3	127 02.8	49.9	179 00.4	16.0
	21	295 36.5	130 50.5	. . 02.0	317 20.1	. . 06.4	142 04.8	. . 49.7	194 02.6	. . 16.0
	22	310 39.0	145 50.0	8 00.8	332 23.2	06.6	157 06.7	49.5	209 04.9	16.0
	23	325 41.4	160 49.6	7 59.6	347 26.3	06.7	172 08.7	49.4	224 07.1	16.0
	2 00	340 43.9	175 49.2	N 7 58.4	2 29.4	S16 06.9	187 10.6	N11 49.2	239 09.3	N22 15.9
	01	355 46.4	190 48.8	57.2	17 32.6	07.0	202 12.5	49.0	254 11.6	15.9
	02	10 48.8	205 48.3	56.0	32 35.7	07.2	217 14.5	48.8	269 13.8	15.9
	03	25 51.3	220 47.9	. . 54.8	47 38.8	. . 07.3	232 16.4	. . 48.6	284 16.0	. . 15.9
	04	40 53.8	235 47.5	53.6	62 41.9	07.5	247 18.4	48.4	299 18.2	15.9
	05	55 56.2	250 47.1	52.5	77 45.0	07.6	262 20.3	48.2	314 20.5	15.9
	06	70 58.7	265 46.6	N 7 51.3	92 48.2	S16 07.8	277 22.3	N11 48.0	329 22.7	N22 15.8
	07	86 01.2	280 46.2	50.1	107 51.3	07.9	292 24.2	47.8	344 24.9	15.8
TUESDAY	08	101 03.6	295 45.8	48.9	122 54.4	08.1	307 26.2	47.6	359 27.1	15.8
	09	116 06.1	310 45.4	. . 47.7	137 57.5	. . 08.2	322 28.1	. . 47.5	14 29.4	. . 15.8
	10	131 08.6	325 44.9	46.5	153 00.6	08.4	337 30.0	47.3	29 31.6	15.8
	11	146 11.0	340 44.5	45.3	168 03.7	08.5	352 32.0	47.1	44 33.8	15.8
	12	161 13.5	355 44.1	N 7 44.1	183 06.9	S16 08.7	7 33.9	N11 46.9	59 36.1	N22 15.7
	13	176 15.9	10 43.7	42.9	198 10.0	08.8	22 35.9	46.7	74 38.3	15.7
	14	191 18.4	25 43.3	41.7	213 13.1	09.0	37 37.8	46.5	89 40.5	15.7
	15	206 20.9	40 42.8	. . 40.5	228 16.2	. . 09.1	52 39.8	. . 46.3	104 42.8	. . 15.7
	16	221 23.3	55 42.4	39.3	243 19.3	09.3	67 41.7	46.1	119 45.0	15.7
	17	236 25.8	70 42.0	38.1	258 22.4	09.4	82 43.7	45.9	134 47.2	15.7
	18	251 28.3	85 41.6	N 7 36.9	273 25.6	S16 09.5	97 45.6	N11 45.7	149 49.4	N22 15.6
	19	266 30.7	100 41.1	35.7	288 28.7	09.7	112 47.6	45.6	164 51.7	15.6
	20	281 33.2	115 40.7	34.5	303 31.8	09.8	127 49.5	45.4	179 53.9	15.6
	21	296 35.7	130 40.3	. . 33.3	318 34.9	. . 10.0	142 51.4	. . 45.2	194 56.1	. . 15.6
	22	311 38.1	145 39.9	32.1	333 38.0	10.1	157 53.4	45.0	209 58.4	15.6
	23	326 40.6	160 39.5	30.9	348 41.1	10.3	172 55.3	44.8	225 00.6	15.6
	3 00	341 43.1	175 39.0	N 7 29.7	3 44.2	S16 10.4	187 57.3	N11 44.6	240 02.8	N22 15.5
	01	356 45.5	190 38.6	28.5	18 47.4	10.5	202 59.2	44.4	255 05.1	15.5
	02	11 48.0	205 38.2	27.3	33 50.5	10.7	218 01.2	44.2	270 07.3	15.5
	03	26 50.4	220 37.8	. . 26.1	48 53.6	. . 10.8	233 03.1	. . 44.0	285 09.5	. . 15.5
	04	41 52.9	235 37.4	24.9	63 56.7	11.0	248 05.1	43.8	300 11.8	15.5
	05	56 55.4	250 37.0	23.7	78 59.8	11.1	263 07.0	43.6	315 14.0	15.5
	06	71 57.8	265 36.5	N 7 22.5	94 02.9	S16 11.2	278 09.0	N11 43.5	330 16.2	N22 15.4
WEDNESDAY	07	87 00.3	280 36.1	21.3	109 06.0	11.4	293 10.9	43.3	345 18.4	15.4
	08	102 02.8	295 35.7	20.1	124 09.1	11.5	308 12.9	43.1	0 20.7	15.4
	09	117 05.2	310 35.3	. . 18.9	139 12.2	. . 11.6	323 14.8	. . 42.9	15 22.9	. . 15.4
	10	132 07.7	325 34.9	17.7	154 15.4	11.8	338 16.7	42.7	30 25.1	15.4
	11	147 10.2	340 34.5	16.5	169 18.5	11.9	353 18.7	42.5	45 27.4	15.4
	12	162 12.6	355 34.0	N 7 15.3	184 21.6	S16 12.1	8 20.6	N11 42.3	60 29.6	N22 15.3
	13	177 15.1	10 33.6	14.1	199 24.7	12.2	23 22.6	42.1	75 31.8	15.3
	14	192 17.6	25 33.2	12.9	214 27.8	12.3	38 24.5	41.9	90 34.1	15.3
	15	207 20.0	40 32.8	. . 11.6	229 30.9	. . 12.5	53 26.5	. . 41.7	105 36.3	. . 15.3
	16	222 22.5	55 32.4	10.4	244 34.0	12.6	68 28.4	41.6	120 38.5	15.3
	17	237 24.9	70 32.0	09.2	259 37.1	12.7	83 30.4	41.4	135 40.8	15.3
	18	252 27.4	85 31.5	N 7 08.0	274 40.2	S16 12.9	98 32.3	N11 41.2	150 43.0	N22 15.2
	19	267 29.9	100 31.1	06.8	289 43.3	13.0	113 34.3	41.0	165 45.2	15.2
	20	282 32.3	115 30.7	05.6	304 46.4	13.1	128 36.2	40.8	180 47.5	15.2
	21	297 34.8	130 30.3	. . 04.4	319 49.5	. . 13.2	143 38.2	. . 40.6	195 49.7	. . 15.2
	22	312 37.3	145 29.9	03.2	334 52.6	13.4	158 40.1	40.4	210 51.9	15.2
	23	327 39.7	160 29.5	02.0	349 55.7	13.5	173 42.0	40.2	225 54.2	15.2
	Mer. Pass.	h m 1 16.9	*v* −0.4	*d* 1.2	*v* 3.1	*d* 0.1	*v* 1.9	*d* 0.2	*v* 2.2	*d* 0.0

STARS

Name	SHA ° ′	Dec ° ′
Acamar	315 23.9	S40 17.1
Achernar	335 31.8	S57 12.9
Acrux	173 18.9	S63 07.2
Adhara	255 18.8	S28 58.3
Aldebaran	290 58.3	N16 31.1
Alioth	166 27.6	N55 56.7
Alkaid	153 05.1	N49 18.0
Al Na'ir	27 52.8	S46 56.6
Alnilam	275 54.3	S 1 11.8
Alphard	218 04.0	S 8 40.2
Alphecca	126 17.6	N26 42.4
Alpheratz	357 51.3	N29 06.6
Altair	62 15.5	N 8 52.7
Ankaa	353 22.8	S42 17.1
Antares	112 35.8	S26 26.5
Arcturus	146 02.9	N19 10.0
Atria	107 44.6	S69 02.4
Avior	234 21.9	S59 30.9
Bellatrix	278 40.4	N 6 21.3
Betelgeuse	271 09.8	N 7 24.6
Canopus	263 59.8	S52 41.5
Capella	280 46.0	N46 00.0
Deneb	49 36.4	N45 17.7
Denebola	182 41.8	N14 33.3
Diphda	349 03.3	S17 57.9
Dubhe	194 01.5	N61 44.1
Elnath	278 22.5	N28 36.7
Eltanin	90 49.5	N51 29.5
Enif	33 54.4	N 9 53.5
Fomalhaut	15 32.0	S29 36.2
Gacrux	172 10.3	S57 08.0
Gienah	176 00.6	S17 33.6
Hadar	148 59.5	S60 23.6
Hamal	328 09.3	N23 28.8
Kaus Aust.	83 53.8	S34 23.1
Kochab	137 19.8	N74 08.7
Markab	13 45.7	N15 13.5
Menkar	314 23.0	N 4 06.3
Menkent	148 17.0	S36 23.3
Miaplacidus	221 42.5	S69 43.7
Mirfak	308 51.4	N49 52.3
Nunki	76 07.6	S26 17.6
Peacock	53 30.7	S56 43.6
Pollux	243 37.4	N28 01.2
Procyon	245 08.0	N 5 13.2
Rasalhague	96 13.5	N12 33.6
Regulus	207 52.0	N11 57.2
Rigel	281 19.5	S 8 11.6
Rigil Kent.	140 03.0	S60 51.2
Sabik	102 21.4	S15 43.8
Schedar	349 49.1	N56 33.3
Shaula	96 32.3	S37 06.5
Sirius	258 40.7	S16 43.0
Spica	158 39.7	S11 10.7
Suhail	222 58.6	S43 26.5
Vega	80 44.0	N38 47.4
Zuben'ubi	137 14.2	S16 03.4

	SHA ° ′	Mer. Pass. h m
Venus	195 05.3	12 17
Mars	21 45.5	23 45
Jupiter	206 26.7	11 30
Saturn	258 25.4	8 02

UT d	UT h		SUN GHA ° ′	SUN Dec ° ′	MOON GHA ° ′	MOON v ′	MOON Dec ° ′	MOON d ′	MOON HP ′
1	00		179 56.0	N 8 30.5	126 41.7	10.0	S11 35.1	14.7	59.6
	01		194 56.2	29.6	141 10.7	10.0	11 49.8	14.7	59.6
	02		209 56.4	28.7	155 39.7	10.0	12 04.5	14.6	59.6
	03		224 56.6	. . 27.8	170 08.7	9.9	12 19.1	14.5	59.6
	04		239 56.8	26.9	184 37.6	9.8	12 33.6	14.4	59.6
	05		254 57.0	26.0	199 06.4	9.8	12 48.0	14.4	59.6
	06		269 57.2	N 8 25.1	213 35.2	9.7	S13 02.4	14.3	59.6
	07		284 57.4	24.2	228 03.9	9.6	13 16.7	14.2	59.6
	08		299 57.6	23.3	242 32.5	9.6	13 30.9	14.1	59.6
	09	M	314 57.8	. . 22.4	257 01.1	9.5	13 45.0	14.1	59.6
	10	O	329 58.0	21.5	271 29.6	9.5	13 59.1	13.9	59.6
	11	N	344 58.2	20.6	285 58.1	9.4	14 13.0	13.9	59.6
	12	D	359 58.4	N 8 19.7	300 26.5	9.3	S14 26.9	13.8	59.6
	13	A	14 58.5	18.8	314 54.8	9.3	14 40.7	13.8	59.6
	14	Y	29 58.7	17.9	329 23.1	9.2	14 54.5	13.6	59.6
	15		44 58.9	. . 17.0	343 51.3	9.1	15 08.1	13.5	59.6
	16		59 59.1	16.1	358 19.4	9.1	15 21.6	13.5	59.5
	17		74 59.3	15.1	12 47.5	9.0	15 35.1	13.4	59.5
	18		89 59.5	N 8 14.2	27 15.5	8.9	S15 48.5	13.2	59.5
	19		104 59.7	13.3	41 43.4	8.9	16 01.7	13.2	59.5
	20		119 59.9	12.4	56 11.3	8.7	16 14.9	13.1	59.5
	21		135 00.1	. . 11.5	70 39.0	8.8	16 28.0	13.0	59.5
	22		150 00.3	10.6	85 06.8	8.6	16 41.0	12.8	59.5
	23		165 00.5	09.7	99 34.4	8.6	16 53.8	12.8	59.5
2	00		180 00.7	N 8 08.8	114 02.0	8.5	S17 06.6	12.7	59.5
	01		195 00.9	07.9	128 29.5	8.5	17 19.3	12.6	59.5
	02		210 01.1	07.0	142 57.0	8.3	17 31.9	12.4	59.5
	03		225 01.3	. . 06.1	157 24.3	8.3	17 44.3	12.4	59.5
	04		240 01.5	05.2	171 51.6	8.3	17 56.7	12.2	59.5
	05		255 01.7	04.2	186 18.9	8.1	18 08.9	12.2	59.5
	06		270 01.9	N 8 03.3	200 46.0	8.1	S18 21.1	12.0	59.5
	07		285 02.1	02.4	215 13.1	8.0	18 33.1	11.9	59.5
	08	T	300 02.3	01.5	229 40.1	7.9	18 45.0	11.8	59.4
	09	U	315 02.5	8 00.6	244 07.0	7.9	18 56.8	11.7	59.4
	10	E	330 02.7	7 59.7	258 33.9	7.8	19 08.5	11.6	59.4
	11	S	345 02.9	58.8	273 00.7	7.7	19 20.1	11.4	59.4
	12	D	0 03.1	N 7 57.9	287 27.4	7.7	S19 31.5	11.3	59.4
	13	A	15 03.3	57.0	301 54.1	7.5	19 42.8	11.2	59.4
	14	Y	30 03.5	56.0	316 20.6	7.5	19 54.0	11.1	59.4
	15		45 03.7	. . 55.1	330 47.1	7.5	20 05.1	11.0	59.4
	16		60 03.9	54.2	345 13.6	7.3	20 16.1	10.8	59.4
	17		75 04.1	53.3	359 39.9	7.3	20 26.9	10.7	59.4
	18		90 04.3	N 7 52.4	14 06.2	7.2	S20 37.6	10.5	59.4
	19		105 04.5	51.5	28 32.4	7.2	20 48.1	10.5	59.4
	20		120 04.7	50.6	42 58.6	7.0	20 58.6	10.3	59.3
	21		135 04.9	. . 49.7	57 24.6	7.0	21 08.9	10.2	59.3
	22		150 05.1	48.7	71 50.6	7.0	21 19.1	10.0	59.3
	23		165 05.3	47.8	86 16.6	6.8	21 29.1	9.9	59.3
3	00		180 05.5	N 7 46.9	100 42.4	6.8	S21 39.0	9.8	59.3
	01		195 05.7	46.0	115 08.2	6.7	21 48.8	9.6	59.3
	02		210 05.9	45.1	129 33.9	6.7	21 58.4	9.5	59.3
	03		225 06.1	. . 44.2	143 59.6	6.5	22 07.9	9.3	59.3
	04		240 06.3	43.3	158 25.1	6.5	22 17.2	9.2	59.3
	05		255 06.5	42.3	172 50.6	6.5	22 26.4	9.1	59.3
	06		270 06.7	N 7 41.4	187 16.1	6.4	S22 35.5	8.9	59.2
	07	W	285 06.9	40.5	201 41.5	6.3	22 44.4	8.8	59.2
	08	E	300 07.1	39.6	216 06.8	6.2	22 53.2	8.6	59.2
	09	D	315 07.4	. . 38.7	230 32.0	6.2	23 01.8	8.5	59.2
	10	N	330 07.6	37.8	244 57.2	6.1	23 10.3	8.3	59.2
	11	E	345 07.8	36.9	259 22.3	6.1	23 18.6	8.2	59.2
	12	S	0 08.0	N 7 35.9	273 47.4	5.9	S23 26.8	8.0	59.2
	13	D	15 08.2	35.0	288 12.3	6.0	23 34.8	7.8	59.2
	14	A	30 08.4	34.1	302 37.3	5.9	23 42.6	7.8	59.2
	15	Y	45 08.6	. . 33.2	317 02.2	5.8	23 50.4	7.5	59.1
	16		60 08.8	32.3	331 27.0	5.7	23 57.9	7.4	59.1
	17		75 09.0	31.3	345 51.7	5.7	24 05.3	7.3	59.1
	18		90 09.2	N 7 30.4	0 16.4	5.7	S24 12.6	7.0	59.1
	19		105 09.4	29.5	14 41.1	5.5	24 19.6	7.0	59.1
	20		120 09.6	28.6	29 05.6	5.6	24 26.6	6.7	59.1
	21		135 09.8	. . 27.7	43 30.2	5.5	24 33.3	6.6	59.1
	22		150 10.0	26.8	57 54.7	5.4	24 39.9	6.5	59.1
	23		165 10.2	25.8	72 19.1	5.4	S24 46.4	6.2	59.0
			SD 15.9	d 0.9	SD 16.2		16.2		16.1

Lat.	Twilight Naut.	Twilight Civil	Sunrise	Moonrise 1	Moonrise 2	Moonrise 3	Moonrise 4
°	h m	h m	h m	h m	h m	h m	h m
N 72	////	02 35	04 04	13 44	■	■	■
N 70	////	03 02	04 17	13 08	■	■	■
68	01 36	03 23	04 28	12 43	15 23	■	■
66	02 13	03 38	04 37	12 24	14 39	■	■
64	02 38	03 51	04 44	12 08	14 09	16 23	■
62	02 57	04 02	04 51	11 55	13 47	15 42	17 31
60	03 13	04 11	04 56	11 45	13 30	15 15	16 50
N 58	03 26	04 19	05 01	11 35	13 15	14 53	16 22
56	03 36	04 26	05 05	11 27	13 02	14 36	16 00
54	03 46	04 32	05 09	11 20	12 51	14 21	15 42
52	03 54	04 37	05 13	11 14	12 42	14 08	15 27
50	04 01	04 42	05 16	11 08	12 33	13 57	15 14
45	04 16	04 52	05 22	10 56	12 15	13 33	14 47
N 40	04 27	05 00	05 28	10 45	12 00	13 15	14 26
35	04 37	05 07	05 33	10 37	11 48	12 59	14 08
30	04 44	05 13	05 37	10 29	11 37	12 46	13 53
20	04 56	05 22	05 44	10 16	11 19	12 23	13 27
N 10	05 05	05 30	05 51	10 05	11 03	12 03	13 05
0	05 11	05 36	05 57	09 54	10 48	11 45	12 44
S 10	05 16	05 41	06 02	09 44	10 34	11 27	12 24
20	05 20	05 46	06 08	09 33	10 18	11 07	12 02
30	05 23	05 51	06 15	09 21	10 00	10 45	11 37
35	05 24	05 53	06 18	09 14	09 50	10 32	11 22
40	05 24	05 56	06 23	09 05	09 38	10 17	11 05
45	05 24	05 58	06 28	08 56	09 25	10 00	10 44
S 50	05 24	06 01	06 33	08 45	09 08	09 38	10 18
52	05 23	06 02	06 36	08 40	09 00	09 28	10 06
54	05 23	06 03	06 39	08 34	08 52	09 16	09 52
56	05 22	06 05	06 42	08 28	08 42	09 03	09 35
58	05 21	06 06	06 46	08 21	08 31	08 48	09 15
S 60	05 20	06 08	06 50	08 13	08 19	08 30	08 51

Lat.	Sunset	Twilight Civil	Twilight Naut.	Moonset 1	Moonset 2	Moonset 3	Moonset 4
°	h m	h m	h m	h m	h m	h m	h m
N 72	19 52	21 19	////	18 04	■	■	■
N 70	19 39	20 53	23 26	18 42	■	■	■
68	19 29	20 34	22 16	19 09	18 23	■	■
66	19 21	20 18	21 42	19 30	19 08	■	■
64	19 13	20 06	21 18	19 47	19 39	19 25	■
62	19 07	19 55	20 59	20 01	20 02	20 06	20 22
60	19 02	19 47	20 44	20 13	20 20	20 34	21 03
N 58	18 57	19 39	20 32	20 23	20 36	20 56	21 32
56	18 53	19 32	20 21	20 32	20 49	21 14	21 54
54	18 49	19 26	20 12	20 40	21 00	21 29	22 12
52	18 46	19 21	20 04	20 47	21 11	21 43	22 27
50	18 43	19 16	19 57	20 54	21 20	21 54	22 41
45	18 36	19 06	19 43	21 08	21 39	22 19	23 08
N 40	18 31	18 58	19 31	21 20	21 55	22 38	23 30
35	18 26	18 52	19 22	21 30	22 09	22 54	23 48
30	18 22	18 46	19 15	21 38	22 20	23 08	24 03
20	18 15	18 37	19 03	21 54	22 41	23 33	24 29
N 10	18 09	18 30	18 55	22 07	22 58	23 53	24 52
0	18 03	18 24	18 48	22 20	23 15	24 13	00 13
S 10	17 58	18 19	18 43	22 32	23 31	24 33	00 33
20	17 52	18 14	18 40	22 46	23 49	24 54	00 54
30	17 45	18 09	18 37	23 01	24 10	00 10	01 18
35	17 42	18 07	18 36	23 11	24 22	00 22	01 32
40	17 37	18 05	18 36	23 21	24 36	00 36	01 49
45	17 33	18 02	18 36	23 33	24 52	00 52	02 09
S 50	17 27	18 00	18 37	23 48	25 13	01 13	02 34
52	17 24	17 58	18 37	23 55	25 23	01 23	02 46
54	17 22	17 57	18 38	24 03	00 03	01 34	03 00
56	17 18	17 56	18 39	24 12	00 12	01 46	03 16
58	17 15	17 55	18 40	24 22	00 22	02 01	03 36
S 60	17 11	17 53	18 41	24 33	00 33	02 19	04 00

Day	SUN Eqn. of Time 00^h	SUN Eqn. of Time 12^h	SUN Mer. Pass.	MOON Mer. Pass. Upper	MOON Mer. Pass. Lower	MOON Age	MOON Phase
d	m s	m s	h m	h m	h m	d	%
1	00 16	00 07	12 00	16 07	03 41	05	28
2	00 02	00 12	12 00	17 01	04 34	06	38
3	00 22	00 31	11 59	17 59	05 30	07	50

	UT	ARIES	VENUS −3.9		MARS −2.8		JUPITER −1.7		SATURN +0.1	
		GHA	GHA	Dec	GHA	Dec	GHA	Dec	GHA	Dec
	d h	° ′	° ′	° ′	° ′	° ′	° ′	° ′	° ′	° ′
	4 00	342 42.2	175 29.1	N 7 00.8	4 58.8	S16 13.6	188 44.0	N11 40.0	240 56.4	N22 15.1
	01	357 44.7	190 28.6	6 59.6	20 01.9	13.8	203 45.9	39.8	255 58.6	15.1
	02	12 47.1	205 28.2	58.4	35 05.1	13.9	218 47.9	39.7	271 00.9	15.1
	03	27 49.6	220 27.8	. . 57.2	50 08.2	. . 14.0	233 49.8	. . 39.5	286 03.1	. . 15.1
	04	42 52.0	235 27.4	55.9	65 11.3	14.2	248 51.8	39.3	301 05.3	15.1
	05	57 54.5	250 27.0	54.7	80 14.4	14.3	263 53.7	39.1	316 07.6	15.1
	06	72 57.0	265 26.6	N 6 53.5	95 17.5	S16 14.4	278 55.7	N11 38.9	331 09.8	N22 15.0
	07	87 59.4	280 26.2	52.3	110 20.6	14.5	293 57.6	38.7	346 12.1	15.0
T	08	103 01.9	295 25.8	51.1	125 23.7	14.7	308 59.6	38.5	1 14.3	15.0
H	09	118 04.4	310 25.4	. . 49.9	140 26.8	. . 14.8	324 01.5	. . 38.3	16 16.5	. . 15.0
U	10	133 06.8	325 24.9	48.7	155 29.9	14.9	339 03.5	38.1	31 18.8	15.0
R	11	148 09.3	340 24.5	47.5	170 33.0	15.0	354 05.4	37.9	46 21.0	15.0
S	12	163 11.8	355 24.1	N 6 46.2	185 36.1	S16 15.1	9 07.4	N11 37.7	61 23.2	N22 15.0
D	13	178 14.2	10 23.7	45.0	200 39.2	15.3	24 09.3	37.6	76 25.5	14.9
A	14	193 16.7	25 23.3	43.8	215 42.3	15.4	39 11.3	37.4	91 27.7	14.9
Y	15	208 19.2	40 22.9	. . 42.6	230 45.4	. . 15.5	54 13.2	. . 37.2	106 29.9	. . 14.9
	16	223 21.6	55 22.5	41.4	245 48.5	15.6	69 15.1	37.0	121 32.2	14.9
	17	238 24.1	70 22.1	40.2	260 51.6	15.8	84 17.1	36.8	136 34.4	14.9
	18	253 26.5	85 21.7	N 6 39.0	275 54.6	S16 15.9	99 19.0	N11 36.6	151 36.7	N22 14.9
	19	268 29.0	100 21.3	37.7	290 57.7	16.0	114 21.0	36.4	166 38.9	14.8
	20	283 31.5	115 20.9	36.5	306 00.8	16.1	129 22.9	36.2	181 41.1	14.8
	21	298 33.9	130 20.4	. . 35.3	321 03.9	. . 16.2	144 24.9	. . 36.0	196 43.4	. . 14.8
	22	313 36.4	145 20.0	34.1	336 07.0	16.4	159 26.8	35.8	211 45.6	14.8
	23	328 38.9	160 19.6	32.9	351 10.1	16.5	174 28.8	35.7	226 47.8	14.8
	5 00	343 41.3	175 19.2	N 6 31.7	6 13.2	S16 16.6	189 30.7	N11 35.5	241 50.1	N22 14.8
	01	358 43.8	190 18.8	30.4	21 16.3	16.7	204 32.7	35.3	256 52.3	14.7
	02	13 46.3	205 18.4	29.2	36 19.4	16.8	219 34.6	35.1	271 54.6	14.7
	03	28 48.7	220 18.0	. . 28.0	51 22.5	. . 16.9	234 36.6	. . 34.9	286 56.8	. . 14.7
	04	43 51.2	235 17.6	26.8	66 25.6	17.0	249 38.5	34.7	301 59.0	14.7
	05	58 53.7	250 17.2	25.6	81 28.7	17.2	264 40.5	34.5	317 01.3	14.7
	06	73 56.1	265 16.8	N 6 24.4	96 31.8	S16 17.3	279 42.4	N11 34.3	332 03.5	N22 14.7
	07	88 58.6	280 16.4	23.1	111 34.9	17.4	294 44.4	34.1	347 05.7	14.6
	08	104 01.0	295 16.0	21.9	126 37.9	17.5	309 46.3	33.9	2 08.0	14.6
F	09	119 03.5	310 15.6	. . 20.7	141 41.0	. . 17.6	324 48.3	. . 33.7	17 10.2	. . 14.6
R	10	134 06.0	325 15.2	19.5	156 44.1	17.7	339 50.2	33.6	32 12.5	14.6
I	11	149 08.4	340 14.8	18.3	171 47.2	17.8	354 52.2	33.4	47 14.7	14.6
D	12	164 10.9	355 14.4	N 6 17.0	186 50.3	S16 17.9	9 54.1	N11 33.2	62 16.9	N22 14.6
A	13	179 13.4	10 13.9	15.8	201 53.4	18.1	24 56.0	33.0	77 19.2	14.5
Y	14	194 15.8	25 13.5	14.6	216 56.5	18.2	39 58.0	32.8	92 21.4	14.5
	15	209 18.3	40 13.1	. . 13.4	231 59.6	. . 18.3	54 59.9	. . 32.6	107 23.7	. . 14.5
	16	224 20.8	55 12.7	12.2	247 02.6	18.4	70 01.9	32.4	122 25.9	14.5
	17	239 23.2	70 12.3	10.9	262 05.7	18.5	85 03.8	32.2	137 28.1	14.5
	18	254 25.7	85 11.9	N 6 09.7	277 08.8	S16 18.6	100 05.8	N11 32.0	152 30.4	N22 14.5
	19	269 28.2	100 11.5	08.5	292 11.9	18.7	115 07.7	31.8	167 32.6	14.4
	20	284 30.6	115 11.1	07.3	307 15.0	18.8	130 09.7	31.7	182 34.9	14.4
	21	299 33.1	130 10.7	. . 06.0	322 18.1	. . 18.9	145 11.6	. . 31.5	197 37.1	. . 14.4
	22	314 35.5	145 10.3	04.8	337 21.2	19.0	160 13.6	31.3	212 39.3	14.4
	23	329 38.0	160 09.9	03.6	352 24.2	19.1	175 15.5	31.1	227 41.6	14.4
	6 00	344 40.5	175 09.5	N 6 02.4	7 27.3	S16 19.2	190 17.5	N11 30.9	242 43.8	N22 14.4
	01	359 42.9	190 09.1	6 01.1	22 30.4	19.3	205 19.4	30.7	257 46.1	14.3
	02	14 45.4	205 08.7	5 59.9	37 33.5	19.4	220 21.4	30.5	272 48.3	14.3
	03	29 47.9	220 08.3	. . 58.7	52 36.6	. . 19.5	235 23.3	. . 30.3	287 50.5	. . 14.3
	04	44 50.3	235 07.9	57.5	67 39.6	19.6	250 25.3	30.1	302 52.8	14.3
	05	59 52.8	250 07.5	56.2	82 42.7	19.8	265 27.2	29.9	317 55.0	14.3
	06	74 55.3	265 07.1	N 5 55.0	97 45.8	S16 19.9	280 29.2	N11 29.7	332 57.3	N22 14.3
	07	89 57.7	280 06.7	53.8	112 48.9	20.0	295 31.1	29.6	347 59.5	14.3
S	08	105 00.2	295 06.3	52.6	127 51.9	20.1	310 33.1	29.4	3 01.8	14.2
A	09	120 02.7	310 05.9	. . 51.3	142 55.0	. . 20.2	325 35.0	. . 29.2	18 04.0	. . 14.2
T	10	135 05.1	325 05.5	50.1	157 58.1	20.3	340 37.0	29.0	33 06.2	14.2
U	11	150 07.6	340 05.1	48.9	173 01.2	20.3	355 38.9	28.8	48 08.5	14.2
R	12	165 10.0	355 04.7	N 5 47.7	188 04.2	S16 20.4	10 40.9	N11 28.6	63 10.7	N22 14.2
D	13	180 12.5	10 04.3	46.4	203 07.3	20.5	25 42.8	28.4	78 13.0	14.2
A	14	195 15.0	25 03.9	45.2	218 10.4	20.6	40 44.8	28.2	93 15.2	14.1
Y	15	210 17.4	40 03.5	. . 44.0	233 13.5	. . 20.7	55 46.7	. . 28.0	108 17.5	. . 14.1
	16	225 19.9	55 03.1	42.7	248 16.5	20.8	70 48.7	27.8	123 19.7	14.1
	17	240 22.4	70 02.7	41.5	263 19.6	20.9	85 50.6	27.7	138 21.9	14.1
	18	255 24.8	85 02.3	N 5 40.3	278 22.7	S16 21.0	100 52.6	N11 27.5	153 24.2	N22 14.1
	19	270 27.3	100 01.9	39.0	293 25.7	21.1	115 54.5	27.3	168 26.4	14.1
	20	285 29.8	115 01.5	37.8	308 28.8	21.2	130 56.5	27.1	183 28.7	14.0
	21	300 32.2	130 01.1	. . 36.6	323 31.9	. . 21.3	145 58.4	. . 26.9	198 30.9	. . 14.0
	22	315 34.7	145 00.7	35.4	338 35.0	21.4	161 00.4	26.7	213 33.2	14.0
	23	330 37.1	160 00.3	34.1	353 38.0	21.5	176 02.3	26.5	228 35.4	14.0
	Mer. Pass.	h m 1 05.1	v −0.4	d 1.2	v 3.1	d 0.1	v 1.9	d 0.2	v 2.2	d 0.0

STARS		
Name	SHA	Dec
	° ′	° ′
Acamar	315 23.9	S40 17.1
Achernar	335 31.8	S57 12.9
Acrux	173 18.9	S63 07.2
Adhara	255 18.8	S28 58.3
Aldebaran	290 58.2	N16 31.1
Alioth	166 27.7	N55 56.7
Alkaid	153 05.1	N49 18.0
Al Na'ir	27 52.7	S46 56.7
Alnilam	275 54.3	S 1 11.8
Alphard	218 04.0	S 8 40.2
Alphecca	126 17.6	N26 42.4
Alpheratz	357 51.2	N29 06.6
Altair	62 15.5	N 8 52.7
Ankaa	353 22.7	S42 17.1
Antares	112 35.8	S26 26.5
Arcturus	146 02.9	N19 10.0
Atria	107 44.6	S69 02.4
Avior	234 21.8	S59 30.9
Bellatrix	278 40.3	N 6 21.3
Betelgeuse	271 09.7	N 7 24.6
Canopus	263 59.8	S52 41.5
Capella	280 45.9	N46 00.0
Deneb	49 36.5	N45 17.7
Denebola	182 41.8	N14 33.3
Diphda	349 03.3	S17 57.9
Dubhe	194 01.4	N61 44.0
Elnath	278 22.4	N28 36.7
Eltanin	90 49.6	N51 29.5
Enif	33 54.4	N 9 53.5
Fomalhaut	15 32.0	S29 36.2
Gacrux	172 10.3	S57 08.0
Gienah	176 00.6	S17 33.6
Hadar	148 59.6	S60 23.6
Hamal	328 09.3	N23 28.8
Kaus Aust.	83 53.9	S34 23.1
Kochab	137 19.8	N74 08.7
Markab	13 45.7	N15 13.5
Menkar	314 23.0	N 4 06.4
Menkent	148 17.0	S36 23.3
Miaplacidus	221 42.5	S69 43.7
Mirfak	308 51.4	N49 52.3
Nunki	76 07.6	S26 17.6
Peacock	53 30.7	S56 43.6
Pollux	243 37.4	N28 01.2
Procyon	245 08.0	N 5 13.2
Rasalhague	96 13.5	N12 33.6
Regulus	207 52.0	N11 57.2
Rigel	281 19.5	S 8 11.6
Rigil Kent.	140 03.0	S60 51.2
Sabik	102 21.4	S15 43.8
Schedar	349 49.1	N56 33.3
Shaula	96 32.3	S37 06.5
Sirius	258 40.7	S16 43.0
Spica	158 39.7	S11 10.7
Suhail	222 58.6	S43 26.5
Vega	80 44.0	N38 47.4
Zuben'ubi	137 14.2	S16 03.4
	SHA	Mer. Pass
	° ′	h m
Venus	191 37.9	12 19
Mars	22 31.9	23 30
Jupiter	205 49.4	11 20
Saturn	258 08.7	7 51

Day	UT (d h)	SUN GHA (° ′)	SUN Dec (° ′)	MOON GHA (° ′)	v (′)	MOON Dec (° ′)	d (′)	HP (′)
THURSDAY	4 00	180 10.4	N 7 24.9	86 43.5	5.3	S24 52.6	6.2	59.0
	01	195 10.6	24.0	101 07.8	5.3	24 58.8	5.9	59.0
	02	210 10.8	23.1	115 32.1	5.3	25 04.7	5.8	59.0
	03	225 11.0	. . 22.2	129 56.4	5.2	25 10.5	5.6	59.0
	04	240 11.2	21.2	144 20.6	5.1	25 16.1	5.5	59.0
	05	255 11.4	20.3	158 44.7	5.1	25 21.6	5.2	59.0
	06	270 11.6	N 7 19.4	173 08.8	5.1	S25 26.8	5.2	58.9
	07	285 11.8	18.5	187 32.9	5.1	25 32.0	4.9	58.9
	08	300 12.0	17.6	201 57.0	5.0	25 36.9	4.8	58.9
	09	315 12.3	. . 16.6	216 21.0	5.0	25 41.7	4.6	58.9
	10	330 12.5	15.7	230 45.0	4.9	25 46.3	4.4	58.9
	11	345 12.7	14.8	245 08.9	4.9	25 50.7	4.3	58.9
	12	0 12.9	N 7 13.9	259 32.8	4.9	S25 55.0	4.1	58.9
	13	15 13.1	13.0	273 56.7	4.9	25 59.1	3.9	58.9
	14	30 13.3	12.0	288 20.6	4.8	26 03.0	3.7	58.8
	15	45 13.5	. . 11.1	302 44.4	4.8	26 06.7	3.6	58.8
	16	60 13.7	10.2	317 08.2	4.8	26 10.3	3.4	58.8
	17	75 13.9	09.3	331 32.0	4.8	26 13.7	3.2	58.8
	18	90 14.1	N 7 08.3	345 55.8	4.7	S26 16.9	3.1	58.8
	19	105 14.3	07.4	0 19.5	4.8	26 20.0	2.9	58.8
	20	120 14.5	06.5	14 43.3	4.7	26 22.9	2.7	58.7
	21	135 14.7	. . 05.6	29 07.0	4.7	26 25.6	2.5	58.7
	22	150 14.9	04.7	43 30.7	4.7	26 28.1	2.3	58.7
	23	165 15.1	03.7	57 54.4	4.7	26 30.4	2.2	58.7
FRIDAY	5 00	180 15.4	N 7 02.8	72 18.1	4.7	S26 32.6	2.0	58.7
	01	195 15.6	01.9	86 41.8	4.7	26 34.6	1.8	58.7
	02	210 15.8	01.0	101 05.5	4.7	26 36.4	1.7	58.7
	03	225 16.0	7 00.0	115 29.2	4.7	26 38.1	1.4	58.6
	04	240 16.2	6 59.1	129 52.9	4.7	26 39.5	1.3	58.6
	05	255 16.4	58.2	144 16.6	4.7	26 40.8	1.1	58.6
	06	270 16.6	N 6 57.3	158 40.3	4.7	S26 41.9	1.0	58.6
	07	285 16.8	56.3	173 04.0	4.7	26 42.9	0.8	58.6
	08	300 17.0	55.4	187 27.7	4.7	26 43.7	0.5	58.6
	09	315 17.2	. . 54.5	201 51.4	4.8	26 44.2	0.5	58.6
	10	330 17.4	53.6	216 15.2	4.7	26 44.7	0.2	58.5
	11	345 17.6	52.6	230 38.9	4.8	26 44.9	0.1	58.5
	12	0 17.9	N 6 51.7	245 02.7	4.8	S26 45.0	0.2	58.5
	13	15 18.1	50.8	259 26.5	4.8	26 44.8	0.2	58.5
	14	30 18.3	49.9	273 50.3	4.8	26 44.6	0.5	58.5
	15	45 18.5	. . 48.9	288 14.1	4.9	26 44.1	0.6	58.5
	16	60 18.7	48.0	302 38.0	4.9	26 43.5	0.9	58.4
	17	75 18.9	47.1	317 01.9	4.9	26 42.6	0.9	58.4
	18	90 19.1	N 6 46.1	331 25.8	4.9	S26 41.7	1.2	58.4
	19	105 19.3	45.2	345 49.7	5.0	26 40.5	1.3	58.4
	20	120 19.5	44.3	0 13.7	5.0	26 39.2	1.5	58.4
	21	135 19.7	. . 43.4	14 37.7	5.1	26 37.7	1.7	58.4
	22	150 19.9	42.4	29 01.8	5.1	26 36.0	1.8	58.3
	23	165 20.2	41.5	43 25.9	5.1	26 34.2	2.1	58.3
SATURDAY	6 00	180 20.4	N 6 40.6	57 50.0	5.2	S26 32.1	2.1	58.3
	01	195 20.6	39.7	72 14.2	5.2	26 30.0	2.4	58.3
	02	210 20.8	38.7	86 38.4	5.3	26 27.6	2.5	58.3
	03	225 21.0	. . 37.8	101 02.7	5.3	26 25.1	2.7	58.3
	04	240 21.2	36.9	115 27.0	5.4	26 22.4	2.9	58.2
	05	255 21.4	35.9	129 51.4	5.4	26 19.5	3.0	58.2
	06	270 21.6	N 6 35.0	144 15.8	5.5	S26 16.5	3.2	58.2
	07	285 21.8	34.1	158 40.3	5.5	26 13.3	3.3	58.2
	08	300 22.1	33.1	173 04.8	5.6	26 10.0	3.5	58.2
	09	315 22.3	. . 32.2	187 29.4	5.7	26 06.5	3.7	58.1
	10	330 22.5	31.3	201 54.1	5.7	26 02.8	3.9	58.1
	11	345 22.7	30.4	216 18.8	5.7	25 58.9	4.0	58.1
	12	0 22.9	N 6 29.4	230 43.5	5.9	S25 54.9	4.1	58.1
	13	15 23.1	28.5	245 08.4	5.9	25 50.8	4.4	58.1
	14	30 23.3	27.6	259 33.3	6.0	25 46.4	4.4	58.1
	15	45 23.5	. . 26.6	273 58.3	6.0	25 42.0	4.7	58.0
	16	60 23.7	25.7	288 23.3	6.1	25 37.3	4.8	58.0
	17	75 24.0	24.8	302 48.4	6.2	25 32.5	4.9	58.0
	18	90 24.2	N 6 23.8	317 13.6	6.2	S25 27.6	5.1	58.0
	19	105 24.4	22.9	331 38.8	6.3	25 22.5	5.3	58.0
	20	120 24.6	22.0	346 04.1	6.4	25 17.2	5.4	58.0
	21	135 24.8	. . 21.0	0 29.5	6.5	25 11.8	5.5	57.9
	22	150 25.0	20.1	14 55.0	6.6	25 06.3	5.8	57.9
	23	165 25.2	19.2	29 20.6	6.6	S25 00.5	5.8	57.9
		SD 15.9	d 0.9	SD 16.0		15.9		15.8

Lat. (°)	Twilight Naut. (h m)	Twilight Civil (h m)	Sunrise (h m)	Moonrise 4 (h m)	Moonrise 5 (h m)	Moonrise 6 (h m)	Moonrise 7 (h m)
N 72	////	02 55	04 19	■	■	■	■
N 70	01 11	03 18	04 30	■	■	■	■
68	02 00	03 36	04 39	■	■	■	■
66	02 30	03 50	04 47	■	■	■	20 26
64	02 52	04 01	04 53	■	■	20 04	19 44
62	03 08	04 11	04 58	17 31	18 44	19 09	19 16
60	03 22	04 19	05 03	16 50	17 59	18 36	18 54
N 58	03 34	04 26	05 07	16 22	17 30	18 12	18 36
56	03 44	04 32	05 11	16 00	17 07	17 52	18 21
54	03 52	04 38	05 14	15 42	16 48	17 36	18 08
52	04 00	04 42	05 17	15 27	16 33	17 21	17 56
50	04 06	04 47	05 20	15 14	16 19	17 09	17 46
45	04 20	04 56	05 26	14 47	15 51	16 43	17 24
N 40	04 31	05 03	05 31	14 26	15 29	16 23	17 07
35	04 39	05 09	05 35	14 08	15 11	16 06	16 52
30	04 46	05 15	05 39	13 53	14 55	15 51	16 40
20	04 57	05 23	05 45	13 27	14 29	15 26	16 18
N 10	05 05	05 29	05 51	13 05	14 06	15 04	15 59
0	05 11	05 35	05 56	12 44	13 45	14 44	15 41
S 10	05 15	05 39	06 00	12 24	13 23	14 24	15 23
20	05 18	05 43	06 05	12 02	13 01	14 02	15 04
30	05 19	05 47	06 11	11 37	12 34	13 37	14 41
35	05 20	05 49	06 14	11 22	12 19	13 22	14 28
40	05 20	05 51	06 18	11 05	12 01	13 05	14 13
45	05 19	05 53	06 22	10 44	11 39	12 44	13 55
S 50	05 17	05 55	06 27	10 18	11 11	12 18	13 33
52	05 17	05 55	06 29	10 06	10 58	12 05	13 22
54	05 16	05 56	06 32	09 52	10 43	11 50	13 09
56	05 14	05 57	06 34	09 35	10 24	11 33	12 55
58	05 13	05 58	06 37	09 15	10 02	11 12	12 39
S 60	05 11	05 59	06 41	08 51	09 33	10 46	12 19

Lat. (°)	Sunset (h m)	Twilight Civil (h m)	Twilight Naut. (h m)	Moonset 4 (h m)	Moonset 5 (h m)	Moonset 6 (h m)	Moonset 7 (h m)
N 72	19 36	20 58	////	■	■	■	■
N 70	19 25	20 35	22 35	■	■	■	■
68	19 16	20 19	21 52	■	■	■	■
66	19 09	20 05	21 23	■	■	■	23 35
64	19 03	19 54	21 03	■	■	22 00	24 15
62	18 58	19 45	20 46	20 22	21 16	22 54	24 43
60	18 53	19 37	20 33	21 03	22 01	23 27	25 04
N 58	18 49	19 30	20 22	21 32	22 30	23 51	25 22
56	18 45	19 24	20 12	21 54	22 53	24 10	00 10
54	18 42	19 19	20 04	22 12	23 11	24 26	00 26
52	18 39	19 14	19 56	22 27	23 27	24 40	00 40
50	18 36	19 10	19 50	22 41	23 41	24 52	00 52
45	18 31	19 01	19 37	23 08	24 08	00 08	01 17
N 40	18 26	18 54	19 26	23 30	24 30	00 30	01 37
35	18 22	18 48	19 18	23 48	24 48	00 48	01 53
30	18 18	18 43	19 11	24 03	00 03	01 04	02 07
20	18 12	18 34	19 00	24 29	00 29	01 30	02 32
N 10	18 07	18 28	18 53	24 52	00 52	01 52	02 52
0	18 02	18 23	18 47	00 13	01 13	02 13	03 12
S 10	17 57	18 18	18 43	00 33	01 34	02 34	03 31
20	17 52	18 15	18 40	00 54	01 57	02 57	03 51
30	17 47	18 11	18 39	01 18	02 23	03 23	04 15
35	17 44	18 09	18 38	01 32	02 39	03 38	04 29
40	17 40	18 07	18 39	01 49	02 57	03 55	04 44
45	17 36	18 06	18 40	02 09	03 18	04 17	05 03
S 50	17 31	18 04	18 41	02 34	03 46	04 43	05 27
52	17 29	18 03	18 42	02 46	03 59	04 56	05 38
54	17 27	18 02	18 43	03 00	04 14	05 11	05 51
56	17 24	18 02	18 45	03 16	04 33	05 29	06 05
58	17 21	18 01	18 46	03 36	04 55	05 50	06 22
S 60	17 18	18 00	18 48	04 00	05 23	06 16	06 43

Day	SUN Eqn. of Time 00^h	SUN Eqn. of Time 12^h	SUN Mer. Pass.	MOON Mer. Pass. Upper	MOON Mer. Pass. Lower	MOON Age	MOON Phase
d	m s	m s	h m	h m	h m	d %	
4	00 41	00 51	11 59	18 59	06 29	08 61	
5	01 01	01 11	11 59	19 59	07 29	09 72	
6	01 21	01 31	11 58	20 58	08 29	10 81	

	UT	ARIES	VENUS −3.9		MARS −2.7		JUPITER −1.7		SATURN +0.1		STARS		
		GHA	GHA	Dec	GHA	Dec	GHA	Dec	GHA	Dec	Name	SHA	Dec
	d h	° ′	° ′	° ′	° ′	° ′	° ′	° ′	° ′	° ′		° ′	° ′
	7 00	345 39.6	174 59.9	N 5 32.9	8 41.1	S16 21.6	191 04.3	N11 26.3	243 37.7	N22 14.0	Acamar	315 23.9	S40 17.1
	01	0 42.1	189 59.5	31.7	23 44.2	21.7	206 06.2	26.1	258 39.9	14.0	Achernar	335 31.8	S57 12.9
	02	15 44.5	204 59.1	30.4	38 47.2	21.8	221 08.2	25.9	273 42.1	13.9	Acrux	173 18.9	S63 07.1
	03	30 47.0	219 58.8	29.2	53 50.3	21.9	236 10.1	25.7	288 44.4	13.9	Adhara	255 18.8	S28 58.3
	04	45 49.5	234 58.4	28.0	68 53.4	21.9	251 12.1	25.6	303 46.6	13.9	Aldebaran	290 58.2	N16 31.1
	05	60 51.9	249 58.0	26.7	83 56.4	22.0	266 14.0	25.4	318 48.9	13.9			
	06	75 54.4	264 57.6	N 5 25.5	98 59.5	S16 22.1	281 15.9	N11 25.2	333 51.1	N22 13.9	Alioth	166 27.7	N55 56.7
	07	90 56.9	279 57.2	24.3	114 02.5	22.2	296 17.9	25.0	348 53.4	13.9	Alkaid	153 05.1	N49 18.0
	08	105 59.3	294 56.8	23.0	129 05.6	22.3	311 19.8	24.8	3 55.6	13.8	Al Na'ir	27 52.7	S46 56.7
S	09	121 01.8	309 56.4	21.8	144 08.7	22.4	326 21.8	24.6	18 57.9	13.8	Alnilam	275 54.2	S 1 11.8
U	10	136 04.3	324 56.0	20.6	159 11.7	22.5	341 23.7	24.4	34 00.1	13.8	Alphard	218 04.0	S 8 40.2
N	11	151 06.7	339 55.6	19.3	174 14.8	22.6	356 25.7	24.2	49 02.4	13.8			
D	12	166 09.2	354 55.2	N 5 18.1	189 17.8	S16 22.6	11 27.6	N11 24.0	64 04.6	N22 13.8	Alphecca	126 17.6	N26 42.4
A	13	181 11.6	9 54.8	16.9	204 20.9	22.7	26 29.6	23.8	79 06.9	13.8	Alpheratz	357 51.2	N29 06.6
Y	14	196 14.1	24 54.4	15.6	219 24.0	22.8	41 31.5	23.7	94 09.1	13.8	Altair	62 15.5	N 8 52.7
	15	211 16.6	39 54.0	14.4	234 27.0	22.9	56 33.5	23.5	109 11.3	13.7	Ankaa	353 22.7	S42 17.1
	16	226 19.0	54 53.6	13.2	249 30.1	23.0	71 35.4	23.3	124 13.6	13.7	Antares	112 35.8	S26 26.5
	17	241 21.5	69 53.2	11.9	264 33.1	23.1	86 37.4	23.1	139 15.8	13.7			
	18	256 24.0	84 52.8	N 5 10.7	279 36.2	S16 23.1	101 39.3	N11 22.9	154 18.1	N22 13.7	Arcturus	146 02.9	N19 10.0
	19	271 26.4	99 52.4	09.5	294 39.2	23.2	116 41.3	22.7	169 20.3	13.7	Atria	107 44.7	S69 02.4
	20	286 28.9	114 52.0	08.2	309 42.3	23.3	131 43.2	22.5	184 22.6	13.7	Avior	234 21.8	S59 30.9
	21	301 31.4	129 51.7	07.0	324 45.4	23.4	146 45.2	22.3	199 24.8	13.6	Bellatrix	278 40.3	N 6 21.3
	22	316 33.8	144 51.3	05.7	339 48.4	23.5	161 47.2	22.1	214 27.1	13.6	Betelgeuse	271 09.7	N 7 24.6
	23	331 36.3	159 50.9	04.5	354 51.5	23.5	176 49.1	21.9	229 29.3	13.6			
	8 00	346 38.8	174 50.5	N 5 03.3	9 54.5	S16 23.6	191 51.1	N11 21.7	244 31.6	N22 13.6	Canopus	263 59.8	S52 41.4
	01	1 41.2	189 50.1	02.0	24 57.6	23.7	206 53.0	21.6	259 33.8	13.6	Capella	280 45.9	N46 00.0
	02	16 43.7	204 49.7	5 00.8	40 00.6	23.8	221 55.0	21.4	274 36.1	13.6	Deneb	49 36.5	N45 17.7
	03	31 46.1	219 49.3	4 59.6	55 03.7	23.8	236 56.9	21.2	289 38.3	13.5	Denebola	182 41.8	N14 33.3
	04	46 48.6	234 48.9	58.3	70 06.7	23.9	251 58.9	21.0	304 40.6	13.5	Diphda	349 03.2	S17 57.9
	05	61 51.1	249 48.5	57.1	85 09.8	24.0	267 00.8	20.8	319 42.8	13.5			
	06	76 53.5	264 48.1	N 4 55.8	100 12.8	S16 24.1	282 02.8	N11 20.6	334 45.1	N22 13.5	Dubhe	194 01.4	N61 44.0
	07	91 56.0	279 47.7	54.6	115 15.9	24.1	297 04.7	20.4	349 47.3	13.5	Elnath	278 22.4	N28 36.7
	08	106 58.5	294 47.4	53.4	130 18.9	24.2	312 06.7	20.2	4 49.6	13.5	Eltanin	90 49.6	N51 29.5
M	09	122 00.9	309 47.0	52.1	145 21.9	24.3	327 08.6	20.0	19 51.8	13.4	Enif	33 54.4	N 9 53.5
O	10	137 03.4	324 46.6	50.9	160 25.0	24.4	342 10.6	19.8	34 54.1	13.4	Fomalhaut	15 32.0	S29 36.2
N	11	152 05.9	339 46.2	49.6	175 28.0	24.4	357 12.5	19.7	49 56.3	13.4			
D	12	167 08.3	354 45.8	N 4 48.4	190 31.1	S16 24.5	12 14.5	N11 19.5	64 58.6	N22 13.4	Gacrux	172 10.3	S57 08.0
A	13	182 10.8	9 45.4	47.2	205 34.1	24.6	27 16.4	19.3	80 00.8	13.4	Gienah	176 00.6	S17 33.6
Y	14	197 13.2	24 45.0	45.9	220 37.2	24.7	42 18.4	19.1	95 03.1	13.4	Hadar	148 59.6	S60 23.5
	15	212 15.7	39 44.6	44.7	235 40.2	24.7	57 20.3	18.9	110 05.3	13.4	Hamal	328 09.3	N23 28.8
	16	227 18.2	54 44.2	43.4	250 43.2	24.8	72 22.3	18.7	125 07.6	13.3	Kaus Aust.	83 53.9	S34 23.1
	17	242 20.6	69 43.8	42.2	265 46.3	24.9	87 24.2	18.5	140 09.8	13.3			
	18	257 23.1	84 43.5	N 4 41.0	280 49.3	S16 24.9	102 26.2	N11 18.3	155 12.1	N22 13.3	Kochab	137 19.9	N74 08.7
	19	272 25.6	99 43.1	39.7	295 52.4	25.0	117 28.1	18.1	170 14.3	13.3	Markab	13 45.7	N15 13.5
	20	287 28.0	114 42.7	38.5	310 55.4	25.1	132 30.1	17.9	185 16.6	13.3	Menkar	314 23.0	N 4 06.4
	21	302 30.5	129 42.3	37.2	325 58.4	25.1	147 32.0	17.8	200 18.8	13.3	Menkent	148 17.1	S36 23.3
	22	317 33.0	144 41.9	36.0	341 01.5	25.2	162 34.0	17.6	215 21.1	13.2	Miaplacidus	221 42.5	S69 43.6
	23	332 35.4	159 41.5	34.7	356 04.5	25.3	177 35.9	17.4	230 23.3	13.2			
	9 00	347 37.9	174 41.1	N 4 33.5	11 07.5	S16 25.3	192 37.9	N11 17.2	245 25.6	N22 13.2	Mirfak	308 51.3	N49 52.3
	01	2 40.4	189 40.7	32.3	26 10.6	25.4	207 39.8	17.0	260 27.8	13.2	Nunki	76 07.7	S26 17.6
	02	17 42.8	204 40.4	31.0	41 13.6	25.5	222 41.8	16.8	275 30.1	13.2	Peacock	53 30.7	S56 43.6
	03	32 45.3	219 40.0	29.8	56 16.6	25.5	237 43.7	16.6	290 32.3	13.2	Pollux	243 37.3	N28 01.2
	04	47 47.7	234 39.6	28.5	71 19.7	25.6	252 45.7	16.4	305 34.6	13.1	Procyon	245 08.0	N 5 13.2
	05	62 50.2	249 39.2	27.3	86 22.7	25.6	267 47.6	16.2	320 36.8	13.1			
	06	77 52.7	264 38.8	N 4 26.0	101 25.7	S16 25.7	282 49.6	N11 16.0	335 39.1	N22 13.1	Rasalhague	96 13.5	N12 33.6
	07	92 55.1	279 38.4	24.8	116 28.8	25.8	297 51.5	15.8	350 41.3	13.1	Regulus	207 52.0	N11 57.2
T	08	107 57.6	294 38.0	23.5	131 31.8	25.8	312 53.5	15.7	5 43.6	13.1	Rigel	281 19.5	S 8 11.6
U	09	123 00.1	309 37.6	22.3	146 34.8	25.9	327 55.4	15.5	20 45.9	13.1	Rigil Kent.	140 03.0	S60 51.1
E	10	138 02.5	324 37.3	21.1	161 37.8	25.9	342 57.4	15.3	35 48.1	13.1	Sabik	102 21.4	S15 43.8
S	11	153 05.0	339 36.9	19.8	176 40.9	26.0	357 59.3	15.1	50 50.4	13.0			
D	12	168 07.5	354 36.5	N 4 18.6	191 43.9	S16 26.1	13 01.3	N11 14.9	65 52.6	N22 13.0	Schedar	349 49.1	N56 33.3
A	13	183 09.9	9 36.1	17.3	206 46.9	26.1	28 03.2	14.7	80 54.9	13.0	Shaula	96 32.3	S37 06.5
Y	14	198 12.4	24 35.7	16.1	221 49.9	26.2	43 05.2	14.5	95 57.1	13.0	Sirius	258 40.7	S16 42.9
	15	213 14.9	39 35.3	14.8	236 53.0	26.2	58 07.2	14.3	110 59.4	13.0	Spica	158 39.7	S11 10.7
	16	228 17.3	54 34.9	13.6	251 56.0	26.3	73 09.1	14.1	126 01.6	13.0	Suhail	222 58.6	S43 26.5
	17	243 19.8	69 34.6	12.3	266 59.0	26.3	88 11.1	13.9	141 03.9	12.9			
	18	258 22.2	84 34.2	N 4 11.1	282 02.0	S16 26.4	103 13.0	N11 13.8	156 06.1	N22 12.9	Vega	80 44.0	N38 47.4
	19	273 24.7	99 33.8	09.8	297 05.1	26.5	118 15.0	13.6	171 08.4	12.9	Zuben'ubi	137 14.2	S16 03.4
	20	288 27.2	114 33.4	08.6	312 08.1	26.5	133 16.9	13.4	186 10.7	12.9		SHA	Mer. Pass.
	21	303 29.6	129 33.0	07.3	327 11.1	26.6	148 18.9	13.2	201 12.9	12.9		° ′	h m
	22	318 32.1	144 32.6	06.1	342 14.1	26.6	163 20.8	13.0	216 15.2	12.9	Venus	188 11.7	12 21
	23	333 34.6	159 32.3	04.8	357 17.1	26.7	178 22.8	12.8	231 17.4	12.9	Mars	23 15.8	23 16
		h m									Jupiter	205 12.3	11 11
	Mer. Pass.	0 53.3	*v* −0.4	*d* 1.2	*v* 3.0	*d* 0.1	*v* 2.0	*d* 0.2	*v* 2.3	*d* 0.0	Saturn	257 52.8	7 41

UT	SUN GHA	SUN Dec	MOON GHA	v	MOON Dec	d	HP
d h	° ′	° ′	° ′	′	° ′	′	′
7 00	180 25.4	N 6 18.2	43 46.2	6.7	S24 54.7	6.0	57.9
01	195 25.7	17.3	58 11.9	6.8	24 48.7	6.2	57.9
02	210 25.9	16.4	72 37.7	6.9	24 42.5	6.3	57.8
03	225 26.1	. . 15.4	87 03.6	6.9	24 36.2	6.4	57.8
04	240 26.3	14.5	101 29.5	7.1	24 29.8	6.6	57.8
05	255 26.5	13.6	115 55.6	7.1	24 23.2	6.7	57.8
06	270 26.7	N 6 12.6	130 21.7	7.2	S24 16.5	6.9	57.8
07	285 26.9	11.7	144 47.9	7.3	24 09.6	7.0	57.7
08	300 27.1	10.8	159 14.2	7.4	24 02.6	7.1	57.7
09	315 27.4	. . 09.8	173 40.6	7.5	23 55.5	7.3	57.7
10	330 27.6	08.9	188 07.1	7.5	23 48.2	7.4	57.7
11	345 27.8	08.0	202 33.6	7.7	23 40.8	7.6	57.7
12	0 28.0	N 6 07.0	217 00.3	7.7	S23 33.2	7.6	57.6
13	15 28.2	06.1	231 27.0	7.9	23 25.6	7.8	57.6
14	30 28.4	05.2	245 53.9	7.9	23 17.8	8.0	57.6
15	45 28.6	. . 04.2	260 20.8	8.0	23 09.8	8.0	57.6
16	60 28.8	03.3	274 47.8	8.1	23 01.8	8.2	57.6
17	75 29.1	02.4	289 14.9	8.2	22 53.6	8.3	57.6
18	90 29.3	N 6 01.4	303 42.1	8.3	S22 45.3	8.5	57.5
19	105 29.5	6 00.5	318 09.4	8.4	22 36.8	8.5	57.5
20	120 29.7	5 59.6	332 36.8	8.5	22 28.3	8.7	57.5
21	135 29.9	. . 58.6	347 04.3	8.6	22 19.6	8.8	57.5
22	150 30.1	57.7	1 31.9	8.6	22 10.8	9.0	57.5
23	165 30.3	56.7	15 59.5	8.8	22 01.8	9.0	57.4
8 00	180 30.6	N 5 55.8	30 27.3	8.9	S21 52.8	9.2	57.4
01	195 30.8	54.9	44 55.2	8.9	21 43.6	9.2	57.4
02	210 31.0	53.9	59 23.1	9.1	21 34.4	9.4	57.4
03	225 31.2	. . 53.0	73 51.2	9.1	21 25.0	9.5	57.4
04	240 31.4	52.1	88 19.3	9.3	21 15.5	9.6	57.3
05	255 31.6	51.1	102 47.6	9.3	21 05.9	9.7	57.3
06	270 31.9	N 5 50.2	117 15.9	9.5	S20 56.2	9.8	57.3
07	285 32.1	49.2	131 44.4	9.5	20 46.4	9.9	57.3
08	300 32.3	48.3	146 12.9	9.6	20 36.5	10.1	57.3
09	315 32.5	. . 47.4	160 41.5	9.8	20 26.4	10.1	57.2
10	330 32.7	46.4	175 10.3	9.8	20 16.3	10.2	57.2
11	345 32.9	45.5	189 39.1	9.9	20 06.1	10.4	57.2
12	0 33.1	N 5 44.6	204 08.0	10.0	S19 55.7	10.4	57.2
13	15 33.4	43.6	218 37.0	10.1	19 45.3	10.5	57.2
14	30 33.6	42.7	233 06.1	10.2	19 34.8	10.6	57.1
15	45 33.8	. . 41.7	247 35.3	10.3	19 24.2	10.7	57.1
16	60 34.0	40.8	262 04.6	10.4	19 13.5	10.8	57.1
17	75 34.2	39.9	276 34.0	10.5	19 02.7	10.9	57.1
18	90 34.4	N 5 38.9	291 03.5	10.6	S18 51.8	11.0	57.0
19	105 34.7	38.0	305 33.1	10.7	18 40.8	11.1	57.0
20	120 34.9	37.0	320 02.8	10.7	18 29.7	11.2	57.0
21	135 35.1	. . 36.1	334 32.5	10.9	18 18.5	11.2	57.0
22	150 35.3	35.2	349 02.4	11.0	18 07.3	11.3	57.0
23	165 35.5	34.2	3 32.4	11.0	17 56.0	11.4	56.9
9 00	180 35.7	N 5 33.3	18 02.4	11.1	S17 44.6	11.5	56.9
01	195 36.0	32.3	32 32.5	11.3	17 33.1	11.6	56.9
02	210 36.2	31.4	47 02.8	11.3	17 21.5	11.7	56.9
03	225 36.4	. . 30.5	61 33.1	11.4	17 09.8	11.7	56.9
04	240 36.6	29.5	76 03.5	11.5	16 58.1	11.8	56.8
05	255 36.8	28.6	90 34.0	11.6	16 46.3	11.9	56.8
06	270 37.0	N 5 27.6	105 04.6	11.7	S16 34.4	11.9	56.8
07	285 37.3	26.7	119 35.3	11.7	16 22.5	12.0	56.8
08	300 37.5	25.7	134 06.0	11.9	16 10.5	12.1	56.8
09	315 37.7	. . 24.8	148 36.9	11.9	15 58.4	12.2	56.7
10	330 37.9	23.9	163 07.8	12.0	15 46.2	12.2	56.7
11	345 38.1	22.9	177 38.8	12.1	15 34.0	12.3	56.7
12	0 38.3	N 5 22.0	192 09.9	12.2	S15 21.7	12.4	56.7
13	15 38.6	21.0	206 41.1	12.3	15 09.3	12.4	56.6
14	30 38.8	20.1	221 12.4	12.4	14 56.9	12.5	56.6
15	45 39.0	. . 19.1	235 43.8	12.4	14 44.4	12.5	56.6
16	60 39.2	18.2	250 15.2	12.5	14 31.9	12.6	56.6
17	75 39.4	17.3	264 46.7	12.6	14 19.3	12.7	56.6
18	90 39.6	N 5 16.3	279 18.3	12.7	S14 06.6	12.7	56.5
19	105 39.9	15.4	293 50.0	12.7	13 53.9	12.7	56.5
20	120 40.1	14.4	308 21.7	12.9	13 41.2	12.9	56.5
21	135 40.3	. . 13.5	322 53.6	12.9	13 28.3	12.8	56.5
22	150 40.5	12.5	337 25.5	13.0	13 15.5	13.0	56.5
23	165 40.7	11.6	351 57.5	13.1	S13 02.5	12.9	56.4
	SD 15.9	*d* 0.9	SD 15.7		15.6		15.4

Lat.	Twilight Naut.	Twilight Civil	Sunrise	Moonrise 7	Moonrise 8	Moonrise 9	Moonrise 10
°	h m	h m	h m	h m	h m	h m	h m
N 72	00 19	03 14	04 33	■	■	20 42	20 00
N 70	01 44	03 33	04 42	■	21 07	20 16	19 47
68	02 20	03 49	04 50	■	20 25	19 57	19 37
66	02 45	04 01	04 56	20 26	19 57	19 41	19 29
64	03 04	04 11	05 02	19 44	19 35	19 28	19 22
62	03 19	04 19	05 06	19 16	19 17	19 17	19 16
60	03 32	04 27	05 10	18 54	19 03	19 07	19 10
N 58	03 42	04 33	05 14	18 36	18 50	18 59	19 05
56	03 51	04 38	05 17	18 21	18 39	18 52	19 01
54	03 59	04 43	05 20	18 08	18 29	18 45	18 57
52	04 06	04 48	05 22	17 56	18 21	18 39	18 54
50	04 12	04 52	05 25	17 46	18 13	18 34	18 51
45	04 24	05 00	05 30	17 24	17 56	18 22	18 44
N 40	04 34	05 06	05 34	17 07	17 43	18 12	18 38
35	04 42	05 12	05 37	16 52	17 31	18 04	18 33
30	04 48	05 16	05 40	16 40	17 21	17 57	18 28
20	04 58	05 24	05 46	16 18	17 03	17 44	18 21
N 10	05 05	05 29	05 50	15 59	16 48	17 33	18 14
0	05 10	05 34	05 55	15 41	16 33	17 22	18 07
S 10	05 13	05 38	05 59	15 23	16 19	17 11	18 01
20	05 15	05 41	06 03	15 04	16 03	17 00	17 54
30	05 16	05 44	06 08	14 41	15 45	16 47	17 46
35	05 16	05 45	06 10	14 28	15 35	16 39	17 42
40	05 15	05 46	06 13	14 13	15 23	16 31	17 36
45	05 13	05 47	06 17	13 55	15 08	16 21	17 30
S 50	05 11	05 48	06 21	13 33	14 51	16 08	17 23
52	05 10	05 49	06 22	13 22	14 42	16 02	17 20
54	05 08	05 49	06 24	13 09	14 33	15 56	17 16
56	05 07	05 50	06 27	12 55	14 23	15 49	17 12
58	05 05	05 50	06 29	12 39	14 11	15 41	17 07
S 60	05 02	05 50	06 32	12 19	13 57	15 32	17 02

Lat.	Sunset	Twilight Civil	Twilight Naut.	Moonset 7	Moonset 8	Moonset 9	Moonset 10
°	h m	h m	h m	h m	h m	h m	h m
N 72	19 20	20 37	23 07	■	■	■	02 50
N 70	19 11	20 19	22 04	■	■	00 44	03 14
68	19 03	20 04	21 30	■	■	01 24	03 31
66	18 57	19 52	21 07	23 35	25 51	01 51	03 46
64	18 52	19 42	20 48	24 15	00 15	02 12	03 57
62	18 48	19 34	20 34	24 43	00 43	02 28	04 07
60	18 44	19 27	20 22	25 04	01 04	02 42	04 15
N 58	18 40	19 21	20 12	25 22	01 22	02 54	04 23
56	18 37	19 16	20 03	00 10	01 36	03 04	04 29
54	18 35	19 11	19 55	00 26	01 49	03 13	04 35
52	18 32	19 07	19 49	00 40	02 00	03 21	04 40
50	18 30	19 03	19 43	00 52	02 10	03 28	04 44
45	18 25	18 55	19 30	01 17	02 30	03 43	04 54
N 40	18 21	18 49	19 21	01 37	02 46	03 56	05 03
35	18 18	18 43	19 13	01 53	03 00	04 06	05 10
30	18 15	18 39	19 07	02 07	03 12	04 15	05 16
20	18 10	18 32	18 58	02 32	03 32	04 31	05 26
N 10	18 05	18 26	18 51	02 52	03 50	04 44	05 36
0	18 01	18 22	18 46	03 12	04 06	04 57	05 44
S 10	17 57	18 18	18 43	03 31	04 23	05 10	05 53
20	17 53	18 15	18 41	03 51	04 40	05 23	06 02
30	17 48	18 12	18 40	04 15	05 00	05 38	06 12
35	17 46	18 11	18 41	04 29	05 11	05 47	06 18
40	17 43	18 10	18 41	04 44	05 24	05 57	06 24
45	17 40	18 09	18 43	05 03	05 40	06 08	06 32
S 50	17 36	18 08	18 46	05 27	05 58	06 22	06 41
52	17 34	18 08	18 47	05 38	06 07	06 29	06 45
54	17 32	18 07	18 48	05 51	06 17	06 36	06 50
56	17 30	18 07	18 50	06 05	06 28	06 44	06 55
58	17 28	18 07	18 52	06 22	06 41	06 53	07 01
S 60	17 25	18 07	18 55	06 43	06 56	07 03	07 07

Day	SUN Eqn. of Time 00^h	SUN Eqn. of Time 12^h	SUN Mer. Pass.	MOON Mer. Pass. Upper	MOON Mer. Pass. Lower	MOON Age	MOON Phase
d	m s	m s	h m	h m	h m	d %	
7	01 41	01 52	11 58	21 54	09 26	11 89	
8	02 02	02 12	11 58	22 45	10 20	12 95	○
9	02 23	02 33	11 57	23 33	11 10	13 98	

UT d	UT h	ARIES GHA	VENUS −3.9 GHA	VENUS Dec	MARS −2.7 GHA	MARS Dec	JUPITER −1.7 GHA	JUPITER Dec	SATURN +0.1 GHA	SATURN Dec
		° ′	° ′	° ′	° ′	° ′	° ′	° ′	° ′	° ′
10	00	348 37.0	174 31.9	N 4 03.6	12 20.1	S16 26.7	193 24.7	N11 12.6	246 19.7	N22 12.8
	01	3 39.5	189 31.5	02.3	27 23.2	26.8	208 26.7	12.4	261 21.9	12.8
	02	18 42.0	204 31.1	4 01.1	42 26.2	26.8	223 28.6	12.2	276 24.2	12.8
	03	33 44.4	219 30.7	3 59.8	57 29.2	26.9	238 30.6	12.0	291 26.4	12.8
	04	48 46.9	234 30.3	58.6	72 32.2	26.9	253 32.5	11.9	306 28.7	12.8
	05	63 49.3	249 30.0	57.3	87 35.2	27.0	268 34.5	11.7	321 31.0	12.8
W	06	78 51.8	264 29.6	N 3 56.1	102 38.2	S16 27.0	283 36.4	N11 11.5	336 33.2	N22 12.7
E	07	93 54.3	279 29.2	54.8	117 41.2	27.1	298 38.4	11.3	351 35.5	12.7
D	08	108 56.7	294 28.8	53.6	132 44.2	27.1	313 40.3	11.1	6 37.7	12.7
N	09	123 59.2	309 28.4	52.3	147 47.3	27.2	328 42.3	10.9	21 40.0	12.7
E	10	139 01.7	324 28.0	51.1	162 50.3	27.2	343 44.3	10.7	36 42.2	12.7
S	11	154 04.1	339 27.7	49.8	177 53.3	27.2	358 46.2	10.5	51 44.5	12.7
D	12	169 06.6	354 27.3	N 3 48.6	192 56.3	S16 27.3	13 48.2	N11 10.3	66 46.8	N22 12.6
A	13	184 09.1	9 26.9	47.3	207 59.3	27.3	28 50.1	10.1	81 49.0	12.6
Y	14	199 11.5	24 26.5	46.1	223 02.3	27.4	43 52.1	10.0	96 51.3	12.6
	15	214 14.0	39 26.1	44.8	238 05.3	27.4	58 54.0	09.8	111 53.5	12.6
	16	229 16.5	54 25.7	43.6	253 08.3	27.5	73 56.0	09.6	126 55.8	12.6
	17	244 18.9	69 25.4	42.3	268 11.3	27.5	88 57.9	09.4	141 58.0	12.6
	18	259 21.4	84 25.0	N 3 41.1	283 14.3	S16 27.5	103 59.9	N11 09.2	157 00.3	N22 12.6
	19	274 23.8	99 24.6	39.8	298 17.3	27.6	119 01.8	09.0	172 02.6	12.5
	20	289 26.3	114 24.2	38.6	313 20.3	27.6	134 03.8	08.8	187 04.8	12.5
	21	304 28.8	129 23.8	37.3	328 23.3	27.7	149 05.7	08.6	202 07.1	12.5
	22	319 31.2	144 23.5	36.1	343 26.3	27.7	164 07.7	08.4	217 09.3	12.5
	23	334 33.7	159 23.1	34.8	358 29.3	27.7	179 09.6	08.2	232 11.6	12.5
11	00	349 36.2	174 22.7	N 3 33.6	13 32.3	S16 27.8	194 11.6	N11 08.1	247 13.9	N22 12.5
	01	4 38.6	189 22.3	32.3	28 35.3	27.8	209 13.6	07.9	262 16.1	12.4
	02	19 41.1	204 21.9	31.1	43 38.3	27.9	224 15.5	07.7	277 18.4	12.4
	03	34 43.6	219 21.6	29.8	58 41.3	27.9	239 17.5	07.5	292 20.6	12.4
	04	49 46.0	234 21.2	28.5	73 44.3	27.9	254 19.4	07.3	307 22.9	12.4
	05	64 48.5	249 20.8	27.3	88 47.3	28.0	269 21.4	07.1	322 25.2	12.4
	06	79 50.9	264 20.4	N 3 26.0	103 50.2	S16 28.0	284 23.3	N11 06.9	337 27.4	N22 12.4
T	07	94 53.4	279 20.0	24.8	118 53.2	28.0	299 25.3	06.7	352 29.7	12.4
H	08	109 55.9	294 19.7	23.5	133 56.2	28.1	314 27.2	06.5	7 31.9	12.3
U	09	124 58.3	309 19.3	22.3	148 59.2	28.1	329 29.2	06.3	22 34.2	12.3
R	10	140 00.8	324 18.9	21.0	164 02.2	28.1	344 31.1	06.1	37 36.5	12.3
S	11	155 03.3	339 18.5	19.8	179 05.2	28.2	359 33.1	06.0	52 38.7	12.3
D	12	170 05.7	354 18.1	N 3 18.5	194 08.2	S16 28.2	14 35.1	N11 05.8	67 41.0	N22 12.3
A	13	185 08.2	9 17.8	17.3	209 11.2	28.2	29 37.0	05.6	82 43.2	12.3
Y	14	200 10.7	24 17.4	16.0	224 14.1	28.2	44 39.0	05.4	97 45.5	12.2
	15	215 13.1	39 17.0	14.7	239 17.1	28.3	59 40.9	05.2	112 47.8	12.2
	16	230 15.6	54 16.6	13.5	254 20.1	28.3	74 42.9	05.0	127 50.0	12.2
	17	245 18.1	69 16.2	12.2	269 23.1	28.3	89 44.8	04.8	142 52.3	12.2
	18	260 20.5	84 15.9	N 3 11.0	284 26.1	S16 28.4	104 46.8	N11 04.6	157 54.5	N22 12.2
	19	275 23.0	99 15.5	09.7	299 29.0	28.4	119 48.7	04.4	172 56.8	12.2
	20	290 25.4	114 15.1	08.5	314 32.0	28.4	134 50.7	04.2	187 59.1	12.2
	21	305 27.9	129 14.7	07.2	329 35.0	28.4	149 52 6	04.1	203 01.3	12.1
	22	320 30.4	144 14.4	05.9	344 38.0	28.5	164 54.6	03.9	218 03.6	12.1
	23	335 32.8	159 14.0	04.7	359 41.0	28.5	179 56.6	03.7	233 05.9	12.1
12	00	350 35.3	174 13.6	N 3 03.4	14 43.9	S16 28.5	194 58.5	N11 03.5	248 08.1	N22 12.1
	01	5 37.8	189 13.2	02.2	29 46.9	28.5	210 00.5	03.3	263 10.4	12.1
	02	20 40.2	204 12.8	3 00.9	44 49.9	28.6	225 02.4	03.1	278 12.7	12.1
	03	35 42.7	219 12.5	2 59.6	59 52.9	28.6	240 04.4	02.9	293 14.9	12.0
	04	50 45.2	234 12.1	58.4	74 55.8	28.6	255 06.3	02.7	308 17.2	12.0
	05	65 47.6	249 11.7	57.1	89 58.8	28.6	270 08.3	02.5	323 19.4	12.0
	06	80 50.1	264 11.3	N 2 55.9	105 01.8	S16 28.6	285 10.2	N11 02.3	338 21.7	N22 12.0
	07	95 52.5	279 11.0	54.6	120 04.7	28.7	300 12.2	02.2	353 24.0	12.0
	08	110 55.0	294 10.6	53.4	135 07.7	28.7	315 14.1	02.0	8 26.2	12.0
F	09	125 57.5	309 10.2	52.1	150 10.7	28.7	330 16.1	01.8	23 28.5	12.0
R	10	140 59.9	324 09.8	50.8	165 13.6	28.7	345 18.1	01.6	38 30.8	11.9
I	11	156 02.4	339 09.5	49.6	180 16.6	28.7	0 20.0	01.4	53 33.0	11.9
D	12	171 04.9	354 09.1	N 2 48.3	195 19.6	S16 28.8	15 22.0	N11 01.2	68 35.3	N22 11.9
A	13	186 07.3	9 08.7	47.1	210 22.5	28.8	30 23.9	01.0	83 37.6	11.9
Y	14	201 09.8	24 08.3	45.8	225 25.5	28.8	45 25.9	00.8	98 39.8	11.9
	15	216 12.3	39 08.0	44.5	240 28.5	28.8	60 27.8	00.6	113 42.1	11.9
	16	231 14.7	54 07.6	43.3	255 31.4	28.8	75 29.8	00.4	128 44.4	11.8
	17	246 17.2	69 07.2	42.0	270 34.4	28.8	90 31.7	00.3	143 46.6	11.8
	18	261 19.7	84 06.8	N 2 40.8	285 37.3	S16 28.8	105 33.7	N11 00.1	158 48.9	N22 11.8
	19	276 22.1	99 06.4	39.5	300 40.3	28.9	120 35.7	10 59.9	173 51.2	11.8
	20	291 24.6	114 06.1	38.2	315 43.2	28.9	135 37.6	59.7	188 53.4	11.8
	21	306 27.0	129 05.7	37.0	330 46.2	28.9	150 39.6	59.5	203 55.7	11.8
	22	321 29.5	144 05.3	35.7	345 49.2	28.9	165 41.5	59.3	218 58.0	11.8
	23	336 32.0	159 04.9	34.4	0 52.1	28.9	180 43.5	59.1	234 00.2	11.7
Mer. Pass.		h m 0 41.5	v −0.4	d 1.3	v 3.0	d 0.0	v 2.0	d 0.2	v 2.3	d 0.0

STARS

Name	SHA	Dec
	° ′	° ′
Acamar	315 23.8	S40 17.1
Achernar	335 31.8	S57 12.[illegible]
Acrux	173 18.9	S63 07.1
Adhara	255 18.8	S28 58.1
Aldebaran	290 58.2	N16 31.1
Alioth	166 27.7	N55 56.6
Alkaid	153 05.1	N49 18.0
Al Na'ir	27 52.7	S46 56.7
Alnilam	275 54.2	S 1 11.8
Alphard	218 04.0	S 8 40.2
Alphecca	126 17.6	N26 42.4
Alpheratz	357 51.2	N29 06.6
Altair	62 15.5	N 8 52.7
Ankaa	353 22.7	S42 17.1
Antares	112 35.8	S26 26.5
Arcturus	146 02.9	N19 10.6
Atria	107 44.7	S69 02.4
Avior	234 21.8	S59 30.9
Bellatrix	278 40.3	N 6 21.3
Betelgeuse	271 09.7	N 7 24.6
Canopus	263 59.7	S52 41.4
Capella	280 45.9	N46 00.6
Deneb	49 36.5	N45 17.7
Denebola	182 41.8	N14 33.3
Diphda	349 03.2	S17 57.9
Dubhe	194 01.4	N61 44.0
Elnath	278 22.4	N28 36.7
Eltanin	90 49.6	N51 29.5
Enif	33 54.4	N 9 53.5
Fomalhaut	15 32.0	S29 36.2
Gacrux	172 10.3	S57 08.0
Gienah	176 00.6	S17 33.6
Hadar	148 59.6	S60 23.5
Hamal	328 09.3	N23 28.8
Kaus Aust.	83 53.9	S34 23.1
Kochab	137 19.9	N74 08.7
Markab	13 45.7	N15 13.5
Menkar	314 22.9	N 4 06.4
Menkent	148 17.1	S36 23.3
Miaplacidus	221 42.5	S69 43.6
Mirfak	308 51.3	N49 52.3
Nunki	76 07.7	S26 17.6
Peacock	53 30.7	S56 43.6
Pollux	243 37.3	N28 01.2
Procyon	245 08.0	N 5 13.2
Rasalhague	96 13.6	N12 33.6
Regulus	207 52.0	N11 57.2
Rigel	281 19.5	S 8 11.6
Rigil Kent.	140 03.0	S60 51.1
Sabik	102 21.4	S15 43.8
Schedar	349 49.1	N56 33.4
Shaula	96 32.3	S37 06.5
Sirius	258 40.7	S16 42.9
Spica	158 39.7	S11 10.7
Suhail	222 58.6	S43 26.5
Vega	80 44.0	N38 47.4
Zuben'ubi	137 14.2	S16 03.4

	SHA	Mer. Pass.
	° ′	h m
Venus	184 46.5	12 23
Mars	23 56.1	23 01
Jupiter	204 35.4	11 02
Saturn	257 37.7	7 30

UT	SUN GHA	SUN Dec	MOON GHA	v	MOON Dec	d	HP
d h	° ′	° ′	° ′	′	° ′	′	′
10 00	180 41.0	N 5 10.7	6 29.6	13.1	S12 49.6	13.1	56.4
01	195 41.2	09.7	21 01.7	13.2	12 36.5	13.1	56.4
02	210 41.4	08.8	35 33.9	13.3	12 23.4	13.1	56.4
03	225 41.6	. . 07.8	50 06.2	13.4	12 10.3	13.1	56.3
04	240 41.8	06.9	64 38.6	13.4	11 57.2	13.2	56.3
05	255 42.0	05.9	79 11.0	13.5	11 44.0	13.3	56.3
06	270 42.3	N 5 05.0	93 43.5	13.6	S11 30.7	13.3	56.3
07	285 42.5	04.0	108 16.1	13.6	11 17.4	13.3	56.3
08	300 42.7	03.1	122 48.7	13.7	11 04.1	13.4	56.2
09	315 42.9	. . 02.1	137 21.4	13.8	10 50.7	13.4	56.2
10	330 43.1	01.2	151 54.2	13.8	10 37.3	13.5	56.2
11	345 43.4	5 00.3	166 27.0	14.0	10 23.8	13.4	56.2
12	0 43.6	N 4 59.3	181 00.0	13.9	S10 10.4	13.6	56.2
13	15 43.8	58.4	195 32.9	14.1	9 56.8	13.5	56.1
14	30 44.0	57.4	210 06.0	14.1	9 43.3	13.6	56.1
15	45 44.2	. . 56.5	224 39.1	14.1	9 29.7	13.6	56.1
16	60 44.4	55.5	239 12.2	14.2	9 16.1	13.6	56.1
17	75 44.7	54.6	253 45.4	14.3	9 02.5	13.7	56.0
18	90 44.9	N 4 53.6	268 18.7	14.3	S 8 48.8	13.7	56.0
19	105 45.1	52.7	282 52.0	14.4	8 35.1	13.7	56.0
20	120 45.3	51.7	297 25.4	14.5	8 21.4	13.8	56.0
21	135 45.5	. . 50.8	311 58.9	14.5	8 07.6	13.8	56.0
22	150 45.8	49.8	326 32.4	14.6	7 53.8	13.7	55.9
23	165 46.0	48.9	341 06.0	14.6	7 40.1	13.9	55.9
11 00	180 46.2	N 4 47.9	355 39.6	14.6	S 7 26.2	13.8	55.9
01	195 46.4	47.0	10 13.2	14.8	7 12.4	13.9	55.9
02	210 46.6	46.0	24 47.0	14.7	6 58.5	13.8	55.9
03	225 46.9	. . 45.1	39 20.7	14.8	6 44.7	13.9	55.8
04	240 47.1	44.2	53 54.5	14.9	6 30.8	13.9	55.8
05	255 47.3	43.2	68 28.4	14.9	6 16.9	14.0	55.8
06	270 47.5	N 4 42.3	83 02.3	15.0	S 6 02.9	13.9	55.8
07	285 47.7	41.3	97 36.3	15.0	5 49.0	14.0	55.7
08	300 48.0	40.4	112 10.3	15.0	5 35.0	13.9	55.7
09	315 48.2	. . 39.4	126 44.3	15.1	5 21.1	14.0	55.7
10	330 48.4	38.5	141 18.4	15.2	5 07.1	14.0	55.7
11	345 48.6	37.5	155 52.6	15.2	4 53.1	14.0	55.7
12	0 48.8	N 4 36.6	170 26.8	15.2	S 4 39.1	14.0	55.6
13	15 49.1	35.6	185 01.0	15.2	4 25.1	14.0	55.6
14	30 49.3	34.7	199 35.2	15.3	4 11.1	14.0	55.6
15	45 49.5	. . 33.7	214 09.5	15.4	3 57.1	14.1	55.6
16	60 49.7	32.8	228 43.9	15.3	3 43.0	14.0	55.6
17	75 49.9	31.8	243 18.2	15.4	3 29.0	14.0	55.5
18	90 50.2	N 4 30.9	257 52.6	15.5	S 3 15.0	14.1	55.5
19	105 50.4	29.9	272 27.1	15.5	3 00.9	14.0	55.5
20	120 50.6	29.0	287 01.6	15.5	2 46.9	14.1	55.5
21	135 50.8	. . 28.0	301 36.1	15.5	2 32.8	14.0	55.5
22	150 51.0	27.1	316 10.6	15.6	2 18.8	14.0	55.4
23	165 51.3	26.1	330 45.2	15.6	2 04.8	14.1	55.4
12 00	180 51.5	N 4 25.2	345 19.8	15.6	S 1 50.7	14.0	55.4
01	195 51.7	24.2	359 54.4	15.7	1 36.7	14.0	55.4
02	210 51.9	23.2	14 29.1	15.6	1 22.7	14.1	55.4
03	225 52.2	. . 22.3	29 03.7	15.7	1 08.6	14.0	55.3
04	240 52.4	21.3	43 38.4	15.8	0 54.6	14.0	55.3
05	255 52.6	20.4	58 13.2	15.7	0 40.6	14.0	55.3
06	270 52.8	N 4 19.4	72 47.9	15.8	S 0 26.6	14.0	55.3
07	285 53.0	18.5	87 22.7	15.8	S 0 12.6	14.0	55.3
08	300 53.3	17.5	101 57.5	15.8	N 0 01.4	13.9	55.2
09	315 53.5	. . 16.6	116 32.3	15.9	0 15.3	14.0	55.2
10	330 53.7	15.6	131 07.2	15.8	0 29.3	14.0	55.2
11	345 53.9	14.7	145 42.0	15.9	0 43.3	13.9	55.2
12	0 54.1	N 4 13.7	160 16.9	15.9	N 0 57.2	13.9	55.2
13	15 54.4	12.8	174 51.8	15.9	1 11.1	13.9	55.2
14	30 54.6	11.8	189 26.7	15.9	1 25.0	13.9	55.1
15	45 54.8	. . 10.9	204 01.6	16.0	1 38.9	13.9	55.1
16	60 55.0	09.9	218 36.6	15.9	1 52.8	13.9	55.1
17	75 55.2	09.0	233 11.5	16.0	2 06.7	13.8	55.1
18	90 55.5	N 4 08.0	247 46.5	15.9	N 2 20.5	13.9	55.1
19	105 55.7	07.1	262 21.4	16.0	2 34.4	13.8	55.0
20	120 55.9	06.1	276 56.4	16.0	2 48.2	13.8	55.0
21	135 56.1	. . 05.1	291 31.4	16.0	3 02.0	13.7	55.0
22	150 56.4	04.2	306 06.4	16.0	3 15.7	13.8	55.0
23	165 56.6	03.2	320 41.4	16.0	N 3 29.5	13.7	55.0
	SD 15.9	d 0.9	SD 15.3		15.2		15.0

Lat.	Twilight Naut.	Twilight Civil	Sunrise	Moonrise 10	Moonrise 11	Moonrise 12	Moonrise 13
°	h m	h m	h m	h m	h m	h m	h m
N 72	01 22	03 31	04 47	20 00	19 28	19 00	18 30
N 70	02 09	03 48	04 54	19 47	19 24	19 03	18 42
68	02 38	04 01	05 01	19 37	19 21	19 06	18 51
66	02 59	04 11	05 06	19 29	19 18	19 08	18 58
64	03 16	04 20	05 10	19 22	19 16	19 10	19 05
62	03 29	04 28	05 14	19 16	19 14	19 12	19 10
60	03 40	04 34	05 17	19 10	19 12	19 14	19 15
N 58	03 50	04 40	05 20	19 05	19 10	19 15	19 20
56	03 58	04 45	05 23	19 01	19 09	19 16	19 23
54	04 05	04 49	05 25	18 57	19 08	19 17	19 27
52	04 11	04 53	05 27	18 54	19 06	19 18	19 30
50	04 17	04 56	05 29	18 51	19 05	19 19	19 33
45	04 28	05 03	05 33	18 44	19 03	19 21	19 39
N 40	04 37	05 09	05 37	18 38	19 01	19 23	19 45
35	04 44	05 14	05 40	18 33	18 59	19 24	19 49
30	04 50	05 18	05 42	18 28	18 58	19 26	19 53
20	04 58	05 24	05 46	18 21	18 55	19 28	20 01
N 10	05 05	05 29	05 50	18 14	18 53	19 30	20 07
0	05 09	05 33	05 53	18 07	18 50	19 32	20 13
S 10	05 11	05 36	05 57	18 01	18 48	19 34	20 19
20	05 13	05 38	06 00	17 54	18 46	19 36	20 26
30	05 12	05 40	06 04	17 46	18 43	19 38	20 33
35	05 11	05 41	06 06	17 42	18 42	19 40	20 37
40	05 10	05 41	06 08	17 36	18 40	19 41	20 42
45	05 08	05 42	06 11	17 30	18 38	19 43	20 48
S 50	05 05	05 42	06 14	17 23	18 35	19 46	20 55
52	05 03	05 42	06 16	17 20	18 34	19 47	20 58
54	05 01	05 42	06 17	17 16	18 33	19 48	21 02
56	04 59	05 42	06 19	17 12	18 32	19 49	21 06
58	04 56	05 42	06 21	17 07	18 30	19 50	21 10
S 60	04 53	05 41	06 23	17 02	18 28	19 52	21 15

Lat.	Sunset	Twilight Civil	Twilight Naut.	Moonset 10	Moonset 11	Moonset 12	Moonset 13
°	h m	h m	h m	h m	h m	h m	h m
N 72	19 04	20 19	22 21	02 50	05 06	07 07	09 05
N 70	18 57	20 02	21 39	03 14	05 16	07 08	08 57
68	18 51	19 50	21 11	03 31	05 23	07 08	08 50
66	18 46	19 40	20 51	03 46	05 30	07 08	08 44
64	18 42	19 31	20 35	03 57	05 35	07 08	08 39
62	18 38	19 24	20 22	04 07	05 40	07 09	08 35
60	18 35	19 18	20 11	04 15	05 44	07 09	08 32
N 58	18 32	19 12	20 02	04 23	05 47	07 09	08 29
56	18 30	19 07	19 54	04 29	05 50	07 09	08 26
54	18 27	19 03	19 47	04 35	05 53	07 09	08 23
52	18 25	19 00	19 41	04 40	05 56	07 09	08 21
50	18 23	18 56	19 35	04 44	05 58	07 09	08 19
45	18 19	18 49	19 24	04 54	06 03	07 09	08 15
N 40	18 16	18 43	19 16	05 03	06 07	07 10	08 11
35	18 13	18 39	19 09	05 10	06 11	07 10	08 08
30	18 11	18 35	19 03	05 16	06 14	07 10	08 05
20	18 07	18 29	18 55	05 26	06 19	07 10	08 00
N 10	18 03	18 24	18 49	05 36	06 24	07 10	07 56
0	18 00	18 21	18 45	05 44	06 28	07 10	07 52
S 10	17 57	18 18	18 42	05 53	06 33	07 10	07 48
20	17 54	18 16	18 41	06 02	06 37	07 11	07 43
30	17 50	18 14	18 42	06 12	06 42	07 11	07 39
35	17 48	18 13	18 43	06 18	06 45	07 11	07 36
40	17 46	18 13	18 44	06 24	06 48	07 11	07 33
45	17 43	18 13	18 47	06 32	06 52	07 11	07 29
S 50	17 40	18 13	18 50	06 41	06 57	07 11	07 25
52	17 39	18 13	18 52	06 45	06 59	07 11	07 23
54	17 37	18 13	18 54	06 50	07 01	07 11	07 21
56	17 36	18 13	18 56	06 55	07 04	07 11	07 18
58	17 34	18 13	18 59	07 01	07 06	07 11	07 16
S 60	17 32	18 13	19 02	07 07	07 09	07 11	07 13

Day	SUN Eqn. of Time 00^h	SUN Eqn. of Time 12^h	SUN Mer. Pass.	MOON Mer. Pass. Upper	MOON Mer. Pass. Lower	MOON Age	MOON Phase
d	m s	m s	h m	h m	h m	d %	
10	02 43	02 54	11 57	24 18	11 56	14 100	○
11	03 04	03 15	11 57	00 18	12 39	15 99	
12	03 26	03 36	11 56	01 00	13 21	16 97	

	UT	ARIES	VENUS −3.9		MARS −2.6		JUPITER −1.7		SATURN +0.1	
		GHA	GHA	Dec	GHA	Dec	GHA	Dec	GHA	Dec
	d h	° ′	° ′	° ′	° ′	° ′	° ′	° ′	° ′	° ′
	13 00	351 34.4	174 04.6	N 2 33.2	15 55.1	S16 28.9	195 45.4	N10 58.9	249 02.5	N22 11.7
	01	6 36.9	189 04.2	31.9	30 58.0	28.9	210 47.4	58.7	264 04.8	11.7
	02	21 39.4	204 03.8	30.7	46 01.0	28.9	225 49.4	58.6	279 07.0	11.7
	03	36 41.8	219 03.4	. . 29.4	61 03.9	. . 28.9	240 51.3	. . 58.4	294 09.3	. . 11.7
	04	51 44.3	234 03.1	28.1	76 06.9	28.9	255 53.3	58.2	309 11.6	11.7
	05	66 46.8	249 02.7	26.9	91 09.8	29.0	270 55.2	58.0	324 13.8	11.7
	06	81 49.2	264 02.3	N 2 25.6	106 12.8	S16 29.0	285 57.2	N10 57.8	339 16.1	N22 11.6
	07	96 51.7	279 01.9	24.3	121 15.7	29.0	300 59.1	57.6	354 18.4	11.6
S	08	111 54.2	294 01.6	23.1	136 18.7	29.0	316 01.1	57.4	9 20.6	11.6
A	09	126 56.6	309 01.2	. . 21.8	151 21.6	. . 29.0	331 03.0	. . 57.2	24 22.9	. . 11.6
T	10	141 59.1	324 00.8	20.6	166 24.5	29.0	346 05.0	57.0	39 25.2	11.6
U	11	157 01.5	339 00.5	19.3	181 27.5	29.0	1 07.0	56.8	54 27.4	11.6
R	12	172 04.0	354 00.1	N 2 18.0	196 30.4	S16 29.0	16 08.9	N10 56.7	69 29.7	N22 11.5
D	13	187 06.5	8 59.7	16.8	211 33.4	29.0	31 10.9	56.5	84 32.0	11.5
A	14	202 08.9	23 59.3	15.5	226 36.3	29.0	46 12.8	56.3	99 34.2	11.5
Y	15	217 11.4	38 59.0	. . 14.2	241 39.2	. . 29.0	61 14.8	. . 56.1	114 36.5	. . 11.5
	16	232 13.9	53 58.6	13.0	256 42.2	29.0	76 16.7	55.9	129 38.8	11.5
	17	247 16.3	68 58.2	11.7	271 45.1	29.0	91 18.7	55.7	144 41.0	11.5
	18	262 18.8	83 57.8	N 2 10.4	286 48.1	S16 29.0	106 20.7	N10 55.5	159 43.3	N22 11.5
	19	277 21.3	98 57.5	09.2	301 51.0	29.0	121 22.6	55.3	174 45.6	11.4
	20	292 23.7	113 57.1	07.9	316 53.9	29.0	136 24.6	55.1	189 47.9	11.4
	21	307 26.2	128 56.7	. . 06.7	331 56.9	. . 29.0	151 26.5	. . 54.9	204 50.1	. . 11.4
	22	322 28.6	143 56.3	05.4	346 59.8	29.0	166 28.5	54.8	219 52.4	11.4
	23	337 31.1	158 56.0	04.1	2 02.7	29.0	181 30.4	54.6	234 54.7	11.4
	14 00	352 33.6	173 55.6	N 2 02.9	17 05.6	S16 29.0	196 32.4	N10 54.4	249 56.9	N22 11.4
	01	7 36.0	188 55.2	01.6	32 08.6	29.0	211 34.4	54.2	264 59.2	11.4
	02	22 38.5	203 54.9	2 00.3	47 11.5	29.0	226 36.3	54.0	280 01.5	11.3
	03	37 41.0	218 54.5	1 59.1	62 14.4	. . 29.0	241 38.3	. . 53.8	295 03.8	. . 11.3
	04	52 43.4	233 54.1	57.8	77 17.4	29.0	256 40.2	53.6	310 06.0	11.3
	05	67 45.9	248 53.7	56.5	92 20.3	29.0	271 42.2	53.4	325 08.3	11.3
	06	82 48.4	263 53.4	N 1 55.3	107 23.2	S16 28.9	286 44.2	N10 53.2	340 10.6	N22 11.3
	07	97 50.8	278 53.0	54.0	122 26.1	28.9	301 46.1	53.0	355 12.8	11.3
	08	112 53.3	293 52.6	52.7	137 29.0	28.9	316 48.1	52.9	10 15.1	11.2
S	09	127 55.8	308 52.2	. . 51.5	152 32.0	. . 28.9	331 50.0	. . 52.7	25 17.4	. . 11.2
U	10	142 58.2	323 51.9	50.2	167 34.9	28.9	346 52.0	52.5	40 19.7	11.2
N	11	158 00.7	338 51.5	48.9	182 37.8	28.9	1 53.9	52.3	55 21.9	11.2
D	12	173 03.1	353 51.1	N 1 47.7	197 40.7	S16 28.9	16 55.9	N10 52.1	70 24.2	N22 11.2
A	13	188 05.6	8 50.8	46.4	212 43.6	28.9	31 57.9	51.9	85 26.5	11.2
Y	14	203 08.1	23 50.4	45.1	227 46.5	28.9	46 59.8	51.7	100 28.7	11.2
	15	218 10.5	38 50.0	. . 43.9	242 49.5	. . 28.9	62 01.8	. . 51.5	115 31.0	. . 11.1
	16	233 13.0	53 49.6	42.6	257 52.4	28.8	77 03.7	51.3	130 33.3	11.1
	17	248 15.5	68 49.3	41.3	272 55.3	28.8	92 05.7	51.1	145 35.6	11.1
	18	263 17.9	83 48.9	N 1 40.1	287 58.2	S16 28.8	107 07.6	N10 51.0	160 37.8	N22 11.1
	19	278 20.4	98 48.5	38.8	303 01.1	28.8	122 09.6	50.8	175 40.1	11.1
	20	293 22.9	113 48.2	37.5	318 04.0	28.8	137 11.6	50.6	190 42.4	11.1
	21	308 25.3	128 47.8	. . 36.3	333 06.9	. . 28.8	152 13.5	. . 50.4	205 44.7	. . 11.1
	22	323 27.8	143 47.4	35.0	348 09.8	28.7	167 15.5	50.2	220 46.9	11.0
	23	338 30.2	158 47.0	33.7	3 12.7	28.7	182 17.4	50.0	235 49.2	11.0
	15 00	353 32.7	173 46.7	N 1 32.5	18 15.6	S16 28.7	197 19.4	N10 49.8	250 51.5	N22 11.0
	01	8 35.2	188 46.3	31.2	33 18.5	28.7	212 21.4	49.6	265 53.8	11.0
	02	23 37.6	203 45.9	29.9	48 21.5	28.7	227 23.3	49.4	280 56.0	11.0
	03	38 40.1	218 45.6	. . 28.7	63 24.4	. . 28.7	242 25.3	. . 49.3	295 58.3	. . 11.0
	04	53 42.6	233 45.2	27.4	78 27.3	28.6	257 27.2	49.1	311 00.6	11.0
	05	68 45.0	248 44.8	26.1	93 30.2	28.6	272 29.2	48.9	326 02.9	10.9
	06	83 47.5	263 44.4	N 1 24.9	108 33.1	S16 28.6	287 31.2	N10 48.7	341 05.1	N22 10.9
	07	98 50.0	278 44.1	23.6	123 35.9	28.6	302 33.1	48.5	356 07.4	10.9
	08	113 52.4	293 43.7	22.3	138 38.8	28.5	317 35.1	48.3	11 09.7	10.9
M	09	128 54.9	308 43.3	. . 21.0	153 41.7	. . 28.5	332 37.0	. . 48.1	26 12.0	. . 10.9
O	10	143 57.4	323 43.0	19.8	168 44.6	28.5	347 39.0	47.9	41 14.2	10.9
N	11	158 59.8	338 42.6	18.5	183 47.5	28.5	2 40.9	47.7	56 16.5	10.8
D	12	174 02.3	353 42.2	N 1 17.2	198 50.4	S16 28.5	17 42.9	N10 47.5	71 18.8	N22 10.8
A	13	189 04.7	8 41.8	16.0	213 53.3	28.4	32 44.9	47.4	86 21.1	10.8
Y	14	204 07.2	23 41.5	14.7	228 56.2	28.4	47 46.8	47.2	101 23.3	10.8
	15	219 09.7	38 41.1	. . 13.4	243 59.1	. . 28.4	62 48.8	. . 47.0	116 25.6	. . 10.8
	16	234 12.1	53 40.7	12.2	259 02.0	28.3	77 50.7	46.8	131 27.9	10.8
	17	249 14.6	68 40.4	10.9	274 04.9	28.3	92 52.7	46.6	146 30.2	10.8
	18	264 17.1	83 40.0	N 1 09.6	289 07.8	S16 28.3	107 54.7	N10 46.4	161 32.5	N22 10.7
	19	279 19.5	98 39.6	08.3	304 10.6	28.3	122 56.6	46.2	176 34.7	10.7
	20	294 22.0	113 39.2	07.1	319 13.5	28.2	137 58.6	46.0	191 37.0	10.7
	21	309 24.5	128 38.9	. . 05.8	334 16.4	. . 28.2	153 00.5	. . 45.8	206 39.3	. . 10.7
	22	324 26.9	143 38.5	04.5	349 19.3	28.2	168 02.5	45.7	221 41.6	10.7
	23	339 29.4	158 38.1	03.3	4 22.2	28.1	183 04.5	45.5	236 43.8	10.7
	Mer. Pass.	h m 0 29.7	*v* −0.4	*d* 1.3	*v* 2.9	*d* 0.0	*v* 2.0	*d* 0.2	*v* 2.3	*d* 0.0

STARS

Name	SHA	Dec
	° ′	° ′
Acamar	315 23.8	S40 17.1
Achernar	335 31.7	S57 12.9
Acrux	173 19.0	S63 07.1
Adhara	255 18.7	S28 58.3
Aldebaran	290 58.2	N16 31.1
Alioth	166 27.7	N55 56.6
Alkaid	153 05.2	N49 18.0
Al Na'ir	27 52.7	S46 56.7
Alnilam	275 54.2	S 1 11.8
Alphard	218 03.9	S 8 40.2
Alphecca	126 17.6	N26 42.4
Alpheratz	357 51.2	N29 06.7
Altair	62 15.6	N 8 52.7
Ankaa	353 22.7	S42 17.1
Antares	112 35.8	S26 26.5
Arcturus	146 03.0	N19 10.6
Atria	107 44.8	S69 02.4
Avior	234 21.8	S59 30.9
Bellatrix	278 40.3	N 6 21.4
Betelgeuse	271 09.7	N 7 24.6
Canopus	263 59.7	S52 41.4
Capella	280 45.9	N46 00.0
Deneb	49 36.5	N45 17.7
Denebola	182 41.8	N14 33.3
Diphda	349 03.2	S17 57.9
Dubhe	194 01.4	N61 44.6
Elnath	278 22.4	N28 36.7
Eltanin	90 49.6	N51 29.5
Enif	33 54.4	N 9 53.5
Fomalhaut	15 32.0	S29 36.2
Gacrux	172 10.3	S57 08.0
Gienah	176 00.6	S17 33.6
Hadar	148 59.6	S60 23.5
Hamal	328 09.3	N23 28.8
Kaus Aust.	83 53.9	S34 23.1
Kochab	137 20.0	N74 08.7
Markab	13 45.7	N15 13.5
Menkar	314 22.9	N 4 06.4
Menkent	148 17.1	S36 23.3
Miaplacidus	221 42.4	S69 43.6
Mirfak	308 51.3	N49 52.4
Nunki	76 07.7	S26 17.6
Peacock	53 30.7	S56 43.[illegible]
Pollux	243 37.3	N28 01.3
Procyon	245 07.9	N 5 13.2
Rasalhague	96 13.6	N12 33.6
Regulus	207 52.0	N11 57.[illegible]
Rigel	281 19.4	S 8 11.6
Rigil Kent.	140 03.0	S60 51.1
Sabik	102 21.4	S15 43.8
Schedar	349 49.1	N56 33.4
Shaula	96 32.4	S37 06.5
Sirius	258 40.7	S16 42.9
Spica	158 39.7	S11 10.7
Suhail	222 58.6	S43 26.5
Vega	80 44.1	N38 47.4
Zuben'ubi	137 14.2	S16 03.4

	SHA	Mer. Pass
	° ′	h m
Venus	181 22.0	12 25
Mars	24 32.1	22 47
Jupiter	203 58.8	10 52
Saturn	257 23.4	7 19

	UT	SUN GHA	SUN Dec	MOON GHA	v	MOON Dec	d	HP
	d h	° ′	° ′	° ′	′	° ′	′	′
13	00	180 56.8	N 4 02.3	335 16.4	16.0	N 3 43.2	13.7	55.0
	01	195 57.0	01.3	349 51.4	16.1	3 56.9	13.7	54.9
	02	210 57.2	4 00.4	4 26.5	16.0	4 10.6	13.6	54.9
	03	225 57.5	3 59.4	19 01.5	16.0	4 24.2	13.7	54.9
	04	240 57.7	58.5	33 36.5	16.0	4 37.9	13.6	54.9
	05	255 57.9	57.5	48 11.5	16.1	4 51.5	13.5	54.9
	06	270 58.1	N 3 56.6	62 46.6	16.0	N 5 05.0	13.6	54.9
	07	285 58.4	55.6	77 21.6	16.0	5 18.6	13.5	54.8
S	08	300 58.6	54.6	91 56.6	16.0	5 32.1	13.5	54.8
A	09	315 58.8	. . 53.7	106 31.6	16.0	5 45.6	13.4	54.8
T	10	330 59.0	52.7	121 06.6	16.1	5 59.0	13.5	54.8
U	11	345 59.2	51.8	135 41.7	16.0	6 12.5	13.4	54.8
R	12	0 59.5	N 3 50.8	150 16.7	16.0	N 6 25.9	13.3	54.8
D	13	15 59.7	49.9	164 51.7	16.0	6 39.2	13.3	54.7
A	14	30 59.9	48.9	179 26.7	15.9	6 52.5	13.3	54.7
Y	15	46 00.1	. . 48.0	194 01.6	16.0	7 05.8	13.3	54.7
	16	61 00.3	47.0	208 36.6	16.0	7 19.1	13.2	54.7
	17	76 00.6	46.0	223 11.6	15.9	7 32.3	13.2	54.7
	18	91 00.8	N 3 45.1	237 46.5	16.0	N 7 45.5	13.1	54.7
	19	106 01.0	44.1	252 21.5	15.9	7 58.6	13.2	54.7
	20	121 01.2	43.2	266 56.4	15.9	8 11.8	13.0	54.6
	21	136 01.5	. . 42.2	281 31.3	15.9	8 24.8	13.1	54.6
	22	151 01.7	41.3	296 06.2	15.9	8 37.9	13.0	54.6
	23	166 01.9	40.3	310 41.1	15.9	8 50.9	12.9	54.6
14	00	181 02.1	N 3 39.3	325 16.0	15.8	N 9 03.8	12.9	54.6
	01	196 02.4	38.4	339 50.8	15.9	9 16.7	12.9	54.6
	02	211 02.6	37.4	354 25.7	15.8	9 29.6	12.8	54.6
	03	226 02.8	. . 36.5	9 00.5	15.8	9 42.4	12.8	54.5
	04	241 03.0	35.5	23 35.3	15.8	9 55.2	12.7	54.5
	05	256 03.2	34.6	38 10.1	15.7	10 07.9	12.7	54.5
	06	271 03.5	N 3 33.6	52 44.8	15.7	N10 20.6	12.7	54.5
	07	286 03.7	32.6	67 19.5	15.8	10 33.3	12.6	54.5
	08	301 03.9	31.7	81 54.3	15.6	10 45.9	12.5	54.5
S	09	316 04.1	. . 30.7	96 28.9	15.7	10 58.4	12.5	54.5
U	10	331 04.4	29.8	111 03.6	15.6	11 10.9	12.5	54.5
N	11	346 04.6	28.8	125 38.2	15.6	11 23.4	12.4	54.5
D	12	1 04.8	N 3 27.8	140 12.8	15.6	N11 35.8	12.3	54.4
A	13	16 05.0	26.9	154 47.4	15.6	11 48.1	12.3	54.4
Y	14	31 05.2	25.9	169 22.0	15.5	12 00.4	12.3	54.4
	15	46 05.5	. . 25.0	183 56.5	15.5	12 12.7	12.2	54.4
	16	61 05.7	24.0	198 31.0	15.5	12 24.9	12.1	54.4
	17	76 05.9	23.1	213 05.5	15.4	12 37.0	12.1	54.4
	18	91 06.1	N 3 22.1	227 39.9	15.4	N12 49.1	12.0	54.4
	19	106 06.4	21.1	242 14.3	15.4	13 01.1	12.0	54.4
	20	121 06.6	20.2	256 48.7	15.3	13 13.1	12.0	54.4
	21	136 06.8	. . 19.2	271 23.0	15.3	13 25.1	11.8	54.4
	22	151 07.0	18.3	285 57.3	15.2	13 36.9	11.8	54.3
	23	166 07.3	17.3	300 31.5	15.3	13 48.7	11.8	54.3
15	00	181 07.5	N 3 16.3	315 05.8	15.2	N14 00.5	11.7	54.3
	01	196 07.7	15.4	329 40.0	15.1	14 12.2	11.6	54.3
	02	211 07.9	14.4	344 14.1	15.1	14 23.8	11.6	54.3
	03	226 08.1	. . 13.5	358 48.2	15.1	14 35.4	11.5	54.3
	04	241 08.4	12.5	13 22.3	15.0	14 46.9	11.4	54.3
	05	256 08.6	11.5	27 56.3	15.0	14 58.3	11.4	54.3
	06	271 08.8	N 3 10.6	42 30.3	15.0	N15 09.7	11.4	54.3
	07	286 09.0	09.6	57 04.3	14.9	15 21.1	11.2	54.3
	08	301 09.3	08.6	71 38.2	14.9	15 32.3	11.2	54.3
M	09	316 09.5	. . 07.7	86 12.1	14.8	15 43.5	11.2	54.3
O	10	331 09.7	06.7	100 45.9	14.8	15 54.7	11.0	54.3
N	11	346 09.9	05.8	115 19.7	14.7	16 05.7	11.0	54.2
D	12	1 10.1	N 3 04.8	129 53.4	14.7	N16 16.7	10.9	54.2
A	13	16 10.4	03.8	144 27.1	14.7	16 27.6	10.9	54.2
	14	31 10.6	02.9	159 00.8	14.6	16 38.5	10.8	54.2
Y	15	46 10.8	. . 01.9	173 34.4	14.5	16 49.3	10.7	54.2
	16	61 11.0	01.0	188 07.9	14.5	17 00.0	10.7	54.2
	17	76 11.3	3 00.0	202 41.4	14.5	17 10.7	10.6	54.2
	18	91 11.5	N 2 59.0	217 14.9	14.4	N17 21.3	10.5	54.2
	19	106 11.7	58.1	231 48.3	14.3	17 31.8	10.4	54.2
	20	121 11.9	57.1	246 21.6	14.3	17 42.2	10.4	54.2
	21	136 12.2	. . 56.1	260 54.9	14.3	17 52.6	10.2	54.2
	22	151 12.4	55.2	275 28.2	14.2	18 02.8	10.3	54.2
	23	166 12.6	54.2	290 01.4	14.2	N18 13.1	10.1	54.2
		SD 15.9	*d* 1.0	SD 14.9		14.8		14.8

Lat.	Twilight Naut.	Twilight Civil	Sunrise	Moonrise 13	Moonrise 14	Moonrise 15	Moonrise 16
°	h m	h m	h m	h m	h m	h m	h m
N 72	01 54	03 47	05 00	18 30	17 54	16 46	▭
N 70	02 30	04 01	05 06	18 42	18 16	17 37	▭
68	02 54	04 13	05 11	18 51	18 33	18 10	17 25
66	03 13	04 22	05 15	18 58	18 47	18 34	18 14
64	03 27	04 30	05 19	19 05	18 59	18 53	18 45
62	03 39	04 36	05 22	19 10	19 09	19 08	19 08
60	03 49	04 42	05 24	19 15	19 17	19 21	19 27
N 58	03 57	04 47	05 27	19 20	19 25	19 32	19 43
56	04 05	04 51	05 29	19 23	19 32	19 42	19 56
54	04 11	04 55	05 30	19 27	19 38	19 51	20 08
52	04 17	04 58	05 32	19 30	19 43	19 59	20 18
50	04 22	05 01	05 34	19 33	19 48	20 06	20 28
45	04 32	05 07	05 37	19 39	19 59	20 21	20 47
N 40	04 40	05 12	05 39	19 45	20 08	20 34	21 03
35	04 46	05 16	05 42	19 49	20 16	20 44	21 17
30	04 52	05 20	05 44	19 53	20 22	20 54	21 28
20	04 59	05 25	05 47	20 01	20 34	21 10	21 49
N 10	05 04	05 29	05 50	20 07	20 45	21 24	22 06
0	05 08	05 32	05 52	20 13	20 55	21 38	22 23
S 10	05 10	05 34	05 55	20 19	21 05	21 51	22 40
20	05 10	05 35	05 57	20 26	21 15	22 06	22 58
30	05 09	05 36	06 00	20 33	21 28	22 23	23 19
35	05 07	05 36	06 02	20 37	21 35	22 33	23 31
40	05 05	05 36	06 03	20 42	21 43	22 44	23 45
45	05 02	05 36	06 05	20 48	21 52	22 57	24 02
S 50	04 58	05 35	06 08	20 55	22 04	23 14	24 23
52	04 56	05 35	06 09	20 58	22 10	23 21	24 33
54	04 54	05 35	06 10	21 02	22 16	23 30	24 44
56	04 51	05 34	06 11	21 06	22 22	23 40	24 57
58	04 48	05 33	06 12	21 10	22 30	23 51	25 12
S 60	04 44	05 32	06 14	21 15	22 38	24 03	00 03

Lat.	Sunset	Twilight Civil	Twilight Naut.	Moonset 13	Moonset 14	Moonset 15	Moonset 16
°	h m	h m	h m	h m	h m	h m	h m
N 72	18 48	20 00	21 49	09 05	11 10	13 47	▭
N 70	18 43	19 47	21 16	08 57	10 50	12 57	▭
68	18 38	19 36	20 53	08 50	10 34	12 26	14 43
66	18 34	19 27	20 36	08 44	10 22	12 03	13 56
64	18 31	19 20	20 21	08 39	10 11	11 46	13 25
62	18 28	19 13	20 10	08 35	10 02	11 31	13 03
60	18 26	19 08	20 00	08 32	09 55	11 19	12 44
N 58	18 24	19 03	19 52	08 29	09 48	11 08	12 29
56	18 22	18 59	19 45	08 26	09 42	10 59	12 17
54	18 20	18 56	19 39	08 23	09 37	10 51	12 05
52	18 18	18 52	19 33	08 21	09 32	10 44	11 56
50	18 17	18 49	19 28	08 19	09 28	10 37	11 47
45	18 14	18 43	19 18	08 15	09 19	10 24	11 28
N 40	18 11	18 38	19 10	08 11	09 11	10 12	11 13
35	18 09	18 34	19 04	08 08	09 05	10 03	11 01
30	18 07	18 31	18 59	08 05	08 59	09 54	10 50
20	18 04	18 26	18 52	08 00	08 49	09 40	10 31
N 10	18 01	18 22	18 47	07 56	08 41	09 27	10 15
0	17 59	18 20	18 44	07 52	08 33	09 15	09 59
S 10	17 57	18 18	18 42	07 48	08 25	09 04	09 44
20	17 54	18 16	18 42	07 43	08 16	08 51	09 28
30	17 52	18 16	18 43	07 39	08 07	08 37	09 10
35	17 50	18 15	18 45	07 36	08 01	08 29	08 59
40	17 49	18 16	18 47	07 33	07 55	08 19	08 47
45	17 47	18 16	18 50	07 29	07 48	08 08	08 32
S 50	17 45	18 17	18 54	07 25	07 39	07 55	08 15
52	17 44	18 17	18 57	07 23	07 35	07 49	08 06
54	17 43	18 18	18 59	07 21	07 31	07 43	07 57
56	17 42	18 19	19 02	07 18	07 26	07 35	07 47
58	17 40	18 19	19 05	07 16	07 21	07 27	07 35
S 60	17 39	18 20	19 09	07 13	07 15	07 17	07 22

Day	SUN Eqn. of Time 00^h	SUN Eqn. of Time 12^h	SUN Mer. Pass.	MOON Mer. Pass. Upper	MOON Mer. Pass. Lower	MOON Age	MOON Phase
d	m s	m s	h m	h m	h m	d %	
13	03 47	03 57	11 56	01 42	14 02	17 92	
14	04 08	04 19	11 56	02 23	14 44	18 86	
15	04 29	04 40	11 55	03 05	15 26	19 79	

UT		ARIES	VENUS −3.9		MARS −2.5		JUPITER −1.7		SATURN +0.1	
		GHA	GHA	Dec	GHA	Dec	GHA	Dec	GHA	Dec
d	h	° ′	° ′	° ′	° ′	° ′	° ′	° ′	° ′	° ′
16	00	354 31.9	173 37.8	N 1 02.0	19 25.0	S16 28.1	198 06.4	N10 45.3	251 46.1	N22 10.7
	01	9 34.3	188 37.4	1 00.7	34 27.9	28.1	213 08.4	45.1	266 48.4	10.6
	02	24 36.8	203 37.0	0 59.5	49 30.8	28.0	228 10.3	44.9	281 50.7	10.6
	03	39 39.2	218 36.7	. . 58.2	64 33.7	. . 28.0	243 12.3	. . 44.7	296 53.0	. . 10.6
	04	54 41.7	233 36.3	56.9	79 36.6	28.0	258 14.3	44.5	311 55.2	10.6
	05	69 44.2	248 35.9	55.6	94 39.4	27.9	273 16.2	44.3	326 57.5	10.6
	06	84 46.6	263 35.5	N 0 54.4	109 42.3	S16 27.9	288 18.2	N10 44.1	341 59.8	N22 10.6
	07	99 49.1	278 35.2	53.1	124 45.2	27.9	303 20.1	43.9	357 02.1	10.6
TUESDAY	08	114 51.6	293 34.8	51.8	139 48.0	27.8	318 22.1	43.8	12 04.4	10.5
	09	129 54.0	308 34.4	. . 50.6	154 50.9	. . 27.8	333 24.1	. . 43.6	27 06.6	. . 10.5
	10	144 56.5	323 34.1	49.3	169 53.8	27.8	348 26.0	43.4	42 08.9	10.5
	11	159 59.0	338 33.7	48.0	184 56.6	27.7	3 28.0	43.2	57 11.2	10.5
	12	175 01.4	353 33.3	N 0 46.7	199 59.5	S16 27.7	18 30.0	N10 43.0	72 13.5	N22 10.5
	13	190 03.9	8 33.0	45.5	215 02.4	27.6	33 31.9	42.8	87 15.8	10.5
	14	205 06.3	23 32.6	44.2	230 05.2	27.6	48 33.9	42.6	102 18.0	10.5
	15	220 08.8	38 32.2	. . 42.9	245 08.1	. . 27.6	63 35.8	. . 42.4	117 20.3	. . 10.4
	16	235 11.3	53 31.8	41.7	260 11.0	27.5	78 37.8	42.2	132 22.6	10.4
	17	250 13.7	68 31.5	40.4	275 13.8	27.5	93 39.8	42.1	147 24.9	10.4
	18	265 16.2	83 31.1	N 0 39.1	290 16.7	S16 27.4	108 41.7	N10 41.9	162 27.2	N22 10.4
	19	280 18.7	98 30.7	37.8	305 19.5	27.4	123 43.7	41.7	177 29.4	10.4
	20	295 21.1	113 30.4	36.6	320 22.4	27.4	138 45.6	41.5	192 31.7	10.4
	21	310 23.6	128 30.0	. . 35.3	335 25.3	. . 27.3	153 47.6	. . 41.3	207 34.0	. . 10.4
	22	325 26.1	143 29.6	34.0	350 28.1	27.3	168 49.6	41.1	222 36.3	10.3
	23	340 28.5	158 29.3	32.8	5 31.0	27.2	183 51.5	40.9	237 38.6	10.3
17	00	355 31.0	173 28.9	N 0 31.5	20 33.8	S16 27.2	198 53.5	N10 40.7	252 40.9	N22 10.3
	01	10 33.5	188 28.5	30.2	35 36.7	27.1	213 55.4	40.5	267 43.1	10.3
	02	25 35.9	203 28.2	28.9	50 39.5	27.1	228 57.4	40.4	282 45.4	10.3
	03	40 38.4	218 27.8	. . 27.7	65 42.4	. . 27.0	243 59.4	. . 40.2	297 47.7	. . 10.3
	04	55 40.8	233 27.4	26.4	80 45.2	27.0	259 01.3	40.0	312 50.0	10.3
	05	70 43.3	248 27.0	25.1	95 48.1	26.9	274 03.3	39.8	327 52.3	10.2
	06	85 45.8	263 26.7	N 0 23.8	110 50.9	S16 26.9	289 05.3	N10 39.6	342 54.6	N22 10.2
WEDNESDAY	07	100 48.2	278 26.3	22.6	125 53.8	26.8	304 07.2	39.4	357 56.8	10.2
	08	115 50.7	293 25.9	21.3	140 56.6	26.8	319 09.2	39.2	12 59.1	10.2
	09	130 53.2	308 25.6	. . 20.0	155 59.4	. . 26.7	334 11.1	. . 39.0	28 01.4	. . 10.2
	10	145 55.6	323 25.2	18.8	171 02.3	26.7	349 13.1	38.8	43 03.7	10.2
	11	160 58.1	338 24.8	17.5	186 05.1	26.6	4 15.1	38.7	58 06.0	10.2
	12	176 00.6	353 24.5	N 0 16.2	201 08.0	S16 26.6	19 17.0	N10 38.5	73 08.3	N22 10.1
	13	191 03.0	8 24.1	14.9	216 10.8	26.5	34 19.0	38.3	88 10.5	10.1
	14	206 05.5	23 23.7	13.7	231 13.6	26.5	49 21.0	38.1	103 12.8	10.1
	15	221 08.0	38 23.4	. . 12.4	246 16.5	. . 26.4	64 22.9	. . 37.9	118 15.1	. . 10.1
	16	236 10.4	53 23.0	11.1	261 19.3	26.4	79 24.9	37.7	133 17.4	10.1
	17	251 12.9	68 22.6	09.8	276 22.1	26.3	94 26.8	37.5	148 19.7	10.1
	18	266 15.3	83 22.2	N 0 08.6	291 25.0	S16 26.2	109 28.8	N10 37.3	163 22.0	N22 10.1
	19	281 17.8	98 21.9	07.3	306 27.8	26.2	124 30.8	37.1	178 24.3	10.0
	20	296 20.3	113 21.5	06.0	321 30.6	26.1	139 32.7	36.9	193 26.5	10.0
	21	311 22.7	128 21.1	. . 04.8	336 33.5	. . 26.1	154 34.7	. . 36.8	208 28.8	. . 10.0
	22	326 25.2	143 20.8	03.5	351 36.3	26.0	169 36.7	36.6	223 31.1	10.0
	23	341 27.7	158 20.4	02.2	6 39.1	26.0	184 38.6	36.4	238 33.4	10.0
18	00	356 30.1	173 20.0	N 0 00.9	21 41.9	S16 25.9	199 40.6	N10 36.2	253 35.7	N22 10.0
	01	11 32.6	188 19.7	S 00.3	36 44.8	25.8	214 42.5	36.0	268 38.0	10.0
	02	26 35.1	203 19.3	01.6	51 47.6	25.8	229 44.5	35.8	283 40.3	09.9
	03	41 37.5	218 18.9	. . 02.9	66 50.4	. . 25.7	244 46.5	. . 35.6	298 42.5	. . 09.9
	04	56 40.0	233 18.5	04.2	81 53.2	25.7	259 48.4	35.4	313 44.8	09.9
	05	71 42.4	248 18.2	05.4	96 56.0	25.6	274 50.4	35.2	328 47.1	09.9
	06	86 44.9	263 17.8	S 0 06.7	111 58.9	S16 25.5	289 52.4	N10 35.1	343 49.4	N22 09.9
	07	101 47.4	278 17.4	08.0	127 01.7	25.5	304 54.3	34.9	358 51.7	09.9
THURSDAY	08	116 49.8	293 17.1	09.3	142 04.5	25.4	319 56.3	34.7	13 54.0	09.9
	09	131 52.3	308 16.7	. . 10.5	157 07.3	. . 25.3	334 58.2	. . 34.5	28 56.3	. . 09.8
	10	146 54.8	323 16.3	11.8	172 10.1	25.3	350 00.2	34.3	43 58.6	09.8
	11	161 57.2	338 16.0	13.1	187 12.9	25.2	5 02.2	34.1	59 00.8	09.8
	12	176 59.7	353 15.6	S 0 14.4	202 15.8	S16 25.1	20 04.1	N10 33.9	74 03.1	N22 09.8
	13	192 02.2	8 15.2	15.6	217 18.6	25.1	35 06.1	33.7	89 05.4	09.8
	14	207 04.6	23 14.9	16.9	232 21.4	25.0	50 08.1	33.5	104 07.7	09.8
	15	222 07.1	38 14.5	. . 18.2	247 24.2	. . 24.9	65 10.0	. . 33.4	119 10.0	. . 09.8
	16	237 09.6	53 14.1	19.5	262 27.0	24.9	80 12.0	33.2	134 12.3	09.7
	17	252 12.0	68 13.7	20.7	277 29.8	24.8	95 14.0	33.0	149 14.6	09.7
	18	267 14.5	83 13.4	S 0 22.0	292 32.6	S16 24.7	110 15.9	N10 32.8	164 16.9	N22 09.7
	19	282 16.9	98 13.0	23.3	307 35.4	24.7	125 17.9	32.6	179 19.2	09.7
	20	297 19.4	113 12.6	24.6	322 38.2	24.6	140 19.8	32.4	194 21.4	09.7
	21	312 21.9	128 12.3	. . 25.8	337 41.0	. . 24.5	155 21.8	. . 32.2	209 23.7	. . 09.7
	22	327 24.3	143 11.9	27.1	352 43.8	24.4	170 23.8	32.0	224 26.0	09.7
	23	342 26.8	158 11.5	28.4	7 46.6	24.4	185 25.7	31.9	239 28.3	09.6
Mer. Pass.		h m 0 17.9	v −0.4	d 1.3	v 2.8	d 0.1	v 2.0	d 0.2	v 2.3	d 0.0

STARS		
Name	SHA	Dec
	° ′	°
Acamar	315 23.8	S40 17
Achernar	335 31.7	S57 12
Acrux	173 19.0	S63 07
Adhara	255 18.7	S28 58
Aldebaran	290 58.1	N16 31
Alioth	166 27.7	N55 56
Alkaid	153 05.2	N49 17
Al Na'ir	27 52.8	S46 56
Alnilam	275 54.2	S 1 11
Alphard	218 03.9	S 8 40
Alphecca	126 17.6	N26 42
Alpheratz	357 51.2	N29 06
Altair	62 15.6	N 8 52
Ankaa	353 22.7	S42 17
Antares	112 35.8	S26 26
Arcturus	146 03.0	N19 10
Atria	107 44.8	S69 02
Avior	234 21.7	S59 30
Bellatrix	278 40.3	N 6 21
Betelgeuse	271 09.7	N 7 24
Canopus	263 59.7	S52 41
Capella	280 45.8	N46 00
Deneb	49 36.5	N45 17
Denebola	182 41.8	N14 33
Diphda	349 03.2	S17 57
Dubhe	194 01.4	N61 44
Elnath	278 22.3	N28 36
Eltanin	90 49.7	N51 29
Enif	33 54.4	N 9 53
Fomalhaut	15 32.0	S29 36
Gacrux	172 10.3	S57 07
Gienah	176 00.6	S17 33
Hadar	148 59.6	S60 23
Hamal	328 09.2	N23 28
Kaus Aust.	83 53.9	S34 23
Kochab	137 20.0	N74 08
Markab	13 45.7	N15 13
Menkar	314 22.9	N 4 06
Menkent	148 17.1	S36 23
Miaplacidus	221 42.4	S69 43
Mirfak	308 51.3	N49 52
Nunki	76 07.7	S26 17
Peacock	53 30.8	S56 43
Pollux	243 37.3	N28 01
Procyon	245 07.9	N 5 13
Rasalhague	96 13.6	N12 33
Regulus	207 52.0	N11 57
Rigel	281 19.4	S 8 11
Rigil Kent.	140 03.1	S60 51
Sabik	102 21.4	S15 43
Schedar	349 49.0	N56 33
Shaula	96 32.4	S37 06
Sirius	258 40.6	S16 42
Spica	158 39.7	S11 10
Suhail	222 58.6	S43 26
Vega	80 44.1	N38 47
Zuben'ubi	137 14.2	S16 03

	SHA	Mer. Pas
	° ′	h m
Venus	177 57.9	12 26
Mars	25 02.8	22 33
Jupiter	203 22.5	10 43
Saturn	257 09.9	7 08

UT d h	SUN GHA ° ′	SUN Dec ° ′	MOON GHA ° ′	MOON *v* ′	MOON Dec ° ′	MOON *d* ′	MOON HP ′
16 00	181 12.8	N 2 53.3	304 34.6	14.1	N18 23.2	10.1	54.2
01	196 13.1	52.3	319 07.7	14.0	18 33.3	9.9	54.2
02	211 13.3	51.3	333 40.7	14.0	18 43.2	10.0	54.2
03	226 13.5	. . 50.4	348 13.7	14.0	18 53.2	9.8	54.2
04	241 13.7	49.4	2 46.7	13.8	19 03.0	9.7	54.2
05	256 13.9	48.4	17 19.5	13.9	19 12.7	9.7	54.2
06	271 14.2	N 2 47.5	31 52.4	13.8	N19 22.4	9.6	54.2
07	286 14.4	46.5	46 25.2	13.7	19 32.0	9.5	54.2
T 08	301 14.6	45.6	60 57.9	13.7	19 41.5	9.4	54.2
U 09	316 14.8	. . 44.6	75 30.6	13.6	19 50.9	9.4	54.2
E 10	331 15.1	43.6	90 03.2	13.5	20 00.3	9.2	54.2
S 11	346 15.3	42.7	104 35.7	13.5	20 09.5	9.2	54.2
D 12	1 15.5	N 2 41.7	119 08.2	13.5	N20 18.7	9.1	54.2
A 13	16 15.7	40.7	133 40.7	13.3	20 27.8	9.0	54.2
Y 14	31 16.0	39.8	148 13.0	13.4	20 36.8	8.9	54.2
15	46 16.2	. . 38.8	162 45.4	13.2	20 45.7	8.8	54.2
16	61 16.4	37.9	177 17.6	13.2	20 54.5	8.7	54.2
17	76 16.6	36.9	191 49.8	13.2	21 03.2	8.7	54.2
18	91 16.8	N 2 35.9	206 22.0	13.1	N21 11.9	8.5	54.2
19	106 17.1	35.0	220 54.1	13.0	21 20.4	8.5	54.2
20	121 17.3	34.0	235 26.1	12.9	21 28.9	8.4	54.2
21	136 17.5	. . 33.0	249 58.0	13.0	21 37.3	8.3	54.2
22	151 17.7	32.1	264 30.0	12.8	21 45.6	8.2	54.2
23	166 18.0	31.1	279 01.8	12.8	21 53.8	8.1	54.2
17 00	181 18.2	N 2 30.1	293 33.6	12.7	N22 01.9	8.0	54.2
01	196 18.4	29.2	308 05.3	12.7	22 09.9	7.9	54.2
02	211 18.6	28.2	322 37.0	12.6	22 17.8	7.8	54.2
03	226 18.9	. . 27.2	337 08.6	12.5	22 25.6	7.7	54.2
04	241 19.1	26.3	351 40.1	12.5	22 33.3	7.6	54.2
05	256 19.3	25.3	6 11.6	12.4	22 40.9	7.5	54.2
06	271 19.5	N 2 24.3	20 43.0	12.3	N22 48.4	7.4	54.2
W 07	286 19.7	23.4	35 14.3	12.3	22 55.8	7.4	54.2
E 08	301 20.0	22.4	49 45.6	12.3	23 03.2	7.2	54.3
D 09	316 20.2	. . 21.5	64 16.9	12.1	23 10.4	7.1	54.3
N 10	331 20.4	20.5	78 48.0	12.1	23 17.5	7.0	54.3
E 11	346 20.6	19.5	93 19.1	12.1	23 24.5	6.9	54.3
S 12	1 20.9	N 2 18.6	107 50.2	12.0	N23 31.4	6.8	54.3
D 13	16 21.1	17.6	122 21.2	11.9	23 38.2	6.7	54.3
A 14	31 21.3	16.6	136 52.1	11.8	23 44.9	6.6	54.3
Y 15	46 21.5	. . 15.7	151 22.9	11.8	23 51.5	6.5	54.3
16	61 21.8	14.7	165 53.7	11.8	23 58.0	6.4	54.3
17	76 22.0	13.7	180 24.5	11.6	24 04.4	6.3	54.3
18	91 22.2	N 2 12.8	194 55.1	11.6	N24 10.7	6.2	54.3
19	106 22.4	11.8	209 25.7	11.6	24 16.9	6.0	54.3
20	121 22.7	10.8	223 56.3	11.5	24 22.9	6.0	54.4
21	136 22.9	. . 09.9	238 26.8	11.4	24 28.9	5.8	54.4
22	151 23.1	08.9	252 57.2	11.4	24 34.7	5.8	54.4
23	166 23.3	07.9	267 27.6	11.3	24 40.5	5.6	54.4
18 00	181 23.5	N 2 07.0	281 57.9	11.2	N24 46.1	5.5	54.4
01	196 23.8	06.0	296 28.1	11.2	24 51.6	5.4	54.4
02	211 24.0	05.0	310 58.3	11.1	24 57.0	5.3	54.4
03	226 24.2	. . 04.1	325 28.4	11.1	25 02.3	5.1	54.4
04	241 24.4	03.1	339 58.5	11.0	25 07.4	5.1	54.4
05	256 24.7	02.1	354 28.5	10.9	25 12.5	4.9	54.5
06	271 24.9	N 2 01.2	8 58.4	10.9	N25 17.4	4.8	54.5
T 07	286 25.1	2 00.2	23 28.3	10.8	25 22.2	4.7	54.5
H 08	301 25.3	1 59.2	37 58.1	10.8	25 26.9	4.6	54.5
U 09	316 25.6	. . 58.3	52 27.9	10.7	25 31.5	4.5	54.5
R 10	331 25.8	57.3	66 57.6	10.7	25 36.0	4.3	54.5
S 11	346 26.0	56.3	81 27.3	10.6	25 40.3	4.2	54.5
D 12	1 26.2	N 1 55.4	95 56.9	10.5	N25 44.5	4.1	54.5
A 13	16 26.4	54.4	110 26.4	10.5	25 48.6	4.0	54.6
Y 14	31 26.7	53.4	124 55.9	10.4	25 52.6	3.9	54.6
15	46 26.9	. . 52.5	139 25.3	10.4	25 56.5	3.7	54.6
16	61 27.1	51.5	153 54.7	10.3	26 00.2	3.6	54.6
17	76 27.3	50.5	168 24.0	10.3	26 03.8	3.5	54.6
18	91 27.6	N 1 49.5	182 53.3	10.2	N26 07.3	3.4	54.6
19	106 27.8	48.6	197 22.5	10.2	26 10.7	3.2	54.7
20	121 28.0	47.6	211 51.7	10.1	26 13.9	3.1	54.7
21	136 28.2	. . 46.6	226 20.8	10.1	26 17.0	3.0	54.7
22	151 28.4	45.7	240 49.9	10.0	26 20.0	2.9	54.7
23	166 28.7	44.7	255 18.9	10.0	N26 22.9	2.7	54.7
	SD 15.9	*d* 1.0	SD 14.8		14.8		14.9

Lat. °	Twilight Naut. h m	Twilight Civil h m	Sunrise h m	Moonrise 16 h m	Moonrise 17 h m	Moonrise 18 h m	Moonrise 19 h m
N 72	02 20	04 03	05 14	▭	▭	▭	▭
N 70	02 48	04 15	05 18	▭	▭	▭	▭
68	03 09	04 24	05 22	17 25	▭	▭	▭
66	03 25	04 32	05 25	18 14	17 06	▭	▭
64	03 38	04 39	05 27	18 45	18 34	▭	▭
62	03 48	04 44	05 29	19 08	19 12	19 23	19 56
60	03 57	04 49	05 31	19 27	19 38	20 00	20 41
N 58	04 05	04 53	05 33	19 43	19 59	20 26	21 10
56	04 12	04 57	05 34	19 56	20 17	20 47	21 33
54	04 17	05 00	05 36	20 08	20 31	21 04	21 51
52	04 22	05 03	05 37	20 18	20 44	21 19	22 07
50	04 27	05 05	05 38	20 28	20 55	21 32	22 20
45	04 36	05 11	05 40	20 47	21 19	21 59	22 48
N 40	04 43	05 15	05 42	21 03	21 38	22 20	23 10
35	04 49	05 19	05 44	21 17	21 54	22 37	23 28
30	04 53	05 21	05 45	21 28	22 08	22 52	23 43
20	05 00	05 26	05 48	21 49	22 31	23 18	24 10
N 10	05 04	05 29	05 50	22 06	22 52	23 40	24 32
0	05 07	05 31	05 51	22 23	23 11	24 01	00 01
S 10	05 08	05 32	05 53	22 40	23 30	24 22	00 22
20	05 07	05 33	05 55	22 58	23 51	24 45	00 45
30	05 05	05 33	05 56	23 19	24 15	00 15	01 11
35	05 03	05 32	05 57	23 31	24 29	00 29	01 26
40	05 00	05 31	05 58	23 45	24 46	00 46	01 44
45	04 56	05 30	06 00	24 02	00 02	01 05	02 06
S 50	04 51	05 29	06 01	24 23	00 23	01 30	02 33
52	04 49	05 28	06 02	24 33	00 33	01 42	02 46
54	04 46	05 27	06 02	24 44	00 44	01 56	03 02
56	04 43	05 26	06 03	24 57	00 57	02 12	03 20
58	04 39	05 25	06 04	25 12	01 12	02 31	03 43
S 60	04 35	05 23	06 05	00 03	01 30	02 55	04 12

Lat. °	Sunset h m	Twilight Civil h m	Twilight Naut. h m	Moonset 16 h m	Moonset 17 h m	Moonset 18 h m	Moonset 19 h m
N 72	18 33	19 43	21 23	▭	▭	▭	▭
N 70	18 29	19 32	20 56	▭	▭	▭	▭
68	18 25	19 22	20 36	14 43	▭	▭	▭
66	18 23	19 15	20 21	13 56	16 42	▭	▭
64	18 20	19 08	20 09	13 25	15 14	▭	▭
62	18 18	19 03	19 58	13 03	14 37	16 09	17 23
60	18 17	18 59	19 50	12 44	14 10	15 32	16 39
N 58	18 15	18 55	19 42	12 29	13 50	15 06	16 10
56	18 14	18 51	19 36	12 17	13 33	14 45	15 47
54	18 12	18 48	19 30	12 05	13 19	14 28	15 29
52	18 11	18 45	19 26	11 56	13 06	14 13	15 13
50	18 10	18 43	19 21	11 47	12 55	14 01	15 00
45	18 08	18 37	19 12	11 28	12 33	13 35	14 32
N 40	18 06	18 33	19 05	11 13	12 14	13 14	14 10
35	18 05	18 30	19 00	11 01	11 59	12 57	13 52
30	18 03	18 27	18 55	10 50	11 46	12 42	13 37
20	18 01	18 23	18 49	10 31	11 23	12 17	13 10
N 10	17 59	18 20	18 45	10 15	11 04	11 55	12 48
0	17 58	18 19	18 43	09 59	10 46	11 35	12 26
S 10	17 56	18 17	18 42	09 44	10 28	11 15	12 05
20	17 55	18 17	18 42	09 28	10 09	10 53	11 42
30	17 53	18 17	18 45	09 10	09 46	10 28	11 16
35	17 52	18 18	18 47	08 59	09 33	10 14	11 00
40	17 51	18 18	18 50	08 47	09 19	09 57	10 42
45	17 50	18 20	18 54	08 32	09 01	09 36	10 20
S 50	17 49	18 21	18 59	08 15	08 39	09 11	09 53
52	17 49	18 22	19 02	08 06	08 29	08 59	09 39
54	17 48	18 23	19 05	07 57	08 17	08 44	09 24
56	17 47	18 24	19 08	07 47	08 04	08 28	09 05
58	17 47	18 26	19 12	07 35	07 48	08 09	08 43
S 60	17 46	18 27	19 16	07 22	07 30	07 44	08 13

Day	SUN Eqn. of Time 00^h	SUN Eqn. of Time 12^h	SUN Mer. Pass.	MOON Mer. Pass. Upper	MOON Mer. Pass. Lower	MOON Age	MOON Phase
d	m s	m s	h m	h m	h m	d %	
16	04 51	05 02	11 55	03 49	16 11	20 71	
17	05 12	05 23	11 55	04 34	16 58	21 62	
18	05 34	05 44	11 54	05 23	17 48	22 53	

	UT	ARIES	VENUS −3.9		MARS −2.4		JUPITER −1.7		SATURN +0.1		STARS		
		GHA	GHA	Dec	GHA	Dec	GHA	Dec	GHA	Dec	Name	SHA	Dec
	d h	° ′	° ′	° ′	° ′	° ′	° ′	° ′	° ′	° ′		° ′	° ′
	19 00	357 29.3	173 11.2	S 0 29.6	22 49.4	S16 24.3	200 27.7	N10 31.7	254 30.6	N22 09.6	Acamar	315 23.8	S40 17.
	01	12 31.7	188 10.8	30.9	37 52.2	24.2	215 29.7	31.5	269 32.9	09.6	Achernar	335 31.7	S57 12.
	02	27 34.2	203 10.4	32.2	52 55.0	24.1	230 31.6	31.3	284 35.2	09.6	Acrux	173 18.9	S63 07.
	03	42 36.7	218 10.1	. . 33.5	67 57.8	. . 24.1	245 33.6	. . 31.1	299 37.5	. . 09.6	Adhara	255 18.7	S28 58.
	04	57 39.1	233 09.7	34.7	83 00.6	24.0	260 35.6	30.9	314 39.8	09.6	Aldebaran	290 58.1	N16 31.
	05	72 41.6	248 09.3	36.0	98 03.4	23.9	275 37.5	30.7	329 42.1	09.6			
	06	87 44.1	263 08.9	S 0 37.3	113 06.2	S16 23.8	290 39.5	N10 30.5	344 44.4	N22 09.6	Alioth	166 27.7	N55 56.
	07	102 46.5	278 08.6	38.6	128 08.9	23.8	305 41.5	30.3	359 46.6	09.5	Alkaid	153 05.2	N49 17.
	08	117 49.0	293 08.2	39.8	143 11.7	23.7	320 43.4	30.2	14 48.9	09.5	Al Na'ir	27 52.8	S46 56.
F	09	132 51.4	308 07.8	. . 41.1	158 14.5	. . 23.6	335 45.4	. . 30.0	29 51.2	. . 09.5	Alnilam	275 54.2	S 1 11.
R	10	147 53.9	323 07.5	42.4	173 17.3	23.5	350 47.4	29.8	44 53.5	09.5	Alphard	218 03.9	S 8 40.
I	11	162 56.4	338 07.1	43.7	188 20.1	23.4	5 49.3	29.6	59 55.8	09.5			
D	12	177 58.8	353 06.7	S 0 44.9	203 22.9	S16 23.4	20 51.3	N10 29.4	74 58.1	N22 09.5	Alphecca	126 17.7	N26 42.
A	13	193 01.3	8 06.4	46.2	218 25.6	23.3	35 53.2	29.2	90 00.4	09.5	Alpheratz	357 51.2	N29 06.
Y	14	208 03.8	23 06.0	47.5	233 28.4	23.2	50 55.2	29.0	105 02.7	09.4	Altair	62 15.6	N 8 52.
	15	223 06.2	38 05.6	. . 48.8	248 31.2	. . 23.1	65 57.2	. . 28.8	120 05.0	. . 09.4	Ankaa	353 22.7	S42 17.
	16	238 08.7	53 05.2	50.0	263 34.0	23.0	80 59.1	28.6	135 07.3	09.4	Antares	112 35.9	S26 26.
	17	253 11.2	68 04.9	51.3	278 36.8	23.0	96 01.1	28.5	150 09.6	09.4			
	18	268 13.6	83 04.5	S 0 52.6	293 39.5	S16 22.9	111 03.1	N10 28.3	165 11.9	N22 09.4	Arcturus	146 03.0	N19 10.
	19	283 16.1	98 04.1	53.9	308 42.3	22.8	126 05.0	28.1	180 14.2	09.4	Atria	107 44.8	S69 02.
	20	298 18.6	113 03.8	55.1	323 45.1	22.7	141 07.0	27.9	195 16.5	09.4	Avior	234 21.7	S59 30.
	21	313 21.0	128 03.4	. . 56.4	338 47.9	. . 22.6	156 09.0	. . 27.7	210 18.8	. . 09.3	Bellatrix	278 40.2	N 6 21.
	22	328 23.5	143 03.0	57.7	353 50.6	22.5	171 10.9	27.5	225 21.0	09.3	Betelgeuse	271 09.6	N 7 24.
	23	343 25.9	158 02.7	0 59.0	8 53.4	22.4	186 12.9	27.3	240 23.3	09.3			
	20 00	358 28.4	173 02.3	S 1 00.2	23 56.2	S16 22.4	201 14.9	N10 27.1	255 25.6	N22 09.3	Canopus	263 59.7	S52 41.
	01	13 30.9	188 01.9	01.5	38 58.9	22.3	216 16.8	26.9	270 27.9	09.3	Capella	280 45.8	N46 00.
	02	28 33.3	203 01.6	02.8	54 01.7	22.2	231 18.8	26.8	285 30.2	09.3	Deneb	49 36.5	N45 17.
	03	43 35.8	218 01.2	. . 04.1	69 04.5	. . 22.1	246 20.8	. . 26.6	300 32.5	. . 09.3	Denebola	182 41.8	N14 33.
	04	58 38.3	233 00.8	05.3	84 07.2	22.0	261 22.7	26.4	315 34.8	09.3	Diphda	349 03.2	S17 57.
	05	73 40.7	248 00.4	06.6	99 10.0	21.9	276 24.7	26.2	330 37.1	09.2			
	06	88 43.2	263 00.1	S 1 07.9	114 12.7	S16 21.8	291 26.7	N10 26.0	345 39.4	N22 09.2	Dubhe	194 01.4	N61 44.
S	07	103 45.7	277 59.7	09.2	129 15.5	21.7	306 28.6	25.8	0 41.7	09.2	Elnath	278 22.3	N28 36.
	08	118 48.1	292 59.3	10.4	144 18.3	21.6	321 30.6	25.6	15 44.0	09.2	Eltanin	90 49.7	N51 29.
A	09	133 50.6	307 59.0	. . 11.7	159 21.0	. . 21.5	336 32.6	. . 25.4	30 46.3	. . 09.2	Enif	33 54.4	N 9 53.
T	10	148 53.0	322 58.6	13.0	174 23.8	21.4	351 34.5	25.3	45 48.6	09.2	Fomalhaut	15 32.0	S29 36.
U	11	163 55.5	337 58.2	14.3	189 26.5	21.4	6 36.5	25.1	60 50.9	09.2			
R	12	178 58.0	352 57.8	S 1 15.5	204 29.3	S16 21.3	21 38.5	N10 24.9	75 53.2	N22 09.1	Gacrux	172 10.3	S57 07.
D	13	194 00.4	7 57.5	16.8	219 32.0	21.2	36 40.4	24.7	90 55.5	09.1	Gienah	176 00.6	S17 33
A	14	209 02.9	22 57.1	18.1	234 34.8	21.1	51 42.4	24.5	105 57.8	09.1	Hadar	148 59.6	S60 23.
Y	15	224 05.4	37 56.7	. . 19.4	249 37.5	. . 21.0	66 44.4	. . 24.3	121 00.1	. . 09.1	Hamal	328 09.2	N23 28.
	16	239 07.8	52 56.4	20.6	264 40.3	20.9	81 46.3	24.1	136 02.4	09.1	Kaus Aust.	83 53.9	S34 23.
	17	254 10.3	67 56.0	21.9	279 43.0	20.8	96 48.3	23.9	151 04.7	09.1			
	18	269 12.8	82 55.6	S 1 23.2	294 45.8	S16 20.7	111 50.3	N10 23.7	166 07.0	N22 09.1	Kochab	137 20.1	N74 08.
	19	284 15.2	97 55.3	24.5	309 48.5	20.6	126 52.2	23.6	181 09.3	09.1	Markab	13 45.7	N15 13.
	20	299 17.7	112 54.9	25.7	324 51.3	20.5	141 54.2	23.4	196 11.6	09.0	Menkar	314 22.9	N 4 06.
	21	314 20.2	127 54.5	. . 27.0	339 54.0	. . 20.4	156 56.2	. . 23.2	211 13.9	. . 09.0	Menkent	148 17.1	S36 23.
	22	329 22.6	142 54.1	28.3	354 56.7	20.3	171 58.1	23.0	226 16.2	09.0	Miaplacidus	221 42.4	S69 43.
	23	344 25.1	157 53.8	29.6	9 59.5	20.2	187 00.1	22.8	241 18.5	09.0			
	21 00	359 27.5	172 53.4	S 1 30.8	25 02.2	S16 20.1	202 02.1	N10 22.6	256 20.8	N22 09.0	Mirfak	308 51.2	N49 52.
	01	14 30.0	187 53.0	32.1	40 05.0	20.0	217 04.0	22.4	271 23.1	09.0	Nunki	76 07.7	S26 17.
	02	29 32.5	202 52.7	33.4	55 07.7	19.9	232 06.0	22.2	286 25.4	09.0	Peacock	53 30.8	S56 43
	03	44 34.9	217 52.3	. . 34.7	70 10.4	. . 19.8	247 08.0	. . 22.1	301 27.7	. . 08.9	Pollux	243 37.3	N28 01.
	04	59 37.4	232 51.9	35.9	85 13.2	19.7	262 09.9	21.9	316 30.0	08.9	Procyon	245 07.9	N 5 13.
	05	74 39.9	247 51.5	37.2	100 15.9	19.6	277 11.9	21.7	331 32.3	08.9			
	06	89 42.3	262 51.2	S 1 38.5	115 18.6	S16 19.5	292 13.9	N10 21.5	346 34.6	N22 08.9	Rasalhague	96 13.6	N12 33
	07	104 44.8	277 50.8	39.8	130 21.4	19.4	307 15.8	21.3	1 36.9	08.9	Regulus	207 51.9	N11 57.
	08	119 47.3	292 50.4	41.0	145 24.1	19.3	322 17.8	21.1	16 39.2	08.9	Rigel	281 19.4	S 8 11.
S	09	134 49.7	307 50.1	. . 42.3	160 26.8	. . 19.2	337 19.8	. . 20.9	31 41.5	. . 08.9	Rigil Kent.	140 03.1	S60 51.
U	10	149 52.2	322 49.7	43.6	175 29.5	19.0	352 21.8	20.7	46 43.8	08.9	Sabik	102 21.4	S15 43.
N	11	164 54.7	337 49.3	44.9	190 32.3	18.9	7 23.7	20.6	61 46.1	08.8			
D	12	179 57.1	352 48.9	S 1 46.1	205 35.0	S16 18.8	22 25.7	N10 20.4	76 48.4	N22 08.8	Schedar	349 49.0	N56 33.
A	13	194 59.6	7 48.6	47.4	220 37.7	18.7	37 27.7	20.2	91 50.7	08.8	Shaula	96 32.4	S37 06.
Y	14	210 02.0	22 48.2	48.7	235 40.4	18.6	52 29.6	20.0	106 53.0	08.8	Sirius	258 40.6	S16 42.
	15	225 04.5	37 47.8	. . 50.0	250 43.1	. . 18.5	67 31.6	. . 19.8	121 55.3	. . 08.8	Spica	158 39.7	S11 10.
	16	240 07.0	52 47.5	51.2	265 45.9	18.4	82 33.6	19.6	136 57.6	08.8	Suhail	222 58.5	S43 26.
	17	255 09.4	67 47.1	52.5	280 48.6	18.3	97 35.5	19.4	151 59.9	08.8			
	18	270 11.9	82 46.7	S 1 53.8	295 51.3	S16 18.2	112 37.5	N10 19.2	167 02.2	N22 08.7	Vega	80 44.1	N38 47.
	19	285 14.4	97 46.3	55.1	310 54.0	18.1	127 39.5	19.1	182 04.5	08.7	Zuben'ubi	137 14.2	S16 03.
	20	300 16.8	112 46.0	56.3	325 56.7	18.0	142 41.4	18.9	197 06.8	08.7		SHA	Mer. Pas
	21	315 19.3	127 45.6	. . 57.6	340 59.4	. . 17.8	157 43.4	. . 18.7	212 09.1	. . 08.7		° ′	h m
	22	330 21.8	142 45.2	1 58.9	356 02.1	17.7	172 45.4	18.5	227 11.4	08.7	Venus	174 33.9	12 28
	23	345 24.2	157 44.8	S 2 00.2	11 04.9	17.6	187 47.3	18.3	242 13.7	08.7	Mars	25 27.8	22 20
		h m									Jupiter	202 46.5	10 34
	Mer. Pass.	0 06.1	v −0.4	d 1.3	v 2.8	d 0.1	v 2.0	d 0.2	v 2.3	d 0.0	Saturn	256 57.2	6 57

JT	SUN GHA	SUN Dec	MOON GHA	v	MOON Dec	d	HP
d h	° ′	° ′	° ′	′	° ′	′	′
9 00	181 28.9	N 1 43.7	269 47.9	9.9	N26 25.6	2.6	54.7
01	196 29.1	42.8	284 16.8	9.9	26 28.2	2.5	54.8
02	211 29.3	41.8	298 45.7	9.8	26 30.7	2.3	54.8
03	226 29.6	. . 40.8	313 14.5	9.8	26 33.0	2.2	54.8
04	241 29.8	39.9	327 43.3	9.7	26 35.2	2.1	54.8
05	256 30.0	38.9	342 12.0	9.7	26 37.3	2.0	54.8
06	271 30.2	N 1 37.9	356 40.7	9.6	N26 39.3	1.8	54.9
07	286 30.5	37.0	11 09.3	9.6	26 41.1	1.7	54.9
08	301 30.7	36.0	25 37.9	9.6	26 42.8	1.5	54.9
09	316 30.9	. . 35.0	40 06.5	9.5	26 44.3	1.5	54.9
10	331 31.1	34.0	54 35.0	9.5	26 45.8	1.2	54.9
11	346 31.3	33.1	69 03.5	9.4	26 47.0	1.2	55.0
12	1 31.6	N 1 32.1	83 31.9	9.4	N26 48.2	1.0	55.0
13	16 31.8	31.1	98 00.3	9.4	26 49.2	0.9	55.0
14	31 32.0	30.2	112 28.7	9.3	26 50.1	0.7	55.0
15	46 32.2	. . 29.2	126 57.0	9.3	26 50.8	0.6	55.1
16	61 32.5	28.2	141 25.3	9.3	26 51.4	0.5	55.1
17	76 32.7	27.3	155 53.6	9.2	26 51.9	0.4	55.1
18	91 32.9	N 1 26.3	170 21.8	9.2	N26 52.3	0.2	55.1
19	106 33.1	25.3	184 50.0	9.2	26 52.5	0.0	55.1
20	121 33.3	24.4	199 18.2	9.1	26 52.5	0.1	55.2
21	136 33.6	. . 23.4	213 46.3	9.1	26 52.4	0.2	55.2
22	151 33.8	22.4	228 14.4	9.1	26 52.2	0.3	55.2
23	166 34.0	21.4	242 42.5	9.0	26 51.9	0.5	55.2
0 00	181 34.2	N 1 20.5	257 10.5	9.0	N26 51.4	0.7	55.3
01	196 34.5	19.5	271 38.5	9.0	26 50.7	0.7	55.3
02	211 34.7	18.5	286 06.5	9.0	26 50.0	1.0	55.3
03	226 34.9	. . 17.6	300 34.5	8.9	26 49.0	1.0	55.3
04	241 35.1	16.6	315 02.4	8.9	26 48.0	1.2	55.4
05	256 35.3	15.6	329 30.3	8.9	26 46.8	1.3	55.4
06	271 35.6	N 1 14.7	343 58.2	8.9	N26 45.5	1.5	55.4
07	286 35.8	13.7	358 26.1	8.9	26 44.0	1.6	55.4
08	301 36.0	12.7	12 54.0	8.8	26 42.4	1.8	55.5
09	316 36.2	. . 11.7	27 21.8	8.9	26 40.6	1.9	55.5
10	331 36.5	10.8	41 49.7	8.8	26 38.7	2.0	55.5
11	346 36.7	09.8	56 17.5	8.8	26 36.7	2.2	55.6
12	1 36.9	N 1 08.8	70 45.3	8.7	N26 34.5	2.4	55.6
13	16 37.1	07.9	85 13.0	8.8	26 32.1	2.4	55.6
14	31 37.3	06.9	99 40.8	8.8	26 29.7	2.7	55.6
15	46 37.6	. . 05.9	114 08.6	8.7	26 27.0	2.7	55.7
16	61 37.8	04.9	128 36.3	8.8	26 24.3	2.9	55.7
17	76 38.0	04.0	143 04.1	8.7	26 21.4	3.1	55.7
18	91 38.2	N 1 03.0	157 31.8	8.7	N26 18.3	3.1	55.8
19	106 38.5	02.0	171 59.5	8.7	26 15.2	3.4	55.8
20	121 38.7	01.1	186 27.2	8.7	26 11.8	3.4	55.8
21	136 38.9	1 00.1	200 54.9	8.7	26 08.4	3.7	55.8
22	151 39.1	0 59.1	215 22.6	8.7	26 04.7	3.7	55.9
23	166 39.3	58.1	229 50.3	8.7	26 01.0	3.9	55.9
1 00	181 39.6	N 0 57.2	244 18.0	8.8	N25 57.1	4.1	55.9
01	196 39.8	56.2	258 45.8	8.7	25 53.0	4.2	56.0
02	211 40.0	55.2	273 13.5	8.7	25 48.8	4.3	56.0
03	226 40.2	. . 54.3	287 41.2	8.7	25 44.5	4.5	56.0
04	241 40.4	53.3	302 08.9	8.7	25 40.0	4.6	56.1
05	256 40.7	52.3	316 36.6	8.7	25 35.4	4.7	56.1
06	271 40.9	N 0 51.3	331 04.3	8.7	N25 30.7	4.9	56.1
07	286 41.1	50.4	345 32.0	8.7	25 25.8	5.1	56.1
08	301 41.3	49.4	359 59.7	8.8	25 20.7	5.2	56.2
09	316 41.5	. . 48.4	14 27.5	8.7	25 15.5	5.3	56.2
10	331 41.8	47.5	28 55.2	8.7	25 10.2	5.5	56.2
11	346 42.0	46.5	43 22.9	8.8	25 04.7	5.6	56.3
12	1 42.2	N 0 45.5	57 50.7	8.8	N24 59.1	5.7	56.3
13	16 42.4	44.5	72 18.5	8.8	24 53.4	5.9	56.3
14	31 42.7	43.6	86 46.3	8.7	24 47.5	6.1	56.4
15	46 42.9	. . 42.6	101 14.0	8.9	24 41.4	6.1	56.4
16	61 43.1	41.6	115 41.9	8.8	24 35.3	6.4	56.4
17	76 43.3	40.7	130 09.7	8.8	24 28.9	6.4	56.5
18	91 43.5	N 0 39.7	144 37.5	8.9	N24 22.5	6.6	56.5
19	106 43.8	38.7	159 05.4	8.8	24 15.9	6.7	56.5
20	121 44.0	37.7	173 33.2	8.9	24 09.2	6.9	56.6
21	136 44.2	. . 36.8	188 01.1	8.9	24 02.3	7.0	56.6
22	151 44.4	35.8	202 29.0	8.9	23 55.3	7.2	56.6
23	166 44.6	34.8	216 56.9	9.0	N23 48.1	7.3	56.7
	SD 16.0	d 1.0	SD 15.0		15.1		15.3

Lat.	Twilight Naut.	Twilight Civil	Sunrise	Moonrise 19	Moonrise 20	Moonrise 21	Moonrise 22
°	h m	h m	h m	h m	h m	h m	h m
N 72	02 42	04 18	05 27	▭	▭	▭	▭
N 70	03 05	04 28	05 30	▭	▭	▭	▭
68	03 23	04 36	05 32	▭	▭	▭	23 02
66	03 37	04 42	05 34	▭	▭	20 43	23 45
64	03 48	04 48	05 36	▭	▭	22 12	24 14
62	03 58	04 52	05 37	19 56	21 10	22 49	24 35
60	04 06	04 56	05 38	20 41	21 48	23 16	24 53
N 58	04 12	05 00	05 39	21 10	22 14	23 36	25 07
56	04 18	05 03	05 40	21 33	22 35	23 53	25 20
54	04 23	05 06	05 41	21 51	22 53	24 07	00 07
52	04 28	05 08	05 42	22 07	23 07	24 20	00 20
50	04 32	05 10	05 42	22 20	23 20	24 31	00 31
45	04 40	05 15	05 44	22 48	23 47	24 54	00 54
N 40	04 46	05 18	05 45	23 10	24 07	00 07	01 12
35	04 51	05 21	05 46	23 28	24 25	00 25	01 27
30	04 55	05 23	05 47	23 43	24 40	00 40	01 40
20	05 01	05 26	05 48	24 10	00 10	01 05	02 03
N 10	05 04	05 28	05 49	24 32	00 32	01 27	02 22
0	05 06	05 30	05 50	00 01	00 54	01 47	02 41
S 10	05 06	05 30	05 51	00 22	01 15	02 07	02 59
20	05 04	05 30	05 52	00 45	01 38	02 29	03 18
30	05 01	05 29	05 53	01 11	02 04	02 54	03 40
35	04 58	05 28	05 53	01 26	02 20	03 09	03 53
40	04 55	05 27	05 53	01 44	02 38	03 26	04 08
45	04 51	05 25	05 54	02 06	03 00	03 47	04 26
S 50	04 44	05 22	05 54	02 33	03 28	04 12	04 48
52	04 42	05 21	05 55	02 46	03 41	04 25	04 58
54	04 38	05 20	05 55	03 02	03 57	04 39	05 10
56	04 34	05 18	05 55	03 20	04 16	04 56	05 23
58	04 30	05 16	05 55	03 43	04 38	05 15	05 38
S 60	04 25	05 14	05 56	04 12	05 08	05 40	05 57

Lat.	Sunset	Twilight Civil	Twilight Naut.	Moonset 19	Moonset 20	Moonset 21	Moonset 22
°	h m	h m	h m	h m	h m	h m	h m
N 72	18 17	19 26	21 00	▭	▭	▭	▭
N 70	18 15	19 17	20 37	▭	▭	▭	▭
68	18 13	19 09	20 21	▭	▭	▭	19 53
66	18 11	19 03	20 07	▭	▭	20 21	19 09
64	18 10	18 57	19 56	▭	▭	18 51	18 40
62	18 08	18 53	19 47	17 23	18 01	18 13	18 17
60	18 07	18 49	19 40	16 39	17 23	17 46	17 59
N 58	18 06	18 46	19 33	16 10	16 56	17 25	17 44
56	18 06	18 43	19 27	15 47	16 35	17 08	17 31
54	18 05	18 40	19 22	15 29	16 17	16 53	17 19
52	18 04	18 38	19 18	15 13	16 02	16 40	17 09
50	18 04	18 36	19 14	15 00	15 49	16 29	17 00
45	18 02	18 32	19 06	14 32	15 23	16 05	16 41
N 40	18 01	18 28	19 00	14 10	15 01	15 46	16 25
35	18 00	18 26	18 55	13 52	14 44	15 30	16 12
30	18 00	18 24	18 51	13 37	14 29	15 17	16 00
20	17 59	18 21	18 46	13 10	14 03	14 53	15 40
N 10	17 58	18 19	18 43	12 48	13 40	14 32	15 22
0	17 57	18 17	18 41	12 26	13 19	14 13	15 06
S 10	17 56	18 17	18 41	12 05	12 58	13 53	14 49
20	17 55	18 17	18 43	11 42	12 36	13 33	14 31
30	17 55	18 19	18 47	11 16	12 10	13 08	14 11
35	17 55	18 20	18 49	11 00	11 54	12 54	13 58
40	17 54	18 21	18 53	10 42	11 36	12 37	13 44
45	17 54	18 23	18 58	10 20	11 14	12 17	13 28
S 50	17 54	18 26	19 04	09 53	10 47	11 52	13 07
52	17 53	18 27	19 07	09 39	10 33	11 40	12 57
54	17 53	18 29	19 10	09 24	10 18	11 26	12 46
56	17 53	18 30	19 14	09 05	09 59	11 10	12 33
58	17 53	18 32	19 19	08 43	09 37	10 51	12 18
S 60	17 53	18 34	19 24	08 13	09 07	10 26	12 00

Day	SUN Eqn. of Time 00^h	SUN Eqn. of Time 12^h	SUN Mer. Pass.	MOON Mer. Pass. Upper	MOON Mer. Pass. Lower	MOON Age	MOON Phase
d	m s	m s	h m	h m	h m	d %	
19	05 55	06 06	11 54	06 14	18 40	23 43	
20	06 16	06 27	11 54	07 06	19 33	24 34	
21	06 38	06 48	11 53	08 00	20 27	25 25	

UT	ARIES	VENUS −3.9		MARS −2.3		JUPITER −1.7		SATURN +0.1		STARS		
	GHA	GHA	Dec	GHA	Dec	GHA	Dec	GHA	Dec	Name	SHA	Dec
d h	° ′	° ′	° ′	° ′	° ′	° ′	° ′	° ′	° ′		° ′	°
22 00	0 26.7	172 44.5	S 2 01.4	26 07.6	S16 17.5	202 49.3	N10 18.1	257 16.0	N22 08.7	Acamar	315 23.8	S40 17
01	15 29.2	187 44.1	02.7	41 10.3	17.4	217 51.3	17.9	272 18.3	08.7	Achernar	335 31.7	S57 12
02	30 31.6	202 43.7	04.0	56 13.0	17.3	232 53.3	17.7	287 20.6	08.6	Acrux	173 19.0	S63 07
03	45 34.1	217 43.4	. . 05.2	71 15.7	. . 17.1	247 55.2	. . 17.5	302 22.9	. . 08.6	Adhara	255 18.7	S28 58
04	60 36.5	232 43.0	06.5	86 18.4	17.0	262 57.2	17.4	317 25.2	08.6	Aldebaran	290 58.1	N16 31
05	75 39.0	247 42.6	07.8	101 21.1	16.9	277 59.2	17.2	332 27.5	08.6			
06	90 41.5	262 42.2	S 2 09.1	116 23.8	S16 16.8	293 01.1	N10 17.0	347 29.8	N22 08.6	Alioth	166 27.7	N55 56
07	105 43.9	277 41.9	10.3	131 26.5	16.7	308 03.1	16.8	2 32.1	08.6	Alkaid	153 05.2	N49 17
08	120 46.4	292 41.5	11.6	146 29.2	16.6	323 05.1	16.6	17 34.4	08.6	Al Na'ir	27 52.8	S46 56
M 09	135 48.9	307 41.1	. . 12.9	161 31.9	. . 16.4	338 07.0	. . 16.4	32 36.7	. . 08.6	Alnilam	275 54.1	S 1 11
O 10	150 51.3	322 40.7	14.2	176 34.6	16.3	353 09.0	16.2	47 39.0	08.5	Alphard	218 03.9	S 8 40
N 11	165 53.8	337 40.4	15.4	191 37.3	16.2	8 11.0	16.0	62 41.3	08.5			
D 12	180 56.3	352 40.0	S 2 16.7	206 40.0	S16 16.1	23 12.9	N10 15.9	77 43.6	N22 08.5	Alphecca	126 17.7	N26 42
A 13	195 58.7	7 39.6	18.0	221 42.7	16.0	38 14.9	15.7	92 45.9	08.5	Alpheratz	357 51.2	N29 06
Y 14	211 01.2	22 39.2	19.3	236 45.3	15.8	53 16.9	15.5	107 48.2	08.5	Altair	62 15.6	N 8 52
15	226 03.6	37 38.9	. . 20.5	251 48.0	. . 15.7	68 18.9	. . 15.3	122 50.5	. . 08.5	Ankaa	353 22.7	S42 17
16	241 06.1	52 38.5	21.8	266 50.7	15.6	83 20.8	15.1	137 52.9	08.5	Antares	112 35.9	S26 26
17	256 08.6	67 38.1	23.1	281 53.4	15.5	98 22.8	14.9	152 55.2	08.5			
18	271 11.0	82 37.8	S 2 24.4	296 56.1	S16 15.3	113 24.8	N10 14.7	167 57.5	N22 08.4	Arcturus	146 03.0	N19 10
19	286 13.5	97 37.4	25.6	311 58.8	15.2	128 26.7	14.5	182 59.8	08.4	Atria	107 44.9	S69 02
20	301 16.0	112 37.0	26.9	327 01.5	15.1	143 28.7	14.4	198 02.1	08.4	Avior	234 21.7	S59 30
21	316 18.4	127 36.6	. . 28.2	342 04.1	. . 15.0	158 30.7	. . 14.2	213 04.4	. . 08.4	Bellatrix	278 40.2	N 6 21
22	331 20.9	142 36.3	29.4	357 06.8	14.8	173 32.7	14.0	228 06.7	08.4	Betelgeuse	271 09.6	N 7 24
23	346 23.4	157 35.9	30.7	12 09.5	14.7	188 34.6	13.8	243 09.0	08.4			
23 00	1 25.8	172 35.5	S 2 32.0	27 12.2	S16 14.6	203 36.6	N10 13.6	258 11.3	N22 08.4	Canopus	263 59.6	S52 41
01	16 28.3	187 35.1	33.3	42 14.8	14.5	218 38.6	13.4	273 13.6	08.3	Capella	280 45.7	N46 00
02	31 30.8	202 34.8	34.5	57 17.5	14.3	233 40.5	13.2	288 15.9	08.3	Deneb	49 36.5	N45 17
03	46 33.2	217 34.4	. . 35.8	72 20.2	. . 14.2	248 42.5	. . 13.1	303 18.2	. . 08.3	Denebola	182 41.8	N14 33
04	61 35.7	232 34.0	37.1	87 22.9	14.1	263 44.5	12.9	318 20.5	08.3	Diphda	349 03.2	S17 57
05	76 38.1	247 33.6	38.4	102 25.5	13.9	278 46.4	12.7	333 22.8	08.3			
06	91 40.6	262 33.3	S 2 39.6	117 28.2	S16 13.8	293 48.4	N10 12.5	348 25.2	N22 08.3	Dubhe	194 01.4	N61 43
07	106 43.1	277 32.9	40.9	132 30.9	13.7	308 50.4	12.3	3 27.5	08.3	Elnath	278 22.3	N28 36
T 08	121 45.5	292 32.5	42.2	147 33.5	13.5	323 52.4	12.1	18 29.8	08.3	Eltanin	90 49.7	N51 29
U 09	136 48.0	307 32.1	. . 43.4	162 36.2	. . 13.4	338 54.3	. . 11.9	33 32.1	. . 08.2	Enif	33 54.4	N 9 53
E 10	151 50.5	322 31.8	44.7	177 38.9	13.3	353 56.3	11.7	48 34.4	08.2	Fomalhaut	15 32.0	S29 36
S 11	166 52.9	337 31.4	46.0	192 41.5	13.1	8 58.3	11.6	63 36.7	08.2			
D 12	181 55.4	352 31.0	S 2 47.3	207 44.2	S16 13.0	24 00.2	N10 11.4	78 39.0	N22 08.2	Gacrux	172 10.3	S57 07
A 13	196 57.9	7 30.6	48.5	222 46.9	12.9	39 02.2	11.2	93 41.3	08.2	Gienah	176 00.6	S17 33
Y 14	212 00.3	22 30.3	49.8	237 49.5	12.7	54 04.2	11.0	108 43.6	08.2	Hadar	148 59.7	S60 23
15	227 02.8	37 29.9	. . 51.1	252 52.2	. . 12.6	69 06.2	. . 10.8	123 45.9	. . 08.2	Hamal	328 09.2	N23 28
16	242 05.3	52 29.5	52.4	267 54.8	12.5	84 08.1	10.6	138 48.2	08.2	Kaus Aust.	83 53.9	S34 23
17	257 07.7	67 29.1	53.6	282 57.5	12.3	99 10.1	10.4	153 50.6	08.1			
18	272 10.2	82 28.7	S 2 54.9	298 00.1	S16 12.2	114 12.1	N10 10.2	168 52.9	N22 08.1	Kochab	137 20.1	N74 08
19	287 12.6	97 28.4	56.2	313 02.8	12.0	129 14.1	10.1	183 55.2	08.1	Markab	13 45.7	N15 13
20	302 15.1	112 28.0	57.4	328 05.4	11.9	144 16.0	09.9	198 57.5	08.1	Menkar	314 22.9	N 4 06
21	317 17.6	127 27.6	2 58.7	343 08.1	. . 11.8	159 18.0	. . 09.7	213 59.8	. . 08.1	Menkent	148 17.1	S36 23
22	332 20.0	142 27.2	3 00.0	358 10.7	11.6	174 20.0	09.5	229 02.1	08.1	Miaplacidus	221 42.3	S69 43
23	347 22.5	157 26.9	01.3	13 13.4	11.5	189 21.9	09.3	244 04.4	08.1			
24 00	2 25.0	172 26.5	S 3 02.5	28 16.0	S16 11.3	204 23.9	N10 09.1	259 06.7	N22 08.1	Mirfak	308 51.2	N49 52
01	17 27.4	187 26.1	03.8	43 18.7	11.2	219 25.9	08.9	274 09.0	08.0	Nunki	76 07.7	S26 17
02	32 29.9	202 25.7	05.1	58 21.3	11.1	234 27.9	08.7	289 11.4	08.0	Peacock	53 30.8	S56 43
03	47 32.4	217 25.4	. . 06.3	73 24.0	. . 10.9	249 29.8	. . 08.6	304 13.7	. . 08.0	Pollux	243 37.2	N28 01
04	62 34.8	232 25.0	07.6	88 26.6	10.8	264 31.8	08.4	319 16.0	08.0	Procyon	245 07.9	N 5 13
05	77 37.3	247 24.6	08.9	103 29.2	10.6	279 33.8	08.2	334 18.3	08.0			
06	92 39.7	262 24.2	S 3 10.1	118 31.9	S16 10.5	294 35.8	N10 08.0	349 20.6	N22 08.0	Rasalhague	96 13.6	N12 33
W 07	107 42.2	277 23.8	11.4	133 34.5	10.3	309 37.7	07.8	4 22.9	08.0	Regulus	207 51.9	N11 57
E 08	122 44.7	292 23.5	12.7	148 37.1	10.2	324 39.7	07.6	19 25.2	08.0	Rigel	281 19.4	S 8 11
D 09	137 47.1	307 23.1	. . 14.0	163 39.8	. . 10.0	339 41.7	. . 07.4	34 27.5	. . 07.9	Rigil Kent.	140 03.1	S60 51
N 10	152 49.6	322 22.7	15.2	178 42.4	09.9	354 43.6	07.3	49 29.9	07.9	Sabik	102 21.4	S15 43
E 11	167 52.1	337 22.3	16.5	193 45.0	09.8	9 45.6	07.1	64 32.2	07.9			
S 12	182 54.5	352 22.0	S 3 17.8	208 47.7	S16 09.6	24 47.6	N10 06.9	79 34.5	N22 07.9	Schedar	349 49.0	N56 33
D 13	197 57.0	7 21.6	19.0	223 50.3	09.5	39 49.6	06.7	94 36.8	07.9	Shaula	96 32.4	S37 06
A 14	212 59.5	22 21.2	20.3	238 52.9	09.3	54 51.5	06.5	109 39.1	07.9	Sirius	258 40.6	S16 42
Y 15	228 01.9	37 20.8	. . 21.6	253 55.6	. . 09.2	69 53.5	. . 06.3	124 41.4	. . 07.9	Spica	158 39.7	S11 10
16	243 04.4	52 20.4	22.9	268 58.2	09.0	84 55.5	06.1	139 43.7	07.9	Suhail	222 58.5	S43 26
17	258 06.9	67 20.1	24.1	284 00.8	08.9	99 57.5	05.9	154 46.1	07.9			
18	273 09.3	82 19.7	S 3 25.4	299 03.4	S16 08.7	114 59.4	N10 05.8	169 48.4	N22 07.8	Vega	80 44.1	N38 47
19	288 11.8	97 19.3	26.7	314 06.0	08.6	130 01.4	05.6	184 50.7	07.8	Zuben'ubi	137 14.2	S16 03
20	303 14.2	112 18.9	27.9	329 08.7	08.4	145 03.4	05.4	199 53.0	07.8			
21	318 16.7	127 18.5	. . 29.2	344 11.3	. . 08.2	160 05.4	. . 05.2	214 55.3	. . 07.8			
22	333 19.2	142 18.2	30.5	359 13.9	08.1	175 07.3	05.0	229 57.6	07.8			
23	348 21.6	157 17.8	31.7	14 16.5	07.9	190 09.3	04.8	244 59.9	07.8			
Mer. Pass.	h m 23 50.4	v −0.4	d 1.3	v 2.7	d 0.1	v 2.0	d 0.2	v 2.3	d 0.0			

	SHA	Mer. Pas
	° ′	h m
Venus	171 09.7	12 30
Mars	25 46.3	22 07
Jupiter	202 10.8	10 24
Saturn	256 45.5	6 46

UT	SUN GHA	SUN Dec	MOON GHA	v	MOON Dec	d	HP
d h	° ′	° ′	° ′	′	° ′	′	′
22 00	181 44.9	N 0 33.8	231 24.9	9.0	N23 40.8	7.4	56.7
01	196 45.1	32.9	245 52.9	8.9	23 33.4	7.5	56.7
02	211 45.3	31.9	260 20.8	9.0	23 25.9	7.7	56.8
03	226 45.5	. . 30.9	274 48.8	9.1	23 18.2	7.9	56.8
04	241 45.7	30.0	289 16.9	9.0	23 10.3	7.9	56.8
05	256 46.0	29.0	303 44.9	9.1	23 02.4	8.1	56.9
06	271 46.2	N 0 28.0	318 13.0	9.1	N22 54.3	8.3	56.9
07	286 46.4	27.0	332 41.1	9.1	22 46.0	8.3	57.0
08	301 46.6	26.1	347 09.2	9.2	22 37.7	8.5	57.0
09	316 46.8	. . 25.1	1 37.4	9.2	22 29.2	8.7	57.0
10	331 47.1	24.1	16 05.6	9.2	22 20.5	8.7	57.1
11	346 47.3	23.2	30 33.8	9.2	22 11.8	8.9	57.1
12	1 47.5	N 0 22.2	45 02.0	9.2	N22 02.9	9.1	57.1
13	16 47.7	21.2	59 30.2	9.3	21 53.8	9.1	57.2
14	31 47.9	20.2	73 58.5	9.3	21 44.7	9.3	57.2
15	46 48.2	. . 19.3	88 26.8	9.4	21 35.4	9.4	57.2
16	61 48.4	18.3	102 55.2	9.4	21 26.0	9.6	57.3
17	76 48.6	17.3	117 23.6	9.4	21 16.4	9.6	57.3
18	91 48.8	N 0 16.3	131 52.0	9.4	N21 06.8	9.8	57.3
19	106 49.0	15.4	146 20.4	9.4	20 57.0	10.0	57.4
20	121 49.3	14.4	160 48.8	9.5	20 47.0	10.0	57.4
21	136 49.5	. . 13.4	175 17.3	9.5	20 37.0	10.2	57.4
22	151 49.7	12.4	189 45.8	9.6	20 26.8	10.3	57.5
23	166 49.9	11.5	204 14.4	9.6	20 16.5	10.4	57.5
23 00	181 50.1	N 0 10.5	218 43.0	9.6	N20 06.1	10.5	57.6
01	196 50.4	09.5	233 11.6	9.6	19 55.6	10.7	57.6
02	211 50.6	08.6	247 40.2	9.7	19 44.9	10.8	57.6
03	226 50.8	. . 07.6	262 08.9	9.7	19 34.1	10.9	57.7
04	241 51.0	06.6	276 37.6	9.7	19 23.2	11.0	57.7
05	256 51.2	05.6	291 06.3	9.8	19 12.2	11.1	57.7
06	271 51.4	N 0 04.7	305 35.1	9.8	N19 01.1	11.3	57.8
07	286 51.7	03.7	320 03.9	9.8	18 49.8	11.4	57.8
08	301 51.9	02.7	334 32.7	9.9	18 38.4	11.5	57.8
09	316 52.1	. . 01.7	349 01.6	9.9	18 26.9	11.6	57.9
10	331 52.3	N 00.8	3 30.5	9.9	18 15.3	11.7	57.9
11	346 52.5	S 00.2	17 59.4	10.0	18 03.6	11.8	57.9
12	1 52.8	S 0 01.2	32 28.4	10.0	N17 51.8	11.9	58.0
13	16 53.0	02.2	46 57.4	10.0	17 39.9	12.1	58.0
14	31 53.2	03.1	61 26.4	10.0	17 27.8	12.1	58.1
15	46 53.4	. . 04.1	75 55.4	10.1	17 15.7	12.3	58.1
16	61 53.6	05.1	90 24.5	10.1	17 03.4	12.3	58.1
17	76 53.9	06.0	104 53.6	10.2	16 51.1	12.5	58.2
18	91 54.1	S 0 07.0	119 22.8	10.1	N16 38.6	12.6	58.2
19	106 54.3	08.0	133 51.9	10.2	16 26.0	12.7	58.2
20	121 54.5	09.0	148 21.1	10.3	16 13.3	12.8	58.3
21	136 54.7	. . 09.9	162 50.4	10.2	16 00.5	12.8	58.3
22	151 54.9	10.9	177 19.6	10.3	15 47.7	13.0	58.3
23	166 55.2	11.9	191 48.9	10.3	15 34.7	13.1	58.4
24 00	181 55.4	S 0 12.9	206 18.2	10.4	N15 21.6	13.2	58.4
01	196 55.6	13.8	220 47.6	10.4	15 08.4	13.3	58.4
02	211 55.8	14.8	235 17.0	10.4	14 55.1	13.4	58.5
03	226 56.0	. . 15.8	249 46.4	10.4	14 41.7	13.4	58.5
04	241 56.3	16.8	264 15.8	10.5	14 28.3	13.6	58.5
05	256 56.5	17.7	278 45.3	10.4	14 14.7	13.7	58.6
06	271 56.7	S 0 18.7	293 14.7	10.5	N14 01.0	13.7	58.6
07	286 56.9	19.7	307 44.2	10.6	13 47.3	13.8	58.6
08	301 57.1	20.7	322 13.8	10.5	13 33.5	14.0	58.7
09	316 57.3	. . 21.6	336 43.3	10.6	13 19.5	14.0	58.7
10	331 57.6	22.6	351 12.9	10.6	13 05.5	14.1	58.7
11	346 57.8	23.6	5 42.5	10.7	12 51.4	14.2	58.8
12	1 58.0	S 0 24.6	20 12.2	10.6	N12 37.2	14.2	58.8
13	16 58.2	25.5	34 41.8	10.7	12 23.0	14.4	58.8
14	31 58.4	26.5	49 11.5	10.7	12 08.6	14.4	58.9
15	46 58.6	. . 27.5	63 41.2	10.7	11 54.2	14.5	58.9
16	61 58.9	28.5	78 10.9	10.8	11 39.7	14.6	58.9
17	76 59.1	29.4	92 40.7	10.7	11 25.1	14.7	59.0
18	91 59.3	S 0 30.4	107 10.4	10.8	N11 10.4	14.7	59.0
19	106 59.5	31.4	121 40.2	10.8	10 55.7	14.8	59.0
20	121 59.7	32.3	136 10.0	10.8	10 40.9	14.9	59.1
21	136 59.9	. . 33.3	150 39.8	10.9	10 26.0	15.0	59.1
22	152 00.2	34.3	165 09.7	10.8	10 11.0	15.0	59.1
23	167 00.4	35.3	179 39.5	10.9	N 9 56.0	15.1	59.2
	SD 16.0	*d* 1.0	SD	15.6		15.8	16.0

Lat.	Twilight Naut.	Twilight Civil	Sunrise	Moonrise 22	23	24	25
°	h m	h m	h m	h m	h m	h m	h m
N 72	03 01	04 32	05 41	▭	▭	00 31	03 18
N 70	03 21	04 40	05 42	▭	▭	01 08	03 32
68	03 36	04 47	05 43	23 02	25 34	01 34	03 43
66	03 48	04 52	05 44	23 45	25 53	01 53	03 52
64	03 58	04 57	05 44	24 14	00 14	02 09	04 00
62	04 07	05 00	05 45	24 35	00 35	02 22	04 07
60	04 14	05 04	05 45	24 53	00 53	02 32	04 12
N 58	04 19	05 06	05 46	25 07	01 07	02 42	04 17
56	04 25	05 09	05 46	25 20	01 20	02 50	04 22
54	04 29	05 11	05 46	00 07	01 30	02 57	04 26
52	04 33	05 13	05 47	00 20	01 40	03 04	04 29
50	04 37	05 15	05 47	00 31	01 48	03 10	04 32
45	04 44	05 18	05 48	00 54	02 06	03 22	04 39
N 40	04 49	05 21	05 48	01 12	02 21	03 32	04 45
35	04 54	05 23	05 48	01 27	02 33	03 41	04 50
30	04 57	05 25	05 49	01 40	02 44	03 49	04 54
20	05 01	05 27	05 49	02 03	03 02	04 02	05 02
N 10	05 04	05 28	05 49	02 22	03 18	04 14	05 08
0	05 05	05 29	05 49	02 41	03 33	04 24	05 14
S 10	05 04	05 28	05 49	02 59	03 48	04 35	05 21
20	05 02	05 27	05 49	03 18	04 04	04 46	05 27
30	04 57	05 25	05 49	03 40	04 22	04 59	05 34
35	04 54	05 24	05 49	03 53	04 32	05 07	05 39
40	04 50	05 22	05 49	04 08	04 44	05 15	05 44
45	04 45	05 19	05 48	04 26	04 58	05 25	05 49
S 50	04 38	05 16	05 48	04 48	05 15	05 37	05 56
52	04 34	05 14	05 48	04 58	05 23	05 42	05 59
54	04 30	05 12	05 47	05 10	05 32	05 48	06 02
56	04 26	05 10	05 47	05 23	05 42	05 55	06 06
58	04 21	05 08	05 47	05 38	05 53	06 02	06 10
S 60	04 15	05 05	05 46	05 57	06 06	06 11	06 14

Lat.	Sunset	Twilight Civil	Twilight Naut.	Moonset 22	23	24	25
°	h m	h m	h m	h m	h m	h m	h m
N 72	18 02	19 10	20 39	▭	20 16	19 16	18 39
N 70	18 01	19 02	20 20	▭	19 37	18 59	18 32
68	18 00	18 56	20 05	19 53	19 10	18 46	18 27
66	18 00	18 51	19 54	19 09	18 49	18 35	18 23
64	17 59	18 46	19 44	18 40	18 32	18 25	18 19
62	17 59	18 43	19 36	18 17	18 18	18 17	18 16
60	17 58	18 40	19 30	17 59	18 06	18 10	18 13
N 58	17 58	18 37	19 24	17 44	17 56	18 04	18 10
56	17 58	18 35	19 19	17 31	17 46	17 58	18 08
54	17 57	18 33	19 14	17 19	17 38	17 53	18 06
52	17 57	18 31	19 11	17 09	17 31	17 49	18 04
50	17 57	18 29	19 07	17 00	17 24	17 45	18 02
45	17 57	18 26	19 00	16 41	17 10	17 36	17 59
N 40	17 56	18 23	18 55	16 25	16 59	17 28	17 55
35	17 56	18 21	18 51	16 12	16 48	17 22	17 53
30	17 56	18 20	18 48	16 00	16 40	17 16	17 50
20	17 56	18 18	18 43	15 40	16 24	17 06	17 46
N 10	17 56	18 17	18 41	15 22	16 11	16 57	17 42
0	17 56	18 16	18 40	15 06	15 58	16 48	17 38
S 10	17 56	18 17	18 41	14 49	15 45	16 40	17 34
20	17 56	18 18	18 44	14 31	15 31	16 31	17 30
30	17 57	18 20	18 48	14 11	15 15	16 20	17 26
35	17 57	18 22	18 52	13 58	15 06	16 14	17 23
40	17 57	18 24	18 56	13 44	14 55	16 07	17 20
45	17 58	18 27	19 01	13 28	14 42	15 59	17 16
S 50	17 58	18 30	19 09	13 07	14 27	15 49	17 12
52	17 58	18 32	19 12	12 57	14 19	15 44	17 10
54	17 59	18 34	19 16	12 46	14 11	15 39	17 08
56	17 59	18 36	19 21	12 33	14 02	15 33	17 05
58	17 59	18 39	19 26	12 18	13 52	15 27	17 03
S 60	18 00	18 42	19 32	12 00	13 39	15 20	16 59

Day	SUN Eqn. of Time 00^h	SUN Eqn. of Time 12^h	SUN Mer. Pass.	MOON Mer. Pass. Upper	MOON Mer. Pass. Lower	MOON Age	MOON Phase
d	m s	m s	h m	h m	h m	d	%
22	06 59	07 10	11 53	08 53	21 20	26	16
23	07 20	07 31	11 52	09 45	22 11	27	9
24	07 41	07 52	11 52	10 36	23 01	28	4

	UT	ARIES	VENUS −3.9		MARS −2.2		JUPITER −1.7		SATURN +0.1	
	d h	GHA ° ′	GHA ° ′	Dec ° ′	GHA ° ′	Dec ° ′	GHA ° ′	Dec ° ′	GHA ° ′	Dec ° ′
	25 00	3 24.1	172 17.4	S 3 33.0	29 19.1	S16 07.8	205 11.3	N10 04.6	260 02.3	N22 07.8
	01	18 26.6	187 17.0	34.3	44 21.7	07.6	220 13.3	04.5	275 04.6	07.8
	02	33 29.0	202 16.6	35.5	59 24.3	07.5	235 15.2	04.3	290 06.9	07.7
	03	48 31.5	217 16.3	. . 36.8	74 27.0	. . 07.3	250 17.2	. . 04.1	305 09.2	. . 07.7
	04	63 34.0	232 15.9	38.1	89 29.6	07.2	265 19.2	03.9	320 11.5	07.7
	05	78 36.4	247 15.5	39.3	104 32.2	07.0	280 21.2	03.7	335 13.8	07.7
	06	93 38.9	262 15.1	S 3 40.6	119 34.8	S16 06.8	295 23.1	N10 03.5	350 16.2	N22 07.7
	07	108 41.3	277 14.7	41.9	134 37.4	06.7	310 25.1	03.3	5 18.5	07.7
T	08	123 43.8	292 14.4	43.2	149 40.0	06.5	325 27.1	03.2	20 20.8	07.7
H	09	138 46.3	307 14.0	. . 44.4	164 42.6	. . 06.4	340 29.1	. . 03.0	35 23.1	. . 07.7
U	10	153 48.7	322 13.6	45.7	179 45.2	06.2	355 31.0	02.8	50 25.4	07.6
R	11	168 51.2	337 13.2	47.0	194 47.8	06.1	10 33.0	02.6	65 27.7	07.6
S	12	183 53.7	352 12.8	S 3 48.2	209 50.4	S16 05.9	25 35.0	N10 02.4	80 30.1	N22 07.6
D	13	198 56.1	7 12.5	49.5	224 53.0	05.7	40 37.0	02.2	95 32.4	07.6
A	14	213 58.6	22 12.1	50.8	239 55.6	05.6	55 38.9	02.0	110 34.7	07.6
Y	15	229 01.1	37 11.7	. . 52.0	254 58.2	. . 05.4	70 40.9	. . 01.8	125 37.0	. . 07.6
	16	244 03.5	52 11.3	53.3	270 00.8	05.2	85 42.9	01.7	140 39.3	07.6
	17	259 06.0	67 10.9	54.6	285 03.4	05.1	100 44.9	01.5	155 41.7	07.6
	18	274 08.5	82 10.6	S 3 55.8	300 05.9	S16 04.9	115 46.8	N10 01.3	170 44.0	N22 07.6
	19	289 10.9	97 10.2	57.1	315 08.5	04.8	130 48.8	01.1	185 46.3	07.5
	20	304 13.4	112 09.8	58.4	330 11.1	04.6	145 50.8	00.9	200 48.6	07.5
	21	319 15.8	127 09.4	3 59.6	345 13.7	. . 04.4	160 52.8	. . 00.7	215 50.9	. . 07.5
	22	334 18.3	142 09.0	4 00.9	0 16.3	04.3	175 54.7	00.5	230 53.3	07.5
	23	349 20.8	157 08.6	02.2	15 18.9	04.1	190 56.7	00.4	245 55.6	07.5
	26 00	4 23.2	172 08.3	S 4 03.4	30 21.5	S16 03.9	205 58.7	N10 00.2	260 57.9	N22 07.5
	01	19 25.7	187 07.9	04.7	45 24.0	03.8	221 00.7	10 00.0	276 00.2	07.5
	02	34 28.2	202 07.5	06.0	60 26.6	03.6	236 02.6	9 59.8	291 02.5	07.5
	03	49 30.6	217 07.1	. . 07.2	75 29.2	. . 03.4	251 04.6	. . 59.6	306 04.9	. . 07.4
	04	64 33.1	232 06.7	08.5	90 31.8	03.2	266 06.6	59.4	321 07.2	07.4
	05	79 35.6	247 06.3	09.8	105 34.3	03.1	281 08.6	59.2	336 09.5	07.4
	06	94 38.0	262 06.0	S 4 11.0	120 36.9	S16 02.9	296 10.5	N 9 59.1	351 11.8	N22 07.4
	07	109 40.5	277 05.6	12.3	135 39.5	02.7	311 12.5	58.9	6 14.1	07.4
	08	124 42.9	292 05.2	13.6	150 42.1	02.6	326 14.5	58.7	21 16.5	07.4
F	09	139 45.4	307 04.8	. . 14.8	165 44.6	. . 02.4	341 16.5	. . 58.5	36 18.8	. . 07.4
R	10	154 47.9	322 04.4	16.1	180 47.2	02.2	356 18.5	58.3	51 21.1	07.4
I	11	169 50.3	337 04.0	17.3	195 49.8	02.0	11 20.4	58.1	66 23.4	07.3
D	12	184 52.8	352 03.6	S 4 18.6	210 52.3	S16 01.9	26 22.4	N 9 57.9	81 25.8	N22 07.3
A	13	199 55.3	7 03.3	19.9	225 54.9	01.7	41 24.4	57.8	96 28.1	07.3
Y	14	214 57.7	22 02.9	21.1	240 57.5	01.5	56 26.4	57.6	111 30.4	07.3
	15	230 00.2	37 02.5	. . 22.4	256 00.0	. . 01.3	71 28.3	. . 57.4	126 32.7	. . 07.3
	16	245 02.7	52 02.1	23.7	271 02.6	01.2	86 30.3	57.2	141 35.0	07.3
	17	260 05.1	67 01.7	24.9	286 05.1	01.0	101 32.3	57.0	156 37.4	07.3
	18	275 07.6	82 01.3	S 4 26.2	301 07.7	S16 00.8	116 34.3	N 9 56.8	171 39.7	N22 07.3
	19	290 10.1	97 01.0	27.5	316 10.3	00.6	131 36.3	56.6	186 42.0	07.3
	20	305 12.5	112 00.6	28.7	331 12.8	00.5	146 38.2	56.5	201 44.3	07.2
	21	320 15.0	127 00.2	. . 30.0	346 15.4	. . 00.3	161 40.2	. . 56.3	216 46.7	. . 07.2
	22	335 17.4	141 59.8	31.3	1 17.9	16 00.1	176 42.2	56.1	231 49.0	07.2
	23	350 19.9	156 59.4	32.5	16 20.5	15 59.9	191 44.2	55.9	246 51.3	07.2
	27 00	5 22.4	171 59.0	S 4 33.8	31 23.0	S15 59.7	206 46.1	N 9 55.7	261 53.6	N22 07.2
	01	20 24.8	186 58.6	35.0	46 25.6	59.6	221 48.1	55.5	276 56.0	07.2
	02	35 27.3	201 58.2	36.3	61 28.1	59.4	236 50.1	55.3	291 58.3	07.2
	03	50 29.8	216 57.9	. . 37.6	76 30.6	. . 59.2	251 52.1	. . 55.2	307 00.6	. . 07.2
	04	65 32.2	231 57.5	38.8	91 33.2	59.0	266 54.1	55.0	322 02.9	07.2
	05	80 34.7	246 57.1	40.1	106 35.7	58.8	281 56.0	54.8	337 05.3	07.1
	06	95 37.2	261 56.7	S 4 41.4	121 38.3	S15 58.6	296 58.0	N 9 54.6	352 07.6	N22 07.1
	07	110 39.6	276 56.3	42.6	136 40.8	58.5	312 00.0	54.4	7 09.9	07.1
S	08	125 42.1	291 55.9	43.9	151 43.4	58.3	327 02.0	54.2	22 12.2	07.1
A	09	140 44.5	306 55.5	. . 45.1	166 45.9	. . 58.1	342 04.0	. . 54.0	37 14.6	. . 07.1
T	10	155 47.0	321 55.1	46.4	181 48.4	57.9	357 05.9	53.9	52 16.9	07.1
U	11	170 49.5	336 54.8	47.7	196 51.0	57.7	12 07.9	53.7	67 19.2	07.1
R	12	185 51.9	351 54.4	S 4 48.9	211 53.5	S15 57.5	27 09.9	N 9 53.5	82 21.5	N22 07.1
D	13	200 54.4	6 54.0	50.2	226 56.0	57.3	42 11.9	53.3	97 23.9	07.1
A	14	215 56.9	21 53.6	51.4	241 58.6	57.2	57 13.8	53.1	112 26.2	07.0
Y	15	230 59.3	36 53.2	. . 52.7	257 01.1	. . 57.0	72 15.8	. . 52.9	127 28.5	. . 07.0
	16	246 01.8	51 52.8	54.0	272 03.6	56.8	87 17.8	52.8	142 30.9	07.0
	17	261 04.3	66 52.4	55.2	287 06.1	56.6	102 19.8	52.6	157 33.2	07.0
	18	276 06.7	81 52.0	S 4 56.5	302 08.7	S15 56.4	117 21.8	N 9 52.4	172 35.5	N22 07.0
	19	291 09.2	96 51.6	57.7	317 11.2	56.2	132 23.7	52.2	187 37.8	07.0
	20	306 11.7	111 51.3	4 59.0	332 13.7	56.0	147 25.7	52.0	202 40.2	07.0
	21	321 14.1	126 50.9	5 00.3	347 16.2	. . 55.8	162 27.7	. . 51.8	217 42.5	. . 07.0
	22	336 16.6	141 50.5	01.5	2 18.8	55.6	177 29.7	51.6	232 44.8	06.9
	23	351 19.0	156 50.1	02.8	17 21.3	55.4	192 31.7	51.5	247 47.2	06.9
	Mer. Pass.	h m 23 38.6	v −0.4	d 1.3	v 2.6	d 0.2	v 2.0	d 0.2	v 2.3	d 0.0

STARS

Name	SHA ° ′	Dec ° ′
Acamar	315 23.7	S40 17.2
Achernar	335 31.7	S57 12.9
Acrux	173 19.0	S63 07.1
Adhara	255 18.7	S28 58.2
Aldebaran	290 58.1	N16 31.1
Alioth	166 27.7	N55 56.6
Alkaid	153 05.2	N49 17.9
Al Na'ir	27 52.8	S46 56.7
Alnilam	275 54.1	S 1 11.8
Alphard	218 03.9	S 8 40.2
Alphecca	126 17.7	N26 42.3
Alpheratz	357 51.2	N29 06.7
Altair	62 15.6	N 8 52.7
Ankaa	353 22.7	S42 17.1
Antares	112 35.9	S26 26.5
Arcturus	146 03.0	N19 10.0
Atria	107 44.9	S69 02.3
Avior	234 21.7	S59 30.9
Bellatrix	278 40.2	N 6 21.4
Betelgeuse	271 09.6	N 7 24.6
Canopus	263 59.6	S52 41.4
Capella	280 45.7	N46 00.1
Deneb	49 36.6	N45 17.8
Denebola	182 41.8	N14 33.3
Diphda	349 03.2	S17 57.9
Dubhe	194 01.4	N61 43.9
Elnath	278 22.3	N28 36.7
Eltanin	90 49.7	N51 29.5
Enif	33 54.4	N 9 53.6
Fomalhaut	15 32.0	S29 36.2
Gacrux	172 10.3	S57 07.9
Gienah	176 00.6	S17 33.5
Hadar	148 59.7	S60 23.5
Hamal	328 09.2	N23 28.8
Kaus Aust.	83 54.0	S34 23.1
Kochab	137 20.1	N74 08.6
Markab	13 45.7	N15 13.6
Menkar	314 22.9	N 4 06.4
Menkent	148 17.1	S36 23.2
Miaplacidus	221 42.3	S69 43.6
Mirfak	308 51.2	N49 52.4
Nunki	76 07.7	S26 17.6
Peacock	53 30.8	S56 43.7
Pollux	243 37.2	N28 01.1
Procyon	245 07.9	N 5 13.2
Rasalhague	96 13.6	N12 33.6
Regulus	207 51.9	N11 57.2
Rigel	281 19.3	S 8 11.6
Rigil Kent.	140 03.1	S60 51.1
Sabik	102 21.4	S15 43.8
Schedar	349 49.0	N56 33.4
Shaula	96 32.4	S37 06.5
Sirius	258 40.6	S16 42.[illegible]
Spica	158 39.7	S11 10.7
Suhail	222 58.5	S43 26.5
Vega	80 44.1	N38 47.4
Zuben'ubi	137 14.2	S16 03.4

	SHA ° ′	Mer. Pass h m
Venus	167 45.0	12 32
Mars	25 58.2	21 55
Jupiter	201 35.5	10 15
Saturn	256 34.7	6 35

UT d h	SUN GHA ° ′	SUN Dec ° ′	MOON GHA ° ′	v ′	MOON Dec ° ′	d ′	HP ′
5 00	182 00.6	S 0 36.2	194 09.4	10.9	N 9 40.9	15.2	59.2
01	197 00.8	37.2	208 39.3	10.9	9 25.7	15.2	59.2
02	212 01.0	38.2	223 09.2	10.9	9 10.5	15.3	59.2
03	227 01.2	. . 39.2	237 39.1	10.9	8 55.2	15.4	59.3
04	242 01.5	40.1	252 09.0	10.9	8 39.8	15.4	59.3
05	257 01.7	41.1	266 38.9	11.0	8 24.4	15.5	59.3
06	272 01.9	S 0 42.1	281 08.9	10.9	N 8 08.9	15.5	59.4
07	287 02.1	43.1	295 38.8	11.0	7 53.4	15.6	59.4
08	302 02.3	44.0	310 08.8	11.0	7 37.8	15.7	59.4
09	317 02.5	. . 45.0	324 38.8	11.0	7 22.1	15.7	59.4
10	332 02.7	46.0	339 08.8	10.9	7 06.4	15.8	59.5
11	347 03.0	47.0	353 38.7	11.0	6 50.6	15.8	59.5
12	2 03.2	S 0 47.9	8 08.7	11.0	N 6 34.8	15.8	59.5
13	17 03.4	48.9	22 38.7	11.0	6 19.0	15.9	59.5
14	32 03.6	49.9	37 08.7	11.1	6 03.1	16.0	59.6
15	47 03.8	. . 50.9	51 38.8	11.0	5 47.1	16.0	59.6
16	62 04.0	51.8	66 08.8	11.0	5 31.1	16.0	59.6
17	77 04.3	52.8	80 38.8	11.0	5 15.1	16.1	59.6
18	92 04.5	S 0 53.8	95 08.8	11.0	N 4 59.0	16.2	59.7
19	107 04.7	54.8	109 38.8	11.0	4 42.8	16.1	59.7
20	122 04.9	55.7	124 08.8	11.0	4 26.7	16.2	59.7
21	137 05.1	. . 56.7	138 38.8	11.0	4 10.5	16.3	59.7
22	152 05.3	57.7	153 08.8	11.0	3 54.2	16.2	59.8
23	167 05.5	58.7	167 38.8	11.0	3 38.0	16.4	59.8
6 00	182 05.8	S 0 59.6	182 08.8	11.0	N 3 21.6	16.3	59.8
01	197 06.0	1 00.6	196 38.8	11.0	3 05.3	16.4	59.8
02	212 06.2	01.6	211 08.8	11.0	2 48.9	16.3	59.9
03	227 06.4	. . 02.5	225 38.8	10.9	2 32.6	16.5	59.9
04	242 06.6	03.5	240 08.7	11.0	2 16.1	16.4	59.9
05	257 06.8	04.5	254 38.7	10.9	1 59.7	16.5	59.9
06	272 07.0	S 1 05.5	269 08.6	10.9	N 1 43.2	16.4	59.9
07	287 07.3	06.4	283 38.5	11.0	1 26.8	16.6	60.0
08	302 07.5	07.4	298 08.5	10.9	1 10.2	16.5	60.0
09	317 07.7	. . 08.4	312 38.4	10.8	0 53.7	16.5	60.0
10	332 07.9	09.4	327 08.2	10.9	0 37.2	16.6	60.0
11	347 08.1	10.3	341 38.1	10.9	0 20.6	16.5	60.0
12	2 08.3	S 1 11.3	356 08.0	10.8	N 0 04.1	16.6	60.1
13	17 08.5	12.3	10 37.8	10.8	S 0 12.5	16.6	60.1
14	32 08.8	13.3	25 07.6	10.8	0 29.1	16.6	60.1
15	47 09.0	. . 14.2	39 37.4	10.8	0 45.7	16.6	60.1
16	62 09.2	15.2	54 07.2	10.7	1 02.3	16.5	60.1
17	77 09.4	16.2	68 36.9	10.8	1 18.8	16.6	60.1
18	92 09.6	S 1 17.2	83 06.7	10.7	S 1 35.4	16.6	60.2
19	107 09.8	18.1	97 36.4	10.6	1 52.0	16.6	60.2
20	122 10.0	19.1	112 06.0	10.7	2 08.6	16.6	60.2
21	137 10.2	. . 20.1	126 35.7	10.6	2 25.2	16.6	60.2
22	152 10.5	21.1	141 05.3	10.6	2 41.8	16.6	60.2
23	167 10.7	22.0	155 34.9	10.6	2 58.4	16.6	60.2
7 00	182 10.9	S 1 23.0	170 04.5	10.5	S 3 15.0	16.5	60.2
01	197 11.1	24.0	184 34.0	10.5	3 31.5	16.6	60.3
02	212 11.3	25.0	199 03.5	10.5	3 48.1	16.5	60.3
03	227 11.5	. . 25.9	213 33.0	10.4	4 04.6	16.5	60.3
04	242 11.7	26.9	228 02.4	10.4	4 21.1	16.5	60.3
05	257 12.0	27.9	242 31.8	10.4	4 37.6	16.5	60.3
06	272 12.2	S 1 28.9	257 01.2	10.3	S 4 54.1	16.4	60.3
07	287 12.4	29.8	271 30.5	10.3	5 10.5	16.5	60.3
08	302 12.6	30.8	285 59.8	10.2	5 27.0	16.4	60.3
09	317 12.8	. . 31.8	300 29.0	10.2	5 43.4	16.4	60.3
10	332 13.0	32.7	314 58.2	10.2	5 59.8	16.3	60.3
11	347 13.2	33.7	329 27.4	10.1	6 16.1	16.3	60.4
12	2 13.4	S 1 34.7	343 56.5	10.1	S 6 32.4	16.3	60.4
13	17 13.6	35.7	358 25.6	10.0	6 48.7	16.3	60.4
14	32 13.9	36.6	12 54.6	10.0	7 05.0	16.2	60.4
15	47 14.1	. . 37.6	27 23.6	10.0	7 21.2	16.2	60.4
16	62 14.3	38.6	41 52.6	9.9	7 37.4	16.1	60.4
17	77 14.5	39.6	56 21.5	9.8	7 53.5	16.1	60.4
18	92 14.7	S 1 40.5	70 50.3	9.8	S 8 09.6	16.0	60.4
19	107 14.9	41.5	85 19.1	9.7	8 25.6	16.0	60.4
20	122 15.1	42.5	99 47.8	9.7	8 41.6	16.0	60.4
21	137 15.3	. . 43.5	114 16.5	9.7	8 57.6	15.9	60.4
22	152 15.5	44.4	128 45.2	9.6	9 13.5	15.8	60.4
23	167 15.8	45.4	143 13.8	9.5	S 9 29.3	15.8	60.4
	SD 16.0	*d* 1.0	SD	16.2		16.4	16.4

Lat.	Twilight Naut.	Twilight Civil	Sunrise	Moonrise 25	26	27	28
°	h m	h m	h m	h m	h m	h m	h m
N 72	03 19	04 46	05 54	03 18	05 43	08 07	10 48
N 70	03 36	04 53	05 54	03 32	05 45	07 59	10 23
68	03 49	04 58	05 53	03 43	05 47	07 52	10 04
66	03 59	05 02	05 53	03 52	05 49	07 46	09 49
64	04 08	05 05	05 53	04 00	05 50	07 41	09 37
62	04 15	05 08	05 53	04 07	05 51	07 37	09 27
60	04 21	05 11	05 52	04 12	05 52	07 34	09 18
N 58	04 27	05 13	05 52	04 17	05 53	07 31	09 10
56	04 31	05 15	05 52	04 22	05 54	07 28	09 04
54	04 35	05 17	05 52	04 26	05 55	07 25	08 58
52	04 38	05 18	05 52	04 29	05 55	07 23	08 53
50	04 41	05 19	05 52	04 32	05 56	07 21	08 48
45	04 48	05 22	05 51	04 39	05 57	07 17	08 37
N 40	04 52	05 24	05 51	04 45	05 58	07 13	08 29
35	04 56	05 25	05 51	04 50	05 59	07 10	08 22
30	04 59	05 26	05 50	04 54	06 00	07 07	08 15
20	05 02	05 28	05 50	05 02	06 02	07 02	08 04
N 10	05 04	05 28	05 49	05 08	06 03	06 58	07 55
0	05 04	05 28	05 48	05 14	06 04	06 54	07 46
S 10	05 02	05 26	05 47	05 21	06 05	06 51	07 38
20	04 59	05 24	05 46	05 27	06 07	06 47	07 28
30	04 53	05 21	05 45	05 34	06 08	06 42	07 18
35	04 50	05 19	05 44	05 39	06 09	06 40	07 12
40	04 45	05 17	05 44	05 44	06 10	06 37	07 05
45	04 39	05 13	05 42	05 49	06 11	06 34	06 58
S 50	04 31	05 09	05 41	05 56	06 13	06 30	06 48
52	04 27	05 07	05 41	05 59	06 13	06 28	06 44
54	04 22	05 05	05 40	06 02	06 14	06 26	06 39
56	04 17	05 02	05 39	06 06	06 15	06 24	06 34
58	04 12	04 59	05 38	06 10	06 16	06 22	06 29
S 60	04 05	04 56	05 37	06 14	06 17	06 19	06 22

Lat.	Sunset	Twilight Civil	Twilight Naut.	Moonset 25	26	27	28
°	h m	h m	h m	h m	h m	h m	h m
N 72	17 46	18 54	20 19	18 39	18 06	17 30	16 41
N 70	17 47	18 48	20 03	18 32	18 08	17 42	17 08
68	17 48	18 43	19 51	18 27	18 10	17 52	17 29
66	17 48	18 39	19 41	18 23	18 12	18 00	17 46
64	17 48	18 36	19 33	18 19	18 13	18 07	18 00
62	17 49	18 33	19 26	18 16	18 14	18 13	18 11
60	17 49	18 31	19 20	18 13	18 15	18 18	18 21
N 58	17 49	18 28	19 15	18 10	18 16	18 22	18 30
56	17 50	18 27	19 10	18 08	18 17	18 27	18 38
54	17 50	18 25	19 07	18 06	18 18	18 30	18 45
52	17 50	18 24	19 03	18 04	18 19	18 34	18 51
50	17 50	18 23	19 00	18 02	18 19	18 37	18 57
45	17 51	18 20	18 54	17 59	18 21	18 43	19 09
N 40	17 51	18 18	18 50	17 55	18 22	18 49	19 19
35	17 52	18 17	18 46	17 53	18 23	18 54	19 28
30	17 52	18 16	18 44	17 50	18 24	18 58	19 35
20	17 53	18 15	18 41	17 46	18 25	19 06	19 49
N 10	17 54	18 15	18 39	17 42	18 27	19 12	20 01
0	17 55	18 15	18 39	17 38	18 28	19 19	20 12
S 10	17 56	18 17	18 41	17 34	18 29	19 25	20 23
20	17 57	18 19	18 44	17 30	18 30	19 32	20 35
30	17 58	18 22	18 50	17 26	18 32	19 39	20 48
35	17 59	18 24	18 54	17 23	18 33	19 44	20 56
40	18 00	18 27	18 59	17 20	18 34	19 49	21 06
45	18 01	18 31	19 05	17 16	18 35	19 55	21 16
S 50	18 03	18 35	19 13	17 12	18 36	20 02	21 29
52	18 03	18 37	19 17	17 10	18 37	20 05	21 35
54	18 04	18 40	19 22	17 08	18 37	20 09	21 42
56	18 05	18 42	19 27	17 05	18 38	20 13	21 50
58	18 06	18 45	19 33	17 03	18 39	20 17	21 58
S 60	18 07	18 49	19 40	16 59	18 40	20 22	22 08

Day	SUN Eqn. of Time 00^h	SUN Eqn. of Time 12^h	SUN Mer. Pass.	MOON Mer. Pass. Upper	MOON Mer. Pass. Lower	Age	Phase
d	m s	m s	h m	h m	h m	d	%
25	08 02	08 12	11 52	11 26	23 51	29	1
26	08 23	08 33	11 51	12 16	24 41	00	0
27	08 43	08 53	11 51	13 07	00 41	01	3

UT	ARIES	VENUS −3.9		MARS −2.2		JUPITER −1.7		SATURN +0.1		STARS		
	GHA	GHA	Dec	GHA	Dec	GHA	Dec	GHA	Dec	Name	SHA	Dec
d h	° ′	° ′	° ′	° ′	° ′	° ′	° ′	° ′	° ′		° ′	° ′
28 00 SUNDAY	6 21.5	171 49.7	S 5 04.0	32 23.8	S15 55.3	207 33.6	N 9 51.3	262 49.5	N22 06.9	Acamar	315 23.7	S40 17.
01	21 24.0	186 49.3	05.3	47 26.3	55.1	222 35.6	51.1	277 51.8	06.9	Achernar	335 31.7	S57 13.
02	36 26.4	201 48.9	06.6	62 28.8	54.9	237 37.6	50.9	292 54.1	06.9	Acrux	173 19.0	S63 07.
03	51 28.9	216 48.5	07.8	77 31.3	54.7	252 39.6	50.7	307 56.5	06.9	Adhara	255 18.6	S28 58.
04	66 31.4	231 48.1	09.1	92 33.8	54.5	267 41.6	50.5	322 58.8	06.9	Aldebaran	290 58.1	N16 31.
05	81 33.8	246 47.7	10.3	107 36.3	54.3	282 43.5	50.3	338 01.1	06.9			
06	96 36.3	261 47.3	S 5 11.6	122 38.9	S15 54.1	297 45.5	N 9 50.2	353 03.5	N22 06.9	Alioth	166 27.7	N55 56.
07	111 38.8	276 47.0	12.9	137 41.4	53.9	312 47.5	50.0	8 05.8	06.8	Alkaid	153 05.2	N49 17.
08	126 41.2	291 46.6	14.1	152 43.9	53.7	327 49.5	49.8	23 08.1	06.8	Al Na'ir	27 52.8	S46 56.
09	141 43.7	306 46.2	15.4	167 46.4	53.5	342 51.5	49.6	38 10.5	06.8	Alnilam	275 54.1	S 1 11.
10	156 46.2	321 45.8	16.6	182 48.9	53.3	357 53.5	49.4	53 12.8	06.8	Alphard	218 03.9	S 8 40.
11	171 48.6	336 45.4	17.9	197 51.4	53.1	12 55.4	49.2	68 15.1	06.8			
12	186 51.1	351 45.0	S 5 19.1	212 53.9	S15 52.9	27 57.4	N 9 49.1	83 17.5	N22 06.8	Alphecca	126 17.7	N26 42.
13	201 53.5	6 44.6	20.4	227 56.4	52.7	42 59.4	48.9	98 19.8	06.8	Alpheratz	357 51.2	N29 06.
14	216 56.0	21 44.2	21.7	242 58.9	52.5	58 01.4	48.7	113 22.1	06.8	Altair	62 15.6	N 8 52.
15	231 58.5	36 43.8	22.9	258 01.4	52.3	73 03.4	48.5	128 24.4	06.8	Ankaa	353 22.7	S42 17.
16	247 00.9	51 43.4	24.2	273 03.9	52.1	88 05.3	48.3	143 26.8	06.7	Antares	112 35.9	S26 26.
17	262 03.4	66 43.0	25.4	288 06.4	51.9	103 07.3	48.1	158 29.1	06.7			
18	277 05.9	81 42.6	S 5 26.7	303 08.9	S15 51.7	118 09.3	N 9 47.9	173 31.4	N22 06.7	Arcturus	146 03.0	N19 10.
19	292 08.3	96 42.2	27.9	318 11.3	51.5	133 11.3	47.8	188 33.8	06.7	Atria	107 45.0	S69 02.
20	307 10.8	111 41.8	29.2	333 13.8	51.3	148 13.3	47.6	203 36.1	06.7	Avior	234 21.6	S59 30.
21	322 13.3	126 41.5	30.4	348 16.3	51.1	163 15.3	47.4	218 38.4	06.7	Bellatrix	278 40.2	N 6 21.
22	337 15.7	141 41.1	31.7	3 18.8	50.9	178 17.2	47.2	233 40.8	06.7	Betelgeuse	271 09.6	N 7 24
23	352 18.2	156 40.7	33.0	18 21.3	50.7	193 19.2	47.0	248 43.1	06.7			
29 00 MONDAY	7 20.6	171 40.3	S 5 34.2	33 23.8	S15 50.5	208 21.2	N 9 46.8	263 45.4	N22 06.7	Canopus	263 59.6	S52 41.
01	22 23.1	186 39.9	35.5	48 26.3	50.3	223 23.2	46.7	278 47.8	06.6	Capella	280 45.7	N46 00.
02	37 25.6	201 39.5	36.7	63 28.7	50.1	238 25.2	46.5	293 50.1	06.6	Deneb	49 36.6	N45 17.
03	52 28.0	216 39.1	38.0	78 31.2	49.8	253 27.1	46.3	308 52.4	06.6	Denebola	182 41.8	N14 33.
04	67 30.5	231 38.7	39.2	93 33.7	49.6	268 29.1	46.1	323 54.8	06.6	Diphda	349 03.2	S17 57.
05	82 33.0	246 38.3	40.5	108 36.2	49.4	283 31.1	45.9	338 57.1	06.6			
06	97 35.4	261 37.9	S 5 41.7	123 38.7	S15 49.2	298 33.1	N 9 45.7	353 59.4	N22 06.6	Dubhe	194 01.3	N61 43.
07	112 37.9	276 37.5	43.0	138 41.1	49.0	313 35.1	45.5	9 01.8	06.6	Elnath	278 22.2	N28 36.
08	127 40.4	291 37.1	44.2	153 43.6	48.8	328 37.1	45.4	24 04.1	06.6	Eltanin	90 49.8	N51 29.
09	142 42.8	306 36.7	45.5	168 46.1	48.6	343 39.0	45.2	39 06.5	06.6	Enif	33 54.4	N 9 53.
10	157 45.3	321 36.3	46.8	183 48.5	48.4	358 41.0	45.0	54 08.8	06.6	Fomalhaut	15 32.0	S29 36.
11	172 47.8	336 35.9	48.0	198 51.0	48.2	13 43.0	44.8	69 11.1	06.5			
12	187 50.2	351 35.5	S 5 49.3	213 53.5	S15 48.0	28 45.0	N 9 44.6	84 13.5	N22 06.5	Gacrux	172 10.3	S57 07.
13	202 52.7	6 35.1	50.5	228 55.9	47.7	43 47.0	44.4	99 15.8	06.5	Gienah	176 00.6	S17 33.
14	217 55.1	21 34.7	51.8	243 58.4	47.5	58 49.0	44.3	114 18.1	06.5	Hadar	148 59.7	S60 23.
15	232 57.6	36 34.3	53.0	259 00.9	47.3	73 50.9	44.1	129 20.5	06.5	Hamal	328 09.2	N23 28
16	248 00.1	51 33.9	54.3	274 03.3	47.1	88 52.9	43.9	144 22.8	06.5	Kaus Aust.	83 54.0	S34 23.
17	263 02.5	66 33.5	55.5	289 05.8	46.9	103 54.9	43.7	159 25.1	06.5			
18	278 05.0	81 33.1	S 5 56.8	304 08.3	S15 46.7	118 56.9	N 9 43.5	174 27.5	N22 06.5	Kochab	137 20.2	N74 08.
19	293 07.5	96 32.7	58.0	319 10.7	46.5	133 58.9	43.3	189 29.8	06.5	Markab	13 45.7	N15 13.
20	308 09.9	111 32.3	5 59.3	334 13.2	46.2	149 00.9	43.2	204 32.2	06.4	Menkar	314 22.8	N 4 06.
21	323 12.4	126 31.9	6 00.5	349 15.6	46.0	164 02.8	43.0	219 34.5	06.4	Menkent	148 17.1	S36 23.
22	338 14.9	141 31.5	01.8	4 18.1	45.8	179 04.8	42.8	234 36.8	06.4	Miaplacidus	221 42.3	S69 43.
23	353 17.3	156 31.1	03.0	19 20.5	45.6	194 06.8	42.6	249 39.2	06.4			
30 00 TUESDAY	8 19.8	171 30.7	S 6 04.3	34 23.0	S15 45.4	209 08.8	N 9 42.4	264 41.5	N22 06.4	Mirfak	308 51.2	N49 52.
01	23 22.3	186 30.3	05.5	49 25.4	45.2	224 10.8	42.2	279 43.8	06.4	Nunki	76 07.8	S26 17.
02	38 24.7	201 29.9	06.8	64 27.9	44.9	239 12.8	42.1	294 46.2	06.4	Peacock	53 30.8	S56 43.
03	53 27.2	216 29.5	08.0	79 30.3	44.7	254 14.8	41.9	309 48.5	06.4	Pollux	243 37.2	N28 01.
04	68 29.6	231 29.1	09.3	94 32.8	44.5	269 16.7	41.7	324 50.9	06.4	Procyon	245 07.8	N 5 13.
05	83 32.1	246 28.7	10.5	109 35.2	44.3	284 18.7	41.5	339 53.2	06.4			
06	98 34.6	261 28.3	S 6 11.8	124 37.7	S15 44.1	299 20.7	N 9 41.3	354 55.5	N22 06.3	Rasalhague	96 13.6	N12 33
07	113 37.0	276 27.9	13.0	139 40.1	43.8	314 22.7	41.1	9 57.9	06.3	Regulus	207 51.9	N11 57
08	128 39.5	291 27.5	14.3	154 42.5	43.6	329 24.7	40.9	25 00.2	06.3	Rigel	281 19.3	S 8 11
09	143 42.0	306 27.1	15.5	169 45.0	43.4	344 26.7	40.8	40 02.6	06.3	Rigil Kent.	140 03.1	S60 51
10	158 44.4	321 26.7	16.8	184 47.4	43.2	359 28.6	40.6	55 04.9	06.3	Sabik	102 21.5	S15 43
11	173 46.9	336 26.3	18.0	199 49.9	42.9	14 30.6	40.4	70 07.2	06.3			
12	188 49.4	351 25.9	S 6 19.3	214 52.3	S15 42.7	29 32.6	N 9 40.2	85 09.6	N22 06.3	Schedar	349 49.0	N56 33
13	203 51.8	6 25.5	20.5	229 54.7	42.5	44 34.6	40.0	100 11.9	06.3	Shaula	96 32.4	S37 06
14	218 54.3	21 25.1	21.7	244 57.2	42.3	59 36.6	39.8	115 14.3	06.3	Sirius	258 40.5	S16 42
15	233 56.7	36 24.7	23.0	259 59.6	42.1	74 38.6	39.7	130 16.6	06.2	Spica	158 39.7	S11 10
16	248 59.2	51 24.3	24.2	275 02.0	41.8	89 40.6	39.5	145 18.9	06.2	Suhail	222 58.5	S43 26
17	264 01.7	66 23.9	25.5	290 04.4	41.6	104 42.5	39.3	160 21.3	06.2			
18	279 04.1	81 23.5	S 6 26.7	305 06.9	S15 41.4	119 44.5	N 9 39.1	175 23.6	N22 06.2	Vega	80 44.2	N38 47
19	294 06.6	96 23.1	28.0	320 09.3	41.1	134 46.5	38.9	190 26.0	06.2	Zuben'ubi	137 14.2	S16 03
20	309 09.1	111 22.7	29.2	335 11.7	40.9	149 48.5	38.7	205 28.3	06.2		SHA	Mer. Pas
21	324 11.5	126 22.3	30.5	350 14.1	40.7	164 50.5	38.6	220 30.6	06.2		° ′	h m
22	339 14.0	141 21.9	31.7	5 16.6	40.5	179 52.5	38.4	235 33.0	06.2	Venus	164 19.6	12 34
23	354 16.5	156 21.5	33.0	20 19.0	40.2	194 54.5	38.2	250 35.3	06.2	Mars	26 03.1	21 43
	h m									Jupiter	201 00.5	10 05
Mer. Pass.	23 26.8	v −0.4	d 1.3	v 2.5	d 0.2	v 2.0	d 0.2	v 2.3	d 0.0	Saturn	256 24.8	6 24

UT	SUN GHA	SUN Dec	MOON GHA	v	MOON Dec	d	HP
d h	° ′	° ′	° ′	′	° ′	′	′
28 00	182 16.0	S 1 46.4	157 42.3	9.5	S 9 45.1	15.8	60.4
01	197 16.2	47.4	172 10.8	9.4	10 00.9	15.7	60.4
02	212 16.4	48.3	186 39.2	9.4	10 16.6	15.6	60.4
03	227 16.6	. . 49.3	201 07.6	9.3	10 32.2	15.6	60.4
04	242 16.8	50.3	215 35.9	9.3	10 47.8	15.5	60.4
05	257 17.0	51.3	230 04.2	9.2	11 03.3	15.4	60.4
06	272 17.2	S 1 52.2	244 32.4	9.1	S11 18.7	15.4	60.4
07	287 17.4	53.2	259 00.5	9.1	11 34.1	15.3	60.4
08	302 17.7	54.2	273 28.6	9.1	11 49.4	15.2	60.4
09	317 17.9	. . 55.1	287 56.7	8.9	12 04.6	15.2	60.4
10	332 18.1	56.1	302 24.6	8.9	12 19.8	15.1	60.4
11	347 18.3	57.1	316 52.5	8.8	12 34.9	15.0	60.4
12	2 18.5	S 1 58.1	331 20.3	8.8	S12 49.9	15.0	60.4
13	17 18.7	1 59.0	345 48.1	8.7	13 04.9	14.8	60.4
14	32 18.9	2 00.0	0 15.8	8.7	13 19.7	14.8	60.4
15	47 19.1	. . 01.0	14 43.5	8.5	13 34.5	14.7	60.4
16	62 19.3	02.0	29 11.0	8.6	13 49.2	14.6	60.4
17	77 19.5	02.9	43 38.6	8.4	14 03.8	14.5	60.4
18	92 19.7	S 2 03.9	58 06.0	8.4	S14 18.3	14.4	60.4
19	107 20.0	04.9	72 33.4	8.3	14 32.7	14.4	60.4
20	122 20.2	05.9	87 00.7	8.2	14 47.1	14.2	60.4
21	137 20.4	. . 06.8	101 27.9	8.2	15 01.3	14.2	60.4
22	152 20.6	07.8	115 55.1	8.1	15 15.5	14.1	60.4
23	167 20.8	08.8	130 22.2	8.0	15 29.6	13.9	60.4
29 00	182 21.0	S 2 09.7	144 49.2	8.0	S15 43.5	13.9	60.4
01	197 21.2	10.7	159 16.2	7.9	15 57.4	13.7	60.4
02	212 21.4	11.7	173 43.1	7.8	16 11.1	13.7	60.4
03	227 21.6	. . 12.7	188 09.9	7.8	16 24.8	13.6	60.3
04	242 21.8	13.6	202 36.7	7.7	16 38.4	13.4	60.3
05	257 22.0	14.6	217 03.4	7.6	16 51.8	13.3	60.3
06	272 22.2	S 2 15.6	231 30.0	7.5	S17 05.1	13.3	60.3
07	287 22.5	16.6	245 56.5	7.5	17 18.4	13.1	60.3
08	302 22.7	17.5	260 23.0	7.4	17 31.5	13.0	60.3
09	317 22.9	. . 18.5	274 49.4	7.3	17 44.5	12.9	60.3
10	332 23.1	19.5	289 15.7	7.3	17 57.4	12.8	60.3
11	347 23.3	20.4	303 42.0	7.1	18 10.2	12.6	60.3
12	2 23.5	S 2 21.4	318 08.1	7.2	S18 22.8	12.5	60.3
13	17 23.7	22.4	332 34.3	7.0	18 35.3	12.4	60.2
14	32 23.9	23.4	347 00.3	7.0	18 47.7	12.3	60.2
15	47 24.1	. . 24.3	1 26.3	6.9	19 00.0	12.2	60.2
16	62 24.3	25.3	15 52.2	6.8	19 12.2	12.0	60.2
17	77 24.5	26.3	30 18.0	6.7	19 24.2	11.9	60.2
18	92 24.7	S 2 27.3	44 43.7	6.7	S19 36.1	11.8	60.2
19	107 24.9	28.2	59 09.4	6.6	19 47.9	11.7	60.2
20	122 25.1	29.2	73 35.0	6.6	19 59.6	11.5	60.2
21	137 25.4	. . 30.2	88 00.6	6.4	20 11.1	11.3	60.1
22	152 25.6	31.1	102 26.0	6.4	20 22.4	11.3	60.1
23	167 25.8	32.1	116 51.4	6.4	20 33.7	11.1	60.1
30 00	182 26.0	S 2 33.1	131 16.8	6.2	S20 44.8	10.9	60.1
01	197 26.2	34.1	145 42.0	6.2	20 55.7	10.9	60.1
02	212 26.4	35.0	160 07.2	6.1	21 06.6	10.7	60.1
03	227 26.6	. . 36.0	174 32.3	6.1	21 17.3	10.5	60.0
04	242 26.8	37.0	188 57.4	6.0	21 27.8	10.4	60.0
05	257 27.0	37.9	203 22.4	5.9	21 38.2	10.2	60.0
06	272 27.2	S 2 38.9	217 47.3	5.9	S21 48.4	10.1	60.0
07	287 27.4	39.9	232 12.2	5.8	21 58.5	10.0	60.0
08	302 27.6	40.9	246 37.0	5.7	22 08.5	9.8	60.0
09	317 27.8	. . 41.8	261 01.7	5.7	22 18.3	9.6	59.9
10	332 28.0	42.8	275 26.4	5.6	22 27.9	9.5	59.9
11	347 28.2	43.8	289 51.0	5.5	22 37.4	9.3	59.9
12	2 28.4	S 2 44.7	304 15.5	5.5	S22 46.7	9.2	59.9
13	17 28.6	45.7	318 40.0	5.4	22 55.9	9.1	59.9
14	32 28.8	46.7	333 04.4	5.4	23 05.0	8.8	59.9
15	47 29.0	. . 47.7	347 28.8	5.3	23 13.8	8.7	59.8
16	62 29.3	48.6	1 53.1	5.3	23 22.5	8.6	59.8
17	77 29.5	49.6	16 17.4	5.2	23 31.1	8.4	59.8
18	92 29.7	S 2 50.6	30 41.6	5.1	S23 39.5	8.2	59.8
19	107 29.9	51.5	45 05.7	5.1	23 47.7	8.0	59.8
20	122 30.1	52.5	59 29.8	5.1	23 55.7	7.9	59.7
21	137 30.3	. . 53.5	73 53.9	5.0	24 03.6	7.8	59.7
22	152 30.5	54.5	88 17.9	4.9	24 11.4	7.5	59.7
23	167 30.7	55.4	102 41.8	4.9	S24 18.9	7.4	59.7
	SD 16.0	d 1.0	SD 16.5		16.4		16.3

Lat.	Twilight Naut.	Twilight Civil	Sunrise	Moonrise 28	Moonrise 29	Moonrise 30	Moonrise 1
°	h m	h m	h m	h m	h m	h m	h m
N 72	03 36	05 00	06 08	10 48	■	■	■
N 70	03 50	05 05	06 06	10 23	13 30	■	■
68	04 01	05 09	06 04	10 04	12 36	■	■
66	04 10	05 12	06 03	09 49	12 03	14 54	■
64	04 18	05 14	06 01	09 37	11 40	13 54	■
62	04 24	05 16	06 00	09 27	11 21	13 20	15 18
60	04 29	05 18	06 00	09 18	11 06	12 56	14 39
N 58	04 33	05 20	05 59	09 10	10 53	12 36	14 12
56	04 37	05 21	05 58	09 04	10 42	12 20	13 51
54	04 41	05 22	05 57	08 58	10 32	12 06	13 33
52	04 44	05 23	05 57	08 53	10 24	11 54	13 18
50	04 46	05 24	05 56	08 48	10 16	11 43	13 05
45	04 51	05 26	05 55	08 37	10 00	11 21	12 39
N 40	04 55	05 27	05 54	08 29	09 46	11 04	12 18
35	04 58	05 28	05 53	08 22	09 35	10 49	12 00
30	05 00	05 28	05 52	08 15	09 25	10 36	11 45
20	05 03	05 28	05 50	08 04	09 09	10 14	11 20
N 10	05 03	05 28	05 49	07 55	08 54	09 55	10 58
0	05 02	05 27	05 47	07 46	08 41	09 38	10 38
S 10	05 00	05 25	05 46	07 38	08 27	09 21	10 18
20	04 56	05 22	05 44	07 28	08 13	09 02	09 56
30	04 50	05 18	05 41	07 18	07 57	08 41	09 32
35	04 45	05 15	05 40	07 12	07 48	08 29	09 17
40	04 40	05 12	05 39	07 05	07 37	08 15	09 00
45	04 33	05 07	05 37	06 58	07 25	07 59	08 40
S 50	04 24	05 02	05 35	06 48	07 10	07 38	08 15
52	04 19	05 00	05 34	06 44	07 03	07 29	08 03
54	04 14	04 57	05 32	06 39	06 56	07 18	07 50
56	04 09	04 54	05 31	06 34	06 47	07 06	07 34
58	04 02	04 50	05 30	06 29	06 38	06 52	07 15
S 60	03 55	04 46	05 28	06 22	06 27	06 36	06 52

Lat.	Sunset	Twilight Civil	Twilight Naut.	Moonset 28	Moonset 29	Moonset 30	Moonset 1
°	h m	h m	h m	h m	h m	h m	h m
N 72	17 31	18 38	20 01	16 41	■	■	■
N 70	17 33	18 34	19 48	17 08	15 57	■	■
68	17 35	18 30	19 37	17 29	16 53	■	■
66	17 37	18 27	19 28	17 46	17 26	16 37	■
64	17 38	18 25	19 21	18 00	17 51	17 38	■
62	17 39	18 23	19 15	18 11	18 11	18 13	18 22
60	17 40	18 21	19 10	18 21	18 27	18 38	19 01
N 58	17 41	18 20	19 06	18 30	18 41	18 58	19 28
56	17 42	18 19	19 02	18 38	18 53	19 15	19 50
54	17 43	18 18	18 59	18 45	19 03	19 29	20 07
52	17 43	18 17	18 56	18 51	19 13	19 42	20 23
50	17 44	18 16	18 54	18 57	19 21	19 53	20 36
45	17 45	18 14	18 49	19 09	19 39	20 16	21 03
N 40	17 46	18 13	18 45	19 19	19 53	20 35	21 24
35	17 48	18 13	18 42	19 28	20 06	20 50	21 42
30	17 49	18 12	18 40	19 35	20 17	21 04	21 58
20	17 50	18 12	18 38	19 49	20 36	21 27	22 24
N 10	17 52	18 13	18 37	20 01	20 52	21 47	22 46
0	17 54	18 14	18 38	20 12	21 08	22 06	23 07
S 10	17 55	18 16	18 41	20 23	21 23	22 25	23 28
20	17 57	18 20	18 45	20 35	21 40	22 46	23 51
30	18 00	18 24	18 52	20 48	21 59	23 09	24 17
35	18 01	18 27	18 56	20 56	22 10	23 23	24 32
40	18 03	18 30	19 02	21 06	22 23	23 39	24 50
45	18 05	18 34	19 09	21 16	22 38	23 59	25 12
S 50	18 07	18 40	19 19	21 29	22 57	24 23	00 23
52	18 08	18 42	19 23	21 35	23 06	24 34	00 34
54	18 10	18 45	19 28	21 42	23 17	24 48	00 48
56	18 11	18 48	19 34	21 50	23 28	25 03	01 03
58	18 12	18 52	19 40	21 58	23 41	25 21	01 21
S 60	18 14	18 56	19 48	22 08	23 57	25 44	01 44

Day	SUN Eqn. of Time 00^h	SUN Eqn. of Time 12^h	SUN Mer. Pass.	MOON Mer. Pass. Upper	MOON Mer. Pass. Lower	MOON Age	MOON Phase
d	m s	m s	h m	h m	h m	d %	
28	09 03	09 14	11 51	13 59	01 32	02 8	
29	09 24	09 34	11 50	14 54	02 26	03 15	
30	09 43	09 53	11 50	15 52	03 23	04 25	

UT	ARIES	VENUS −3.9		MARS −2.1		JUPITER −1.8		SATURN +0.1		STARS		
	GHA	GHA	Dec	GHA	Dec	GHA	Dec	GHA	Dec	Name	SHA	Dec
d h	° ′	° ′	° ′	° ′	° ′	° ′	° ′	° ′	° ′		° ′	°
1 00	9 18.9	171 21.1	S 6 34.2	35 21.4	S15 40.0	209 56.4	N 9 38.0	265 37.7	N22 06.2	Acamar	315 23.7	S40 17
01	24 21.4	186 20.7	35.4	50 23.8	39.8	224 58.4	37.8	280 40.0	06.1	Achernar	335 31.6	S57 13
02	39 23.9	201 20.3	36.7	65 26.2	39.5	240 00.4	37.6	295 42.4	06.1	Acrux	173 18.9	S63 07
03	54 26.3	216 19.8	37.9	80 28.7	39.3	255 02.4	37.5	310 44.7	06.1	Adhara	255 18.6	S28 58
04	69 28.8	231 19.4	39.2	95 31.1	39.1	270 04.4	37.3	325 47.0	06.1	Aldebaran	290 58.0	N16 31
05	84 31.2	246 19.0	40.4	110 33.5	38.8	285 06.4	37.1	340 49.4	06.1			
W 06	99 33.7	261 18.6	S 6 41.7	125 35.9	S15 38.6	300 08.4	N 9 36.9	355 51.7	N22 06.1	Alioth	166 27.7	N55 56
E 07	114 36.2	276 18.2	42.9	140 38.3	38.4	315 10.4	36.7	10 54.1	06.1	Alkaid	153 05.2	N49 17
D 08	129 38.6	291 17.8	44.2	155 40.7	38.1	330 12.3	36.5	25 56.4	06.1	Al Na'ir	27 52.8	S46 56
N 09	144 41.1	306 17.4	45.4	170 43.1	37.9	345 14.3	36.4	40 58.8	06.1	Alnilam	275 54.1	S 1 11
E 10	159 43.6	321 17.0	46.6	185 45.5	37.7	0 16.3	36.2	56 01.1	06.1	Alphard	218 03.8	S 8 40
S 11	174 46.0	336 16.6	47.9	200 47.9	37.4	15 18.3	36.0	71 03.5	06.0			
D 12	189 48.5	351 16.2	S 6 49.1	215 50.3	S15 37.2	30 20.3	N 9 35.8	86 05.8	N22 06.0	Alphecca	126 17.7	N26 42
A 13	204 51.0	6 15.8	50.4	230 52.7	37.0	45 22.3	35.6	101 08.1	06.0	Alpheratz	357 51.2	N29 06
Y 14	219 53.4	21 15.4	51.6	245 55.1	36.7	60 24.3	35.5	116 10.5	06.0	Altair	62 15.6	N 8 52
15	234 55.9	36 15.0	52.8	260 57.5	36.5	75 26.3	35.3	131 12.8	06.0	Ankaa	353 22.7	S42 17
16	249 58.4	51 14.5	54.1	275 59.9	36.3	90 28.2	35.1	146 15.2	06.0	Antares	112 35.9	S26 26
17	265 00.8	66 14.1	55.3	291 02.3	36.0	105 30.2	34.9	161 17.5	06.0			
18	280 03.3	81 13.7	S 6 56.6	306 04.7	S15 35.8	120 32.2	N 9 34.7	176 19.9	N22 06.0	Arcturus	146 03.0	N19 10
19	295 05.7	96 13.3	57.8	321 07.1	35.5	135 34.2	34.5	191 22.2	06.0	Atria	107 45.0	S69 02
20	310 08.2	111 12.9	6 59.0	336 09.5	35.3	150 36.2	34.4	206 24.6	06.0	Avior	234 21.6	S59 30
21	325 10.7	126 12.5	7 00.3	351 11.9	35.1	165 38.2	34.2	221 26.9	05.9	Bellatrix	278 40.1	N 6 21
22	340 13.1	141 12.1	01.5	6 14.3	34.8	180 40.2	34.0	236 29.3	05.9	Betelgeuse	271 09.5	N 7 24
23	355 15.6	156 11.7	02.8	21 16.7	34.6	195 42.2	33.8	251 31.6	05.9			
2 00	10 18.1	171 11.3	S 7 04.0	36 19.1	S15 34.3	210 44.2	N 9 33.6	266 34.0	N22 05.9	Canopus	263 59.5	S52 41
01	25 20.5	186 10.9	05.2	51 21.4	34.1	225 46.1	33.4	281 36.3	05.9	Capella	280 45.7	N46 00
02	40 23.0	201 10.4	06.5	66 23.8	33.9	240 48.1	33.3	296 38.7	05.9	Deneb	49 36.6	N45 17
03	55 25.5	216 10.0	07.7	81 26.2	33.6	255 50.1	33.1	311 41.0	05.9	Denebola	182 41.7	N14 33
04	70 27.9	231 09.6	09.0	96 28.6	33.4	270 52.1	32.9	326 43.3	05.9	Diphda	349 03.2	S17 57
05	85 30.4	246 09.2	10.2	111 31.0	33.1	285 54.1	32.7	341 45.7	05.9			
06	100 32.9	261 08.8	S 7 11.4	126 33.3	S15 32.9	300 56.1	N 9 32.5	356 48.0	N22 05.9	Dubhe	194 01.3	N61 43
T 07	115 35.3	276 08.4	12.7	141 35.7	32.6	315 58.1	32.3	11 50.4	05.8	Elnath	278 22.2	N28 36
H 08	130 37.8	291 08.0	13.9	156 38.1	32.4	331 00.1	32.2	26 52.7	05.8	Eltanin	90 49.8	N51 29
U 09	145 40.2	306 07.6	15.1	171 40.5	32.1	346 02.1	32.0	41 55.1	05.8	Enif	33 51.4	N 9 53
R 10	160 42.7	321 07.1	16.4	186 42.8	31.9	1 04.0	31.8	56 57.4	05.8	Fomalhaut	15 32.0	S29 36
S 11	175 45.2	336 06.7	17.6	201 45.2	31.6	16 06.0	31.6	71 59.8	05.8			
D 12	190 47.6	351 06.3	S 7 18.8	216 47.6	S15 31.4	31 08.0	N 9 31.4	87 02.1	N22 05.8	Gacrux	172 10.3	S57 07
A 13	205 50.1	6 05.9	20.1	231 50.0	31.2	46 10.0	31.3	102 04.5	05.8	Gienah	176 00.6	S17 33
Y 14	220 52.6	21 05.5	21.3	246 52.3	30.9	61 12.0	31.1	117 06.8	05.8	Hadar	148 59.7	S60 23
15	235 55.0	36 05.1	22.6	261 54.7	30.7	76 14.0	30.9	132 09.2	05.8	Hamal	328 09.2	N23 28
16	250 57.5	51 04.7	23.8	276 57.1	30.4	91 16.0	30.7	147 11.5	05.8	Kaus Aust.	83 54.0	S34 23
17	266 00.0	66 04.2	25.0	291 59.4	30.2	106 18.0	30.5	162 13.9	05.8			
18	281 02.4	81 03.8	S 7 26.3	307 01.8	S15 29.9	121 20.0	N 9 30.3	177 16.2	N22 05.7	Kochab	137 20.2	N74 08
19	296 04.9	96 03.4	27.5	322 04.1	29.7	136 22.0	30.2	192 18.6	05.7	Markab	13 45.7	N15 13
20	311 07.4	111 03.0	28.7	337 06.5	29.4	151 23.9	30.0	207 20.9	05.7	Menkar	314 22.8	N 4 06
21	326 09.8	126 02.6	30.0	352 08.9	29.2	166 25.9	29.8	222 23.3	05.7	Menkent	148 17.1	S36 23
22	341 12.3	141 02.2	31.2	7 11.2	28.9	181 27.9	29.6	237 25.6	05.7	Miaplacidus	221 42.2	S69 43
23	356 14.7	156 01.8	32.4	22 13.6	28.6	196 29.9	29.4	252 28.0	05.7			
3 00	11 17.2	171 01.3	S 7 33.7	37 15.9	S15 28.4	211 31.9	N 9 29.2	267 30.3	N22 05.7	Mirfak	308 51.1	N49 52
01	26 19.7	186 00.9	34.9	52 18.3	28.1	226 33.9	29.1	282 32.7	05.7	Nunki	76 07.8	S26 17
02	41 22.1	201 00.5	36.1	67 20.6	27.9	241 35.9	28.9	297 35.1	05.7	Peacock	53 30.9	S56 43
03	56 24.6	216 00.1	37.3	82 23.0	27.6	256 37.9	28.7	312 37.4	05.7	Pollux	243 37.2	N28 01
04	71 27.1	230 59.7	38.6	97 25.3	27.4	271 39.9	28.5	327 39.8	05.6	Procyon	245 07.8	N 5 13
05	86 29.5	245 59.2	39.8	112 27.7	27.1	286 41.9	28.3	342 42.1	05.6			
06	101 32.0	260 58.8	S 7 41.0	127 30.0	S15 26.9	301 43.9	N 9 28.2	357 44.5	N22 05.6	Rasalhague	96 13.7	N12 33
07	116 34.5	275 58.4	42.3	142 32.4	26.6	316 45.8	28.0	12 46.8	05.6	Regulus	207 51.9	N11 57
08	131 36.9	290 58.0	43.5	157 34.7	26.4	331 47.8	27.8	27 49.2	05.6	Rigel	281 19.3	S 8 11
F 09	146 39.4	305 57.6	44.7	172 37.1	26.1	346 49.8	27.6	42 51.5	05.6	Rigil Kent.	140 03.1	S60 51
R 10	161 41.8	320 57.2	46.0	187 39.4	25.8	1 51.8	27.4	57 53.9	05.6	Sabik	102 21.5	S15 43
I 11	176 44.3	335 56.7	47.2	202 41.7	25.6	16 53.8	27.2	72 56.2	05.6			
D 12	191 46.8	350 56.3	S 7 48.4	217 44.1	S15 25.3	31 55.8	N 9 27.1	87 58.6	N22 05.6	Schedar	349 49.0	N56 33
A 13	206 49.2	5 55.9	49.6	232 46.4	25.1	46 57.8	26.9	103 00.9	05.6	Shaula	96 32.4	S37 06
Y 14	221 51.7	20 55.5	50.9	247 48.7	24.8	61 59.8	26.7	118 03.3	05.6	Sirius	258 40.5	S16 42
15	236 54.2	35 55.1	52.1	262 51.1	24.5	77 01.8	26.5	133 05.6	05.5	Spica	158 39.7	S11 10
16	251 56.6	50 54.6	53.3	277 53.4	24.3	92 03.8	26.3	148 08.0	05.5	Suhail	222 58.5	S43 26
17	266 59.1	65 54.2	54.6	292 55.7	24.0	107 05.8	26.2	163 10.4	05.5			
18	282 01.6	80 53.8	S 7 55.8	307 58.1	S15 23.8	122 07.8	N 9 26.0	178 12.7	N22 05.5	Vega	80 44.2	N38 47
19	297 04.0	95 53.4	57.0	323 00.4	23.5	137 09.8	25.8	193 15.1	05.5	Zuben'ubi	137 14.2	S16 03
20	312 06.5	110 52.9	58.2	338 02.7	23.2	152 11.7	25.6	208 17.4	05.5		SHA	Mer. Pas
21	327 09.0	125 52.5	7 59.5	353 05.1	23.0	167 13.7	25.4	223 19.8	05.5		° ′	h m
22	342 11.4	140 52.1	8 00.7	8 07.4	22.7	182 15.7	25.3	238 22.1	05.5	Venus	160 53.2	12 36
23	357 13.9	155 51.7	S 8 01.9	23 09.7	22.4	197 17.7	25.1	253 24.5	05.5	Mars	26 01.0	21 31
Mer. Pass.	h m 23 15.0	v −0.4	d 1.2	v 2.4	d 0.2	v 2.0	d 0.2	v 2.3	d 0.0	Jupiter	200 26.1	9 56
										Saturn	256 15.9	6 13

	UT	SUN GHA	SUN Dec	MOON GHA	v	MOON Dec	d	HP
	d h	° ′	° ′	° ′	′	° ′	′	′
	1 00	182 30.9	S 2 56.4	117 05.7	4.9	S24 26.3	7.3	59.7
	01	197 31.1	57.4	131 29.6	4.8	24 33.6	7.0	59.6
	02	212 31.3	58.3	145 53.4	4.7	24 40.6	6.9	59.6
	03	227 31.5	2 59.3	160 17.1	4.8	24 47.5	6.7	59.6
	04	242 31.7	3 00.3	174 40.9	4.7	24 54.2	6.6	59.6
	05	257 31.9	01.3	189 04.6	4.6	25 00.8	6.3	59.5
	06	272 32.1	S 3 02.2	203 28.2	4.6	S25 07.1	6.2	59.5
W	07	287 32.3	03.2	217 51.8	4.6	25 13.3	6.1	59.5
E	08	302 32.5	04.2	232 15.4	4.5	25 19.4	5.8	59.5
D	09	317 32.7	. . 05.1	246 38.9	4.6	25 25.2	5.7	59.5
N	10	332 32.9	06.1	261 02.5	4.5	25 30.9	5.5	59.4
E	11	347 33.1	07.1	275 26.0	4.4	25 36.4	5.3	59.4
S	12	2 33.3	S 3 08.0	289 49.4	4.5	S25 41.7	5.1	59.4
D	13	17 33.5	09.0	304 12.9	4.4	25 46.8	5.0	59.4
A	14	32 33.7	10.0	318 36.3	4.4	25 51.8	4.8	59.3
Y	15	47 33.9	. . 11.0	332 59.7	4.3	25 56.6	4.6	59.3
	16	62 34.1	11.9	347 23.0	4.4	26 01.2	4.4	59.3
	17	77 34.3	12.9	1 46.4	4.3	26 05.6	4.3	59.3
	18	92 34.5	S 3 13.9	16 09.7	4.3	S26 09.9	4.1	59.3
	19	107 34.7	14.8	30 33.0	4.3	26 14.0	3.8	59.2
	20	122 34.9	15.8	44 56.3	4.3	26 17.8	3.8	59.2
	21	137 35.1	. . 16.8	59 19.6	4.3	26 21.6	3.5	59.2
	22	152 35.3	17.7	73 42.9	4.3	26 25.1	3.3	59.2
	23	167 35.5	18.7	88 06.2	4.3	26 28.4	3.2	59.1
	2 00	182 35.7	S 3 19.7	102 29.5	4.3	S26 31.6	3.0	59.1
	01	197 35.9	20.7	116 52.8	4.3	26 34.6	2.8	59.1
	02	212 36.1	21.6	131 16.1	4.3	26 37.4	2.6	59.1
	03	227 36.3	. . 22.6	145 39.4	4.2	26 40.0	2.5	59.0
	04	242 36.5	23.6	160 02.6	4.3	26 42.5	2.2	59.0
	05	257 36.7	24.5	174 25.9	4.3	26 44.7	2.1	59.0
	06	272 36.9	S 3 25.5	188 49.2	4.4	S26 46.8	1.9	59.0
	07	287 37.1	26.5	203 12.6	4.3	26 48.7	1.7	58.9
T	08	302 37.3	27.4	217 35.9	4.3	26 50.4	1.6	58.9
H	09	317 37.5	. . 28.4	231 59.2	4.4	26 52.0	1.3	58.9
U	10	332 37.7	29.4	246 22.6	4.4	26 53.3	1.2	58.9
R	11	347 37.9	30.3	260 46.0	4.4	26 54.5	1.0	58.8
S	12	2 38.1	S 3 31.3	275 09.4	4.4	S26 55.5	0.8	58.8
D	13	17 38.3	32.3	289 32.8	4.4	26 56.3	0.7	58.8
A	14	32 38.5	33.2	303 56.2	4.5	26 57.0	0.4	58.8
Y	15	47 38.7	. . 34.2	318 19.7	4.5	26 57.4	0.3	58.7
	16	62 38.9	35.2	332 43.2	4.6	26 57.7	0.1	58.7
	17	77 39.1	36.1	347 06.8	4.5	26 57.8	0.1	58.7
	18	92 39.3	S 3 37.1	1 30.3	4.6	S26 57.7	0.2	58.7
	19	107 39.5	38.1	15 53.9	4.7	26 57.5	0.4	58.6
	20	122 39.7	39.1	30 17.6	4.7	26 57.1	0.7	58.6
	21	137 39.9	. . 40.0	44 41.3	4.7	26 56.4	0.7	58.6
	22	152 40.1	41.0	59 05.0	4.7	26 55.7	1.0	58.6
	23	167 40.3	42.0	73 28.7	4.9	26 54.7	1.1	58.5
	3 00	182 40.5	S 3 42.9	87 52.6	4.8	S26 53.6	1.3	58.5
	01	197 40.7	43.9	102 16.4	4.9	26 52.3	1.5	58.5
	02	212 40.9	44.9	116 40.3	5.0	26 50.8	1.7	58.5
	03	227 41.1	. . 45.8	131 04.3	5.0	26 49.1	1.8	58.4
	04	242 41.3	46.8	145 28.3	5.1	26 47.3	2.0	58.4
	05	257 41.5	47.8	159 52.4	5.1	26 45.3	2.1	58.4
	06	272 41.7	S 3 48.7	174 16.5	5.2	S26 43.2	2.4	58.4
	07	287 41.9	49.7	188 40.7	5.2	26 40.8	2.5	58.3
	08	302 42.1	50.7	203 04.9	5.3	26 38.3	2.7	58.3
F	09	317 42.3	. . 51.6	217 29.2	5.4	26 35.6	2.8	58.3
R	10	332 42.5	52.6	231 53.6	5.4	26 32.8	3.0	58.2
I	11	347 42.7	53.6	246 18.0	5.5	26 29.8	3.2	58.2
D	12	2 42.9	S 3 54.5	260 42.5	5.6	S26 26.6	3.3	58.2
A	13	17 43.1	55.5	275 07.1	5.6	26 23.3	3.5	58.2
Y	14	32 43.3	56.5	289 31.7	5.8	26 19.8	3.7	58.1
	15	47 43.5	. . 57.4	303 56.5	5.7	26 16.1	3.8	58.1
	16	62 43.7	58.4	318 21.2	5.9	26 12.3	4.0	58.1
	17	77 43.9	3 59.4	332 46.1	5.9	26 08.3	4.1	58.1
	18	92 44.0	S 4 00.3	347 11.0	6.0	S26 04.2	4.3	58.0
	19	107 44.2	01.3	1 36.0	6.1	25 59.9	4.5	58.0
	20	122 44.4	02.3	16 01.1	6.2	25 55.4	4.6	58.0
	21	137 44.6	. . 03.2	30 26.3	6.2	25 50.8	4.7	58.0
	22	152 44.8	04.2	44 51.5	6.4	25 46.1	5.0	57.9
	23	167 45.0	05.2	59 16.9	6.4	S25 41.1	5.0	57.9
		SD 16.0	*d* 1.0	SD 16.2		16.0		15.9

Lat.	Twilight Naut.	Twilight Civil	Sunrise	Moonrise 1	Moonrise 2	Moonrise 3	Moonrise 4
°	h m	h m	h m	h m	h m	h m	h m
N 72	03 52	05 14	06 21	■	■	■	■
N 70	04 04	05 17	06 18	■	■	■	■
68	04 13	05 19	06 15	■	■	■	■
66	04 21	05 21	06 12	■	■	■	19 07
64	04 27	05 23	06 10	■	■	■	18 05
62	04 32	05 24	06 08	15 18	16 48	17 22	17 30
60	04 36	05 25	06 07	14 39	15 59	16 44	17 05
N 58	04 40	05 26	06 05	14 12	15 28	16 17	16 44
56	04 44	05 27	06 04	13 51	15 04	15 56	16 28
54	04 46	05 28	06 03	13 33	14 45	15 38	16 13
52	04 49	05 28	06 02	13 18	14 29	15 23	16 01
50	04 51	05 28	06 01	13 05	14 15	15 10	15 50
45	04 55	05 29	05 59	12 39	13 47	14 43	15 27
N 40	04 58	05 30	05 57	12 18	13 25	14 22	15 08
35	05 00	05 30	05 55	12 00	13 06	14 04	14 52
30	05 02	05 30	05 54	11 45	12 50	13 49	14 39
20	05 03	05 29	05 51	11 20	12 24	13 23	14 16
N 10	05 03	05 28	05 49	10 58	12 01	13 00	13 55
0	05 01	05 26	05 46	10 38	11 39	12 39	13 37
S 10	04 58	05 23	05 44	10 18	11 18	12 18	13 18
20	04 53	05 19	05 41	09 56	10 55	11 56	12 57
30	04 46	05 14	05 38	09 32	10 28	11 30	12 34
35	04 41	05 11	05 36	09 17	10 13	11 15	12 20
40	04 34	05 07	05 34	09 00	09 55	10 57	12 04
45	04 27	05 02	05 31	08 40	09 33	10 35	11 45
S 50	04 17	04 55	05 28	08 15	09 05	10 08	11 21
52	04 12	04 53	05 27	08 03	08 51	09 54	11 09
54	04 06	04 49	05 25	07 50	08 36	09 39	10 56
56	04 00	04 46	05 23	07 34	08 18	09 21	10 40
58	03 53	04 41	05 21	07 15	07 55	08 59	10 22
S 60	03 44	04 37	05 19	06 52	07 26	08 31	10 00

Lat.	Sunset	Twilight Civil	Twilight Naut.	Moonset 1	Moonset 2	Moonset 3	Moonset 4
°	h m	h m	h m	h m	h m	h m	h m
N 72	17 16	18 22	19 44	■	■	■	■
N 70	17 19	18 20	19 32	■	■	■	■
68	17 22	18 18	19 23	■	■	■	■
66	17 25	18 16	19 16	■	■	■	20 44
64	17 27	18 14	19 10	■	■	■	21 46
62	17 29	18 13	19 05	18 22	18 59	20 30	22 20
60	17 31	18 12	19 01	19 01	19 49	21 08	22 45
N 58	17 33	18 11	18 57	19 28	20 20	21 35	23 04
56	17 34	18 11	18 54	19 50	20 43	21 56	23 20
54	17 35	18 10	18 51	20 07	21 02	22 13	23 34
52	17 36	18 10	18 49	20 23	21 18	22 28	23 46
50	17 37	18 09	18 47	20 36	21 32	22 41	23 57
45	17 40	18 09	18 43	21 03	22 01	23 07	24 19
N 40	17 42	18 09	18 40	21 24	22 23	23 28	24 37
35	17 43	18 09	18 38	21 42	22 41	23 45	24 52
30	17 45	18 09	18 36	21 58	22 57	24 00	00 00
20	17 48	18 10	18 35	22 24	23 24	24 25	00 25
N 10	17 50	18 11	18 36	22 46	23 47	24 47	00 47
0	17 53	18 13	18 37	23 07	24 08	00 08	01 07
S 10	17 55	18 16	18 41	23 28	24 30	00 30	01 27
20	17 58	18 20	18 46	23 51	24 52	00 52	01 49
30	18 02	18 26	18 54	24 17	00 17	01 19	02 13
35	18 04	18 29	18 59	24 32	00 32	01 35	02 28
40	18 06	18 33	19 05	24 50	00 50	01 53	02 44
45	18 09	18 38	19 13	25 12	01 12	02 14	03 04
S 50	18 12	18 45	19 24	00 23	01 39	02 42	03 29
52	18 13	18 48	19 29	00 34	01 53	02 55	03 41
54	18 15	18 51	19 34	00 48	02 08	03 11	03 55
56	18 17	18 55	19 41	01 03	02 26	03 29	04 10
58	18 19	18 59	19 48	01 21	02 49	03 51	04 29
S 60	18 21	19 04	19 57	01 44	03 17	04 20	04 52

Day	SUN Eqn. of Time 00^h	SUN Eqn. of Time 12^h	SUN Mer. Pass.	MOON Mer. Pass. Upper	MOON Mer. Pass. Lower	Age	Phase
d	m s	m s	h m	h m	h m	d %	
1	10 03	10 13	11 50	16 53	04 22	05 36	
2	10 23	10 32	11 49	17 54	05 23	06 47	◑
3	10 42	10 51	11 49	18 53	06 24	07 58	

UT	ARIES	VENUS −3.9		MARS −2.0		JUPITER −1.8		SATURN +0.1	
	GHA	GHA	Dec	GHA	Dec	GHA	Dec	GHA	Dec
d h	° ′	° ′	° ′	° ′	° ′	° ′	° ′	° ′	° ′
4 00	12 16.3	170 51.3	S 8 03.1	38 12.0	S15 22.2	212 19.7	N 9 24.9	268 26.8	N22 05.5
01	27 18.8	185 50.8	04.4	53 14.4	21.9	227 21.7	24.7	283 29.2	05.5
02	42 21.3	200 50.4	05.6	68 16.7	21.7	242 23.7	24.5	298 31.6	05.4
03	57 23.7	215 50.0	. . 06.8	83 19.0	. . 21.4	257 25.7	. . 24.3	313 33.9	. . 05.4
04	72 26.2	230 49.6	08.0	98 21.3	21.1	272 27.7	24.2	328 36.3	05.4
05	87 28.7	245 49.1	09.3	113 23.6	20.9	287 29.7	24.0	343 38.6	05.4
06	102 31.1	260 48.7	S 8 10.5	128 25.9	S15 20.6	302 31.7	N 9 23.8	358 41.0	N22 05.4
07	117 33.6	275 48.3	11.7	143 28.3	20.3	317 33.7	23.6	13 43.4	05.4
S 08	132 36.1	290 47.9	12.9	158 30.6	20.1	332 35.7	23.4	28 45.7	05.4
A 09	147 38.5	305 47.4	. . 14.2	173 32.9	. . 19.8	347 37.7	. . 23.3	43 48.1	. . 05.4
T 10	162 41.0	320 47.0	15.4	188 35.2	19.5	2 39.7	23.1	58 50.4	05.4
U 11	177 43.5	335 46.6	16.6	203 37.5	19.2	17 41.7	22.9	73 52.8	05.4
R 12	192 45.9	350 46.1	S 8 17.8	218 39.8	S15 19.0	32 43.6	N 9 22.7	88 55.1	N22 05.4
D 13	207 48.4	5 45.7	19.0	233 42.1	18.7	47 45.6	22.5	103 57.5	05.3
A 14	222 50.8	20 45.3	20.3	248 44.4	18.4	62 47.6	22.4	118 59.9	05.3
Y 15	237 53.3	35 44.9	. . 21.5	263 46.7	. . 18.2	77 49.6	. . 22.2	134 02.2	. . 05.3
16	252 55.8	50 44.4	22.7	278 49.0	17.9	92 51.6	22.0	149 04.6	05.3
17	267 58.2	65 44.0	23.9	293 51.3	17.6	107 53.6	21.8	164 06.9	05.3
18	283 00.7	80 43.6	S 8 25.1	308 53.6	S15 17.3	122 55.6	N 9 21.6	179 09.3	N22 05.3
19	298 03.2	95 43.2	26.4	323 55.9	17.1	137 57.6	21.5	194 11.7	05.3
20	313 05.6	110 42.7	27.6	338 58.2	16.8	152 59.6	21.3	209 14.0	05.3
21	328 08.1	125 42.3	. . 28.8	354 00.5	. . 16.5	168 01.6	. . 21.1	224 16.4	. . 05.3
22	343 10.6	140 41.9	30.0	9 02.8	16.3	183 03.6	20.9	239 18.7	05.3
23	358 13.0	155 41.4	31.2	24 05.1	16.0	198 05.6	20.7	254 21.1	05.3
5 00	13 15.5	170 41.0	S 8 32.5	39 07.4	S15 15.7	213 07.6	N 9 20.6	269 23.5	N22 05.3
01	28 18.0	185 40.6	33.7	54 09.7	15.4	228 09.6	20.4	284 25.8	05.2
02	43 20.4	200 40.1	34.9	69 12.0	15.1	243 11.6	20.2	299 28.2	05.2
03	58 22.9	215 39.7	. . 36.1	84 14.3	. . 14.9	258 13.6	. . 20.0	314 30.5	. . 05.2
04	73 25.3	230 39.3	37.3	99 16.5	14.6	273 15.6	19.8	329 32.9	05.2
05	88 27.8	245 38.8	38.5	114 18.8	14.3	288 17.6	19.6	344 35.3	05.2
06	103 30.3	260 38.4	S 8 39.8	129 21.1	S15 14.0	303 19.6	N 9 19.5	359 37.6	N22 05.2
07	118 32.7	275 38.0	41.0	144 23.4	13.8	318 21.6	19.3	14 40.0	05.2
08	133 35.2	290 37.6	42.2	159 25.7	13.5	333 23.6	19.1	29 42.4	05.2
S 09	148 37.7	305 37.1	. . 43.4	174 28.0	. . 13.2	348 25.6	. . 18.9	44 44.7	. . 05.2
U 10	163 40.1	320 36.7	44.6	189 30.2	12.9	3 27.5	18.7	59 47.1	05.2
N 11	178 42.6	335 36.3	45.8	204 32.5	12.6	18 29.5	18.6	74 49.4	05.2
D 12	193 45.1	350 35.8	S 8 47.0	219 34.8	S15 12.4	33 31.5	N 9 18.4	89 51.8	N22 05.2
A 13	208 47.5	5 35.4	48.3	234 37.1	12.1	48 33.5	18.2	104 54.2	05.1
Y 14	223 50.0	20 35.0	49.5	249 39.3	11.8	63 35.5	18.0	119 56.5	05.1
15	238 52.4	35 34.5	. . 50.7	264 41.6	. . 11.5	78 37.5	. . 17.8	134 58.9	. . 05.1
16	253 54.9	50 34.1	51.9	279 43.9	11.2	93 39.5	17.7	150 01.3	05.1
17	268 57.4	65 33.6	53.1	294 46.1	11.0	108 41.5	17.5	165 03.6	05.1
18	283 59.8	80 33.2	S 8 54.3	309 48.4	S15 10.7	123 43.5	N 9 17.3	180 06.0	N22 05.1
19	299 02.3	95 32.8	55.5	324 50.7	10.4	138 45.5	17.1	195 08.4	05.1
20	314 04.8	110 32.3	56.7	339 52.9	10.1	153 47.5	16.9	210 10.7	05.1
21	329 07.2	125 31.9	. . 57.9	354 55.2	. . 09.8	168 49.5	. . 16.8	225 13.1	. . 05.1
22	344 09.7	140 31.5	8 59.2	9 57.5	09.5	183 51.5	16.6	240 15.5	05.1
23	359 12.2	155 31.0	9 00.4	24 59.7	09.2	198 53.5	16.4	255 17.8	05.1
6 00	14 14.6	170 30.6	S 9 01.6	40 02.0	S15 09.0	213 55.5	N 9 16.2	270 20.2	N22 05.1
01	29 17.1	185 30.2	02.8	55 04.3	08.7	228 57.5	16.0	285 22.6	05.0
02	44 19.6	200 29.7	04.0	70 06.5	08.4	243 59.5	15.9	300 24.9	05.0
03	59 22.0	215 29.3	. . 05.2	85 08.8	. . 08.1	259 01.5	. . 15.7	315 27.3	. . 05.0
04	74 24.5	230 28.8	06.4	100 11.0	07.8	274 03.5	15.5	330 29.7	05.0
05	89 26.9	245 28.4	07.6	115 13.3	07.5	289 05.5	15.3	345 32.0	05.0
06	104 29.4	260 28.0	S 9 08.8	130 15.5	S15 07.2	304 07.5	N 9 15.2	0 34.4	N22 05.0
07	119 31.9	275 27.5	10.0	145 17.8	06.9	319 09.5	15.0	15 36.8	05.0
08	134 34.3	290 27.1	11.2	160 20.0	06.7	334 11.5	14.8	30 39.1	05.0
M 09	149 36.8	305 26.6	. . 12.4	175 22.3	. . 06.4	349 13.5	. . 14.6	45 41.5	. . 05.0
O 10	164 39.3	320 26.2	13.7	190 24.5	06.1	4 15.5	14.4	60 43.9	05.0
N 11	179 41.7	335 25.8	14.9	205 26.8	05.8	19 17.5	14.3	75 46.2	05.0
D 12	194 44.2	350 25.3	S 9 16.1	220 29.0	S15 05.5	34 19.5	N 9 14.1	90 48.6	N22 05.0
A 13	209 46.7	5 24.9	17.3	235 31.3	05.2	49 21.5	13.9	105 51.0	04.9
Y 14	224 49.1	20 24.4	18.5	250 33.5	04.9	64 23.5	13.7	120 53.3	04.9
15	239 51.6	35 24.0	. . 19.7	265 35.8	. . 04.6	79 25.5	. . 13.5	135 55.7	. . 04.9
16	254 54.1	50 23.5	20.9	280 38.0	04.3	94 27.5	13.4	150 58.1	04.9
17	269 56.5	65 23.1	22.1	295 40.2	04.0	109 29.5	13.2	166 00.4	04.9
18	284 59.0	80 22.7	S 9 23.3	310 42.5	S15 03.7	124 31.5	N 9 13.0	181 02.8	N22 04.9
19	300 01.4	95 22.2	24.5	325 44.7	03.4	139 33.5	12.8	196 05.2	04.9
20	315 03.9	110 21.8	25.7	340 46.9	03.2	154 35.5	12.6	211 07.5	04.9
21	330 06.4	125 21.3	. . 26.9	355 49.2	. . 02.9	169 37.5	. . 12.5	226 09.9	. . 04.9
22	345 08.8	140 20.9	28.1	10 51.4	02.6	184 39.5	12.3	241 12.3	04.9
23	0 11.3	155 20.4	29.3	25 53.6	02.3	199 41.5	12.1	256 14.7	04.9
Mer. Pass.	h m 23 03.2	*v* −0.4	*d* 1.2	*v* 2.3	*d* 0.3	*v* 2.0	*d* 0.2	*v* 2.4	*d* 0.0

STARS Name	SHA	Dec
	° ′	° ′
Acamar	315 23.7	S40 17.2
Achernar	335 31.6	S57 13.0
Acrux	173 18.9	S63 07.0
Adhara	255 18.6	S28 58.3
Aldebaran	290 58.0	N16 31.1
Alioth	166 27.7	N55 56.5
Alkaid	153 05.2	N49 17.9
Al Na'ir	27 52.8	S46 56.8
Alnilam	275 54.0	S 1 11.8
Alphard	218 03.8	S 8 40.2
Alphecca	126 17.7	N26 42.3
Alpheratz	357 51.2	N29 06.7
Altair	62 15.6	N 8 52.8
Ankaa	353 22.7	S42 17.1
Antares	112 35.9	S26 26.5
Arcturus	146 03.0	N19 10.0
Atria	107 45.0	S69 02.3
Avior	234 21.6	S59 30.9
Bellatrix	278 40.1	N 6 21.4
Betelgeuse	271 09.5	N 7 24.6
Canopus	263 59.5	S52 41.4
Capella	280 45.6	N46 00.1
Deneb	49 36.6	N45 17.8
Denebola	182 41.7	N14 33.3
Diphda	349 03.2	S17 57.9
Dubhe	194 01.3	N61 43.9
Elnath	278 22.2	N28 36.7
Eltanin	90 49.8	N51 29.5
Enif	33 54.4	N 9 53.6
Fomalhaut	15 32.0	S29 36.2
Gacrux	172 10.3	S57 07.9
Gienah	176 00.6	S17 33.5
Hadar	148 59.7	S60 23.4
Hamal	328 09.2	N23 28.9
Kaus Aust.	83 54.0	S34 23.1
Kochab	137 20.3	N74 08.6
Markab	13 45.7	N15 13.6
Menkar	314 22.8	N 4 06.4
Menkent	148 17.1	S36 23.2
Miaplacidus	221 42.2	S69 43.5
Mirfak	308 51.1	N49 52.4
Nunki	76 07.8	S26 17.6
Peacock	53 30.9	S56 43.7
Pollux	243 37.1	N28 01.1
Procyon	245 07.8	N 5 13.1
Rasalhague	96 13.7	N12 33.6
Regulus	207 51.9	N11 57.1
Rigel	281 19.3	S 8 11.6
Rigil Kent.	140 03.1	S60 51.1
Sabik	102 21.5	S15 43.8
Schedar	349 49.0	N56 33.5
Shaula	96 32.5	S37 06.5
Sirius	258 40.5	S16 42.9
Spica	158 39.7	S11 10.7
Suhail	222 58.4	S43 26.5
Vega	80 44.2	N38 47.4
Zuben'ubi	137 14.2	S16 03.4

	SHA	Mer. Pass.
	° ′	h m
Venus	157 25.5	12 38
Mars	25 51.9	21 20
Jupiter	199 52.1	9 46
Saturn	256 08.0	6 01

UT	SUN GHA	SUN Dec	MOON GHA	v	MOON Dec	d	HP
d h	° ′	° ′	° ′	′	° ′	′	′
4 00 (SATURDAY)	182 45.2	S 4 06.1	73 42.3	6.5	S25 36.1	5.2	57.9
01	197 45.4	07.1	88 07.8	6.6	25 30.9	5.4	57.9
02	212 45.6	08.1	102 33.4	6.6	25 25.5	5.5	57.8
03	227 45.8	. . 09.0	116 59.0	6.8	25 20.0	5.7	57.8
04	242 46.0	10.0	131 24.8	6.8	25 14.3	5.8	57.8
05	257 46.2	11.0	145 50.6	7.0	25 08.5	6.0	57.8
06	272 46.4	S 4 11.9	160 16.6	7.0	S25 02.5	6.1	57.7
07	287 46.6	12.9	174 42.6	7.1	24 56.4	6.2	57.7
08	302 46.8	13.8	189 08.7	7.2	24 50.2	6.4	57.7
09	317 47.0	. . 14.8	203 34.9	7.3	24 43.8	6.5	57.7
10	332 47.2	15.8	218 01.2	7.4	24 37.3	6.7	57.6
11	347 47.3	16.7	232 27.6	7.5	24 30.6	6.8	57.6
12	2 47.5	S 4 17.7	246 54.1	7.5	S24 23.8	6.9	57.6
13	17 47.7	18.7	261 20.6	7.7	24 16.9	7.1	57.6
14	32 47.9	19.6	275 47.3	7.8	24 09.8	7.2	57.5
15	47 48.1	. . 20.6	290 14.1	7.9	24 02.6	7.3	57.5
16	62 48.3	21.6	304 41.0	7.9	23 55.3	7.5	57.5
17	77 48.5	22.5	319 07.9	8.1	23 47.8	7.6	57.5
18	92 48.7	S 4 23.5	333 35.0	8.1	S23 40.2	7.7	57.4
19	107 48.9	24.5	348 02.1	8.3	23 32.5	7.9	57.4
20	122 49.1	25.4	2 29.4	8.3	23 24.6	8.0	57.4
21	137 49.3	. . 26.4	16 56.7	8.5	23 16.6	8.0	57.4
22	152 49.5	27.3	31 24.2	8.5	23 08.6	8.3	57.3
23	167 49.7	28.3	45 51.7	8.7	23 00.3	8.3	57.3
5 00 (SUNDAY)	182 49.8	S 4 29.3	60 19.4	8.7	S22 52.0	8.5	57.3
01	197 50.0	30.2	74 47.1	8.9	22 43.5	8.6	57.3
02	212 50.2	31.2	89 15.0	8.9	22 34.9	8.6	57.2
03	227 50.4	. . 32.2	103 42.9	9.0	22 26.3	8.9	57.2
04	242 50.6	33.1	118 10.9	9.2	22 17.4	8.9	57.2
05	257 50.8	34.1	132 39.1	9.2	22 08.5	9.0	57.2
06	272 51.0	S 4 35.0	147 07.3	9.4	S21 59.5	9.2	57.1
07	287 51.2	36.0	161 35.7	9.4	21 50.3	9.2	57.1
08	302 51.4	37.0	176 04.1	9.5	21 41.1	9.4	57.1
09	317 51.6	. . 37.9	190 32.6	9.7	21 31.7	9.5	57.1
10	332 51.7	38.9	205 01.3	9.7	21 22.2	9.5	57.1
11	347 51.9	39.9	219 30.0	9.8	21 12.7	9.7	57.0
12	2 52.1	S 4 40.8	233 58.8	10.0	S21 03.0	9.8	57.0
13	17 52.3	41.8	248 27.8	10.0	20 53.2	9.9	57.0
14	32 52.5	42.7	262 56.8	10.1	20 43.3	10.0	57.0
15	47 52.7	. . 43.7	277 25.9	10.3	20 33.3	10.1	56.9
16	62 52.9	44.7	291 55.2	10.3	20 23.2	10.2	56.9
17	77 53.1	45.6	306 24.5	10.4	20 13.0	10.2	56.9
18	92 53.3	S 4 46.6	320 53.9	10.5	S20 02.8	10.4	56.9
19	107 53.4	47.6	335 23.4	10.7	19 52.4	10.5	56.8
20	122 53.6	48.5	349 53.1	10.7	19 41.9	10.5	56.8
21	137 53.8	. . 49.5	4 22.8	10.8	19 31.4	10.7	56.8
22	152 54.0	50.4	18 52.6	10.9	19 20.7	10.7	56.8
23	167 54.2	51.4	33 22.5	11.0	19 10.0	10.8	56.8
6 00 (MONDAY)	182 54.4	S 4 52.4	47 52.5	11.1	S18 59.2	11.0	56.7
01	197 54.6	53.3	62 22.6	11.1	18 48.2	11.0	56.7
02	212 54.8	54.3	76 52.7	11.3	18 37.2	11.0	56.7
03	227 54.9	. . 55.2	91 23.0	11.4	18 26.2	11.2	56.7
04	242 55.1	56.2	105 53.4	11.4	18 15.0	11.2	56.6
05	257 55.3	57.2	120 23.8	11.6	18 03.8	11.4	56.6
06	272 55.5	S 4 58.1	134 54.4	11.6	S17 52.4	11.4	56.6
07	287 55.7	4 59.1	149 25.0	11.8	17 41.0	11.5	56.6
08	302 55.9	5 00.0	163 55.8	11.8	17 29.5	11.5	56.5
09	317 56.1	. . 01.0	178 26.6	11.9	17 18.0	11.6	56.5
10	332 56.3	02.0	192 57.5	12.0	17 06.4	11.7	56.5
11	347 56.4	02.9	207 28.5	12.1	16 54.7	11.8	56.5
12	2 56.6	S 5 03.9	221 59.6	12.2	S16 42.9	11.9	56.5
13	17 56.8	04.8	236 30.8	12.2	16 31.0	11.9	56.4
14	32 57.0	05.8	251 02.0	12.4	16 19.1	12.0	56.4
15	47 57.2	. . 06.8	265 33.4	12.4	16 07.1	12.0	56.4
16	62 57.4	07.7	280 04.8	12.5	15 55.1	12.1	56.4
17	77 57.5	08.7	294 36.3	12.6	15 43.0	12.2	56.4
18	92 57.7	S 5 09.6	309 07.9	12.7	S15 30.8	12.3	56.3
19	107 57.9	10.6	323 39.6	12.7	15 18.5	12.3	56.3
20	122 58.1	11.6	338 11.3	12.8	15 06.2	12.3	56.3
21	137 58.3	. . 12.5	352 43.1	13.0	14 53.9	12.5	56.3
22	152 58.5	13.5	7 15.1	13.0	14 41.4	12.4	56.2
23	167 58.7	14.4	21 47.1	13.0	S14 29.0	12.6	56.2
	SD 16.0	*d* 1.0	SD 15.7		15.5		15.4

Lat.	Twilight Naut.	Twilight Civil	Sunrise	Moonrise 4	Moonrise 5	Moonrise 6	Moonrise 7
°	h m	h m	h m	h m	h m	h m	h m
N 72	04 07	05 27	06 35	■	■	19 17	18 26
N 70	04 17	05 29	06 30	■	20 07	18 43	18 10
68	04 24	05 30	06 25	■	18 56	18 19	17 57
66	04 31	05 31	06 22	19 07	18 19	17 59	17 46
64	04 36	05 31	06 19	18 05	17 52	17 44	17 37
62	04 40	05 32	06 16	17 30	17 31	17 31	17 29
60	04 44	05 32	06 14	17 05	17 14	17 19	17 22
N 58	04 47	05 33	06 12	16 44	17 00	17 10	17 16
56	04 50	05 33	06 10	16 28	16 48	17 01	17 11
54	04 52	05 33	06 08	16 13	16 37	16 54	17 06
52	04 54	05 33	06 07	16 01	16 27	16 47	17 02
50	04 56	05 33	06 05	15 50	16 19	16 41	16 58
45	04 59	05 33	06 02	15 27	16 00	16 27	16 49
N 40	05 01	05 33	06 00	15 08	15 45	16 16	16 42
35	05 03	05 32	05 57	14 52	15 33	16 07	16 36
30	05 04	05 31	05 55	14 39	15 21	15 58	16 30
20	05 04	05 30	05 52	14 16	15 02	15 44	16 21
N 10	05 03	05 27	05 48	13 55	14 46	15 31	16 12
0	05 00	05 25	05 45	13 37	14 30	15 19	16 04
S 10	04 56	05 21	05 42	13 18	14 14	15 07	15 56
20	04 50	05 16	05 38	12 57	13 57	14 54	15 48
30	04 42	05 10	05 34	12 34	13 37	14 39	15 38
35	04 36	05 06	05 32	12 20	13 26	14 30	15 33
40	04 29	05 02	05 29	12 04	13 13	14 21	15 26
45	04 21	04 56	05 26	11 45	12 57	14 09	15 19
S 50	04 09	04 49	05 22	11 21	12 38	13 55	15 09
52	04 04	04 45	05 20	11 09	12 29	13 48	15 05
54	03 58	04 42	05 18	10 56	12 18	13 41	15 01
56	03 51	04 37	05 15	10 40	12 06	13 32	14 55
58	03 43	04 33	05 13	10 22	11 53	13 23	14 50
S 60	03 34	04 27	05 10	10 00	11 37	13 12	14 43

Lat.	Sunset	Twilight Civil	Twilight Naut.	Moonset 4	Moonset 5	Moonset 6	Moonset 7
°	h m	h m	h m	h m	h m	h m	h m
N 72	17 00	18 07	19 27	■	■	■	00 07
N 70	17 06	18 06	19 18	■	21 35	24 39	00 39
68	17 10	18 05	19 10	■	22 45	25 02	01 02
66	17 14	18 05	19 04	20 44	23 21	25 20	01 20
64	17 17	18 04	18 59	21 46	23 46	25 34	01 34
62	17 20	18 04	18 55	22 20	24 06	00 06	01 46
60	17 22	18 03	18 52	22 45	24 22	00 22	01 56
N 58	17 24	18 03	18 49	23 04	24 36	00 36	02 05
56	17 26	18 03	18 46	23 20	24 47	00 47	02 12
54	17 28	18 03	18 44	23 34	24 58	00 58	02 19
52	17 29	18 03	18 42	23 46	25 07	01 07	02 25
50	17 31	18 03	18 40	23 57	25 15	01 15	02 31
45	17 34	18 03	18 37	24 19	00 19	01 32	02 43
N 40	17 37	18 04	18 35	24 37	00 37	01 45	02 52
35	17 39	18 04	18 34	24 52	00 52	01 57	03 01
30	17 41	18 05	18 33	00 00	01 04	02 07	03 08
20	17 45	18 07	18 33	00 25	01 26	02 25	03 20
N 10	17 48	18 10	18 34	00 47	01 45	02 40	03 31
0	17 52	18 13	18 37	01 07	02 03	02 54	03 41
S 10	17 55	18 16	18 41	01 27	02 20	03 08	03 51
20	17 59	18 21	18 47	01 49	02 38	03 22	04 02
30	18 03	18 27	18 56	02 13	03 00	03 39	04 14
35	18 06	18 31	19 02	02 28	03 12	03 49	04 20
40	18 09	18 36	19 09	02 44	03 26	04 00	04 28
45	18 12	18 42	19 18	03 04	03 43	04 13	04 37
S 50	18 17	18 49	19 29	03 29	04 03	04 29	04 48
52	18 19	18 53	19 34	03 41	04 13	04 36	04 53
54	18 21	18 57	19 41	03 55	04 24	04 44	04 59
56	18 23	19 01	19 48	04 10	04 36	04 53	05 05
58	18 26	19 06	19 56	04 29	04 50	05 03	05 11
S 60	18 29	19 12	20 05	04 52	05 07	05 15	05 19

Day	SUN Eqn. of Time 00^h	SUN Eqn. of Time 12^h	SUN Mer. Pass.	MOON Mer. Pass. Upper	MOON Mer. Pass. Lower	MOON Age	MOON Phase
d	m s	m s	h m	h m	h m	d %	
4	11 00	11 10	11 49	19 50	07 22	08 68	
5	11 19	11 28	11 49	20 42	08 16	09 78	
6	11 37	11 46	11 48	21 30	09 06	10 86	

	UT	ARIES	VENUS −3.9		MARS −1.9		JUPITER −1.8		SATURN +0.1	
		GHA	GHA	Dec	GHA	Dec	GHA	Dec	GHA	Dec
	d h	° ′	° ′	° ′	° ′	° ′	° ′	° ′	° ′	° ′
	7 00	15 13.8	170 20.0	S 9 30.5	40 55.9	S15 02.0	214 43.5	N 9 11.9	271 17.0	N22 04.9
	01	30 16.2	185 19.5	31.7	55 58.1	01.7	229 45.5	11.7	286 19.4	04.9
	02	45 18.7	200 19.1	32.9	71 00.3	01.4	244 47.5	11.6	301 21.8	04.8
	03	60 21.2	215 18.7	. . 34.1	86 02.6	. . 01.1	259 49.5	. . 11.4	316 24.1	. . 04.8
	04	75 23.6	230 18.2	35.3	101 04.8	00.8	274 51.5	11.2	331 26.5	04.8
	05	90 26.1	245 17.8	36.5	116 07.0	00.5	289 53.5	11.0	346 28.9	04.8
	06	105 28.5	260 17.3	S 9 37.7	131 09.2	S15 00.2	304 55.5	N 9 10.9	1 31.3	N22 04.8
	07	120 31.0	275 16.9	38.9	146 11.4	14 59.9	319 57.5	10.7	16 33.6	04.8
T	08	135 33.5	290 16.4	40.1	161 13.7	59.6	334 59.5	10.5	31 36.0	04.8
U	09	150 35.9	305 16.0	. . 41.3	176 15.9	. . 59.3	350 01.5	. . 10.3	46 38.4	. . 04.8
	10	165 38.4	320 15.5	42.5	191 18.1	59.0	5 03.5	10.1	61 40.7	04.8
E	11	180 40.9	335 15.1	43.7	206 20.3	58.7	20 05.5	10.0	76 43.1	04.8
S	12	195 43.3	350 14.6	S 9 44.9	221 22.5	S14 58.4	35 07.5	N 9 09.8	91 45.5	N22 04.8
D	13	210 45.8	5 14.2	46.1	236 24.7	58.1	50 09.5	09.6	106 47.9	04.8
A	14	225 48.3	20 13.7	47.3	251 27.0	57.8	65 11.5	09.4	121 50.2	04.8
Y	15	240 50.7	35 13.3	. . 48.5	266 29.2	. . 57.5	80 13.5	. . 09.2	136 52.6	. . 04.7
	16	255 53.2	50 12.8	49.7	281 31.4	57.2	95 15.5	09.1	151 55.0	04.7
	17	270 55.7	65 12.4	50.9	296 33.6	56.9	110 17.5	08.9	166 57.4	04.7
	18	285 58.1	80 11.9	S 9 52.1	311 35.8	S14 56.6	125 19.5	N 9 08.7	181 59.7	N22 04.7
	19	301 00.6	95 11.5	53.3	326 38.0	56.3	140 21.5	08.5	197 02.1	04.7
	20	316 03.0	110 11.0	54.4	341 40.2	56.0	155 23.5	08.4	212 04.5	04.7
	21	331 05.5	125 10.6	. . 55.6	356 42.4	. . 55.6	170 25.5	. . 08.2	227 06.9	. . 04.7
	22	346 08.0	140 10.1	56.8	11 44.6	55.3	185 27.5	08.0	242 09.2	04.7
	23	1 10.4	155 09.7	58.0	26 46.8	55.0	200 29.5	07.8	257 11.6	04.7
	8 00	16 12.9	170 09.2	S 9 59.2	41 49.0	S14 54.7	215 31.5	N 9 07.6	272 14.0	N22 04.7
	01	31 15.4	185 08.8	10 00.4	56 51.2	54.4	230 33.5	07.5	287 16.4	04.7
	02	46 17.8	200 08.3	01.6	71 53.4	54.1	245 35.5	07.3	302 18.7	04.7
	03	61 20.3	215 07.8	. . 02.8	86 55.6	. . 53.8	260 37.5	. . 07.1	317 21.1	. . 04.7
	04	76 22.8	230 07.4	04.0	101 57.8	53.5	275 39.5	06.9	332 23.5	04.6
	05	91 25.2	245 06.9	05.2	117 00.0	53.2	290 41.5	06.7	347 25.9	04.6
	06	106 27.7	260 06.5	S10 06.4	132 02.2	S14 52.9	305 43.6	N 9 06.6	2 28.2	N22 04.6
W	07	121 30.1	275 06.0	07.5	147 04.4	52.6	320 45.6	06.4	17 30.6	04.6
E	08	136 32.6	290 05.6	08.7	162 06.6	52.3	335 47.6	06.2	32 33.0	04.6
D	09	151 35.1	305 05.1	. . 09.9	177 08.8	. . 52.0	350 49.6	. . 06.0	47 35.4	. . 04.6
N	10	166 37.5	320 04.7	11.1	192 11.0	51.6	5 51.6	05.9	62 37.7	04.6
	11	181 40.0	335 04.2	12.3	207 13.1	51.3	20 53.6	05.7	77 40.1	04.6
E	12	196 42.5	350 03.7	S10 13.5	222 15.3	S14 51.0	35 55.6	N 9 05.5	92 42.5	N22 04.6
S	13	211 44.9	5 03.3	14.7	237 17.5	50.7	50 57.6	05.3	107 44.9	04.6
D	14	226 47.4	20 02.8	15.9	252 19.7	50.4	65 59.6	05.1	122 47.3	04.6
A	15	241 49.9	35 02.4	. . 17.0	267 21.9	. . 50.1	81 01.6	. . 05.0	137 49.6	. . 04.6
Y	16	256 52.3	50 01.9	18.2	282 24.1	49.8	96 03.6	04.8	152 52.0	04.6
	17	271 54.8	65 01.4	19.4	297 26.2	49.5	111 05.6	04.6	167 54.4	04.6
	18	286 57.3	80 01.0	S10 20.6	312 28.4	S14 49.1	126 07.6	N 9 04.4	182 56.8	N22 04.5
	19	301 59.7	95 00.5	21.8	327 30.6	48.8	141 09.6	04.3	197 59.1	04.5
	20	317 02.2	110 00.1	23.0	342 32.8	48.5	156 11.6	04.1	213 01.5	04.5
	21	332 04.6	124 59.6	. . 24.2	357 34.9	. . 48.2	171 13.6	. . 03.9	228 03.9	. . 04.5
	22	347 07.1	139 59.1	25.3	12 37.1	47.9	186 15.6	03.7	243 06.3	04.5
	23	2 09.6	154 58.7	26.5	27 39.3	47.6	201 17.6	03.6	258 08.7	04.5
	9 00	17 12.0	169 58.2	S10 27.7	42 41.5	S14 47.3	216 19.6	N 9 03.4	273 11.0	N22 04.5
	01	32 14.5	184 57.8	28.9	57 43.6	46.9	231 21.6	03.2	288 13.4	04.5
	02	47 17.0	199 57.3	30.1	72 45.8	46.6	246 23.6	03.0	303 15.8	04.5
	03	62 19.4	214 56.8	. . 31.2	87 48.0	. . 46.3	261 25.6	. . 02.8	318 18.2	. . 04.5
	04	77 21.9	229 56.4	32.4	102 50.1	46.0	276 27.7	02.7	333 20.6	04.5
	05	92 24.4	244 55.9	33.6	117 52.3	45.7	291 29.7	02.5	348 22.9	04.5
	06	107 26.8	259 55.4	S10 34.8	132 54.5	S14 45.3	306 31.7	N 9 02.3	3 25.3	N22 04.5
	07	122 29.3	274 55.0	36.0	147 56.6	45.0	321 33.7	02.1	18 27.7	04.5
T	08	137 31.7	289 54.5	37.1	162 58.8	44.7	336 35.7	02.0	33 30.1	04.4
H	09	152 34.2	304 54.1	. . 38.3	178 00.9	. . 44.4	351 37.7	. . 01.8	48 32.5	. . 04.4
U	10	167 36.7	319 53.6	39.5	193 03.1	44.1	6 39.7	01.6	63 34.9	04.4
R	11	182 39.1	334 53.1	40.7	208 05.3	43.7	21 41.7	01.4	78 37.2	04.4
S	12	197 41.6	349 52.7	S10 41.9	223 07.4	S14 43.4	36 43.7	N 9 01.2	93 39.6	N22 04.4
D	13	212 44.1	4 52.2	43.0	238 09.6	43.1	51 45.7	01.1	108 42.0	04.4
A	14	227 46.5	19 51.7	44.2	253 11.7	42.8	66 47.7	00.9	123 44.4	04.4
Y	15	242 49.0	34 51.3	. . 45.4	268 13.9	. . 42.5	81 49.7	. . 00.7	138 46.8	. . 04.4
	16	257 51.5	49 50.8	46.6	283 16.0	42.1	96 51.7	00.5	153 49.1	04.4
	17	272 53.9	64 50.3	47.7	298 18.2	41.8	111 53.7	00.4	168 51.5	04.4
	18	287 56.4	79 49.9	S10 48.9	313 20.3	S14 41.5	126 55.7	N 9 00.2	183 53.9	N22 04.4
	19	302 58.9	94 49.4	50.1	328 22.5	41.2	141 57.8	9 00.0	198 56.3	04.4
	20	318 01.3	109 48.9	51.3	343 24.6	40.8	156 59.8	8 59.8	213 58.7	04.4
	21	333 03.8	124 48.4	. . 52.4	358 26.8	. . 40.5	172 01.8	. . 59.7	229 01.1	. . 04.4
	22	348 06.2	139 48.0	53.6	13 28.9	40.2	187 03.8	59.5	244 03.5	04.4
	23	3 08.7	154 47.5	54.8	28 31.1	39.9	202 05.8	59.3	259 05.8	04.3
	Mer. Pass.	h m 22 51.4	*v* −0.5	*d* 1.2	*v* 2.2	*d* 0.3	*v* 2.0	*d* 0.2	*v* 2.4	*d* 0.0

STARS

Name	SHA	Dec
	° ′	° ′
Acamar	315 23.7	S40 17.2
Achernar	335 31.6	S57 13.0
Acrux	173 18.9	S63 07.0
Adhara	255 18.6	S28 58.3
Aldebaran	290 58.0	N16 31.1
Alioth	166 27.7	N55 56.5
Alkaid	153 05.2	N49 17.8
Al Na'ir	27 52.8	S46 56.8
Alnilam	275 54.0	S 1 11.8
Alphard	218 03.8	S 8 40.2
Alphecca	126 17.7	N26 42.3
Alpheratz	357 51.2	N29 06.7
Altair	62 15.6	N 8 52.8
Ankaa	353 22.7	S42 17.2
Antares	112 35.9	S26 26.5
Arcturus	146 03.0	N19 09.9
Atria	107 45.0	S69 02.3
Avior	234 21.5	S59 30.8
Bellatrix	278 40.1	N 6 21.4
Betelgeuse	271 09.5	N 7 24.6
Canopus	263 59.5	S52 41.4
Capella	280 45.6	N46 00.1
Deneb	49 36.6	N45 17.8
Denebola	182 41.7	N14 33.3
Diphda	349 03.2	S17 57.9
Dubhe	194 01.3	N61 43.9
Elnath	278 22.2	N28 36.7
Eltanin	90 49.8	N51 29.5
Enif	33 54.5	N 9 53.6
Fomalhaut	15 32.0	S29 36.2
Gacrux	172 10.3	S57 07.9
Gienah	176 00.6	S17 33.5
Hadar	148 59.7	S60 23.4
Hamal	328 09.1	N23 28.9
Kaus Aust.	83 54.0	S34 23.1
Kochab	137 20.3	N74 08.6
Markab	13 45.7	N15 13.6
Menkar	314 22.8	N 4 06.4
Menkent	148 17.1	S36 23.2
Miaplacidus	221 42.1	S69 43.5
Mirfak	308 51.1	N49 52.4
Nunki	76 07.8	S26 17.7
Peacock	53 30.9	S56 43.7
Pollux	243 37.1	N28 01.1
Procyon	245 07.8	N 5 13.1
Rasalhague	96 13.7	N12 33.5
Regulus	207 51.8	N11 57.1
Rigel	281 19.3	S 8 11.6
Rigil Kent.	140 03.2	S60 51.0
Sabik	102 21.5	S15 43.8
Schedar	349 49.0	N56 33.5
Shaula	96 32.5	S37 06.5
Sirius	258 40.5	S16 42.9
Spica	158 39.7	S11 10.7
Suhail	222 58.4	S43 26.5
Vega	80 44.2	N38 47.4
Zuben'ubi	137 14.2	S16 03.3

	SHA	Mer. Pass.
	° ′	h m
Venus	153 56.3	12 40
Mars	25 36.1	21 10
Jupiter	199 18.6	9 37
Saturn	256 01.1	5 50

UT d	UT h	SUN GHA	SUN Dec	MOON GHA	v	MOON Dec	d	HP
d	h	° ′	° ′	° ′	′	° ′	′	′
7	00	182 58.8	S 5 15.4	36 19.1	13.2	S14 16.4	12.6	56.2
	01	197 59.0	16.3	50 51.3	13.2	14 03.8	12.6	56.2
	02	212 59.2	17.3	65 23.5	13.3	13 51.2	12.7	56.2
	03	227 59.4	. . 18.3	79 55.8	13.4	13 38.5	12.8	56.1
	04	242 59.6	19.2	94 28.2	13.4	13 25.7	12.8	56.1
	05	257 59.8	20.2	109 00.6	13.5	13 12.9	12.9	56.1
	06	272 59.9	S 5 21.1	123 33.1	13.6	S13 00.0	12.9	56.1
	07	288 00.1	22.1	138 05.7	13.7	12 47.1	12.9	56.1
T	08	303 00.3	23.1	152 38.4	13.7	12 34.2	13.0	56.0
U	09	318 00.5	. . 24.0	167 11.1	13.8	12 21.2	13.1	56.0
E	10	333 00.7	25.0	181 43.9	13.9	12 08.1	13.1	56.0
S	11	348 00.8	25.9	196 16.8	13.9	11 55.0	13.1	56.0
D	12	3 01.0	S 5 26.9	210 49.7	14.0	S11 41.9	13.2	56.0
A	13	18 01.2	27.8	225 22.7	14.1	11 28.7	13.2	55.9
Y	14	33 01.4	28.8	239 55.8	14.1	11 15.5	13.3	55.9
	15	48 01.6	. . 29.7	254 28.9	14.2	11 02.2	13.2	55.9
	16	63 01.7	30.7	269 02.1	14.2	10 49.0	13.4	55.9
	17	78 01.9	31.7	283 35.3	14.4	10 35.6	13.4	55.9
	18	93 02.1	S 5 32.6	298 08.7	14.3	S10 22.2	13.4	55.8
	19	108 02.3	33.6	312 42.0	14.5	10 08.8	13.4	55.8
	20	123 02.5	34.5	327 15.5	14.5	9 55.4	13.5	55.8
	21	138 02.7	. . 35.5	341 49.0	14.5	9 41.9	13.5	55.8
	22	153 02.8	36.4	356 22.5	14.6	9 28.4	13.5	55.8
	23	168 03.0	37.4	10 56.1	14.7	9 14.9	13.6	55.7
8	00	183 03.2	S 5 38.4	25 29.8	14.7	S 9 01.3	13.6	55.7
	01	198 03.4	39.3	40 03.5	14.8	8 47.7	13.6	55.7
	02	213 03.5	40.3	54 37.3	14.8	8 34.1	13.6	55.7
	03	228 03.7	. . 41.2	69 11.1	14.8	8 20.5	13.7	55.7
	04	243 03.9	42.2	83 44.9	15.0	8 06.8	13.7	55.6
	05	258 04.1	43.1	98 18.9	14.9	7 53.1	13.8	55.6
	06	273 04.3	S 5 44.1	112 52.8	15.0	S 7 39.3	13.7	55.6
W	07	288 04.4	45.0	127 26.8	15.1	7 25.6	13.8	55.6
E	08	303 04.6	46.0	142 00.9	15.1	7 11.8	13.8	55.6
D	09	318 04.8	. . 46.9	156 35.0	15.2	6 58.0	13.8	55.6
N	10	333 05.0	47.9	171 09.2	15.2	6 44.2	13.8	55.5
E	11	348 05.2	48.9	185 43.4	15.2	6 30.4	13.9	55.5
S	12	3 05.3	S 5 49.8	200 17.6	15.3	S 6 16.5	13.8	55.5
D	13	18 05.5	50.8	214 51.9	15.3	6 02.7	13.9	55.5
A	14	33 05.7	51.7	229 26.2	15.4	5 48.8	13.9	55.5
Y	15	48 05.9	. . 52.7	244 00.6	15.4	5 34.9	13.9	55.4
	16	63 06.0	53.6	258 35.0	15.4	5 21.0	13.9	55.4
	17	78 06.2	54.6	273 09.4	15.5	5 07.1	14.0	55.4
	18	93 06.4	S 5 55.5	287 43.9	15.5	S 4 53.1	13.9	55.4
	19	108 06.6	56.5	302 18.4	15.6	4 39.2	14.0	55.4
	20	123 06.7	57.4	316 53.0	15.6	4 25.2	13.9	55.4
	21	138 06.9	. . 58.4	331 27.6	15.6	4 11.3	14.0	55.3
	22	153 07.1	5 59.3	346 02.2	15.7	3 57.3	14.0	55.3
	23	168 07.3	6 00.3	0 36.9	15.6	3 43.3	14.0	55.3
9	00	183 07.4	S 6 01.2	15 11.5	15.7	S 3 29.3	14.0	55.3
	01	198 07.6	02.2	29 46.2	15.8	3 15.3	14.0	55.3
	02	213 07.8	03.1	44 21.0	15.8	3 01.3	14.0	55.2
	03	228 08.0	. . 04.1	58 55.8	15.8	2 47.3	14.0	55.2
	04	243 08.1	05.0	73 30.6	15.8	2 33.3	14.0	55.2
	05	258 08.3	06.0	88 05.4	15.8	2 19.3	14.0	55.2
	06	273 08.5	S 6 06.9	102 40.2	15.9	S 2 05.3	14.0	55.2
T	07	288 08.7	07.9	117 15.1	15.9	1 51.3	14.0	55.2
H	08	303 08.8	08.8	131 50.0	15.9	1 37.3	14.1	55.1
U	09	318 09.0	. . 09.8	146 24.9	15.9	1 23.2	14.0	55.1
R	10	333 09.2	10.8	160 59.8	16.0	1 09.2	14.0	55.1
S	11	348 09.4	11.7	175 34.8	16.0	0 55.2	14.0	55.1
D	12	3 09.5	S 6 12.7	190 09.8	16.0	S 0 41.2	14.0	55.1
A	13	18 09.7	13.6	204 44.8	16.0	0 27.2	13.9	55.1
Y	14	33 09.9	14.6	219 19.8	16.0	S 0 13.3	14.0	55.0
	15	48 10.0	. . 15.5	233 54.8	16.1	N 0 00.7	14.0	55.0
	16	63 10.2	16.5	248 29.9	16.0	0 14.7	14.0	55.0
	17	78 10.4	17.4	263 04.9	16.1	0 28.7	13.9	55.0
	18	93 10.6	S 6 18.4	277 40.0	16.1	N 0 42.6	14.0	55.0
	19	108 10.7	19.3	292 15.1	16.1	0 56.6	13.9	55.0
	20	123 10.9	20.3	306 50.2	16.1	1 10.5	13.9	55.0
	21	138 11.1	. . 21.2	321 25.3	16.1	1 24.4	13.9	54.9
	22	153 11.2	22.1	336 00.4	16.1	1 38.3	13.9	54.9
	23	168 11.4	23.1	350 35.5	16.2	N 1 52.2	13.9	54.9
		SD 16.0	*d* 1.0	SD 15.2		15.1		15.0

Lat.	Twilight Naut.	Twilight Civil	Sunrise	Moonrise 7	8	9	10
°	h m	h m	h m	h m	h m	h m	h m
N 72	04 21	05 41	06 49	18 26	17 52	17 23	16 54
N 70	04 29	05 41	06 42	18 10	17 45	17 23	17 02
68	04 35	05 40	06 36	17 57	17 39	17 24	17 08
66	04 41	05 40	06 32	17 46	17 35	17 24	17 14
64	04 45	05 40	06 28	17 37	17 31	17 25	17 19
62	04 48	05 40	06 24	17 29	17 27	17 25	17 23
60	04 51	05 39	06 21	17 22	17 24	17 25	17 26
N 58	04 54	05 39	06 18	17 16	17 21	17 25	17 30
56	04 56	05 39	06 16	17 11	17 19	17 26	17 33
54	04 57	05 38	06 14	17 06	17 17	17 26	17 35
52	04 59	05 38	06 12	17 02	17 15	17 26	17 38
50	05 00	05 38	06 10	16 58	17 13	17 26	17 40
45	05 03	05 37	06 06	16 49	17 09	17 27	17 44
N 40	05 04	05 36	06 03	16 42	17 05	17 27	17 49
35	05 05	05 34	06 00	16 36	17 02	17 27	17 52
30	05 05	05 33	05 57	16 30	17 00	17 28	17 55
20	05 05	05 30	05 53	16 21	16 55	17 28	18 00
N 10	05 03	05 27	05 48	16 12	16 51	17 28	18 05
0	04 59	05 24	05 44	16 04	16 47	17 29	18 10
S 10	04 55	05 19	05 40	15 56	16 44	17 29	18 14
20	04 48	05 14	05 36	15 48	16 40	17 30	18 19
30	04 38	05 06	05 31	15 38	16 35	17 30	18 25
35	04 32	05 02	05 28	15 33	16 32	17 31	18 28
40	04 24	04 57	05 24	15 26	16 29	17 31	18 32
45	04 15	04 50	05 20	15 19	16 26	17 32	18 36
S 50	04 02	04 42	05 15	15 09	16 22	17 32	18 41
52	03 56	04 38	05 13	15 05	16 20	17 32	18 44
54	03 50	04 34	05 10	15 01	16 18	17 33	18 46
56	03 42	04 29	05 07	14 55	16 15	17 33	18 49
58	03 33	04 24	05 04	14 50	16 13	17 33	18 53
S 60	03 23	04 18	05 01	14 43	16 10	17 34	18 56

Lat.	Sunset	Twilight Civil	Twilight Naut.	Moonset 7	8	9	10
°	h m	h m	h m	h m	h m	h m	h m
N 72	16 45	17 52	19 11	00 07	02 32	04 36	06 34
N 70	16 52	17 53	19 03	00 39	02 46	04 39	06 29
68	16 57	17 53	18 58	01 02	02 57	04 42	06 24
66	17 02	17 53	18 53	01 20	03 06	04 45	06 21
64	17 06	17 54	18 49	01 34	03 13	04 47	06 18
62	17 10	17 54	18 45	01 46	03 19	04 49	06 16
60	17 13	17 55	18 43	01 56	03 25	04 50	06 13
N 58	17 16	17 55	18 40	02 05	03 30	04 52	06 11
56	17 18	17 55	18 38	02 12	03 34	04 53	06 10
54	17 21	17 56	18 37	02 19	03 38	04 54	06 08
52	17 23	17 56	18 35	02 25	03 41	04 55	06 07
50	17 24	17 57	18 34	02 31	03 44	04 56	06 06
45	17 29	17 58	18 32	02 43	03 51	04 58	06 03
N 40	17 32	17 59	18 30	02 52	03 57	04 59	06 00
35	17 35	18 00	18 30	03 01	04 02	05 01	05 58
30	17 38	18 02	18 29	03 08	04 06	05 02	05 57
20	17 42	18 05	18 30	03 20	04 13	05 04	05 54
N 10	17 47	18 08	18 32	03 31	04 19	05 06	05 51
0	17 51	18 12	18 36	03 41	04 25	05 07	05 49
S 10	17 55	18 16	18 41	03 51	04 31	05 09	05 46
20	18 00	18 22	18 48	04 02	04 37	05 11	05 43
30	18 05	18 29	18 58	04 14	04 44	05 13	05 40
35	18 08	18 34	19 04	04 20	04 48	05 14	05 39
40	18 12	18 39	19 12	04 28	04 53	05 15	05 37
45	18 16	18 46	19 22	04 37	04 58	05 17	05 34
S 50	18 21	18 54	19 34	04 48	05 04	05 18	05 32
52	18 24	18 58	19 40	04 53	05 07	05 19	05 30
54	18 26	19 03	19 47	04 59	05 10	05 20	05 29
56	18 29	19 08	19 55	05 05	05 13	05 21	05 28
58	18 32	19 13	20 04	05 11	05 17	05 22	05 26
S 60	18 36	19 20	20 15	05 19	05 21	05 23	05 24

Day	SUN Eqn. of Time 00^h	SUN Eqn. of Time 12^h	SUN Mer. Pass.	MOON Mer. Pass. Upper	MOON Mer. Pass. Lower	Age	Phase
d	m s	m s	h m	h m	h m	d %	
7	11 55	12 04	11 48	22 15	09 53	11 92	
8	12 12	12 21	11 48	22 57	10 36	12 97	
9	12 29	12 38	11 47	23 39	11 18	13 99	

UT (d h)	ARIES GHA ° ′	VENUS −3.9 GHA ° ′	VENUS Dec ° ′	MARS −1.8 GHA ° ′	MARS Dec ° ′	JUPITER −1.8 GHA ° ′	JUPITER Dec ° ′	SATURN +0.1 GHA ° ′	SATURN Dec ° ′
10 00 (FRIDAY)	18 11.2	169 47.0	S10 55.9	43 33.2	S14 39.5	217 07.8	N 8 59.1	274 08.2	N22 04.3
01	33 13.6	184 46.6	57.1	58 35.3	39.2	232 09.8	59.0	289 10.6	04.3
02	48 16.1	199 46.1	58.3	73 37.5	38.9	247 11.8	58.8	304 13.0	04.3
03	63 18.6	214 45.6	10 59.5	88 39.6	. . 38.6	262 13.8	. . 58.6	319 15.4	. . 04.3
04	78 21.0	229 45.1	11 00.6	103 41.8	38.2	277 15.8	58.4	334 17.8	04.3
05	93 23.5	244 44.7	01.8	118 43.9	37.9	292 17.8	58.2	349 20.1	04.3
06	108 26.0	259 44.2	S11 03.0	133 46.0	S14 37.6	307 19.8	N 8 58.1	4 22.5	N22 04.3
07	123 28.4	274 43.7	04.1	148 48.2	37.2	322 21.9	57.9	19 24.9	04.3
08	138 30.9	289 43.3	05.3	163 50.3	36.9	337 23.9	57.7	34 27.3	04.3
09	153 33.4	304 42.8	. . 06.5	178 52.4	. . 36.6	352 25.9	. . 57.5	49 29.7	. . 04.3
10	168 35.8	319 42.3	07.6	193 54.6	36.3	7 27.9	57.4	64 32.1	04.3
11	183 38.3	334 41.8	08.8	208 56.7	35.9	22 29.9	57.2	79 34.5	04.3
12	198 40.7	349 41.4	S11 10.0	223 58.8	S14 35.6	37 31.9	N 8 57.0	94 36.9	N22 04.3
13	213 43.2	4 40.9	11.1	239 00.9	35.3	52 33.9	56.8	109 39.2	04.3
14	228 45.7	19 40.4	12.3	254 03.1	34.9	67 35.9	56.7	124 41.6	04.2
15	243 48.1	34 39.9	. . 13.5	269 05.2	. . 34.6	82 37.9	. . 56.5	139 44.0	. . 04.2
16	258 50.6	49 39.5	14.6	284 07.3	34.3	97 39.9	56.3	154 46.4	04.2
17	273 53.1	64 39.0	15.8	299 09.4	33.9	112 41.9	56.1	169 48.8	04.2
18	288 55.5	79 38.5	S11 17.0	314 11.5	S14 33.6	127 44.0	N 8 56.0	184 51.2	N22 04.2
19	303 58.0	94 38.0	18.1	329 13.7	33.3	142 46.0	55.8	199 53.6	04.2
20	319 00.5	109 37.5	19.3	344 15.8	32.9	157 48.0	55.6	214 56.0	04.2
21	334 02.9	124 37.1	. . 20.5	359 17.9	. . 32.6	172 50.0	. . 55.4	229 58.3	. . 04.2
22	349 05.4	139 36.6	21.6	14 20.0	32.3	187 52.0	55.3	245 00.7	04.2
23	4 07.8	154 36.1	22.8	29 22.1	31.9	202 54.0	55.1	260 03.1	04.2
11 00 (SATURDAY)	19 10.3	169 35.6	S11 23.9	44 24.2	S14 31.6	217 56.0	N 8 54.9	275 05.5	N22 04.2
01	34 12.8	184 35.1	25.1	59 26.4	31.3	232 58.0	54.7	290 07.9	04.2
02	49 15.2	199 34.7	26.3	74 28.5	30.9	248 00.0	54.6	305 10.3	04.2
03	64 17.7	214 34.2	. . 27.4	89 30.6	. . 30.6	263 02.0	. . 54.4	320 12.7	. . 04.2
04	79 20.2	229 33.7	28.6	104 32.7	30.3	278 04.1	54.2	335 15.1	04.2
05	94 22.6	244 33.2	29.7	119 34.8	29.9	293 06.1	54.0	350 17.5	04.2
06	109 25.1	259 32.7	S11 30.9	134 36.9	S14 29.6	308 08.1	N 8 53.8	5 19.9	N22 04.1
07	124 27.6	274 32.3	32.1	149 39.0	29.2	323 10.1	53.7	20 22.2	04.1
08	139 30.0	289 31.8	33.2	164 41.1	28.9	338 12.1	53.5	35 24.6	04.1
09	154 32.5	304 31.3	. . 34.4	179 43.2	. . 28.6	353 14.1	. . 53.3	50 27.0	. . 04.1
10	169 35.0	319 30.8	35.5	194 45.3	28.2	8 16.1	53.1	65 29.4	04.1
11	184 37.4	334 30.3	36.7	209 47.4	27.9	23 18.1	53.0	80 31.8	04.1
12	199 39.9	349 29.8	S11 37.8	224 49.5	S14 27.5	38 20.2	N 8 52.8	95 34.2	N22 04.1
13	214 42.3	4 29.3	39.0	239 51.6	27.2	53 22.2	52.6	110 36.6	04.1
14	229 44.8	19 28.9	40.2	254 53.7	26.9	68 24.2	52.4	125 39.0	04.1
15	244 47.3	34 28.4	. . 41.3	269 55.8	. . 26.5	83 26.2	. . 52.3	140 41.4	. . 04.1
16	259 49.7	49 27.9	42.5	284 57.9	26.2	98 28.2	52.1	155 43.8	04.1
17	274 52.2	64 27.4	43.6	300 00.0	25.8	113 30.2	51.9	170 46.2	04.1
18	289 54.7	79 26.9	S11 44.8	315 02.1	S14 25.5	128 32.2	N 8 51.7	185 48.6	N22 04.1
19	304 57.1	94 26.4	45.9	330 04.2	25.2	143 34.2	51.6	200 51.0	04.1
20	319 59.6	109 25.9	47.1	345 06.3	24.8	158 36.3	51.4	215 53.3	04.1
21	335 02.1	124 25.5	. . 48.2	0 08.4	. . 24.5	173 38.3	. . 51.2	230 55.7	. . 04.1
22	350 04.5	139 25.0	49.4	15 10.5	24.1	188 40.3	51.0	245 58.1	04.1
23	5 07.0	154 24.5	50.5	30 12.5	23.8	203 42.3	50.9	261 00.5	04.0
12 00 (SUNDAY)	20 09.4	169 24.0	S11 51.7	45 14.6	S14 23.4	218 44.3	N 8 50.7	276 02.9	N22 04.0
01	35 11.9	184 23.5	52.8	60 16.7	23.1	233 46.3	50.5	291 05.3	04.0
02	50 14.4	199 23.0	54.0	75 18.8	22.7	248 48.3	50.3	306 07.7	04.0
03	65 16.8	214 22.5	. . 55.1	90 20.9	. . 22.4	263 50.3	. . 50.2	321 10.1	. . 04.0
04	80 19.3	229 22.0	56.3	105 23.0	22.1	278 52.4	50.0	336 12.5	04.0
05	95 21.8	244 21.5	57.4	120 25.0	21.7	293 54.4	49.8	351 14.9	04.0
06	110 24.2	259 21.0	S11 58.6	135 27.1	S14 21.4	308 56.4	N 8 49.7	6 17.3	N22 04.0
07	125 26.7	274 20.5	11 59.7	150 29.2	21.0	323 58.4	49.5	21 19.7	04.0
08	140 29.2	289 20.1	12 00.9	165 31.3	20.7	339 00.4	49.3	36 22.1	04.0
09	155 31.6	304 19.6	. . 02.0	180 33.3	. . 20.3	354 02.4	. . 49.1	51 24.5	. . 04.0
10	170 34.1	319 19.1	03.1	195 35.4	20.0	9 04.4	49.0	66 26.9	04.0
11	185 36.6	334 18.6	04.3	210 37.5	19.6	24 06.5	48.8	81 29.3	04 0
12	200 39.0	349 18.1	S12 05.4	225 39.6	S14 19.3	39 08.5	N 8 48.6	96 31.7	N22 04.0
13	215 41.5	4 17.6	06.6	240 41.6	18.9	54 10.5	48.4	111 34.1	04.0
14	230 43.9	19 17.1	07.7	255 43.7	18.6	69 12.5	48.3	126 36.5	04.0
15	245 46.4	34 16.6	. . 08.9	270 45.8	. . 18.2	84 14.5	. . 48.1	141 38.9	. . 04.0
16	260 48.9	49 16.1	10.0	285 47.8	17.9	99 16.5	47.9	156 41.3	04.0
17	275 51.3	64 15.6	11.2	300 49.9	17.5	114 18.5	47.7	171 43.7	03.9
18	290 53.8	79 15.1	S12 12.3	315 52.0	S14 17.2	129 20.6	N 8 47.6	186 46.1	N22 03.9
19	305 56.3	94 14.6	13.4	330 54.0	16.8	144 22.6	47.4	201 48.5	03.9
20	320 58.7	109 14.1	14.6	345 56.1	16.5	159 24.6	47.2	216 50.9	03.9
21	336 01.2	124 13.6	. . 15.7	0 58.2	. . 16.1	174 26.6	. . 47.0	231 53.3	. . 03.9
22	351 03.7	139 13.1	16.9	16 00.2	15.8	189 28.6	46.9	246 55.7	03.9
23	6 06.1	154 12.6	18.0	31 02.3	15.4	204 30.6	46.7	261 58.1	03.9
Mer. Pass.	h m 22 39.6	*v* −0.5	*d* 1.2	*v* 2.1	*d* 0.3	*v* 2.0	*d* 0.2	*v* 2.4	*d* 0.0

STARS

Name	SHA ° ′	Dec ° ′
Acamar	315 23.7	S40 17.
Achernar	335 31.6	S57 13.
Acrux	173 18.9	S63 07.
Adhara	255 18.5	S28 58.
Aldebaran	290 58.0	N16 31.
Alioth	166 27.7	N55 56.
Alkaid	153 05.2	N49 17.
Al Na'ir	27 52.8	S46 56.
Alnilam	275 54.0	S 1 11.
Alphard	218 03.8	S 8 40.
Alphecca	126 17.7	N26 42.
Alpheratz	357 51.2	N29 06.
Altair	62 15.7	N 8 52.
Ankaa	353 22.7	S42 17.
Antares	112 35.9	S26 26.
Arcturus	146 03.0	N19 09.
Atria	107 45.1	S69 02.
Avior	234 21.5	S59 30.
Bellatrix	278 40.1	N 6 21.
Betelgeuse	271 09.5	N 7 24.
Canopus	263 59.4	S52 41.
Capella	280 45.6	N46 00.
Deneb	49 36.6	N45 17.
Denebola	182 41.7	N14 33.
Diphda	349 03.2	S17 57.
Dubhe	194 01.3	N61 43.
Elnath	278 22.1	N28 36.
Eltanin	90 49.9	N51 29.
Enif	33 54.5	N 9 53.
Fomalhaut	15 32.0	S29 36.
Gacrux	172 10.3	S57 07.
Gienah	176 00.6	S17 33.
Hadar	148 59.7	S60 23.
Hamal	328 09.1	N23 28.
Kaus Aust.	83 54.0	S34 23.
Kochab	137 20.3	N74 08.
Markab	13 45.7	N15 13.
Menkar	314 22.8	N 4 06.
Menkent	148 17.1	S36 23.
Miaplacidus	221 42.1	S69 43.
Mirfak	308 51.1	N49 52.
Nunki	76 07.8	S26 17.
Peacock	53 30.9	S56 43.
Pollux	243 37.1	N28 01.
Procyon	245 07.8	N 5 13.
Rasalhague	96 13.7	N12 33.
Regulus	207 51.8	N11 57.
Rigel	281 19.2	S 8 11.
Rigil Kent.	140 03.2	S60 51.
Sabik	102 21.5	S15 43.
Schedar	349 49.0	N56 33.
Shaula	96 32.5	S37 06.
Sirius	258 40.5	S16 42.
Spica	158 39.7	S11 10.
Suhail	222 58.4	S43 26.
Vega	80 44.2	N38 47.
Zuben'ubi	137 14.2	S16 03.

	SHA ° ′	Mer. Pass h m
Venus	150 25.3	12 42
Mars	25 13.9	20 59
Jupiter	198 45.7	9 27
Saturn	255 55.2	5 39

	UT	SUN GHA	SUN Dec	MOON GHA	*v*	MOON Dec	*d*	HP
	d h	° ′	° ′	° ′	′	° ′	′	′
	10 00	183 11.6	S 6 24.0	5 10.7	16.1	N 2 06.1	13.9	54.9
	01	198 11.8	25.0	19 45.8	16.1	2 20.0	13.8	54.9
	02	213 11.9	25.9	34 20.9	16.2	2 33.8	13.9	54.9
	03	228 12.1	. . 26.9	48 56.1	16.1	2 47.7	13.8	54.8
	04	243 12.3	27.8	63 31.2	16.2	3 01.5	13.8	54.8
	05	258 12.4	28.8	78 06.4	16.2	3 15.3	13.8	54.8
	06	273 12.6	S 6 29.7	92 41.6	16.1	N 3 29.1	13.7	54.8
	07	288 12.8	30.7	107 16.7	16.2	3 42.8	13.8	54.8
	08	303 12.9	31.6	121 51.9	16.1	3 56.6	13.7	54.8
FRIDAY	09	318 13.1	. . 32.6	136 27.0	16.2	4 10.3	13.7	54.8
	10	333 13.3	33.5	151 02.2	16.2	4 24.0	13.6	54.7
	11	348 13.4	34.5	165 37.4	16.1	4 37.6	13.7	54.7
	12	3 13.6	S 6 35.4	180 12.5	16.2	N 4 51.3	13.6	54.7
	13	18 13.8	36.4	194 47.7	16.1	5 04.9	13.6	54.7
	14	33 13.9	37.3	209 22.8	16.1	5 18.5	13.6	54.7
	15	48 14.1	. . 38.3	223 57.9	16.2	5 32.1	13.5	54.7
	16	63 14.3	39.2	238 33.1	16.1	5 45.6	13.5	54.7
	17	78 14.5	40.1	253 08.2	16.1	5 59.1	13.5	54.7
	18	93 14.6	S 6 41.1	267 43.3	16.1	N 6 12.6	13.5	54.6
	19	108 14.8	42.0	282 18.4	16.1	6 26.1	13.4	54.6
	20	123 15.0	43.0	296 53.5	16.1	6 39.5	13.4	54.6
	21	138 15.1	. . 43.9	311 28.6	16.1	6 52.9	13.3	54.6
	22	153 15.3	44.9	326 03.7	16.0	7 06.2	13.4	54.6
	23	168 15.4	45.8	340 38.7	16.1	7 19.6	13.3	54.6
	11 00	183 15.6	S 6 46.8	355 13.8	16.0	N 7 32.9	13.2	54.6
	01	198 15.8	47.7	9 48.8	16.0	7 46.1	13.2	54.5
	02	213 15.9	48.7	24 23.8	16.0	7 59.3	13.2	54.5
	03	228 16.1	. . 49.6	38 58.8	16.0	8 12.5	13.2	54.5
	04	243 16.3	50.5	53 33.8	15.9	8 25.7	13.1	54.5
	05	258 16.4	51.5	68 08.7	16.0	8 38.8	13.0	54.5
	06	273 16.6	S 6 52.4	82 43.7	15.9	N 8 51.8	13.1	54.5
	07	288 16.8	53.4	97 18.6	15.9	9 04.9	13.0	54.5
SATURDAY	08	303 16.9	54.3	111 53.5	15.9	9 17.9	12.9	54.5
	09	318 17.1	. . 55.3	126 28.4	15.8	9 30.8	12.9	54.5
	10	333 17.3	56.2	141 03.2	15.9	9 43.7	12.9	54.4
	11	348 17.4	57.2	155 38.1	15.8	9 56.6	12.8	54.4
	12	3 17.6	S 6 58.1	170 12.9	15.8	N10 09.4	12.8	54.4
	13	18 17.7	6 59.0	184 47.7	15.7	10 22.2	12.7	54.4
	14	33 17.9	7 00.0	199 22.4	15.8	10 34.9	12.7	54.4
	15	48 18.1	. . 00.9	213 57.2	15.7	10 47.6	12.7	54.4
	16	63 18.2	01.9	228 31.9	15.7	11 00.3	12.6	54.4
	17	78 18.4	02.8	243 06.6	15.6	11 12.9	12.5	54.4
	18	93 18.6	S 7 03.8	257 41.2	15.6	N11 25.4	12.5	54.4
	19	108 18.7	04.7	272 15.8	15.6	11 37.9	12.5	54.3
	20	123 18.9	05.6	286 50.4	15.6	11 50.4	12.4	54.3
	21	138 19.0	. . 06.6	301 25.0	15.5	12 02.8	12.3	54.3
	22	153 19.2	07.5	315 59.5	15.5	12 15.1	12.3	54.3
	23	168 19.4	08.5	330 34.0	15.5	12 27.4	12.3	54.3
	12 00	183 19.5	S 7 09.4	345 08.5	15.4	N12 39.7	12.2	54.3
	01	198 19.7	10.3	359 42.9	15.4	12 51.9	12.1	54.3
	02	213 19.8	11.3	14 17.3	15.4	13 04.0	12.1	54.3
	03	228 20.0	. . 12.2	28 51.7	15.3	13 16.1	12.0	54.3
	04	243 20.2	13.2	43 26.0	15.3	13 28.1	12.0	54.3
	05	258 20.3	14.1	58 00.3	15.3	13 40.1	11.9	54.3
	06	273 20.5	S 7 15.0	72 34.6	15.2	N13 52.0	11.8	54.2
	07	288 20.6	16.0	87 08.8	15.2	14 03.8	11.8	54.2
	08	303 20.8	16.9	101 43.0	15.1	14 15.6	11.8	54.2
SUNDAY	09	318 21.0	. . 17.9	116 17.1	15.1	14 27.4	11.6	54.2
	10	333 21.1	18.8	130 51.2	15.1	14 39.0	11.6	54.2
	11	348 21.3	19.7	145 25.3	15.0	14 50.6	11.6	54.2
	12	3 21.4	S 7 20.7	159 59.3	15.0	N15 02.2	11.5	54.2
	13	18 21.6	21.6	174 33.3	14.9	15 13.7	11.4	54.2
	14	33 21.7	22.6	189 07.2	14.9	15 25.1	11.4	54.2
	15	48 21.9	. . 23.5	203 41.1	14.8	15 36.5	11.3	54.2
	16	63 22.1	24.4	218 14.9	14.8	15 47.8	11.2	54.2
	17	78 22.2	25.4	232 48.7	14.8	15 59.0	11.2	54.2
	18	93 22.4	S 7 26.3	247 22.5	14.7	N16 10.2	11.1	54.2
	19	108 22.5	27.2	261 56.2	14.7	16 21.3	11.0	54.2
	20	123 22.7	28.2	276 29.9	14.6	16 32.3	10.9	54.1
	21	138 22.8	. . 29.1	291 03.5	14.6	16 43.2	10.9	54.1
	22	153 23.0	30.1	305 37.1	14.5	16 54.1	10.9	54.1
	23	168 23.2	31.0	320 10.6	14.5	N17 05.0	10.7	54.1
		SD 16.0	*d* 0.9	SD 14.9		14.8		14.8

Lat.	Twilight Naut.	Twilight Civil	Sunrise	Moonrise 10	11	12	13
°	h m	h m	h m	h m	h m	h m	h m
N 72	04 35	05 54	07 03	16 54	16 20	15 29	▭
N 70	04 41	05 52	06 54	17 02	16 37	16 05	14 50
68	04 46	05 51	06 47	17 08	16 51	16 30	15 56
66	04 50	05 50	06 41	17 14	17 03	16 50	16 32
64	04 53	05 48	06 36	17 19	17 13	17 06	16 58
62	04 56	05 47	06 32	17 23	17 21	17 19	17 19
60	04 58	05 46	06 28	17 26	17 28	17 31	17 35
N 58	05 00	05 46	06 25	17 30	17 34	17 41	17 49
56	05 02	05 45	06 22	17 33	17 40	17 49	18 02
54	05 03	05 44	06 19	17 35	17 45	17 57	18 12
52	05 04	05 43	06 17	17 38	17 50	18 04	18 22
50	05 05	05 42	06 15	17 40	17 54	18 10	18 30
45	05 06	05 40	06 10	17 44	18 03	18 24	18 48
N 40	05 07	05 39	06 06	17 49	18 11	18 35	19 03
35	05 07	05 37	06 02	17 52	18 18	18 45	19 16
30	05 07	05 35	05 59	17 55	18 23	18 54	19 27
20	05 06	05 31	05 53	18 00	18 34	19 08	19 46
N 10	05 03	05 27	05 48	18 05	18 43	19 21	20 02
0	04 59	05 23	05 44	18 10	18 51	19 34	20 18
S 10	04 53	05 17	05 39	18 14	19 00	19 46	20 34
20	04 45	05 11	05 33	18 19	19 09	19 59	20 51
30	04 34	05 03	05 27	18 25	19 19	20 14	21 10
35	04 27	04 58	05 23	18 28	19 26	20 23	21 22
40	04 19	04 52	05 19	18 32	19 33	20 34	21 35
45	04 09	04 44	05 14	18 36	19 41	20 46	21 51
S 50	03 55	04 35	05 09	18 41	19 51	21 00	22 10
52	03 49	04 31	05 06	18 44	19 55	21 07	22 19
54	03 41	04 26	05 03	18 46	20 00	21 15	22 30
56	03 33	04 21	05 00	18 49	20 06	21 24	22 41
58	03 23	04 15	04 56	18 53	20 12	21 33	22 55
S 60	03 12	04 08	04 52	18 56	20 20	21 45	23 11

Lat.	Sunset	Twilight Civil	Twilight Naut.	Moonset 10	11	12	13
°	h m	h m	h m	h m	h m	h m	h m
N 72	16 29	17 38	18 56	06 34	08 35	10 54	▭
N 70	16 38	17 39	18 50	06 29	08 20	10 20	13 06
68	16 45	17 41	18 45	06 24	08 07	09 56	12 01
66	16 51	17 42	18 41	06 21	07 58	09 38	11 26
64	16 56	17 44	18 38	06 18	07 49	09 23	11 01
62	17 00	17 45	18 36	06 16	07 42	09 11	10 42
60	17 04	17 46	18 34	06 13	07 36	09 00	10 26
N 58	17 08	17 47	18 32	06 11	07 31	08 51	10 12
56	17 11	17 48	18 31	06 10	07 26	08 43	10 01
54	17 13	17 49	18 30	06 08	07 22	08 36	09 51
52	17 16	17 50	18 29	06 07	07 18	08 30	09 42
50	17 18	17 50	18 28	06 06	07 15	08 24	09 34
45	17 23	17 53	18 26	06 03	07 07	08 12	09 17
N 40	17 27	17 54	18 26	06 00	07 01	08 02	09 03
35	17 31	17 56	18 26	05 58	06 56	07 53	08 52
30	17 34	17 58	18 26	05 57	06 51	07 46	08 41
20	17 40	18 02	18 28	05 54	06 43	07 33	08 24
N 10	17 45	18 06	18 31	05 51	06 36	07 22	08 09
0	17 50	18 11	18 35	05 49	06 30	07 11	07 55
S 10	17 55	18 16	18 41	05 46	06 23	07 01	07 41
20	18 01	18 23	18 49	05 43	06 16	06 50	07 26
30	18 07	18 31	19 00	05 40	06 08	06 37	07 09
35	18 11	18 37	19 07	05 39	06 04	06 30	06 59
40	18 15	18 43	19 16	05 37	05 58	06 22	06 48
45	18 20	18 50	19 26	05 34	05 53	06 12	06 34
S 50	18 26	18 59	19 40	05 32	05 45	06 01	06 18
52	18 29	19 04	19 47	05 30	05 42	05 55	06 11
54	18 32	19 09	19 54	05 29	05 39	05 49	06 03
56	18 35	19 14	20 03	05 28	05 35	05 43	05 53
58	18 39	19 20	20 13	05 26	05 30	05 36	05 43
S 60	18 43	19 28	20 24	05 24	05 26	05 28	05 31

Day	SUN Eqn. of Time 00^h	SUN Eqn. of Time 12^h	SUN Mer. Pass.	MOON Mer. Pass. Upper	MOON Mer. Pass. Lower	MOON Age	MOON %	Phase
d	m s	m s	h m	h m	h m	d	%	
10	12 46	12 54	11 47	24 20	11 59	14	100	○
11	13 02	13 10	11 47	00 20	12 40	15	99	
12	13 18	13 25	11 47	01 01	13 22	16	96	

UT (d h)	ARIES GHA ° ′	VENUS −3.9 GHA ° ′	VENUS Dec ° ′	MARS −1.7 GHA ° ′	MARS Dec ° ′	JUPITER −1.8 GHA ° ′	JUPITER Dec ° ′	SATURN +0.0 GHA ° ′	SATURN Dec ° ′
13 00 (MONDAY)	21 08.6	169 12.1	S12 19.1	46 04.3	S14 15.1	219 32.7	N 8 46.5	277 00.5	N22 03.9
01	36 11.0	184 11.6	20.3	61 06.4	14.7	234 34.7	46.3	292 02.9	03.9
02	51 13.5	199 11.1	21.4	76 08.4	14.3	249 36.7	46.2	307 05.2	03.9
03	66 16.0	214 10.6	. . 22.5	91 10.5	. . 14.0	264 38.7	. . 46.0	322 07.6	. . 03.9
04	81 18.4	229 10.1	23.7	106 12.6	13.6	279 40.7	45.8	337 10.0	03.9
05	96 20.9	244 09.6	24.8	121 14.6	13.3	294 42.7	45.6	352 12.4	03.9
06	111 23.4	259 09.1	S12 25.9	136 16.7	S14 12.9	309 44.8	N 8 45.5	7 14.9	N22 03.9
07	126 25.8	274 08.6	27.1	151 18.7	12.6	324 46.8	45.3	22 17.3	03.9
08	141 28.3	289 08.1	28.2	166 20.8	12.2	339 48.8	45.1	37 19.7	03.9
09	156 30.8	304 07.6	. . 29.4	181 22.8	. . 11.9	354 50.8	. . 45.0	52 22.1	. . 03.9
10	171 33.2	319 07.1	30.5	196 24.9	11.5	9 52.8	44.8	67 24.5	03.9
11	186 35.7	334 06.6	31.6	211 26.9	11.1	24 54.8	44.6	82 26.9	03.9
12	201 38.2	349 06.1	S12 32.7	226 28.9	S14 10.8	39 56.9	N 8 44.4	97 29.3	N22 03.8
13	216 40.6	4 05.6	33.9	241 31.0	10.4	54 58.9	44.3	112 31.7	03.8
14	231 43.1	19 05.1	35.0	256 33.0	10.1	70 00.9	44.1	127 34.1	03.8
15	246 45.5	34 04.6	. . 36.1	271 35.1	. . 09.7	85 02.9	. . 43.9	142 36.5	. . 03.8
16	261 48.0	49 04.1	37.3	286 37.1	09.3	100 04.9	43.7	157 38.9	03.8
17	276 50.5	64 03.6	38.4	301 39.2	09.0	115 06.9	43.6	172 41.3	03.8
18	291 52.9	79 03.1	S12 39.5	316 41.2	S14 08.6	130 09.0	N 8 43.4	187 43.7	N22 03.8
19	306 55.4	94 02.5	40.7	331 43.2	08.3	145 11.0	43.2	202 46.1	03.8
20	321 57.9	109 02.0	41.8	346 45.3	07.9	160 13.0	43.0	217 48.5	03.8
21	337 00.3	124 01.5	. . 42.9	1 47.3	. . 07.5	175 15.0	. . 42.9	232 50.9	. . 03.8
22	352 02.8	139 01.0	44.0	16 49.3	07.2	190 17.0	42.7	247 53.3	03.8
23	7 05.3	154 00.5	45.2	31 51.4	06.8	205 19.1	42.5	262 55.7	03.8
14 00 (TUESDAY)	22 07.7	169 00.0	S12 46.3	46 53.4	S14 06.5	220 21.1	N 8 42.4	277 58.1	N22 03.8
01	37 10.2	183 59.5	47.4	61 55.4	06.1	235 23.1	42.2	293 00.5	03.8
02	52 12.7	198 59.0	48.5	76 57.5	05.7	250 25.1	42.0	308 02.9	03.8
03	67 15.1	213 58.5	. . 49.7	91 59.5	. . 05.4	265 27.1	. . 41.8	323 05.3	. . 03.8
04	82 17.6	228 58.0	50.8	107 01.5	05.0	280 29.2	41.7	338 07.7	03.8
05	97 20.0	243 57.5	51.9	122 03.5	04.6	295 31.2	41.5	353 10.1	03.8
06	112 22.5	258 56.9	S12 53.0	137 05.6	S14 04.3	310 33.2	N 8 41.3	8 12.5	N22 03.8
07	127 25.0	273 56.4	54.2	152 07.6	03.9	325 35.2	41.1	23 14.9	03.8
08	142 27.4	288 55.9	55.3	167 09.6	03.5	340 37.2	41.0	38 17.3	03.7
09	157 29.9	303 55.4	. . 56.4	182 11.6	. . 03.2	355 39.2	. . 40.8	53 19.7	. . 03.7
10	172 32.4	318 54.9	57.5	197 13.7	02.8	10 41.3	40.6	68 22.1	03.7
11	187 34.8	333 54.4	58.6	212 15.7	02.4	25 43.3	40.5	83 24.6	03.7
12	202 37.3	348 53.9	S12 59.8	227 17.7	S14 02.1	40 45.3	N 8 40.3	98 27.0	N22 03.7
13	217 39.8	3 53.3	13 00.9	242 19.7	01.7	55 47.3	40.1	113 29.4	03.7
14	232 42.2	18 52.8	02.0	257 21.7	01.3	70 49.3	39.9	128 31.8	03.7
15	247 44.7	33 52.3	. . 03.1	272 23.7	. . 01.0	85 51.4	. . 39.8	143 34.2	. . 03.7
16	262 47.2	48 51.8	04.2	287 25.8	00.6	100 53.4	39.6	158 36.6	03.7
17	277 49.6	63 51.3	05.4	302 27.8	14 00.2	115 55.4	39.4	173 39.0	03.7
18	292 52.1	78 50.8	S13 06.5	317 29.8	S13 59.9	130 57.4	N 8 39.3	188 41.4	N22 03.7
19	307 54.5	93 50.2	07.6	332 31.8	59.5	145 59.5	39.1	203 43.8	03.7
20	322 57.0	108 49.7	08.7	347 33.8	59.1	161 01.5	38.9	218 46.2	03.7
21	337 59.5	123 49.2	. . 09.8	2 35.8	. . 58.8	176 03.5	. . 38.7	233 48.6	. . 03.7
22	353 01.9	138 48.7	10.9	17 37.8	58.4	191 05.5	38.6	248 51.0	03.7
23	8 04.4	153 48.2	12.0	32 39.8	58.0	206 07.5	38.4	263 53.4	03.7
15 00 (WEDNESDAY)	23 06.9	168 47.6	S13 13.2	47 41.8	S13 57.7	221 09.6	N 8 38.2	278 55.9	N22 03.7
01	38 09.3	183 47.1	14.3	62 43.8	57.3	236 11.6	38.0	293 58.3	03.7
02	53 11.8	198 46.6	15.4	77 45.8	56.9	251 13.6	37.9	309 00.7	03.7
03	68 14.3	213 46.1	. . 16.5	92 47.8	. . 56.5	266 15.6	. . 37.7	324 03.1	. . 03.7
04	83 16.7	228 45.6	17.6	107 49.8	56.2	281 17.6	37.5	339 05.5	03.7
05	98 19.2	243 45.0	18.7	122 51.8	55.8	296 19.7	37.4	354 07.9	03.7
06	113 21.6	258 44.5	S13 19.8	137 53.8	S13 55.4	311 21.7	N 8 37.2	9 10.3	N22 03.6
07	128 24.1	273 44.0	20.9	152 55.8	55.0	326 23.7	37.0	24 12.7	03.6
08	143 26.6	288 43.5	22.0	167 57.8	54.7	341 25.7	36.8	39 15.1	03.6
09	158 29.0	303 42.9	. . 23.1	182 59.8	. . 54.3	356 27.8	. . 36.7	54 17.5	. . 03.6
10	173 31.5	318 42.4	24.3	198 01.8	53.9	11 29.8	36.5	69 20.0	03.6
11	188 34.0	333 41.9	25.4	213 03.8	53.6	26 31.8	36.3	84 22.4	03.6
12	203 36.4	348 41.4	S13 26.5	228 05.8	S13 53.2	41 33.8	N 8 36.2	99 24.8	N22 03.6
13	218 38.9	3 40.8	27.6	243 07.8	52.8	56 35.9	36.0	114 27.2	03.6
14	233 41.4	18 40.3	28.7	258 09.8	52.4	71 37.9	35.8	129 29.6	03.6
15	248 43.8	33 39.8	. . 29.8	273 11.8	. . 52.1	86 39.9	. . 35.6	144 32.0	. . 03.6
16	263 46.3	48 39.3	30.9	288 13.8	51.7	101 41.9	35.5	159 34.4	03.6
17	278 48.8	63 38.7	32.0	303 15.7	51.3	116 43.9	35.3	174 36.8	03.6
18	293 51.2	78 38.2	S13 33.1	318 17.7	S13 50.9	131 46.0	N 8 35.1	189 39.2	N22 03.6
19	308 53.7	93 37.7	34.2	333 19.7	50.5	146 48.0	35.0	204 41.7	03.6
20	323 56.1	108 37.1	35.3	348 21.7	50.2	161 50.0	34.8	219 44.1	03.6
21	338 58.6	123 36.6	. . 36.4	3 23.7	. . 49.8	176 52.0	. . 34.6	234 46.5	. . 03.6
22	354 01.1	138 36.1	37.5	18 25.7	49.4	191 54.1	34.4	249 48.9	03.6
23	9 03.5	153 35.6	38.6	33 27.6	49.0	206 56.1	34.3	264 51.3	03.6
Mer. Pass. (h m)	22 27.8	v −0.5	d 1.1	v 2.0	d 0.4	v 2.0	d 0.2	v 2.4	d 0.0

STARS Name	SHA ° ′	Dec ° ′
Acamar	315 23.6	S40 17.2
Achernar	335 31.6	S57 13.0
Acrux	173 18.9	S63 07.0
Adhara	255 18.5	S28 58.3
Aldebaran	290 58.0	N16 31.1
Alioth	166 27.7	N55 56.5
Alkaid	153 05.2	N49 17.8
Al Na'ir	27 52.8	S46 56.8
Alnilam	275 54.0	S 1 11.8
Alphard	218 03.8	S 8 40.2
Alphecca	126 17.7	N26 42.3
Alpheratz	357 51.2	N29 06.8
Altair	62 15.7	N 8 52.8
Ankaa	353 22.7	S42 17.2
Antares	112 35.9	S26 26.4
Arcturus	146 03.0	N19 09.9
Atria	107 45.1	S69 02.3
Avior	234 21.5	S59 30.8
Bellatrix	278 40.1	N 6 21.4
Betelgeuse	271 09.5	N 7 24.6
Canopus	263 59.4	S52 41.4
Capella	280 45.5	N46 00.1
Deneb	49 36.7	N45 17.8
Denebola	182 41.7	N14 33.2
Diphda	349 03.2	S17 57.9
Dubhe	194 01.2	N61 43.8
Elnath	278 22.1	N28 36.7
Eltanin	90 49.9	N51 29.5
Enif	33 54.5	N 9 53.6
Fomalhaut	15 32.0	S29 36.2
Gacrux	172 10.3	S57 07.8
Gienah	176 00.6	S17 33.5
Hadar	148 59.7	S60 23.4
Hamal	328 09.1	N23 28.9
Kaus Aust.	83 54.0	S34 23.1
Kochab	137 20.4	N74 08.5
Markab	13 45.7	N15 13.6
Menkar	314 22.8	N 4 06.4
Menkent	148 17.1	S36 23.2
Miaplacidus	221 42.0	S69 43.5
Mirfak	308 51.0	N49 52.5
Nunki	76 07.8	S26 17.7
Peacock	53 30.9	S56 43.7
Pollux	243 37.1	N28 01.1
Procyon	245 07.7	N 5 13.1
Rasalhague	96 13.7	N12 33.5
Regulus	207 51.8	N11 57.1
Rigel	281 19.2	S 8 11.6
Rigil Kent.	140 03.2	S60 51.0
Sabik	102 21.5	S15 43.8
Schedar	349 49.0	N56 33.5
Shaula	96 32.5	S37 06.5
Sirius	258 40.4	S16 42.9
Spica	158 39.7	S11 10.7
Suhail	222 58.4	S43 26.4
Vega	80 44.3	N38 47.4
Zuben'ubi	137 14.2	S16 03.3
	SHA ° ′	**Mer. Pass.** h m
Venus	146 52.3	12 44
Mars	24 45.7	20 50
Jupiter	198 13.4	9 17
Saturn	255 50.4	5 27

UT	SUN GHA	SUN Dec	MOON GHA	v	MOON Dec	d	HP
d h	° ′	° ′	° ′	′	° ′	′	′
13 00	183 23.3	S 7 31.9	334 44.1	14.4	N17 15.7	10.7	54.1
01	198 23.5	32.9	349 17.5	14.4	17 26.4	10.6	54.1
02	213 23.6	33.8	3 50.9	14.4	17 37.0	10.5	54.1
03	228 23.8	. . 34.7	18 24.3	14.2	17 47.5	10.5	54.1
04	243 23.9	35.7	32 57.5	14.3	17 58.0	10.3	54.1
05	258 24.1	36.6	47 30.8	14.2	18 08.3	10.3	54.1
06	273 24.2	S 7 37.6	62 04.0	14.1	N18 18.6	10.3	54.1
07	288 24.4	38.5	76 37.1	14.1	18 28.9	10.1	54.1
08	303 24.5	39.4	91 10.2	14.0	18 39.0	10.1	54.1
09	318 24.7	. . 40.4	105 43.2	14.0	18 49.1	10.0	54.1
10	333 24.9	41.3	120 16.2	13.9	18 59.1	9.9	54.1
11	348 25.0	42.2	134 49.1	13.9	19 09.0	9.8	54.1
12	3 25.2	S 7 43.2	149 22.0	13.8	N19 18.8	9.8	54.1
13	18 25.3	44.1	163 54.8	13.8	19 28.6	9.6	54.1
14	33 25.5	45.0	178 27.6	13.7	19 38.2	9.6	54.1
15	48 25.6	. . 46.0	193 00.3	13.7	19 47.8	9.5	54.1
16	63 25.8	46.9	207 33.0	13.6	19 57.3	9.4	54.1
17	78 25.9	47.8	222 05.6	13.5	20 06.7	9.4	54.1
18	93 26.1	S 7 48.8	236 38.1	13.5	N20 16.1	9.2	54.1
19	108 26.2	49.7	251 10.6	13.5	20 25.3	9.2	54.1
20	123 26.4	50.6	265 43.1	13.3	20 34.5	9.1	54.1
21	138 26.5	. . 51.6	280 15.4	13.4	20 43.6	8.9	54.1
22	153 26.7	52.5	294 47.8	13.2	20 52.5	8.9	54.1
23	168 26.8	53.4	309 20.0	13.3	21 01.4	8.8	54.1
14 00	183 27.0	S 7 54.4	323 52.3	13.1	N21 10.2	8.8	54.0
01	198 27.1	55.3	338 24.4	13.1	21 19.0	8.6	54.0
02	213 27.3	56.2	352 56.5	13.1	21 27.6	8.5	54.0
03	228 27.4	. . 57.2	7 28.6	13.0	21 36.1	8.4	54.0
04	243 27.6	58.1	22 00.6	12.9	21 44.5	8.4	54.0
05	258 27.7	7 59.0	36 32.5	12.9	21 52.9	8.2	54.0
06	273 27.9	S 8 00.0	51 04.4	12.8	N22 01.1	8.2	54.1
07	288 28.0	00.9	65 36.2	12.8	22 09.3	8.0	54.1
08	303 28.2	01.8	80 08.0	12.7	22 17.3	8.0	54.1
09	318 28.3	. . 02.8	94 39.7	12.6	22 25.3	7.9	54.1
10	333 28.5	03.7	109 11.3	12.6	22 33.2	7.7	54.1
11	348 28.6	04.6	123 42.9	12.5	22 40.9	7.7	54.1
12	3 28.7	S 8 05.5	138 14.4	12.5	N22 48.6	7.6	54.1
13	18 28.9	06.5	152 45.9	12.4	22 56.2	7.4	54.1
14	33 29.0	07.4	167 17.3	12.4	23 03.6	7.4	54.1
15	48 29.2	. . 08.3	181 48.7	12.3	23 11.0	7.3	54.1
16	63 29.3	09.3	196 20.0	12.2	23 18.3	7.1	54.1
17	78 29.5	10.2	210 51.2	12.2	23 25.4	7.1	54.1
18	93 29.6	S 8 11.1	225 22.4	12.1	N23 32.5	7.0	54.1
19	108 29.8	12.1	239 53.5	12.1	23 39.5	6.8	54.1
20	123 29.9	13.0	254 24.6	12.0	23 46.3	6.8	54.1
21	138 30.1	. . 13.9	268 55.6	12.0	23 53.1	6.6	54.1
22	153 30.2	14.8	283 26.6	11.9	23 59.7	6.5	54.1
23	168 30.3	15.8	297 57.5	11.9	24 06.2	6.4	54.1
15 00	183 30.5	S 8 16.7	312 28.4	11.7	N24 12.6	6.4	54.1
01	198 30.6	17.6	326 59.1	11.8	24 19.0	6.2	54.1
02	213 30.8	18.5	341 29.9	11.7	24 25.2	6.1	54.1
03	228 30.9	. . 19.5	356 00.6	11.6	24 31.3	6.0	54.1
04	243 31.1	20.4	10 31.2	11.6	24 37.3	5.8	54.1
05	258 31.2	21.3	25 01.8	11.5	24 43.1	5.8	54.1
06	273 31.4	S 8 22.3	39 32.3	11.5	N24 48.9	5.7	54.1
07	288 31.5	23.2	54 02.8	11.4	24 54.6	5.5	54.1
08	303 31.6	24.1	68 33.2	11.3	25 00.1	5.4	54.2
09	318 31.8	. . 25.0	83 03.5	11.3	25 05.5	5.3	54.2
10	333 31.9	26.0	97 33.8	11.3	25 10.8	5.2	54.2
11	348 32.1	26.9	112 04.1	11.2	25 16.0	5.1	54.2
12	3 32.2	S 8 27.8	126 34.3	11.1	N25 21.1	5.0	54.2
13	18 32.3	28.7	141 04.4	11.1	25 26.1	4.8	54.2
14	33 32.5	29.7	155 34.5	11.1	25 30.9	4.8	54.2
15	48 32.6	. . 30.6	170 04.6	11.0	25 35.7	4.6	54.2
16	63 32.8	31.5	184 34.6	10.9	25 40.3	4.5	54.2
17	78 32.9	32.4	199 04.5	10.9	25 44.8	4.4	54.2
18	93 33.0	S 8 33.4	213 34.4	10.9	N25 49.2	4.2	54.2
19	108 33.2	34.3	228 04.3	10.8	25 53.4	4.2	54.2
20	123 33.3	35.2	242 34.1	10.7	25 57.6	4.0	54.3
21	138 33.5	. . 36.1	257 03.8	10.7	26 01.6	3.9	54.3
22	153 33.6	37.1	271 33.5	10.7	26 05.5	3.7	54.3
23	168 33.7	38.0	286 03.2	10.6	N26 09.2	3.7	54.3
	SD 16.1	*d* 0.9	SD 14.7		14.7		14.8

Lat.	Twilight Naut.	Twilight Civil	Sunrise	Moonrise 13	Moonrise 14	Moonrise 15	Moonrise 16
°	h m	h m	h m	h m	h m	h m	h m
N 72	04 49	06 07	07 17	▭	▭	▭	▭
N 70	04 53	06 04	07 07	14 50	▭	▭	▭
68	04 57	06 01	06 58	15 56	▭	▭	▭
66	05 00	05 59	06 51	16 32	15 57	▭	▭
64	05 02	05 57	06 45	16 58	16 48	16 25	▭
62	05 04	05 55	06 40	17 19	17 20	17 26	17 46
60	05 05	05 54	06 36	17 35	17 44	17 59	18 31
N 58	05 07	05 52	06 32	17 49	18 03	18 24	19 00
56	05 08	05 51	06 28	18 02	18 19	18 44	19 23
54	05 08	05 49	06 25	18 12	18 32	19 01	19 41
52	05 09	05 48	06 22	18 22	18 45	19 15	19 57
50	05 09	05 47	06 19	18 30	18 55	19 28	20 11
45	05 10	05 44	06 14	18 48	19 18	19 54	20 38
N 40	05 10	05 42	06 09	19 03	19 36	20 14	21 00
35	05 10	05 39	06 05	19 16	19 51	20 31	21 19
30	05 09	05 37	06 01	19 27	20 04	20 46	21 34
20	05 06	05 32	05 54	19 46	20 27	21 12	22 01
N 10	05 03	05 27	05 48	20 02	20 46	21 33	22 24
0	04 58	05 22	05 43	20 18	21 05	21 54	22 45
S 10	04 51	05 16	05 37	20 34	21 23	22 14	23 06
20	04 42	05 09	05 31	20 51	21 43	22 37	23 29
30	04 31	04 59	05 24	21 10	22 07	23 02	23 56
35	04 23	04 54	05 19	21 22	22 20	23 17	24 12
40	04 14	04 47	05 15	21 35	22 36	23 35	24 30
45	04 03	04 39	05 09	21 51	22 55	23 56	24 53
S 50	03 48	04 29	05 02	22 10	23 19	24 23	00 23
52	03 41	04 24	04 59	22 19	23 30	24 36	00 36
54	03 33	04 19	04 56	22 30	23 43	24 52	00 52
56	03 24	04 13	04 52	22 41	23 58	25 10	01 10
58	03 13	04 06	04 48	22 55	24 16	00 16	01 31
S 60	03 00	03 58	04 43	23 11	24 38	00 38	01 59

Lat.	Sunset	Twilight Civil	Twilight Naut.	Moonset 13	Moonset 14	Moonset 15	Moonset 16
°	h m	h m	h m	h m	h m	h m	h m
N 72	16 13	17 23	18 41	▭	▭	▭	▭
N 70	16 24	17 26	18 37	13 06	▭	▭	▭
68	16 32	17 29	18 33	12 01	▭	▭	▭
66	16 40	17 32	18 31	11 26	13 37	▭	▭
64	16 46	17 34	18 28	11 01	12 47	14 51	▭
62	16 51	17 36	18 27	10 42	12 16	13 51	15 16
60	16 55	17 37	18 25	10 26	11 53	13 17	14 31
N 58	16 59	17 39	18 24	10 12	11 34	12 52	14 02
56	17 03	17 40	18 23	10 01	11 18	12 33	13 39
54	17 06	17 42	18 23	09 51	11 05	12 16	13 21
52	17 09	17 43	18 22	09 42	10 53	12 02	13 05
50	17 12	17 44	18 22	09 34	10 43	11 50	12 51
45	17 18	17 47	18 21	09 17	10 22	11 25	12 24
N 40	17 23	17 50	18 21	09 03	10 04	11 05	12 02
35	17 27	17 53	18 22	08 52	09 50	10 48	11 44
30	17 31	17 55	18 23	08 41	09 37	10 34	11 28
20	17 38	18 00	18 25	08 24	09 16	10 09	11 02
N 10	17 44	18 05	18 29	08 09	08 58	09 48	10 40
0	17 49	18 10	18 35	07 55	08 40	09 28	10 18
S 10	17 55	18 17	18 41	07 41	08 23	09 09	09 57
20	18 02	18 24	18 50	07 26	08 05	08 48	09 35
30	18 09	18 33	19 02	07 09	07 44	08 24	09 09
35	18 13	18 39	19 10	06 59	07 32	08 09	08 53
40	18 18	18 46	19 19	06 48	07 18	07 53	08 35
45	18 24	18 54	19 31	06 34	07 01	07 33	08 13
S 50	18 31	19 05	19 46	06 18	06 40	07 09	07 46
52	18 34	19 09	19 53	06 11	06 31	06 57	07 33
54	18 38	19 15	20 01	06 03	06 20	06 44	07 17
56	18 42	19 21	20 11	05 53	06 08	06 28	06 59
58	18 46	19 28	20 22	05 43	05 53	06 10	06 37
S 60	18 51	19 36	20 34	05 31	05 37	05 48	06 09

Day	SUN Eqn. of Time 00^h	SUN Eqn. of Time 12^h	SUN Mer. Pass.	MOON Mer. Pass. Upper	MOON Mer. Pass. Lower	MOON Age	MOON Phase
d	m s	m s	h m	h m	h m	d %	
13	13 33	13 40	11 46	01 44	14 06	17 91	
14	13 48	13 55	11 46	02 29	14 53	18 85	
15	14 02	14 09	11 46	03 17	15 41	19 78	

	UT	ARIES	VENUS −3.9		MARS −1.6		JUPITER −1.8		SATURN +0.0		STARS		
	d h	GHA ° ′	GHA ° ′	Dec ° ′	GHA ° ′	Dec ° ′	GHA ° ′	Dec ° ′	GHA ° ′	Dec ° ′	Name	SHA ° ′	Dec ° ′
	16 00	24 06.0	168 35.0	S13 39.7	48 29.6	S13 48.7	221 58.1	N 8 34.1	279 53.7	N22 03.6	Acamar	315 23.6	S40 17.2
	01	39 08.5	183 34.5	40.8	63 31.6	48.3	237 00.1	33.9	294 56.1	03.6	Achernar	335 31.6	S57 13.0
	02	54 10.9	198 34.0	41.9	78 33.6	47.9	252 02.2	33.8	309 58.6	03.6	Acrux	173 18.9	S63 07.0
	03	69 13.4	213 33.4	. . 43.0	93 35.5	. . 47.5	267 04.2	. . 33.6	325 01.0	. . 03.6	Adhara	255 18.5	S28 58.3
	04	84 15.9	228 32.9	44.1	108 37.5	47.1	282 06.2	33.4	340 03.4	03.6	Aldebaran	290 57.9	N16 31.1
	05	99 18.3	243 32.4	45.2	123 39.5	46.7	297 08.2	33.3	355 05.8	03.6			
	06	114 20.8	258 31.8	S13 46.3	138 41.5	S13 46.4	312 10.3	N 8 33.1	10 08.2	N22 03.5	Alioth	166 27.7	N55 56.4
	07	129 23.3	273 31.3	47.4	153 43.4	46.0	327 12.3	32.9	25 10.6	03.5	Alkaid	153 05.2	N49 17.8
T	08	144 25.7	288 30.8	48.5	168 45.4	45.6	342 14.3	32.7	40 13.0	03.5	Al Na'ir	27 52.8	S46 56.8
H	09	159 28.2	303 30.2	. . 49.6	183 47.4	. . 45.2	357 16.3	. . 32.6	55 15.5	. . 03.5	Alnilam	275 54.0	S 1 11.8
U	10	174 30.6	318 29.7	50.7	198 49.3	44.8	12 18.4	32.4	70 17.9	03.5	Alphard	218 03.8	S 8 40.2
R	11	189 33.1	333 29.1	51.8	213 51.3	44.5	27 20.4	32.2	85 20.3	03.5			
S	12	204 35.6	348 28.6	S13 52.9	228 53.3	S13 44.1	42 22.4	N 8 32.1	100 22.7	N22 03.5	Alphecca	126 17.7	N26 42.3
D	13	219 38.0	3 28.1	53.9	243 55.2	43.7	57 24.4	31.9	115 25.1	03.5	Alpheratz	357 51.2	N29 06.8
A	14	234 40.5	18 27.5	55.0	258 57.2	43.3	72 26.5	31.7	130 27.5	03.5	Altair	62 15.7	N 8 52.8
Y	15	249 43.0	33 27.0	. . 56.1	273 59.2	. . 42.9	87 28.5	. . 31.5	145 30.0	. . 03.5	Ankaa	353 22.7	S42 17.2
	16	264 45.4	48 26.5	57.2	289 01.1	42.5	102 30.5	31.4	160 32.4	03.5	Antares	112 35.9	S26 26.4
	17	279 47.9	63 25.9	58.3	304 03.1	42.2	117 32.5	31.2	175 34.8	03.5			
	18	294 50.4	78 25.4	S13 59.4	319 05.1	S13 41.8	132 34.6	N 8 31.0	190 37.2	N22 03.5	Arcturus	146 03.0	N19 09.9
	19	309 52.8	93 24.8	14 00.5	334 07.0	41.4	147 36.6	30.9	205 39.6	03.5	Atria	107 45.1	S69 02.3
	20	324 55.3	108 24.3	01.6	349 09.0	41.0	162 38.6	30.7	220 42.0	03.5	Avior	234 21.4	S59 30.8
	21	339 57.8	123 23.8	. . 02.7	4 10.9	. . 40.6	177 40.7	. . 30.5	235 44.5	. . 03.5	Bellatrix	278 40.0	N 6 21.3
	22	355 00.2	138 23.2	03.8	19 12.9	40.2	192 42.7	30.4	250 46.9	03.5	Betelgeuse	271 09.4	N 7 24.6
	23	10 02.7	153 22.7	04.8	34 14.8	39.8	207 44.7	30.2	265 49.3	03.5			
	17 00	25 05.1	168 22.1	S14 05.9	49 16.8	S13 39.4	222 46.7	N 8 30.0	280 51.7	N22 03.5	Canopus	263 59.4	S52 41.5
	01	40 07.6	183 21.6	07.0	64 18.8	39.1	237 48.8	29.8	295 54.1	03.5	Capella	280 45.5	N46 00.1
	02	55 10.1	198 21.0	08.1	79 20.7	38.7	252 50.8	29.7	310 56.6	03.5	Deneb	49 36.7	N45 17.8
	03	70 12.5	213 20.5	. . 09.2	94 22.7	. . 38.3	267 52.8	. . 29.5	325 59.0	. . 03.5	Denebola	182 41.7	N14 33.2
	04	85 15.0	228 20.0	10.3	109 24.6	37.9	282 54.8	29.3	341 01.4	03.5	Diphda	349 03.2	S17 57.9
	05	100 17.5	243 19.4	11.3	124 26.6	37.5	297 56.9	29.2	356 03.8	03.5			
	06	115 19.9	258 18.9	S14 12.4	139 28.5	S13 37.1	312 58.9	N 8 29.0	11 06.2	N22 03.5	Dubhe	194 01.2	N61 43.8
	07	130 22.4	273 18.3	13.5	154 30.4	36.7	328 00.9	28.8	26 08.7	03.5	Elnath	278 22.1	N28 36.7
	08	145 24.9	288 17.8	14.6	169 32.4	36.3	343 03.0	28.7	41 11.1	03.5	Eltanin	90 49.9	N51 29.5
F	09	160 27.3	303 17.2	. . 15.7	184 34.3	. . 35.9	358 05.0	. . 28.5	56 13.5	. . 03.4	Enif	33 54.5	N 9 53.6
R	10	175 29.8	318 16.7	16.7	199 36.3	35.6	13 07.0	28.3	71 15.9	03.4	Fomalhaut	15 32.0	S29 36.2
I	11	190 32.2	333 16.1	17.8	214 38.2	35.2	28 09.0	28.2	86 18.3	03.4			
D	12	205 34.7	348 15.6	S14 18.9	229 40.2	S13 34.8	43 11.1	N 8 28.0	101 20.8	N22 03.4	Gacrux	172 10.3	S57 07.8
A	13	220 37.2	3 15.0	20.0	244 42.1	34.4	58 13.1	27.8	116 23.2	03.4	Gienah	176 00.5	S17 33.5
Y	14	235 39.6	18 14.5	21.1	259 44.0	34.0	73 15.1	27.6	131 25.6	03.4	Hadar	148 59.7	S60 23.4
	15	250 42.1	33 13.9	. . 22.1	274 46.0	. . 33.6	88 17.2	. . 27.5	146 28.0	. . 03.4	Hamal	328 09.1	N23 28.9
	16	265 44.6	48 13.4	23.2	289 47.9	33.2	103 19.2	27.3	161 30.5	03.4	Kaus Aust.	83 54.1	S34 23.1
	17	280 47.0	63 12.8	24.3	304 49.9	32.8	118 21.2	27.1	176 32.9	03.4			
	18	295 49.5	78 12.3	S14 25.4	319 51.8	S13 32.4	133 23.3	N 8 27.0	191 35.3	N22 03.4	Kochab	137 20.4	N74 08.5
	19	310 52.0	93 11.7	26.4	334 53.7	32.0	148 25.3	26.8	206 37.7	03.4	Markab	13 45.7	N15 13.6
	20	325 54.4	108 11.2	27.5	349 55.7	31.6	163 27.3	26.6	221 40.1	03.4	Menkar	314 22.8	N 4 06.4
	21	340 56.9	123 10.6	. . 28.6	4 57.6	. . 31.2	178 29.3	. . 26.5	236 42.6	. . 03.4	Menkent	148 17.1	S36 23.2
	22	355 59.4	138 10.1	29.7	19 59.5	30.8	193 31.4	26.3	251 45.0	03.4	Miaplacidus	221 42.0	S69 43.5
	23	11 01.8	153 09.5	30.7	35 01.4	30.4	208 33.4	26.1	266 47.4	03.4			
	18 00	26 04.3	168 09.0	S14 31.8	50 03.4	S13 30.1	223 35.4	N 8 26.0	281 49.8	N22 03.4	Mirfak	308 51.0	N49 52.5
	01	41 06.7	183 08.4	32.9	65 05.3	29.7	238 37.5	25.8	296 52.3	03.4	Nunki	76 07.8	S26 17.7
	02	56 09.2	198 07.9	33.9	80 07.2	29.3	253 39.5	25.6	311 54.7	03.4	Peacock	53 31.0	S56 43.7
	03	71 11.7	213 07.3	. . 35.0	95 09.2	. . 28.9	268 41.5	. . 25.4	326 57.1	. . 03.4	Pollux	243 37.0	N28 01.1
	04	86 14.1	228 06.8	36.1	110 11.1	28.5	283 43.6	25.3	341 59.5	03.4	Procyon	245 07.7	N 5 13.1
	05	101 16.6	243 06.2	37.1	125 13.0	28.1	298 45.6	25.1	357 02.0	03.4			
	06	116 19.1	258 05.6	S14 38.2	140 14.9	S13 27.7	313 47.6	N 8 24.9	12 04.4	N22 03.4	Rasalhague	96 13.7	N12 33.5
	07	131 21.5	273 05.1	39.3	155 16.8	27.3	328 49.6	24.8	27 06.8	03.4	Regulus	207 51.8	N11 57.1
S	08	146 24.0	288 04.5	40.3	170 18.8	26.9	343 51.7	24.6	42 09.2	03.4	Rigel	281 19.2	S 8 11.6
A	09	161 26.5	303 04.0	. . 41.4	185 20.7	. . 26.5	358 53.7	. . 24.4	57 11.7	. . 03.4	Rigil Kent.	140 03.2	S60 51.0
T	10	176 28.9	318 03.4	42.5	200 22.6	26.1	13 55.7	24.3	72 14.1	03.4	Sabik	102 21.5	S15 43.8
U	11	191 31.4	333 02.9	43.5	215 24.5	25.7	28 57.8	24.1	87 16.5	03.4			
R	12	206 33.9	348 02.3	S14 44.6	230 26.4	S13 25.3	43 59.8	N 8 23.9	102 18.9	N22 03.4	Schedar	349 49.0	N56 33.6
D	13	221 36.3	3 01.7	45.7	245 28.4	24.9	59 01.8	23.8	117 21.4	03.4	Shaula	96 32.5	S37 06.5
A	14	236 38.8	18 01.2	46.7	260 30.3	24.5	74 03.9	23.6	132 23.8	03.4	Sirius	258 40.4	S16 43.0
Y	15	251 41.2	33 00.6	. . 47.8	275 32.2	. . 24.1	89 05.9	. . 23.4	147 26.2	. . 03.4	Spica	158 39.7	S11 10.7
	16	266 43.7	48 00.1	48.9	290 34.1	23.7	104 07.9	23.3	162 28.6	03.4	Suhail	222 58.3	S43 26.4
	17	281 46.2	62 59.5	49.9	305 36.0	23.3	119 10.0	23.1	177 31.1	03.4			
	18	296 48.6	77 58.9	S14 51.0	320 37.9	S13 22.9	134 12.0	N 8 22.9	192 33.5	N22 03.3	Vega	80 44.3	N38 47.4
	19	311 51.1	92 58.4	52.0	335 39.8	22.5	149 14.0	22.8	207 35.9	03.3	Zuben'ubi	137 14.2	S16 03.3
	20	326 53.6	107 57.8	53.1	350 41.7	22.1	164 16.1	22.6	222 38.4	03.3			
	21	341 56.0	122 57.2	. . 54.2	5 43.6	. . 21.7	179 18.1	. . 22.4	237 40.8	. . 03.3			
	22	356 58.5	137 56.7	55.2	20 45.6	21.3	194 20.1	22.3	252 43.2	03.3			
	23	12 01.0	152 56.1	56.3	35 47.5	20.9	209 22.2	22.1	267 45.6	03.3			
	Mer. Pass.	h m 22 16.0	*v* −0.5	*d* 1.1	*v* 1.9	*d* 0.4	*v* 2.0	*d* 0.2	*v* 2.4	*d* 0.0			

	SHA ° ′	Mer. Pass h m
Venus	143 17.0	12 47
Mars	24 11.7	20 40
Jupiter	197 41.6	9 08
Saturn	255 46.6	5 16

UT d	UT h	SUN GHA	SUN Dec	MOON GHA	MOON v	MOON Dec	MOON d	MOON HP
		° ′	° ′	° ′	′	° ′	′	′
16 THURSDAY	00	183 33.9	S 8 38.9	300 32.8	10.6	N26 12.9	3.5	54.3
	01	198 34.0	39.8	315 02.4	10.5	26 16.4	3.4	54.3
	02	213 34.1	40.8	329 31.9	10.5	26 19.8	3.3	54.3
	03	228 34.3	. . 41.7	344 01.4	10.4	26 23.1	3.1	54.3
	04	243 34.4	42.6	358 30.8	10.4	26 26.2	3.1	54.3
	05	258 34.6	43.5	13 00.2	10.4	26 29.3	2.9	54.4
	06	273 34.7	S 8 44.4	27 29.6	10.3	N26 32.2	2.7	54.4
	07	288 34.8	45.4	41 58.9	10.3	26 34.9	2.7	54.4
	08	303 35.0	46.3	56 28.2	10.2	26 37.6	2.5	54.4
	09	318 35.1	. . 47.2	70 57.4	10.2	26 40.1	2.4	54.4
	10	333 35.2	48.1	85 26.6	10.1	26 42.5	2.3	54.4
	11	348 35.4	49.0	99 55.7	10.2	26 44.8	2.1	54.4
	12	3 35.5	S 8 50.0	114 24.9	10.0	N26 46.9	2.0	54.5
	13	18 35.6	50.9	128 53.9	10.1	26 48.9	1.9	54.5
	14	33 35.8	51.8	143 23.0	10.0	26 50.8	1.7	54.5
	15	48 35.9	. . 52.7	157 52.0	10.0	26 52.5	1.7	54.5
	16	63 36.0	53.6	172 21.0	9.9	26 54.2	1.4	54.5
	17	78 36.2	54.6	186 49.9	10.0	26 55.6	1.4	54.5
	18	93 36.3	S 8 55.5	201 18.9	9.9	N26 57.0	1.2	54.5
	19	108 36.4	56.4	215 47.8	9.8	26 58.2	1.1	54.6
	20	123 36.6	57.3	230 16.6	9.8	26 59.3	1.0	54.6
	21	138 36.7	. . 58.2	244 45.4	9.8	27 00.3	0.8	54.6
	22	153 36.8	8 59.2	259 14.2	9.8	27 01.1	0.7	54.6
	23	168 37.0	9 00.1	273 43.0	9.8	27 01.8	0.6	54.6
17 FRIDAY	00	183 37.1	S 9 01.0	288 11.8	9.7	N27 02.4	0.4	54.6
	01	198 37.2	01.9	302 40.5	9.7	27 02.8	0.3	54.7
	02	213 37.4	02.8	317 09.2	9.7	27 03.1	0.1	54.7
	03	228 37.5	. . 03.7	331 37.9	9.6	27 03.2	0.1	54.7
	04	243 37.6	04.7	346 06.5	9.6	27 03.3	0.2	54.7
	05	258 37.8	05.6	0 35.1	9.6	27 03.1	0.2	54.7
	06	273 37.9	S 9 06.5	15 03.7	9.6	N27 02.9	0.4	54.8
	07	288 38.0	07.4	29 32.3	9.6	27 02.5	0.5	54.8
	08	303 38.2	08.3	44 00.9	9.6	27 02.0	0.7	54.8
	09	318 38.3	. . 09.2	58 29.5	9.5	27 01.3	0.7	54.8
	10	333 38.4	10.2	72 58.0	9.5	27 00.6	1.0	54.8
	11	348 38.5	11.1	87 26.5	9.5	26 59.6	1.0	54.9
	12	3 38.7	S 9 12.0	101 55.0	9.5	N26 58.6	1.2	54.9
	13	18 38.8	12.9	116 23.5	9.5	26 57.4	1.4	54.9
	14	33 38.9	13.8	130 52.0	9.5	26 56.0	1.4	54.9
	15	48 39.1	. . 14.7	145 20.5	9.4	26 54.6	1.6	54.9
	16	63 39.2	15.7	159 48.9	9.5	26 53.0	1.8	55.0
	17	78 39.3	16.6	174 17.4	9.4	26 51.2	1.9	55.0
	18	93 39.4	S 9 17.5	188 45.8	9.4	N26 49.3	2.0	55.0
	19	108 39.6	18.4	203 14.2	9.4	26 47.3	2.1	55.0
	20	123 39.7	19.3	217 42.6	9.5	26 45.2	2.3	55.1
	21	138 39.8	. . 20.2	232 11.1	9.4	26 42.9	2.5	55.1
	22	153 39.9	21.1	246 39.5	9.4	26 40.4	2.5	55.1
	23	168 40.1	22.0	261 07.9	9.4	26 37.9	2.7	55.1
18 SATURDAY	00	183 40.2	S 9 23.0	275 36.3	9.4	N26 35.2	2.9	55.2
	01	198 40.3	23.9	290 04.7	9.4	26 32.3	3.0	55.2
	02	213 40.4	24.8	304 33.1	9.4	26 29.3	3.1	55.2
	03	228 40.6	. . 25.7	319 01.5	9.4	26 26.2	3.2	55.2
	04	243 40.7	26.6	333 29.9	9.4	26 23.0	3.4	55.3
	05	258 40.8	27.5	347 58.3	9.4	26 19.6	3.6	55.3
	06	273 40.9	S 9 28.4	2 26.7	9.4	N26 16.0	3.6	55.3
	07	288 41.1	29.3	16 55.1	9.4	26 12.4	3.8	55.3
	08	303 41.2	30.3	31 23.5	9.4	26 08.6	4.0	55.4
	09	318 41.3	. . 31.2	45 51.9	9.4	26 04.6	4.0	55.4
	10	333 41.4	32.1	60 20.3	9.5	26 00.6	4.3	55.4
	11	348 41.6	33.0	74 48.8	9.4	25 56.3	4.3	55.4
	12	3 41.7	S 9 33.9	89 17.2	9.4	N25 52.0	4.5	55.5
	13	18 41.8	34.8	103 45.6	9.5	25 47.5	4.6	55.5
	14	33 41.9	35.7	118 14.1	9.5	25 42.9	4.8	55.5
	15	48 42.1	. . 36.6	132 42.6	9.4	25 38.1	4.9	55.6
	16	63 42.2	37.5	147 11.0	9.5	25 33.2	5.0	55.6
	17	78 42.3	38.4	161 39.5	9.5	25 28.2	5.2	55.6
	18	93 42.4	S 9 39.3	176 08.0	9.5	N25 23.0	5.3	55.6
	19	108 42.5	40.3	190 36.5	9.6	25 17.7	5.4	55.7
	20	123 42.7	41.2	205 05.1	9.5	25 12.3	5.6	55.7
	21	138 42.8	. . 42.1	219 33.6	9.5	25 06.7	5.7	55.7
	22	153 42.9	43.0	234 02.1	9.6	25 01.0	5.9	55.8
	23	168 43.0	43.9	248 30.7	9.6	N24 55.1	6.0	55.8
		SD 16.1	*d* 0.9	SD 14.8		15.0		15.1

Lat.	Twilight Naut.	Twilight Civil	Sunrise	Moonrise 16	Moonrise 17	Moonrise 18	Moonrise 19
°	h m	h m	h m	h m	h m	h m	h m
N 72	05 02	06 20	07 32	▭	▭	▭	▭
N 70	05 05	06 16	07 19	▭	▭	▭	▭
68	05 07	06 12	07 09	▭	▭	▭	▭
66	05 09	06 08	07 01	▭	▭	▭	20 47
64	05 10	06 06	06 54	▭	▭	19 21	21 28
62	05 12	06 03	06 48	17 46	18 42	20 14	21 56
60	05 12	06 01	06 43	18 31	19 26	20 45	22 18
N 58	05 13	05 59	06 38	19 00	19 55	21 09	22 35
56	05 13	05 57	06 34	19 23	20 18	21 28	22 50
54	05 14	05 55	06 31	19 41	20 36	21 44	23 02
52	05 14	05 53	06 27	19 57	20 51	21 58	23 13
50	05 14	05 52	06 24	20 11	21 05	22 10	23 23
45	05 14	05 48	06 18	20 38	21 32	22 35	23 44
N 40	05 13	05 45	06 12	21 00	21 54	22 55	24 00
35	05 12	05 42	06 07	21 19	22 12	23 11	24 14
30	05 11	05 39	06 03	21 34	22 27	23 25	24 26
20	05 07	05 33	05 55	22 01	22 54	23 49	24 47
N 10	05 03	05 27	05 49	22 24	23 16	24 10	00 10
0	04 57	05 21	05 42	22 45	23 37	24 30	00 30
S 10	04 49	05 14	05 36	23 06	23 58	24 49	00 49
20	04 40	05 06	05 29	23 29	24 21	00 21	01 10
30	04 27	04 56	05 20	23 56	24 47	00 47	01 33
35	04 19	04 50	05 16	24 12	00 12	01 02	01 47
40	04 09	04 42	05 10	24 30	00 30	01 20	02 03
45	03 57	04 33	05 04	24 53	00 53	01 42	02 23
S 50	03 41	04 22	04 56	00 23	01 21	02 09	02 47
52	03 33	04 17	04 53	00 36	01 35	02 22	02 58
54	03 24	04 11	04 49	00 52	01 51	02 37	03 11
56	03 14	04 05	04 44	01 10	02 10	02 55	03 26
58	03 03	03 57	04 40	01 31	02 33	03 17	03 44
S 60	02 49	03 49	04 34	01 59	03 04	03 44	04 05

Lat.	Sunset	Twilight Civil	Twilight Naut.	Moonset 16	Moonset 17	Moonset 18	Moonset 19
°	h m	h m	h m	h m	h m	h m	h m
N 72	15 57	17 09	18 26	▭	▭	▭	▭
N 70	16 10	17 13	18 24	▭	▭	▭	▭
68	16 20	17 17	18 22	▭	▭	▭	▭
66	16 28	17 21	18 20	▭	▭	▭	17 41
64	16 35	17 24	18 19	▭	▭	17 18	16 59
62	16 41	17 27	18 18	15 16	16 08	16 25	16 30
60	16 47	17 29	18 17	14 31	15 24	15 53	16 08
N 58	16 51	17 31	18 17	14 02	14 54	15 29	15 50
56	16 56	17 33	18 16	13 39	14 32	15 10	15 35
54	16 59	17 35	18 16	13 21	14 13	14 53	15 22
52	17 03	17 37	18 16	13 05	13 58	14 39	15 11
50	17 06	17 38	18 16	12 51	13 44	14 27	15 00
45	17 13	17 42	18 16	12 24	13 17	14 01	14 39
N 40	17 18	17 46	18 17	12 02	12 55	13 41	14 21
35	17 23	17 49	18 18	11 44	12 37	13 24	14 07
30	17 28	17 52	18 20	11 28	12 21	13 09	13 54
20	17 35	17 58	18 23	11 02	11 54	12 44	13 32
N 10	17 42	18 03	18 28	10 40	11 31	12 23	13 12
0	17 49	18 10	18 34	10 18	11 10	12 02	12 54
S 10	17 55	18 17	18 42	09 57	10 49	11 42	12 36
20	18 03	18 25	18 51	09 35	10 26	11 20	12 17
30	18 11	18 35	19 04	09 09	09 59	10 55	11 54
35	18 16	18 42	19 13	08 53	09 43	10 40	11 41
40	18 21	18 49	19 23	08 35	09 25	10 22	11 25
45	18 28	18 58	19 35	08 13	09 03	10 01	11 07
S 50	18 36	19 10	19 52	07 46	08 35	09 34	10 44
52	18 39	19 15	19 59	07 33	08 21	09 21	10 33
54	18 43	19 21	20 08	07 17	08 05	09 06	10 20
56	18 48	19 28	20 19	06 59	07 46	08 49	10 06
58	18 53	19 36	20 31	06 37	07 22	08 27	09 48
S 60	18 58	19 44	20 45	06 09	06 51	08 00	09 27

Day	SUN Eqn. of Time 00^h	SUN Eqn. of Time 12^h	SUN Mer. Pass.	MOON Mer. Pass. Upper	MOON Mer. Pass. Lower	MOON Age	MOON Phase
d	m s	m s	h m	h m	h m	d %	
16	14 15	14 22	11 46	04 06	16 32	20 69	◑
17	14 28	14 34	11 45	04 58	17 24	21 60	
18	14 41	14 46	11 45	05 50	18 16	22 50	

	UT	ARIES	VENUS −3.9		MARS −1.5		JUPITER −1.8		SATURN +0.0	
	d h	GHA ° ′	GHA ° ′	Dec ° ′	GHA ° ′	Dec ° ′	GHA ° ′	Dec ° ′	GHA ° ′	Dec ° ′
	19 00	27 03.4	167 55.5	S14 57.3	50 49.4	S13 20.5	224 24.2	N 8 21.9	282 48.1	N22 03.3
	01	42 05.9	182 55.0	58.4	65 51.3	20.1	239 26.2	21.8	297 50.5	03.3
	02	57 08.4	197 54.4	14 59.4	80 53.2	19.7	254 28.3	21.6	312 52.9	03.3
	03	72 10.8	212 53.8	15 00.5	95 55.1 . .	19.3	269 30.3 . .	21.4	327 55.4 . .	03.3
	04	87 13.3	227 53.3	01.5	110 57.0	18.9	284 32.3	21.2	342 57.8	03.3
	05	102 15.7	242 52.7	02.6	125 58.9	18.4	299 34.4	21.1	358 00.2	03.3
	06	117 18.2	257 52.1	S15 03.6	141 00.8	S13 18.0	314 36.4	N 8 20.9	13 02.6	N22 03.3
	07	132 20.7	272 51.6	04.7	156 02.7	17.6	329 38.4	20.7	28 05.1	03.3
	08	147 23.1	287 51.0	05.7	171 04.6	17.2	344 40.5	20.6	43 07.5	03.3
S	09	162 25.6	302 50.4 . .	06.8	186 06.5 . .	16.8	359 42.5 . .	20.4	58 09.9 . .	03.3
U	10	177 28.1	317 49.9	07.8	201 08.3	16.4	14 44.5	20.2	73 12.4	03.3
N	11	192 30.5	332 49.3	08.9	216 10.2	16.0	29 46.6	20.1	88 14.8	03.3
D	12	207 33.0	347 48.7	S15 09.9	231 12.1	S13 15.6	44 48.6	N 8 19.9	103 17.2	N22 03.3
A	13	222 35.5	2 48.1	11.0	246 14.0	15.2	59 50.6	19.7	118 19.7	03.3
Y	14	237 37.9	17 47.6	12.0	261 15.9	14.8	74 52.7	19.6	133 22.1	03.3
	15	252 40.4	32 47.0 . .	13.1	276 17.8 . .	14.4	89 54.7 . .	19.4	148 24.5 . .	03.3
	16	267 42.8	47 46.4	14.1	291 19.7	14.0	104 56.8	19.2	163 27.0	03.3
	17	282 45.3	62 45.9	15.2	306 21.6	13.6	119 58.8	19.1	178 29.4	03.3
	18	297 47.8	77 45.3	S15 16.2	321 23.5	S13 13.2	135 00.8	N 8 18.9	193 31.8	N22 03.3
	19	312 50.2	92 44.7	17.3	336 25.4	12.7	150 02.9	18.7	208 34.3	03.3
	20	327 52.7	107 44.1	18.3	351 27.2	12.3	165 04.9	18.6	223 36.7	03.3
	21	342 55.2	122 43.6 . .	19.4	6 29.1 . .	11.9	180 06.9 . .	18.4	238 39.1 . .	03.3
	22	357 57.6	137 43.0	20.4	21 31.0	11.5	195 09.0	18.2	253 41.6	03.3
	23	13 00.1	152 42.4	21.4	36 32.9	11.1	210 11.0	18.1	268 44.0	03.3
	20 00	28 02.6	167 41.8	S15 22.5	51 34.8	S13 10.7	225 13.0	N 8 17.9	283 46.4	N22 03.3
	01	43 05.0	182 41.2	23.5	66 36.6	10.3	240 15.1	17.7	298 48.9	03.3
	02	58 07.5	197 40.7	24.6	81 38.5	09.9	255 17.1	17.6	313 51.3	03.3
	03	73 10.0	212 40.1 . .	25.6	96 40.4 . .	09.5	270 19.2 . .	17.4	328 53.7 . .	03.3
	04	88 12.4	227 39.5	26.6	111 42.3	09.0	285 21.2	17.2	343 56.2	03.3
	05	103 14.9	242 38.9	27.7	126 44.2	08.6	300 23.2	17.1	358 58.6	03.3
	06	118 17.3	257 38.3	S15 28.7	141 46.0	S13 08.2	315 25.3	N 8 16.9	14 01.0	N22 03.3
	07	133 19.8	272 37.8	29.7	156 47.9	07.8	330 27.3	16.7	29 03.5	03.3
	08	148 22.3	287 37.2	30.8	171 49.8	07.4	345 29.3	16.6	44 05.9	03.3
M	09	163 24.7	302 36.6 . .	31.8	186 51.6 . .	07.0	0 31.4 . .	16.4	59 08.3 . .	03.3
O	10	178 27.2	317 36.0	32.8	201 53.5	06.6	15 33.4	16.2	74 10.8	03.3
N	11	193 29.7	332 35.4	33.9	216 55.4	06.2	30 35.5	16.1	89 13.2	03.3
D	12	208 32.1	347 34.9	S15 34.9	231 57.3	S13 05.7	45 37.5	N 8 15.9	104 15.6	N22 03.3
A	13	223 34.6	2 34.3	35.9	246 59.1	05.3	60 39.5	15.7	119 18.1	03.3
Y	14	238 37.1	17 33.7	37.0	262 01.0	04.9	75 41.6	15.6	134 20.5	03.2
	15	253 39.5	32 33.1 . .	38.0	277 02.9 . .	04.5	90 43.6 . .	15.4	149 23.0 . .	03.2
	16	268 42.0	47 32.5	39.0	292 04.7	04.1	105 45.6	15.3	164 25.4	03.2
	17	283 44.5	62 31.9	40.1	307 06.6	03.7	120 47.7	15.1	179 27.8	03.2
	18	298 46.9	77 31.3	S15 41.1	322 08.4	S13 03.2	135 49.7	N 8 14.9	194 30.3	N22 03.2
	19	313 49.4	92 30.8	42.1	337 10.3	02.8	150 51.8	14.8	209 32.7	03.2
	20	328 51.8	107 30.2	43.1	352 12.2	02.4	165 53.8	14.6	224 35.1	03.2
	21	343 54.3	122 29.6 . .	44.2	7 14.0 . .	02.0	180 55.8 . .	14.4	239 37.6 . .	03.2
	22	358 56.8	137 29.0	45.2	22 15.9	01.6	195 57.9	14.3	254 40.0	03.2
	23	13 59.2	152 28.4	46.2	37 17.7	01.2	210 59.9	14.1	269 42.5	03.2
	21 00	29 01.7	167 27.8	S15 47.2	52 19.6	S13 00.7	226 02.0	N 8 13.9	284 44.9	N22 03.2
	01	44 04.2	182 27.2	48.3	67 21.5	13 00.3	241 04.0	13.8	299 47.3	03.2
	02	59 06.6	197 26.6	49.3	82 23.3	12 59.9	256 06.0	13.6	314 49.8	03.2
	03	74 09.1	212 26.0 . .	50.3	97 25.2 . .	59.5	271 08.1 . .	13.4	329 52.2 . .	03.2
	04	89 11.6	227 25.5	51.3	112 27.0	59.1	286 10.1	13.3	344 54.7	03.2
	05	104 14.0	242 24.9	52.4	127 28.9	58.6	301 12.2	13.1	359 57.1	03.2
	06	119 16.5	257 24.3	S15 53.4	142 30.7	S12 58.2	316 14.2	N 8 12.9	14 59.5	N22 03.2
	07	134 18.9	272 23.7	54.4	157 32.6	57.8	331 16.2	12.8	30 02.0	03.2
T	08	149 21.4	287 23.1	55.4	172 34.4	57.4	346 18.3	12.6	45 04.4	03.2
U	09	164 23.9	302 22.5 . .	56.4	187 36.3 . .	57.0	1 20.3 . .	12.4	60 06.9 . .	03.2
E	10	179 26.3	317 21.9	57.5	202 38.1	56.5	16 22.4	12.3	75 09.3	03.2
S	11	194 28.8	332 21.3	58.5	217 40.0	56.1	31 24.4	12.1	90 11.7	03.2
D	12	209 31.3	347 20.7	S15 59.5	232 41.8	S12 55.7	46 26.5	N 8 11.9	105 14.2	N22 03.2
A	13	224 33.7	2 20.1	16 00.5	247 43.7	55.3	61 28.5	11.8	120 16.6	03.2
Y	14	239 36.2	17 19.5	01.5	262 45.5	54.9	76 30.5	11.6	135 19.1	03.2
	15	254 38.7	32 18.9 . .	02.5	277 47.3 . .	54.4	91 32.6 . .	11.5	150 21.5 . .	03.2
	16	269 41.1	47 18.3	03.5	292 49.2	54.0	106 34.6	11.3	165 23.9	03.2
	17	284 43.6	62 17.7	04.6	307 51.0	53.6	121 36.7	11.1	180 26.4	03.2
	18	299 46.1	77 17.1	S16 05.6	322 52.9	S12 53.2	136 38.7	N 8 11.0	195 28.8	N22 03.2
	19	314 48.5	92 16.5	06.6	337 54.7	52.7	151 40.7	10.8	210 31.3	03.2
	20	329 51.0	107 15.9	07.6	352 56.5	52.3	166 42.8	10.6	225 33.7	03.2
	21	344 53.4	122 15.3 . .	08.6	7 58.4 . .	51.9	181 44.8 . .	10.5	240 36.2 . .	03.2
	22	359 55.9	137 14.7	09.6	23 00.2	51.5	196 46.9	10.3	255 38.6	03.2
	23	14 58.4	152 14.1	10.6	38 02.0	51.0	211 48.9	10.1	270 41.0	03.2
	Mer. Pass.	h m 22 04.2	v −0.6	d 1.0	v 1.9	d 0.4	v 2.0	d 0.2	v 2.4	d 0.0

STARS

Name	SHA ° ′	Dec ° ′
Acamar	315 23.6	S40 17.2
Achernar	335 31.6	S57 13.1
Acrux	173 18.9	S63 07.0
Adhara	255 18.5	S28 58.3
Aldebaran	290 57.9	N16 31.1
Alioth	166 27.7	N55 56.4
Alkaid	153 05.2	N49 17.8
Al Na'ir	27 52.8	S46 56.8
Alnilam	275 53.9	S 1 11.8
Alphard	218 03.7	S 8 40.2
Alphecca	126 17.7	N26 42.3
Alpheratz	357 51.2	N29 06.8
Altair	62 15.7	N 8 52.8
Ankaa	353 22.7	S42 17.2
Antares	112 36.0	S26 26.4
Arcturus	146 03.0	N19 09.9
Atria	107 45.2	S69 02.3
Avior	234 21.4	S59 30.8
Bellatrix	278 40.0	N 6 21.3
Betelgeuse	271 09.4	N 7 24.6
Canopus	263 59.3	S52 41.5
Capella	280 45.5	N46 00.1
Deneb	49 36.7	N45 17.8
Denebola	182 41.7	N14 33.2
Diphda	349 03.2	S17 57.9
Dubhe	194 01.2	N61 43.8
Elnath	278 22.1	N28 36.7
Eltanin	90 49.9	N51 29.5
Enif	33 54.5	N 9 53.0
Fomalhaut	15 32.0	S29 36.2
Gacrux	172 10.3	S57 07.8
Gienah	176 00.5	S17 33.5
Hadar	148 59.7	S60 23.4
Hamal	328 09.1	N23 28.9
Kaus Aust.	83 54.1	S34 23.1
Kochab	137 20.4	N74 08.5
Markab	13 45.7	N15 13.6
Menkar	314 22.7	N 4 06.4
Menkent	148 17.1	S36 23.2
Miaplacidus	221 42.0	S69 43.5
Mirfak	308 51.0	N49 52.5
Nunki	76 07.8	S26 17.7
Peacock	53 31.0	S56 43.7
Pollux	243 37.0	N28 01.1
Procyon	245 07.7	N 5 13.1
Rasalhague	96 13.7	N12 33.5
Regulus	207 51.8	N11 57.1
Rigel	281 19.2	S 8 11.6
Rigil Kent.	140 03.2	S60 51.0
Sabik	102 21.5	S15 43.8
Schedar	349 49.0	N56 33.6
Shaula	96 32.5	S37 06.5
Sirius	258 40.4	S16 43.0
Spica	158 39.7	S11 10.7
Suhail	222 58.3	S43 26.4
Vega	80 44.3	N38 47.4
Zuben'ubi	137 14.2	S16 03.

	SHA ° ′	Mer. Pass. h m
Venus	139 39.3	12 50
Mars	23 32.2	20 31
Jupiter	197 10.5	8 58
Saturn	255 43.9	5 04

UT (d h)	SUN GHA ° ′	SUN Dec ° ′	MOON GHA ° ′	v ′	MOON Dec ° ′	d ′	HP ′
19 00 (SUNDAY)	183 43.1	S 9 44.8	262 59.3	9.6	N24 49.1	6.1	55.8
01	198 43.3	45.7	277 27.9	9.6	24 43.0	6.2	55.9
02	213 43.4	46.6	291 56.5	9.7	24 36.8	6.4	55.9
03	228 43.5	. . 47.5	306 25.2	9.6	24 30.4	6.5	55.9
04	243 43.6	48.4	320 53.8	9.7	24 23.9	6.7	56.0
05	258 43.7	49.3	335 22.5	9.7	24 17.2	6.7	56.0
06	273 43.8	S 9 50.2	349 51.2	9.7	N24 10.5	6.9	56.0
07	288 44.0	51.1	4 19.9	9.7	24 03.6	7.1	56.0
08	303 44.1	52.0	18 48.6	9.8	23 56.5	7.2	56.1
09	318 44.2	. . 52.9	33 17.4	9.8	23 49.3	7.3	56.1
10	333 44.3	53.8	47 46.2	9.8	23 42.0	7.4	56.1
11	348 44.4	54.8	62 15.0	9.8	23 34.6	7.6	56.2
12	3 44.5	S 9 55.7	76 43.8	9.9	N23 27.0	7.6	56.2
13	18 44.7	56.6	91 12.7	9.8	23 19.4	7.9	56.2
14	33 44.8	57.5	105 41.5	9.9	23 11.5	7.9	56.3
15	48 44.9	. . 58.4	120 10.4	10.0	23 03.6	8.1	56.3
16	63 45.0	9 59.3	134 39.4	9.9	22 55.5	8.2	56.4
17	78 45.1	10 00.2	149 08.3	10.0	22 47.3	8.3	56.4
18	93 45.2	S10 01.1	163 37.3	10.0	N22 39.0	8.5	56.4
19	108 45.3	02.0	178 06.3	10.0	22 30.5	8.6	56.5
20	123 45.5	02.9	192 35.3	10.1	22 21.9	8.7	56.5
21	138 45.6	. . 03.8	207 04.4	10.0	22 13.2	8.8	56.5
22	153 45.7	04.7	221 33.4	10.1	22 04.4	9.0	56.6
23	168 45.8	05.6	236 02.5	10.2	21 55.4	9.1	56.6
20 00 (MONDAY)	183 45.9	S10 06.5	250 31.7	10.1	N21 46.3	9.2	56.6
01	198 46.0	07.4	265 00.8	10.2	21 37.1	9.3	56.7
02	213 46.1	08.3	279 30.0	10.2	21 27.8	9.5	56.7
03	228 46.2	. . 09.2	293 59.2	10.3	21 18.3	9.6	56.7
04	243 46.4	10.1	308 28.5	10.2	21 08.7	9.7	56.8
05	258 46.5	11.0	322 57.7	10.3	20 59.0	9.8	56.8
06	273 46.6	S10 11.9	337 27.0	10.3	N20 49.2	9.9	56.9
07	288 46.7	12.8	351 56.3	10.4	20 39.3	10.1	56.9
08	303 46.8	13.7	6 25.7	10.4	20 29.2	10.1	56.9
09	318 46.9	. . 14.6	20 55.1	10.4	20 19.1	10.3	57.0
10	333 47.0	15.5	35 24.5	10.4	20 08.8	10.4	57.0
11	348 47.1	16.4	49 53.9	10.4	19 58.4	10.6	57.0
12	3 47.2	S10 17.3	64 23.3	10.5	N19 47.8	10.6	57.1
13	18 47.3	18.2	78 52.8	10.5	19 37.2	10.8	57.1
14	33 47.5	19.1	93 22.3	10.6	19 26.4	10.8	57.2
15	48 47.6	. . 20.0	107 51.9	10.5	19 15.6	11.0	57.2
16	63 47.7	20.9	122 21.4	10.6	19 04.6	11.1	57.2
17	78 47.8	21.8	136 51.0	10.7	18 53.5	11.2	57.3
18	93 47.9	S10 22.7	151 20.7	10.6	N18 42.3	11.3	57.3
19	108 48.0	23.6	165 50.3	10.7	18 31.0	11.5	57.3
20	123 48.1	24.4	180 20.0	10.7	18 19.5	11.5	57.4
21	138 48.2	. . 25.3	194 49.7	10.7	18 08.0	11.7	57.4
22	153 48.3	26.2	209 19.4	10.7	17 56.3	11.7	57.5
23	168 48.4	27.1	223 49.1	10.8	17 44.6	11.9	57.5
21 00 (TUESDAY)	183 48.5	S10 28.0	238 18.9	10.8	N17 32.7	12.0	57.5
01	198 48.6	28.9	252 48.7	10.8	17 20.7	12.0	57.6
02	213 48.7	29.8	267 18.5	10.9	17 08.7	12.2	57.6
03	228 48.8	. . 30.7	281 48.4	10.9	16 56.5	12.3	57.7
04	243 48.9	31.6	296 18.3	10.9	16 44.2	12.4	57.7
05	258 49.0	32.5	310 48.2	10.9	16 31.8	12.5	57.7
06	273 49.1	S10 33.4	325 18.1	10.9	N16 19.3	12.6	57.8
07	288 49.3	34.3	339 48.0	11.0	16 06.7	12.7	57.8
08	303 49.4	35.2	354 18.0	11.0	15 54.0	12.8	57.9
09	318 49.5	. . 36.1	8 48.0	11.0	15 41.2	12.8	57.9
10	333 49.6	37.0	23 18.0	11.0	15 28.4	13.0	57.9
11	348 49.7	37.9	37 48.0	11.0	15 15.4	13.1	58.0
12	3 49.8	S10 38.7	52 18.0	11.1	N15 02.3	13.2	58.0
13	18 49.9	39.6	66 48.1	11.1	14 49.1	13.3	58.1
14	33 50.0	40.5	81 18.2	11.1	14 35.8	13.3	58.1
15	48 50.1	. . 41.4	95 48.3	11.1	14 22.5	13.5	58.1
16	63 50.2	42.3	110 18.4	11.1	14 09.0	13.6	58.2
17	78 50.3	43.2	124 48.5	11.2	13 55.4	13.6	58.2
18	93 50.4	S10 44.1	139 18.7	11.2	N13 41.8	13.7	58.3
19	108 50.5	45.0	153 48.9	11.2	13 28.1	13.9	58.3
20	123 50.6	45.9	168 19.1	11.2	13 14.2	13.9	58.3
21	138 50.7	. . 46.8	182 49.3	11.2	13 00.3	14.0	58.4
22	153 50.8	47.6	197 19.5	11.2	12 46.3	14.0	58.4
23	168 50.9	48.5	211 49.7	11.2	N12 32.3	14.2	58.5
	SD 16.1	*d* 0.9	SD	15.3		15.6	15.8

Lat. °	Twilight Naut. h m	Twilight Civil h m	Sunrise h m	Moonrise 19 h m	Moonrise 20 h m	Moonrise 21 h m	Moonrise 22 h m
N 72	05 15	06 33	07 47	▭	▭	▭	00 10
N 70	05 16	06 27	07 32	▭	21 54	24 32	00 32
68	05 18	06 22	07 21	▭	22 35	24 50	00 50
66	05 18	06 18	07 11	20 47	23 03	25 04	01 04
64	05 19	06 14	07 03	21 28	23 24	25 15	01 15
62	05 19	06 11	06 56	21 56	23 41	25 25	01 25
60	05 19	06 08	06 51	22 18	23 55	25 33	01 33
N 58	05 19	06 05	06 45	22 35	24 07	00 07	01 40
56	05 19	06 03	06 41	22 50	24 17	00 17	01 47
54	05 19	06 00	06 36	23 02	24 26	00 26	01 52
52	05 19	05 58	06 33	23 13	24 34	00 34	01 57
50	05 19	05 56	06 29	23 23	24 41	00 41	02 02
45	05 18	05 52	06 22	23 44	24 57	00 57	02 12
N 40	05 16	05 48	06 15	24 00	00 00	01 09	02 20
35	05 15	05 44	06 10	24 14	00 14	01 20	02 27
30	05 13	05 41	06 05	24 26	00 26	01 29	02 33
20	05 08	05 34	05 56	24 47	00 47	01 45	02 43
N 10	05 03	05 27	05 49	00 10	01 05	01 59	02 53
0	04 56	05 21	05 42	00 30	01 21	02 12	03 01
S 10	04 48	05 13	05 34	00 49	01 38	02 25	03 10
20	04 37	05 04	05 26	01 10	01 55	02 38	03 19
30	04 24	04 53	05 17	01 33	02 16	02 54	03 29
35	04 15	04 46	05 12	01 47	02 27	03 03	03 35
40	04 04	04 38	05 06	02 03	02 41	03 13	03 42
45	03 51	04 28	04 59	02 23	02 57	03 25	03 50
S 50	03 34	04 16	04 50	02 47	03 16	03 39	03 59
52	03 25	04 10	04 46	02 58	03 25	03 46	04 03
54	03 16	04 04	04 42	03 11	03 35	03 54	04 08
56	03 05	03 56	04 37	03 26	03 47	04 02	04 13
58	02 52	03 48	04 31	03 44	04 00	04 11	04 19
S 60	02 37	03 39	04 25	04 05	04 16	04 22	04 25

Lat. °	Sunset h m	Twilight Civil h m	Twilight Naut. h m	Moonset 19 h m	Moonset 20 h m	Moonset 21 h m	Moonset 22 h m
N 72	15 41	16 54	18 12	▭	▭	17 55	17 10
N 70	15 56	17 01	18 11	▭	18 23	17 29	16 58
68	16 07	17 06	18 10	▭	17 40	17 10	16 49
66	16 17	17 10	18 10	17 41	17 11	16 54	16 41
64	16 25	17 14	18 09	16 59	16 49	16 42	16 35
62	16 32	17 18	18 09	16 30	16 31	16 31	16 29
60	16 38	17 21	18 09	16 08	16 16	16 21	16 24
N 58	16 43	17 24	18 09	15 50	16 04	16 13	16 20
56	16 48	17 26	18 09	15 35	15 53	16 06	16 16
54	16 52	17 29	18 09	15 22	15 43	15 59	16 12
52	16 56	17 31	18 10	15 11	15 34	15 53	16 09
50	17 00	17 33	18 10	15 00	15 26	15 48	16 06
45	17 08	17 37	18 11	14 39	15 10	15 36	15 59
N 40	17 14	17 41	18 13	14 21	14 56	15 26	15 54
35	17 20	17 45	18 15	14 07	14 44	15 18	15 49
30	17 24	17 49	18 17	13 54	14 34	15 10	15 45
20	17 33	17 55	18 21	13 32	14 16	14 58	15 37
N 10	17 41	18 02	18 27	13 12	14 00	14 46	15 31
0	17 48	18 09	18 34	12 54	13 45	14 35	15 25
S 10	17 56	18 17	18 42	12 36	13 31	14 25	15 18
20	18 04	18 26	18 53	12 17	13 15	14 13	15 12
30	18 13	18 38	19 07	11 54	12 56	14 00	15 04
35	18 18	18 45	19 16	11 41	12 45	13 52	14 59
40	18 25	18 53	19 27	11 25	12 33	13 43	14 54
45	18 32	19 03	19 40	11 07	12 18	13 32	14 48
S 50	18 41	19 15	19 58	10 44	12 00	13 19	14 41
52	18 45	19 21	20 06	10 33	11 51	13 13	14 37
54	18 49	19 28	20 16	10 20	11 42	13 07	14 34
56	18 54	19 35	20 27	10 06	11 31	12 59	14 30
58	19 00	19 43	20 40	09 48	11 18	12 51	14 25
S 60	19 06	19 53	20 56	09 27	11 03	12 41	14 20

Day	SUN Eqn. of Time 00^h	SUN Eqn. of Time 12^h	SUN Mer. Pass.	MOON Mer. Pass. Upper	MOON Mer. Pass. Lower	Age	Phase
d	m s	m s	h m	h m	h m	d %	
19	14 52	14 58	11 45	06 42	19 08	23 40	
20	15 03	15 09	11 45	07 33	19 59	24 30	
21	15 14	15 19	11 45	08 24	20 48	25 21	

UT d	UT h	ARIES GHA	VENUS −3.9 GHA	VENUS Dec	MARS −1.4 GHA	MARS Dec	JUPITER −1.8 GHA	JUPITER Dec	SATURN +0.0 GHA	SATURN Dec	STARS Name	SHA	Dec
		° ′	° ′	° ′	° ′	° ′	° ′	° ′	° ′	° ′		° ′	° ′
22	00	30 00.8	167 13.5	S16 11.6	53 03.9	S12 50.6	226 51.0	N 8 10.0	285 43.5	N22 03.2	Acamar	315 23.6	S40 17.2
	01	45 03.3	182 12.9	12.6	68 05.7	50.2	241 53.0	09.8	300 45.9	03.2	Achernar	335 31.6	S57 13.1
	02	60 05.8	197 12.3	13.6	83 07.5	49.8	256 55.0	09.6	315 48.4	03.2	Acrux	173 18.9	S63 07.0
	03	75 08.2	212 11.7	. . 14.6	98 09.4	. . 49.3	271 57.1	. . 09.5	330 50.8	. . 03.2	Adhara	255 18.4	S28 58.3
	04	90 10.7	227 11.1	15.6	113 11.2	48.9	286 59.1	09.3	345 53.3	03.2	Aldebaran	290 57.9	N16 31.1
	05	105 13.2	242 10.5	16.6	128 13.0	48.5	302 01.2	09.2	0 55.7	03.2			
W	06	120 15.6	257 09.9	S16 17.7	143 14.9	S12 48.0	317 03.2	N 8 09.0	15 58.2	N22 03.2	Alioth	166 27.6	N55 56.4
E	07	135 18.1	272 09.3	18.7	158 16.7	47.6	332 05.3	08.8	31 00.6	03.2	Alkaid	153 05.2	N49 17.8
D	08	150 20.6	287 08.7	19.7	173 18.5	47.2	347 07.3	08.7	46 03.0	03.2	Al Na'ir	27 52.9	S46 56.8
N	09	165 23.0	302 08.1	. . 20.7	188 20.3	. . 46.8	2 09.4	. . 08.5	61 05.5	. . 03.2	Alnilam	275 53.9	S 1 11.8
E	10	180 25.5	317 07.5	21.7	203 22.2	46.3	17 11.4	08.3	76 07.9	03.2	Alphard	218 03.7	S 8 40.2
S	11	195 27.9	332 06.9	22.7	218 24.0	45.9	32 13.4	08.2	91 10.4	03.2			
D	12	210 30.4	347 06.2	S16 23.7	233 25.8	S12 45.5	47 15.5	N 8 08.0	106 12.8	N22 03.2	Alphecca	126 17.8	N26 42.3
A	13	225 32.9	2 05.6	24.7	248 27.6	45.0	62 17.5	07.8	121 15.3	03.2	Alpheratz	357 51.2	N29 06.8
Y	14	240 35.3	17 05.0	25.7	263 29.5	44.6	77 19.6	07.7	136 17.7	03.2	Altair	62 15.7	N 8 52.7
	15	255 37.8	32 04.4	. . 26.6	278 31.3	. . 44.2	92 21.6	. . 07.5	151 20.2	. . 03.2	Ankaa	353 22.7	S42 17.2
	16	270 40.3	47 03.8	27.6	293 33.1	43.8	107 23.7	07.4	166 22.6	03.2	Antares	112 36.0	S26 26.4
	17	285 42.7	62 03.2	28.6	308 34.9	43.3	122 25.7	07.2	181 25.1	03.2			
	18	300 45.2	77 02.6	S16 29.6	323 36.7	S12 42.9	137 27.8	N 8 07.0	196 27.5	N22 03.2	Arcturus	146 03.0	N19 09.9
	19	315 47.7	92 02.0	30.6	338 38.5	42.5	152 29.8	06.9	211 30.0	03.2	Atria	107 45.2	S69 02.3
	20	330 50.1	107 01.4	31.6	353 40.4	42.0	167 31.8	06.7	226 32.4	03.2	Avior	234 21.3	S59 30.8
	21	345 52.6	122 00.8	. . 32.6	8 42.2	. . 41.6	182 33.9	. . 06.5	241 34.9	. . 03.2	Bellatrix	278 40.0	N 6 21.3
	22	0 55.0	137 00.1	33.6	23 44.0	41.2	197 35.9	06.4	256 37.3	03.2	Betelgeuse	271 09.4	N 7 24.6
	23	15 57.5	151 59.5	34.6	38 45.8	40.7	212 38.0	06.2	271 39.8	03.2			
23	00	31 00.0	166 58.9	S16 35.6	53 47.6	S12 40.3	227 40.0	N 8 06.1	286 42.2	N22 03.2	Canopus	263 59.3	S52 41.5
	01	46 02.4	181 58.3	36.6	68 49.4	39.9	242 42.1	05.9	301 44.6	03.2	Capella	280 45.5	N46 00.1
	02	61 04.9	196 57.7	37.6	83 51.2	39.4	257 44.1	05.7	316 47.1	03.2	Deneb	49 36.7	N45 17.8
	03	76 07.4	211 57.1	. . 38.6	98 53.0	. . 39.0	272 46.2	. . 05.6	331 49.5	. . 03.2	Denebola	182 41.7	N14 33.2
	04	91 09.8	226 56.4	39.5	113 54.8	38.6	287 48.2	05.4	346 52.0	03.2	Diphda	349 03.2	S17 57.9
	05	106 12.3	241 55.8	40.5	128 56.6	38.1	302 50.3	05.2	1 54.4	03.2			
	06	121 14.8	256 55.2	S16 41.5	143 58.5	S12 37.7	317 52.3	N 8 05.1	16 56.9	N22 03.2	Dubhe	194 01.1	N61 43.8
T	07	136 17.2	271 54.6	42.5	159 00.3	37.3	332 54.4	04.9	31 59.3	03.2	Elnath	278 22.0	N28 36.7
H	08	151 19.7	286 54.0	43.5	174 02.1	36.8	347 56.4	04.8	47 01.8	03.2	Eltanin	90 50.0	N51 29.5
U	09	166 22.2	301 53.4	. . 44.5	189 03.9	. . 36.4	2 58.5	. . 04.6	62 04.2	. . 03.2	Enif	33 54.5	N 9 53.6
R	10	181 24.6	316 52.7	45.5	204 05.7	36.0	18 00.5	04.4	77 06.7	03.2	Fomalhaut	15 32.0	S29 36.3
S	11	196 27.1	331 52.1	46.4	219 07.5	35.5	33 02.6	04.3	92 09.2	03.2			
D	12	211 29.5	346 51.5	S16 47.4	234 09.3	S12 35.1	48 04.6	N 8 04.1	107 11.6	N22 03.2	Gacrux	172 10.3	S57 07.8
A	13	226 32.0	1 50.9	48.4	249 11.1	34.6	63 06.7	03.9	122 14.1	03.2	Gienah	176 00.5	S17 33.5
Y	14	241 34.5	16 50.3	49.4	264 12.9	34.2	78 08.7	03.8	137 16.5	03.2	Hadar	148 59.7	S60 23.4
	15	256 36.9	31 49.6	. . 50.4	279 14.7	. . 33.8	93 10.7	. . 03.6	152 19.0	. . 03.2	Hamal	328 09.1	N23 28.9
	16	271 39.4	46 49.0	51.3	294 16.5	33.3	108 12.8	03.5	167 21.4	03.2	Kaus Aust.	83 54.1	S34 23.1
	17	286 41.9	61 48.4	52.3	309 18.3	32.9	123 14.8	03.3	182 23.9	03.2			
	18	301 44.3	76 47.8	S16 53.3	324 20.1	S12 32.5	138 16.9	N 8 03.1	197 26.3	N22 03.2	Kochab	137 20.4	N74 08.5
	19	316 46.8	91 47.1	54.3	339 21.8	32.0	153 18.9	03.0	212 28.8	03.2	Markab	13 45.8	N15 13.6
	20	331 49.3	106 46.5	55.2	354 23.6	31.6	168 21.0	02.8	227 31.2	03.2	Menkar	314 22.7	N 4 06.4
	21	346 51.7	121 45.9	. . 56.2	9 25.4	. . 31.1	183 23.0	. . 02.6	242 33.7	. . 03.2	Menkent	148 17.1	S36 23.2
	22	1 54.2	136 45.3	57.2	24 27.2	30.7	198 25.1	02.5	257 36.1	03.2	Miaplacidus	221 41.9	S69 43.5
	23	16 56.6	151 44.6	58.2	39 29.0	30.3	213 27.1	02.3	272 38.6	03.2			
24	00	31 59.1	166 44.0	S16 59.1	54 30.8	S12 29.8	228 29.2	N 8 02.2	287 41.0	N22 03.2	Mirfak	308 51.0	N49 52.5
	01	47 01.6	181 43.4	17 00.1	69 32.6	29.4	243 31.2	02.0	302 43.5	03.2	Nunki	76 07.9	S26 17.7
	02	62 04.0	196 42.8	01.1	84 34.4	28.9	258 33.3	01.8	317 45.9	03.2	Peacock	53 31.0	S56 43.7
	03	77 06.5	211 42.1	. . 02.0	99 36.2	. . 28.5	273 35.3	. . 01.7	332 48.4	. . 03.2	Pollux	243 37.0	N28 01.1
	04	92 09.0	226 41.5	03.0	114 37.9	28.1	288 37.4	01.5	347 50.9	03.2	Procyon	245 07.7	N 5 13.1
	05	107 11.4	241 40.9	04.0	129 39.7	27.6	303 39.4	01.4	2 53.3	03.2			
	06	122 13.9	256 40.2	S17 05.0	144 41.5	S12 27.2	318 41.5	N 8 01.2	17 55.8	N22 03.2	Rasalhague	96 13.7	N12 33.5
	07	137 16.4	271 39.6	05.9	159 43.3	26.7	333 43.5	01.0	32 58.2	03.2	Regulus	207 51.8	N11 57.1
	08	152 18.8	286 39.0	06.9	174 45.1	26.3	348 45.6	00.9	48 00.7	03.2	Rigel	281 19.2	S 8 11.6
F	09	167 21.3	301 38.3	. . 07.9	189 46.9	. . 25.8	3 47.6	. . 00.7	63 03.1	. . 03.2	Rigil Kent.	140 03.2	S60 51.0
R	10	182 23.8	316 37.7	08.8	204 48.6	25.4	18 49.7	00.5	78 05.6	03.2	Sabik	102 21.5	S15 43.8
I	11	197 26.2	331 37.1	09.8	219 50.4	25.0	33 51.7	00.4	93 08.0	03.2			
D	12	212 28.7	346 36.4	S17 10.7	234 52.2	S12 24.5	48 53.8	N 8 00.2	108 10.5	N22 03.2	Schedar	349 49.0	N56 33.6
A	13	227 31.1	1 35.8	11.7	249 54.0	24.1	63 55.9	8 00.1	123 13.0	03.2	Shaula	96 32.5	S37 06.5
Y	14	242 33.6	16 35.2	12.7	264 55.8	23.6	78 57.9	7 59.9	138 15.4	03.2	Sirius	258 40.4	S16 43.0
	15	257 36.1	31 34.5	. . 13.6	279 57.5	. . 23.2	94 00.0	. . 59.7	153 17.9	. . 03.2	Spica	158 39.7	S11 10.7
	16	272 38.5	46 33.9	14.6	294 59.3	22.7	109 02.0	59.6	168 20.3	03.2	Suhail	222 58.3	S43 26.4
	17	287 41.0	61 33.3	15.5	310 01.1	22.3	124 04.1	59.4	183 22.8	03.2			
	18	302 43.5	76 32.6	S17 16.5	325 02.8	S12 21.8	139 06.1	N 7 59.3	198 25.2	N22 03.2	Vega	80 44.3	N38 47.4
	19	317 45.9	91 32.0	17.5	340 04.6	21.4	154 08.2	59.1	213 27.7	03.2	Zuben'ubi	137 14.2	S16 03.3
	20	332 48.4	106 31.4	18.4	355 06.4	21.0	169 10.2	58.9	228 30.2	03.2		SHA	Mer. Pass
	21	347 50.9	121 30.7	. . 19.4	10 08.2	. . 20.5	184 12.3	. . 58.8	243 32.6	. . 03.2		° ′	h m
	22	2 53.3	136 30.1	20.3	25 09.9	20.1	199 14.3	58.6	258 35.1	03.2	Venus	135 58.9	12 53
	23	17 55.8	151 29.4	21.3	40 11.7	19.6	214 16.4	58.5	273 37.5	03.2	Mars	22 47.6	20 22
		h m									Jupiter	196 40.1	8 48
Mer. Pass.		21 52.4	*v* −0.6	*d* 1.0	*v* 1.8	*d* 0.4	*v* 2.0	*d* 0.2	*v* 2.5	*d* 0.0	Saturn	255 42.2	4 52

UT d h	SUN GHA ° ′	SUN Dec ° ′	MOON GHA ° ′	MOON *v* ′	MOON Dec ° ′	MOON *d* ′	MOON HP ′
22 00 (WEDNESDAY)	183 51.0	S10 49.4	226 19.9	11.3	N12 18.1	14.3	58.5
01	198 51.1	50.3	240 50.2	11.3	12 03.8	14.3	58.5
02	213 51.2	51.2	255 20.5	11.2	11 49.5	14.4	58.6
03	228 51.3	. . 52.1	269 50.7	11.3	11 35.1	14.5	58.6
04	243 51.4	53.0	284 21.0	11.3	11 20.6	14.6	58.7
05	258 51.5	53.9	298 51.3	11.3	11 06.0	14.6	58.7
06	273 51.6	S10 54.7	313 21.6	11.3	N10 51.4	14.8	58.7
07	288 51.6	55.6	327 51.9	11.3	10 36.6	14.8	58.8
08	303 51.7	56.5	342 22.2	11.3	10 21.8	14.8	58.8
09	318 51.8	. . 57.4	356 52.5	11.4	10 07.0	15.0	58.9
10	333 51.9	58.3	11 22.9	11.3	9 52.0	15.0	58.9
11	348 52.0	10 59.2	25 53.2	11.3	9 37.0	15.1	58.9
12	3 52.1	S11 00.1	40 23.5	11.4	N 9 21.9	15.2	59.0
13	18 52.2	00.9	54 53.9	11.3	9 06.7	15.2	59.0
14	33 52.3	01.8	69 24.2	11.3	8 51.5	15.3	59.0
15	48 52.4	. . 02.7	83 54.5	11.4	8 36.2	15.3	59.1
16	63 52.5	03.6	98 24.9	11.3	8 20.9	15.5	59.1
17	78 52.6	04.5	112 55.2	11.3	8 05.4	15.4	59.2
18	93 52.7	S11 05.4	127 25.5	11.4	N 7 50.0	15.6	59.2
19	108 52.8	06.2	141 55.9	11.3	7 34.4	15.6	59.2
20	123 52.9	07.1	156 26.2	11.3	7 18.8	15.7	59.3
21	138 53.0	. . 08.0	170 56.5	11.3	7 03.1	15.7	59.3
22	153 53.1	08.9	185 26.8	11.3	6 47.4	15.8	59.3
23	168 53.2	09.8	199 57.1	11.3	6 31.6	15.8	59.4
23 00 (THURSDAY)	183 53.2	S11 10.7	214 27.4	11.3	N 6 15.8	15.9	59.4
01	198 53.3	11.5	228 57.7	11.3	5 59.9	16.0	59.5
02	213 53.4	12.4	243 28.0	11.2	5 43.9	16.0	59.5
03	228 53.5	. . 13.3	257 58.2	11.3	5 27.9	16.0	59.5
04	243 53.6	14.2	272 28.5	11.2	5 11.9	16.1	59.6
05	258 53.7	15.1	286 58.7	11.3	4 55.8	16.1	59.6
06	273 53.8	S11 15.9	301 29.0	11.2	N 4 39.7	16.2	59.6
07	288 53.9	16.8	315 59.2	11.2	4 23.5	16.2	59.7
08	303 54.0	17.7	330 29.4	11.1	4 07.3	16.3	59.7
09	318 54.1	. . 18.6	344 59.5	11.2	3 51.0	16.3	59.7
10	333 54.1	19.5	359 29.7	11.1	3 34.7	16.3	59.8
11	348 54.2	20.3	13 59.8	11.2	3 18.4	16.4	59.8
12	3 54.3	S11 21.2	28 30.0	11.1	N 3 02.0	16.5	59.8
13	18 54.4	22.1	43 00.1	11.0	2 45.5	16.4	59.9
14	33 54.5	23.0	57 30.1	11.1	2 29.1	16.5	59.9
15	48 54.6	. . 23.8	72 00.2	11.0	2 12.6	16.5	59.9
16	63 54.7	24.7	86 30.2	11.0	1 56.1	16.6	60.0
17	78 54.8	25.6	101 00.2	11.0	1 39.5	16.5	60.0
18	93 54.8	S11 26.5	115 30.2	10.9	N 1 23.0	16.6	60.0
19	108 54.9	27.3	130 00.1	10.9	1 06.4	16.6	60.1
20	123 55.0	28.2	144 30.0	10.9	0 49.8	16.7	60.1
21	138 55.1	. . 29.1	158 59.9	10.8	0 33.1	16.7	60.1
22	153 55.2	30.0	173 29.7	10.9	N 0 16.4	16.6	60.2
23	168 55.3	30.8	187 59.6	10.7	S 0 00.2	16.7	60.2
24 00 (FRIDAY)	183 55.3	S11 31.7	202 29.3	10.8	S 0 16.9	16.8	60.2
01	198 55.4	32.6	216 59.1	10.7	0 33.7	16.7	60.3
02	213 55.5	33.5	231 28.8	10.7	0 50.4	16.7	60.3
03	228 55.6	. . 34.3	245 58.5	10.6	1 07.1	16.8	60.3
04	243 55.7	35.2	260 28.1	10.6	1 23.9	16.8	60.3
05	258 55.8	36.1	274 57.7	10.5	1 40.7	16.7	60.4
06	273 55.8	S11 37.0	289 27.2	10.6	S 1 57.4	16.8	60.4
07	288 55.9	37.8	303 56.8	10.4	2 14.2	16.8	60.4
08	303 56.0	38.7	318 26.2	10.4	2 31.0	16.8	60.4
09	318 56.1	. . 39.6	332 55.6	10.4	2 47.8	16.8	60.5
10	333 56.2	40.5	347 25.0	10.4	3 04.6	16.7	60.5
11	348 56.3	41.3	1 54.4	10.2	3 21.3	16.8	60.5
12	3 56.3	S11 42.2	16 23.6	10.3	S 3 38.1	16.8	60.5
13	18 56.4	43.1	30 52.9	10.2	3 54.9	16.8	60.6
14	33 56.5	43.9	45 22.1	10.1	4 11.7	16.7	60.6
15	48 56.6	. . 44.8	59 51.2	10.1	4 28.4	16.8	60.6
16	63 56.7	45.7	74 20.3	10.0	4 45.2	16.7	60.6
17	78 56.7	46.5	88 49.3	10.0	5 01.9	16.7	60.7
18	93 56.8	S11 47.4	103 18.3	9.9	S 5 18.6	16.7	60.7
19	108 56.9	48.3	117 47.2	9.9	5 35.3	16.7	60.7
20	123 57.0	49.1	132 16.1	9.8	5 52.0	16.6	60.7
21	138 57.0	. . 50.0	146 44.9	9.7	6 08.6	16.7	60.8
22	153 57.1	50.9	161 13.6	9.7	6 25.3	16.6	60.8
23	168 57.2	51.8	175 42.3	9.6	S 6 41.9	16.5	60.8
	SD 16.1	*d* 0.9	SD 16.1		16.3		16.5

Lat. °	Twilight Naut. h m	Twilight Civil h m	Sunrise h m	Moonrise 22 h m	Moonrise 23 h m	Moonrise 24 h m	Moonrise 25 h m
N 72	05 28	06 47	08 02	00 10	02 40	05 03	07 35
N 70	05 28	06 39	07 46	00 32	02 48	05 01	07 20
68	05 28	06 33	07 33	00 50	02 55	04 58	07 08
66	05 27	06 27	07 22	01 04	03 00	04 57	06 58
64	05 27	06 23	07 13	01 15	03 05	04 55	06 50
62	05 27	06 18	07 05	01 25	03 09	04 54	06 43
60	05 26	06 15	06 58	01 33	03 12	04 53	06 37
N 58	05 26	06 11	06 52	01 40	03 15	04 52	06 32
56	05 25	06 08	06 47	01 47	03 18	04 51	06 27
54	05 25	06 06	06 42	01 52	03 20	04 50	06 23
52	05 24	06 03	06 38	01 57	03 22	04 49	06 19
50	05 23	06 01	06 34	02 02	03 24	04 49	06 16
45	05 21	05 56	06 26	02 12	03 28	04 47	06 08
N 40	05 19	05 51	06 18	02 20	03 32	04 46	06 02
35	05 17	05 47	06 12	02 27	03 35	04 45	05 57
30	05 15	05 43	06 07	02 33	03 38	04 44	05 52
20	05 09	05 35	05 57	02 43	03 42	04 43	05 45
N 10	05 03	05 28	05 49	02 53	03 47	04 41	05 38
0	04 55	05 20	05 41	03 01	03 50	04 40	05 32
S 10	04 46	05 11	05 33	03 10	03 54	04 39	05 25
20	04 35	05 02	05 24	03 19	03 58	04 38	05 19
30	04 20	04 49	05 14	03 29	04 03	04 36	05 11
35	04 11	04 42	05 08	03 35	04 06	04 36	05 07
40	03 59	04 33	05 01	03 42	04 09	04 35	05 02
45	03 45	04 23	04 54	03 50	04 12	04 34	04 57
S 50	03 26	04 09	04 44	03 59	04 16	04 33	04 50
52	03 17	04 03	04 40	04 03	04 18	04 32	04 47
54	03 07	03 56	04 35	04 08	04 20	04 32	04 44
56	02 55	03 48	04 29	04 13	04 22	04 31	04 41
58	02 41	03 39	04 23	04 19	04 25	04 31	04 37
S 60	02 24	03 29	04 17	04 25	04 28	04 30	04 32

Lat. °	Sunset h m	Twilight Civil h m	Twilight Naut. h m	Moonset 22 h m	Moonset 23 h m	Moonset 24 h m	Moonset 25 h m
N 72	15 25	16 40	17 59	17 10	16 35	16 01	15 20
N 70	15 41	16 48	17 59	16 58	16 33	16 08	15 38
68	15 55	16 55	17 59	16 49	16 31	16 13	15 53
66	16 06	17 00	18 00	16 41	16 30	16 18	16 05
64	16 15	17 05	18 00	16 35	16 28	16 22	16 15
62	16 23	17 09	18 01	16 29	16 27	16 25	16 23
60	16 30	17 13	18 01	16 24	16 26	16 28	16 31
N 58	16 36	17 16	18 02	16 20	16 25	16 31	16 38
56	16 41	17 19	18 03	16 16	16 24	16 33	16 43
54	16 46	17 22	18 03	16 12	16 24	16 36	16 49
52	16 50	17 25	18 04	16 09	16 23	16 38	16 54
50	16 54	17 27	18 05	16 06	16 22	16 39	16 58
45	17 03	17 33	18 07	15 59	16 21	16 43	17 07
N 40	17 10	17 37	18 09	15 54	16 20	16 47	17 15
35	17 16	17 42	18 11	15 49	16 19	16 50	17 22
30	17 21	17 46	18 14	15 45	16 18	16 52	17 28
20	17 31	17 53	18 19	15 37	16 17	16 57	17 39
N 10	17 40	18 01	18 26	15 31	16 15	17 00	17 48
0	17 48	18 09	18 33	15 25	16 14	17 04	17 57
S 10	17 56	18 17	18 43	15 18	16 12	17 08	18 06
20	18 05	18 28	18 54	15 12	16 11	17 12	18 15
30	18 15	18 40	19 09	15 04	16 09	17 16	18 26
35	18 21	18 47	19 19	14 59	16 08	17 19	18 32
40	18 28	18 56	19 31	14 54	16 07	17 22	18 39
45	18 36	19 07	19 45	14 48	16 06	17 25	18 48
S 50	18 46	19 21	20 04	14 41	16 04	17 29	18 58
52	18 50	19 27	20 13	14 37	16 03	17 31	19 02
54	18 55	19 34	20 24	14 34	16 02	17 33	19 08
56	19 01	19 42	20 36	14 30	16 01	17 36	19 13
58	19 07	19 51	20 50	14 25	16 00	17 38	19 20
S 60	19 14	20 02	21 08	14 20	15 59	17 41	19 27

Day	SUN Eqn. of Time 00^h	SUN Eqn. of Time 12^h	SUN Mer. Pass.	MOON Mer. Pass. Upper	MOON Mer. Pass. Lower	MOON Age	MOON Phase
d	m s	m s	h m	h m	h m	d %	
22	15 24	15 28	11 45	09 13	21 37	26 13	●
23	15 33	15 37	11 44	10 02	22 27	27 6	
24	15 41	15 45	11 44	10 52	23 18	28 2	

	UT	ARIES	VENUS −3.9		MARS −1.3		JUPITER −1.8		SATURN +0.0		STARS		
		GHA	GHA	Dec	GHA	Dec	GHA	Dec	GHA	Dec	Name	SHA	Dec
	d h	° ′	° ′	° ′	° ′	° ′	° ′	° ′	° ′	° ′		° ′	° ′
	25 00	32 58.2	166 28.8	S17 22.2	55 13.5	S12 19.2	229 18.4	N 7 58.3	288 40.0	N22 03.2	Acamar	315 23.6	S40 17.3
	01	48 00.7	181 28.2	23.2	70 15.2	18.7	244 20.5	58.1	303 42.5	03.2	Achernar	335 31.6	S57 13.1
	02	63 03.2	196 27.5	24.1	85 17.0	18.3	259 22.5	58.0	318 44.9	03.2	Acrux	173 18.8	S63 06.9
	03	78 05.6	211 26.9	. . 25.1	100 18.8	. . 17.8	274 24.6	. . 57.8	333 47.4	. . 03.2	Adhara	255 18.4	S28 58.3
	04	93 08.1	226 26.2	26.1	115 20.5	17.4	289 26.6	57.7	348 49.8	03.2	Aldebaran	290 57.9	N16 31.1
	05	108 10.6	241 25.6	27.0	130 22.3	16.9	304 28.7	57.5	3 52.3	03.2			
	06	123 13.0	256 24.9	S17 28.0	145 24.1	S12 16.5	319 30.8	N 7 57.3	18 54.8	N22 03.2	Alioth	166 27.6	N55 56.4
	07	138 15.5	271 24.3	28.9	160 25.8	16.0	334 32.8	57.2	33 57.2	03.2	Alkaid	153 05.2	N49 17.7
S	08	153 18.0	286 23.7	29.8	175 27.6	15.6	349 34.9	57.0	48 59.7	03.2	Al Na'ir	27 52.9	S46 56.8
A	09	168 20.4	301 23.0	. . 30.8	190 29.3	. . 15.1	4 36.9	. . 56.9	64 02.1	. . 03.2	Alnilam	275 53.9	S 1 11.8
T	10	183 22.9	316 22.4	31.7	205 31.1	14.7	19 39.0	56.7	79 04.6	03.2	Alphard	218 03.7	S 8 40.2
U	11	198 25.4	331 21.7	32.7	220 32.8	14.2	34 41.0	56.5	94 07.1	03.2			
R	12	213 27.8	346 21.1	S17 33.6	235 34.6	S12 13.8	49 43.1	N 7 56.4	109 09.5	N22 03.2	Alphecca	126 17.8	N26 42.2
D	13	228 30.3	1 20.4	34.6	250 36.4	13.3	64 45.1	56.2	124 12.0	03.2	Alpheratz	357 51.2	N29 06.8
A	14	243 32.7	16 19.8	35.5	265 38.1	12.9	79 47.2	56.1	139 14.4	03.2	Altair	62 15.7	N 8 52.7
Y	15	258 35.2	31 19.1	. . 36.5	280 39.9	. . 12.4	94 49.2	. . 55.9	154 16.9	. . 03.2	Ankaa	353 22.7	S42 17.2
	16	273 37.7	46 18.5	37.4	295 41.6	12.0	109 51.3	55.7	169 19.4	03.2	Antares	112 36.0	S26 26.4
	17	288 40.1	61 17.8	38.3	310 43.4	11.5	124 53.4	55.6	184 21.8	03.2			
	18	303 42.6	76 17.2	S17 39.3	325 45.1	S12 11.1	139 55.4	N 7 55.4	199 24.3	N22 03.2	Arcturus	146 03.0	N19 09.9
	19	318 45.1	91 16.5	40.2	340 46.9	10.6	154 57.5	55.3	214 26.8	03.2	Atria	107 45.2	S69 02.3
	20	333 47.5	106 15.9	41.2	355 48.6	10.2	169 59.5	55.1	229 29.2	03.2	Avior	234 21.3	S59 30.8
	21	348 50.0	121 15.2	. . 42.1	10 50.4	. . 09.7	185 01.6	. . 54.9	244 31.7	. . 03.2	Bellatrix	278 40.0	N 6 21.3
	22	3 52.5	136 14.6	43.0	25 52.1	09.3	200 03.6	54.8	259 34.1	03.2	Betelgeuse	271 09.4	N 7 24.6
	23	18 54.9	151 13.9	44.0	40 53.9	08.8	215 05.7	54.6	274 36.6	03.2			
	26 00	33 57.4	166 13.3	S17 44.9	55 55.6	S12 08.3	230 07.8	N 7 54.5	289 39.1	N22 03.2	Canopus	263 59.3	S52 41.5
	01	48 59.9	181 12.6	45.8	70 57.4	07.9	245 09.8	54.3	304 41.5	03.2	Capella	280 45.4	N46 00.1
	02	64 02.3	196 12.0	46.8	85 59.1	07.4	260 11.9	54.2	319 44.0	03.2	Deneb	49 36.7	N45 17.8
	03	79 04.8	211 11.3	. . 47.7	101 00.8	. . 07.0	275 13.9	. . 54.0	334 46.5	. . 03.2	Denebola	182 41.7	N14 33.2
	04	94 07.2	226 10.7	48.6	116 02.6	06.5	290 16.0	53.8	349 48.9	03.2	Diphda	349 03.2	S17 58.0
	05	109 09.7	241 10.0	49.6	131 04.3	06.1	305 18.0	53.7	4 51.4	03.2			
	06	124 12.2	256 09.4	S17 50.5	146 06.1	S12 05.6	320 20.1	N 7 53.5	19 53.9	N22 03.2	Dubhe	194 01.1	N61 43.8
	07	139 14.6	271 08.7	51.4	161 07.8	05.2	335 22.2	53.4	34 56.3	03.2	Elnath	278 22.0	N28 36.7
	08	154 17.1	286 08.0	52.4	176 09.6	04.7	350 24.2	53.2	49 58.8	03.2	Eltanin	90 50.0	N51 29.5
S	09	169 19.6	301 07.4	. . 53.3	191 11.3	. . 04.2	5 26.3	. . 53.0	65 01.3	. . 03.2	Enif	33 54.5	N 9 53.6
U	10	184 22.0	316 06.7	54.2	206 13.0	03.8	20 28.3	52.9	80 03.7	03.2	Fomalhaut	15 32.0	S29 36.3
N	11	199 24.5	331 06.1	55.1	221 14.8	03.3	35 30.4	52.7	95 06.2	03.2			
D	12	214 27.0	346 05.4	S17 56.1	236 16.5	S12 02.9	50 32.4	N 7 52.6	110 08.7	N22 03.2	Gacrux	172 10.2	S57 07.8
A	13	229 29.4	1 04.8	57.0	251 18.2	02.4	65 34.5	52.4	125 11.1	03.2	Gienah	176 00.5	S17 33.5
Y	14	244 31.9	16 04.1	57.9	266 20.0	02.0	80 36.6	52.2	140 13.6	03.2	Hadar	148 59.7	S60 23.4
	15	259 34.3	31 03.4	. . 58.8	281 21.7	. . 01.5	95 38.6	. . 52.1	155 16.1	. . 03.2	Hamal	328 09.1	N23 28.9
	16	274 36.8	46 02.8	17 59.8	296 23.4	01.0	110 40.7	51.9	170 18.5	03.2	Kaus Aust.	83 54.1	S34 23.1
	17	289 39.3	61 02.1	18 00.7	311 25.2	00.6	125 42.7	51.8	185 21.0	03.2			
	18	304 41.7	76 01.4	S18 01.6	326 26.9	S12 00.1	140 44.8	N 7 51.6	200 23.5	N22 03.2	Kochab	137 20.4	N74 08.5
	19	319 44.2	91 00.8	02.5	341 28.6	11 59.7	155 46.9	51.5	215 25.9	03.2	Markab	13 45.8	N15 13.6
	20	334 46.7	106 00.1	03.4	356 30.4	59.2	170 48.9	51.3	230 28.4	03.2	Menkar	314 22.7	N 4 06.4
	21	349 49.1	120 59.5	. . 04.4	11 32.1	. . 58.7	185 51.0	. . 51.1	245 30.9	. . 03.2	Menkent	148 17.1	S36 23.2
	22	4 51.6	135 58.8	05.3	26 33.8	58.3	200 53.0	51.0	260 33.3	03.2	Miaplacidus	221 41.9	S69 43.5
	23	19 54.1	150 58.1	06.2	41 35.5	57.8	215 55.1	50.8	275 35.8	03.2			
	27 00	34 56.5	165 57.5	S18 07.1	56 37.3	S11 57.4	230 57.2	N 7 50.7	290 38.3	N22 03.2	Mirfak	308 51.0	N49 52.5
	01	49 59.0	180 56.8	08.0	71 39.0	56.9	245 59.2	50.5	305 40.7	03.2	Nunki	76 07.9	S26 17.7
	02	65 01.5	195 56.1	08.9	86 40.7	56.4	261 01.3	50.4	320 43.2	03.2	Peacock	53 31.0	S56 43.7
	03	80 03.9	210 55.5	. . 09.8	101 42.4	. . 56.0	276 03.3	. . 50.2	335 45.7	. . 03.2	Pollux	243 37.0	N28 01.1
	04	95 06.4	225 54.8	10.8	116 44.2	55.5	291 05.4	50.0	350 48.1	03.2	Procyon	245 07.6	N 5 13.1
	05	110 08.8	240 54.1	11.7	131 45.9	55.1	306 07.5	49.9	5 50.6	03.2			
	06	125 11.3	255 53.5	S18 12.6	146 47.6	S11 54.6	321 09.5	N 7 49.7	20 53.1	N22 03.2	Rasalhague	96 13.7	N12 33.5
	07	140 13.8	270 52.8	13.5	161 49.3	54.1	336 11.6	49.6	35 55.6	03.2	Regulus	207 51.7	N11 57.1
	08	155 16.2	285 52.1	14.4	176 51.0	53.7	351 13.7	49.4	50 58.0	03.2	Rigel	281 19.1	S 8 11.6
M	09	170 18.7	300 51.4	. . 15.3	191 52.8	. . 53.2	6 15.7	. . 49.3	66 00.5	. . 03.2	Rigil Kent.	140 03.2	S60 51.0
O	10	185 21.2	315 50.8	16.2	206 54.5	52.7	21 17.8	49.1	81 03.0	03.2	Sabik	102 21.6	S15 43.8
N	11	200 23.6	330 50.1	17.1	221 56.2	52.3	36 19.8	48.9	96 05.4	03.2			
D	12	215 26.1	345 49.4	S18 18.0	236 57.9	S11 51.8	51 21.9	N 7 48.8	111 07.9	N22 03.2	Schedar	349 49.0	N56 33.6
A	13	230 28.6	0 48.8	18.9	251 59.6	51.3	66 24.0	48.6	126 10.4	03.2	Shaula	96 32.6	S37 06.5
Y	14	245 31.0	15 48.1	19.8	267 01.3	50.9	81 26.0	48.5	141 12.9	03.2	Sirius	258 40.3	S16 43.0
	15	260 33.5	30 47.4	. . 20.7	282 03.0	. . 50.4	96 28.1	. . 48.3	156 15.3	. . 03.2	Spica	158 39.7	S11 10.7
	16	275 36.0	45 46.7	21.6	297 04.8	50.0	111 30.1	48.2	171 17.8	03.2	Suhail	222 58.3	S43 26.4
	17	290 38.4	60 46.1	22.5	312 06.5	49.5	126 32.2	48.0	186 20.3	03.2			
	18	305 40.9	75 45.4	S18 23.4	327 08.2	S11 49.0	141 34.3	N 7 47.8	201 22.7	N22 03.2	Vega	80 44.3	N38 47.4
	19	320 43.3	90 44.7	24.3	342 09.9	48.6	156 36.3	47.7	216 25.2	03.2	Zuben'ubi	137 14.2	S16 03.3
	20	335 45.8	105 44.0	25.2	357 11.6	48.1	171 38.4	47.5	231 27.7	03.2		SHA	Mer. Pass
	21	350 48.3	120 43.4	. . 26.1	12 13.3	. . 47.6	186 40.5	. . 47.4	246 30.2	. . 03.2		° ′	h m
	22	5 50.7	135 42.7	27.0	27 15.0	47.2	201 42.5	47.2	261 32.6	03.2	Venus	132 15.9	12 56
	23	20 53.2	150 42.0	27.9	42 16.7	46.7	216 44.6	47.1	276 35.1	03.2	Mars	21 58.2	20 14
		h m									Jupiter	196 10.4	8 38
	Mer. Pass.	21 40.6	*v* −0.7	*d* 0.9	*v* 1.7	*d* 0.5	*v* 2.1	*d* 0.2	*v* 2.5	*d* 0.0	Saturn	255 41.7	4 41

UT	SUN GHA	SUN Dec	MOON GHA	v	MOON Dec	d	HP
d h	° ′	° ′	° ′	′	° ′	′	′
25 00	183 57.3	S11 52.6	190 10.9	9.6	S 6 58.4	16.6	60.8
01	198 57.4	53.5	204 39.5	9.5	7 15.0	16.5	60.8
02	213 57.4	54.4	219 08.0	9.4	7 31.5	16.5	60.8
03	228 57.5	. . 55.2	233 36.4	9.4	7 48.0	16.4	60.9
04	243 57.6	56.1	248 04.8	9.3	8 04.4	16.5	60.9
05	258 57.7	56.9	262 33.1	9.2	8 20.9	16.3	60.9
06	273 57.7	S11 57.8	277 01.3	9.2	S 8 37.2	16.4	60.9
07	288 57.8	58.7	291 29.5	9.1	8 53.6	16.2	60.9
S 08	303 57.9	11 59.5	305 57.6	9.0	9 09.8	16.3	60.9
A 09	318 58.0	12 00.4	320 25.6	9.0	9 26.1	16.2	61.0
T 10	333 58.0	01.3	334 53.6	8.9	9 42.3	16.1	61.0
U 11	348 58.1	02.1	349 21.5	8.8	9 58.4	16.1	61.0
R 12	3 58.2	S12 03.0	3 49.3	8.8	S10 14.5	16.0	61.0
D 13	18 58.2	03.9	18 17.1	8.6	10 30.5	16.0	61.0
A 14	33 58.3	04.7	32 44.7	8.6	10 46.5	15.9	61.0
Y 15	48 58.4	. . 05.6	47 12.3	8.6	11 02.4	15.9	61.0
16	63 58.5	06.4	61 39.9	8.4	11 18.3	15.8	61.1
17	78 58.5	07.3	76 07.3	8.4	11 34.1	15.7	61.1
18	93 58.6	S12 08.2	90 34.7	8.3	S11 49.8	15.7	61.1
19	108 58.7	09.0	105 02.0	8.2	12 05.5	15.6	61.1
20	123 58.7	09.9	119 29.2	8.1	12 21.1	15.5	61.1
21	138 58.8	. . 10.8	133 56.3	8.0	12 36.6	15.5	61.1
22	153 58.9	11.6	148 23.3	8.0	12 52.1	15.4	61.1
23	168 59.0	12.5	162 50.3	7.9	13 07.5	15.3	61.1
26 00	183 59.0	S12 13.3	177 17.2	7.8	S13 22.8	15.2	61.1
01	198 59.1	14.2	191 44.0	7.7	13 38.0	15.1	61.1
02	213 59.2	15.0	206 10.7	7.7	13 53.1	15.1	61.1
03	228 59.2	. . 15.9	220 37.4	7.5	14 08.2	14.9	61.1
04	243 59.3	16.8	235 03.9	7.5	14 23.1	14.9	61.1
05	258 59.4	17.6	249 30.4	7.4	14 38.0	14.8	61.1
06	273 59.4	S12 18.5	263 56.8	7.3	S14 52.8	14.7	61.1
07	288 59.5	19.3	278 23.1	7.2	15 07.5	14.6	61.2
08	303 59.6	20.2	292 49.3	7.2	15 22.1	14.5	61.2
S 09	318 59.6	. . 21.1	307 15.5	7.0	15 36.6	14.4	61.2
U 10	333 59.7	21.9	321 41.5	7.0	15 51.0	14.3	61.2
N 11	348 59.8	22.8	336 07.5	6.8	16 05.3	14.2	61.2
D 12	3 59.8	S12 23.6	350 33.3	6.8	S16 19.5	14.1	61.2
A 13	18 59.9	24.5	4 59.1	6.7	16 33.6	14.0	61.2
Y 14	34 00.0	25.3	19 24.8	6.6	16 47.6	13.8	61.2
15	49 00.0	. . 26.2	33 50.4	6.6	17 01.4	13.8	61.2
16	64 00.1	27.0	48 16.0	6.4	17 15.2	13.6	61.2
17	79 00.2	27.9	62 41.4	6.4	17 28.8	13.6	61.1
18	94 00.2	S12 28.7	77 06.8	6.2	S17 42.4	13.4	61.1
19	109 00.3	29.6	91 32.0	6.2	17 55.8	13.3	61.1
20	124 00.3	30.4	105 57.2	6.1	18 09.1	13.1	61.1
21	139 00.4	. . 31.3	120 22.3	6.0	18 22.2	13.1	61.1
22	154 00.5	32.2	134 47.3	5.9	18 35.3	12.9	61.1
23	169 00.5	33.0	149 12.2	5.8	18 48.2	12.8	61.1
27 00	184 00.6	S12 33.9	163 37.0	5.8	S19 01.0	12.7	61.1
01	199 00.7	34.7	178 01.8	5.6	19 13.7	12.5	61.1
02	214 00.7	35.6	192 26.4	5.6	19 26.2	12.4	61.1
03	229 00.8	. . 36.4	206 51.0	5.5	19 38.6	12.2	61.1
04	244 00.8	37.3	221 15.5	5.4	19 50.8	12.2	61.1
05	259 00.9	38.1	235 39.9	5.3	20 03.0	11.9	61.1
06	274 01.0	S12 39.0	250 04.2	5.3	S20 14.9	11.9	61.1
07	289 01.0	39.8	264 28.5	5.1	20 26.8	11.7	61.1
08	304 01.1	40.7	278 52.6	5.1	20 38.5	11.5	61.0
M 09	319 01.1	. . 41.5	293 16.7	5.0	20 50.0	11.4	61.0
O 10	334 01.2	42.4	307 40.7	4.9	21 01.4	11.3	61.0
N 11	349 01.3	43.2	322 04.6	4.8	21 12.7	11.1	61.0
D 12	4 01.3	S12 44.0	336 28.4	4.8	S21 23.8	10.9	61.0
A 13	19 01.4	44.9	350 52.2	4.7	21 34.7	10.8	61.0
Y 14	34 01.4	45.7	5 15.9	4.6	21 45.5	10.6	61.0
15	49 01.5	. . 46.6	19 39.5	4.5	21 56.1	10.5	61.0
16	64 01.5	47.4	34 03.0	4.5	22 06.6	10.3	60.9
17	79 01.6	48.3	48 26.5	4.3	22 16.9	10.2	60.9
18	94 01.7	S12 49.1	62 49.8	4.3	S22 27.1	10.0	60.9
19	109 01.7	50.0	77 13.1	4.3	22 37.1	9.8	60.9
20	124 01.8	50.8	91 36.4	4.2	22 46.9	9.6	60.9
21	139 01.8	. . 51.7	105 59.6	4.0	22 56.5	9.5	60.9
22	154 01.9	52.5	120 22.6	4.1	23 06.0	9.4	60.8
23	169 01.9	53.3	134 45.7	3.9	S23 15.4	9.1	60.8
	SD 16.1	*d* 0.9	SD	16.6		16.7	16.6

Lat.	Twilight Naut.	Twilight Civil	Sunrise	Moonrise 25	Moonrise 26	Moonrise 27	Moonrise 28
°	h m	h m	h m	h m	h m	h m	h m
N 72	05 40	07 00	08 19	07 35	10 49	■■■■	■■■■
N 70	05 39	06 51	08 00	07 20	10 02	■■■■	■■■■
68	05 38	06 43	07 44	07 08	09 32	13 00	■■■■
66	05 36	06 37	07 32	06 58	09 10	11 43	■■■■
64	05 35	06 31	07 22	06 50	08 52	11 06	13 44
62	05 34	06 26	07 13	06 43	08 38	10 40	12 46
60	05 33	06 22	07 06	06 37	08 26	10 20	12 12
N 58	05 32	06 18	06 59	06 32	08 16	10 03	11 48
56	05 31	06 14	06 53	06 27	08 07	09 49	11 28
54	05 30	06 11	06 48	06 23	07 59	09 37	11 12
52	05 29	06 08	06 43	06 19	07 52	09 26	10 58
50	05 28	06 06	06 39	06 16	07 46	09 17	10 46
45	05 25	05 59	06 30	06 08	07 32	08 57	10 21
N 40	05 22	05 54	06 22	06 02	07 21	08 41	10 01
35	05 19	05 49	06 15	05 57	07 12	08 28	09 44
30	05 17	05 45	06 09	05 52	07 03	08 16	09 30
20	05 10	05 36	05 59	05 45	06 49	07 57	09 05
N 10	05 03	05 28	05 49	05 38	06 37	07 40	08 44
0	04 55	05 19	05 41	05 32	06 26	07 24	08 25
S 10	04 45	05 10	05 32	05 25	06 15	07 08	08 06
20	04 33	04 59	05 22	05 19	06 03	06 51	07 45
30	04 17	04 46	05 11	05 11	05 49	06 32	07 22
35	04 07	04 38	05 05	05 07	05 42	06 21	07 08
40	03 54	04 29	04 57	05 02	05 33	06 09	06 52
45	03 39	04 17	04 49	04 57	05 23	05 54	06 33
S 50	03 19	04 03	04 38	04 50	05 11	05 36	06 10
52	03 10	03 56	04 34	04 47	05 05	05 28	05 59
54	02 58	03 49	04 28	04 44	04 59	05 18	05 46
56	02 45	03 40	04 22	04 41	04 52	05 08	05 32
58	02 30	03 31	04 16	04 37	04 45	04 56	05 15
S 60	02 11	03 19	04 08	04 32	04 36	04 42	04 55

Lat.	Sunset	Twilight Civil	Twilight Naut.	Moonset 25	Moonset 26	Moonset 27	Moonset 28
°	h m	h m	h m	h m	h m	h m	h m
N 72	15 08	16 26	17 46	15 20	14 03	■■■■	■■■■
N 70	15 27	16 36	17 47	15 38	14 52	■■■■	■■■■
68	15 42	16 43	17 49	15 53	15 24	13 59	■■■■
66	15 55	16 50	17 50	16 05	15 48	15 17	■■■■
64	16 05	16 56	17 51	16 15	16 07	15 55	15 27
62	16 14	17 01	17 53	16 23	16 22	16 22	16 26
60	16 21	17 05	17 54	16 31	16 35	16 43	17 00
N 58	16 28	17 09	17 55	16 38	16 47	17 00	17 25
56	16 34	17 13	17 56	16 43	16 56	17 15	17 45
54	16 39	17 16	17 57	16 49	17 05	17 28	18 01
52	16 44	17 19	17 58	16 54	17 13	17 39	18 16
50	16 48	17 22	17 59	16 58	17 20	17 49	18 28
45	16 58	17 28	18 02	17 07	17 35	18 10	18 54
N 40	17 06	17 33	18 05	17 15	17 48	18 27	19 15
35	17 13	17 38	18 08	17 22	17 59	18 42	19 32
30	17 19	17 43	18 11	17 28	18 08	18 54	19 47
20	17 29	17 52	18 18	17 39	18 25	19 16	20 13
N 10	17 39	18 00	18 25	17 48	18 39	19 35	20 35
0	17 47	18 09	18 33	17 57	18 53	19 53	20 55
S 10	17 56	18 18	18 43	18 06	19 06	20 10	21 16
20	18 06	18 29	18 56	18 15	19 21	20 29	21 38
30	18 17	18 42	19 12	18 26	19 38	20 51	22 03
35	18 24	18 50	19 22	18 32	19 48	21 04	22 18
40	18 31	19 00	19 35	18 39	19 59	21 19	22 36
45	18 40	19 12	19 50	18 48	20 12	21 37	22 57
S 50	18 51	19 26	20 10	18 58	20 29	21 59	23 24
52	18 56	19 33	20 20	19 02	20 36	22 10	23 37
54	19 01	19 41	20 32	19 08	20 45	22 22	23 52
56	19 07	19 49	20 45	19 13	20 55	22 36	24 09
58	19 14	19 59	21 01	19 20	21 06	22 52	24 31
S 60	19 22	20 11	21 21	19 27	21 18	23 12	24 58

Day	SUN Eqn. of Time 00^h	SUN Eqn. of Time 12^h	SUN Mer. Pass.	MOON Mer. Pass. Upper	MOON Mer. Pass. Lower	MOON Age	MOON Phase
d	m s	m s	h m	h m	h m	d	%
25	15 49	15 53	11 44	11 44	24 11	29	0
26	15 56	15 59	11 44	12 39	00 11	01	1
27	16 02	16 05	11 44	13 38	01 08	02	6

	UT	ARIES	VENUS −3.9		MARS −1.3		JUPITER −1.8		SATURN +0.0	
		GHA	GHA	Dec	GHA	Dec	GHA	Dec	GHA	Dec
	d h	° ′	° ′	° ′	° ′	° ′	° ′	° ′	° ′	° ′
TUESDAY	28 00	35 55.7	165 41.3	S18 28.8	57 18.4	S11 46.2	231 46.7	N 7 46.9	291 37.6	N22 03.2
	01	50 58.1	180 40.7	29.7	72 20.1	45.8	246 48.7	46.8	306 40.1	03.2
	02	66 00.6	195 40.0	30.6	87 21.8	45.3	261 50.8	46.6	321 42.5	03.2
	03	81 03.1	210 39.3	. . 31.5	102 23.5	. . 44.8	276 52.9	. . 46.4	336 45.0	. . 03.2
	04	96 05.5	225 38.6	32.4	117 25.2	44.3	291 54.9	46.3	351 47.5	03.2
	05	111 08.0	240 37.9	33.3	132 26.9	43.9	306 57.0	46.1	6 50.0	03.2
	06	126 10.5	255 37.2	S18 34.2	147 28.6	S11 43.4	321 59.0	N 7 46.0	21 52.4	N22 03.2
	07	141 12.9	270 36.6	35.1	162 30.3	42.9	337 01.1	45.8	36 54.9	03.2
	08	156 15.4	285 35.9	36.0	177 32.0	42.5	352 03.2	45.7	51 57.4	03.2
	09	171 17.8	300 35.2	. . 36.8	192 33.7	. . 42.0	7 05.2	. . 45.5	66 59.9	. . 03.2
	10	186 20.3	315 34.5	37.7	207 35.4	41.5	22 07.3	45.3	82 02.3	03.2
	11	201 22.8	330 33.8	38.6	222 37.1	41.1	37 09.4	45.2	97 04.8	03.2
	12	216 25.2	345 33.1	S18 39.5	237 38.8	S11 40.6	52 11.4	N 7 45.0	112 07.3	N22 03.3
	13	231 27.7	0 32.5	40.4	252 40.5	40.1	67 13.5	44.9	127 09.8	03.3
	14	246 30.2	15 31.8	41.3	267 42.2	39.6	82 15.6	44.7	142 12.2	03.3
	15	261 32.6	30 31.1	. . 42.2	282 43.9	. . 39.2	97 17.6	. . 44.6	157 14.7	. . 03.3
	16	276 35.1	45 30.4	43.0	297 45.6	38.7	112 19.7	44.4	172 17.2	03.3
	17	291 37.6	60 29.7	43.9	312 47.3	38.2	127 21.8	44.3	187 19.7	03.3
	18	306 40.0	75 29.0	S18 44.8	327 49.0	S11 37.8	142 23.8	N 7 44.1	202 22.2	N22 03.3
	19	321 42.5	90 28.3	45.7	342 50.7	37.3	157 25.9	44.0	217 24.6	03.3
	20	336 44.9	105 27.7	46.6	357 52.3	36.8	172 28.0	43.8	232 27.1	03.3
	21	351 47.4	120 27.0	. . 47.4	12 54.0	. . 36.3	187 30.0	. . 43.6	247 29.6	. . 03.3
	22	6 49.9	135 26.3	48.3	27 55.7	35.9	202 32.1	43.5	262 32.1	03.3
	23	21 52.3	150 25.6	49.2	42 57.4	35.4	217 34.2	43.3	277 34.5	03.3
WEDNESDAY	29 00	36 54.8	165 24.9	S18 50.1	57 59.1	S11 34.9	232 36.2	N 7 43.2	292 37.0	N22 03.3
	01	51 57.3	180 24.2	50.9	73 00.8	34.4	247 38.3	43.0	307 39.5	03.3
	02	66 59.7	195 23.5	51.8	88 02.5	34.0	262 40.4	42.9	322 42.0	03.3
	03	82 02.2	210 22.8	. . 52.7	103 04.1	. . 33.5	277 42.4	. . 42.7	337 44.5	. . 03.3
	04	97 04.7	225 22.1	53.6	118 05.8	33.0	292 44.5	42.6	352 46.9	03.3
	05	112 07.1	240 21.4	54.4	133 07.5	32.5	307 46.6	42.4	7 49.4	03.3
	06	127 09.6	255 20.7	S18 55.3	148 09.2	S11 32.1	322 48.7	N 7 42.2	22 51.9	N22 03.3
	07	142 12.1	270 20.0	56.2	163 10.9	31.6	337 50.7	42.1	37 54.4	03.3
	08	157 14.5	285 19.3	57.0	178 12.5	31.1	352 52.8	41.9	52 56.9	03.3
	09	172 17.0	300 18.6	. . 57.9	193 14.2	. . 30.6	7 54.9	. . 41.8	67 59.3	. . 03.3
	10	187 19.4	315 17.9	58.8	208 15.9	30.2	22 56.9	41.6	83 01.8	03.3
	11	202 21.9	330 17.3	18 59.6	223 17.6	29.7	37 59.0	41.5	98 04.3	03.3
	12	217 24.4	345 16.6	S19 00.5	238 19.3	S11 29.2	53 01.1	N 7 41.3	113 06.8	N22 03.3
	13	232 26.8	0 15.9	01.4	253 20.9	28.7	68 03.1	41.2	128 09.3	03.3
	14	247 29.3	15 15.2	02.2	268 22.6	28.3	83 05.2	41.0	143 11.8	03.3
	15	262 31.8	30 14.5	. . 03.1	283 24.3	. . 27.8	98 07.3	. . 40.9	158 14.2	. . 03.3
	16	277 34.2	45 13.8	03.9	298 25.9	27.3	113 09.3	40.7	173 16.7	03.3
	17	292 36.7	60 13.1	04.8	313 27.6	26.8	128 11.4	40.6	188 19.2	03.3
	18	307 39.2	75 12.4	S19 05.7	328 29.3	S11 26.3	143 13.5	N 7 40.4	203 21.7	N22 03.3
	19	322 41.6	90 11.7	06.5	343 31.0	25.9	158 15.6	40.2	218 24.2	03.3
	20	337 44.1	105 11.0	07.4	358 32.6	25.4	173 17.6	40.1	233 26.7	03.3
	21	352 46.6	120 10.2	. . 08.2	13 34.3	. . 24.9	188 19.7	. . 39.9	248 29.1	. . 03.3
	22	7 49.0	135 09.5	09.1	28 36.0	24.4	203 21.8	39.8	263 31.6	03.3
	23	22 51.5	150 08.8	09.9	43 37.6	23.9	218 23.8	39.6	278 34.1	03.3
THURSDAY	30 00	37 53.9	165 08.1	S19 10.8	58 39.3	S11 23.5	233 25.9	N 7 39.5	293 36.6	N22 03.3
	01	52 56.4	180 07.4	11.7	73 41.0	23.0	248 28.0	39.3	308 39.1	03.3
	02	67 58.9	195 06.7	12.5	88 42.6	22.5	263 30.1	39.2	323 41.6	03.3
	03	83 01.3	210 06.0	. . 13.4	103 44.3	. . 22.0	278 32.1	. . 39.0	338 44.0	. . 03.3
	04	98 03.8	225 05.3	14.2	118 46.0	21.5	293 34.2	38.9	353 46.5	03.3
	05	113 06.3	240 04.6	15.1	133 47.6	21.1	308 36.3	38.7	8 49.0	03.3
	06	128 08.7	255 03.9	S19 15.9	148 49.3	S11 20.6	323 38.4	N 7 38.6	23 51.5	N22 03.4
	07	143 11.2	270 03.2	16.7	163 50.9	20.1	338 40.4	38.4	38 54.0	03.4
	08	158 13.7	285 02.5	17.6	178 52.6	19.6	353 42.5	38.3	53 56.5	03.4
	09	173 16.1	300 01.8	. . 18.4	193 54.3	. . 19.1	8 44.6	. . 38.1	68 59.0	. . 03.4
	10	188 18.6	315 01.1	19.3	208 55.9	18.6	23 46.6	38.0	84 01.4	03.4
	11	203 21.1	330 00.4	20.1	223 57.6	18.2	38 48.7	37.8	99 03.9	03.4
	12	218 23.5	344 59.6	S19 21.0	238 59.2	S11 17.7	53 50.8	N 7 37.6	114 06.4	N22 03.4
	13	233 26.0	359 58.9	21.8	254 00.9	17.2	68 52.9	37.5	129 08.9	03.4
	14	248 28.4	14 58.2	22.7	269 02.5	16.7	83 54.9	37.3	144 11.4	03.4
	15	263 30.9	29 57.5	. . 23.5	284 04.2	. . 16.2	98 57.0	. . 37.2	159 13.9	. . 03.4
	16	278 33.4	44 56.8	24.3	299 05.8	15.7	113 59.1	37.0	174 16.4	03.4
	17	293 35.8	59 56.1	25.2	314 07.5	15.3	129 01.2	36.9	189 18.9	03.4
	18	308 38.3	74 55.4	S19 26.0	329 09.2	S11 14.8	144 03.2	N 7 36.7	204 21.3	N22 03.4
	19	323 40.8	89 54.7	26.8	344 10.8	14.3	159 05.3	36.6	219 23.8	03.4
	20	338 43.2	104 53.9	27.7	359 12.5	13.8	174 07.4	36.4	234 26.3	03.4
	21	353 45.7	119 53.2	. . 28.5	14 14.1	. . 13.3	189 09.5	. . 36.3	249 28.8	. . 03.4
	22	8 48.2	134 52.5	29.4	29 15.8	12.8	204 11.5	36.1	264 31.3	03.4
	23	23 50.6	149 51.8	30.2	44 17.4	12.3	219 13.6	36.0	279 33.8	03.4
	Mer. Pass.	h m 21 28.8	v −0.7	d 0.9	v 1.7	d 0.5	v 2.1	d 0.2	v 2.5	d 0.0

STARS

Name	SHA	Dec
	° ′	° ′
Acamar	315 23.6	S40 17.
Achernar	335 31.6	S57 13.
Acrux	173 18.8	S63 06.
Adhara	255 18.4	S28 58.
Aldebaran	290 57.9	N16 31.
Alioth	166 27.6	N55 56.
Alkaid	153 05.2	N49 17.
Al Na'ir	27 52.9	S46 56.
Alnilam	275 53.9	S 1 11.
Alphard	218 03.7	S 8 40.
Alphecca	126 17.8	N26 42.
Alpheratz	357 51.2	N29 06.
Altair	62 15.7	N 8 52.
Ankaa	353 22.7	S42 17.
Antares	112 36.0	S26 26.
Arcturus	146 03.0	N19 09.
Atria	107 45.2	S69 02.
Avior	234 21.3	S59 30.
Bellatrix	278 39.9	N 6 21.
Betelgeuse	271 09.3	N 7 24.
Canopus	263 59.3	S52 41.
Capella	280 45.4	N46 00.
Deneb	49 36.8	N45 17.
Denebola	182 41.6	N14 33.
Diphda	349 03.2	S17 58.
Dubhe	194 01.1	N61 43.
Elnath	278 22.0	N28 36.
Eltanin	90 50.0	N51 29.
Enif	33 54.5	N 9 53.
Fomalhaut	15 32.0	S29 36.
Gacrux	172 10.2	S57 07
Gienah	176 00.5	S17 33.
Hadar	148 59.7	S60 23.
Hamal	328 09.1	N23 28.
Kaus Aust.	83 54.1	S34 23.
Kochab	137 20.5	N74 08.
Markab	13 45.8	N15 13.
Menkar	314 22.7	N 4 06
Menkent	148 17.1	S36 23
Miaplacidus	221 41.8	S69 43
Mirfak	308 50.9	N49 52
Nunki	76 07.9	S26 17
Peacock	53 31.1	S56 43
Pollux	243 36.9	N28 01
Procyon	245 07.6	N 5 13
Rasalhague	96 13.8	N12 33
Regulus	207 51.7	N11 57
Rigel	281 19.1	S 8 11
Rigil Kent.	140 03.2	S60 51
Sabik	102 21.6	S15 43
Schedar	349 49.0	N56 33
Shaula	96 32.6	S37 06
Sirius	258 40.3	S16 43
Spica	158 39.6	S11 10
Suhail	222 58.2	S43 26
Vega	80 44.3	N38 47
Zuben'ubi	137 14.2	S16 03

	SHA	Mer. Pas
	° ′	h m
Venus	128 30.1	12 59
Mars	21 04.3	20 06
Jupiter	195 41.4	8 28
Saturn	255 42.2	4 29

	UT	SUN GHA	SUN Dec	MOON GHA	MOON v	MOON Dec	MOON d	MOON HP
	d h	° ′	° ′	° ′	′	° ′	′	′
	28 00	184 02.0	S12 54.2	149 08.6	4.0	S23 24.5	9.0	60.8
	01	199 02.0	55.0	163 31.6	3.8	23 33.5	8.8	60.8
	02	214 02.1	55.9	177 54.4	3.8	23 42.3	8.7	60.8
	03	229 02.1	. . 56.7	192 17.2	3.7	23 51.0	8.4	60.7
	04	244 02.2	57.6	206 39.9	3.7	23 59.4	8.3	60.7
	05	259 02.3	58.4	221 02.6	3.6	24 07.7	8.1	60.7
	06	274 02.3	S12 59.2	235 25.2	3.5	S24 15.8	7.9	60.7
	07	289 02.4	13 00.1	249 47.7	3.5	24 23.7	7.8	60.7
T	08	304 02.4	00.9	264 10.2	3.5	24 31.5	7.5	60.6
U	09	319 02.5	. . 01.8	278 32.7	3.4	24 39.0	7.4	60.6
E	10	334 02.5	02.6	292 55.1	3.4	24 46.4	7.2	60.6
S	11	349 02.6	03.4	307 17.5	3.3	24 53.6	7.0	60.6
D	12	4 02.6	S13 04.3	321 39.8	3.3	S25 00.6	6.9	60.6
A	13	19 02.7	05.1	336 02.1	3.2	25 07.5	6.6	60.5
Y	14	34 02.7	06.0	350 24.3	3.2	25 14.1	6.4	60.5
	15	49 02.8	. . 06.8	4 46.5	3.2	25 20.5	6.3	60.5
	16	64 02.8	07.6	19 08.7	3.1	25 26.8	6.1	60.5
	17	79 02.9	08.5	33 30.8	3.1	25 32.9	5.9	60.4
	18	94 02.9	S13 09.3	47 52.9	3.1	S25 38.8	5.7	60.4
	19	109 03.0	10.1	62 15.0	3.0	25 44.5	5.5	60.4
	20	124 03.0	11.0	76 37.0	3.0	25 50.0	5.3	60.4
	21	139 03.0	. . 11.8	90 59.0	3.0	25 55.3	5.1	60.3
	22	154 03.1	12.6	105 21.0	3.0	26 00.4	4.9	60.3
	23	169 03.1	13.5	119 43.0	3.0	26 05.3	4.8	60.3
	29 00	184 03.2	S13 14.3	134 05.0	2.9	S26 10.1	4.5	60.3
	01	199 03.2	15.2	148 26.9	3.0	26 14.6	4.4	60.2
	02	214 03.3	16.0	162 48.9	2.9	26 19.0	4.1	60.2
	03	229 03.3	. . 16.8	177 10.8	2.9	26 23.1	4.0	60.2
	04	244 03.4	17.7	191 32.7	3.0	26 27.1	3.8	60.1
	05	259 03.4	18.5	205 54.7	2.9	26 30.9	3.5	60.1
	06	274 03.5	S13 19.3	220 16.6	2.9	S26 34.4	3.4	60.1
W	07	289 03.5	20.2	234 38.5	2.9	26 37.8	3.2	60.1
E	08	304 03.5	21.0	249 00.4	3.0	26 41.0	3.0	60.0
D	09	319 03.6	. . 21.8	263 22.4	2.9	26 44.0	2.8	60.0
N	10	334 03.6	22.6	277 44.3	3.0	26 46.8	2.6	60.0
E	11	349 03.7	23.5	292 06.3	2.9	26 49.4	2.4	59.9
S	12	4 03.7	S13 24.3	306 28.2	3.0	S26 51.8	2.2	59.9
D	13	19 03.8	25.1	320 50.2	3.0	26 54.0	2.0	59.9
A	14	34 03.8	26.0	335 12.2	3.0	26 56.0	1.8	59.8
Y	15	49 03.8	. . 26.8	349 34.2	3.1	26 57.8	1.6	59.8
	16	64 03.9	27.6	3 56.3	3.0	26 59.4	1.5	59.8
	17	79 03.9	28.5	18 18.3	3.1	27 00.9	1.2	59.7
	18	94 04.0	S13 29.3	32 40.4	3.2	S27 02.1	1.1	59.7
	19	109 04.0	30.1	47 02.6	3.1	27 03.2	0.8	59.7
	20	124 04.1	30.9	61 24.7	3.2	27 04.0	0.7	59.7
	21	139 04.1	. . 31.8	75 46.9	3.3	27 04.7	0.4	59.6
	22	154 04.1	32.6	90 09.2	3.3	27 05.1	0.3	59.6
	23	169 04.2	33.4	104 31.5	3.3	27 05.4	0.1	59.6
	30 00	184 04.2	S13 34.2	118 53.8	3.4	S27 05.5	0.1	59.5
	01	199 04.2	35.1	133 16.2	3.4	27 05.4	0.3	59.5
	02	214 04.3	35.9	147 38.6	3.5	27 05.1	0.5	59.5
	03	229 04.3	. . 36.7	162 01.1	3.5	27 04.6	0.6	59.4
	04	244 04.4	37.5	176 23.6	3.6	27 04.0	0.9	59.4
	05	259 04.4	38.4	190 46.2	3.6	27 03.1	1.0	59.4
	06	274 04.4	S13 39.2	205 08.8	3.7	S27 02.1	1.2	59.3
	07	289 04.5	40.0	219 31.5	3.8	27 00.9	1.4	59.3
T	08	304 04.5	40.8	233 54.3	3.8	26 59.5	1.6	59.3
H	09	319 04.5	. . 41.7	248 17.1	3.9	26 57.9	1.8	59.2
U	10	334 04.6	42.5	262 40.0	4.0	26 56.1	2.0	59.2
R	11	349 04.6	43.3	277 03.0	4.1	26 54.1	2.1	59.2
S	12	4 04.6	S13 44.1	291 26.1	4.1	S26 52.0	2.3	59.1
D	13	19 04.7	44.9	305 49.2	4.2	26 49.7	2.5	59.1
A	14	34 04.7	45.8	320 12.4	4.2	26 47.2	2.7	59.1
Y	15	49 04.7	. . 46.6	334 35.6	4.4	26 44.5	2.8	59.0
	16	64 04.8	47.4	348 59.0	4.4	26 41.7	3.0	59.0
	17	79 04.8	48.2	3 22.4	4.5	26 38.7	3.2	59.0
	18	94 04.8	S13 49.0	17 45.9	4.6	S26 35.5	3.4	58.9
	19	109 04.9	49.9	32 09.5	4.7	26 32.1	3.5	58.9
	20	124 04.9	50.7	46 33.2	4.7	26 28.6	3.7	58.8
	21	139 04.9	. . 51.5	60 56.9	4.9	26 24.9	3.9	58.8
	22	154 05.0	52.3	75 20.8	4.9	26 21.0	4.1	58.8
	23	169 05.0	53.1	89 44.7	5.1	S26 16.9	4.2	58.7
		SD 16.1	*d* 0.8	SD	16.5	16.3		16.1

Lat.	Twilight Naut.	Twilight Civil	Sunrise	Moonrise 28	Moonrise 29	Moonrise 30	Moonrise 31
°	h m	h m	h m	h m	h m	h m	h m
N 72	05 52	07 14	08 36	■	■	■	■
N 70	05 50	07 03	08 14	■	■	■	■
68	05 47	06 54	07 57	■	■	■	■
66	05 45	06 46	07 43	■	■	■	■
64	05 43	06 40	07 31	13 44	■	■	16 28
62	05 42	06 34	07 22	12 46	14 38	15 32	15 44
60	05 40	06 29	07 13	12 12	13 48	14 47	15 14
N 58	05 38	06 24	07 06	11 48	13 17	14 17	14 51
56	05 37	06 20	07 00	11 28	12 53	13 55	14 33
54	05 35	06 17	06 54	11 12	12 34	13 36	14 17
52	05 34	06 13	06 49	10 58	12 18	13 20	14 04
50	05 32	06 10	06 44	10 46	12 04	13 07	13 52
45	05 29	06 03	06 34	10 21	11 36	12 39	13 27
N 40	05 25	05 57	06 25	10 01	11 14	12 16	13 07
35	05 22	05 52	06 18	09 44	10 55	11 58	12 51
30	05 18	05 47	06 11	09 30	10 39	11 42	12 37
20	05 11	05 37	06 00	09 05	10 13	11 16	12 12
N 10	05 03	05 28	05 50	08 44	09 50	10 53	11 51
0	04 54	05 19	05 40	08 25	09 28	10 31	11 31
S 10	04 44	05 09	05 31	08 06	09 07	10 10	11 11
20	04 31	04 58	05 20	07 45	08 44	09 47	10 50
30	04 14	04 43	05 08	07 22	08 18	09 20	10 25
35	04 03	04 35	05 01	07 08	08 03	09 04	10 11
40	03 50	04 25	04 53	06 52	07 45	08 46	09 54
45	03 34	04 12	04 44	06 33	07 23	08 24	09 34
S 50	03 12	03 57	04 33	06 10	06 56	07 56	09 08
52	03 02	03 50	04 27	05 59	06 43	07 42	08 56
54	02 50	03 42	04 22	05 46	06 27	07 26	08 41
56	02 36	03 32	04 15	05 32	06 10	07 08	08 25
58	02 19	03 22	04 08	05 15	05 48	06 45	08 05
S 60	01 57	03 09	04 00	04 55	05 21	06 15	07 40

Lat.	Sunset	Twilight Civil	Twilight Naut.	Moonset 28	Moonset 29	Moonset 30	Moonset 31
°	h m	h m	h m	h m	h m	h m	h m
N 72	14 50	16 12	17 33	■	■	■	■
N 70	15 12	16 23	17 36	■	■	■	■
68	15 30	16 33	17 39	■	■	■	■
66	15 44	16 40	17 41	■	■	■	■
64	15 55	16 47	17 43	15 27	■	■	19 11
62	16 05	16 53	17 45	16 26	16 47	18 04	19 55
60	16 13	16 58	17 47	17 00	17 37	18 49	20 24
N 58	16 21	17 02	17 48	17 25	18 08	19 18	20 46
56	16 27	17 06	17 50	17 45	18 32	19 41	21 04
54	16 33	17 10	17 51	18 01	18 51	19 59	21 20
52	16 38	17 13	17 53	18 16	19 07	20 15	21 33
50	16 43	17 17	17 54	18 28	19 21	20 28	21 44
45	16 53	17 24	17 58	18 54	19 50	20 56	22 08
N 40	17 02	17 30	18 02	19 15	20 12	21 18	22 27
35	17 09	17 35	18 05	19 32	20 31	21 35	22 43
30	17 16	17 40	18 09	19 47	20 47	21 51	22 57
20	17 27	17 50	18 16	20 13	21 14	22 17	23 20
N 10	17 38	17 59	18 24	20 35	21 37	22 40	23 40
0	17 47	18 08	18 33	20 55	21 59	23 00	23 58
S 10	17 57	18 19	18 44	21 16	22 20	23 21	24 17
20	18 07	18 30	18 57	21 38	22 43	23 43	24 36
30	18 20	18 45	19 14	22 03	23 10	24 09	00 09
35	18 27	18 53	19 25	22 18	23 26	24 24	00 24
40	18 35	19 04	19 39	22 36	23 44	24 42	00 42
45	18 44	19 16	19 55	22 57	24 07	00 07	01 02
S 50	18 56	19 32	20 17	23 24	24 35	00 35	01 29
52	19 01	19 39	20 27	23 37	24 49	00 49	01 41
54	19 07	19 47	20 40	23 52	25 04	01 04	01 56
56	19 14	19 57	20 54	24 09	00 09	01 23	02 13
58	19 21	20 08	21 12	24 31	00 31	01 46	02 33
S 60	19 30	20 20	21 34	24 58	00 58	02 17	02 58

Day	SUN Eqn. of Time 00^h	SUN Eqn. of Time 12^h	SUN Mer. Pass.	MOON Mer. Pass. Upper	MOON Mer. Pass. Lower	MOON Age	MOON Phase
d	m s	m s	h m	h m	h m	d %	
28	16 08	16 10	11 44	14 40	02 09	03 13	
29	16 13	16 15	11 44	15 44	03 12	04 22	
30	16 17	16 19	11 44	16 46	04 15	05 32	

2003 OCT. 31, NOV. 1, 2 (FRI., SAT., SUN.)

	UT	ARIES	VENUS −3.9		MARS −1.2		JUPITER −1.8		SATURN −0.1		STARS		
		GHA	GHA	Dec	GHA	Dec	GHA	Dec	GHA	Dec	Name	SHA	Dec
	d h	° ′	° ′	° ′	° ′	° ′	° ′	° ′	° ′	° ′		° ′	° ′
	31 00	38 53.1	164 51.1	S19 31.0	59 19.1	S11 11.9	234 15.7	N 7 35.8	294 36.3	N22 03.4	Acamar	315 23.6	S40 17.
	01	53 55.6	179 50.4	31.8	74 20.7	11.4	249 17.8	35.7	309 38.8	03.4	Achernar	335 31.6	S57 13.
	02	68 58.0	194 49.6	32.7	89 22.3	10.9	264 19.8	35.5	324 41.2	03.4	Acrux	173 18.8	S63 06.
	03	84 00.5	209 48.9	. . 33.5	104 24.0	. . 10.4	279 21.9	. . 35.4	339 43.7	. . 03.4	Adhara	255 18.4	S28 58.
	04	99 02.9	224 48.2	34.3	119 25.6	09.9	294 24.0	35.2	354 46.2	03.4	Aldebaran	290 57.8	N16 31.
	05	114 05.4	239 47.5	35.2	134 27.3	09.4	309 26.1	35.1	9 48.7	03.4			
	06	129 07.9	254 46.8	S19 36.0	149 28.9	S11 08.9	324 28.1	N 7 34.9	24 51.2	N22 03.4	Alioth	166 27.6	N55 56.
	07	144 10.3	269 46.0	36.8	164 30.6	08.4	339 30.2	34.8	39 53.7	03.4	Alkaid	153 05.2	N49 17.
	08	159 12.8	284 45.3	37.6	179 32.2	08.0	354 32.3	34.6	54 56.2	03.4	Al Na'ir	27 52.9	S46 56
F	09	174 15.3	299 44.6	. . 38.5	194 33.8	. . 07.5	9 34.4	. . 34.5	69 58.7	. . 03.4	Alnilam	275 53.9	S 1 11.
R	10	189 17.7	314 43.9	39.3	209 35.5	07.0	24 36.4	34.3	85 01.2	03.4	Alphard	218 03.6	S 8 40.
I	11	204 20.2	329 43.2	40.1	224 37.1	06.5	39 38.5	34.2	100 03.7	03.4			
D	12	219 22.7	344 42.4	S19 40.9	239 38.8	S11 06.0	54 40.6	N 7 34.0	115 06.2	N22 03.4	Alphecca	126 17.8	N26 42.
A	13	234 25.1	359 41.7	41.8	254 40.4	05.5	69 42.7	33.9	130 08.6	03.4	Alpheratz	357 51.2	N29 06.
Y	14	249 27.6	14 41.0	42.6	269 42.0	05.0	84 44.8	33.7	145 11.1	03.5	Altair	62 15.7	N 8 52.
	15	264 30.1	29 40.3	. . 43.4	284 43.7	. . 04.5	99 46.8	. . 33.6	160 13.6	. . 03.5	Ankaa	353 22.7	S42 17.
	16	279 32.5	44 39.5	44.2	299 45.3	04.0	114 48.9	33.4	175 16.1	03.5	Antares	112 36.0	S26 26.
	17	294 35.0	59 38.8	45.0	314 46.9	03.5	129 51.0	33.3	190 18.6	03.5			
	18	309 37.4	74 38.1	S19 45.8	329 48.6	S11 03.1	144 53.1	N 7 33.1	205 21.1	N22 03.5	Arcturus	146 03.0	N19 09.
	19	324 39.9	89 37.4	46.7	344 50.2	02.6	159 55.2	33.0	220 23.6	03.5	Atria	107 45.2	S69 02.
	20	339 42.4	104 36.6	47.5	359 51.8	02.1	174 57.2	32.8	235 26.1	03.5	Avior	234 21.2	S59 30.
	21	354 44.8	119 35.9	. . 48.3	14 53.5	. . 01.6	189 59.3	. . 32.7	250 28.6	. . 03.5	Bellatrix	278 39.9	N 6 21.
	22	9 47.3	134 35.2	49.1	29 55.1	01.1	205 01.4	32.5	265 31.1	03.5	Betelgeuse	271 09.3	N 7 24.
	23	24 49.8	149 34.4	49.9	44 56.7	00.6	220 03.5	32.4	280 33.6	03.5			
	1 00	39 52.2	164 33.7	S19 50.7	59 58.4	S11 00.1	235 05.5	N 7 32.2	295 36.1	N22 03.5	Canopus	263 59.2	S52 41
	01	54 54.7	179 33.0	51.5	75 00.0	10 59.6	250 07.6	32.1	310 38.6	03.5	Capella	280 45.4	N46 00
	02	69 57.2	194 32.3	52.3	90 01.6	59.1	265 09.7	31.9	325 41.1	03.5	Deneb	49 36.8	N45 17
	03	84 59.6	209 31.5	. . 53.1	105 03.2	. . 58.6	280 11.8	. . 31.8	340 43.6	. . 03.5	Denebola	182 41.6	N14 33
	04	100 02.1	224 30.8	53.9	120 04.9	58.1	295 13.9	31.6	355 46.0	03.5	Diphda	349 03.2	S17 58
	05	115 04.5	239 30.1	54.7	135 06.5	57.6	310 15.9	31.5	10 48.5	03.5			
	06	130 07.0	254 29.3	S19 55.6	150 08.1	S10 57.1	325 18.0	N 7 31.3	25 51.0	N22 03.5	Dubhe	194 01.0	N61 43.
	07	145 09.5	269 28.6	56.4	165 09.7	56.6	340 20.1	31.2	40 53.5	03.5	Elnath	278 22.0	N28 36.
S	08	160 11.9	284 27.9	57.2	180 11.4	56.2	355 22.2	31.0	55 56.0	03.5	Eltanin	90 50.0	N51 29.
A	09	175 14.4	299 27.1	. . 58.0	195 13.0	. . 55.7	10 24.3	. . 30.9	70 58.5	. . 03.5	Enif	33 54.5	N 9 53
T	10	190 16.9	314 26.4	58.8	210 14.6	55.2	25 26.3	30.7	86 01.0	03.5	Fomalhaut	15 32.0	S29 36
U	11	205 19.3	329 25.7	19 59.6	225 16.2	54.7	40 28.4	30.6	101 03.5	03.5			
R	12	220 21.8	344 24.9	S20 00.4	240 17.9	S10 54.2	55 30.5	N 7 30.4	116 06.0	N22 03.5	Gacrux	172 10.2	S57 07
D	13	235 24.3	359 24.2	01.2	255 19.5	53.7	70 32.6	30.3	131 08.5	03.5	Gienah	176 00.5	S17 33
A	14	250 26.7	14 23.4	02.0	270 21.1	53.2	85 34.7	30.1	146 11.0	03.5	Hadar	148 59.6	S60 23
Y	15	265 29.2	29 22.7	. . 02.7	285 22.7	. . 52.7	100 36.8	. . 30.0	161 13.5	. . 03.5	Hamal	328 09.1	N23 28.
	16	280 31.7	44 22.0	03.5	300 24.3	52.2	115 38.8	29.8	176 16.0	03.5	Kaus Aust.	83 54.1	S34 23
	17	295 34.1	59 21.2	04.3	315 26.0	51.7	130 40.9	29.7	191 18.5	03.6			
	18	310 36.6	74 20.5	S20 05.1	330 27.6	S10 51.2	145 43.0	N 7 29.5	206 21.0	N22 03.6	Kochab	137 20.5	N74 08
	19	325 39.0	89 19.8	05.9	345 29.2	50.7	160 45.1	29.4	221 23.5	03.6	Markab	13 45.8	N15 13
	20	340 41.5	104 19.0	06.7	0 30.8	50.2	175 47.2	29.2	236 26.0	03.6	Menkar	314 22.7	N 4 06
	21	355 44.0	119 18.3	. . 07.5	15 32.4	. . 49.7	190 49.3	. . 29.1	251 28.5	. . 03.6	Menkent	148 17.1	S36 23
	22	10 46.4	134 17.5	08.3	30 34.0	49.2	205 51.3	28.9	266 31.0	03.6	Miaplacidus	221 41.7	S69 43
	23	25 48.9	149 16.8	09.1	45 35.6	48.7	220 53.4	28.8	281 33.5	03.6			
	2 00	40 51.4	164 16.1	S20 09.9	60 37.2	S10 48.2	235 55.5	N 7 28.6	296 36.0	N22 03.6	Mirfak	308 50.9	N49 52
	01	55 53.8	179 15.3	10.7	75 38.9	47.7	250 57.6	28.5	311 38.5	03.6	Nunki	76 07.9	S26 17
	02	70 56.3	194 14.6	11.4	90 40.5	47.2	265 59.7	28.3	326 41.0	03.6	Peacock	53 31.1	S56 43
	03	85 58.8	209 13.8	. . 12.2	105 42.1	. . 46.7	281 01.8	. . 28.2	341 43.5	. . 03.6	Pollux	243 36.9	N28 01
	04	101 01.2	224 13.1	13.0	120 43.7	46.2	296 03.8	28.0	356 46.0	03.6	Procyon	245 07.6	N 5 13
	05	116 03.7	239 12.3	13.8	135 45.3	45.7	311 05.9	27.9	11 48.5	03.6			
	06	131 06.2	254 11.6	S20 14.6	150 46.9	S10 45.2	326 08.0	N 7 27.7	26 51.0	N22 03.6	Rasalhague	96 13.8	N12 33
	07	146 08.6	269 10.8	15.3	165 48.5	44.7	341 10.1	27.6	41 53.5	03.6	Regulus	207 51.7	N11 57
	08	161 11.1	284 10.1	16.1	180 50.1	44.2	356 12.2	27.4	56 56.0	03.6	Rigel	281 19.1	S 8 11
S	09	176 13.5	299 09.3	. . 16.9	195 51.7	. . 43.7	11 14.3	. . 27.3	71 58.5	. . 03.6	Rigil Kent.	140 03.2	S60 51
U	10	191 16.0	314 08.6	17.7	210 53.3	43.2	26 16.3	27.1	87 01.0	03.6	Sabik	102 21.6	S15 43
N	11	206 18.5	329 07.9	18.5	225 54.9	42.7	41 18.4	27.0	102 03.5	03.6			
D	12	221 20.9	344 07.1	S20 19.2	240 56.5	S10 42.2	56 20.5	N 7 26.8	117 06.0	N22 03.6	Schedar	349 49.0	N56 33
A	13	236 23.4	359 06.4	20.0	255 58.1	41.7	71 22.6	26.7	132 08.5	03.6	Shaula	96 32.6	S37 06
Y	14	251 25.9	14 05.6	20.8	270 59.7	41.2	86 24.7	26.5	147 11.0	03.6	Sirius	258 40.3	S16 43
	15	266 28.3	29 04.9	. . 21.6	286 01.3	. . 40.7	101 26.8	. . 26.4	162 13.5	. . 03.6	Spica	158 39.6	S11 10
	16	281 30.8	44 04.1	22.3	301 02.9	40.2	116 28.9	26.2	177 16.0	03.6	Suhail	222 58.2	S43 26
	17	296 33.3	59 03.4	23.1	316 04.5	39.7	131 30.9	26.1	192 18.5	03.7			
	18	311 35.7	74 02.6	S20 23.9	331 06.1	S10 39.2	146 33.0	N 7 26.0	207 21.0	N22 03.7	Vega	80 44.4	N38 47
	19	326 38.2	89 01.8	24.6	346 07.7	38.7	161 35.1	25.8	222 23.5	03.7	Zuben'ubi	137 14.2	S16 03
	20	341 40.6	104 01.1	25.4	1 09.3	38.2	176 37.2	25.7	237 26.0	03.7		SHA	Mer. Pas
	21	356 43.1	119 00.3	. . 26.2	16 10.9	. . 37.7	191 39.3	. . 25.5	252 28.5	. . 03.7		° ′	h m
	22	11 45.6	133 59.6	26.9	31 12.5	37.2	206 41.4	25.4	267 31.0	03.7	Venus	124 41.5	13 02
	23	26 48.0	148 58.8	27.7	46 14.1	36.7	221 43.5	25.2	282 33.5	03.7	Mars	20 06.1	19 58
		h m									Jupiter	195 13.3	8 18
	Mer. Pass.	21 17.0	*v* −0.7	*d* 0.8	*v* 1.6	*d* 0.5	*v* 2.1	*d* 0.1	*v* 2.5	*d* 0.0	Saturn	255 43.8	4 17

UT	SUN GHA	SUN Dec	MOON GHA	v	MOON Dec	d	HP
d h	° ′	° ′	° ′	′	° ′	′	′
31 00	184 05.0	S13 53.9	104 08.8	5.1	S26 12.7	4.3	58.7
01	199 05.1	54.8	118 32.9	5.2	26 08.4	4.6	58.7
02	214 05.1	55.6	132 57.1	5.4	26 03.8	4.7	58.6
03	229 05.1	. . 56.4	147 21.5	5.4	25 59.1	4.8	58.6
04	244 05.2	57.2	161 45.9	5.5	25 54.3	5.1	58.6
05	259 05.2	58.0	176 10.4	5.6	25 49.2	5.1	58.5
06	274 05.2	S13 58.8	190 35.0	5.8	S25 44.1	5.4	58.5
07	289 05.2	13 59.7	204 59.8	5.8	25 38.7	5.4	58.5
08	304 05.3	14 00.5	219 24.6	5.9	25 33.3	5.7	58.4
09	319 05.3	. . 01.3	233 49.5	6.1	25 27.6	5.8	58.4
10	334 05.3	02.1	248 14.6	6.1	25 21.8	5.9	58.4
11	349 05.4	02.9	262 39.7	6.2	25 15.9	6.1	58.3
12	4 05.4	S14 03.7	277 04.9	6.4	S25 09.8	6.2	58.3
13	19 05.4	04.5	291 30.3	6.5	25 03.6	6.4	58.3
14	34 05.4	05.3	305 55.8	6.5	24 57.2	6.5	58.2
15	49 05.5	. . 06.1	320 21.3	6.7	24 50.7	6.7	58.2
16	64 05.5	07.0	334 47.0	6.8	24 44.0	6.8	58.2
17	79 05.5	07.8	349 12.8	6.9	24 37.2	7.0	58.1
18	94 05.5	S14 08.6	3 38.7	7.0	S24 30.2	7.1	58.1
19	109 05.6	09.4	18 04.7	7.2	24 23.1	7.2	58.1
20	124 05.6	10.2	32 30.9	7.2	24 15.9	7.3	58.0
21	139 05.6	. . 11.0	46 57.1	7.4	24 08.6	7.5	58.0
22	154 05.6	11.8	61 23.5	7.4	24 01.1	7.6	58.0
23	169 05.7	12.6	75 49.9	7.6	23 53.5	7.8	57.9
1 00	184 05.7	S14 13.4	90 16.5	7.7	S23 45.7	7.9	57.9
01	199 05.7	14.2	104 43.2	7.8	23 37.8	8.0	57.8
02	214 05.7	15.0	119 10.0	7.9	23 29.8	8.1	57.8
03	229 05.7	. . 15.8	133 36.9	8.1	23 21.7	8.3	57.8
04	244 05.8	16.7	148 04.0	8.1	23 13.4	8.4	57.7
05	259 05.8	17.5	162 31.1	8.3	23 05.0	8.5	57.7
06	274 05.8	S14 18.3	176 58.4	8.4	S22 56.5	8.6	57.7
07	289 05.8	19.1	191 25.8	8.5	22 47.9	8.7	57.6
08	304 05.8	19.9	205 53.3	8.6	22 39.2	8.9	57.6
09	319 05.9	. . 20.7	220 20.9	8.7	22 30.3	8.9	57.6
10	334 05.9	21.5	234 48.6	8.9	22 21.4	9.1	57.5
11	349 05.9	22.3	249 16.5	8.9	22 12.3	9.2	57.5
12	4 05.9	S14 23.1	263 44.4	9.1	S22 03.1	9.3	57.5
13	19 05.9	23.9	278 12.5	9.2	21 53.8	9.4	57.4
14	34 06.0	24.7	292 40.7	9.3	21 44.4	9.5	57.4
15	49 06.0	. . 25.5	307 09.0	9.4	21 34.9	9.6	57.4
16	64 06.0	26.3	321 37.4	9.5	21 25.3	9.8	57.4
17	79 06.0	27.1	336 05.9	9.6	21 15.5	9.8	57.3
18	94 06.0	S14 27.9	350 34.5	9.8	S21 05.7	9.9	57.3
19	109 06.0	28.7	5 03.3	9.8	20 55.8	10.0	57.3
20	124 06.1	29.5	19 32.1	10.0	20 45.8	10.2	57.2
21	139 06.1	. . 30.3	34 01.1	10.1	20 35.6	10.2	57.2
22	154 06.1	31.1	48 30.2	10.2	20 25.4	10.3	57.2
23	169 06.1	31.9	62 59.4	10.3	20 15.1	10.4	57.1
2 00	184 06.1	S14 32.7	77 28.7	10.4	S20 04.7	10.5	57.1
01	199 06.1	33.5	91 58.1	10.5	19 54.2	10.6	57.1
02	214 06.1	34.3	106 27.6	10.6	19 43.6	10.7	57.0
03	229 06.2	. . 35.1	120 57.2	10.8	19 32.9	10.7	57.0
04	244 06.2	35.9	135 27.0	10.8	19 22.2	10.9	57.0
05	259 06.2	36.7	149 56.8	11.0	19 11.3	10.9	56.9
06	274 06.2	S14 37.5	164 26.8	11.0	S19 00.4	11.0	56.9
07	289 06.2	38.2	178 56.8	11.2	18 49.4	11.1	56.9
08	304 06.2	39.0	193 27.0	11.2	18 38.3	11.2	56.8
09	319 06.2	. . 39.8	207 57.2	11.4	18 27.1	11.3	56.8
10	334 06.2	40.6	222 27.6	11.5	18 15.8	11.3	56.8
11	349 06.3	41.4	236 58.1	11.5	18 04.5	11.4	56.8
12	4 06.3	S14 42.2	251 28.6	11.7	S17 53.1	11.5	56.7
13	19 06.3	43.0	265 59.3	11.8	17 41.6	11.6	56.7
14	34 06.3	43.8	280 30.1	11.8	17 30.0	11.6	56.7
15	49 06.3	. . 44.6	295 00.9	12.0	17 18.4	11.7	56.6
16	64 06.3	45.4	309 31.9	12.0	17 06.7	11.8	56.6
17	79 06.3	46.2	324 02.9	12.2	16 54.9	11.8	56.6
18	94 06.3	S14 47.0	338 34.1	12.2	S16 43.1	11.9	56.6
19	109 06.3	47.8	353 05.3	12.4	16 31.2	12.0	56.5
20	124 06.3	48.5	7 36.7	12.4	16 19.2	12.0	56.5
21	139 06.3	. . 49.3	22 08.1	12.6	16 07.2	12.1	56.5
22	154 06.4	50.1	36 39.7	12.6	15 55.1	12.2	56.4
23	169 06.4	50.9	51 11.3	12.7	S15 42.9	12.2	56.4
	SD 16.1	d 0.8	SD 15.9		15.7		15.5

Lat.	Twilight Naut.	Twilight Civil	Sunrise	Moonrise 31	Moonrise 1	Moonrise 2	Moonrise 3
°	h m	h m	h m	h m	h m	h m	h m
N 72	06 04	07 27	08 54	■	■	18 00	16 53
N 70	06 00	07 14	08 28	■	■	17 13	16 32
68	05 57	07 04	08 09	■	17 36	16 42	16 16
66	05 54	06 55	07 53	■	16 43	16 18	16 03
64	05 51	06 48	07 41	16 28	16 10	16 00	15 52
62	05 49	06 42	07 30	15 44	15 45	15 45	15 43
60	05 46	06 36	07 21	15 14	15 26	15 32	15 35
N 58	05 44	06 31	07 13	14 51	15 10	15 21	15 28
56	05 42	06 26	07 06	14 33	14 56	15 11	15 22
54	05 40	06 22	07 00	14 17	14 44	15 03	15 16
52	05 39	06 18	06 54	14 04	14 34	14 55	15 11
50	05 37	06 15	06 49	13 52	14 24	14 48	15 06
45	05 32	06 07	06 38	13 27	14 04	14 33	14 56
N 40	05 28	06 00	06 28	13 07	13 48	14 20	14 47
35	05 24	05 54	06 21	12 51	13 34	14 10	14 40
30	05 21	05 49	06 13	12 37	13 22	14 01	14 34
20	05 12	05 39	06 01	12 12	13 01	13 44	14 23
N 10	05 04	05 29	05 50	11 51	12 43	13 30	14 13
0	04 54	05 19	05 40	11 31	12 27	13 17	14 04
S 10	04 43	05 08	05 30	11 11	12 10	13 04	13 54
20	04 29	04 56	05 19	10 50	11 51	12 50	13 44
30	04 11	04 41	05 06	10 25	11 30	12 33	13 33
35	03 59	04 31	04 58	10 11	11 18	12 23	13 26
40	03 45	04 21	04 50	09 54	11 04	12 13	13 19
45	03 28	04 08	04 40	09 34	10 47	12 00	13 10
S 50	03 05	03 51	04 27	09 08	10 26	11 44	12 59
52	02 54	03 43	04 22	08 56	10 16	11 36	12 54
54	02 41	03 35	04 15	08 41	10 04	11 28	12 49
56	02 26	03 25	04 08	08 25	09 51	11 18	12 42
58	02 07	03 13	04 00	08 05	09 36	11 08	12 36
S 60	01 42	03 00	03 51	07 40	09 18	10 55	12 28

Lat.	Sunset	Twilight Civil	Twilight Naut.	Moonset 31	Moonset 1	Moonset 2	Moonset 3
°	h m	h m	h m	h m	h m	h m	h m
N 72	14 32	15 58	17 21	■	■	21 19	24 02
N 70	14 58	16 11	17 25	■	■	22 05	24 20
68	15 17	16 22	17 29	■	19 58	22 35	24 34
66	15 33	16 31	17 32	■	20 50	22 57	24 46
64	15 45	16 38	17 35	19 11	21 22	23 14	24 55
62	15 56	16 45	17 37	19 55	21 46	23 28	25 03
60	16 05	16 50	17 40	20 24	22 04	23 40	25 10
N 58	16 13	16 55	17 42	20 46	22 20	23 50	25 16
56	16 20	17 00	17 44	21 04	22 33	23 59	25 21
54	16 27	17 04	17 46	21 20	22 44	24 07	00 07
52	16 32	17 08	17 48	21 33	22 54	24 14	00 14
50	16 38	17 12	17 50	21 44	23 03	24 20	00 20
45	16 49	17 19	17 54	22 08	23 22	24 34	00 34
N 40	16 58	17 26	17 58	22 27	23 37	24 45	00 45
35	17 06	17 32	18 02	22 43	23 50	24 54	00 54
30	17 13	17 38	18 06	22 57	24 01	00 01	01 02
20	17 26	17 48	18 15	23 20	24 20	00 20	01 16
N 10	17 37	17 58	18 23	23 40	24 36	00 36	01 28
0	17 47	18 08	18 33	23 58	24 51	00 51	01 40
S 10	17 58	18 19	18 45	24 17	00 17	01 06	01 51
20	18 09	18 32	18 59	24 36	00 36	01 22	02 03
30	18 22	18 47	19 17	00 09	00 59	01 41	02 17
35	18 30	18 56	19 29	00 24	01 12	01 51	02 24
40	18 38	19 07	19 43	00 42	01 27	02 04	02 33
45	18 48	19 21	20 00	01 02	01 45	02 18	02 43
S 50	19 01	19 37	20 23	01 29	02 07	02 35	02 56
52	19 07	19 45	20 35	01 41	02 18	02 43	03 02
54	19 13	19 54	20 48	01 56	02 30	02 52	03 08
56	19 20	20 04	21 04	02 13	02 43	03 02	03 15
58	19 28	20 16	21 24	02 33	02 59	03 14	03 23
S 60	19 38	20 30	21 49	02 58	03 18	03 27	03 31

Day	SUN Eqn. of Time 00^h	SUN Eqn. of Time 12^h	SUN Mer. Pass.	MOON Mer. Pass. Upper	MOON Mer. Pass. Lower	Age	Phase
d	m s	m s	h m	h m	h m	d %	
31	16 20	16 21	11 44	17 45	05 16	06 43	
1	16 23	16 24	11 44	18 39	06 13	07 54	
2	16 24	16 25	11 44	19 29	07 04	08 64	

UT d h	ARIES GHA	VENUS −3.9 GHA	VENUS Dec	MARS −1.1 GHA	MARS Dec	JUPITER −1.9 GHA	JUPITER Dec	SATURN −0.1 GHA	SATURN Dec
	° ′	° ′	° ′	° ′	° ′	° ′	° ′	° ′	° ′
3 00	41 50.5	163 58.1	S20 28.5	61 15.7	S10 36.2	236 45.6	N 7 25.1	297 36.0	N22 03.7
01	56 53.0	178 57.3	29.2	76 17.3	35.7	251 47.6	24.9	312 38.5	03.7
02	71 55.4	193 56.6	30.0	91 18.9	35.2	266 49.7	24.8	327 41.0	03.7
03	86 57.9	208 55.8	. . 30.8	106 20.5	. . 34.7	281 51.8	. . 24.6	342 43.5	. . 03.7
04	102 00.4	223 55.1	31.5	121 22.1	34.2	296 53.9	24.5	357 46.0	03.7
05	117 02.8	238 54.3	32.3	136 23.7	33.7	311 56.0	24.3	12 48.6	03.7
06	132 05.3	253 53.5	S20 33.0	151 25.3	S10 33.1	326 58.1	N 7 24.2	27 51.1	N22 03.7
M 07	147 07.8	268 52.8	33.8	166 26.9	32.6	342 00.2	24.0	42 53.6	03.7
O 08	162 10.2	283 52.0	34.5	181 28.5	32.1	357 02.3	23.9	57 56.1	03.7
N 09	177 12.7	298 51.3	. . 35.3	196 30.0	. . 31.6	12 04.3	. . 23.8	72 58.6	. . 03.7
D 10	192 15.1	313 50.5	36.1	211 31.6	31.1	27 06.4	23.6	88 01.1	03.7
A 11	207 17.6	328 49.7	36.8	226 33.2	30.6	42 08.5	23.5	103 03.6	03.7
Y 12	222 20.1	343 49.0	S20 37.6	241 34.8	S10 30.1	57 10.6	N 7 23.3	118 06.1	N22 03.7
13	237 22.5	358 48.2	38.3	256 36.4	29.6	72 12.7	23.2	133 08.6	03.7
14	252 25.0	13 47.5	39.1	271 38.0	29.1	87 14.8	23.0	148 11.1	03.8
15	267 27.5	28 46.7	. . 39.8	286 39.6	. . 28.6	102 16.9	. . 22.9	163 13.6	. . 03.8
16	282 29.9	43 45.9	40.6	301 41.1	28.1	117 19.0	22.7	178 16.1	03.8
17	297 32.4	58 45.2	41.3	316 42.7	27.6	132 21.1	22.6	193 18.6	03.8
18	312 34.9	73 44.4	S20 42.0	331 44.3	S10 27.1	147 23.2	N 7 22.4	208 21.1	N22 03.8
19	327 37.3	88 43.6	42.8	346 45.9	26.6	162 25.2	22.3	223 23.6	03.8
20	342 39.8	103 42.9	43.5	1 47.5	26.0	177 27.3	22.1	238 26.2	03.8
21	357 42.3	118 42.1	. . 44.3	16 49.1	. . 25.5	192 29.4	. . 22.0	253 28.7	. . 03.8
22	12 44.7	133 41.3	45.0	31 50.6	25.0	207 31.5	21.9	268 31.2	03.8
23	27 47.2	148 40.6	45.8	46 52.2	24.5	222 33.6	21.7	283 33.7	03.8
4 00	42 49.6	163 39.8	S20 46.5	61 53.8	S10 24.0	237 35.7	N 7 21.6	298 36.2	N22 03.8
01	57 52.1	178 39.0	47.2	76 55.4	23.5	252 37.8	21.4	313 38.7	03.8
02	72 54.6	193 38.3	48.0	91 56.9	23.0	267 39.9	21.3	328 41.2	03.8
03	87 57.0	208 37.5	. . 48.7	106 58.5	. . 22.5	282 42.0	. . 21.1	343 43.7	. . 03.8
04	102 59.5	223 36.7	49.5	122 00.1	22.0	297 44.1	21.0	358 46.2	03.8
05	118 02.0	238 36.0	50.2	137 01.7	21.5	312 46.2	20.8	13 48.7	03.8
06	133 04.4	253 35.2	S20 50.9	152 03.2	S10 20.9	327 48.3	N 7 20.7	28 51.2	N22 03.8
07	148 06.9	268 34.4	51.7	167 04.8	20.4	342 50.3	20.6	43 53.8	03.8
T 08	163 09.4	283 33.7	52.4	182 06.4	19.9	357 52.4	20.4	58 56.3	03.8
U 09	178 11.8	298 32.9	. . 53.1	197 08.0	. . 19.4	12 54.5	. . 20.3	73 58.8	. . 03.8
E 10	193 14.3	313 32.1	53.9	212 09.5	18.9	27 56.6	20.1	89 01.3	03.9
S 11	208 16.7	328 31.3	54.6	227 11.1	18.4	42 58.7	20.0	104 03.8	03.9
D 12	223 19.2	343 30.6	S20 55.3	242 12.7	S10 17.9	58 00.8	N 7 19.8	119 06.3	N22 03.9
A 13	238 21.7	358 29.8	56.0	257 14.3	17.4	73 02.9	19.7	134 08.8	03.9
Y 14	253 24.1	13 29.0	56.8	272 15.8	16.9	88 05.0	19.5	149 11.3	03.9
15	268 26.6	28 28.2	. . 57.5	287 17.4	. . 16.3	103 07.1	. . 19.4	164 13.8	. . 03.9
16	283 29.1	43 27.5	58.2	302 19.0	15.8	118 09.2	19.3	179 16.4	03.9
17	298 31.5	58 26.7	58.9	317 20.5	15.3	133 11.3	19.1	194 18.9	03.9
18	313 34.0	73 25.9	S20 59.7	332 22.1	S10 14.8	148 13.4	N 7 19.0	209 21.4	N22 03.9
19	328 36.5	88 25.1	21 00.4	347 23.7	14.3	163 15.5	18.8	224 23.9	03.9
20	343 38.9	103 24.4	01.1	2 25.2	13.8	178 17.6	18.7	239 26.4	03.9
21	358 41.4	118 23.6	. . 01.8	17 26.8	. . 13.3	193 19.7	. . 18.5	254 28.9	. . 03.9
22	13 43.9	133 22.8	02.5	32 28.3	12.7	208 21.8	18.4	269 31.4	03.9
23	28 46.3	148 22.0	03.2	47 29.9	12.2	223 23.8	18.2	284 34.0	03.9
5 00	43 48.8	163 21.2	S21 04.0	62 31.5	S10 11.7	238 25.9	N 7 18.1	299 36.5	N22 03.9
01	58 51.2	178 20.5	04.7	77 33.0	11.2	253 28.0	18.0	314 39.0	03.9
02	73 53.7	193 19.7	05.4	92 34.6	10.7	268 30.1	17.8	329 41.5	03.9
03	88 56.2	208 18.9	. . 06.1	107 36.2	. . 10.2	283 32.2	. . 17.7	344 44.0	. . 03.9
04	103 58.6	223 18.1	06.8	122 37.7	09.7	298 34.3	17.5	359 46.5	04.0
05	119 01.1	238 17.3	07.5	137 39.3	09.1	313 36.4	17.4	14 49.0	04.0
06	134 03.6	253 16.6	S21 08.2	152 40.8	S10 08.6	328 38.5	N 7 17.2	29 51.6	N22 04.0
W 07	149 06.0	268 15.8	08.9	167 42.4	08.1	343 40.6	17.1	44 54.1	04.0
E 08	164 08.5	283 15.0	09.7	182 44.0	07.6	358 42.7	17.0	59 56.6	04.0
D 09	179 11.0	298 14.2	. . 10.4	197 45.5	. . 07.1	13 44.8	. . 16.8	74 59.1	. . 04.0
N 10	194 13.4	313 13.4	11.1	212 47.1	06.6	28 46.9	16.7	90 01.6	04.0
E 11	209 15.9	328 12.6	11.8	227 48.6	06.0	43 49.0	16.5	105 04.1	04.0
S 12	224 18.3	343 11.9	S21 12.5	242 50.2	S10 05.5	58 51.1	N 7 16.4	120 06.6	N22 04.0
D 13	239 20.8	358 11.1	13.2	257 51.7	05.0	73 53.2	16.2	135 09.2	04.0
A 14	254 23.3	13 10.3	13.9	272 53.3	04.5	88 55.3	16.1	150 11.7	04.0
Y 15	269 25.7	28 09.5	. . 14.6	287 54.8	. . 04.0	103 57.4	. . 16.0	165 14.2	. . 04.0
16	284 28.2	43 08.7	15.3	302 56.4	03.4	118 59.5	15.8	180 16.7	04.0
17	299 30.7	58 07.9	16.0	317 57.9	02.9	134 01.6	15.7	195 19.2	04.0
18	314 33.1	73 07.1	S21 16.7	332 59.5	S10 02.4	149 03.7	N 7 15.5	210 21.7	N22 04.0
19	329 35.6	88 06.3	17.4	348 01.0	01.9	164 05.8	15.4	225 24.3	04.0
20	344 38.1	103 05.6	18.1	3 02.6	01.4	179 07.9	15.2	240 26.8	04.0
21	359 40.5	118 04.8	. . 18.8	18 04.1	. . 00.8	194 10.0	. . 15.1	255 29.3	. . 04.1
22	14 43.0	133 04.0	19.5	33 05.7	10 00.3	209 12.1	15.0	270 31.8	04.1
23	29 45.5	148 03.2	20.1	48 07.2	S 9 59.8	224 14.2	14.8	285 34.3	04.1
Mer. Pass. (h m)	21 05.2	v −0.8	d 0.7	v 1.6	d 0.5	v 2.1	d 0.1	v 2.5	d 0.0

STARS

Name	SHA	Dec
	° ′	° ′
Acamar	315 23.6	S40 17.3
Achernar	335 31.6	S57 13.1
Acrux	173 18.7	S63 06.9
Adhara	255 18.3	S28 58.3
Aldebaran	290 57.8	N16 31.1
Alioth	166 27.6	N55 56.3
Alkaid	153 05.2	N49 17.7
Al Na'ir	27 52.9	S46 56.8
Alnilam	275 53.8	S 1 11.8
Alphard	218 03.6	S 8 40.2
Alphecca	126 17.8	N26 42.2
Alpheratz	357 51.2	N29 06.8
Altair	62 15.8	N 8 52.7
Ankaa	353 22.7	S42 17.3
Antares	112 36.0	S26 26.4
Arcturus	146 03.0	N19 09.9
Atria	107 45.2	S69 02.2
Avior	234 21.2	S59 30.9
Bellatrix	278 39.9	N 6 21.3
Betelgeuse	271 09.3	N 7 24.6
Canopus	263 59.2	S52 41.5
Capella	280 45.3	N46 00.1
Deneb	49 36.8	N45 17.8
Denebola	182 41.6	N14 33.2
Diphda	349 03.2	S17 58.0
Dubhe	194 01.0	N61 43.7
Elnath	278 22.0	N28 36.7
Eltanin	90 50.0	N51 29.5
Enif	33 54.5	N 9 53.6
Fomalhaut	15 32.1	S29 36.3
Gacrux	172 10.2	S57 07.8
Gienah	176 00.5	S17 33.5
Hadar	148 59.6	S60 23.3
Hamal	328 09.1	N23 28.9
Kaus Aust.	83 54.1	S34 23.1
Kochab	137 20.5	N74 08.4
Markab	13 45.8	N15 13.6
Menkar	314 22.7	N 4 06.4
Menkent	148 17.1	S36 23.2
Miaplacidus	221 41.7	S69 43.5
Mirfak	308 50.9	N49 52.5
Nunki	76 07.9	S26 17.7
Peacock	53 31.1	S56 43.7
Pollux	243 36.9	N28 01.1
Procyon	245 07.6	N 5 13.1
Rasalhague	96 13.8	N12 33.5
Regulus	207 51.7	N11 57.1
Rigel	281 19.1	S 8 11.6
Rigil Kent.	140 03.2	S60 50.9
Sabik	102 21.6	S15 43.8
Schedar	349 49.0	N56 33.6
Shaula	96 32.6	S37 06.5
Sirius	258 40.3	S16 43.0
Spica	158 39.6	S11 10.7
Suhail	222 58.2	S43 26.5
Vega	80 44.4	N38 47.4
Zuben'ubi	137 14.2	S16 03.4

	SHA	Mer. Pass
	° ′	h m
Venus	120 50.2	13 06
Mars	19 04.1	19 50
Jupiter	194 46.1	8 08
Saturn	255 46.5	4 05

UT	SUN GHA	SUN Dec	MOON GHA	v	MOON Dec	d	HP
d h	° ′	° ′	° ′	′	° ′	′	′
3 00	184 06.4	S14 51.7	65 43.0	12.8	S15 30.7	12.3	56.4
01	199 06.4	52.5	80 14.8	12.9	15 18.4	12.3	56.4
02	214 06.4	53.3	94 46.7	12.9	15 06.1	12.4	56.3
03	229 06.4	. . 54.1	109 18.6	13.1	14 53.7	12.5	56.3
04	244 06.4	54.8	123 50.7	13.2	14 41.2	12.5	56.3
05	259 06.4	55.6	138 22.9	13.2	14 28.7	12.5	56.2
06	274 06.4	S14 56.4	152 55.1	13.3	S14 16.2	12.7	56.2
07	289 06.4	57.2	167 27.4	13.4	14 03.5	12.6	56.2
08	304 06.4	58.0	181 59.8	13.5	13 50.9	12.7	56.2
09	319 06.4	. . 58.8	196 32.3	13.5	13 38.2	12.8	56.1
10	334 06.4	14 59.5	211 04.8	13.7	13 25.4	12.8	56.1
11	349 06.4	15 00.3	225 37.5	13.7	13 12.6	12.9	56.1
12	4 06.4	S15 01.1	240 10.2	13.8	S12 59.7	12.9	56.1
13	19 06.4	01.9	254 43.0	13.8	12 46.8	12.9	56.0
14	34 06.4	02.7	269 15.8	14.0	12 33.9	13.0	56.0
15	49 06.4	. . 03.5	283 48.8	14.0	12 20.9	13.0	56.0
16	64 06.4	04.2	298 21.8	14.1	12 07.9	13.1	56.0
17	79 06.4	05.0	312 54.9	14.1	11 54.8	13.1	55.9
18	94 06.4	S15 05.8	327 28.0	14.3	S11 41.7	13.2	55.9
19	109 06.4	06.6	342 01.3	14.3	11 28.5	13.2	55.9
20	124 06.4	07.4	356 34.6	14.3	11 15.3	13.2	55.9
21	139 06.4	. . 08.1	11 07.9	14.4	11 02.1	13.2	55.8
22	154 06.4	08.9	25 41.3	14.5	10 48.9	13.3	55.8
23	169 06.4	09.7	40 14.8	14.6	10 35.6	13.4	55.8
4 00	184 06.4	S15 10.5	54 48.4	14.6	S10 22.2	13.3	55.8
01	199 06.4	11.2	69 22.0	14.7	10 08.9	13.4	55.7
02	214 06.4	12.0	83 55.7	14.8	9 55.5	13.4	55.7
03	229 06.4	. . 12.8	98 29.5	14.8	9 42.1	13.5	55.7
04	244 06.4	13.6	113 03.3	14.8	9 28.6	13.5	55.7
05	259 06.4	14.3	127 37.1	15.0	9 15.1	13.5	55.6
06	274 06.4	S15 15.1	142 11.1	14.9	S 9 01.6	13.5	55.6
07	289 06.4	15.9	156 45.0	15.1	8 48.1	13.6	55.6
08	304 06.4	16.7	171 19.1	15.1	8 34.5	13.6	55.6
09	319 06.4	. . 17.4	185 53.2	15.1	8 20.9	13.6	55.6
10	334 06.4	18.2	200 27.3	15.2	8 07.3	13.6	55.5
11	349 06.4	19.0	215 01.5	15.2	7 53.7	13.7	55.5
12	4 06.4	S15 19.8	229 35.7	15.3	S 7 40.0	13.7	55.5
13	19 06.4	20.5	244 10.0	15.4	7 26.3	13.7	55.5
14	34 06.4	21.3	258 44.4	15.4	7 12.6	13.7	55.4
15	49 06.3	. . 22.1	273 18.8	15.4	6 58.9	13.7	55.4
16	64 06.3	22.8	287 53.2	15.5	6 45.2	13.8	55.4
17	79 06.3	23.6	302 27.7	15.5	6 31.4	13.7	55.4
18	94 06.3	S15 24.4	317 02.2	15.6	S 6 17.7	13.8	55.4
19	109 06.3	25.1	331 36.8	15.6	6 03.9	13.8	55.3
20	124 06.3	25.9	346 11.4	15.7	5 50.1	13.8	55.3
21	139 06.3	. . 26.7	0 46.1	15.7	5 36.3	13.9	55.3
22	154 06.3	27.5	15 20.8	15.7	5 22.4	13.8	55.3
23	169 06.3	28.2	29 55.5	15.8	5 08.6	13.9	55.3
5 00	184 06.3	S15 29.0	44 30.3	15.8	S 4 54.7	13.8	55.2
01	199 06.3	29.8	59 05.1	15.8	4 40.9	13.9	55.2
02	214 06.2	30.5	73 39.9	15.9	4 27.0	13.9	55.2
03	229 06.2	. . 31.3	88 14.8	15.9	4 13.1	13.9	55.2
04	244 06.2	32.0	102 49.7	15.9	3 59.2	13.9	55.2
05	259 06.2	32.8	117 24.6	16.0	3 45.3	13.9	55.1
06	274 06.2	S15 33.6	131 59.6	16.0	S 3 31.4	13.9	55.1
07	289 06.2	34.3	146 34.6	16.0	3 17.5	13.9	55.1
08	304 06.2	35.1	161 09.6	16.1	3 03.6	13.9	55.1
09	319 06.2	. . 35.9	175 44.7	16.1	2 49.7	13.9	55.1
10	334 06.1	36.6	190 19.8	16.1	2 35.8	13.9	55.1
11	349 06.1	37.4	204 54.9	16.1	2 21.9	14.0	55.0
12	4 06.1	S15 38.2	219 30.0	16.2	S 2 07.9	13.9	55.0
13	19 06.1	38.9	234 05.2	16.2	1 54.0	13.9	55.0
14	34 06.1	39.7	248 40.4	16.2	1 40.1	13.9	55.0
15	49 06.1	. . 40.4	263 15.6	16.2	1 26.2	13.9	55.0
16	64 06.1	41.2	277 50.8	16.2	1 12.3	13.9	55.0
17	79 06.0	42.0	292 26.0	16.3	0 58.4	14.0	54.9
18	94 06.0	S15 42.7	307 01.3	16.3	S 0 44.4	13.9	54.9
19	109 06.0	43.5	321 36.6	16.3	0 30.5	13.9	54.9
20	124 06.0	44.2	336 11.9	16.3	0 16.6	13.9	54.9
21	139 06.0	. . 45.0	350 47.2	16.3	S 0 02.7	13.8	54.9
22	154 06.0	45.7	5 22.5	16.3	N 0 11.1	13.9	54.9
23	169 05.9	46.5	19 57.8	16.4	N 0 25.0	13.9	54.8
	SD 16.2	*d* 0.8	SD 15.3		15.1		15.0

Lat.	Twilight Naut.	Twilight Civil	Sunrise	Moonrise 3	Moonrise 4	Moonrise 5	Moonrise 6
°	h m	h m	h m	h m	h m	h m	h m
N 72	06 16	07 41	09 13	16 53	16 15	15 45	15 16
N 70	06 11	07 26	08 44	16 32	16 06	15 43	15 22
68	06 06	07 14	08 22	16 16	15 58	15 42	15 26
66	06 03	07 05	08 04	16 03	15 51	15 41	15 30
64	05 59	06 56	07 50	15 52	15 46	15 40	15 33
62	05 56	06 49	07 39	15 43	15 41	15 39	15 36
60	05 53	06 43	07 29	15 35	15 37	15 38	15 39
N 58	05 50	06 37	07 20	15 28	15 33	15 37	15 41
56	05 48	06 32	07 12	15 22	15 30	15 36	15 43
54	05 46	06 28	07 06	15 16	15 26	15 36	15 45
52	05 43	06 23	06 59	15 11	15 24	15 35	15 46
50	05 41	06 20	06 54	15 06	15 21	15 35	15 48
45	05 36	06 11	06 42	14 56	15 16	15 34	15 51
N 40	05 32	06 04	06 32	14 47	15 11	15 33	15 54
35	05 27	05 57	06 23	14 40	15 07	15 32	15 56
30	05 23	05 51	06 16	14 34	15 04	15 31	15 58
20	05 14	05 40	06 03	14 23	14 57	15 30	16 02
N 10	05 04	05 29	05 51	14 13	14 52	15 29	16 06
0	04 54	05 19	05 40	14 04	14 47	15 28	16 09
S 10	04 42	05 07	05 29	13 54	14 42	15 27	16 12
20	04 27	04 54	05 17	13 44	14 36	15 26	16 15
30	04 08	04 38	05 03	13 33	14 30	15 25	16 19
35	03 56	04 28	04 55	13 26	14 26	15 24	16 21
40	03 41	04 17	04 46	13 19	14 22	15 24	16 24
45	03 23	04 03	04 35	13 10	14 17	15 23	16 27
S 50	02 59	03 45	04 22	12 59	14 12	15 22	16 31
52	02 47	03 37	04 16	12 54	14 09	15 21	16 32
54	02 32	03 28	04 09	12 49	14 06	15 21	16 34
56	02 16	03 17	04 02	12 42	14 03	15 20	16 36
58	01 54	03 05	03 53	12 36	13 59	15 20	16 39
S 60	01 26	02 50	03 43	12 28	13 55	15 19	16 41

Lat.	Sunset	Twilight Civil	Twilight Naut.	Moonset 3	Moonset 4	Moonset 5	Moonset 6
°	h m	h m	h m	h m	h m	h m	h m
N 72	14 12	15 45	17 09	24 02	00 02	02 09	04 07
N 70	14 42	15 59	17 15	24 20	00 20	02 16	04 05
68	15 04	16 11	17 19	24 34	00 34	02 21	04 03
66	15 22	16 21	17 23	24 46	00 46	02 26	04 02
64	15 36	16 30	17 27	24 55	00 55	02 29	04 00
62	15 47	16 37	17 30	25 03	01 03	02 33	03 59
60	15 58	16 43	17 33	25 10	01 10	02 35	03 58
N 58	16 06	16 49	17 36	25 16	01 16	02 38	03 58
56	16 14	16 54	17 38	25 21	01 21	02 40	03 57
54	16 21	16 59	17 41	00 07	01 26	02 42	03 56
52	16 27	17 03	17 43	00 14	01 30	02 44	03 56
50	16 33	17 07	17 45	00 20	01 34	02 46	03 55
45	16 45	17 16	17 50	00 34	01 43	02 49	03 54
N 40	16 55	17 23	17 55	00 45	01 49	02 52	03 53
35	17 03	17 30	18 00	00 54	01 55	02 54	03 52
30	17 11	17 36	18 04	01 02	02 01	02 57	03 51
20	17 24	17 47	18 13	01 16	02 10	03 00	03 50
N 10	17 36	17 58	18 23	01 28	02 17	03 04	03 49
0	17 47	18 08	18 33	01 40	02 25	03 07	03 48
S 10	17 58	18 20	18 46	01 51	02 32	03 10	03 47
20	18 10	18 33	19 01	02 03	02 39	03 13	03 45
30	18 24	18 50	19 20	02 17	02 48	03 17	03 44
35	18 32	19 00	19 32	02 24	02 53	03 19	03 43
40	18 42	19 11	19 47	02 33	02 58	03 21	03 42
45	18 53	19 25	20 06	02 43	03 05	03 24	03 41
S 50	19 06	19 43	20 30	02 56	03 12	03 27	03 40
52	19 12	19 51	20 43	03 02	03 16	03 28	03 40
54	19 19	20 01	20 57	03 08	03 20	03 30	03 39
56	19 27	20 12	21 14	03 15	03 24	03 32	03 38
58	19 36	20 25	21 36	03 23	03 29	03 33	03 37
S 60	19 46	20 40	22 06	03 31	03 34	03 36	03 37

Day	SUN Eqn. of Time 00^h	SUN Eqn. of Time 12^h	SUN Mer. Pass.	MOON Mer. Pass. Upper	MOON Mer. Pass. Lower	MOON Age	MOON Phase
d	m s	m s	h m	h m	h m	d %	
3	16 25	16 26	11 44	20 14	07 52	09 73	
4	16 26	16 25	11 44	20 57	08 36	10 82	
5	16 25	16 25	11 44	21 38	09 18	11 89	

UT	ARIES	VENUS −3.9		MARS −1.0		JUPITER −1.9		SATURN −0.1		STARS		
	GHA	GHA	Dec	GHA	Dec	GHA	Dec	GHA	Dec	Name	SHA	Dec
d h	° ′	° ′	° ′	° ′	° ′	° ′	° ′	° ′	° ′		° ′	° ′
6 00 (THURSDAY)	44 47.9	163 02.4	S21 20.8	63 08.8	S 9 59.3	239 16.3	N 7 14.7	300 36.9	N22 04.1	Acamar	315 23.6	S40 17.3
01	59 50.4	178 01.6	21.5	78 10.3	58.8	254 18.4	14.5	315 39.4	04.1	Achernar	335 31.6	S57 13.
02	74 52.8	193 00.8	22.2	93 11.9	58.2	269 20.5	14.4	330 41.9	04.1	Acrux	173 18.7	S63 06.9
03	89 55.3	208 00.0 . .	22.9	108 13.4 . .	57.7	284 22.6 . .	14.2	345 44.4 . .	04.1	Adhara	255 18.3	S28 58.3
04	104 57.8	222 59.2	23.6	123 15.0	57.2	299 24.7	14.1	0 46.9	04.1	Aldebaran	290 57.8	N16 31.
05	120 00.2	237 58.4	24.3	138 16.5	56.7	314 26.8	14.0	15 49.5	04.1			
06	135 02.7	252 57.6	S21 25.0	153 18.0	S 9 56.2	329 28.9	N 7 13.8	30 52.0	N22 04.1	Alioth	166 27.6	N55 56.
07	150 05.2	267 56.8	25.6	168 19.6	55.6	344 31.0	13.7	45 54.5	04.1	Alkaid	153 05.2	N49 17.7
08	165 07.6	282 56.0	26.3	183 21.1	55.1	359 33.1	13.5	60 57.0	04.1	Al Na'ir	27 52.9	S46 56.8
09	180 10.1	297 55.2 . .	27.0	198 22.7 . .	54.6	14 35.2 . .	13.4	75 59.5 . .	04.1	Alnilam	275 53.8	S 1 11.8
10	195 12.6	312 54.5	27.7	213 24.2	54.1	29 37.3	13.3	91 02.1	04.1	Alphard	218 03.6	S 8 40.2
11	210 15.0	327 53.7	28.4	228 25.8	53.6	44 39.4	13.1	106 04.6	04.1			
12	225 17.5	342 52.9	S21 29.1	243 27.3	S 9 53.0	59 41.5	N 7 13.0	121 07.1	N22 04.1	Alphecca	126 17.8	N26 42.
13	240 20.0	357 52.1	29.7	258 28.8	52.5	74 43.6	12.8	136 09.6	04.1	Alpheratz	357 51.2	N29 06.8
14	255 22.4	12 51.3	30.4	273 30.4	52.0	89 45.7	12.7	151 12.1	04.2	Altair	62 15.8	N 8 52.7
15	270 24.9	27 50.5 . .	31.1	288 31.9 . .	51.5	104 47.8 . .	12.6	166 14.7 . .	04.2	Ankaa	353 22.7	S42 17.
16	285 27.3	42 49.7	31.8	303 33.4	50.9	119 49.9	12.4	181 17.2	04.2	Antares	112 36.0	S26 26.
17	300 29.8	57 48.9	32.4	318 35.0	50.4	134 52.0	12.3	196 19.7	04.2			
18	315 32.3	72 48.1	S21 33.1	333 36.5	S 9 49.9	149 54.1	N 7 12.1	211 22.2	N22 04.2	Arcturus	146 03.0	N19 09.
19	330 34.7	87 47.3	33.8	348 38.1	49.4	164 56.2	12.0	226 24.8	04.2	Atria	107 45.3	S69 02.
20	345 37.2	102 46.5	34.4	3 39.6	48.8	179 58.3	11.8	241 27.3	04.2	Avior	234 21.2	S59 30.
21	0 39.7	117 45.7 . .	35.1	18 41.1 . .	48.3	195 00.4 . .	11.7	256 29.8 . .	04.2	Bellatrix	278 39.9	N 6 21.
22	15 42.1	132 44.9	35.8	33 42.7	47.8	210 02.5	11.6	271 32.3	04.2	Betelgeuse	271 09.3	N 7 24.
23	30 44.6	147 44.1	36.4	48 44.2	47.3	225 04.6	11.4	286 34.8	04.2			
7 00 (FRIDAY)	45 47.1	162 43.3	S21 37.1	63 45.7	S 9 46.7	240 06.7	N 7 11.3	301 37.4	N22 04.2	Canopus	263 59.2	S52 41.
01	60 49.5	177 42.4	37.8	78 47.3	46.2	255 08.8	11.1	316 39.9	04.2	Capella	280 45.3	N46 00.
02	75 52.0	192 41.6	38.4	93 48.8	45.7	270 10.9	11.0	331 42.4	04.2	Deneb	49 36.8	N45 17.
03	90 54.4	207 40.8 . .	39.1	108 50.3 . .	45.2	285 13.1 . .	10.9	346 44.9 . .	04.2	Denebola	182 41.6	N14 33.
04	105 56.9	222 40.0	39.8	123 51.8	44.6	300 15.2	10.7	1 47.5	04.2	Diphda	349 03.2	S17 58.
05	120 59.4	237 39.2	40.4	138 53.4	44.1	315 17.3	10.6	16 50.0	04.3			
06	136 01.8	252 38.4	S21 41.1	153 54.9	S 9 43.6	330 19.4	N 7 10.4	31 52.5	N22 04.3	Dubhe	194 01.0	N61 43.
07	151 04.3	267 37.6	41.7	168 56.4	43.1	345 21.5	10.3	46 55.0	04.3	Elnath	278 21.9	N28 36.
08	166 06.8	282 36.8	42.4	183 58.0	42.5	0 23.6	10.2	61 57.6	04.3	Eltanin	90 50.1	N51 29.
09	181 09.2	297 36.0 . .	43.1	198 59.5 . .	42.0	15 25.7 . .	10.0	77 00.1 . .	04.3	Enif	33 54.6	N 9 53.
10	196 11.7	312 35.2	43.7	214 01.0	41.5	30 27.8	09.9	92 02.6	04.3	Fomalhaut	15 32.1	S29 36.
11	211 14.2	327 34.4	44.4	229 02.5	40.9	45 29.9	09.8	107 05.1	04.3			
12	226 16.6	342 33.6	S21 45.0	244 04.1	S 9 40.4	60 32.0	N 7 09.6	122 07.7	N22 04.3	Gacrux	172 10.1	S57 07.
13	241 19.1	357 32.8	45.7	259 05.6	39.9	75 34.1	09.5	137 10.2	04.3	Gienah	176 00.4	S17 33.
14	256 21.6	12 32.0	46.3	274 07.1	39.4	90 36.2	09.3	152 12.7	04.3	Hadar	148 59.6	S60 23.
15	271 24.0	27 31.1 . .	47.0	289 08.6 . .	38.8	105 38.3 . .	09.2	167 15.2 . .	04.3	Hamal	328 09.1	N23 28.
16	286 26.5	42 30.3	47.6	304 10.1	38.3	120 40.4	09.1	182 17.8	04.3	Kaus Aust.	83 54.1	S34 23.
17	301 28.9	57 29.5	48.3	319 11.7	37.8	135 42.5	08.9	197 20.3	04.3			
18	316 31.4	72 28.7	S21 48.9	334 13.2	S 9 37.2	150 44.6	N 7 08.8	212 22.8	N22 04.3	Kochab	137 20.5	N74 08.
19	331 33.9	87 27.9	49.6	349 14.7	36.7	165 46.7	08.6	227 25.4	04.3	Markab	13 45.8	N15 13.
20	346 36.3	102 27.1	50.2	4 16.2	36.2	180 48.9	08.5	242 27.9	04.4	Menkar	314 22.7	N 4 06.
21	1 38.8	117 26.3 . .	50.9	19 17.7 . .	35.7	195 51.0 . .	08.4	257 30.4 . .	04.4	Menkent	148 17.1	S36 23.
22	16 41.3	132 25.5	51.5	34 19.3	35.1	210 53.1	08.2	272 32.9	04.4	Miaplacidus	221 41.7	S69 43.
23	31 43.7	147 24.6	52.1	49 20.8	34.6	225 55.2	08.1	287 35.5	04.4			
8 00 (SATURDAY)	46 46.2	162 23.8	S21 52.8	64 22.3	S 9 34.1	240 57.3	N 7 07.9	302 38.0	N22 04.4	Mirfak	308 50.9	N49 52.
01	61 48.7	177 23.0	53.4	79 23.8	33.5	255 59.4	07.8	317 40.5	04.4	Nunki	76 07.9	S26 17.
02	76 51.1	192 22.2	54.1	94 25.3	33.0	271 01.5	07.7	332 43.1	04.4	Peacock	53 31.1	S56 43.
03	91 53.6	207 21.4 . .	54.7	109 26.9 . .	32.5	286 03.6 . .	07.5	347 45.6 . .	04.4	Pollux	243 36.8	N28 01.
04	106 56.0	222 20.6	55.3	124 28.4	31.9	301 05.7	07.4	2 48.1	04.4	Procyon	245 07.5	N 5 13.
05	121 58.5	237 19.7	56.0	139 29.9	31.4	316 07.8	07.3	17 50.6	04.4			
06	137 01.0	252 18.9	S21 56.6	154 31.4	S 9 30.9	331 09.9	N 7 07.1	32 53.2	N22 04.4	Rasalhague	96 13.8	N12 33.
07	152 03.4	267 18.1	57.2	169 32.9	30.4	346 12.1	07.0	47 55.7	04.4	Regulus	207 51.6	N11 57.
08	167 05.9	282 17.3	57.9	184 34.4	29.8	1 14.2	06.8	62 58.2	04.4	Rigel	281 19.1	S 8 11.
09	182 08.4	297 16.5 . .	58.5	199 35.9 . .	29.3	16 16.3 . .	06.7	78 00.8 . .	04.4	Rigil Kent.	140 03.1	S60 50.
10	197 10.8	312 15.7	59.1	214 37.4	28.8	31 18.4	06.6	93 03.3	04.5	Sabik	102 21.6	S15 43.
11	212 13.3	327 14.8	21 59.8	229 39.0	28.2	46 20.5	06.4	108 05.8	04.5			
12	227 15.8	342 14.0	S22 00.4	244 40.5	S 9 27.7	61 22.6	N 7 06.3	123 08.4	N22 04.5	Schedar	349 49.0	N56 33.
13	242 18.2	357 13.2	01.0	259 42.0	27.2	76 24.7	06.2	138 10.9	04.5	Shaula	96 32.6	S37 06.
14	257 20.7	12 12.4	01.6	274 43.5	26.6	91 26.8	06.0	153 13.4	04.5	Sirius	258 40.3	S16 43.
15	272 23.2	27 11.5 . .	02.3	289 45.0 . .	26.1	106 28.9 . .	05.9	168 15.9 . .	04.5	Spica	158 39.6	S11 10.
16	287 25.6	42 10.7	02.9	304 46.5	25.6	121 31.0	05.7	183 18.5	04.5	Suhail	222 58.2	S43 26.
17	302 28.1	57 09.9	03.5	319 48.0	25.0	136 33.2	05.6	198 21.0	04.5			
18	317 30.5	72 09.1	S22 04.1	334 49.5	S 9 24.5	151 35.3	N 7 05.5	213 23.5	N22 04.5	Vega	80 44.4	N38 47.
19	332 33.0	87 08.3	04.7	349 51.0	24.0	166 37.4	05.3	228 26.1	04.5	Zuben'ubi	137 14.2	S16 03
20	347 35.5	102 07.4	05.4	4 52.5	23.4	181 39.5	05.2	243 28.6	04.5		SHA	Mer. Pas
21	2 37.9	117 06.6 . .	06.0	19 54.0 . .	22.9	196 41.6 . .	05.1	258 31.1 . .	04.5		° ′	h m
22	17 40.4	132 05.8	06.6	34 55.5	22.4	211 43.7	04.9	273 33.7	04.5	Venus	116 56.2	13 10
23	32 42.9	147 05.0	07.2	49 57.0	21.8	226 45.8	04.8	288 36.2	04.5	Mars	17 58.7	19 43
	h m									Jupiter	194 19.7	7 58
Mer. Pass.	20 53.4	*v* −0.8	*d* 0.7	*v* 1.5	*d* 0.5	*v* 2.1	*d* 0.1	*v* 2.5	*d* 0.0	Saturn	255 50.3	3 53

UT	SUN GHA	SUN Dec	MOON GHA	MOON *v*	MOON Dec	MOON *d*	MOON HP
d h	° ′	° ′	° ′	′	° ′	′	′
6 00	184 05.9	S15 47.3	34 33.2	16.3	N 0 38.9	13.9	54.8
01	199 05.9	48.0	49 08.5	16.4	0 52.8	13.8	54.8
02	214 05.9	48.8	63 43.9	16.4	1 06.6	13.9	54.8
03	229 05.9	. . 49.5	78 19.3	16.3	1 20.5	13.8	54.8
04	244 05.8	50.3	92 54.6	16.4	1 34.3	13.8	54.8
05	259 05.8	51.0	107 30.0	16.4	1 48.1	13.8	54.7
06	274 05.8	S15 51.8	122 05.4	16.4	N 2 01.9	13.8	54.7
07	289 05.8	52.5	136 40.8	16.4	2 15.7	13.8	54.7
08	304 05.8	53.3	151 16.2	16.4	2 29.5	13.7	54.7
09	319 05.7	. . 54.0	165 51.6	16.5	2 43.2	13.8	54.7
10	334 05.7	54.8	180 27.1	16.4	2 57.0	13.7	54.7
11	349 05.7	55.5	195 02.5	16.4	3 10.7	13.7	54.7
12	4 05.7	S15 56.3	209 37.9	16.4	N 3 24.4	13.7	54.6
13	19 05.6	57.0	224 13.3	16.4	3 38.1	13.7	54.6
14	34 05.6	57.8	238 48.7	16.4	3 51.8	13.6	54.6
15	49 05.6	. . 58.5	253 24.1	16.4	4 05.4	13.7	54.6
16	64 05.6	15 59.3	267 59.5	16.4	4 19.1	13.6	54.6
17	79 05.5	16 00.0	282 34.9	16.4	4 32.7	13.6	54.6
18	94 05.5	S16 00.8	297 10.3	16.4	N 4 46.3	13.5	54.6
19	109 05.5	01.5	311 45.7	16.4	4 59.8	13.6	54.5
20	124 05.5	02.3	326 21.1	16.3	5 13.4	13.5	54.5
21	139 05.4	. . 03.0	340 56.4	16.4	5 26.9	13.5	54.5
22	154 05.4	03.8	355 31.8	16.3	5 40.4	13.4	54.5
23	169 05.4	04.5	10 07.1	16.4	5 53.8	13.5	54.5
7 00	184 05.4	S16 05.3	24 42.5	16.3	N 6 07.3	13.4	54.5
01	199 05.3	06.0	39 17.8	16.3	6 20.7	13.4	54.5
02	214 05.3	06.7	53 53.1	16.3	6 34.1	13.3	54.5
03	229 05.3	. . 07.5	68 28.4	16.3	6 47.4	13.3	54.5
04	244 05.2	08.2	83 03.7	16.2	7 00.7	13.3	54.4
05	259 05.2	09.0	97 38.9	16.3	7 14.0	13.3	54.4
06	274 05.2	S16 09.7	112 14.2	16.2	N 7 27.3	13.2	54.4
07	289 05.2	10.5	126 49.4	16.2	7 40.5	13.2	54.4
08	304 05.1	11.2	141 24.6	16.2	7 53.7	13.2	54.4
09	319 05.1	. . 11.9	155 59.8	16.2	8 06.9	13.1	54.4
10	334 05.1	12.7	170 35.0	16.1	8 20.0	13.1	54.4
11	349 05.0	13.4	185 10.1	16.2	8 33.1	13.0	54.4
12	4 05.0	S16 14.2	199 45.3	16.1	N 8 46.1	13.1	54.4
13	19 05.0	14.9	214 20.4	16.1	8 59.2	12.9	54.3
14	34 04.9	15.6	228 55.5	16.0	9 12.1	13.0	54.3
15	49 04.9	. . 16.4	243 30.5	16.0	9 25.1	12.9	54.3
16	64 04.9	17.1	258 05.5	16.0	9 38.0	12.8	54.3
17	79 04.8	17.8	272 40.5	16.0	9 50.8	12.8	54.3
18	94 04.8	S16 18.6	287 15.5	16.0	N10 03.6	12.8	54.3
19	109 04.8	19.3	301 50.5	15.9	10 16.4	12.8	54.3
20	124 04.7	20.1	316 25.4	15.9	10 29.2	12.6	54.3
21	139 04.7	. . 20.8	331 00.3	15.9	10 41.8	12.7	54.3
22	154 04.7	21.5	345 35.2	15.8	10 54.5	12.6	54.3
23	169 04.6	22.3	0 10.0	15.8	11 07.1	12.6	54.2
8 00	184 04.6	S16 23.0	14 44.8	15.8	N11 19.7	12.5	54.2
01	199 04.6	23.7	29 19.6	15.7	11 32.2	12.4	54.2
02	214 04.5	24.5	43 54.3	15.7	11 44.6	12.4	54.2
03	229 04.5	. . 25.2	58 29.0	15.7	11 57.0	12.4	54.2
04	244 04.4	25.9	73 03.7	15.6	12 09.4	12.3	54.2
05	259 04.4	26.6	87 38.3	15.6	12 21.7	12.3	54.2
06	274 04.4	S16 27.4	102 12.9	15.5	N12 34.0	12.2	54.2
07	289 04.3	28.1	116 47.4	15.6	12 46.2	12.2	54.2
08	304 04.3	28.8	131 22.0	15.4	12 58.4	12.1	54.2
09	319 04.3	. . 29.6	145 56.4	15.5	13 10.5	12.0	54.2
10	334 04.2	30.3	160 30.9	15.4	13 22.5	12.0	54.2
11	349 04.2	31.0	175 05.3	15.3	13 34.5	12.0	54.1
12	4 04.1	S16 31.8	189 39.6	15.4	N13 46.5	11.9	54.1
13	19 04.1	32.5	204 14.0	15.2	13 58.4	11.8	54.1
14	34 04.0	33.2	218 48.2	15.3	14 10.2	11.8	54.1
15	49 04.0	. . 33.9	233 22.5	15.2	14 22.0	11.7	54.1
16	64 04.0	34.7	247 56.7	15.1	14 33.7	11.6	54.1
17	79 03.9	35.4	262 30.8	15.1	14 45.3	11.6	54.1
18	94 03.9	S16 36.1	277 04.9	15.1	N14 56.9	11.6	54.1
19	109 03.8	36.8	291 39.0	15.0	15 08.5	11.5	54.1
20	124 03.8	37.6	306 13.0	15.0	15 20.0	11.4	54.1
21	139 03.8	. . 38.3	320 47.0	14.9	15 31.4	11.3	54.1
22	154 03.7	39.0	335 20.9	14.9	15 42.7	11.3	54.1
23	169 03.7	39.7	349 54.8	14.8	N15 54.0	11.2	54.1
	SD 16.2	*d* 0.7	SD 14.9		14.8		14.8

Lat.	Twilight Naut.	Twilight Civil	Sunrise	Moonrise 6	Moonrise 7	Moonrise 8	Moonrise 9
°	h m	h m	h m	h m	h m	h m	h m
N 72	06 28	07 55	09 35	15 16	14 45	14 03	▭
N 70	06 21	07 38	08 59	15 22	14 59	14 30	13 41
68	06 16	07 25	08 35	15 26	15 10	14 51	14 23
66	06 11	07 14	08 15	15 30	15 19	15 07	14 51
64	06 07	07 05	08 00	15 33	15 27	15 21	15 13
62	06 03	06 57	07 47	15 36	15 34	15 32	15 31
60	05 59	06 50	07 36	15 39	15 40	15 42	15 46
N 58	05 56	06 44	07 27	15 41	15 45	15 51	15 58
56	05 53	06 38	07 19	15 43	15 50	15 58	16 09
54	05 51	06 33	07 11	15 45	15 54	16 05	16 19
52	05 48	06 28	07 05	15 46	15 58	16 11	16 27
50	05 46	06 24	06 59	15 48	16 02	16 17	16 35
45	05 40	06 15	06 46	15 51	16 09	16 29	16 52
N 40	05 35	06 07	06 35	15 54	16 16	16 39	17 05
35	05 30	06 00	06 26	15 56	16 21	16 48	17 17
30	05 25	05 53	06 18	15 58	16 26	16 55	17 27
20	05 15	05 41	06 04	16 02	16 35	17 09	17 45
N 10	05 05	05 30	05 52	16 06	16 42	17 20	18 00
0	04 54	05 19	05 40	16 09	16 49	17 31	18 15
S 10	04 41	05 06	05 28	16 12	16 57	17 42	18 30
20	04 25	04 53	05 16	16 15	17 04	17 54	18 45
30	04 05	04 36	05 01	16 19	17 13	18 08	19 03
35	03 52	04 25	04 53	16 21	17 18	18 16	19 14
40	03 37	04 13	04 43	16 24	17 24	18 25	19 26
45	03 18	03 58	04 31	16 27	17 31	18 36	19 41
S 50	02 52	03 40	04 17	16 31	17 39	18 49	19 59
52	02 39	03 31	04 11	16 32	17 43	18 55	20 07
54	02 24	03 21	04 03	16 34	17 48	19 02	20 16
56	02 05	03 09	03 55	16 36	17 52	19 09	20 27
58	01 42	02 56	03 46	16 39	17 58	19 18	20 39
S 60	01 08	02 40	03 35	16 41	18 04	19 28	20 54

Lat.	Sunset	Twilight Civil	Twilight Naut.	Moonset 6	Moonset 7	Moonset 8	Moonset 9
°	h m	h m	h m	h m	h m	h m	h m
N 72	13 51	15 31	16 58	04 07	06 05	08 15	▭
N 70	14 27	15 48	17 05	04 05	05 54	07 49	10 08
68	14 52	16 01	17 10	04 03	05 45	07 30	09 28
66	15 11	16 12	17 15	04 02	05 37	07 15	09 00
64	15 26	16 22	17 19	04 00	05 31	07 03	08 39
62	15 39	16 30	17 23	03 59	05 25	06 52	08 22
60	15 50	16 37	17 27	03 58	05 21	06 44	08 09
N 58	16 00	16 43	17 30	03 58	05 16	06 36	07 57
56	16 08	16 48	17 33	03 57	05 13	06 29	07 47
54	16 15	16 54	17 36	03 56	05 09	06 23	07 38
52	16 22	16 58	17 39	03 56	05 06	06 18	07 30
50	16 28	17 02	17 41	03 55	05 04	06 13	07 22
45	16 41	17 12	17 47	03 54	04 58	06 02	07 07
N 40	16 52	17 20	17 52	03 53	04 53	05 54	06 55
35	17 01	17 27	17 57	03 52	04 49	05 46	06 44
30	17 09	17 34	18 02	03 51	04 45	05 40	06 35
20	17 23	17 46	18 12	03 50	04 39	05 28	06 19
N 10	17 35	17 57	18 22	03 49	04 33	05 18	06 05
0	17 47	18 09	18 34	03 48	04 28	05 09	05 52
S 10	17 59	18 21	18 47	03 47	04 23	05 00	05 39
20	18 12	18 35	19 02	03 45	04 17	04 51	05 26
30	18 27	18 52	19 23	03 44	04 11	04 40	05 10
35	18 35	19 03	19 36	03 43	04 08	04 33	05 01
40	18 45	19 15	19 51	03 42	04 04	04 26	04 51
45	18 57	19 30	20 11	03 41	03 59	04 18	04 39
S 50	19 11	19 49	20 37	03 40	03 53	04 08	04 24
52	19 18	19 58	20 50	03 40	03 51	04 03	04 18
54	19 25	20 08	21 06	03 39	03 48	03 58	04 10
56	19 34	20 20	21 25	03 38	03 45	03 52	04 02
58	19 43	20 33	21 50	03 37	03 42	03 46	03 52
S 60	19 54	20 50	22 25	03 37	03 38	03 39	03 42

Day	SUN Eqn. of Time 00^h	SUN Eqn. of Time 12^h	SUN Mer. Pass.	MOON Mer. Pass. Upper	MOON Mer. Pass. Lower	MOON Age	MOON Phase
d	m s	m s	h m	h m	h m	d %	
6	16 24	16 23	11 44	22 18	09 58	12 94	
7	16 21	16 20	11 44	22 59	10 39	13 98	○
8	16 18	16 17	11 44	23 42	11 20	14 100	

UT	ARIES	VENUS −3.9		MARS −0.9		JUPITER −1.9		SATURN −0.1		STARS		
	GHA	GHA	Dec	GHA	Dec	GHA	Dec	GHA	Dec	Name	SHA	Dec
d h	° ′	° ′	° ′	° ′	° ′	° ′	° ′	° ′	° ′		° ′	°
9 00	47 45.3	162 04.1	S22 07.8	64 58.5	S 9 21.3	241 47.9	N 7 04.6	303 38.7	N22 04.6	Acamar	315 23.6	S40 17
01	62 47.8	177 03.3	08.4	80 00.0	20.7	256 50.1	04.5	318 41.3	04.6	Achernar	335 31.6	S57 13
02	77 50.3	192 02.5	09.0	95 01.5	20.2	271 52.2	04.4	333 43.8	04.6	Acrux	173 18.7	S63 06
03	92 52.7	207 01.6	. . 09.7	110 03.1	. . 19.7	286 54.3	. . 04.2	348 46.3	. . 04.6	Adhara	255 18.3	S28 58
04	107 55.2	222 00.8	10.3	125 04.6	19.1	301 56.4	04.1	3 48.9	04.6	Aldebaran	290 57.8	N16 31
05	122 57.7	237 00.0	10.9	140 06.1	18.6	316 58.5	04.0	18 51.4	04.6			
06	138 00.1	251 59.2	S22 11.5	155 07.6	S 9 18.1	332 00.6	N 7 03.8	33 53.9	N22 04.6	Alioth	166 27.6	N55 56
07	153 02.6	266 58.3	12.1	170 09.1	17.5	347 02.7	03.7	48 56.5	04.6	Alkaid	153 05.2	N49 17
08	168 05.0	281 57.5	12.7	185 10.5	17.0	2 04.9	03.6	63 59.0	04.6	Al Na'ir	27 53.0	S46 56
S 09	183 07.5	296 56.7	. . 13.3	200 12.0	. . 16.5	17 07.0	. . 03.4	79 01.5	. . 04.6	Alnilam	275 53.8	S 1 11
U 10	198 10.0	311 55.8	13.9	215 13.5	15.9	32 09.1	03.3	94 04.1	04.6	Alphard	218 03.6	S 8 40
N 11	213 12.4	326 55.0	14.5	230 15.0	15.4	47 11.2	03.1	109 06.6	04.6			
D 12	228 14.9	341 54.2	S22 15.1	245 16.5	S 9 14.8	62 13.3	N 7 03.0	124 09.2	N22 04.6	Alphecca	126 17.8	N26 42
A 13	243 17.4	356 53.3	15.7	260 18.0	14.3	77 15.4	02.9	139 11.7	04.7	Alpheratz	357 51.2	N29 06
Y 14	258 19.8	11 52.5	16.3	275 19.5	13.8	92 17.5	02.7	154 14.2	04.7	Altair	62 15.8	N 8 52
15	273 22.3	26 51.7	. . 16.9	290 21.0	. . 13.2	107 19.7	. . 02.6	169 16.8	. . 04.7	Ankaa	353 22.7	S42 17
16	288 24.8	41 50.8	17.5	305 22.5	12.7	122 21.8	02.5	184 19.3	04.7	Antares	112 36.0	S26 26
17	303 27.2	56 50.0	18.1	320 24.0	12.2	137 23.9	02.3	199 21.8	04.7			
18	318 29.7	71 49.2	S22 18.7	335 25.5	S 9 11.6	152 26.0	N 7 02.2	214 24.4	N22 04.7	Arcturus	146 03.0	N19 09
19	333 32.2	86 48.3	19.3	350 27.0	11.1	167 28.1	02.1	229 26.9	04.7	Atria	107 45.3	S69 02
20	348 34.6	101 47.5	19.9	5 28.5	10.5	182 30.2	01.9	244 29.4	04.7	Avior	234 21.1	S59 30
21	3 37.1	116 46.7	. . 20.5	20 30.0	. . 10.0	197 32.4	. . 01.8	259 32.0	. . 04.7	Bellatrix	278 39.9	N 6 21
22	18 39.5	131 45.8	21.1	35 31.5	09.5	212 34.5	01.7	274 34.5	04.7	Betelgeuse	271 09.3	N 7 24
23	33 42.0	146 45.0	21.6	50 33.0	08.9	227 36.6	01.5	289 37.1	04.7			
10 00	48 44.5	161 44.2	S22 22.2	65 34.4	S 9 08.4	242 38.7	N 7 01.4	304 39.6	N22 04.7	Canopus	263 59.2	S52 41
01	63 46.9	176 43.3	22.8	80 35.9	07.8	257 40.8	01.2	319 42.1	04.7	Capella	280 45.3	N46 00
02	78 49.4	191 42.5	23.4	95 37.4	07.3	272 42.9	01.1	334 44.7	04.8	Deneb	49 36.8	N45 17
03	93 51.9	206 41.7	. . 24.0	110 38.9	. . 06.8	287 45.1	. . 01.0	349 47.2	. . 04.8	Denebola	182 41.6	N14 33
04	108 54.3	221 40.8	24.6	125 40.4	06.2	302 47.2	00.8	4 49.8	04.8	Diphda	349 03.2	S17 58
05	123 56.8	236 40.0	25.2	140 41.9	05.7	317 49.3	00.7	19 52.3	04.8			
06	138 59.3	251 39.1	S22 25.7	155 43.4	S 9 05.1	332 51.4	N 7 00.6	34 54.8	N22 04.8	Dubhe	194 00.9	N61 43
07	154 01.7	266 38.3	26.3	170 44.9	04.6	347 53.5	00.4	49 57.4	04.8	Elnath	278 21.9	N28 36
08	169 04.2	281 37.5	26.9	185 46.3	04.1	2 55.6	00.3	64 59.9	04.8	Eltanin	90 50.1	N51 29
M 09	184 06.6	296 36.6	. . 27.5	200 47.8	. . 03.5	17 57.8	. . 00.2	80 02.4	. . 04.8	Enif	33 54.6	N 9 53
O 10	199 09.1	311 35.8	28.1	215 49.3	03.0	32 59.9	7 00.0	95 05.0	04.8	Fomalhaut	15 32.1	S29 36
N 11	214 11.6	326 34.9	28.6	230 50.8	02.4	48 02.0	6 59.9	110 07.5	04.8			
D 12	229 14.0	341 34.1	S22 29.2	245 52.3	S 9 01.9	63 04.1	N 6 59.8	125 10.1	N22 04.8	Gacrux	172 10.1	S57 07
A 13	244 16.5	356 33.2	29.8	260 53.8	01.3	78 06.2	59.6	140 12.6	04.8	Gienah	176 00.4	S17 33
Y 14	259 19.0	11 32.4	30.3	275 55.2	00.8	93 08.4	59.5	155 15.1	04.8	Hadar	148 59.6	S60 23
15	274 21.4	26 31.6	. . 30.9	290 56.7	9 00.3	108 10.5	. . 59.4	170 17.7	. . 04.9	Hamal	328 09.1	N23 28
16	289 23.9	41 30.7	31.5	305 58.2	8 59.7	123 12.6	59.2	185 20.2	04.9	Kaus Aust.	83 54.1	S34 23
17	304 26.4	56 29.9	32.1	320 59.7	59.2	138 14.7	59.1	200 22.8	04.9			
18	319 28.8	71 29.0	S22 32.6	336 01.2	S 8 58.6	153 16.8	N 6 59.0	215 25.3	N22 04.9	Kochab	137 20.5	N74 08
19	334 31.3	86 28.2	33.2	351 02.6	58.1	168 19.0	58.8	230 27.9	04.9	Markab	13 45.8	N15 13
20	349 33.8	101 27.3	33.8	6 04.1	57.5	183 21.1	58.7	245 30.4	04.9	Menkar	314 22.7	N 4 06
21	4 36.2	116 26.5	. . 34.3	21 05.6	. . 57.0	198 23.2	. . 58.6	260 32.9	. . 04.9	Menkent	148 17.1	S36 23
22	19 38.7	131 25.6	34.9	36 07.1	56.5	213 25.3	58.4	275 35.5	04.9	Miaplacidus	221 41.6	S69 43
23	34 41.1	146 24.8	35.4	51 08.6	55.9	228 27.5	58.3	290 38.0	04.9			
11 00	49 43.6	161 23.9	S22 36.0	66 10.0	S 8 55.4	243 29.6	N 6 58.2	305 40.6	N22 04.9	Mirfak	308 50.9	N49 52
01	64 46.1	176 23.1	36.6	81 11.5	54.8	258 31.7	58.0	320 43.1	04.9	Nunki	76 07.9	S26 17
02	79 48.5	191 22.2	37.1	96 13.0	54.3	273 33.8	57.9	335 45.7	04.9	Peacock	53 31.1	S56 41
03	94 51.0	206 21.4	. . 37.7	111 14.5	. . 53.7	288 35.9	. . 57.8	350 48.2	. . 05.0	Pollux	243 36.8	N28 01
04	109 53.5	221 20.5	38.2	126 15.9	53.2	303 38.1	57.6	5 50.7	05.0	Procyon	245 07.5	N 5 11
05	124 55.9	236 19.7	38.8	141 17.4	52.6	318 40.2	57.5	20 53.3	05.0			
06	139 58.4	251 18.8	S22 39.4	156 18.9	S 8 52.1	333 42.3	N 6 57.4	35 55.8	N22 05.0	Rasalhague	96 13.8	N12 33
07	155 00.9	266 18.0	39.9	171 20.3	51.6	348 44.4	57.2	50 58.4	05.0	Regulus	207 51.6	N11 57
T 08	170 03.3	281 17.1	40.5	186 21.8	51.0	3 46.6	57.1	66 00.9	05.0	Rigel	281 19.1	S 8 11
U 09	185 05.8	296 16.3	. . 41.0	201 23.3	. . 50.5	18 48.7	. . 57.0	81 03.5	. . 05.0	Rigil Kent.	140 03.1	S60 50
E 10	200 08.3	311 15.4	41.6	216 24.8	49.9	33 50.8	56.8	96 06.0	05.0	Sabik	102 21.6	S15 43
S 11	215 10.7	326 14.6	42.1	231 26.2	49.4	48 52.9	56.7	111 08.6	05.0			
D 12	230 13.2	341 13.7	S22 42.7	246 27.7	S 8 48.8	63 55.0	N 6 56.6	126 11.1	N22 05.0	Schedar	349 49.0	N56 31
A 13	245 15.6	356 12.9	43.2	261 29.2	48.3	78 57.2	56.4	141 13.6	05.0	Shaula	96 32.6	S37 06
Y 14	260 18.1	11 12.0	43.7	276 30.6	47.7	93 59.3	56.3	156 16.2	05.0	Sirius	258 40.2	S16 43
15	275 20.6	26 11.2	. . 44.3	291 32.1	. . 47.2	109 01.4	. . 56.2	171 18.7	. . 05.1	Spica	158 39.6	S11 10
16	290 23.0	41 10.3	44.8	306 33.6	46.6	124 03.5	56.1	186 21.3	05.1	Suhail	222 58.1	S43 26
17	305 25.5	56 09.5	45.4	321 35.0	46.1	139 05.7	55.9	201 23.8	05.1			
18	320 28.0	71 08.6	S22 45.9	336 36.5	S 8 45.5	154 07.8	N 6 55.8	216 26.4	N22 05.1	Vega	80 44.4	N38 47
19	335 30.4	86 07.7	46.5	351 38.0	45.0	169 09.9	55.7	231 28.9	05.1	Zuben'ubi	137 14.2	S16 01
20	350 32.9	101 06.9	47.0	6 39.4	44.4	184 12.0	55.5	246 31.5	05.1		SHA	Mer.Pa
21	5 35.4	116 06.0	. . 47.5	21 40.9	. . 43.9	199 14.2	. . 55.4	261 34.0	. . 05.1		° ′	h m
22	20 37.8	131 05.2	48.1	36 42.4	43.3	214 16.3	55.3	276 36.6	05.1	Venus	112 59.7	13 14
23	35 40.3	146 04.3	48.6	51 43.8	42.8	229 18.4	55.1	291 39.1	05.1	Mars	16 50.0	19 36
	h m									Jupiter	193 54.2	7 48
Mer. Pass.	20 41.6	*v* −0.8	*d* 0.6	*v* 1.5	*d* 0.5	*v* 2.1	*d* 0.1	*v* 2.5	*d* 0.0	Saturn	255 55.1	3 41

UT		SUN GHA	SUN Dec	MOON GHA	v	MOON Dec	d	HP
d h		° ′	° ′	° ′	′	° ′	′	′
9 00		184 03.6	S16 40.4	4 28.6	14.8	N16 05.2	11.2	54.1
01		199 03.6	41.2	19 02.4	14.7	16 16.4	11.1	54.1
02		214 03.5	41.9	33 36.1	14.7	16 27.5	11.0	54.1
03		229 03.5	. . 42.6	48 09.8	14.6	16 38.5	10.9	54.1
04		244 03.4	43.3	62 43.4	14.6	16 49.4	10.9	54.0
05		259 03.4	44.1	77 17.0	14.6	17 00.3	10.8	54.0
06		274 03.3	S16 44.8	91 50.6	14.4	N17 11.1	10.8	54.0
07		289 03.3	45.5	106 24.0	14.5	17 21.9	10.6	54.0
08		304 03.2	46.2	120 57.5	14.4	17 32.5	10.6	54.0
09	S	319 03.2	. . 46.9	135 30.9	14.3	17 43.1	10.6	54.0
10	U	334 03.1	47.6	150 04.2	14.3	17 53.7	10.4	54.0
11	N	349 03.1	48.4	164 37.5	14.2	18 04.1	10.4	54.0
12	D	4 03.1	S16 49.1	179 10.7	14.1	N18 14.5	10.3	54.0
13	A	19 03.0	49.8	193 43.8	14.2	18 24.8	10.2	54.0
14	Y	34 03.0	50.5	208 17.0	14.0	18 35.0	10.2	54.0
15		49 02.9	. . 51.2	222 50.0	14.0	18 45.2	10.0	54.0
16		64 02.9	51.9	237 23.0	14.0	18 55.2	10.0	54.0
17		79 02.8	52.6	251 56.0	13.9	19 05.2	9.9	54.0
18		94 02.7	S16 53.4	266 28.9	13.8	N19 15.1	9.9	54.0
19		109 02.7	54.1	281 01.7	13.8	19 25.0	9.7	54.0
20		124 02.6	54.8	295 34.5	13.7	19 34.7	9.7	54.0
21		139 02.6	. . 55.5	310 07.2	13.7	19 44.4	9.6	54.0
22		154 02.5	56.2	324 39.9	13.6	19 54.0	9.5	54.0
23		169 02.5	56.9	339 12.5	13.5	20 03.5	9.4	54.0
10 00		184 02.4	S16 57.6	353 45.0	13.5	N20 12.9	9.3	54.0
01		199 02.4	58.3	8 17.5	13.5	20 22.2	9.3	54.0
02		214 02.3	59.0	22 50.0	13.4	20 31.5	9.1	54.0
03		229 02.3	16 59.8	37 22.4	13.3	20 40.6	9.1	54.0
04		244 02.2	17 00.5	51 54.7	13.3	20 49.7	9.0	54.0
05		259 02.2	01.2	66 27.0	13.2	20 58.7	8.9	54.0
06		274 02.1	S17 01.9	80 59.2	13.1	N21 07.6	8.8	54.0
07		289 02.0	02.6	95 31.3	13.1	21 16.4	8.7	54.0
08		304 02.0	03.3	110 03.4	13.0	21 25.1	8.6	54.0
09	M	319 01.9	. . 04.0	124 35.4	13.0	21 33.7	8.5	54.0
10	O	334 01.9	04.7	139 07.4	12.9	21 42.2	8.4	54.0
11	N	349 01.8	05.4	153 39.3	12.9	21 50.6	8.4	54.0
12	D	4 01.8	S17 06.1	168 11.2	12.8	N21 59.0	8.2	54.0
13	A	19 01.7	06.8	182 43.0	12.8	22 07.2	8.2	54.0
14	Y	34 01.6	07.5	197 14.8	12.6	22 15.4	8.0	54.0
15		49 01.6	. . 08.2	211 46.4	12.7	22 23.4	8.0	54.0
16		64 01.5	08.9	226 18.1	12.6	22 31.4	7.8	54.0
17		79 01.5	09.6	240 49.7	12.5	22 39.2	7.8	54.0
18		94 01.4	S17 10.3	255 21.2	12.4	N22 47.0	7.6	54.0
19		109 01.3	11.0	269 52.6	12.4	22 54.6	7.6	54.0
20		124 01.3	11.7	284 24.0	12.4	23 02.2	7.4	54.0
21		139 01.2	. . 12.4	298 55.4	12.2	23 09.6	7.4	54.0
22		154 01.2	13.1	313 26.6	12.3	23 17.0	7.2	54.0
23		169 01.1	13.8	327 57.9	12.1	23 24.2	7.2	54.0
11 00		184 01.0	S17 14.5	342 29.0	12.2	N23 31.4	7.0	54.0
01		199 01.0	15.2	357 00.2	12.0	23 38.4	6.9	54.0
02		214 00.9	15.9	11 31.2	12.0	23 45.3	6.9	54.0
03		229 00.8	. . 16.6	26 02.2	12.0	23 52.2	6.7	54.0
04		244 00.8	17.3	40 33.2	11.9	23 58.9	6.6	54.0
05		259 00.7	18.0	55 04.1	11.8	24 05.5	6.5	54.0
06		274 00.6	S17 18.7	69 34.9	11.8	N24 12.0	6.4	54.0
07		289 00.6	19.4	84 05.7	11.7	24 18.4	6.3	54.0
08	T	304 00.5	20.1	98 36.4	11.7	24 24.7	6.2	54.0
09	U	319 00.5	. . 20.8	113 07.1	11.6	24 30.9	6.0	54.0
10	E	334 00.4	21.5	127 37.7	11.5	24 36.9	6.0	54.0
11	S	349 00.3	22.2	142 08.2	11.6	24 42.9	5.8	54.0
12	D	4 00.3	S17 22.9	156 38.8	11.4	N24 48.7	5.8	54.0
13	A	19 00.2	23.5	171 09.2	11.4	24 54.5	5.6	54.0
14	Y	34 00.1	24.2	185 39.6	11.4	25 00.1	5.5	54.0
15		49 00.1	. . 24.9	200 10.0	11.3	25 05.6	5.4	54.0
16		64 00.0	25.6	214 40.3	11.3	25 11.0	5.3	54.0
17		78 59.9	26.3	229 10.6	11.2	25 16.3	5.1	54.0
18		93 59.8	S17 27.0	243 40.8	11.1	N25 21.4	5.1	54.0
19		108 59.8	27.7	258 10.9	11.1	25 26.5	4.9	54.0
20		123 59.7	28.4	272 41.0	11.1	25 31.4	4.8	54.1
21		138 59.6	. . 29.1	287 11.1	11.0	25 36.2	4.7	54.1
22		153 59.6	29.7	301 41.1	11.0	25 40.9	4.5	54.1
23		168 59.5	30.4	316 11.1	10.9	N25 45.4	4.5	54.1
		SD 16.2	*d* 0.7	SD 14.7		14.7		14.7

Lat.	Twilight Naut.	Twilight Civil	Sunrise	Moonrise 9	Moonrise 10	Moonrise 11	Moonrise 12
°	h m	h m	h m	h m	h m	h m	h m
N 72	06 39	08 09	09 59	□	□	□	□
N 70	06 31	07 50	09 16	13 41	□	□	□
68	06 25	07 35	08 48	14 23	12 44	□	□
66	06 19	07 23	08 27	14 51	14 25	□	□
64	06 14	07 13	08 10	15 13	15 04	14 48	□
62	06 10	07 04	07 56	15 31	15 31	15 34	15 47
60	06 06	06 57	07 44	15 46	15 52	16 05	16 29
N 58	06 02	06 50	07 34	15 58	16 10	16 28	16 58
56	05 59	06 44	07 25	16 09	16 24	16 46	17 20
54	05 56	06 38	07 17	16 19	16 37	17 02	17 38
52	05 53	06 33	07 10	16 27	16 48	17 16	17 53
50	05 50	06 29	07 04	16 35	16 58	17 28	18 07
45	05 44	06 19	06 50	16 52	17 19	17 53	18 34
N 40	05 38	06 10	06 39	17 05	17 36	18 12	18 56
35	05 32	06 02	06 29	17 17	17 50	18 29	19 14
30	05 27	05 55	06 21	17 27	18 03	18 43	19 29
20	05 16	05 43	06 06	17 45	18 25	19 08	19 56
N 10	05 05	05 31	05 53	18 00	18 43	19 29	20 18
0	04 54	05 19	05 40	18 15	19 01	19 49	20 40
S 10	04 40	05 06	05 28	18 30	19 19	20 09	21 01
20	04 24	04 51	05 15	18 45	19 38	20 31	21 24
30	04 03	04 33	04 59	19 03	20 00	20 56	21 50
35	03 49	04 22	04 50	19 14	20 13	21 11	22 06
40	03 33	04 10	04 40	19 26	20 28	21 28	22 24
45	03 13	03 54	04 27	19 41	20 46	21 48	22 47
S 50	02 45	03 35	04 12	19 59	21 08	22 14	23 15
52	02 31	03 25	04 05	20 07	21 19	22 27	23 29
54	02 15	03 14	03 58	20 16	21 31	22 42	23 45
56	01 55	03 02	03 49	20 27	21 45	22 59	24 04
58	01 28	02 48	03 39	20 39	22 01	23 20	24 27
S 60	00 46	02 30	03 28	20 54	22 21	23 46	24 58

Lat.	Sunset	Twilight Civil	Twilight Naut.	Moonset 9	Moonset 10	Moonset 11	Moonset 12
°	h m	h m	h m	h m	h m	h m	h m
N 72	13 28	15 18	16 47	□	□	□	□
N 70	14 11	15 36	16 55	10 08	□	□	□
68	14 39	15 51	17 02	09 28	12 40	□	□
66	15 00	16 04	17 07	09 00	11 00	□	□
64	15 17	16 14	17 12	08 39	10 22	12 18	□
62	15 31	16 23	17 17	08 22	09 56	11 32	13 03
60	15 43	16 30	17 21	08 09	09 35	11 02	12 21
N 58	15 53	16 37	17 25	07 57	09 19	10 39	11 52
56	16 02	16 43	17 28	07 47	09 04	10 21	11 31
54	16 10	16 49	17 31	07 38	08 52	10 05	11 13
52	16 17	16 54	17 34	07 30	08 42	09 52	10 57
50	16 23	16 58	17 37	07 22	08 32	09 40	10 44
45	16 37	17 08	17 44	07 07	08 12	09 16	10 17
N 40	16 49	17 17	17 50	06 55	07 56	08 57	09 55
35	16 58	17 25	17 55	06 44	07 42	08 41	09 38
30	17 07	17 32	18 01	06 35	07 31	08 27	09 22
20	17 22	17 45	18 11	06 19	07 10	08 03	08 57
N 10	17 35	17 57	18 22	06 05	06 53	07 43	08 34
0	17 47	18 09	18 34	05 52	06 37	07 24	08 13
S 10	18 00	18 22	18 48	05 39	06 21	07 05	07 53
20	18 13	18 37	19 04	05 26	06 03	06 45	07 30
30	18 29	18 55	19 26	05 10	05 44	06 22	07 05
35	18 38	19 06	19 39	05 01	05 32	06 08	06 50
40	18 49	19 19	19 56	04 51	05 19	05 52	06 32
45	19 01	19 35	20 16	04 39	05 04	05 34	06 11
S 50	19 16	19 54	20 44	04 24	04 45	05 10	05 44
52	19 23	20 04	20 58	04 18	04 36	04 59	05 31
54	19 31	20 15	21 15	04 10	04 26	04 47	05 17
56	19 40	20 27	21 36	04 02	04 14	04 32	04 59
58	19 50	20 42	22 04	03 52	04 01	04 15	04 38
S 60	20 02	21 00	22 50	03 42	03 46	03 55	04 12

Day	SUN Eqn. of Time 00^h	SUN Eqn. of Time 12^h	SUN Mer. Pass.	MOON Mer. Pass. Upper	MOON Mer. Pass. Lower	MOON Age	MOON %	Phase
d	m s	m s	h m	h m	h m	d	%	
9	16 15	16 12	11 44	24 26	12 03	15	100	
10	16 10	16 07	11 44	00 26	12 49	16	98	○
11	16 04	16 01	11 44	01 12	13 37	17	95	

UT	ARIES	VENUS −3.9		MARS −0.8		JUPITER −1.9		SATURN −0.1	
	GHA	GHA	Dec	GHA	Dec	GHA	Dec	GHA	Dec
d h	° ′	° ′	° ′	° ′	° ′	° ′	° ′	° ′	° ′
12 00	50 42.8	161 03.5	S22 49.1	66 45.3	S 8 42.3	244 20.5	N 6 55.0	306 41.7	N22 05.1
01	65 45.2	176 02.6	49.7	81 46.8	41.7	259 22.7	54.9	321 44.2	05.1
02	80 47.7	191 01.7	50.2	96 48.2	41.2	274 24.8	54.7	336 46.7	05.2
03	95 50.1	206 00.9 . .	50.7	111 49.7 . .	40.6	289 26.9 . .	54.6	351 49.3 . .	05.2
04	110 52.6	221 00.0	51.3	126 51.1	40.1	304 29.1	54.5	6 51.8	05.2
05	125 55.1	235 59.2	51.8	141 52.6	39.5	319 31.2	54.3	21 54.4	05.2
06	140 57.5	250 58.3	S22 52.3	156 54.1	S 8 39.0	334 33.3	N 6 54.2	36 56.9	N22 05.2
W 07	156 00.0	265 57.4	52.8	171 55.5	38.4	349 35.4	54.1	51 59.5	05.2
E 08	171 02.5	280 56.6	53.4	186 57.0	37.9	4 37.6	54.0	67 02.0	05.2
D 09	186 04.9	295 55.7 . .	53.9	201 58.4 . .	37.3	19 39.7 . .	53.8	82 04.6 . .	05.2
N 10	201 07.4	310 54.9	54.4	216 59.9	36.8	34 41.8	53.7	97 07.1	05.2
E 11	216 09.9	325 54.0	54.9	232 01.4	36.2	49 44.0	53.6	112 09.7	05.2
S 12	231 12.3	340 53.1	S22 55.4	247 02.8	S 8 35.7	64 46.1	N 6 53.4	127 12.2	N22 05.2
D 13	246 14.8	355 52.3	56.0	262 04.3	35.1	79 48.2	53.3	142 14.8	05.3
A 14	261 17.2	10 51.4	56.5	277 05.7	34.5	94 50.3	53.2	157 17.3	05.3
Y 15	276 19.7	25 50.5 . .	57.0	292 07.2 . .	34.0	109 52.5 . .	53.0	172 19.9 . .	05.3
16	291 22.2	40 49.7	57.5	307 08.6	33.4	124 54.6	52.9	187 22.4	05.3
17	306 24.6	55 48.8	58.0	322 10.1	32.9	139 56.7	52.8	202 25.0	05.3
18	321 27.1	70 47.9	S22 58.5	337 11.5	S 8 32.3	154 58.9	N 6 52.7	217 27.5	N22 05.3
19	336 29.6	85 47.1	59.1	352 13.0	31.8	170 01.0	52.5	232 30.1	05.3
20	351 32.0	100 46.2	22 59.6	7 14.4	31.2	185 03.1	52.4	247 32.6	05.3
21	6 34.5	115 45.3	23 00.1	22 15.9 . .	30.7	200 05.2 . .	52.3	262 35.2 . .	05.3
22	21 37.0	130 44.5	00.6	37 17.4	30.1	215 07.4	52.1	277 37.7	05.3
23	36 39.4	145 43.6	01.1	52 18.8	29.6	230 09.5	52.0	292 40.3	05.3
13 00	51 41.9	160 42.7	S23 01.6	67 20.3	S 8 29.0	245 11.6	N 6 51.9	307 42.9	N22 05.4
01	66 44.4	175 41.9	02.1	82 21.7	28.5	260 13.8	51.7	322 45.4	05.4
02	81 46.8	190 41.0	02.6	97 23.2	27.9	275 15.9	51.6	337 48.0	05.4
03	96 49.3	205 40.1 . .	03.1	112 24.6 . .	27.4	290 18.0 . .	51.5	352 50.5 . .	05.4
04	111 51.7	220 39.3	03.6	127 26.0	26.8	305 20.2	51.4	7 53.1	05.4
05	126 54.2	235 38.4	04.1	142 27.5	26.3	320 22.3	51.2	22 55.6	05.4
06	141 56.7	250 37.5	S23 04.6	157 28.9	S 8 25.7	335 24.4	N 6 51.1	37 58.2	N22 05.4
07	156 59.1	265 36.7	05.1	172 30.4	25.1	350 26.6	51.0	53 00.7	05.4
T 08	172 01.6	280 35.8	05.6	187 31.8	24.6	5 28.7	50.8	68 03.3	05.4
H 09	187 04.1	295 34.9 . .	06.1	202 33.3 . .	24.0	20 30.8 . .	50.7	83 05.8 . .	05.4
U 10	202 06.5	310 34.0	06.6	217 34.7	23.5	35 33.0	50.6	98 08.4	05.4
R 11	217 09.0	325 33.2	07.1	232 36.2	22.9	50 35.1	50.5	113 10.9	05.5
S 12	232 11.5	340 32.3	S23 07.6	247 37.6	S 8 22.4	65 37.2	N 6 50.3	128 13.5	N22 05.5
D 13	247 13.9	355 31.4	08.1	262 39.1	21.8	80 39.4	50.2	143 16.0	05.5
A 14	262 16.4	10 30.5	08.6	277 40.5	21.3	95 41.5	50.1	158 18.6	05.5
Y 15	277 18.9	25 29.7 . .	09.0	292 41.9 . .	20.7	110 43.6 . .	49.9	173 21.2 . .	05.5
16	292 21.3	40 28.8	09.5	307 43.4	20.1	125 45.8	49.8	188 23.7	05.5
17	307 23.8	55 27.9	10.0	322 44.8	19.6	140 47.9	49.7	203 26.3	05.5
18	322 26.2	70 27.0	S23 10.5	337 46.3	S 8 19.0	155 50.0	N 6 49.6	218 28.8	N22 05.5
19	337 28.7	85 26.2	11.0	352 47.7	18.5	170 52.2	49.4	233 31.4	05.5
20	352 31.2	100 25.3	11.5	7 49.2	17.9	185 54.3	49.3	248 33.9	05.5
21	7 33.6	115 24.4 . .	11.9	22 50.6 . .	17.4	200 56.4 . .	49.2	263 36.5 . .	05.6
22	22 36.1	130 23.5	12.4	37 52.0	16.8	215 58.6	49.0	278 39.1	05.6
23	37 38.6	145 22.7	12.9	52 53.5	16.3	231 00.7	48.9	293 41.6	05.6
14 00	52 41.0	160 21.8	S23 13.4	67 54.9	S 8 15.7	246 02.8	N 6 48.8	308 44.2	N22 05.6
01	67 43.5	175 20.9	13.9	82 56.3	15.1	261 05.0	48.7	323 46.7	05.6
02	82 46.0	190 20.0	14.3	97 57.8	14.6	276 07.1	48.5	338 49.3	05.6
03	97 48.4	205 19.2 . .	14.8	112 59.2 . .	14.0	291 09.2 . .	48.4	353 51.8 . .	05.6
04	112 50.9	220 18.3	15.3	128 00.7	13.5	306 11.4	48.3	8 54.4	05.6
05	127 53.4	235 17.4	15.8	143 02.1	12.9	321 13.5	48.2	23 57.0	05.6
06	142 55.8	250 16.5	S23 16.2	158 03.5	S 8 12.3	336 15.7	N 6 48.0	38 59.5	N22 05.6
07	157 58.3	265 15.6	16.7	173 05.0	11.8	351 17.8	47.9	54 02.1	05.6
08	173 00.7	280 14.8	17.2	188 06.4	11.2	6 19.9	47.8	69 04.6	05.7
F 09	188 03.2	295 13.9 . .	17.6	203 07.8 . .	10.7	21 22.1 . .	47.7	84 07.2 . .	05.7
R 10	203 05.7	310 13.0	18.1	218 09.3	10.1	36 24.2	47.5	99 09.7	05.7
I 11	218 08.1	325 12.1	18.6	233 10.7	09.6	51 26.3	47.4	114 12.3	05.7
D 12	233 10.6	340 11.2	S23 19.0	248 12.1	S 8 09.0	66 28.5	N 6 47.3	129 14.9	N22 05.7
A 13	248 13.1	355 10.3	19.5	263 13.6	08.4	81 30.6	47.1	144 17.4	05.7
Y 14	263 15.5	10 09.5	20.0	278 15.0	07.9	96 32.8	47.0	159 20.0	05.7
15	278 18.0	25 08.6 . .	20.4	293 16.4 . .	07.3	111 34.9 . .	46.9	174 22.5 . .	05.7
16	293 20.5	40 07.7	20.9	308 17.9	06.8	126 37.0	46.8	189 25.1	05.7
17	308 22.9	55 06.8	21.3	323 19.3	06.2	141 39.2	46.6	204 27.7	05.7
18	323 25.4	70 05.9	S23 21.8	338 20.7	S 8 05.6	156 41.3	N 6 46.5	219 30.2	N22 05.8
19	338 27.9	85 05.0	22.2	353 22.1	05.1	171 43.4	46.4	234 32.8	05.8
20	353 30.3	100 04.2	22.7	8 23.6	04.5	186 45.6	46.3	249 35.3	05.8
21	8 32.8	115 03.3 . .	23.1	23 25.0 . .	03.9	201 47.7 . .	46.1	264 37.9 . .	05.8
22	23 35.2	130 02.4	23.6	38 26.4	03.4	216 49.9	46.0	279 40.5	05.8
23	38 37.7	145 01.5	24.0	53 27.9	02.8	231 52.0	45.9	294 43.0	05.8
Mer. Pass.	h m 20 29.8	v −0.9	d 0.5	v 1.4	d 0.6	v 2.1	d 0.1	v 2.6	d 0.0

STARS Name	SHA	Dec
	° ′	° ′
Acamar	315 23.6	S40 17.3
Achernar	335 31.6	S57 13.2
Acrux	173 18.7	S63 06.9
Adhara	255 18.3	S28 58.
Aldebaran	290 57.8	N16 31.
Alioth	166 27.5	N55 56.
Alkaid	153 05.1	N49 17.6
Al Na'ir	27 53.0	S46 56.
Alnilam	275 53.8	S 1 11.
Alphard	218 03.6	S 8 40.
Alphecca	126 17.8	N26 42.2
Alpheratz	357 51.2	N29 06.
Altair	62 15.8	N 8 52.
Ankaa	353 22.7	S42 17.
Antares	112 36.0	S26 26.
Arcturus	146 02.9	N19 09.
Atria	107 45.3	S69 02.
Avior	234 21.1	S59 30.
Bellatrix	278 39.9	N 6 21.
Betelgeuse	271 09.2	N 7 24.
Canopus	263 59.1	S52 41.
Capella	280 45.3	N46 00.
Deneb	49 36.9	N45 17.
Denebola	182 41.5	N14 33.
Diphda	349 03.2	S17 58.
Dubhe	194 00.9	N61 43.
Elnath	278 21.9	N28 36.
Eltanin	90 50.1	N51 29.
Enif	33 54.6	N 9 53.
Fomalhaut	15 32.1	S29 36.
Gacrux	172 10.1	S57 07.
Gienah	176 00.4	S17 33.
Hadar	148 59.6	S60 23.
Hamal	328 09.1	N23 28.
Kaus Aust.	83 54.1	S34 23.
Kochab	137 20.5	N74 08.
Markab	13 45.8	N15 13.
Menkar	314 22.7	N 4 06.
Menkent	148 17.0	S36 23.
Miaplacidus	221 41.5	S69 43.
Mirfak	308 50.9	N49 52.
Nunki	76 07.9	S26 17.
Peacock	53 31.2	S56 43.
Pollux	243 36.8	N28 01.
Procyon	245 07.5	N 5 13.
Rasalhague	96 13.8	N12 33.
Regulus	207 51.6	N11 57.
Rigel	281 19.0	S 8 11.
Rigil Kent.	140 03.1	S60 50
Sabik	102 21.6	S15 43
Schedar	349 49.0	N56 33
Shaula	96 32.6	S37 06
Sirius	258 40.2	S16 43
Spica	158 39.6	S11 10
Suhail	222 58.1	S43 26
Vega	80 44.4	N38 47
Zuben'ubi	137 14.2	S16 03

	SHA	Mer. Pas
	° ′	h m
Venus	109 00.8	13 18
Mars	15 38.4	19 29
Jupiter	193 29.7	7 38
Saturn	256 01.0	3 29

UT	SUN GHA	SUN Dec	MOON GHA	v	MOON Dec	d	HP
d h	° ′	° ′	° ′	′	° ′	′	′
12 00	183 59.4	S17 31.1	330 41.0	10.9	N25 49.9	4.3	54.1
01	198 59.3	31.8	345 10.9	10.8	25 54.2	4.2	54.1
02	213 59.3	32.5	359 40.7	10.8	25 58.4	4.1	54.1
03	228 59.2	. . 33.2	14 10.5	10.7	26 02.5	4.0	54.1
04	243 59.1	33.8	28 40.2	10.7	26 06.5	3.8	54.1
05	258 59.1	34.5	43 09.9	10.7	26 10.3	3.7	54.1
06	273 59.0	S17 35.2	57 39.6	10.6	N26 14.0	3.6	54.1
W 07	288 58.9	35.9	72 09.2	10.6	26 17.6	3.5	54.1
E 08	303 58.8	36.6	86 38.8	10.6	26 21.1	3.3	54.1
D 09	318 58.8	. . 37.3	101 08.4	10.5	26 24.4	3.3	54.1
N 10	333 58.7	37.9	115 37.9	10.4	26 27.7	3.0	54.1
E 11	348 58.6	38.6	130 07.3	10.5	26 30.7	3.0	54.2
S 12	3 58.5	S17 39.3	144 36.8	10.4	N26 33.7	2.9	54.2
D 13	18 58.5	40.0	159 06.2	10.3	26 36.6	2.7	54.2
A 14	33 58.4	40.7	173 35.5	10.3	26 39.3	2.6	54.2
Y 15	48 58.3	. . 41.3	188 04.8	10.3	26 41.9	2.4	54.2
16	63 58.2	42.0	202 34.1	10.3	26 44.3	2.4	54.2
17	78 58.1	42.7	217 03.4	10.2	26 46.7	2.2	54.2
18	93 58.1	S17 43.4	231 32.6	10.2	N26 48.9	2.0	54.2
19	108 58.0	44.0	246 01.8	10.2	26 50.9	2.0	54.2
20	123 57.9	44.7	260 31.0	10.2	26 52.9	1.8	54.2
21	138 57.8	. . 45.4	275 00.2	10.1	26 54.7	1.7	54.3
22	153 57.8	46.1	289 29.3	10.1	26 56.4	1.5	54.3
23	168 57.7	46.7	303 58.4	10.0	26 57.9	1.5	54.3
13 00	183 57.6	S17 47.4	318 27.4	10.1	N26 59.4	1.3	54.3
01	198 57.5	48.1	332 56.5	10.0	27 00.7	1.1	54.3
02	213 57.4	48.8	347 25.5	10.0	27 01.8	1.1	54.3
03	228 57.4	. . 49.4	1 54.5	10.0	27 02.9	0.9	54.3
04	243 57.3	50.1	16 23.5	9.9	27 03.8	0.7	54.3
05	258 57.2	50.8	30 52.4	10.0	27 04.5	0.7	54.3
06	273 57.1	S17 51.4	45 21.4	9.9	N27 05.2	0.5	54.4
T 07	288 57.0	52.1	59 50.3	9.9	27 05.7	0.3	54.4
08	303 56.9	52.8	74 19.2	9.9	27 06.0	0.3	54.4
H 09	318 56.9	. . 53.4	88 48.1	9.8	27 06.3	0.1	54.4
U 10	333 56.8	54.1	103 16.9	9.9	27 06.4	0.1	54.4
R 11	348 56.7	54.8	117 45.8	9.8	27 06.3	0.1	54.4
S 12	3 56.6	S17 55.4	132 14.6	9.9	N27 06.2	0.3	54.4
D 13	18 56.5	56.1	146 43.5	9.8	27 05.9	0.4	54.4
A 14	33 56.4	56.8	161 12.3	9.8	27 05.5	0.6	54.5
Y 15	48 56.3	. . 57.4	175 41.1	9.8	27 04.9	0.7	54.5
16	63 56.3	58.1	190 09.9	9.8	27 04.2	0.8	54.5
17	78 56.2	58.8	204 38.7	9.8	27 03.4	1.0	54.5
18	93 56.1	S17 59.4	219 07.5	9.8	N27 02.4	1.1	54.5
19	108 56.0	18 00.1	233 36.3	9.8	27 01.3	1.2	54.5
20	123 55.9	00.8	248 05.1	9.8	27 00.1	1.4	54.6
21	138 55.8	. . 01.4	262 33.9	9.7	26 58.7	1.5	54.6
22	153 55.7	02.1	277 02.6	9.8	26 57.2	1.6	54.6
23	168 55.6	02.7	291 31.4	9.8	26 55.6	1.8	54.6
14 00	183 55.6	S18 03.4	306 00.2	9.8	N26 53.8	1.9	54.6
01	198 55.5	04.1	320 29.0	9.7	26 51.9	2.0	54.6
02	213 55.4	04.7	334 57.7	9.8	26 49.9	2.2	54.6
03	228 55.3	. . 05.4	349 26.5	9.8	26 47.7	2.3	54.7
04	243 55.2	06.0	3 55.3	9.8	26 45.4	2.4	54.7
05	258 55.1	06.7	18 24.1	9.8	26 43.0	2.6	54.7
06	273 55.0	S18 07.3	32 52.9	9.8	N26 40.4	2.7	54.7
07	288 54.9	08.0	47 21.7	9.8	26 37.7	2.9	54.7
08	303 54.8	08.7	61 50.5	9.8	26 34.8	2.9	54.8
09	318 54.7	. . 09.3	76 19.3	9.8	26 31.9	3.1	54.8
10	333 54.6	10.0	90 48.1	9.9	26 28.8	3.3	54.8
11	348 54.5	10.6	105 17.0	9.8	26 25.5	3.4	54.8
12	3 54.4	S18 11.3	119 45.8	9.9	N26 22.1	3.5	54.8
13	18 54.4	11.9	134 14.7	9.8	26 18.6	3.6	54.9
14	33 54.3	12.6	148 43.5	9.9	26 15.0	3.8	54.9
15	48 54.2	. . 13.2	163 12.4	9.9	26 11.2	3.9	54.9
16	63 54.1	13.9	177 41.3	9.9	26 07.3	4.0	54.9
17	78 54.0	14.5	192 10.2	10.0	26 03.3	4.2	54.9
18	93 53.9	S18 15.2	206 39.2	9.9	N25 59.1	4.3	55.0
19	108 53.8	15.8	221 08.1	10.0	25 54.8	4.4	55.0
20	123 53.7	16.5	235 37.1	10.0	25 50.4	4.6	55.0
21	138 53.6	. . 17.1	250 06.1	10.0	25 45.8	4.7	55.0
22	153 53.5	17.8	264 35.1	10.0	25 41.1	4.8	55.0
23	168 53.4	18.4	279 04.1	10.1	N25 36.3	5.0	55.1
	SD 16.2	*d* 0.7	SD 14.8		14.8		14.9

Lat.	Twilight Naut.	Twilight Civil	Sunrise	Moonrise 12	Moonrise 13	Moonrise 14	Moonrise 15
°	h m	h m	h m	h m	h m	h m	h m
N 72	06 50	08 23	10 28	□	□	□	□
N 70	06 41	08 02	09 34	□	□	□	□
68	06 34	07 46	09 01	□	□	□	□
66	06 27	07 32	08 38	□	□	□	17 47
64	06 22	07 21	08 19	□	□	□	18 53
62	06 17	07 12	08 04	15 47	16 28	17 49	19 28
60	06 12	07 03	07 52	16 29	17 15	18 26	19 53
N 58	06 08	06 56	07 41	16 58	17 45	18 52	20 13
56	06 04	06 50	07 31	17 20	18 08	19 13	20 29
54	06 01	06 44	07 23	17 38	18 27	19 30	20 43
52	05 57	06 38	07 15	17 53	18 43	19 45	20 56
50	05 54	06 33	07 09	18 07	18 57	19 57	21 06
45	05 47	06 23	06 54	18 34	19 25	20 23	21 29
N 40	05 41	06 13	06 42	18 56	19 47	20 44	21 47
35	05 35	06 05	06 32	19 14	20 05	21 01	22 02
30	05 29	05 58	06 23	19 29	20 20	21 16	22 15
20	05 18	05 44	06 07	19 56	20 47	21 41	22 37
N 10	05 06	05 32	05 54	20 18	21 10	22 03	22 56
0	04 54	05 19	05 41	20 40	21 31	22 23	23 14
S 10	04 39	05 05	05 28	21 01	21 53	22 43	23 31
20	04 22	04 50	05 14	21 24	22 15	23 05	23 50
30	04 00	04 31	04 57	21 50	22 42	23 29	24 12
35	03 46	04 20	04 48	22 06	22 58	23 44	24 25
40	03 29	04 07	04 37	22 24	23 16	24 01	00 01
45	03 08	03 50	04 24	22 47	23 38	24 21	00 21
S 50	02 39	03 30	04 08	23 15	24 06	00 06	00 46
52	02 24	03 20	04 01	23 29	24 19	00 19	00 59
54	02 06	03 08	03 52	23 45	24 35	00 35	01 13
56	01 44	02 55	03 43	24 04	00 04	00 54	01 29
58	01 13	02 39	03 32	24 27	00 27	01 17	01 48
S 60	00 12	02 21	03 20	24 58	00 58	01 46	02 12

Lat.	Sunset	Twilight Civil	Twilight Naut.	Moonset 12	Moonset 13	Moonset 14	Moonset 15
°	h m	h m	h m	h m	h m	h m	h m
N 72	12 59	15 04	16 37	□	□	□	□
N 70	13 54	15 25	16 46	□	□	□	□
68	14 26	15 42	16 53	□	□	□	□
66	14 50	15 55	17 00	□	□	□	16 25
64	15 08	16 06	17 06	□	□	□	15 18
62	15 23	16 16	17 11	13 03	14 09	14 35	14 43
60	15 36	16 24	17 16	12 21	13 22	13 58	14 17
N 58	15 47	16 32	17 20	11 52	12 51	13 32	13 57
56	15 56	16 38	17 24	11 31	12 28	13 11	13 40
54	16 05	16 44	17 27	11 13	12 09	12 53	13 25
52	16 12	16 49	17 30	10 57	11 54	12 39	13 13
50	16 19	16 54	17 34	10 44	11 40	12 26	13 02
45	16 34	17 05	17 41	10 17	11 12	11 59	12 38
N 40	16 46	17 15	17 47	09 55	10 50	11 38	12 20
35	16 56	17 23	17 53	09 38	10 31	11 20	12 04
30	17 05	17 30	17 59	09 22	10 16	11 05	11 50
20	17 21	17 44	18 11	08 57	09 49	10 39	11 27
N 10	17 35	17 57	18 22	08 34	09 26	10 17	11 07
0	17 48	18 09	18 35	08 13	09 05	09 56	10 47
S 10	18 01	18 23	18 49	07 53	08 43	09 35	10 28
20	18 15	18 39	19 06	07 30	08 20	09 13	10 08
30	18 32	18 58	19 29	07 05	07 53	08 47	09 44
35	18 41	19 09	19 43	06 50	07 37	08 31	09 30
40	18 52	19 23	20 00	06 32	07 19	08 13	09 13
45	19 05	19 39	20 22	06 11	06 57	07 52	08 54
S 50	19 21	20 00	20 51	05 44	06 29	07 24	08 29
52	19 29	20 10	21 06	05 31	06 15	07 10	08 17
54	19 37	20 22	21 25	05 17	05 59	06 55	08 04
56	19 47	20 35	21 48	04 59	05 40	06 36	07 48
58	19 58	20 51	22 20	04 38	05 16	06 14	07 29
S 60	20 10	21 11	////	04 12	04 45	05 44	07 05

Day	SUN Eqn. of Time 00^h	SUN Eqn. of Time 12^h	SUN Mer. Pass.	MOON Mer. Pass. Upper	MOON Mer. Pass. Lower	Age	Phase
d	m s	m s	h m	h m	h m	d %	
12	15 58	15 54	11 44	02 01	14 27	18 90	
13	15 51	15 47	11 44	02 52	15 18	19 83	◑
14	15 42	15 38	11 44	03 44	16 10	20 76	

UT		ARIES	VENUS −3.9		MARS −0.8		JUPITER −1.9		SATURN −0.2		STARS		
		GHA	GHA	Dec	GHA	Dec	GHA	Dec	GHA	Dec	Name	SHA	Dec
d	h	° ′	° ′	° ′	° ′	° ′	° ′	° ′	° ′	° ′		° ′	°
15	00	53 40.2	160 00.6	S23 24.5	68 29.3	S 8 02.3	246 54.1	N 6 45.8	309 45.6	N22 05.8	Acamar	315 23.5	S40 17
	01	68 42.6	174 59.7	24.9	83 30.7	01.7	261 56.3	45.6	324 48.1	05.8	Achernar	335 31.6	S57 13
	02	83 45.1	189 58.8	25.4	98 32.1	01.1	276 58.4	45.5	339 50.7	05.8	Acrux	173 18.6	S63 06
	03	98 47.6	204 57.9	. . 25.8	113 33.6	. . 00.6	292 00.6	. . 45.4	354 53.3	. . 05.8	Adhara	255 18.3	S28 58
	04	113 50.0	219 57.1	26.3	128 35.0	8 00.0	307 02.7	45.3	9 55.8	05.9	Aldebaran	290 57.8	N16 31
	05	128 52.5	234 56.2	26.7	143 36.4	7 59.5	322 04.9	45.1	24 58.4	05.9			
	06	143 55.0	249 55.3	S23 27.2	158 37.8	S 7 58.9	337 07.0	N 6 45.0	40 01.0	N22 05.9	Alioth	166 27.5	N55 56
	07	158 57.4	264 54.4	27.6	173 39.2	58.3	352 09.1	44.9	55 03.5	05.9	Alkaid	153 05.1	N49 17
S	08	173 59.9	279 53.5	28.1	188 40.7	57.8	7 11.3	44.8	70 06.1	05.9	Al Na'ir	27 53.0	S46 56
A	09	189 02.3	294 52.6	. . 28.5	203 42.1	. . 57.2	22 13.4	. . 44.6	85 08.6	. . 05.9	Alnilam	275 53.8	S 1 11
T	10	204 04.8	309 51.7	28.9	218 43.5	56.6	37 15.6	44.5	100 11.2	05.9	Alphard	218 03.5	S 8 40
U	11	219 07.3	324 50.8	29.4	233 44.9	56.1	52 17.7	44.4	115 13.8	05.9			
R	12	234 09.7	339 49.9	S23 29.8	248 46.4	S 7 55.5	67 19.8	N 6 44.3	130 16.3	N22 05.9	Alphecca	126 17.7	N26 42
D	13	249 12.2	354 49.0	30.2	263 47.8	54.9	82 22.0	44.1	145 18.9	06.0	Alpheratz	357 51.2	N29 06
A	14	264 14.7	9 48.2	30.7	278 49.2	54.4	97 24.1	44.0	160 21.5	06.0	Altair	62 15.8	N 8 52
Y	15	279 17.1	24 47.3	. . 31.1	293 50.6	. . 53.8	112 26.3	. . 43.9	175 24.0	. . 06.0	Ankaa	353 22.7	S42 17
	16	294 19.6	39 46.4	31.5	308 52.0	53.3	127 28.4	43.8	190 26.6	06.0	Antares	112 36.0	S26 26
	17	309 22.1	54 45.5	32.0	323 53.4	52.7	142 30.6	43.6	205 29.2	06.0			
	18	324 24.5	69 44.6	S23 32.4	338 54.9	S 7 52.1	157 32.7	N 6 43.5	220 31.7	N22 06.0	Arcturus	146 02.9	N19 09
	19	339 27.0	84 43.7	32.8	353 56.3	51.6	172 34.9	43.4	235 34.3	06.0	Atria	107 45.3	S69 02
	20	354 29.5	99 42.8	33.2	8 57.7	51.0	187 37.0	43.3	250 36.9	06.0	Avior	234 21.0	S59 30
	21	9 31.9	114 41.9	. . 33.7	23 59.1	. . 50.4	202 39.1	. . 43.2	265 39.4	. . 06.0	Bellatrix	278 39.8	N 6 21
	22	24 34.4	129 41.0	34.1	39 00.5	49.9	217 41.3	43.0	280 42.0	06.0	Betelgeuse	271 09.2	N 7 24
	23	39 36.8	144 40.1	34.5	54 01.9	49.3	232 43.4	42.9	295 44.6	06.1			
16	00	54 39.3	159 39.2	S23 34.9	69 03.4	S 7 48.7	247 45.6	N 6 42.8	310 47.1	N22 06.1	Canopus	263 59.1	S52 41
	01	69 41.8	174 38.3	35.3	84 04.8	48.2	262 47.7	42.7	325 49.7	06.1	Capella	280 45.2	N46 00
	02	84 44.2	189 37.4	35.8	99 06.2	47.6	277 49.9	42.5	340 52.3	06.1	Deneb	49 36.9	N45 17
	03	99 46.7	204 36.5	. . 36.2	114 07.6	. . 47.0	292 52.0	. . 42.4	355 54.8	. . 06.1	Denebola	182 41.5	N14 33
	04	114 49.2	219 35.6	36.6	129 09.0	46.5	307 54.2	42.3	10 57.4	06.1	Diphda	349 03.2	S17 58
	05	129 51.6	234 34.7	37.0	144 10.4	45.9	322 56.3	42.2	26 00.0	06.1			
	06	144 54.1	249 33.8	S23 37.4	159 11.8	S 7 45.3	337 58.5	N 6 42.0	41 02.5	N22 06.1	Dubhe	194 00.9	N61 4
	07	159 56.6	264 32.9	37.8	174 13.2	44.8	353 00.6	41.9	56 05.1	06.1	Elnath	278 21.9	N28 3
	08	174 59.0	279 32.0	38.2	189 14.7	44.2	8 02.7	41.8	71 07.7	06.2	Eltanin	90 50.1	N51 2
S	09	190 01.5	294 31.1	. . 38.7	204 16.1	. . 43.6	23 04.9	. . 41.7	86 10.2	. . 06.2	Enif	33 54.6	N 9 5
U	10	205 04.0	309 30.2	39.1	219 17.5	43.1	38 07.0	41.6	101 12.8	06.2	Fomalhaut	15 32.1	S29 3
N	11	220 06.4	324 29.3	39.5	234 18.9	42.5	53 09.2	41.4	116 15.4	06.2			
D	12	235 08.9	339 28.4	S23 39.9	249 20.3	S 7 41.9	68 11.3	N 6 41.3	131 17.9	N22 06.2	Gacrux	172 10.1	S57 0
A	13	250 11.3	354 27.5	40.3	264 21.7	41.4	83 13.5	41.2	146 20.5	06.2	Gienah	176 00.4	S17 3
Y	14	265 13.8	9 26.6	40.7	279 23.1	40.8	98 15.6	41.1	161 23.1	06.2	Hadar	148 59.6	S60 2
	15	280 16.3	24 25.7	. . 41.1	294 24.5	. . 40.2	113 17.8	. . 40.9	176 25.6	. . 06.2	Hamal	328 09.1	N23 2
	16	295 18.7	39 24.8	41.5	309 25.9	39.7	128 19.9	40.8	191 28.2	06.2	Kaus Aust.	83 54.2	S34 2
	17	310 21.2	54 23.9	41.9	324 27.3	39.1	143 22.1	40.7	206 30.8	06.2			
	18	325 23.7	69 23.0	S23 42.3	339 28.7	S 7 38.5	158 24.2	N 6 40.6	221 33.3	N22 06.3	Kochab	137 20.5	N74 0
	19	340 26.1	84 22.1	42.7	354 30.1	38.0	173 26.4	40.5	236 35.9	06.3	Markab	13 45.8	N15 1
	20	355 28.6	99 21.2	43.1	9 31.5	37.4	188 28.5	40.3	251 38.5	06.3	Menkar	314 22.7	N 4 0
	21	10 31.1	114 20.3	. . 43.5	24 33.0	. . 36.8	203 30.7	. . 40.2	266 41.0	. . 06.3	Menkent	148 17.0	S36 2
	22	25 33.5	129 19.4	43.9	39 34.4	36.2	218 32.8	40.1	281 43.6	06.3	Miaplacidus	221 41.5	S69 4
	23	40 36.0	144 18.5	44.3	54 35.8	35.7	233 35.0	40.0	296 46.2	06.3			
17	00	55 38.5	159 17.6	S23 44.7	69 37.2	S 7 35.1	248 37.1	N 6 39.8	311 48.8	N22 06.3	Mirfak	308 50.8	N49 5
	01	70 40.9	174 16.7	45.0	84 38.6	34.5	263 39.3	39.7	326 51.3	06.3	Nunki	76 07.9	S26 1
	02	85 43.4	189 15.8	45.4	99 40.0	34.0	278 41.4	39.6	341 53.9	06.3	Peacock	53 31.2	S56 4
	03	100 45.8	204 14.9	. . 45.8	114 41.4	. . 33.4	293 43.6	. . 39.5	356 56.5	. . 06.4	Pollux	243 36.8	N28 0
	04	115 48.3	219 14.0	46.2	129 42.8	32.8	308 45.7	39.4	11 59.0	06.4	Procyon	245 07.5	N 5 1
	05	130 50.8	234 13.1	46.6	144 44.2	32.3	323 47.9	39.2	27 01.6	06.4			
	06	145 53.2	249 12.2	S23 47.0	159 45.6	S 7 31.7	338 50.0	N 6 39.1	42 04.2	N22 06.4	Rasalhague	96 13.8	N12 3
	07	160 55.7	264 11.3	47.4	174 47.0	31.1	353 52.2	39.0	57 06.8	06.4	Regulus	207 51.6	N11 5
	08	175 58.2	279 10.4	47.7	189 48.4	30.5	8 54.3	38.9	72 09.3	06.4	Rigel	281 19.0	S 8 1
M	09	191 00.6	294 09.5	. . 48.1	204 49.8	. . 30.0	23 56.5	. . 38.8	87 11.9	. . 06.4	Rigil Kent.	140 03.1	S60 5
O	10	206 03.1	309 08.6	48.5	219 51.2	29.4	38 58.6	38.6	102 14.5	06.4	Sabik	102 21.6	S15 4
N	11	221 05.6	324 07.7	48.9	234 52.6	28.8	54 00.8	38.5	117 17.0	06.4			
D	12	236 08.0	339 06.8	S23 49.3	249 54.0	S 7 28.3	69 02.9	N 6 38.4	132 19.6	N22 06.5	Schedar	349 49.0	N56 3
A	13	251 10.5	354 05.9	49.6	264 55.4	27.7	84 05.1	38.3	147 22.2	06.5	Shaula	96 32.6	S37 0
Y	14	266 12.9	9 04.9	50.0	279 56.8	27.1	99 07.2	38.2	162 24.8	06.5	Sirius	258 40.2	S16 4
	15	281 15.4	24 04.0	. . 50.4	294 58.2	. . 26.5	114 09.4	. . 38.0	177 27.3	. . 06.5	Spica	158 39.6	S11 1
	16	296 17.9	39 03.1	50.8	309 59.5	26.0	129 11.6	37.9	192 29.9	06.5	Suhail	222 58.1	S43 2
	17	311 20.3	54 02.2	51.1	325 00.9	25.4	144 13.7	37.8	207 32.5	06.5			
	18	326 22.8	69 01.3	S23 51.5	340 02.3	S 7 24.8	159 15.9	N 6 37.7	222 35.1	N22 06.5	Vega	80 44.4	N38 4
	19	341 25.3	84 00.4	51.9	355 03.7	24.3	174 18.0	37.6	237 37.6	06.5	Zuben'ubi	137 14.2	S16 0
	20	356 27.7	98 59.5	52.2	10 05.1	23.7	189 20.2	37.4	252 40.2	06.5		SHA	Mer.Pa
	21	11 30.2	113 58.6	. . 52.6	25 06.5	. . 23.1	204 22.3	. . 37.3	267 42.8	. . 06.6		° ′	h
	22	26 32.7	128 57.7	53.0	40 07.9	22.5	219 24.5	37.2	282 45.4	06.6	Venus	104 59.9	13 2
	23	41 35.1	143 56.8	53.3	55 09.3	22.0	234 26.6	37.1	297 47.9	06.6	Mars	14 24.0	19 2
Mer. Pass.		h m 20 18.0	v −0.9	d 0.4	v 1.4	d 0.6	v 2.1	d 0.1	v 2.6	d 0.0	Jupiter	193 06.3	7 2
											Saturn	256 07.8	3 1

UT	SUN GHA	SUN Dec	MOON GHA	v	MOON Dec	d	HP
d h	° ′	° ′	° ′	′	° ′	′	′
5 00	183 53.3	S18 19.1	293 33.2	10.0	N25 31.3	5.0	55.1
01	198 53.2	19.7	308 02.2	10.1	25 26.3	5.2	55.1
02	213 53.1	20.4	322 31.3	10.2	25 21.1	5.4	55.1
03	228 53.0	. . 21.0	337 00.5	10.1	25 15.7	5.5	55.1
04	243 52.9	21.7	351 29.6	10.2	25 10.2	5.6	55.2
05	258 52.8	22.3	5 58.8	10.2	25 04.6	5.7	55.2
06	273 52.7	S18 22.9	20 28.0	10.2	N24 58.9	5.8	55.2
07	288 52.6	23.6	34 57.2	10.2	24 53.1	6.0	55.2
08	303 52.5	24.2	49 26.4	10.3	24 47.1	6.1	55.3
09	318 52.4	. . 24.9	63 55.7	10.3	24 41.0	6.3	55.3
10	333 52.3	25.5	78 25.0	10.3	24 34.7	6.3	55.3
11	348 52.2	26.2	92 54.3	10.4	24 28.4	6.5	55.3
12	3 52.1	S18 26.8	107 23.7	10.3	N24 21.9	6.6	55.4
13	18 52.0	27.4	121 53.0	10.4	24 15.3	6.8	55.4
14	33 51.9	28.1	136 22.4	10.5	24 08.5	6.9	55.4
15	48 51.8	. . 28.7	150 51.9	10.5	24 01.6	6.9	55.4
16	63 51.7	29.3	165 21.4	10.5	23 54.7	7.2	55.5
17	78 51.6	30.0	179 50.9	10.5	23 47.5	7.2	55.5
18	93 51.5	S18 30.6	194 20.4	10.5	N23 40.3	7.3	55.5
19	108 51.4	31.3	208 49.9	10.6	23 33.0	7.5	55.5
20	123 51.2	31.9	223 19.5	10.7	23 25.5	7.6	55.6
21	138 51.1	. . 32.5	237 49.2	10.6	23 17.9	7.7	55.6
22	153 51.0	33.2	252 18.8	10.7	23 10.2	7.9	55.6
23	168 50.9	33.8	266 48.5	10.7	23 02.3	8.0	55.7
6 00	183 50.8	S18 34.4	281 18.2	10.8	N22 54.3	8.0	55.7
01	198 50.7	35.1	295 48.0	10.8	22 46.3	8.2	55.7
02	213 50.6	35.7	310 17.8	10.8	22 38.1	8.4	55.7
03	228 50.5	. . 36.3	324 47.6	10.9	22 29.7	8.4	55.8
04	243 50.4	37.0	339 17.5	10.8	22 21.3	8.5	55.8
05	258 50.3	37.6	353 47.3	11.0	22 12.8	8.7	55.8
06	273 50.2	S18 38.2	8 17.3	10.9	N22 04.1	8.8	55.9
07	288 50.1	38.8	22 47.2	11.0	21 55.3	8.9	55.9
08	303 49.9	39.5	37 17.2	11.0	21 46.4	9.0	55.9
09	318 49.8	. . 40.1	51 47.2	11.1	21 37.4	9.2	56.0
10	333 49.7	40.7	66 17.3	11.1	21 28.2	9.2	56.0
11	348 49.6	41.4	80 47.4	11.1	21 19.0	9.4	56.0
12	3 49.5	S18 42.0	95 17.5	11.2	N21 09.6	9.4	56.0
13	18 49.4	42.6	109 47.7	11.2	21 00.2	9.6	56.1
14	33 49.3	43.2	124 17.9	11.2	20 50.6	9.7	56.1
15	48 49.2	. . 43.9	138 48.1	11.2	20 40.9	9.8	56.1
16	63 49.1	44.5	153 18.3	11.3	20 31.1	9.9	56.2
17	78 48.9	45.1	167 48.6	11.4	20 21.2	10.1	56.2
18	93 48.8	S18 45.7	182 19.0	11.3	N20 11.1	10.1	56.2
19	108 48.7	46.4	196 49.3	11.4	20 01.0	10.2	56.3
20	123 48.6	47.0	211 19.7	11.4	19 50.8	10.4	56.3
21	138 48.5	. . 47.6	225 50.1	11.5	19 40.4	10.4	56.3
22	153 48.4	48.2	240 20.6	11.5	19 30.0	10.6	56.4
23	168 48.2	48.8	254 51.1	11.5	19 19.4	10.7	56.4
7 00	183 48.1	S18 49.5	269 21.6	11.5	N19 08.7	10.7	56.4
01	198 48.0	50.1	283 52.1	11.6	18 58.0	10.9	56.5
02	213 47.9	50.7	298 22.7	11.6	18 47.1	11.0	56.5
03	228 47.8	. . 51.3	312 53.3	11.7	18 36.1	11.1	56.5
04	243 47.7	51.9	327 24.0	11.6	18 25.0	11.1	56.6
05	258 47.5	52.5	341 54.6	11.7	18 13.9	11.3	56.6
06	273 47.4	S18 53.2	356 25.3	11.8	N18 02.6	11.4	56.6
07	288 47.3	53.8	10 56.1	11.7	17 51.2	11.5	56.7
08	303 47.2	54.4	25 26.8	11.8	17 39.7	11.6	56.7
09	318 47.1	. . 55.0	39 57.6	11.8	17 28.1	11.7	56.7
10	333 46.9	55.6	54 28.4	11.9	17 16.4	11.7	56.8
11	348 46.8	56.2	68 59.3	11.8	17 04.7	11.9	56.8
12	3 46.7	S18 56.8	83 30.1	11.9	N16 52.8	12.0	56.8
13	18 46.6	57.5	98 01.0	12.0	16 40.8	12.0	56.9
14	33 46.5	58.1	112 32.0	11.9	16 28.8	12.2	56.9
15	48 46.3	. . 58.7	127 02.9	12.0	16 16.6	12.2	57.0
16	63 46.2	59.3	141 33.9	12.0	16 04.4	12.4	57.0
17	78 46.1	18 59.9	156 04.9	12.0	15 52.0	12.4	57.0
18	93 46.0	S19 00.5	170 35.9	12.0	N15 39.6	12.6	57.1
19	108 45.9	01.1	185 06.9	12.1	15 27.0	12.6	57.1
20	123 45.7	01.7	199 38.0	12.1	15 14.4	12.7	57.1
21	138 45.6	. . 02.3	214 09.1	12.1	15 01.7	12.8	57.2
22	153 45.5	02.9	228 40.2	12.1	14 48.9	12.9	57.2
23	168 45.4	03.5	243 11.3	12.2	N14 36.0	12.9	57.3
	SD 16.2	*d* 0.6	SD 15.1		15.3		15.5

Lat.	Twilight Naut.	Twilight Civil	Sunrise	Moonrise 15	Moonrise 16	Moonrise 17	Moonrise 18
°	h m	h m	h m	h m	h m	h m	h m
N 72	07 01	08 37	11 15	▭	▭	21 04	23 42
N 70	06 51	08 14	09 53	▭	▭	21 39	23 56
68	06 42	07 56	09 15	▭	19 43	22 04	24 07
66	06 35	07 41	08 49	17 47	20 22	22 23	24 16
64	06 29	07 29	08 29	18 53	20 49	22 38	24 24
62	06 23	07 19	08 13	19 28	21 10	22 51	24 31
60	06 18	07 10	07 59	19 53	21 26	23 01	24 36
N 58	06 14	07 02	07 48	20 13	21 41	23 10	24 41
56	06 09	06 55	07 38	20 29	21 53	23 18	24 46
54	06 06	06 49	07 29	20 43	22 03	23 26	24 50
52	06 02	06 43	07 21	20 56	22 12	23 32	24 53
50	05 59	06 38	07 14	21 06	22 21	23 38	24 56
45	05 51	06 26	06 58	21 29	22 38	23 50	25 03
N 40	05 44	06 17	06 46	21 47	22 52	24 00	00 00
35	05 37	06 08	06 35	22 02	23 05	24 09	00 09
30	05 31	06 00	06 25	22 15	23 15	24 16	00 16
20	05 19	05 46	06 09	22 37	23 33	24 29	00 29
N 10	05 07	05 33	05 55	22 56	23 49	24 41	00 41
0	04 54	05 19	05 41	23 14	24 03	00 03	00 51
S 10	04 39	05 05	05 27	23 31	24 18	00 18	01 02
20	04 21	04 49	05 13	23 50	24 33	00 33	01 13
30	03 58	04 30	04 56	24 12	00 12	00 51	01 26
35	03 44	04 18	04 46	24 25	00 25	01 01	01 33
40	03 26	04 04	04 34	00 01	00 39	01 13	01 41
45	03 03	03 47	04 21	00 21	00 57	01 26	01 51
S 50	02 33	03 25	04 04	00 46	01 18	01 43	02 03
52	02 17	03 14	03 56	00 59	01 28	01 50	02 08
54	01 58	03 02	03 47	01 13	01 40	01 59	02 14
56	01 33	02 48	03 37	01 29	01 53	02 09	02 21
58	00 56	02 31	03 26	01 48	02 08	02 20	02 28
S 60	////	02 11	03 13	02 12	02 25	02 32	02 36

Lat.	Sunset	Twilight Civil	Twilight Naut.	Moonset 15	Moonset 16	Moonset 17	Moonset 18
°	h m	h m	h m	h m	h m	h m	h m
N 72	12 13	14 51	16 27	▭	▭	16 38	15 41
N 70	13 35	15 15	16 37	▭	▭	16 01	15 24
68	14 13	15 33	16 46	▭	16 14	15 34	15 11
66	14 39	15 47	16 53	16 25	15 34	15 14	15 00
64	14 59	15 59	17 00	15 18	15 06	14 57	14 50
62	15 16	16 10	17 05	14 43	14 44	14 44	14 42
60	15 29	16 19	17 10	14 17	14 27	14 32	14 35
N 58	15 41	16 26	17 15	13 57	14 12	14 22	14 29
56	15 51	16 34	17 19	13 40	13 59	14 13	14 24
54	16 00	16 40	17 23	13 25	13 48	14 05	14 19
52	16 08	16 46	17 27	13 13	13 38	13 58	14 14
50	16 15	16 51	17 30	13 02	13 29	13 51	14 10
45	16 31	17 02	17 38	12 38	13 11	13 38	14 01
N 40	16 43	17 12	17 45	12 20	12 55	13 26	13 53
35	16 54	17 21	17 52	12 04	12 42	13 16	13 47
30	17 04	17 29	17 58	11 50	12 31	13 07	13 41
20	17 20	17 43	18 10	11 27	12 11	12 52	13 31
N 10	17 35	17 57	18 22	11 07	11 54	12 39	13 22
0	17 48	18 10	18 35	10 47	11 38	12 26	13 14
S 10	18 02	18 24	18 50	10 28	11 21	12 14	13 06
20	18 17	18 41	19 09	10 08	11 04	12 00	12 56
30	18 34	19 00	19 32	09 44	10 44	11 44	12 46
35	18 44	19 12	19 46	09 30	10 32	11 35	12 40
40	18 56	19 26	20 04	09 13	10 18	11 25	12 33
45	19 10	19 44	20 27	08 54	10 02	11 12	12 25
S 50	19 26	20 06	20 58	08 29	09 41	10 57	12 15
52	19 34	20 16	21 14	08 17	09 32	10 50	12 10
54	19 43	20 29	21 34	08 04	09 21	10 42	12 05
56	19 53	20 43	22 00	07 48	09 08	10 33	12 00
58	20 05	21 00	22 39	07 29	08 54	10 23	11 53
S 60	20 18	21 21	////	07 05	08 37	10 11	11 46

Day	SUN Eqn. of Time 00^h	SUN Eqn. of Time 12^h	SUN Mer. Pass.	MOON Mer. Pass. Upper	MOON Mer. Pass. Lower	Age	Phase
d	m s	m s	h m	h m	h m	d %	
15	15 33	15 29	11 45	04 35	17 01	21 67	
16	15 24	15 18	11 45	05 26	17 50	22 57	
17	15 13	15 07	11 45	06 15	18 39	23 47	

UT	ARIES	VENUS −3.9		MARS −0.7		JUPITER −1.9		SATURN −0.2		STARS		
d h	GHA ° ′	GHA ° ′	Dec ° ′	GHA ° ′	Dec ° ′	GHA ° ′	Dec ° ′	GHA ° ′	Dec ° ′	Name	SHA ° ′	Dec ° ′
18 00 TUESDAY	56 37.6	158 55.8	S23 53.7	70 10.7	S 7 21.4	249 28.8	N 6 37.0	312 50.5	N22 06.6	Acamar	315 23.5	S40 17.
01	71 40.1	173 54.9	54.0	85 12.1	20.8	264 30.9	36.8	327 53.1	06.6	Achernar	335 31.6	S57 13.
02	86 42.5	188 54.0	54.4	100 13.5	20.2	279 33.1	36.7	342 55.7	06.6	Acrux	173 18.6	S63 06.
03	101 45.0	203 53.1	54.8	115 14.9	19.7	294 35.3	36.6	357 58.2	06.6	Adhara	255 18.2	S28 58.
04	116 47.4	218 52.2	55.1	130 16.3	19.1	309 37.4	36.5	13 00.8	06.6	Aldebaran	290 57.8	N16 31.
05	131 49.9	233 51.3	55.5	145 17.6	18.5	324 39.6	36.4	28 03.4	06.7			
06	146 52.4	248 50.4	S23 55.8	160 19.0	S 7 17.9	339 41.7	N 6 36.2	43 06.0	N22 06.7	Alioth	166 27.5	N55 56.
07	161 54.8	263 49.5	56.2	175 20.4	17.4	354 43.9	36.1	58 08.5	06.7	Alkaid	153 05.1	N49 17
08	176 57.3	278 48.5	56.5	190 21.8	16.8	9 46.0	36.0	73 11.1	06.7	Al Na'ir	27 53.0	S46 56.
09	191 59.8	293 47.6	56.9	205 23.2	16.2	24 48.2	35.9	88 13.7	06.7	Alnilam	275 53.7	S 1 11
10	207 02.2	308 46.7	57.2	220 24.6	15.6	39 50.3	35.8	103 16.3	06.7	Alphard	218 03.5	S 8 40
11	222 04.7	323 45.8	57.6	235 26.0	15.1	54 52.5	35.7	118 18.8	06.7			
12	237 07.2	338 44.9	S23 57.9	250 27.4	S 7 14.5	69 54.7	N 6 35.5	133 21.4	N22 06.7	Alphecca	126 17.7	N26 42
13	252 09.6	353 44.0	58.3	265 28.7	13.9	84 56.8	35.4	148 24.0	06.7	Alpheratz	357 51.2	N29 06
14	267 12.1	8 43.1	58.6	280 30.1	13.3	99 59.0	35.3	163 26.6	06.8	Altair	62 15.8	N 8 52
15	282 14.6	23 42.1	59.0	295 31.5	12.8	115 01.1	35.2	178 29.2	06.8	Ankaa	353 22.7	S42 17
16	297 17.0	38 41.2	59.3	310 32.9	12.2	130 03.3	35.1	193 31.7	06.8	Antares	112 36.0	S26 26
17	312 19.5	53 40.3	23 59.7	325 34.3	11.6	145 05.5	34.9	208 34.3	06.8			
18	327 21.9	68 39.4	S24 00.0	340 35.7	S 7 11.0	160 07.6	N 6 34.8	223 36.9	N22 06.8	Arcturus	146 02.9	N19 09
19	342 24.4	83 38.5	00.3	355 37.1	10.5	175 09.8	34.7	238 39.5	06.8	Atria	107 45.3	S69 02
20	357 26.9	98 37.6	00.7	10 38.4	09.9	190 11.9	34.6	253 42.1	06.8	Avior	234 21.0	S59 30
21	12 29.3	113 36.6	01.0	25 39.8	09.3	205 14.1	34.5	268 44.6	06.8	Bellatrix	278 39.8	N 6 21
22	27 31.8	128 35.7	01.3	40 41.2	08.7	220 16.3	34.4	283 47.2	06.8	Betelgeuse	271 09.2	N 7 24
23	42 34.3	143 34.8	01.7	55 42.6	08.2	235 18.4	34.2	298 49.8	06.9			
19 00 WEDNESDAY	57 36.7	158 33.9	S24 02.0	70 44.0	S 7 07.6	250 20.6	N 6 34.1	313 52.4	N22 06.9	Canopus	263 59.1	S52 41
01	72 39.2	173 33.0	02.3	85 45.3	07.0	265 22.7	34.0	328 54.9	06.9	Capella	280 45.2	N46 00
02	87 41.7	188 32.1	02.7	100 46.7	06.4	280 24.9	33.9	343 57.5	06.9	Deneb	49 36.9	N45 17
03	102 44.1	203 31.1	03.0	115 48.1	05.8	295 27.1	33.8	359 00.1	06.9	Denebola	182 41.5	N14 33
04	117 46.6	218 30.2	03.3	130 49.5	05.3	310 29.2	33.7	14 02.7	06.9	Diphda	349 03.2	S17 58
05	132 49.0	233 29.3	03.7	145 50.9	04.7	325 31.4	33.5	29 05.3	06.9			
06	147 51.5	248 28.4	S24 04.0	160 52.2	S 7 04.1	340 33.5	N 6 33.4	44 07.9	N22 06.9	Dubhe	194 00.8	N61 43
07	162 54.0	263 27.5	04.3	175 53.6	03.5	355 35.7	33.3	59 10.4	07.0	Elnath	278 21.9	N28 36
08	177 56.4	278 26.5	04.6	190 55.0	03.0	10 37.9	33.2	74 13.0	07.0	Eltanin	90 50.1	N51 29
09	192 58.9	293 25.6	04.9	205 56.4	02.4	25 40.0	33.1	89 15.6	07.0	Enif	33 54.6	N 9 53
10	208 01.4	308 24.7	05.3	220 57.7	01.8	40 42.2	33.0	104 18.2	07.0	Fomalhaut	15 32.1	S29 36
11	223 03.8	323 23.8	05.6	235 59.1	01.2	55 44.3	32.8	119 20.8	07.0			
12	238 06.3	338 22.9	S24 05.9	251 00.5	S 7 00.6	70 46.5	N 6 32.7	134 23.3	N22 07.0	Gacrux	172 10.0	S57 07
13	253 08.8	353 21.9	06.2	266 01.9	7 00.1	85 48.7	32.6	149 25.9	07.0	Gienah	176 00.4	S17 33
14	268 11.2	8 21.0	06.5	281 03.2	6 59.5	100 50.8	32.5	164 28.5	07.0	Hadar	148 59.5	S60 23
15	283 13.7	23 20.1	06.8	296 04.6	58.9	115 53.0	32.4	179 31.1	07.1	Hamal	328 09.1	N23 28
16	298 16.2	38 19.2	07.2	311 06.0	58.3	130 55.2	32.3	194 33.7	07.1	Kaus Aust.	83 54.2	S34 23
17	313 18.6	53 18.2	07.5	326 07.4	57.7	145 57.3	32.2	209 36.3	07.1			
18	328 21.1	68 17.3	S24 07.8	341 08.7	S 6 57.2	160 59.5	N 6 32.0	224 38.8	N22 07.1	Kochab	137 20.5	N74 08
19	343 23.5	83 16.4	08.1	356 10.1	56.6	176 01.7	31.9	239 41.4	07.1	Markab	13 45.8	N15 13
20	358 26.0	98 15.5	08.4	11 11.5	56.0	191 03.8	31.8	254 44.0	07.1	Menkar	314 22.7	N 4 06
21	13 28.5	113 14.5	08.7	26 12.9	55.4	206 06.0	31.7	269 46.6	07.1	Menkent	148 17.0	S36 23
22	28 30.9	128 13.6	09.0	41 14.2	54.8	221 08.2	31.6	284 49.2	07.1	Miaplacidus	221 41.4	S69 43
23	43 33.4	143 12.7	09.3	56 15.6	54.3	236 10.3	31.5	299 51.8	07.1			
20 00 THURSDAY	58 35.9	158 11.8	S24 09.6	71 17.0	S 6 53.7	251 12.5	N 6 31.3	314 54.3	N22 07.2	Mirfak	308 50.8	N49 52
01	73 38.3	173 10.8	09.9	86 18.3	53.1	266 14.6	31.2	329 56.9	07.2	Nunki	76 07.9	S26 17
02	88 40.8	188 09.9	10.2	101 19.7	52.5	281 16.8	31.1	344 59.5	07.2	Peacock	53 31.2	S56 43
03	103 43.3	203 09.0	10.5	116 21.1	51.9	296 19.0	31.0	0 02.1	07.2	Pollux	243 36.7	N28 01
04	118 45.7	218 08.1	10.8	131 22.4	51.4	311 21.1	30.9	15 04.7	07.2	Procyon	245 07.5	N 5 11
05	133 48.2	233 07.1	11.1	146 23.8	50.8	326 23.3	30.8	30 07.3	07.2			
06	148 50.6	248 06.2	S24 11.4	161 25.2	S 6 50.2	341 25.5	N 6 30.7	45 09.8	N22 07.2	Rasalhague	96 13.8	N12 3
07	163 53.1	263 05.3	11.7	176 26.6	49.6	356 27.6	30.5	60 12.4	07.2	Regulus	207 51.5	N11 5
08	178 55.6	278 04.4	12.0	191 27.9	49.0	11 29.8	30.4	75 15.0	07.3	Rigel	281 19.0	S 8 1
09	193 58.0	293 03.4	12.3	206 29.3	48.4	26 32.0	30.3	90 17.6	07.3	Rigil Kent.	140 03.1	S60 5
10	209 00.5	308 02.5	12.6	221 30.6	47.9	41 34.1	30.2	105 20.2	07.3	Sabik	102 21.6	S15 4
11	224 03.0	323 01.6	12.9	236 32.0	47.3	56 36.3	30.1	120 22.8	07.3			
12	239 05.4	338 00.7	S24 13.1	251 33.4	S 6 46.7	71 38.5	N 6 30.0	135 25.4	N22 07.3	Schedar	349 49.0	N56 3
13	254 07.9	352 59.7	13.4	266 34.7	46.1	86 40.6	29.9	150 27.9	07.3	Shaula	96 32.6	S37 0
14	269 10.4	7 58.8	13.7	281 36.1	45.5	101 42.8	29.7	165 30.5	07.3	Sirius	258 40.2	S16 4
15	284 12.8	22 57.9	14.0	296 37.5	44.9	116 45.0	29.6	180 33.1	07.3	Spica	158 39.5	S11 1
16	299 15.3	37 56.9	14.3	311 38.8	44.4	131 47.2	29.5	195 35.7	07.4	Suhail	222 58.0	S43 2
17	314 17.8	52 56.0	14.6	326 40.2	43.8	146 49.3	29.4	210 38.3	07.4			
18	329 20.2	67 55.1	S24 14.8	341 41.6	S 6 43.2	161 51.5	N 6 29.3	225 40.9	N22 07.4	Vega	80 44.4	N38 4
19	344 22.7	82 54.2	15.1	356 42.9	42.6	176 53.7	29.2	240 43.5	07.4	Zuben'ubi	137 14.2	S16 0
20	359 25.1	97 53.2	15.4	11 44.3	42.0	191 55.8	29.1	255 46.1	07.4		SHA	Mer. Pa
21	14 27.6	112 52.3	15.7	26 45.6	41.4	206 58.0	29.0	270 48.6	07.4		° ′	h
22	29 30.1	127 51.4	15.9	41 47.0	40.9	222 00.2	28.8	285 51.2	07.4	Venus	100 57.2	13 2
23	44 32.5	142 50.4	16.2	56 48.4	40.3	237 02.3	28.7	300 53.8	07.4	Mars	13 07.2	19 1
Mer. Pass.	h m 20 06.2	*v* −0.9	*d* 0.3	*v* 1.4	*d* 0.6	*v* 2.2	*d* 0.1	*v* 2.6	*d* 0.0	Jupiter	192 43.8	7 1
										Saturn	256 15.6	3 0

UT	SUN GHA	SUN Dec	MOON GHA	*v*	MOON Dec	*d*	HP
d h	° ′	° ′	° ′	′	° ′	′	′
18 00	183 45.2	S19 04.2	257 42.5	12.1	N14 23.1	13.1	57.3
01	198 45.1	04.8	272 13.6	12.2	14 10.0	13.1	57.3
02	213 45.0	05.4	286 44.8	12.2	13 56.9	13.3	57.4
03	228 44.9	. . 06.0	301 16.0	12.2	13 43.6	13.3	57.4
04	243 44.7	06.6	315 47.2	12.2	13 30.3	13.4	57.4
05	258 44.6	07.2	330 18.4	12.3	13 16.9	13.4	57.5
06	273 44.5	S19 07.8	344 49.7	12.3	N13 03.5	13.6	57.5
07	288 44.3	08.4	359 21.0	12.2	12 49.9	13.6	57.6
T 08	303 44.2	09.0	13 52.2	12.3	12 36.3	13.7	57.6
U 09	318 44.1	. . 09.6	28 23.5	12.3	12 22.6	13.8	57.6
E 10	333 44.0	10.2	42 54.8	12.3	12 08.8	13.9	57.7
S 11	348 43.8	10.8	57 26.1	12.3	11 54.9	13.9	57.7
D 12	3 43.7	S19 11.4	71 57.4	12.4	N11 41.0	14.1	57.8
A 13	18 43.6	12.0	86 28.8	12.3	11 26.9	14.1	57.8
Y 14	33 43.4	12.6	101 00.1	12.3	11 12.8	14.1	57.8
15	48 43.3	. . 13.2	115 31.4	12.4	10 58.7	14.3	57.9
16	63 43.2	13.8	130 02.8	12.3	10 44.4	14.3	57.9
17	78 43.0	14.4	144 34.1	12.4	10 30.1	14.4	57.9
18	93 42.9	S19 15.0	159 05.5	12.3	N10 15.7	14.4	58.0
19	108 42.8	15.5	173 36.8	12.4	10 01.3	14.6	58.0
20	123 42.6	16.1	188 08.2	12.4	9 46.7	14.5	58.1
21	138 42.5	. . 16.7	202 39.6	12.3	9 32.2	14.7	58.1
22	153 42.4	17.3	217 10.9	12.4	9 17.5	14.7	58.1
23	168 42.2	17.9	231 42.3	12.3	9 02.8	14.8	58.2
19 00	183 42.1	S19 18.5	246 13.6	12.4	N 8 48.0	14.9	58.2
01	198 42.0	19.1	260 45.0	12.4	8 33.1	14.9	58.3
02	213 41.8	19.7	275 16.4	12.3	8 18.2	15.0	58.3
03	228 41.7	. . 20.3	289 47.7	12.3	8 03.2	15.0	58.3
04	243 41.6	20.9	304 19.0	12.4	7 48.2	15.1	58.4
05	258 41.4	21.5	318 50.4	12.3	7 33.1	15.2	58.4
06	273 41.3	S19 22.0	333 21.7	12.3	N 7 17.9	15.2	58.5
W 07	288 41.2	22.6	347 53.0	12.3	7 02.7	15.2	58.5
E 08	303 41.0	23.2	2 24.3	12.3	6 47.5	15.4	58.5
D 09	318 40.9	. . 23.8	16 55.6	12.3	6 32.1	15.3	58.6
N 10	333 40.8	24.4	31 26.9	12.3	6 16.8	15.5	58.6
E 11	348 40.6	25.0	45 58.2	12.2	6 01.3	15.5	58.7
S 12	3 40.5	S19 25.6	60 29.4	12.2	N 5 45.8	15.5	58.7
D 13	18 40.3	26.1	75 00.6	12.3	5 30.3	15.6	58.7
A 14	33 40.2	26.7	89 31.9	12.2	5 14.7	15.6	58.8
Y 15	48 40.1	. . 27.3	104 03.1	12.1	4 59.1	15.7	58.8
16	63 39.9	27.9	118 34.2	12.2	4 43.4	15.7	58.9
17	78 39.8	28.5	133 05.4	12.1	4 27.7	15.8	58.9
18	93 39.6	S19 29.1	147 36.5	12.2	N 4 11.9	15.8	58.9
19	108 39.5	29.6	162 07.7	12.1	3 56.1	15.8	59.0
20	123 39.4	30.2	176 38.8	12.0	3 40.3	15.9	59.0
21	138 39.2	. . 30.8	191 09.8	12.1	3 24.4	16.0	59.1
22	153 39.1	31.4	205 40.9	12.0	3 08.4	15.9	59.1
23	168 38.9	31.9	220 11.9	11.9	2 52.5	16.0	59.1
20 00	183 38.8	S19 32.5	234 42.8	12.0	N 2 36.5	16.1	59.2
01	198 38.6	33.1	249 13.8	11.9	2 20.4	16.1	59.2
02	213 38.5	33.7	263 44.7	11.9	2 04.3	16.1	59.3
03	228 38.4	. . 34.3	278 15.6	11.8	1 48.2	16.1	59.3
04	243 38.2	34.8	292 46.4	11.9	1 32.1	16.2	59.3
05	258 38.1	35.4	307 17.3	11.7	1 15.9	16.2	59.4
06	273 37.9	S19 36.0	321 48.0	11.8	N 0 59.7	16.2	59.4
T 07	288 37.8	36.5	336 18.8	11.7	0 43.5	16.3	59.5
H 08	303 37.6	37.1	350 49.5	11.6	0 27.2	16.2	59.5
U 09	318 37.5	. . 37.7	5 20.1	11.6	N 0 11.0	16.3	59.5
R 10	333 37.3	38.3	19 50.7	11.6	S 0 05.3	16.4	59.6
S 11	348 37.2	38.8	34 21.3	11.5	0 21.7	16.3	59.6
D 12	3 37.0	S19 39.4	48 51.8	11.5	S 0 38.0	16.4	59.6
A 13	18 36.9	40.0	63 22.3	11.5	0 54.4	16.3	59.7
Y 14	33 36.8	40.5	77 52.8	11.3	1 10.7	16.4	59.7
15	48 36.6	. . 41.1	92 23.1	11.4	1 27.1	16.4	59.8
16	63 36.5	41.7	106 53.5	11.3	1 43.5	16.5	59.8
17	78 36.3	42.2	121 23.8	11.2	2 00.0	16.4	59.8
18	93 36.2	S19 42.8	135 54.0	11.2	S 2 16.4	16.4	59.9
19	108 36.0	43.4	150 24.2	11.1	2 32.8	16.5	59.9
20	123 35.9	43.9	164 54.3	11.1	2 49.3	16.4	59.9
21	138 35.7	. . 44.5	179 24.4	11.0	3 05.7	16.5	60.0
22	153 35.6	45.1	193 54.4	10.9	3 22.2	16.4	60.0
23	168 35.4	45.6	208 24.3	10.9	S 3 38.6	16.5	60.0
	SD 16.2	*d* 0.6	SD 15.7		16.0		16.3

Lat.	Twilight Naut.	Twilight Civil	Sunrise	Moonrise 18	19	20	21
°	h m	h m	h m	h m	h m	h m	h m
N 72	07 12	08 51	■	23 42	26 02	02 02	04 24
N 70	07 00	08 25	10 15	23 56	26 04	02 04	04 16
68	06 51	08 06	09 30	24 07	00 07	02 07	04 09
66	06 43	07 50	09 01	24 16	00 16	02 09	04 04
64	06 36	07 37	08 39	24 24	00 24	02 10	03 59
62	06 29	07 26	08 21	24 31	00 31	02 11	03 56
60	06 24	07 16	08 07	24 36	00 36	02 13	03 52
N 58	06 19	07 08	07 54	24 41	00 41	02 14	03 49
56	06 14	07 01	07 44	24 46	00 46	02 15	03 47
54	06 10	06 54	07 34	24 50	00 50	02 16	03 44
52	06 06	06 48	07 26	24 53	00 53	02 16	03 42
50	06 03	06 42	07 18	24 56	00 56	02 17	03 40
45	05 54	06 30	07 02	25 03	01 03	02 19	03 36
N 40	05 47	06 20	06 49	00 00	01 09	02 20	03 33
35	05 40	06 11	06 38	00 09	01 14	02 21	03 30
30	05 33	06 02	06 28	00 16	01 19	02 22	03 27
20	05 21	05 48	06 11	00 29	01 26	02 24	03 23
N 10	05 08	05 34	05 56	00 41	01 33	02 25	03 19
0	04 54	05 20	05 42	00 51	01 39	02 26	03 16
S 10	04 39	05 05	05 27	01 02	01 45	02 28	03 12
20	04 20	04 48	05 12	01 13	01 51	02 29	03 08
30	03 56	04 28	04 54	01 26	01 59	02 31	03 04
35	03 41	04 16	04 44	01 33	02 03	02 32	03 02
40	03 23	04 01	04 32	01 41	02 08	02 33	02 59
45	02 59	03 43	04 18	01 51	02 13	02 35	02 56
S 50	02 27	03 20	04 00	02 03	02 20	02 36	02 53
52	02 10	03 09	03 52	02 08	02 23	02 37	02 51
54	01 49	02 56	03 43	02 14	02 26	02 38	02 49
56	01 21	02 42	03 32	02 21	02 30	02 39	02 47
58	00 35	02 24	03 20	02 28	02 34	02 40	02 45
S 60	////	02 01	03 06	02 36	02 39	02 41	02 43

Lat.	Sunset	Twilight Civil	Twilight Naut.	Moonset 18	19	20	21
°	h m	h m	h m	h m	h m	h m	h m
N 72	■	14 38	16 17	15 41	15 04	14 31	13 56
N 70	13 15	15 04	16 29	15 24	14 58	14 33	14 07
68	14 00	15 24	16 39	15 11	14 52	14 35	14 16
66	14 29	15 40	16 47	15 00	14 48	14 36	14 24
64	14 51	15 53	16 54	14 50	14 44	14 37	14 30
62	15 09	16 04	17 00	14 42	14 40	14 38	14 36
60	15 23	16 13	17 06	14 35	14 37	14 39	14 41
N 58	15 36	16 22	17 11	14 29	14 35	14 40	14 46
56	15 46	16 29	17 15	14 24	14 32	14 41	14 50
54	15 56	16 36	17 20	14 19	14 30	14 41	14 53
52	16 04	16 42	17 24	14 14	14 28	14 42	14 56
50	16 12	16 48	17 27	14 10	14 26	14 42	14 59
45	16 28	17 00	17 36	14 01	14 22	14 44	15 06
N 40	16 41	17 10	17 43	13 53	14 19	14 44	15 11
35	16 52	17 20	17 50	13 47	14 16	14 45	15 16
30	17 02	17 28	17 57	13 41	14 14	14 46	15 20
20	17 20	17 43	18 10	13 31	14 09	14 47	15 27
N 10	17 35	17 57	18 23	13 22	14 05	14 48	15 34
0	17 49	18 11	18 36	13 14	14 01	14 49	15 40
S 10	18 03	18 26	18 52	13 06	13 57	14 51	15 46
20	18 19	18 43	19 11	12 56	13 53	14 52	15 52
30	18 37	19 03	19 35	12 46	13 49	14 53	16 00
35	18 47	19 16	19 50	12 40	13 46	14 53	16 04
40	18 59	19 30	20 09	12 33	13 43	14 54	16 09
45	19 14	19 48	20 32	12 25	13 39	14 55	16 14
S 50	19 31	20 11	21 05	12 15	13 34	14 56	16 21
52	19 40	20 23	21 23	12 10	13 32	14 57	16 24
54	19 49	20 36	21 44	12 05	13 30	14 57	16 28
56	20 00	20 51	22 13	12 00	13 28	14 58	16 32
58	20 12	21 09	23 04	11 53	13 25	14 58	16 36
S 60	20 26	21 32	////	11 46	13 22	14 59	16 41

Day	SUN Eqn. of Time 00^h	SUN Eqn. of Time 12^h	SUN Mer. Pass.	MOON Mer. Pass. Upper	MOON Mer. Pass. Lower	Age	Phase
d	m s	m s	h m	h m	h m	d %	
18	15 01	14 55	11 45	07 03	19 26	24 36	
19	14 49	14 42	11 45	07 50	20 14	25 26	
20	14 35	14 28	11 46	08 38	21 02	26 17	

UT		ARIES	VENUS −3.9		MARS −0.6		JUPITER −2.0		SATURN −0.2	
d h		GHA ° ′	GHA ° ′	Dec ° ′	GHA ° ′	Dec ° ′	GHA ° ′	Dec ° ′	GHA ° ′	Dec ° ′
21 00		59 35.0	157 49.5	S24 16.5	71 49.7	S 6 39.7	252 04.5	N 6 28.6	315 56.4	N22 07.5
01		74 37.5	172 48.6	16.8	86 51.1	39.1	267 06.7	28.5	330 59.0	07.5
02		89 39.9	187 47.6	17.0	101 52.4	38.5	282 08.9	28.4	346 01.6	07.5
03		104 42.4	202 46.7	17.3	116 53.8	37.9	297 11.0	28.3	1 04.2	07.5
04		119 44.9	217 45.8	17.6	131 55.2	37.3	312 13.2	28.2	16 06.8	07.5
05		134 47.3	232 44.9	17.8	146 56.5	36.8	327 15.4	28.1	31 09.3	07.5
06		149 49.8	247 43.9	S24 18.1	161 57.9	S 6 36.2	342 17.5	N 6 27.9	46 11.9	N22 07.5
07		164 52.3	262 43.0	18.4	176 59.2	35.6	357 19.7	27.8	61 14.5	07.5
08		179 54.7	277 42.1	18.6	192 00.6	35.0	12 21.9	27.7	76 17.1	07.6
09	F	194 57.2	292 41.1	18.9	207 01.9	34.4	27 24.1	27.6	91 19.7	07.6
10	R	209 59.6	307 40.2	19.1	222 03.3	33.8	42 26.2	27.5	106 22.3	07.6
11	I	225 02.1	322 39.3	19.4	237 04.7	33.2	57 28.4	27.4	121 24.9	07.6
12	D	240 04.6	337 38.3	S24 19.7	252 06.0	S 6 32.7	72 30.6	N 6 27.3	136 27.5	N22 07.6
13	A	255 07.0	352 37.4	19.9	267 07.4	32.1	87 32.8	27.2	151 30.1	07.6
14	Y	270 09.5	7 36.5	20.2	282 08.7	31.5	102 34.9	27.0	166 32.7	07.6
15		285 12.0	22 35.5	20.4	297 10.1	30.9	117 37.1	26.9	181 35.3	07.6
16		300 14.4	37 34.6	20.7	312 11.4	30.3	132 39.3	26.8	196 37.8	07.7
17		315 16.9	52 33.7	20.9	327 12.8	29.7	147 41.4	26.7	211 40.4	07.7
18		330 19.4	67 32.7	S24 21.2	342 14.1	S 6 29.1	162 43.6	N 6 26.6	226 43.0	N22 07.7
19		345 21.8	82 31.8	21.4	357 15.5	28.6	177 45.8	26.5	241 45.6	07.7
20		0 24.3	97 30.9	21.7	12 16.8	28.0	192 48.0	26.4	256 48.2	07.7
21		15 26.7	112 29.9	21.9	27 18.2	27.4	207 50.1	26.3	271 50.8	07.7
22		30 29.2	127 29.0	22.2	42 19.5	26.8	222 52.3	26.2	286 53.4	07.7
23		45 31.7	142 28.0	22.4	57 20.9	26.2	237 54.5	26.0	301 56.0	07.7
22 00		60 34.1	157 27.1	S24 22.6	72 22.2	S 6 25.6	252 56.7	N 6 25.9	316 58.6	N22 07.8
01		75 36.6	172 26.2	22.9	87 23.6	25.0	267 58.8	25.8	332 01.2	07.8
02		90 39.1	187 25.2	23.1	102 24.9	24.4	283 01.0	25.7	347 03.8	07.8
03		105 41.5	202 24.3	23.4	117 26.3	23.8	298 03.2	25.6	2 06.4	07.8
04		120 44.0	217 23.4	23.6	132 27.6	23.3	313 05.4	25.5	17 08.9	07.8
05		135 46.5	232 22.4	23.8	147 29.0	22.7	328 07.6	25.4	32 11.5	07.8
06		150 48.9	247 21.5	S24 24.1	162 30.3	S 6 22.1	343 09.7	N 6 25.3	47 14.1	N22 07.8
07		165 51.4	262 20.6	24.3	177 31.7	21.5	358 11.9	25.2	62 16.7	07.9
08	S	180 53.9	277 19.6	24.5	192 33.0	20.9	13 14.1	25.1	77 19.3	07.9
09	A	195 56.3	292 18.7	24.8	207 34.4	20.3	28 16.3	24.9	92 21.9	07.9
10	T	210 58.8	307 17.7	25.0	222 35.7	19.7	43 18.4	24.8	107 24.5	07.9
11	U	226 01.2	322 16.8	25.2	237 37.0	19.1	58 20.6	24.7	122 27.1	07.9
12	R	241 03.7	337 15.9	S24 25.4	252 38.4	S 6 18.5	73 22.8	N 6 24.6	137 29.7	N22 07.9
13	D	256 06.2	352 14.9	25.7	267 39.7	18.0	88 25.0	24.5	152 32.3	07.9
14	A	271 08.6	7 14.0	25.9	282 41.1	17.4	103 27.2	24.4	167 34.9	07.9
15	Y	286 11.1	22 13.1	26.1	297 42.4	16.8	118 29.3	24.3	182 37.5	08.0
16		301 13.6	37 12.1	26.3	312 43.8	16.2	133 31.5	24.2	197 40.1	08.0
17		316 16.0	52 11.2	26.6	327 45.1	15.6	148 33.7	24.1	212 42.7	08.0
18		331 18.5	67 10.2	S24 26.8	342 46.5	S 6 15.0	163 35.9	N 6 24.0	227 45.3	N22 08.0
19		346 21.0	82 09.3	27.0	357 47.8	14.4	178 38.1	23.9	242 47.9	08.0
20		1 23.4	97 08.4	27.2	12 49.1	13.8	193 40.2	23.7	257 50.5	08.0
21		16 25.9	112 07.4	27.4	27 50.5	13.2	208 42.4	23.6	272 53.1	08.0
22		31 28.4	127 06.5	27.6	42 51.8	12.6	223 44.6	23.5	287 55.7	08.1
23		46 30.8	142 05.5	27.9	57 53.2	12.0	238 46.8	23.4	302 58.3	08.1
23 00		61 33.3	157 04.6	S24 28.1	72 54.5	S 6 11.4	253 49.0	N 6 23.3	318 00.9	N22 08.1
01		76 35.7	172 03.7	28.3	87 55.8	10.9	268 51.1	23.2	333 03.4	08.1
02		91 38.2	187 02.7	28.5	102 57.2	10.3	283 53.3	23.1	348 06.0	08.1
03		106 40.7	202 01.8	28.7	117 58.5	09.7	298 55.5	23.0	3 08.6	08.1
04		121 43.1	217 00.8	28.9	132 59.8	09.1	313 57.7	22.9	18 11.2	08.1
05		136 45.6	231 59.9	29.1	148 01.2	08.5	328 59.9	22.8	33 13.8	08.1
06		151 48.1	246 58.9	S24 29.3	163 02.5	S 6 07.9	344 02.1	N 6 22.7	48 16.4	N22 08.2
07		166 50.5	261 58.0	29.5	178 03.9	07.3	359 04.2	22.6	63 19.0	08.2
08		181 53.0	276 57.1	29.7	193 05.2	06.7	14 06.4	22.5	78 21.6	08.2
09	S	196 55.5	291 56.1	29.9	208 06.5	06.1	29 08.6	22.3	93 24.2	08.2
10	U	211 57.9	306 55.2	30.1	223 07.9	05.5	44 10.8	22.2	108 26.8	08.2
11	N	227 00.4	321 54.2	30.3	238 09.2	04.9	59 13.0	22.1	123 29.4	08.2
12	D	242 02.8	336 53.3	S24 30.5	253 10.5	S 6 04.3	74 15.2	N 6 22.0	138 32.0	N22 08.2
13	A	257 05.3	351 52.4	30.7	268 11.9	03.7	89 17.3	21.9	153 34.6	08.3
14	Y	272 07.8	6 51.4	30.9	283 13.2	03.1	104 19.5	21.8	168 37.2	08.3
15		287 10.2	21 50.5	31.1	298 14.5	02.6	119 21.7	21.7	183 39.8	08.3
16		302 12.7	36 49.5	31.3	313 15.9	02.0	134 23.9	21.6	198 42.4	08.3
17		317 15.2	51 48.6	31.5	328 17.2	01.4	149 26.1	21.5	213 45.0	08.3
18		332 17.6	66 47.6	S24 31.6	343 18.5	S 6 00.8	164 28.3	N 6 21.4	228 47.6	N22 08.3
19		347 20.1	81 46.7	31.8	358 19.9	6 00.2	179 30.4	21.3	243 50.2	08.3
20		2 22.6	96 45.8	32.0	13 21.2	5 59.6	194 32.6	21.2	258 52.8	08.3
21		17 25.0	111 44.8	32.2	28 22.5	59.0	209 34.8	21.1	273 55.4	08.4
22		32 27.5	126 43.9	32.4	43 23.9	58.4	224 37.0	21.0	288 58.0	08.4
23		47 30.0	141 42.9	32.6	58 25.2	57.8	239 39.2	20.9	304 00.6	08.4
Mer. Pass.		h m 19 54.5	*v* −0.9	*d* 0.2	*v* 1.3	*d* 0.6	*v* 2.2	*d* 0.1	*v* 2.6	*d* 0.0

STARS

Name	SHA ° ′	Dec ° ′
Acamar	315 23.5	S40 17.
Achernar	335 31.7	S57 13.
Acrux	173 18.5	S63 06.
Adhara	255 18.2	S28 58.
Aldebaran	290 57.7	N16 31.
Alioth	166 27.5	N55 56.
Alkaid	153 05.1	N49 17.
Al Na'ir	27 53.0	S46 56.
Alnilam	275 53.7	S 1 11.
Alphard	218 03.5	S 8 40.
Alphecca	126 17.7	N26 42.
Alpheratz	357 51.2	N29 06.
Altair	62 15.8	N 8 52.
Ankaa	353 22.7	S42 17.
Antares	112 36.0	S26 26.
Arcturus	146 02.9	N19 09.
Atria	107 45.3	S69 02.
Avior	234 21.0	S59 30.
Bellatrix	278 39.8	N 6 21.
Betelgeuse	271 09.2	N 7 24.
Canopus	263 59.1	S52 41.
Capella	280 45.2	N46 00.
Deneb	49 36.9	N45 17.
Denebola	182 41.5	N14 33.
Diphda	349 03.2	S17 58.
Dubhe	194 00.8	N61 43.
Elnath	278 21.8	N28 36.
Eltanin	90 50.1	N51 29.
Enif	33 54.6	N 9 53.
Fomalhaut	15 32.1	S29 36.
Gacrux	172 10.0	S57 07.
Gienah	176 00.3	S17 33.
Hadar	148 59.5	S60 23.
Hamal	328 09.1	N23 28.
Kaus Aust.	83 54.2	S34 23.
Kochab	137 20.4	N74 08.
Markab	13 45.8	N15 13.
Menkar	314 22.7	N 4 06.
Menkent	148 17.0	S36 23.
Miaplacidus	221 41.4	S69 43.
Mirfak	308 50.8	N49 52.
Nunki	76 07.9	S26 17.
Peacock	53 31.2	S56 43.
Pollux	243 36.7	N28 01.
Procyon	245 07.4	N 5 13.
Rasalhague	96 13.8	N12 33.
Regulus	207 51.5	N11 57.
Rigel	281 19.0	S 8 11.
Rigil Kent.	140 03.1	S60 50.
Sabik	102 21.6	S15 43.
Schedar	349 49.1	N56 33.
Shaula	96 32.6	S37 06.
Sirius	258 40.2	S16 43.
Spica	158 39.5	S11 10.
Suhail	222 58.0	S43 26.
Vega	80 44.4	N38 47.
Zuben'ubi	137 14.2	S16 03.

	SHA ° ′	Mer. Pas h m
Venus	96 53.0	13 31
Mars	11 48.1	19 09
Jupiter	192 22.5	7 07
Saturn	256 24.4	2 52

UT d h	SUN GHA ° ′	SUN Dec ° ′	MOON GHA ° ′	v ′	MOON Dec ° ′	d ′	HP ′
21 00 FRIDAY	183 35.3	S19 46.2	222 54.2	10.9	S 3 55.1	16.4	60.1
01	198 35.1	46.7	237 24.1	10.7	4 11.5	16.5	60.1
02	213 34.9	47.3	251 53.8	10.7	4 28.0	16.5	60.1
03	228 34.8	. . 47.9	266 23.5	10.7	4 44.5	16.4	60.2
04	243 34.6	48.4	280 53.2	10.6	5 00.9	16.4	60.2
05	258 34.5	49.0	295 22.8	10.5	5 17.3	16.5	60.2
06	273 34.3	S19 49.5	309 52.3	10.4	S 5 33.8	16.4	60.3
07	288 34.2	50.1	324 21.7	10.4	5 50.2	16.4	60.3
08	303 34.0	50.7	338 51.1	10.3	6 06.6	16.4	60.3
09	318 33.9	. . 51.2	353 20.4	10.2	6 23.0	16.3	60.4
10	333 33.7	51.8	7 49.6	10.1	6 39.3	16.4	60.4
11	348 33.6	52.3	22 18.7	10.1	6 55.7	16.3	60.4
12	3 33.4	S19 52.9	36 47.8	10.0	S 7 12.0	16.3	60.5
13	18 33.3	53.4	51 16.8	9.9	7 28.3	16.3	60.5
14	33 33.1	54.0	65 45.7	9.8	7 44.6	16.3	60.5
15	48 32.9	. . 54.5	80 14.5	9.8	8 00.9	16.2	60.6
16	63 32.8	55.1	94 43.3	9.7	8 17.1	16.2	60.6
17	78 32.6	55.6	109 12.0	9.6	8 33.3	16.1	60.6
18	93 32.5	S19 56.2	123 40.6	9.5	S 8 49.4	16.2	60.6
19	108 32.3	56.7	138 09.1	9.4	9 05.6	16.1	60.7
20	123 32.1	57.3	152 37.5	9.4	9 21.7	16.0	60.7
21	138 32.0	. . 57.8	167 05.9	9.2	9 37.7	16.0	60.7
22	153 31.8	58.4	181 34.1	9.2	9 53.7	16.0	60.8
23	168 31.7	58.9	196 02.3	9.1	10 09.7	15.9	60.8
22 00 SATURDAY	183 31.5	S19 59.5	210 30.4	8.9	S10 25.6	15.9	60.8
01	198 31.3	20 00.0	224 58.3	8.9	10 41.5	15.8	60.8
02	213 31.2	00.6	239 26.2	8.8	10 57.3	15.8	60.9
03	228 31.0	. . 01.1	253 54.0	8.8	11 13.1	15.8	60.9
04	243 30.9	01.7	268 21.8	8.6	11 28.9	15.6	60.9
05	258 30.7	02.2	282 49.4	8.5	11 44.5	15.6	60.9
06	273 30.5	S20 02.8	297 16.9	8.4	S12 00.1	15.6	61.0
07	288 30.4	03.3	311 44.3	8.4	12 15.7	15.5	61.0
08	303 30.2	03.8	326 11.7	8.2	12 31.2	15.4	61.0
09	318 30.1	. . 04.4	340 38.9	8.1	12 46.6	15.4	61.0
10	333 29.9	04.9	355 06.0	8.1	13 02.0	15.3	61.0
11	348 29.7	05.5	9 33.1	7.9	13 17.3	15.2	61.1
12	3 29.6	S20 06.0	24 00.0	7.8	S13 32.5	15.1	61.1
13	18 29.4	06.5	38 26.8	7.7	13 47.6	15.1	61.1
14	33 29.2	07.1	52 53.5	7.7	14 02.7	15.0	61.1
15	48 29.1	. . 07.6	67 20.2	7.5	14 17.7	14.9	61.1
16	63 28.9	08.2	81 46.7	7.4	14 32.6	14.8	61.2
17	78 28.7	08.7	96 13.1	7.3	14 47.4	14.8	61.2
18	93 28.6	S20 09.2	110 39.4	7.2	S15 02.2	14.6	61.2
19	108 28.4	09.8	125 05.6	7.2	15 16.8	14.6	61.2
20	123 28.2	10.3	139 31.8	7.0	15 31.4	14.5	61.2
21	138 28.1	. . 10.8	153 57.8	6.8	15 45.9	14.4	61.2
22	153 27.9	11.4	168 23.6	6.8	16 00.3	14.3	61.3
23	168 27.7	11.9	182 49.4	6.7	16 14.6	14.2	61.3
23 00 SUNDAY	183 27.6	S20 12.4	197 15.1	6.6	S16 28.8	14.1	61.3
01	198 27.4	13.0	211 40.7	6.5	16 42.9	14.0	61.3
02	213 27.2	13.5	226 06.2	6.3	16 56.9	13.8	61.3
03	228 27.1	. . 14.0	240 31.5	6.3	17 10.7	13.8	61.3
04	243 26.9	14.5	254 56.8	6.1	17 24.5	13.7	61.3
05	258 26.7	15.1	269 21.9	6.0	17 38.2	13.5	61.4
06	273 26.5	S20 15.6	283 46.9	6.0	S17 51.7	13.5	61.4
07	288 26.4	16.1	298 11.9	5.8	18 05.2	13.3	61.4
08	303 26.2	16.7	312 36.7	5.7	18 18.5	13.2	61.4
09	318 26.0	. . 17.2	327 01.4	5.6	18 31.7	13.1	61.4
10	333 25.9	17.7	341 26.0	5.5	18 44.8	12.9	61.4
11	348 25.7	18.2	355 50.5	5.3	18 57.7	12.9	61.4
12	3 25.5	S20 18.8	10 14.8	5.3	S19 10.6	12.7	61.4
13	18 25.3	19.3	24 39.1	5.2	19 23.3	12.6	61.4
14	33 25.2	19.8	39 03.3	5.1	19 35.9	12.4	61.4
15	48 25.0	. . 20.3	53 27.4	4.9	19 48.3	12.3	61.4
16	63 24.8	20.8	67 51.3	4.9	20 00.6	12.2	61.4
17	78 24.6	21.4	82 15.2	4.7	S20 12.8	12.0	61.4
18	93 24.5	S20 21.9					
19	108 24.3	22.4					
20	123 24.1	22.9					
21	138 23.9	. . 23.4					
22	153 23.8	24.0					
23	168 23.6	24.5					
	SD 16.2	*d* 0.5	SD 16.5		16.6		16.7

A total eclipse of the Sun occurs on this date. See page 5.

Lat. °	Twilight Naut. h m	Twilight Civil h m	Sunrise h m	Moonrise 21 h m	Moonrise 22 h m	Moonrise 23 h m	Moonrise 24 h m
N 72	07 22	09 05	■	04 24	07 05	■	■
N 70	07 09	08 37	10 41	04 16	06 40	09 55	■
68	06 59	08 15	09 45	04 09	06 22	09 00	■
66	06 50	07 59	09 12	04 04	06 08	08 27	11 39
64	06 42	07 45	08 48	03 59	05 56	08 03	10 27
62	06 36	07 33	08 29	03 56	05 46	07 44	09 51
60	06 30	07 23	08 14	03 52	05 37	07 29	09 25
N 58	06 24	07 14	08 01	03 49	05 30	07 16	09 05
56	06 19	07 06	07 50	03 47	05 23	07 05	08 49
54	06 15	06 59	07 40	03 44	05 18	06 55	08 34
52	06 11	06 53	07 31	03 42	05 12	06 47	08 22
50	06 07	06 47	07 23	03 40	05 08	06 39	08 11
45	05 58	06 34	07 06	03 36	04 58	06 22	07 49
N 40	05 50	06 23	06 52	03 33	04 49	06 09	07 31
35	05 43	06 13	06 41	03 30	04 42	05 58	07 16
30	05 36	06 05	06 30	03 27	04 36	05 48	07 03
20	05 22	05 49	06 13	03 23	04 25	05 31	06 41
N 10	05 09	05 35	05 57	03 19	04 16	05 17	06 22
0	04 55	05 20	05 42	03 16	04 08	05 03	06 04
S 10	04 39	05 05	05 28	03 12	03 59	04 50	05 46
20	04 20	04 48	05 12	03 08	03 50	04 36	05 28
30	03 55	04 27	04 53	03 04	03 40	04 20	05 06
35	03 39	04 14	04 42	03 02	03 34	04 11	04 54
40	03 20	03 59	04 30	02 59	03 28	04 00	04 40
45	02 55	03 40	04 15	02 56	03 20	03 48	04 23
S 50	02 21	03 16	03 57	02 53	03 11	03 33	04 02
52	02 03	03 05	03 48	02 51	03 07	03 26	03 53
54	01 41	02 51	03 38	02 49	03 02	03 19	03 42
56	01 09	02 35	03 27	02 47	02 57	03 10	03 30
58	////	02 16	03 15	02 45	02 52	03 01	03 15
S 60	////	01 52	03 00	02 43	02 45	02 50	02 59

Lat. °	Sunset h m	Twilight Civil h m	Twilight Naut. h m	Moonset 21 h m	Moonset 22 h m	Moonset 23 h m	Moonset 24 h m
N 72	■	14 26	16 09	13 56	13 06	■	■
N 70	12 51	14 54	16 22	14 07	13 33	12 17	■
68	13 46	15 16	16 32	14 16	13 53	13 14	■
66	14 19	15 33	16 41	14 24	14 09	13 49	12 45
64	14 43	15 47	16 49	14 30	14 23	14 14	13 58
62	15 02	15 58	16 56	14 36	14 34	14 33	14 35
60	15 17	16 09	17 02	14 41	14 44	14 50	15 01
N 58	15 31	16 18	17 07	14 46	14 53	15 04	15 22
56	15 42	16 25	17 12	14 50	15 01	15 16	15 39
54	15 52	16 33	17 17	14 53	15 07	15 26	15 54
52	16 01	16 39	17 21	14 56	15 13	15 36	16 07
50	16 09	16 45	17 25	14 59	15 19	15 44	16 18
45	16 25	16 58	17 34	15 06	15 31	16 02	16 42
N 40	16 39	17 09	17 42	15 11	15 41	16 17	17 01
35	16 51	17 18	17 49	15 16	15 50	16 29	17 17
30	17 01	17 27	17 56	15 20	15 58	16 40	17 31
20	17 19	17 43	18 10	15 27	16 11	16 59	17 54
N 10	17 35	17 57	18 23	15 34	16 22	17 16	18 15
0	17 50	18 12	18 37	15 40	16 33	17 31	18 34
S 10	18 05	18 27	18 53	15 46	16 44	17 47	18 53
20	18 21	18 45	19 13	15 52	16 56	18 04	19 14
30	18 39	19 06	19 38	16 00	17 10	18 23	19 38
35	18 50	19 19	19 53	16 04	17 18	18 35	19 52
40	19 03	19 34	20 13	16 09	17 27	18 48	20 09
45	19 18	19 53	20 38	16 14	17 37	19 03	20 28
S 50	19 36	20 17	21 12	16 21	17 50	19 22	20 53
52	19 45	20 29	21 31	16 24	17 56	19 31	21 05
54	19 55	20 42	21 54	16 28	18 03	19 42	21 19
56	20 06	20 58	22 27	16 32	18 11	19 53	21 34
58	20 19	21 18	////	16 36	18 19	20 07	21 53
S 60	20 34	21 43	////	16 41	18 29	20 23	22 17

Day	SUN Eqn. of Time 00^h m s	SUN Eqn. of Time 12^h m s	SUN Mer. Pass. h m	MOON Mer. Pass. Upper h m	MOON Mer. Pass. Lower h m	MOON Age d	MOON Phase %
21	14 21	14 14	11 46	09 28	21 54	27	9
22	14 06	13 59	11 46	10 20	22 48	28	3
23	13 51	13 42	11 46	11 17	23 48	29	0

UT	ARIES	VENUS −3.9		MARS −0.5		JUPITER −2.0		SATURN −0.2	
	GHA	GHA	Dec	GHA	Dec	GHA	Dec	GHA	Dec
d h	° ′	° ′	° ′	° ′	° ′	° ′	° ′	° ′	° ′
24 00 (Monday)	62 32.4	156 42.0	S24 32.7	73 26.5	S 5 57.2	254 41.4	N 6 20.7	319 03.2	N22 08.4
01	77 34.9	171 41.0	32.9	88 27.8	56.6	269 43.6	20.6	334 05.8	08.4
02	92 37.3	186 40.1	33.1	103 29.2	56.0	284 45.8	20.5	349 08.4	08.4
03	107 39.8	201 39.1	33.3	118 30.5	55.4	299 47.9	20.4	4 11.0	08.4
04	122 42.3	216 38.2	33.4	133 31.8	54.8	314 50.1	20.3	19 13.6	08.5
05	137 44.7	231 37.3	33.6	148 33.2	54.2	329 52.3	20.2	34 16.2	08.5
06	152 47.2	246 36.3	S24 33.8	163 34.5	S 5 53.6	344 54.5	N 6 20.1	49 18.8	N22 08.5
07	167 49.7	261 35.4	34.0	178 35.8	53.0	359 56.7	20.0	64 21.4	08.5
08	182 52.1	276 34.4	34.1	193 37.1	52.4	14 58.9	19.9	79 24.0	08.5
09	197 54.6	291 33.5	34.3	208 38.5	51.8	30 01.1	19.8	94 26.6	08.5
10	212 57.1	306 32.5	34.5	223 39.8	51.2	45 03.3	19.7	109 29.2	08.5
11	227 59.5	321 31.6	34.6	238 41.1	50.6	60 05.4	19.6	124 31.8	08.6
12	243 02.0	336 30.6	S24 34.8	253 42.4	S 5 50.0	75 07.6	N 6 19.5	139 34.5	N22 08.6
13	258 04.5	351 29.7	35.0	268 43.8	49.4	90 09.8	19.4	154 37.1	08.6
14	273 06.9	6 28.7	35.1	283 45.1	48.9	105 12.0	19.3	169 39.7	08.6
15	288 09.4	21 27.8	35.3	298 46.4	48.3	120 14.2	19.2	184 42.3	08.6
16	303 11.8	36 26.9	35.4	313 47.7	47.7	135 16.4	19.1	199 44.9	08.6
17	318 14.3	51 25.9	35.6	328 49.1	47.1	150 18.6	19.0	214 47.5	08.6
18	333 16.8	66 25.0	S24 35.8	343 50.4	S 5 46.5	165 20.8	N 6 18.9	229 50.1	N22 08.6
19	348 19.2	81 24.0	35.9	358 51.7	45.9	180 23.0	18.8	244 52.7	08.7
20	3 21.7	96 23.1	36.1	13 53.0	45.3	195 25.2	18.6	259 55.3	08.7
21	18 24.2	111 22.1	36.2	28 54.3	44.7	210 27.3	18.5	274 57.9	08.7
22	33 26.6	126 21.2	36.4	43 55.7	44.1	225 29.5	18.4	290 00.5	08.7
23	48 29.1	141 20.2	36.5	58 57.0	43.5	240 31.7	18.3	305 03.1	08.7
25 00 (Tuesday)	63 31.6	156 19.3	S24 36.7	73 58.3	S 5 42.9	255 33.9	N 6 18.2	320 05.7	N22 08.7
01	78 34.0	171 18.3	36.8	88 59.6	42.3	270 36.1	18.1	335 08.3	08.7
02	93 36.5	186 17.4	37.0	104 00.9	41.7	285 38.3	18.0	350 10.9	08.8
03	108 39.0	201 16.4	37.1	119 02.3	41.1	300 40.5	17.9	5 13.5	08.8
04	123 41.4	216 15.5	37.3	134 03.6	40.5	315 42.7	17.8	20 16.1	08.8
05	138 43.9	231 14.5	37.4	149 04.9	39.9	330 44.9	17.7	35 18.7	08.8
06	153 46.3	246 13.6	S24 37.5	164 06.2	S 5 39.3	345 47.1	N 6 17.6	50 21.3	N22 08.8
07	168 48.8	261 12.6	37.7	179 07.5	38.7	0 49.3	17.5	65 23.9	08.8
08	183 51.3	276 11.7	37.8	194 08.9	38.1	15 51.5	17.4	80 26.5	08.8
09	198 53.7	291 10.7	38.0	209 10.2	37.5	30 53.7	17.3	95 29.2	08.9
10	213 56.2	306 09.8	38.1	224 11.5	36.9	45 55.9	17.2	110 31.8	08.9
11	228 58.7	321 08.8	38.2	239 12.8	36.3	60 58.1	17.1	125 34.4	08.9
12	244 01.1	336 07.9	S24 38.4	254 14.1	S 5 35.7	76 00.2	N 6 17.0	140 37.0	N22 08.9
13	259 03.6	351 07.0	38.5	269 15.4	35.1	91 02.4	16.9	155 39.6	08.9
14	274 06.1	6 06.0	38.6	284 16.7	34.5	106 04.6	16.8	170 42.2	08.9
15	289 08.5	21 05.1	38.8	299 18.1	33.9	121 06.8	16.7	185 44.8	08.9
16	304 11.0	36 04.1	38.9	314 19.4	33.3	136 09.0	16.6	200 47.4	09.0
17	319 13.5	51 03.2	39.0	329 20.7	32.7	151 11.2	16.5	215 50.0	09.0
18	334 15.9	66 02.2	S24 39.1	344 22.0	S 5 32.1	166 13.4	N 6 16.4	230 52.6	N22 09.0
19	349 18.4	81 01.3	39.3	359 23.3	31.5	181 15.6	16.3	245 55.2	09.0
20	4 20.8	96 00.3	39.4	14 24.6	30.9	196 17.8	16.2	260 57.8	09.0
21	19 23.3	110 59.4	39.5	29 25.9	30.3	211 20.0	16.1	276 00.4	09.0
22	34 25.8	125 58.4	39.6	44 27.3	29.7	226 22.2	16.0	291 03.1	09.0
23	49 28.2	140 57.5	39.8	59 28.6	29.1	241 24.4	15.9	306 05.7	09.1
26 00 (Wednesday)	64 30.7	155 56.5	S24 39.9	74 29.9	S 5 28.5	256 26.6	N 6 15.8	321 08.3	N22 09.1
01	79 33.2	170 55.6	40.0	89 31.2	27.9	271 28.8	15.7	336 10.9	09.1
02	94 35.6	185 54.6	40.1	104 32.5	27.3	286 31.0	15.6	351 13.5	09.1
03	109 38.1	200 53.7	40.2	119 33.8	26.7	301 33.2	15.5	6 16.1	09.1
04	124 40.6	215 52.7	40.3	134 35.1	26.1	316 35.4	15.4	21 18.7	09.1
05	139 43.0	230 51.8	40.4	149 36.4	25.5	331 37.6	15.3	36 21.3	09.1
06	154 45.5	245 50.8	S24 40.6	164 37.7	S 5 24.9	346 39.8	N 6 15.2	51 23.9	N22 09.2
07	169 48.0	260 49.9	40.7	179 39.0	24.2	1 42.0	15.1	66 26.5	09.2
08	184 50.4	275 48.9	40.8	194 40.3	23.6	16 44.2	15.0	81 29.1	09.2
09	199 52.9	290 48.0	40.9	209 41.7	23.0	31 46.4	14.9	96 31.8	09.2
10	214 55.3	305 47.0	41.0	224 43.0	22.4	46 48.6	14.8	111 34.4	09.2
11	229 57.8	320 46.1	41.1	239 44.3	21.8	61 50.8	14.7	126 37.0	09.2
12	245 00.3	335 45.1	S24 41.2	254 45.6	S 5 21.2	76 53.0	N 6 14.6	141 39.6	N22 09.2
13	260 02.7	350 44.2	41.3	269 46.9	20.6	91 55.2	14.5	156 42.2	09.3
14	275 05.2	5 43.2	41.4	284 48.2	20.0	106 57.4	14.4	171 44.8	09.3
15	290 07.7	20 42.3	41.5	299 49.5	19.4	121 59.6	14.3	186 47.4	09.3
16	305 10.1	35 41.3	41.6	314 50.8	18.8	137 01.8	14.2	201 50.0	09.3
17	320 12.6	50 40.3	41.7	329 52.1	18.2	152 04.0	14.1	216 52.6	09.3
18	335 15.1	65 39.4	S24 41.8	344 53.4	S 5 17.6	167 06.2	N 6 14.0	231 55.3	N22 09.3
19	350 17.5	80 38.4	41.9	359 54.7	17.0	182 08.4	13.9	246 57.9	09.3
20	5 20.0	95 37.5	42.0	14 56.0	16.4	197 10.6	13.8	262 00.5	09.4
21	20 22.5	110 36.5	42.1	29 57.3	15.8	212 12.8	13.7	277 03.1	09.4
22	35 24.9	125 35.6	42.1	44 58.6	15.2	227 15.0	13.6	292 05.7	09.4
23	50 27.4	140 34.6	42.2	59 59.9	14.6	242 17.2	13.5	307 08.3	09.4
Mer. Pass.	h m 19 42.7	*v* −0.9	*d* 0.1	*v* 1.3	*d* 0.6	*v* 2.2	*d* 0.1	*v* 2.6	*d* 0.0

STARS

Name	SHA	Dec
	° ′	° ′
Acamar	315 23.5	S40 17.4
Achernar	335 31.7	S57 13.2
Acrux	173 18.5	S63 06.9
Adhara	255 18.2	S28 58.4
Aldebaran	290 57.7	N16 31.1
Alioth	166 27.4	N55 56.2
Alkaid	153 05.1	N49 17.6
Al Na'ir	27 53.0	S46 56.9
Alnilam	275 53.7	S 1 11.8
Alphard	218 03.5	S 8 40.3
Alphecca	126 17.7	N26 42.1
Alpheratz	357 51.2	N29 06.8
Altair	62 15.8	N 8 52.7
Ankaa	353 22.7	S42 17.3
Antares	112 36.0	S26 26.4
Arcturus	146 02.9	N19 09.8
Atria	107 45.3	S69 02.1
Avior	234 20.9	S59 30.9
Bellatrix	278 39.8	N 6 21.3
Betelgeuse	271 09.2	N 7 24.6
Canopus	263 59.0	S52 41.6
Capella	280 45.2	N46 00.2
Deneb	49 36.9	N45 17.8
Denebola	182 41.4	N14 33.1
Diphda	349 03.2	S17 58.0
Dubhe	194 00.7	N61 43.6
Elnath	278 21.8	N28 36.7
Eltanin	90 50.1	N51 29.4
Enif	33 54.6	N 9 53.6
Fomalhaut	15 32.1	S29 36.3
Gacrux	172 10.0	S57 07.7
Gienah	176 00.3	S17 33.6
Hadar	148 59.5	S60 23.3
Hamal	328 09.1	N23 28.9
Kaus Aust.	83 54.2	S34 23.1
Kochab	137 20.4	N74 08.3
Markab	13 45.8	N15 13.6
Menkar	314 22.6	N 4 06.4
Menkent	148 17.0	S36 23.1
Miaplacidus	221 41.3	S69 43.5
Mirfak	308 50.8	N49 52.6
Nunki	76 07.9	S26 17.6
Peacock	53 31.2	S56 43.7
Pollux	243 36.7	N28 01.0
Procyon	245 07.4	N 5 13.0
Rasalhague	96 13.8	N12 33.4
Regulus	207 51.5	N11 57.0
Rigel	281 19.0	S 8 11.7
Rigil Kent.	140 03.0	S60 50.9
Sabik	102 21.6	S15 43.8
Schedar	349 49.1	N56 33.7
Shaula	96 32.6	S37 06.5
Sirius	258 40.1	S16 43.1
Spica	158 39.5	S11 10.7
Suhail	222 58.0	S43 26.5
Vega	80 44.5	N38 47.3
Zuben'ubi	137 14.2	S16 03.4

	SHA	Mer. Pass.
	° ′	h m
Venus	92 47.7	13 36
Mars	10 26.7	19 02
Jupiter	192 02.4	6 57
Saturn	256 34.1	2 39

UT	SUN GHA	SUN Dec	MOON GHA	v	MOON Dec	d	HP
d h	° ′	° ′	° ′	′	° ′	′	′
24 00 (MONDAY)	183 23.4	S20 25.0	182 59.2	4.0	S21 33.9	11.0	61.5
01	198 23.2	25.5	197 22.2	3.9	21 44.9	10.8	61.5
02	213 23.1	26.0	211 45.1	3.9	21 55.7	10.7	61.5
03	228 22.9	. . 26.5	226 08.0	3.7	22 06.4	10.5	61.5
04	243 22.7	27.0	240 30.7	3.6	22 16.9	10.4	61.4
05	258 22.5	27.6	254 53.3	3.6	22 27.3	10.2	61.4
06	273 22.4	S20 28.1	269 15.9	3.4	S22 37.5	10.0	61.4
07	288 22.2	28.6	283 38.3	3.4	22 47.5	9.9	61.4
08	303 22.0	29.1	298 00.7	3.3	22 57.4	9.6	61.4
09	318 21.8	. . 29.6	312 23.0	3.2	23 07.0	9.5	61.4
10	333 21.6	30.1	326 45.2	3.1	23 16.5	9.4	61.4
11	348 21.5	30.6	341 07.3	3.0	23 25.9	9.1	61.4
12	3 21.3	S20 31.1	355 29.3	2.9	S23 35.0	9.0	61.4
13	18 21.1	31.6	9 51.2	2.9	23 44.0	8.8	61.4
14	33 20.9	32.1	24 13.1	2.7	23 52.8	8.7	61.4
15	48 20.7	. . 32.7	38 34.8	2.7	24 01.5	8.4	61.4
16	63 20.5	33.2	52 56.5	2.7	24 09.9	8.3	61.4
17	78 20.4	33.7	67 18.2	2.5	24 18.2	8.0	61.4
18	93 20.2	S20 34.2	81 39.7	2.5	S24 26.2	7.9	61.3
19	108 20.0	34.7	96 01.2	2.4	24 34.1	7.7	61.3
20	123 19.8	35.2	110 22.6	2.4	24 41.8	7.5	61.3
21	138 19.6	. . 35.7	124 44.0	2.2	24 49.3	7.3	61.3
22	153 19.4	36.2	139 05.2	2.3	24 56.6	7.2	61.3
23	168 19.3	36.7	153 26.5	2.1	25 03.8	6.9	61.3
25 00 (TUESDAY)	183 19.1	S20 37.2	167 47.6	2.1	S25 10.7	6.7	61.3
01	198 18.9	37.7	182 08.7	2.1	25 17.4	6.6	61.3
02	213 18.7	38.2	196 29.8	2.0	25 24.0	6.3	61.2
03	228 18.5	. . 38.7	210 50.8	1.9	25 30.3	6.2	61.2
04	243 18.3	39.2	225 11.7	1.9	25 36.5	5.9	61.2
05	258 18.2	39.7	239 32.6	1.9	25 42.4	5.7	61.2
06	273 18.0	S20 40.2	253 53.5	1.8	S25 48.1	5.6	61.2
07	288 17.8	40.7	268 14.3	1.8	25 53.7	5.3	61.2
08	303 17.6	41.2	282 35.1	1.8	25 59.0	5.2	61.1
09	318 17.4	. . 41.7	296 55.9	1.7	26 04.2	4.9	61.1
10	333 17.2	42.1	311 16.6	1.7	26 09.1	4.7	61.1
11	348 17.0	42.6	325 37.3	1.7	26 13.8	4.6	61.1
12	3 16.8	S20 43.1	339 58.0	1.6	S26 18.4	4.3	61.1
13	18 16.7	43.6	354 18.6	1.6	26 22.7	4.1	61.0
14	33 16.5	44.1	8 39.2	1.7	26 26.8	3.9	61.0
15	48 16.3	. . 44.6	22 59.9	1.6	26 30.7	3.7	61.0
16	63 16.1	45.1	37 20.5	1.5	26 34.4	3.5	61.0
17	78 15.9	45.6	51 41.0	1.6	26 37.9	3.3	60.9
18	93 15.7	S20 46.1	66 01.6	1.6	S26 41.2	3.1	60.9
19	108 15.5	46.6	80 22.2	1.6	26 44.3	2.8	60.9
20	123 15.3	47.0	94 42.8	1.6	26 47.1	2.7	60.9
21	138 15.1	. . 47.5	109 03.4	1.5	26 49.8	2.5	60.8
22	153 14.9	48.0	123 23.9	1.6	26 52.3	2.2	60.8
23	168 14.7	48.5	137 44.5	1.6	26 54.5	2.1	60.8
26 00 (WEDNESDAY)	183 14.6	S20 49.0	152 05.1	1.7	S26 56.6	1.8	60.8
01	198 14.4	49.5	166 25.8	1.6	26 58.4	1.6	60.7
02	213 14.2	50.0	180 46.4	1.7	27 00.0	1.4	60.7
03	228 14.0	. . 50.4	195 07.1	1.6	27 01.4	1.3	60.7
04	243 13.8	50.9	209 27.7	1.8	27 02.7	1.0	60.7
05	258 13.6	51.4	223 48.5	1.7	27 03.7	0.8	60.6
06	273 13.4	S20 51.9	238 09.2	1.8	S27 04.5	0.5	60.6
07	288 13.2	52.4	252 30.0	1.8	27 05.0	0.4	60.6
08	303 13.0	52.8	266 50.8	1.9	27 05.4	0.2	60.5
09	318 12.8	. . 53.3	281 11.7	1.9	27 05.6	0.0	60.5
10	333 12.6	53.8	295 32.6	1.9	27 05.6	0.2	60.5
11	348 12.4	54.3	309 53.5	2.0	27 05.4	0.5	60.5
12	3 12.2	S20 54.7	324 14.5	2.0	S27 04.9	0.6	60.4
13	18 12.0	55.2	338 35.5	2.1	27 04.3	0.8	60.4
14	33 11.8	55.7	352 56.6	2.2	27 03.5	1.0	60.4
15	48 11.6	. . 56.2	7 17.8	2.2	27 02.5	1.3	60.3
16	63 11.4	56.6	21 39.0	2.3	27 01.2	1.4	60.3
17	78 11.2	57.1	36 00.3	2.4	26 59.8	1.6	60.3
18	93 11.0	S20 57.6	50 21.7	2.4	S26 58.2	1.9	60.2
19	108 10.8	58.1	64 43.1	2.5	26 56.3	2.0	60.2
20	123 10.6	58.5	79 04.6	2.5	26 54.3	2.2	60.2
21	138 10.4	. . 59.0	93 26.1	2.7	26 52.1	2.4	60.1
22	153 10.2	59.5	107 47.8	2.7	26 49.7	2.6	60.1
23	168 10.0	59.9	122 09.5	2.8	S26 47.1	2.8	60.1
	SD 16.2	*d* 0.5	SD 16.7		16.6		16.5

Lat.	Twilight Naut.	Twilight Civil	Sunrise	Moonrise 24	25	26	27
°	h m	h m	h m	h m	h m	h m	h m
N 72	07 32	09 20	■	■	■	■	■
N 70	07 18	08 48	11 21	■	■	■	■
68	07 07	08 25	10 00	■	■	■	■
66	06 57	08 07	09 23	11 39	■	■	■
64	06 49	07 52	08 57	10 27	■	■	15 03
62	06 41	07 39	08 37	09 51	11 58	13 29	13 53
60	06 35	07 29	08 21	09 25	11 16	12 37	13 18
N 58	06 29	07 19	08 07	09 05	10 47	12 05	12 52
56	06 24	07 11	07 55	08 49	10 25	11 42	12 31
54	06 19	07 04	07 45	08 34	10 07	11 22	12 14
52	06 15	06 57	07 36	08 22	09 52	11 06	11 59
50	06 11	06 51	07 28	08 11	09 39	10 52	11 47
45	06 01	06 37	07 10	07 49	09 11	10 23	11 20
N 40	05 53	06 26	06 56	07 31	08 50	10 01	10 59
35	05 45	06 16	06 44	07 16	08 32	09 42	10 42
30	05 38	06 07	06 33	07 03	08 17	09 26	10 27
20	05 24	05 51	06 15	06 41	07 51	08 59	10 01
N 10	05 10	05 36	05 58	06 22	07 29	08 36	09 39
0	04 55	05 21	05 43	06 04	07 08	08 14	09 18
S 10	04 39	05 05	05 28	05 46	06 48	07 52	08 57
20	04 19	04 47	05 11	05 28	06 26	07 29	08 35
30	03 54	04 26	04 52	05 06	06 01	07 02	08 09
35	03 37	04 12	04 41	04 54	05 46	06 47	07 54
40	03 17	03 57	04 28	04 40	05 29	06 28	07 36
45	02 52	03 37	04 13	04 23	05 08	06 06	07 15
S 50	02 16	03 13	03 54	04 02	04 43	05 38	06 48
52	01 57	03 00	03 45	03 53	04 30	05 24	06 35
54	01 32	02 46	03 35	03 42	04 16	05 08	06 19
56	00 55	02 30	03 23	03 30	04 00	04 50	06 02
58	////	02 09	03 10	03 15	03 41	04 27	05 40
S 60	////	01 42	02 54	02 59	03 17	03 57	05 12

Lat.	Sunset	Twilight Civil	Twilight Naut.	Moonset 24	25	26	27
°	h m	h m	h m	h m	h m	h m	h m
N 72	■	14 13	16 01	■	■	■	■
N 70	12 12	14 45	16 15	■	■	■	■
68	13 33	15 08	16 26	■	■	■	■
66	14 10	15 26	16 36	12 45	■	■	■
64	14 36	15 41	16 44	13 58	■	■	16 10
62	14 56	15 54	16 52	14 35	14 44	15 32	17 20
60	15 12	16 04	16 58	15 01	15 26	16 23	17 55
N 58	15 26	16 14	17 04	15 22	15 55	16 55	18 20
56	15 38	16 22	17 09	15 39	16 18	17 19	18 40
54	15 48	16 30	17 14	15 54	16 36	17 38	18 57
52	15 57	16 36	17 18	16 07	16 52	17 54	19 11
50	16 06	16 42	17 23	16 18	17 05	18 08	19 24
45	16 23	16 56	17 32	16 42	17 33	18 37	19 50
N 40	16 38	17 07	17 41	17 01	17 55	18 59	20 10
35	16 50	17 17	17 48	17 17	18 13	19 18	20 27
30	17 01	17 26	17 56	17 31	18 29	19 33	20 42
20	17 19	17 43	18 10	17 54	18 55	20 00	21 06
N 10	17 35	17 58	18 24	18 15	19 18	20 23	21 27
0	17 51	18 13	18 38	18 34	19 39	20 45	21 47
S 10	18 06	18 29	18 55	18 53	20 01	21 06	22 07
20	18 23	18 47	19 15	19 14	20 24	21 29	22 28
30	18 42	19 08	19 41	19 38	20 50	21 56	22 52
35	18 53	19 22	19 57	19 52	21 06	22 11	23 06
40	19 06	19 38	20 17	20 09	21 24	22 29	23 22
45	19 22	19 57	20 43	20 28	21 46	22 51	23 41
S 50	19 41	20 22	21 19	20 53	22 14	23 19	24 05
52	19 50	20 34	21 39	21 05	22 28	23 32	24 17
54	20 00	20 49	22 04	21 19	22 44	23 47	24 30
56	20 12	21 06	22 43	21 34	23 02	24 05	00 05
58	20 25	21 27	////	21 53	23 25	24 27	00 27
S 60	20 41	21 54	////	22 17	23 55	24 56	00 56

Day	SUN Eqn. of Time 00^h	SUN Eqn. of Time 12^h	SUN Mer. Pass.	MOON Mer. Pass. Upper	MOON Mer. Pass. Lower	MOON Age	MOON Phase
d	m s	m s	h m	h m	h m	d	%
24	13 34	13 25	11 47	12 19	24 51	01	0
25	13 17	13 08	11 47	13 24	00 51	02	4
26	12 59	12 49	11 47	14 29	01 57	03	10

UT	ARIES	VENUS −3.9		MARS −0.5		JUPITER −2.0		SATURN −0.2		STARS		
	GHA	GHA	Dec	GHA	Dec	GHA	Dec	GHA	Dec	Name	SHA	Dec
d h	° ′	° ′	° ′	° ′	° ′	° ′	° ′	° ′	° ′		° ′	° ′
27 00	65 29.8	155 33.7	S24 42.3	75 01.2	S 5 14.0	257 19.4	N 6 13.4	322 10.9	N22 09.4	Acamar	315 23.5	S40 17.
01	80 32.3	170 32.7	42.4	90 02.5	13.4	272 21.6	13.3	337 13.5	09.4	Achernar	335 31.7	S57 13.
02	95 34.8	185 31.8	42.5	105 03.8	12.8	287 23.8	13.2	352 16.2	09.5	Acrux	173 18.5	S63 06.
03	110 37.2	200 30.8	42.6	120 05.1	12.2	302 26.0	13.1	7 18.8	09.5	Adhara	255 18.2	S28 58.
04	125 39.7	215 29.9	42.6	135 06.4	11.6	317 28.2	13.0	22 21.4	09.5	Aldebaran	290 57.7	N16 31.
05	140 42.2	230 28.9	42.7	150 07.7	11.0	332 30.4	12.9	37 24.0	09.5			
06	155 44.6	245 28.0	S24 42.8	165 09.0	S 5 10.3	347 32.7	N 6 12.8	52 26.6	N22 09.5	Alioth	166 27.4	N55 56.
07	170 47.1	260 27.0	42.9	180 10.3	09.7	2 34.9	12.7	67 29.2	09.5	Alkaid	153 05.0	N49 17.
T 08	185 49.6	275 26.1	43.0	195 11.6	09.1	17 37.1	12.6	82 31.8	09.5	Al Na'ir	27 53.0	S46 56.
H 09	200 52.0	290 25.1	43.0	210 12.9	08.5	32 39.3	12.5	97 34.5	09.6	Alnilam	275 53.7	S 1 11.
U 10	215 54.5	305 24.2	43.1	225 14.2	07.9	47 41.5	12.4	112 37.1	09.6	Alphard	218 03.4	S 8 40.
R 11	230 57.0	320 23.2	43.2	240 15.5	07.3	62 43.7	12.3	127 39.7	09.6			
S 12	245 59.4	335 22.3	S24 43.3	255 16.8	S 5 06.7	77 45.9	N 6 12.2	142 42.3	N22 09.6	Alphecca	126 17.7	N26 42.
D 13	261 01.9	350 21.3	43.3	270 18.1	06.1	92 48.1	12.1	157 44.9	09.6	Alpheratz	357 51.2	N29 06.
A 14	276 04.3	5 20.4	43.4	285 19.4	05.5	107 50.3	12.0	172 47.5	09.6	Altair	62 15.8	N 8 52.
Y 15	291 06.8	20 19.4	43.5	300 20.7	04.9	122 52.5	11.9	187 50.1	09.6	Ankaa	353 22.8	S42 17.
16	306 09.3	35 18.5	43.5	315 22.0	04.3	137 54.7	11.8	202 52.8	09.7	Antares	112 35.9	S26 26.
17	321 11.7	50 17.5	43.6	330 23.3	03.7	152 56.9	11.7	217 55.4	09.7			
18	336 14.2	65 16.6	S24 43.7	345 24.6	S 5 03.1	167 59.1	N 6 11.6	232 58.0	N22 09.7	Arcturus	146 02.9	N19 09.
19	351 16.7	80 15.6	43.7	0 25.9	02.5	183 01.3	11.5	248 00.6	09.7	Atria	107 45.2	S69 02.
20	6 19.1	95 14.6	43.8	15 27.2	01.9	198 03.5	11.4	263 03.2	09.7	Avior	234 20.9	S59 30.
21	21 21.6	110 13.7	43.8	30 28.5	01.2	213 05.8	11.3	278 05.8	09.7	Bellatrix	278 39.8	N 6 21.
22	36 24.1	125 12.7	43.9	45 29.8	00.6	228 08.0	11.2	293 08.5	09.8	Betelgeuse	271 09.1	N 7 24.
23	51 26.5	140 11.8	44.0	60 31.1	5 00.0	243 10.2	11.1	308 11.1	09.8			
28 00	66 29.0	155 10.8	S24 44.0	75 32.4	S 4 59.4	258 12.4	N 6 11.0	323 13.7	N22 09.8	Canopus	263 59.0	S52 41.
01	81 31.5	170 09.9	44.1	90 33.6	58.8	273 14.6	10.9	338 16.3	09.8	Capella	280 45.2	N46 00.
02	96 33.9	185 08.9	44.1	105 34.9	58.2	288 16.8	10.8	353 18.9	09.8	Deneb	49 36.9	N45 17.
03	111 36.4	200 08.0	44.2	120 36.2	57.6	303 19.0	10.7	8 21.5	09.8	Denebola	182 41.4	N14 33.
04	126 38.8	215 07.0	44.2	135 37.5	57.0	318 21.2	10.6	23 24.2	09.8	Diphda	349 03.2	S17 58.
05	141 41.3	230 06.1	44.3	150 38.8	56.4	333 23.4	10.6	38 26.8	09.9			
06	156 43.8	245 05.1	S24 44.3	165 40.1	S 4 55.8	348 25.6	N 6 10.5	53 29.4	N22 09.9	Dubhe	194 00.7	N61 43.
07	171 46.2	260 04.2	44.4	180 41.4	55.2	3 27.8	10.4	68 32.0	09.9	Elnath	278 21.8	N28 36.
08	186 48.7	275 03.2	44.4	195 42.7	54.6	18 30.1	10.3	83 34.6	09.9	Eltanin	90 50.2	N51 29.
F 09	201 51.2	290 02.3	44.4	210 44.0	53.9	33 32.3	10.2	98 37.3	09.9	Enif	33 54.6	N 9 53.
R 10	216 53.6	305 01.3	44.5	225 45.3	53.3	48 34.5	10.1	113 39.9	09.9	Fomalhaut	15 32.1	S29 36.
I 11	231 56.1	320 00.4	44.5	240 46.5	52.7	63 36.7	10.0	128 42.5	09.9			
D 12	246 58.6	334 59.4	S24 44.6	255 47.8	S 4 52.1	78 38.9	N 6 09.9	143 45.1	N22 10.0	Gacrux	172 09.9	S57 07.
A 13	262 01.0	349 58.5	44.6	270 49.1	51.5	93 41.1	09.8	158 47.7	10.0	Gienah	176 00.3	S17 33.
Y 14	277 03.5	4 57.5	44.6	285 50.4	50.9	108 43.3	09.7	173 50.3	10.0	Hadar	148 59.4	S60 23.
15	292 05.9	19 56.6	44.7	300 51.7	50.3	123 45.5	09.6	188 53.0	10.0	Hamal	328 09.0	N23 28.
16	307 08.4	34 55.6	44.7	315 53.0	49.7	138 47.8	09.5	203 55.6	10.0	Kaus Aust.	83 54.2	S34 23.
17	322 10.9	49 54.6	44.8	330 54.3	49.1	153 50.0	09.4	218 58.2	10.0			
18	337 13.3	64 53.7	S24 44.8	345 55.6	S 4 48.5	168 52.2	N 6 09.3	234 00.8	N22 10.1	Kochab	137 20.4	N74 08.
19	352 15.8	79 52.7	44.8	0 56.8	47.8	183 54.4	09.2	249 03.4	10.1	Markab	13 45.8	N15 13.
20	7 18.3	94 51.8	44.8	15 58.1	47.2	198 56.6	09.1	264 06.1	10.1	Menkar	314 22.6	N 4 06.
21	22 20.7	109 50.8	44.9	30 59.4	46.6	213 58.8	09.0	279 08.7	10.1	Menkent	148 16.9	S36 23.
22	37 23.2	124 49.9	44.9	46 00.7	46.0	229 01.0	08.9	294 11.3	10.1	Miaplacidus	221 41.3	S69 43.
23	52 25.7	139 48.9	44.9	61 02.0	45.4	244 03.3	08.8	309 13.9	10.1			
29 00	67 28.1	154 48.0	S24 44.9	76 03.3	S 4 44.8	259 05.5	N 6 08.8	324 16.5	N22 10.1	Mirfak	308 50.8	N49 52.
01	82 30.6	169 47.0	45.0	91 04.6	44.2	274 07.7	08.7	339 19.2	10.2	Nunki	76 07.9	S26 17.
02	97 33.1	184 46.1	45.0	106 05.8	43.6	289 09.9	08.6	354 21.8	10.2	Peacock	53 31.2	S56 43.
03	112 35.5	199 45.1	45.0	121 07.1	43.0	304 12.1	08.5	9 24.4	10.2	Pollux	243 36.7	N28 01.
04	127 38.0	214 44.2	45.0	136 08.4	42.3	319 14.3	08.4	24 27.0	10.2	Procyon	245 07.4	N 5 13.
05	142 40.4	229 43.2	45.0	151 09.7	41.7	334 16.5	08.3	39 29.6	10.2			
06	157 42.9	244 42.3	S24 45.1	166 11.0	S 4 41.1	349 18.8	N 6 08.2	54 32.3	N22 10.2	Rasalhague	96 13.8	N12 33.
07	172 45.4	259 41.3	45.1	181 12.3	40.5	4 21.0	08.1	69 34.9	10.3	Regulus	207 51.5	N11 57.
S 08	187 47.8	274 40.4	45.1	196 13.5	39.9	19 23.2	08.0	84 37.5	10.3	Rigel	281 19.0	S 8 11.
A 09	202 50.3	289 39.4	45.1	211 14.8	39.3	34 25.4	07.9	99 40.1	10.3	Rigil Kent.	140 03.0	S60 50.
T 10	217 52.8	304 38.5	45.1	226 16.1	38.7	49 27.6	07.8	114 42.8	10.3	Sabik	102 21.6	S15 43.
U 11	232 55.2	319 37.5	45.1	241 17.4	38.1	64 29.8	07.7	129 45.4	10.3			
R 12	247 57.7	334 36.5	S24 45.1	256 18.7	S 4 37.5	79 32.1	N 6 07.6	144 48.0	N22 10.3	Schedar	349 49.1	N56 33.
D 13	263 00.2	349 35.6	45.1	271 19.9	36.8	94 34.3	07.5	159 50.6	10.3	Shaula	96 32.6	S37 06.
A 14	278 02.6	4 34.6	45.1	286 21.2	36.2	109 36.5	07.4	174 53.2	10.4	Sirius	258 40.1	S16 43.
Y 15	293 05.1	19 33.7	45.2	301 22.5	35.6	124 38.7	07.4	189 55.9	10.4	Spica	158 39.5	S11 10.
16	308 07.6	34 32.7	45.2	316 23.8	35.0	139 40.9	07.3	204 58.5	10.4	Suhail	222 57.9	S43 26.
17	323 10.0	49 31.8	45.2	331 25.1	34.4	154 43.2	07.2	220 01.1	10.4			
18	338 12.5	64 30.8	S24 45.2	346 26.3	S 4 33.8	169 45.4	N 6 07.1	235 03.7	N22 10.4	Vega	80 44.5	N38 47.
19	353 14.9	79 29.9	45.2	1 27.6	33.2	184 47.6	07.0	250 06.4	10.4	Zuben'ubi	137 14.1	S16 03.
20	8 17.4	94 28.9	45.2	16 28.9	32.5	199 49.8	06.9	265 09.0	10.5		SHA	Mer. Pas
21	23 19.9	109 28.0	45.2	31 30.2	31.9	214 52.0	06.8	280 11.6	10.5		° ′	h m
22	38 22.3	124 27.0	45.1	46 31.4	31.3	229 54.3	06.7	295 14.2	10.5	Venus	88 41.9	13 40
23	53 24.8	139 26.1	45.1	61 32.7	30.7	244 56.5	06.6	310 16.9	10.5	Mars	9 03.4	18 56
	h m									Jupiter	191 43.4	6 46
Mer. Pass.	19 30.9	v −1.0	d 0.0	v 1.3	d 0.6	v 2.2	d 0.1	v 2.6	d 0.0	Saturn	256 44.7	2 27

	UT	SUN GHA	SUN Dec	MOON GHA	v	MOON Dec	d	HP
	d h	° ′	° ′	° ′	′	° ′	′	′
	27 00	183 09.8	S21 00.4	136 31.3	2.9	S26 44.3	3.0	60.0
	01	198 09.6	00.9	150 53.2	3.0	26 41.3	3.1	60.0
	02	213 09.4	01.3	165 15.2	3.1	26 38.2	3.4	60.0
	03	228 09.2	. . 01.8	179 37.3	3.1	26 34.8	3.5	59.9
	04	243 09.0	02.3	193 59.4	3.3	26 31.3	3.7	59.9
	05	258 08.8	02.7	208 21.7	3.3	26 27.6	3.9	59.8
	06	273 08.6	S21 03.2	222 44.0	3.5	S26 23.7	4.1	59.8
	07	288 08.4	03.7	237 06.5	3.5	26 19.6	4.3	59.8
T	08	303 08.2	04.1	251 29.0	3.7	26 15.3	4.4	59.7
H	09	318 08.0	. . 04.6	265 51.7	3.7	26 10.9	4.6	59.7
U	10	333 07.8	05.0	280 14.4	3.9	26 06.3	4.8	59.7
R	11	348 07.6	05.5	294 37.3	3.9	26 01.5	5.0	59.6
S	12	3 07.4	S21 06.0	309 00.2	4.1	S25 56.5	5.1	59.6
D	13	18 07.2	06.4	323 23.3	4.2	25 51.4	5.3	59.6
A	14	33 07.0	06.9	337 46.5	4.3	25 46.1	5.5	59.5
Y	15	48 06.8	. . 07.3	352 09.8	4.4	25 40.6	5.7	59.5
	16	63 06.6	07.8	6 33.2	4.5	25 34.9	5.8	59.4
	17	78 06.4	08.3	20 56.7	4.7	25 29.1	5.9	59.4
	18	93 06.2	S21 08.7	35 20.4	4.7	S25 23.2	6.2	59.4
	19	108 06.0	09.2	49 44.1	4.9	25 17.0	6.3	59.3
	20	123 05.8	09.6	64 08.0	5.0	25 10.7	6.4	59.3
	21	138 05.6	. . 10.1	78 32.0	5.2	25 04.3	6.6	59.2
	22	153 05.4	10.5	92 56.2	5.2	24 57.7	6.8	59.2
	23	168 05.2	11.0	107 20.4	5.4	24 50.9	6.9	59.2
	28 00	183 04.9	S21 11.4	121 44.8	5.5	S24 44.0	7.0	59.1
	01	198 04.7	11.9	136 09.3	5.6	24 37.0	7.2	59.1
	02	213 04.5	12.3	150 33.9	5.7	24 29.8	7.4	59.1
	03	228 04.3	. . 12.8	164 58.6	5.9	24 22.4	7.5	59.0
	04	243 04.1	13.2	179 23.5	6.0	24 14.9	7.6	59.0
	05	258 03.9	13.7	193 48.5	6.1	24 07.3	7.8	58.9
	06	273 03.7	S21 14.1	208 13.6	6.3	S23 59.5	7.9	58.9
	07	288 03.5	14.6	222 38.9	6.4	23 51.6	8.1	58.9
	08	303 03.3	15.0	237 04.3	6.5	23 43.5	8.2	58.8
F	09	318 03.1	. . 15.5	251 29.8	6.6	23 35.3	8.3	58.8
R	10	333 02.9	15.9	265 55.4	6.8	23 27.0	8.5	58.7
I	11	348 02.6	16.3	280 21.2	6.9	23 18.5	8.5	58.7
D	12	3 02.4	S21 16.8	294 47.1	7.0	S23 10.0	8.8	58.7
A	13	18 02.2	17.2	309 13.1	7.2	23 01.2	8.8	58.6
Y	14	33 02.0	17.7	323 39.3	7.3	22 52.4	9.0	58.6
	15	48 01.8	. . 18.1	338 05.6	7.4	22 43.4	9.0	58.5
	16	63 01.6	18.6	352 32.0	7.6	22 34.4	9.3	58.5
	17	78 01.4	19.0	6 58.6	7.7	22 25.1	9.3	58.5
	18	93 01.2	S21 19.4	21 25.3	7.8	S22 15.8	9.4	58.4
	19	108 00.9	19.9	35 52.1	8.0	22 06.4	9.6	58.4
	20	123 00.7	20.3	50 19.1	8.0	21 56.8	9.7	58.3
	21	138 00.5	. . 20.7	64 46.1	8.2	21 47.1	9.7	58.3
	22	153 00.3	21.2	79 13.3	8.4	21 37.4	9.9	58.3
	23	168 00.1	21.6	93 40.7	8.5	21 27.5	10.0	58.2
	29 00	182 59.9	S21 22.0	108 08.2	8.6	S21 17.5	10.1	58.2
	01	197 59.7	22.5	122 35.8	8.7	21 07.4	10.2	58.1
	02	212 59.4	22.9	137 03.5	8.8	20 57.2	10.4	58.1
	03	227 59.2	. . 23.3	151 31.3	9.0	20 46.8	10.4	58.1
	04	242 59.0	23.8	165 59.3	9.2	20 36.4	10.5	58.0
	05	257 58.8	24.2	180 27.5	9.2	20 25.9	10.6	58.0
	06	272 58.6	S21 24.6	194 55.7	9.4	S20 15.3	10.7	57.9
S	07	287 58.4	25.1	209 24.1	9.5	20 04.6	10.8	57.9
	08	302 58.2	25.5	223 52.6	9.6	19 53.8	10.9	57.9
A	09	317 57.9	. . 25.9	238 21.2	9.7	19 42.9	10.9	57.8
T	10	332 57.7	26.4	252 49.9	9.9	19 32.0	11.1	57.8
U	11	347 57.5	26.8	267 18.8	10.0	19 20.9	11.2	57.7
R	12	2 57.3	S21 27.2	281 47.8	10.1	S19 09.7	11.2	57.7
D	13	17 57.1	27.6	296 16.9	10.2	18 58.5	11.3	57.7
A	14	32 56.8	28.1	310 46.1	10.4	18 47.2	11.4	57.6
Y	15	47 56.6	. . 28.5	325 15.5	10.4	18 35.8	11.5	57.6
	16	62 56.4	28.9	339 44.9	10.6	18 24.3	11.6	57.5
	17	77 56.2	29.3	354 14.5	10.7	18 12.7	11.6	57.5
	18	92 56.0	S21 29.7	8 44.2	10.9	S18 01.1	11.7	57.5
	19	107 55.7	30.2	23 14.1	10.9	17 49.4	11.8	57.4
	20	122 55.5	30.6	37 44.0	11.1	17 37.6	11.9	57.4
	21	137 55.3	. . 31.0	52 14.1	11.1	17 25.7	11.9	57.4
	22	152 55.1	31.4	66 44.2	11.3	17 13.8	12.0	57.3
	23	167 54.9	31.8	81 14.5	11.4	S17 01.8	12.1	57.3
		SD 16.2	d 0.4	SD 16.2		16.0		15.7

Lat.	Twilight Naut.	Twilight Civil	Sunrise	Moonrise 27	Moonrise 28	Moonrise 29	Moonrise 30
°	h m	h m	h m	h m	h m	h m	h m
N 72	07 41	09 34	■	■	■	■	15 22
N 70	07 26	08 59	■	■	■	15 50	14 56
68	07 14	08 34	10 17	■	■	15 07	14 36
66	07 04	08 15	09 35	■	15 12	14 38	14 20
64	06 55	07 59	09 06	15 03	14 27	14 16	14 07
62	06 47	07 46	08 45	13 53	13 58	13 58	13 56
60	06 40	07 35	08 28	13 18	13 35	13 43	13 47
N 58	06 34	07 25	08 13	12 52	13 16	13 30	13 38
56	06 29	07 16	08 01	12 31	13 01	13 19	13 31
54	06 24	07 08	07 50	12 14	12 47	13 09	13 24
52	06 19	07 01	07 41	11 59	12 36	13 00	13 18
50	06 14	06 55	07 32	11 47	12 25	12 52	13 13
45	06 05	06 41	07 14	11 20	12 03	12 36	13 01
N 40	05 56	06 29	06 59	10 59	11 46	12 22	12 51
35	05 48	06 19	06 46	10 42	11 31	12 10	12 43
30	05 40	06 10	06 35	10 27	11 18	12 00	12 35
20	05 26	05 53	06 16	10 01	10 55	11 42	12 23
N 10	05 11	05 37	06 00	09 39	10 36	11 26	12 11
0	04 56	05 22	05 44	09 18	10 18	11 12	12 01
S 10	04 39	05 06	05 28	08 57	09 59	10 57	11 50
20	04 19	04 47	05 11	08 35	09 40	10 41	11 39
30	03 53	04 25	04 52	08 09	09 17	10 23	11 25
35	03 36	04 11	04 40	07 54	09 04	10 12	11 18
40	03 15	03 55	04 27	07 36	08 48	10 00	11 09
45	02 49	03 35	04 11	07 15	08 30	09 46	10 59
S 50	02 12	03 09	03 51	06 48	08 07	09 28	10 46
52	01 51	02 57	03 42	06 35	07 56	09 19	10 41
54	01 24	02 42	03 31	06 19	07 43	09 10	10 34
56	00 40	02 24	03 19	06 02	07 29	08 59	10 27
58	////	02 02	03 05	05 40	07 12	08 47	10 19
S 60	////	01 33	02 49	05 12	06 51	08 33	10 09

Lat.	Sunset	Twilight Civil	Twilight Naut.	Moonset 27	Moonset 28	Moonset 29	Moonset 30
°	h m	h m	h m	h m	h m	h m	h m
N 72	■	14 01	15 54	■	■	■	21 27
N 70	■	14 36	16 09	■	■	19 17	21 51
68	13 18	15 01	16 21	■	■	19 59	22 09
66	14 01	15 20	16 31	■	18 03	20 27	22 23
64	14 29	15 36	16 40	16 10	18 47	20 48	22 35
62	14 50	15 49	16 48	17 20	19 16	21 05	22 45
60	15 07	16 01	16 55	17 55	19 39	21 19	22 53
N 58	15 22	16 10	17 01	18 20	19 56	21 31	23 00
56	15 34	16 19	17 07	18 40	20 11	21 41	23 07
54	15 45	16 27	17 12	18 57	20 24	21 50	23 13
52	15 55	16 34	17 16	19 11	20 35	21 58	23 18
50	16 03	16 40	17 21	19 24	20 45	22 06	23 22
45	16 22	16 54	17 31	19 50	21 06	22 21	23 32
N 40	16 36	17 06	17 40	20 10	21 23	22 34	23 41
35	16 49	17 17	17 48	20 27	21 37	22 44	23 48
30	17 00	17 26	17 55	20 42	21 49	22 53	23 54
20	17 19	17 43	18 10	21 06	22 10	23 09	24 05
N 10	17 36	17 58	18 24	21 27	22 28	23 23	24 14
0	17 51	18 14	18 40	21 47	22 44	23 36	24 23
S 10	18 07	18 30	18 57	22 07	23 01	23 48	24 31
20	18 24	18 49	19 17	22 28	23 18	24 02	00 02
30	18 44	19 11	19 44	22 52	23 38	24 17	00 17
35	18 56	19 25	20 00	23 06	23 50	24 26	00 26
40	19 09	19 41	20 21	23 22	24 03	00 03	00 36
45	19 25	20 01	20 48	23 41	24 19	00 19	00 47
S 50	19 45	20 27	21 26	24 05	00 05	00 38	01 01
52	19 55	20 40	21 47	24 17	00 17	00 47	01 08
54	20 05	20 55	22 14	24 30	00 30	00 57	01 15
56	20 17	21 13	23 01	00 05	00 45	01 08	01 23
58	20 32	21 35	////	00 27	01 02	01 21	01 32
S 60	20 48	22 05	////	00 56	01 24	01 36	01 42

Day	SUN Eqn. of Time 00^h	SUN Eqn. of Time 12^h	SUN Mer. Pass.	MOON Mer. Pass. Upper	MOON Mer. Pass. Lower	Age	Phase
d	m s	m s	h m	h m	h m	d %	
27	12 40	12 30	11 47	15 33	03 02	04 18	
28	12 20	12 10	11 48	16 31	04 03	05 27	
29	12 00	11 50	11 48	17 24	04 58	06 37	

	UT	ARIES	VENUS −3.9		MARS −0.4		JUPITER −2.0		SATURN −0.3		STARS		
		GHA	GHA	Dec	GHA	Dec	GHA	Dec	GHA	Dec	Name	SHA	Dec
	d h	° ′	° ′	° ′	° ′	° ′	° ′	° ′	° ′	° ′		° ′	° ′
	30 00	68 27.3	154 25.1	S24 45.1	76 34.0	S 4 30.1	259 58.7	N 6 06.5	325 19.5	N22 10.5	Acamar	315 23.5	S40 17.4
	01	83 29.7	169 24.2	45.1	91 35.3	29.5	275 00.9	06.4	340 22.1	10.5	Achernar	335 31.7	S57 13.2
	02	98 32.2	184 23.2	45.1	106 36.5	28.9	290 03.1	06.3	355 24.7	10.6	Acrux	173 18.4	S63 06.9
	03	113 34.7	199 22.3	. . 45.1	121 37.8	. . 28.3	305 05.4	. . 06.3	10 27.4	. . 10.6	Adhara	255 18.2	S28 58.4
	04	128 37.1	214 21.3	45.1	136 39.1	27.6	320 07.6	06.2	25 30.0	10.6	Aldebaran	290 57.7	N16 31.1
	05	143 39.6	229 20.4	45.1	151 40.4	27.0	335 09.8	06.1	40 32.6	10.6			
	06	158 42.1	244 19.4	S24 45.1	166 41.6	S 4 26.4	350 12.0	N 6 06.0	55 35.2	N22 10.6	Alioth	166 27.4	N55 56.2
	07	173 44.5	259 18.5	45.0	181 42.9	25.8	5 14.3	05.9	70 37.9	10.6	Alkaid	153 05.0	N49 17.5
	08	188 47.0	274 17.5	45.0	196 44.2	25.2	20 16.5	05.8	85 40.5	10.6	Al Na'ir	27 53.1	S46 56.9
S	09	203 49.4	289 16.6	. . 45.0	211 45.5	. . 24.6	35 18.7	. . 05.7	100 43.1	. . 10.7	Alnilam	275 53.7	S 1 11.8
U	10	218 51.9	304 15.6	45.0	226 46.7	24.0	50 20.9	05.6	115 45.7	10.7	Alphard	218 03.4	S 8 40.3
N	11	233 54.4	319 14.6	45.0	241 48.0	23.3	65 23.1	05.5	130 48.4	10.7			
D	12	248 56.8	334 13.7	S24 44.9	256 49.3	S 4 22.7	80 25.4	N 6 05.4	145 51.0	N22 10.7	Alphecca	126 17.7	N26 42.1
A	13	263 59.3	349 12.7	44.9	271 50.6	22.1	95 27.6	05.3	160 53.6	10.7	Alpheratz	357 51.3	N29 06.8
Y	14	279 01.8	4 11.8	44.9	286 51.8	21.5	110 29.8	05.3	175 56.2	10.7	Altair	62 15.8	N 8 52.7
	15	294 04.2	19 10.8	. . 44.9	301 53.1	. . 20.9	125 32.0	. . 05.2	190 58.9	. . 10.8	Ankaa	353 22.8	S42 17.3
	16	309 06.7	34 09.9	44.8	316 54.4	20.3	140 34.3	05.1	206 01.5	10.8	Antares	112 35.9	S26 26.4
	17	324 09.2	49 08.9	44.8	331 55.6	19.6	155 36.5	05.0	221 04.1	10.8			
	18	339 11.6	64 08.0	S24 44.8	346 56.9	S 4 19.0	170 38.7	N 6 04.9	236 06.7	N22 10.8	Arcturus	146 02.9	N19 09.7
	19	354 14.1	79 07.0	44.7	1 58.2	18.4	185 40.9	04.8	251 09.4	10.8	Atria	107 45.2	S69 02.1
	20	9 16.5	94 06.1	44.7	16 59.5	17.8	200 43.2	04.7	266 12.0	10.8	Avior	234 20.9	S59 30.9
	21	24 19.0	109 05.1	. . 44.7	32 00.7	. . 17.2	215 45.4	. . 04.6	281 14.6	. . 10.9	Bellatrix	278 39.8	N 6 21.3
	22	39 21.5	124 04.2	44.6	47 02.0	16.6	230 47.6	04.5	296 17.3	10.9	Betelgeuse	271 09.1	N 7 24.6
	23	54 23.9	139 03.2	44.6	62 03.3	15.9	245 49.8	04.5	311 19.9	10.9			
	1 00	69 26.4	154 02.3	S24 44.6	77 04.5	S 4 15.3	260 52.1	N 6 04.4	326 22.5	N22 10.9	Canopus	263 59.0	S52 41.6
	01	84 28.9	169 01.3	44.5	92 05.8	14.7	275 54.3	04.3	341 25.1	10.9	Capella	280 45.1	N46 00.2
	02	99 31.3	184 00.4	44.5	107 07.1	14.1	290 56.5	04.2	356 27.8	10.9	Deneb	49 37.0	N45 17.8
	03	114 33.8	198 59.4	. . 44.4	122 08.3	. . 13.5	305 58.8	. . 04.1	11 30.4	. . 11.0	Denebola	182 41.4	N14 33.1
	04	129 36.3	213 58.5	44.4	137 09.6	12.9	321 01.0	04.0	26 33.0	11.0	Diphda	349 03.2	S17 58.0
	05	144 38.7	228 57.5	44.4	152 10.9	12.2	336 03.2	03.9	41 35.6	11.0			
	06	159 41.2	243 56.6	S24 44.3	167 12.1	S 4 11.6	351 05.4	N 6 03.8	56 38.3	N22 11.0	Dubhe	194 00.6	N61 43.6
	07	174 43.7	258 55.6	44.3	182 13.4	11.0	6 07.7	03.7	71 40.9	11.0	Elnath	278 21.8	N28 36.7
	08	189 46.1	273 54.7	44.2	197 14.7	10.4	21 09.9	03.7	86 43.5	11.0	Eltanin	90 50.2	N51 29.3
M	09	204 48.6	288 53.7	. . 44.2	212 15.9	. . 09.8	36 12.1	. . 03.6	101 46.2	. . 11.0	Enif	33 54.6	N 9 53.6
O	10	219 51.0	303 52.8	44.1	227 17.2	09.2	51 14.4	03.5	116 48.8	11.1	Fomalhaut	15 32.2	S29 36.3
N	11	234 53.5	318 51.8	44.1	242 18.5	08.5	66 16.6	03.4	131 51.4	11.1			
D	12	249 56.0	333 50.9	S24 44.0	257 19.7	S 4 07.9	81 18.8	N 6 03.3	146 54.1	N22 11.1	Gacrux	172 09.9	S57 07.7
A	13	264 58.4	348 49.9	43.9	272 21.0	07.3	96 21.0	03.2	161 56.7	11.1	Gienah	176 00.3	S17 33.6
Y	14	280 00.9	3 49.0	43.9	287 22.2	06.7	111 23.3	03.1	176 59.3	11.1	Hadar	148 59.4	S60 23.3
	15	295 03.4	18 48.0	. . 43.8	302 23.5	. . 06.1	126 25.5	. . 03.0	192 01.9	. . 11.1	Hamal	328 09.1	N23 29.0
	16	310 05.8	33 47.1	43.8	317 24.8	05.5	141 27.7	03.0	207 04.6	11.2	Kaus Aust.	83 54.2	S34 23.1
	17	325 08.3	48 46.1	43.7	332 26.0	04.8	156 30.0	02.9	222 07.2	11.2			
	18	340 10.8	63 45.2	S24 43.6	347 27.3	S 4 04.2	171 32.2	N 6 02.8	237 09.8	N22 11.2	Kochab	137 20.4	N74 08.2
	19	355 13.2	78 44.2	43.6	2 28.6	03.6	186 34.4	02.7	252 12.5	11.2	Markab	13 45.9	N15 13.6
	20	10 15.7	93 43.3	43.5	17 29.8	03.0	201 36.7	02.6	267 15.1	11.2	Menkar	314 22.6	N 4 06.4
	21	25 18.2	108 42.3	. . 43.4	32 31.1	. . 02.4	216 38.9	. . 02.5	282 17.7	. . 11.2	Menkent	148 16.9	S36 23.2
	22	40 20.6	123 41.4	43.4	47 32.3	01.7	231 41.1	02.4	297 20.4	11.3	Miaplacidus	221 41.2	S69 43.6
	23	55 23.1	138 40.4	43.3	62 33.6	01.1	246 43.4	02.4	312 23.0	11.3			
	2 00	70 25.5	153 39.5	S24 43.2	77 34.9	S 4 00.5	261 45.6	N 6 02.3	327 25.6	N22 11.3	Mirfak	308 50.8	N49 52.6
	01	85 28.0	168 38.5	43.2	92 36.1	3 59.9	276 47.8	02.2	342 28.2	11.3	Nunki	76 07.9	S26 17.6
	02	100 30.5	183 37.6	43.1	107 37.4	59.3	291 50.1	02.1	357 30.9	11.3	Peacock	53 31.3	S56 43.7
	03	115 32.9	198 36.6	. . 43.0	122 38.6	. . 58.7	306 52.3	. . 02.0	12 33.5	. . 11.3	Pollux	243 36.6	N28 01.0
	04	130 35.4	213 35.7	42.9	137 39.9	58.0	321 54.5	01.9	27 36.1	11.4	Procyon	245 07.4	N 5 13.0
	05	145 37.9	228 34.7	42.9	152 41.2	57.4	336 56.8	01.8	42 38.8	11.4			
	06	160 40.3	243 33.8	S24 42.8	167 42.4	S 3 56.8	351 59.0	N 6 01.7	57 41.4	N22 11.4	Rasalhague	96 13.8	N12 33.4
	07	175 42.8	258 32.8	42.7	182 43.7	56.2	7 01.2	01.7	72 44.0	11.4	Regulus	207 51.4	N11 57.0
	08	190 45.3	273 31.9	42.6	197 44.9	55.6	22 03.5	01.6	87 46.7	11.4	Rigel	281 19.0	S 8 11.7
T	09	205 47.7	288 30.9	. . 42.6	212 46.2	. . 54.9	37 05.7	. . 01.5	102 49.3	. . 11.4	Rigil Kent.	140 03.0	S60 50.9
U	10	220 50.2	303 30.0	42.5	227 47.5	54.3	52 07.9	01.4	117 51.9	11.5	Sabik	102 21.5	S15 43.8
E	11	235 52.6	318 29.1	42.4	242 48.7	53.7	67 10.2	01.3	132 54.6	11.5			
S	12	250 55.1	333 28.1	S24 42.3	257 50.0	S 3 53.1	82 12.4	N 6 01.2	147 57.2	N22 11.5	Schedar	349 49.1	N56 33.7
D	13	265 57.6	348 27.2	42.2	272 51.2	52.5	97 14.6	01.2	162 59.8	11.5	Shaula	96 32.6	S37 06.5
A	14	281 00.0	3 26.2	42.1	287 52.5	51.8	112 16.9	01.1	178 02.5	11.5	Sirius	258 40.1	S16 43.1
Y	15	296 02.5	18 25.3	. . 42.0	302 53.7	. . 51.2	127 19.1	. . 01.0	193 05.1	. . 11.5	Spica	158 39.5	S11 10.8
	16	311 05.0	33 24.3	41.9	317 55.0	50.6	142 21.3	00.9	208 07.7	11.6	Suhail	222 57.9	S43 26.5
	17	326 07.4	48 23.4	41.9	332 56.2	50.0	157 23.6	00.8	223 10.4	11.6			
	18	341 09.9	63 22.4	S24 41.8	347 57.5	S 3 49.4	172 25.8	N 6 00.7	238 13.0	N22 11.6	Vega	80 44.5	N38 47.3
	19	356 12.4	78 21.5	41.7	2 58.8	48.7	187 28.1	00.6	253 15.6	11.6	Zuben'ubi	137 14.1	S16 03.4
	20	11 14.8	93 20.5	41.6	18 00.0	48.1	202 30.3	00.6	268 18.3	11.6		SHA	Mer. Pass.
	21	26 17.3	108 19.6	. . 41.5	33 01.3	. . 47.5	217 32.5	. . 00.5	283 20.9	. . 11.6		° ′	h m
	22	41 19.8	123 18.6	41.4	48 02.5	46.9	232 34.8	00.4	298 23.5	11.7	Venus	84 35.9	13 45
	23	56 22.2	138 17.7	41.3	63 03.8	46.2	247 37.0	00.3	313 26.2	11.7	Mars	7 38.1	18 50
	Mer. Pass.	h m 19 19.1	*v* −0.9	*d* 0.1	*v* 1.3	*d* 0.6	*v* 2.2	*d* 0.1	*v* 2.6	*d* 0.0	Jupiter	191 25.7	6 36
											Saturn	256 56.1	2 14

	UT	SUN GHA	SUN Dec	MOON GHA	*v*	MOON Dec	*d*	HP
	d h	° ′	° ′	° ′	′	° ′	′	′
	30 00	182 54.6	S21 32.3	95 44.9	11.5	S16 49.7	12.1	57.2
	01	197 54.4	32.7	110 15.4	11.6	16 37.6	12.2	57.2
	02	212 54.2	33.1	124 46.0	11.8	16 25.4	12.3	57.2
	03	227 54.0	. . 33.5	139 16.8	11.8	16 13.1	12.3	57.1
	04	242 53.8	33.9	153 47.6	11.9	16 00.8	12.4	57.1
	05	257 53.5	34.3	168 18.5	12.1	15 48.4	12.5	57.1
	06	272 53.3	S21 34.8	182 49.6	12.1	S15 35.9	12.5	57.0
	07	287 53.1	35.2	197 20.7	12.3	15 23.4	12.5	57.0
	08	302 52.9	35.6	211 52.0	12.3	15 10.9	12.7	56.9
S	09	317 52.6	. . 36.0	226 23.3	12.5	14 58.2	12.6	56.9
U	10	332 52.4	36.4	240 54.8	12.5	14 45.6	12.8	56.9
N	11	347 52.2	36.8	255 26.3	12.7	14 32.8	12.7	56.8
D	12	2 52.0	S21 37.2	269 58.0	12.7	S14 20.1	12.9	56.8
A	13	17 51.7	37.6	284 29.7	12.9	14 07.2	12.8	56.8
Y	14	32 51.5	38.0	299 01.6	12.9	13 54.4	13.0	56.7
	15	47 51.3	. . 38.4	313 33.5	13.0	13 41.4	12.9	56.7
	16	62 51.1	38.8	328 05.5	13.2	13 28.5	13.0	56.7
	17	77 50.8	39.3	342 37.7	13.2	13 15.5	13.1	56.6
	18	92 50.6	S21 39.7	357 09.9	13.3	S13 02.4	13.1	56.6
	19	107 50.4	40.1	11 42.2	13.4	12 49.3	13.1	56.6
	20	122 50.2	40.5	26 14.6	13.5	12 36.2	13.2	56.5
	21	137 49.9	. . 40.9	40 47.1	13.5	12 23.0	13.2	56.5
	22	152 49.7	41.3	55 19.6	13.7	12 09.8	13.3	56.5
	23	167 49.5	41.7	69 52.3	13.7	11 56.5	13.3	56.4
	1 00	182 49.2	S21 42.1	84 25.0	13.9	S11 43.2	13.3	56.4
	01	197 49.0	42.5	98 57.9	13.9	11 29.9	13.4	56.4
	02	212 48.8	42.9	113 30.8	13.9	11 16.5	13.4	56.3
	03	227 48.6	. . 43.3	128 03.7	14.1	11 03.1	13.4	56.3
	04	242 48.3	43.7	142 36.8	14.1	10 49.7	13.5	56.3
	05	257 48.1	44.1	157 09.9	14.2	10 36.2	13.4	56.2
	06	272 47.9	S21 44.5	171 43.1	14.3	S10 22.8	13.6	56.2
	07	287 47.6	44.8	186 16.4	14.4	10 09.2	13.5	56.2
	08	302 47.4	45.2	200 49.8	14.4	9 55.7	13.6	56.1
M	09	317 47.2	. . 45.6	215 23.2	14.5	9 42.1	13.6	56.1
O	10	332 46.9	46.0	229 56.7	14.6	9 28.5	13.6	56.1
N	11	347 46.7	46.4	244 30.3	14.6	9 14.9	13.7	56.0
D	12	2 46.5	S21 46.8	259 03.9	14.7	S 9 01.2	13.6	56.0
A	13	17 46.2	47.2	273 37.6	14.8	8 47.6	13.7	56.0
Y	14	32 46.0	47.6	288 11.4	14.8	8 33.9	13.7	55.9
	15	47 45.8	. . 48.0	302 45.2	14.9	8 20.2	13.8	55.9
	16	62 45.6	48.4	317 19.1	15.0	8 06.4	13.7	55.9
	17	77 45.3	48.8	331 53.1	15.0	7 52.7	13.8	55.9
	18	92 45.1	S21 49.1	346 27.1	15.1	S 7 38.9	13.8	55.8
	19	107 44.9	49.5	1 01.2	15.2	7 25.1	13.8	55.8
	20	122 44.6	49.9	15 35.4	15.1	7 11.3	13.8	55.8
	21	137 44.4	. . 50.3	30 09.5	15.3	6 57.5	13.8	55.7
	22	152 44.1	50.7	44 43.8	15.3	6 43.7	13.9	55.7
	23	167 43.9	51.1	59 18.1	15.4	6 29.8	13.8	55.7
	2 00	182 43.7	S21 51.5	73 52.5	15.4	S 6 16.0	13.9	55.7
	01	197 43.4	51.8	88 26.9	15.4	6 02.1	13.9	55.6
	02	212 43.2	52.2	103 01.3	15.5	5 48.2	13.9	55.6
	03	227 43.0	. . 52.6	117 35.8	15.6	5 34.3	13.9	55.6
	04	242 42.7	53.0	132 10.4	15.6	5 20.4	13.9	55.6
	05	257 42.5	53.4	146 45.0	15.7	5 06.5	13.9	55.5
	06	272 42.3	S21 53.7	161 19.7	15.6	S 4 52.6	13.9	55.5
	07	287 42.0	54.1	175 54.3	15.8	4 38.7	14.0	55.5
T	08	302 41.8	54.5	190 29.1	15.8	4 24.7	13.9	55.4
U	09	317 41.6	. . 54.9	205 03.9	15.8	4 10.8	13.9	55.4
E	10	332 41.3	55.2	219 38.7	15.8	3 56.9	14.0	55.4
S	11	347 41.1	55.6	234 13.5	15.9	3 42.9	13.9	55.4
D	12	2 40.8	S21 56.0	248 48.4	16.0	S 3 29.0	14.0	55.3
A	13	17 40.6	56.4	263 23.4	15.9	3 15.0	13.9	55.3
Y	14	32 40.4	56.7	277 58.3	16.0	3 01.1	14.0	55.3
	15	47 40.1	. . 57.1	292 33.3	16.1	2 47.1	13.9	55.3
	16	62 39.9	57.5	307 08.4	16.0	2 33.2	13.9	55.2
	17	77 39.6	57.9	321 43.4	16.1	2 19.3	14.0	55.2
	18	92 39.4	S21 58.2	336 18.5	16.1	S 2 05.3	13.9	55.2
	19	107 39.2	58.6	350 53.6	16.2	1 51.4	14.0	55.2
	20	122 38.9	59.0	5 28.8	16.2	1 37.4	13.9	55.2
	21	137 38.7	. . 59.3	20 04.0	16.2	1 23.5	13.9	55.1
	22	152 38.4	21 59.7	34 39.2	16.2	1 09.6	13.9	55.1
	23	167 38.2	S22 00.1	49 14.4	16.3	S 0 55.7	13.9	55.1
		SD 16.2	*d* 0.4	SD 15.5		15.3		15.1

Lat.	Twilight Naut.	Twilight Civil	Sunrise	Moonrise 30	Moonrise 1	Moonrise 2	Moonrise 3
°	h m	h m	h m	h m	h m	h m	h m
N 72	07 50	09 48	■	15 22	14 38	14 07	13 38
N 70	07 34	09 09	■	14 56	14 26	14 03	13 41
68	07 21	08 43	10 34	14 36	14 16	13 59	13 44
66	07 10	08 22	09 45	14 20	14 07	13 56	13 46
64	07 00	08 05	09 15	14 07	14 00	13 54	13 48
62	06 52	07 52	08 52	13 56	13 54	13 52	13 49
60	06 45	07 40	08 34	13 47	13 49	13 50	13 51
N 58	06 39	07 30	08 19	13 38	13 44	13 48	13 52
56	06 33	07 21	08 06	13 31	13 40	13 47	13 53
54	06 28	07 13	07 55	13 24	13 36	13 45	13 54
52	06 23	07 05	07 45	13 18	13 32	13 44	13 55
50	06 18	06 59	07 36	13 13	13 29	13 43	13 56
45	06 08	06 44	07 17	13 01	13 22	13 41	13 58
N 40	05 59	06 32	07 02	12 51	13 16	13 38	14 00
35	05 50	06 21	06 49	12 43	13 11	13 37	14 01
30	05 42	06 12	06 38	12 35	13 07	13 35	14 02
20	05 27	05 55	06 18	12 23	12 59	13 32	14 04
N 10	05 13	05 39	06 01	12 11	12 52	13 30	14 06
0	04 57	05 23	05 45	12 01	12 46	13 28	14 08
S 10	04 40	05 06	05 29	11 50	12 39	13 25	14 10
20	04 19	04 47	05 12	11 39	12 32	13 23	14 12
30	03 52	04 24	04 51	11 25	12 24	13 20	14 14
35	03 35	04 10	04 39	11 18	12 20	13 19	14 16
40	03 14	03 54	04 26	11 09	12 14	13 17	14 17
45	02 46	03 33	04 09	10 59	12 08	13 15	14 19
S 50	02 07	03 07	03 49	10 46	12 01	13 12	14 21
52	01 45	02 53	03 39	10 41	11 57	13 11	14 22
54	01 16	02 38	03 28	10 34	11 54	13 10	14 23
56	00 21	02 19	03 16	10 27	11 50	13 08	14 25
58	////	01 56	03 01	10 19	11 45	13 07	14 26
S 60	////	01 24	02 44	10 09	11 40	13 05	14 27

Lat.	Sunset	Twilight Civil	Twilight Naut.	Moonset 30	Moonset 1	Moonset 2	Moonset 3
°	h m	h m	h m	h m	h m	h m	h m
N 72	■	13 49	15 47	21 27	23 43	25 43	01 43
N 70	■	14 28	16 03	21 51	23 52	25 43	01 43
68	13 03	14 55	16 17	22 09	24 00	00 00	01 44
66	13 52	15 15	16 27	22 23	24 07	00 07	01 44
64	14 22	15 32	16 37	22 35	24 12	00 12	01 44
62	14 45	15 46	16 45	22 45	24 17	00 17	01 45
60	15 03	15 57	16 52	22 53	24 21	00 21	01 45
N 58	15 18	16 08	16 59	23 00	24 25	00 25	01 45
56	15 31	16 17	17 04	23 07	24 28	00 28	01 45
54	15 43	16 25	17 10	23 13	24 31	00 31	01 45
52	15 52	16 32	17 15	23 18	24 33	00 33	01 45
50	16 01	16 39	17 19	23 22	24 35	00 35	01 46
45	16 20	16 53	17 30	23 32	24 40	00 40	01 46
N 40	16 35	17 05	17 39	23 41	24 45	00 45	01 46
35	16 48	17 16	17 47	23 48	24 48	00 48	01 46
30	17 00	17 26	17 55	23 54	24 51	00 51	01 46
20	17 19	17 43	18 10	24 05	00 05	00 57	01 47
N 10	17 36	17 59	18 25	24 14	00 14	01 02	01 47
0	17 53	18 15	18 41	24 23	00 23	01 06	01 47
S 10	18 09	18 32	18 58	24 31	00 31	01 10	01 47
20	18 26	18 51	19 19	00 02	00 40	01 15	01 48
30	18 47	19 14	19 46	00 17	00 50	01 20	01 48
35	18 59	19 28	20 04	00 26	00 56	01 23	01 48
40	19 13	19 45	20 25	00 36	01 03	01 26	01 48
45	19 29	20 05	20 52	00 47	01 11	01 30	01 48
S 50	19 49	20 32	21 32	01 01	01 20	01 35	01 48
52	19 59	20 45	21 54	01 08	01 24	01 37	01 48
54	20 10	21 01	22 25	01 15	01 29	01 39	01 48
56	20 23	21 20	23 27	01 23	01 34	01 42	01 49
58	20 38	21 44	////	01 32	01 39	01 44	01 49
S 60	20 55	22 17	////	01 42	01 46	01 48	01 49

Day	SUN Eqn. of Time 00^h	SUN Eqn. of Time 12^h	SUN Mer. Pass.	MOON Mer. Pass. Upper	MOON Mer. Pass. Lower	MOON Age	MOON Phase
d	m s	m s	h m	h m	h m	d %	
30	11 39	11 28	11 49	18 12	05 48	07 48	
1	11 17	11 06	11 49	18 56	06 34	08 58	◐
2	10 55	10 44	11 49	19 37	07 17	09 68	

Day	UT (d h)	ARIES GHA	VENUS −3.9 GHA	VENUS Dec	MARS −0.3 GHA	MARS Dec	JUPITER −2.0 GHA	JUPITER Dec	SATURN −0.3 GHA	SATURN Dec
		° ′	° ′	° ′	° ′	° ′	° ′	° ′	° ′	° ′
WEDNESDAY	3 00	71 24.7	153 16.7	S24 41.2	78 05.0	S 3 45.6	262 39.2	N 6 00.2	328 28.8	N22 11.7
	01	86 27.1	168 15.8	41.1	93 06.3	45.0	277 41.5	00.1	343 31.4	11.7
	02	101 29.6	183 14.9	41.0	108 07.5	44.4	292 43.7	00.1	358 34.1	11.7
	03	116 32.1	198 13.9	40.9	123 08.8	43.8	307 46.0	6 00.0	13 36.7	11.7
	04	131 34.5	213 13.0	40.7	138 10.0	43.1	322 48.2	5 59.9	28 39.3	11.8
	05	146 37.0	228 12.0	40.6	153 11.3	42.5	337 50.4	59.8	43 42.0	11.8
	06	161 39.5	243 11.1	S24 40.5	168 12.5	S 3 41.9	352 52.7	N 5 59.7	58 44.6	N22 11.8
	07	176 41.9	258 10.1	40.4	183 13.8	41.3	7 54.9	59.6	73 47.3	11.8
	08	191 44.4	273 09.2	40.3	198 15.0	40.7	22 57.2	59.6	88 49.9	11.8
	09	206 46.9	288 08.2	40.2	213 16.3	40.0	37 59.4	59.5	103 52.5	11.8
	10	221 49.3	303 07.3	40.1	228 17.5	39.4	53 01.6	59.4	118 55.2	11.9
	11	236 51.8	318 06.3	40.0	243 18.8	38.8	68 03.9	59.3	133 57.8	11.9
	12	251 54.2	333 05.4	S24 39.8	258 20.0	S 3 38.2	83 06.1	N 5 59.2	149 00.4	N22 11.9
	13	266 56.7	348 04.5	39.7	273 21.3	37.5	98 08.4	59.1	164 03.1	11.9
	14	281 59.2	3 03.5	39.6	288 22.5	36.9	113 10.6	59.1	179 05.7	11.9
	15	297 01.6	18 02.6	39.5	303 23.8	36.3	128 12.9	59.0	194 08.3	11.9
	16	312 04.1	33 01.6	39.4	318 25.0	35.7	143 15.1	58.9	209 11.0	12.0
	17	327 06.6	48 00.7	39.2	333 26.3	35.0	158 17.3	58.8	224 13.6	12.0
	18	342 09.0	62 59.7	S24 39.1	348 27.5	S 3 34.4	173 19.6	N 5 58.7	239 16.3	N22 12.0
	19	357 11.5	77 58.8	39.0	3 28.8	33.8	188 21.8	58.7	254 18.9	12.0
	20	12 14.0	92 57.8	38.9	18 30.0	33.2	203 24.1	58.6	269 21.5	12.0
	21	27 16.4	107 56.9	38.7	33 31.3	32.6	218 26.3	58.5	284 24.2	12.0
	22	42 18.9	122 56.0	38.6	48 32.5	31.9	233 28.6	58.4	299 26.8	12.1
	23	57 21.4	137 55.0	38.5	63 33.8	31.3	248 30.8	58.3	314 29.4	12.1
THURSDAY	4 00	72 23.8	152 54.1	S24 38.3	78 35.0	S 3 30.7	263 33.0	N 5 58.2	329 32.1	N22 12.1
	01	87 26.3	167 53.1	38.2	93 36.2	30.1	278 35.3	58.2	344 34.7	12.1
	02	102 28.7	182 52.2	38.1	108 37.5	29.4	293 37.5	58.1	359 37.4	12.1
	03	117 31.2	197 51.2	37.9	123 38.7	28.8	308 39.8	58.0	14 40.0	12.1
	04	132 33.7	212 50.3	37.8	138 40.0	28.2	323 42.0	57.9	29 42.6	12.2
	05	147 36.1	227 49.4	37.7	153 41.2	27.6	338 44.3	57.8	44 45.3	12.2
	06	162 38.6	242 48.4	S24 37.5	168 42.5	S 3 26.9	353 46.5	N 5 57.8	59 47.9	N22 12.2
	07	177 41.1	257 47.5	37.4	183 43.7	26.3	8 48.8	57.7	74 50.6	12.2
	08	192 43.5	272 46.5	37.2	198 45.0	25.7	23 51.0	57.6	89 53.2	12.2
	09	207 46.0	287 45.6	37.1	213 46.2	25.1	38 53.3	57.5	104 55.8	12.2
	10	222 48.5	302 44.7	36.9	228 47.4	24.4	53 55.5	57.4	119 58.5	12.3
	11	237 50.9	317 43.7	36.8	243 48.7	23.8	68 57.8	57.4	135 01.1	12.3
	12	252 53.4	332 42.8	S24 36.6	258 49.9	S 3 23.2	84 00.0	N 5 57.3	150 03.7	N22 12.3
	13	267 55.9	347 41.8	36.5	273 51.2	22.6	99 02.2	57.2	165 06.4	12.3
	14	282 58.3	2 40.9	36.3	288 52.4	21.9	114 04.5	57.1	180 09.0	12.3
	15	298 00.8	17 40.0	36.2	303 53.7	21.3	129 06.7	57.0	195 11.7	12.3
	16	313 03.2	32 39.0	36.0	318 54.9	20.7	144 09.0	57.0	210 14.3	12.4
	17	328 05.7	47 38.1	35.9	333 56.1	20.1	159 11.2	56.9	225 16.9	12.4
	18	343 08.2	62 37.1	S24 35.7	348 57.4	S 3 19.4	174 13.5	N 5 56.8	240 19.6	N22 12.4
	19	358 10.6	77 36.2	35.6	3 58.6	18.8	189 15.7	56.7	255 22.2	12.4
	20	13 13.1	92 35.3	35.4	18 59.9	18.2	204 18.0	56.6	270 24.9	12.4
	21	28 15.6	107 34.3	35.2	34 01.1	17.6	219 20.2	56.6	285 27.5	12.4
	22	43 18.0	122 33.4	35.1	49 02.3	16.9	234 22.5	56.5	300 30.2	12.5
	23	58 20.5	137 32.4	34.9	64 03.6	16.3	249 24.7	56.4	315 32.8	12.5
FRIDAY	5 00	73 23.0	152 31.5	S24 34.8	79 04.8	S 3 15.7	264 27.0	N 5 56.3	330 35.4	N22 12.5
	01	88 25.4	167 30.6	34.6	94 06.0	15.1	279 29.2	56.3	345 38.1	12.5
	02	103 27.9	182 29.6	34.4	109 07.3	14.4	294 31.5	56.2	0 40.7	12.5
	03	118 30.3	197 28.7	34.3	124 08.5	13.8	309 33.7	56.1	15 43.4	12.5
	04	133 32.8	212 27.7	34.1	139 09.8	13.2	324 36.0	56.0	30 46.0	12.6
	05	148 35.3	227 26.8	33.9	154 11.0	12.6	339 38.3	55.9	45 48.6	12.6
	06	163 37.7	242 25.9	S24 33.7	169 12.2	S 3 11.9	354 40.5	N 5 55.9	60 51.3	N22 12.6
	07	178 40.2	257 24.9	33.6	184 13.5	11.3	9 42.8	55.8	75 53.9	12.6
	08	193 42.7	272 24.0	33.4	199 14.7	10.7	24 45.0	55.7	90 56.6	12.6
	09	208 45.1	287 23.1	33.2	214 15.9	10.1	39 47.3	55.6	105 59.2	12.6
	10	223 47.6	302 22.1	33.0	229 17.2	09.4	54 49.5	55.6	121 01.9	12.7
	11	238 50.1	317 21.2	32.9	244 18.4	08.8	69 51.8	55.5	136 04.5	12.7
	12	253 52.5	332 20.2	S24 32.7	259 19.7	S 3 08.2	84 54.0	N 5 55.4	151 07.1	N22 12.7
	13	268 55.0	347 19.3	32.5	274 20.9	07.5	99 56.3	55.3	166 09.8	12.7
	14	283 57.5	2 18.4	32.3	289 22.1	06.9	114 58.5	55.2	181 12.4	12.7
	15	298 59.9	17 17.4	32.1	304 23.4	06.3	130 00.8	55.2	196 15.1	12.8
	16	314 02.4	32 16.5	32.0	319 24.6	05.7	145 03.0	55.1	211 17.7	12.8
	17	329 04.8	47 15.6	31.8	334 25.8	05.0	160 05.3	55.0	226 20.4	12.8
	18	344 07.3	62 14.6	S24 31.6	349 27.1	S 3 04.4	175 07.6	N 5 54.9	241 23.0	N22 12.8
	19	359 09.8	77 13.7	31.4	4 28.3	03.8	190 09.8	54.9	256 25.6	12.8
	20	14 12.2	92 12.8	31.2	19 29.5	03.2	205 12.1	54.8	271 28.3	12.8
	21	29 14.7	107 11.8	31.0	34 30.8	02.5	220 14.3	54.7	286 30.9	12.9
	22	44 17.2	122 10.9	30.8	49 32.0	01.9	235 16.6	54.6	301 33.6	12.9
	23	59 19.6	137 10.0	30.6	64 33.2	01.3	250 18.8	54.6	316 36.2	12.9
	Mer. Pass.	h m 19 07.3	*v* −0.9	*d* 0.1	*v* 1.2	*d* 0.6	*v* 2.2	*d* 0.1	*v* 2.6	*d* 0.0

STARS

Name	SHA	Dec
	° ′	° ′
Acamar	315 23.5	S40 17.4
Achernar	335 31.7	S57 13.3
Acrux	173 18.4	S63 06.9
Adhara	255 18.1	S28 58.4
Aldebaran	290 57.7	N16 31.1
Alioth	166 27.3	N55 56.2
Alkaid	153 05.0	N49 17.5
Al Na'ir	27 53.1	S46 56.9
Alnilam	275 53.7	S 1 11.9
Alphard	218 03.4	S 8 40.3
Alphecca	126 17.7	N26 42.1
Alpheratz	357 51.3	N29 06.8
Altair	62 15.8	N 8 52.7
Ankaa	353 22.8	S42 17.3
Antares	112 35.9	S26 26.4
Arcturus	146 02.8	N19 09.7
Atria	107 45.2	S69 02.1
Avior	234 20.8	S59 31.0
Bellatrix	278 39.8	N 6 21.3
Betelgeuse	271 09.1	N 7 24.6
Canopus	263 59.0	S52 41.6
Capella	280 45.1	N46 00.2
Deneb	49 37.0	N45 17.8
Denebola	182 41.4	N14 33.1
Diphda	349 03.2	S17 58.0
Dubhe	194 00.6	N61 43.6
Elnath	278 21.8	N28 36.7
Eltanin	90 50.2	N51 29.3
Enif	33 54.6	N 9 53.6
Fomalhaut	15 32.2	S29 36.3
Gacrux	172 09.9	S57 07.7
Gienah	176 00.2	S17 33.6
Hadar	148 59.4	S60 23.3
Hamal	328 09.1	N23 29.0
Kaus Aust.	83 54.2	S34 23.1
Kochab	137 20.4	N74 08.2
Markab	13 45.9	N15 13.6
Menkar	314 22.6	N 4 06.4
Menkent	148 16.9	S36 23.2
Miaplacidus	221 41.2	S69 43.6
Mirfak	308 50.8	N49 52.6
Nunki	76 07.9	S26 17.6
Peacock	53 31.3	S56 43.7
Pollux	243 36.6	N28 01.0
Procyon	245 07.3	N 5 13.0
Rasalhague	96 13.8	N12 33.4
Regulus	207 51.4	N11 57.0
Rigel	281 19.0	S 8 11.7
Rigil Kent.	140 02.9	S60 50.8
Sabik	102 21.5	S15 43.8
Schedar	349 49.1	N56 33.7
Shaula	96 32.6	S37 06.5
Sirius	258 40.1	S16 43.1
Spica	158 39.4	S11 10.8
Suhail	222 57.9	S43 26.5
Vega	80 44.5	N38 47.3
Zuben'ubi	137 14.1	S16 03.4

	SHA	Mer. Pass.
	° ′	h m
Venus	80 30.3	13 49
Mars	6 11.2	18 44
Jupiter	191 09.2	6 25
Saturn	257 08.3	2 02

	UT	SUN GHA	SUN Dec	MOON GHA	v	MOON Dec	d	HP
	d h	° ′	° ′	° ′	′	° ′	′	′
	3 00	182 38.0	S22 00.4	63 49.7	16.3	S 0 41.8	13.9	55.1
	01	197 37.7	00.8	78 25.0	16.3	0 27.9	13.9	55.0
	02	212 37.5	01.2	93 00.3	16.3	0 14.0	13.9	55.0
	03	227 37.2	. . 01.5	107 35.6	16.3	S 0 00.1	13.9	55.0
	04	242 37.0	01.9	122 10.9	16.4	N 0 13.8	13.9	55.0
	05	257 36.8	02.2	136 46.3	16.3	0 27.7	13.8	55.0
	06	272 36.5	S22 02.6	151 21.6	16.4	N 0 41.5	13.9	54.9
WEDNESDAY	07	287 36.3	03.0	165 57.0	16.4	0 55.4	13.8	54.9
	08	302 36.0	03.3	180 32.4	16.4	1 09.2	13.8	54.9
	09	317 35.8	. . 03.7	195 07.8	16.5	1 23.0	13.8	54.9
	10	332 35.5	04.0	209 43.3	16.4	1 36.8	13.8	54.9
	11	347 35.3	04.4	224 18.7	16.5	1 50.6	13.8	54.8
	12	2 35.0	S22 04.8	238 54.2	16.4	N 2 04.4	13.7	54.8
	13	17 34.8	05.1	253 29.6	16.5	2 18.1	13.8	54.8
	14	32 34.6	05.5	268 05.1	16.4	2 31.9	13.7	54.8
	15	47 34.3	. . 05.8	282 40.5	16.5	2 45.6	13.7	54.8
	16	62 34.1	06.2	297 16.0	16.5	2 59.3	13.7	54.7
	17	77 33.8	06.5	311 51.5	16.5	3 13.0	13.7	54.7
	18	92 33.6	S22 06.9	326 27.0	16.5	N 3 26.7	13.6	54.7
	19	107 33.3	07.2	341 02.5	16.5	3 40.3	13.6	54.7
	20	122 33.1	07.6	355 38.0	16.4	3 53.9	13.6	54.7
	21	137 32.8	. . 07.9	10 13.4	16.5	4 07.5	13.6	54.7
	22	152 32.6	08.3	24 48.9	16.5	4 21.1	13.6	54.6
	23	167 32.3	08.6	39 24.4	16.5	4 34.7	13.5	54.6
	4 00	182 32.1	S22 09.0	53 59.9	16.5	N 4 48.2	13.5	54.6
	01	197 31.8	09.3	68 35.4	16.4	5 01.7	13.5	54.6
	02	212 31.6	09.7	83 10.8	16.5	5 15.2	13.4	54.6
	03	227 31.4	. . 10.0	97 46.3	16.5	5 28.6	13.5	54.6
	04	242 31.1	10.4	112 21.8	16.4	5 42.1	13.4	54.5
	05	257 30.9	10.7	126 57.2	16.5	5 55.5	13.3	54.5
	06	272 30.6	S22 11.0	141 32.7	16.4	N 6 08.8	13.4	54.5
THURSDAY	07	287 30.4	11.4	156 08.1	16.4	6 22.2	13.3	54.5
	08	302 30.1	11.7	170 43.5	16.4	6 35.5	13.3	54.5
	09	317 29.9	. . 12.1	185 18.9	16.4	6 48.8	13.2	54.5
	10	332 29.6	12.4	199 54.3	16.4	7 02.0	13.3	54.5
	11	347 29.4	12.7	214 29.7	16.3	7 15.3	13.2	54.4
	12	2 29.1	S22 13.1	229 05.0	16.4	N 7 28.5	13.1	54.4
	13	17 28.9	13.4	243 40.4	16.3	7 41.6	13.1	54.4
	14	32 28.6	13.8	258 15.7	16.3	7 54.7	13.1	54.4
	15	47 28.4	. . 14.1	272 51.0	16.3	8 07.8	13.1	54.4
	16	62 28.1	14.4	287 26.3	16.3	8 20.9	13.0	54.4
	17	77 27.9	14.8	302 01.6	16.2	8 33.9	13.0	54.4
	18	92 27.6	S22 15.1	316 36.8	16.3	N 8 46.9	12.9	54.4
	19	107 27.4	15.4	331 12.1	16.2	8 59.8	12.9	54.3
	20	122 27.1	15.8	345 47.3	16.1	9 12.7	12.9	54.3
	21	137 26.8	. . 16.1	0 22.4	16.2	9 25.6	12.8	54.3
	22	152 26.6	16.4	14 57.6	16.1	9 38.4	12.8	54.3
	23	167 26.3	16.8	29 32.7	16.1	9 51.2	12.8	54.3
	5 00	182 26.1	S22 17.1	44 07.8	16.1	N10 04.0	12.7	54.3
	01	197 25.8	17.4	58 42.9	16.1	10 16.7	12.6	54.3
	02	212 25.6	17.7	73 18.0	16.0	10 29.3	12.7	54.3
	03	227 25.3	. . 18.1	87 53.0	16.0	10 42.0	12.5	54.3
	04	242 25.1	18.4	102 28.0	15.9	10 54.5	12.6	54.2
	05	257 24.8	18.7	117 02.9	15.9	11 07.1	12.5	54.2
	06	272 24.6	S22 19.0	131 37.8	15.9	N11 19.6	12.4	54.2
	07	287 24.3	19.4	146 12.7	15.9	11 32.0	12.4	54.2
	08	302 24.1	19.7	160 47.6	15.8	11 44.4	12.3	54.2
FRIDAY	09	317 23.8	. . 20.0	175 22.4	15.8	11 56.7	12.3	54.2
	10	332 23.5	20.3	189 57.2	15.8	12 09.0	12.3	54.2
	11	347 23.3	20.7	204 32.0	15.7	12 21.3	12.2	54.2
	12	2 23.0	S22 21.0	219 06.7	15.6	N12 33.5	12.1	54.2
	13	17 22.8	21.3	233 41.3	15.7	12 45.6	12.1	54.2
	14	32 22.5	21.6	248 16.0	15.6	12 57.7	12.1	54.1
	15	47 22.3	. . 21.9	262 50.6	15.5	13 09.8	12.0	54.1
	16	62 22.0	22.3	277 25.1	15.6	13 21.8	11.9	54.1
	17	77 21.7	22.6	291 59.7	15.4	13 33.7	11.9	54.1
	18	92 21.5	S22 22.9	306 34.1	15.5	N13 45.6	11.8	54.1
	19	107 21.2	23.2	321 08.6	15.4	13 57.4	11.8	54.1
	20	122 21.0	23.5	335 43.0	15.3	14 09.2	11.7	54.1
	21	137 20.7	. . 23.8	350 17.3	15.3	14 20.9	11.7	54.1
	22	152 20.5	24.1	4 51.6	15.3	14 32.6	11.6	54.1
	23	167 20.2	24.5	19 25.9	15.2	N14 44.2	11.5	54.1
		SD 16.3	*d* 0.3	SD 14.9		14.8		14.8

Lat.	Twilight Naut.	Twilight Civil	Sunrise	Moonrise 3	Moonrise 4	Moonrise 5	Moonrise 6
°	h m	h m	h m	h m	h m	h m	h m
N 72	07 58	10 02	■	13 38	13 08	12 31	11 25
N 70	07 41	09 19	■	13 41	13 19	12 53	12 14
68	07 27	08 51	10 53	13 44	13 28	13 10	12 46
66	07 15	08 29	09 56	13 46	13 35	13 24	13 10
64	07 06	08 12	09 23	13 48	13 42	13 35	13 28
62	06 57	07 57	08 59	13 49	13 47	13 45	13 44
60	06 50	07 45	08 40	13 51	13 52	13 54	13 57
N 58	06 43	07 34	08 24	13 52	13 56	14 01	14 08
56	06 37	07 25	08 11	13 53	14 00	14 08	14 18
54	06 31	07 17	07 59	13 54	14 04	14 14	14 26
52	06 26	07 09	07 49	13 55	14 07	14 19	14 34
50	06 21	07 02	07 40	13 56	14 10	14 24	14 41
45	06 11	06 47	07 21	13 58	14 16	14 35	14 56
N 40	06 01	06 35	07 05	14 00	14 21	14 44	15 09
35	05 53	06 24	06 52	14 01	14 26	14 51	15 19
30	05 44	06 14	06 40	14 02	14 30	14 58	15 29
20	05 29	05 56	06 20	14 04	14 37	15 10	15 45
N 10	05 14	05 40	06 03	14 06	14 43	15 20	15 59
0	04 58	05 24	05 46	14 08	14 49	15 30	16 13
S 10	04 40	05 07	05 30	14 10	14 55	15 40	16 26
20	04 19	04 48	05 12	14 12	15 01	15 50	16 41
30	03 51	04 24	04 51	14 14	15 08	16 02	16 57
35	03 34	04 10	04 39	14 16	15 13	16 09	17 07
40	03 12	03 53	04 25	14 17	15 17	16 18	17 18
45	02 44	03 32	04 08	14 19	15 23	16 27	17 32
S 50	02 04	03 04	03 47	14 21	15 30	16 38	17 48
52	01 40	02 51	03 37	14 22	15 33	16 44	17 56
54	01 08	02 35	03 26	14 23	15 36	16 50	18 04
56	////	02 15	03 13	14 25	15 40	16 56	18 14
58	////	01 50	02 58	14 26	15 44	17 04	18 25
S 60	////	01 15	02 40	14 27	15 49	17 12	18 37

Lat.	Sunset	Twilight Civil	Twilight Naut.	Moonset 3	Moonset 4	Moonset 5	Moonset 6
°	h m	h m	h m	h m	h m	h m	h m
N 72	■	13 37	15 42	01 43	03 39	05 43	08 18
N 70	■	14 20	15 59	01 43	03 31	05 23	07 30
68	12 47	14 49	16 13	01 44	03 24	05 08	06 59
66	13 44	15 11	16 24	01 44	03 19	04 55	06 37
64	14 17	15 28	16 34	01 44	03 14	04 45	06 19
62	14 41	15 43	16 43	01 45	03 10	04 36	06 05
60	15 00	15 55	16 50	01 45	03 07	04 29	05 53
N 58	15 15	16 05	16 57	01 45	03 04	04 22	05 42
56	15 29	16 15	17 03	01 45	03 01	04 17	05 33
54	15 40	16 23	17 08	01 45	02 59	04 12	05 25
52	15 51	16 31	17 14	01 45	02 56	04 07	05 18
50	16 00	16 37	17 18	01 46	02 54	04 03	05 12
45	16 19	16 52	17 29	01 46	02 50	03 54	04 58
N 40	16 35	17 05	17 39	01 46	02 46	03 46	04 47
35	16 48	17 16	17 47	01 46	02 43	03 40	04 37
30	17 00	17 26	17 55	01 46	02 40	03 34	04 29
20	17 20	17 44	18 11	01 47	02 36	03 25	04 14
N 10	17 37	18 00	18 26	01 47	02 32	03 16	04 02
0	17 54	18 16	18 42	01 47	02 28	03 08	03 50
S 10	18 10	18 33	19 00	01 47	02 24	03 00	03 39
20	18 28	18 53	19 22	01 48	02 20	02 52	03 26
30	18 49	19 16	19 49	01 48	02 15	02 43	03 12
35	19 01	19 31	20 07	01 48	02 12	02 37	03 04
40	19 16	19 48	20 28	01 48	02 09	02 31	02 55
45	19 32	20 09	20 57	01 48	02 06	02 24	02 44
S 50	19 53	20 37	21 38	01 48	02 02	02 15	02 31
52	20 04	20 50	22 01	01 48	02 00	02 11	02 25
54	20 15	21 07	22 35	01 48	01 58	02 07	02 18
56	20 28	21 26	////	01 49	01 55	02 02	02 11
58	20 43	21 51	////	01 49	01 53	01 57	02 03
S 60	21 01	22 27	////	01 49	01 50	01 51	01 53

Day	SUN Eqn. of Time 00^h	SUN Eqn. of Time 12^h	SUN Mer. Pass.	MOON Mer. Pass. Upper	MOON Mer. Pass. Lower	MOON Age	MOON Phase
d	m s	m s	h m	h m	h m	d %	
3	10 32	10 21	11 50	20 18	07 58	10 76	
4	10 09	09 57	11 50	20 58	08 38	11 84	◐
5	09 45	09 33	11 50	21 40	09 19	12 90	

	UT	ARIES	VENUS −3.9		MARS −0.3		JUPITER −2.0		SATURN −0.3		STARS		
		GHA	GHA	Dec	GHA	Dec	GHA	Dec	GHA	Dec	Name	SHA	Dec
	d h	° ′	° ′	° ′	° ′	° ′	° ′	° ′	° ′	° ′		° ′	° ′
	6 00	74 22.1	152 09.0	S24 30.4	79 34.5	S 3 00.6	265 21.1	N 5 54.5	331 38.9	N22 12.9	Acamar	315 23.6	S40 17.4
	01	89 24.6	167 08.1	30.2	94 35.7	3 00.0	280 23.3	54.4	346 41.5	12.9	Achernar	335 31.7	S57 13.3
	02	104 27.0	182 07.2	30.0	109 36.9	2 59.4	295 25.6	54.3	1 44.1	12.9	Acrux	173 18.3	S63 06.9
	03	119 29.5	197 06.2	. . 29.8	124 38.1	. . 58.8	310 27.9	. . 54.3	16 46.8	. . 13.0	Adhara	255 18.1	S28 58.4
	04	134 32.0	212 05.3	29.6	139 39.4	58.1	325 30.1	54.2	31 49.4	13.0	Aldebaran	290 57.7	N16 31.1
	05	149 34.4	227 04.4	29.4	154 40.6	57.5	340 32.4	54.1	46 52.1	13.0			
	06	164 36.9	242 03.4	S24 29.2	169 41.8	S 2 56.9	355 34.6	N 5 54.0	61 54.7	N22 13.0	Alioth	166 27.3	N55 56.1
	07	179 39.3	257 02.5	29.0	184 43.1	56.2	10 36.9	54.0	76 57.4	13.0	Alkaid	153 05.0	N49 17.5
S	08	194 41.8	272 01.6	28.8	199 44.3	55.6	25 39.2	53.9	92 00.0	13.0	Al Na'ir	27 53.1	S46 56.9
A	09	209 44.3	287 00.6	. . 28.6	214 45.5	. . 55.0	40 41.4	. . 53.8	107 02.7	. . 13.1	Alnilam	275 53.7	S 1 11.9
T	10	224 46.7	301 59.7	28.4	229 46.8	54.4	55 43.7	53.7	122 05.3	13.1	Alphard	218 03.4	S 8 40.3
U	11	239 49.2	316 58.8	28.2	244 48.0	53.7	70 45.9	53.7	137 08.0	13.1			
R	12	254 51.7	331 57.8	S24 28.0	259 49.2	S 2 53.1	85 48.2	N 5 53.6	152 10.6	N22 13.1	Alphecca	126 17.7	N26 42.0
D	13	269 54.1	346 56.9	27.8	274 50.4	52.5	100 50.5	53.5	167 13.2	13.1	Alpheratz	357 51.3	N29 06.8
A	14	284 56.6	1 56.0	27.6	289 51.7	51.8	115 52.7	53.4	182 15.9	13.1	Altair	62 15.8	N 8 52.7
Y	15	299 59.1	16 55.0	. . 27.4	304 52.9	. . 51.2	130 55.0	. . 53.4	197 18.5	. . 13.2	Ankaa	353 22.8	S42 17.4
	16	315 01.5	31 54.1	27.1	319 54.1	50.6	145 57.2	53.3	212 21.2	13.2	Antares	112 35.9	S26 26.4
	17	330 04.0	46 53.2	26.9	334 55.3	50.0	160 59.5	53.2	227 23.8	13.2			
	18	345 06.4	61 52.3	S24 26.7	349 56.6	S 2 49.3	176 01.8	N 5 53.1	242 26.5	N22 13.2	Arcturus	146 02.8	N19 09.7
	19	0 08.9	76 51.3	26.5	4 57.8	48.7	191 04.0	53.1	257 29.1	13.2	Atria	107 45.2	S69 02.1
	20	15 11.4	91 50.4	26.3	19 59.0	48.1	206 06.3	53.0	272 31.8	13.3	Avior	234 20.8	S59 31.0
	21	30 13.8	106 49.5	. . 26.0	35 00.3	. . 47.4	221 08.5	. . 52.9	287 34.4	. . 13.3	Bellatrix	278 39.7	N 6 21.3
	22	45 16.3	121 48.5	25.8	50 01.5	46.8	236 10.8	52.8	302 37.1	13.3	Betelgeuse	271 09.1	N 7 24.6
	23	60 18.8	136 47.6	25.6	65 02.7	46.2	251 13.1	52.8	317 39.7	13.3			
	7 00	75 21.2	151 46.7	S24 25.4	80 03.9	S 2 45.5	266 15.3	N 5 52.7	332 42.4	N22 13.3	Canopus	263 59.0	S52 41.7
	01	90 23.7	166 45.8	25.1	95 05.2	44.9	281 17.6	52.6	347 45.0	13.3	Capella	280 45.1	N46 00.2
	02	105 26.2	181 44.8	24.9	110 06.4	44.3	296 19.9	52.5	2 47.7	13.4	Deneb	49 37.0	N45 17.8
	03	120 28.6	196 43.9	. . 24.7	125 07.6	. . 43.7	311 22.1	. . 52.5	17 50.3	. . 13.4	Denebola	182 41.4	N14 33.0
	04	135 31.1	211 43.0	24.4	140 08.8	43.0	326 24.4	52.4	32 52.9	13.4	Diphda	349 03.2	S17 58.0
	05	150 33.6	226 42.0	24.2	155 10.1	42.4	341 26.7	52.3	47 55.6	13.4			
	06	165 36.0	241 41.1	S24 24.0	170 11.3	S 2 41.8	356 28.9	N 5 52.3	62 58.2	N22 13.4	Dubhe	194 00.6	N61 43.6
	07	180 38.5	256 40.2	23.7	185 12.5	41.1	11 31.2	52.2	78 00.9	13.4	Elnath	278 21.8	N28 36.7
	08	195 40.9	271 39.3	23.5	200 13.7	40.5	26 33.5	52.1	93 03.5	13.5	Eltanin	90 50.2	N51 29.3
S	09	210 43.4	286 38.3	. . 23.3	215 14.9	. . 39.9	41 35.7	. . 52.0	108 06.2	. . 13.5	Enif	33 54.6	N 9 53.6
U	10	225 45.9	301 37.4	23.0	230 16.2	39.2	56 38.0	52.0	123 08.8	13.5	Fomalhaut	15 32.2	S29 36.3
N	11	240 48.3	316 36.5	22.8	245 17.4	38.6	71 40.2	51.9	138 11.5	13.5			
D	12	255 50.8	331 35.6	S24 22.6	260 18.6	S 2 38.0	86 42.5	N 5 51.8	153 14.1	N22 13.5	Gacrux	172 09.8	S57 07.7
A	13	270 53.3	346 34.6	22.3	275 19.8	37.4	101 44.8	51.7	168 16.8	13.6	Gienah	176 00.2	S17 33.6
Y	14	285 55.7	1 33.7	22.1	290 21.1	36.7	116 47.0	51.7	183 19.4	13.6	Hadar	148 59.4	S60 23.2
	15	300 58.2	16 32.8	. . 21.8	305 22.3	. . 36.1	131 49.3	. . 51.6	198 22.1	. . 13.6	Hamal	328 09.1	N23 29.0
	16	316 00.7	31 31.9	21.6	320 23.5	35.5	146 51.6	51.5	213 24.7	13.6	Kaus Aust.	83 54.2	S34 23.1
	17	331 03.1	46 30.9	21.3	335 24.7	34.8	161 53.9	51.5	228 27.4	13.6			
	18	346 05.6	61 30.0	S24 21.1	350 25.9	S 2 34.2	176 56.1	N 5 51.4	243 30.0	N22 13.6	Kochab	137 20.3	N74 08.2
	19	1 08.1	76 29.1	20.8	5 27.2	33.6	191 58.4	51.3	258 32.7	13.7	Markab	13 45.9	N15 13.6
	20	16 10.5	91 28.2	20.6	20 28.4	32.9	207 00.7	51.3	273 35.3	13.7	Menkar	314 22.6	N 4 06.3
	21	31 13.0	106 27.2	. . 20.3	35 29.6	. . 32.3	222 02.9	. . 51.2	288 38.0	. . 13.7	Menkent	148 16.9	S36 23.2
	22	46 15.4	121 26.3	20.1	50 30.8	31.7	237 05.2	51.1	303 40.6	13.7	Miaplacidus	221 41.1	S69 43.6
	23	61 17.9	136 25.4	19.8	65 32.0	31.0	252 07.5	51.0	318 43.3	13.7			
	8 00	76 20.4	151 24.5	S24 19.6	80 33.2	S 2 30.4	267 09.7	N 5 51.0	333 45.9	N22 13.7	Mirfak	308 50.8	N49 52.6
	01	91 22.8	166 23.6	19.3	95 34.5	29.8	282 12.0	50.9	348 48.6	13.8	Nunki	76 07.9	S26 17.6
	02	106 25.3	181 22.6	19.0	110 35.7	29.1	297 14.3	50.8	3 51.2	13.8	Peacock	53 31.3	S56 43.7
	03	121 27.8	196 21.7	. . 18.8	125 36.9	. . 28.5	312 16.5	. . 50.8	18 53.9	. . 13.8	Pollux	243 36.6	N28 01.0
	04	136 30.2	211 20.8	18.5	140 38.1	27.9	327 18.8	50.7	33 56.5	13.8	Procyon	245 07.3	N 5 13.0
	05	151 32.7	226 19.9	18.3	155 39.3	27.2	342 21.1	50.6	48 59.2	13.8			
	06	166 35.2	241 18.9	S24 18.0	170 40.6	S 2 26.6	357 23.4	N 5 50.5	64 01.8	N22 13.9	Rasalhague	96 13.8	N12 33.4
	07	181 37.6	256 18.0	17.7	185 41.8	26.0	12 25.6	50.5	79 04.5	13.9	Regulus	207 51.4	N11 57.6
	08	196 40.1	271 17.1	17.5	200 43.0	25.4	27 27.9	50.4	94 07.1	13.9	Rigel	281 18.9	S 8 11.7
M	09	211 42.6	286 16.2	. . 17.2	215 44.2	. . 24.7	42 30.2	. . 50.3	109 09.8	. . 13.9	Rigil Kent.	140 02.9	S60 50.8
O	10	226 45.0	301 15.3	16.9	230 45.4	24.1	57 32.4	50.3	124 12.4	13.9	Sabik	102 21.5	S15 43.8
N	11	241 47.5	316 14.3	16.6	245 46.6	23.5	72 34.7	50.2	139 15.1	13.9			
D	12	256 49.9	331 13.4	S24 16.4	260 47.8	S 2 22.8	87 37.0	N 5 50.1	154 17.7	N22 14.0	Schedar	349 49.1	N56 33.7
A	13	271 52.4	346 12.5	16.1	275 49.1	22.2	102 39.3	50.1	169 20.4	14.0	Shaula	96 32.6	S37 06.5
Y	14	286 54.9	1 11.6	15.8	290 50.3	21.6	117 41.5	50.0	184 23.0	14.0	Sirius	258 40.1	S16 43.1
	15	301 57.3	16 10.7	. . 15.6	305 51.5	. . 20.9	132 43.8	. . 49.9	199 25.7	. . 14.0	Spica	158 39.4	S11 10.8
	16	316 59.8	31 09.8	15.3	320 52.7	20.3	147 46.1	49.9	214 28.3	14.0	Suhail	222 57.9	S43 26.6
	17	332 02.3	46 08.8	15.0	335 53.9	19.7	162 48.4	49.8	229 31.0	14.0			
	18	347 04.7	61 07.9	S24 14.7	350 55.1	S 2 19.0	177 50.6	N 5 49.7	244 33.7	N22 14.1	Vega	80 44.5	N38 47.1
	19	2 07.2	76 07.0	14.4	5 56.3	18.4	192 52.9	49.6	259 36.3	14.1	Zuben'ubi	137 14.1	S16 03.4
	20	17 09.7	91 06.1	14.2	20 57.6	17.8	207 55.2	49.6	274 39.0	14.1		SHA	Mer. Pass.
	21	32 12.1	106 05.2	. . 13.9	35 58.8	. . 17.1	222 57.5	. . 49.5	289 41.6	. . 14.1		° ′	h m
	22	47 14.6	121 04.3	13.6	51 00.0	16.5	237 59.7	49.4	304 44.3	14.1	Venus	76 25.4	13 54
	23	62 17.1	136 03.3	13.3	66 01.2	15.9	253 02.0	49.4	319 46.9	14.2	Mars	4 42.7	18 38
		h m									Jupiter	190 54.1	6 14
	Mer. Pass.	18 55.5	*v* −0.9	*d* 0.2	*v* 1.2	*d* 0.6	*v* 2.3	*d* 0.1	*v* 2.6	*d* 0.0	Saturn	257 21.1	1 49

	UT	SUN GHA	SUN Dec	MOON GHA	v	MOON Dec	d	HP
	d h	° ′	° ′	° ′	′	° ′	′	′
SATURDAY	6 00	182 19.9	S22 24.8	34 00.1	15.2	N14 55.7	11.5	54.1
	01	197 19.7	25.1	48 34.3	15.1	15 07.2	11.4	54.1
	02	212 19.4	25.4	63 08.4	15.1	15 18.6	11.4	54.1
	03	227 19.2	. . 25.7	77 42.5	15.0	15 30.0	11.3	54.1
	04	242 18.9	26.0	92 16.5	15.0	15 41.3	11.2	54.1
	05	257 18.6	26.3	106 50.5	14.9	15 52.5	11.2	54.1
	06	272 18.4	S22 26.6	121 24.4	14.9	N16 03.7	11.1	54.0
	07	287 18.1	26.9	135 58.3	14.8	16 14.8	11.1	54.0
	08	302 17.9	27.2	150 32.1	14.8	16 25.9	10.9	54.0
	09	317 17.6	. . 27.5	165 05.9	14.7	16 36.8	11.0	54.0
	10	332 17.3	27.8	179 39.6	14.6	16 47.8	10.8	54.0
	11	347 17.1	28.1	194 13.2	14.7	16 58.6	10.8	54.0
	12	2 16.8	S22 28.5	208 46.9	14.5	N17 09.4	10.7	54.0
	13	17 16.6	28.8	223 20.4	14.5	17 20.1	10.6	54.0
	14	32 16.3	29.1	237 53.9	14.5	17 30.7	10.6	54.0
	15	47 16.0	. . 29.4	252 27.4	14.4	17 41.3	10.5	54.0
	16	62 15.8	29.7	267 00.8	14.3	17 51.8	10.4	54.0
	17	77 15.5	30.0	281 34.1	14.3	18 02.2	10.3	54.0
	18	92 15.2	S22 30.2	296 07.4	14.2	N18 12.5	10.3	54.0
	19	107 15.0	30.5	310 40.6	14.2	18 22.8	10.2	54.0
	20	122 14.7	30.8	325 13.8	14.1	18 33.0	10.1	54.0
	21	137 14.4	. . 31.1	339 46.9	14.1	18 43.1	10.1	54.0
	22	152 14.2	31.4	354 20.0	14.0	18 53.2	9.9	54.0
	23	167 13.9	31.7	8 53.0	13.9	19 03.1	9.9	54.0
SUNDAY	7 00	182 13.7	S22 32.0	23 25.9	13.9	N19 13.0	9.8	54.0
	01	197 13.4	32.3	37 58.8	13.8	19 22.8	9.8	54.0
	02	212 13.1	32.6	52 31.6	13.8	19 32.6	9.6	54.0
	03	227 12.9	. . 32.9	67 04.4	13.7	19 42.2	9.6	54.0
	04	242 12.6	33.2	81 37.1	13.7	19 51.8	9.5	54.0
	05	257 12.3	33.5	96 09.8	13.5	20 01.3	9.4	54.0
	06	272 12.1	S22 33.8	110 42.3	13.6	N20 10.7	9.3	54.0
	07	287 11.8	34.1	125 14.9	13.4	20 20.0	9.2	54.0
	08	302 11.5	34.3	139 47.3	13.4	20 29.2	9.2	54.0
	09	317 11.3	. . 34.6	154 19.7	13.4	20 38.4	9.0	54.0
	10	332 11.0	34.9	168 52.1	13.3	20 47.4	9.0	54.0
	11	347 10.7	35.2	183 24.4	13.2	20 56.4	8.9	54.0
	12	2 10.5	S22 35.5	197 56.6	13.1	N21 05.3	8.8	54.0
	13	17 10.2	35.8	212 28.7	13.1	21 14.1	8.7	54.0
	14	32 09.9	36.0	227 00.8	13.1	21 22.8	8.6	54.0
	15	47 09.7	. . 36.3	241 32.9	12.9	21 31.4	8.5	54.0
	16	62 09.4	36.6	256 04.8	12.9	21 39.9	8.4	54.0
	17	77 09.1	36.9	270 36.7	12.9	21 48.3	8.4	54.0
	18	92 08.9	S22 37.2	285 08.6	12.8	N21 56.7	8.2	54.0
	19	107 08.6	37.4	299 40.4	12.7	22 04.9	8.1	54.0
	20	122 08.3	37.7	314 12.1	12.7	22 13.0	8.1	54.0
	21	137 08.1	. . 38.0	328 43.8	12.6	22 21.1	7.9	54.0
	22	152 07.8	38.3	343 15.4	12.5	22 29.0	7.9	54.0
	23	167 07.5	38.6	357 46.9	12.5	22 36.9	7.8	54.0
MONDAY	8 00	182 07.2	S22 38.8	12 18.4	12.4	N22 44.7	7.6	54.0
	01	197 07.0	39.1	26 49.8	12.4	22 52.3	7.6	54.0
	02	212 06.7	39.4	41 21.2	12.3	22 59.9	7.4	54.0
	03	227 06.4	. . 39.6	55 52.5	12.2	23 07.3	7.4	54.0
	04	242 06.2	39.9	70 23.7	12.2	23 14.7	7.2	54.0
	05	257 05.9	40.2	84 54.9	12.1	23 21.9	7.2	54.0
	06	272 05.6	S22 40.5	99 26.0	12.0	N23 29.1	7.0	54.0
	07	287 05.4	40.7	113 57.0	12.0	23 36.1	7.0	54.0
	08	302 05.1	41.0	128 28.0	12.0	23 43.1	6.8	54.0
	09	317 04.8	. . 41.3	142 59.0	11.8	23 49.9	6.7	54.0
	10	332 04.5	41.5	157 29.8	11.9	23 56.6	6.6	54.0
	11	347 04.3	41.8	172 00.7	11.7	24 03.2	6.6	54.0
	12	2 04.0	S22 42.1	186 31.4	11.7	N24 09.8	6.4	54.0
	13	17 03.7	42.3	201 02.1	11.7	24 16.2	6.3	54.0
	14	32 03.4	42.6	215 32.8	11.6	24 22.5	6.1	54.0
	15	47 03.2	. . 42.9	230 03.4	11.5	24 28.6	6.1	54.0
	16	62 02.9	43.1	244 33.9	11.5	24 34.7	6.0	54.0
	17	77 02.6	43.4	259 04.4	11.4	24 40.7	5.8	54.0
	18	92 02.4	S22 43.6	273 34.8	11.3	N24 46.5	5.8	54.0
	19	107 02.1	43.9	288 05.1	11.3	24 52.3	5.6	54.0
	20	122 01.8	44.2	302 35.4	11.3	24 57.9	5.5	54.0
	21	137 01.5	. . 44.4	317 05.7	11.2	25 03.4	5.4	54.1
	22	152 01.3	44.7	331 35.9	11.1	25 08.8	5.3	54.1
	23	167 01.0	44.9	346 06.0	11.1	N25 14.1	5.1	54.1
		SD 16.3	d 0.3	SD 14.7		14.7		14.7

Lat.	Twilight Naut.	Twilight Civil	Sunrise	Moonrise 6	Moonrise 7	Moonrise 8	Moonrise 9
°	h m	h m	h m	h m	h m	h m	h m
N 72	08 05	10 15	■	11 25	□	□	□
N 70	07 47	09 28	■	12 14	□	□	□
68	07 33	08 58	11 15	12 46	12 01	□	□
66	07 21	08 35	10 05	13 10	12 49	□	□
64	07 10	08 17	09 30	13 28	13 20	13 08	□
62	07 02	08 02	09 05	13 44	13 43	13 46	13 54
60	06 54	07 50	08 45	13 57	14 02	14 12	14 32
N 58	06 47	07 39	08 29	14 08	14 18	14 33	14 59
56	06 41	07 29	08 15	14 18	14 31	14 51	15 20
54	06 35	07 20	08 03	14 26	14 43	15 06	15 38
52	06 29	07 13	07 53	14 34	14 53	15 18	15 53
50	06 25	07 06	07 44	14 41	15 02	15 30	16 06
45	06 13	06 50	07 24	14 56	15 22	15 53	16 32
N 40	06 04	06 38	07 08	15 09	15 38	16 12	16 54
35	05 55	06 26	06 54	15 19	15 51	16 28	17 11
30	05 47	06 16	06 42	15 29	16 03	16 42	17 26
20	05 31	05 58	06 22	15 45	16 23	17 06	17 52
N 10	05 15	05 42	06 04	15 59	16 41	17 26	18 15
0	04 59	05 25	05 48	16 13	16 58	17 46	18 36
S 10	04 41	05 08	05 31	16 26	17 15	18 05	18 57
20	04 19	04 48	05 12	16 41	17 33	18 26	19 19
30	03 51	04 24	04 51	16 57	17 53	18 50	19 45
35	03 33	04 10	04 39	17 07	18 06	19 04	20 01
40	03 11	03 52	04 25	17 18	18 20	19 21	20 19
45	02 42	03 31	04 07	17 32	18 37	19 40	20 41
S 50	02 00	03 03	03 46	17 48	18 58	20 05	21 08
52	01 36	02 48	03 36	17 56	19 08	20 18	21 22
54	01 01	02 32	03 24	18 04	19 19	20 31	21 37
56	////	02 12	03 11	18 14	19 32	20 47	21 56
58	////	01 46	02 55	18 25	19 47	21 07	22 19
S 60	////	01 07	02 37	18 37	20 05	21 31	22 48

Lat.	Sunset	Twilight Civil	Twilight Naut.	Moonset 6	Moonset 7	Moonset 8	Moonset 9
°	h m	h m	h m	h m	h m	h m	h m
N 72	■	13 27	15 37	08 18	□	□	□
N 70	■	14 14	15 55	07 30	□	□	□
68	12 27	14 44	16 10	06 59	09 17	□	□
66	13 37	15 07	16 22	06 37	08 30	□	□
64	14 12	15 25	16 32	06 19	07 59	09 49	□
62	14 37	15 40	16 41	06 05	07 37	09 12	10 46
60	14 57	15 53	16 48	05 53	07 19	08 46	10 08
N 58	15 13	16 04	16 55	05 42	07 04	08 25	09 42
56	15 27	16 13	17 02	05 33	06 51	08 08	09 21
54	15 39	16 22	17 08	05 25	06 40	07 54	09 04
52	15 49	16 30	17 13	05 18	06 30	07 41	08 49
50	15 59	16 37	17 18	05 12	06 21	07 30	08 36
45	16 19	16 52	17 29	04 58	06 03	07 08	08 10
N 40	16 35	17 05	17 39	04 47	05 48	06 49	07 49
35	16 48	17 16	17 48	04 37	05 35	06 34	07 32
30	17 00	17 26	17 56	04 29	05 24	06 21	07 17
20	17 20	17 44	18 12	04 14	05 06	05 58	06 52
N 10	17 38	18 01	18 27	04 02	04 49	05 39	06 30
0	17 55	18 17	18 43	03 50	04 34	05 21	06 10
S 10	18 12	18 35	19 02	03 39	04 19	05 03	05 49
20	18 30	18 55	19 24	03 26	04 03	04 43	05 28
30	18 51	19 19	19 52	03 12	03 45	04 21	05 03
35	19 04	19 33	20 10	03 04	03 34	04 08	04 48
40	19 18	19 51	20 32	02 55	03 22	03 53	04 31
45	19 36	20 12	21 01	02 44	03 07	03 36	04 10
S 50	19 57	20 41	21 43	02 31	02 50	03 14	03 45
52	20 07	20 55	22 08	02 25	02 42	03 03	03 32
54	20 19	21 12	22 44	02 18	02 33	02 52	03 18
56	20 32	21 32	////	02 11	02 22	02 38	03 02
58	20 48	21 59	////	02 03	02 11	02 23	02 42
S 60	21 07	22 38	////	01 53	01 57	02 04	02 18

Day	SUN Eqn. of Time 00^h	SUN Eqn. of Time 12^h	SUN Mer. Pass.	MOON Mer. Pass. Upper	MOON Mer. Pass. Lower	MOON Age	MOON Phase
d	m s	m s	h m	h m	h m	d %	
6	09 20	09 08	11 51	22 23	10 01	13 95	○
7	08 55	08 42	11 51	23 09	10 46	14 98	
8	08 30	08 17	11 52	23 58	11 33	15 100	

UT	ARIES	VENUS −3.9		MARS −0.2		JUPITER −2.1		SATURN −0.3		STARS		
	GHA	GHA	Dec	GHA	Dec	GHA	Dec	GHA	Dec	Name	SHA	Dec
d h	° ′	° ′	° ′	° ′	° ′	° ′	° ′	° ′	° ′		° ′	° ′
9 00	77 19.5	151 02.4	S24 13.0	81 02.4	S 2 15.2	268 04.3	N 5 49.3	334 49.6	N22 14.2	Acamar	315 23.6	S40 17.4
01	92 22.0	166 01.5	12.7	96 03.6	14.6	283 06.6	49.2	349 52.2	14.2	Achernar	335 31.7	S57 13.3
02	107 24.4	181 00.6	12.4	111 04.8	14.0	298 08.8	49.2	4 54.9	14.2	Acrux	173 18.3	S63 06.9
03	122 26.9	195 59.7	12.1	126 06.0	13.3	313 11.1	49.1	19 57.5	14.2	Adhara	255 18.1	S28 58.5
04	137 29.4	210 58.8	11.8	141 07.3	12.7	328 13.4	49.0	35 00.2	14.2	Aldebaran	290 57.7	N16 31.1
05	152 31.8	225 57.8	11.6	156 08.5	12.1	343 15.7	49.0	50 02.8	14.3			
T 06	167 34.3	240 56.9	S24 11.3	171 09.7	S 2 11.4	358 17.9	N 5 48.9	65 05.5	N22 14.3	Alioth	166 27.3	N55 56.1
U 07	182 36.8	255 56.0	11.0	186 10.9	10.8	13 20.2	48.8	80 08.1	14.3	Alkaid	153 05.0	N49 17.5
E 08	197 39.2	270 55.1	10.7	201 12.1	10.2	28 22.5	48.8	95 10.8	14.3	Al Na'ir	27 53.1	S46 56.8
S 09	212 41.7	285 54.2	10.4	216 13.3	09.5	43 24.8	48.7	110 13.5	14.3	Alnilam	275 53.7	S 1 11.9
D 10	227 44.2	300 53.3	10.1	231 14.5	08.9	58 27.1	48.6	125 16.1	14.4	Alphard	218 03.3	S 8 40.4
A 11	242 46.6	315 52.4	09.8	246 15.7	08.2	73 29.3	48.6	140 18.8	14.4			
Y 12	257 49.1	330 51.5	S24 09.5	261 16.9	S 2 07.6	88 31.6	N 5 48.5	155 21.4	N22 14.4	Alphecca	126 17.7	N26 42.0
13	272 51.5	345 50.5	09.2	276 18.1	07.0	103 33.9	48.4	170 24.1	14.4	Alpheratz	357 51.3	N29 06.8
14	287 54.0	0 49.6	08.9	291 19.4	06.3	118 36.2	48.4	185 26.7	14.4	Altair	62 15.8	N 8 52.7
15	302 56.5	15 48.7	08.5	306 20.6	05.7	133 38.4	48.3	200 29.4	14.4	Ankaa	353 22.8	S42 17.4
16	317 58.9	30 47.8	08.2	321 21.8	05.1	148 40.7	48.2	215 32.0	14.5	Antares	112 35.9	S26 26.4
17	333 01.4	45 46.9	07.9	336 23.0	04.4	163 43.0	48.2	230 34.7	14.5			
18	348 03.9	60 46.0	S24 07.6	351 24.2	S 2 03.8	178 45.3	N 5 48.1	245 37.3	N22 14.5	Arcturus	146 02.8	N19 09.7
19	3 06.3	75 45.1	07.3	6 25.4	03.2	193 47.6	48.0	260 40.0	14.5	Atria	107 45.2	S69 02.1
20	18 08.8	90 44.2	07.0	21 26.6	02.5	208 49.9	48.0	275 42.7	14.5	Avior	234 20.8	S59 31.0
21	33 11.3	105 43.3	06.7	36 27.8	01.9	223 52.1	47.9	290 45.3	14.6	Bellatrix	278 39.7	N 6 21.3
22	48 13.7	120 42.4	06.4	51 29.0	01.3	238 54.4	47.8	305 48.0	14.6	Betelgeuse	271 09.1	N 7 24.6
23	63 16.2	135 41.4	06.1	66 30.2	00.6	253 56.7	47.8	320 50.6	14.6			
10 00	78 18.7	150 40.5	S24 05.7	81 31.4	S 2 00.0	268 59.0	N 5 47.7	335 53.3	N22 14.6	Canopus	263 59.0	S52 41.7
01	93 21.1	165 39.6	05.4	96 32.6	1 59.4	284 01.3	47.6	350 55.9	14.6	Capella	280 45.1	N46 00.2
02	108 23.6	180 38.7	05.1	111 33.8	58.7	299 03.5	47.6	5 58.6	14.6	Deneb	49 37.0	N45 17.8
03	123 26.0	195 37.8	04.8	126 35.0	58.1	314 05.8	47.5	21 01.2	14.7	Denebola	182 41.3	N14 33.0
04	138 28.5	210 36.9	04.5	141 36.2	57.5	329 08.1	47.5	36 03.9	14.7	Diphda	349 03.2	S17 58.0
05	153 31.0	225 36.0	04.1	156 37.5	56.8	344 10.4	47.4	51 06.6	14.7			
06	168 33.4	240 35.1	S24 03.8	171 38.7	S 1 56.2	359 12.7	N 5 47.3	66 09.2	N22 14.7	Dubhe	194 00.5	N61 43.6
W 07	183 35.9	255 34.2	03.5	186 39.9	55.5	14 15.0	47.3	81 11.9	14.7	Elnath	278 21.7	N28 36.7
E 08	198 38.4	270 33.3	03.1	201 41.1	54.9	29 17.2	47.2	96 14.5	14.8	Eltanin	90 50.2	N51 29.3
D 09	213 40.8	285 32.4	02.8	216 42.3	54.3	44 19.5	47.1	111 17.2	14.8	Enif	33 54.7	N 9 53.5
N 10	228 43.3	300 31.5	02.5	231 43.5	53.6	59 21.8	47.1	126 19.8	14.8	Fomalhaut	15 32.2	S29 36.3
E 11	243 45.8	315 30.6	02.2	246 44.7	53.0	74 24.1	47.0	141 22.5	14.8			
S 12	258 48.2	330 29.7	S24 01.8	261 45.9	S 1 52.4	89 26.4	N 5 46.9	156 25.2	N22 14.8	Gacrux	172 09.8	S57 07.7
D 13	273 50.7	345 28.8	01.5	276 47.1	51.7	104 28.7	46.9	171 27.8	14.8	Gienah	176 00.2	S17 33.6
A 14	288 53.2	0 27.9	01.2	291 48.3	51.1	119 31.0	46.8	186 30.5	14.9	Hadar	148 59.3	S60 23.2
Y 15	303 55.6	15 27.0	00.8	306 49.5	50.5	134 33.2	46.8	201 33.1	14.9	Hamal	328 09.1	N23 29.0
16	318 58.1	30 26.1	00.5	321 50.7	49.8	149 35.5	46.7	216 35.8	14.9	Kaus Aust.	83 54.2	S34 23.1
17	334 00.5	45 25.1	24 00.1	336 51.9	49.2	164 37.8	46.6	231 38.4	14.9			
18	349 03.0	60 24.2	S23 59.8	351 53.1	S 1 48.6	179 40.1	N 5 46.6	246 41.1	N22 14.9	Kochab	137 20.3	N74 08.2
19	4 05.5	75 23.3	59.5	6 54.3	47.9	194 42.4	46.5	261 43.8	15.0	Markab	13 45.9	N15 13.6
20	19 07.9	90 22.4	59.1	21 55.5	47.3	209 44.7	46.4	276 46.4	15.0	Menkar	314 22.6	N 4 06.3
21	34 10.4	105 21.5	58.8	36 56.7	46.6	224 47.0	46.4	291 49.1	15.0	Menkent	148 16.9	S36 23.2
22	49 12.9	120 20.6	58.4	51 57.9	46.0	239 49.3	46.3	306 51.7	15.0	Miaplacidus	221 41.1	S69 43.6
23	64 15.3	135 19.7	58.1	66 59.1	45.4	254 51.5	46.2	321 54.4	15.0			
11 00	79 17.8	150 18.8	S23 57.7	82 00.3	S 1 44.7	269 53.8	N 5 46.2	336 57.0	N22 15.0	Mirfak	308 50.8	N49 52.7
01	94 20.3	165 17.9	57.4	97 01.5	44.1	284 56.1	46.1	351 59.7	15.1	Nunki	76 07.9	S26 17.6
02	109 22.7	180 17.0	57.0	112 02.7	43.5	299 58.4	46.1	7 02.4	15.1	Peacock	53 31.3	S56 43.7
03	124 25.2	195 16.1	56.7	127 03.9	42.8	315 00.7	46.0	22 05.0	15.1	Pollux	243 36.6	N28 01.0
04	139 27.7	210 15.2	56.3	142 05.1	42.2	330 03.0	45.9	37 07.7	15.1	Procyon	245 07.3	N 5 13.0
05	154 30.1	225 14.3	56.0	157 06.3	41.5	345 05.3	45.9	52 10.3	15.1			
06	169 32.6	240 13.4	S23 55.6	172 07.5	S 1 40.9	0 07.6	N 5 45.8	67 13.0	N22 15.2	Rasalhague	96 13.8	N12 33.4
07	184 35.0	255 12.5	55.3	187 08.7	40.3	15 09.9	45.8	82 15.7	15.2	Regulus	207 51.4	N11 56.9
T 08	199 37.5	270 11.6	54.9	202 09.9	39.6	30 12.2	45.7	97 18.3	15.2	Rigel	281 18.9	S 8 11.7
H 09	214 40.0	285 10.8	54.5	217 11.1	39.0	45 14.4	45.6	112 21.0	15.2	Rigil Kent.	140 02.9	S60 50.8
U 10	229 42.4	300 09.9	54.2	232 12.3	38.4	60 16.7	45.6	127 23.6	15.2	Sabik	102 21.5	S15 43.8
R 11	244 44.9	315 09.0	53.8	247 13.5	37.7	75 19.0	45.5	142 26.3	15.2			
S 12	259 47.4	330 08.1	S23 53.5	262 14.7	S 1 37.1	90 21.3	N 5 45.4	157 29.0	N22 15.3	Schedar	349 49.1	N56 33.7
D 13	274 49.8	345 07.2	53.1	277 15.9	36.4	105 23.6	45.4	172 31.6	15.3	Shaula	96 32.6	S37 06.5
A 14	289 52.3	0 06.3	52.7	292 17.1	35.8	120 25.9	45.3	187 34.3	15.3	Sirius	258 40.1	S16 43.1
Y 15	304 54.8	15 05.4	52.4	307 18.3	35.2	135 28.2	45.3	202 36.9	15.3	Spica	158 39.4	S11 10.8
16	319 57.2	30 04.5	52.0	322 19.5	34.5	150 30.5	45.2	217 39.6	15.3	Suhail	222 57.8	S43 26.6
17	334 59.7	45 03.6	51.6	337 20.6	33.9	165 32.8	45.1	232 42.3	15.4			
18	350 02.2	60 02.7	S23 51.3	352 21.8	S 1 33.3	180 35.1	N 5 45.1	247 44.9	N22 15.4	Vega	80 44.5	N38 47.2
19	5 04.6	75 01.8	50.9	7 23.0	32.6	195 37.4	45.0	262 47.6	15.4	Zuben'ubi	137 14.1	S16 03.4
20	20 07.1	90 00.9	50.5	22 24.2	32.0	210 39.7	45.0	277 50.2	15.4		SHA	Mer. Pass
21	35 09.5	105 00.0	50.1	37 25.4	31.3	225 42.0	44.9	292 52.9	15.4		° ′	h m
22	50 12.0	119 59.1	49.8	52 26.6	30.7	240 44.3	44.8	307 55.6	15.4	Venus	72 21.9	13 58
23	65 14.5	134 58.2	49.4	67 27.8	30.1	255 46.6	44.8	322 58.2	15.5	Mars	3 12.8	18 32
	h m									Jupiter	190 40.3	6 03
Mer. Pass.	18 43.7	*v* −0.9	*d* 0.3	*v* 1.2	*d* 0.6	*v* 2.3	*d* 0.1	*v* 2.7	*d* 0.0	Saturn	257 34.6	1 36

UT	SUN GHA	SUN Dec	MOON GHA	v	MOON Dec	d	HP
d h	° ′	° ′	° ′	′	° ′	′	′
9 00 (TUESDAY)	182 00.7	S22 45.2	0 36.1	11.1	N25 19.2	5.1	54.1
01	197 00.4	45.4	15 06.2	11.0	25 24.3	4.9	54.1
02	212 00.2	45.7	29 36.2	10.9	25 29.2	4.8	54.1
03	226 59.9	. . 46.0	44 06.1	10.9	25 34.0	4.7	54.1
04	241 59.6	46.2	58 36.0	10.8	25 38.7	4.5	54.1
05	256 59.3	46.5	73 05.8	10.8	25 43.2	4.5	54.1
06	271 59.1	S22 46.7	87 35.6	10.8	N25 47.7	4.3	54.1
07	286 58.8	47.0	102 05.4	10.7	25 52.0	4.2	54.1
08	301 58.5	47.2	116 35.1	10.6	25 56.2	4.1	54.1
09	316 58.2	. . 47.5	131 04.7	10.6	26 00.3	4.0	54.1
10	331 58.0	47.7	145 34.3	10.6	26 04.3	3.8	54.1
11	346 57.7	48.0	160 03.9	10.5	26 08.1	3.7	54.1
12	1 57.4	S22 48.2	174 33.4	10.5	N26 11.8	3.6	54.1
13	16 57.1	48.5	189 02.9	10.4	26 15.4	3.4	54.1
14	31 56.8	48.7	203 32.3	10.4	26 18.8	3.4	54.2
15	46 56.6	. . 48.9	218 01.7	10.4	26 22.2	3.2	54.2
16	61 56.3	49.2	232 31.1	10.3	26 25.4	3.1	54.2
17	76 56.0	49.4	247 00.4	10.3	26 28.5	2.9	54.2
18	91 55.7	S22 49.7	261 29.7	10.3	N26 31.4	2.9	54.2
19	106 55.5	49.9	275 59.0	10.2	26 34.3	2.7	54.2
20	121 55.2	50.2	290 28.2	10.2	26 37.0	2.5	54.2
21	136 54.9	. . 50.4	304 57.4	10.1	26 39.5	2.5	54.2
22	151 54.6	50.6	319 26.5	10.1	26 42.0	2.3	54.2
23	166 54.3	50.9	333 55.6	10.1	26 44.3	2.2	54.2
10 00 (WEDNESDAY)	181 54.1	S22 51.1	348 24.7	10.0	N26 46.5	2.0	54.2
01	196 53.8	51.4	2 53.7	10.1	26 48.5	1.9	54.2
02	211 53.5	51.6	17 22.8	10.0	26 50.4	1.8	54.3
03	226 53.2	. . 51.8	31 51.8	9.9	26 52.2	1.7	54.3
04	241 52.9	52.1	46 20.7	10.0	26 53.9	1.5	54.3
05	256 52.7	52.3	60 49.7	9.9	26 55.4	1.4	54.3
06	271 52.4	S22 52.5	75 18.6	9.9	N26 56.8	1.3	54.3
07	286 52.1	52.8	89 47.5	9.9	26 58.1	1.1	54.3
08	301 51.8	53.0	104 16.4	9.8	26 59.2	1.0	54.3
09	316 51.5	. . 53.2	118 45.2	9.8	27 00.2	0.9	54.3
10	331 51.3	53.4	133 14.0	9.9	27 01.1	0.7	54.3
11	346 51.0	53.7	147 42.9	9.8	27 01.8	0.6	54.3
12	1 50.7	S22 53.9	162 11.7	9.7	N27 02.4	0.5	54.3
13	16 50.4	54.1	176 40.4	9.8	27 02.9	0.3	54.4
14	31 50.1	54.4	191 09.2	9.7	27 03.2	0.2	54.4
15	46 49.9	. . 54.6	205 37.9	9.8	27 03.4	0.1	54.4
16	61 49.6	54.8	220 06.7	9.7	27 03.5	0.1	54.4
17	76 49.3	55.0	234 35.4	9.7	27 03.4	0.2	54.4
18	91 49.0	S22 55.3	249 04.1	9.7	N27 03.2	0.3	54.4
19	106 48.7	55.5	263 32.8	9.7	27 02.9	0.5	54.4
20	121 48.4	55.7	278 01.5	9.7	27 02.4	0.6	54.4
21	136 48.2	. . 55.9	292 30.2	9.7	27 01.8	0.7	54.4
22	151 47.9	56.1	306 58.9	9.7	27 01.1	0.9	54.5
23	166 47.6	56.4	321 27.6	9.7	27 00.2	1.0	54.5
11 00 (THURSDAY)	181 47.3	S22 56.6	335 56.3	9.7	N26 59.2	1.2	54.5
01	196 47.0	56.8	350 25.0	9.6	26 58.0	1.3	54.5
02	211 46.7	57.0	4 53.6	9.7	26 56.7	1.4	54.5
03	226 46.5	. . 57.2	19 22.3	9.7	26 55.3	1.5	54.5
04	241 46.2	57.5	33 51.0	9.7	26 53.8	1.7	54.5
05	256 45.9	57.7	48 19.7	9.7	26 52.1	1.8	54.5
06	271 45.6	S22 57.9	62 48.4	9.7	N26 50.3	2.0	54.6
07	286 45.3	58.1	77 17.1	9.7	26 48.3	2.1	54.6
08	301 45.0	58.3	91 45.8	9.7	26 46.2	2.2	54.6
09	316 44.7	. . 58.5	106 14.5	9.7	26 44.0	2.4	54.6
10	331 44.5	58.7	120 43.2	9.7	26 41.6	2.5	54.6
11	346 44.2	58.9	135 11.9	9.7	26 39.1	2.6	54.6
12	1 43.9	S22 59.1	149 40.6	9.8	N26 36.5	2.8	54.6
13	16 43.6	59.4	164 09.4	9.8	26 33.7	2.9	54.6
14	31 43.3	59.6	178 38.2	9.7	26 30.8	3.0	54.7
15	46 43.0	22 59.8	193 06.9	9.8	26 27.8	3.1	54.7
16	61 42.7	23 00.0	207 35.7	9.8	26 24.7	3.3	54.7
17	76 42.5	00.2	222 04.5	9.9	26 21.4	3.5	54.7
18	91 42.2	S23 00.4	236 33.4	9.8	N26 17.9	3.5	54.7
19	106 41.9	00.6	251 02.2	9.9	26 14.4	3.7	54.7
20	121 41.6	00.8	265 31.1	9.8	26 10.7	3.8	54.8
21	136 41.3	. . 01.0	279 59.9	9.9	26 06.9	4.0	54.8
22	151 41.0	01.2	294 28.8	10.0	26 02.9	4.1	54.8
23	166 40.7	01.4	308 57.8	9.9	N25 58.8	4.2	54.8
	SD 16.3	*d* 0.2	SD 14.8		14.8		14.9

Lat.	Twilight Naut.	Twilight Civil	Sunrise	Moonrise 9	10	11	12
°	h m	h m	h m	h m	h m	h m	h m
N 72	08 11	10 28	■	□	□	□	□
N 70	07 52	09 36	■	□	□	□	□
68	07 37	09 04	■	□	□	□	□
66	07 25	08 41	10 14	□	□	□	□
64	07 15	08 22	09 37	□	□	□	16 28
62	07 05	08 07	09 11	13 54	14 24	15 34	17 10
60	06 57	07 54	08 50	14 32	15 11	16 15	17 38
N 58	06 50	07 42	08 33	14 59	15 41	16 42	17 59
56	06 44	07 33	08 19	15 20	16 04	17 04	18 17
54	06 38	07 24	08 07	15 38	16 23	17 22	18 32
52	06 32	07 16	07 56	15 53	16 38	17 37	18 45
50	06 27	07 09	07 47	16 06	16 52	17 50	18 57
45	06 16	06 53	07 27	16 32	17 20	18 17	19 20
N 40	06 06	06 40	07 10	16 54	17 42	18 38	19 39
35	05 57	06 29	06 57	17 11	18 00	18 55	19 55
30	05 49	06 18	06 45	17 26	18 16	19 11	20 08
20	05 33	06 00	06 24	17 52	18 43	19 36	20 31
N 10	05 17	05 43	06 06	18 15	19 06	19 58	20 51
0	05 00	05 26	05 49	18 36	19 27	20 19	21 10
S 10	04 42	05 09	05 32	18 57	19 49	20 40	21 29
20	04 20	04 49	05 13	19 19	20 12	21 02	21 48
30	03 51	04 24	04 52	19 45	20 38	21 27	22 11
35	03 33	04 10	04 39	20 01	20 54	21 42	22 25
40	03 11	03 52	04 24	20 19	21 12	22 00	22 40
45	02 41	03 30	04 07	20 41	21 35	22 20	22 58
S 50	01 58	03 01	03 45	21 08	22 03	22 47	23 21
52	01 33	02 47	03 35	21 22	22 16	22 59	23 32
54	00 54	02 30	03 23	21 37	22 32	23 14	23 44
56	////	02 09	03 09	21 56	22 51	23 31	23 58
58	////	01 42	02 53	22 19	23 15	23 51	24 14
S 60	////	01 00	02 34	22 48	23 45	24 17	00 17

Lat.	Sunset	Twilight Civil	Twilight Naut.	Moonset 9	10	11	12
°	h m	h m	h m	h m	h m	h m	h m
N 72	■	13 17	15 34	□	□	□	□
N 70	■	14 09	15 52	□	□	□	□
68	■	14 41	16 07	□	□	□	□
66	13 31	15 04	16 20	□	□	□	□
64	14 08	15 23	16 30	□	□	□	13 35
62	14 34	15 38	16 39	10 46	12 03	12 41	12 52
60	14 55	15 51	16 47	10 08	11 16	12 00	12 24
N 58	15 12	16 03	16 55	09 42	10 46	11 32	12 02
56	15 26	16 12	17 01	09 21	10 23	11 11	11 43
54	15 38	16 21	17 07	09 04	10 05	10 53	11 28
52	15 49	16 29	17 13	08 49	09 49	10 38	11 15
50	15 58	16 36	17 18	08 36	09 35	10 24	11 03
45	16 18	16 52	17 29	08 10	09 07	09 57	10 39
N 40	16 35	17 05	17 39	07 49	08 45	09 36	10 19
35	16 48	17 16	17 48	07 32	08 27	09 18	10 03
30	17 00	17 27	17 57	07 17	08 11	09 02	09 49
20	17 21	17 45	18 13	06 52	07 45	08 36	09 24
N 10	17 39	18 02	18 28	06 30	07 22	08 13	09 03
0	17 56	18 19	18 45	06 10	07 00	07 52	08 44
S 10	18 14	18 37	19 03	05 49	06 39	07 31	08 24
20	18 32	18 57	19 26	05 28	06 16	07 08	08 03
30	18 54	19 21	19 54	05 03	05 50	06 42	07 38
35	19 06	19 36	20 12	04 48	05 34	06 26	07 24
40	19 21	19 54	20 35	04 31	05 16	06 08	07 07
45	19 39	20 16	21 05	04 10	04 54	05 46	06 46
S 50	20 00	20 44	21 48	03 45	04 26	05 18	06 21
52	20 11	20 59	22 14	03 32	04 12	05 04	06 08
54	20 23	21 16	22 53	03 18	03 56	04 48	05 54
56	20 37	21 37	////	03 02	03 38	04 30	05 37
58	20 53	22 05	////	02 42	03 15	04 06	05 17
S 60	21 12	22 48	////	02 18	02 45	03 36	04 51

Day	SUN Eqn. of Time 00^h	SUN Eqn. of Time 12^h	SUN Mer. Pass.	MOON Mer. Pass. Upper	MOON Mer. Pass. Lower	Age	Phase
d	m s	m s	h m	h m	h m	d %	
9	08 03	07 50	11 52	24 48	12 23	16 100	○
10	07 37	07 23	11 53	00 48	13 14	17 97	
11	07 10	06 56	11 53	01 40	14 06	18 94	

	UT	ARIES	VENUS −4.0		MARS −0.1		JUPITER −2.1		SATURN −0.3		STARS		
		GHA	GHA	Dec	GHA	Dec	GHA	Dec	GHA	Dec	Name	SHA	Dec
	d h	° ′	° ′	° ′	° ′	° ′	° ′	° ′	° ′	° ′		° ′	° ′
	12 00	80 16.9	149 57.3	S23 49.0	82 29.0	S 1 29.4	270 48.8	N 5 44.7	338 00.9	N22 15.5	Acamar	315 23.6	S40 17.5
	01	95 19.4	164 56.4	48.6	97 30.2	28.8	285 51.1	44.7	353 03.5	15.5	Achernar	335 31.8	S57 13.3
	02	110 21.9	179 55.6	48.3	112 31.4	28.2	300 53.4	44.6	8 06.2	15.5	Acrux	173 18.2	S63 06.9
	03	125 24.3	194 54.7	. . 47.9	127 32.6	. . 27.5	315 55.7	. . 44.5	23 08.9	. . 15.5	Adhara	255 18.1	S28 58.5
	04	140 26.8	209 53.8	47.5	142 33.8	26.9	330 58.0	44.5	38 11.5	15.6	Aldebaran	290 57.7	N16 31.1
	05	155 29.3	224 52.9	47.1	157 35.0	26.2	346 00.3	44.4	53 14.2	15.6			
	06	170 31.7	239 52.0	S23 46.7	172 36.2	S 1 25.6	1 02.6	N 5 44.4	68 16.8	N22 15.6	Alioth	166 27.2	N55 56.1
	07	185 34.2	254 51.1	46.3	187 37.4	25.0	16 04.9	44.3	83 19.5	15.6	Alkaid	153 04.9	N49 17.5
	08	200 36.7	269 50.2	45.9	202 38.6	24.3	31 07.2	44.3	98 22.2	15.6	Al Na'ir	27 53.1	S46 56.8
F	09	215 39.1	284 49.3	. . 45.6	217 39.7	. . 23.7	46 09.5	. . 44.2	113 24.8	. . 15.7	Alnilam	275 53.6	S 1 11.9
R	10	230 41.6	299 48.4	45.2	232 40.9	23.0	61 11.8	44.1	128 27.5	15.7	Alphard	218 03.3	S 8 40.4
I	11	245 44.0	314 47.6	44.8	247 42.1	22.4	76 14.1	44.1	143 30.2	15.7			
D	12	260 46.5	329 46.7	S23 44.4	262 43.3	S 1 21.8	91 16.4	N 5 44.0	158 32.8	N22 15.7	Alphecca	126 17.6	N26 42.0
A	13	275 49.0	344 45.8	44.0	277 44.5	21.1	106 18.7	44.0	173 35.5	15.7	Alpheratz	357 51.3	N29 06.8
Y	14	290 51.4	359 44.9	43.6	292 45.7	20.5	121 21.0	43.9	188 38.1	15.7	Altair	62 15.8	N 8 52.7
	15	305 53.9	14 44.0	. . 43.2	307 46.9	. . 19.8	136 23.3	. . 43.8	203 40.8	. . 15.8	Ankaa	353 22.8	S42 17.4
	16	320 56.4	29 43.1	42.8	322 48.1	19.2	151 25.6	43.8	218 43.5	15.8	Antares	112 35.9	S26 26.4
	17	335 58.8	44 42.2	42.4	337 49.3	18.6	166 27.9	43.7	233 46.1	15.8			
	18	351 01.3	59 41.4	S23 42.0	352 50.5	S 1 17.9	181 30.2	N 5 43.7	248 48.8	N22 15.8	Arcturus	146 02.8	N19 09.7
	19	6 03.8	74 40.5	41.6	7 51.6	17.3	196 32.5	43.6	263 51.5	15.8	Atria	107 45.2	S69 02.1
	20	21 06.2	89 39.6	41.2	22 52.8	16.7	211 34.8	43.6	278 54.1	15.9	Avior	234 20.8	S59 31.0
	21	36 08.7	104 38.7	. . 40.8	37 54.0	. . 16.0	226 37.1	. . 43.5	293 56.8	. . 15.9	Bellatrix	278 39.7	N 6 21.3
	22	51 11.2	119 37.8	40.4	52 55.2	15.4	241 39.4	43.4	308 59.4	15.9	Betelgeuse	271 09.1	N 7 24.6
	23	66 13.6	134 36.9	40.0	67 56.4	14.7	256 41.7	43.4	324 02.1	15.9			
	13 00	81 16.1	149 36.0	S23 39.6	82 57.6	S 1 14.1	271 44.0	N 5 43.3	339 04.8	N22 15.9	Canopus	263 58.9	S52 41.7
	01	96 18.5	164 35.2	39.2	97 58.8	13.5	286 46.3	43.3	354 07.4	16.0	Capella	280 45.1	N46 00.2
	02	111 21.0	179 34.3	38.8	113 00.0	12.8	301 48.6	43.2	9 10.1	16.0	Deneb	49 37.0	N45 17.7
	03	126 23.5	194 33.4	. . 38.3	128 01.1	. . 12.2	316 50.9	. . 43.2	24 12.8	. . 16.0	Denebola	182 41.3	N14 33.0
	04	141 25.9	209 32.5	37.9	143 02.3	11.5	331 53.2	43.1	39 15.4	16.0	Diphda	349 03.2	S17 58.0
	05	156 28.4	224 31.6	37.5	158 03.5	10.9	346 55.5	43.1	54 18.1	16.0			
	06	171 30.9	239 30.8	S23 37.1	173 04.7	S 1 10.3	1 57.8	N 5 43.0	69 20.8	N22 16.0	Dubhe	194 00.5	N61 43.6
	07	186 33.3	254 29.9	36.7	188 05.9	09.6	17 00.1	42.9	84 23.4	16.1	Elnath	278 21.7	N28 36.7
S	08	201 35.8	269 29.0	36.3	203 07.1	09.0	32 02.4	42.9	99 26.1	16.1	Eltanin	90 50.2	N51 29.3
A	09	216 38.3	284 28.1	. . 35.9	218 08.3	. . 08.3	47 04.7	. . 42.8	114 28.7	. . 16.1	Enif	33 54.7	N 9 53.5
T	10	231 40.7	299 27.2	35.4	233 09.5	07.7	62 07.1	42.8	129 31.4	16.1	Fomalhaut	15 32.2	S29 36.3
U	11	246 43.2	314 26.4	35.0	248 10.6	07.1	77 09.4	42.7	144 34.1	16.1			
R	12	261 45.6	329 25.5	S23 34.6	263 11.8	S 1 06.4	92 11.7	N 5 42.7	159 36.7	N22 16.2	Gacrux	172 09.7	S57 07.8
D	13	276 48.1	344 24.6	34.2	278 13.0	05.8	107 14.0	42.6	174 39.4	16.2	Gienah	176 00.2	S17 33.6
A	14	291 50.6	359 23.7	33.7	293 14.2	05.1	122 16.3	42.6	189 42.1	16.2	Hadar	148 59.3	S60 23.2
Y	15	306 53.0	14 22.9	. . 33.3	308 15.4	. . 04.5	137 18.6	. . 42.5	204 44.7	. . 16.2	Hamal	328 09.1	N23 29.0
	16	321 55.5	29 22.0	32.9	323 16.6	03.9	152 20.9	42.4	219 47.4	16.2	Kaus Aust.	83 54.1	S34 23.1
	17	336 58.0	44 21.1	32.5	338 17.7	03.2	167 23.2	42.4	234 50.1	16.3			
	18	352 00.4	59 20.2	S23 32.0	353 18.9	S 1 02.6	182 25.5	N 5 42.3	249 52.7	N22 16.3	Kochab	137 20.3	N74 08.2
	19	7 02.9	74 19.4	31.6	8 20.1	01.9	197 27.8	42.3	264 55.4	16.3	Markab	13 45.9	N15 13.6
	20	22 05.4	89 18.5	31.2	23 21.3	01.3	212 30.1	42.2	279 58.1	16.3	Menkar	314 22.6	N 4 06.3
	21	37 07.8	104 17.6	. . 30.7	38 22.5	. . 00.6	227 32.4	. . 42.2	295 00.7	. . 16.3	Menkent	148 16.8	S36 23.2
	22	52 10.3	119 16.7	30.3	53 23.7	1 00.0	242 34.7	42.1	310 03.4	16.3	Miaplacidus	221 41.1	S69 43.6
	23	67 12.8	134 15.9	29.9	68 24.8	0 59.4	257 37.0	42.1	325 06.1	16.4			
	14 00	82 15.2	149 15.0	S23 29.4	83 26.0	S 0 58.7	272 39.3	N 5 42.0	340 08.7	N22 16.4	Mirfak	308 50.8	N49 52.7
	01	97 17.7	164 14.1	29.0	98 27.2	58.1	287 41.7	42.0	355 11.4	16.4	Nunki	76 07.9	S26 17.6
	02	112 20.1	179 13.2	28.6	113 28.4	57.4	302 44.0	41.9	10 14.1	16.4	Peacock	53 31.3	S56 43.6
	03	127 22.6	194 12.4	. . 28.1	128 29.6	. . 56.8	317 46.3	. . 41.9	25 16.7	. . 16.4	Pollux	243 36.5	N28 01.0
	04	142 25.1	209 11.5	27.7	143 30.7	56.2	332 48.6	41.8	40 19.4	16.5	Procyon	245 07.3	N 5 13.0
	05	157 27.5	224 10.6	27.2	158 31.9	55.5	347 50.9	41.7	55 22.1	16.5			
	06	172 30.0	239 09.8	S23 26.8	173 33.1	S 0 54.9	2 53.2	N 5 41.7	70 24.7	N22 16.5	Rasalhague	96 13.8	N12 33.4
	07	187 32.5	254 08.9	26.3	188 34.3	54.2	17 55.5	41.6	85 27.4	16.5	Regulus	207 51.3	N11 56.9
	08	202 34.9	269 08.0	25.9	203 35.5	53.6	32 57.8	41.6	100 30.1	16.5	Rigel	281 18.9	S 8 11.7
S	09	217 37.4	284 07.1	. . 25.4	218 36.7	. . 53.0	48 00.1	. . 41.5	115 32.7	. . 16.6	Rigil Kent.	140 02.8	S60 50.8
U	10	232 39.9	299 06.3	25.0	233 37.8	52.3	63 02.4	41.5	130 35.4	16.6	Sabik	102 21.5	S15 43.8
N	11	247 42.3	314 05.4	24.6	248 39.0	51.7	78 04.8	41.4	145 38.1	16.6			
D	12	262 44.8	329 04.5	S23 24.1	263 40.2	S 0 51.0	93 07.1	N 5 41.4	160 40.7	N22 16.6	Schedar	349 49.2	N56 33.8
A	13	277 47.3	344 03.7	23.6	278 41.4	50.4	108 09.4	41.3	175 43.4	16.6	Shaula	96 32.5	S37 06.5
Y	14	292 49.7	359 02.8	23.2	293 42.5	49.7	123 11.7	41.3	190 46.1	16.6	Sirius	258 40.0	S16 43.1
	15	307 52.2	14 01.9	. . 22.7	308 43.7	. . 49.1	138 14.0	. . 41.2	205 48.7	. . 16.7	Spica	158 39.4	S11 10.8
	16	322 54.6	29 01.1	22.3	323 44.9	48.5	153 16.3	41.2	220 51.4	16.7	Suhail	222 57.8	S43 26.6
	17	337 57.1	44 00.2	21.8	338 46.1	47.8	168 18.6	41.1	235 54.1	16.7			
	18	352 59.6	58 59.3	S23 21.4	353 47.3	S 0 47.2	183 21.0	N 5 41.1	250 56.7	N22 16.7	Vega	80 44.5	N38 47.2
	19	8 02.0	73 58.5	20.9	8 48.4	46.5	198 23.3	41.0	265 59.4	16.7	Zuben'ubi	137 14.0	S16 03.4
	20	23 04.5	88 57.6	20.4	23 49.6	45.9	213 25.6	41.0	281 02.1	16.8			
	21	38 07.0	103 56.7	. . 20.0	38 50.8	. . 45.3	228 27.9	. . 40.9	296 04.7	. . 16.8			
	22	53 09.4	118 55.9	19.5	53 52.0	44.6	243 30.2	40.9	311 07.4	16.8			
	23	68 11.9	133 55.0	19.1	68 53.1	44.0	258 32.5	40.8	326 10.1	16.8			
	Mer. Pass.	h m 18 31.9	v −0.9	d 0.4	v 1.2	d 0.6	v 2.3	d 0.1	v 2.7	d 0.0			

	SHA	Mer. Pass
	° ′	h m
Venus	68 20.0	14 02
Mars	1 41.5	18 27
Jupiter	190 27.9	5 52
Saturn	257 48.7	1 23

UT d	UT h	SUN GHA ° ′	SUN Dec ° ′	MOON GHA ° ′	v ′	MOON Dec ° ′	d ′	HP ′
12 FRIDAY	00	181 40.4	S23 01.6	323 26.7	10.0	N25 54.6	4.3	54.8
	01	196 40.2	01.8	337 55.7	10.0	25 50.3	4.5	54.8
	02	211 39.9	02.0	352 24.7	10.0	25 45.8	4.6	54.8
	03	226 39.6	. . 02.2	6 53.7	10.1	25 41.2	4.8	54.9
	04	241 39.3	02.4	21 22.8	10.1	25 36.4	4.8	54.9
	05	256 39.0	02.6	35 51.9	10.1	25 31.6	5.0	54.9
	06	271 38.7	S23 02.8	50 21.0	10.1	N25 26.6	5.2	54.9
	07	286 38.4	03.0	64 50.1	10.2	25 21.4	5.2	54.9
	08	301 38.1	03.2	79 19.3	10.2	25 16.2	5.4	54.9
	09	316 37.8	. . 03.4	93 48.5	10.2	25 10.8	5.5	55.0
	10	331 37.6	03.6	108 17.7	10.3	25 05.3	5.6	55.0
	11	346 37.3	03.8	122 47.0	10.3	24 59.7	5.8	55.0
	12	1 37.0	S23 03.9	137 16.3	10.3	N24 53.9	5.9	55.0
	13	16 36.7	04.1	151 45.6	10.4	24 48.0	6.0	55.0
	14	31 36.4	04.3	166 15.0	10.4	24 42.0	6.1	55.0
	15	46 36.1	. . 04.5	180 44.4	10.4	24 35.9	6.3	55.1
	16	61 35.8	04.7	195 13.8	10.5	24 29.6	6.3	55.1
	17	76 35.5	04.9	209 43.3	10.5	24 23.3	6.5	55.1
	18	91 35.2	S23 05.1	224 12.8	10.5	N24 16.8	6.7	55.1
	19	106 34.9	05.3	238 42.3	10.6	24 10.1	6.7	55.1
	20	121 34.7	05.4	253 11.9	10.6	24 03.4	6.9	55.2
	21	136 34.4	. . 05.6	267 41.5	10.7	23 56.5	7.0	55.2
	22	151 34.1	05.8	282 11.2	10.7	23 49.5	7.1	55.2
	23	166 33.8	06.0	296 40.9	10.7	23 42.4	7.2	55.2
13 SATURDAY	00	181 33.5	S23 06.2	311 10.6	10.8	N23 35.2	7.4	55.2
	01	196 33.2	06.3	325 40.4	10.8	23 27.8	7.4	55.3
	02	211 32.9	06.5	340 10.2	10.9	23 20.4	7.6	55.3
	03	226 32.6	. . 06.7	354 40.1	10.9	23 12.8	7.7	55.3
	04	241 32.3	06.9	9 10.0	11.0	23 05.1	7.8	55.3
	05	256 32.0	07.1	23 40.0	10.9	22 57.3	8.0	55.3
	06	271 31.7	S23 07.2	38 09.9	11.1	N22 49.3	8.0	55.4
	07	286 31.4	07.4	52 40.0	11.1	22 41.3	8.2	55.4
	08	301 31.2	07.6	67 10.1	11.1	22 33.1	8.2	55.4
	09	316 30.9	. . 07.8	81 40.2	11.1	22 24.9	8.4	55.4
	10	331 30.6	07.9	96 10.3	11.2	22 16.5	8.5	55.4
	11	346 30.3	08.1	110 40.5	11.3	22 08.0	8.6	55.5
	12	1 30.0	S23 08.3	125 10.8	11.3	N21 59.4	8.8	55.5
	13	16 29.7	08.4	139 41.1	11.3	21 50.6	8.8	55.5
	14	31 29.4	08.6	154 11.4	11.4	21 41.8	8.9	55.5
	15	46 29.1	. . 08.8	168 41.8	11.4	21 32.9	9.1	55.5
	16	61 28.8	09.0	183 12.2	11.4	21 23.8	9.2	55.6
	17	76 28.5	09.1	197 42.6	11.5	21 14.6	9.2	55.6
	18	91 28.2	S23 09.3	212 13.1	11.6	N21 05.4	9.4	55.6
	19	106 27.9	09.5	226 43.7	11.6	20 56.0	9.5	55.6
	20	121 27.6	09.6	241 14.3	11.6	20 46.5	9.6	55.7
	21	136 27.3	. . 09.8	255 44.9	11.7	20 36.9	9.6	55.7
	22	151 27.0	09.9	270 15.6	11.7	20 27.3	9.8	55.7
	23	166 26.7	10.1	284 46.3	11.8	20 17.5	9.9	55.7
14 SUNDAY	00	181 26.4	S23 10.3	299 17.1	11.8	N20 07.6	10.0	55.8
	01	196 26.1	10.4	313 47.9	11.8	19 57.6	10.1	55.8
	02	211 25.9	10.6	328 18.7	11.9	19 47.5	10.2	55.8
	03	226 25.6	. . 10.8	342 49.6	12.0	19 37.3	10.3	55.8
	04	241 25.3	10.9	357 20.6	11.9	19 27.0	10.4	55.9
	05	256 25.0	11.1	11 51.5	12.0	19 16.6	10.5	55.9
	06	271 24.7	S23 11.2	26 22.5	12.1	N19 06.1	10.6	55.9
	07	286 24.4	11.4	40 53.6	12.1	18 55.5	10.7	55.9
	08	301 24.1	11.5	55 24.7	12.1	18 44.8	10.8	56.0
	09	316 23.8	. . 11.7	69 55.8	12.2	18 34.0	10.9	56.0
	10	331 23.5	11.8	84 27.0	12.2	18 23.1	10.9	56.0
	11	346 23.2	12.0	98 58.2	12.3	18 12.2	11.1	56.0
	12	1 22.9	S23 12.2	113 29.5	12.3	N18 01.1	11.2	56.1
	13	16 22.6	12.3	128 00.8	12.3	17 49.9	11.2	56.1
	14	31 22.3	12.5	142 32.1	12.4	17 38.7	11.4	56.1
	15	46 22.0	. . 12.6	157 03.5	12.4	17 27.3	11.4	56.1
	16	61 21.7	12.8	171 34.9	12.5	17 15.9	11.5	56.2
	17	76 21.4	12.9	186 06.4	12.4	17 04.4	11.7	56.2
	18	91 21.1	S23 13.0	200 37.8	12.6	N16 52.7	11.7	56.2
	19	106 20.8	13.2	215 09.4	12.5	16 41.0	11.8	56.2
	20	121 20.5	13.3	229 40.9	12.6	16 29.2	11.8	56.3
	21	136 20.2	. . 13.5	244 12.5	12.6	16 17.4	12.0	56.3
	22	151 19.9	13.6	258 44.1	12.7	16 05.4	12.0	56.3
	23	166 19.6	13.8	273 15.8	12.7	N15 53.4	12.2	56.4
		SD 16.3	*d* 0.2	SD	15.0	15.1		15.3

Lat.	Twilight Naut.	Twilight Civil	Sunrise	Moonrise 12	13	14	15
°	h m	h m	h m	h m	h m	h m	h m
N 72	08 17	10 39	■	□	□	18 06	21 01
N 70	07 57	09 43	■	□	□	19 00	21 20
68	07 42	09 10	■	□	16 56	19 32	21 36
66	07 29	08 45	10 22	□	17 53	19 56	21 48
64	07 18	08 26	09 42	16 28	18 26	20 14	21 58
62	07 09	08 10	09 15	17 10	18 50	20 29	22 06
60	07 01	07 57	08 54	17 38	19 09	20 42	22 14
N 58	06 53	07 46	08 37	17 59	19 25	20 52	22 20
56	06 47	07 36	08 23	18 17	19 38	21 02	22 26
54	06 41	07 27	08 10	18 32	19 50	21 10	22 31
52	06 35	07 19	07 59	18 45	20 00	21 17	22 35
50	06 30	07 11	07 50	18 57	20 09	21 24	22 40
45	06 19	06 56	07 29	19 20	20 28	21 38	22 48
N 40	06 08	06 42	07 13	19 39	20 43	21 49	22 56
35	05 59	06 31	06 59	19 55	20 56	21 59	23 02
30	05 51	06 20	06 47	20 08	21 08	22 08	23 08
20	05 34	06 02	06 26	20 31	21 27	22 22	23 17
N 10	05 18	05 45	06 07	20 51	21 44	22 35	23 25
0	05 02	05 28	05 50	21 10	21 59	22 47	23 33
S 10	04 43	05 10	05 33	21 29	22 15	22 59	23 41
20	04 21	04 50	05 14	21 48	22 32	23 12	23 49
30	03 52	04 25	04 52	22 11	22 51	23 26	23 59
35	03 33	04 10	04 40	22 25	23 02	23 34	24 04
40	03 11	03 52	04 25	22 40	23 14	23 44	24 10
45	02 41	03 30	04 07	22 58	23 29	23 55	24 17
S 50	01 56	03 01	03 45	23 21	23 47	24 08	00 08
52	01 30	02 46	03 34	23 32	23 56	24 14	00 14
54	00 48	02 29	03 22	23 44	24 05	00 05	00 21
56	////	02 07	03 08	23 58	24 16	00 16	00 28
58	////	01 39	02 52	24 14	00 14	00 28	00 37
S 60	////	00 53	02 32	00 17	00 33	00 42	00 46

Lat.	Sunset	Twilight Civil	Twilight Naut.	Moonset 12	13	14	15
°	h m	h m	h m	h m	h m	h m	h m
N 72	■	13 09	15 31	□	□	15 25	14 09
N 70	■	14 05	15 50	□	□	14 29	13 47
68	■	14 38	16 06	□	14 52	13 56	13 30
66	13 26	15 02	16 19	□	13 54	13 31	13 16
64	14 05	15 22	16 29	13 35	13 20	13 12	13 04
62	14 33	15 37	16 39	12 52	12 55	12 56	12 54
60	14 54	15 51	16 47	12 24	12 36	12 42	12 46
N 58	15 11	16 02	16 54	12 02	12 19	12 31	12 38
56	15 25	16 12	17 01	11 43	12 05	12 21	12 32
54	15 37	16 21	17 07	11 28	11 53	12 12	12 26
52	15 48	16 29	17 13	11 15	11 43	12 04	12 20
50	15 58	16 36	17 18	11 03	11 33	11 56	12 15
45	16 18	16 52	17 29	10 39	11 13	11 41	12 05
N 40	16 35	17 06	17 40	10 19	10 56	11 28	11 56
35	16 49	17 17	17 49	10 03	10 42	11 17	11 48
30	17 01	17 28	17 57	09 49	10 30	11 08	11 41
20	17 22	17 46	18 14	09 24	10 09	10 51	11 30
N 10	17 40	18 03	18 30	09 03	09 51	10 36	11 19
0	17 58	18 20	18 46	08 44	09 34	10 22	11 09
S 10	18 15	18 38	19 05	08 24	09 17	10 09	10 59
20	18 34	18 58	19 28	08 03	08 58	09 54	10 49
30	18 56	19 23	19 56	07 38	08 37	09 36	10 36
35	19 08	19 38	20 15	07 24	08 24	09 26	10 29
40	19 23	19 56	20 38	07 07	08 10	09 15	10 21
45	19 41	20 18	21 08	06 46	07 52	09 01	10 11
S 50	20 03	20 48	21 52	06 21	07 30	08 44	10 00
52	20 14	21 02	22 19	06 08	07 20	08 36	09 54
54	20 26	21 20	23 02	05 54	07 08	08 28	09 48
56	20 40	21 42	////	05 37	06 55	08 18	09 42
58	20 57	22 10	////	05 17	06 39	08 06	09 34
S 60	21 17	22 57	////	04 51	06 20	07 53	09 25

Day	SUN Eqn. of Time 00^h	12^h	Mer. Pass.	MOON Mer. Pass. Upper	Lower	Age	Phase
d	m s	m s	h m	h m	h m	d %	
12	06 42	06 28	11 54	02 31	14 57	19 88	
13	06 15	06 00	11 54	03 22	15 47	20 81	
14	05 46	05 32	11 54	04 11	16 35	21 73	

UT	ARIES	VENUS −4.0		MARS −0.1		JUPITER −2.1		SATURN −0.4	
	GHA	GHA	Dec	GHA	Dec	GHA	Dec	GHA	Dec
d h	° ′	° ′	° ′	° ′	° ′	° ′	° ′	° ′	° ′
15 00	83 14.4	148 54.2	S23 18.6	83 54.3	S 0 43.3	273 34.8	N 5 40.8	341 12.7	N22 16.8
01	98 16.8	163 53.3	18.1	98 55.5	42.7	288 37.2	40.7	356 15.4	16.9
02	113 19.3	178 52.4	17.7	113 56.7	42.0	303 39.5	40.7	11 18.1	16.9
03	128 21.8	193 51.6	. . 17.2	128 57.9	. . 41.4	318 41.8	. . 40.6	26 20.7	. . 16.9
04	143 24.2	208 50.7	16.7	143 59.0	40.8	333 44.1	40.6	41 23.4	16.9
05	158 26.7	223 49.9	16.2	159 00.2	40.1	348 46.4	40.5	56 26.1	16.9
06	173 29.1	238 49.0	S23 15.8	174 01.4	S 0 39.5	3 48.7	N 5 40.5	71 28.7	N22 17.0
07	188 31.6	253 48.1	15.3	189 02.6	38.8	18 51.1	40.4	86 31.4	17.0
08	203 34.1	268 47.3	14.8	204 03.7	38.2	33 53.4	40.4	101 34.1	17.0
M 09	218 36.5	283 46.4	. . 14.3	219 04.9	. . 37.5	48 55.7	. . 40.3	116 36.7	. . 17.0
O 10	233 39.0	298 45.6	13.9	234 06.1	36.9	63 58.0	40.3	131 39.4	17.0
N 11	248 41.5	313 44.7	13.4	249 07.2	36.3	79 00.3	40.2	146 42.1	17.0
D 12	263 43.9	328 43.8	S23 12.9	264 08.4	S 0 35.6	94 02.6	N 5 40.2	161 44.7	N22 17.1
A 13	278 46.4	343 43.0	12.4	279 09.6	35.0	109 05.0	40.1	176 47.4	17.1
Y 14	293 48.9	358 42.1	11.9	294 10.8	34.3	124 07.3	40.1	191 50.1	17.1
15	308 51.3	13 41.3	. . 11.5	309 11.9	. . 33.7	139 09.6	. . 40.0	206 52.8	. . 17.1
16	323 53.8	28 40.4	11.0	324 13.1	33.0	154 11.9	40.0	221 55.4	17.1
17	338 56.2	43 39.6	10.5	339 14.3	32.4	169 14.2	39.9	236 58.1	17.2
18	353 58.7	58 38.7	S23 10.0	354 15.5	S 0 31.8	184 16.6	N 5 39.9	252 00.8	N22 17.2
19	9 01.2	73 37.8	09.5	9 16.6	31.1	199 18.9	39.8	267 03.4	17.2
20	24 03.6	88 37.0	09.0	24 17.8	30.5	214 21.2	39.8	282 06.1	17.2
21	39 06.1	103 36.1	. . 08.5	39 19.0	. . 29.8	229 23.5	. . 39.7	297 08.8	. . 17.2
22	54 08.6	118 35.3	08.0	54 20.1	29.2	244 25.8	39.7	312 11.4	17.3
23	69 11.0	133 34.4	07.6	69 21.3	28.5	259 28.2	39.6	327 14.1	17.3
16 00	84 13.5	148 33.6	S23 07.1	84 22.5	S 0 27.9	274 30.5	N 5 39.6	342 16.8	N22 17.3
01	99 16.0	163 32.7	06.6	99 23.7	27.3	289 32.8	39.5	357 19.5	17.3
02	114 18.4	178 31.9	06.1	114 24.8	26.6	304 35.1	39.5	12 22.1	17.3
03	129 20.9	193 31.0	. . 05.6	129 26.0	. . 26.0	319 37.5	. . 39.4	27 24.8	. . 17.4
04	144 23.4	208 30.2	05.1	144 27.2	25.3	334 39.8	39.4	42 27.5	17.4
05	159 25.8	223 29.3	04.6	159 28.3	24.7	349 42.1	39.3	57 30.1	17.4
06	174 28.3	238 28.5	S23 04.1	174 29.5	S 0 24.0	4 44.4	N 5 39.3	72 32.8	N22 17.4
07	189 30.7	253 27.6	03.6	189 30.7	23.4	19 46.8	39.2	87 35.5	17.4
T 08	204 33.2	268 26.8	03.1	204 31.8	22.8	34 49.1	39.2	102 38.1	17.4
U 09	219 35.7	283 25.9	. . 02.6	219 33.0	. . 22.1	49 51.4	. . 39.1	117 40.8	. . 17.5
E 10	234 38.1	298 25.1	02.1	234 34.2	21.5	64 53.7	39.1	132 43.5	17.5
S 11	249 40.6	313 24.2	01.5	249 35.3	20.8	79 56.1	39.1	147 46.2	17.5
D 12	264 43.1	328 23.4	S23 01.0	264 36.5	S 0 20.2	94 58.4	N 5 39.0	162 48.8	N22 17.5
A 13	279 45.5	343 22.5	00.5	279 37.7	19.5	110 00.7	39.0	177 51.5	17.5
Y 14	294 48.0	358 21.7	23 00.0	294 38.9	18.9	125 03.0	38.9	192 54.2	17.6
15	309 50.5	13 20.9	22 59.5	309 40.0	. . 18.2	140 05.4	. . 38.9	207 56.8	. . 17.6
16	324 52.9	28 20.0	59.0	324 41.2	17.6	155 07.7	38.8	222 59.5	17.6
17	339 55.4	43 19.2	58.5	339 42.4	17.0	170 10.0	38.8	238 02.2	17.6
18	354 57.8	58 18.3	S22 58.0	354 43.5	S 0 16.3	185 12.3	N 5 38.7	253 04.9	N22 17.6
19	10 00.3	73 17.5	57.4	9 44.7	15.7	200 14.7	38.7	268 07.5	17.7
20	25 02.8	88 16.6	56.9	24 45.9	15.0	215 17.0	38.6	283 10.2	17.7
21	40 05.2	103 15.8	. . 56.4	39 47.0	. . 14.4	230 19.3	. . 38.6	298 12.9	. . 17.7
22	55 07.7	118 14.9	55.9	54 48.2	13.7	245 21.6	38.5	313 15.5	17.7
23	70 10.2	133 14.1	55.4	69 49.4	13.1	260 24.0	38.5	328 18.2	17.7
17 00	85 12.6	148 13.3	S22 54.8	84 50.5	S 0 12.4	275 26.3	N 5 38.5	343 20.9	N22 17.8
01	100 15.1	163 12.4	54.3	99 51.7	11.8	290 28.6	38.4	358 23.6	17.8
02	115 17.6	178 11.6	53.8	114 52.8	11.2	305 31.0	38.4	13 26.2	17.8
03	130 20.0	193 10.7	. . 53.3	129 54.0	. . 10.5	320 33.3	. . 38.3	28 28.9	. . 17.8
04	145 22.5	208 09.9	52.7	144 55.2	09.9	335 35.6	38.3	43 31.6	17.8
05	160 25.0	223 09.1	52.2	159 56.3	09.2	350 38.0	38.2	58 34.3	17.8
06	175 27.4	238 08.2	S22 51.7	174 57.5	S 0 08.6	5 40.3	N 5 38.2	73 36.9	N22 17.9
W 07	190 29.9	253 07.4	51.1	189 58.7	07.9	20 42.6	38.1	88 39.6	17.9
E 08	205 32.3	268 06.6	50.6	204 59.8	07.3	35 44.9	38.1	103 42.3	17.9
D 09	220 34.8	283 05.7	. . 50.1	220 01.0	. . 06.6	50 47.3	. . 38.1	118 44.9	. . 17.9
N 10	235 37.3	298 04.9	49.5	235 02.2	06.0	65 49.6	38.0	133 47.6	17.9
E 11	250 39.7	313 04.0	49.0	250 03.3	05.4	80 51.9	38.0	148 50.3	18.0
S 12	265 42.2	328 03.2	S22 48.5	265 04.5	S 0 04.7	95 54.3	N 5 37.9	163 53.0	N22 18.0
D 13	280 44.7	343 02.4	47.9	280 05.6	04.1	110 56.6	37.9	178 55.6	18.0
A 14	295 47.1	358 01.5	47.4	295 06.8	03.4	125 58.9	37.8	193 58.3	18.0
Y 15	310 49.6	13 00.7	. . 46.9	310 08.0	. . 02.8	141 01.3	. . 37.8	209 01.0	. . 18.0
16	325 52.1	27 59.9	46.3	325 09.1	02.1	156 03.6	37.8	224 03.7	18.1
17	340 54.5	42 59.0	45.8	340 10.3	01.5	171 05.9	37.7	239 06.3	18.1
18	355 57.0	57 58.2	S22 45.2	355 11.5	S 0 00.8	186 08.3	N 5 37.7	254 09.0	N22 18.1
19	10 59.5	72 57.4	44.7	10 12.6	S 00.2	201 10.6	37.6	269 11.7	18.1
20	26 01.9	87 56.5	44.1	25 13.8	N 00.4	216 12.9	37.6	284 14.3	18.1
21	41 04.4	102 55.7	. . 43.6	40 14.9	. . 01.1	231 15.3	. . 37.5	299 17.0	. . 18.2
22	56 06.8	117 54.9	43.0	55 16.1	01.7	246 17.6	37.5	314 19.7	18.2
23	71 09.3	132 54.1	42.5	70 17.3	02.4	261 20.0	37.5	329 22.4	18.2
Mer. Pass.	h m 18 20.1	v −0.8	d 0.5	v 1.2	d 0.6	v 2.3	d 0.0	v 2.7	d 0.0

STARS		
Name	SHA	Dec
	° ′	° ′
Acamar	315 23.6	S40 17.5
Achernar	335 31.8	S57 13.
Acrux	173 18.2	S63 06.
Adhara	255 18.1	S28 58.5
Aldebaran	290 57.7	N16 31.
Alioth	166 27.2	N55 56.
Alkaid	153 04.9	N49 17.
Al Na'ir	27 53.1	S46 56.
Alnilam	275 53.6	S 1 11.
Alphard	218 03.3	S 8 40.
Alphecca	126 17.6	N26 42.
Alpheratz	357 51.3	N29 06.
Altair	62 15.8	N 8 52.
Ankaa	353 22.8	S42 17.
Antares	112 35.9	S26 26.
Arcturus	146 02.8	N19 09.
Atria	107 45.1	S69 02.
Avior	234 20.7	S59 31.
Bellatrix	278 39.7	N 6 21.
Betelgeuse	271 09.1	N 7 24.
Canopus	263 58.9	S52 41.
Capella	280 45.1	N46 00.
Deneb	49 37.0	N45 17.
Denebola	182 41.3	N14 33.
Diphda	349 03.2	S17 58.
Dubhe	194 00.4	N61 43.
Elnath	278 21.7	N28 36.
Eltanin	90 50.2	N51 29.
Enif	33 54.7	N 9 53.
Fomalhaut	15 32.2	S29 36.
Gacrux	172 09.7	S57 07.
Gienah	176 00.1	S17 33.
Hadar	148 59.2	S60 23.
Hamal	328 09.1	N23 29.
Kaus Aust.	83 54.1	S34 23.
Kochab	137 20.2	N74 08.
Markab	13 45.9	N15 13.
Menkar	314 22.6	N 4 06.
Menkent	148 16.8	S36 23.
Miaplacidus	221 41.0	S69 43.
Mirfak	308 50.8	N49 52.
Nunki	76 07.9	S26 17.
Peacock	53 31.3	S56 43.
Pollux	243 36.5	N28 01.
Procyon	245 07.3	N 5 13.
Rasalhague	96 13.8	N12 33.
Regulus	207 51.3	N11 56.
Rigel	281 18.9	S 8 11.
Rigil Kent.	140 02.8	S60 50.
Sabik	102 21.5	S15 43.
Schedar	349 49.2	N56 33.
Shaula	96 32.5	S37 06.
Sirius	258 40.0	S16 43.
Spica	158 39.3	S11 10.
Suhail	222 57.8	S43 26.
Vega	80 44.5	N38 47.
Zuben'ubi	137 14.0	S16 03.

	SHA	Mer. Pass.
	° ′	h m
Venus	64 20.1	14 07
Mars	0 09.0	18 21
Jupiter	190 17.0	5 41
Saturn	258 03.3	1 11

UT	SUN GHA	SUN Dec	MOON GHA	v	MOON Dec	d	HP
d h	° ′	° ′	° ′	′	° ′	′	′
15 00	181 19.3	S23 13.9	287 47.5	12.7	N15 41.2	12.2	56.4
01	196 19.0	14.1	302 19.2	12.8	15 29.0	12.3	56.4
02	211 18.7	14.2	316 51.0	12.7	15 16.7	12.4	56.4
03	226 18.4	. . 14.3	331 22.7	12.9	15 04.3	12.4	56.5
04	241 18.1	14.5	345 54.6	12.8	14 51.9	12.6	56.5
05	256 17.8	14.6	0 26.4	12.9	14 39.3	12.6	56.5
06	271 17.5	S23 14.8	14 58.3	12.9	N14 26.7	12.7	56.6
07	286 17.2	14.9	29 30.2	12.9	14 14.0	12.7	56.6
08	301 16.9	15.0	44 02.1	13.0	14 01.3	12.9	56.6
09	316 16.6	. . 15.2	58 34.1	13.0	13 48.4	12.9	56.6
10	331 16.3	15.3	73 06.1	13.0	13 35.5	13.0	56.7
11	346 16.0	15.4	87 38.1	13.0	13 22.5	13.1	56.7
12	1 15.7	S23 15.6	102 10.1	13.1	N13 09.4	13.1	56.7
13	16 15.4	15.7	116 42.2	13.0	12 56.3	13.3	56.8
14	31 15.1	15.8	131 14.2	13.1	12 43.0	13.2	56.8
15	46 14.8	. . 16.0	145 46.3	13.2	12 29.8	13.4	56.8
16	61 14.5	16.1	160 18.5	13.1	12 16.4	13.4	56.8
17	76 14.2	16.2	174 50.6	13.2	12 03.0	13.5	56.9
18	91 13.9	S23 16.3	189 22.8	13.1	N11 49.5	13.6	56.9
19	106 13.6	16.5	203 54.9	13.2	11 35.9	13.6	56.9
20	121 13.3	16.6	218 27.1	13.3	11 22.3	13.7	57.0
21	136 13.0	. . 16.7	232 59.4	13.2	11 08.6	13.8	57.0
22	151 12.7	16.9	247 31.6	13.2	10 54.8	13.8	57.0
23	166 12.4	17.0	262 03.8	13.3	10 41.0	13.9	57.1
16 00	181 12.1	S23 17.1	276 36.1	13.2	N10 27.1	14.0	57.1
01	196 11.8	17.2	291 08.3	13.3	10 13.1	14.0	57.1
02	211 11.5	17.3	305 40.6	13.3	9 59.1	14.1	57.2
03	226 11.2	. . 17.5	320 12.9	13.3	9 45.0	14.1	57.2
04	241 10.9	17.6	334 45.2	13.3	9 30.9	14.2	57.2
05	256 10.6	17.7	349 17.5	13.3	9 16.7	14.3	57.3
06	271 10.3	S23 17.8	3 49.8	13.4	N 9 02.4	14.3	57.3
07	286 10.0	17.9	18 22.2	13.3	8 48.1	14.3	57.3
08	301 09.7	18.1	32 54.5	13.3	8 33.8	14.5	57.4
09	316 09.4	. . 18.2	47 26.8	13.3	8 19.3	14.4	57.4
10	331 09.1	18.3	61 59.1	13.4	8 04.9	14.6	57.4
11	346 08.8	18.4	76 31.5	13.3	7 50.3	14.5	57.5
12	1 08.5	S23 18.5	91 03.8	13.3	N 7 35.8	14.7	57.5
13	16 08.2	18.6	105 36.1	13.4	7 21.1	14.7	57.5
14	31 07.9	18.7	120 08.5	13.3	7 06.4	14.7	57.6
15	46 07.6	. . 18.9	134 40.8	13.3	6 51.7	14.8	57.6
16	61 07.3	19.0	149 13.1	13.4	6 36.9	14.8	57.6
17	76 07.0	19.1	163 45.5	13.3	6 22.1	14.9	57.7
18	91 06.7	S23 19.2	178 17.8	13.3	N 6 07.2	14.9	57.7
19	106 06.4	19.3	192 50.1	13.3	5 52.3	15.0	57.7
20	121 06.1	19.4	207 22.4	13.3	5 37.3	15.0	57.8
21	136 05.8	. . 19.5	221 54.7	13.2	5 22.3	15.0	57.8
22	151 05.5	19.6	236 26.9	13.3	5 07.3	15.1	57.8
23	166 05.1	19.7	250 59.2	13.3	4 52.2	15.2	57.9
17 00	181 04.8	S23 19.8	265 31.5	13.2	N 4 37.0	15.1	57.9
01	196 04.5	19.9	280 03.7	13.2	4 21.9	15.2	57.9
02	211 04.2	20.0	294 35.9	13.2	4 06.7	15.3	58.0
03	226 03.9	. . 20.1	309 08.1	13.2	3 51.4	15.3	58.0
04	241 03.6	20.2	323 40.3	13.2	3 36.1	15.3	58.0
05	256 03.3	20.3	338 12.5	13.1	3 20.8	15.3	58.1
06	271 03.0	S23 20.4	352 44.6	13.2	N 3 05.5	15.4	58.1
07	286 02.7	20.5	7 16.8	13.1	2 50.1	15.5	58.1
08	301 02.4	20.6	21 48.9	13.0	2 34.6	15.4	58.2
09	316 02.1	. . 20.7	36 20.9	13.1	2 19.2	15.5	58.2
10	331 01.8	20.8	50 53.0	13.0	2 03.7	15.5	58.2
11	346 01.5	20.9	65 25.0	13.0	1 48.2	15.5	58.3
12	1 01.2	S23 21.0	79 57.0	13.0	N 1 32.7	15.6	58.3
13	16 00.9	21.1	94 29.0	12.9	1 17.1	15.6	58.3
14	31 00.6	21.2	109 00.9	12.9	1 01.5	15.6	58.4
15	46 00.3	. . 21.3	123 32.8	12.9	0 45.9	15.6	58.4
16	61 00.0	21.4	138 04.7	12.8	0 30.3	15.7	58.5
17	75 59.7	21.5	152 36.5	12.8	N 0 14.6	15.7	58.5
18	90 59.4	S23 21.6	167 08.3	12.7	S 0 01.1	15.7	58.5
19	105 59.0	21.6	181 40.0	12.8	0 16.8	15.7	58.6
20	120 58.7	21.7	196 11.8	12.6	0 32.5	15.7	58.6
21	135 58.4	. . 21.8	210 43.4	12.7	0 48.2	15.7	58.6
22	150 58.1	21.9	225 15.1	12.6	1 03.9	15.8	58.7
23	165 57.8	22.0	239 46.7	12.5	S 1 19.7	15.8	58.7
	SD 16.3	*d* 0.1	SD		15.5	15.7	15.9

Lat.	Twilight Naut.	Twilight Civil	Sunrise	Moonrise 15	Moonrise 16	Moonrise 17	Moonrise 18
°	h m	h m	h m	h m	h m	h m	h m
N 72	08 21	10 48	■■■	21 01	23 19	25 32	01 32
N 70	08 01	09 49	■■■	21 20	23 26	25 29	01 29
68	07 45	09 14	■■■	21 36	23 31	25 27	01 27
66	07 32	08 49	10 28	21 48	23 36	25 25	01 25
64	07 21	08 30	09 47	21 58	23 40	25 23	01 23
62	07 12	08 14	09 19	22 06	23 43	25 21	01 21
60	07 04	08 00	08 58	22 14	23 46	25 20	01 20
N 58	06 56	07 49	08 40	22 20	23 49	25 19	01 19
56	06 49	07 38	08 26	22 26	23 51	25 18	01 18
54	06 43	07 29	08 13	22 31	23 53	25 17	01 17
52	06 38	07 21	08 02	22 35	23 55	25 16	01 16
50	06 32	07 14	07 52	22 40	23 56	25 15	01 15
45	06 21	06 58	07 32	22 48	24 00	00 00	01 14
N 40	06 10	06 44	07 15	22 56	24 03	00 03	01 12
35	06 01	06 33	07 01	23 02	24 06	00 06	01 11
30	05 52	06 22	06 49	23 08	24 08	00 08	01 10
20	05 36	06 03	06 27	23 17	24 12	00 12	01 08
N 10	05 20	05 46	06 09	23 25	24 16	00 16	01 07
0	05 03	05 29	05 52	23 33	24 19	00 19	01 05
S 10	04 44	05 11	05 34	23 41	24 22	00 22	01 04
20	04 22	04 51	05 15	23 49	24 26	00 26	01 03
30	03 53	04 26	04 53	23 59	24 30	00 30	01 01
35	03 34	04 11	04 40	24 04	00 04	00 32	01 00
40	03 11	03 53	04 25	24 10	00 10	00 35	00 59
45	02 41	03 30	04 08	24 17	00 17	00 38	00 58
S 50	01 56	03 01	03 45	00 08	00 25	00 41	00 57
52	01 28	02 46	03 34	00 14	00 29	00 43	00 56
54	00 44	02 28	03 22	00 21	00 34	00 45	00 55
56	////	02 06	03 08	00 28	00 38	00 47	00 55
58	////	01 37	02 51	00 37	00 43	00 49	00 54
S 60	////	00 49	02 31	00 46	00 49	00 51	00 53

Lat.	Sunset	Twilight Civil	Twilight Naut.	Moonset 15	Moonset 16	Moonset 17	Moonset 18
°	h m	h m	h m	h m	h m	h m	h m
N 72	■■■	13 02	15 30	14 09	13 29	12 56	12 24
N 70	■■■	14 02	15 50	13 47	13 19	12 55	12 31
68	■■■	14 37	16 05	13 30	13 10	12 53	12 36
66	13 22	15 02	16 18	13 16	13 04	12 52	12 41
64	14 04	15 21	16 29	13 04	12 58	12 51	12 45
62	14 32	15 37	16 39	12 54	12 53	12 51	12 48
60	14 53	15 50	16 47	12 46	12 48	12 50	12 52
N 58	15 10	16 02	16 55	12 38	12 44	12 49	12 54
56	15 25	16 12	17 01	12 32	12 41	12 49	12 57
54	15 38	16 21	17 07	12 26	12 37	12 48	12 59
52	15 49	16 29	17 13	12 20	12 35	12 48	13 01
50	15 58	16 37	17 18	12 15	12 32	12 47	13 03
45	16 19	16 53	17 30	12 05	12 26	12 46	13 07
N 40	16 36	17 06	17 40	11 56	12 21	12 45	13 10
35	16 50	17 18	17 50	11 48	12 17	12 45	13 13
30	17 02	17 29	17 58	11 41	12 13	12 44	13 16
20	17 23	17 47	18 15	11 30	12 06	12 43	13 20
N 10	17 42	18 05	18 31	11 19	12 01	12 42	13 24
0	17 59	18 22	18 48	11 09	11 55	12 41	13 28
S 10	18 17	18 40	19 07	10 59	11 49	12 40	13 32
20	18 36	19 00	19 29	10 49	11 43	12 39	13 36
30	18 58	19 25	19 58	10 36	11 37	12 38	13 40
35	19 11	19 40	20 17	10 29	11 33	12 37	13 43
40	19 26	19 58	20 40	10 21	11 28	12 36	13 46
45	19 43	20 21	21 10	10 11	11 23	12 35	13 50
S 50	20 06	20 50	21 56	10 00	11 16	12 34	13 54
52	20 17	21 05	22 23	09 54	11 13	12 33	13 56
54	20 29	21 23	23 08	09 48	11 10	12 33	13 58
56	20 43	21 45	////	09 42	11 06	12 32	14 00
58	21 00	22 15	////	09 34	11 02	12 31	14 03
S 60	21 20	23 04	////	09 25	10 57	12 30	14 06

Day	SUN Eqn. of Time 00^h	SUN Eqn. of Time 12^h	SUN Mer. Pass.	MOON Mer. Pass. Upper	MOON Mer. Pass. Lower	Age	Phase
d	m s	m s	h m	h m	h m	d %	
15	05 18	05 03	11 55	04 58	17 21	22 63	
16	04 49	04 35	11 55	05 44	18 07	23 53	
17	04 20	04 05	11 56	06 30	18 53	24 42	

UT		ARIES	VENUS −4.0		MARS +0.0		JUPITER −2.1		SATURN −0.4		STARS		
		GHA	GHA	Dec	GHA	Dec	GHA	Dec	GHA	Dec	Name	SHA	Dec
d	h	° ′	° ′	° ′	° ′	° ′	° ′	° ′	° ′	° ′		° ′	° ′
18	00	86 11.8	147 53.2	S22 41.9	85 18.4	N 0 03.0	276 22.3	N 5 37.4	344 25.0	N22 18.2	Acamar	315 23.6	S40 17.
	01	101 14.2	162 52.4	41.4	100 19.6	03.7	291 24.6	37.4	359 27.7	18.2	Achernar	335 31.8	S57 13.
	02	116 16.7	177 51.6	40.8	115 20.7	04.3	306 27.0	37.3	14 30.4	18.3	Acrux	173 18.2	S63 06.
	03	131 19.2	192 50.7	40.3	130 21.9	05.0	321 29.3	37.3	29 33.1	18.3	Adhara	255 18.1	S28 58.
	04	146 21.6	207 49.9	39.7	145 23.0	05.6	336 31.6	37.2	44 35.7	18.3	Aldebaran	290 57.7	N16 31.
	05	161 24.1	222 49.1	39.2	160 24.2	06.3	351 34.0	37.2	59 38.4	18.3			
	06	176 26.6	237 48.3	S22 38.6	175 25.4	N 0 06.9	6 36.3	N 5 37.2	74 41.1	N22 18.3	Alioth	166 27.2	N55 56.
	07	191 29.0	252 47.4	38.1	190 26.5	07.5	21 38.6	37.1	89 43.8	18.3	Alkaid	153 04.9	N49 17.
T	08	206 31.5	267 46.6	37.5	205 27.7	08.2	36 41.0	37.1	104 46.4	18.4	Al Na'ir	27 53.2	S46 56.
H	09	221 33.9	282 45.8	36.9	220 28.8	08.8	51 43.3	37.0	119 49.1	18.4	Alnilam	275 53.6	S 1 11.
U	10	236 36.4	297 45.0	36.4	235 30.0	09.5	66 45.7	37.0	134 51.8	18.4	Alphard	218 03.3	S 8 40.
R	11	251 38.9	312 44.1	35.8	250 31.2	10.1	81 48.0	37.0	149 54.5	18.4			
S	12	266 41.3	327 43.3	S22 35.2	265 32.3	N 0 10.8	96 50.3	N 5 36.9	164 57.1	N22 18.4	Alphecca	126 17.6	N26 42.
D	13	281 43.8	342 42.5	34.7	280 33.5	11.4	111 52.7	36.9	179 59.8	18.5	Alpheratz	357 51.3	N29 06.
A	14	296 46.3	357 41.7	34.1	295 34.6	12.1	126 55.0	36.8	195 02.5	18.5	Altair	62 15.8	N 8 52.
Y	15	311 48.7	12 40.8	33.5	310 35.8	12.7	141 57.4	36.8	210 05.2	18.5	Ankaa	353 22.8	S42 17.
	16	326 51.2	27 40.0	33.0	325 36.9	13.4	156 59.7	36.8	225 07.8	18.5	Antares	112 35.9	S26 26.
	17	341 53.7	42 39.2	32.4	340 38.1	14.0	172 02.0	36.7	240 10.5	18.5			
	18	356 56.1	57 38.4	S22 31.8	355 39.2	N 0 14.7	187 04.4	N 5 36.7	255 13.2	N22 18.6	Arcturus	146 02.7	N19 09.
	19	11 58.6	72 37.6	31.3	10 40.4	15.3	202 06.7	36.6	270 15.9	18.6	Atria	107 45.1	S69 02.
	20	27 01.1	87 36.7	30.7	25 41.6	15.9	217 09.1	36.6	285 18.5	18.6	Avior	234 20.7	S59 31.
	21	42 03.5	102 35.9	30.1	40 42.7	16.6	232 11.4	36.6	300 21.2	18.6	Bellatrix	278 39.7	N 6 21.
	22	57 06.0	117 35.1	29.5	55 43.9	17.2	247 13.8	36.5	315 23.9	18.6	Betelgeuse	271 09.1	N 7 24.
	23	72 08.4	132 34.3	29.0	70 45.0	17.9	262 16.1	36.5	330 26.6	18.7			
19	00	87 10.9	147 33.5	S22 28.4	85 46.2	N 0 18.5	277 18.4	N 5 36.4	345 29.2	N22 18.7	Canopus	263 58.9	S52 41.
	01	102 13.4	162 32.6	27.8	100 47.3	19.2	292 20.8	36.4	0 31.9	18.7	Capella	280 45.1	N46 00.
	02	117 15.8	177 31.8	27.2	115 48.5	19.8	307 23.1	36.4	15 34.6	18.7	Deneb	49 37.0	N45 17.
	03	132 18.3	192 31.0	26.6	130 49.6	20.5	322 25.5	36.3	30 37.3	18.7	Denebola	182 41.2	N14 33.
	04	147 20.8	207 30.2	26.1	145 50.8	21.1	337 27.8	36.3	45 39.9	18.8	Diphda	349 03.3	S17 58.
	05	162 23.2	222 29.4	25.5	160 51.9	21.8	352 30.2	36.2	60 42.6	18.8			
	06	177 25.7	237 28.6	S22 24.9	175 53.1	N 0 22.4	7 32.5	N 5 36.2	75 45.3	N22 18.8	Dubhe	194 00.4	N61 43.
	07	192 28.2	252 27.7	24.3	190 54.3	23.1	22 34.8	36.2	90 48.0	18.8	Elnath	278 21.7	N28 36.
	08	207 30.6	267 26.9	23.7	205 55.4	23.7	37 37.2	36.1	105 50.6	18.8	Eltanin	90 50.2	N51 29.
F	09	222 33.1	282 26.1	23.1	220 56.6	24.4	52 39.5	36.1	120 53.3	18.9	Enif	33 54.7	N 9 53.
R	10	237 35.6	297 25.3	22.5	235 57.7	25.0	67 41.9	36.1	135 56.0	18.9	Fomalhaut	15 32.2	S29 36.
I	11	252 38.0	312 24.5	22.0	250 58.9	25.6	82 44.2	36.0	150 58.7	18.9			
D	12	267 40.5	327 23.7	S22 21.4	266 00.0	N 0 26.3	97 46.6	N 5 36.0	166 01.3	N22 18.9	Gacrux	172 09.7	S57 07.
A	13	282 42.9	342 22.9	20.8	281 01.2	26.9	112 48.9	35.9	181 04.0	18.9	Gienah	176 00.1	S17 33.
Y	14	297 45.4	357 22.1	20.2	296 02.3	27.6	127 51.3	35.9	196 06.7	18.9	Hadar	148 59.2	S60 23.
	15	312 47.9	12 21.3	19.6	311 03.5	28.2	142 53.6	35.9	211 09.4	19.0	Hamal	328 09.1	N23 29.
	16	327 50.3	27 20.4	19.0	326 04.6	28.9	157 56.0	35.8	226 12.1	19.0	Kaus Aust.	83 54.1	S34 23.
	17	342 52.8	42 19.6	18.4	341 05.8	29.5	172 58.3	35.8	241 14.7	19.0			
	18	357 55.3	57 18.8	S22 17.8	356 06.9	N 0 30.2	188 00.7	N 5 35.8	256 17.4	N22 19.0	Kochab	137 20.2	N74 08.
	19	12 57.7	72 18.0	17.2	11 08.1	30.8	203 03.0	35.7	271 20.1	19.0	Markab	13 45.9	N15 13.
	20	28 00.2	87 17.2	16.6	26 09.2	31.5	218 05.4	35.7	286 22.8	19.1	Menkar	314 22.6	N 4 06.
	21	43 02.7	102 16.4	16.0	41 10.4	32.1	233 07.7	35.6	301 25.4	19.1	Menkent	148 16.8	S36 23.
	22	58 05.1	117 15.6	15.4	56 11.5	32.8	248 10.1	35.6	316 28.1	19.1	Miaplacidus	221 41.0	S69 43.
	23	73 07.6	132 14.8	14.8	71 12.7	33.4	263 12.4	35.6	331 30.8	19.1			
20	00	88 10.0	147 14.0	S22 14.2	86 13.8	N 0 34.1	278 14.8	N 5 35.5	346 33.5	N22 19.1	Mirfak	308 50.8	N49 52.
	01	103 12.5	162 13.2	13.6	101 15.0	34.7	293 17.1	35.5	1 36.1	19.2	Nunki	76 07.9	S26 17.
	02	118 15.0	177 12.4	13.0	116 16.1	35.3	308 19.5	35.5	16 38.8	19.2	Peacock	53 31.3	S56 43.
	03	133 17.4	192 11.6	12.3	131 17.3	36.0	323 21.8	35.4	31 41.5	19.2	Pollux	243 36.5	N28 01.
	04	148 19.9	207 10.8	11.7	146 18.4	36.6	338 24.2	35.4	46 44.2	19.2	Procyon	245 07.3	N 5 13.
	05	163 22.4	222 10.0	11.1	161 19.6	37.3	353 26.5	35.4	61 46.9	19.2			
	06	178 24.8	237 09.2	S22 10.5	176 20.7	N 0 37.9	8 28.9	N 5 35.3	76 49.5	N22 19.3	Rasalhague	96 13.8	N12 33.
	07	193 27.3	252 08.4	09.9	191 21.9	38.6	23 31.2	35.3	91 52.2	19.3	Regulus	207 51.3	N11 56.
S	08	208 29.8	267 07.6	09.3	206 23.0	39.2	38 33.6	35.3	106 54.9	19.3	Rigel	281 18.9	S 8 11.
A	09	223 32.2	282 06.8	08.7	221 24.1	39.9	53 35.9	35.2	121 57.6	19.3	Rigil Kent.	140 02.8	S60 50.
T	10	238 34.7	297 06.0	08.1	236 25.3	40.5	68 38.3	35.2	137 00.2	19.3	Sabik	102 21.5	S15 43.
U	11	253 37.2	312 05.2	07.4	251 26.4	41.2	83 40.6	35.2	152 02.9	19.4			
R	12	268 39.6	327 04.4	S22 06.8	266 27.6	N 0 41.8	98 43.0	N 5 35.1	167 05.6	N22 19.4	Schedar	349 49.2	N56 33.
D	13	283 42.1	342 03.6	06.2	281 28.7	42.5	113 45.3	35.1	182 08.3	19.4	Shaula	96 32.5	S37 06.
A	14	298 44.5	357 02.8	05.6	296 29.9	43.1	128 47.7	35.0	197 11.0	19.4	Sirius	258 40.0	S16 43.
Y	15	313 47.0	12 02.0	05.0	311 31.0	43.8	143 50.0	35.0	212 13.6	19.4	Spica	158 39.3	S11 10.
	16	328 49.5	27 01.2	04.3	326 32.2	44.4	158 52.4	35.0	227 16.3	19.5	Suhail	222 57.8	S43 26.
	17	343 51.9	42 00.4	03.7	341 33.3	45.1	173 54.7	34.9	242 19.0	19.5			
	18	358 54.4	56 59.6	S22 03.1	356 34.5	N 0 45.7	188 57.1	N 5 34.9	257 21.7	N22 19.5	Vega	80 44.5	N38 47.
	19	13 56.9	71 58.8	02.5	11 35.6	46.4	203 59.5	34.9	272 24.4	19.5	Zuben'ubi	137 14.0	S16 03.
	20	28 59.3	86 58.0	01.8	26 36.7	47.0	219 01.8	34.8	287 27.0	19.5		SHA	Mer. Pas
	21	44 01.8	101 57.2	01.2	41 37.9	47.7	234 04.2	34.8	302 29.7	19.5		° ′	h m
	22	59 04.3	116 56.4	22 00.6	56 39.0	48.3	249 06.5	34.8	317 32.4	19.6	Venus	60 22.5	14 11
	23	74 06.7	131 55.6	S21 59.9	71 40.2	49.0	264 08.9	34.7	332 35.1	19.6	Mars	358 35.3	18 16
		h m									Jupiter	190 07.5	5 30
Mer. Pass.		18 08.3	v −0.8	d 0.6	v 1.2	d 0.6	v 2.3	d 0.0	v 2.7	d 0.0	Saturn	258 18.3	0 58

UT		SUN GHA	SUN Dec	MOON GHA	v	MOON Dec	d	HP
d	h	° ′	° ′	° ′	′	° ′	′	′
18	00	180 57.5	S23 22.1	254 18.2	12.5	S 1 35.5	15.8	58.7
	01	195 57.2	22.2	268 49.7	12.5	1 51.3	15.8	58.8
	02	210 56.9	22.2	283 21.2	12.4	2 07.1	15.8	58.8
	03	225 56.6	. . 22.3	297 52.6	12.3	2 22.9	15.8	58.8
	04	240 56.3	22.4	312 23.9	12.3	2 38.7	15.8	58.9
	05	255 56.0	22.5	326 55.2	12.3	2 54.5	15.8	58.9
	06	270 55.7	S23 22.6	341 26.5	12.1	S 3 10.3	15.8	58.9
	07	285 55.4	22.6	355 57.6	12.2	3 26.1	15.9	59.0
T	08	300 55.1	22.7	10 28.8	12.1	3 42.0	15.8	59.0
H	09	315 54.8	. . 22.8	24 59.9	12.0	3 57.8	15.8	59.0
U	10	330 54.4	22.9	39 30.9	11.9	4 13.6	15.9	59.1
R	11	345 54.1	23.0	54 01.8	11.9	4 29.5	15.8	59.1
S	12	0 53.8	S23 23.0	68 32.7	11.9	S 4 45.3	15.8	59.2
D	13	15 53.5	23.1	83 03.6	11.8	5 01.1	15.8	59.2
A	14	30 53.2	23.2	97 34.4	11.7	5 16.9	15.8	59.2
Y	15	45 52.9	. . 23.2	112 05.1	11.6	5 32.7	15.8	59.3
	16	60 52.6	23.3	126 35.7	11.6	5 48.5	15.8	59.3
	17	75 52.3	23.4	141 06.3	11.5	6 04.3	15.8	59.3
	18	90 52.0	S23 23.5	155 36.8	11.4	S 6 20.1	15.7	59.4
	19	105 51.7	23.5	170 07.2	11.4	6 35.8	15.8	59.4
	20	120 51.4	23.6	184 37.6	11.3	6 51.6	15.7	59.4
	21	135 51.1	. . 23.7	199 07.9	11.2	7 07.3	15.7	59.5
	22	150 50.8	23.7	213 38.1	11.1	7 23.0	15.7	59.5
	23	165 50.4	23.8	228 08.2	11.1	7 38.7	15.7	59.5
19	00	180 50.1	S23 23.9	242 38.3	10.9	S 7 54.4	15.6	59.6
	01	195 49.8	23.9	257 08.2	10.9	8 10.0	15.6	59.6
	02	210 49.5	24.0	271 38.1	10.8	8 25.6	15.6	59.6
	03	225 49.2	. . 24.1	286 07.9	10.8	8 41.2	15.6	59.7
	04	240 48.9	24.1	300 37.7	10.6	8 56.8	15.5	59.7
	05	255 48.6	24.2	315 07.3	10.6	9 12.3	15.5	59.7
	06	270 48.3	S23 24.2	329 36.9	10.5	S 9 27.8	15.5	59.8
	07	285 48.0	24.3	344 06.4	10.3	9 43.3	15.4	59.8
	08	300 47.7	24.4	358 35.7	10.3	9 58.7	15.4	59.8
F	09	315 47.4	. . 24.4	13 05.0	10.2	10 14.1	15.4	59.8
R	10	330 47.0	24.5	27 34.2	10.2	10 29.5	15.3	59.9
I	11	345 46.7	24.5	42 03.4	10.0	10 44.8	15.3	59.9
D	12	0 46.4	S23 24.6	56 32.4	9.9	S11 00.1	15.2	59.9
A	13	15 46.1	24.6	71 01.3	9.8	11 15.3	15.2	60.0
Y	14	30 45.8	24.7	85 30.1	9.8	11 30.5	15.1	60.0
	15	45 45.5	. . 24.7	99 58.9	9.6	11 45.6	15.1	60.0
	16	60 45.2	24.8	114 27.5	9.5	12 00.7	15.0	60.1
	17	75 44.9	24.8	128 56.0	9.5	12 15.7	15.0	60.1
	18	90 44.6	S23 24.9	143 24.5	9.3	S12 30.7	14.9	60.1
	19	105 44.3	24.9	157 52.8	9.2	12 45.6	14.9	60.1
	20	120 44.0	25.0	172 21.0	9.2	13 00.5	14.8	60.2
	21	135 43.6	. . 25.0	186 49.2	9.0	13 15.3	14.7	60.2
	22	150 43.3	25.1	201 17.2	8.9	13 30.0	14.7	60.2
	23	165 43.0	25.1	215 45.1	8.8	13 44.7	14.6	60.3
20	00	180 42.7	S23 25.2	230 12.9	8.7	S13 59.3	14.5	60.3
	01	195 42.4	25.2	244 40.6	8.6	14 13.8	14.5	60.3
	02	210 42.1	25.3	259 08.2	8.5	14 28.3	14.4	60.3
	03	225 41.8	. . 25.3	273 35.7	8.4	14 42.7	14.3	60.4
	04	240 41.5	25.4	288 03.1	8.2	14 57.0	14.3	60.4
	05	255 41.2	25.4	302 30.3	8.2	15 11.3	14.2	60.4
	06	270 40.9	S23 25.4	316 57.5	8.0	S15 25.5	14.0	60.4
S	07	285 40.5	25.5	331 24.5	7.9	15 39.5	14.1	60.5
A	08	300 40.2	25.5	345 51.4	7.8	15 53.6	13.9	60.5
T	09	315 39.9	. . 25.5	0 18.2	7.7	16 07.5	13.8	60.5
U	10	330 39.6	25.6	14 44.9	7.6	16 21.3	13.8	60.5
R	11	345 39.3	25.6	29 11.5	7.5	16 35.1	13.6	60.6
D	12	0 39.0	S23 25.7	43 38.0	7.3	S16 48.7	13.6	60.6
A	13	15 38.7	25.7	58 04.3	7.2	17 02.3	13.4	60.6
Y	14	30 38.4	25.7	72 30.5	7.1	17 15.7	13.4	60.6
	15	45 38.1	. . 25.8	86 56.6	7.0	17 29.1	13.3	60.7
	16	60 37.7	25.8	101 22.6	6.9	17 42.4	13.2	60.7
	17	75 37.4	25.8	115 48.5	6.8	17 55.6	13.0	60.7
	18	90 37.1	S23 25.9	130 14.3	6.6	S18 08.6	13.0	60.7
	19	105 36.8	25.9	144 39.9	6.5	18 21.6	12.8	60.8
	20	120 36.5	25.9	159 05.4	6.4	18 34.4	12.8	60.8
	21	135 36.2	. . 25.9	173 30.8	6.3	18 47.2	12.6	60.8
	22	150 35.9	26.0	187 56.1	6.1	18 59.8	12.5	60.8
	23	165 35.6	26.0	202 21.2	6.1	S19 12.3	12.4	60.8
		SD 16.3	d 0.1	SD	16.1	16.3		16.5

Lat.	Twilight Naut.	Twilight Civil	Sunrise	Moonrise 18	Moonrise 19	Moonrise 20	Moonrise 21
°	h m	h m	h m	h m	h m	h m	h m
N 72	08 24	10 55	■	01 32	03 55	06 57	■
N 70	08 04	09 52	■	01 29	03 40	06 14	■
68	07 48	09 17	■	01 27	03 29	05 46	08 55
66	07 35	08 52	10 33	01 25	03 19	05 25	07 53
64	07 24	08 32	09 50	01 23	03 11	05 08	07 18
62	07 14	08 16	09 22	01 21	03 04	04 54	06 54
60	07 06	08 03	09 00	01 20	02 58	04 43	06 34
N 58	06 58	07 51	08 43	01 19	02 53	04 33	06 18
56	06 52	07 41	08 28	01 18	02 48	04 24	06 05
54	06 45	07 32	08 15	01 17	02 44	04 16	05 53
52	06 40	07 23	08 04	01 16	02 41	04 10	05 42
50	06 34	07 16	07 54	01 15	02 37	04 03	05 33
45	06 23	07 00	07 34	01 14	02 30	03 50	05 14
N 40	06 12	06 46	07 17	01 12	02 24	03 39	04 58
35	06 03	06 35	07 03	01 11	02 19	03 30	04 45
30	05 54	06 24	06 50	01 10	02 14	03 22	04 34
20	05 38	06 05	06 29	01 08	02 07	03 09	04 15
N 10	05 21	05 48	06 11	01 07	02 00	02 57	03 58
0	05 04	05 31	05 53	01 05	01 54	02 46	03 42
S 10	04 45	05 12	05 36	01 04	01 48	02 35	03 27
20	04 23	04 52	05 17	01 03	01 41	02 23	03 11
30	03 54	04 27	04 55	01 01	01 34	02 10	02 52
35	03 35	04 12	04 42	01 00	01 30	02 03	02 41
40	03 12	03 54	04 26	00 59	01 25	01 54	02 29
45	02 41	03 31	04 08	00 58	01 20	01 44	02 15
S 50	01 56	03 01	03 46	00 57	01 13	01 32	01 57
52	01 28	02 46	03 35	00 56	01 10	01 27	01 49
54	00 41	02 28	03 23	00 55	01 07	01 21	01 40
56	////	02 06	03 08	00 55	01 03	01 14	01 30
58	////	01 36	02 52	00 54	01 00	01 07	01 18
S 60	////	00 46	02 31	00 53	00 55	00 59	01 05

Lat.	Sunset	Twilight Civil	Twilight Naut.	Moonset 18	Moonset 19	Moonset 20	Moonset 21
°	h m	h m	h m	h m	h m	h m	h m
N 72	■	12 59	15 30	12 24	11 45	10 34	■
N 70	■	14 01	15 50	12 31	12 02	11 19	■
68	■	14 36	16 06	12 36	12 17	11 49	10 41
66	13 21	15 02	16 19	12 41	12 28	12 12	11 44
64	14 03	15 21	16 30	12 45	12 38	12 30	12 19
62	14 32	15 37	16 39	12 48	12 47	12 45	12 45
60	14 53	15 51	16 48	12 52	12 54	12 58	13 05
N 58	15 11	16 03	16 55	12 54	13 00	13 09	13 22
56	15 26	16 13	17 02	12 57	13 06	13 18	13 37
54	15 38	16 22	17 08	12 59	13 11	13 27	13 49
52	15 49	16 30	17 14	13 01	13 16	13 35	14 00
50	15 59	16 38	17 19	13 03	13 20	13 42	14 10
45	16 20	16 54	17 31	13 07	13 30	13 56	14 30
N 40	16 37	17 07	17 42	13 10	13 37	14 09	14 47
35	16 51	17 19	17 51	13 13	13 44	14 19	15 01
30	17 03	17 30	18 00	13 16	13 50	14 29	15 14
20	17 25	17 49	18 16	13 20	14 00	14 45	15 35
N 10	17 43	18 06	18 32	13 24	14 09	14 59	15 54
0	18 01	18 23	18 49	13 28	14 18	15 12	16 11
S 10	18 18	18 41	19 08	13 32	14 27	15 25	16 28
20	18 37	19 02	19 31	13 36	14 36	15 40	16 47
30	18 59	19 27	20 00	13 40	14 46	15 56	17 09
35	19 12	19 42	20 19	13 43	14 52	16 06	17 22
40	19 27	20 00	20 42	13 46	14 59	16 17	17 36
45	19 45	20 23	21 13	13 50	15 08	16 30	17 54
S 50	20 08	20 53	21 58	13 54	15 17	16 45	18 16
52	20 19	21 08	22 26	13 56	15 22	16 53	18 26
54	20 31	21 26	23 13	13 58	15 27	17 01	18 38
56	20 45	21 48	////	14 00	15 33	17 11	18 52
58	21 02	22 18	////	14 03	15 39	17 21	19 07
S 60	21 23	23 09	////	14 06	15 46	17 34	19 27

Day	SUN Eqn. of Time 00^h	SUN Eqn. of Time 12^h	SUN Mer. Pass.	MOON Mer. Pass. Upper	MOON Mer. Pass. Lower	MOON Age	MOON Phase
d	m s	m s	h m	h m	h m	d	%
18	03 51	03 36	11 56	07 17	19 41	25	31
19	03 21	03 06	11 57	08 06	20 32	26	21
20	02 51	02 37	11 57	08 59	21 27	27	12

	UT	ARIES	VENUS −4.0		MARS +0.0		JUPITER −2.1		SATURN −0.4		STARS		
	d h	GHA	GHA	Dec	GHA	Dec	GHA	Dec	GHA	Dec	Name	SHA	Dec
		° ′	° ′	° ′	° ′	° ′	° ′	° ′	° ′	° ′		° ′	° ′
	21 00	89 09.2	146 54.8	S21 59.3	86 41.3	N 0 49.6	279 11.2	N 5 34.7	347 37.7	N22 19.6	Acamar	315 23.6	S40 17.5
	01	104 11.7	161 54.0	58.7	101 42.5	50.2	294 13.6	34.7	2 40.4	19.6	Achernar	335 31.8	S57 13.3
	02	119 14.1	176 53.2	58.0	116 43.6	50.9	309 16.0	34.6	17 43.1	19.6	Acrux	173 18.1	S63 06.9
	03	134 16.6	191 52.4	. . 57.4	131 44.8	. . 51.5	324 18.3	. . 34.6	32 45.8	. . 19.7	Adhara	255 18.0	S28 58.5
	04	149 19.0	206 51.7	56.8	146 45.9	52.2	339 20.7	34.6	47 48.5	19.7	Aldebaran	290 57.7	N16 31.1
	05	164 21.5	221 50.9	56.1	161 47.0	52.8	354 23.0	34.5	62 51.1	19.7			
	06	179 24.0	236 50.1	S21 55.5	176 48.2	N 0 53.5	9 25.4	N 5 34.5	77 53.8	N22 19.7	Alioth	166 27.1	N55 56.1
	07	194 26.4	251 49.3	54.8	191 49.3	54.1	24 27.7	34.5	92 56.5	19.7	Alkaid	153 04.8	N49 17.4
	08	209 28.9	266 48.5	54.2	206 50.5	54.8	39 30.1	34.5	107 59.2	19.8	Al Na'ir	27 53.2	S46 56.8
S	09	224 31.4	281 47.7	. . 53.6	221 51.6	. . 55.4	54 32.5	. . 34.4	123 01.9	. . 19.8	Alnilam	275 53.6	S 1 11.9
U	10	239 33.8	296 46.9	52.9	236 52.7	56.1	69 34.8	34.4	138 04.5	19.8	Alphard	218 03.2	S 8 40.4
N	11	254 36.3	311 46.1	52.3	251 53.9	56.7	84 37.2	34.4	153 07.2	19.8			
D	12	269 38.8	326 45.3	S21 51.6	266 55.0	N 0 57.4	99 39.5	N 5 34.3	168 09.9	N22 19.8	Alphecca	126 17.6	N26 42.0
A	13	284 41.2	341 44.6	51.0	281 56.2	58.0	114 41.9	34.3	183 12.6	19.9	Alpheratz	357 51.3	N29 06.8
Y	14	299 43.7	356 43.8	50.3	296 57.3	58.7	129 44.3	34.3	198 15.3	19.9	Altair	62 15.8	N 8 52.6
	15	314 46.2	11 43.0	. . 49.7	311 58.4	0 59.3	144 46.6	. . 34.2	213 17.9	. . 19.9	Ankaa	353 22.9	S42 17.4
	16	329 48.6	26 42.2	49.0	326 59.6	1 00.0	159 49.0	34.2	228 20.6	19.9	Antares	112 35.8	S26 26.4
	17	344 51.1	41 41.4	48.4	342 00.7	00.6	174 51.4	34.2	243 23.3	19.9			
	18	359 53.5	56 40.6	S21 47.7	357 01.9	N 1 01.3	189 53.7	N 5 34.1	258 26.0	N22 20.0	Arcturus	146 02.7	N19 09.6
	19	14 56.0	71 39.9	47.1	12 03.0	01.9	204 56.1	34.1	273 28.7	20.0	Atria	107 45.1	S69 02.0
	20	29 58.5	86 39.1	46.4	27 04.1	02.6	219 58.4	34.1	288 31.3	20.0	Avior	234 20.7	S59 31.1
	21	45 00.9	101 38.3	. . 45.8	42 05.3	. . 03.2	235 00.8	. . 34.0	303 34.0	. . 20.0	Bellatrix	278 39.7	N 6 21.3
	22	60 03.4	116 37.5	45.1	57 06.4	03.9	250 03.2	34.0	318 36.7	20.0	Betelgeuse	271 09.1	N 7 24.5
	23	75 05.9	131 36.7	44.5	72 07.6	04.5	265 05.5	34.0	333 39.4	20.1			
	22 00	90 08.3	146 36.0	S21 43.8	87 08.7	N 1 05.2	280 07.9	N 5 34.0	348 42.1	N22 20.1	Canopus	263 58.9	S52 41.8
	01	105 10.8	161 35.2	43.1	102 09.8	05.8	295 10.3	33.9	3 44.7	20.1	Capella	280 45.1	N46 00.2
	02	120 13.3	176 34.4	42.5	117 11.0	06.5	310 12.6	33.9	18 47.4	20.1	Deneb	49 37.1	N45 17.7
	03	135 15.7	191 33.6	. . 41.8	132 12.1	. . 07.1	325 15.0	. . 33.9	33 50.1	. . 20.1	Denebola	182 41.2	N14 33.0
	04	150 18.2	206 32.8	41.1	147 13.3	07.8	340 17.4	33.8	48 52.8	20.2	Diphda	349 03.3	S17 58.1
	05	165 20.7	221 32.1	40.5	162 14.4	08.4	355 19.7	33.8	63 55.5	20.2			
	06	180 23.1	236 31.3	S21 39.8	177 15.5	N 1 09.1	10 22.1	N 5 33.8	78 58.1	N22 20.2	Dubhe	194 00.3	N61 43.6
	07	195 25.6	251 30.5	39.2	192 16.7	09.7	25 24.5	33.8	94 00.8	20.2	Elnath	278 21.7	N28 36.7
	08	210 28.0	266 29.7	38.5	207 17.8	10.4	40 26.8	33.7	109 03.5	20.2	Eltanin	90 50.2	N51 29.2
M	09	225 30.5	281 29.0	. . 37.8	222 18.9	. . 11.0	55 29.2	. . 33.7	124 06.2	. . 20.3	Enif	33 54.7	N 9 53.5
O	10	240 33.0	296 28.2	37.1	237 20.1	11.7	70 31.6	33.7	139 08.9	20.3	Fomalhaut	15 32.2	S29 36.3
N	11	255 35.4	311 27.4	36.5	252 21.2	12.3	85 33.9	33.6	154 11.5	20.3			
D	12	270 37.9	326 26.6	S21 35.8	267 22.3	N 1 13.0	100 36.3	N 5 33.6	169 14.2	N22 20.3	Gacrux	172 09.6	S57 07.8
A	13	285 40.4	341 25.9	35.1	282 23.5	13.6	115 38.7	33.6	184 16.9	20.3	Gienah	176 00.1	S17 33.7
Y	14	300 42.8	356 25.1	34.5	297 24.6	14.2	130 41.0	33.5	199 19.6	20.3	Hadar	148 59.2	S60 23.2
	15	315 45.3	11 24.3	. . 33.8	312 25.7	. . 14.9	145 43.4	. . 33.5	214 22.3	. . 20.4	Hamal	328 09.1	N23 29.0
	16	330 47.8	26 23.6	33.1	327 26.9	15.5	160 45.8	33.5	229 24.9	20.4	Kaus Aust.	83 54.1	S34 23.1
	17	345 50.2	41 22.8	32.4	342 28.0	16.2	175 48.1	33.5	244 27.6	20.4			
	18	0 52.7	56 22.0	S21 31.8	357 29.2	N 1 16.8	190 50.5	N 5 33.4	259 30.3	N22 20.4	Kochab	137 20.1	N74 08.1
	19	15 55.2	71 21.2	31.1	12 30.3	17.5	205 52.9	33.4	274 33.0	20.4	Markab	13 45.9	N15 13.6
	20	30 57.6	86 20.5	30.4	27 31.4	18.1	220 55.2	33.4	289 35.7	20.5	Menkar	314 22.6	N 4 06.3
	21	46 00.1	101 19.7	. . 29.7	42 32.6	. . 18.8	235 57.6	. . 33.4	304 38.3	. . 20.5	Menkent	148 16.7	S36 23.2
	22	61 02.5	116 18.9	29.0	57 33.7	19.4	251 00.0	33.3	319 41.0	20.5	Miaplacidus	221 40.9	S69 43.7
	23	76 05.0	131 18.2	28.3	72 34.8	20.1	266 02.4	33.3	334 43.7	20.5			
	23 00	91 07.5	146 17.4	S21 27.7	87 36.0	N 1 20.7	281 04.7	N 5 33.3	349 46.4	N22 20.5	Mirfak	308 50.8	N49 52.7
	01	106 09.9	161 16.6	27.0	102 37.1	21.4	296 07.1	33.2	4 49.1	20.6	Nunki	76 07.9	S26 17.6
	02	121 12.4	176 15.9	26.3	117 38.2	22.0	311 09.5	33.2	19 51.8	20.6	Peacock	53 31.3	S56 43.6
	03	136 14.9	191 15.1	. . 25.6	132 39.4	. . 22.7	326 11.8	. . 33.2	34 54.4	. . 20.6	Pollux	243 36.5	N28 01.0
	04	151 17.3	206 14.3	24.9	147 40.5	23.3	341 14.2	33.2	49 57.1	20.6	Procyon	245 07.2	N 5 13.0
	05	166 19.8	221 13.6	24.2	162 41.6	24.0	356 16.6	33.1	64 59.8	20.6			
	06	181 22.3	236 12.8	S21 23.5	177 42.7	N 1 24.6	11 19.0	N 5 33.1	80 02.5	N22 20.7	Rasalhague	96 13.7	N12 33.4
	07	196 24.7	251 12.1	22.8	192 43.9	25.3	26 21.3	33.1	95 05.2	20.7	Regulus	207 51.3	N11 56.9
T	08	211 27.2	266 11.3	22.1	207 45.0	25.9	41 23.7	33.1	110 07.8	20.7	Rigel	281 18.9	S 8 11.8
U	09	226 29.7	281 10.5	. . 21.5	222 46.1	. . 26.6	56 26.1	. . 33.0	125 10.5	. . 20.7	Rigil Kent.	140 02.7	S60 50.8
E	10	241 32.1	296 09.8	20.8	237 47.3	27.2	71 28.5	33.0	140 13.2	20.7	Sabik	102 21.5	S15 43.8
S	11	256 34.6	311 09.0	20.1	252 48.4	27.9	86 30.8	33.0	155 15.9	20.8			
D	12	271 37.0	326 08.3	S21 19.4	267 49.5	N 1 28.5	101 33.2	N 5 33.0	170 18.6	N22 20.8	Schedar	349 49.2	N56 33.8
A	13	286 39.5	341 07.5	18.7	282 50.7	29.2	116 35.6	32.9	185 21.3	20.8	Shaula	96 32.5	S37 06.4
Y	14	301 42.0	356 06.7	18.0	297 51.8	29.8	131 38.0	32.9	200 23.9	20.8	Sirius	258 40.0	S16 43.2
	15	316 44.4	11 06.0	. . 17.3	312 52.9	. . 30.5	146 40.3	. . 32.9	215 26.6	. . 20.8	Spica	158 39.3	S11 10.8
	16	331 46.9	26 05.2	16.6	327 54.1	31.1	161 42.7	32.9	230 29.3	20.9	Suhail	222 57.7	S43 26.6
	17	346 49.4	41 04.5	15.9	342 55.2	31.8	176 45.1	32.8	245 32.0	20.9			
	18	1 51.8	56 03.7	S21 15.2	357 56.3	N 1 32.4	191 47.5	N 5 32.8	260 34.7	N22 20.9	Vega	80 44.5	N38 47.2
	19	16 54.3	71 03.0	14.5	12 57.4	33.1	206 49.8	32.8	275 37.3	20.9	Zuben'ubi	137 14.0	S16 03.4
	20	31 56.8	86 02.2	13.7	27 58.6	33.7	221 52.2	32.8	290 40.0	20.9			
	21	46 59.2	101 01.4	. . 13.0	42 59.7	. . 34.4	236 54.6	. . 32.7	305 42.7	. . 21.0			
	22	62 01.7	116 00.7	12.3	58 00.8	35.0	251 57.0	32.7	320 45.4	21.0			
	23	77 04.2	130 59.9	11.6	73 02.0	35.7	266 59.3	32.7	335 48.1	21.0			
	Mer. Pass.	h m 17 56.5	*v* −0.8	*d* 0.7	*v* 1.1	*d* 0.6	*v* 2.4	*d* 0.0	*v* 2.7	*d* 0.0			

	SHA	Mer. Pass
	° ′	h m
Venus	56 27.6	14 14
Mars	357 00.4	18 10
Jupiter	189 59.6	5 19
Saturn	258 33.7	0 45

UT		SUN GHA	SUN Dec	MOON GHA	v	MOON Dec	d	HP
	d h	° ′	° ′	° ′	′	° ′	′	′
	21 00	180 35.3	S23 26.0	216 46.3	5.9	S19 24.7	12.2	60.8
	01	195 35.0	26.0	231 11.2	5.8	19 36.9	12.2	60.9
	02	210 34.6	26.1	245 36.0	5.6	19 49.1	12.0	60.9
	03	225 34.3	. . 26.1	260 00.6	5.6	20 01.1	11.9	60.9
	04	240 34.0	26.1	274 25.2	5.4	20 13.0	11.8	60.9
	05	255 33.7	26.1	288 49.6	5.3	20 24.8	11.6	60.9
	06	270 33.4	S23 26.2	303 13.9	5.2	S20 36.4	11.5	61.0
	07	285 33.1	26.2	317 38.1	5.1	20 47.9	11.4	61.0
	08	300 32.8	26.2	332 02.2	5.0	20 59.3	11.2	61.0
S	09	315 32.5	. . 26.2	346 26.2	4.8	21 10.5	11.1	61.0
U	10	330 32.1	26.2	0 50.0	4.7	21 21.6	10.9	61.0
N	11	345 31.8	26.3	15 13.7	4.7	21 32.5	10.8	61.0
D	12	0 31.5	S23 26.3	29 37.4	4.4	S21 43.3	10.7	61.0
A	13	15 31.2	26.3	44 00.8	4.4	21 54.0	10.5	61.0
Y	14	30 30.9	26.3	58 24.2	4.3	22 04.5	10.4	61.1
	15	45 30.6	. . 26.3	72 47.5	4.1	22 14.9	10.2	61.1
	16	60 30.3	26.3	87 10.6	4.1	22 25.1	10.0	61.1
	17	75 30.0	26.3	101 33.7	3.9	22 35.1	9.9	61.1
	18	90 29.7	S23 26.3	115 56.6	3.8	S22 45.0	9.7	61.1
	19	105 29.3	26.4	130 19.4	3.7	22 54.7	9.6	61.1
	20	120 29.0	26.4	144 42.1	3.6	23 04.3	9.4	61.1
	21	135 28.7	. . 26.4	159 04.7	3.5	23 13.7	9.2	61.1
	22	150 28.4	26.4	173 27.2	3.4	23 22.9	9.1	61.1
	23	165 28.1	26.4	187 49.6	3.3	23 32.0	8.9	61.1
	22 00	180 27.8	S23 26.4	202 11.9	3.2	S23 40.9	8.8	61.2
	01	195 27.5	26.4	216 34.1	3.1	23 49.7	8.5	61.2
	02	210 27.2	26.4	230 56.2	3.0	23 58.2	8.4	61.2
	03	225 26.8	. . 26.4	245 18.2	2.9	24 06.6	8.2	61.2
	04	240 26.5	26.4	259 40.1	2.8	24 14.8	8.0	61.2
	05	255 26.2	26.4	274 01.9	2.7	24 22.8	7.9	61.2
	06	270 25.9	S23 26.4	288 23.6	2.6	S24 30.7	7.7	61.2
	07	285 25.6	26.4	302 45.2	2.6	24 38.4	7.4	61.2
	08	300 25.3	26.4	317 06.8	2.4	24 45.8	7.3	61.2
M	09	315 25.0	. . 26.4	331 28.2	2.4	24 53.1	7.2	61.2
O	10	330 24.7	26.4	345 49.6	2.3	25 00.3	6.9	61.2
N	11	345 24.4	26.4	0 10.9	2.2	25 07.2	6.7	61.2
D	12	0 24.0	S23 26.4	14 32.1	2.1	S25 13.9	6.6	61.2
A	13	15 23.7	26.4	28 53.2	2.0	25 20.5	6.3	61.2
Y	14	30 23.4	26.4	43 14.2	2.0	25 26.8	6.2	61.2
	15	45 23.1	. . 26.4	57 35.2	1.9	25 33.0	5.9	61.2
	16	60 22.8	26.4	71 56.1	1.9	25 38.9	5.8	61.2
	17	75 22.5	26.4	86 17.0	1.8	25 44.7	5.6	61.2
	18	90 22.2	S23 26.4	100 37.8	1.7	S25 50.3	5.3	61.2
	19	105 21.9	26.4	114 58.5	1.7	25 55.6	5.2	61.2
	20	120 21.5	26.4	129 19.2	1.6	26 00.8	4.9	61.2
	21	135 21.2	. . 26.3	143 39.8	1.5	26 05.7	4.8	61.2
	22	150 20.9	26.3	158 00.3	1.6	26 10.5	4.6	61.2
	23	165 20.6	26.3	172 20.9	1.4	26 15.1	4.3	61.2
	23 00	180 20.3	S23 26.3	186 41.3	1.4	S26 19.4	4.2	61.2
	01	195 20.0	26.3	201 01.7	1.4	26 23.6	3.9	61.1
	02	210 19.7	26.3	215 22.1	1.4	26 27.5	3.7	61.1
	03	225 19.4	. . 26.3	229 42.5	1.3	26 31.2	3.6	61.1
	04	240 19.0	26.2	244 02.8	1.3	26 34.8	3.3	61.1
	05	255 18.7	26.2	258 23.1	1.2	26 38.1	3.1	61.1
	06	270 18.4	S23 26.2	272 43.3	1.3	S26 41.2	2.9	61.1
	07	285 18.1	26.2	287 03.6	1.2	26 44.1	2.7	61.1
T	08	300 17.8	26.2	301 23.8	1.2	26 46.8	2.5	61.1
U	09	315 17.5	. . 26.1	315 44.0	1.2	26 49.3	2.2	61.1
E	10	330 17.2	26.1	330 04.2	1.1	26 51.5	2.1	61.1
S	11	345 16.9	26.1	344 24.3	1.2	26 53.6	1.8	61.0
D	12	0 16.5	S23 26.1	358 44.5	1.2	S26 55.4	1.7	61.0
A	13	15 16.2	26.1	13 04.7	1.1	26 57.1	1.4	61.0
Y	14	30 15.9	26.0	27 24.8	1.2	26 58.5	1.2	61.0
	15	45 15.6	. . 26.0	41 45.0	1.2	26 59.7	1.0	61.0
	16	60 15.3	26.0	56 05.2	1.2	27 00.7	0.8	61.0
	17	75 15.0	26.0	70 25.4	1.2	27 01.5	0.6	61.0
	18	90 14.7	S23 25.9	84 45.6	1.2	S27 02.1	0.4	60.9
	19	105 14.4	25.9	99 05.8	1.2	27 02.5	0.1	60.9
	20	120 14.0	25.9	113 26.0	1.3	27 02.6	0.0	60.9
	21	135 13.7	. . 25.8	127 46.3	1.3	27 02.6	0.3	60.9
	22	150 13.4	25.8	142 06.6	1.3	27 02.3	0.5	60.9
	23	165 13.1	25.8	156 26.9	1.4	S27 01.8	0.6	60.8
		SD 16.3	d 0.0	SD 16.6		16.7		16.6

Lat.	Twilight Naut.	Twilight Civil	Sunrise	Moonrise 21	Moonrise 22	Moonrise 23	Moonrise 24
°	h m	h m	h m	h m	h m	h m	h m
N 72	08 26	10 58	▬	▬	▬	▬	▬
N 70	08 06	09 55	▬	▬	▬	▬	▬
68	07 50	09 19	▬	08 55	▬	▬	▬
66	07 37	08 54	10 35	07 53	▬	▬	▬
64	07 26	08 34	09 52	07 18	09 54	▬	▬
62	07 16	08 18	09 24	06 54	09 01	10 58	11 55
60	07 08	08 04	09 02	06 34	08 28	10 09	11 11
N 58	07 00	07 53	08 45	06 18	08 04	09 38	10 41
56	06 53	07 42	08 30	06 05	07 45	09 14	10 19
54	06 47	07 33	08 17	05 53	07 29	08 55	10 00
52	06 41	07 25	08 06	05 42	07 15	08 39	09 45
50	06 36	07 18	07 56	05 33	07 03	08 25	09 31
45	06 24	07 02	07 35	05 14	06 38	07 57	09 03
N 40	06 14	06 48	07 19	04 58	06 18	07 35	08 41
35	06 04	06 36	07 04	04 45	06 02	07 16	08 23
30	05 56	06 26	06 52	04 34	05 48	07 00	08 07
20	05 39	06 07	06 31	04 15	05 24	06 34	07 40
N 10	05 23	05 49	06 12	03 58	05 03	06 11	07 17
0	05 06	05 32	05 55	03 42	04 44	05 49	06 56
S 10	04 47	05 14	05 37	03 27	04 25	05 28	06 34
20	04 24	04 53	05 18	03 11	04 05	05 05	06 11
30	03 55	04 28	04 56	02 52	03 41	04 39	05 45
35	03 36	04 13	04 43	02 41	03 28	04 24	05 29
40	03 13	03 55	04 28	02 29	03 12	04 06	05 11
45	02 42	03 32	04 10	02 15	02 53	03 44	04 49
S 50	01 57	03 02	03 47	01 57	02 30	03 17	04 21
52	01 29	02 47	03 36	01 49	02 20	03 04	04 07
54	00 41	02 29	03 24	01 40	02 07	02 49	03 51
56	////	02 07	03 10	01 30	01 53	02 31	03 33
58	////	01 37	02 53	01 18	01 36	02 10	03 10
S 60	////	00 45	02 32	01 05	01 17	01 43	02 40

Lat.	Sunset	Twilight Civil	Twilight Naut.	Moonset 21	Moonset 22	Moonset 23	Moonset 24
°	h m	h m	h m	h m	h m	h m	h m
N 72	▬	12 59	15 31	▬	▬	▬	▬
N 70	▬	14 02	15 51	▬	▬	▬	▬
68	▬	14 37	16 07	10 41	▬	▬	▬
66	13 22	15 03	16 20	11 44	▬	▬	▬
64	14 04	15 23	16 31	12 19	11 55	▬	▬
62	14 33	15 39	16 41	12 45	12 49	13 10	14 32
60	14 54	15 52	16 49	13 05	13 22	14 00	15 16
N 58	15 12	16 04	16 57	13 22	13 46	14 31	15 45
56	15 27	16 14	17 04	13 37	14 06	14 54	16 08
54	15 40	16 23	17 10	13 49	14 22	15 14	16 26
52	15 51	16 32	17 15	14 00	14 36	15 30	16 42
50	16 01	16 39	17 21	14 10	14 49	15 44	16 55
45	16 21	16 55	17 33	14 30	15 15	16 12	17 23
N 40	16 38	17 09	17 43	14 47	15 35	16 35	17 44
35	16 52	17 21	17 52	15 01	15 52	16 53	18 02
30	17 05	17 31	18 01	15 14	16 07	17 09	18 18
20	17 26	17 50	18 18	15 35	16 32	17 36	18 44
N 10	17 45	18 08	18 34	15 54	16 54	17 59	19 06
0	18 02	18 25	18 51	16 11	17 15	18 21	19 27
S 10	18 20	18 43	19 10	16 28	17 35	18 43	19 47
20	18 39	19 03	19 33	16 47	17 57	19 06	20 10
30	19 01	19 28	20 02	17 09	18 23	19 33	20 35
35	19 14	19 44	20 20	17 22	18 38	19 48	20 50
40	19 29	20 02	20 44	17 36	18 55	20 07	21 07
45	19 47	20 25	21 14	17 54	19 16	20 29	21 28
S 50	20 10	20 54	22 00	18 16	19 43	20 57	21 54
52	20 21	21 09	22 28	18 26	19 55	21 11	22 07
54	20 33	21 28	23 15	18 38	20 10	21 27	22 21
56	20 47	21 50	////	18 52	20 28	21 46	22 38
58	21 04	22 20	////	19 07	20 49	22 08	22 57
S 60	21 25	23 11	////	19 27	21 16	22 38	23 22

Day	SUN Eqn. of Time 00^h	SUN Eqn. of Time 12^h	SUN Mer. Pass.	MOON Mer. Pass. Upper	MOON Mer. Pass. Lower	MOON Age	MOON Phase
d	m s	m s	h m	h m	h m	d	%
21	02 22	02 07	11 58	09 57	22 27	28	5
22	01 52	01 37	11 58	10 59	23 32	29	1
23	01 22	01 07	11 59	12 05	24 39	00	0

UT		ARIES	VENUS −4.0		MARS +0.1		JUPITER −2.2		SATURN −0.4	
		GHA	GHA	Dec	GHA	Dec	GHA	Dec	GHA	Dec
d	h	° ′	° ′	° ′	° ′	° ′	° ′	° ′	° ′	° ′
24	00	92 06.6	145 59.2	S21 10.9	88 03.1	N 1 36.3	282 01.7	N 5 32.7	350 50.8	N22 21.0
	01	107 09.1	160 58.4	10.2	103 04.2	37.0	297 04.1	32.6	5 53.4	21.0
	02	122 11.5	175 57.7	09.5	118 05.3	37.6	312 06.5	32.6	20 56.1	21.1
	03	137 14.0	190 56.9	08.8	133 06.5	38.3	327 08.9	32.6	35 58.8	21.1
	04	152 16.5	205 56.2	08.1	148 07.6	38.9	342 11.2	32.6	51 01.5	21.1
	05	167 18.9	220 55.4	07.3	163 08.7	39.6	357 13.6	32.5	66 04.2	21.1
	06	182 21.4	235 54.7	S21 06.6	178 09.8	N 1 40.2	12 16.0	N 5 32.5	81 06.9	N22 21.1
WEDNESDAY	07	197 23.9	250 53.9	05.9	193 11.0	40.9	27 18.4	32.5	96 09.5	21.2
	08	212 26.3	265 53.2	05.2	208 12.1	41.5	42 20.8	32.5	111 12.2	21.2
	09	227 28.8	280 52.4	04.5	223 13.2	42.2	57 23.1	32.5	126 14.9	21.2
	10	242 31.3	295 51.7	03.8	238 14.3	42.8	72 25.5	32.4	141 17.6	21.2
	11	257 33.7	310 50.9	03.0	253 15.5	43.5	87 27.9	32.4	156 20.3	21.2
	12	272 36.2	325 50.2	S21 02.3	268 16.6	N 1 44.1	102 30.3	N 5 32.4	171 23.0	N22 21.2
	13	287 38.7	340 49.5	01.6	283 17.7	44.8	117 32.7	32.4	186 25.6	21.3
	14	302 41.1	355 48.7	00.9	298 18.8	45.4	132 35.1	32.3	201 28.3	21.3
	15	317 43.6	10 48.0	21 00.1	313 20.0	46.1	147 37.4	32.3	216 31.0	21.3
	16	332 46.0	25 47.2	20 59.4	328 21.1	46.7	162 39.8	32.3	231 33.7	21.3
	17	347 48.5	40 46.5	58.7	343 22.2	47.4	177 42.2	32.3	246 36.4	21.3
	18	2 51.0	55 45.7	S20 58.0	358 23.3	N 1 48.0	192 44.6	N 5 32.3	261 39.0	N22 21.4
	19	17 53.4	70 45.0	57.2	13 24.5	48.7	207 47.0	32.2	276 41.7	21.4
	20	32 55.9	85 44.3	56.5	28 25.6	49.3	222 49.4	32.2	291 44.4	21.4
	21	47 58.4	100 43.5	55.8	43 26.7	50.0	237 51.7	32.2	306 47.1	21.4
	22	63 00.8	115 42.8	55.0	58 27.8	50.6	252 54.1	32.2	321 49.8	21.4
	23	78 03.3	130 42.0	54.3	73 29.0	51.3	267 56.5	32.2	336 52.5	21.5
25	00	93 05.8	145 41.3	S20 53.6	88 30.1	N 1 51.9	282 58.9	N 5 32.1	351 55.1	N22 21.5
	01	108 08.2	160 40.6	52.8	103 31.2	52.6	298 01.3	32.1	6 57.8	21.5
	02	123 10.7	175 39.8	52.1	118 32.3	53.2	313 03.7	32.1	22 00.5	21.5
	03	138 13.1	190 39.1	51.3	133 33.5	53.9	328 06.1	32.1	37 03.2	21.5
	04	153 15.6	205 38.4	50.6	148 34.6	54.5	343 08.4	32.1	52 05.9	21.6
	05	168 18.1	220 37.6	49.9	163 35.7	55.2	358 10.8	32.0	67 08.6	21.6
	06	183 20.5	235 36.9	S20 49.1	178 36.8	N 1 55.9	13 13.2	N 5 32.0	82 11.3	N22 21.6
	07	198 23.0	250 36.1	48.4	193 37.9	56.5	28 15.6	32.0	97 13.9	21.6
THURSDAY	08	213 25.5	265 35.4	47.6	208 39.1	57.2	43 18.0	32.0	112 16.6	21.6
	09	228 27.9	280 34.7	46.9	223 40.2	57.8	58 20.4	32.0	127 19.3	21.7
	10	243 30.4	295 33.9	46.2	238 41.3	58.5	73 22.8	31.9	142 22.0	21.7
	11	258 32.9	310 33.2	45.4	253 42.4	59.1	88 25.2	31.9	157 24.7	21.7
	12	273 35.3	325 32.5	S20 44.7	268 43.5	N 1 59.8	103 27.6	N 5 31.9	172 27.4	N22 21.7
	13	288 37.8	340 31.8	43.9	283 44.7	2 00.4	118 29.9	31.9	187 30.0	21.7
	14	303 40.3	355 31.0	43.2	298 45.8	01.1	133 32.3	31.9	202 32.7	21.8
	15	318 42.7	10 30.3	42.4	313 46.9	01.7	148 34.7	31.8	217 35.4	21.8
	16	333 45.2	25 29.6	41.7	328 48.0	02.4	163 37.1	31.8	232 38.1	21.8
	17	348 47.6	40 28.8	40.9	343 49.1	03.0	178 39.5	31.8	247 40.8	21.8
	18	3 50.1	55 28.1	S20 40.2	358 50.3	N 2 03.7	193 41.9	N 5 31.8	262 43.5	N22 21.8
	19	18 52.6	70 27.4	39.4	13 51.4	04.3	208 44.3	31.8	277 46.1	21.9
	20	33 55.0	85 26.7	38.6	28 52.5	05.0	223 46.7	31.7	292 48.8	21.9
	21	48 57.5	100 25.9	37.9	43 53.6	05.6	238 49.1	31.7	307 51.5	21.9
	22	64 00.0	115 25.2	37.1	58 54.7	06.3	253 51.5	31.7	322 54.2	21.9
	23	79 02.4	130 24.5	36.4	73 55.9	06.9	268 53.9	31.7	337 56.9	21.9
26	00	94 04.9	145 23.8	S20 35.6	88 57.0	N 2 07.6	283 56.3	N 5 31.7	352 59.6	N22 22.0
	01	109 07.4	160 23.0	34.9	103 58.1	08.2	298 58.6	31.7	8 02.2	22.0
	02	124 09.8	175 22.3	34.1	118 59.2	08.9	314 01.0	31.6	23 04.9	22.0
	03	139 12.3	190 21.6	33.3	134 00.3	09.5	329 03.4	31.6	38 07.6	22.0
	04	154 14.8	205 20.9	32.6	149 01.4	10.2	344 05.8	31.6	53 10.3	22.0
	05	169 17.2	220 20.1	31.8	164 02.6	10.8	359 08.2	31.6	68 13.0	22.1
	06	184 19.7	235 19.4	S20 31.0	179 03.7	N 2 11.5	14 10.6	N 5 31.6	83 15.7	N22 22.1
	07	199 22.1	250 18.7	30.3	194 04.8	12.1	29 13.0	31.6	98 18.4	22.1
	08	214 24.6	265 18.0	29.5	209 05.9	12.8	44 15.4	31.5	113 21.0	22.1
FRIDAY	09	229 27.1	280 17.3	28.7	224 07.0	13.4	59 17.8	31.5	128 23.7	22.1
	10	244 29.5	295 16.5	28.0	239 08.1	14.1	74 20.2	31.5	143 26.4	22.2
	11	259 32.0	310 15.8	27.2	254 09.3	14.7	89 22.6	31.5	158 29.1	22.2
	12	274 34.5	325 15.1	S20 26.4	269 10.4	N 2 15.4	104 25.0	N 5 31.5	173 31.8	N22 22.2
	13	289 36.9	340 14.4	25.6	284 11.5	16.0	119 27.4	31.5	188 34.5	22.2
	14	304 39.4	355 13.7	24.9	299 12.6	16.7	134 29.8	31.4	203 37.1	22.2
	15	319 41.9	10 13.0	24.1	314 13.7	17.3	149 32.2	31.4	218 39.8	22.3
	16	334 44.3	25 12.3	23.3	329 14.8	18.0	164 34.6	31.4	233 42.5	22.3
	17	349 46.8	40 11.5	22.5	344 15.9	18.6	179 37.0	31.4	248 45.2	22.3
	18	4 49.3	55 10.8	S20 21.8	359 17.1	N 2 19.3	194 39.4	N 5 31.4	263 47.9	N22 22.3
	19	19 51.7	70 10.1	21.0	14 18.2	19.9	209 41.8	31.4	278 50.6	22.3
	20	34 54.2	85 09.4	20.2	29 19.3	20.6	224 44.2	31.3	293 53.3	22.4
	21	49 56.6	100 08.7	19.4	44 20.4	21.2	239 46.6	31.3	308 55.9	22.4
	22	64 59.1	115 08.0	18.6	59 21.5	21.9	254 49.0	31.3	323 58.6	22.4
	23	80 01.6	130 07.3	17.9	74 22.6	22.5	269 51.4	31.3	339 01.3	22.4
Mer. Pass.		h m 17 44.7	v −0.7	d 0.7	v 1.1	d 0.7	v 2.4	d 0.0	v 2.7	d 0.0

STARS		
Name	SHA	Dec
	° ′	° ′
Acamar	315 23.6	S40 17.5
Achernar	335 31.8	S57 13.3
Acrux	173 18.1	S63 06.9
Adhara	255 18.0	S28 58.5
Aldebaran	290 57.6	N16 31.1
Alioth	166 27.1	N55 56.1
Alkaid	153 04.8	N49 17.4
Al Na'ir	27 53.2	S46 56.8
Alnilam	275 53.6	S 1 11.9
Alphard	218 03.2	S 8 40.4
Alphecca	126 17.6	N26 42.0
Alpheratz	357 51.3	N29 06.8
Altair	62 15.8	N 8 52.6
Ankaa	353 22.9	S42 17.4
Antares	112 35.8	S26 26.4
Arcturus	146 02.7	N19 09.6
Atria	107 45.0	S69 02.0
Avior	234 20.7	S59 31.1
Bellatrix	278 39.7	N 6 21.3
Betelgeuse	271 09.0	N 7 24.5
Canopus	263 58.9	S52 41.8
Capella	280 45.0	N46 00.2
Deneb	49 37.1	N45 17.7
Denebola	182 41.2	N14 33.6
Diphda	349 03.3	S17 58.
Dubhe	194 00.3	N61 43.6
Elnath	278 21.7	N28 36.7
Eltanin	90 50.2	N51 29.2
Enif	33 54.7	N 9 53.5
Fomalhaut	15 32.2	S29 36.
Gacrux	172 09.6	S57 07.8
Gienah	176 00.1	S17 33.7
Hadar	148 59.1	S60 23.
Hamal	328 09.1	N23 29.
Kaus Aust.	83 54.1	S34 23.
Kochab	137 20.1	N74 08.
Markab	13 45.9	N15 13.
Menkar	314 22.6	N 4 06.
Menkent	148 16.7	S36 23.
Miaplacidus	221 40.9	S69 43.
Mirfak	308 50.8	N49 52.
Nunki	76 07.9	S26 17.
Peacock	53 31.3	S56 43.
Pollux	243 36.5	N28 01.
Procyon	245 07.2	N 5 13.
Rasalhague	96 13.7	N12 33.
Regulus	207 51.2	N11 56.
Rigel	281 18.9	S 8 11.
Rigil Kent.	140 02.7	S60 50.
Sabik	102 21.5	S15 43.
Schedar	349 49.2	N56 33.
Shaula	96 32.5	S37 06.
Sirius	258 40.0	S16 43.
Spica	158 39.3	S11 10.
Suhail	222 57.7	S43 26.
Vega	80 44.5	N38 47.
Zuben'ubi	137 14.0	S16 03.

	SHA	Mer. Pas
	° ′	h m
Venus	52 35.5	14 18
Mars	355 24.3	18 05
Jupiter	189 53.1	5 07
Saturn	258 49.4	0 32

UT	SUN GHA	SUN Dec	MOON GHA	v	MOON Dec	d	HP
d h	° ′	° ′	° ′	′	° ′	′	′
24 00	180 12.8	S23 25.7	170 47.3	1.4	S27 01.2	0.9	60.8
01	195 12.5	25.7	185 07.7	1.4	27 00.3	1.1	60.8
02	210 12.2	25.7	199 28.1	1.5	26 59.2	1.3	60.8
03	225 11.9	. . 25.6	213 48.6	1.5	26 57.9	1.5	60.8
04	240 11.5	25.6	228 09.1	1.6	26 56.4	1.7	60.7
05	255 11.2	25.6	242 29.7	1.6	26 54.7	2.0	60.7
06	270 10.9	S23 25.5	256 50.3	1.7	S26 52.7	2.1	60.7
07	285 10.6	25.5	271 11.0	1.8	26 50.6	2.3	60.7
08	300 10.3	25.4	285 31.8	1.8	26 48.3	2.5	60.7
09	315 10.0	. . 25.4	299 52.6	1.9	26 45.8	2.8	60.6
10	330 09.7	25.4	314 13.5	2.0	26 43.0	2.9	60.6
11	345 09.4	25.3	328 34.5	2.0	26 40.1	3.1	60.6
12	0 09.1	S23 25.3	342 55.5	2.1	S26 37.0	3.3	60.6
13	15 08.7	25.2	357 16.6	2.2	26 33.7	3.6	60.5
14	30 08.4	25.2	11 37.8	2.2	26 30.1	3.7	60.5
15	45 08.1	. . 25.1	25 59.0	2.4	26 26.4	3.9	60.5
16	60 07.8	25.1	40 20.4	2.4	26 22.5	4.1	60.5
17	75 07.5	25.1	54 41.8	2.5	26 18.4	4.3	60.4
18	90 07.2	S23 25.0	69 03.3	2.6	S26 14.1	4.4	60.4
19	105 06.9	25.0	83 24.9	2.7	26 09.7	4.7	60.4
20	120 06.6	24.9	97 46.6	2.8	26 05.0	4.9	60.3
21	135 06.2	. . 24.9	112 08.4	2.9	26 00.1	5.0	60.3
22	150 05.9	24.8	126 30.3	3.0	25 55.1	5.2	60.3
23	165 05.6	24.8	140 52.3	3.1	25 49.9	5.4	60.3
25 00	180 05.3	S23 24.7	155 14.4	3.2	S25 44.5	5.6	60.2
01	195 05.0	24.6	169 36.6	3.3	25 38.9	5.8	60.2
02	210 04.7	24.6	183 58.9	3.5	25 33.1	5.9	60.2
03	225 04.4	. . 24.5	198 21.4	3.5	25 27.2	6.1	60.1
04	240 04.1	24.5	212 43.9	3.6	25 21.1	6.3	60.1
05	255 03.8	24.4	227 06.5	3.8	25 14.8	6.4	60.1
06	270 03.4	S23 24.4	241 29.3	3.9	S25 08.4	6.6	60.0
07	285 03.1	24.3	255 52.2	4.0	25 01.8	6.8	60.0
08	300 02.8	24.2	270 15.2	4.1	24 55.0	7.0	60.0
09	315 02.5	. . 24.2	284 38.3	4.2	24 48.0	7.1	59.9
10	330 02.2	24.1	299 01.5	4.3	24 40.9	7.3	59.9
11	345 01.9	24.1	313 24.8	4.5	24 33.6	7.4	59.9
12	0 01.6	S23 24.0	327 48.3	4.6	S24 26.2	7.6	59.8
13	15 01.3	23.9	342 11.9	4.7	24 18.6	7.7	59.8
14	30 01.0	23.9	356 35.6	4.9	24 10.9	7.9	59.8
15	45 00.6	. . 23.8	10 59.5	5.0	24 03.0	8.1	59.7
16	60 00.3	23.7	25 23.5	5.1	23 54.9	8.2	59.7
17	75 00.0	23.7	39 47.6	5.2	23 46.7	8.3	59.7
18	89 59.7	S23 23.6	54 11.8	5.4	S23 38.4	8.5	59.6
19	104 59.4	23.5	68 36.2	5.5	23 29.9	8.7	59.6
20	119 59.1	23.5	83 00.7	5.6	23 21.2	8.7	59.6
21	134 58.8	. . 23.4	97 25.3	5.8	23 12.5	9.0	59.5
22	149 58.5	23.3	111 50.1	5.9	23 03.5	9.0	59.5
23	164 58.2	23.3	126 15.0	6.0	22 54.5	9.2	59.4
26 00	179 57.8	S23 23.2	140 40.0	6.2	S22 45.3	9.3	59.4
01	194 57.5	23.1	155 05.2	6.3	22 36.0	9.5	59.4
02	209 57.2	23.0	169 30.5	6.4	22 26.5	9.6	59.3
03	224 56.9	. . 23.0	183 55.9	6.6	22 16.9	9.7	59.3
04	239 56.6	22.9	198 21.5	6.7	22 07.2	9.8	59.3
05	254 56.3	22.8	212 47.2	6.8	21 57.4	9.9	59.2
06	269 56.0	S23 22.7	227 13.0	7.0	S21 47.5	10.1	59.2
07	284 55.7	22.7	241 39.0	7.1	21 37.4	10.2	59.1
08	299 55.4	22.6	256 05.1	7.2	21 27.2	10.3	59.1
09	314 55.0	. . 22.5	270 31.3	7.4	21 16.9	10.5	59.1
10	329 54.7	22.4	284 57.7	7.5	21 06.4	10.5	59.0
11	344 54.4	22.3	299 24.2	7.7	20 55.9	10.6	59.0
12	359 54.1	S23 22.3	313 50.9	7.8	S20 45.3	10.8	58.9
13	14 53.8	22.2	328 17.7	7.9	20 34.5	10.9	58.9
14	29 53.5	22.1	342 44.6	8.0	20 23.6	10.9	58.9
15	44 53.2	. . 22.0	357 11.6	8.2	20 12.7	11.1	58.8
16	59 52.9	21.9	11 38.8	8.3	20 01.6	11.2	58.8
17	74 52.6	21.8	26 06.1	8.5	19 50.4	11.2	58.8
18	89 52.3	S23 21.8	40 33.6	8.6	S19 39.2	11.4	58.7
19	104 51.9	21.7	55 01.2	8.7	19 27.8	11.5	58.7
20	119 51.6	21.6	69 28.9	8.8	19 16.3	11.5	58.6
21	134 51.3	. . 21.5	83 56.7	9.0	19 04.8	11.6	58.6
22	149 51.0	21.4	98 24.7	9.1	18 53.2	11.8	58.6
23	164 50.7	21.3	112 52.8	9.3	S18 41.4	11.8	58.5
	SD 16.3	*d* 0.1	SD		16.5	16.3	16.1

Lat.	Twilight Naut.	Twilight Civil	Sunrise	Moonrise 24	25	26	27
°	h m	h m	h m	h m	h m	h m	h m
N 72	08 27	10 58	■	■	■	■	14 00
N 70	08 07	09 55	■	■	■	■	13 24
68	07 51	09 20	■	■	■	13 42	12 58
66	07 38	08 55	10 36	■	■	13 00	12 37
64	07 27	08 35	09 53	■	12 47	12 31	12 21
62	07 17	08 19	09 25	11 55	12 07	12 08	12 08
60	07 09	08 06	09 03	11 11	11 38	11 50	11 56
N 58	07 01	07 54	08 46	10 41	11 17	11 35	11 46
56	06 55	07 44	08 31	10 19	10 59	11 22	11 37
54	06 48	07 35	08 18	10 00	10 43	11 11	11 29
52	06 43	07 26	08 07	09 45	10 30	11 01	11 22
50	06 37	07 19	07 57	09 31	10 19	10 52	11 16
45	06 26	07 03	07 37	09 03	09 54	10 33	11 02
N 40	06 15	06 49	07 20	08 41	09 35	10 17	10 50
35	06 06	06 38	07 06	08 23	09 18	10 04	10 41
30	05 57	06 27	06 53	08 07	09 04	09 52	10 32
20	05 41	06 08	06 32	07 40	08 40	09 32	10 17
N 10	05 24	05 51	06 14	07 17	08 19	09 15	10 04
0	05 07	05 34	05 56	06 56	08 00	08 58	09 52
S 10	04 48	05 15	05 39	06 34	07 40	08 42	09 39
20	04 26	04 55	05 20	06 11	07 19	08 25	09 26
30	03 57	04 30	04 58	05 45	06 55	08 04	09 11
35	03 38	04 15	04 44	05 29	06 40	07 52	09 02
40	03 15	03 57	04 29	05 11	06 23	07 38	08 51
45	02 44	03 34	04 11	04 49	06 03	07 22	08 39
S 50	01 59	03 04	03 49	04 21	05 38	07 02	08 25
52	01 31	02 49	03 38	04 07	05 26	06 52	08 18
54	00 44	02 31	03 26	03 51	05 12	06 41	08 10
56	////	02 09	03 11	03 33	04 56	06 29	08 02
58	////	01 39	02 55	03 10	04 36	06 15	07 52
S 60	////	00 48	02 34	02 40	04 12	05 57	07 41

Lat.	Sunset	Twilight Civil	Twilight Naut.	Moonset 24	25	26	27
°	h m	h m	h m	h m	h m	h m	h m
N 72	■	13 02	15 33	■	■	■	18 32
N 70	■	14 05	15 53	■	■	■	19 06
68	■	14 40	16 09	■	■	16 59	19 30
66	13 24	15 05	16 22	■	■	17 40	19 49
64	14 07	15 25	16 33	■	15 52	18 08	20 04
62	14 35	15 41	16 43	14 32	16 32	18 29	20 16
60	14 56	15 54	16 51	15 16	17 00	18 46	20 27
N 58	15 14	16 06	16 59	15 45	17 21	19 01	20 36
56	15 29	16 16	17 05	16 08	17 38	19 13	20 44
54	15 41	16 25	17 11	16 26	17 53	19 24	20 51
52	15 52	16 33	17 17	16 42	18 06	19 33	20 57
50	16 02	16 41	17 22	16 55	18 17	19 41	21 03
45	16 23	16 57	17 34	17 23	18 40	19 59	21 15
N 40	16 40	17 10	17 45	17 44	18 59	20 14	21 25
35	16 54	17 22	17 54	18 02	19 14	20 26	21 34
30	17 06	17 33	18 03	18 18	19 28	20 36	21 41
20	17 28	17 52	18 19	18 44	19 51	20 54	21 54
N 10	17 46	18 09	18 35	19 06	20 10	21 10	22 05
0	18 04	18 26	18 52	19 27	20 28	21 24	22 15
S 10	18 21	18 44	19 11	19 47	20 46	21 39	22 25
20	18 40	19 05	19 34	20 10	21 06	21 54	22 36
30	19 02	19 30	20 03	20 35	21 28	22 12	22 48
35	19 15	19 45	20 22	20 50	21 41	22 22	22 55
40	19 30	20 03	20 45	21 07	21 55	22 33	23 03
45	19 48	20 26	21 15	21 28	22 13	22 46	23 13
S 50	20 11	20 55	22 01	21 54	22 34	23 03	23 24
52	20 22	21 11	22 29	22 07	22 44	23 10	23 29
54	20 34	21 28	23 15	22 21	22 56	23 19	23 34
56	20 48	21 51	////	22 38	23 09	23 28	23 41
58	21 05	22 21	////	22 57	23 24	23 39	23 48
S 60	21 25	23 11	////	23 22	23 42	23 51	23 55

Day	SUN Eqn. of Time 00^h	Eqn. of Time 12^h	Mer. Pass.	MOON Mer. Pass. Upper	Mer. Pass. Lower	Age	Phase
d	m s	m s	h m	h m	h m	d %	
24	00 52	00 37	11 59	13 11	00 39	01 2	●
25	00 22	00 07	12 00	14 14	01 43	02 6	
26	00 08	00 23	12 00	15 12	02 44	03 13	

UT		ARIES	VENUS −4.0		MARS +0.2		JUPITER −2.2		SATURN −0.4	
		GHA	GHA	Dec	GHA	Dec	GHA	Dec	GHA	Dec
d	h	° ′	° ′	° ′	° ′	° ′	° ′	° ′	° ′	° ′
27	00	95 04.0	145 06.6	S20 17.1	89 23.7	N 2 23.2	284 53.8	N 5 31.3	354 04.0	N22 22.4
	01	110 06.5	160 05.8	16.3	104 24.9	23.8	299 56.2	31.3	9 06.7	22.4
	02	125 09.0	175 05.1	15.5	119 26.0	24.5	314 58.6	31.3	24 09.4	22.5
	03	140 11.4	190 04.4	. . 14.7	134 27.1	. . 25.1	330 01.0	. . 31.2	39 12.1	. . 22.5
	04	155 13.9	205 03.7	13.9	149 28.2	25.8	345 03.4	31.2	54 14.7	22.5
	05	170 16.4	220 03.0	13.1	164 29.3	26.4	0 05.8	31.2	69 17.4	22.5
	06	185 18.8	235 02.3	S20 12.4	179 30.4	N 2 27.1	15 08.2	N 5 31.2	84 20.1	N22 22.5
	07	200 21.3	250 01.6	11.6	194 31.5	27.7	30 10.6	31.2	99 22.8	22.6
S	08	215 23.8	265 00.9	10.8	209 32.6	28.4	45 13.0	31.2	114 25.5	22.6
A	09	230 26.2	280 00.2	. . 10.0	224 33.7	. . 29.0	60 15.4	. . 31.2	129 28.2	. . 22.6
T	10	245 28.7	294 59.5	09.2	239 34.9	29.7	75 17.8	31.2	144 30.8	22.6
U	11	260 31.1	309 58.8	08.4	254 36.0	30.4	90 20.2	31.1	159 33.5	22.6
R	12	275 33.6	324 58.1	S20 07.6	269 37.1	N 2 31.0	105 22.6	N 5 31.1	174 36.2	N22 22.7
D	13	290 36.1	339 57.4	06.8	284 38.2	31.7	120 25.0	31.1	189 38.9	22.7
A	14	305 38.5	354 56.7	06.0	299 39.3	32.3	135 27.4	31.1	204 41.6	22.7
Y	15	320 41.0	9 56.0	. . 05.2	314 40.4	. . 33.0	150 29.8	. . 31.1	219 44.3	. . 22.7
	16	335 43.5	24 55.3	04.4	329 41.5	33.6	165 32.2	31.1	234 47.0	22.7
	17	350 45.9	39 54.6	03.6	344 42.6	34.3	180 34.6	31.1	249 49.6	22.8
	18	5 48.4	54 53.9	S20 02.8	359 43.7	N 2 34.9	195 37.0	N 5 31.1	264 52.3	N22 22.8
	19	20 50.9	69 53.2	02.0	14 44.8	35.6	210 39.4	31.0	279 55.0	22.8
	20	35 53.3	84 52.5	01.2	29 46.0	36.2	225 41.8	31.0	294 57.7	22.8
	21	50 55.8	99 51.8	20 00.4	44 47.1	. . 36.9	240 44.3	. . 31.0	310 00.4	. . 22.8
	22	65 58.2	114 51.1	19 59.6	59 48.2	37.5	255 46.7	31.0	325 03.1	22.9
	23	81 00.7	129 50.4	58.8	74 49.3	38.2	270 49.1	31.0	340 05.8	22.9
28	00	96 03.2	144 49.7	S19 58.0	89 50.4	N 2 38.8	285 51.5	N 5 31.0	355 08.4	N22 22.9
	01	111 05.6	159 49.0	57.2	104 51.5	39.5	300 53.9	31.0	10 11.1	22.9
	02	126 08.1	174 48.3	56.4	119 52.6	40.1	315 56.3	31.0	25 13.8	22.9
	03	141 10.6	189 47.6	. . 55.6	134 53.7	. . 40.8	330 58.7	. . 31.0	40 16.5	. . 23.0
	04	156 13.0	204 46.9	54.7	149 54.8	41.4	346 01.1	30.9	55 19.2	23.0
	05	171 15.5	219 46.2	53.9	164 55.9	42.1	1 03.5	30.9	70 21.9	23.0
	06	186 18.0	234 45.5	S19 53.1	179 57.0	N 2 42.7	16 05.9	N 5 30.9	85 24.6	N22 23.0
	07	201 20.4	249 44.9	52.3	194 58.1	43.4	31 08.3	30.9	100 27.2	23.0
	08	216 22.9	264 44.2	51.5	209 59.2	44.0	46 10.7	30.9	115 29.9	23.1
S	09	231 25.4	279 43.5	. . 50.7	225 00.4	. . 44.7	61 13.2	. . 30.9	130 32.6	. . 23.1
U	10	246 27.8	294 42.8	49.9	240 01.5	45.3	76 15.6	30.9	145 35.3	23.1
N	11	261 30.3	309 42.1	49.0	255 02.6	46.0	91 18.0	30.9	160 38.0	23.1
D	12	276 32.7	324 41.4	S19 48.2	270 03.7	N 2 46.6	106 20.4	N 5 30.9	175 40.7	N22 23.1
A	13	291 35.2	339 40.7	47.4	285 04.8	47.3	121 22.8	30.8	190 43.4	23.2
Y	14	306 37.7	354 40.0	46.6	300 05.9	47.9	136 25.2	30.8	205 46.0	23.2
	15	321 40.1	9 39.4	. . 45.8	315 07.0	. . 48.6	151 27.6	. . 30.8	220 48.7	. . 23.2
	16	336 42.6	24 38.7	44.9	330 08.1	49.2	166 30.0	30.8	235 51.4	23.2
	17	351 45.1	39 38.0	44.1	345 09.2	49.9	181 32.5	30.8	250 54.1	23.2
	18	6 47.5	54 37.3	S19 43.3	0 10.3	N 2 50.5	196 34.9	N 5 30.8	265 56.8	N22 23.3
	19	21 50.0	69 36.6	42.5	15 11.4	51.2	211 37.3	30.8	280 59.5	23.3
	20	36 52.5	84 35.9	41.6	30 12.5	51.8	226 39.7	30.8	296 02.2	23.3
	21	51 54.9	99 35.3	. . 40.8	45 13.6	. . 52.5	241 42.1	. . 30.8	311 04.8	. . 23.3
	22	66 57.4	114 34.6	40.0	60 14.7	53.1	256 44.5	30.8	326 07.5	23.3
	23	81 59.9	129 33.9	39.2	75 15.8	53.8	271 46.9	30.8	341 10.2	23.4
29	00	97 02.3	144 33.2	S19 38.3	90 16.9	N 2 54.4	286 49.4	N 5 30.8	356 12.9	N22 23.4
	01	112 04.8	159 32.5	37.5	105 18.0	55.1	301 51.8	30.7	11 15.6	23.4
	02	127 07.2	174 31.9	36.7	120 19.1	55.8	316 54.2	30.7	26 18.3	23.4
	03	142 09.7	189 31.2	. . 35.8	135 20.2	. . 56.4	331 56.6	. . 30.7	41 21.0	. . 23.4
	04	157 12.2	204 30.5	35.0	150 21.3	57.1	346 59.0	30.7	56 23.6	23.5
	05	172 14.6	219 29.8	34.2	165 22.4	57.7	2 01.4	30.7	71 26.3	23.5
	06	187 17.1	234 29.1	S19 33.3	180 23.5	N 2 58.4	17 03.9	N 5 30.7	86 29.0	N22 23.5
	07	202 19.6	249 28.5	32.5	195 24.6	59.0	32 06.3	30.7	101 31.7	23.5
	08	217 22.0	264 27.8	31.6	210 25.7	2 59.7	47 08.7	30.7	116 34.4	23.5
M	09	232 24.5	279 27.1	. . 30.8	225 26.9	3 00.3	62 11.1	. . 30.7	131 37.1	. . 23.6
O	10	247 27.0	294 26.4	30.0	240 28.0	01.0	77 13.5	30.7	146 39.8	23.6
N	11	262 29.4	309 25.8	29.1	255 29.1	01.6	92 15.9	30.7	161 42.4	23.6
D	12	277 31.9	324 25.1	S19 28.3	270 30.2	N 3 02.3	107 18.4	N 5 30.7	176 45.1	N22 23.6
A	13	292 34.3	339 24.4	27.4	285 31.3	02.9	122 20.8	30.7	191 47.8	23.6
Y	14	307 36.8	354 23.8	26.6	300 32.4	03.6	137 23.2	30.7	206 50.5	23.6
	15	322 39.3	9 23.1	. . 25.8	315 33.5	. . 04.2	152 25.6	. . 30.6	221 53.2	. . 23.7
	16	337 41.7	24 22.4	24.9	330 34.6	04.9	167 28.0	30.6	236 55.9	23.7
	17	352 44.2	39 21.7	24.1	345 35.7	05.5	182 30.5	30.6	251 58.6	23.7
	18	7 46.7	54 21.1	S19 23.2	0 36.8	N 3 06.2	197 32.9	N 5 30.6	267 01.3	N22 23.7
	19	22 49.1	69 20.4	22.4	15 37.9	06.8	212 35.3	30.6	282 03.9	23.7
	20	37 51.6	84 19.7	21.5	30 39.0	07.5	227 37.7	30.6	297 06.6	23.8
	21	52 54.1	99 19.1	. . 20.7	45 40.1	. . 08.1	242 40.1	. . 30.6	312 09.3	. . 23.8
	22	67 56.5	114 18.4	19.8	60 41.2	08.8	257 42.6	30.6	327 12.0	23.8
	23	82 59.0	129 17.7	19.0	75 42.3	09.4	272 45.0	30.6	342 14.7	23.8
Mer. Pass.		h m 17 32.9	*v* −0.7	*d* 0.8	*v* 1.1	*d* 0.7	*v* 2.4	*d* 0.0	*v* 2.7	*d* 0.0

STARS		
Name	SHA	Dec
	° ′	° ′
Acamar	315 23.6	S40 17.5
Achernar	335 31.9	S57 13.3
Acrux	173 18.0	S63 06.9
Adhara	255 18.0	S28 58.5
Aldebaran	290 57.6	N16 31.1
Alioth	166 27.0	N55 56.1
Alkaid	153 04.8	N49 17.4
Al Na'ir	27 53.2	S46 56.8
Alnilam	275 53.6	S 1 11.9
Alphard	218 03.2	S 8 40.4
Alphecca	126 17.6	N26 41.9
Alpheratz	357 51.3	N29 06.8
Altair	62 15.8	N 8 52.6
Ankaa	353 22.9	S42 17.4
Antares	112 35.8	S26 26.4
Arcturus	146 02.7	N19 09.6
Atria	107 45.0	S69 02.0
Avior	234 20.6	S59 31.1
Bellatrix	278 39.7	N 6 21.3
Betelgeuse	271 09.0	N 7 24.5
Canopus	263 58.9	S52 41.8
Capella	280 45.0	N46 00.2
Deneb	49 37.1	N45 17.
Denebola	182 41.2	N14 33.0
Diphda	349 03.3	S17 58.1
Dubhe	194 00.2	N61 43.6
Elnath	278 21.7	N28 36.
Eltanin	90 50.2	N51 29.2
Enif	33 54.7	N 9 53.5
Fomalhaut	15 32.2	S29 36.
Gacrux	172 09.5	S57 07.
Gienah	176 00.0	S17 33.
Hadar	148 59.1	S60 23.
Hamal	328 09.1	N23 29.
Kaus Aust.	83 54.1	S34 23.
Kochab	137 20.0	N74 08.
Markab	13 45.9	N15 13.
Menkar	314 22.6	N 4 06.
Menkent	148 16.7	S36 23.
Miaplacidus	221 40.9	S69 43.
Mirfak	308 50.8	N49 52.
Nunki	76 07.9	S26 17.
Peacock	53 31.3	S56 43.
Pollux	243 36.4	N28 01.
Procyon	245 07.2	N 5 13.
Rasalhague	96 13.7	N12 33.
Regulus	207 51.2	N11 56.
Rigel	281 18.9	S 8 11.
Rigil Kent.	140 02.6	S60 50.
Sabik	102 21.4	S15 43.
Schedar	349 49.3	N56 33.
Shaula	96 32.5	S37 06.
Sirius	258 40.0	S16 43.
Spica	158 39.2	S11 10.
Suhail	222 57.7	S43 26.
Vega	80 44.5	N38 47
Zuben'ubi	137 13.9	S16 03
	SHA	Mer. Pas
	° ′	h m
Venus	48 46.5	14 21
Mars	353 47.2	17 59
Jupiter	189 48.3	4 56
Saturn	259 05.3	0 19

ALTITUDE CORRECTION TABLES 10°-90°—SUN, STARS, PLANETS

OCT.—MAR. **SUN**			APR.—SEPT.			**STARS AND PLANETS**	
App. Alt.	Lower Limb	Upper Limb	App. Alt.	Lower Limb	Upper Limb	App Alt.	Corr^{n}
° ′	′	′	° ′	′	′	° ′	′
9 34			9 39			9 56	
	+10.8	−21.5		+10.6	−21.2		−5.3
9 45			9 51			10 08	
	+10.9	−21.4		+10.7	−21.1		−5.2
9 56			10 03			10 20	
	+11.0	−21.3		+10.8	−21.0		−5.1
10 08			10 15			10 33	
	+11.1	−21.2		+10.9	−20.9		−5.0
10 21			10 27			10 46	
	+11.2	−21.1		+11.0	−20.8		−4.9
10 34			10 40			11 00	
	+11.3	−21.0		+11.1	−20.7		−4.8
10 47			10 54			11 14	
	+11.4	−20.9		+11.2	−20.6		−4.7
11 01			11 08			11 29	
	+11.5	−20.8		+11.3	−20.5		−4.6
11 15			11 23			11 45	
	+11.6	−20.7		+11.4	−20.4		−4.5
11 30			11 38			12 01	
	+11.7	−20.6		+11.5	−20.3		−4.4
11 46			11 54			12 18	
	+11.8	−20.5		+11.6	−20.2		−4.3
12 02			12 10			12 35	
	+11.9	−20.4		+11.7	−20.1		−4.2
12 19			12 28			12 54	
	+12.0	−20.3		+11.8	−20.0		−4.1
12 37			12 46			13 13	
	+12.1	−20.2		+11.9	−19.9		−4.0
12 55			13 05			13 33	
	+12.2	−20.1		+12.0	−19.8		−3.9
13 14			13 24			13 54	
	+12.3	−20.0		+12.1	−19.7		−3.8
13 35			13 45			14 16	
	+12.4	−19.9		+12.2	−19.6		−3.7
13 56			14 07			14 40	
	+12.5	−19.8		+12.3	−19.5		−3.6
14 18			14 30			15 04	
	+12.6	−19.7		+12.4	−19.4		−3.5
14 42			14 54			15 30	
	+12.7	−19.6		+12.5	−19.3		−3.4
15 06			15 19			15 57	
	+12.8	−19.5		+12.6	−19.2		−3.3
15 32			15 46			16 26	
	+12.9	−19.4		+12.7	−19.1		−3.2
15 59			16 14			16 56	
	+13.0	−19.3		+12.8	−19.0		−3.1
16 28			16 44			17 28	
	+13.1	−19.2		+12.9	−18.9		−3.0
16 59			17 15			18 02	
	+13.2	−19.1		+13.0	−18.8		−2.9
17 32			17 48			18 38	
	+13.3	−19.0		+13.1	−18.7		−2.8
18 06			18 24			19 17	
	+13.4	−18.9		+13.2	−18.6		−2.7
18 42			19 01			19 58	
	+13.5	−18.8		+13.3	−18.5		−2.6
19 21			19 42			20 42	
	+13.6	−18.7		+13.4	−18.4		−2.5
20 03			20 25			21 28	
	+13.7	−18.6		+13.5	−18.3		−2.4
20 48			21 11			22 19	
	+13.8	−18.5		+13.6	−18.2		−2.3
21 35			22 00			23 13	
	+13.9	−18.4		+13.7	−18.1		−2.2
22 26			22 54			24 11	
	+14.0	−18.3		+13.8	−18.0		−2.1
23 22			23 51			25 14	
	+14.1	−18.2		+13.9	−17.9		−2.0
24 21			24 53			26 22	
	+14.2	−18.1		+14.0	−17.8		−1.9
25 26			26 00			27 36	
	+14.3	−18.0		+14.1	−17.7		−1.8
26 36			27 13			28 56	
	+14.4	−17.9		+14.2	−17.6		−1.7
27 52			28 33			30 24	
	+14.5	−17.8		+14.3	−17.5		−1.6
29 15			30 00			32 00	
	+14.6	−17.7		+14.4	−17.4		−1.5
30 46			31 35			33 45	
	+14.7	−17.6		+14.5	−17.3		−1.4
32 26			33 20			35 40	
	+14.8	−17.5		+14.6	−17.2		−1.3
34 17			35 17			37 48	
	+14.9	−17.4		+14.7	−17.1		−1.2
36 20			37 26			40 08	
	+15.0	−17.3		+14.8	−17.0		−1.1
38 36			39 50			42 44	
	+15.1	−17.2		+14.9	−16.9		−1.0
41 08			42 31			45 36	
	+15.2	−17.1		+15.0	−16.8		−0.9
43 59			45 31			48 47	
	+15.3	−17.0		+15.1	−16.7		−0.8
47 10			48 55			52 18	
	+15.4	−16.9		+15.2	−16.6		−0.7
50 46			52 44			56 11	
	+15.5	−16.8		+15.3	−16.5		−0.6
54 49			57 02			60 28	
	+15.6	−16.7		+15.4	−16.4		−0.5
59 23			61 51			65 08	
	+15.7	−16.6		+15.5	−16.3		−0.4
64 30			67 17			70 11	
	+15.8	−16.5		+15.6	−16.2		−0.3
70 12			73 16			75 34	
	+15.9	−16.4		+15.7	−16.1		−0.2
76 26			79 43			81 13	
	+16.0	−16.3		+15.8	−16.0		−0.1
83 05			86 32			87 03	
	+16.1	−16.2		+15.9	−15.9		0.0
90 00			90 00			90 00	

STARS AND PLANETS — App. Alt. / Additional Corr^{n}

2003

VENUS

Jan. 1–Feb. 20	
°	′
0	
	+0.2
41	
	+0.1
76	

Feb. 21–Dec. 31	
°	′
0	
	+0.1
60	

MARS

Jan. 1–May 2 Dec. 17–Dec. 31	
°	′
0	
	+0.1
60	

May 3–June 26 Oct. 26–Dec. 16	
°	′
0	
	+0.2
41	
	+0.1
76	

June 27–Aug. 1 Sept. 23–Oct. 25	
°	′
0	
	+0.3
34	
	+0.2
60	
	+0.1
80	

Aug. 2–Sept. 22	
°	′
0	
	+0.4
29	
	+0.3
51	
	+0.2
68	
	+0.1
83	

DIP

Ht. of Eye	Corr^{n}	Ht. of Eye
m	′	ft.
2.4		8.0
	−2.8	
2.6		8.6
	−2.9	
2.8		9.2
	−3.0	
3.0		9.8
	−3.1	
3.2		10.5
	−3.2	
3.4		11.2
	−3.3	
3.6		11.9
	−3.4	
3.8		12.6
	−3.5	
4.0		13.3
	−3.6	
4.3		14.1
	−3.7	
4.5		14.9
	−3.8	
4.7		15.7
	−3.9	
5.0		16.5
	−4.0	
5.2		17.4
	−4.1	
5.5		18.3
	−4.2	
5.8		19.1
	−4.3	
6.1		20.1
	−4.4	
6.3		21.0
	−4.5	
6.6		22.0
	−4.6	
6.9		22.9
	−4.7	
7.2		23.9
	−4.8	
7.5		24.9
	−4.9	
7.9		26.0
	−5.0	
8.2		27.1
	−5.1	
8.5		28.1
	−5.2	
8.8		29.2
	−5.3	
9.2		30.4
	−5.4	
9.5		31.5
	−5.5	
9.9		32.7
	−5.6	
10.3		33.9
	−5.7	
10.6		35.1
	−5.8	
11.0		36.3
	−5.9	
11.4		37.6
	−6.0	
11.8		38.9
	−6.1	
12.2		40.1
	−6.2	
12.6		41.5
	−6.3	
13.0		42.8
	−6.4	
13.4		44.2
	−6.5	
13.8		45.5
	−6.6	
14.2		46.9
	−6.7	
14.7		48.4
	−6.8	
15.1		49.8
	−6.9	
15.5		51.3
	−7.0	
16.0		52.8
	−7.1	
16.5		54.3
	−7.2	
16.9		55.8
	−7.3	
17.4		57.4
	−7.4	
17.9		58.9
	−7.5	
18.4		60.5
	−7.6	
18.8		62.1
	−7.7	
19.3		63.8
	−7.8	
19.8		65.4
	−7.9	
20.4		67.1
	−8.0	
20.9		68.8
	−8.1	
21.4		70.5

Ht. of Eye	Corr^{n}
m	′
1.0	−1.8
1.5	−2.2
2.0	−2.5
2.5	−2.8
3.0	−3.0
See table ←	
m	′
20	−7.9
22	−8.3
24	−8.6
26	−9.0
28	−9.3
30	−9.6
32	−10.0
34	−10.3
36	−10.6
38	−10.8
40	−11.1
42	−11.4
44	−11.7
46	−11.9
48	−12.2
ft.	′
2	−1.4
4	−1.9
6	−2.4
8	−2.7
10	−3.1
See table ←	
ft.	′
70	−8.1
75	−8.4
80	−8.7
85	−8.9
90	−9.2
95	−9.5
100	−9.7
105	−9.9
110	−10.2
115	−10.4
120	−10.6
125	−10.8
130	−11.1
135	−11.3
140	−11.5
145	−11.7
150	−11.9
155	−12.1

App. Alt. = Apparent altitude = Sextant altitude corrected for index error and dip.

INDEX TO SELECTED STARS, 2003

Name	No	Mag	SHA	Dec
			°	°
Acamar	**7**	3·1	315	S 40
Achernar	**5**	0·6	336	S 57
Acrux	**30**	1·1	173	S 63
Adhara	**19**	1·6	255	S 29
Aldebaran	**10**	1·1	291	N 17
Alioth	**32**	1·7	166	N 56
Alkaid	**34**	1·9	153	N 49
Al Na'ir	**55**	2·2	28	S 47
Alnilam	**15**	1·8	276	S 1
Alphard	**25**	2·2	218	S 9
Alphecca	**41**	2·3	126	N 27
Alpheratz	**1**	2·2	358	N 29
Altair	**51**	0·9	62	N 9
Ankaa	**2**	2·4	353	S 42
Antares	**42**	1·2	113	S 26
Arcturus	**37**	0·2	146	N 19
Atria	**43**	1·9	108	S 69
Avior	**22**	1·7	234	S 60
Bellatrix	**13**	1·7	279	N 6
Betelgeuse	**16**	Var.*	271	N 7
Canopus	**17**	−0·9	264	S 53
Capella	**12**	0·2	281	N 46
Deneb	**53**	1·3	50	N 45
Denebola	**28**	2·2	183	N 15
Diphda	**4**	2·2	349	S 18
Dubhe	**27**	2·0	194	N 62
Elnath	**14**	1·8	278	N 29
Eltanin	**47**	2·4	91	N 51
Enif	**54**	2·5	34	N 10
Fomalhaut	**56**	1·3	16	S 30
Gacrux	**31**	1·6	172	S 57
Gienah	**29**	2·8	176	S 18
Hadar	**35**	0·9	149	S 60
Hamal	**6**	2·2	328	N 23
Kaus Australis	**48**	2·0	84	S 34
Kochab	**40**	2·2	137	N 74
Markab	**57**	2·6	14	N 15
Menkar	**8**	2·8	314	N 4
Menkent	**36**	2·3	148	S 36
Miaplacidus	**24**	1·8	222	S 70
Mirfak	**9**	1·9	309	N 50
Nunki	**50**	2·1	76	S 26
Peacock	**52**	2·1	54	S 57
Pollux	**21**	1·2	244	N 28
Procyon	**20**	0·5	245	N 5
Rasalhague	**46**	2·1	96	N 13
Regulus	**26**	1·3	208	N 12
Rigel	**11**	0·3	281	S 8
Rigil Kentaurus	**38**	0·1	140	S 61
Sabik	**44**	2·6	102	S 16
Schedar	**3**	2·5	350	N 57
Shaula	**45**	1·7	97	S 37
Sirius	**18**	−1·6	259	S 17
Spica	**33**	1·2	159	S 11
Suhail	**23**	2·2	223	S 43
Vega	**49**	0·1	81	N 39
Zubenelgenubi	**39**	2·9	137	S 16

No	Name	Mag	SHA	Dec
			°	°
1	*Alpheratz*	2·2	358	N 29
2	*Ankaa*	2·4	353	S 42
3	*Schedar*	2·5	350	N 57
4	*Diphda*	2·2	349	S 18
5	*Achernar*	0·6	336	S 57
6	*Hamal*	2·2	328	N 23
7	*Acamar*	3·1	315	S 40
8	*Menkar*	2·8	314	N 4
9	*Mirfak*	1·9	309	N 50
10	*Aldebaran*	1·1	291	N 17
11	*Rigel*	0·3	281	S 8
12	*Capella*	0·2	281	N 46
13	*Bellatrix*	1·7	279	N 6
14	*Elnath*	1·8	278	N 29
15	*Alnilam*	1·8	276	S 1
16	*Betelgeuse*	Var.*	271	N 7
17	*Canopus*	−0·9	264	S 53
18	*Sirius*	−1·6	259	S 17
19	*Adhara*	1·6	255	S 29
20	*Procyon*	0·5	245	N 5
21	*Pollux*	1·2	244	N 28
22	*Avior*	1·7	234	S 60
23	*Suhail*	2·2	223	S 43
24	*Miaplacidus*	1·8	222	S 70
25	*Alphard*	2·2	218	S 9
26	*Regulus*	1·3	208	N 12
27	*Dubhe*	2·0	194	N 62
28	*Denebola*	2·2	183	N 15
29	*Gienah*	2·8	176	S 18
30	*Acrux*	1·1	173	S 63
31	*Gacrux*	1·6	172	S 57
32	*Alioth*	1·7	166	N 56
33	*Spica*	1·2	159	S 11
34	*Alkaid*	1·9	153	N 49
35	*Hadar*	0·9	149	S 60
36	*Menkent*	2·3	148	S 36
37	*Arcturus*	0·2	146	N 19
38	*Rigil Kentaurus*	0·1	140	S 61
39	*Zubenelgenubi*	2·9	137	S 16
40	*Kochab*	2·2	137	N 74
41	*Alphecca*	2·3	126	N 27
42	*Antares*	1·2	113	S 26
43	*Atria*	1·9	108	S 69
44	*Sabik*	2·6	102	S 16
45	*Shaula*	1·7	97	S 37
46	*Rasalhague*	2·1	96	N 13
47	*Eltanin*	2·4	91	N 51
48	*Kaus Australis*	2·0	84	S 34
49	*Vega*	0·1	81	N 39
50	*Nunki*	2·1	76	S 26
51	*Altair*	0·9	62	N 9
52	*Peacock*	2·1	54	S 57
53	*Deneb*	1·3	50	N 45
54	*Enif*	2·5	34	N 10
55	*Al Na'ir*	2·2	28	S 47
56	*Fomalhaut*	1·3	16	S 30
57	*Markab*	2·6	14	N 15

*0·1 — 1·2

d	h	SUN GHA ° ′	SUN Dec ° ′	MOON GHA ° ′	v ′	MOON Dec ° ′	d ′	HP ′
27	00	179 50.4	S23 21.2	127 21.1	9.3	S18 29.6	11.9	58.5
	01	194 50.1	21.1	141 49.4	9.5	18 17.7	12.0	58.4
	02	209 49.8	21.0	156 17.9	9.6	18 05.7	12.1	58.4
	03	224 49.5	. . 20.9	170 46.5	9.8	17 53.6	12.1	58.4
	04	239 49.2	20.8	185 15.3	9.9	17 41.5	12.2	58.3
	05	254 48.9	20.7	199 44.2	10.0	17 29.3	12.3	58.3
	06	269 48.5	S23 20.6	214 13.2	10.1	S17 17.0	12.4	58.2
	07	284 48.2	20.5	228 42.3	10.2	17 04.6	12.5	58.2
S	08	299 47.9	20.4	243 11.5	10.4	16 52.1	12.5	58.2
A	09	314 47.6	. . 20.3	257 40.9	10.5	16 39.6	12.6	58.1
T	10	329 47.3	20.2	272 10.4	10.6	16 27.0	12.6	58.1
U	11	344 47.0	20.1	286 40.0	10.7	16 14.4	12.7	58.0
R	12	359 46.7	S23 20.0	301 09.7	10.8	S16 01.7	12.8	58.0
D	13	14 46.4	19.9	315 39.5	11.0	15 48.9	12.9	57.9
A	14	29 46.1	19.8	330 09.5	11.1	15 36.0	12.9	57.9
Y	15	44 45.8	. . 19.7	344 39.6	11.2	15 23.1	12.9	57.9
	16	59 45.5	19.6	359 09.8	11.3	15 10.2	13.1	57.8
	17	74 45.1	19.5	13 40.1	11.4	14 57.1	13.1	57.8
	18	89 44.8	S23 19.4	28 10.5	11.5	S14 44.0	13.1	57.7
	19	104 44.5	19.3	42 41.0	11.6	14 30.9	13.2	57.7
	20	119 44.2	19.2	57 11.6	11.8	14 17.7	13.2	57.7
	21	134 43.9	. . 19.1	71 42.4	11.8	14 04.5	13.3	57.6
	22	149 43.6	19.0	86 13.2	12.0	13 51.2	13.3	57.6
	23	164 43.3	18.9	100 44.2	12.0	13 37.9	13.4	57.6
28	00	179 43.0	S23 18.8	115 15.2	12.2	S13 24.5	13.5	57.5
	01	194 42.7	18.7	129 46.4	12.3	13 11.0	13.4	57.5
	02	209 42.4	18.5	144 17.7	12.3	12 57.6	13.5	57.4
	03	224 42.1	. . 18.4	158 49.0	12.5	12 44.1	13.6	57.4
	04	239 41.8	18.3	173 20.5	12.6	12 30.5	13.6	57.4
	05	254 41.5	18.2	187 52.1	12.6	12 16.9	13.6	57.3
	06	269 41.1	S23 18.1	202 23.7	12.8	S12 03.3	13.7	57.3
	07	284 40.8	18.0	216 55.5	12.8	11 49.6	13.7	57.2
	08	299 40.5	17.8	231 27.3	13.0	11 35.9	13.7	57.2
S	09	314 40.2	. . 17.7	245 59.3	13.0	11 22.2	13.8	57.2
U	10	329 39.9	17.6	260 31.3	13.1	11 08.4	13.8	57.1
N	11	344 39.6	17.5	275 03.4	13.3	10 54.6	13.9	57.1
D	12	359 39.3	S23 17.4	289 35.7	13.3	S10 40.7	13.8	57.0
A	13	14 39.0	17.2	304 08.0	13.4	10 26.9	13.9	57.0
Y	14	29 38.7	17.1	318 40.4	13.4	10 13.0	13.9	57.0
	15	44 38.4	. . 17.0	333 12.8	13.6	9 59.1	14.0	56.9
	16	59 38.1	16.9	347 45.4	13.6	9 45.1	14.0	56.9
	17	74 37.8	16.7	2 18.0	13.8	9 31.1	13.9	56.9
	18	89 37.5	S23 16.6	16 50.8	13.8	S 9 17.2	14.1	56.8
	19	104 37.2	16.5	31 23.6	13.9	9 03.1	14.0	56.8
	20	119 36.9	16.4	45 56.5	13.9	8 49.1	14.0	56.7
	21	134 36.5	. . 16.2	60 29.4	14.1	8 35.1	14.1	56.7
	22	149 36.2	16.1	75 02.5	14.1	8 21.0	14.1	56.7
	23	164 35.9	16.0	89 35.6	14.2	8 06.9	14.1	56.6
29	00	179 35.6	S23 15.8	104 08.8	14.2	S 7 52.8	14.1	56.6
	01	194 35.3	15.7	118 42.0	14.4	7 38.7	14.2	56.6
	02	209 35.0	15.6	133 15.4	14.4	7 24.5	14.1	56.5
	03	224 34.7	. . 15.4	147 48.8	14.4	7 10.4	14.2	56.5
	04	239 34.4	15.3	162 22.2	14.6	6 56.2	14.2	56.5
	05	254 34.1	15.2	176 55.8	14.6	6 42.0	14.1	56.4
	06	269 33.8	S23 15.0	191 29.4	14.6	S 6 27.9	14.2	56.4
	07	284 33.5	14.9	206 03.0	14.7	6 13.7	14.2	56.4
	08	299 33.2	14.8	220 36.7	14.8	5 59.5	14.2	56.3
M	09	314 32.9	. . 14.6	235 10.5	14.9	5 45.3	14.2	56.3
O	10	329 32.6	14.5	249 44.4	14.9	5 31.1	14.3	56.2
N	11	344 32.3	14.4	264 18.3	14.9	5 16.8	14.2	56.2
D	12	359 32.0	S23 14.2	278 52.2	15.1	S 5 02.6	14.2	56.2
A	13	14 31.7	14.1	293 26.3	15.0	4 48.4	14.2	56.1
Y	14	29 31.4	13.9	308 00.3	15.2	4 34.2	14.3	56.1
	15	44 31.1	. . 13.8	322 34.5	15.1	4 19.9	14.2	56.1
	16	59 30.7	13.6	337 08.6	15.3	4 05.7	14.2	56.0
	17	74 30.4	13.5	351 42.9	15.2	3 51.5	14.2	56.0
	18	89 30.1	S23 13.4	6 17.1	15.4	S 3 37.3	14.3	56.0
	19	104 29.8	13.2	20 51.5	15.3	3 23.0	14.2	55.9
	20	119 29.5	13.1	35 25.8	15.5	3 08.8	14.2	55.9
	21	134 29.2	. . 12.9	50 00.3	15.4	2 54.6	14.2	55.9
	22	149 28.9	12.8	64 34.7	15.5	2 40.4	14.2	55.9
	23	164 28.6	12.6	79 09.2	15.6	S 2 26.2	14.2	55.8
		SD 16.3	*d* 0.1	SD 15.8		15.5		15.3

Lat. °	Twilight Naut. h m	Twilight Civil h m	Sunrise h m	Moonrise 27 h m	Moonrise 28 h m	Moonrise 29 h m	Moonrise 30 h m
N 72	08 27	10 54	■	14 00	13 04	12 28	11 59
N 70	08 07	09 54	■	13 24	12 47	12 22	12 00
68	07 51	09 20	■	12 58	12 34	12 16	12 00
66	07 38	08 55	10 34	12 37	12 23	12 11	12 01
64	07 27	08 36	09 53	12 21	12 14	12 07	12 01
62	07 18	08 20	09 25	12 08	12 06	12 04	12 01
60	07 09	08 06	09 04	11 56	11 59	12 01	12 02
N 58	07 02	07 55	08 46	11 46	11 53	11 58	12 02
56	06 55	07 44	08 32	11 37	11 47	11 55	12 02
54	06 49	07 35	08 19	11 29	11 43	11 53	12 03
52	06 44	07 27	08 08	11 22	11 38	11 51	12 03
50	06 38	07 20	07 58	11 16	11 34	11 49	12 03
45	06 27	07 04	07 38	11 02	11 25	11 45	12 03
N 40	06 16	06 50	07 21	10 50	11 18	11 42	12 04
35	06 07	06 39	07 07	10 41	11 12	11 39	12 04
30	05 58	06 28	06 55	10 32	11 06	11 36	12 04
20	05 42	06 09	06 33	10 17	10 56	11 32	12 05
N 10	05 26	05 52	06 15	10 04	10 48	11 28	12 05
0	05 09	05 35	05 58	09 52	10 40	11 24	12 06
S 10	04 50	05 17	05 40	09 39	10 32	11 20	12 06
20	04 27	04 57	05 21	09 26	10 23	11 16	12 07
30	03 59	04 32	04 59	09 11	10 13	11 12	12 07
35	03 40	04 17	04 46	09 02	10 07	11 09	12 08
40	03 17	03 59	04 31	08 51	10 01	11 06	12 08
45	02 46	03 36	04 13	08 39	09 53	11 02	12 09
S 50	02 01	03 07	03 51	08 25	09 44	10 58	12 09
52	01 34	02 52	03 40	08 18	09 39	10 56	12 10
54	00 49	02 34	03 28	08 10	09 35	10 54	12 10
56	////	02 12	03 14	08 02	09 29	10 52	12 10
58	////	01 42	02 57	07 52	09 23	10 49	12 11
S 60	////	00 54	02 37	07 41	09 17	10 46	12 11

Lat. °	Sunset h m	Twilight Civil h m	Twilight Naut. h m	Moonset 27 h m	Moonset 28 h m	Moonset 29 h m	Moonset 30 h m
N 72	■	13 09	15 36	18 32	21 06	23 13	25 12
N 70	■	14 09	15 56	19 06	21 20	23 16	25 06
68	■	14 43	16 12	19 30	21 31	23 19	25 02
66	13 29	15 08	16 25	19 49	21 40	23 22	24 59
64	14 10	15 27	16 36	20 04	21 48	23 24	24 56
62	14 38	15 43	16 45	20 16	21 54	23 26	24 53
60	14 59	15 57	16 53	20 27	22 00	23 27	24 51
N 58	15 17	16 08	17 01	20 36	22 05	23 28	24 49
56	15 31	16 18	17 08	20 44	22 09	23 30	24 47
54	15 44	16 27	17 14	20 51	22 13	23 31	24 46
52	15 55	16 36	17 19	20 57	22 17	23 32	24 44
50	16 05	16 43	17 25	21 03	22 20	23 33	24 43
45	16 25	16 59	17 36	21 15	22 27	23 35	24 40
N 40	16 42	17 12	17 46	21 25	22 32	23 36	24 38
35	16 56	17 24	17 56	21 34	22 37	23 37	24 36
30	17 08	17 35	18 04	21 41	22 41	23 39	24 34
20	17 29	17 53	18 21	21 54	22 49	23 41	24 31
N 10	17 48	18 11	18 37	22 05	22 55	23 43	24 28
0	18 05	18 28	18 54	22 15	23 01	23 44	24 25
S 10	18 23	18 46	19 13	22 25	23 07	23 46	24 23
20	18 41	19 06	19 35	22 36	23 13	23 48	24 20
30	19 03	19 31	20 04	22 48	23 20	23 49	24 17
35	19 16	19 46	20 23	22 55	23 24	23 51	24 15
40	19 31	20 04	20 46	23 03	23 29	23 52	24 13
45	19 49	20 26	21 16	23 13	23 34	23 53	24 11
S 50	20 12	20 56	22 01	23 24	23 40	23 55	24 08
52	20 22	21 11	22 28	23 29	23 43	23 56	24 07
54	20 34	21 29	23 12	23 34	23 46	23 56	24 06
56	20 49	21 51	////	23 41	23 50	23 57	24 04
58	21 05	22 20	////	23 48	23 54	23 58	24 03
S 60	21 25	23 08	////	23 55	23 58	23 59	24 01

Day	SUN Eqn. of Time 00^h	SUN Eqn. of Time 12^h	SUN Mer. Pass.	MOON Mer. Pass. Upper	MOON Mer. Pass. Lower	MOON Age	MOON Phase
d	m s	m s	h m	h m	h m	d	%
27	00 38	00 53	12 01	16 03	03 38	04	22
28	01 07	01 22	12 01	16 51	04 27	05	31
29	01 37	01 52	12 02	17 34	05 13	06	41

UT		ARIES	VENUS −4.0		MARS +0.2		JUPITER −2.2		SATURN −0.4	
		GHA	GHA	Dec	GHA	Dec	GHA	Dec	GHA	Dec
d h		° ′	° ′	° ′	° ′	° ′	° ′	° ′	° ′	° ′
30 00		98 01.5	144 17.1	S19 18.1	90 43.4	N 3 10.1	287 47.4	N 5 30.6	357 17.4	N22 23.8
01		113 03.9	159 16.4	17.3	105 44.5	10.7	302 49.8	30.6	12 20.1	23.9
02		128 06.4	174 15.8	16.4	120 45.6	11.4	317 52.3	30.6	27 22.7	23.9
03		143 08.8	189 15.1	. . 15.6	135 46.7	. . 12.0	332 54.7	. . 30.6	42 25.4	. . 23.9
04		158 11.3	204 14.4	14.7	150 47.8	12.7	347 57.1	30.6	57 28.1	23.9
05		173 13.8	219 13.8	13.8	165 48.8	13.3	2 59.5	30.6	72 30.8	23.9
06		188 16.2	234 13.1	S19 13.0	180 49.9	N 3 14.0	18 01.9	N 5 30.6	87 33.5	N22 24.0
07		203 18.7	249 12.4	12.1	195 51.0	14.6	33 04.4	30.6	102 36.2	24.0
08	T	218 21.2	264 11.8	11.3	210 52.1	15.3	48 06.8	30.6	117 38.9	24.0
09	U	233 23.6	279 11.1	. . 10.4	225 53.2	. . 15.9	63 09.2	. . 30.6	132 41.5	. . 24.0
10	E	248 26.1	294 10.5	09.5	240 54.3	16.6	78 11.6	30.6	147 44.2	24.0
11		263 28.6	309 09.8	08.7	255 55.4	17.2	93 14.1	30.5	162 46.9	24.1
12	S	278 31.0	324 09.1	S19 07.8	270 56.5	N 3 17.9	108 16.5	N 5 30.5	177 49.6	N22 24.1
13	D	293 33.5	339 08.5	07.0	285 57.6	18.5	123 18.9	30.5	192 52.3	24.1
14	A	308 36.0	354 07.8	06.1	300 58.7	19.2	138 21.4	30.5	207 55.0	24.1
15	Y	323 38.4	9 07.2	. . 05.2	315 59.8	. . 19.8	153 23.8	. . 30.5	222 57.7	. . 24.1
16		338 40.9	24 06.5	04.4	331 00.9	20.5	168 26.2	30.5	238 00.4	24.2
17		353 43.3	39 05.9	03.5	346 02.0	21.2	183 28.6	30.5	253 03.0	24.2
18		8 45.8	54 05.2	S19 02.6	1 03.1	N 3 21.8	198 31.1	N 5 30.5	268 05.7	N22 24.2
19		23 48.3	69 04.6	01.7	16 04.2	22.5	213 33.5	30.5	283 08.4	24.2
20		38 50.7	84 03.9	00.9	31 05.3	23.1	228 35.9	30.5	298 11.1	24.2
21		53 53.2	99 03.3	19 00.0	46 06.4	. . 23.8	243 38.3	. . 30.5	313 13.8	. . 24.3
22		68 55.7	114 02.6	18 59.1	61 07.5	24.4	258 40.8	30.5	328 16.5	24.3
23		83 58.1	129 02.0	58.3	76 08.6	25.1	273 43.2	30.5	343 19.2	24.3
31 00		99 00.6	144 01.3	S18 57.4	91 09.7	N 3 25.7	288 45.6	N 5 30.5	358 21.8	N22 24.3
01		114 03.1	159 00.7	56.5	106 10.8	26.4	303 48.1	30.5	13 24.5	24.3
02		129 05.5	174 00.0	55.6	121 11.9	27.0	318 50.5	30.5	28 27.2	24.4
03		144 08.0	188 59.4	. . 54.8	136 13.0	. . 27.7	333 52.9	. . 30.5	43 29.9	. . 24.4
04		159 10.4	203 58.7	53.9	151 14.1	28.3	348 55.4	30.5	58 32.6	24.4
05		174 12.9	218 58.1	53.0	166 15.1	29.0	3 57.8	30.5	73 35.3	24.4
06		189 15.4	233 57.4	S18 52.1	181 16.2	N 3 29.6	19 00.2	N 5 30.5	88 38.0	N22 24.4
07	W	204 17.8	248 56.8	51.2	196 17.3	30.3	34 02.7	30.5	103 40.7	24.5
08	E	219 20.3	263 56.1	50.4	211 18.4	30.9	49 05.1	30.5	118 43.3	24.5
09	D	234 22.8	278 55.5	. . 49.5	226 19.5	. . 31.6	64 07.5	. . 30.5	133 46.0	. . 24.5
10	N	249 25.2	293 54.8	48.6	241 20.6	32.2	79 10.0	30.5	148 48.7	24.5
11	E	264 27.7	308 54.2	47.7	256 21.7	32.9	94 12.4	30.5	163 51.4	24.5
12	S	279 30.2	323 53.6	S18 46.8	271 22.8	N 3 33.5	109 14.8	N 5 30.5	178 54.1	N22 24.5
13	D	294 32.6	338 52.9	45.9	286 23.9	34.2	124 17.3	30.5	193 56.8	24.6
14	A	309 35.1	353 52.3	45.0	301 25.0	34.8	139 19.7	30.5	208 59.5	24.6
15	Y	324 37.6	8 51.6	. . 44.2	316 26.1	. . 35.5	154 22.1	. . 30.5	224 02.1	. . 24.6
16		339 40.0	23 51.0	43.3	331 27.2	36.1	169 24.6	30.5	239 04.8	24.6
17		354 42.5	38 50.4	42.4	346 28.3	36.8	184 27.0	30.5	254 07.5	24.6
18		9 44.9	53 49.7	S18 41.5	1 29.3	N 3 37.4	199 29.4	N 5 30.5	269 10.2	N22 24.7
19		24 47.4	68 49.1	40.6	16 30.4	38.1	214 31.9	30.5	284 12.9	24.7
20		39 49.9	83 48.4	39.7	31 31.5	38.7	229 34.3	30.5	299 15.6	24.7
21		54 52.3	98 47.8	. . 38.8	46 32.6	. . 39.4	244 36.7	. . 30.5	314 18.3	. . 24.7
22		69 54.8	113 47.2	37.9	61 33.7	40.0	259 39.2	30.5	329 20.9	24.7
23		84 57.3	128 46.5	37.0	76 34.8	40.7	274 41.6	30.5	344 23.6	24.8
1 00		99 59.7	143 45.9	S18 36.1	91 35.9	N 3 41.3	289 44.0	N 5 30.5	359 26.3	N22 24.8
01		115 02.2	158 45.3	35.2	106 37.0	42.0	304 46.5	30.5	14 29.0	24.8
02		130 04.7	173 44.6	34.3	121 38.1	42.6	319 48.9	30.5	29 31.7	24.8
03		145 07.1	188 44.0	. . 33.4	136 39.2	. . 43.3	334 51.4	. . 30.5	44 34.4	. . 24.8
04		160 09.6	203 43.4	32.5	151 40.2	43.9	349 53.8	30.5	59 37.1	24.9
05		175 12.1	218 42.7	31.6	166 41.3	44.6	4 56.2	30.5	74 39.8	24.9
06		190 14.5	233 42.1	S18 30.7	181 42.4	N 3 45.2	19 58.7	N 5 30.5	89 42.4	N22 24.9
07		205 17.0	248 41.5	29.8	196 43.5	45.9	35 01.1	30.5	104 45.1	24.9
08	T	220 19.4	263 40.8	28.9	211 44.6	46.5	50 03.6	30.5	119 47.8	24.9
09	H	235 21.9	278 40.2	. . 28.0	226 45.7	. . 47.2	65 06.0	. . 30.5	134 50.5	. . 25.0
10	U	250 24.4	293 39.6	27.1	241 46.8	47.8	80 08.4	30.5	149 53.2	25.0
11	R	265 26.8	308 39.0	26.2	256 47.9	48.5	95 10.9	30.5	164 55.9	25.0
12	S	280 29.3	323 38.3	S18 25.3	271 49.0	N 3 49.1	110 13.3	N 5 30.5	179 58.6	N22 25.0
13	D	295 31.8	338 37.7	24.4	286 50.0	49.8	125 15.8	30.5	195 01.2	25.0
14	A	310 34.2	353 37.1	23.5	301 51.1	50.4	140 18.2	30.5	210 03.9	25.1
15	Y	325 36.7	8 36.5	. . 22.6	316 52.2	. . 51.1	155 20.6	. . 30.5	225 06.6	. . 25.1
16		340 39.2	23 35.8	21.7	331 53.3	51.7	170 23.1	30.5	240 09.3	25.1
17		355 41.6	38 35.2	20.8	346 54.4	52.4	185 25.5	30.5	255 12.0	25.1
18		10 44.1	53 34.6	S18 19.8	1 55.5	N 3 53.0	200 28.0	N 5 30.6	270 14.7	N22 25.1
19		25 46.5	68 34.0	18.9	16 56.6	53.7	215 30.4	30.6	285 17.4	25.2
20		40 49.0	83 33.3	18.0	31 57.7	54.3	230 32.9	30.6	300 20.1	25.2
21		55 51.5	98 32.7	. . 17.1	46 58.7	. . 55.0	245 35.3	. . 30.6	315 22.7	. . 25.2
22		70 53.9	113 32.1	16.2	61 59.8	55.6	260 37.7	30.6	330 25.4	25.2
23		85 56.4	128 31.5	15.3	77 00.9	56.3	275 40.2	30.6	345 28.1	25.2
Mer. Pass.		h m 17 21.1	v −0.6	d 0.9	v 1.1	d 0.7	v 2.4	d 0.0	v 2.7	d 0.0

STARS		
Name	SHA	Dec
	° ′	° ′
Acamar	315 23.6	S40 17.
Achernar	335 31.9	S57 13.
Acrux	173 18.0	S63 06.
Adhara	255 18.0	S28 58.
Aldebaran	290 57.6	N16 31.
Alioth	166 27.0	N55 56.
Alkaid	153 04.7	N49 17.
Al Na'ir	27 53.2	S46 56.
Alnilam	275 53.6	S 1 11.
Alphard	218 03.2	S 8 40.
Alphecca	126 17.5	N26 41.
Alpheratz	357 51.4	N29 06.
Altair	62 15.8	N 8 52.
Ankaa	353 22.9	S42 17.
Antares	112 35.8	S26 26.
Arcturus	146 02.6	N19 09.
Atria	107 45.0	S69 02.
Avior	234 20.6	S59 31.
Bellatrix	278 39.7	N 6 21.
Betelgeuse	271 09.0	N 7 24.
Canopus	263 58.9	S52 41.
Capella	280 45.0	N46 00.
Deneb	49 37.1	N45 17.
Denebola	182 41.1	N14 33.
Diphda	349 03.3	S17 58.
Dubhe	194 00.2	N61 43.
Elnath	278 21.7	N28 36.
Eltanin	90 50.1	N51 29
Enif	33 54.7	N 9 53.
Fomalhaut	15 32.2	S29 36.
Gacrux	172 09.5	S57 07.
Gienah	176 00.0	S17 33.
Hadar	148 59.0	S60 23.
Hamal	328 09.1	N23 29.
Kaus Aust.	83 54.1	S34 23.
Kochab	137 20.0	N74 08.
Markab	13 45.9	N15 13
Menkar	314 22.7	N 4 06.
Menkent	148 16.7	S36 23.
Miaplacidus	221 40.8	S69 43.
Mirfak	308 50.8	N49 52.
Nunki	76 07.9	S26 17.
Peacock	53 31.3	S56 43
Pollux	243 36.4	N28 01
Procyon	245 07.2	N 5 13
Rasalhague	96 13.7	N12 33
Regulus	207 51.2	N11 56
Rigel	281 18.9	S 8 11
Rigil Kent.	140 02.6	S60 50
Sabik	102 21.4	S15 43
Schedar	349 49.3	N56 33
Shaula	96 32.5	S37 06
Sirius	258 40.0	S16 43
Spica	158 39.2	S11 10
Suhail	222 57.7	S43 26
Vega	80 44.5	N38 47
Zuben'ubi	137 13.9	S16 03

	SHA	Mer. Pas
	° ′	h m
Venus	45 00.7	14 25
Mars	352 09.1	17 54
Jupiter	189 45.0	4 44
Saturn	259 21.3	0 07

UT	SUN GHA	SUN Dec	MOON GHA	v	MOON Dec	d	HP
d h	° ′	° ′	° ′	′	° ′	′	′
0 00	179 28.3	S23 12.5	93 43.8	15.6	S 2 12.0	14.2	55.8
01	194 28.0	12.3	108 18.4	15.6	1 57.8	14.2	55.8
02	209 27.7	12.2	122 53.0	15.6	1 43.6	14.2	55.7
03	224 27.4	. . 12.0	137 27.6	15.7	1 29.4	14.2	55.7
04	239 27.1	11.9	152 02.3	15.8	1 15.2	14.1	55.7
05	254 26.8	11.7	166 37.1	15.7	1 01.1	14.2	55.6
06	269 26.5	S23 11.5	181 11.8	15.8	S 0 46.9	14.1	55.6
07	284 26.2	11.4	195 46.6	15.9	0 32.8	14.1	55.6
08	299 25.9	11.2	210 21.5	15.8	0 18.7	14.1	55.6
09	314 25.6	. . 11.1	224 56.3	15.9	S 0 04.6	14.1	55.5
10	329 25.3	10.9	239 31.2	15.9	N 0 09.5	14.1	55.5
11	344 25.0	10.8	254 06.1	16.0	0 23.6	14.1	55.5
12	359 24.7	S23 10.6	268 41.1	15.9	N 0 37.7	14.0	55.4
13	14 24.4	10.4	283 16.0	16.0	0 51.7	14.0	55.4
14	29 24.1	10.3	297 51.0	16.0	1 05.7	14.1	55.4
15	44 23.8	. . 10.1	312 26.0	16.1	1 19.8	13.9	55.4
16	59 23.5	09.9	327 01.1	16.0	1 33.7	14.0	55.3
17	74 23.2	09.8	341 36.1	16.1	1 47.7	14.0	55.3
18	89 22.9	S23 09.6	356 11.2	16.1	N 2 01.7	13.9	55.3
19	104 22.6	09.5	10 46.3	16.1	2 15.6	13.9	55.3
20	119 22.3	09.3	25 21.4	16.1	2 29.5	13.9	55.2
21	134 22.0	. . 09.1	39 56.5	16.2	2 43.4	13.9	55.2
22	149 21.7	08.9	54 31.7	16.1	2 57.3	13.8	55.2
23	164 21.4	08.8	69 06.8	16.2	3 11.1	13.8	55.2
1 00	179 21.1	S23 08.6	83 42.0	16.1	N 3 24.9	13.8	55.1
01	194 20.8	08.4	98 17.1	16.2	3 38.7	13.8	55.1
02	209 20.5	08.3	112 52.3	16.2	3 52.5	13.8	55.1
03	224 20.2	. . 08.1	127 27.5	16.2	4 06.3	13.7	55.1
04	239 19.9	07.9	142 02.7	16.2	4 20.0	13.7	55.0
05	254 19.6	07.8	156 37.9	16.2	4 33.7	13.6	55.0
06	269 19.3	S23 07.6	171 13.1	16.2	N 4 47.3	13.6	55.0
07	284 19.0	07.4	185 48.3	16.2	5 00.9	13.6	55.0
08	299 18.7	07.2	200 23.5	16.3	5 14.5	13.6	54.9
09	314 18.4	. . 07.0	214 58.8	16.2	5 28.1	13.6	54.9
10	329 18.1	06.9	229 34.0	16.2	5 41.7	13.5	54.9
11	344 17.8	06.7	244 09.2	16.2	5 55.2	13.4	54.9
12	359 17.5	S23 06.5	258 44.4	16.2	N 6 08.6	13.5	54.9
13	14 17.2	06.3	273 19.6	16.2	6 22.1	13.4	54.8
14	29 16.9	06.2	287 54.8	16.2	6 35.5	13.4	54.8
15	44 16.6	. . 06.0	302 30.0	16.2	6 48.9	13.3	54.8
16	59 16.3	05.8	317 05.2	16.2	7 02.2	13.3	54.8
17	74 16.0	05.6	331 40.4	16.2	7 15.5	13.3	54.8
18	89 15.7	S23 05.4	346 15.6	16.1	N 7 28.8	13.2	54.7
19	104 15.4	05.2	0 50.7	16.2	7 42.0	13.2	54.7
20	119 15.1	05.0	15 25.9	16.1	7 55.2	13.2	54.7
21	134 14.8	. . 04.9	30 01.0	16.2	8 08.4	13.1	54.7
22	149 14.5	04.7	44 36.2	16.1	8 21.5	13.1	54.7
23	164 14.2	04.5	59 11.3	16.1	8 34.6	13.0	54.6
00	179 13.9	S23 04.3	73 46.4	16.1	N 8 47.6	13.0	54.6
01	194 13.6	04.1	88 21.5	16.0	9 00.6	13.0	54.6
02	209 13.3	03.9	102 56.5	16.1	9 13.6	12.9	54.6
03	224 13.0	. . 03.7	117 31.6	16.0	9 26.5	12.9	54.6
04	239 12.7	03.5	132 06.6	16.0	9 39.4	12.8	54.6
05	254 12.4	03.3	146 41.6	16.0	9 52.2	12.8	54.5
06	269 12.1	S23 03.1	161 16.6	16.0	N10 05.0	12.7	54.5
07	284 11.8	03.0	175 51.6	15.9	10 17.7	12.7	54.5
08	299 11.5	02.8	190 26.5	16.0	10 30.4	12.6	54.5
09	314 11.2	. . 02.6	205 01.5	15.9	10 43.0	12.7	54.5
10	329 10.9	02.4	219 36.4	15.8	10 55.7	12.5	54.5
11	344 10.6	02.2	234 11.2	15.9	11 08.2	12.5	54.4
12	359 10.3	S23 02.0	248 46.1	15.8	N11 20.7	12.5	54.4
13	14 10.0	01.8	263 20.9	15.8	11 33.2	12.4	54.4
14	29 09.7	01.6	277 55.7	15.7	11 45.6	12.3	54.4
15	44 09.4	. . 01.4	292 30.4	15.8	11 57.9	12.3	54.4
16	59 09.2	01.2	307 05.2	15.7	12 10.2	12.3	54.4
17	74 08.9	01.0	321 39.9	15.6	12 22.5	12.2	54.4
18	89 08.6	S23 00.8	336 14.5	15.7	N12 34.7	12.2	54.4
19	104 08.3	00.6	350 49.2	15.6	12 46.9	12.1	54.3
20	119 08.0	00.4	5 23.8	15.5	12 59.0	12.0	54.3
21	134 07.7	23 00.1	19 58.3	15.6	13 11.0	12.0	54.3
22	149 07.4	22 59.9	34 32.9	15.4	13 23.0	11.9	54.3
23	164 07.1	S22 59.7	49 07.3	15.5	N13 34.9	11.9	54.3
	SD 16.3	*d* 0.2	SD 15.1		14.9		14.8

Lat.	Twilight Naut.	Twilight Civil	Sunrise	Moonrise 30	Moonrise 31	Moonrise 1	Moonrise 2
°	h m	h m	h m	h m	h m	h m	h m
N 72	08 25	10 47	▬	11 59	11 30	10 56	10 05
N 70	08 06	09 52	▬	12 00	11 38	11 13	10 41
68	07 51	09 18	▬	12 00	11 45	11 27	11 06
66	07 38	08 54	10 30	12 01	11 50	11 39	11 26
64	07 27	08 35	09 51	12 01	11 55	11 49	11 42
62	07 18	08 19	09 24	12 01	11 59	11 57	11 56
60	07 10	08 06	09 03	12 02	12 03	12 04	12 07
N 58	07 02	07 55	08 46	12 02	12 06	12 11	12 17
56	06 56	07 45	08 32	12 02	12 09	12 17	12 26
54	06 50	07 36	08 19	12 03	12 12	12 22	12 33
52	06 44	07 28	08 08	12 03	12 14	12 26	12 40
50	06 39	07 20	07 59	12 03	12 16	12 31	12 47
45	06 27	07 04	07 38	12 03	12 21	12 40	13 00
N 40	06 17	06 51	07 22	12 04	12 25	12 47	13 12
35	06 08	06 40	07 08	12 04	12 29	12 54	13 21
30	05 59	06 29	06 56	12 04	12 32	13 00	13 30
20	05 43	06 11	06 35	12 05	12 37	13 10	13 45
N 10	05 27	05 54	06 16	12 05	12 42	13 19	13 58
0	05 10	05 37	05 59	12 06	12 47	13 28	14 10
S 10	04 52	05 19	05 42	12 06	12 51	13 36	14 22
20	04 29	04 58	05 23	12 07	12 56	13 46	14 36
30	04 01	04 34	05 01	12 07	13 02	13 56	14 51
35	03 42	04 19	04 48	12 08	13 05	14 02	15 00
40	03 19	04 01	04 34	12 08	13 09	14 09	15 10
45	02 49	03 39	04 16	12 09	13 13	14 18	15 22
S 50	02 05	03 09	03 54	12 09	13 19	14 28	15 37
52	01 38	02 55	03 43	12 10	13 21	14 32	15 44
54	00 56	02 37	03 31	12 10	13 24	14 37	15 51
56	////	02 15	03 17	12 10	13 27	14 43	16 00
58	////	01 47	03 01	12 11	13 30	14 49	16 10
S 60	////	01 01	02 41	12 11	13 34	14 57	16 21

Lat.	Sunset	Twilight Civil	Twilight Naut.	Moonset 30	Moonset 31	Moonset 1	Moonset 2
°	h m	h m	h m	h m	h m	h m	h m
N 72	▬	13 19	15 41	25 12	01 12	03 13	05 32
N 70	▬	14 14	16 00	25 06	01 06	02 57	04 57
68	▬	14 48	16 15	25 02	01 02	02 45	04 33
66	13 36	15 12	16 28	24 59	00 59	02 35	04 15
64	14 15	15 31	16 39	24 56	00 56	02 27	04 00
62	14 42	15 47	16 48	24 53	00 53	02 20	03 47
60	15 03	16 00	16 56	24 51	00 51	02 14	03 37
N 58	15 20	16 11	17 04	24 49	00 49	02 08	03 28
56	15 34	16 21	17 10	24 47	00 47	02 03	03 20
54	15 47	16 30	17 16	24 46	00 46	01 59	03 13
52	15 58	16 38	17 22	24 44	00 44	01 55	03 07
50	16 07	16 46	17 27	24 43	00 43	01 52	03 01
45	16 28	17 01	17 38	24 40	00 40	01 44	02 49
N 40	16 44	17 15	17 49	24 38	00 38	01 38	02 39
35	16 58	17 26	17 58	24 36	00 36	01 33	02 30
30	17 10	17 37	18 06	24 34	00 34	01 28	02 23
20	17 31	17 55	18 23	24 31	00 31	01 20	02 10
N 10	17 49	18 12	18 38	24 28	00 28	01 13	01 58
0	18 07	18 29	18 55	24 25	00 25	01 06	01 48
S 10	18 24	18 47	19 14	24 23	00 23	01 00	01 37
20	18 43	19 07	19 36	24 20	00 20	00 53	01 26
30	19 04	19 32	20 05	24 17	00 17	00 45	01 14
35	19 17	19 47	20 23	24 15	00 15	00 40	01 06
40	19 32	20 05	20 46	24 13	00 13	00 35	00 58
45	19 50	20 27	21 16	24 11	00 11	00 29	00 49
S 50	20 12	20 56	22 00	24 08	00 08	00 22	00 37
52	20 22	21 11	22 26	24 07	00 07	00 19	00 32
54	20 34	21 28	23 08	24 06	00 06	00 15	00 26
56	20 48	21 49	////	24 04	00 04	00 11	00 19
58	21 05	22 18	////	24 03	00 03	00 07	00 12
S 60	21 24	23 02	////	24 01	00 01	00 02	00 04

Day	SUN Eqn. of Time 00^h	SUN Eqn. of Time 12^h	SUN Mer. Pass.	MOON Mer. Pass. Upper	MOON Mer. Pass. Lower	Age	Phase
d	m s	m s	h m	h m	h m	d	%
30	02 06	02 21	12 02	18 16	05 55	07	51
31	02 35	02 50	12 03	18 57	06 36	08	61
1	03 04	03 18	12 03	19 38	07 17	09	70

EXPLANATION

PRINCIPLE AND ARRANGEMENT

1. *Object.* The object of this Almanac is to provide, in a convenient form, the data required for the practice of astronomical navigation at sea.

2. *Principle.* The main contents of the Almanac consist of data from which the *Greenwich Hour Angle* (GHA) and the *Declination* (Dec) of all the bodies used for navigation can be obtained for any instant of *Universal Time* (UT), or *Greenwich Mean Time* (GMT). The *Local Hour Angle* (LHA) can then be obtained by means of the formula:

$$\text{LHA} = \text{GHA} \begin{smallmatrix} -\text{ west} \\ +\text{ east} \end{smallmatrix} \text{ longitude}$$

The remaining data consist of: times of rising and setting of the Sun and Moon, and times of twilight; miscellaneous calendarial and planning data and auxiliary tables, including a list of Standard Times; corrections to be applied to observed altitude.

For the Sun, Moon, and planets the GHA and Dec are tabulated directly for each hour of UT throughout the year. For the stars the *Sidereal Hour Angle* (SHA) is given, and the GHA is obtained from:

$$\text{GHA Star} = \text{GHA Aries} + \text{SHA Star}$$

The SHA and Dec of the stars change slowly and may be regarded as constant over periods of several days. GHA Aries, or the Greenwich Hour Angle of the first point of Aries (the Vernal Equinox), is tabulated for each hour. Permanent tables give the appropriate increments and corrections to the tabulated hourly values of GHA and Dec for the minutes and seconds of UT.

The six-volume series of *Sight Reduction Tables for Marine Navigation* (published in U.S.A. as Pub. No. 229 and in U.K. as N.P. 401) has been designed for the solution of the navigational triangle and is intended for use with *The Nautical Almanac*.

Two alternative procedures for sight reduction are described on pages 277–318. The first requires the use of programmable calculators or computers, while the second uses a set of concise tables that is given on pages 286–317.

The tabular accuracy is 0′.1 throughout. The time argument on the daily pages of this Almanac is 12^h + the Greenwich Hour Angle of the mean sun and is here denoted by UT, although it is also known as GMT. This scale may differ from the broadcast time signals (UTC) by an amount which, if ignored, will introduce an error of up to 0′.2 in longitude determined from astronomical observations. (The difference arises because the time argument depends on the variable rate of rotation of the Earth while the broadcast time signals are now based on an atomic time-scale.) Step adjustments of exactly one second are made to the time signals as required (normally at 24^h on December 31 and June 30) so that the difference between the time signals and UT, as used in this Almanac, may not exceed $0^s.9$. Those who require to reduce observations to a precision of better than 1^s must therefore obtain the correction (DUT1) to the time signals from coding in the signal, or from other sources; the required time is given by UT1=UTC+DUT1 to a precision of $0^s.1$. Alternatively, the longitude, when determined from astronomical observations, may be corrected by the corresponding amount shown in the following table:

Correction to time signals	Correction to longitude
$-0^s.9$ to $-0^s.7$	0′.2 to east
$-0^s.6$ to $-0^s.3$	0′.1 to east
$-0^s.2$ to $+0^s.2$	no correction
$+0^s.3$ to $+0^s.6$	0′.1 to west
$+0^s.7$ to $+0^s.9$	0′.2 to west

3. *Lay-out.* The ephemeral data for three days are presented on an opening of two pages: the left-hand page contains the data for the planets and stars; the right-hand page contains the data for the Sun and Moon, together with times of twilight, sunrise, sunset, moonrise and moonset.

The remaining contents are arranged as follows: for ease of reference the altitude-correction tables are given on pages A2, A3, A4, xxxiv and xxxv; calendar, Moon's phases, eclipses, and planet notes (i.e. data of general interest) precede the main tabulations. The Explanation is followed by information on standard times, star charts and list of star positions, sight reduction procedures and concise sight reduction tables, tables of increments and corrections and other auxiliary tables that are frequently used.

MAIN DATA

4. *Daily pages.* The daily pages give the GHA of Aries, the GHA and Dec of the Sun, Moon, and the four navigational planets, for each hour of UT. For the Moon, values of v and d are also tabulated for each hour to facilitate the correction of GHA and Dec to intermediate times; v and d for the Sun and planets change so slowly that they are given, at the foot of the appropriate columns, once only on the page; v is zero for Aries and negligible for the Sun, and is omitted. The SHA and Dec of the 57 selected stars, arranged in alphabetical order of proper name, are also given.

5. *Stars.* The SHA and Dec of 173 stars, including the 57 selected stars, are tabulated for each month on pages 268–273; no interpolation is required and the data can be used in precisely the same way as those for the selected stars on the daily pages. The stars are arranged in order of SHA.

The list of 173 includes all stars down to magnitude 3·0, together with a few fainter ones to fill the larger gaps. The 57 selected stars have been chosen from amongst these on account of brightness and distribution in the sky; they will suffice for the majority of observations.

The 57 selected stars are known by their proper names, but they are also numbered in descending order of SHA. In the list of 173 stars, the constellation names are always given on the left-hand page; on the facing page proper names are given where well-known names exist. Numbers for the selected stars are given in both columns.

An index to the selected stars, containing lists in both alphabetical and numerical order, is given on page xxxiii and is also reprinted on the bookmark.

6. *Increments and corrections.* The tables printed on tinted paper (pages ii–xxxi) at the back of the Almanac provide the increments and corrections for minutes and seconds to be applied to the hourly values of GHA and Dec. They consist of sixty tables, one for each minute, separated into two parts: increments to GHA for Sun and planets, Aries, and Moon for every minute and second; and, for each minute, corrections to be applied to GHA and Dec corresponding to the values of v and d given on the daily pages.

The increments are based on the following adopted hourly rates of increase of the GHA: Sun and planets, 15° precisely; Aries, 15° 02′.46; Moon, 14° 19′.0. The values of v on the daily pages are the excesses of the actual hourly motions over the adopted values; they are generally positive, except for Venus. The tabulated hourly values of the Sun's GHA have been adjusted to reduce to a minimum the error caused by treating v as negligible. The values of d on the daily pages are the hourly differences of the Dec. For the Moon, the true values of v and d are given for each hour; otherwise mean values are given for the three days on the page.

7. *Method of entry.* The UT of an observation is expressed as a day and hour, followed by a number of minutes and seconds. The tabular values of GHA and Dec, and, where necessary, the corresponding values of v and d, are taken directly from the daily pages for the day and hour of UT; this hour is always *before* the time of observation. SHA and Dec of the selected stars are also taken from the daily pages.

The table of Increments and Corrections for the minute of UT is then selected. For the GHA, the increment for minutes and seconds is taken from the appropriate column opposite the seconds of UT; the v-correction is taken from the second part of the same table opposite the value of v as given on the daily pages. Both increment and v-correction are to be added to the GHA, except for Venus when v is prefixed by a minus sign and the v-correction is to be subtracted. For the Dec there is no increment, but a d-correction is applied in the same way as the v-correction; d is given without sign on the daily pages and the sign of the correction is to be supplied by inspection of the Dec column. In many cases the correction may be applied mentally.

8. *Examples.* (a) Sun and Moon. Required the GHA and Dec of the Sun and Moon on 2003 December 1 at $15^h\ 47^m\ 13^s$ UT.

		SUN			MOON			
		GHA	Dec	d	GHA	v	Dec	d
		° ′	° ′	′	° ′	′	° ′	′
Daily page, December $1^d\ 15^h$		47 45·8	S 21 48·0	0·4	302 45·2	14·9	S 8 20·2	13·8
Increments for	$47^m\ 13^s$	11 48·3			11 16·0			
v or d corrections for	47^m		+0·3		+11·8		−10·9	
Sum for December $1^d\ 15^h\ 47^m\ 13^s$		59 34·1	S 21 48·3		314 13·0		S 8 09·3	

(b) Planets. Required the LHA and Dec of (i) Venus on 2003 December 1 at $13^h\ 20^m\ 02^s$ UT in longitude E 89° 20′; (ii) Jupiter on 2003 December 1 at $7^h\ 18^m\ 32^s$ UT in longitude W 34° 43′.

		VENUS					JUPITER			
		GHA	v	Dec	d		GHA	v	Dec	d
		° ′	′	° ′	′		° ′	′	° ′	′
Daily page, Dec. 1^d	(13^h)	348 49·9	−0·9	S24 43·9	0·1	(7^h)	6 07·7	2·2	N6 03·7	0·1
Increments (planets)	($20^m\ 02^s$)	5 00·5				($18^m\ 32^s$)	4 38·0			
v or d corrections	(20^m)	−0·3		+0·0		(18^m)	+0·7		+0·0	
Sum = GHA and Dec.		353 50·1		S24 43·9			10 46·4		N6 03·7	
Longitude	(east)	+ 89 20·0				(west)	− 34 43·0			
Multiples of 360°		−360					+360			
LHA planet		83 10·1					336 03·4			

(c) Stars. Required the GHA and Dec of (i) *Sirius* on 2003 December 1 at $7^h\ 18^m\ 32^s$ UT; (ii) *Vega* on 2003 December 1 at $20^h\ 48^m\ 32^s$ UT.

		Sirius			*Vega*	
		GHA	Dec		GHA	Dec
		° ′	° ′		° ′	° ′
Daily page (SHA and Dec)		258 40·1	S 16 43.1		80 44·5	N 38 47.3
Daily page (GHA Aries)	(7^h)	174 43·7		(20^h)	10 15·7	
Increments (Aries)	($18^m\ 32^s$)	4 38·8		($48^m\ 32^s$)	12 10·0	
Sum = GHA star		438 02·6			103 10·2	
Multiples of 360°		−360				
GHA star		78 02·6			103 10·2	

9. *Polaris (Pole Star) tables.* The tables on pages 274–276 provide means by which the latitude can be deduced from an observed altitude of *Polaris*, and they also give its azimuth; their use is explained and illustrated on those pages. They are based on the following formula:

$$\text{Latitude} - H_O = -p\cos h + \tfrac{1}{2}p\ \sin p\ \sin^2 h\ \tan(\text{latitude})$$

where H_O = Apparent altitude (corrected for refraction)

p = polar distance of *Polaris* = 90° − Dec

h = local hour angle of *Polaris* = LHA Aries + SHA

a_0, which is a function of LHA Aries only, is the value of both terms of the above formula calculated for mean values of the SHA (321° 12′) and Dec (N 89° 16′·8) of *Polaris*, for a mean latitude of 50°, and adjusted by the addition of a constant (58′·8).

a_1, which is a function of LHA Aries and latitude, is the excess of the value of the second term over its mean value for latitude 50°, increased by a constant (0′.6) to make it always positive. a_2, which is a function of LHA Aries and date, is the correction to the first term for the variation of *Polaris* from its adopted mean position; it is increased by a constant (0′.6) to make it positive. The sum of the added constants is 1°, so that:

$$\text{Latitude} = \text{Apparent altitude (corrected for refraction)} - 1^\circ + a_0 + a_1 + a_2$$

RISING AND SETTING PHENOMENA

10. *General.* On the right-hand daily pages are given the times of sunrise and sunset, of the beginning and end of civil and nautical twilights, and of moonrise and moonset for a range of latitudes from N 72° to S 60°. These times, which are given to the nearest minute, are strictly the UT of the phenomena on the Greenwich meridian; they are given for every day for moonrise and moonset, but only for the middle day of the three on each page for the solar phenomena.

They are approximately the Local Mean Times (LMT) of the corresponding phenomena on other meridians; they can be formally interpolated if desired. The UT of a phenomenon is obtained from the LMT by:

$$\text{UT} = \text{LMT} \begin{matrix} + \text{ west} \\ - \text{ east} \end{matrix} \text{ longitude}$$

in which the longitude must first be converted to time by the table on page i or otherwise.

Interpolation for latitude can be done mentally or with the aid of Table I on page xxxii.

The following symbols are used to indicate the conditions under which, in high latitudes, some of the phenomena do not occur:

□ Sun or Moon remains continuously above the horizon;

■ Sun or Moon remains continuously below the horizon;

//// twilight lasts all night.

Basis of the tabulations. At sunrise and sunset 16′ is allowed for semi-diameter and 34′ for horizontal refraction, so that at the times given the Sun's upper limb is on the visible horizon; all times refer to phenomena as seen from sea level with a clear horizon.

At the times given for the beginning and end of twilight, the Sun's zenith distance is 96° for civil, and 102° for nautical twilight. The degree of illumination at the times given for civil twilight (in good conditions and in the absence of other illumination) is such that the brightest stars are visible and the horizon is clearly defined. At the times given for nautical twilight the horizon is in general not visible, and it is too dark for observation with a marine sextant.

Times corresponding to other depressions of the Sun may be obtained by interpolation or, for depressions of more than 12°, less reliably, by extrapolation; times so obtained will be subject to considerable uncertainty near extreme conditions.

At moonrise and moonset allowance is made for semi-diameter, parallax, and refraction (34′), so that at the times given the Moon's upper limb is on the visible horizon as seen from sea level.

11. *Sunrise, sunset, twilight.* The tabulated times may be regarded, without serious error, as the LMT of the phenomena on any of the three days on the page and in any longitude. Precise times may normally be obtained by interpolating the tabular values for latitude and to the correct day and longitude, the latter being expressed as a fraction of a day by dividing it by 360°, positive for west and negative for east longitudes. In the extreme conditions near □, ■ or //// interpolation may not be possible in one direction, but accurate times are of little value in these circumstances.

Examples. Required the UT of (a) the beginning of morning twilights and sunrise on 2003 January 22 for latitude S 48° 55′, longitude E 75° 18′; (b) sunset and the end of evening twilights on 2003 January 24 for latitude N 67° 10′, longitude W 168° 05′.

	(a)	Twilight Nautical	Twilight Civil	Sunrise	(b)	Sunset	Twilight Civil	Twilight Nautical
From p. 25		d h m	d h m	d h m		d h m	d h m	d h m
LMT for Lat	S 45°	22 03 24	22 04 08	22 04 43	N 66°	24 14 56	24 16 06	24 17 14
Corr. to (p. xxxii, Table I)	S 48° 55′	−27	−20	−16	N 67° 10′	−17	−9	−6
Long (p. i)	E 75° 18′	−5 01	−5 01	−5 01	W 168° 05′	+11 12	+11 12	+11 12
UT		21 21 56	21 22 47	21 23 26		25 01 51	25 03 09	25 04 20

The LMT are strictly for January 23 (middle date on page) and 0° longitude; for more precise times it is necessary to interpolate, but rounding errors may accumulate to about 2^{m}.

(a) to January $22^{d} - 75°/360° =$ Jan. 21ᵈ8, i.e. $\frac{1}{3}(1{\cdot}2) = 0{\cdot}4$ backwards towards the data for the same latitude interpolated similarly from page 23; the corrections are -2^{m} to nautical twilight, -2^{m} to civil twilight and -2^{m} to sunrise.

(b) to January $24^{d} + 168°/360° =$ Jan. 24ᵈ5, i.e. $\frac{1}{3}(1{\cdot}5) = 0{\cdot}5$ forwards towards the data for the same latitude interpolated similarly from page 27; the corrections are $+7^{m}$ to sunset, $+5^{m}$ to civil twilight, and $+4^{m}$ to nautical twilight.

12. *Moonrise, moonset.* Precise times of moonrise and moonset are rarely needed; a glance at the tables will generally give sufficient indication of whether the Moon is available for observation and of the hours of rising and setting. If needed, precise times may be obtained as follows. Interpolate for latitude, using Table I on page xxxii, on the day wanted and also on the preceding day in east longitudes or the following day in west longitudes; take the difference between these times and interpolate for longitude by applying to the time for the day wanted the correction from Table II on page xxxii, so that the resulting time is between the two times used. In extreme conditions near ☐ or ■ interpolation for latitude or longitude may be possible only in one direction; accurate times are of little value in these circumstances.

To facilitate this interpolation the times of moonrise and moonset are given for four days on each page; where no phenomenon occurs during a particular day (as happens once a month) the time of the phenomenon on the following day, increased by 24^{h}, is given; extra care must be taken when interpolating between two values, when one of those values exceeds 24^{h}. In practice it suffices to use the daily difference between the times for the nearest tabular latitude, and generally, to enter Table II with the nearest tabular arguments as in the examples below.

Examples. Required the UT of moonrise and moonset in latitude S 47° 10′, longitudes E 124° 00′ and W 78° 31′ on 2003 January 8.

	Longitude E 124° 00′ Moonrise	Longitude E 124° 00′ Moonset	Longitude W 78° 31′ Moonrise	Longitude W 78° 31′ Moonset
	d h m	d h m	d h m	d h m
LMT for Lat. S 45°	8 10 11	8 23 14	8 10 11	8 23 14
Lat correction (p. xxxii, Table I)	−02	+02	−02	+02
Long correction (p. xxxii, Table II)	−23	−07	+13	+04
Correct LMT	8 09 46	8 23 09	8 10 22	8 23 20
Longitude (p. i)	−8 16	−8 16	+5 14	+5 14
UT	8 01 30	8 14 53	8 15 36	9 04 34

ALTITUDE CORRECTION TABLES

13. *General.* In general two corrections are given for application to altitudes observed with a marine sextant; additional corrections are required for Venus and Mars and also for very low altitudes.

Tables of the correction for dip of the horizon, due to height of eye above sea level, are given on pages A2 and xxxiv. Strictly this correction should be applied first and subtracted from the sextant altitude to give apparent altitude, which is the correct argument for the other tables.

Separate tables are given of the second correction for the Sun, for stars and planets (on pages A2 and A3), and for the Moon (on pages xxxiv and xxxv). For the Sun, values are given for both lower and upper limbs, for two periods of the year. The star tables are used for the planets, but additional corrections for parallax (page A2) are required for Venus and Mars. The Moon tables are in two parts: the main correction is a function of apparent altitude only and is tabulated for the lower limb (30′ must be subtracted to obtain the correction for the upper limb); the other, which is given for both lower and upper limbs, depends also on the horizontal parallax, which has to be taken from the daily pages.

An additional correction, given on page A4, is required for the change in the refraction, due to variations of pressure and temperature from the adopted standard conditions; it may generally be ignored for altitudes greater than 10°, except possibly in extreme conditions. The correction tables for the Sun, stars, and planets are in two parts; only those for altitudes greater than 10° are reprinted on the bookmark.

14. *Critical tables.* Some of the altitude correction tables are arranged as critical tables. In these an interval of apparent altitude (or height of eye) corresponds to a single value of the correction; no interpolation is required. At a "critical" entry the upper of the two possible values of the correction is to be taken. For example, in the table of dip, a correction of −4′.1 corresponds to all values of the height of eye from 5·3 to 5·5 metres (17·5 to 18·3 feet) inclusive.

15. *Examples.* The following examples illustrate the use of the altitude correction tables; the sextant altitudes given are assumed to be taken on 2003 December 1 with a marine sextant at height 5·4 metres (18 feet), temperature −3°C and pressure 982 mb, the Moon sights being taken at about 10ʰ UT.

	SUN lower limb	SUN upper limb	MOON lower limb	MOON upper limb	VENUS	*Polaris*
	° ′	° ′	° ′	° ′	° ′	° ′
Sextant altitude	21 19·7	3 20·2	33 27·6	26 06·7	4 32·6	49 36·5
Dip, height 5·4 metres (18 feet)	−4·1	−4·1	−4·1	−4·1	−4·1	−4·1
Main correction	+13·8	−29·6	+57·4	+60·5	−10·8	−0·8
−30′ for upper limb (Moon)	—	—	—	−30·0	—	—
L, U correction for Moon	—	—	+3·2	+2·6	—	—
Additional correction for Venus	—	—	—	—	+0·1	—
Additional refraction correction	−0·1	−0·3	0·0	−0·1	−0·3	0·0
Corrected sextant altitude	21 29·3	2 46·2	34 24·1	26 35·6	4 17·5	49 31·6

The main corrections have been taken out with apparent altitude (sextant altitude corrected for dip) as argument, interpolating where possible. These refinements are rarely necessary.

16. *Composition of the Corrections.* The table for the dip of the sea horizon is based on the formula:

$$\text{Correction for dip} = -1'.76\sqrt{(\text{height of eye in metres})} = -0'.97\sqrt{(\text{height of eye in feet})}$$

The correction table for the Sun includes the effects of semi-diameter, parallax and mean refraction.

The correction tables for the stars and planets allow for the effect of mean refraction.

The phase correction for Venus has been incorporated in the tabulations for GHA and Dec, and no correction for phase is required. The additional corrections for Venus and Mars allow for parallax. Alternatively, the correction for parallax may be calculated from $p \cos H$, where p is the parallax and H is the altitude. In 2003 the values for p are:

	Jan. 1		Feb. 20		Dec. 31
Venus		0′.2		0′.1	

	Jan. 1		May 2		June 26		Aug. 1		Sept. 22		Oct. 25		Dec. 16		Dec. 31
Mars		0′.1		0′.2		0′.3		0′.4		0′.3		0′.2		0′.1	

The correction table for the Moon includes the effect of semi-diameter, parallax, augmentation and mean refraction.

Mean refraction is calculated for a temperature of 10°C (50°F) and a pressure of 1010 mb (29·83 inches).

17. *Bubble sextant observations.* When observing with a bubble sextant no correction is necessary for dip, semi-diameter, or augmentation. The altitude corrections for the stars and planets on page A2 and on the bookmark should be used for the Sun as well as for the stars and planets; for the Moon it is easiest to take the mean of the corrections for lower and upper limbs and subtract 15′ from the altitude; the correction for dip must not be applied.

AUXILIARY AND PLANNING DATA

18. *Sun and Moon.* On the daily pages are given: hourly values of the horizontal parallax of the Moon; the semi-diameters and the times of meridian passage of both Sun and Moon over the Greenwich meridian; the equation of time; the age of the Moon, the percent (%) illuminated and a symbol indicating the phase. The times of the phases of the Moon are given in UT on page 4. For the Moon, the semi-diameters for each of the three days are given at the foot of the column; for the Sun a single value is sufficient. Table II on page xxxii may be used for interpolating the time of the Moon's meridian passage for longitude. The equation of time is given daily at 00^{h} and 12^{h} UT. The sign is *positive* for unshaded values and *negative* for shaded values. To obtain apparent time add the equation of time to mean time when the sign is *positive*. Subtract the equation of time from mean time when the sign is *negative*. At 12^{h} UT, when the sign is *positive*, meridian passage of the Sun occurs *before* 12^{h} UT, otherwise it occurs *after* 12^{h} UT.

19. *Planets.* The magnitudes of the planets are given immediately following their names in the headings on the daily pages; also given, for the middle day of the three on the page, are their SHA at 00^{h} UT and their times of meridian passage.

The planet notes and diagram on pages 8 and 9 provide descriptive information as to the suitability of the planets for observation during the year, and of their positions and movements.

20. *Stars.* The time of meridian passage of the first point of Aries over the Greenwich meridian is given on the daily pages, for the middle day of the three on the page, to $0^{m}\!.1$. The interval between successive meridian passages is $23^{h}\ 56^{m}\!.1$ (24^{h} less $3^{m}\!.9$) so that times for intermediate days and other meridians can readily be derived. If a precise time is required it may be obtained by finding the UT at which LHA Aries is zero.

The meridian passage of a star occurs when its LHA is zero, that is when LHA Aries + SHA = 360°. An approximate time can be obtained from the planet diagram on page 9.

The star charts on pages 266 and 267 are intended to assist identification. They show the relative positions of the stars in the sky as seen from the Earth and include all 173 stars used in the Almanac, together with a few others to complete the main constellation configurations. The local meridian at any time may be located on the chart by means of its SHA which is 360° − LHA Aries, or west longitude − GHA Aries.

21. *Star globe.* To set a star globe on which is printed a scale of LHA Aries, first set the globe for latitude and then rotate about the polar axis until the scale under the edge of the meridian circle reads LHA Aries.

To mark the positions of the Sun, Moon, and planets on the star globe, take the difference GHA Aries − GHA body and use this along the LHA Aries scale, in conjunction with the declination, to plot the position. GHA Aries − GHA body is most conveniently found by taking the difference when the GHA of the body is small (less than 15°), which happens once a day.

22. *Calendar.* On page 4 are given lists of ecclesiastical festivals, and of the principal anniversaries and holidays in the United Kingdom and the United States of America. The calendar on page 5 includes the day of the year as well as the day of the week.

Brief particulars are given, at the foot of page 5, of the solar and lunar eclipses occurring during the year; the times given are in UT. The principal features of the more important solar eclipses are shown on the maps on pages 6 and 7.

23. *Standard times.* The lists on pages 262–265 give the standard times used in most countries. In general no attempt is made to give details of the beginning and end of summer time, since they are liable to frequent changes at short notice. For the latest information consult Admiralty List of Radio Signals Volume 2 (NP 282) corrected by Section VI of the weekly edition of Admiralty Notices to Mariners.

The Date or Calendar Line is an arbitrary line, on either side of which the date differs by one day; when crossing this line on a westerly course, the date must be advanced one day; when crossing it on an easterly course, the date must be put back one day. The line is a modification of the line of the 180th meridian, and is drawn so as to include, as far as possible, islands of any one group, etc., on the same side of the line. It may be traced by starting at the South Pole and joining up to the following positions:

	°	°	°	°	°	°	°
Lat	S 51·0	S 45·0	S 15·0	S 5·0	N 48·0	N 53·0	N 65·5
Long	180·0	W 172·5	W 172·5	180·0	180·0	E 170·0	W 169·0

thence through the middle of the Diomede Islands to Lat N 68°0, Long W 169°0, passing east of Ostrov Vrangelya (Wrangel Island) to Lat N 75°0, Long 180°0, and thence to the North Pole.

ACCURACY

24. *Main data.* The quantities tabulated in this Almanac are generally correct to the nearest 0′1; the exception is the Sun's GHA which is deliberately adjusted by up to 0′15 to reduce the error due to ignoring the v-correction. The GHA and Dec at intermediate times cannot be obtained to this precision, since at least two quantities must be added; moreover, the v- and d-corrections are based on mean values of v and d and are taken from tables for the whole minute only. The largest error that can occur in the GHA or Dec of any body other than the Sun or Moon is less than 0′2; it may reach 0′25 for the GHA of the Sun and 0′3 for that of the Moon.

In practice it may be expected that only one third of the values of GHA and Dec taken out will have errors larger than 0′05 and less than one tenth will have errors larger than 0′1.

25. *Altitude corrections.* The errors in the altitude corrections are nominally of the same order as those in GHA and Dec, as they result from the addition of several quantities each correctly rounded off to 0′1. But the actual values of the dip and of the refraction at low altitudes may, in extreme atmospheric conditions, differ considerably from the mean values used in the tables.

USE OF THIS ALMANAC IN 2004

This Almanac may be used for the Sun and stars in 2004 in the following manner.

For the Sun, take out the GHA and Dec for the same date but, for January and February (for February 29 use March 1), for a time $5^h\ 48^m\ 00^s$ *earlier* and, for March to December, for a time $18^h\ 12^m\ 00^s$ *later* than the UT of observation; in both cases *add* 87° 00′ to the GHA so obtained. The error, mainly due to planetary perturbations of the Earth, is unlikely to exceed 0′4.

For the stars, calculate the GHA and Dec for the same date and the same time, but for January and February (for February 29 use March 1) *subtract* 15′1 and for March to December *add* 44′0 to the GHA so found. The error, due to incomplete correction for precession and nutation, is unlikely to exceed 0′4. If preferred, the same result can be obtained by using a time $5^h\ 48^m\ 00^s$ earlier for January and February (for February 29 use March 1) and $18^h\ 12^m\ 00^s$ later for March to December, than the UT of observation (as for the Sun) and adding 86° 59′2 to the GHA (or adding 87° as for the Sun and subtracting 0′8, for precession, from the SHA of the star).

The Almanac cannot be so used for the Moon or planets.

LIST I — PLACES FAST ON UTC (mainly those EAST OF GREENWICH)

The times given below should be } *added* to UTC to give Standard Time / *subtracted* from Standard Time to give UTC.

Place	h	m
Admiralty Islands	10	
Afghanistan	04	30
Albania*	01	
Algeria	01	
Amirante Islands	04	
Andaman Islands	05	30
Angola	01	
Armenia*	04	
Australia		
Australian Capital Territory*	10	
New South Wales[1]*	10	
Northern Territory	09	30
Queensland	10	
South Australia*	09	30
Tasmania*	10	
Victoria*	10	
Western Australia	08	
Whitsunday Islands	10	
Austria*†	01	
Azerbaijan*	04	
Bahrain	03	
Balearic Islands*	01	
Bangladesh	06	
Belarus*	02	
Belgium*†	01	
Benin	01	
Bosnia and Herzegovina*	01	
Botswana, Republic of	02	
Brunei	08	
Bulgaria*	02	
Burma (Myanmar)	06	30
Burundi	02	
Cambodia	07	
Cameroon Republic	01	
Caroline Islands[2]	10	
Central African Republic	01	
Chad	01	
Chagos Archipelago	05	
Chatham Islands*	12	45
China, People's Republic of	08	
Christmas Island, Indian Ocean	07	
Cocos (Keeling) Islands	06	30
Comoro Islands (Comoros)	03	
Congo, Democratic Republic		
Kinshasa, Mbandaka	01	
Haut-Zaire, Kasai, Kivu, Shaba	02	
Congo Republic	01	
Corsica*	01	
Crete*	02	
Croatia*	01	
Cyprus: Ercan*, Larnaca*	02	
Czech Republic*	01	

Place	h	m
Denmark*†	01	
Djibouti	03	
East Timor	09	
Egypt, Arab Republic of*	02	
Equatorial Guinea, Republic of	01	
Eritrea	03	
Estonia	02	
Ethiopia	03	
Fiji	12	
Finland*†	02	
France*†	01	
Gabon	01	
Georgia*	04	
Germany*†	01	
Gibraltar*	01	
Greece*†	02	
Guam	10	
Hong Kong	08	
Hungary*	01	
India	05	30
Indonesia, Republic of		
Bangka, Billiton, Java, West and Central Kalimantan, Madura, Sumatra	07	
Bali, Flores, South and East Kalimantan, Lombok, Sulawesi, Sumba, Sumbawa, West Timor	08	
Aru, Irian Jaya, Kai, Moluccas, Tanimbar	09	
Iran*	03	30
Iraq*	03	
Israel*	02	
Italy*†	01	
Jan Mayen Island*	01	
Japan	09	
Jordan*	02	
Kazakhstan*		
Western, Aktau, Uralsk, Atyrau	04	
Central, Aktyubinsk, Kzyl-Orda	05	
Eastern	06	
Kenya	03	
Kiribati Republic		
Gilbert Islands	12	
Phoenix Islands[3]	13	
Line Islands[3]	14	
Korea, North,	09	
Republic of (South)	09	
Kuril Islands	11	
Kuwait	03	
Kyrgyzstan*	05	
Laccadive Islands	05	30
Laos	07	
Latvia*	02	

* Daylight-saving time may be kept in these places. † For Summer time dates see List II footnotes.

1 Except Broken Hill Area which keeps $09^h\ 30^m$.

2 Except Pohnpei, Pingelap and Kosrae which keep 11^h.

3 The Line and Phoenix Is. not part of the Kiribati Republic keep 10^h and 11^h, respectively, slow on UTC.

LIST I — (*continued*)

	h m
Lebanon*	02
Lesotho	02
Libya	02
Liechtenstein*	01
Lithuania	02
Lord Howe Island*	10 30
Luxembourg*†	01
Macau	08
Macedonia*, former Yugoslav Republic	01
Macias Nguema (Fernando Póo)	01
Madagascar, Democratic Republic of	03
Malawi	02
Malaysia, Malaya, Sabah, Sarawak	08
Maldives, Republic of The	05
Malta*	01
Mariana Islands	10
Marshall Islands[1]	12
Mauritius	04
Moldova*	02
Monaco*	01
Mongolia*[3]	09
Mozambique	02
Namibia*	01
Nauru	12
Nepal	05 45
Netherlands, The*†	01
New Caledonia	11
New Zealand*	12
Nicobar Islands	05 30
Niger	01
Nigeria, Republic of	01
Norfolk Island	11 30
Norway*	01
Novaya Zemlya	03
Okinawa	09
Oman	04
Pagalu (Annobon Islands)	01
Pakistan	05
Palau Islands	09
Papua New Guinea	10
Pescadores Islands	08
Philippine Republic	08
Poland*	01
Qatar	03
Reunion	04
Romania*	02
Russia[2]*	
Zone 1 Kaliningrad	02
Zone 2 Moscow, St Petersburg, Arkhangelsk, Astrakhan	03
Zone 3 Samara, Izhevsk	04
Zone 4 Perm, Amderna, Novyy Port	05
Zone 5 Omsk, Novosibirsk	06
Zone 6 Norilsk, Kyzyl, Dikson	07
Russia (*continued*)	
Zone 7 Bratsk, Irkutsk, Ulan-Ude	08
Zone 8 Yakutsk, Chita, Tiksi	09
Zone 9 Vladivostok, Khabarovsk, Okhotsk	10
Zone 10 Magadan	11
Zone 11 Petropavlovsk, Pevek	12
Rwanda	02
Ryukyu Islands	09
Sakhalin Island*	10
Santa Cruz Islands	11
Sardinia*	01
Saudi Arabia	03
Schouten Islands	10
Serbia*	01
Seychelles	04
Sicily*	01
Singapore	08
Slovakia*	01
Slovenia*	01
Socotra	03
Solomon Islands	11
Somalia Republic	03
South Africa, Republic of	02
Spain*†	01
Spanish Possessions in North Africa*	01
Spitsbergen (Svalbard)*	01
Sri Lanka	06
Sudan, Republic of	03
Swaziland	02
Sweden*†	01
Switzerland*	01
Syria (Syrian Arab Republic)*	02
Taiwan	08
Tajikistan	05
Tanzania	03
Thailand	07
Tonga*	13
Tunisia	01
Turkey*	02
Turkmenistan	05
Tuvalu	12
Uganda	03
Ukraine*	02
United Arab Emirates	04
Uzbekistan	05
Vanuatu, Republic of	11
Vietnam, Socialist Republic of	07
Yemen	03
Yugoslavia*, Federal Republic of	01
Zambia, Republic of	02
Zimbabwe	02

* Daylight-saving time may be kept in these places. † For Summer time dates see List II footnotes.

1 Except the Ebon Atol which keeps time 24^h slow on that of the rest of the islands.

2 The boundaries between the zones are irregular; listed are chief towns in each zone.

3 Western and eastern regions of Mongolia keep 08^h and 10^h, respectively.

LIST II — PLACES NORMALLY KEEPING UTC

Ascension Island	Ghana	Irish Republic*†	Morocco	Sierra Leone
Burkina-Faso	Great Britain†	Ivory Coast	Portugal*†	Togo Republic
Canary Islands*	Guinea-Bissau	Liberia	Principe	Tristan da Cunha
Channel Islands†	Guinea Republic	Madeira*	St. Helena	
Faeroes*, The	Iceland	Mali	São Tomé	
Gambia, The	Ireland, Northern†	Mauritania	Senegal	

* Daylight-saving time may be kept in these places.

† Summer time (daylight-saving time), one hour in advance of UTC, will be kept from 2002 March 31^{d} 01^{h} to October 27^{d} 01^{h} UTC (Ninth Summer Time Directive of the European Union). Ratification by member countries has not been verified.

LIST III — PLACES SLOW ON UTC (WEST OF GREENWICH)

The times given below should be *subtracted* from UTC to give Standard Time / *added* to Standard Time to give UTC.

Place	h	m
Argentina	03	
Austral (Tubuai) Islands[1]	10	
Azores*	01	
Bahamas*	05	
Barbados	04	
Belize	06	
Bermuda*	04	
Bolivia	04	
Brazil		
Oceanic Islands	02	
S and E coastal states, Bahia, Goiás, Tocantins, Brasilia*	03	
N and NE coastal states, Eastern Para	03	
Mato Grosso, Mato Grosso do Sul*	04	
Central states, western Para, E and Central Amazonas	04	
Acre, western Amazonas	05	
British Antarctic Territory[2]	03	
Canada		
Alberta*	07	
British Columbia[3]*	08	
Labrador[3]*	04	
Manitoba*	06	
New Brunswick*	04	
Newfoundland*	03	30
Northwest Territories[3]*		
east of long. W. 85°	05	
long. W. 85° to W. 102°	06	
west of long. W. 102°	07	
Nova Scotia*	04	
Ontario, east of long. W. 90°*	05	
west of long. W. 90°	05	
Prince Edward Island*	04	
Canada (*continued*)		
Quebec, east of long. W. 63°[3]	04	
west of long. W. 63°*	05	
Saskatchewan[3]	06	
Yukon*	08	
Cape Verde Islands	01	
Cayman Islands	05	
Chile*	04	
Colombia	05	
Cook Islands	10	
Costa Rica	06	
Cuba*	05	
Curaçao Island	04	
Dominican Republic	04	
Easter Island (I. de Pascua)*	06	
Ecuador	05	
El Salvador	06	
Falkland Islands*	04	
Fanning Island	10	
Fernando de Noronha Island	02	
French Guiana	03	
Galápagos Islands	06	
Greenland[4]		
General*	03	
Scoresby Sound*	01	
Thule area*	04	
Grenada	04	
Guadeloupe	04	
Guatemala	06	
Guyana, Republic of	04	
Haiti	05	
Honduras	06	

* Daylight-saving time may be kept in these places.

1 This is the legal standard time, but local mean time is generally used.

2 Most stations use UTC.

3 Some areas may keep another time zone.

4 Mesters Vig and Danmarkshavn keep UTC.

LIST III — (*continued*)

	h m
Jamaica	05
Johnston Island	10
Juan Fernandez Islands*	04
Leeward Islands	04
Marquesas Islands	09 30
Martinique	04
Mexico*[1]	06
Midway Islands	11
Nicaragua	06
Niue	11
Panama, Republic of	05
Paraguay*	04
Peru	05
Pitcairn Island	08
Puerto Rico	04
St. Pierre and Miquelon*	03
Samoa	11
Society Islands	10
South Georgia	02
Suriname	03
Trindade Island, South Atlantic	02
Trinidad and Tobago	04
Tuamotu Archipelago	10
Tubuai (Austral) Islands	10
Turks and Caicos Islands*	05
United States of America[2]	
Alabama	06
Alaska	09
Aleutian Islands, east of W. 169° 30′	09
Aleutian Islands, west of W. 169° 30′	10
Arizona[3]	07
Arkansas	06
California	08
Colorado	07
Connecticut	05
Delaware	05
District of Columbia	05
Florida[4]	05
Georgia	05
Hawaii[3]	10
Idaho[4]	07
Illinois	06
Indiana[3,4]	05

	h m
United States of America[2] (*continued*)	
Iowa	06
Kansas[4]	06
Kentucky, eastern part	05
western part	06
Louisiana	06
Maine	05
Maryland	05
Massachusetts	05
Michigan[4]	05
Minnesota	06
Mississippi	06
Missouri	06
Montana	07
Nebraska, eastern part	06
western part	07
Nevada	08
New Hampshire	05
New Jersey	05
New Mexico	07
New York	05
North Carolina	05
North Dakota[4]	06
Ohio	05
Oklahoma	06
Oregon[4]	08
Pennsylvania	05
Rhode Island	05
South Carolina	05
South Dakota, eastern part	06
western part	07
Tennessee, eastern part	05
western part	06
Texas[4]	06
Utah	07
Vermont	05
Virginia	05
Washington D.C.	05
Washington	08
West Virginia	05
Wisconsin	06
Wyoming	07
Uruguay	03
Venezuela	04
Virgin Islands	04
Windward Islands	04

* Daylight-saving time may be kept in these places.

1 Except the states of Sonora, Sinaloa*, Nayarit*, Chihuahua* and the Southern District of Lower California* which keep 07^h, and the Northern District of Lower California* which keeps 08^h.

2 Daylight-saving (Summer) time, one hour fast on the time given, is kept from the first Sunday in April to the last Sunday in October, changing at $02^h\ 00^m$ local clock time.

3 Exempt from keeping daylight-saving time.

4 A small portion of the state is in another time zone.

NORTHERN STARS

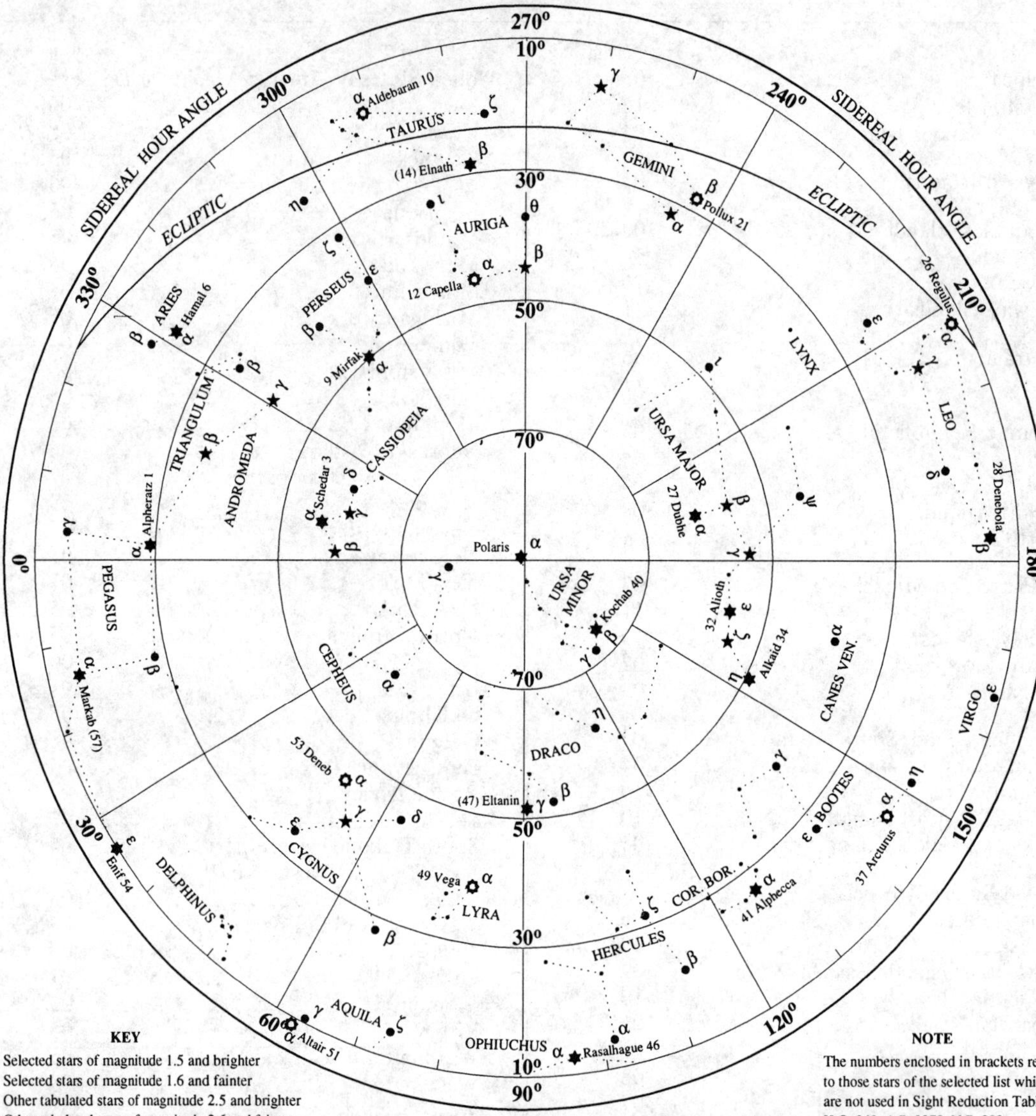

KEY

- Selected stars of magnitude 1.5 and brighter
- Selected stars of magnitude 1.6 and fainter
- Other tabulated stars of magnitude 2.5 and brighter
- Other tabulated stars of magnitude 2.6 and fainter
- Untabulated stars

NOTE

The numbers enclosed in brackets refer to those stars of the selected list which are not used in Sight Reduction Tables H.O. 249, A.P. 3270, N.P. 303.

EQUATORIAL STARS (SHA 0° to 180°)

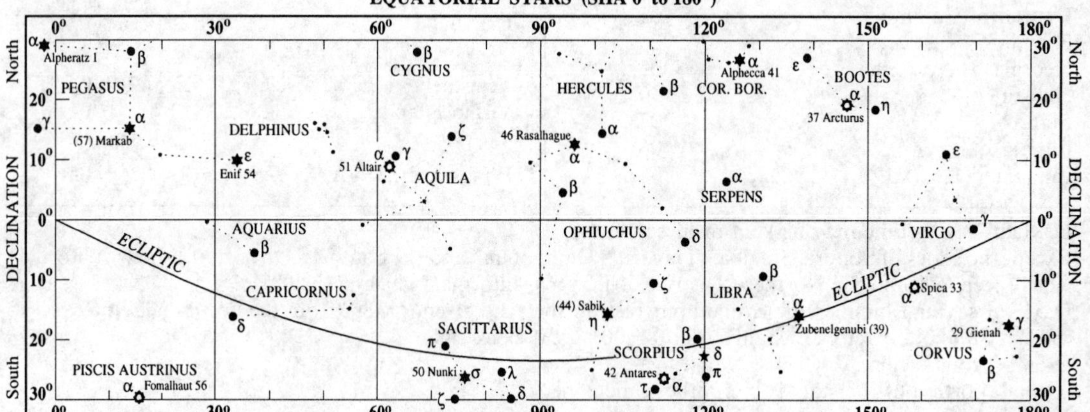

SIDEREAL HOUR ANGLE

SOUTHERN STARS

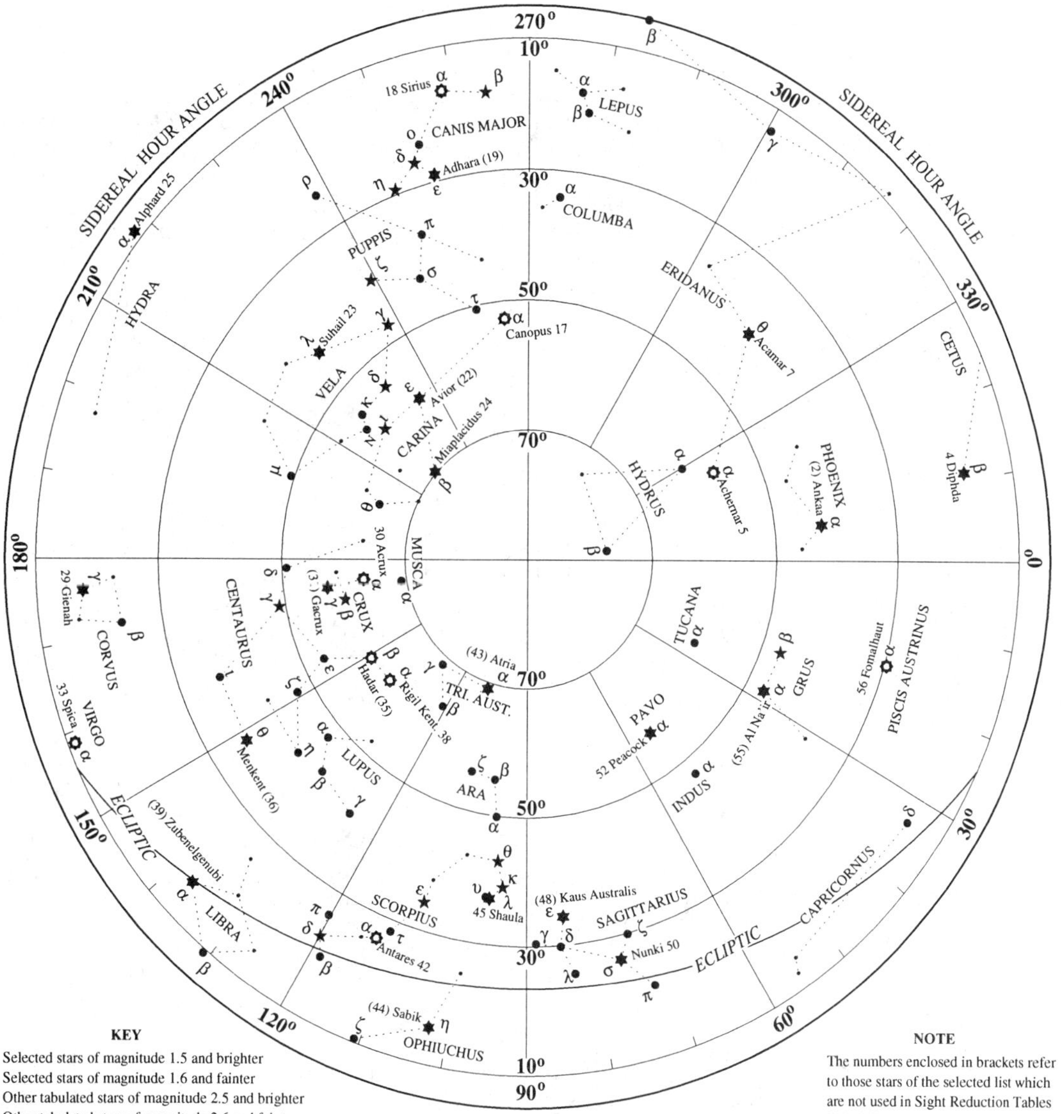

KEY

- Selected stars of magnitude 1.5 and brighter
- Selected stars of magnitude 1.6 and fainter
- Other tabulated stars of magnitude 2.5 and brighter
- Other tabulated stars of magnitude 2.6 and fainter
- Untabulated stars

NOTE

The numbers enclosed in brackets refer to those stars of the selected list which are not used in Sight Reduction Tables H.O. 249, A.P. 3270, N.P. 303.

EQUATORIAL STARS (SHA 180° to 360°)

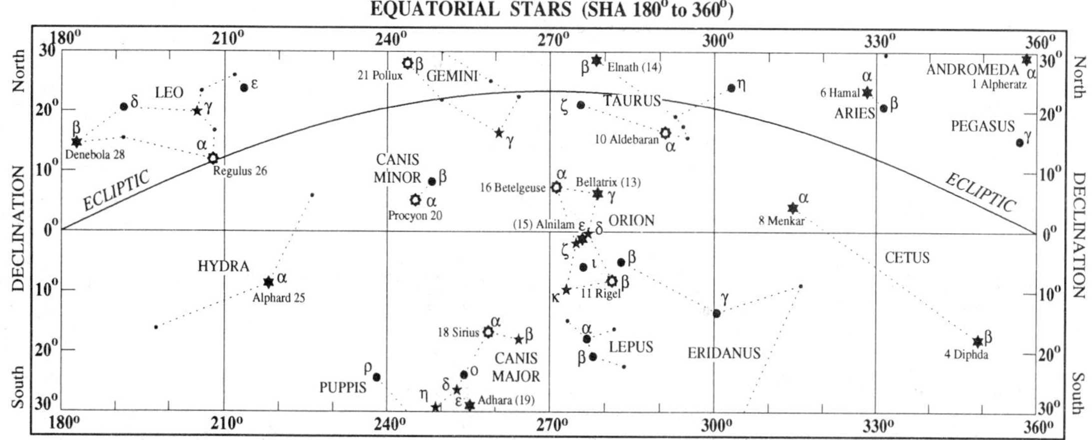

Mag.	Name and Number		SHA							Declination						
				JAN.	FEB.	MAR.	APR.	MAY	JUNE		JAN.	FEB.	MAR.	APR.	MAY	JUNE
			°	′	′	′	′	′	′	°	′	′	′	′	′	′
3·4	γ Cephei		**5**	09·1	09·6	09·8	09·6	09·0	08·2	N **77**	39·1	39·0	38·9	38·8	38·7	38·7
2·6	α Pegasi	57	**13**	46·8	46·8	46·8	46·6	46·4	46·2	N **15**	13·2	13·1	13·1	13·1	13·1	13·2
2·6	β Pegasi		**14**	01·6	01·7	01·7	01·5	01·3	01·1	N **28**	05·9	05·8	05·8	05·7	05·7	05·8
1·3	α Piscis Aust.	56	**15**	33·2	33·2	33·1	33·0	32·8	32·5	S **29**	36·7	36·6	36·5	36·4	36·3	36·2
2·2	β Gruis		**19**	17·8	17·8	17·8	17·6	17·3	17·0	S **46**	52·5	52·4	52·2	52·1	52·0	51·9
2·9	α Tucanæ		**25**	20·1	20·1	20·0	19·7	19·4	18·9	S **60**	15·0	14·9	14·7	14·6	14·5	14·4
2·2	α Gruis	55	**27**	54·3	54·2	54·1	53·9	53·6	53·3	S **46**	57·1	57·0	56·9	56·7	56·6	56·6
3·0	δ Capricorni		**33**	12·4	12·4	12·3	12·1	11·9	11·6	S **16**	07·0	07·0	07·0	06·9	06·8	06·7
2·5	ε Pegasi	54	**33**	55·5	55·5	55·4	55·2	55·0	54·7	N **9**	53·2	53·1	53·1	53·1	53·2	53·3
3·1	β Aquarii		**37**	04·7	04·7	04·6	04·4	04·2	03·9	S **5**	33·6	33·7	33·7	33·6	33·6	33·5
2·6	α Cephei		**40**	21·1	21·1	21·0	20·6	20·3	19·9	N **62**	35·9	35·8	35·6	35·5	35·5	35·6
2·6	ε Cygni		**48**	25·6	25·5	25·4	25·2	24·9	24·7	N **33**	58·8	58·7	58·6	58·6	58·6	58·7
1·3	α Cygni	53	**49**	37·6	37·6	37·4	37·2	36·9	36·6	N **45**	17·4	17·3	17·1	17·1	17·1	17·3
3·2	α Indi		**50**	34·1	33·9	33·7	33·4	33·1	32·8	S **47**	17·1	17·0	16·9	16·8	16·7	16·7
2·1	α Pavonis	52	**53**	32·6	32·5	32·2	31·8	31·4	31·1	S **56**	43·7	43·6	43·5	43·4	43·4	43·4
2·3	γ Cygni		**54**	25·6	25·5	25·3	25·1	24·8	24·6	N **40**	15·9	15·8	15·7	15·6	15·7	15·8
0·9	α Aquilæ	51	**62**	16·6	16·5	16·3	16·1	15·9	15·7	N **8**	52·4	52·4	52·3	52·3	52·4	52·5
2·8	γ Aquilæ		**63**	24·5	24·4	24·2	24·0	23·8	23·6	N **10**	37·1	37·0	37·0	37·0	37·1	37·2
3·0	δ Cygni		**63**	44·6	44·5	44·3	44·0	43·7	43·5	N **45**	08·2	08·0	07·9	07·9	08·0	08·1
3·2	β Cygni		**67**	17·9	17·8	17·6	17·4	17·1	16·9	N **27**	57·8	57·7	57·6	57·6	57·7	57·8
3·0	π Sagittarii		**72**	31·5	31·3	31·1	30·8	30·6	30·4	S **21**	01·2	01·2	01·2	01·2	01·2	01·1
3·0	ζ Aquilæ		**73**	37·3	37·2	37·0	36·8	36·5	36·4	N **13**	51·9	51·9	51·8	51·8	51·9	52·0
2·7	ζ Sagittarii		**74**	18·6	18·4	18·2	17·9	17·7	17·4	S **29**	52·7	52·6	52·6	52·6	52·6	52·6
2·1	σ Sagittarii	50	**76**	08·8	08·7	08·4	08·2	07·9	07·7	S **26**	17·7	17·7	17·6	17·6	17·6	17·6
0·1	α Lyræ	49	**80**	45·0	44·8	44·6	44·3	44·1	43·9	N **38**	47·0	46·9	46·8	46·8	46·9	47·1
2·9	λ Sagittarii		**82**	58·3	58·1	57·8	57·6	57·3	57·1	S **25**	25·3	25·3	25·3	25·3	25·2	25·2
2·0	ε Sagittarii	48	**83**	55·1	54·9	54·6	54·3	54·1	53·9	S **34**	23·0	23·0	23·0	23·0	23·0	23·0
2·8	δ Sagittarii		**84**	42·8	42·6	42·4	42·1	41·9	41·7	S **29**	49·7	49·6	49·6	49·6	49·6	49·6
3·1	γ Sagittarii		**88**	30·6	30·4	30·1	29·9	29·6	29·4	S **30**	25·5	25·5	25·5	25·5	25·5	25·5
2·4	γ Draconis	47	**90**	50·5	50·2	50·0	49·6	49·4	49·3	N **51**	29·1	29·0	28·9	28·9	29·1	29·2
2·9	β Ophiuchi		**94**	06·2	06·0	05·8	05·6	05·4	05·2	N **4**	33·9	33·8	33·7	33·7	33·8	33·9
2·5	κ Scorpii		**94**	20·2	20·0	19·7	19·4	19·2	19·0	S **39**	01·9	01·9	01·9	01·9	01·9	02·0
2·0	θ Scorpii		**95**	37·7	37·4	37·1	36·8	36·5	36·3	S **42**	59·9	59·9	59·9	60·0	60·0	60·1
2·1	α Ophiuchi	46	**96**	14·4	14·2	14·0	13·7	13·6	13·4	N **12**	33·3	33·2	33·2	33·2	33·3	33·4
1·7	λ Scorpii	45	**96**	33·4	33·2	32·9	32·6	32·4	32·2	S **37**	06·3	06·3	06·3	06·4	06·4	06·4
3·0	α Aræ		**96**	59·7	59·3	59·0	58·7	58·4	58·2	S **49**	52·6	52·6	52·6	52·7	52·7	52·8
2·8	υ Scorpii		**97**	16·1	15·9	15·6	15·3	15·1	14·9	S **37**	17·9	17·9	17·9	17·9	17·9	18·0
3·0	β Draconis		**97**	23·1	22·8	22·5	22·2	22·0	21·9	N **52**	17·7	17·6	17·6	17·6	17·7	17·9
2·8	β Aræ		**98**	37·6	37·2	36·9	36·5	36·2	36·0	S **55**	31·9	31·8	31·8	31·9	31·9	32·0
Var.‡	α Herculis		**101**	18·7	18·5	18·3	18·0	17·9	17·8	N **14**	23·1	23·0	22·9	23·0	23·0	23·1
2·6	η Ophiuchi	44	**102**	22·3	22·0	21·8	21·6	21·4	21·3	S **15**	43·7	43·8	43·8	43·8	43·8	43·8
3·1	ζ Aræ		**105**	17·8	17·4	17·0	16·7	16·4	16·2	S **55**	59·6	59·5	59·6	59·6	59·7	59·8
2·4	ε Scorpii		**107**	25·2	24·9	24·7	24·4	24·2	24·1	S **34**	17·9	17·9	17·9	18·0	18·0	18·0
1·9	α Triang. Aust.	43	**107**	46·4	45·9	45·3	44·7	44·3	44·1	S **69**	01·8	01·8	01·8	01·9	02·0	02·1
3·0	ζ Herculis		**109**	39·4	39·2	39·0	38·7	38·6	38·5	N **31**	35·7	35·6	35·5	35·6	35·7	35·8
2·7	ζ Ophiuchi		**110**	40·6	40·4	40·1	39·9	39·8	39·7	S **10**	34·4	34·5	34·5	34·5	34·5	34·5
2·9	τ Scorpii		**110**	59·5	59·2	59·0	58·7	58·6	58·4	S **28**	13·3	13·3	13·4	13·4	13·4	13·5
2·8	β Herculis		**112**	25·2	25·0	24·7	24·5	24·4	24·3	N **21**	28·8	28·7	28·7	28·7	28·8	28·9
1·2	α Scorpii	42	**112**	36·6	36·4	36·1	35·9	35·7	35·6	S **26**	26·3	26·3	26·4	26·4	26·4	26·4
2·9	η Draconis		**113**	60·0	59·7	59·3	59·0	58·8	58·8	N **61**	30·2	30·1	30·1	30·1	30·3	30·4
3·0	δ Ophiuchi		**116**	22·9	22·6	22·4	22·2	22·1	22·0	S **3**	42·2	42·2	42·3	42·3	42·3	42·2
2·8	β Scorpii		**118**	36·3	36·0	35·8	35·6	35·4	35·4	S **19**	48·8	48·8	48·9	48·9	48·9	49·0
2·5	δ Scorpii		**119**	52·8	52·5	52·3	52·1	52·0	51·9	S **22**	37·8	37·8	37·9	37·9	37·9	38·0
3·0	π Scorpii		**120**	15·0	14·7	14·5	14·2	14·1	14·0	S **26**	07·3	07·3	07·4	07·5	07·5	07·5
3·0	β Trianguli Aust.		**121**	09·7	09·2	08·8	08·4	08·1	08·0	S **63**	26·2	26·2	26·2	26·3	26·5	26·6
2·8	α Serpentis		**123**	54·1	53·9	53·7	53·5	53·4	53·3	N **6**	24·9	24·8	24·8	24·8	24·8	24·9
3·0	γ Lupi		**126**	10·4	10·1	09·8	09·6	09·4	09·4	S **41**	10·4	10·5	10·6	10·7	10·7	10·8
2·3	α Coronæ Bor.	41	**126**	18·1	17·9	17·6	17·5	17·4	17·3	N **26**	42·1	42·0	42·0	42·0	42·1	42·3

‡ 3·0 — 3·7

Mag.	Name and Number		SHA							Declination						
				JULY	AUG.	SEPT.	OCT.	NOV.	DEC.		JULY	AUG.	SEPT.	OCT.	NOV.	DEC.
			°	′	′	′	′	′	′	°	′	′	′	′	′	′
3·4	γ Cephei		**5**	07·4	06·9	06·6	06·8	07·2	07·9	N **77**	38·8	38·9	39·1	39·3	39·4	39·5
2·6	*Markab*	57	**13**	46·0	45·8	45·7	45·7	45·8	45·9	N **15**	13·3	13·4	13·5	13·6	13·6	13·6
2·6	*Scheat*		**14**	00·8	00·6	00·6	00·6	00·7	00·8	N **28**	05·9	06·1	06·2	06·3	06·3	06·3
1·3	*Fomalhaut*	56	**15**	32·3	32·1	32·0	32·0	32·1	32·2	S **29**	36·1	36·1	36·2	36·2	36·3	36·3
2·2	β Gruis		**19**	16·7	16·5	16·4	16·4	16·6	16·7	S **46**	51·9	51·9	52·0	52·1	52·2	52·2
2·9	α Tucanæ		**25**	18·5	18·3	18·2	18·3	18·6	18·8	S **60**	14·4	14·5	14·6	14·7	14·8	14·8
2·2	*Al Na'ir*	55	**27**	53·0	52·8	52·7	52·8	53·0	53·1	S **46**	56·5	56·6	56·7	56·8	56·8	56·8
3·0	δ Capricorni		**33**	11·4	11·3	11·2	11·3	11·4	11·5	S **16**	06·7	06·7	06·7	06·7	06·7	06·7
2·5	*Enif*	54	**33**	54·5	54·4	54·4	54·5	54·6	54·7	N **9**	53·4	53·5	53·5	53·6	53·6	53·5
3·1	β Aquarii		**37**	03·7	03·6	03·6	03·7	03·8	03·9	S **5**	33·4	33·3	33·3	33·3	33·3	33·4
2·6	*Alderamin*		**40**	19·6	19·5	19·7	19·9	20·2	20·5	N **62**	35·8	36·0	36·2	36·3	36·3	36·3
2·6	ε Cygni		**48**	24·5	24·4	24·5	24·6	24·8	24·9	N **33**	58·9	59·0	59·1	59·2	59·2	59·1
1·3	*Deneb*	53	**49**	36·5	36·4	36·5	36·7	36·9	37·0	N **45**	17·4	17·6	17·7	17·8	17·8	17·7
3·2	α Indi		**50**	32·5	32·4	32·5	32·6	32·8	32·9	S **47**	16·7	16·8	16·9	17·0	17·0	16·9
2·1	*Peacock*	52	**53**	30·8	30·7	30·8	30·9	31·2	31·3	S **56**	43·4	43·5	43·6	43·7	43·7	43·6
2·3	γ Cygni		**54**	24·4	24·4	24·5	24·7	24·8	25·0	N **40**	16·0	16·1	16·2	16·3	16·3	16·2
0·9	*Altair*	51	**62**	15·5	15·5	15·6	15·7	15·8	15·8	N **8**	52·6	52·7	52·7	52·8	52·7	52·7
2·8	γ Aquilæ		**63**	23·4	23·4	23·5	23·6	23·7	23·8	N **10**	37·3	37·3	37·4	37·4	37·4	37·3
3·0	δ Cygni		**63**	43·4	43·4	43·6	43·8	44·0	44·1	N **45**	08·3	08·4	08·5	08·6	08·6	08·5
3·2	*Albireo*		**67**	16·8	16·8	16·9	17·1	17·2	17·3	N **27**	58·0	58·1	58·2	58·2	58·2	58·1
3·0	π Sagittarii		**72**	30·3	30·2	30·3	30·5	30·6	30·6	S **21**	01·1	01·1	01·1	01·2	01·2	01·2
3·0	ζ Aquilæ		**73**	36·3	36·3	36·4	36·5	36·6	36·6	N **13**	52·1	52·2	52·2	52·2	52·2	52·1
2·7	ζ Sagittarii		**74**	17·3	17·3	17·4	17·5	17·6	17·7	S **29**	52·6	52·6	52·6	52·6	52·6	52·6
2·1	*Nunki*	50	**76**	07·6	07·6	07·7	07·8	07·9	07·9	S **26**	17·6	17·6	17·6	17·7	17·6	17·6
0·1	*Vega*	49	**80**	43·8	43·9	44·1	44·3	44·4	44·5	N **38**	47·2	47·3	47·4	47·4	47·3	47·2
2·9	λ Sagittarii		**82**	57·0	57·1	57·2	57·3	57·4	57·4	S **25**	25·2	25·3	25·3	25·3	25·3	25·3
2·0	*Kaus Australis*	48	**83**	53·8	53·8	53·9	54·1	54·1	54·1	S **34**	23·1	23·1	23·1	23·1	23·1	23·1
2·8	δ Sagittarii		**84**	41·6	41·6	41·7	41·8	41·9	41·9	S **29**	49·7	49·7	49·7	49·7	49·7	49·7
3·1	γ Sagittarii		**88**	29·4	29·4	29·5	29·6	29·7	29·7	S **30**	25·5	25·5	25·6	25·6	25·5	25·5
2·4	*Eltanin*	47	**90**	49·3	49·4	49·6	49·9	50·1	50·2	N **51**	29·4	29·5	29·5	29·5	29·4	29·3
2·9	β Ophiuchi		**94**	05·2	05·2	05·3	05·5	05·5	05·5	N **4**	34·0	34·0	34·0	34·0	34·0	33·9
2·5	κ Scorpii		**94**	18·9	18·9	19·1	19·3	19·3	19·3	S **39**	02·0	02·1	02·1	02·1	02·0	02·0
2·0	θ Scorpii		**95**	36·3	36·3	36·5	36·7	36·8	36·7	S **43**	00·1	00·2	00·2	00·2	00·1	00·1
2·1	*Rasalhague*	46	**96**	13·4	13·5	13·6	13·7	13·8	13·8	N **12**	33·5	33·5	33·6	33·5	33·5	33·4
1·7	*Shaula*	45	**96**	32·2	32·2	32·4	32·5	32·6	32·5	S **37**	06·5	06·5	06·5	06·5	06·5	06·5
3·0	α Aræ		**96**	58·1	58·2	58·4	58·6	58·7	58·6	S **49**	52·9	52·9	53·0	52·9	52·9	52·8
2·8	υ Scorpii		**97**	14·8	14·9	15·0	15·2	15·3	15·2	S **37**	18·0	18·1	18·1	18·1	18·0	18·0
3·0	β Draconis		**97**	21·9	22·1	22·3	22·6	22·8	22·8	N **52**	18·0	18·1	18·2	18·1	18·0	17·8
2·8	β Aræ		**98**	35·9	36·0	36·2	36·5	36·6	36·5	S **55**	32·1	32·2	32·2	32·2	32·1	32·1
Var.‡	α Herculis		**101**	17·7	17·8	17·9	18·1	18·1	18·1	N **14**	23·2	23·3	23·3	23·3	23·2	23·1
2·6	*Sabik*	44	**102**	21·2	21·3	21·4	21·5	21·6	21·5	S **15**	43·8	43·8	43·8	43·8	43·8	43·8
3·1	ζ Aræ		**105**	16·2	16·3	16·5	16·7	16·8	16·7	S **55**	59·9	60·0	60·0	60·0	59·9	59·8
2·4	ε Scorpii		**107**	24·1	24·1	24·3	24·4	24·5	24·4	S **34**	18·1	18·1	18·1	18·1	18·0	18·0
1·9	*Atria*	43	**107**	44·1	44·4	44·8	45·1	45·3	45·1	S **69**	02·2	02·3	02·4	02·3	02·2	02·1
3·0	ζ Herculis		**109**	38·5	38·6	38·8	39·0	39·0	39·0	N **31**	35·9	36·0	36·0	35·9	35·8	35·7
2·7	ζ Ophiuchi		**110**	39·7	39·7	39·9	40·0	40·0	39·9	S **10**	34·5	34·5	34·4	34·4	34·5	34·5
2·9	τ Scorpii		**110**	58·4	58·5	58·6	58·8	58·8	58·7	S **28**	13·5	13·5	13·5	13·5	13·4	13·4
2·8	β Herculis		**112**	24·3	24·4	24·6	24·7	24·7	24·7	N **21**	29·0	29·1	29·1	29·0	28·9	28·8
1·2	*Antares*	42	**112**	35·6	35·7	35·8	35·9	36·0	35·9	S **26**	26·5	26·5	26·5	26·4	26·4	26·4
2·9	η Draconis		**113**	58·9	59·2	59·5	59·8	60·0	60·0	N **61**	30·6	30·6	30·6	30·5	30·4	30·2
3·0	δ Ophiuchi		**116**	22·0	22·1	22·2	22·3	22·3	22·2	S **3**	42·2	42·2	42·2	42·2	42·2	42·3
2·8	β Scorpii		**118**	35·4	35·4	35·6	35·7	35·7	35·6	S **19**	49·0	49·0	48·9	48·9	48·9	48·9
2·5	*Dschubba*		**119**	51·9	52·0	52·1	52·2	52·2	52·1	S **22**	38·0	38·0	37·9	37·9	37·9	37·9
3·0	π Scorpii		**120**	14·0	14·1	14·3	14·4	14·4	14·2	S **26**	07·5	07·5	07·5	07·5	07·5	07·5
3·0	β Trianguli Aust.		**121**	08·1	08·4	08·7	08·9	08·9	08·7	S **63**	26·7	26·8	26·7	26·7	26·5	26·4
2·8	α Serpentis		**123**	53·4	53·4	53·6	53·7	53·7	53·5	N **6**	24·9	25·0	25·0	25·0	24·9	24·8
3·0	γ Lupi		**126**	09·4	09·5	09·7	09·8	09·8	09·6	S **41**	10·9	10·9	10·9	10·8	10·7	10·7
2·3	*Alphecca*	41	**126**	17·4	17·5	17·6	17·7	17·7	17·6	N **26**	42·3	42·4	42·4	42·3	42·2	42·0

‡ 3·0 — 3·7

Mag.	Name and Number		SHA							Declination						
				JAN.	FEB.	MAR.	APR.	MAY	JUNE		JAN.	FEB.	MAR.	APR.	MAY	JUNE
			°	′	′	′	′	′	′	°	′	′	′	′	′	′
3·1	γ Ursæ Minoris		**129**	49·5	48·9	48·4	48·0	47·9	48·1	**N 71**	49·1	49·0	49·1	49·2	49·3	49·5
3·1	γ Trianguli Aust.		**130**	13·0	12·4	11·9	11·5	11·3	11·2	**S 68**	41·1	41·2	41·3	41·4	41·5	41·7
2·7	β Libræ		**130**	42·8	42·6	42·4	42·2	42·1	42·1	**S 9**	23·6	23·7	23·8	23·8	23·8	23·8
2·8	β Lupi		**135**	19·5	19·2	18·9	18·7	18·6	18·6	**S 43**	08·6	08·6	08·7	08·8	08·9	09·0
2·9	α Libræ	39	**137**	14·7	14·4	14·2	14·1	14·0	13·9	**S 16**	03·2	03·3	03·3	03·4	03·4	03·4
2·2	β Ursæ Minoris	40	**137**	19·8	19·1	18·6	18·2	18·1	18·4	**N 74**	08·3	08·2	08·3	08·4	08·6	08·7
2·6	ε Bootis		**138**	43·5	43·2	43·0	42·9	42·8	42·8	**N 27**	03·5	03·4	03·4	03·5	03·6	03·7
2·9	α Lupi		**139**	28·5	28·2	27·9	27·7	27·6	27·6	**S 47**	23·8	23·9	24·0	24·1	24·2	24·3
0·1	α Centauri	38	**140**	03·4	03·0	02·7	02·4	02·3	02·4	**S 60**	50·5	50·6	50·7	50·9	51·0	51·1
2·6	η Centauri		**141**	04·9	04·6	04·4	04·2	04·1	04·1	**S 42**	10·0	10·1	10·2	10·3	10·4	10·5
3·0	γ Bootis		**141**	57·3	57·0	56·7	56·6	56·5	56·6	**N 38**	17·5	17·4	17·4	17·5	17·7	17·8
0·2	α Bootis	37	**146**	03·3	03·0	02·8	02·7	02·7	02·7	**N 19**	09·9	09·8	09·8	09·8	09·9	10·0
2·3	θ Centauri	36	**148**	17·4	17·1	16·9	16·8	16·7	16·8	**S 36**	22·9	23·0	23·1	23·2	23·3	23·3
0·9	β Centauri	35	**148**	59·9	59·5	59·2	59·0	58·9	59·0	**S 60**	22·9	23·0	23·2	23·3	23·5	23·6
3·1	ζ Centauri		**151**	04·4	04·1	03·9	03·7	03·7	03·7	**S 47**	17·9	18·0	18·1	18·3	18·4	18·5
2·8	η Bootis		**151**	17·8	17·5	17·4	17·3	17·2	17·3	**N 18**	22·9	22·8	22·8	22·8	22·9	23·0
1·9	η Ursæ Majoris	34	**153**	05·3	04·9	04·7	04·6	04·6	04·7	**N 49**	17·7	17·6	17·7	17·8	17·9	18·0
2·6	ε Centauri		**154**	59·2	58·8	58·6	58·4	58·4	58·5	**S 53**	28·6	28·7	28·8	29·0	29·1	29·2
1·2	α Virginis	33	**158**	39·9	39·7	39·5	39·4	39·4	39·5	**S 11**	10·5	10·6	10·7	10·8	10·8	10·8
2·2	ζ Ursæ Majoris		**158**	59·4	59·0	58·8	58·7	58·7	58·9	**N 54**	54·3	54·3	54·4	54·5	54·7	54·7
2·9	ι Centauri		**159**	48·7	48·5	48·3	48·2	48·2	48·2	**S 36**	43·5	43·6	43·7	43·8	43·9	44·0
3·0	ε Virginis		**164**	25·2	25·0	24·9	24·8	24·8	24·9	**N 10**	56·5	56·5	56·4	56·5	56·5	56·6
2·9	α Canum Venat.		**165**	57·6	57·3	57·1	57·0	57·1	57·2	**N 38**	17·9	17·9	18·0	18·1	18·2	18·2
1·7	ε Ursæ Majoris	32	**166**	27·6	27·2	27·0	26·9	27·0	27·2	**N 55**	56·4	56·4	56·5	56·6	56·7	56·8
1·5	β Crucis		**168**	01·7	01·4	01·2	01·1	01·1	01·3	**S 59**	42·0	42·1	42·3	42·4	42·6	42·7
2·9	γ Virginis		**169**	32·9	32·7	32·6	32·5	32·6	32·6	**S 1**	27·9	28·0	28·1	28·1	28·1	28·0
2·4	γ Centauri		**169**	34·9	34·6	34·5	34·4	34·4	34·5	**S 48**	58·3	58·4	58·6	58·7	58·8	58·9
2·9	α Muscæ		**170**	39·7	39·2	39·0	38·9	39·0	39·3	**S 69**	08·8	08·9	09·1	09·3	09·4	09·5
2·8	β Corvi		**171**	22·0	21·7	21·6	21·5	21·6	21·6	**S 23**	24·6	24·8	24·9	25·0	25·0	25·0
1·6	γ Crucis	31	**172**	10·1	09·8	09·6	09·6	09·7	09·8	**S 57**	07·5	07·6	07·8	07·9	08·1	08·1
1·1	α Crucis	30	**173**	18·6	18·2	18·0	18·0	18·1	18·3	**S 63**	06·6	06·7	06·9	07·1	07·2	07·3
2·8	γ Corvi	29	**176**	00·7	00·5	00·4	00·3	00·4	00·4	**S 17**	33·4	33·5	33·6	33·7	33·7	33·7
2·9	δ Centauri		**177**	52·4	52·1	51·9	51·9	52·0	52·2	**S 50**	44·1	44·2	44·4	44·5	44·6	44·7
2·5	γ Ursæ Majoris		**181**	30·1	29·8	29·6	29·6	29·7	29·9	**N 53**	40·5	40·5	40·6	40·7	40·8	40·9
2·2	β Leonis	28	**182**	41·8	41·6	41·5	41·5	41·6	41·7	**N 14**	33·3	33·2	33·2	33·2	33·3	33·3
2·6	δ Leonis		**191**	26·0	25·8	25·7	25·7	25·8	25·9	**N 20**	30·4	30·3	30·4	30·4	30·5	30·5
3·2	ψ Ursæ Majoris		**192**	32·4	32·2	32·1	32·1	32·3	32·4	**N 44**	28·8	28·8	28·9	29·0	29·1	29·1
2·0	α Ursæ Majoris	27	**194**	01·0	00·7	00·6	00·7	00·9	01·2	**N 61**	43·9	44·0	44·1	44·2	44·3	44·3
2·4	β Ursæ Majoris		**194**	29·5	29·2	29·1	29·2	29·3	29·5	**N 56**	21·8	21·9	22·0	22·1	22·2	22·2
2·8	μ Velorum		**198**	16·4	16·2	16·2	16·3	16·4	16·6	**S 49**	25·9	26·1	26·2	26·4	26·4	26·4
3·0	θ Carinæ		**199**	13·8	13·5	13·5	13·7	14·0	14·3	**S 64**	24·3	24·5	24·7	24·8	24·9	24·9
2·3	γ Leonis		**204**	57·9	57·7	57·7	57·7	57·8	57·9	**N 19**	49·6	49·5	49·5	49·6	49·6	49·7
1·3	α Leonis	26	**207**	52·0	51·8	51·8	51·8	51·9	52·0	**N 11**	57·1	57·1	57·1	57·1	57·2	57·2
3·1	ε Leonis		**213**	29·6	29·4	29·4	29·5	29·6	29·7	**N 23**	45·6	45·6	45·6	45·7	45·7	45·7
3·0	N Velorum		**217**	09·9	09·9	09·9	10·1	10·4	10·6	**S 57**	02·6	02·8	03·0	03·1	03·1	03·1
2·2	α Hydræ	25	**218**	03·9	03·8	03·8	03·9	04·0	04·0	**S 8**	40·2	40·3	40·4	40·4	40·4	40·4
2·6	κ Velorum		**219**	26·6	26·5	26·6	26·8	27·0	27·2	**S 55**	01·2	01·4	01·6	01·7	01·7	01·7
2·2	ι Carinæ		**220**	42·0	42·0	42·1	42·3	42·6	42·8	**S 59**	17·1	17·3	17·4	17·5	17·6	17·5
1·8	β Carinæ	24	**221**	40·8	40·8	41·0	41·4	41·8	42·2	**S 69**	43·6	43·7	43·9	44·0	44·1	44·0
2·2	λ Velorum	23	**222**	58·1	58·1	58·1	58·3	58·5	58·6	**S 43**	26·5	26·7	26·8	26·9	26·9	26·9
3·1	ι Ursæ Majoris		**225**	08·6	08·5	08·5	08·6	08·8	08·9	**N 48**	01·7	01·8	01·9	02·0	02·0	02·0
2·0	δ Velorum		**228**	47·8	47·8	47·9	48·1	48·4	48·6	**S 54**	43·0	43·2	43·3	43·4	43·4	43·4
1·7	ε Carinæ	22	**234**	20·8	20·9	21·1	21·4	21·7	21·9	**S 59**	31·0	31·2	31·3	31·4	31·4	31·3
1·9	γ Velorum		**237**	35·3	35·3	35·4	35·6	35·8	36·0	**S 47**	20·6	20·8	20·9	21·0	21·0	20·9
2·9	ρ Puppis		**238**	04·7	04·7	04·8	04·9	05·1	05·1	**S 24**	18·7	18·8	18·9	19·0	18·9	18·9
2·3	ζ Puppis		**239**	04·4	04·4	04·5	04·7	04·8	05·0	**S 40**	00·6	00·8	00·9	00·9	00·9	00·8
1·2	β Geminorum	21	**243**	37·4	37·3	37·4	37·5	37·6	37·7	**N 28**	01·2	01·2	01·2	01·2	01·2	01·2
0·5	α Canis Minoris	20	**245**	08·0	08·0	08·0	08·2	08·3	08·3	**N 5**	13·1	13·0	13·0	13·0	13·0	13·1

Mag.	Name and Number		SHA							Declination						
				JULY	AUG.	SEPT.	OCT.	NOV.	DEC.		JULY	AUG.	SEPT.	OCT.	NOV.	DEC.
			°	′	′	′	′	′	′	°	′	′	′	′	′	′
3·1	γ Ursæ Minoris		**129**	48·4	48·9	49·4	49·8	50·0	49·9	N **71**	49·6	49·6	49·5	49·4	49·2	49·0
3·1	γ Trianguli Aust.		**130**	11·4	11·7	12·1	12·4	12·3	12·1	S **68**	41·8	41·8	41·8	41·7	41·6	41·5
2·7	β Libræ		**130**	42·1	42·2	42·3	42·4	42·3	42·2	S **9**	23·7	23·7	23·7	23·7	23·7	23·8
2·8	β Lupi		**135**	18·6	18·8	19·0	19·0	19·0	18·8	S **43**	09·0	09·0	09·0	08·9	08·9	08·8
2·9	*Zubenelgenubi*	39	**137**	14·0	14·1	14·2	14·2	14·2	14·0	S **16**	03·4	03·4	03·4	03·3	03·4	03·4
2·2	*Kochab*	40	**137**	18·9	19·4	20·0	20·4	20·5	20·2	N **74**	08·8	08·8	08·7	08·5	08·3	08·1
2·6	ε Bootis		**138**	42·9	43·0	43·1	43·2	43·2	43·0	N **27**	03·8	03·8	03·8	03·7	03·5	03·4
2·9	α Lupi		**139**	27·7	27·8	28·0	28·1	28·0	27·8	S **47**	24·4	24·4	24·3	24·2	24·1	24·1
0·1	*Rigil Kent.*	38	**140**	02·5	02·8	03·1	03·2	03·1	02·8	S **60**	51·2	51·2	51·1	51·0	50·9	50·8
2·6	η Centauri		**141**	04·2	04·3	04·5	04·5	04·5	04·3	S **42**	10·5	10·5	10·5	10·4	10·3	10·3
3·0	γ Bootis		**141**	56·7	56·8	57·0	57·1	57·0	56·9	N **38**	17·8	17·8	17·8	17·7	17·5	17·3
0·2	*Arcturus*	37	**146**	02·7	02·8	03·0	03·0	02·9	02·8	N **19**	10·0	10·0	10·0	09·9	09·8	09·7
2·3	*Menkent*	36	**148**	16·8	17·0	17·1	17·1	17·0	16·8	S **36**	23·4	23·3	23·3	23·2	23·2	23·2
0·9	*Hadar*	35	**148**	59·1	59·4	59·6	59·7	59·6	59·3	S **60**	23·6	23·6	23·5	23·4	23·3	23·2
3·1	ζ Centauri		**151**	03·8	04·0	04·1	04·2	04·1	03·8	S **47**	18·5	18·5	18·4	18·3	18·2	18·2
2·8	η Bootis		**151**	17·3	17·4	17·5	17·5	17·5	17·3	N **18**	23·0	23·0	23·0	22·9	22·8	22·6
1·9	*Alkaid*	34	**153**	04·8	05·0	05·2	05·2	05·1	04·9	N **49**	18·1	18·0	18·0	17·8	17·6	17·4
2·6	ε Centauri		**154**	58·6	58·8	59·0	59·0	58·9	58·6	S **53**	29·2	29·2	29·1	29·0	28·9	28·9
1·2	*Spica*	33	**158**	39·5	39·6	39·7	39·7	39·6	39·4	S **11**	10·7	10·7	10·7	10·7	10·7	10·8
2·2	*Mizar*		**158**	59·0	59·2	59·4	59·4	59·3	59·0	N **54**	54·8	54·7	54·6	54·5	54·3	54·1
2·9	ι Centauri		**159**	48·3	48·5	48·6	48·6	48·4	48·2	S **36**	44·0	43·9	43·8	43·8	43·7	43·7
3·0	ε Virginis		**164**	24·9	25·0	25·1	25·0	24·9	24·7	N **10**	56·6	56·6	56·6	56·5	56·4	56·3
2·9	*Cor Caroli*		**165**	57·3	57·4	57·5	57·5	57·3	57·1	N **38**	18·3	18·2	18·1	18·0	17·8	17·7
1·7	*Alioth*	32	**166**	27·4	27·6	27·7	27·7	27·5	27·2	N **55**	56·8	56·7	56·6	56·5	56·3	56·1
1·5	*Mimosa*		**168**	01·5	01·8	01·9	01·9	01·7	01·3	S **59**	42·7	42·6	42·5	42·4	42·3	42·3
2·9	γ Virginis		**169**	32·7	32·7	32·8	32·8	32·6	32·4	S **1**	28·0	28·0	28·0	28·0	28·1	28·2
2·4	*Muhlifain*		**169**	34·7	34·9	35·0	34·9	34·7	34·4	S **48**	58·9	58·8	58·7	58·6	58·5	58·5
2·9	α Muscæ		**170**	39·7	40·0	40·2	40·2	39·9	39·4	S **69**	09·5	09·4	09·3	09·2	09·1	09·0
2·8	β Corvi		**171**	21·7	21·8	21·8	21·8	21·6	21·4	S **23**	25·0	24·9	24·9	24·8	24·8	24·9
1·6	*Gacrux*	31	**172**	10·0	10·2	10·3	10·3	10·1	09·7	S **57**	08·1	08·1	08·0	07·8	07·8	07·8
1·1	*Acrux*	30	**173**	18·5	18·8	19·0	18·9	18·6	18·2	S **63**	07·3	07·2	07·1	07·0	06·9	06·9
2·8	*Gienah*	29	**176**	00·5	00·6	00·6	00·6	00·4	00·1	S **17**	33·7	33·6	33·6	33·5	33·6	33·6
2·9	δ Centauri		**177**	52·3	52·5	52·5	52·5	52·2	51·9	S **50**	44·7	44·6	44·5	44·3	44·3	44·3
2·5	*Phecda*		**181**	30·1	30·2	30·2	30·1	29·9	29·6	N **53**	40·9	40·8	40·7	40·5	40·3	40·2
2·2	*Denebola*	28	**182**	41·7	41·8	41·8	41·7	41·5	41·3	N **14**	33·4	33·3	33·3	33·2	33·1	33·0
2·6	δ Leonis		**191**	25·9	26·0	25·9	25·8	25·6	25·4	N **20**	30·5	30·5	30·4	30·3	30·2	30·1
3·2	ψ Ursæ Majoris		**192**	32·5	32·5	32·5	32·4	32·1	31·8	N **44**	29·1	29·0	28·9	28·7	28·6	28·5
2·0	*Dubhe*	27	**194**	01·3	01·4	01·4	01·2	00·9	00·4	N **61**	44·3	44·2	44·0	43·8	43·7	43·6
2·4	*Merak*		**194**	29·7	29·7	29·7	29·5	29·2	28·8	N **56**	22·2	22·1	21·9	21·7	21·6	21·5
2·8	μ Velorum		**198**	16·7	16·8	16·8	16·7	16·4	16·1	S **49**	26·4	26·3	26·1	26·0	26·0	26·1
3·0	θ Carinæ		**199**	14·5	14·7	14·7	14·5	14·1	13·7	S **64**	24·9	24·8	24·6	24·5	24·4	24·5
2·3	*Algeiba*		**204**	57·9	57·9	57·9	57·7	57·5	57·2	N **19**	49·7	49·6	49·6	49·5	49·4	49·3
1·3	*Regulus*	26	**207**	52·1	52·0	52·0	51·8	51·6	51·3	N **11**	57·2	57·2	57·2	57·1	57·0	56·9
3·1	ε Leonis		**213**	29·7	29·7	29·6	29·4	29·2	28·9	N **23**	45·7	45·7	45·6	45·5	45·4	45·4
3·0	*N* Velorum		**217**	10·8	10·8	10·7	10·5	10·1	09·8	S **57**	03·0	02·9	02·7	02·6	02·6	02·7
2·2	*Alphard*	25	**218**	04·1	04·0	03·9	03·8	03·5	03·3	S **8**	40·3	40·2	40·2	40·2	40·3	40·4
2·6	κ Velorum		**219**	27·4	27·4	27·3	27·0	26·7	26·3	S **55**	01·5	01·4	01·3	01·2	01·2	01·3
2·2	ι Carinæ		**220**	43·0	43·0	42·9	42·6	42·3	41·9	S **59**	17·4	17·3	17·1	17·0	17·0	17·1
1·8	*Miaplacidus*	24	**221**	42·5	42·6	42·4	42·0	41·5	41·0	S **69**	43·9	43·8	43·6	43·5	43·5	43·6
2·2	*Suhail*	23	**222**	58·7	58·7	58·6	58·4	58·1	57·8	S **43**	26·8	26·6	26·5	26·4	26·5	26·6
3·1	ι Ursæ Majoris		**225**	08·9	08·9	08·7	08·4	08·1	07·7	N **48**	01·9	01·8	01·7	01·6	01·5	01·5
2·0	δ Velorum		**228**	48·7	48·7	48·5	48·2	47·9	47·6	S **54**	43·2	43·1	42·9	42·9	42·9	43·0
1·7	*Avior*	22	**234**	22·0	22·0	21·8	21·4	21·1	20·7	S **59**	31·2	31·0	30·9	30·8	30·9	31·0
1·9	γ Velorum		**237**	36·0	36·0	35·8	35·5	35·2	35·0	S **47**	20·7	20·6	20·5	20·4	20·5	20·6
2·9	ρ Puppis		**238**	05·1	05·0	04·9	04·7	04·4	04·2	S **24**	18·7	18·6	18·6	18·5	18·6	18·7
2·3	ζ Puppis		**239**	05·0	04·9	04·7	04·5	04·2	04·0	S **40**	00·7	00·5	00·4	00·4	00·5	00·6
1·2	*Pollux*	21	**243**	37·6	37·5	37·3	37·1	36·8	36·5	N **28**	01·2	01·2	01·1	01·1	01·1	01·0
0·5	*Procyon*	20	**245**	08·2	08·1	07·9	07·7	07·5	07·3	N **5**	13·1	13·1	13·2	13·1	13·1	13·0

Mag.	Name and Number		SHA							Declination						
				JAN.	FEB.	MAR.	APR.	MAY	JUNE		JAN.	FEB.	MAR.	APR.	MAY	JUNE
			°	′	′	′	′	′	′	°	′	′	′	′	′	′
1·6	α Geminorum		**246**	18·0	17·9	18·0	18·2	18·3	18·3	N **31**	52·9	53·0	53·0	53·0	53·0	53·0
3·3	σ Puppis		**247**	39·8	39·8	40·0	40·2	40·4	40·5	S **43**	18·4	18·6	18·7	18·7	18·6	18·5
3·1	β Canis Minoris		**248**	10·2	10·1	10·2	10·3	10·4	10·5	N **8**	17·0	17·0	17·0	17·0	17·0	17·0
2·4	η Canis Majoris		**248**	56·6	56·6	56·7	56·9	57·0	57·1	S **29**	18·5	18·6	18·7	18·7	18·7	18·6
2·7	π Puppis		**250**	41·0	41·0	41·2	41·4	41·5	41·6	S **37**	06·2	06·3	06·4	06·4	06·3	06·2
2·0	δ Canis Majoris		**252**	52·1	52·1	52·2	52·4	52·5	52·6	S **26**	23·9	24·0	24·1	24·1	24·0	23·9
3·1	o Canis Majoris		**254**	12·5	12·6	12·7	12·8	13·0	13·0	S **23**	50·3	50·4	50·4	50·4	50·4	50·3
1·6	ε Canis Majoris	19	**255**	18·6	18·6	18·8	18·9	19·1	19·1	S **28**	58·6	58·7	58·8	58·8	58·7	58·6
2·8	τ Puppis		**257**	29·4	29·5	29·7	30·0	30·2	30·3	S **50**	37·1	37·3	37·3	37·3	37·3	37·1
−1·6	α Canis Majoris	18	**258**	40·7	40·7	40·8	40·9	41·0	41·1	S **16**	43·2	43·3	43·3	43·3	43·3	43·2
1·9	γ Geminorum		**260**	31·6	31·6	31·7	31·8	31·9	31·9	N **16**	23·8	23·8	23·8	23·8	23·8	23·9
−0·9	α Carinæ	17	**263**	59·3	59·4	59·6	59·9	60·1	60·2	S **52**	41·9	42·0	42·1	42·1	42·0	41·8
2·0	β Canis Majoris		**264**	17·4	17·4	17·5	17·7	17·8	17·8	S **17**	57·5	57·6	57·6	57·6	57·5	57·4
2·7	θ Aurigæ		**270**	01·0	01·0	01·1	01·3	01·4	01·4	N **37**	12·8	12·9	12·9	12·9	12·9	12·8
2·1	β Aurigæ		**270**	03·6	03·7	03·8	04·0	04·1	04·1	N **44**	57·0	57·0	57·1	57·0	57·0	56·9
Var.‡	α Orionis	16	**271**	09·9	09·9	10·0	10·2	10·2	10·2	N **7**	24·5	24·4	24·4	24·4	24·5	24·5
2·2	κ Orionis		**273**	01·4	01·4	01·6	01·7	01·8	01·7	S **9**	40·2	40·2	40·2	40·2	40·2	40·1
1·9	ζ Orionis		**274**	46·2	46·3	46·4	46·5	46·6	46·6	S **1**	56·5	56·5	56·5	56·5	56·5	56·4
2·8	α Columbæ		**275**	03·4	03·5	03·7	03·8	03·9	03·9	S **34**	04·5	04·5	04·6	04·5	04·4	04·3
3·0	ζ Tauri		**275**	32·5	32·6	32·7	32·8	32·9	32·9	N **21**	08·7	08·7	08·7	08·7	08·7	08·7
1·8	ε Orionis	15	**275**	54·4	54·5	54·6	54·7	54·8	54·7	S **1**	12·0	12·1	12·1	12·1	12·0	12·0
2·9	ι Orionis		**276**	06·2	06·2	06·4	06·5	06·6	06·5	S **5**	54·5	54·6	54·6	54·6	54·5	54·4
2·7	α Leporis		**276**	46·9	47·0	47·1	47·3	47·3	47·3	S **17**	49·3	49·3	49·4	49·3	49·3	49·2
2·5	δ Orionis		**276**	57·5	57·6	57·7	57·8	57·9	57·8	S **0**	17·8	17·9	17·9	17·9	17·8	17·8
3·0	β Leporis		**277**	54·2	54·3	54·4	54·6	54·7	54·6	S **20**	45·5	45·6	45·6	45·6	45·5	45·4
1·8	β Tauri	14	**278**	22·7	22·7	22·9	23·0	23·1	23·0	N **28**	36·7	36·7	36·7	36·7	36·7	36·6
1·7	γ Orionis	13	**278**	40·5	40·6	40·7	40·8	40·9	40·8	N **6**	21·1	21·1	21·1	21·1	21·1	21·2
0·2	α Aurigæ	12	**280**	46·2	46·3	46·5	46·7	46·7	46·6	N **46**	00·2	00·2	00·3	00·2	00·2	00·1
0·3	β Orionis	11	**281**	19·7	19·7	19·9	20·0	20·0	20·0	S **8**	11·9	12·0	12·0	12·0	11·9	11·8
2·9	β Eridani		**283**	00·0	00·0	00·1	00·3	00·3	00·3	S **5**	05·0	05·0	05·1	05·0	05·0	04·9
2·9	ι Aurigæ		**285**	42·1	42·2	42·3	42·5	42·5	42·4	N **33**	10·3	10·4	10·4	10·3	10·3	10·3
1·1	α Tauri	10	**290**	58·6	58·6	58·8	58·9	58·9	58·8	N **16**	30·9	30·9	30·9	30·9	30·9	30·9
3·2	γ Eridani		**300**	27·4	27·5	27·6	27·7	27·8	27·7	S **13**	30·1	30·2	30·2	30·1	30·0	29·9
3·0	ε Persei		**300**	29·2	29·3	29·4	29·6	29·6	29·4	N **40**	01·3	01·3	01·3	01·2	01·1	01·1
2·9	ζ Persei		**301**	25·2	25·3	25·4	25·5	25·5	25·4	N **31**	53·6	53·6	53·6	53·6	53·6	53·5
3·0	η Tauri		**303**	05·1	05·2	05·3	05·4	05·4	05·3	N **24**	06·9	06·9	06·9	06·9	06·9	06·9
1·9	α Persei	9	**308**	52·0	52·2	52·3	52·5	52·4	52·2	N **49**	52·5	52·5	52·5	52·4	52·3	52·2
Var.§	β Persei		**312**	54·6	54·8	54·9	55·0	55·0	54·8	N **40**	58·2	58·2	58·1	58·1	58·0	58·0
2·8	α Ceti	8	**314**	23·5	23·6	23·7	23·8	23·7	23·6	N **4**	06·0	06·0	06·0	06·0	06·1	06·1
3·1	θ Eridani	7	**315**	24·3	24·5	24·6	24·7	24·7	24·6	S **40**	17·8	17·9	17·8	17·7	17·5	17·4
2·1	α Ursæ Minoris		**321**	12·6	24·6	35·1	40·8	39·0	31·0	N **89**	16·9	16·9	16·9	16·7	16·6	16·5
3·1	β Trianguli		**327**	34·3	34·5	34·6	34·6	34·5	34·3	N **35**	00·2	00·2	00·1	00·0	00·0	00·0
2·2	α Arietis	6	**328**	10·0	10·1	10·2	10·3	10·2	10·0	N **23**	28·6	28·6	28·6	28·5	28·5	28·6
2·2	γ Andromedæ		**328**	58·9	59·1	59·2	59·2	59·2	58·9	N **42**	20·8	20·8	20·7	20·6	20·5	20·5
3·0	α Hydri		**330**	17·0	17·3	17·5	17·6	17·5	17·3	S **61**	33·7	33·7	33·5	33·4	33·2	33·0
2·7	β Arietis		**331**	18·1	18·2	18·3	18·3	18·2	18·0	N **20**	49·4	49·3	49·3	49·3	49·3	49·3
0·6	α Eridani	5	**335**	32·7	32·9	33·0	33·1	33·0	32·7	S **57**	13·7	13·6	13·5	13·3	13·1	13·0
2·8	δ Cassiopeiæ		**338**	30·1	30·4	30·6	30·6	30·4	30·0	N **60**	15·3	15·2	15·1	15·0	14·9	14·9
2·4	β Andromedæ		**342**	31·8	31·9	32·0	32·0	31·8	31·6	N **35**	38·3	38·2	38·1	38·1	38·0	38·1
Var.‖	γ Cassiopeiæ		**345**	47·1	47·3	47·4	47·4	47·2	46·8	N **60**	44·2	44·1	44·0	43·9	43·8	43·8
2·2	β Ceti	4	**349**	04·1	04·2	04·3	04·2	04·1	03·9	S **17**	58·4	58·4	58·4	58·3	58·2	58·1
2·5	α Cassiopeiæ	3	**349**	50·2	50·4	50·5	50·5	50·3	49·9	N **56**	33·4	33·3	33·2	33·1	33·0	33·0
2·4	α Phœnicis	2	**353**	23·8	23·9	23·9	23·9	23·7	23·4	S **42**	17·7	17·7	17·6	17·4	17·3	17·1
2·9	β Hydri		**353**	31·8	32·3	32·5	32·4	32·0	31·3	S **77**	14·7	14·5	14·4	14·2	14·0	13·9
2·9	γ Pegasi		**356**	39·5	39·5	39·6	39·5	39·3	39·1	N **15**	12·0	11·9	11·9	11·9	11·9	12·0
2·4	β Cassiopeiæ		**357**	40·4	40·6	40·7	40·6	40·4	40·0	N **59**	10·1	10·1	09·9	09·8	09·7	09·7
2·2	α Andromedæ	1	**357**	52·2	52·3	52·3	52·3	52·1	51·8	N **29**	06·4	06·4	06·3	06·2	06·2	06·3

‡ 0·1 — 1·2 § 2·3 — 3·5 ‖ Irregular variable; 2001 mag. 2·2

Mag.	Name and Number		SHA							Declination						
				JULY	AUG.	SEPT.	OCT.	NOV.	DEC.		JULY	AUG.	SEPT.	OCT.	NOV.	DEC.
			°	′	′	′	′	′	′	°	′	′	′	′	′	′
1·6	*Castor*		**246**	18·2	18·1	17·9	17·6	17·3	17·1	N **31**	53·0	53·0	52·9	52·9	52·8	52·8
3·3	σ Puppis		**247**	40·5	40·4	40·2	39·9	39·6	39·4	S **43**	18·4	18·2	18·1	18·1	18·2	18·3
3·1	β Canis Minoris		**248**	10·4	10·3	10·1	09·9	09·6	09·4	N **8**	17·1	17·1	17·1	17·1	17·0	17·0
2·4	η Canis Majoris		**248**	57·0	56·9	56·7	56·5	56·3	56·0	S **29**	18·5	18·3	18·3	18·2	18·3	18·4
2·7	π Puppis		**250**	41·6	41·4	41·2	41·0	40·7	40·5	S **37**	06·1	06·0	05·9	05·9	05·9	06·1
2·0	*Wezen*		**252**	52·5	52·4	52·2	52·0	51·7	51·5	S **26**	23·8	23·7	23·6	23·6	23·7	23·8
3·1	ο Canis Majoris		**254**	13·0	12·8	12·6	12·4	12·2	12·0	S **23**	50·2	50·0	50·0	50·0	50·1	50·2
1·6	*Adhara*	19	**255**	19·1	18·9	18·7	18·5	18·3	18·1	S **28**	58·5	58·3	58·3	58·3	58·3	58·5
2·8	τ Puppis		**257**	30·3	30·1	29·9	29·6	29·3	29·1	S **50**	37·0	36·8	36·7	36·7	36·8	37·0
−1·6	*Sirius*	18	**258**	41·0	40·8	40·7	40·4	40·2	40·0	S **16**	43·1	43·0	42·9	42·9	43·0	43·1
1·9	*Alhena*		**260**	31·8	31·7	31·4	31·2	31·0	30·8	N **16**	23·9	23·9	23·9	23·9	23·9	23·8
−0·9	*Canopus*	17	**263**	60·2	60·0	59·7	59·4	59·1	58·9	S **52**	41·7	41·5	41·4	41·4	41·5	41·7
2·0	*Mirzam*		**264**	17·7	17·5	17·3	17·1	16·9	16·7	S **17**	57·3	57·2	57·1	57·2	57·2	57·3
2·7	θ Aurigæ		**270**	01·2	01·0	00·7	00·4	00·2	00·0	N **37**	12·8	12·8	12·8	12·8	12·8	12·8
2·1	*Menkalinan*		**270**	03·9	03·7	03·4	03·0	02·7	02·5	N **44**	56·9	56·8	56·8	56·8	56·9	56·9
Var.‡	*Betelgeuse*	16	**271**	10·1	09·9	09·7	09·4	09·2	09·1	N **7**	24·6	24·6	24·6	24·6	24·6	24·6
2·2	κ Orionis		**273**	01·6	01·5	01·2	01·0	00·8	00·7	S **9**	40·0	39·9	39·8	39·8	39·9	40·0
1·9	*Alnitak*		**274**	46·4	46·3	46·0	45·8	45·6	45·5	S **1**	56·3	56·3	56·2	56·2	56·3	56·3
2·8	*Phact*		**275**	03·8	03·6	03·4	03·2	03·0	02·8	S **34**	04·2	04·0	04·0	04·0	04·1	04·2
3·0	ζ Tauri		**275**	32·7	32·5	32·3	32·0	31·8	31·6	N **21**	08·7	08·8	08·8	08·8	08·8	08·8
1·8	*Alnilam*	15	**275**	54·6	54·4	54·2	54·0	53·8	53·6	S **1**	11·9	11·8	11·8	11·8	11·8	11·9
2·9	ι Orionis		**276**	06·4	06·2	06·0	05·8	05·6	05·4	S **5**	54·3	54·2	54·2	54·2	54·3	54·4
2·7	α Leporis		**276**	47·2	47·0	46·8	46·6	46·4	46·2	S **17**	49·0	48·9	48·9	48·9	49·0	49·1
2·5	δ Orionis		**276**	57·7	57·5	57·3	57·1	56·9	56·7	S **0**	17·7	17·6	17·6	17·6	17·6	17·7
3·0	β Leporis		**277**	54·5	54·3	54·1	53·9	53·7	53·6	S **20**	45·2	45·1	45·1	45·1	45·2	45·3
1·8	*Elnath*	14	**278**	22·9	22·6	22·4	22·1	21·9	21·7	N **28**	36·6	36·7	36·7	36·7	36·7	36·7
1·7	*Bellatrix*	13	**278**	40·7	40·5	40·3	40·0	39·8	39·7	N **6**	21·3	21·3	21·4	21·3	21·3	21·3
0·2	*Capella*	12	**280**	46·4	46·2	45·8	45·5	45·3	45·1	N **46**	00·0	00·0	00·0	00·1	00·1	00·2
0·3	*Rigel*	11	**281**	19·9	19·7	19·4	19·2	19·0	18·9	S **8**	11·7	11·6	11·6	11·6	11·7	11·8
2·9	β Eridani		**282**	60·1	59·9	59·7	59·5	59·3	59·2	S **5**	04·8	04·7	04·7	04·7	04·7	04·8
2·9	ι Aurigæ		**285**	42·2	42·0	41·7	41·4	41·2	41·1	N **33**	10·3	10·3	10·3	10·3	10·4	10·4
1·1	*Aldebaran*	10	**290**	58·6	58·4	58·2	58·0	57·8	57·7	N **16**	31·0	31·1	31·1	31·1	31·1	31·1
3·2	γ Eridani		**300**	27·5	27·2	27·0	26·8	26·7	26·6	S **13**	29·8	29·7	29·7	29·7	29·7	29·8
3·0	ε Persei		**300**	29·2	28·9	28·6	28·4	28·2	28·1	N **40**	01·1	01·1	01·2	01·3	01·3	01·4
2·9	ζ Persei		**301**	25·2	24·9	24·7	24·4	24·3	24·2	N **31**	53·6	53·6	53·7	53·7	53·8	53·8
3·0	*Alcyone*		**303**	05·0	04·8	04·5	04·3	04·2	04·1	N **24**	06·9	07·0	07·0	07·1	07·1	07·1
1·9	*Mirfak*	9	**308**	51·9	51·6	51·3	51·0	50·8	50·8	N **49**	52·2	52·3	52·4	52·5	52·6	52·7
Var.§	*Algol*		**312**	54·5	54·2	53·9	53·7	53·6	53·5	N **40**	58·0	58·0	58·1	58·2	58·3	58·4
2·8	*Menkar*	8	**314**	23·4	23·1	22·9	22·8	22·7	22·6	N **4**	06·2	06·3	06·4	06·4	06·4	06·3
3·1	*Acamar*	7	**315**	24·3	24·1	23·8	23·6	23·5	23·6	S **40**	17·2	17·1	17·1	17·2	17·3	17·5
2·1	*Polaris*		**320**	78·9	65·4	53·3	45·0	42·1	46·5	N **89**	16·4	16·5	16·6	16·8	16·9	17·1
3·1	β Trianguli		**327**	34·0	33·8	33·5	33·4	33·3	33·3	N **35**	00·1	00·1	00·3	00·4	00·4	00·5
2·2	*Hamal*	6	**328**	09·7	09·5	09·3	09·1	09·1	09·1	N **23**	28·6	28·7	28·8	28·9	28·9	29·0
2·2	*Almak*		**328**	58·6	58·3	58·1	57·9	57·8	57·9	N **42**	20·6	20·7	20·8	20·9	21·0	21·1
3·0	α Hydri		**330**	16·9	16·5	16·1	16·0	16·0	16·1	S **61**	32·9	32·9	32·9	33·0	33·2	33·3
2·7	*Sheratan*		**331**	17·8	17·5	17·3	17·2	17·1	17·1	N **20**	49·4	49·5	49·6	49·7	49·7	49·7
0·6	*Achernar*	5	**335**	32·4	32·0	31·7	31·6	31·6	31·8	S **57**	12·9	12·8	12·9	13·0	13·2	13·3
2·8	*Ruchbah*		**338**	29·6	29·3	29·0	28·8	28·8	28·9	N **60**	14·9	15·0	15·2	15·3	15·5	15·6
2·4	*Mirach*		**342**	31·3	31·0	30·8	30·7	30·7	30·8	N **35**	38·1	38·3	38·4	38·5	38·6	38·6
Var.‖	γ Cassiopeiæ		**345**	46·4	46·1	45·8	45·7	45·7	45·9	N **60**	43·8	43·9	44·1	44·3	44·4	44·5
2·2	*Diphda*	4	**349**	03·6	03·4	03·2	03·2	03·2	03·2	S **17**	58·0	57·9	57·9	57·9	58·0	58·1
2·5	*Schedar*	3	**349**	49·6	49·3	49·1	49·0	49·0	49·2	N **56**	33·1	33·2	33·4	33·5	33·7	33·8
2·4	*Ankaa*	2	**353**	23·1	22·9	22·7	22·7	22·7	22·8	S **42**	17·0	17·0	17·1	17·2	17·3	17·4
2·9	β Hydri		**353**	30·5	29·7	29·2	29·2	29·5	30·1	S **77**	13·8	13·9	14·0	14·1	14·3	14·3
2·9	*Algenib*		**356**	38·8	38·6	38·5	38·5	38·5	38·6	N **15**	12·1	12·2	12·3	12·3	12·4	12·4
2·4	*Caph*		**357**	39·6	39·3	39·1	39·1	39·2	39·4	N **59**	09·8	10·0	10·1	10·3	10·4	10·5
2·2	*Alpheratz*	1	**357**	51·6	51·3	51·2	51·2	51·2	51·3	N **29**	06·4	06·5	06·7	06·8	06·8	06·8

‡ 0·1 — 1·2 § 2·3 — 3·5 ‖ Irregular variable; 2001 mag. 2·2

POLARIS (POLE STAR) TABLES, 2003

FOR DETERMINING LATITUDE FROM SEXTANT ALTITUDE AND FOR AZIMUTH

LHA ARIES	0° – 9°	10° – 19°	20° – 29°	30° – 39°	40° – 49°	50° – 59°	60° – 69°	70° – 79°	80° – 89°	90° – 99°	100° – 109°	110° – 119°
	a_0	a_0	a_0	a_0	a_0	a_0	a_0	a_0	a_0	a_0	a_0	a_0
°	° ′	° ′	° ′	° ′	° ′	° ′	° ′	° ′	° ′	° ′	° ′	° ′
0	0 25·3	0 21·0	0 17·9	0 16·1	0 15·6	0 16·4	0 18·6	0 21·9	0 26·4	0 31·9	0 38·2	0 45·2
1	24·8	20·7	17·7	16·0	15·6	16·6	18·8	22·3	26·9	32·5	38·9	45·9
2	24·3	20·3	17·5	15·9	15·7	16·8	19·1	22·7	27·5	33·1	39·6	46·6
3	23·9	20·0	17·3	15·8	15·7	16·9	19·5	23·2	28·0	33·7	40·3	47·3
4	23·4	19·6	17·1	15·8	15·8	17·1	19·8	23·6	28·5	34·4	40·9	48·1
5	0 23·0	0 19·3	0 16·9	0 15·7	0 15·9	0 17·3	0 20·1	0 24·1	0 29·1	0 35·0	0 41·6	0 48·8
6	22·6	19·0	16·7	15·7	15·9	17·6	20·4	24·5	29·6	35·6	42·3	49·5
7	22·2	18·7	16·5	15·6	16·0	17·8	20·8	25·0	30·2	36·3	43·0	50·3
8	21·8	18·5	16·4	15·6	16·2	18·0	21·2	25·5	30·8	36·9	43·7	51·0
9	21·4	18·2	16·2	15·6	16·3	18·3	21·5	25·9	31·3	37·6	44·5	51·8
10	0 21·0	0 17·9	0 16·1	0 15·6	0 16·4	0 18·6	0 21·9	0 26·4	0 31·9	0 38·2	0 45·2	0 52·5
Lat.	a_1	a_1	a_1	a_1	a_1	a_1	a_1	a_1	a_1	a_1	a_1	a_1
°	′	′	′	′	′	′	′	′	′	′	′	′
0	0·5	0·5	0·6	0·6	0·6	0·6	0·5	0·5	0·4	0·4	0·3	0·3
10	·5	·6	·6	·6	·6	·6	·5	·5	·5	·4	·4	·3
20	·5	·6	·6	·6	·6	·6	·6	·5	·5	·4	·4	·4
30	·5	·6	·6	·6	·6	·6	·6	·5	·5	·5	·5	·4
40	0·6	0·6	0·6	0·6	0·6	0·6	0·6	0·6	0·6	0·5	0·5	0·5
45	·6	·6	·6	·6	·6	·6	·6	·6	·6	·6	·6	·6
50	·6	·6	·6	·6	·6	·6	·6	·6	·6	·6	·6	·6
55	·6	·6	·6	·6	·6	·6	·6	·6	·6	·6	·7	·7
60	·6	·6	·6	·6	·6	·6	·6	·7	·7	·7	·7	·7
62	0·7	0·6	0·6	0·6	0·6	0·6	0·6	0·7	0·7	0·7	0·8	0·8
64	·7	·6	·6	·6	·6	·6	·6	·7	·7	·8	·8	·8
66	·7	·6	·6	·6	·6	·6	·7	·7	·7	·8	·8	·9
68	0·7	0·7	0·6	0·6	0·6	0·6	0·7	0·7	0·8	0·8	0·9	0·9
Month	a_2	a_2	a_2	a_2	a_2	a_2	a_2	a_2	a_2	a_2	a_2	a_2
	′	′	′	′	′	′	′	′	′	′	′	′
Jan.	0·7	0·7	0·7	0·7	0·7	0·7	0·7	0·7	0·7	0·7	0·7	0·6
Feb.	·6	·7	·7	·7	·8	·8	·8	·8	·8	·8	·8	·8
Mar.	·5	·5	·6	·6	·7	·7	·8	·8	·9	·9	·9	·9
Apr.	0·3	0·4	0·5	0·5	0·6	0·6	0·7	0·8	0·8	0·9	0·9	0·9
May	·2	·3	·3	·4	·4	·5	·6	·6	·7	·8	·8	·9
June	·2	·2	·2	·3	·3	·4	·4	·5	·6	·6	·7	·8
July	0·3	0·2	0·2	0·2	0·3	0·3	0·3	0·4	0·4	0·5	0·5	0·6
Aug.	·4	·3	·3	·3	·3	·3	·3	·3	·3	·4	·4	·4
Sept.	·6	·5	·5	·4	·4	·3	·3	·3	·3	·3	·3	·3
Oct.	0·8	0·7	0·6	0·6	0·5	0·5	0·4	0·4	0·3	0·3	0·3	0·3
Nov.	0·9	0·9	0·8	·8	·7	·6	·6	·5	·4	·4	·3	·3
Dec.	1·0	1·0	1·0	0·9	0·9	0·8	0·7	0·7	0·6	0·5	0·4	0·4
Lat.	AZIMUTH											
°	°	°	°	°	°	°	°	°	°	°	°	°
0	0·4	0·3	0·2	0·0	359·9	359·8	359·7	359·6	359·5	359·4	359·3	359·3
20	0·4	0·3	0·2	0·1	359·9	359·8	359·7	359·5	359·4	359·4	359·3	359·3
40	0·5	0·4	0·2	0·1	359·9	359·7	359·6	359·4	359·3	359·2	359·1	359·1
50	0·6	0·5	0·3	0·1	359·9	359·7	359·5	359·3	359·2	359·1	359·0	358·9
55	0·7	0·5	0·3	0·1	359·9	359·6	359·4	359·2	359·1	358·9	358·8	358·8
60	0·8	0·6	0·4	0·1	359·8	359·6	359·4	359·1	358·9	358·8	358·7	358·6
65	1·0	0·7	0·4	0·1	359·8	359·5	359·2	359·0	358·7	358·6	358·4	358·3

$$\text{Latitude} = \text{Apparent altitude (corrected for refraction)} - 1^\circ + a_0 + a_1 + a_2$$

The table is entered with LHA Aries to determine the column to be used; each column refers to a range of 10°. a_0 is taken, with mental interpolation, from the upper table with the units of LHA Aries in degrees as argument; a_1, a_2 are taken, without interpolation, from the second and third tables with arguments latitude and month respectively. a_0, a_1, a_2, are always positive. The final table gives the azimuth of *Polaris*.

POLARIS (POLE STAR) TABLES, 2003

FOR DETERMINING LATITUDE FROM SEXTANT ALTITUDE AND FOR AZIMUTH

LHA ARIES	120°– 129°	130°– 139°	140°– 149°	150°– 159°	160°– 169°	170°– 179°	180°– 189°	190°– 199°	200°– 209°	210°– 219°	220°– 229°	230°– 239°
	a_0	a_0	a_0	a_0	a_0	a_0	a_0	a_0	a_0	a_0	a_0	a_0
°	° ′	° ′	° ′	° ′	° ′	° ′	° ′	° ′	° ′	° ′	° ′	° ′
0	0 52·5	1 00·0	1 07·5	1 14·7	1 21·4	1 27·4	1 32·6	1 36·7	1 39·7	1 41·5	1 42·0	1 41·2
1	53·3	00·8	08·2	15·4	22·1	28·0	33·1	37·1	40·0	41·6	42·0	41·0
2	54·0	01·5	09·0	16·1	22·7	28·5	33·5	37·4	40·2	41·7	41·9	40·9
3	54·8	02·3	09·7	16·8	23·3	29·1	33·9	37·8	40·4	41·8	41·9	40·7
4	55·5	03·0	10·4	17·5	23·9	29·6	34·4	38·1	40·6	41·9	41·8	40·5
5	0 56·3	1 03·8	1 11·2	1 18·1	1 24·5	1 30·1	1 34·8	1 38·4	1 40·8	1 41·9	1 41·8	1 40·3
6	57·0	04·5	11·9	18·8	25·1	30·6	35·2	38·7	40·9	41·9	41·7	40·1
7	57·8	05·3	12·6	19·5	25·7	31·1	35·6	39·0	41·1	42·0	41·6	39·9
8	58·5	06·0	13·3	20·1	26·3	31·6	36·0	39·2	41·2	42·0	41·5	39·6
9	0 59·3	06·8	14·0	20·8	26·9	32·1	36·4	39·5	41·4	42·0	41·3	39·4
10	1 00·0	1 07·5	1 14·7	1 21·4	1 27·4	1 32·6	1 36·7	1 39·7	1 41·5	1 42·0	1 41·2	1 39·1
Lat.	a_1	a_1	a_1	a_1	a_1	a_1	a_1	a_1	a_1	a_1	a_1	a_1
°	′	′	′	′	′	′	′	′	′	′	′	′
0	0·3	0·3	0·3	0·3	0·4	0·4	0·5	0·5	0·6	0·6	0·6	0·6
10	·3	·3	·3	·4	·4	·5	·5	·6	·6	·6	·6	·6
20	·4	·4	·4	·4	·5	·5	·5	·6	·6	·6	·6	·6
30	·4	·4	·4	·5	·5	·5	·5	·6	·6	·6	·6	·6
40	0·5	0·5	0·5	0·5	0·5	0·6	0·6	0·6	0·6	0·6	0·6	0·6
45	·5	·5	·6	·6	·6	·6	·6	·6	·6	·6	·6	·6
50	·6	·6	·6	·6	·6	·6	·6	·6	·6	·6	·6	·6
55	·7	·7	·7	·7	·6	·6	·6	·6	·6	·6	·6	·6
60	·7	·7	·7	·7	·7	·7	·6	·6	·6	·6	·6	·6
62	0·8	0·8	0·8	0·8	0·7	0·7	0·7	0·6	0·6	0·6	0·6	0·6
64	·8	·8	·8	·8	·8	·7	·7	·6	·6	·6	·6	·6
66	·9	·9	·9	·8	·8	·7	·7	·6	·6	·6	·6	·6
68	0·9	0·9	0·9	0·9	0·8	0·8	0·7	0·7	0·6	0·6	0·6	0·6
Month	a_2	a_2	a_2	a_2	a_2	a_2	a_2	a_2	a_2	a_2	a_2	a_2
	′	′	′	′	′	′	′	′	′	′	′	′
Jan.	0·6	0·6	0·6	0·6	0·5	0·5	0·5	0·5	0·5	0·5	0·5	0·5
Feb.	·8	·7	·7	·7	·6	·6	·6	·5	·5	·5	·4	·4
Mar.	0·9	0·9	0·9	0·8	·8	·8	·7	·7	·6	·6	·5	·5
Apr.	1·0	1·0	1·0	1·0	0·9	0·9	0·9	0·8	0·7	0·7	0·6	0·6
May	0·9	1·0	1·0	1·0	1·0	1·0	1·0	0·9	0·9	·8	·8	·7
June	·8	0·9	0·9	1·0	1·0	1·0	1·0	1·0	1·0	0·9	·9	·8
July	0·7	0·7	0·8	0·8	0·9	0·9	0·9	1·0	1·0	1·0	0·9	0·9
Aug.	·5	·6	·6	·7	·7	·8	·8	0·9	0·9	0·9	·9	·9
Sept.	·4	·4	·4	·5	·5	·6	·6	·7	·7	·8	·8	·9
Oct.	0·3	0·3	0·3	0·3	0·4	0·4	0·4	0·5	0·6	0·6	0·7	0·7
Nov.	·2	·2	·2	·2	·2	·2	·3	·3	·4	·4	·5	·6
Dec.	0·3	0·2	0·2	0·2	0·2	0·2	0·2	0·2	0·2	0·3	0·3	0·4
Lat.	AZIMUTH											
°	°	°	°	°	°	°	°	°	°	°	°	°
0	359·3	359·3	359·3	359·4	359·4	359·5	359·6	359·7	359·8	0·0	0·1	0·2
20	359·2	359·2	359·3	359·3	359·4	359·5	359·6	359·7	359·8	359·9	0·1	0·2
40	359·1	359·1	359·1	359·2	359·2	359·4	359·5	359·6	359·8	359·9	0·1	0·3
50	358·9	358·9	358·9	359·0	359·1	359·2	359·4	359·6	359·7	359·9	0·1	0·3
55	358·7	358·8	358·8	358·9	359·0	359·1	359·3	359·5	359·7	359·9	0·1	0·3
60	358·6	358·6	358·6	358·7	358·9	359·0	359·2	359·4	359·7	359·9	0·2	0·4
65	358·3	358·3	358·4	358·5	358·6	358·8	359·1	359·3	359·6	359·9	0·2	0·5

ILLUSTRATION

On 2003 April 21 at 23h 18m 56s UT in longitude W 37° 14′ the apparent altitude (corrected for refraction), H_O, of *Polaris* was 49° 31′.6

From the daily pages:	° ′		° ′
GHA Aries (23h)	194 35·9	H_O	49 31·6
Increment (18m 56s)	4 44·8	a_0 (argument 162° 07′)	1 22·8
Longitude (west)	−37 14	a_1 (Lat 50° approx.)	0·6
		a_2 (April)	0·9
LHA Aries	162 07	Sum − 1° = Lat =	49 55·9

FOR DETERMINING LATITUDE FROM SEXTANT ALTITUDE AND FOR AZIMUTH

LHA ARIES	240° – 249°	250° – 259°	260° – 269°	270° – 279°	280° – 289°	290° – 299°	300° – 309°	310° – 319°	320° – 329°	330° – 339°	340° – 349°	350° – 359°
	a_0	a_0	a_0	a_0	a_0	a_0	a_0	a_0	a_0	a_0	a_0	a_0
°	° ′	° ′	° ′	° ′	° ′	° ′	° ′	° ′	° ′	° ′	° ′	° ′
0	1 39·1	1 35·8	1 31·4	1 26·1	1 19·9	1 13·0	1 05·7	0 58·2	0 50·7	0 43·5	0 36·7	0 30·[illegible]
1	38·8	35·4	30·9	25·5	19·2	12·3	05·0	57·5	50·0	42·8	36·0	30·[illegible]
2	38·6	35·0	30·4	24·9	18·5	11·6	04·2	56·7	49·2	42·1	35·4	29·[illegible]
3	38·3	34·6	29·9	24·3	17·9	10·9	03·5	56·0	48·5	41·4	34·7	28·[illegible]
4	37·9	34·2	29·4	23·7	17·2	10·1	02·7	55·2	47·8	40·7	34·1	28·[illegible]
5	1 37·6	1 33·8	1 28·9	1 23·1	1 16·5	1 09·4	1 02·0	0 54·5	0 47·0	0 40·0	0 33·5	0 27·[illegible]
6	37·3	33·3	28·3	22·4	15·8	08·7	01·2	53·7	46·3	39·3	32·9	27·[illegible]
7	36·9	32·9	27·8	21·8	15·1	07·9	1 00·5	53·0	45·6	38·6	32·3	26·[illegible]
8	36·6	32·4	27·2	21·2	14·4	07·2	0 59·7	52·2	44·9	38·0	31·7	26·[illegible]
9	36·2	31·9	26·6	20·5	13·7	06·5	59·0	51·5	44·2	37·3	31·1	25·[illegible]
10	1 35·8	1 31·4	1 26·1	1 19·9	1 13·0	1 05·7	0 58·2	0 50·7	0 43·5	0 36·7	0 30·5	0 25·[illegible]
Lat.	a_1	a_1	a_1	a_1	a_1	a_1	a_1	a_1	a_1	a_1	a_1	a_1
°	′	′	′	′	′	′	′	′	′	′	′	′
0	0·5	0·5	0·4	0·4	0·3	0·3	0·3	0·3	0·3	0·3	0·4	0·4
10	·5	·5	·5	·4	·4	·3	·3	·3	·3	·4	·4	·5
20	·6	·5	·5	·4	·4	·4	·4	·4	·4	·4	·5	·5
30	·6	·5	·5	·5	·5	·4	·4	·4	·4	·5	·5	·5
40	0·6	0·6	0·6	0·5	0·5	0·5	0·5	0·5	0·5	0·5	0·5	0·6
45	·6	·6	·6	·6	·6	·6	·5	·5	·6	·6	·6	·6
50	·6	·6	·6	·6	·6	·6	·6	·6	·6	·6	·6	·6
55	·6	·6	·6	·6	·7	·7	·7	·7	·7	·7	·6	·6
60	·6	·7	·7	·7	·7	·7	·7	·7	·7	·7	·7	·7
62	0·6	0·7	0·7	0·7	0·8	0·8	0·8	0·8	0·8	0·8	0·7	0·7
64	·6	·7	·7	·8	·8	·8	·8	·8	·8	·8	·8	·7
66	·7	·7	·7	·8	·8	·9	·9	·9	·9	·8	·8	·7
68	0·7	0·7	0·8	0·8	0·9	0·9	0·9	0·9	0·9	0·9	0·8	0·8
Month	a_2	a_2	a_2	a_2	a_2	a_2	a_2	a_2	a_2	a_2	a_2	a_2
	′	′	′	′	′	′	′	′	′	′	′	′
Jan.	0·5	0·5	0·5	0·5	0·5	0·6	0·6	0·6	0·6	0·6	0·7	0·7
Feb.	·4	·4	·4	·4	·4	·4	·4	·5	·5	·5	·6	·6
Mar.	·4	·4	·3	·3	·3	·3	·3	·3	·3	·4	·4	·4
Apr.	0·5	0·4	0·4	0·3	0·3	0·3	0·2	0·2	0·2	0·2	0·3	0·3
May	·6	·6	·5	·4	·4	·3	·3	·2	·2	·2	·2	·2
June	·8	·7	·6	·6	·5	·4	·4	·3	·3	·2	·2	·2
July	0·9	0·8	0·8	0·7	0·7	0·6	0·5	0·5	0·4	0·4	0·3	0·3
Aug.	·9	·9	·9	·8	·8	·8	·7	·6	·6	·5	·5	·4
Sept.	·9	·9	·9	·9	·9	·9	·8	·8	·8	·7	·7	·6
Oct.	0·8	0·8	0·9	0·9	0·9	0·9	0·9	0·9	0·9	0·9	0·8	0·8
Nov.	·6	·7	·8	·8	·9	·9	1·0	1·0	1·0	1·0	1·0	1·0
Dec.	0·5	0·5	0·6	0·7	0·8	0·8	0·9	1·0	1·0	1·0	1·0	1·0
Lat.	AZIMUTH											
°	°	°	°	°	°	°	°	°	°	°	°	°
0	0·3	0·4	0·5	0·6	0·7	0·7	0·7	0·7	0·7	0·6	0·6	0·5
20	0·3	0·5	0·6	0·6	0·7	0·7	0·8	0·8	0·7	0·7	0·6	0·5
40	0·4	0·6	0·7	0·8	0·9	0·9	0·9	0·9	0·9	0·8	0·8	0·7
50	0·5	0·7	0·8	0·9	1·0	1·1	1·1	1·1	1·1	1·0	0·9	0·8
55	0·5	0·7	0·9	1·0	1·1	1·2	1·3	1·3	1·2	1·1	1·0	0·9
60	0·6	0·8	1·0	1·2	1·3	1·4	1·4	1·4	1·4	1·3	1·2	1·0
65	0·7	1·0	1·2	1·4	1·5	1·6	1·7	1·7	1·6	1·5	1·4	1·2

$$\text{Latitude} = \text{Apparent altitude (corrected for refraction)} - 1^\circ + a_0 + a_1 + a_2$$

The table is entered with LHA Aries to determine the column to be used; each column refers to a range o[illegible] 10°. a_0 is taken, with mental interpolation, from the upper table with the units of LHA Aries in degrees a[illegible] argument; a_1, a_2 are taken, without interpolation, from the second and third tables with arguments latitude an[illegible] month respectively. a_0, a_1, a_2, are always positive. The final table gives the azimuth of *Polaris*.

SIGHT REDUCTION PROCEDURES

METHODS AND FORMULAE FOR DIRECT COMPUTATION

1. *Introduction.* In this section formulae and methods are provided for *calculating* position at sea from observed altitudes taken with a marine sextant using a computer or programmable calculator.

The method uses analogous concepts and similar terminology as that used in *manual* methods of astro-navigation, where position is found by plotting position lines from their intercept and azimuth on a marine chart.

The algorithms are presented in standard algebra suitable for translating into the programming language of the user's computer. The basic ephemeris data may be taken directly from the main tabular pages of a current version of *The Nautical Almanac*. Formulae are given for calculating altitude and azimuth from the *GHA* and *Dec* of a body, and the estimated position of the observer. Formulae are also given for reducing sextant observations to observed altitudes by applying the corrections for dip, refraction, parallax and semi-diameter.

The intercept and azimuth obtained from each observation determines a position line, and the observer should lie on or close to each position line. The method of least squares is used to calculate the fix by finding the position where the sum of the squares of the distances from the position lines is a minimum. The use of least squares has other advantages. For example it is possible to improve the estimated position at the time of fix by repeating the calculation. It is also possible to include more observations in the solution and to reject doubtful ones.

2. *Notation.*

GHA = Greenwich hour angle. The range of *GHA* is from 0° to 360° starting at 0° on the Greenwich meridian increasing to the west, back to 360° on the Greenwich meridian.

SHA = sidereal hour angle. The range is 0° to 360°.

Dec = declination. The sign convention for declination is north is positive, south is negative. The range is from −90° at the south celestial pole to +90° at the north celestial pole.

Long = longitude. The sign convention is east is positive, west is negative. The range is −180° to +180°.

Lat = latitude. The sign convention is north is positive, south is negative. The range is from −90° to +90°.

LHA = *GHA* + *Long* = local hour angle. The *LHA* increases to the west from 0° on the local meridian to 360°.

H_C = calculated altitude. Above the horizon is positive, below the horizon is negative. The range is from −90° in the nadir to +90° in the zenith.

H_S = sextant altitude.

H = apparent altitude = sextant altitude corrected for instrumental error and dip.

H_O = observed altitude = apparent altitude corrected for refraction and, in appropriate cases, corrected for parallax and semi-diameter.

Z = Z_n = true azimuth. *Z* is measured from true north through east, south, west and back to north. The range is from 0° to 360°.

I = sextant index error.

D = dip of horizon.

R = atmospheric refraction.

HP = horizontal parallax of the Sun, Moon, Venus or Mars.
PA = parallax in altitude of the Sun, Moon, Venus or Mars.
S = semi-diameter of the Sun or Moon.
p = intercept $= H_O - H_C$. Towards is positive, away is negative.
T = course or track, measured as for azimuth from the north.
V = speed in knots.

3. *Entering Basic Data.* When quantities such as *GHA* are entered, which in *The Nautical Almanac* are given in degrees and minutes, convert them to degrees and decimals of a degree by dividing the minutes by 60 and adding to the degrees; for example, if $GHA = 123° \; 45'\!.6$, enter the two numbers 123 and 45·6 into the memory and set $GHA = 123 + 45{\cdot}6/60 = 123^{\circ}\!.7600$. Although four decimal places of a degree are shown in the examples, it is assumed that full precision is maintained in the calculations.

When using a computer or programmable calculator, write a subroutine to convert degrees and minutes to degrees and decimals. Scientific calculators usually have a special key for this purpose. For quantities like *Dec* which require a minus sign for southern declination, change the sign from plus to minus after the value has been converted to degrees and decimals, *e.g.* $Dec = \text{S}\,0° \; 12'\!.3 = \text{S}\,0^{\circ}\!.2050 = -0^{\circ}\!.2050$. Other quantities which require conversion are semi-diameter, horizontal parallax, longitude and latitude.

4. *Interpolation of GHA and Dec.* The *GHA* and *Dec* of the Sun, Moon and planets are interpolated to the time of observation by direct calculation as follows: If the universal time is $a^{\text{h}} \; b^{\text{m}} \; c^{\text{s}}$, form the interpolation factor $x = b/60 + c/3600$. Enter the tabular value GHA_0 for the preceding hour (a) and the tabular value GHA_1 for the following hour ($a + 1$) then the interpolated value *GHA* is given by

$$GHA = GHA_0 + x(GHA_1 - GHA_0)$$

If the *GHA* passes through 360° between tabular values add 360° to GHA_1 before interpolation. If the interpolated value exceeds 360°, subtract 360° from *GHA*.

Similarly for declination, enter the tabular value Dec_0 for the preceding hour (a) and the tabular value Dec_1 for the following hour ($a + 1$), then the interpolated value *Dec* is given by

$$Dec = Dec_0 + x(Dec_1 - Dec_0)$$

5. *Example.* (a) Find the *GHA* and *Dec* of the Sun on 2003 December 1 at $19^{\text{h}} \; 57^{\text{m}} \; 44^{\text{s}}$ UT.

The interpolation factor $x = 57/60 + 44/3600 = 0^{\text{h}}\!.9622$

page 233 $19^{\text{h}} \; GHA_0 = 107° \; 44'\!.9 = 107^{\circ}\!.7483$

$20^{\text{h}} \; GHA_1 = 122° \; 44'\!.6 = 122^{\circ}\!.7433$

$19^{\text{h}}\!.9622 \; GHA = 107{\cdot}7483 + 0{\cdot}9622(122{\cdot}7433 - 107{\cdot}7483) = 122^{\circ}\!.1769$

$19^{\text{h}} \; Dec_0 = \text{S}\,21° \; 49'\!.5 = -21^{\circ}\!.8250$

$20^{\text{h}} \; Dec_1 = \text{S}\,21° \; 49'\!.9 = -21^{\circ}\!.8317$

$19^{\text{h}}\!.9622 \; Dec = -21{\cdot}8250 + 0{\cdot}9622(-21{\cdot}8317 + 21{\cdot}8250) = -21^{\circ}\!.8314$

GHA Aries is interpolated in the same way as *GHA* of a body. For a star the *SHA* and *Dec* are taken from the tabular page and do not require interpolation, then

$$GHA = GHA \text{ Aries} + SHA$$

where *GHA* Aries is interpolated to the time of observation.

(b) Find the *GHA* and *Dec* of *Vega* on 2003 December 1 at $19^h\ 57^m\ 44^s$ UT.

The interpolation factor $x = 0^h\!.9622$ as in the previous example

$$\begin{aligned}
\text{page 232} \quad 19^h\ GHA\ \text{Aries}_0 &= 355°\ 13'\!.2 = 355°\!.2200 \\
20^h\ GHA\ \text{Aries}_1 &= 10°\ 15'\!.7 = 370°\!.2617 \quad (360°\ \text{added}) \\
19^h\!.9622\ GHA\ \text{Aries} &= 355{\cdot}2200 + 0{\cdot}9622(370{\cdot}2617 - 355{\cdot}2200) = 369°\!.6934 \\
SHA &= 80°\ 44'\!.5 = 80°\!.7417 \\
GHA = GHA\ \text{Aries} + SHA &= 90°\!.4351 \quad (\text{multiple of } 360°\ \text{removed}) \\
Dec &= \text{N}\,38°\ 47'\!.3 = +38°\!.7883
\end{aligned}$$

6. *The calculated altitude and azimuth.* The calculated altitude H_C and true azimuth Z are determined from the *GHA* and *Dec* interpolated to the time of observation and from the *Long* and *Lat* estimated at the time of observation as follows:

Step 1. Calculate the local hour angle

$$LHA = GHA + Long$$

Add or subtract multiples of 360° to set *LHA* in the range 0° to 360°.

Step 2. Calculate S, C and the altitude H_C from

$$\begin{aligned}
S &= \sin Dec \\
C &= \cos Dec \cos LHA \\
H_C &= \sin^{-1}(S \sin Lat + C \cos Lat)
\end{aligned}$$

where $\sin^{-1}$ is the inverse function of sine.

Step 3. Calculate X and A from

$$\begin{aligned}
X &= (S \cos Lat - C \sin Lat)/\cos H_C \\
&\text{If } X > +1 \quad \text{set} \quad X = +1 \\
&\text{If } X < -1 \quad \text{set} \quad X = -1 \\
A &= \cos^{-1} X
\end{aligned}$$

where $\cos^{-1}$ is the inverse function of cosine.

Step 4. Determine the azimuth Z

$$\begin{aligned}
&\text{If } LHA > 180° \quad \text{then} \quad Z = A \\
&\text{Otherwise} \quad Z = 360° - A
\end{aligned}$$

7. *Example.* Find the calculated altitude H_C and azimuth Z when

$$GHA = 53° \quad Dec = \text{S}\,15° \quad Lat = \text{N}\,32° \quad Long = \text{W}\,16°$$

For the calculation

$$GHA = 53°\!.0000 \quad Dec = -15°\!.0000 \quad Lat = +32°\!.0000 \quad Long = -16°\!.0000$$

Step 1. $LHA = 53{\cdot}0000 - 16{\cdot}0000 = 37{\cdot}0000$

Step 2.

$$\begin{aligned}
S &= -0{\cdot}2588 \\
C &= +0{\cdot}9659 \times 0{\cdot}7986 = 0{\cdot}7714 \\
\sin H_C &= -0{\cdot}2588 \times 0{\cdot}5299 + 0{\cdot}7714 \times 0{\cdot}8480 = 0{\cdot}5171 \\
H_C &= 31°\!.1346
\end{aligned}$$

Step 3. $X = (-0{\cdot}2588 \times 0{\cdot}8480 - 0{\cdot}7714 \times 0{\cdot}5299)/0{\cdot}8560 = -0{\cdot}7340$

$A = 137^{\circ}{\cdot}2239$

Step 4. Since $LHA \leq 180^{\circ}$ then $Z = 360^{\circ} - A = 222^{\circ}{\cdot}7761$

8. *Reduction from sextant altitude to observed altitude.* The sextant altitude H_S is corrected for both dip and index error to produce the apparent altitude. The observed altitude H_O is calculated by applying a correction for refraction. For the Sun, Moon, Venus and Mars a correction for parallax is also applied to H, and for the Sun and Moon a further correction for semi-diameter is required. The corrections are calculated as follows:

Step 1. Calculate dip

$$D = 0^{\circ}{\cdot}0293\sqrt{h}$$

where h is the height of eye above the horizon in metres.

Step 2. Calculate apparent altitude

$$H = H_S + I - D$$

where I is the sextant index error.

Step 3. Calculate refraction (R) at a standard temperature of 10° Celsius (C) and pressure of 1010 millibars (mb)

$$R_0 = 0^{\circ}{\cdot}0167/\tan(H + 7{\cdot}31/(H + 4{\cdot}4))$$

If the temperature $T°$ C and pressure P mb are known calculate the refraction from

$$R = fR_0 \quad \text{where} \quad f = 0{\cdot}28P/(T + 273)$$

otherwise set $R = R_0$

Step 4. Calculate the parallax in altitude (PA) from the horizontal parallax (HP) and the apparent altitude (H) for the Sun, Moon, Venus and Mars as follows:

$$PA = HP\cos H$$

For the Sun $HP = 0^{\circ}{\cdot}0024$. This correction is very small and could be ignored.

For the Moon HP is taken for the nearest hour from the main tabular page and converted to degrees.

For Venus and Mars the HP is taken from the critical table at the bottom of page 259 and converted to degrees.

For the navigational stars and the remaining planets, Jupiter and Saturn set $PA = 0$.

If an error of 0'·2 is significant the expression for the parallax in altitude for the Moon should include a small correction OB for the oblateness of the Earth as follows:

$$PA = HP\cos H + OB$$

$$\text{where} \quad OB = -0^{\circ}{\cdot}0032\sin^2 Lat\,\cos H + 0^{\circ}{\cdot}0032\sin(2Lat)\cos Z\,\sin H$$

At mid-latitudes and for altitudes of the Moon below 60° a simple approximation to OB is

$$OB = -0^{\circ}{\cdot}0017\cos H$$

Step 5. Calculate the semi-diameter for the Sun and Moon as follows:

Sun: S is taken from the main tabular page and converted to degrees.

Moon: $S = 0^\circ.2724HP$ where HP is taken for the nearest hour from the main tabular page and converted to degrees.

Step 6. Calculate the observed altitude

$$H_O = H - R + PA \pm S$$

where the plus sign is used if the lower limb of the Sun or Moon was observed and the minus sign if the upper limb was observed.

9. *Example.* The following example illustrates how to use a calculator to reduce the sextant altitude (H_S) to observed altitude (H_O); the sextant altitudes given are assumed to be taken on 2003 December 1 with a marine sextant, zero index error, at height 5·4 m, temperature −3° C and pressure 982 mb, the Moon sights are assumed to be taken at 10^h UT.

Body limb	Sun lower	Sun upper	Moon lower	Moon upper	Venus —	*Polaris* —
Sextant altitude:						
H_S	21·3283	3·3367	33·4600	26·1117	4·5433	49·6083
Step 1. Dip:						
$D = 0{\cdot}0293\sqrt{h}$	0·0681	0·0681	0·0681	0·0681	0·0681	0·0681
Step 2. Apparent altitude:						
$H = H_S + I - D$	21·2602	3·2686	33·3919	26·0436	4·4752	49·5402
Step 3. Refraction:						
R_0	0·0423	0·2262	0·0251	0·0338	0·1801	0·0142
f	1·0184	1·0184	1·0184	1·0184	1·0184	1·0184
$R = fR_0$	0·0431	0·2304	0·0256	0·0344	0·1834	0·0144
Step 4. Parallax:			(56′.1)	(56′.1)	(0′.1)	
HP	0·0024	0·0024	0·9350	0·9350	0·0017	—
Parallax in altitude:						
$PA = HP\cos H$	0·0022	0·0024	0·7807	0·8401	0·0017	—
Step 5. Semi-diameter:						
Sun : $S = 16{\cdot}2/60$	0·2700	0·2700	—	—	—	—
Moon : $S = 0{\cdot}2724HP$	—	—	0·2547	0·2547	—	—
Step 6. Observed altitude:						
$H_O = H - R + PA \pm S$	21·4894	2·7706	34·4017	26·5945	4·2935	49·5258

Note that for the Moon the correction for the oblateness of the Earth of about $-0^\circ.0017\cos H$, which equals $-0^\circ.0014$ for the lower limb and $-0^\circ.0015$ for the upper limb, has been ignored in the above calculation.

10. *Position from intercept and azimuth using a chart.* An estimate is made of the position at the adopted time of fix. The position at the time of observation is then calculated by dead reckoning from the time of fix. For example if the course (track) T and the speed V (in knots) of the observer are constant then *Long* and *Lat* at the time of observation are calculated from

$$Long = L_F + t(V/60)\sin T/\cos B_F$$
$$Lat = B_F + t(V/60)\cos T$$

where L_F and B_F are the estimated longitude and latitude at the time of fix and t is the time interval in hours from the time of fix to the time of observation, t is positive if the time of observation is after the time of fix and negative if it was before.

The position line of an observation is plotted on a chart using the intercept

$$p = H_O - H_C$$

and azimuth Z with origin at the calculated position (*Long*, *Lat*) at the time of observation, where H_C and Z are calculated using the method in section 6, page 279. Starting from this calculated position a line is drawn on the chart along the direction of the azimuth to the body. Convert p to nautical miles by multiplying by 60. The position line is drawn at right angles to the azimuth line, distance p from (*Long*, *Lat*) towards the body if p is positive and distance p away from the body if p is negative. Provided there are no gross errors the navigator should be somewhere on or near the position line at the time of observation. Two or more position lines are required to determine a fix.

11. *Position from intercept and azimuth by calculation.* The position of the fix may be calculated from two or more sextant observations as follows.

If p_1, Z_1, are the intercept and azimuth of the first observation, p_2, Z_2, of the second observation and so on, form the summations

$$A = \cos^2 Z_1 + \cos^2 Z_2 + \cdots$$
$$B = \cos Z_1 \sin Z_1 + \cos Z_2 \sin Z_2 + \cdots$$
$$C = \sin^2 Z_1 + \sin^2 Z_2 + \cdots$$
$$D = p_1 \cos Z_1 + p_2 \cos Z_2 + \cdots$$
$$E = p_1 \sin Z_1 + p_2 \sin Z_2 + \cdots$$

where the number of terms in each summation is equal to the number of observations.

With $G = AC - B^2$, an improved estimate of the position at the time of fix (L_I, B_I) is given by

$$L_I = L_F + (AE - BD)/(G\cos B_F), \qquad B_I = B_F + (CD - BE)/G$$

Calculate the distance d between the initial estimated position (L_F, B_F) at the time of fix and the improved estimated position (L_I, B_I) in nautical miles from

$$d = 60\sqrt{((L_I - L_F)^2 \cos^2 B_F + (B_I - B_F)^2)}$$

If d exceeds about 20 nautical miles set $L_F = L_I$, $B_F = B_I$ and repeat the calculation until d, the distance between the position at the previous estimate and the improved estimate, is less than about 20 nautical miles.

12. *Example of direct computation.* Using the method described above, calculate the position of a ship on 2003 June 25 at $21^h\ 00^m\ 00^s$ UT from the marine sextant observations of the three stars *Regulus* (No. 26) at $20^h\ 39^m\ 23^s$ UT, *Antares* (No. 42) at $20^h\ 45^m\ 47^s$ UT and *Kochab* (No. 40) at $21^h\ 10^m\ 34^s$ UT, where the observed altitudes of the three stars corrected for the effects of refraction, dip and instrumental error, are 34°7411, 21°9385 and 47°3760 respectively. The ship was travelling at a constant speed of 20 knots on a course of 325° during the period of observation, and the position of the ship at the time of fix $21^h\ 00^m\ 00^s$ UT is only known to the nearest whole degree W 15°, N 32°.

Intermediate values for the first iteration are shown in the table. *GHA* Aries was interpolated from the nearest tabular values on page 126. For the first iteration set $L_F = -15^\circ 0000$, $B_F = +32^\circ 0000$ at the time of fix at $21^h\ 00^m\ 00^s$ UT.

	First Iteration		
Body	*Regulus*	*Antares*	*Kochab*
No.	26	42	40
time of observation	$20^h\ 39^m\ 23^s$	$20^h\ 45^m\ 47^s$	$21^h\ 10^m\ 34^s$
H_O	34·7411	21·9385	47·3760
interpolation factor	0·6564	0·7631	0·1761
GHA Aries	223·4165	225·0210	231·2337
SHA (page 126)	207·8667	112·5933	137·3083
GHA	71·2832	337·6143	8·5420
Dec (page 126)	+11·9533	−26·4417	+74·1450
t	−0·3436	−0·2369	+0·1761
Long	−14·9225	−14·9466	−15·0397
Lat	+31·9062	+31·9353	+32·0481
Z	262·2660	144·2577	2·6368
H_C	34·7180	21·6327	47·7761
p	+0·0231	+0·3058	−0·4001

$$A = 1{\cdot}6748 \quad B = -0{\cdot}2948 \quad C = 1{\cdot}3252 \quad D = -0{\cdot}6510 \quad E = 0{\cdot}1373 \quad G = 2{\cdot}1325$$
$$(A\,E - B\,D)/(G\cos B_F) = +0{\cdot}0211, \qquad (C\,D - B\,E)/G = -0{\cdot}3856$$

An improved estimate of the position at the time of fix is

$$L_I = L_F + 0{\cdot}0211 = -14{\cdot}9789 \quad \text{and} \quad B_I = B_F - 0{\cdot}3856 = +31{\cdot}6144$$

Since the distance between the previous estimated position and the improved estimate $d = 23{\cdot}2$ nautical miles set $L_F = -14{\cdot}9789$, and $B_F = +31{\cdot}6144$ and repeat the calculation. The table shows the intermediate values of the calculation for the second iteration. In each iteration the quantities H_O, *GHA*, *Dec* and *t* do not change.

	Second Iteration		
Body	*Regulus*	*Antares*	*Kochab*
No.	26	42	40
Long	−14·9018	−14·9257	−15·0185
Lat	+31·5206	+31·5497	+31·6625
Z	262·5435	144·1843	2·6090
H_C	34·7514	21·9559	47·3918
p	−0·0104	−0·0174	−0·0158

$$A = 1{\cdot}6723 \quad B = -0{\cdot}3004 \quad C = 1{\cdot}3277 \quad D = -0{\cdot}0003 \quad E = -0{\cdot}0006 \quad G = 2{\cdot}1301$$
$$(A\,E - B\,D)/(G\cos B_F) = -0{\cdot}0006, \qquad (C\,D - B\,E)/G = -0{\cdot}0003$$

An improved estimate of the position at the time of fix is

$$L_I = L_F - 0{\cdot}0006 = -14{\cdot}9795 \quad \text{and} \quad B_I = B_F - 0{\cdot}0003 = +31{\cdot}6142$$

The distance between the previous estimated position and the improved estimated position $d = 0{\cdot}04$ nautical miles is so small that a third iteration would produce a negligible improvement to the estimate of the position.

USE OF CONCISE SIGHT REDUCTION TABLES

1. *Introduction.* The concise sight reduction tables given on pages 286 to 317 are intended for use when neither more extensive tables nor electronic computing aids are available. These "NAO sight reduction tables" provide for the reduction of the local hour angle and declination of a celestial object to azimuth and altitude, referred to an assumed position on the Earth, for use in the intercept method of celestial navigation which is now standard practice.

2. *Form of tables.* Entries in the reduction table are at a fixed interval of one degree for all latitudes and hour angles. A compact arrangement results from division of the navigational triangle into two right spherical triangles, so that the table has to be entered twice. Assumed latitude and local hour angle are the arguments for the first entry. The reduction table responds with the intermediate arguments A, B, and Z_1, where A is used as one of the arguments for the second entry to the table, B has to be incremented by the declination to produce the quantity F, and Z_1 is a component of the azimuth angle. The reduction table is then reentered with A and F and yields H, P, and Z_2 where H is the altitude, P is the complement of the parallactic angle, and Z_2 is the second component of the azimuth angle. It is usually necessary to adjust the tabular altitude for the fractional parts of the intermediate entering arguments to derive computed altitude, and an auxiliary table is provided for the purpose. Rules governing signs of the quantities which must be added or subtracted are given in the instructions and summarized on each tabular page. Azimuth angle is the sum of two components and is converted to true azimuth by familiar rules, repeated at the bottom of the tabular pages.

Tabular altitude and intermediate quantities are given to the nearest minute of arc, although errors of 2′ in computed altitude may accrue during adjustment for the minutes parts of entering arguments. Components of azimuth angle are stated to 0°.1; for derived true azimuth, only whole degrees are warranted. Since objects near the zenith are difficult to observe with a marine sextant, they should be avoided; altitudes greater than about 80° are not suited to reduction by this method.

In many circumstances the accuracy provided by these tables is sufficient. However, to maintain the full accuracy (0′.1) of the ephemeral data in the almanac throughout their reduction to altitude and azimuth, more extensive tables or a calculator should be used.

3. *Use of Tables.*

Step 1. Determine the Greenwich hour angle (*GHA*) and Declination (*Dec*) of the body from the almanac. Select an assumed latitude (*Lat*) of integral degrees nearest to the estimated latitude. Choose an assumed longitude nearest to the estimated longitude such that the local hour angle

$$LHA = GHA \begin{array}{c} -\text{ west} \\ +\text{ east} \end{array} \text{longitude}$$

has integral degrees.

Step 2. Enter the reduction table with *Lat* and *LHA* as arguments. Record the quantities A, B and Z_1. Apply the rules for the sign of B and Z_1: B is minus if $90° < LHA < 270°$: Z_1 has the same sign as B. Set $A°$ = nearest whole degree of A and A' = minutes part of A. This step may be repeated for all reductions before leaving the latitude opening of the table.

Step 3. Record the declination *Dec*. Apply the rules for the sign of *Dec*: *Dec* is minus if the name of *Dec* (*i.e.* N or S) is contrary to latitude. Add B and *Dec* algebraically to produce F. If F is negative, the object is below the horizon (in sight reduction, this can occur when the objects are close to the horizon). Regard F as positive until step 7. Set $F°$ = nearest whole degree of F and F' = minutes part of F.

Step 4. Enter the reduction table a second time with A° and F° as arguments and record H, P, and Z_2. Set P° = nearest whole degree of P and $Z_2{}^\circ$ = nearest whole degree of Z_2.

Step 5. Enter the auxiliary table with F' and P° as arguments to obtain $corr_1$ to H for F'. Apply the rule for the sign of $corr_1$: $corr_1$ is minus if $F < 90^\circ$ and $F' > 29'$ or if $F > 90^\circ$ and $F' < 30'$, otherwise $corr_1$ is plus.

Step 6. Enter the auxiliary table with A' and $Z_2{}^\circ$ as arguments to obtain $corr_2$ to H for A'. Apply the rule for the sign of $corr_2$: $corr_2$ is minus if $A' < 30'$, otherwise $corr_2$ is plus.

Step 7. Calculate the computed altitude H_C as the sum of H, $corr_1$ and $corr_2$. Apply the rule for the sign of H_C: H_C is minus if F is negative.

Step 8. Apply the rule for the sign of Z_2: Z_2 is minus if $F > 90^\circ$. If F is negative, replace Z_2 by $180^\circ - Z_2$. Set the azimuth angle Z equal to the algebraic sum of Z_1 and Z_2 and ignore the resulting sign. Obtain the true azimuth Z_n from the rules

For N latitude, if	$LHA > 180^\circ$	$Z_n = Z$
if	$LHA < 180^\circ$	$Z_n = 360^\circ - Z$
For S latitude, if	$LHA > 180^\circ$	$Z_n = 180^\circ - Z$
if	$LHA < 180^\circ$	$Z_n = 180^\circ + Z$

Observed altitude H_O is compared with H_C to obtain the altitude difference, which, with Z_n, is used to plot the position line.

4. *Example.* (a) Required the altitude and azimuth of *Schedar* on 2003 February 5 at UT 06[h] 30[m] from the estimated position 5° east, 53° north.

1. Assumed latitude	$Lat =$ 53° N		
From the almanac	$GHA =$ 222° 20′		
Assumed longitude	4° 40′ E		
Local hour angle	$LHA =$ 227		
2. Reduction table, 1st entry $(Lat, LHA) = (53, 227)$	$A =$ 26 07	$A^\circ = 26$, $A' = 7$	
	$B = -27$ 12	$Z_1 = -49{\cdot}4$,	$90^\circ < LHA < 270^\circ$
3. From the almanac	$Dec = +56$ 33		*Lat* and *Dec* same
Sum $= B + Dec$	$F = +29$ 21	$F^\circ = 29$, $F' = 21$	
4. Reduction table, 2nd entry $(A^\circ, F^\circ) = (26, 29)$	$H =$ 25 50	$P^\circ = 61$	
		$Z_2 = 76{\cdot}3$	
5. Auxiliary table, 1st entry $(F', P^\circ) = (21, 61)$	$corr_1 =$ +18		$F < 90^\circ$, $F' < 29'$
Sum	26 08		
6. Auxiliary table, 2nd entry $(A', Z_2{}^\circ) = (7, 76)$	$corr_2 =$ −2		$A' < 30'$
7. Sum = computed altitude	$H_C = +26^\circ$ 06′		$F > 0^\circ$
8. Azimuth, first component	$Z_1 = -49{\cdot}4$		same sign as B
second component	$Z_2 = +76{\cdot}3$		$F < 90^\circ$, $F > 0^\circ$
Sum = azimuth angle	$Z =$ 26·9		
True azimuth	$Z_n =$ 027°		N *Lat*, $LHA > 180^\circ$

continued on page 318

SIGHT REDUCTION TABLE

B: (–) for 90°< LHA < 270°
Dec: (–) for Lat. contrary name

Z_1: same sign as B
Z_2: (–) for F > 90°

Lat. / A		0°			1°			2°			3°			4°			5°			Lat. / A	
LHA / F		A / H	B / P	Z_1 / Z_2	A / H	B / P	Z_1 / Z_2	A / H	B / P	Z_1 / Z_2	A / H	B / P	Z_1 / Z_2	A / H	B / P	Z_1 / Z_2	A / H	B / P	Z_1 / Z_2	LHA	
°	°	° ′	° ′	°	° ′	° ′	°	° ′	° ′	°	° ′	° ′	°	° ′	° ′	°	° ′	° ′	°	°	°
0	180	0 00	90 00	90.0	0 00	89 00	90.0	0 00	88 00	90.0	0 00	87 00	90.0	0 00	86 00	90.0	0 00	85 00	90.0	180	360
1	179	1 00	90 00	90.0	1 00	89 00	90.0	1 00	88 00	90.0	1 00	87 00	89.9	1 00	86 00	89.9	1 00	85 00	89.9	181	359
2	178	2 00	90 00	90.0	2 00	89 00	90.0	2 00	88 00	89.9	2 00	87 00	89.9	2 00	86 00	89.9	2 00	85 00	89.8	182	358
3	177	3 00	90 00	90.0	3 00	89 00	89.9	3 00	88 00	89.9	3 00	87 00	89.8	3 00	86 00	89.8	2 59	85 00	89.7	183	357
4	176	4 00	90 00	90.0	4 00	89 00	89.9	4 00	88 00	89.9	4 00	87 00	89.8	3 59	85 59	89.7	3 59	84 59	89.7	184	356
5	175	5 00	90 00	90.0	5 00	89 00	89.9	5 00	88 00	89.8	5 00	86 59	89.7	4 59	85 59	89.7	4 59	84 59	89.6	185	355
6	174	6 00	90 00	90.0	6 00	89 00	89.9	6 00	87 59	89.8	6 00	86 59	89.7	5 59	85 59	89.6	5 59	84 58	89.5	186	354
7	173	7 00	90 00	90.0	7 00	89 00	89.9	7 00	87 59	89.8	6 59	86 59	89.6	6 59	85 58	89.5	6 58	84 58	89.4	187	353
8	172	8 00	90 00	90.0	8 00	88 59	89.9	8 00	87 59	89.7	7 59	86 58	89.6	7 59	85 58	89.4	7 58	84 57	89.3	188	352
9	171	9 00	90 00	90.0	9 00	88 59	89.8	9 00	87 59	89.7	8 59	86 58	89.5	8 59	85 57	89.4	8 58	84 56	89.2	189	351
10	170	10 00	90 00	90.0	10 00	88 59	89.8	10 00	87 58	89.6	9 59	86 57	89.5	9 59	85 56	89.3	9 58	84 55	89.1	190	350
11	169	11 00	90 00	90.0	11 00	88 59	89.8	11 00	87 58	89.6	10 59	86 57	89.4	10 58	85 56	89.2	10 57	84 54	89.0	191	349
12	168	12 00	90 00	90.0	12 00	88 59	89.8	12 00	87 57	89.6	11 59	86 56	89.4	11 58	85 55	89.2	11 57	84 53	88.9	192	348
13	167	13 00	90 00	90.0	13 00	88 58	89.8	13 00	87 57	89.5	12 59	86 55	89.3	12 58	85 54	89.1	12 57	84 52	88.8	193	347
14	166	14 00	90 00	90.0	14 00	88 58	89.8	13 59	87 56	89.5	13 59	86 54	89.3	13 58	85 53	89.0	13 57	84 51	88.8	194	346
15	165	15 00	90 00	90.0	15 00	88 58	89.7	14 59	87 56	89.5	14 59	86 54	89.2	14 58	85 52	88.9	14 56	84 49	88.7	195	345
16	164	16 00	90 00	90.0	16 00	88 58	89.7	15 59	87 55	89.4	15 59	86 53	89.1	15 58	85 50	88.9	15 56	84 48	88.6	196	344
17	163	17 00	90 00	90.0	17 00	88 57	89.7	16 59	87 55	89.4	16 59	86 52	89.1	16 57	85 49	88.8	16 56	84 46	88.5	197	343
18	162	18 00	90 00	90.0	18 00	88 57	89.7	17 59	87 54	89.4	17 58	86 51	89.0	17 57	85 48	88.7	17 56	84 45	88.4	198	342
19	161	19 00	90 00	90.0	19 00	88 57	89.7	18 59	87 53	89.3	18 58	86 50	89.0	18 57	85 46	88.6	18 55	84 43	88.3	199	341
20	160	20 00	90 00	90.0	20 00	88 56	89.6	19 59	87 52	89.3	19 58	86 48	88.9	19 57	85 45	88.5	19 55	84 41	88.2	200	340
21	159	21 00	90 00	90.0	21 00	88 56	89.6	20 59	87 51	89.2	20 58	86 47	88.8	20 57	85 43	88.5	20 55	84 39	88.1	201	339
22	158	22 00	90 00	90.0	22 00	88 55	89.6	21 59	87 51	89.2	21 58	86 46	88.8	21 57	85 41	88.4	21 55	84 37	88.0	202	338
23	157	23 00	90 00	90.0	23 00	88 55	89.6	22 59	87 50	89.2	22 58	86 44	88.7	22 56	85 39	88.3	22 54	84 34	87.9	203	337
24	156	24 00	90 00	90.0	24 00	88 54	89.6	23 59	87 49	89.1	23 58	86 43	88.7	23 56	85 37	88.2	23 54	84 32	87.8	204	336
25	155	25 00	90 00	90.0	25 00	88 54	89.5	24 59	87 48	89.1	24 58	86 41	88.6	24 56	85 35	88.1	24 54	84 29	87.7	205	335
26	154	26 00	90 00	90.0	26 00	88 53	89.5	25 59	87 46	89.0	25 58	86 40	88.5	25 56	85 33	88.1	25 54	84 26	87.6	206	334
27	153	27 00	90 00	90.0	27 00	88 53	89.5	26 59	87 45	89.0	26 58	86 38	88.5	26 56	85 31	88.0	26 53	84 24	87.5	207	333
28	152	28 00	90 00	90.0	28 00	88 52	89.5	27 59	87 44	88.9	27 57	86 36	88.4	27 56	85 28	87.9	27 53	84 20	87.3	208	332
29	151	29 00	90 00	90.0	29 00	88 51	89.4	28 59	87 43	88.9	28 57	86 34	88.3	28 55	85 26	87.8	28 53	84 17	87.2	209	331
30	150	30 00	90 00	90.0	30 00	88 51	89.4	29 59	87 41	88.8	29 57	86 32	88.3	29 55	85 23	87.7	29 52	84 14	87.1	210	330
31	149	31 00	90 00	90.0	31 00	88 50	89.4	30 59	87 40	88.8	30 57	86 30	88.2	30 55	85 20	87.6	30 52	84 10	87.0	211	329
32	148	32 00	90 00	90.0	32 00	88 49	89.4	31 59	87 39	88.8	31 57	86 28	88.1	31 55	85 17	87.5	31 52	84 07	86.9	212	328
33	147	33 00	90 00	90.0	33 00	88 48	89.4	32 59	87 37	88.7	32 57	86 25	88.1	32 55	85 14	87.4	32 52	84 03	86.8	213	327
34	146	34 00	90 00	90.0	34 00	88 48	89.3	33 59	87 35	88.7	33 57	86 23	88.0	33 54	85 11	87.3	33 51	83 59	86.6	214	326
35	145	35 00	90 00	90.0	35 00	88 47	89.3	34 59	87 34	88.6	34 57	86 20	87.9	34 54	85 07	87.2	34 51	83 54	86.5	215	325
36	144	36 00	90 00	90.0	36 00	88 46	89.3	35 58	87 32	88.5	35 57	86 18	87.8	35 54	85 04	87.1	35 51	83 50	86.4	216	324
37	143	37 00	90 00	90.0	37 00	88 45	89.2	36 58	87 30	88.5	36 56	86 15	87.7	36 54	85 00	87.0	36 50	83 45	86.2	217	323
38	142	38 00	90 00	90.0	38 00	88 44	89.2	37 58	87 28	88.4	37 56	86 12	87.7	37 53	84 56	86.9	37 50	83 40	86.1	218	322
39	141	39 00	90 00	90.0	39 00	88 43	89.2	38 58	87 26	88.4	38 56	86 09	87.6	38 53	84 52	86.8	38 49	83 35	86.0	219	321
40	140	40 00	90 00	90.0	40 00	88 42	89.2	39 58	87 23	88.3	39 56	86 05	87.5	39 53	84 47	86.7	39 49	83 29	85.8	220	320
41	139	41 00	90 00	90.0	41 00	88 40	89.1	40 58	87 21	88.3	40 56	86 02	87.4	40 53	84 42	86.5	40 49	83 23	85.7	221	319
42	138	42 00	90 00	90.0	42 00	88 39	89.1	41 58	87 19	88.2	41 56	85 58	87.3	41 52	84 37	86.4	41 48	83 17	85.5	222	318
43	137	43 00	90 00	90.0	43 00	88 38	89.1	42 58	87 16	88.1	42 56	85 54	87.2	42 52	84 32	86.3	42 48	83 11	85.4	223	317
44	136	44 00	90 00	90.0	43 59	88 37	89.0	43 58	87 13	88.1	43 55	85 50	87.1	43 52	84 27	86.1	43 47	83 04	85.2	224	316
45	135	45 00	90 00	90.0	44 59	88 35	89.0	44 58	87 10	88.0	44 55	85 46	87.0	44 52	84 21	86.0	44 47	82 57	85.0	225	315

Lat. / A		0°			1°			2°			3°			4°			5°			Lat. / A	
LHA / F		A / H	B / P	Z_1 / Z_2	A / H	B / P	Z_1 / Z_2	A / H	B / P	Z_1 / Z_2	A / H	B / P	Z_1 / Z_2	A / H	B / P	Z_1 / Z_2	A / H	B / P	Z_1 / Z_2	LHA	
°	°	° ′	° ′	°	° ′	° ′	°	° ′	° ′	°	° ′	° ′	°	° ′	° ′	°	° ′	° ′	°	°	°
45	135	45 00	90 00	90.0	44 59	88 35	89.0	44 58	87 10	88.0	44 55	85 46	87.0	44 52	84 21	86.0	44 47	82 57	85.0	225	315
46	134	46 00	90 00	90.0	45 59	88 34	89.0	45 58	87 07	87.9	45 55	85 41	86.9	45 51	84 15	85.9	45 46	82 49	84.8	226	314
47	133	47 00	90 00	90.0	46 59	88 32	88.9	46 58	87 04	87.9	46 55	85 36	86.8	46 51	84 09	85.7	46 46	82 41	84.7	227	313
48	132	48 00	90 00	90.0	47 59	88 30	88.9	47 58	87 01	87.8	47 55	85 31	86.7	47 51	84 02	85.6	47 46	82 33	84.5	228	312
49	131	49 00	90 00	90.0	48 59	88 29	88.8	48 58	86 57	87.7	48 55	85 26	86.6	48 50	83 55	85.4	48 45	82 24	84.3	229	311
50	130	50 00	90 00	90.0	49 59	88 27	88.8	49 58	86 53	87.6	49 54	85 20	86.4	49 50	83 47	85.2	49 44	82 15	84.1	230	310
51	129	51 00	90 00	90.0	50 59	88 25	88.8	50 57	86 49	87.5	50 54	85 14	86.3	50 50	83 40	85.1	50 44	82 05	83.9	231	309
52	128	52 00	90 00	90.0	51 59	88 23	88.7	51 57	86 45	87.4	51 54	85 08	86.2	51 49	83 31	84.9	51 43	81 55	83.6	232	308
53	127	53 00	90 00	90.0	52 59	88 20	88.7	52 57	86 41	87.3	52 54	85 01	86.0	52 49	83 22	84.7	52 43	81 44	83.4	233	307
54	126	54 00	90 00	90.0	53 59	88 18	88.6	53 57	86 36	87.2	53 54	84 54	85.9	53 49	83 13	84.5	53 42	81 32	83.2	234	306
55	125	55 00	90 00	90.0	54 59	88 15	88.6	54 57	86 31	87.1	54 53	84 47	85.7	54 48	83 03	84.3	54 41	81 20	82.9	235	305
56	124	56 00	90 00	90.0	55 59	88 13	88.5	55 57	86 26	87.0	55 53	84 39	85.6	55 48	82 52	84.1	55 41	81 06	82.6	236	304
57	123	57 00	90 00	90.0	56 59	88 10	88.5	56 57	86 20	86.9	56 53	84 30	85.4	56 47	82 41	83.9	56 40	80 52	82.4	237	303
58	122	58 00	90 00	90.0	57 59	88 07	88.4	57 57	86 14	86.8	57 52	84 21	85.2	57 47	82 29	83.6	57 39	80 38	82.1	238	302
59	121	59 00	90 00	90.0	58 59	88 04	88.3	58 57	86 07	86.7	58 52	84 11	85.0	58 46	82 16	83.4	58 38	80 22	81.7	239	301
60	120	60 00	90 00	90.0	59 59	88 00	88.3	59 56	86 00	86.5	59 52	84 01	84.8	59 46	82 02	83.1	59 37	80 04	81.4	240	300
61	119	61 00	90 00	90.0	60 59	87 56	88.2	60 56	85 53	86.4	60 52	83 50	84.6	60 45	81 48	82.8	60 37	79 46	81.1	241	299
62	118	62 00	90 00	90.0	61 59	87 52	88.1	61 56	85 45	86.2	61 51	83 38	84.4	61 44	81 32	82.5	61 36	79 27	80.7	242	298
63	117	63 00	90 00	90.0	62 59	87 48	88.0	62 56	85 36	86.1	62 51	83 25	84.1	62 44	81 15	82.2	62 35	79 06	80.3	243	297
64	116	64 00	90 00	90.0	63 59	87 43	88.0	63 56	85 27	85.9	63 50	83 11	83.9	63 43	80 56	81.9	63 33	78 43	79.9	244	296
65	115	65 00	90 00	90.0	64 59	87 38	87.9	64 56	85 17	85.7	64 50	82 56	83.6	64 42	80 36	81.5	64 32	78 18	79.4	245	295
66	114	66 00	90 00	90.0	65 59	87 33	87.8	65 55	85 06	85.5	65 49	82 39	83.3	65 41	80 15	81.1	65 31	77 52	78.9	246	294
67	113	67 00	90 00	90.0	66 59	87 26	87.6	66 55	84 54	85.3	66 49	82 22	83.0	66 40	79 51	80.7	66 29	77 23	78.4	247	293
68	112	68 00	90 00	90.0	67 59	87 20	87.5	67 55	84 40	85.1	67 48	82 02	82.6	67 39	79 26	80.2	67 28	76 51	77.8	248	292
69	111	69 00	90 00	90.0	68 59	87 13	87.4	68 55	84 26	84.8	68 48	81 41	82.2	68 38	78 57	79.7	68 26	76 17	77.2	249	291
70	110	70 00	90 00	90.0	69 59	87 05	87.3	69 54	84 10	84.5	69 47	81 17	81.8	69 37	78 27	79.2	69 25	75 39	76.5	250	290
71	109	71 00	90 00	90.0	70 58	86 56	87.1	70 54	83 53	84.2	70 46	80 51	81.4	70 36	77 53	78.5	70 23	74 57	75.8	251	289
72	108	72 00	90 00	90.0	71 58	86 46	86.9	71 54	83 33	83.9	71 46	80 22	80.8	71 35	77 15	77.9	71 20	74 12	75.0	252	288
73	107	73 00	90 00	90.0	72 58	86 35	86.7	72 53	83 11	83.5	72 45	79 50	80.3	72 33	76 33	77.1	72 18	73 20	74.1	253	287
74	106	74 00	90 00	90.0	73 58	86 22	86.5	73 53	82 47	83.1	73 44	79 14	79.7	73 31	75 46	76.3	73 15	72 23	73.1	254	286
75	105	75 00	90 00	90.0	74 58	86 08	86.3	74 52	82 19	82.6	74 43	78 33	78.9	74 29	74 53	75.4	74 12	71 19	72.0	255	285
76	104	76 00	90 00	90.0	75 58	85 52	86.0	75 52	81 47	82.0	75 41	77 47	78.1	75 27	73 53	74.4	75 09	70 07	70.7	256	284
77	103	77 00	90 00	90.0	76 58	85 34	85.7	76 51	81 11	81.4	76 40	76 53	77.2	76 25	72 44	73.2	76 05	68 45	69.3	257	283
78	102	78 00	90 00	90.0	77 58	85 12	85.3	77 50	80 28	80.7	77 38	75 51	76.2	77 22	71 25	71.8	77 01	67 11	67.7	258	282
79	101	79 00	90 00	90.0	78 57	84 46	84.9	78 49	79 38	79.8	78 36	74 38	74.9	78 18	69 52	70.3	77 56	65 22	65.8	259	281
80	100	80 00	90 00	90.0	79 57	84 16	84.3	79 48	78 38	78.8	79 34	73 12	73.5	79 14	68 04	68.4	78 50	63 16	63.7	260	280
81	99	81 00	90 00	90.0	80 57	83 38	83.7	80 47	77 25	77.6	80 31	71 29	71.7	80 09	65 55	66.2	79 43	60 47	61.2	261	279
82	98	82 00	90 00	90.0	81 56	82 51	82.9	81 45	75 55	76.1	81 28	69 22	69.6	81 04	63 19	63.6	80 34	57 51	58.2	262	278
83	97	83 00	90 00	90.0	82 56	81 51	81.9	82 43	74 01	74.1	82 23	66 44	66.9	81 57	60 09	60.4	81 24	54 20	54.6	263	277
84	96	84 00	90 00	90.0	83 55	80 31	80.6	83 41	71 32	71.6	83 18	63 22	63.5	82 48	56 13	56.4	82 12	50 04	50.3	264	276
85	95	85 00	90 00	90.0	84 54	78 40	78.7	84 37	68 10	68.3	84 10	58 59	59.1	83 36	51 16	51.4	82 56	44 53	45.1	265	275
86	94	86 00	90 00	90.0	85 53	75 57	76.0	85 32	63 24	63.5	85 00	53 05	53.2	84 21	44 56	45.1	83 36	38 34	38.7	266	274
87	93	87 00	90 00	90.0	86 50	71 33	71.6	86 24	56 17	56.3	85 45	44 58	45.0	85 00	36 49	36.9	84 10	30 53	31.0	267	273
88	92	88 00	90 00	90.0	87 46	63 26	63.4	87 10	44 59	45.0	86 24	33 40	33.7	85 32	26 31	26.6	84 37	21 45	21.8	268	272
89	91	89 00	90 00	90.0	88 35	45 00	45.0	87 46	26 33	26.6	86 50	18 25	18.4	85 53	14 01	14.0	84 54	11 17	11.3	269	271
90	90	90 00	0 00	0.0	89 00	0 00	0.0	88 00	0 00	0.0	87 00	0 00	0.0	86 00	0 00	0.0	85 00	0 00	0.0	270	270

N. Lat.: for LHA > 180°... $Z_n = Z$
for LHA < 180°... $Z_n = 360° - Z$

S. Lat.: for LHA > 180°... $Z_n = 180° - Z$
for LHA < 180°... $Z_n = 180° + Z$

SIGHT REDUCTION TABLE

B: (–) for 90° < LHA < 270°
Dec: (–) for Lat. contrary name

Z_1 : same sign as B
Z_2 : (–) for F > 90°

Lat. / A		6°			7°			8°			9°			10°			11°			Lat. / A	
LHA / F		A / H	B / P	Z_1 / Z_2	A / H	B / P	Z_1 / Z_2	A / H	B / P	Z_1 / Z_2	A / H	B / P	Z_1 / Z_2	A / H	B / P	Z_1 / Z_2	A / H	B / P	Z_1 / Z_2	LHA	
°	°	° ′	° ′	°	° ′	° ′	°	° ′	° ′	°	° ′	° ′	°	° ′	° ′	°	° ′	° ′	°	°	°
0	180	0 00	84 00	90.0	0 00	83 00	90.0	0 00	82 00	90.0	0 00	81 00	90.0	0 00	80 00	90.0	0 00	79 00	90.0	180	360
1	179	1 00	84 00	89.9	1 00	83 00	89.9	0 59	82 00	89.9	0 59	81 00	89.8	0 59	80 00	89.8	0 59	79 00	89.8	181	359
2	178	1 59	84 00	89.8	1 59	83 00	89.8	1 59	82 00	89.7	1 59	81 00	89.7	1 58	80 00	89.7	1 58	79 00	89.6	182	358
3	177	2 59	84 00	89.7	2 59	82 59	89.6	2 58	81 59	89.6	2 58	80 59	89.5	2 57	79 59	89.5	2 57	78 59	89.4	183	357
4	176	3 59	83 59	89.6	3 58	82 59	89.5	3 58	81 59	89.4	3 57	80 59	89.4	3 56	79 59	89.3	3 56	78 58	89.2	184	356
5	175	4 58	83 59	89.5	4 58	82 58	89.4	4 57	81 58	89.3	4 56	80 58	89.2	4 55	79 58	89.1	4 54	78 58	89.0	185	355
6	174	5 58	83 58	89.4	5 57	82 58	89.3	5 56	81 57	89.2	5 56	80 57	89.1	5 55	79 57	89.0	5 53	78 56	88.9	186	354
7	173	6 58	83 57	89.3	6 57	82 57	89.1	6 56	81 56	89.0	6 55	80 56	88.9	6 54	79 56	88.8	6 52	78 55	88.7	187	353
8	172	7 57	83 56	89.2	7 56	82 56	89.0	7 55	81 55	88.9	7 54	80 55	88.7	7 53	79 54	88.6	7 51	78 54	88.5	188	352
9	171	8 57	83 56	89.1	8 56	82 55	88.9	8 55	81 54	88.7	8 53	80 53	88.6	8 52	79 53	88.4	8 50	78 52	88.3	189	351
10	170	9 57	83 54	88.9	9 55	82 54	88.8	9 54	81 53	88.6	9 53	80 52	88.4	9 51	79 51	88.2	9 49	78 50	88.1	190	350
11	169	10 56	83 53	88.8	10 55	82 52	88.6	10 53	81 51	88.5	10 52	80 50	88.3	10 50	79 49	88.1	10 48	78 48	87.9	191	349
12	168	11 56	83 52	88.7	11 55	82 51	88.5	11 53	81 49	88.3	11 51	80 48	88.1	11 49	79 47	87.9	11 47	78 46	87.7	192	348
13	167	12 56	83 51	88.6	12 54	82 49	88.4	12 52	81 48	88.2	12 50	80 46	87.9	12 48	79 45	87.7	12 45	78 43	87.5	193	347
14	166	13 55	83 49	88.5	13 54	82 47	88.3	13 52	81 46	88.0	13 49	80 44	87.8	13 47	79 42	87.5	13 44	78 40	87.3	194	346
15	165	14 55	83 47	88.4	14 53	82 45	88.1	14 51	81 43	87.9	14 49	80 41	87.6	14 46	79 39	87.3	14 43	78 37	87.1	195	345
16	164	15 55	83 46	88.3	15 53	82 43	88.0	15 50	81 41	87.7	15 48	80 39	87.4	15 45	79 36	87.1	15 42	78 34	86.9	196	344
17	163	16 54	83 44	88.2	16 52	82 41	87.9	16 50	81 38	87.6	16 47	80 36	87.3	16 44	79 33	87.0	16 41	78 31	86.7	197	343
18	162	17 54	83 42	88.1	17 52	82 39	87.7	17 49	81 36	87.4	17 46	80 33	87.1	17 43	79 30	86.8	17 39	78 27	86.5	198	342
19	161	18 54	83 39	87.9	18 51	82 36	87.6	18 48	81 33	87.3	18 45	80 29	86.9	18 42	79 26	86.6	18 38	78 23	86.2	199	341
20	160	19 53	83 37	87.8	19 51	82 33	87.5	19 48	81 30	87.1	19 45	80 26	86.7	19 41	79 22	86.4	19 37	78 19	86.0	200	340
21	159	20 53	83 35	87.7	20 50	82 30	87.3	20 47	81 26	86.9	20 44	80 22	86.6	20 40	79 18	86.2	20 36	78 14	85.8	201	339
22	158	21 52	83 32	87.6	21 50	82 27	87.2	21 46	81 23	86.8	21 43	80 18	86.4	21 39	79 14	86.0	21 35	78 10	85.6	202	338
23	157	22 52	83 29	87.5	22 49	82 24	87.0	22 46	81 19	86.6	22 42	80 14	86.2	22 38	79 09	85.8	22 33	78 05	85.4	203	337
24	156	23 52	83 26	87.3	23 49	82 21	86.9	23 45	81 15	86.5	23 41	80 10	86.0	23 37	79 05	85.6	23 32	77 59	85.1	204	336
25	155	24 51	83 23	87.2	24 48	82 17	86.7	24 44	81 11	86.3	24 40	80 05	85.8	24 36	78 59	85.4	24 31	77 54	84.9	205	335
26	154	25 51	83 20	87.1	25 48	82 13	86.6	25 44	81 07	86.1	25 39	80 00	85.6	25 35	78 54	85.2	25 29	77 48	84.7	206	334
27	153	26 50	83 16	87.0	26 47	82 09	86.4	26 43	81 02	85.9	26 38	79 55	85.4	26 33	78 48	84.9	26 28	77 42	84.4	207	333
28	152	27 50	83 13	86.8	27 46	82 05	86.3	27 42	80 57	85.8	27 38	79 50	85.2	27 32	78 42	84.7	27 27	77 35	84.2	208	332
29	151	28 50	83 09	86.7	28 46	82 01	86.1	28 41	80 52	85.6	28 37	79 44	85.0	28 31	78 36	84.5	28 25	77 28	84.0	209	331
30	150	29 49	83 05	86.5	29 45	81 56	86.0	29 41	80 47	85.4	29 36	79 38	84.8	29 30	78 29	84.3	29 24	77 21	83.7	210	330
31	149	30 49	83 01	86.4	30 45	81 51	85.8	30 40	80 41	85.2	30 35	79 32	84.6	30 29	78 23	84.0	30 22	77 13	83.5	211	329
32	148	31 48	82 56	86.3	31 44	81 46	85.6	31 39	80 35	85.0	31 34	79 25	84.4	31 27	78 15	83.8	31 21	77 05	83.2	212	328
33	147	32 48	82 51	86.1	32 43	81 40	85.5	32 38	80 29	84.8	32 33	79 18	84.2	32 26	78 08	83.6	32 19	76 57	82.9	213	327
34	146	33 47	82 46	86.0	33 43	81 35	85.3	33 37	80 23	84.6	33 32	79 11	84.0	33 25	78 00	83.3	33 18	76 48	82.7	214	326
35	145	34 47	82 41	85.8	34 42	81 29	85.1	34 37	80 16	84.4	34 30	79 03	83.7	34 24	77 51	83.1	34 16	76 39	82.4	215	325
36	144	35 46	82 36	85.7	35 41	81 22	84.9	35 36	80 09	84.2	35 29	78 55	83.5	35 22	77 42	82.8	35 14	76 29	82.1	216	324
37	143	36 46	82 30	85.5	36 41	81 16	84.8	36 35	80 01	84.0	36 28	78 47	83.3	36 21	77 33	82.5	36 13	76 19	81.8	217	323
38	142	37 45	82 24	85.3	37 40	81 09	84.6	37 34	79 53	83.8	37 27	78 38	83.0	37 19	77 23	82.3	37 11	76 09	81.5	218	322
39	141	38 45	82 18	85.2	38 39	81 01	84.4	38 33	79 45	83.6	38 26	78 29	82.8	38 18	77 13	82.0	38 09	75 57	81.2	219	321
40	140	39 44	82 11	85.0	39 39	80 54	84.2	39 32	79 36	83.3	39 25	78 19	82.5	39 16	77 02	81.7	39 07	75 46	80.9	220	320
41	139	40 44	82 04	84.8	40 38	80 46	84.0	40 31	79 27	83.1	40 23	78 09	82.3	40 15	76 51	81.4	40 05	75 33	80.6	221	319
42	138	41 43	81 57	84.6	41 37	80 37	83.7	41 30	79 17	82.9	41 22	77 58	82.0	41 13	76 39	81.1	41 04	75 21	80.3	222	318
43	137	42 42	81 49	84.4	42 36	80 28	83.5	42 29	79 07	82.6	42 21	77 47	81.7	42 12	76 27	80.8	42 02	75 07	79.9	223	317
44	136	43 42	81 41	84.2	43 35	80 19	83.3	43 28	78 57	82.3	43 19	77 35	81.4	43 10	76 14	80.5	43 00	74 53	79.6	224	316

Lat. / A		6°			7°			8°			9°			10°			11°			Lat. / A	
LHA / F		A / H	B / P	Z_1 / Z_2	A / H	B / P	Z_1 / Z_2	A / H	B / P	Z_1 / Z_2	A / H	B / P	Z_1 / Z_2	A / H	B / P	Z_1 / Z_2	A / H	B / P	Z_1 / Z_2	LHA	
°	°	° ′	° ′	°	° ′	° ′	°	° ′	° ′	°	° ′	° ′	°	° ′	° ′	°	° ′	° ′	°	°	°
45	135	44 41	81 33	84.0	44 34	80 09	83.1	44 27	78 46	82.1	44 18	77 22	81.1	44 08	76 00	80.1	43 57	74 38	79.2	225	315
46	134	45 41	81 24	83.8	45 34	79 59	82.8	45 26	78 34	81.8	45 16	77 09	80.8	45 06	75 45	79.8	44 55	74 22	78.8	226	314
47	133	46 40	81 14	83.6	46 33	79 48	82.6	46 24	78 21	81.5	46 15	76 56	80.5	46 04	75 30	79.5	45 53	74 05	78.4	227	313
48	132	47 39	81 04	83.4	47 32	79 36	82.3	47 23	78 08	81.2	47 13	76 41	80.1	47 03	75 14	79.1	46 51	73 48	78.0	228	312
49	131	48 38	80 54	83.1	48 31	79 24	82.0	48 22	77 55	80.9	48 12	76 26	79.8	48 01	74 57	78.7	47 48	73 30	77.6	229	311
50	130	49 38	80 43	82.9	49 30	79 11	81.7	49 20	77 40	80.6	49 10	76 09	79.4	48 58	74 40	78.3	48 46	73 10	77.2	230	310
51	129	50 37	80 31	82.6	50 29	78 58	81.4	50 19	77 25	80.2	50 08	75 52	79.1	49 56	74 21	77.9	49 43	72 50	76.7	231	309
52	128	51 36	80 19	82.4	51 27	78 43	81.1	51 18	77 08	79.9	51 06	75 34	78.7	50 54	74 01	77.5	50 40	72 29	76.3	232	308
53	127	52 35	80 06	82.1	52 26	78 28	80.8	52 16	76 51	79.5	52 04	75 15	78.3	51 52	73 40	77.0	51 37	72 06	75.8	233	307
54	126	53 34	79 52	81.8	53 25	78 12	80.5	53 14	76 33	79.2	53 02	74 55	77.8	52 49	73 18	76.6	52 35	71 42	75.3	234	306
55	125	54 33	79 37	81.5	54 24	77 55	80.1	54 13	76 14	78.8	54 00	74 34	77.4	53 47	72 55	76.1	53 31	71 17	74.8	235	305
56	124	55 32	79 21	81.2	55 22	77 37	79.8	55 11	75 54	78.3	54 58	74 11	76.9	54 44	72 30	75.6	54 28	70 50	74.2	236	304
57	123	56 31	79 05	80.9	56 21	77 18	79.4	56 09	75 32	77.9	55 56	73 47	76.5	55 41	72 04	75.0	55 25	70 22	73.6	237	303
58	122	57 30	78 47	80.5	57 19	76 57	79.0	57 07	75 09	77.4	56 53	73 22	75.9	56 38	71 36	74.5	56 21	69 51	73.0	238	302
59	121	58 29	78 28	80.1	58 18	76 35	78.5	58 05	74 44	77.0	57 51	72 54	75.4	57 35	71 06	73.9	57 17	69 19	72.4	239	301
60	120	59 28	78 08	79.7	59 16	76 12	78.1	59 03	74 18	76.4	58 48	72 25	74.8	58 32	70 34	73.3	58 13	68 45	71.7	240	300
61	119	60 26	77 46	79.3	60 14	75 47	77.6	60 01	73 50	75.9	59 45	71 54	74.2	59 28	70 01	72.6	59 09	68 09	71.0	241	299
62	118	61 25	77 23	78.9	61 12	75 21	77.1	60 58	73 20	75.3	60 42	71 21	73.6	60 24	69 25	71.9	60 05	67 30	70.3	242	298
63	117	62 23	76 58	78.4	62 10	74 52	76.5	61 56	72 48	74.7	61 39	70 46	72.9	61 20	68 46	71.2	61 00	66 49	69.5	243	297
64	116	63 22	76 31	77.9	63 08	74 21	76.0	62 53	72 13	74.1	62 35	70 08	72.2	62 16	68 05	70.4	61 55	66 05	68.6	244	296
65	115	64 20	76 02	77.4	64 06	73 48	75.4	63 50	71 36	73.4	63 32	69 27	71.5	63 12	67 21	69.6	62 50	65 18	67.7	245	295
66	114	65 18	75 31	76.8	65 03	73 12	74.7	64 47	70 56	72.6	64 28	68 43	70.6	64 07	66 34	68.7	63 44	64 27	66.8	246	294
67	113	66 16	74 57	76.2	66 01	72 33	74.0	65 43	70 13	71.8	65 23	67 56	69.8	65 02	65 43	67.8	64 38	63 33	65.8	247	293
68	112	67 14	74 20	75.5	66 58	71 51	73.2	66 40	69 26	71.0	66 19	67 05	68.8	65 56	64 48	66.7	65 32	62 35	64.7	248	292
69	111	68 12	73 39	74.8	67 55	71 05	72.4	67 36	68 35	70.1	67 14	66 09	67.8	66 50	63 48	65.7	66 25	61 31	63.6	249	291
70	110	69 09	72 55	74.0	68 51	70 15	71.5	68 31	67 40	69.1	68 09	65 09	66.7	67 44	62 44	64.5	67 17	60 23	62.3	250	290
71	109	70 07	72 06	73.1	69 48	69 20	70.5	69 27	66 39	68.0	69 03	64 03	65.6	68 37	61 34	63.2	68 09	59 10	61.0	251	289
72	108	71 03	71 13	72.2	70 44	68 20	69.4	70 21	65 33	66.8	69 57	62 52	64.3	69 29	60 17	61.9	69 00	57 50	59.6	252	288
73	107	72 00	70 14	71.1	71 39	67 13	68.3	71 16	64 20	65.5	70 50	61 33	62.9	70 21	58 54	60.4	69 50	56 23	58.0	253	287
74	106	72 56	69 08	70.0	72 34	65 59	67.0	72 09	62 59	64.1	71 42	60 07	61.4	71 12	57 24	58.8	70 40	54 48	56.4	254	286
75	105	73 52	67 54	68.7	73 29	64 37	65.5	73 03	61 30	62.6	72 34	58 32	59.7	72 02	55 44	57.1	71 28	53 06	54.5	255	285
76	104	74 48	66 31	67.3	74 23	63 05	64.0	73 55	59 51	60.8	73 24	56 47	57.9	72 51	53 55	55.1	72 16	51 13	52.6	256	284
77	103	75 42	64 57	65.6	75 16	61 22	62.2	74 46	58 00	58.9	74 14	54 51	55.9	73 39	51 55	53.1	73 02	49 10	50.4	257	283
78	102	76 36	63 11	63.8	76 08	59 26	60.2	75 37	55 57	56.8	75 02	52 42	53.6	74 26	49 42	50.8	73 47	46 56	48.1	258	282
79	101	77 29	61 09	61.7	76 59	57 14	57.9	76 26	53 38	54.4	75 49	50 18	51.2	75 11	47 16	48.2	74 30	44 28	45.5	259	281
80	100	78 21	58 49	59.3	77 49	54 44	55.3	77 13	51 01	51.7	76 35	47 38	48.4	75 54	44 34	45.4	75 11	41 47	42.7	260	280
81	99	79 12	56 06	56.6	78 37	51 52	52.4	77 59	48 04	48.7	77 18	44 39	45.4	76 35	41 35	42.4	75 49	38 50	39.7	261	279
82	98	80 01	52 56	53.4	79 23	48 35	49.1	78 42	44 43	45.3	77 59	41 18	41.9	77 13	38 17	39.0	76 26	35 36	36.4	262	278
83	97	80 47	49 13	49.6	80 07	44 47	45.2	79 23	40 56	41.4	78 37	37 35	38.1	77 49	34 39	35.3	76 59	32 05	32.8	263	277
84	96	81 31	44 51	45.2	80 47	40 24	40.8	80 01	36 38	37.1	79 12	33 25	33.9	78 21	30 40	31.2	77 29	28 16	28.8	264	276
85	95	82 12	39 40	39.9	81 24	35 22	35.7	80 34	31 48	32.2	79 43	28 49	29.2	78 50	26 18	26.7	77 56	24 09	24.6	265	275
86	94	82 48	33 34	33.8	81 57	29 36	29.8	81 04	26 24	26.7	80 09	23 46	24.1	79 14	21 35	21.9	78 18	19 44	20.1	266	274
87	93	83 18	26 28	26.6	82 23	23 05	23.3	81 28	20 25	20.6	80 31	18 17	18.5	79 34	16 32	16.8	78 36	15 04	15.4	267	273
88	92	83 41	18 22	18.5	82 43	15 52	16.0	81 45	13 57	14.1	80 47	12 26	12.6	79 48	11 12	11.4	78 49	10 11	10.4	268	272
89	91	83 55	9 26	9.5	82 56	8 05	8.2	81 56	7 05	7.1	80 57	6 17	6.4	79 57	5 39	5.7	78 57	5 08	5.2	269	271
90	90	84 00	0 00	0.0	83 00	0 00	0.0	82 00	0 00	0.0	81 00	0 00	0.0	80 00	0 00	0.0	79 00	0 00	0.0	270	270

N. Lat.: for LHA > 180°... $Z_n = Z$
for LHA < 180°... $Z_n = 360° - Z$

S. Lat.: for LHA > 180°... $Z_n = 180° - Z$
for LHA < 180°... $Z_n = 180° + Z$

SIGHT REDUCTION TABLE

B: (–) for 90° < LHA < 270°
Dec: (–) for Lat. contrary name

Z_1 : same sign as B
Z_2 : (–) for F > 90°

Lat. / A		12°			13°			14°			15°			16°			17°			Lat. / A	
LHA / F		A / H	B / P	Z_1 / Z_2	A / H	B / P	Z_1 / Z_2	A / H	B / P	Z_1 / Z_2	A / H	B / P	Z_1 / Z_2	A / H	B / P	Z_1 / Z_2	A / H	B / P	Z_1 / Z_2	LHA	
°	°	° ′	° ′	°	° ′	° ′	°	° ′	° ′	°	° ′	° ′	°	° ′	° ′	°	° ′	° ′	°	°	°
0	180	0 00	78 00	90.0	0 00	77 00	90.0	0 00	76 00	90.0	0 00	75 00	90.0	0 00	74 00	90.0	0 00	73 00	90.0	180	360
1	179	0 59	78 00	89.8	0 58	77 00	89.8	0 58	76 00	89.8	0 58	75 00	89.7	0 58	74 00	89.7	0 57	73 00	89.7	181	359
2	178	1 57	78 00	89.6	1 57	77 00	89.5	1 56	76 00	89.5	1 56	74 59	89.5	1 55	73 59	89.4	1 55	72 59	89.4	182	358
3	177	2 56	77 59	89.4	2 55	76 59	89.3	2 55	75 59	89.3	2 54	74 59	89.2	2 53	73 59	89.2	2 52	72 59	89.1	183	357
4	176	3 55	77 58	89.2	3 54	76 58	89.1	3 53	75 58	89.0	3 52	74 58	89.0	3 51	73 58	88.9	3 49	72 58	88.8	184	356
5	175	4 53	77 57	89.0	4 52	76 57	88.9	4 51	75 57	88.8	4 50	74 57	88.7	4 48	73 57	88.6	4 47	72 56	88.5	185	355
6	174	5 52	77 56	88.7	5 51	76 56	88.6	5 49	75 56	88.5	5 48	74 55	88.4	5 46	73 55	88.3	5 44	72 55	88.2	186	354
7	173	6 51	77 55	88.5	6 49	76 54	88.4	6 47	75 54	88.3	6 46	74 54	88.2	6 44	73 53	88.1	6 42	72 53	87.9	187	353
8	172	7 49	77 53	88.3	7 48	76 53	88.2	7 46	75 52	88.1	7 44	74 52	87.9	7 41	73 51	87.8	7 39	72 51	87.6	188	352
9	171	8 48	77 51	88.1	8 46	76 51	88.0	8 44	75 50	87.8	8 41	74 49	87.7	8 39	73 49	87.5	8 36	72 48	87.3	189	351
10	170	9 47	77 49	87.9	9 44	76 48	87.7	9 42	75 48	87.6	9 39	74 47	87.4	9 37	73 46	87.2	9 34	72 45	87.0	190	350
11	169	10 45	77 47	87.7	10 43	76 46	87.5	10 40	75 45	87.3	10 37	74 44	87.1	10 34	73 43	86.9	10 31	72 42	86.7	191	349
12	168	11 44	77 44	87.5	11 41	76 43	87.3	11 38	75 42	87.1	11 35	74 41	86.9	11 32	73 40	86.6	11 28	72 39	86.4	192	348
13	167	12 43	77 42	87.3	12 40	76 40	87.0	12 36	75 39	86.8	12 33	74 37	86.6	12 29	73 36	86.4	12 25	72 35	86.1	193	347
14	166	13 41	77 39	87.0	13 38	76 37	86.8	13 35	75 35	86.5	13 31	74 34	86.3	13 27	73 32	86.1	13 23	72 31	85.8	194	346
15	165	14 40	77 35	86.8	14 36	76 33	86.6	14 33	75 32	86.3	14 29	74 30	86.0	14 24	73 28	85.8	14 20	72 26	85.5	195	345
16	164	15 38	77 32	86.6	15 35	76 30	86.3	15 31	75 28	86.0	15 26	74 25	85.8	15 22	73 23	85.5	15 17	72 21	85.2	196	344
17	163	16 37	77 28	86.4	16 33	76 26	86.1	16 29	75 23	85.8	16 24	74 21	85.5	16 19	73 19	85.2	16 14	72 16	84.9	197	343
18	162	17 36	77 24	86.1	17 31	76 21	85.8	17 27	75 19	85.5	17 22	74 16	85.2	17 17	73 13	84.9	17 11	72 11	84.6	198	342
19	161	18 34	77 20	85.9	18 30	76 17	85.6	18 25	75 14	85.2	18 20	74 11	84.9	18 14	73 08	84.6	18 08	72 05	84.3	199	341
20	160	19 33	77 15	85.7	19 28	76 12	85.3	19 23	75 08	85.0	19 17	74 05	84.6	19 12	73 02	84.3	19 05	71 59	83.9	200	340
21	159	20 31	77 10	85.4	20 26	76 07	85.1	20 21	75 03	84.7	20 15	73 59	84.3	20 09	72 56	84.0	20 03	71 52	83.6	201	339
22	158	21 30	77 05	85.2	21 24	76 01	84.8	21 19	74 57	84.4	21 13	73 53	84.0	21 06	72 49	83.6	21 00	71 45	83.3	202	338
23	157	22 28	77 00	85.0	22 23	75 55	84.5	22 17	74 51	84.1	22 10	73 46	83.7	22 04	72 42	83.3	21 56	71 38	82.9	203	337
24	156	23 27	76 54	84.7	23 21	75 49	84.3	23 15	74 44	83.9	23 08	73 39	83.4	23 01	72 34	83.0	22 53	71 30	82.6	204	336
25	155	24 25	76 48	84.5	24 19	75 43	84.0	24 13	74 37	83.6	24 06	73 32	83.1	23 58	72 27	82.7	23 50	71 22	82.2	205	335
26	154	25 23	76 42	84.2	25 17	75 36	83.7	25 10	74 30	83.3	25 03	73 24	82.8	24 55	72 18	82.3	24 47	71 13	81.9	206	334
27	153	26 22	76 35	84.0	26 15	75 28	83.5	26 08	74 22	83.0	26 01	73 16	82.5	25 52	72 10	82.0	25 44	71 04	81.5	207	333
28	152	27 20	76 28	83.7	27 13	75 21	83.2	27 06	74 14	82.7	26 58	73 07	82.2	26 50	72 00	81.7	26 41	70 54	81.2	208	332
29	151	28 18	76 20	83.4	28 11	75 13	82.9	28 04	74 05	82.4	27 55	72 58	81.8	27 47	71 51	81.3	27 37	70 44	80.8	209	331
30	150	29 17	76 13	83.2	29 09	75 04	82.6	29 01	73 56	82.0	28 53	72 48	81.5	28 44	71 41	81.0	28 34	70 33	80.4	210	330
31	149	30 15	76 04	82.9	30 07	74 56	82.3	29 59	73 47	81.7	29 50	72 38	81.2	29 41	71 30	80.6	29 30	70 22	80.0	211	329
32	148	31 13	75 56	82.6	31 05	74 46	82.0	30 57	73 37	81.4	30 47	72 28	80.8	30 37	71 19	80.2	30 27	70 11	79.6	212	328
33	147	32 11	75 47	82.3	32 03	74 37	81.7	31 54	73 27	81.1	31 44	72 17	80.5	31 34	71 07	79.9	31 23	69 58	79.2	213	327
34	146	33 10	75 37	82.0	33 01	74 26	81.4	32 52	73 16	80.7	32 42	72 05	80.1	32 31	70 55	79.5	32 20	69 45	78.8	214	326
35	145	34 08	75 27	81.7	33 59	74 16	81.0	33 49	73 04	80.4	33 39	71 53	79.7	33 28	70 42	79.1	33 16	69 32	78.4	215	325
36	144	35 06	75 17	81.4	34 56	74 04	80.7	34 46	72 52	80.0	34 36	71 40	79.4	34 24	70 29	78.7	34 12	69 18	78.0	216	324
37	143	36 04	75 06	81.1	35 54	73 53	80.4	35 44	72 40	79.7	35 33	71 27	79.0	35 21	70 15	78.3	35 08	69 03	77.6	217	323
38	142	37 02	74 54	80.8	36 52	73 40	80.0	36 41	72 27	79.3	36 29	71 13	78.6	36 17	70 00	77.8	36 04	68 48	77.1	218	322
39	141	38 00	74 42	80.4	37 49	73 27	79.7	37 38	72 13	78.9	37 26	70 59	78.2	37 13	69 45	77.4	37 00	68 32	76.7	219	321
40	140	38 57	74 30	80.1	38 47	73 14	79.3	38 35	71 58	78.5	38 23	70 43	77.7	38 10	69 29	77.0	37 56	68 15	76.2	220	320
41	139	39 55	74 16	79.8	39 44	72 59	78.9	39 32	71 43	78.1	39 19	70 27	77.3	39 06	69 12	76.5	38 51	67 57	75.7	221	319
42	138	40 53	74 02	79.4	40 41	72 45	78.5	40 29	71 27	77.7	40 16	70 10	76.9	40 02	68 54	76.1	39 47	67 38	75.3	222	318
43	137	41 51	73 48	79.0	41 39	72 29	78.2	41 26	71 11	77.3	41 12	69 53	76.4	40 58	68 35	75.6	40 42	67 19	74.7	223	317
44	136	42 48	73 32	78.6	42 36	72 12	77.7	42 23	70 53	76.9	42 09	69 34	76.0	41 54	68 16	75.1	41 38	66 58	74.2	224	316
45	135	43 46	73 16	78.3	43 33	71 55	77.3	43 19	70 35	76.4	43 05	69 15	75.5	42 49	67 56	74.6	42 33	66 37	73.7	225	315

Lat. / A		12°			13°			14°			15°			16°			17°			Lat. / A	
LHA / F		A / H	B / P	Z_1 / Z_2	A / H	B / P	Z_1 / Z_2	A / H	B / P	Z_1 / Z_2	A / H	B / P	Z_1 / Z_2	A / H	B / P	Z_1 / Z_2	A / H	B / P	Z_1 / Z_2	LHA	
°	°	° ′	° ′	°	° ′	° ′	°	° ′	° ′	°	° ′	° ′	°	° ′	° ′	°	° ′	° ′	°	°	°
45	135	43 46	73 16	78.3	43 33	71 55	77.3	43 19	70 35	76.4	43 05	69 15	75.5	42 49	67 56	74.6	42 33	66 37	73.7	225	315
46	134	44 43	72 59	77.8	44 30	71 37	76.9	44 16	70 15	75.9	44 01	68 54	75.0	43 45	67 34	74.1	43 28	66 15	73.2	226	314
47	133	45 40	72 41	77.4	45 27	71 18	76.4	45 12	69 55	75.5	44 57	68 33	74.5	44 40	67 12	73.5	44 23	65 51	72.6	227	313
48	132	46 38	72 23	77.0	46 24	70 58	76.0	46 09	69 34	75.0	45 53	68 11	74.0	45 35	66 48	73.0	45 17	65 27	72.0	228	312
49	131	47 35	72 03	76.5	47 20	70 37	75.5	47 05	69 11	74.4	46 48	67 47	73.4	46 30	66 23	72.4	46 12	65 01	71.4	229	311
50	130	48 32	71 42	76.1	48 17	70 15	75.0	48 01	68 48	73.9	47 44	67 22	72.9	47 25	65 58	71.8	47 06	64 34	70.8	230	310
51	129	49 29	71 20	75.6	49 13	69 51	74.5	48 57	68 23	73.4	48 39	66 56	72.3	48 20	65 30	71.2	48 00	64 05	70.1	231	309
52	128	50 25	70 57	75.1	50 09	69 27	73.9	49 52	67 57	72.8	49 34	66 29	71.7	49 15	65 02	70.6	48 54	63 35	69.5	232	308
53	127	51 22	70 33	74.6	51 06	69 01	73.4	50 48	67 30	72.2	50 29	66 00	71.0	50 09	64 31	69.9	49 48	63 04	68.8	233	307
54	126	52 19	70 07	74.0	52 02	68 33	72.8	51 43	67 01	71.6	51 24	65 30	70.4	51 03	64 00	69.2	50 41	62 31	68.1	234	306
55	125	53 15	69 40	73.5	52 57	68 04	72.2	52 38	66 30	70.9	52 18	64 58	69.7	51 57	63 26	68.5	51 34	61 56	67.3	235	305
56	124	54 11	69 11	72.9	53 53	67 34	71.6	53 33	65 58	70.3	53 12	64 24	69.0	52 50	62 51	67.8	52 27	61 20	66.6	236	304
57	123	55 07	68 41	72.2	54 48	67 02	70.9	54 28	65 24	69.6	54 06	63 48	68.3	53 43	62 14	67.0	53 19	60 42	65.8	237	303
58	122	56 03	68 09	71.6	55 43	66 28	70.2	55 22	64 48	68.8	55 00	63 11	67.5	54 36	61 35	66.2	54 12	60 01	64.9	238	302
59	121	56 59	67 34	70.9	56 38	65 51	69.5	56 16	64 10	68.1	55 53	62 31	66.7	55 29	60 54	65.4	55 03	59 18	64.1	239	301
60	120	57 54	66 58	70.2	57 33	65 13	68.7	57 10	63 30	67.3	56 46	61 49	65.9	56 21	60 10	64.5	55 55	58 33	63.1	240	300
61	119	58 49	66 20	69.4	58 27	64 32	67.9	58 04	62 47	66.4	57 39	61 04	65.0	57 13	59 24	63.6	56 46	57 46	62.2	241	299
62	118	59 44	65 38	68.6	59 21	63 49	67.1	58 57	62 02	65.5	58 31	60 17	64.0	58 05	58 35	62.6	57 36	56 56	61.2	242	298
63	117	60 38	64 55	67.8	60 15	63 03	66.2	59 50	61 13	64.6	59 23	59 27	63.1	58 55	57 43	61.6	58 26	56 03	60.2	243	297
64	116	61 32	64 08	66.9	61 08	62 14	65.2	60 42	60 22	63.6	60 15	58 34	62.0	59 46	56 49	60.5	59 16	55 06	59.1	244	296
65	115	62 26	63 18	66.0	62 01	61 21	64.2	61 34	59 28	62.6	61 06	57 37	61.0	60 36	55 51	59.4	60 05	54 07	57.9	245	295
66	114	63 20	62 25	65.0	62 53	60 25	63.2	62 26	58 30	61.5	61 56	56 37	59.8	61 25	54 49	58.2	60 53	53 04	56.7	246	294
67	113	64 13	61 27	63.9	63 45	59 25	62.1	63 16	57 27	60.3	62 46	55 34	58.6	62 14	53 44	57.0	61 41	51 57	55.4	247	293
68	112	65 05	60 26	62.8	64 37	58 21	60.9	64 07	56 21	59.1	63 35	54 25	57.4	63 02	52 34	55.7	62 27	50 47	54.1	248	292
69	111	65 57	59 20	61.6	65 27	57 13	59.6	64 56	55 10	57.8	64 23	53 13	56.0	63 49	51 20	54.3	63 14	49 32	52.7	249	291
70	110	66 48	58 08	60.3	66 18	55 59	58.3	65 45	53 55	56.4	65 11	51 55	54.6	64 36	50 01	52.9	63 59	48 12	51.2	250	290
71	109	67 39	56 52	58.9	67 07	54 40	56.8	66 33	52 33	54.9	65 58	50 33	53.1	65 21	48 38	51.3	64 43	46 48	49.7	251	289
72	108	68 29	55 29	57.4	67 55	53 14	55.3	67 20	51 06	53.3	66 44	49 04	51.5	66 06	47 08	49.7	65 26	45 18	48.0	252	288
73	107	69 18	53 59	55.8	68 43	51 42	53.7	68 07	49 33	51.6	67 29	47 30	49.8	66 49	45 33	48.0	66 08	43 43	46.3	253	287
74	106	70 06	52 22	54.1	69 30	50 03	51.9	68 52	47 52	49.8	68 12	45 49	47.9	67 31	43 52	46.1	66 49	42 02	44.4	254	286
75	105	70 53	50 36	52.2	70 15	48 16	50.0	69 36	46 04	47.9	68 55	44 00	46.0	68 12	42 04	44.2	67 29	40 15	42.5	255	285
76	104	71 38	48 42	50.2	70 59	46 20	47.9	70 18	44 08	45.9	69 36	42 05	43.9	68 52	40 09	42.1	68 07	38 21	40.5	256	284
77	103	72 23	46 37	48.0	71 42	44 15	45.7	70 59	42 03	43.7	70 15	40 01	41.7	69 30	38 07	39.9	68 43	36 21	38.3	257	283
78	102	73 06	44 22	45.6	72 23	42 00	43.4	71 38	39 49	41.3	70 53	37 49	39.4	70 06	35 57	37.6	69 18	34 13	36.0	258	282
79	101	73 47	41 55	43.1	73 02	39 34	40.8	72 16	37 26	38.8	71 28	35 27	36.9	70 40	33 38	35.2	69 50	31 58	33.6	259	281
80	100	74 26	39 15	40.3	73 39	36 57	38.1	72 51	34 51	36.1	72 02	32 57	34.3	71 12	31 12	32.6	70 21	29 36	31.1	260	280
81	99	75 02	36 21	37.3	74 14	34 07	35.1	73 24	32 06	33.2	72 34	30 17	31.5	71 42	28 37	29.9	70 50	27 06	28.4	261	279
82	98	75 37	33 13	34.1	74 46	31 05	32.0	73 55	29 10	30.2	73 03	27 27	28.5	72 09	25 53	27.0	71 16	24 29	25.7	262	278
83	97	76 08	29 50	30.6	75 16	27 50	28.6	74 23	26 03	26.9	73 29	24 27	25.4	72 34	23 02	24.0	71 39	21 44	22.8	263	277
84	96	76 36	26 11	26.8	75 42	24 22	25.0	74 48	22 45	23.5	73 52	21 19	22.1	72 56	20 02	20.9	72 00	18 53	19.8	264	276
85	95	77 01	22 18	22.8	76 05	20 41	21.3	75 09	19 16	19.9	74 12	18 01	18.7	73 15	16 54	17.6	72 18	15 55	16.7	265	275
86	94	77 22	18 10	18.6	76 25	16 49	17.3	75 27	15 38	16.1	74 29	14 36	15.1	73 31	13 40	14.2	72 33	12 51	13.5	266	274
87	93	77 38	13 50	14.1	76 40	12 46	13.1	75 41	11 51	12.2	74 43	11 03	11.4	73 44	10 21	10.8	72 45	9 43	10.2	267	273
88	92	77 50	9 19	9.5	76 51	8 36	8.8	75 52	7 58	8.2	74 52	7 25	7.7	73 53	6 56	7.2	72 53	6 31	6.8	268	272
89	91	77 58	4 42	4.8	76 58	4 19	4.4	75 58	4 00	4.1	74 58	3 44	3.9	73 58	3 29	3.6	72 58	3 16	3.4	269	271
90	90	78 00	0 00	0.0	77 00	0 00	0.0	76 00	0 00	0.0	75 00	0 00	0.0	74 00	0 00	0.0	73 00	0 00	0.0	270	270

N. Lat.: for LHA > 180°... $Z_n = Z$
for LHA < 180°... $Z_n = 360° - Z$

S. Lat.: for LHA > 180°... $Z_n = 180° - Z$
for LHA < 180°... $Z_n = 180° + Z$

SIGHT REDUCTION TABLE

B: (–) for 90°< LHA < 270°
Dec: (–) for Lat. contrary name

Z_1 : same sign as B
Z_2 : (–) for F > 90°

Lat. / A		18°			19°			20°			21°			22°			23°			Lat. / A	
LHA / F		A / H	B / P	Z_1 / Z_2	A / H	B / P	Z_1 / Z_2	A / H	B / P	Z_1 / Z_2	A / H	B / P	Z_1 / Z_2	A / H	B / P	Z_1 / Z_2	A / H	B / P	Z_1 / Z_2	LHA	
°	°	° ′	° ′	°	° ′	° ′	°	° ′	° ′	°	° ′	° ′	°	° ′	° ′	°	° ′	° ′	°	°	°
0	180	0 00	72 00	90.0	0 00	71 00	90.0	0 00	70 00	90.0	0 00	69 00	90.0	0 00	68 00	90.0	0 00	67 00	90.0	180	360
1	179	0 57	72 00	89.7	0 57	71 00	89.7	0 56	70 00	89.7	0 56	69 00	89.6	0 56	68 00	89.6	0 55	67 00	89.6	181	359
2	178	1 54	71 59	89.4	1 53	70 59	89.3	1 53	69 59	89.3	1 52	68 59	89.3	1 51	67 59	89.3	1 50	66 59	89.2	182	358
3	177	2 51	71 59	89.1	2 50	70 59	89.0	2 49	69 58	89.0	2 48	68 58	88.9	2 47	67 58	88.9	2 46	66 58	88.8	183	357
4	176	3 48	71 58	88.8	3 47	70 57	88.7	3 46	69 57	88.6	3 44	68 57	88.6	3 42	67 57	88.5	3 41	66 57	88.4	184	356
5	175	4 45	71 56	88.5	4 44	70 56	88.4	4 42	69 56	88.3	4 40	68 56	88.2	4 38	67 55	88.1	4 36	66 55	88.0	185	355
6	174	5 42	71 54	88.1	5 40	70 54	88.0	5 38	69 54	87.9	5 36	68 54	87.8	5 34	67 53	87.7	5 31	66 53	87.6	186	354
7	173	6 39	71 52	87.8	6 37	70 52	87.7	6 35	69 52	87.6	6 32	68 51	87.5	6 29	67 51	87.4	6 26	66 51	87.3	187	353
8	172	7 36	71 50	87.5	7 34	70 50	87.4	7 31	69 49	87.2	7 28	68 49	87.1	7 25	67 48	87.0	7 22	66 48	86.9	188	352
9	171	8 33	71 47	87.2	8 30	70 47	87.0	8 27	69 46	86.9	8 24	68 46	86.8	8 20	67 45	86.6	8 17	66 45	86.5	189	351
10	170	9 30	71 44	86.9	9 27	70 44	86.7	9 23	69 43	86.5	9 20	68 42	86.4	9 16	67 42	86.2	9 12	66 41	86.1	190	350
11	169	10 27	71 41	86.6	10 24	70 40	86.4	10 20	69 39	86.2	10 16	68 39	86.0	10 11	67 38	85.8	10 07	66 37	85.7	191	349
12	168	11 24	71 37	86.2	11 20	70 36	86.0	11 16	69 35	85.8	11 12	68 34	85.6	11 07	67 33	85.4	11 02	66 32	85.3	192	348
13	167	12 21	71 33	85.9	12 17	70 32	85.7	12 12	69 31	85.5	12 07	68 30	85.3	12 02	67 29	85.1	11 57	66 28	84.8	193	347
14	166	13 18	71 29	85.6	13 13	70 28	85.4	13 08	69 26	85.1	13 03	68 25	84.9	12 58	67 24	84.7	12 52	66 22	84.4	194	346
15	165	14 15	71 24	85.3	14 10	70 23	85.0	14 05	69 21	84.8	13 59	68 20	84.5	13 53	67 18	84.3	13 47	66 17	84.0	195	345
16	164	15 12	71 19	84.9	15 06	70 18	84.7	15 01	69 16	84.4	14 55	68 14	84.1	14 48	67 12	83.9	14 42	66 10	83.6	196	344
17	163	16 09	71 14	84.6	16 03	70 12	84.3	15 57	69 10	84.0	15 50	68 08	83.7	15 44	67 06	83.5	15 37	66 04	83.2	197	343
18	162	17 05	71 08	84.3	16 59	70 06	84.0	16 53	69 03	83.7	16 46	68 01	83.4	16 39	66 59	83.1	16 32	65 57	82.8	198	342
19	161	18 02	71 02	83.9	17 56	69 59	83.6	17 49	68 57	83.3	17 42	67 54	83.0	17 34	66 52	82.7	17 26	65 49	82.3	199	341
20	160	18 59	70 56	83.6	18 52	69 53	83.2	18 45	68 50	82.9	18 37	67 47	82.6	18 29	66 44	82.2	18 21	65 41	81.9	200	340
21	159	19 56	70 49	83.2	19 48	69 45	82.9	19 41	68 42	82.5	19 33	67 39	82.2	19 24	66 36	81.8	19 16	65 33	81.5	201	339
22	158	20 52	70 41	82.9	20 45	69 38	82.5	20 37	68 34	82.1	20 28	67 31	81.8	20 19	66 27	81.4	20 10	65 24	81.0	202	338
23	157	21 49	70 33	82.5	21 41	69 29	82.1	21 32	68 26	81.7	21 24	67 22	81.4	21 14	66 18	81.0	21 05	65 15	80.6	203	337
24	156	22 45	70 25	82.2	22 37	69 21	81.8	22 28	68 17	81.3	22 19	67 12	80.9	22 09	66 09	80.5	21 59	65 05	80.1	204	336
25	155	23 42	70 17	81.8	23 33	69 12	81.4	23 24	68 07	80.9	23 14	67 03	80.5	23 04	65 58	80.1	22 54	64 54	79.7	205	335
26	154	24 38	70 07	81.4	24 29	69 02	81.0	24 20	67 57	80.5	24 09	66 52	80.1	23 59	65 48	79.6	23 48	64 43	79.2	206	334
27	153	25 35	69 58	81.1	25 25	68 52	80.6	25 15	67 47	80.1	25 05	66 42	79.7	24 54	65 36	79.2	24 42	64 32	78.7	207	333
28	152	26 31	69 48	80.7	26 21	68 42	80.2	26 11	67 36	79.7	26 00	66 30	79.2	25 48	65 25	78.7	25 36	64 19	78.3	208	332
29	151	27 27	69 37	80.3	27 17	68 31	79.8	27 06	67 24	79.3	26 55	66 18	78.8	26 43	65 12	78.3	26 30	64 07	77.8	209	331
30	150	28 24	69 26	79.9	28 13	68 19	79.4	28 01	67 12	78.8	27 50	66 06	78.3	27 37	64 59	77.8	27 24	63 53	77.3	210	330
31	149	29 20	69 14	79.5	29 09	68 07	78.9	28 57	67 00	78.4	28 44	65 53	77.8	28 31	64 46	77.3	28 18	63 39	76.8	211	329
32	148	30 16	69 02	79.1	30 04	67 54	78.5	29 52	66 46	77.9	29 39	65 39	77.4	29 26	64 32	76.8	29 12	63 25	76.3	212	328
33	147	31 12	68 49	78.7	31 00	67 41	78.1	30 47	66 32	77.5	30 34	65 24	76.9	30 20	64 17	76.3	30 05	63 09	75.8	213	327
34	146	32 08	68 36	78.2	31 55	67 27	77.6	31 42	66 18	77.0	31 28	65 09	76.4	31 14	64 01	75.8	30 59	62 53	75.2	214	326
35	145	33 04	68 22	77.8	32 51	67 12	77.2	32 37	66 03	76.5	32 23	64 54	75.9	32 08	63 45	75.3	31 52	62 36	74.7	215	325
36	144	33 59	68 07	77.3	33 46	66 57	76.7	33 32	65 47	76.0	33 17	64 37	75.4	33 01	63 28	74.8	32 45	62 19	74.2	216	324
37	143	34 55	67 52	76.9	34 41	66 41	76.2	34 26	65 30	75.5	34 11	64 20	74.9	33 55	63 10	74.2	33 38	62 01	73.6	217	323
38	142	35 50	67 36	76.4	35 36	66 24	75.7	35 21	65 13	75.0	35 05	64 02	74.4	34 48	62 51	73.7	34 31	61 41	73.0	218	322
39	141	36 46	67 19	76.0	36 31	66 06	75.2	36 15	64 54	74.5	35 59	63 43	73.8	35 42	62 32	73.1	35 24	61 21	72.4	219	321
40	140	37 41	67 01	75.5	37 26	65 48	74.7	37 10	64 35	74.0	36 53	63 23	73.3	36 35	62 12	72.6	36 17	61 01	71.8	220	320
41	139	38 36	66 42	75.0	38 20	65 29	74.2	38 04	64 15	73.4	37 46	63 02	72.7	37 28	61 50	72.0	37 09	60 39	71.2	221	319
42	138	39 31	66 23	74.5	39 15	65 08	73.7	38 58	63 54	72.9	38 40	62 41	72.1	38 21	61 28	71.4	38 01	60 16	70.6	222	318
43	137	40 26	66 03	73.9	40 09	64 47	73.1	39 51	63 33	72.3	39 33	62 18	71.5	39 13	61 05	70.7	38 53	59 52	70.0	223	317
44	136	41 21	65 42	73.4	41 03	64 25	72.5	40 45	63 10	71.7	40 26	61 55	70.9	40 06	60 41	70.1	39 45	59 27	69.3	224	316
45	135	42 16	65 19	72.8	41 57	64 02	72.0	41 38	62 46	71.1	41 19	61 30	70.3	40 58	60 15	69.5	40 37	59 01	68.7	225	315

Lat. / A		18°			19°			20°			21°			22°			23°			Lat. / A	
LHA / F		A / H	B / P	Z_1 / Z_2	A / H	B / P	Z_1 / Z_2	A / H	B / P	Z_1 / Z_2	A / H	B / P	Z_1 / Z_2	A / H	B / P	Z_1 / Z_2	A / H	B / P	Z_1 / Z_2	LHA	
°	°	° ′	° ′	°	° ′	° ′	°	° ′	° ′	°	° ′	° ′	°	° ′	° ′	°	° ′	° ′	°	°	°
45	135	42 16	65 19	72.8	41 57	64 02	72.0	41 38	62 46	71.1	41 19	61 30	70.3	40 58	60 15	69.5	40 37	59 01	68.7	225	315
46	134	43 10	64 56	72.3	42 51	63 38	71.4	42 32	62 21	70.5	42 11	61 05	69.6	41 50	59 49	68.8	41 28	58 34	68.0	226	314
47	133	44 04	64 32	71.7	43 45	63 13	70.8	43 25	61 55	69.9	43 04	60 38	69.0	42 42	59 21	68.1	42 19	58 06	67.3	227	313
48	132	44 58	64 06	71.1	44 38	62 46	70.1	44 18	61 27	69.2	43 56	60 09	68.3	43 33	58 53	67.4	43 10	57 37	66.5	228	312
49	131	45 52	63 39	70.4	45 32	62 18	69.5	45 10	60 59	68.5	44 48	59 40	67.6	44 24	58 22	66.7	44 00	57 06	65.8	229	311
50	130	46 46	63 11	69.8	46 25	61 49	68.8	46 03	60 29	67.8	45 39	59 09	66.9	45 15	57 51	65.9	44 50	56 34	65.0	230	310
51	129	47 39	62 42	69.1	47 17	61 19	68.1	46 55	59 57	67.1	46 31	58 37	66.1	46 06	57 18	65.2	45 40	56 00	64.2	231	309
52	128	48 33	62 11	68.4	48 10	60 47	67.4	47 46	59 25	66.4	47 22	58 03	65.4	46 56	56 44	64.4	46 30	55 25	63.4	232	308
53	127	49 25	61 38	67.7	49 02	60 13	66.6	48 38	58 50	65.6	48 13	57 28	64.6	47 46	56 07	63.6	47 19	54 48	62.6	233	307
54	126	50 18	61 04	67.0	49 54	59 38	65.9	49 29	58 14	64.8	49 03	56 51	63.7	48 36	55 30	62.7	48 08	54 10	61.7	234	306
55	125	51 10	60 28	66.2	50 46	59 01	65.1	50 20	57 36	64.0	49 53	56 12	62.9	49 25	54 50	61.9	48 56	53 30	60.8	235	305
56	124	52 03	59 50	65.4	51 37	58 23	64.2	51 10	56 56	63.1	50 43	55 32	62.0	50 14	54 09	61.0	49 44	52 48	59.9	236	304
57	123	52 54	59 11	64.6	52 28	57 42	63.4	52 00	56 15	62.2	51 32	54 49	61.1	51 02	53 26	60.0	50 32	52 04	59.0	237	303
58	122	53 46	58 29	63.7	53 18	56 59	62.5	52 50	55 31	61.3	52 21	54 05	60.2	51 50	52 41	59.1	51 19	51 18	58.0	238	302
59	121	54 37	57 45	62.8	54 08	56 14	61.5	53 39	54 45	60.4	53 09	53 18	59.2	52 38	51 53	58.1	52 06	50 30	57.0	239	301
60	120	55 27	56 59	61.8	54 58	55 27	60.6	54 28	53 57	59.4	53 57	52 29	58.2	53 25	51 04	57.0	52 52	49 40	55.9	240	300
61	119	56 17	56 10	60.9	55 47	54 37	59.6	55 16	53 06	58.3	54 44	51 38	57.1	54 11	50 12	55.9	53 37	48 48	54.8	241	299
62	118	57 07	55 19	59.8	56 36	53 45	58.5	56 04	52 13	57.2	55 31	50 44	56.0	54 57	49 17	54.8	54 22	47 53	53.7	242	298
63	117	57 56	54 25	58.8	57 24	52 49	57.4	56 51	51 17	56.1	56 17	49 47	54.9	55 42	48 20	53.7	55 06	46 55	52.5	243	297
64	116	58 44	53 27	57.6	58 12	51 51	56.3	57 38	50 18	55.0	57 03	48 48	53.7	56 27	47 20	52.5	55 50	45 55	51.3	244	296
65	115	59 32	52 27	56.5	58 58	50 50	55.1	58 23	49 16	53.7	57 47	47 45	52.5	57 10	46 17	51.2	56 32	44 52	50.0	245	295
66	114	60 19	51 23	55.2	59 45	49 45	53.8	59 09	48 11	52.5	58 32	46 39	51.2	57 53	45 11	49.9	57 14	43 47	48.7	246	294
67	113	61 06	50 15	53.9	60 30	48 37	52.5	59 53	47 02	51.1	59 15	45 30	49.8	58 36	44 02	48.6	57 55	42 38	47.4	247	293
68	112	61 52	49 04	52.6	61 15	47 25	51.1	60 36	45 50	49.8	59 57	44 18	48.4	59 17	42 50	47.2	58 36	41 26	46.0	248	292
69	111	62 37	47 48	51.2	61 58	46 09	49.7	61 19	44 33	48.3	60 39	43 02	47.0	59 57	41 34	45.7	59 15	40 10	44.5	249	291
70	110	63 21	46 28	49.7	62 41	44 48	48.2	62 01	43 13	46.8	61 19	41 42	45.4	60 36	40 15	44.2	59 53	38 52	43.0	250	290
71	109	64 04	45 03	48.1	63 23	43 24	46.6	62 41	41 49	45.2	61 58	40 18	43.9	61 15	38 52	42.6	60 30	37 29	41.4	251	289
72	108	64 45	43 34	46.4	64 04	41 54	44.9	63 21	40 20	43.5	62 37	38 50	42.2	61 52	37 25	40.9	61 06	36 03	39.7	252	288
73	107	65 26	41 59	44.7	64 43	40 20	43.2	63 59	38 46	41.8	63 14	37 18	40.5	62 27	35 53	39.2	61 41	34 34	38.0	253	287
74	106	66 06	40 19	42.9	65 21	38 41	41.4	64 36	37 08	40.0	63 49	35 41	38.7	63 02	34 18	37.4	62 14	33 00	36.3	254	286
75	105	66 44	38 32	40.9	65 58	36 56	39.5	65 11	35 25	38.1	64 23	33 59	36.8	63 35	32 39	35.6	62 46	31 22	34.4	255	285
76	104	67 20	36 40	38.9	66 33	35 05	37.4	65 45	33 37	36.1	64 56	32 13	34.8	64 07	30 55	33.6	63 16	29 41	32.5	256	284
77	103	67 55	34 42	36.8	67 07	33 09	35.3	66 18	31 43	34.0	65 27	30 22	32.8	64 37	29 06	31.6	63 45	27 55	30.6	257	283
78	102	68 29	32 37	34.5	67 39	31 07	33.1	66 48	29 44	31.9	65 57	28 26	30.7	65 05	27 14	29.6	64 13	26 06	28.5	258	282
79	101	69 00	30 25	32.2	68 09	29 00	30.8	67 17	27 40	29.6	66 25	26 26	28.5	65 32	25 17	27.4	64 38	24 12	26.4	259	281
80	100	69 29	28 07	29.7	68 37	26 46	28.4	67 44	25 30	27.3	66 50	24 20	26.2	65 56	23 15	25.2	65 02	22 15	24.3	260	280
81	99	69 57	25 43	27.1	69 03	24 26	25.9	68 09	23 15	24.8	67 14	22 10	23.8	66 19	21 10	22.9	65 23	20 14	22.1	261	279
82	98	70 21	23 11	24.5	69 27	22 00	23.3	68 31	20 56	22.3	67 36	19 56	21.4	66 40	19 00	20.6	65 43	18 09	19.8	262	278
83	97	70 44	20 34	21.7	69 48	19 29	20.7	68 51	18 31	19.7	67 55	17 37	18.9	66 58	16 47	18.1	66 01	16 01	17.4	263	277
84	96	71 03	17 50	18.8	70 07	16 53	17.9	69 09	16 01	17.1	68 12	15 14	16.3	67 14	14 30	15.7	66 16	13 50	15.1	264	276
85	95	71 20	15 01	15.8	70 23	14 12	15.0	69 25	13 28	14.3	68 26	12 48	13.7	67 28	12 10	13.1	66 29	11 36	12.6	265	275
86	94	71 35	12 07	12.8	70 36	11 27	12.1	69 37	10 51	11.6	68 38	10 18	11.0	67 39	9 48	10.6	66 40	9 20	10.1	266	274
87	93	71 46	9 09	9.6	70 46	8 39	9.1	69 47	8 11	8.7	68 48	7 46	8.3	67 48	7 23	8.0	66 49	7 02	7.6	267	273
88	92	71 54	6 08	6.4	70 54	5 47	6.1	69 54	5 29	5.8	68 55	5 12	5.6	67 55	4 56	5.3	66 55	4 42	5.1	268	272
89	91	71 58	3 04	3.2	70 58	2 54	3.1	69 59	2 45	2.9	68 59	2 36	2.8	67 59	2 28	2.7	66 59	2 21	2.6	269	271
90	90	72 00	0 00	0.0	71 00	0 00	0.0	70 00	0 00	0.0	69 00	0 00	0.0	68 00	0 00	0.0	67 00	0 00	0.0	270	270

N. Lat.: for LHA > 180°... $Z_n = Z$
for LHA < 180°... $Z_n = 360° - Z$

S. Lat.: for LHA > 180°... $Z_n = 180° - Z$
for LHA < 180°... $Z_n = 180° + Z$

SIGHT REDUCTION TABLE

B: (–) for 90°< LHA < 270°
Dec: (–) for Lat. contrary name

Z_1 : same sign as B
Z_2 : (–) for F > 90°

Lat. / A		24°			25°			26°			27°			28°			29°			Lat. / A	
LHA / F		A / H	B / P	Z_1 / Z_2	A / H	B / P	Z_1 / Z_2	A / H	B / P	Z_1 / Z_2	A / H	B / P	Z_1 / Z_2	A / H	B / P	Z_1 / Z_2	A / H	B / P	Z_1 / Z_2	LHA	
°	°	° ′	° ′	°	° ′	° ′	°	° ′	° ′	°	° ′	° ′	°	° ′	° ′	°	° ′	° ′	°	°	°
0	180	0 00	66 00	90.0	0 00	65 00	90.0	0 00	64 00	90.0	0 00	63 00	90.0	0 00	62 00	90.0	0 00	61 00	90.0	180	360
1	179	0 55	66 00	89.6	0 54	65 00	89.6	0 54	64 00	89.6	0 53	63 00	89.5	0 53	62 00	89.5	0 52	61 00	89.5	181	359
2	178	1 50	65 59	89.2	1 49	64 59	89.2	1 48	63 59	89.1	1 47	62 59	89.1	1 46	61 59	89.1	1 45	60 59	89.0	182	358
3	177	2 44	65 58	88.8	2 43	64 58	88.7	2 42	63 58	88.7	2 40	62 58	88.6	2 39	61 58	88.6	2 37	60 58	88.5	183	357
4	176	3 39	65 57	88.4	3 37	64 57	88.3	3 36	63 57	88.2	3 34	62 57	88.2	3 32	61 57	88.1	3 30	60 56	88.1	184	356
5	175	4 34	65 55	88.0	4 32	64 55	87.9	4 30	63 55	87.8	4 27	62 55	87.7	4 25	61 55	87.6	4 22	60 54	87.6	185	355
6	174	5 29	65 53	87.6	5 26	64 53	87.5	5 23	63 53	87.4	5 21	62 52	87.3	5 18	61 52	87.2	5 15	60 52	87.1	186	354
7	173	6 24	65 50	87.1	6 20	64 50	87.0	6 17	63 50	86.9	6 14	62 50	86.8	6 11	61 49	86.7	6 07	60 49	86.6	187	353
8	172	7 18	65 47	86.7	7 15	64 47	86.6	7 11	63 47	86.5	7 07	62 46	86.3	7 04	61 46	86.2	6 59	60 46	86.1	188	352
9	171	8 13	65 44	86.3	8 09	64 44	86.2	8 05	63 43	86.0	8 01	62 43	85.9	7 56	61 42	85.7	7 52	60 42	85.6	189	351
10	170	9 08	65 40	85.9	9 03	64 40	85.7	8 59	63 39	85.6	8 54	62 39	85.4	8 49	61 38	85.3	8 44	60 38	85.1	190	350
11	169	10 02	65 36	85.5	9 57	64 35	85.3	9 52	63 35	85.1	9 47	62 34	85.0	9 42	61 33	84.8	9 36	60 33	84.6	191	349
12	168	10 57	65 32	85.1	10 52	64 31	84.9	10 46	63 30	84.7	10 41	62 29	84.5	10 35	61 28	84.3	10 29	60 28	84.1	192	348
13	167	11 52	65 27	84.6	11 46	64 26	84.4	11 40	63 25	84.2	11 34	62 24	84.0	11 27	61 23	83.8	11 21	60 22	83.6	193	347
14	166	12 46	65 21	84.2	12 40	64 20	84.0	12 34	63 19	83.8	12 27	62 18	83.5	12 20	61 17	83.3	12 13	60 16	83.1	194	346
15	165	13 41	65 15	83.8	13 34	64 14	83.5	13 27	63 13	83.3	13 20	62 11	83.1	13 13	61 10	82.8	13 05	60 09	82.6	195	345
16	164	14 35	65 09	83.3	14 28	64 07	83.1	14 21	63 06	82.8	14 13	62 04	82.6	14 05	61 03	82.3	13 57	60 02	82.1	196	344
17	163	15 29	65 02	82.9	15 22	64 00	82.6	15 14	62 59	82.4	15 06	61 57	82.1	14 58	60 56	81.8	14 49	59 54	81.6	197	343
18	162	16 24	64 55	82.5	16 16	63 53	82.2	16 08	62 51	81.9	15 59	61 49	81.6	15 50	60 47	81.3	15 41	59 46	81.0	198	342
19	161	17 18	64 47	82.0	17 10	63 45	81.7	17 01	62 43	81.4	16 52	61 41	81.1	16 42	60 39	80.8	16 33	59 37	80.5	199	341
20	160	18 12	64 39	81.6	18 03	63 36	81.3	17 54	62 34	80.9	17 45	61 32	80.6	17 35	60 30	80.3	17 24	59 28	80.0	200	340
21	159	19 07	64 30	81.1	18 57	63 28	80.8	18 47	62 25	80.4	18 37	61 23	80.1	18 27	60 20	79.8	18 16	59 18	79.5	201	339
22	158	20 01	64 21	80.7	19 51	63 18	80.3	19 41	62 15	80.0	19 30	61 13	79.6	19 19	60 10	79.3	19 08	59 08	78.9	202	338
23	157	20 55	64 11	80.2	20 44	63 08	79.8	20 34	62 05	79.5	20 22	61 02	79.1	20 11	59 59	78.7	19 59	58 57	78.4	203	337
24	156	21 49	64 01	79.7	21 38	62 58	79.3	21 27	61 54	79.0	21 15	60 51	78.6	21 03	59 48	78.2	20 50	58 45	77.8	204	336
25	155	22 43	63 50	79.3	22 31	62 46	78.9	22 19	61 43	78.4	22 07	60 39	78.0	21 55	59 36	77.7	21 42	58 33	77.3	205	335
26	154	23 36	63 39	78.8	23 25	62 35	78.4	23 12	61 31	77.9	22 59	60 27	77.5	22 46	59 24	77.1	22 33	58 20	76.7	206	334
27	153	24 30	63 27	78.3	24 18	62 22	77.8	24 05	61 18	77.4	23 52	60 14	77.0	23 38	59 10	76.5	23 24	58 07	76.1	207	333
28	152	25 24	63 14	77.8	25 11	62 10	77.3	24 57	61 05	76.9	24 44	60 01	76.4	24 29	58 57	76.0	24 15	57 53	75.5	208	332
29	151	26 17	63 01	77.3	26 04	61 56	76.8	25 50	60 51	76.3	25 36	59 47	75.9	25 21	58 42	75.4	25 05	57 38	75.0	209	331
30	150	27 11	62 48	76.8	26 57	61 42	76.3	26 42	60 37	75.8	26 27	59 32	75.3	26 12	58 27	74.8	25 56	57 23	74.4	210	330
31	149	28 04	62 33	76.3	27 50	61 27	75.8	27 35	60 22	75.2	27 19	59 16	74.7	27 03	58 11	74.2	26 46	57 07	73.8	211	329
32	148	28 57	62 18	75.7	28 42	61 12	75.2	28 27	60 06	74.7	28 10	59 00	74.2	27 54	57 55	73.7	27 37	56 50	73.1	212	328
33	147	29 50	62 02	75.2	29 35	60 56	74.7	29 19	59 49	74.1	29 02	58 43	73.6	28 45	57 38	73.0	28 27	56 32	72.5	213	327
34	146	30 43	61 46	74.7	30 27	60 39	74.1	30 10	59 32	73.5	29 53	58 26	73.0	29 35	57 20	72.4	29 17	56 14	71.9	214	326
35	145	31 36	61 28	74.1	31 19	60 21	73.5	31 02	59 14	72.9	30 44	58 07	72.4	30 26	57 01	71.8	30 07	55 55	71.2	215	325
36	144	32 29	61 10	73.5	32 11	60 02	72.9	31 53	58 55	72.3	31 35	57 48	71.7	31 16	56 41	71.2	30 56	55 35	70.6	216	324
37	143	33 21	60 52	73.0	33 03	59 43	72.3	32 45	58 35	71.7	32 26	57 28	71.1	32 06	56 21	70.5	31 46	55 14	69.9	217	323
38	142	34 13	60 32	72.4	33 55	59 23	71.7	33 36	58 15	71.1	33 16	57 07	70.5	32 56	55 59	69.9	32 35	54 53	69.3	218	322
39	141	35 06	60 11	71.8	34 47	59 02	71.1	34 27	57 53	70.5	34 06	56 45	69.8	33 45	55 37	69.2	33 24	54 30	68.6	219	321
40	140	35 58	59 50	71.2	35 38	58 40	70.5	35 17	57 31	69.8	34 56	56 22	69.1	34 35	55 14	68.5	34 12	54 07	67.9	220	320
41	139	36 49	59 28	70.5	36 29	58 17	69.8	36 08	57 08	69.1	35 46	55 59	68.5	35 24	54 50	67.8	35 01	53 42	67.1	221	319
42	138	37 41	59 04	69.9	37 20	57 54	69.2	36 58	56 43	68.5	36 36	55 34	67.8	36 13	54 25	67.1	35 49	53 17	66.4	222	318
43	137	38 32	58 40	69.2	38 11	57 29	68.5	37 48	56 18	67.8	37 25	55 08	67.1	37 02	53 59	66.4	36 37	52 50	65.7	223	317
44	136	39 23	58 15	68.6	39 01	57 03	67.8	38 38	55 52	67.1	38 14	54 41	66.3	37 50	53 32	65.6	37 25	52 23	64.9	224	316
45	135	40 14	57 48	67.9	39 51	56 36	67.1	39 28	55 24	66.3	39 03	54 13	65.6	38 38	53 04	64.9	38 12	51 54	64.1	225	315

Lat. / A		24°			25°			26°			27°			28°			29°			Lat. / A	
LHA / F		A / H	B / P	Z_1 / Z_2	A / H	B / P	Z_1 / Z_2	A / H	B / P	Z_1 / Z_2	A / H	B / P	Z_1 / Z_2	A / H	B / P	Z_1 / Z_2	A / H	B / P	Z_1 / Z_2	LHA	
°	°	° ′	° ′	°	° ′	° ′	°	° ′	° ′	°	° ′	° ′	°	° ′	° ′	°	° ′	° ′	°	°	°
45	135	40 14	57 48	67.9	39 51	56 36	67.1	39 28	55 24	66.3	39 03	54 13	65.6	38 38	53 04	64.9	38 12	51 54	64.1	225	315
46	134	41 05	57 21	67.2	40 41	56 08	66.4	40 17	54 56	65.6	39 52	53 44	64.8	39 26	52 34	64.1	38 59	51 25	63.3	226	314
47	133	41 55	56 52	66.4	41 31	55 38	65.6	41 06	54 26	64.8	40 40	53 14	64.0	40 13	52 04	63.3	39 46	50 54	62.5	227	313
48	132	42 45	56 22	65.7	42 20	55 08	64.9	41 54	53 55	64.0	41 28	52 43	63.2	41 00	51 32	62.5	40 32	50 22	61.7	228	312
49	131	43 35	55 50	64.9	43 09	54 36	64.1	42 43	53 22	63.2	42 15	52 10	62.4	41 47	50 59	61.6	41 18	49 48	60.9	229	311
50	130	44 25	55 17	64.1	43 58	54 02	63.3	43 31	52 49	62.4	43 03	51 36	61.6	42 34	50 24	60.8	42 04	49 14	60.0	230	310
51	129	45 14	54 43	63.3	44 47	53 28	62.4	44 18	52 13	61.6	43 49	51 00	60.7	43 20	49 48	59.9	42 49	48 38	59.1	231	309
52	128	46 03	54 08	62.5	45 35	52 52	61.6	45 06	51 37	60.7	44 36	50 23	59.8	44 05	49 11	59.0	43 34	48 00	58.2	232	308
53	127	46 51	53 30	61.6	46 22	52 14	60.7	45 52	50 59	59.8	45 22	49 45	58.9	44 51	48 32	58.1	44 18	47 21	57.2	233	307
54	126	47 39	52 51	60.8	47 09	51 34	59.8	46 39	50 19	58.9	46 07	49 05	58.0	45 35	47 52	57.1	45 02	46 41	56.3	234	306
55	125	48 27	52 11	59.8	47 56	50 53	58.9	47 25	49 37	58.0	46 53	48 23	57.0	46 19	47 10	56.2	45 46	45 59	55.3	235	305
56	124	49 14	51 28	58.9	48 43	50 11	57.9	48 10	48 54	57.0	47 37	47 40	56.1	47 03	46 27	55.2	46 29	45 15	54.3	236	304
57	123	50 01	50 44	57.9	49 28	49 26	56.9	48 55	48 09	56.0	48 21	46 54	55.0	47 46	45 41	54.1	47 11	44 30	53.3	237	303
58	122	50 47	49 58	56.9	50 14	48 39	55.9	49 40	47 22	54.9	49 05	46 07	54.0	48 29	44 54	53.1	47 53	43 43	52.2	238	302
59	121	51 33	49 09	55.9	50 58	47 51	54.9	50 23	46 34	53.9	49 48	45 18	52.9	49 11	44 05	52.0	48 34	42 54	51.1	239	301
60	120	52 18	48 19	54.8	51 43	47 00	53.8	51 07	45 43	52.8	50 30	44 28	51.8	49 53	43 14	50.9	49 14	42 03	50.0	240	300
61	119	53 02	47 26	53.7	52 26	46 07	52.7	51 49	44 50	51.7	51 12	43 35	50.7	50 33	42 22	49.7	49 54	41 10	48.8	241	299
62	118	53 46	46 31	52.6	53 09	45 12	51.5	52 31	43 54	50.5	51 53	42 39	49.5	51 13	41 27	48.6	50 33	40 16	47.6	242	298
63	117	54 29	45 33	51.4	53 51	44 14	50.3	53 13	42 57	49.3	52 33	41 42	48.3	51 53	40 30	47.3	51 12	39 19	46.4	243	297
64	116	55 12	44 33	50.2	54 33	43 14	49.1	53 53	41 57	48.1	53 13	40 42	47.1	52 31	39 30	46.1	51 49	38 20	45.2	244	296
65	115	55 53	43 30	48.9	55 13	42 11	47.8	54 33	40 55	46.8	53 51	39 40	45.8	53 09	38 29	44.8	52 26	37 19	43.9	245	295
66	114	56 34	42 25	47.6	55 53	41 06	46.5	55 12	39 50	45.4	54 29	38 36	44.4	53 46	37 25	43.5	53 02	36 16	42.6	246	294
67	113	57 14	41 16	46.2	56 32	39 58	45.1	55 50	38 42	44.1	55 06	37 29	43.1	54 22	36 19	42.1	53 37	35 11	41.2	247	293
68	112	57 53	40 05	44.8	57 10	38 47	43.7	56 27	37 32	42.7	55 42	36 19	41.7	54 57	35 10	40.7	54 11	34 03	39.8	248	292
69	111	58 32	38 50	43.3	57 47	37 33	42.2	57 03	36 18	41.2	56 17	35 07	40.2	55 31	33 59	39.3	54 44	32 53	38.4	249	291
70	110	59 09	37 32	41.8	58 23	36 16	40.7	57 38	35 02	39.7	56 51	33 52	38.7	56 04	32 45	37.8	55 16	31 41	36.9	250	290
71	109	59 45	36 11	40.2	58 58	34 55	39.2	58 12	33 43	38.1	57 24	32 35	37.2	56 36	31 29	36.3	55 47	30 26	35.4	251	289
72	108	60 19	34 46	38.6	59 32	33 32	37.6	58 44	32 21	36.5	57 56	31 14	35.6	57 07	30 10	34.7	56 17	29 08	33.8	252	288
73	107	60 53	33 18	36.9	60 05	32 05	35.9	59 16	30 56	34.9	58 26	29 51	34.0	57 36	28 48	33.1	56 46	27 49	32.2	253	287
74	106	61 25	31 46	35.2	60 36	30 35	34.2	59 46	29 28	33.2	58 55	28 25	32.3	58 05	27 24	31.4	57 13	26 26	30.6	254	286
75	105	61 56	30 10	33.4	61 06	29 02	32.4	60 15	27 57	31.4	59 23	26 56	30.5	58 31	25 57	29.7	57 39	25 02	28.9	255	285
76	104	62 26	28 31	31.5	61 34	27 25	30.5	60 42	26 23	29.6	59 50	25 24	28.8	58 57	24 28	28.0	58 04	23 35	27.2	256	284
77	103	62 53	26 48	29.6	62 01	25 45	28.6	61 08	24 46	27.8	60 15	23 49	27.0	59 21	22 56	26.2	58 27	22 05	25.5	257	283
78	102	63 20	25 02	27.6	62 26	24 02	26.7	61 32	23 05	25.9	60 38	22 12	25.1	59 44	21 21	24.4	58 49	20 34	23.7	258	282
79	101	63 44	23 12	25.5	62 50	22 15	24.7	61 55	21 22	23.9	61 00	20 32	23.2	60 05	19 44	22.5	59 09	19 00	21.8	259	281
80	100	64 07	21 18	23.4	63 12	20 25	22.6	62 16	19 36	21.9	61 20	18 49	21.2	60 24	18 05	20.6	59 28	17 24	20.0	260	280
81	99	64 28	19 22	21.3	63 32	18 33	20.5	62 35	17 47	19.9	61 39	17 04	19.2	60 42	16 24	18.6	59 45	15 46	18.1	261	279
82	98	64 47	17 22	19.1	63 50	16 37	18.4	62 53	15 56	17.8	61 56	15 17	17.2	60 58	14 40	16.7	60 01	14 06	16.2	262	278
83	97	65 03	15 18	16.8	64 06	14 39	16.2	63 08	14 02	15.6	62 10	13 27	15.1	61 12	12 55	14.7	60 14	12 24	14.2	263	277
84	96	65 18	13 13	14.5	64 20	12 38	14.0	63 22	12 06	13.5	62 23	11 36	13.0	61 25	11 07	12.6	60 26	10 41	12.2	264	276
85	95	65 31	11 05	12.1	64 32	10 35	11.7	63 33	10 08	11.3	62 35	9 42	10.9	61 36	9 19	10.6	60 37	8 56	10.2	265	275
86	94	65 41	8 54	9.8	64 42	8 30	9.4	63 43	8 08	9.1	62 44	7 48	8.8	61 44	7 28	8.5	60 45	7 10	8.2	266	274
87	93	65 49	6 42	7.3	64 50	6 24	7.1	63 50	6 07	6.8	62 51	5 52	6.6	61 51	5 37	6.4	60 52	5 24	6.2	267	273
88	92	65 55	4 29	4.9	64 56	4 17	4.7	63 56	4 06	4.6	62 56	3 55	4.4	61 56	3 45	4.3	60 56	3 36	4.1	268	272
89	91	65 59	2 15	2.5	64 59	2 09	2.4	63 59	2 03	2.3	62 59	1 58	2.2	61 59	1 53	2.1	60 59	1 48	2.1	269	271
90	90	66 00	0 00	0.0	65 00	0 00	0.0	64 00	0 00	0.0	63 00	0 00	0.0	62 00	0 00	0.0	61 00	0 00	0.0	270	270

N. Lat.: for LHA > 180°... $Z_n = Z$
for LHA < 180°... $Z_n = 360° - Z$

S. Lat.: for LHA > 180°... $Z_n = 180° - Z$
for LHA < 180°... $Z_n = 180° + Z$

SIGHT REDUCTION TABLE

B: (–) for 90°< LHA < 270°
Dec: (–) for Lat. contrary name

Z_1 : same sign as B
Z_2 : (–) for F > 90°

Lat. / A		30°			31°			32°			33°			34°			35°			Lat. / A	
LHA / F		A / H	B / P	Z_1 / Z_2	A / H	B / P	Z_1 / Z_2	A / H	B / P	Z_1 / Z_2	A / H	B / P	Z_1 / Z_2	A / H	B / P	Z_1 / Z_2	A / H	B / P	Z_1 / Z_2	LHA	
°	°	° ′	° ′	°	° ′	° ′	°	° ′	° ′	°	° ′	° ′	°	° ′	° ′	°	° ′	° ′	°	°	°
0	180	0 00	60 00	90.0	0 00	59 00	90.0	0 00	58 00	90.0	0 00	57 00	90.0	0 00	56 00	90.0	0 00	55 00	90.0	180	360
1	179	0 52	60 00	89.5	0 51	59 00	89.5	0 51	58 00	89.5	0 50	57 00	89.5	0 50	56 00	89.4	0 49	55 00	89.4	181	359
2	178	1 44	59 59	89.0	1 43	58 59	89.0	1 42	57 59	88.9	1 41	56 59	88.9	1 39	55 59	88.9	1 38	54 59	88.9	182	358
3	177	2 36	59 58	88.5	2 34	58 58	88.5	2 33	57 58	88.4	2 31	56 58	88.4	2 29	55 58	88.3	2 27	54 58	88.3	183	357
4	176	3 28	59 56	88.0	3 26	58 56	87.9	3 23	57 56	87.9	3 21	56 56	87.8	3 19	55 56	87.8	3 17	54 56	87.7	184	356
5	175	4 20	59 54	87.5	4 17	58 54	87.4	4 14	57 54	87.3	4 12	56 54	87.3	4 09	55 54	87.2	4 06	54 54	87.1	185	355
6	174	5 12	59 52	87.0	5 08	58 52	86.9	5 05	57 52	86.8	5 02	56 51	86.7	4 58	55 51	86.6	4 55	54 51	86.6	186	354
7	173	6 04	59 49	86.5	6 00	58 49	86.4	5 56	57 48	86.3	5 52	56 48	86.2	5 48	55 48	86.1	5 44	54 48	86.0	187	353
8	172	6 55	59 45	86.0	6 51	58 45	85.9	6 47	57 45	85.7	6 42	56 45	85.6	6 38	55 44	85.5	6 33	54 44	85.4	188	352
9	171	7 47	59 42	85.5	7 42	58 41	85.3	7 37	57 41	85.2	7 32	56 40	85.1	7 27	55 40	84.9	7 22	54 40	84.8	189	351
10	170	8 39	59 37	85.0	8 34	58 37	84.8	8 28	57 36	84.7	8 22	56 36	84.5	8 17	55 36	84.4	8 11	54 35	84.2	190	350
11	169	9 31	59 32	84.4	9 25	58 32	84.3	9 19	57 31	84.1	9 13	56 31	84.0	9 06	55 30	83.8	9 00	54 30	83.6	191	349
12	168	10 22	59 27	83.9	10 16	58 26	83.8	10 09	57 26	83.6	10 03	56 25	83.4	9 56	55 25	83.2	9 48	54 24	83.0	192	348
13	167	11 14	59 21	83.4	11 07	58 20	83.2	11 00	57 20	83.0	10 52	56 19	82.8	10 45	55 18	82.6	10 37	54 18	82.5	193	347
14	166	12 06	59 15	82.9	11 58	58 14	82.7	11 50	57 13	82.5	11 42	56 12	82.3	11 34	55 12	82.1	11 26	54 11	81.9	194	346
15	165	12 57	59 08	82.4	12 49	58 07	82.1	12 41	57 06	81.9	12 32	56 05	81.7	12 23	55 04	81.5	12 14	54 04	81.3	195	345
16	164	13 49	59 01	81.8	13 40	57 59	81.6	13 31	56 58	81.4	13 22	55 57	81.1	13 13	54 57	80.9	13 03	53 56	80.7	196	344
17	163	14 40	58 53	81.3	14 31	57 51	81.1	14 21	56 50	80.8	14 12	55 49	80.5	14 02	54 48	80.3	13 51	53 47	80.1	197	343
18	162	15 31	58 44	80.8	15 22	57 43	80.5	15 12	56 42	80.2	15 01	55 40	80.0	14 51	54 39	79.7	14 40	53 38	79.4	198	342
19	161	16 23	58 35	80.2	16 12	57 34	79.9	16 02	56 32	79.7	15 51	55 31	79.4	15 40	54 30	79.1	15 28	53 29	78.8	199	341
20	160	17 14	58 26	79.7	17 03	57 24	79.4	16 52	56 23	79.1	16 40	55 21	78.8	16 28	54 20	78.5	16 16	53 19	78.2	200	340
21	159	18 05	58 16	79.1	17 53	57 14	78.8	17 42	56 12	78.5	17 29	55 11	78.2	17 17	54 09	77.9	17 04	53 08	77.6	201	339
22	158	18 56	58 05	78.6	18 44	57 03	78.2	18 31	56 01	77.9	18 19	55 00	77.6	18 06	53 58	77.3	17 52	52 56	77.0	202	338
23	157	19 47	57 54	78.0	19 34	56 52	77.7	19 21	55 50	77.3	19 08	54 48	77.0	18 54	53 46	76.6	18 40	52 44	76.3	203	337
24	156	20 37	57 42	77.4	20 24	56 40	77.1	20 11	55 38	76.7	19 57	54 36	76.4	19 42	53 34	76.0	19 28	52 32	75.7	204	336
25	155	21 28	57 30	76.9	21 14	56 27	76.5	21 00	55 25	76.1	20 46	54 23	75.7	20 31	53 21	75.4	20 15	52 19	75.0	205	335
26	154	22 19	57 17	76.3	22 04	56 14	75.9	21 49	55 12	75.5	21 34	54 09	75.1	21 19	53 07	74.7	21 03	52 05	74.4	206	334
27	153	23 09	57 03	75.7	22 54	56 00	75.3	22 39	54 57	74.9	22 23	53 55	74.5	22 07	52 52	74.1	21 50	51 50	73.7	207	333
28	152	23 59	56 49	75.1	23 44	55 46	74.7	23 28	54 43	74.3	23 11	53 40	73.8	22 54	52 37	73.4	22 37	51 35	73.0	208	332
29	151	24 50	56 34	74.5	24 33	55 31	74.1	24 17	54 27	73.6	23 59	53 24	73.2	23 42	52 22	72.8	23 24	51 19	72.4	209	331
30	150	25 40	56 19	73.9	25 23	55 15	73.4	25 05	54 11	73.0	24 48	53 08	72.5	24 29	52 05	72.1	24 11	51 03	71.7	210	330
31	149	26 29	56 02	73.3	26 12	54 58	72.8	25 54	53 54	72.3	25 35	52 51	71.9	25 17	51 48	71.4	24 57	50 45	71.0	211	329
32	148	27 19	55 45	72.6	27 01	54 41	72.2	26 42	53 37	71.7	26 23	52 33	71.2	26 04	51 30	70.7	25 44	50 27	70.3	212	328
33	147	28 09	55 27	72.0	27 50	54 23	71.5	27 31	53 19	71.0	27 11	52 15	70.5	26 50	51 12	70.0	26 30	50 08	69.6	213	327
34	146	28 58	55 09	71.4	28 38	54 04	70.8	28 19	53 00	70.3	27 58	51 56	69.8	27 37	50 52	69.3	27 16	49 49	68.8	214	326
35	145	29 47	54 49	70.7	29 27	53 44	70.2	29 06	52 40	69.6	28 45	51 36	69.1	28 24	50 32	68.6	28 01	49 29	68.1	215	325
36	144	30 36	54 29	70.0	30 15	53 24	69.5	29 54	52 19	68.9	29 32	51 15	68.4	29 10	50 11	67.9	28 47	49 07	67.4	216	324
37	143	31 25	54 08	69.4	31 03	53 03	68.8	30 41	51 58	68.2	30 19	50 53	67.7	29 56	49 49	67.2	29 32	48 45	66.6	217	323
38	142	32 13	53 46	68.7	31 51	52 40	68.1	31 28	51 35	67.5	31 05	50 30	66.9	30 41	49 26	66.4	30 17	48 23	65.9	218	322
39	141	33 02	53 23	68.0	32 39	52 17	67.4	32 15	51 12	66.8	31 51	50 07	66.2	31 27	49 03	65.6	31 02	47 59	65.1	219	321
40	140	33 50	53 00	67.2	33 26	51 53	66.6	33 02	50 48	66.0	32 37	49 43	65.4	32 12	48 38	64.9	31 46	47 34	64.3	220	320
41	139	34 37	52 35	66.5	34 13	51 29	65.9	33 48	50 23	65.3	33 23	49 17	64.7	32 57	48 13	64.1	32 30	47 09	63.5	221	319
42	138	35 25	52 09	65.8	35 00	51 03	65.1	34 34	49 56	64.5	34 08	48 51	63.9	33 42	47 46	63.3	33 14	46 42	62.7	222	318
43	137	36 12	51 43	65.0	35 46	50 36	64.3	35 20	49 29	63.7	34 53	48 24	63.1	34 26	47 19	62.5	33 58	46 15	61.9	223	317
44	136	36 59	51 15	64.2	36 33	50 08	63.6	36 06	49 01	62.9	35 38	47 55	62.3	35 10	46 51	61.6	34 41	45 46	61.0	224	316
45	135	37 46	50 46	63.4	37 19	49 39	62.7	36 51	48 32	62.1	36 22	47 26	61.4	35 53	46 21	60.8	35 24	45 17	60.2	225	315

Lat. / A		30°			31°			32°			33°			34°			35°			Lat. / A	
LHA / F		A / H	B / P	Z_1 / Z_2	A / H	B / P	Z_1 / Z_2	A / H	B / P	Z_1 / Z_2	A / H	B / P	Z_1 / Z_2	A / H	B / P	Z_1 / Z_2	A / H	B / P	Z_1 / Z_2	LHA	
°	°	° ′	° ′	°	° ′	° ′	°	° ′	° ′	°	° ′	° ′	°	° ′	° ′	°	° ′	° ′	°	°	°
45	135	37 46	50 46	63.4	37 19	49 39	62.7	36 51	48 32	62.1	36 22	47 26	61.4	35 53	46 21	60.8	35 24	45 17	60.2	225	315
46	134	38 32	50 16	62.6	38 04	49 08	61.9	37 36	48 02	61.2	37 06	46 56	60.6	36 37	45 51	59.9	36 06	44 46	59.3	226	314
47	133	39 18	49 45	61.8	38 49	48 37	61.1	38 20	47 30	60.4	37 50	46 24	59.7	37 19	45 19	59.1	36 48	44 15	58.4	227	313
48	132	40 04	49 13	61.0	39 34	48 05	60.2	39 04	46 58	59.5	38 33	45 51	58.8	38 02	44 46	58.2	37 30	43 42	57.5	228	312
49	131	40 49	48 39	60.1	40 19	47 31	59.4	39 48	46 24	58.6	39 16	45 18	57.9	38 44	44 12	57.2	38 11	43 08	56.6	229	311
50	130	41 34	48 04	59.2	41 03	46 56	58.5	40 31	45 49	57.7	39 59	44 42	57.0	39 26	43 37	56.3	38 52	42 33	55.6	230	310
51	129	42 18	47 28	58.3	41 46	46 20	57.5	41 14	45 12	56.8	40 41	44 06	56.1	40 07	43 01	55.4	39 32	41 57	54.7	231	309
52	128	43 02	46 50	57.4	42 29	45 42	56.6	41 56	44 34	55.9	41 22	43 28	55.1	40 47	42 23	54.4	40 12	41 19	53.7	232	308
53	127	43 46	46 11	56.4	43 12	45 03	55.6	42 38	43 55	54.9	42 03	42 49	54.1	41 28	41 44	53.4	40 52	40 41	52.7	233	307
54	126	44 29	45 31	55.5	43 54	44 22	54.7	43 19	43 15	53.9	42 44	42 09	53.1	42 07	41 04	52.4	41 30	40 01	51.7	234	306
55	125	45 11	44 49	54.5	44 36	43 40	53.7	44 00	42 33	52.9	43 24	41 27	52.1	42 46	40 23	51.4	42 09	39 19	50.7	235	305
56	124	45 53	44 05	53.5	45 17	42 57	52.6	44 40	41 50	51.8	44 03	40 44	51.1	43 25	39 40	50.3	42 46	38 37	49.6	236	304
57	123	46 35	43 20	52.4	45 58	42 11	51.6	45 20	41 05	50.8	44 42	39 59	50.0	44 03	38 55	49.3	43 24	37 53	48.5	237	303
58	122	47 16	42 33	51.3	46 38	41 25	50.5	45 59	40 18	49.7	45 20	39 13	48.9	44 40	38 09	48.2	44 00	37 07	47.5	238	302
59	121	47 56	41 44	50.2	47 17	40 36	49.4	46 38	39 30	48.6	45 58	38 25	47.8	45 17	37 22	47.1	44 36	36 20	46.3	239	301
60	120	48 35	40 54	49.1	47 56	39 46	48.3	47 16	38 40	47.5	46 35	37 36	46.7	45 53	36 33	45.9	45 11	35 32	45.2	240	300
61	119	49 14	40 01	47.9	48 34	38 54	47.1	47 53	37 48	46.3	47 11	36 45	45.5	46 29	35 42	44.7	45 46	34 42	44.0	241	299
62	118	49 53	39 07	46.8	49 11	38 00	45.9	48 29	36 55	45.1	47 46	35 52	44.3	47 03	34 50	43.6	46 19	33 50	42.8	242	298
63	117	50 30	38 11	45.5	49 48	37 04	44.7	49 05	36 00	43.9	48 21	34 57	43.1	47 37	33 57	42.3	46 53	32 57	41.6	243	297
64	116	51 07	37 13	44.3	50 23	36 07	43.4	49 40	35 03	42.6	48 55	34 01	41.8	48 10	33 01	41.1	47 25	32 03	40.4	244	296
65	115	51 43	36 12	43.0	50 58	35 07	42.2	50 14	34 04	41.3	49 28	33 03	40.6	48 43	32 04	39.8	47 56	31 07	39.1	245	295
66	114	52 18	35 10	41.7	51 33	34 06	40.8	50 47	33 04	40.0	50 01	32 04	39.3	49 14	31 05	38.5	48 27	30 09	37.8	246	294
67	113	52 52	34 05	40.3	52 06	33 02	39.5	51 19	32 01	38.7	50 32	31 02	37.9	49 44	30 05	37.2	48 56	29 10	36.5	247	293
68	112	53 25	32 59	38.9	52 38	31 56	38.1	51 50	30 57	37.3	51 02	29 59	36.6	50 14	29 03	35.8	49 25	28 09	35.2	248	292
69	111	53 57	31 50	37.5	53 09	30 49	36.7	52 21	29 50	35.9	51 32	28 53	35.2	50 43	27 59	34.5	49 53	27 06	33.8	249	291
70	110	54 28	30 39	36.1	53 39	29 39	35.2	52 50	28 42	34.5	52 00	27 46	33.8	51 10	26 53	33.1	50 20	26 02	32.4	250	290
71	109	54 58	29 25	34.6	54 08	28 27	33.8	53 18	27 31	33.0	52 28	26 38	32.3	51 37	25 46	31.6	50 46	24 56	31.0	251	289
72	108	55 27	28 09	33.0	54 37	27 13	32.2	53 46	26 19	31.5	52 54	25 27	30.8	52 03	24 37	30.2	51 10	23 49	29.5	252	288
73	107	55 55	26 51	31.4	55 03	25 57	30.7	54 12	25 04	30.0	53 19	24 14	29.3	52 27	23 26	28.7	51 34	22 40	28.1	253	287
74	106	56 21	25 31	29.8	55 29	24 39	29.1	54 36	23 48	28.4	53 43	23 00	27.8	52 50	22 14	27.1	51 57	21 29	26.6	254	286
75	105	56 46	24 09	28.2	55 53	23 18	27.5	55 00	22 30	26.8	54 06	21 44	26.2	53 12	21 00	25.6	52 18	20 17	25.0	255	285
76	104	57 10	22 44	26.5	56 16	21 56	25.8	55 22	21 10	25.2	54 28	20 26	24.6	53 33	19 44	24.0	52 38	19 04	23.5	256	284
77	103	57 33	21 17	24.8	56 38	20 31	24.1	55 43	19 48	23.5	54 48	19 06	23.0	53 53	18 27	22.4	52 57	17 49	21.9	257	283
78	102	57 54	19 48	23.0	56 59	19 05	22.4	56 03	18 24	21.9	55 07	17 45	21.3	54 11	17 08	20.8	53 15	16 32	20.3	258	282
79	101	58 13	18 17	21.2	57 17	17 37	20.7	56 21	16 59	20.1	55 25	16 22	19.6	54 28	15 48	19.2	53 31	15 15	18.7	259	281
80	100	58 32	16 44	19.4	57 35	16 07	18.9	56 38	15 32	18.4	55 41	14 58	17.9	54 44	14 26	17.5	53 47	13 56	17.1	260	280
81	99	58 48	15 10	17.6	57 51	14 36	17.1	56 53	14 03	16.6	55 56	13 33	16.2	54 58	13 03	15.8	54 00	12 36	15.4	261	279
82	98	59 03	13 33	15.7	58 05	13 02	15.3	57 07	12 33	14.9	56 09	12 06	14.5	55 11	11 40	14.1	54 13	11 14	13.8	262	278
83	97	59 16	11 55	13.8	58 18	11 28	13.4	57 19	11 02	13.0	56 21	10 38	12.7	55 22	10 14	12.4	54 24	9 52	12.1	263	277
84	96	59 28	10 16	11.9	58 29	9 52	11.5	57 30	9 30	11.2	56 31	9 09	10.9	55 32	8 49	10.6	54 33	8 29	10.4	264	276
85	95	59 37	8 35	9.9	58 38	8 15	9.6	57 39	7 56	9.4	56 40	7 39	9.1	55 41	7 22	8.9	54 41	7 06	8.7	265	275
86	94	59 46	6 53	8.0	58 46	6 37	7.7	57 47	6 22	7.5	56 47	6 08	7.3	55 48	5 54	7.1	54 48	5 41	7.0	266	274
87	93	59 52	5 11	6.0	58 52	4 59	5.8	57 52	4 47	5.6	56 53	4 36	5.5	55 53	4 26	5.4	54 53	4 16	5.2	267	273
88	92	59 56	3 28	4.0	58 57	3 19	3.9	57 57	3 12	3.8	56 57	3 05	3.7	55 57	2 58	3.6	54 57	2 51	3.5	268	272
89	91	59 59	1 44	2.0	58 59	1 40	1.9	57 59	1 36	1.9	56 59	1 32	1.8	55 59	1 29	1.8	54 59	1 26	1.7	269	271
90	90	60 00	0 00	0.0	59 00	0 00	0.0	58 00	0 00	0.0	57 00	0 00	0.0	56 00	0 00	0.0	55 00	0 00	0.0	270	270

N. Lat.: for LHA > 180°... $Z_n = Z$
for LHA < 180°... $Z_n = 360° - Z$

S. Lat.: for LHA > 180°... $Z_n = 180° - Z$
for LHA < 180°... $Z_n = 180° + Z$

B: (–) for 90° < LHA < 270°
Dec: (–) for Lat. contrary name

SIGHT REDUCTION TABLE

Z_1 : same sign as B
Z_2 : (–) for F > 90°

Lat. / A		36°			37°			38°			39°			40°			41°			Lat. / A	
LHA / F		A / H	B / P	Z_1 / Z_2	A / H	B / P	Z_1 / Z_2	A / H	B / P	Z_1 / Z_2	A / H	B / P	Z_1 / Z_2	A / H	B / P	Z_1 / Z_2	A / H	B / P	Z_1 / Z_2	LHA	
°	°	° ′	° ′	°	° ′	° ′	°	° ′	° ′	°	° ′	° ′	°	° ′	° ′	°	° ′	° ′	°	°	°
0	180	0 00	54 00	90.0	0 00	53 00	90.0	0 00	52 00	90.0	0 00	51 00	90.0	0 00	50 00	90.0	0 00	49 00	90.0	180	360
1	179	0 49	54 00	89.4	0 48	53 00	89.4	0 47	52 00	89.4	0 47	51 00	89.4	0 46	50 00	89.4	0 45	49 00	89.3	181	359
2	178	1 37	53 59	88.8	1 36	52 59	88.8	1 35	51 59	88.8	1 33	50 59	88.7	1 32	49 59	88.7	1 31	48 59	88.7	182	358
3	177	2 26	53 58	88.2	2 24	52 58	88.2	2 22	51 58	88.2	2 20	50 58	88.1	2 18	49 58	88.1	2 16	48 58	88.0	183	357
4	176	3 14	53 56	87.6	3 12	52 56	87.6	3 09	51 56	87.5	3 06	50 56	87.5	3 04	49 56	87.4	3 01	48 56	87.4	184	356
5	175	4 03	53 54	87.1	3 59	52 54	87.0	3 56	51 54	86.9	3 53	50 54	86.8	3 50	49 54	86.8	3 46	48 54	86.7	185	355
6	174	4 51	53 51	86.5	4 47	52 51	86.4	4 43	51 51	86.3	4 40	50 51	86.2	4 36	49 51	86.1	4 31	48 51	86.1	186	354
7	173	5 39	53 48	85.9	5 35	52 48	85.8	5 31	51 48	85.7	5 26	50 47	85.6	5 21	49 47	85.5	5 17	48 47	85.4	187	353
8	172	6 28	53 44	85.3	6 23	52 44	85.2	6 18	51 44	85.1	6 13	50 44	84.9	6 07	49 43	84.8	6 02	48 43	84.7	188	352
9	171	7 16	53 40	84.7	7 11	52 39	84.6	7 05	51 39	84.4	6 59	50 39	84.3	6 53	49 39	84.2	6 47	48 39	84.1	189	351
10	170	8 05	53 35	84.1	7 58	52 35	83.9	7 52	51 34	83.8	7 45	50 34	83.7	7 39	49 34	83.5	7 32	48 34	83.4	190	350
11	169	8 53	53 30	83.5	8 46	52 29	83.3	8 39	51 29	83.2	8 32	50 29	83.0	8 24	49 29	82.9	8 17	48 28	82.7	191	349
12	168	9 41	53 24	82.9	9 33	52 23	82.7	9 26	51 23	82.5	9 18	50 23	82.4	9 10	49 23	82.2	9 02	48 22	82.1	192	348
13	167	10 29	53 17	82.3	10 21	52 17	82.1	10 13	51 17	81.9	10 04	50 16	81.7	9 55	49 16	81.6	9 46	48 16	81.4	193	347
14	166	11 17	53 10	81.7	11 08	52 10	81.5	10 59	51 10	81.3	10 50	50 09	81.1	10 41	49 09	80.9	10 31	48 09	80.7	194	346
15	165	12 05	53 03	81.0	11 56	52 02	80.8	11 46	51 02	80.6	11 36	50 02	80.4	11 26	49 01	80.2	11 16	48 01	80.0	195	345
16	164	12 53	52 55	80.4	12 43	51 54	80.2	12 33	50 54	80.0	12 22	49 53	79.8	12 11	48 53	79.6	12 00	47 53	79.3	196	344
17	163	13 41	52 46	79.8	13 30	51 46	79.6	13 19	50 45	79.3	13 08	49 45	79.1	12 57	48 44	78.9	12 45	47 44	78.7	197	343
18	162	14 29	52 37	79.2	14 17	51 37	78.9	14 06	50 36	78.7	13 54	49 35	78.4	13 42	48 35	78.2	13 29	47 34	78.0	198	342
19	161	15 16	52 28	78.6	15 04	51 27	78.3	14 52	50 26	78.0	14 39	49 25	77.8	14 27	48 25	77.5	14 13	47 24	77.3	199	341
20	160	16 04	52 17	77.9	15 51	51 16	77.6	15 38	50 16	77.4	15 25	49 15	77.1	15 11	48 14	76.8	14 58	47 14	76.6	200	340
21	159	16 51	52 07	77.3	16 38	51 05	77.0	16 24	50 05	76.7	16 10	49 04	76.4	15 56	48 03	76.1	15 42	47 03	75.9	201	339
22	158	17 39	51 55	76.6	17 24	50 54	76.3	17 10	49 53	76.0	16 56	48 52	75.7	16 41	47 51	75.4	16 25	46 51	75.2	202	338
23	157	18 26	51 43	76.0	18 11	50 42	75.7	17 56	49 41	75.4	17 41	48 40	75.0	17 25	47 39	74.7	17 09	46 38	74.4	203	337
24	156	19 13	51 30	75.3	18 57	50 29	75.0	18 42	49 28	74.7	18 26	48 27	74.3	18 09	47 26	74.0	17 53	46 25	73.7	204	336
25	155	20 00	51 17	74.7	19 44	50 15	74.3	19 27	49 14	74.0	19 10	48 13	73.6	18 53	47 12	73.3	18 36	46 12	73.0	205	335
26	154	20 46	51 03	74.0	20 30	50 01	73.6	20 13	49 00	73.3	19 55	47 59	72.9	19 37	46 58	72.6	19 19	45 57	72.3	206	334
27	153	21 33	50 48	73.3	21 15	49 47	73.0	20 58	48 45	72.6	20 40	47 44	72.2	20 21	46 43	71.9	20 02	45 42	71.5	207	333
28	152	22 19	50 33	72.6	22 01	49 31	72.3	21 43	48 30	71.9	21 24	47 28	71.5	21 05	46 28	71.1	20 45	45 27	70.8	208	332
29	151	23 06	50 17	72.0	22 47	49 15	71.6	22 28	48 14	71.2	22 08	47 12	70.8	21 48	46 11	70.4	21 28	45 11	70.0	209	331
30	150	23 52	50 00	71.3	23 32	48 58	70.8	23 12	47 57	70.4	22 52	46 55	70.0	22 31	45 54	69.6	22 10	44 54	69.3	210	330
31	149	24 37	49 43	70.5	24 17	48 41	70.1	23 57	47 39	69.7	23 36	46 38	69.3	23 14	45 37	68.9	22 52	44 36	68.5	211	329
32	148	25 23	49 25	69.8	25 02	48 23	69.4	24 41	47 21	69.0	24 19	46 19	68.5	23 57	45 18	68.1	23 34	44 17	67.7	212	328
33	147	26 09	49 06	69.1	25 47	48 04	68.7	25 25	47 02	68.2	25 02	46 00	67.8	24 40	44 59	67.3	24 16	43 58	66.9	213	327
34	146	26 54	48 46	68.4	26 32	47 44	67.9	26 09	46 42	67.4	25 45	45 40	67.0	25 22	44 39	66.6	24 58	43 39	66.1	214	326
35	145	27 39	48 26	67.6	27 16	47 23	67.1	26 52	46 21	66.7	26 28	45 20	66.2	26 04	44 19	65.8	25 39	43 18	65.3	215	325
36	144	28 24	48 04	66.9	28 00	47 02	66.4	27 36	46 00	65.9	27 11	44 58	65.4	26 46	43 57	65.0	26 20	42 57	64.5	216	324
37	143	29 08	47 42	66.1	28 44	46 40	65.6	28 19	45 38	65.1	27 53	44 36	64.6	27 27	43 35	64.2	27 01	42 34	63.7	217	323
38	142	29 52	47 19	65.3	29 27	46 17	64.8	29 01	45 15	64.3	28 35	44 13	63.8	28 08	43 12	63.3	27 41	42 12	62.9	218	322
39	141	30 36	46 56	64.5	30 10	45 53	64.0	29 44	44 51	63.5	29 17	43 49	63.0	28 49	42 48	62.5	28 21	41 48	62.0	219	321
40	140	31 20	46 31	63.7	30 53	45 28	63.2	30 26	44 26	62.7	29 58	43 25	62.2	29 30	42 24	61.7	29 01	41 23	61.2	220	320
41	139	32 03	46 05	62.9	31 36	45 03	62.4	31 08	44 01	61.8	30 39	42 59	61.3	30 10	41 58	60.8	29 41	40 58	60.3	221	319
42	138	32 46	45 39	62.1	32 18	44 36	61.5	31 49	43 34	61.0	31 20	42 33	60.5	30 50	41 32	59.9	30 20	40 32	59.4	222	318
43	137	33 29	45 11	61.3	33 00	44 09	60.7	32 30	43 07	60.1	32 00	42 05	59.6	31 30	41 05	59.1	30 59	40 04	58.5	223	317
44	136	34 12	44 43	60.4	33 42	43 40	59.8	33 11	42 38	59.3	32 40	41 37	58.7	32 09	40 36	58.2	31 37	39 36	57.6	224	316
45	135	34 54	44 13	59.6	34 23	43 11	59.0	33 52	42 09	58.4	33 20	41 08	57.8	32 48	40 07	57.3	32 15	39 08	56.7	225	315

Lat. / A		36°			37°			38°			39°			40°			41°			Lat. / A	
LHA / F		A / H	B / P	Z_1 / Z_2	A / H	B / P	Z_1 / Z_2	A / H	B / P	Z_1 / Z_2	A / H	B / P	Z_1 / Z_2	A / H	B / P	Z_1 / Z_2	A / H	B / P	Z_1 / Z_2	LHA	
°	°	° ′	° ′	°	° ′	° ′	°	° ′	° ′	°	° ′	° ′	°	° ′	° ′	°	° ′	° ′	°	°	°
45	135	34 54	44 13	59.6	34 23	43 11	59.0	33 52	42 09	58.4	33 20	41 08	57.8	32 48	40 07	57.3	32 15	39 08	56.7	225	315
46	134	35 35	43 43	58.7	35 04	42 40	58.1	34 32	41 38	57.5	33 59	40 37	56.9	33 26	39 37	56.4	32 53	38 38	55.8	226	314
47	133	36 17	43 11	57.8	35 44	42 09	57.2	35 12	41 07	56.6	34 38	40 06	56.0	34 04	39 06	55.4	33 30	38 07	54.9	227	313
48	132	36 57	42 39	56.9	36 24	41 36	56.2	35 51	40 35	55.6	35 17	39 34	55.0	34 42	38 34	54.5	34 07	37 35	53.9	228	312
49	131	37 38	42 05	55.9	37 04	41 03	55.3	36 30	40 01	54.7	35 55	39 01	54.1	35 19	38 01	53.5	34 43	37 03	53.0	229	311
50	130	38 18	41 30	55.0	37 43	40 28	54.4	37 08	39 27	53.7	36 32	38 27	53.1	35 56	37 27	52.5	35 19	36 29	52.0	230	310
51	129	38 57	40 54	54.0	38 22	39 52	53.4	37 46	38 51	52.8	37 09	37 51	52.1	36 32	36 52	51.6	35 55	35 54	51.0	231	309
52	128	39 36	40 17	53.0	39 00	39 15	52.4	38 23	38 14	51.8	37 46	37 15	51.1	37 08	36 16	50.6	36 30	35 18	50.0	232	308
53	127	40 15	39 38	52.0	39 38	38 37	51.4	39 00	37 36	50.8	38 22	36 37	50.1	37 43	35 39	49.5	37 04	34 42	49.0	233	307
54	126	40 53	38 58	51.0	40 15	37 57	50.4	39 36	36 57	49.7	38 57	35 58	49.1	38 18	35 01	48.5	37 38	34 04	47.9	234	306
55	125	41 30	38 17	50.0	40 52	37 17	49.3	40 12	36 17	48.7	39 32	35 19	48.1	38 52	34 21	47.4	38 11	33 25	46.9	235	305
56	124	42 07	37 35	48.9	41 28	36 35	48.3	40 47	35 36	47.6	40 07	34 38	47.0	39 26	33 41	46.4	38 44	32 45	45.8	236	304
57	123	42 44	36 51	47.9	42 03	35 51	47.2	41 22	34 53	46.5	40 41	33 55	45.9	39 59	32 59	45.3	39 16	32 04	44.7	237	303
58	122	43 19	36 06	46.8	42 38	35 07	46.1	41 56	34 09	45.4	41 14	33 12	44.8	40 31	32 16	44.2	39 48	31 22	43.6	238	302
59	121	43 54	35 20	45.6	43 12	34 21	45.0	42 29	33 24	44.3	41 46	32 27	43.7	41 03	31 32	43.1	40 19	30 39	42.5	239	301
60	120	44 29	34 32	44.5	43 46	33 34	43.8	43 02	32 37	43.2	42 18	31 42	42.5	41 34	30 47	41.9	40 49	29 54	41.3	240	300
61	119	45 02	33 43	43.3	44 18	32 45	42.6	43 34	31 49	42.0	42 49	30 55	41.4	42 04	30 01	40.8	41 18	29 09	40.2	241	299
62	118	45 35	32 52	42.1	44 51	31 55	41.5	44 05	31 00	40.8	43 20	30 06	40.2	42 34	29 14	39.6	41 47	28 22	39.0	242	298
63	117	46 07	32 00	40.9	45 22	31 04	40.3	44 36	30 10	39.6	43 49	29 17	39.0	43 03	28 25	38.4	42 15	27 35	37.8	243	297
64	116	46 39	31 06	39.7	45 52	30 11	39.0	45 06	29 18	38.4	44 18	28 26	37.8	43 31	27 35	37.2	42 43	26 46	36.6	244	296
65	115	47 09	30 11	38.4	46 22	29 17	37.8	45 35	28 25	37.1	44 47	27 34	36.5	43 58	26 44	36.0	43 09	25 56	35.4	245	295
66	114	47 39	29 14	37.1	46 51	28 21	36.5	46 03	27 30	35.9	45 14	26 40	35.3	44 25	25 52	34.7	43 35	25 04	34.2	246	294
67	113	48 08	28 16	35.8	47 19	27 24	35.2	46 30	26 34	34.6	45 40	25 45	34.0	44 50	24 58	33.4	44 00	24 12	32.9	247	293
68	112	48 36	27 17	34.5	47 46	26 26	33.9	46 56	25 37	33.3	46 06	24 50	32.7	45 15	24 03	32.2	44 24	23 19	31.6	248	292
69	111	49 03	26 15	33.1	48 13	25 26	32.5	47 22	24 38	31.9	46 31	23 52	31.4	45 39	23 08	30.8	44 48	22 24	30.3	249	291
70	110	49 29	25 13	31.8	48 38	24 25	31.2	47 46	23 39	30.6	46 55	22 54	30.0	46 03	22 11	29.5	45 10	21 29	29.0	250	290
71	109	49 54	24 08	30.4	49 02	23 22	29.8	48 10	22 37	29.2	47 17	21 54	28.7	46 25	21 12	28.2	45 32	20 32	27.7	251	289
72	108	50 18	23 02	28.9	49 25	22 18	28.4	48 33	21 35	27.8	47 39	20 53	27.3	46 46	20 13	26.8	45 52	19 34	26.3	252	288
73	107	50 41	21 55	27.5	49 48	21 12	26.9	48 54	20 31	26.4	48 00	19 51	25.9	47 06	19 13	25.4	46 12	18 35	25.0	253	287
74	106	51 03	20 47	26.0	50 09	20 06	25.5	49 15	19 26	25.0	48 20	18 48	24.5	47 25	18 11	24.0	46 30	17 36	23.6	254	286
75	105	51 24	19 36	24.5	50 29	18 57	24.0	49 34	18 20	23.5	48 39	17 43	23.1	47 44	17 09	22.6	46 48	16 35	22.2	255	285
76	104	51 43	18 25	23.0	50 48	17 48	22.5	49 52	17 12	22.0	48 57	16 38	21.6	48 01	16 05	21.2	47 05	15 33	20.8	256	284
77	103	52 02	17 12	21.4	51 06	16 37	21.0	50 09	16 04	20.6	49 13	15 31	20.1	48 17	15 00	19.8	47 20	14 31	19.4	257	283
78	102	52 19	15 58	19.9	51 22	15 25	19.5	50 25	14 54	19.0	49 29	14 24	18.7	48 32	13 55	18.3	47 35	13 27	18.0	258	282
79	101	52 35	14 43	18.3	51 37	14 13	17.9	50 40	13 43	17.5	49 43	13 16	17.2	48 46	12 49	16.8	47 48	12 23	16.5	259	281
80	100	52 49	13 27	16.7	51 52	12 59	16.3	50 54	12 32	16.0	49 56	12 06	15.7	48 58	11 42	15.3	48 01	11 18	15.0	260	280
81	99	53 02	12 09	15.1	52 04	11 44	14.7	51 06	11 19	14.4	50 08	10 56	14.1	49 10	10 34	13.8	48 12	10 12	13.6	261	279
82	98	53 14	10 51	13.4	52 16	10 28	13.1	51 18	10 06	12.9	50 19	9 45	12.6	49 20	9 25	12.3	48 22	9 06	12.1	262	278
83	97	53 25	9 31	11.8	52 26	9 11	11.5	51 27	8 52	11.3	50 29	8 34	11.0	49 30	8 16	10.8	48 31	7 59	10.6	263	277
84	96	53 34	8 11	10.1	52 35	7 54	9.9	51 36	7 37	9.7	50 37	7 21	9.5	49 38	7 06	9.3	48 38	6 51	9.1	264	276
85	95	53 42	6 50	8.5	52 43	6 36	8.3	51 43	6 22	8.1	50 44	6 09	7.9	49 44	5 56	7.8	48 45	5 44	7.6	265	275
86	94	53 49	5 29	6.8	52 49	5 17	6.6	51 49	5 06	6.5	50 50	4 55	6.3	49 50	4 45	6.2	48 50	4 35	6.1	266	274
87	93	53 54	4 07	5.1	52 54	3 58	5.0	51 54	3 50	4.9	50 54	3 42	4.8	49 54	3 34	4.7	48 55	3 27	4.6	267	273
88	92	53 57	2 45	3.4	52 57	2 39	3.3	51 57	2 33	3.2	50 57	2 28	3.2	49 58	2 23	3.1	48 58	2 18	3.0	268	272
89	91	53 59	1 23	1.7	52 59	1 20	1.7	51 59	1 17	1.6	50 59	1 14	1.6	49 59	1 11	1.6	48 59	1 09	1.5	269	271
90	90	54 00	0 00	0.0	53 00	0 00	0.0	52 00	0 00	0.0	51 00	0 00	0.0	50 00	0 00	0.0	49 00	0 00	0.0	270	270

N. Lat.: for LHA > 180°... $Z_n = Z$
for LHA < 180°... $Z_n = 360° - Z$

S. Lat.: for LHA > 180°... $Z_n = 180° - Z$
for LHA < 180°... $Z_n = 180° + Z$

SIGHT REDUCTION TABLE

B: (–) for 90° < LHA < 270°
Dec: (–) for Lat. contrary name

Z_1 : same sign as B
Z_2 : (–) for F > 90°

Lat. / A		42°			43°			44°			45°			46°			47°			Lat. / A	
LHA / F		A / H	B / P	Z_1 / Z_2	A / H	B / P	Z_1 / Z_2	A / H	B / P	Z_1 / Z_2	A / H	B / P	Z_1 / Z_2	A / H	B / P	Z_1 / Z_2	A / H	B / P	Z_1 / Z_2	LHA	
°	°	° ′	° ′	°	° ′	° ′	°	° ′	° ′	°	° ′	° ′	°	° ′	° ′	°	° ′	° ′	°	°	°
0	180	0 00	48 00	90.0	0 00	47 00	90.0	0 00	46 00	90.0	0 00	45 00	90.0	0 00	44 00	90.0	0 00	43 00	90.0	180	360
1	179	0 45	48 00	89.3	0 44	47 00	89.3	0 43	46 00	89.3	0 42	45 00	89.3	0 42	44 00	89.3	0 41	43 00	89.3	181	359
2	178	1 29	47 59	88.7	1 28	46 59	88.6	1 26	45 59	88.6	1 25	44 59	88.6	1 23	43 59	88.6	1 22	42 59	88.5	182	358
3	177	2 14	47 58	88.0	2 12	46 58	88.0	2 09	45 58	87.9	2 07	44 58	87.9	2 05	43 58	87.8	2 03	42 58	87.8	183	357
4	176	2 58	47 56	87.3	2 55	46 56	87.3	2 53	45 56	87.2	2 50	44 56	87.2	2 47	43 56	87.1	2 44	42 56	87.1	184	356
5	175	3 43	47 53	86.6	3 39	46 53	86.6	3 36	45 53	86.5	3 32	44 53	86.5	3 28	43 53	86.4	3 24	42 53	86.3	185	355
6	174	4 27	47 51	86.0	4 23	46 51	85.9	4 19	45 51	85.8	4 14	44 51	85.7	4 10	43 51	85.7	4 05	42 51	85.6	186	354
7	173	5 12	47 47	85.3	5 07	46 47	85.2	5 02	45 47	85.1	4 57	44 47	85.0	4 51	43 47	85.0	4 46	42 47	84.9	187	353
8	172	5 56	47 43	84.6	5 51	46 43	84.5	5 45	45 43	84.4	5 39	44 43	84.3	5 33	43 43	84.2	5 27	42 43	84.1	188	352
9	171	6 41	47 39	84.0	6 34	46 39	83.8	6 28	45 39	83.7	6 21	44 39	83.6	6 14	43 39	83.5	6 07	42 39	83.4	189	351
10	170	7 25	47 34	83.3	7 18	46 34	83.1	7 11	45 34	83.0	7 03	44 34	82.9	6 56	43 34	82.8	6 48	42 34	82.7	190	350
11	169	8 09	47 28	82.6	8 01	46 28	82.4	7 53	45 28	82.3	7 45	44 28	82.2	7 37	43 28	82.0	7 29	42 28	81.9	191	349
12	168	8 53	47 22	81.9	8 45	46 22	81.8	8 36	45 22	81.6	8 27	44 22	81.5	8 18	43 22	81.3	8 09	42 22	81.2	192	348
13	167	9 37	47 16	81.2	9 28	46 15	81.1	9 19	45 15	80.9	9 09	44 15	80.7	8 59	43 15	80.6	8 49	42 16	80.4	193	347
14	166	10 21	47 08	80.5	10 11	46 08	80.3	10 01	45 08	80.2	9 51	44 08	80.0	9 40	43 08	79.8	9 30	42 08	79.7	194	346
15	165	11 05	47 01	79.8	10 55	46 00	79.6	10 44	45 00	79.5	10 33	44 00	79.3	10 21	43 00	79.1	10 10	42 01	78.9	195	345
16	164	11 49	46 52	79.1	11 38	45 52	78.9	11 26	44 52	78.7	11 14	43 52	78.5	11 02	42 52	78.3	10 50	41 52	78.2	196	344
17	163	12 33	46 43	78.4	12 21	45 43	78.2	12 08	44 43	78.0	11 56	43 43	77.8	11 43	42 43	77.6	11 30	41 44	77.4	197	343
18	162	13 17	46 34	77.7	13 04	45 34	77.5	12 51	44 34	77.3	12 37	43 34	77.1	12 24	42 34	76.8	12 10	41 34	76.6	198	342
19	161	14 00	46 24	77.0	13 46	45 24	76.8	13 33	44 24	76.5	13 19	43 24	76.3	13 04	42 24	76.1	12 50	41 24	75.9	199	341
20	160	14 43	46 13	76.3	14 29	45 13	76.1	14 15	44 13	75.8	14 00	43 13	75.6	13 45	42 13	75.3	13 29	41 14	75.1	200	340
21	159	15 27	46 02	75.6	15 12	45 02	75.3	14 56	44 02	75.1	14 41	43 02	74.8	14 25	42 02	74.6	14 09	41 03	74.3	201	339
22	158	16 10	45 50	74.9	15 54	44 50	74.6	15 38	43 50	74.3	15 22	42 50	74.1	15 05	41 50	73.8	14 48	40 51	73.5	202	338
23	157	16 53	45 38	74.1	16 36	44 38	73.9	16 19	43 38	73.6	16 02	42 38	73.3	15 45	41 38	73.0	15 27	40 39	72.8	203	337
24	156	17 36	45 25	73.4	17 18	44 25	73.1	17 01	43 25	72.8	16 43	42 25	72.5	16 25	41 25	72.2	16 06	40 26	72.0	204	336
25	155	18 18	45 11	72.7	18 00	44 11	72.4	17 42	43 11	72.1	17 23	42 11	71.8	17 04	41 12	71.5	16 45	40 12	71.2	205	335
26	154	19 01	44 57	71.9	18 42	43 57	71.6	18 23	42 57	71.3	18 03	41 57	71.0	17 44	40 57	70.7	17 24	39 58	70.4	206	334
27	153	19 43	44 42	71.2	19 24	43 42	70.8	19 04	42 42	70.5	18 43	41 42	70.2	18 23	40 43	69.9	18 02	39 43	69.6	207	333
28	152	20 25	44 26	70.4	20 05	43 26	70.1	19 44	42 26	69.7	19 23	41 27	69.4	19 02	40 27	69.1	18 40	39 28	68.8	208	332
29	151	21 07	44 10	69.6	20 46	43 10	69.3	20 25	42 10	68.9	20 03	41 10	68.6	19 41	40 11	68.3	19 18	39 12	67.9	209	331
30	150	21 49	43 53	68.9	21 27	42 53	68.5	21 05	41 53	68.1	20 42	40 54	67.8	20 19	39 54	67.4	19 56	38 55	67.1	210	330
31	149	22 30	43 35	68.1	22 08	42 35	67.7	21 45	41 36	67.3	21 21	40 36	67.0	20 58	39 37	66.6	20 34	38 38	66.3	211	329
32	148	23 11	43 17	67.3	22 48	42 17	66.9	22 24	41 17	66.5	22 00	40 18	66.2	21 36	39 19	65.8	21 11	38 20	65.4	212	328
33	147	23 53	42 58	66.5	23 28	41 58	66.1	23 04	40 58	65.7	22 39	39 59	65.3	22 14	39 00	65.0	21 48	38 02	64.6	213	327
34	146	24 33	42 38	65.7	24 08	41 38	65.3	23 43	40 39	64.9	23 17	39 40	64.5	22 51	38 41	64.1	22 25	37 42	63.7	214	326
35	145	25 14	42 18	64.9	24 48	41 18	64.5	24 22	40 18	64.1	23 56	39 19	63.7	23 29	38 21	63.3	23 02	37 23	62.9	215	325
36	144	25 54	41 56	64.1	25 28	40 57	63.6	25 01	39 57	63.2	24 34	38 58	62.8	24 06	38 00	62.4	23 38	37 02	62.0	216	324
37	143	26 34	41 34	63.2	26 07	40 35	62.8	25 39	39 35	62.4	25 11	38 37	61.9	24 43	37 38	61.5	24 14	36 41	61.1	217	323
38	142	27 14	41 11	62.4	26 46	40 12	61.9	26 17	39 13	61.5	25 48	38 14	61.1	25 19	37 16	60.7	24 50	36 19	60.3	218	322
39	141	27 53	40 48	61.5	27 24	39 48	61.1	26 55	38 50	60.6	26 25	37 51	60.2	25 55	36 53	59.8	25 25	35 56	59.4	219	321
40	140	28 32	40 23	60.7	28 02	39 24	60.2	27 32	38 25	59.8	27 02	37 27	59.3	26 31	36 30	58.9	26 00	35 32	58.5	220	320
41	139	29 11	39 58	59.8	28 40	38 59	59.3	28 10	38 01	58.9	27 38	37 03	58.4	27 07	36 05	58.0	26 35	35 08	57.6	221	319
42	138	29 49	39 32	58.9	29 18	38 33	58.4	28 46	37 35	58.0	28 14	36 37	57.5	27 42	35 40	57.1	27 09	34 43	56.6	222	318
43	137	30 27	39 05	58.0	29 55	38 06	57.5	29 23	37 08	57.1	28 50	36 11	56.6	28 17	35 14	56.1	27 43	34 18	55.7	223	317
44	136	31 05	38 37	57.1	30 32	37 39	56.6	29 59	36 41	56.1	29 25	35 44	55.7	28 51	34 47	55.2	28 17	33 51	54.8	224	316
45	135	31 42	38 09	56.2	31 08	37 10	55.7	30 34	36 13	55.2	30 00	35 16	54.7	29 25	34 20	54.3	28 50	33 24	53.8	225	315

Lat. / A		42°			43°			44°			45°			46°			47°			Lat. / A	
LHA / F		A / H	B / P	Z_1 / Z_2	A / H	B / P	Z_1 / Z_2	A / H	B / P	Z_1 / Z_2	A / H	B / P	Z_1 / Z_2	A / H	B / P	Z_1 / Z_2	A / H	B / P	Z_1 / Z_2	LHA	
°	°	° ′	° ′	°	° ′	° ′	°	° ′	° ′	°	° ′	° ′	°	° ′	° ′	°	° ′	° ′	°	°	°
45	135	31 42	38 09	56.2	31 08	37 10	55.7	30 34	36 13	55.2	30 00	35 16	54.7	29 25	34 20	54.3	28 50	33 24	53.8	225	315
46	134	32 19	37 39	55.3	31 45	36 41	54.8	31 10	35 44	54.3	30 34	34 47	53.8	29 59	33 51	53.3	29 23	32 56	52.9	226	314
47	133	32 55	37 08	54.3	32 20	36 11	53.8	31 45	35 14	53.3	31 08	34 18	52.8	30 32	33 22	52.4	29 55	32 27	51.9	227	313
48	132	33 31	36 37	53.4	32 55	35 40	52.9	32 19	34 43	52.3	31 42	33 47	51.9	31 05	32 52	51.4	30 27	31 58	50.9	228	312
49	131	34 07	36 05	52.4	33 30	35 08	51.9	32 53	34 11	51.4	32 15	33 16	50.9	31 37	32 21	50.4	30 59	31 27	49.9	229	311
50	130	34 42	35 31	51.4	34 04	34 35	50.9	33 26	33 39	50.4	32 48	32 44	49.9	32 09	31 50	49.4	31 30	30 56	48.9	230	310
51	129	35 17	34 57	50.4	34 38	34 01	49.9	33 59	33 05	49.4	33 20	32 11	48.9	32 40	31 17	48.4	32 00	30 24	47.9	231	309
52	128	35 51	34 22	49.4	35 12	33 26	48.9	34 32	32 31	48.4	33 52	31 37	47.9	33 11	30 44	47.4	32 30	29 52	46.9	232	308
53	127	36 24	33 45	48.4	35 44	32 50	47.9	35 04	31 56	47.3	34 23	31 02	46.8	33 42	30 10	46.3	33 00	29 18	45.9	233	307
54	126	36 57	33 08	47.4	36 17	32 13	46.8	35 35	31 20	46.3	34 54	30 27	45.8	34 12	29 35	45.3	33 29	28 44	44.8	234	306
55	125	37 30	32 30	46.3	36 48	31 36	45.8	36 06	30 43	45.2	35 24	29 50	44.7	34 41	28 59	44.2	33 58	28 08	43.8	235	305
56	124	38 02	31 51	45.2	37 19	30 57	44.7	36 37	30 04	44.2	35 53	29 13	43.6	35 10	28 22	43.2	34 26	27 32	42.7	236	304
57	123	38 33	31 10	44.1	37 50	30 17	43.6	37 06	29 25	43.1	36 22	28 34	42.6	35 38	27 45	42.1	34 53	26 56	41.6	237	303
58	122	39 04	30 29	43.0	38 20	29 36	42.5	37 36	28 45	42.0	36 51	27 55	41.5	36 06	27 06	41.0	35 20	26 18	40.5	238	302
59	121	39 34	29 46	41.9	38 49	28 55	41.4	38 04	28 04	40.9	37 19	27 15	40.4	36 33	26 27	39.9	35 46	25 39	39.4	239	301
60	120	40 04	29 03	40.8	39 18	28 12	40.2	38 32	27 22	39.7	37 46	26 34	39.2	36 59	25 46	38.8	36 12	25 00	38.3	240	300
61	119	40 32	28 18	39.6	39 46	27 28	39.1	38 59	26 39	38.6	38 12	25 52	38.1	37 25	25 05	37.6	36 37	24 20	37.2	241	299
62	118	41 00	27 32	38.5	40 13	26 43	37.9	39 26	25 56	37.4	38 38	25 09	36.9	37 50	24 23	36.5	37 02	23 39	36.0	242	298
63	117	41 28	26 45	37.3	40 40	25 58	36.8	39 52	25 11	36.3	39 03	24 25	35.8	38 14	23 40	35.3	37 25	22 57	34.9	243	297
64	116	41 54	25 58	36.1	41 06	25 11	35.6	40 17	24 25	35.1	39 28	23 40	34.6	38 38	22 57	34.1	37 48	22 14	33.7	244	296
65	115	42 20	25 09	34.9	41 31	24 23	34.4	40 41	23 38	33.9	39 51	22 55	33.4	39 01	22 12	33.0	38 11	21 31	32.5	245	295
66	114	42 45	24 19	33.6	41 55	23 34	33.1	41 05	22 50	32.7	40 14	22 08	32.2	39 23	21 27	31.8	38 32	20 46	31.3	246	294
67	113	43 10	23 28	32.4	42 19	22 44	31.9	41 28	22 02	31.4	40 37	21 21	31.0	39 45	20 40	30.5	38 53	20 01	30.1	247	293
68	112	43 33	22 35	31.1	42 42	21 53	30.6	41 50	21 12	30.2	40 58	20 32	29.7	40 06	19 53	29.3	39 13	19 15	28.9	248	292
69	111	43 56	21 42	29.8	43 04	21 01	29.4	42 11	20 22	28.9	41 19	19 43	28.5	40 26	19 05	28.1	39 33	18 29	27.7	249	291
70	110	44 18	20 48	28.5	43 25	20 08	28.1	42 32	19 30	27.7	41 38	18 53	27.2	40 45	18 17	26.8	39 51	17 41	26.5	250	290
71	109	44 38	19 53	27.2	43 45	19 15	26.8	42 51	18 38	26.4	41 57	18 02	26.0	41 03	17 27	25.6	40 09	16 53	25.2	251	289
72	108	44 58	18 57	25.9	44 04	18 20	25.5	43 10	17 45	25.1	42 16	17 10	24.7	41 21	16 37	24.3	40 26	16 05	24.0	252	288
73	107	45 17	17 59	24.6	44 23	17 24	24.1	43 28	16 51	23.8	42 33	16 18	23.4	41 38	15 46	23.0	40 42	15 15	22.7	253	287
74	106	45 35	17 01	23.2	44 40	16 28	22.8	43 45	15 56	22.4	42 49	15 25	22.1	41 54	14 54	21.7	40 58	14 25	21.4	254	286
75	105	45 53	16 02	21.8	44 57	15 31	21.4	44 01	15 00	21.1	43 05	14 31	20.8	42 09	14 02	20.4	41 12	13 34	20.1	255	285
76	104	46 09	15 02	20.4	45 12	14 33	20.1	44 16	14 04	19.7	43 19	13 36	19.4	42 23	13 09	19.1	41 26	12 43	18.8	256	284
77	103	46 24	14 02	19.0	45 27	13 34	18.7	44 30	13 07	18.4	43 33	12 41	18.1	42 36	12 15	17.8	41 39	11 51	17.5	257	283
78	102	46 38	13 00	17.6	45 40	12 34	17.3	44 43	12 09	17.0	43 46	11 45	16.7	42 48	11 21	16.5	41 51	10 58	16.2	258	282
79	101	46 51	11 58	16.2	45 53	11 34	15.9	44 55	11 11	15.6	43 57	10 48	15.4	43 00	10 26	15.1	42 02	10 05	14.9	259	281
80	100	47 03	10 55	14.8	46 04	10 33	14.5	45 06	10 12	14.2	44 08	9 51	14.0	43 10	9 31	13.8	42 12	9 12	13.6	260	280
81	99	47 13	9 51	13.3	46 15	9 31	13.1	45 16	9 12	12.8	44 18	8 53	12.6	43 19	8 35	12.4	42 21	8 18	12.2	261	279
82	98	47 23	8 47	11.9	46 24	8 29	11.6	45 26	8 12	11.4	44 27	7 55	11.2	43 28	7 39	11.1	42 29	7 24	10.9	262	278
83	97	47 32	7 42	10.4	46 33	7 27	10.2	45 34	7 12	10.0	44 34	6 57	9.9	43 35	6 43	9.7	42 36	6 29	9.5	263	277
84	96	47 39	6 37	8.9	46 40	6 24	8.8	45 41	6 11	8.6	44 41	5 58	8.5	43 42	5 46	8.3	42 42	5 34	8.2	264	276
85	95	47 46	5 32	7.4	46 46	5 20	7.3	45 46	5 09	7.2	44 47	4 59	7.1	43 47	4 49	6.9	42 48	4 39	6.8	265	275
86	94	47 51	4 26	6.0	46 51	4 17	5.9	45 51	4 08	5.7	44 52	3 59	5.6	43 52	3 51	5.6	42 52	3 43	5.5	266	274
87	93	47 55	3 20	4.5	46 55	3 13	4.4	45 55	3 06	4.3	44 55	3 00	4.2	43 55	2 54	4.2	42 56	2 48	4.1	267	273
88	92	47 58	2 13	3.0	46 58	2 09	2.9	45 58	2 04	2.9	44 58	2 00	2.8	43 58	1 56	2.8	42 58	1 52	2.7	268	272
89	91	47 59	1 07	1.5	46 59	1 04	1.5	45 59	1 02	1.4	44 59	1 00	1.4	43 59	0 58	1.4	43 00	0 56	1.4	269	271
90	90	48 00	0 00	0.0	47 00	0 00	0.0	46 00	0 00	0.0	45 00	0 00	0.0	44 00	0 00	0.0	43 00	0 00	0.0	270	270

N. Lat.: for LHA > 180°… $Z_n = Z$
for LHA < 180°… $Z_n = 360° - Z$

S. Lat.: for LHA > 180°… $Z_n = 180° - Z$
for LHA < 180°… $Z_n = 180° + Z$

SIGHT REDUCTION TABLE

B: (–) for 90° < LHA < 270°
Dec: (–) for Lat. contrary name

Z_1 : same sign as B
Z_2 : (–) for F > 90°

Lat. / A		48°			49°			50°			51°			52°			53°			Lat. / A	
LHA / F		A / H	B / P	Z_1 / Z_2	A / H	B / P	Z_1 / Z_2	A / H	B / P	Z_1 / Z_2	A / H	B / P	Z_1 / Z_2	A / H	B / P	Z_1 / Z_2	A / H	B / P	Z_1 / Z_2	LHA	
°	°	° ′	° ′	°	° ′	° ′	°	° ′	° ′	°	° ′	° ′	°	° ′	° ′	°	° ′	° ′	°	°	°
0	180	0 00	42 00	90.0	0 00	41 00	90.0	0 00	40 00	90.0	0 00	39 00	90.0	0 00	38 00	90.0	0 00	37 00	90.0	180	360
1	179	0 40	42 00	89.3	0 39	41 00	89.2	0 39	40 00	89.2	0 38	39 00	89.2	0 37	38 00	89.2	0 36	37 00	89.2	181	359
2	178	1 20	41 59	88.5	1 19	40 59	88.5	1 17	39 59	88.5	1 16	38 59	88.4	1 14	37 59	88.4	1 12	36 59	88.4	182	358
3	177	2 00	41 58	87.8	1 58	40 58	87.7	1 56	39 58	87.7	1 53	38 58	87.7	1 51	37 58	87.6	1 48	36 58	87.6	183	357
4	176	2 41	41 56	87.0	2 37	40 56	87.0	2 34	39 56	86.9	2 31	38 56	86.9	2 28	37 56	86.8	2 24	36 56	86.8	184	356
5	175	3 21	41 53	86.3	3 17	40 54	86.2	3 13	39 54	86.2	3 09	38 54	86.1	3 05	37 54	86.1	3 00	36 54	86.0	185	355
6	174	4 01	41 51	85.5	3 56	40 51	85.5	3 51	39 51	85.4	3 46	38 51	85.3	3 41	37 51	85.3	3 36	36 51	85.2	186	354
7	173	4 41	41 47	84.8	4 35	40 47	84.7	4 30	39 47	84.6	4 24	38 47	84.5	4 18	37 48	84.5	4 12	36 48	84.4	187	353
8	172	5 21	41 43	84.0	5 14	40 43	83.9	5 08	39 43	83.9	5 01	38 44	83.8	4 55	37 44	83.7	4 48	36 44	83.6	188	352
9	171	6 01	41 39	83.3	5 53	40 39	83.2	5 46	39 39	83.1	5 39	38 39	83.0	5 32	37 39	82.9	5 24	36 40	82.8	189	351
10	170	6 40	41 34	82.5	6 32	40 34	82.4	6 25	39 34	82.3	6 16	38 34	82.2	6 08	37 35	82.1	6 00	36 35	82.0	190	350
11	169	7 20	41 28	81.8	7 11	40 28	81.7	7 03	39 29	81.5	6 54	38 29	81.4	6 45	37 29	81.3	6 36	36 29	81.2	191	349
12	168	8 00	41 22	81.0	7 50	40 22	80.9	7 41	39 23	80.8	7 31	38 23	80.6	7 21	37 23	80.5	7 11	36 24	80.4	192	348
13	167	8 39	41 16	80.3	8 29	40 16	80.1	8 19	39 16	80.0	8 08	38 16	79.8	7 58	37 17	79.7	7 47	36 17	79.6	193	347
14	166	9 19	41 09	79.5	9 08	40 09	79.3	8 57	39 09	79.2	8 45	38 09	79.0	8 34	37 10	78.9	8 22	36 10	78.7	194	346
15	165	9 58	41 01	78.7	9 47	40 01	78.6	9 35	39 02	78.4	9 22	38 02	78.2	9 10	37 02	78.1	8 58	36 03	77.9	195	345
16	164	10 38	40 53	78.0	10 25	39 53	77.8	10 12	38 53	77.6	9 59	37 54	77.4	9 46	36 54	77.3	9 33	35 55	77.1	196	344
17	163	11 17	40 44	77.2	11 04	39 44	77.0	10 50	38 45	76.8	10 36	37 45	76.6	10 22	36 46	76.5	10 08	35 47	76.3	197	343
18	162	11 56	40 34	76.4	11 42	39 35	76.2	11 27	38 35	76.0	11 13	37 36	75.8	10 58	36 37	75.6	10 43	35 38	75.5	198	342
19	161	12 35	40 25	75.6	12 20	39 25	75.4	12 05	38 26	75.2	11 49	37 26	75.0	11 34	36 27	74.8	11 18	35 28	74.6	199	341
20	160	13 14	40 14	74.9	12 58	39 15	74.6	12 42	38 15	74.4	12 26	37 16	74.2	12 09	36 17	74.0	11 53	35 18	73.8	200	340
21	159	13 52	40 03	74.1	13 36	39 04	73.8	13 19	38 04	73.6	13 02	37 05	73.4	12 45	36 06	73.2	12 27	35 08	73.0	201	339
22	158	14 31	39 51	73.3	14 14	38 52	73.0	13 56	37 53	72.8	13 38	36 54	72.6	13 20	35 55	72.3	13 02	34 56	72.1	202	338
23	157	15 09	39 39	72.5	14 51	38 40	72.2	14 33	37 41	72.0	14 14	36 42	71.7	13 55	35 43	71.5	13 36	34 45	71.3	203	337
24	156	15 48	39 26	71.7	15 29	38 27	71.4	15 09	37 28	71.2	14 50	36 30	70.9	14 30	35 31	70.7	14 10	34 33	70.4	204	336
25	155	16 26	39 13	70.9	16 06	38 14	70.6	15 46	37 15	70.3	15 25	36 17	70.1	15 05	35 18	69.8	14 44	34 20	69.6	205	335
26	154	17 03	38 59	70.1	16 43	38 00	69.8	16 22	37 01	69.5	16 01	36 03	69.2	15 39	35 05	69.0	15 18	34 07	68.7	206	334
27	153	17 41	38 44	69.3	17 20	37 46	69.0	16 58	36 47	68.7	16 36	35 49	68.4	16 14	34 51	68.1	15 51	33 53	67.9	207	333
28	152	18 19	38 29	68.4	17 56	37 30	68.1	17 34	36 32	67.8	17 11	35 34	67.5	16 48	34 36	67.3	16 25	33 38	67.0	208	332
29	151	18 56	38 13	67.6	18 33	37 15	67.3	18 09	36 16	67.0	17 46	35 18	66.7	17 22	34 21	66.4	16 58	33 23	66.1	209	331
30	150	19 33	37 57	66.8	19 09	36 58	66.5	18 45	36 00	66.1	18 20	35 03	65.8	17 56	34 05	65.5	17 31	33 08	65.2	210	330
31	149	20 10	37 40	65.9	19 45	36 41	65.6	19 20	35 44	65.3	18 55	34 46	65.0	18 29	33 49	64.7	18 03	32 52	64.4	211	329
32	148	20 46	37 22	65.1	20 21	36 24	64.8	19 55	35 26	64.4	19 29	34 29	64.1	19 02	33 32	63.8	18 36	32 35	63.5	212	328
33	147	21 22	37 03	64.2	20 56	36 06	63.9	20 30	35 08	63.6	20 03	34 11	63.2	19 35	33 14	62.9	19 08	32 18	62.6	213	327
34	146	21 58	36 44	63.4	21 31	35 47	63.0	21 04	34 49	62.7	20 36	33 53	62.3	20 08	32 56	62.0	19 40	32 00	61.7	214	326
35	145	22 34	36 25	62.5	22 06	35 27	62.1	21 38	34 30	61.8	21 10	33 33	61.4	20 41	32 37	61.1	20 12	31 41	60.8	215	325
36	144	23 10	36 04	61.6	22 41	35 07	61.3	22 12	34 10	60.9	21 43	33 14	60.5	21 13	32 18	60.2	20 43	31 22	59.9	216	324
37	143	23 45	35 43	60.8	23 15	34 46	60.4	22 45	33 50	60.0	22 15	32 54	59.6	21 45	31 58	59.3	21 14	31 02	59.0	217	323
38	142	24 20	35 21	59.9	23 49	34 25	59.5	23 19	33 28	59.1	22 48	32 33	58.7	22 16	31 37	58.4	21 45	30 42	58.0	218	322
39	141	24 54	34 59	59.0	24 23	34 02	58.6	23 52	33 07	58.2	23 20	32 11	57.8	22 48	31 16	57.5	22 15	30 21	57.1	219	321
40	140	25 28	34 36	58.1	24 57	33 40	57.7	24 24	32 44	57.3	23 52	31 49	56.9	23 19	30 54	56.5	22 45	30 00	56.2	220	320
41	139	26 02	34 12	57.1	25 30	33 16	56.7	24 57	32 21	56.3	24 23	31 26	56.0	23 49	30 32	55.6	23 15	29 38	55.2	221	319
42	138	26 36	33 47	56.2	26 02	32 52	55.8	25 28	31 57	55.4	24 54	31 02	55.0	24 20	30 08	54.6	23 45	29 15	54.3	222	318
43	137	27 09	33 22	55.3	26 35	32 27	54.9	26 00	31 32	54.5	25 25	30 38	54.1	24 50	29 45	53.7	24 14	28 52	53.3	223	317
44	136	27 42	32 56	54.3	27 07	32 01	53.9	26 31	31 07	53.5	25 55	30 13	53.1	25 19	29 20	52.7	24 43	28 28	52.4	224	316
45	135	28 14	32 29	53.4	27 38	31 35	53.0	27 02	30 41	52.5	26 25	29 48	52.1	25 48	28 55	51.8	25 11	28 03	51.4	225	315

Lat. / A		48°			49°			50°			51°			52°			53°			Lat. / A	
LHA / F		A / H	B / P	Z_1/Z_2	A / H	B / P	Z_1/Z_2	A / H	B / P	Z_1/Z_2	A / H	B / P	Z_1/Z_2	A / H	B / P	Z_1/Z_2	A / H	B / P	Z_1/Z_2	LHA	
°	°	° ′	° ′	°	° ′	° ′	°	° ′	° ′	°	° ′	° ′	°	° ′	° ′	°	° ′	° ′	°	°	°
45	135	28 14	32 29	53.4	27 38	31 35	53.0	27 02	30 41	52.5	26 25	29 48	52.1	25 48	28 55	51.8	25 11	28 03	51.4	225	315
46	134	28 46	32 01	52.4	28 10	31 08	52.0	27 32	30 14	51.6	26 55	29 22	51.2	26 17	28 29	50.8	25 39	27 38	50.4	226	314
47	133	29 18	31 33	51.4	28 40	30 40	51.0	28 02	29 47	50.6	27 24	28 55	50.2	26 46	28 03	49.8	26 07	27 12	49.4	227	313
48	132	29 49	31 04	50.5	29 11	30 11	50.0	28 32	29 19	49.6	27 53	28 27	49.2	27 14	27 36	48.8	26 34	26 46	48.4	228	312
49	131	30 20	30 34	49.5	29 41	29 42	49.0	29 01	28 50	48.6	28 21	27 59	48.2	27 41	27 08	47.8	27 01	26 18	47.4	229	311
50	130	30 50	30 04	48.5	30 10	29 12	48.0	29 30	28 20	47.6	28 49	27 30	47.2	28 08	26 40	46.8	27 27	25 51	46.4	230	310
51	129	31 20	29 32	47.5	30 39	28 41	47.0	29 58	27 50	46.6	29 17	27 00	46.2	28 35	26 11	45.8	27 53	25 22	45.4	231	309
52	128	31 49	29 00	46.4	31 08	28 09	46.0	30 26	27 19	45.6	29 44	26 30	45.2	29 01	25 41	44.8	28 19	24 53	44.4	232	308
53	127	32 18	28 27	45.4	31 36	27 37	45.0	30 53	26 48	44.5	30 10	25 59	44.1	29 27	25 11	43.7	28 44	24 24	43.3	233	307
54	126	32 46	27 53	44.4	32 03	27 04	43.9	31 20	26 15	43.5	30 36	25 27	43.1	29 52	24 40	42.7	29 08	23 53	42.3	234	306
55	125	33 14	27 19	43.3	32 30	26 30	42.9	31 46	25 42	42.4	31 02	24 55	42.0	30 17	24 08	41.6	29 32	23 23	41.2	235	305
56	124	33 42	26 44	42.2	32 57	25 55	41.8	32 12	25 08	41.4	31 27	24 22	41.0	30 41	23 36	40.6	29 56	22 51	40.2	236	304
57	123	34 08	26 07	41.1	33 23	25 20	40.7	32 37	24 34	40.3	31 51	23 48	39.9	31 05	23 03	39.5	30 19	22 19	39.1	237	303
58	122	34 34	25 30	40.1	33 48	24 44	39.6	33 02	23 58	39.2	32 15	23 14	38.8	31 28	22 29	38.4	30 41	21 46	38.0	238	302
59	121	35 00	24 53	39.0	34 13	24 07	38.5	33 26	23 22	38.1	32 39	22 38	37.7	31 51	21 55	37.3	31 03	21 13	37.0	239	301
60	120	35 25	24 14	37.8	34 37	23 30	37.4	33 50	22 46	37.0	33 02	22 03	36.6	32 13	21 20	36.2	31 25	20 39	35.9	240	300
61	119	35 49	23 35	36.7	35 01	22 51	36.3	34 12	22 08	35.9	33 24	21 26	35.5	32 35	20 45	35.1	31 46	20 04	34.8	241	299
62	118	36 13	22 55	35.6	35 24	22 12	35.2	34 35	21 30	34.8	33 45	20 49	34.4	32 56	20 09	34.0	32 06	19 29	33.7	242	298
63	117	36 36	22 14	34.4	35 46	21 32	34.0	34 56	20 51	33.6	34 06	20 11	33.3	33 16	19 32	32.9	32 26	18 53	32.5	243	297
64	116	36 58	21 32	33.3	36 08	20 52	32.9	35 17	20 12	32.5	34 27	19 33	32.1	33 36	18 54	31.8	32 45	18 17	31.4	244	296
65	115	37 20	20 50	32.1	36 29	20 10	31.7	35 38	19 32	31.3	34 47	18 54	31.0	33 55	18 16	30.6	33 03	17 40	30.3	245	295
66	114	37 41	20 07	30.9	36 49	19 28	30.5	35 58	18 51	30.2	35 06	18 14	29.8	34 13	17 38	29.5	33 21	17 02	29.1	246	294
67	113	38 01	19 23	29.7	37 09	18 46	29.4	36 17	18 09	29.0	35 24	17 33	28.6	34 31	16 59	28.3	33 38	16 24	28.0	247	293
68	112	38 21	18 38	28.5	37 28	18 02	28.2	36 35	17 27	27.8	35 42	16 53	27.5	34 48	16 19	27.1	33 55	15 46	26.8	248	292
69	111	38 40	17 53	27.3	37 46	17 18	27.0	36 53	16 44	26.6	35 59	16 11	26.3	35 05	15 38	26.0	34 11	15 07	25.7	249	291
70	110	38 58	17 07	26.1	38 04	16 33	25.7	37 10	16 01	25.4	36 15	15 29	25.1	35 21	14 58	24.8	34 26	14 27	24.5	250	290
71	109	39 15	16 20	24.9	38 20	15 48	24.5	37 26	15 17	24.2	36 31	14 46	23.9	35 36	14 16	23.6	34 41	13 47	23.3	251	289
72	108	39 31	15 33	23.6	38 36	15 02	23.3	37 41	14 32	23.0	36 46	14 03	22.7	35 50	13 34	22.4	34 55	13 07	22.1	252	288
73	107	39 47	14 45	22.4	38 51	14 16	22.1	37 56	13 47	21.8	37 00	13 19	21.5	36 04	12 52	21.2	35 08	12 25	20.9	253	287
74	106	40 02	13 56	21.1	39 06	13 28	20.8	38 10	13 01	20.5	37 13	12 35	20.3	36 17	12 09	20.0	35 21	11 44	19.8	254	286
75	105	40 16	13 07	19.8	39 19	12 41	19.5	38 23	12 15	19.3	37 26	11 50	19.0	36 29	11 26	18.8	35 33	11 02	18.5	255	285
76	104	40 29	12 17	18.5	39 32	11 53	18.3	38 35	11 28	18.0	37 38	11 05	17.8	36 41	10 42	17.6	35 44	10 20	17.3	256	284
77	103	40 41	11 27	17.3	39 44	11 04	17.0	38 47	10 41	16.8	37 49	10 19	16.5	36 52	9 58	16.3	35 54	9 37	16.1	257	283
78	102	40 53	10 36	16.0	39 55	10 15	15.7	38 57	9 54	15.5	38 00	9 33	15.3	37 02	9 14	15.1	36 04	8 54	14.9	258	282
79	101	41 04	9 45	14.7	40 05	9 25	14.4	39 07	9 06	14.2	38 09	8 47	14.0	37 11	8 29	13.9	36 13	8 11	13.7	259	281
80	100	41 13	8 53	13.3	40 15	8 35	13.2	39 16	8 17	13.0	38 18	8 00	12.8	37 19	7 44	12.6	36 21	7 27	12.5	260	280
81	99	41 22	8 01	12.0	40 23	7 45	11.9	39 25	7 29	11.7	38 26	7 13	11.5	37 27	6 58	11.4	36 28	6 43	11.2	261	279
82	98	41 30	7 09	10.7	40 31	6 54	10.5	39 32	6 40	10.4	38 33	6 26	10.3	37 34	6 12	10.1	36 35	5 59	10.0	262	278
83	97	41 37	6 16	9.4	40 38	6 03	9.2	39 39	5 50	9.1	38 39	5 38	9.0	37 40	5 26	8.9	36 41	5 15	8.7	263	277
84	96	41 43	5 23	8.1	40 44	5 12	7.9	39 44	5 01	7.8	38 45	4 50	7.7	37 45	4 40	7.6	36 46	4 30	7.5	264	276
85	95	41 48	4 29	6.7	40 49	4 20	6.6	39 49	4 11	6.5	38 49	4 02	6.4	37 50	3 54	6.3	36 50	3 45	6.3	265	275
86	94	41 52	3 36	5.4	40 53	3 28	5.3	39 53	3 21	5.2	38 53	3 14	5.1	37 53	3 07	5.1	36 54	3 01	5.0	266	274
87	93	41 56	2 42	4.0	40 56	2 36	4.0	39 56	2 31	3.9	38 56	2 26	3.9	37 56	2 20	3.8	36 56	2 16	3.8	267	273
88	92	41 58	1 48	2.7	40 58	1 44	2.6	39 58	1 41	2.6	38 58	1 37	2.6	37 58	1 34	2.5	36 58	1 30	2.5	268	272
89	91	42 00	0 54	1.3	41 00	0 52	1.3	40 00	0 50	1.3	39 00	0 49	1.3	38 00	0 47	1.3	37 00	0 45	1.3	269	271
90	90	42 00	0 00	0.0	41 00	0 00	0.0	40 00	0 00	0.0	39 00	0 00	0.0	38 00	0 00	0.0	37 00	0 00	0.0	270	270

N. Lat.: for LHA > 180°... $Z_n = Z$
for LHA < 180°... $Z_n = 360° - Z$

S. Lat.: for LHA > 180°... $Z_n = 180° - Z$
for LHA < 180°... $Z_n = 180° + Z$

SIGHT REDUCTION TABLE

B: (–) for 90°< LHA < 270°
Dec: (–) for Lat. contrary name

Z_1 : same sign as B
Z_2 : (–) for F > 90°

Lat. / A		54°			55°			56°			57°			58°			59°			Lat. / A	
LHA / F		A / H	B / P	Z_1 / Z_2	A / H	B / P	Z_1 / Z_2	A / H	B / P	Z_1 / Z_2	A / H	B / P	Z_1 / Z_2	A / H	B / P	Z_1 / Z_2	A / H	B / P	Z_1 / Z_2	LHA	
°	°	° ′	° ′	°	° ′	° ′	°	° ′	° ′	°	° ′	° ′	°	° ′	° ′	°	° ′	° ′	°	°	°
0	180	0 00	36 00	90.0	0 00	35 00	90.0	0 00	34 00	90.0	0 00	33 00	90.0	0 00	32 00	90.0	0 00	31 00	90.0	180	360
1	179	0 35	36 00	89.2	0 34	35 00	89.2	0 34	34 00	89.2	0 33	33 00	89.2	0 32	32 00	89.2	0 31	31 00	89.1	181	359
2	178	1 11	35 59	88.4	1 09	34 59	88.4	1 07	33 59	88.3	1 05	32 59	88.3	1 04	31 59	88.3	1 02	30 59	88.3	182	358
3	177	1 46	35 58	87.6	1 43	34 58	87.5	1 41	33 58	87.5	1 38	32 58	87.5	1 35	31 58	87.5	1 33	30 58	87.4	183	357
4	176	2 21	35 56	86.8	2 18	34 56	86.7	2 14	33 56	86.7	2 11	32 56	86.6	2 07	31 56	86.6	2 04	30 56	86.6	184	356
5	175	2 56	35 54	86.0	2 52	34 54	85.9	2 48	33 54	85.9	2 43	32 54	85.8	2 39	31 54	85.8	2 34	30 54	85.7	185	355
6	174	3 31	35 51	85.1	3 26	34 51	85.1	3 21	33 51	85.0	3 16	32 51	85.0	3 11	31 52	84.9	3 05	30 52	84.9	186	354
7	173	4 06	35 48	84.3	4 00	34 48	84.3	3 54	33 48	84.2	3 48	32 48	84.1	3 42	31 48	84.1	3 36	30 49	84.0	187	353
8	172	4 42	35 44	83.5	4 35	34 44	83.4	4 28	33 44	83.4	4 21	32 45	83.3	4 14	31 45	83.2	4 07	30 45	83.1	188	352
9	171	5 17	35 40	82.7	5 09	34 40	82.6	5 01	33 40	82.5	4 53	32 41	82.4	4 45	31 41	82.3	4 37	30 41	82.3	189	351
10	170	5 51	35 35	81.9	5 43	34 35	81.8	5 34	33 36	81.7	5 26	32 36	81.6	5 17	31 36	81.5	5 08	30 37	81.4	190	350
11	169	6 26	35 30	81.1	6 17	34 30	81.0	6 08	33 31	80.8	5 58	32 31	80.7	5 48	31 31	80.6	5 38	30 32	80.5	191	349
12	168	7 01	35 24	80.2	6 51	34 24	80.1	6 41	33 25	80.0	6 30	32 25	79.9	6 20	31 26	79.8	6 09	30 27	79.7	192	348
13	167	7 36	35 18	79.4	7 25	34 18	79.3	7 14	33 19	79.2	7 02	32 19	79.0	6 51	31 20	78.9	6 39	30 21	78.8	193	347
14	166	8 11	35 11	78.6	7 59	34 12	78.5	7 46	33 12	78.3	7 34	32 13	78.2	7 22	31 14	78.1	7 09	30 15	77.9	194	346
15	165	8 45	35 04	77.8	8 32	34 04	77.6	8 19	33 05	77.5	8 06	32 06	77.3	7 53	31 07	77.2	7 40	30 08	77.1	195	345
16	164	9 19	34 56	76.9	9 06	33 57	76.8	8 52	32 58	76.6	8 38	31 58	76.5	8 24	31 00	76.3	8 10	30 01	76.2	196	344
17	163	9 54	34 47	76.1	9 39	33 48	75.9	9 25	32 49	75.8	9 10	31 50	75.6	8 55	30 52	75.5	8 40	29 53	75.3	197	343
18	162	10 28	34 39	75.3	10 13	33 40	75.1	9 57	32 41	74.9	9 41	31 42	74.8	9 25	30 43	74.6	9 09	29 45	74.4	198	342
19	161	11 02	34 29	74.4	10 46	33 30	74.2	10 29	32 32	74.1	10 13	31 33	73.9	9 56	30 35	73.7	9 39	29 36	73.6	199	341
20	160	11 36	34 19	73.6	11 19	33 21	73.4	11 02	32 22	73.2	10 44	31 24	73.0	10 27	30 25	72.8	10 09	29 27	72.7	200	340
21	159	12 10	34 09	72.7	11 52	33 10	72.5	11 34	32 12	72.3	11 15	31 14	72.2	10 57	30 15	72.0	10 38	29 17	71.8	201	339
22	158	12 43	33 58	71.9	12 24	33 00	71.7	12 06	32 01	71.5	11 46	31 03	71.3	11 27	30 05	71.1	11 07	29 07	70.9	202	338
23	157	13 17	33 46	71.0	12 57	32 48	70.8	12 37	31 50	70.6	12 17	30 52	70.4	11 57	29 54	70.2	11 37	28 57	70.0	203	337
24	156	13 50	33 34	70.2	13 29	32 36	70.0	13 09	31 38	69.7	12 48	30 41	69.5	12 27	29 43	69.3	12 06	28 46	69.1	204	336
25	155	14 23	33 22	69.3	14 02	32 24	69.1	13 40	31 26	68.9	13 18	30 29	68.6	12 56	29 31	68.4	12 34	28 34	68.2	205	335
26	154	14 56	33 09	68.5	14 34	32 11	68.2	14 11	31 14	68.0	13 49	30 16	67.8	13 26	29 19	67.5	13 03	28 22	67.3	206	334
27	153	15 29	32 55	67.6	15 06	31 58	67.3	14 42	31 00	67.1	14 19	30 03	66.9	13 55	29 06	66.6	13 31	28 10	66.4	207	333
28	152	16 01	32 41	66.7	15 37	31 44	66.5	15 13	30 47	66.2	14 49	29 50	66.0	14 24	28 53	65.7	14 00	27 57	65.5	208	332
29	151	16 33	32 26	65.8	16 09	31 29	65.6	15 44	30 32	65.3	15 19	29 36	65.1	14 53	28 39	64.8	14 28	27 43	64.6	209	331
30	150	17 05	32 11	65.0	16 40	31 14	64.7	16 14	30 17	64.4	15 48	29 21	64.2	15 22	28 25	63.9	14 55	27 29	63.7	210	330
31	149	17 37	31 55	64.1	17 11	30 58	63.8	16 44	30 02	63.5	16 17	29 06	63.3	15 50	28 10	63.0	15 23	27 15	62.7	211	329
32	148	18 09	31 38	63.2	17 42	30 42	62.9	17 14	29 46	62.6	16 47	28 51	62.3	16 19	27 55	62.1	15 50	27 00	61.8	212	328
33	147	18 40	31 21	62.3	18 12	30 25	62.0	17 44	29 30	61.7	17 15	28 34	61.4	16 47	27 39	61.2	16 17	26 45	60.9	213	327
34	146	19 11	31 04	61.4	18 42	30 08	61.1	18 13	29 13	60.8	17 44	28 18	60.5	17 14	27 23	60.2	16 44	26 29	60.0	214	326
35	145	19 42	30 46	60.5	19 12	29 50	60.2	18 42	28 55	59.9	18 12	28 01	59.6	17 42	27 06	59.3	17 11	26 12	59.0	215	325
36	144	20 13	30 27	59.6	19 42	29 32	59.2	19 11	28 37	58.9	18 40	27 43	58.6	18 09	26 49	58.4	17 37	25 55	58.1	216	324
37	143	20 43	30 07	58.6	20 12	29 13	58.3	19 40	28 19	58.0	19 08	27 25	57.7	18 36	26 31	57.4	18 03	25 38	57.1	217	323
38	142	21 13	29 48	57.7	20 41	28 53	57.4	20 08	27 59	57.1	19 35	27 06	56.8	19 02	26 13	56.5	18 29	25 20	56.2	218	322
39	141	21 43	29 27	56.8	21 10	28 33	56.4	20 36	27 40	56.1	20 03	26 47	55.8	19 29	25 54	55.5	18 55	25 02	55.2	219	321
40	140	22 12	29 06	55.8	21 38	28 13	55.5	21 04	27 20	55.2	20 30	26 27	54.9	19 55	25 35	54.6	19 20	24 43	54.3	220	320
41	139	22 41	28 44	54.9	22 06	27 51	54.5	21 31	26 59	54.2	20 56	26 07	53.9	20 21	25 15	53.6	19 45	24 24	53.3	221	319
42	138	23 10	28 22	53.9	22 34	27 29	53.6	21 58	26 37	53.3	21 22	25 46	52.9	20 46	24 55	52.6	20 10	24 04	52.3	222	318
43	137	23 38	27 59	53.0	23 02	27 07	52.6	22 25	26 15	52.3	21 48	25 24	52.0	21 11	24 34	51.7	20 34	23 43	51.4	223	317
44	136	24 06	27 36	52.0	23 29	26 44	51.7	22 51	25 53	51.3	22 14	25 02	51.0	21 36	24 12	50.7	20 58	23 23	50.4	224	316
45	135	24 34	27 11	51.0	23 56	26 20	50.7	23 17	25 30	50.3	22 39	24 40	50.0	22 00	23 50	49.7	21 21	23 01	49.4	225	315

Lat. / A		54°			55°			56°			57°			58°			59°			Lat. / A	
LHA / F		A / H	B / P	Z_1 / Z_2	A / H	B / P	Z_1 / Z_2	A / H	B / P	Z_1 / Z_2	A / H	B / P	Z_1 / Z_2	A / H	B / P	Z_1 / Z_2	A / H	B / P	Z_1 / Z_2	LHA	
°	°	° ′	° ′	°	° ′	° ′	°	° ′	° ′	°	° ′	° ′	°	° ′	° ′	°	° ′	° ′	°	°	°
45	135	24 34	27 11	51.0	23 56	26 20	50.7	23 17	25 30	50.3	22 39	24 40	50.0	22 00	23 50	49.7	21 21	23 01	49.4	225	315
46	134	25 01	26 47	50.0	24 22	25 56	49.7	23 43	25 06	49.4	23 04	24 17	49.0	22 24	23 28	48.7	21 45	22 39	48.4	226	314
47	133	25 28	26 22	49.1	24 48	25 32	48.7	24 08	24 42	48.4	23 28	23 53	48.0	22 48	23 05	47.7	22 08	22 17	47.4	227	313
48	132	25 54	25 56	48.1	25 14	25 06	47.7	24 33	24 17	47.4	23 53	23 29	47.0	23 11	22 41	46.7	22 30	21 54	46.4	228	312
49	131	26 20	25 29	47.1	25 39	24 40	46.7	24 58	23 52	46.4	24 16	23 05	46.0	23 34	22 17	45.7	22 52	21 31	45.4	229	311
50	130	26 46	25 02	46.0	26 04	24 14	45.7	25 22	23 26	45.3	24 40	22 39	45.0	23 57	21 53	44.7	23 14	21 07	44.4	230	310
51	129	27 11	24 34	45.0	26 28	23 47	44.7	25 45	23 00	44.3	25 02	22 14	44.0	24 19	21 28	43.7	23 36	20 43	43.4	231	309
52	128	27 36	24 06	44.0	26 52	23 19	43.6	26 09	22 33	43.3	25 25	21 48	43.0	24 41	21 03	42.7	23 57	20 18	42.3	232	308
53	127	28 00	23 37	43.0	27 16	22 51	42.6	26 32	22 06	42.3	25 47	21 21	41.9	25 02	20 37	41.6	24 17	19 53	41.3	233	307
54	126	28 24	23 07	41.9	27 39	22 22	41.6	26 54	21 38	41.2	26 09	20 54	40.9	25 23	20 10	40.6	24 37	19 27	40.3	234	306
55	125	28 47	22 37	40.9	28 01	21 53	40.5	27 16	21 09	40.2	26 30	20 26	39.9	25 44	19 43	39.5	24 57	19 01	39.2	235	305
56	124	29 10	22 07	39.8	28 24	21 23	39.5	27 37	20 40	39.1	26 50	19 57	38.8	26 04	19 16	38.5	25 17	18 34	38.2	236	304
57	123	29 32	21 35	38.8	28 45	20 52	38.4	27 58	20 10	38.1	27 11	19 29	37.8	26 23	18 48	37.4	25 35	18 07	37.1	237	303
58	122	29 54	21 03	37.7	29 06	20 21	37.3	28 19	19 40	37.0	27 31	18 59	36.7	26 42	18 19	36.4	25 54	17 40	36.1	238	302
59	121	30 15	20 31	36.6	29 27	19 50	36.3	28 38	19 09	35.9	27 50	18 30	35.6	27 01	17 50	35.3	26 12	17 12	35.0	239	301
60	120	30 36	19 58	35.5	29 47	19 18	35.2	28 58	18 38	34.9	28 09	17 59	34.5	27 19	17 21	34.2	26 29	16 43	34.0	240	300
61	119	30 56	19 24	34.4	30 07	18 45	34.1	29 17	18 06	33.8	28 27	17 29	33.5	27 37	16 51	33.2	26 46	16 14	32.9	241	299
62	118	31 16	18 50	33.3	30 26	18 12	33.0	29 35	17 34	32.7	28 45	16 57	32.4	27 54	16 21	32.1	27 03	15 45	31.8	242	298
63	117	31 35	18 15	32.2	30 44	17 38	31.9	29 53	17 02	31.6	29 02	16 26	31.3	28 10	15 50	31.0	27 19	15 15	30.7	243	297
64	116	31 53	17 40	31.1	31 02	17 04	30.8	30 10	16 28	30.5	29 19	15 53	30.2	28 27	15 19	29.9	27 35	14 45	29.6	244	296
65	115	32 11	17 04	30.0	31 19	16 29	29.7	30 27	15 55	29.4	29 35	15 21	29.1	28 42	14 48	28.8	27 50	14 15	28.5	245	295
66	114	32 29	16 28	28.8	31 36	15 54	28.5	30 43	15 20	28.2	29 50	14 48	28.0	28 57	14 16	27.7	28 04	13 44	27.4	246	294
67	113	32 45	15 51	27.7	31 52	15 18	27.4	30 59	14 46	27.1	30 05	14 14	26.8	29 12	13 43	26.6	28 18	13 13	26.3	247	293
68	112	33 01	15 14	26.5	32 08	14 42	26.3	31 14	14 11	26.0	30 20	13 40	25.7	29 26	13 10	25.5	28 31	12 41	25.2	248	292
69	111	33 17	14 36	25.4	32 23	14 05	25.1	31 28	13 35	24.8	30 34	13 06	24.6	29 39	12 37	24.4	28 44	12 09	24.1	249	291
70	110	33 32	13 57	24.2	32 37	13 28	24.0	31 42	12 59	23.7	30 47	12 31	23.5	29 52	12 04	23.2	28 57	11 37	23.0	250	290
71	109	33 46	13 18	23.1	32 51	12 51	22.8	31 55	12 23	22.6	31 00	11 56	22.3	30 04	11 30	22.1	29 09	11 04	21.9	251	289
72	108	33 59	12 39	21.9	33 04	12 13	21.6	32 08	11 46	21.4	31 12	11 21	21.2	30 16	10 56	21.0	29 20	10 31	20.8	252	288
73	107	34 12	12 00	20.7	33 16	11 34	20.5	32 20	11 09	20.2	31 23	10 45	20.0	30 27	10 21	19.8	29 30	9 58	19.6	253	287
74	106	34 24	11 19	19.5	33 28	10 55	19.3	32 31	10 32	19.1	31 34	10 09	18.9	30 37	9 46	18.7	29 41	9 24	18.5	254	286
75	105	34 36	10 39	18.3	33 39	10 16	18.1	32 42	9 54	17.9	31 44	9 32	17.7	30 47	9 11	17.5	29 50	8 50	17.4	255	285
76	104	34 46	9 58	17.1	33 49	9 37	16.9	32 52	9 16	16.7	31 54	8 56	16.6	30 57	8 36	16.4	29 59	8 16	16.2	256	284
77	103	34 56	9 17	15.9	33 59	8 57	15.7	33 01	8 38	15.6	32 03	8 19	15.4	31 05	8 00	15.2	30 07	7 42	15.1	257	283
78	102	35 06	8 35	14.7	34 08	8 17	14.5	33 10	7 59	14.4	32 11	7 41	14.2	31 13	7 24	14.1	30 15	7 07	13.9	258	282
79	101	35 14	7 54	13.5	34 16	7 37	13.3	33 18	7 20	13.2	32 19	7 04	13.0	31 21	6 48	12.9	30 22	6 32	12.8	259	281
80	100	35 22	7 11	12.3	34 24	6 56	12.1	33 25	6 41	12.0	32 26	6 26	11.9	31 27	6 12	11.7	30 29	5 57	11.6	260	280
81	99	35 29	6 29	11.1	34 30	6 15	10.9	33 32	6 01	10.8	32 33	5 48	10.7	31 34	5 35	10.6	30 35	5 22	10.5	261	279
82	98	35 36	5 46	9.9	34 37	5 34	9.7	33 37	5 22	9.6	32 38	5 10	9.5	31 39	4 58	9.4	30 40	4 47	9.3	262	278
83	97	35 41	5 04	8.6	34 42	4 53	8.5	33 43	4 42	8.4	32 43	4 32	8.3	31 44	4 21	8.2	30 45	4 11	8.2	263	277
84	96	35 46	4 21	7.4	34 47	4 11	7.3	33 47	4 02	7.2	32 48	3 53	7.1	31 48	3 44	7.1	30 49	3 36	7.0	264	276
85	95	35 51	3 37	6.2	34 51	3 30	6.1	33 51	3 22	6.0	32 52	3 14	6.0	31 52	3 07	5.9	30 52	3 00	5.8	265	275
86	94	35 54	2 54	4.9	34 54	2 48	4.9	33 54	2 42	4.8	32 55	2 36	4.8	31 55	2 30	4.7	30 55	2 24	4.7	266	274
87	93	35 57	2 11	3.7	34 57	2 06	3.7	33 57	2 01	3.6	32 57	1 57	3.6	31 57	1 52	3.5	30 57	1 48	3.5	267	273
88	92	35 58	1 27	2.5	34 59	1 24	2.4	33 59	1 21	2.4	32 59	1 18	2.4	31 59	1 15	2.4	30 59	1 12	2.3	268	272
89	91	36 00	0 44	1.2	35 00	0 42	1.2	34 00	0 40	1.2	33 00	0 39	1.2	32 00	0 37	1.2	31 00	0 36	1.2	269	271
90	90	36 00	0 00	0.0	35 00	0 00	0.0	34 00	0 00	0.0	33 00	0 00	0.0	32 00	0 00	0.0	31 00	0 00	0.0	270	270

N. Lat.: for LHA > 180°... $Z_n = Z$
for LHA < 180°... $Z_n = 360° - Z$

S. Lat.: for LHA > 180°... $Z_n = 180° - Z$
for LHA < 180°... $Z_n = 180° + Z$

SIGHT REDUCTION TABLE

B: (–) for 90°< LHA < 270°
Dec: (–) for Lat. contrary name

Z_1 : same sign as B
Z_2 : (–) for F > 90°

Lat. / A		60°			61°			62°			63°			64°			65°			Lat. / A	
LHA / F		A / H	B / P	Z_1 / Z_2	A / H	B / P	Z_1 / Z_2	A / H	B / P	Z_1 / Z_2	A / H	B / P	Z_1 / Z_2	A / H	B / P	Z_1 / Z_2	A / H	B / P	Z_1 / Z_2	LHA	
°	°	° ′	° ′	°	° ′	° ′	°	° ′	° ′	°	° ′	° ′	°	° ′	° ′	°	° ′	° ′	°	°	°
0	180	0 00	30 00	90.0	0 00	29 00	90.0	0 00	28 00	90.0	0 00	27 00	90.0	0 00	26 00	90.0	0 00	25 00	90.0	180	360
1	179	0 30	30 00	89.1	0 29	29 00	89.1	0 28	28 00	89.1	0 27	27 00	89.1	0 26	26 00	89.1	0 25	25 00	89.1	181	359
2	178	1 00	29 59	88.3	0 58	28 59	88.3	0 56	27 59	88.2	0 54	26 59	88.2	0 53	25 59	88.2	0 51	24 59	88.2	182	358
3	177	1 30	29 58	87.4	1 27	28 58	87.4	1 24	27 58	87.4	1 22	26 58	87.3	1 19	25 58	87.3	1 16	24 58	87.3	183	357
4	176	2 00	29 56	86.5	1 56	28 56	86.5	1 53	27 57	86.5	1 49	26 57	86.4	1 45	25 57	86.4	1 41	24 57	86.4	184	356
5	175	2 30	29 54	85.7	2 25	28 54	85.6	2 21	27 55	85.6	2 16	26 55	85.5	2 11	25 55	85.5	2 07	24 55	85.5	185	355
6	174	3 00	29 52	84.8	2 54	28 52	84.7	2 49	27 52	84.7	2 43	26 52	84.6	2 38	25 53	84.6	2 32	24 53	84.6	186	354
7	173	3 30	29 49	83.9	3 23	28 49	83.9	3 17	27 49	83.8	3 10	26 50	83.8	3 04	25 50	83.7	2 57	24 50	83.7	187	353
8	172	3 59	29 45	83.1	3 52	28 46	83.0	3 45	27 46	82.9	3 37	26 46	82.9	3 30	25 47	82.8	3 22	24 47	82.7	188	352
9	171	4 29	29 42	82.2	4 21	28 42	82.1	4 13	27 42	82.0	4 04	26 43	82.0	3 56	25 43	81.9	3 47	24 44	81.8	189	351
10	170	4 59	29 37	81.3	4 50	28 38	81.2	4 41	27 38	81.2	4 31	26 39	81.1	4 22	25 39	81.0	4 13	24 40	80.9	190	350
11	169	5 28	29 33	80.4	5 18	28 33	80.4	5 08	27 34	80.3	4 58	26 34	80.2	4 48	25 35	80.1	4 38	24 36	80.0	191	349
12	168	5 58	29 27	79.6	5 47	28 28	79.5	5 36	27 29	79.4	5 25	26 29	79.3	5 14	25 30	79.2	5 02	24 31	79.1	192	348
13	167	6 27	29 22	78.7	6 16	28 22	78.6	6 04	27 23	78.5	5 52	26 24	78.4	5 40	25 25	78.3	5 27	24 26	78.2	193	347
14	166	6 57	29 15	77.8	6 44	28 16	77.7	6 31	27 17	77.6	6 18	26 18	77.5	6 05	25 20	77.4	5 52	24 21	77.3	194	346
15	165	7 26	29 09	76.9	7 13	28 10	76.8	6 59	27 11	76.7	6 45	26 12	76.6	6 31	25 14	76.5	6 17	24 15	76.4	195	345
16	164	7 55	29 02	76.1	7 41	28 03	75.9	7 26	27 04	75.8	7 11	26 06	75.7	6 56	25 07	75.5	6 41	24 09	75.4	196	344
17	163	8 24	28 54	75.2	8 09	27 56	75.0	7 53	26 57	74.9	7 38	25 59	74.8	7 22	25 00	74.6	7 06	24 02	74.5	197	343
18	162	8 53	28 46	74.3	8 37	27 48	74.1	8 20	26 50	74.0	8 04	25 51	73.9	7 47	24 53	73.7	7 30	23 55	73.6	198	342
19	161	9 22	28 38	73.4	9 05	27 40	73.2	8 48	26 41	73.1	8 30	25 43	72.9	8 12	24 45	72.8	7 55	23 48	72.7	199	341
20	160	9 51	28 29	72.5	9 33	27 31	72.3	9 14	26 33	72.2	8 56	25 35	72.0	8 37	24 37	71.9	8 19	23 40	71.7	200	340
21	159	10 19	28 19	71.6	10 00	27 22	71.4	9 41	26 24	71.3	9 22	25 26	71.1	9 02	24 29	71.0	8 43	23 32	70.8	201	339
22	158	10 48	28 10	70.7	10 28	27 12	70.5	10 08	26 15	70.4	9 48	25 17	70.2	9 27	24 20	70.0	9 07	23 23	69.9	202	338
23	157	11 16	27 59	69.8	10 55	27 02	69.6	10 34	26 05	69.5	10 13	25 08	69.3	9 52	24 11	69.1	9 30	23 14	69.0	203	337
24	156	11 44	27 49	68.9	11 22	26 51	68.7	11 00	25 54	68.5	10 38	24 58	68.4	10 16	24 01	68.2	9 54	23 04	68.0	204	336
25	155	12 12	27 37	68.0	11 49	26 40	67.8	11 27	25 44	67.6	11 04	24 47	67.4	10 41	23 51	67.3	10 17	22 55	67.1	205	335
26	154	12 40	27 26	67.1	12 16	26 29	66.9	11 53	25 33	66.7	11 29	24 36	66.5	11 05	23 40	66.3	10 41	22 44	66.2	206	334
27	153	13 07	27 13	66.2	12 43	26 17	66.0	12 18	25 21	65.8	11 54	24 25	65.6	11 29	23 29	65.4	11 04	22 34	65.2	207	333
28	152	13 35	27 01	65.3	13 09	26 05	65.1	12 44	25 09	64.9	12 18	24 13	64.7	11 53	23 18	64.5	11 27	22 23	64.3	208	332
29	151	14 02	26 48	64.4	13 36	25 52	64.1	13 09	24 56	63.9	12 43	24 01	63.7	12 16	23 06	63.5	11 49	22 11	63.3	209	331
30	150	14 29	26 34	63.4	14 02	25 39	63.2	13 35	24 43	63.0	13 07	23 49	62.8	12 40	22 54	62.6	12 12	21 59	62.4	210	330
31	149	14 55	26 20	62.5	14 28	25 25	62.3	14 00	24 30	62.1	13 31	23 36	61.8	13 03	22 41	61.6	12 34	21 47	61.4	211	329
32	148	15 22	26 05	61.6	14 53	25 11	61.3	14 24	24 16	61.1	13 55	23 22	60.9	13 26	22 28	60.7	12 56	21 35	60.5	212	328
33	147	15 48	25 50	60.6	15 19	24 56	60.4	14 49	24 02	60.2	14 19	23 08	59.9	13 49	22 15	59.7	13 18	21 22	59.5	213	327
34	146	16 14	25 35	59.7	15 44	24 41	59.5	15 13	23 47	59.2	14 42	22 54	59.0	14 11	22 01	58.8	13 40	21 08	58.6	214	326
35	145	16 40	25 19	58.8	16 09	24 25	58.5	15 37	23 32	58.3	15 06	22 39	58.0	14 34	21 47	57.8	14 02	20 54	57.6	215	325
36	144	17 05	25 02	57.8	16 33	24 09	57.6	16 01	23 17	57.3	15 29	22 24	57.1	14 56	21 32	56.9	14 23	20 40	56.6	216	324
37	143	17 31	24 45	56.9	16 58	23 53	56.6	16 25	23 00	56.4	15 51	22 09	56.1	15 18	21 17	55.9	14 44	20 26	55.7	217	323
38	142	17 56	24 28	55.9	17 22	23 36	55.7	16 48	22 44	55.4	16 14	21 53	55.2	15 39	21 01	54.9	15 05	20 11	54.7	218	322
39	141	18 20	24 10	55.0	17 46	23 18	54.7	17 11	22 27	54.4	16 36	21 36	54.2	16 01	20 46	54.0	15 25	19 55	53.7	219	321
40	140	18 45	23 52	54.0	18 09	23 00	53.7	17 34	22 10	53.5	16 58	21 19	53.2	16 22	20 29	53.0	15 46	19 39	52.7	220	320
41	139	19 09	23 33	53.0	18 33	22 42	52.8	17 56	21 52	52.5	17 20	21 02	52.2	16 43	20 13	52.0	16 06	19 23	51.8	221	319
42	138	19 33	23 13	52.1	18 56	22 23	51.8	18 19	21 34	51.5	17 41	20 44	51.3	17 03	19 55	51.0	16 26	19 07	50.8	222	318
43	137	19 56	22 54	51.1	19 18	22 04	50.8	18 40	21 15	50.5	18 02	20 26	50.3	17 24	19 38	50.0	16 45	18 50	49.8	223	317
44	136	20 19	22 33	50.1	19 41	21 44	49.8	19 02	20 56	49.5	18 23	20 08	49.3	17 44	19 20	49.0	17 04	18 33	48.8	224	316
45	135	20 42	22 12	49.1	20 03	21 24	48.8	19 23	20 36	48.6	18 43	19 49	48.3	18 03	19 02	48.1	17 23	18 15	47.8	225	315

Lat. / A		60°			61°			62°			63°			64°			65°			Lat. / A	
LHA / F		A / H	B / P	Z_1 / Z_2	A / H	B / P	Z_1 / Z_2	A / H	B / P	Z_1 / Z_2	A / H	B / P	Z_1 / Z_2	A / H	B / P	Z_1 / Z_2	A / H	B / P	Z_1 / Z_2	LHA	
°	°	° ′	° ′	°	° ′	° ′	°	° ′	° ′	°	° ′	° ′	°	° ′	° ′	°	° ′	° ′	°	°	°
45	135	20 42	22 12	49.1	20 03	21 24	48.8	19 23	20 36	48.6	18 43	19 49	48.3	18 03	19 02	48.1	17 23	18 15	47.8	225	315
46	134	21 05	21 51	48.1	20 25	21 04	47.8	19 44	20 16	47.6	19 04	19 29	47.3	18 23	18 43	47.1	17 42	17 57	46.8	226	314
47	133	21 27	21 30	47.1	20 46	20 43	46.8	20 05	19 56	46.6	19 24	19 10	46.3	18 42	18 24	46.1	18 00	17 39	45.8	227	313
48	132	21 49	21 07	46.1	21 07	20 21	45.8	20 25	19 35	45.6	19 43	18 50	45.3	19 01	18 04	45.1	18 18	17 20	44.8	228	312
49	131	22 10	20 45	45.1	21 28	19 59	44.8	20 45	19 14	44.6	20 02	18 29	44.3	19 19	17 45	44.0	18 36	17 01	43.8	229	311
50	130	22 31	20 22	44.1	21 48	19 37	43.8	21 05	18 52	43.5	20 21	18 08	43.3	19 37	17 24	43.0	18 53	16 41	42.8	230	310
51	129	22 52	19 58	43.1	22 08	19 14	42.8	21 24	18 30	42.5	20 40	17 47	42.3	19 55	17 04	42.0	19 10	16 21	41.8	231	309
52	128	23 12	19 34	42.1	22 28	18 51	41.8	21 43	18 08	41.5	20 58	17 25	41.2	20 13	16 43	41.0	19 27	16 01	40.8	232	308
53	127	23 32	19 10	41.0	22 47	18 27	40.7	22 01	17 45	40.5	21 15	17 03	40.2	20 30	16 21	40.0	19 44	15 41	39.7	233	307
54	126	23 52	18 45	40.0	23 06	18 03	39.7	22 19	17 21	39.4	21 33	16 40	39.2	20 46	16 00	39.0	20 00	15 20	38.7	234	306
55	125	24 11	18 19	39.0	23 24	17 38	38.7	22 37	16 58	38.4	21 50	16 17	38.2	21 03	15 38	37.9	20 15	14 58	37.7	235	305
56	124	24 29	17 54	37.9	23 42	17 13	37.6	22 54	16 34	37.4	22 07	15 54	37.1	21 19	15 15	36.9	20 31	14 37	36.7	236	304
57	123	24 48	17 27	36.9	23 59	16 48	36.6	23 11	16 09	36.3	22 23	15 31	36.1	21 34	14 53	35.8	20 46	14 15	35.6	237	303
58	122	25 05	17 01	35.8	24 17	16 22	35.5	23 28	15 44	35.3	22 39	15 07	35.0	21 49	14 29	34.8	21 00	13 53	34.6	238	302
59	121	25 23	16 34	34.8	24 33	15 56	34.5	23 44	15 19	34.2	22 54	14 42	34.0	22 04	14 06	33.8	21 14	13 30	33.5	239	301
60	120	25 40	16 06	33.7	24 50	15 29	33.4	23 59	14 53	33.2	23 09	14 18	32.9	22 19	13 42	32.7	21 28	13 07	32.5	240	300
61	119	25 56	15 38	32.6	25 05	15 03	32.4	24 15	14 27	32.1	23 24	13 53	31.9	22 33	13 18	31.7	21 42	12 44	31.5	241	299
62	118	26 12	15 10	31.5	25 21	14 35	31.3	24 29	14 01	31.1	23 38	13 27	30.8	22 46	12 54	30.6	21 55	12 21	30.4	242	298
63	117	26 27	14 41	30.5	25 36	14 08	30.2	24 44	13 34	30.0	23 52	13 01	29.8	22 59	12 29	29.5	22 07	11 57	29.3	243	297
64	116	26 42	14 12	29.4	25 50	13 39	29.1	24 57	13 07	28.9	24 05	12 35	28.7	23 12	12 04	28.5	22 19	11 33	28.3	244	296
65	115	26 57	13 43	28.3	26 04	13 11	28.1	25 11	12 40	27.8	24 18	12 09	27.6	23 25	11 39	27.4	22 31	11 09	27.2	245	295
66	114	27 11	13 13	27.2	26 17	12 42	27.0	25 24	12 12	26.8	24 30	11 43	26.6	23 36	11 13	26.4	22 43	10 44	26.2	246	294
67	113	27 24	12 43	26.1	26 30	12 13	25.9	25 36	11 44	25.7	24 42	11 16	25.5	23 48	10 47	25.3	22 54	10 20	25.1	247	293
68	112	27 37	12 12	25.0	26 43	11 44	24.8	25 48	11 16	24.6	24 54	10 48	24.4	23 59	10 21	24.2	23 04	9 55	24.0	248	292
69	111	27 50	11 41	23.9	26 55	11 14	23.7	26 00	10 47	23.5	25 05	10 21	23.3	24 09	9 55	23.1	23 14	9 29	23.0	249	291
70	110	28 01	11 10	22.8	27 06	10 44	22.6	26 11	10 18	22.4	25 15	9 53	22.2	24 20	9 28	22.0	23 24	9 04	21.9	250	290
71	109	28 13	10 39	21.7	27 17	10 14	21.5	26 21	9 49	21.3	25 25	9 25	21.1	24 29	9 01	21.0	23 33	8 38	20.8	251	289
72	108	28 24	10 07	20.6	27 27	9 43	20.4	26 31	9 20	20.2	25 35	8 57	20.0	24 38	8 34	19.9	23 42	8 12	19.7	252	288
73	107	28 34	9 35	19.4	27 37	9 12	19.3	26 41	8 50	19.1	25 44	8 28	18.9	24 47	8 07	18.8	23 50	7 46	18.6	253	287
74	106	28 44	9 03	18.3	27 47	8 41	18.2	26 50	8 20	18.0	25 52	8 00	17.8	24 55	7 39	17.7	23 58	7 19	17.6	254	286
75	105	28 53	8 30	17.2	27 55	8 10	17.0	26 58	7 50	16.9	26 01	7 31	16.7	25 03	7 12	16.6	24 06	6 53	16.5	255	285
76	104	29 01	7 57	16.1	28 04	7 38	15.9	27 06	7 20	15.8	26 08	7 02	15.6	25 10	6 44	15.5	24 13	6 26	15.4	256	284
77	103	29 09	7 24	14.9	28 11	7 06	14.8	27 13	6 49	14.7	26 15	6 32	14.5	25 17	6 16	14.4	24 19	5 59	14.3	257	283
78	102	29 17	6 51	13.8	28 18	6 34	13.7	27 20	6 19	13.5	26 22	6 03	13.4	25 23	5 47	13.3	24 25	5 32	13.2	258	282
79	101	29 24	6 17	12.7	28 25	6 02	12.5	27 27	5 48	12.4	26 28	5 33	12.3	25 29	5 19	12.2	24 31	5 05	12.1	259	281
80	100	29 30	5 44	11.5	28 31	5 30	11.4	27 32	5 17	11.3	26 33	5 03	11.2	25 35	4 50	11.1	24 36	4 38	11.0	260	280
81	99	29 36	5 10	10.4	28 37	4 57	10.3	27 38	4 45	10.2	26 38	4 33	10.1	25 39	4 22	10.0	24 40	4 10	9.9	261	279
82	98	29 41	4 36	9.2	28 41	4 25	9.1	27 42	4 14	9.0	26 43	4 03	9.0	25 44	3 53	8.9	24 44	3 43	8.8	262	278
83	97	29 45	4 01	8.1	28 46	3 52	8.0	27 46	3 42	7.9	26 47	3 33	7.8	25 48	3 24	7.8	24 48	3 15	7.7	263	277
84	96	29 49	3 27	6.9	28 50	3 19	6.9	27 50	3 11	6.8	26 50	3 03	6.7	25 51	2 55	6.7	24 51	2 47	6.6	264	276
85	95	29 52	2 53	5.8	28 53	2 46	5.7	27 53	2 39	5.7	26 53	2 33	5.6	25 54	2 26	5.6	24 54	2 20	5.5	265	275
86	94	29 55	2 18	4.6	28 55	2 13	4.6	27 56	2 07	4.5	26 56	2 02	4.5	25 56	1 57	4.4	24 56	1 52	4.4	266	274
87	93	29 57	1 44	3.5	28 57	1 40	3.4	27 57	1 36	3.4	26 58	1 32	3.4	25 58	1 28	3.3	24 58	1 24	3.3	267	273
88	92	29 59	1 09	2.3	28 59	1 06	2.3	27 59	1 04	2.3	26 59	1 01	2.2	25 59	0 59	2.2	24 59	0 56	2.2	268	272
89	91	30 00	0 35	1.2	29 00	0 33	1.1	28 00	0 32	1.1	27 00	0 31	1.1	26 00	0 29	1.1	25 00	0 28	1.1	269	271
90	90	30 00	0 00	0.0	29 00	0 00	0.0	28 00	0 00	0.0	27 00	0 00	0.0	26 00	0 00	0.0	25 00	0 00	0.0	270	270

N. Lat.: for LHA > 180°... $Z_n = Z$
for LHA < 180°... $Z_n = 360° - Z$

S. Lat.: for LHA > 180°... $Z_n = 180° - Z$
for LHA < 180°... $Z_n = 180° + Z$

SIGHT REDUCTION TABLE

B: (–) for 90°< LHA < 270°
Dec: (–) for Lat. contrary name

Z_1 : same sign as B
Z_2 : (–) for F > 90°

Lat. / A		66°			67°			68°			69°			70°			71°			Lat. / A	
LHA / F		A / H	B / P	Z_1 / Z_2	A / H	B / P	Z_1 / Z_2	A / H	B / P	Z_1 / Z_2	A / H	B / P	Z_1 / Z_2	A / H	B / P	Z_1 / Z_2	A / H	B / P	Z_1 / Z_2	LHA	
°	°	° ′	° ′	°	° ′	° ′	°	° ′	° ′	°	° ′	° ′	°	° ′	° ′	°	° ′	° ′	°	°	°
0	180	0 00	24 00	90.0	0 00	23 00	90.0	0 00	22 00	90.0	0 00	21 00	90.0	0 00	20 00	90.0	0 00	19 00	90.0	180	360
1	179	0 24	24 00	89.1	0 23	23 00	89.1	0 22	22 00	89.1	0 22	21 00	89.1	0 21	20 00	89.1	0 20	19 00	89.1	181	359
2	178	0 49	23 59	88.2	0 47	22 59	88.2	0 45	21 59	88.1	0 43	20 59	88.1	0 41	19 59	88.1	0 39	18 59	88.1	182	358
3	177	1 13	23 58	87.3	1 10	22 58	87.2	1 07	21 58	87.2	1 04	20 58	87.2	1 02	19 58	87.2	0 59	18 59	87.2	183	357
4	176	1 38	23 57	86.3	1 34	22 57	86.3	1 30	21 57	86.3	1 26	20 57	86.3	1 22	19 57	86.2	1 18	18 57	86.2	184	356
5	175	2 02	23 55	85.4	1 57	22 55	85.4	1 52	21 55	85.4	1 47	20 56	85.3	1 42	19 56	85.3	1 38	18 56	85.3	185	355
6	174	2 26	23 53	84.5	2 20	22 53	84.5	2 15	21 53	84.4	2 09	20 54	84.4	2 03	19 54	84.4	1 57	18 54	84.3	186	354
7	173	2 50	23 50	83.6	2 44	22 51	83.6	2 37	21 51	83.5	2 30	20 51	83.5	2 23	19 52	83.4	2 16	18 52	83.4	187	353
8	172	3 15	23 48	82.7	3 07	22 48	82.6	2 59	21 48	82.6	2 52	20 49	82.5	2 44	19 49	82.5	2 36	18 50	82.4	188	352
9	171	3 39	23 44	81.8	3 30	22 45	81.7	3 22	21 45	81.6	3 13	20 46	81.6	3 04	19 46	81.5	2 55	18 47	81.5	189	351
10	170	4 03	23 41	80.8	3 53	22 41	80.8	3 44	21 42	80.7	3 34	20 42	80.7	3 24	19 43	80.6	3 14	18 44	80.5	190	350
11	169	4 27	23 36	79.9	4 17	22 37	79.9	4 06	21 38	79.8	3 55	20 39	79.7	3 45	19 40	79.6	3 34	18 41	79.6	191	349
12	168	4 51	23 32	79.0	4 40	22 33	78.9	4 28	21 34	78.9	4 16	20 35	78.8	4 05	19 36	78.7	3 53	18 37	78.6	192	348
13	167	5 15	23 27	78.1	5 03	22 28	78.0	4 50	21 29	77.9	4 37	20 30	77.8	4 25	19 32	77.8	4 12	18 33	77.7	193	347
14	166	5 39	23 22	77.2	5 25	22 23	77.1	5 12	21 24	77.0	4 58	20 26	76.9	4 45	19 27	76.8	4 31	18 28	76.7	194	346
15	165	6 03	23 16	76.2	5 48	22 18	76.1	5 34	21 19	76.0	5 19	20 21	76.0	5 05	19 22	75.9	4 50	18 24	75.8	195	345
16	164	6 26	23 10	75.3	6 11	22 12	75.2	5 56	21 13	75.1	5 40	20 15	75.0	5 25	19 17	74.9	5 09	18 19	74.8	196	344
17	163	6 50	23 04	74.4	6 34	22 06	74.3	6 17	21 08	74.2	6 01	20 09	74.1	5 44	19 11	74.0	5 28	18 14	73.9	197	343
18	162	7 13	22 57	73.5	6 56	21 59	73.3	6 39	21 01	73.2	6 21	20 03	73.1	6 04	19 06	73.0	5 46	18 08	72.9	198	342
19	161	7 37	22 50	72.5	7 19	21 52	72.4	7 00	20 54	72.3	6 42	19 57	72.2	6 24	18 59	72.1	6 05	18 02	72.0	199	341
20	160	8 00	22 42	71.6	7 41	21 45	71.5	7 22	20 47	71.4	7 02	19 50	71.2	6 43	18 53	71.1	6 24	17 56	71.0	200	340
21	159	8 23	22 34	70.7	8 03	21 37	70.5	7 43	20 40	70.4	7 23	19 43	70.3	7 02	18 46	70.2	6 42	17 49	70.1	201	339
22	158	8 46	22 26	69.7	8 25	21 29	69.6	8 04	20 32	69.5	7 43	19 35	69.3	7 22	18 39	69.2	7 00	17 42	69.1	202	338
23	157	9 09	22 17	68.8	8 47	21 21	68.7	8 25	20 24	68.5	8 03	19 28	68.4	7 41	18 31	68.3	7 19	17 35	68.1	203	337
24	156	9 31	22 08	67.9	9 09	21 12	67.7	8 46	20 16	67.6	8 23	19 19	67.4	8 00	18 24	67.3	7 37	17 28	67.2	204	336
25	155	9 54	21 58	66.9	9 30	21 03	66.8	9 07	20 07	66.6	8 43	19 11	66.5	8 19	18 15	66.3	7 55	17 20	66.2	205	335
26	154	10 16	21 49	66.0	9 52	20 53	65.8	9 27	19 57	65.7	9 02	19 02	65.5	8 37	18 07	65.4	8 12	17 12	65.2	206	334
27	153	10 38	21 38	65.0	10 13	20 43	64.9	9 48	19 48	64.7	9 22	18 53	64.6	8 56	17 58	64.4	8 30	17 03	64.3	207	333
28	152	11 00	21 28	64.1	10 34	20 33	63.9	10 08	19 38	63.8	9 41	18 43	63.6	9 14	17 49	63.5	8 48	16 55	63.3	208	332
29	151	11 22	21 17	63.1	10 55	20 22	63.0	10 28	19 28	62.8	10 00	18 34	62.6	9 33	17 39	62.5	9 05	16 46	62.3	209	331
30	150	11 44	21 05	62.2	11 16	20 11	62.0	10 48	19 17	61.8	10 19	18 23	61.7	9 51	17 30	61.5	9 22	16 36	61.4	210	330
31	149	12 06	20 53	61.2	11 37	20 00	61.1	11 07	19 06	60.9	10 38	18 13	60.7	10 09	17 20	60.5	9 39	16 27	60.4	211	329
32	148	12 27	20 41	60.3	11 57	19 48	60.1	11 27	18 55	59.9	10 57	18 02	59.7	10 27	17 09	59.6	9 56	16 17	59.4	212	328
33	147	12 48	20 29	59.3	12 17	19 36	59.1	11 46	18 43	58.9	11 15	17 51	58.8	10 44	16 58	58.6	10 13	16 06	58.4	213	327
34	146	13 09	20 16	58.4	12 37	19 23	58.2	12 06	18 31	58.0	11 34	17 39	57.8	11 02	16 47	57.6	10 29	15 56	57.5	214	326
35	145	13 29	20 02	57.4	12 57	19 10	57.2	12 24	18 19	57.0	11 52	17 27	56.8	11 19	16 36	56.7	10 46	15 45	56.5	215	325
36	144	13 50	19 49	56.4	13 17	18 57	56.2	12 43	18 06	56.0	12 10	17 15	55.9	11 36	16 24	55.7	11 02	15 34	55.5	216	324
37	143	14 10	19 34	55.5	13 36	18 44	55.3	13 02	17 53	55.1	12 27	17 03	54.9	11 53	16 12	54.7	11 18	15 23	54.5	217	323
38	142	14 30	19 20	54.5	13 55	18 30	54.3	13 20	17 40	54.1	12 45	16 50	53.9	12 09	16 00	53.7	11 34	15 11	53.5	218	322
39	141	14 50	19 05	53.5	14 14	18 15	53.3	13 38	17 26	53.1	13 02	16 37	52.9	12 26	15 48	52.7	11 49	14 59	52.6	219	321
40	140	15 09	18 50	52.5	14 33	18 01	52.3	13 56	17 12	52.1	13 19	16 23	51.9	12 42	15 35	51.7	12 05	14 47	51.6	220	320
41	139	15 29	18 34	51.5	14 51	17 46	51.3	14 14	16 57	51.1	13 36	16 09	50.9	12 58	15 22	50.8	12 20	14 34	50.6	221	319
42	138	15 48	18 18	50.6	15 09	17 30	50.3	14 31	16 43	50.1	13 52	15 55	49.9	13 14	15 08	49.8	12 35	14 21	49.6	222	318
43	137	16 06	18 02	49.6	15 27	17 15	49.4	14 48	16 28	49.2	14 09	15 41	49.0	13 29	14 54	48.8	12 50	14 08	48.6	223	317
44	136	16 25	17 46	48.6	15 45	16 59	48.4	15 05	16 12	48.2	14 25	15 26	48.0	13 45	14 40	47.8	13 04	13 55	47.6	224	316
45	135	16 43	17 29	47.6	16 02	16 42	47.4	15 22	15 57	47.2	14 41	15 11	47.0	14 00	14 26	46.8	13 19	13 41	46.6	225	315

Lat. / A		66°			67°			68°			69°			70°			71°			Lat. / A	
LHA / F		A / H	B / P	Z_1 / Z_2	A / H	B / P	Z_1 / Z_2	A / H	B / P	Z_1 / Z_2	A / H	B / P	Z_1 / Z_2	A / H	B / P	Z_1 / Z_2	A / H	B / P	Z_1 / Z_2	LHA	
°	°	° ′	° ′	°	° ′	° ′	°	° ′	° ′	°	° ′	° ′	°	° ′	° ′	°	° ′	° ′	°	°	°
45	135	16 43	17 29	47.6	16 02	16 42	47.4	15 22	15 57	47.2	14 41	15 11	47.0	14 00	14 26	46.8	13 19	13 41	46.6	225	315
46	134	17 01	17 11	46.6	16 19	16 26	46.4	15 38	15 41	46.2	14 56	14 56	46.0	14 15	14 11	45.8	13 33	13 27	45.6	226	314
47	133	17 18	16 53	45.6	16 36	16 09	45.4	15 54	15 24	45.2	15 12	14 40	45.0	14 29	13 56	44.8	13 46	13 13	44.6	227	313
48	132	17 36	16 35	44.6	16 53	15 51	44.4	16 10	15 08	44.2	15 27	14 24	44.0	14 43	13 41	43.8	14 00	12 58	43.6	228	312
49	131	17 53	16 17	43.6	17 09	15 34	43.4	16 25	14 51	43.2	15 42	14 08	43.0	14 58	13 26	42.8	14 13	12 44	42.6	229	311
50	130	18 09	15 58	42.6	17 25	15 16	42.4	16 41	14 33	42.1	15 56	13 52	41.9	15 11	13 10	41.8	14 27	12 29	41.6	230	310
51	129	18 26	15 39	41.6	17 41	14 57	41.3	16 56	14 16	41.1	16 10	13 35	40.9	15 25	12 54	40.8	14 39	12 14	40.6	231	309
52	128	18 42	15 20	40.5	17 56	14 39	40.3	17 10	13 58	40.1	16 24	13 18	39.9	15 38	12 38	39.7	14 52	11 58	39.6	232	308
53	127	18 57	15 00	39.5	18 11	14 20	39.3	17 24	13 40	39.1	16 38	13 00	38.9	15 51	12 21	38.7	15 04	11 42	38.6	233	307
54	126	19 13	14 40	38.5	18 26	14 01	38.3	17 39	13 22	38.1	16 51	12 43	37.9	16 04	12 05	37.7	15 16	11 26	37.5	234	306
55	125	19 28	14 20	37.5	18 40	13 41	37.3	17 52	13 03	37.1	17 04	12 25	36.9	16 16	11 48	36.7	15 28	11 10	36.5	235	305
56	124	19 42	13 59	36.4	18 54	13 21	36.2	18 06	12 44	36.0	17 17	12 07	35.8	16 28	11 30	35.7	15 40	10 54	35.5	236	304
57	123	19 57	13 38	35.4	19 08	13 01	35.2	18 19	12 25	35.0	17 29	11 49	34.8	16 40	11 13	34.6	15 51	10 37	34.5	237	303
58	122	20 11	13 17	34.4	19 21	12 41	34.2	18 31	12 05	34.0	17 42	11 30	33.8	16 52	10 55	33.6	16 02	10 20	33.5	238	302
59	121	20 24	12 55	33.3	19 34	12 20	33.1	18 44	11 45	32.9	17 53	11 11	32.8	17 03	10 37	32.6	16 12	10 03	32.4	239	301
60	120	20 37	12 33	32.3	19 47	11 59	32.1	18 56	11 25	31.9	18 05	10 52	31.7	17 14	10 19	31.6	16 23	9 46	31.4	240	300
61	119	20 50	12 11	31.2	19 59	11 38	31.1	19 08	11 05	30.9	18 16	10 33	30.7	17 24	10 00	30.5	16 33	9 29	30.4	241	299
62	118	21 03	11 48	30.2	20 11	11 16	30.0	19 19	10 44	29.8	18 27	10 13	29.7	17 35	9 42	29.5	16 42	9 11	29.4	242	298
63	117	21 15	11 26	29.2	20 22	10 54	29.0	19 30	10 24	28.8	18 37	9 53	28.6	17 45	9 23	28.5	16 52	8 53	28.3	243	297
64	116	21 27	11 03	28.1	20 34	10 32	27.9	19 41	10 03	27.7	18 47	9 33	27.6	17 54	9 04	27.4	17 01	8 35	27.3	244	296
65	115	21 38	10 39	27.0	20 44	10 10	26.9	19 51	9 41	26.7	18 57	9 13	26.5	18 03	8 45	26.4	17 10	8 17	26.3	245	295
66	114	21 49	10 16	26.0	20 55	9 48	25.8	20 01	9 20	25.7	19 07	8 52	25.5	18 12	8 25	25.4	17 18	7 58	25.2	246	294
67	113	21 59	9 52	24.9	21 05	9 25	24.8	20 10	8 58	24.6	19 16	8 32	24.5	18 21	8 06	24.3	17 26	7 40	24.2	247	293
68	112	22 09	9 28	23.9	21 14	9 02	23.7	20 19	8 36	23.5	19 24	8 11	23.4	18 29	7 46	23.3	17 34	7 21	23.1	248	292
69	111	22 19	9 04	22.8	21 24	8 39	22.6	20 28	8 14	22.5	19 33	7 50	22.4	18 37	7 26	22.2	17 42	7 02	22.1	249	291
70	110	22 28	8 40	21.7	21 32	8 16	21.6	20 37	7 52	21.4	19 41	7 29	21.3	18 45	7 06	21.2	17 49	6 43	21.1	250	290
71	109	22 37	8 15	20.7	21 41	7 52	20.5	20 45	7 30	20.4	19 48	7 07	20.2	18 52	6 45	20.1	17 56	6 24	20.0	251	289
72	108	22 45	7 50	19.6	21 49	7 28	19.4	20 52	7 07	19.3	19 56	6 46	19.2	18 59	6 25	19.1	18 02	6 04	19.0	252	288
73	107	22 53	7 25	18.5	21 56	7 04	18.4	21 00	6 44	18.2	20 03	6 24	18.1	19 05	6 04	18.0	18 08	5 45	17.9	253	287
74	106	23 01	7 00	17.4	22 04	6 40	17.3	21 06	6 21	17.2	20 09	6 02	17.1	19 12	5 44	17.0	18 14	5 25	16.9	254	286
75	105	23 08	6 34	16.3	22 10	6 16	16.2	21 13	5 58	16.1	20 15	5 40	16.0	19 17	5 23	15.9	18 20	5 06	15.8	255	285
76	104	23 15	6 09	15.3	22 17	5 52	15.2	21 19	5 35	15.1	20 21	5 18	15.0	19 23	5 02	14.9	18 25	4 46	14.8	256	284
77	103	23 21	5 43	14.2	22 23	5 27	14.1	21 24	5 12	14.0	20 26	4 56	13.9	19 28	4 41	13.8	18 30	4 26	13.7	257	283
78	102	23 27	5 17	13.1	22 28	5 03	13.0	21 30	4 48	12.9	20 31	4 34	12.8	19 33	4 20	12.7	18 34	4 06	12.7	258	282
79	101	23 32	4 51	12.0	22 33	4 38	11.9	21 35	4 24	11.8	20 36	4 11	11.8	19 37	3 58	11.7	18 38	3 46	11.6	259	281
80	100	23 37	4 25	10.9	22 38	4 13	10.8	21 39	4 01	10.8	20 40	3 49	10.7	19 41	3 37	10.6	18 42	3 25	10.6	260	280
81	99	23 41	3 59	9.8	22 42	3 48	9.8	21 43	3 37	9.7	20 44	3 26	9.6	19 45	3 16	9.6	18 45	3 05	9.5	261	279
82	98	23 45	3 33	8.7	22 46	3 23	8.7	21 46	3 13	8.6	20 47	3 03	8.6	19 48	2 54	8.5	18 48	2 45	8.5	262	278
83	97	23 49	3 06	7.7	22 49	2 58	7.6	21 50	2 49	7.5	20 50	2 41	7.5	19 51	2 32	7.4	18 51	2 24	7.4	263	277
84	96	23 52	2 40	6.6	22 52	2 32	6.5	21 52	2 25	6.5	20 53	2 18	6.4	19 53	2 11	6.4	18 54	2 04	6.3	264	276
85	95	23 54	2 13	5.5	22 54	2 07	5.4	21 55	2 01	5.4	20 55	1 55	5.4	19 55	1 49	5.3	18 55	1 43	5.3	265	275
86	94	23 56	1 47	4.4	22 56	1 42	4.3	21 57	1 37	4.3	20 57	1 32	4.3	19 57	1 27	4.3	18 57	1 23	4.2	266	274
87	93	23 58	1 20	3.3	22 58	1 16	3.3	21 58	1 13	3.2	20 58	1 09	3.2	19 58	1 05	3.2	18 58	1 02	3.2	267	273
88	92	23 59	0 53	2.2	22 59	0 51	2.2	21 59	0 48	2.2	20 59	0 46	2.1	19 59	0 44	2.1	18 59	0 41	2.1	268	272
89	91	24 00	0 27	1.1	23 00	0 25	1.1	22 00	0 24	1.1	21 00	0 23	1.1	20 00	0 22	1.1	19 00	0 21	1.1	269	271
90	90	24 00	0 00	0.0	23 00	0 00	0.0	22 00	0 00	0.0	21 00	0 00	0.0	20 00	0 00	0.0	19 00	0 00	0.0	270	270

N. Lat.: for LHA > 180°... $Z_n = Z$
for LHA < 180°... $Z_n = 360° - Z$

S. Lat.: for LHA > 180°... $Z_n = 180° - Z$
for LHA < 180°... $Z_n = 180° + Z$

SIGHT REDUCTION TABLE

B: (–) for 90° < LHA < 270°
Dec: (–) for Lat. contrary name

Z_1 : same sign as B
Z_2 : (–) for F > 90°

Lat. / A		72°			73°			74°			75°			76°			77°			Lat. / A	
LHA / F		A / H	B / P	Z_1 / Z_2	A / H	B / P	Z_1 / Z_2	A / H	B / P	Z_1 / Z_2	A / H	B / P	Z_1 / Z_2	A / H	B / P	Z_1 / Z_2	A / H	B / P	Z_1 / Z_2	LHA	
°	°	° ′	° ′	°	° ′	° ′	°	° ′	° ′	°	° ′	° ′	°	° ′	° ′	°	° ′	° ′	°	°	°
0	180	0 00	18 00	90.0	0 00	17 00	90.0	0 00	16 00	90.0	0 00	15 00	90.0	0 00	14 00	90.0	0 00	13 00	90.0	180	360
1	179	0 19	18 00	89.0	0 18	17 00	89.0	0 17	16 00	89.0	0 16	15 00	89.0	0 15	14 00	89.0	0 13	13 00	89.0	181	359
2	178	0 37	17 59	88.1	0 35	16 59	88.1	0 33	15 59	88.1	0 31	14 59	88.1	0 29	14 00	88.1	0 27	13 00	88.1	182	358
3	177	0 56	17 59	87.1	0 53	16 59	87.1	0 50	15 59	87.1	0 47	14 59	87.1	0 44	13 59	87.1	0 40	12 59	87.1	183	357
4	176	1 14	17 58	86.2	1 10	16 58	86.2	1 06	15 58	86.2	1 02	14 58	86.1	0 58	13 58	86.1	0 54	12 58	86.1	184	356
5	175	1 33	17 56	85.2	1 28	16 56	85.2	1 23	15 57	85.2	1 18	14 57	85.2	1 12	13 57	85.1	1 07	12 57	85.1	185	355
6	174	1 51	17 54	84.3	1 45	16 55	84.3	1 39	15 55	84.2	1 33	14 55	84.2	1 27	13 56	84.2	1 21	12 56	84.2	186	354
7	173	2 09	17 52	83.3	2 03	16 53	83.3	1 56	15 53	83.3	1 48	14 54	83.2	1 41	13 54	83.2	1 34	12 54	83.2	187	353
8	172	2 28	17 50	82.4	2 20	16 51	82.3	2 12	15 51	82.3	2 04	14 52	82.3	1 56	13 52	82.2	1 48	12 53	82.2	188	352
9	171	2 46	17 48	81.4	2 37	16 48	81.4	2 28	15 49	81.3	2 19	14 49	81.3	2 10	13 50	81.3	2 01	12 51	81.2	189	351
10	170	3 05	17 45	80.5	2 55	16 45	80.4	2 45	15 46	80.4	2 35	14 47	80.3	2 24	13 48	80.3	2 14	12 49	80.3	190	350
11	169	3 23	17 41	79.5	3 12	16 42	79.5	3 01	15 43	79.4	2 50	14 44	79.4	2 39	13 45	79.3	2 28	12 46	79.3	191	349
12	168	3 41	17 38	78.6	3 29	16 39	78.5	3 17	15 40	78.5	3 05	14 41	78.4	2 53	13 42	78.3	2 41	12 44	78.3	192	348
13	167	3 59	17 34	77.6	3 46	16 35	77.5	3 33	15 37	77.5	3 20	14 38	77.4	3 07	13 39	77.4	2 54	12 41	77.3	193	347
14	166	4 17	17 30	76.7	4 03	16 31	76.6	3 49	15 33	76.5	3 35	14 34	76.5	3 21	13 36	76.4	3 07	12 38	76.3	194	346
15	165	4 35	17 25	75.7	4 20	16 27	75.6	4 05	15 29	75.6	3 50	14 31	75.5	3 35	13 32	75.4	3 20	12 34	75.4	195	345
16	164	4 53	17 21	74.7	4 37	16 23	74.7	4 21	15 25	74.6	4 05	14 27	74.5	3 49	13 29	74.5	3 33	12 31	74.4	196	344
17	163	5 11	17 16	73.8	4 54	16 18	73.7	4 37	15 20	73.6	4 20	14 22	73.5	4 03	13 25	73.5	3 46	12 27	73.4	197	343
18	162	5 29	17 10	72.8	5 11	16 13	72.7	4 53	15 15	72.7	4 35	14 18	72.6	4 17	13 20	72.5	3 59	12 23	72.4	198	342
19	161	5 46	17 05	71.9	5 28	16 07	71.8	5 09	15 10	71.7	4 50	14 13	71.6	4 31	13 16	71.5	4 12	12 19	71.5	199	341
20	160	6 04	16 59	70.9	5 44	16 02	70.8	5 25	15 05	70.7	5 05	14 08	70.6	4 45	13 11	70.5	4 25	12 14	70.5	200	340
21	159	6 21	16 52	69.9	6 01	15 56	69.8	5 40	14 59	69.7	5 19	14 03	69.7	4 58	13 06	69.6	4 37	12 10	69.5	201	339
22	158	6 39	16 46	69.0	6 17	15 50	68.9	5 56	14 53	68.8	5 34	13 57	68.7	5 12	13 01	68.6	4 50	12 05	68.5	202	338
23	157	6 56	16 39	68.0	6 34	15 43	67.9	6 11	14 47	67.8	5 48	13 51	67.7	5 25	12 56	67.6	5 03	12 00	67.5	203	337
24	156	7 13	16 32	67.1	6 50	15 36	66.9	6 26	14 41	66.8	6 03	13 45	66.7	5 39	12 50	66.6	5 15	11 55	66.5	204	336
25	155	7 30	16 25	66.1	7 06	15 29	66.0	6 41	14 34	65.9	6 17	13 39	65.8	5 52	12 44	65.7	5 27	11 49	65.6	205	335
26	154	7 47	16 17	65.1	7 22	15 22	65.0	6 56	14 27	64.9	6 31	13 32	64.8	6 05	12 38	64.7	5 40	11 43	64.6	206	334
27	153	8 04	16 09	64.1	7 38	15 14	64.0	7 11	14 20	63.9	6 45	13 26	63.8	6 18	12 32	63.7	5 52	11 37	63.6	207	333
28	152	8 20	16 00	63.2	7 53	15 06	63.0	7 26	14 12	62.9	6 59	13 19	62.8	6 31	12 25	62.7	6 04	11 31	62.6	208	332
29	151	8 37	15 52	62.2	8 09	14 58	62.1	7 41	14 05	61.9	7 13	13 11	61.8	6 44	12 18	61.7	6 16	11 25	61.6	209	331
30	150	8 53	15 43	61.2	8 24	14 50	61.1	7 55	13 57	61.0	7 26	13 04	60.9	6 57	12 11	60.7	6 27	11 18	60.6	210	330
31	149	9 09	15 34	60.3	8 40	14 41	60.1	8 10	13 49	60.0	7 40	12 56	59.9	7 09	12 04	59.8	6 39	11 12	59.7	211	329
32	148	9 25	15 24	59.3	8 55	14 32	59.1	8 24	13 40	59.0	7 53	12 48	58.9	7 22	11 56	58.8	6 51	11 05	58.7	212	328
33	147	9 41	15 15	58.3	9 10	14 23	58.2	8 38	13 31	58.0	8 06	12 40	57.9	7 34	11 49	57.8	7 02	10 57	57.7	213	327
34	146	9 57	15 05	57.3	9 25	14 13	57.2	8 52	13 22	57.0	8 19	12 31	56.9	7 46	11 41	56.8	7 14	10 50	56.7	214	326
35	145	10 13	14 54	56.3	9 39	14 04	56.2	9 06	13 13	56.1	8 32	12 23	55.9	7 59	11 33	55.8	7 25	10 43	55.7	215	325
36	144	10 28	14 44	55.4	9 54	13 54	55.2	9 19	13 04	55.1	8 45	12 14	54.9	8 11	11 24	54.8	7 36	10 35	54.7	216	324
37	143	10 43	14 33	54.4	10 08	13 43	54.2	9 33	12 54	54.1	8 58	12 05	53.9	8 22	11 16	53.8	7 47	10 27	53.7	217	323
38	142	10 58	14 22	53.4	10 22	13 33	53.2	9 46	12 44	53.1	9 10	11 55	53.0	8 34	11 07	52.8	7 58	10 19	52.7	218	322
39	141	11 13	14 10	52.4	10 36	13 22	52.2	9 59	12 34	52.1	9 22	11 46	52.0	8 45	10 58	51.8	8 08	10 10	51.7	219	321
40	140	11 27	13 59	51.4	10 50	13 11	51.3	10 12	12 23	51.1	9 35	11 36	51.0	8 57	10 49	50.8	8 19	10 02	50.7	220	320
41	139	11 42	13 47	50.4	11 04	13 00	50.3	10 25	12 13	50.1	9 47	11 26	50.0	9 08	10 39	49.9	8 29	9 53	49.7	221	319
42	138	11 56	13 34	49.4	11 17	12 48	49.3	10 38	12 02	49.1	9 58	11 16	49.0	9 19	10 30	48.9	8 39	9 44	48.7	222	318
43	137	12 10	13 22	48.4	11 30	12 36	48.3	10 50	11 51	48.1	10 10	11 05	48.0	9 30	10 20	47.9	8 49	9 35	47.7	223	317
44	136	12 24	13 09	47.4	11 43	12 24	47.3	11 02	11 39	47.1	10 21	10 55	47.0	9 40	10 10	46.9	8 59	9 26	46.7	224	316
45	135	12 37	12 56	46.4	11 56	12 12	46.3	11 14	11 28	46.1	10 33	10 44	46.0	9 51	10 00	45.9	9 09	9 16	45.7	225	315

Lat. / A		72°			73°			74°			75°			76°			77°			Lat. / A	
LHA / F		A / H	B / P	Z_1 / Z_2	A / H	B / P	Z_1 / Z_2	A / H	B / P	Z_1 / Z_2	A / H	B / P	Z_1 / Z_2	A / H	B / P	Z_1 / Z_2	A / H	B / P	Z_1 / Z_2	LHA	
°	°	° ′	° ′	°	° ′	° ′	°	° ′	° ′	°	° ′	° ′	°	° ′	° ′	°	° ′	° ′	°	°	°
45	135	12 37	12 56	46.4	11 56	12 12	46.3	11 14	11 28	46.1	10 33	10 44	46.0	9 51	10 00	45.9	9 09	9 16	45.7	225	315
46	134	12 51	12 43	45.4	12 08	11 59	45.3	11 26	11 16	45.1	10 44	10 33	45.0	10 01	9 50	44.9	9 19	9 07	44.7	226	314
47	133	13 04	12 30	44.4	12 21	11 47	44.3	11 38	11 04	44.1	10 55	10 21	44.0	10 11	9 39	43.9	9 28	8 57	43.7	227	313
48	132	13 17	12 16	43.4	12 33	11 34	43.3	11 49	10 52	43.1	11 05	10 10	43.0	10 21	9 28	42.9	9 37	8 47	42.7	228	312
49	131	13 29	12 02	42.4	12 45	11 21	42.3	12 00	10 39	42.1	11 16	9 58	42.0	10 31	9 17	41.9	9 46	8 37	41.7	229	311
50	130	13 42	11 48	41.4	12 57	11 07	41.3	12 11	10 27	41.1	11 26	9 46	41.0	10 41	9 06	40.9	9 55	8 26	40.7	230	310
51	129	13 54	11 33	40.4	13 08	10 53	40.3	12 22	10 14	40.1	11 36	9 34	40.0	10 50	8 55	39.8	10 04	8 16	39.7	231	309
52	128	14 06	11 19	39.4	13 19	10 40	39.2	12 33	10 01	39.1	11 46	9 22	39.0	10 59	8 44	38.8	10 13	8 05	38.7	232	308
53	127	14 17	11 04	38.4	13 30	10 26	38.2	12 43	9 47	38.1	11 56	9 10	38.0	11 08	8 32	37.8	10 21	7 55	37.7	233	307
54	126	14 29	10 49	37.4	13 41	10 11	37.2	12 53	9 34	37.1	12 05	8 57	36.9	11 17	8 20	36.8	10 29	7 44	36.7	234	306
55	125	14 40	10 33	36.4	13 51	9 57	36.2	13 03	9 20	36.1	12 14	8 44	35.9	11 26	8 08	35.8	10 37	7 33	35.7	235	305
56	124	14 51	10 18	35.3	14 02	9 42	35.2	13 13	9 07	35.1	12 23	8 31	34.9	11 34	7 56	34.8	10 45	7 21	34.7	236	304
57	123	15 01	10 02	34.3	14 12	9 27	34.2	13 22	8 53	34.0	12 32	8 18	33.9	11 42	7 44	33.8	10 52	7 10	33.7	237	303
58	122	15 12	9 46	33.3	14 21	9 12	33.2	13 31	8 38	33.0	12 41	8 05	32.9	11 50	7 32	32.8	11 00	6 59	32.7	238	302
59	121	15 22	9 30	32.3	14 31	8 57	32.1	13 40	8 24	32.0	12 49	7 51	31.9	11 58	7 19	31.8	11 07	6 47	31.7	239	301
60	120	15 31	9 14	31.3	14 40	8 41	31.1	13 49	8 10	31.0	12 57	7 38	30.9	12 06	7 06	30.8	11 14	6 35	30.6	240	300
61	119	15 41	8 57	30.2	14 49	8 26	30.1	13 57	7 55	30.0	13 05	7 24	29.8	12 13	6 54	29.7	11 21	6 23	29.6	241	299
62	118	15 50	8 40	29.2	14 58	8 10	29.1	14 05	7 40	28.9	13 13	7 10	28.8	12 20	6 41	28.7	11 27	6 11	28.6	242	298
63	117	15 59	8 23	28.2	15 06	7 54	28.0	14 13	7 25	27.9	13 20	6 56	27.8	12 27	6 27	27.7	11 34	5 59	27.6	243	297
64	116	16 08	8 06	27.2	15 14	7 38	27.0	14 21	7 10	26.9	13 27	6 42	26.8	12 34	6 14	26.7	11 40	5 47	26.6	244	296
65	115	16 16	7 49	26.1	15 22	7 22	26.0	14 28	6 55	25.9	13 34	6 28	25.8	12 40	6 01	25.7	11 46	5 34	25.6	245	295
66	114	16 24	7 32	25.1	15 29	7 05	25.0	14 35	6 39	24.9	13 41	6 13	24.7	12 46	5 47	24.6	11 52	5 22	24.6	246	294
67	113	16 32	7 14	24.1	15 37	6 49	23.9	14 42	6 24	23.8	13 47	5 59	23.7	12 52	5 34	23.6	11 57	5 09	23.5	247	293
68	112	16 39	6 56	23.0	15 44	6 32	22.9	14 48	6 08	22.8	13 53	5 44	22.7	12 58	5 20	22.6	12 02	4 57	22.5	248	292
69	111	16 46	6 39	22.0	15 50	6 15	21.9	14 55	5 52	21.8	13 59	5 29	21.7	13 03	5 06	21.6	12 07	4 44	21.5	249	291
70	110	16 53	6 20	20.9	15 57	5 58	20.8	15 01	5 36	20.7	14 05	5 14	20.6	13 08	4 52	20.6	12 12	4 31	20.5	250	290
71	109	16 59	6 02	19.9	16 03	5 41	19.8	15 06	5 20	19.7	14 10	4 59	19.6	13 13	4 38	19.5	12 17	4 18	19.5	251	289
72	108	17 05	5 44	18.9	16 09	5 24	18.8	15 12	5 04	18.7	14 15	4 44	18.6	13 18	4 24	18.5	12 21	4 05	18.4	252	288
73	107	17 11	5 26	17.8	16 14	5 06	17.7	15 17	4 48	17.6	14 20	4 29	17.6	13 23	4 10	17.5	12 25	3 52	17.4	253	287
74	106	17 17	5 07	16.8	16 19	4 49	16.7	15 22	4 31	16.6	14 24	4 13	16.5	13 27	3 56	16.5	12 29	3 38	16.4	254	286
75	105	17 22	4 48	15.7	16 24	4 31	15.7	15 26	4 15	15.6	14 29	3 58	15.5	13 31	3 42	15.4	12 33	3 25	15.4	255	285
76	104	17 27	4 30	14.7	16 29	4 14	14.6	15 31	3 58	14.5	14 33	3 43	14.5	13 35	3 27	14.4	12 36	3 12	14.4	256	284
77	103	17 31	4 11	13.6	16 33	3 56	13.6	15 35	3 41	13.5	14 36	3 27	13.4	13 38	3 13	13.4	12 40	2 58	13.3	257	283
78	102	17 36	3 52	12.6	16 37	3 38	12.5	15 38	3 25	12.5	14 40	3 11	12.4	13 41	2 58	12.4	12 43	2 45	12.3	258	282
79	101	17 39	3 33	11.6	16 41	3 20	11.5	15 42	3 08	11.4	14 43	2 56	11.4	13 44	2 43	11.3	12 45	2 31	11.3	259	281
80	100	17 43	3 14	10.5	16 44	3 02	10.4	15 45	2 51	10.4	14 46	2 40	10.3	13 47	2 29	10.3	12 48	2 18	10.3	260	280
81	99	17 46	2 55	9.5	16 47	2 44	9.4	15 48	2 34	9.4	14 49	2 24	9.3	13 49	2 14	9.3	12 50	2 04	9.2	261	279
82	98	17 49	2 35	8.4	16 50	2 26	8.4	15 50	2 17	8.3	14 51	2 08	8.3	13 52	1 59	8.2	12 52	1 50	8.2	262	278
83	97	17 52	2 16	7.4	16 52	2 08	7.3	15 53	2 00	7.3	14 53	1 52	7.2	13 54	1 44	7.2	12 54	1 37	7.2	263	277
84	96	17 54	1 57	6.3	16 54	1 50	6.3	15 55	1 43	6.2	14 55	1 36	6.2	13 55	1 30	6.2	12 56	1 23	6.2	264	276
85	95	17 56	1 37	5.3	16 56	1 32	5.2	15 56	1 26	5.2	14 56	1 20	5.2	13 57	1 15	5.2	12 57	1 09	5.1	265	275
86	94	17 57	1 18	4.2	16 57	1 13	4.2	15 58	1 09	4.2	14 58	1 04	4.1	13 58	1 00	4.1	12 58	0 55	4.1	266	274
87	93	17 58	0 58	3.2	16 59	0 55	3.1	15 59	0 52	3.1	14 59	0 48	3.1	13 59	0 45	3.1	12 59	0 42	3.1	267	273
88	92	17 59	0 39	2.1	16 59	0 37	2.1	15 59	0 34	2.1	14 59	0 32	2.1	13 59	0 30	2.1	13 00	0 28	2.1	268	272
89	91	18 00	0 19	1.1	17 00	0 18	1.0	16 00	0 17	1.0	15 00	0 16	1.0	14 00	0 15	1.0	13 00	0 14	1.0	269	271
90	90	18 00	0 00	0.0	17 00	0 00	0.0	16 00	0 00	0.0	15 00	0 00	0.0	14 00	0 00	0.0	13 00	0 00	0.0	270	270

N. Lat.: for LHA > 180°... $Z_n = Z$
for LHA < 180°... $Z_n = 360° - Z$

S. Lat.: for LHA > 180°... $Z_n = 180° - Z$
for LHA < 180°... $Z_n = 180° + Z$

SIGHT REDUCTION TABLE

B: (–) for 90° < LHA < 270°
Dec: (–) for Lat. contrary name

Z_1 : same sign as B
Z_2 : (–) for F > 90°

Lat. / A		78°			79°			80°			81°			82°			83°			Lat. / A	
LHA / F		A / H	B / P	Z_1 / Z_2	A / H	B / P	Z_1 / Z_2	A / H	B / P	Z_1 / Z_2	A / H	B / P	Z_1 / Z_2	A / H	B / P	Z_1 / Z_2	A / H	B / P	Z_1 / Z_2	LHA	
°	°	° ′	° ′	°	° ′	° ′	°	° ′	° ′	°	° ′	° ′	°	° ′	° ′	°	° ′	° ′	°	°	°
0	180	0 00	12 00	90.0	0 00	11 00	90.0	0 00	10 00	90.0	0 00	9 00	90.0	0 00	8 00	90.0	0 00	7 00	90.0	180	360
1	179	0 12	12 00	89.0	0 11	11 00	89.0	0 10	10 00	89.0	0 09	9 00	89.0	0 08	8 00	89.0	0 07	7 00	89.0	181	359
2	178	0 25	12 00	88.0	0 23	11 00	88.0	0 21	10 00	88.0	0 19	9 00	88.0	0 17	8 00	88.0	0 15	7 00	88.0	182	358
3	177	0 37	11 59	87.1	0 34	10 59	87.1	0 31	9 59	87.0	0 28	8 59	87.0	0 25	7 59	87.0	0 22	6 59	87.0	183	357
4	176	0 50	11 58	86.1	0 46	10 58	86.1	0 42	9 59	86.1	0 38	8 59	86.0	0 33	7 59	86.0	0 29	6 59	86.0	184	356
5	175	1 02	11 57	85.1	0 57	10 58	85.1	0 52	9 58	85.1	0 47	8 58	85.1	0 42	7 58	85.0	0 37	6 58	85.0	185	355
6	174	1 15	11 56	84.1	1 09	10 56	84.1	1 02	9 57	84.1	0 56	8 57	84.1	0 50	7 57	84.1	0 44	6 58	84.0	186	354
7	173	1 27	11 55	83.2	1 20	10 55	83.1	1 13	9 56	83.1	1 06	8 56	83.1	0 58	7 56	83.1	0 51	6 57	83.1	187	353
8	172	1 39	11 53	82.2	1 31	10 54	82.1	1 23	9 54	82.1	1 15	8 55	82.1	1 07	7 55	82.1	0 58	6 56	82.1	188	352
9	171	1 52	11 51	81.2	1 43	10 52	81.2	1 33	9 53	81.1	1 24	8 53	81.1	1 15	7 54	81.1	1 06	6 55	81.1	189	351
10	170	2 04	11 49	80.2	1 54	10 50	80.2	1 44	9 51	80.1	1 33	8 52	80.1	1 23	7 53	80.1	1 13	6 54	80.1	190	350
11	169	2 16	11 47	79.2	2 05	10 48	79.2	1 54	9 49	79.2	1 43	8 50	79.1	1 31	7 51	79.1	1 20	6 52	79.1	191	349
12	168	2 29	11 45	78.3	2 16	10 46	78.2	2 04	9 47	78.2	1 52	8 48	78.1	1 39	7 50	78.1	1 27	6 51	78.1	192	348
13	167	2 41	11 42	77.3	2 28	10 43	77.2	2 14	9 45	77.2	2 01	8 46	77.2	1 48	7 48	77.1	1 34	6 49	77.1	193	347
14	166	2 53	11 39	76.3	2 39	10 41	76.2	2 24	9 43	76.2	2 10	8 44	76.2	1 56	7 46	76.1	1 41	6 48	76.1	194	346
15	165	3 05	11 36	75.3	2 50	10 38	75.3	2 35	9 40	75.2	2 19	8 42	75.2	2 04	7 44	75.1	1 48	6 46	75.1	195	345
16	164	3 17	11 33	74.3	3 01	10 35	74.3	2 45	9 37	74.2	2 28	8 39	74.2	2 12	7 42	74.1	1 56	6 44	74.1	196	344
17	163	3 29	11 29	73.4	3 12	10 32	73.3	2 55	9 34	73.2	2 37	8 37	73.2	2 20	7 39	73.2	2 03	6 42	73.1	197	343
18	162	3 41	11 26	72.4	3 23	10 28	72.3	3 05	9 31	72.3	2 46	8 34	72.2	2 28	7 37	72.2	2 09	6 40	72.1	198	342
19	161	3 53	11 22	71.4	3 34	10 25	71.3	3 14	9 28	71.3	2 55	8 31	71.2	2 36	7 34	71.2	2 16	6 37	71.1	199	341
20	160	4 05	11 18	70.4	3 45	10 21	70.3	3 24	9 24	70.3	3 04	8 28	70.2	2 44	7 31	70.2	2 23	6 35	70.1	200	340
21	159	4 16	11 13	69.4	3 55	10 17	69.4	3 34	9 21	69.3	3 13	8 25	69.2	2 52	7 28	69.2	2 30	6 32	69.1	201	339
22	158	4 28	11 09	68.4	4 06	10 13	68.4	3 44	9 17	68.3	3 22	8 21	68.2	2 59	7 25	68.2	2 37	6 30	68.1	202	338
23	157	4 40	11 04	67.5	4 17	10 09	67.4	3 53	9 13	67.3	3 30	8 18	67.3	3 07	7 22	67.2	2 44	6 27	67.2	203	337
24	156	4 51	10 59	66.5	4 27	10 04	66.4	4 03	9 09	66.3	3 39	8 14	66.3	3 15	7 19	66.2	2 50	6 24	66.2	204	336
25	155	5 02	10 54	65.5	4 38	9 59	65.4	4 13	9 05	65.3	3 47	8 10	65.3	3 22	7 16	65.2	2 57	6 21	65.2	205	335
26	154	5 14	10 49	64.5	4 48	9 55	64.4	4 22	9 00	64.3	3 56	8 06	64.3	3 30	7 12	64.2	3 04	6 18	64.2	206	334
27	153	5 25	10 43	63.5	4 58	9 50	63.4	4 31	8 56	63.4	4 04	8 02	63.3	3 37	7 08	63.2	3 10	6 15	63.2	207	333
28	152	5 36	10 38	62.5	5 08	9 44	62.4	4 41	8 51	62.4	4 13	7 58	62.3	3 45	7 04	62.2	3 17	6 11	62.2	208	332
29	151	5 47	10 32	61.5	5 18	9 39	61.4	4 50	8 46	61.4	4 21	7 53	61.3	3 52	7 00	61.2	3 23	6 08	61.2	209	331
30	150	5 58	10 26	60.5	5 28	9 33	60.5	4 59	8 41	60.4	4 29	7 49	60.3	3 59	6 56	60.2	3 30	6 04	60.2	210	330
31	149	6 09	10 20	59.6	5 38	9 28	59.5	5 08	8 36	59.4	4 37	7 44	59.3	4 07	6 52	59.2	3 36	6 00	59.2	211	329
32	148	6 20	10 13	58.6	5 48	9 22	58.5	5 17	8 30	58.4	4 45	7 39	58.3	4 14	6 48	58.3	3 42	5 57	58.2	212	328
33	147	6 30	10 06	57.6	5 58	9 16	57.5	5 26	8 25	57.4	4 53	7 34	57.3	4 21	6 43	57.3	3 48	5 53	57.2	213	327
34	146	6 41	10 00	56.6	6 08	9 09	56.5	5 34	8 19	56.4	5 01	7 29	56.3	4 28	6 39	56.3	3 54	5 49	56.2	214	326
35	145	6 51	9 53	55.6	6 17	9 03	55.5	5 43	8 13	55.4	5 09	7 24	55.3	4 35	6 34	55.3	4 00	5 45	55.2	215	325
36	144	7 01	9 45	54.6	6 26	8 56	54.5	5 51	8 07	54.4	5 17	7 18	54.3	4 42	6 29	54.3	4 06	5 40	54.2	216	324
37	143	7 11	9 38	53.6	6 36	8 49	53.5	6 00	8 01	53.4	5 24	7 13	53.3	4 48	6 24	53.3	4 12	5 36	53.2	217	323
38	142	7 21	9 31	52.6	6 45	8 43	52.5	6 08	7 55	52.4	5 32	7 07	52.3	4 55	6 19	52.3	4 18	5 32	52.2	218	322
39	141	7 31	9 23	51.6	6 54	8 35	51.5	6 16	7 48	51.4	5 39	7 01	51.3	5 01	6 14	51.3	4 24	5 27	51.2	219	321
40	140	7 41	9 15	50.6	7 03	8 28	50.5	6 25	7 42	50.4	5 46	6 55	50.3	5 08	6 09	50.3	4 30	5 22	50.2	220	320
41	139	7 50	9 07	49.6	7 11	8 21	49.5	6 32	7 35	49.4	5 53	6 49	49.4	5 14	6 03	49.3	4 35	5 18	49.2	221	319
42	138	8 00	8 59	48.6	7 20	8 13	48.5	6 40	7 28	48.4	6 01	6 43	48.4	5 21	5 58	48.3	4 41	5 13	48.2	222	318
43	137	8 09	8 50	47.6	7 29	8 05	47.5	6 48	7 21	47.4	6 07	6 36	47.4	5 27	5 52	47.3	4 46	5 08	47.2	223	317
44	136	8 18	8 42	46.6	7 37	7 58	46.5	6 56	7 14	46.4	6 14	6 30	46.4	5 33	5 46	46.3	4 51	5 03	46.2	224	316
45	135	8 27	8 33	45.6	7 45	7 50	45.5	7 03	7 06	45.4	6 21	6 23	45.4	5 39	5 41	45.3	4 57	4 58	45.2	225	315

Lat. / A		78°			79°			80°			81°			82°			83°			Lat. / A	
LHA / F		A / H	B / P	Z_1 / Z_2	A / H	B / P	Z_1 / Z_2	A / H	B / P	Z_1 / Z_2	A / H	B / P	Z_1 / Z_2	A / H	B / P	Z_1 / Z_2	A / H	B / P	Z_1 / Z_2	LHA	
°	°	° ′	° ′	°	° ′	° ′	°	° ′	° ′	°	° ′	° ′	°	° ′	° ′	°	° ′	° ′	°	°	°
45	135	8 27	8 33	45.6	7 45	7 50	45.5	7 03	7 06	45.4	6 21	6 23	45.4	5 39	5 41	45.3	4 57	4 58	45.2	225	315
46	134	8 36	8 24	44.6	7 53	7 41	44.5	7 11	6 59	44.4	6 28	6 17	44.4	5 45	5 35	44.3	5 02	4 53	44.2	226	314
47	133	8 45	8 15	43.6	8 01	7 33	43.5	7 18	6 51	43.4	6 34	6 10	43.4	5 51	5 29	43.3	5 07	4 47	43.2	227	313
48	132	8 53	8 06	42.6	8 09	7 25	42.5	7 25	6 44	42.4	6 41	6 03	42.4	5 56	5 22	42.3	5 12	4 42	42.2	228	312
49	131	9 02	7 56	41.6	8 17	7 16	41.5	7 32	6 36	41.4	6 47	5 56	41.4	6 02	5 16	41.3	5 17	4 36	41.2	229	311
50	130	9 10	7 47	40.6	8 24	7 07	40.5	7 39	6 28	40.4	6 53	5 49	40.3	6 07	5 10	40.3	5 21	4 31	40.2	230	310
51	129	9 18	7 37	39.6	8 32	6 58	39.5	7 45	6 20	39.4	6 59	5 42	39.3	6 13	5 03	39.3	5 26	4 25	39.2	231	309
52	128	9 26	7 27	38.6	8 39	6 49	38.5	7 52	6 12	38.4	7 05	5 34	38.3	6 18	4 57	38.3	5 31	4 19	38.2	232	308
53	127	9 33	7 17	37.6	8 46	6 40	37.5	7 58	6 03	37.4	7 11	5 27	37.3	6 23	4 50	37.3	5 35	4 14	37.2	233	307
54	126	9 41	7 07	36.6	8 53	6 31	36.5	8 05	5 55	36.4	7 16	5 19	36.3	6 28	4 43	36.3	5 39	4 08	36.2	234	306
55	125	9 48	6 57	35.6	9 00	6 22	35.5	8 11	5 47	35.4	7 22	5 11	35.3	6 33	4 37	35.3	5 44	4 02	35.2	235	305
56	124	9 56	6 47	34.6	9 06	6 12	34.5	8 17	5 38	34.4	7 27	5 04	34.3	6 38	4 30	34.3	5 48	3 56	34.2	236	304
57	123	10 03	6 36	33.6	9 13	6 03	33.5	8 22	5 29	33.4	7 32	4 56	33.3	6 42	4 23	33.3	5 52	3 50	33.2	237	303
58	122	10 09	6 26	32.6	9 19	5 53	32.5	8 28	5 20	32.4	7 37	4 48	32.3	6 47	4 16	32.3	5 56	3 43	32.2	238	302
59	121	10 16	6 15	31.6	9 25	5 43	31.5	8 34	5 11	31.4	7 42	4 40	31.3	6 51	4 08	31.2	6 00	3 37	31.2	239	301
60	120	10 22	6 04	30.6	9 31	5 33	30.5	8 39	5 02	30.4	7 47	4 32	30.3	6 55	4 01	30.2	6 04	3 31	30.2	240	300
61	119	10 29	5 53	29.5	9 36	5 23	29.5	8 44	4 53	29.4	7 52	4 23	29.3	6 59	3 54	29.2	6 07	3 24	29.2	241	299
62	118	10 35	5 42	28.5	9 42	5 13	28.4	8 49	4 44	28.4	7 56	4 15	28.3	7 04	3 46	28.2	6 11	3 18	28.2	242	298
63	117	10 41	5 31	27.5	9 47	5 03	27.4	8 54	4 35	27.4	8 01	4 07	27.3	7 07	3 39	27.2	6 14	3 11	27.2	243	297
64	116	10 46	5 19	26.5	9 52	4 52	26.4	8 59	4 25	26.3	8 05	3 58	26.3	7 11	3 32	26.2	6 17	3 05	26.2	244	296
65	115	10 52	5 08	25.5	9 57	4 42	25.4	9 03	4 16	25.3	8 09	3 50	25.3	7 15	3 24	25.2	6 20	2 58	25.2	245	295
66	114	10 57	4 56	24.5	10 02	4 31	24.4	9 08	4 06	24.3	8 13	3 41	24.3	7 18	3 16	24.2	6 24	2 52	24.2	246	294
67	113	11 02	4 45	23.5	10 07	4 21	23.4	9 12	3 56	23.3	8 17	3 32	23.3	7 22	3 09	23.2	6 26	2 45	23.2	247	293
68	112	11 07	4 33	22.4	10 11	4 10	22.4	9 16	3 47	22.3	8 20	3 24	22.2	7 25	3 01	22.2	6 29	2 38	22.1	248	292
69	111	11 12	4 21	21.4	10 16	3 59	21.4	9 20	3 37	21.3	8 24	3 15	21.2	7 28	2 53	21.2	6 32	2 31	21.1	249	291
70	110	11 16	4 09	20.4	10 20	3 48	20.3	9 23	3 27	20.3	8 27	3 06	20.2	7 31	2 45	20.2	6 35	2 24	20.1	250	290
71	109	11 20	3 58	19.4	10 24	3 37	19.3	9 27	3 17	19.3	8 30	2 57	19.2	7 34	2 37	19.2	6 37	2 17	19.1	251	289
72	108	11 24	3 45	18.4	10 27	3 26	18.3	9 30	3 07	18.3	8 33	2 48	18.2	7 36	2 29	18.2	6 39	2 10	18.1	252	288
73	107	11 28	3 33	17.4	10 31	3 15	17.3	9 34	2 57	17.2	8 36	2 39	17.2	7 39	2 21	17.2	6 42	2 03	17.1	253	287
74	106	11 32	3 21	16.3	10 34	3 04	16.3	9 37	2 47	16.2	8 39	2 30	16.2	7 41	2 13	16.1	6 44	1 56	16.1	254	286
75	105	11 35	3 09	15.3	10 37	2 53	15.3	9 39	2 37	15.2	8 41	2 21	15.2	7 44	2 05	15.1	6 46	1 49	15.1	255	285
76	104	11 38	2 57	14.3	10 40	2 42	14.3	9 42	2 27	14.2	8 44	2 12	14.2	7 46	1 57	14.1	6 47	1 42	14.1	256	284
77	103	11 41	2 44	13.3	10 43	2 30	13.2	9 44	2 16	13.2	8 46	2 02	13.2	7 48	1 49	13.1	6 49	1 35	13.1	257	283
78	102	11 44	2 32	12.3	10 45	2 19	12.2	9 47	2 06	12.2	8 48	1 53	12.1	7 49	1 40	12.1	6 51	1 28	12.1	258	282
79	101	11 47	2 19	11.2	10 48	2 07	11.2	9 49	1 56	11.2	8 50	1 44	11.1	7 51	1 32	11.1	6 52	1 21	11.1	259	281
80	100	11 49	2 07	10.2	10 50	1 56	10.2	9 51	1 45	10.2	8 52	1 35	10.1	7 53	1 24	10.1	6 54	1 13	10.1	260	280
81	99	11 51	1 54	9.2	10 52	1 45	9.2	9 53	1 35	9.1	8 53	1 25	9.1	7 54	1 16	9.1	6 55	1 06	9.1	261	279
82	98	11 53	1 42	8.2	10 53	1 33	8.1	9 54	1 24	8.1	8 55	1 16	8.1	7 55	1 07	8.1	6 56	0 59	8.1	262	278
83	97	11 55	1 29	7.2	10 55	1 21	7.1	9 55	1 14	7.1	8 56	1 06	7.1	7 56	0 59	7.1	6 57	0 51	7.1	263	277
84	96	11 56	1 16	6.1	10 56	1 10	6.1	9 57	1 03	6.1	8 57	0 57	6.1	7 57	0 50	6.1	6 58	0 44	6.0	264	276
85	95	11 57	1 04	5.1	10 57	0 58	5.1	9 58	0 53	5.1	8 58	0 47	5.1	7 58	0 42	5.0	6 58	0 37	5.0	265	275
86	94	11 58	0 51	4.1	10 58	0 47	4.1	9 59	0 42	4.1	8 59	0 38	4.0	7 59	0 34	4.0	6 59	0 29	4.0	266	274
87	93	11 59	0 38	3.1	10 59	0 35	3.1	9 59	0 32	3.0	8 59	0 28	3.0	7 59	0 25	3.0	6 59	0 22	3.0	267	273
88	92	12 00	0 26	2.0	11 00	0 23	2.0	10 00	0 21	2.0	9 00	0 19	2.0	8 00	0 17	2.0	7 00	0 15	2.0	268	272
89	91	12 00	0 13	1.0	11 00	0 12	1.0	10 00	0 11	1.0	9 00	0 10	1.0	8 00	0 08	1.0	7 00	0 07	1.0	269	271
90	90	12 00	0 00	0.0	11 00	0 00	0.0	10 00	0 00	0.0	9 00	0 00	0.0	8 00	0 00	0.0	7 00	0 00	0.0	270	270

N. Lat.: for LHA > 180°... $Z_n = Z$
for LHA < 180°... $Z_n = 360° - Z$

S. Lat.: for LHA > 180°... $Z_n = 180° - Z$
for LHA < 180°... $Z_n = 180° + Z$

SIGHT REDUCTION TABLE

B: (–) for 90° < LHA < 270°
Dec: (–) for Lat. contrary name

Z_1 : same sign as B
Z_2 : (–) for F > 90°

Lat. / A		84°			85°			86°			87°			88°			89°			Lat. / A	
LHA / F		A / H	B / P	Z_1 / Z_2	A / H	B / P	Z_1 / Z_2	A / H	B / P	Z_1 / Z_2	A / H	B / P	Z_1 / Z_2	A / H	B / P	Z_1 / Z_2	A / H	B / P	Z_1 / Z_2	LHA	
°	°	° ′	° ′	°	° ′	° ′	°	° ′	° ′	°	° ′	° ′	°	° ′	° ′	°	° ′	° ′	°	°	°
0	180	0 00	6 00	90.0	0 00	5 00	90.0	0 00	4 00	90.0	0 00	3 00	90.0	0 00	2 00	90.0	0 00	1 00	90.0	180	360
1	179	0 06	6 00	89.0	0 05	5 00	89.0	0 04	4 00	89.0	0 03	3 00	89.0	0 02	2 00	89.0	0 01	1 00	89.0	181	359
2	178	0 13	6 00	88.0	0 10	5 00	88.0	0 08	4 00	88.0	0 06	3 00	88.0	0 04	2 00	88.0	0 02	1 00	88.0	182	358
3	177	0 19	6 00	87.0	0 16	5 00	87.0	0 13	4 00	87.0	0 09	3 00	87.0	0 06	2 00	87.0	0 03	1 00	87.0	183	357
4	176	0 25	5 59	86.0	0 21	4 59	86.0	0 17	3 59	86.0	0 13	3 00	86.0	0 08	2 00	86.0	0 04	1 00	86.0	184	356
5	175	0 31	5 59	85.0	0 26	4 59	85.0	0 21	3 59	85.0	0 16	2 59	85.0	0 10	2 00	85.0	0 05	1 00	85.0	185	355
6	174	0 38	5 58	84.0	0 31	4 58	84.0	0 25	3 59	84.0	0 19	2 59	84.0	0 13	1 59	84.0	0 06	1 00	84.0	186	354
7	173	0 44	5 57	83.0	0 37	4 58	83.0	0 29	3 58	83.0	0 22	2 59	83.0	0 15	1 59	83.0	0 07	1 00	83.0	187	353
8	172	0 50	5 57	82.0	0 42	4 57	82.0	0 33	3 58	82.0	0 25	2 58	82.0	0 17	1 59	82.0	0 08	0 59	82.0	188	352
9	171	0 56	5 56	81.0	0 47	4 56	81.0	0 38	3 57	81.0	0 28	2 58	81.0	0 19	1 59	81.0	0 09	0 59	81.0	189	351
10	170	1 02	5 55	80.1	0 52	4 55	80.0	0 42	3 56	80.0	0 31	2 57	80.0	0 21	1 58	80.0	0 10	0 59	80.0	190	350
11	169	1 09	5 53	79.1	0 57	4 55	79.0	0 46	3 56	79.0	0 34	2 57	79.0	0 23	1 58	79.0	0 11	0 59	79.0	191	349
12	168	1 15	5 52	78.1	1 02	4 53	78.0	0 50	3 55	78.0	0 37	2 56	78.0	0 25	1 57	78.0	0 12	0 59	78.0	192	348
13	167	1 21	5 51	77.1	1 07	4 52	77.0	0 54	3 54	77.0	0 40	2 55	77.0	0 27	1 57	77.0	0 13	0 58	77.0	193	347
14	166	1 27	5 49	76.1	1 12	4 51	76.1	0 58	3 53	76.0	0 44	2 55	76.0	0 29	1 56	76.0	0 15	0 58	76.0	194	346
15	165	1 33	5 48	75.1	1 18	4 50	75.1	1 02	3 52	75.0	0 47	2 54	75.0	0 31	1 56	75.0	0 16	0 58	75.0	195	345
16	164	1 39	5 46	74.1	1 23	4 48	74.1	1 06	3 51	74.0	0 50	2 53	74.0	0 33	1 55	74.0	0 17	0 58	74.0	196	344
17	163	1 45	5 44	73.1	1 28	4 47	73.1	1 10	3 50	73.0	0 53	2 52	73.0	0 35	1 55	73.0	0 18	0 57	73.0	197	343
18	162	1 51	5 43	72.1	1 33	4 45	72.1	1 14	3 48	72.0	0 56	2 51	72.0	0 37	1 54	72.0	0 19	0 57	72.0	198	342
19	161	1 57	5 41	71.1	1 38	4 44	71.1	1 18	3 47	71.0	0 59	2 50	71.0	0 39	1 53	71.0	0 20	0 57	71.0	199	341
20	160	2 03	5 38	70.1	1 42	4 42	70.1	1 22	3 46	70.0	1 02	2 49	70.0	0 41	1 53	70.0	0 21	0 56	70.0	200	340
21	159	2 09	5 36	69.1	1 47	4 40	69.1	1 26	3 44	69.0	1 04	2 48	69.0	0 43	1 52	69.0	0 22	0 56	69.0	201	339
22	158	2 15	5 34	68.1	1 52	4 38	68.1	1 30	3 43	68.0	1 07	2 47	68.0	0 45	1 51	68.0	0 22	0 56	68.0	202	338
23	157	2 20	5 32	67.1	1 57	4 36	67.1	1 34	3 41	67.1	1 10	2 46	67.0	0 47	1 50	67.0	0 23	0 55	67.0	203	337
24	156	2 26	5 29	66.1	2 02	4 34	66.1	1 38	3 39	66.1	1 13	2 44	66.0	0 49	1 50	66.0	0 24	0 55	66.0	204	336
25	155	2 32	5 26	65.1	2 07	4 32	65.1	1 41	3 38	65.1	1 16	2 43	65.0	0 51	1 49	65.0	0 25	0 54	65.0	205	335
26	154	2 38	5 24	64.1	2 11	4 30	64.1	1 45	3 36	64.1	1 19	2 42	64.0	0 53	1 48	64.0	0 26	0 54	64.0	206	334
27	153	2 43	5 21	63.1	2 16	4 27	63.1	1 49	3 34	63.1	1 22	2 40	63.0	0 54	1 47	63.0	0 27	0 53	63.0	207	333
28	152	2 49	5 18	62.1	2 21	4 25	62.1	1 53	3 32	62.1	1 24	2 39	62.0	0 56	1 46	62.0	0 28	0 53	62.0	208	332
29	151	2 54	5 15	61.1	2 25	4 23	61.1	1 56	3 30	61.1	1 27	2 37	61.0	0 58	1 45	61.0	0 29	0 52	61.0	209	331
30	150	3 00	5 12	60.1	2 30	4 20	60.1	2 00	3 28	60.1	1 30	2 36	60.0	1 00	1 44	60.0	0 30	0 52	60.0	210	330
31	149	3 05	5 09	59.1	2 34	4 17	59.1	2 04	3 26	59.1	1 33	2 34	59.0	1 02	1 43	59.0	0 31	0 51	59.0	211	329
32	148	3 11	5 06	58.1	2 39	4 15	58.1	2 07	3 24	58.1	1 35	2 33	58.0	1 04	1 42	58.0	0 32	0 51	58.0	212	328
33	147	3 16	5 02	57.1	2 43	4 12	57.1	2 11	3 21	57.1	1 38	2 31	57.0	1 05	1 41	57.0	0 33	0 50	57.0	213	327
34	146	3 21	4 59	56.1	2 48	4 09	56.1	2 14	3 19	56.1	1 41	2 29	56.0	1 07	1 39	56.0	0 34	0 50	56.0	214	326
35	145	3 26	4 55	55.1	2 52	4 06	55.1	2 18	3 17	55.1	1 43	2 27	55.0	1 09	1 38	55.0	0 34	0 49	55.0	215	325
36	144	3 31	4 52	54.1	2 56	4 03	54.1	2 21	3 14	54.1	1 46	2 26	54.0	1 11	1 37	54.0	0 35	0 49	54.0	216	324
37	143	3 36	4 48	53.2	3 00	4 00	53.1	2 24	3 12	53.1	1 48	2 24	53.0	1 12	1 36	53.0	0 36	0 48	53.0	217	323
38	142	3 41	4 44	52.2	3 05	3 57	52.1	2 28	3 09	52.1	1 51	2 22	52.0	1 14	1 35	52.0	0 37	0 47	52.0	218	322
39	141	3 46	4 40	51.2	3 09	3 53	51.1	2 31	3 07	51.1	1 53	2 20	51.0	1 16	1 33	51.0	0 38	0 47	51.0	219	321
40	140	3 51	4 36	50.2	3 13	3 50	50.1	2 34	3 04	50.1	1 56	2 18	50.0	1 17	1 32	50.0	0 39	0 46	50.0	220	320
41	139	3 56	4 32	49.2	3 17	3 47	49.1	2 37	3 01	49.1	1 58	2 16	49.0	1 19	1 31	49.0	0 39	0 45	49.0	221	319
42	138	4 01	4 28	48.2	3 21	3 43	48.1	2 41	2 58	48.1	2 00	2 14	48.0	1 20	1 29	48.0	0 40	0 45	48.0	222	318
43	137	4 05	4 24	47.2	3 24	3 40	47.1	2 44	2 56	47.1	2 03	2 12	47.0	1 22	1 28	47.0	0 41	0 44	47.0	223	317
44	136	4 10	4 19	46.2	3 28	3 36	46.1	2 47	2 53	46.1	2 05	2 10	46.0	1 23	1 26	46.0	0 42	0 43	46.0	224	316
45	135	4 14	4 15	45.2	3 32	3 32	45.1	2 50	2 50	45.1	2 07	2 07	45.0	1 25	1 25	45.0	0 42	0 42	45.0	225	315

Lat. / A		84°			85°			86°			87°			88°			89°			Lat. / A	
LHA / F		A / H	B / P	Z_1 / Z_2	A / H	B / P	Z_1 / Z_2	A / H	B / P	Z_1 / Z_2	A / H	B / P	Z_1 / Z_2	A / H	B / P	Z_1 / Z_2	A / H	B / P	Z_1 / Z_2	LHA	
°	°	° ′	° ′	°	° ′	° ′	°	° ′	° ′	°	° ′	° ′	°	° ′	° ′	°	° ′	° ′	°	°	°
45	135	4 14	4 15	45.2	3 32	3 32	45.1	2 50	2 50	45.1	2 07	2 07	45.0	1 25	1 25	45.0	0 42	0 42	45.0	225	315
46	134	4 19	4 11	44.2	3 36	3 29	44.1	2 53	2 47	44.1	2 09	2 05	44.0	1 26	1 23	44.0	0 43	0 42	44.0	226	314
47	133	4 23	4 06	43.2	3 39	3 25	43.1	2 55	2 44	43.1	2 12	2 03	43.0	1 28	1 22	43.0	0 44	0 41	43.0	227	313
48	132	4 27	4 01	42.2	3 43	3 21	42.1	2 58	2 41	42.1	2 14	2 01	42.0	1 29	1 20	42.0	0 45	0 40	42.0	228	312
49	131	4 31	3 57	41.2	3 46	3 17	41.1	3 01	2 38	41.1	2 16	1 58	41.0	1 31	1 19	41.0	0 45	0 39	41.0	229	311
50	130	4 36	3 52	40.2	3 50	3 13	40.1	3 04	2 34	40.1	2 18	1 56	40.0	1 32	1 17	40.0	0 46	0 39	40.0	230	310
51	129	4 40	3 47	39.2	3 53	3 09	39.1	3 06	2 31	39.1	2 20	1 53	39.0	1 33	1 16	39.0	0 47	0 38	39.0	231	309
52	128	4 43	3 42	38.2	3 56	3 05	38.1	3 09	2 28	38.1	2 22	1 51	38.0	1 35	1 14	38.0	0 47	0 37	38.0	232	308
53	127	4 47	3 37	37.2	3 59	3 01	37.1	3 12	2 25	37.1	2 24	1 48	37.0	1 36	1 12	37.0	0 48	0 36	37.0	233	307
54	126	4 51	3 32	36.1	4 03	2 57	36.1	3 14	2 21	36.1	2 26	1 46	36.0	1 37	1 11	36.0	0 49	0 35	36.0	234	306
55	125	4 55	3 27	35.1	4 06	2 52	35.1	3 17	2 18	35.1	2 27	1 43	35.0	1 38	1 09	35.0	0 49	0 34	35.0	235	305
56	124	4 58	3 22	34.1	4 09	2 48	34.1	3 19	2 14	34.1	2 29	1 41	34.0	1 39	1 07	34.0	0 50	0 34	34.0	236	304
57	123	5 02	3 17	33.1	4 12	2 44	33.1	3 21	2 11	33.1	2 31	1 38	33.0	1 41	1 05	33.0	0 50	0 33	33.0	237	303
58	122	5 05	3 11	32.1	4 14	2 39	32.1	3 23	2 07	32.1	2 33	1 35	32.0	1 42	1 04	32.0	0 51	0 32	32.0	238	302
59	121	5 08	3 06	31.1	4 17	2 35	31.1	3 26	2 04	31.1	2 34	1 33	31.0	1 43	1 02	31.0	0 51	0 31	31.0	239	301
60	120	5 12	3 00	30.1	4 20	2 30	30.1	3 28	2 00	30.1	2 36	1 30	30.0	1 44	1 00	30.0	0 52	0 30	30.0	240	300
61	119	5 15	2 55	29.1	4 22	2 26	29.1	3 30	1 56	29.1	2 37	1 27	29.0	1 45	0 58	29.0	0 52	0 29	29.0	241	299
62	118	5 18	2 49	28.1	4 25	2 21	28.1	3 32	1 53	28.1	2 39	1 25	28.0	1 46	0 56	28.0	0 53	0 28	28.0	242	298
63	117	5 21	2 44	27.1	4 27	2 16	27.1	3 34	1 49	27.1	2 40	1 22	27.0	1 47	0 54	27.0	0 53	0 27	27.0	243	297
64	116	5 23	2 38	26.1	4 30	2 12	26.1	3 36	1 45	26.1	2 42	1 19	26.0	1 48	0 53	26.0	0 54	0 26	26.0	244	296
65	115	5 26	2 33	25.1	4 32	2 07	25.1	3 37	1 42	25.1	2 43	1 16	25.0	1 49	0 51	25.0	0 54	0 25	25.0	245	295
66	114	5 29	2 27	24.1	4 34	2 02	24.1	3 39	1 38	24.1	2 44	1 13	24.0	1 50	0 49	24.0	0 55	0 24	24.0	246	294
67	113	5 31	2 21	23.1	4 36	1 57	23.1	3 41	1 34	23.1	2 46	1 10	23.0	1 50	0 47	23.0	0 55	0 23	23.0	247	293
68	112	5 34	2 15	22.1	4 38	1 53	22.1	3 42	1 30	22.0	2 47	1 07	22.0	1 51	0 45	22.0	0 56	0 22	22.0	248	292
69	111	5 36	2 09	21.1	4 40	1 48	21.1	3 44	1 26	21.0	2 48	1 05	21.0	1 52	0 43	21.0	0 56	0 22	21.0	249	291
70	110	5 38	2 04	20.1	4 42	1 43	20.1	3 46	1 22	20.0	2 49	1 02	20.0	1 53	0 41	20.0	0 56	0 21	20.0	250	290
71	109	5 40	1 58	19.1	4 44	1 38	19.1	3 47	1 18	19.0	2 50	0 59	19.0	1 53	0 39	19.0	0 57	0 20	19.0	251	289
72	108	5 42	1 52	18.1	4 45	1 33	18.1	3 48	1 14	18.0	2 51	0 56	18.0	1 54	0 37	18.0	0 57	0 19	18.0	252	288
73	107	5 44	1 46	17.1	4 47	1 28	17.1	3 49	1 10	17.0	2 52	0 53	17.0	1 55	0 35	17.0	0 57	0 18	17.0	253	287
74	106	5 46	1 40	16.1	4 48	1 23	16.1	3 51	1 06	16.0	2 53	0 50	16.0	1 55	0 33	16.0	0 58	0 17	16.0	254	286
75	105	5 48	1 33	15.1	4 50	1 18	15.1	3 52	1 02	15.0	2 54	0 47	15.0	1 56	0 31	15.0	0 58	0 16	15.0	255	285
76	104	5 49	1 27	14.1	4 51	1 13	14.1	3 53	0 58	14.0	2 55	0 44	14.0	1 56	0 29	14.0	0 58	0 15	14.0	256	284
77	103	5 51	1 21	13.1	4 52	1 08	13.0	3 54	0 54	13.0	2 55	0 41	13.0	1 57	0 27	13.0	0 58	0 13	13.0	257	283
78	102	5 52	1 15	12.1	4 53	1 03	12.0	3 55	0 50	12.0	2 56	0 37	12.0	1 57	0 25	12.0	0 59	0 12	12.0	258	282
79	101	5 53	1 09	11.1	4 54	0 57	11.0	3 56	0 46	11.0	2 57	0 34	11.0	1 58	0 23	11.0	0 59	0 11	11.0	259	281
80	100	5 55	1 03	10.1	4 55	0 52	10.0	3 56	0 42	10.0	2 57	0 31	10.0	1 58	0 21	10.0	0 59	0 10	10.0	260	280
81	99	5 56	0 57	9.0	4 56	0 47	9.0	3 57	0 38	9.0	2 58	0 28	9.0	1 59	0 19	9.0	0 59	0 09	9.0	261	279
82	98	5 56	0 50	8.0	4 57	0 42	8.0	3 58	0 33	8.0	2 58	0 25	8.0	1 59	0 17	8.0	0 59	0 08	8.0	262	278
83	97	5 57	0 44	7.0	4 58	0 37	7.0	3 58	0 29	7.0	2 59	0 22	7.0	1 59	0 15	7.0	1 00	0 07	7.0	263	277
84	96	5 58	0 38	6.0	4 58	0 31	6.0	3 59	0 25	6.0	2 59	0 19	6.0	1 59	0 13	6.0	1 00	0 06	6.0	264	276
85	95	5 59	0 31	5.0	4 59	0 26	5.0	3 59	0 21	5.0	2 59	0 16	5.0	2 00	0 10	5.0	1 00	0 05	5.0	265	275
86	94	5 59	0 25	4.0	4 59	0 21	4.0	3 59	0 17	4.0	3 00	0 13	4.0	2 00	0 08	4.0	1 00	0 04	4.0	266	274
87	93	6 00	0 19	3.0	5 00	0 16	3.0	4 00	0 13	3.0	3 00	0 09	3.0	2 00	0 06	3.0	1 00	0 03	3.0	267	273
88	92	6 00	0 13	2.0	5 00	0 10	2.0	4 00	0 08	2.0	3 00	0 06	2.0	2 00	0 04	2.0	1 00	0 02	2.0	268	272
89	91	6 00	0 06	1.0	5 00	0 05	1.0	4 00	0 04	1.0	3 00	0 03	1.0	2 00	0 02	1.0	1 00	0 01	1.0	269	271
90	90	6 00	0 00	0.0	5 00	0 00	0.0	4 00	0 00	0.0	3 00	0 00	0.0	2 00	0 00	0.0	1 00	0 00	0.0	270	270

N. Lat.: for LHA > 180°... $Z_n = Z$
for LHA < 180°... $Z_n = 360° - Z$

S. Lat.: for LHA > 180°... $Z_n = 180° - Z$
for LHA < 180°... $Z_n = 180° + Z$

F < 90° and F′ > 29′ : (–) corr.
F > 90° and F′ < 30′ : (–) corr.

AUXILIARY TABLE

A′ < 30′ : (–) corr.

F′ / A′	1′ 59	2′ 58	3′ 57	4′ 56	5′ 55	6′ 54	7′ 53	8′ 52	9′ 51	10′ 50	11′ 49	12′ 48	13′ 47	14′ 46	15′ 45	16′ 44	17′ 43	18′ 42	19′ 41	20′ 40	21′ 39	22′ 38	23′ 37	24′ 36	25′ 35	26′ 34	27′ 33	28′ 32	29′ 31	30′ 30	F′ / A′
P °	′	′	′	′	′	′	′	′	′	′	′	′	′	′	′	′	′	′	′	′	′	′	′	′	′	′	′	′	′	′	Z_2 °
1	0	0	0	0	0	0	0	0	0	0	0	0	0	0	0	0	0	0	0	0	0	0	0	0	0	0	0	0	1	1	89
2	0	0	0	0	0	0	0	0	0	0	0	0	0	0	1	1	1	1	1	1	1	1	1	1	1	1	1	1	1	1	88
3	0	0	0	0	0	0	0	0	0	1	1	1	1	1	1	1	1	1	1	1	1	1	1	1	1	1	1	1	2	2	87
4	0	0	0	0	0	0	0	1	1	1	1	1	1	1	1	1	1	1	1	1	1	2	2	2	2	2	2	2	2	2	86
5	0	0	0	0	0	1	1	1	1	1	1	1	1	1	1	1	1	2	2	2	2	2	2	2	2	2	2	2	3	3	85
6	0	0	0	0	1	1	1	1	1	1	1	1	1	1	2	2	2	2	2	2	2	2	2	3	3	3	3	3	3	3	84
7	0	0	0	0	1	1	1	1	1	1	1	1	2	2	2	2	2	2	2	2	3	3	3	3	3	3	3	3	4	4	83
8	0	0	0	1	1	1	1	1	1	1	2	2	2	2	2	2	2	3	3	3	3	3	3	3	3	4	4	4	4	4	82
9	0	0	0	1	1	1	1	1	1	2	2	2	2	2	2	3	3	3	3	3	3	3	4	4	4	4	4	4	5	5	81
10	0	0	1	1	1	1	1	1	2	2	2	2	2	2	3	3	3	3	3	3	4	4	4	4	4	5	5	5	5	5	80
11	0	0	1	1	1	1	1	2	2	2	2	2	2	3	3	3	3	3	4	4	4	4	4	5	5	5	5	5	6	6	79
12	0	0	1	1	1	1	1	2	2	2	2	2	3	3	3	3	4	4	4	4	4	5	5	5	5	5	6	6	6	6	78
13	0	0	1	1	1	1	2	2	2	2	2	3	3	3	3	4	4	4	4	4	5	5	5	5	6	6	6	6	7	7	77
14	0	0	1	1	1	1	2	2	2	2	3	3	3	3	4	4	4	4	5	5	5	5	6	6	6	6	7	7	7	7	76
15	0	1	1	1	1	2	2	2	2	3	3	3	3	4	4	4	4	5	5	5	5	6	6	6	6	7	7	7	8	8	75
16	0	1	1	1	1	2	2	2	2	3	3	3	4	4	4	4	5	5	5	6	6	6	6	7	7	7	7	8	8	8	74
17	0	1	1	1	1	2	2	2	3	3	3	4	4	4	4	5	5	5	6	6	6	6	7	7	7	8	8	8	8	9	73
18	0	1	1	1	2	2	2	2	3	3	3	4	4	4	5	5	5	6	6	6	6	7	7	7	8	8	8	9	9	9	72
19	0	1	1	1	2	2	2	3	3	3	4	4	4	5	5	5	6	6	6	7	7	7	7	8	8	8	9	9	9	10	71
20	0	1	1	1	2	2	2	3	3	3	4	4	4	5	5	5	6	6	6	7	7	8	8	8	9	9	9	10	10	10	70
21	0	1	1	1	2	2	3	3	3	4	4	4	5	5	5	6	6	6	7	7	8	8	8	9	9	9	10	10	10	11	69
22	0	1	1	1	2	2	3	3	3	4	4	4	5	5	6	6	6	7	7	7	8	8	9	9	9	10	10	10	11	11	68
23	0	1	1	2	2	2	3	3	4	4	4	5	5	5	6	6	7	7	7	8	8	9	9	9	10	10	11	11	11	12	67
24	0	1	1	2	2	2	3	3	4	4	4	5	5	6	6	7	7	7	8	8	9	9	9	10	10	11	11	11	12	12	66
25	0	1	1	2	2	3	3	3	4	4	5	5	5	6	6	7	7	8	8	8	9	9	10	10	11	11	11	12	12	13	65
26	0	1	1	2	2	3	3	4	4	4	5	5	6	6	7	7	7	8	8	9	9	10	10	11	11	11	12	12	13	13	64
27	0	1	1	2	2	3	3	4	4	5	5	5	6	6	7	7	8	8	9	9	10	10	10	11	11	12	12	13	13	14	63
28	0	1	1	2	2	3	3	4	4	5	5	6	6	7	7	8	8	8	9	9	10	10	11	11	12	12	13	13	14	14	62
29	0	1	1	2	2	3	3	4	4	5	5	6	6	7	7	8	8	9	9	10	10	11	11	12	12	13	13	14	14	15	61
30	0	1	1	2	2	3	3	4	4	5	5	6	6	7	7	8	8	9	9	10	10	11	11	12	12	13	13	14	14	15	60
31	1	1	2	2	3	3	4	4	5	5	6	6	7	7	8	8	9	9	10	10	11	11	12	12	13	13	14	14	15	15	59
32	1	1	2	2	3	3	4	4	5	5	6	6	7	7	8	8	9	10	10	11	11	12	12	13	13	14	14	15	15	16	58
33	1	1	2	2	3	3	4	4	5	5	6	7	7	8	8	9	9	10	10	11	11	12	13	13	14	14	15	15	16	16	57
34	1	1	2	2	3	3	4	4	5	6	6	7	7	8	8	9	10	10	11	11	12	12	13	13	14	15	15	16	16	17	56
35	1	1	2	2	3	3	4	5	5	6	6	7	7	8	9	9	10	10	11	11	12	13	13	14	14	15	15	16	17	17	55
36	1	1	2	2	3	4	4	5	5	6	6	7	8	8	9	9	10	11	11	12	12	13	14	14	15	15	16	16	17	18	54
37	1	1	2	2	3	4	4	5	5	6	7	7	8	8	9	10	10	11	11	12	13	13	14	14	15	16	16	17	17	18	53
38	1	1	2	2	3	4	4	5	6	6	7	7	8	9	9	10	10	11	12	12	13	14	14	15	15	16	17	17	18	18	52
39	1	1	2	3	3	4	4	5	6	6	7	8	8	9	9	10	11	11	12	13	13	14	14	15	16	16	17	18	18	19	51
40	1	1	2	3	3	4	4	5	6	6	7	8	8	9	10	10	11	12	12	13	13	14	15	15	16	17	17	18	19	19	50

F′ / A′	1′ 59	2′ 58	3′ 57	4′ 56	5′ 55	6′ 54	7′ 53	8′ 52	9′ 51	10′ 50	11′ 49	12′ 48	13′ 47	14′ 46	15′ 45	16′ 44	17′ 43	18′ 42	19′ 41	20′ 40	21′ 39	22′ 38	23′ 37	24′ 36	25′ 35	26′ 34	27′ 33	28′ 32	29′ 31	30′ 30	F′ / A′
P °	′	′	′	′	′	′	′	′	′	′	′	′	′	′	′	′	′	′	′	′	′	′	′	′	′	′	′	′	′	′	Z_2 °
41	1	1	2	3	3	4	5	5	6	7	7	8	9	9	10	10	11	12	12	13	14	14	15	16	16	17	18	18	19	20	49
42	1	1	2	3	3	4	5	5	6	7	7	8	9	9	10	11	11	12	13	13	14	15	15	16	17	17	18	19	19	20	48
43	1	1	2	3	3	4	5	5	6	7	8	8	9	10	10	11	12	12	13	14	14	15	16	16	17	18	18	19	20	20	47
44	1	1	2	3	3	4	5	6	6	7	8	8	9	10	10	11	12	13	13	14	15	15	16	17	17	18	19	19	20	21	46
45	1	1	2	3	4	4	5	6	6	7	8	8	9	10	11	11	12	13	13	14	15	16	16	17	18	18	19	20	21	21	45
46	1	1	2	3	4	4	5	6	6	7	8	9	9	10	11	12	12	13	14	14	15	16	17	17	18	19	19	20	21	22	44
47	1	1	2	3	4	4	5	6	7	7	8	9	10	10	11	12	12	13	14	15	15	16	17	18	18	19	20	20	21	22	43
48	1	1	2	3	4	4	5	6	7	7	8	9	10	10	11	12	13	13	14	15	16	16	17	18	19	19	20	21	22	22	42
49	1	2	2	3	4	5	5	6	7	8	8	9	10	11	11	12	13	14	14	15	16	17	17	18	19	20	20	21	22	23	41
50	1	2	2	3	4	5	5	6	7	8	8	9	10	11	11	12	13	14	15	15	16	17	18	18	19	20	21	21	22	23	40
51	1	2	2	3	4	5	5	6	7	8	9	9	10	11	12	12	13	14	15	16	16	17	18	19	19	20	21	22	23	23	39
52	1	2	2	3	4	5	6	6	7	8	9	9	10	11	12	13	13	14	15	16	17	17	18	19	20	20	21	22	23	24	38
53	1	2	2	3	4	5	6	6	7	8	9	10	10	11	12	13	14	14	15	16	17	18	18	19	20	21	22	22	23	24	37
54	1	2	2	3	4	5	6	6	7	8	9	10	11	11	12	13	14	15	15	16	17	18	19	19	20	21	22	23	23	24	36
55	1	2	2	3	4	5	6	7	7	8	9	10	11	11	12	13	14	15	16	16	17	18	19	20	20	21	22	23	24	25	35
56	1	2	2	3	4	5	6	7	7	8	9	10	11	12	12	13	14	15	16	17	17	18	19	20	21	22	22	23	24	25	34
57	1	2	3	3	4	5	6	7	8	8	9	10	11	12	13	13	14	15	16	17	18	18	19	20	21	22	23	23	24	25	33
58	1	2	3	3	4	5	6	7	8	8	9	10	11	12	13	14	14	15	16	17	18	19	20	20	21	22	23	24	25	25	32
59	1	2	3	3	4	5	6	7	8	9	9	10	11	12	13	14	15	15	16	17	18	19	20	21	21	22	23	24	25	26	31
60	1	2	3	3	4	5	6	7	8	9	10	10	11	12	13	14	15	16	16	17	18	19	20	21	22	23	23	24	25	26	30
61	1	2	3	3	4	5	6	7	8	9	10	10	11	12	13	14	15	16	17	17	18	19	20	21	22	23	24	24	25	26	29
62	1	2	3	4	4	5	6	7	8	9	10	11	11	12	13	14	15	16	17	18	19	19	20	21	22	23	24	25	26	26	28
63	1	2	3	4	4	5	6	7	8	9	10	11	12	12	13	14	15	16	17	18	19	20	20	21	22	23	24	25	26	27	27
64	1	2	3	4	4	5	6	7	8	9	10	11	12	13	13	14	15	16	17	18	19	20	21	22	22	23	24	25	26	27	26
65	1	2	3	4	5	5	6	7	8	9	10	11	12	13	14	15	15	16	17	18	19	20	21	22	23	24	24	25	26	27	25
66	1	2	3	4	5	5	6	7	8	9	10	11	12	13	14	15	16	16	17	18	19	20	21	22	23	24	25	26	26	27	24
67	1	2	3	4	5	6	6	7	8	9	10	11	12	13	14	15	16	17	17	18	19	20	21	22	23	24	25	26	27	28	23
68	1	2	3	4	5	6	6	7	8	9	10	11	12	13	14	15	16	17	18	19	19	20	21	22	23	24	25	26	27	28	22
69	1	2	3	4	5	6	7	7	8	9	10	11	12	13	14	15	16	17	18	19	20	21	21	22	23	24	25	26	27	28	21
70	1	2	3	4	5	6	7	8	8	9	10	11	12	13	14	15	16	17	18	19	20	21	22	23	23	24	25	26	27	28	20
71	1	2	3	4	5	6	7	8	9	9	10	11	12	13	14	15	16	17	18	19	20	21	22	23	24	25	26	26	27	28	19
72	1	2	3	4	5	6	7	8	9	10	10	11	12	13	14	15	16	17	18	19	20	21	22	23	24	25	26	27	28	29	18
73	1	2	3	4	5	6	7	8	9	10	11	11	12	13	14	15	16	17	18	19	20	21	22	23	24	25	26	27	28	29	17
74	1	2	3	4	5	6	7	8	9	10	11	12	12	13	14	15	16	17	18	19	20	21	22	23	24	25	26	27	28	29	16
75	1	2	3	4	5	6	7	8	9	10	11	12	13	14	14	15	16	17	18	19	20	21	22	23	24	25	26	27	28	29	15
76	1	2	3	4	5	6	7	8	9	10	11	12	13	14	15	16	16	17	18	19	20	21	22	23	24	25	26	27	28	29	14
77	1	2	3	4	5	6	7	8	9	10	11	12	13	14	15	16	17	18	19	19	20	21	22	23	24	25	26	27	28	29	13
78	1	2	3	4	5	6	7	8	9	10	11	12	13	14	15	16	17	18	19	20	21	22	22	23	24	25	26	27	28	29	12
79	1	2	3	4	5	6	7	8	9	10	11	12	13	14	15	16	17	18	19	20	21	22	23	24	25	26	27	27	28	29	11
80	1	2	3	4	5	6	7	8	9	10	11	12	13	14	15	16	17	18	19	20	21	22	23	24	25	26	27	28	29	30	10

For P > 80°, use 80°.

For Z_2 < 10°, use 10°.

USE OF CONCISE SIGHT REDUCTION TABLES (continued)

4. *Example.* (b) Required the altitude and azimuth of *Vega* on 2003 July 29 at UT $04^h\ 51^m$ from the estimated position 152° west, 15° south.

1. Assumed latitude	$Lat =$	15° S		
From the almanac	$GHA =$	99° 55′		
Assumed longitude		151° 55′ W		
Local hour angle	$LHA =$	308		
2. Reduction table, 1st entry $(Lat, LHA) = (15, 308)$	$A =$	49 34	$A° = 50$, $A' = 34$	
	$B =$	+66 29	$Z_1 = +71{\cdot}7$,	$LHA > 270°$
3. From the almanac	$Dec =$	−38 47		*Lat* and *Dec* contrary
Sum $= B + Dec$	$F =$	+27 42 ,	$F° = 28$, $F' = 42$	
4. Reduction table, 2nd entry $(A°, F°) = (50, 28)$	$H =$	17 34	$P° = 37$	
			$Z_2 = 67{\cdot}8$	
5. Auxiliary table, 1st entry $(F', P°) = (42, 37)$	$corr_1 =$	−11		$F < 90°$, $F' > 29'$
Sum		17 23		
6. Auxiliary table, 2nd entry $(A', Z_2°) = (34, 68)$	$corr_2 =$	+10		$A' > 30'$
7. Sum = computed altitude	$H_C =$	+17° 33′		$F > 0°$
8. Azimuth, first component	$Z_1 =$	+71·7		same sign as B
second component	$Z_2 =$	+67·8		$F < 90°$, $F > 0°$
Sum = azimuth angle	$Z =$	139·5		
True azimuth	$Z_n =$	041°		S *Lat*, $LHA > 180°$

5. *Form for use with the Concise Sight Reduction Tables.* The form on the following page lays out the procedure explained on pages 284-285. Each step is shown, with notes and rules to ensure accuracy, rather than speed, throughout the calculation. The form is mainly intended for the calculation of star positions. It therefore includes the formation of the Greenwich hour of Aries (*GHA* Aries), and thus the Greenwich hour angle of the star (*GHA*) from its tabular sidereal hour angle (*SHA*). These calculations, included in step 1 of the form, can easily be replaced by the interpolation of *GHA* and *Dec* for the Sun, Moon or planets.

The form may be freely copied, however, acknowledgment of the source is requested.

Date & UT of observation h m s	**Body**	**Estimated Latitude & Longitude** ° ′ ° ′

Step	**Calculate Altitude & Azimuth**		**Summary of Rules & Notes**
Assumed latitude	*Lat* = °		Nearest estimated latitude, integral number of degrees.
Assumed longitude	*Long* = ° ′		Choose *Long* so that *LHA* has integral number of degrees.
1. From the almanac:	*Dec* = ° ′		Record the *Dec* for use in Step 3.
GHA Aries h	= ° ′		Needed if using *SHA*. Tabular value.
Increment m s	= ° ′		for minutes and seconds of time.
SHA	*SHA* = ° ′		
$GHA = GHA\ Aries + SHA$	*GHA* = ° ′		Remove multiples of 360°.
Assumed longitude	*Long* = ° ′		West longitudes are negative.
$LHA = GHA + Long$	*LHA* = °		Remove multiples of 360°.
2. Reduction table, 1st entry $(Lat, LHA) = ($ °, °$)$	*A* = ° ′	$A^\circ =$ °	nearest whole degree of *A*.
record *A*, *B* and Z_1.		$A' =$ ′	minutes part of *A*.
	B = ° ′		*B* is minus if $90° < LHA < 270°$.
		$Z_1 =$ °	Z_1 has the same sign as *B*.
3. From step 1	*Dec* = ° ′		*Dec* is minus if contrary to *Lat*.
$F = B + Dec$	*F* = ° ′		Regard *F* as positive until step 7.
		$F^\circ =$ °	nearest whole degree of *F*.
		$F' =$ ′	minutes part of *F*.
4. Reduction table, 2nd entry $(A^\circ, F^\circ) = ($ °, °$)$	*H* = ° ′	$P^\circ =$ °	nearest whole degree of *P*.
record *H*, *P* and Z_2.		$Z_2 =$ °	
5. Auxiliary table, 1st entry $(F', P^\circ) = ($ ′, °$)$ record $corr_1$	$corr_1 =$ ′		$corr_1$ is minus if $F < 90°$ & $F' > 29'$, or if $F > 90°$ & $F' < 30'$.
6. Auxiliary table, 2nd entry $(A', Z_2^\circ) = ($ ′, °$)$ record $corr_2$	$corr_2 =$ ′		Z_2° nearest whole degree of Z_2. $corr_2$ is minus if $A' < 30'$.
7. Calculated altitude = $H_C = H + corr_1 + corr_2$	$H_C =$ ° ′		H_C is minus if *F* is negative, and object is below the horizon.
8. Azimuth, 1st component	$Z_1 =$ °		Z_1 has the same sign as *B*.
2nd component	$Z_2 =$ °		Z_2 is minus if $F > 90°$. If *F* is negative, $Z_2 = 180° - Z_2$
$Z = Z_1 + Z_2$	*Z* = °		Ignore the sign of *Z*.
		N *Lat*:	If $LHA > 180°$, $Z_n = Z$, or if $LHA < 180°$, $Z_n = 360° - Z$,
		S *Lat*:	If $LHA > 180°$, $Z_n = 180° - Z$, or if $LHA < 180°$, $Z_n = 180° + Z$.
True azimuth	$Z_n =$ °		

For use with *The Nautical Almanac's* Concise Sight Reduction Tables pages 284-318.

CONVERSION OF ARC TO TIME

0°–59°		60°–119°		120°–179°		180°–239°		240°–299°		300°–359°			0′.00	0′.25	0′.50	0′.75
°	h m	°	h m	°	h m	°	h m	°	h m	°	h m	′	m s	m s	m s	m s
0	0 00	**60**	4 00	**120**	8 00	**180**	12 00	**240**	16 00	**300**	20 00	**0**	0 00	0 01	0 02	0 03
1	0 04	**61**	4 04	**121**	8 04	**181**	12 04	**241**	16 04	**301**	20 04	**1**	0 04	0 05	0 06	0 07
2	0 08	**62**	4 08	**122**	8 08	**182**	12 08	**242**	16 08	**302**	20 08	**2**	0 08	0 09	0 10	0 11
3	0 12	**63**	4 12	**123**	8 12	**183**	12 12	**243**	16 12	**303**	20 12	**3**	0 12	0 13	0 14	0 15
4	0 16	**64**	4 16	**124**	8 16	**184**	12 16	**244**	16 16	**304**	20 16	**4**	0 16	0 17	0 18	0 19
5	0 20	**65**	4 20	**125**	8 20	**185**	12 20	**245**	16 20	**305**	20 20	**5**	0 20	0 21	0 22	0 23
6	0 24	**66**	4 24	**126**	8 24	**186**	12 24	**246**	16 24	**306**	20 24	**6**	0 24	0 25	0 26	0 27
7	0 28	**67**	4 28	**127**	8 28	**187**	12 28	**247**	16 28	**307**	20 28	**7**	0 28	0 29	0 30	0 31
8	0 32	**68**	4 32	**128**	8 32	**188**	12 32	**248**	16 32	**308**	20 32	**8**	0 32	0 33	0 34	0 35
9	0 36	**69**	4 36	**129**	8 36	**189**	12 36	**249**	16 36	**309**	20 36	**9**	0 36	0 37	0 38	0 39
10	0 40	**70**	4 40	**130**	8 40	**190**	12 40	**250**	16 40	**310**	20 40	**10**	0 40	0 41	0 42	0 43
11	0 44	**71**	4 44	**131**	8 44	**191**	12 44	**251**	16 44	**311**	20 44	**11**	0 44	0 45	0 46	0 47
12	0 48	**72**	4 48	**132**	8 48	**192**	12 48	**252**	16 48	**312**	20 48	**12**	0 48	0 49	0 50	0 51
13	0 52	**73**	4 52	**133**	8 52	**193**	12 52	**253**	16 52	**313**	20 52	**13**	0 52	0 53	0 54	0 55
14	0 56	**74**	4 56	**134**	8 56	**194**	12 56	**254**	16 56	**314**	20 56	**14**	0 56	0 57	0 58	0 59
15	1 00	**75**	5 00	**135**	9 00	**195**	13 00	**255**	17 00	**315**	21 00	**15**	1 00	1 01	1 02	1 03
16	1 04	**76**	5 04	**136**	9 04	**196**	13 04	**256**	17 04	**316**	21 04	**16**	1 04	1 05	1 06	1 07
17	1 08	**77**	5 08	**137**	9 08	**197**	13 08	**257**	17 08	**317**	21 08	**17**	1 08	1 09	1 10	1 11
18	1 12	**78**	5 12	**138**	9 12	**198**	13 12	**258**	17 12	**318**	21 12	**18**	1 12	1 13	1 14	1 15
19	1 16	**79**	5 16	**139**	9 16	**199**	13 16	**259**	17 16	**319**	21 16	**19**	1 16	1 17	1 18	1 19
20	1 20	**80**	5 20	**140**	9 20	**200**	13 20	**260**	17 20	**320**	21 20	**20**	1 20	1 21	1 22	1 23
21	1 24	**81**	5 24	**141**	9 24	**201**	13 24	**261**	17 24	**321**	21 24	**21**	1 24	1 25	1 26	1 27
22	1 28	**82**	5 28	**142**	9 28	**202**	13 28	**262**	17 28	**322**	21 28	**22**	1 28	1 29	1 30	1 31
23	1 32	**83**	5 32	**143**	9 32	**203**	13 32	**263**	17 32	**323**	21 32	**23**	1 32	1 33	1 34	1 35
24	1 36	**84**	5 36	**144**	9 36	**204**	13 36	**264**	17 36	**324**	21 36	**24**	1 36	1 37	1 38	1 39
25	1 40	**85**	5 40	**145**	9 40	**205**	13 40	**265**	17 40	**325**	21 40	**25**	1 40	1 41	1 42	1 43
26	1 44	**86**	5 44	**146**	9 44	**206**	13 44	**266**	17 44	**326**	21 44	**26**	1 44	1 45	1 46	1 47
27	1 48	**87**	5 48	**147**	9 48	**207**	13 48	**267**	17 48	**327**	21 48	**27**	1 48	1 49	1 50	1 51
28	1 52	**88**	5 52	**148**	9 52	**208**	13 52	**268**	17 52	**328**	21 52	**28**	1 52	1 53	1 54	1 55
29	1 56	**89**	5 56	**149**	9 56	**209**	13 56	**269**	17 56	**329**	21 56	**29**	1 56	1 57	1 58	1 59
30	2 00	**90**	6 00	**150**	10 00	**210**	14 00	**270**	18 00	**330**	22 00	**30**	2 00	2 01	2 02	2 03
31	2 04	**91**	6 04	**151**	10 04	**211**	14 04	**271**	18 04	**331**	22 04	**31**	2 04	2 05	2 06	2 07
32	2 08	**92**	6 08	**152**	10 08	**212**	14 08	**272**	18 08	**332**	22 08	**32**	2 08	2 09	2 10	2 11
33	2 12	**93**	6 12	**153**	10 12	**213**	14 12	**273**	18 12	**333**	22 12	**33**	2 12	2 13	2 14	2 15
34	2 16	**94**	6 16	**154**	10 16	**214**	14 16	**274**	18 16	**334**	22 16	**34**	2 16	2 17	2 18	2 19
35	2 20	**95**	6 20	**155**	10 20	**215**	14 20	**275**	18 20	**335**	22 20	**35**	2 20	2 21	2 22	2 23
36	2 24	**96**	6 24	**156**	10 24	**216**	14 24	**276**	18 24	**336**	22 24	**36**	2 24	2 25	2 26	2 27
37	2 28	**97**	6 28	**157**	10 28	**217**	14 28	**277**	18 28	**337**	22 28	**37**	2 28	2 29	2 30	2 31
38	2 32	**98**	6 32	**158**	10 32	**218**	14 32	**278**	18 32	**338**	22 32	**38**	2 32	2 33	2 34	2 35
39	2 36	**99**	6 36	**159**	10 36	**219**	14 36	**279**	18 36	**339**	22 36	**39**	2 36	2 37	2 38	2 39
40	2 40	**100**	6 40	**160**	10 40	**220**	14 40	**280**	18 40	**340**	22 40	**40**	2 40	2 41	2 42	2 43
41	2 44	**101**	6 44	**161**	10 44	**221**	14 44	**281**	18 44	**341**	22 44	**41**	2 44	2 45	2 46	2 47
42	2 48	**102**	6 48	**162**	10 48	**222**	14 48	**282**	18 48	**342**	22 48	**42**	2 48	2 49	2 50	2 51
43	2 52	**103**	6 52	**163**	10 52	**223**	14 52	**283**	18 52	**343**	22 52	**43**	2 52	2 53	2 54	2 55
44	2 56	**104**	6 56	**164**	10 56	**224**	14 56	**284**	18 56	**344**	22 56	**44**	2 56	2 57	2 58	2 59
45	3 00	**105**	7 00	**165**	11 00	**225**	15 00	**285**	19 00	**345**	23 00	**45**	3 00	3 01	3 02	3 03
46	3 04	**106**	7 04	**166**	11 04	**226**	15 04	**286**	19 04	**346**	23 04	**46**	3 04	3 05	3 06	3 07
47	3 08	**107**	7 08	**167**	11 08	**227**	15 08	**287**	19 08	**347**	23 08	**47**	3 08	3 09	3 10	3 11
48	3 12	**108**	7 12	**168**	11 12	**228**	15 12	**288**	19 12	**348**	23 12	**48**	3 12	3 13	3 14	3 15
49	3 16	**109**	7 16	**169**	11 16	**229**	15 16	**289**	19 16	**349**	23 16	**49**	3 16	3 17	3 18	3 19
50	3 20	**110**	7 20	**170**	11 20	**230**	15 20	**290**	19 20	**350**	23 20	**50**	3 20	3 21	3 22	3 23
51	3 24	**111**	7 24	**171**	11 24	**231**	15 24	**291**	19 24	**351**	23 24	**51**	3 24	3 25	3 26	3 27
52	3 28	**112**	7 28	**172**	11 28	**232**	15 28	**292**	19 28	**352**	23 28	**52**	3 28	3 29	3 30	3 31
53	3 32	**113**	7 32	**173**	11 32	**233**	15 32	**293**	19 32	**353**	23 32	**53**	3 32	3 33	3 34	3 35
54	3 36	**114**	7 36	**174**	11 36	**234**	15 36	**294**	19 36	**354**	23 36	**54**	3 36	3 37	3 38	3 39
55	3 40	**115**	7 40	**175**	11 40	**235**	15 40	**295**	19 40	**355**	23 40	**55**	3 40	3 41	3 42	3 43
56	3 44	**116**	7 44	**176**	11 44	**236**	15 44	**296**	19 44	**356**	23 44	**56**	3 44	3 45	3 46	3 47
57	3 48	**117**	7 48	**177**	11 48	**237**	15 48	**297**	19 48	**357**	23 48	**57**	3 48	3 49	3 50	3 51
58	3 52	**118**	7 52	**178**	11 52	**238**	15 52	**298**	19 52	**358**	23 52	**58**	3 52	3 53	3 54	3 55
59	3 56	**119**	7 56	**179**	11 56	**239**	15 56	**299**	19 56	**359**	23 56	**59**	3 56	3 57	3 58	3 59

The above table is for converting expressions in arc to their equivalent in time; its main use in this Almanac is for the conversion of longitude for application to LMT (*added* if *west*, *subtracted* if *east*) to give UT or vice versa, particularly in the case of sunrise, sunset, etc.

0m	SUN PLANETS	ARIES	MOON	v or d	Corrn	v or d	Corrn	v or d	Corrn
s	° ′	° ′	° ′	′	′	′	′	′	′
00	0 00·0	0 00·0	0 00·0	0·0	0·0	6·0	0·1	12·0	0·1
01	0 00·3	0 00·3	0 00·2	0·1	0·0	6·1	0·1	12·1	0·1
02	0 00·5	0 00·5	0 00·5	0·2	0·0	6·2	0·1	12·2	0·1
03	0 00·8	0 00·8	0 00·7	0·3	0·0	6·3	0·1	12·3	0·1
04	0 01·0	0 01·0	0 01·0	0·4	0·0	6·4	0·1	12·4	0·1
05	0 01·3	0 01·3	0 01·2	0·5	0·0	6·5	0·1	12·5	0·1
06	0 01·5	0 01·5	0 01·4	0·6	0·0	6·6	0·1	12·6	0·1
07	0 01·8	0 01·8	0 01·7	0·7	0·0	6·7	0·1	12·7	0·1
08	0 02·0	0 02·0	0 01·9	0·8	0·0	6·8	0·1	12·8	0·1
09	0 02·3	0 02·3	0 02·1	0·9	0·0	6·9	0·1	12·9	0·1
10	0 02·5	0 02·5	0 02·4	1·0	0·0	7·0	0·1	13·0	0·1
11	0 02·8	0 02·8	0 02·6	1·1	0·0	7·1	0·1	13·1	0·1
12	0 03·0	0 03·0	0 02·9	1·2	0·0	7·2	0·1	13·2	0·1
13	0 03·3	0 03·3	0 03·1	1·3	0·0	7·3	0·1	13·3	0·1
14	0 03·5	0 03·5	0 03·3	1·4	0·0	7·4	0·1	13·4	0·1
15	0 03·8	0 03·8	0 03·6	1·5	0·0	7·5	0·1	13·5	0·1
16	0 04·0	0 04·0	0 03·8	1·6	0·0	7·6	0·1	13·6	0·1
17	0 04·3	0 04·3	0 04·1	1·7	0·0	7·7	0·1	13·7	0·1
18	0 04·5	0 04·5	0 04·3	1·8	0·0	7·8	0·1	13·8	0·1
19	0 04·8	0 04·8	0 04·5	1·9	0·0	7·9	0·1	13·9	0·1
20	0 05·0	0 05·0	0 04·8	2·0	0·0	8·0	0·1	14·0	0·1
21	0 05·3	0 05·3	0 05·0	2·1	0·0	8·1	0·1	14·1	0·1
22	0 05·5	0 05·5	0 05·2	2·2	0·0	8·2	0·1	14·2	0·1
23	0 05·8	0 05·8	0 05·5	2·3	0·0	8·3	0·1	14·3	0·1
24	0 06·0	0 06·0	0 05·7	2·4	0·0	8·4	0·1	14·4	0·1
25	0 06·3	0 06·3	0 06·0	2·5	0·0	8·5	0·1	14·5	0·1
26	0 06·5	0 06·5	0 06·2	2·6	0·0	8·6	0·1	14·6	0·1
27	0 06·8	0 06·8	0 06·4	2·7	0·0	8·7	0·1	14·7	0·1
28	0 07·0	0 07·0	0 06·7	2·8	0·0	8·8	0·1	14·8	0·1
29	0 07·3	0 07·3	0 06·9	2·9	0·0	8·9	0·1	14·9	0·1
30	0 07·5	0 07·5	0 07·2	3·0	0·0	9·0	0·1	15·0	0·1
31	0 07·8	0 07·8	0 07·4	3·1	0·0	9·1	0·1	15·1	0·1
32	0 08·0	0 08·0	0 07·6	3·2	0·0	9·2	0·1	15·2	0·1
33	0 08·3	0 08·3	0 07·9	3·3	0·0	9·3	0·1	15·3	0·1
34	0 08·5	0 08·5	0 08·1	3·4	0·0	9·4	0·1	15·4	0·1
35	0 08·8	0 08·8	0 08·4	3·5	0·0	9·5	0·1	15·5	0·1
36	0 09·0	0 09·0	0 08·6	3·6	0·0	9·6	0·1	15·6	0·1
37	0 09·3	0 09·3	0 08·8	3·7	0·0	9·7	0·1	15·7	0·1
38	0 09·5	0 09·5	0 09·1	3·8	0·0	9·8	0·1	15·8	0·1
39	0 09·8	0 09·8	0 09·3	3·9	0·0	9·9	0·1	15·9	0·1
40	0 10·0	0 10·0	0 09·5	4·0	0·0	10·0	0·1	16·0	0·1
41	0 10·3	0 10·3	0 09·8	4·1	0·0	10·1	0·1	16·1	0·1
42	0 10·5	0 10·5	0 10·0	4·2	0·0	10·2	0·1	16·2	0·1
43	0 10·8	0 10·8	0 10·3	4·3	0·0	10·3	0·1	16·3	0·1
44	0 11·0	0 11·0	0 10·5	4·4	0·0	10·4	0·1	16·4	0·1
45	0 11·3	0 11·3	0 10·7	4·5	0·0	10·5	0·1	16·5	0·1
46	0 11·5	0 11·5	0 11·0	4·6	0·0	10·6	0·1	16·6	0·1
47	0 11·8	0 11·8	0 11·2	4·7	0·0	10·7	0·1	16·7	0·1
48	0 12·0	0 12·0	0 11·5	4·8	0·0	10·8	0·1	16·8	0·1
49	0 12·3	0 12·3	0 11·7	4·9	0·0	10·9	0·1	16·9	0·1
50	0 12·5	0 12·5	0 11·9	5·0	0·0	11·0	0·1	17·0	0·1
51	0 12·8	0 12·8	0 12·2	5·1	0·0	11·1	0·1	17·1	0·1
52	0 13·0	0 13·0	0 12·4	5·2	0·0	11·2	0·1	17·2	0·1
53	0 13·3	0 13·3	0 12·6	5·3	0·0	11·3	0·1	17·3	0·1
54	0 13·5	0 13·5	0 12·9	5·4	0·0	11·4	0·1	17·4	0·1
55	0 13·8	0 13·8	0 13·1	5·5	0·0	11·5	0·1	17·5	0·1
56	0 14·0	0 14·0	0 13·4	5·6	0·0	11·6	0·1	17·6	0·1
57	0 14·3	0 14·3	0 13·6	5·7	0·0	11·7	0·1	17·7	0·1
58	0 14·5	0 14·5	0 13·8	5·8	0·0	11·8	0·1	17·8	0·1
59	0 14·8	0 14·8	0 14·1	5·9	0·0	11·9	0·1	17·9	0·1
60	0 15·0	0 15·0	0 14·3	6·0	0·1	12·0	0·1	18·0	0·2

1m	SUN PLANETS	ARIES	MOON	v or d	Corrn	v or d	Corrn	v or d	Corr
s	° ′	° ′	° ′	′	′	′	′	′	′
00	0 15·0	0 15·0	0 14·3	0·0	0·0	6·0	0·2	12·0	0·
01	0 15·3	0 15·3	0 14·6	0·1	0·0	6·1	0·2	12·1	0·
02	0 15·5	0 15·5	0 14·8	0·2	0·0	6·2	0·2	12·2	0·
03	0 15·8	0 15·8	0 15·0	0·3	0·0	6·3	0·2	12·3	0·
04	0 16·0	0 16·0	0 15·3	0·4	0·0	6·4	0·2	12·4	0·
05	0 16·3	0 16·3	0 15·5	0·5	0·0	6·5	0·2	12·5	0·
06	0 16·5	0 16·5	0 15·7	0·6	0·0	6·6	0·2	12·6	0·
07	0 16·8	0 16·8	0 16·0	0·7	0·0	6·7	0·2	12·7	0·
08	0 17·0	0 17·0	0 16·2	0·8	0·0	6·8	0·2	12·8	0·
09	0 17·3	0 17·3	0 16·5	0·9	0·0	6·9	0·2	12·9	0·
10	0 17·5	0 17·5	0 16·7	1·0	0·0	7·0	0·2	13·0	0·
11	0 17·8	0 17·8	0 16·9	1·1	0·0	7·1	0·2	13·1	0·
12	0 18·0	0 18·0	0 17·2	1·2	0·0	7·2	0·2	13·2	0·
13	0 18·3	0 18·3	0 17·4	1·3	0·0	7·3	0·2	13·3	0·
14	0 18·5	0 18·6	0 17·7	1·4	0·0	7·4	0·2	13·4	0·
15	0 18·8	0 18·8	0 17·9	1·5	0·0	7·5	0·2	13·5	0·
16	0 19·0	0 19·1	0 18·1	1·6	0·0	7·6	0·2	13·6	0·
17	0 19·3	0 19·3	0 18·4	1·7	0·0	7·7	0·2	13·7	0·
18	0 19·5	0 19·6	0 18·6	1·8	0·0	7·8	0·2	13·8	0·
19	0 19·8	0 19·8	0 18·9	1·9	0·0	7·9	0·2	13·9	0·
20	0 20·0	0 20·1	0 19·1	2·0	0·1	8·0	0·2	14·0	0·
21	0 20·3	0 20·3	0 19·3	2·1	0·1	8·1	0·2	14·1	0·
22	0 20·5	0 20·6	0 19·6	2·2	0·1	8·2	0·2	14·2	0·
23	0 20·8	0 20·8	0 19·8	2·3	0·1	8·3	0·2	14·3	0·
24	0 21·0	0 21·1	0 20·0	2·4	0·1	8·4	0·2	14·4	0·
25	0 21·3	0 21·3	0 20·3	2·5	0·1	8·5	0·2	14·5	0·
26	0 21·5	0 21·6	0 20·5	2·6	0·1	8·6	0·2	14·6	0·
27	0 21·8	0 21·8	0 20·8	2·7	0·1	8·7	0·2	14·7	0·
28	0 22·0	0 22·1	0 21·0	2·8	0·1	8·8	0·2	14·8	0·
29	0 22·3	0 22·3	0 21·2	2·9	0·1	8·9	0·2	14·9	0·
30	0 22·5	0 22·6	0 21·5	3·0	0·1	9·0	0·2	15·0	0·
31	0 22·8	0 22·8	0 21·7	3·1	0·1	9·1	0·2	15·1	0·
32	0 23·0	0 23·1	0 22·0	3·2	0·1	9·2	0·2	15·2	0·
33	0 23·3	0 23·3	0 22·2	3·3	0·1	9·3	0·2	15·3	0·
34	0 23·5	0 23·6	0 22·4	3·4	0·1	9·4	0·2	15·4	0·
35	0 23·8	0 23·8	0 22·7	3·5	0·1	9·5	0·2	15·5	0·
36	0 24·0	0 24·1	0 22·9	3·6	0·1	9·6	0·2	15·6	0·
37	0 24·3	0 24·3	0 23·1	3·7	0·1	9·7	0·2	15·7	0·
38	0 24·5	0 24·6	0 23·4	3·8	0·1	9·8	0·2	15·8	0·
39	0 24·8	0 24·8	0 23·6	3·9	0·1	9·9	0·2	15·9	0
40	0 25·0	0 25·1	0 23·9	4·0	0·1	10·0	0·3	16·0	0·
41	0 25·3	0 25·3	0 24·1	4·1	0·1	10·1	0·3	16·1	0·
42	0 25·5	0 25·6	0 24·3	4·2	0·1	10·2	0·3	16·2	0·
43	0 25·8	0 25·8	0 24·6	4·3	0·1	10·3	0·3	16·3	0·
44	0 26·0	0 26·1	0 24·8	4·4	0·1	10·4	0·3	16·4	0·
45	0 26·3	0 26·3	0 25·1	4·5	0·1	10·5	0·3	16·5	0·
46	0 26·5	0 26·6	0 25·3	4·6	0·1	10·6	0·3	16·6	0·
47	0 26·8	0 26·8	0 25·5	4·7	0·1	10·7	0·3	16·7	0·
48	0 27·0	0 27·1	0 25·8	4·8	0·1	10·8	0·3	16·8	0·
49	0 27·3	0 27·3	0 26·0	4·9	0·1	10·9	0·3	16·9	0
50	0 27·5	0 27·6	0 26·2	5·0	0·1	11·0	0·3	17·0	0
51	0 27·8	0 27·8	0 26·5	5·1	0·1	11·1	0·3	17·1	0·
52	0 28·0	0 28·1	0 26·7	5·2	0·1	11·2	0·3	17·2	0·
53	0 28·3	0 28·3	0 27·0	5·3	0·1	11·3	0·3	17·3	0
54	0 28·5	0 28·6	0 27·2	5·4	0·1	11·4	0·3	17·4	0
55	0 28·8	0 28·8	0 27·4	5·5	0·1	11·5	0·3	17·5	0·
56	0 29·0	0 29·1	0 27·7	5·6	0·1	11·6	0·3	17·6	0·
57	0 29·3	0 29·3	0 27·9	5·7	0·1	11·7	0·3	17·7	0
58	0 29·5	0 29·6	0 28·2	5·8	0·1	11·8	0·3	17·8	0
59	0 29·8	0 29·8	0 28·4	5·9	0·1	11·9	0·3	17·9	0
60	0 30·0	0 30·1	0 28·6	6·0	0·2	12·0	0·3	18·0	0

m 2	SUN PLANETS	ARIES	MOON	v or d	Corr^{n}	v or d	Corr^{n}	v or d	Corr^{n}
s	° ′	° ′	° ′	′	′	′	′	′	′
00	0 30·0	0 30·1	0 28·6	0·0	0·0	6·0	0·3	12·0	0·5
01	0 30·3	0 30·3	0 28·9	0·1	0·0	6·1	0·3	12·1	0·5
02	0 30·5	0 30·6	0 29·1	0·2	0·0	6·2	0·3	12·2	0·5
03	0 30·8	0 30·8	0 29·3	0·3	0·0	6·3	0·3	12·3	0·5
04	0 31·0	0 31·1	0 29·6	0·4	0·0	6·4	0·3	12·4	0·5
05	0 31·3	0 31·3	0 29·8	0·5	0·0	6·5	0·3	12·5	0·5
06	0 31·5	0 31·6	0 30·1	0·6	0·0	6·6	0·3	12·6	0·5
07	0 31·8	0 31·8	0 30·3	0·7	0·0	6·7	0·3	12·7	0·5
08	0 32·0	0 32·1	0 30·5	0·8	0·0	6·8	0·3	12·8	0·5
09	0 32·3	0 32·3	0 30·8	0·9	0·0	6·9	0·3	12·9	0·5
10	0 32·5	0 32·6	0 31·0	1·0	0·0	7·0	0·3	13·0	0·5
11	0 32·8	0 32·8	0 31·3	1·1	0·0	7·1	0·3	13·1	0·5
12	0 33·0	0 33·1	0 31·5	1·2	0·1	7·2	0·3	13·2	0·6
13	0 33·3	0 33·3	0 31·7	1·3	0·1	7·3	0·3	13·3	0·6
14	0 33·5	0 33·6	0 32·0	1·4	0·1	7·4	0·3	13·4	0·6
15	0 33·8	0 33·8	0 32·2	1·5	0·1	7·5	0·3	13·5	0·6
16	0 34·0	0 34·1	0 32·5	1·6	0·1	7·6	0·3	13·6	0·6
17	0 34·3	0 34·3	0 32·7	1·7	0·1	7·7	0·3	13·7	0·6
18	0 34·5	0 34·6	0 32·9	1·8	0·1	7·8	0·3	13·8	0·6
19	0 34·8	0 34·8	0 33·2	1·9	0·1	7·9	0·3	13·9	0·6
20	0 35·0	0 35·1	0 33·4	2·0	0·1	8·0	0·3	14·0	0·6
21	0 35·3	0 35·3	0 33·6	2·1	0·1	8·1	0·3	14·1	0·6
22	0 35·5	0 35·6	0 33·9	2·2	0·1	8·2	0·3	14·2	0·6
23	0 35·8	0 35·8	0 34·1	2·3	0·1	8·3	0·3	14·3	0·6
24	0 36·0	0 36·1	0 34·4	2·4	0·1	8·4	0·4	14·4	0·6
25	0 36·3	0 36·3	0 34·6	2·5	0·1	8·5	0·4	14·5	0·6
26	0 36·5	0 36·6	0 34·8	2·6	0·1	8·6	0·4	14·6	0·6
27	0 36·8	0 36·9	0 35·1	2·7	0·1	8·7	0·4	14·7	0·6
28	0 37·0	0 37·1	0 35·3	2·8	0·1	8·8	0·4	14·8	0·6
29	0 37·3	0 37·4	0 35·6	2·9	0·1	8·9	0·4	14·9	0·6
30	0 37·5	0 37·6	0 35·8	3·0	0·1	9·0	0·4	15·0	0·6
31	0 37·8	0 37·9	0 36·0	3·1	0·1	9·1	0·4	15·1	0·6
32	0 38·0	0 38·1	0 36·3	3·2	0·1	9·2	0·4	15·2	0·6
33	0 38·3	0 38·4	0 36·5	3·3	0·1	9·3	0·4	15·3	0·6
34	0 38·5	0 38·6	0 36·7	3·4	0·1	9·4	0·4	15·4	0·6
35	0 38·8	0 38·9	0 37·0	3·5	0·1	9·5	0·4	15·5	0·6
36	0 39·0	0 39·1	0 37·2	3·6	0·2	9·6	0·4	15·6	0·7
37	0 39·3	0 39·4	0 37·5	3·7	0·2	9·7	0·4	15·7	0·7
38	0 39·5	0 39·6	0 37·7	3·8	0·2	9·8	0·4	15·8	0·7
39	0 39·8	0 39·9	0 37·9	3·9	0·2	9·9	0·4	15·9	0·7
40	0 40·0	0 40·1	0 38·2	4·0	0·2	10·0	0·4	16·0	0·7
41	0 40·3	0 40·4	0 38·4	4·1	0·2	10·1	0·4	16·1	0·7
42	0 40·5	0 40·6	0 38·7	4·2	0·2	10·2	0·4	16·2	0·7
43	0 40·8	0 40·9	0 38·9	4·3	0·2	10·3	0·4	16·3	0·7
44	0 41·0	0 41·1	0 39·1	4·4	0·2	10·4	0·4	16·4	0·7
45	0 41·3	0 41·4	0 39·4	4·5	0·2	10·5	0·4	16·5	0·7
46	0 41·5	0 41·6	0 39·6	4·6	0·2	10·6	0·4	16·6	0·7
47	0 41·8	0 41·9	0 39·8	4·7	0·2	10·7	0·4	16·7	0·7
48	0 42·0	0 42·1	0 40·1	4·8	0·2	10·8	0·5	16·8	0·7
49	0 42·3	0 42·4	0 40·3	4·9	0·2	10·9	0·5	16·9	0·7
50	0 42·5	0 42·6	0 40·6	5·0	0·2	11·0	0·5	17·0	0·7
51	0 42·8	0 42·9	0 40·8	5·1	0·2	11·1	0·5	17·1	0·7
52	0 43·0	0 43·1	0 41·0	5·2	0·2	11·2	0·5	17·2	0·7
53	0 43·3	0 43·4	0 41·3	5·3	0·2	11·3	0·5	17·3	0·7
54	0 43·5	0 43·6	0 41·5	5·4	0·2	11·4	0·5	17·4	0·7
55	0 43·8	0 43·9	0 41·8	5·5	0·2	11·5	0·5	17·5	0·7
56	0 44·0	0 44·1	0 42·0	5·6	0·2	11·6	0·5	17·6	0·7
57	0 44·3	0 44·4	0 42·2	5·7	0·2	11·7	0·5	17·7	0·7
58	0 44·5	0 44·6	0 42·5	5·8	0·2	11·8	0·5	17·8	0·7
59	0 44·8	0 44·9	0 42·7	5·9	0·2	11·9	0·5	17·9	0·7
60	0 45·0	0 45·1	0 43·0	6·0	0·3	12·0	0·5	18·0	0·8

m 3	SUN PLANETS	ARIES	MOON	v or d	Corr^{n}	v or d	Corr^{n}	v or d	Corr^{n}
s	° ′	° ′	° ′	′	′	′	′	′	′
00	0 45·0	0 45·1	0 43·0	0·0	0·0	6·0	0·4	12·0	0·7
01	0 45·3	0 45·4	0 43·2	0·1	0·0	6·1	0·4	12·1	0·7
02	0 45·5	0 45·6	0 43·4	0·2	0·0	6·2	0·4	12·2	0·7
03	0 45·8	0 45·9	0 43·7	0·3	0·0	6·3	0·4	12·3	0·7
04	0 46·0	0 46·1	0 43·9	0·4	0·0	6·4	0·4	12·4	0·7
05	0 46·3	0 46·4	0 44·1	0·5	0·0	6·5	0·4	12·5	0·7
06	0 46·5	0 46·6	0 44·4	0·6	0·0	6·6	0·4	12·6	0·7
07	0 46·8	0 46·9	0 44·6	0·7	0·0	6·7	0·4	12·7	0·7
08	0 47·0	0 47·1	0 44·9	0·8	0·0	6·8	0·4	12·8	0·7
09	0 47·3	0 47·4	0 45·1	0·9	0·1	6·9	0·4	12·9	0·8
10	0 47·5	0 47·6	0 45·3	1·0	0·1	7·0	0·4	13·0	0·8
11	0 47·8	0 47·9	0 45·6	1·1	0·1	7·1	0·4	13·1	0·8
12	0 48·0	0 48·1	0 45·8	1·2	0·1	7·2	0·4	13·2	0·8
13	0 48·3	0 48·4	0 46·1	1·3	0·1	7·3	0·4	13·3	0·8
14	0 48·5	0 48·6	0 46·3	1·4	0·1	7·4	0·4	13·4	0·8
15	0 48·8	0 48·9	0 46·5	1·5	0·1	7·5	0·4	13·5	0·8
16	0 49·0	0 49·1	0 46·8	1·6	0·1	7·6	0·4	13·6	0·8
17	0 49·3	0 49·4	0 47·0	1·7	0·1	7·7	0·4	13·7	0·8
18	0 49·5	0 49·6	0 47·2	1·8	0·1	7·8	0·5	13·8	0·8
19	0 49·8	0 49·9	0 47·5	1·9	0·1	7·9	0·5	13·9	0·8
20	0 50·0	0 50·1	0 47·7	2·0	0·1	8·0	0·5	14·0	0·8
21	0 50·3	0 50·4	0 48·0	2·1	0·1	8·1	0·5	14·1	0·8
22	0 50·5	0 50·6	0 48·2	2·2	0·1	8·2	0·5	14·2	0·8
23	0 50·8	0 50·9	0 48·4	2·3	0·1	8·3	0·5	14·3	0·8
24	0 51·0	0 51·1	0 48·7	2·4	0·1	8·4	0·5	14·4	0·8
25	0 51·3	0 51·4	0 48·9	2·5	0·1	8·5	0·5	14·5	0·8
26	0 51·5	0 51·6	0 49·2	2·6	0·2	8·6	0·5	14·6	0·9
27	0 51·8	0 51·9	0 49·4	2·7	0·2	8·7	0·5	14·7	0·9
28	0 52·0	0 52·1	0 49·6	2·8	0·2	8·8	0·5	14·8	0·9
29	0 52·3	0 52·4	0 49·9	2·9	0·2	8·9	0·5	14·9	0·9
30	0 52·5	0 52·6	0 50·1	3·0	0·2	9·0	0·5	15·0	0·9
31	0 52·8	0 52·9	0 50·3	3·1	0·2	9·1	0·5	15·1	0·9
32	0 53·0	0 53·1	0 50·6	3·2	0·2	9·2	0·5	15·2	0·9
33	0 53·3	0 53·4	0 50·8	3·3	0·2	9·3	0·5	15·3	0·9
34	0 53·5	0 53·6	0 51·1	3·4	0·2	9·4	0·5	15·4	0·9
35	0 53·8	0 53·9	0 51·3	3·5	0·2	9·5	0·6	15·5	0·9
36	0 54·0	0 54·1	0 51·5	3·6	0·2	9·6	0·6	15·6	0·9
37	0 54·3	0 54·4	0 51·8	3·7	0·2	9·7	0·6	15·7	0·9
38	0 54·5	0 54·6	0 52·0	3·8	0·2	9·8	0·6	15·8	0·9
39	0 54·8	0 54·9	0 52·3	3·9	0·2	9·9	0·6	15·9	0·9
40	0 55·0	0 55·2	0 52·5	4·0	0·2	10·0	0·6	16·0	0·9
41	0 55·3	0 55·4	0 52·7	4·1	0·2	10·1	0·6	16·1	0·9
42	0 55·5	0 55·7	0 53·0	4·2	0·2	10·2	0·6	16·2	0·9
43	0 55·8	0 55·9	0 53·2	4·3	0·3	10·3	0·6	16·3	1·0
44	0 56·0	0 56·2	0 53·4	4·4	0·3	10·4	0·6	16·4	1·0
45	0 56·3	0 56·4	0 53·7	4·5	0·3	10·5	0·6	16·5	1·0
46	0 56·5	0 56·7	0 53·9	4·6	0·3	10·6	0·6	16·6	1·0
47	0 56·8	0 56·9	0 54·2	4·7	0·3	10·7	0·6	16·7	1·0
48	0 57·0	0 57·2	0 54·4	4·8	0·3	10·8	0·6	16·8	1·0
49	0 57·3	0 57·4	0 54·6	4·9	0·3	10·9	0·6	16·9	1·0
50	0 57·5	0 57·7	0 54·9	5·0	0·3	11·0	0·6	17·0	1·0
51	0 57·8	0 57·9	0 55·1	5·1	0·3	11·1	0·6	17·1	1·0
52	0 58·0	0 58·2	0 55·4	5·2	0·3	11·2	0·7	17·2	1·0
53	0 58·3	0 58·4	0 55·6	5·3	0·3	11·3	0·7	17·3	1·0
54	0 58·5	0 58·7	0 55·8	5·4	0·3	11·4	0·7	17·4	1·0
55	0 58·8	0 58·9	0 56·1	5·5	0·3	11·5	0·7	17·5	1·0
56	0 59·0	0 59·2	0 56·3	5·6	0·3	11·6	0·7	17·6	1·0
57	0 59·3	0 59·4	0 56·6	5·7	0·3	11·7	0·7	17·7	1·0
58	0 59·5	0 59·7	0 56·8	5·8	0·3	11·8	0·7	17·8	1·0
59	0 59·8	0 59·9	0 57·0	5·9	0·3	11·9	0·7	17·9	1·0
60	1 00·0	1 00·2	0 57·3	6·0	0·4	12·0	0·7	18·0	1·1

m 4 s	SUN PLANETS ° ′	ARIES ° ′	MOON ° ′	v or d ′	Corrⁿ ′	v or d ′	Corrⁿ ′	v or d ′	Corrⁿ ′
00	1 00·0	1 00·2	0 57·3	0·0	0·0	6·0	0·5	12·0	0·9
01	1 00·3	1 00·4	0 57·5	0·1	0·0	6·1	0·5	12·1	0·9
02	1 00·5	1 00·7	0 57·7	0·2	0·0	6·2	0·5	12·2	0·9
03	1 00·8	1 00·9	0 58·0	0·3	0·0	6·3	0·5	12·3	0·9
04	1 01·0	1 01·2	0 58·2	0·4	0·0	6·4	0·5	12·4	0·9
05	1 01·3	1 01·4	0 58·5	0·5	0·0	6·5	0·5	12·5	0·9
06	1 01·5	1 01·7	0 58·7	0·6	0·0	6·6	0·5	12·6	0·9
07	1 01·8	1 01·9	0 58·9	0·7	0·1	6·7	0·5	12·7	1·0
08	1 02·0	1 02·2	0 59·2	0·8	0·1	6·8	0·5	12·8	1·0
09	1 02·3	1 02·4	0 59·4	0·9	0·1	6·9	0·5	12·9	1·0
10	1 02·5	1 02·7	0 59·7	1·0	0·1	7·0	0·5	13·0	1·0
11	1 02·8	1 02·9	0 59·9	1·1	0·1	7·1	0·5	13·1	1·0
12	1 03·0	1 03·2	1 00·1	1·2	0·1	7·2	0·5	13·2	1·0
13	1 03·3	1 03·4	1 00·4	1·3	0·1	7·3	0·5	13·3	1·0
14	1 03·5	1 03·7	1 00·6	1·4	0·1	7·4	0·6	13·4	1·0
15	1 03·8	1 03·9	1 00·8	1·5	0·1	7·5	0·6	13·5	1·0
16	1 04·0	1 04·2	1 01·1	1·6	0·1	7·6	0·6	13·6	1·0
17	1 04·3	1 04·4	1 01·3	1·7	0·1	7·7	0·6	13·7	1·0
18	1 04·5	1 04·7	1 01·6	1·8	0·1	7·8	0·6	13·8	1·0
19	1 04·8	1 04·9	1 01·8	1·9	0·1	7·9	0·6	13·9	1·0
20	1 05·0	1 05·2	1 02·0	2·0	0·2	8·0	0·6	14·0	1·1
21	1 05·3	1 05·4	1 02·3	2·1	0·2	8·1	0·6	14·1	1·1
22	1 05·5	1 05·7	1 02·5	2·2	0·2	8·2	0·6	14·2	1·1
23	1 05·8	1 05·9	1 02·8	2·3	0·2	8·3	0·6	14·3	1·1
24	1 06·0	1 06·2	1 03·0	2·4	0·2	8·4	0·6	14·4	1·1
25	1 06·3	1 06·4	1 03·2	2·5	0·2	8·5	0·6	14·5	1·1
26	1 06·5	1 06·7	1 03·5	2·6	0·2	8·6	0·6	14·6	1·1
27	1 06·8	1 06·9	1 03·7	2·7	0·2	8·7	0·7	14·7	1·1
28	1 07·0	1 07·2	1 03·9	2·8	0·2	8·8	0·7	14·8	1·1
29	1 07·3	1 07·4	1 04·2	2·9	0·2	8·9	0·7	14·9	1·1
30	1 07·5	1 07·7	1 04·4	3·0	0·2	9·0	0·7	15·0	1·1
31	1 07·8	1 07·9	1 04·7	3·1	0·2	9·1	0·7	15·1	1·1
32	1 08·0	1 08·2	1 04·9	3·2	0·2	9·2	0·7	15·2	1·1
33	1 08·3	1 08·4	1 05·1	3·3	0·2	9·3	0·7	15·3	1·1
34	1 08·5	1 08·7	1 05·4	3·4	0·3	9·4	0·7	15·4	1·2
35	1 08·8	1 08·9	1 05·6	3·5	0·3	9·5	0·7	15·5	1·2
36	1 09·0	1 09·2	1 05·9	3·6	0·3	9·6	0·7	15·6	1·2
37	1 09·3	1 09·4	1 06·1	3·7	0·3	9·7	0·7	15·7	1·2
38	1 09·5	1 09·7	1 06·3	3·8	0·3	9·8	0·7	15·8	1·2
39	1 09·8	1 09·9	1 06·6	3·9	0·3	9·9	0·7	15·9	1·2
40	1 10·0	1 10·2	1 06·8	4·0	0·3	10·0	0·8	16·0	1·2
41	1 10·3	1 10·4	1 07·0	4·1	0·3	10·1	0·8	16·1	1·2
42	1 10·5	1 10·7	1 07·3	4·2	0·3	10·2	0·8	16·2	1·2
43	1 10·8	1 10·9	1 07·5	4·3	0·3	10·3	0·8	16·3	1·2
44	1 11·0	1 11·2	1 07·8	4·4	0·3	10·4	0·8	16·4	1·2
45	1 11·3	1 11·4	1 08·0	4·5	0·3	10·5	0·8	16·5	1·2
46	1 11·5	1 11·7	1 08·2	4·6	0·3	10·6	0·8	16·6	1·2
47	1 11·8	1 11·9	1 08·5	4·7	0·4	10·7	0·8	16·7	1·3
48	1 12·0	1 12·2	1 08·7	4·8	0·4	10·8	0·8	16·8	1·3
49	1 12·3	1 12·4	1 09·0	4·9	0·4	10·9	0·8	16·9	1·3
50	1 12·5	1 12·7	1 09·2	5·0	0·4	11·0	0·8	17·0	1·3
51	1 12·8	1 12·9	1 09·4	5·1	0·4	11·1	0·8	17·1	1·3
52	1 13·0	1 13·2	1 09·7	5·2	0·4	11·2	0·8	17·2	1·3
53	1 13·3	1 13·5	1 09·9	5·3	0·4	11·3	0·8	17·3	1·3
54	1 13·5	1 13·7	1 10·2	5·4	0·4	11·4	0·9	17·4	1·3
55	1 13·8	1 14·0	1 10·4	5·5	0·4	11·5	0·9	17·5	1·3
56	1 14·0	1 14·2	1 10·6	5·6	0·4	11·6	0·9	17·6	1·3
57	1 14·3	1 14·5	1 10·9	5·7	0·4	11·7	0·9	17·7	1·3
58	1 14·5	1 14·7	1 11·1	5·8	0·4	11·8	0·9	17·8	1·3
59	1 14·8	1 15·0	1 11·3	5·9	0·4	11·9	0·9	17·9	1·3
60	1 15·0	1 15·2	1 11·6	6·0	0·5	12·0	0·9	18·0	1·4

m 5 s	SUN PLANETS ° ′	ARIES ° ′	MOON ° ′	v or d ′	Corrⁿ ′	v or d ′	Corrⁿ ′	v or d ′	Corrⁿ ′
00	1 15·0	1 15·2	1 11·6	0·0	0·0	6·0	0·6	12·0	1·1
01	1 15·3	1 15·5	1 11·8	0·1	0·0	6·1	0·6	12·1	1·1
02	1 15·5	1 15·7	1 12·1	0·2	0·0	6·2	0·6	12·2	1·1
03	1 15·8	1 16·0	1 12·3	0·3	0·0	6·3	0·6	12·3	1·1
04	1 16·0	1 16·2	1 12·5	0·4	0·0	6·4	0·6	12·4	1·1
05	1 16·3	1 16·5	1 12·8	0·5	0·0	6·5	0·6	12·5	1·1
06	1 16·5	1 16·7	1 13·0	0·6	0·1	6·6	0·6	12·6	1·2
07	1 16·8	1 17·0	1 13·3	0·7	0·1	6·7	0·6	12·7	1·2
08	1 17·0	1 17·2	1 13·5	0·8	0·1	6·8	0·6	12·8	1·2
09	1 17·3	1 17·5	1 13·7	0·9	0·1	6·9	0·6	12·9	1·2
10	1 17·5	1 17·7	1 14·0	1·0	0·1	7·0	0·6	13·0	1·2
11	1 17·8	1 18·0	1 14·2	1·1	0·1	7·1	0·7	13·1	1·2
12	1 18·0	1 18·2	1 14·4	1·2	0·1	7·2	0·7	13·2	1·2
13	1 18·3	1 18·5	1 14·7	1·3	0·1	7·3	0·7	13·3	1·2
14	1 18·5	1 18·7	1 14·9	1·4	0·1	7·4	0·7	13·4	1·2
15	1 18·8	1 19·0	1 15·2	1·5	0·1	7·5	0·7	13·5	1·2
16	1 19·0	1 19·2	1 15·4	1·6	0·1	7·6	0·7	13·6	1·2
17	1 19·3	1 19·5	1 15·6	1·7	0·2	7·7	0·7	13·7	1·3
18	1 19·5	1 19·7	1 15·9	1·8	0·2	7·8	0·7	13·8	1·3
19	1 19·8	1 20·0	1 16·1	1·9	0·2	7·9	0·7	13·9	1·3
20	1 20·0	1 20·2	1 16·4	2·0	0·2	8·0	0·7	14·0	1·3
21	1 20·3	1 20·5	1 16·6	2·1	0·2	8·1	0·7	14·1	1·3
22	1 20·5	1 20·7	1 16·8	2·2	0·2	8·2	0·8	14·2	1·3
23	1 20·8	1 21·0	1 17·1	2·3	0·2	8·3	0·8	14·3	1·3
24	1 21·0	1 21·2	1 17·3	2·4	0·2	8·4	0·8	14·4	1·3
25	1 21·3	1 21·5	1 17·5	2·5	0·2	8·5	0·8	14·5	1·3
26	1 21·5	1 21·7	1 17·8	2·6	0·2	8·6	0·8	14·6	1·3
27	1 21·8	1 22·0	1 18·0	2·7	0·2	8·7	0·8	14·7	1·3
28	1 22·0	1 22·2	1 18·3	2·8	0·3	8·8	0·8	14·8	1·4
29	1 22·3	1 22·5	1 18·5	2·9	0·3	8·9	0·8	14·9	1·4
30	1 22·5	1 22·7	1 18·7	3·0	0·3	9·0	0·8	15·0	1·4
31	1 22·8	1 23·0	1 19·0	3·1	0·3	9·1	0·8	15·1	1·4
32	1 23·0	1 23·2	1 19·2	3·2	0·3	9·2	0·8	15·2	1·4
33	1 23·3	1 23·5	1 19·5	3·3	0·3	9·3	0·9	15·3	1·4
34	1 23·5	1 23·7	1 19·7	3·4	0·3	9·4	0·9	15·4	1·4
35	1 23·8	1 24·0	1 19·9	3·5	0·3	9·5	0·9	15·5	1·4
36	1 24·0	1 24·2	1 20·2	3·6	0·3	9·6	0·9	15·6	1·4
37	1 24·3	1 24·5	1 20·4	3·7	0·3	9·7	0·9	15·7	1·4
38	1 24·5	1 24·7	1 20·7	3·8	0·3	9·8	0·9	15·8	1·4
39	1 24·8	1 25·0	1 20·9	3·9	0·4	9·9	0·9	15·9	1·5
40	1 25·0	1 25·2	1 21·1	4·0	0·4	10·0	0·9	16·0	1·5
41	1 25·3	1 25·5	1 21·4	4·1	0·4	10·1	0·9	16·1	1·5
42	1 25·5	1 25·7	1 21·6	4·2	0·4	10·2	0·9	16·2	1·5
43	1 25·8	1 26·0	1 21·8	4·3	0·4	10·3	0·9	16·3	1·5
44	1 26·0	1 26·2	1 22·1	4·4	0·4	10·4	1·0	16·4	1·5
45	1 26·3	1 26·5	1 22·3	4·5	0·4	10·5	1·0	16·5	1·5
46	1 26·5	1 26·7	1 22·6	4·6	0·4	10·6	1·0	16·6	1·5
47	1 26·8	1 27·0	1 22·8	4·7	0·4	10·7	1·0	16·7	1·5
48	1 27·0	1 27·2	1 23·0	4·8	0·4	10·8	1·0	16·8	1·5
49	1 27·3	1 27·5	1 23·3	4·9	0·4	10·9	1·0	16·9	1·5
50	1 27·5	1 27·7	1 23·5	5·0	0·5	11·0	1·0	17·0	1·6
51	1 27·8	1 28·0	1 23·8	5·1	0·5	11·1	1·0	17·1	1·6
52	1 28·0	1 28·2	1 24·0	5·2	0·5	11·2	1·0	17·2	1·6
53	1 28·3	1 28·5	1 24·2	5·3	0·5	11·3	1·0	17·3	1·6
54	1 28·5	1 28·7	1 24·5	5·4	0·5	11·4	1·0	17·4	1·6
55	1 28·8	1 29·0	1 24·7	5·5	0·5	11·5	1·1	17·5	1·6
56	1 29·0	1 29·2	1 24·9	5·6	0·5	11·6	1·1	17·6	1·6
57	1 29·3	1 29·5	1 25·2	5·7	0·5	11·7	1·1	17·7	1·6
58	1 29·5	1 29·7	1 25·4	5·8	0·5	11·8	1·1	17·8	1·6
59	1 29·8	1 30·0	1 25·7	5·9	0·5	11·9	1·1	17·9	1·6
60	1 30·0	1 30·2	1 25·9	6·0	0·6	12·0	1·1	18·0	1·7

m 6	SUN PLANETS	ARIES	MOON	v or d	Corrn	v or d	Corrn	v or d	Corrn
s	° ′	° ′	° ′	′	′	′	′	′	′
00	1 30·0	1 30·2	1 25·9	0·0	0·0	6·0	0·7	12·0	1·3
01	1 30·3	1 30·5	1 26·1	0·1	0·0	6·1	0·7	12·1	1·3
02	1 30·5	1 30·7	1 26·4	0·2	0·0	6·2	0·7	12·2	1·3
03	1 30·8	1 31·0	1 26·6	0·3	0·0	6·3	0·7	12·3	1·3
04	1 31·0	1 31·2	1 26·9	0·4	0·0	6·4	0·7	12·4	1·3
05	1 31·3	1 31·5	1 27·1	0·5	0·1	6·5	0·7	12·5	1·4
06	1 31·5	1 31·8	1 27·3	0·6	0·1	6·6	0·7	12·6	1·4
07	1 31·8	1 32·0	1 27·6	0·7	0·1	6·7	0·7	12·7	1·4
08	1 32·0	1 32·3	1 27·8	0·8	0·1	6·8	0·7	12·8	1·4
09	1 32·3	1 32·5	1 28·0	0·9	0·1	6·9	0·7	12·9	1·4
10	1 32·5	1 32·8	1 28·3	1·0	0·1	7·0	0·8	13·0	1·4
11	1 32·8	1 33·0	1 28·5	1·1	0·1	7·1	0·8	13·1	1·4
12	1 33·0	1 33·3	1 28·8	1·2	0·1	7·2	0·8	13·2	1·4
13	1 33·3	1 33·5	1 29·0	1·3	0·1	7·3	0·8	13·3	1·4
14	1 33·5	1 33·8	1 29·2	1·4	0·2	7·4	0·8	13·4	1·5
15	1 33·8	1 34·0	1 29·5	1·5	0·2	7·5	0·8	13·5	1·5
16	1 34·0	1 34·3	1 29·7	1·6	0·2	7·6	0·8	13·6	1·5
17	1 34·3	1 34·5	1 30·0	1·7	0·2	7·7	0·8	13·7	1·5
18	1 34·5	1 34·8	1 30·2	1·8	0·2	7·8	0·8	13·8	1·5
19	1 34·8	1 35·0	1 30·4	1·9	0·2	7·9	0·9	13·9	1·5
20	1 35·0	1 35·3	1 30·7	2·0	0·2	8·0	0·9	14·0	1·5
21	1 35·3	1 35·5	1 30·9	2·1	0·2	8·1	0·9	14·1	1·5
22	1 35·5	1 35·8	1 31·1	2·2	0·2	8·2	0·9	14·2	1·5
23	1 35·8	1 36·0	1 31·4	2·3	0·2	8·3	0·9	14·3	1·5
24	1 36·0	1 36·3	1 31·6	2·4	0·3	8·4	0·9	14·4	1·6
25	1 36·3	1 36·5	1 31·9	2·5	0·3	8·5	0·9	14·5	1·6
26	1 36·5	1 36·8	1 32·1	2·6	0·3	8·6	0·9	14·6	1·6
27	1 36·8	1 37·0	1 32·3	2·7	0·3	8·7	0·9	14·7	1·6
28	1 37·0	1 37·3	1 32·6	2·8	0·3	8·8	1·0	14·8	1·6
29	1 37·3	1 37·5	1 32·8	2·9	0·3	8·9	1·0	14·9	1·6
30	1 37·5	1 37·8	1 33·1	3·0	0·3	9·0	1·0	15·0	1·6
31	1 37·8	1 38·0	1 33·3	3·1	0·3	9·1	1·0	15·1	1·6
32	1 38·0	1 38·3	1 33·5	3·2	0·3	9·2	1·0	15·2	1·6
33	1 38·3	1 38·5	1 33·8	3·3	0·4	9·3	1·0	15·3	1·7
34	1 38·5	1 38·8	1 34·0	3·4	0·4	9·4	1·0	15·4	1·7
35	1 38·8	1 39·0	1 34·3	3·5	0·4	9·5	1·0	15·5	1·7
36	1 39·0	1 39·3	1 34·5	3·6	0·4	9·6	1·0	15·6	1·7
37	1 39·3	1 39·5	1 34·7	3·7	0·4	9·7	1·1	15·7	1·7
38	1 39·5	1 39·8	1 35·0	3·8	0·4	9·8	1·1	15·8	1·7
39	1 39·8	1 40·0	1 35·2	3·9	0·4	9·9	1·1	15·9	1·7
40	1 40·0	1 40·3	1 35·4	4·0	0·4	10·0	1·1	16·0	1·7
41	1 40·3	1 40·5	1 35·7	4·1	0·4	10·1	1·1	16·1	1·7
42	1 40·5	1 40·8	1 35·9	4·2	0·5	10·2	1·1	16·2	1·8
43	1 40·8	1 41·0	1 36·2	4·3	0·5	10·3	1·1	16·3	1·8
44	1 41·0	1 41·3	1 36·4	4·4	0·5	10·4	1·1	16·4	1·8
45	1 41·3	1 41·5	1 36·6	4·5	0·5	10·5	1·1	16·5	1·8
46	1 41·5	1 41·8	1 36·9	4·6	0·5	10·6	1·1	16·6	1·8
47	1 41·8	1 42·0	1 37·1	4·7	0·5	10·7	1·2	16·7	1·8
48	1 42·0	1 42·3	1 37·4	4·8	0·5	10·8	1·2	16·8	1·8
49	1 42·3	1 42·5	1 37·6	4·9	0·5	10·9	1·2	16·9	1·8
50	1 42·5	1 42·8	1 37·8	5·0	0·5	11·0	1·2	17·0	1·8
51	1 42·8	1 43·0	1 38·1	5·1	0·6	11·1	1·2	17·1	1·9
52	1 43·0	1 43·3	1 38·3	5·2	0·6	11·2	1·2	17·2	1·9
53	1 43·3	1 43·5	1 38·5	5·3	0·6	11·3	1·2	17·3	1·9
54	1 43·5	1 43·8	1 38·8	5·4	0·6	11·4	1·2	17·4	1·9
55	1 43·8	1 44·0	1 39·0	5·5	0·6	11·5	1·2	17·5	1·9
56	1 44·0	1 44·3	1 39·3	5·6	0·6	11·6	1·3	17·6	1·9
57	1 44·3	1 44·5	1 39·5	5·7	0·6	11·7	1·3	17·7	1·9
58	1 44·5	1 44·8	1 39·7	5·8	0·6	11·8	1·3	17·8	1·9
59	1 44·8	1 45·0	1 40·0	5·9	0·6	11·9	1·3	17·9	1·9
60	1 45·0	1 45·3	1 40·2	6·0	0·7	12·0	1·3	18·0	2·0

m 7	SUN PLANETS	ARIES	MOON	v or d	Corrn	v or d	Corrn	v or d	Corrn
s	° ′	° ′	° ′	′	′	′	′	′	′
00	1 45·0	1 45·3	1 40·2	0·0	0·0	6·0	0·8	12·0	1·5
01	1 45·3	1 45·5	1 40·5	0·1	0·0	6·1	0·8	12·1	1·5
02	1 45·5	1 45·8	1 40·7	0·2	0·0	6·2	0·8	12·2	1·5
03	1 45·8	1 46·0	1 40·9	0·3	0·0	6·3	0·8	12·3	1·5
04	1 46·0	1 46·3	1 41·2	0·4	0·1	6·4	0·8	12·4	1·6
05	1 46·3	1 46·5	1 41·4	0·5	0·1	6·5	0·8	12·5	1·6
06	1 46·5	1 46·8	1 41·6	0·6	0·1	6·6	0·8	12·6	1·6
07	1 46·8	1 47·0	1 41·9	0·7	0·1	6·7	0·8	12·7	1·6
08	1 47·0	1 47·3	1 42·1	0·8	0·1	6·8	0·9	12·8	1·6
09	1 47·3	1 47·5	1 42·4	0·9	0·1	6·9	0·9	12·9	1·6
10	1 47·5	1 47·8	1 42·6	1·0	0·1	7·0	0·9	13·0	1·6
11	1 47·8	1 48·0	1 42·8	1·1	0·1	7·1	0·9	13·1	1·6
12	1 48·0	1 48·3	1 43·1	1·2	0·2	7·2	0·9	13·2	1·7
13	1 48·3	1 48·5	1 43·3	1·3	0·2	7·3	0·9	13·3	1·7
14	1 48·5	1 48·8	1 43·6	1·4	0·2	7·4	0·9	13·4	1·7
15	1 48·8	1 49·0	1 43·8	1·5	0·2	7·5	0·9	13·5	1·7
16	1 49·0	1 49·3	1 44·0	1·6	0·2	7·6	1·0	13·6	1·7
17	1 49·3	1 49·5	1 44·3	1·7	0·2	7·7	1·0	13·7	1·7
18	1 49·5	1 49·8	1 44·5	1·8	0·2	7·8	1·0	13·8	1·7
19	1 49·8	1 50·1	1 44·8	1·9	0·2	7·9	1·0	13·9	1·7
20	1 50·0	1 50·3	1 45·0	2·0	0·3	8·0	1·0	14·0	1·8
21	1 50·3	1 50·6	1 45·2	2·1	0·3	8·1	1·0	14·1	1·8
22	1 50·5	1 50·8	1 45·5	2·2	0·3	8·2	1·0	14·2	1·8
23	1 50·8	1 51·1	1 45·7	2·3	0·3	8·3	1·0	14·3	1·8
24	1 51·0	1 51·3	1 45·9	2·4	0·3	8·4	1·1	14·4	1·8
25	1 51·3	1 51·6	1 46·2	2·5	0·3	8·5	1·1	14·5	1·8
26	1 51·5	1 51·8	1 46·4	2·6	0·3	8·6	1·1	14·6	1·8
27	1 51·8	1 52·1	1 46·7	2·7	0·3	8·7	1·1	14·7	1·8
28	1 52·0	1 52·3	1 46·9	2·8	0·4	8·8	1·1	14·8	1·9
29	1 52·3	1 52·6	1 47·1	2·9	0·4	8·9	1·1	14·9	1·9
30	1 52·5	1 52·8	1 47·4	3·0	0·4	9·0	1·1	15·0	1·9
31	1 52·8	1 53·1	1 47·6	3·1	0·4	9·1	1·1	15·1	1·9
32	1 53·0	1 53·3	1 47·9	3·2	0·4	9·2	1·2	15·2	1·9
33	1 53·3	1 53·6	1 48·1	3·3	0·4	9·3	1·2	15·3	1·9
34	1 53·5	1 53·8	1 48·3	3·4	0·4	9·4	1·2	15·4	1·9
35	1 53·8	1 54·1	1 48·6	3·5	0·4	9·5	1·2	15·5	1·9
36	1 54·0	1 54·3	1 48·8	3·6	0·5	9·6	1·2	15·6	2·0
37	1 54·3	1 54·6	1 49·0	3·7	0·5	9·7	1·2	15·7	2·0
38	1 54·5	1 54·8	1 49·3	3·8	0·5	9·8	1·2	15·8	2·0
39	1 54·8	1 55·1	1 49·5	3·9	0·5	9·9	1·2	15·9	2·0
40	1 55·0	1 55·3	1 49·8	4·0	0·5	10·0	1·3	16·0	2·0
41	1 55·3	1 55·6	1 50·0	4·1	0·5	10·1	1·3	16·1	2·0
42	1 55·5	1 55·8	1 50·2	4·2	0·5	10·2	1·3	16·2	2·0
43	1 55·8	1 56·1	1 50·5	4·3	0·5	10·3	1·3	16·3	2·0
44	1 56·0	1 56·3	1 50·7	4·4	0·6	10·4	1·3	16·4	2·1
45	1 56·3	1 56·6	1 51·0	4·5	0·6	10·5	1·3	16·5	2·1
46	1 56·5	1 56·8	1 51·2	4·6	0·6	10·6	1·3	16·6	2·1
47	1 56·8	1 57·1	1 51·4	4·7	0·6	10·7	1·3	16·7	2·1
48	1 57·0	1 57·3	1 51·7	4·8	0·6	10·8	1·4	16·8	2·1
49	1 57·3	1 57·6	1 51·9	4·9	0·6	10·9	1·4	16·9	2·1
50	1 57·5	1 57·8	1 52·1	5·0	0·6	11·0	1·4	17·0	2·1
51	1 57·8	1 58·1	1 52·4	5·1	0·6	11·1	1·4	17·1	2·1
52	1 58·0	1 58·3	1 52·6	5·2	0·7	11·2	1·4	17·2	2·2
53	1 58·3	1 58·6	1 52·9	5·3	0·7	11·3	1·4	17·3	2·2
54	1 58·5	1 58·8	1 53·1	5·4	0·7	11·4	1·4	17·4	2·2
55	1 58·8	1 59·1	1 53·3	5·5	0·7	11·5	1·4	17·5	2·2
56	1 59·0	1 59·3	1 53·6	5·6	0·7	11·6	1·5	17·6	2·2
57	1 59·3	1 59·6	1 53·8	5·7	0·7	11·7	1·5	17·7	2·2
58	1 59·5	1 59·8	1 54·1	5·8	0·7	11·8	1·5	17·8	2·2
59	1 59·8	2 00·1	1 54·3	5·9	0·7	11·9	1·5	17·9	2·2
60	2 00·0	2 00·3	1 54·5	6·0	0·8	12·0	1·5	18·0	2·3

m 8 s	SUN PLANETS ° ′	ARIES ° ′	MOON ° ′	v or d ′	Corr^n ′	v or d ′	Corr^n ′	v or d ′	Corr^n ′
00	2 00·0	2 00·3	1 54·5	0·0	0·0	6·0	0·9	12·0	1·7
01	2 00·3	2 00·6	1 54·8	0·1	0·0	6·1	0·9	12·1	1·7
02	2 00·5	2 00·8	1 55·0	0·2	0·0	6·2	0·9	12·2	1·7
03	2 00·8	2 01·1	1 55·2	0·3	0·0	6·3	0·9	12·3	1·7
04	2 01·0	2 01·3	1 55·5	0·4	0·1	6·4	0·9	12·4	1·8
05	2 01·3	2 01·6	1 55·7	0·5	0·1	6·5	0·9	12·5	1·8
06	2 01·5	2 01·8	1 56·0	0·6	0·1	6·6	0·9	12·6	1·8
07	2 01·8	2 02·1	1 56·2	0·7	0·1	6·7	0·9	12·7	1·8
08	2 02·0	2 02·3	1 56·4	0·8	0·1	6·8	1·0	12·8	1·8
09	2 02·3	2 02·6	1 56·7	0·9	0·1	6·9	1·0	12·9	1·8
10	2 02·5	2 02·8	1 56·9	1·0	0·1	7·0	1·0	13·0	1·8
11	2 02·8	2 03·1	1 57·2	1·1	0·2	7·1	1·0	13·1	1·9
12	2 03·0	2 03·3	1 57·4	1·2	0·2	7·2	1·0	13·2	1·9
13	2 03·3	2 03·6	1 57·6	1·3	0·2	7·3	1·0	13·3	1·9
14	2 03·5	2 03·8	1 57·9	1·4	0·2	7·4	1·0	13·4	1·9
15	2 03·8	2 04·1	1 58·1	1·5	0·2	7·5	1·1	13·5	1·9
16	2 04·0	2 04·3	1 58·4	1·6	0·2	7·6	1·1	13·6	1·9
17	2 04·3	2 04·6	1 58·6	1·7	0·2	7·7	1·1	13·7	1·9
18	2 04·5	2 04·8	1 58·8	1·8	0·3	7·8	1·1	13·8	2·0
19	2 04·8	2 05·1	1 59·1	1·9	0·3	7·9	1·1	13·9	2·0
20	2 05·0	2 05·3	1 59·3	2·0	0·3	8·0	1·1	14·0	2·0
21	2 05·3	2 05·6	1 59·5	2·1	0·3	8·1	1·1	14·1	2·0
22	2 05·5	2 05·8	1 59·8	2·2	0·3	8·2	1·2	14·2	2·0
23	2 05·8	2 06·1	2 00·0	2·3	0·3	8·3	1·2	14·3	2·0
24	2 06·0	2 06·3	2 00·3	2·4	0·3	8·4	1·2	14·4	2·0
25	2 06·3	2 06·6	2 00·5	2·5	0·4	8·5	1·2	14·5	2·1
26	2 06·5	2 06·8	2 00·7	2·6	0·4	8·6	1·2	14·6	2·1
27	2 06·8	2 07·1	2 01·0	2·7	0·4	8·7	1·2	14·7	2·1
28	2 07·0	2 07·3	2 01·2	2·8	0·4	8·8	1·2	14·8	2·1
29	2 07·3	2 07·6	2 01·5	2·9	0·4	8·9	1·3	14·9	2·1
30	2 07·5	2 07·8	2 01·7	3·0	0·4	9·0	1·3	15·0	2·1
31	2 07·8	2 08·1	2 01·9	3·1	0·4	9·1	1·3	15·1	2·1
32	2 08·0	2 08·4	2 02·2	3·2	0·5	9·2	1·3	15·2	2·2
33	2 08·3	2 08·6	2 02·4	3·3	0·5	9·3	1·3	15·3	2·2
34	2 08·5	2 08·9	2 02·6	3·4	0·5	9·4	1·3	15·4	2·2
35	2 08·8	2 09·1	2 02·9	3·5	0·5	9·5	1·3	15·5	2·2
36	2 09·0	2 09·4	2 03·1	3·6	0·5	9·6	1·4	15·6	2·2
37	2 09·3	2 09·6	2 03·4	3·7	0·5	9·7	1·4	15·7	2·2
38	2 09·5	2 09·9	2 03·6	3·8	0·5	9·8	1·4	15·8	2·2
39	2 09·8	2 10·1	2 03·8	3·9	0·6	9·9	1·4	15·9	2·3
40	2 10·0	2 10·4	2 04·1	4·0	0·6	10·0	1·4	16·0	2·3
41	2 10·3	2 10·6	2 04·3	4·1	0·6	10·1	1·4	16·1	2·3
42	2 10·5	2 10·9	2 04·6	4·2	0·6	10·2	1·4	16·2	2·3
43	2 10·8	2 11·1	2 04·8	4·3	0·6	10·3	1·5	16·3	2·3
44	2 11·0	2 11·4	2 05·0	4·4	0·6	10·4	1·5	16·4	2·3
45	2 11·3	2 11·6	2 05·3	4·5	0·6	10·5	1·5	16·5	2·3
46	2 11·5	2 11·9	2 05·5	4·6	0·7	10·6	1·5	16·6	2·4
47	2 11·8	2 12·1	2 05·7	4·7	0·7	10·7	1·5	16·7	2·4
48	2 12·0	2 12·4	2 06·0	4·8	0·7	10·8	1·5	16·8	2·4
49	2 12·3	2 12·6	2 06·2	4·9	0·7	10·9	1·5	16·9	2·4
50	2 12·5	2 12·9	2 06·5	5·0	0·7	11·0	1·6	17·0	2·4
51	2 12·8	2 13·1	2 06·7	5·1	0·7	11·1	1·6	17·1	2·4
52	2 13·0	2 13·4	2 06·9	5·2	0·7	11·2	1·6	17·2	2·4
53	2 13·3	2 13·6	2 07·2	5·3	0·8	11·3	1·6	17·3	2·5
54	2 13·5	2 13·9	2 07·4	5·4	0·8	11·4	1·6	17·4	2·5
55	2 13·8	2 14·1	2 07·7	5·5	0·8	11·5	1·6	17·5	2·5
56	2 14·0	2 14·4	2 07·9	5·6	0·8	11·6	1·6	17·6	2·5
57	2 14·3	2 14·6	2 08·1	5·7	0·8	11·7	1·7	17·7	2·5
58	2 14·5	2 14·9	2 08·4	5·8	0·8	11·8	1·7	17·8	2·5
59	2 14·8	2 15·1	2 08·6	5·9	0·8	11·9	1·7	17·9	2·5
60	2 15·0	2 15·4	2 08·9	6·0	0·9	12·0	1·7	18·0	2·6

m 9 s	SUN PLANETS ° ′	ARIES ° ′	MOON ° ′	v or d ′	Corr^n ′	v or d ′	Corr^n ′	v or d ′	Corr^n ′
00	2 15·0	2 15·4	2 08·9	0·0	0·0	6·0	1·0	12·0	1·9
01	2 15·3	2 15·6	2 09·1	0·1	0·0	6·1	1·0	12·1	1·9
02	2 15·5	2 15·9	2 09·3	0·2	0·0	6·2	1·0	12·2	1·9
03	2 15·8	2 16·1	2 09·6	0·3	0·0	6·3	1·0	12·3	1·9
04	2 16·0	2 16·4	2 09·8	0·4	0·1	6·4	1·0	12·4	2·0
05	2 16·3	2 16·6	2 10·0	0·5	0·1	6·5	1·0	12·5	2·0
06	2 16·5	2 16·9	2 10·3	0·6	0·1	6·6	1·0	12·6	2·0
07	2 16·8	2 17·1	2 10·5	0·7	0·1	6·7	1·1	12·7	2·0
08	2 17·0	2 17·4	2 10·8	0·8	0·1	6·8	1·1	12·8	2·0
09	2 17·3	2 17·6	2 11·0	0·9	0·1	6·9	1·1	12·9	2·0
10	2 17·5	2 17·9	2 11·2	1·0	0·2	7·0	1·1	13·0	2·1
11	2 17·8	2 18·1	2 11·5	1·1	0·2	7·1	1·1	13·1	2·1
12	2 18·0	2 18·4	2 11·7	1·2	0·2	7·2	1·1	13·2	2·1
13	2 18·3	2 18·6	2 12·0	1·3	0·2	7·3	1·2	13·3	2·1
14	2 18·5	2 18·9	2 12·2	1·4	0·2	7·4	1·2	13·4	2·1
15	2 18·8	2 19·1	2 12·4	1·5	0·2	7·5	1·2	13·5	2·1
16	2 19·0	2 19·4	2 12·7	1·6	0·3	7·6	1·2	13·6	2·2
17	2 19·3	2 19·6	2 12·9	1·7	0·3	7·7	1·2	13·7	2·2
18	2 19·5	2 19·9	2 13·1	1·8	0·3	7·8	1·2	13·8	2·2
19	2 19·8	2 20·1	2 13·4	1·9	0·3	7·9	1·3	13·9	2·2
20	2 20·0	2 20·4	2 13·6	2·0	0·3	8·0	1·3	14·0	2·2
21	2 20·3	2 20·6	2 13·9	2·1	0·3	8·1	1·3	14·1	2·2
22	2 20·5	2 20·9	2 14·1	2·2	0·3	8·2	1·3	14·2	2·2
23	2 20·8	2 21·1	2 14·3	2·3	0·4	8·3	1·3	14·3	2·3
24	2 21·0	2 21·4	2 14·6	2·4	0·4	8·4	1·3	14·4	2·3
25	2 21·3	2 21·6	2 14·8	2·5	0·4	8·5	1·3	14·5	2·3
26	2 21·5	2 21·9	2 15·1	2·6	0·4	8·6	1·4	14·6	2·3
27	2 21·8	2 22·1	2 15·3	2·7	0·4	8·7	1·4	14·7	2·3
28	2 22·0	2 22·4	2 15·5	2·8	0·4	8·8	1·4	14·8	2·3
29	2 22·3	2 22·6	2 15·8	2·9	0·5	8·9	1·4	14·9	2·4
30	2 22·5	2 22·9	2 16·0	3·0	0·5	9·0	1·4	15·0	2·4
31	2 22·8	2 23·1	2 16·2	3·1	0·5	9·1	1·4	15·1	2·4
32	2 23·0	2 23·4	2 16·5	3·2	0·5	9·2	1·5	15·2	2·4
33	2 23·3	2 23·6	2 16·7	3·3	0·5	9·3	1·5	15·3	2·4
34	2 23·5	2 23·9	2 17·0	3·4	0·5	9·4	1·5	15·4	2·4
35	2 23·8	2 24·1	2 17·2	3·5	0·6	9·5	1·5	15·5	2·5
36	2 24·0	2 24·4	2 17·4	3·6	0·6	9·6	1·5	15·6	2·5
37	2 24·3	2 24·6	2 17·7	3·7	0·6	9·7	1·5	15·7	2·5
38	2 24·5	2 24·9	2 17·9	3·8	0·6	9·8	1·6	15·8	2·5
39	2 24·8	2 25·1	2 18·2	3·9	0·6	9·9	1·6	15·9	2·5
40	2 25·0	2 25·4	2 18·4	4·0	0·6	10·0	1·6	16·0	2·5
41	2 25·3	2 25·6	2 18·6	4·1	0·6	10·1	1·6	16·1	2·5
42	2 25·5	2 25·9	2 18·9	4·2	0·7	10·2	1·6	16·2	2·6
43	2 25·8	2 26·1	2 19·1	4·3	0·7	10·3	1·6	16·3	2·6
44	2 26·0	2 26·4	2 19·3	4·4	0·7	10·4	1·6	16·4	2·6
45	2 26·3	2 26·7	2 19·6	4·5	0·7	10·5	1·7	16·5	2·6
46	2 26·5	2 26·9	2 19·8	4·6	0·7	10·6	1·7	16·6	2·6
47	2 26·8	2 27·2	2 20·1	4·7	0·7	10·7	1·7	16·7	2·6
48	2 27·0	2 27·4	2 20·3	4·8	0·8	10·8	1·7	16·8	2·7
49	2 27·3	2 27·7	2 20·5	4·9	0·8	10·9	1·7	16·9	2·7
50	2 27·5	2 27·9	2 20·8	5·0	0·8	11·0	1·7	17·0	2·7
51	2 27·8	2 28·2	2 21·0	5·1	0·8	11·1	1·8	17·1	2·7
52	2 28·0	2 28·4	2 21·3	5·2	0·8	11·2	1·8	17·2	2·7
53	2 28·3	2 28·7	2 21·5	5·3	0·8	11·3	1·8	17·3	2·7
54	2 28·5	2 28·9	2 21·7	5·4	0·9	11·4	1·8	17·4	2·8
55	2 28·8	2 29·2	2 22·0	5·5	0·9	11·5	1·8	17·5	2·8
56	2 29·0	2 29·4	2 22·2	5·6	0·9	11·6	1·8	17·6	2·8
57	2 29·3	2 29·7	2 22·5	5·7	0·9	11·7	1·9	17·7	2·8
58	2 29·5	2 29·9	2 22·7	5·8	0·9	11·8	1·9	17·8	2·8
59	2 29·8	2 30·2	2 22·9	5·9	0·9	11·9	1·9	17·9	2·8
60	2 30·0	2 30·4	2 23·2	6·0	1·0	12·0	1·9	18·0	2·9

m 10 s	SUN PLANETS ° ′	ARIES ° ′	MOON ° ′	v or d ′	Corrⁿ ′	v or d ′	Corrⁿ ′	v or d ′	Corrⁿ ′
00	2 30·0	2 30·4	2 23·2	0·0	0·0	6·0	1·1	12·0	2·1
01	2 30·3	2 30·7	2 23·4	0·1	0·0	6·1	1·1	12·1	2·1
02	2 30·5	2 30·9	2 23·6	0·2	0·0	6·2	1·1	12·2	2·1
03	2 30·8	2 31·2	2 23·9	0·3	0·1	6·3	1·1	12·3	2·2
04	2 31·0	2 31·4	2 24·1	0·4	0·1	6·4	1·1	12·4	2·2
05	2 31·3	2 31·7	2 24·4	0·5	0·1	6·5	1·1	12·5	2·2
06	2 31·5	2 31·9	2 24·6	0·6	0·1	6·6	1·2	12·6	2·2
07	2 31·8	2 32·2	2 24·8	0·7	0·1	6·7	1·2	12·7	2·2
08	2 32·0	2 32·4	2 25·1	0·8	0·1	6·8	1·2	12·8	2·2
09	2 32·3	2 32·7	2 25·3	0·9	0·2	6·9	1·2	12·9	2·3
10	2 32·5	2 32·9	2 25·6	1·0	0·2	7·0	1·2	13·0	2·3
11	2 32·8	2 33·2	2 25·8	1·1	0·2	7·1	1·2	13·1	2·3
12	2 33·0	2 33·4	2 26·0	1·2	0·2	7·2	1·3	13·2	2·3
13	2 33·3	2 33·7	2 26·3	1·3	0·2	7·3	1·3	13·3	2·3
14	2 33·5	2 33·9	2 26·5	1·4	0·2	7·4	1·3	13·4	2·3
15	2 33·8	2 34·2	2 26·7	1·5	0·3	7·5	1·3	13·5	2·4
16	2 34·0	2 34·4	2 27·0	1·6	0·3	7·6	1·3	13·6	2·4
17	2 34·3	2 34·7	2 27·2	1·7	0·3	7·7	1·3	13·7	2·4
18	2 34·5	2 34·9	2 27·5	1·8	0·3	7·8	1·4	13·8	2·4
19	2 34·8	2 35·2	2 27·7	1·9	0·3	7·9	1·4	13·9	2·4
20	2 35·0	2 35·4	2 27·9	2·0	0·4	8·0	1·4	14·0	2·5
21	2 35·3	2 35·7	2 28·2	2·1	0·4	8·1	1·4	14·1	2·5
22	2 35·5	2 35·9	2 28·4	2·2	0·4	8·2	1·4	14·2	2·5
23	2 35·8	2 36·2	2 28·7	2·3	0·4	8·3	1·5	14·3	2·5
24	2 36·0	2 36·4	2 28·9	2·4	0·4	8·4	1·5	14·4	2·5
25	2 36·3	2 36·7	2 29·1	2·5	0·4	8·5	1·5	14·5	2·5
26	2 36·5	2 36·9	2 29·4	2·6	0·5	8·6	1·5	14·6	2·6
27	2 36·8	2 37·2	2 29·6	2·7	0·5	8·7	1·5	14·7	2·6
28	2 37·0	2 37·4	2 29·8	2·8	0·5	8·8	1·5	14·8	2·6
29	2 37·3	2 37·7	2 30·1	2·9	0·5	8·9	1·6	14·9	2·6
30	2 37·5	2 37·9	2 30·3	3·0	0·5	9·0	1·6	15·0	2·6
31	2 37·8	2 38·2	2 30·6	3·1	0·5	9·1	1·6	15·1	2·6
32	2 38·0	2 38·4	2 30·8	3·2	0·6	9·2	1·6	15·2	2·7
33	2 38·3	2 38·7	2 31·0	3·3	0·6	9·3	1·6	15·3	2·7
34	2 38·5	2 38·9	2 31·3	3·4	0·6	9·4	1·6	15·4	2·7
35	2 38·8	2 39·2	2 31·5	3·5	0·6	9·5	1·7	15·5	2·7
36	2 39·0	2 39·4	2 31·8	3·6	0·6	9·6	1·7	15·6	2·7
37	2 39·3	2 39·7	2 32·0	3·7	0·6	9·7	1·7	15·7	2·7
38	2 39·5	2 39·9	2 32·2	3·8	0·7	9·8	1·7	15·8	2·8
39	2 39·8	2 40·2	2 32·5	3·9	0·7	9·9	1·7	15·9	2·8
40	2 40·0	2 40·4	2 32·7	4·0	0·7	10·0	1·8	16·0	2·8
41	2 40·3	2 40·7	2 32·9	4·1	0·7	10·1	1·8	16·1	2·8
42	2 40·5	2 40·9	2 33·2	4·2	0·7	10·2	1·8	16·2	2·8
43	2 40·8	2 41·2	2 33·4	4·3	0·8	10·3	1·8	16·3	2·9
44	2 41·0	2 41·4	2 33·7	4·4	0·8	10·4	1·8	16·4	2·9
45	2 41·3	2 41·7	2 33·9	4·5	0·8	10·5	1·8	16·5	2·9
46	2 41·5	2 41·9	2 34·1	4·6	0·8	10·6	1·9	16·6	2·9
47	2 41·8	2 42·2	2 34·4	4·7	0·8	10·7	1·9	16·7	2·9
48	2 42·0	2 42·4	2 34·6	4·8	0·8	10·8	1·9	16·8	2·9
49	2 42·3	2 42·7	2 34·9	4·9	0·9	10·9	1·9	16·9	3·0
50	2 42·5	2 42·9	2 35·1	5·0	0·9	11·0	1·9	17·0	3·0
51	2 42·8	2 43·2	2 35·3	5·1	0·9	11·1	1·9	17·1	3·0
52	2 43·0	2 43·4	2 35·6	5·2	0·9	11·2	2·0	17·2	3·0
53	2 43·3	2 43·7	2 35·8	5·3	0·9	11·3	2·0	17·3	3·0
54	2 43·5	2 43·9	2 36·1	5·4	0·9	11·4	2·0	17·4	3·0
55	2 43·8	2 44·2	2 36·3	5·5	1·0	11·5	2·0	17·5	3·1
56	2 44·0	2 44·4	2 36·5	5·6	1·0	11·6	2·0	17·6	3·1
57	2 44·3	2 44·7	2 36·8	5·7	1·0	11·7	2·0	17·7	3·1
58	2 44·5	2 45·0	2 37·0	5·8	1·0	11·8	2·1	17·8	3·1
59	2 44·8	2 45·2	2 37·2	5·9	1·0	11·9	2·1	17·9	3·1
60	2 45·0	2 45·5	2 37·5	6·0	1·1	12·0	2·1	18·0	3·2

m 11 s	SUN PLANETS ° ′	ARIES ° ′	MOON ° ′	v or d ′	Corrⁿ ′	v or d ′	Corrⁿ ′	v or d ′	Corrⁿ ′
00	2 45·0	2 45·5	2 37·5	0·0	0·0	6·0	1·2	12·0	2·3
01	2 45·3	2 45·7	2 37·7	0·1	0·0	6·1	1·2	12·1	2·3
02	2 45·5	2 46·0	2 38·0	0·2	0·0	6·2	1·2	12·2	2·3
03	2 45·8	2 46·2	2 38·2	0·3	0·1	6·3	1·2	12·3	2·4
04	2 46·0	2 46·5	2 38·4	0·4	0·1	6·4	1·2	12·4	2·4
05	2 46·3	2 46·7	2 38·7	0·5	0·1	6·5	1·2	12·5	2·4
06	2 46·5	2 47·0	2 38·9	0·6	0·1	6·6	1·3	12·6	2·4
07	2 46·8	2 47·2	2 39·2	0·7	0·1	6·7	1·3	12·7	2·4
08	2 47·0	2 47·5	2 39·4	0·8	0·2	6·8	1·3	12·8	2·5
09	2 47·3	2 47·7	2 39·6	0·9	0·2	6·9	1·3	12·9	2·5
10	2 47·5	2 48·0	2 39·9	1·0	0·2	7·0	1·3	13·0	2·5
11	2 47·8	2 48·2	2 40·1	1·1	0·2	7·1	1·4	13·1	2·5
12	2 48·0	2 48·5	2 40·3	1·2	0·2	7·2	1·4	13·2	2·5
13	2 48·3	2 48·7	2 40·6	1·3	0·2	7·3	1·4	13·3	2·5
14	2 48·5	2 49·0	2 40·8	1·4	0·3	7·4	1·4	13·4	2·6
15	2 48·8	2 49·2	2 41·1	1·5	0·3	7·5	1·4	13·5	2·6
16	2 49·0	2 49·5	2 41·3	1·6	0·3	7·6	1·5	13·6	2·6
17	2 49·3	2 49·7	2 41·5	1·7	0·3	7·7	1·5	13·7	2·6
18	2 49·5	2 50·0	2 41·8	1·8	0·3	7·8	1·5	13·8	2·6
19	2 49·8	2 50·2	2 42·0	1·9	0·4	7·9	1·5	13·9	2·7
20	2 50·0	2 50·5	2 42·3	2·0	0·4	8·0	1·5	14·0	2·7
21	2 50·3	2 50·7	2 42·5	2·1	0·4	8·1	1·6	14·1	2·7
22	2 50·5	2 51·0	2 42·7	2·2	0·4	8·2	1·6	14·2	2·7
23	2 50·8	2 51·2	2 43·0	2·3	0·4	8·3	1·6	14·3	2·7
24	2 51·0	2 51·5	2 43·2	2·4	0·5	8·4	1·6	14·4	2·8
25	2 51·3	2 51·7	2 43·4	2·5	0·5	8·5	1·6	14·5	2·8
26	2 51·5	2 52·0	2 43·7	2·6	0·5	8·6	1·6	14·6	2·8
27	2 51·8	2 52·2	2 43·9	2·7	0·5	8·7	1·7	14·7	2·8
28	2 52·0	2 52·5	2 44·2	2·8	0·5	8·8	1·7	14·8	2·8
29	2 52·3	2 52·7	2 44·4	2·9	0·6	8·9	1·7	14·9	2·9
30	2 52·5	2 53·0	2 44·6	3·0	0·6	9·0	1·7	15·0	2·9
31	2 52·8	2 53·2	2 44·9	3·1	0·6	9·1	1·7	15·1	2·9
32	2 53·0	2 53·5	2 45·1	3·2	0·6	9·2	1·8	15·2	2·9
33	2 53·3	2 53·7	2 45·4	3·3	0·6	9·3	1·8	15·3	2·9
34	2 53·5	2 54·0	2 45·6	3·4	0·7	9·4	1·8	15·4	3·0
35	2 53·8	2 54·2	2 45·8	3·5	0·7	9·5	1·8	15·5	3·0
36	2 54·0	2 54·5	2 46·1	3·6	0·7	9·6	1·8	15·6	3·0
37	2 54·3	2 54·7	2 46·3	3·7	0·7	9·7	1·9	15·7	3·0
38	2 54·5	2 55·0	2 46·6	3·8	0·7	9·8	1·9	15·8	3·0
39	2 54·8	2 55·2	2 46·8	3·9	0·7	9·9	1·9	15·9	3·0
40	2 55·0	2 55·5	2 47·0	4·0	0·8	10·0	1·9	16·0	3·1
41	2 55·3	2 55·7	2 47·3	4·1	0·8	10·1	1·9	16·1	3·1
42	2 55·5	2 56·0	2 47·5	4·2	0·8	10·2	2·0	16·2	3·1
43	2 55·8	2 56·2	2 47·7	4·3	0·8	10·3	2·0	16·3	3·1
44	2 56·0	2 56·5	2 48·0	4·4	0·8	10·4	2·0	16·4	3·1
45	2 56·3	2 56·7	2 48·2	4·5	0·9	10·5	2·0	16·5	3·2
46	2 56·5	2 57·0	2 48·5	4·6	0·9	10·6	2·0	16·6	3·2
47	2 56·8	2 57·2	2 48·7	4·7	0·9	10·7	2·1	16·7	3·2
48	2 57·0	2 57·5	2 48·9	4·8	0·9	10·8	2·1	16·8	3·2
49	2 57·3	2 57·7	2 49·2	4·9	0·9	10·9	2·1	16·9	3·2
50	2 57·5	2 58·0	2 49·4	5·0	1·0	11·0	2·1	17·0	3·3
51	2 57·8	2 58·2	2 49·7	5·1	1·0	11·1	2·1	17·1	3·3
52	2 58·0	2 58·5	2 49·9	5·2	1·0	11·2	2·1	17·2	3·3
53	2 58·3	2 58·7	2 50·1	5·3	1·0	11·3	2·2	17·3	3·3
54	2 58·5	2 59·0	2 50·4	5·4	1·0	11·4	2·2	17·4	3·3
55	2 58·8	2 59·2	2 50·6	5·5	1·1	11·5	2·2	17·5	3·4
56	2 59·0	2 59·5	2 50·8	5·6	1·1	11·6	2·2	17·6	3·4
57	2 59·3	2 59·7	2 51·1	5·7	1·1	11·7	2·2	17·7	3·4
58	2 59·5	3 00·0	2 51·3	5·8	1·1	11·8	2·3	17·8	3·4
59	2 59·8	3 00·2	2 51·6	5·9	1·1	11·9	2·3	17·9	3·4
60	3 00·0	3 00·5	2 51·8	6·0	1·2	12·0	2·3	18·0	3·5

12ᵐ	SUN PLANETS	ARIES	MOON	v or d	Corrⁿ	v or d	Corrⁿ	v or d	Corrⁿ
s	° ′	° ′	° ′	′	′	′	′	′	′
00	3 00·0	3 00·5	2 51·8	0·0	0·0	6·0	1·3	12·0	2·5
01	3 00·3	3 00·7	2 52·0	0·1	0·0	6·1	1·3	12·1	2·5
02	3 00·5	3 01·0	2 52·3	0·2	0·0	6·2	1·3	12·2	2·5
03	3 00·8	3 01·2	2 52·5	0·3	0·1	6·3	1·3	12·3	2·6
04	3 01·0	3 01·5	2 52·8	0·4	0·1	6·4	1·3	12·4	2·6
05	3 01·3	3 01·7	2 53·0	0·5	0·1	6·5	1·4	12·5	2·6
06	3 01·5	3 02·0	2 53·2	0·6	0·1	6·6	1·4	12·6	2·6
07	3 01·8	3 02·2	2 53·5	0·7	0·1	6·7	1·4	12·7	2·6
08	3 02·0	3 02·5	2 53·7	0·8	0·2	6·8	1·4	12·8	2·7
09	3 02·3	3 02·7	2 53·9	0·9	0·2	6·9	1·4	12·9	2·7
10	3 02·5	3 03·0	2 54·2	1·0	0·2	7·0	1·5	13·0	2·7
11	3 02·8	3 03·3	2 54·4	1·1	0·2	7·1	1·5	13·1	2·7
12	3 03·0	3 03·5	2 54·7	1·2	0·3	7·2	1·5	13·2	2·8
13	3 03·3	3 03·8	2 54·9	1·3	0·3	7·3	1·5	13·3	2·8
14	3 03·5	3 04·0	2 55·1	1·4	0·3	7·4	1·5	13·4	2·8
15	3 03·8	3 04·3	2 55·4	1·5	0·3	7·5	1·6	13·5	2·8
16	3 04·0	3 04·5	2 55·6	1·6	0·3	7·6	1·6	13·6	2·8
17	3 04·3	3 04·8	2 55·9	1·7	0·4	7·7	1·6	13·7	2·9
18	3 04·5	3 05·0	2 56·1	1·8	0·4	7·8	1·6	13·8	2·9
19	3 04·8	3 05·3	2 56·3	1·9	0·4	7·9	1·6	13·9	2·9
20	3 05·0	3 05·5	2 56·6	2·0	0·4	8·0	1·7	14·0	2·9
21	3 05·3	3 05·8	2 56·8	2·1	0·4	8·1	1·7	14·1	2·9
22	3 05·5	3 06·0	2 57·0	2·2	0·5	8·2	1·7	14·2	3·0
23	3 05·8	3 06·3	2 57·3	2·3	0·5	8·3	1·7	14·3	3·0
24	3 06·0	3 06·5	2 57·5	2·4	0·5	8·4	1·8	14·4	3·0
25	3 06·3	3 06·8	2 57·8	2·5	0·5	8·5	1·8	14·5	3·0
26	3 06·5	3 07·0	2 58·0	2·6	0·5	8·6	1·8	14·6	3·0
27	3 06·8	3 07·3	2 58·2	2·7	0·6	8·7	1·8	14·7	3·1
28	3 07·0	3 07·5	2 58·5	2·8	0·6	8·8	1·8	14·8	3·1
29	3 07·3	3 07·8	2 58·7	2·9	0·6	8·9	1·9	14·9	3·1
30	3 07·5	3 08·0	2 59·0	3·0	0·6	9·0	1·9	15·0	3·1
31	3 07·8	3 08·3	2 59·2	3·1	0·6	9·1	1·9	15·1	3·1
32	3 08·0	3 08·5	2 59·4	3·2	0·7	9·2	1·9	15·2	3·2
33	3 08·3	3 08·8	2 59·7	3·3	0·7	9·3	1·9	15·3	3·2
34	3 08·5	3 09·0	2 59·9	3·4	0·7	9·4	2·0	15·4	3·2
35	3 08·8	3 09·3	3 00·2	3·5	0·7	9·5	2·0	15·5	3·2
36	3 09·0	3 09·5	3 00·4	3·6	0·8	9·6	2·0	15·6	3·3
37	3 09·3	3 09·8	3 00·6	3·7	0·8	9·7	2·0	15·7	3·3
38	3 09·5	3 10·0	3 00·9	3·8	0·8	9·8	2·0	15·8	3·3
39	3 09·8	3 10·3	3 01·1	3·9	0·8	9·9	2·1	15·9	3·3
40	3 10·0	3 10·5	3 01·3	4·0	0·8	10·0	2·1	16·0	3·3
41	3 10·3	3 10·8	3 01·6	4·1	0·9	10·1	2·1	16·1	3·4
42	3 10·5	3 11·0	3 01·8	4·2	0·9	10·2	2·1	16·2	3·4
43	3 10·8	3 11·3	3 02·1	4·3	0·9	10·3	2·1	16·3	3·4
44	3 11·0	3 11·5	3 02·3	4·4	0·9	10·4	2·2	16·4	3·4
45	3 11·3	3 11·8	3 02·5	4·5	0·9	10·5	2·2	16·5	3·4
46	3 11·5	3 12·0	3 02·8	4·6	1·0	10·6	2·2	16·6	3·5
47	3 11·8	3 12·3	3 03·0	4·7	1·0	10·7	2·2	16·7	3·5
48	3 12·0	3 12·5	3 03·3	4·8	1·0	10·8	2·3	16·8	3·5
49	3 12·3	3 12·8	3 03·5	4·9	1·0	10·9	2·3	16·9	3·5
50	3 12·5	3 13·0	3 03·7	5·0	1·0	11·0	2·3	17·0	3·5
51	3 12·8	3 13·3	3 04·0	5·1	1·1	11·1	2·3	17·1	3·6
52	3 13·0	3 13·5	3 04·2	5·2	1·1	11·2	2·3	17·2	3·6
53	3 13·3	3 13·8	3 04·4	5·3	1·1	11·3	2·4	17·3	3·6
54	3 13·5	3 14·0	3 04·7	5·4	1·1	11·4	2·4	17·4	3·6
55	3 13·8	3 14·3	3 04·9	5·5	1·1	11·5	2·4	17·5	3·6
56	3 14·0	3 14·5	3 05·2	5·6	1·2	11·6	2·4	17·6	3·7
57	3 14·3	3 14·8	3 05·4	5·7	1·2	11·7	2·4	17·7	3·7
58	3 14·5	3 15·0	3 05·6	5·8	1·2	11·8	2·5	17·8	3·7
59	3 14·8	3 15·3	3 05·9	5·9	1·2	11·9	2·5	17·9	3·7
60	3 15·0	3 15·5	3 06·1	6·0	1·3	12·0	2·5	18·0	3·8

13ᵐ	SUN PLANETS	ARIES	MOON	v or d	Corrⁿ	v or d	Corrⁿ	v or d	Corrⁿ
s	° ′	° ′	° ′	′	′	′	′	′	′
00	3 15·0	3 15·5	3 06·1	0·0	0·0	6·0	1·4	12·0	2·7
01	3 15·3	3 15·8	3 06·4	0·1	0·0	6·1	1·4	12·1	2·7
02	3 15·5	3 16·0	3 06·6	0·2	0·0	6·2	1·4	12·2	2·7
03	3 15·8	3 16·3	3 06·8	0·3	0·1	6·3	1·4	12·3	2·8
04	3 16·0	3 16·5	3 07·1	0·4	0·1	6·4	1·4	12·4	2·8
05	3 16·3	3 16·8	3 07·3	0·5	0·1	6·5	1·5	12·5	2·8
06	3 16·5	3 17·0	3 07·5	0·6	0·1	6·6	1·5	12·6	2·8
07	3 16·8	3 17·3	3 07·8	0·7	0·2	6·7	1·5	12·7	2·9
08	3 17·0	3 17·5	3 08·0	0·8	0·2	6·8	1·5	12·8	2·9
09	3 17·3	3 17·8	3 08·3	0·9	0·2	6·9	1·6	12·9	2·9
10	3 17·5	3 18·0	3 08·5	1·0	0·2	7·0	1·6	13·0	2·9
11	3 17·8	3 18·3	3 08·7	1·1	0·2	7·1	1·6	13·1	2·9
12	3 18·0	3 18·5	3 09·0	1·2	0·3	7·2	1·6	13·2	3·0
13	3 18·3	3 18·8	3 09·2	1·3	0·3	7·3	1·6	13·3	3·0
14	3 18·5	3 19·0	3 09·5	1·4	0·3	7·4	1·7	13·4	3·0
15	3 18·8	3 19·3	3 09·7	1·5	0·3	7·5	1·7	13·5	3·0
16	3 19·0	3 19·5	3 09·9	1·6	0·4	7·6	1·7	13·6	3·1
17	3 19·3	3 19·8	3 10·2	1·7	0·4	7·7	1·7	13·7	3·1
18	3 19·5	3 20·0	3 10·4	1·8	0·4	7·8	1·8	13·8	3·1
19	3 19·8	3 20·3	3 10·7	1·9	0·4	7·9	1·8	13·9	3·1
20	3 20·0	3 20·5	3 10·9	2·0	0·5	8·0	1·8	14·0	3·2
21	3 20·3	3 20·8	3 11·1	2·1	0·5	8·1	1·8	14·1	3·2
22	3 20·5	3 21·0	3 11·4	2·2	0·5	8·2	1·8	14·2	3·2
23	3 20·8	3 21·3	3 11·6	2·3	0·5	8·3	1·9	14·3	3·2
24	3 21·0	3 21·6	3 11·8	2·4	0·5	8·4	1·9	14·4	3·2
25	3 21·3	3 21·8	3 12·1	2·5	0·6	8·5	1·9	14·5	3·3
26	3 21·5	3 22·1	3 12·3	2·6	0·6	8·6	1·9	14·6	3·3
27	3 21·8	3 22·3	3 12·6	2·7	0·6	8·7	2·0	14·7	3·3
28	3 22·0	3 22·6	3 12·8	2·8	0·6	8·8	2·0	14·8	3·3
29	3 22·3	3 22·8	3 13·0	2·9	0·7	8·9	2·0	14·9	3·4
30	3 22·5	3 23·1	3 13·3	3·0	0·7	9·0	2·0	15·0	3·4
31	3 22·8	3 23·3	3 13·5	3·1	0·7	9·1	2·0	15·1	3·4
32	3 23·0	3 23·6	3 13·8	3·2	0·7	9·2	2·1	15·2	3·4
33	3 23·3	3 23·8	3 14·0	3·3	0·7	9·3	2·1	15·3	3·4
34	3 23·5	3 24·1	3 14·2	3·4	0·8	9·4	2·1	15·4	3·5
35	3 23·8	3 24·3	3 14·5	3·5	0·8	9·5	2·1	15·5	3·5
36	3 24·0	3 24·6	3 14·7	3·6	0·8	9·6	2·2	15·6	3·5
37	3 24·3	3 24·8	3 14·9	3·7	0·8	9·7	2·2	15·7	3·5
38	3 24·5	3 25·1	3 15·2	3·8	0·9	9·8	2·2	15·8	3·6
39	3 24·8	3 25·3	3 15·4	3·9	0·9	9·9	2·2	15·9	3·6
40	3 25·0	3 25·6	3 15·7	4·0	0·9	10·0	2·3	16·0	3·6
41	3 25·3	3 25·8	3 15·9	4·1	0·9	10·1	2·3	16·1	3·6
42	3 25·5	3 26·1	3 16·1	4·2	0·9	10·2	2·3	16·2	3·6
43	3 25·8	3 26·3	3 16·4	4·3	1·0	10·3	2·3	16·3	3·7
44	3 26·0	3 26·6	3 16·6	4·4	1·0	10·4	2·3	16·4	3·7
45	3 26·3	3 26·8	3 16·9	4·5	1·0	10·5	2·4	16·5	3·7
46	3 26·5	3 27·1	3 17·1	4·6	1·0	10·6	2·4	16·6	3·7
47	3 26·8	3 27·3	3 17·3	4·7	1·1	10·7	2·4	16·7	3·8
48	3 27·0	3 27·6	3 17·6	4·8	1·1	10·8	2·4	16·8	3·8
49	3 27·3	3 27·8	3 17·8	4·9	1·1	10·9	2·5	16·9	3·8
50	3 27·5	3 28·1	3 18·0	5·0	1·1	11·0	2·5	17·0	3·8
51	3 27·8	3 28·3	3 18·3	5·1	1·1	11·1	2·5	17·1	3·8
52	3 28·0	3 28·6	3 18·5	5·2	1·2	11·2	2·5	17·2	3·9
53	3 28·3	3 28·8	3 18·8	5·3	1·2	11·3	2·5	17·3	3·9
54	3 28·5	3 29·1	3 19·0	5·4	1·2	11·4	2·6	17·4	3·9
55	3 28·8	3 29·3	3 19·2	5·5	1·2	11·5	2·6	17·5	3·9
56	3 29·0	3 29·6	3 19·5	5·6	1·3	11·6	2·6	17·6	4·0
57	3 29·3	3 29·8	3 19·7	5·7	1·3	11·7	2·6	17·7	4·0
58	3 29·5	3 30·1	3 20·0	5·8	1·3	11·8	2·7	17·8	4·0
59	3 29·8	3 30·3	3 20·2	5·9	1·3	11·9	2·7	17·9	4·0
60	3 30·0	3 30·6	3 20·4	6·0	1·4	12·0	2·7	18·0	4·1

14m s	SUN PLANETS ° ′	ARIES ° ′	MOON ° ′	v or d ′	Corrn ′	v or d ′	Corrn ′	v or d ′	Corrn ′
00	3 30·0	3 30·6	3 20·4	0·0	0·0	6·0	1·5	12·0	2·9
01	3 30·3	3 30·8	3 20·7	0·1	0·0	6·1	1·5	12·1	2·9
02	3 30·5	3 31·1	3 20·9	0·2	0·0	6·2	1·5	12·2	2·9
03	3 30·8	3 31·3	3 21·1	0·3	0·1	6·3	1·5	12·3	3·0
04	3 31·0	3 31·6	3 21·4	0·4	0·1	6·4	1·5	12·4	3·0
05	3 31·3	3 31·8	3 21·6	0·5	0·1	6·5	1·6	12·5	3·0
06	3 31·5	3 32·1	3 21·9	0·6	0·1	6·6	1·6	12·6	3·0
07	3 31·8	3 32·3	3 22·1	0·7	0·2	6·7	1·6	12·7	3·1
08	3 32·0	3 32·6	3 22·3	0·8	0·2	6·8	1·6	12·8	3·1
09	3 32·3	3 32·8	3 22·6	0·9	0·2	6·9	1·7	12·9	3·1
10	3 32·5	3 33·1	3 22·8	1·0	0·2	7·0	1·7	13·0	3·1
11	3 32·8	3 33·3	3 23·1	1·1	0·3	7·1	1·7	13·1	3·2
12	3 33·0	3 33·6	3 23·3	1·2	0·3	7·2	1·7	13·2	3·2
13	3 33·3	3 33·8	3 23·5	1·3	0·3	7·3	1·8	13·3	3·2
14	3 33·5	3 34·1	3 23·8	1·4	0·3	7·4	1·8	13·4	3·2
15	3 33·8	3 34·3	3 24·0	1·5	0·4	7·5	1·8	13·5	3·3
16	3 34·0	3 34·6	3 24·3	1·6	0·4	7·6	1·8	13·6	3·3
17	3 34·3	3 34·8	3 24·5	1·7	0·4	7·7	1·9	13·7	3·3
18	3 34·5	3 35·1	3 24·7	1·8	0·4	7·8	1·9	13·8	3·3
19	3 34·8	3 35·3	3 25·0	1·9	0·5	7·9	1·9	13·9	3·4
20	3 35·0	3 35·6	3 25·2	2·0	0·5	8·0	1·9	14·0	3·4
21	3 35·3	3 35·8	3 25·4	2·1	0·5	8·1	2·0	14·1	3·4
22	3 35·5	3 36·1	3 25·7	2·2	0·5	8·2	2·0	14·2	3·4
23	3 35·8	3 36·3	3 25·9	2·3	0·6	8·3	2·0	14·3	3·5
24	3 36·0	3 36·6	3 26·2	2·4	0·6	8·4	2·0	14·4	3·5
25	3 36·3	3 36·8	3 26·4	2·5	0·6	8·5	2·1	14·5	3·5
26	3 36·5	3 37·1	3 26·6	2·6	0·6	8·6	2·1	14·6	3·5
27	3 36·8	3 37·3	3 26·9	2·7	0·7	8·7	2·1	14·7	3·6
28	3 37·0	3 37·6	3 27·1	2·8	0·7	8·8	2·1	14·8	3·6
29	3 37·3	3 37·8	3 27·4	2·9	0·7	8·9	2·2	14·9	3·6
30	3 37·5	3 38·1	3 27·6	3·0	0·7	9·0	2·2	15·0	3·6
31	3 37·8	3 38·3	3 27·8	3·1	0·7	9·1	2·2	15·1	3·6
32	3 38·0	3 38·6	3 28·1	3·2	0·8	9·2	2·2	15·2	3·7
33	3 38·3	3 38·8	3 28·3	3·3	0·8	9·3	2·2	15·3	3·7
34	3 38·5	3 39·1	3 28·5	3·4	0·8	9·4	2·3	15·4	3·7
35	3 38·8	3 39·3	3 28·8	3·5	0·8	9·5	2·3	15·5	3·7
36	3 39·0	3 39·6	3 29·0	3·6	0·9	9·6	2·3	15·6	3·8
37	3 39·3	3 39·9	3 29·3	3·7	0·9	9·7	2·3	15·7	3·8
38	3 39·5	3 40·1	3 29·5	3·8	0·9	9·8	2·4	15·8	3·8
39	3 39·8	3 40·4	3 29·7	3·9	0·9	9·9	2·4	15·9	3·8
40	3 40·0	3 40·6	3 30·0	4·0	1·0	10·0	2·4	16·0	3·9
41	3 40·3	3 40·9	3 30·2	4·1	1·0	10·1	2·4	16·1	3·9
42	3 40·5	3 41·1	3 30·5	4·2	1·0	10·2	2·5	16·2	3·9
43	3 40·8	3 41·4	3 30·7	4·3	1·0	10·3	2·5	16·3	3·9
44	3 41·0	3 41·6	3 30·9	4·4	1·1	10·4	2·5	16·4	4·0
45	3 41·3	3 41·9	3 31·2	4·5	1·1	10·5	2·5	16·5	4·0
46	3 41·5	3 42·1	3 31·4	4·6	1·1	10·6	2·6	16·6	4·0
47	3 41·8	3 42·4	3 31·6	4·7	1·1	10·7	2·6	16·7	4·0
48	3 42·0	3 42·6	3 31·9	4·8	1·2	10·8	2·6	16·8	4·1
49	3 42·3	3 42·9	3 32·1	4·9	1·2	10·9	2·6	16·9	4·1
50	3 42·5	3 43·1	3 32·4	5·0	1·2	11·0	2·7	17·0	4·1
51	3 42·8	3 43·4	3 32·6	5·1	1·2	11·1	2·7	17·1	4·1
52	3 43·0	3 43·6	3 32·8	5·2	1·3	11·2	2·7	17·2	4·2
53	3 43·3	3 43·9	3 33·1	5·3	1·3	11·3	2·7	17·3	4·2
54	3 43·5	3 44·1	3 33·3	5·4	1·3	11·4	2·8	17·4	4·2
55	3 43·8	3 44·4	3 33·6	5·5	1·3	11·5	2·8	17·5	4·2
56	3 44·0	3 44·6	3 33·8	5·6	1·4	11·6	2·8	17·6	4·3
57	3 44·3	3 44·9	3 34·0	5·7	1·4	11·7	2·8	17·7	4·3
58	3 44·5	3 45·1	3 34·3	5·8	1·4	11·8	2·9	17·8	4·3
59	3 44·8	3 45·4	3 34·5	5·9	1·4	11·9	2·9	17·9	4·3
60	3 45·0	3 45·6	3 34·8	6·0	1·5	12·0	2·9	18·0	4·4

15m s	SUN PLANETS ° ′	ARIES ° ′	MOON ° ′	v or d ′	Corrn ′	v or d ′	Corrn ′	v or d ′	Corrn ′
00	3 45·0	3 45·6	3 34·8	0·0	0·0	6·0	1·6	12·0	3·1
01	3 45·3	3 45·9	3 35·0	0·1	0·0	6·1	1·6	12·1	3·1
02	3 45·5	3 46·1	3 35·2	0·2	0·1	6·2	1·6	12·2	3·2
03	3 45·8	3 46·4	3 35·5	0·3	0·1	6·3	1·6	12·3	3·2
04	3 46·0	3 46·6	3 35·7	0·4	0·1	6·4	1·7	12·4	3·2
05	3 46·3	3 46·9	3 35·9	0·5	0·1	6·5	1·7	12·5	3·2
06	3 46·5	3 47·1	3 36·2	0·6	0·2	6·6	1·7	12·6	3·3
07	3 46·8	3 47·4	3 36·4	0·7	0·2	6·7	1·7	12·7	3·3
08	3 47·0	3 47·6	3 36·7	0·8	0·2	6·8	1·8	12·8	3·3
09	3 47·3	3 47·9	3 36·9	0·9	0·2	6·9	1·8	12·9	3·3
10	3 47·5	3 48·1	3 37·1	1·0	0·3	7·0	1·8	13·0	3·4
11	3 47·8	3 48·4	3 37·4	1·1	0·3	7·1	1·8	13·1	3·4
12	3 48·0	3 48·6	3 37·6	1·2	0·3	7·2	1·9	13·2	3·4
13	3 48·3	3 48·9	3 37·9	1·3	0·3	7·3	1·9	13·3	3·4
14	3 48·5	3 49·1	3 38·1	1·4	0·4	7·4	1·9	13·4	3·5
15	3 48·8	3 49·4	3 38·3	1·5	0·4	7·5	1·9	13·5	3·5
16	3 49·0	3 49·6	3 38·6	1·6	0·4	7·6	2·0	13·6	3·5
17	3 49·3	3 49·9	3 38·8	1·7	0·4	7·7	2·0	13·7	3·5
18	3 49·5	3 50·1	3 39·0	1·8	0·5	7·8	2·0	13·8	3·6
19	3 49·8	3 50·4	3 39·3	1·9	0·5	7·9	2·0	13·9	3·6
20	3 50·0	3 50·6	3 39·5	2·0	0·5	8·0	2·1	14·0	3·6
21	3 50·3	3 50·9	3 39·8	2·1	0·5	8·1	2·1	14·1	3·6
22	3 50·5	3 51·1	3 40·0	2·2	0·6	8·2	2·1	14·2	3·7
23	3 50·8	3 51·4	3 40·2	2·3	0·6	8·3	2·1	14·3	3·7
24	3 51·0	3 51·6	3 40·5	2·4	0·6	8·4	2·2	14·4	3·7
25	3 51·3	3 51·9	3 40·7	2·5	0·6	8·5	2·2	14·5	3·7
26	3 51·5	3 52·1	3 41·0	2·6	0·7	8·6	2·2	14·6	3·8
27	3 51·8	3 52·4	3 41·2	2·7	0·7	8·7	2·2	14·7	3·8
28	3 52·0	3 52·6	3 41·4	2·8	0·7	8·8	2·3	14·8	3·8
29	3 52·3	3 52·9	3 41·7	2·9	0·7	8·9	2·3	14·9	3·8
30	3 52·5	3 53·1	3 41·9	3·0	0·8	9·0	2·3	15·0	3·9
31	3 52·8	3 53·4	3 42·1	3·1	0·8	9·1	2·4	15·1	3·9
32	3 53·0	3 53·6	3 42·4	3·2	0·8	9·2	2·4	15·2	3·9
33	3 53·3	3 53·9	3 42·6	3·3	0·9	9·3	2·4	15·3	4·0
34	3 53·5	3 54·1	3 42·9	3·4	0·9	9·4	2·4	15·4	4·0
35	3 53·8	3 54·4	3 43·1	3·5	0·9	9·5	2·5	15·5	4·0
36	3 54·0	3 54·6	3 43·3	3·6	0·9	9·6	2·5	15·6	4·0
37	3 54·3	3 54·9	3 43·6	3·7	1·0	9·7	2·5	15·7	4·1
38	3 54·5	3 55·1	3 43·8	3·8	1·0	9·8	2·5	15·8	4·1
39	3 54·8	3 55·4	3 44·1	3·9	1·0	9·9	2·6	15·9	4·1
40	3 55·0	3 55·6	3 44·3	4·0	1·0	10·0	2·6	16·0	4·1
41	3 55·3	3 55·9	3 44·5	4·1	1·1	10·1	2·6	16·1	4·2
42	3 55·5	3 56·1	3 44·8	4·2	1·1	10·2	2·6	16·2	4·2
43	3 55·8	3 56·4	3 45·0	4·3	1·1	10·3	2·7	16·3	4·2
44	3 56·0	3 56·6	3 45·2	4·4	1·1	10·4	2·7	16·4	4·2
45	3 56·3	3 56·9	3 45·5	4·5	1·2	10·5	2·7	16·5	4·3
46	3 56·5	3 57·1	3 45·7	4·6	1·2	10·6	2·7	16·6	4·3
47	3 56·8	3 57·4	3 46·0	4·7	1·2	10·7	2·8	16·7	4·3
48	3 57·0	3 57·6	3 46·2	4·8	1·2	10·8	2·8	16·8	4·3
49	3 57·3	3 57·9	3 46·4	4·9	1·3	10·9	2·8	16·9	4·4
50	3 57·5	3 58·2	3 46·7	5·0	1·3	11·0	2·8	17·0	4·4
51	3 57·8	3 58·4	3 46·9	5·1	1·3	11·1	2·9	17·1	4·4
52	3 58·0	3 58·7	3 47·2	5·2	1·3	11·2	2·9	17·2	4·4
53	3 58·3	3 58·9	3 47·4	5·3	1·4	11·3	2·9	17·3	4·5
54	3 58·5	3 59·2	3 47·6	5·4	1·4	11·4	2·9	17·4	4·5
55	3 58·8	3 59·4	3 47·9	5·5	1·4	11·5	3·0	17·5	4·5
56	3 59·0	3 59·7	3 48·1	5·6	1·4	11·6	3·0	17·6	4·5
57	3 59·3	3 59·9	3 48·4	5·7	1·5	11·7	3·0	17·7	4·6
58	3 59·5	4 00·2	3 48·6	5·8	1·5	11·8	3·0	17·8	4·6
59	3 59·8	4 00·4	3 48·8	5·9	1·5	11·9	3·1	17·9	4·6
60	4 00·0	4 00·7	3 49·1	6·0	1·6	12·0	3·1	18·0	4·7

16^m	SUN PLANETS	ARIES	MOON	v or d	Corr^n	v or d	Corr^n	v or d	Corr^n
s	° ′	° ′	° ′	′	′	′	′	′	′
00	4 00·0	4 00·7	3 49·1	0·0	0·0	6·0	1·7	12·0	3·3
01	4 00·3	4 00·9	3 49·3	0·1	0·0	6·1	1·7	12·1	3·3
02	4 00·5	4 01·2	3 49·5	0·2	0·1	6·2	1·7	12·2	3·4
03	4 00·8	4 01·4	3 49·8	0·3	0·1	6·3	1·7	12·3	3·4
04	4 01·0	4 01·7	3 50·0	0·4	0·1	6·4	1·8	12·4	3·4
05	4 01·3	4 01·9	3 50·3	0·5	0·1	6·5	1·8	12·5	3·4
06	4 01·5	4 02·2	3 50·5	0·6	0·2	6·6	1·8	12·6	3·5
07	4 01·8	4 02·4	3 50·7	0·7	0·2	6·7	1·8	12·7	3·5
08	4 02·0	4 02·7	3 51·0	0·8	0·2	6·8	1·9	12·8	3·5
09	4 02·3	4 02·9	3 51·2	0·9	0·2	6·9	1·9	12·9	3·5
10	4 02·5	4 03·2	3 51·5	1·0	0·3	7·0	1·9	13·0	3·6
11	4 02·8	4 03·4	3 51·7	1·1	0·3	7·1	2·0	13·1	3·6
12	4 03·0	4 03·7	3 51·9	1·2	0·3	7·2	2·0	13·2	3·6
13	4 03·3	4 03·9	3 52·2	1·3	0·4	7·3	2·0	13·3	3·7
14	4 03·5	4 04·2	3 52·4	1·4	0·4	7·4	2·0	13·4	3·7
15	4 03·8	4 04·4	3 52·6	1·5	0·4	7·5	2·1	13·5	3·7
16	4 04·0	4 04·7	3 52·9	1·6	0·4	7·6	2·1	13·6	3·7
17	4 04·3	4 04·9	3 53·1	1·7	0·5	7·7	2·1	13·7	3·8
18	4 04·5	4 05·2	3 53·4	1·8	0·5	7·8	2·1	13·8	3·8
19	4 04·8	4 05·4	3 53·6	1·9	0·5	7·9	2·2	13·9	3·8
20	4 05·0	4 05·7	3 53·8	2·0	0·6	8·0	2·2	14·0	3·9
21	4 05·3	4 05·9	3 54·1	2·1	0·6	8·1	2·2	14·1	3·9
22	4 05·5	4 06·2	3 54·3	2·2	0·6	8·2	2·3	14·2	3·9
23	4 05·8	4 06·4	3 54·6	2·3	0·6	8·3	2·3	14·3	3·9
24	4 06·0	4 06·7	3 54·8	2·4	0·7	8·4	2·3	14·4	4·0
25	4 06·3	4 06·9	3 55·0	2·5	0·7	8·5	2·3	14·5	4·0
26	4 06·5	4 07·2	3 55·3	2·6	0·7	8·6	2·4	14·6	4·0
27	4 06·8	4 07·4	3 55·5	2·7	0·7	8·7	2·4	14·7	4·0
28	4 07·0	4 07·7	3 55·7	2·8	0·8	8·8	2·4	14·8	4·1
29	4 07·3	4 07·9	3 56·0	2·9	0·8	8·9	2·4	14·9	4·1
30	4 07·5	4 08·2	3 56·2	3·0	0·8	9·0	2·5	15·0	4·1
31	4 07·8	4 08·4	3 56·5	3·1	0·9	9·1	2·5	15·1	4·2
32	4 08·0	4 08·7	3 56·7	3·2	0·9	9·2	2·5	15·2	4·2
33	4 08·3	4 08·9	3 56·9	3·3	0·9	9·3	2·6	15·3	4·2
34	4 08·5	4 09·2	3 57·2	3·4	0·9	9·4	2·6	15·4	4·2
35	4 08·8	4 09·4	3 57·4	3·5	1·0	9·5	2·6	15·5	4·3
36	4 09·0	4 09·7	3 57·7	3·6	1·0	9·6	2·6	15·6	4·3
37	4 09·3	4 09·9	3 57·9	3·7	1·0	9·7	2·7	15·7	4·3
38	4 09·5	4 10·2	3 58·1	3·8	1·0	9·8	2·7	15·8	4·3
39	4 09·8	4 10·4	3 58·4	3·9	1·1	9·9	2·7	15·9	4·4
40	4 10·0	4 10·7	3 58·6	4·0	1·1	10·0	2·8	16·0	4·4
41	4 10·3	4 10·9	3 58·8	4·1	1·1	10·1	2·8	16·1	4·4
42	4 10·5	4 11·2	3 59·1	4·2	1·2	10·2	2·8	16·2	4·5
43	4 10·8	4 11·4	3 59·3	4·3	1·2	10·3	2·8	16·3	4·5
44	4 11·0	4 11·7	3 59·6	4·4	1·2	10·4	2·9	16·4	4·5
45	4 11·3	4 11·9	3 59·8	4·5	1·2	10·5	2·9	16·5	4·5
46	4 11·5	4 12·2	4 00·0	4·6	1·3	10·6	2·9	16·6	4·6
47	4 11·8	4 12·4	4 00·3	4·7	1·3	10·7	2·9	16·7	4·6
48	4 12·0	4 12·7	4 00·5	4·8	1·3	10·8	3·0	16·8	4·6
49	4 12·3	4 12·9	4 00·8	4·9	1·3	10·9	3·0	16·9	4·6
50	4 12·5	4 13·2	4 01·0	5·0	1·4	11·0	3·0	17·0	4·7
51	4 12·8	4 13·4	4 01·2	5·1	1·4	11·1	3·1	17·1	4·7
52	4 13·0	4 13·7	4 01·5	5·2	1·4	11·2	3·1	17·2	4·7
53	4 13·3	4 13·9	4 01·7	5·3	1·5	11·3	3·1	17·3	4·8
54	4 13·5	4 14·2	4 02·0	5·4	1·5	11·4	3·1	17·4	4·8
55	4 13·8	4 14·4	4 02·2	5·5	1·5	11·5	3·2	17·5	4·8
56	4 14·0	4 14·7	4 02·4	5·6	1·5	11·6	3·2	17·6	4·8
57	4 14·3	4 14·9	4 02·7	5·7	1·6	11·7	3·2	17·7	4·9
58	4 14·5	4 15·2	4 02·9	5·8	1·6	11·8	3·2	17·8	4·9
59	4 14·8	4 15·4	4 03·1	5·9	1·6	11·9	3·3	17·9	4·9
60	4 15·0	4 15·7	4 03·4	6·0	1·7	12·0	3·3	18·0	5·0

17^m	SUN PLANETS	ARIES	MOON	v or d	Corr^n	v or d	Corr^n	v or d	Corr^n
s	° ′	° ′	° ′	′	′	′	′	′	′
00	4 15·0	4 15·7	4 03·4	0·0	0·0	6·0	1·8	12·0	3·5
01	4 15·3	4 15·9	4 03·6	0·1	0·0	6·1	1·8	12·1	3·5
02	4 15·5	4 16·2	4 03·9	0·2	0·1	6·2	1·8	12·2	3·6
03	4 15·8	4 16·5	4 04·1	0·3	0·1	6·3	1·8	12·3	3·6
04	4 16·0	4 16·7	4 04·3	0·4	0·1	6·4	1·9	12·4	3·6
05	4 16·3	4 17·0	4 04·6	0·5	0·1	6·5	1·9	12·5	3·6
06	4 16·5	4 17·2	4 04·8	0·6	0·2	6·6	1·9	12·6	3·7
07	4 16·8	4 17·5	4 05·1	0·7	0·2	6·7	2·0	12·7	3·7
08	4 17·0	4 17·7	4 05·3	0·8	0·2	6·8	2·0	12·8	3·7
09	4 17·3	4 18·0	4 05·5	0·9	0·3	6·9	2·0	12·9	3·8
10	4 17·5	4 18·2	4 05·8	1·0	0·3	7·0	2·0	13·0	3·8
11	4 17·8	4 18·5	4 06·0	1·1	0·3	7·1	2·1	13·1	3·8
12	4 18·0	4 18·7	4 06·2	1·2	0·4	7·2	2·1	13·2	3·9
13	4 18·3	4 19·0	4 06·5	1·3	0·4	7·3	2·1	13·3	3·9
14	4 18·5	4 19·2	4 06·7	1·4	0·4	7·4	2·2	13·4	3·9
15	4 18·8	4 19·5	4 07·0	1·5	0·4	7·5	2·2	13·5	3·9
16	4 19·0	4 19·7	4 07·2	1·6	0·5	7·6	2·2	13·6	4·0
17	4 19·3	4 20·0	4 07·4	1·7	0·5	7·7	2·2	13·7	4·0
18	4 19·5	4 20·2	4 07·7	1·8	0·5	7·8	2·3	13·8	4·0
19	4 19·8	4 20·5	4 07·9	1·9	0·6	7·9	2·3	13·9	4·1
20	4 20·0	4 20·7	4 08·2	2·0	0·6	8·0	2·3	14·0	4·1
21	4 20·3	4 21·0	4 08·4	2·1	0·6	8·1	2·4	14·1	4·1
22	4 20·5	4 21·2	4 08·6	2·2	0·6	8·2	2·4	14·2	4·1
23	4 20·8	4 21·5	4 08·9	2·3	0·7	8·3	2·4	14·3	4·2
24	4 21·0	4 21·7	4 09·1	2·4	0·7	8·4	2·5	14·4	4·2
25	4 21·3	4 22·0	4 09·3	2·5	0·7	8·5	2·5	14·5	4·2
26	4 21·5	4 22·2	4 09·6	2·6	0·8	8·6	2·5	14·6	4·3
27	4 21·8	4 22·5	4 09·8	2·7	0·8	8·7	2·5	14·7	4·3
28	4 22·0	4 22·7	4 10·1	2·8	0·8	8·8	2·6	14·8	4·3
29	4 22·3	4 23·0	4 10·3	2·9	0·8	8·9	2·6	14·9	4·3
30	4 22·5	4 23·2	4 10·5	3·0	0·9	9·0	2·6	15·0	4·4
31	4 22·8	4 23·5	4 10·8	3·1	0·9	9·1	2·7	15·1	4·4
32	4 23·0	4 23·7	4 11·0	3·2	0·9	9·2	2·7	15·2	4·4
33	4 23·3	4 24·0	4 11·3	3·3	1·0	9·3	2·7	15·3	4·5
34	4 23·5	4 24·2	4 11·5	3·4	1·0	9·4	2·7	15·4	4·5
35	4 23·8	4 24·5	4 11·7	3·5	1·0	9·5	2·8	15·5	4·5
36	4 24·0	4 24·7	4 12·0	3·6	1·1	9·6	2·8	15·6	4·6
37	4 24·3	4 25·0	4 12·2	3·7	1·1	9·7	2·8	15·7	4·6
38	4 24·5	4 25·2	4 12·5	3·8	1·1	9·8	2·9	15·8	4·6
39	4 24·8	4 25·5	4 12·7	3·9	1·1	9·9	2·9	15·9	4·6
40	4 25·0	4 25·7	4 12·9	4·0	1·2	10·0	2·9	16·0	4·7
41	4 25·3	4 26·0	4 13·2	4·1	1·2	10·1	2·9	16·1	4·7
42	4 25·5	4 26·2	4 13·4	4·2	1·2	10·2	3·0	16·2	4·7
43	4 25·8	4 26·5	4 13·6	4·3	1·3	10·3	3·0	16·3	4·8
44	4 26·0	4 26·7	4 13·9	4·4	1·3	10·4	3·0	16·4	4·8
45	4 26·3	4 27·0	4 14·1	4·5	1·3	10·5	3·1	16·5	4·8
46	4 26·5	4 27·2	4 14·4	4·6	1·3	10·6	3·1	16·6	4·8
47	4 26·8	4 27·5	4 14·6	4·7	1·4	10·7	3·1	16·7	4·9
48	4 27·0	4 27·7	4 14·8	4·8	1·4	10·8	3·2	16·8	4·9
49	4 27·3	4 28·0	4 15·1	4·9	1·4	10·9	3·2	16·9	4·9
50	4 27·5	4 28·2	4 15·3	5·0	1·5	11·0	3·2	17·0	5·0
51	4 27·8	4 28·5	4 15·6	5·1	1·5	11·1	3·2	17·1	5·0
52	4 28·0	4 28·7	4 15·8	5·2	1·5	11·2	3·3	17·2	5·0
53	4 28·3	4 29·0	4 16·0	5·3	1·5	11·3	3·3	17·3	5·0
54	4 28·5	4 29·2	4 16·3	5·4	1·6	11·4	3·3	17·4	5·1
55	4 28·8	4 29·5	4 16·5	5·5	1·6	11·5	3·4	17·5	5·1
56	4 29·0	4 29·7	4 16·7	5·6	1·6	11·6	3·4	17·6	5·1
57	4 29·3	4 30·0	4 17·0	5·7	1·7	11·7	3·4	17·7	5·2
58	4 29·5	4 30·2	4 17·2	5·8	1·7	11·8	3·4	17·8	5·2
59	4 29·8	4 30·5	4 17·5	5·9	1·7	11·9	3·5	17·9	5·2
60	4 30·0	4 30·7	4 17·7	6·0	1·8	12·0	3·5	18·0	5·3

m 18	SUN PLANETS	ARIES	MOON	v or d	Corrⁿ	v or d	Corrⁿ	v or d	Corrⁿ
s	° ′	° ′	° ′	′	′	′	′	′	′
00	4 30·0	4 30·7	4 17·7	0·0	0·0	6·0	1·9	12·0	3·7
01	4 30·3	4 31·0	4 17·9	0·1	0·0	6·1	1·9	12·1	3·7
02	4 30·5	4 31·2	4 18·2	0·2	0·1	6·2	1·9	12·2	3·8
03	4 30·8	4 31·5	4 18·4	0·3	0·1	6·3	1·9	12·3	3·8
04	4 31·0	4 31·7	4 18·7	0·4	0·1	6·4	2·0	12·4	3·8
05	4 31·3	4 32·0	4 18·9	0·5	0·2	6·5	2·0	12·5	3·9
06	4 31·5	4 32·2	4 19·1	0·6	0·2	6·6	2·0	12·6	3·9
07	4 31·8	4 32·5	4 19·4	0·7	0·2	6·7	2·1	12·7	3·9
08	4 32·0	4 32·7	4 19·6	0·8	0·2	6·8	2·1	12·8	3·9
09	4 32·3	4 33·0	4 19·8	0·9	0·3	6·9	2·1	12·9	4·0
10	4 32·5	4 33·2	4 20·1	1·0	0·3	7·0	2·2	13·0	4·0
11	4 32·8	4 33·5	4 20·3	1·1	0·3	7·1	2·2	13·1	4·0
12	4 33·0	4 33·7	4 20·6	1·2	0·4	7·2	2·2	13·2	4·1
13	4 33·3	4 34·0	4 20·8	1·3	0·4	7·3	2·3	13·3	4·1
14	4 33·5	4 34·2	4 21·0	1·4	0·4	7·4	2·3	13·4	4·1
15	4 33·8	4 34·5	4 21·3	1·5	0·5	7·5	2·3	13·5	4·2
16	4 34·0	4 34·8	4 21·5	1·6	0·5	7·6	2·3	13·6	4·2
17	4 34·3	4 35·0	4 21·8	1·7	0·5	7·7	2·4	13·7	4·2
18	4 34·5	4 35·3	4 22·0	1·8	0·6	7·8	2·4	13·8	4·3
19	4 34·8	4 35·5	4 22·2	1·9	0·6	7·9	2·4	13·9	4·3
20	4 35·0	4 35·8	4 22·5	2·0	0·6	8·0	2·5	14·0	4·3
21	4 35·3	4 36·0	4 22·7	2·1	0·6	8·1	2·5	14·1	4·3
22	4 35·5	4 36·3	4 22·9	2·2	0·7	8·2	2·5	14·2	4·4
23	4 35·8	4 36·5	4 23·2	2·3	0·7	8·3	2·6	14·3	4·4
24	4 36·0	4 36·8	4 23·4	2·4	0·7	8·4	2·6	14·4	4·4
25	4 36·3	4 37·0	4 23·7	2·5	0·8	8·5	2·6	14·5	4·5
26	4 36·5	4 37·3	4 23·9	2·6	0·8	8·6	2·7	14·6	4·5
27	4 36·8	4 37·5	4 24·1	2·7	0·8	8·7	2·7	14·7	4·5
28	4 37·0	4 37·8	4 24·4	2·8	0·9	8·8	2·7	14·8	4·6
29	4 37·3	4 38·0	4 24·6	2·9	0·9	8·9	2·7	14·9	4·6
30	4 37·5	4 38·3	4 24·9	3·0	0·9	9·0	2·8	15·0	4·6
31	4 37·8	4 38·5	4 25·1	3·1	1·0	9·1	2·8	15·1	4·7
32	4 38·0	4 38·8	4 25·3	3·2	1·0	9·2	2·8	15·2	4·7
33	4 38·3	4 39·0	4 25·6	3·3	1·0	9·3	2·9	15·3	4·7
34	4 38·5	4 39·3	4 25·8	3·4	1·0	9·4	2·9	15·4	4·7
35	4 38·8	4 39·5	4 26·1	3·5	1·1	9·5	2·9	15·5	4·8
36	4 39·0	4 39·8	4 26·3	3·6	1·1	9·6	3·0	15·6	4·8
37	4 39·3	4 40·0	4 26·5	3·7	1·1	9·7	3·0	15·7	4·8
38	4 39·5	4 40·3	4 26·8	3·8	1·2	9·8	3·0	15·8	4·9
39	4 39·8	4 40·5	4 27·0	3·9	1·2	9·9	3·1	15·9	4·9
40	4 40·0	4 40·8	4 27·2	4·0	1·2	10·0	3·1	16·0	4·9
41	4 40·3	4 41·0	4 27·5	4·1	1·3	10·1	3·1	16·1	5·0
42	4 40·5	4 41·3	4 27·7	4·2	1·3	10·2	3·1	16·2	5·0
43	4 40·8	4 41·5	4 28·0	4·3	1·3	10·3	3·2	16·3	5·0
44	4 41·0	4 41·8	4 28·2	4·4	1·4	10·4	3·2	16·4	5·1
45	4 41·3	4 42·0	4 28·4	4·5	1·4	10·5	3·2	16·5	5·1
46	4 41·5	4 42·3	4 28·7	4·6	1·4	10·6	3·3	16·6	5·1
47	4 41·8	4 42·5	4 28·9	4·7	1·4	10·7	3·3	16·7	5·1
48	4 42·0	4 42·8	4 29·2	4·8	1·5	10·8	3·3	16·8	5·2
49	4 42·3	4 43·0	4 29·4	4·9	1·5	10·9	3·4	16·9	5·2
50	4 42·5	4 43·3	4 29·6	5·0	1·5	11·0	3·4	17·0	5·2
51	4 42·8	4 43·5	4 29·9	5·1	1·6	11·1	3·4	17·1	5·3
52	4 43·0	4 43·8	4 30·1	5·2	1·6	11·2	3·5	17·2	5·3
53	4 43·3	4 44·0	4 30·3	5·3	1·6	11·3	3·5	17·3	5·3
54	4 43·5	4 44·3	4 30·6	5·4	1·7	11·4	3·5	17·4	5·4
55	4 43·8	4 44·5	4 30·8	5·5	1·7	11·5	3·5	17·5	5·4
56	4 44·0	4 44·8	4 31·1	5·6	1·7	11·6	3·6	17·6	5·4
57	4 44·3	4 45·0	4 31·3	5·7	1·8	11·7	3·6	17·7	5·5
58	4 44·5	4 45·3	4 31·5	5·8	1·8	11·8	3·6	17·8	5·5
59	4 44·8	4 45·5	4 31·8	5·9	1·8	11·9	3·7	17·9	5·5
60	4 45·0	4 45·8	4 32·0	6·0	1·9	12·0	3·7	18·0	5·6

m 19	SUN PLANETS	ARIES	MOON	v or d	Corrⁿ	v or d	Corrⁿ	v or d	Corrⁿ
s	° ′	° ′	° ′	′	′	′	′	′	′
00	4 45·0	4 45·8	4 32·0	0·0	0·0	6·0	2·0	12·0	3·9
01	4 45·3	4 46·0	4 32·3	0·1	0·0	6·1	2·0	12·1	3·9
02	4 45·5	4 46·3	4 32·5	0·2	0·1	6·2	2·0	12·2	4·0
03	4 45·8	4 46·5	4 32·7	0·3	0·1	6·3	2·0	12·3	4·0
04	4 46·0	4 46·8	4 33·0	0·4	0·1	6·4	2·1	12·4	4·0
05	4 46·3	4 47·0	4 33·2	0·5	0·2	6·5	2·1	12·5	4·1
06	4 46·5	4 47·3	4 33·4	0·6	0·2	6·6	2·1	12·6	4·1
07	4 46·8	4 47·5	4 33·7	0·7	0·2	6·7	2·2	12·7	4·1
08	4 47·0	4 47·8	4 33·9	0·8	0·3	6·8	2·2	12·8	4·2
09	4 47·3	4 48·0	4 34·2	0·9	0·3	6·9	2·2	12·9	4·2
10	4 47·5	4 48·3	4 34·4	1·0	0·3	7·0	2·3	13·0	4·2
11	4 47·8	4 48·5	4 34·6	1·1	0·4	7·1	2·3	13·1	4·3
12	4 48·0	4 48·8	4 34·9	1·2	0·4	7·2	2·3	13·2	4·3
13	4 48·3	4 49·0	4 35·1	1·3	0·4	7·3	2·4	13·3	4·3
14	4 48·5	4 49·3	4 35·4	1·4	0·5	7·4	2·4	13·4	4·4
15	4 48·8	4 49·5	4 35·6	1·5	0·5	7·5	2·4	13·5	4·4
16	4 49·0	4 49·8	4 35·8	1·6	0·5	7·6	2·5	13·6	4·4
17	4 49·3	4 50·0	4 36·1	1·7	0·6	7·7	2·5	13·7	4·5
18	4 49·5	4 50·3	4 36·3	1·8	0·6	7·8	2·5	13·8	4·5
19	4 49·8	4 50·5	4 36·6	1·9	0·6	7·9	2·6	13·9	4·5
20	4 50·0	4 50·8	4 36·8	2·0	0·7	8·0	2·6	14·0	4·6
21	4 50·3	4 51·0	4 37·0	2·1	0·7	8·1	2·6	14·1	4·6
22	4 50·5	4 51·3	4 37·3	2·2	0·7	8·2	2·7	14·2	4·6
23	4 50·8	4 51·5	4 37·5	2·3	0·7	8·3	2·7	14·3	4·6
24	4 51·0	4 51·8	4 37·7	2·4	0·8	8·4	2·7	14·4	4·7
25	4 51·3	4 52·0	4 38·0	2·5	0·8	8·5	2·8	14·5	4·7
26	4 51·5	4 52·3	4 38·2	2·6	0·8	8·6	2·8	14·6	4·7
27	4 51·8	4 52·5	4 38·5	2·7	0·9	8·7	2·8	14·7	4·8
28	4 52·0	4 52·8	4 38·7	2·8	0·9	8·8	2·9	14·8	4·8
29	4 52·3	4 53·1	4 38·9	2·9	0·9	8·9	2·9	14·9	4·8
30	4 52·5	4 53·3	4 39·2	3·0	1·0	9·0	2·9	15·0	4·9
31	4 52·8	4 53·6	4 39·4	3·1	1·0	9·1	3·0	15·1	4·9
32	4 53·0	4 53·8	4 39·7	3·2	1·0	9·2	3·0	15·2	4·9
33	4 53·3	4 54·1	4 39·9	3·3	1·1	9·3	3·0	15·3	5·0
34	4 53·5	4 54·3	4 40·1	3·4	1·1	9·4	3·1	15·4	5·0
35	4 53·8	4 54·6	4 40·4	3·5	1·1	9·5	3·1	15·5	5·0
36	4 54·0	4 54·8	4 40·6	3·6	1·2	9·6	3·1	15·6	5·1
37	4 54·3	4 55·1	4 40·8	3·7	1·2	9·7	3·2	15·7	5·1
38	4 54·5	4 55·3	4 41·1	3·8	1·2	9·8	3·2	15·8	5·1
39	4 54·8	4 55·6	4 41·3	3·9	1·3	9·9	3·2	15·9	5·2
40	4 55·0	4 55·8	4 41·6	4·0	1·3	10·0	3·3	16·0	5·2
41	4 55·3	4 56·1	4 41·8	4·1	1·3	10·1	3·3	16·1	5·2
42	4 55·5	4 56·3	4 42·0	4·2	1·4	10·2	3·3	16·2	5·3
43	4 55·8	4 56·6	4 42·3	4·3	1·4	10·3	3·3	16·3	5·3
44	4 56·0	4 56·8	4 42·5	4·4	1·4	10·4	3·4	16·4	5·3
45	4 56·3	4 57·1	4 42·8	4·5	1·5	10·5	3·4	16·5	5·4
46	4 56·5	4 57·3	4 43·0	4·6	1·5	10·6	3·4	16·6	5·4
47	4 56·8	4 57·6	4 43·2	4·7	1·5	10·7	3·5	16·7	5·4
48	4 57·0	4 57·8	4 43·5	4·8	1·6	10·8	3·5	16·8	5·5
49	4 57·3	4 58·1	4 43·7	4·9	1·6	10·9	3·5	16·9	5·5
50	4 57·5	4 58·3	4 43·9	5·0	1·6	11·0	3·6	17·0	5·5
51	4 57·8	4 58·6	4 44·2	5·1	1·7	11·1	3·6	17·1	5·6
52	4 58·0	4 58·8	4 44·4	5·2	1·7	11·2	3·6	17·2	5·6
53	4 58·3	4 59·1	4 44·7	5·3	1·7	11·3	3·7	17·3	5·6
54	4 58·5	4 59·3	4 44·9	5·4	1·8	11·4	3·7	17·4	5·7
55	4 58·8	4 59·6	4 45·1	5·5	1·8	11·5	3·7	17·5	5·7
56	4 59·0	4 59·8	4 45·4	5·6	1·8	11·6	3·8	17·6	5·7
57	4 59·3	5 00·1	4 45·6	5·7	1·9	11·7	3·8	17·7	5·8
58	4 59·5	5 00·3	4 45·9	5·8	1·9	11·8	3·8	17·8	5·8
59	4 59·8	5 00·6	4 46·1	5·9	1·9	11·9	3·9	17·9	5·8
60	5 00·0	5 00·8	4 46·3	6·0	2·0	12·0	3·9	18·0	5·9

m 20 s	SUN PLANETS ° ′	ARIES ° ′	MOON ° ′	v or d ′	Corrn ′	v or d ′	Corrn ′	v or d ′	Corrn ′
00	5 00·0	5 00·8	4 46·3	0·0	0·0	6·0	2·1	12·0	4·1
01	5 00·3	5 01·1	4 46·6	0·1	0·0	6·1	2·1	12·1	4·1
02	5 00·5	5 01·3	4 46·8	0·2	0·1	6·2	2·1	12·2	4·2
03	5 00·8	5 01·6	4 47·0	0·3	0·1	6·3	2·2	12·3	4·2
04	5 01·0	5 01·8	4 47·3	0·4	0·1	6·4	2·2	12·4	4·2
05	5 01·3	5 02·1	4 47·5	0·5	0·2	6·5	2·2	12·5	4·3
06	5 01·5	5 02·3	4 47·8	0·6	0·2	6·6	2·3	12·6	4·3
07	5 01·8	5 02·6	4 48·0	0·7	0·2	6·7	2·3	12·7	4·3
08	5 02·0	5 02·8	4 48·2	0·8	0·3	6·8	2·3	12·8	4·4
09	5 02·3	5 03·1	4 48·5	0·9	0·3	6·9	2·4	12·9	4·4
10	5 02·5	5 03·3	4 48·7	1·0	0·3	7·0	2·4	13·0	4·4
11	5 02·8	5 03·6	4 49·0	1·1	0·4	7·1	2·4	13·1	4·5
12	5 03·0	5 03·8	4 49·2	1·2	0·4	7·2	2·5	13·2	4·5
13	5 03·3	5 04·1	4 49·4	1·3	0·4	7·3	2·5	13·3	4·5
14	5 03·5	5 04·3	4 49·7	1·4	0·5	7·4	2·5	13·4	4·6
15	5 03·8	5 04·6	4 49·9	1·5	0·5	7·5	2·6	13·5	4·6
16	5 04·0	5 04·8	4 50·2	1·6	0·5	7·6	2·6	13·6	4·6
17	5 04·3	5 05·1	4 50·4	1·7	0·6	7·7	2·6	13·7	4·7
18	5 04·5	5 05·3	4 50·6	1·8	0·6	7·8	2·7	13·8	4·7
19	5 04·8	5 05·6	4 50·9	1·9	0·6	7·9	2·7	13·9	4·7
20	5 05·0	5 05·8	4 51·1	2·0	0·7	8·0	2·7	14·0	4·8
21	5 05·3	5 06·1	4 51·3	2·1	0·7	8·1	2·8	14·1	4·8
22	5 05·5	5 06·3	4 51·6	2·2	0·8	8·2	2·8	14·2	4·9
23	5 05·8	5 06·6	4 51·8	2·3	0·8	8·3	2·8	14·3	4·9
24	5 06·0	5 06·8	4 52·1	2·4	0·8	8·4	2·9	14·4	4·9
25	5 06·3	5 07·1	4 52·3	2·5	0·9	8·5	2·9	14·5	5·0
26	5 06·5	5 07·3	4 52·5	2·6	0·9	8·6	2·9	14·6	5·0
27	5 06·8	5 07·6	4 52·8	2·7	0·9	8·7	3·0	14·7	5·0
28	5 07·0	5 07·8	4 53·0	2·8	1·0	8·8	3·0	14·8	5·1
29	5 07·3	5 08·1	4 53·3	2·9	1·0	8·9	3·0	14·9	5·1
30	5 07·5	5 08·3	4 53·5	3·0	1·0	9·0	3·1	15·0	5·1
31	5 07·8	5 08·6	4 53·7	3·1	1·1	9·1	3·1	15·1	5·2
32	5 08·0	5 08·8	4 54·0	3·2	1·1	9·2	3·1	15·2	5·2
33	5 08·3	5 09·1	4 54·2	3·3	1·1	9·3	3·2	15·3	5·2
34	5 08·5	5 09·3	4 54·4	3·4	1·2	9·4	3·2	15·4	5·3
35	5 08·8	5 09·6	4 54·7	3·5	1·2	9·5	3·2	15·5	5·3
36	5 09·0	5 09·8	4 54·9	3·6	1·2	9·6	3·3	15·6	5·3
37	5 09·3	5 10·1	4 55·2	3·7	1·3	9·7	3·3	15·7	5·4
38	5 09·5	5 10·3	4 55·4	3·8	1·3	9·8	3·3	15·8	5·4
39	5 09·8	5 10·6	4 55·6	3·9	1·3	9·9	3·4	15·9	5·4
40	5 10·0	5 10·8	4 55·9	4·0	1·4	10·0	3·4	16·0	5·5
41	5 10·3	5 11·1	4 56·1	4·1	1·4	10·1	3·5	16·1	5·5
42	5 10·5	5 11·4	4 56·4	4·2	1·4	10·2	3·5	16·2	5·5
43	5 10·8	5 11·6	4 56·6	4·3	1·5	10·3	3·5	16·3	5·6
44	5 11·0	5 11·9	4 56·8	4·4	1·5	10·4	3·6	16·4	5·6
45	5 11·3	5 12·1	4 57·1	4·5	1·5	10·5	3·6	16·5	5·6
46	5 11·5	5 12·4	4 57·3	4·6	1·6	10·6	3·6	16·6	5·7
47	5 11·8	5 12·6	4 57·5	4·7	1·6	10·7	3·7	16·7	5·7
48	5 12·0	5 12·9	4 57·8	4·8	1·6	10·8	3·7	16·8	5·7
49	5 12·3	5 13·1	4 58·0	4·9	1·7	10·9	3·7	16·9	5·8
50	5 12·5	5 13·4	4 58·3	5·0	1·7	11·0	3·8	17·0	5·8
51	5 12·8	5 13·6	4 58·5	5·1	1·7	11·1	3·8	17·1	5·8
52	5 13·0	5 13·9	4 58·7	5·2	1·8	11·2	3·8	17·2	5·9
53	5 13·3	5 14·1	4 59·0	5·3	1·8	11·3	3·9	17·3	5·9
54	5 13·5	5 14·4	4 59·2	5·4	1·8	11·4	3·9	17·4	5·9
55	5 13·8	5 14·6	4 59·5	5·5	1·9	11·5	3·9	17·5	6·0
56	5 14·0	5 14·9	4 59·7	5·6	1·9	11·6	4·0	17·6	6·0
57	5 14·3	5 15·1	4 59·9	5·7	1·9	11·7	4·0	17·7	6·0
58	5 14·5	5 15·4	5 00·2	5·8	2·0	11·8	4·0	17·8	6·1
59	5 14·8	5 15·6	5 00·4	5·9	2·0	11·9	4·1	17·9	6·1
60	5 15·0	5 15·9	5 00·7	6·0	2·1	12·0	4·1	18·0	6·2

m 21 s	SUN PLANETS ° ′	ARIES ° ′	MOON ° ′	v or d ′	Corrn ′	v or d ′	Corrn ′	v or d ′	Corrn ′
00	5 15·0	5 15·9	5 00·7	0·0	0·0	6·0	2·2	12·0	4·3
01	5 15·3	5 16·1	5 00·9	0·1	0·0	6·1	2·2	12·1	4·3
02	5 15·5	5 16·4	5 01·1	0·2	0·1	6·2	2·2	12·2	4·4
03	5 15·8	5 16·6	5 01·4	0·3	0·1	6·3	2·3	12·3	4·4
04	5 16·0	5 16·9	5 01·6	0·4	0·1	6·4	2·3	12·4	4·4
05	5 16·3	5 17·1	5 01·8	0·5	0·2	6·5	2·3	12·5	4·5
06	5 16·5	5 17·4	5 02·1	0·6	0·2	6·6	2·4	12·6	4·5
07	5 16·8	5 17·6	5 02·3	0·7	0·3	6·7	2·4	12·7	4·6
08	5 17·0	5 17·9	5 02·6	0·8	0·3	6·8	2·4	12·8	4·6
09	5 17·3	5 18·1	5 02·8	0·9	0·3	6·9	2·5	12·9	4·6
10	5 17·5	5 18·4	5 03·0	1·0	0·4	7·0	2·5	13·0	4·7
11	5 17·8	5 18·6	5 03·3	1·1	0·4	7·1	2·5	13·1	4·7
12	5 18·0	5 18·9	5 03·5	1·2	0·4	7·2	2·6	13·2	4·7
13	5 18·3	5 19·1	5 03·8	1·3	0·5	7·3	2·6	13·3	4·8
14	5 18·5	5 19·4	5 04·0	1·4	0·5	7·4	2·7	13·4	4·8
15	5 18·8	5 19·6	5 04·2	1·5	0·5	7·5	2·7	13·5	4·8
16	5 19·0	5 19·9	5 04·5	1·6	0·6	7·6	2·7	13·6	4·9
17	5 19·3	5 20·1	5 04·7	1·7	0·6	7·7	2·8	13·7	4·9
18	5 19·5	5 20·4	5 04·9	1·8	0·6	7·8	2·8	13·8	4·9
19	5 19·8	5 20·6	5 05·2	1·9	0·7	7·9	2·8	13·9	5·0
20	5 20·0	5 20·9	5 05·4	2·0	0·7	8·0	2·9	14·0	5·0
21	5 20·3	5 21·1	5 05·7	2·1	0·8	8·1	2·9	14·1	5·1
22	5 20·5	5 21·4	5 05·9	2·2	0·8	8·2	2·9	14·2	5·1
23	5 20·8	5 21·6	5 06·1	2·3	0·8	8·3	3·0	14·3	5·1
24	5 21·0	5 21·9	5 06·4	2·4	0·9	8·4	3·0	14·4	5·2
25	5 21·3	5 22·1	5 06·6	2·5	0·9	8·5	3·0	14·5	5·2
26	5 21·5	5 22·4	5 06·9	2·6	0·9	8·6	3·1	14·6	5·2
27	5 21·8	5 22·6	5 07·1	2·7	1·0	8·7	3·1	14·7	5·3
28	5 22·0	5 22·9	5 07·3	2·8	1·0	8·8	3·2	14·8	5·3
29	5 22·3	5 23·1	5 07·6	2·9	1·0	8·9	3·2	14·9	5·3
30	5 22·5	5 23·4	5 07·8	3·0	1·1	9·0	3·2	15·0	5·4
31	5 22·8	5 23·6	5 08·0	3·1	1·1	9·1	3·3	15·1	5·4
32	5 23·0	5 23·9	5 08·3	3·2	1·1	9·2	3·3	15·2	5·4
33	5 23·3	5 24·1	5 08·5	3·3	1·2	9·3	3·3	15·3	5·5
34	5 23·5	5 24·4	5 08·8	3·4	1·2	9·4	3·4	15·4	5·5
35	5 23·8	5 24·6	5 09·0	3·5	1·3	9·5	3·4	15·5	5·6
36	5 24·0	5 24·9	5 09·2	3·6	1·3	9·6	3·4	15·6	5·6
37	5 24·3	5 25·1	5 09·5	3·7	1·3	9·7	3·5	15·7	5·6
38	5 24·5	5 25·4	5 09·7	3·8	1·4	9·8	3·5	15·8	5·7
39	5 24·8	5 25·6	5 10·0	3·9	1·4	9·9	3·5	15·9	5·7
40	5 25·0	5 25·9	5 10·2	4·0	1·4	10·0	3·6	16·0	5·7
41	5 25·3	5 26·1	5 10·4	4·1	1·5	10·1	3·6	16·1	5·8
42	5 25·5	5 26·4	5 10·7	4·2	1·5	10·2	3·7	16·2	5·8
43	5 25·8	5 26·6	5 10·9	4·3	1·5	10·3	3·7	16·3	5·8
44	5 26·0	5 26·9	5 11·1	4·4	1·6	10·4	3·7	16·4	5·9
45	5 26·3	5 27·1	5 11·4	4·5	1·6	10·5	3·8	16·5	5·9
46	5 26·5	5 27·4	5 11·6	4·6	1·6	10·6	3·8	16·6	5·9
47	5 26·8	5 27·6	5 11·9	4·7	1·7	10·7	3·8	16·7	6·0
48	5 27·0	5 27·9	5 12·1	4·8	1·7	10·8	3·9	16·8	6·0
49	5 27·3	5 28·1	5 12·3	4·9	1·8	10·9	3·9	16·9	6·1
50	5 27·5	5 28·4	5 12·6	5·0	1·8	11·0	3·9	17·0	6·1
51	5 27·8	5 28·6	5 12·8	5·1	1·8	11·1	4·0	17·1	6·1
52	5 28·0	5 28·9	5 13·1	5·2	1·9	11·2	4·0	17·2	6·2
53	5 28·3	5 29·1	5 13·3	5·3	1·9	11·3	4·0	17·3	6·2
54	5 28·5	5 29·4	5 13·5	5·4	1·9	11·4	4·1	17·4	6·2
55	5 28·8	5 29·7	5 13·8	5·5	2·0	11·5	4·1	17·5	6·3
56	5 29·0	5 29·9	5 14·0	5·6	2·0	11·6	4·2	17·6	6·3
57	5 29·3	5 30·2	5 14·3	5·7	2·0	11·7	4·2	17·7	6·3
58	5 29·5	5 30·4	5 14·5	5·8	2·1	11·8	4·2	17·8	6·4
59	5 29·8	5 30·7	5 14·7	5·9	2·1	11·9	4·3	17·9	6·4
60	5 30·0	5 30·9	5 15·0	6·0	2·2	12·0	4·3	18·0	6·5

22m	SUN PLANETS	ARIES	MOON	v or d	Corrⁿ	v or d	Corrⁿ	v or d	Corrⁿ
s	° ′	° ′	° ′	′	′	′	′	′	′
00	5 30·0	5 30·9	5 15·0	0·0	0·0	6·0	2·3	12·0	4·5
01	5 30·3	5 31·2	5 15·2	0·1	0·0	6·1	2·3	12·1	4·5
02	5 30·5	5 31·4	5 15·4	0·2	0·1	6·2	2·3	12·2	4·6
03	5 30·8	5 31·7	5 15·7	0·3	0·1	6·3	2·4	12·3	4·6
04	5 31·0	5 31·9	5 15·9	0·4	0·2	6·4	2·4	12·4	4·7
05	5 31·3	5 32·2	5 16·2	0·5	0·2	6·5	2·4	12·5	4·7
06	5 31·5	5 32·4	5 16·4	0·6	0·2	6·6	2·5	12·6	4·7
07	5 31·8	5 32·7	5 16·6	0·7	0·3	6·7	2·5	12·7	4·8
08	5 32·0	5 32·9	5 16·9	0·8	0·3	6·8	2·6	12·8	4·8
09	5 32·3	5 33·2	5 17·1	0·9	0·3	6·9	2·6	12·9	4·8
10	5 32·5	5 33·4	5 17·4	1·0	0·4	7·0	2·6	13·0	4·9
11	5 32·8	5 33·7	5 17·6	1·1	0·4	7·1	2·7	13·1	4·9
12	5 33·0	5 33·9	5 17·8	1·2	0·5	7·2	2·7	13·2	5·0
13	5 33·3	5 34·2	5 18·1	1·3	0·5	7·3	2·7	13·3	5·0
14	5 33·5	5 34·4	5 18·3	1·4	0·5	7·4	2·8	13·4	5·0
15	5 33·8	5 34·7	5 18·5	1·5	0·6	7·5	2·8	13·5	5·1
16	5 34·0	5 34·9	5 18·8	1·6	0·6	7·6	2·9	13·6	5·1
17	5 34·3	5 35·2	5 19·0	1·7	0·6	7·7	2·9	13·7	5·1
18	5 34·5	5 35·4	5 19·3	1·8	0·7	7·8	2·9	13·8	5·2
19	5 34·8	5 35·7	5 19·5	1·9	0·7	7·9	3·0	13·9	5·2
20	5 35·0	5 35·9	5 19·7	2·0	0·8	8·0	3·0	14·0	5·3
21	5 35·3	5 36·2	5 20·0	2·1	0·8	8·1	3·0	14·1	5·3
22	5 35·5	5 36·4	5 20·2	2·2	0·8	8·2	3·1	14·2	5·3
23	5 35·8	5 36·7	5 20·5	2·3	0·9	8·3	3·1	14·3	5·4
24	5 36·0	5 36·9	5 20·7	2·4	0·9	8·4	3·2	14·4	5·4
25	5 36·3	5 37·2	5 20·9	2·5	0·9	8·5	3·2	14·5	5·4
26	5 36·5	5 37·4	5 21·2	2·6	1·0	8·6	3·2	14·6	5·5
27	5 36·8	5 37·7	5 21·4	2·7	1·0	8·7	3·3	14·7	5·5
28	5 37·0	5 37·9	5 21·6	2·8	1·0	8·8	3·3	14·8	5·6
29	5 37·3	5 38·2	5 21·9	2·9	1·1	8·9	3·3	14·9	5·6
30	5 37·5	5 38·4	5 22·1	3·0	1·1	9·0	3·4	15·0	5·6
31	5 37·8	5 38·7	5 22·4	3·1	1·2	9·1	3·4	15·1	5·7
32	5 38·0	5 38·9	5 22·6	3·2	1·2	9·2	3·5	15·2	5·7
33	5 38·3	5 39·2	5 22·8	3·3	1·2	9·3	3·5	15·3	5·7
34	5 38·5	5 39·4	5 23·1	3·4	1·3	9·4	3·5	15·4	5·8
35	5 38·8	5 39·7	5 23·3	3·5	1·3	9·5	3·6	15·5	5·8
36	5 39·0	5 39·9	5 23·6	3·6	1·4	9·6	3·6	15·6	5·9
37	5 39·3	5 40·2	5 23·8	3·7	1·4	9·7	3·6	15·7	5·9
38	5 39·5	5 40·4	5 24·0	3·8	1·4	9·8	3·7	15·8	5·9
39	5 39·8	5 40·7	5 24·3	3·9	1·5	9·9	3·7	15·9	6·0
40	5 40·0	5 40·9	5 24·5	4·0	1·5	10·0	3·8	16·0	6·0
41	5 40·3	5 41·2	5 24·7	4·1	1·5	10·1	3·8	16·1	6·0
42	5 40·5	5 41·4	5 25·0	4·2	1·6	10·2	3·8	16·2	6·1
43	5 40·8	5 41·7	5 25·2	4·3	1·6	10·3	3·9	16·3	6·1
44	5 41·0	5 41·9	5 25·5	4·4	1·7	10·4	3·9	16·4	6·1
45	5 41·3	5 42·2	5 25·7	4·5	1·7	10·5	3·9	16·5	6·2
46	5 41·5	5 42·4	5 25·9	4·6	1·7	10·6	4·0	16·6	6·2
47	5 41·8	5 42·7	5 26·2	4·7	1·8	10·7	4·0	16·7	6·3
48	5 42·0	5 42·9	5 26·4	4·8	1·8	10·8	4·1	16·8	6·3
49	5 42·3	5 43·2	5 26·7	4·9	1·8	10·9	4·1	16·9	6·3
50	5 42·5	5 43·4	5 26·9	5·0	1·9	11·0	4·1	17·0	6·4
51	5 42·8	5 43·7	5 27·1	5·1	1·9	11·1	4·2	17·1	6·4
52	5 43·0	5 43·9	5 27·4	5·2	2·0	11·2	4·2	17·2	6·5
53	5 43·3	5 44·2	5 27·6	5·3	2·0	11·3	4·2	17·3	6·5
54	5 43·5	5 44·4	5 27·9	5·4	2·0	11·4	4·3	17·4	6·5
55	5 43·8	5 44·7	5 28·1	5·5	2·1	11·5	4·3	17·5	6·6
56	5 44·0	5 44·9	5 28·3	5·6	2·1	11·6	4·4	17·6	6·6
57	5 44·3	5 45·2	5 28·6	5·7	2·1	11·7	4·4	17·7	6·6
58	5 44·5	5 45·4	5 28·8	5·8	2·2	11·8	4·4	17·8	6·7
59	5 44·8	5 45·7	5 29·0	5·9	2·2	11·9	4·5	17·9	6·7
60	5 45·0	5 45·9	5 29·3	6·0	2·3	12·0	4·5	18·0	6·8

23m	SUN PLANETS	ARIES	MOON	v or d	Corrⁿ	v or d	Corrⁿ	v or d	Corrⁿ
s	° ′	° ′	° ′	′	′	′	′	′	′
00	5 45·0	5 45·9	5 29·3	0·0	0·0	6·0	2·4	12·0	4·7
01	5 45·3	5 46·2	5 29·5	0·1	0·0	6·1	2·4	12·1	4·7
02	5 45·5	5 46·4	5 29·8	0·2	0·1	6·2	2·4	12·2	4·8
03	5 45·8	5 46·7	5 30·0	0·3	0·1	6·3	2·5	12·3	4·8
04	5 46·0	5 46·9	5 30·2	0·4	0·2	6·4	2·5	12·4	4·9
05	5 46·3	5 47·2	5 30·5	0·5	0·2	6·5	2·5	12·5	4·9
06	5 46·5	5 47·4	5 30·7	0·6	0·2	6·6	2·6	12·6	4·9
07	5 46·8	5 47·7	5 31·0	0·7	0·3	6·7	2·6	12·7	5·0
08	5 47·0	5 48·0	5 31·2	0·8	0·3	6·8	2·7	12·8	5·0
09	5 47·3	5 48·2	5 31·4	0·9	0·4	6·9	2·7	12·9	5·1
10	5 47·5	5 48·5	5 31·7	1·0	0·4	7·0	2·7	13·0	5·1
11	5 47·8	5 48·7	5 31·9	1·1	0·4	7·1	2·8	13·1	5·1
12	5 48·0	5 49·0	5 32·1	1·2	0·5	7·2	2·8	13·2	5·2
13	5 48·3	5 49·2	5 32·4	1·3	0·5	7·3	2·9	13·3	5·2
14	5 48·5	5 49·5	5 32·6	1·4	0·5	7·4	2·9	13·4	5·2
15	5 48·8	5 49·7	5 32·9	1·5	0·6	7·5	2·9	13·5	5·3
16	5 49·0	5 50·0	5 33·1	1·6	0·6	7·6	3·0	13·6	5·3
17	5 49·3	5 50·2	5 33·3	1·7	0·7	7·7	3·0	13·7	5·4
18	5 49·5	5 50·5	5 33·6	1·8	0·7	7·8	3·1	13·8	5·4
19	5 49·8	5 50·7	5 33·8	1·9	0·7	7·9	3·1	13·9	5·4
20	5 50·0	5 51·0	5 34·1	2·0	0·8	8·0	3·1	14·0	5·5
21	5 50·3	5 51·2	5 34·3	2·1	0·8	8·1	3·2	14·1	5·5
22	5 50·5	5 51·5	5 34·5	2·2	0·9	8·2	3·2	14·2	5·6
23	5 50·8	5 51·7	5 34·8	2·3	0·9	8·3	3·3	14·3	5·6
24	5 51·0	5 52·0	5 35·0	2·4	0·9	8·4	3·3	14·4	5·6
25	5 51·3	5 52·2	5 35·2	2·5	1·0	8·5	3·3	14·5	5·7
26	5 51·5	5 52·5	5 35·5	2·6	1·0	8·6	3·4	14·6	5·7
27	5 51·8	5 52·7	5 35·7	2·7	1·1	8·7	3·4	14·7	5·8
28	5 52·0	5 53·0	5 36·0	2·8	1·1	8·8	3·4	14·8	5·8
29	5 52·3	5 53·2	5 36·2	2·9	1·1	8·9	3·5	14·9	5·8
30	5 52·5	5 53·5	5 36·4	3·0	1·2	9·0	3·5	15·0	5·9
31	5 52·8	5 53·7	5 36·7	3·1	1·2	9·1	3·6	15·1	5·9
32	5 53·0	5 54·0	5 36·9	3·2	1·3	9·2	3·6	15·2	6·0
33	5 53·3	5 54·2	5 37·2	3·3	1·3	9·3	3·6	15·3	6·0
34	5 53·5	5 54·5	5 37·4	3·4	1·3	9·4	3·7	15·4	6·0
35	5 53·8	5 54·7	5 37·6	3·5	1·4	9·5	3·7	15·5	6·1
36	5 54·0	5 55·0	5 37·9	3·6	1·4	9·6	3·8	15·6	6·1
37	5 54·3	5 55·2	5 38·1	3·7	1·4	9·7	3·8	15·7	6·1
38	5 54·5	5 55·5	5 38·4	3·8	1·5	9·8	3·8	15·8	6·2
39	5 54·8	5 55·7	5 38·6	3·9	1·5	9·9	3·9	15·9	6·2
40	5 55·0	5 56·0	5 38·8	4·0	1·6	10·0	3·9	16·0	6·3
41	5 55·3	5 56·2	5 39·1	4·1	1·6	10·1	4·0	16·1	6·3
42	5 55·5	5 56·5	5 39·3	4·2	1·6	10·2	4·0	16·2	6·3
43	5 55·8	5 56·7	5 39·5	4·3	1·7	10·3	4·0	16·3	6·4
44	5 56·0	5 57·0	5 39·8	4·4	1·7	10·4	4·1	16·4	6·4
45	5 56·3	5 57·2	5 40·0	4·5	1·8	10·5	4·1	16·5	6·5
46	5 56·5	5 57·5	5 40·3	4·6	1·8	10·6	4·2	16·6	6·5
47	5 56·8	5 57·7	5 40·5	4·7	1·8	10·7	4·2	16·7	6·5
48	5 57·0	5 58·0	5 40·7	4·8	1·9	10·8	4·2	16·8	6·6
49	5 57·3	5 58·2	5 41·0	4·9	1·9	10·9	4·3	16·9	6·6
50	5 57·5	5 58·5	5 41·2	5·0	2·0	11·0	4·3	17·0	6·7
51	5 57·8	5 58·7	5 41·5	5·1	2·0	11·1	4·3	17·1	6·7
52	5 58·0	5 59·0	5 41·7	5·2	2·0	11·2	4·4	17·2	6·7
53	5 58·3	5 59·2	5 41·9	5·3	2·1	11·3	4·4	17·3	6·8
54	5 58·5	5 59·5	5 42·2	5·4	2·1	11·4	4·5	17·4	6·8
55	5 58·8	5 59·7	5 42·4	5·5	2·2	11·5	4·5	17·5	6·9
56	5 59·0	6 00·0	5 42·6	5·6	2·2	11·6	4·5	17·6	6·9
57	5 59·3	6 00·2	5 42·9	5·7	2·2	11·7	4·6	17·7	6·9
58	5 59·5	6 00·5	5 43·1	5·8	2·3	11·8	4·6	17·8	7·0
59	5 59·8	6 00·7	5 43·4	5·9	2·3	11·9	4·7	17·9	7·0
60	6 00·0	6 01·0	5 43·6	6·0	2·4	12·0	4·7	18·0	7·1

m 24	SUN PLANETS	ARIES	MOON	v or d	Corrn	v or d	Corrn	v or d	Corrn
s	° ′	° ′	° ′	′	′	′	′	′	′
00	6 00·0	6 01·0	5 43·6	0·0	0·0	6·0	2·5	12·0	4·9
01	6 00·3	6 01·2	5 43·8	0·1	0·0	6·1	2·5	12·1	4·9
02	6 00·5	6 01·5	5 44·1	0·2	0·1	6·2	2·5	12·2	5·0
03	6 00·8	6 01·7	5 44·3	0·3	0·1	6·3	2·6	12·3	5·0
04	6 01·0	6 02·0	5 44·6	0·4	0·2	6·4	2·6	12·4	5·1
05	6 01·3	6 02·2	5 44·8	0·5	0·2	6·5	2·7	12·5	5·1
06	6 01·5	6 02·5	5 45·0	0·6	0·2	6·6	2·7	12·6	5·1
07	6 01·8	6 02·7	5 45·3	0·7	0·3	6·7	2·7	12·7	5·2
08	6 02·0	6 03·0	5 45·5	0·8	0·3	6·8	2·8	12·8	5·2
09	6 02·3	6 03·2	5 45·7	0·9	0·4	6·9	2·8	12·9	5·3
10	6 02·5	6 03·5	5 46·0	1·0	0·4	7·0	2·9	13·0	5·3
11	6 02·8	6 03·7	5 46·2	1·1	0·4	7·1	2·9	13·1	5·3
12	6 03·0	6 04·0	5 46·5	1·2	0·5	7·2	2·9	13·2	5·4
13	6 03·3	6 04·2	5 46·7	1·3	0·5	7·3	3·0	13·3	5·4
14	6 03·5	6 04·5	5 46·9	1·4	0·6	7·4	3·0	13·4	5·5
15	6 03·8	6 04·7	5 47·2	1·5	0·6	7·5	3·1	13·5	5·5
16	6 04·0	6 05·0	5 47·4	1·6	0·7	7·6	3·1	13·6	5·6
17	6 04·3	6 05·2	5 47·7	1·7	0·7	7·7	3·1	13·7	5·6
18	6 04·5	6 05·5	5 47·9	1·8	0·7	7·8	3·2	13·8	5·6
19	6 04·8	6 05·7	5 48·1	1·9	0·8	7·9	3·2	13·9	5·7
20	6 05·0	6 06·0	5 48·4	2·0	0·8	8·0	3·3	14·0	5·7
21	6 05·3	6 06·3	5 48·6	2·1	0·9	8·1	3·3	14·1	5·8
22	6 05·5	6 06·5	5 48·8	2·2	0·9	8·2	3·3	14·2	5·8
23	6 05·8	6 06·8	5 49·1	2·3	0·9	8·3	3·4	14·3	5·8
24	6 06·0	6 07·0	5 49·3	2·4	1·0	8·4	3·4	14·4	5·9
25	6 06·3	6 07·3	5 49·6	2·5	1·0	8·5	3·5	14·5	5·9
26	6 06·5	6 07·5	5 49·8	2·6	1·1	8·6	3·5	14·6	6·0
27	6 06·8	6 07·8	5 50·0	2·7	1·1	8·7	3·6	14·7	6·0
28	6 07·0	6 08·0	5 50·3	2·8	1·1	8·8	3·6	14·8	6·0
29	6 07·3	6 08·3	5 50·5	2·9	1·2	8·9	3·6	14·9	6·1
30	6 07·5	6 08·5	5 50·8	3·0	1·2	9·0	3·7	15·0	6·1
31	6 07·8	6 08·8	5 51·0	3·1	1·3	9·1	3·7	15·1	6·2
32	6 08·0	6 09·0	5 51·2	3·2	1·3	9·2	3·8	15·2	6·2
33	6 08·3	6 09·3	5 51·5	3·3	1·3	9·3	3·8	15·3	6·2
34	6 08·5	6 09·5	5 51·7	3·4	1·4	9·4	3·8	15·4	6·3
35	6 08·8	6 09·8	5 52·0	3·5	1·4	9·5	3·9	15·5	6·3
36	6 09·0	6 10·0	5 52·2	3·6	1·5	9·6	3·9	15·6	6·4
37	6 09·3	6 10·3	5 52·4	3·7	1·5	9·7	4·0	15·7	6·4
38	6 09·5	6 10·5	5 52·7	3·8	1·6	9·8	4·0	15·8	6·5
39	6 09·8	6 10·8	5 52·9	3·9	1·6	9·9	4·0	15·9	6·5
40	6 10·0	6 11·0	5 53·1	4·0	1·6	10·0	4·1	16·0	6·5
41	6 10·3	6 11·3	5 53·4	4·1	1·7	10·1	4·1	16·1	6·6
42	6 10·5	6 11·5	5 53·6	4·2	1·7	10·2	4·2	16·2	6·6
43	6 10·8	6 11·8	5 53·9	4·3	1·8	10·3	4·2	16·3	6·7
44	6 11·0	6 12·0	5 54·1	4·4	1·8	10·4	4·2	16·4	6·7
45	6 11·3	6 12·3	5 54·3	4·5	1·8	10·5	4·3	16·5	6·7
46	6 11·5	6 12·5	5 54·6	4·6	1·9	10·6	4·3	16·6	6·8
47	6 11·8	6 12·8	5 54·8	4·7	1·9	10·7	4·4	16·7	6·8
48	6 12·0	6 13·0	5 55·1	4·8	2·0	10·8	4·4	16·8	6·9
49	6 12·3	6 13·3	5 55·3	4·9	2·0	10·9	4·5	16·9	6·9
50	6 12·5	6 13·5	5 55·5	5·0	2·0	11·0	4·5	17·0	6·9
51	6 12·8	6 13·8	5 55·8	5·1	2·1	11·1	4·5	17·1	7·0
52	6 13·0	6 14·0	5 56·0	5·2	2·1	11·2	4·6	17·2	7·0
53	6 13·3	6 14·3	5 56·2	5·3	2·2	11·3	4·6	17·3	7·1
54	6 13·5	6 14·5	5 56·5	5·4	2·2	11·4	4·7	17·4	7·1
55	6 13·8	6 14·8	5 56·7	5·5	2·2	11·5	4·7	17·5	7·1
56	6 14·0	6 15·0	5 57·0	5·6	2·3	11·6	4·7	17·6	7·2
57	6 14·3	6 15·3	5 57·2	5·7	2·3	11·7	4·8	17·7	7·2
58	6 14·5	6 15·5	5 57·4	5·8	2·4	11·8	4·8	17·8	7·3
59	6 14·8	6 15·8	5 57·7	5·9	2·4	11·9	4·9	17·9	7·3
60	6 15·0	6 16·0	5 57·9	6·0	2·5	12·0	4·9	18·0	7·4

m 25	SUN PLANETS	ARIES	MOON	v or d	Corrn	v or d	Corrn	v or d	Corrn
s	° ′	° ′	° ′	′	′	′	′	′	′
00	6 15·0	6 16·0	5 57·9	0·0	0·0	6·0	2·6	12·0	5·1
01	6 15·3	6 16·3	5 58·2	0·1	0·0	6·1	2·6	12·1	5·1
02	6 15·5	6 16·5	5 58·4	0·2	0·1	6·2	2·6	12·2	5·2
03	6 15·8	6 16·8	5 58·6	0·3	0·1	6·3	2·7	12·3	5·2
04	6 16·0	6 17·0	5 58·9	0·4	0·2	6·4	2·7	12·4	5·3
05	6 16·3	6 17·3	5 59·1	0·5	0·2	6·5	2·8	12·5	5·3
06	6 16·5	6 17·5	5 59·3	0·6	0·3	6·6	2·8	12·6	5·4
07	6 16·8	6 17·8	5 59·6	0·7	0·3	6·7	2·8	12·7	5·4
08	6 17·0	6 18·0	5 59·8	0·8	0·3	6·8	2·9	12·8	5·4
09	6 17·3	6 18·3	6 00·1	0·9	0·4	6·9	2·9	12·9	5·5
10	6 17·5	6 18·5	6 00·3	1·0	0·4	7·0	3·0	13·0	5·5
11	6 17·8	6 18·8	6 00·5	1·1	0·5	7·1	3·0	13·1	5·6
12	6 18·0	6 19·0	6 00·8	1·2	0·5	7·2	3·1	13·2	5·6
13	6 18·3	6 19·3	6 01·0	1·3	0·6	7·3	3·1	13·3	5·7
14	6 18·5	6 19·5	6 01·3	1·4	0·6	7·4	3·1	13·4	5·7
15	6 18·8	6 19·8	6 01·5	1·5	0·6	7·5	3·2	13·5	5·7
16	6 19·0	6 20·0	6 01·7	1·6	0·7	7·6	3·2	13·6	5·8
17	6 19·3	6 20·3	6 02·0	1·7	0·7	7·7	3·3	13·7	5·8
18	6 19·5	6 20·5	6 02·2	1·8	0·8	7·8	3·3	13·8	5·9
19	6 19·8	6 20·8	6 02·5	1·9	0·8	7·9	3·4	13·9	5·9
20	6 20·0	6 21·0	6 02·7	2·0	0·9	8·0	3·4	14·0	6·0
21	6 20·3	6 21·3	6 02·9	2·1	0·9	8·1	3·4	14·1	6·0
22	6 20·5	6 21·5	6 03·2	2·2	0·9	8·2	3·5	14·2	6·0
23	6 20·8	6 21·8	6 03·4	2·3	1·0	8·3	3·5	14·3	6·1
24	6 21·0	6 22·0	6 03·6	2·4	1·0	8·4	3·6	14·4	6·1
25	6 21·3	6 22·3	6 03·9	2·5	1·1	8·5	3·6	14·5	6·2
26	6 21·5	6 22·5	6 04·1	2·6	1·1	8·6	3·7	14·6	6·2
27	6 21·8	6 22·8	6 04·4	2·7	1·1	8·7	3·7	14·7	6·2
28	6 22·0	6 23·0	6 04·6	2·8	1·2	8·8	3·7	14·8	6·3
29	6 22·3	6 23·3	6 04·8	2·9	1·2	8·9	3·8	14·9	6·3
30	6 22·5	6 23·5	6 05·1	3·0	1·3	9·0	3·8	15·0	6·4
31	6 22·8	6 23·8	6 05·3	3·1	1·3	9·1	3·9	15·1	6·4
32	6 23·0	6 24·0	6 05·6	3·2	1·4	9·2	3·9	15·2	6·5
33	6 23·3	6 24·3	6 05·8	3·3	1·4	9·3	4·0	15·3	6·5
34	6 23·5	6 24·5	6 06·0	3·4	1·4	9·4	4·0	15·4	6·5
35	6 23·8	6 24·8	6 06·3	3·5	1·5	9·5	4·0	15·5	6·6
36	6 24·0	6 25·1	6 06·5	3·6	1·5	9·6	4·1	15·6	6·6
37	6 24·3	6 25·3	6 06·7	3·7	1·6	9·7	4·1	15·7	6·7
38	6 24·5	6 25·6	6 07·0	3·8	1·6	9·8	4·2	15·8	6·7
39	6 24·8	6 25·8	6 07·2	3·9	1·7	9·9	4·2	15·9	6·8
40	6 25·0	6 26·1	6 07·5	4·0	1·7	10·0	4·3	16·0	6·8
41	6 25·3	6 26·3	6 07·7	4·1	1·7	10·1	4·3	16·1	6·8
42	6 25·5	6 26·6	6 07·9	4·2	1·8	10·2	4·3	16·2	6·9
43	6 25·8	6 26·8	6 08·2	4·3	1·8	10·3	4·4	16·3	6·9
44	6 26·0	6 27·1	6 08·4	4·4	1·9	10·4	4·4	16·4	7·0
45	6 26·3	6 27·3	6 08·7	4·5	1·9	10·5	4·5	16·5	7·0
46	6 26·5	6 27·6	6 08·9	4·6	2·0	10·6	4·5	16·6	7·1
47	6 26·8	6 27·8	6 09·1	4·7	2·0	10·7	4·5	16·7	7·1
48	6 27·0	6 28·1	6 09·4	4·8	2·0	10·8	4·6	16·8	7·1
49	6 27·3	6 28·3	6 09·6	4·9	2·1	10·9	4·6	16·9	7·2
50	6 27·5	6 28·6	6 09·8	5·0	2·1	11·0	4·7	17·0	7·2
51	6 27·8	6 28·8	6 10·1	5·1	2·2	11·1	4·7	17·1	7·3
52	6 28·0	6 29·1	6 10·3	5·2	2·2	11·2	4·8	17·2	7·3
53	6 28·3	6 29·3	6 10·6	5·3	2·3	11·3	4·8	17·3	7·4
54	6 28·5	6 29·6	6 10·8	5·4	2·3	11·4	4·8	17·4	7·4
55	6 28·8	6 29·8	6 11·0	5·5	2·3	11·5	4·9	17·5	7·4
56	6 29·0	6 30·1	6 11·3	5·6	2·4	11·6	4·9	17·6	7·5
57	6 29·3	6 30·3	6 11·5	5·7	2·4	11·7	5·0	17·7	7·5
58	6 29·5	6 30·6	6 11·8	5·8	2·5	11·8	5·0	17·8	7·6
59	6 29·8	6 30·8	6 12·0	5·9	2·5	11·9	5·1	17·9	7·6
60	6 30·0	6 31·1	6 12·2	6·0	2·6	12·0	5·1	18·0	7·7

26m	SUN PLANETS	ARIES	MOON	v or d	Corrn	v or d	Corrn	v or d	Corrn
s	° ′	° ′	° ′	′	′	′	′	′	′
00	6 30·0	6 31·1	6 12·2	0·0	0·0	6·0	2·7	12·0	5·3
01	6 30·3	6 31·3	6 12·5	0·1	0·0	6·1	2·7	12·1	5·3
02	6 30·5	6 31·6	6 12·7	0·2	0·1	6·2	2·7	12·2	5·4
03	6 30·8	6 31·8	6 12·9	0·3	0·1	6·3	2·8	12·3	5·4
04	6 31·0	6 32·1	6 13·2	0·4	0·2	6·4	2·8	12·4	5·5
05	6 31·3	6 32·3	6 13·4	0·5	0·2	6·5	2·9	12·5	5·5
06	6 31·5	6 32·6	6 13·7	0·6	0·3	6·6	2·9	12·6	5·6
07	6 31·8	6 32·8	6 13·9	0·7	0·3	6·7	3·0	12·7	5·6
08	6 32·0	6 33·1	6 14·1	0·8	0·4	6·8	3·0	12·8	5·7
09	6 32·3	6 33·3	6 14·4	0·9	0·4	6·9	3·0	12·9	5·7
10	6 32·5	6 33·6	6 14·6	1·0	0·4	7·0	3·1	13·0	5·7
11	6 32·8	6 33·8	6 14·9	1·1	0·5	7·1	3·1	13·1	5·8
12	6 33·0	6 34·1	6 15·1	1·2	0·5	7·2	3·2	13·2	5·8
13	6 33·3	6 34·3	6 15·3	1·3	0·6	7·3	3·2	13·3	5·9
14	6 33·5	6 34·6	6 15·6	1·4	0·6	7·4	3·3	13·4	5·9
15	6 33·8	6 34·8	6 15·8	1·5	0·7	7·5	3·3	13·5	6·0
16	6 34·0	6 35·1	6 16·1	1·6	0·7	7·6	3·4	13·6	6·0
17	6 34·3	6 35·3	6 16·3	1·7	0·8	7·7	3·4	13·7	6·1
18	6 34·5	6 35·6	6 16·5	1·8	0·8	7·8	3·4	13·8	6·1
19	6 34·8	6 35·8	6 16·8	1·9	0·8	7·9	3·5	13·9	6·1
20	6 35·0	6 36·1	6 17·0	2·0	0·9	8·0	3·5	14·0	6·2
21	6 35·3	6 36·3	6 17·2	2·1	0·9	8·1	3·6	14·1	6·2
22	6 35·5	6 36·6	6 17·5	2·2	1·0	8·2	3·6	14·2	6·3
23	6 35·8	6 36·8	6 17·7	2·3	1·0	8·3	3·7	14·3	6·3
24	6 36·0	6 37·1	6 18·0	2·4	1·1	8·4	3·7	14·4	6·4
25	6 36·3	6 37·3	6 18·2	2·5	1·1	8·5	3·8	14·5	6·4
26	6 36·5	6 37·6	6 18·4	2·6	1·1	8·6	3·8	14·6	6·4
27	6 36·8	6 37·8	6 18·7	2·7	1·2	8·7	3·8	14·7	6·5
28	6 37·0	6 38·1	6 18·9	2·8	1·2	8·8	3·9	14·8	6·5
29	6 37·3	6 38·3	6 19·2	2·9	1·3	8·9	3·9	14·9	6·6
30	6 37·5	6 38·6	6 19·4	3·0	1·3	9·0	4·0	15·0	6·6
31	6 37·8	6 38·8	6 19·6	3·1	1·4	9·1	4·0	15·1	6·7
32	6 38·0	6 39·1	6 19·9	3·2	1·4	9·2	4·1	15·2	6·7
33	6 38·3	6 39·3	6 20·1	3·3	1·5	9·3	4·1	15·3	6·8
34	6 38·5	6 39·6	6 20·3	3·4	1·5	9·4	4·2	15·4	6·8
35	6 38·8	6 39·8	6 20·6	3·5	1·5	9·5	4·2	15·5	6·8
36	6 39·0	6 40·1	6 20·8	3·6	1·6	9·6	4·2	15·6	6·9
37	6 39·3	6 40·3	6 21·1	3·7	1·6	9·7	4·3	15·7	6·9
38	6 39·5	6 40·6	6 21·3	3·8	1·7	9·8	4·3	15·8	7·0
39	6 39·8	6 40·8	6 21·5	3·9	1·7	9·9	4·4	15·9	7·0
40	6 40·0	6 41·1	6 21·8	4·0	1·8	10·0	4·4	16·0	7·1
41	6 40·3	6 41·3	6 22·0	4·1	1·8	10·1	4·5	16·1	7·1
42	6 40·5	6 41·6	6 22·3	4·2	1·9	10·2	4·5	16·2	7·2
43	6 40·8	6 41·8	6 22·5	4·3	1·9	10·3	4·5	16·3	7·2
44	6 41·0	6 42·1	6 22·7	4·4	1·9	10·4	4·6	16·4	7·2
45	6 41·3	6 42·3	6 23·0	4·5	2·0	10·5	4·6	16·5	7·3
46	6 41·5	6 42·6	6 23·2	4·6	2·0	10·6	4·7	16·6	7·3
47	6 41·8	6 42·8	6 23·4	4·7	2·1	10·7	4·7	16·7	7·4
48	6 42·0	6 43·1	6 23·7	4·8	2·1	10·8	4·8	16·8	7·4
49	6 42·3	6 43·4	6 23·9	4·9	2·2	10·9	4·8	16·9	7·5
50	6 42·5	6 43·6	6 24·2	5·0	2·2	11·0	4·9	17·0	7·5
51	6 42·8	6 43·9	6 24·4	5·1	2·3	11·1	4·9	17·1	7·6
52	6 43·0	6 44·1	6 24·6	5·2	2·3	11·2	4·9	17·2	7·6
53	6 43·3	6 44·4	6 24·9	5·3	2·3	11·3	5·0	17·3	7·6
54	6 43·5	6 44·6	6 25·1	5·4	2·4	11·4	5·0	17·4	7·7
55	6 43·8	6 44·9	6 25·4	5·5	2·4	11·5	5·1	17·5	7·7
56	6 44·0	6 45·1	6 25·6	5·6	2·5	11·6	5·1	17·6	7·8
57	6 44·3	6 45·4	6 25·8	5·7	2·5	11·7	5·2	17·7	7·8
58	6 44·5	6 45·6	6 26·1	5·8	2·6	11·8	5·2	17·8	7·9
59	6 44·8	6 45·9	6 26·3	5·9	2·6	11·9	5·3	17·9	7·9
60	6 45·0	6 46·1	6 26·6	6·0	2·7	12·0	5·3	18·0	8·0

27m	SUN PLANETS	ARIES	MOON	v or d	Corrn	v or d	Corrn	v or d	Corrn
s	° ′	° ′	° ′	′	′	′	′	′	′
00	6 45·0	6 46·1	6 26·6	0·0	0·0	6·0	2·8	12·0	5·5
01	6 45·3	6 46·4	6 26·8	0·1	0·0	6·1	2·8	12·1	5·5
02	6 45·5	6 46·6	6 27·0	0·2	0·1	6·2	2·8	12·2	5·6
03	6 45·8	6 46·9	6 27·3	0·3	0·1	6·3	2·9	12·3	5·6
04	6 46·0	6 47·1	6 27·5	0·4	0·2	6·4	2·9	12·4	5·7
05	6 46·3	6 47·4	6 27·7	0·5	0·2	6·5	3·0	12·5	5·7
06	6 46·5	6 47·6	6 28·0	0·6	0·3	6·6	3·0	12·6	5·8
07	6 46·8	6 47·9	6 28·2	0·7	0·3	6·7	3·1	12·7	5·8
08	6 47·0	6 48·1	6 28·5	0·8	0·4	6·8	3·1	12·8	5·9
09	6 47·3	6 48·4	6 28·7	0·9	0·4	6·9	3·2	12·9	5·9
10	6 47·5	6 48·6	6 28·9	1·0	0·5	7·0	3·2	13·0	6·0
11	6 47·8	6 48·9	6 29·2	1·1	0·5	7·1	3·3	13·1	6·0
12	6 48·0	6 49·1	6 29·4	1·2	0·6	7·2	3·3	13·2	6·1
13	6 48·3	6 49·4	6 29·7	1·3	0·6	7·3	3·3	13·3	6·1
14	6 48·5	6 49·6	6 29·9	1·4	0·6	7·4	3·4	13·4	6·1
15	6 48·8	6 49·9	6 30·1	1·5	0·7	7·5	3·4	13·5	6·2
16	6 49·0	6 50·1	6 30·4	1·6	0·7	7·6	3·5	13·6	6·2
17	6 49·3	6 50·4	6 30·6	1·7	0·8	7·7	3·5	13·7	6·3
18	6 49·5	6 50·6	6 30·8	1·8	0·8	7·8	3·6	13·8	6·3
19	6 49·8	6 50·9	6 31·1	1·9	0·9	7·9	3·6	13·9	6·4
20	6 50·0	6 51·1	6 31·3	2·0	0·9	8·0	3·7	14·0	6·4
21	6 50·3	6 51·4	6 31·6	2·1	1·0	8·1	3·7	14·1	6·5
22	6 50·5	6 51·6	6 31·8	2·2	1·0	8·2	3·8	14·2	6·5
23	6 50·8	6 51·9	6 32·0	2·3	1·1	8·3	3·8	14·3	6·6
24	6 51·0	6 52·1	6 32·3	2·4	1·1	8·4	3·9	14·4	6·6
25	6 51·3	6 52·4	6 32·5	2·5	1·1	8·5	3·9	14·5	6·6
26	6 51·5	6 52·6	6 32·8	2·6	1·2	8·6	3·9	14·6	6·7
27	6 51·8	6 52·9	6 33·0	2·7	1·2	8·7	4·0	14·7	6·7
28	6 52·0	6 53·1	6 33·2	2·8	1·3	8·8	4·0	14·8	6·8
29	6 52·3	6 53·4	6 33·5	2·9	1·3	8·9	4·1	14·9	6·8
30	6 52·5	6 53·6	6 33·7	3·0	1·4	9·0	4·1	15·0	6·9
31	6 52·8	6 53·9	6 33·9	3·1	1·4	9·1	4·2	15·1	6·9
32	6 53·0	6 54·1	6 34·2	3·2	1·5	9·2	4·2	15·2	7·0
33	6 53·3	6 54·4	6 34·4	3·3	1·5	9·3	4·3	15·3	7·0
34	6 53·5	6 54·6	6 34·7	3·4	1·6	9·4	4·3	15·4	7·1
35	6 53·8	6 54·9	6 34·9	3·5	1·6	9·5	4·4	15·5	7·1
36	6 54·0	6 55·1	6 35·1	3·6	1·7	9·6	4·4	15·6	7·2
37	6 54·3	6 55·4	6 35·4	3·7	1·7	9·7	4·4	15·7	7·2
38	6 54·5	6 55·6	6 35·6	3·8	1·7	9·8	4·5	15·8	7·2
39	6 54·8	6 55·9	6 35·9	3·9	1·8	9·9	4·5	15·9	7·3
40	6 55·0	6 56·1	6 36·1	4·0	1·8	10·0	4·6	16·0	7·3
41	6 55·3	6 56·4	6 36·3	4·1	1·9	10·1	4·6	16·1	7·4
42	6 55·5	6 56·6	6 36·6	4·2	1·9	10·2	4·7	16·2	7·4
43	6 55·8	6 56·9	6 36·8	4·3	2·0	10·3	4·7	16·3	7·5
44	6 56·0	6 57·1	6 37·0	4·4	2·0	10·4	4·8	16·4	7·5
45	6 56·3	6 57·4	6 37·3	4·5	2·1	10·5	4·8	16·5	7·6
46	6 56·5	6 57·6	6 37·5	4·6	2·1	10·6	4·9	16·6	7·6
47	6 56·8	6 57·9	6 37·8	4·7	2·2	10·7	4·9	16·7	7·7
48	6 57·0	6 58·1	6 38·0	4·8	2·2	10·8	5·0	16·8	7·7
49	6 57·3	6 58·4	6 38·2	4·9	2·2	10·9	5·0	16·9	7·7
50	6 57·5	6 58·6	6 38·5	5·0	2·3	11·0	5·0	17·0	7·8
51	6 57·8	6 58·9	6 38·7	5·1	2·3	11·1	5·1	17·1	7·8
52	6 58·0	6 59·1	6 39·0	5·2	2·4	11·2	5·1	17·2	7·9
53	6 58·3	6 59·4	6 39·2	5·3	2·4	11·3	5·2	17·3	7·9
54	6 58·5	6 59·6	6 39·4	5·4	2·5	11·4	5·2	17·4	8·0
55	6 58·8	6 59·9	6 39·7	5·5	2·5	11·5	5·3	17·5	8·0
56	6 59·0	7 00·1	6 39·9	5·6	2·6	11·6	5·3	17·6	8·1
57	6 59·3	7 00·4	6 40·2	5·7	2·6	11·7	5·4	17·7	8·1
58	6 59·5	7 00·6	6 40·4	5·8	2·7	11·8	5·4	17·8	8·2
59	6 59·8	7 00·9	6 40·6	5·9	2·7	11·9	5·5	17·9	8·2
60	7 00·0	7 01·1	6 40·9	6·0	2·8	12·0	5·5	18·0	8·3

28m s	SUN PLANETS ° ′	ARIES ° ′	MOON ° ′	v or d ′	Corrⁿ ′	v or d ′	Corrⁿ ′	v or d ′	Corrⁿ ′
00	7 00·0	7 01·1	6 40·9	0·0	0·0	6·0	2·9	12·0	5·7
01	7 00·3	7 01·4	6 41·1	0·1	0·0	6·1	2·9	12·1	5·7
02	7 00·5	7 01·7	6 41·3	0·2	0·1	6·2	2·9	12·2	5·8
03	7 00·8	7 01·9	6 41·6	0·3	0·1	6·3	3·0	12·3	5·8
04	7 01·0	7 02·2	6 41·8	0·4	0·2	6·4	3·0	12·4	5·9
05	7 01·3	7 02·4	6 42·1	0·5	0·2	6·5	3·1	12·5	5·9
06	7 01·5	7 02·7	6 42·3	0·6	0·3	6·6	3·1	12·6	6·0
07	7 01·8	7 02·9	6 42·5	0·7	0·3	6·7	3·2	12·7	6·0
08	7 02·0	7 03·2	6 42·8	0·8	0·4	6·8	3·2	12·8	6·1
09	7 02·3	7 03·4	6 43·0	0·9	0·4	6·9	3·3	12·9	6·1
10	7 02·5	7 03·7	6 43·3	1·0	0·5	7·0	3·3	13·0	6·2
11	7 02·8	7 03·9	6 43·5	1·1	0·5	7·1	3·4	13·1	6·2
12	7 03·0	7 04·2	6 43·7	1·2	0·6	7·2	3·4	13·2	6·3
13	7 03·3	7 04·4	6 44·0	1·3	0·6	7·3	3·5	13·3	6·3
14	7 03·5	7 04·7	6 44·2	1·4	0·7	7·4	3·5	13·4	6·4
15	7 03·8	7 04·9	6 44·4	1·5	0·7	7·5	3·6	13·5	6·4
16	7 04·0	7 05·2	6 44·7	1·6	0·8	7·6	3·6	13·6	6·5
17	7 04·3	7 05·4	6 44·9	1·7	0·8	7·7	3·7	13·7	6·5
18	7 04·5	7 05·7	6 45·2	1·8	0·9	7·8	3·7	13·8	6·6
19	7 04·8	7 05·9	6 45·4	1·9	0·9	7·9	3·8	13·9	6·6
20	7 05·0	7 06·2	6 45·6	2·0	1·0	8·0	3·8	14·0	6·7
21	7 05·3	7 06·4	6 45·9	2·1	1·0	8·1	3·8	14·1	6·7
22	7 05·5	7 06·7	6 46·1	2·2	1·0	8·2	3·9	14·2	6·7
23	7 05·8	7 06·9	6 46·4	2·3	1·1	8·3	3·9	14·3	6·8
24	7 06·0	7 07·2	6 46·6	2·4	1·1	8·4	4·0	14·4	6·8
25	7 06·3	7 07·4	6 46·8	2·5	1·2	8·5	4·0	14·5	6·9
26	7 06·5	7 07·7	6 47·1	2·6	1·2	8·6	4·1	14·6	6·9
27	7 06·8	7 07·9	6 47·3	2·7	1·3	8·7	4·1	14·7	7·0
28	7 07·0	7 08·2	6 47·5	2·8	1·3	8·8	4·2	14·8	7·0
29	7 07·3	7 08·4	6 47·8	2·9	1·4	8·9	4·2	14·9	7·1
30	7 07·5	7 08·7	6 48·0	3·0	1·4	9·0	4·3	15·0	7·1
31	7 07·8	7 08·9	6 48·3	3·1	1·5	9·1	4·3	15·1	7·2
32	7 08·0	7 09·2	6 48·5	3·2	1·5	9·2	4·4	15·2	7·2
33	7 08·3	7 09·4	6 48·7	3·3	1·6	9·3	4·4	15·3	7·3
34	7 08·5	7 09·7	6 49·0	3·4	1·6	9·4	4·5	15·4	7·3
35	7 08·8	7 09·9	6 49·2	3·5	1·7	9·5	4·5	15·5	7·4
36	7 09·0	7 10·2	6 49·5	3·6	1·7	9·6	4·6	15·6	7·4
37	7 09·3	7 10·4	6 49·7	3·7	1·8	9·7	4·6	15·7	7·5
38	7 09·5	7 10·7	6 49·9	3·8	1·8	9·8	4·7	15·8	7·5
39	7 09·8	7 10·9	6 50·2	3·9	1·9	9·9	4·7	15·9	7·6
40	7 10·0	7 11·2	6 50·4	4·0	1·9	10·0	4·8	16·0	7·6
41	7 10·3	7 11·4	6 50·6	4·1	1·9	10·1	4·8	16·1	7·6
42	7 10·5	7 11·7	6 50·9	4·2	2·0	10·2	4·8	16·2	7·7
43	7 10·8	7 11·9	6 51·1	4·3	2·0	10·3	4·9	16·3	7·7
44	7 11·0	7 12·2	6 51·4	4·4	2·1	10·4	4·9	16·4	7·8
45	7 11·3	7 12·4	6 51·6	4·5	2·1	10·5	5·0	16·5	7·8
46	7 11·5	7 12·7	6 51·8	4·6	2·2	10·6	5·0	16·6	7·9
47	7 11·8	7 12·9	6 52·1	4·7	2·2	10·7	5·1	16·7	7·9
48	7 12·0	7 13·2	6 52·3	4·8	2·3	10·8	5·1	16·8	8·0
49	7 12·3	7 13·4	6 52·6	4·9	2·3	10·9	5·2	16·9	8·0
50	7 12·5	7 13·7	6 52·8	5·0	2·4	11·0	5·2	17·0	8·1
51	7 12·8	7 13·9	6 53·0	5·1	2·4	11·1	5·3	17·1	8·1
52	7 13·0	7 14·2	6 53·3	5·2	2·5	11·2	5·3	17·2	8·2
53	7 13·3	7 14·4	6 53·5	5·3	2·5	11·3	5·4	17·3	8·2
54	7 13·5	7 14·7	6 53·8	5·4	2·6	11·4	5·4	17·4	8·3
55	7 13·8	7 14·9	6 54·0	5·5	2·6	11·5	5·5	17·5	8·3
56	7 14·0	7 15·2	6 54·2	5·6	2·7	11·6	5·5	17·6	8·4
57	7 14·3	7 15·4	6 54·5	5·7	2·7	11·7	5·6	17·7	8·4
58	7 14·5	7 15·7	6 54·7	5·8	2·8	11·8	5·6	17·8	8·5
59	7 14·8	7 15·9	6 54·9	5·9	2·8	11·9	5·7	17·9	8·5
60	7 15·0	7 16·2	6 55·2	6·0	2·9	12·0	5·7	18·0	8·6

29m s	SUN PLANETS ° ′	ARIES ° ′	MOON ° ′	v or d ′	Corrⁿ ′	v or d ′	Corrⁿ ′	v or d ′	Corrⁿ ′
00	7 15·0	7 16·2	6 55·2	0·0	0·0	6·0	3·0	12·0	5·9
01	7 15·3	7 16·4	6 55·4	0·1	0·0	6·1	3·0	12·1	5·9
02	7 15·5	7 16·7	6 55·7	0·2	0·1	6·2	3·0	12·2	6·0
03	7 15·8	7 16·9	6 55·9	0·3	0·1	6·3	3·1	12·3	6·0
04	7 16·0	7 17·2	6 56·1	0·4	0·2	6·4	3·1	12·4	6·1
05	7 16·3	7 17·4	6 56·4	0·5	0·2	6·5	3·2	12·5	6·1
06	7 16·5	7 17·7	6 56·6	0·6	0·3	6·6	3·2	12·6	6·2
07	7 16·8	7 17·9	6 56·9	0·7	0·3	6·7	3·3	12·7	6·2
08	7 17·0	7 18·2	6 57·1	0·8	0·4	6·8	3·3	12·8	6·3
09	7 17·3	7 18·4	6 57·3	0·9	0·4	6·9	3·4	12·9	6·3
10	7 17·5	7 18·7	6 57·6	1·0	0·5	7·0	3·4	13·0	6·4
11	7 17·8	7 18·9	6 57·8	1·1	0·5	7·1	3·5	13·1	6·4
12	7 18·0	7 19·2	6 58·0	1·2	0·6	7·2	3·5	13·2	6·5
13	7 18·3	7 19·4	6 58·3	1·3	0·6	7·3	3·6	13·3	6·5
14	7 18·5	7 19·7	6 58·5	1·4	0·7	7·4	3·6	13·4	6·6
15	7 18·8	7 20·0	6 58·8	1·5	0·7	7·5	3·7	13·5	6·6
16	7 19·0	7 20·2	6 59·0	1·6	0·8	7·6	3·7	13·6	6·7
17	7 19·3	7 20·5	6 59·2	1·7	0·8	7·7	3·8	13·7	6·7
18	7 19·5	7 20·7	6 59·5	1·8	0·9	7·8	3·8	13·8	6·8
19	7 19·8	7 21·0	6 59·7	1·9	0·9	7·9	3·9	13·9	6·8
20	7 20·0	7 21·2	7 00·0	2·0	1·0	8·0	3·9	14·0	6·9
21	7 20·3	7 21·5	7 00·2	2·1	1·0	8·1	4·0	14·1	6·9
22	7 20·5	7 21·7	7 00·4	2·2	1·1	8·2	4·0	14·2	7·0
23	7 20·8	7 22·0	7 00·7	2·3	1·1	8·3	4·1	14·3	7·0
24	7 21·0	7 22·2	7 00·9	2·4	1·2	8·4	4·1	14·4	7·1
25	7 21·3	7 22·5	7 01·1	2·5	1·2	8·5	4·2	14·5	7·1
26	7 21·5	7 22·7	7 01·4	2·6	1·3	8·6	4·2	14·6	7·2
27	7 21·8	7 23·0	7 01·6	2·7	1·3	8·7	4·3	14·7	7·2
28	7 22·0	7 23·2	7 01·9	2·8	1·4	8·8	4·3	14·8	7·3
29	7 22·3	7 23·5	7 02·1	2·9	1·4	8·9	4·4	14·9	7·3
30	7 22·5	7 23·7	7 02·3	3·0	1·5	9·0	4·4	15·0	7·4
31	7 22·8	7 24·0	7 02·6	3·1	1·5	9·1	4·5	15·1	7·4
32	7 23·0	7 24·2	7 02·8	3·2	1·6	9·2	4·5	15·2	7·5
33	7 23·3	7 24·5	7 03·1	3·3	1·6	9·3	4·6	15·3	7·5
34	7 23·5	7 24·7	7 03·3	3·4	1·7	9·4	4·6	15·4	7·6
35	7 23·8	7 25·0	7 03·5	3·5	1·7	9·5	4·7	15·5	7·6
36	7 24·0	7 25·2	7 03·8	3·6	1·8	9·6	4·7	15·6	7·7
37	7 24·3	7 25·5	7 04·0	3·7	1·8	9·7	4·8	15·7	7·7
38	7 24·5	7 25·7	7 04·3	3·8	1·9	9·8	4·8	15·8	7·8
39	7 24·8	7 26·0	7 04·5	3·9	1·9	9·9	4·9	15·9	7·8
40	7 25·0	7 26·2	7 04·7	4·0	2·0	10·0	4·9	16·0	7·9
41	7 25·3	7 26·5	7 05·0	4·1	2·0	10·1	5·0	16·1	7·9
42	7 25·5	7 26·7	7 05·2	4·2	2·1	10·2	5·0	16·2	8·0
43	7 25·8	7 27·0	7 05·4	4·3	2·1	10·3	5·1	16·3	8·0
44	7 26·0	7 27·2	7 05·7	4·4	2·2	10·4	5·1	16·4	8·1
45	7 26·3	7 27·5	7 05·9	4·5	2·2	10·5	5·2	16·5	8·1
46	7 26·5	7 27·7	7 06·2	4·6	2·3	10·6	5·2	16·6	8·2
47	7 26·8	7 28·0	7 06·4	4·7	2·3	10·7	5·3	16·7	8·2
48	7 27·0	7 28·2	7 06·6	4·8	2·4	10·8	5·3	16·8	8·3
49	7 27·3	7 28·5	7 06·9	4·9	2·4	10·9	5·4	16·9	8·3
50	7 27·5	7 28·7	7 07·1	5·0	2·5	11·0	5·4	17·0	8·4
51	7 27·8	7 29·0	7 07·4	5·1	2·5	11·1	5·5	17·1	8·4
52	7 28·0	7 29·2	7 07·6	5·2	2·6	11·2	5·5	17·2	8·5
53	7 28·3	7 29·5	7 07·8	5·3	2·6	11·3	5·6	17·3	8·5
54	7 28·5	7 29·7	7 08·1	5·4	2·7	11·4	5·6	17·4	8·6
55	7 28·8	7 30·0	7 08·3	5·5	2·7	11·5	5·7	17·5	8·6
56	7 29·0	7 30·2	7 08·5	5·6	2·8	11·6	5·7	17·6	8·7
57	7 29·3	7 30·5	7 08·8	5·7	2·8	11·7	5·8	17·7	8·7
58	7 29·5	7 30·7	7 09·0	5·8	2·9	11·8	5·8	17·8	8·8
59	7 29·8	7 31·0	7 09·3	5·9	2·9	11·9	5·9	17·9	8·8
60	7 30·0	7 31·2	7 09·5	6·0	3·0	12·0	5·9	18·0	8·9

30m	SUN PLANETS	ARIES	MOON	v or d	Corrⁿ	v or d	Corrⁿ	v or d	Corrⁿ
s	° ′	° ′	° ′	′	′	′	′	′	′
00	7 30·0	7 31·2	7 09·5	0·0	0·0	6·0	3·1	12·0	6·1
01	7 30·3	7 31·5	7 09·7	0·1	0·1	6·1	3·1	12·1	6·2
02	7 30·5	7 31·7	7 10·0	0·2	0·1	6·2	3·2	12·2	6·2
03	7 30·8	7 32·0	7 10·2	0·3	0·2	6·3	3·2	12·3	6·3
04	7 31·0	7 32·2	7 10·5	0·4	0·2	6·4	3·3	12·4	6·3
05	7 31·3	7 32·5	7 10·7	0·5	0·3	6·5	3·3	12·5	6·4
06	7 31·5	7 32·7	7 10·9	0·6	0·3	6·6	3·4	12·6	6·4
07	7 31·8	7 33·0	7 11·2	0·7	0·4	6·7	3·4	12·7	6·5
08	7 32·0	7 33·2	7 11·4	0·8	0·4	6·8	3·5	12·8	6·5
09	7 32·3	7 33·5	7 11·6	0·9	0·5	6·9	3·5	12·9	6·6
10	7 32·5	7 33·7	7 11·9	1·0	0·5	7·0	3·6	13·0	6·6
11	7 32·8	7 34·0	7 12·1	1·1	0·6	7·1	3·6	13·1	6·7
12	7 33·0	7 34·2	7 12·4	1·2	0·6	7·2	3·7	13·2	6·7
13	7 33·3	7 34·5	7 12·6	1·3	0·7	7·3	3·7	13·3	6·8
14	7 33·5	7 34·7	7 12·8	1·4	0·7	7·4	3·8	13·4	6·8
15	7 33·8	7 35·0	7 13·1	1·5	0·8	7·5	3·8	13·5	6·9
16	7 34·0	7 35·2	7 13·3	1·6	0·8	7·6	3·9	13·6	6·9
17	7 34·3	7 35·5	7 13·6	1·7	0·9	7·7	3·9	13·7	7·0
18	7 34·5	7 35·7	7 13·8	1·8	0·9	7·8	4·0	13·8	7·0
19	7 34·8	7 36·0	7 14·0	1·9	1·0	7·9	4·0	13·9	7·1
20	7 35·0	7 36·2	7 14·3	2·0	1·0	8·0	4·1	14·0	7·1
21	7 35·3	7 36·5	7 14·5	2·1	1·1	8·1	4·1	14·1	7·2
22	7 35·5	7 36·7	7 14·7	2·2	1·1	8·2	4·2	14·2	7·2
23	7 35·8	7 37·0	7 15·0	2·3	1·2	8·3	4·2	14·3	7·3
24	7 36·0	7 37·2	7 15·2	2·4	1·2	8·4	4·3	14·4	7·3
25	7 36·3	7 37·5	7 15·5	2·5	1·3	8·5	4·3	14·5	7·4
26	7 36·5	7 37·7	7 15·7	2·6	1·3	8·6	4·4	14·6	7·4
27	7 36·8	7 38·0	7 15·9	2·7	1·4	8·7	4·4	14·7	7·5
28	7 37·0	7 38·3	7 16·2	2·8	1·4	8·8	4·5	14·8	7·5
29	7 37·3	7 38·5	7 16·4	2·9	1·5	8·9	4·5	14·9	7·6
30	7 37·5	7 38·8	7 16·7	3·0	1·5	9·0	4·6	15·0	7·6
31	7 37·8	7 39·0	7 16·9	3·1	1·6	9·1	4·6	15·1	7·7
32	7 38·0	7 39·3	7 17·1	3·2	1·6	9·2	4·7	15·2	7·7
33	7 38·3	7 39·5	7 17·4	3·3	1·7	9·3	4·7	15·3	7·8
34	7 38·5	7 39·8	7 17·6	3·4	1·7	9·4	4·8	15·4	7·8
35	7 38·8	7 40·0	7 17·9	3·5	1·8	9·5	4·8	15·5	7·9
36	7 39·0	7 40·3	7 18·1	3·6	1·8	9·6	4·9	15·6	7·9
37	7 39·3	7 40·5	7 18·3	3·7	1·9	9·7	4·9	15·7	8·0
38	7 39·5	7 40·8	7 18·6	3·8	1·9	9·8	5·0	15·8	8·0
39	7 39·8	7 41·0	7 18·8	3·9	2·0	9·9	5·0	15·9	8·1
40	7 40·0	7 41·3	7 19·0	4·0	2·0	10·0	5·1	16·0	8·1
41	7 40·3	7 41·5	7 19·3	4·1	2·1	10·1	5·1	16·1	8·2
42	7 40·5	7 41·8	7 19·5	4·2	2·1	10·2	5·2	16·2	8·2
43	7 40·8	7 42·0	7 19·8	4·3	2·2	10·3	5·2	16·3	8·3
44	7 41·0	7 42·3	7 20·0	4·4	2·2	10·4	5·3	16·4	8·3
45	7 41·3	7 42·5	7 20·2	4·5	2·3	10·5	5·3	16·5	8·4
46	7 41·5	7 42·8	7 20·5	4·6	2·3	10·6	5·4	16·6	8·4
47	7 41·8	7 43·0	7 20·7	4·7	2·4	10·7	5·4	16·7	8·5
48	7 42·0	7 43·3	7 21·0	4·8	2·4	10·8	5·5	16·8	8·5
49	7 42·3	7 43·5	7 21·2	4·9	2·5	10·9	5·5	16·9	8·6
50	7 42·5	7 43·8	7 21·4	5·0	2·5	11·0	5·6	17·0	8·6
51	7 42·8	7 44·0	7 21·7	5·1	2·6	11·1	5·6	17·1	8·7
52	7 43·0	7 44·3	7 21·9	5·2	2·6	11·2	5·7	17·2	8·7
53	7 43·3	7 44·5	7 22·1	5·3	2·7	11·3	5·7	17·3	8·8
54	7 43·5	7 44·8	7 22·4	5·4	2·7	11·4	5·8	17·4	8·8
55	7 43·8	7 45·0	7 22·6	5·5	2·8	11·5	5·8	17·5	8·9
56	7 44·0	7 45·3	7 22·9	5·6	2·8	11·6	5·9	17·6	8·9
57	7 44·3	7 45·5	7 23·1	5·7	2·9	11·7	5·9	17·7	9·0
58	7 44·5	7 45·8	7 23·3	5·8	2·9	11·8	6·0	17·8	9·0
59	7 44·8	7 46·0	7 23·6	5·9	3·0	11·9	6·0	17·9	9·1
60	7 45·0	7 46·3	7 23·8	6·0	3·1	12·0	6·1	18·0	9·2

31m	SUN PLANETS	ARIES	MOON	v or d	Corrⁿ	v or d	Corrⁿ	v or d	Corrⁿ
s	° ′	° ′	° ′	′	′	′	′	′	′
00	7 45·0	7 46·3	7 23·8	0·0	0·0	6·0	3·2	12·0	6·3
01	7 45·3	7 46·5	7 24·1	0·1	0·1	6·1	3·2	12·1	6·4
02	7 45·5	7 46·8	7 24·3	0·2	0·1	6·2	3·3	12·2	6·4
03	7 45·8	7 47·0	7 24·5	0·3	0·2	6·3	3·3	12·3	6·5
04	7 46·0	7 47·3	7 24·8	0·4	0·2	6·4	3·4	12·4	6·5
05	7 46·3	7 47·5	7 25·0	0·5	0·3	6·5	3·4	12·5	6·6
06	7 46·5	7 47·8	7 25·2	0·6	0·3	6·6	3·5	12·6	6·6
07	7 46·8	7 48·0	7 25·5	0·7	0·4	6·7	3·5	12·7	6·7
08	7 47·0	7 48·3	7 25·7	0·8	0·4	6·8	3·6	12·8	6·7
09	7 47·3	7 48·5	7 26·0	0·9	0·5	6·9	3·6	12·9	6·8
10	7 47·5	7 48·8	7 26·2	1·0	0·5	7·0	3·7	13·0	6·8
11	7 47·8	7 49·0	7 26·4	1·1	0·6	7·1	3·7	13·1	6·9
12	7 48·0	7 49·3	7 26·7	1·2	0·6	7·2	3·8	13·2	6·9
13	7 48·3	7 49·5	7 26·9	1·3	0·7	7·3	3·8	13·3	7·0
14	7 48·5	7 49·8	7 27·2	1·4	0·7	7·4	3·9	13·4	7·0
15	7 48·8	7 50·0	7 27·4	1·5	0·8	7·5	3·9	13·5	7·1
16	7 49·0	7 50·3	7 27·6	1·6	0·8	7·6	4·0	13·6	7·1
17	7 49·3	7 50·5	7 27·9	1·7	0·9	7·7	4·0	13·7	7·2
18	7 49·5	7 50·8	7 28·1	1·8	0·9	7·8	4·1	13·8	7·2
19	7 49·8	7 51·0	7 28·4	1·9	1·0	7·9	4·1	13·9	7·3
20	7 50·0	7 51·3	7 28·6	2·0	1·1	8·0	4·2	14·0	7·4
21	7 50·3	7 51·5	7 28·8	2·1	1·1	8·1	4·3	14·1	7·4
22	7 50·5	7 51·8	7 29·1	2·2	1·2	8·2	4·3	14·2	7·5
23	7 50·8	7 52·0	7 29·3	2·3	1·2	8·3	4·4	14·3	7·5
24	7 51·0	7 52·3	7 29·5	2·4	1·3	8·4	4·4	14·4	7·6
25	7 51·3	7 52·5	7 29·8	2·5	1·3	8·5	4·5	14·5	7·6
26	7 51·5	7 52·8	7 30·0	2·6	1·4	8·6	4·5	14·6	7·7
27	7 51·8	7 53·0	7 30·3	2·7	1·4	8·7	4·6	14·7	7·7
28	7 52·0	7 53·3	7 30·5	2·8	1·5	8·8	4·6	14·8	7·8
29	7 52·3	7 53·5	7 30·7	2·9	1·5	8·9	4·7	14·9	7·8
30	7 52·5	7 53·8	7 31·0	3·0	1·6	9·0	4·7	15·0	7·9
31	7 52·8	7 54·0	7 31·2	3·1	1·6	9·1	4·8	15·1	7·9
32	7 53·0	7 54·3	7 31·5	3·2	1·7	9·2	4·8	15·2	8·0
33	7 53·3	7 54·5	7 31·7	3·3	1·7	9·3	4·9	15·3	8·0
34	7 53·5	7 54·8	7 31·9	3·4	1·8	9·4	4·9	15·4	8·1
35	7 53·8	7 55·0	7 32·2	3·5	1·8	9·5	5·0	15·5	8·1
36	7 54·0	7 55·3	7 32·4	3·6	1·9	9·6	5·0	15·6	8·2
37	7 54·3	7 55·5	7 32·6	3·7	1·9	9·7	5·1	15·7	8·2
38	7 54·5	7 55·8	7 32·9	3·8	2·0	9·8	5·1	15·8	8·3
39	7 54·8	7 56·0	7 33·1	3·9	2·0	9·9	5·2	15·9	8·3
40	7 55·0	7 56·3	7 33·4	4·0	2·1	10·0	5·3	16·0	8·4
41	7 55·3	7 56·6	7 33·6	4·1	2·2	10·1	5·3	16·1	8·5
42	7 55·5	7 56·8	7 33·8	4·2	2·2	10·2	5·4	16·2	8·5
43	7 55·8	7 57·1	7 34·1	4·3	2·3	10·3	5·4	16·3	8·6
44	7 56·0	7 57·3	7 34·3	4·4	2·3	10·4	5·5	16·4	8·6
45	7 56·3	7 57·6	7 34·6	4·5	2·4	10·5	5·5	16·5	8·7
46	7 56·5	7 57·8	7 34·8	4·6	2·4	10·6	5·6	16·6	8·7
47	7 56·8	7 58·1	7 35·0	4·7	2·5	10·7	5·6	16·7	8·8
48	7 57·0	7 58·3	7 35·3	4·8	2·5	10·8	5·7	16·8	8·8
49	7 57·3	7 58·6	7 35·5	4·9	2·6	10·9	5·7	16·9	8·9
50	7 57·5	7 58·8	7 35·7	5·0	2·6	11·0	5·8	17·0	8·9
51	7 57·8	7 59·1	7 36·0	5·1	2·7	11·1	5·8	17·1	9·0
52	7 58·0	7 59·3	7 36·2	5·2	2·7	11·2	5·9	17·2	9·0
53	7 58·3	7 59·6	7 36·5	5·3	2·8	11·3	5·9	17·3	9·1
54	7 58·5	7 59·8	7 36·7	5·4	2·8	11·4	6·0	17·4	9·1
55	7 58·8	8 00·1	7 36·9	5·5	2·9	11·5	6·0	17·5	9·2
56	7 59·0	8 00·3	7 37·2	5·6	2·9	11·6	6·1	17·6	9·2
57	7 59·3	8 00·6	7 37·4	5·7	3·0	11·7	6·1	17·7	9·3
58	7 59·5	8 00·8	7 37·7	5·8	3·0	11·8	6·2	17·8	9·3
59	7 59·8	8 01·1	7 37·9	5·9	3·1	11·9	6·2	17·9	9·4
60	8 00·0	8 01·3	7 38·1	6·0	3·2	12·0	6·3	18·0	9·5

32m s	SUN PLANETS ° ′	ARIES ° ′	MOON ° ′	v or d ′	Corrn ′	v or d ′	Corrn ′	v or d ′	Corrn ′
00	8 00·0	8 01·3	7 38·1	0·0	0·0	6·0	3·3	12·0	6·5
01	8 00·3	8 01·6	7 38·4	0·1	0·1	6·1	3·3	12·1	6·6
02	8 00·5	8 01·8	7 38·6	0·2	0·1	6·2	3·4	12·2	6·6
03	8 00·8	8 02·1	7 38·8	0·3	0·2	6·3	3·4	12·3	6·7
04	8 01·0	8 02·3	7 39·1	0·4	0·2	6·4	3·5	12·4	6·7
05	8 01·3	8 02·6	7 39·3	0·5	0·3	6·5	3·5	12·5	6·8
06	8 01·5	8 02·8	7 39·6	0·6	0·3	6·6	3·6	12·6	6·8
07	8 01·8	8 03·1	7 39·8	0·7	0·4	6·7	3·6	12·7	6·9
08	8 02·0	8 03·3	7 40·0	0·8	0·4	6·8	3·7	12·8	6·9
09	8 02·3	8 03·6	7 40·3	0·9	0·5	6·9	3·7	12·9	7·0
10	8 02·5	8 03·8	7 40·5	1·0	0·5	7·0	3·8	13·0	7·0
11	8 02·8	8 04·1	7 40·8	1·1	0·6	7·1	3·8	13·1	7·1
12	8 03·0	8 04·3	7 41·0	1·2	0·7	7·2	3·9	13·2	7·2
13	8 03·3	8 04·6	7 41·2	1·3	0·7	7·3	4·0	13·3	7·2
14	8 03·5	8 04·8	7 41·5	1·4	0·8	7·4	4·0	13·4	7·3
15	8 03·8	8 05·1	7 41·7	1·5	0·8	7·5	4·1	13·5	7·3
16	8 04·0	8 05·3	7 42·0	1·6	0·9	7·6	4·1	13·6	7·4
17	8 04·3	8 05·6	7 42·2	1·7	0·9	7·7	4·2	13·7	7·4
18	8 04·5	8 05·8	7 42·4	1·8	1·0	7·8	4·2	13·8	7·5
19	8 04·8	8 06·1	7 42·7	1·9	1·0	7·9	4·3	13·9	7·5
20	8 05·0	8 06·3	7 42·9	2·0	1·1	8·0	4·3	14·0	7·6
21	8 05·3	8 06·6	7 43·1	2·1	1·1	8·1	4·4	14·1	7·6
22	8 05·5	8 06·8	7 43·4	2·2	1·2	8·2	4·4	14·2	7·7
23	8 05·8	8 07·1	7 43·6	2·3	1·2	8·3	4·5	14·3	7·7
24	8 06·0	8 07·3	7 43·9	2·4	1·3	8·4	4·6	14·4	7·8
25	8 06·3	8 07·6	7 44·1	2·5	1·4	8·5	4·6	14·5	7·9
26	8 06·5	8 07·8	7 44·3	2·6	1·4	8·6	4·7	14·6	7·9
27	8 06·8	8 08·1	7 44·6	2·7	1·5	8·7	4·7	14·7	8·0
28	8 07·0	8 08·3	7 44·8	2·8	1·5	8·8	4·8	14·8	8·0
29	8 07·3	8 08·6	7 45·1	2·9	1·6	8·9	4·8	14·9	8·1
30	8 07·5	8 08·8	7 45·3	3·0	1·6	9·0	4·9	15·0	8·1
31	8 07·8	8 09·1	7 45·5	3·1	1·7	9·1	4·9	15·1	8·2
32	8 08·0	8 09·3	7 45·8	3·2	1·7	9·2	5·0	15·2	8·2
33	8 08·3	8 09·6	7 46·0	3·3	1·8	9·3	5·0	15·3	8·3
34	8 08·5	8 09·8	7 46·2	3·4	1·8	9·4	5·1	15·4	8·3
35	8 08·8	8 10·1	7 46·5	3·5	1·9	9·5	5·1	15·5	8·4
36	8 09·0	8 10·3	7 46·7	3·6	2·0	9·6	5·2	15·6	8·5
37	8 09·3	8 10·6	7 47·0	3·7	2·0	9·7	5·3	15·7	8·5
38	8 09·5	8 10·8	7 47·2	3·8	2·1	9·8	5·3	15·8	8·6
39	8 09·8	8 11·1	7 47·4	3·9	2·1	9·9	5·4	15·9	8·6
40	8 10·0	8 11·3	7 47·7	4·0	2·2	10·0	5·4	16·0	8·7
41	8 10·3	8 11·6	7 47·9	4·1	2·2	10·1	5·5	16·1	8·7
42	8 10·5	8 11·8	7 48·2	4·2	2·3	10·2	5·5	16·2	8·8
43	8 10·8	8 12·1	7 48·4	4·3	2·3	10·3	5·6	16·3	8·8
44	8 11·0	8 12·3	7 48·6	4·4	2·4	10·4	5·6	16·4	8·9
45	8 11·3	8 12·6	7 48·9	4·5	2·4	10·5	5·7	16·5	8·9
46	8 11·5	8 12·8	7 49·1	4·6	2·5	10·6	5·7	16·6	9·0
47	8 11·8	8 13·1	7 49·3	4·7	2·5	10·7	5·8	16·7	9·0
48	8 12·0	8 13·3	7 49·6	4·8	2·6	10·8	5·9	16·8	9·1
49	8 12·3	8 13·6	7 49·8	4·9	2·7	10·9	5·9	16·9	9·2
50	8 12·5	8 13·8	7 50·1	5·0	2·7	11·0	6·0	17·0	9·2
51	8 12·8	8 14·1	7 50·3	5·1	2·8	11·1	6·0	17·1	9·3
52	8 13·0	8 14·3	7 50·5	5·2	2·8	11·2	6·1	17·2	9·3
53	8 13·3	8 14·6	7 50·8	5·3	2·9	11·3	6·1	17·3	9·4
54	8 13·5	8 14·9	7 51·0	5·4	2·9	11·4	6·2	17·4	9·4
55	8 13·8	8 15·1	7 51·3	5·5	3·0	11·5	6·2	17·5	9·5
56	8 14·0	8 15·4	7 51·5	5·6	3·0	11·6	6·3	17·6	9·5
57	8 14·3	8 15·6	7 51·7	5·7	3·1	11·7	6·3	17·7	9·6
58	8 14·5	8 15·9	7 52·0	5·8	3·1	11·8	6·4	17·8	9·6
59	8 14·8	8 16·1	7 52·2	5·9	3·2	11·9	6·4	17·9	9·7
60	8 15·0	8 16·4	7 52·5	6·0	3·3	12·0	6·5	18·0	9·8

33m s	SUN PLANETS ° ′	ARIES ° ′	MOON ° ′	v or d ′	Corrn ′	v or d ′	Corrn ′	v or d ′	Corrn ′
00	8 15·0	8 16·4	7 52·5	0·0	0·0	6·0	3·4	12·0	6·7
01	8 15·3	8 16·6	7 52·7	0·1	0·1	6·1	3·4	12·1	6·8
02	8 15·5	8 16·9	7 52·9	0·2	0·1	6·2	3·5	12·2	6·8
03	8 15·8	8 17·1	7 53·2	0·3	0·2	6·3	3·5	12·3	6·9
04	8 16·0	8 17·4	7 53·4	0·4	0·2	6·4	3·6	12·4	6·9
05	8 16·3	8 17·6	7 53·6	0·5	0·3	6·5	3·6	12·5	7·0
06	8 16·5	8 17·9	7 53·9	0·6	0·3	6·6	3·7	12·6	7·0
07	8 16·8	8 18·1	7 54·1	0·7	0·4	6·7	3·7	12·7	7·1
08	8 17·0	8 18·4	7 54·4	0·8	0·4	6·8	3·8	12·8	7·1
09	8 17·3	8 18·6	7 54·6	0·9	0·5	6·9	3·9	12·9	7·2
10	8 17·5	8 18·9	7 54·8	1·0	0·6	7·0	3·9	13·0	7·3
11	8 17·8	8 19·1	7 55·1	1·1	0·6	7·1	4·0	13·1	7·3
12	8 18·0	8 19·4	7 55·3	1·2	0·7	7·2	4·0	13·2	7·4
13	8 18·3	8 19·6	7 55·6	1·3	0·7	7·3	4·1	13·3	7·4
14	8 18·5	8 19·9	7 55·8	1·4	0·8	7·4	4·1	13·4	7·5
15	8 18·8	8 20·1	7 56·0	1·5	0·8	7·5	4·2	13·5	7·5
16	8 19·0	8 20·4	7 56·3	1·6	0·9	7·6	4·2	13·6	7·6
17	8 19·3	8 20·6	7 56·5	1·7	0·9	7·7	4·3	13·7	7·6
18	8 19·5	8 20·9	7 56·7	1·8	1·0	7·8	4·4	13·8	7·7
19	8 19·8	8 21·1	7 57·0	1·9	1·1	7·9	4·4	13·9	7·8
20	8 20·0	8 21·4	7 57·2	2·0	1·1	8·0	4·5	14·0	7·8
21	8 20·3	8 21·6	7 57·5	2·1	1·2	8·1	4·5	14·1	7·9
22	8 20·5	8 21·9	7 57·7	2·2	1·2	8·2	4·6	14·2	7·9
23	8 20·8	8 22·1	7 57·9	2·3	1·3	8·3	4·6	14·3	8·0
24	8 21·0	8 22·4	7 58·2	2·4	1·3	8·4	4·7	14·4	8·0
25	8 21·3	8 22·6	7 58·4	2·5	1·4	8·5	4·7	14·5	8·1
26	8 21·5	8 22·9	7 58·7	2·6	1·5	8·6	4·8	14·6	8·2
27	8 21·8	8 23·1	7 58·9	2·7	1·5	8·7	4·9	14·7	8·2
28	8 22·0	8 23·4	7 59·1	2·8	1·6	8·8	4·9	14·8	8·3
29	8 22·3	8 23·6	7 59·4	2·9	1·6	8·9	5·0	14·9	8·3
30	8 22·5	8 23·9	7 59·6	3·0	1·7	9·0	5·0	15·0	8·4
31	8 22·8	8 24·1	7 59·8	3·1	1·7	9·1	5·1	15·1	8·4
32	8 23·0	8 24·4	8 00·1	3·2	1·8	9·2	5·1	15·2	8·5
33	8 23·3	8 24·6	8 00·3	3·3	1·8	9·3	5·2	15·3	8·5
34	8 23·5	8 24·9	8 00·6	3·4	1·9	9·4	5·2	15·4	8·6
35	8 23·8	8 25·1	8 00·8	3·5	2·0	9·5	5·3	15·5	8·7
36	8 24·0	8 25·4	8 01·0	3·6	2·0	9·6	5·4	15·6	8·7
37	8 24·3	8 25·6	8 01·3	3·7	2·1	9·7	5·4	15·7	8·8
38	8 24·5	8 25·9	8 01·5	3·8	2·1	9·8	5·5	15·8	8·8
39	8 24·8	8 26·1	8 01·8	3·9	2·2	9·9	5·5	15·9	8·9
40	8 25·0	8 26·4	8 02·0	4·0	2·2	10·0	5·6	16·0	8·9
41	8 25·3	8 26·6	8 02·2	4·1	2·3	10·1	5·6	16·1	9·0
42	8 25·5	8 26·9	8 02·5	4·2	2·3	10·2	5·7	16·2	9·0
43	8 25·8	8 27·1	8 02·7	4·3	2·4	10·3	5·8	16·3	9·1
44	8 26·0	8 27·4	8 02·9	4·4	2·5	10·4	5·8	16·4	9·2
45	8 26·3	8 27·6	8 03·2	4·5	2·5	10·5	5·9	16·5	9·2
46	8 26·5	8 27·9	8 03·4	4·6	2·6	10·6	5·9	16·6	9·3
47	8 26·8	8 28·1	8 03·7	4·7	2·6	10·7	6·0	16·7	9·3
48	8 27·0	8 28·4	8 03·9	4·8	2·7	10·8	6·0	16·8	9·4
49	8 27·3	8 28·6	8 04·1	4·9	2·7	10·9	6·1	16·9	9·4
50	8 27·5	8 28·9	8 04·4	5·0	2·8	11·0	6·1	17·0	9·5
51	8 27·8	8 29·1	8 04·6	5·1	2·8	11·1	6·2	17·1	9·5
52	8 28·0	8 29·4	8 04·9	5·2	2·9	11·2	6·3	17·2	9·6
53	8 28·3	8 29·6	8 05·1	5·3	3·0	11·3	6·3	17·3	9·7
54	8 28·5	8 29·9	8 05·3	5·4	3·0	11·4	6·4	17·4	9·7
55	8 28·8	8 30·1	8 05·6	5·5	3·1	11·5	6·4	17·5	9·8
56	8 29·0	8 30·4	8 05·8	5·6	3·1	11·6	6·5	17·6	9·8
57	8 29·3	8 30·6	8 06·1	5·7	3·2	11·7	6·5	17·7	9·9
58	8 29·5	8 30·9	8 06·3	5·8	3·2	11·8	6·6	17·8	9·9
59	8 29·8	8 31·1	8 06·5	5·9	3·3	11·9	6·6	17·9	10·0
60	8 30·0	8 31·4	8 06·8	6·0	3·4	12·0	6·7	18·0	10·1

34^m	SUN PLANETS	ARIES	MOON	v or d	Corr^n	v or d	Corr^n	v or d	Corr^n
s	° ′	° ′	° ′	′	′	′	′	′	′
00	8 30·0	8 31·4	8 06·8	0·0	0·0	6·0	3·5	12·0	6·9
01	8 30·3	8 31·6	8 07·0	0·1	0·1	6·1	3·5	12·1	7·0
02	8 30·5	8 31·9	8 07·2	0·2	0·1	6·2	3·6	12·2	7·0
03	8 30·8	8 32·1	8 07·5	0·3	0·2	6·3	3·6	12·3	7·1
04	8 31·0	8 32·4	8 07·7	0·4	0·2	6·4	3·7	12·4	7·1
05	8 31·3	8 32·6	8 08·0	0·5	0·3	6·5	3·7	12·5	7·2
06	8 31·5	8 32·9	8 08·2	0·6	0·3	6·6	3·8	12·6	7·2
07	8 31·8	8 33·2	8 08·4	0·7	0·4	6·7	3·9	12·7	7·3
08	8 32·0	8 33·4	8 08·7	0·8	0·5	6·8	3·9	12·8	7·4
09	8 32·3	8 33·7	8 08·9	0·9	0·5	6·9	4·0	12·9	7·4
10	8 32·5	8 33·9	8 09·2	1·0	0·6	7·0	4·0	13·0	7·5
11	8 32·8	8 34·2	8 09·4	1·1	0·6	7·1	4·1	13·1	7·5
12	8 33·0	8 34·4	8 09·6	1·2	0·7	7·2	4·1	13·2	7·6
13	8 33·3	8 34·7	8 09·9	1·3	0·7	7·3	4·2	13·3	7·6
14	8 33·5	8 34·9	8 10·1	1·4	0·8	7·4	4·3	13·4	7·7
15	8 33·8	8 35·2	8 10·3	1·5	0·9	7·5	4·3	13·5	7·8
16	8 34·0	8 35·4	8 10·6	1·6	0·9	7·6	4·4	13·6	7·8
17	8 34·3	8 35·7	8 10·8	1·7	1·0	7·7	4·4	13·7	7·9
18	8 34·5	8 35·9	8 11·1	1·8	1·0	7·8	4·5	13·8	7·9
19	8 34·8	8 36·2	8 11·3	1·9	1·1	7·9	4·5	13·9	8·0
20	8 35·0	8 36·4	8 11·5	2·0	1·2	8·0	4·6	14·0	8·1
21	8 35·3	8 36·7	8 11·8	2·1	1·2	8·1	4·7	14·1	8·1
22	8 35·5	8 36·9	8 12·0	2·2	1·3	8·2	4·7	14·2	8·2
23	8 35·8	8 37·2	8 12·3	2·3	1·3	8·3	4·8	14·3	8·2
24	8 36·0	8 37·4	8 12·5	2·4	1·4	8·4	4·8	14·4	8·3
25	8 36·3	8 37·7	8 12·7	2·5	1·4	8·5	4·9	14·5	8·3
26	8 36·5	8 37·9	8 13·0	2·6	1·5	8·6	4·9	14·6	8·4
27	8 36·8	8 38·2	8 13·2	2·7	1·6	8·7	5·0	14·7	8·5
28	8 37·0	8 38·4	8 13·4	2·8	1·6	8·8	5·1	14·8	8·5
29	8 37·3	8 38·7	8 13·7	2·9	1·7	8·9	5·1	14·9	8·6
30	8 37·5	8 38·9	8 13·9	3·0	1·7	9·0	5·2	15·0	8·6
31	8 37·8	8 39·2	8 14·2	3·1	1·8	9·1	5·2	15·1	8·7
32	8 38·0	8 39·4	8 14·4	3·2	1·8	9·2	5·3	15·2	8·7
33	8 38·3	8 39·7	8 14·6	3·3	1·9	9·3	5·3	15·3	8·8
34	8 38·5	8 39·9	8 14·9	3·4	2·0	9·4	5·4	15·4	8·9
35	8 38·8	8 40·2	8 15·1	3·5	2·0	9·5	5·5	15·5	8·9
36	8 39·0	8 40·4	8 15·4	3·6	2·1	9·6	5·5	15·6	9·0
37	8 39·3	8 40·7	8 15·6	3·7	2·1	9·7	5·6	15·7	9·0
38	8 39·5	8 40·9	8 15·8	3·8	2·2	9·8	5·6	15·8	9·1
39	8 39·8	8 41·2	8 16·1	3·9	2·2	9·9	5·7	15·9	9·1
40	8 40·0	8 41·4	8 16·3	4·0	2·3	10·0	5·8	16·0	9·2
41	8 40·3	8 41·7	8 16·5	4·1	2·4	10·1	5·8	16·1	9·3
42	8 40·5	8 41·9	8 16·8	4·2	2·4	10·2	5·9	16·2	9·3
43	8 40·8	8 42·2	8 17·0	4·3	2·5	10·3	5·9	16·3	9·4
44	8 41·0	8 42·4	8 17·3	4·4	2·5	10·4	6·0	16·4	9·4
45	8 41·3	8 42·7	8 17·5	4·5	2·6	10·5	6·0	16·5	9·5
46	8 41·5	8 42·9	8 17·7	4·6	2·6	10·6	6·1	16·6	9·5
47	8 41·8	8 43·2	8 18·0	4·7	2·7	10·7	6·2	16·7	9·6
48	8 42·0	8 43·4	8 18·2	4·8	2·8	10·8	6·2	16·8	9·7
49	8 42·3	8 43·7	8 18·5	4·9	2·8	10·9	6·3	16·9	9·7
50	8 42·5	8 43·9	8 18·7	5·0	2·9	11·0	6·3	17·0	9·8
51	8 42·8	8 44·2	8 18·9	5·1	2·9	11·1	6·4	17·1	9·8
52	8 43·0	8 44·4	8 19·2	5·2	3·0	11·2	6·4	17·2	9·9
53	8 43·3	8 44·7	8 19·4	5·3	3·0	11·3	6·5	17·3	9·9
54	8 43·5	8 44·9	8 19·7	5·4	3·1	11·4	6·6	17·4	10·0
55	8 43·8	8 45·2	8 19·9	5·5	3·2	11·5	6·6	17·5	10·1
56	8 44·0	8 45·4	8 20·1	5·6	3·2	11·6	6·7	17·6	10·1
57	8 44·3	8 45·7	8 20·4	5·7	3·3	11·7	6·7	17·7	10·2
58	8 44·5	8 45·9	8 20·6	5·8	3·3	11·8	6·8	17·8	10·2
59	8 44·8	8 46·2	8 20·8	5·9	3·4	11·9	6·8	17·9	10·3
60	8 45·0	8 46·4	8 21·1	6·0	3·5	12·0	6·9	18·0	10·4

35^m	SUN PLANETS	ARIES	MOON	v or d	Corr^n	v or d	Corr^n	v or d	Corr^n
s	° ′	° ′	° ′	′	′	′	′	′	′
00	8 45·0	8 46·4	8 21·1	0·0	0·0	6·0	3·6	12·0	7·1
01	8 45·3	8 46·7	8 21·3	0·1	0·1	6·1	3·6	12·1	7·2
02	8 45·5	8 46·9	8 21·6	0·2	0·1	6·2	3·7	12·2	7·2
03	8 45·8	8 47·2	8 21·8	0·3	0·2	6·3	3·7	12·3	7·3
04	8 46·0	8 47·4	8 22·0	0·4	0·2	6·4	3·8	12·4	7·3
05	8 46·3	8 47·7	8 22·3	0·5	0·3	6·5	3·8	12·5	7·4
06	8 46·5	8 47·9	8 22·5	0·6	0·4	6·6	3·9	12·6	7·5
07	8 46·8	8 48·2	8 22·8	0·7	0·4	6·7	4·0	12·7	7·5
08	8 47·0	8 48·4	8 23·0	0·8	0·5	6·8	4·0	12·8	7·6
09	8 47·3	8 48·7	8 23·2	0·9	0·5	6·9	4·1	12·9	7·6
10	8 47·5	8 48·9	8 23·5	1·0	0·6	7·0	4·1	13·0	7·7
11	8 47·8	8 49·2	8 23·7	1·1	0·7	7·1	4·2	13·1	7·8
12	8 48·0	8 49·4	8 23·9	1·2	0·7	7·2	4·3	13·2	7·8
13	8 48·3	8 49·7	8 24·2	1·3	0·8	7·3	4·3	13·3	7·9
14	8 48·5	8 49·9	8 24·4	1·4	0·8	7·4	4·4	13·4	7·9
15	8 48·8	8 50·2	8 24·7	1·5	0·9	7·5	4·4	13·5	8·0
16	8 49·0	8 50·4	8 24·9	1·6	0·9	7·6	4·5	13·6	8·0
17	8 49·3	8 50·7	8 25·1	1·7	1·0	7·7	4·6	13·7	8·1
18	8 49·5	8 50·9	8 25·4	1·8	1·1	7·8	4·6	13·8	8·2
19	8 49·8	8 51·2	8 25·6	1·9	1·1	7·9	4·7	13·9	8·2
20	8 50·0	8 51·5	8 25·9	2·0	1·2	8·0	4·7	14·0	8·3
21	8 50·3	8 51·7	8 26·1	2·1	1·2	8·1	4·8	14·1	8·3
22	8 50·5	8 52·0	8 26·3	2·2	1·3	8·2	4·9	14·2	8·4
23	8 50·8	8 52·2	8 26·6	2·3	1·4	8·3	4·9	14·3	8·5
24	8 51·0	8 52·5	8 26·8	2·4	1·4	8·4	5·0	14·4	8·5
25	8 51·3	8 52·7	8 27·0	2·5	1·5	8·5	5·0	14·5	8·6
26	8 51·5	8 53·0	8 27·3	2·6	1·5	8·6	5·1	14·6	8·6
27	8 51·8	8 53·2	8 27·5	2·7	1·6	8·7	5·1	14·7	8·7
28	8 52·0	8 53·5	8 27·8	2·8	1·7	8·8	5·2	14·8	8·8
29	8 52·3	8 53·7	8 28·0	2·9	1·7	8·9	5·3	14·9	8·8
30	8 52·5	8 54·0	8 28·2	3·0	1·8	9·0	5·3	15·0	8·9
31	8 52·8	8 54·2	8 28·5	3·1	1·8	9·1	5·4	15·1	8·9
32	8 53·0	8 54·5	8 28·7	3·2	1·9	9·2	5·4	15·2	9·0
33	8 53·3	8 54·7	8 29·0	3·3	2·0	9·3	5·5	15·3	9·1
34	8 53·5	8 55·0	8 29·2	3·4	2·0	9·4	5·6	15·4	9·1
35	8 53·8	8 55·2	8 29·4	3·5	2·1	9·5	5·6	15·5	9·2
36	8 54·0	8 55·5	8 29·7	3·6	2·1	9·6	5·7	15·6	9·2
37	8 54·3	8 55·7	8 29·9	3·7	2·2	9·7	5·7	15·7	9·3
38	8 54·5	8 56·0	8 30·2	3·8	2·2	9·8	5·8	15·8	9·3
39	8 54·8	8 56·2	8 30·4	3·9	2·3	9·9	5·9	15·9	9·4
40	8 55·0	8 56·5	8 30·6	4·0	2·4	10·0	5·9	16·0	9·5
41	8 55·3	8 56·7	8 30·9	4·1	2·4	10·1	6·0	16·1	9·5
42	8 55·5	8 57·0	8 31·1	4·2	2·5	10·2	6·0	16·2	9·6
43	8 55·8	8 57·2	8 31·3	4·3	2·5	10·3	6·1	16·3	9·6
44	8 56·0	8 57·5	8 31·6	4·4	2·6	10·4	6·2	16·4	9·7
45	8 56·3	8 57·7	8 31·8	4·5	2·7	10·5	6·2	16·5	9·8
46	8 56·5	8 58·0	8 32·1	4·6	2·7	10·6	6·3	16·6	9·8
47	8 56·8	8 58·2	8 32·3	4·7	2·8	10·7	6·3	16·7	9·9
48	8 57·0	8 58·5	8 32·5	4·8	2·8	10·8	6·4	16·8	9·9
49	8 57·3	8 58·7	8 32·8	4·9	2·9	10·9	6·4	16·9	10·0
50	8 57·5	8 59·0	8 33·0	5·0	3·0	11·0	6·5	17·0	10·1
51	8 57·8	8 59·2	8 33·3	5·1	3·0	11·1	6·6	17·1	10·1
52	8 58·0	8 59·5	8 33·5	5·2	3·1	11·2	6·6	17·2	10·2
53	8 58·3	8 59·7	8 33·7	5·3	3·1	11·3	6·7	17·3	10·2
54	8 58·5	9 00·0	8 34·0	5·4	3·2	11·4	6·7	17·4	10·3
55	8 58·8	9 00·2	8 34·2	5·5	3·3	11·5	6·8	17·5	10·4
56	8 59·0	9 00·5	8 34·4	5·6	3·3	11·6	6·9	17·6	10·4
57	8 59·3	9 00·7	8 34·7	5·7	3·4	11·7	6·9	17·7	10·5
58	8 59·5	9 01·0	8 34·9	5·8	3·4	11·8	7·0	17·8	10·5
59	8 59·8	9 01·2	8 35·2	5·9	3·5	11·9	7·0	17·9	10·6
60	9 00·0	9 01·5	8 35·4	6·0	3·6	12·0	7·1	18·0	10·7

36m s	SUN PLANETS ° ′	ARIES ° ′	MOON ° ′	v or d ′	Corrⁿ ′	v or d ′	Corrⁿ ′	v or d ′	Corrⁿ ′
00	9 00·0	9 01·5	8 35·4	0·0	0·0	6·0	3·7	12·0	7·3
01	9 00·3	9 01·7	8 35·6	0·1	0·1	6·1	3·7	12·1	7·4
02	9 00·5	9 02·0	8 35·9	0·2	0·1	6·2	3·8	12·2	7·4
03	9 00·8	9 02·2	8 36·1	0·3	0·2	6·3	3·8	12·3	7·5
04	9 01·0	9 02·5	8 36·4	0·4	0·2	6·4	3·9	12·4	7·5
05	9 01·3	9 02·7	8 36·6	0·5	0·3	6·5	4·0	12·5	7·6
06	9 01·5	9 03·0	8 36·8	0·6	0·4	6·6	4·0	12·6	7·7
07	9 01·8	9 03·2	8 37·1	0·7	0·4	6·7	4·1	12·7	7·7
08	9 02·0	9 03·5	8 37·3	0·8	0·5	6·8	4·1	12·8	7·8
09	9 02·3	9 03·7	8 37·5	0·9	0·5	6·9	4·2	12·9	7·8
10	9 02·5	9 04·0	8 37·8	1·0	0·6	7·0	4·3	13·0	7·9
11	9 02·8	9 04·2	8 38·0	1·1	0·7	7·1	4·3	13·1	8·0
12	9 03·0	9 04·5	8 38·3	1·2	0·7	7·2	4·4	13·2	8·0
13	9 03·3	9 04·7	8 38·5	1·3	0·8	7·3	4·4	13·3	8·1
14	9 03·5	9 05·0	8 38·7	1·4	0·9	7·4	4·5	13·4	8·2
15	9 03·8	9 05·2	8 39·0	1·5	0·9	7·5	4·6	13·5	8·2
16	9 04·0	9 05·5	8 39·2	1·6	1·0	7·6	4·6	13·6	8·3
17	9 04·3	9 05·7	8 39·5	1·7	1·0	7·7	4·7	13·7	8·3
18	9 04·5	9 06·0	8 39·7	1·8	1·1	7·8	4·7	13·8	8·4
19	9 04·8	9 06·2	8 39·9	1·9	1·2	7·9	4·8	13·9	8·5
20	9 05·0	9 06·5	8 40·2	2·0	1·2	8·0	4·9	14·0	8·5
21	9 05·3	9 06·7	8 40·4	2·1	1·3	8·1	4·9	14·1	8·6
22	9 05·5	9 07·0	8 40·6	2·2	1·3	8·2	5·0	14·2	8·6
23	9 05·8	9 07·2	8 40·9	2·3	1·4	8·3	5·0	14·3	8·7
24	9 06·0	9 07·5	8 41·1	2·4	1·5	8·4	5·1	14·4	8·8
25	9 06·3	9 07·7	8 41·4	2·5	1·5	8·5	5·2	14·5	8·8
26	9 06·5	9 08·0	8 41·6	2·6	1·6	8·6	5·2	14·6	8·9
27	9 06·8	9 08·2	8 41·8	2·7	1·6	8·7	5·3	14·7	8·9
28	9 07·0	9 08·5	8 42·1	2·8	1·7	8·8	5·4	14·8	9·0
29	9 07·3	9 08·7	8 42·3	2·9	1·8	8·9	5·4	14·9	9·1
30	9 07·5	9 09·0	8 42·6	3·0	1·8	9·0	5·5	15·0	9·1
31	9 07·8	9 09·2	8 42·8	3·1	1·9	9·1	5·5	15·1	9·2
32	9 08·0	9 09·5	8 43·0	3·2	1·9	9·2	5·6	15·2	9·2
33	9 08·3	9 09·8	8 43·3	3·3	2·0	9·3	5·7	15·3	9·3
34	9 08·5	9 10·0	8 43·5	3·4	2·1	9·4	5·7	15·4	9·4
35	9 08·8	9 10·3	8 43·8	3·5	2·1	9·5	5·8	15·5	9·4
36	9 09·0	9 10·5	8 44·0	3·6	2·2	9·6	5·8	15·6	9·5
37	9 09·3	9 10·8	8 44·2	3·7	2·3	9·7	5·9	15·7	9·6
38	9 09·5	9 11·0	8 44·5	3·8	2·3	9·8	6·0	15·8	9·6
39	9 09·8	9 11·3	8 44·7	3·9	2·4	9·9	6·0	15·9	9·7
40	9 10·0	9 11·5	8 44·9	4·0	2·4	10·0	6·1	16·0	9·7
41	9 10·3	9 11·8	8 45·2	4·1	2·5	10·1	6·1	16·1	9·8
42	9 10·5	9 12·0	8 45·4	4·2	2·6	10·2	6·2	16·2	9·9
43	9 10·8	9 12·3	8 45·7	4·3	2·6	10·3	6·3	16·3	9·9
44	9 11·0	9 12·5	8 45·9	4·4	2·7	10·4	6·3	16·4	10·0
45	9 11·3	9 12·8	8 46·1	4·5	2·7	10·5	6·4	16·5	10·0
46	9 11·5	9 13·0	8 46·4	4·6	2·8	10·6	6·4	16·6	10·1
47	9 11·8	9 13·3	8 46·6	4·7	2·9	10·7	6·5	16·7	10·2
48	9 12·0	9 13·5	8 46·9	4·8	2·9	10·8	6·6	16·8	10·2
49	9 12·3	9 13·8	8 47·1	4·9	3·0	10·9	6·6	16·9	10·3
50	9 12·5	9 14·0	8 47·3	5·0	3·0	11·0	6·7	17·0	10·3
51	9 12·8	9 14·3	8 47·6	5·1	3·1	11·1	6·8	17·1	10·4
52	9 13·0	9 14·5	8 47·8	5·2	3·2	11·2	6·8	17·2	10·5
53	9 13·3	9 14·8	8 48·0	5·3	3·2	11·3	6·9	17·3	10·5
54	9 13·5	9 15·0	8 48·3	5·4	3·3	11·4	6·9	17·4	10·6
55	9 13·8	9 15·3	8 48·5	5·5	3·3	11·5	7·0	17·5	10·6
56	9 14·0	9 15·5	8 48·8	5·6	3·4	11·6	7·1	17·6	10·7
57	9 14·3	9 15·8	8 49·0	5·7	3·5	11·7	7·1	17·7	10·8
58	9 14·5	9 16·0	8 49·2	5·8	3·5	11·8	7·2	17·8	10·8
59	9 14·8	9 16·3	8 49·5	5·9	3·6	11·9	7·2	17·9	10·9
60	9 15·0	9 16·5	8 49·7	6·0	3·7	12·0	7·3	18·0	11·0

37m s	SUN PLANETS ° ′	ARIES ° ′	MOON ° ′	v or d ′	Corrⁿ ′	v or d ′	Corrⁿ ′	v or d ′	Corrⁿ ′
00	9 15·0	9 16·5	8 49·7	0·0	0·0	6·0	3·8	12·0	7·5
01	9 15·3	9 16·8	8 50·0	0·1	0·1	6·1	3·8	12·1	7·6
02	9 15·5	9 17·0	8 50·2	0·2	0·1	6·2	3·9	12·2	7·6
03	9 15·8	9 17·3	8 50·4	0·3	0·2	6·3	3·9	12·3	7·7
04	9 16·0	9 17·5	8 50·7	0·4	0·3	6·4	4·0	12·4	7·8
05	9 16·3	9 17·8	8 50·9	0·5	0·3	6·5	4·1	12·5	7·8
06	9 16·5	9 18·0	8 51·1	0·6	0·4	6·6	4·1	12·6	7·9
07	9 16·8	9 18·3	8 51·4	0·7	0·4	6·7	4·2	12·7	7·9
08	9 17·0	9 18·5	8 51·6	0·8	0·5	6·8	4·3	12·8	8·0
09	9 17·3	9 18·8	8 51·9	0·9	0·6	6·9	4·3	12·9	8·1
10	9 17·5	9 19·0	8 52·1	1·0	0·6	7·0	4·4	13·0	8·1
11	9 17·8	9 19·3	8 52·3	1·1	0·7	7·1	4·4	13·1	8·2
12	9 18·0	9 19·5	8 52·6	1·2	0·8	7·2	4·5	13·2	8·3
13	9 18·3	9 19·8	8 52·8	1·3	0·8	7·3	4·6	13·3	8·3
14	9 18·5	9 20·0	8 53·1	1·4	0·9	7·4	4·6	13·4	8·4
15	9 18·8	9 20·3	8 53·3	1·5	0·9	7·5	4·7	13·5	8·4
16	9 19·0	9 20·5	8 53·5	1·6	1·0	7·6	4·8	13·6	8·5
17	9 19·3	9 20·8	8 53·8	1·7	1·1	7·7	4·8	13·7	8·6
18	9 19·5	9 21·0	8 54·0	1·8	1·1	7·8	4·9	13·8	8·6
19	9 19·8	9 21·3	8 54·3	1·9	1·2	7·9	4·9	13·9	8·7
20	9 20·0	9 21·5	8 54·5	2·0	1·3	8·0	5·0	14·0	8·8
21	9 20·3	9 21·8	8 54·7	2·1	1·3	8·1	5·1	14·1	8·8
22	9 20·5	9 22·0	8 55·0	2·2	1·4	8·2	5·1	14·2	8·9
23	9 20·8	9 22·3	8 55·2	2·3	1·4	8·3	5·2	14·3	8·9
24	9 21·0	9 22·5	8 55·4	2·4	1·5	8·4	5·3	14·4	9·0
25	9 21·3	9 22·8	8 55·7	2·5	1·6	8·5	5·3	14·5	9·1
26	9 21·5	9 23·0	8 55·9	2·6	1·6	8·6	5·4	14·6	9·1
27	9 21·8	9 23·3	8 56·2	2·7	1·7	8·7	5·4	14·7	9·2
28	9 22·0	9 23·5	8 56·4	2·8	1·8	8·8	5·5	14·8	9·3
29	9 22·3	9 23·8	8 56·6	2·9	1·8	8·9	5·6	14·9	9·3
30	9 22·5	9 24·0	8 56·9	3·0	1·9	9·0	5·6	15·0	9·4
31	9 22·8	9 24·3	8 57·1	3·1	1·9	9·1	5·7	15·1	9·4
32	9 23·0	9 24·5	8 57·4	3·2	2·0	9·2	5·8	15·2	9·5
33	9 23·3	9 24·8	8 57·6	3·3	2·1	9·3	5·8	15·3	9·6
34	9 23·5	9 25·0	8 57·8	3·4	2·1	9·4	5·9	15·4	9·6
35	9 23·8	9 25·3	8 58·1	3·5	2·2	9·5	5·9	15·5	9·7
36	9 24·0	9 25·5	8 58·3	3·6	2·3	9·6	6·0	15·6	9·8
37	9 24·3	9 25·8	8 58·5	3·7	2·3	9·7	6·1	15·7	9·8
38	9 24·5	9 26·0	8 58·8	3·8	2·4	9·8	6·1	15·8	9·9
39	9 24·8	9 26·3	8 59·0	3·9	2·4	9·9	6·2	15·9	9·9
40	9 25·0	9 26·5	8 59·3	4·0	2·5	10·0	6·3	16·0	10·0
41	9 25·3	9 26·8	8 59·5	4·1	2·6	10·1	6·3	16·1	10·1
42	9 25·5	9 27·0	8 59·7	4·2	2·6	10·2	6·4	16·2	10·1
43	9 25·8	9 27·3	9 00·0	4·3	2·7	10·3	6·4	16·3	10·2
44	9 26·0	9 27·5	9 00·2	4·4	2·8	10·4	6·5	16·4	10·3
45	9 26·3	9 27·8	9 00·5	4·5	2·8	10·5	6·6	16·5	10·3
46	9 26·5	9 28·1	9 00·7	4·6	2·9	10·6	6·6	16·6	10·4
47	9 26·8	9 28·3	9 00·9	4·7	2·9	10·7	6·7	16·7	10·4
48	9 27·0	9 28·6	9 01·2	4·8	3·0	10·8	6·8	16·8	10·5
49	9 27·3	9 28·8	9 01·4	4·9	3·1	10·9	6·8	16·9	10·6
50	9 27·5	9 29·1	9 01·6	5·0	3·1	11·0	6·9	17·0	10·6
51	9 27·8	9 29·3	9 01·9	5·1	3·2	11·1	6·9	17·1	10·7
52	9 28·0	9 29·6	9 02·1	5·2	3·3	11·2	7·0	17·2	10·8
53	9 28·3	9 29·8	9 02·4	5·3	3·3	11·3	7·1	17·3	10·8
54	9 28·5	9 30·1	9 02·6	5·4	3·4	11·4	7·1	17·4	10·9
55	9 28·8	9 30·3	9 02·8	5·5	3·4	11·5	7·2	17·5	10·9
56	9 29·0	9 30·6	9 03·1	5·6	3·5	11·6	7·3	17·6	11·0
57	9 29·3	9 30·8	9 03·3	5·7	3·6	11·7	7·3	17·7	11·1
58	9 29·5	9 31·1	9 03·6	5·8	3·6	11·8	7·4	17·8	11·1
59	9 29·8	9 31·3	9 03·8	5·9	3·7	11·9	7·4	17·9	11·2
60	9 30·0	9 31·6	9 04·0	6·0	3·8	12·0	7·5	18·0	11·3

38m s	SUN PLANETS ° ′	ARIES ° ′	MOON ° ′	v or d ′	Corrn ′	v or d ′	Corrn ′	v or d ′	Corrn ′
00	9 30·0	9 31·6	9 04·0	0·0	0·0	6·0	3·9	12·0	7·7
01	9 30·3	9 31·8	9 04·3	0·1	0·1	6·1	3·9	12·1	7·8
02	9 30·5	9 32·1	9 04·5	0·2	0·1	6·2	4·0	12·2	7·8
03	9 30·8	9 32·3	9 04·7	0·3	0·2	6·3	4·0	12·3	7·9
04	9 31·0	9 32·6	9 05·0	0·4	0·3	6·4	4·1	12·4	8·0
05	9 31·3	9 32·8	9 05·2	0·5	0·3	6·5	4·2	12·5	8·0
06	9 31·5	9 33·1	9 05·5	0·6	0·4	6·6	4·2	12·6	8·1
07	9 31·8	9 33·3	9 05·7	0·7	0·4	6·7	4·3	12·7	8·1
08	9 32·0	9 33·6	9 05·9	0·8	0·5	6·8	4·4	12·8	8·2
09	9 32·3	9 33·8	9 06·2	0·9	0·6	6·9	4·4	12·9	8·3
10	9 32·5	9 34·1	9 06·4	1·0	0·6	7·0	4·5	13·0	8·3
11	9 32·8	9 34·3	9 06·7	1·1	0·7	7·1	4·6	13·1	8·4
12	9 33·0	9 34·6	9 06·9	1·2	0·8	7·2	4·6	13·2	8·5
13	9 33·3	9 34·8	9 07·1	1·3	0·8	7·3	4·7	13·3	8·5
14	9 33·5	9 35·1	9 07·4	1·4	0·9	7·4	4·7	13·4	8·6
15	9 33·8	9 35·3	9 07·6	1·5	1·0	7·5	4·8	13·5	8·7
16	9 34·0	9 35·6	9 07·9	1·6	1·0	7·6	4·9	13·6	8·7
17	9 34·3	9 35·8	9 08·1	1·7	1·1	7·7	4·9	13·7	8·8
18	9 34·5	9 36·1	9 08·3	1·8	1·2	7·8	5·0	13·8	8·9
19	9 34·8	9 36·3	9 08·6	1·9	1·2	7·9	5·1	13·9	8·9
20	9 35·0	9 36·6	9 08·8	2·0	1·3	8·0	5·1	14·0	9·0
21	9 35·3	9 36·8	9 09·0	2·1	1·3	8·1	5·2	14·1	9·0
22	9 35·5	9 37·1	9 09·3	2·2	1·4	8·2	5·3	14·2	9·1
23	9 35·8	9 37·3	9 09·5	2·3	1·5	8·3	5·3	14·3	9·2
24	9 36·0	9 37·6	9 09·8	2·4	1·5	8·4	5·4	14·4	9·2
25	9 36·3	9 37·8	9 10·0	2·5	1·6	8·5	5·5	14·5	9·3
26	9 36·5	9 38·1	9 10·2	2·6	1·7	8·6	5·5	14·6	9·4
27	9 36·8	9 38·3	9 10·5	2·7	1·7	8·7	5·6	14·7	9·4
28	9 37·0	9 38·6	9 10·7	2·8	1·8	8·8	5·6	14·8	9·5
29	9 37·3	9 38·8	9 11·0	2·9	1·9	8·9	5·7	14·9	9·6
30	9 37·5	9 39·1	9 11·2	3·0	1·9	9·0	5·8	15·0	9·6
31	9 37·8	9 39·3	9 11·4	3·1	2·0	9·1	5·8	15·1	9·7
32	9 38·0	9 39·6	9 11·7	3·2	2·1	9·2	5·9	15·2	9·8
33	9 38·3	9 39·8	9 11·9	3·3	2·1	9·3	6·0	15·3	9·8
34	9 38·5	9 40·1	9 12·1	3·4	2·2	9·4	6·0	15·4	9·9
35	9 38·8	9 40·3	9 12·4	3·5	2·2	9·5	6·1	15·5	9·9
36	9 39·0	9 40·6	9 12·6	3·6	2·3	9·6	6·2	15·6	10·0
37	9 39·3	9 40·8	9 12·9	3·7	2·4	9·7	6·2	15·7	10·1
38	9 39·5	9 41·1	9 13·1	3·8	2·4	9·8	6·3	15·8	10·1
39	9 39·8	9 41·3	9 13·3	3·9	2·5	9·9	6·4	15·9	10·2
40	9 40·0	9 41·6	9 13·6	4·0	2·6	10·0	6·4	16·0	10·3
41	9 40·3	9 41·8	9 13·8	4·1	2·6	10·1	6·5	16·1	10·3
42	9 40·5	9 42·1	9 14·1	4·2	2·7	10·2	6·5	16·2	10·4
43	9 40·8	9 42·3	9 14·3	4·3	2·8	10·3	6·6	16·3	10·5
44	9 41·0	9 42·6	9 14·5	4·4	2·8	10·4	6·7	16·4	10·5
45	9 41·3	9 42·8	9 14·8	4·5	2·9	10·5	6·7	16·5	10·6
46	9 41·5	9 43·1	9 15·0	4·6	3·0	10·6	6·8	16·6	10·7
47	9 41·8	9 43·3	9 15·2	4·7	3·0	10·7	6·9	16·7	10·7
48	9 42·0	9 43·6	9 15·5	4·8	3·1	10·8	6·9	16·8	10·8
49	9 42·3	9 43·8	9 15·7	4·9	3·1	10·9	7·0	16·9	10·8
50	9 42·5	9 44·1	9 16·0	5·0	3·2	11·0	7·1	17·0	10·9
51	9 42·8	9 44·3	9 16·2	5·1	3·3	11·1	7·1	17·1	11·0
52	9 43·0	9 44·6	9 16·4	5·2	3·3	11·2	7·2	17·2	11·0
53	9 43·3	9 44·8	9 16·7	5·3	3·4	11·3	7·3	17·3	11·1
54	9 43·5	9 45·1	9 16·9	5·4	3·5	11·4	7·3	17·4	11·2
55	9 43·8	9 45·3	9 17·2	5·5	3·5	11·5	7·4	17·5	11·2
56	9 44·0	9 45·6	9 17·4	5·6	3·6	11·6	7·4	17·6	11·3
57	9 44·3	9 45·8	9 17·6	5·7	3·7	11·7	7·5	17·7	11·4
58	9 44·5	9 46·1	9 17·9	5·8	3·7	11·8	7·6	17·8	11·4
59	9 44·8	9 46·4	9 18·1	5·9	3·8	11·9	7·6	17·9	11·5
60	9 45·0	9 46·6	9 18·4	6·0	3·9	12·0	7·7	18·0	11·6

39m s	SUN PLANETS ° ′	ARIES ° ′	MOON ° ′	v or d ′	Corrn ′	v or d ′	Corrn ′	v or d ′	Corrn ′
00	9 45·0	9 46·6	9 18·4	0·0	0·0	6·0	4·0	12·0	7·9
01	9 45·3	9 46·9	9 18·6	0·1	0·1	6·1	4·0	12·1	8·0
02	9 45·5	9 47·1	9 18·8	0·2	0·1	6·2	4·1	12·2	8·0
03	9 45·8	9 47·4	9 19·1	0·3	0·2	6·3	4·1	12·3	8·1
04	9 46·0	9 47·6	9 19·3	0·4	0·3	6·4	4·2	12·4	8·2
05	9 46·3	9 47·9	9 19·5	0·5	0·3	6·5	4·3	12·5	8·2
06	9 46·5	9 48·1	9 19·8	0·6	0·4	6·6	4·3	12·6	8·3
07	9 46·8	9 48·4	9 20·0	0·7	0·5	6·7	4·4	12·7	8·4
08	9 47·0	9 48·6	9 20·3	0·8	0·5	6·8	4·5	12·8	8·4
09	9 47·3	9 48·9	9 20·5	0·9	0·6	6·9	4·5	12·9	8·5
10	9 47·5	9 49·1	9 20·7	1·0	0·7	7·0	4·6	13·0	8·6
11	9 47·8	9 49·4	9 21·0	1·1	0·7	7·1	4·7	13·1	8·6
12	9 48·0	9 49·6	9 21·2	1·2	0·8	7·2	4·7	13·2	8·7
13	9 48·3	9 49·9	9 21·5	1·3	0·9	7·3	4·8	13·3	8·8
14	9 48·5	9 50·1	9 21·7	1·4	0·9	7·4	4·9	13·4	8·8
15	9 48·8	9 50·4	9 21·9	1·5	1·0	7·5	4·9	13·5	8·9
16	9 49·0	9 50·6	9 22·2	1·6	1·1	7·6	5·0	13·6	9·0
17	9 49·3	9 50·9	9 22·4	1·7	1·1	7·7	5·1	13·7	9·0
18	9 49·5	9 51·1	9 22·6	1·8	1·2	7·8	5·1	13·8	9·1
19	9 49·8	9 51·4	9 22·9	1·9	1·3	7·9	5·2	13·9	9·2
20	9 50·0	9 51·6	9 23·1	2·0	1·3	8·0	5·3	14·0	9·2
21	9 50·3	9 51·9	9 23·4	2·1	1·4	8·1	5·3	14·1	9·3
22	9 50·5	9 52·1	9 23·6	2·2	1·4	8·2	5·4	14·2	9·3
23	9 50·8	9 52·4	9 23·8	2·3	1·5	8·3	5·5	14·3	9·4
24	9 51·0	9 52·6	9 24·1	2·4	1·6	8·4	5·5	14·4	9·5
25	9 51·3	9 52·9	9 24·3	2·5	1·6	8·5	5·6	14·5	9·5
26	9 51·5	9 53·1	9 24·6	2·6	1·7	8·6	5·7	14·6	9·6
27	9 51·8	9 53·4	9 24·8	2·7	1·8	8·7	5·7	14·7	9·7
28	9 52·0	9 53·6	9 25·0	2·8	1·8	8·8	5·8	14·8	9·7
29	9 52·3	9 53·9	9 25·3	2·9	1·9	8·9	5·9	14·9	9·8
30	9 52·5	9 54·1	9 25·5	3·0	2·0	9·0	5·9	15·0	9·9
31	9 52·8	9 54·4	9 25·7	3·1	2·0	9·1	6·0	15·1	9·9
32	9 53·0	9 54·6	9 26·0	3·2	2·1	9·2	6·1	15·2	10·0
33	9 53·3	9 54·9	9 26·2	3·3	2·2	9·3	6·1	15·3	10·1
34	9 53·5	9 55·1	9 26·5	3·4	2·2	9·4	6·2	15·4	10·1
35	9 53·8	9 55·4	9 26·7	3·5	2·3	9·5	6·3	15·5	10·2
36	9 54·0	9 55·6	9 26·9	3·6	2·4	9·6	6·3	15·6	10·3
37	9 54·3	9 55·9	9 27·2	3·7	2·4	9·7	6·4	15·7	10·3
38	9 54·5	9 56·1	9 27·4	3·8	2·5	9·8	6·5	15·8	10·4
39	9 54·8	9 56·4	9 27·7	3·9	2·6	9·9	6·5	15·9	10·5
40	9 55·0	9 56·6	9 27·9	4·0	2·6	10·0	6·6	16·0	10·5
41	9 55·3	9 56·9	9 28·1	4·1	2·7	10·1	6·6	16·1	10·6
42	9 55·5	9 57·1	9 28·4	4·2	2·8	10·2	6·7	16·2	10·7
43	9 55·8	9 57·4	9 28·6	4·3	2·8	10·3	6·8	16·3	10·7
44	9 56·0	9 57·6	9 28·8	4·4	2·9	10·4	6·8	16·4	10·8
45	9 56·3	9 57·9	9 29·1	4·5	3·0	10·5	6·9	16·5	10·9
46	9 56·5	9 58·1	9 29·3	4·6	3·0	10·6	7·0	16·6	10·9
47	9 56·8	9 58·4	9 29·6	4·7	3·1	10·7	7·0	16·7	11·0
48	9 57·0	9 58·6	9 29·8	4·8	3·2	10·8	7·1	16·8	11·1
49	9 57·3	9 58·9	9 30·0	4·9	3·2	10·9	7·2	16·9	11·1
50	9 57·5	9 59·1	9 30·3	5·0	3·3	11·0	7·2	17·0	11·2
51	9 57·8	9 59·4	9 30·5	5·1	3·4	11·1	7·3	17·1	11·3
52	9 58·0	9 59·6	9 30·8	5·2	3·4	11·2	7·4	17·2	11·3
53	9 58·3	9 59·9	9 31·0	5·3	3·5	11·3	7·4	17·3	11·4
54	9 58·5	10 00·1	9 31·2	5·4	3·6	11·4	7·5	17·4	11·5
55	9 58·8	10 00·4	9 31·5	5·5	3·6	11·5	7·6	17·5	11·5
56	9 59·0	10 00·6	9 31·7	5·6	3·7	11·6	7·6	17·6	11·6
57	9 59·3	10 00·9	9 32·0	5·7	3·8	11·7	7·7	17·7	11·7
58	9 59·5	10 01·1	9 32·2	5·8	3·8	11·8	7·8	17·8	11·7
59	9 59·8	10 01·4	9 32·4	5·9	3·9	11·9	7·8	17·9	11·8
60	10 00·0	10 01·6	9 32·7	6·0	4·0	12·0	7·9	18·0	11·9

40m s	SUN PLANETS ° ′	ARIES ° ′	MOON ° ′	v or d ′	Corrn ′	v or d ′	Corrn ′	v or d ′	Corrn ′
00	10 00·0	10 01·6	9 32·7	0·0	0·0	6·0	4·1	12·0	8·1
01	10 00·3	10 01·9	9 32·9	0·1	0·1	6·1	4·1	12·1	8·2
02	10 00·5	10 02·1	9 33·1	0·2	0·1	6·2	4·2	12·2	8·2
03	10 00·8	10 02·4	9 33·4	0·3	0·2	6·3	4·3	12·3	8·3
04	10 01·0	10 02·6	9 33·6	0·4	0·3	6·4	4·3	12·4	8·4
05	10 01·3	10 02·9	9 33·9	0·5	0·3	6·5	4·4	12·5	8·4
06	10 01·5	10 03·1	9 34·1	0·6	0·4	6·6	4·5	12·6	8·5
07	10 01·8	10 03·4	9 34·3	0·7	0·5	6·7	4·5	12·7	8·6
08	10 02·0	10 03·6	9 34·6	0·8	0·5	6·8	4·6	12·8	8·6
09	10 02·3	10 03·9	9 34·8	0·9	0·6	6·9	4·7	12·9	8·7
10	10 02·5	10 04·1	9 35·1	1·0	0·7	7·0	4·7	13·0	8·8
11	10 02·8	10 04·4	9 35·3	1·1	0·7	7·1	4·8	13·1	8·8
12	10 03·0	10 04·7	9 35·5	1·2	0·8	7·2	4·9	13·2	8·9
13	10 03·3	10 04·9	9 35·8	1·3	0·9	7·3	4·9	13·3	9·0
14	10 03·5	10 05·2	9 36·0	1·4	0·9	7·4	5·0	13·4	9·0
15	10 03·8	10 05·4	9 36·2	1·5	1·0	7·5	5·1	13·5	9·1
16	10 04·0	10 05·7	9 36·5	1·6	1·1	7·6	5·1	13·6	9·2
17	10 04·3	10 05·9	9 36·7	1·7	1·1	7·7	5·2	13·7	9·2
18	10 04·5	10 06·2	9 37·0	1·8	1·2	7·8	5·3	13·8	9·3
19	10 04·8	10 06·4	9 37·2	1·9	1·3	7·9	5·3	13·9	9·4
20	10 05·0	10 06·7	9 37·4	2·0	1·4	8·0	5·4	14·0	9·5
21	10 05·3	10 06·9	9 37·7	2·1	1·4	8·1	5·5	14·1	9·5
22	10 05·5	10 07·2	9 37·9	2·2	1·5	8·2	5·5	14·2	9·6
23	10 05·8	10 07·4	9 38·2	2·3	1·6	8·3	5·6	14·3	9·7
24	10 06·0	10 07·7	9 38·4	2·4	1·6	8·4	5·7	14·4	9·7
25	10 06·3	10 07·9	9 38·6	2·5	1·7	8·5	5·7	14·5	9·8
26	10 06·5	10 08·2	9 38·9	2·6	1·8	8·6	5·8	14·6	9·9
27	10 06·8	10 08·4	9 39·1	2·7	1·8	8·7	5·9	14·7	9·9
28	10 07·0	10 08·7	9 39·3	2·8	1·9	8·8	5·9	14·8	10·0
29	10 07·3	10 08·9	9 39·6	2·9	2·0	8·9	6·0	14·9	10·1
30	10 07·5	10 09·2	9 39·8	3·0	2·0	9·0	6·1	15·0	10·1
31	10 07·8	10 09·4	9 40·1	3·1	2·1	9·1	6·1	15·1	10·2
32	10 08·0	10 09·7	9 40·3	3·2	2·2	9·2	6·2	15·2	10·3
33	10 08·3	10 09·9	9 40·5	3·3	2·2	9·3	6·3	15·3	10·3
34	10 08·5	10 10·2	9 40·8	3·4	2·3	9·4	6·3	15·4	10·4
35	10 08·8	10 10·4	9 41·0	3·5	2·4	9·5	6·4	15·5	10·5
36	10 09·0	10 10·7	9 41·3	3·6	2·4	9·6	6·5	15·6	10·5
37	10 09·3	10 10·9	9 41·5	3·7	2·5	9·7	6·5	15·7	10·6
38	10 09·5	10 11·2	9 41·7	3·8	2·6	9·8	6·6	15·8	10·7
39	10 09·8	10 11·4	9 42·0	3·9	2·6	9·9	6·7	15·9	10·7
40	10 10·0	10 11·7	9 42·2	4·0	2·7	10·0	6·8	16·0	10·8
41	10 10·3	10 11·9	9 42·4	4·1	2·8	10·1	6·8	16·1	10·9
42	10 10·5	10 12·2	9 42·7	4·2	2·8	10·2	6·9	16·2	10·9
43	10 10·8	10 12·4	9 42·9	4·3	2·9	10·3	7·0	16·3	11·0
44	10 11·0	10 12·7	9 43·2	4·4	3·0	10·4	7·0	16·4	11·1
45	10 11·3	10 12·9	9 43·4	4·5	3·0	10·5	7·1	16·5	11·1
46	10 11·5	10 13·2	9 43·6	4·6	3·1	10·6	7·2	16·6	11·2
47	10 11·8	10 13·4	9 43·9	4·7	3·2	10·7	7·2	16·7	11·3
48	10 12·0	10 13·7	9 44·1	4·8	3·2	10·8	7·3	16·8	11·3
49	10 12·3	10 13·9	9 44·4	4·9	3·3	10·9	7·4	16·9	11·4
50	10 12·5	10 14·2	9 44·6	5·0	3·4	11·0	7·4	17·0	11·5
51	10 12·8	10 14·4	9 44·8	5·1	3·4	11·1	7·5	17·1	11·5
52	10 13·0	10 14·7	9 45·1	5·2	3·5	11·2	7·6	17·2	11·6
53	10 13·3	10 14·9	9 45·3	5·3	3·6	11·3	7·6	17·3	11·7
54	10 13·5	10 15·2	9 45·6	5·4	3·6	11·4	7·7	17·4	11·7
55	10 13·8	10 15·4	9 45·8	5·5	3·7	11·5	7·8	17·5	11·8
56	10 14·0	10 15·7	9 46·0	5·6	3·8	11·6	7·8	17·6	11·9
57	10 14·3	10 15·9	9 46·3	5·7	3·8	11·7	7·9	17·7	11·9
58	10 14·5	10 16·2	9 46·5	5·8	3·9	11·8	8·0	17·8	12·0
59	10 14·8	10 16·4	9 46·7	5·9	4·0	11·9	8·0	17·9	12·1
60	10 15·0	10 16·7	9 47·0	6·0	4·1	12·0	8·1	18·0	12·2

41m s	SUN PLANETS ° ′	ARIES ° ′	MOON ° ′	v or d ′	Corrn ′	v or d ′	Corrn ′	v or d ′	Corrn ′
00	10 15·0	10 16·7	9 47·0	0·0	0·0	6·0	4·2	12·0	8·3
01	10 15·3	10 16·9	9 47·2	0·1	0·1	6·1	4·2	12·1	8·4
02	10 15·5	10 17·2	9 47·5	0·2	0·1	6·2	4·3	12·2	8·4
03	10 15·8	10 17·4	9 47·7	0·3	0·2	6·3	4·4	12·3	8·5
04	10 16·0	10 17·7	9 47·9	0·4	0·3	6·4	4·4	12·4	8·6
05	10 16·3	10 17·9	9 48·2	0·5	0·3	6·5	4·5	12·5	8·6
06	10 16·5	10 18·2	9 48·4	0·6	0·4	6·6	4·6	12·6	8·7
07	10 16·8	10 18·4	9 48·7	0·7	0·5	6·7	4·6	12·7	8·8
08	10 17·0	10 18·7	9 48·9	0·8	0·6	6·8	4·7	12·8	8·9
09	10 17·3	10 18·9	9 49·1	0·9	0·6	6·9	4·8	12·9	8·9
10	10 17·5	10 19·2	9 49·4	1·0	0·7	7·0	4·8	13·0	9·0
11	10 17·8	10 19·4	9 49·6	1·1	0·8	7·1	4·9	13·1	9·1
12	10 18·0	10 19·7	9 49·8	1·2	0·8	7·2	5·0	13·2	9·1
13	10 18·3	10 19·9	9 50·1	1·3	0·9	7·3	5·0	13·3	9·2
14	10 18·5	10 20·2	9 50·3	1·4	1·0	7·4	5·1	13·4	9·3
15	10 18·8	10 20·4	9 50·6	1·5	1·0	7·5	5·2	13·5	9·3
16	10 19·0	10 20·7	9 50·8	1·6	1·1	7·6	5·3	13·6	9·4
17	10 19·3	10 20·9	9 51·0	1·7	1·2	7·7	5·3	13·7	9·5
18	10 19·5	10 21·2	9 51·3	1·8	1·2	7·8	5·4	13·8	9·5
19	10 19·8	10 21·4	9 51·5	1·9	1·3	7·9	5·5	13·9	9·6
20	10 20·0	10 21·7	9 51·8	2·0	1·4	8·0	5·5	14·0	9·7
21	10 20·3	10 21·9	9 52·0	2·1	1·5	8·1	5·6	14·1	9·8
22	10 20·5	10 22·2	9 52·2	2·2	1·5	8·2	5·7	14·2	9·8
23	10 20·8	10 22·4	9 52·5	2·3	1·6	8·3	5·7	14·3	9·9
24	10 21·0	10 22·7	9 52·7	2·4	1·7	8·4	5·8	14·4	10·0
25	10 21·3	10 23·0	9 52·9	2·5	1·7	8·5	5·9	14·5	10·0
26	10 21·5	10 23·2	9 53·2	2·6	1·8	8·6	5·9	14·6	10·1
27	10 21·8	10 23·5	9 53·4	2·7	1·9	8·7	6·0	14·7	10·2
28	10 22·0	10 23·7	9 53·7	2·8	1·9	8·8	6·1	14·8	10·2
29	10 22·3	10 24·0	9 53·9	2·9	2·0	8·9	6·2	14·9	10·3
30	10 22·5	10 24·2	9 54·1	3·0	2·1	9·0	6·2	15·0	10·4
31	10 22·8	10 24·5	9 54·4	3·1	2·1	9·1	6·3	15·1	10·4
32	10 23·0	10 24·7	9 54·6	3·2	2·2	9·2	6·4	15·2	10·5
33	10 23·3	10 25·0	9 54·9	3·3	2·3	9·3	6·4	15·3	10·6
34	10 23·5	10 25·2	9 55·1	3·4	2·4	9·4	6·5	15·4	10·7
35	10 23·8	10 25·5	9 55·3	3·5	2·4	9·5	6·6	15·5	10·7
36	10 24·0	10 25·7	9 55·6	3·6	2·5	9·6	6·6	15·6	10·8
37	10 24·3	10 26·0	9 55·8	3·7	2·6	9·7	6·7	15·7	10·9
38	10 24·5	10 26·2	9 56·1	3·8	2·6	9·8	6·8	15·8	10·9
39	10 24·8	10 26·5	9 56·3	3·9	2·7	9·9	6·8	15·9	11·0
40	10 25·0	10 26·7	9 56·5	4·0	2·8	10·0	6·9	16·0	11·1
41	10 25·3	10 27·0	9 56·8	4·1	2·8	10·1	7·0	16·1	11·1
42	10 25·5	10 27·2	9 57·0	4·2	2·9	10·2	7·1	16·2	11·2
43	10 25·8	10 27·5	9 57·2	4·3	3·0	10·3	7·1	16·3	11·3
44	10 26·0	10 27·7	9 57·5	4·4	3·0	10·4	7·2	16·4	11·3
45	10 26·3	10 28·0	9 57·7	4·5	3·1	10·5	7·3	16·5	11·4
46	10 26·5	10 28·2	9 58·0	4·6	3·2	10·6	7·3	16·6	11·5
47	10 26·8	10 28·5	9 58·2	4·7	3·3	10·7	7·4	16·7	11·6
48	10 27·0	10 28·7	9 58·4	4·8	3·3	10·8	7·5	16·8	11·6
49	10 27·3	10 29·0	9 58·7	4·9	3·4	10·9	7·5	16·9	11·7
50	10 27·5	10 29·2	9 58·9	5·0	3·5	11·0	7·6	17·0	11·8
51	10 27·8	10 29·5	9 59·2	5·1	3·5	11·1	7·7	17·1	11·8
52	10 28·0	10 29·7	9 59·4	5·2	3·6	11·2	7·7	17·2	11·9
53	10 28·3	10 30·0	9 59·6	5·3	3·7	11·3	7·8	17·3	12·0
54	10 28·5	10 30·2	9 59·9	5·4	3·7	11·4	7·9	17·4	12·0
55	10 28·8	10 30·5	10 00·1	5·5	3·8	11·5	8·0	17·5	12·1
56	10 29·0	10 30·7	10 00·3	5·6	3·9	11·6	8·0	17·6	12·2
57	10 29·3	10 31·0	10 00·6	5·7	3·9	11·7	8·1	17·7	12·2
58	10 29·5	10 31·2	10 00·8	5·8	4·0	11·8	8·2	17·8	12·3
59	10 29·8	10 31·5	10 01·1	5·9	4·1	11·9	8·2	17·9	12·4
60	10 30·0	10 31·7	10 01·3	6·0	4·2	12·0	8·3	18·0	12·5

42m	SUN PLANETS	ARIES	MOON	v or d	Corrn	v or d	Corrn	v or d	Corrn
s	° ′	° ′	° ′	′	′	′	′	′	′
00	10 30·0	10 31·7	10 01·3	0·0	0·0	6·0	4·3	12·0	8·5
01	10 30·3	10 32·0	10 01·5	0·1	0·1	6·1	4·3	12·1	8·6
02	10 30·5	10 32·2	10 01·8	0·2	0·1	6·2	4·4	12·2	8·6
03	10 30·8	10 32·5	10 02·0	0·3	0·2	6·3	4·5	12·3	8·7
04	10 31·0	10 32·7	10 02·3	0·4	0·3	6·4	4·5	12·4	8·8
05	10 31·3	10 33·0	10 02·5	0·5	0·4	6·5	4·6	12·5	8·9
06	10 31·5	10 33·2	10 02·7	0·6	0·4	6·6	4·7	12·6	8·9
07	10 31·8	10 33·5	10 03·0	0·7	0·5	6·7	4·7	12·7	9·0
08	10 32·0	10 33·7	10 03·2	0·8	0·6	6·8	4·8	12·8	9·1
09	10 32·3	10 34·0	10 03·4	0·9	0·6	6·9	4·9	12·9	9·1
10	10 32·5	10 34·2	10 03·7	1·0	0·7	7·0	5·0	13·0	9·2
11	10 32·8	10 34·5	10 03·9	1·1	0·8	7·1	5·0	13·1	9·3
12	10 33·0	10 34·7	10 04·2	1·2	0·9	7·2	5·1	13·2	9·4
13	10 33·3	10 35·0	10 04·4	1·3	0·9	7·3	5·2	13·3	9·4
14	10 33·5	10 35·2	10 04·6	1·4	1·0	7·4	5·2	13·4	9·5
15	10 33·8	10 35·5	10 04·9	1·5	1·1	7·5	5·3	13·5	9·6
16	10 34·0	10 35·7	10 05·1	1·6	1·1	7·6	5·4	13·6	9·6
17	10 34·3	10 36·0	10 05·4	1·7	1·2	7·7	5·5	13·7	9·7
18	10 34·5	10 36·2	10 05·6	1·8	1·3	7·8	5·5	13·8	9·8
19	10 34·8	10 36·5	10 05·8	1·9	1·3	7·9	5·6	13·9	9·8
20	10 35·0	10 36·7	10 06·1	2·0	1·4	8·0	5·7	14·0	9·9
21	10 35·3	10 37·0	10 06·3	2·1	1·5	8·1	5·7	14·1	10·0
22	10 35·5	10 37·2	10 06·5	2·2	1·6	8·2	5·8	14·2	10·1
23	10 35·8	10 37·5	10 06·8	2·3	1·6	8·3	5·9	14·3	10·1
24	10 36·0	10 37·7	10 07·0	2·4	1·7	8·4	6·0	14·4	10·2
25	10 36·3	10 38·0	10 07·3	2·5	1·8	8·5	6·0	14·5	10·3
26	10 36·5	10 38·2	10 07·5	2·6	1·8	8·6	6·1	14·6	10·3
27	10 36·8	10 38·5	10 07·7	2·7	1·9	8·7	6·2	14·7	10·4
28	10 37·0	10 38·7	10 08·0	2·8	2·0	8·8	6·2	14·8	10·5
29	10 37·3	10 39·0	10 08·2	2·9	2·1	8·9	6·3	14·9	10·6
30	10 37·5	10 39·2	10 08·5	3·0	2·1	9·0	6·4	15·0	10·6
31	10 37·8	10 39·5	10 08·7	3·1	2·2	9·1	6·4	15·1	10·7
32	10 38·0	10 39·7	10 08·9	3·2	2·3	9·2	6·5	15·2	10·8
33	10 38·3	10 40·0	10 09·2	3·3	2·3	9·3	6·6	15·3	10·8
34	10 38·5	10 40·2	10 09·4	3·4	2·4	9·4	6·7	15·4	10·9
35	10 38·8	10 40·5	10 09·7	3·5	2·5	9·5	6·7	15·5	11·0
36	10 39·0	10 40·7	10 09·9	3·6	2·6	9·6	6·8	15·6	11·1
37	10 39·3	10 41·0	10 10·1	3·7	2·6	9·7	6·9	15·7	11·1
38	10 39·5	10 41·3	10 10·4	3·8	2·7	9·8	6·9	15·8	11·2
39	10 39·8	10 41·5	10 10·6	3·9	2·8	9·9	7·0	15·9	11·3
40	10 40·0	10 41·8	10 10·8	4·0	2·8	10·0	7·1	16·0	11·3
41	10 40·3	10 42·0	10 11·1	4·1	2·9	10·1	7·2	16·1	11·4
42	10 40·5	10 42·3	10 11·3	4·2	3·0	10·2	7·2	16·2	11·5
43	10 40·8	10 42·5	10 11·6	4·3	3·0	10·3	7·3	16·3	11·5
44	10 41·0	10 42·8	10 11·8	4·4	3·1	10·4	7·4	16·4	11·6
45	10 41·3	10 43·0	10 12·0	4·5	3·2	10·5	7·4	16·5	11·7
46	10 41·5	10 43·3	10 12·3	4·6	3·3	10·6	7·5	16·6	11·8
47	10 41·8	10 43·5	10 12·5	4·7	3·3	10·7	7·6	16·7	11·8
48	10 42·0	10 43·8	10 12·8	4·8	3·4	10·8	7·7	16·8	11·9
49	10 42·3	10 44·0	10 13·0	4·9	3·5	10·9	7·7	16·9	12·0
50	10 42·5	10 44·3	10 13·2	5·0	3·5	11·0	7·8	17·0	12·0
51	10 42·8	10 44·5	10 13·5	5·1	3·6	11·1	7·9	17·1	12·1
52	10 43·0	10 44·8	10 13·7	5·2	3·7	11·2	7·9	17·2	12·2
53	10 43·3	10 45·0	10 13·9	5·3	3·8	11·3	8·0	17·3	12·3
54	10 43·5	10 45·3	10 14·2	5·4	3·8	11·4	8·1	17·4	12·3
55	10 43·8	10 45·5	10 14·4	5·5	3·9	11·5	8·1	17·5	12·4
56	10 44·0	10 45·8	10 14·7	5·6	4·0	11·6	8·2	17·6	12·5
57	10 44·3	10 46·0	10 14·9	5·7	4·0	11·7	8·3	17·7	12·5
58	10 44·5	10 46·3	10 15·1	5·8	4·1	11·8	8·4	17·8	12·6
59	10 44·8	10 46·5	10 15·4	5·9	4·2	11·9	8·4	17·9	12·7
60	10 45·0	10 46·8	10 15·6	6·0	4·3	12·0	8·5	18·0	12·8

43m	SUN PLANETS	ARIES	MOON	v or d	Corrn	v or d	Corrn	v or d	Corrn
s	° ′	° ′	° ′	′	′	′	′	′	′
00	10 45·0	10 46·8	10 15·6	0·0	0·0	6·0	4·4	12·0	8·7
01	10 45·3	10 47·0	10 15·9	0·1	0·1	6·1	4·4	12·1	8·8
02	10 45·5	10 47·3	10 16·1	0·2	0·1	6·2	4·5	12·2	8·8
03	10 45·8	10 47·5	10 16·3	0·3	0·2	6·3	4·6	12·3	8·9
04	10 46·0	10 47·8	10 16·6	0·4	0·3	6·4	4·6	12·4	9·0
05	10 46·3	10 48·0	10 16·8	0·5	0·4	6·5	4·7	12·5	9·1
06	10 46·5	10 48·3	10 17·0	0·6	0·4	6·6	4·8	12·6	9·1
07	10 46·8	10 48·5	10 17·3	0·7	0·5	6·7	4·9	12·7	9·2
08	10 47·0	10 48·8	10 17·5	0·8	0·6	6·8	4·9	12·8	9·3
09	10 47·3	10 49·0	10 17·8	0·9	0·7	6·9	5·0	12·9	9·4
10	10 47·5	10 49·3	10 18·0	1·0	0·7	7·0	5·1	13·0	9·4
11	10 47·8	10 49·5	10 18·2	1·1	0·8	7·1	5·1	13·1	9·5
12	10 48·0	10 49·8	10 18·5	1·2	0·9	7·2	5·2	13·2	9·6
13	10 48·3	10 50·0	10 18·7	1·3	0·9	7·3	5·3	13·3	9·6
14	10 48·5	10 50·3	10 19·0	1·4	1·0	7·4	5·4	13·4	9·7
15	10 48·8	10 50·5	10 19·2	1·5	1·1	7·5	5·4	13·5	9·8
16	10 49·0	10 50·8	10 19·4	1·6	1·2	7·6	5·5	13·6	9·9
17	10 49·3	10 51·0	10 19·7	1·7	1·2	7·7	5·6	13·7	9·9
18	10 49·5	10 51·3	10 19·9	1·8	1·3	7·8	5·7	13·8	10·0
19	10 49·8	10 51·5	10 20·2	1·9	1·4	7·9	5·7	13·9	10·1
20	10 50·0	10 51·8	10 20·4	2·0	1·5	8·0	5·8	14·0	10·2
21	10 50·3	10 52·0	10 20·6	2·1	1·5	8·1	5·9	14·1	10·2
22	10 50·5	10 52·3	10 20·9	2·2	1·6	8·2	5·9	14·2	10·3
23	10 50·8	10 52·5	10 21·1	2·3	1·7	8·3	6·0	14·3	10·4
24	10 51·0	10 52·8	10 21·3	2·4	1·7	8·4	6·1	14·4	10·4
25	10 51·3	10 53·0	10 21·6	2·5	1·8	8·5	6·2	14·5	10·5
26	10 51·5	10 53·3	10 21·8	2·6	1·9	8·6	6·2	14·6	10·6
27	10 51·8	10 53·5	10 22·1	2·7	2·0	8·7	6·3	14·7	10·7
28	10 52·0	10 53·8	10 22·3	2·8	2·0	8·8	6·4	14·8	10·7
29	10 52·3	10 54·0	10 22·5	2·9	2·1	8·9	6·5	14·9	10·8
30	10 52·5	10 54·3	10 22·8	3·0	2·2	9·0	6·5	15·0	10·9
31	10 52·8	10 54·5	10 23·0	3·1	2·2	9·1	6·6	15·1	10·9
32	10 53·0	10 54·8	10 23·3	3·2	2·3	9·2	6·7	15·2	11·0
33	10 53·3	10 55·0	10 23·5	3·3	2·4	9·3	6·7	15·3	11·1
34	10 53·5	10 55·3	10 23·7	3·4	2·5	9·4	6·8	15·4	11·2
35	10 53·8	10 55·5	10 24·0	3·5	2·5	9·5	6·9	15·5	11·2
36	10 54·0	10 55·8	10 24·2	3·6	2·6	9·6	7·0	15·6	11·3
37	10 54·3	10 56·0	10 24·4	3·7	2·7	9·7	7·0	15·7	11·4
38	10 54·5	10 56·3	10 24·7	3·8	2·8	9·8	7·1	15·8	11·5
39	10 54·8	10 56·5	10 24·9	3·9	2·8	9·9	7·2	15·9	11·5
40	10 55·0	10 56·8	10 25·2	4·0	2·9	10·0	7·3	16·0	11·6
41	10 55·3	10 57·0	10 25·4	4·1	3·0	10·1	7·3	16·1	11·7
42	10 55·5	10 57·3	10 25·6	4·2	3·0	10·2	7·4	16·2	11·7
43	10 55·8	10 57·5	10 25·9	4·3	3·1	10·3	7·5	16·3	11·8
44	10 56·0	10 57·8	10 26·1	4·4	3·2	10·4	7·5	16·4	11·9
45	10 56·3	10 58·0	10 26·4	4·5	3·3	10·5	7·6	16·5	12·0
46	10 56·5	10 58·3	10 26·6	4·6	3·3	10·6	7·7	16·6	12·0
47	10 56·8	10 58·5	10 26·8	4·7	3·4	10·7	7·8	16·7	12·1
48	10 57·0	10 58·8	10 27·1	4·8	3·5	10·8	7·8	16·8	12·2
49	10 57·3	10 59·0	10 27·3	4·9	3·6	10·9	7·9	16·9	12·3
50	10 57·5	10 59·3	10 27·5	5·0	3·6	11·0	8·0	17·0	12·3
51	10 57·8	10 59·6	10 27·8	5·1	3·7	11·1	8·0	17·1	12·4
52	10 58·0	10 59·8	10 28·0	5·2	3·8	11·2	8·1	17·2	12·5
53	10 58·3	11 00·1	10 28·3	5·3	3·8	11·3	8·2	17·3	12·5
54	10 58·5	11 00·3	10 28·5	5·4	3·9	11·4	8·3	17·4	12·6
55	10 58·8	11 00·6	10 28·7	5·5	4·0	11·5	8·3	17·5	12·7
56	10 59·0	11 00·8	10 29·0	5·6	4·1	11·6	8·4	17·6	12·8
57	10 59·3	11 01·1	10 29·2	5·7	4·1	11·7	8·5	17·7	12·8
58	10 59·5	11 01·3	10 29·5	5·8	4·2	11·8	8·6	17·8	12·9
59	10 59·8	11 01·6	10 29·7	5·9	4·3	11·9	8·6	17·9	13·0
60	11 00·0	11 01·8	10 29·9	6·0	4·4	12·0	8·7	18·0	13·1

44m	SUN PLANETS	ARIES	MOON	v or d	Corrn	v or d	Corrn	v or d	Corrn
s	° ′	° ′	° ′	′	′	′	′	′	′
00	11 00·0	11 01·8	10 29·9	0·0	0·0	6·0	4·5	12·0	8·9
01	11 00·3	11 02·1	10 30·2	0·1	0·1	6·1	4·5	12·1	9·0
02	11 00·5	11 02·3	10 30·4	0·2	0·1	6·2	4·6	12·2	9·0
03	11 00·8	11 02·6	10 30·6	0·3	0·2	6·3	4·7	12·3	9·1
04	11 01·0	11 02·8	10 30·9	0·4	0·3	6·4	4·7	12·4	9·2
05	11 01·3	11 03·1	10 31·1	0·5	0·4	6·5	4·8	12·5	9·3
06	11 01·5	11 03·3	10 31·4	0·6	0·4	6·6	4·9	12·6	9·3
07	11 01·8	11 03·6	10 31·6	0·7	0·5	6·7	5·0	12·7	9·4
08	11 02·0	11 03·8	10 31·8	0·8	0·6	6·8	5·0	12·8	9·5
09	11 02·3	11 04·1	10 32·1	0·9	0·7	6·9	5·1	12·9	9·6
10	11 02·5	11 04·3	10 32·3	1·0	0·7	7·0	5·2	13·0	9·6
11	11 02·8	11 04·6	10 32·6	1·1	0·8	7·1	5·3	13·1	9·7
12	11 03·0	11 04·8	10 32·8	1·2	0·9	7·2	5·3	13·2	9·8
13	11 03·3	11 05·1	10 33·0	1·3	1·0	7·3	5·4	13·3	9·9
14	11 03·5	11 05·3	10 33·3	1·4	1·0	7·4	5·5	13·4	9·9
15	11 03·8	11 05·6	10 33·5	1·5	1·1	7·5	5·6	13·5	10·0
16	11 04·0	11 05·8	10 33·8	1·6	1·2	7·6	5·6	13·6	10·1
17	11 04·3	11 06·1	10 34·0	1·7	1·3	7·7	5·7	13·7	10·2
18	11 04·5	11 06·3	10 34·2	1·8	1·3	7·8	5·8	13·8	10·2
19	11 04·8	11 06·6	10 34·5	1·9	1·4	7·9	5·9	13·9	10·3
20	11 05·0	11 06·8	10 34·7	2·0	1·5	8·0	5·9	14·0	10·4
21	11 05·3	11 07·1	10 34·9	2·1	1·6	8·1	6·0	14·1	10·5
22	11 05·5	11 07·3	10 35·2	2·2	1·6	8·2	6·1	14·2	10·5
23	11 05·8	11 07·6	10 35·4	2·3	1·7	8·3	6·2	14·3	10·6
24	11 06·0	11 07·8	10 35·7	2·4	1·8	8·4	6·2	14·4	10·7
25	11 06·3	11 08·1	10 35·9	2·5	1·9	8·5	6·3	14·5	10·8
26	11 06·5	11 08·3	10 36·1	2·6	1·9	8·6	6·4	14·6	10·8
27	11 06·8	11 08·6	10 36·4	2·7	2·0	8·7	6·5	14·7	10·9
28	11 07·0	11 08·8	10 36·6	2·8	2·1	8·8	6·5	14·8	11·0
29	11 07·3	11 09·1	10 36·9	2·9	2·2	8·9	6·6	14·9	11·1
30	11 07·5	11 09·3	10 37·1	3·0	2·2	9·0	6·7	15·0	11·1
31	11 07·8	11 09·6	10 37·3	3·1	2·3	9·1	6·7	15·1	11·2
32	11 08·0	11 09·8	10 37·6	3·2	2·4	9·2	6·8	15·2	11·3
33	11 08·3	11 10·1	10 37·8	3·3	2·4	9·3	6·9	15·3	11·3
34	11 08·5	11 10·3	10 38·0	3·4	2·5	9·4	7·0	15·4	11·4
35	11 08·8	11 10·6	10 38·3	3·5	2·6	9·5	7·0	15·5	11·5
36	11 09·0	11 10·8	10 38·5	3·6	2·7	9·6	7·1	15·6	11·6
37	11 09·3	11 11·1	10 38·8	3·7	2·7	9·7	7·2	15·7	11·6
38	11 09·5	11 11·3	10 39·0	3·8	2·8	9·8	7·3	15·8	11·7
39	11 09·8	11 11·6	10 39·2	3·9	2·9	9·9	7·3	15·9	11·8
40	11 10·0	11 11·8	10 39·5	4·0	3·0	10·0	7·4	16·0	11·9
41	11 10·3	11 12·1	10 39·7	4·1	3·0	10·1	7·5	16·1	11·9
42	11 10·5	11 12·3	10 40·0	4·2	3·1	10·2	7·6	16·2	12·0
43	11 10·8	11 12·6	10 40·2	4·3	3·2	10·3	7·6	16·3	12·1
44	11 11·0	11 12·8	10 40·4	4·4	3·3	10·4	7·7	16·4	12·2
45	11 11·3	11 13·1	10 40·7	4·5	3·3	10·5	7·8	16·5	12·2
46	11 11·5	11 13·3	10 40·9	4·6	3·4	10·6	7·9	16·6	12·3
47	11 11·8	11 13·6	10 41·1	4·7	3·5	10·7	7·9	16·7	12·4
48	11 12·0	11 13·8	10 41·4	4·8	3·6	10·8	8·0	16·8	12·5
49	11 12·3	11 14·1	10 41·6	4·9	3·6	10·9	8·1	16·9	12·5
50	11 12·5	11 14·3	10 41·9	5·0	3·7	11·0	8·2	17·0	12·6
51	11 12·8	11 14·6	10 42·1	5·1	3·8	11·1	8·2	17·1	12·7
52	11 13·0	11 14·8	10 42·3	5·2	3·9	11·2	8·3	17·2	12·8
53	11 13·3	11 15·1	10 42·6	5·3	3·9	11·3	8·4	17·3	12·8
54	11 13·5	11 15·3	10 42·8	5·4	4·0	11·4	8·5	17·4	12·9
55	11 13·8	11 15·6	10 43·1	5·5	4·1	11·5	8·5	17·5	13·0
56	11 14·0	11 15·8	10 43·3	5·6	4·2	11·6	8·6	17·6	13·1
57	11 14·3	11 16·1	10 43·5	5·7	4·2	11·7	8·7	17·7	13·1
58	11 14·5	11 16·3	10 43·8	5·8	4·3	11·8	8·8	17·8	13·2
59	11 14·8	11 16·6	10 44·0	5·9	4·4	11·9	8·8	17·9	13·3
60	11 15·0	11 16·8	10 44·3	6·0	4·5	12·0	8·9	18·0	13·4

45m	SUN PLANETS	ARIES	MOON	v or d	Corrn	v or d	Corrn	v or d	Corrn
s	° ′	° ′	° ′	′	′	′	′	′	′
00	11 15·0	11 16·8	10 44·3	0·0	0·0	6·0	4·6	12·0	9·1
01	11 15·3	11 17·1	10 44·5	0·1	0·1	6·1	4·6	12·1	9·2
02	11 15·5	11 17·3	10 44·7	0·2	0·2	6·2	4·7	12·2	9·3
03	11 15·8	11 17·6	10 45·0	0·3	0·2	6·3	4·8	12·3	9·3
04	11 16·0	11 17·9	10 45·2	0·4	0·3	6·4	4·9	12·4	9·4
05	11 16·3	11 18·1	10 45·4	0·5	0·4	6·5	4·9	12·5	9·5
06	11 16·5	11 18·4	10 45·7	0·6	0·5	6·6	5·0	12·6	9·6
07	11 16·8	11 18·6	10 45·9	0·7	0·5	6·7	5·1	12·7	9·6
08	11 17·0	11 18·9	10 46·2	0·8	0·6	6·8	5·2	12·8	9·7
09	11 17·3	11 19·1	10 46·4	0·9	0·7	6·9	5·2	12·9	9·8
10	11 17·5	11 19·4	10 46·6	1·0	0·8	7·0	5·3	13·0	9·9
11	11 17·8	11 19·6	10 46·9	1·1	0·8	7·1	5·4	13·1	9·9
12	11 18·0	11 19·9	10 47·1	1·2	0·9	7·2	5·5	13·2	10·0
13	11 18·3	11 20·1	10 47·4	1·3	1·0	7·3	5·5	13·3	10·1
14	11 18·5	11 20·4	10 47·6	1·4	1·1	7·4	5·6	13·4	10·2
15	11 18·8	11 20·6	10 47·8	1·5	1·1	7·5	5·7	13·5	10·2
16	11 19·0	11 20·9	10 48·1	1·6	1·2	7·6	5·8	13·6	10·3
17	11 19·3	11 21·1	10 48·3	1·7	1·3	7·7	5·8	13·7	10·4
18	11 19·5	11 21·4	10 48·5	1·8	1·4	7·8	5·9	13·8	10·5
19	11 19·8	11 21·6	10 48·8	1·9	1·4	7·9	6·0	13·9	10·5
20	11 20·0	11 21·9	10 49·0	2·0	1·5	8·0	6·1	14·0	10·6
21	11 20·3	11 22·1	10 49·3	2·1	1·6	8·1	6·1	14·1	10·7
22	11 20·5	11 22·4	10 49·5	2·2	1·7	8·2	6·2	14·2	10·8
23	11 20·8	11 22·6	10 49·7	2·3	1·7	8·3	6·3	14·3	10·8
24	11 21·0	11 22·9	10 50·0	2·4	1·8	8·4	6·4	14·4	10·9
25	11 21·3	11 23·1	10 50·2	2·5	1·9	8·5	6·4	14·5	11·0
26	11 21·5	11 23·4	10 50·5	2·6	2·0	8·6	6·5	14·6	11·1
27	11 21·8	11 23·6	10 50·7	2·7	2·0	8·7	6·6	14·7	11·1
28	11 22·0	11 23·9	10 50·9	2·8	2·1	8·8	6·7	14·8	11·2
29	11 22·3	11 24·1	10 51·2	2·9	2·2	8·9	6·7	14·9	11·3
30	11 22·5	11 24·4	10 51·4	3·0	2·3	9·0	6·8	15·0	11·4
31	11 22·8	11 24·6	10 51·6	3·1	2·4	9·1	6·9	15·1	11·5
32	11 23·0	11 24·9	10 51·9	3·2	2·4	9·2	7·0	15·2	11·5
33	11 23·3	11 25·1	10 52·1	3·3	2·5	9·3	7·1	15·3	11·6
34	11 23·5	11 25·4	10 52·4	3·4	2·6	9·4	7·1	15·4	11·7
35	11 23·8	11 25·6	10 52·6	3·5	2·7	9·5	7·2	15·5	11·8
36	11 24·0	11 25·9	10 52·8	3·6	2·7	9·6	7·3	15·6	11·8
37	11 24·3	11 26·1	10 53·1	3·7	2·8	9·7	7·4	15·7	11·9
38	11 24·5	11 26·4	10 53·3	3·8	2·9	9·8	7·4	15·8	12·0
39	11 24·8	11 26·6	10 53·6	3·9	3·0	9·9	7·5	15·9	12·1
40	11 25·0	11 26·9	10 53·8	4·0	3·0	10·0	7·6	16·0	12·1
41	11 25·3	11 27·1	10 54·0	4·1	3·1	10·1	7·7	16·1	12·2
42	11 25·5	11 27·4	10 54·3	4·2	3·2	10·2	7·7	16·2	12·3
43	11 25·8	11 27·6	10 54·5	4·3	3·3	10·3	7·8	16·3	12·4
44	11 26·0	11 27·9	10 54·7	4·4	3·3	10·4	7·9	16·4	12·4
45	11 26·3	11 28·1	10 55·0	4·5	3·4	10·5	8·0	16·5	12·5
46	11 26·5	11 28·4	10 55·2	4·6	3·5	10·6	8·0	16·6	12·6
47	11 26·8	11 28·6	10 55·5	4·7	3·6	10·7	8·1	16·7	12·7
48	11 27·0	11 28·9	10 55·7	4·8	3·6	10·8	8·2	16·8	12·7
49	11 27·3	11 29·1	10 55·9	4·9	3·7	10·9	8·3	16·9	12·8
50	11 27·5	11 29·4	10 56·2	5·0	3·8	11·0	8·3	17·0	12·9
51	11 27·8	11 29·6	10 56·4	5·1	3·9	11·1	8·4	17·1	13·0
52	11 28·0	11 29·9	10 56·7	5·2	3·9	11·2	8·5	17·2	13·0
53	11 28·3	11 30·1	10 56·9	5·3	4·0	11·3	8·6	17·3	13·1
54	11 28·5	11 30·4	10 57·1	5·4	4·1	11·4	8·6	17·4	13·2
55	11 28·8	11 30·6	10 57·4	5·5	4·2	11·5	8·7	17·5	13·3
56	11 29·0	11 30·9	10 57·6	5·6	4·2	11·6	8·8	17·6	13·3
57	11 29·3	11 31·1	10 57·9	5·7	4·3	11·7	8·9	17·7	13·4
58	11 29·5	11 31·4	10 58·1	5·8	4·4	11·8	8·9	17·8	13·5
59	11 29·8	11 31·6	10 58·3	5·9	4·5	11·9	9·0	17·9	13·6
60	11 30·0	11 31·9	10 58·6	6·0	4·6	12·0	9·1	18·0	13·7

46m s	SUN PLANETS ° ′	ARIES ° ′	MOON ° ′	v or d ′	Corrn ′	v or d ′	Corrn ′	v or d ′	Corrn ′
00	11 30·0	11 31·9	10 58·6	0·0	0·0	6·0	4·7	12·0	9·3
01	11 30·3	11 32·1	10 58·8	0·1	0·1	6·1	4·7	12·1	9·4
02	11 30·5	11 32·4	10 59·0	0·2	0·2	6·2	4·8	12·2	9·5
03	11 30·8	11 32·6	10 59·3	0·3	0·2	6·3	4·9	12·3	9·5
04	11 31·0	11 32·9	10 59·5	0·4	0·3	6·4	5·0	12·4	9·6
05	11 31·3	11 33·1	10 59·8	0·5	0·4	6·5	5·0	12·5	9·7
06	11 31·5	11 33·4	11 00·0	0·6	0·5	6·6	5·1	12·6	9·8
07	11 31·8	11 33·6	11 00·2	0·7	0·5	6·7	5·2	12·7	9·8
08	11 32·0	11 33·9	11 00·5	0·8	0·6	6·8	5·3	12·8	9·9
09	11 32·3	11 34·1	11 00·7	0·9	0·7	6·9	5·3	12·9	10·0
10	11 32·5	11 34·4	11 01·0	1·0	0·8	7·0	5·4	13·0	10·1
11	11 32·8	11 34·6	11 01·2	1·1	0·9	7·1	5·5	13·1	10·2
12	11 33·0	11 34·9	11 01·4	1·2	0·9	7·2	5·6	13·2	10·2
13	11 33·3	11 35·1	11 01·7	1·3	1·0	7·3	5·7	13·3	10·3
14	11 33·5	11 35·4	11 01·9	1·4	1·1	7·4	5·7	13·4	10·4
15	11 33·8	11 35·6	11 02·1	1·5	1·2	7·5	5·8	13·5	10·5
16	11 34·0	11 35·9	11 02·4	1·6	1·2	7·6	5·9	13·6	10·5
17	11 34·3	11 36·2	11 02·6	1·7	1·3	7·7	6·0	13·7	10·6
18	11 34·5	11 36·4	11 02·9	1·8	1·4	7·8	6·0	13·8	10·7
19	11 34·8	11 36·7	11 03·1	1·9	1·5	7·9	6·1	13·9	10·8
20	11 35·0	11 36·9	11 03·3	2·0	1·6	8·0	6·2	14·0	10·9
21	11 35·3	11 37·2	11 03·6	2·1	1·6	8·1	6·3	14·1	10·9
22	11 35·5	11 37·4	11 03·8	2·2	1·7	8·2	6·4	14·2	11·0
23	11 35·8	11 37·7	11 04·1	2·3	1·8	8·3	6·4	14·3	11·1
24	11 36·0	11 37·9	11 04·3	2·4	1·9	8·4	6·5	14·4	11·2
25	11 36·3	11 38·2	11 04·5	2·5	1·9	8·5	6·6	14·5	11·2
26	11 36·5	11 38·4	11 04·8	2·6	2·0	8·6	6·7	14·6	11·3
27	11 36·8	11 38·7	11 05·0	2·7	2·1	8·7	6·7	14·7	11·4
28	11 37·0	11 38·9	11 05·2	2·8	2·2	8·8	6·8	14·8	11·5
29	11 37·3	11 39·2	11 05·5	2·9	2·2	8·9	6·9	14·9	11·5
30	11 37·5	11 39·4	11 05·7	3·0	2·3	9·0	7·0	15·0	11·6
31	11 37·8	11 39·7	11 06·0	3·1	2·4	9·1	7·1	15·1	11·7
32	11 38·0	11 39·9	11 06·2	3·2	2·5	9·2	7·1	15·2	11·8
33	11 38·3	11 40·2	11 06·4	3·3	2·6	9·3	7·2	15·3	11·9
34	11 38·5	11 40·4	11 06·7	3·4	2·6	9·4	7·3	15·4	11·9
35	11 38·8	11 40·7	11 06·9	3·5	2·7	9·5	7·4	15·5	12·0
36	11 39·0	11 40·9	11 07·2	3·6	2·8	9·6	7·4	15·6	12·1
37	11 39·3	11 41·2	11 07·4	3·7	2·9	9·7	7·5	15·7	12·2
38	11 39·5	11 41·4	11 07·6	3·8	2·9	9·8	7·6	15·8	12·2
39	11 39·8	11 41·7	11 07·9	3·9	3·0	9·9	7·7	15·9	12·3
40	11 40·0	11 41·9	11 08·1	4·0	3·1	10·0	7·8	16·0	12·4
41	11 40·3	11 42·2	11 08·3	4·1	3·2	10·1	7·8	16·1	12·5
42	11 40·5	11 42·4	11 08·6	4·2	3·3	10·2	7·9	16·2	12·6
43	11 40·8	11 42·7	11 08·8	4·3	3·3	10·3	8·0	16·3	12·6
44	11 41·0	11 42·9	11 09·1	4·4	3·4	10·4	8·1	16·4	12·7
45	11 41·3	11 43·2	11 09·3	4·5	3·5	10·5	8·1	16·5	12·8
46	11 41·5	11 43·4	11 09·5	4·6	3·6	10·6	8·2	16·6	12·9
47	11 41·8	11 43·7	11 09·8	4·7	3·6	10·7	8·3	16·7	12·9
48	11 42·0	11 43·9	11 10·0	4·8	3·7	10·8	8·4	16·8	13·0
49	11 42·3	11 44·2	11 10·3	4·9	3·8	10·9	8·4	16·9	13·1
50	11 42·5	11 44·4	11 10·5	5·0	3·9	11·0	8·5	17·0	13·2
51	11 42·8	11 44·7	11 10·7	5·1	4·0	11·1	8·6	17·1	13·3
52	11 43·0	11 44·9	11 11·0	5·2	4·0	11·2	8·7	17·2	13·3
53	11 43·3	11 45·2	11 11·2	5·3	4·1	11·3	8·8	17·3	13·4
54	11 43·5	11 45·4	11 11·5	5·4	4·2	11·4	8·8	17·4	13·5
55	11 43·8	11 45·7	11 11·7	5·5	4·3	11·5	8·9	17·5	13·6
56	11 44·0	11 45·9	11 11·9	5·6	4·3	11·6	9·0	17·6	13·6
57	11 44·3	11 46·2	11 12·2	5·7	4·4	11·7	9·1	17·7	13·7
58	11 44·5	11 46·4	11 12·4	5·8	4·5	11·8	9·1	17·8	13·8
59	11 44·8	11 46·7	11 12·6	5·9	4·6	11·9	9·2	17·9	13·9
60	11 45·0	11 46·9	11 12·9	6·0	4·7	12·0	9·3	18·0	14·0

47m s	SUN PLANETS ° ′	ARIES ° ′	MOON ° ′	v or d ′	Corrn ′	v or d ′	Corrn ′	v or d ′	Corrn ′
00	11 45·0	11 46·9	11 12·9	0·0	0·0	6·0	4·8	12·0	9·5
01	11 45·3	11 47·2	11 13·1	0·1	0·1	6·1	4·8	12·1	9·6
02	11 45·5	11 47·4	11 13·4	0·2	0·2	6·2	4·9	12·2	9·7
03	11 45·8	11 47·7	11 13·6	0·3	0·2	6·3	5·0	12·3	9·7
04	11 46·0	11 47·9	11 13·8	0·4	0·3	6·4	5·1	12·4	9·8
05	11 46·3	11 48·2	11 14·1	0·5	0·4	6·5	5·1	12·5	9·9
06	11 46·5	11 48·4	11 14·3	0·6	0·5	6·6	5·2	12·6	10·0
07	11 46·8	11 48·7	11 14·6	0·7	0·6	6·7	5·3	12·7	10·1
08	11 47·0	11 48·9	11 14·8	0·8	0·6	6·8	5·4	12·8	10·1
09	11 47·3	11 49·2	11 15·0	0·9	0·7	6·9	5·5	12·9	10·2
10	11 47·5	11 49·4	11 15·3	1·0	0·8	7·0	5·5	13·0	10·3
11	11 47·8	11 49·7	11 15·5	1·1	0·9	7·1	5·6	13·1	10·4
12	11 48·0	11 49·9	11 15·7	1·2	1·0	7·2	5·7	13·2	10·5
13	11 48·3	11 50·2	11 16·0	1·3	1·0	7·3	5·8	13·3	10·5
14	11 48·5	11 50·4	11 16·2	1·4	1·1	7·4	5·9	13·4	10·6
15	11 48·8	11 50·7	11 16·5	1·5	1·2	7·5	5·9	13·5	10·7
16	11 49·0	11 50·9	11 16·7	1·6	1·3	7·6	6·0	13·6	10·8
17	11 49·3	11 51·2	11 16·9	1·7	1·3	7·7	6·1	13·7	10·8
18	11 49·5	11 51·4	11 17·2	1·8	1·4	7·8	6·2	13·8	10·9
19	11 49·8	11 51·7	11 17·4	1·9	1·5	7·9	6·3	13·9	11·0
20	11 50·0	11 51·9	11 17·7	2·0	1·6	8·0	6·3	14·0	11·1
21	11 50·3	11 52·2	11 17·9	2·1	1·7	8·1	6·4	14·1	11·2
22	11 50·5	11 52·4	11 18·1	2·2	1·7	8·2	6·5	14·2	11·2
23	11 50·8	11 52·7	11 18·4	2·3	1·8	8·3	6·6	14·3	11·3
24	11 51·0	11 52·9	11 18·6	2·4	1·9	8·4	6·7	14·4	11·4
25	11 51·3	11 53·2	11 18·8	2·5	2·0	8·5	6·7	14·5	11·5
26	11 51·5	11 53·4	11 19·1	2·6	2·1	8·6	6·8	14·6	11·6
27	11 51·8	11 53·7	11 19·3	2·7	2·1	8·7	6·9	14·7	11·6
28	11 52·0	11 53·9	11 19·6	2·8	2·2	8·8	7·0	14·8	11·7
29	11 52·3	11 54·2	11 19·8	2·9	2·3	8·9	7·0	14·9	11·8
30	11 52·5	11 54·5	11 20·0	3·0	2·4	9·0	7·1	15·0	11·9
31	11 52·8	11 54·7	11 20·3	3·1	2·5	9·1	7·2	15·1	12·0
32	11 53·0	11 55·0	11 20·5	3·2	2·5	9·2	7·3	15·2	12·0
33	11 53·3	11 55·2	11 20·8	3·3	2·6	9·3	7·4	15·3	12·1
34	11 53·5	11 55·5	11 21·0	3·4	2·7	9·4	7·4	15·4	12·2
35	11 53·8	11 55·7	11 21·2	3·5	2·8	9·5	7·5	15·5	12·3
36	11 54·0	11 56·0	11 21·5	3·6	2·9	9·6	7·6	15·6	12·4
37	11 54·3	11 56·2	11 21·7	3·7	2·9	9·7	7·7	15·7	12·4
38	11 54·5	11 56·5	11 22·0	3·8	3·0	9·8	7·8	15·8	12·5
39	11 54·8	11 56·7	11 22·2	3·9	3·1	9·9	7·8	15·9	12·6
40	11 55·0	11 57·0	11 22·4	4·0	3·2	10·0	7·9	16·0	12·7
41	11 55·3	11 57·2	11 22·7	4·1	3·2	10·1	8·0	16·1	12·7
42	11 55·5	11 57·5	11 22·9	4·2	3·3	10·2	8·1	16·2	12·8
43	11 55·8	11 57·7	11 23·1	4·3	3·4	10·3	8·2	16·3	12·9
44	11 56·0	11 58·0	11 23·4	4·4	3·5	10·4	8·2	16·4	13·0
45	11 56·3	11 58·2	11 23·6	4·5	3·6	10·5	8·3	16·5	13·1
46	11 56·5	11 58·5	11 23·9	4·6	3·6	10·6	8·4	16·6	13·1
47	11 56·8	11 58·7	11 24·1	4·7	3·7	10·7	8·5	16·7	13·2
48	11 57·0	11 59·0	11 24·3	4·8	3·8	10·8	8·6	16·8	13·3
49	11 57·3	11 59·2	11 24·6	4·9	3·9	10·9	8·6	16·9	13·4
50	11 57·5	11 59·5	11 24·8	5·0	4·0	11·0	8·7	17·0	13·5
51	11 57·8	11 59·7	11 25·1	5·1	4·0	11·1	8·8	17·1	13·5
52	11 58·0	12 00·0	11 25·3	5·2	4·1	11·2	8·9	17·2	13·6
53	11 58·3	12 00·2	11 25·5	5·3	4·2	11·3	8·9	17·3	13·7
54	11 58·5	12 00·5	11 25·8	5·4	4·3	11·4	9·0	17·4	13·8
55	11 58·8	12 00·7	11 26·0	5·5	4·4	11·5	9·1	17·5	13·9
56	11 59·0	12 01·0	11 26·2	5·6	4·4	11·6	9·2	17·6	13·9
57	11 59·3	12 01·2	11 26·5	5·7	4·5	11·7	9·3	17·7	14·0
58	11 59·5	12 01·5	11 26·7	5·8	4·6	11·8	9·3	17·8	14·1
59	11 59·8	12 01·7	11 27·0	5·9	4·7	11·9	9·4	17·9	14·2
60	12 00·0	12 02·0	11 27·2	6·0	4·8	12·0	9·5	18·0	14·3

48 m	SUN PLANETS	ARIES	MOON	v or d	Corrn	v or d	Corrn	v or d	Corrn
s	° ′	° ′	° ′	′	′	′	′	′	′
00	12 00·0	12 02·0	11 27·2	0·0	0·0	6·0	4·9	12·0	9·7
01	12 00·3	12 02·2	11 27·4	0·1	0·1	6·1	4·9	12·1	9·8
02	12 00·5	12 02·5	11 27·7	0·2	0·2	6·2	5·0	12·2	9·9
03	12 00·8	12 02·7	11 27·9	0·3	0·2	6·3	5·1	12·3	9·9
04	12 01·0	12 03·0	11 28·2	0·4	0·3	6·4	5·2	12·4	10·0
05	12 01·3	12 03·2	11 28·4	0·5	0·4	6·5	5·3	12·5	10·1
06	12 01·5	12 03·5	11 28·6	0·6	0·5	6·6	5·3	12·6	10·2
07	12 01·8	12 03·7	11 28·9	0·7	0·6	6·7	5·4	12·7	10·3
08	12 02·0	12 04·0	11 29·1	0·8	0·6	6·8	5·5	12·8	10·3
09	12 02·3	12 04·2	11 29·3	0·9	0·7	6·9	5·6	12·9	10·4
10	12 02·5	12 04·5	11 29·6	1·0	0·8	7·0	5·7	13·0	10·5
11	12 02·8	12 04·7	11 29·8	1·1	0·9	7·1	5·7	13·1	10·6
12	12 03·0	12 05·0	11 30·1	1·2	1·0	7·2	5·8	13·2	10·7
13	12 03·3	12 05·2	11 30·3	1·3	1·1	7·3	5·9	13·3	10·8
14	12 03·5	12 05·5	11 30·5	1·4	1·1	7·4	6·0	13·4	10·8
15	12 03·8	12 05·7	11 30·8	1·5	1·2	7·5	6·1	13·5	10·9
16	12 04·0	12 06·0	11 31·0	1·6	1·3	7·6	6·1	13·6	11·0
17	12 04·3	12 06·2	11 31·3	1·7	1·4	7·7	6·2	13·7	11·1
18	12 04·5	12 06·5	11 31·5	1·8	1·5	7·8	6·3	13·8	11·2
19	12 04·8	12 06·7	11 31·7	1·9	1·5	7·9	6·4	13·9	11·2
20	12 05·0	12 07·0	11 32·0	2·0	1·6	8·0	6·5	14·0	11·3
21	12 05·3	12 07·2	11 32·2	2·1	1·7	8·1	6·5	14·1	11·4
22	12 05·5	12 07·5	11 32·4	2·2	1·8	8·2	6·6	14·2	11·5
23	12 05·8	12 07·7	11 32·7	2·3	1·9	8·3	6·7	14·3	11·6
24	12 06·0	12 08·0	11 32·9	2·4	1·9	8·4	6·8	14·4	11·6
25	12 06·3	12 08·2	11 33·2	2·5	2·0	8·5	6·9	14·5	11·7
26	12 06·5	12 08·5	11 33·4	2·6	2·1	8·6	7·0	14·6	11·8
27	12 06·8	12 08·7	11 33·6	2·7	2·2	8·7	7·0	14·7	11·9
28	12 07·0	12 09·0	11 33·9	2·8	2·3	8·8	7·1	14·8	12·0
29	12 07·3	12 09·2	11 34·1	2·9	2·3	8·9	7·2	14·9	12·0
30	12 07·5	12 09·5	11 34·4	3·0	2·4	9·0	7·3	15·0	12·1
31	12 07·8	12 09·7	11 34·6	3·1	2·5	9·1	7·4	15·1	12·2
32	12 08·0	12 10·0	11 34·8	3·2	2·6	9·2	7·4	15·2	12·3
33	12 08·3	12 10·2	11 35·1	3·3	2·7	9·3	7·5	15·3	12·4
34	12 08·5	12 10·5	11 35·3	3·4	2·7	9·4	7·6	15·4	12·4
35	12 08·8	12 10·7	11 35·6	3·5	2·8	9·5	7·7	15·5	12·5
36	12 09·0	12 11·0	11 35·8	3·6	2·9	9·6	7·8	15·6	12·6
37	12 09·3	12 11·2	11 36·0	3·7	3·0	9·7	7·8	15·7	12·7
38	12 09·5	12 11·5	11 36·3	3·8	3·1	9·8	7·9	15·8	12·8
39	12 09·8	12 11·7	11 36·5	3·9	3·2	9·9	8·0	15·9	12·9
40	12 10·0	12 12·0	11 36·7	4·0	3·2	10·0	8·1	16·0	12·9
41	12 10·3	12 12·2	11 37·0	4·1	3·3	10·1	8·2	16·1	13·0
42	12 10·5	12 12·5	11 37·2	4·2	3·4	10·2	8·2	16·2	13·1
43	12 10·8	12 12·8	11 37·5	4·3	3·5	10·3	8·3	16·3	13·2
44	12 11·0	12 13·0	11 37·7	4·4	3·6	10·4	8·4	16·4	13·3
45	12 11·3	12 13·3	11 37·9	4·5	3·6	10·5	8·5	16·5	13·3
46	12 11·5	12 13·5	11 38·2	4·6	3·7	10·6	8·6	16·6	13·4
47	12 11·8	12 13·8	11 38·4	4·7	3·8	10·7	8·6	16·7	13·5
48	12 12·0	12 14·0	11 38·7	4·8	3·9	10·8	8·7	16·8	13·6
49	12 12·3	12 14·3	11 38·9	4·9	4·0	10·9	8·8	16·9	13·7
50	12 12·5	12 14·5	11 39·1	5·0	4·0	11·0	8·9	17·0	13·7
51	12 12·8	12 14·8	11 39·4	5·1	4·1	11·1	9·0	17·1	13·8
52	12 13·0	12 15·0	11 39·6	5·2	4·2	11·2	9·1	17·2	13·9
53	12 13·3	12 15·3	11 39·8	5·3	4·3	11·3	9·1	17·3	14·0
54	12 13·5	12 15·5	11 40·1	5·4	4·4	11·4	9·2	17·4	14·1
55	12 13·8	12 15·8	11 40·3	5·5	4·4	11·5	9·3	17·5	14·1
56	12 14·0	12 16·0	11 40·6	5·6	4·5	11·6	9·4	17·6	14·2
57	12 14·3	12 16·3	11 40·8	5·7	4·6	11·7	9·5	17·7	14·3
58	12 14·5	12 16·5	11 41·0	5·8	4·7	11·8	9·5	17·8	14·4
59	12 14·8	12 16·8	11 41·3	5·9	4·8	11·9	9·6	17·9	14·5
60	12 15·0	12 17·0	11 41·5	6·0	4·9	12·0	9·7	18·0	14·6

49 m	SUN PLANETS	ARIES	MOON	v or d	Corrn	v or d	Corrn	v or d	Corrn
s	° ′	° ′	° ′	′	′	′	′	′	′
00	12 15·0	12 17·0	11 41·5	0·0	0·0	6·0	5·0	12·0	9·9
01	12 15·3	12 17·3	11 41·8	0·1	0·1	6·1	5·0	12·1	10·0
02	12 15·5	12 17·5	11 42·0	0·2	0·2	6·2	5·1	12·2	10·1
03	12 15·8	12 17·8	11 42·2	0·3	0·2	6·3	5·2	12·3	10·1
04	12 16·0	12 18·0	11 42·5	0·4	0·3	6·4	5·3	12·4	10·2
05	12 16·3	12 18·3	11 42·7	0·5	0·4	6·5	5·4	12·5	10·3
06	12 16·5	12 18·5	11 42·9	0·6	0·5	6·6	5·4	12·6	10·4
07	12 16·8	12 18·8	11 43·2	0·7	0·6	6·7	5·5	12·7	10·5
08	12 17·0	12 19·0	11 43·4	0·8	0·7	6·8	5·6	12·8	10·6
09	12 17·3	12 19·3	11 43·7	0·9	0·7	6·9	5·7	12·9	10·6
10	12 17·5	12 19·5	11 43·9	1·0	0·8	7·0	5·8	13·0	10·7
11	12 17·8	12 19·8	11 44·1	1·1	0·9	7·1	5·9	13·1	10·8
12	12 18·0	12 20·0	11 44·4	1·2	1·0	7·2	5·9	13·2	10·9
13	12 18·3	12 20·3	11 44·6	1·3	1·1	7·3	6·0	13·3	11·0
14	12 18·5	12 20·5	11 44·9	1·4	1·2	7·4	6·1	13·4	11·1
15	12 18·8	12 20·8	11 45·1	1·5	1·2	7·5	6·2	13·5	11·1
16	12 19·0	12 21·0	11 45·3	1·6	1·3	7·6	6·3	13·6	11·2
17	12 19·3	12 21·3	11 45·6	1·7	1·4	7·7	6·4	13·7	11·3
18	12 19·5	12 21·5	11 45·8	1·8	1·5	7·8	6·4	13·8	11·4
19	12 19·8	12 21·8	11 46·1	1·9	1·6	7·9	6·5	13·9	11·5
20	12 20·0	12 22·0	11 46·3	2·0	1·7	8·0	6·6	14·0	11·6
21	12 20·3	12 22·3	11 46·5	2·1	1·7	8·1	6·7	14·1	11·6
22	12 20·5	12 22·5	11 46·8	2·2	1·8	8·2	6·8	14·2	11·7
23	12 20·8	12 22·8	11 47·0	2·3	1·9	8·3	6·8	14·3	11·8
24	12 21·0	12 23·0	11 47·2	2·4	2·0	8·4	6·9	14·4	11·9
25	12 21·3	12 23·3	11 47·5	2·5	2·1	8·5	7·0	14·5	12·0
26	12 21·5	12 23·5	11 47·7	2·6	2·1	8·6	7·1	14·6	12·0
27	12 21·8	12 23·8	11 48·0	2·7	2·2	8·7	7·2	14·7	12·1
28	12 22·0	12 24·0	11 48·2	2·8	2·3	8·8	7·3	14·8	12·2
29	12 22·3	12 24·3	11 48·4	2·9	2·4	8·9	7·3	14·9	12·3
30	12 22·5	12 24·5	11 48·7	3·0	2·5	9·0	7·4	15·0	12·4
31	12 22·8	12 24·8	11 48·9	3·1	2·6	9·1	7·5	15·1	12·5
32	12 23·0	12 25·0	11 49·2	3·2	2·6	9·2	7·6	15·2	12·5
33	12 23·3	12 25·3	11 49·4	3·3	2·7	9·3	7·7	15·3	12·6
34	12 23·5	12 25·5	11 49·6	3·4	2·8	9·4	7·8	15·4	12·7
35	12 23·8	12 25·8	11 49·9	3·5	2·9	9·5	7·8	15·5	12·8
36	12 24·0	12 26·0	11 50·1	3·6	3·0	9·6	7·9	15·6	12·9
37	12 24·3	12 26·3	11 50·3	3·7	3·1	9·7	8·0	15·7	13·0
38	12 24·5	12 26·5	11 50·6	3·8	3·1	9·8	8·1	15·8	13·0
39	12 24·8	12 26·8	11 50·8	3·9	3·2	9·9	8·2	15·9	13·1
40	12 25·0	12 27·0	11 51·1	4·0	3·3	10·0	8·3	16·0	13·2
41	12 25·3	12 27·3	11 51·3	4·1	3·4	10·1	8·3	16·1	13·3
42	12 25·5	12 27·5	11 51·5	4·2	3·5	10·2	8·4	16·2	13·4
43	12 25·8	12 27·8	11 51·8	4·3	3·5	10·3	8·5	16·3	13·4
44	12 26·0	12 28·0	11 52·0	4·4	3·6	10·4	8·6	16·4	13·5
45	12 26·3	12 28·3	11 52·3	4·5	3·7	10·5	8·7	16·5	13·6
46	12 26·5	12 28·5	11 52·5	4·6	3·8	10·6	8·7	16·6	13·7
47	12 26·8	12 28·8	11 52·7	4·7	3·9	10·7	8·8	16·7	13·8
48	12 27·0	12 29·0	11 53·0	4·8	4·0	10·8	8·9	16·8	13·9
49	12 27·3	12 29·3	11 53·2	4·9	4·0	10·9	9·0	16·9	13·9
50	12 27·5	12 29·5	11 53·4	5·0	4·1	11·0	9·1	17·0	14·0
51	12 27·8	12 29·8	11 53·7	5·1	4·2	11·1	9·2	17·1	14·1
52	12 28·0	12 30·0	11 53·9	5·2	4·3	11·2	9·2	17·2	14·2
53	12 28·3	12 30·3	11 54·2	5·3	4·4	11·3	9·3	17·3	14·3
54	12 28·5	12 30·5	11 54·4	5·4	4·5	11·4	9·4	17·4	14·4
55	12 28·8	12 30·8	11 54·6	5·5	4·5	11·5	9·5	17·5	14·4
56	12 29·0	12 31·1	11 54·9	5·6	4·6	11·6	9·6	17·6	14·5
57	12 29·3	12 31·3	11 55·1	5·7	4·7	11·7	9·7	17·7	14·6
58	12 29·5	12 31·6	11 55·4	5·8	4·8	11·8	9·7	17·8	14·7
59	12 29·8	12 31·8	11 55·6	5·9	4·9	11·9	9·8	17·9	14·8
60	12 30·0	12 32·1	11 55·8	6·0	5·0	12·0	9·9	18·0	14·9

50m	SUN PLANETS	ARIES	MOON	v or d	Corrⁿ	v or d	Corrⁿ	v or d	Corrⁿ
s	° ′	° ′	° ′	′	′	′	′	′	′
00	12 30·0	12 32·1	11 55·8	0·0	0·0	6·0	5·1	12·0	10·1
01	12 30·3	12 32·3	11 56·1	0·1	0·1	6·1	5·1	12·1	10·2
02	12 30·5	12 32·6	11 56·3	0·2	0·2	6·2	5·2	12·2	10·3
03	12 30·8	12 32·8	11 56·5	0·3	0·3	6·3	5·3	12·3	10·4
04	12 31·0	12 33·1	11 56·8	0·4	0·3	6·4	5·4	12·4	10·4
05	12 31·3	12 33·3	11 57·0	0·5	0·4	6·5	5·5	12·5	10·5
06	12 31·5	12 33·6	11 57·3	0·6	0·5	6·6	5·6	12·6	10·6
07	12 31·8	12 33·8	11 57·5	0·7	0·6	6·7	5·6	12·7	10·7
08	12 32·0	12 34·1	11 57·7	0·8	0·7	6·8	5·7	12·8	10·8
09	12 32·3	12 34·3	11 58·0	0·9	0·8	6·9	5·8	12·9	10·9
10	12 32·5	12 34·6	11 58·2	1·0	0·8	7·0	5·9	13·0	10·9
11	12 32·8	12 34·8	11 58·5	1·1	0·9	7·1	6·0	13·1	11·0
12	12 33·0	12 35·1	11 58·7	1·2	1·0	7·2	6·1	13·2	11·1
13	12 33·3	12 35·3	11 58·9	1·3	1·1	7·3	6·1	13·3	11·2
14	12 33·5	12 35·6	11 59·2	1·4	1·2	7·4	6·2	13·4	11·3
15	12 33·8	12 35·8	11 59·4	1·5	1·3	7·5	6·3	13·5	11·4
16	12 34·0	12 36·1	11 59·7	1·6	1·3	7·6	6·4	13·6	11·4
17	12 34·3	12 36·3	11 59·9	1·7	1·4	7·7	6·5	13·7	11·5
18	12 34·5	12 36·6	12 00·1	1·8	1·5	7·8	6·6	13·8	11·6
19	12 34·8	12 36·8	12 00·4	1·9	1·6	7·9	6·6	13·9	11·7
20	12 35·0	12 37·1	12 00·6	2·0	1·7	8·0	6·7	14·0	11·8
21	12 35·3	12 37·3	12 00·8	2·1	1·8	8·1	6·8	14·1	11·9
22	12 35·5	12 37·6	12 01·1	2·2	1·9	8·2	6·9	14·2	12·0
23	12 35·8	12 37·8	12 01·3	2·3	1·9	8·3	7·0	14·3	12·0
24	12 36·0	12 38·1	12 01·6	2·4	2·0	8·4	7·1	14·4	12·1
25	12 36·3	12 38·3	12 01·8	2·5	2·1	8·5	7·2	14·5	12·2
26	12 36·5	12 38·6	12 02·0	2·6	2·2	8·6	7·2	14·6	12·3
27	12 36·8	12 38·8	12 02·3	2·7	2·3	8·7	7·3	14·7	12·4
28	12 37·0	12 39·1	12 02·5	2·8	2·4	8·8	7·4	14·8	12·5
29	12 37·3	12 39·3	12 02·8	2·9	2·4	8·9	7·5	14·9	12·5
30	12 37·5	12 39·6	12 03·0	3·0	2·5	9·0	7·6	15·0	12·6
31	12 37·8	12 39·8	12 03·2	3·1	2·6	9·1	7·7	15·1	12·7
32	12 38·0	12 40·1	12 03·5	3·2	2·7	9·2	7·7	15·2	12·8
33	12 38·3	12 40·3	12 03·7	3·3	2·8	9·3	7·8	15·3	12·9
34	12 38·5	12 40·6	12 03·9	3·4	2·9	9·4	7·9	15·4	13·0
35	12 38·8	12 40·8	12 04·2	3·5	2·9	9·5	8·0	15·5	13·0
36	12 39·0	12 41·1	12 04·4	3·6	3·0	9·6	8·1	15·6	13·1
37	12 39·3	12 41·3	12 04·7	3·7	3·1	9·7	8·2	15·7	13·2
38	12 39·5	12 41·6	12 04·9	3·8	3·2	9·8	8·2	15·8	13·3
39	12 39·8	12 41·8	12 05·1	3·9	3·3	9·9	8·3	15·9	13·4
40	12 40·0	12 42·1	12 05·4	4·0	3·4	10·0	8·4	16·0	13·5
41	12 40·3	12 42·3	12 05·6	4·1	3·5	10·1	8·5	16·1	13·6
42	12 40·5	12 42·6	12 05·9	4·2	3·5	10·2	8·6	16·2	13·6
43	12 40·8	12 42·8	12 06·1	4·3	3·6	10·3	8·7	16·3	13·7
44	12 41·0	12 43·1	12 06·3	4·4	3·7	10·4	8·8	16·4	13·8
45	12 41·3	12 43·3	12 06·6	4·5	3·8	10·5	8·8	16·5	13·9
46	12 41·5	12 43·6	12 06·8	4·6	3·9	10·6	8·9	16·6	14·0
47	12 41·8	12 43·8	12 07·0	4·7	4·0	10·7	9·0	16·7	14·1
48	12 42·0	12 44·1	12 07·3	4·8	4·0	10·8	9·1	16·8	14·1
49	12 42·3	12 44·3	12 07·5	4·9	4·1	10·9	9·2	16·9	14·2
50	12 42·5	12 44·6	12 07·8	5·0	4·2	11·0	9·3	17·0	14·3
51	12 42·8	12 44·8	12 08·0	5·1	4·3	11·1	9·3	17·1	14·4
52	12 43·0	12 45·1	12 08·2	5·2	4·4	11·2	9·4	17·2	14·5
53	12 43·3	12 45·3	12 08·5	5·3	4·5	11·3	9·5	17·3	14·6
54	12 43·5	12 45·6	12 08·7	5·4	4·5	11·4	9·6	17·4	14·6
55	12 43·8	12 45·8	12 09·0	5·5	4·6	11·5	9·7	17·5	14·7
56	12 44·0	12 46·1	12 09·2	5·6	4·7	11·6	9·8	17·6	14·8
57	12 44·3	12 46·3	12 09·4	5·7	4·8	11·7	9·8	17·7	14·9
58	12 44·5	12 46·6	12 09·7	5·8	4·9	11·8	9·9	17·8	15·0
59	12 44·8	12 46·8	12 09·9	5·9	5·0	11·9	10·0	17·9	15·1
60	12 45·0	12 47·1	12 10·2	6·0	5·1	12·0	10·1	18·0	15·2

51m	SUN PLANETS	ARIES	MOON	v or d	Corrⁿ	v or d	Corrⁿ	v or d	Corrⁿ
s	° ′	° ′	° ′	′	′	′	′	′	′
00	12 45·0	12 47·1	12 10·2	0·0	0·0	6·0	5·2	12·0	10·3
01	12 45·3	12 47·3	12 10·4	0·1	0·1	6·1	5·2	12·1	10·4
02	12 45·5	12 47·6	12 10·6	0·2	0·2	6·2	5·3	12·2	10·5
03	12 45·8	12 47·8	12 10·9	0·3	0·3	6·3	5·4	12·3	10·6
04	12 46·0	12 48·1	12 11·1	0·4	0·3	6·4	5·5	12·4	10·6
05	12 46·3	12 48·3	12 11·3	0·5	0·4	6·5	5·6	12·5	10·7
06	12 46·5	12 48·6	12 11·6	0·6	0·5	6·6	5·7	12·6	10·8
07	12 46·8	12 48·8	12 11·8	0·7	0·6	6·7	5·8	12·7	10·9
08	12 47·0	12 49·1	12 12·1	0·8	0·7	6·8	5·8	12·8	11·0
09	12 47·3	12 49·4	12 12·3	0·9	0·8	6·9	5·9	12·9	11·1
10	12 47·5	12 49·6	12 12·5	1·0	0·9	7·0	6·0	13·0	11·2
11	12 47·8	12 49·9	12 12·8	1·1	0·9	7·1	6·1	13·1	11·2
12	12 48·0	12 50·1	12 13·0	1·2	1·0	7·2	6·2	13·2	11·3
13	12 48·3	12 50·4	12 13·3	1·3	1·1	7·3	6·3	13·3	11·4
14	12 48·5	12 50·6	12 13·5	1·4	1·2	7·4	6·4	13·4	11·5
15	12 48·8	12 50·9	12 13·7	1·5	1·3	7·5	6·4	13·5	11·6
16	12 49·0	12 51·1	12 14·0	1·6	1·4	7·6	6·5	13·6	11·7
17	12 49·3	12 51·4	12 14·2	1·7	1·5	7·7	6·6	13·7	11·8
18	12 49·5	12 51·6	12 14·4	1·8	1·5	7·8	6·7	13·8	11·8
19	12 49·8	12 51·9	12 14·7	1·9	1·6	7·9	6·8	13·9	11·9
20	12 50·0	12 52·1	12 14·9	2·0	1·7	8·0	6·9	14·0	12·0
21	12 50·3	12 52·4	12 15·2	2·1	1·8	8·1	7·0	14·1	12·1
22	12 50·5	12 52·6	12 15·4	2·2	1·9	8·2	7·0	14·2	12·2
23	12 50·8	12 52·9	12 15·6	2·3	2·0	8·3	7·1	14·3	12·3
24	12 51·0	12 53·1	12 15·9	2·4	2·1	8·4	7·2	14·4	12·4
25	12 51·3	12 53·4	12 16·1	2·5	2·1	8·5	7·3	14·5	12·4
26	12 51·5	12 53·6	12 16·4	2·6	2·2	8·6	7·4	14·6	12·5
27	12 51·8	12 53·9	12 16·6	2·7	2·3	8·7	7·5	14·7	12·6
28	12 52·0	12 54·1	12 16·8	2·8	2·4	8·8	7·6	14·8	12·7
29	12 52·3	12 54·4	12 17·1	2·9	2·5	8·9	7·6	14·9	12·8
30	12 52·5	12 54·6	12 17·3	3·0	2·6	9·0	7·7	15·0	12·9
31	12 52·8	12 54·9	12 17·5	3·1	2·7	9·1	7·8	15·1	13·0
32	12 53·0	12 55·1	12 17·8	3·2	2·7	9·2	7·9	15·2	13·0
33	12 53·3	12 55·4	12 18·0	3·3	2·8	9·3	8·0	15·3	13·1
34	12 53·5	12 55·6	12 18·3	3·4	2·9	9·4	8·1	15·4	13·2
35	12 53·8	12 55·9	12 18·5	3·5	3·0	9·5	8·2	15·5	13·3
36	12 54·0	12 56·1	12 18·7	3·6	3·1	9·6	8·2	15·6	13·4
37	12 54·3	12 56·4	12 19·0	3·7	3·2	9·7	8·3	15·7	13·5
38	12 54·5	12 56·6	12 19·2	3·8	3·3	9·8	8·4	15·8	13·6
39	12 54·8	12 56·9	12 19·5	3·9	3·3	9·9	8·5	15·9	13·6
40	12 55·0	12 57·1	12 19·7	4·0	3·4	10·0	8·6	16·0	13·7
41	12 55·3	12 57·4	12 19·9	4·1	3·5	10·1	8·7	16·1	13·8
42	12 55·5	12 57·6	12 20·2	4·2	3·6	10·2	8·8	16·2	13·9
43	12 55·8	12 57·9	12 20·4	4·3	3·7	10·3	8·8	16·3	14·0
44	12 56·0	12 58·1	12 20·6	4·4	3·8	10·4	8·9	16·4	14·1
45	12 56·3	12 58·4	12 20·9	4·5	3·9	10·5	9·0	16·5	14·2
46	12 56·5	12 58·6	12 21·1	4·6	3·9	10·6	9·1	16·6	14·2
47	12 56·8	12 58·9	12 21·4	4·7	4·0	10·7	9·2	16·7	14·3
48	12 57·0	12 59·1	12 21·6	4·8	4·1	10·8	9·3	16·8	14·4
49	12 57·3	12 59·4	12 21·8	4·9	4·2	10·9	9·4	16·9	14·5
50	12 57·5	12 59·6	12 22·1	5·0	4·3	11·0	9·4	17·0	14·6
51	12 57·8	12 59·9	12 22·3	5·1	4·4	11·1	9·5	17·1	14·7
52	12 58·0	13 00·1	12 22·6	5·2	4·5	11·2	9·6	17·2	14·8
53	12 58·3	13 00·4	12 22·8	5·3	4·5	11·3	9·7	17·3	14·8
54	12 58·5	13 00·6	12 23·0	5·4	4·6	11·4	9·8	17·4	14·9
55	12 58·8	13 00·9	12 23·3	5·5	4·7	11·5	9·9	17·5	15·0
56	12 59·0	13 01·1	12 23·5	5·6	4·8	11·6	10·0	17·6	15·1
57	12 59·3	13 01·4	12 23·8	5·7	4·9	11·7	10·0	17·7	15·2
58	12 59·5	13 01·6	12 24·0	5·8	5·0	11·8	10·1	17·8	15·3
59	12 59·8	13 01·9	12 24·2	5·9	5·1	11·9	10·2	17·9	15·4
60	13 00·0	13 02·1	12 24·5	6·0	5·2	12·0	10·3	18·0	15·5

52m	SUN PLANETS	ARIES	MOON	v or d	Corrn	v or d	Corrn	v or d	Corrn
s	° ′	° ′	° ′	′	′	′	′	′	′
00	13 00·0	13 02·1	12 24·5	0·0	0·0	6·0	5·3	12·0	10·5
01	13 00·3	13 02·4	12 24·7	0·1	0·1	6·1	5·3	12·1	10·6
02	13 00·5	13 02·6	12 24·9	0·2	0·2	6·2	5·4	12·2	10·7
03	13 00·8	13 02·9	12 25·2	0·3	0·3	6·3	5·5	12·3	10·8
04	13 01·0	13 03·1	12 25·4	0·4	0·4	6·4	5·6	12·4	10·9
05	13 01·3	13 03·4	12 25·7	0·5	0·4	6·5	5·7	12·5	10·9
06	13 01·5	13 03·6	12 25·9	0·6	0·5	6·6	5·8	12·6	11·0
07	13 01·8	13 03·9	12 26·1	0·7	0·6	6·7	5·9	12·7	11·1
08	13 02·0	13 04·1	12 26·4	0·8	0·7	6·8	6·0	12·8	11·2
09	13 02·3	13 04·4	12 26·6	0·9	0·8	6·9	6·0	12·9	11·3
10	13 02·5	13 04·6	12 26·9	1·0	0·9	7·0	6·1	13·0	11·4
11	13 02·8	13 04·9	12 27·1	1·1	1·0	7·1	6·2	13·1	11·5
12	13 03·0	13 05·1	12 27·3	1·2	1·1	7·2	6·3	13·2	11·6
13	13 03·3	13 05·4	12 27·6	1·3	1·1	7·3	6·4	13·3	11·6
14	13 03·5	13 05·6	12 27·8	1·4	1·2	7·4	6·5	13·4	11·7
15	13 03·8	13 05·9	12 28·0	1·5	1·3	7·5	6·6	13·5	11·8
16	13 04·0	13 06·1	12 28·3	1·6	1·4	7·6	6·7	13·6	11·9
17	13 04·3	13 06·4	12 28·5	1·7	1·5	7·7	6·7	13·7	12·0
18	13 04·5	13 06·6	12 28·8	1·8	1·6	7·8	6·8	13·8	12·1
19	13 04·8	13 06·9	12 29·0	1·9	1·7	7·9	6·9	13·9	12·2
20	13 05·0	13 07·1	12 29·2	2·0	1·8	8·0	7·0	14·0	12·3
21	13 05·3	13 07·4	12 29·5	2·1	1·8	8·1	7·1	14·1	12·3
22	13 05·5	13 07·7	12 29·7	2·2	1·9	8·2	7·2	14·2	12·4
23	13 05·8	13 07·9	12 30·0	2·3	2·0	8·3	7·3	14·3	12·5
24	13 06·0	13 08·2	12 30·2	2·4	2·1	8·4	7·4	14·4	12·6
25	13 06·3	13 08·4	12 30·4	2·5	2·2	8·5	7·4	14·5	12·7
26	13 06·5	13 08·7	12 30·7	2·6	2·3	8·6	7·5	14·6	12·8
27	13 06·8	13 08·9	12 30·9	2·7	2·4	8·7	7·6	14·7	12·9
28	13 07·0	13 09·2	12 31·1	2·8	2·5	8·8	7·7	14·8	13·0
29	13 07·3	13 09·4	12 31·4	2·9	2·5	8·9	7·8	14·9	13·0
30	13 07·5	13 09·7	12 31·6	3·0	2·6	9·0	7·9	15·0	13·1
31	13 07·8	13 09·9	12 31·9	3·1	2·7	9·1	8·0	15·1	13·2
32	13 08·0	13 10·2	12 32·1	3·2	2·8	9·2	8·0	15·2	13·3
33	13 08·3	13 10·4	12 32·3	3·3	2·9	9·3	8·1	15·3	13·4
34	13 08·5	13 10·7	12 32·6	3·4	3·0	9·4	8·2	15·4	13·5
35	13 08·8	13 10·9	12 32·8	3·5	3·1	9·5	8·3	15·5	13·6
36	13 09·0	13 11·2	12 33·1	3·6	3·2	9·6	8·4	15·6	13·7
37	13 09·3	13 11·4	12 33·3	3·7	3·2	9·7	8·5	15·7	13·7
38	13 09·5	13 11·7	12 33·5	3·8	3·3	9·8	8·6	15·8	13·8
39	13 09·8	13 11·9	12 33·8	3·9	3·4	9·9	8·7	15·9	13·9
40	13 10·0	13 12·2	12 34·0	4·0	3·5	10·0	8·8	16·0	14·0
41	13 10·3	13 12·4	12 34·2	4·1	3·6	10·1	8·8	16·1	14·1
42	13 10·5	13 12·7	12 34·5	4·2	3·7	10·2	8·9	16·2	14·2
43	13 10·8	13 12·9	12 34·7	4·3	3·8	10·3	9·0	16·3	14·3
44	13 11·0	13 13·2	12 35·0	4·4	3·9	10·4	9·1	16·4	14·3
45	13 11·3	13 13·4	12 35·2	4·5	3·9	10·5	9·2	16·5	14·4
46	13 11·5	13 13·7	12 35·4	4·6	4·0	10·6	9·3	16·6	14·5
47	13 11·8	13 13·9	12 35·7	4·7	4·1	10·7	9·4	16·7	14·6
48	13 12·0	13 14·2	12 35·9	4·8	4·2	10·8	9·5	16·8	14·7
49	13 12·3	13 14·4	12 36·2	4·9	4·3	10·9	9·5	16·9	14·8
50	13 12·5	13 14·7	12 36·4	5·0	4·4	11·0	9·6	17·0	14·9
51	13 12·8	13 14·9	12 36·6	5·1	4·5	11·1	9·7	17·1	15·0
52	13 13·0	13 15·2	12 36·9	5·2	4·6	11·2	9·8	17·2	15·1
53	13 13·3	13 15·4	12 37·1	5·3	4·6	11·3	9·9	17·3	15·1
54	13 13·5	13 15·7	12 37·4	5·4	4·7	11·4	10·0	17·4	15·2
55	13 13·8	13 15·9	12 37·6	5·5	4·8	11·5	10·1	17·5	15·3
56	13 14·0	13 16·2	12 37·8	5·6	4·9	11·6	10·2	17·6	15·4
57	13 14·3	13 16·4	12 38·1	5·7	5·0	11·7	10·2	17·7	15·5
58	13 14·5	13 16·7	12 38·3	5·8	5·1	11·8	10·3	17·8	15·6
59	13 14·8	13 16·9	12 38·5	5·9	5·2	11·9	10·4	17·9	15·7
60	13 15·0	13 17·2	12 38·8	6·0	5·3	12·0	10·5	18·0	15·8

53m	SUN PLANETS	ARIES	MOON	v or d	Corrn	v or d	Corrn	v or d	Corrn
s	° ′	° ′	° ′	′	′	′	′	′	′
00	13 15·0	13 17·2	12 38·8	0·0	0·0	6·0	5·4	12·0	10·7
01	13 15·3	13 17·4	12 39·0	0·1	0·1	6·1	5·4	12·1	10·8
02	13 15·5	13 17·7	12 39·3	0·2	0·2	6·2	5·5	12·2	10·9
03	13 15·8	13 17·9	12 39·5	0·3	0·3	6·3	5·6	12·3	11·0
04	13 16·0	13 18·2	12 39·7	0·4	0·4	6·4	5·7	12·4	11·1
05	13 16·3	13 18·4	12 40·0	0·5	0·4	6·5	5·8	12·5	11·1
06	13 16·5	13 18·7	12 40·2	0·6	0·5	6·6	5·9	12·6	11·2
07	13 16·8	13 18·9	12 40·5	0·7	0·6	6·7	6·0	12·7	11·3
08	13 17·0	13 19·2	12 40·7	0·8	0·7	6·8	6·1	12·8	11·4
09	13 17·3	13 19·4	12 40·9	0·9	0·8	6·9	6·2	12·9	11·5
10	13 17·5	13 19·7	12 41·2	1·0	0·9	7·0	6·2	13·0	11·6
11	13 17·8	13 19·9	12 41·4	1·1	1·0	7·1	6·3	13·1	11·7
12	13 18·0	13 20·2	12 41·6	1·2	1·1	7·2	6·4	13·2	11·8
13	13 18·3	13 20·4	12 41·9	1·3	1·2	7·3	6·5	13·3	11·9
14	13 18·5	13 20·7	12 42·1	1·4	1·2	7·4	6·6	13·4	11·9
15	13 18·8	13 20·9	12 42·4	1·5	1·3	7·5	6·7	13·5	12·0
16	13 19·0	13 21·2	12 42·6	1·6	1·4	7·6	6·8	13·6	12·1
17	13 19·3	13 21·4	12 42·8	1·7	1·5	7·7	6·9	13·7	12·2
18	13 19·5	13 21·7	12 43·1	1·8	1·6	7·8	7·0	13·8	12·3
19	13 19·8	13 21·9	12 43·3	1·9	1·7	7·9	7·0	13·9	12·4
20	13 20·0	13 22·2	12 43·6	2·0	1·8	8·0	7·1	14·0	12·5
21	13 20·3	13 22·4	12 43·8	2·1	1·9	8·1	7·2	14·1	12·6
22	13 20·5	13 22·7	12 44·0	2·2	2·0	8·2	7·3	14·2	12·7
23	13 20·8	13 22·9	12 44·3	2·3	2·1	8·3	7·4	14·3	12·8
24	13 21·0	13 23·2	12 44·5	2·4	2·1	8·4	7·5	14·4	12·8
25	13 21·3	13 23·4	12 44·7	2·5	2·2	8·5	7·6	14·5	12·9
26	13 21·5	13 23·7	12 45·0	2·6	2·3	8·6	7·7	14·6	13·0
27	13 21·8	13 23·9	12 45·2	2·7	2·4	8·7	7·8	14·7	13·1
28	13 22·0	13 24·2	12 45·5	2·8	2·5	8·8	7·8	14·8	13·2
29	13 22·3	13 24·4	12 45·7	2·9	2·6	8·9	7·9	14·9	13·3
30	13 22·5	13 24·7	12 45·9	3·0	2·7	9·0	8·0	15·0	13·4
31	13 22·8	13 24·9	12 46·2	3·1	2·8	9·1	8·1	15·1	13·5
32	13 23·0	13 25·2	12 46·4	3·2	2·9	9·2	8·2	15·2	13·6
33	13 23·3	13 25·4	12 46·7	3·3	2·9	9·3	8·3	15·3	13·6
34	13 23·5	13 25·7	12 46·9	3·4	3·0	9·4	8·4	15·4	13·7
35	13 23·8	13 26·0	12 47·1	3·5	3·1	9·5	8·5	15·5	13·8
36	13 24·0	13 26·2	12 47·4	3·6	3·2	9·6	8·6	15·6	13·9
37	13 24·3	13 26·5	12 47·6	3·7	3·3	9·7	8·6	15·7	14·0
38	13 24·5	13 26·7	12 47·9	3·8	3·4	9·8	8·7	15·8	14·1
39	13 24·8	13 27·0	12 48·1	3·9	3·5	9·9	8·8	15·9	14·2
40	13 25·0	13 27·2	12 48·3	4·0	3·6	10·0	8·9	16·0	14·3
41	13 25·3	13 27·5	12 48·6	4·1	3·7	10·1	9·0	16·1	14·4
42	13 25·5	13 27·7	12 48·8	4·2	3·7	10·2	9·1	16·2	14·4
43	13 25·8	13 28·0	12 49·0	4·3	3·8	10·3	9·2	16·3	14·5
44	13 26·0	13 28·2	12 49·3	4·4	3·9	10·4	9·3	16·4	14·6
45	13 26·3	13 28·5	12 49·5	4·5	4·0	10·5	9·4	16·5	14·7
46	13 26·5	13 28·7	12 49·8	4·6	4·1	10·6	9·5	16·6	14·8
47	13 26·8	13 29·0	12 50·0	4·7	4·2	10·7	9·5	16·7	14·9
48	13 27·0	13 29·2	12 50·2	4·8	4·3	10·8	9·6	16·8	15·0
49	13 27·3	13 29·5	12 50·5	4·9	4·4	10·9	9·7	16·9	15·1
50	13 27·5	13 29·7	12 50·7	5·0	4·5	11·0	9·8	17·0	15·2
51	13 27·8	13 30·0	12 51·0	5·1	4·5	11·1	9·9	17·1	15·2
52	13 28·0	13 30·2	12 51·2	5·2	4·6	11·2	10·0	17·2	15·3
53	13 28·3	13 30·5	12 51·4	5·3	4·7	11·3	10·1	17·3	15·4
54	13 28·5	13 30·7	12 51·7	5·4	4·8	11·4	10·2	17·4	15·5
55	13 28·8	13 31·0	12 51·9	5·5	4·9	11·5	10·3	17·5	15·6
56	13 29·0	13 31·2	12 52·1	5·6	5·0	11·6	10·3	17·6	15·7
57	13 29·3	13 31·5	12 52·4	5·7	5·1	11·7	10·4	17·7	15·8
58	13 29·5	13 31·7	12 52·6	5·8	5·2	11·8	10·5	17·8	15·9
59	13 29·8	13 32·0	12 52·9	5·9	5·3	11·9	10·6	17·9	16·0
60	13 30·0	13 32·2	12 53·1	6·0	5·4	12·0	10·7	18·0	16·1

54m s	SUN PLANETS ° ′	ARIES ° ′	MOON ° ′	v or d ′	Corrn ′	v or d ′	Corrn ′	v or d ′	Corrn ′
00	13 30·0	13 32·2	12 53·1	0·0	0·0	6·0	5·5	12·0	10·9
01	13 30·3	13 32·5	12 53·3	0·1	0·1	6·1	5·5	12·1	11·0
02	13 30·5	13 32·7	12 53·6	0·2	0·2	6·2	5·6	12·2	11·1
03	13 30·8	13 33·0	12 53·8	0·3	0·3	6·3	5·7	12·3	11·2
04	13 31·0	13 33·2	12 54·1	0·4	0·4	6·4	5·8	12·4	11·3
05	13 31·3	13 33·5	12 54·3	0·5	0·5	6·5	5·9	12·5	11·4
06	13 31·5	13 33·7	12 54·5	0·6	0·5	6·6	6·0	12·6	11·4
07	13 31·8	13 34·0	12 54·8	0·7	0·6	6·7	6·1	12·7	11·5
08	13 32·0	13 34·2	12 55·0	0·8	0·7	6·8	6·2	12·8	11·6
09	13 32·3	13 34·5	12 55·2	0·9	0·8	6·9	6·3	12·9	11·7
10	13 32·5	13 34·7	12 55·5	1·0	0·9	7·0	6·4	13·0	11·8
11	13 32·8	13 35·0	12 55·7	1·1	1·0	7·1	6·4	13·1	11·9
12	13 33·0	13 35·2	12 56·0	1·2	1·1	7·2	6·5	13·2	12·0
13	13 33·3	13 35·5	12 56·2	1·3	1·2	7·3	6·6	13·3	12·1
14	13 33·5	13 35·7	12 56·4	1·4	1·3	7·4	6·7	13·4	12·2
15	13 33·8	13 36·0	12 56·7	1·5	1·4	7·5	6·8	13·5	12·3
16	13 34·0	13 36·2	12 56·9	1·6	1·5	7·6	6·9	13·6	12·4
17	13 34·3	13 36·5	12 57·2	1·7	1·5	7·7	7·0	13·7	12·4
18	13 34·5	13 36·7	12 57·4	1·8	1·6	7·8	7·1	13·8	12·5
19	13 34·8	13 37·0	12 57·6	1·9	1·7	7·9	7·2	13·9	12·6
20	13 35·0	13 37·2	12 57·9	2·0	1·8	8·0	7·3	14·0	12·7
21	13 35·3	13 37·5	12 58·1	2·1	1·9	8·1	7·4	14·1	12·8
22	13 35·5	13 37·7	12 58·3	2·2	2·0	8·2	7·4	14·2	12·9
23	13 35·8	13 38·0	12 58·6	2·3	2·1	8·3	7·5	14·3	13·0
24	13 36·0	13 38·2	12 58·8	2·4	2·2	8·4	7·6	14·4	13·1
25	13 36·3	13 38·5	12 59·1	2·5	2·3	8·5	7·7	14·5	13·2
26	13 36·5	13 38·7	12 59·3	2·6	2·4	8·6	7·8	14·6	13·3
27	13 36·8	13 39·0	12 59·5	2·7	2·5	8·7	7·9	14·7	13·4
28	13 37·0	13 39·2	12 59·8	2·8	2·5	8·8	8·0	14·8	13·4
29	13 37·3	13 39·5	13 00·0	2·9	2·6	8·9	8·1	14·9	13·5
30	13 37·5	13 39·7	13 00·3	3·0	2·7	9·0	8·2	15·0	13·6
31	13 37·8	13 40·0	13 00·5	3·1	2·8	9·1	8·3	15·1	13·7
32	13 38·0	13 40·2	13 00·7	3·2	2·9	9·2	8·4	15·2	13·8
33	13 38·3	13 40·5	13 01·0	3·3	3·0	9·3	8·4	15·3	13·9
34	13 38·5	13 40·7	13 01·2	3·4	3·1	9·4	8·5	15·4	14·0
35	13 38·8	13 41·0	13 01·5	3·5	3·2	9·5	8·6	15·5	14·1
36	13 39·0	13 41·2	13 01·7	3·6	3·3	9·6	8·7	15·6	14·2
37	13 39·3	13 41·5	13 01·9	3·7	3·4	9·7	8·8	15·7	14·3
38	13 39·5	13 41·7	13 02·2	3·8	3·5	9·8	8·9	15·8	14·4
39	13 39·8	13 42·0	13 02·4	3·9	3·5	9·9	9·0	15·9	14·4
40	13 40·0	13 42·2	13 02·6	4·0	3·6	10·0	9·1	16·0	14·5
41	13 40·3	13 42·5	13 02·9	4·1	3·7	10·1	9·2	16·1	14·6
42	13 40·5	13 42·7	13 03·1	4·2	3·8	10·2	9·3	16·2	14·7
43	13 40·8	13 43·0	13 03·4	4·3	3·9	10·3	9·4	16·3	14·8
44	13 41·0	13 43·2	13 03·6	4·4	4·0	10·4	9·4	16·4	14·9
45	13 41·3	13 43·5	13 03·8	4·5	4·1	10·5	9·5	16·5	15·0
46	13 41·5	13 43·7	13 04·1	4·6	4·2	10·6	9·6	16·6	15·1
47	13 41·8	13 44·0	13 04·3	4·7	4·3	10·7	9·7	16·7	15·2
48	13 42·0	13 44·3	13 04·6	4·8	4·4	10·8	9·8	16·8	15·3
49	13 42·3	13 44·5	13 04·8	4·9	4·5	10·9	9·9	16·9	15·4
50	13 42·5	13 44·8	13 05·0	5·0	4·5	11·0	10·0	17·0	15·4
51	13 42·8	13 45·0	13 05·3	5·1	4·6	11·1	10·1	17·1	15·5
52	13 43·0	13 45·3	13 05·5	5·2	4·7	11·2	10·2	17·2	15·6
53	13 43·3	13 45·5	13 05·7	5·3	4·8	11·3	10·3	17·3	15·7
54	13 43·5	13 45·8	13 06·0	5·4	4·9	11·4	10·4	17·4	15·8
55	13 43·8	13 46·0	13 06·2	5·5	5·0	11·5	10·4	17·5	15·9
56	13 44·0	13 46·3	13 06·5	5·6	5·1	11·6	10·5	17·6	16·0
57	13 44·3	13 46·5	13 06·7	5·7	5·2	11·7	10·6	17·7	16·1
58	13 44·5	13 46·8	13 06·9	5·8	5·3	11·8	10·7	17·8	16·2
59	13 44·8	13 47·0	13 07·2	5·9	5·4	11·9	10·8	17·9	16·3
60	13 45·0	13 47·3	13 07·4	6·0	5·5	12·0	10·9	18·0	16·4

55m s	SUN PLANETS ° ′	ARIES ° ′	MOON ° ′	v or d ′	Corrn ′	v or d ′	Corrn ′	v or d ′	Corrn ′
00	13 45·0	13 47·3	13 07·4	0·0	0·0	6·0	5·6	12·0	11·1
01	13 45·3	13 47·5	13 07·7	0·1	0·1	6·1	5·6	12·1	11·2
02	13 45·5	13 47·8	13 07·9	0·2	0·2	6·2	5·7	12·2	11·3
03	13 45·8	13 48·0	13 08·1	0·3	0·3	6·3	5·8	12·3	11·4
04	13 46·0	13 48·3	13 08·4	0·4	0·4	6·4	5·9	12·4	11·5
05	13 46·3	13 48·5	13 08·6	0·5	0·5	6·5	6·0	12·5	11·6
06	13 46·5	13 48·8	13 08·8	0·6	0·6	6·6	6·1	12·6	11·7
07	13 46·8	13 49·0	13 09·1	0·7	0·6	6·7	6·2	12·7	11·7
08	13 47·0	13 49·3	13 09·3	0·8	0·7	6·8	6·3	12·8	11·8
09	13 47·3	13 49·5	13 09·6	0·9	0·8	6·9	6·4	12·9	11·9
10	13 47·5	13 49·8	13 09·8	1·0	0·9	7·0	6·5	13·0	12·0
11	13 47·8	13 50·0	13 10·0	1·1	1·0	7·1	6·6	13·1	12·1
12	13 48·0	13 50·3	13 10·3	1·2	1·1	7·2	6·7	13·2	12·2
13	13 48·3	13 50·5	13 10·5	1·3	1·2	7·3	6·8	13·3	12·3
14	13 48·5	13 50·8	13 10·8	1·4	1·3	7·4	6·8	13·4	12·4
15	13 48·8	13 51·0	13 11·0	1·5	1·4	7·5	6·9	13·5	12·5
16	13 49·0	13 51·3	13 11·2	1·6	1·5	7·6	7·0	13·6	12·6
17	13 49·3	13 51·5	13 11·5	1·7	1·6	7·7	7·1	13·7	12·7
18	13 49·5	13 51·8	13 11·7	1·8	1·7	7·8	7·2	13·8	12·8
19	13 49·8	13 52·0	13 12·0	1·9	1·8	7·9	7·3	13·9	12·9
20	13 50·0	13 52·3	13 12·2	2·0	1·9	8·0	7·4	14·0	13·0
21	13 50·3	13 52·5	13 12·4	2·1	1·9	8·1	7·5	14·1	13·0
22	13 50·5	13 52·8	13 12·7	2·2	2·0	8·2	7·6	14·2	13·1
23	13 50·8	13 53·0	13 12·9	2·3	2·1	8·3	7·7	14·3	13·2
24	13 51·0	13 53·3	13 13·1	2·4	2·2	8·4	7·8	14·4	13·3
25	13 51·3	13 53·5	13 13·4	2·5	2·3	8·5	7·9	14·5	13·4
26	13 51·5	13 53·8	13 13·6	2·6	2·4	8·6	8·0	14·6	13·5
27	13 51·8	13 54·0	13 13·9	2·7	2·5	8·7	8·0	14·7	13·6
28	13 52·0	13 54·3	13 14·1	2·8	2·6	8·8	8·1	14·8	13·7
29	13 52·3	13 54·5	13 14·3	2·9	2·7	8·9	8·2	14·9	13·8
30	13 52·5	13 54·8	13 14·6	3·0	2·8	9·0	8·3	15·0	13·9
31	13 52·8	13 55·0	13 14·8	3·1	2·9	9·1	8·4	15·1	14·0
32	13 53·0	13 55·3	13 15·1	3·2	3·0	9·2	8·5	15·2	14·1
33	13 53·3	13 55·5	13 15·3	3·3	3·1	9·3	8·6	15·3	14·2
34	13 53·5	13 55·8	13 15·5	3·4	3·1	9·4	8·7	15·4	14·2
35	13 53·8	13 56·0	13 15·8	3·5	3·2	9·5	8·8	15·5	14·3
36	13 54·0	13 56·3	13 16·0	3·6	3·3	9·6	8·9	15·6	14·4
37	13 54·3	13 56·5	13 16·2	3·7	3·4	9·7	9·0	15·7	14·5
38	13 54·5	13 56·8	13 16·5	3·8	3·5	9·8	9·1	15·8	14·6
39	13 54·8	13 57·0	13 16·7	3·9	3·6	9·9	9·2	15·9	14·7
40	13 55·0	13 57·3	13 17·0	4·0	3·7	10·0	9·3	16·0	14·8
41	13 55·3	13 57·5	13 17·2	4·1	3·8	10·1	9·3	16·1	14·9
42	13 55·5	13 57·8	13 17·4	4·2	3·9	10·2	9·4	16·2	15·0
43	13 55·8	13 58·0	13 17·7	4·3	4·0	10·3	9·5	16·3	15·1
44	13 56·0	13 58·3	13 17·9	4·4	4·1	10·4	9·6	16·4	15·2
45	13 56·3	13 58·5	13 18·2	4·5	4·2	10·5	9·7	16·5	15·3
46	13 56·5	13 58·8	13 18·4	4·6	4·3	10·6	9·8	16·6	15·4
47	13 56·8	13 59·0	13 18·6	4·7	4·3	10·7	9·9	16·7	15·4
48	13 57·0	13 59·3	13 18·9	4·8	4·4	10·8	10·0	16·8	15·5
49	13 57·3	13 59·5	13 19·1	4·9	4·5	10·9	10·1	16·9	15·6
50	13 57·5	13 59·8	13 19·3	5·0	4·6	11·0	10·2	17·0	15·7
51	13 57·8	14 00·0	13 19·6	5·1	4·7	11·1	10·3	17·1	15·8
52	13 58·0	14 00·3	13 19·8	5·2	4·8	11·2	10·4	17·2	15·9
53	13 58·3	14 00·5	13 20·1	5·3	4·9	11·3	10·5	17·3	16·0
54	13 58·5	14 00·8	13 20·3	5·4	5·0	11·4	10·5	17·4	16·1
55	13 58·8	14 01·0	13 20·5	5·5	5·1	11·5	10·6	17·5	16·2
56	13 59·0	14 01·3	13 20·8	5·6	5·2	11·6	10·7	17·6	16·3
57	13 59·3	14 01·5	13 21·0	5·7	5·3	11·7	10·8	17·7	16·4
58	13 59·5	14 01·8	13 21·3	5·8	5·4	11·8	10·9	17·8	16·5
59	13 59·8	14 02·0	13 21·5	5·9	5·5	11·9	11·0	17·9	16·6
60	14 00·0	14 02·3	13 21·7	6·0	5·6	12·0	11·1	18·0	16·7

56^{m} s	SUN PLANETS ° ′	ARIES ° ′	MOON ° ′	v or d ′	Corr^{n} ′	v or d ′	Corr^{n} ′	v or d ′	Corr^{n} ′
00	14 00·0	14 02·3	13 21·7	0·0	0·0	6·0	5·7	12·0	11·3
01	14 00·3	14 02·6	13 22·0	0·1	0·1	6·1	5·7	12·1	11·4
02	14 00·5	14 02·8	13 22·2	0·2	0·2	6·2	5·8	12·2	11·5
03	14 00·8	14 03·1	13 22·4	0·3	0·3	6·3	5·9	12·3	11·6
04	14 01·0	14 03·3	13 22·7	0·4	0·4	6·4	6·0	12·4	11·7
05	14 01·3	14 03·6	13 22·9	0·5	0·5	6·5	6·1	12·5	11·8
06	14 01·5	14 03·8	13 23·2	0·6	0·6	6·6	6·2	12·6	11·9
07	14 01·8	14 04·1	13 23·4	0·7	0·7	6·7	6·3	12·7	12·0
08	14 02·0	14 04·3	13 23·6	0·8	0·8	6·8	6·4	12·8	12·1
09	14 02·3	14 04·6	13 23·9	0·9	0·8	6·9	6·5	12·9	12·1
10	14 02·5	14 04·8	13 24·1	1·0	0·9	7·0	6·6	13·0	12·2
11	14 02·8	14 05·1	13 24·4	1·1	1·0	7·1	6·7	13·1	12·3
12	14 03·0	14 05·3	13 24·6	1·2	1·1	7·2	6·8	13·2	12·4
13	14 03·3	14 05·6	13 24·8	1·3	1·2	7·3	6·9	13·3	12·5
14	14 03·5	14 05·8	13 25·1	1·4	1·3	7·4	7·0	13·4	12·6
15	14 03·8	14 06·1	13 25·3	1·5	1·4	7·5	7·1	13·5	12·7
16	14 04·0	14 06·3	13 25·6	1·6	1·5	7·6	7·2	13·6	12·8
17	14 04·3	14 06·6	13 25·8	1·7	1·6	7·7	7·3	13·7	12·9
18	14 04·5	14 06·8	13 26·0	1·8	1·7	7·8	7·3	13·8	13·0
19	14 04·8	14 07·1	13 26·3	1·9	1·8	7·9	7·4	13·9	13·1
20	14 05·0	14 07·3	13 26·5	2·0	1·9	8·0	7·5	14·0	13·2
21	14 05·3	14 07·6	13 26·7	2·1	2·0	8·1	7·6	14·1	13·3
22	14 05·5	14 07·8	13 27·0	2·2	2·1	8·2	7·7	14·2	13·4
23	14 05·8	14 08·1	13 27·2	2·3	2·2	8·3	7·8	14·3	13·5
24	14 06·0	14 08·3	13 27·5	2·4	2·3	8·4	7·9	14·4	13·6
25	14 06·3	14 08·6	13 27·7	2·5	2·4	8·5	8·0	14·5	13·7
26	14 06·5	14 08·8	13 27·9	2·6	2·4	8·6	8·1	14·6	13·7
27	14 06·8	14 09·1	13 28·2	2·7	2·5	8·7	8·2	14·7	13·8
28	14 07·0	14 09·3	13 28·4	2·8	2·6	8·8	8·3	14·8	13·9
29	14 07·3	14 09·6	13 28·7	2·9	2·7	8·9	8·4	14·9	14·0
30	14 07·5	14 09·8	13 28·9	3·0	2·8	9·0	8·5	15·0	14·1
31	14 07·8	14 10·1	13 29·1	3·1	2·9	9·1	8·6	15·1	14·2
32	14 08·0	14 10·3	13 29·4	3·2	3·0	9·2	8·7	15·2	14·3
33	14 08·3	14 10·6	13 29·6	3·3	3·1	9·3	8·8	15·3	14·4
34	14 08·5	14 10·8	13 29·8	3·4	3·2	9·4	8·9	15·4	14·5
35	14 08·8	14 11·1	13 30·1	3·5	3·3	9·5	8·9	15·5	14·6
36	14 09·0	14 11·3	13 30·3	3·6	3·4	9·6	9·0	15·6	14·7
37	14 09·3	14 11·6	13 30·6	3·7	3·5	9·7	9·1	15·7	14·8
38	14 09·5	14 11·8	13 30·8	3·8	3·6	9·8	9·2	15·8	14·9
39	14 09·8	14 12·1	13 31·0	3·9	3·7	9·9	9·3	15·9	15·0
40	14 10·0	14 12·3	13 31·3	4·0	3·8	10·0	9·4	16·0	15·1
41	14 10·3	14 12·6	13 31·5	4·1	3·9	10·1	9·5	16·1	15·2
42	14 10·5	14 12·8	13 31·8	4·2	4·0	10·2	9·6	16·2	15·3
43	14 10·8	14 13·1	13 32·0	4·3	4·0	10·3	9·7	16·3	15·3
44	14 11·0	14 13·3	13 32·2	4·4	4·1	10·4	9·8	16·4	15·4
45	14 11·3	14 13·6	13 32·5	4·5	4·2	10·5	9·9	16·5	15·5
46	14 11·5	14 13·8	13 32·7	4·6	4·3	10·6	10·0	16·6	15·6
47	14 11·8	14 14·1	13 32·9	4·7	4·4	10·7	10·1	16·7	15·7
48	14 12·0	14 14·3	13 33·2	4·8	4·5	10·8	10·2	16·8	15·8
49	14 12·3	14 14·6	13 33·4	4·9	4·6	10·9	10·3	16·9	15·9
50	14 12·5	14 14·8	13 33·7	5·0	4·7	11·0	10·4	17·0	16·0
51	14 12·8	14 15·1	13 33·9	5·1	4·8	11·1	10·5	17·1	16·1
52	14 13·0	14 15·3	13 34·1	5·2	4·9	11·2	10·5	17·2	16·2
53	14 13·3	14 15·6	13 34·4	5·3	5·0	11·3	10·6	17·3	16·3
54	14 13·5	14 15·8	13 34·6	5·4	5·1	11·4	10·7	17·4	16·4
55	14 13·8	14 16·1	13 34·9	5·5	5·2	11·5	10·8	17·5	16·5
56	14 14·0	14 16·3	13 35·1	5·6	5·3	11·6	10·9	17·6	16·6
57	14 14·3	14 16·6	13 35·3	5·7	5·4	11·7	11·0	17·7	16·7
58	14 14·5	14 16·8	13 35·6	5·8	5·5	11·8	11·1	17·8	16·8
59	14 14·8	14 17·1	13 35·8	5·9	5·6	11·9	11·2	17·9	16·9
60	14 15·0	14 17·3	13 36·1	6·0	5·7	12·0	11·3	18·0	17·0

57^{m} s	SUN PLANETS ° ′	ARIES ° ′	MOON ° ′	v or d ′	Corr^{n} ′	v or d ′	Corr^{n} ′	v or d ′	Corr^{n} ′
00	14 15·0	14 17·3	13 36·1	0·0	0·0	6·0	5·8	12·0	11·5
01	14 15·3	14 17·6	13 36·3	0·1	0·1	6·1	5·8	12·1	11·6
02	14 15·5	14 17·8	13 36·5	0·2	0·2	6·2	5·9	12·2	11·7
03	14 15·8	14 18·1	13 36·8	0·3	0·3	6·3	6·0	12·3	11·8
04	14 16·0	14 18·3	13 37·0	0·4	0·4	6·4	6·1	12·4	11·9
05	14 16·3	14 18·6	13 37·2	0·5	0·5	6·5	6·2	12·5	12·0
06	14 16·5	14 18·8	13 37·5	0·6	0·6	6·6	6·3	12·6	12·1
07	14 16·8	14 19·1	13 37·7	0·7	0·7	6·7	6·4	12·7	12·2
08	14 17·0	14 19·3	13 38·0	0·8	0·8	6·8	6·5	12·8	12·3
09	14 17·3	14 19·6	13 38·2	0·9	0·9	6·9	6·6	12·9	12·4
10	14 17·5	14 19·8	13 38·4	1·0	1·0	7·0	6·7	13·0	12·5
11	14 17·8	14 20·1	13 38·7	1·1	1·1	7·1	6·8	13·1	12·6
12	14 18·0	14 20·3	13 38·9	1·2	1·2	7·2	6·9	13·2	12·7
13	14 18·3	14 20·6	13 39·2	1·3	1·2	7·3	7·0	13·3	12·7
14	14 18·5	14 20·9	13 39·4	1·4	1·3	7·4	7·1	13·4	12·8
15	14 18·8	14 21·1	13 39·6	1·5	1·4	7·5	7·2	13·5	12·9
16	14 19·0	14 21·4	13 39·9	1·6	1·5	7·6	7·3	13·6	13·0
17	14 19·3	14 21·6	13 40·1	1·7	1·6	7·7	7·4	13·7	13·1
18	14 19·5	14 21·9	13 40·3	1·8	1·7	7·8	7·5	13·8	13·2
19	14 19·8	14 22·1	13 40·6	1·9	1·8	7·9	7·6	13·9	13·3
20	14 20·0	14 22·4	13 40·8	2·0	1·9	8·0	7·7	14·0	13·4
21	14 20·3	14 22·6	13 41·1	2·1	2·0	8·1	7·8	14·1	13·5
22	14 20·5	14 22·9	13 41·3	2·2	2·1	8·2	7·9	14·2	13·6
23	14 20·8	14 23·1	13 41·5	2·3	2·2	8·3	8·0	14·3	13·7
24	14 21·0	14 23·4	13 41·8	2·4	2·3	8·4	8·1	14·4	13·8
25	14 21·3	14 23·6	13 42·0	2·5	2·4	8·5	8·1	14·5	13·9
26	14 21·5	14 23·9	13 42·3	2·6	2·5	8·6	8·2	14·6	14·0
27	14 21·8	14 24·1	13 42·5	2·7	2·6	8·7	8·3	14·7	14·1
28	14 22·0	14 24·4	13 42·7	2·8	2·7	8·8	8·4	14·8	14·2
29	14 22·3	14 24·6	13 43·0	2·9	2·8	8·9	8·5	14·9	14·3
30	14 22·5	14 24·9	13 43·2	3·0	2·9	9·0	8·6	15·0	14·4
31	14 22·8	14 25·1	13 43·4	3·1	3·0	9·1	8·7	15·1	14·5
32	14 23·0	14 25·4	13 43·7	3·2	3·1	9·2	8·8	15·2	14·6
33	14 23·3	14 25·6	13 43·9	3·3	3·2	9·3	8·9	15·3	14·7
34	14 23·5	14 25·9	13 44·2	3·4	3·3	9·4	9·0	15·4	14·8
35	14 23·8	14 26·1	13 44·4	3·5	3·4	9·5	9·1	15·5	14·9
36	14 24·0	14 26·4	13 44·6	3·6	3·5	9·6	9·2	15·6	15·0
37	14 24·3	14 26·6	13 44·9	3·7	3·5	9·7	9·3	15·7	15·0
38	14 24·5	14 26·9	13 45·1	3·8	3·6	9·8	9·4	15·8	15·1
39	14 24·8	14 27·1	13 45·4	3·9	3·7	9·9	9·5	15·9	15·2
40	14 25·0	14 27·4	13 45·6	4·0	3·8	10·0	9·6	16·0	15·3
41	14 25·3	14 27·6	13 45·8	4·1	3·9	10·1	9·7	16·1	15·4
42	14 25·5	14 27·9	13 46·1	4·2	4·0	10·2	9·8	16·2	15·5
43	14 25·8	14 28·1	13 46·3	4·3	4·1	10·3	9·9	16·3	15·6
44	14 26·0	14 28·4	13 46·5	4·4	4·2	10·4	10·0	16·4	15·7
45	14 26·3	14 28·6	13 46·8	4·5	4·3	10·5	10·1	16·5	15·8
46	14 26·5	14 28·9	13 47·0	4·6	4·4	10·6	10·2	16·6	15·9
47	14 26·8	14 29·1	13 47·3	4·7	4·5	10·7	10·3	16·7	16·0
48	14 27·0	14 29·4	13 47·5	4·8	4·6	10·8	10·4	16·8	16·1
49	14 27·3	14 29·6	13 47·7	4·9	4·7	10·9	10·4	16·9	16·2
50	14 27·5	14 29·9	13 48·0	5·0	4·8	11·0	10·5	17·0	16·3
51	14 27·8	14 30·1	13 48·2	5·1	4·9	11·1	10·6	17·1	16·4
52	14 28·0	14 30·4	13 48·5	5·2	5·0	11·2	10·7	17·2	16·5
53	14 28·3	14 30·6	13 48·7	5·3	5·1	11·3	10·8	17·3	16·6
54	14 28·5	14 30·9	13 48·9	5·4	5·2	11·4	10·9	17·4	16·7
55	14 28·8	14 31·1	13 49·2	5·5	5·3	11·5	11·0	17·5	16·8
56	14 29·0	14 31·4	13 49·4	5·6	5·4	11·6	11·1	17·6	16·9
57	14 29·3	14 31·6	13 49·7	5·7	5·5	11·7	11·2	17·7	17·0
58	14 29·5	14 31·9	13 49·9	5·8	5·6	11·8	11·3	17·8	17·1
59	14 29·8	14 32·1	13 50·1	5·9	5·7	11·9	11·4	17·9	17·2
60	14 30·0	14 32·4	13 50·4	6·0	5·8	12·0	11·5	18·0	17·3

58^m	SUN PLANETS	ARIES	MOON	v or d	Corr^n	v or d	Corr^n	v or d	Corr^n
s	° ′	° ′	° ′	′	′	′	′	′	′
00	14 30·0	14 32·4	13 50·4	0·0	0·0	6·0	5·9	12·0	11·7
01	14 30·3	14 32·6	13 50·6	0·1	0·1	6·1	5·9	12·1	11·8
02	14 30·5	14 32·9	13 50·8	0·2	0·2	6·2	6·0	12·2	11·9
03	14 30·8	14 33·1	13 51·1	0·3	0·3	6·3	6·1	12·3	12·0
04	14 31·0	14 33·4	13 51·3	0·4	0·4	6·4	6·2	12·4	12·1
05	14 31·3	14 33·6	13 51·6	0·5	0·5	6·5	6·3	12·5	12·2
06	14 31·5	14 33·9	13 51·8	0·6	0·6	6·6	6·4	12·6	12·3
07	14 31·8	14 34·1	13 52·0	0·7	0·7	6·7	6·5	12·7	12·4
08	14 32·0	14 34·4	13 52·3	0·8	0·8	6·8	6·6	12·8	12·5
09	14 32·3	14 34·6	13 52·5	0·9	0·9	6·9	6·7	12·9	12·6
10	14 32·5	14 34·9	13 52·8	1·0	1·0	7·0	6·8	13·0	12·7
11	14 32·8	14 35·1	13 53·0	1·1	1·1	7·1	6·9	13·1	12·8
12	14 33·0	14 35·4	13 53·2	1·2	1·2	7·2	7·0	13·2	12·9
13	14 33·3	14 35·6	13 53·5	1·3	1·3	7·3	7·1	13·3	13·0
14	14 33·5	14 35·9	13 53·7	1·4	1·4	7·4	7·2	13·4	13·1
15	14 33·8	14 36·1	13 53·9	1·5	1·5	7·5	7·3	13·5	13·2
16	14 34·0	14 36·4	13 54·2	1·6	1·6	7·6	7·4	13·6	13·3
17	14 34·3	14 36·6	13 54·4	1·7	1·7	7·7	7·5	13·7	13·4
18	14 34·5	14 36·9	13 54·7	1·8	1·8	7·8	7·6	13·8	13·5
19	14 34·8	14 37·1	13 54·9	1·9	1·9	7·9	7·7	13·9	13·6
20	14 35·0	14 37·4	13 55·1	2·0	2·0	8·0	7·8	14·0	13·7
21	14 35·3	14 37·6	13 55·4	2·1	2·0	8·1	7·9	14·1	13·7
22	14 35·5	14 37·9	13 55·6	2·2	2·1	8·2	8·0	14·2	13·8
23	14 35·8	14 38·1	13 55·9	2·3	2·2	8·3	8·1	14·3	13·9
24	14 36·0	14 38·4	13 56·1	2·4	2·3	8·4	8·2	14·4	14·0
25	14 36·3	14 38·6	13 56·3	2·5	2·4	8·5	8·3	14·5	14·1
26	14 36·5	14 38·9	13 56·6	2·6	2·5	8·6	8·4	14·6	14·2
27	14 36·8	14 39·2	13 56·8	2·7	2·6	8·7	8·5	14·7	14·3
28	14 37·0	14 39·4	13 57·0	2·8	2·7	8·8	8·6	14·8	14·4
29	14 37·3	14 39·7	13 57·3	2·9	2·8	8·9	8·7	14·9	14·5
30	14 37·5	14 39·9	13 57·5	3·0	2·9	9·0	8·8	15·0	14·6
31	14 37·8	14 40·2	13 57·8	3·1	3·0	9·1	8·9	15·1	14·7
32	14 38·0	14 40·4	13 58·0	3·2	3·1	9·2	9·0	15·2	14·8
33	14 38·3	14 40·7	13 58·2	3·3	3·2	9·3	9·1	15·3	14·9
34	14 38·5	14 40·9	13 58·5	3·4	3·3	9·4	9·2	15·4	15·0
35	14 38·8	14 41·2	13 58·7	3·5	3·4	9·5	9·3	15·5	15·1
36	14 39·0	14 41·4	13 59·0	3·6	3·5	9·6	9·4	15·6	15·2
37	14 39·3	14 41·7	13 59·2	3·7	3·6	9·7	9·5	15·7	15·3
38	14 39·5	14 41·9	13 59·4	3·8	3·7	9·8	9·6	15·8	15·4
39	14 39·8	14 42·2	13 59·7	3·9	3·8	9·9	9·7	15·9	15·5
40	14 40·0	14 42·4	13 59·9	4·0	3·9	10·0	9·8	16·0	15·6
41	14 40·3	14 42·7	14 00·1	4·1	4·0	10·1	9·8	16·1	15·7
42	14 40·5	14 42·9	14 00·4	4·2	4·1	10·2	9·9	16·2	15·8
43	14 40·8	14 43·2	14 00·6	4·3	4·2	10·3	10·0	16·3	15·9
44	14 41·0	14 43·4	14 00·9	4·4	4·3	10·4	10·1	16·4	16·0
45	14 41·3	14 43·7	14 01·1	4·5	4·4	10·5	10·2	16·5	16·1
46	14 41·5	14 43·9	14 01·3	4·6	4·5	10·6	10·3	16·6	16·2
47	14 41·8	14 44·2	14 01·6	4·7	4·6	10·7	10·4	16·7	16·3
48	14 42·0	14 44·4	14 01·8	4·8	4·7	10·8	10·5	16·8	16·4
49	14 42·3	14 44·7	14 02·1	4·9	4·8	10·9	10·6	16·9	16·5
50	14 42·5	14 44·9	14 02·3	5·0	4·9	11·0	10·7	17·0	16·6
51	14 42·8	14 45·2	14 02·5	5·1	5·0	11·1	10·8	17·1	16·7
52	14 43·0	14 45·4	14 02·8	5·2	5·1	11·2	10·9	17·2	16·8
53	14 43·3	14 45·7	14 03·0	5·3	5·2	11·3	11·0	17·3	16·9
54	14 43·5	14 45·9	14 03·3	5·4	5·3	11·4	11·1	17·4	17·0
55	14 43·8	14 46·2	14 03·5	5·5	5·4	11·5	11·2	17·5	17·1
56	14 44·0	14 46·4	14 03·7	5·6	5·5	11·6	11·3	17·6	17·2
57	14 44·3	14 46·7	14 04·0	5·7	5·6	11·7	11·4	17·7	17·3
58	14 44·5	14 46·9	14 04·2	5·8	5·7	11·8	11·5	17·8	17·4
59	14 44·8	14 47·2	14 04·4	5·9	5·8	11·9	11·6	17·9	17·5
60	14 45·0	14 47·4	14 04·7	6·0	5·9	12·0	11·7	18·0	17·6

59^m	SUN PLANETS	ARIES	MOON	v or d	Corr^n	v or d	Corr^n	v or d	Corr^n
s	° ′	° ′	° ′	′	′	′	′	′	′
00	14 45·0	14 47·4	14 04·7	0·0	0·0	6·0	6·0	12·0	11·9
01	14 45·3	14 47·7	14 04·9	0·1	0·1	6·1	6·0	12·1	12·0
02	14 45·5	14 47·9	14 05·2	0·2	0·2	6·2	6·1	12·2	12·1
03	14 45·8	14 48·2	14 05·4	0·3	0·3	6·3	6·2	12·3	12·2
04	14 46·0	14 48·4	14 05·6	0·4	0·4	6·4	6·3	12·4	12·3
05	14 46·3	14 48·7	14 05·9	0·5	0·5	6·5	6·4	12·5	12·4
06	14 46·5	14 48·9	14 06·1	0·6	0·6	6·6	6·5	12·6	12·5
07	14 46·8	14 49·2	14 06·4	0·7	0·7	6·7	6·6	12·7	12·6
08	14 47·0	14 49·4	14 06·6	0·8	0·8	6·8	6·7	12·8	12·7
09	14 47·3	14 49·7	14 06·8	0·9	0·9	6·9	6·8	12·9	12·8
10	14 47·5	14 49·9	14 07·1	1·0	1·0	7·0	6·9	13·0	12·9
11	14 47·8	14 50·2	14 07·3	1·1	1·1	7·1	7·0	13·1	13·0
12	14 48·0	14 50·4	14 07·5	1·2	1·2	7·2	7·1	13·2	13·1
13	14 48·3	14 50·7	14 07·8	1·3	1·3	7·3	7·2	13·3	13·2
14	14 48·5	14 50·9	14 08·0	1·4	1·4	7·4	7·3	13·4	13·3
15	14 48·8	14 51·2	14 08·3	1·5	1·5	7·5	7·4	13·5	13·4
16	14 49·0	14 51·4	14 08·5	1·6	1·6	7·6	7·5	13·6	13·5
17	14 49·3	14 51·7	14 08·7	1·7	1·7	7·7	7·6	13·7	13·6
18	14 49·5	14 51·9	14 09·0	1·8	1·8	7·8	7·7	13·8	13·7
19	14 49·8	14 52·2	14 09·2	1·9	1·9	7·9	7·8	13·9	13·8
20	14 50·0	14 52·4	14 09·5	2·0	2·0	8·0	7·9	14·0	13·9
21	14 50·3	14 52·7	14 09·7	2·1	2·1	8·1	8·0	14·1	14·0
22	14 50·5	14 52·9	14 09·9	2·2	2·2	8·2	8·1	14·2	14·1
23	14 50·8	14 53·2	14 10·2	2·3	2·3	8·3	8·2	14·3	14·2
24	14 51·0	14 53·4	14 10·4	2·4	2·4	8·4	8·3	14·4	14·3
25	14 51·3	14 53·7	14 10·6	2·5	2·5	8·5	8·4	14·5	14·4
26	14 51·5	14 53·9	14 10·9	2·6	2·6	8·6	8·5	14·6	14·5
27	14 51·8	14 54·2	14 11·1	2·7	2·7	8·7	8·6	14·7	14·6
28	14 52·0	14 54·4	14 11·4	2·8	2·8	8·8	8·7	14·8	14·7
29	14 52·3	14 54·7	14 11·6	2·9	2·9	8·9	8·8	14·9	14·8
30	14 52·5	14 54·9	14 11·8	3·0	3·0	9·0	8·9	15·0	14·9
31	14 52·8	14 55·2	14 12·1	3·1	3·1	9·1	9·0	15·1	15·0
32	14 53·0	14 55·4	14 12·3	3·2	3·2	9·2	9·1	15·2	15·1
33	14 53·3	14 55·7	14 12·6	3·3	3·3	9·3	9·2	15·3	15·2
34	14 53·5	14 55·9	14 12·8	3·4	3·4	9·4	9·3	15·4	15·3
35	14 53·8	14 56·2	14 13·0	3·5	3·5	9·5	9·4	15·5	15·4
36	14 54·0	14 56·4	14 13·3	3·6	3·6	9·6	9·5	15·6	15·5
37	14 54·3	14 56·7	14 13·5	3·7	3·7	9·7	9·6	15·7	15·6
38	14 54·5	14 56·9	14 13·8	3·8	3·8	9·8	9·7	15·8	15·7
39	14 54·8	14 57·2	14 14·0	3·9	3·9	9·9	9·8	15·9	15·8
40	14 55·0	14 57·5	14 14·2	4·0	4·0	10·0	9·9	16·0	15·9
41	14 55·3	14 57·7	14 14·5	4·1	4·1	10·1	10·0	16·1	16·0
42	14 55·5	14 58·0	14 14·7	4·2	4·2	10·2	10·1	16·2	16·1
43	14 55·8	14 58·2	14 14·9	4·3	4·3	10·3	10·2	16·3	16·2
44	14 56·0	14 58·5	14 15·2	4·4	4·4	10·4	10·3	16·4	16·3
45	14 56·3	14 58·7	14 15·4	4·5	4·5	10·5	10·4	16·5	16·4
46	14 56·5	14 59·0	14 15·7	4·6	4·6	10·6	10·5	16·6	16·5
47	14 56·8	14 59·2	14 15·9	4·7	4·7	10·7	10·6	16·7	16·6
48	14 57·0	14 59·5	14 16·1	4·8	4·8	10·8	10·7	16·8	16·7
49	14 57·3	14 59·7	14 16·4	4·9	4·9	10·9	10·8	16·9	16·8
50	14 57·5	15 00·0	14 16·6	5·0	5·0	11·0	10·9	17·0	16·9
51	14 57·8	15 00·2	14 16·9	5·1	5·1	11·1	11·0	17·1	17·0
52	14 58·0	15 00·5	14 17·1	5·2	5·2	11·2	11·1	17·2	17·1
53	14 58·3	15 00·7	14 17·3	5·3	5·3	11·3	11·2	17·3	17·2
54	14 58·5	15 01·0	14 17·6	5·4	5·4	11·4	11·3	17·4	17·3
55	14 58·8	15 01·2	14 17·8	5·5	5·5	11·5	11·4	17·5	17·4
56	14 59·0	15 01·5	14 18·0	5·6	5·6	11·6	11·5	17·6	17·5
57	14 59·3	15 01·7	14 18·3	5·7	5·7	11·7	11·6	17·7	17·6
58	14 59·5	15 02·0	14 18·5	5·8	5·8	11·8	11·7	17·8	17·7
59	14 59·8	15 02·2	14 18·8	5·9	5·9	11·9	11·8	17·9	17·8
60	15 00·0	15 02·5	14 19·0	6·0	6·0	12·0	11·9	18·0	17·9

TABLES FOR INTERPOLATING SUNRISE, MOONRISE, ETC.

TABLE I—FOR LATITUDE

Tabular Interval 10°	5°	2°	Difference between the times for consecutive latitudes 5^m	10^m	15^m	20^m	25^m	30^m	35^m	40^m	45^m	50^m	55^m	60^m	1^h 05^m	1^h 10^m	1^h 15^m	1^h 20^m
° ′	° ′	° ′	m	m	m	m	m	m	m	m	m	m	m	m	h m	h m	h m	h m
0 30	**0 15**	**0 06**	0	0	1	1	1	1	1	2	2	2	2	2	0 02	0 02	0 02	0 02
1 00	**0 30**	**0 12**	0	1	1	2	2	3	3	3	4	4	4	5	05	05	05	05
1 30	**0 45**	**0 18**	1	1	2	3	3	4	4	5	5	6	7	7	07	07	07	07
2 00	**1 00**	**0 24**	1	2	3	4	5	5	6	7	7	8	9	10	10	10	10	10
2 30	**1 15**	**0 30**	1	2	4	5	6	7	8	9	9	10	11	12	12	13	13	13
3 00	**1 30**	**0 36**	1	3	4	6	7	8	9	10	11	12	13	14	0 15	0 15	0 16	0 16
3 30	**1 45**	**0 42**	2	3	5	7	8	10	11	12	13	14	16	17	18	18	19	19
4 00	**2 00**	**0 48**	2	4	6	8	9	11	13	14	15	16	18	19	20	21	22	22
4 30	**2 15**	**0 54**	2	4	7	9	11	13	15	16	18	19	21	22	23	24	25	26
5 00	**2 30**	**1 00**	2	5	7	10	12	14	16	18	20	22	23	25	26	27	28	29
5 30	**2 45**	**1 06**	3	5	8	11	13	16	18	20	22	24	26	28	0 29	0 30	0 31	0 32
6 00	**3 00**	**1 12**	3	6	9	12	14	17	20	22	24	26	29	31	32	33	34	36
6 30	**3 15**	**1 18**	3	6	10	13	16	19	22	24	26	29	31	34	36	37	38	40
7 00	**3 30**	**1 24**	3	7	10	14	17	20	23	26	29	31	34	37	39	41	42	44
7 30	**3 45**	**1 30**	4	7	11	15	18	22	25	28	31	34	37	40	43	44	46	48
8 00	**4 00**	**1 36**	4	8	12	16	20	23	27	30	34	37	41	44	0 47	0 48	0 51	0 53
8 30	**4 15**	**1 42**	4	8	13	17	21	25	29	33	36	40	44	48	0 51	0 53	0 56	0 58
9 00	**4 30**	**1 48**	4	9	13	18	22	27	31	35	39	43	47	52	0 55	0 58	1 01	1 04
9 30	**4 45**	**1 54**	5	9	14	19	24	28	33	38	42	47	51	56	1 00	1 04	1 08	1 12
10 00	**5 00**	**2 00**	5	10	15	20	25	30	35	40	45	50	55	60	1 05	1 10	1 15	1 20

Table I is for interpolating the LMT of sunrise, twilight, moonrise, etc., for latitude. It is to be entered, in the appropriate column on the left, with the difference between true latitude and the nearest tabular latitude which is *less* than the true latitude; and with the argument at the top which is the nearest value of the difference between the times for the tabular latitude and the next higher one; the correction so obtained is applied to the time for the tabular latitude; the sign of the correction can be seen by inspection. It is to be noted that the interpolation is not linear, so that when using this table it is essential to take out the tabular phenomenon for the latitude *less* than the true latitude.

TABLE II—FOR LONGITUDE

Long. East or West	Difference between the times for given date and preceding date (for east longitude) or for given date and following date (for west longitude) 10^m	20^m	30^m	40^m	50^m	60^m	1^h + 10^m	20^m	30^m	1^h + 40^m	50^m	60^m	2^h 10^m	2^h 20^m	2^h 30^m	2^h 40^m	2^h 50^m	3^h 00^m
°	m	m	m	m	m	m	m	m	m	m	m	m	h m	h m	h m	h m	h m	h m
0	0	0	0	0	0	0	0	0	0	0	0	0	0 00	0 00	0 00	0 00	0 00	0 00
10	0	1	1	1	1	2	2	2	2	3	3	3	04	04	04	04	05	05
20	1	1	2	2	3	3	4	4	5	6	6	7	07	08	08	09	09	10
30	1	2	2	3	4	5	6	7	7	8	9	10	11	12	12	13	14	15
40	1	2	3	4	6	7	8	9	10	11	12	13	14	16	17	18	19	20
50	1	3	4	6	7	8	10	11	12	14	15	17	0 18	0 19	0 21	0 22	0 24	0 25
60	2	3	5	7	8	10	12	13	15	17	18	20	22	23	25	27	28	30
70	2	4	6	8	10	12	14	16	17	19	21	23	25	27	29	31	33	35
80	2	4	7	9	11	13	16	18	20	22	24	27	29	31	33	36	38	40
90	2	5	7	10	12	15	17	20	22	25	27	30	32	35	37	40	42	45
100	3	6	8	11	14	17	19	22	25	28	31	33	0 36	0 39	0 42	0 44	0 47	0 50
110	3	6	9	12	15	18	21	24	27	31	34	37	40	43	46	49	0 52	0 55
120	3	7	10	13	17	20	23	27	30	33	37	40	43	47	50	53	0 57	1 00
130	4	7	11	14	18	22	25	29	32	36	40	43	47	51	54	0 58	1 01	1 05
140	4	8	12	16	19	23	27	31	35	39	43	47	51	54	0 58	1 02	1 06	1 10
150	4	8	13	17	21	25	29	33	38	42	46	50	0 54	0 58	1 03	1 07	1 11	1 15
160	4	9	13	18	22	27	31	36	40	44	49	53	0 58	1 02	1 07	1 11	1 16	1 20
170	5	9	14	19	24	28	33	38	42	47	52	57	1 01	1 06	1 11	1 16	1 20	1 25
180	5	10	15	20	25	30	35	40	45	50	55	60	1 05	1 10	1 15	1 20	1 25	1 30

Table II is for interpolating the LMT of moonrise, moonset and the Moon's meridian passage for longitude. It is entered with longitude and with the difference between the times for the given date and for the preceding date (in east longitudes) or following date (in west longitudes). The correction is normally *added* for west longitudes and *subtracted* for east longitudes, but if, as occasionally happens, the times become earlier each day instead of later, the signs of the corrections must be reversed.

INDEX TO SELECTED STARS, 2003

Name	No	Mag	SHA	Dec
			°	°
Acamar	**7**	3·1	315	S 40
Achernar	**5**	0·6	336	S 57
Acrux	**30**	1·1	173	S 63
Adhara	**19**	1·6	255	S 29
Aldebaran	**10**	1·1	291	N 17
Alioth	**32**	1·7	166	N 56
Alkaid	**34**	1·9	153	N 49
Al Na'ir	**55**	2·2	28	S 47
Alnilam	**15**	1·8	276	S 1
Alphard	**25**	2·2	218	S 9
Alphecca	**41**	2·3	126	N 27
Alpheratz	**1**	2·2	358	N 29
Altair	**51**	0·9	62	N 9
Ankaa	**2**	2·4	353	S 42
Antares	**42**	1·2	113	S 26
Arcturus	**37**	0·2	146	N 19
Atria	**43**	1·9	108	S 69
Avior	**22**	1·7	234	S 60
Bellatrix	**13**	1·7	279	N 6
Betelgeuse	**16**	Var.*	271	N 7
Canopus	**17**	−0·9	264	S 53
Capella	**12**	0·2	281	N 46
Deneb	**53**	1·3	50	N 45
Denebola	**28**	2·2	183	N 15
Diphda	**4**	2·2	349	S 18
Dubhe	**27**	2·0	194	N 62
Elnath	**14**	1·8	278	N 29
Eltanin	**47**	2·4	91	N 51
Enif	**54**	2·5	34	N 10
Fomalhaut	**56**	1·3	16	S 30
Gacrux	**31**	1·6	172	S 57
Gienah	**29**	2·8	176	S 18
Hadar	**35**	0·9	149	S 60
Hamal	**6**	2·2	328	N 23
Kaus Australis	**48**	2·0	84	S 34
Kochab	**40**	2·2	137	N 74
Markab	**57**	2·6	14	N 15
Menkar	**8**	2·8	314	N 4
Menkent	**36**	2·3	148	S 36
Miaplacidus	**24**	1·8	222	S 70
Mirfak	**9**	1·9	309	N 50
Nunki	**50**	2·1	76	S 26
Peacock	**52**	2·1	54	S 57
Pollux	**21**	1·2	244	N 28
Procyon	**20**	0·5	245	N 5
Rasalhague	**46**	2·1	96	N 13
Regulus	**26**	1·3	208	N 12
Rigel	**11**	0·3	281	S 8
Rigil Kentaurus	**38**	0·1	140	S 61
Sabik	**44**	2·6	102	S 16
Schedar	**3**	2·5	350	N 57
Shaula	**45**	1·7	97	S 37
Sirius	**18**	−1·6	259	S 17
Spica	**33**	1·2	159	S 11
Suhail	**23**	2·2	223	S 43
Vega	**49**	0·1	81	N 39
Zubenelgenubi	**39**	2·9	137	S 16

No	Name	Mag	SHA	Dec
			°	°
1	*Alpheratz*	2·2	358	N 29
2	*Ankaa*	2·4	353	S 42
3	*Schedar*	2·5	350	N 57
4	*Diphda*	2·2	349	S 18
5	*Achernar*	0·6	336	S 57
6	*Hamal*	2·2	328	N 23
7	*Acamar*	3·1	315	S 40
8	*Menkar*	2·8	314	N 4
9	*Mirfak*	1·9	309	N 50
10	*Aldebaran*	1·1	291	N 17
11	*Rigel*	0·3	281	S 8
12	*Capella*	0·2	281	N 46
13	*Bellatrix*	1·7	279	N 6
14	*Elnath*	1·8	278	N 29
15	*Alnilam*	1·8	276	S 1
16	*Betelgeuse*	Var.*	271	N 7
17	*Canopus*	−0·9	264	S 53
18	*Sirius*	−1·6	259	S 17
19	*Adhara*	1·6	255	S 29
20	*Procyon*	0·5	245	N 5
21	*Pollux*	1·2	244	N 28
22	*Avior*	1·7	234	S 60
23	*Suhail*	2·2	223	S 43
24	*Miaplacidus*	1·8	222	S 70
25	*Alphard*	2·2	218	S 9
26	*Regulus*	1·3	208	N 12
27	*Dubhe*	2·0	194	N 62
28	*Denebola*	2·2	183	N 15
29	*Gienah*	2·8	176	S 18
30	*Acrux*	1·1	173	S 63
31	*Gacrux*	1·6	172	S 57
32	*Alioth*	1·7	166	N 56
33	*Spica*	1·2	159	S 11
34	*Alkaid*	1·9	153	N 49
35	*Hadar*	0·9	149	S 60
36	*Menkent*	2·3	148	S 36
37	*Arcturus*	0·2	146	N 19
38	*Rigil Kentaurus*	0·1	140	S 61
39	*Zubenelgenubi*	2·9	137	S 16
40	*Kochab*	2·2	137	N 74
41	*Alphecca*	2·3	126	N 27
42	*Antares*	1·2	113	S 26
43	*Atria*	1·9	108	S 69
44	*Sabik*	2·6	102	S 16
45	*Shaula*	1·7	97	S 37
46	*Rasalhague*	2·1	96	N 13
47	*Eltanin*	2·4	91	N 51
48	*Kaus Australis*	2·0	84	S 34
49	*Vega*	0·1	81	N 39
50	*Nunki*	2·1	76	S 26
51	*Altair*	0·9	62	N 9
52	*Peacock*	2·1	54	S 57
53	*Deneb*	1·3	50	N 45
54	*Enif*	2·5	34	N 10
55	*Al Na'ir*	2·2	28	S 47
56	*Fomalhaut*	1·3	16	S 30
57	*Markab*	2·6	14	N 15

*0·1 — 1·2

ALTITUDE CORRECTION TABLES 0°–35°— MOON

App. Alt.	0°–4° Corrⁿ	5°–9° Corrⁿ	10°–14° Corrⁿ	15°–19° Corrⁿ	20°–24° Corrⁿ	25°–29° Corrⁿ	30°–34° Corrⁿ	App. Alt.
′	° ′	° ′	° ′	° ′	° ′	° ′	° ′	′
00	**0** 33·8	**5** 58·2	**10** 62·1	**15** 62·8	**20** 62·2	**25** 60·8	**30** 58·9	**00**
10	35·9	58·5	62·2	62·8	62·1	60·8	58·8	**10**
20	37·8	58·7	62·2	62·8	62·1	60·7	58·8	**20**
30	39·6	58·9	62·3	62·8	62·1	60·7	58·7	**30**
40	41·2	59·1	62·3	62·8	62·0	60·6	58·6	**40**
50	42·6	59·3	62·4	62·7	62·0	60·6	58·5	**50**
00	**1** 44·0	**6** 59·5	**11** 62·4	**16** 62·7	**21** 62·0	**26** 60·5	**31** 58·5	**00**
10	45·2	59·7	62·4	62·7	61·9	60·4	58·4	**10**
20	46·3	59·9	62·5	62·7	61·9	60·4	58·3	**20**
30	47·3	60·0	62·5	62·7	61·9	60·3	58·2	**30**
40	48·3	60·2	62·5	62·7	61·8	60·3	58·2	**40**
50	49·2	60·3	62·6	62·7	61·8	60·2	58·1	**50**
00	**2** 50·0	**7** 60·5	**12** 62·6	**17** 62·7	**22** 61·7	**27** 60·1	**32** 58·0	**00**
10	50·8	60·6	62·6	62·6	61·7	60·1	57·9	**10**
20	51·4	60·7	62·6	62·6	61·6	60·0	57·8	**20**
30	52·1	60·9	62·7	62·6	61·6	59·9	57·8	**30**
40	52·7	61·0	62·7	62·6	61·5	59·9	57·7	**40**
50	53·3	61·1	62·7	62·6	61·5	59·8	57·6	**50**
00	**3** 53·8	**8** 61·2	**13** 62·7	**18** 62·5	**23** 61·5	**28** 59·7	**33** 57·5	**00**
10	54·3	61·3	62·7	62·5	61·4	59·7	57·4	**10**
20	54·8	61·4	62·7	62·5	61·4	59·6	57·4	**20**
30	55·2	61·5	62·8	62·5	61·3	59·6	57·3	**30**
40	55·6	61·6	62·8	62·4	61·3	59·5	57·2	**40**
50	56·0	61·6	62·8	62·4	61·2	59·4	57·1	**50**
00	**4** 56·4	**9** 61·7	**14** 62·8	**19** 62·4	**24** 61·2	**29** 59·3	**34** 57·0	**00**
10	56·7	61·8	62·8	62·3	61·1	59·3	56·9	**10**
20	57·1	61·9	62·8	62·3	61·1	59·2	56·9	**20**
30	57·4	61·9	62·8	62·3	61·0	59·1	56·8	**30**
40	57·7	62·0	62·8	62·2	60·9	59·1	56·7	**40**
50	57·9	62·1	62·8	62·2	60·9	59·0	56·6	**50**

HP	L	U	L	U	L	U	L	U	L	U	L	U	L	U	HP
′	′	′	′	′	′	′	′	′	′	′	′	′	′	′	′
54·0	0·3	0·9	0·3	0·9	0·4	1·0	0·5	1·1	0·6	1·2	0·7	1·3	0·9	1·5	**54·0**
54·3	0·7	1·1	0·7	1·2	0·7	1·2	0·8	1·3	0·9	1·4	1·1	1·5	1·2	1·7	**54·3**
54·6	1·1	1·4	1·1	1·4	1·1	1·4	1·2	1·5	1·3	1·6	1·4	1·7	1·5	1·8	**54·6**
54·9	1·4	1·6	1·5	1·6	1·5	1·6	1·6	1·7	1·6	1·8	1·8	1·9	1·9	2·0	**54·9**
55·2	1·8	1·8	1·8	1·8	1·9	1·9	1·9	1·9	2·0	2·0	2·1	2·1	2·2	2·2	**55·2**
55·5	2·2	2·0	2·2	2·0	2·3	2·1	2·3	2·1	2·4	2·2	2·4	2·3	2·5	2·4	**55·5**
55·8	2·6	2·2	2·6	2·2	2·6	2·3	2·7	2·3	2·7	2·4	2·8	2·4	2·9	2·5	**55·8**
56·1	3·0	2·4	3·0	2·5	3·0	2·5	3·0	2·5	3·1	2·6	3·1	2·6	3·2	2·7	**56·1**
56·4	3·4	2·7	3·4	2·7	3·4	2·7	3·4	2·7	3·4	2·8	3·5	2·8	3·5	2·9	**56·4**
56·7	3·7	2·9	3·7	2·9	3·8	2·9	3·8	2·9	3·8	3·0	3·8	3·0	3·9	3·0	**56·7**
57·0	4·1	3·1	4·1	3·1	4·1	3·1	4·1	3·1	4·2	3·1	4·2	3·2	4·2	3·2	**57·0**
57·3	4·5	3·3	4·5	3·3	4·5	3·3	4·5	3·3	4·5	3·3	4·5	3·4	4·6	3·4	**57·3**
57·6	4·9	3·5	4·9	3·5	4·9	3·5	4·9	3·5	4·9	3·5	4·9	3·5	4·9	3·6	**57·6**
57·9	5·3	3·8	5·3	3·8	5·2	3·8	5·2	3·7	5·2	3·7	5·2	3·7	5·2	3·7	**57·9**
58·2	5·6	4·0	5·6	4·0	5·6	4·0	5·6	4·0	5·6	3·9	5·6	3·9	5·6	3·9	**58·2**
58·5	6·0	4·2	6·0	4·2	6·0	4·2	6·0	4·2	6·0	4·1	5·9	4·1	5·9	4·1	**58·5**
58·8	6·4	4·4	6·4	4·4	6·4	4·4	6·3	4·4	6·3	4·3	6·3	4·3	6·2	4·2	**58·8**
59·1	6·8	4·6	6·8	4·6	6·7	4·6	6·7	4·6	6·7	4·5	6·6	4·5	6·6	4·4	**59·1**
59·4	7·2	4·8	7·1	4·8	7·1	4·8	7·1	4·8	7·0	4·7	7·0	4·7	6·9	4·6	**59·4**
59·7	7·5	5·1	7·5	5·0	7·5	5·0	7·5	5·0	7·4	4·9	7·3	4·8	7·2	4·7	**59·7**
60·0	7·9	5·3	7·9	5·3	7·9	5·2	7·8	5·2	7·8	5·1	7·7	5·0	7·6	4·9	**60·0**
60·3	8·3	5·5	8·3	5·5	8·2	5·4	8·2	5·4	8·1	5·3	8·0	5·2	7·9	5·1	**60·3**
60·6	8·7	5·7	8·7	5·7	8·6	5·7	8·6	5·6	8·5	5·5	8·4	5·4	8·2	5·3	**60·6**
60·9	9·1	5·9	9·0	5·9	9·0	5·9	8·9	5·8	8·8	5·7	8·7	5·6	8·6	5·4	**60·9**
61·2	9·5	6·2	9·4	6·1	9·4	6·1	9·3	6·0	9·2	5·9	9·1	5·8	8·9	5·6	**61·2**
61·5	9·8	6·4	9·8	6·3	9·7	6·3	9·7	6·2	9·5	6·1	9·4	5·9	9·2	5·8	**61·5**

DIP					
Ht. of Eye	Corrⁿ	Ht. of Eye	Ht. of Eye	Corrⁿ	Ht. of Eye
m	′	ft.	m	′	ft.
2·4		8·0	9·5		31·5
	−2·8			−5·5	
2·6		8·6	9·9		32·7
	−2·9			−5·6	
2·8		9·2	10·3		33·9
	−3·0			−5·7	
3·0		9·8	10·6		35·1
	−3·1			−5·8	
3·2		10·5	11·0		36·3
	−3·2			−5·9	
3·4		11·2	11·4		37·6
	−3·3			−6·0	
3·6		11·9	11·8		38·9
	−3·4			−6·1	
3·8		12·6	12·2		40·1
	−3·5			−6·2	
4·0		13·3	12·6		41·5
	−3·6			−6·3	
4·3		14·1	13·0		42·8
	−3·7			−6·4	
4·5		14·9	13·4		44·2
	−3·8			−6·5	
4·7		15·7	13·8		45·5
	−3·9			−6·6	
5·0		16·5	14·2		46·9
	−4·0			−6·7	
5·2		17·4	14·7		48·4
	−4·1			−6·8	
5·5		18·3	15·1		49·8
	−4·2			−6·9	
5·8		19·1	15·5		51·3
	−4·3			−7·0	
6·1		20·1	16·0		52·8
	−4·4			−7·1	
6·3		21·0	16·5		54·3
	−4·5			−7·2	
6·6		22·0	16·9		55·8
	−4·6			−7·3	
6·9		22·9	17·4		57·4
	−4·7			−7·4	
7·2		23·9	17·9		58·9
	−4·8			−7·5	
7·5		24·9	18·4		60·5
	−4·9			−7·6	
7·9		26·0	18·8		62·1
	−5·0			−7·7	
8·2		27·1	19·3		63·8
	−5·1			−7·8	
8·5		28·1	19·8		65·4
	−5·2			−7·9	
8·8		29·2	20·4		67·1
	−5·3			−8·0	
9·2		30·4	20·9		68·8
	−5·4			−8·1	
9·5		31·5	21·4		70·5

MOON CORRECTION TABLE

The correction is in two parts; the first correction is taken from the upper part of the table with argument apparent altitude, and the second from the lower part, with argument HP, in the same column as that from which the first correction was taken. Separate corrections are given in the lower part for lower (L) and upper (U) limbs. All corrections are to be **added** to apparent altitude, *but 30′ is to be subtracted from the altitude of the upper limb.*

For corrections for pressure and temperature see page A4.

For bubble sextant observations ignore dip, take the mean of upper and lower limb corrections and subtract 15′ from the altitude.

App. Alt. = Apparent altitude = Sextant altitude corrected for index error and dip.